MW01105688

C. Yeh and F. Shimabukuro

From Analytic to Numerical Electromagnetics

Contributions by C. Yeh and His Collaborators

California Advanced Studies Press, Los Angeles, California, USA

C. Yeh

California Advanced Studies
2432 Nalin Drive
Los Angeles
CA 90077
USA
Cavour.Yeh@gmail.com

F. I. Shimabukuro

California Advanced Studies
2432 Nalin Drive
Los Angeles
CA 90077
USA
fredshima@gmail.com

ISBN 978-145753-625-0

Library of Congress Control Number: 2015913876

Printed on acid-free paper

9 8 7 6 5 4 3 2 1

California Advanced Studies

To Our Families
Vivian, John, and Evelyn; Siblings—Dorothy, Richard, and Vicky
Karen, Susan, and Lee

PREFACE

It is our responsibility as scientists, knowing the great progress which comes from a satisfactory philosophy of ignorance, the great progress which is the fruit of freedom of thought, to proclaim the value of this freedom, to teach how doubt is not to be feared but welcomed and discussed; and to demand this freedom as our duty to all coming generations.

— *Richard Feynman, 1955*

...to boldly go where no man has gone before...

—*Star Trek, 1966*

When writing a book on a classic subject—electromagnetics (EM)—it is prudent to ask the question "What's new?" This book will address mostly original material that has not appeared in book form elsewhere and contains original research carried out at the California Institute of Technology (Caltech), the University of Southern California (USC), the University of California, Los Angeles (UCLA), the Hughes Research laboratory, and the Jet Propulsion Laboratory (JPL) by Yeh and his collaborators. Even with this single-source limitation, the material given here contains many necessary fundamental topics that can complement any modern EM textbook and can provide ideas for further research. It also leads the way to show the pioneering development of several numerical approaches to the solutions of complex EM problems. In other words, three major categories of new EM material are given here:

(1) Original applied EM theory that cannot be found in any EM textbook (specifics given in Chapter 2)

(2) Original canonical solutions that reinforce the foundation of applied EM theory and that provide the necessary stepping-stones for the development of numerical approaches and other research ideas (examples given in Chapter 1)

(3) Original pioneering development of several important numerical approaches that have become the standard-bearers of some modern numerical EM simulations (specifics given in Chapter 1)

The varied topics are divided into nine chapters. Chapter 1 gives an introduction of the contents of this book. Chapter 2 provides a number of articles of general interest in EM theory. Scattering and diffraction of waves is discussed in Chapter 3. Chapter 4 gives articles on guided waves, while Chapter 5 gives articles on antennas, radiation, and propagation. Chapter 6 delves into the interaction of EM waves with relativistically moving medium. Classical radiation from charged particles in the presence of material medium is treated in Chapter 7. The remaining two chapters deal with related subjects, such as high-speed fiber networks and acoustic waves. To present this large body of material in a most economic, clear, and efficient manner, we chose to present them in a format containing reprints of original articles. The merit of having all the papers collected in a self-contained book should not be overlooked. This way, all reading and studying can be done without resorting to distracting Internet searches. This book can be used in a senior or graduate seminar course, as a reference material for researchers, or as a supplement for an advanced EM course. Papers that are most useful for instructional purposes are indicated with an asterisk (*) on the listings given in the Contents.

Innovations in research are usually built upon prior work. Proper citation of prior work becomes the backbone of research integrity. It is curious to note that a number of popular references (with many citations in the literature) may not represent the papers published by the originators of the concepts. We have taken special care not to follow this erroneous path.

It is with deep gratitude and great pleasure for one of us (Yeh) to acknowledge the significant contributions by his collaborators (coauthors). Their names are listed on a separate page in this book. We are very grateful to Professors C. H. Papas, J. R. Whinnery, and R. W. Gould for their kind guidance and encouragement. We also wish to express our deepest thanks and gratitude to our colleagues who provided much positive advice and encouragement throughout our professional careers.

C. Yeh
F. I. Shimabukuro

Los Angeles, CA

C. Yeh's Collaborators
(Paper number in brackets)

J. Armstrong *(3-6) (5-2-3)*

P. Barber *(3-5) (3-7-4)*
L. Bergman *(4-6-1) (4-6-2) (4-6-3) (4-6-4) (4-6-5) (4-6-6)*
W. P. Brown *(2-2) (4-4-2) (4-5-2) (4-9) (5-2-1) (5-2-2)*

K. F. Casey *(2-3) (4-2-1) (4-2-2) (6-1-3) (6-4) (7-1) (7-2) (7-4)*
L. Casperson *(2-2) (3-7-1) (3-7-2) (3-7-3) (4-5-1)*
J. Chu *(4-11-1)*
S. Colak *(3-7-3) (3-7-4) (3-7-5)*
T. Cwik *(4-14)*

S. B. Dong *(4-4-1) (4-4-2) (9-7)*

C. Elachi *(4-2-3) (4-2-4) (4-2-5) (4-2-6) (4-2-7) (4-13)*
G. Evans *(4-2-7)*

M. Gerla *(8-1) (8-2) (8-3) (8-4) (8-5)*

K. Ha *(4-4-2)*
D. B. Hall *(4-10)*
S. Hamada *(2-2)*

W. A. Imbriale *(2-5) (4-11-2)*
A. Ishimaru *(3-6) (3-8-1) (3-8-2) (3-8-3c)*

D. L. Jaggard *(4-2-8)*
V. Jamnejad *(4-11-2)*
A. Johnston *(2-3) (4-15)*

O. Kaldirim *(8-6)*
Z. A. Kaprielian *(3-3-2) (4-2-1) (5-1-2) (7-1)*
C. S. Kim *(3-1-1) (3-1-4)*
S. G. Knorr *(8-6)*
H. H. Kuehl *(4-12)*

D. Lesselier *(3-8-2) (3-8-3)*

M. Lin *(8-1)*
G. Lindgren *(4-3)*

F. Manshadi *(2-3) (4-5-4) (4-9) (4-11-2)*
G. E. Mariki *(4-8) (4-9)*
R. Matthes *(4-2-2)*
S. Monacos *(4-6-2) (4-6-5)*
J. Morookian *(4-6-2) (4-6-4) (4-6-5)*

W. Oliver *(4-4-1) (9-7)*
T. Y. Otoshi *(2-5) (2-7)*

J. E. Pearson *(5-2-1) (5-2-2)*
R. Pogorzelski *(7-3) (7-4) (7-5)*

P. Rodrigues *(8-1) (8-2) (8-3) (8-5)*
W. V. T. Rusch *(3-3-3) (3-4) (5-1-7)*

W. E. Salmond *(4-1-4a) (4-4-4b)*
F. I. Shimabukuro *(4-11-1) (4-11-2) (4-11-3) (4-11-4)*
P. H. Siegel *(4-11-3)*
P. Stanton *(4-11-2)*
G. E. Stewart *(4-1c)*
B. Szejn *(4-5-1) (4-5-2)*

P. K. C. Wang *(3-3-1)*
K. E. Wilson *(5-2-3)*
R. Woo *(3-6)*

W. F. Yeung *(3-7-1)*

Vitae

C. Yeh has a BS (with honors), an MS, and a PhD from Cal Tech in electrical engineering. He is a fellow of IEEE (Institute of Electrical and Electronics Engineers), a fellow of OSA (Optical Society of America), and a senior scientist at California Advanced Studies. He has more than forty years of teaching and research experience as a professor (UCLA) and associate professor (USC), as a consultant at the Jet Propulsion Laboratory, Hughes Research Laboratory, and others. He gained four US patents and published more than 150 papers in optical and microwave-guided waves, dielectric waveguides, scattering and diffraction of waves, antennas and radiation, optical fibers, WDM solitons, fiber networks, numerical techniques, THz waveguides, moving medium,; a chapter in two books; an article on dynamic equations in the American Institute of Physics handbook; and a book entitled *The Essence of Dielectric Waveguides* (507 pp., Springer Book Co., New York, NY, 2008).

F. I. Shimabukuro has a BS and MS from MIT and PhD from Cal Tech in electrical engineering. He is a senior scientist at California Advanced Studies, and has more than forty years of research experience at the Hughes Aircraft Co. and the Aerospace Corp. He has gained one US patent and has published more than sixty papers in radio astronomy, atmospheric physics, guided waves, millimeter wave technology, THz waveguides, plasma waveguides, and a book entitled *The Essence of Dielectric Waveguides* (507 pp., Springer Book Co., New York, NY, 2008).

CONTENTS

Preface iv

List of C. Yeh's Collaborators vi

Vitae viii

1. Introduction 1

2. Applied Electromagnetic Theory 11

*2.1 α vs. Q (Paper 2-1, C. Yeh, Proc. IRE, 2145, Oct.,1962) 15

*2.2 Scalar-wave approach for single-mode inhomogeneous fiber problems. (Paper 2-2, C. Yeh, L. Casperson, and W. P. Brown, Appl. Phys. Lett. 34(7), 460, Apr., 1979) 18

*2.3 Accuracy of directional coupler theory in fiber or integrated Optics applications. (Paper 2-3, C. Yeh, F. Manshadi, K. F. Casey, and A. Johnston, J. Opt. Soc. Am., Vol. 68, No.8, 1079,) 21

*2.4 Boundary conditions in electromagnetics. (Paper 2-4, C. Yeh, Phy. Rev. E, Vol. 48, No. 2, 1426, Aug., 1993; C. Yeh and F. Shimabukuro, "The essence of dielectric waveguides," pps 20-27, Springer Book Co.,New York, NY (2008) 26

*2.5 Power loss for multimode waveguides and its application to beam- waveguide system. (Paper 2-5, W. A. Imbriale, T. Y. Otoshi, and C. Yeh, IEEE Tran. on Micro. Theory and Tech., Vol. 46, No. 5, 523, May, 1998) 35

*2.6 Computer reconstruction of near-zone field from given far-zone data of two-dimensional objects. (Paper 2-6, S. Hamada and C. Yeh, IEEE Tran. on Ant. and Prop., Vol. AP-25, No. 3, 304, May, 1977) 41

*2.7 Noise temperature of a lossy plate for elliptically polarized wave. (Paper 2-7, T. Y. Otoshi and C. Yeh, IEEE Trans. on Micro.Theory and Tech., Vol. 48, No. 9, 1588, Sept. 2000). 49

*2.8 Excitation of higher order modes by a step discontinuity of a circular waveguide. (Paper 2-8, C. Yeh, JPL Technical Report No. 32-496, The Jet Propulsion Laboratory, California Institute of Technology, Pasadena, California, Feb. 1, 1964) 53

3. Scattering and Diffraction 67

3.1	Scattering by elliptical dielectric cylinders	
	*3.1.1 Normal incident case. (Paper 3-1-1, C. Yeh, J. Math. Phys, Vol. 4, No. 1, 65, Jan. 1963)	76
	*3.1.2 Oblique incident case. (Paper 3-1-2, C. Yeh, J. Opt. Soc. Am., Vol. 54, No. 10, 1227, Oct. 1964)	83
	3.1.3 Exact solution to a lens problem. (Paper 3-1-3, C. Yeh, J. Opt. Soc. Am., Vol. 55, No. 7, 860, July 1965)	88
	3.1.4 Layered elliptical cylinder. (Paper 3-1-4, C. S. Kim and C. Yeh, Radio Science, Vol. 26, No. 5, 1165, Sept. 1991) .	96
*3.2	Scattering by parabolic dielectric cylinders. (Paper 3-2a, C. Yeh, J. Opt. Soc. Am., Vol.57, No. 2, 195, Feb. 1967; Paper 3-2b, C. Yeh, J. Math. and Phys., Vol. XLV, No. 2, 231, June, 1966)	108
3.3	Scattering from layered structures	
	*3.3.1 Scattering of obliquely incident waves by a radially inhomogeneous dielectric cylinder. (Paper 3-3-1, C. Yeh and P. K. C. Wang, I. Appl. Phys., Vol. 43, No. 10, 3999, Oct. 1972)	117
	3.3.2 Scattering from a cylinder coated with an inhomogeneous dielectric sheath. (Paper 3-3-2, C. Yeh and Z. A. Kaprielian, Can. J. Phys., Vol. 41, 143, Jan., 1963).	125
	*3.3.3 Scattering by an infinite cylinder coated with an inhomogeneous and anisotropic plasma sheath. (Paper 3-3-3,V. T. Rusch and C. Yeh, IEEE Trans Ant. and Prop.,Vol. AP-15, No., 3, 452, May, 1967).	134
*3.4	Interaction of microwaves with an inhomogeneous and anisotropic plasma column. (Paper 3-4, C. Yeh and W. V. T. Rusch, J. Appl. Phys., Vol. 36, No. 7, 2302, July, 1965).	140
*3.5	Scattering by arbitrarily shaped dielectric bodies. (Paper 3-5, P. Barber and C. Yeh, Appl. Opt., Vol. 14, 2864, Dec., 1975).	145
3.6	Scattering by Pruppacher-Pitter raindrops at 30 GHz (Paper 3-6, C. Yeh, R. Woo, J. W. Armstrong and A. Ishimaru, Radio Science, Vol. 17, No. 4, 757, July, 1982)	154
3.7	Scattering by focused beams.	
	*3.7.1 Single particle scattering with focused laser beams. (Paper 3-7-1, L. W. Casperson, C. Yeh, W. F. Yeung, Appl. Opt., Vol. 16, 1104, Apr., 1977)	163
	3.7.2 Rayleigh-Debye scattering with focused laser beams. (Paper 3-7-2, L. W. Casperson and C. Yeh, Appl. Opt., Vol. 17, 1637, May, 1978)	167
	*3.7.3 Scattering of focused beams by tenuous particles. (Paper 3-7-3, S. Colak, C. Yeh, and L. W. Casperson, Appl. Opt., Vol. 18, 294, Feb., 1979)	174
	*3.7.4 Scattering of sharply focused beams by arbitrarily shaped dielectric particles: an exact solution.	

(Paper 3-7-4, C. Yeh, S. Colak, and P. Barber, Appl. Opt., Vol. 21, No. 24, 4426, Dec. 1982) 183

3.7.5 Scattering of a focused beam by moving particles. (Paper 3-7-5, S. Colak and C. Yeh, Appl. Opt., Vol. 19, 256, Jan., 1980). 191

3.8 Radiative transfer approach to multiple scattering problems.

3.8.1 Matrix representation of the vector radiative-transfer heory for randomly distributed nonspherical particles. (Paper 3-8-1, A. Ishimaru and C. Yeh, J. Opt. Soc. Am. A, Vol. 1, 359, May, 1984) 198

*3.8.2 Multiple scattering calculations for nonspherical particles based on the vector radiative transfer theory. (Paper 3-8-2, A. Ishimaru, D. Lesselier, and C. Yeh, Radio Science, Vol. 19, No. 5, 1356, Sept.,1984). 204

*3.8.3 First-order multiple scattering theory for nonspherical particles. (Paper 3-8-3, A. Ishimaru, C. Yeh, and D. Lesselier, Appl. Opt.,Vol. 23, 4132, Nov., 1984). 214

3.9 Perturbation approach to the diffraction of electromagnetic waves by arbitrarily shaped dielectric obstacles.

3.9.1 3-D object. (Paper 3-9-1, C. Yeh, Phys. Rev., Vol. 135, No. 5A, A1193, Aug., 1964). 222

3.9.2 2-D object. (Paper 3-9-2, C. Yeh, J. Math. Phys., Vol. 6, No. 12, 2008, Dec., 1965). 231

4. Guided waves and waveguides. 237

4.1 Elliptical dielectric waveguides

*4.1.1 Exact solutions. (Paper 4-1-1,C. Yeh, J. Appl. Phys., Vol. 33, No. 11, 3235, Nov. 1962) 251

*4.1.2 Attenuation calculation. (Paper 4-1-2, C. Yeh, IEEE Trans. Ant. And Prop., Vol. AP-11, No. 2, 177, Mar. 1963). 261

4.1.3 Weakly guiding case. (Paper 4-1-3, C. Yeh, Opt. and Quant. Elec., Vol. 8, 43, July 1976). 269

*4.1.4 Ferrite-filled elliptical waveguides. (Paper 4-1-4a, W. E. Salmond and C. Yeh, J. Appl. Phys., Vol. 41, No. 8, 3210, July, 1970; Paper 4-1-4b, W. E. Salmond and C. Yeh, J. Appl. Phys., Vol. 41, No. 8, 3221, July, 1970). 274

4.2 Wave guidance in periodic dielectric/plasma medium.

*4.2.1 Sinusoidally stratified dielectric medium. (Paper 4-2-1, C. Yeh, K. F. Casey, and Z. A. Kaprielian, IEEE Trans. on Micro. Theo. and Tech., Vol. MTT-13, No. 3, 297, May, 1965) 291

*4.2.2 Sinusoidally stratified plasma medium . (Paper 4-2-2,

K. F. Casey, J. R. Matthes, and C. Yeh, J. Math. Phys., Vol. 10, No. 5, 891, May, 1969)....................... 297

*4.2.3 Periodic structure in integrated optics. (Paper 4-2-3, C. Elachi and C. Yeh, J. Appl. Phys., Vol. 44, 3146, Aug., 1973)....................................... 304

4.2.4 Frequency selective coupler for integrated optics systems. (Paper 4-2-4, C. Elachi and C. Yeh, Opt. Comm., Vol. 7, No. 3, 201, Mar., 1973)....................... 311

4.2.5 Stop bands for optical wave propagation in cholesteric liquid crystals. (Paper 4-2-5, C. Elachi and C. Yeh, J. Opt. Soc. Am., Vol. 63, No. 7, 840, July, 1973)........ 315

4.2.6 Mode conversion in periodically disturbed thin-film waveguides. (Paper 4-2-6, C. Elachi and C. Yeh, J. Appl. Phys., Vol. 45, No. 8, 3494, Aug. 1974)......... 318

*4.2.7 Transversely bounded DFB lasers. (Paper 4-2-7, C. Elachi, G. Evans, and C. Yeh, J. Opt. Soc. Am., Vol.65, No. 4, 404, Apr., 1973)...................... 324

4.2.8 Transients in a periodic slab: coupled waves approach. (Paper 4-2-8, C. Elachi, D. L. Jaggard, and C. Yeh, IEEE Trans. Ant. and Prop., Vol. AP-23, No. 3, 352, May, 1975)....................................... 333

*4.3 An efficient 4 x 4 matrix method to compute the propagation characteristics of radially stratified fibers. (Paper 4-3, C. Yeh and G. Lindgren, Appl. Opt., Vol. 16, 483, Feb., 1977, submitted for publication, Feb. 27, 1976)..................... 340

4.4 Finite element approach

*4.4.1 Arbitrarily shaped inhomogeneous optical fiber or integrated optical waveguides (Paper 4-4-1, C. Yeh, S. B. Dong, and W. Oliver, J. Appl. Phys., Vol. 46, No. 5, 2125, May, 1975)........................... 351

*4.4.2 Single-mode optical waveguides. (Paper 4-4-2, C. Yeh, K. Ha, S. B. Dong, and W. P. Brown, Appl. Opt., Vol.18, No. 10, 1490, May, 1979)........... 356

4.5 Beam propagation method (BPM) or Scalar wave FFT method

*4.5.1 Propagation of truncated Gaussian beams in multimode fiber guides. (Paper 4-5-1, C. Yeh, L. Casperson, and B. Szejn, J. Opt. Soc. Am., Vol. 68, No. 7, 989, July, 1978, submitted for publication Jan 21, 1977)....... 371

*4.5.2 Multimode or single mode inhomogeneous fiber couplers. (Paper 4-5-2, C. Yeh, W. P. Brown, and R. Szejn, Appl. Opt., Vol. 18, 489, Feb., 1979).......... 375

*4.5.3 On multimode or single mode fiber couplers, tapers, and horns. (Paper 4-5-3, C. Yeh, J. of Ceramics, 1980)....... 382

*4.5.4 On weakly guiding single-mode optical waveguides (Paper 4-5-4, C. Yeh and F. Manshadi, J. Light. Tech., Vol. LT-3, No. 1, 199, Feb., 1985).................... 394

4.6 Nonlinear optical fibers

*4.6.1 Pulse shepherding in nonlinear fiber optics. (Paper 4-6-1, C. Yeh and L. Bergman, J. Appl. Phys. Vol. 80, No. 6, 3174, Sept. 1996) 401

4.6.2 Experimental verification of the pulse shepherding concept in dispersion shifted single-mode fiber for bit-parallel wavelength links. (Paper 4-6-2, L. Bergman, J. Morookian, C. Yeh, and S.Monacos, Proc. Inter. Conf. Massively Parallel Proc. Using Opt. Interconnections, IEEE Computer Soc., Montreal, Can., p 25, June 22-24, 1997). 406

*4.6.3 Enhanced pulse compression in a nonlinear fiber by wavelength division multiplexed optical pulse, (Paper 4-6-3, C. Yeh and L. Bergman, Phys. Rev. E, Vol. 57, No. 2, 2398, Feb., 1998). 412

*4.6.4 An all-optical long-distance multi-Gbytes/s bit-parallel WDM single-fiber link. (Paper 4-6-4, L. Bergman, J. Morookian, and C. Yeh, J. Light. Tech., Vol. 16, No. 9, 1577, Sept., 1998) 419

*4.6.5 Generation of time-aligned picoseconds pulses on wavelength division multiplexed beams in a nonlinear fiber. (Paper 4-6-5, C. Yeh, L. Bergman, J. Morookian, and S. Monacos, Phys. Rev. E, Vol. 57, No. 5, 5135, May, 1998). 425

*4.6.6 Existence of optical solitons on wavelength division multiplexed beams in a nonlinear fiber .(Paper 4-6-6, C. Yeh and L. Bergman, Phys. Rev. E, Vol. 60, No. 2, 2306, Aug., 1999). 430

4.7 Optical waveguide theory .(Paper 4-7, C. Yeh, IEEE Trans. Circuits and Systems, Vol. CAS-26, No. 12, 1011, Dec. 1979) 433

*4.8 Dynamic three-dimensional TLM analysis of microstripline on anisotropic substrate. (Paper 4-8, G. E. Mariki and C. Yeh, IEEE Trans. Micro. Theo. And Tech., Vol. MTT-33, No. 9, 789 Sept., 1985). 442

4.9 Modeling of star-shaped and parallel wire carrier distribution systems. (Paper 4-9, C. Yeh, F. Manshadi and G. Mariki, and W. P. Brown, J. The Franklin Inst., Vol. 305, No. 2, 67, Feb., 1978) . . 453

*4.10 Leaky waves in a heteroepitaxial film. (Paper 4-10, D. B. Hall and C. Yeh, J. Appl. Phys., Vol. 44, No. 5, 2271, May, 1973). 464

4.11 Terahertz and millimeter dielectric waveguides.

*4.11.1 Dielectric waveguides: An optimum configuration for ultra-low loss millimeter/submillimeter dielectric waveguide. (Paper 4-11-1, C. Yeh , F. I. Shimabukuro, and J. Chu, IEEE Trans. Micro. Theo. And Tech., Vol. 38, No. 6, 691, June, 1990) 468

*4.11.2 Communication at millimetre-submillimetre wavelengths

using a ceramic ribbon. ((Paper 4-11-2, C. Yeh, F. Shimabukuro, P. Stanton, V. Jamnejad, W. Imbriale, and F. Manshadi, Nature, Vol. 404, 584, Apr., 2000). 480

*4.11.3 Low-loss terahertz ribbon waveguides. (Paper 4-11-3, C. Yeh, F. Shimabukuro, and P. H. Siegel, Appl. Opt., Vol. 44, No. 28, 5937, Oct., 2005) 485

*4.11.4 Attenuation measurement of very low loss dielectric waveguides by the cavity resonator method applicable in the millimeter/submillimeter wavelength range. (Paper 4-11-4, F. Shimabukuro and C. Yeh, IEEE Trans. Micro. Theo. And Tech., MTT-36, 1160, July, 1988). 495

4.12 Wave propagation in hot plasma waveguides with infinite magnetostatic fields. (Paper 4-12, H. H. Kuehl, G. E. Stewart, and C. Yeh, Phys. Fluids, Vol. 8, No. 4, 723, Apr., 1965) 502

4.13 Distribution networks and electrically controllable couplers for integrated optics. (Paper 4-13, C. Elachi and C. Yeh, Appl. Opt., Vol.13, 1372, June, 1974). 508

*4.14 Highly sensitive quantum well infrared photodetectors (Paper 4-14, T. Cwik and C. Yeh, J. Appl. Phys., Vol. 86, No. 5, 2779, Sept., 1999). 512

4.15 How does one induce leakage in an optical fiber link? (Paper 4-15, C. Yeh and A. Johnston, Agard Conferences No. AGARD-CP-219, 26-1, May, 1977) 518

5. Antennas, Radiators, and Propagation 525

5.1 Antennas and radiators

*5.1.1 Dielectric coated prolate spheroidal antennas. (Paper 5-1-1, C. Yeh, J. Math. and Phys., Vol. XLII, No. 1, 68, Mar., 1963) 530

5.1.2 Radiation from an axially slotted infinite cylinder coated with an inhomogeneous dielectric sheath. (Paper 5-1-2, C. Yeh and Z. A. Kaprielian, Brit. J. Appl. Phys., Vol. 14, 677, Oct., 1963). 540

*5.1.3 Dyadic Green's function for a radially inhomogeneous spherical medium. (Paper 5-1-3, C. Yeh, Phys. Rev. Vol. 131, No. 5, 2350, Sept. 1963) 545

*5.1.4 External field produced by an arbitrary slot on a ribbon coated with a penetrable sheath. (Paper 5-1-4, C. Yeh, Appl. Sci. Res., Sec. B, Vol. 10, 417, June, 1963). 549

*5.1.5 Application of Sommerfeld's complex-order wavefunctions to an antenna problem. (Paper 5-1-5, C. Yeh, J. Math. Phys., Vol. 5, No. 3, 344, Mar., 1964). . . . 567

*5.1.6 Electromagnetic radiation from an arbitrary slot on a conducting cylinder coated with a uniform cold plasma

sheath with an axial static magnetic field. (Paper 5-1-6, C. Yeh, Can. J. Phys, Vol.42, 1369, July, 1964).......... 574

*5.1.7 Radiation through an inhomogeneous, magnetized plasma sheath. (Paper 5-1-7, C. Yeh and W. V. T. Rusch, IEEE Trans. Ant. and Prop., Vol. AP-15, No. 2, 328, Mar., 1967).. 587

5.2 Propagation

*5.2.1 Enhanced focal-plane irradiance in the presence of thermal blooming. (Paper 5-2-1, C. Yeh, J. E. Pearson, and W. P. Brown, Appl. Opt., Vol.15, 2913, Nov., 1976)........... 588

*5.2.2 Propagation of laser beams having an on-axis null in the presence of thermal blooming. (Paper 5-2-2, J. E. Pearson, C. Yeh, and W. P. Brown, J. Opt. Soc. Am., Vol. 66, No. 12, 1384, Dec. 1976)........................... 592

*5.2.3 Earth-to-deep-space optical communication system with adaptive tilt and scintillation correction by use of near-Earth relay mirrors. (Paper 5-2-3, J. W. Armstrong, C. Yeh, and K. E. Wilson, Opt. Lett., Vol. 23, No. 14, 1087, July, 1998).. 597

6. Moving medium 601

6.1 Moving dielectric half-space or slabs

*6.1.1 Reflection and transmission of electromagnetic waves by a moving dielectric medium. (Paper 6-1-1, C. Yeh, J. Appl. Phys., Vol. 36, No. 11, 3513, Nov. 1965)....... 608

*6.1.2 Brewster angle for a dielectric medium moving at relativistic speed. (Paper 6-1-2, C. Yeh, J. Appl. Phys., Vol. 38, No. 13, 5194, Dec., 1967).................... 613

*6.1.3 Reflection and transmission of electromagnetic waves by a moving dielectric slab. (Paper 6-1-3, C. Yeh and K. F. Casey, Phys. Rev., Vol.144, No. 2, 665, Apr. 1966).. 620

*6.1.4 Reflection and transmission of electromagnetic waves by a moving dielectric slab II, Parallel polarizations. (Paper 6-1-4, C. Yeh, Phys. Rev., Vol. 167, No. 3, 875, Mar.,1968).. 625

*6.1.5 Reflection and transmission of electromagnetic waves by a moving plasma medium. (Paper 6-1-5, C. Yeh, J. Appl. Phys., Vol.37, No. 8, 3079, July, 1966)................. 628

*6.1.6 Reflection and transmission of electromagnetic waves by a moving plasma medium II, Parallel polarizations. (Paper 6-1-6, C. Yeh, J. Appl. Phys., Vol.38, No. 7, 2817, June, 1967).. 632

*6.1.7 Reflection from a dielectric-coated moving mirror. (Paper 6-1-7, C. Yeh, J. Opt. Soc. Am., Vol.57, No. 5,

657, May, 1967) 635
6.2 Guided wave on moving structure.
*6.2.1 Propagation along moving dielectric waveguides. (Paper 6-2-1, C. Yeh, J. Opt. Soc. Am., Vol.58, No. 6, 767, June, 1968). 640
*6.2.2 Wave propagation on a moving plasma column. (Paper 6-2-2, C. Yeh, J. Appl. Phys., Vol. 39, No. 13, 6112, Dec., 1968) 644
6.3 Scattering by moving plasma column or sheath.
*6.3.1 Scattering of obliquely incident microwaves by a moving plasma column. (Paper 6-3-1, C. Yeh, J. Appl. Phys., Vol. 40, No. 13, 5066, Dec., 1969). 646
*6.3.2 Diffraction of waves by a conducting cylinder coated with a moving plasma sheath. (Paper 6-3-2, C. Yeh, J. Math. Phys., Vol. 11, No. 1, 99, Jan., 1970). 655
*6.4 Radiation from an aperture in a conducting cylinder coated with a moving plasma sheath. (Paper 6-4, K. F. Casey and C. Yeh IEEE Trans. Ant. and Prop., Vol. AP-17, No. 6, Nov., 1969) 661
6.5 A proposed method of shifting the frequency of light. (Paper 6-5 C. Yeh, Appl. Phys. Lett., Vol. 9, No. 5, 184, Sept., 1966) 666

7. **Classical radiation from charged** particles. 669

*7.1 Čerenkov radiation in inhomogeneous periodic media. (Paper 7-1, K. F. Casey, C. Yeh, and Z. A. Kaprielian, Phys. Rev., Vol.140, No. 3B, B768, Nov., 1965). 672
*7.2 Transition radiation in a periodically stratified plasma. (Paper 7-2, K. F. Casey and C. Yeh, Phy. Rev. A, Vol. 2, No. 3, No. 3, 810Sept., 1970) 680
*7.3 Diffraction radiation from a charged particle moving through a penetrable sphere. (Paper 7-3, R. Pogorzelski and C. Yeh, Phys. Rev. A, Vol. 8, No. 1, 137, July, 1973). 689
*7.4 On the Čerenkov threshold associated with synchrotron radiation in a dielectric media. (Paper 7-4, R. J. Pogorzelski, C. Yeh, and K. F. Casey, J. Appl. Phys., Vol. 45, No. 12, 5251, Dec., 1974). 697
*7.5 Synchrotron-diffraction radiation spectra in the presence of a penetrable sphere. (Paper 7-5, R. J. Pogorzelski and C. Yeh, J. Appl. Phys., Vol. 46, No., 2, 643, Feb. 1975). 702

8. **High data-rate fiber networks and low-noise optics** 709

*8.1 RATO-Net: a random-access protocol for unidirectional ultra-high

speed optical fiber network. (Paper 8-1, C. Yeh, M. Lin, M. Gerla, and P. Rodrigues, J. Lightwave Tech., Vol. 8, No. 1, Jan., 1990). 712

8.2 U-Net: a unidirectional fiber bus network. (Paper 8-2, M. Gerla, P. Rodrigues, and C. Yeh, FOC-LAN 84, Sept., 1984). 724

*8.3 Token-based protocols for high-speed optical fiber networks. (Paper 8-3, M. Gerla, P. Rodrigues, and C. Yeh, J. Lightwave Tech., Vol. LT-3, No. 3, 449, June, 1985). 729

*8.4 Interconnection of fiber optics local area networks (FOLAN). (Paper 8-4, M. Gerla and C. Yeh, SPIE O-E/Fibers '87, Symp. On Fiber Optics and Integrated Optoelectronics, San Diego, 835 53, May, 1987). 749

8.5 Buzz-net: a hybrid random access/virtual token local network. (Paper 8-5, M. Gerla, P. Rodrigues, and C. Yeh, June, 1983) 760

8.6 Low-noise fiber optics receiver with super-beta bipolar transistors. (Paper 8-6, S. G. Knorr, O. Kaldirim, and C. Yeh, Fiber and Int. Opt., Vol. 1, No. 4, 76, Dec., 1978) 764

9. Acoustic waves 783

*9.1 Scattering by elliptical fluid cylinders. (Paper 9-1, C. Yeh, J. Acoust. Soc. Am., Vol. 54, No. 3, 760, Jan., 1973) 786

*9.2 Diffraction of sound waves by penetrable oblate spheroid. (Paper 9-2, C. Yeh, Ann. Physik, 7. Folge, Bd. 13, Heft 1-2, 53, Jan., 1964) 797

*9.3 Scattering by penetrable prolate spheroid. (Paper 9-3, C. Yeh, J. Acoust. Soc. Am., Vol. 42, No. 2, 518, Aug., 1967) 806

9.4 Scattering by liquid-coated prolate spheroids. (Paper 9-4, C. Yeh, J. Acoust. Soc. Am., Vol. 46, No. 3 (Part 2), 797, Sept., 1969) 810

*9.5 Reflection and transmission of sound waves by a moving fluid layer. (Paper 9-5, C. Yeh, J. Acoust. Soc. Am., Vol. 43, No. 6, 1454, June, 1968) 815

*9.6 Diffraction by sound waves by a moving fluid cylinder. (Paper 9-6, C. Yeh, J. Acoust. Soc. Am., Vol. 44, No. 5, 1216, Nov., 1968) 817

*9.7 Finite element approach to solve acoustic waveguide problems. (Paper 9-7, W. A. Oliver, S. B. Dong, and C. Yeh, J. Acoust. Soc. Am., Vol. 69, No. 1145, Jan.,1981) 821

1

INTRODUCTION

The advent of increasing computing power in the past fifty years has revolutionized the way scientific and technical problems are solved. Prior to that, problems were usually solved by the exact, perturbation, approximation, variation, or asymptotic approach. Now, most technical problems, including those in electromagnetics, are solved routinely by numerical means with the use of computers. This book presents the contributions of Yeh and his collaborators (Y&C) in this transition period in which numerical approaches became increasingly utilized in solving electromagnetic problems.

It is of interest to learn how and when this transition took place and to understand what were the foundations on which the numerical approaches were based. The foundations were the analytical solutions of canonical problems. This is not to minimize the importance of using numerical means to solve complicated problems, however.

The research described here also went through a similar transition. The importance of finding analytic solutions for canonical problems persists even today and for the foreseeable future; thus, a great deal of the research effort was spent in finding analytic solutions to canonical problems. Initiation of our research is usually based on our desire to answer certain relevant and intriguing questions, to understand and explore new phenomena, or to find innovative ways to solve problems. In addition to canonical analytic solutions, Y&C were also instrumental in being the very first to develop some of today's very popular and useful numerical techniques in electromagnetic waves.

Below are listed a few of the highlights.

On classic canonical problems

- Elliptical dielectric waveguides (Papers 4-1-1 through 4-1-4)—These papers yield the analytic solution to a classic fundamental problem on noncircular dielectric waveguides. (Since the angular wave functions in spherical or circular cylindrical coordinates are only functions of the angular coordinate, boundary conditions can be satisfied in a straightforward manner at the bounding surfaces in these coordinate systems. In contrast, the angular wave functions in elliptical or parabolic coordinates are functions not only of the angular coordinate but also of the constitutive parameters of the medium to which they apply. A unique innovative way must be found.)

- Scattering by elliptical dielectric cylinders (Papers 3-1-1 through 3-1-4)—These papers yield the analytic solution to a classic fundamental problem on scattering by noncircular dielectric cylinders. (The same method that was used to match the boundary conditions for the elliptical dielectric waveguide case was used to resolve the difficulty encountered here.)

- Reflection and refraction of a plane wave by a moving dielectric or plasma slab (Papers 6-1-1 through 6-1-7)—These papers provide the first application of special relativity to solve the classic problem of the reflection and transmission of an incident electromagnetic wave by moving plane dielectric slabs. (Although this problem was known to be a classic one, it was surprising to learn that its solution could not be found in the literature or in any textbooks.)

- Wave guidance in periodic dielectric/plasma medium (Papers 4-2-1 through 4-2-8)—These papers provide the exact analytic solution to the classic fundamental problem of wave propagation in a periodic medium. (Brillioun was a pioneer in the study of wave propagation in periodic structures or medium. Here, for the first time, full vector wave solutions for the TM modes were given. Applications can be found in modern research on photonic structures.)

- Scattering by parabolic dielectric cylinders (Papers 3-2a and b)—These papers give an analytic solution to another classic noncircular cylinder problem. (This is another classic canonical problem. The same mathematical technique that we developed for the elliptical dielectric cylinder was used successfully to find the analytic solution to this problem.)

- Scattering by focused beam on an isolated single particle (Papers 3-7-1 through 3-7-5)—These papers provide the exact solution to the problem of scattering of an arbitrarily shaped dielectric particle by a focused beam. (To obtain the scattering signature of a certain particle or to maneuver that particle in an ensemble of particles in their natural environment is a problem of interest in many disciplines. The scattering signature may be used to identify the particle while maneuvering it may be used to steer the particle. A focused beam could be used to attain this goal. Using this setup, one may calculate the forces acting on the particle, which has direct applications on particle trapping and on predicting the action of an optical tweezer.)

- Radiation from a dipole in a radial inhomogeneous spherical medium (Paper 5-1-3)—This paper gives the solution for a classic problem of a radiator in a radial inhomogeneous medium. (The solution of dipole radiation in a homogeneous spherical medium is a well-known classic textbook topic. Here, we find the solution for the problem of dipole radiation in a radial inhomogeneous spherical medium.)

- Dielectric coated prolate spheroidal antenna (Paper 5-1-1)—This paper gives the solution to a classic fundamental problem. (Representation of conducting prolate spheroidal antenna as a linear antenna is a well-known method of solving this classic fundamental problem. What is unknown is the effect of a dielectric coating on this linear antenna. A dielectric coated prolate spheroid can be used to approximate a dielectric coated linear antenna. The difficulty in matching the boundary conditions at the boundary interface because the angular function is a function of the angular coordinate and the constitutive parameters of the medium in which the function applies can be resolved by the same technique that we developed earlier for the elliptical dielectric waveguide case.)

- Scattering of sound waves by elliptical fluid cylinder (Paper 9-1)—This paper gives the solution to a classic fundamental problem. (The same technique we introduced earlier in solving the elliptical dielectric waveguide problem can be used to solve this problem.)

- Scattering of acoustic waves by a prolate or an oblate liquid spheroid (Papers 9-2 through 9-4)—These papers give the solution to a classic fundamental problem. (The same technique we introduced earlier in solving the elliptical dielectric waveguide problem can be used to solve this problem.)

On Applied Electromagnetic Wave Theory

- Conditions for the validity of scalar-wave approach (Paper 2-2)—This paper gives the fundamental limitations of the scalar-wave approach in solving the guided wave problems. (It is much easier to deal with scalar-wave equations rather than vector wave equations. We must, however, know the conditions under which this scalar-wave approach is valid.)

- A new proof of electromagnetic boundary conditions (Paper 2-4)—This paper provides the fundamental proof for the important boundary conditions that state that, for time-varying fields, satisfaction of the continuity condition for the tangential **E** and **H** fields across the boundary of two dielectrics means the satisfaction of the continuity condition for the normal **D** and **B** fields, but not the converse, and, for static fields, the boundary conditions for the tangential **E** and normal **D** for dielectric material or for the tangential **H** and normal **B** for magnetic material, must be satisfied simultaneously. (It was surprising to learn that this fundamental proof, which should have been taught in the same way the usual boundary conditions were taught, was nowhere to be found, even after more than 100 years since Maxwell's equations were established.)

- Accuracy of coupled-mode theory for guided waves (Paper 2-3)—This paper provides the fundamental limitations of the coupled-mode approximation. (The coupled-mode theory, based on the perturbation theory, had been used extensively, yet, the exact fundamental limitations for this approximation were unknown. It was important to learn these limitations.)

- Correct way to calculate attenuation for multimode waveguides (Paper 2-5)—This paper fulfills a needed presentation on an important topic missing in all well-known textbooks on microwaves. (The traditional way to find the loss of multimode propagation in a waveguide is simply to sum up the loss of each mode proportional to the amount of power each one carries. That this was not always correct was highlighted by a measurement made for a dual-mode horn at JPL. The correct solution for this problem was sought and is given here.)

On Nonlinear Optical Fibers

- Pulse shepherding in nonlinear fiber (Papers 4-6-1 through 4-6-5)—These papers show that, using a shepherd pulse, it is possible to align, compress, or generate co-propagating WDM pulses in flight. (It is known that the value of a nonlinear fiber's constitutive parameter depends on the strength of the propagating pulses. This interaction will then modify the shape of these pulses.)

- Discovery of WDM optical solitons in fiber (Paper 4-6-6)—This paper provides the ultimate data capacity of an optical fiber. (It is known that the nonlinear cross-phase modulation effect can distort and contaminate WDM data pulses in a single fiber. The question is whether there can be WDM optical solitons.)

On Classical Radiation from Moving Charged Particles

- Radiation from a fast-moving particle in an inhomogeneous periodic medium (Papers 7-1 through 7-5)—These papers provide the classic fundamental solutions for the generation of Čerenkov and transition radiation for a moving charged particle in a continuously inhomogeneous periodic medium. (Although the four distinct methods of generating electromagnetic radiation by classic—not quantum—means were well known, the complex radiation that could be generated when two or more of these situations existed simultaneously had not been studied. Very interesting and fundamental results were found.)

On Low-Loss Terahertz Waveguides

- Low-loss millimeter-submillimeter (terahertz) waveguides (Papers 4-11-1 through 4-11-4)—These papers describe the means to guide terahertz waves at a loss factor that is 100 times less than that for the next best guide. (Unlike the optical waveguide case, the luxury of eliminating unwanted contaminants from the guiding material to attain ultra-low-loss guiding does not exist in this frequency band. This is because for terahertz waves, the intrinsic molecular vibrational absorption characteristics of the material cannot be eliminated. A new way must be found to make a low-loss waveguide for terahertz wave.)

On Laser Beam Propagation in the Atmosphere

- High-energy laser beam propagation with thermal blooming in the atmosphere (Papers 5-2-1 and 5-2-2)—These papers show a new way to significantly increase the target intensity of a laser beam propagating through an atmosphere with turbulence and thermal blooming. (The presence of thermal blooming in the atmosphere tends to limit the intensity of power that could be delivered to a target. It is of interest to find an innovative way to overcome this.)

- Earth-to-deep-space optical communication with turbulence in the atmosphere (Paper 5-2-3)—This paper provides a new way to communicate with a spacecraft in space or on Mars using adaptive optics on Earth. (It is of paramount importance that a high-data-rate communication link be made available between Earth and spacecraft or between Earth and the other planets. Optical link is a viable solution; however, Earth's turbulent atmosphere is of concern. Learning how to overcome this is a worthwhile pursuit.)

On Interaction of Waves with Moving or Stationary Plasma Column

- Scattering of waves by an inhomogeneous and anisotropic plasma cylinder (Paper 3-4)—This paper gives the only known fundamental solution to this problem. (Only a few cases exist in which an analytic solution can be found for the vector wave equation in a complex medium. This was one of them.)

- Scattering from moving plasma column (Papers 6-3-1 and 6-3-2)—These papers give the first relativistic solution for scattering from moving plasma cylinder. (It is always of great intellectual interest to solve problems involving relativity. This problem could be solved only when the special theory of relativity was used.)

- Moving dielectric or plasma waveguides (Papers 6-2-1 and 6-2-2)—These papers give the first relativistic guided wave solution for moving dielectric waveguide. (It is curious to learn how a wave can be guided by a moving material column. What would happen to the guided wave when the material column was moving faster than the velocity of light in the medium?)

On Acoustic Wave Interaction with a Moving Fluid Cylinder

- Scattering of acoustic waves by a moving fluid cylinder (Paper 9-6)—This paper gives the solution to a classic problem. (What are the scattering characteristics of acoustic waves by a jet exhaust or by streaming fluid from a nozzle? How is the scattered field affected by the movement of the material?)

On High-Speed Fiber Network

- Gigabit rate, random access fiber network (Papers 8-1 and 8-2)—This paper provides a workable random-access, fair, and bounded-delay-access-to-all-stations protocol that could replace the popular Ethernet at ultra-high- speed transmission. (As the demand for ultra-high data rate is increased, Ethernet protocol fails when the signal transition time between stations is longer than the packet data length because data collision can occur. New protocols and architectures must be found.)

In spite of the importance of canonical solutions in providing fundamental understanding, one is nevertheless limited by the availability of analytic techniques to solve the more complicated and more realistic problems. One must resort to using numerical techniques. Successful development and application of certain numerical techniques can lead the way to solving problems that could not be solved by traditional analytic means. Being a pioneer in successfully developing a given numerical technique is therefore important because one finds a way to explore a new family of unsolved problems. Listed below are the techniques that were first developed by Y&C:

On the Development of Numerical Methods for Electromagnetic Waves

- The 4 x 4 matrix approach to solve layered structures (Paper 4-3)—This paper shows how a $4n$ x $4n$ matrix operation for an n-layered problem can be reduced to an n(4 x 4) matrices operation, thereby enabling the solution of an n-layered problem without any restriction on the number of layers due to finite computing time and computing capacity. This was demonstrated by an n-layered cylindrical structure. Of course, the simpler n-layered planar structure that yields a $2n$ x $2n$ matrix can be done in the same manner. This technique was first introduced by Y&C. (The usefulness and importance of this approach can be appreciated for its application to reflection and refraction from multilayered planar, cylindrical, or spherical structures in electromagnetics, mechanics, hydrodynamics, earthquake analyses, etc. .)

- Beam propagation method to treat linear or nonlinear guided wave problems (Papers 4-5-1 through 4-5-4)—These papers show how to obtain the evolution of a scalar field as it propagates along a linear or nonlinear dielectric structure with direct applications in WDM beams in optical fiber, integrated optics, etc.. (The fast Fourier transform scalar-wave method, otherwise called the beam propagation method, had been used successfully to calculate laser beam propagation in the atmosphere for many years. In mid-1975, Y&C were the first to successfully adopt this technique to wave propagation in fiber structure. The special advantage of this technique was that it provided the evolution of the complete fields in the structure, rather than the traditional method of dealing with each propagating mode. Thus, the single-mode structure or multimode structure can be treated in the same manner and the structure and medium may vary along the direction of propagation.)

- Finite-element method to solve arbitrarily shaped and inhomogeneous dielectric waveguide problems (Papers 4-4-1 and 4-4-2)—These papers provide an exact numerical solution to the vector wave equation with specific application to arbitrarily shaped and inhomogeneous fiber or integrated optic guide. (In the structural engineering field and civil engineering field, the finite-element method had been used successfully for many years to analyze very complex mechanical or earthen structures. Y&C were the first to adopt the finite-element method to solve problems dealing with electromagnetic dielectric waveguide structures.)

- Transmission line method (TLM) to treat guided wave problems (Paper 4-8)—This paper shows a powerful numerical technique to solve guided wave problems in inhomogeneous and/or anisotropic medium, with applications in optical fibers, integrated optics, and microstrip lines. (Schelkunoff first derived a set of transmission line equations to represent Maxwell's equations. The representation was an exact one. By numerically solving this set of transmission line equations, one may obtain all fields of Maxwell's equations. This approach may be used to numerically solve many guided wave problems.)

- Scattering of electromagnetic wave by arbitrarily shaped dielectric bodies (the extended boundary condition method, EBCM) (Papers 3-5 and 3-7-4)—These papers provide the exact full wave solutions to the scattering problems dealing with dielectric bodies that are spherical or nonspherical shapes. (Scattering by spherical structure is the only 3-D problem that possesses a complete analytic solution, i.e., the Mie solution. The numerical method must be used to solve problems involving other 3-D shapes. The EBCM was first used by Waterman to solve the problem of scattering by 3-D-conducting bodies. Y&C's solution on 3-D dielectric bodies was an extension of his work.)

- Two-point boundary condition method to treat scattering of waves by radially inhomogeneous dielectric cylinders or spheres (Paper 3-3-1)—This paper yields the full wave solution to a problem that could not be solved analytically. (Although this problem could be solved by dividing the cylindrical inhomogeneous medium into layers of cylindrical homogeneous medium whose analytic solution was well known, formulating this problem in terms of two-point boundary condition method could provide a simplified way to solve this problem.)

The successful development of these efficient and versatile numerical techniques can provide a much expanded horizon for analyzing electromagnetic problems. In fact, computer simulation of an actually complex system can now be performed. Some of these numerical-technique programs have been successfully marketed as commercial products by others.

As noted in the preface, this book is divided into 9 chapters:

Chapter 1. Introduction
Chapter 2. Applied Electromagnetic Theory
Chapter 3. Scattering and Diffraction
Chapter 4. Guided Waves and Waveguides
Chapter 5. Antennas, Radiation, and Propagation
Chapter 6. Moving Medium
Chapter 7. Classical Radiation from Moving Particles
Chapter 8. High-Data-Rate Fiber Networks and Low-Noise Fiberoptics Receivers
Chapter 9. Acoustic waves

Thus, this book provides a glimpse at the transition from analytical to numerical electromagnetics through the contributions by Y&C.

For the reader to have easy access to these contributions, original papers are included in this book. The fact that most of the work was focused on fundamental canonical solutions is important. Finding the varied and broad applications related to electromagnetic waves, as shown here, reveals the future potential and importance of electromagnetics research. Recent specific examples are the intense research activities in plasmons for nanotechnology and wave interaction with metamaterials.

2

APPLIED ELECTROMAGNETIC THEORY

To gain new and useful insight into the theory of electromagnetic waves, a number of specific investigations were carried out. These results yielded further understanding of the theory and provided more efficient means for solving electromagnetic problems.

While teaching the derivation of boundary conditions in elecromagnetics, using "pill-box" and "line-integrals" one frequently states that, for a boundary between two dielectrics, satisfying the continuity conditions for the tangential electric and magnetic fields at the boundary implies that the boundary conditions for the normal component of the electric displacement field and the normal component of the magnetic induction flux field are automatically satisfied whereas the converse is not true. Furthermore, the above statement is true only for time-varying fields and not for static fields. Why? The proof cannot be found anywhere except in Yeh's paper and our textbook. This proof is addressed in Paper 2-4.

It is much easier to solve the scalar wave equation than the vector wave equation. Before one attempts to find the solutions for the scalar wave equation, it is important for one to know the conditions under which these solutions may be applied to practical fiber problems. These conditions are clarified in Paper 2-2.

When applying the perturbation approach of coupled mode theory for fiber waveguide couplers, no discussion on the regions of validity for this approximate method could be found and so this is discussed in Paper 2-3.

While working on beam waveguide systems, the question of how to calculate the attenuation of waves in a multimode oversize waveguide was asked. The simple and rather obvious answer was that because the total power was equal to the sum of the power for each different propagating mode, then the total power loss was also the sum of the loss for each various mode, with the loss of each mode calculated according to the attenuation constant for each mode propagating separately. This answer was wrong, and the correct answer is given in Paper 2-5.

It is known that a complete set of near-zone field that contains reactive as well as real power can be used to reconstruct the far-zone field, which contains only real radiation power. Can one reconstruct the near-zone field from a complete set of the far-zone field? The question is answered in Paper 2-6.

When the waveguide resonant cavity method was used to measure the attenuation factor, α, of the dominant hybrid mode on a dielectric waveguide, it was necessary to use the Q factor to calculate α for the hybrid mode. This formula had to be derived. Paper 2-1 gives the relationship between α and Q for the hybrid mode.

In some situations, it is necessary to calculate noise temperature of a lossy flat-plate reflector. Reflector losses can be due to metallic surface resistivity and multilayer dielectric sheets, including thin layers of plating, paint, and primer on the reflector surface. The incident wave is elliptically polarized, which is general enough to include linear and circular polarizations as well. The derivations show that the noise temperature for the circularly polarized wave case is simply the average of those for perpendicular and parallel polarizations. This is discussed in Paper 2-7.

It was found that exciting an antenna dish with an appropriate multimode horn can produce a radiation pattern with lower side lobes. How to excite this multimode horn is the subject of investigation in Paper 2-8.

2.1 A relation between *α* and *Q*. (Paper 2-1)

A relation between α and Q is derived. The symbol α is the attenuation constant of a single mode propagating in a waveguide, and Q is the quality factor of a resonance by shorting both ends of this waveguide. This derived relation is valid for TEM, TE. TM, or HE modes.

2.2 Scalar-wave approach for single-mode inhomogeneous fiber problems.(Paper 2-2)

It has generally been accepted that accurate results may be obtained using the scalar-wave approach to solve problems dealing with inhomogeneous single-mode guided wave structures. The problem of the applicability of the scalar-wave approach to obtain wave-propagation characteristics in single-mode fiber or integrated optical circuit guides with inhomogeneous index profiles is examined in this paper. It is shown that if certain limiting conditions are satisfied, the scalar-wave approach will yield valid results for single-mode structures. These limiting conditions are usually satisfied by many practical single-mode inhomogeneous fibers of integrated optical circuits (IOC) structures.

2.3 Accuracy of directional coupler theory in fiber or integrated optics applications. (Paper 2-3)

Based on the solution of a basic canonical problem, the possible margin of error on the predicted coupling length according to coupled-mode theory as applied to fiber and integrated optical guides is inferred. It was found that coupling length obtained according to the coupled-mode theory is usually accurate to within 20% of the actual value, provided that the frequency of operation is above the cutoff frequency of the antisymmetric mode of the coupled structure.

2.4 Boundary conditions in electromagnetics. (Paper 2-4)

A proof is given to show that, for time-varying fields, satisfying the boundary conditions on the tangential electric and magnetic fields across the boundary surface of two different material media means that the boundary conditions on the normal magnetic-induction vector and electric-displacement vector are automatically satisfied while the converse is not true. This statement is not true for static fields.

2.5 Power loss for multimode waveguides and its application to beam-waveguide system. (Paper 2-5)

The conventional way of expressing power loss in decibels per meter for a multimode waveguide system with finite-wall conductivity (such as a beam-waveguide (BWG) system with a protective shroud) can be incorrect and misleading. The power loss (in decibels) for a multimode waveguiding system is, in general, not linearly proportional to the length of the waveguide. New power-loss formulas for multimode systems are derived in this paper for arbitrarily shaped conducting waveguide tubes. In these formulas, there are factors such as $[\exp(jx) - 1]/(jx)$, where $x = (\beta_a - \beta_b)l$, with β_a and β_b being the propagation constants of the different propagating modes and l being the distance from the source plane to the plane of interest along the guide. For a large BWG supporting many propagating modes, β_as are quite close to β_bs; thus, the mode-coupling terms remain important for a very long distance from the source plane. The multimode power-loss formula for a large circular conducting tube has been verified by experiments. The formula was also used to calculate the additional noise temperature contributions due to the presence of a protective shroud surrounding a millimeter-wave BWG system.

2.6 Computer reconstruction of near-zone field from given far-zone data of two-dimensional objects. (Paper 2-6)

Using a complete set of cylindrical vector wave functions, it is shown that an accurate near-zone electromagnetic field can be easily reconstructed from given far-zone data of two-dimensional objects. The validity of this approach is shown by the comparison of the reconstructed results with those obtained according to the exact solution of a two-dimensional slit problem.

2.7 Noise temperature of a lossy flat-plate reflector for the elliptically polarized wave-case. (Paper 2-7)

This short paper presents the derivation of equations necessary to calculate noise temperature of a lossy flat-plate reflector. Reflector losses can be due to metallic surface resistivity and multilayer dielectric sheets, including thin layers of plating, paint, and primer on the reflector surface. The incident wave is elliptically polarized, which is general enough to include linear and circular polarizations as well. The derivations show that the noise temperature for the circularly polarized wave case is simply the average of those for perpendicular and parallel polarizations.

2-8 Excitation of higher order modes by a step discontinuity of a circular waveguide. (Paper 2-8)

The problem of excitation of higher-order modes by a step discontinuity in a circular waveguide with an incident dominant H_{11} mode is considered. An approximate method is used to solve this problem. In this method, it is assumed that the tangential electric field at the discontinuity is zero everywhere except in the aperture, where it is equal to the incident tangential electric field. Relative amplitudes of these excited modes are given and discussed.

A Relation Between α and Q*

In 1944, Davidson and Simmonds[1] derived a relation between the Q of a cavity composed of a uniform transmission line with short-circuiting ends and the attenuation constant α of such a transmission line. Later, in 1950, Barlow and Cullin[2] rederived this relation. These authors showed that this relation is quite general and is applicable to arbitrary cross-section uniform metal tube waveguides. Since then one of the standard techniques for the measurements of the attenuation constant α has become the use of the cavity method.[3] This method offers an excellent way of measuring the attenuation constant of a waveguide when the loss is quite small. Later on this method was generalized and applied to open waveguides,[4] such as the single-wire transmission line and the dielectric waveguide, by various authors.

However, it is noted that the formula by Davidson, etc., was derived under the assumption that there exists a single equivalent transmission line for the mode under consideration. This assumption is true for a pure TE, TM, or TEM mode, but it is not clear that such a single equivalent transmission line exists for a hybrid wave. This suspicion originates from the fact that 1) the TE and TM waves are intimately coupled to each other, and 2) the characteristic impedance defined by Schelkunoff[5] is not constant with respect to the transverse coordinates. It is, therefore, very difficult to conceive the possibility that there exists a single equivalent transmission line for this hybrid wave; at best the hybrid wave may be represented by a set of transmission lines coupled tightly with one another. Hence the formula by Davidson, etc., may not be applicable to a hybrid wave.

A more general relation[6] between Q and α can be obtained without using the transmission line equivalent circuit, provided that α is very small compared with the phase constant β and that the loss contributed by the short-circuiting end plates is negligible compared with the total loss of the waveguide section under consideration. The propagation constant of a guided wave with a small attenuation constant at ω_0 is

$$\Gamma(\omega_0) = \alpha(\omega_0) + i\beta(\omega_0). \qquad (1)$$

At resonance, the following relation is true:

$$\Gamma(\omega_0) + \frac{\partial\Gamma}{\partial\omega}\Delta\omega \approx i\beta(\omega_0). \qquad (2)$$

Combining (1) and (2) gives

$$\alpha(\omega_0) = -i\frac{\partial\beta}{\partial\omega}\Delta\omega. \qquad (3)$$

Since the group velocity v_g and Q are given by the relations

$$v_g = \frac{\partial\omega}{\partial\beta},$$

and

$$Q = \frac{\omega_0}{2(\Delta\omega/i)},$$

one arrives at the relation

$$\alpha = \frac{\omega_0}{2Qv_g} = \frac{v_p}{v_g}\frac{\beta}{2Q} \qquad (4)$$

where v_p is the phase velocity of the wave. This is the general relation that is sought.

Substituting the values of v_p/v_g for a TE, TM, or TEM wave into (4), one gets the relations derived by Davidson, etc. For a TM or TE wave in a metal waveguide,

$$\frac{v_p}{v_g} = \frac{1}{1 - \left(\frac{\lambda}{\lambda_c}\right)^2}, \qquad (5)$$

where λ_c is the cutoff wavelength of the wave under consideration. Thus

$$\alpha_{TE,TM} = \frac{1}{1 - \left(\frac{\lambda}{\lambda_c}\right)^2}\frac{\beta}{2Q}. \qquad (6)$$

For a TEM wave,

$$\frac{v_p}{v_g} = 1. \qquad (7)$$

Hence

$$\alpha_{TEM} = \frac{\beta}{2Q}. \qquad (8)$$

For a hybrid wave, v_g and v_p are not simply related. They may be obtained graphically from the ω-β diagram. However, for a dominant hybrid wave on a dielectric rod at very low frequencies or at very high frequencies, the relation (8) is a good approximation since at these frequencies $v_p \approx v_g$.[7]

C. Yeh
Elec. Engrg. Dept.
Univ. So. Calif.
Los Angeles, Calif.
Formerly with
Calif. Inst. Tech.
Pasadena, Calif.

* Received April 9, 1962.

[1] C. F. Davidson and J. C. Simmonds, "Cylindrical cavity resonators," *Wireless Engr.*, vol. 21, pp. 420–424; September, 1944.

[2] H. M. Barlow and A. L. Cullen, "Microwave Measurements," Constable and Company, Ltd., London, England; 1950.

[3] E. L. Ginzton, "Microwave Measurements," McGraw-Hill Book Co., Inc., New York, N. Y.; 1957.

[4] C. H. Chandler, "An investigation of dielectric rod as waveguides," *J. Appl. Phys.*, vol. 20, pp. 1188–1192; December, 1949.

E. H. Sheibe, B. G. King, and D. L. Van Zeeland, "Loss measurements of surface wave transmission lines," *J. Appl. Phys.*, vol. 25, pp. 790–797; June, 1954.

[5] S. A. Schelkunoff, "The impedance concept and its application to problems of reflection, refraction, shielding and power absorption," *Bell Sys. Tech. J.*, vol. 17, pp. 17–48; January, 1938.

[6] Private communication with Prof. R. W. Gould, Calif. Inst. Tech., Pasadena, Calif.

[7] C. Yeh, "Electromagnetic surface wave propagation along a dielectric cylinder of elliptical cross section," Ph.D. thesis, Calif. Inst. Tech., Pasadena, 1962.

Reprinted from:
PROCEEDINGS OF THE IRE
October 1962

An alternate derivation of the relation between α and Q is given below.

Relation between α and Q

For a swept input signal in the vicinity of resonant wavelength, λ_g, through a low loss single mode waveguide of length d ($d = n\lambda_g/2$), shorted at both ends with miniscule coupling at the input and output shorting sections, the output is a narrow transmitted resonance at ω_0 with a half-power bandwidth $2\Delta\omega$ (See Fig. 1). The transmitted power is given by

$$P_t \sim K\left|\frac{e^{-\Gamma d}}{1 - R^2 e^{-2\Gamma d}}\right|^2 = K\left|\frac{1}{e^{\Gamma d} - R^2 e^{-\Gamma d}}\right|^2 \qquad (1)$$

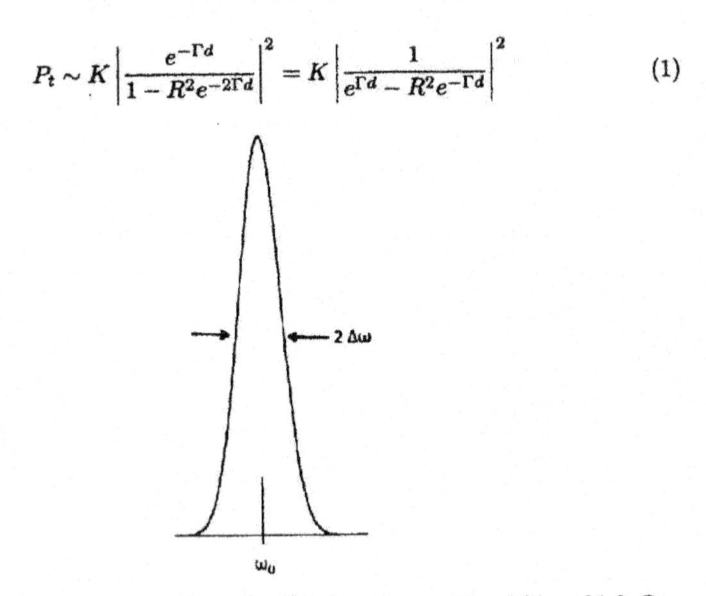

Fig. 1. Transmitted power through a low loss waveguide within a high Q cavity.

where

- K is a constant dependent of the incident power and the coupling coefiicients
- R is the reflection coefficient at the shorted ends
- d is the cavity length
- n is an integer
- $\Gamma = j\beta + \alpha$
- β is the propagation constant ($= 2\pi/\lambda_g$)
- λ_g is the guide wavelength
- α is the attenuation constant

For a low loss waveguide, the coupling and reflectivity losses at the input and output are small compared to the waveguide losses. For this case $R \approx 1$ At the given resonance, $\omega = \omega_0$, $\beta_0 = 2\pi/\lambda_g = \pi n/d$, the transmission is at a maximum and is given by

$$P_t(\omega = \omega_0) \sim \left|\frac{1}{e^{\alpha d} - e^{-\alpha d}}\right|^2 \tag{2}$$

At the frequency, $\omega = \omega_0 \pm \Delta\omega$, where the power output is one half that of the maximum.

$$P_t(\omega = \omega_0 \pm \Delta\omega) \sim \left|\frac{1}{e^{\Gamma d} - e^{-\Gamma d}}\right|^2 = \frac{1}{2}\left|\frac{1}{e^{\alpha d} - e^{-\alpha d}}\right|^2 \tag{3}$$

Expanding Γ in a power series about ω_0,

$$\Gamma = j\beta_0 + j\frac{\partial\beta}{\partial\omega}\Delta\omega + ...\alpha + \frac{\partial\alpha}{\partial\omega}\Delta\omega + . \tag{4}$$

Assuming α does not change over the interval $\Delta\omega$, for a low loss waveguide (high Q), expanding the exponentials to first order, the relation in (3) can be written as

$$\begin{aligned}
\left|2\alpha d + 2j\frac{\partial\beta}{\partial\omega}\Delta\omega d\right|^2 &= 2(2\alpha d)^2 \\
(2\alpha d)^2 + \left(2\frac{\partial\beta}{\partial\omega}\Delta\omega d\right)^2 &= 2(2\alpha d)^2 \\
\alpha &= \frac{\partial\beta}{\partial\omega}\Delta\omega
\end{aligned} \tag{5}$$

Since the phase velocity, v_{ph}, is equal to ω/β, the group velocity, v_g is equal to $\partial\omega/\partial\beta$, and $Q = \omega/2\Delta\omega$, (5) becomes

$$\alpha = \frac{v_{ph}}{v_g}\frac{\beta}{2Q} \tag{6}$$

Scalar-wave approach for single-mode inhomogeneous fiber problems[a]

C. Yeh and L. Casperson

Electrical Sciences and Engineering Department, University of California Los Angeles, Los Angeles, California 90024

W. P. Brown

Hughes Research Laboratories, Malibu, California 90265

(Received 19 December 1978; accepted for publication 8 January 1979)

It has generally been accepted that accurate results may be obtained using the scalar-wave approach to solve problems dealing with inhomogeneous multimode guided-wave structures. The problem of the applicability of the scalar-wave approach to obtain wave propagation characteristics in single-mode fiber or integrated optical circuit guides with inhomogeneous index profiles is examined in this paper. It is shown that if certain limiting conditions are satisfied, the scalar-wave approach will yield valid results for single-mode structures. These limiting conditions are usually satisfied by many practical single-mode inhomogeneous fibers or IOC structures.

PACS numbers: 42.65.Cq, 42.40. – i, 42.55.Rz, 42.60.He

It has been universally accepted that the full set of Maxwell equations resulting in the vector-wave equation must be used to treat waveguides supporting single mode. This requirement confines the analytical treatment to only a few simple structures. The advent of optical-fiber guides as viable communication links as well as the dawn of small high-bandwidth integrated optical circuits demand that the analytical horizon be expanded. It is recognized that many more problems can be solved if only the scalar-wave equation needs be considered. The purpose of this paper is to investigate in depth the conditions under which the scalar wave equation can be used instead of the vector wave equation and to demonstrate, with concrete examples, the validity of the scalar-wave approach[1] in providing accurate results for the graded index fiber guide.

Starting with the vector-wave equation for the electric field vector $\mathbf{E}$ in the fiber structure,

$$\nabla\times\nabla\times\mathbf{E} - \omega^2\mu_0\epsilon\mathbf{E} = 0, \tag{1}$$

where ω is the frequency of the wave, μ_0 is the permeability, and $\epsilon(\mathbf{r})$ is the inhomogenous permittivity of the structure; and making use of the vector identity

$$\nabla\times\nabla\times\mathbf{E} = \nabla(\nabla\cdot\mathbf{E}) - \nabla^2\mathbf{E} \tag{2}$$

and the relation

$$\nabla\cdot\mathbf{E} = -\epsilon^{-1}\nabla\epsilon\cdot\mathbf{E}, \tag{3}$$

one has

$$\nabla^2\mathbf{E} + \omega^2\mu_0\epsilon\mathbf{E} - \nabla(\epsilon^{-1}\nabla\epsilon\cdot\mathbf{E}) = 0. \tag{4}$$

Rewriting Eq. (4) gives

$$\nabla^2\mathbf{E} + \omega^2\mu_0\epsilon_0\left\{\frac{\epsilon}{\epsilon_0}\mathbf{E} - \left[\frac{1}{\omega^2\mu_0\epsilon_0}\nabla\left(\frac{1}{\epsilon}\nabla\epsilon\cdot\mathbf{E}\right)\right]\right\} = 0. \tag{5}$$

The relative importance of the terms within the curly brackets can be determined from the following:

$$\frac{\epsilon}{\epsilon_0}\mathbf{E} = O\left(\frac{\epsilon}{\epsilon_0}\right)\mathbf{E}, \tag{6}$$

$$\frac{1}{\omega^2\mu\epsilon_0}\nabla\left(\frac{1}{\epsilon}\nabla\epsilon\cdot\mathbf{E}\right) = \frac{1}{k_0^2}O\left(\frac{\nabla\epsilon}{\epsilon}\nabla\mathbf{E}\right) = O\left(\frac{\epsilon/\epsilon_0}{k_0 l}\mathbf{E}\right), \tag{7}$$

where O means the "order of magnitude" and l is the smaller of the distances over which ϵ/ϵ_0 and E change appreciably. For single-model fiber structures, the values of ϵ/ϵ_0 and $k_0 l$ are typically in the range

$$\epsilon/\epsilon_0 = O(2), \tag{8}$$

$$k_0 l \simeq (2\pi/\lambda)l = O(10^2 \text{ or } 10^3),$$

$$l \simeq O(10\text{–}100\ \mu), \quad \lambda \simeq O(1\ \mu) \tag{9}$$

It follows that the second term within the curly brackets in Eq. (5) is several orders of magnitude smaller than the first term, $(\epsilon/\epsilon_0)E$. It is therefore justifiable to neglect the second term and write Eq. (5) in the form

$$\nabla^2\mathbf{E} + k_0^2(\epsilon/\epsilon_0)\mathbf{E} = 0 \tag{10}$$

The physical significance of replacing Eq. (5) by Eq. (10) is this. By discarding the term $\nabla(\epsilon^{-1}\nabla\epsilon\cdot\mathbf{E})$, we are neglecting any depolarization effects that may occur. This means that the wave retains the linear polarization it has at the source, which is evidenced by the fact that Eq. (10) can be reduced to a scalar equation by writing $\mathbf{E}(\mathbf{x})$ in the form

$$\mathbf{E}(\mathbf{x}) = \hat{e}_p u(\mathbf{x}), \tag{11}$$

where $\hat{e}_p$ is a unit vector in the direction of the initial polarization of the wave. Substituting Eq. (11) in Eq. (10), we find that $u(\mathbf{x})$ satisfies the scalar-wave equation

$$\nabla^2 u + k_0^2(\epsilon/\epsilon_0)u = 0. \tag{12}$$

This equation with the boundary condition on the initial sur-

[a] Partially supported by Hughes IR & D.

 0003-6951/79/070460-03$00.50

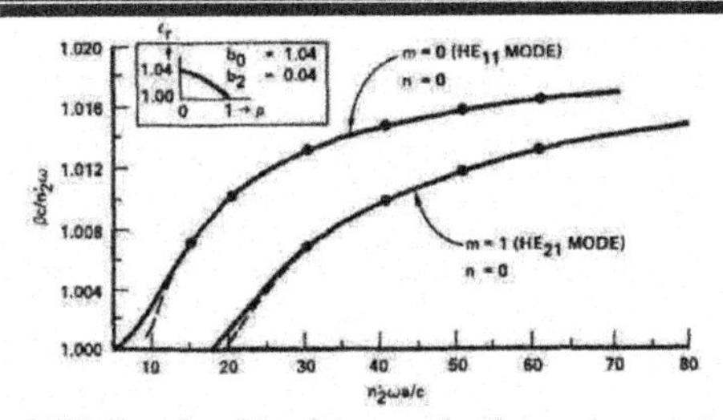

FIG. 1. Comparison of the scalar-wave results with vector-wave results for the dispersion characteristics of HE_{11} and HE_{21} modes. The data points are scalar-wave results and the solid curves are exact vector-wave results. The core index variations are given by $n_r = n_2'\epsilon_r^{1/2} = n_2'[b_0 - b_2\rho^2]^{1/2}$, where $\rho = r/a$, n_2' is approximately the index of the cladding and a is the core radius.

face, and the radiation condition at infinity, completely specifies $u(x)$, from which we can then obtain the electromagnetic field vectors **E** and **H**.

To verify the fact that the scalar-wave approach may be used to obtain accurate results for the case of wave propagation along single-mode inhomogeneous fiber structures, we shall now consider the special case of the dominant mode propagating in a fiber with a parabolic radial index profile. This case was chosen because exact vector-wave solutions exist for this problem. By comparing our results based on the scalar-wave approach with the exact results, one may verify the applicability of this scalar-wave approach to single-mode problems.

Vector-wave solutions exist for radially inhomogeneous fibers.[2,3] Two methods can usually be used to obtain the propagation characteristics of dominant modes in these types of fibers: (1) the radially inhomogeneous cylinder is subdivided into a number of concentric layers and the problem is solved by matching the solution for each homogeneous layer at the subdivided boundaries;[3] (2) the problem for the radially inhomogeneous cylinder is formulated in terms of a set of four coupled first-order differential equations for the transverse field components, and direct numerical integration is then performed to obtain the propagation constants of the lower-order modes.[3] Both methods have been used to obtain the dispersion characteristics of the dominant HE_{11} mode for a parabolic index profile fiber. Results are shown in Fig. 1. The normalized propagation constant $\beta c/n_2'\omega$ is plotted against the normalized frequency $n_2'\omega a/c$ in Fig. 1 for the following index profile:

$$n = n_2' b_0^{1/2}\left(1 - \frac{1}{2}\frac{b_2}{b_0}\rho^2\right), \quad \text{for} \quad \rho < 1, \qquad (13)$$

$$n = n_2' b_0^{1/2}\left(1 - \frac{1}{2}\frac{b_2}{b_0}\right), \quad \text{for} \quad \rho > 1, \qquad (14)$$

with $\rho = r/a$, a is the radius of the inhomogeneous core, ω is the frequency of the wave, c is the speed of light in vaccum, and n_2', b_0, and b_2 are given constants. Data given in Fig. 1 will be used to check the accuracy of the scalar-wave results.

The scalar-wave solution for the propagating wave in a parabolic index profile medium can be written in the form[4]

$$u = u_0(x,y)\exp\left[-i\frac{2\pi n_0}{\lambda} - (m+n+1)(n_2/n_0)^{1/2}\right]z, \qquad (15)$$

where the propagating field is assumed to be linearly polarized, $u_0(x,y)$ is a real Hermite-Gaussian function, $\lambda = 2\pi c/\omega$, m and n are integers (0, 1, 2,...), and the index profile is given in the standard Gaussian beam notation

$$n = n_0[1 - \tfrac{1}{2}(n_2/n_0)a^2\rho^2]. \qquad (16)$$

The values of n_0 and n_2 in this expression are related to n_2', b_0, and b_2 given previously by the relationships

$$n_0 = n_2' b_0^{1/2}, \qquad (17)$$

$$(n_2/n_0)a^2 = b_2/b_0. \qquad (18)$$

The propagation constant β for the field can be obtained readily from Eq. (15),

$$\beta = 2\pi n_0/\lambda - (m+n+1)(n_2/n_0)^{1/2} \qquad (19)$$

or

$$\frac{\beta c}{n_2'\omega} = \sqrt{b_0} - (m+n+1)\left(1 - \frac{1}{b_0}\right)^{1/2}\left(\frac{c}{n_2'\omega a}\right). \qquad (20)$$

This is the analytic result for the normalized propagation constant as a function of the normalized frequency for various modes ($m, n = 0, 1,\cdots$) in a parabolic index guide based on the scalar-wave equation. Numerical results calculated according to this equation are also displayed in Fig. 1 by the dashed curves. It is clearly seen that excellent agreement

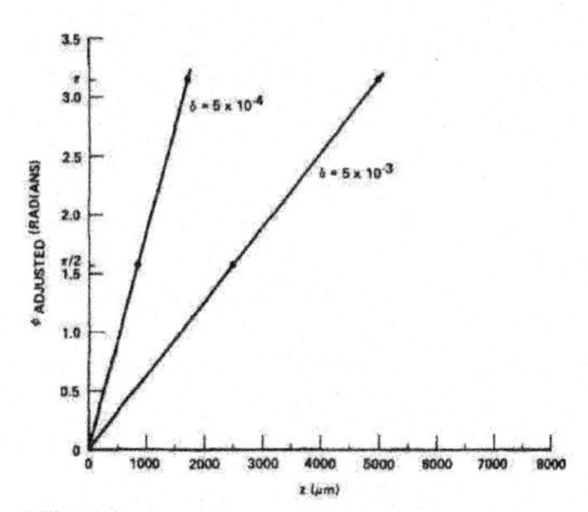

FIG. 2. Comparison of the analytic scalar-wave results with the numerical FFT scalar-wave results for the adjusted phase shift of a Gaussian beam propagation along a parabolic index profile fiber with index $n(r) = n_0[1 - \delta(r/a)]^2$. The data points are numerical FFT scalar-wave results and the solid lines are analytic scalar-wave results. $a = 50\,\mu$m and $n_0 = 1.5$. Computations were carried out for various free-space wavelengths $\lambda = 0.6$, 0.8, 1.0, and 1.2 μm and the corresponding beam radii $b = 10.03$, 11.58, 12.95, and 14.20 μm for $\delta = 5\times10^{-4}$ case and $b = 5.64$, 6.51, 7.28, and 7.97 μm for the $\delta = 5\times10^{-3}$ case [$\delta = (n_2/n_0)a^2 1/2$)]

between the scalar-wave results and the vector-wave results is obtained. Only when the operating frequencies are near the cutoff frequency of the mode does any difference exist. In fact, the difference is *not* caused by the inadequacy or inaccuracy of the scalar-wave approach, but rather by the difference in our choice of index profiles for the vector and the scalar cases. In the vector case the index profile is no longer parabolic for $\rho > 1$, while in the scalar case it is always parabolic. By extending the parabolic index profile beyond $\rho = 1$ for the vector case, we can show (according to our computations) that the difference becomes very small indeed. It appears, therefore, that if the conditions described by Eqs. (8) and (9) are satisfied, excellent results may be obtained using the scalar-wave equation.

Let us now consider the case of the numerical solution of the scalar-wave equation by the FFT (fast Fourier transform) technique.[1] Of course, one must recognize that the important feature of this approach is in its ability in solving problems with nonparabolic index profiles; but to determine whether this approach may be used to treat single-mode problems, we shall compare the numerical results obtained according to this FFT approach for the parabolic index profile case with those obtained according to the analytic scalar-wave solution. Since we have already shown that the analytic scalar-wave solution is as good as the vector-wave solution for the situation under consideration, we can safely conclude that if the numerical FFT scalar wave solution checks out with the analytic scalar-wave solution, then this numerical FFT scalar-wave approach may be used with confidence for single-mode fibers with arbitrary index profiles as long as the limiting conditions specified earlier are satisfied.

Figure 2 shows the "adjusted phase shift" of the beam as it propagates down a parabolic index profile fiber. Adjusted phase shift is defined as the difference between the phase shift of a plane wave in an n_0 medium, $(2\pi n_0/\lambda)z$, and the actual phase shift of the beam; i.e., $\Phi_{\text{adjusted}} = (2\pi n_0/\lambda)z - \beta z$. The expression for the index profile is given by Eq. (16) with $n_0 = 1.5$ and $(n_2/n_0)a^2 = 0.005$. The data points refer to the numerical results obtained by the numerical FFT scalar-wave approach while the solid lines were calculated according to the formula

$$\Phi_{\text{adjusted}} = (2\pi n_0/\lambda)z - \beta z = (m + n + 1)(n_2/n_0)^{1/2}z, \tag{21}$$

which is the analytic expression [Eq. (15)] obtained according to the scalar-wave solution discussed earlier. For the dominant HE_{11} mode, $m = 0$ and $n = 0$. So, $\Phi_{\text{adjusted}} = (n_2/n_0)^{1/2}z$. It can be seen that excellent agreement was achieved.

In conclusion we can safely state that our numerical FFT scalar-wave approach can be used with confidence to solve single-mode problems dealing with fiber or IOC structures with general inhomogeneous index profiles provided that the following conditions are satisfied:

(a) The depolarization effects are negligible, or

$$\left|\frac{1}{\omega^2\mu\epsilon_0}\nabla\frac{1}{\epsilon}\nabla\epsilon\cdot\mathbf{E}\right| \ll \left|\frac{\epsilon}{\epsilon_0}\mathbf{E}\right|, \tag{22}$$

where ϵ is the dielectric permittivity, $\mathbf{E}$ is the electric field vector, μ and ϵ_0 are, respectively, the free-space permeability and permittivity, and ω is the frequency of the wave. Physically, Eq. (22) implies that the index profile of the fiber varies little over distances of the order of the wavelength.

(b) The paraxial ray approximation may be used, or the factor $(\partial^2 A/\partial z^2)(\mathbf{x},z)$ is negligible compared to the factor

$$\left(2ikn_0\frac{\partial}{\partial z} + \nabla_t^2 + k^2(n^2(\mathbf{x},z) - n_0^2)\right)A(\mathbf{x},z),$$

where

$$u = \exp(ikn_0z)A(\mathbf{x},z).$$

This condition means that the complex amplitude $A(\mathbf{x},z)$ varies much more rapidly transverse to the direction of propagation than it does along the direction of propagation, which is satisfied for fields propagating at small angles to the z axis.

The authors wish to thank R. Szejn for carrying out the computer programming task.

[1]C. Yeh, L. Casperson, and R. Szejn, J. Opt. Soc. Am. 68, 989 (1978).
[2]C. Yeh and G. Lindgren, Appl. Opt. 16, 483 (1977).
[3]J.G. Dil and H. Blok, Opt. Electron 5, 415 (1973).
[4]L. Casperson, Appl. Opt. 12, 2434 (1973).

Accuracy of directional coupler theory in fiber or integrated optics applications

C. Yeh, F. Manshadi, K. F. Casey*, and A. Johnston†

Electrical Sciences and Engineering Department, University of California, Los Angeles, Los Angeles, California 90024

(Received 23 January 1978)

Based on the solution of a simplified canonical problem, the possible margin of error on the predicted coupling length according to the coupled-mode theory as applied to fiber and integrated optical guides is inferred. It was found that coupling length obtained according to the coupled-mode theory is usually accurate to within 20% of the actual value provided that the frequency of operation is above the cutoff frequency of the antisymmetric mode of the coupled structure.

Coupled-mode theory has been used extensively in recent years in the analysis of many important problems in integrated optics and fiber optics.[1–3] It is therefore of extreme interest to learn the region of validity for which the coupled-mode theory will yield accurate results. Since exact solutions for most practical problems are not available, we shall therefore deal specifically with a canonical problem whose exact solution, as well as approximate solution based on the coupled-mode theory, is obtainable. Comparison of the results may then provide an indication of the accuracy of the coupled-mode theory approach. It is shown that when the coupled-mode theory is used properly, the prediction of coupling distances is surprisingly accurate even when the separation between the guides is relatively small. However, as far as the field configurations of coupled guides are concerned, significant inaccuracy is observed when the separation distance becomes small.

The geometry of the problem is shown in Fig. 1. Two identical dielectric slab waveguides are located in the regions $d/2 \le |x| \le a + d/2$. The permittivity of the slab guides is ϵ_1 and that of the regions outside the guides is ϵ_2; furthermore, $\epsilon_1 > \epsilon_2$. The permeability is assumed to be μ_0 everywhere. A perfectly conducting plane covers the surface $z = 0$ except for the region $d/2 \le x \le a + d/2$. In that aperture, the tangential electric field is $E_0\vec{a}_x$. All electromagnetic field quantities are independent of the coordinate y and the time dependence $\exp(j\omega t)$ is assumed and suppressed.

The electromagnetic field has components E_x, E_z, and H_y, which are related as

$$E_x = \frac{-1}{j\omega\epsilon}\frac{\partial H_y}{\partial z}, \tag{1a}$$

$$E_z = \frac{1}{j\omega\epsilon}\frac{\partial H_y}{\partial x}. \tag{1b}$$

H_y satisfies

$$\frac{\partial^2 H_y}{\partial x^2} + \frac{\partial^2 H_y}{\partial z^2} + k_i^2 H_y = 0, \tag{2}$$

where $k_i^2 = \omega^2\mu_0\epsilon_i$, $i = 1,2$. Now it is necessary to specify that all field components have the same z dependence, i.e., $\exp(-j\beta z)$. Thus if $E_x(x,z) = \hat{E}_x(x)\exp(-j\beta z)$ and similarly for E_z and H_y, we have

$$\hat{E}_x = (\beta/\omega\epsilon)\hat{H}_y, \tag{3a}$$

$$\hat{E}_z = \frac{1}{j\omega\epsilon}\frac{d\hat{H}_y}{dx}, \tag{3b}$$

$$\frac{d^2\hat{H}_y}{dx^2} + (k_i^2 - \beta^2)\hat{H}_y = 0 \tag{3c}$$

We are concerned with the excitation by the aperture field of the guided modes on this structure; the guided-wave field will be the dominant portion of the total field for large z. Solutions of Eq. (3c) corresponding to the guided-wave field are

$$-\infty \le x \le -a - d/z:\quad \hat{H}_{yn}^{(1)}(x) = A_n \exp[v_n(x + a + d/2)]; \tag{4a}$$

$$-a - d/2 \le x \le -d/2:\quad \hat{H}_{yn}^{(2)}(x) = B_n \sin u_n(x + d/2) + C_n \cos u_n(x + d/2); \tag{4b}$$

$$-d/2 \le x \le d/2:\quad \hat{H}_{yn}^{(3)}(x) = D_n \sinh v_n x + E_n \cosh v_n x; \tag{4c}$$

$$d/2 \le x \le a + d/2:\quad \hat{H}_{yn}^{(4)}(x) = F_n \sin u_n(x - d/2) + G_n \cos u_n(x - d/2); \tag{4d}$$

$$a + d/2 \le x \le \infty:\quad \hat{H}_{yn}^{(5)}(x) = H_n \exp[-v_n(x - a - d/2)]; \tag{4e}$$

in which $u_n^2 = k_1^2 - \beta_n^2$, $v_n^2 = \beta_n^2 - k_2^2$, and the coefficients A_n – H_n are to be determined. The n subscripts refer to the nth guided-wave mode.

The characteristic equation from which the values of β_n may be determined is found by forcing the tangential field components E_z and H_y to be continuous at the four dielectric interfaces. By virtue of the symmetry of the configuration about $x = 0$, the electromagnetic field may be separated into two parts whose axial electric field E_z has either even or odd symmetry about $x = 0$. For the even modes, $H_n = -A_n$, $G_n = -C_n$, $F_n = B_n$, and $E_n = 0$; for the odd modes, $H_n = A_n$, $G_n = C_n$, $F_n = -B_n$, and $D_n = 0$. Thus considering the two cases separately, one matches the boundary conditions at $x = d/2$ and $x = a + d/2$, and obtains

$$\left[\tan u_n a - \frac{2qu_n v_n}{q^2u_n^2 - v_n^2 + (q^2u_n^2 + v_n^2)\exp(-v_n d)}\right] \times \left[\tan u_n a - \frac{2qu_n v_n}{q^2u_n^2 - v_n^2 - (q^2u_n^2 + v_n^2 + v_n^2)\exp(-v_n d)}\right] = 0 \tag{5}$$

in which $q = \epsilon_2/\epsilon_1$. The roots of the first factor in square

 0030-3941/78/6808-1079$00.50

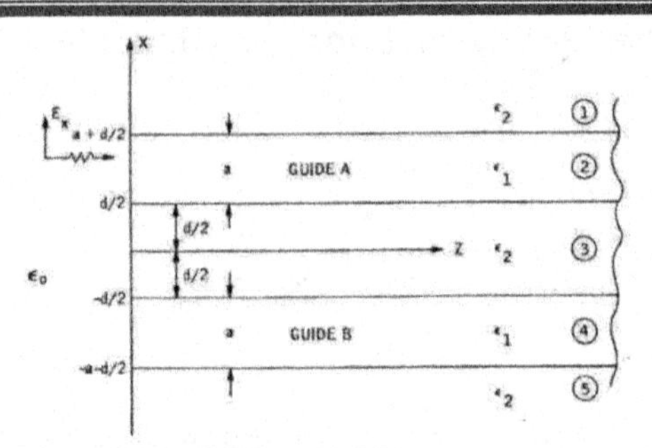

FIG. 1. Geometry of the canonical problem.

brackets yield the even modes; those of the second factor, the odd modes.

We may also evaluate the coefficients $A_n - H_n$ in terms of any single one of them, say A_n. The relationship is expressed as $B_n = b_n A_n, \ldots, H_n = h_n A_n$ in which $b_n - h_n$ are given by

even modes:

$$b_n^e = f_n^e = (1/\Delta_n^e)(v_n/\epsilon_2)\cosh(v_n d/2); \quad (6a)$$

$$c_n^e = -g_n^e = -(1/\Delta_n^e)(u_n/\epsilon_1)\sinh(v_n d/2); \quad (6b)$$

$$d_n^e = \frac{1}{\Delta_n^e}\frac{u_n}{\epsilon_1}, \quad e_n^e = 0, \quad h_n^e = -1; \quad (6c)$$

$$\Delta_n^e = (-v_n/\epsilon_2)\cosh(v_n d/2)\sin u_n a - (u_n/\epsilon_1)\sinh(v_n d/2)\cos u_n a. \quad (6d)$$

odd modes:

$$b_n^o = -f_n^o = (1/\epsilon_n^o)(v_n/\epsilon_2)\sinh(v_n d/2); \quad (7a)$$

$$c_n^o = g_n^o = -(1/\Delta_n^o)(u_n/\epsilon_1)\cosh(v_n d/2); \quad (7b)$$

$$d_n^o = 0, \quad e_n^o = -\frac{1}{\Delta_n^o}\frac{u_n}{\epsilon_1}, \quad h_n^o = 1; \quad (7c)$$

$$\Delta_n^o = -(v_n/\epsilon_2)\sinh(v_n d/2)\sin u_n a - (u_n/\epsilon_1)\cosh(v_n d/2)\cos u_n a. \quad (7d)$$

This completes the formal analysis of the guided-mode electromagnetic field on the dual slab waveguide structure.

We now consider the boundary condition at $z = 0$, i.e., that $E_x(x,0) = E_0$ for $d/2 \le x \le a + d/2$ and $E_x(x,0) = 0$ elsewhere. Equating the excitation field to the total field E_x for $z \ge 0$ at $z = 0$ yields

$$E_0[U(x - d/2) - U(x - a - d/2)] = \sum_n \frac{-\beta_n}{\omega\epsilon(x)}\hat{H}_{yn}(x) + (\text{radiated field})|_{x=0}, \quad (8)$$

where U is the unit step function. The sum is taken over the guided modes for which β_n is a proper root, i.e., a root for which $\mathrm{Re}(v_n) \ge 0$, of Eq. (5). By virtue of the orthogonality relation

$$\int_{-\infty}^{\infty}\frac{1}{\epsilon(x)}\hat{H}_y(x;\beta)\hat{H}_y^*(x;\beta')dx = 0 \quad (\beta \ne \beta'), \quad (9)$$

we readily obtain

$$-\frac{\omega}{\beta_n}E_0\int_{d/2}^{a+d/2}\hat{H}_{yn}^{(4)*}(x)\,dx = \frac{1}{\epsilon_2}\int_{-\infty}^{-a-d/2}|\hat{H}_{yn}^{(1)}(x)|^2dx + \frac{1}{\epsilon_1}\int_{-a-d/2}^{-d/2}|\hat{H}_{yn}^{(2)}(x)|^2dx + \frac{1}{\epsilon_2}\int_{-d/2}^{d/2}|\hat{H}_{yn}^{(3)}(x)|^2dx + \frac{1}{\epsilon_1}\int_{d/2}^{a+d/2}|\hat{H}_{yn}^{(4)}(x)|^2dx + \frac{1}{\epsilon_2}\int_{a+d/2}^{\infty}|\hat{H}_{yn}^{(5)}(x)|^2dx + (\text{radiated fields}). \quad (10)$$

The integrals in Eq. (10) are easily evaluated and yield a relation between E_o and A_n. We find that

$$A_n = \frac{-\omega E_o}{u_n\beta_n}[f_n^*(1 - \cos u_n a) + g_n^*\sin u_n a] \times \left\{\frac{1}{v_n\epsilon_2} + \frac{1}{u_n\epsilon_1}[|g_n|^2(u_n a + \tfrac{1}{2}\sin 2u_n a) + |f_n|^2(u_n a - \tfrac{1}{2}\sin 2u_n a) + (f_n g_n^* + f_n^* g_n)\sin^2 u_n a]\right.$$

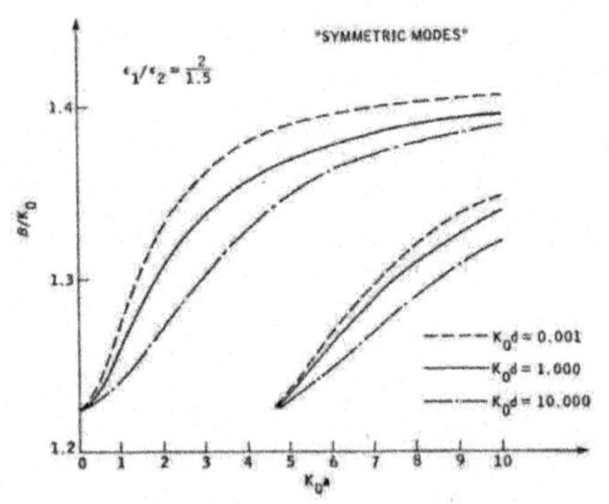

FIG. 2. Normalized propagation constants as a function of normalized frequencies. The lowest-order mode $n = 1$ has zero cutoff frequency.

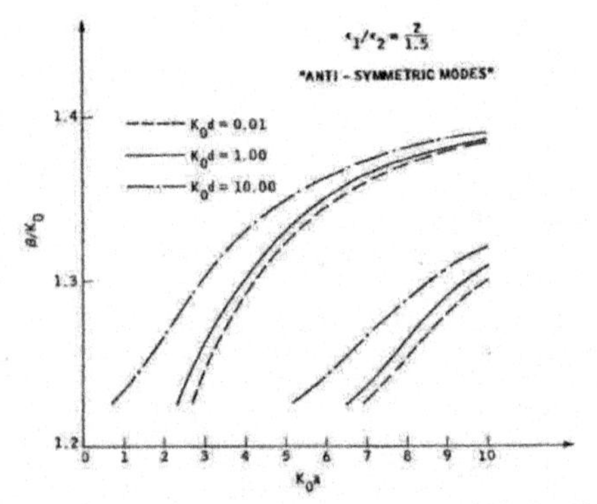

FIG. 3. Normalized propagation constants as a function of normalized frequencies. All antisymmetric modes have cutoff frequencies.

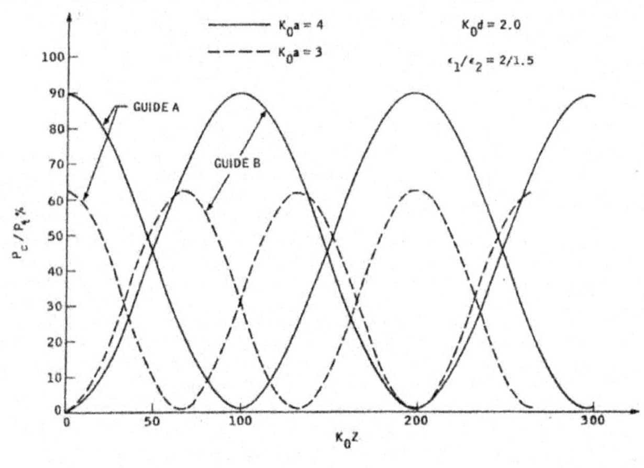

FIG. 4. Normalized power in the core region of guides A and B as a function of the normalized longitudinal distance. The coupling length (defined as the length for which complete exchange of power in the cores of guide A and guide B occurs) is longer for less tightly bounded fields.

$$+\frac{u_n^2}{2\Delta_n^2 v_n \epsilon_1^2 \epsilon_2}(\sinh v_n d \pm v_n d)\Bigg]^{-1} \quad (11)$$

in which the lower (−) sign is taken for the even modes, and the upper (+) sign for the odd modes. Δ_n^2 is either Δ_n^{e2} or Δ_n^{o2} as appropriate, as are f_n and g_n. This completes the formal solution for the excitation of the guided-wave field by the aperture electric field at $z = 0$. Here we have assumed that the contribution of the exciting field to the radiated field is negligible as compared with the guided field.

We now consider the special case which is of particular interest for the purposes of this paper, that in which only a single mode is possible on each of the dielectric slabs in isolation. We shall further require that the lowest-order even and odd modes will propagate on the dual-guide structure. Thus

$$(2a/qd) < ua \tan ua < \infty,$$

where $0 < ua < \pi/2$. We now evaluate the z component of the Poynting vector in each of the slab waveguides under the assumption that only these lowest-order modes exist; their propagation constants are denoted β_e and β_o, and the associated fields $\hat{H}_y(x)$ are denoted $\hat{H}_{ye}(x)$ and $\hat{H}_{yo}(x)$, respectively. In the "upper" guide ($d/2 < x < a + d/2$) or "guide A", we have

$$P_{z\mathrm{A}}(x,z) = \frac{1}{2\omega}\mathrm{Re}\Big\{\frac{1}{\epsilon_1}[\beta_e|\hat{H}_{ye}^{(4)}(x)|^2 + \beta_o|\hat{H}_{yo}^{(4)}(x)|^2 + \beta_e\hat{H}_{ye}^{(4)}(x)\hat{H}_{yo}^{(4)*}(x)e^{-j(\beta_e-\beta_o^*)z} + \beta_o\hat{H}_{yo}^{(4)}(x)\hat{H}_{ye}^{(4)*}(x)e^{j(\beta_e^*-\beta_o)z}]\Big\}. \quad (12)$$

In the "lower" guide, or "guide B,"

$$P_{z\mathrm{B}}(x,z) = \frac{1}{2\omega}\mathrm{Re}\Big\{\frac{1}{\epsilon_1}[\beta_e|\hat{H}_{ye}^{(2)}(x)|^2 + \beta_o|\hat{H}_{yo}^{(2)}(x)|^2 + \beta_e\hat{H}_{ye}^{(2)}(x)\hat{H}_{yo}^{(2)*}(x)e^{-j(\beta_e-\beta_o^*)z} + \beta_o\hat{H}_{yo}^{(2)}(x)\hat{H}_{ye}^{(2)*}(x)e^{-j(\beta_e^*-\beta_o)z}]\Big\}. \quad (13)$$

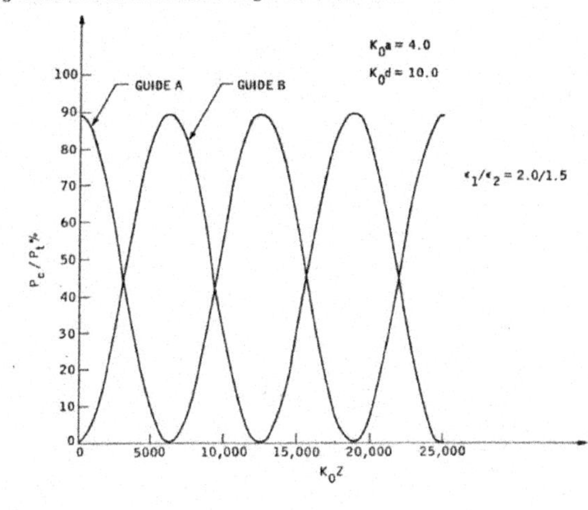

FIG. 5. Normalized power in the core region of guides A and B as a function of the normalized longitudinal distance. P_C is the power in the core region of guide A or guide B as appropriate. P_T is the total guided power.

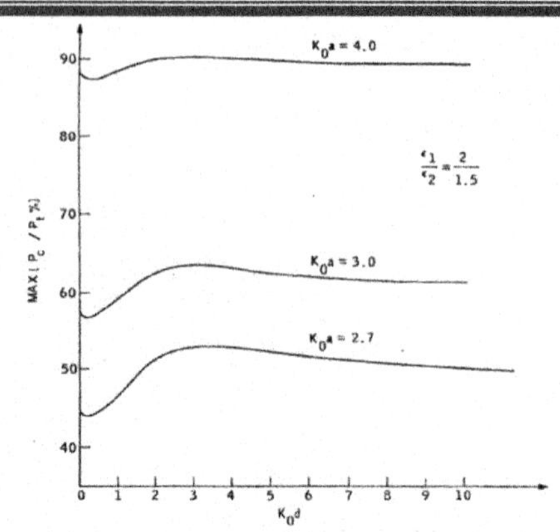

FIG. 6. Maximum normalized power in the core region of guide A as a function of separation distance of the two guides.

We may now calculate the total power per unit width carried in each of the slab waveguides (The power carried outside the slab guides is not included in the calculation. Thus P_A and P_B will not reduce to the total power carried in a single slab waveguide mode in the limit $d \to \infty$.) by integrating P_{zA} from $x = d/2$ to $x = a + d/2$ and P_{zB} from $x = -a - d/2$ to $x = -d/2$. Denoting these powers by P_A and P_B, we obtain, assuming that ϵ_1, ϵ_2, β_e, and β_o are purely real,

$$P_A(z) = \int_{d/2}^{a+d/2} P_{zA}(x,z)\, dx,$$

$$P_B(z) = \int_{-a-d/2}^{-d/2} P_{zB}(x,z)\, dx.$$

Now we are in a position to compute exactly how the input power is transferred from one guide to other. (The only assumption is that no radiated power is present.) We must first determine the propagation constants of the guided modes along the slabs from Eq. (5). Results for the two lowest-order modes are displayed in Fig. 2 for the symmetric modes and in Fig. 3 for the antisymmetric modes. In these figures, the normalized propagation constant β/k_0 is plotted against the normalized thickness of the slab k_0a for various values of the normalized separation of the slabs k_0d. The constant k_0, the free-space wave number, is defined as $\omega(\mu_0\epsilon_0)^{1/2}$. It can be seen that there exists no cutoff frequency for the lowest-order symmetric mode while a cutoff frequency does exist for the lowest-order antisymmetric mode. Of course, all higher-order symmetric and antisymmetric modes possess cutoff frequencies.

To illustrate how the guided power exchanges between the two guiding structures, Figs. 4 and 5 are introduced. The operating frequency is so selected that only the dominant symmetric mode and the lowest-order antisymmetric mode may exist along the guiding structure. It is interesting to note that although the initial exciting field exists only at the en-

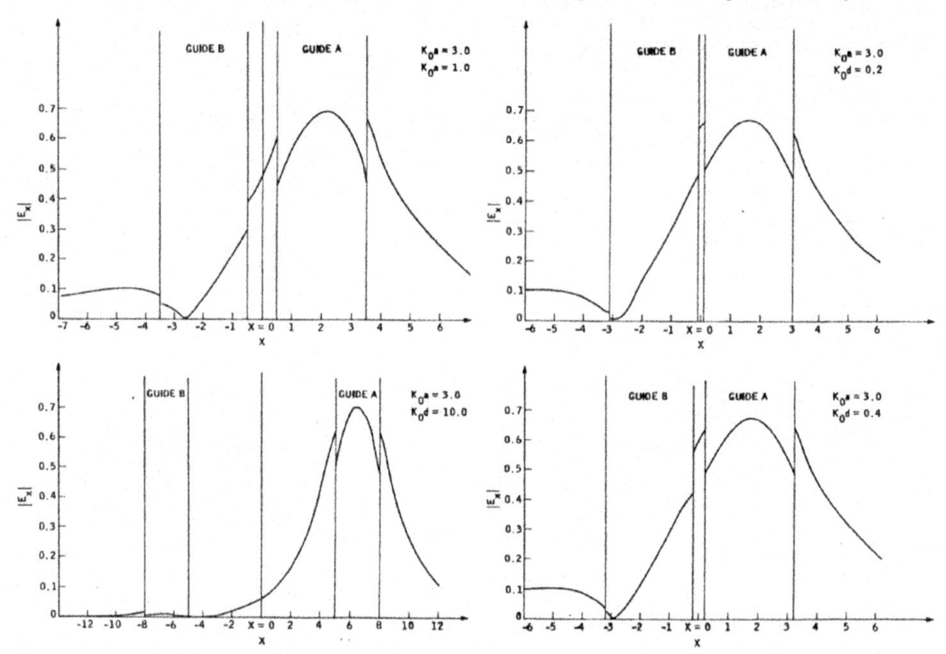

FIG. 7. Transverse electric field distribution across the two coupled guides.

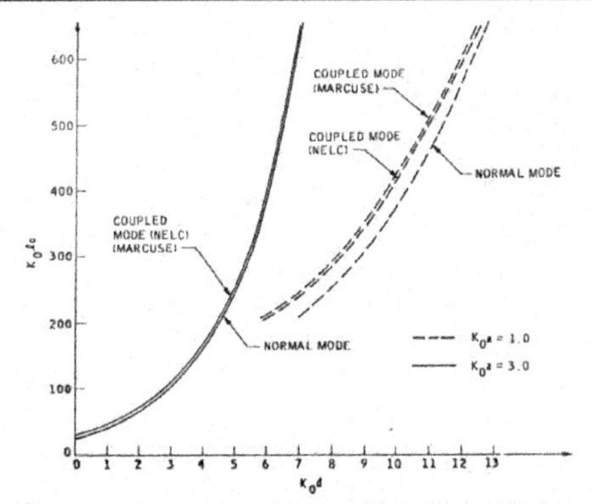

FIG. 8. Normalized coupling length as a function of the normalized separation distance. Note that the coupling length ceases to exist for $k_0d < 7.0$ when $k_0a = 1.0$ according to the normal-mode theory.

trance of the core region of guide A, according to our computed results there exists a small amount of guided power in guide B. The reason for this is that to satisfy the initial given field configuration at the entrance of the guiding structure, radiation mode as well as guided modes must be taken into account. Since we have *a priori* ignored the radiation mode in our calculation, we can only satisfy approximately the given field. The quantity P_B at $z = 0$ represents the power calculated from that field that extends from guide A into the core region of guide B. It can also be seen from these figures that guided power exchanges from one guide to the other in a periodic fashion as expected. The distance for which maximum guided power is transferred from one guide to the other is called the coupling length. It is noted that the coupling length becomes shorter as the separation distance between the guides is shortened. To correlate the maximum power contained in the core region of guide A with the normalized separation distance k_0d, we have performed the computation at the entrance of the coupled guide. Results are shown in Fig. 6. It can be seen that the normalized maximum power in the core

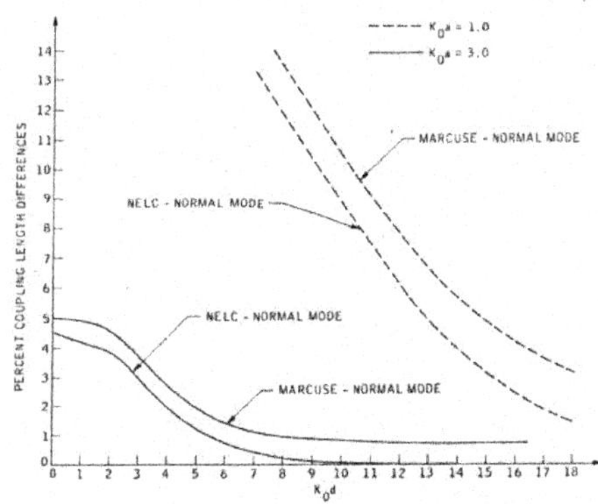

FIG. 9. Percent coupling length differences between normal-mode theory and coupled-mode theory as a function of normalized separation distance.

region of guide A varies in a rather unusual fashion for small separation distances. At large k_0d, the maximum power contained in the core region approaches that for the case of an isolated slab guide, as expected. To further understand the behavior of the transverse fields in the guides when the separation distance is small, we have plotted in Fig. 7 the quantity $|E_x|$ versus the tranverse distance. The complexity of the evolution of the transverse electric field as the separation distances are changed indicates the complex nature of the curves in Fig. 6 when the separation distances are small.

Recall that the primary purpose of our investigation is to determine how accurate the coupled-mode theory is in its treatment of the coupled dielectric waveguide problem. We have carried out the cases above according to the coupled-mode theory described by Marcuse[1] and by NELC researchers.[2] The "exact" normal-mode results are then compared with those obtained according to the various coupled-mode theories. Displayed in Fig. 8 are the curves for the coupling length as a function of the normalized separation distance between two parallel dielectric guides as shown for two different k_0a values in Fig. 1. One notes that as the separation distance is increased, the agreement between the results based on coupled-mode theories and our normal-mode results becomes better, and that closer agreement is obtained for larger k_0a values or when more power is confined within the core of the guide. This is because large k_0a corresponds to more tightly bounded field; hence, the coupling field is weaker and the coupled-mode theory which is based on perturbation concept tends to be more accurate. It is worthwhile to point out that for certain combinations of k_0a and k_0d values such as for $k_0a = 1$ and $k_0d \leq 7.0$ only the dominant symmetric mode exists, so according to the exact normal-mode theory, no back and forth exchange of propagating power takes place between guide A and guide B. On the other hand, the approximate coupled-mode theory continues to predict the power exchange phenomenon. Another way of expressing the differences for the results based on different theories is shown in Fig. 9 where the percent differences between different coupled-mode theories and the normal-mode theory are plotted against the normalized separation distances.

It can be seen that the coupled-mode theory is surprisingly good (within 20%) in predicting the coupling length of two parallel dielectric slab guides even when the separation distance is relatively small and the confinement of guided power is relatively weak. Extrapolating our present results to other geometrics involving optical fibers or integrated optical guides, it is inferred that the coupling distances predicted according to the coupled-mode theory are accurate to within 20% of the actual values if the symmetric and antisymmetric modes are both above cutoff. Finally, it should be noted that when the separation distance is small the transverse field configurations of the coupled guides (see Fig. 7) are significantly different than those assumed in the couple-mode theory.

*K. F. Casey is with the Department of Electrical Engineering, Kansas State University, Manhattan, Kans.
†A. Johnston is with the Jet Propulsion Laboratories, Pasadena, Calif. 91103.

[1]D. Marcuse, *Light Transmission Optics*, (Van Nostrand-Reinhold, New York, 1972, pp. 417–421.
[2]D. B. Hall, "Frequency Selective Coupling between Planar Waveguides," NELC Technical Note TN-2583 (1974), Naval Electronics Laboratory Center, San Diego, Calif. 92152.
[3]S. E. Miller, Bell Syst. Tech. J. **33**, 661 (1954).

PHYSICAL REVIEW E VOLUME 48, NUMBER 2 AUGUST 1993

Boundary conditions in electromagnetics

C. Yeh
2432 Nalin Drive, Los Angeles, California 90077
(Received 29 March 1993)

A proof is given to show that satisfying the boundary conditions on the tangential electric and magnetic fields across the boundary surface of two different material media means that the boundary conditions on the normal magnetic-induction vector and electric-displacement vector are automatically satisfied while the converse is not true.

PACS number(s): 41.20.−q

Boundary conditions for electromagnetic fields are the cornerstones for classical theory of electrodynamics. Their derivations are standard textbook material [1–5]. For example, the boundary conditions for electromagnetic fields at the boundary of two distinct dielectric media, derived through the application of Stoke's theorem to Maxwell's curl equations over a rectangular area which borders the two boundary media, are

$$\mathbf{n}\times\mathbf{E}_1=\mathbf{n}\times\mathbf{E}_2\ , \tag{1}$$

$$\mathbf{n}\times\mathbf{H}_1=\mathbf{n}\times\mathbf{H}_2\ , \tag{2}$$

where $\mathbf{n}$ is a unit vector normal to the boundary, $\mathbf{E}_1$ and $\mathbf{E}_2$ are, respectively, the electric field in medium 1 and in medium 2 at the boundary, and $\mathbf{H}_1$ and $\mathbf{H}_2$ are, respectively, the magnetic field in medium 1 and in medium 2 at the boundary; the boundary conditions derived through the application of divergence theorem to Maxwell's divergence equations over a pill box which borders the two boundary media, are

$$\mathbf{n}\cdot\mathbf{D}_1=\mathbf{n}\cdot\mathbf{D}_2\ , \tag{3}$$

$$\mathbf{n}\cdot\mathbf{B}_1=\mathbf{n}\cdot\mathbf{B}_2\ , \tag{4}$$

where $\mathbf{D}_1$ and $\mathbf{D}_2$ are, respectively, the displacement vector in medium 1 and in medium 2 at the boundary, while $\mathbf{B}_1$ and $\mathbf{B}_2$ are, respectively, the magnetic flux density in medium 1 and in medium 2 at the boundary.

It is important to note that the boundary conditions on the normal components of $\mathbf{D}$ and $\mathbf{B}$ [Eqs. (3) and (4)] are redundant, since, according to the uniqueness theorem [1], the boundary conditions on the tangential components of $\mathbf{E}$ and $\mathbf{H}$ [Eqs. (1) and (2)] are necessary and sufficient boundary conditions. In other words, satisfying Eqs. (1) and (2) implies that Eqs. (3) and (4) are automatically satisfied while the converse is not true. Although this fact is well known, no proof has been found in the standard literature. It is the purpose of this paper to present this proof.

Across the boundary, as shown in Fig. 1, let us introduce two small parallel surface areas of rectangular shape that are mirror images of each other. The top rectangle, parallel to the interface, is located in medium 1, characterized by $(\epsilon_1,\mu_1,\sigma_1)$ where ϵ_1, μ_1, and σ_1 are, respectively, the permittivity, the permeability, and the conductivity of medium 1, while the bottom rectangle, also parallel to the interface, is located in medium 2, characterized by $(\epsilon_2,\mu_2,\sigma_2)$ where ϵ_2,μ_2, and σ_2 are, respectively, the permittivity, the permeability, and the conductivity of medium 2. The small rectangle has sides Δs_1 and Δs_2. The unit vectors $\mathbf{n}_1$ and $\mathbf{n}_2$ are normal to the rectangular surfaces as shown in Fig. 1. The vectors $\mathbf{e}_x$, $\mathbf{e}_y$, and $\mathbf{e}_z$ are the three unit vectors in the x,y,z directions, respectively.

The source-free Maxwells equations are

$$\nabla\times\mathbf{E}=-\frac{\partial\mathbf{B}}{\partial t}\ , \tag{5}$$

$$\nabla\times\mathbf{H}=\frac{\partial\mathbf{D}}{\partial t}\ , \tag{6}$$

where $\mathbf{E}$, $\mathbf{H}$, $\mathbf{B}$, $\mathbf{D}$ refer to the field quantities in the medium in which they apply. Integrating Eq. (5) over the rectangular area ΔS_1 in region 1 yields

$$\int_{\Delta S_1}(\nabla\times\mathbf{E}_1)\cdot\mathbf{n}_1 dS=-\frac{\partial}{\partial t}\int_{\Delta S_1}\mathbf{B}_1\cdot\mathbf{n}_1 dS\ , \tag{7}$$

and integrating Eq. (5) over the rectangular area ΔS_2 in region 2 yields

$$\int_{\Delta S_2}(\nabla\times\mathbf{E}_2)\cdot\mathbf{n}_2 dS=-\frac{\partial}{\partial t}\int_{\Delta S_2}\mathbf{B}_2\cdot\mathbf{n}_2 dS\ . \tag{8}$$

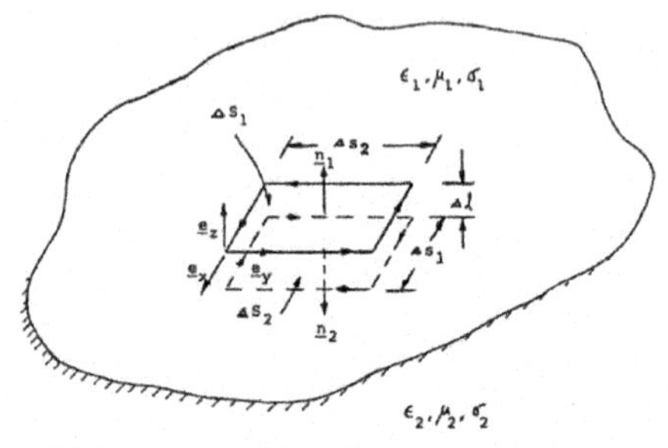

FIG. 1. Geometry of the problem. Rectangular area ΔS_1 is parallel to rectangular area ΔS_2. Sides of rectangle are Δs_1 and Δs_2. Δl is the separation between the two rectangles.

Application of Stoke's theorem to Eqs. (7) and (8) and adding the resultant equations yield

$$\int_{c_1} \mathbf{E}_1\cdot d\mathbf{s}_1 + \int_{c_2} \mathbf{E}_2\cdot d\mathbf{s}_2 = -\frac{\partial}{\partial t}\left[\int_{\Delta S_1} \mathbf{B}_1\cdot\mathbf{n}_1 dS + \int_{\Delta S_2} \mathbf{B}_2\cdot\mathbf{n}_2 dS\right], \quad (9)$$

where c_1 and c_2 are, respectively, the circumferences of the rectangular areas ΔS_1 and ΔS_2. In rectangular coordinates, with $\Delta s_1 \to 0$ and $\Delta s_2 \to 0$, E has a constant value along each side. Allowing the separation Δl between the two parallel rectangular areas in medium 1 and in medium 2 to approach zero, one has

$$(E_{1x}-E_{2x})2\Delta x + (E_{1y}-E_{2y})2\Delta y = -\frac{\partial}{\partial t}(B_{1z}-B_{2z})\Delta x\,\Delta y\ . \quad (10)$$

From Fig. 1, one may immediately identify the following:

$$\Delta s_1 = \Delta x,\quad \Delta s_2 = \Delta y\ ,$$

$$\mathbf{s}_1 = \mathbf{e}_x,\quad \mathbf{s}_2 = \mathbf{e}_y\ ,$$

$$\mathbf{n}_1 = -\mathbf{n}_2 = \mathbf{n} = \mathbf{e}_z\ .$$

The quantities E_{1x} and E_{2x} are, respectively, the x-directed electric field tangential to the boundary in region 1 and in region 2, the quantities E_{1y} and E_{2y} are, respectively, the y-directed electric field tangential to the boundary in region 1 and in region 2, and the quantities B_{1z} and B_{2z} are, respectively, the z-directed **B** field normal to the boundary surface in region 1 and in region 2.

Equation (10) shows that if the condition that the tangential components of the electric field are continuous across the boundary is satisfied, i.e., $E_{1x}=E_{2x}$ *and* $E_{1y}=E_{2y}$, then it follows that $B_{1z}=B_{2z}$, i.e., the condition that the normal component of the **B** field is continuous across the boundary is automatically satisfied. But, if the condition that the normal component of **B** field is continuous across the boundary is satisfied, i.e., $B_{1z}=B_{2z}$, according to Eq. (10) the two boundary conditions ($E_{1x}=E_{2x}$, $E_{1y}=E_{2y}$) on the two components of the tangential electric field are *not necessarily* satisfied.

A similar proof can be made for the tangential components of **H** and the normal component of **D**, using Eq. (6).

It is therefore proven that satisfying the boundary conditions on the tangential electric and magnetic fields across the boundary surface implies that the boundary conditions on the normal magnetic induction vector and the normal electric displacement vector are satisfied while the converse is not true.

The author wishes to thank Tom Otoshi of JPL for reading this paper.

[1] J. A. Stratton, *Electromagnetic Theory* (McGraw-Hill, New York, 1942), pp. 483 and 486.
[2] S. Ramo, J. Whinnery, and T. Van Duzer, *Fields and Waves in Communication Electronics* (Wiley, New York, 1965).
[3] J. D. Jackson, *Classical Electrodynamics,* 2nd ed. (Wiley, New York, 1975), p. 17.
[4] C. H. Papas, *Theory of Electromagnetic Wave Propagation* (McGraw-Hill, New York, 1965).
[5] C. Yeh, in *Dynamic Fields, American Institute of Physics Handbook,* 3rd ed. edited by D. E. Gray (McGraw-Hill, New York, 1972).

Another more detailed version is given in the book entitled *The Essence of Dielectric Waveguides* (Springer, New York, N Y, 2008).

2.3.1. Boundary Conditions

Electromagnetic fields across a given boundary between two distinct media must satisfy a set of boundary conditions. Let P be a smooth surface separating two media, 1 and 2; let the unit vector normal to the boundary be $\mathbf{n}$, pointing from medium 1 into medium 2. Consider the following three cases: (a) medium 1 and 2 are dielectrics; (b) medium 1 is a perfect conductor and medium 2 is a dielectric; and (c) medium 1 is an imperfect conductor and medium 2 is a dielectric.

Case a. Media 1 and 2 are dielectrics having constitutive parameters $(\epsilon_1, \mu_1, \sigma_1)$, **and** $(\epsilon_2, \mu_2, \sigma_2)$, **respectively**

To thoroughly understand the boundary conditions for the field quantities $\mathbf{E}$, $\mathbf{D}$, $\mathbf{H}$, and $\mathbf{B}$, let us consider the following scenarios:

Scenario 1

Across the boundary, as shown in Fig. 2.1, let us introduce two small parallel surface areas of rectangular shape that are mirror images of each other.

The top rectangle, parallel to the interface, is located in medium 1, while the bottom rectangle, also parallel to the interface, is located in medium 2. The small rectangle has sides Δs_1 and Δs_2 .

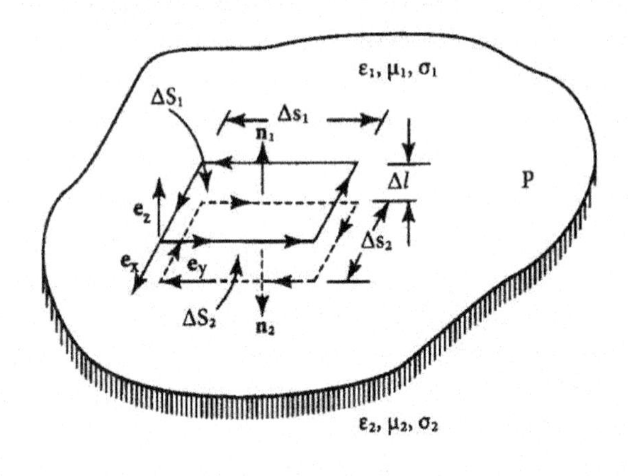

Figure 1: Geometry for scenario 1, showing the relationship between the tangential electric fields and the normal magnetic fluxes across a dielectric boundary, as well as the tangential magnetic fields and the normal displacement vectors. The rectangular area ΔS_1 is parallel to the rectangular area ΔS_2. The sides of the rectangle are Δs_1 and Δs_2 and the separation between the two rectangles is Δl. P is the plane that separates medium 1 and medium 2.

The unit vectors $\mathbf{n}_1$ and $\mathbf{n}_2$ are normal to the rectangular surfaces. The vectors $\mathbf{e}_x$, $\mathbf{e}_y$, and $\mathbf{e}_z$ are the three unit vectors in the x, y, z directions, respectively. P is the plane that separates medium 1 from medium 2. The source-free Maxwell equations are

$$\nabla \times \mathbf{E}(\mathbf{r},t) = -\frac{\partial \mathbf{B}(\mathbf{r},t)}{\partial t} \tag{2.36}$$

$$\nabla \times \mathbf{H}(\mathbf{r},t) = \mathbf{J}(\mathbf{r},t) + \frac{\partial \mathbf{D}(\mathbf{r},t)}{\partial t} \tag{2.37}$$

Integrating Eq. (2.36) over the rectangular area ΔS_1 in region 1 yields

$$\int_{\Delta S_1} (\nabla \times \mathbf{E}_1) \cdot \mathbf{n}_1 dS = -\frac{\partial}{\partial t}\left(\int_{\Delta S_1} \mathbf{B}_1 \cdot \mathbf{n}_1 dS\right) \tag{2.38}$$

and integrating Eq. (2.36) over the rectangular area ΔS_2 in region 2 yields

$$\int_{\Delta S_2} (\nabla \times \mathbf{E}_2) \cdot \mathbf{n}_2 dS = -\frac{\partial}{\partial t}\left(\int_{\Delta S_2} \mathbf{B}_2 \cdot \mathbf{n}_2 dS\right) \tag{2.39}$$

Application of Stokes theorem to Eqs. (2.38) and (2.39) and adding the resultant equations yield

$$\int_{c_1} \mathbf{E}_1 \cdot d\mathbf{s}_1 + \int_{c_2} \mathbf{E}_2 \cdot d\mathbf{s}_2 = -\left[\frac{\partial}{\partial t}\left(\int_{\Delta S_1} \mathbf{B}_1 \cdot \mathbf{n}_1 dS\right) + \frac{\partial}{\partial t}\left(\int_{\Delta S_2} \mathbf{B}_2 \cdot \mathbf{n}_2 dS\right)\right] \tag{2.40}$$

where c_1 and c_2 are, respectively, the circumferences of the rectangular areas ΔS_1 and ΔS_2. In rectangular coordinates, with $\Delta s_1 \to 0$ and $\Delta s_2 \to 0$, $\mathbf{E}$ has a constant value along each side. Allowing the separation Δl between the two parallel rectangular areas in medium 1 and in medium 2 to approach zero, one has,

$$(E_{1x} - E_{2x})2\Delta x + (E_{1y} - E_{2y})2\Delta y = -\frac{\partial}{\partial t}(B_{1z} - B_{2z})\Delta x \Delta y \tag{2.41}$$

where $\Delta s_1 = \Delta x$, $\Delta s_2 = \Delta y$, $\mathbf{s}_1 = \mathbf{e}_x$, $\mathbf{s}_2 = \mathbf{e}_y$, $\mathbf{n}_1 = - \mathbf{n}_2 = \mathbf{n} = \mathbf{e}_z$, E_{1x} and E_{2x} are, respectively, the x-directed electric field tangential to the boundary in region 1 and in region 2, E_{1y} and E_{2y} are, respectively, the y-directed electric field tangential to the boundary in region 1 and in region 2, B_{1z} and B_{2z} are, respectively, the z-directed $\mathbf{B}$ field normal to the boundary surface in region 1 and in region 2.

In a similar manner, one may derive the following relation from Eq. (2.37) for the tangential components of $\mathbf{H}$ and the normal component of $\mathbf{D}$ on the boundary surface:

$$(H_{1x} - H_{2x})2\Delta x + (H_{1y} - H_{2y})2\Delta y = \frac{\partial}{\partial t}(D_{1z} - D_{2z})\Delta x \Delta y + (J_{1z} - J_{2z})\Delta x \Delta y \tag{2.42}$$

The significance of Eqs. (2.41) and (2.42) will be discussed later.

Scenario 2

Let us consider a rectangular path c crossing through the boundary as shown in Fig. 2.2.

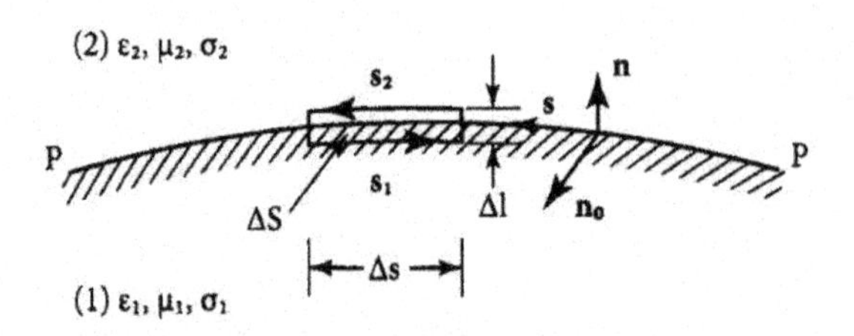

Figure 2: Geometry for scenario 2, showing the continuity of the tangential electric and magnetic fields across a dielectric boundary plane P.

Two sides with length Δs of this rectangle are parallel to the boundary P and the height is Δl. The unit vector $\mathbf{n}_0$ is normal to the rectangular area ΔS. The unit vectors $\mathbf{s}_1$and $\mathbf{s}_2$ are parallel to the boundary interface and are normal to $\mathbf{n}_0$ and $\mathbf{n}$, where $\mathbf{n}$ is the unit vector normal to the boundary surface. Also, $\mathbf{s}_1 = -\ \mathbf{s}_2$, $\mathbf{s}_1 = \mathbf{n}_0 \times\ \mathbf{n}$, and $\mathbf{s}_2 = -\mathbf{n}_0 \times\ \mathbf{n}$. Integrating Eq. (2.36) over the rectangular area yields

$$\int_{\Delta S} (\nabla \times \mathbf{E}) \cdot \mathbf{n}_0 dS = -\frac{\partial}{\partial t} \left(\int_{\Delta S} \mathbf{B} \cdot \mathbf{n}_0 dS \right) \tag{2.43}$$

Application of Stoke's theorem gives

$$\int_c \mathbf{E} \cdot d\mathbf{s} = -\frac{\partial}{\partial t} \left(\int_{\Delta S} \mathbf{B} \cdot \mathbf{n}_0 dS \right) \tag{2.44}$$

which for a very small rectangle with $\Delta s \to 0$ and $\Delta l \to 0$, $\mathbf{E}$ has a constant value along each side, giving,

$$\begin{aligned} &(\mathbf{s}_1 \cdot \mathbf{E}_1 + \mathbf{s}_2 \cdot \mathbf{E}_2)\Delta s + \text{contribution from ends} \\ &= -\frac{\partial \mathbf{B}}{\partial t} \cdot \mathbf{n}_0 dS \\ &= -\frac{\partial \mathbf{B}}{\partial t} \cdot \mathbf{n}_0 \Delta s \Delta l \end{aligned} \tag{2.45}$$

Taking the limit $\Delta s \to 0$ and $\Delta l \to 0$, one has

$$(\mathbf{n}_0 \times \mathbf{n} \cdot \mathbf{E}_1 - \mathbf{n}_0 \times \mathbf{n} \cdot \mathbf{E}_2) = -\underset{\Delta l \to 0}{Lim} \left(\frac{\partial \mathbf{B}}{\partial t} \cdot \mathbf{n}_0 \Delta l \right) \tag{2.46}$$

or

$$\mathbf{n}_0 \cdot [\mathbf{n} \times (\mathbf{E}_1 - \mathbf{E}_2)] = \underset{\Delta l \to 0}{Lim} \left(\mathbf{n}_0 \cdot \frac{\partial \mathbf{B}}{\partial t} \right) \Delta l \tag{2.47}$$

The contribution from ends on the left side Eq. (2.45) is zero as $\Delta l \to 0$. So,

$$\mathbf{n} \times (\mathbf{E}_1 - \mathbf{E}_2) = 0 \tag{2.48}$$

In a similar manner, using Eq. (2.37), one may derive the relation

$$\mathbf{n} \times (\mathbf{H}_1 - \mathbf{H}_2) = \mathbf{J}_s \tag{2.49}$$

where $\mathbf{J}_s$ is the surface current density at the boundary surface. It is shown that the tangential components of the electric and magnetic fields must be continuous across the dielectric boundary. That these boundary conditions on the tangential components of the electric and magnetic fields at the boundary of two dissimilar dielectric media are necessary and sufficient boundary conditions will be shown later.

Scenario 3

Let us consider a small right circular cylinder with an end area Δa and height Δl, situated across the boundary interface as shown in Fig. 2.3. The unit vectors

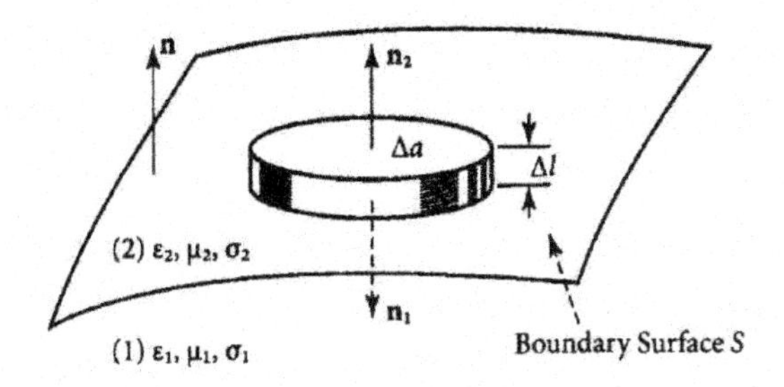

Figure 3: Geometry for scenario 3, showing the continuity of normal magnetic fluxes and the continuity of normal displacement vectors across a dielectric boundary.

$\mathbf{n}_1$ and $\mathbf{n}_2$ are normal to the end surfaces which are normal to the boundary. The volume of this cylinder is ΔV. The boundary surface S separates medium 1 and medium 2. Integrating Eq. (2.5) over this enclosed volume gives

$$\int_{\Delta V} \nabla \cdot \mathbf{B} \, dV = 0 \tag{2.50}$$

Applying the divergence theorem yields

$$\int \mathbf{B} \cdot \mathbf{n}\, da = 0 \tag{2.51}$$

where the area integration is over the walls and ends of the cylinder. For a very small cylinder with $\Delta a \to 0$ and $\Delta l \to 0$, $\mathbf{B}$ has a constant value over each end surface, and

$$(\mathbf{B}_1 \cdot \mathbf{n}_1 + \mathbf{B}_2 \cdot \mathbf{n}_2)\, \Delta a + \text{contribution of the side walls} = 0 \tag{2.52}$$

Here $\mathbf{B}_1$ is the $\mathbf{B}$ field in region 1 and $\mathbf{B}_1$ is the $\mathbf{B}$ field in region 2. Since $\mathbf{n}_2 = -\mathbf{n}_1 = -\mathbf{n}$, Eq. (2.52) becomes

$$(\mathbf{B}_1 - \mathbf{B}_2) \cdot \mathbf{n} = 0 \tag{2.53}$$

which is the needed boundary condition on $\mathbf{B}$. Similarly, using Eq. (2.8) for the source free boundary, one may derive the relation

$$(\mathbf{D}_1 - \mathbf{D}_2) \cdot \mathbf{n} = \rho_s \tag{2.54}$$

where ρ_s is the surface charge density at the boundary surface. These equations indicate that the normal components of $\mathbf{B}$ and $\mathbf{D}$ must be continuous across the boundary of dissimilar dielectric media. That these boundary conditions are necessary but not sufficient boundary conditions will be shown in the following.

From the above discussion, according to the three scenarios, one may restate the following boundary conditions at the boundary of two dielectrics:

$$(E_{1x} - E_{2x})2\Delta x + (E_{1y} - E_{2y})2\Delta y = -\frac{\partial}{\partial t}(B_{1z} - B_{2z})\Delta x \Delta y \tag{2.41}$$

$$\begin{aligned}(H_{1x} - H_{2x})2\Delta x &+ (H_{1y} - H_{2y})2\Delta y \\ &= \frac{\partial}{\partial t}(D_{1z} - D_{2z})\Delta x \Delta y + (J_{1z} - J_{2z})\Delta x \Delta y\end{aligned} \tag{2.42}$$

$$\mathbf{n} \times (\mathbf{E}_1 - \mathbf{E}_2) = 0 \tag{2.48}$$

$$\mathbf{n} \times (\mathbf{H}_1 - \mathbf{H}_2) = \mathbf{J}_s \tag{2.49}$$

$$(\mathbf{B}_1 - \mathbf{B}_2) \cdot \mathbf{n} = 0 \tag{2.53}$$

$$(\mathbf{D}_1 - \mathbf{D}_2) \cdot \mathbf{n} = \rho_s \tag{2.54}$$

The discussion on the necessary and sufficient boundary conditions at the boundary of two dielectrics was first given by Yeh in 1993 [20]. Referring to Eqs. (2.41) and (2.42), for time-harmonic fields, we may replace $\partial/\partial t$ by $j\omega$ resulting in

$$(E_{1x} - E_{2x})2\triangle x + (E_{1y} - E_{2y})2\triangle y = -j\omega(B_{1z} - B_{2z})\triangle x\triangle y \tag{2.55}$$

and

$$(H_{1x} - H_{2x})2\triangle x + (H_{1y} - H_{2y})2\triangle y$$

$$= j\omega(D_{1z} - D_{2z})\triangle x\triangle y + (J_{1z} - J_{2z})\triangle x\triangle y \tag{2.56}$$

Here, ω is the frequency of the time-harmonic fields. All field symbols are now time-independent phasors. These equations are valid at the interface boundary. Equation (2.55) shows that if the boundary conditions on the tangential electric field are satisfied, i.e., $E_{1x} = E_{2x}$ and $E_{1y} = E_{2y}$, then the left hand side of the equation is identically zero. Since, for time-varying fields, ω is non-zero, and $\triangle x\triangle y \neq 0$, then $B_{1z} - B_{2z} = 0$; or $B_{1z} = B_{2z}$, which is the boundary condition on the normal component of **B**. This is proof that satisfying the boundary conditions on the tangential components of electric field means that the boundary condition on the normal component of **B** is satisfied. On the other hand, if the boundary condition on the normal component of **B** is satisfied, i.e., if the right hand side of Eq. (2.55) is zero, it only means that

$$(E_{1x} - E_{2x})2\triangle x + (E_{1y} - E_{2y})2\triangle y = 0 \tag{2.57}$$

It is not possible to conclude that the boundary conditions on the tangential components of **E** are satisfied. Equation (2.57) only indicates that the combined terms of $(E_{1x} - E_{2x})2\triangle x$ and $(E_{1y} - E_{2y})2\triangle y$ must be zero and not necessarily each term must be zero. Similar conclusion can be reached from Eq. (2.56), i.e., satisfying the boundary conditions on the tangential components of magnetic field **H**, Eq. (2.49), means that the boundary condition on the normal component of **D**, Eq. (2.54) is satisfied, while the converse is not true.

To summarize, the necessary and sufficient boundary conditions on the time-varying electromagnetic fields across two distinct dielectric media are that the tangential electric fields must be continuous across the boundary and the tangential magnetic fields must be discontinuous by the surface current density, i.e., Eq.(2.48) and Eq.(2.49). Satisfying the boundary conditions for the normal components of **D** *and* **B**, i.e., *Eq.(2.53) and Eq.(2.54) does not guarantee that all the necessary boundary conditions are satisfied.*

Since the constitutive relations were never used to arrive at the above proof [20], the proof is equally applicable to isotropic or anisotropic, dispersive or non-dispersive, moving or stationary, linear or nonlinear, and left-handed (metamaterial), or right-handed media.

Let us now investigate the special case of static fields. Here, ω is identically zero. In that case, Eqs. (2.55) and (2.56) can be read as follows:

$$(E_{1x} - E_{2x})2\triangle x + (E_{1y} - E_{2y})2\triangle y = 0 \tag{2.55a}$$

$$(H_{1x} - H_{2x})2\triangle x + (H_{1y} - H_{2y})2\triangle y = (J_{1z} - J_{2z})\triangle x\triangle y \tag{2.56a}$$

There is no connection between electric and magnetic fields. Hence, unlike the time-dependent case, satisfying the boundary conditions on the tangential electric or magnetic fields says nothing about the satisfaction of the boundary conditions on the normal component of **D** and **B**. Furthermore, Eq. (2.55a) only shows that, on the boundary, the conditions $E_{1x} = E_{2x}$ and $E_{1y} = E_{2y}$ must be satisfied simultaneously. Indeed, for electrostatic problems, the complete boundary conditions require the satisfaction of the continuity of the tangential components of **E** across the boundary as well as the condition that the normal **D** is discontinuous by the surface charge density across the boundary. Using the same argument on Eq. (2.56a) for the magnetostatic case, one may reach similar conclusion, i.e., the complete magnetostatic boundary conditions require the condition that the tangential components of **H** be discontinuous by the surface current density as well as the continuity of normal **B** across the boundary.

The necessary and sufficient boundary condition on the static electric and magnetic fields across two distinct dielectric media are that (a) for the electrostatic fields, the tangential component of **E** must be continuous and the normal component of **D** must be discontinuous by the surface charge density at the boundary, and, (b) for the magnetostatic fields, the tangential component of **H** must be discontinuous by the surface current density and the normal component of **B** must be continuous at the boundary.

IEEE TRANSACTIONS ON MICROWAVE THEORY AND TECHNIQUES, VOL. 46, NO. 5, MAY 1998 523

Power Loss for Multimode Waveguides and Its Application to Beam–Waveguide System

W. A. Imbriale, *Fellow, IEEE,* T. Y. Otoshi, *Life Fellow, IEEE,* and C. Yeh, *Fellow, IEEE*

***Abstract*—The conventional way of expressing power loss in decibels/meter for a multimode waveguiding system with finite wall conductivity (such as a beam–waveguide (BWG) system with protective shroud) can be incorrect and misleading. The power loss (in decibels) for a multimode waveguiding system is, in general, not linearly proportional to the length of the waveguide. New power-loss formulas for multimode system are derived in this paper for arbitrarily shaped conducting waveguide tubes. In these formulas, there are factors such as $[\exp(jx)-1]/(jx)$, where $x = (\beta_a - \beta_b)\ell$, with β_a and β_b being the propagation constants of the different propagating modes and ℓ being the distance from the source plane to the plane of interest along the guide. For a large BWG supporting many propagating modes, β_a's are quite close to β_b's, thus the mode coupling terms remain important for a very long distance from the source plane. The multimode power-loss formula for a large circular conducting tube has been verified by experiments. This formula was also used to calculate the additional noise temperature contribution due to the presence of a protective shroud surrounding a millimeter-wave BWG system.**

***Index Terms*—Beam waveguides, cylindrical waveguide, noise measurement, waveguide theory.**

I. Introduction and the Consideration of a Fundamental Concept

IN TEXTBOOKS on electromagnetics and guided waves, the perturbation technique is used to calculate the attenuation factor of a given propagating mode in a slightly lossy and highly conducting hollow metallic waveguide. Based on this technique, the attenuation constant for the mth mode $\alpha^{(m)}$ due to conductor loss in a general cylindrical hollow metallic waveguide is found to be [1]–[6]

$$\alpha^{(m)} \simeq \frac{R \oint_c \left[\underline{H}^{(m)} \cdot \underline{H}^{(m)*}\right] d\ell}{2\mathrm{Re} \iint_s \left[\underline{E}^{(m)} \times \underline{H}^{(m)*}\right] \cdot \underline{e}_z \, dA} \tag{1}$$

where R denotes the surface resistance of the metal walls, $\underline{E}^{(m)}$ and $\underline{H}^{(m)}$ are the unperturbed electric and magnetic fields for the mth propagating mode in this waveguide with perfectly conducting walls, Re denotes the real part of the integral, $\underline{e}_z$ is the unit vector in the z-propagating direction, * denotes the complex conjugate of the integral, C is the contour around the cross section of the waveguide, and S is the cross-sectional area of the waveguide. Here, $\alpha^{(m)}$ expressed in nepers/meter is the attenuation constant for the mth propagating mode per unit length of the waveguide.

Manuscript received January 22, 1997; revised February 13, 1998. This work was supported by the National Aeronautics and Space Administration under Contract NAS 71260.

The authors are with the Jet Propulsion Laboratory (JPL), California Institute of Technology, Pasadena, CA 91109 USA.

Publisher Item Identifier S 0018-9480(98)03165-2.

A more accurate determination of the attenuation constant $\alpha^{(m)}$ can be obtained through the boundary-value-problem approach. Here, the fields in different regions [i.e., the metal region characterized by (ϵ, μ, σ) and the vacuum region characterized by (ϵ_0, μ_0)] of the waveguide are matched at the boundary, yielding a dispersion relation from which the complex propagation constant for each mode may be determined. For this approach, in general, all field components must be assumed to be present. In other words, for a hollow circular metal pipe, the field components (E_z, E_r, E_ϕ, H_z, H_r, H_ϕ) will all be present when circular symmetry of the mode is not present. Here, the circular cylindrical coordinates (r, ϕ, z) are assumed. This was the approach (called the hybrid-mode approach) used by Chou and Lee to calculate modal attenuation in multilayered coated waveguides [7]. Other improved versions of the perturbation formula of (1) for the attenuation constant of a single mode were given by Gustincic [8] and Collin [2].

The intent of this paper is not to improve the power-loss calculation for a single mode. One notes that, for the small loss case, this improvement is negligible. The intent of this paper is to provide a correct way of finding the power loss for the multimode case.

In all of the above considerations, the power loss has always been expressed by $\alpha^{(m)}$ for each mth mode in nepers/meter.

It is the limitation of this way of expressing power loss that we wish to address in the following sections.

When a single mode, say the mth mode, is propagating in this hollow waveguide, the following expression is normally used to represent the power carried by this mode along this waveguide structure:

$$P^{(m)}(z) = P_0^{(m)} e^{-2\alpha^{(m)} z} \approx P_0^{(m)}[1 - 2\alpha^{(m)} z] \tag{2}$$

where $P_0^{(m)}$ is the initial input power of the mth mode and z is the distance along the guide. That this expression is valid if and only if a single mode is propagating alone in this waveguide is usually glossed over in the textbooks. Furthermore, (1) and (2) offer the impression that the power loss in a given waveguide may be expressed by the attenuation constant $\alpha^{(m)}$ in nepers/meter. From (2), for small attenuation, the power loss $P_L^{(m)}$ is

$$\begin{aligned} P_L^{(m)} &= P_0^{(m)} \text{ (Power Input)} - P^{(m)}(z) \text{ (Power Output)} \\ &\simeq P_0^{(m)} 2\alpha^{(m)} z. \end{aligned}$$

Consequently, one may obtain the mistaken impression that since the modes are orthogonal, the total loss is additive when more than one mode is present simultaneously in the waveguide; after all, we know that the total power is additive. For the multimode propagation case, the total power loss should not be expressed through an attenuation constant as certain nepers/meter (or decibels/meter). Indeed, due to the contributions of the cross-product terms in $\underline{J} \cdot \underline{J}^*$ where $\underline{J}$ is the total surface current, and the total power loss in the multimode case is no longer a linear function of the length of the guide, as in the single-mode case.

For example, assume that a given source in an infinitely long hollow conducting waveguide excites two equal amplitude lowest order propagating modes. Further assume that the waveguide can only support these two lowest order propagating modes. The walls of the waveguide are made with highly conducting (but not perfectly conducting) metal. Let us find the total power loss at a distance d from the source plane.

According to the classical textbook formula (1), the attenuation constant for each mode can be calculated using this formula. Say the answer for mode 1 is $\alpha_1 = 0.001$ (nepers/meter) and for mode 2 is $\alpha_2 = 0.002$. (Even if we use the more exact way of calculating the attenuation constant by the boundary-value-problem approach (or the hybrid-mode approach) described in [1], due to the highly conducting nature of the walls, the attenuation constants for these two modes would not deviate much from the given values). Let P_0 be the input power for mode 1 as well as for mode 2. Thus, the power of mode 1 after propagating for a distance z in the waveguide is $P_0 \exp[-2\alpha^{(1)}z]$ and for mode 2 is $P_0 \exp[-2\alpha^{(2)}z]$. Since the power is additive, the total power loss is

$$\begin{aligned} P_{\text{Total Loss}} &= P_{\text{Input}} - P_{\text{Output}} \\ &= 2P_0 - P_0\{\exp[-2\alpha^{(1)}z] + \exp[-2\alpha^{(2)}z]\} \\ &= 2P_0[\alpha^{(1)} + \alpha^{(2)}]z. \end{aligned} \tag{3}$$

Extending this concept to n modes would yield

$$P_{\text{Total Loss}} = 2P_0(\alpha^{(1)} + \alpha^{(2)} + \cdots + \alpha^{(n)})z.$$

Thus, according to the above formula, no matter how small z or α is, $P_{\text{Total Loss}}$ is proportional to z. To demonstrate that this concept is incorrect, consider the following: a high-gain horn radiating inside a waveguide with boresite along the axis of the waveguide. If the modes are considered to be uncoupled, then the loss for each mode can be independently computed and summed. Therefore, the power loss per unit length would be independent of the position in the waveguide. A simple thought experiment should be sufficient to conclude that if the diameter became larger and larger, one would certainly expect the loss per unit length in a region near the plane of the horn aperture (where there is virtually no radiation from the horn) to be quite different from the loss at a distance where the radiation pattern from the horn would illuminate the waveguide walls. This is very similar to the experiment described later in this paper. A correct interpretation of the perturbation theory would be to apply it to the total tangential H field on the waveguide wall. Since the power loss formula uses tangential H squared, the modal fields are thus coupled through this term resulting in a waveguide loss which varies as a function of the axial dimension. The results are precisely what one would expect for the horn example, i.e., very little loss very near the aperture plane and increasing significantly when the radiation pattern of the horn intersects the waveguide wall. Using this approach, the resultant theoretical/numerical data compares very favorably with the measured experimental data.

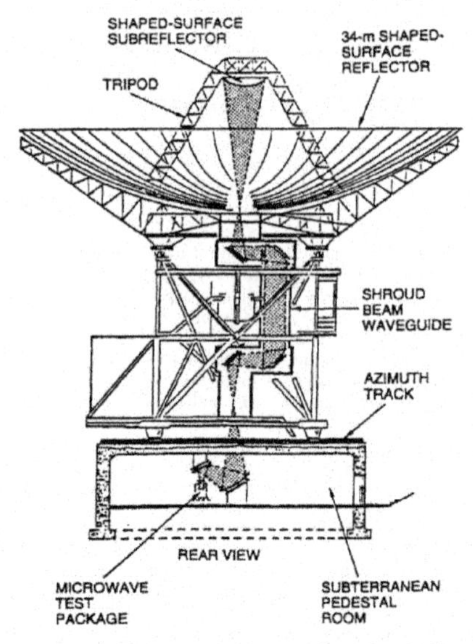

Fig. 1. View of Jet Propulsion Laboratory's (JPL's) DSS-13 BWG antenna.

Therefore, the purpose of this paper is to address the power-loss problem when more than one mode is simultaneously present in the waveguide. This effort is motivated by our desire to verify the measured data for a millimeter-wave beam–waveguide (BWG) with a protective shroud consisting of sections of a round conducting tube, as shown in Fig. 1. Solution of this problem is of great importance in optimizing the design to yield minimum noise temperature for the NASA/Deep Space Network's low-noise microwave receiving system [9].

II. Formulation of the Problem and Formal Solution

Shown in Fig. 2 is the geometry of the canonical problem. A uniform conducting waveguide of arbitrary cross section with its axis aligned in the z-direction has a length ℓ. In the $z = 0$ plane, the transverse electric field $\underline{E}_t(x, y)$ is assumed to be given. Thus, the amplitudes of all the modes (propagating and evanescent modes) can be calculated [1]–[5] and are assumed to be known. We wish to calculate the power loss of the fields due to the imperfect conductivity of the wall with intrinsic wave resistance (surface resistance) R.

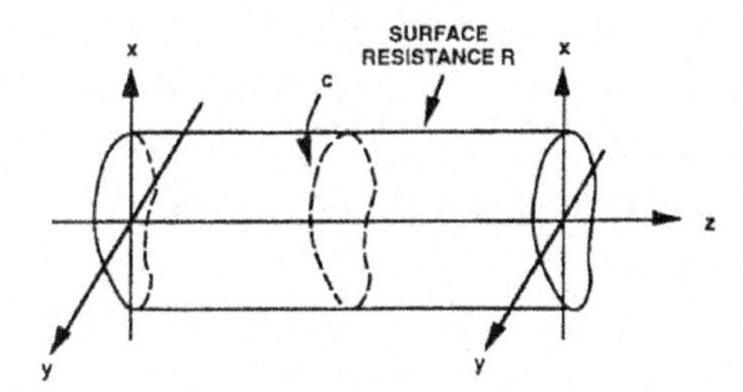

Fig. 2. Geometry of the problem.

From Ohm's Law and Poynting's vector theorem, the power loss is given by [1]–[6]

$$P_L = \frac{1}{2} R \iint_A (\underline{J}_s \cdot \underline{J}_s^*)\, dA \tag{4}$$

where

$\underline{J}_s$ $\underline{n} \times \underline{H}$ = surface current density on the wall;
$\underline{n}$ unit vector normal to the wall surface;
$\underline{H}$ total magnetic field in the waveguide;
A surface area of the wall

or

$$P_L = \frac{1}{2} R \iint_A (\underline{H}_\tau \cdot \underline{H}_\tau^*)\, dA \tag{5}$$

where $\underline{H}_t$ is the component of the total magnetic field which is tangential to the wall surface. It is known that in a hollow arbitrarily shaped uniform waveguide with a conducting wall there can exist two sets of eigenmodes: [1]–[6] transverse electric (TE) modes and transverse magnetic (TM) modes with a specific propagation constant for each mode. The total fields for TE modes are

$$\underline{E}^{(TE)}(x, y, z) = \sum_{m=1}^{\infty} A_m^{(\mathrm{TE})} \underline{E}_{mt}^{(\mathrm{TE})}(x, y) e^{j\beta_m^{(\mathrm{TE})} z} \tag{6}$$

$$\underline{H}^{(\mathrm{TE})}(x, y, z) = \sum_{m=1}^{\infty} A_m^{(\mathrm{TE})} \Big[\underline{H}_{mt}^{(\mathrm{TE})}(x, y) + H_{mz}^{(\mathrm{TE})}(x, y) \underline{e}_z \Big] e^{j\beta_m^{(\mathrm{TE})} z} \tag{7}$$

where $\underline{E}_{mt}^{(\mathrm{TE})}$ and $(\underline{H}_{mt}^{(\mathrm{TE})} + H_{mz}^{(\mathrm{TE})} \underline{e}_z)$ are connected through the Maxwell's equations and $\beta_m^{(\mathrm{TE})}$ is the propagation constant of the mth TE eigenmode, and the total fields for TM modes are

$$\underline{E}^{(\mathrm{TM})}(x, y, z) = \sum_{m=1}^{\infty} A_m^{(\mathrm{TM})} \Big[\underline{E}_{mt}^{(\mathrm{TM})}(x, y) + E_{mz}^{(\mathrm{TM})}(x, y) \underline{e}_z \Big] e^{j\beta_m^{(\mathrm{TM})} z} \tag{8}$$

$$\underline{H}^{(\mathrm{TM})}(x, y, z) = \sum_{m=1}^{\infty} A_m^{(\mathrm{TM})} \underline{H}_{mt}^{(\mathrm{TM})}(x, y)\, e^{j\beta_m^{(\mathrm{TM})} z} \tag{9}$$

where $\underline{H}_{mt}^{(\mathrm{TM})}$ and $(\underline{E}_{mt}^{(\mathrm{TM})} + E_{mz}^{(\mathrm{TM})} \underline{e}_z)$ are connected through Maxwell's equations and $\beta_m^{(\mathrm{TM})}$ is the propagation constant of the mth TM eigenmode. $A_m^{(\mathrm{TE})}$ and $A_m^{(\mathrm{TM})}$ are arbitrary amplitude coefficients for TE and TM modes. The subscript t indicates the transverse components of the field (transverse to the z-direction). The index m is used to tally the modes—it does not necessarily correspond to mode order. One notes that β_m may take on negative values, indicating modes propagating in the opposite direction.

Substituting (6)–(9) into (5) yields

$$\begin{aligned} P_L = \frac{1}{2} R \Bigg\{ & \sum_{m=1}^{M} \sum_{n=1}^{M} A_m^{(\mathrm{TE})} A_n^{(\mathrm{TE})*} \oint_c [H_{mc}^{(\mathrm{TE})} H_{nc}^{(\mathrm{TE})*} + H_{mz}^{(\mathrm{TE})} H_{nz}^{(\mathrm{TE})*}]\, dc \int_0^{\ell} e^{j(\beta_m^{(\mathrm{TE})} - \beta_n^{(\mathrm{TE})}) z}\, dz \\ & + \sum_{m'=1}^{M'} \sum_{n=1}^{M} A_{m'}^{(\mathrm{TM})} A_n^{(\mathrm{TE})*} \oint_c [H_{m'c}^{(\mathrm{TM})} H_{nc}^{(\mathrm{TE})*}]\, dc \cdot \int_0^{\ell} e^{j(\beta_{m'}^{(\mathrm{TM})} - \beta_n^{(\mathrm{TE})}) z}\, dz \\ & + \sum_{m=1}^{M} \sum_{n'=1}^{M'} A_m^{(\mathrm{TE})} A_{n'}^{(\mathrm{TM})*} \oint_c [H_{mc}^{(\mathrm{TE})} H_{n'c}^{(\mathrm{TM})*}]\, dc \cdot \int_0^{\ell} e^{j(\beta_m^{(\mathrm{TE})} - \beta_{n'}^{(\mathrm{TM})}) z}\, dz \\ & + \sum_{m'=1}^{M'} \sum_{n'=1}^{M'} A_{m'}^{(\mathrm{TM})} A_{n'}^{(\mathrm{TM})*} \oint_c [H_{m'c}^{(\mathrm{TM})} H_{n'c}^{(\mathrm{TM})*}]\, dc \cdot \int_0^{\ell} e^{j(\beta_{m'}^{(\mathrm{TM})} - \beta_{n'}^{(\mathrm{TM})}) z}\, dz \Bigg\}. \end{aligned} \tag{10}$$

Here, c is the contour around the inner surface of the waveguide, which is also normal to the z-axis (see Fig. 2). The subscript c represents the component of the transverse field that is tangential to the contour c, M is the number of TE propagating modes, M' is the number of TM propagating modes and m, m', n, n' are mode indices. Simplifying (10) gives

$$P_L = [\text{Part 1}] + [\text{Part 2}] \tag{11}$$

with (12) and (13), shown at the bottom of the following page, where

$$\begin{aligned} I_m^{(\mathrm{TE})} &= \oint_c [|H_{mc}^{(\mathrm{TE})}|^2 + |H_{mz}^{(\mathrm{TE})}|^2]\, dc \\ I_{m'}^{(\mathrm{TM})} &= \oint_c |H_{m'c}^{(\mathrm{TM})}|^2\, dc \\ I_{mn}^{(\mathrm{TE})} &= \oint_c [H_{mc}^{(\mathrm{TE})} H_{nc}^{(\mathrm{TE})*} + H_{mz}^{(\mathrm{TE})} H_{nz}^{(\mathrm{TE})*}]\, dc \\ I_{m'n'}^{(\mathrm{TM})} &= \oint_c [H_{m'c}^{(\mathrm{TM})} H_{n'c}^{(\mathrm{TM})*}]\, dc \\ I_{m'n}^{(\mathrm{TM})(\mathrm{TE})} &= \oint_c [H_{m'c}^{(\mathrm{TM})} H_{nc}^{(\mathrm{TE})*}]\, dc \\ I_{mn'}^{(\mathrm{TE})(\mathrm{TM})} &= \oint_c [H_{mc}^{(\mathrm{TE})} H_{n'c}^{(\mathrm{TM})*}]\, dc. \end{aligned} \tag{14}$$

It should be noted that P_L is always purely real.

One should point out that all the field components used in the above expressions are assumed to be the field components for a perfectly conducting waveguide. The use of surface resistance R and Ohm's Law to calculate the total power loss is an application of the perturbation technique.

Using the orthogonality properties of these field components, one can show that the total power carried in a multimode waveguide is the sum of the power carried by each propagating mode in this multimode waveguide. On the other hand, (11) shows that power losses or attenuation of different simultaneously existing modes are not simply additive, as indicated by the first bracketed term [Part 1]. The correct expression must include the second bracketed term [Part 2], which shows the cross-product terms. Indeed, the use of an attenuation constant to describe power loss in a waveguide should be limited to the single-mode unidirectional propagation case only, because only for this case is the power loss linearly dependent on the length of the guide. For the multimode propagation case, the power loss varies with the length of the guide in a rather complicated manner, as shown in (11). Equation (11) vividly demonstrates the importance of the modal coupling term. Since the factor

$$f(x) = [\exp(jx) - 1]/(jx) \tag{15}$$

(where $x = (\beta_1 - \beta_2)\ell$, β_1, and β_2 are the propagation constants for the coupling modes, and ℓ is the distance from the entrance of the waveguide to the point of interest along the guide) determines the importance of the coupling term, let us now examine this factor closely. The function $|f(x)|$ is largest when $x \rightarrow 0$ and begins to diminish and approaches zero when x increases. This means that the cross-product terms in (11) (i.e., [Part 2]) are important when the difference between the propagation constants of the propagating modes that are excited in the waveguide is small and/or when ℓ is small, such that the product $(\beta_1 - \beta_2)\ell$ is small. This condition is particularly true when the transverse dimensions of the waveguide are very large, such as in the BWG case that we considered. It is also noted that under certain conditions, [Part 2] can be negative. This means that the total power loss can be less than that given by [Part 1], the part representing only the additive aspect of power loss by each mode in a multimode waveguide.

When $|x| \gg 1$, then $f(x) \rightarrow 0$, and the coupling terms in (11) [Part 2] approach zero. This means that when $\ell \rightarrow \infty$, [Part 2] $\rightarrow 0$, and the usual decoupled result given by [Part 1] in (11) becomes valid. Thus, as $\ell \rightarrow \infty$, the power loss for each mode in the multimode waveguide is additive.

III. Application to BWG Noise Temperature Computations

We shall now apply the above theory to calculate the conductivity loss (power loss) in a large BWG tube. The noise temperature contributed by the conductivity loss in a BWG can then be easily computed. Computed results are compared with measured data from an experiment, validating the theory.

Large BWG-type ground-station antennas are generally designed with metallic tubes enclosing the BWG mirrors. The scattered field from a BWG mirror is obtained by the use of a physical optics integration procedure with a Green's function appropriate to the circular waveguide geometry [10]. In this manner, the coefficients $A_m^{(\mathrm{TE}),(\mathrm{TM})}$ of the circular waveguide modes that are propagating in the oversized waveguides are determined.

Knowing the coefficients $A_m^{(\mathrm{TE}),(\mathrm{TM})}$, one may calculate the tangential magnetic fields for the TE and TM modes from (7) and (9). The total tangential magnetic field is the sum of these tangential magnetic fields. Substituting the total tangential magnetic field into (5) and carrying out the integral in (5) numerically, one may readily obtain the total power loss P_L. This numerical technique is quite general; it can be applied to a metal tube waveguide of arbitrary shape. Another way may also be used: knowing $A_m^{(\mathrm{TE}),(\mathrm{TM})}$ for the modes in a circular metal tube (sleeve) waveguide or in a rectangular

$$[\text{Part 1}] = \tfrac{1}{2}R\ell\left[\sum_{m=1}^{M} |A_m^{(\mathrm{TE})}|^2 I_m^{(\mathrm{TE})} + \sum_{m'=1}^{M'} |A_{m'}^{(\mathrm{TM})}|^2 I_{m'}^{(\mathrm{TM})}\right] \tag{12}$$

$$\begin{aligned}[\text{Part 2}] = \tfrac{1}{2}R\ell\Bigg(&\sum_{m=1}^{M}\sum_{\substack{n=1\\ n\neq m}}^{M} A_m^{(\mathrm{TE})}A_n^{(\mathrm{TE})*} I_{mn}^{(\mathrm{TE})}\left\{\frac{e^{j[\beta_m^{(\mathrm{TE})}-\beta_n^{(\mathrm{TE})}]\ell}-1}{j[\beta_m^{(\mathrm{TE})}-\beta_n^{(\mathrm{TE})}]\ell}\right\}\\
&+\sum_{m'=1}^{M'}\sum_{\substack{n'=1,\\ n'\neq m'}}^{M'} A_{m'}^{(\mathrm{TM})}A_{n'}^{(\mathrm{TM})*} I_{m'n'}^{(\mathrm{TM})}\left\{\frac{e^{j[\beta_{m'}^{(\mathrm{TM})}-\beta_{n'}^{(\mathrm{TM})}]\ell}-1}{j[\beta_{m'}^{(\mathrm{TM})}-\beta_{n'}^{(\mathrm{TM})}]\ell}\right\}\\
&+\sum_{m'=1}^{M'}\sum_{n=1}^{M} A_{m'}^{(\mathrm{TM})}A_n^{(\mathrm{TE})*} I_{m'n}^{(\mathrm{TM})(\mathrm{TE})}\left\{\frac{e^{j[\beta_{m'}^{(\mathrm{TM})}-\beta_n^{(\mathrm{TE})}]\ell}-1}{j[\beta_{m'}^{(\mathrm{TM})}-\beta_n^{(\mathrm{TE})}]\ell}\right\}\\
&+\sum_{m=1}^{M}\sum_{n'=1}^{M'} A_m^{(\mathrm{TE})}A_{n'}^{(\mathrm{TM})*} I_{mn'}^{(\mathrm{TE})(\mathrm{TM})}\left\{\frac{e^{j[\beta_m^{(\mathrm{TE})}-\beta_{n'}^{(\mathrm{TM})}]\ell}-1}{j[\beta_m^{(\mathrm{TE})}-\beta_{n'}^{(\mathrm{TM})}]\ell}\right\}\Bigg)\end{aligned} \tag{13}$$

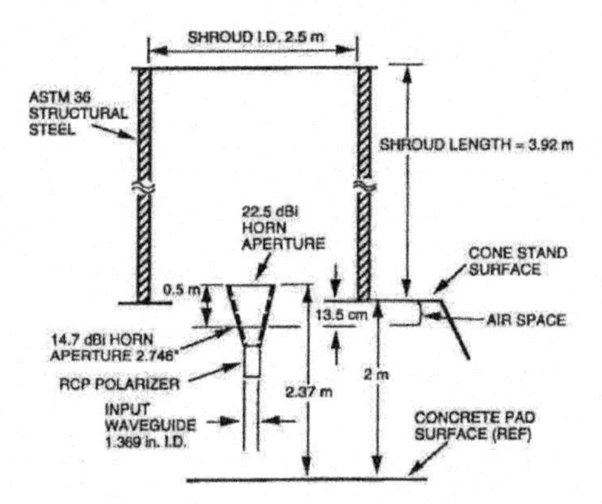

Fig. 3. Experimental setup.

metal tube waveguide, one may derive analytic expressions for the total power loss P_L.

The above numerical approach was used to calculate the conductivity loss of a short length of BWG tube [11]. The experiment utilized a 3.92-m-long 2.5-m-diameter structure steel tube and a very sensitive noise temperature measuring radiometer (see Figs. 3 and 4). Noise temperature comparisons were made between several different horns radiating in free space and radiating into the BWG tube. The experiment also included measurements with the steel tube and the tube lined with aluminum sheets. Utilizing the measured conductivity of the aluminum and steel [12] (see Table I) and the computed modes in the BWG tube, a conductivity loss was computed and converted into a noise temperature prediction. For the 14.7-dBi gain horn, the TE_{1p} and TM_{1p} modes to $p = 50$ were included, and for the 22.5 dBi gain horn, modes to $p = 22$ were included. The following formula was used for the conversion:

$$\text{Noise Temperature in K} = (P_L/P_T)T_0 \tag{16}$$

where P_L is the total power loss, P_T is the given total input power, and T_0 is the ambient temperature in K (for room temperature, $T_0 = 293.1$ K). A comparison of the measurement with both the new theory (11) and the textbook theory (3) is shown in Table II. The most dramatic difference was with the higher gain (22.5-dBi) horn. It was this experimental result which showed that the result obtained according to (3) was incorrect. The measurement was 0.1 K ± 0.1 K and there was no question that the calculation of 2.6 K from (3) was significantly outside the range measurement uncertainty. The explanation can be seen in Fig. 5, which plots the attenuation loss as a function of tube size. Because the high-gain horn does not "illuminate" the wall until further down the tube from its aperture plane, there is only a very small loss near the aperture. This clearly demonstrates the fact that the power loss is not linearly dependent on z, and thus validates the analysis.

(a)

(b)

Fig. 4. Measurement setup. (a) Horn in free space. (b) Horn with BWG tube.

IV. Conclusions

The concept of expressing power loss along a given uniform waveguide in nepers/meter must be used with caution. This concept is only generally true for single-mode unidirectional propagation. When more than one mode exists simultaneously, the power loss is no longer linearly proportional to the length of waveguide. Depending on the differences for the

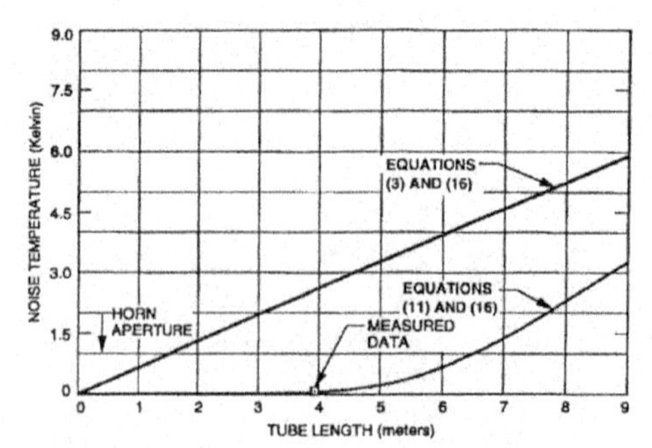

Fig. 5. Noise temperature versus tube length for 22.5-dBi gain horn.

TABLE I
ELECTRICAL CONDUCTIVITIES OF SHROUD MATERIALS [12]

Material	Effective Conductivity mhos/meter
BWG antenna shroud ASTM A36 steel	0.003×10^7
0.064 in. thick 6061 aluminum sheet	2.2×10^7
0.024 in. thick galvanized steel	1.2×10^7
High-conductivity copper	5.66×10^7

TABLE II
COMPARISON OF EXPERIMENTAL AND THEORETICAL RESULTS. THEORETICAL RESULTS ARE CALCULATED FROM (11) AND (16) AND FROM (3) AND (16)

	Measured, K	Calculated	
		New Method (11), K	Textbook Method (3), K
22.5-dBi gain horn with steel tube	0.1 ± 0.1	0.1	2.6
14.7-dBi gain horn with steel tube	2.5 ± 0.4	2.3	3.0
14.7-dBi gain horn with aluminum tube	0.2 ± 0.1	0.09	0.11

propagation constants of the coexisting propagating modes and the length of the waveguide, the total power loss may be more than, equal to, or less than the proportional sum of the power losses for each mode propagating separately, as shown in (11).

Accurate formulas for the total power loss by taking mode coupling into account can be derived for an arbitrarily shaped conducting tube, circular conducting tube, and rectangular conducting tube. The factor $f(x)$ [see (15)]—where $x = (\beta_a - \beta_b)\ell$, with $(\beta_a - \beta_b)$ being the differences between the propagation constants of various modes propagating simultaneously in the conducting tube, and ℓ being the length of the waveguide from the source plane to the plane of interest along the guide—appears to be the governing factor that controls the importance of mode coupling between mode a and mode b in affecting the total power loss calculation. Since the factor $\sin x/x$ approaches zero as x approaches infinity, the effect of the term containing this factor approaches zero, indicating the diminishing effect of mode coupling on the total power-loss calculation. Since $x = (\beta_a - \beta_b)\ell$ in order that x may approach a large value quickly, two possibilities exist.

1) If β_a is close to β_b, as in the case of a very large guide, then ℓ must be very long in order that x may be large, indicating that the mode-coupling effect can affect the total loss calculation for a very long distance from the source plane.
2) If β_a is not close to β_b, as in the case of a smaller guide, then ℓ can be relatively short for x to be large enough so that the $f(x)$ term may be negligible, indicating that the mode coupling term only affects the total loss calculation for a relatively short distance from the source plane.

When applied to the JPL millimeter-wave BWG case, one notes that β_a is very close to β_b. The newly developed loss formula for an oversized circular conducting tube was thus used to calculate the additional noise temperature contribution due to the presence of a protective shroud surrounding a millimeter-wave BWG.

ACKNOWLEDGMENT

The authors wish to thank Dr. V. Jamnejad of the JPL, Pasadena, CA, and the reviewer for their careful reading of this paper and their very helpful comments. They also wish to thank M. Franco of the JPL for his assistance in obtaining the experimental results.

REFERENCES

[1] J. A. Stratton, *Electromagnetic Theory*. New York: McGraw-Hill, 1941, pp. 543–544.
[2] R. E. Collin, *Field Theory of Guided Waves*. New York: McGraw-Hill, 1960, p. 341.
[3] C. Yeh, *Dynamic Fields, American Institute of Physics Handbook*, 3rd ed., D. E. Gray, Ed. New York: McGraw-Hill, 1972.
[4] S. Ramo, J. R. Whinnery, and Van Duzer, *Fields and Waves in Modern Communications*. New York: Wiley, 1967, pp. 413, 417.
[5] R. F. Harrington, *Time Harmonic Electromagnetic Field*. New York: McGraw-Hill Book, 1961, pp. 73, 255, 376.
[6] N. Marcuvitz, *Waveguide Handbook* (MIT Radiation Lab. Series), New York: McGraw-Hill, 1951, vol. 10, p. 25.
[7] R. C. Chou and S. W. Lee, "Modal attenuation in multilayered coated waveguides," *IEEE Trans. Microwave Theory Tech.*, vol. 36, pp. 1167–1176, July 1988.
[8] J. J. Gustincic, "A general power loss method for attenuation of cavities and waveguides," *IEEE Trans. Microwave Theory Tech.*, vol. MTT-11, pp. 83–87, Jan. 1963.
[9] D. A. Bathker, W. Veruttipong, T. Y. Otoshi, and P. W. Cramer, "Beam-waveguide antenna performance predictions with comparisons to experimental results," *IEEE Trans. Microwave Theory Tech.*, vol. 40, pp. 1274–1285, June 1992.
[10] A. G. Cha and W. A. Imbriale, "A new analysis of beam-waveguide antennas considering the presence of a metal enclosure," *IEEE Trans. Antennas Propagat.*, vol. 40, pp. 1041–1046, Sept. 1992.
[11] W. A. Imbriale, T. Y. Otoshi, and C. Yeh, "An analytic technique for computing the conductivity loss in the walls of a beam waveguide system," presented at the URSI Int. Conf., Kyoto, Japan, Aug. 25, 1992.
[12] T. Y. Otoshi and M. M. Franco, "The electrical conductivities of steel and other candidate materials for shrouds in a beam-waveguide antenna system," *IEEE Trans. Instrum. Meas.*, vol. 45, pp. 77–83, Feb. 1996.

304 IEEE TRANSACTIONS ON ANTENNAS AND PROPAGATION, VOL. AP-25, NO. 3, MAY 1977

Computer Reconstruction of Near-Zone Field from Given Far-Zone Data of Two-Dimensional Objects

SHINOBU HAMADA, MEMBER, IEEE, AND C. YEH, MEMBER, IEEE

***Abstract*—Using a complete set of cylindrical vector wave functions, it is shown that accurate near-zone electromagnetic field can be easily reconstructed from given far-zone data of two-dimensional objects. The validity of this approach was shown by the comparison of the reconstructed results with those obtained according to the exact solution of a two-dimensional slit problem.**

I. INTRODUCTION

IT HAS BEEN drilled into the minds of many electromagnetics students that the near-zone field of an antenna contains both the radiated power (the real part of $E \times H$) and the reactive power (from the imaginary part of $E \times H$), and the far-zone field contains basically the radiated power. Hence the logical conclusion of this premise is that one may obtain the far-zone field from the knowledge of near-zone field as was illustrated by many authors [1]-[4]. It seems reasonable to *assume* that one will not be able to derive the near-zone field from the knowledge of the far-zone field (i.e., one should not be expected to get something (the near-zone reactive power and radiated power) for nothing (the far-zone radiated power)). Curiously in recent years another school of researchers [5], [6] has provided some evidence that one *can* derive substantial knowledge (including reactive power) of near-zone field from far-field data. For example, Ludwig [6] has shown that if he expands the far-zone measured field of a horn antenna in terms of a complete set of spherical vector wave functions which satisfy the three-dimensional Maxwell equations, he can then obtain the expansion coefficients of the set of spherical vector wave functions. Using these expansion coefficients which remain unchanged whether one has far-field or near-field expansions, he calculates the near-zone field of the horn antenna. Excellent agreement was obtained when he compares his results with measured data. However some doubts still remain because the far-zone and near-zone data were obtained by measurements (probe-perturbation complications), and more rigorous verification is not possible because the exact theoretical result for a horn does not exist.

Realizing that exact solution to the problem of the scattering by a two-dimensional slit exists and that a complete set of cylindrical vector wave functions exists for the two-dimensional Maxwell equations, we, therefore, initiated the investigation of this two-dimensional problem whose results and conclusions are given here. Several specific problems were considered.

Manuscript received April 7, 1975; revised April 15, 1976.

S. J. Hamada was with the Electrical Sciences and Engineering Department, University of California, Los Angeles, CA. He is now with TRW Defense and Space Systems Group, Redondo Beach, CA.

C. Yeh is with the Electrical Sciences and Engineering Department, University of California, Los Angeles, CA 90024.

a) Using the exact far-field data which are calculated from exact solution of the slit problem, we are able to obtain the expansion coefficients of the cylindrical wave functions and using these same expansion coefficients we can reconstruct the near-zone field which are then compared with exact data calculated from the exact solution of the slit problem. Excellent agreements for the amplitude and phase of the field were obtained as close as 2λ from the center of a 3λ wide slit. (λ is the free-space wavelength).

b) We wish to investigate how consistent is GTD (geometrical theory of diffraction) in providing the far-zone and near-zone results. Again expansion coefficients of complete cylindrical wave were obtained using far-zone GTD results. The near-zone cylindrical wave results are compared with the exact slit results. Excellent agreement was obtained indicating the selfconsistency of GTD technique.

c) It is known that the far-zone data obtained according to the scalar Huygens-Fresnal principle is quite accurate for slit width $\geqslant 3\lambda$. We wish to investigate here whether these far-zone results can be used to produce the near-zone field. We will show that the near-zone field constructed using these far-zone results according to the cylindrical wave expansion is unreliable. Hence the scalar Huygens-Fresnel formula is not self-consistent.

d) Since the vector Kirchoff integral formula can be derived from the vector wave equation, we expect this technique to be self-consistent. This is shown using our cylindrical wave expansion.

The implication of our investigation is very clear. It means that the near-zone field (including the reactive component) may be reconstructed from far-zone field if one believes that the complete set of spherical or cylindrical vector wave functions represents the unique and complete solution of a three- or two-dimensional source-free vector wave equation, respectively. Some computational limitations do exist. They are discussed in this paper.

II. ANALYTICAL FORMULATION

The homogeneous two-dimensional vector wave equation in free space is solved to obtain the field in cylindrical coordinates. The radiated field is expanded in a series of cylindrical wave functions. We assume a time dependence of $e^{-i\omega t}$ which will be suppressed throughout. These equations can be reduced to the more familiar scalar wave equations by separating the electromagnetic field into the transverse electric wave (TE) and the transverse magnetic wave (TM). Since we are concerned with a two-dimensional problem, it will be assumed that all fields are independent of z. In the TM

Printed in U.S.A. Annals No. 705AP003

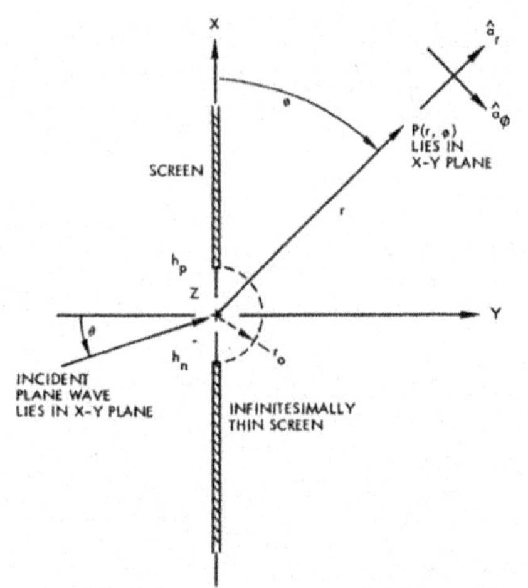

Fig. 1. Classical infinite slit, incident wave, and coordinates system.

case, the electric vector is parallel to the z axis, and in the TE case, the magnetic vector is parallel to the z axis.

The magnetic field of the TE wave may be derived from a scalar Helmholtz equations as follows:

$$H(r, \phi) = \sum_{n=-\infty}^{\infty} a_n H_n^{(1)}(kr) e^{in\phi} \mathbf{a}_z, \qquad r > r_0 \tag{1}$$

where $k = \omega\sqrt{\mu_0 \epsilon_0}$ is the free space wave number. We have assumed that the source is surrounded completely by a cylindrical surface of radius r_0 as shown in Fig. 1. $\mathbf{a}_r$, $\mathbf{a}_\phi$, and $\mathbf{a}_z$ are unit vectors in cylindrical coordinates.

To illustrate how we may obtain the near-zone field from a known far-zone field, let us first consider the case of the TE wave. A similar approach may be used for TM wave. The far-zone magnetic field expression for the TE wave can be obtained by using the asymptotic expression for the Hankel function of the first kind. For $r = A \gg 4r_0^2/\lambda$, we have

$$H_z^{\mathrm{far}}(A, \phi) = \sqrt{\frac{2}{\pi k A}}\, e^{i(kA - \pi/4)} \sum_{n=-\infty}^{\infty} a_n e^{in\left(\phi - \frac{\pi}{2}\right)}. \tag{2}$$

It is noted that the modal coefficients a_n remain unchanged for far-zone or near-zone expressions. Hence, if a_n can be determined from a known tangential magnetic field H_z^{far}, the same coefficients can be used to compute the near-zone field.

A Fourier analysis of (2) gives

$$a_n = \sqrt{\frac{kA}{8\pi}}\, e^{-i\left(kA - \frac{(2n+1)\pi}{4}\right)} \int_0^{2\pi} H_z^{\mathrm{far}}(A, \phi) e^{-in\phi}\, d\phi \tag{3}$$

where $n = 0, \pm 1, \pm 2 \cdots \infty$.

For numerical computation, it is necessary to truncate the summation in (2) so that the sum is carried out for a finite number of terms N_0, i.e.,

$$a_n = 0, \qquad \text{when } n > \frac{N_0}{2} \tag{4}$$

$$\frac{3}{2} kr_1 \geqslant \frac{N_0}{2} > kr_0. \tag{5}$$

This constraint for $N_0/2$ is derived empirically which will be discussed in detail in the following section. Here r_1 represents the radial distance at which the near-zone field is reconstructed. If this problem is to be solved by discrete Fourier transformation (DFT) [7], the number of computed sample points L of the exact far-field must be equal to the number of unknown coefficients N_0, i.e.,

$$L = N_0. \tag{6}$$

The Fourier coefficients of the far-field D_n^{far} can be represented in the discrete Fourier transform as follows:

$$D_n^{\mathrm{far}} = \sum_{m=-\frac{L}{2}+1}^{m=\frac{L}{2}} H_z^{\mathrm{far}}(A, \phi_m) e^{-in\phi_m} \tag{7}$$

where

$$\phi = \phi_m = m\Delta\phi, \quad \Delta\phi = \frac{2\pi}{L}$$

$$m = -\frac{L}{2} + 1, -\frac{L}{2} + 2, \cdots, -1, 0, 1, \cdots, \frac{L}{2}.$$

The discrete inverse Fourier transform is then given by

$$H_z^{\mathrm{far}}(A, \phi_m) = \frac{1}{L} \sum_{n=-\frac{L}{2}+1}^{n=\frac{L}{2}} D_n^{\mathrm{far}} e^{in\phi_m}. \tag{8}$$

So the modal coefficient a_n is

$$a_n = \frac{D_n^{\mathrm{far}}}{L} \sqrt{\frac{\pi k A}{2}}\, e^{-i\left(kA - \frac{(2n+1)\pi}{4}\right)}. \tag{9}$$

Using (1) and (9), the near-zone field at $r = r_1$ may be reconstructed as follows:

$$H_z^{\mathrm{nf}}(r_1, \phi_m) = \sum_{n=-\frac{L}{2}+1}^{n=\frac{L}{2}} a_n H_n^{(1)}(kr_1) e^{in\phi_m}, \qquad r_1 > r_0. \tag{10}$$

Expressing the near-zone field in a DFT form yields

$$H_z^{nf}(r_1, \phi_m) = \frac{1}{L} \sum_{n=-\frac{L}{2}+1}^{n=\frac{L}{2}} D_n^{nf} e^{in\phi_m}, \qquad r_1 > r_0 \quad (11)$$

where the near-field Fourier coefficients D_n^{nf} are

$$D_n^{nf} = La_n H_n^{(1)}(kr_1), \qquad r_1 > r_0. \qquad (12)$$

This means that the mathematical connection between the far-zone field and the near-zone field is given by a very simple relationship between the respective Fourier coefficients D_n^{far} and D_n^{nf} as follows:

$$D_n^{nf} = D_n^{far} H_n^{(1)}(kr_1) \sqrt{\frac{\pi KA}{2}}\, e^{-i\left(kA - \frac{(2n+1)}{4}\pi\right)}. \qquad (13)$$

The determination of the near-field from the known far-field is thus obtainable by a double discrete Fourier transformation. The coefficients D_n^{far} are first calculated from a Fourier transform of the given far-zone field; then the coefficients D_n^{nf} are determined from (13). A Fourier transformation of the D_n^{nf} according to (11) gives the required near-zone field.

A similar analysis can be carried out for the TM wave. The results are identical to those for the TE wave given above except H_z should be replaced by E_z and a_n by b_n throughout.

III. NUMERICAL RESULTS

Now we are in a position to determine how well the analytical technique described in the previous section performs in obtaining the near-zone field from given far-zone field data and to test the constraints on the truncation of the infinite summation. The canonical two-dimensional problem we have selected to use for our computer experiment is the classical infinite slit diffraction problem as shown in Fig. 1. Complete solution of this problem in terms of Mathieu functions is well known [8]. Exact numerical results for far-zone as well as near-zone fields can be computed from this Mathieu function solution. We shall carry out the computer experiment as follows:

a) The exact far-zone fields for an incident TE plane wave and for an incident TM plane wave are first calculated from the Mathieu function solutions at a far-zone distance of $A = 10\,000\lambda$ using several sample points: $L = 16, 32$, and 64. Two different slit widths are used: 3λ and 10λ. Accordingly, the expansion coefficients for the complete set of cylindrical waves are computed using these various sample points. The near-zone fields are then reconstructed at a distance of $r_1 = 2\lambda$ using the fast Fourier transform (FFT) technique according to the algorithm described in Section II. The reconstructed near-zone field intensity and phase for TE wave are shown in Fig. 2 for the following truncating number of the harmonics: $N_0/2 = 8, 16, 32$. The diamond points in Fig. 2 refer to the case $N_0/2 = 8$, the circle points refer to the case $N_0/2 = 16$, and the triangle points refer to the case $N_0/2 = 32$. The solid curves represent the exact near-field results obtained from

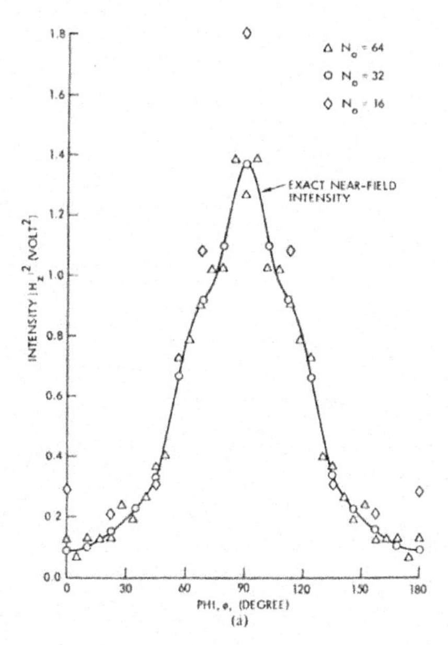

(a)

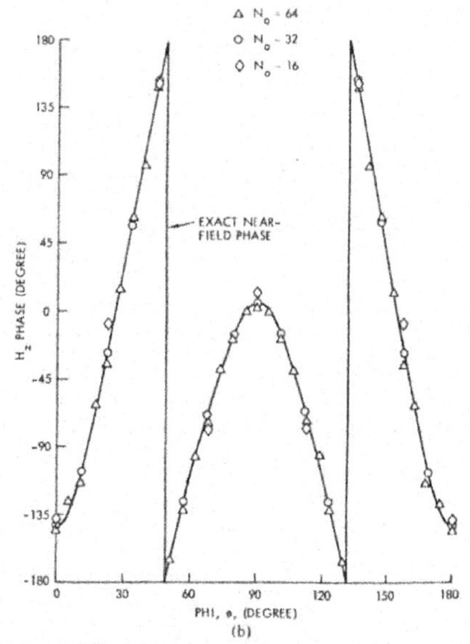

(b)

Fig. 2. (a) Effect of truncation number for harmonics on near-field intensity computations. Slit width = 3λ, $r_1 = 2\lambda$, $r_0 = 10\,000\lambda$. TE case. (b) Effect of truncation number for harmonics on near-field phase computations. Slit width = 3λ, $r_1 = 2\lambda$, $r_0 = 10\,000\lambda$. TE case.

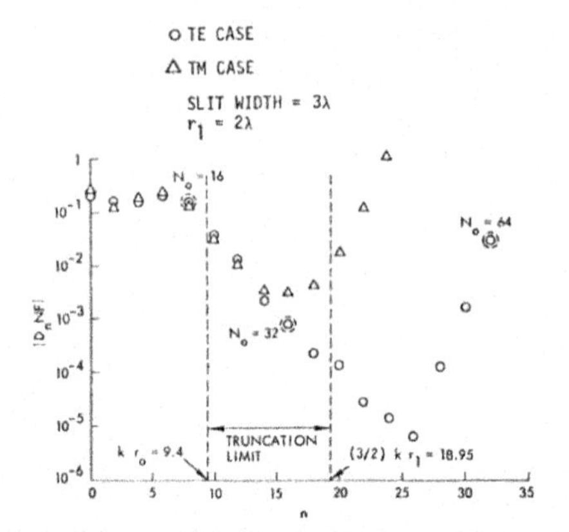

Fig. 3. Magnitudes of near-field coefficients as function of truncation number n.

the Mathieu function solution at $r_1 = 2\lambda$. It can be seen that with $kr_0 = 9.42$ and $3kr_1/2 = 18.85$, only the $N_0/2 = 16$ satisfies the empirical conditions given by (4) and (5). This observation is confirmed by Fig. 2 Figs. 2(a) and 2(b) show that if $N_0/2$ is chosen outside the constraint given by (5), such as for the case $N_0/2 = 8$ or for the case $N_0/2 = 32$, errors are introduced for the reconstructed near-zone field. Only if $N_0/2$ is chosen within the constraint (5), will good agreement be obtainable between the reconstructed near-zone field and the exact near-zone field. Further illustration is given by Fig. 3 in which the magnitude of the Fourier coefficients of the near-zone field is plotted as a function of the number of harmonics n. Fig. 3 indicates that the first truncation number $N_0/2 = 8$ was not big enough. The magnitude of near-field Fourier coefficients D_n^{nf} decreases as a function of the number of harmonics until n has reached 26 for the TE case (18 for the TM case), then it starts to increase very rapidly beyond this number. This phenomenon can be understood with the help of (13). For a fixed argument kr_1, the Hankel function $H_n^{(1)}(kr_1)$ increases very rapidly with increasing n, while D_n^{far} is not decaying fast enough with increasing n.

Further comparisons between the exact and reconstructed near-zone fields are shown in Table I and Figs. 4 and 5 for the normally incident TE case and in Table II and Fig. 6 for the obliquely incident TE case. It can be seen that when the empirical formulas ((4) and (5)) are obeyed, excellent agreement between the exact and the reconstructed near-zone fields was achieved.

Similar computations were carried out for the TM case. Again if the empirical relations were satisfied, excellent agreement between the exact and the reconstructed near-zone fields was found. Fig. 7 illustrates the situation.

It is not a coincidence that the reconstructed near-zone field as close as 2λ from the aperture agrees so well with the exact results. The argument that this near-zone region contains negligible reactive field is certainly unjustified. The reason that one obtains such excellent near-zone reconstruction is due to the use of a complete set of wave functions to represent the far-zone as well as the near-zone fields. It is anticipated that this technique can be used to obtain the near-zone fields of mutually coupled slot radiators or dipole antennas from their far-zone fields [10].

TABLE I
COMPARISON OF EXACT NEAR-FIELDS AND RECONSTRUCTED NEAR-FIELDS USING CYLINDRICAL WAVE EXPANSIONS

PHI, ϕ (DEG)	EXACT NEAR FIELDS		RECONSTRUCTED NEAR FIELDS	
	INTENSITY, $\|H_z\|^2$	PHASE (DEG)	INTENSITY, $\|H_z\|^2$	PHASE (DEG)
0.00	0.0915	-143.27	0.0915	-143.45
11.25	0.1022	-107.40	0.1025	-107.31
22.50	0.1512	-28.59	0.1510	-28.56
33.75	0.2259	57.65	0.2259	57.65
45.00	0.3260	149.90	0.3260	149.91
56.25	0.6581	-130.31	0.6579	-130.33
67.50	0.9089	-69.13	0.9096	-69.12
78.75	1.1002	-16.08	1.0996	-16.06
90.00	1.3779	5.46	1.3733	5.44
101.75	1.1002	-16.08	1.0996	-16.06
112.50	0.9089	-69.13	0.9096	-69.12
123.75	0.6581	-130.31	0.6579	-130.33
135.00	0.3260	149.90	0.3260	149.91
146.25	0.2259	57.65	0.2259	57.65
157.50	0.1512	-28.59	0.1510	-28.56
168.75	0.1022	-107.40	0.1025	-107.31
180.00	0.0915	-143.27	0.0915	-143.45

Condition: (TE case, normal incidence, slit width = 3λ and $r_1 = 2\lambda$)

TABLE II
COMPARISON OF EXACT NEAR-FIELDS AND RECONSTRUCTED NEAR-FIELDS USING CYLINDRICAL WAVE EXPANSIONS

PHI, ϕ (DEG)	EXACT NEAR FIELDS		RECONSTRUCTED NEAR FIELDS	
	INTENSITY, $\|H_z\|^2$	PHASE (DEG)	INTENSITY, $\|H_z\|^2$	PHASE (DEG)
0.00	0.0699	11.93	0.0699	11.63
11.25	0.0705	48.79	0.0708	48.96
22.50	0.0803	133.44	0.0801	133.52
33.75	0.1036	-131.25	0.1034	-131.26
45.00	0.1160	-29.86	0.1159	-29.87
56.25	0.2000	68.53	0.1999	68.50
67.50	0.2712	152.22	0.2715	152.21
78.75	0.4545	-124.85	0.4547	-124.83
90.00	0.7886	-68.89	0.7880	-68.89
101.25	0.9191	-23.20	0.9196	-23.22
112.50	1.3449	1.32	1.3446	1.33
123.75	1.3251	-13.79	1.3255	-13.81
135.00	0.8252	-63.94	0.8246	-63.93
146.25	0.5605	-132.85	0.5609	-132.84
157.50	0.3369	158.24	0.3367	158.22
168.75	0.1763	92.87	0.1764	92.94
180.00	0.1338	60.88	0.1339	60.77

Condition: (TE case, oblique angle of incidence, slit width = 3λ and $r_1 = 2\lambda$)

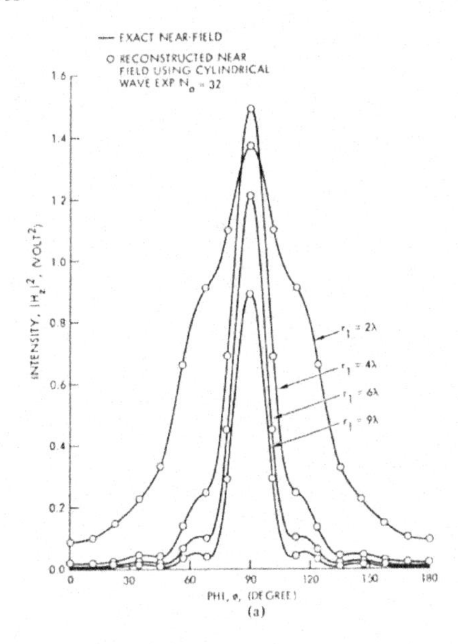

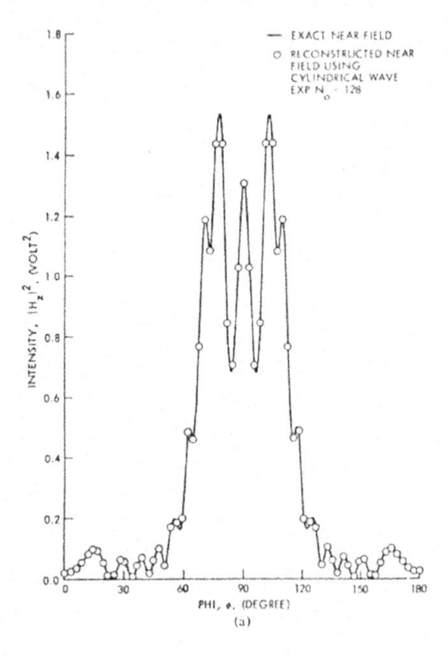

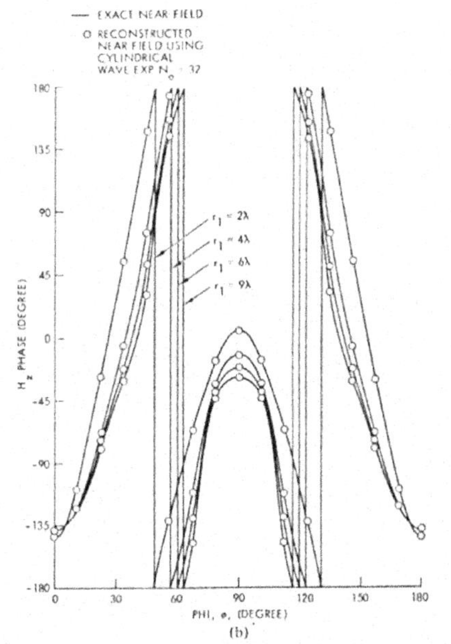

Fig. 4. (a) Comparison of exact and reconstructed near-field intensity of three-wavelength slit at $r_1 = 2\lambda$, 4λ, 6λ, and 9λ. TE case, normal incidence. (b) Comparison of exact and reconstructed near-field phase of three-wavelength slit at $r_1 = 2\lambda$, 4λ, 6λ, and 9λ. TE case, normal incidence.

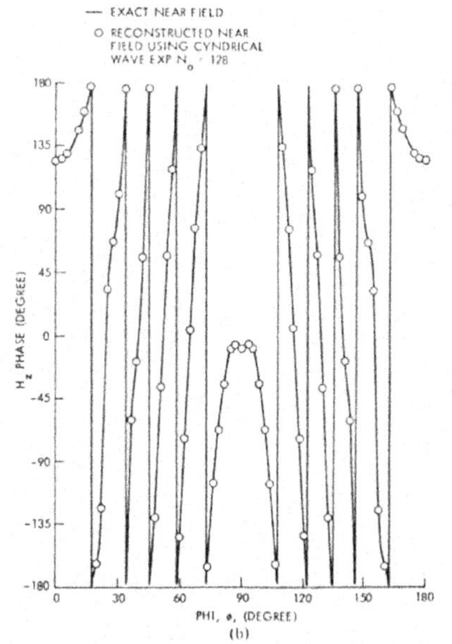

Fig. 5. (a) Comparison of exact and reconstructed near-field intensity of ten-wavelength slit. TE case, normal incidence. (b) Comparison of exact and reconstructed near-field phase of ten-wavelength slit. TE case, normal incidence.

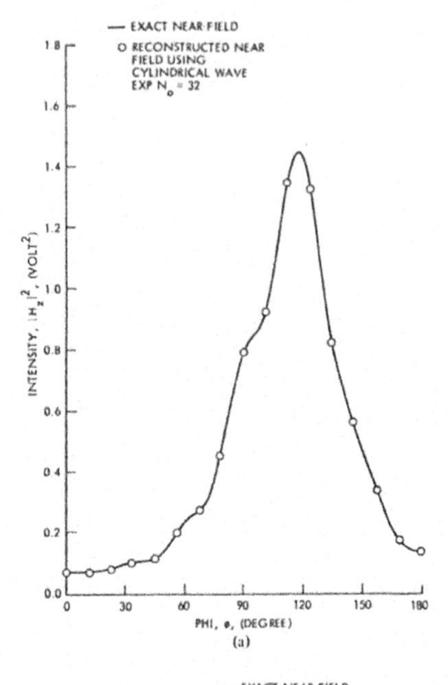

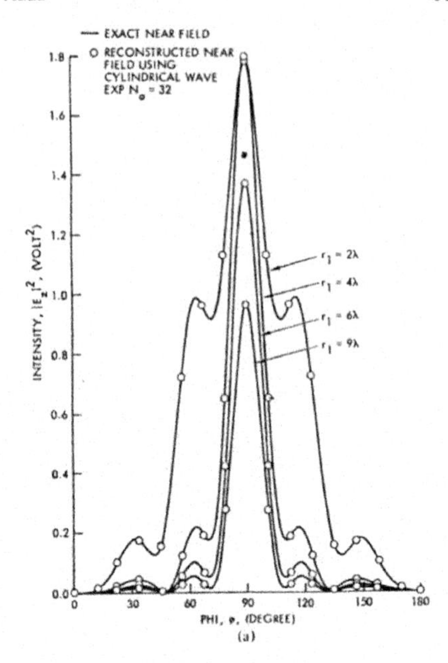

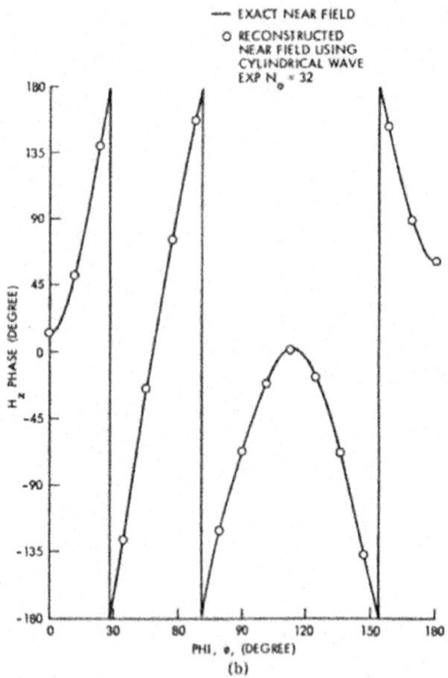

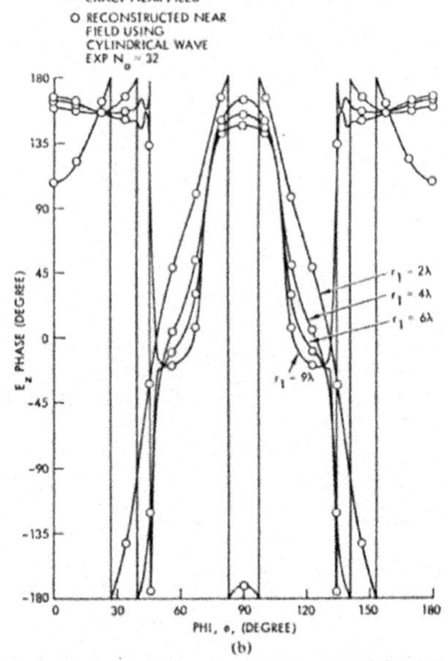

Fig. 6. (a) Comparison of exact and reconstructed near-field intensity of three-wavelength slit. TE case, oblique angle of incidence (θ = 22.5°). (b) Comparison of exact and reconstructed near-field phase of three-wavelength slit. TE case, oblique angle of incidence (θ = 22.5°).

Fig. 7. (a) Comparison of exact and reconstructed near-field intensity of three-wavelength slit at r_1 = 2λ, 4λ, 6λ, and 9λ. TM case, normal incidence. (b) Comparison of exact and near-field phase of three-wavelength slit at r_1 = 2λ, 4λ, 6λ, and 9λ. TM case, normal incidence.

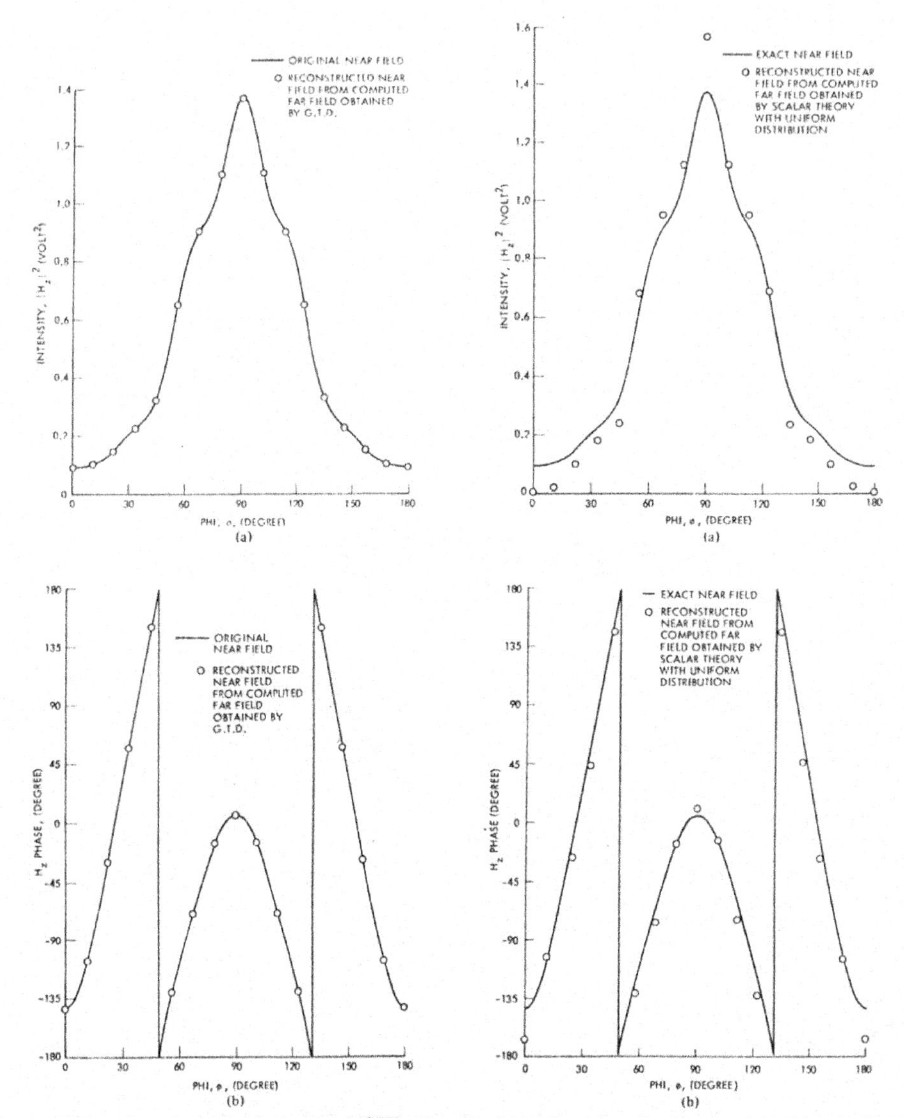

Fig. 8. (a) Comparison of original and reconstructed (GTD) near-field intensity of three-wavelength slit. TE case, normal incidence. (b) Comparison of original and reconstructed (GTD) near-field phase of three-wavelength slit. TE case, normal incidence.

Fig. 9. (a) Comparison of exact and reconstructed (scalar theory) near-field intensity of three-wavelength slit. TE case, normal incidence. (b) Comparison of exact and reconstructed (scalar theory) near-field phase of three wavelength slit. TE case, normal incidence.

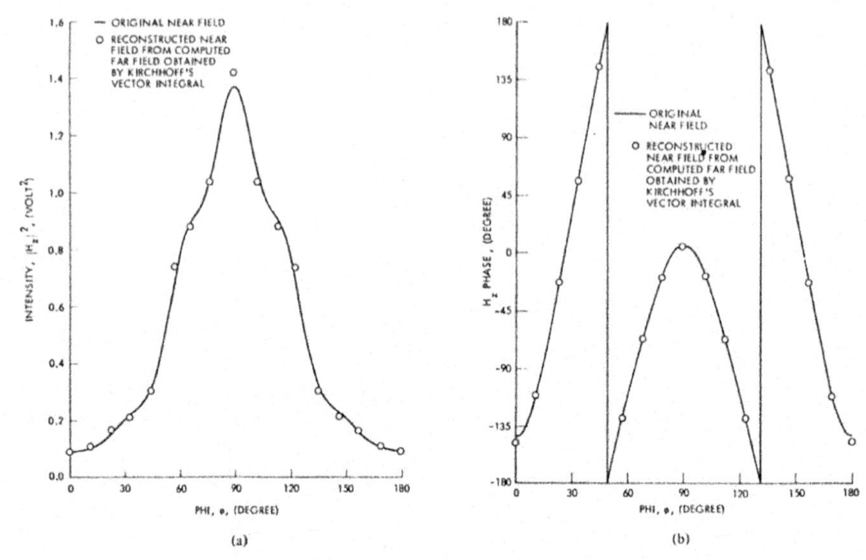

Fig. 10. (a) Comparison of original and reconstructed (Kirchhoff's vector integral) near-field intensity of three-wavelength slit. TE case, normal incidence. (b) Comparison of original and reconstructed (Kirchhoff's vector integral) near-field phase of three-wavelength slit. TE case, normal incidence.

b) The secondary purpose of this paper is to investigate the self-consistency of various classical techniques in predicting the far-zone radiated fields using a simple assumed aperture field or using the geometrical theory of diffraction. What we do to check the self-consistency of these techniques was to obtain the expansion coefficients for cylindrical wave functions from the far-zone data supplied by these techniques and then to calculate the near-zone fields using the same expansion coefficients and to compare with the exact results obtained according to the Mathieu function solution. Figs. 8–10 illustrate the comparisons. They confirm our understanding that both GTD and vector Kirchhoff integral formula are self-consistent while the scalar Huggens-Fresnal formula is not self-consistent.

In conclusion we can say that the near-zone field of a radiator can be generated if the far-zone field is known. Our computational experience shows that even if the far-zone field is not known completely, a reasonably good approximate near-zone field can still be obtained using this cylindrical vector wave function expansion for the two-dimensional problem and using the spherical vector wave function expansion for the three-dimensional problem.

REFERENCES

[1] H. G. Booker and P. C. Clemmow, "The Concept of an Angular Spectrum of Plane Waves and its Relation to that of Polar Diagram and Aperture Distribution," *J. IEE*, Vol. 97, Part III, Paper No. 922R, pp. 11–17, Jan. 1950.

[2] M. A. K. Hamid, "The Radiation Pattern of an Antenna from Near-Field Correlation Measurements," *IEEE Trans. Antennas Propagat.*, Vol. AP-16, No. 3, pp. 351–353, May 1968.

[3] D. T. Paris, "Digital Computer Analysis of Aperture Antennas," *IEEE Trans. Antennas Propagat.*, Vol. AP-16, No. 2, pp. 262–264, March 1968.

[4] J. Brown and E. V. Jull, "The Prediction of Aerial Radiation Patterns from Near-Field Measurements," *Proc. Inst. Elec. Eng.*, Vol. 108B, pp. 635–644, Nov. 1961; E. V. Jull, "The Prediction of Aerial Radiation Patterns from Limited Near-Field Measurements," *Proc. Inst. Eng.*, Vol. 110, pp. 501–506, March 1963; F. Jensen, *Electromagnetic Near-Field Far-Field Correlations*, Ph.D. dissertation, Tech. Univ. Denmark, Lyngby, Denmark, July 1970.

[5] D. R. Wilton and R. Mittra, "A New Numerical Approach to the Calculation of Electromagnetic Scattering Properties of Two-Dimensional Bodies of Arbitrary Cross Section," *IEEE Trans. Antennas Propagat.*, Vol. AP-20, No. 3, pp. 310–317, May 1972.

[6] A. C. Ludwig, "Near-Field, Far-Field Transformations Using Spherical-Wave Expansions," *IEEE Trans. Antennas Propagat.*, Vol. AP-19, No. 2, pp. 214–220, March 1971.

[7] E. O. Brigham and R. E. Morrow, "The Fast Fourier Transform," *IEEE Spectrum*, Vol. 4, No. 12, pp. 63–70, December 1967.

[8] P. M. Morse and P. J. Rubenstein, "The Diffraction of Waves by Ribbons and Slits," *Physical Review*, Vol. 54, pp. 895–898, December 1938.

[9] S. J. Hamada, "Theoretical Predictions of Electromagnetic Near-Field Distribution of Two-Dimensional Objects from Known Far-Field Data," Ph.D. dissertation, UCLA, Los Angeles (1973).

[10] ——, unpublished data, TRW.

1588 IEEE TRANSACTIONS ON MICROWAVE THEORY AND TECHNIQUES, VOL. 48, NO. 9, SEPTEMBER 2000

Noise Temperature of a Lossy Flat-Plate Reflector for the Elliptically Polarized Wave-Case

T. Y. Otoshi and C. Yeh

***Abstract*—This short paper presents the derivation of equations necessary to calculate noise temperature of a lossy flat-plate reflector. Reflector losses can be due to metallic surface resistivity and multilayer dielectric sheets, including thin layers of plating, paint, and primer on the reflector surface. The incident wave is elliptically polarized, which is general enough to include linear and circular polarizations as well. The derivations show that the noise temperature for the circularly polarized incident wave case is simply the average of those for perpendicular and parallel polarizations.**

***Index Terms*—Elliptical polarization, noise temperature, point, primer, reflector.**

I. INTRODUCTION

Although equations for power in an incident and reflected elliptically polarized wave can be derived in a straightforward manner, the equations for the associated noise temperatures are not well known nor, to the authors' knowledge, can they be found in published literature. It is especially of interest to know what the relations are when expressed in terms of perpendicular and parallel polarizations and the corresponding reflection coefficients. The following presents the derivations of noise-temperature equations for three cases of interest.

II. THEORY

A. *Power Relationships*

For the coordinate system geometry shown in Fig. 1, the fields for an incident elliptically polarized plane wave at the reflection point are [1], [2]

$$\overline{E}_i = E_{xi}\hat{\mathbf{a}}_{xi} + E_{yi}\hat{\mathbf{a}}_{yi} \quad (1)$$

$$\overline{H}_i = H_{xi}\hat{\mathbf{a}}_{xi} + H_{yi}\hat{\mathbf{a}}_{yi} \quad (2)$$

where

$$E_{xi} = E_1 e^{j(\omega t - k z_i)} \quad (3)$$

$$E_{yi} = E_2 e^{j(\omega t - k z_i + \delta)} \quad (4)$$

$$H_{xi} = -\frac{E_{yi}}{\eta} \quad (5)$$

$$H_{yi} = \frac{E_{xi}}{\eta} \quad (6)$$

Manuscript received December 16, 1999. This work was carried out by the Jet Propulsion Laboratory, California Institute of Technology, under a contract with the National Aeronautics and Space Administration.

The authors are with the Jet Propulsion Laboratory (JPL), California Institute of Technology, Pasadena, CA 91109 USA.

Publisher Item Identifier S 0018-9480(00)07404-4.

IEEE TRANSACTIONS ON MICROWAVE THEORY AND TECHNIQUES, VOL. 48, NO. 9, SEPTEMBER 2000 1589

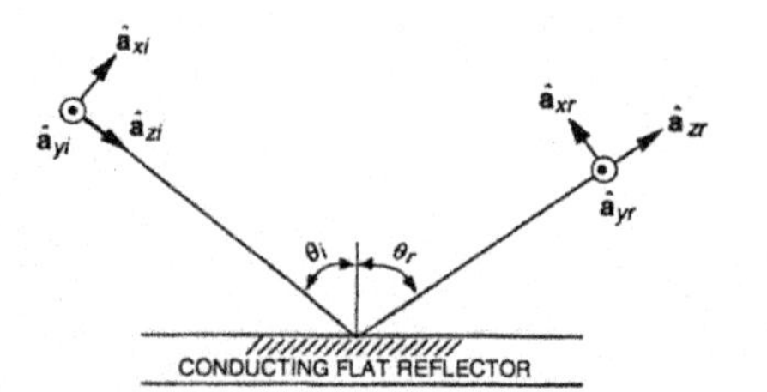

Fig. 1. Coordinate system for incident and reflected plane waves. The symbols with boldface $\hat{\mathbf{a}}$ are unit vectors and θ_i and θ_r are angles of incidence and reflection, respectively. The plane of incidence is the plane of this page.

where ω is the angular frequency, t is time, η is the characteristic impedance of free space, k is the free-space wavenumber, and z_i is the distance from an arbitrarily chosen source point on the incident wave ray path to the reflection point on the reflector surface (Fig. 1). In (3) and (4), it is important to note that E_1 and E_2 are scalar magnitudes and δ is the phase difference between E_{xi} and E_{yi}.

The Poynting vector [1] for the incident wave is expressed as

$$\overline{P}_i = \frac{1}{2}\mathrm{Re}\left(\overline{E}_i \times \overline{H}_i^*\right) \tag{7}$$

where $\times$, $*$, and Re denote cross product, complex conjugate, and real part, respectively.

Then assuming all of the incident power travels through an area A in the direction of Poynting vector, the total incident wave power is

$$P_{Ti} = \int \left(\overline{P}_i \cdot \hat{\mathbf{a}}_{zi}\right) dA \tag{8}$$

where $\cdot$ denotes the dot product. Substitutions of (1)–(7) into (8) result in

$$P_{Ti} = \frac{1}{2\eta}\left(E_1^2 + E_2^2\right) A. \tag{9}$$

The equations for the reflected wave are obtained by replacing the subscript i with r in all of the equations for the incident wave except for (3) and (4). From Fig. 1, it can be seen that the expressions for E_{xr} and E_{yr} are

$$E_{xr} = \Gamma_{\|} E_{xi} e^{-jkz_r} \tag{10}$$

$$E_{yr} = \Gamma_{\perp} E_{yi} e^{-jkz_r} \tag{11}$$

where

$\Gamma_{\|}$ is the voltage reflection coefficient for parallel polarization at the reflection point and is a function of incidence angle θ_i (see Fig. 1)

$\Gamma_{\perp}$ is the voltage reflection coefficient for perpendicular polarization at the reflection point and is a function of incidence angle θ_i

and z_r is the distance from the reflection point on the reflector surface to an arbitrary observation point along the reflected ray path.

Then following steps similar to those used to obtain (9), the total power for the reflected wave, can be derived as

$$P_{Tr} = \frac{1}{2\eta}\left[\left|\Gamma_{\|}\right|^2 E_1^2 + \left|\Gamma_{\perp}\right|^2 E_2^2\right] A. \tag{12}$$

It is assumed that the lossy flat-plate reflector in Fig. 1 has sufficient thickness so that no power is transmitted out the bottom side. Then the dissipated power is

$$P_d = P_{Ti} - P_{Tr}. \tag{13}$$

B. Noise-Temperature Relationships

For the geometry of Fig. 1, the equation for the noise temperature of the lossy flat-plate reflector can be derived from

$$T_n = \left(\frac{P_d}{P_{Ti}}\right) T_p \tag{14}$$

where T_p is the physical temperature of the reflector in units of kelvin. For example, if the lossy conductor is at a physical temperature of 20°C, then T_p = 293.16K. Use of (9), (12), and (13) in (14) gives

$$T_n = \left(1 - \left|\Gamma_{ep}\right|^2\right) T_p \tag{15}$$

where

$$\left|\Gamma_{ep}\right|^2 = \frac{\left|\Gamma_{\|}\right|^2 E_1^2 + \left|\Gamma_{\perp}\right|^2 E_2^2}{E_1^2 + E_2^2}. \tag{16}$$

Equation (15) is the elliptically polarized wave noise-temperature equation that is general enough to apply to linear and circular polarizations as well. In the following, the noise-temperature expressions for three different polarization cases are derived.

Case 1: If the incident wave is linearly polarized with the E-field perpendicular to the plane of incidence, then $E_1 = 0$ and (15) becomes

$$T_n = (T_n)_{\perp} = \left(1 - \left|\Gamma_{\perp}\right|^2\right) T_p. \tag{17}$$

Case 2: If the incident wave is linearly polarized with the E-field parallel to the plane of incidence, then $E_2 = 0$ and (15) becomes

$$T_n = (T_n)_{\|} = \left(1 - \left|\Gamma_{\|}\right|^2\right) T_p. \tag{18}$$

Case 3: If the incident wave is circularly polarized, then $E_1 = E_2$ and

$$T_n = (T_n)_{cp} = \left[1 - \frac{\left(\left|\Gamma_{\|}\right|^2 + \left|\Gamma_{\perp}\right|^2\right)}{2}\right] T_p. \tag{19}$$

Note then that $(T_n)_{cp}$ is also just the average of $(T_n)_{\perp}$ and $(T_n)_{\|}$ or

$$(T_n)_{cp} = \frac{1}{2}\left[(T_n)_{\perp} + (T_n)_{\|}\right]. \tag{20}$$

The reader is reminded that, since the reflection coefficients are functions of incidence angle θ_i, the noise temperatures are also functions of θ_i as well as polarization.

C. Excess Noise-Temperature Relationships

It is of interest to see what the relationship is for excess noise temperature as well. For painted reflector noise-temperature analyses [3], it is convenient to use the term excess noise temperature (ENT). It is defined in [3] as the total noise temperature of a painted reflector minus the noise temperature of the reflector (bare metal) without paint. Mathematically, it is expressed as

$$\Delta T_n = T_{n2} - T_{n1} = \left(1 - \left|\Gamma_2\right|^2\right) T_p - \left(1 - \left|\Gamma_1\right|^2\right) T_p \tag{21}$$

IEEE TRANSACTIONS ON MICROWAVE THEORY AND TECHNIQUES, VOL. 48, NO. 9, SEPTEMBER 2000 1589

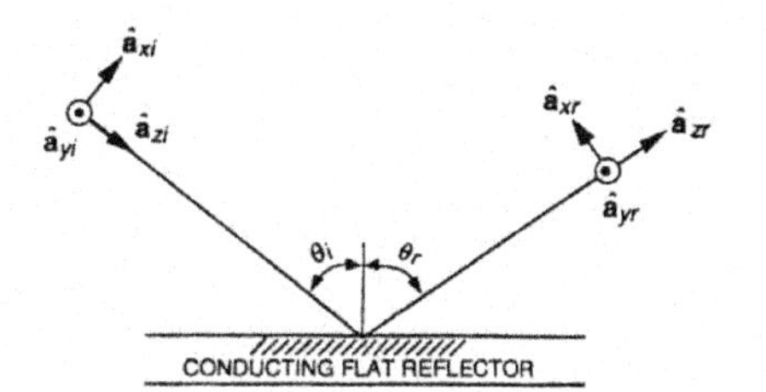

Fig. 1. Coordinate system for incident and reflected plane waves. The symbols with boldface $\hat{\mathbf{a}}$ are unit vectors and θ_i and θ_r are angles of incidence and reflection, respectively. The plane of incidence is the plane of this page.

where ω is the angular frequency, t is time, η is the characteristic impedance of free space, k is the free-space wavenumber, and z_i is the distance from an arbitrarily chosen source point on the incident wave ray path to the reflection point on the reflector surface (Fig. 1). In (3) and (4), it is important to note that E_1 and E_2 are scalar magnitudes and δ is the phase difference between E_{xi} and E_{yi}.

The Poynting vector [1] for the incident wave is expressed as

$$\overline{P}_i = \frac{1}{2}\mathrm{Re}\left(\overline{E}_i \times \overline{H}_i^*\right) \tag{7}$$

where $\times$, $*$, and Re denote cross product, complex conjugate, and real part, respectively.

Then assuming all of the incident power travels through an area A in the direction of Poynting vector, the total incident wave power is

$$P_{Ti} = \int \left(\overline{P}_i \cdot \hat{\mathbf{a}}_{zi}\right) dA \tag{8}$$

where $\cdot$ denotes the dot product. Substitutions of (1)–(7) into (8) result in

$$P_{Ti} = \frac{1}{2\eta}\left(E_1^2 + E_2^2\right) A. \tag{9}$$

The equations for the reflected wave are obtained by replacing the subscript i with r in all of the equations for the incident wave except for (3) and (4). From Fig. 1, it can be seen that the expressions for E_{xr} and E_{yr} are

$$E_{xr} = \Gamma_{\parallel} E_{xi} e^{-jkz_r} \tag{10}$$

$$E_{yr} = \Gamma_{\perp} E_{yi} e^{-jkz_r} \tag{11}$$

where

$\Gamma_{\parallel}$ is the voltage reflection coefficient for parallel polarization at the reflection point and is a function of incidence angle θ_i (see Fig. 1)

$\Gamma_{\perp}$ is the voltage reflection coefficient for perpendicular polarization at the reflection point and is a function of incidence angle θ_i

and z_r is the distance from the reflection point on the reflector surface to an arbitrary observation point along the reflected ray path.

Then following steps similar to those used to obtain (9), the total power for the reflected wave, can be derived as

$$P_{Tr} = \frac{1}{2\eta}\left[\left|\Gamma_{\parallel}\right|^2 E_1^2 + \left|\Gamma_{\perp}\right|^2 E_2^2\right] A. \tag{12}$$

It is assumed that the lossy flat-plate reflector in Fig. 1 has sufficient thickness so that no power is transmitted out the bottom side. Then the dissipated power is

$$P_d = P_{Ti} - P_{Tr}. \tag{13}$$

B. Noise-Temperature Relationships

For the geometry of Fig. 1, the equation for the noise temperature of the lossy flat-plate reflector can be derived from

$$T_n = \left(\frac{P_d}{P_{Ti}}\right) T_p \tag{14}$$

where T_p is the physical temperature of the reflector in units of kelvin. For example, if the lossy conductor is at a physical temperature of 20°C, then T_p = 293.16K. Use of (9), (12), and (13) in (14) gives

$$T_n = \left(1 - \left|\Gamma_{ep}\right|^2\right) T_p \tag{15}$$

where

$$\left|\Gamma_{ep}\right|^2 = \frac{\left|\Gamma_{\parallel}\right|^2 E_1^2 + \left|\Gamma_{\perp}\right|^2 E_2^2}{E_1^2 + E_2^2}. \tag{16}$$

Equation (15) is the elliptically polarized wave noise-temperature equation that is general enough to apply to linear and circular polarizations as well. In the following, the noise-temperature expressions for three different polarization cases are derived.

Case 1: If the incident wave is linearly polarized with the E-field perpendicular to the plane of incidence, then $E_1 = 0$ and (15) becomes

$$T_n = (T_n)_{\perp} = \left(1 - \left|\Gamma_{\perp}\right|^2\right) T_p. \tag{17}$$

Case 2: If the incident wave is linearly polarized with the E-field parallel to the plane of incidence, then $E_2 = 0$ and (15) becomes

$$T_n = (T_n)_{\parallel} = \left(1 - \left|\Gamma_{\parallel}\right|^2\right) T_p. \tag{18}$$

Case 3: If the incident wave is circularly polarized, then $E_1 = E_2$ and

$$T_n = (T_n)_{cp} = \left[1 - \frac{\left(\left|\Gamma_{\parallel}\right|^2 + \left|\Gamma_{\perp}\right|^2\right)}{2}\right] T_p. \tag{19}$$

Note then that $(T_n)_{cp}$ is also just the average of $(T_n)_{\perp}$ and $(T_n)_{\parallel}$ or

$$(T_n)_{cp} = \frac{1}{2}\left[(T_n)_{\perp} + (T_n)_{\parallel}\right]. \tag{20}$$

The reader is reminded that, since the reflection coefficients are functions of incidence angle θ_i, the noise temperatures are also functions of θ_i as well as polarization.

C. Excess Noise-Temperature Relationships

It is of interest to see what the relationship is for excess noise temperature as well. For painted reflector noise-temperature analyses [3], it is convenient to use the term excess noise temperature (ENT). It is defined in [3] as the total noise temperature of a painted reflector minus the noise temperature of the reflector (bare metal) without paint. Mathematically, it is expressed as

$$\Delta T_n = T_{n2} - T_{n1} = \left(1 - \left|\Gamma_2\right|^2\right) T_p - \left(1 - \left|\Gamma_1\right|^2\right) T_p \tag{21}$$

IEEE TRANSACTIONS ON MICROWAVE THEORY AND TECHNIQUES, VOL. 48, NO. 9, SEPTEMBER 2000 1591

normal incidence case either (17) or (18) can be used to calculate the noise temperature (of the particular wet reflector configuration under study) as a function of the equivalent load reflection coefficient phase angle.

IV. Concluding Remarks

In the short paper, noise-temperature equations were derived from power equations for the incident and reflected wave. The relationships between noise temperatures of the different polarized wave cases were not obvious to the authors until the equations were derived from basic theoretical considerations. Hence, this paper serves to document the relationships and derivations. These noise-temperature formulas have proven to be useful for painted reflector studies [3] and will be useful for studies of plating [6]–[8] on reflector surfaces as well.

References

[1] S. Ramo and J. R. Whinnery, *Fields and Waves in Modern Radio*. New York: Wiley, 1953.

[2] J. A. Stratton, *Electromagnetic Theory*. New York: McGraw-Hill, 1941.

[3] T. Y. Otoshi, Y. Rhamat-Samii, R. Cirillo, and J. Sosnowski, "Noise-temperature and gain loss due to paints and primers on DSN antenna reflector surfaces," The Telecommunications and Mission Operations Progress Rep. 42-140, [Online] Available: http://tmo.jpl.nasa.gov/tmo/progress_report/42-140/140F.pdf, Feb. 15, 2000.

[4] H.-P. Ip and Y. Rahmat-Samii, "Analysis and characterization of multi-layered reflector antennas: Rain/snow accumulation and deployable membrane," *IEEE Trans. Antennas Propagat.*, vol. 46, pp. 1593–1605, Nov. 1998.

[5] T. Y. Otoshi, "Maximum and minimum return losses from a passive two-port network terminated with a mismatched load," *IEEE Trans. Microwave Theory Tech.*, vol. 42, pp. 787–792, May 1994.

[6] E. H. Thom and T. Y. Otoshi, "Surface resistivity measurements of candidate subreflector surfaces," The Telecommunications and Data Acquisition Progress Rep. 42-65, Jet Propulsion Lab., Pasadena, CA, Oct. 15, 1981.

[7] T. Y. Otoshi and M. M. Franco, "The electrical conductivities of steel and other candidate material for shrouds in a beam-waveguide antenna system," *IEEE Trans. Instrum. Meas.*, vol. 45, pp. 73–83, Feb. 1996.

[8] ——, "Correction to 'The electrical conductivities of steel and other candidate material for shrouds in a beam-waveguide antenna system'," *IEEE Trans. Instrum. Meas.*, vol. 45, p. 839, Aug. 1996.

Technical Report No. 32-496

EXCITATION OF HIGHER ORDER MODES BY A STEP DISCONTINUITY OF A CIRCULAR WAVEGUIDE

Cavour Yeh

R. Stevens

Robertson Stevens, *Chief*
Communications Element Research Section

JET PROPULSION LABORATORY
California Institute of Technology
Pasadena, California
February 1, 1964

CONTENTS

FIGURES

ABSTRACT

The problem of excitation of higher order modes by a step discontinuity in a circular waveguide with an incident dominant H_{11} mode is considered. An approximate method is used to solve this problem. In this method, it is assumed that the tangential electric field at the discontinuity is zero everywhere except in the aperture where it is equal to the incident tangential electric field. Relative amplitudes of these excited modes are given and discussed.

I. INTRODUCTION

The use of a horn as a radiator has been considered either theoretically or experimentally by many investigators (Ref. 1-4). However, most previous work has been concerned chiefly with the radiation characteristics of a horn radiator excited by a single waveguide mode. In order to suppress side lobes of the radiation pattern and still retain an equal beamwidth for the main lobe, Potter (Ref. 5) recently investigated the use of a horn radiator excited by a combination of two waveguide modes. He has shown that a significant reduction of the side lobe amplitude is achieved if the E_{11} circular waveguide mode is properly used in combination with the commonly used dominant H_{11} circular waveguide mode (i. e., the relative amplitude and phase of the E_{11} and H_{11} modes must be properly chosen). One of the simplest ways of generating these modes is to make an abrupt discontinuity of the cross section of the waveguide. The purpose of this report is to investigate this excitation problem.

The problem of an obstacle or discontinuity in waveguides has been treated by Schwinger (Ref. 6) and his followers (Ref. 7-9) using the integral equation and variational techniques. However, when more than one propagating mode can exist in the waveguide, the solution of the resultant integral equation becomes very involved. Hence, an approximate method is introduced to consider the present problem. In this method it is assumed that the tangential electric field at the discontinuity is a known quantity. Namely, it is the tangential electric field of the incident wave at the aperture and is otherwise zero at the discontinuity. The excited electromagnetic waves are expanded in terms of the orthonormal modes of the waveguide. The expansion coefficients are then obtained by matching the tangential electric field at the discontinuity. Relative amplitudes of the excited waves are plotted. Results will be discussed.

II. FORMULATION OF THE PROBLEM

One end of a circular waveguide of radius a, called waveguide 1, is connected with the other end of a circular waveguide of radius b, called waveguide 2, as shown in Fig. 1. It is assumed that $b > a$. Only the dominant H_{11} mode may propagate in waveguide 1; all other modes are evanescent. The only propagating modes in waveguide 2 are the H_{11}, H_{21}, H_{01}, E_{01}, and E_{11} modes. Because of the symmetrical characteristics of the discontinuity, the azimuthal dependence of all excited modes will be either sin θ or cos θ if the incident wave is the dominant H_{11} wave.

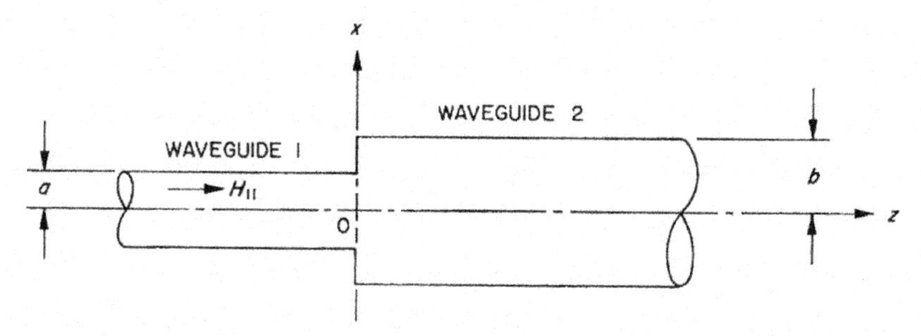

Fig. 1. Geometrical configuration

The tangential electric and longitudinal magnetic field components of the incident H_{11} wave in a circular waveguide may be given by[1]

$$H_z^{(i)} = H_0 \, \alpha_1^i \, J_1(\gamma_{11}\rho) \cos\theta \, e^{ih_{11}z} \tag{1a}$$

$$E_\theta^{(i)} = -\frac{i\omega\mu_0}{\gamma_{11}} H_0 \, \alpha_1^{(i)} J_1'(\gamma_{11}\rho) \cos\theta \, e^{ih_{11}z} \tag{1b}$$

$$E_\rho^{(i)} = -\frac{i\omega\mu_0}{\gamma_{11}^2} H_0 \, \alpha_1^{(i)} \frac{1}{\rho} J_1(\gamma_{11}\rho) \sin\theta \, e^{ih_{11}z} \tag{1c}$$

[1] The nomenclature used in this Report follows that used in Ref. 10.

where

$$a_1^{(i)} = \sqrt{\frac{2}{\pi}} \frac{\gamma_{11}^2}{\sqrt{(\gamma_{11}a)^2 - 1}} \frac{1}{J_1(\gamma_{11}a)}, \quad \gamma_{11}a = 1.8412, \; h_{11}^2 = k_0^2 - \gamma_{11}^2$$

$$k_0^2 = \omega^2 \mu \epsilon_0$$

and H_0 is the amplitude of the incident wave. The cylindrical coordinates (ρ, θ, z) are used. The harmonic time dependence, $e^{-i\omega t}$, is assumed and suppressed throughout. The appropriate expression for the transvers electric field components of the transmitted wave in waveguide 2 is (Ref. 10):

$$\mathbf{E}_t^{(t)} = \sum_p A_p \mathbf{E}_{tp}^m + B_p \mathbf{E}_{tp}^e \tag{2}$$

Where

$$\mathbf{E}_{tp}^m = a_p^m \left[-\frac{i\omega\mu_0}{\gamma_{1p}^m} J_1'(\gamma_{1p}^m \rho) \cos\theta \, \mathbf{e}_\theta - \frac{i\omega\mu_0}{\gamma_{1p}^{m\,2}} \frac{1}{\rho} J_1(\gamma_{1p}^m \rho) \sin\theta \, \mathbf{e}_\rho \right] e^{ih_{1p}^m z}$$

$$\mathbf{E}_{tp}^e = a_p^e \left[\frac{ih_{1p}^e}{\gamma_{1p}^{e\,2}} \frac{1}{\rho} J_1(\gamma_{1p}^e \rho) \cos\theta \, \mathbf{e}_\theta + \frac{ih_{1p}^e}{\gamma_{1p}^e} J_1'(\gamma_{1p}^e \rho) \sin\theta \, \mathbf{e}_\rho \right] e^{ih_{1p}^e z},$$

with

$$h_{1p}^m = \sqrt{k^2 - \gamma_{1p}^{m\,2}}, \qquad h_{1p}^e = \sqrt{k^2 - \gamma_{1p}^{e\,2}}$$

$$J_1'(\gamma_{1p}^m b) = 0, \qquad J_1(\gamma_{1p}^e b) = 0$$

$$a_p^m = \sqrt{\frac{2}{\pi}} \frac{\gamma_{1p}^{m\,2}}{\sqrt{(\gamma_{1p}^m b)^2 - 1}} \frac{1}{J_1(\gamma_{1p}^m b)}, \qquad a_p^e = \sqrt{\frac{2}{\pi}} \frac{\gamma_{1p}^e}{b} \frac{1}{J_2(\gamma_{1p}^e b)}$$

where the prime signifies the derivative of the function with respect to its argument; $\mathbf{e}_\theta$ and $\mathbf{e}_\rho$ are respectively the unit vector in θ and ρ directions. A_p and B_p are yet unknown coefficients. They may be found by matching the fields at the discontinuity.

III. MATCHING AT THE DISCONTINUITY

At the discontinuity, i.e., at $z = 0$, the transverse components of the transmitted electric field must be identical with the aperture electric field $\mathbf{E}_{ap}$ in the aperture and must be zero on the perfectly conducting obstacle, i.e.,

$$\begin{aligned} \mathbf{E}_t^{(t)} &= \mathbf{E}_{ap} && \text{for } 0 \le \rho \le a \\ &= 0 && \text{for } a \le \rho \le b \end{aligned} \tag{3}$$

at $z = 0$. $\mathbf{E}_t^{(t)}$ is given by Eq. (2) with $z = 0$.

Applying the orthogonality relations (Ref. 10)

$$\int_0^{2\pi} \int_0^b (\mathbf{E}_{tp}^e \cdot \mathbf{E}_{tq}^m)\, \rho\, d\rho\, d\theta = 0 \qquad \text{for all p and } q \tag{4}$$

$$\begin{aligned} \int_0^{2\pi} \int_0^b (\mathbf{E}_{tp}^e \cdot \mathbf{E}_{tq}^e)\, \rho\, d\rho\, d\theta &= 0 && \text{for p} \neq q \\ &= h_{1p}^{e2} && \text{for p} = q \end{aligned} \tag{5}$$

$$\begin{aligned} \int_0^{2\pi} \int_0^b (\mathbf{E}_{tp}^m \cdot \mathbf{E}_{tq}^m)\, \rho\, d\rho\, d\theta &= 0 && \text{for p} \neq q \\ &= \omega^2 \mu_0^2 && \text{for p} = q \end{aligned} \tag{6}$$

to Eq. (3), one obtains

$$A_p = \frac{1}{\omega^2 \mu^2} \int_0^{2\pi} \int_0^a (\mathbf{E}_{ap} \cdot \mathbf{E}_{tp}^m)\, \rho\, d\rho\, d\theta \tag{7}$$

$$B_p = \frac{1}{h_{1p}^{e2}} \int_0^{2\pi} \int_0^a (\mathbf{E}_{ap} \cdot \mathbf{E}_{tp}^e)\, \rho\, d\rho\, d\theta \tag{8}$$

It is noted that the tangential aperture electric field $\mathbf{E}_{ap}$ is still an unknown quantity. If one follows the Schwinger (Ref. 6) formulation, one then obtains a Schwinger-type integral equation in which $\mathbf{E}_{ap}$ is the unknown function. The solution for $\mathbf{E}_{ap}$ is very involved and tedious if more than one propagating mode is allowed in waveguide 2 as in the present case. To overcome this difficulty, we shall assume that the tangential aperture electric field $\mathbf{E}_{ap}$ is identical with the tangential components of the incident electric field at $z = 0$ (see Eq. 1), as a first-order approximation. In other words,

$$\mathbf{E}_{ap} = H_0 a_1^{(i)} \left[-\frac{i\omega\mu_0}{\gamma_{11}} J_1'(\gamma_{11}\rho) \cos\theta \, \mathbf{e}_\theta - \frac{i\omega\mu_0}{\gamma_{11}^2} \frac{1}{\rho} J_1(\gamma_{11}\rho) \sin\theta \, \mathbf{e}_\rho \right] \tag{9}$$

The limitations of this approximation are similar to the ones made for the slot antenna (Ref. 11). A discussion of slots in waveguides has been given by Stevenson (Ref. 12). Substituting Eq. (9) into Eq. (7) and (8), and carrying out the integration gives

$$A_p = H_0 \frac{2(\gamma_{11} a)\, \gamma_{1p}^{m\,2}}{(\gamma_{11}^2 - \gamma_{1p}^{m\,2})} \cdot \frac{1}{\sqrt{[(\gamma_{11} a)^2 - 1]\,[(\gamma_{1p}^m b)^2 - 1]}} \frac{J_1(\gamma_{1p}^m a)}{J_1(\gamma_{1p}^m b)} \tag{10}$$

$$B_p = -H_0 \frac{\omega\mu_0 \gamma_{11}^2}{h_{1p}^e(\gamma_{11} b)\sqrt{(\gamma_{11} a)^2 - 1}} \frac{X}{(\gamma_{11}^2 - \gamma_{1p}^{e\,2}) J_1(\gamma_{11} a) J_2(\gamma_{1p}^e b)} \tag{11}$$

with

$$X = \gamma_{11} a \,[J_2(\gamma_{1p}^e a) J_2'(\gamma_{11} a) - J_0(\gamma_{1p}^e a) J_0'(\gamma_{11} a)]$$
$$+ \gamma_{1p}^e a \,[J_0(\gamma_{11} a) J_0'(\gamma_{1p}^e a) - J_2(\gamma_{11} a) J_2'(\gamma_{1p}^e a)] \tag{12}$$

One obtains the complete expressions for the transverse components of the transmitted electric field by substituting the expressions for A_p and B_p into Eq. (2):

$$E_\rho^{(t)} = -\frac{i\omega\mu_0}{a} H_0 \, 2\sqrt{\frac{2}{\pi}} \, C \sum_p \left[P_p \frac{1}{\gamma_{1p}^m \rho} J_1(\gamma_{1p}^m \rho) \sin\theta \, e^{ih_{1p}^m z} + Q_p J_1'(\gamma_{1p}^e \rho) \sin\theta \, e^{ih_{1p}^e z} \right] \tag{13}$$

$$E_\theta^{(t)} = -\frac{i\omega\mu_0}{a} H_0 \, 2\sqrt{\frac{2}{\pi}} \, C \sum_p \left[P_p J_1'(\gamma_{1p}^m \rho) \cos\theta \, e^{ih_{1p}^m z} + Q_p \frac{1}{\gamma_{1p}^e \rho} J_1(\gamma_{1p}^e \rho) \cos\theta \, e^{ih_{1p}^e z} \right] \tag{14}$$

where

$$C = \frac{\gamma_{11} a}{\sqrt{(\gamma_{11} a)^2 - 1}} \; \frac{\gamma_{11}^{m\,3} a^3}{(\gamma_{11}^2 a^2 + \gamma_{11}^{m\,2} a^2)\,[(\gamma_{11}^m b)^2 - 1]} \; \frac{J_1(\gamma_{11}^m a)}{J_1(\gamma_{11}^m b)} \tag{15}$$

$$P_p = \left(\frac{\gamma_{11}^2 a^2 - \gamma_{11}^{m\,2} a^2}{\gamma_{11}^2 a^2 - \gamma_{1p}^{m\,2} a^2} \right) \frac{\sqrt{(\gamma_{11}^m b)^2 - 1}}{\sqrt{(\gamma_{1p}^m b)^2 - 1}} \; \frac{J_1(\gamma_{1p}^m a)}{J_1(\gamma_{11}^m a)} \; \frac{J_1(\gamma_{11}^m b)}{J_1(\gamma_{1p}^m b)} \; \frac{\gamma_{1p}^{m\,3} a^3}{\gamma_{11}^{m\,3} a^3} \; \frac{\sqrt{(\gamma_{11}^m b)^2 - 1}}{\sqrt{(\gamma_{1p}^m b)^2 - 1}} \tag{16}$$

$$Q_p = \frac{1}{2} \; \frac{(\gamma_{11}^2 a^2 - \gamma_{11}^{m\,2} a^2)\,[(\gamma_{11}^m b)^2 - 1]}{\gamma_{11}^{m\,3} a b^2} \; \frac{J_1(\gamma_{11}^m b)}{J_1(\gamma_{11}^m a)} \; \frac{X}{(\gamma_{11}^2 a^2 - \gamma_{1p}^{e\,2} a^2)\, J_1(\gamma_{11} a)\, J_2^2(\gamma_{1p}^e b)} \tag{17}$$

and X is given by Eq. (12)

Numerical computations of C, P_p, and Q_p were carried out using the IBM 7090 computer for various values of $k_0 a$ and $k_0 b$. As mention in Sec. II, we assume that the size of waveguide 1 is such that only the dominant H_{11} mode may propagate and that the size of waveguide 2 is such that only H_{11} and E_{11} modes may propagate.[2] Hence, the range of $k_0\,a$ and $k_0\,b$, that we will consider, are respectively $1.841 \leq k_0\,a \leq 5.331$ and $1.841 \leq k_0\,b \leq 5.331$. P_p and Q_p for various values of $k_0\,a$ and $k_0\,b$ are plotted in Fig. 2 through 5. It can be seen that, as expected, if $k_0\,b \approx k_0\,a$, the only mode that is strongly excited is the H_{11} wave. As the difference between $k_0\,a$ and $k_0\,b$ becomes larger, other modes may be more strongly excited. It is also interesting to note that for some specific values of $k_0\,a$, the amplitudes of certain higher order modes are zero. It should be recalled that the analysis given here is an approximate one. Hence, the results are applicable only when the assumption, that the tangential aperture electric field is the incident electric field at the discontinuity, may be used.

[2] Because of the symmetrical nature of the discontinuity, H_{01}, E_{01}, and H_{21} propagating modes in waveguide 2 are not excited.

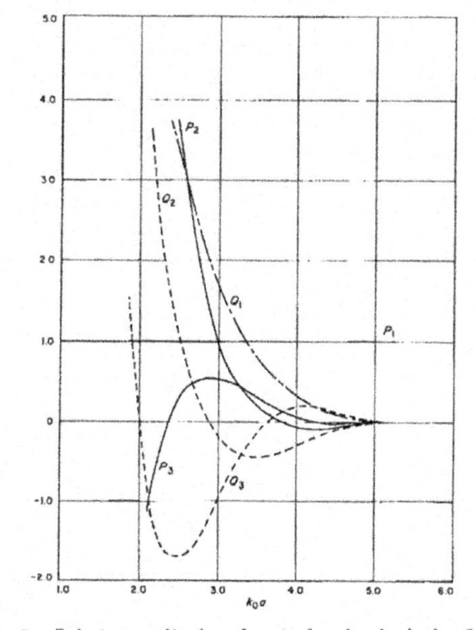

Fig. 2. Relative amplitudes of excited modes for $k_0\,b = 5.251$

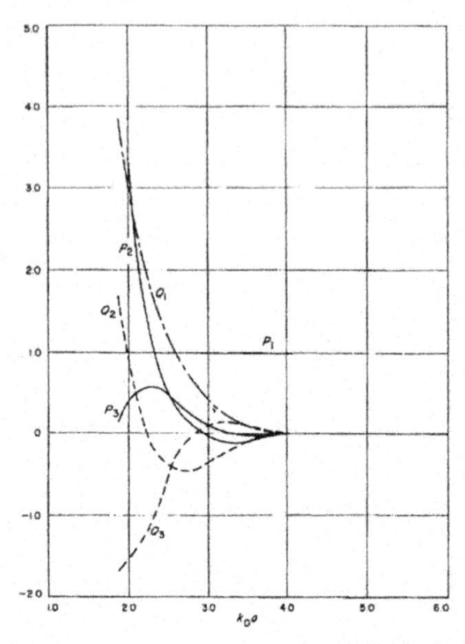

Fig. 3. Relative amplitudes of excited modes for $k_0 b = 4.051$

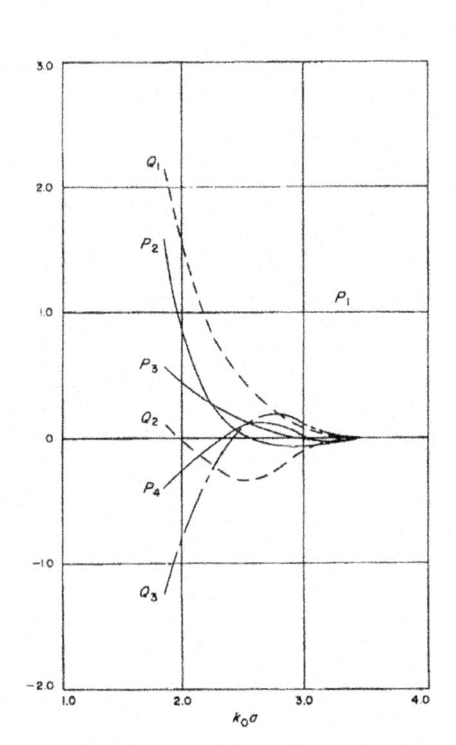

Fig. 4. Relative amplitudes of excited modes for $k_0\,b = 3.151$

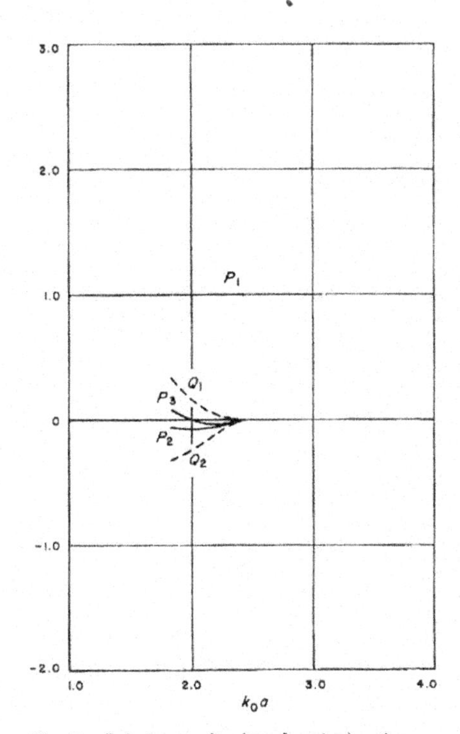

Fig. 5. Relative amplitudes of excited modes for $k_0\,b = 2.451$

ACKNOWLEDGEMENT

The author wishes to thank Mr. Philip D. Potter of the Jet Propulsion Laboratory, Pasadena, California, for many valuable discussions.

3

SCATTERING AND DIFFRACTION

Separation of variables is a very powerful mathematical technique that enables us to find analytical solutions of vector wave equations for several canonical structures, such as planar surfaces, spheres, circular cylinders, elliptical cylinders, and parabolic cylinders. Results from the scattering by dielectric/conducting spheres or circular cylinders, and the reflection, refraction, and transmission for planar/conducting slabs or layers have become the backbone of many important developments dealing with electromagnetic waves. By the early 1960s, we realized that the scattering problems for dielectric elliptical/parabolic cylinders had not been solved, although the analytic solutions for the vector wave equations in these coordinate systems were well known. It turned out that there was a complication, i.e., unlike the spherical or circular cylindrical cases, the angular functions in these other coordinates were not only functions of the angular coordinates but also of the constitutive parameters of the medium to which they applied. Consequently, when matching the boundary conditions for the fields, the angular function of a given index in one medium cannot be cancelled by the angular function with the same index number in the other medium, resulting in an equation dependent on the angular function of both media, with infinite sums of indices. This complexity prevented the finding of the scattering solutions for the elliptical/parabolic dielectric cylinders in the same way as was done for the dielectric sphere or dielectric circular cylinder case. C. Yeh discovered a way to resolve this problem, and the canonical solutions for the scattering by elliptical/parabolic dielectric cylinders were found. These solutions were presented in a series of fundamental papers (Papers 3-1-1, 3-2a, and 3-2b).

The advent of increasing computational power in the 1960s presented us with an opportunity to investigate the use of numerical techniques to solve problems that could not be solved using traditional analytical means. Subsequently, we solved the problem of plane wave scattering by (a) radially inhomogeneous dielectric cylinder (Papers 3-3-1 and 3-3-2), (b) a radially inhomogeneous and anisotropic plasma cylinder (Papers 3-3-3 and 3-4), and (c) an arbitrarily shaped dielectric object (Paper 3-5).

The invention of lasers in 1960 meant that optical waves could be strongly focused to a diffraction-limited spot size. In 1975, we initiated the study on the problem of the scattering of a dielectric object by a focused beam (Papers 3-7-1 through 3-7-4). This focused-beam-scattering research was eventually developed into a very fruitful area of optical tweezers, or optical confinement of very small individual particles or even atoms.

When multiple scattering bodies are in the vicinity of each other, multiple scattering effects must be taken into consideration, making most analyses intractable. Using the radiative transfer approach, we investigated the scattering by randomly distributed nonspherical particles (Papers 3-8-1 through 3-8-3).

A perturbation approach to handling the diffraction of electromagnetic waves by arbitrarily shaped dielectric obstacles is given for 3-D objects (Paper 3-9-1) and for 2-D objects (Paper 3-9-2).

Abstracts of all papers in this chapter appear below.

3.1 Scattering by elliptical dielectric cylinders

3.1.1 Normal incident case. (Paper 3-1-1) (The reprint here now includes the corrected numerical results. See C. Yeh and C. S. Kim, "Erratum: The diffraction of waves by a penetrable ribbon," J. Math. Phys., Vol. 29, No. 3, 721, 1988.)

The exact solution of the diffraction of waves by a dielectric ribbon or by an elliptical dielectric cylinder is obtained. Results are given in terms of Mathieu and modified Mathieu functions. It is found that each expansion coefficient of the scattered or transmitted wave is coupled to all coefficients of the series expansion for the incident wave, except when the elliptical cylinder degenerates into a circular one. Both polarizations of the incident wave are considered: one with the incident electric field vector in the axial direction and the other with the incident magnetic vector in the axial direction. The technique used in this paper to satisfy the boundary conditions may be applied to similar types of problems, such as the plasma-coated ribbon radiator and the corresponding acoustic problems.

3.1.2 Oblique incident case. (Paper 3-1-2)

The exact solution of the scattering of obliquely incident plane waves by an elliptic dielectric cylinder is obtained. Each expansion coefficient of the scattered or transmitted wave is coupled to all coefficients of the series expansion for the incident wave except when the elliptical cylinder degenerates into a circular one. Both polarizations of the incident wave are considered: one with the incident electric field vector in the axial direction and the other with the incident magnetic vector in the axial direction. In the general case of oblique incidence, the scattered field contains a significant cross-polarized component that vanishes at normal incidence.

3.1.3 Exact solution to a lens problem. (Paper 3-1-3)

Exact solution is obtained for the diffraction of a plane electromagnetic wave by an elliptic cylindrical lens surrounding an infinitely long parallel slit in a perfectly conducting plane. Results are expressed in terms of Mathieu functions. Approximate expressions for the expansion coefficients for the diffracted waves, the scattered intensity at infinity, and the transmission coefficients are also obtained in the long-wavelength (Rayleigh scattering) limit. Possible numerical investigations are discussed.

3.1.4 Layered elliptical cylinder. (Paper 3-1-4)

The scattering problem of obliquely incident waves on a multilayered elliptical cylinder is considered. The cylinder is assumed homogeneous and isotropic but lossy. Both polarizations of the incident wave, the ***E*** wave, and the ***H*** wave are considered. In the theoretical analysis, the number of layers are unlimited, but in the numerical analysis, we limit the number of layers so we can avoid excessively long computer running time. The numerical results include the bistatic radar cross sections, the specular differential cross sections, and the efficiency factors for absorption, scattering, and extinction. The computed results show the dependence of the scattering on the size, shape, and material, and on the angle of the incident wave.

3.2 Scattering by parabolic dielectric cylinders. (Papers 3-2a, 3-2b)

The exact solution of the diffraction of waves by a dielectric parabolic cylinder is obtained. Results are given in terms of parabolic-cylinder functions. It is found that each expansion coefficient of the scattered or transmitted wave is coupled to all coefficients of the series expansion of the incident wave, even for very thin cylinders. Both polarizations of the incident wave are considered: one with the incident electric vector in the axial direction and the other with the magnetic field vector in the axial direction. Numerical computations are carried out for the backscattering cross section and the radiation pattern for the scattered wave.

3.3 Scattering from layered structures.

3.3.1 Scattering of obliquely incident waves by a radially inhomogeneous dielectric cylinder. (Paper 3-3-1)

The problem of the interaction of obliquely incident electromagnetic waves with a radially inhomogeneous fiber cylinder is treated analytically. By formulating this problem in terms of the two-point boundary value problem, we eliminate the need to evaluate untabulated functions or to invert large matrices. Numerical results for the different scattering cross sections are obtained for the case of an inhomogeneous fiber whose radial dielectric variation may be described by the Luneburg profile. It is found that the induced cross-polarized scattered fields as well as the characteristics of the total scattered fields are influenced significantly by the inhomogeneity of the dielectric medium.

3.3.2 Scattering from a cylinder coated with an inhomogeneous dielectric sheath. (Paper 3-3-2)

As a space vehicle reenters the atmosphere, a plasma sheath, surrounding the vehicle, is generated. It is well known that the sheath is inhomogeneous; however, to make this problem suitable for theoretical analysis, most investigators make the assumption that the sheath is homogeneous. To investigate the validity of this assumption, the idealized problem of the scattering of plane waves by a conducting cylinder coated with a stratified dielectric sheath is considered. The wave equation is separated using the vector function method of Hansen and Stratton. It is then applied to the plane-wave scattering problem. The backscattering cross section is defined and obtained. Analytical expressions for the scattering coefficients of a thin inhomogeneous sheath are also given.Numerical computations are carried out for a specific variation of the dielectric sheath, i.e., $\epsilon(r) = \epsilon_0\ \alpha/k_0 r$, where α is a constant, $k_0^2 = \omega^2 \mu \epsilon_0$, and ϵ_0 is the free space dielectric constant. Results are compared with the homogeneous sheath problem; the dielectric constant of the homogeneous sheath is taken to be the average value of that for the inhomogeneous sheath. It is found that in general, rather distinct differences are observed except when the sheath is very thin.

3.3.3 Scattering by an infinite cylinder coated with an inhomogeneous and anisotropic plasma sheath. (Paper 3-3-3)

The problem of the interaction of an incident ***H*** wave with an infinite conducting cylinder coated with an inhomogeneous and anisotropic plasma sheath is treated analytically. Numerical results of the far-field pattern of the scattered field as well as the backscattering cross sections are presented for various interesting ranges of the parameters involved. Detailed discussion of the effects of the impressed static magnetic field, the sheath thickness, and the density and profile of the plasma sheath on the scattered field is also presented. A parabolic electron density profile is assumed.

3.4 Interaction of microwaves with an inhomogeneous and anisotropic plasma column. (Paper 3-4)

The problem of the interaction of microwaves with a radially inhomogeneous plasma column confined by an impressed axial static magnetic field is treated analytically. Extensive numerical results for the backscattering cross sections are presented for various interesting ranges of the parameters involved. It is found that the often-used homogeneous model for the interpretation of experimental data can sometimes lead to ambiguous results. The ambiguity arises because of the interdependence of the average plasma frequency, the gyrofrequency, and the inhomogeneity of the plasma density distribution. Further measurements are suggested to distinguish these effects.

3.5 Scattering of electromagnetic waves by arbitrarily shaped dielectric bodies (Paper 3-5)

The differential scattering characteristics of closed three-dimensional dielectric objects are theoretically investigated. The scattering problem is solved in a spherical basis by the extended boundary condition method (EBCM), which results in a system of linear equations for the expansion coefficients of the scattered field in terms of the incident field coefficients. The equations are solved numerically for dielectric spheres, spheroids, and finite cylinders to study the dependence of the differential scattering on the size, shape, and index of refraction of the scattering object. The method developed here appears to be most applicable to objects whose physical size is on the order of the wavelength of the incident radiation.

3.6 Scattering by Pruppacher-Pitter raindrops at 30 GHz (Paper 3-6)

Optimum design of modern ground-satellite communication systems requires the knowledge of rain-induced differential attenuation, differential phase shift, and cross-polarization factors. After a comprehensive assessment of different available analytical techniques, an efficient technique is chosen to yield the desired scattering results. Tabulation of the scattered fields for Pruppacher-Pitter raindrops with sizes ranging from 0.25 mm to 3.5 mm at 20°C and 30 GHz for several incidence angles is given.

3.7 Scattering by focused beams

3.7.1 Single particle scattering with focused laser beams. (Paper 3-7-1)

A scattering technique is described in which the incident laser beam is tightly focused to isolate the effects of a single particle. In this way, the individual particles may be studied in their natural environment, and experiments with latex spheres are in agreement with the theory.

3.7.2 Rayleigh-Debye scattering with focused laser beams. (Paper 3-7-2)

A focused beam technique has been developed for diagnosing the characteristics of individual particles in a polydisperse ensemble. In the Rayleigh-Debye approximation, the scattered fields are related to the orientation and properties of a scatterer by means of explicit analytical formulas. The results simplify when the particle size is small compared to the minimum beam diameter.

3.7.3 Scattering of focused beams by tenuous particles. (Paper 3-7-3)

This paper deals with the problem of the scattering of focused laser beams by tenuous particles using an iterative technique. The results are shown to be accurate, provided that (1) the polarizability of the particle medium is small and (2) the phase shift of the central ray is less than 2. It was found that when the size of the incident beam waist is close to that of the scatterer, the scattered field deviates significantly from that for the incident plane-wave case. Specific examples are given.

3.7.4 Scattering of sharply focused beams by arbitrarily shaped dielectric particles: an exact solution. (Paper 3-7-4)

By expanding the incident focused beam field in terms of its plane-wave spectrum and by using the technique we developed earlier to treat the problem of the scattering of plane waves by arbitrarily shaped dielectric obstacles, we have been successful in solving the problem of the scattering of sharply focused beams by arbitrarily shaped dielectric particles. It is found that the presence of the curvature of the incident wave front and the nonuniformity of the incident wave intensity affect greatly the scattering characteristics.

3.7.5 Scattering of a focused beam by moving particles. (Paper 3-7-5)

The scattering of a focused laser beam by moving particles is studied. A solution is given for a spherical scatterer that is small compared to the wavelength. Several numerical examples are given. It is concluded that for particles that have a short time of flight within the laser beam, the bandwidth and frequency spectrum of the scattered field is critically dependent on the laser beam shape. This fact is particularly important when information on the motion of small particles or macromolecules is to be inferred from the measured bandwidth and spectrum of the scattered field due to an incident wave that is not a plane wave.

3.8 Radiative transfer approach to multiple scattering problems

3.8.1 Matrix representations of the vector radiative-transfer theory for randomly distributed nonspherical particles. (Paper 3-8-1)

General matrix representation of the vector radiative-transfer equation for randomly distributed nonspherical particles is given with compact representation of the extinction matrix and the Mueller matrix. The propagation of the coherent Stokes vector and the coherency matrix in such a medium is expressed in matrix form. The extinction matrix is related to the generalized optical theorem for partially polarized waves. The first-order scattering solution of the Stokes vector is given in matrix form, and discussions of the Fourier expansion of the equation of transfer and the limitation of the equation of transfer are given.

3.8.2 Multiple scattering calculations for nonspherical particles based on the vector radiative transfer theory. (Paper 3-8-2)

On the basis of the vector radiative transfer equation, multiple scattering calculations were performed for an obliquely incident linearly polarized wave upon a plane-parallel slab consisting of arrays of vertically oriented, uniformly distributed nonspherical spheroidal particles of identical size. Results are given in terms of the Stokes parameters for the incoherent field. Owing to the symmetry of the particles and their orientation, decoupling of the Fourier expansion coefficients in the solution of the radiative transfer equation occurs. Each Fourier component satisfies a single equation, which is then solved by the Gaussian quadrature and eigenvalue and eigenvector technique. The behavior of the forward scattered incoherent field is investigated for low-loss as well as high-loss particles, for different densities and shapes, and for various angles of incidence. Comparison with the results from the approximate first-order theory shows good quantitative agreement for thin and sparsely populated layers.

3.8.3 First-order multiple scattering theory for nonspherical particles. (Paper 3-8-3)

Using the vector radiative transfer equation, we have obtained the first-order solution for the problem dealing with multiple scattering from an ensemble of nonspherical particles. Representative results for low-absorbent disks (oblate spheroids) as well as high-absorbent disks were obtained. In general, it was found that multiple scattering effects were more prominent for low-absorbent particles.

3.9 Perturbation approach to the diffraction of electromagnetic waves by arbitrarily shaped dielectric obstacles

3.9.1 3-D object. (Paper 3-9-1)

A perturbation method is developed to consider the problem of the diffraction of electromagnetic waves by an arbitrarily shaped dielectric obstacle whose boundary may be expressed in the general form, in spherical coordinates, $r_p = r_0[1 + \delta f_1(\theta, \varphi) + \delta^2 f_2(\theta, \varphi) + \cdots)]$, where r_0 is the radius of the unperturbed sphere and $f_n(\theta, \varphi)$ are arbitrary, single-valued analytic functions. δ is chosen such that

$$\sum_{n=1}^{\infty} |\delta^n f_n(\theta, \varphi)| < 1, \ 0 \le \theta \le \pi, \ 0 \le \varphi \le 2\pi$$

Detailed analysis is carried out to the first order in δ. Procedures to obtain higher-order terms are also indicated. The perturbation solutions are valid for the near-zone region of the obstacle as well as for the far-zone region, and they are applicable for all frequencies. Possible applications of this perturbation technique to elementary-particle scattering problems and other electromagnetic problems are noted.

3.9.2 2-D object. (Paper 3-9-2)

The perturbation technique based on a Taylor series expansion of the boundary conditions at the perturbed boundary is extended to consider the problem of the diffraction of waves by a dielectric object with perturbed boundary. Since this approach attacks the complete boundary-value problem, the result is valid for the near zone as well as for the far zone and is valid for all frequencies. By way of illustration, the problem of the diffraction of electromagnetic waves by a dielectric cylinder with perturbed boundary is treated. A specific example on the scattering of plane waves by a dielectric elliptic cylinder with small eccentricity is given. Numerical results are also computed for this specific example and are compared with those obtained from the exact solution.

Reprinted from *Printed in U.S.A.*
JOURNAL OF MATHEMATICAL PHYSICS VOLUME 4, NUMBER 1 JANUARY 1963

The Diffraction of Waves By a Penetrable Ribbon*

C. YEH
Electrical Engineering Department, University of Southern California, Los Angeles, California
(Received 18 May 1962)

The exact solution of the diffraction of waves by a dielectric ribbon or by an elliptical dielectric cylinder is obtained. Results are given in terms of Mathieu and modified Mathieu functions. It is found that each expansion coefficient of the scattered or transmitted wave is coupled to all coefficients of the series expansion for the incident wave, except when the elliptical cylinder degenerates to a circular one. Both polarizations of the incident wave are considered: one with the incident electric vector in the axial direction and the other with the incident magnetic vector in the axial direction. It is noted that the technique used in this paper to satisfy the boundary conditions may be applied to similar types of problems; such as the plasma-coated ribbon radiator and the corresponding acoustical problems.

I. INTRODUCTION

THE problems of scattering of waves by a circular cylinder have been considered by many authors.[1-3] The exact solution of the problem of the diffraction of waves by a perfectly conducting elliptical cylinder, or by a ribbon, has been obtained by Sieger,[4] and Morse and Rubenstein.[5] However, the corresponding solution for the diffraction of waves by a dielectric elliptical cylinder or by a dielectric ribbon has not been found. It is the purpose of this paper to present the exact solution of this problem. It is shown that certain mathematical difficulties can be overcome by separating the wave equation in elliptic cylinder coordinates and by applying the orthogonality properties of the Mathieu functions.

II. FORMULATION OF THE PROBLEM

To analyze this problem, the elliptical cylinder coordinated (ξ, η, z), as shown in Fig. 1, are introduced. In terms of the rectangular coordinates (x, y, z), the elliptical cylinder coordinates are defined by the following relations:

$$x = q \cosh \xi \cos \eta,$$
$$y = q \sinh \xi \sin \eta, \qquad (1)$$
$$z = z,$$
$$(0 \leq \xi < \infty, 0 \leq \eta \leq 2\pi),$$

where q is the semi-focal length of the ellipse. The contour surfaces of constant ξ are confocal elliptic cylinders, and those of constant η are confocal hyperbolic cylinders. One of the confocal elliptic cylinders with $\xi = \xi_0$ is assumed to coincide with the boundary of the solid dielectric cylinder, and z axis concides with its longitudinal axis. We shall consider waves whose propagation vector is in the x-y plane, so that the z coordinate may be omitted from the discussion. A possible solution of the wave equation is then $R(\xi)\Theta(\eta)e^{-i\omega t}$, where R and Θ satisfy the differential equations

$$d^2R/d\xi^2 - (c - 2\gamma^2 \cosh 2\xi)R = 0 \qquad (2)$$

$$d^2\Theta/d\eta^2 + (c - 2\gamma^2 \cos 2\eta)\Theta = 0, \qquad (3)$$

where c is the separation constant and $\gamma^2 = k^2q^2/4$, k being the wave number. Equations (2) and (3) are, respectively, the modified Mathieu and Mathieu differential equations.

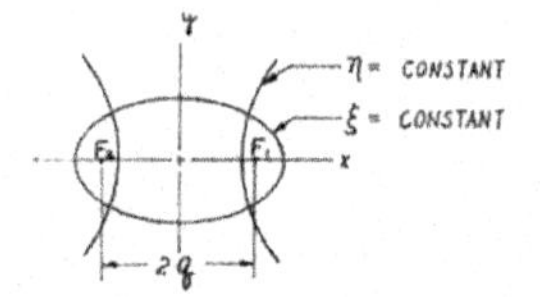

FIG. 1. The Elliptical cylinder coordinates. F_1 and F_2 are the foci of the ellipse. The distance between foci is the focal distance $2q$.

The periodic solutions of the equation in η are of two types: even about $\eta = 0$, and odd about $\eta = 0$. They are possible only for certain characteristic values of c. The even and odd functions are, respectively, denoted by $ce_n(\eta, \gamma^2)$ and $se_n(\eta, \gamma^2)$, with the sequence in n according to increasing values of c. It is noted that these functions are orthogonal functions. The solutions of (2) corresponding to the even function $ce_n(\eta, \gamma^2)$ having the same characteristic values of c are $Ce_n(\xi, \gamma^2)$ and $Fey_n(\xi, \gamma^2)$,

* This work was supported by Air Force Cambridge Research Center.
[1] P. Debye, Physik. Z. 9, 775 (1908).
[2] V. Fock, Doklady Akad. Nauk. S. S. S. R. 109, 477 (1956).
[3] R. Kind and T. T. Wu, *The Scattering and Diffraction of Waves* (Harvard University Press, Cambridge, Massachusetts, 1959).
[4] B. Sieger, Ann. Physik (Liepzig) 27, 626 (1908).
[5] P. M. Morse and P. L. Rubenstein, Phys. Rev. 54, 895 (1938).

and those corresponding to the odd function $se_n(\eta, \gamma^2)$ are $Se_n(\xi, \gamma^2)$ and $Gey_n(\xi, \gamma^2)$.[6]

The proper choice of these functions to represent the electromagnetic fields depends upon the boundary conditions. For the region within the dielectric cylinder, all field components must be finite. All field components for the scattered wave must satisfy the Sommerfeld's radiation condition at infinity. Consequently, the appropriate solutions of the wave equation for the region inside the dielectric cylinder are

$$\begin{Bmatrix} Ce_n(\xi, \gamma_1^2) & ce_n(\eta, \gamma_1^2) \\ Se_n(\xi, \gamma_1^2) & se_n(\eta, \gamma_1^2) \end{Bmatrix}, \tag{4}$$

and those for the scattered wave are

$$\begin{Bmatrix} Me_n^{(1),(2)}(\xi, \gamma_0^2) & ce_n(\eta, \gamma_0^2) \\ Ne_n^{(1),(2)}(\xi, \gamma_0^2) & se_n(\eta, \gamma_0^2) \end{Bmatrix}, \tag{5}$$

where

$$Me_n^{(1),(2)}(\xi, \gamma_0^2) = Ce_n(\xi, \gamma_0^2) \pm iFey_n(\xi, \gamma_0^2) \tag{6}$$

$$Ne_n^{(1),(2)}(\xi, \gamma_0^2) = Se_n(\xi, \gamma_0^2) \pm iGey_n(\xi, \gamma_0^2), \tag{7}$$

with

$$\gamma_0^2 = k_0^2 q^2/4 \quad \text{and} \quad \gamma_1^2 = k_1^2 q^2/4.$$

$$k_0^2 = \omega^2 \mu \epsilon_0 \quad \text{and} \quad k_1^2 = \omega^2 \mu \epsilon_1,$$

where ϵ_0 and ϵ_1 are, respectively, the dielectric constant of the free-space and of the elliptical cylinder.

III. SCATTERING OF A NORMALLY INCIDENT PLANE WAVE

Two types of normally incident waves are possible. One, called an E wave, is defined by $H_z = 0$, and the other, called an H wave, is defined by $E_z = 0$.

To simplify the notations for the Mathieu and modified Mathieu functions without any ambiguities, the following abbreviations are used:

$$\begin{aligned} ce_n(\eta) &= ce_n(\eta, \gamma_0^2), & se_n(\eta) &= se_n(\eta, \gamma_0^2), \\ ce_n^*(\eta) &= ce_n(\eta, \gamma_1^2), & se_n^*(\eta) &= se_n(\eta, \gamma_1^2), \\ Ce_n(\xi) &= Ce_n(\xi, \gamma_0^2), & Se_n(\xi) &= Se_n(\xi, \gamma_0^2), \\ Ce_n^*(\xi) &= Ce_n(\xi, \gamma_1^2), & Se_n^*(\xi) &= Se_n(\xi, \gamma_1^2), \end{aligned} \tag{8}$$
$$Me_n^{(1),(2)}(\xi) = Me_n^{(1),(2)}(\xi, \gamma_0^2),$$
$$Ne_n^{(1),(2)}(\xi) = Ne_n^{(1),(2)}(\xi, \gamma_0^2).$$

[6] We follow the notations adopted by Ince, [N. McLachlan, *Theory and Application of Mathieu Functions* (Oxford University Press, New York, 1951)].

It can be shown that the incident field of a plane wave with propagation vector in the x-y plane and at an angle θ with the x axis is proportional to the factor

$$e^{ik_0(x\cos\theta + y\sin\theta)} = e^{ik_0 q(\cosh\xi\cos\eta\cos\theta + \sinh\xi\sin\eta\sin\theta)}$$
$$= 2\sum_{n=0}^{\infty}\Big[\frac{1}{p_{2n}} Ce_{2n}(\xi)ce_{2n}(\eta)ce_{2n}(\theta)$$
$$+ (1/s_{2n+2})Se_{2n+2}(\xi)se_{2n+2}(\eta)se_{2n+2}(\theta)$$
$$+ (i/p_{2n+1})Ce_{2n+1}(\xi)ce_{2n+1}(\eta)ce_{2n+1}(\theta)$$
$$+ (i/s_{2n+1})Se_{2n+1}(\xi)se_{2n+1}(\eta)se_{2n+1}(\theta)\Big]. \tag{9}$$

where p_{2n}, p_{2n+1}, s_{2n+1}, and s_{2n+2} are joining factors.[6]

E wave. For an E wave, the field components of an incident wave are:

$$E_z^i = E_0 \,\{\text{the right-hand side of Eq. (9)}\}, \tag{10}$$

$$H_\eta^i = (1/k_0^2 p)\{+i\omega\epsilon_0\, \partial E_z^i/\partial\xi\}. \tag{11}$$

$$H_\xi^i = (1/k_0^2 p)\{-i\omega\epsilon_0\, \partial E_z^i/\partial\eta\}, \tag{12}$$

where $p = q\,(\sinh^2\xi + \sin^2\eta)^{1/2}$. Referring to (4) and (5), we see that the scattered field and the transmitted field inside the dielectric cylinder must be of the form

$$E_z^s = 2E_0 \sum_{n=0}^{\infty}\Big[\frac{A_{2n}}{p_{2n}} Me_{2n}^{(1)}(\xi)ce_{2n}(\eta)ce_{2n}(\theta)$$
$$+ (B_{2n+2}/s_{2n+2})Ne_{2n+2}^{(1)}(\xi)se_{2n+2}(\eta)se_{2n+2}(\theta)$$
$$+ i(A_{2n+1}/p_{2n+1})Me_{2n+1}^{(1)}(\xi)ce_{2n+1}(\eta)ce_{2n+1}(\theta)$$
$$+ i(B_{2n+1}/s_{2n+1})Ne_{2n+1}^{(1)}(\xi)se_{2n+1}(\eta)se_{2n+1}(\theta)\Big], \tag{13}$$

$$H_\eta^s = (1/k_0^2 p)\{+i\omega\epsilon_0\, \partial E_z^s/\partial\xi\}, \tag{14}$$

$$H_\xi^s = (1/k_0^2 p)\{-i\omega\epsilon_0\, \partial E_z^s/\partial\eta\}, \tag{15}$$

and

$$E_z^t = 2E_0 \sum_{n=0}^{\infty}\Big[\frac{C_{2n}}{p_{2n}^*} Ce_{2n}^*(\xi)ce_{2n}^*(\eta)ce_{2n}(\theta)$$
$$+ (D_{2n+2}/s_{2n+2}^*)Se_{2n+2}^*(\xi)se_{2n+2}^*(\eta)se_{2n+2}(\theta)$$
$$+ i(C_{2n+1}/p_{2n+1}^*)Ce_{2n+1}^*(\xi)ce_{2n+1}^*(\eta)ce_{2n+1}(\theta)$$
$$+ i(D_{2n+1}/s_{2n+1}^*)Se_{2n+1}^*(\xi)se_{2n+1}^*(\eta)se_{2n+1}(\theta)\Big], \tag{16}$$

$$H_\eta^t = (1/k_1^2 p)\{+i\omega\epsilon_1\, \partial E_z^t/\partial\xi\}, \tag{17}$$

$$H_\xi^t = (1/k_1^2 p)\{-i\omega\epsilon_1\, \partial E_z^t/\partial\eta\}, \tag{18}$$

where A_{2n}, A_{2n+1}, B_{2n+2}, B_{2n+1}, C_{2n}, C_{2n+1}, D_{2n+2}, and D_{2n+1} are arbitrary unknown coefficients that can be determined by applying the boundary conditions. p^*_{2n}, p^*_{2n+1}, s^*_{2n+2}, and s^*_{2n+1} are joining factors.[6]

The boundary conditions require the continuity of the tangential components of the electric and magnetic field at the boundary surface $\xi = \xi_0$; i.e.,

$$\sum_{n=0}^{\infty} [(g_{2n} + A_{2n}a_{2n})ce_{2n}(\eta) + (h_{2n+2} + B_{2n+2}b_{2n+2})se_{2n+2}(\eta) + i(g_{2n+1} + A_{2n+1}a_{2n+1})ce_{2n+1}(\eta) + i(h_{2n+1} + B_{2n+1}b_{2n+1})se_{2n+1}(\eta)]$$
$$= \sum_{n=0}^{\infty} \Big[C_{2n}c_{2n} \sum_{m=0}^{\infty} \alpha_{2n,2m}ce_{2m}(\eta) + D_{2n+2}\, d_{2n+2} \sum_{m=0}^{\infty} \beta_{2n+2,2m+2}se_{2m+2}(\eta) + iC_{2n+1}c_{2n+1} \sum_{m=0}^{\infty} \alpha_{2n+1,2m+1}ce_{2m+1}(\eta) + i\, D_{2n+1}\, d_{2n+1} \sum_{m=0}^{\infty} \beta_{2n+1,2m+1}se_{2m+1}(\eta) \Big], \quad (19)$$

$$\sum_{n=0}^{\infty} [(g'_{2n} + A_{2n}a'_{2n})ce_{2n}(\eta) + (h'_{2n+2} + B_{2n+2}b'_{2n+2})se_{2n+2}(\eta) + i(g'_{2n+1} + A_{2n+1}a'_{2n+1})ce_{2n+1}(\eta) + i(h'_{2n+1} + B_{2n+1}b'_{2n+1})se_{2n+1}(\eta)]$$
$$= \sum_{n=0}^{\infty} \Big[C_{2n}c'_{2n} \sum_{m=0}^{\infty} \alpha_{2n,2m}ce_{2m}(\eta) + D_{2n+2}\, d'_{2n+2} \sum_{m=0}^{\infty} \beta_{2n+2,2m+2}se_{2m+2}(\eta) + iC_{2n+1}c'_{2n+1} \sum_{m=0}^{\infty} \alpha_{2n+1,2m+1}ce_{2m+1}(\eta) + i\, D_{2n+1}\, d'_{2n+1} \sum_{m=0}^{\infty} \beta_{2n+1,2m+1}se_{2m+1}(\eta) \Big], \quad (20)$$

in which the relations

$$ce^*_{2n}(\eta) = \sum_{m=0}^{\infty} \alpha_{2n,2m}ce_{2m}(\eta),$$
$$se^*_{2n+2}(\eta) = \sum_{m=0}^{\infty} \beta_{2n+2,2m+2}se_{2m+2}(\eta),$$
$$ce^*_{2n+1}(\eta) = \sum_{m=0}^{\infty} \alpha_{2n+1,2m+1}ce_{2m+1}(\eta),$$
$$se^*_{2n+1}(\eta) = \sum_{m=0}^{\infty} \beta_{2n+1,2m+1}se_{2m+1}(\eta), \quad (21)$$

and the abbreviations

$$h_l = (1/s_l)Se_l(\xi_0)se_l(\theta),$$
$$h'_l = (1/s_l)Se'_l(\xi_0)se_l(\theta),$$
$$g_l = (1/p_l)Ce_l(\xi_0)ce_l(\theta),$$
$$g'_l = (1/p_l)Ce'_l(\xi_0)ce_l(\theta),$$
$$a_l = (1/p_l)Me^{(1)}_l(\xi_0)ce_l(\theta),$$
$$a'_l = (1/p_l)Me^{(1)\prime}_l(\xi_0)ce_l(\theta),$$
$$b_l = (1/s_l)Ne^{(1)}_l(\xi_0)se_l(\theta), \quad (22)$$
$$b'_l = (1/s_l)Ne^{(1)\prime}_l(\xi_0)se_l(\theta),$$
$$c_l = (1/p^*_l)Ce^*_l(\xi_0)ce_l(\theta),$$
$$c'_l = (1/p^*_l)Ce^{*\prime}_l(\xi_0)ce_l(\theta),$$
$$d_l = (1/s^*_l)Se^*_l(\xi_0)se_l(\theta),$$
$$d'_l = (1/s^*_l)Se^{*\prime}_l(\xi_0)se_l(\theta),$$
$$(l = 0, 1, 2, \cdots)$$

have been used. The primes denote differentiation with respect to ξ_0. $\alpha_{2n,2m}$, $\alpha_{2n+1,2m+1}$, $\beta_{2n+2,2m+2}$, and $\beta_{2n+1,2m+1}$ are given in the Appendix. Applying the orthogonality relations of the Mathieu functions to Eq. (19) gives the following expressions:

$$g_{2j} + A_{2j}a_{2j} = \sum_{r=0}^{\infty} c_{2r}\alpha_{2r,2j}C_{2r} \quad (23a)$$
$$g_{2j+1} + A_{2j+1}a_{2j+1} = \sum_{r=0}^{\infty} c_{2r+1}\alpha_{2r+1,2j+1}C_{2r+1} \quad (23b)$$
$$h_{2j+2} + B_{2j+2}b_{2j+2} = \sum_{r=0}^{\infty} d_{2r+2}\beta_{2r+2,2j+2}\, D_{2r+2} \quad (23c)$$
$$h_{2j+1} + B_{2j+1}b_{2j+1} = \sum_{r=0}^{\infty} d_{2r+1}\beta_{2r+1,2j+1}\, D_{2r+1} \quad (23d)$$
$$(j = 0, 1, 2 \cdots).$$

Similarly; from Eq. (20) one contains the following expressions:

$$g'_{2j} + A_{2j}a'_{2j} = \sum_{r=0}^{\infty} c'_{2r}\alpha_{2r,2j}C_{2r}, \quad (24a)$$
$$g'_{2j+1} + A_{2j+1}a'_{2j+1} = \sum_{r=0}^{\infty} c'_{2r+1}\alpha_{2r+1,2j+1}C_{2r+1} \quad (24b)$$
$$h'_{2j+2} + B_{2j+2}b'_{2j+2} = \sum_{r=0}^{\infty} d'_{2r+2}\beta_{2r+2,2j+2}\, D_{2r+2}, \quad (24c)$$
$$h'_{2j+1} + B_{2j+1}b'_{2j+1} = \sum_{r=0}^{\infty} d'_{2r+1}\beta_{2r+1,2j+1}\, D_{2r+1} \quad (24d)$$
$$(j = 0, 1, 2 \cdots).$$

Solving these equations for the arbitrary constants C_{2r}, C_{2r+1}, D_{2r+1} and D_{2r+2}, one obtains in matrix notations,

$$C_{2r} = R^{-1}_{2r,2j}G_{2j}, \quad (25a)$$

$$C_{2r+1} = R^{-1}_{2r+1,2i+1}G_{2i+1}, \quad (25b)$$

$$D_{2r+1} = Q^{-1}_{2r+1,2i+1}H_{2i+1}, \quad (25c)$$

$$D_{2r+2} = Q^{-1}_{2r+2,2i+2}H_{2i+2}, \quad (25d)$$

where $R^{-1}_{2r,2i}$ is the inverse of the matrix

$$R_{2i,2r} = (c_{2r} - c'_{2r}a_{2i}/a'_{2i})\alpha_{2r,2i};$$

$R^{-1}_{2r+1,2i+1}$ is the inverse of the matrix

$$R_{2i+1,2r+1} = (c_{2r+1} - c'_{2r+1}a_{2i+1}/a'_{2i+1})\alpha_{2r+1,2i+1};$$

$Q^{-1}_{2r+1,2i+1}$ is the inverse of the matrix

$$Q_{2i+1,2r+1} = (d_{2r+1} - d'_{2r+1}b_{2i+1}/b'_{2i+1})\beta_{2r+1,2i+1};$$

$Q^{-1}_{2r+2,2i+2}$ is the inverse of the matrix

$$Q_{2i+2,2r+2} = (d_{2r+2} - d'_{2r+2}b_{2i+2}/b'_{2i+2})\beta_{2r+2,2i+2};$$

$G_{2i} = g_{2i} - g'_{2i}\, a_{2i}/a'_{2i}$ is a column matrix; $G_{2i+1} = g_{2i+1} - g'_{2i+1}\, a_{2i+1}/a'_{2i+1}$ is a column matrix; $H_{2i+1} = h_{2i+1} - h'_{2i+1}\, b_{2i+1}/b'_{2i+1}$ is a column matrix; and

$$H'_{2i+2} = h_{2i+2} - h'_{2i+2}b_{2i+2}/b'_{2i+2}$$

is a column matrix. The other arbitrary constants can be easily obtained from Eq. (23) or (24); they are

$$A_{2i} = \frac{1}{a_{2i}}\left[-g_{2i} + \sum_{r=0}^{\infty} c_{2r}\alpha_{2r,2i}C_{2r}\right], \quad (25e)$$

$$A_{2i+1} = \frac{1}{a_{2i+1}}\left[-g_{2i+1} + \sum_{r=0}^{\infty} c_{2r+1}\alpha_{2r+1,2i+1}C_{2r+1}\right], \quad (25f)$$

$$B_{2i+1} = \frac{1}{b_{2i+1}}\left[-h_{2i+1} + \sum_{r=0}^{\infty} d_{2r+1}\beta_{2r+1,2i+1}\, D_{2r+1}\right], \quad (25g)$$

$$B_{2i+2} = \frac{1}{b_{2i+2}}\left[-h_{2i+2} + \sum_{r=0}^{\infty} d_{2r+2}\beta_{2r+2,2i+2}\, D_{2r+2}\right]. \quad (25h)$$

H wave. The expressions for the field components of an incident H wave are:

$$H^i_z = H_0\,\{\text{the right-hand side of Eq. (9)}\}, \quad (26)$$

$$E^i_\eta = (1/k_0^2p)\{-i\omega\mu\; \partial H^i_z/\partial\xi\}, \quad (27)$$

$$E^i_\xi = (1/k_0^2p)\{i\omega\mu\; \partial H^i_z/\partial\eta\}. \quad (28)$$

The scattered field and the transmitted field inside the dielectric cylinder are of the form

$$H^s_z = 2H_0 \sum_{n=0}^{\infty}\left[\frac{U_{2n}}{p_{2n}} Me^{(1)}_{2n}(\xi)ce_{2n}(\eta)ce_{2n}(\theta) + (V_{2n+2}/s_{2n+2})Ne^{(1)}_{2n+2}(\xi)se_{2n+2}(\eta)se_{2n+2}(\theta) + i(U_{2n+1}/p_{2n+1})Me^{(1)}_{2n+1}(\xi)ce_{2n+1}(\eta)ce_{2n+1}(\theta) + i(V_{2n+1}/s_{2n+1})Ne^{(1)}_{2n+1}(\xi)se_{2n+1}(\eta)se_{2n+1}(\theta)\right], \quad (29)$$

$$E^s_\eta = (1/k_0^2p)\{-i\omega\mu\; \partial H^s_z/\partial\xi\}, \quad (30)$$

$$E^s_\xi = (1/k_0^2p)\{i\omega\mu\; \partial H^s_z/\partial\eta\}, \quad (31)$$

and,

$$H^t_z = 2H_0 \sum_{n=0}^{\infty}\left[\frac{W_{2n}}{p^*_{2n}} Ce^*_{2n}(\xi)ce^*_{2n}(\eta)ce_{2n}(\theta) + (X_{2n+2}/s^*_{2n+2})Se^*_{2n+2}(\xi)se^*_{2n+2}(\eta)se_{2n+2}(\theta) + i(W_{2n+1}/p^*_{2n+1})Ce^*_{2n+1}(\xi)ce^*_{2n+1}(\eta)ce_{2n+1}(\theta) + i(X_{2n+1}/s^*_{2n+1})Se^*_{2n+1}(\xi)se^*_{2n+1}(\eta)se_{2n+1}(\theta)\right]. \quad (32)$$

$$E^t_\eta = (1/k_1^2p)\{-i\omega\mu\; \partial H^t_z/\partial\xi\}, \quad (33)$$

$$E^t_\xi = (1/k_1^2p)\{i\omega\mu\; \partial H^t_z/\partial\eta\}. \quad (34)$$

The unknown coefficients U_{2n}, U_{2n+1}, V_{2n+2}, V_{2n+1}, W_{2n}, W_{2n+1}, X_{2n+2}, and X_{2n+1} are to be determined by applying the boundary conditions that the tangential components of the magnetic and electric fields are continuous at the boundary surface $\xi = \xi_0$. Applying the similar procedures as described for the E wave, one obtains the following expressions for the arbitrary constants (in matrix notations):

$$W_{2r} = P^{-1}_{2r,2i}L_{2i}, \quad (35a)$$

$$W_{2r+1} = P^{-1}_{2r+1,2i+1}L_{2i+1}, \quad (35b)$$

$$X_{2r+1} = S^{-1}_{2r+1,2i+1}M_{2i+1}, \quad (35c)$$

$$X_{2r+2} = S^{-1}_{2r+2,2i+2}M_{2i+2}, \quad (35d)$$

$$U_{2i} = \frac{1}{a_{2i}}\left[-g_{2i} + \sum_{r=0}^{\infty} c_{2r}\alpha_{2r,2i}W_{2r}\right], \quad (35e)$$

$$U_{2i+1} = \frac{1}{a_{2i+1}}\left[-g_{2i+1} + \sum_{r=0}^{\infty} c_{2r+1}\alpha_{2r+1,2i+1}W_{2r+1}\right], \quad (35f)$$

$$V_{2i+1} = \frac{1}{b_{2i+1}}\left[-h_{2i+1} + \sum_{r=0}^{\infty} d_{2r+1}\beta_{2r+1,2i+1}X_{2r+1}\right], \quad (35g)$$

$$V_{2i+2} = \frac{1}{b_{2i+2}}\left[-h_{2i+2} + \sum_{r=0}^{\infty} d_{2r+2}\beta_{2r+2,2i+2}X_{2r+2}\right], \quad (35h)$$

where $P^{-1}_{2r,2i}$ is the inverse of the matrix

$$P_{2i,2r} = \left(c_{2r} - c'_{2r}\,\frac{a_{2i}}{a'_{2i}}\frac{\epsilon_0}{\epsilon_1}\right)\alpha_{2r,2i};$$

$P^{-1}_{2r+1,2i+1}$ is the inverse of the matrix

$$P_{2i+1,2r+1} = \left(c_{2r+1} - c'_{2r+1}\,\frac{a_{2i+1}}{a'_{2i+1}}\frac{\epsilon_0}{\epsilon_1}\right)\alpha_{2r+1,2i+1};$$

$S^{-1}_{2r+1,2j+1}$ is the inverse of the matrix

$$S_{2j+1,2r+1} = \left(d_{2r+1} - d'_{2r+1}\frac{b_{2j+1}}{b'_{2j+1}}\frac{\epsilon_0}{\epsilon_1}\right)\beta_{2r+1,2j+1};$$

$S^{-1}_{2r+2,2j+2}$ is the inverse of the matrix

$$S_{2j+2,2r+2} = \left(d_{2r+2} - d'_{2r+2}\frac{b_{2r+2}}{b'_{2r+2}}\frac{\epsilon_0}{\epsilon_1}\right)\beta_{2r+2,2j+2};$$

$L_{2j} = g_{2j} - g'_{2j}\,a_{2j}/a'_{2j}$ is a column matrix; $L_{2j+1} = g_{2j+1} - g'_{2j+1}\,a_{2j+1}/a'_{2j+1}$ is a column matrix; $M_{2j+1} = h_{2j+1} - h'_{2j+1}\,b_{2j+1}/b'_{2j+1}$ is a column matrix; and $M_{2j+2} = h_{2j+2} - h'_{2j+2}\,b_{2j+2}/b'_{2j+2}$ is a column matrix.

At large distance from the dielectric cylinder, the confocal ellipses are now sensibly concentric circles, and it is permissible to use the following asymptotic expressions for the radial Mathieu functions when $k_0r \gg 1$ and $k_0r \gg M$, where M is the order of the Mathieu function and $\frac{1}{4}k_0^2q^2\cosh^2\xi \approx \frac{1}{4}k_0^2r^2$:

$$Me^{(1)}_{2n}(\xi) \sim -ip_{2n}(2/\pi k_0r)^{1/2}e^{[i(k_0r+\pi/4)]}, \tag{36}$$

$$Me^{(1)}_{2n+1}(\xi) \sim -p_{2n+1}(2/\pi k_0r)^{1/2}e^{[i(k_0r+\pi/4)]}, \tag{37}$$

$$Ne^{(1)}_{2n+1}(\xi) \sim -s_{2n+1}(2/\pi k_0r)^{1/2}e^{[i(k_0r+\pi/4)]}, \tag{38}$$

$$Ne^{(1)}_{2n+2}(\xi) \sim -is_{2n+2}(2/\pi k_0r)^{1/2}e^{[i(k_0r+\pi/4)]}. \tag{39}$$

Using the above equations, one obtains the expressions for the far-zone scattered field:

$$H^s_z(E\text{ wave}) \approx 2E_0\left[(-i)\left(\frac{\epsilon_0}{\mu}\right)^{1/2}\right]\left(\frac{2}{\pi k_0r}\right)^{1/2}e^{i(k_0r+\pi/4)} \times \sum_{n=0}^{\infty}[A_{2n}ce_{2n}(\eta)ce_{2n}(\theta) + B_{2n+2}se_{2n+2}(\eta)se_{2n+2}(\theta) + A_{2n+1}ce_{2n+1}(\eta)ce_{2n+1}(\theta) + B_{2n+1}se_{2n+1}(\eta)se_{2n+1}(\theta)], \tag{40}$$

$$E^s_z(H\text{ wave}) \approx 2H_0\left[(-i)\left(\frac{\mu}{\epsilon_0}\right)^{1/2}\right]\left(\frac{2}{\pi k_0r}\right)^{1/2}e^{i(k_0r+\pi/4)} \times \sum_{n=0}^{\infty}[U_{2n}ce_{2n}(\eta)ce_{2n}(\theta) + V_{2n+2}se_{2n+2}(\eta)se_{2n+2}(\theta) + U_{2n+1}ce_{2n+1}(\eta)ce_{2n+1}(\theta) + V_{2n+1}se_{2n+1}(\eta)se_{2n+1}(\theta)]. \tag{41}$$

The ribbon corresponds to the limiting case of a cylinder of very small thickness, $\xi_0 \to 0$.

As an ellipse degenerates to a circle its semifocal length q tends to zero while ξ_0 approaches infinity such that the product $q\cosh\xi_0$ or $q\sinh\xi_0$ or $\frac{1}{2}qe^{\xi_0}$ tends to a constant r_0, which is the radius of the degenerated circle. Using the degenerated forms of Mathieu and modified Mathieu functions,[6] the arbitrary constants for the E wave become

$$C_s = D_s = \frac{H^{(1)\prime}_s(k_0r_0)J_s(k_0r_0) - J'_s(k_0r_0)H^{(1)}_s(k_0r_0)}{H^{(1)\prime}_s(k_0r_0)J_s(k_1r_0) - (\epsilon_1/\epsilon_0)^{1/2}J'_s(k_1r_0)H^{(1)}_s(k_0r_0)}, \tag{42a}$$

$$A_s = B_s = \frac{(\epsilon_1/\epsilon_0)^{1/2}J_s(k_0r_0)J'_s(k_1r_0) - J'_s(k_0r_0)J_s(k_1r_0)}{H^{(1)\prime}_s(k_0r_0)J_s(k_1r_0) - (\epsilon_1/\epsilon_0)^{1/2}J'_s(k_1r_0)H^{(1)}_s(k_0r_0)}, \tag{42b}$$

and the arbitrary constants for the H wave become

$$W_s = X_s = \frac{H^{(1)\prime}_s(k_0r_0)J_s(k_0r_0) - J'_s(k_0r_0)H^{(1)}_s(k_0r_0)}{H^{(1)\prime}_s(k_0r_0)J_s(k_1r_0) - J'_s(k_1r_0)H^{(1)}_s(k_0r_0)(\epsilon_0/\epsilon_1)^{1/2}}, \tag{43a}$$

$$U_s = V_s = \frac{(\epsilon_0/\epsilon_1)^{1/2}J_s(k_0r_0)J'_s(k_1r_0) - J'_s(k_0r_0)J_s(k_1r_0)}{H^{(1)\prime}_s(k_0r_0)J_s(k_1r_0) - J'_s(k_1r_0)H^{(1)}_s(k_0r_0)(\epsilon_0/\epsilon_1)^{1/2}}. \tag{43b}$$

These are well-known expressions for the circular dielectric cylinder. This completes the derivation of the fundamental formulae involved in the diffraction of waves by an elliptical dielectric cylinder.

IV. CONCLUSION

The exact solution of the problem of the diffraction of waves by an elliptical dielectric cylinder, or by a dielectric ribbon is obtained. It is interesting to note that unlike the case for a circular dielectric cylinder, or the case for a perfectly conducting elliptical cylinder, each expansion coefficient of the scattered or transmitted wave for the dielectric elliptical cylinder is coupled to all coefficients of the series expansion for the incident wave. This characteristic is also found in the problem of surface wave propagation along an elliptical dielectric cylinder.[7] To qualitatively illustrate how the solutions behave, the radiation patterns of the scattered fields for the two polarizations of the incident wave are

[7] C. Yeh, Ph.D. Thesis, California Institute of Technology, Pasadena, California, 1962; J. Appl. Phys. 33, 3235 (1962).

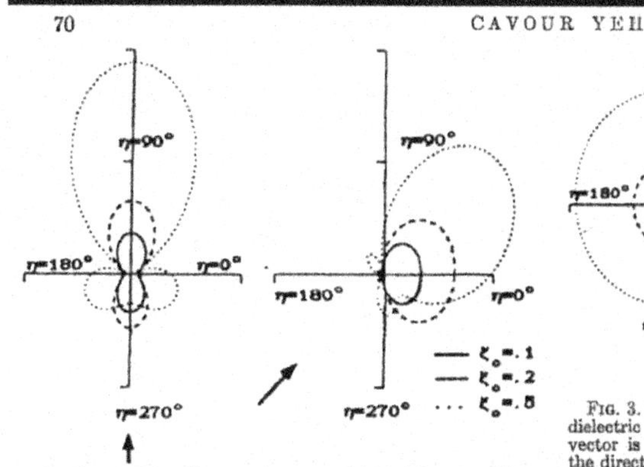

FIG. 2. Polar diagrams for waves ($|H_y^s|$) scattered by a dielectric ribbon with $k_0^2q^2 = 10$. The incident electric vector is polarized in the axial direction. (Arrows indicate the direction of incident waves.)

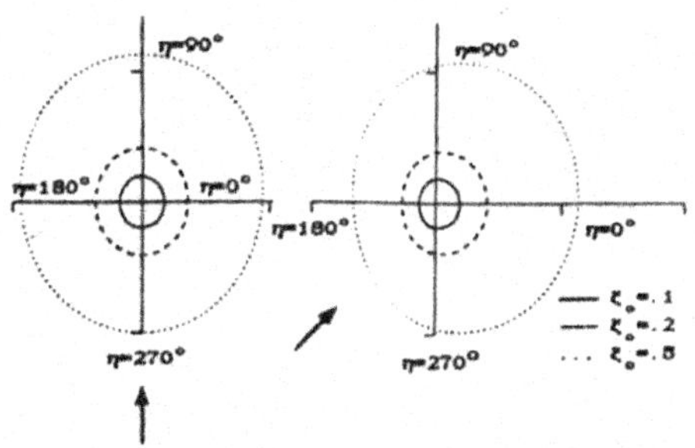

FIG. 3. Polar diagrams for waves ($|H_y^s|$) scattered by a dielectric ribbon with $k_0^2 q^2 = 1.0$. The incident electric vector is polarized in the axial direction. (Arrows indicate the direction of incident waves.)

computed. Numerical computations are carried out using the available tables on the expansion coefficients of Mathieu functions,[8] and using the high-speed IBM 7090 computer. Two cases of k_0q are considered: one with $k_0q = 1$, and the other with $k_0q = (10)^{1/2}$. Various values of ξ_0 are used. It is assumed that $\epsilon_1/\epsilon_0 = 2.0$. Results are shown in Figs. 2 through 5. Numerical investigation shows that the infinite series representing these expansion coefficients for the scattered or transmitted wave converge quite rapidly for small values of $\gamma_1 = k_1q/2$; only the first few terms of the infinite series are needed as long as γ_1 is less than 4. The rate of convergence can best be illustrated by the example in Table I, where A_{2n}, A_{2n+1}, B_{2n+2}, and B_{2n+1} are

TABLE I. The rate of convergence for $\xi_0 = 0.2$, $k_0q = (10)^{1/2}$, and $\theta = 90°$.

	A_{2n}			A_{2n+1}		
n	$m=2$	$m=3$	$m=4$	$m=2$	$m=3$	$m=4$
0	-0.151	-0.167	-0.166	-0.254	-0.261	-0.261
	$+0.328i$	$+0.333i$	$+0.333i$	$+0.318i$	$+0.326i$	$+0.326i$
1	$+0.137\times10^{-1}$	$+0.678\times10^{-1}$	$+0.675\times10^{-1}$	$+0.400\times10^{-1}$	$+0.408\times10^{-1}$	$+0.408\times10^{-1}$
	$+0.212i$	$+0.126i$	$+0.127i$	$-0.142\times10^{-1}i$	$-0.205\times10^{-1}i$	$-0.205\times10^{-1}i$
2		-0.107×10^{-2}	-0.105×10^{-2}		-0.307×10^{-3}	-0.307×10^{-3}
		$-0.109\times10^{-1}i$	$-0.111\times10^{-1}i$		$-0.261\times10^{-3}i$	$-0.265\times10^{-3}i$
3			$+0.330\times10^{-5}$			$+0.443\times10^{-6}$
			$+0.271\times10^{-4}i$			$+0.791\times10^{-7}i$
	B_{2n+2}			B_{2n+1}		
n	$m=2$	$m=3$	$m=4$	$m=2$	$m=3$	$m=4$
0	-0.405×10^{-3}	-0.405×10^{-3}	-0.405×10^{-3}	-0.375×10^{-2}	-0.375×10^{-2}	-0.375×10^{-2}
	$+0.192\times10^{-1}i$	$+0.192\times10^{-1}i$	$+0.192\times10^{-1}i$	$+0.593\times10^{-1}i$	$+0.594\times10^{-1}i$	$+0.594\times10^{-1}i$
1	$+0.152\times10^{-4}$	$+0.153\times10^{-4}$	$+0.153\times10^{-4}$	$+0.373\times10^{-3}$	$+0.375\times10^{-3}$	$+0.375\times10^{-3}$
	$-0.440\times10^{-3}i$	$-0.469\times10^{-3}i$	$-0.469\times10^{-3}i$	$-0.216\times10^{-2}i$	$-0.255\times10^{-2}i$	$-0.255\times10^{-2}i$
2		-0.421×10^{-7}	-0.421×10^{-7}		-0.365×10^{-5}	-0.365×10^{-5}
		$-0.172\times10^{-5}i$	$-0.175\times10^{-5}i$		$-0.427\times10^{-4}i$	$-0.438\times10^{-4}i$
3			$+0.266\times10^{-10}$			$+0.534\times10^{-8}$
			$-0.105\times10^{-8}i$			$-0.571\times10^{-7}i$

[8] National Bureau of Standards, *Tables Relating to Mathieu Functions* (Columbia University Press, New York, 1951).

the expansions coefficients for the scattered field H_y^s [see Eq. (40)]. $m \times m$ is the size of the matrix used. It is observed that the infinite series converge faster for smaller values of k_0q and ξ_0. For example, when $k_0q = 1.0$, $\xi_0 = 0.2$, only three terms of the infinite series (i.e., $m = 3$) are required.

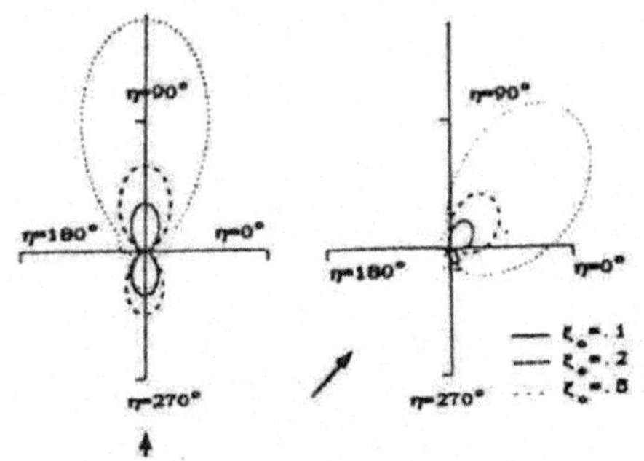

FIG. 4. Polar diagrams for waves ($|E_y^s|$) scattered by a dielectric ribbon with $k_0^2 q^2 = 10$. The incident magnetic vector is polarized in the axial direction. (Arrows indicate the direction of incident waves.)

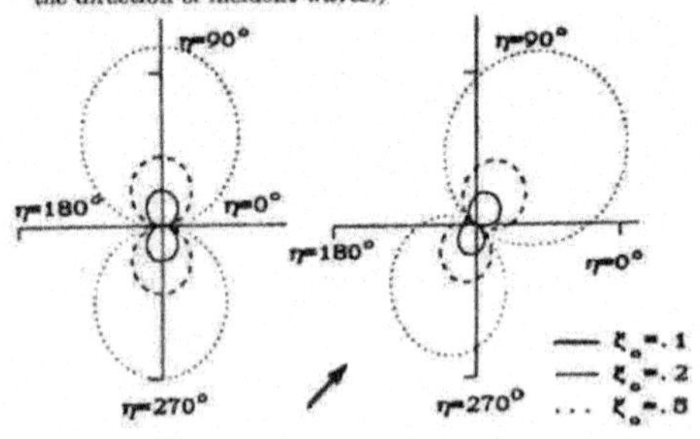

FIG. 5. Polar diagrams for waves ($|E_y^s|$) scattered by a dielectric ribbon with $k_0^2 q^2 = 1.0$. The incident magnetic vector is polarized in the axial direction. (Arrows indicate the direction of incident waves.)

It is interesting to note that the solutions with a slight modification are applicable to the problem of the diffraction of waves from a plasma ribbon, and that the method used in analyzing this problem are applicable to problems involving dielectric coated elliptical cylindrical radiators. Corresponding acoustical problems may also be analyzed in a similar manner.

ACKNOWLEDGMENTS

I would like to thank Miss Linda LaBella for typing this manuscript. I also wish to thank Mrs. G. Knudson and Mrs. K. K. Higa of the Hughes Aircraft Company in Culver City for providing the subroutines to compute Bessel functions and to invert matrix. The use of the computing facilities at the Western Data Processing Center at UCLA is gratefully acknowledged.

APPENDIX

FORMULAS FOR $\alpha_{2n,2m}$, $\alpha_{2n+1,2m+1}$, $\beta_{2n+2,2m+2}$, AND $\beta_{2n+1,2m+1}$

It can readily be shown, from the theory of Mathieu functions that

$$\alpha_{2n,2m} = \frac{\int_0^{2\pi} ce_{2n}^*(\eta) ce_{2m}(\eta)\, d\eta}{\int_0^{2\pi} ce_{2m}^2(\eta)\, d\eta} = \frac{2A_0^{*(2n)}A_0^{(2m)} + \sum_{r=1}^{\infty}[A_{2r}^{*(2n)}A_{2r}^{(2m)}]}{2[A_0^{(2m)}]^2 + \sum_{r=1}^{\infty}[A_{2r}^{(2m)}]^2},$$

$$\alpha_{2n+1,2m+1} = \frac{\int_0^{2\pi} ce_{2n+1}^*(\eta) ce_{2m+1}(\eta)\, d\eta}{\int_0^{2\pi} ce_{2m+1}^2(\eta)\, d\eta} = \frac{\sum_{r=0}^{\infty}[A_{2r+1}^{*(2n+1)}A_{2r+1}^{(2m+1)}]}{\sum_{r=0}^{\infty}[A_{2r+1}^{(2m+1)}]^2},$$

$$\beta_{2n+2,2m+2} = \frac{\int_0^{2\pi} se_{2n+2}^*(\eta) se_{2m+2}(\eta)\, d\eta}{\int_0^{2\pi} se_{2m+2}^2(\eta)\, d\eta} = \frac{\sum_{r=0}^{\infty}[B_{2r+2}^{*(2n+2)}B_{2r+2}^{(2m+2)}]}{\sum_{r=0}^{\infty}[B_{2r+2}^{(2m+2)}]^2},$$

$$\beta_{2n+1,2m+1} = \frac{\int_0^{2\pi} se_{2n+1}^*(\eta) se_{2m+1}(\eta)\, d\eta}{\int_0^{2\pi} se_{2m+1}^2(\eta)\, d\eta} = \frac{\sum_{r=0}^{\infty}[B_{2r+1}^{*(2n+1)}B_{2r+1}^{(2m+1)}]}{\sum_{r=0}^{\infty}[B_{2r+1}^{(2m+1)}]^2},$$

where

$$A_{2r}^{(2n)},\quad A_{2r}^{*(2n)},\quad A_{2r+1}^{(2n+1)},\quad A_{2r+1}^{*(2n+1)},\quad B_{2r+2}^{(2n+2)},\quad B_{2r+2}^{*(2n+2)},\quad B_{2r+1}^{(2n+1)},$$

and $B_{2r+1}^{*(2n+1)}$ are, respectively, the expansion coefficients for

$$ce_{2n}(\eta),\quad ce_{2n}^*(\eta),\quad ce_{2n+1}(\eta),\quad ce_{2n+1}^*(\eta),\quad se_{2n+2}(\eta),\quad se_{2n+2}^*(\eta),\quad se_{2n+1}(\eta),$$

and $se_{2n+1}^*(\eta)$ (see reference 6).

Reprinted from JOURNAL OF THE OPTICAL SOCIETY OF AMERICA, Vol. 54, No. 10, 1227–1231, October 1964
Printed in U. S. A.

Scattering of Obliquely Incident Light Waves by Elliptical Fibers*

C. YEH
Electrical Engineering Department, University of Southern California, Los Angeles, California 90007

(Received 15 April 1963; revision received 16 June 1964)

The exact solution of the scattering of obliquely incident plane waves by an elliptical dielectric cylinder is obtained. Each expansion coefficient of the scattered or transmitted wave is coupled to all coefficients of the series expansion for the incident wave except when the elliptical cylinder degenerates to a circular one. Both polarizations of the incident wave are considered: one with the incident electric vector in the axial direction, and the other with the incident magnetic vector in the axial direction. In the general case of oblique incidence, the scattered field contains a significant cross-polarized component which vanishes at normal incidence.

I. INTRODUCTION

THE problems of scattering of waves by a circular dielectric cylinder have been considered by many authors.[1–3] Most recently, the exact solution of the problem of diffraction of normally incident plane waves by a dielectric elliptical cylinder was obtained.[4] However, the general case for oblique incidence has not been considered. It is the purpose of this paper to present a complete solution for this general problem. It is hoped that the results will be applicable to the problem of scattering of light by noncircular fibers.

Since this work is an extension of Ref. 4, the notations used there will be carried over.

II. SCATTERING OF AN OBLIQUELY INCIDENT PLANE WAVE

The geometry of this problem is shown in Fig. 1. It is assumed that the elliptical dielectric cylinder, which has a permittivity ϵ_1, a permeability μ_1, and a conductivity of zero, is embedded in a homogeneous perfect dielectric medium (ϵ_0, μ_0, $\sigma_0=0$). A time dependence of $e^{-i\omega t}$ has been assumed and suppressed throughout.

Two types of incident waves are possible. The one, called an **E** wave is defined by $\mathbf{H}_z=0$, and the other called an **H** wave is defined by $\mathbf{E}_z=0$.

---

* Supported by the U. S. Air Force Cambridge Research Laboratory.

[1] P. Debye, Physik Z. 9, 775 (1908).

[2] J. R. Wait, Can. J. Phys. 33, 189 (1955).

[3] H. C. Van de Hulst, *Light Scattering by Small Particles* (John Wiley & Sons, Inc., New York, 1957).

[4] C. Yeh, J. Math. Phys. 4, 65 (1963).

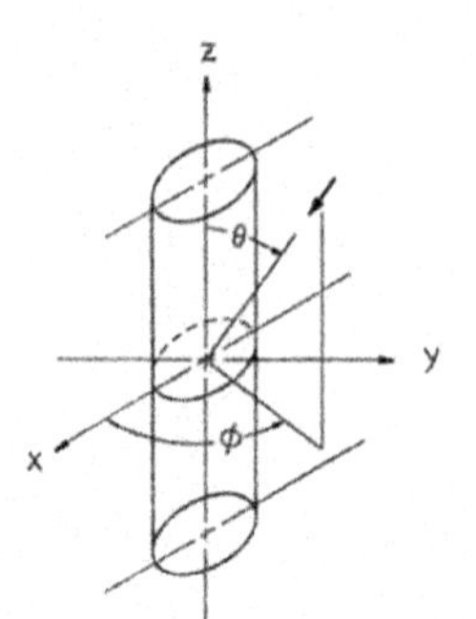

FIG. 1. Plane wave incident on an elliptical dielectric cylinder. Arrow indicates the direction of incident wave.

A plane wave with its direction of propagation defined by the angles ϕ and θ (see Fig. 1) is given by

$$\exp[ik_0(x\cos\phi\sin\theta+y\sin\phi\sin\theta+z\cos\theta)]$$
$$=\exp\{ik_0[q\sin\theta(\cosh\xi\cos\eta\cos\phi+\sinh\xi\sin\eta\sin\phi)+z\cos\theta]\}$$
$$=2\sum_{n=0}^{\infty}\Bigg[\frac{1}{p_{2n}}Ce_{2n}(\xi,\gamma_0^2)ce_{2n}(\eta,\gamma_0^2)ce_{2n}(\phi,\gamma_0^2)+\frac{1}{s_{2n+2}}Se_{2n+2}(\xi,\gamma_0^2)se_{2n+2}(\eta,\gamma_0^2)se_{2n+2}(\phi,\gamma_0^2)+\frac{i}{p_{2n+1}}Ce_{2n+1}(\xi,\gamma_0^2)ce_{2n+1}(\eta,\gamma_0^2)ce_{2n+1}(\phi,\gamma_0^2)+\frac{i}{s_{2n+1}}Se_{2n+1}(\xi,\gamma_0^2)se_{2n+1}(\eta,\gamma_0^2)se_{2n+1}(\phi,\gamma_0^2)\Bigg]\times e^{ik_0z\cos\theta},\quad(1)$$

where $\gamma_0^2=k_0^2q^2\sin^2\theta/4$, $k_0^2=2\pi/\lambda_0$, and λ_0 is the free-space wavelength. p_{2n}, p_{2n+1}, s_{2n+2}, and s_{2n+1} are joining factors.[5]

In order to simplify the notations for the Mathieu and modified Mathieu functions without any ambiguities, the following abbreviations are used:

$$\begin{aligned}Ce_n(\xi,\gamma_0^2)&=Ce_n(\xi), & ce_n(\eta,\gamma_0^2)&=ce_n(\eta),\\ Se_n(\xi,\gamma_0^2)&=Se_n(\xi), & se_n(\eta,\gamma_0^2)&=se_n(\eta),\\ Ce_n(\xi,\gamma_1^2)&=Ce_n^*(\xi), & ce_n(\eta,\gamma_1^2)&=ce_n^*(\eta),\\ Se_n(\xi,\gamma_1^2)&=Se_n^*(\xi), & se_n(\eta,\gamma_1^2)&=se_n^*(\eta),\end{aligned}$$
$$Me_n^{(1),(2)}(\xi,\gamma_0^2)=Me_n^{(1),(2)}(\xi),$$
$$Ne_n^{(1),(2)}(\xi,\gamma_0^2)=Ne_n^{(1),(2)}(\xi),\quad(2)$$

with $\gamma_1^2=(k_1^2-k_0^2\cos^2\theta)q^2/4$ where $k_1^2=\omega^2\mu_1\epsilon_1$.

[5] N. McLachlan, *Theory and Application of Mathieu Function* (Oxford University Press, London 1951).

E Wave

The axial components of an incident **E** wave are

$$\mathbf{E}_z^i=\mathbf{E}_0\,[\text{the right-hand side of Eq. (1)}]\quad(3)$$
$$\mathbf{H}_z^i=0.\quad(4)$$

Unlike the case for a normally incident wave or for a perfectly conducting cylinder at oblique incidence, the boundary conditions for the present general case of oblique incidence on a dielectric elliptical cylinder cannot be satisfied if the z component of the scattered or transmitted magnetic field is taken to be zero for an incident **E** wave, or if the z component of the scattered or transmitted electric field is taken to be zero for an incident **H** wave. Hence, we see that the axial components of the scattered field and of the transmitted field inside the dielectric cylinder must be of the form

$$\mathbf{E}_z^s=2\mathbf{E}_0\sum_{n=0}^{\infty}\Bigg[\frac{A_{2n}}{p_{2n}}Me_{2n}^{(1)}(\xi)ce_{2n}(\eta)ce_{2n}(\phi)+\frac{B_{2n+2}}{s_{2n+2}}Ne_{2n+2}^{(1)}(\xi)se_{2n+2}(\eta)se_{2n+2}(\phi)+i\frac{A_{2n+1}}{p_{2n+1}}Me_{2n+1}^{(1)}(\xi)ce_{2n+1}(\eta)ce_{2n+1}(\phi)+i\frac{B_{2n+1}}{s_{2n+1}}Ne_{2n+1}^{(1)}(\xi)se_{2n+1}(\eta)se_{2n+1}(\phi)\Bigg]\times e^{ik_0z\cos\theta},\quad(5)$$

$$\mathbf{H}_z^s=2\mathbf{E}_0\sum_{n=0}^{\infty}\Bigg[\frac{C_{2n}}{p_{2n}}Me_{2n}^{(1)}(\xi)ce_{2n}(\eta)ce_{2n}(\phi)+\frac{D_{2n+2}}{s_{2n+1}}Ne_{2n+2}^{(1)}(\xi)se_{2n+2}(\eta)se_{2n+2}(\phi)+i\frac{C_{2n+1}}{p_{2n+1}}Me_{2n+1}^{(1)}(\xi)ce_{2n+1}(\eta)ce_{2n+1}(\phi)+i\frac{D_{2n+1}}{s_{2n+1}}Ne_{2n+1}^{(1)}(\xi)se_{2n+1}(\eta)se_{2n+1}(\phi)\Bigg]\times e^{ik_0z\cos\theta},\quad(6)$$

and

$$\mathbf{E}_z^t=2\mathbf{E}_0\sum_{n=0}^{\infty}\Bigg[\frac{F_{2n}}{p_{2n}^*}Ce_{2n}^*(\xi)ce_{2n}^*(\eta)ce_{2n}(\phi)+\frac{G_{2n+2}}{s_{2n+2}^*}Se_{2n+2}^*(\xi)se_{2n+2}^*(\eta)se_{2n+2}(\phi)+i\frac{F_{2n+1}}{p_{2n+1}^*}Ce_{2n+1}^*(\xi)ce_{2n+1}^*(\eta)ce_{2n+1}(\phi)+i\frac{G_{2n+1}}{s_{2n+1}^*}Se_{2n+1}^*(\xi)se_{2n+1}^*(\eta)se_{2n+1}(\phi)\Bigg]\times e^{ik_0z\cos\theta},\quad(7)$$

$$\mathbf{H}_z{}^t=2\mathbf{E}_0\sum_{n=0}^{\infty}\Big[\frac{P_{2n}}{p_{2n}{}^*}Ce_{2n}{}^*(\xi)ce_{2n}{}^*(\eta)ce_{2n}(\phi)$$
$$+\frac{Q_{2n+2}}{s_{2n+2}{}^*}Se_{2n+2}{}^*(\xi)se_{2n+2}{}^*(\eta)se_{2n+2}(\phi)$$
$$+i\frac{P_{2n+1}}{p_{2n+1}{}^*}Ce_{2n+1}{}^*(\xi)ce_{2n+1}{}^*(\eta)ce_{2n+1}(\phi)$$
$$+i\frac{Q_{2n+1}}{s_{2n+1}{}^*}Se_{2n+1}{}^*(\xi)se_{2n+1}{}^*(\eta)se_{2n+1}(\phi)\Big]\times e^{ik_0z\cos\theta}, \quad (8)$$

where A_n, B_n, C_n, D_n, F_n, G_n, P_n, and Q_n are arbitrary unknown coefficients that can be determined by applying the boundary conditions. $p_{2n}{}^*$, $p_{2n+1}{}^*$, $s_{2n+2}{}^*$, and $s_{2n+1}{}^*$ are joining factors. All transverse fields can be derived from Maxwell's equations with the knowledge of the axial fields.

The boundary conditions require the continuity of the tangential components of the electric and magnetic field at the boundary surface $\xi=\xi_0$; i.e.,

$$[\text{r.h.s. of (3) with } \xi=\xi_0]+[\text{r.h.s. of (5) with } \xi=\xi_0]=[\text{r.h.s. of (7) with } \xi=\xi_0], \quad (9)$$

$$[\text{r.h.s. of (6) with } \xi=\xi_0]=[\text{r.h.s. of (8) with } \xi=\xi_0], \quad (10)$$

$$\frac{1}{k_0^2\sin^2\theta}\Big\{ik_0\cos\theta\frac{\partial}{\partial\eta}[\text{r.h.s. of (3) with } \xi=\xi_0]+ik_0\cos\theta\frac{\partial}{\partial\eta}[\text{r.h.s. of (5) with } \xi=\xi_0]-i\omega\mu_0\frac{\partial}{\partial\xi_0}[\text{r.h.s. of (6) with } \xi=\xi_0]\Big\}$$
$$=\frac{1}{(k_1^2-k_0^2\cos^2\theta)}\Big\{ik_0\cos\theta\frac{\partial}{\partial\eta}\times[\text{r.h.s. of (7) with } \xi=\xi_0]-i\omega\mu_1\frac{\partial}{\partial\xi_0}[\text{r.h.s. of (8) with } \xi=\xi_0]\Big\}, \quad (11)$$

$$\frac{1}{k_0^2\sin^2\theta}\Big\{-i\omega\epsilon_0\frac{\partial}{\partial\xi_0}[\text{r.h.s. of (3) with } \xi=\xi_0]-i\omega\epsilon_0\frac{\partial}{\partial\xi_0}[\text{r.h.s. of (5) with } \xi=\xi_0]-ik_0\cos\theta\frac{\partial}{\partial\eta}[\text{r.h.s. of (6) with } \xi=\xi_0]\Big\}$$
$$=\frac{1}{(k_1^2-k_0^2\cos^2\theta)}\Big\{-i\omega\epsilon_1\frac{\partial}{\partial\xi_0}\times[\text{r.h.s. of (7) with } \xi=\xi_0]-ik_0\cos\theta\frac{\partial}{\partial\eta}[\text{r.h.s. of (8) with } \xi=\xi_0]\Big\}, \quad (12)$$

where r.h.s. means the right-hand side. In contrast with the spherical or circular cylinder case, the angular functions in the elliptical cylinder cases are functions not only of the angular component but also of the characteristics of the medium. Consequently, the summation signs and the angular functions in the above equations may not be omitted. However, it will be shown that this difficulty may be overcome by the orthogonality properties of Mathieu functions. Substituting the expansions

$$ce_m{}^*(\eta)=\sum_{n=0}^{\infty}{}'\alpha_{m,n}ce_n(\eta), \quad (13a)$$

$$se_m{}^*(\eta)=\sum_{n=0}^{\infty}{}'\beta_{m,n}se_n(\eta), \quad (13b)$$

$$\frac{d}{d\eta}ce_m(\eta)=\sum_{n=0}^{\infty}{}'\gamma_{m,n}se_n(\eta), \quad (13c)$$

$$\frac{d}{d\eta}se_m(\eta)=\sum_{n=0}^{\infty}{}'\chi_{m,n}ce_n(\eta), \quad (13d)$$

into Eqs. (9)–(12), and applying the orthogonality relations of Mathieu functions, leads to the following expressions:

$$\frac{1}{p_n}[Ce_n(\xi_0)+A_nMe_n{}^{(1)}(\xi_0)]ce_n(\phi)=\sum_{m=0}^{\infty}{}'\frac{F_m}{p_m{}^*}Ce_m{}^*(\xi_0)ce_m(\phi)\alpha_{m,n}, \quad (14)$$

$$\frac{1}{s_n}[Se_n(\xi_0)+B_nNe_n{}^{(1)}(\xi_0)]se_n(\phi)=\sum_{m=0}^{\infty}{}'\frac{G_m}{s_m{}^*}Se_m{}^*(\xi_0)se_m(\phi)\beta_{m,n}, \quad (15)$$

$$\frac{C_n}{p_n}Me_n{}^{(1)}(\xi_0)ce_n(\phi)=\sum_{m=0}^{\infty}{}'\frac{P_m}{p_m{}^*}Ce_m{}^*(\xi_0)ce_m(\phi)\alpha_{m,n}, \quad (16)$$

$$\frac{D_n}{s_n}Ne_n{}^{(1)}(\xi_0)se_n(\phi)=\sum_{m=0}^{\infty}{}'\frac{Q_m}{s_m{}^*}Se_m{}^*(\xi_0)se_m(\phi)\beta_{m,n}, \quad (17)$$

$$\left(\frac{\epsilon_0}{\mu_0}\right)^{\frac{1}{2}}\cos\theta\left(1-\frac{k_0^2\sin^2\theta}{k_1^2-k_0^2\cos^2\theta}\right)\left\{\sum_{m=0}^{\infty}{}'\frac{1}{s_m}[Se_m(\xi_0)+B_mNe_m^{(1)}(\xi_0)]se_m(\phi)\chi_{m,n}\right\}-\frac{C_n}{p_n}Me_n^{(1)\prime}(\xi_0)ce_n(\phi)$$
$$=\left(\frac{\mu_1}{\mu_0}\right)\left(\frac{k_0^2\sin^2\theta}{k_1^2-k_0^2\cos^2\theta}\right)\sum_{m=0}^{\infty}{}'\frac{P_m}{p_m^*}Ce_m^{*\prime}(\xi_0)ce_m(\phi)\alpha_{m,n}, \quad (18)$$

$$\left(\frac{\epsilon_0}{\mu_0}\right)^{\frac{1}{2}}\cos\theta\left(1-\frac{k_0^2\sin^2\theta}{k_1^2-k_0^2\cos^2\theta}\right)\left\{\sum_{m=0}^{\infty}{}'\frac{1}{p_m}[Ce_m(\xi_0)+A_mMe_m^{(1)}(\xi_0)]ce_m(\phi)\gamma_{m,n}\right\}-\frac{D_n}{s_n}Ne_n^{(1)\prime}(\xi_0)se_n(\phi)$$
$$=\left(\frac{\mu_1}{\mu_0}\right)\frac{k_0^2\sin^2\theta}{k_1^2-k_0^2\cos^2\theta}\sum_{m=0}^{\infty}{}'\frac{Q_m}{s_m^*}Se_m^{*\prime}(\xi_0)se_m(\phi)\beta_{m,n}, \quad (19)$$

$$\frac{1}{s_n}[Se_n'(\xi_0)+B_nNe_n^{(1)\prime}(\xi_0)]se_n(\phi)+\left(\frac{\mu_0}{\epsilon_0}\right)^{\frac{1}{2}}\cos\theta\left(1-\frac{k_0^2\sin^2\theta}{k_1^2-k_0^2\cos^2\theta}\right)\sum_{m=0}^{\infty}{}'\frac{C_m}{p_m}Me_m^{(1)}(\xi_0)ce_m(\phi)\gamma_{m,n}$$
$$=\frac{k_0^2\sin^2\theta}{k_1^2-k_0^2\cos^2\theta}\frac{\epsilon_1}{\epsilon_0}\sum_{m=0}^{\infty}{}'\frac{G_m}{s_m^*}Se_m^{*\prime}(\xi_0)se_m(\phi)\beta_{m,n}, \quad (20)$$

$$\frac{1}{p_n}[Ce_n'(\xi_0)+A_nMe_n^{(1)\prime}(\xi_0)]ce_n(\phi)+\left(\frac{\mu_0}{\epsilon_0}\right)^{\frac{1}{2}}\cos\theta\left(1-\frac{k_0^2\sin^2\theta}{k_1^2-k_0^2\cos^2\theta}\right)\sum_{m=0}^{\infty}{}'\frac{D_m}{s_m}Ne_m^{(1)}(\xi_0)se_m(\phi)\chi_{m,n}$$
$$=\frac{k_0^2\sin^2\theta}{k_1^2-k_0^2\cos^2\theta}\frac{\epsilon_1}{\epsilon_0}\sum_{m=0}^{\infty}{}'\frac{F_m}{p_m^*}Ce_m^{*\prime}(\xi_0)ce_m(\phi)\alpha_{m,n}, \quad (21)$$

$$(n=0,1,2,3,4\cdots).$$

$\alpha_{m,n}$, $\beta_{m,n}$, $\gamma_{m,n}$, and $\chi_{m,n}$ are given in the Appendix. The prime on the summation sign means that when n is odd, the above series are summed over all odd values of m, and when n is even, the series are summed over all even values of m. The primes on the modified Mathieu functions denote differentiation with respect to ξ_0. The unknown coefficients A_n, B_n, C_n, D_n, F_n, G_n, P_n, and Q_n can now be obtained from the above equations. Combining Eqs. (14), (17), (19), and (21) gives

$$\sum_{m=0}^{\infty}{}' F_m s_{mn}^e+Q_m t_{mn}^e=0, \quad (22)$$

$$\sum_{m=0}^{\infty}{}' F_m u_{mn}^e+Q_m v_{mn}^e=w_n^e, \quad (23)$$

$$(n=0,1,2,3\cdots),$$

and combining Eqs. (15), (16), (18), and (20) gives

$$\sum_{m=0}^{\infty}{}' G_m s_{mn}^0+P_m t_{mn}^0=0, \quad (24)$$

$$\sum_{m=0}^{\infty}{}' G_m u_{mn}^0+P_m v_{mn}^0=w_n^0, \quad (25)$$

$$(n=0,1,2,3\cdots),$$

where

$$s_{mn}^e=\left(\frac{\epsilon_0}{\mu_0}\right)^{\frac{1}{2}}\cos\theta(1-x^2)f_m\sum_{r=0}^{\infty}{}'\alpha_{mr}\gamma_{rn},$$

$$s_{mn}^0=\left(\frac{\epsilon_0}{\mu_0}\right)^{\frac{1}{2}}\cos\theta(1-x^2)g_m\sum_{r=0}^{\infty}{}'\beta_{mr}\chi_{rn},$$

$$t_{mn}^e=\left[\frac{\mu_1}{\mu_0}x^2g_m'-\frac{b_n'}{b_n}g_m\right]\beta_{mn},$$

$$t_{mn}^0=\left[\frac{\mu_1}{\mu_0}x^2h_m'-\frac{a_n'}{a_n}h_m\right]\alpha_{mn},$$

$$u_{mn}^e=\left[f_m\frac{a_n'}{a_n}-x^2\frac{\epsilon_1}{\epsilon_0}f_m'\right]\alpha_{mn}, \quad (26)$$

$$u_{mn}^0=\left[g_m\frac{b_n'}{b_n}-x^2\frac{\epsilon_1}{\epsilon_0}g_m'\right]\beta_{mn},$$

$$v_{mn}^e=\left(\frac{\mu_0}{\epsilon_0}\right)^{\frac{1}{2}}\cos\theta(1-x^2)g_m\sum_{r=0}^{\infty}{}'\beta_{mr}\chi_{rn},$$

$$v_{mn}^0=\left(\frac{\mu_0}{\epsilon_0}\right)^{\frac{1}{2}}\cos\theta(1-x^2)f_m\sum_{r=0}^{\infty}{}'\alpha_{mr}\gamma_{rn},$$

$$w_n^o = c_n \frac{a_n'}{a_n} - c_n',$$

$$w_n^e = d_n \frac{b_n'}{b_n} - d_n'.$$

The following abbreviations have been used:

$$\begin{aligned}
c_n &= (1/p_n)Ce_n(\xi_0)ce_n(\phi),\\
c_n' &= (1/p_n)Ce_n'(\xi_0)ce_n(\phi),\\
d_n &= (1/s_n)Se_n(\xi_0)se_n(\phi),\\
d_n' &= (1/s_n)Se_n'(\xi_0)se_n(\phi),\\
a_n &= (1/p_n)Me_n^{(1)}(\xi_0)ce_n(\phi),\\
a_n' &= (1/p_n)Me_n^{(1)\prime}(\xi_0)ce_n(\phi),\\
b_n &= (1/s_n)Ne_n^{(1)}(\xi_0)se_n(\phi),\\
b_n' &= (1/s_n)Ne_n^{(1)\prime}(\xi_0)se_n(\phi),\\
f_n &= (1/p_n^*)Ce_n^*(\xi_0)ce_n(\phi),\\
f_n' &= (1/p_n^*)Ce_n^{*\prime}(\xi_0)ce_n(\phi),\\
g_n &= (1/s_n^*)Se_n^*(\xi_0)se_n(\phi),\\
g_n' &= (1/s_n^*)Se_n^{*\prime}(\xi_0)se_n(\phi),\\
\kappa^2 &= k_0^2\sin^2\theta/(k_1^2 - k_0^2\cos^2\theta).
\end{aligned} \tag{27}$$

The coefficients F_m and Q_m, and G_m and P_m can be obtained readily from Eqs. (22) and (23), and Eqs. (24) and (25). Knowing F_m, Q_m, G_m, and P_m, the coefficients A_n, B_n, C_n, and D_n can be found from Eqs. (14)–(17).

The roots of the determinant of Eqs. (22) and (23) provide the propagation constants of a set of surface waves along an elliptical dielectric cylinder.[6] Owing to the asymmetry of the elliptic cylinder, it is possible to have two orientations for the field configurations. The propagation constants of the other set of surface waves are obtained from the roots of the determinant of Eqs. (24) and (25). This surface-wave has a direct application in the guiding of light waves along a flat fiber (i.e., the fiber optics problem).[6]

The transmitted fields in the dielectric cylinder and the scattered fields in free space due to an incident **E** wave are now completely determined.

H Wave

The corresponding result for the case of an incident **H** wave is of the same form as above if **E** is replaced by **H**, **H** is replaced by $-$**E**, ϵ is replaced by μ, and μ is replaced by ϵ, throughout.

[6] C. Yeh, J. Appl. Phys. 33, 3235 (1962).

III. CONCLUSIONS

The exact solution of the problem of the scattering of an obliquely incident plane wave by an elliptical dielectric cylinder, or by a dielectric ribbon is obtained. Unlike the case for a circular dielectric cylinder, each expansion coefficient of the scattered or transmitted wave for the elliptical dielectric cylinder is coupled to all coefficients of the series expansion for the incident wave. Hence, the results are much more involved. Numerical investigation for the case of normally incident plane waves shows that the infinite determinants representing these expansion coefficients for the scattered or transmitted wave converge quite rapidly for small values of $k_0q/2$; only the first few terms in the determinants are needed as long as $k_0q/2$ is less than 5.[4] This range of $k_0q/2$ covers, however, most of the range not convered by the usual approximate diffraction theory. The results presented here are particularly useful in studying the scattering of light by thin fiber ribbons.

APPENDIX

Formulas for $\alpha_{m,n}$, $\beta_{m,n}$, $\gamma_{m,n}$, and $\chi_{m,n}$

It can readily be shown from the theory of Mathieu functions that

$$\alpha_{m,n} = \int_0^{2\pi} ce_m^*(\eta)ce_n(\eta)d\eta \Big/ \int_0^{2\pi} ce_n^2(\eta)d\eta,$$

$$\beta_{m,n} = \int_0^{2\pi} ce_m^*(\eta)ce_n(\eta)d\eta \Big/ \int_0^{2\pi} ce_n^2(\eta)d\eta,$$

$$\gamma_{m,n} = \int_0^{2\pi} ce_m'(\eta)se_n(\eta)d\eta \Big/ \int_0^{2\pi} se_n^2(\eta)d\eta,$$

$$\chi_{m,n} = \int_0^{2\pi} se_m'(\eta)ce_n(\eta)d\eta \Big/ \int_0^{2\pi} ce_n^2(\eta)d\eta,$$

where the prime signifies the derivative of the function with respect to its argument. The above integrals can easily be integrated using the series expansions of Mathieu functions in terms of trigonometric functions.[5]

Reprinted from JOURNAL OF THE OPTICAL SOCIETY OF AMERICA, Vol. 55, No. 7, 860–867, July 1965
Printed in U. S. A.

Exact Solution to a Lens Problem*

C. YEH
Electrical Engineering Department, University of Southern California, Los Angeles, California 90007
(Received 20 March 1965)

Exact solution is obtained for the diffraction of plane electromagnetic waves by an elliptic cylindrical lens surrounding an infinitely long parallel slit in a perfectly conducting plane. Results are expressed in terms of Mathieu functions. Approximate expressions for the expansion coefficients for the diffracted waves, the scattered intensity at infinity, and the transmission coefficients are also obtained in the long-wavelength (Rayleigh scattering) limit. Possible numerical investigations are discussed.

I. INTRODUCTION

ONE of the classical diffraction problems is the diffraction of electromagnetic waves by an infinitely long parallel slit in a perfectly conducting plane embedded in a homogeneous medium.[1] This problem was first treated by Sieger[2]; extensive numerical results were given by Morse and Rubenstein,[3] and by Skavlem.[4] A modification of this problem, in which the medium on one side of the conducting plane differs in dielectric constant and magnetic permeability from the medium on the other side, has been considered by Meixner[5] and recently by Barakat.[6] Another variation is to include the presence of an elliptical lens surrounding the slit (see Fig. 1). This problem is important in that it offers an exact solution to the problem of transmission and focusing of electromagnetic waves through lenses with noncircular cylindrical boundary shape. Elliptical lenses are also good approximations to actual two-dimensional lenses. It is hoped that the solution of this problem will provide further understanding of the behavior of electromagnetic fields in the focal region of the lens.

II. FORMULATION OF THE PROBLEM

To analyze this problem, the elliptical-cylinder coordinates (ξ,η,z) are introduced. In terms of the rectangular coordinates (x,y,z), the elliptical-cylinder coordinates are defined by the following relations:

$$\begin{aligned} x&=q\cosh\xi\cos\eta,\\ y&=q\sinh\xi\sin\eta,\\ z&=z,\\ &(0\leq\xi<\infty,\ 0\leq\eta\leq2\pi), \end{aligned} \tag{1}$$

where q is the semifocal length of the ellipse. The contour surfaces of constant ξ are confocal elliptic cylinders, and those of constant η are confocal hyperbolic cylinders. The geometry of this lens structure is shown in Fig. 1. In the half-plane $y>0$, the medium in the region bounded by $\xi_1\leq\xi<\infty$ is characterized by a dielectric constant ϵ_1 and permeability μ_1 and the medium bounded by $0\leq\xi\leq\xi_1$ is characterized by ϵ_2 and μ_2. In the half-plane $y<0$, the medium in the region bounded by $0\leq\xi\leq\xi_2$ is characterized by a dielectric constant ϵ_2 and permeability μ_2 and the medium bounded by $\xi_2\leq\xi<\infty$ is characterized by ϵ_3 and μ_3. There is a perfectly conducting plane at $y=0$ with a slit of width $2q$. It should be noted that $\xi=\xi_1$ and $\xi=\xi_2$ are boundaries of confocal elliptical cylinders with a focal length of $2q$. We consider waves whose propagation vector is in the x,y plane, so that the z coordinate may be omitted from the discussion. Furthermore, the usual harmonic time-dependence factor $e^{i\omega t}$ of all field components is assumed and suppressed throughout.

Two types of incident plane waves are possible. The one called an E wave is defined by $H_z=0$, and the other called an H wave is defined by $E_z=0$.

III. FORMAL SOLUTION FOR ELECTRIC VECTOR PARALLEL TO SLIT

The expansion of an incident E wave of unit amplitude making an angle β with the x axis, when expressed in terms of Mathieu functions, is[7,8]

$$\begin{aligned} E_z^i=2\sum_{m=0}^{\infty}(i)^m Mc_m^{(1)}(\xi,h_1)ce_m(\eta,h_1^2)ce_m(\beta,h_1^2)\\ +2\sum_{m=1}^{\infty}(i)^m Ms_m^{(1)}(\xi,h_1)se_m(\eta,h_1^2)se_m(\beta,h_1^2)\\ (0\leq\eta\leq\pi), \end{aligned} \tag{2}$$

where $h_1=\frac{1}{2}k_1q$ and $k_1=\omega(\mu_1\epsilon_1)^{\frac{1}{2}}$. In the absence of the slit and the lens, the reflected wave is

$$\begin{aligned} E_z^r=-2\sum_{m=0}^{\infty}(i)^m Mc_m^{(1)}(\xi,h_1)ce_m(\eta,h_1^2)ce_m(\beta,h_1^2)\\ +2\sum_{m=1}^{\infty}(i)^m Ms_m^{(1)}(\xi,h_1)se_m(\eta,h_1^2)se_m(\beta,h_1^2)\\ (0\leq\eta\leq\pi). \end{aligned} \tag{3}$$

* This work was supported by the Joint Services Electronics Program (U. S. Army, U. S. Navy, and U. S. Air Force) under grant No. AF-AFOSR-496-64.

[1] H. Hönl, A. W. Maue, and K. Westpfahl, "Theorie der Beugung," in *Handbuch der Physik*, S. Flügge, ed. (Springer-Verlag, Berlin 1961), Vol. 25, Chap. 1.
[2] B. Sieger, Ann. Physik 27, 626 (1908).
[3] P. M. Morse and P. J. Rubenstein, Phys. Rev. 54, 895 (1938).
[4] S. Skavlem, Arch. Math. Naturvidenskab **B51**, 61 (1950).
[5] J. Meixner, "Diffraction of Electromagnetic Waves by a Slit in a Conducting Plane Between Different Media," New York University, Institute of Mathematical Sciences, Division of Electromagnetic Research, Research Rept. No. EM-68 (1954).
[6] R. Barakat, J. Opt. Soc. Am. 53, 1231 (1963).
[7] J. Meixner and F. W. Schäfke, *Mathieusche Funktionen und Sphäroidfunktionen* (Springer-Verlag, Berlin, 1954).
[8] *Tables Relating to Mathieu Functions* (Columbia University Press, New York, 1951); G. Blanch and D. S. Clemm, "Tables Relating to the Radial Mathieu Functions, Vol. 1, Functions of the First Kind" (Aeronautical Research Laboratories, Office of Aerospace Research, U. S. Air Force, 1961).

In order that the diffracted field may satisfy Sommerfeld's radiation condition at infinity, Bouwkamp–Meixner's edge condition[9] at the slit edges and the condition that the tangential electric field must be identically zero on the conducting screen, the diffracted field must be of the following form:

for the region $y\geq 0$, $\xi_1\leq\xi<\infty$, $0\leq\eta\leq\pi$,

$$ {}_1E_z{}^d=\sum_{m=1}^{\infty}{}_1V_m Ms_m{}^{(4)}(\xi,h_1)se_m(\eta,h_1{}^2); \quad (4)$$

for the region $y\geq 0$, $0\leq\xi\leq\xi_2$, $0\leq\eta\leq\pi$,

$$ {}_2E_z{}^d=\sum_{m=1}^{\infty}{}_2V_m{}^{(1)}Ms_m{}^{(1)}(\xi,h_2)se_m(\eta,h_2{}^2) + {}_2V_m{}^{(2)}Ms_m{}^{(2)}(\xi,h_2)se_m(\eta,h_2{}^2); \quad (5)$$

for the region $y\leq 0$, $0\leq\xi\leq\xi_2$, $-\pi\leq\eta\leq 0$,

$$ {}_3E_z{}^d=\sum_{m=1}^{\infty}{}_3V_m{}^{(1)}Ms_m{}^{(1)}(\xi,h_2)se_m(\eta,h_2{}^2) + {}_3V_m{}^{(2)}Ms_m{}^{(2)}(\xi,h_2)se_m(\eta,h_2{}^2); \quad (6)$$

and for the region $y\leq 0$, $\xi_2\leq\xi<\infty$, $-\pi\leq\eta\leq 0$,

$$ {}_4E_z{}^d=\sum_{m=1}^{\infty}{}_4V_m Ms_m{}^{(4)}(\xi,h_3)se_m(\eta,h_3{}^2), \quad (7)$$

where $h_2=\frac{1}{2}k_2q$, $h_3=\frac{1}{2}k_3q$, $k_2=\omega(\mu_2\epsilon_2)^{\frac{1}{2}}$, $k_3=\omega(\mu_3\epsilon_3)^{\frac{1}{2}}$; ${}_1V_m$, ${}_2V_m{}^{(1)}$, ${}_2V_m{}^{(2)}$, ${}_3V_m{}^{(1)}$, ${}_3V_m{}^{(2)}$, and ${}_4V_m$ are arbitrary constants to be determined by the boundary conditions.

The boundary conditions require the continuity of the tangential electric and magnetic fields at the boundary surfaces. We have

(a) on the elliptical cylinder $\xi=\xi_1$,

$$E_z{}^i+E_z{}^r+{}_1E_z{}^d={}_2E_z{}^d|_{\xi=\xi_1,0\leq\eta\leq\pi}, \quad (8)$$

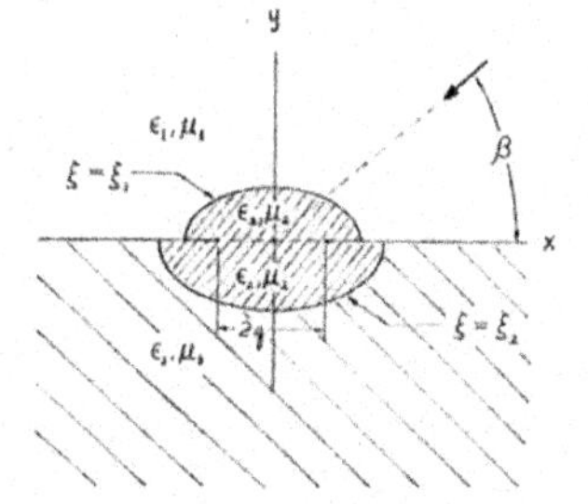

FIG. 1. Geometry of the lens. The arrow indicates the direction of incident wave.

$$\frac{1}{\mu_1}\frac{\partial}{\partial\xi}[E_z{}^i+E_z{}^r+{}_1E_z{}^d]=\frac{1}{\mu_2}\frac{\partial}{\partial\xi}[{}_2E_z{}^d]|_{\xi=\xi_1,0\leq\eta\leq\pi}; \quad (9)$$

(b) in the slit $\xi=0$,

$$ {}_2E_z{}^d|_{\xi=0,0\leq\eta\leq\pi}={}_3E_z{}^d|_{\xi=0,-\pi\leq\eta\leq 0}, \quad (10)$$

$$\frac{\partial}{\partial y}[{}_2E_z{}^d]|_{\xi=0,0\leq\eta\leq\pi}=\frac{\partial}{\partial y}[{}_3E_z{}^d]|_{\xi=0,-\pi\leq\eta\leq 0}; \quad (11)$$

(c) on the elliptical cylinder $\xi=\xi_2$,

$$ {}_3E_z{}^d={}_4E_z{}^d|_{\xi=\xi_2,-\pi\leq\eta\leq 0}, \quad (12)$$

$$\frac{1}{\mu_2}\frac{\partial}{\partial\xi}[{}_3E_z{}^d]=\frac{1}{\mu_3}\frac{\partial}{\partial\xi}[{}_4E_z{}^d]|_{\xi=\xi_2,-\pi\leq\eta\leq 0}. \quad (13)$$

It is noted that

$$\frac{\partial}{\partial y}=\frac{1}{q\sin\eta}\frac{\partial}{\partial\xi} \quad \text{for} \quad \xi=0;$$

$$se_m(-\eta,h^2)=-se_m(\eta,h^2).$$

Substituting Eqs. (2)–(7) into Eqs. (8)–(13) gives, for $0\leq\eta\leq\pi$,

$$\sum_{m=1}^{\infty}[4(i)^m Ms_m{}^{(1)}(\xi_1,h_1)se_m(\beta,h_1{}^2)+{}_1V_m Ms_m{}^{(4)}(\xi_1,h_1)]se_m(\eta,h_1{}^2) = \sum_{m=1}^{\infty}[{}_2V_m{}^{(1)}Ms_m{}^{(1)}(\xi_1,h_2)+{}_2V_m{}^{(2)}Ms_m{}^{(2)}(\xi_1,h_2)]se_m(\eta,h_2{}^2), \quad (14)$$

$$\sum_{m=1}^{\infty}[4(i)^m Ms_m{}^{(1)\prime}(\xi_1,h_1)se_m(\beta,h_1{}^2)+{}_1V_m Ms_m{}^{(4)\prime}(\xi_1,h_1)]se_m(\eta,h_1{}^2) = \frac{\mu_1}{\mu_2}\sum_{m=1}^{\infty}[{}_2V_m{}^{(1)}Ms_m{}^{(1)\prime}(\xi_1,h_2)+{}_2V_m{}^{(2)}Ms_m{}^{(2)\prime}(\xi_1,h_2)]se_m(\eta,h_2{}^2), \quad (15)$$

$$\sum_{m=1}^{\infty}{}_2V_m{}^{(2)}Ms_m{}^{(2)}(0,h_2)se_m(\eta,h_2{}^2)=-\sum_{m=1}^{\infty}{}_3V_m{}^{(2)}Ms_m{}^{(2)}(0,h_2)se_m(\eta,h_2{}^2), \quad (16)$$

$$\sum_{m=1}^{\infty}[{}_2V_m{}^{(1)}Ms_m{}^{(1)\prime}(0,h_2)+{}_2V_m{}^{(2)}Ms_m{}^{(2)\prime}(0,h_2)]se_m(\eta,h_2{}^2) = \sum_{m=1}^{\infty}[{}_3V_m{}^{(1)}Ms_m{}^{(1)\prime}(0,h_2)+{}_3V_m{}^{(2)}Ms_m{}^{(2)\prime}(0,h_2)]se_m(\eta,h_2{}^2), \quad (17)$$

[9] C. J. Bouwkamp, Physica 12, 467 (1946); J. Meixner, Ann. Physik 6, 2 (1949).

$$\sum_{m=1}^{\infty}[{}_3V_m^{(1)}Ms_m^{(1)}(\xi_2,h_2)+{}_3V_m^{(2)}Ms_m^{(2)}(\xi_2,h_2)]se_m(\eta,h_2^2)=\sum_{m=1}^{\infty}{}_4V_mMs_m^{(4)}(\xi_2,h_3)se_m(\eta,h_3^2), \tag{18}$$

$$\sum_{m=1}^{\infty}[{}_3V_m^{(1)}Ms_m^{(1)\prime}(\xi_2,h_2)+{}_3V_m^{(2)}Ms_m^{(2)\prime}(\xi_2,h_2)]se_m(\eta,h_2^2)=\frac{\mu_2}{\mu_3}\sum_{m=1}^{\infty}{}_4V_mMs_m^{(4)\prime}(\xi_2,h_3)se_m(\eta,h_3^2), \tag{19}$$

where the prime signifies the derivative of the function with respect to ξ. These equations must be valid for all values of η between $\eta=0$ and $\eta=\pi$. Let us multiply both sides of the above equations by $se_n(\eta,h_2^2)$ and integrate with respect to η from $\eta=0$ to $\eta=\pi$. Using the orthogonality relations of the angular Mathieu function and introducing the following relations:

$$\alpha_{nm}=\frac{2}{\pi}\int_0^{\pi}se_n(\eta,h_2^2)se_m(\eta,h_1^2)d\eta \tag{20}$$

and

$$\gamma_{nm}=\frac{2}{\pi}\int_0^{\pi}se_n(\eta,h_2^2)se_m(\eta,h_3^2)d\eta, \tag{21}$$

we obtain

$$\sum_{m=1}^{\infty}[4(i)^mMs_m^{(1)}(\xi_1,h_1)se_m(\beta,h_1^2)+{}_1V_mMs_m^{(4)}(\xi_1,h_1)]\alpha_{nm}={}_2V_n^{(1)}Ms_n^{(1)}(\xi_1,h_2)+{}_2V_n^{(2)}Ms_n^{(2)}(\xi_1,h_2), \tag{22}$$

$$\sum_{m=1}^{\infty}[4(i)^mMs_m^{(1)\prime}(\xi_1,h_1)se_m(\beta,h_1^2)+{}_1V_mMs_m^{(4)\prime}(\xi_1,h_1)]\alpha_{nm}=\frac{\mu_1}{\mu_2}[{}_2V_n^{(1)}Ms_n^{(1)\prime}(\xi_1,h_2)+{}_2V_n^{(2)}Ms_n^{(2)\prime}(\xi_1,h_2)], \tag{23}$$

$${}_2V_n^{(2)}Ms_n^{(2)}(0,h_2)=-{}_3V_n^{(2)}Ms_n^{(2)}(0,h_2), \tag{24}$$

$${}_2V_n^{(1)}Ms_n^{(1)\prime}(0,h_2)+{}_2V_n^{(2)}Ms_n^{(2)\prime}(0,h_2)={}_3V_n^{(1)}Ms_n^{(1)\prime}(0,h_2)+{}_3V_n^{(2)}Ms_n^{(2)\prime}(0,h_2), \tag{25}$$

$$\sum_{m=1}^{\infty}{}_4V_mMs_m^{(4)}(\xi_2,h_3)\gamma_{nm}={}_3V_n^{(1)}Ms_n^{(1)}(\xi_2,h_2)+{}_3V_n^{(2)}Ms_n^{(2)}(\xi_2,h_2), \tag{26}$$

$$\sum_{m=1}^{\infty}\frac{\mu_2}{\mu_3}{}_4V_mMs_m^{(4)\prime}(\xi_2,h_3)\gamma_{nm}={}_3V_n^{(1)}Ms_n^{(1)\prime}(\xi_2,h_2)+{}_3V_n^{(2)}Ms_n^{(2)\prime}(\xi_2,h_2),\quad(n=1,2,3\cdots). \tag{27}$$

Eliminating ${}_2V_n^{(1)}$, ${}_2V_n^{(2)}$, ${}_3V_n^{(1)}$, ${}_3V_n^{(2)}$ from the above equations and simplifying gives

$$\sum_{m=1}^{\infty}{}_1V_ma_{n,m}^{(1)}+{}_4V_ma_{n,m}^{(4)}=b_n, \tag{28}$$

$$\sum_{m=1}^{\infty}{}_1V_mc_{n,m}^{(1)}+{}_4V_mc_{n,m}^{(4)}=d_n,\quad(n=1,2,3\cdots), \tag{29}$$

where

$$a_{n,m}^{(1)}=-Ms_m^{(4)}(\xi_1,h_1)\alpha_{n,m} \tag{30a}$$

$$\begin{aligned}a_{n,m}^{(4)}=(1/\Delta_n)\gamma_{n,m}\Bigg\{&Ms_n^{(1)}(\xi_1,h_2)[Ms_m^{(4)}(\xi_2,h_3)Ms_n^{(2)\prime}(\xi_2,h_2)-(\mu_2/\mu_3)Ms_m^{(4)\prime}(\xi_2,h_3)Ms_n^{(2)}(\xi_2,h_2)]\\&+\left[2Ms_n^{(1)}(\xi_1,h_2)\frac{Ms_n^{(2)\prime}(0,h_2)}{Ms_n^{(1)\prime}(0,h_2)}-Ms_n^{(2)}(\xi_1,h_2)\right][(\mu_2/\mu_3)Ms_m^{(4)\prime}(\xi_2,h_3)Ms_m^{(1)}(\xi_2,h_2)\\&-Ms_m^{(4)}(\xi_2,h_3)Ms_n^{(1)\prime}(\xi_2,h_2)]\Bigg\},\end{aligned} \tag{30b}$$

$$c_{n,m}^{(1)}=-Ms_m^{(4)\prime}(\xi_1,h_1)\alpha_{n,m}, \tag{30c}$$

$$c_{n,m}^{(4)}=(\mu_1/\mu_2)(1/\Delta_n)\gamma_{n,m}\left\{Ms_n^{(1)\prime}(\xi_1,h_2)[Ms_m^{(4)}(\xi_2,h_3)Ms_n^{(2)\prime}(\xi_2,h_2)-(\mu_2/\mu_3)Ms_m^{(4)\prime}(\xi_2,h_3)Ms_n^{(2)}(\xi_2,h_2)]\right.$$
$$+\left[2Ms_n^{(1)\prime}(\xi_1,h_2)\frac{Ms_n^{(2)\prime}(0,h_2)}{Ms_n^{(1)\prime}(0,h_2)}-Ms_n^{(2)\prime}(\xi_1,h_2)\right][(\mu_2/\mu_3)Ms_m^{(4)\prime}(\xi_2,h_3)Ms_n^{(1)}(\xi_2,h_2)$$
$$\left.-Ms_m^{(4)}(\xi_2,h_3)Ms_n^{(1)\prime}(\xi_2,h_2)]\right\}, \quad (30d)$$

$$b_n=\sum_{m=1}^{\infty}4(i)^mMs_m^{(1)}(\xi_1,h_1)se_m(\beta,h_1^2)\alpha_{n,m}, \quad (30e)$$

$$d_n=\sum_{m=1}^{\infty}4(i)^mMs_m^{(1)\prime}(\xi_1,h_1)se_m(\beta,h_1^2)\alpha_{n,m}, \quad (30f)$$

with

$$\Delta_n=Ms_n^{(1)}(\xi_2,h_2)Ms_n^{(2)\prime}(\xi_2,h_2)-Ms_n^{(2)}(\xi_2,h_2)Ms_n^{(1)\prime}(\xi_2,h_2).$$

The expansion coefficients for the diffracted wave, ${}_1V_m$ and ${}_4V_m$, can therefore be found from Eqs. (28) and (29). Thus the diffracted wave is completely determined. The above solution satisfies the edge condition.[9]

When $\xi_1=\xi_2=0$, $\mu_2\to\mu_1$ and $\epsilon_2\to\epsilon_1$ or $\mu_2\to\mu_3$ and $\epsilon_2\to\epsilon_3$ as appropriate, the above results reduce to those given by Meixner[5] and by Barakat[6] for the case of diffraction of plane wave by a slit between two different media.

IV. FORMAL SOLUTION FOR MAGNETIC VECTOR PARALLEL TO SLIT

The expression in terms of Mathieu functions for an incident H wave of unit amplitude making an angle β with the x axis is

$$H_z^i=2\sum_{m=0}^{\infty}(i)^mMc_m^{(1)}(\xi,h_1)ce_m(\eta,h_1^2)ce_m(\beta,h_1^2)$$
$$+2\sum_{m=1}^{\infty}(i)^mMs_m^{(1)}(\xi,h_1)se_m(\eta,h_1^2)se_m(\beta,h_1^2),$$
$$(0\leq\eta\leq\pi). \quad (31)$$

The expression for the reflected wave in the absence of the slit and the lens is

$$H_z^r=2\sum_{m=0}^{\infty}(i)^mMc_m^{(1)}(\xi,h_1)ce_m(\eta,h_1^2)ce_m(\beta,h_1^2)$$
$$-2\sum_{m=1}^{\infty}(i)^mMs_m^{(1)}(\xi,h_1)se_m(\eta,h_1^2)se_m(\beta,h_1^2)$$
$$(0\leq\eta\leq\pi). \quad (32)$$

The proper expressions for the diffracted fields are:

for the region $y\geq0$, $\xi_1\leq\xi<\infty$, $0\leq\eta\leq\pi$,

$${}_1H_z^d=\sum_{m=0}^{\infty}{}_1U_mMc_m^{(4)}(\xi,h_1)ce_m(\eta,h_1^2); \quad (33)$$

for the region $y\geq0$, $0\leq\xi\leq\xi_1$, $0\leq\eta\leq\pi$,

$${}_2H_z^d=\sum_{m=0}^{\infty}{}_2U_m^{(1)}Mc_m^{(1)}(\xi,h_2)ce_m(\eta,h_2^2)$$
$$+{}_2U_m^{(2)}Mc_m^{(2)}(\xi,h_2)ce_m(\eta,h_2^2); \quad (34)$$

for the region $y\leq0$, $0\leq\xi\leq\xi_2$, $-\pi\leq\eta\leq0$,

$${}_3H_z^d=\sum_{m=0}^{\infty}{}_3U_m^{(1)}Mc_m^{(1)}(\xi,h_2)ce_m(\eta,h_2^2)$$
$$+{}_3U_m^{(2)}Mc_m^{(2)}(\xi,h_2)ce_m(\eta,h_2^2); \quad (35)$$

and for the region $y\leq0$, $\xi_2\leq\xi<\infty$, $-\pi\leq\eta\leq0$,

$${}_4H_z^d=\sum_{m=1}^{\infty}{}_4U_mMc_m^{(4)}(\xi,h_3)ce_m(\eta,h_3^2), \quad (36)$$

where ${}_1U_m$, ${}_2U_m^{(1)}$, ${}_2U_m^{(2)}$, ${}_3U_m^{(1)}$, ${}_3U_m^{(2)}$, and ${}_4U_m$ are arbitrary constants to be determined by the boundary conditions.

The appropriate boundary conditions are:

(a) on the elliptical cylinder $\xi=\xi_1$,

$$H_z^i+H_z^r+{}_1H_z^d={}_2H_z^d|_{\xi=\xi_1,\,0\leq\eta\leq\pi}, \quad (37)$$

$$\frac{1}{\epsilon_1}\frac{\partial}{\partial\xi}[H_z^i+H_z^r+{}_1H_z^d]=\frac{1}{\epsilon_2}\frac{\partial}{\partial\xi}[{}_2H_z^d]|_{\xi=\xi_1,\,0\leq\eta\leq\pi}; \quad (38)$$

(b) in the slit $\xi=0$,

$${}_2H_z^d|_{\xi=0,\,0\leq\eta\leq\pi}={}_3H_z^d|_{\xi=0,\,-\pi\leq\eta\leq0}, \quad (39)$$

$$\frac{\partial}{\partial y}[{}_2H_z^d]|_{\xi=0,\,0\leq\eta\leq\pi}=\frac{\partial}{\partial y}[{}_3H_z^d]|_{\xi=0,\,-\pi\leq\eta\leq0}; \quad (40)$$

(c) on the elliptical cylinder $\xi=\xi_2$,

$${}_3H_z^d={}_4H_z^d|_{\xi=\xi_2,\,-\pi\leq\eta\leq0}, \quad (41)$$

$$\frac{1}{\epsilon_2}\frac{\partial}{\partial\xi}[{}_3H_z^d]=\frac{1}{\epsilon_3}\frac{\partial}{\partial\xi}[{}_4H_z^d]|_{\xi=\xi_2,\,-\pi\leq\eta\leq0}. \quad (42)$$

Putting Eqs. (31)–(36) into Eqs. (37)–(42), multiplying both sides of the resultant equations by $ce_n(\eta,h_2^2)$ and integrating with respect to η from π to 0 yields

$$\sum_{m=0}^{\infty}[4(i)^m Mc_m^{(1)}(\xi_1,h_1)ce_m(\beta,h_1^2)+{}_1U_m Mc_m^{(4)}(\xi_1,h_1)]\beta_{nm}={}_2U_n^{(1)}Mc_n^{(1)}(\xi_1,h_2)+{}_2U_n^{(2)}Mc_n^{(2)}(\xi_1,h_2), \quad (43)$$

$$\sum_{m=0}^{\infty}[4(i)^m Mc_m^{(1)\prime}(\xi_1,h_1)ce_m(\beta,h_1^2)+{}_1U_m Mc_m^{(4)\prime}(\xi_1,h_1)]\beta_{nm}=\frac{\epsilon_1}{\epsilon_2}[{}_2U_n^{(1)}Mc_n^{(1)\prime}(\xi_1,h_2)+{}_2U_n^{(2)}Mc_n^{(2)\prime}(\xi_1,h_2)], \quad (44)$$

$${}_2U_n^{(1)}Mc_n^{(1)}(0,h_2)+{}_2U_n^{(2)}Mc_n^{(2)}(0,h_2)={}_3U_n^{(1)}Mc_n^{(1)}(0,h_2)+{}_3U_n^{(2)}Mc_n^{(2)}(0,h_2), \quad (45)$$

$${}_2U_n^{(2)}Mc_n^{(2)\prime}(0,h_2)=-{}_3U_n^{(2)}Mc_n^{(2)\prime}(0,h_2), \quad (46)$$

$$\sum_{m=0}^{\infty}{}_4U_m Mc_m^{(4)}(\xi_2,h_3)\chi_{nm}={}_3U_n^{(1)}Mc_n^{(1)}(\xi_2,h_2)+{}_3U_n^{(2)}Mc_n^{(2)}(\xi_2,h_2), \quad (47)$$

$$\sum_{m=0}^{\infty}\frac{\epsilon_2}{\epsilon_3}{}_4U_m Mc_m^{(4)\prime}(\xi_2,h_3)\chi_{nm}={}_3U_n^{(1)}Mc_n^{(1)\prime}(\xi_2,h_2)+{}_3U_n^{(2)}Mc_n^{(2)\prime}(\xi_2,h_2), \quad (n=0,1,2,3\cdots), \quad (48)$$

with

$$\beta_{nm}=\frac{2}{\pi}\int_0^{\pi}ce_n(\eta,h_2^2)ce_m(\eta,h_1^2)d\eta, \quad (49)$$

$$\chi_{nm}=\frac{2}{\pi}\int_0^{\pi}ce_n(\eta,h_2^2)ce_m(\eta,h_3^2)d\eta. \quad (50)$$

The prime signifies the derivative of the function with respect to ξ. Eliminating ${}_2U_n^{(1)}$, ${}_2U_n^{(2)}$, ${}_3U_n^{(1)}$, ${}_3U_n^{(2)}$ from the above equations and simplifying gives

$$\sum_{m=0}^{\infty}{}_1U_m f_{n,m}^{(1)}+{}_4U_m f_{n,m}^{(4)}=p_n, \quad (51)$$

$$\sum_{m=0}^{\infty}{}_1U_m s_{n,m}^{(1)}+{}_4U_m s_{n,m}^{(4)}=w_n, \quad (n=0,1,2,3\cdots), \quad (52)$$

where

$$f_{n,m}^{(1)}=-Mc_m^{(4)}(\xi_1,h_1)\beta_{nm}, \quad (53a)$$

$$f_{n,m}^{(4)}=\frac{1}{\lambda_n}\chi_{nm}\Bigg\{Mc_n^{(1)}(\xi_1,h_2)\left[Mc_m^{(4)}(\xi_2,h_3)Mc_n^{(2)\prime}(\xi_2,h_2)-\frac{\epsilon_2}{\epsilon_3}Mc_m^{(4)\prime}(\xi_2,h_3)Mc_n^{(2)}(\xi_2,h_2)\right]$$
$$+\left[2\frac{Mc_n^{(2)}(0,h_2)}{Mc_n^{(1)}(0,h_2)}Mc_n^{(1)}(\xi_1,h_2)-Mc_n^{(2)}(\xi_1,h_2)\right]\Bigg[\frac{\epsilon_2}{\epsilon_3}Mc_m^{(4)\prime}(\xi_2,h_3)Mc_n^{(1)}(\xi_2,h_2)$$
$$-Mc_m^{(4)}(\xi_2,h_3)Mc_n^{(1)\prime}(\xi_2,h_2)\Bigg]\Bigg\}, \quad (53b)$$

$$s_{n,m}^{(1)}=-Mc_m^{(4)\prime}(\xi_1,h_1)\beta_{nm}, \quad (53c)$$

$$s_{n,m}^{(4)}=\frac{1}{\lambda_n}\chi_{nm}\frac{\epsilon_1}{\epsilon_2}\Bigg\{Mc_n^{(1)\prime}(\xi_1,h_2)\left[Mc_m^{(4)}(\xi_2,h_3)Mc_n^{(2)\prime}(\xi_2,h_2)-\frac{\epsilon_2}{\epsilon_3}Mc_m^{(4)\prime}(\xi_2,h_3)Mc_n^{(2)}(\xi_2,h_2)\right]$$
$$+\left[2\frac{Mc_n^{(2)}(0,h_2)}{Mc_n^{(1)}(0,h_2)}Mc_n^{(1)\prime}(\xi_1,h_2)-Mc_n^{(2)\prime}(\xi_1,h_2)\right]\Bigg[\frac{\epsilon_2}{\epsilon_3}Mc_m^{(4)\prime}(\xi_2,h_3)Mc_n^{(1)}(\xi_2,h_2)$$
$$-Mc_m^{(4)}(\xi_2,h_3)Mc_n^{(1)\prime}(\xi_2,h_2)\Bigg]\Bigg\}, \quad (53d)$$

$$p_n = \sum_{m=0}^{\infty} 4(i)^m Mc_m^{(1)}(\xi_1,h_1)ce_m(\beta,h_1^2)\beta_{nm}, \tag{53e}$$

$$w_n = \sum_{m=0}^{\infty} 4(i)^m Mc_m^{(1)\prime}(\xi_1,h_1)ce_m(\beta,h_1^2)\beta_{nm}, \tag{53f}$$

with

$$\lambda_n = Mc_n^{(1)}(\xi_2,h_2)Mc_n^{(2)\prime}(\xi_2,h_2) - Mc_n^{(2)}(\xi_2,h_2)Mc_n^{(1)\prime}(\xi_2,h_2).$$

Hence, the expansion coefficients for the diffracted wave, ${}_1U_m$ and ${}_4U_m$, can be found from Eqs. (51) and (52). The edge condition is also satisfied by the above solution.[9]

When $\xi_1 = \xi_2 = 0$, $\mu_2 \to \mu_1$ and $\epsilon_2 \to \epsilon_1$ or $\mu_2 \to \mu_3$ and $\epsilon_2 \to \epsilon_3$ as appropriate, the above results reduce to those given by Meixner[5] and by Barakat.[6]

V. SCATTERED INTENSITY IN THE FAR ZONE AND TRANSMISSION COEFFICIENTS

The scattered intensity in the far zone of the slit is defined as follows:

$$I^E(y<0) = \lim_{\xi\to\infty} |{}_4E_z^d|^2$$

$$= \frac{q}{h_3 r} \sum_{n=1}^{\infty}\sum_{m=1}^{\infty} {}_4V_n\, {}_4V_m^* se_n(\eta,h_3^2)se_m(\eta,h_3^2), \tag{54}$$

$$I^E(y>0) = \lim_{\xi\to\infty} |{}_1E_z^d|^2$$

$$= \frac{q}{h_1 r} \sum_{n=1}^{\infty}\sum_{m=1}^{\infty} {}_1V_n\, {}_1V_m^* se_n(\eta,h_1^2)se_m(\eta,h_1^2) \tag{55}$$

for an incident E wave;

$$I^H(y<0) = \lim_{\xi\to\infty} |{}_4H_z^d|^2$$

$$= \frac{q}{h_3 r} \sum_{n=0}^{\infty}\sum_{m=0}^{\infty} {}_4U_n\, {}_4U_m^* ce_n(\eta,h_3^2)ce_m(\eta,h_3^2), \tag{56}$$

$$I^H(y>0) = \lim_{\xi\to\infty} |{}_1H_z^d|^2$$

$$= \frac{q}{h_1 r} \sum_{n=0}^{\infty}\sum_{m=0}^{\infty} {}_1U_n\, {}_1U_m^* ce_n(\eta,h_3^2)ce_m(\eta,h_3^2) \tag{57}$$

for an incident H wave. r is the distance from the center of the slit to the field point of interest. The asymptotic expressions[7] for $Ms_m(\xi,h)$ and $Mc_m(\xi,h)$ have been used to derive the above equations.

The transmission coefficient is defined as the total flux penetrating the slit per unit flux incident on the slit; i.e.,

$$T^{E,H} = \frac{1}{(1/r)} \int_{\pi}^{2\pi} [I^{E,H}(y<0)]dy. \tag{58}$$

Hence the quantity $T^{E,H}$ is the ratio of the flux actually transmitted through the lens to the flux which geometric optics predicts would be transmitted at normal incidence. Substituting (7) and (36) into Eq. (58) and carrying out the integration gives:

$$T^E = \frac{q\pi}{2h_3} \sum_{n=1}^{\infty} |{}_4V_n|^2, \tag{59}$$

$$T^H = \frac{q\pi}{2h_3} \sum_{n=0}^{\infty} |{}_4U_n|^2. \tag{60}$$

VI. RAYLEIGH APPROXIMATIONS

If the wavelength of the incident wave is large compared to the characteristic dimension of the diffraction object, i.e., h_1, h_2, $h_3 \ll 1$, the expressions for the expansion coefficients for the diffracted fields given in Secs. III and IV can be greatly simplified. We consider the case (the Rayleigh case[10]) in which the frequency is assumed to be so low that only the first expansion coefficient is important.

Expansions of Mathieu and modified Mathieu functions in powers of h for small values of h are known.[7] They are

$$ce_0(\eta,h^2) = 2^{-\frac{1}{2}}[1 - \tfrac{1}{4}h^2 - \tfrac{1}{2}h^2\cos 2\eta] + O(h^4),$$

$$ce_1(\eta,h^2) = \cos\eta + O(h^2),$$

$$se_1(\eta,h^2) = \sin\eta + O(h^2),$$

$$Mc_0^{(1)}(\xi,h) = 1 - \tfrac{1}{2}h^2\cosh 2\xi + \tfrac{1}{32}h^4(6+\cosh 4\xi) + O(h^6), \tag{61}$$

$$Mc_1^{(1)}(\xi,h) = h\cosh\xi + O(h^3),$$

$$Ms_1^{(1)}(\xi,h) = h\sinh\xi + O(h^3),$$

$$Mc_0^{(2)}(\xi,h) = (2/\pi)[\ln(\tfrac{1}{2}\gamma h e^{\xi})]Mc_0^{(1)}(\xi,h) + \pi^{-1}[h^2\sinh 2\xi - h^4(\tfrac{1}{4}+\tfrac{3}{32}\sinh 4\xi)] + O(h^3),$$

$$Mc_1^{(2)}(\xi,h) \approx Ms_1^{(2)}(\xi,h) \approx -(2/\pi h)e^{-\xi},$$

provided that ξ is not too large. $\gamma = 1.781$. Furthermore,

$$\alpha_{11} \approx \gamma_{11} = 1 + O(h^2),$$

$$\beta_{00} \approx \chi_{00} = 1 + O(h^2). \tag{62}$$

[10] Lord Rayleigh, Proc. Roy. Soc. (London) A89, 194 (1913).

Substituting Eqs. (61) and (62) into Eq. (30) and simplifying, gives

$$a_{11}^{(1)} \approx -h_1 \sinh\xi_1 + i(2/\pi h_1)e^{-\xi_1},$$
$$a_{11}^{(4)} \approx h_2\nu_1(\xi_1,\xi_2,\mu_2/\mu_3) + (i/h_3)\nu_2(\xi_1,\xi_2,\mu_2/\mu_3),$$
$$c_{11}^{(1)} \approx -h_1 \cosh\xi_1 - i(2/\pi h_1)e^{-\xi_1},$$
$$c_{11}^{(4)} \approx (\mu_1/\mu_2)[h_3\nu_3(\xi_1,\xi_2,\mu_2/\mu_3) + (i/h_3)\nu_4(\xi_1,\xi_2,\mu_2/\mu_3)], \quad (63)$$
$$b_1 \approx 4ih_1 \sinh\xi_1 \sin\beta,$$
$$d_1 \approx 4ih_1 \cosh\xi_1 \sin\beta,$$

where

$$\begin{aligned}
\nu_1(\xi_1,\xi_2,\mu_2/\mu_3) &= \sinh\xi_1 e^{-\xi_2}[\sinh\xi_2 + (\mu_2/\mu_3)\cosh\xi_2] + e^{\xi_1}\sinh\xi_2\cosh\xi_2[(\mu_2/\mu_3)-1],\\
\nu_2(\xi_1,\xi_2,\mu_2/\mu_3) &= (2/\pi)e^{-\xi_2}\{e^{-\xi_2}\sinh\xi_1[1-(\mu_2/\mu_3)] - e^{\xi_1}[(\mu_2/\mu_3)\sinh\xi_2 - \cosh\xi_2]\},\\
\nu_3(\xi_1,\xi_2,\mu_2/\mu_3) &= \cosh\xi_1 e^{-\xi_2}[\sinh\xi_2 + (\mu_2/\mu_3)\cosh\xi_2] + e^{\xi_1}\sinh\xi_2\cosh\xi_2[(\mu_2/\mu_3)-1],\\
\nu_4(\xi_1,\xi_2,\mu_2/\mu_3) &= (2/\pi)e^{-\xi_2}\{e^{-\xi_2}\cosh\xi_1[1-(\mu_2/\mu_3)] - e^{\xi_1}[(\mu_2/\mu_3)\sinh\xi_2 - \cosh\xi_2]\}.
\end{aligned} \quad (64)$$

Inserting (63) into Eqs. (28) and (29) and solving for ${}_1V_1$ and ${}_4V_1$, we obtain

$${}_1V_1 = 4h_1^2 \sin\beta\left(\left[\left(\nu_2\cosh\xi_1 - \frac{\mu_1}{\mu_2}\nu_4\sinh\xi_1\right) + ih_3^2\left(\frac{\mu_1}{\mu_2}\nu_3\sinh\xi_1 - \nu_1\cosh\xi_1\right)\right] \Big/ \left\{-\frac{2}{\pi}e^{-\xi_1}\left(\nu_1+\nu_4\frac{\mu_1}{\mu_2}\right) + i\left[\frac{2}{\pi}h_3^2 e^{-\xi_1}\left(\nu_1+\frac{\mu_1}{\mu_2}\nu_3\right) + h_1^2\left(\nu_2\cosh\xi_1 - \frac{\mu_1}{\mu_2}\sinh\xi_1\nu_4\right)\right]\right\}\right), \quad (65)$$

$${}_4V_1 = (2h_1h_3\pi\sin\beta) \Big/ \left\{\frac{2}{\pi}e^{-\xi_1}\left(\nu_1+\nu_4\frac{\mu_1}{\mu_2}\right) - i\left[\frac{2}{\pi}h_3^2e^{-\xi_1}\left(\nu_1+\frac{\mu_1}{\mu_2}\nu_3\right) + h_1^2\left(\nu_2\cosh\xi_1 - \frac{\mu_1}{\mu_2}\nu_4\sinh\xi_1\right)\right]\right\}. \quad (66)$$

The magnitudes of ${}_1V_2$ and ${}_4V_2$ are all of $O(h_1^4)$ in comparison with ${}_1V_1$ and ${}_4V_1$ which were of $O(h_1^2)$ and hence higher-order expansion coefficients can be neglected. It is interesting to note that the diffracted field for an incident E wave in the Rayleigh approximation is not a function of ϵ_2 although it is still a function of ξ_1 and ξ_2, the boundary of the lens. It can also be shown that if $\mu_1/\mu_2 = \mu_2/\mu_3 = 1$, the diffracted field for an incident E wave in the Rayleigh approximation is not a function of ϵ_2 and ξ_2.

The behavior of the expansion coefficients for an incident H wave is very different from the behavior exhibited by the expansion coefficients of an incident E wave, as shown below. Substituting Eq. (61) into Eq. (53) and carrying out the algebraic manipulation to $O(1)$ gives

$$\begin{aligned}
f_{00}^{(1)} &\approx -1 + i(2/\pi)\ln(\tfrac{1}{2}h_1\gamma e^{\xi_1}),\\
f_{00}^{(4)} &\approx 1 + i(2/\pi)[(\epsilon_2/\epsilon_3)(\xi_1+\xi_2) - \ln(\tfrac{1}{2}h_3\gamma e^{\xi_2})],\\
s_{00}^{(1)} &\approx i(2/\pi),\\
s_{00}^{(4)} &\approx -i(2/\pi)(\epsilon_1/\epsilon_2),\\
p_0 &\approx 2\sqrt{2},\\
w_0 &\approx 0.
\end{aligned} \quad (67)$$

Inserting (67) into Eqs. (51) and (52) and solving for ${}_1U_0$ and ${}_4U_0$, we obtain

$${}_1U_0 \approx 2\sqrt{2}\frac{\epsilon_2}{\epsilon_3} \Big/ \left\{\left(1+\frac{\epsilon_1}{\epsilon_3}\right) + i\frac{2}{\pi}\left[\frac{\epsilon_2}{\epsilon_3}(\xi_1+\xi_2) - \frac{\epsilon_1}{\epsilon_3}\ln(\tfrac{1}{2}h_1\gamma e^{\xi_1}) - \ln(\tfrac{1}{2}h_3\gamma e^{\xi_2})\right]\right\}, \quad (68)$$

$${}_4U_0 \approx -(\epsilon_3/\epsilon_2)\,{}_1U_0. \quad (69)$$

${}_1U_1$ and ${}_4U_1$ are all of $O(h^2 \ln h)$ in comparison with ${}_1U_0$ and ${}_4U_0$ which were of $O(1)$; hence higher-order expansion coefficients can be neglected. We note from the above results that, in the Rayleigh approximation, the diffracted field for an incident H wave is independent of the angle of incidence and is quite insensitive to the slitwidth/wavelength ratio since k_0q occurs only in the logarithmic term. The presence of the lens seems to modify quite significantly the diffracted field of an H wave, even in the Rayleigh limit.

The expressions for the scattered intensities in the far zone, in the Rayleigh limit, can now be given. They are

$$I^E(y<0) \approx \frac{q\sin^2\eta}{r}\left\{(h_1^2h_3\pi^4\sin^2\beta) \Big/ \left[e^{-2\xi_1}\left(\nu_1+\nu_4\frac{\mu_1}{\mu_2}\right)^2\right]\right\}, \quad (70)$$

$$I^E(y>0) = \frac{q\sin^2\eta}{r}\left\{\left[(h_1^4/h_3)\sin^2\beta\pi^2\left(\nu_2\cosh\xi_1 - \frac{\mu_1}{\mu_2}\nu_4\sinh\xi_1\right)^2\right] \Big/ \left[e^{-2\xi_1}\left(\nu_1+\nu_4\frac{\mu_1}{\mu_2}\right)^2\right]\right\} \quad (71)$$

for an incident E wave;

$$I^H(y<0)=\frac{q}{rh_3}\left(4\Big/\left\{\left[\left(1+\frac{\epsilon_1}{\epsilon_3}\right)\right]^2+\frac{4}{\pi^2}\left[\frac{\epsilon_2}{\epsilon_3}(\xi_1+\xi_2)-\frac{\epsilon_1}{\epsilon_3}\ln(\tfrac{1}{2}h_1\gamma e^{\xi_1})-\ln(\tfrac{1}{2}h_3\gamma e^{\xi_2})\right]^2\right\}\right), \quad (72)$$

$$I^H(y>0)=\frac{q}{rh_3}\left((4\epsilon_2/\epsilon_3)\Big/\left\{\left(1+\frac{\epsilon_1}{\epsilon_3}\right)^2+\frac{4}{\pi^2}\left[\frac{\epsilon_2}{\epsilon_3}(\xi_1+\xi_2)-\frac{\epsilon_1}{\epsilon_3}\ln(\tfrac{1}{2}h_1\gamma e^{\xi_1})-\ln(\tfrac{1}{2}h_3\gamma e^{\xi_2})\right]^2\right\}\right) \quad (73)$$

for an incident H wave. The transmission coefficients in the Rayleigh approximation are

$$T^E\approx(q\pi^3h_1^2h_3\sin^2\beta)/\{2e^{-2\xi_1}[\nu_1+\nu_4(\mu_1/\mu_2)]^2\}, \quad (74)$$

$$T^H\approx(4q\pi/h_3)\Big/\left\{\left(1+\frac{\epsilon_1}{\epsilon_3}\right)^2+\frac{4}{\pi^2}\left[\frac{\epsilon_2}{\epsilon_3}(\xi_1+\xi_2)-\frac{\epsilon_1}{\epsilon_3}\ln(\tfrac{1}{2}h_1\gamma e^{\xi_1})-\ln(\tfrac{1}{2}h_3\gamma e^{\xi_2})\right]^2\right\}. \quad (75)$$

ν_1 and ν_4 are given in Eq. (64).

VII. CONCLUSIONS

By use of the orthogonality properties of Mathieu functions, the exact solution of the problem of transmission of electromagnetic wave through a dielectric lens whose boundary surface coincides with an elliptical cylinder is obtained. Also given are the approximate analytical expressions for the transmission coefficients and expansion coefficients in the Rayleigh (long-wavelength) limit. Initial numerical investigation shows that, for small values of h_1 and h_2, it is only necessary to use the first few equations of the set of infinite simultaneous equations as represented by Eqs. (28) and (29) or Eqs. (51) and (52). For example, for $h_1\leq 1$, $\epsilon_3/\epsilon_2=2$, $\epsilon_3/\epsilon_1=1$, $\mu_2/\mu_1=\mu_3/\mu_1=1$, and $\xi_1=\xi_2\leq 1$, less than six simultaneous equations are required to give an accuracy of three significant figures for the expansion coefficients of the diffracted waves; for $h_1\leq 2.5$, $\epsilon_2/\epsilon_1=2$, $\epsilon_3/\epsilon_1=1$, $\mu_2/\mu_1=\mu_3/\mu_1=1$, and $\xi_1=\xi_2\leq 1$, less than 14 simultaneous equations are required to give the same accuracy. For the quasi-optical case in which $h_1\approx 10$, $\epsilon_2/\epsilon_1=2$, $\epsilon_3/\epsilon_1=1$, $\mu_2/\mu_1=\mu_3/\mu_1=1$, and $\xi_1=\xi_2=1$, it is estimated that 40 simultaneous equations will suffice. This is well within the capability of modern high-speed computers. It is particularly interesting in the quasi-optical case to compute the behavior of the diffracted fields at and in the neighborhood of the "optical focal region."[11] Also noted is the fact that the geometry of the present problem approximates closely the shape of an actual two-dimensional lens. Boundary-perturbation technique[12] may be used to treat problems with boundary shapes slightly different from the elliptical ones considered here. A presentation of the numerical results is complicated and lengthy and will therefore be reserved for future communications.

ACKNOWLEDGMENTS

The author wishes to express his sincere appreciation to Professor J. Meixner for suggesting this problem.

[11] Although incident parallel rays can *not* be focused by a lens with elliptic cylindrical boundary to a point in the two-dimensional space, nevertheless there exists a region in which the rays are concentrated. This region is the "optical focal region" of interest. This "optical focal region" may be in the near zone of the elliptic lens.

[12] C. Yeh, Phys. Rev. 135, A1193 (1964).

Radio Science, Volume 26, Number 5, Pages 1165–1176, September–October 1991

Scattering of an obliquely incident wave by a multilayered elliptical lossy dielectric cylinder

Chang S. Kim

Submarine Electromagnetic Systems Department, Naval Underwater Systems Center, New London, Connecticut

C. Yeh

Department of Electrical Engineering, University of California, Los Angeles

(Received July 20, 1989; revised May 28, 1991; accepted May 30, 1991.)

The scattering problem of obliquely incident waves on a multilayered elliptical cylinder is considered. The cylinder is assumed homogeneous and isotropic but lossy. Both polarization of the incident wave, the *E* wave, and the *H* wave are considered. In the theoretical analysis the number of layers are unlimited, but in the numerical analysis we limit the number of layers so we can avoid excessively long computer running time. The numerical results include the bistatic radar cross sections, the specular differential cross sections, and the efficiency factors for absorption, scattering, and extinction. The computed results show the dependence of the scattering on the size, shape, material, and the angle of the incident.

1. INTRODUCTION

Rayleigh [1881] first used the Maxwell's equation [*Maxwell*, 1864] to solve the problem of scattering of an electromagnetic wave normally incident on a circular lossless dielectric cylinder by using the separation of variables method. *Van de Hulst* [1949] solved the problem of the scattering of a normally incident wave by a circular lossy dielectric cylinder, and *Wait* [1955] did the scattering of obliquely incident wave by a circular lossless dielectric cylinder theoretically and numerically by using the separation of variables method. For a multilayered circular lossless dielectric cylinder. *Evans et al.* [1964] gave first the complete solutions of the scattered wave incident normally, and many interesting numerical results were shown [*Datta and Som*, 1975]. Both used the separation of variables method. *Bussey and Richmond* [1975] did the same analysis for a lossy layered cylinder. Compared to the circular cylinder, the scattering problem of an elliptical cylinder was solved much later because of its complexity. *Morse and Rubenstein* [1938] obtained the exact solution and numerical results of the scattering problem by the elliptical conducting cylinder after a table of the Mathieu functions was published, and *Yeh* [1963, 1965] solved the scattering problem theoretically and numerically when the wave is incident normally on a homogeneous dielectric elliptical cylinder. *Yeh* [1964] also did the oblique incidence case only theoretically because of the complexity of the numerical calculations. Morse and Yeh used the separation of variables method, but *Alexopoulous and Tadler* [1975] used the impedance boundary conditions method for the solution of the scattering of normally incident wave by an elliptical lossy dielectric cylinder. *Morita* [1979] used the extended boundary conditions method, and *Goel and Jain* [1981] used the integral equations for a lossless dielectric cylinder when the wave is incident normally. Lately, *Richmond* [1988] investigated the scattering properties for a single layered elliptical conducting cylinder. He used Galerkin's method but limited investigations only for a lossless layer and normal incident case.

The purpose of this paper is to approach the scattering problem more generally. An infinitely long elliptical cylinder is chosen as a scattering object because the shape of the cross section is changeable from a flat ribbon to a circular shape. Also, for a more general approach to the scattering problem the analyses have been done for lossy and lossless. Finally, some number of layers are added to the elliptical cylinder because altering the thickness, material, and number of layers is a very effective way to change the scattering patterns. For

Paper number 91RS01706.
0048-6604/91/91RS-01706$08.00

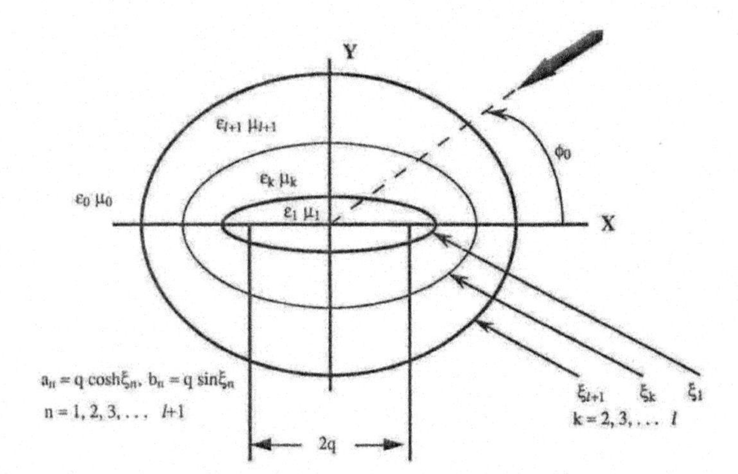

Fig. 1. Scattering geometry in two dimensional view. An infinitely long multilayered elliptical dielectric cylinder is illuminated by a plane wave with its direction at an angle ϕ_0 with the X axis in the X-Y plane and θ_0 with the Z axis. There are $l + 1$ boundary surfaces where the boundary conditions are applied.

the theoretical analysis of a multilayered elliptical dielectric cylinder the number of layers are unlimited. The numerical results are based on the theoretical analyses but the number of layers is limited to avoid excessively long computer running time.

2. FORMULATION OF THE PROBLEM

The scattering geometry of this report is shown in Figure 1. The direction of the incident plane wave is at angle ϕ_0 with the X axis on the X-Y plane and θ_0 with Z axis. The medium of each layer, whose permittivity and permeability are ε_k and $\mu_k (k = 1, 2, \cdots, l + 1)$ respectively, is assumed homogeneous and isotropic but lossy or lossless and the medium outside the cylinder is free space whose permittivity and permeability are denoted by ε_0 and μ_0, respectively. All the dielectric interfaces in Figure 1 are confocal ellipses and each interface coincides with the surface of constant ξ of the elliptical cylinder coordinates (ξ, η, z). Hence these coordinates are used to find the solutions of the wave equations.

The elliptical cylinder coordinates are related to the rectangular coordinates (x, y, z) through the following relationship [*McLachlan*, 1951]:

$$x = q \cosh \xi \cos \eta \tag{1a}$$

$$y = q \sinh \xi \sin \eta \tag{1b}$$

$$z = z \tag{1c}$$

where $0 \le \xi \le \infty$ and $0 \le \xi \le 2\pi$. The distance between the two foci of the ellipse is $2q$. Two types of plane waves, the E wave and H wave, are possible. The E wave is defined by $H_z = 0$, and the H wave is defined by $E_z = 0$.

To simplify the notation for the radial and angular Mathieu functions, the following abbreviations are used:

$$ce_n(\eta) = ce_n(\eta, \gamma_0^2),\ ce_n^{(k)}(\eta) = ce_n(\eta, \gamma_k^2) \tag{2a}$$

$$se_n(\eta) = se_n(\eta, \gamma_0^2),\ se_n^{(k)}(\eta) = se_n(\eta, \gamma_k^2) \tag{2b}$$

$$Ce_n(\xi) = Ce_n(\xi, \gamma_0^2),\ Ce_n^{(k)}(\xi) = Ce_n(\xi, \gamma_k^2) \tag{2c}$$

$$Se_n(\xi) = Se_n(\xi, \gamma_0^2),\ Se_n^{(k)}(\xi) = Se_n(\xi, \gamma_k^2) \tag{2d}$$

$$Fey_n(\xi) = Fey_n(\xi, \gamma_0^2),\ Fey_n^{(k)}(\xi) = Fey_n(\xi, \gamma_k^2) \tag{2e}$$

$$Gey_n(\xi) = Gey_n(\xi, \gamma_0^2),\ Gey_n^{(k)}(\xi) = Gey_n(\xi, \gamma_k^2) \tag{2f}$$

with

$$p = q\sqrt{\sinh^2 \xi + \sin^2 \eta} \qquad k = 0, 1, 2, \cdots \tag{3a}$$

$$k_k^2 = \omega^2 \epsilon_k \mu_k \qquad k = 0, 1, 2 \cdots \tag{3b}$$

$$\kappa_k^2 = k_k^2 - k_0^2 \cos^2\theta_k \qquad k = 0, 1, 2, \cdots \tag{3c}$$

$$\gamma_k^2 = \kappa_k^2 q^2/4 \qquad k = 0, 1, 2, \cdots \tag{3d}$$

where $Ce_n(\xi, \gamma_k^2)$ and $Se_n(\xi, \gamma_k^2)$ are the even and odd radial Mathieu functions of the first kind, $Fey_n(\xi, \gamma_k^2)$ and $Gey_n(\xi, \gamma_k^2)$ are the even and odd radial Mathieu functions of the second kind, $ce_n(\eta, \gamma_k^2)$ and $se_n(\eta, \gamma_k^2)$ are the even and odd angular Mathieu functions, respectively, and ω is the angular frequency of the incident wave.

2.1. *E wave incident*

The z component of the incident electric field E_z^i is simply defined by in the elliptical cylinder coordinates [*Yeh*, 1963]:

$$E_z^i = E_0 \exp\{ik_0[q \sin\theta_0(\cosh\xi \cos\eta \cos\phi_0 + \sinh\xi \sin\eta \sin\phi_0) + z\cos\theta_0]\}$$

$$= 2E_0 e^{jk_0 z \cos\theta_0} \sum_{n=0}^{\infty} \left[\frac{1}{p_{2n}} Ce_{2n}(\xi) ce_{2n}(\eta) ce_{2n}(\phi_0) + \frac{1}{s_{2n+2}} Se_{2n+2}(\xi) se_{2n+2}(\eta) se_{2n+2}(\phi_0) + \frac{j}{p_{2n+1}} Ce_{2n+1}(\xi) ce_{2n+1}(\eta) ce_{2n+1}(\phi_0) + \frac{j}{s_{2n+1}} Se_{2n+1}(\xi) se_{2n+1}(\eta) se_{2n+1}(\phi_0)\right] \tag{4}$$

where E_0 is a complex amplitude of E_z^i and p_n and s_n are joining factors which are given in [*McLachlan*, 1951]. The harmonic time dependence of $e^{-j\omega t}$ is assumed and suppressed through out. The z component of the incident magnetic field is given by

$$H_z^i = 0 \tag{5}$$

The scattered fields should satisfy Sommerfeld's radiation conditions [*Stratton*, 1941] at infinity because the source of the scattering is confined at the region of the elliptical cylinder. Therefore the solutions of the scattered waves are expressed by using the following functions defined by

$$Me_n(\xi) = Ce_n(\xi) + jFey_n(\xi) \tag{6}$$

for even solutions and

$$Ne_n(\xi) = Se_n(\xi) + jGey_n(\xi) \tag{7}$$

for odd solutions.

The boundary condition for the case of oblique incidence requires the z component of the scattered magnetic field be nonzero. Therefore the scattered fields are expressed as a superposition of the E wave and the H wave even though the z component of the incident magnetic field is zero.

$$E_z^s = 2E_0 e^{jk_0 z \cos\theta_0} \cdot \sum_{n=0}^{\infty} \left[\frac{1}{p_{2n}} A_{2n} Me_{2n}(\xi) ce_{2n}(\eta) ce_{2n}(\phi_0) + \frac{1}{s_{2n+2}} B_{2n+2} Ne_{2n+2}(\xi) se_{2n+2}(\eta) se_{2n+2}(\phi_0) + \frac{j}{p_{2n+1}} A_{2n+1} Me_{2n+1}(\xi) ce_{2n+1}(\eta) ce_{2n+1}(\phi_0) + \frac{j}{s_{2n+1}} B_{2n+1} Ne_{2n+1}(\xi) se_{2n+1}(\eta) se_{2n+1}(\phi_0)\right] \tag{8}$$

and the z component of the scattered magnetic field

$$H_z^s = 2E_0 e^{jk_0 z \cos\theta_0} \cdot \sum_{n=0}^{\infty} \left[\frac{1}{p_{2n}} C_{2n} Me_{2n}(\xi) ce_{2n}(\eta) ce_{2n}(\phi_0) + \frac{1}{s_{2n+2}} D_{2n+2} Ne_{2n+2}(\xi) se_{2n+2}(\eta) se_{2n+2}(\phi_0) + \frac{j}{p_{2n+1}} C_{2n+1} Me_{2n+1}(\xi) ce_{2n+1}(\eta) ce_{2n+1}(\phi_0) + \frac{j}{s_{2n+1}} D_{2n+1} Ne_{2n+1}(\xi) se_{2n+1}(\eta) se_{2n+1}(\phi_0)\right] \tag{9}$$

where A_n, B_n, C_n, and D_n are the unknown expansion coefficients of the scattered wave.

The transmitted waves into each layer are finite and should satisfy the boundary conditions at the boundary surface. Therefore the linear combinations of the first and the second modified Mathieu functions are used for the field expression as follows:

$$E^t_{z,k} = 2E_0 e^{jk_0 z \cos\theta_0} \sum_{n=0}^{\infty} \left[\frac{1}{p^{(k)}_{2n}} \{F^{(k)}_{2n} Ce^{(k)}_{2n}(\xi) + R^{(k)}_{2n} Fey^{(k)}_{2n}(\xi)\} ce^{(k)}_{2n}(\eta) ce^{(k)}_{2n}(\phi_0) + \frac{1}{s^{(k)}_{2n+2}} \{G^{(k)}_{2n+2} Se^{(k)}_{2n+2}(\xi) + T^{(k)}_{2n+2} Gey^{(k)}_{2n+2}(\xi)\} se^{(k)}_{2n+2}(\eta) se^{(k)}_{2n+2}(\phi_0) + \frac{j}{p^{(k)}_{2n+1}} \{F^{(k)}_{2n+1} Ce^{(k)}_{2n+1}(\xi) + R^{(k)}_{2n+1} Fey^{(k)}_{2n+1}(\xi)\} ce^{(k)}_{2n+1}(\eta) ce^{(k)}_{2n+1}(\phi_0) + \frac{j}{s^{(k)}_{2n+1}} \{G^{(k)}_{2n+1} Se^{(k)}_{2n+1}(\xi) + T^{(k)}_{2n+1} Gey^{(k)}_{2n+1}(\xi)\} se^{(k)}_{2n+1}(\eta) se_{2n+1}(\phi_0) \right] \quad (10)$$

$$H^t_{z,k} = 2E_0 e^{jk_0 z \cos\theta_0} \sum_{n=0}^{\infty} \left[\frac{1}{p^{(k)}_{2n}} \{P^{(k)}_{2n} Ce^{(k)}_{2n}(\xi) + X^{(k)}_{2n} Fey^{(k)}_{2n}(\xi)\} ce^{(k)}_{2n}(\eta) ce^{(k)}_{2n}(\phi_0) + \frac{1}{s^{(k)}_{2n+2}} \{Q^{(k)}_{2n+2} Se^{(k)}_{2n+2}(\xi) + Y^{(k)}_{2n+2} Gey^{(k)}_{2n+2}(\xi)\} se^{(k)}_{2n+2}(\eta) se^{(k)}_{2n+2}(\phi_0) + \frac{j}{p^{(k)}_{2n+1}} \{P^{(k)}_{2n+1} Ce^{(k)}_{2n+1}(\xi) + X^{(k)}_{2n+1} Fey^{(k)}_{2n+1}(\xi)\} ce^{(k)}_{2n+1}(\eta) ce^{(k)}_{2n+1}(\phi_0) + \frac{j}{s^{(k)}_{2n+1}} \{Q^{(k)}_{2n+1} Se^{(k)}_{2n+1}(\xi) + Y^{(k)}_{2n+1} Gey^{(k)}_{2n+1}(\xi)\} se^{(k)}_{2n+1}(\eta) se_{2n+1}(\phi_0) \right] \quad (11)$$

where $F^{(k)}_n$, $G^{(k)}_n$, $P^{(k)}_n$, $Q^{(k)}_n$, $R^{(k)}_n$, $T^{(k)}_n$, $X^{(k)}_n$, and $Y^{(k)}_n$ are the unknown expansion coefficients of the transmitted wave and $p^{(k)}_n$ and $s^{(k)}_n$ are joining factors and given by *Kim* [1989].

Equations (10) and (11) represent the components of the transmitted wave into each layer ($k = 1, 2, \cdots, l+1$), but the following coefficients, $R^{(1)}_n$, $T^{(1)}_n$, $X^{(1)}_n$, and $Y^{(1)}_n$ vanish in the innermost layer because the radial Mathieu functions of the second kind have infinite values at $\xi = 0$. Other components of electric and magnetic fields are expressed as written in Appendix A.1.

To find the unknown expansion coefficients of the scattered and the transmitted waves, the boundary conditions are applied at each dielectric interface. At $\xi = \xi_k$ ($k = 1, 2, \cdots, l$) the following equations are formed:

$$E^t_{z,k} = E^t_{z,k+1} \quad (12a)$$

$$H^t_{z,k} = H^t_{z,k+1} \quad (12b)$$

$$E^t_{\eta,k} = E^t_{\eta,k+1} \quad (12c)$$

$$H^t_{\eta,k} = H^t_{\eta,k+1} \quad (12d)$$

At $\xi = \xi_{l+1}$ the boundary conditions are stated as follows:

$$E^i_z + E^s_z = E^t_{z,l+1} \quad (13a)$$

$$H^s_z = H^t_{z,l+1} \quad (13b)$$

$$E^i_\eta + E^s_\eta = E^t_{\eta,l+1} \quad (13c)$$

$$H^i_\eta + H^s_\eta = H^t_{\eta,l+1} \quad (13d)$$

In the actual computations the infinite series in (4) and (8) through (11) should be truncated to finite sums and every element of each unknown coefficient set becomes a member of its family. Equations (12) and (13) give four families of equations at each interface, although we have eight families of unknown coefficients at every boundary surface. However, if the orthogonality properties of the angular Mathieu functions in Appendix A.2 are substituted into (12) and (13), we can separate each equation into two independent equations [*Yeh*, 1964; *Kim*, 1989]. Finally, we have ($8l + 8$) families of equations for ($8l + 8$) families of unknowns for ($l + 1$) interfaces as indicated in Figure 1.

2.2. *H wave incident*

The corresponding result for the case of an incident *H* wave can be obtained by replacing **E** with **H**, **H** with $-$**E**, ε with μ, and μ with ε.

2.3. *Cross sections and efficiency factors*

From the expansion coefficients obtained the electromagnetic waves can be found at any location either inside or outside the cylinder. Now, it is possible to find the power density distribution at any point and the total amount of power absorbed and scattered by the cylinder.

2.3.1. *Differential cross section (or bistatic radar cross section).* The differential scattering cross section (σ_d) represents the density of the power scattered by cylinder in various directions at the region far from the cylinder and defined by [*Yeh and Wang*, 1972]:

$$\sigma_d(\varphi, \phi_0) = \lim_{r\to\infty} 2\pi r \frac{\mathrm{Re}\,[(\mathbf{E}^s \times \mathbf{H}^{s*}) \cdot \hat{e}_r]}{\mathrm{Re}\,|\mathbf{E}^i \times \mathbf{H}^{i*}|} \tag{14}$$

where the asterisk indicates complex conjugates and $\hat{e}_r$ is the unit vector in the radial direction.

At large distances from the cylinder the confocal ellipses are now sensibly concentric circles and the fields can be expressed more simply by using the asymptotic expressions of the radial Mathieu functions [*McLachlan*, 1951], when $k_0 r \gg M$, where M is the order of the Mathieu function and $r \approx q$ cosh $\xi \approx q$ sinh ξ and $\varphi \approx \eta$. For the E wave incidence case the z components of the scattered waved at a region far from the cylinder are expressed as

$$E_z^s \approx j2E_0 \sqrt{\frac{2}{\pi \kappa_0 r}} \exp\{jk_0 z \cos\theta_0 \cdot \exp(j\kappa_0 r + (\pi/4)\} T_1(\varphi, \phi_0) \tag{15a}$$

$$H_z^s \approx j2E_0 \sqrt{\frac{2}{\pi \kappa_0 r}} \exp\{jk_0 z \cos\theta_0 \cdot \exp(j\kappa_0 r + (\pi/4)\} T_2(\varphi, \phi_0) \tag{15b}$$

where

$$T_1(\varphi, \phi_0) = -\sum_{n=0}^{\infty} A_n ce_n(\varphi) ce_n(\phi_0) - \sum_{m=1}^{\infty} B_m se_m(\varphi) se_m(\phi_0) \tag{16a}$$

$$T_2(\varphi, \phi_0) = -\sum_{n=0}^{\infty} C_n ce_n(\varphi) ce_n(\phi_0) - \sum_{m=1}^{\infty} D_m se_m(\varphi) se_m(\phi_0) \tag{16b}$$

For the H wave incidence case the z components of the scattered wave are expressed by replacing E_0 with H_0.

Using the formulas in Appendix A.1 and substituting (15) into (14) give

$$\sigma_d(\varphi, \phi_0) = \frac{16}{\kappa_0} \left[|T_1(\varphi, \phi_0)|^2 + \frac{\mu_0}{\varepsilon_0} |T_2(\varphi, \phi_0)|^2 \right] \tag{17}$$

for the E wave incidence case. Similarly, the following equation is formed for the H wave incidence case.

$$\sigma_d(\varphi, \phi_0) = \frac{16}{\kappa_0} \left[\frac{\varepsilon_0}{\mu_0} |T_1(\varphi, \phi_0)|^2 + |T_2(\varphi, \phi_0)|^2 \right] \tag{18}$$

The power scattered in the backward direction is very interesting in the radar detection. Let the specular differential scattering cross section be defined by

$$\sigma_b(\phi_0) = \sigma_d(-\phi_0, \phi_0) \tag{19}$$

σ_b becomes the monostatic radar cross section if the plane wave is incident normally into the cylinder.

2.3.2. *Absorption cross section.* When the medium of the cylinder is lossy, part of the transmitted wave is absorbed. The total power absorbed by the cylinder is equivalent to the power of the incident wave that falls on a certain area. This area is called the absorption cross section [*Bohren and Huffman*, 1983] and expressed by

$$C_{\mathrm{abs}} = \frac{\text{total power absorbed by the cylinder per unit length}}{\text{incident power flux density}} \tag{20a}$$

$$C_{\mathrm{abs}} = \frac{-\mathrm{Re}\left[\int_0^{2\pi} (\mathbf{E} \times \mathbf{H}^*) \cdot \hat{e}_\xi p \, d\eta\right]}{\mathrm{Re}\,|\mathbf{E}^i \times \mathbf{H}^{i*}|} \tag{20b}$$

where $\mathbf{E}$ and $\mathbf{H}$ are total electric and magnetic fields and expressed as

$$\mathbf{E} = \mathbf{E}^i + \mathbf{E}^s \tag{21a}$$

$$\mathbf{H} = \mathbf{H}^i + \mathbf{H}^s \tag{21b}$$

2.3.3. *Scattering cross section.* The total power scattered by the cylinder is equivalent to the

$k_0a_1 = 0.9425$, $\varepsilon_1/\varepsilon_0 = 67.0 + j43.0$, $\mu_1/\mu_0 = 1.0$
$k_0a_2 = 1.2566$, $\varepsilon_2/\varepsilon_0 = 6.00 + j0.50$, $\mu_2/\mu_0 = 1.0$

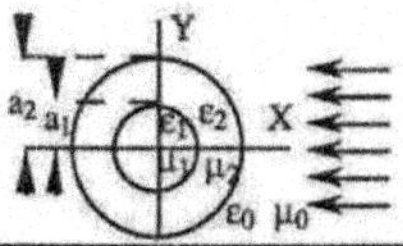

E-Wave		H-Wave	
Kim & Yeh	Bussey & Richmond	Kim & Yeh	Bussey & Richmond
-0.82524 + j0.01109	-0.82522 + j0.01107	-0.24856 + j0.26672	-0.24860 + j0.26670
-0.24859 + j0.26672	-0.24860 + j0.26670	-0.36530 - j0.33502	-0.36538 - j0.33501
-0.03665 + j0.03370	-0.03666 + j0.03371	-0.08624 - j0.19621	-0.08629 - j0.19627
-0.00222 + j0.00047	-0.00222 + j0.00047	-0.00118 - j0.01224	-0.00118 - j0.01224

(a)

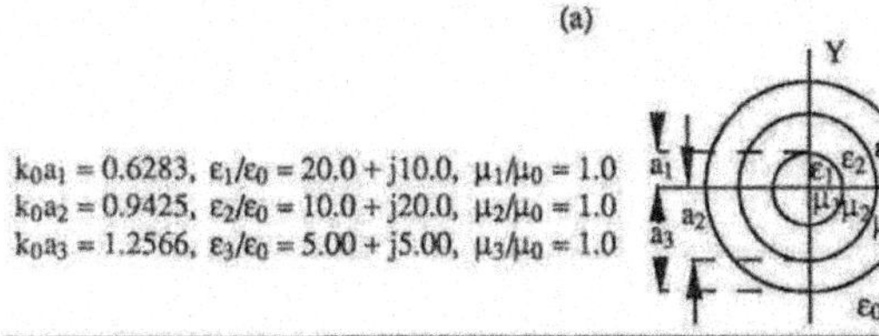

$k_0a_1 = 0.6283$, $\varepsilon_1/\varepsilon_0 = 20.0 + j10.0$, $\mu_1/\mu_0 = 1.0$
$k_0a_2 = 0.9425$, $\varepsilon_2/\varepsilon_0 = 10.0 + j20.0$, $\mu_2/\mu_0 = 1.0$
$k_0a_3 = 1.2566$, $\varepsilon_3/\varepsilon_0 = 5.00 + j5.00$, $\mu_3/\mu_0 = 1.0$

E-Wave		H-Wave	
Kim & Yeh	Bussey & Richmond	Kim & Yeh	Bussey & Richmond
-0.75139 - j0.05412	-0.75138 - j0.05418	-0.30702 + j0.25491	-0.30707 + j0.25491
-0.30704 + j0.25491	-0.30707 + j0.25491	-0.34855 - j0.25229	-0.34857 - j0.25228
-0.05949 - j0.04259	-0.05950 - j0.04261	-0.11967 - j0.17025	-0.11972 - j0.17028
-0.00364 + j0.00022	-0.00364 + j0.00022	-0.00340 - j0.01271	-0.00340 - j0.01272

(b)

Fig. 2. Numerical calculations of expansion coefficients of the scattered waves. Bussey and Richmond calculated Bessel functions for the circular cylindrical multilayer [*Bussey and Richmond*, 1975].

power of the incident wave that falls on the scattering cross section (C_{sct}). C_{sct} is expressed as [*Bohren and Huffman*, 1983]:

$$C_{sct} = \frac{\text{total power scattered by the cylinder per unit length}}{\text{incident power flux density}} \quad (22a)$$

$$C_{sct} = \frac{\operatorname{Re}\left[\int_0^{2\pi} (\mathbf{E}^s \times \mathbf{H}^{s*}) \cdot \hat{e}_\xi p \, d\eta\right]}{\operatorname{Re}|\mathbf{E}^i \times \mathbf{H}^{i*}|} \quad (22b)$$

2.3.4. *Extinction cross section*. The total power attenuated by the cylinder is equivalent to the sum of the total power scattered and total power absorbed. The extinction cross section C_{ext} is defined by [*Kerker*, 1969]:

$$C_{ext} = C_{abs} + C_{sct} \tag{23}$$

2.3.5. *Efficiency factors.* The efficiency factors for extinction, scattering, and absorption are dimensionless constants as defined by [*Kerker*, 1969]:

$$Q_{ext} = C_{ext}/G \tag{24a}$$

$$Q_{sct} = C_{sct}/G \tag{24b}$$

$$Q_{abs} = C_{abs}/G \tag{24c}$$

where G is the geometrical cross section.

3. NUMERICAL RESULTS

In this section it is investigated how the scattering characteristics are dependent on the size, the number of layers, the material of the object, and the angle of the incident wave.

In the previous section all waves are expressed with the angular and radial Mathieu functions. It is very important to find the accurate values of these functions for the numerical results. The accuracy of Mathieu functions are checked with following processes. First, the trigonometric functions, J-Bessel and Y-Bessel functions with real and complex arguments are calculated and checked with the published tables [*Abramowitz and Stegun*, 1972; *National Bureau of Standards*, (NBS), 1947, 1950], since the angular and radial Mathieu functions are expressed in terms of infinites series of those functions with unknown series expansion coefficients [*McLahlan*, 1951]. Second, the eigenvalues of the Mathieu differential equations for real and complex parameters are calculated and checked with the published tables [*NBS*, 1945; *Blanch and Clemm*, 1969]. Third, unknown expansion series coefficients are calculated [*Kim*, 1989]. Numerical value of each series term is carefully investigated, and the truncated point is decided for each Mathieu function.

With confidence in the numerical values of the Mathieu functions, numerical analyses are carried out to solve the scattering problem for a multilayered elliptical lossy dielectric cylinder. The expansion coefficients of the scattered waves are calculated and compared with the published data [*Bussey and Richmond*, 1975] in Figure 2. The numerical values of coefficients are agreed well. We choose

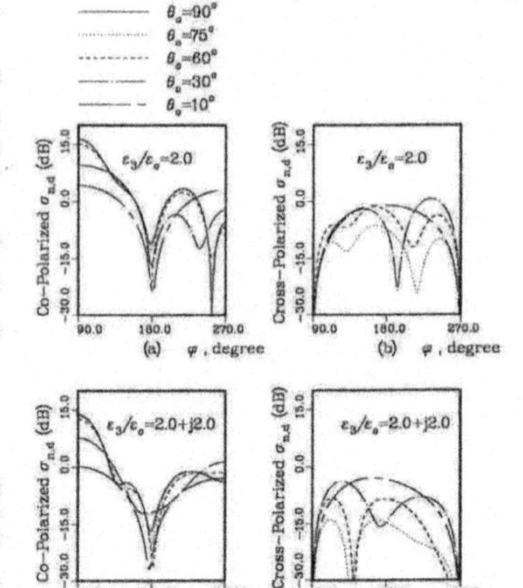

Fig. 3. Comparison of the copolarized and cross-polarized components as a function of the azimuthal angle (φ) for multilayered cylinder scattering, H wave incident, k_0a_1 = 3.48, a_1/b_1 = 10.0, k_0a_2 = 3.58, a_2/b_2 = 4.0, k_0a_3 = 4.00, a_3/b_3 = 2.0, $\mu_1/\mu_0 = \mu_2/\mu_0 = \mu_3/\mu_0$ = 1.0, $\varepsilon_1/\varepsilon_0$ = 8.0, $\varepsilon_2/\varepsilon_0$ = 4.0, ϕ_0 = 90°.

$\xi > 5$ in order to make the confocal ellipses sensibly concentric circles. Bussey and Richmond directly calculated Bessel functions with complex parameters for the circular cylinder.

The computer program also gives the numerical results of the normalized bistatic radar cross section ($\sigma_{n,d}$) and the normalized specular differential cross section ($\sigma_{n,b}$) which are defined by

$$\sigma_{n,d} = \frac{\sigma_d}{k_0/4} \tag{25a}$$

$$\sigma_{n,b} = \frac{\sigma_b}{k_0/4} \tag{25b}$$

Also, efficiencies for absorption, scattering and extinction are given.

Figure 3 shows the cross polarized $\sigma_{n,d}$ for the resonant frequency case when the H wave is incident obliquely. At low obliquity the copolarized components for the lossless and for the lossy case

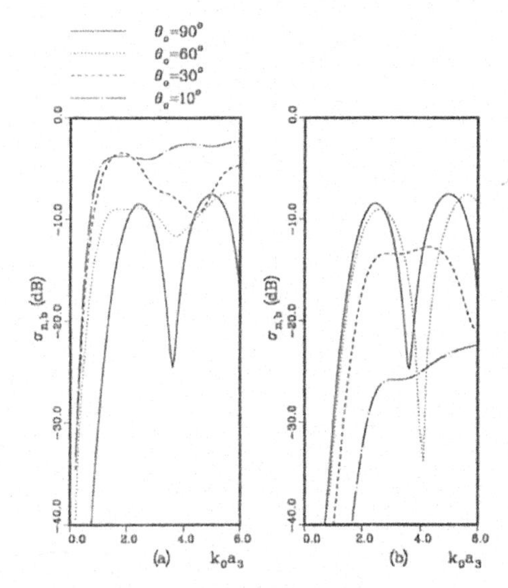

Fig. 4. Normalized specular cross section as a function of frequency for multilayered dielectric cylinder scattering, H wave incident, $a_1/b_1 = 10.0$, $a_2/b_2 = 5.0$, $a_3/b_3 = 2.0$, $a_1/a_2 = 0.98$, $a_2/a_3 = 0.88$, $\mu_1/\mu_0 = \mu_2/\mu_0 = \mu_3/\mu_0 = 1.0$, $\phi_0 = 90°$: (*a*) $\varepsilon_1/\varepsilon_0 = \varepsilon_2/\varepsilon_0 = 4.0$, $\varepsilon_3/\varepsilon_0 = 2.0 + j2.0$; (*b*) $\varepsilon_1/\varepsilon_0 = 4.2$, $\varepsilon_2/\varepsilon_0 = 4.0$, $\varepsilon_3/\varepsilon_0 = 2.0 + j2.0$.

are quite different, but as the obliquity increase, they become very similar as shown in Figures 3*a* and 3*c* and resembles that of Rayleigh scattering of the circular cylinder at normal incidence case [*Van de Hulst*, 1957]. At high obliquity, Figures 3*b* and 3*d* shown that the cross-polarized curves have very broad maxima which are also shown in the problem of the scattering by the circular cylinder [*Kerker*, 1969].

In Figures 4 and 5, $\sigma_{n,b}$ are compared between a single layered cylinder ($\varepsilon_1 = \varepsilon_2 \neq \varepsilon_3$) and a double layered cylinder ($\varepsilon_1 \neq \varepsilon_2 \neq \varepsilon_3$) when the H wave is incident with $\phi_0 = 90°$. Figure 4 shows that both of $\sigma_{n,b}$ curves are very similar at normal incidence case ($\theta_0 = 90°$) when ε_3 is complex. This explains that the transmitted waves into the cylinder are absorbed in the outermost layer and the waves cannot reach to the inner interface. However, as θ_0 is decreased, $\sigma_{n,b}$ curves become very different and for the double layered cylinder the powers scattered in the backward direction are much smaller than those of a single layered cylinder. These show that as the obliquity is increased, the wave transmits more deeply inside and the double layered cylinder acts like a waveguide and some portion of the transmitted power is trapped inside the cylinder.

If the inner layer is lossy and the outer layer is lossless, Figure 5 illustrates that the backscattering curves are similar for the normal incidence for the both cylinders when the size of the cylinder is small ($k_0a_3 < 3.0$), but for the large size of the cylinder the curves are different even in the normal incidence case. However, if you compare the curves with those in Figure 4, you realize that the curves in Figure 5 are much smoother around the minimum. As θ_0 is decreased, the backscattering is very smaller for a large double layered cylinder. This indicates that the combination of materials in the layers in Figure 4 can reduce the backscattering much below that in Figure 4.

In Figure 6, waveguide effect can be observed clearly. The solid line is the backscattering curve of

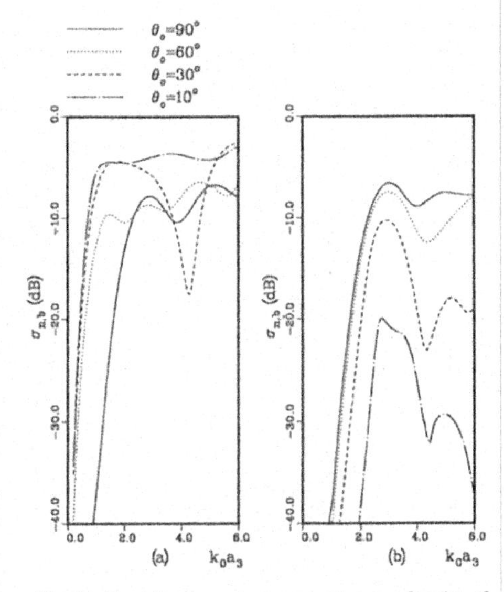

Fig. 5. Normalized specular cross section as a function of frequency for multilayered dielectric cylinder scattering, H wave incident, $a_1/b_1 = 10.0$, $a_2/b_2 = 5.0$, $a_3/b_3 = 2.0$, $a_1/a_2 = 0.98$, $a_2/a_3 = 0.88$, $\mu_1/\mu_0 = \mu_2/\mu_0 = \mu_3/\mu_0 = 1.0$, $\phi_0 = 90°$: (*a*) $\varepsilon_1/\varepsilon_0 = \varepsilon_2/\varepsilon_0 = 4.0 + j4.0$, $\varepsilon_3/\varepsilon_0 = 2.0$; (*b*) $\varepsilon_1/\varepsilon_0 = 4.2$, $\varepsilon_2/\varepsilon_0 = 4.0 + j4.0$, $\varepsilon_3/\varepsilon_0 = 2.0$.

a dielectric cylinder without layers. The dotted lines are for a single layered cylinder but the thickness of the layer is different. Finally, the chain dotted line is for the double layered cylinder. Figure 6*a* shows that the dielectric cylinder has the minimum backscattering for normal incidence because the penetrated waves are reduced from the absorption. But for the oblique incidence case, the double layered cylinder has minimum backscattering powers because some portion of the transmitted wave is trapped into the inner cylinder and propagates through the z axis. For a single layered cylinder a thicker lossy layer can reduce the backscattering more.

The efficiency factors for the multilayered dielectric cylinders are shown in Figures 7 and 8. First, we can observe that Q_{ext} of the lossy cylinder is much smoother than that of the lossless cylinder. This indicates that the interferences between partial waves are reduced in the lossy cylinder due to absorption. For the E wave, Q_{abs} and Q_{sct} are very similar for normal incidence as given in Figure 7,

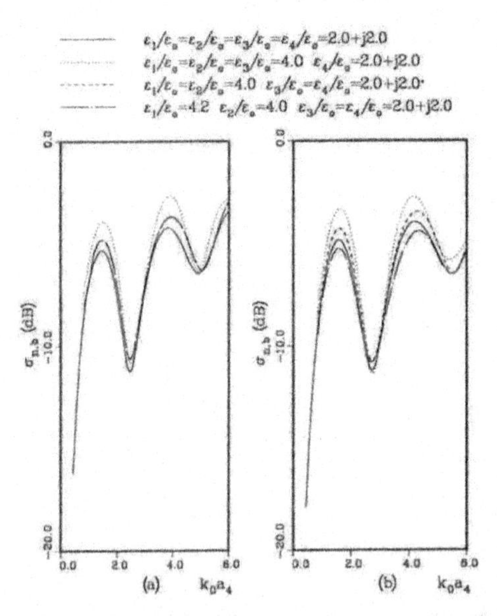

Fig. 6. Normalized specular cross section as a function of frequency for multilayered dielectric cylinder scattering, E wave incident, $a_1/b_1 = 20.0$, $a_2/b_2 = 10.0$, $a_3/b_3 = 5.0$, $a_4/b_4 = 2.0$, $a_1/a_2 = 0.99$, $a_2/a_3 = 0.98$, $a_3/a_4 = 0.88$, $\mu_1/\mu_0 = \mu_2/\mu_0 = \mu_3/\mu_0 = \mu_4/\mu_0 = 1.0$, $\phi_0 = 90°$; (*a*) $\theta_0 = 90°$; (*b*) $\theta_0 = 45°$.

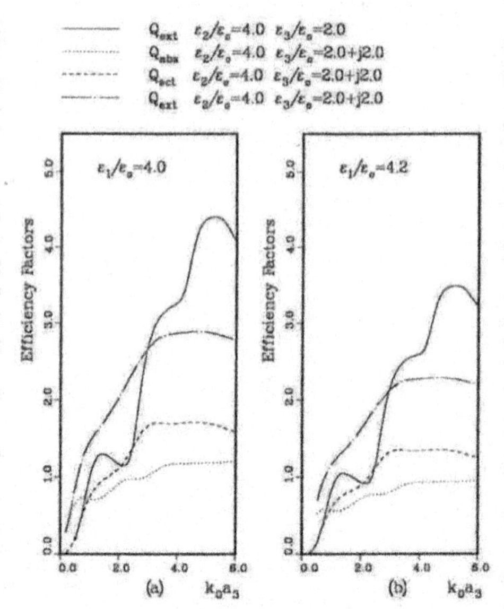

Fig. 7. Efficiency factors as a function of frequency for multilayered dielectric cylinder scattering, E wave incident, $a_1/b_1 = 10.0$, $a_2/b_2 = 5.0$, $a_3/b_3 = 2.0$, $a_1/a_2 = 0.98$, $a_2/a_3 = 0.88$, $\mu_1/\mu_0 = \mu_2/\mu_0 = \mu_3/\mu_0 = 1.0$, $\phi_0 = 90°$, $\theta_0 = 90°$; (*a*) $\varepsilon_1/\varepsilon_0 = 4.0$; (*b*) $\varepsilon_1/\varepsilon_0 = 4.2$.

but they are different for oblique incidence as shown in Figure 8. However, the curves of Q_{ext} are very similar for lossless and lossy material, but the value of Q_{ext} is much lower for a double layered cylinder than that for a single layered cylinder like the H wave incidence case. Therefore we know that Q_{ext} can be reduced by adding layers in the cylinder.

4. CONCLUSION

The scattering of electromagnetic waves by a multilayered elliptical cylinder has been investigated theoretically and numerically. The scattering problem has been solved by using the separation of variables method. The numerical analysis gives us very interesting results as follows: (1) The interference is reduced in lossy materials. (2) The cross-polarization components are hardly changed between the lossless and the lossy materials at high obliquity. (3) Using the principle of a waveguide, $\sigma_{n,b}$ can be reduced by a multilayered elliptical

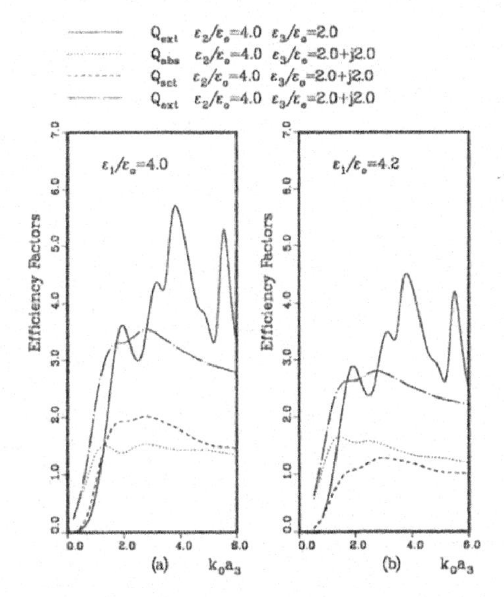

Fig. 8. Efficiency factors as a function of frequency for multilayered dielectric cylinder scattering, E wave incident, $a_1/b_1 = 10.0$, $a_2/b_2 = 5.0$, $a_3/b_3 = 2.0$, $a_1/a_2 = 0.98$, $a_2/a_3 = 0.88$, $\mu_1/\mu_0 = \mu_2/\mu_0 = \mu_3/\mu_0 = 1.0$, $\phi_0 = 90°$, $\theta_0 = 30°$: (*a*) $\varepsilon_1/\varepsilon_0 = 4.0$; (*b*) $\varepsilon_1/\varepsilon_0 = 4.2$.

cylinder. From those results described above the separation of variables method works well for the present scattering problems. However, the exact method is impractical in the geometrical optics region because a computer cannot handle the enormous dimension of matrix equations and the values of radial Mathieu functions. For the high-frequency region, many approximate methods are available, but no matter what method is used, the numerical results should be checked with the correct answers at the region where the exact solution is available. Therefore this research provides the ground base for the scattering of a high-frequency incident wave by an elliptical cylinder, because the separation of variables method can provide the correct numerical result around the beginning of geometrical optics region. With a proper approximate technique for high frequency the scattering problem can be solved completely without limitations of size, shape, material, and incident angle.

In extending this research the theoretical analysis and the computer program may be useful for the solution of the scattering problems of inhomogeneous or anisotropic elliptical cylinder as those for a homogeneous and isotropic circular cylinder have been used for the solutions of the scattering by an inhomogeneous or anisotropic circular cylinder [*Rusch and Yeh*, 1967; *Gallawa*, 1970; *Yeh and Wang*, 1972].

APPENDIX

A.1. Formulas for E_ξ, H_ξ, E_η and H_η in terms of E_z and H_z

At the region characterized with the parameter ε_k and μ_k,

$$E_{\xi,k} = \frac{jk_0 \cos\theta_0}{\kappa_k^2 p}\frac{\partial E_{z,k}}{\partial \xi} + \frac{j\omega\mu_k}{\kappa_k^2 p}\frac{\partial H_{z,k}}{\partial \eta} \tag{A1}$$

$$H_{\xi,k} = \frac{jk_0 \cos\theta_0}{\kappa_k^2 p}\frac{\partial H_{z,k}}{\partial \xi} - \frac{j\omega\varepsilon_k}{\kappa_k^2 p}\frac{\partial E_{z,k}}{\partial \eta} \tag{A2}$$

$$E_{\eta,k} = \frac{jk_0 \cos\theta_0}{\kappa_k^2 p}\frac{\partial E_{z,k}}{\partial \eta} - \frac{j\omega\mu_k}{\kappa_k^2 p}\frac{\partial H_{z,k}}{\partial \xi} \tag{A3}$$

$$H_{\eta,k} = \frac{jk_0 \cos\theta_0}{\kappa_k^2 p}\frac{\partial H_{z,k}}{\partial \eta} + \frac{j\omega\varepsilon_k}{\kappa_k^2 p}\frac{\partial E_{z,k}}{\partial \xi} \tag{A4}$$

A.2. Orthogonality properties of the Mathieu functions

Angular Mathieu functions have the following orthogonality relations:

$$\int_0^{2\pi} ce_m(\eta, \gamma_k^2)ce_n(\eta, \gamma_k^2)\, d\eta = 0, \tag{A5}$$

$$m \neq n,\ k = 0, 1, 2, \cdots$$

$$\int_0^{2\pi} se_m(\eta, \gamma_k^2)se_n(\eta, \gamma_k^2)\, d\eta = 0, \tag{A6}$$

$$m \neq n,\ k = 0, 1, 2, \cdots$$

$$\int_0^{2\pi} ce_m(\eta, \gamma_k^2)se_n(\eta, \gamma_k^2)\, d\eta = 0, \tag{A7}$$

$$\text{any } m, n, \text{ and } k = 0, 1, 2, \cdots$$

where m and n are nonnegative integers.

The orthogonality properties of angular Mathieu functions make it possible to express their derivatives or angular Mathieu functions with different parameters as expansions of angular Mathieu functions [*Yeh*, 1964]:

$$ce_n'(\eta) = \frac{\partial ce_n(\eta)}{\partial \eta} = \sum_{m=1}^{\infty} \nu_{n,m} se_m(\eta) \tag{A8}$$

$$se_n'(\eta) = \frac{\partial se_n(\eta)}{\partial \eta} = \sum_{m=0}^{\infty} \chi_{n,m} ce_m(\eta) \tag{A9}$$

$$ce_n^{(k)}(\eta) = \sum_{m=0}^{\infty} \alpha_{n,m}^{(k)} ce_m(\eta) \tag{A10}$$

$$se_n^{(k)}(\eta) = \sum_{m=1}^{\infty} \beta_{n,m}^{(k)} se_m(\eta) \tag{A11}$$

$$\frac{\partial ce_n^{(k)}(\eta)}{\partial \eta} = \sum_{m=0}^{\infty} \alpha_{n,m}^{(k)} \sum_{p=1}^{\infty} \nu_{m,p} se_p(\eta) \tag{A12}$$

$$\frac{\partial se_n^{(k)}(\eta)}{\partial \eta} = \sum_{m=1}^{\infty} \beta_{n,m}^{(k)} \sum_{p=0}^{\infty} \chi_{m,p} ce_p(\eta) \tag{A13}$$

Inversely, the coefficients can be expressed as follows:

$$\alpha_{n,m}^{(k)} = \frac{\int_0^{2\pi} ce_n^{(k)}(\eta) ce_m(\eta)\, d\eta}{\int_0^{2\pi} ce_m^2(\eta)\, d\eta} \tag{A14}$$

$$\beta_{n,m}^{(k)} = \frac{\int_0^{2\pi} se_n^{(k)}(\eta) se_m(\eta)\, d\eta}{\int_0^{2\pi} se_m^2(\eta)\, d\eta} \tag{A15}$$

$$\chi_{n,m} = \frac{\int_0^{2\pi} se_n'(\eta) ce_m(\eta)\, d\eta}{\int_0^{2\pi} ce_m^2(\eta)\, d\eta} \tag{A16}$$

$$\nu_{n,m} = \frac{\int_0^{2\pi} ce_n'(\eta) se_m(\eta)\, d\eta}{\int_0^{2\pi} se_m^2(\eta)\, d\eta} \tag{A17}$$

REFERENCES

Abramowitz, M., and Stegun, I. (Eds.), *Handbook of Mathematical Functions*, U. S. Government Printing Office, Washington, D. C., 1972.

Alexopoulous, N., and G. Tadler, Electromagnetic scattering from an elliptic cylinder loaded by continuous and discontinuous surface impedances, *J. Appl. Phys.*, *46*(3), 1128–1134, 1975.

Blanch, G., and D. Clemm, *Table of the Characteristic Values of Mathieu Differential Equations for Complex Parameters*, Aerospace Research Laboratory, Washington, D. C., 1969.

Bohren, C., and D. Huffman, *Absorption and Scattering of Light by Small Particles*, 530 pp., John Wiley, New York, 1983.

Bussey, H., and J. Richmond, Scattering by a lossy dielectric circular cylindrical multilayer, *IEEE Trans. Antennas Propag.*, *23*(5), 723–725, 1975.

Datta, A., and S. Som, Numerical study of the scattered electromagnetic field inside a hollow dielectric cylinder, 1, Scattering of a single beam, *Appl. Opt.*, *14*(7), 1516–1523, 1975.

Evans, L., J. Chen, and S. Churchill, Scattering of electromagnetic radiation by infinitely long, hollow and coated cylinder, *J. Opt. Soc. Am.*, *54*(8), 1004–1007, 1964.

Gallawa, R., Scattering from graded cylindrical media, *IEEE Trans. Antennas Propag.*, *18*(1), 136–139, 1970.

Goel, G., and D. Jain, Scattering of plane waves by a penetrable elliptic cylinder, *J. Acoust. Soc. Am.*, *69*(2), 371–379, 1981.

Kerker, M., *The Scattering of Light and Other Electromagnetic Radiation*, 666 pp., Academic, San Diego, Calif., 1969.

Kim, C., Scattering of an Obliquely Incident Wave by a Multi-Layered Elliptical Lossy Cylinder, 202 pp., Ph.D. dissertation, Univ. of Calif., Los Angeles, 1989.

Maxwell, J., A dynamical theory of the electromagnetic field, *Proc. R. Soc., London*, *43*, 531–536, 1864.

McLachlan, N. W., *Theory and Application of Mathieu Function*, 401 pp., Oxford University Press, London, 1951.

Morita, N., Another method of extending the boundary condition for the problem of scattering by dielectric cylinders, *IEEE Trans. Antennas Propag.*, *27*(1), 1979.

Morse, P., and P. Rubenstein, The diffraction of waves by ribbons and by slits, *Phys. Rev.*, *54*(12), 895–898, 1938.

National Bureau of Standards (NBS), *Table of the Characteristic Values of Mathieu Differential Equations*, U. S. Government Printing Office, Washington, D. C., 1945.

NBS, *Table of the Bessel Functions J0(z) and J1(z) for Complex Arguments*, Columbia University Press, New York, 1947.

NBS, *Table of the Bessel Functions J0(z) and J1(z) for Complex Arguments*, Columbia University Press, New York, 1950.

Rayleigh, L., On the electromagnetic theory of light, *Philos. Mag.*, *12*, 81–101, 1881.

Richmond, J., Scattering by a conducting elliptical cylinder with dielectric coating, *Radio Sci.*, *23*(6), 1061–1066, 1988.

Rusch, W., and C. Yeh, Scattering by an infinite cylinder coated with an inhomogeneous and anisotropic plasma sheath, *IEEE Trans. Antennas Propag.*, *15*(5), 452–459, 1967.

Stratton, J., *Electromagnetic Theory*, 615 pp., McGraw-Hill, New York, 1941.

Van de Hulst, H., On the attenuation of plane waves by obstacles of arbitrary size and form, *Physica*, *15*, 740–746, 1949.

Van de Hulst, H., *Light Scattering By Small Particle*, 470 pp., John Wiley, New York, 1957.

Wait, J., Scattering of a plane wave from a circular dielectric cylinder at oblique incidence, *Can. J. Phys.*, *33*(2), 189–195, 1955.

Yeh, C., The diffraction of waves by a penetrable ribbon, *J. Math. Phys.*, *4*(1), 65–71, 1963.

Yeh, C., Scattering of obliquely incident light waves by elliptical fibers, *J. Opt. Soc. Am.*, *54*(10), 1227–1231, 1964.

Yeh, C., Backscattering cross section of a dielectric elliptical cylinder, *J. Opt. Soc. Am.*, *55*(3), 309–314, 1965.

Yeh, C., and P. Wang, Scattering of obliquely incident waves by inhomogeneous fibers, *J. Appl. Phys.*, *43*(10), 3999–4006, 1972.

C. S. Kim, Submarine Electromagnetic Systems Department, Naval Underwater Systems Center, New London, CT 06320.

C. Yeh, Department of Electrical Engineering, University of California, Los Angeles, CA 90024.

Reprinted from JOURNAL OF THE OPTICAL SOCIETY OF AMERICA, Vol. 57, No. 2, 195-201, February 1967
Printed in U. S. A.

Diffraction of Waves by a Dielectric Parabolic Cylinder*

C. YEH
Department of Electrical Engineering, University of Southern California, Los Angeles, California 90007
(Received 20 May 1966)

The exact solution of the diffraction of waves by a dielectric parabolic cylinder is obtained. Results are given in terms of parabolic-cylinder functions. It is found that each expansion coefficient of the scattered or transmitted wave is coupled to all coefficients of the series expansion for the incident wave, even for very thin cylinders. Both polarizations of the incident wave are considered: one with the incident electric vector in the axial direction and the other with the incident magnetic vector in the axial direction. Numerical computations are carried out for the backscattering cross section and the radiation patterns of the scattered wave.
INDEX HEADINGS: Diffraction; Polarization; Scattering.

THE classic problem of the diffraction of waves by a perfectly conducting and infinitely thin half-plane was solved by Sommerfeld[1] by the use of a contour-integral representation many years ago. The corresponding problem for the diffraction by a finite-conducting half-plane or by a finite-conducting parabolic cylinder was considered by Epstein.[2] However, very few qualitative results were given by him. In recent years Fock[3] and Rice[4] obtained an asymptotic solution for a large perfectly conducting parabolic cylinder. Fock used an integral-equation approach while Rice made use of the Watson transformation technique.

If a thin half-plane is transparent, i.e., a dielectric half-plane, the behaviors of the diffracted wave and the penetrating wave are expected to be quite different than for a perfectly conducting or a highly conducting half-plane. The purpose of this study is to examine this dielectric half-plane problem. The exact solution of the diffraction of a plane electromagnetic wave by a dielectric parabolic cylinder is given in terms of series of parabolic-cylinder functions. The special case of a very thin parabolic cylinder is treated in detail. Numerical computations are carried out for the backscattering cross section as well as for the radiation pattern of the scattered wave.

The solutions with a slight modification are applicable to the problem of the diffraction of waves from a thin plasma half-plane, and the method of analysis is applicable to problems involving dielectric or plasma-coated parabolic cylinders.

I. FORMULATION OF THE PROBLEM

The parabolic-cylinder coordinates (ξ,η,z), which are shown in Fig. 1, are related to the rectangular coordinates (x,y,z) by the following expressions:

$$\begin{aligned} x&=\xi\eta,\\ y&=\tfrac{1}{2}(\eta^2-\xi^2),\\ z&=z,\\ (-\infty\leq\xi&\leq\infty,\ 0\leq\eta<\infty). \end{aligned}\tag{1}$$

The surfaces of constant η are confocal parabolic cylinders having their focus on the z axis. One of the confocal parabolic cylinders with $\eta=\eta_0$ is assumed to coincide with the boundary of the solid dielectric cylinder. The lines $\xi=$constant are halves of upward-curving confocal parabolas having their common focus on the z axis. ξ is positive in the half-plane $x>0$ and negative in $x<0$. We consider waves whose propagation vectors are in the x-y plane, so that the z coordinate may be omitted from the discussion. A possible solution of the wave equation is then

$$\exp(ik\eta^2/2)\cdot R_m(i^{-\frac{1}{2}}k^{\frac{1}{2}}\eta)\cdot\exp(-ik\xi^2/2)\cdot\Theta_m(i^{\frac{1}{2}}k^{\frac{1}{2}}\xi)\cdot\exp(i\omega t),$$

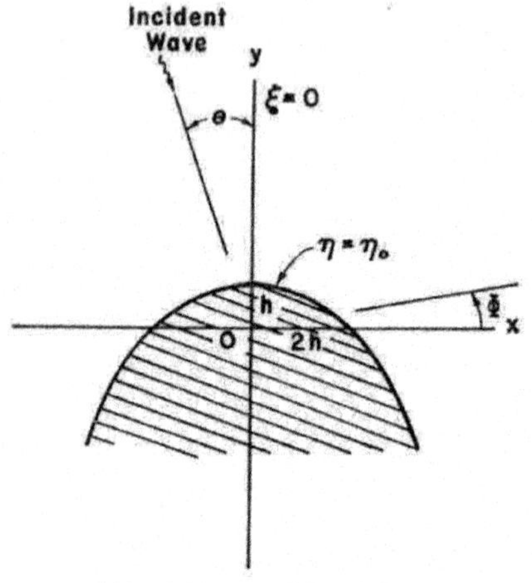

FIG. 1. Geometry of the problem.

* Work partly supported by the Joint Services Electronics Program and partly supported by the Naval Ordnance Test Station.
[1] A. Sommerfeld, Math. Ann. 47, 317 (1896).
[2] P. S. Epstein, dissertation, Munich (1914).
[3] V. A. Fock, J. Phys. (USSR) 10, 130 (1949); 10, 399 (1949).
[4] S. O. Rice, Bell System Tech. J. 33, 417 (1954).

where $R_m(z)$ and $\Theta_m(z)$ satisfy the differential equation

$$\frac{d^2T_m}{dz^2}-2z\cdot\frac{dT_m}{dz}+2mT_m=0, \tag{2}$$

where m are integers and k is the wave number.

The solutions of Eq. (2) have been investigated quite thoroughly by Epstein[3] and Rice.[4] The notations adopted by Rice[4] are followed. The proper choice of the solutions to represent the electromagnetic fields depends upon the boundary conditions. For the region within the dielectric cylinder, all field components must be finite. All field components for the scattered wave must satisfy the Sommerfeld radiation condition at infinity. Consequently, the appropriate solution of the wave equation for the region inside the dielectric cylinder is

$$\{S_m^{(1)}(\eta_1')S_m^{(1)}(\xi_1')\} \tag{3}$$

and that for the scattered wave is

$$\{S_m^{(3)}(\eta_0')S_m^{(1)}(\xi_0')\}, \tag{4}$$

where

$$S_m^{(1)}(z)=\exp(-z^2/2)\cdot U_m(z)$$

and

$$S_m^{(3)}(z)=\exp(-z^2/2)\cdot W_m(z);$$

$U_m(z)$ and $W_m(z)$ are defined in Ref. 4; $\eta_1'=(k_1/i)^{\frac{1}{2}}\eta$ and $\xi_0'=(ik_0)^{\frac{1}{2}}\xi$, with $k_0^2=\omega^2\mu\epsilon_0$ and $k_1^2=\omega^2\mu\epsilon_1$; ϵ_0 and ϵ_1 are, respectively, the dielectric constant of the free space and of the parabolic cylinder.

II. SCATTERING OF A NORMALLY INCIDENT PLANE WAVE

Two types of normally incident waves are possible. One, called an E wave, is defined by $H_z=0$, and the other, called an H wave, is defined by $E_z=0$.

It can be shown that the incident field of a plane wave with propagation vector in the x–y plane and at an angle θ with the y axis is proportional to the factor

$$\begin{aligned}&\exp[-ik_0(x\sin\theta-y\cos\theta)]\\&=\exp\{-ik_0[\xi\eta\sin\theta-\tfrac{1}{2}(\eta^2-\xi^2)\cos\theta]\}\\&=\sec\frac{\theta}{2}\sum_{m=0}^{\infty}m!\left(-\frac{i}{2}\tan\frac{\theta}{2}\right)^m S_m^{(1)}(\eta_0')S_m^{(1)}(\xi_0').\end{aligned} \tag{5}$$

The above series converges when $|\theta|<\pi/2$.

E Wave

The axial components of an incident E wave are:

$$E_z^i=[\text{the right-hand side of Eq. (5)}] \tag{6}$$

$$H_z^i=0. \tag{7}$$

Referring to (3) and (4), we see that the scattered axial fields and the transmitted axial fields inside the dielectric cylinder must be of the form

$$E_z^s=\sum_{m=0}^{\infty}A_mS_m^{(3)}(\eta_0')S_m^{(1)}(\xi_0') \tag{8}$$

$$H_z^s=0, \tag{9}$$

and

$$E_z^t=\sum_{m=0}^{\infty}B_mS_m^{(1)}(\eta_1')S_m^{(1)}(\xi_1') \tag{10}$$

$$H_z^t=0, \tag{11}$$

where A_m and B_m are arbitrary unknown coefficients that can be determined by applying the boundary conditions.

The boundary conditions require the continuity of the tangential components of the electric and magnetic field at the boundary surface $\eta=\eta_0$; i.e.,

$$\sum_{m=0}^{\infty}{}'\,m!\sec\left(\frac{\theta}{2}\right)\left(-\frac{i}{2}\tan\frac{\theta}{2}\right)^m S_m^{(1)}(\eta_0')\alpha_{nm}+A_mS_m^{(3)}(\eta_0')\alpha_{nm}=B_nS_n^{(1)}(\eta_1') \tag{12}$$

$$\sum_{m=0}^{\infty}{}'\,m!\sec\left(\frac{\theta}{2}\right)\left(-\frac{i}{2}\tan\frac{\theta}{2}\right)^m i^{-\frac{1}{2}}\frac{d}{d\eta_0'}[S_m^{(1)}(\eta_0')]\alpha_{nm}+A_mi^{-\frac{1}{2}}\frac{d}{d\eta_0'}[S_m^{(3)}(\eta_0')]\alpha_{nm}=B_n\left(\frac{k_1}{k_0}\right)^{\frac{1}{2}}i^{-\frac{1}{2}}\frac{d}{d\eta_1'}[S_n^{(1)}(\eta_1')] \tag{13}$$

in which the orthogonality relation of the function $S_n^{(1)}(\xi_1')$ has been used. α_{nm} is given in Appendix A. The prime on the summation sign indicates that only odd integers of m are summed if n is odd and only even integers of m are summed if n is even. Eliminating the constant B_n from the above equations gives

$$\sum_{m+0}^{\infty}{}'\,P_{nm}A_m=Q_n,\quad(n=0,1,\cdots), \tag{14}$$

where

$$P_{nm}=\alpha_{nm}\left\{S_n^{(1)}(\eta_1')\frac{d}{d\eta}[S_m^{(3)}(\eta_0')]-S_m^{(3)}(\eta_0')\frac{d}{d\eta}[S_n^{(1)}(\eta_1')]\right\}$$

$$Q_n=\sum_{m=0}^{\infty}{}'\,m!\sec\left(\frac{\theta}{2}\right)\left(-\frac{i}{2}\tan\frac{\theta}{2}\right)^m\alpha_{nm}\left\{S_m^{(1)}(\eta_0')\frac{d}{d\eta}[S_n^{(1)}(\eta_1')]-S_n^{(1)}(\eta_1')\frac{d}{d\eta}[S_m^{(1)}(\eta_0')]\right\}. \tag{15}$$

Equation (14) is an infinite set of linear algebraic equations from which A_m may be obtained by the method of successive approximation.

H Wave

A similar approach may be carried out for an incident H wave

$$\sum_{m=0}^{\infty}{}' G_{nm}C_m = K_n \quad (n=0, 1, 2 \cdots), \tag{16}$$

where

$$G_{nm} = \alpha_{nm}\left\{S_n^{(1)}(\eta_1')\frac{d}{d\eta}[S_m^{(3)}(\eta_0')] - \frac{\epsilon_0}{\epsilon_1}S_m^{(3)}(\eta_0')\frac{d}{d\eta}[S_n^{(1)}(\eta_1')]\right\}$$

$$K_n = \sum_{m=0}^{\infty}{}' m!\sec\left(\frac{\theta}{2}\right)\left(-\frac{i}{2}\tan\frac{\theta}{2}\right)^m \alpha_{mn}\left\{\frac{\epsilon_0}{\epsilon_1}S_m^{(1)}(\eta_0')\frac{d}{d\eta}[S_n^{(1)}(\eta_1')] - S_n^{(1)}(\eta_1')\frac{d}{d\eta}[S_m^{(1)}(\eta_0')]\right\}, \tag{17}$$

where C_m are the expansion coefficients for the scattered field.

At a large distance from the dielectric cylinder, it is permissible to use the following asymptotic expressions for the parabolic-cylinder functions when $k_0 r \gg 1$ and $k_0 r \gg M$, where M is the order of the parabolic-cylinder functions[4]

$$W_m(\eta_0') \sim i(\eta_0')^{-m-1}\exp(\eta_0'^2)/2\pi^{\frac{1}{2}} \tag{18}$$

$$U_m(\xi_0') \sim 2^m \xi_0'^m/m!. \tag{19}$$

Using the above equations, we obtain the expressions for the far-zone scattered field

$$E_z^s(E\text{ wave}) \approx \frac{i}{2}\left(\frac{i}{2\pi k_0 r}\right)^{\frac{1}{2}}\exp(-ik_0 r)\left[\sum_{m=0}^{\infty} A_m \frac{(2i)^m \cot^m(\frac{1}{2}\varphi+\frac{1}{4}\pi)}{m!\ \sin(\frac{1}{2}\varphi+\frac{1}{4}\pi)}\right], \tag{20}$$

$$H_z^s(H\text{ wave}) \approx \frac{i}{2}\left(\frac{i}{2\pi k_0 r}\right)^{\frac{1}{2}}\exp(-ik_0 r)\left[\sum_{m=0}^{\infty} C_m \frac{(2i)^m \cot^m(\frac{1}{2}\varphi+\frac{1}{4}\pi)}{m!\ \sin(\frac{1}{2}\varphi+\frac{1}{4}\pi)}\right]. \tag{21}$$

This completes the derivation of the fundamental formulas involved in the diffraction of waves by a parabolic dielectric cylinder.

III. SCATTERING BY A THIN PENETRABLE HALF-PLANE

The thin penetrable half-plane corresponds to the limiting case of a parabolic cylinder of very small thickness, $\eta_0 \to 0$.

According to the theory of parabolic-cylinder functions[4] we find that if $k_0^{\frac{1}{2}}\eta_0 \ll 1$ and $k_1^{\frac{1}{2}}\eta_0 \ll 1$,

$$U_n(q) \approx \frac{1}{\Gamma(1+\frac{1}{2}n)}\cos\left(\frac{n\pi}{2}\right) + \frac{2q}{\Gamma[(1+n)/2]}\sin\left(\frac{n\pi}{2}\right); \tag{22}$$

$$W_n(q) \approx \frac{-i^n}{2\Gamma(1+\frac{1}{2}n)} - \frac{qi^{n-1}}{\Gamma[(1+n)/2]} \tag{23}$$

$$\frac{d}{dq}U_n(q) \approx \frac{2}{\Gamma[(1+n)/2]}\sin\left(\frac{n\pi}{2}\right) - \frac{nq}{\Gamma(1+\frac{1}{2}n)}\cos\left(\frac{n\pi}{2}\right) \tag{24}$$

$$\frac{d}{dq}W_n(q) \approx -\frac{i^{n-1}}{\Gamma[(1+n)/2]} + \frac{nqi^n}{\Gamma(1+\frac{1}{2}n)}, \tag{25}$$

where q may be $i^{-\frac{1}{2}}k_0^{\frac{1}{2}}\eta_0$ or $i^{-\frac{1}{2}}k_1^{\frac{1}{2}}\eta_0$. Substituting the above expressions into Eqs. (15) and (17) and carrying out the algebraic manipulation, we have

$$P_{nm}\approx i^{-\frac{1}{2}}k_0^{\frac{1}{2}}\alpha_{nm}i^m\left[\frac{-i^{-1}\cos(n\pi/2)}{\Gamma(1+\frac{1}{2}n)\Gamma[(1+m)/2]}-\left(\frac{k_1}{k_0}\right)^{\frac{1}{2}}\frac{\sin(n\pi/2)}{\Gamma[(1+n)/2]\Gamma(1+\frac{1}{2}m)}\right]$$

$$+i^{-\frac{1}{2}}k_0^{\frac{1}{2}}\eta_0\left[\frac{\cos(n\pi/2)}{\Gamma(1+\frac{1}{2}n)\Gamma(1+\frac{1}{2}m)}\left((1+m)-\frac{k_1}{k_0}\frac{(1+n)}{2}\right)+\frac{2i^{-1}\sin(n\pi/2)}{\Gamma[(1+m)/2]\Gamma[(1+n)/2]}\left(1-\frac{k_1^{\frac{1}{2}}}{k_0^{\frac{1}{2}}}\right)\right]$$

$$Q_n=i^{-\frac{1}{2}}k_0^{\frac{1}{2}}\eta_0\sum_{m=0}^{\infty}{}'\,m!\sec\frac{\theta}{2}\left(-\frac{i}{2}\tan\frac{\theta}{2}\right)^m\alpha_{nm}\left\{\frac{\cos(n\pi/2)\cos(m\pi/2)}{\Gamma(1+\frac{1}{2}n)(1+\frac{1}{2}m)}\left[(1+m)-(1+n)\frac{k_1}{k_0}\right]\right\}+0(k_0\eta_0^2) \quad (26)$$

and

$$G_{nm}=i^{-\frac{1}{2}}k_0^{\frac{1}{2}}\alpha_{nm}i^m\left(\left[\frac{-i^{-1}\cos(n\pi/2)}{\Gamma(1+\frac{1}{2}n)\Gamma[(1+m)/2]}-\left(\frac{\epsilon_0}{\epsilon_1}\right)^{\frac{1}{2}}\frac{\sin(n\pi/2)}{\Gamma[(1+n)/2]\Gamma(1+\frac{1}{2}m)}\right]+i^{-\frac{1}{2}}k_0^{\frac{1}{2}}\eta_0\right.$$

$$\left.\times\left\{\frac{\cos(n\pi/2)}{\Gamma(1+\frac{1}{2}n)\Gamma(1+\frac{1}{2}m)}\left[(1+m)-\left(\frac{\epsilon_0}{\epsilon_1}\right)^{\frac{1}{2}}\frac{(1+n)}{2}\right]+\frac{2i^{-1}\sin(n\pi/2)}{\Gamma[(1+m)/2]\Gamma[(1+n)/2]}\left[1-\left(\frac{\epsilon_0}{\epsilon_1}\right)^{\frac{1}{2}}\right]\right\}\right)$$

$$K_n=i^{-\frac{1}{2}}k_0^{\frac{1}{2}}(i^{-\frac{1}{2}}k_0^{\frac{1}{2}}\eta_0)\sum_{m=0}^{\infty}{}'\,m!\sec\frac{\theta}{2}\left(-\frac{i}{2}\tan\frac{\theta}{2}\right)^m\alpha_{nm}$$

$$\times\left\{\frac{\cos(n\pi/2)\cos(m\pi/2)}{\Gamma(1+\frac{1}{2}n)\Gamma(1+\frac{1}{2}m)}\left[(1+m)-\left(\frac{\epsilon_0}{\epsilon_1}\right)^{\frac{1}{2}}(1+n)\right]+\frac{4(\epsilon_1/\epsilon_0)^{\frac{1}{2}}\sin(m\pi/2)\sin(n\pi/2)}{\Gamma[(1+n)/2]\Gamma[(1+m)/2]}\left[\frac{\epsilon_0}{\epsilon_1}-1\right]\right\}+0(k_0\eta_0^2). \quad (27)$$

The coefficients A_m for the E wave and C_m for the H wave can be obtained, respectively, by solving Eqs. (14) and (16) using Eqs. (26) and (27). Very little simplification is achieved by considering this a special case of a thin penetrable half-plane. Unlike the case of a perfectly conducting half-plane (i.e., Sommerfeld's problem), it does not seem possible to express the solution for the present penetrable case in a closed form. However, we may deduce from the above expressions the orders of magnitude of the expansion coefficients for the scattered wave. For example, for an incident E wave,

$$P_{nm}=0(1),\quad Q_n=0(k_0^{\frac{1}{2}}\eta_0)$$

for $n=0, 2, 4 \cdots$ and $m=0, 2, 4 \cdots$, thus

$$A_m=0(k_0^{\frac{1}{2}}\eta_0)\quad \text{with}\quad m=0, 2, 4 \cdots; \quad (28)$$

and

$$P_{nm}=0(1),\quad Q_n=0(k_0\eta_0^2),$$

for $n=1, 3, 5 \cdots$ and $m=1, 3, 5 \cdots$, thus

$$A_m=0(k_0\eta_0^2)\quad \text{with}\quad m=1, 3, 5, \cdots. \quad (29)$$

For an incident H wave, $G_{nm}=0(1)$ and $K_n=0(k_0^{\frac{1}{2}}\eta_0)$ for $n=0, 1, 2, 3 \cdots$ and $m=0, 1, 2, 3 \cdots$, thus

$$C_m=0(k_0^{\frac{1}{2}}\eta_0)\quad \text{with}\quad m=0, 1, 2 \cdots. \quad (30)$$

If $\eta_0=0$ (i.e., zero thickness) or if $\epsilon_1/\epsilon_0=1$, the scattering coefficients A_m for E wave and C_m for H wave are identically zero, as expected.

Because of the complicated interaction between the incident electromagnetic wave with the penetrable parabolic cylinder and the involved expressions for the scattered waves even for the special case of a very thin parabolic cylinder, numerical computations are reported in the following section.

IV. NUMERICAL COMPUTATIONS

To illustrate qualitatively how the solutions behave, the radiation patterns of the scattered fields as well as the backscattering cross section of a dielectric parabolic cylinder are computed. The backscattering cross section is defined as the ratio of the total power scattered, by a fictitious isotropic scatterer which scatters energy in all directions with intensity equal to that scattered directly back toward the source by the actual scattering object, to the power incident per unit area on the scatterer,[8] i.e.,

$$\sigma_B=2\pi r(P_B^s/S^i), \quad (31)$$

where P_B^s is the power density of the far-zone scattered field in the direction of source and S^i is the incident power density. For an incident E wave

$$\sigma_B^E=\frac{1}{4k_0}\left|\sum_{m=0}^{\infty}A_m\frac{(2i)^m\cot^m(\frac{1}{2}\theta+\frac{1}{4}\pi)}{m!\ \sin(\frac{1}{2}\theta+\frac{1}{4}\pi)}\right|^2, \quad (32)$$

and for an incident H wave

$$\sigma_B^H=\frac{1}{4k_0}\left|\sum_{m=0}^{\infty}C_m\frac{(2i)^m\cot^m(\frac{1}{2}\theta+\frac{1}{4}\pi)}{m!\ \sin(\frac{1}{2}\theta+\frac{1}{4}\pi)}\right|, \quad (33)$$

where θ has been defined in Fig. 1.

[8] R. King and T. T. Wu, *The Scattering and Diffraction of Waves* (Harvard University Press, Cambridge, Massachusetts, 1959).

TABLE I. Tabulation of $(\{[1+(\epsilon_0/\epsilon_1)^{\frac{1}{2}}]/2\}^{\frac{1}{2}}2^n/n!)\alpha_{nm}$ with $\epsilon_1/\epsilon_0=2.0$.

$n \backslash m$	0	2	4	6	8	10	12
0	1.000	−0.1716	0.1472×10^{-1}	-0.8418×10^{-2}	0.3611×10^{-4}	-0.1239×10^{-5}	0.3543×10^{-7}
2	0.1716	1.912	−0.3305	0.2843×10^{-1}	-0.1628×10^{-2}	0.6987×10^{-4}	-0.2399×10^{-5}
4	0.1472×10^{-1}	0.3305	0.5711	−0.1029	0.8963×10^{-2}	-0.5166×10^{-3}	0.2226×10^{-4}
6	0.8418×10^{-2}	0.2843×10^{-1}	0.1029	0.6320×10^{-1}	-0.1238×10^{-1}	0.1106×10^{-2}	-0.6455×10^{-4}
8	0.3611×10^{-4}	0.1628×10^{-2}	0.8963×10^{-2}	0.1238×10^{-1}	0.3376×10^{-2}	-0.7691×10^{-3}	0.7152×10^{-4}
10	0.1239×10^{-5}	0.6987×10^{-4}	0.5166×10^{-3}	0.1106×10^{-2}	0.7691×10^{-3}	0.9434×10^{-4}	-0.2846×10^{-4}
12	0.3543×10^{-7}	0.2399×10^{-5}	0.2226×10^{-4}	0.6455×10^{-4}	0.7152×10^{-4}	0.2846×10^{-4}	0.1157×10^{-5}

$n \backslash m$	1	3	5	7	9	11	13
1	1.970	−0.3381	0.2900×10^{-1}	-0.1659×10^{-2}	0.7114×10^{-4}	-0.2441×10^{-5}	0.6981×10^{-7}
3	0.3381	1.217	−0.2138	0.1848×10^{-1}	-0.1061×10^{-2}	0.4561×10^{-4}	-0.1568×10^{-5}
5	0.2900×10^{-1}	0.2138	0.2104	-0.3296×10^{-1}	0.3459×10^{-2}	-0.2005×10^{-3}	0.8665×10^{-5}
7	0.1659×10^{-2}	0.1848×10^{-1}	0.3926×10^{-1}	0.1586×10^{-1}	-0.3316×10^{-2}	0.3016×10^{-3}	-0.1775×10^{-4}
9	0.7114×10^{-4}	0.1061×10^{-2}	0.3459×10^{-2}	0.3316×10^{-2}	0.6128×10^{-3}	-0.1569×10^{-3}	0.1498×10^{-4}
11	0.2441×10^{-5}	0.4561×10^{-4}	0.2005×10^{-3}	0.3016×10^{-3}	0.1569×10^{-3}	0.1200×10^{-4}	-0.4630×10^{-5}
13	0.6981×10^{-7}	0.1568×10^{-5}	0.8665×10^{-5}	0.1775×10^{-4}	0.1498×10^{-4}	0.4630×10^{-5}	0.5321×10^{-7}

Numerical values for the functions $U_n(z)$ and $W_n(z)$ are computed from the series expression given in Ref. 4. α_{nm} are calculated from Eq. (A2). Some representative values for α_{nm} with $\epsilon_1/\epsilon_0=2.0$ are given in Table I. It can be seen that α_{nm} gets smaller as $|n-m|$ becomes larger. The expansion coefficients for the scattered wave, A_m and C_m, are computed from Eqs. (14) and (16), respectively, by the method of successive approximation.[6] This method, described in Ref. 6, may be used provided that the matrix coefficients, i.e., P_{nm} or G_{nm}, satisfy the conditions $\sum_{n,m}|P_{n,m}|^2<\infty$ and $\sum_{n,m}|G_{nm}|^2<\infty$. With the help of the asymptotic expressions for the parabolic cylinder functions given by Rice,[4] the above conditions on the matrix coefficients can be verified. The success of the method of successive approximation in solving the two sets of infinite simultaneous equations [i.e., Eqs. (14) and (16)] can be demonstrated numerically. The first 3 simultaneous equations were used to calculate, say A_m, and then $\sigma_B{}^E$. Then the first 4 simultaneous equations were used and then the first 5 equations, etc., until the desired accuracy for $\sigma_B{}^E$ was reached. For the range of $k_0{}^{\frac{1}{2}}\eta_0$ considered, at no time were more than ten simultaneous equations required to achieve an accuracy of three significant figures. Numerical investigation shows that the infinite series representing the scattered wave, i.e., Eqs. (20) and (21), converge quite rapidly for $0\leq\varphi\leq\pi$; only the first few terms of the series are needed. It appears that the rate of convergence of the series depends, to a great extent, on the direction of the point of observation. More and more terms of the series are required as φ decreases from 0° to −90° or as φ increases from 180° to 270°. Referring to Eqs. (20) and (21), we see that if $0\leq\varphi\leq\pi$, then $|\cot(\frac{1}{2}\varphi+\frac{1}{4}\pi)|\leq1$ and if $-\frac{1}{2}\pi<\varphi\leq0$ or if $\pi\leq\varphi<\frac{3}{2}\pi$, then $|\cot(\frac{1}{2}\varphi+\frac{1}{4}\pi)|\geq1$. Consequently, the quantity $|\cot^m(\frac{1}{2}\varphi+\frac{1}{4}\pi)|$ decreases as m increases when $0\leq\varphi\leq\pi$, thus helping the convergence of the

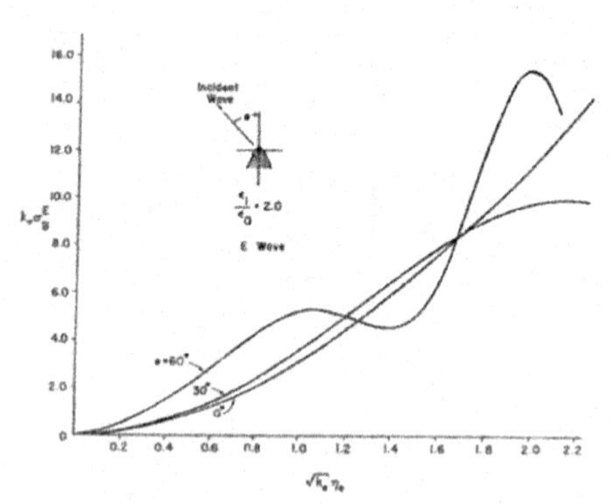

FIG. 2. Normalized backscattering cross section for an incident E wave.

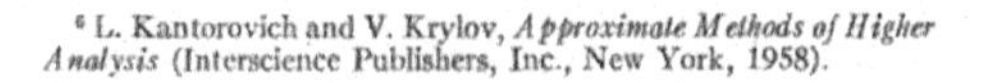

[6] L. Kantorovich and V. Krylov, *Approximate Methods of Higher Analysis* (Interscience Publishers, Inc., New York, 1958).

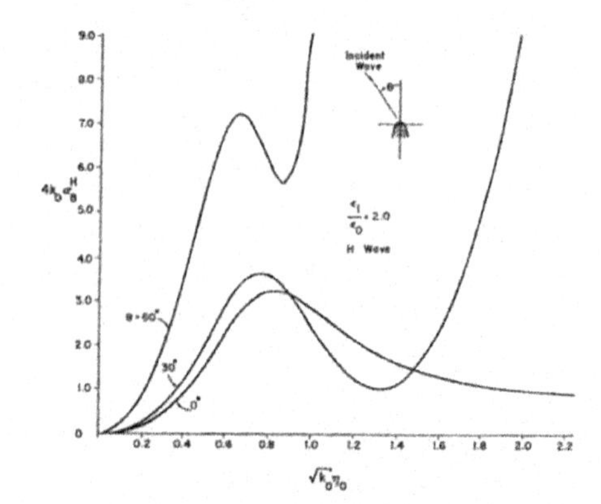

FIG. 3. Normalized backscattering cross section for an incident H wave.

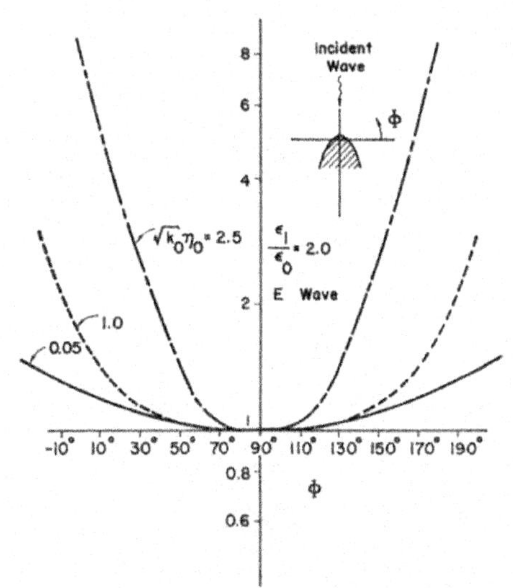

FIG. 4. Normalized magnitude of the scattered wave for an incident E wave at an angle of $\theta=0°$.

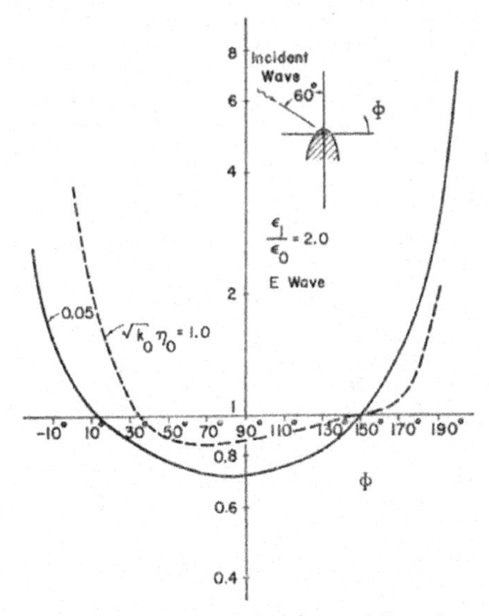

FIG. 6. Normalized magnitude of the scattered wave for an incident E wave at an angle of $\theta=60°$.

series. On the other hand when $-\pi/2<\varphi\leq 0$ or when $\pi\leq\varphi<3\pi/2$, the quantity $|\cot^m(\frac{1}{2}\varphi+\frac{1}{4}\pi)|$ increases as m increases, thus slowing the rate of convergence of the series.

The backscattering cross sections as a function of $k_0^{\frac{1}{2}}\eta_0$ for the two polarizations of the incident waves have been computed for various angles of incidence. It was assumed that $\epsilon_1/\epsilon_0=2.0$. Results are presented in Figs. 2–3 for three directions of incidence: $\theta=0°$, $30°$, $60°$.

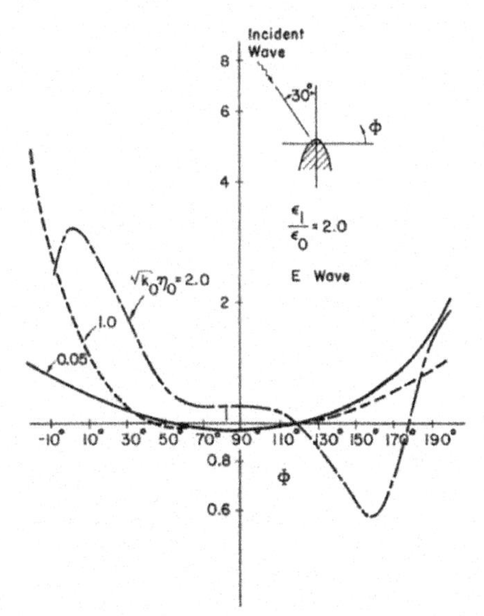

FIG. 5. Normalized magnitude of the scattered wave for an incident E wave at an angle of $\theta=30°$.

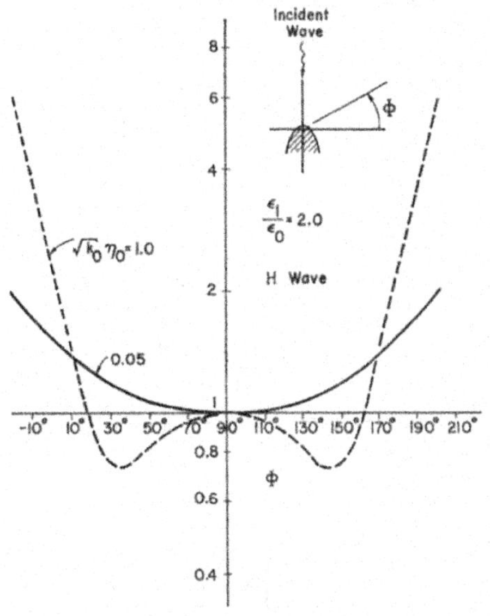

FIG. 7. Normalized magnitude of the scattered wave for an incident H wave at an angle of $\theta=0°$.

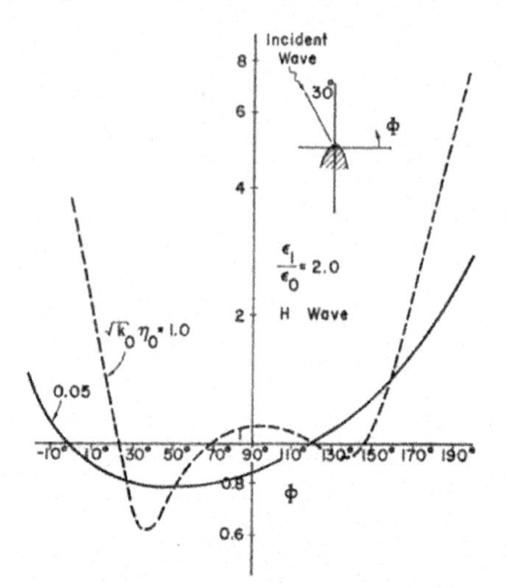

FIG. 8. Normalized magnitude of the scattered wave for an incident H wave at an angle of $\theta=30°$.

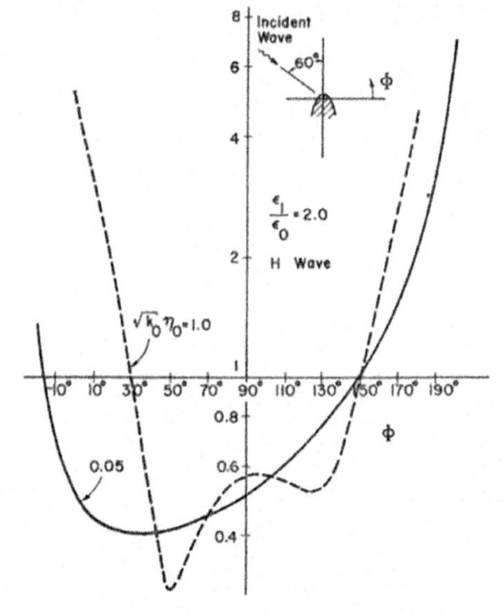

FIG. 9. Normalized magnitude of the scattered wave for an incident H wave at an angle of $\theta=60°$.

These figures show that when $k_0^{\frac{1}{2}}\eta_0 \ll 1$, the normalzied backscattering cross section, $4k_0\sigma_B{}^E$ or $4k_0\sigma_B{}^H$, is directly proportional to $k_0\eta_0{}^2$ as expected from the analysis on the scattering by a thin dielectric half-plane given in Sec. III. At low frequencies, the backscattering cross section increases as the angle of incidence θ increases from 0° which is the head-on incidence case. This is probably due to an increase of cross-sectional view as "seen" by the incident wave. As frequency increases, the general oscillatory behavior of the backscattering cross section of dielectric obstacles can be seen from these figures. Resonances are also shown for the backscattering cross section at smaller values of $k_0^{\frac{1}{2}}\eta_0$ as θ increases from 0°.

Figs. 4–9 show the radiation patterns of the scattered fields for the two polarizations of the incident waves. Three directions of incidence are considered: $\theta=0°$, 30°, 60°. Various values of $k_0^{\frac{1}{2}}\eta_0$ are used. It is again assumed that $\epsilon_1/\epsilon_0=2.0$. For the head-on incidence cases, the relative magnitude of the scattered wave becomes greater as the direction of the point of observation goes away from the head-on direction; and $|E_z|$ or $|H_z|$ grows greater faster as $k_0^{\frac{1}{2}}\eta_0$ increases. The evidence of trapped surface waves along the interface of the parabolic dielectric cylinder with considerable leakage is quite clear.

For $\theta\neq 0°$ cases (i.e., for the non-head-on incidence cases) the radiation patterns are no longer symmetric about the y axis. However, as expected, the radiation patterns are still relatively symmetric about the y axis and insensitive to the direction of incident wave when $k_0^{\frac{1}{2}}\eta_0 \ll 1$, i.e., for very thin dielectric half-plane.

APPENDIX A

Formula for α_{nm}

α_{nm} is defined as follows:

$$\alpha_{nm}=\frac{1}{N_n}\int_{-\infty}^{\infty} S_m{}^{(1)}(\xi_0')S_n{}^{(1)}(\xi_1')d\xi_1', \qquad \text{(A1)}$$

where

$$N_n=\int_{-\infty}^{\infty}[S_n{}^{(1)}(\xi_1')]^2 d\xi_1'=2^n\pi^{\frac{1}{2}}/n!.$$

The integral in Eq. (A1) can be evaluated using the technique described in Ref. 7. We have

$$\alpha_{nm}=\frac{n!}{2^n}\left[\frac{2}{1+(k_0/k_1)}\right]^{\frac{1}{2}}\sum_{r=0}^{(n,m)}\left[1\Big/\left(\frac{n-r}{2}\right)!\left(\frac{m-r}{2}\right)!r!\right]$$
$$\times\alpha^{(n-r)/2}\beta^{(m-r)/2}\gamma, \qquad \text{(A2)}$$

$\alpha=(1-c)/(1+c)$, $\beta=(c-1)/(c+1)$, $\gamma=4c^{\frac{1}{2}}/(c+1)$, and $c=k_0/k_1$; the summation sign indicates that if n is even and m is even then even r is summed from 0 to n if $n<m$ or from 0 to m if $m<n$; if n is odd and m is odd then odd r is summed from 1 to n if $n<m$ or from 1 to m if $m<n$; if n is odd and m is even or if n is even and m is odd, $\alpha_{nm}=0$.

[7] C. Yeh, J. Math. & Phys. 45, 231 (1966).

Reprinted from JOURNAL OF MATHEMATICS AND PHYSICS
Vol. XLV, No. 2, June, 1966
Printed in U.S.A.

A NOTE ON INTEGRALS INVOLVING PARABOLIC CYLINDER FUNCTIONS*

BY C. YEH

The purpose of this note is to present an evaluation of the integral

$$I_{nm}(a, b) = \int_{-\infty}^{\infty} U_n(z) U_m(az) e^{-bz^2} \, dz \tag{1}$$

where a and b are constants and n and m are non-negative integers. $U_n(z)$ is related to the parabolic cylinder function $D_n(z)$[1,2] through the equation

$$U_n(z) = 2^{n/2} e^{z^2/2} D_n(2^{\frac{1}{2}} z)/\Gamma(n + 1); \tag{2}$$

$U_n(z)$ is also related to Hermite's polynomial $H_n(z)$[1,2] through the equation

$$U_n(z) = H_n(z)/n! \tag{3}$$

and $H_n(z)$ satisfies the differential equation

$$\frac{d^2 U_n(z)}{dz^2} - 2z \frac{dU_n(z)}{dz} + 2nU_n(z) = 0. \tag{4}$$

The integral I_{nm} occurs in problems involving parabolic cylindrical boundaries, whether in potential theory, or in acoustic or electromagnetic wave diffraction theory. For example, in the problem of diffraction of waves by penetrable parabolic cylinder,[3] it is necessary to expand one set of parabolic cylinder function in terms of another set, i.e.,

$$U_n(z) e^{-z^2/2} = \textstyle\sum_{m=0}^{\infty} \alpha_{nm} U_m(az) e^{-a^2 z^2/2} \tag{5}$$

where

$$\alpha_{nm} = \frac{I_{nm}[a, \frac{1}{2}(1 + a^2)]a}{I_{mm}(1, 1)}. \tag{6}$$

Making use of the generating function for Hermite's polynomial,[1,2] we have

$$e^{-s^2+2sz} = \textstyle\sum_{n=0}^{\infty} U_n(z) s^n, \qquad e^{-t^2+2atz} = \textstyle\sum_{m=0}^{\infty} U_m(az) t^m. \tag{7, 8}$$

Combining Eqs. (7) and (8) gives

$$e^{-(s^2+t^2)+2(s+at)z} = \textstyle\sum_{n=0}^{\infty} \sum_{m=0}^{\infty} U_n(z) U_m(az) s^n t^m. \tag{9}$$

Multiplying both sides of (9) by $e^{-bz^2}\,dz$ and integrating with respect to z from $-\infty$ to $+\infty$, one obtains

$$\int_{-\infty}^{\infty} e^{-(s^2+t^2)+2(s+at)z-bz^2} \, dz = \sum_{n=0}^{\infty} \sum_{m=0}^{\infty} s^n t^m \int_{-\infty}^{\infty} U_n(z) U_m(az) e^{-bz^2} \, dz. \tag{10}$$

The integral on the left hand side of Eq. (10) can be evaluated immediately;

* Supported by the Naval Ordnance Test Station.

it is

$$\sqrt{\frac{\pi}{b}} \exp\left\{\frac{1}{b}[s^2(1-b)+t^2(a^2-b)+2ast]\right\}$$
$$= \sqrt{\frac{\pi}{b}} \sum_{p=0}^{\infty}\sum_{q=0}^{\infty}\sum_{r=0}^{\infty} \frac{1}{p!\,q!\,r!}\left\{\left(\frac{1-b}{b}\right)^p \left(\frac{a^2-b}{b}\right)^q \left(\frac{2a}{b}\right)^r s^{2p}t^{2q}(st)^r\right\}. \tag{11}$$

If equal powers of s and t are equated in the series on the right sides of Eqs. (10) and (11), we obtain

$$I_{nm}(a,b) = \int_{-\infty}^{\infty} U_n(z)U_m(az)e^{-bz^2dz}$$
$$= \sum_{r=0,1}^{(n,m)} \frac{\sqrt{\pi/b}}{[\frac{1}{2}(n-r)]![\frac{1}{2}(m-r)]!\,r!}\left[\left(\frac{1-b}{b}\right)^{\frac{1}{2}(n-r)}\left(\frac{a^2-b}{b}\right)^{\frac{1}{2}(m-r)}\left(\frac{2a}{b}\right)^r\right] \tag{12}$$

where the summation sign means that if n is even and m is even then even r is summed from 0 to n if $n < m$ and from 0 to m if $m < n$; if n is odd and m is odd then odd r is summed from 1 to n if $n < m$ and from 1 to m if $m < n$; if n is odd and m is even or if n is even and m is odd the integral $I_{nm}(a,b) = 0$.

Referring to (6), one has

$$\alpha_{nm} = \frac{m!}{\sqrt{\pi}2^m}\sum_{r=0,1}^{(n,m)} \frac{\sqrt{2\pi/1+a^2}\,a}{[\frac{1}{2}(n-r)]![\frac{1}{2}(m-r)]!\,r!}\left[\left(\frac{1-a^2}{1+a^2}\right)^{\frac{1}{2}(n-r)} \cdot \left(\frac{a^2-1}{a^2+1}\right)^{\frac{1}{2}(m-r)}\left(\frac{4a}{1+a^2}\right)^r\right]. \tag{13}$$

REFERENCES

1. E. T. WHITTAKER AND G. N. WATSON, "Modern Analysis", Cambridge Univ. Press (1927).
2. A. ERDELYI, "Higher Transcendental Functions", McGraw-Hill Book Co., Inc., New York (1953).
3. C. YEH, "Diffraction of Electromagnetic Waves by Penetrable Parabolic Cylinders" USCEC Rept., Elec. Eng. Dept., Univ. of So. Calif., Jan., 1965.

ELECTRICAL ENGINEERING DEPARTMENT
UNIVERSITY OF SOUTHERN CALIFORNIA
LOS ANGELES, CALIFORNIA

(Received January 27, 1965)

Scattering of obliquely incident waves by inhomogeneous fibers*

C. Yeh and P. K. C. Wang
School of Engineering and Applied Sciences, University of California, Los Angeles, California 90024
(Received 21 January 1972)

The problem of the interaction of obliquely incident electromagnetic waves with a radially inhomogeneous fiber cylinder is treated analytically. By formulating this problem in terms of the two-point boundary-value problem, we eliminate the need to evaluate untabulated functions or to invert large matrices. Numerical results for the differential scattering cross sections are obtained for the case of an inhomogeneous fiber whose radial dielectric variation may be described by the Luneberg profile. It is found that the induced cross-polarized scattered field as well as the characteristics of the total scattered fields are influenced significantly by the inhomogeneity of the dielectric medium.

I. INTRODUCTION

During the past decade, a great deal of work has been carried out by various investigators on the problem of the interaction of electromagnetic waves with an inhomogeneous dielectric or plasma column.[1-6] Applications include the reflection of microwaves by meteor trails, reentry communication, plasma diagnostics, and the scattering of light by optical fibers. For a radially inhomogeneous cylinder, the solution of the wave equation even for the normal incidence case is a very difficult task; it usually involves infinite summations of untabulated functions, or of solutions to an infinite set of second-order differential equations.[7] For the oblique incidence case, the wave equation reduces to two coupled second-order differential equations whose solutions are usually very difficult to obtain even with the help of computers.

A useful approximate approach for an analytical solution of electromagnetic problems involving a radially inhomogeneous column is to subdivide it into thin homogeneous layers, and to solve an easier problem in each layer.[8] The fields in each layer are expanded in appropriate eigenfunctions and the expansion coefficients determined by matching boundary conditions. However this straightforward approach becomes much too tedious, and the number of simultaneous equations to be solved tends to be prohibitively large as the number of layers increase. It is therefore quite apparent that a different approach must be taken.

The purpose of this paper is to present the formulation of the problem of the scattering of an obliquely incident electromagnetic wave by a radially inhomogeneous fiber cylinder in terms of the two-point boundary-value problem. Solution of this two-point boundary-value problem will then be discussed. Numerical results are obtained for a specific radial dielectric variation, i.e., the Luneberg profile. It will be shown that the method described here greatly simplifies the computational procedure and that this technique can be readily adapted to

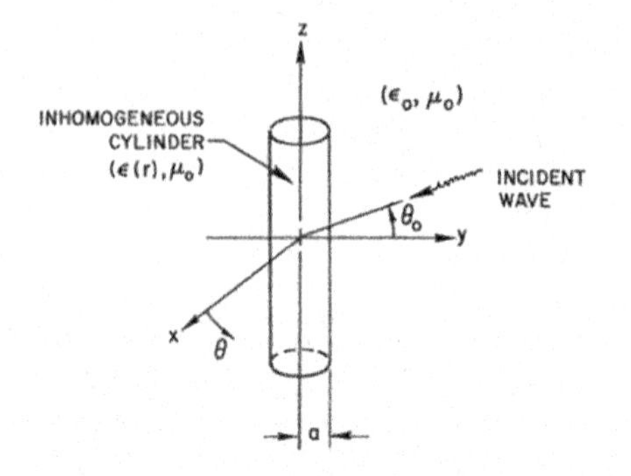

FIG. 1. Geometry of the problem.

solve other important physical problems, such as the scattering of waves by inhomogeneous plasma spheres or slabs, or by anisotropic and inhomogeneous cylinders.

II. FORMULATION OF THE PROBLEM

An infinite inhomogeneous dielectric cylinder of radius a immersed in free space is coaxial with the x axis (Fig. 1). It is assumed that the inhomogeneous cylinder has a permeability μ_0 and a permittivity $\epsilon(r)$, which is a function only of the radial distance r from the axis of the cylinder. Let a plane wave be incident obliquely on this structure. The propagation vector of the incident wave, which is assumed to be in the yz plane, makes an angle θ_0 with the positive y axis. Thus the axial components of the incident plane wave take the form

$$E_z^i = E_0 \sin\gamma \cos\theta_0 \exp(-ik_0 \cos\theta_0 y + ik_0 \sin\theta_0 z), \tag{1}$$

$$H_z^i = -(\epsilon_0/\mu_0)^{1/2} E_0 \cos\gamma \cos\theta_0 \exp(-ik_0 \cos\theta_0 y + ik_0 \sin\theta_0 z), \tag{2}$$

where E_0 and ω are, respectively, the amplitude and the frequency of the incident wave and $k_0 = \omega(\mu_0\epsilon_0)^{1/2}$. γ is the angle between the incident $\mathbf{E}^i$ vector and the x axis.

The time-dependence factor $\exp(-i\omega t)$ is assumed and suppressed throughout. Writing Eqs. (1) and (2) in cylindrical coordinates (r, θ, z) gives

$$E_z^i = E_0 \sin\gamma \cos\theta_0 \sum_{n=-\infty}^{\infty} (-1)^n J_n(k_0 r \cos\theta_0) \times \exp(in\theta) \exp(ik_0 \sin\theta_0 z), \tag{3}$$

$$H_z^i = -(k_0/\omega\mu_0) E_0 \cos\gamma \cos\theta_0 \sum_{n=-\infty}^{\infty} (-1)^n J_n(k_0 r \cos\theta_0) \times \exp(in\theta) \exp(ik_0 \sin\theta_0 z), \tag{4}$$

where $J_n(p)$ is the Bessel function of order n and argument p. The scattered wave in free space takes the form

$$E_z^s = \sum_{n=-\infty}^{\infty} (-1)^n A_n H_n^{(1)}(k_0 r \cos\theta_0) \exp(in\theta) \exp(ik_0 \sin\theta_0 z) \cos\theta_0, \tag{5}$$

$$H_z^s = \sum_{n=-\infty}^{\infty} (-1)^n B_n i(\epsilon_0/\mu_0)^{1/2} H_n^{(1)}(k_0 r \cos\theta_0) \times \exp(in\theta) \exp(ik_0 \sin\theta_0 z) \cos\theta_0, \tag{6}$$

where $H_n^{(1)}(p)$ is the Hankel function and A_n and B_n are as yet unknown arbitrary constants.

Unlike the case for the scattering by homogeneous dielectric cylinders, the field components within the inhomogeneous cylinder for the present problem cannot be expressed simply in terms of infinite series of Bessel functions. Furthermore, the wave equations for E_z^p and H_z^p within the inhomogeneous cylinder are coupled to each other. From physical grounds one may assume that all field components in a radially inhomogeneous cylinder must be periodic with respect to the angular coordinate θ; for example, the tangential components of the penetrated wave may take the following form:

$$E_z^p = \sum_{n=-\infty}^{\infty} (-1)^n E_{zn} \exp(in\theta) \exp(ik_0 \sin\theta_0 z) \cos\theta_0, \tag{7}$$

$$H_z^p = \sum_{n=-\infty}^{\infty} (-1)^n H_{zn} \exp(in\theta) \exp(ik_0 \sin\theta_0 z) \cos\theta_0, \tag{8}$$

$$E_\theta^p = \sum_{n=-\infty}^{\infty} (-1)^n E_{\theta n} \exp(in\theta) \exp(ik_0 \sin\theta_0 z) \cos\theta_0, \tag{9}$$

$$H_\theta^p = \sum_{n=-\infty}^{\infty} (-1)^n H_{\theta n} \exp(in\theta) \exp(ik_0 \sin\theta_0 z) \cos\theta_0, \tag{10}$$

where E_{zn}, H_{zn}, $E_{\theta n}$, and $H_{\theta n}$ are yet unknown functions of r, the radial coordinates. Substituting Eqs. (7)–(10) into Maxwell's equations gives, in matrix form,

$$\frac{d}{d\rho}\begin{bmatrix} E_{zn} \\ \rho E_{\theta n} \\ \left(\frac{\mu_0}{\epsilon_0}\right)^{1/2} H_{zn} \\ \rho\left(\frac{\mu_0}{\epsilon_0}\right)^{1/2} H_{\theta n} \end{bmatrix} = \begin{bmatrix} 0 & 0 & \frac{n\sin\theta_0}{i\rho\epsilon/\epsilon_0} & \frac{1}{i\rho}\left(1-\frac{\sin\theta_0}{\epsilon/\epsilon_0}\right) \\ 0 & 0 & i\left(\rho-\frac{n^2}{\rho\epsilon/\epsilon_0}\right) & \frac{in\sin\theta_0}{\rho\epsilon/\epsilon_0} \\ \frac{in\sin\theta_0}{\rho} & \frac{i}{\rho}\left(\frac{\epsilon}{\epsilon_0}-\sin^2\theta_0\right) & 0 & 0 \\ i\left(\frac{n^2}{\rho}-\rho\frac{\epsilon}{\epsilon_0}\right) & \frac{n\sin\theta_0}{i\rho} & 0 & 0 \end{bmatrix} \begin{bmatrix} E_{zn} \\ \rho E_{\theta n} \\ \left(\frac{\mu_0}{\epsilon_0}\right)^{1/2} H_{zn} \\ \rho\left(\frac{\mu_0}{\epsilon_0}\right)^{1/2} H_{\theta n} \end{bmatrix}, \tag{11}$$

with $\rho = k_0 r$. ϵ/ϵ_0 is the inhomogeneous radial dielectric variation within the cylinder.

III. BOUNDARY CONDITIONS

To obtain the arbitrary constant A_n and B_n for the scattered fields and the appropriate expressions for E_{zn}, $E_{\theta n}$, H_{zn}, and $H_{\theta n}$, we must make use of the boundary conditions for this problem. The boundary conditions require the continuity of the tangential electric and magnetic fields at the boundary $\rho = \rho_a$, where $\rho_a = k_0 a$ and a is the radius of the inhomogeneous cylinder. In order that Eq. (11) may be integrated numerically, we shall assume that there exists an inner homogeneous dielectric cylindrical core of radius r_0 which can be made arbitrarily small. The dielectric constant of the inner core is assumed to be ϵ_1. Matching the boundary conditions at $\rho = \rho_a$ and at $\rho = \rho_0$ $(\rho_0 = k_0 r_0)$ gives

$$E_0^i J_n(\rho_a \cos\theta_0) + A_n H_n^{(1)}(\rho_a \cos\theta_0) = E_{zn}(\rho_a), \tag{12a}$$

$$\left(\frac{\mu_0}{\epsilon_0}\right)^{1/2} H_0' J_n(\rho_a \cos\theta_0) + iB_n H_n^{(1)}(\rho_a \cos\theta_0) = \left(\frac{\mu_0}{\epsilon_0}\right)^{1/2} H_{zn}(\rho_a), \tag{12b}$$

$$-\frac{\sin\theta_0 n E_0'}{\cos^2\theta_0} J_n(\rho_a \cos\theta_0) - \frac{n\sin\theta_0}{\cos^2\theta_0} A_n H_n^{(1)}(\rho_a\cos\theta_0) - \frac{i}{\cos\theta_0}\left(\frac{\mu_0}{\epsilon_0}\right)^{1/2} H_0' \rho_a J_n'(\rho_a\cos\theta_0) + \frac{1}{\cos\theta_0} B_n \rho_a H_n^{(1)\prime}(\rho_a\cos\theta_0) = \rho_a E_{\theta n}(\rho_a), \tag{12c}$$

$$-\frac{i\rho_a}{\cos\theta_0} E_0' J_n'(\rho_a\cos\theta_0) - \frac{in}{\cos\theta_0}\sin\theta_0 B_n H_n^{(1)}(\rho_a\cos\theta_0) + \left(\frac{\mu_0}{\epsilon_0}\right)^{1/2}\frac{\sin\theta_0}{\cos^2\theta_0} n H_0' J_n(\rho_a\cos\theta_0) + \frac{i\rho_a}{\cos\theta_0} A_n H_n^{(1)\prime}(\rho_a\cos\theta_0) = \left(\frac{\mu_0}{\epsilon_0}\right)^{1/2}\rho_a H_n(\rho_a), \tag{12d}$$

and

$$C_n J_n(\lambda\rho_0) = E_{zn}(\rho_0), \tag{13a}$$

$$D_n J_n(\lambda\rho_0) = (\mu_0/\epsilon_0)^{1/2} H_{zn}(\rho_0), \tag{13b}$$

$$-\frac{n\sin\theta_0}{\lambda^2} C_n J_n(\lambda\rho_0) - \frac{i}{\lambda} D_n \rho_0 J_n'(\lambda\rho_0) = \rho_0 E_{\theta n}(\rho_0), \tag{13c}$$

$$\frac{n\sin\theta_0}{\lambda^2} D_n J_n(\lambda\rho_0) - i\frac{\epsilon_1}{\epsilon_0}\frac{1}{\lambda} C_n \rho_0 J_n'(\lambda\rho_0) = \left(\frac{\mu_0}{\epsilon_0}\right)^{1/2}\rho_0 H_{\theta n}(\rho_0), \tag{13d}$$

where the prime signifies the derivative of the function with respect to its argument and $\lambda^2 = [(\epsilon_1/\epsilon_0) - \sin^2\theta_0)]$. C_n and D_n are arbitrary expansion coefficients for the axial fields within the homogeneous inner core. Eliminating A_n and B_n from Eq. (12) and C_n and D_n from Eq. (13), and combining the resultant equations gives, in matrix form,

$$\begin{bmatrix} n\lambda^{-2}\sin\theta_0 & 1 & \frac{i\rho_0}{\lambda}\frac{J_n'(\lambda\rho_0)}{J_n(\lambda\rho_0)} & 0 \\ -\frac{i\rho_0}{\lambda}\left(\frac{\epsilon_1}{\epsilon_0}\right)\frac{J_n'(\lambda\rho_0)}{J_n(\lambda\rho_0)} & 0 & \lambda^{-2} n\sin\theta_0 & 1 \\ 0 & 0 & 0 & 0 \\ 0 & 0 & 0 & 0 \end{bmatrix} \begin{bmatrix} E_{zn}(\rho_0) \\ \rho_0 E_{\theta n}(\rho_0) \\ \left(\frac{\mu_0}{\epsilon_0}\right)^{1/2} H_{zn}(\rho_0) \\ \rho_0\left(\frac{\mu_0}{\epsilon_0}\right)^{1/2} H_{\theta n}(\rho_0) \end{bmatrix}$$

$$+\begin{bmatrix} 0 & 0 & 0 & 0 \\ 0 & 0 & 0 & 0 \\ in\sin\theta_0 H_n^{(1)}(v) & iH_n^{(1)}(v)\cos^2\theta_0 & -\rho_a\cos\theta_0 H_n^{(1)\prime}(v) & 0 \\ \rho_a\cos\theta_0 H_n^{(1)\prime}(v) & 0 & in\sin\theta_0 H_n^{(1)}(v) & iH_n^{(1)}(v)\cos^2\theta_0 \end{bmatrix} \begin{bmatrix} E_{zn}(\rho_a) \\ \rho_a E_{\theta n}(\rho_a) \\ \left(\frac{\mu_0}{\epsilon_0}\right)^{1/2} H_{zn}(\rho_a) \\ \rho_a\left(\frac{\mu_0}{\epsilon_0}\right)^{1/2} H_{\theta n}(\rho_a) \end{bmatrix} = \begin{bmatrix} 0 \\ 0 \\ -\frac{2i}{\pi}\left(\frac{\mu_0}{\epsilon_0}\right)^{1/2} H_0' \\ \frac{2i}{\pi} E_0' \end{bmatrix}, \tag{14}$$

with

$$v = \rho_a\cos\theta_0, \quad E_0' = E_0\sin\gamma, \quad H_0' = -(\epsilon_0/\mu_0)^{1/2} E_0\cos\gamma.$$

Hence, we have completed the formulation of the present scattering problem as a two-point boundary-value problem; the differential equations are given by Eq. (11) while the two-point boundary conditions are given by Eq. (14).

IV. REDUCTION TO CANONICAL FORM AND RESCALING OF EQS. (11) AND (14)

In order to achieve uniform accuracy, step sizes used in the numerical integration of Eq. (11) must be varied according to how fast the function changes as ρ is varied. To circumvent this difficulty for the present problem, we shall rescale the radial coordinate according to the following scheme:

$$\rho' = n^2 ln\rho, \quad \rho = \exp(\rho'/n^2), \quad \rho_a' = ln\rho_a, \quad d\rho' = n^2\frac{d\rho}{\rho}, \quad d\rho = n^{-2}\rho\, d\rho', \quad \rho_0' = ln\rho_0. \tag{15}$$

Equation (11) then reduces to

$$\frac{d}{d\rho'}\begin{bmatrix}\mathbf{x}_R \\ \mathbf{x}_I\end{bmatrix} = \begin{bmatrix} 0_4 & -\tilde{A} \\ \tilde{A} & 0_4 \end{bmatrix}\begin{bmatrix}\mathbf{x}_R \\ \mathbf{x}_I\end{bmatrix}, \tag{16}$$

where

$$x_R = \mathrm{Re}\begin{bmatrix} E_{zn} \\ \rho E_{\theta n} \\ (\mu_0/\epsilon_0)^{1/2} H_{zn} \\ \rho(\mu_0/\epsilon_0)^{1/2} H_{\theta n} \end{bmatrix}, \qquad x_I = \mathrm{Im}\begin{bmatrix} E_{zn} \\ \rho E_{\theta n} \\ (\mu_0/\epsilon_0)^{1/2} H_{zn} \\ \rho(\mu_0/\epsilon_0)^{1/2} H_{\theta n} \end{bmatrix},$$

$$\tilde{A} = \left[\begin{array}{cc|cc} \multicolumn{2}{c|}{\multirow{2}{*}{0_2}} & -n^{-1}\left(\frac{\epsilon}{\epsilon_0}\right)^{-1}\sin\theta_0 & n^{-2}\left(\frac{\epsilon_0}{\epsilon}\sin^2\theta_0 - 1\right) \\ & & n^{-2}\exp\left(\frac{2\rho'}{n^2}\right) - \frac{\epsilon_0}{\epsilon} & n^{-1}\left(\frac{\epsilon_0}{\epsilon}\right)\sin\theta_0 \\ \hline n^{-1}\sin\theta_0 & n^{-2}\left(\frac{\epsilon}{\epsilon_0} - \sin^2\theta_0\right) & \multicolumn{2}{c}{\multirow{2}{*}{0_2}} \\ 1 - \frac{\epsilon}{\epsilon_0 n^2}\exp\left(\frac{2\rho'}{n^2}\right) & -n^{-1}\sin\theta_0 & & \end{array}\right],$$

and 0_4 is a 4×4 zero matrix. Similarily, the boundary conditions, Eq. (14), become

$$\left[\begin{array}{c|c} \tilde{B}_R & -\tilde{B}_I \\ \hline \tilde{B}_I & \tilde{B}_R \end{array}\right] \left[\begin{array}{c} \mathbf{x}_R(\rho_0') \\ \hline \mathbf{x}_I(\rho_0') \end{array}\right] + \left[\begin{array}{c|c} \tilde{C}_R & -\tilde{C}_I \\ \hline \tilde{C}_I & \tilde{C}_R \end{array}\right] \left[\begin{array}{c} \mathbf{x}_R(\rho_a') \\ \mathbf{x}_I(\rho_a') \end{array}\right] = \left[\begin{array}{c} 0_4 \\ R_I \end{array}\right], \tag{17}$$

where

$$\tilde{B}_R = \left[\begin{array}{cc|cc} n\lambda^{-2}\sin\theta_0 & 1 & 0 & 0 \\ 0 & 0 & n\lambda^{-2}\sin\theta_0 & 1 \\ \hline \multicolumn{2}{c|}{0_2} & \multicolumn{2}{c}{0_2} \end{array}\right],$$

$$\tilde{B}_I = \left[\begin{array}{cc|cc} 0 & 0 & \frac{\rho_0}{\lambda}\frac{J_n'(\lambda\rho_0)}{J_n(\lambda\rho_0)} & 0 \\ -\frac{\rho_0\epsilon}{\lambda\epsilon_0}\frac{J_n'(\lambda\rho_0)}{J_n(\lambda\rho_0)} & 0 & 0 & 0 \\ \hline \multicolumn{2}{c|}{0_2} & \multicolumn{2}{c}{0_2} \end{array}\right],$$

$$\tilde{C}_R = \left[\begin{array}{cc|cc} \multicolumn{2}{c|}{0_2} & \multicolumn{2}{c}{0_2} \\ \hline -n\sin\theta_0 Y_n(v) & -Y_n(v)\cos^2\theta_0 & -vJ_n'(v) & 0 \\ vJ_n'(v) & 0 & -n\sin\theta_0 Y_n'(v) & -Y_n(v)\cos^2\theta_0 \end{array}\right],$$

$$\tilde{C}_I = \left[\begin{array}{cc|cc} \multicolumn{2}{c|}{0_2} & \multicolumn{2}{c}{0_2} \\ \hline n\sin\theta_0 J_n(v) & \cos^2\theta_0 J_n(v) & -vY_n'(v) & 0 \\ vY_n'(v) & 0 & n\sin\theta_0 J_n(v) & \cos^2\theta_0 J_n(v) \end{array}\right],$$

$$\mathbf{R}_I = \left[\begin{array}{c} 0 \\ 0 \\ -\frac{2}{\pi}\left(\frac{\mu_0}{\epsilon_0}\right)^{1/2} H_0' \\ \frac{2}{\pi}E_0' \end{array}\right].$$

Equations (16) and (17) are now in the canonical form of the two-point boundary-value problem.

V. SOLUTION OF THE LINEAR TWO-POINT BOUNDARY-VALUE PROBLEM

It is assumed that $\mathbf{x}$ is an n-dimensional vector, $\mathbf{d}$ is also an n-dimensional vector, and $A(\rho)$, B, and C are $n\times n$ matrices. (For the present problem $n=8$.) Equations (16) and (17) can now be written in the following form

$$\frac{d\mathbf{x}}{d\rho} = A(\rho)\mathbf{x}, \tag{18}$$

$$B\mathbf{x}(\rho_0) + C\mathbf{x}(\rho_a) = \mathbf{d}, \tag{19}$$

where ρ is the independent variable and $\rho_a > \rho_0$. The basic idea is to first determine $\mathbf{x}(\rho_0)$ or $\mathbf{x}(\rho_a)$ completely. Then $\mathbf{x}(\rho)$ can be found by integrating Eq. (18) either forward or backward in ρ with the initial value $\mathbf{x}(\rho_0)$ or $\mathbf{x}(\rho_a)$.

To determine $\mathbf{x}(\rho_0)$, we shall make use of the adjoint equation which is

$$\frac{d\mathbf{z}^b(\rho)}{d\rho} = -A^T(\rho)\mathbf{z}^b(\rho), \tag{20}$$

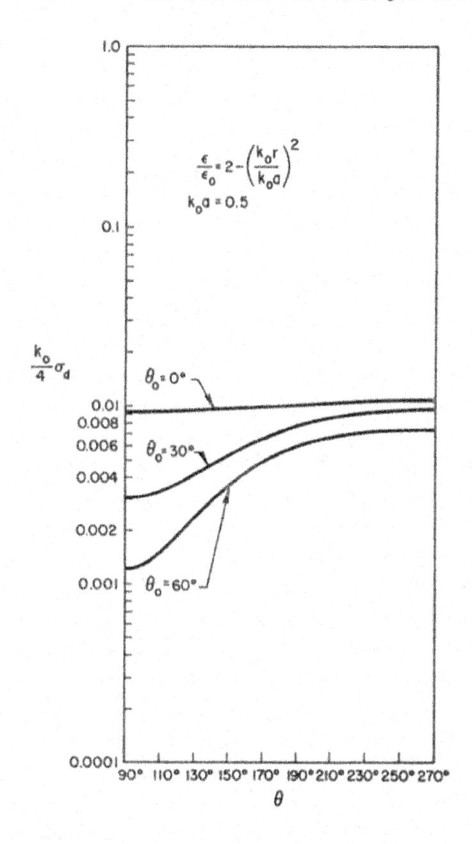

FIG. 2. Normalized differential scattering cross section as a function of the azimuthal angle for $k_0a = 0.5$. θ_0 is the angle of incidence.

where $A^T(\rho)$ is the transpose of $A(\rho)$ and where $\mathbf{z}^p(\rho)$ is the solution of this adjoint equation with final value

$$\mathbf{z}^p(\rho_a) = \mathbf{e}_p = \begin{bmatrix} 0 \\ \vdots \\ 0 \\ 1 \\ 0 \\ \vdots \\ 0 \end{bmatrix}, \quad 1 \leq p \leq n. \tag{21}$$

Here, the pth component of the vector $\mathbf{e}_p$ is unity while all other components of $\mathbf{e}_p$ are zero.

Let us now consider the expression

$$\frac{d}{d\rho}\{[\mathbf{z}^p(\rho)]^T\mathbf{x}(\rho)\} = \left(\frac{d}{d\rho}\mathbf{z}^p(\rho)\right)^T\mathbf{x}(\rho) + [\mathbf{z}^p(\rho)]^T\left[\frac{d}{d\rho}\mathbf{x}(\rho)\right]. \tag{22}$$

Substituting Eqs. (18) and (20) into the above equation gives

$$\frac{d}{d\rho}\{[\mathbf{z}^p(\rho)]^T\mathbf{x}(\rho)\} = -[A^T(\rho)\mathbf{z}^p(\rho)]^T\mathbf{x}(\rho) + [\mathbf{z}^p(\rho)]^T A(\rho)\mathbf{x}(\rho)$$

The right-hand side of this equation is identically zero; hence,

$$\frac{d}{d\rho}\{[\mathbf{z}^p(\rho)]^T\mathbf{x}(\rho)\} = 0$$

or

$$[\mathbf{z}^p(\rho_a)]^T\mathbf{x}(\rho_a) = [\mathbf{z}^p(\rho_0)]^T\mathbf{x}(\rho_0), \tag{23}$$

which is a relation between $\mathbf{x}(\rho_a)$ and $\mathbf{x}(\rho_0)$. Now, if we integrate Eq. (20) with final value $\mathbf{z}^p(\rho_a)$ given by Eq. (21) backward in ρ for all $1 \leq p \leq n$, then, by using Eq. (23), we find

$$\mathbf{x}(\rho_a) = \begin{bmatrix} [\mathbf{z}^1(\rho_0)]^T \\ \vdots \\ [\mathbf{z}^n(\rho_0)]^T \end{bmatrix} \mathbf{x}(\rho_0) = \Phi(\rho_a, \rho_0)\mathbf{x}(\rho_0). \tag{24}$$

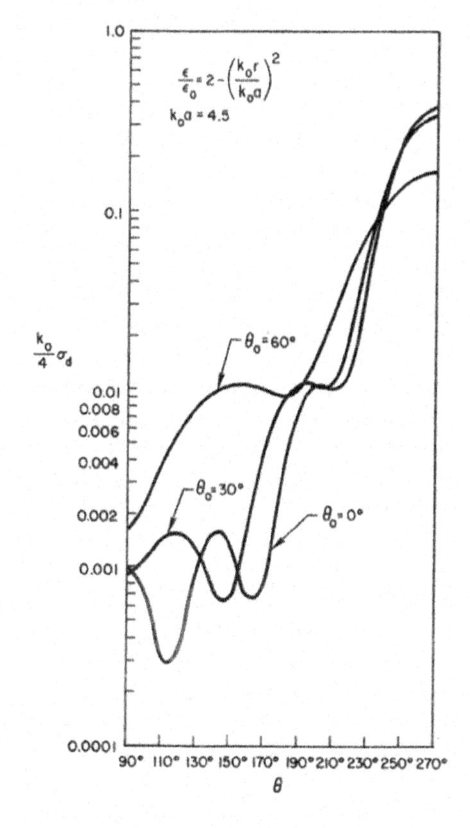

FIG. 3. Normalized differential scattering cross section as a function of the azimuthal angle for $k_0a = 4.5$.

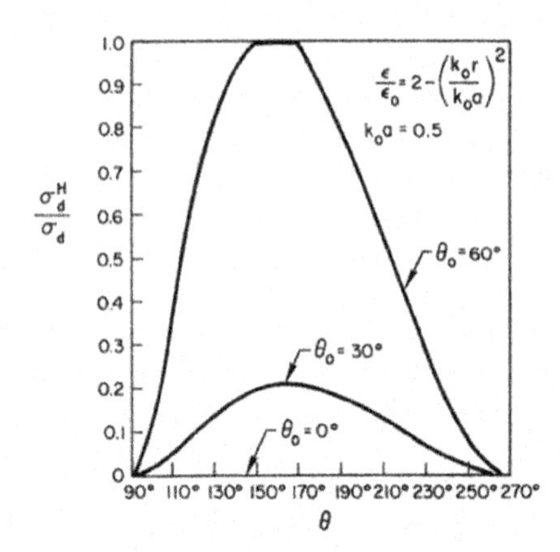

FIG. 4. Ratio of the cross-polarized component of the differential scattering cross section and the total differential scattering cross section vs the azimuthal angle for $k_0a = 0.5$.

Substituting Eq. (24) into the two-point boundary condition, Eq. (19) gives

$$\mathbf{x}(\rho_a) = [B\Phi(\rho_a, \rho_0)^{-1} + C]^{-1}\mathbf{d}$$

and

$$\mathbf{x}(\rho_0) = [B + C\Phi(\rho_a, \rho_0)]^{-1}\mathbf{d}. \tag{25}$$

Therefore the initial condition $\mathbf{x}(\rho_0)$ for Eq. (18) is completely determined. We can now proceed to find $\mathbf{x}(\rho)$ by integrating Eq. (18) forward with the initial conditions given by Eq. (25). Knowing $\mathbf{x}(\rho_a)$, i.e., knowing $E_{zn}(\rho_a)$, $E_{\theta n}(\rho_a)$, $(\mu_0/\epsilon_0)^{1/2} H_{zn}(\rho_a)$, and $(\mu_0/\epsilon_0)^{1/2}\rho_a H_{\theta n}(\rho_a)$, it would be a simple matter to find the expansion coefficients A_n and B_n for the scattered fields from Eq. (12).

VI. DIFFERENTIAL SCATTERING CROSS SECTION

Of special interest to many physical problems is the differential scattering cross section which represents the amount of energy scattered by the cylinder in various directions. The differential scattering cross section per unit length is defined by

$$\sigma_d = \lim_{r\to\infty} 2\pi r \frac{\mathbf{S}^s \cdot \mathbf{e}_r}{|\mathbf{S}^i|}, \tag{26}$$

where $\mathbf{S}^i$ and $\mathbf{S}^s$ are, respectively, the time-averaged incident and scattered Poynting vectors

$$\mathbf{S}^i = \tfrac{1}{2}\mathrm{Re}(\mathbf{E}^i \times \mathbf{H}^{i*}),$$

$$\mathbf{S}^s = \tfrac{1}{2}\mathrm{Re}(\mathbf{E}^s \times \mathbf{H}^{s*}).$$

The $*$ represents the complex conjugate of the function. Substituting Eqs. (3)–(6) into Eq. (26) gives

$$\sigma_d = \sigma_d^E + \sigma_d^H \tag{27}$$

with

$$\sigma_d^E = \frac{4}{k_0} \left| \sum_{n=-\infty}^{\infty} i^n A_n \exp(in\theta) \right|^2, \tag{28}$$

$$\sigma_d^H = \frac{4}{k_0} \left| \sum_{n=-\infty}^{\infty} i^n B_n \exp(in\theta) \right|^2. \tag{29}$$

VII. NUMERICAL EXAMPLES

Computer programs were prepared according to the method discussed above. As a specific example, we shall choose the Luneburg lens profile as the dielectric variation within the cylinder; in other words

$$\epsilon(\rho)/\epsilon_0 = 2 - (\rho/\rho_a)^2. \tag{30}$$

Extensive computations were carried out for the case of an obliquely incident wave whose electric vector is polarized in the z direction. Two representative cases were chosen: the low-frequency case with $\rho_a = k_0 a = 0.5$ and the resonant-frequency case with $\rho_a = 4.5$. In numerical computation, the number of terms (modes) taken in the summation [as given in Eqs. (28) and (29)] is determined by the size of the argument in the Bessel and Hankel functions. Computations are carried out to include all terms which contribute more than 0.01% of the sum. Based on this criterion, no more than nine terms of the series are needed and no more than 10^5 steps were used to integrate the differential equations for the range of ρ_a considered in this paper.

The differential scattering cross section as a function of the azimuthal angle for various angles of incidence is plotted in Fig. 2 for the low-frequency case ($k_0 a = 0.5$) and in Fig. 3 for the resonant-frequency case ($k_0 a = 0.5$). Due to the symmetry of the problem, only the region $90° \leq \theta \leq 270°$ was included in all figures. The curves for region $270° \leq \theta \leq 90°$ are mirror images about $\theta = 270°$ of the given curves. It can be seen from Fig. 2 that the scattered pattern becomes more directional as the angle of incidence departs from the normal axis, i.e., as θ_0 increases. At normal incidence ($\theta_0 = 0°$) the scattered field is almost omnidirectional. It appears that higher multipoles must be used to represent the induced field as θ_0 becomes larger. However, the magnitudes of the

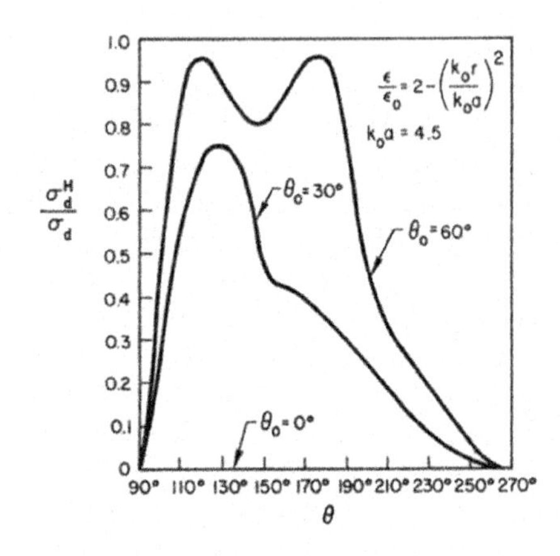

FIG. 5. σ_d^H/σ_d vs the azimuthal angle for $k_0a = 4.5$.

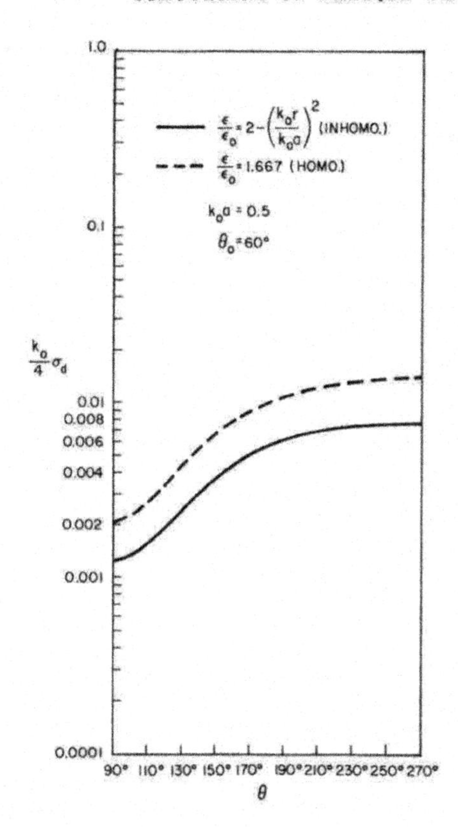

FIG. 6. Comparison of the normalized differential scattering cross section for inhomogeneous and equivalent homogeneous fibers. $k_0a = 0.5$.

scattered energy are larger for a normally incident wave, indicating that perhaps better matching of the incident wave at oblique incidence is achieved. Much more fluctuations for the differential scattering cross section vs the azimuthal angle curves are observed for the resonant-frequency case ($k_0a = 4.5$) from Fig. 3. This is due to the diffraction phenomena. There are fewer valleys and peaks for larger θ_0 indicating again that better matching of the incident wave at oblique incidence is achieved. It is important to note that the scattering patterns for the obliquely incident wave are very much different than that for the normally incident wave. Furthermore a pure E (with $H_z = 0$) or H (with $E_z = 0$) incident wave will give rise to a cross-polarized component and result in a combination of coupled E and H waves.[9] To see how much scattered energy is distributed in the cross-polarized field (i.e., the H wave) for an incident E wave, Figs. 4 and 5 are introduced. As expected no cross-polarized wave was excited for the normally incident wave (i.e., the $\theta_0 = 0°$ curves); coupling of a cross-polarized wave becomes more pronounced at larger angles of oblique incidence. At a large angle of oblique incidence even 100% coupling of a cross-polarized wave was observed in certain scattering directions; for example, when $k_0a = 0.5$, $\theta_0 = 60°$, 100% coupling occurs for $150° < \theta < 170°$. It should be noted that no coupling occurs for the forward as well as backward scattered wave for any angles of oblique incidence.

For purposes of comparison, the normalized differential scattering cross section has been computed for a homogeneous dielectric cylinder. The equivalent homogeneous cylinder has a uniform dielectric within the cylinder whose dielectric constant is equal to 1.6667. Thus the homogeneous and inhomogeneous cylinders being compared have the same average dielectric constant, but the inhomogeneous model has a continuous dielectric variation at the edge (although the gradient is discontinuous) while the homogeneous model has a discontinuous dielectric constant at the edge. Figures 6 and 7 compare the scattering pattern from inhomogeneous and equivalent

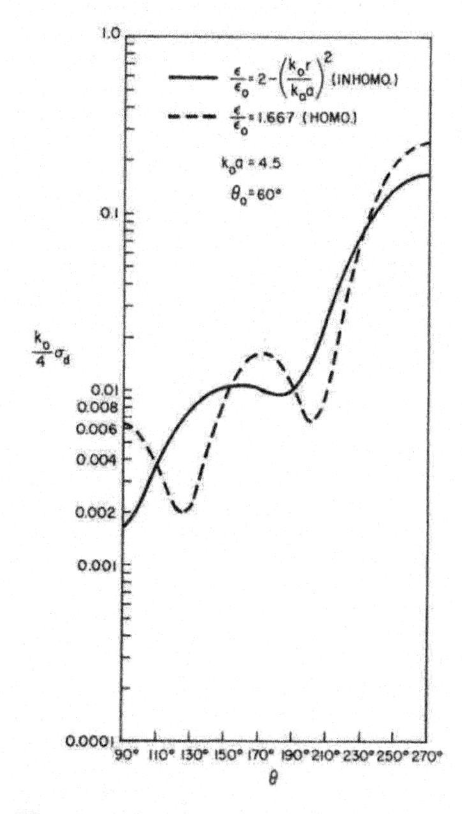

FIG. 7. Comparison of the normalized differential scattering cross section for inhomogeneous and equivalent homogeneous fibers. $k_0a = 4.5$.

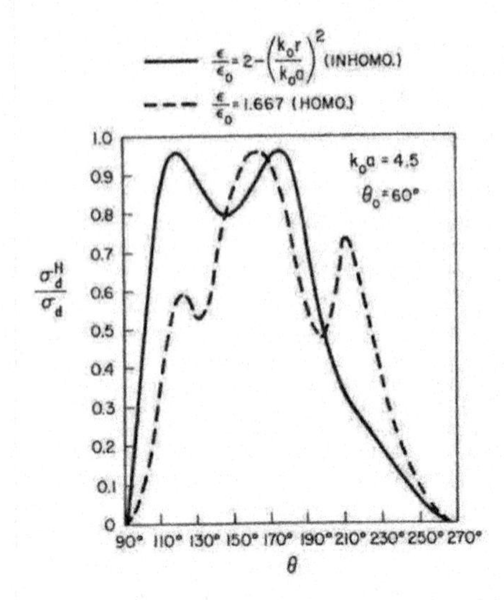

FIG. 8. Comparison of σ_d^H/σ_d for inhomogeneous and equivalent homogeneous fibers. $k_0a = 4.5$.

homogeneous cylinders for $\theta_0 = 60°$ and $k_0a = 0.5$ and 4.5. For the low-frequency case $k_0a = 0.5$, the angular dependence of the differential scattering cross section for the inhomogeneous and homogeneous cylinders is almost identical; the only difference is in the magnitude of the scattered energy. This is because at low frequencies the diffraction characteristics are not strongly dependent on the homogeneity of the dielectric cylinder. It is still apparent, however, that the mismatch between the homogeneous dielectric cylinder and its surrounding free space contributed greater scattered energy than the case involving inhomogeneous cylinders. At moderate frequency ($k_0a = 4.5$), the scattering patterns for the homogeneous and inhomogeneous cylinders are significantly different due to expected diffraction efforts. A smoother pattern for the inhomogeneous case is obtained.

Figure 8 shows that the coupling between cross-polarized fields for the equivalent homogeneous cylinder is very much different from that for the inhomogeneous cylinder.

In conclusion we may state that significant differences are observed between the diffracted energies for an obliquely incident wave and those for a normally incident wave and that the diffraction characteristics for an inhomogeneous cylinder are quite different from those for an equivalent homogeneous cylinder. The need for exact analysis of the inhomogeneous cylinder case with an obliquely incident wave is very apparent. It should be further noted that the technique used to solve this problem may be readily adapted to treat other important physical problems, such as the problems of the interaction of obliquely incident microwaves with an inhomogeneous and anisotropic plasma column.

---

*Supported by the National Science Foundation (C. Yeh) and by AFOSR Grant No. 72-2303 (P. Wang).

[1]M. Kerker, *The Scattering of Light and Other Electromagnetic Radiation* (Academic, New York, 1969).
[2]C. Yeh and W. V. T. Rusch, J. Appl. Phys. **36**, 2302 (1965).
[3]G. H. Keitel, Proc. IRE **43**, 1481 (1955).
[4]P. M. Platzman and H. T. Ozaki, J. Appl. Phys. **31**, 1597 (1960).
[5]C. Yeh, J. Appl. Phys. **40**, 5066 (1969).
[6]V. C. Wong and D. K. Cheng, Proc. Inst. Elec. Eng. **115**, 1446 (1968).
[7]R. Burman, IEEE Trans. Antennas Propag. **13**, 646 (1965).
[8]J. R. Wait, J. Res. Nat. Bur. Stand. B **65**, 137 (1961).
[9]J. R. Wait, Can. J. Phys. **33**, 189 (1955).

SCATTERING FROM A CYLINDER COATED WITH AN INHOMOGENEOUS DIELECTRIC SHEATH*

C. YEH AND Z. A. KAPRIELIAN
Electrical Engineering Department, University of Southern California, Los Angeles, California
Received July 6, 1962

ABSTRACT

As a space vehicle re-enters the atmosphere, a plasma sheath, surrounding the vehicle, is generated. It is well known that the sheath is inhomogeneous. However, to make this problem suitable for theoretical analysis, most investigators make the assumption that the sheath is homogeneous. To investigate the validity of this assumption, the idealized problem of the scattering of plane waves by a conducting cylinder coated with a stratified dielectric sheath is considered. The wave equation is separated using the vector wave-function method of Hansen and Stratton. It is then applied to the plane-wave scattering problem. The backscattering cross section is defined and obtained. Analytical expressions for the scattering coefficients of a thin inhomogeneous sheath are also given. Numerical computations are carried out for a specific variation of the dielectric sheath: i.e., $\epsilon(r) = \epsilon_0\alpha/k_0r$, where α is a constant, $k_0^2 = \omega^2\mu\epsilon_0$, and ϵ_0 is the free-space dielectric constant. Results are compared with the homogeneous sheath problem; the dielectric constant of the homogeneous sheath is taken to be the average value of that for the inhomogeneous sheath. It is found that in general rather distinct differences are observed except when the sheath is very thin.

I. INTRODUCTION

It is well known that the plasma sheath surrounding a re-entering space vehicle is inhomogeneous. To render this problem suitable for theoretical analysis, most investigators make the assumption that the plasma sheath is homogeneous. The purpose of this paper is to investigate the validity of this assumption. To bring out the effects due to the inhomogeneity of the sheath and to avoid unnecessary complicated mathematical operations, the following idealized problem is analyzed. A plane wave is assumed to be impinging normally upon an infinitely long cylinder coated with a radially inhomogeneous dielectric sheath. The backscattering cross section is obtained. It is then compared with the backscattering cross section for the identical structure coated with a homogeneous dielectric sheath whose dielectric constant equals the average value of the dielectric constant of the inhomogeneous dielectric. Numerical computations are carried out for a specific variation of the dielectric constant: i.e., $\epsilon(r)/\epsilon_0 = \alpha/k_0r$, where α is a constant, $k_0^2 = \omega^2\mu\epsilon_0$, ϵ_0 is the free-space dielectric constant, and r is the radial component in the cylindrical co-ordinates. With this choice of dielectric variation, the wave functions can be expressed in terms of familiar functions (i.e., Bessel functions of nonintegral order).

The electromagnetic field associated with the inhomogeneous dielectric medium has been discussed by a number of authors. A rather systematic treatment is given by Marcuvitz (1951). Tai (1958) investigated the electromagnetic field involved in a radially stratified spherical medium and then

*This research was sponsored by the Office of Naval Research.

considered specifically the electromagnetic theory of the spherical Luneberg lens. The theory of the cylindrical or the two-dimensional Luneberg lens has been obtained by Jasik (1953). The scattering of a plane wave normally incident on a meteor trail which has a Gaussian radial distribution of ionization was also investigated by Keitel (1955). In the present work the vector wave-function method of Hansen and Stratton will be extended to represent the fields in a radially stratified cylindrical medium without any restriction upon the exact nature of the dielectric constant. According to this method all field components in this medium can be derived from the scalar quantity $\Phi(r,\theta)$ or $\Psi(r,\theta)$ as follows:

$$\mathbf{E} = \nabla\times(\Phi(r,\theta)\mathbf{e}_r), \tag{1a}$$

$$\mathbf{H} = \frac{i}{\omega\mu_0}\nabla\times\nabla\times(\Phi(r,\theta)\mathbf{e}_r) \tag{1b}$$

for TM waves, and

$$\mathbf{H} = \nabla\times(\Psi(r,\theta)\mathbf{e}_r), \tag{2a}$$

$$\mathbf{E} = -\frac{i}{\omega\epsilon(r)}\nabla\times\nabla\times(\Psi(r,\theta)\mathbf{e}_r) \tag{2b}$$

for TE waves. All field components are assumed to be independent of the axial component.* $\mathbf{e}_r$ is the unit vector in the r direction of the cylindrical co-ordinates (r, θ, z), and the harmonic time-dependence $e^{i\omega t}$ is assumed and suppressed throughout. The above formulation assures the fulfillment of the divergence conditions in Maxwell's equations. It is noted that μ_0 may also be a function of r in this formulation. The solutions for $\Phi(r,\theta)$ and $\Psi(r,\theta)$ may be obtained, respectively, by substituting equations (1) and (2) into Maxwell's equations. One gets

$$\Phi(r,\theta) = \begin{Bmatrix} U_n^{(1)}(kr) \\ U_n^{(2)}(kr) \end{Bmatrix}\{e^{\pm in\theta}\} \tag{3}$$

and

$$\Psi(r,\theta) = \begin{Bmatrix} V_n^{(1)}(kr) \\ V_n^{(2)}(kr) \end{Bmatrix}\{e^{\pm in\theta}\}, \tag{4}$$

where $U_n^{(1),(2)}(kr)$ and $V_n^{(1),(2)}(kr)$ satisfy, respectively, the differential equations

$$\left[\frac{d^2}{dr^2} - \frac{1}{r}\frac{d}{dr} + \left(k_0^2\frac{\epsilon(kr)}{\epsilon_0} + \frac{1-n^2}{r^2}\right)\right]U_n^{(1),(2)}(kr) = 0 \tag{5}$$

and

*The above formulation does not apply when the fields are functions of r, θ, and z. No general formulation seems to be available. One obvious way of approaching this problem is to divide the inhomogeneous medium into many small segments, to treat each segment separately by assuming the dielectric constant as constant within this segment, and to connect the fields in these segments by matching the boundary conditions.

$$(6)\quad \left[\frac{d^2}{dr^2} - \left(1 + \frac{r}{\epsilon(kr)}\frac{d\epsilon(kr)}{dr}\right)\left(\frac{1}{r}\frac{d}{dr}\right) + \left(k_0^2\frac{\epsilon(kr)}{\epsilon_0} + \frac{1}{r\epsilon(kr)}\frac{d\epsilon(kr)}{dr} + \frac{1-n^2}{r^2}\right)\right]V_n^{(1),(2)}(kr) = 0$$

with $k_0^2 = \omega^2\mu\epsilon_0$. $\epsilon(kr)$ is the variation of the inhomogeneous dielectric constant with respect to r.

II. SCATTERING OF NORMALLY INCIDENT PLANE WAVES

Figure 1 shows the geometric arrangement for a plane wave incident in the direction of the positive x axis. The polarization may take an arbitrary

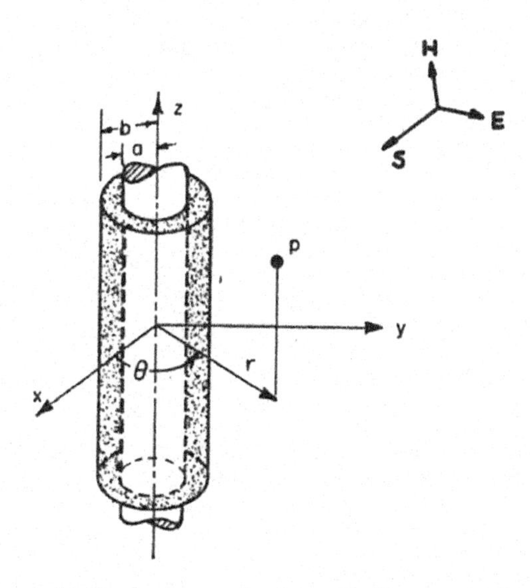

FIG. 1. Plane wave incident on a conducting cylinder coated with an inhomogeneous dielectric sheath.

direction. If the incident electric vector is in the z direction, the incident E_z field may be expressed as

$$(7)\quad E_z^{\text{inc}} = E_0 e^{-ik_0 r\cos\theta} = E_0\left[J_0(k_0 r) + 2\sum_{n=1}^{\infty}(-i)^n J_n(k_0 r)\cos n\theta\right],$$

where E_0 is the amplitude of the incident wave. Then the scattered field must be of the form

$$(8)\quad E_z^{\text{sc}} = A_0 H_0^{(1)}(k_0 r) + 2\sum_{n=1}^{\infty} A_n(-i)^n H_n^{(2)}(k_0 r)\cos n\theta,$$

where $k_0^2 = \omega^2\mu\epsilon_0$, ϵ_0 is the free-space dielectric constant, $H_n^{(2)}(k_0 r)$ are the Hankel functions of the second kind of order n representing the outgoing traveling wave, and A_n are the arbitrary constants to be determined from the boundary conditions. According to equations (1*a*) and (3), the required E_z field within the sheath (i.e., $a \leqslant r \leqslant b$) is

$$(9) \qquad E_z^{\,sh} = \sum_{n=0}^{\infty} B_n(2-\delta_n^0)(i)^n[U_n^{(1)}(kr) + C_n U_n^{(2)}(kr)] \cos n\theta/kr,$$

in which B_n and C_n are the arbitrary constants and δ_n^0 is the Kronecker Delta. Satisfying the appropriate boundary conditions at $r = a$ and $r = b$ gives the following relations:

$$(10a) \qquad A_n = E_0 \frac{J_n(k_0 b)\, U_n'(ka, kb) - J_n'(k_0 b)\, U_n(ka, kb)}{H_n^{(2)\prime}(k_0 b)\, U_n(ka, kb) - H_n^{(2)}(k_0 b)\, U_n'(ka, kb)},$$

$$(10b) \qquad B_n = E_0 \frac{H_n^{(2)\prime}(k_0 b)\, J_n(k_0 b) - H_n^{(2)}(k_0 b)\, J_n'(k_0 b)}{H_n^{(2)\prime}(k_0 b)\, U_n(ka, kb) - H_n^{(2)}(k_0 b)\, U_n'(ka, kb)},$$

$$(10c) \qquad C_n = -\frac{U_n^{(1)}(ka)}{U_n^{(2)}(ka)},$$

where

$$(11) \qquad U_n(ka, kb) = \left[U_n^{(1)}(kb) - \frac{U_n^{(1)}(ka)}{U_n^{(2)}(ka)} U_n^{(2)}(kb)\right] \Big/ kb,$$

$$(12) \qquad \frac{k_0}{k} U_n'(ka, kb) = \left[\frac{d}{d(kb)} \frac{U_n^{(1)}(kb)}{kb}\right] - \frac{U_n^{(1)}(ka)}{U_n^{(2)}(ka)} \left[\frac{d}{d(kb)} \frac{U_n^{(2)}(kb)}{kb}\right],$$

and the prime indicates the derivative of the function with respect to its argument.

If the incident magnetic vector is in the z direction, the incident H_z field, the scattered H_z field, and the H_z field within the sheath may be expressed respectively as

$$(13) \qquad H_z^{\,inc} = H_0 \sum_{n=0}^{\infty} (2-\delta_n^0)(-i)^n J_n(k_0 r) \cos n\theta,$$

$$(14) \qquad H_z^{\,sc} = \sum_{n=0}^{\infty} a_n(2-\delta_n^0)(-i)^n H_n^{(2)}(k_0 r) \cos n\theta,$$

and

$$(15) \qquad H_z^{\,sh} = \sum_{n=0}^{\infty} b_n(2-\delta_n^0)(-i)^n[V_n^{(1)}(kr) + c_n V_n^{(2)}(kr)] \cos n\theta/k_0 r.$$

a_n, b_n, and c_n are the arbitrary constants, and H_0 is the amplitude of the incident wave. Matching the appropriate boundary conditions at $r = a$ and $r = b$ gives

$$(16a) \quad a_n = H_0 \frac{J_n'(k_0 b)\, V_n(ka, kb)[\epsilon(kb)/\epsilon_0] - J_n(k_0 b)\, V_n'(ka, kb)}{H_n^{(2)}(k_0 b)\, V_n'(ka, kb) - [\epsilon(kb)/\epsilon_0] H_n^{(2)\prime}(k_0 b)\, V_n(ka, kb)},$$

$$(16b) \quad b_n = H_0 \frac{\epsilon(kb)}{\epsilon_0} \frac{J_n'(k_0 b)\, H_n^{(2)}(k_0 b) - J_n(k_0 b)\, H_n^{(2)\prime}(k_0 b)}{H_n^{(2)}(k_0 b)\, V_n'(ka, kb) - [\epsilon(kb)/\epsilon_0] H_n^{(2)\prime}(k_0 b)\, V_n(ka, kb)},$$

$$c_n = -\left[\frac{d}{d(ka)} \frac{V_n^{(1)}(ka)}{ka}\right] \Big/ \left[\frac{d}{d(ka)} \frac{V_n^{(2)}(ka)}{ka}\right], \tag{16c}$$

in which

$$\frac{k}{k_0} V_n(ka, kb) = [V_n^{(1)}(kb) + c_n V_n^{(2)}(kb)]/kb, \tag{17}$$

$$V_n'(ka, kb) = \left[\frac{d}{d(kb)} \frac{V_n^{(1)}(kb)}{kb}\right] + c_n \left[\frac{d}{d(kb)} \frac{V_n^{(1)}(kb)}{kb}\right], \tag{18}$$

and the prime indicates the derivative of the function with respect to its argument.

The backscattering cross section is conventionally the convenient quantity to define. It represents the ratio of the total power scattered by a fictitious isotropic scatterer, which scatters energy in all directions with intensity equal to that scattered directly back toward the source by the actual scattering object, to the incident power per unit area on the scatterer: i.e.,

$$\sigma_B = 2\pi r \frac{S_B^{sc}}{S^{inc}}, \tag{19}$$

where S^{inc} is the incident power density and S_B^{sc} is the power density of the far-zone scattered field in the direction of the source. Hence, the backscattering cross section for the TM wave is

$$\sigma_B^{TM} = 2\pi r \frac{|E^{sc}|^2}{|E^{inc}|^2} = \frac{4}{k_0} \left| \sum_{n=0}^{\infty} \frac{A_n}{E_0} (2-\delta_n^0)(-1)^n \right|^2, \tag{20}$$

and the backscattering cross section for the TE wave is

$$\sigma_B^{TE} = 2\pi r \frac{|H^{sc}|^2}{|H^{inc}|^2} = \frac{4}{k_0} \left| \sum_{n=0}^{\infty} \frac{a_n}{H_0} (2-\delta_n^0)(-1)^n \right|^2. \tag{21}$$

III. THIN-SHEATH APPROXIMATION

If the inhomogeneous dielectric sheath is very thin, certain simplifications for the expressions for the expansion coefficients can be made. For instance, when $k(b-a) \ll 1$, then the following power-series expansions can be used:

$$U_n^{(1)}(ka) \approx U_n^{(1)}(kb) + U_n^{(1)\prime}(kb)\, k(a-b), \tag{22}$$

$$U_n^{(2)}(ka) \approx U_n^{(2)}(kb) + U_n^{(2)\prime}(kb)\, k(a-b). \tag{23}$$

Hence

$$U_n(ka, kb) \approx k(a-b) \left[\frac{ka}{U_n^{(2)}(ka)}\right] P, \tag{24}$$

$$\frac{k_0}{k} U_n'(ka, kb) \approx -P \left[\frac{ka}{U_n^{(2)}(ka)}\right], \tag{25}$$

where P is the Wronskian for the expression

$$\left\{U_n^{(1)}(kb)\left[\frac{d}{d(kb)}\frac{U_n^{(2)}(kb)}{kb}\right]-U_n^{(2)}(kb)\left[\frac{d}{d(kb)}\frac{U_n^{(1)}(kb)}{kb}\right]\right\}\frac{1}{kb}.$$

Substituting equations (24) and (25) into equation (10a) and simplifying gives

$$A_n \approx -E_0\frac{J_n(k_0b)+k(a-b)\,J_n'(k_0b)}{H_n^{(2)}(k_0b)+k(a-b)\,H_n^{(2)\prime}(k_0b)}. \tag{26}$$

It is noted that the expansion coefficients for the scattered field for a conducting cylinder coated with a thin homogeneous dielectric sheath are given by

$$A_n^{\text{homo}} \approx -E_0\frac{J_n(k_0b)+k_1(a-b)\,J_n'(k_0b)}{H_n^{(2)}(k_0b)+k_1(a-b)\,H_n^{(2)\prime}(k_0b)}, \tag{27}$$

where $k_1^2 = \omega^2\mu\epsilon_1$ and ϵ_1 is the dielectric constant of the homogeneous sheath. Upon comparing equations (26) and (27), one notes that if $k \approx k_1$, then the scattering characteristics of the cylinder coated with a thin inhomogeneous dielectric sheath are identical with those of the cylinder coated with a thin homogeneous sheath. This effect is demonstrated numerically in Fig. 5. Similar conclusions may be reached for the TE waves.

IV. A SPECIFIC EXAMPLE

To get an idea of how the solutions behave, numerical computations are carried out for a specific variation of the dielectric constant. The inhomogeneous dielectric is taken to be

$$\epsilon(r) = \epsilon_0(\alpha/k_0r), \tag{28}$$

where α is a constant and $k_0^2 = \omega^2\mu\epsilon_0$. The variation of the dielectric constant is chosen for the following reasons. First of all, with this variation, the solutions of equations (5) and (6) may be expressed in terms of well-known functions: i.e., the solutions of equations (5) and (6) are, respectively,

$$U_n^{(1),(2)}(\alpha k_0r) = \alpha k_0r\begin{Bmatrix}J_{2n}(2\sqrt{\alpha k_0r})\\ Y_{2n}(2\sqrt{\alpha k_0r})\end{Bmatrix} \tag{29}$$

and

$$V_n^{(1),(2)}(\alpha k_0r) = \sqrt{\alpha k_0r}\begin{Bmatrix}J_{\sqrt{(1+4n^2)}}(2\sqrt{\alpha k_0r})\\ J_{-\sqrt{(1+4n^2)}}(2\sqrt{\alpha k_0r})\end{Bmatrix}, \tag{30}$$

where $J_\nu(p)$ and $Y_\nu(p)$ are respectively the Bessel function and the Neumann function of order ν and argument p. Secondly, this variation roughly approximates the equivalent dielectric sheath profile of a re-entering vehicle (see Lew and Angelo 1960).

Numerical computations are carried out for the backscattering cross section for the TM wave using equations (29), (10a), and (20). The results are shown in Figs. 2 and 3. It can be seen from these figures that the backscattering behavior of the inhomogeneous dielectric coated cylinder does not present a pattern which can be predicted from a simple model. This rather unpredictable behavior is probably due to the fact that multiple scattering occurs at the boundary surfaces and part of the wave is trapped in the sheath.

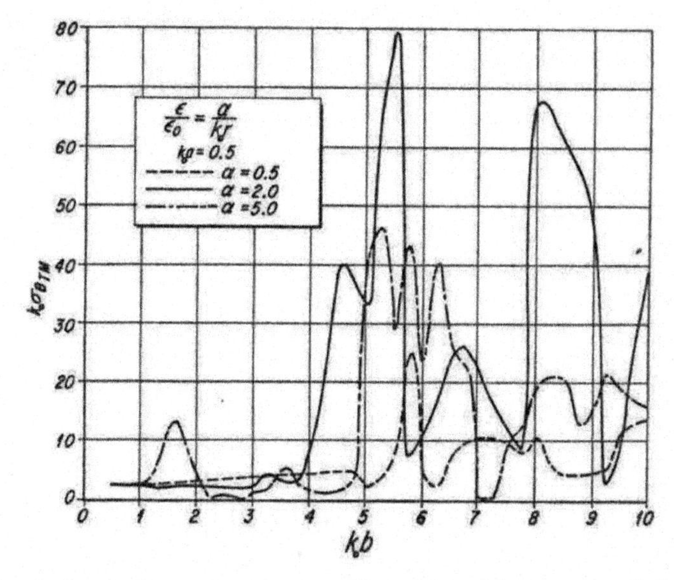

Fig. 2. Backscattering cross section of a cylinder with an inhomogeneous dielectric coating.

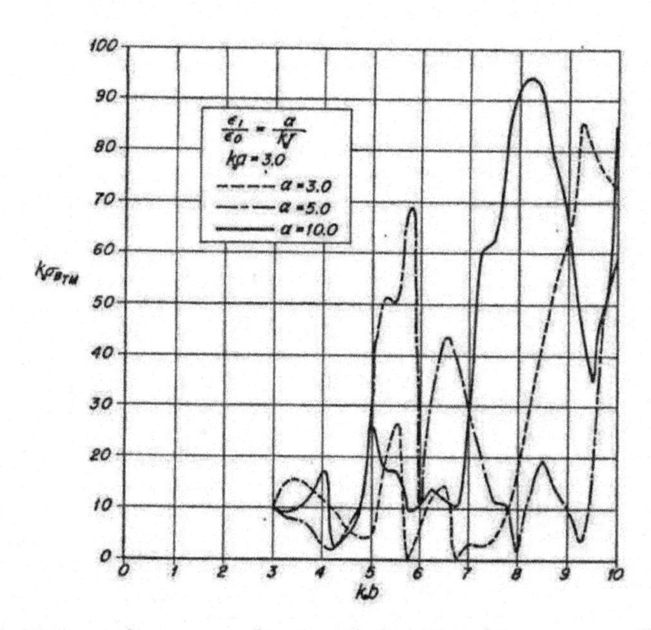

Fig. 3. Backscattering cross section of a cylinder with an inhomogeneous dielectric coating.

The effects of inhomogeneity of the dielectric on the backscattering characteristics of the wave can best be illustrated by comparing the backscattering cross section of a cylinder coated with a homogeneous dielectric with that of the inhomogeneous dielectric coated cylinder. Figures 4 and 5 show this

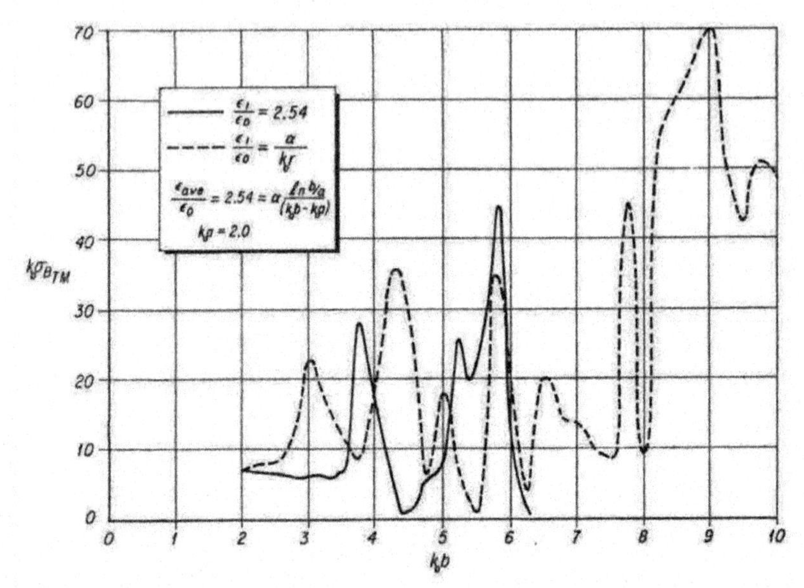

FIG. 4. A comparison between the backscattering cross section of a cylinder coated with an inhomogeneous dielectric sheath with that coated with a homogeneous dielectric sheath.

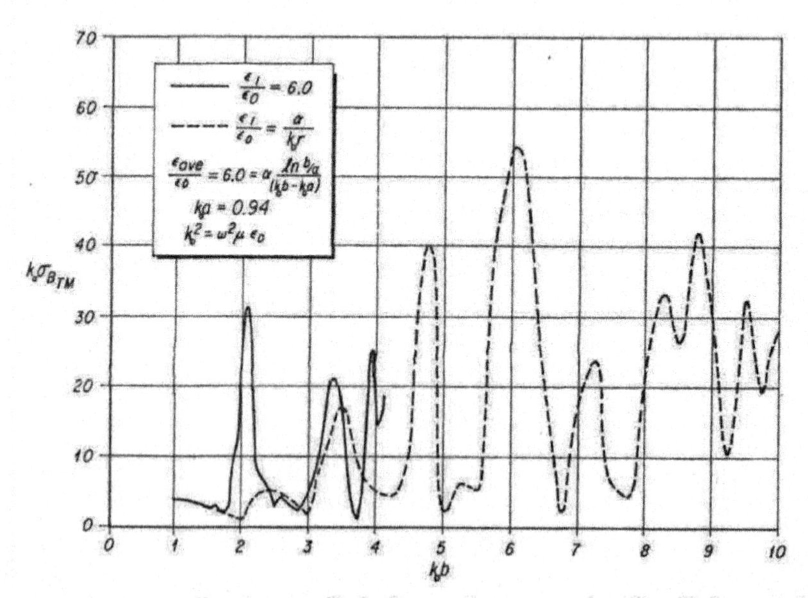

FIG. 5. A comparison between the backscattering cross section of a cylinder coated with an inhomogeneous dielectric sheath with that coated with a homogeneous dielectric sheath.

comparison. The data on the σ_B^{TM} for the homogeneous dielectric sheath case are obtained from a paper by Tang (1957). It is noted that distinct differences are observed except when the thickness of the coating is small and when the condition given in Section III is obeyed. One also notes that the major contribution to the backscattering cross section for a thinly coated cylinder is due to the conducting cylinder. It is learned from this example that caution is required in making the homogeneous sheath assumption for the plasma sheath of a re-entering vehicle, and in generalizing the conclusions obtained from the homogeneous sheath analysis.

In conclusion we shall remark that for obliquely incident plane waves the footnote in Section I should be taken into consideration.

We wish to thank Dick Joe for carrying out the computation. We also gratefully acknowledge the use of the computing facilities of the Western Data Processing Center at U.C.L.A.

REFERENCES

JASICK, H. 1953. Dissertation for Ph.D. degree at Polytechnic Institute of Brooklyn.
KEITEL, G. H. 1955. Proc. I.R.E. **43**, 1481.
LEW, H. G. and ANGELO, V. A. 1960. GE Space Science Lab., Philadelphia, Pa., Tech. Information Sec. No. R60SD356.
MARCUVITZ, N. 1951. Communs. Pure Appl. Math. **4**, 263.
TAI, C. T. 1958. Appl. Sci. Research, B, **7** (2), 113.
TANG, C. 1957. J. Appl. Phys. **25**, 628.

Scattering by an Infinite Cylinder Coated with an Inhomogeneous and Anisotropic Plasma Sheath

WILLARD V. T. RUSCH, MEMBER, IEEE, AND CAVOUR YEH, MEMBER, IEEE

Abstract—The problem of the interaction of an incident H wave with an infinite conducting cylinder coated with an inhomogeneous and anisotropic plasma sheath is treated analytically. Numerical results for the far-field pattern of the scattered field as well as the backscattering cross sections are presented for various interesting ranges of the parameters involved. Detailed discussion of the effects of the impressed static magnetic field, the sheath thickness, and the density and profile of the plasma sheath on the scattered wave is also presented. A parabolic electron density profile is assumed.

Manuscript received July 20, 1966; revised November 14, 1966. This paper was supported by the Joint Services Electronics Program.

W. V. T. Rusch is with Bell Telephone Laboratories, Inc., Holmdel, N. J. He is on leave of absence from the Department of Electrical Engineering, University of Southern California.

C. Yeh is with the Department of Electrical Engineering, University of Southern California, Los Angeles, Calif.

I. INTRODUCTION

THE problem of the scattering of electromagnetic waves by composite cylinders has been treated by various authors [1]–[3]. A rather comprehensive account of the analytical and experimental results has been given in a book by King and Wu [1]. Recently, Kodis [2] calculated the scattering cross section of a dielectric coated conducting cylinder using the geometric optics method. The case of the scattering by a cylinder coated with an inhomogeneous dielectric sheath has also been considered [3]. Due to current interest in the reentry problem [4], investigations have been carried out on the scattering by plasma coated cylinders [5], [6]. Most previous theoretical studies [4]–[6] have treated the scattering of electromagnetic waves by either a cylinder coated with an inhomogeneous plasma sheath or a cylinder coated with an anisotropic plasma sheath, but have not treated both effects simultaneously.

The purpose of this analysis is to consider the problem of the scattering of electromagnetic waves by a conducting cylinder coated with an anisotropic and inhomogeneous plasma sheath. The plasma density profile of the sheath is chosen such that it approximates the actual profile encountered for a reentry vehicle [4]. Numerical computations for the backscattering cross sections as well as the differential cross sections are carried out for various plasma sheath thicknesses and various plasma parameters. Results are discussed.

II. Formulation of the Problem

An infinite, perfectly conducting cylinder of radius b is coaxial with the z axis. The cylinder is coated with a radially inhomogeneous anisotropic plasma sheath. The plasma is anisotropic by virtue of an impressed magnetostatic field in the axial direction. It is assumed that the electromagnetic properties of the plasma may be adequately described in terms of a local macroscopic dielectric tensor [7], i.e.,

$$\epsilon(r) = \begin{bmatrix} \epsilon_1(r) & i\epsilon_2(r) & 0 \\ -i\epsilon_2(r) & \epsilon_1(r) & 0 \\ 0 & 0 & \epsilon_3(r) \end{bmatrix} \tag{1}$$

where

$$\epsilon_1(r) = \epsilon_0\left[1 - \frac{W_p^2(r)}{1 - W_g^2}\right]$$

$$\epsilon_2(r) = -\epsilon_0\left[\frac{W_p^2(r) W_g}{1 - W_g^2}\right]$$

$$\epsilon_3(r) = \epsilon_0[1 - W_p^2(r)] \tag{2}$$

with

$$W_p(r) = \sqrt{\frac{n(r)e^2}{m\epsilon_0\omega^2}} = \frac{\omega_p(r)}{\omega}$$

$$W_g = \frac{eB_0}{m\omega} = \frac{\omega_g}{\omega}. \tag{3}$$

$n(r)$ is the free electron density distribution, e is the electron charge, m is the electron mass, ϵ_0 is the free-space permittivity, and B_0 is the impressed magnetostatic field.

It can be shown from Maxwell's equations that, for fields independent of z, two independent modes exist: an E mode ($E_z \neq 0$; $H_z = 0$) and an H mode ($H_z \neq 0$; $E_z = 0$). The nonzero field components of an E mode are $E_z(r, \theta)$, $H_r(r, \theta)$, and $H_\theta(r, \theta)$. Assuming $E_z(r, \theta) = P(r)e^{\pm in\theta}$, one obtains from Maxwell's equations

$$\frac{d^2P(r)}{dr^2} + \frac{1}{r}\frac{dP(r)}{dr} + \left[\omega^2\mu_0\epsilon_3(r) - \frac{n^2}{r^2}\right]P(r) = 0. \tag{4}$$

The nonzero field components of an H mode are $H_z(r, \theta)$, $E_r(r, \theta)$, and $E_\theta(r, \theta)$. Assuming $H_z(r, \theta) = R(r)e^{\pm in\theta}$, one obtins from Maxwell's equations

$$\frac{d^2R(r)}{dr^2} + \frac{1}{rg(r)}\frac{d}{dr}(rg(r))\frac{dR(r)}{dr} + \frac{1}{g(r)}\left\{\omega^2\mu_0\epsilon_0 \pm \frac{n}{r}\frac{d}{dr}\left[\frac{\epsilon_2(r)}{\epsilon_1(r)}g(r)\right] - \frac{n^2}{r^2}g(r)\right\}R(r) = 0 \tag{5}$$

where

$$g(r) = \frac{\epsilon_0\epsilon_3(r)}{\epsilon_1^2(r) - \epsilon_2^2(r)}. \tag{6}$$

A time dependence of $e^{i\omega t}$ for all field components has been assumed and suppressed throughout. One notes from (4) that the anisotropy of the plasma does not enter into the solution for the E mode. Consequently, the analysis that follows will be restricted to the H mode.

III. Scattering of a Plane Wave

A plane wave with its magnetic vector polarized in the axial direction (i.e., an H wave) is assumed to impinge normally upon a plasma coated cylinder (Fig. 1). The axial component of the incident H wave may be expressed as

$$H_z^i = H_0 e^{ik_0 r\cos\theta} = H_0 \sum_{n=-\infty}^{\infty} (i)^n J_n(k_0 r) e^{in\theta},$$

$$E_z^i = 0. \tag{7}$$

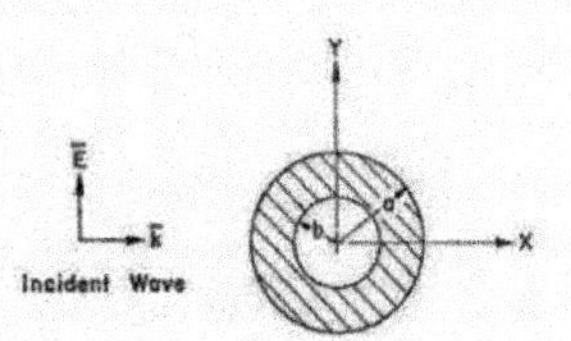

Fig. 1. Geometry.

The axial components of the scattered wave outside the cylinder and the wave within the plasma sheath take the form

$$H_z^s = H_0 \sum_{n=-\infty}^{\infty} B_n H_n^{(1)}(k_0 r) e^{in\theta}$$

$$E_z^s = 0 \tag{8}$$

and

$$H_z^t = H_0 \sum_{n=-\infty}^{\infty} A_n\{R_n^{(1)}(k_0 r) - Q_n(k_0 b) R_n^{(2)}(k_0 r)\} e^{in\theta}$$

$$E_z^t = 0 \tag{9}$$

with

$$Q_n(k_0 b) = \frac{\dfrac{\epsilon_2(b)}{\epsilon_1(b)}\dfrac{n}{k_0 b} R_n^{(1)}(k_0 b) + R_n^{(1)\prime}(k_0 b)}{\dfrac{\epsilon_2(b)}{\epsilon_1(b)}\dfrac{n}{k_0 b} R_n^{(2)}(k_0 b) + R_n^{(2)\prime}(k_0 b)} \tag{10}$$

where the prime signifies the derivative of the function with respect to its argument. A_n and B_n are arbitrary constants that can be determined by satisfying the boundary conditions at $r=a$, where a is the outer radius of the plasma sheath. The boundary condition at the conducting surface of the cylinder, i.e., at $r=b$, has been satisfied by (9). $J_n(k_0r)$ and $H_n^{(1)}(k_0r)$ are, respectively, the Bessel function and the Hankel function of the first kind. $R_n^{(1),(2)}(k_0r)$ are the two independent solutions of (5). Matching the tangential electric and magnetic field at $r=a$ gives

$$A_n = \frac{(i^n)}{\Delta}\left[H_n^{(1)}(k_0a)J_n'(k_0a) - J_n(k_0a)H_n^{(1)\prime}(k_0a)\right] \quad (11)$$

$$B_n = \frac{(i^n)}{\Delta}\left\{P_n(k_0a, k_0b)J_n'(k_0a) + \frac{\epsilon_0 J_n(k_0a)}{[\epsilon_2^2(a)-\epsilon_1^2(a)]} \times\left[\frac{\epsilon_2(a)n}{k_0a}P_n(k_0a, k_0b) + \epsilon_1(a)S_n(k_0a, k_0b)\right]\right\} \quad (12)$$

where

$$\Delta = \frac{-\epsilon_0 H_n^{(1)}(k_0a)}{[\epsilon_2^2(a) - \epsilon_1^2(a)]}\left[\frac{\epsilon_2(a)n}{k_0a}P_n(k_0a,k_0b) + \epsilon_1(a)S_n(k_0a,k_0b)\right] - P_n(k_0a, k_0b)H_n^{(1)}(k_0a),$$

$$P_n(k_0a, k_0b) = R_n^{(1)}(k_0a) - Q_n(k_0b)R_n^{(2)}(k_0a),$$

$$S_n(k_0a, k_0b) = R_n^{(1)\prime}(k_0a) - Q_n(k_0b)R_n^{(2)\prime}(k_0a), \quad (13)$$

where the prime denotes the derivative of the function with respect to its argument and $Q_n(k_0b)$ is given by (10).

The backscattering cross section per unit length is given

$$\sigma_B = \frac{4}{k_0}\left|\sum_{n=-\infty}^{\infty} B_n e^{in\pi/2}\right|^2. \quad (14)$$

The far-field pattern of the scattered wave is proportional to

$$\xi(\theta) = \left|\sum_{n=-\infty}^{\infty} B_n e^{in(\theta-\pi/2)}\right|, \quad (15)$$

where B_n is given by (12).

IV. Numerical Results and Discussion

In order to investigate the characteristics of the scattered field, the following free electron density within the sheath has been chosen:

$$n(r) = n_0\left[1 - \left(\frac{r}{d}\right)^2\right], \qquad b \le r \le a, d \ge a \quad (16)$$

where n_0 determines the magnitude of the electron density, and d is a constant related to the profile shape. Thus if $d=a$, the density decays monotonically to zero at the edge of the sheath ($r=a$); in the other limit, if $d \gg a$, the electron density is essentially constant within the sheath and drops abruptly to zero at the edge. The parabolic plasma density profile of (16) closely approximates the radial distribution of electron density in the plasma sheath surrounding a hypersonic reentry vehicle [4]. The two independent wavefunction solutions for (5) are of the form:

$$R_n^{(1)}(k_0r) = (k_0r)^{|n|}\sum_{s=0}^{\infty} f(n, s)(k_0r)^{2s} \quad (17)$$

$$R_n^{(2)}(k_0r) = [\ln (k_0r)^2]R_n^{(1)}(k_0r) + (k_0r)^{-|n|}\sum_{t=0}^{|n|-1} g(n, t)(k_0r)^{2t} + (k_0r)^{|n|}\sum_{s=1}^{\infty} h(n, s)(k_0r)^{2s}. \quad (18)$$

The series coefficients $f(n, s)$, $g(n, t)$ and $h(n, s)$ are evaluated by inserting (17) and (18) into (5). In order to assure series convergence for $b \le r \le a$, it is necessary for either of the following two conditions to be satisfied:

$$W_y > 1 \quad \text{and} \quad 1 + W_y > 2W_{p0}^2$$
$$W_y < 1 \quad \text{and} \quad 1 - W_y > 2W_{p0}^2$$

where

$$W_{p0}^2 = \frac{n_0 e^2}{m\epsilon_0\omega^2}.$$

The rate of convergence of these series has been discussed elsewhere [8].

In each of the cases considered below the incident plane wave is polarized with its E field normal to its H field parallel to the axis of the plasma-coated cylinder. The sheath thickness is determined by the ratio b/a. Unless specified otherwise, the electron density goes to zero at the edge of the sheath, i.e., $d/a=1.0$. The absolute plasma density is specified by the quantity $k_{p0}a=\omega W_{p0}a/c$, which, unless specified otherwise, will be 0.9. The static axial magnetic field is specified by the quantity $k_ya=\omega W_ya/c$, which will be either 1.5 or 3.0 in the cases considered. The absolute frequency, or size of the sheath-coated cylinder in wavelengths, is specified by the quantity $k_0a=\omega a/c=2\pi(a/\lambda)$ where λ is the free-space wavelength.[1]

The scattering from a relatively thin ($b/a=0.75$), moderately magnetized ($k_ya=1.5$) sheath is considered initially. The scattering cross section is plotted in Fig. 2 for a "low" frequency ($k_0a=1.05$); for a "medium" frequency ($k_0a=1.45$, which is also near the so-called "upper hybrid" resonant condition corresponding to $\omega^2=\omega_c^2+\omega_p^2$ in a homogeneous magnetized plasma [8]); and for a "high" frequency ($k_0a=5.0$). The low-frequency pattern is very nearly symmetric, and quite similar to the pattern for a cylinder of radius $k_0b=0.75\times1.05=0.79$. Hence, the sheath, inasmuch as its thickness is only 4 percent of a free-space wavelength, has virtually no effect. The medium-frequency or "resonant" pattern is considerably more asymmetric; the large forward lobe is more than four times greater than the

[1] Although the numerical results are presented in terms of the normalized parameters k_0a, k_ya, and $k_{p0}a$, the range of values considered may be translated into dimensions, frequencies, electron densities, etc., well within the present "state of the art." For example, if $a=10$ cm, the values $k_0a=2.95$, $k_ya=3.0$, and $k_{p0}a=0.9$ correspond, respectively, to a free-space wavelength of 21.3 cm (1.43 GHz), magnetic flux density of 0.51 kG, and an axial electron density of 0.2×10^{10} electrons/cm^3.

forward lobe of the low-frequency pattern. The high-frequency pattern is again almost indistinguishable from the scattering pattern from an equivalent uncoated cylinder. This is because the plasma sheath is transparent at very high frequencies. The predominant back lobe of the low-frequency pattern has changed to a predominant forward lobe. Only near the resonant condition is the effect of the sheath significant.

The previous calculations were repeated for the same values of $k_{p0}a$ and $k_c a$ but for a thicker plasma coating ($b/a=0.5$). The scattering cross section is plotted in Fig. 3 for a "low" frequency ($k_0 a=1.05$); for a "medium" frequency ($k_0=1.45$); and for a "high" frequency ($k_0 a=5.0$). The low- and high-frequency patterns are very similar to those of the corresponding conducting cylinder without a sheath. However, for the medium frequency, which is near the resonant condition, the effects of the sheath are again quite pronounced: the pattern is extremely asymmetric and considerably enhanced in amplitude, even larger than the high-frequency result.

The transition from a perfectly conducting cylinder to a sheath-coated cylinder to an inhomogeneous plasma column without an inner conductor is illustrated in Fig. 4. The parameters are $k_c a=1.5$; $k_{p0}a=0.9$; $k_0 a=0.9$. The sheath thickness is varied so that $b/a=1.0$ in Fig. 4(a); $b/a=0.75$ in Fig. 4(b); $b/a=0.5$ in Fig. 4(c). The pattern in Fig. 4(d) is for an inhomogeneous plasma column. The three-lobed pattern for the conducting cylinder changes finally into the single-lobed pattern of the plasma column. The transition that occurs illustrates the strong effect of the sheath thickness upon the scattering characteristics.

The calculations were also carried out for a cylinder with the same sheath size ($b/a=0.5$) but with a doubled magnetic field ($k_c a=3.0$). Scattering cross sections are plottted in Fig. 5 for a "low" frequency ($k_0 a=0.95$); a "medium" frequency near resonance ($k_0 a=2.95$); and a "high" frequency ($k_0 a=4.95$). Again, the resonant pattern is extremely large and asymmetric compared to the other two, which correspond closely to the scattering from an uncoated metallic cylinder.

In order to investigate the effects of sheath profile shape, patterns were computed as the shape of the profile was changed by varying c/a. Typical shapes are plotted in Fig. 6. Each profile has the same total number of electrons. In most cases profile shape has virtually no effect on the scattering patterns. No profile dependency could be detected as d/a was varied from 1.0 to ∞ in any of the cases plotted in Figs. 2-4. Only when the sheath exerts a relatively large effect on the scattering (compared to the inner cylinder) can a profile effect be noticed. For frequencies away from the resonant frequency, the metallic cylinder exerts a dominating effect on the scattered wave. Hence this variation of the sheath profile shape offers very little effect on the scattered wave. The asymmetric scattering cross section for a relatively thick sheath ($b/a=0.5$) and a relatively large magnetic field ($k_c a=3.0$) is plotted in Fig. 7 for a nearly resonant frequency ($k_0 a=2.95$). The two patterns in Fig. 7 are plotted for a profile which falls to zero at the edge of the sheath ($d/a=1$), and for a profile which is virtually constant within the sheath and drops to zero at the edge ($d/a=10$). The two patterns are nearly the same. The greatest difference is noted in the back lobe. This difference disappears, however, for other frequencies not near resonance.

The backscattering cross section, as defined in (14), is plotted in Fig. 8 for the previous cylinder ($d/a=0.5$; $k_c a=3.0$). The three curves plotted in Fig. 8 are: the backscattering curve which results if $d/a=1.0$; the backscattering curve which results if $d/a=10$; the backscattering curve which results for an uncoated metallic cylinder. It is evident that in each case the inner cylinder dominates the backscattering. Very little difference can be noted among the three curves, except near the resonant condition. The backscattering does not exhibit the deep nulls which are found for plasma cylinders without an inner metallic cylinder [8]. It should also be noted that a strong profile effect is absent, except near resonance.

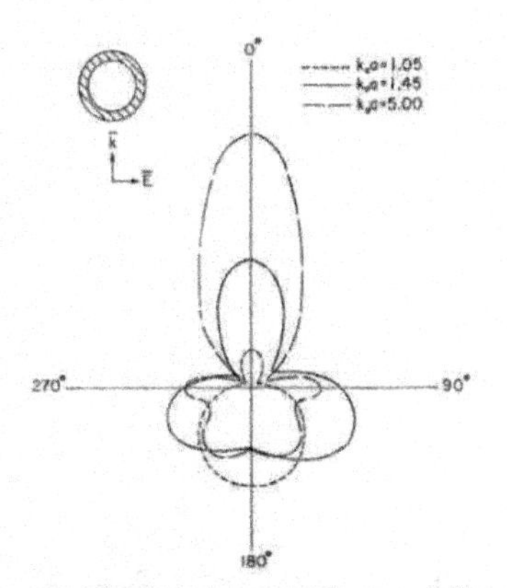

Fig. 2. Normalized scattering cross section vs. angle ($k_{p0}a=0.9$; $k_c a=1.5$; $b/a=0.75$; $d/a=1.0$).

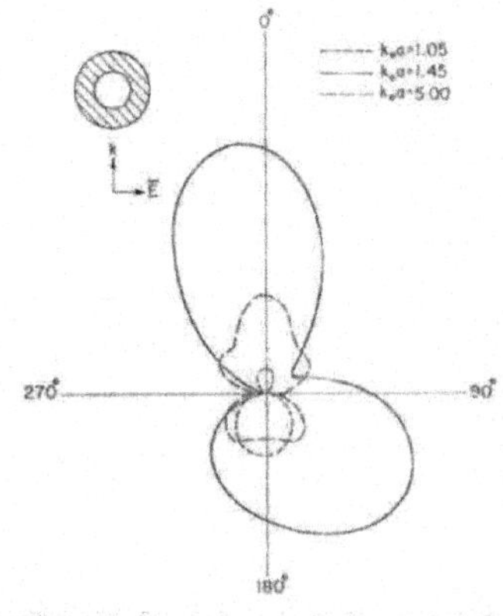

Fig. 3. Normalized scattering cross section vs. angle ($k_{p0}a=0.9$; $k_c a=1.5$; $b/a=0.5$; $d/a=1.0$).

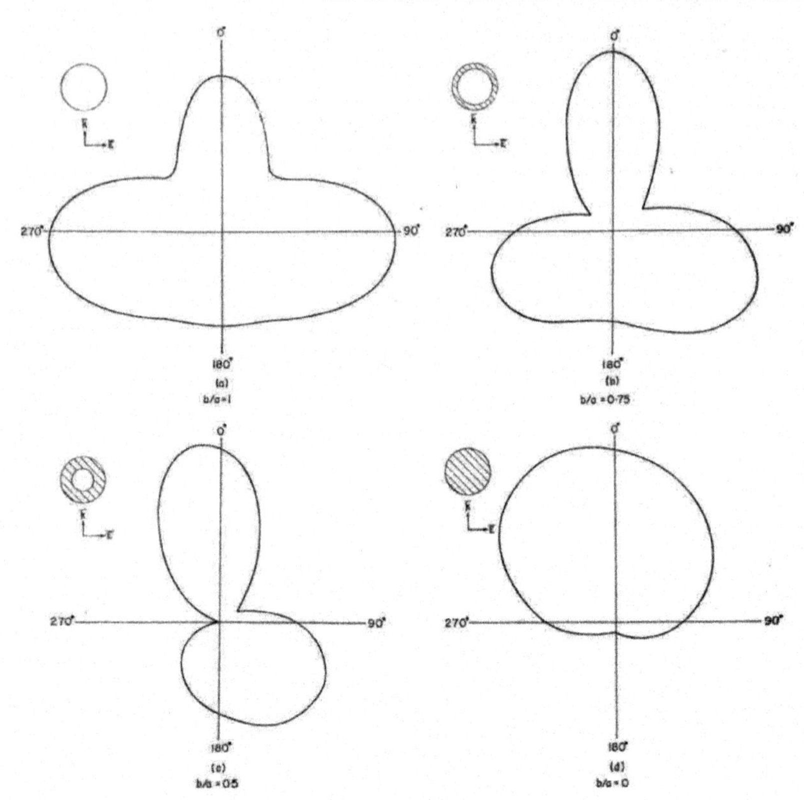

Fig. 4 Normalized scattering cross section vs. angle ($k_{p0}a=0.9$; $k_{p}a=1.5$; $k_{0}a=1.45$; $d/a=1.0$). (a) $b/a=1$; (b) $b/a=0.75$; (c) $b/a=0.5$; (d) $b/a=0$.

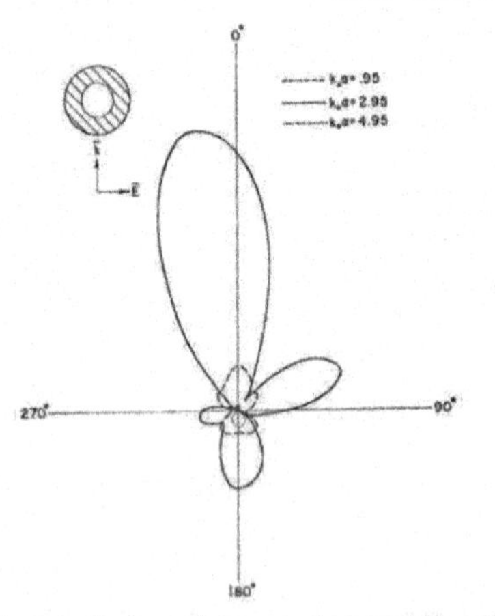

Fig. 5. Normalized scattering cross section vs. angle ($k_{p0}a=0.9$; $k_{i}a=3.0$; $b/a=0.5$; $d/a=1.0$).

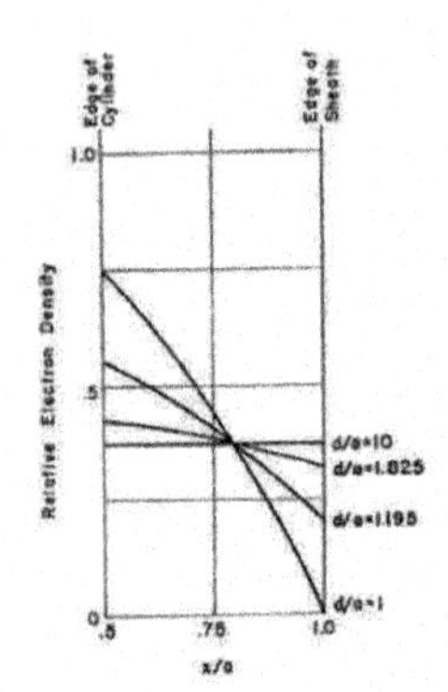

Fig. 6. Electron density profiles.

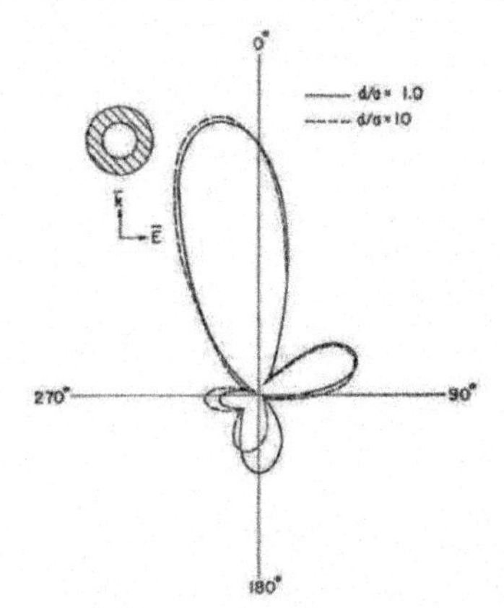

Fig. 7. Normalized scattering cross section vs. angle ($k_{p0}a=0.9$; $k_s a=3.0$; $k_0 a=2.95$; $b/a=0.5$).

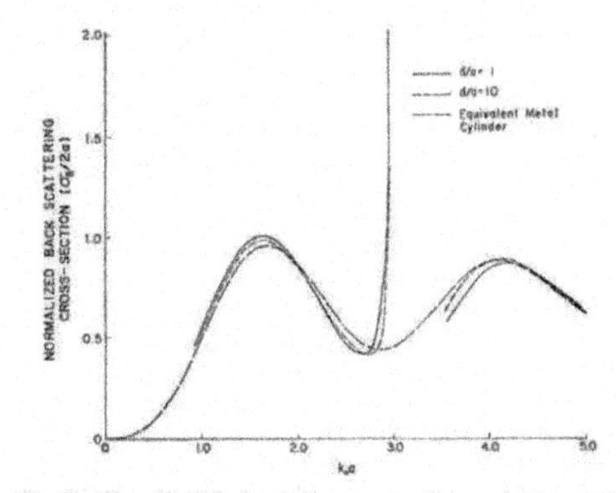

Fig. 8. Normalized backscattering cross section vs. $k_0 a$ ($k_{p0}a=0.9$; $k_s a=3.0$; $b/a=0.5$).

References

[1] R. W. P. King and T. T. Wu, *The Scattering and Diffraction of Waves*. Cambridge, Mass.: Harvard University Press, 1959.

[2] R. D. Kodis, "The scattering cross section of a composite cylinder, geometric optics," *IEEE Trans. on Antennas and Propagation*, vol. AP-11, pp. 86-93, January 1963.

[3] C. Yeh and Z. A. Kaprielian, "Scattering from a cylinder coated with an inhomogeneous dielectric sheath," *Canad. J. Phys.*, vol. 41, pp. 143-151, 1963.

[4] W. Rotman and G. Meltz, "Experimental investigation of the electromagnetic effects of re-entry," USAF Cambridge Research Center, Beford, Mass., Tech. Rept. AFCRL 87, 1961.

[5] H. C. Chen and D. K. Cheng, "Scattering of electromagnetic waves by an anisotropic plasma-coated conducting cylinder," *IEEE Trans. on Antennas and Propagation*, vol. AP-12, pp. 348-353, May 1964; "Scattered electromagnetic and plasma waves of a conducting cylinder coated with a compressible plasma," *Appl. Sci. Res.*, vol. B-11, pp. 442-452, 1965.

[6] J. R. Wait, *Electromagnetic Waves in a Stratified Media*. New York: Pergamon, 1962.

[7] W. P. Allis, S. J. Buchsbaum, and A. Bers, *Waves in Plasmas*. New York: Wiley, 1963.

[8] C. Yeh and W. V. T. Rusch, "Interaction of microwaves with an inhomogeneous and anisotrpic plasma medium," *J. Appl. Phys.*, vol. 36, pp. 2302-2306, 1965.

Reprinted from JOURNAL OF APPLIED PHYSICS, Vol. 36, No. 7, 2302-2306, July, 1965

Interaction of Microwaves with an Inhomogeneous and Anisotropic Plasma Column*

C. YEH AND W. V. T. RUSCH
Electrical Engineering Department, University of Southern California, Los Angeles, California
(Received 23 June 1964; in final form 15 February 1965)

The problem of the interaction of microwaves with a radially inhomogeneous plasma column confined by an impressed axial static magnetic field is treated analytically. Extensive numerical results for the backscattering cross sections are presented for various interesting ranges of the parameters involved. It is found that the often-used homogeneous model for interpretation of experimental data can sometimes lead to ambiguous results. This ambiguity arises because of the interdependence of the average plasma frequency, the gyrofrequency, and the inhomogeneity of the plasma density distribution. Further measurements are suggested to distinguish these effects.

I. INTRODUCTION

THE application of microwave measurement techniques to determine various plasma properties has been rather fruitful in recent years.[1,2] It is frequently the case, however, that the experimental data are interpreted in terms of a homogeneous plasma occupying a well-defined region of space. Although considerable theoretical work[3-5] has been concerned with the effect of departures from this idealized model, these analyses are often approximate or present results of such a general nature that their application to a particular physical situation is extremely difficult, if not impossible.

Wyatt[6] has recently considered the problem of the interaction of electromagnetic waves with a radially inhomogeneous plasma sphere. He found that the differential scattered intensity is extremely sensitive to variations in plasma density and surface diffusivity. The experimental data may lead to ambiguous conclusions about the physical properties of the plasma, depending upon the assumed theoretical model (i.e., whether the assumed plasma density is homogeneous or not).

It is noted that a cylindrical column of plasma is more physically realizable in the laboratory than an unsupported plasma sphere. Such a plasma column might be produced in a discharge tube with an impressed axial magnetic field to confine the plasma. This impressed field will produce anisotropic effects within the plasma. Most previous theoretical studies[3-5,7] have treated the interaction of electromagnetic waves with either an inhomogeneous plasma or an anisotropic plasma, but have not treated both effects simultaneously.

It is, therefore, the purpose of this analysis to treat the problem of the interaction of an electromagnetic wave with a cold, anisotropic, inhomogeneous plasma column. The validity of the cold-plasma model has been discussed quite thoroughly by Stix.[8] Numerical backscattering cross sections are evaluated for a wide range of plasma parameters. The parabolic plasma density profile assumed for the numerical computations closely approximates the actual profile encountered in a discharge tube.[9] In order to evaluate the effect of the inhomogeneity of plasma density on the backscattering cross section, results are compared with those for a homogeneous plasma column with a plasma density which is equal to the average density of the inhomogeneous column. Significant differences are observed.

The results of the analysis indicate that extreme care must be taken in the interpretation of experimental data. Furthermore, there is a need for new experimental evidence to differentiate between effects due to the magnetic field, to the plasma density inhomogeneity, and to the average plasma density.

II. FORMULATION OF THE PROBLEM

An infinite plasma cylinder of radius a is coaxial with the z axis (Fig. 1). It shall be assumed that the electromagnetic properties of the plasma may be adequately described in terms of a local, macroscopic dielectric tensor.[1] The plasma is anisotropic by virtue of an impressed magnetostatic field in the axial direction. It is reasonable to assume that the free electron density within the plasma cylinder is inhomogeneous in the radial direction. Consequently, the plasma frequency, defined by the relation

$$\omega_p^2(r) = n(r)e^2/m\epsilon_0, \tag{1}$$

will be a function of r, the radial coordinate. In Eq. (1) $n(r)$ is the free electron density, e is the electron charge, m is the electron mass, and ϵ_0 is the free-space permittivity.

* The research reported in this paper was supported partly by the Air Force Cambridge Research Laboratories and partly by the Joint Services Electronics Program.

[1] S. C. Brown, *Basic Data Plasma Physics* (John Wiley & Sons, Inc., New York, 1961).
[2] B. Agdur and B. Enander, J. Appl. Phys. 33, 575 (1962); B. Agdur, B. Kerzar, and F. Sellberg, Phys. Rev. 128, 1 (1962).
[3] F. A. Albini and R. G. Jahn, J. Appl. Phys. 32, 75 (1961).
[4] P. A. Clavier, J. Appl. Phys. 32, 570 (1961).
[5] W. P. Allis, S. J. Buchbaum, and A. Bers, *Waves in Plasmas* (John Wiley & Sons, Inc., New York, 1963).
[6] P. J. Wyatt, J. Appl. Phys. 34, 2078 (1963).
[7] P. M. Platzman and H. T. Ozaki, J. Appl. Phys. 31, 1597 (1960).
[8] T. H. Stix, *The Theory of Plasma Waves* (McGraw-Hill Book Company, Inc., New York, 1962).
[9] W. P. Thompson, *An Introduction to Plasma Physics* (Addison-Wesley Publishing Company, Inc., London, 1962).

The dielectric tensor (in cylindrical coordinates) of the cold, inhomogeneous, and anisotropic plasma with the magnetostatic field in the positive z direction is

$$\epsilon(r)=\begin{bmatrix} \epsilon_1(r) & i\epsilon_2(r) & 0 \\ -i\epsilon_2(r) & \epsilon_1(r) & 0 \\ 0 & 0 & \epsilon_3(r) \end{bmatrix}, \quad (2)$$

where

$$\begin{aligned} \epsilon_1(r)&=\epsilon_0\{1-[W_p^2(r)/(1-W_g^2)]\}, \\ \epsilon_2(r)&=-\epsilon_0[W_p^2(r)W_g/(1-W_g^2)], \\ \epsilon_3(r)&=\epsilon_0[1-W_p^2(r)], \end{aligned} \quad (3)$$

and

$$\begin{aligned} W_p(r)&=\omega_p(r)/\omega, \\ W_g&=eB_0/m\omega=\omega_g/\omega. \end{aligned} \quad (4)$$

It is the purpose of this paper to study the solutions of Maxwell's equations within such a plasma medium, with the restriction that none of the field quantities possess an axial variation. In polar coordinates, the z-independent Maxwell's equations may be written in component form (with an assumed $e^{-i\omega t}$ time dependence which is suppressed throughout):

$$(1/r)(\partial H_z/\partial\Theta)=-i\omega[\epsilon_1(r)E_r+i\epsilon_2(r)E_\Theta], \quad (5)$$

$$-(\partial H_z/\partial r)=-i\omega[-i\epsilon_2(r)E_r+\epsilon_1(r)E_\Theta], \quad (6)$$

$$\frac{1}{r}\frac{\partial}{\partial r}(rH_\Theta)-\frac{1}{r}\frac{\partial H_r}{\partial\Theta}=-i\omega\epsilon_3(r)E_z, \quad (7)$$

$$(1/r)(\partial E_z/\partial\Theta)=i\omega\mu_0 H_r, \quad (8)$$

$$-(\partial E_z/\partial r)=i\omega\mu_0 H_\Theta, \quad (9)$$

$$\frac{1}{r}\frac{\partial}{\partial r}(rE_\Theta)-\frac{1}{r}\frac{\partial E_r}{\partial\Theta}=i\omega\mu_0 H_z, \quad (10)$$

where μ_0 is the permeability of free space.

It can be seen from Eqs. (5)–(10) that two independent modes exist: an E mode ($E_z\neq 0$; $H_z=0$) and an H mode ($H_z\neq 0$; $E_z=0$).

E Mode

The nonzero field components of an E mode are $E_z(r,\Theta)$, $H_r(r,\Theta)$, and $H_\Theta(r,\Theta)$. Assuming $E_z(r,\Theta)=P(r)e^{\pm in\Theta}$, and combining Eqs. (7)–(9), one obtains

$$\frac{d^2P(r)}{dr^2}+\frac{1}{r}\frac{dP(r)}{dr}+\left[\omega^2\mu_0\epsilon_3(r)-\frac{n^2}{r^2}\right]P(r)=0. \quad (11)$$

$H_r(r,\Theta)$ and $H_\Theta(r,\Theta)$ can then be obtained from Eqs. (8) and (9) by differentiation of $E_z(r,\Theta)$. It is evident from Eqs. (7)–(10) that the anisotropy of the plasma does not enter into the solution for the E mode, since the electric vector is parallel to the static magnetic field. The solution is then identical to the solution for an isotropic, inhomogeneous plasma, which has been considered elsewhere.[3] Consequently, the analysis that follows will be restricted to the H mode.

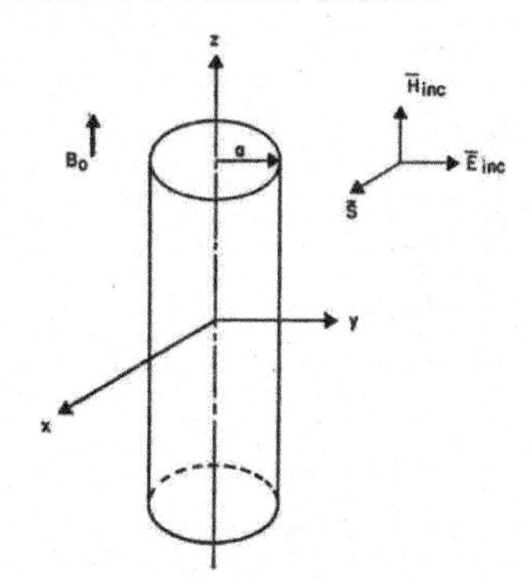

FIG. 1. Geometry of the problem.

H Mode

The nonzero field components of an H mode are $H_z(r,\Theta)$, $E_r(r,\Theta)$, and $E_\Theta(r,\Theta)$. Assuming $H_z(r,\Theta)=R(k_0r)e^{\pm in\Theta}$, and combining Eqs. (5), (6), and (10), one obtains

$$\frac{d^2R(k_0r)}{dr^2}+\frac{1}{rg(r)}\frac{d}{dr}[rg(r)]\frac{dR(k_0r)}{dr}+\frac{1}{g(r)}\times\left\{\omega^2\mu_0\epsilon_0\pm\frac{n}{r}\frac{d}{dr}\left[\frac{\epsilon_2(r)}{\epsilon_1(r)}g(r)\right]-\frac{n^2}{r^2}g(r)\right\}R(k_0r)=0, \quad (12)$$

where

$$g(r)=\epsilon_0\epsilon_1(r)/[\epsilon_1^2(r)-\epsilon_2^2(r)]. \quad (13)$$

Equation (12) can be solved when $W_p(r)$ is specified. $E_r(r,\Theta)$ and $E_\Theta(r,\Theta)$ can then be obtained from Eqs. (5) and (6) by differentiation of $H_z(r,\Theta)$. Since Eq. (7) is the only point at which $\epsilon_3(r)$ enters the field equations, it is evident that the classical plasma resonance which occurs at $W_p(r)=1$ should not necessarily produce a singularity in the H mode solution.

III. BACKSCATTERING FROM AN INHOMOGENEOUS, ANISOTROPIC PLASMA CYLINDER

A plane wave with its magnetic vector polarized in the axial direction (i.e., an H wave) is assumed to impinge normally upon a cold, inhomogeneous plasma cylinder with an impressed magnetostatic field in the z direction (Fig. 1). The nature of the assumed free electron density distribution will be discussed in Sec.

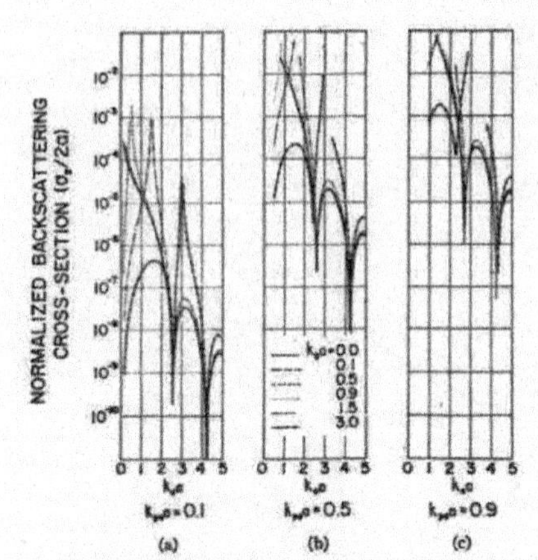

FIG. 2. Normalized backscattering cross section vs k_0a for increasing k_pa. (a) $k_{p0}a=0.1$, (b) $k_{p0}a=0.5$, (c) $k_{p0}a=0.9$.

IV. The axial component of the incident H wave may be expressed as

$$H_z^i = H_0 e^{ik_0 r\cos\theta} = H_0 \sum_{n=-\infty}^{+\infty} (i)^n J_n(k_0 r) e^{in\theta},$$

$$E_z^i = 0. \tag{14}$$

The axial components of the scattered wave outside the cylinder and the wave that has penetrated the cylinder take the form:

$$H_z^r = H_0 \sum_{n=-\infty}^{+\infty} B_n H_n^{(1)}(k_0 r) e^{in\theta},$$

$$E_z^r = 0, \tag{15}$$

and

$$H_z^t = H_0 \sum_{n=-\infty}^{+\infty} A_n R_n(k_0 r) e^{in\theta},$$

$$E_z^t = 0. \tag{16}$$

A_n and B_n are arbitrary constants that can be determined by satisfying the boundary conditions for the tangential fields. $R_n(k_0r)$ is one of the two independent solutions of Eq. (12) that is finite for $0 \leq r \leq a$. $H_n^{(1)}(k_0r)$ is the Hankel function of the nth order and the first kind. Matching the tangential electric and magnetic fields at $r=a$ gives:

$$A_n = [(i)^n/\Delta][H_n^{(1)}(k_0a)J_n'(k_0a) - J_n(k_0a)H_n^{(1)\prime}(k_0a)], \tag{17}$$

$$B_n = \frac{(i)^n}{\Delta}\left\{R_n(k_0a)J_n'(k_0a) + \frac{\epsilon_0 J_n(k_0a)}{[\epsilon_2^2(a)-\epsilon_1^2(a)]} \times\left[\frac{\epsilon_2(a)n}{k_0a}R_n(k_0a)+\epsilon_1(a)R_n'(k_0a)\right]\right\}, \tag{18}$$

where

$$\Delta = \frac{-\epsilon_0 H_n^{(1)}(k_0a)}{[\epsilon_2^2(a)-\epsilon_1^2(a)]}\left[\frac{\epsilon_2(a)n}{k_0a}R_n(k_0a)+\epsilon_1(a)R_n'(k_0a)\right] - R_n(k_0a)H_n^{(1)\prime}(k_0a), \tag{19}$$

where the prime denotes derivative of the function with respect to its argument. The backscattering cross section is then:

$$\sigma_B = \frac{4}{k_0}\left|\sum_{n=-\infty}^{+\infty} B_n e^{in(\pi/2)}\right|^2, \tag{20}$$

where B_n is given in Eq. (18).

IV. SOLUTIONS FOR THE RADIAL WAVEFUNCTIONS

The radial wavefunction solutions of Eq. (12) have been obtained for a radially inhomogeneous plasma region within which the free electron density is given by:

$$n(r) = n_0[1-(r/a)^2], \quad 0 \leq r \leq a, \tag{21}$$

where n_0 is the electron density on axis and a is the outer radius of the cylindrical region. This profile closely approximates the actual one encountered in a discharge tube.[8] The two independent wavefunction solutions are of the form:

$$R_n(k_0r) = (k_0r)^{|n|} \sum_{s=0}^{\infty} f(n,s)(k_0r)^{2s}, \tag{22}$$

$$S_n(k_0r) = [\ln(k_0r)^2]R_n(k_0r) + (k_0r)^{-|n|}\sum_{t=0}^{|n|-1} g(n,t)(k_0r)^{2t} + (k_0r)^{|n|}\sum_{s=1}^{\infty} h(n,s)(k_0r)^{2s}. \tag{23}$$

The series coefficients $f(n,s)$, $g(n,t)$, and $h(n,s)$ are evaluated by inserting Eqs. (22) and (23) into Eq. (12).

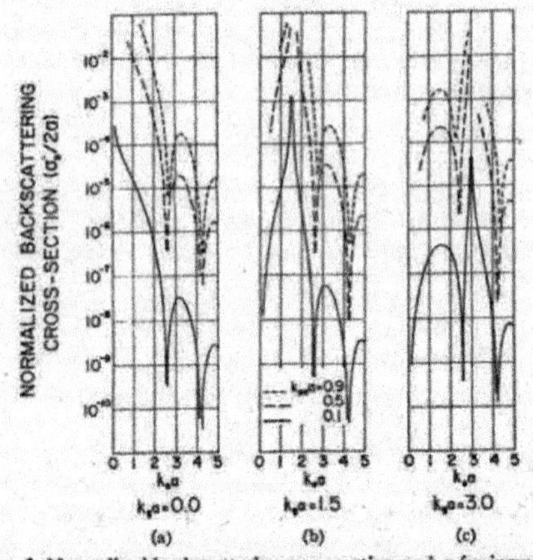

FIG. 3. Normalized backscattering cross section vs k_0a for increasing $k_{p0}a$. (a) $k_ea=0.0$, (b) $k_ea=1.5$, (c) $k_ea=3.0$.

To assure series convergence for $r \leq a$, it is necessary for either of the following two conditions to be satisfied:

$$W_g > 1 \quad \text{and} \quad 1 + W_g > 2W_{p0}^2, \tag{24}$$

$$W_g < 1 \quad \text{and} \quad 1 - W_g > 2W_{p0}^2, \tag{25}$$

where

$$W_{p0}^2 = n_0 e^2 / m \epsilon_0 \omega^2. \tag{26}$$

V. NUMERICAL RESULTS

The backscattering cross sections (per unit length of the cylinder) have been computed as a function of $k_0 a$ for various magnetic field intensities and electron densities. The normalized backscattering cross section, $\sigma_B/2a$, is plotted in Figs. 2-4. The independent variable for these curves (abscissa) is $k_0 a$, and the variable parameters are $k_g a$ and $k_{p0} a$, which are defined by

$$k_g a = \omega_g (\mu_0 \epsilon_0)^{\frac{1}{2}} a = (eB_0/m)(\mu_0 \epsilon_0)^{\frac{1}{2}} a, \tag{27}$$

$$k_{p0} a = \omega_{p0} (\mu_0 \epsilon_0)^{\frac{1}{2}} a = (n_0 e^2 / m \epsilon_0)^{\frac{1}{2}} (\mu_0 \epsilon_0)^{\frac{1}{2}} a. \tag{28}$$

Figures 2(a), 2(b), and 2(c) indicate the variation in normalized backscattering cross section for various values of the electron density: $k_{p0} a = 0.1, 0.5, 0.9$. Certain regions of these curves are not completed, because within these regions the relationship between frequency, magnetic field, and electron density is such that the convergence conditions in Eqs. (24) and (25) are not satisfied. In each case the series convergence fails as the backscattering cross section becomes anomalously large. These resonances are of two types: one is related to the resonance condition

$$\omega^2 = \omega_g^2 + \omega_p^2 \tag{29}$$

that occurs when a wave propagates at right angles to the impressed static magnetic field in a homogeneous plasma and the H field of the wave is parallel to the static field. This resonance condition will be referred to as the "upper hybrid" resonance.[8] The second type of resonance is related to the dipolar resonance[8]

$$\omega^2 = \omega_p^2 / 2 \tag{30}$$

occurring in a homogeneous plasma with no impressed field.

The backscattering cross sections in Fig. 2(a) consist primarily of a series of monotonically decreasing "geometrical" resonances upon which is superimposed a single resonance of the "upper hybrid" type indicated in Eq. (29). As the magnetic field increases, i.e., as $k_g a$ progresses from 0 to 3.0, this resonance occurs at correspondingly larger values of $k_0 a$. A secondary effect with increasing magnetic field is a moderate increase in the magnitude of the geometrical resonances where they are not masked by the dominant "upper hybrid" resonance.

Comparison of Figs. 2(b) and 2(c) with 2(a) indicates that allowing the electron density to increase over a moderate range shifts the entire pattern to a higher level

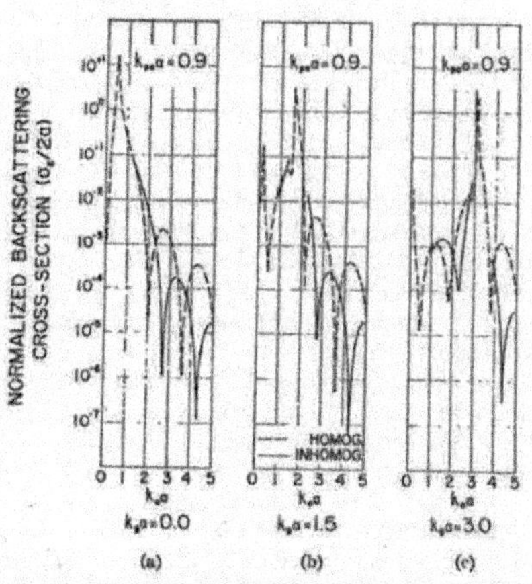

FIG. 4. Comparison of normalized backscattering cross section for inhomogeneous and equivalent homogeneous cylinders. (a) $k_g a = 0.0$, $k_{p0} a = 0.9$; (b) $k_g a = 1.5$, $k_{p0} a = 0.9$; (c) $k_g a = 3.0$, $k_{p0} a = 0.9$.

as the total number of scatterers increases. However, the relative shape and relative magnitudes of the curves are virtually unchanged.

The effect of an increasing electron density on the backscattering is illustrated more clearly in Figs. 3(a), 3(b), and 3(c), each with a constant value of impressed magnetic field ($k_g a = 0$, 1.5, 3.0). The pattern shape remains relatively unchanged as the entire pattern shifts upwards with increasing electron density. Over the range of parameters considered the backscattering cross section is approximately proportional to the square of the electron density.

For purposes of comparison, the normalized backscattering cross section has been computed for a homogeneous, magnetized plasma cylinder. The equivalent homogeneous cylinder has a uniform electron density within the cylinder which is equal to $\frac{1}{2}$ the axial electron density of the inhomogeneous cylinder. Thus the homogeneous and inhomogeneous cylinders being compared have the same average electron density and the same total number of electrons, but the inhomogeneous model has a continuous density at the edge (although the gradient is discontinuous) while the homogeneous model has a discontinuous density at the edge.

Figures 4(a), 4(b), and 4(c) compare the backscattering from inhomogeneous and equivalent homogeneous cylinders for $k_{p0} a = 0.9$ and $k_g a = 0.0$, 1.5, and 3.0. In the homogeneous cases it has been possible to compute the low-frequency "dipolar" resonance of Eq. (30) which could only be approached from the high-frequency side for the inhomogeneous cylinder because of series convergence difficulties. This dipolar resonance occurs at lower frequencies as the magnetic field increases.

Figures 4(b) and 4(c) indicate that the "upper hybrid" resonances occur very nearly at the same frequencies in the homogeneous and inhomogeneous cases. However, the greatest differences are noted in the magnitude and position of the geometrical resonances, which are considerably greater and shifted lower in frequency for the homogeneous cylinder. The differences are especially pronounced in Fig. 4(c), where the magnitudes of the first and third geometrical resonances are virtually the same for the homogeneous cylinder, but which are more than an order of magnitude different for the inhomogeneous cylinder.

To facilitate converting the normalized parameters into specific values, the following example is cited: if the radius of the plasma column is 2 cm, the value of $k_{p0}a=0.9$ corresponds to an axial electron density of $5.7\times10^{+18}$ m^{-3}, $k_c a=3.0$ corresponds to $B_0=2570$ G, and $k_0 a=1$ corresponds to $\lambda=12.6$ cm ($f=2.4$ kMc/sec). These values are applicable to the interpretation of Fig. 4(c).

On the basis of the above calculations, it appears that the often-used homogeneous model is liable to result in ambiguous interpretations when applied to a particular experimental configuration. The backscattering cross section at a particular frequency and a particular magnetic field, in general, is a complex function of *both* the magnitude and also the profile of the plasma distribution, and hence cannot be used for a unique determination of either one. To determine both these quantities, it is suggested that backscattering measurements should be carried out over a relatively wide range of frequencies, as might be carried out for the specific example in the previous paragraph.

ACKNOWLEDGMENTS

We wish to thank Dr. Gordon Stewart for many helpful discussions. The use of computing facilities at the Western Data Processing Center at UCLA is gratefully acknowledged.

Reprinted from APPLIED OPTICS, Vol. 14, page 2864, December 1975

Scattering of electromagnetic waves by arbitrarily shaped dielectric bodies

P. Barber and C. Yeh

The differential scattering characteristics of closed three-dimensional dielectric objects are theoretically investigated. The scattering problem is solved in a spherical basis by the Extended Boundary Condition Method (EBCM) which results in a system of linear equations for the expansion coefficients of the scattered field in terms of the incident field coefficients. The equations are solved numerically for dielectric spheres, spheroids, and finite cylinders to study the dependence of the differential scattering on the size, shape, and index of refraction of the scattering object. The method developed here appears to be most applicable to objects whose physical size is on the order of the wavelength of the incident radiation.

I. Introduction

The calculation of the scattering of electromagnetic waves by dielectric objects is a problem that has received increased attention in recent years. Knowledge of the scattered field is required in many areas, such as in investigations of the scattering of microwaves by raindrops and the scattering of light by small chemical and biological particles. The solution of the scattering problem for spherical objects is well known (the Mie theory) and has been used to great advantage in the study of many physical systems. However, many investigations are concerned with the scattering by nonspherical bodies, and the need for a method to determine rapidly the theoretical scattering by nonspherical objects is clearly indicated. Furthermore, a primary need is for methods that are applicable to objects whose physical size is on the order of the wavelength of the incident radiation (the so-called resonance region). This paper will be concerned with the solution of the scattering problem for nonspherical dielectric bodies. The method is most applicable to objects lying in the electromagnetics resonance region. The theoretical equations will be solved numerically for common solid geometric shapes to study the dependence of the scattering on the size, shape, and index of refraction of the scattering object.

When this work was done, both authors were with the University of California, Electrical Sciences & Engineering Department, School of Engineering & Applied Science, Los Angeles, California 90024. P. Barber is now with the University of Utah, Salt Lake City, Utah 84112.

Received 23 September 1974.

Many techniques have been developed for analyzing scattering and diffraction problems involving dielectric obstacles. Each of the available methods generally has a range of applicability that is determined by the size of the scattering object relative to the wavelength of the incident radation. The scattering by objects that are very small compared to the wavelength can be analyzed by the Rayleigh approximation,[1] and geometrical optics methods can be employed for objects that are electrically large.[2,3] Objects whose size is on the order of the wavelength of the incident radiation lie in the range commonly called the resonance region, and the complete wave nature of the incident radiation must be considered in the solution of the scattering problem.

The classical method of solution in the resonance region utilizes the separation of variables technique. This is only effective for bodies whose bounding surface coincides with one of the coordinate systems for which the vector Helmholtz equation is separable, and therefore its application has been restricted to infinite circular[4] and elliptic cylinders[5] in two dimensions and to the sphere[6] (the Mie theory) in three dimensions. Other methods that have been used to solve dielectric scattering problems in the resonance region include point matching methods and perturbation techniques. The point matching method[7,8] assumes a spherical expansion (or circular cylindrical expansion in two dimensions) of the scattered field is valid on the surface of a nonspherical body. The perturbation technique[9] considers a nonspherical body as a distortion (perturbation) of a perfect sphere. It is most effective for bodies that are only slightly nonspherical. Various integral equation[10,11] formulations have also been quite successful for solving dielectric scattering problems. Based on exact

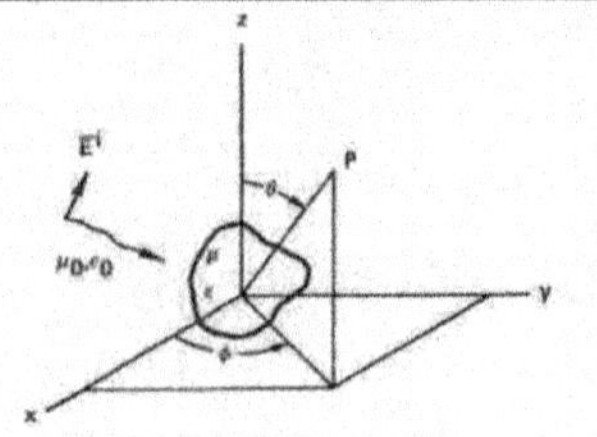
Fig. 1. Scattering geometry.

theory, these methods are quite general and are very amenable to solution by digital computer.

A new matrix formulation of scattering, which could be classified as an integral equation method, has recently been developed by Waterman.[12–15] Originally published for conducting bodies,[12,13] this method, called the Extended Boundary Condition Method, is exact and provides a general formulation for scattering from obstacles of arbitrary size and shape with size ratios from the Rayleigh region to the geometric optics limit. Waterman later extended the theory to dielectric obstacles[14] using the vector Huygen's principle.[16]

The present investigation begins with a theoretical development of the Extended Boundary Condition Method as it applies to scattering by dielectric objects. It will be shown that an alternate, but conceptually similar derivation, results using Schelkunoff's equivalence theorem,[17] rather than the Huygen's principle.

The over-all goal is to determine the scattered field when an arbitrary dielectric body is illuminated by a plane electromagnetic wave as shown in Fig. 1. Specifically, we wish to compute the differential scattering, which is the complete scattering pattern in all directions due to an incident wave in one particular direction. The dielectric body, assumed homogeneous and isotropic, is characterized by the constitutive parameters μ, ϵ, where μ is the permeability, and ϵ is the permittivity of the dielectric material (ϵ may be complex to account for the lossy case). The surrounding medium is considered to be free space with parameters μ_0, ϵ_0. The scattered field must be determined for all locations P on a sphere surrounding the scattering object.

The scattering problem is illustrated schematically in Fig. 2. The total field everywhere is given by the sum of the incident field and the scattered field, where the incident field $\overline{E}^i$, $\overline{H}^i$ is the field present in the absence of the scatterer, and the scattered field $\overline{E}^s$, $\overline{H}^s$ is given by the difference between the field with the object present $(\overline{E}, \overline{H})$ and the incident field, that is,

$$\overline{E}^s = \overline{E} - \overline{E}^i, \overline{H}^s = \overline{H} - \overline{H}^i. \quad (1)$$

This scattered field can be thought of as the field produced by polarization currents within the scattering object.

The goal of the theoretical development is to find a solution for the scattered field in terms of the incident field and the physical characteristics of the scattering object. This is accomplished by an application of the equivalence theorem which casts the scattered field as due to a set of surface currents located coincident with the surface of the scattering object. The remainder of the analysis then consists of finding an expression for these surface currents. The approach followed ultimately finds the internal fields in terms of the incident field, then the surface currents in terms of the internal fields, and finally the scattered field as a function of these surface currents. The entire analysis can conveniently be broken down into the following steps:

(1) The equivalence theorem is applied, which, as far as the external fields are concerned, effectively replaces the scattering object by a set of surface currents over S. These surface currents, which are derived from the tangential components of the external field, are found to be the source of the scattered field outside S, and because the equivalence theorem results in a null field inside S, they radiate a scattered field within S that cancels the incident field.

(2) The problem resulting from the application of the equivalence theorem is analyzed. This problem is called the external problem, because the fields external to S are the same as in the original problem. One expression for the unknown surface currents in terms of the incident field is derived from the fact that these surface currents radiate a scattered field that cancels the incident field throughout the interior volume. These equations are put in a form suitable for numerical solution by expansion of the various field quantities in vector spherical wave functions.

(3) Considering the complete problem again, the fields internal to the dielectric region are now expanded in regular vector spherical wave functions with coefficients to be determined.

(4) The boundary conditions at the surface are applied, and the continuity of the tangential fields leads to the linear system of integral equations for the coefficients of the unknown internal fields in terms of the expansion coefficients of the incident field.

(5) The scattered far field is then determined by evaluating the internal field at the surface and then

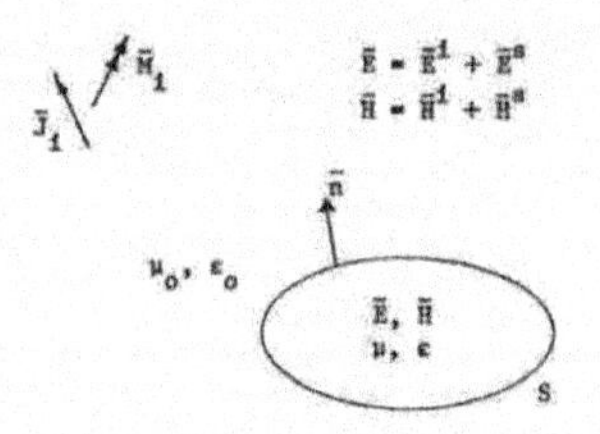
Fig. 2. The scattering problem. $\overline{J}_i$ and $\overline{M}_i$ are the sources of the incident field.

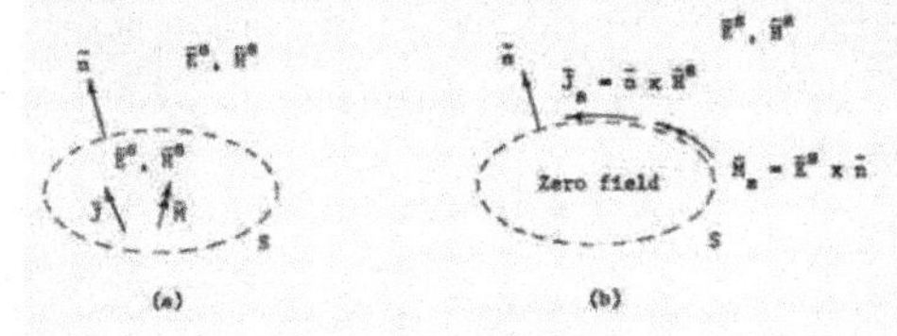

Fig. 3. Application of the equivalence theorem to the scattered field sources.

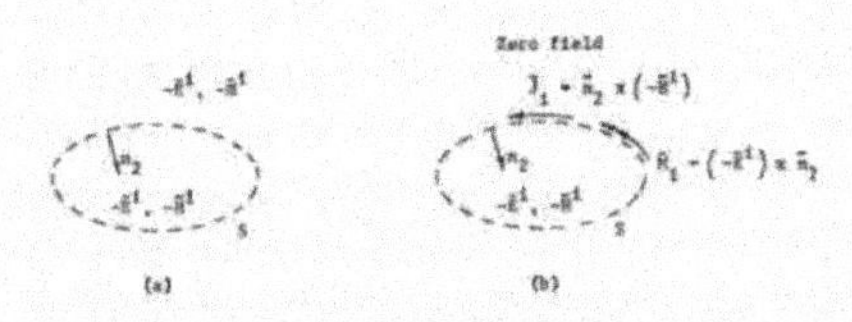

Fig. 4. Application of the equivalence theorem to the negative of the incident field sources.

solving for the scattered field in terms of the surface fields. The differential scattering cross section is proportional to the square of the far field amplitude.

II. Application of the Equivalence Principle

Let a closed surface S separate an isotropic homogenous medium into two regions as shown in Fig. 3(a). All the sources are contained within S so that the region outside S is source-free. Schelkunoff's equivalence theorem states that the field in a source-free region bounded by a surface S could be produced by a distribution of electric and magnetic currents on this surface, and, in this sense, the actual source distribution can be replaced by an equivalent distribution. Furthermore, if the field produced by the original source is $\bar{E}^s$, $\bar{H}^s$, the equivalent sources on S consist of an electric current sheet of density $\hat{n} \times \bar{H}^s$ and a magnetic current sheet of density $\bar{E}^s \times \hat{n}$, where the normal $\hat{n}$ points from the region containing the sources to the source-free region. The application of Schelkunoff's equivalence theorem is shown in Fig. 3(b). Note that the boundary conditions at S indicate that the surface currents produce a null field within S.

We relate the situation shown in Fig. 3 to the scattered field portion of our scattering problem by identifying $\bar{J}$ and $\bar{M}$ in Fig. 3(a) as polarization currents[17] within S (which have been induced by an incident field), which radiate in free space to produce the scattered field $\bar{E}^s$, $\bar{H}^s$. In Fig. 3(b), these polarization currents have been replaced by equivalent surface currents that radiate the scattered field external to S and a null field inside S.

Now we go through some additional transformations to construct the complete scattering problem as a sum of fields due to different sources.

Consider the situation depicted in Fig. 4(a). Here we have a set of sources $-\bar{J}_i$, $-\bar{M}_i$ radiating in free space and producing a field $-\bar{E}^i$, $-\bar{H}^i$. The equivalene theorem can be applied as shown in Fig. 4(b) so that the sources outside S can be replaced by equivalent surface currents on S, which radiate the fields $-\bar{E}^i$, $-\bar{H}^i$ within S and a null field outside S. Relative to the total scattering problem, we identify these fields as the negative of the incident field.

We now apply superposition and add together the sources and fields from Figs. 3(b) and 4(b), obtaining the situation shown in Fig. 5(a). Here we end up with a set of surface currents that radiate the scattered field external to S and the negative of the incident field internal to S. If we now add to Fig. 5(a) a set of sources $\bar{J}_i$, $\bar{M}_i$ producing an incident field $\bar{E}^i$, $\bar{H}^i$, we obtain the configuration shown in Fig. 5(b). $\bar{E}_+$ and $\bar{H}_+$ are the values of the external $\bar{E}$, $\bar{H}$ fields at the surface. It can be seen that external to the surface S, the sources and fields are exactly the same as those existing in the original scattering problem, and we have replaced the scattering object by a set of surface currents over a surface S. Furthermore, these surface currents radiate in unbounded free space to produce the scattered field outside S and the negative of the incident field inside S.

III. External Problem

Referring to Fig. 5(b), the entire region is unbounded, and, therefore, the scattered fields everywhere due to $\bar{J}_+$ and $\bar{M}_+$ can be determined from the vector potentials $\bar{A}$ and $\bar{F}$.[4]

$$\bar{E}^s = -\nabla \times \bar{F} - \frac{1}{j\omega\epsilon_0}(\nabla \times \nabla \times \bar{A}), \tag{2a}$$

$$\bar{H}^s = \nabla \times \bar{A} - \frac{1}{j\omega\mu_0}(\nabla \times \nabla \times \bar{F}), \tag{2b}$$

where

$$\bar{A} = \frac{1}{4\pi}\int_s \frac{\bar{J}_s \exp(jk|\bar{r} - \bar{r}'|)}{|\bar{r} - \bar{r}'|} dS, \bar{J}_s = \bar{n} \times \bar{H}_s \tag{3a}$$

and

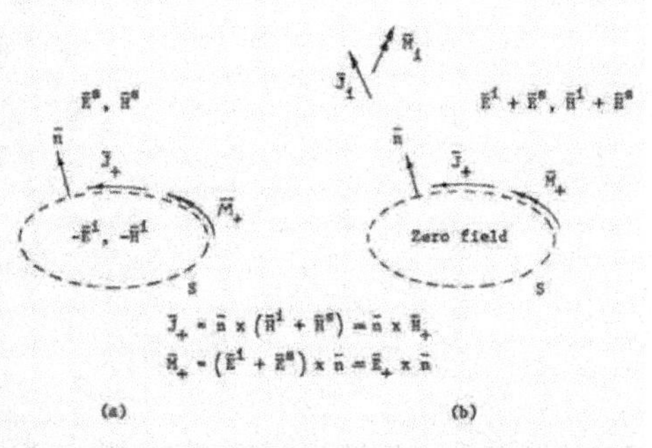

Fig. 5. Summing of sources and fields to obtain the external problem: (a) the sum of Figs. 3(b) and 4(b); (b) adding in the incident sources and fields.

$$\bar{F} = \frac{1}{4\pi}\int_s \frac{\bar{M}_s \exp(jk|\bar{r} - \bar{r}'|)}{|\bar{r} - \bar{r}'|} dS, \bar{M}_s = \bar{E}_s \times \bar{n}; \quad (3b)$$

$\bar{r}'$ and $\bar{r}$ are position vectors from an interior origin to source and field points, respectively, and an $e^{-j\omega t}$ time variation has been assumed. The expression for the scattered electric field can be obtained by substituting Eqs. (3a) and (3b) into Eq. (2a).

$$\bar{E}^s(\bar{r}) = \nabla \times \int_s (\bar{n} \times \bar{E}_s) g(kR) dS - \nabla \times \nabla \times \int_s \frac{1}{j\omega\epsilon_0} (\bar{n} \times \bar{H}_s) g(kR) dS, \quad (4)$$

where $g(kR)$ is the free space Green's function $[\exp(jkR)]/(4\pi R)$

$$R = |r - r'| \text{ and } k = 2\pi/\lambda.$$

The expression for the total field is given by

$$\left.\begin{matrix}\bar{E}(r)\\ 0\end{matrix}\right\} = \bar{E}^i(\bar{r}) + \nabla \times \int_s (\bar{n} \times \bar{E}_s) g(kR) dS - \nabla \times \nabla \times \int_s \frac{1}{j\omega\epsilon_0} (\bar{n} \times \bar{H}_s) g(kR) dS; \bar{r} \begin{cases}\text{outside } S\\ \text{inside } S.\end{cases} \quad (5)$$

It can be seen that for $\bar{R}$ inside S, Eq. (5) partially determines the surface currents because of the requirement that the scattered field must cancel the incident field throughout the interior volume as previously indicated, i.e., for $\bar{r}$ inside S,

$$\nabla \times \int_s (\bar{n} \times \bar{E}_s) g(kR) dS - \nabla \times \nabla \times \int_s \frac{1}{j\omega\epsilon_0} (\bar{n} \times \bar{H}_s) g(kR) dS = -\bar{E}^i(\bar{r}). \quad (6)$$

This equation can be expanded by making use of the spherical vector harmonics $\bar{M}$ and $\bar{N}$ which have been defined by Stratton.[6] These functions, solutions of the vector wave equation, are given by

$$\bar{M}_{\sigma mn}(\bar{r}) = \nabla \times \bar{r} \begin{matrix}\cos m\phi\\ \sin m\phi\end{matrix} P_n^m(\cos\theta) z_n(kr) \quad (7a)$$

$$\bar{N}_{\sigma mn}(\bar{r}) = \frac{1}{k}\nabla \times \bar{M}_{\sigma mn}(\bar{r}), \quad (7b)$$

where

σ = even or odd;
$P_n^m(\cos\theta)$ = associated Legendre function;
$z_n(kr)$ = an appropriate spherical Bessel function.

For solutions of the wave equation that must be finite at $r = 0$, $z_n(kr) = j_n(kr)$, and the resulting vector spherical functions $\bar{M}$ and $\bar{N}$ are known as solutions of the first kind. The other vector spherical functions we will be using are obtained by using the spherical Hankel functions, i.e., $z_n(kr) = h_n^{(1)}(kr) = j_n(kr) + jn_n(kr)$. This solution, representing outgoing waves, is called the solution of the third kind.

Using these spherical vector harmonics, the incident field is given by

$$\bar{E}^i(\bar{r}) = \sum_{\nu=1}^{\infty} D_\nu [a_\nu \bar{M}_\nu^1(k\bar{r}) + b_\nu \bar{N}_\nu^1(k\bar{r})], \quad (8)$$

where ν is a combined index incorporating σ, m, and n. D_ν is a normalization constant, and the expansion coefficients a_ν and b_ν are known for a specified incident field.

$$D_\nu = \epsilon_m \frac{(2n + 1)(n - m)!}{4n(n + 1)(n + m)!}, \epsilon_m = \left.\begin{matrix}1\\ 2\end{matrix}\right\}, m \begin{matrix} = 0\\ > 0.\end{matrix}$$

The terms in the integrals of Eq. (6) are expanded as follows:

$$(\bar{n} \times \bar{E}_s) g(kR) = (\bar{n} \times \bar{E}_s) \cdot \bar{\bar{G}}, \quad (9a)$$

$$(\bar{n} \times \bar{H}_s) g(kR) = (\bar{n} \times \bar{H}_s) \cdot \bar{\bar{G}}, \quad (9b)$$

where $\bar{\bar{G}}(kR)$ is the free space Green's dyadic given by Morse and Feshbach.[18]

$$\bar{\bar{G}}(kr) = \frac{jk}{\pi}\sum_{\nu=1}^{\infty} D_\nu [\bar{M}_\nu^3(k\bar{r}_>)\bar{M}_\nu^1(k\bar{r}_<) + \bar{N}_\nu^3(k\bar{r}_>)\bar{N}_\nu^1(k\bar{r}_<)], \quad (10)$$

$\bar{r}_>$, $\bar{r}_<$ are, respectively, the greater and lesser of $\bar{r}$, $\bar{r}'$. The irrotational terms normally present in Eq. (10) are not specified as they will wash out due to the curl operation in Eq. (6).

Equation (6) is required to hold throughout the entire interior volume, i.e., for all r inside S. Now we wish to substitute the expansions of Eq. (8) through Eq. (10) into Eq. (6). Before this is done, the region of convergence of the expansions of Eqs. (8) and (10) should be investigated. The series expansion for the incident field given in Eq. (8) obviously has no singularities within the interior volume, but the same cannot be said for the Green's function expansion in Eq. (10); $g(kR) = [\exp(jkR)]/(4\pi R)$ has a singularity at $R = 0$, i.e., for $\bar{r} = \bar{r}'$. Therefore, an expansion about an origin in the enclosed volume is valid only within an inscribed sphere (and outside a circumscribed sphere), as shown in Fig. 6.

For the moment, restrict the field point $\bar{r}$ to lie within the inscribed sphere. Substituting Eq. (8) through Eq. (10) into Eq. (6) gives the following set of equations:

$$\frac{jk^2}{\pi}\int_s \left[\bar{N}_\nu^3(k\bar{r}')\cdot(\bar{n} \times \bar{E}_s) + j\left(\frac{\mu_0}{\epsilon_0}\right)^{1/2} \bar{M}_\nu^3(k\bar{r}')\cdot(\bar{n} \times \bar{H}_s)\right] dS = -a_\nu; \quad (11a)$$

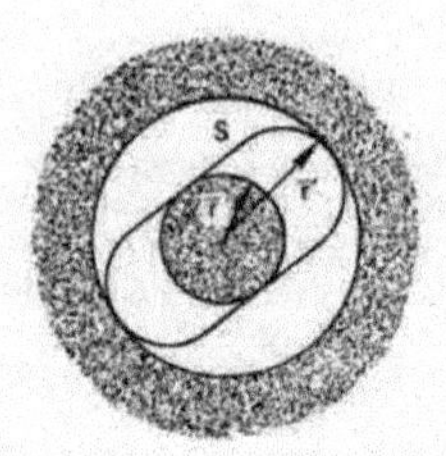

Fig. 6. Region of convergence of the Green's function expansion (shaded).

$$\frac{ik^2}{\pi}\int_S\Big[\overline{M}_\nu^{3}(k\overline{r}')\cdot(\hat{n}\times\overline{E}_+)$$

$$+\,j\left(\frac{\mu_0}{\epsilon_0}\right)^{1/2}\overline{N}_\nu^{3}(k\overline{r}')\cdot(\hat{n}\times\overline{H}_+)\Big]\,ds=-b_\nu;\quad\text{(11b)}$$

where $\nu = 1, 2, 3, \ldots$.

Note that the substitution into Eq. (6) has resulted in two sets of equations. The reason for this is that the coefficient of each regular wave function ($\overline{M}$ or $\overline{N}$) must vanish separately due to the orthogonality of the functions over a spherical surface about the origin.

The solution of Eqs. (11) guarantees that the total field will be zero within the inscribed sphere, but we need to guarantee a null field throughout the entire interior volume. As shown by Waterman,[12] this can be accomplished by using the concept of analytic continuation.

From the theory of complex variables, if an analytic function can be expanded in a region with a finite radius of convergence, that function can also be expanded about any origin within the original region of convergence out to the nearest singularity of the function. This process can be repeated, resulting in the analytic continuation of the function from its original region of definition to other parts of the enclosed volume.[19] The result is that the solution of Eqs. (11), which guarantees zero total field within the inscribed sphere, is sufficient to guarantee that the total field is zero within the entire enclosed volume.

It should be noted that Eqs. (11) can be solved for $\hat{n}\times\overline{E}_+$ and $\hat{n}\times\overline{H}_+$ (the surface currents). These surface currents can then be substituted into Eq. (4) to determine the scattered field, but this scattered field is the result for the external problem alone, which by now is recognized as the problem of scattering by a perfect conductor. The external problem represents only half of the problem for dielectric scattering, and the other half, the internal problem, must now be considered.

IV. Internal Problem

Assume that the field inside the dielectric can be approximated by

$$\overline{E}(k'\overline{r})=\sum_{\mu=1}^{N}[c_\mu\overline{M}_\mu^{1}(k'\overline{r})+d_\mu\overline{N}_\mu^{1}(k'\overline{r})]\,,\quad\text{(12a)}$$

where μ incorporates the indices σ, m, n, and c_μ and d_μ are unknown coefficients. $k'=\omega(\mu\epsilon)^{1/2}=(\mu_r\epsilon_r)^{1/2}k$. The $\overline{H}$ field internal to S is given by

$$\overline{H}(k'\overline{r})=\frac{1}{j\omega\mu}[\nabla\times\overline{E}(k'r)]$$

$$=-j\left(\frac{\epsilon_r}{\mu_r}\right)^{1/2}\left(\frac{\epsilon_0}{\mu_0}\right)^{1/2}\sum_{\mu=1}^{N}[c_\mu\overline{N}_\mu^{1}(k'\overline{r})+d_\mu\overline{M}_\mu^{1}(k'\overline{r})].\quad\text{(12b)}$$

V. Application of the Boundary Conditions at the Surface

The boundary conditions at the surface require that the tangential components of the fields be continuous, i.e.,

$$\overline{n}\times\overline{H}_+=\overline{n}\times\overline{H}_-\quad\text{(13a)}$$

and

$$\overline{n}\times\overline{E}_+=\overline{n}\times\overline{E}_-.\quad\text{(13b)}$$

The plus (+) sign and minus (−) sign subscripts on the surface fields indicate that these fields are the external and internal fields, respectively, evaluated at the surface. From Eq. (12), the tangential components of the internal fields evaluated at the surface are

$$\overline{n}\times\overline{E}_-=\sum_{\mu=1}^{N}[c_\mu\overline{n}\times\overline{M}_\mu^{1}(k'\overline{r}')+d_\mu\overline{n}\times\overline{N}_\mu^{1}(k'\overline{r}')],\quad\text{(14a)}$$

$$\overline{n}\times\overline{H}_-=-j\left(\frac{\epsilon_r}{\mu_r}\right)^{1/2}\left(\frac{\epsilon_0}{\mu_0}\right)^{1/2}\sum_{\mu=1}^{N}[c_\mu\overline{n}\times\overline{N}_\mu^{1}(k'\overline{r}')+d_\mu\overline{n}\times\overline{M}_\mu^{1}(k'\overline{r}')].\quad\text{(14b)}$$

Now, due to the equality of the tangential surface fields in Eqs. (13), Eqs. (14) can now be substituted into the first $2N$ of Eqs. (11), giving the system of equations,

$$\left[K+\left(\frac{\epsilon_r}{\mu_r}\right)^{1/2}J\right]c_\mu+\left[L+\left(\frac{\epsilon_r}{\mu_r}\right)^{1/2}I\right]d_\mu=-ja_\nu$$

$$\nu=1,2,\ldots,N,\quad\text{(15a)}$$

$$\left[I+\left(\frac{\epsilon_r}{\mu_r}\right)^{1/2}L\right]c_\mu+\left[J+\left(\frac{\epsilon_r}{\mu_r}\right)^{1/2}K\right]d_\mu=-jb_\nu,\quad\text{(15b)}$$

where

$$I=\frac{k^2}{\pi}\int_S\overline{n}\cdot\overline{M}_\nu^{3}(k\overline{r}')\times\overline{M}_\mu^{1}(k'\overline{r}')dS,$$

$$J=\frac{k^2}{\pi}\int_S\overline{n}\cdot\overline{M}_\nu^{3}(k\overline{r}')\times\overline{N}_\mu^{1}(k'\overline{r}')dS,$$

$$K=\frac{k^2}{\pi}\int_S\overline{n}\cdot\overline{N}_\nu^{3}(k\overline{r}')\times\overline{M}_\mu^{1}(k'\overline{r}')dS,$$

$$L=\frac{k^2}{\pi}\int_S\overline{n}\cdot\overline{N}_\nu^{3}(k\overline{r}')\times\overline{N}_\mu^{1}(k'\overline{r}')dS.$$

These are a set of simultaneous linear equations that can be solved for the expansion coefficients of the internal field. Subsequent analysis will provide a relationship between these coefficients and the expansion coefficients of the scattered field.

VI. Evaluation of the Scattered Field

The coefficients of the internal field can be obtained by solution of Eqs. (15). These coefficients can then be used in Eqs. (14) to find $\hat{n}\times\overline{E}_-$ and $\hat{n}\times\overline{H}_-$, which when substituted into Eq. (4) will give the expression for the scattered field.

$$\overline{E}^s(k\overline{r})=\sum_{\nu=1}^{N}[p_\nu\overline{M}_\nu^{3}(k\overline{r})+q_\nu\overline{N}_\nu^{3}(k\overline{r})],\quad\text{(16)}$$

where r is outside a circumscribed sphere,

$$p_\nu=-jD_\nu\sum_{\mu=1}^{N}\left\{\left[K'+\left(\frac{\epsilon_r}{\mu_r}\right)^{1/2}J'\right]c_\mu+\left[L'+\left(\frac{\epsilon_r}{\mu_r}\right)^{1/2}I'\right]d_\mu\right\},\quad\text{(17a)}$$

$$q_\nu = -jD_\nu \sum_{\mu=1}^{N} \left\{ \left[I' + \left(\frac{\epsilon_r}{\mu_r}\right)^{1/2} L' \right] c_\mu + \left[J' + \left(\frac{\epsilon_r}{\mu_r}\right)^{1/2} K' \right] d_\mu \right\}, \quad (17b)$$

and

$$I' = \frac{k^2}{\pi} \int_S \bar{n} \cdot \bar{M}_\nu^{1}(k\bar{r}') \times \bar{M}_\mu^{1}(k'\bar{r}') dS,$$

$$J' = \frac{k^2}{\pi} \int_S \bar{n} \cdot \bar{M}_\nu^{1}(k\bar{r}') \times \bar{N}_\mu^{1}(k'\bar{r}') dS,$$

$$K' = \frac{k^2}{\pi} \int_S \bar{n} \cdot \bar{N}_\nu^{1}(k\bar{r}') \times \bar{M}_\mu^{1}(k'\bar{r}') dS,$$

$$L' = \frac{k^2}{\pi} \int_S \bar{n} \cdot \bar{N}_\nu^{1}(k\bar{r}') \times \bar{N}_\mu^{1}(k'\bar{r}') dS.$$

The vector far-field amplitude of the scattered field is defined by

$$\bar{E}^s(k\bar{r}) = \bar{F}(\theta_s, \phi_s/\theta_i, \phi_i) \frac{\exp(jkr)}{r}, kr \to \infty, \quad (18)$$

where $\bar{F}(\theta_s, \phi_s/\theta_i, \phi_i)$ is the vector far-field amplitude in the (θ_s, ϕ_s) direction due to an incident field in a given (θ_i, ϕ_i) direction.

The differential scattering cross section is defined as

$$\sigma_D = \lim_{r \to \infty} \left[4\pi r^2 \frac{S_s(\theta_s, \phi_s)}{S_i(\theta_i, \phi_i)} \right], \quad (19)$$

where $S_s(\theta_s, \phi_s)$ = the scattered power density

$$= \frac{|\bar{F}(\theta_s, \phi_s/\theta_i, \phi_i)|^2}{2Z_0 r^2}, Z_0 = \sqrt{\mu_0/\epsilon_0},$$

and $S_i(\theta_i, \phi_i)$ = the incident power density

$$= \frac{|\bar{E}^i|^2}{2Z_0}.$$

Assuming that the incident electric field $\bar{E}^i$ has unit amplitude and substituting the expression for S_s and S_i into Eq. (19) give

$$\sigma_D(\theta_s, \phi_s/\theta_i, \phi_i) = 4\pi |\bar{F}(\theta_s, \phi_s/\theta_i, \phi_i)|^2. \quad (20)$$

For $(\theta_s, \phi_s) = (\pi - \theta_i, \pi + \phi_i)$, this reduces to the usual definition for backscatter cross section.

VII. Computed Results

The equations of the previous sections have been solved numerically on a digital computer to investigate the dependence of the differential scattering characteristics of closed three-dimensional bodies on the size, shape, and dielectric constant of the scattering object.

The method of numerical solution for a particular scattering problem consists of choosing a value for N, solving the set of equations for the differential scattering cross section, then repeating the calculation for successively larger N values until the final result (the differential scattering cross section) converges to a specified accuracy. The maximum value of N required for a given problem is dependent on the shape (deviation from a sphere), size, and index of refraction. Small spheres with an index of refraction near unity require small values of N, while a large cylinder with relatively high index of refraction would require a much larger N value for solution. This characteristic of the numerical solution accounts for the suitability of this method to resonance-sized objects. The scattering by smaller objects can be solved more simply by the Rayleigh approximation, and, although large objects can be handled by this method, the large number of terms required will in many cases result in a very large mathematical problem that might be better solved by geometrical optics methods.

The validity of the computational procedure was first verified by making calculations for spherical bodies and comparing the results to those obtained by the Mie theory. After satisfactory results for this test were obtained, the validity check was extended to nonspherical shapes by computing the scattering due to a sphere with the origin of the sphere moved off-center. This has the effect of making the sphere appear nonspherical as far as the mathematics is concerned, however, the computer program must still calculate the same scattered field as is obtained with the origin at the center. Further tests were made including compliance with the law of reciprocity. Specifically, scattering calculations were made for a prolate spheroid wherein the scattered return in one direction due to an incident wave from another direction was compared to the scattered return in the original incident field direction due to an incident wave from the original scattered direction. The scattered field obtained from the two cases was identical.

In this paper we will be concerned with the scattering by axisymmetric bodies, i.e., those bodies having the z axis as an axis of revolution. Specifically, the differential scattering will be evaluated for spheres, prolate and oblate spheroids, and finite cylinders. The scattered field will be evaluated over two planes, which can be called the azimuthal plane and the equatorial plane. The incident wave is defined to be in the (θ_i, ϕ_i) direction and the scattered wave in the (θ_s, ϕ_s) direction. Referring to Fig. 1, the azimuthal and equatorial planes are given as follows:

Azimuthal plane—the incident wave is in the (0°, 0°) direction, i.e., traveling along the z axis in the $+z$ direction. The scattered field is in the $(\theta_s, 0°)$ direction where θ_s varies from 0° to 180°. Therefore, the scattered field is evaluated over that half of the x-z plane lying on the $+x$ side of the origin.

Equatorial plane—the incident wave is in the (90°, 0°) direction, i.e., traveling along the x axis in the $+z$ direction. The scattered field is in the $(90°, \phi_s)$ direction where ϕ_s varies from 0° to 180°. Therefore, the scattered field is evaluated over that half of the x-y plane lying on the $+y$ side of the origin.

The curves of differential scattering cross section vs scattering angle will conform to the convention

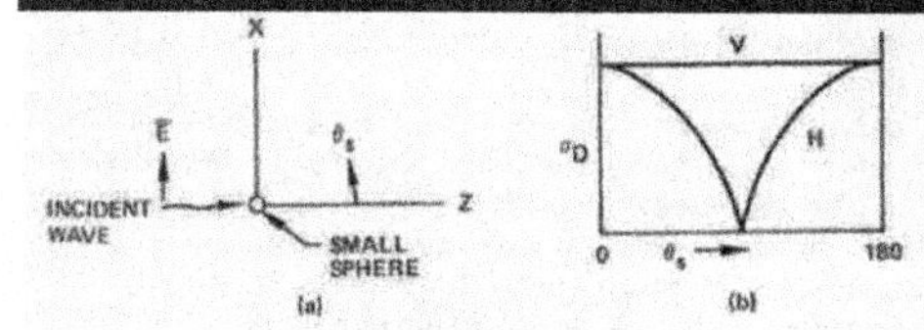

Fig. 7. Rayleigh scattering.

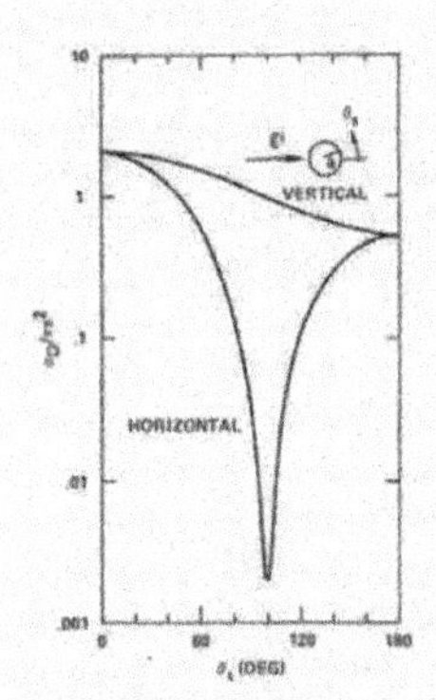

Fig. 8. Sphere scattering, $ka = 1.0$, $\epsilon_r = 4.0$.

used in light scattering studies where the scattering results are plotted from the forward direction to the backward direction, i.e., $\theta_s = 0°$ defines the forward scatter and $\theta_s = 180°$ the backscatter. The scattering angle is plotted along the horizontal axis. The vertical axis is the normalized differential scattering cross section defined as σ_D [see Eq. (20)] divided by πa^2, where a is some characteristic dimension of the scatterer that will be defined in each case. The polarization of the incident and scattered fields will be referred to as either vertical, which is perpendicular to the scattering plane, or horizontal, which is parallel to the scattering plane. It should be noted that for the particular case of the scattering by axisymmetric bodies evaluated over the given planes, symmetry considerations indicate that no cross polarized return will be generated, i.e., a vertically polarized incident wave will generate only a vertically polarized scattered wave and likewise for the horizontally polarized wave case.

Before looking at the results for nonspherical bodies, it will be useful to examine the behavior of the differential scattering curves for lossless dielectric spheres. For spheres lying in the Rayleigh region, the scattered field will be the same as that of a small electric dipole as shown in Fig. 7.

For the horizontally polarized incident wave shown, the differential scattering cross section in the x-z plane is given by curve H in Fig. 7(b). The curve due to a vertically polarized incident wave is a straight line in this plane.

Keeping the dielectric constant the same and increasing the size of the sphere result in a more complex interaction, and the differential scattering curve starts to change in a predictable manner. The curve in Fig. 7 would apply to a sphere with ka (a = the radius) no greater than about 0.157. The curve for $ka = 1.0$, $\epsilon_r = 4.0$, which is given in Fig. 8, shows that for a slightly larger sphere, the scattering starts to peak in the forward direction. This forward peak is a characteristic trait of all larger spheres.

Figure 9 shows the vertical polarization differential scattering curve for spheres of size $ka = 3$ and 5, where the dielectric constant is held to $\epsilon_r = 4.0$. A series of maxima and minima have begun to appear in the backward direction, and, as the size is increased, they move toward the forward direction, and additional maxima and minima arise in the backward direction. Qualitatively, it is observed that the larger the sphere, the more peaks and valleys are evident in the vertical polarization differential scattering curve. A similar effect occurs for the horizontally

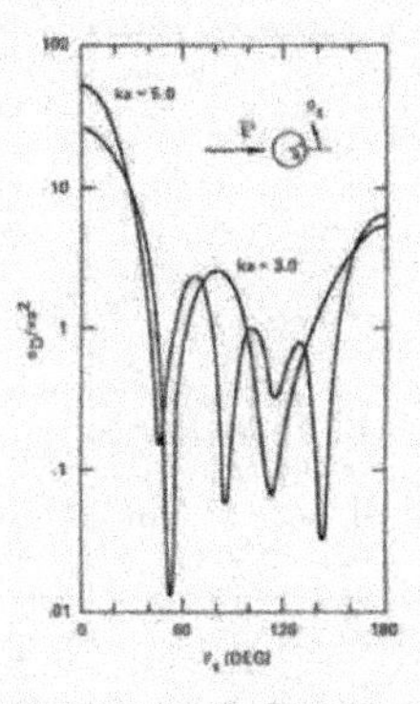

Fig. 9. Sphere scattering, vertical polarization, $ka = 3.0, 5.0$, $\epsilon_r = 4.0$.

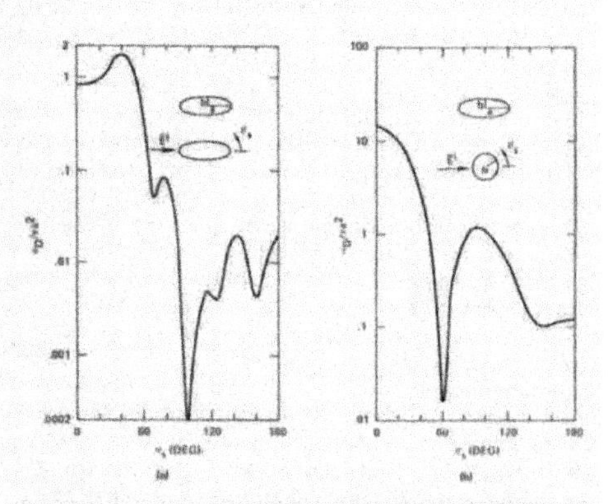

Fig. 10. 3:1 prolate spheroid scattering, vertical polarization, $ka = 7.114$, $\epsilon_r = 5.0$: (a) azimuthal plane; (b) equatorial plane.

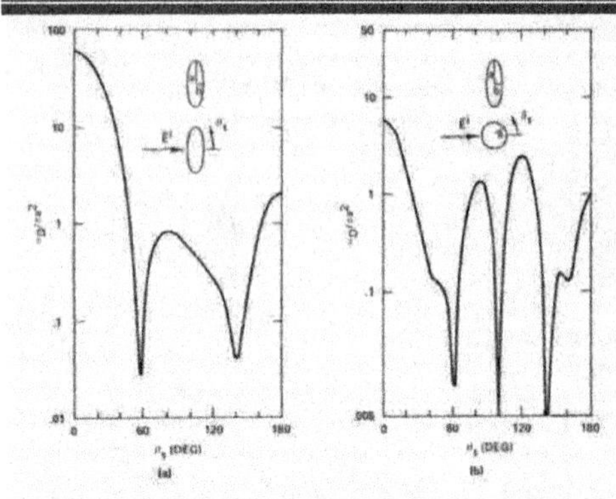

Fig. 11. 3:1 oblate spheroid scattering, vertical polarization, ka = 4.932, ϵ_r = 5.0: (a) azimuthal plane; (b) equatorial plane.

polarized return; however, the amplitude of the oscillation is not as great as in the vertically polarized case, and other effects take place that are not as predictable as in the vertical polarization case. For a different dielectric constant, the behavior of the scattering pattern with increasing size is similar, i.e., more peaks and valleys appear for larger spheres.

The vertical polarization scattering curves for a 3:1 prolate spheroid are given in Figs. 10(a) and 10(b) for the azimuthal and equatorial planes, respectively. One very interesting phenomenon is observed. The azimuthal plane scattering curve [Fig. 10(a)] oscillates rather rapidly with scattering angle and qualitatively is very similar to the return that might be expected from a relatively large sphere. On the other hand, the scattering curve in the equatorial plane [Fig. 10(b)] has fewer oscillations similar to that which one might expect from a relatively smaller sphere. The differential scattering characteristics apparently indicate some measure of size of the scatterer in the direction in which the incident wave is traveling. This behavior was further examined by repeating the calculation for a 3:1 oblate spheroid. These results are shown in Fig. 11. As expected, the curve in the azimuthal plane has fewer oscillations than the curve over the equatorial plane, because the scattering in the azimuthal plane is primarily influenced by the thickness of the spheroid along the minor axis while the equatorial plane scattering is dependent on the major axis thickness. This measurement theory is very interesting, because it indicates a method of experimentally obtaining relative size and shape information for nonspherical particles. (The experimental measurement procedure is a problem unto itself.) Another interesting feature of the scattering behavior is evident in Fig. 10(a). The peak in the curve at 40° is a characteristic that is not expected based on experience with spherical scatterers, where the peak return usually occurs in the forward direction (0°). The peak here is believed due to a specular reflection off the leading edge of the spheroid.

Discussion in the preceding paragraph has concluded that the scattering in a given plane is dependent on the features of the scatterer in that plane. It will be interesting to determine to what extent the scattering in one plane is dependent on the physical features of the scatterer in another plane. Figure 12 shows the differential scattering in the equatorial plane for 2:1 and 3:1 capped cylinders with the same minor diameters, but different lengths. It can be seen that both vertical and horizontal curves are almost identical in shape. Calculations were also made for a shorter cylinder (1.5:1) with the same minor diameter, and the scattering curves, although similar to the 2:1 curves in Fig. 12(a), did not have the almost exact correlation that is evident for the 2:1 and 3:1 cylinders. The conclusion is that the scatter-

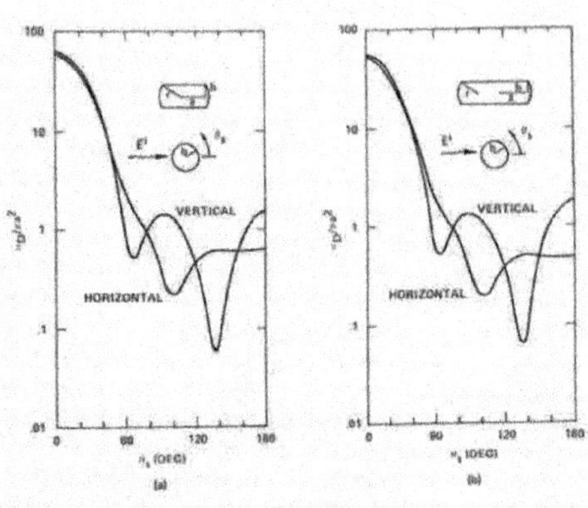

Fig. 12. Cylinder scattering, equatorial plane, ϵ_r = 1.96, r = $2b$: (a) 2:1, ka = 5.288, b/a = 0.577; (b) 3:1, ka = 8.341, b/a = 0.366.

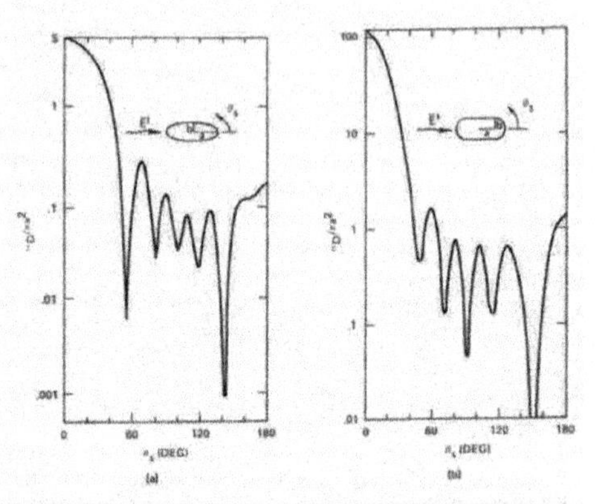

Fig. 13. Scattering by similar bodies, vertical polarization, azimuthal plane, ϵ_r = 2.28: (a) 2:1 prolate spheroid, ka = 9.529, a/b = 2.0; (b) 2:1 cylinder, ka = 4.765, a/b = 1.0.

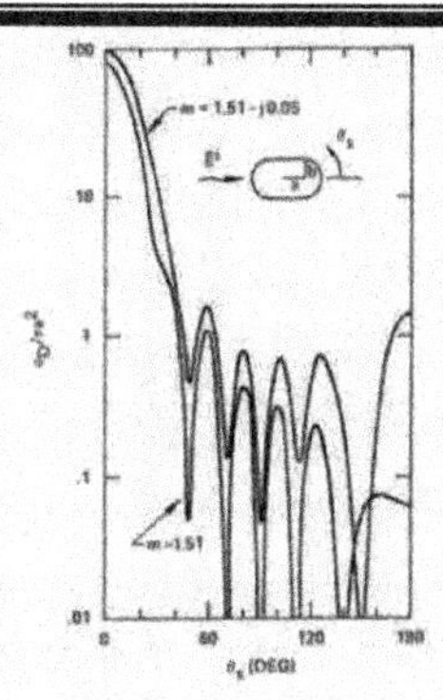

Fig. 14. Cylinder scattering, azimuthal plane, $ka = 4.765$, $a/b = 1.0$, $m = 1.51$, $m = 1.51-j0.05$.

ing in the equatorial plane becomes independent of length for longer cylinders.

To further investigate the dependence of the differential scattering curves on shape, a calculation was made for a prolate spheroid and a capped cylinder, both bodies having the same length and width (and dielectric constant). The prolate spheroid vertical polarization scattering curve for the azimuthal plane is shown in Fig. 13(a), and that for the cylinder is given in Fig. 13(b). It can be seen that even with the same over-all dimensions, small variations in shape cause quite noticeable differences in the differential scattering characteristics.

Figure 14 compares the vertical polarization differential scattering curves for a lossless cylinder (real index of refraction) and for the same cylinder with a small amount of loss added (complex index of refraction). The behavior of the curves is very similar to that which is observed for spherical scatterers. It can be seen that the absorption has caused a change in the differential scattering characteristic that is manifested primarily by a reduction in the scattered energy all the way around the cylinder, although the effect of the internal loss is much more noticeable in the backscatter direction than in the forward direction. The reason for this is probably attributable in part to a diffraction of a large percentage of the incident radiation around the edge of the cylinder. This component is unaffected by the internal features and therefore shows no reduction when the interior becomes absorbing. The small reduction in the forward direction is due to that portion of the forward scattered wave that is refracted through the cylinder, while a major portion of the backscattered energy is due to that portion of the incident wave that is refracted into the interior region and then exists in the backward direction after multiple internal reflections. This component is directly dependent on the internal characteristics of the scattering object and therefore is attenuated when loss is added. The net effect of all this is that the effect of loss in the interior is much more noticeable in the backward than in the forward directions.

This is illustrated very graphically by looking at the ratio of forward to backscatter in the two cases. That ratio is 80 in the lossless case and almost 1500 in the lossy case.

VIII. Conclusion

The differential scattering of electromagnetic plane waves by homogeneous isotropic dielectric bodies has been theoretically investigated. The numerical results obtained here have investigated the dependence of the scattering on the physical characteristics of the scatterer. The analytical method has wide application to many physical problems including those where the scattering by known objects must be determined (the direct scattering problem) and those where theoretical results are needed for comparison to experimental scattering results to determine the characteristics of unknown scatterers (the inverse scattering problem). An example of the former would be in the area of atmospheric physics where the interaction of various types of electromagnetic waves with aerosols is required. The solution of the inverse problem is required in the study of small chemical and biological particles by light scattering methods.

The authors express their sincere appreciation to P. C. Waterman of the Mitre Corporation and P. J. Wyatt of Science Spectrum, Incorporated, for their interest in this project and their valuable comments.

References

1. M. Kerker, *The Scattering of Light and Other Electromagnetic Radiation* (Academic, New York, 1969).
2. R. G. Kouyoumjian, L. Peters, Jr., and D. T. Thomas, IEEE Trans. Antennas Propag. **AP-11**, 690 (1963).
3. J. B. Keller, J. Opt. Soc. Am. **52**, 116 (1962).
4. R. F. Harrington, *Time-Harmonic Electromagnetic Fields* (McGraw-Hill, New York, 1961).
5. C. Yeh, J. Opt. Soc. Am. **55**, 309 (1965).
6. J. A. Stratton, *Electromagnetic Theory* (McGraw-Hill, New York, 1941).
7. R. H. T. Bates, J. R. James, I. N. L. Gallett, and R. F. Millar, Radio Electron. Eng. **43**, 193 (1973).
8. D. R. Wilton and R. Mittra, *IEEE Trans. Antennas Propag.* **AP-20**, 310 (1972).
9. C. Yeh, Phys. Rev. **135**, A1193 (31 August 1964).
10. J. H. Richmond, IEEE Trans. Antennas Propag. **AP-13**, 334 (1965).
11. J. H. Richmond, Proc. IEEE **53**, 796 (1965).
12. P. C. Waterman, Proc. IEEE **53**, 805 (1965).
13. P. C. Waterman and C. V. McCarthy, "Numerical Solution of Electromagnetic Scattering Problems," Mitre Corporation, Bedford, Massachusetts, Report MTP-74 (N69-31912) (June 1968).
14. P. C. Waterman, Alta Freq. **38** (Speciale), 348 (1969).
15. P. C. Waterman, Phys. Rev. D **3**, 825 (1971).
16. H. Honl, A. W. Maue, and K. Westpfahl, *Handbuch der Physik* (Springer Verlag, Berlin, 1961), Vol. 25/1.
17. S. A. Schelkunoff, *Electromagnetic Waves* (D. Von Nostrand, New York, 1943).
18. P. M. Morse and H. Feshbach, *Methods of Theoretical Physics* (McGraw-Hill, New York, 1953).
19. A. Nehari, *Introduction to Complex Analysis* (Allyn and Bacon, Boston, Mass., 1968).

Radio Science, Volume 17, Number 4, pages 757–765, July–August 1982

Scattering by Pruppacher-Pitter raindrops at 30 GHz

C. Yeh

Electrical Engineering Department, University of California
Los Angeles, California 90024

R. Woo and J. W. Armstrong

Jet Propulsion Laboratory, California Institute of Technology
Pasadena, California 91103

A. Ishimaru

Department of Electrical Engineering, University of Washington
Seattle, Washington 98195

(Received October 19, 1981; revised January 25, 1982; accepted January 25, 1982.)

Optimum design of modern ground-satellite communication systems requires the knowledge of rain-induced differential attenuation, differential phase shift, and cross-polarization factors. After a comprehensive assessment of different available analytical techniques, an efficient technique is chosen to yield the desired scattering results. Tabulation of the scattered fields for Pruppacher-Pitter raindrops with sizes ranging from 0.25 mm to 3.5 mm at 20°C and at 30 GHz for several incidence angles is given.

INTRODUCTION

The optimum design of modern ground-satellite communication systems requires detailed knowledge of rain-induced radio propagation effects [*Fang and Lee*, 1978; *Oguchi*, 1977, 1981; *Holt*, 1982; *Morrison and Cross*, 1974; *Bringi and Seliga*, 1977]. Investigation by *Pruppacher and Pitter* [1977] of the shapes of realistic raindrops shows that raindrops take on shapes which deviate significantly from spheres when their sizes exceed about 1 mm. In fact, their shapes even depart noticeably from oblate spheroidal shapes. It is well known according to the electromagnetic scattering theory that the shape of a scatterer is a crucial factor in determining its resonant scattering characteristics. Hence accurate treatment of the scattering by realistic nonspherical raindrops is essential in the determination of rain-induced radio propagation effects which may include the generation of cross-polarized fields [*Oguchi*, 1981] and the introduction of incoherent fields due to the presence of multiple scattering [*Ishimaru*, 1978; *Kattawar and Plass*, 1972].

Various aspects of the rain scattering problem have been treated by a number of investigators using various techniques [*Oguchi*, 1977; *Morrison and Cross*, 1974; *Holt*, 1982]. The purpose of this paper is twofold. We shall first provide a comprehensive assessment of previously used techniques. On the basis of our assessment we shall then select the most attractive method for rain scattering calculation. Then a complete tabulation of scattered fields for Pruppacher-Pitter raindrops with sizes ranging from 0.25 mm to 3.5 mm at 20°C and at 30 GHz for several incidence angles will be given. These extensive single particle scattering results also represent the necessary input for the determination of the distribution of incoherent and coherent intensities due to multiple scattering effects.

AN ASSESSMENT OF AVAILABLE TECHNIQUES FOR THE RAINDROP SCATTERING PROBLEMS

The classical solutions of electromagnetic wave scattering problems are limited to simple objects of separable boundaries such as spheres, cylinders, etc. [*Kerker*, 1969]. Recent developments in scattering

Paper number 2S0184.
0048-6604/82/0708-0184$08.00

analysis for scatterers of arbitrary shapes include the geometrical theory of diffraction for high-frequency scattering [*Keller*, 1962] and the method of moments [*Harrington*, 1968], the perturbation method [*Yeh*, 1964], the point-matching method [*Oguchi*, 1977; *Morrison and Cross*, 1974], the Fredholm equation approach [*Holt*, 1980], the extended boundary condition method [*Barber and Yeh*, 1975], and the finite-element technique for resonant region scattering [*Mei*, 1974; *Morgan*, 1980; *Yeh et al.*, 1975].

To achieve a proper perspective of the state of the art of various analytical/numerical techniques, the following assessment is made.

Geometrical theory of diffraction. This method is mainly a high-frequency technique which treats the smooth part of the scatterer by geometric optics and obtains the scattered fields from edges and grazing surfaces via solutions of canonical problems such as scattering by wedges, tips, etc. The geometrical theory of diffraction may yield reasonable results even at resonant frequencies. However, it is still a technique most conveniently applied to convex and perfectly conducting targets at high frequencies. When the targets are concave or transparent, the necessity for tracing the multiply scattered rays makes the method overly cumbersome to use. Its advantages are that it is easy to learn and is a clear physical concept. Its disadvantages are that it is difficult to apply to dielectric bodies and is of questionable validity at lower frequencies.

Method of moments. This method is based on the solution of an integral equation. The equation is a surface integral equation if the scatterer is a perfectly conducting body, a pair of coupled surface integral equations if the scatterer is a homogeneous dielectric body, and a volume integral equation if the scatterer is an inhomogeneous dielectric body. If the dielectric scatterer has a volume greater than $(0.1\lambda)^3$, the demand of the method of moments for computer memory or time becomes excessive. The advantage of this method is that it is efficient for thin conducting scatterer. The disadvantage is that it is limited to dielectric scatterer of small volume.

Perturbation method. This method is an extension of the Mie scattering solution to scatterers of nonspherical geometry by the boundary perturbation technique. This technique is inherently limited to particle shapes which are small perturbations of a sphere. The advantage of this method is that analytic expressions are available. The disadvantages are that it has uncertain convergence, it is limited to bodies that are small perturbations of a sphere, and analytic expressions for higher-order terms are excessively complicated.

Point-matching method. Field expressions for the interior region of the scatterer and for the exterior region are given in terms of complete sets of spherical harmonics. Boundary conditions are satisfied by point matching on the boundary surface. The advantage of this method is that it is conceptually simple. The disadvantages are that it has uncertain convergence, it demands excessive computer time, and it is limited to bodies that are close to spherical shapes.

The Fredholm integral equation method. The scattering problem is first formulated in the form of the Fredholm integral equation which is then solved by a matrix method. The major limitation of the method is that the matrix elements have to be evaluated analytically, and in practice this is only feasible for a limited number of model shapes. The advantage of this method is that it may be used to treat scatterers of relatively large volume and scatterers with edges. The disadvantages are that it requires separate analysis for each different scatterer leading to different programs for each scatterer and in practice only a few model shapes are feasible.

Extended boundary condition method. Integral representations for the fields can be derived which satisfy the wave equation and all necessary boundary conditions. By expanding the fields in terms of a complete set of spherical harmonics and by making use of analytic continuation techniques, one may reduce the integral representations to a set of linear algebraic equations. The advantages of this method are that it may be used to treat scatterers of large volume and arbitrary shapes and that numerical results can be generated very efficiently. The disadvantage is that it is usually not applicable to inhomogeneous dielectric scatterers.

Global-local finite element method. The global-local finite element method is a numerical analysis technque in which both the contemporary finite element and classical Rayleigh-Ritz approximations are employed. The scattering problem is divided into two regions: the exterior region and the interior region. Without any sacrifice of rigor, it is assumed that the boundary surface between the exterior and interior regions is a sphere for three-dimensional problems and a circular cylinder for two-dimensional problems. It is further assumed that the boundary surface is so chosen that the medium in the exterior region is homogeneous and that the inhomogeneous, irregularly shaped absorbing object is contained within the interior region. Solutions of wave equations in the exterior homogeneous region are well known. Therefore knowing the interior fields at the boundary surface provides the necessary information to solve for the exterior fields by using the boundary conditions at the boundary surface. The finite element technique is used to find the interior field. The advantages of this method are that it is adaptable to scatterers of nonspherical shapes and inhomogeneous material and that results are based on exact Maxwell's equations. The disadvantages are that it requires extensive computer memory and uses complicated computer codes.

On the basis of this assessment it is noted that only two techniques are sufficiently reliable to provide scattering results that possess little uncertainty for Pruppacher-Pitter raindrops: the extended boundary condition method and the finite element technique. The complexity of the finite element formulation for truly three-dimensional problems and its enormous demand for computer memory render this technique less than attractive for extensive computations. The extended boundary condtion technique is therefore preferred.

A BRIEF SUMMARY OF THE EXTENDED BOUNDARY CONDITION METHOD

The geometry of the problem is shown in Figure 1. For a linearly polarized incident plane wave,

$$\mathbf{E}_i(z) = E_0 e^{ikz} \mathbf{e}_i \tag{1}$$

where E_0 is the amplitude of the plane wave, $\mathbf{e}_i$ is a unit vector indicating the polarization direction of the incident wave which is taken to be either the x or y direction, k is the free-space wave number, and the wave is propagating in the $+z$ direction. The scattered field takes on the following form:

$$\mathbf{E}_s(r) = \mathbf{f}(\mathbf{o}, \mathbf{i}) \frac{e^{ikR}}{R} \tag{2}$$

where $\mathbf{i}$ and $\mathbf{o}$ are, respectively, the incident direction and the observation direction. Following the matrix notation of *Ishimaru* [1978] and *Van de Hulst* [1957], the incident field and the far-zone scattered field can be written as follows:

$$\begin{aligned} \mathbf{E}_i &= E_{i\perp} \mathbf{e}_i + E_{i\parallel} \mathbf{e}_{i\parallel} \\ \mathbf{E}_s &= E_{s\perp} \mathbf{e}_s + E_{s\parallel} \mathbf{e}_{s\parallel} \end{aligned} \tag{3}$$

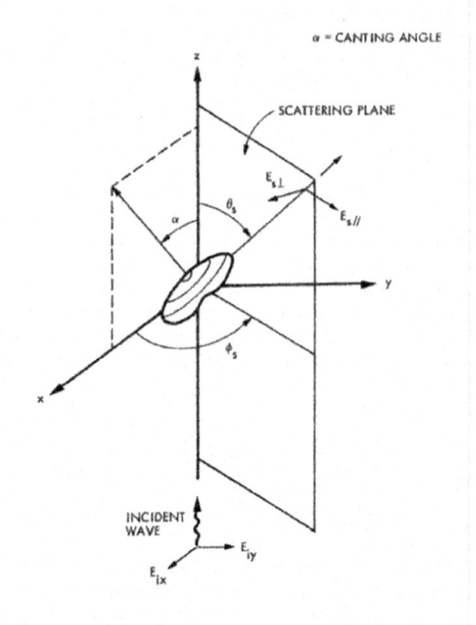

Fig. 1. The geometry of the problem.

where $\mathbf{e}_{i\perp}$ or $\mathbf{e}_{s\perp}$ and $\mathbf{e}_{i\parallel}$ or $\mathbf{e}_{s\parallel}$ are, respectively, the unit vectors perpendicular and parallel to the plane of scattering, which is the plane containing the vectors $\mathbf{i}$ and $\mathbf{0}$. Note that $\mathbf{e}_{i\perp}$ is necessarily equal to $\mathbf{e}_{s\perp}$, but $\mathbf{e}_{i\parallel}$ is not necessarily equal to $\mathbf{e}_{s\parallel}$ and $E_{s\perp}$ and $E_{s\parallel}$ are related to $E_{i\parallel}$ and $E_{i\parallel}$ by the following relation:

$$\begin{bmatrix} E_{s\perp} \\ E_{s\parallel} \end{bmatrix} = \frac{e^{ikR}}{R} \begin{bmatrix} f_{11} & f_{11} \\ f_{21} & f_{22} \end{bmatrix} \begin{bmatrix} E_{i\perp} \\ E_{i\parallel} \end{bmatrix} \tag{4}$$

The scattering functions S_1, S_2, S_3, and S_4 used by Van de Hulst as well as the Stokes parameters can all be obtained from f_{11}, f_{12}, f_{21}, and f_{22}. For example,

$$S_1 = \frac{k}{i} f_{11} \quad S_2 = \frac{k}{i} f_{22} \quad S_3 = \frac{k}{i} f_{21} \quad S_4 = \frac{k}{i} f_{12} \tag{5}$$

and if the incident wave has an arbitrary state of polarization and its Stokes parameters are given by I_{1i}, I_{2i}, U_i, and V_i, then the Stokes parameters I_{1s},

I_{2s}, U_s, and V_s of the scattered wave are given by

$$\begin{bmatrix} I_{1s} \\ I_{2s} \\ U_s \\ V_s \end{bmatrix} = \frac{1}{R^2} \begin{bmatrix} |f_{11}|^2 & |f_{12}|^2 & \mathrm{Re}\,(f_{11}f_{12}^*) & -\mathrm{Im}\,(f_{11}f_{12}^*) \\ |f_{21}|^2 & |f_{22}|^2 & \mathrm{Re}\,(f_{21}f_{22}^*) & -\mathrm{Im}\,(f_{21}f_{22}^*) \\ 2\,\mathrm{Re}\,(f_{11}f_{21}^*) & 2\,\mathrm{Re}\,(f_{12}f_{22}^*) & \mathrm{Re}\,(f_{11}f_{22}^* + f_{12}f_{21}^*) & -\mathrm{Im}\,(f_{11}f_{22}^* + f_{12}f_{21}^*) \\ 2\,\mathrm{Im}\,(f_{11}f_{21}^*) & 2\,\mathrm{Im}\,(f_{12}f_{22}^*) & \mathrm{Im}\,(f_{11}f_{22}^* + f_{12}f_{21}^*) & \mathrm{Re}\,(f_{11}f_{22}^*) \end{bmatrix} \begin{bmatrix} I_{1i} \\ I_{2i} \\ U_i \\ V_i \end{bmatrix} \quad (6)$$

where the asterisk means the complex conjugate.

Other commonly used parameters may also be defined in terms of the function $\mathbf{f}(\mathbf{o}, \mathbf{i})$, such as

Differential scattering cross section

$$\sigma_d(\mathbf{o}, \mathbf{i}) = |f(\mathbf{o}, \mathbf{i})|^2 \quad (7)$$

Bistatic radar cross section

$$\sigma_b = 4\pi |f(\mathbf{o}, \mathbf{i})|^2 \quad (8)$$

Backscattering cross section

$$\sigma_b = 4\pi |f(-\mathbf{i}, \mathbf{i})|^2 \quad (9)$$

Scattering cross section

$$\sigma_s = \int_{4\pi} |f(\mathbf{o}, \mathbf{i})|^2 \, d\Omega \quad (10)$$

where $d\Omega$ is a differential solid angle,

Extinction cross section

$$\sigma_e = \sigma_s + \sigma_a = \frac{4\pi}{k} \mathrm{Im}\,[f(\mathbf{i}, \mathbf{i})] \cdot \mathbf{e}_i \quad (11)$$

where σ_a is an absorption cross section and $\mathbf{e}_i$ is a unit vector in the direction polarization of the incident wave.

Therefore knowing the matrix elements f_{11}, f_{12}, f_{21}, and f_{22}, one may calculate all the desired parameters. These matrix elements are obtained by solving the boundary value problem. We intend to use the extended boundary condition technique to solve this scattering problem.

The scattering body, assumed to be homogeneous and isotropic, is characterized by the constitutive parameters μ_0 and ε where μ_0 is the free-space permeability and ε is the permittivity of the dielectric material (ε may be complex to account for the lossy case). The surrounding medium is considered to be free space with parameters μ_0 and ε_0. For the given incident field, the scattered field must be determined.

The theoretical development for the extended boundary condition technique is as follows:

First, the equivalence principle is applied, breaking the scattering problem into two separate problems, an exterior part and an interior part. One equation for the unknown field quantities is derived from the external problem where it is found that the scattered field due to the surface currents must completely cancel the incident field throughout the interior volume.

The internal problem is then considered, and the fields within the dielectric region are expanded in regular vector spherical wave functions with coefficients to be determined. Superposition is applied, and the boundary conditions at the surface lead to a linear system of integral equations for the coefficients of the unknown surface fields in terms of the incident field.

The scattered far field is then determined by evaluating the internal field at the surface and substituting into the original expression which gives the exterior scattered field in terms of the surface fields.

To avoid writing many lengthy mathematical expressions, a detailed derivation will not be given here. We shall simply state the final results:

$$(K + (\varepsilon_r)^{1/2} J)c_\mu + (L + \varepsilon^{1/2} I)d_\mu = -ja_\nu \quad (12)$$

$$\nu = 1, 2, \cdots, N$$

$$(I + (\varepsilon_r)^{1/2} L)c_\mu + (J + (\varepsilon_r)^{1/2} K)d_\mu = -jb_\nu \quad (13)$$

where

$$I = \frac{k^2}{\pi} \int_S \mathbf{n} \cdot \mathbf{M}_\nu^3(k\mathbf{r}') \times \mathbf{M}_\mu^1(k'\mathbf{r}') \, dS$$

$$J = \frac{k^2}{\pi} \int_S \mathbf{n} \cdot \mathbf{M}_\nu^3(k\mathbf{r}') \times \mathbf{N}_\mu^1(k'\mathbf{r}') \, dS$$

$$K = \frac{k^2}{\pi} \int_S \mathbf{n} \cdot \mathbf{N}_\nu^3(k\mathbf{r}') \times \mathbf{M}_\mu^1(k'\mathbf{r}') \, dS$$

$$L = \frac{k^2}{\pi} \int_S \mathbf{n} \cdot \mathbf{N}_\nu^3(k\mathbf{r}') \times \mathbf{N}_\mu^1(k'\mathbf{r}') \, dS$$

and $\varepsilon_r = (\varepsilon/\varepsilon_0)$; $\mathbf{n}$ is a unit vector normal to the boundary surface S and $\mathbf{r}'$ is the radial vector from the origin to the boundary surface; a_ν and b_ν are expansion coefficients for the incident wave defined as follows:

$$\mathbf{E}_i(r) = \sum_{\nu=1}^{\infty} D_\nu[a_\nu \mathbf{M}_\nu^1(kr) + b_\nu \mathbf{N}_\nu^1(kr)] \quad (14)$$

where ν is a combined index incorporating σ, m, and n. D_ν is a normalization constant

$$D_\nu = \varepsilon_m \frac{(2n+1)(n-m)!}{4n(n+1)(n+m)!}$$

$$\varepsilon_m = 1 \qquad m = 0$$

$$\varepsilon_m = 2 \qquad m > 0$$

The spherical wave harmonics $\mathbf{M}$ and $\mathbf{N}$ have been defined by *Stratton* [1941]. These functions, solutions of the vector wave equation, are given by

$$\mathbf{M}_{\sigma mn}^{1,3}(\mathbf{r}) = \nabla \times \mathbf{r} \begin{matrix} \cos m\phi \\ \sin m\phi \end{matrix} P_n^m(\cos\theta) z_n^{1,3}(kr) \quad (15)$$

$$\mathbf{N}_{\sigma mn}^{1,3}(\mathbf{r}) = \frac{1}{k} \nabla \times \mathbf{M}_{\sigma mn}^{1,3}(\mathbf{r})$$

where

σ even or odd;
$P_n^m(\cos\theta)$ associated Legendre function;
$z_n^{1,3}(kr)$ appropriate spherical Bessel function.

For solutions of the wave equation which must be finite at $r = 0$, $z_n^1(kr) = j_n(kr)$ and the resulting vector spherical functions $\mathbf{M}$ and $\mathbf{N}$ are known as solutions of the first kind. The other vector spherical functions we will be using are obtained by using the sperhical Hankel functions, i.e.,

$$z_n^3(kr) = h_n^{(1)}(kr) = j_n(kr) + jn_n(kr)$$

This solution, representing outgoing waves, is called the solution of the third kind.

The c_μ and d_μ are the expansion coefficients for the field inside the dielectric which are defined as follows:

$$\mathbf{E}(k'\mathbf{r}) = \sum_{\mu=1}^{N} [c_\mu \mathbf{M}_\mu^1(k'\mathbf{r}) + d_\mu \mathbf{N}_\mu^1(k'\mathbf{r})] \quad (16)$$

where μ incorporates the indices σ, m, and n; c_μ and d_μ are unknown coefficients; and $k' = (\varepsilon_r)^{1/2}k$ and $k = \omega(\mu_0\varepsilon_0)^{1/2}$. The H field internal to S is given by

$$\mathbf{H}(k'\mathbf{r}) = \frac{1}{j\omega\mu_0}[\nabla \times \mathbf{E}(k'\mathbf{r})]$$

$$= \frac{1}{j\omega\mu_0}\sum_{\mu=1}^{N}\{c_\mu \nabla \times \mathbf{M}_\mu^1(k'\mathbf{r}) + d_\mu \nabla \times \mathbf{N}_\mu^1(k'\mathbf{r})\}$$

$$= -j\frac{k}{\mu_0}\sum_{\mu=1}^{N}[c_\mu \mathbf{N}_\mu^1(k'\mathbf{r}) + d_\mu \mathbf{M}_\mu^1(k'\mathbf{r})] \quad (17)$$

The coefficients of the scattered fields which are defined as follows:

$$\mathbf{E}_s(k\mathbf{r}) = \sum_{\nu=1}^{N}[p_\nu \mathbf{M}_\nu^3(k\mathbf{r}) + q_\nu \mathbf{N}_\nu^3(k\mathbf{r})] \quad (18)$$

where r is outside a circumscribed sphere, are

$$p_\nu = -jD_\nu \sum_{\mu=1}^{N}\{[K' + (\varepsilon_r)^{1/2}J']c_\mu + [L' + (\varepsilon_r)^{1/2} I']\, d_\mu\}$$

$$q_\nu = -jD_\nu \sum_{\mu=1}^{N}\{[I' + (\varepsilon_r)^{1/2} L']c_\mu + [J' + (\varepsilon_r)^{1/2} K']\, d_\mu\} \quad (19)$$

where

$$I' = \frac{k^2}{\pi}\int_S \mathbf{n} \cdot \mathbf{M}_\nu^3(k\mathbf{r}') \times \mathbf{M}_\mu^1(k'\mathbf{r}')\, dS$$

$$J' = \frac{k^2}{\pi}\int_S \mathbf{n} \cdot \mathbf{M}_\nu^3(k\mathbf{r}') \times \mathbf{N}_\mu^1(k'\mathbf{r}')\, dS$$

$$K' = \frac{k^2}{\pi}\int_S \mathbf{n} \cdot \mathbf{N}_\nu^3(k\mathbf{r}') \times \mathbf{M}_\mu^1(k'\mathbf{r}')\, dS$$

$$L' = \frac{k^2}{\pi}\int_S \mathbf{n} \cdot \mathbf{N}_\nu^3(k\mathbf{r}') \times \mathbf{N}_\mu^1(k'\mathbf{r}')\, dS$$

The exact theoretical problem requires the solution of an infinite set of equations. The equations resulting from the external problem are such a set. The first concession to practical considerations is made in the derivation of the internal problem, where the infinite term spherical expansion for the internal field is truncated to N terms. This finite internal field expansion is then substituted into the first $2N$ equations of the infinite set of (13), and the final set of truncated equations is obtained. The scattering results are obtained numerically by solving the complete set of equations for successive values of N until the final results (such as the differential scattering cross section) converge to a specified accuracy. This insures that enough of the expansion terms have been kept to guarantee the correct final result.

The indices μ and ν each run from 1 to N and are related to the internal field expansions. As indicated previously, they incorporate the indices σ, m, and n where σ is even or odd and m and n are indices of Bessel and Legendre functions. To distinguish between expansions over ν and μ, the indices which they represent are unprimed and primed, respectively, i.e., ν is over σ, m, and n, and μ is over σ', m', and n', where the expansion scheme is as follows:

$$\sigma mn = e01, o01, e11, o11, e02, o02, e12\ o12, e22, o22, \cdots$$

where e is even and o is odd.

There are essentially four main steps in the solution for the differential scattering cross section:

1. Solution of the system of equations (13) for the coefficients of the internal field c_μ and d_μ in terms of the incident field coefficients a_μ and b_μ. Note that the solution of (13) requires a matrix inversion procedure.

2. Substitution of the coefficients c_μ and d_μ into (19) to determine the coefficients of the scattered field.

3. Substitution of the scattered field coefficients into a spherical expansion for the far-field vector amplitude.

$$\mathbf{f}(\theta_s, \phi_s | \theta_i, \phi_i) = \frac{1}{jk} \sum_{\nu=1}^{N} [p_\nu \mathbf{C}_\nu(\theta, \phi) + jq_\nu \mathbf{B}_\nu(\theta, \phi)]$$

4. Computation of the differential scattering cross section

$$\sigma_d(\theta_s, \phi_s) = |f(\theta_s, \phi_s | \theta_i, \phi_i)|^2$$

DISPLAY OF RESULTS AND DISCUSSION

Selected Pruppacher-Pitter raindrop shapes are displayed in Figure 2. It can be seen that as the raindrop size increases, its shape deviates significantly from a sphere, and it even differs noticeably from an oblate spheroid. These Pruppacher-Pitter raindrop shapes were used in our computation and compilation of the scattering results. Specifically, we have obtained the scattering matrix elements f_{11}, f_{12}, f_{21}, and f_{22} for the following parameters: raindrop sizes: 0.25 mm, (0.25 mm), 3.5 mm; frequency: 30 GHz; temperature: 20°C; θ_s: 0°, (10°), 360°; α: 0°, (30°), 90°; ϕ_s: 0°, (30°), 180°.

Results are tabulated in Table 1, which has been condensed into microfiche form.[1]

[1] Table 1 is available with entire article on microfiche. Order from American Geophysical Union, 2000 Florida Avenue, N. W., Washington, D. C. 20009. Document S82-001; $5.00. Payment must accompany order.

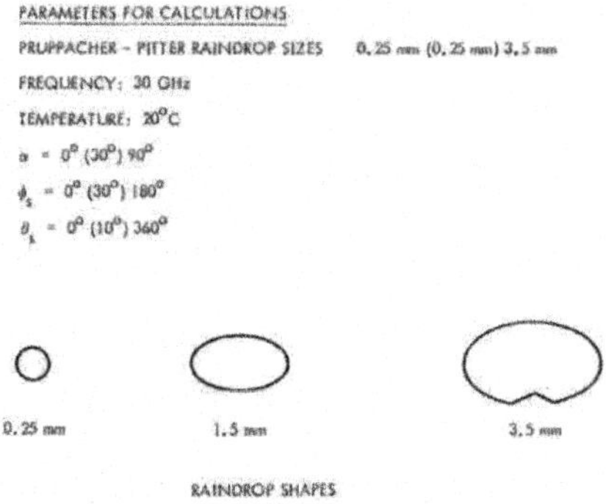

PARAMETERS FOR CALCULATIONS
PRUPPACHER - PITTER RAINDROP SIZES 0.25 mm (0.25 mm) 3.5 mm
FREQUENCY: 30 GHz
TEMPERATURE: 20°C
α = 0° (30°) 90°
ϕ_s = 0° (30°) 180°
θ_s = 0° (10°) 360°

0.25 mm
1.5 mm
3.5 mm

RAINDROP SHAPES

Fig. 2. Pruppacher-Pitter raindrops.

We have, of course, compared our results with those obtained by *Morrison and Cross* [1974] and *Oguchi* [1977], who used the point-matching technique, and by *Morgan* [1980], who used the finite element approach. The results all agree within a few percentage points. These differences may be caused by numerical errors due to finite grid size, finite number of matching points, finite matrix size, finite integration steps, accumulated numerical errors, or the limitation of the particular numerical scheme. On the basis of this comparison one may conclude that the point-matching technique actually provides rather good results in spite of uncertain convergence and the lack of rigor due to the 'Rayleigh phenomenon.' It is also significant to note that our results agree quite well with those obtained according to the finite element approach which is based on the exact Maxwell equations.

It is of interest to note that the scattering results for a Pruppacher-Pitter-shaped raindrop compare very closely to those for an equivalent oblate spheroidal–shaped raindrop even when their shapes are visibly different for large size drops (see Figure 2). It appears that the absorptive characteristics of water at 30 GHz and 20° are responsible for minimizing the differences in the scattering results due to shape differences in the Pruppacher-Pitter and equivalent oblate spheroidal drops.

To illustrate the effect of drop size changes on the scattered field, Figures 3–5 are introduced. In Figure 3 the normalized scattering, absorption, and extinction cross sections are plotted as a function of raindrop sizes. The relatively smooth character of the curves can again be attributed to the absorptive nature of water. Polar plots of scattered fields in the $\phi_s = 90°$ plane for canting angles $\alpha = 0°$ and for

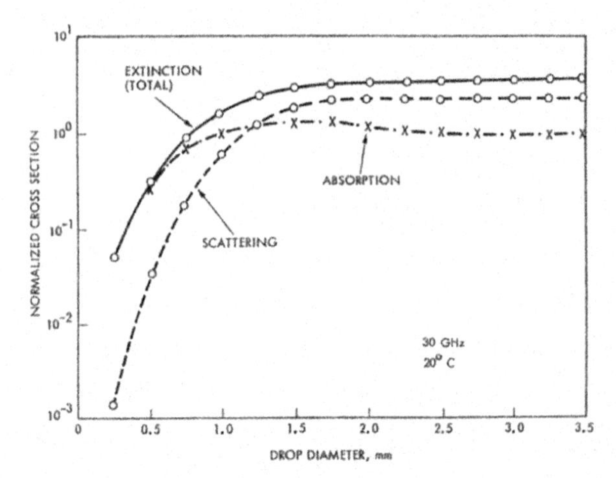

Fig. 3. Normalized cross section as a function of raindrop size.

various drop sizes are shown in Figures 4–6. It can be seen that no cross-polarized field is generated for small drop sizes (with spherical shapes) or for canting angle $\alpha = 0$ (see Figure 4). For large drop sizes (with nonspherical shapes) and for canting angles $\alpha \neq 0$, a cross-polarized field is generated (see Figures 5 and 6). One also notes that the induced cross-polarized field is at least 1 order of magnitude weaker than the non-cross-polarized scattered field. This is again due to the absorptive characteristics of the raindrops under consideration.

In conclusion, one may state that the extended boundary condition method is an efficient method in yielding reliable scattering results for nonspherical

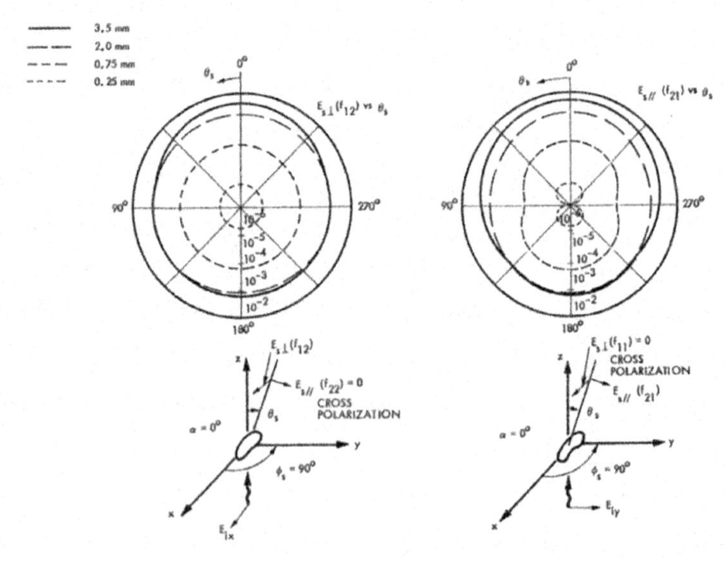

Fig. 4. Polar diagrams for scattered fields with $\alpha = 0°$ (no cross polarization).

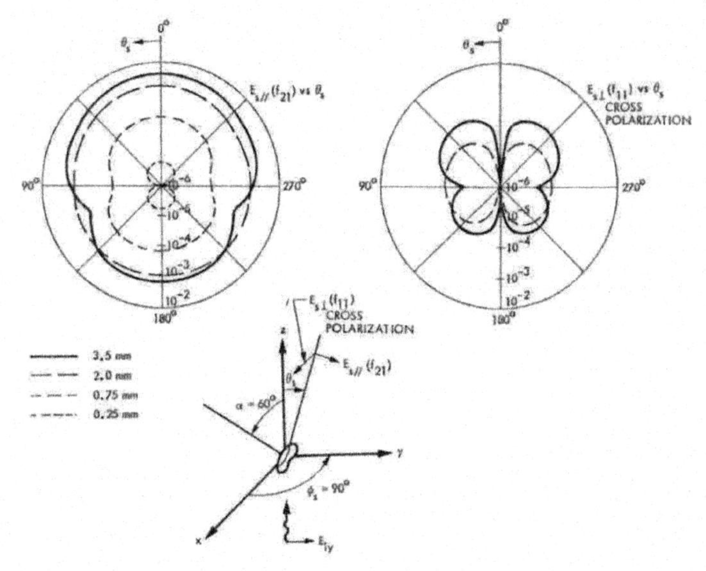

Fig. 5. Polar diagrams for scattered fields with $\alpha = 60°$ (cross polarization present for nonspherical raindrops). Incident field is polarized in the y direction.

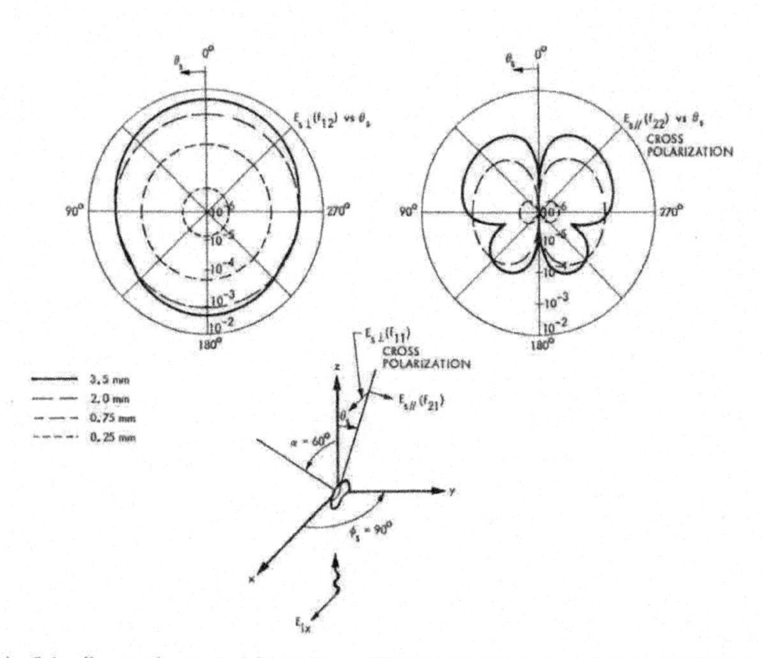

Fig. 6. Polar diagrams for scattered fields with $\alpha = 60°$ (cross polarization present for nonspherical raindrops). Incident field is polarized in the x direction.

Pruppacher-Pitter raindrops. One notes that the large absorption constant of water at 30 GHz and 20°C tends to minimize the shape effect on the scattered field. It is expected that because of the transparent nature of ice at 30 GHz the shape factor for ice particles will play a significant role in giving rise to the generation of cross-polarized field.

Acknowledgments. The authors wish to thank H. R. Pruppacher of UCLA for his advice on raindrop shapes and his interest in this work. The research described in this paper was carried out by the Jet Propulsion Laboratory, California Institute of Technology, under contract with the National Aeronautics and Space Administration.

REFERENCES

Barber, P., and C. Yeh (1975), Scattering of electromagnetic waves by arbitrarily shaped dielectric bodies, *Appl. Opt., 14,* 2864–2872.

Bringi, V. N., and T. Seliga (1977), Scattering from non-spherical hydrometeors, *Ann. Telecommun., 32,* 392–397.

Fang, D. J., and F. J. Lee (1978), Tabulations of raindrop induced forward and backward scattering amplitudes, *COMSAT Tech. Rev., 8,* 455–484.

Harrington, R. F. (1968), *Field Computation by Moment Method,* Macmillan, New York.

Holt, A. R. (1980), The Fredholm integral equation method and comparison with *T*-matrix approach, in *Acoustic, Electromagnetic and Elastic Wave Scattering; Focus on the T-Matrix Approach,* edited by V. K. Varadan and V. V. Varadan, pp. 255–268, Pergamon, New York.

Holt, A. R. (1982), The scattering of electromagnetic waves by single hydrometeors, *Radio Sci,* in press.

Ishimaru, A. (1978), *Wave Propagation and Scattering in Random Media,* vol. I, II, Academic, New York.

Kattawar, G. W., and G. N. Plass (1972), Degree and direction of polarization of multiple scattered light, *Appl. Opt., 11,* 2851–2865.

Keller, J. B. (1962), Geometrical theory of diffraction, *J. Opt. Soc. Am., 52,* 116–130.

Kerker, M. (1969), *The Scattering of Light and Other Electromagnetic Radiations,* Academic, New York.

Mei, K. K. (1974), Unimoment method of solving antenna and scattering problems, *IEEE Trans. Antennas Propag., AP-22,* 760–766.

Morgan, M. A. (1980), Finite element computation of microwave scattering by raindrops, *Radio Sci., 15,* 1109–1119.

Morrison, J. A., and M. J. Cross (1974), Scattering of a plane electromagnetic wave by axisymmetric raindrops, *Bell Syst. Tech. J., 53,* 955–1019.

Oguchi, T. (1977), Scattering properties of Pruppacher-and-Pitter form raindrops and cross polarization due to rain: Calculations at 11, 13, 19.3, and 34.8 GHz, *Radio Sci., 12,* 41–51.

Oguchi, T. (1981), Scattering from hydrometeors: A survey, *Radio Sci., 16,* 691–730.

Pruppacher, H. R., and R. L. Pitter (1971), A semi-empirical determination of the shape of cloud and raindrops, *J. Atmos. Sci., 28,* 86–94.

Stratton, J. A. (1941), *Electromagnetic Theory,* McGraw-Hill, New York.

Van de Hulst, H. C. (1957), *Light Scattering by Small Particles,* John Wiley, New York.

Yeh, C. (1964), Perturbation approach to the diffraction of electromagnetic waves by arbitrarily shaped dielectric obstacles, *Phys. Rev. A, 135,* 1193–1201.

Yeh, C., S. Dong, and W. Oliver (1975), Arbitrarily shaped inhomogeneous optical fiber or integrated optical waveguides, *J. Appl. Phys., 46,* 2125–2129.

Reprinted from APPLIED OPTICS, Vol. 16, page 1104, April 1977

Single particle scattering with focused laser beams

Lee W. Casperson, C. Yeh, and Wing F. Yeung

A scattering technique is described in which the incident laser beam is tightly focused to isolate the effects of a single particle. In this way the individual particles may be studied in their natural environment, and experiments with latex spheres are in agreement with the theory.

I. Introduction

The importance of using light scattering techniques to determine the physical properties of particulates is well recognized.[1-3] Optical methods may be fast, inexpensive, and nondestructive, and under optimum conditions the individual particles can be studied in surroundings approaching their natural environment. When the scattering objects are known to be nearly identical (monodisperse) and spherically symmetric, substantial amounts of information can be obtained from the spatial distribution of light scattered from several particles simultaneously. However, for critical applications where the size or shape of individual particles must be determined within a polydisperse medium, it is essential that in some way the effects of the particles be isolated. A brute force method for accomplishing this isolation involves electrostatically supporting an individual particle while it is illuminated by a laser beam.[4] The simplest application of single particle scattering is in particle counting and sizing. For this purpose gas suspended particles may be made to flow through a small illuminated aperture one at a time, and the resulting flashes of light provide information on the density and size of the particles.

In the previously mentioned single-particle scattering schemes the environment of the scatterer is somewhat artificial. This fact may be of great consequence when one deals with biologically interesting objects such as viruses and bacteria and when one wishes to obtain more precise information on pollutants, since the mere fact of isolation is likely to alter the properties of the species under study. The purpose of this work is to describe a new type of scattering technique which is designed to overcome some of these limitations. The general experimental configuration is shown schematically in Fig. 1. An incident laser beam is tightly focused by means of lenses to yield a focal spot in the scattering material with a diameter on the order of the wavelength of the light. If the beam is kept stationary it will, from time to time, impinge on one of the scatterers which drifts into the focal region. When this occurs, there will be a pulse of scattered light which can be detected to yield information about that specific scatterer. An advantage of this technique is that in many cases a polydisperse ensemble can be studied particle by particle without removing the scatterers from their normal surroundings.

The authors are with University of California, School of Engineering & Applied Science, Los Angeles, California 90024.
Received 19 August 1976.

II. Concentration Limits

As a first step in the analysis we estimate the maximum particle concentration for which this technique is useful. The incident beam is assumed to have the complex Gaussian amplitude distribution

$$E(r,z) = E_o(w_o/w)\exp - i[kz + kr^2/2R - ir^2/w^2 - \tan^{-1}(z\lambda/\pi w_o^2)], \quad (1)$$

where w is the $1/e$ spot size, R is the radius of curvature of the spherical phase fronts, and λ is the wavelength in the medium. The spot size varies with z according to

$$w = w_o[1 + (\lambda z/\pi w_o^2)^2]^{1/2}, \quad (2)$$

where w_o is the spot size at the beam waist. The light will be strongly scattered by any particles located within the spot radius, and the scattering will be relatively unimportant for particles outside of this radius. Also, the intensity falls off rapidly for distances z away from the beam waist. If we define z_m as the maximum distance at which scattering is important, the total active volume can be found from the integral

$$V = \pi \int_{-z_m}^{z_m} w^2(z)dz = \pi w_o^2 \int_{-z_m}^{z_m} [1 + (\lambda z/\pi w_o^2)^2]dz$$
$$= \pi w_o^2[2z_m + 2z_m^3(\lambda/\pi w_o^2)^2/3]. \quad (3)$$

At best one might expect to eliminate from the scattering data events which occur where the intensity is down to half of its maximum value. From Eq. (2) this occurs at the Rayleigh length $z_o = \pi w_o^2/\lambda$, and Eq. (3) reduces to $V = 8\pi^2 w_o^4/3\lambda$. If the data analysis proce-

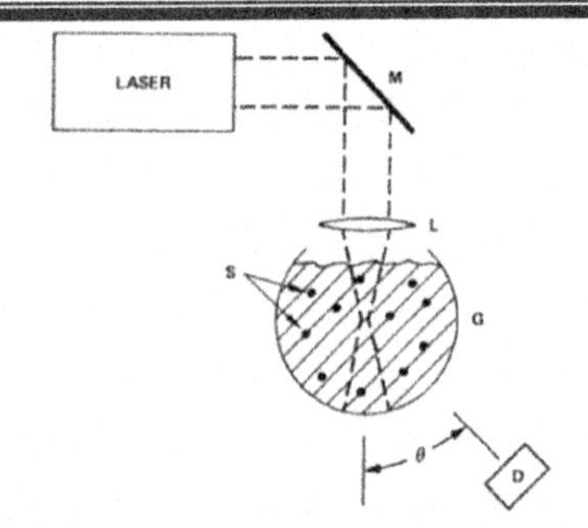

Fig. 1. Schematic drawing of experimental apparatus (not to scale) used in focused beam scattering studies with M, a plane mirror; L, a microscope objective lens; G, a spherical glass envelope containing the scatterers; D, a low noise detector; and S, individual scatterers.

dure can tolerate an average of one particle in this volume, the maximum usable particle density is $\rho_m = 3\lambda/8\pi^2 w_o^4$. Thus the maximum particle density depends critically on the minimum spot size w_o, and it is essential to use high quality low f-number optics. From Eq. (2) the far-field diffraction angle is $w/z = \lambda/\pi w_o$. But the maximum diffraction angle associated with a lens is approximately $(2f)^{-1}$, where f is the f-number. Therefore, the minimum spot size can be written $w_o = 2\lambda f/\pi$, and the maximum density is

$$\rho_m = \frac{3\pi^2}{128\lambda^3 f^4} = \frac{0.23}{\lambda^3 f^4}. \tag{4}$$

For an f value of unity and a wavelength of 5000 Å the maximum density is 1.85×10^{12} cm^{-3}. If one uses a lens having a larger f, or if a smaller number of particle coincidences is required, this maximum concentration would be reduced accordingly. There is no minimum usable density, but with low particle concentrations, data accumulation will, of course, be very slow.

III. Particle Motion

Another aspect of the focused beam scattering technique involves motion of the scattering object. Any object suspended in a gas or liquid exhibits a random drifting behavior known as Brownian motion. This motion is helpful in that it continuously brings new particles into the beam. On the other hand, without feedback and beam tracking methods the Brownian motion places a lower limit on the residence time of each particle in the beam and imposes a limit on the response time of the data recording system.

The distribution function for a particle in a fluid can be written

$$f(\bar r,t) = (4\pi Dt)^{-3/2} \exp\left(\frac{-|\bar r - \bar r_o|^2}{4Dt}\right), \tag{5}$$

where $\bar r_o$ is the initial position vector of the particle at time $t = 0$, and D is the diffusion coefficient.[5] For a spherical particle of radius a in a fluid of viscosity η the diffusion coefficient is

$$D = kT/6\pi a\eta, \tag{6}$$

where k is Boltzmann's constant and T is the temperature. The average time spent by a particle within the beam can be estimated from Eq. (5) by assuming that the particle starts at the axis of the cylindrical beam volume at time $t = 0$. Then, one can calculate the time at which the integral of the distribution function over the beam volume is equal to one half, or

$$\frac{1}{2} = (4\pi Dt)^{-3/2} \int_0^{2\pi}\int_0^{w_o}\int_{-\infty}^{\infty} \exp\left[-\frac{(r^2+z^2)}{4Dt}\right] dz\,dr\,d\theta$$
$$= 1 - \exp(-w_o^2/4Dt). \tag{7}$$

Therefore, the interaction time is about

$$t_{1/2} = \frac{w_o^2}{4D \ln 2} = \frac{\lambda^2 f^2}{\pi^2 D \ln 2}. \tag{8}$$

To illustrate the application of these formulas we consider an example corresponding to the experiment described later in this paper. The viscosity of water at 25°C is approximately 0.90 cP or 0.90×10^{-3} N-sec/m^2.[6] Therefore, from Eq. (6) a particle of radius $a = 0.2405$ µm will have a diffusion coefficient $D = 1.01 \times 10^{-12}$ m^2/sec. The wavelength of the incident beam is $\lambda = 0.5145$ µm, and the focusing lens has an f-number of 1.65 corresponding to a beam waist spot size $w_o = 2\lambda f/\pi = 0.54$ µm. Thus Eq. (8) implies that the residence time of a particle near the focus is about $t_{1/2} = 0.1$ sec. This value agrees with typical pulse lengths observed in our experiments. For very large values of focal intensity the above analysis can be generalized to include the effects of radiation pressure.

IV. Scattered Field Distribution

In the previous sections a general focused-beam technique has been described for determining the detailed characteristics of polydisperse systems of scatterers. In order to apply this technique it is necessary that the scattering properties of the individual particles be understood when the incident light is in the form of a focused Gaussian beam. Most previous treatments of scattering have regarded the light source as being a uniform plane wave, and we cannot *a priori* expect to obtain the same results with a focused beam. Recently the scattered field has been derived for a spherical dielectric particle at the axis of plane wave Gaussian beam for particles which are small compared to the beam diameter.[7] A more general treatment has also been given which applies to on-axis particles of arbitrary size.[8]

As one would expect, the field distribution resulting with an incident Gaussian beam reduces to the familiar plane wave results in the limit that the beam diameter is much larger than the particle diameter. As an example, typical scattering patterns are reproduced in Fig. 2 for the case of a small conducting spherical particle.[8] This figure shows the field distribution that results when a Gaussian beam of spot size $w_o = 4.19\lambda$ propagating in the z direction and polarized in the x direction is incident on a conducting sphere of radius 1.47λ.

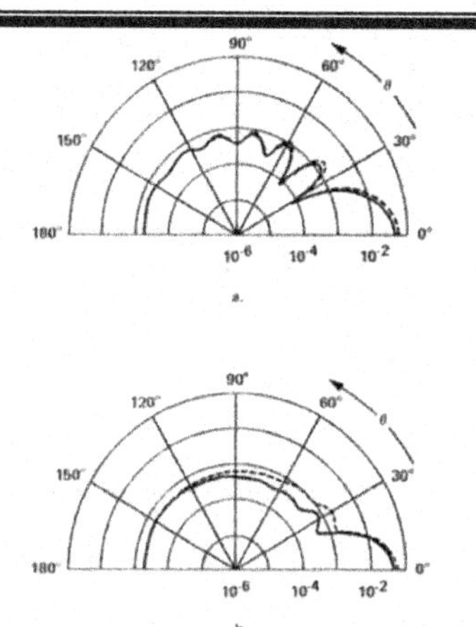

Fig. 2. Scattered field distribution that results when a Gaussian beam of spot size $w_o = 4.19\lambda$ is incident on a conducting sphere of radius $a = 1.47\lambda$. Parts (a) and (b) show, respectively, the $|E_\theta{}^r|^2$ and $|E_\phi{}^r|^2$ values of the scattered field with the exact results given by solid lines and the corresponding plane wave results given by dashed lines (after Ref. 8).

Parts (a) and (b) show, respectively, the $|E_\theta{}^r|^2$ and $|E_\phi{}^r|^2$ values of the scattered field with the exact results given by solid lines and the corresponding plane wave results given by dashed lines. ($E_\theta{}^r$ and $E_\phi{}^r$ are, respectively, the θ component in the x-z plane and the ϕ component in the y-z plane.) These curves confirm the reasonable fact that for particles small compared to the beam diameter, the scattering pattern for a Gaussian beam is essentially identical to the more familiar plane wave scattering data.

For particles which are comparable in size to the beam diameter, the plane wave data cannot be employed. The previously obtained beam solutions[7,8] are also not rigorously applicable in this case because of two complicating features of the problem. First, in our case the particles are not necessarily going to be at the axis of the light beam, and the scattering pattern will not be symmetric about the beam axis. Second, with a very low f-number focusing lens the phase fronts of the beam may be spherical rather than plane when they impinge upon the particle, and this fact will also affect the scattered fields. Because of the desirability of adopting the familiar plane wave scattering results, our initial experiments have employed focusing lenses with intermediate f-numbers and particles which are small compared to the beam diameter. More general theoretical models are in progress, so that in the future the above restrictions will not be necessary. Besides including off-axis spherical particles of arbitrary size and curved phase fronts of the incident beam, we are also adapting a previous analysis which accounts for scattering by particles of arbitrary shape.[9] With these results it will be possible, for example, to distinguish a needle-shaped asbestos fiber in the atmosphere from other harmless particles of comparable volume.

V. Experiment

A general focused-beam technique has been described for determining the characteristics of an ensemble of scattering particles. We have confirmed the validity of this approach by means of experiments using particles of known properties, and the field distribution scattered from latex spheres is found to be in good agreement with the theoretical predictions. The experimental apparatus is shown schematically in Fig. 1. The linearly polarized 5145-Å beam from an argon laser (Lexel model 75) is deflected downward into a cell containing the particles suspended in water. The cell has a spherical glass envelope 42 mm in diameter and is centered on the beam focus. This spherical shape is useful for eliminating refraction effects with liquid-suspended particles. Focusing has been accomplished by means of a microscope objective lens. The particles used in this study are latex spheres having a radius of $a = 0.2405\ \mu m$ and a refractive index of $n = 1.59$.

In an optimized configuration designed for investigating the properties of polydisperse systems of nonspherical particles, the detection scheme should consist of an array of detectors completely surrounding the scattering cell. Information from these detectors would be fed to an appropriate multichannel analysis system to determine the properties of the individual particles as they pass through the beam. Alternatively, the information could be carried from many points surrounding the cell to a remote detector array by means of optical fibers. However, our initial experiments have involved uniform dielectric spheres, and it has been sufficient to use a single detector which can be accurately scanned about the focus as shown in Fig. 1. Several events are recorded for each angular position, and the largest values correspond to particles near the focus. The weaker events result from particles crossing the beam away from the focus and are discarded. The off-focus events can also be largely eliminated if the detector is preceded by a spatial filter.

Typical experimental results obtained at intervals of 15° are summarized in Fig. 3. The beam waist spot size is $w_o = 0.54\ \mu m$, and each point represents approximately twenty measurements. The theoretical curve is a plot of $|E_\theta{}^r|^2$ obtained from Lorenz-Mie scattering theory using the known refractive index of the particles together with the value $ka = 3.6$. This latter number gives a slightly better fit than the theoretical value $ka = 2\pi na/\lambda = 3.9$, and discrepancies of this order are not uncommon in detailed scattering studies.[10] Measurements have also been made of the total power in the transmitted beam. When a particle passes through the focus, the transmitted power decreases, and the magnitude of this decrease can be used to determine the total scattering cross section.

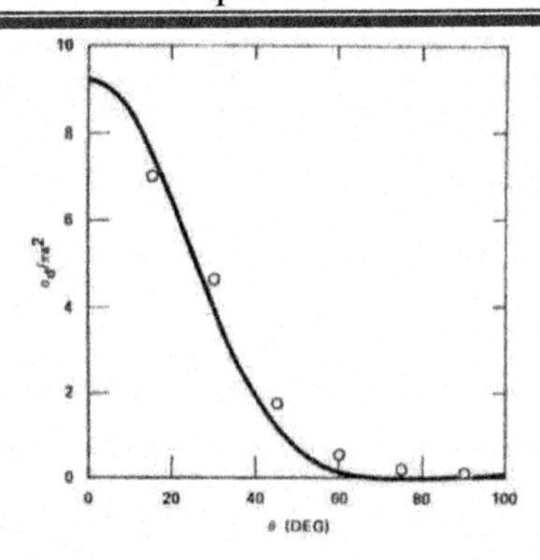

Fig. 3. Angular dependence of the differential scattering cross section σ_d divided by the area πa^2. Solid line is exact result of Lorenz-Mie theory, and points are experimental values.

VI. Conclusion

A focused-beam scattering technique has been developed in which the effects of individual scatterers can be isolated without removing the scatterers from their natural environment. This technique is potentially advantageous for investigations of living organisms or particles which for any reason have characteristics that depend on the host medium. Experimental verification of the basic ideas has been obtained using monodisperse dielectric spheres. Extensions of these concepts will involve detailed scattering calculations and seeing to what extent a unique inversion is possible with more general systems of polydisperse and nonspherical particles.

This work was supported in part by the National Science Foundation.

References

1. M. Kerker, *The Scattering of Light and Other Electromagnetic Radiation* (Academic, New York, 1969).
2. C. Tanford, *Physical Chemistry of Macromolecules* (Wiley, New York, 1961).
3. *Remote Measurement of Pollution* (National Aeronautics and Space Administration, Washington, D.C., 1971).
4. D. T. Phillips, P. J. Wyatt, and R. M. Berkman, J. Colloid Interface Sci. **34**, 159 (1970).
5. P. M. Morse, *Thermal Physics* (Benjamin, New York, 1965), Chap. 15.
6. *Smithsonian Physical Tables*, W. E. Forsythe, Ed. (Smithsonian, Washington, D.C., 1969), p. 319.
7. N. Morita, T. Tanaka, T. Yamasaki, and Y. Nakanishi, IEEE Trans. Antennas Propag. **AP-16**, 724 (1968).
8. W.-C. Tsai and R. J. Pogorzelski, J. Opt. Soc. Am. **65**, 1457 (1975).
9. P. Barber and C. Yeh, Appl. Opt. 14, 2864 (1975).
10. R. G. Pinnick, D. E. Carroll, and D. J. Hofmann, Appl. Opt. 15, 384 (1976).

Reprinted from APPLIED OPTICS, Vol. 17, page 1637, May 15, 1978

Rayleigh-Debye scattering with focused laser beams

Lee W. Casperson and C. Yeh

A focused beam technique has been developed for diagnosing the characteristics of individual particles within a polydisperse ensemble. In the Rayleigh-Debye approximation the scattered fields are related to the orientation and properties of a scatterer by means of explicit analytic formulas. The results simplify when the particle size is small compared to the minimum beam diameter.

I. Introduction

Light scattering techniques have for many years provided an effective means for determining the concentration and properties of small particles suspended in a homogeneous medium. These techniques are much less useful, however, when the particle ensemble is polydisperse, i.e., the properties vary from particle to particle. With polydisperse media it has been necessary to physically isolate individual particles from the ensemble prior to applying an electromagnetic wave. Recently, however, we have described a new scattering technique in which the light source is a focused laser beam.[1] With typical scattering media, at most a single particle is likely to be near the focal region at any instant of time. Therefore, the radiation scattered from the focused beam may be used to deduce the properties of individual particles within an ensemble, and there is no need to remove the particles from their normal surroundings. In order to take full advantage of the focused beam technique, however, the relationships between the scattered fields and the particle properties must be known in detail. The purpose of this work is to develop realistic models for the scattering process which take proper account of the focused beam fields.

In the vast majority of treatments of electromagnetic scattering it is assumed that the unperturbed incident fields are homogeneous plane waves. If the scattering particles are spheres, an exact solution for this case is now commonly known as Lorentz-Mie or Mie scattering.[2] The Lorentz-Mie solutions and other exact results are numerically complex,[3] and several approximation methods have also been developed. The most generally useful approximation is known as Rayleigh-Debye scattering.[4] In this approximation the additional phase shift experienced by a wave traversing the particle is required to be small compared to π. Thus the conditions of Rayleigh-Debye scattering are met if either the particle is small or its complex refractive index is close to unity. Following the work of Rayleigh and Debye this model was also discussed by Gans[5] and Rocard[6] and often goes by the name of Rayleigh-Gans or Rayleigh-Gans-Rocard scattering. The advantage of this method is that one can usually write simple explicit formulas for the scattered fields, even for nonspherical particles. Because of their simplicity such results are commonly applied when the basic phase shift condition is not well satisfied, and it is a remarkable fact that qualitatively useful results are sometimes still obtainable.

The subject of the present study is the application of the Rayleigh-Debye method to the problem of scattering from a focused Gaussian laser beam. In spite of the common usage of Gaussian beams in laboratory research, there have been very few treatments of light scattering by such beams. Pearson *et al.* have considered the scattering of plane wave Gaussian beams from edges.[7] Morita *et al.* have treated the scattering of a plane-wave beam from a small spherical particle at the beam axis.[8] Tsai and Pogorzelski generalized these results for an arbitrary sized spherical particle at the beam axis.[9] Recently, Tam has outlined a possible procedure for treating the scattering of a plane-wave beam from an off-axis sphere.[10] All these analyses have concerned the scattering of a plane wave having a Gaussian amplitude variation in the radial direction. Such a beam mode cannot be an exact solution of Maxwell's equations, and radial phase variations are always present except at the beam waist. A linearly polarized beam is also not possible in a rigorous solution, and minor field components must always be present in the longitudinal direction and in the transverse direction normal to the principal component. These extra field components are of no significance for most laser

The authors are with University of California, School of Engineering & Applied Science, Los Angeles, California 90024.
Received 23 November 1977.
0003-6935/78/0515-1637$0.50/0.

applications. However, with focused-beam scattering spatial resolution is best if the beam is brought to a very sharp focus with low f-number optics. Under these conditions the other field components, especially the longitudinal components, become comparable in magnitude to the principal transverse polarization components. A consideration of these other fields is thus essential for our present investigations. Recently the scattering of uniform converging beams by spheres has also been discussed.[11]

In Sec. II the basic scattering formulas for the Rayleigh-Debye model are derived in a form suitable for an arbitrary spatially dependent incident field and an arbitrarily shaped refracting particle. The field components of a focused Gaussian beam are derived in Sec. III, and general numerical solutions are presented. In Sec. IV a simplified central field model is introduced and discussed. The advantage of this model is that it leads to explicit analytic formulas for the scattered fields.

II. Scattering Formulas

The essential feature of the Rayleigh-Debye approximation is that the electromagnetic field experienced by each volume element in the scatterer is the same as the incident field that would exist in the absence of the scatterer. The scattered field at any point in space is then a coherent superposition of the dipole radiation fields arising from each volume element. The result of this superposition can be inferred from the formulas for a phased array of conventional dipole antennas. This analogy forms the basis for the following discussion.

The complex amplitude of the retarded vector potential corresponding to a small antenna segment can be written[12]

$$A_z = \frac{\mu I dz}{4\pi\rho}\exp(-ik\rho) \tag{1}$$

where I is the peak value of the time-harmonic current flowing in the z-directed antenna segment of differential length dz. The coefficient μ is the permeability of the medium surrounding the antenna, ρ is the distance between the antenna and the detection point, and k is the propagation constant. This antenna segment is basically a small dipole, and the dipole moment can be related to the current by[13]

$$p_z = -iIdz/\omega. \tag{2}$$

Combining these formulas, one finds that the vector potential of the scattered fields is related to the dipole moment $\overline{p}$ by

$$\overline{A}_s = \frac{i\omega\mu\overline{p}}{4\pi\rho}\exp(-ik\rho). \tag{3}$$

If the dipole source is distributed in space, Eq. (3) must be replaced by

$$\overline{A}_s = \frac{i\omega\mu}{4\pi}\int_v \frac{\overline{p}(\overline{r})}{\rho}\exp(-ik\rho)dv, \tag{4}$$

where dv is a volume element in the scatterer.

To complete this formulation, the incident field may be related to the dipole moment, and the scattered field may be related to the vector potential. First, the dipole moment of a small dielectric body is given by[14]

$$\overline{p}(\overline{r})dv = 3\epsilon_2\left(\frac{\epsilon_1-\epsilon_2}{\epsilon_1+\epsilon_2}\right)\overline{E}(\overline{r})dv, \tag{5}$$

where ϵ_1 is the permittivity of the scatterer, ϵ_2 is the permittivity of the surrounding medium, and $\overline{E}(\overline{r})$ is the electric field that would have existed in the absence of the scatterer. Combining Eqs. (4) and (5), the vector potential of the scattered fields is

$$\overline{A}_s = \frac{3ik^2}{4\pi\omega}\left(\frac{n_1^2-n_2^2}{n_1^2+2n_2^2}\right)\int_v \frac{\overline{E}(\overline{r})\exp(-ik\rho)}{\rho}dv. \tag{6}$$

In this expression the propagation constant k is related to the permittivity and vacuum wavelength by

$$k = \omega(\mu\epsilon_2)^{1/2} = (2\pi n_2)/\lambda, \tag{7}$$

and the index of refraction is related to the permittivity by $n_2 = (\epsilon_2/\epsilon_0)^{1/2}$.

The scattered electric field amplitude at a point $\overline{r}'$ is related to the vector potential by the equation[15]

$$\overline{E}_s(\overline{r}') = -i\omega\left\{\overline{A}_s(\overline{r}') + \frac{1}{k^2}\nabla'[\nabla'\cdot\overline{A}_s(\overline{r}')]\right\}, \tag{8}$$

where ∇' represents the gradient with respect to the primed coordinates. The total electric field at the point $\overline{r}'$ may be obtained by adding on the field of the input beam according to

$$\overline{E}_t(\overline{r}') = \overline{E}_s(\overline{r}') + \overline{E}(\overline{r}'). \tag{9}$$

Equations (4)–(9) may be combined to yield the electric field at any point in space, and the result is

$$\overline{E}_t(\overline{r}') = \frac{3k^2}{4\pi}\left(\frac{n_1^2-n_2^2}{n_1^2+2n_2^2}\right)\int_v\left[1+\frac{\nabla'}{k^2}(\nabla'\cdot)\right] \times\left[\frac{\overline{E}(\overline{r})}{|\overline{r}'-\overline{r}|}\exp(-ik|\overline{r}'-\overline{r}|)\right]dv + \overline{E}(\overline{r}'), \tag{10}$$

where the identity $\rho = |\overline{r}' - \overline{r}|$ has been used. By employing the total field at a point inside of the scatterer as the input field for a second iteration, the accuracy of the Rayleigh-Debye method can be increased.[16] For our present purposes, however, the first order solutions are adequate.

Equation (10) can be simplified if we make the reasonable assumption that ρ is always large compared with the wavelength λ/n_2. The result is

$$\overline{E}_t(\overline{r}') = \frac{3k^2}{4\pi}\left(\frac{n_1^2-n_2^2}{n_1^2+2n_2^2}\right)\int_v\left\{\frac{\overline{E}(\overline{r})}{|\overline{r}'-\overline{r}|} - \frac{(\overline{r}'-\overline{r})[(\overline{r}'-\overline{r})\cdot\overline{E}(\overline{r})]}{|\overline{r}'-\overline{r}|^3}\right\} \times\exp(-ik|\overline{r}'-\overline{r}|)dv + \overline{E}(\overline{r}). \tag{11}$$

For the problem of scattering from a small object one can also assume that the distance ρ remains large compared to a typical dimension or displacement of the object. More specifically we require that every point of the object be close to the coordinate reference so the following conditions apply:

$$x' \gg x, \quad y' \gg y, \quad z' \gg z, \qquad \rho \simeq (x'^2 + y'^2 + z'^2)^{1/2}. \tag{12}$$

Now Eq. (11) reduces to

$$\overline{E}_t(\overline{r}') = \frac{3k^2}{4\pi}\left(\frac{n_1^2 - n_2^2}{n_1^2 + 2n_2^2}\right) \cdot \frac{\exp(-ik\rho)}{\rho}\left[\overline{S} - \frac{\overline{r}'(\overline{r}'\cdot\overline{S})}{\rho^2}\right] + \overline{E}(\overline{r}'), \quad (13)$$

where $\overline{S}$ represents the integral

$$\overline{S} = \int_v \overline{E}(\overline{r})\exp(ik\overline{r}\cdot\overline{r}'/\rho)dv. \quad (14)$$

Equation (13) can also be written in the forms

$$E_t(\overline{r}') = -\frac{3k^2}{4\pi}\left(\frac{n_1^2 - n_2^2}{n_1^2 + 2n_2^2}\right)\frac{\exp(-ik\rho)}{\rho^3}[\overline{r}' \times (\overline{r}' \times \overline{S})] + \overline{E}(\overline{r}') \quad (15)$$

$$= \frac{3k^2}{4\pi}\left(\frac{n_1^2 - n_2^2}{n_1^2 + 2n_2^2}\right)\frac{\exp(-ik\rho)}{\rho}(\sin\chi)|\overline{S}|\overline{i}_s + \overline{E}(\overline{r}'), \quad (16)$$

where χ is the angle between the $\overline{S}$ vector and $\overline{r}'$, and $\overline{i}_s$ represents the polarization direction of the scattered field.

Equations (13) and (14) are our basic results for scattering with focused beams, and it is appropriate to compare these formulas with previous treatments. One significant feature here is that the spatial variations of the incident field prevent the easy removal of $\overline{E}(\overline{r})$ from the integral in Eq. (14). If the field had been a homogeneous plane wave, this removal would have been possible, and Eq. (13) would simplify to well known formulas. Analytic techniques for reducing Eq. (14) with a Gaussian input field are discussed in Sec. IV. Another difference from some treatments is that the incident field also is a direct component of the total detected field, and this is due to the large beamwidth of the incident field.

III. Focused Beam Fields

In order to apply the previously derived scattering formulas, explicit procedures are needed for obtaining the components of the incident electromagnetic wave. Unfortunately, there exist no exact analytic solutions of Maxwell's equations for freely propagating beams. Such solutions can only be obtained by numerical integration of the wave equation or by a Fourier expansion in terms of infinite plane waves.[17] It is also reasonable to enquire whether sufficient accuracy might be obtainable using approximate analytical expressions for the field components. The dominant transverse field components can be expressed in terms of polynomial-Gaussian functions.[18] In this section approximate formulas are given for the fundamental Gaussian beam including first-order corrections for nontransverse field components.[19] Detailed solutions are presented, and the significance of the correction terms is explored.

Maxwell's equations for the complex amplitudes of the electric and magnetic fields can be written in the well known form

$$\nabla \times \overline{E} = -i\omega\mu\overline{H}, \quad (17)$$

$$\nabla \times \overline{H} = i\omega\epsilon\overline{E}. \quad (18)$$

In a homogeneous medium these equations may be combined to yield the wave equation

$$\nabla^2\overline{E} + k^2\overline{E} = 0, \quad (19)$$

where the propagation constant is $k = \omega(\mu\epsilon)^{1/2}$. In approximate solutions of Eq. (19) one typically assumes that the beam is nearly a linearly polarized plane wave in the form

$$E_x = \psi(x,y,z)\exp(-ikz), \quad (20)$$

where $\psi(x,y,z)$ is a slowly varying function of the spatial coordinates. When Eq. (20) is substituted into Eq. (19), one obtains the new wave equation

$$\frac{\partial\psi^2}{\partial x^2} + \frac{\partial^2\psi}{\partial y^2} + \frac{\partial^2\psi}{\partial z^2} - 2ik\frac{\partial\psi}{\partial z} = 0. \quad (21)$$

In the paraxial approximation the second derivative of ψ with respect to z is neglected, and several complete orthogonal sets of solutions of the resulting equation are known. Of particular interest here is the fundamental Gaussian mode which may be shown by direct substitution to have the form

$$\psi = \psi_0\exp[-iQ(z)(x^2 + y^2)/2 - iP(z)], \quad (22)$$

where the parameters Q and P have the z dependences

$$Q(z) = Q_0(1 + zQ_0/k)^{-1}, \quad (23)$$

$$P(z) = P_0 - i\ln(1 + zQ_0/k). \quad (24)$$

In these results Q_0 and P_0 represent, respectively, the values of $Q(z)$ and $P(z)$ at the beam waist ($z = 0$). $Q(z)$ can be identified with the phase front curvature $R(z)$ and the $1/e$ amplitude spotsize $w(z)$ using the equation

$$Q(z) = k/R(z) - 2i/w^2(z), \quad (25)$$

and $P(z)$ is the complex phase.

For the present application the beams are strongly focused, and there is no *a priori* reason for assuming that the paraxial approximation will provide a useful description of the focused beam fields. Therefore, we have employed an iterative procedure for obtaining increasingly accurate descriptions of the fields. This procedure involves substituting an approximate expression for the field components into the left-hand sides of Eqs. (17) and (18) in order to obtain new expressions for $\overline{E}$ and $\overline{H}$. As our starting approximation we use the Gaussian beam solution given in Eq. (22). Thus a reasonable set of starting fields is

$$E_x(x,y,z) = \psi(x,y,z)\exp(-ikz), \quad (26)$$

$$H_y(x,y,z) = (\epsilon/\mu)^{1/2}\psi(x,y,z)\exp(-ikz). \quad (27)$$

When Eqs. (26) and (27) are introduced into the left-hand sides of Eqs. (17) and (18), the leading z components are found to be

$$E_z(x,y,z) = -xQ(z)\psi(x,y,z)\exp(-ikz)/k, \quad (28)$$

$$H_z(x,y,z) = -(\epsilon/\mu)^{1/2}yQ(z)\psi(x,y,z)\exp(-ikz)/k. \quad (29)$$

We may assume that x (or y) is at most comparable to w, and near the beam waist w is on the order of $|Q|^{-1/2}$. Therefore, the z components are smaller than the leading transverse components by a factor of about $(kw)^{-1}$.

The orthogonal field components and other corrections can be found by further iterations. In establishing the magnitude of these corrections it suffices to assume that the most rapid z variations occur in the terms $\exp(-ikz)$. Thus self-consistent expressions for the orthogonal components are

$$E_y(x,y,z) = -xyQ^2(z)\psi(x,y,z)\exp(-ikz)/2k^2, \quad (30)$$

$$H_x(x,y,z) = -(\epsilon/u)^{1/2}xyQ^2(z)\psi(x,y,z)\exp(-ikz)/2k^2. \quad (31)$$

Near the beam waist the magnitude of these components is smaller than the dominant transverse components by about $(kw)^{-2}$. This is also the approximate amount by which the higher order corrections to E_x and H_y are smaller than the leading terms, and these corrections are not necessary for the present discussion. The smallness of the corrections is easily confirmed with a specific example. The smallest spot size that one might expect to encounter in practice is about $w = \lambda$, and the correction factor is already as small as $(kw)^{-2} = 0.0025$.

IV. Central Field Approximation

In Sec. II a general formula has been developed for calculating the scattered field distribution that results when an arbitrary electromagnetic beam is incident on a homogeneous particle. In essence the problem is reduced to evaluating the integral in Eq. (14). Unfortunately, this integral can be performed explicitly only for certain simple distributions of the electromagnetic fields. For the Gaussian field distributions of focused beams the integrals cannot be expressed in terms of elementary functions. Numerical integration is always straightforward, and some solutions will be discussed. However, it has proved to be very useful in our work to introduce a simpler model for the electromagnetic fields, and the result may be called a central field approximation. The basic geometry that we have in mind is illustrated in Fig. 1. The incident beam is a Gaussian having its waist at the plane $z = 0$.

Combining Eqs. (14), (26), and (28), we find that the general form for the scattering integral is

$$S_x = \int_v \psi(x,y,z)\exp(i\delta)dv, \quad (32)$$

$$S_z = -\int_v xQ(z)\psi(x,y,z)\exp(i\delta)dv, \quad (33)$$

where the term $\delta = k(\bar{r}\cdot\bar{r}'/\rho - z)$ in the exponent represents the phase shift of a scattered ray with respect to a ray travelling straight from the origin. These integrals cannot be performed explicitly. However, if the particle size is small or at most comparable to the spot size, simplifications are possible. Such a smallness constraint may apply to many reasonable practical configurations and is consistent with the basic phase shift condition underlying the Rayleigh-Debye approximation. In this case the spatial variations of the field in the vicinity of the particle can be accurately described using only a few terms in a power series expansion. The expansion is performed about some typical location in the particle, and for a spherical scatterer the natural expansion point is the particle center. Thus the Gaussian field distributions described in the previous section can be expanded in a Taylor series with respect to the Cartesian coordinates, and some of the integrals represented by Eq. (14) may be performed analytically.

In our own studies the general expansions have often proved to be unnecessarily accurate, and for brevity we only present the simplest limiting cases. The first step is to observe that for a particle small compared to the beam diameter the only z variation of importance in Eqs. (26) and (28) is the $\exp(-ikz)$ plane wave factor. Thus, Eqs. (32) and (33) can be written

$$S_x \int_x\int_y \psi(x,y,z_1)\int_z \exp(i\delta)dzdydx, \quad (34)$$

$$S_z = -\int_x\int_y xQ(z_1)\psi(x,y,z_1)\int_z \exp(i\delta)dzdydx, \quad (35)$$

where z_1 is the coordinate of the center of the particle (or some other characteristic reference). With a change of coordinates the z integration reduces to the integration of a simple exponential.

A slightly less accurate subsequent approximation is to assume that the transverse field variations in the vicinity of the particle are entirely negligible. Thus, Eqs. (34) and (35) reduce to

$$S_x = \psi(x_1,y_1,z_1)\int_v \exp(i\delta)dv, \quad (36)$$

$$S_z = -x_1Q(z_1)\psi(x_1,y_1,z_1)\int_v \exp(i\delta)dv, \quad (37)$$

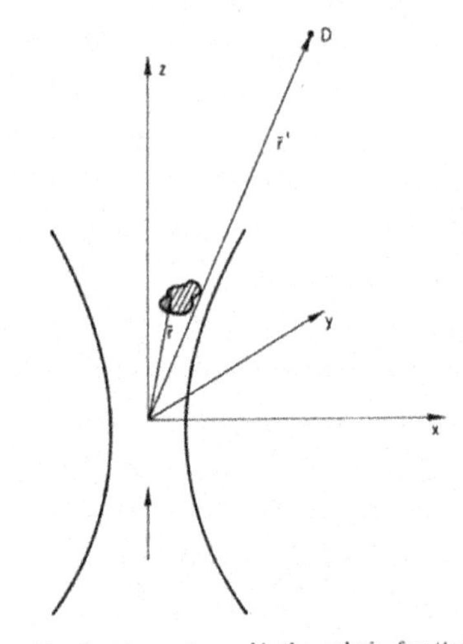

Fig. 1. Coordinate geometry used in the analysis of scattering with focused beams. The beam waist is at $z = 0$, and D represents a detection point.

where x_1 and y_1 are the transverse coordinates of the particle center. These remaining integrations can be performed explicitly for many simple particle shapes. In the case of spheres the result is[20]

$$\int_v \exp(i\delta)dv = vG(u), \tag{38}$$

where v is the volume of the sphere, and the variable u is related to the scattering angle θ and the particle radius a by

$$u = 2ka \sin(\theta/2). \tag{39}$$

The function $G(u)$ is given by

$$G(u) = 3v(\sin u - u \cos u)/u^3. \tag{40}$$

Thus, Eqs. (36) and (37) now have the explicit forms

$$S_x = v\psi(x_1,y_1,z_1)G(u), \tag{41}$$

$$S_z = -x_1Q(z_1)v\psi(x_1,y_1,z_1)G(u). \tag{42}$$

The preceding equations and their obvious extensions to higher orders constitute the essence of the central field approximation. In the simplest limit the field is assumed to be a constant in the vicinity of the particle, and the value of this constant is the actual value of the field at the center of the particle.

It is now reasonable to inquire whether these equations reduce in the appropriate limit to the well known expressions for Rayleigh-Debye scattering of homogeneous plane waves. A Gaussian beam reduces to a plane wave if the spot size w and phase front curvature R are permitted to increase without limit. Therefore, from Eq. (25) the beam parameter Q vanishes and with it any longitudinal or orthogonal field components implied by Eqs. (28)–(31). Equation (13) now reduces to

$$E_t(r') = \frac{3k^2v\psi}{4\pi}\left(\frac{n_1^2 - n_2^2}{n_1^2 + 2n_2^2}\right) G(u) \frac{\exp(-ik\rho)}{\rho} \times \left[\vec{i}_x - \frac{\vec{r}'(\vec{r}' \cdot \vec{i}_x)}{\rho^2}\right] + \overline{E}(r'). \tag{43}$$

The magnitude of the bracketed terms can sometimes be replaced by unity or $\cos\theta$ if the incident radiation is polarized, respectively, perpendicular to the scattering plane or parallel to the scattering plane. These results are in agreement with the conventional Rayleigh-Debye scattering formulas.[4]

An important application of the scattering formalism that has been developed is the determination of the location and properties of small scatterers. For this purpose one would like to be able to obtain a unique inversion of the formulas, and it is desirable to know whether such an inversion is possible. This question can be most easily answered by considering the symmetry characteristics of the scattering formulas themselves. It is assumed that the direct beam $\overline{E}(\vec{r}')$ has been eliminated from the scattering pattern represented by Eq. (13), and this term will be included in the later discussion.

We treat first the case in which the scattering particle is inverted across the z axis in such a way that the dielectric constant at every point (x,y,z) is replaced by the dielectric constant of the corresponding point $(-x,-y,z)$. Simultaneously the detection point (x',y',z') is moved to the point $(-x',-y',z')$. From Eqs. (32) and (33) and the symmetry of the beam it follows that this transformation leaves S_x (and S_y) unchanged while the sign of S_z is reversed. Now an examination of Eq. (13) reveals that E_{tx} and E_{ty} are unchanged, while the sign of E_{tz} is reversed. This last sign change has no effect on the energy density $u_e \alpha \overline{E} \cdot \overline{E}^*$, so the measured intensity is not altered by the transformation. This result is reasonable, since we expect intuitively that a reversal of scatterer and detector should yield no change in the output.

Somewhat more subtle is the question of what happens if the detector is inverted across the z axis while the scatterer is inverted across the $z = 0$ plane. Thus we replace the dielectric constant at every point (x,y,z) by the dielectric constant at the opposite point $(x,y,-z)$, and the detection point (x',y',z') is replaced by the point $(-x',-y',z')$. From Eqs. (23) and (24) the real parts of $Q(z)$ and $P(z)$ change sign (with $P_0 = 0$), and the imaginary parts are unchanged. Then from Eq. (22) (with ψ_0 real) ψ is replaced by its complex conjugate. Equations (26)–(31) imply that E_x and E_y are replaced by their complex conjugates while E_z is replaced by its negative conjugate. Next, Eqs. (32) and (33) imply that S_x (and S_y) are replaced by their conjugates while S_z is replaced by its negative conjugate. Finally, with Eq. (13) one finds again that the observed intensity is unchanged by the transformation.

Combining the previously derived symmetry relations (or performing a third derivation) one discovers that there is a basic degeneracy in the scattering formulas. In particular, the field distribution is unchanged in this Rayleigh-Debye approximation when the scatterer is inverted across the origin (0,0,0). This degeneracy is an unfortunate source of uncertainty unless either the particle shape or its location can be determined by some independent means. If the particles themselves possess inversion symmetry, the degeneracy makes the scattering problem nonunique. For example, the scattering pattern of a sphere centered at the point (x_1,y_1,z_1) is exactly the same as the pattern for the sphere located at the point $(-x_1,-y_1,-z_1)$. Furthermore, a sphere situated anywhere on the x axis somewhat surprisingly yields a scattering pattern which is symmetric in the detection coordinate x'.

An examination of Eq. (13) shows that the degeneracy is removed if the incident field $\overline{E}(\vec{r}')$ is included in the detected radiation. Then the scattering pattern is superimposed with the intense forward beam, and a characteristic pattern of interference fringes occurs where the two signals are comparable in amplitude. The lack of uniqueness can also be overcome by additional complexity in the experimental apparatus. For example, if the focal point of the incident beam is dithered in any direction on a time scale short compared to the transit time of the particle, the response of the scattered fields will eliminate the uniqueness problem. Alternatively, spatial filtering in front of the detectors can reveal additional information about the particle location. To eliminate the degeneracy it would be sufficient, for example, to know whether the particle is above or below the $z = 0$ plane.

V. Results

The implications of the scattering formulas that have been obtained can be most readily illustrated by means of specific examples. The examples chosen correspond to the configurations of certain of our experiments. The assumed radiation wavelength is $\lambda = 0.5145 \times 10^{-6}$ m corresponding to the blue-green line of an argon laser. The scattering particles are latex spheres of 0.24×10^{-6}, m radius and refractive index $n_1 = 1.59$. These spheres are suspended in water with a refractive index of $n_2 = 1.33$, and the beam is focused in the water to a minimum spot size of $w_0 = 0.54 \times 10^{-6}$. The purpose of these calculations is to check on the accuracy of the central field approximation and to see how the scattered intensity depends on the location of the scatterer.

The scattered field distribution in the $z' = 1$-cm plane is plotted in Fig. 2 for the spherical particle at two locations in the $z = 0$ plane. The incident field is polarized primarily in the x direction, and the normalized energy density is $u = \overline{E}_t \cdot \overline{E}_t^*$ with $\psi_0 = 1$. The solid lines are the general Rayleigh-Debye results based upon Eqs. (13), (32), and (33). The incident field $\overline{E}(\overline{r}')$ is omitted from Eq. (13) as discussed previously. The dashed lines are the central field approximation of Eqs. (13), (41), and (42). It is apparent from the figure that even in this extreme case of a particle comparable in size to the beam the central field model is accurate within a few percent. The central field model exaggerates the scattering in Fig. 2(a) because it also exaggerates the incident field near the edge of the particle. The symmetry of the scattering pattern in Fig. 2(b) follows from the relations of the previous section.

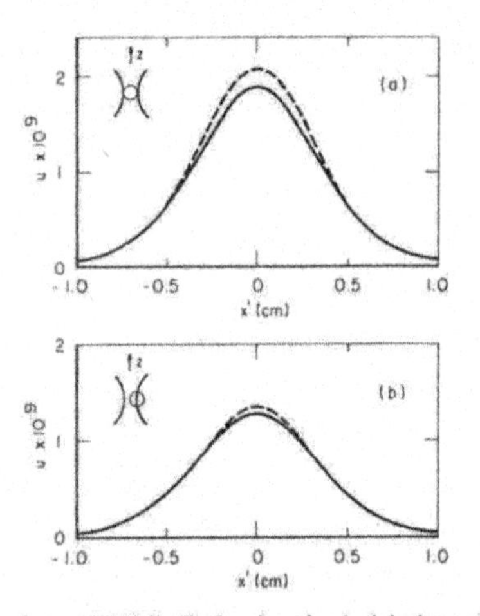

Fig. 2. Scattered field distribution along the x' axis in the $z' = 1$-cm plane for (a) a particle at the origin $x = 0$ and (b) a particle at $x = 0.25 \times 10^{-6}$ m. The solid lines represent the exact Rayleigh-Debye solutions, and the dashed lines are the central field results. The particle and beam properties are described in the text. Note that the scattered field remains symmetric in x' even though the particle in (b) is off axis.

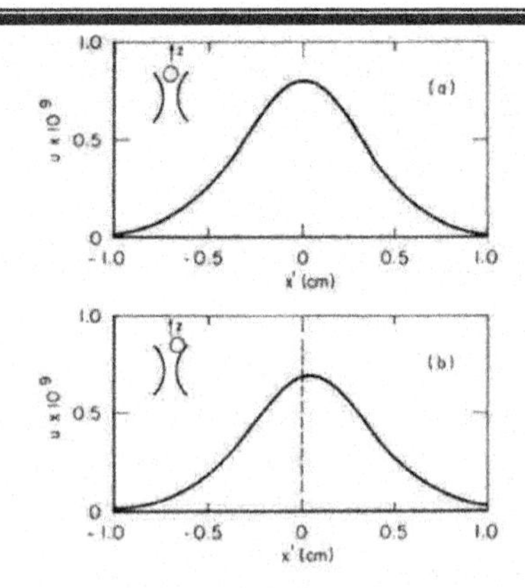

Fig. 3. Scattered field distribution along the x' axis in the $z' = 1$-cm plane for (a) a particle at the coordinate $x = 0, y = 0, z = 3 \times 10^{-6}$ m and (b) a particle at the coordinates $x = 0.25 \times 10^{-6}$ m, $y = 0, z = 3 \times 10^{-6}$ m. Note that in case (b) the scattered fields are slightly asymmetric in x'.

In Fig. 3 are presented the corresponding results for a scattering particle located in the $z = 3 \times 10^{-6}$-m plane. The scattered intensity is reduced because the particle is away from the focus. A small but distinct asymmetry always appears when the particle is away from the z axis (except in the $z = 0$ plane). The agreement between the exact and central field models in this case is too close to be represented in the figure, and both models imply the asymmetry for off-axis particles. This slight deflection of the scattered radiation beam can be qualitatively understood from a consideration of the wavefront orientation at the particle.

VI. Conclusion

A Rayleigh-Debye method has been developed for treating the scattering of focused light beams, and this is the first focused beam treatment which accounts for nontransverse components of the electromagnetic fields. Any focused beam must possess such nontransverse components, and the examples considered here have involved the important Gaussian beams which commonly result with laser sources. Another area of investigation has involved reducing the volume integrals inherent in the Rayleigh-Debye method to explicit analytic formulas. This has been accomplished by a sequence of central field approximations. The result of this study is that even the poorest of these approximations retains the major field asymmetries and differs by at most a few percent from the general Rayleigh-Debye values. It is also worth noting that the general shape of the scattering diagrams is largely insensitive to the position of the particle in the beam and differ in relatively minor respects from the scattering patterns for homogeneous plane waves. This fact should lead to a considerable simplification of the interpretation and inversion of focused beam scattering data.

This work was supported in part by the National Science Foundation.

References

1. L. W. Casperson, C. Yeh, and W. F. Yeung, Appl. Opt. **16**, 1104 (1977).
2. J. A. Stratton, *Electromagnetic Theory* (McGraw-Hill, New York, 1941), p. 563.
3. P. Barber and C. Yeh, Appl. Opt. **14**, 2864 (1975).
4. M. Kerker, *The Scattering of Light and Other Electromagnetic Radiation* (Academic, New York, 1968), Chap. 8.
5. R. Gans, Ann. Phys. **76**, 29 (1925).
6. I. Rocard, Rev. Opt. **9**, 97 (1930).
7. J. E. Pearson, T. C. McGill, S. Kurtin, and A. Yariv, J. Opt. Soc. Am. **59**, 1440 (1969).
8. N. Morita, T. Tanaka, T. Yamasaki, and Y. Nakanishi, IEEE Trans. Antennas Propag. **AP-16**, 724 (1968).
9. W.-C. Tasi and R. J. Pogorzelski, J. Opt. Soc. Am. **65**, 1457 (1975).
10. W. G. Tam, Appl. Opt. **16**, 2016 (1977).
11. H. Chew, M. Kerker, and D. D. Cooke, Opt. Lett. **1**, 138 (1977).
12. S. Ramo, J. R. Whinnery, and T. Van Duzer, *Fields and Waves in Communication Electronics* (Wiley, New York, 1965), p. 643.
13. J. A. Stratton, *Electromagnetic Theory* (McGraw-Hill, New York, 1941), p. 431.
14. Ref. 13, p. 205.
15. Ref. 12, p. 269.
16. K. S. Shifrin, *Scattering of Light in a Turbid Medium* (Moscow, 1951) (NASA Technical Translation TTF-477, 1968), Chap. 8.
17. W. H. Carter, J. Opt. Soc. Am. **62**, 1195 (1972).
18. H. Kogelnik and T. Li, Appl. Opt. **5**, 1550 (1966).
19. L. W. Casperson, Appl. Opt. **12**, 2434 (1973).
20. H. C. Van De Hulst, *Light Scattering by Small Particles* (Wiley, New York, 1957), Chap. 7.

Reprinted from APPLIED OPTICS, Vol. 18, page 294, February 1, 1979

Scattering of focused beams by tenuous particles

S. Colak, C. Yeh, and L. W. Casperson

This paper deals with the problem of the scattering of focused laser beams by tenuous particles using an iterative technique. The results are shown to be accurate provided that (a) the polarizability of the particle medium is small and (b) the phase shift of the central ray is less than 2. It was found that when the size of the incident beam waist is close to that of the scatterer, the scattered field deviates significantly from that for the incident plane wave case. Specific examples are given.

I. Introduction

By strongly focusing a laser beam in a polydisperse ensemble medium, we have shown that[1] it is possible to obtain the scattering characteristics of individual particles in such a medium. In order to make use of this focused beam technique to deduce the physical properties of individual particles within an ensemble, the relationships between the scattered fields and the particle properties must be known in advance. In a previous paper,[2] we have derived an explicit analytical formula for the scattered field of a particle in a focused beam under the Rayleigh-Gans approximation. This approximation requires that the following two criteria be satisfied: (1) the relative refractive index n for the scatterer with respect to its surrounding medium is close to 1, that is, $|n-1| \ll 1$, and (2) the phase shift δ of the central ray through the particle is small, $\delta = 2ka|n-1| \ll 1$, where k is the propagation constant, and a is a typical particle dimension. It should be recognized that these criteria are very restrictive. The purpose of this paper is to extend the iterative technique developed by Shifrin[3] and later applied by Acquista[4] for the treatment of the plane wave scattering problem to the focused beam scattering problem. This iterative technique will yield results which will be valid under two broader criteria: (1) the polarizability α is small, $\alpha = [3(n^2-1)]/[4\pi(n^2+2)] \ll 1$, and (2) the phase shift δ of the central ray is less than 2. Several numerical examples will be given.

The authors are with University of California, Los Angeles, Electrical Sciences & Engineering Department, Los Angeles, California 90024.
Received 14 July 1978.
0003-6935/79/03294-09$00.50/0.

II. Formulation of the Problem

The geometry of the problem is shown in Fig. 1. A tenuous particle whose physical parameters satisfy the necessary broader criteria given in Sec. I is situated in an incident focused Gaussian beam. We are interested in the scattered field.

The transverse electric field of an incident Gaussian beam takes the form[5]

$$\mathbf{E}_b(x,y,z,t) = (A_x\hat{x}_x + A_y\hat{y}_y)f(x,y,z)\exp(i\omega t), \tag{1a}$$

where

$$\begin{aligned} f(x,y,z) &= \frac{W_0}{W(z)}\exp(-ikz)\exp[-ik(x^2+y^2)/2R(z)] \\ &\quad \cdot \exp[-(x^2+y^2)/W^2(z)] \\ &\quad \cdot \exp[-i\arctan(z/z_0)], \end{aligned} \tag{1b}$$

$$W(z) = W_0[1+(z^2/z_0^2)]^{1/2}, \tag{1c}$$

$$R(z) = z[1+(z_0^2/z^2)], \tag{1d}$$

$$z_0 = \pi W_0^2/\lambda. \tag{1e}$$

The quantities $W(z)$ and $R(z)$ are, respectively, the spot size and the radius of curvature of the phase front of the beam at z. A_x or A_y are, respectively, the complex amplitude constant of the electric field vectors polarized in the x or y direction. W_0 is the beam waist spot size, and z_0 is the Rayleigh length of the beam. ω is the frequency of the wave, and $k = \omega(\epsilon_0\mu_0)^{1/2}$ is the wavenumber. The relations given above assume that the longitudinal z component of the beam field is much smaller than the transverse components. This assumption does not cause any error greater than 2.5×10^{-3} in the transverse components given by Eq. (1) even in the most sharply focused beam cases.[2]

Instead of solving the problem for the complicated beam expression of Eq. (1), this work will involve a more precise and practical method, which makes use of the plane wave spectrum (PWS) of the incident beam. The PWS of the beam is obtained as follows: the beam expression given by Eq. (1) can be Fourier transformed

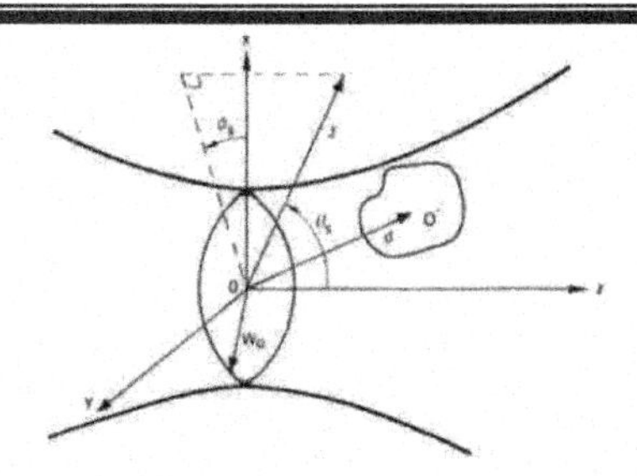

Fig. 1. Geometry of the scattering problem. W_0 is the spot size of the beam which is centered on the O system. The beam propagates along the $+z$ direction. The scatterer is centered on the O' system, which is displaced from the O system by $\mathbf{d} = (d_x, d_y, d_z)$, and it may also be rotated with respect to the O system. θ_s and ϕ_s define the scattering angles.

(spatially) in the $z = 0$ plane to yield the spectrum functions of the beam field components, which are

$$S_x(p,q) = \frac{A_x}{\lambda^2} \iint_{-\infty}^{+\infty} f(x,y,0) \exp[ik(px + qy)]dxdy$$
$$= A_x[(b^2)/\pi] \exp[-b^2(p^2 + q^2)], \quad \text{(2a)}$$
$$S_y(p,q) = \frac{A_y}{\lambda^2} \iint_{-\infty}^{+\infty} f(x,y,0) \exp[-ik(px + qy)]dxdy$$
$$= A_y[(b^2)/\pi] \exp[-b^2(p^2 + q^2)], \quad \text{(2b)}$$

where

$$b = kW_0/2. \quad \text{(2c)}$$

The inverse transform gives the x and y polarized incident beam electric fields at the $z = 0$ plane:

$$E_{bx}(x,y,0) = A_x f(x,y,0) = \iint_{-\infty}^{+\infty} S_x(p,q) \exp[ik(px + qy)]dpdq, \quad \text{(3a)}$$

$$E_{by}(x,y,0) = A_y f(x,y,0) = \iint_{-\infty}^{+\infty} S_y(p,q) \exp[ik(px + qy)]dpdq, \quad \text{(3b)}$$

where the time dependence, $\exp(i\omega t)$, of the fields is omitted. The scalar Helmholtz equation,

$$\nabla^2 E_{bj}(x,y,z) + k^2 E_{bj}(x,y,z) = 0, \quad j = x, \text{ or } y, \quad \text{(4)}$$

together with the boundary conditions given by Eq. (3) are used[6] to find the spectrum of the incident fields everywhere in space. Thus, for a wave propagating in the positive z direction, the following equations for the transverse incident beam components are obtained:

$$E_{bx}(x,y,z) = \iint_{-\infty}^{+\infty} S_x(p,q) \exp[ik(px + qy - sz)]dpdq, \quad \text{(5a)}$$

$$E_{by}(x,y,z) = \iint_{-\infty}^{+\infty} S_y(p,q) \exp[ik(px + qy - sz)]dpdq, \quad \text{(5b)}$$

where

$$s = (1 - p^2 - q^2)^{1/2}. \quad \text{(5c)}$$

The plane wave spectrum of the z component of the incident electric field is found using the divergence condition $\nabla \cdot \mathbf{E}_b = 0$. Finally, the PWS of the total incident electric field is found by summing all three electric field expressions corresponding to three different polarizations and by using Eq. (2). The result is

$$\mathbf{E}_b(x,y,z) = \frac{b^2}{\pi} \iint_{-\infty}^{+\infty} \left(A_x\hat{x} + A_y\hat{y} + \frac{pA_x + qA_y}{s}\hat{z}\right) \cdot \exp[-b^2(p^2 + q^2)] \exp[ik(px + qy - sz)]dpdq, \quad \text{(6)}$$

where $s = (1 - p^2 - q^2)^{1/2}$, and $b = kW_0/2$. Similar derivations have been carried out by Clemmow[7] and Carter.[8] It is seen that the integrand in Eq. (6) can be viewed as a form of plane wave with amplitudes

$$\mathbf{A}_i(p,q) = \left(A_x\hat{x} + A_y\hat{y} + \frac{pA_x + qA_y}{s}\hat{z}\right) \exp[-b^2(p^2 + q^2)] \quad \text{(7)}$$

and propagation vectors

$$\mathbf{k}_i(p,q) = k(p\hat{x} + q\hat{y} - s\hat{z}). \quad \text{(8)}$$

Hence, each plane wave component in the integrand can be represented by the general form

$$\mathbf{E}_i(p,q) = \mathbf{A}_i(p,q) \exp[i\mathbf{k}_i(p,q) \cdot \mathbf{r}]. \quad \text{(9)}$$

The expression of Eq. (6) is called the PWS (plane wave spectrum) of the incident laser beam. The plane wave components of this spectrum have polarization and propagation vectors as defined by Eqs. (7) and (8), and they can be expressed in a compact form as shown in Eq. (9). If the scattering problem is solved for one of these plane wave components, the total scattering field pattern due to the incident laser beam will be found by summing over a continuous superposition of such solutions. Of course, this procedure is true for linear phenomenon only.

As mentioned earlier, Shifrin[3] developed an integro-differential equation, which gives the effective field in the presence of a tenuous scatterer. Later, Acquista,[4] by expanding the effective field and the scattered fields in the power series of the polarizability α and by making use of the Fourier transforms, obtained the scattered field expressions for the first two iterations. They are summarized as follows:

$$\mathbf{E}_s^{(1)}(\mathbf{r}) = \frac{k^2}{r} \exp(ikr)\, \mathbf{E}_{o\perp}\, u(k\hat{r} - \mathbf{k}),$$

$$\mathbf{E}_s^{(2)}(\mathbf{r}) = \frac{2k^2}{(2\pi)^2} \frac{\exp(ikr)}{r} \int_{V_m} \frac{u(\mathbf{m} + k\hat{r})u(\mathbf{m} + \mathbf{k})}{m^2 - k^2} \quad \text{(10a)}$$
$$\cdot \left[\left(\frac{2}{3}k^2 + \frac{1}{3}m^2\right)\mathbf{E}_{o\perp} - (\mathbf{m} \cdot \mathbf{E}_o)\mathbf{m}_\perp\right] d^3m, \quad \text{(10b)}$$

$$\mathbf{E}_s(r) = \alpha \mathbf{E}_s^{(1)}(\mathbf{r}) + \alpha^2 \mathbf{E}_s^{(2)}(\mathbf{r}), \quad \text{(11)}$$

where $\mathbf{E}_s$ is the scattered field, and $\mathbf{E}_s^{(1)}$ and $\mathbf{E}_s^{(2)}$ are the scattered field in the first and second iterations, respectively. Note that the above equations were derived assuming that the incident wave is a single plane wave with vector amplitude $\mathbf{E}_o$ and propagation vector $\mathbf{k}$. The symbols in the above equations are defined as follows: $u(\mathbf{v})$ is the spatial Fourier transform of the 3-D pupil function $u(\mathbf{r})$ of the scatterer, which is given by

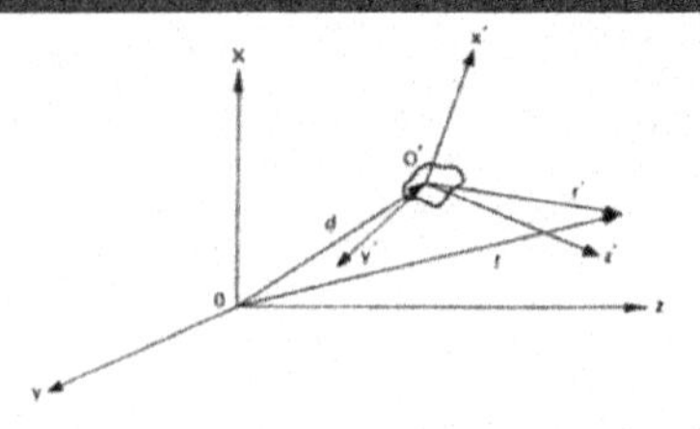

Fig. 2. Placement of the rotated scatterer coordinate system O' with respect to the stationary beam system O. $\mathbf{r}$ and $\mathbf{r}'$ are arbitrary position vectors in the beam and the scatterer system, respectively. $\mathbf{d}$ is the vector defining the displacement of the scatterer from the origin of the beam system.

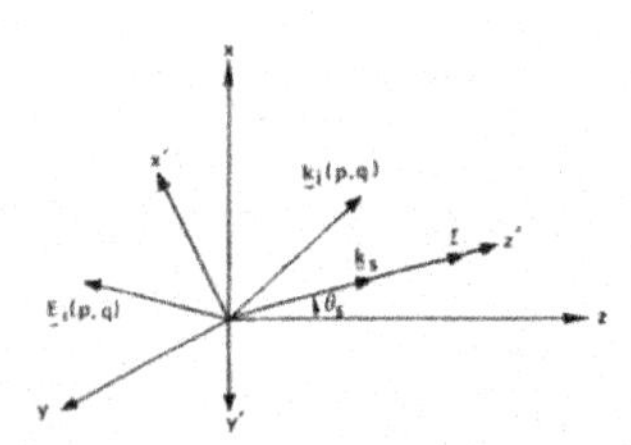

Fig. 3. Diagram for coordinate transformations. $\mathbf{k}_s$, $\mathbf{r}$, and $\hat{z}'$ are vectors all directed to the observation point. The incident field propagation vector $\mathbf{k}_i(p,q)$ is in the $x'z'$ plane. The scattering plane is chosen to be the xz plane, and thus the vector $\mathbf{k}_s$ (also $\mathbf{r}$ and $\hat{z}'$) is in this plane, and it makes a scattering angle θ_s with the z axis of the beam system O.

$$u(\mathbf{v}) = \int_{V_s} \exp(i\mathbf{v}\cdot\mathbf{r})d^3r, \tag{12}$$

where V_s means that the integration is to be carried out over the total scatterer volume; $\mathbf{m}$ and d^3m are, respectively, the vectors and the volume elements of the 3-D spatial frequency domain; V_m indicates that the integration in this domain is to be carried out over all space; $\mathbf{E}_{o\perp}$ is the incident plane wave vector amplitude component perpendicular to the scattered field propagation vector which is along $\mathbf{r}$ as shown in Fig. 1; the polarizability α of the scatterer material is given by

$$\alpha = \frac{3}{4\pi}\left(\frac{n^2-1}{n^2+2}\right), \tag{13}$$

where n is the index of refraction of the scatterer.

Acquista's results given by Eq. (9) were obtained by assuming a rotated coordinate system (scatterer system) whose z' axis was along the vector $\mathbf{r}$, which defined the scattering direction in the laboratory coordinate system (beam system). He also chooses the x' axis of his coordinate system such that the propagation vector of the incident plane wave lies in the $x'z'$ plane of the scatterer system. The relations between the scatterer and the beam systems are summarized in Figs. 2 and 3. In general, the scatterer system, which is centered on an off-centered particle, should be considered as rotated as well as displaced with respect to the beam system. If the terms defined in the scatterer system are denoted by primed quantities, the transformations between the beam and the scatterer systems will be given as

$$\mathbf{r}' = [T](\mathbf{r}+\mathbf{d}), \tag{14a}$$

$$\mathbf{r} = [T]^{-1}\mathbf{r}' - \mathbf{d}, \tag{14b}$$

where $\mathbf{d}$ is the vector defining the displacement of the scatterer from the origin of the beam system, and $[T]$ is the coordinate rotation transformation matrix given as

$$[T] = \begin{bmatrix} a_{11} & a_{12} & a_{13} \\ a_{21} & a_{22} & a_{23} \\ a_{31} & a_{32} & a_{33} \end{bmatrix}, \tag{14c}$$

If the scattering plane is taken to be the xz plane of the beam system, the elements of $[T]$ are given as

$$\left.\begin{aligned}
&a_{31} = \sin\theta_s, \\
&a_{32} = 0, \\
&a_{33} = \cos\theta_s, \\
&a_{21} = -qa_{33}/a_r, \\
&a_{22} = (a_{33}p + sa_{31})/a_r, \\
&a_{23} = qa_{31}/a_r,\ a_r = [(qa_{33})^2 + (pa_{33} + sa_{31})^2 + (qa_{31})^2]^{1/2}, \\
&a_{11} = a_{22}a_{33}, \\
&a_{12} = q/a_r, \\
&a_{13} = -a_{22}a_{31},
\end{aligned}\right\} \tag{14d}$$

and θ_s is the scattering angle which is assumed to be in the xz plane as shown in Fig. 3.

Note that with these coordinate transformations, the plane wave components of the PWS of the laser beam can be transformed to the scatterer system. Applications of such transformations to Eq. (9) give the corresponding plane wave components in the scatterer system,

$$\mathbf{E}_i'(p,q) = \mathbf{A}_i'(p,q)\exp[-i\Delta(p,q)]\exp[i\mathbf{k}_i'(p,q)\cdot\mathbf{r}'], \tag{15a}$$

where

$$\mathbf{A}_i'(p,q) = [T]\mathbf{A}_i(p,q), \tag{15b}$$

$$\mathbf{k}_i'(p,q) = [T]\mathbf{k}_i(p,q), \tag{15c}$$

$$\Delta(p,q) = \mathbf{k}_i(p,q)\cdot\mathbf{d}. \tag{15d}$$

So applying the transformation

$$\left.\begin{aligned}
&\mathbf{E}_0 \to \mathbf{A}_i'(p,q)\exp[-i\Delta(p,q)], \\
&\mathbf{k} \to \mathbf{k}_i'(p,q), \\
&\mathbf{r} \to \mathbf{r}' \quad \text{and} \quad \mathbf{m} \to \mathbf{m}'.
\end{aligned}\right\} \tag{16}$$

In Eq. (10) give the scattered fields due to a single plane wave component of the PWS of the laser beam. These scattered fields are defined in the scatterer system. The corresponding scattered field components in the beam system are then found using the transformations defined in Eq. (14). Finally, the total scattered field due to the incident laser beam is found by superposing these results over the whole p,q plane. The results are[9]

$$\mathbf{E}_{st}^{(1)}(\mathbf{r}) = \frac{k^2}{r}\exp(ikr)\iint_{-\infty}^{+\infty}\mathbf{E}_{i\perp}^{b}(p,q)\exp[-i\Delta(p,q)] \cdot u[k\hat{r}' - \mathbf{k}_i(p,q)]dpdq, \tag{17a}$$

$$\mathbf{E}_{st}^{(2)}(\mathbf{r}) = \frac{2k^2}{(2\pi)^2}\frac{\exp(ikr)}{r}\iint_{-\infty}^{+\infty} \cdot \exp[-i\Delta(p,q)]\int_{V_m}\frac{u(\mathbf{m}' + k\hat{r}')u[\mathbf{m}' + \mathbf{k}_i'(p,q)]}{m^2 - k^2} \cdot \left\{\left(\frac{2}{3}k^2 + \frac{1}{3}m^2\right)\mathbf{E}_{i\perp}^{b}(p,q) - [\mathbf{m}' \cdot \mathbf{E}_i'(p,q)]\mathbf{m}_{\perp}^{b}\right\} d^3m'dpdq, \tag{17b}$$

where

$$\mathbf{E}_i^{b}(p,q) = [T]^{-1}\mathbf{E}_i'(p,q), \tag{17d}$$

$$\mathbf{m}^{b} = [T]^{-1}\mathbf{m}', \tag{17e}$$

and all primed quantities are defined in the scatterer system. $\mathbf{E}_{st}^{(1)}(\mathbf{r})$ and $\mathbf{E}_{st}^{(2)}(\mathbf{r})$ are the complete results of the scattered fields in the first two iterations due to the incident laser beam. The total scattered field due to an incident Gaussian beam is given by

$$\mathbf{E}_{st}(\mathbf{r}) = \alpha\mathbf{E}_{st}^{(1)}(\mathbf{r}) + \alpha^2\mathbf{E}_{st}^{(2)}(\mathbf{r}). \tag{18}$$

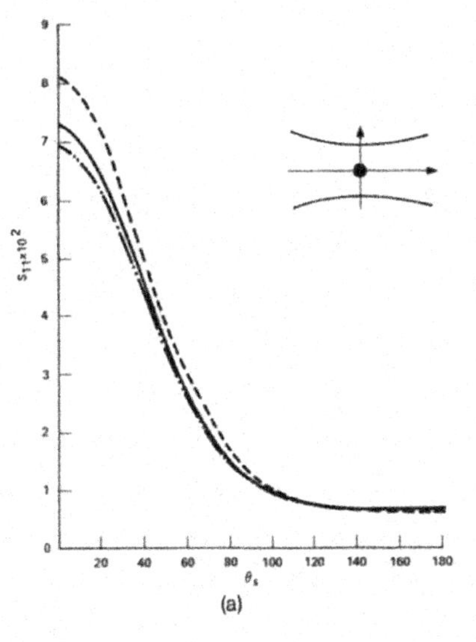

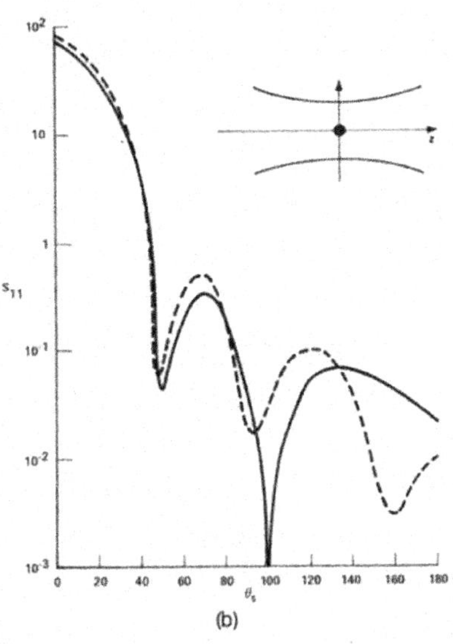

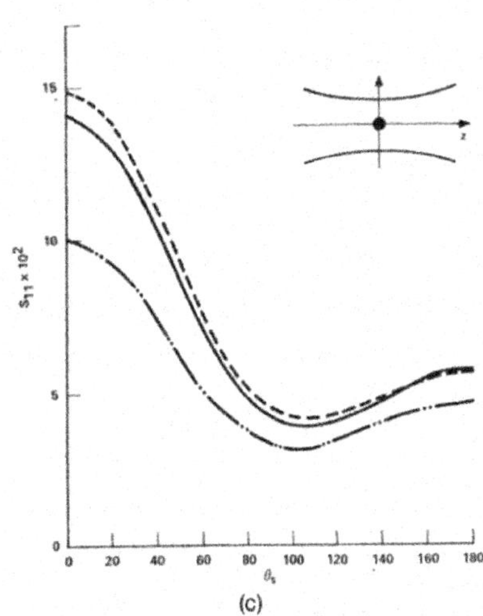

Fig. 4. Beam scattering patterns for centered spheres (solid lines). In each case, the incident laser beam has a large spot size. Dashed lines give the plane wave results for comparison purposes. The dotted lines represent the first iteration results [if $\mathbf{E}_{st}^{(2)}(\mathbf{r})$ is taken to be zero]. For all figures $W_0/a = 5.0$. The scatterer parameters are (a) $ka = 1.60$, $n = 1.10$; (b) $ka = 5.0$, $n = 1.10$; (c) $ka = 1.0$, $n = 1.55$.

Computations may be carried out in terms of the Stokes parameter S_{11}, which is defined as follows:

$$S_{11} = k^2 r^2 \{[I_s(\mathbf{r})]/[I_i(\mathbf{r})]\}. \tag{19}$$

where $I_s(\mathbf{r})$ and $I_i(\mathbf{r})$ are, respectively, the scattered and the incident field intensities, i.e.,

$$I_s(\mathbf{r}) = |\alpha \mathbf{E}_{si}^{(1)}(\mathbf{r}) + \alpha^2 \mathbf{E}_{si}^{(2)}(\mathbf{r})|^2, \tag{20a}$$

$$I_i(\mathbf{r}) = (A_x^2 + A_y^2)^{1/2}, \tag{20b}$$

where $\mathbf{E}_{si}^{(1)}(\mathbf{r})$ and $\mathbf{E}_{si}^{(2)}(\mathbf{r})$ are given by Eq. (17), and A_x and A_y have been defined for a Gaussian beam as shown in Eq. (1). Note that in Eq. (20b) maximum intensity of the laser beam (which occurs at origin) is taken to be the incident intensity. By combining Eqs. (17), (19), and (20) and omitting small terms, the following equation is obtained:

$$S_{11} = k^2 \alpha^2 \int\int\int\int_{-\infty}^{\infty} \cos[\Delta(p,q) - \Delta(p',q')]$$

$$\cdot \Big\{ \hat{\epsilon}_{i\perp}^{b}(p,q) \cdot \hat{\epsilon}_{i\perp}^{b}(p',q') u[kr' - \mathbf{k}_i'(p,q)] u[kr' - \mathbf{k}_i'(p',q')]$$

$$+ \frac{\alpha}{\pi^2} \int_{V_m} u[kr' - \mathbf{k}_i'(p,q)] \frac{u(\mathbf{m}' + kr')u[\mathbf{m}' + \mathbf{k}_i'(p',q')]}{m^2 - k^2}$$

$$\cdot \Big\{ \Big[\Big(\frac{2}{3} k^2 + \frac{1}{3} m^2 \Big) \hat{\epsilon}_{i\perp}^{b}(p,q) \cdot \hat{\epsilon}_{i\perp}^{b}(p',q') - [\hat{\epsilon}_{i\perp}^{b}(p,q) \cdot \mathbf{m}_{\perp}^{b}]$$

$$\cdot [\hat{\epsilon}_i'(p',q') \cdot \mathbf{m}'] d^3 m' \Big\} dp dq dp' dq', \tag{21}$$

where $\hat{\epsilon}_{i\perp}^{b}(p,q)$ and $\hat{\epsilon}_i(p,q)$ are unit vectors along the electric field vectors $\mathbf{E}_{i\perp}^{b}(p,q)$ and $\mathbf{E}_i'(p,q)$, respectively.

Numerical examples have been carried out for various particle positions, incident beam spot sizes, and various ka values for spherical scatterers. The applications can easily be extended to other particle shapes using Eq. (12). For spherical scatterers,

$$u(\mathbf{v}) = V_s f(av), \tag{22a}$$

$$V_s = 4/3\pi a^3, \tag{22b}$$

$$f(av) = \frac{3}{(av)^3} [\sin(av) - (av)\cos(av)], \tag{22c}$$

where a is the radius of the scatterer.

III. Numerical Examples

The first computations were carried out to check if the results obtained in the limit for which the beam waist spot size, W_0's large, were comparable to those for the plane wave scattering case. The comparisons were done for different values of ka and index of refraction n. For the case of large W_0, the scattered field distributions are in close agreement with Acquista's single plane wave scattering results. Figures 4(a), 4(b), and 4(c) show some examples of these comparisons. In Figs. 4(a) and 4(c), first and second iterations are plotted separately to reveal the advantage of the present approach to the Rayleigh-Gans approximation. The first iteration corresponds to the case in which $E_{si}^{(2)}(\mathbf{r})$ is taken to be zero. The second iteration is the total result as defined in Eq. (18).

As can be seen from these figures, the case of the scattering by Gaussian beam with large spot size closely resembles that by plane wave case. The major difference is a slight decrease in the scattered amplitude for the beam case. This effect is more observable as the spot size decreases.

The next calculation is carried out to observe the effect of the size of the scatterer and/or the beam spot size on the scattered field of a centered particle. Different ratios of W_0/a were used. It is observed that as W_0/a is decreased, the scattering pattern shows a shift from $\theta_s = 0°$ to $\theta_s = \pm 180°$. In other words, the dips and the peaks of the scattering pattern move toward $\theta_s = \pm 180°$ direction. As the spot decreases, the forward scattered beam is broadened, and its peak is decreased. This effect is more observable as the value of ka increases. As ka increases to a large value, a dip appears in the forward direction. The pattern shift in such cases is as much as 10 degrees for the first dip, and this corresponds to a spot size decrease by a factor of 10. However, it is seen that the shifts are not linearly dependent on spot size changes. Figures 5(a), 5(b), 5(c), and 5(d) summarize the effects of changing W_0/a ratios on the scattered field. Figures 5(c) and 5(d) are included to show the effects of a beam waist spot size which is less than the scatterer radius, that is, the $(W_0/a) < 1$ case. As seen in Fig. 5(c) the forward peak is relatively flat for $(W_0/a) = 0.65$, and in Fig. 5(d) the forward scattered field begins to show a dip at $\theta_s = 0°$ for $(W_0/a) = 0.50$. In earlier figures these dips did not appear. This is due to the fact that for the parameters chosen the diffraction limit of the beam was exceeded in these cases. The flattening and the dipping of the pattern in the forward scattering direction and the shifting of the scattered energy toward $\theta_s = \pm 180°$ are caused by the interference effects of fields inside the particle. These results show strong resemblance to the fringe pattern changes and shifts obtained from a Fabry-Perot etalon.[10] Using the geometrical optics principles, the pattern shifts or the pattern expansions may be explained by the changes of the incident beam phase front curvature. According to the beam expression given by Eq. (1), one notes that decreasing the spot size W_0 increases the beam phase front curvature, $1/R(a)$, on the particle surface. As this phase curvature increases, it becomes better matched to the particle surface. Consequently fewer fringes are expected. This also causes the scattered beam pattern to expand. Similar results for centered spheres were also found by Tsai and Pogorzelski.[11]

However, an obvious question arises with the appearance of a dip in the forward peak. The above etalon example could also explain this, but only if a change in the round-trip phase delay along the particle axis due to spot size changes is assumed. If the beam expression of Eq. (1) is considered, it is seen that, in fact, the phase of the beam does depend on the waist spot size through the last exponential term in Eq. (1b). A simple derivation shows that for a complete destructive interference along the z axis of the sphere, one should have

$$4k_0 na + 4\tan^{-1}(a/z_0) = (2h + 1)\pi,$$

where h is any integer. Using Eq. (1c) for z_0 and rearranging gives the relation,

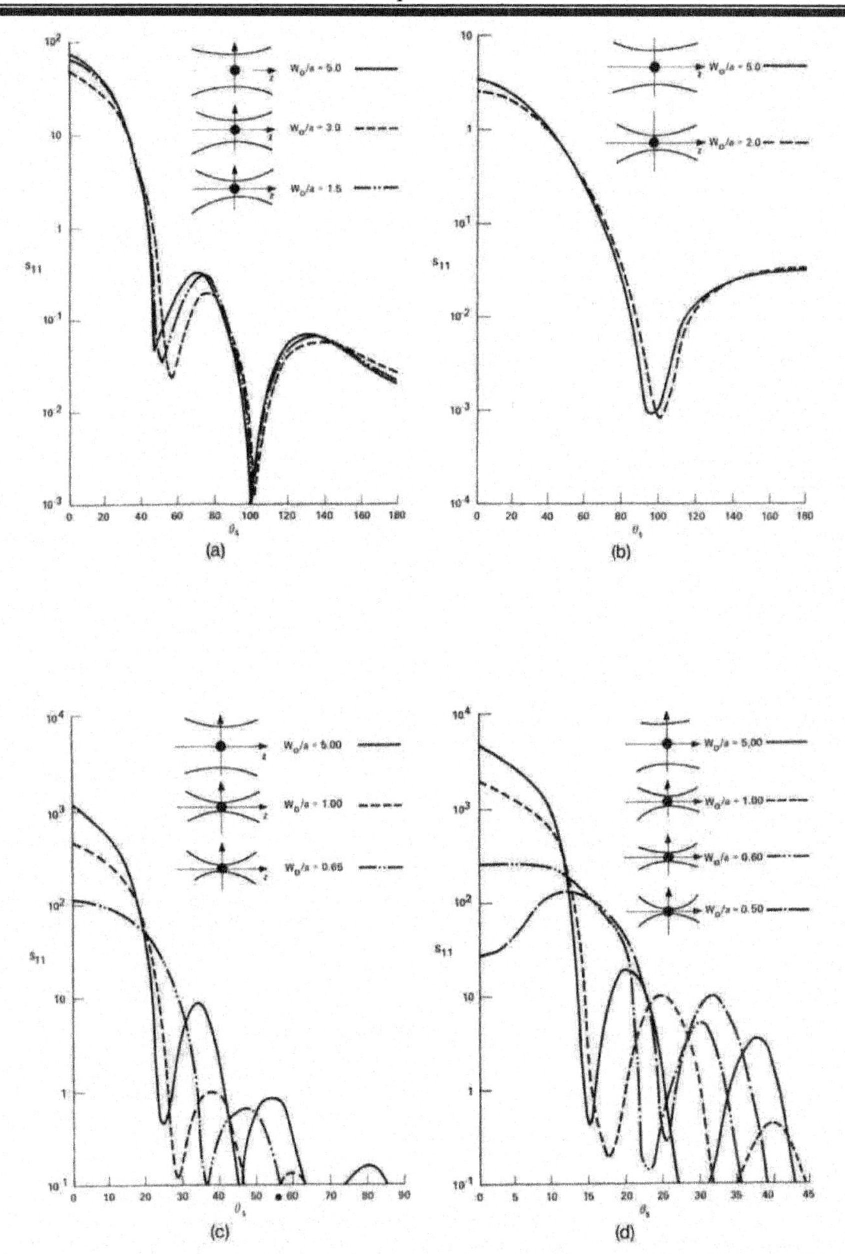

Fig. 5. Beam scattering patterns showing the effect of decreasing beam waist spot size W_0 for centered spherical scatterers. The expansion of the pattern for smaller spot sizes is easily observable. The scatterer parameters are: (a) $ka = 5.0, n = 1.10$; (b) $ka = 3.0, n_r = 1.10$; (c) $ka = 10.0, n = 1.05$; (d) $ka = 15.0, n = 1.03$. The spot sizes are indicated in figures in W_0/a ratios.

$$\frac{W_0}{a} = \left[\frac{2}{nk_0a \tan\left(\frac{2h+1}{4}\pi - nk_0a\right)}\right]^{1/2},$$

using this equation for the parameters of Fig. 5(d) shows that the only value of the (W_0/a) ratio which may cause a complete destructive interference along the z axis within the particle is 0.28. This corresponds to a beam phase front curvature radius at the scatterer surface of about $1.60a$.

The above reasons explain the pattern shift, forward peak flattening, and dipping. The decrease of the over-all magnitude of the pattern is due to the fact that the more the beam is focused, the smaller is the amplitude of the incident field. This is simply caused by taking the peak amplitude of the fields being constant, i.e., unity. In practical experiments, this will not be true. In other words, focusing the laser beam will increase the field magnitudes near the origin, and thus increase the magnitude of the scattering pattern. These effects can be taken into account by a simple proportional increase of the patterns given in this section.

Another set of calculations was done to find the effects of off-centered particles. Some results of this set are shown in Figs. 6. Figures 6(a) and 6(b) show the effect of changing the position of the particle in the $z = 0$ plane. Changing the particle position in this plane reduces the intensity of the scattered fields, but does not change the scattering pattern. The latter effect is due to the Gaussian amplitude distribution of the incident beam. Figure 6(a) shows that as the particle is moved from the origin (0,0,0) to the point (2,1/2,0) or to (4,3,0), the parameter S_{11}, at the forward peak, is reduced by

$$\frac{S_{11,(0,0,0)}}{S_{11,(2,1/2,0)}} = 1.71, \qquad \frac{S_{11,(0,0,0)}}{S_{11,(4,3,0)}} = 8.3$$

for each case. These values support the reasoning given above, because the intensity of the beam at the origin (0,0,0), is greater than its value at the points (2,1/2,0) and (4,3,0) by

$$\frac{I_{\mathrm{inc},(0,0,0)}}{I_{\mathrm{inc},(2,1/2,0)}} = \exp[2(d_x^2 + d_y^2)/W_0^2] = 1.65, \qquad \frac{I_{\mathrm{inc},(0,0,0)}}{I_{\mathrm{inc},(4,3,0)}} = 7.38,$$

which are close to the respective values obtained for S_{11}. Similar calculations for the parameters of Fig. 6(b) show that

$$\frac{S_{11,(0,0,0)}}{S_{11,(2,2,0)}} = 1.88, \qquad \frac{I_{\mathrm{inc},(0,0,0)}}{I_{\mathrm{inc},(2,2,0)}} = 1.90,$$

$$\frac{S_{11,(0,0,0)}}{S_{11,(4,4,0)}} = 12.48, \qquad \frac{I_{\mathrm{inc},(0,0,0)}}{I_{\mathrm{inc},(4,4,0)}} = 12.94.$$

The decrease in the magnitude of S_{11} is nearly equal to the decrease of the incident intensity caused by off-centering the particle.

The scattering pattern caused by an off-centered particle in $z = 0$ plane is found to be exactly symmetrical. Although this is unrealistic because of propagation characteristics of a Gaussian beam, it is somewhat predictable because the present approach of this study is just an iterative approximation for the scattering problem. The asymmetry for $z \neq 0$ case as seen from Fig. 6(c) is relatively small. However, one expects increasing asymmetry as the particle moves parallel to the z axis toward the $z = 0$ plane. This last point, again, can be reasoned from the beam propagation characteristics.

Another result is the change of the scattering pattern caused by a change of position of an off-centered particle on a circular path, which is centered on the origin and lies in the $z = 0$ plane. That is, a particle located at (1,0,0) gives different scattering pattern than a particle at $(\sqrt{2},\sqrt{2},0)$, although they are at equal distances from the origin. This assumes that the other parameters of the scattering are fixed. The above effect is a beam scattering characteristic and cannot be observed in the far-field behavior of the single plane wave scattering.

If the scatterer is off-centered at a value $z \neq 0$, and for nonzero d_x, and d_y displacements, the scattering pattern is asymmetric. This is because the incident field evaluated at the center of the scatterer has a phase front which is directed in a different way than the z axis. This seems to be the cause of the asymmetry of the pattern for displacements away from the $z = 0$ plane. Figure 6(c) gives an example of this. The difference between the S_{11} values for both signs of the scattering angle θ_s from 0° to 180° is plotted on a different scale to show the asymmetric distribution.

The results of our earlier study[1] show that if the displacement (d_x,d_y,d_z) is changed to $(-d_x,-d_y,-d_z)$, the scattering pattern completely repeats itself. However, if (d_x,d_y,d_z) is changed to $(-d_x,-d_y,d_z)$, one gets inverted asymmetrical distribution compared with the original case. If $d_z = 0$, the pattern is the same and symmetrical for all the cases given above. The computations of the present work agreed with these results. This behavior can be predicted by considering the correlation term $\cos[\Delta(p,q) - \Delta(p',q')]$ of Eq. (29). As stated there, $\Delta(p,q)$ is a linear function of displacement given by $\Delta(p,q) = k(pd_x + qd_y - sd_z)$. So a change of (d_x,d_y,d_z) to $(-d_x,-d_y,-d_z)$ does not change the scattering pattern because the cosine function is even. However, as explained before, this behavior is not realistic, and the error comes from the inherent assumptions of Rayleigh type scattering solutions about the behavior of fields inside the scatterer.

The computation of Eq. (21) to obtain scattered field patterns was carried out on an IBM 360 using a Gaussian quadrature scheme. Since the integrand contains a relatively localized Gaussian envelope function, very little computer time was needed. For centered particles the CPU time was about 10 sec, and for off-centered particles the time increased to a maximum of 40 sec/plot.

IV. Conclusions

Results of the previous section showed that if the maximum dimension of the scatterer is a, and the waist spot size of the beam is W_0, under the conditions that the scatterer is centered on the beam waist, and $a \lesssim (W_0/5)$, the scattering patterns obtained have slight differences (less than 5%) from the single plane wave scattering patterns. However, if either of the above

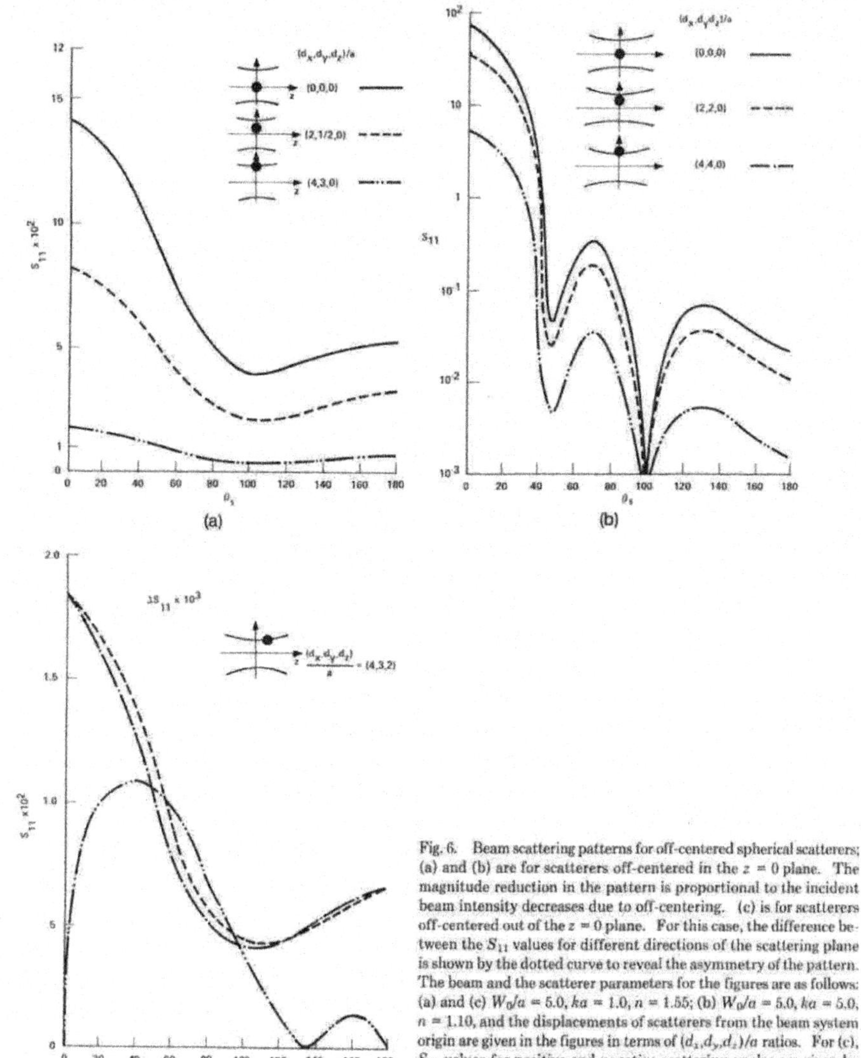

Fig. 6. Beam scattering patterns for off-centered spherical scatterers; (a) and (b) are for scatterers off-centered in the $z = 0$ plane. The magnitude reduction in the pattern is proportional to the incident beam intensity decreases due to off-centering. (c) is for scatterers off-centered out of the $z = 0$ plane. For this case, the difference between the S_{11} values for different directions of the scattering plane is shown by the dotted curve to reveal the asymmetry of the pattern. The beam and the scatterer parameters for the figures are as follows: (a) and (c) $W_0/a = 5.0$, $ka = 1.0$, $n = 1.55$; (b) $W_0/a = 5.0$, $ka = 5.0$, $n = 1.10$, and the displacements of scatterers from the beam system origin are given in the figures in terms of $(d_x,d_y,d_z)/a$ ratios. For (c), S_{11} values for positive and negative scattering angles are given by dashed curves without and with dots, respectively. The difference between these two cases is plotted by the dotted curve.

conditions is not satisfied, the patterns of the beam scattering showed great differences compared with the single plane wave results. These differences are summarized below.

Decreasing the spot size of the beam relative to the dimensions of the scatterer which is centered on the beam waist caused three common changes in the scattering pattern in most cases. First there was the decrease in the over-all magnitude of the scattered field. This is simply because in each case the maximum amplitude of the beam is taken to be unity. So, the sharper the beam is focused, the less is the total energy contained in the beam. Second was the flattening of the forward peak and even an appearance of a dip in the center of the peak for very small spot sizes. This effect was due to the fact that the curvature of the beam becomes a better fit for the scatterer surfaces when the spot size gets smaller. When the curvature becomes a closer fit to the surface of the particle, the fields start interfering within the particle. The calculations done in previous sections showed that the dips in the forward peak occurred close to some specific beam parameters, which satisfied the conditions of a destructive interference on the z axis of the scatterer. The other change in the pattern which is caused by a decrease in the spot size is the expansion of the pattern, i.e., the shift of the dips and the peaks from forward to the backward scattering directions. The explanation of this effect is also found to be the better fit of the phase front curvature of the beam to the scatterer surface. These points are explained in a more detailed way in the previous section.

Off-centered particles created two different situations. First, changing the particle position in the $z = 0$ plane away from the origin caused a smooth decrease in the scattering pattern. The reason for this is the Gaussian amplitude distribution of the incident laser beam (see the previous section for numerical examples). Second, moving an off-centered scatterer out of the $z = 0$ plane parallel to the z axis caused an asymmetry in the scattering pattern.

This work was supported in part by the National Science Foundation.

References

1. L. W. Casperson, C. Yeh, and W. F. Yeung, Appl. Opt. **16**, 1104 (1977).
2. L. W. Casperson and C. Yeh, Appl. Opt. **17**, 1637 (1978).
3. K. S. Shifrin, *Scattering of Light in a Turbid Medium* (Nauka, Moscow, 1951; NASA, Washington, D.C., 1968), Technical Translation TTF-477.
4. C. Acquista, Appl. Opt. 15, 2932 (1976).
5. A. Yariv, *Quantum Electronics* (Wiley, New York, 1975).
6. J. W. Goodman, *Introduction to Fourier Optics* (McGraw-Hill, New York, 1968).
7. P. C. Clemmow, *Plane Wave Spectrum Representation of Electromagnetic Fields* (Pergamon, London, 1966).
8. W. H. Carter, J. Opt. Soc. Am. **62**, 1195 (1972).
9. S. Colak, "Focused Laser Beam Scattering by Stationary and Moving Particles," Ph.D. Thesis, U. California, Los Angeles (1978).
10. A. Yariv, *Introduction to Optical Electronics* (1971).
11. W. C. Tsai and R. J. Pogorzelski, J. Opt. Soc. Am. **65**, 1457 (1975).

Scattering of sharply focused beams by arbitrarily shaped dielectric particles: an exact solution

C. Yeh, S. Colak, and P. Barber

By expanding the incident focused beam field in terms of its plane wave spectrum and by using the technique we developed earlier to treat the problem of the scattering of plane waves by arbitrarily shaped dielectric obstacles, we have been successful in solving the problem of the scattering of sharply focused beams by arbitrarily shaped dielectric particles. It was found that the presence of the curvature of the incident wave front and the nonuniformity of the incident wave intensity affect greatly the scattering characteristics.

I. Introduction

Recently, it was shown experimentally[1] that the scattering characteristics of a single submicron particle in an ensemble of many particles can be measured by the use of a sharply focused laser beam. Refinement of this experimental technique is now being pursued. A parallel effort in calculating the expected theoretical scattering characteristics of a particle due to an incident sharply focused beam has also been carried out. By combining the technique that we developed earlier to treat the problem of the scattering of plane waves by arbitrarily shaped dielectric obstacles[2] and the technique of representing an incident focused beam by its plane wave spectrum,[3] we have successfully obtained the exact solution of the scattering of arbitrarily shaped dielectric particles by a sharply focused beam. It will be shown in this paper that very significant deviations of the scattering characteristics from those for an incident plane wave are observed. Our exact results are also compared with the earlier results obtained according to the Rayleigh-type approximations.[4,5] In spite of the common use of Gaussian beams in laboratory research, there have been very few treatments of light scattering by such beams. Pearson *et al.*[6] have considered the scattering of plane wave Gaussian beams from edges. Morita *et al.*[7] have treated the scattering of a plane wave beam from a small spherical particle at the beam axis. Tsai and Pogorzelski[8] generalized these results for an arbitrarily sized spherical particle at the beam axis. Recently, Tam has outlined a possible procedure for treating the scattering of a plane wave beam from an off-axis sphere,[9,10] and Chew *et al.*[10] considered the problem of scattering of uniform converging beams by spheres. A related problem of the scattering of dipole field by a sphere has also been treated by Chew *et al.*[11]

P. Barber is with Clarkson College, Department of Electrical Engineering, Potsdam, New York 13676; the other authors are with University of California at Los Angeles, Electrical Engineering Department, Los Angeles, California 90024.

Received 12 December 1979; revised manuscript 25 August 1982.
0003-6935/82/244426-08$01.00/0.

II. Method of Solution

The transverse electric field vector of a laser beam is given by the expression[5]

$$\mathbf{E}_b(x,y,z,t) = (A_x\mathbf{e}_x + A_y\mathbf{e}_y)h(x,y,z)\exp(i\omega t), \quad (1a)$$

where

$$h(x,y,z) = \frac{W_0}{W(z)}\exp(-ikz)\exp[-ik(x^2+y^2)/2R(z)]$$

$$\times \exp[-(x^2+y^2)/W(z)]\exp\left[-i\arctan^{-1}\left(\frac{z}{z_0}\right)\right], \quad (1b)$$

where $W(z)$ and $R(z)$ are the spot size, and the radius of curvature of the phase fronts of the beam is given by $W(z) = W_0[1 + (z^2/z_0^2)]^{1/2}$ and $R(z) = z[1 + (z_0^2/z^2)]$. These relations assume that the chosen coordinate system is centered on the beam waist (see Fig. 1). A_x and A_y are complex constant amplitudes of the **x** and **y** polarized field vectors, W_0 is the beam waist spot size, z_0 is the Rayleigh length of the beam given by $z_0 = \pi W_0^2/\lambda$. λ and ω are the wavelength and the frequency of oscillation of the incident beam wave, respectively, and $k = \omega\sqrt{\epsilon_0\mu_0}$ is the wave number. Note that the small longitudinal field component (z directed) is not included in the above expressions. However, the first-order correction to it is taken into account when the plane wave spectrum representation of the beam is derived. This representation has been found in our

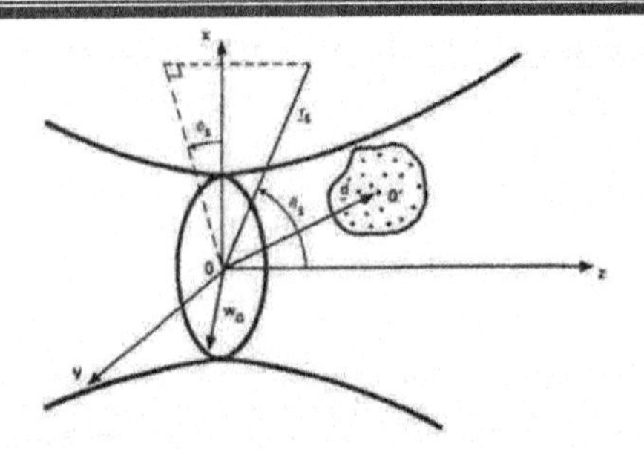

Fig. 1. Geometry for the focused laser beam scattering by arbitrarily shaped and positioned homogeneous particles. W_0 is the beam waist spot size, and $\mathbf{r}_s$ is the scattering direction defined by the scattering angles (θ_s,ϕ_s). $\mathbf{d}$ is the displacement vector to the position of the particle which is centered on the 0′ (primed) coordinate system.

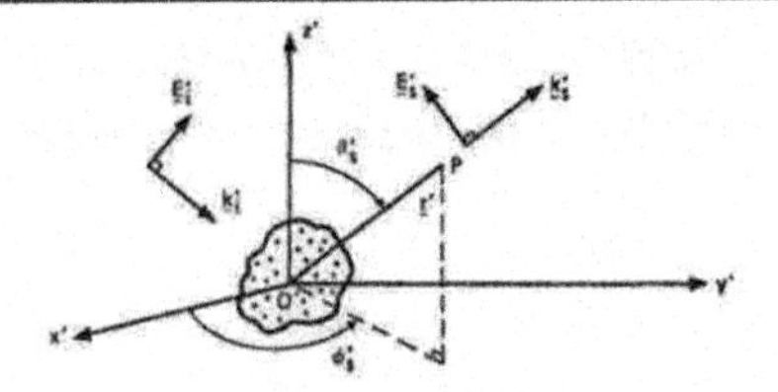

Fig. 2. Single-plane wave scattering geometry used by Barber and Yeh.[2] The terms are primed to indicate that they are defined in a scatterer coordinate system which may be different from the beam system. The scattering angles are (θ_s',ϕ_s'), and the incidence angles of the plane wave are (θ_i',ϕ_i') which are defined by $\mathbf{k}_i'$.

earlier paper,[5] and it is given by the following expression:

$$E_b(x,y,z,t) = \frac{b^2}{\pi}\exp(i\omega t)\iint_{-\infty}^{+\infty}\exp[-b^2(p^2+q^2)] \times\left\{A_x\mathbf{e}_x + A_y\mathbf{e}_y + \frac{pA_x+qA_y}{s}\mathbf{e}_z\right\} \times\exp[ik(px+qy-sz)]dpdq, \quad (2)$$

where, for a homogeneous and isotropic propagation medium, $s=\sqrt{1-p^2-q^2}$ and $b=kW_0/2$. Note that the integrand has the form of a plane wave which can be written

$$\mathbf{E}_i(p,q) = \mathbf{A}_i(p,q)\exp[i\mathbf{k}_i(p,q)\cdot\mathbf{r}], \quad (3a)$$

where

$$\mathbf{A}_i(p,q) = \left\{A_x\mathbf{e}_x + A_y\mathbf{e}_y + \frac{pA_x+qA_y}{s}\mathbf{e}_z\right\}\exp[-b^2(p^2+q^2)], \quad (3b)$$

$$\mathbf{k}_i(p,q) = k(p\mathbf{e}_x + q\mathbf{e}_y - s\mathbf{e}_z) \quad (3c)$$

are the amplitudes and the propagation vectors of such plane wave components. These plane wave components are assumed to be incident on a scatterer. The scattering results for each such plane wave component can be obtained using the Extended Boundary Condition technique described in Ref. 2 for the case of single-plane wave scattering by arbitrarily shaped homogeneous particles. The complete scattering pattern due to the focused laser beam is found using the integral superposition of these results.

The exact solution of single-plane wave scattering by arbitrarily shaped dielectric bodies has been worked out by Barber and Yeh.[2] This solution was achieved by an application of the Extended Boundary Condition Method. The results for the far-field amplitude of the scattered field (for a single plane wave) are

$$\mathbf{E}_s'(k\mathbf{r}) = \mathbf{F}'(\theta_s',\phi_s'|\theta_i',\phi_i')\frac{\exp(ikr')}{r'}, \quad (4)$$

where $\mathbf{F}'(\theta_s',\phi_s'|\theta_i',\phi_i')$ is the vector far-field amplitude for the scattered field in the (θ_s',ϕ_s') direction due to an incident plane wave in the (θ_i',ϕ_i') direction, and it is given by

$$\mathbf{F}'(\theta_s',\phi_s'|\theta_i',\phi_i') = \frac{4}{jk}\sum_{r=1}^{N} D_r^{(-)}[f_r(\theta_i',\phi_i')\mathbf{C}_r(\theta_s',\phi_s') + jg_r(\theta_i',\phi_i')\mathbf{B}_r(\theta_s',\phi_s')]. \quad (5)$$

The geometry of this single-plane wave scattering case is shown in Fig. 2. The combined index r is used to represent the spherical vector harmonic indices, emn or omn, and

$$D_r^{(-)} = (-1)^m\epsilon_m\frac{(2n+1)(n-m)!}{4n(n+1)(n+m)!}, \quad (6)$$

$$\epsilon_m = \begin{cases}1 \text{ for } m=0,\\ 2 \text{ for } m>0.\end{cases}$$

$\mathbf{C}_r(\theta,\phi)$ and $\mathbf{B}_r(\theta,\phi)$ are related to the spherical vector harmonics $\mathbf{M}_r(k\mathbf{r})$ and $\mathbf{N}_r(k\mathbf{r})$. These four functions are listed in the Appendix. The coefficients $f_r(\theta_i',\phi_i')$ and $g_r(\theta_i',\phi_i')$ are given by

$$\begin{bmatrix} f_r(\theta_i',\phi_i')\\ g_r(\theta_i',\phi_i')\end{bmatrix} = [T_{rr'}]\begin{bmatrix}-\mathbf{e}_0'\cdot\mathbf{C}_r(\theta_i',\phi_i')\\ j\mathbf{e}_0'\cdot\mathbf{B}_r(\theta_i',\phi_i')\end{bmatrix}, \quad (7)$$

where $\mathbf{e}_0'$ is the vector amplitude of the incident plane wave; $T_{rr'}$ is the scatterer matrix which depends on the scatterer shape and material, and it is given by

$$[T_{rr'}] = \begin{bmatrix}K'+\sqrt{\epsilon_r}J' & L'+\sqrt{\epsilon_r}I'\\ I'+\sqrt{\epsilon_r}L' & J'+\sqrt{\epsilon_r}K'\end{bmatrix}\cdot\begin{bmatrix}K+\sqrt{\epsilon_r}J & L+\sqrt{\epsilon_r}I\\ I+\sqrt{\epsilon_r}L & J+\sqrt{\epsilon_r}K\end{bmatrix}, \quad (8)$$

where ϵ_r is the relative dielectric permittivity of the scatterer. The terms I', J', K', L' and I, J, K, L are related to the surface integrals of the spherical vector harmonics, $\mathbf{M}_n(k\mathbf{r})$ and $\mathbf{N}_r(k\mathbf{r})$, and their derivations are given in Ref. 2. They are also included in the Appendix.

The numerical computations of the above results are greatly simplified if (1) the scatterer is assumed to be axisymmetric, and (2) a scatterer coordinate system is chosen so that the incident wave in this system stays in the $\phi'=0$ plane. For the beam scattering cases, the

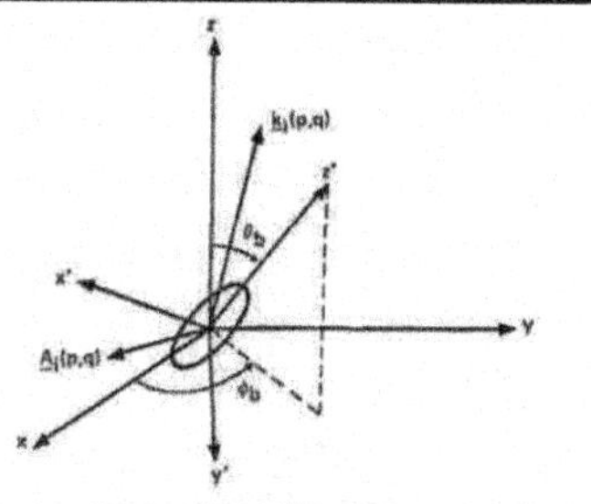

Fig. 3. Diagram showing the relations between the beam and the scatterer system. The symmetry axis of the particle is the z' axis, and it makes the angles (θ_b,ϕ_b) with respect to the beam frame. The x' axis of the scatterer frame is defined so that the incident wave propagation vector $\mathbf{k}_i(p,q)$ stays in the x'-z' plane of this frame.

second assumption requires a new scatterer coordinate system for each plane wave component of the beam. In such a case, the scattered fields should be transformed back to a fixed beam system before the final superposition can be carried out.

As previously discussed, the incident laser beam can be written in its plane wave spectrum representation form as in Eq. (2). Then each plane wave component of the beam will have the form given by Eq. (3). For each such component, a different scatterer coordinate system is chosen so that the corresponding incident propagation vector $\mathbf{k}_i(p,q)$ lies in the $\phi' = 0$ plane of the respective scatterer system. The symmetry axis of the scatterer, which is the z' axis of the scatterer coordinate system, is assumed to be directed so that it makes (θ_b,ϕ_b) angles with respect to the beam coordinate system. This is shown in Fig. 3. With this information, the transformations between the beam system 0 and the scatterer system 0′ are defined by

$$\mathbf{r}' = [C](\mathbf{r} + \mathbf{d}) \qquad \mathbf{r} = [C]^{-1}\mathbf{r}' - \mathbf{d}, \tag{9}$$

where $\mathbf{d}$ is the displacement vector of the scatterer position, and [C] is the coordinate transformation matrix given by

$$[C] = \begin{bmatrix} c_{11} & c_{12} & c_{13} \\ c_{21} & c_{22} & c_{23} \\ c_{31} & c_{32} & c_{33} \end{bmatrix}, \tag{10}$$

where

$$\begin{aligned}
c_{31} &= \sin\theta_b \cos\phi_b, \\
c_{32} &= \sin\theta_b \sin\phi_b, \\
c_{33} &= \cos\theta_b, \\
c_{21} &= -(c_{32}s + pc_{33})/c_r, \\
c_{22} &= (c_{33}q + sc_{31})/c_r, \\
c_{23} &= (c_{31}p - qc_{32})/c_r, \\
c_r &= [(c_{32}s + pc_{33})^2 + (c_{33}q + sc_{31})^2 + (c_{31}p - qc_{32})^2]^{1/2}, \\
c_{11} &= c_{22}c_{33} - c_{32}c_{13}, \\
c_{12} &= c_{23}c_{31} - c_{33}c_{21}, \\
c_{13} &= c_{21}c_{32} - c_{22}c_{31}.
\end{aligned}$$

The application of these transformations to the plane wave components as defined by Eq. (3) gives the corresponding plane wave components in the scatterer system as

$$\mathbf{E}_i'(p,q) = \mathbf{A}_i'(p,q)\exp[-i\Delta(p,q)]\exp[i\mathbf{k}_i'(p,q)\cdot r'], \tag{11a}$$

where

$$\mathbf{A}_i'(p,q) = [c]\mathbf{A}_i(p,q), \tag{11b}$$

$$\mathbf{k}_i'(p,q) = [c]\mathbf{k}_i(p,q), \tag{11c}$$

$$\Delta(p,q) = \mathbf{k}_i(p,q)\cdot \mathbf{d}. \tag{11d}$$

For each plane wave component, the amplitude $\mathbf{A}_i'(p,q)$ defines the corresponding incident field vector amplitude which is represented by $\mathbf{e}_0'$ in Eq. (7). Also the propagation vector $\mathbf{k}_i'(p,q)$ defines the incidence angle (θ_i',ϕ_i') of each respective plane wave component of the beam, so that the terms $\mathbf{C}_r(\theta_i',\phi_i')$ and $\mathbf{B}_r(\theta_i',\phi_i')$ for Eq. (7) can be found. Note that with our choice of the scatterer coordinate system ϕ_i' will always be zero. This is indicated in Fig. 3. The scatterer matrix [C] in Eq. (7) is evaluated only once for each problem because it depends only on the scatterer geometry and material. Then since the right-hand side of Eq. (7) is completely known, it gives the corresponding coefficients $f_r(\theta_i',\phi_i')$ and $g_r(\theta_i',\phi_i')$ for each beam plane wave component. Using these coefficients in Eq. (5) and assuming that the scattering angles (θ_s',ϕ_s') in the scatterer system are also known from the transformation of scattering angles (θ_s,ϕ_s) in the beam system, one gets the far-field vector amplitude $\mathbf{F}'[(\theta_s',\phi_s'|\theta_i',\phi_i'),p,q]$ of the scattered fields in the scatterer system. The p, q dependence of this quantity comes from the fact that it corresponds to a single plane wave component of the beam plane wave spectrum with parameters p and q. The corresponding far-field vector amplitude in the fixed beam system is found by a simple coordinate transformation. Thus

$$\mathbf{F}[(\theta_s,\phi_s|\theta_i,\phi_i),p,q] = [C]^{-1}\mathbf{F}'[(\theta_s',\phi_s'|\theta_i',\phi_i'),p,q]. \tag{12}$$

Superposition of these results is achieved in the fixed beam system. This superposition corresponds to the addition of far-field vector amplitudes for each plane wave component of the beam. The scattered electric field for the laser beam is then given by

$$\mathbf{E}_s(k\mathbf{r}) = \frac{\exp(ikr)}{r}\iint_{-\infty}^{+\infty}\mathbf{F}[(\theta_s,\phi_s|\theta_i,\phi_i),p,q]dpdq. \tag{13}$$

These results are computed in terms of two different quantities. First, there is the normalized differential cross section defined by

$$\sigma_{dn} = \sigma_d/\pi a^2, \tag{14a}$$

where

$$\begin{aligned}
\sigma_d &= \lim_{r\to\infty}\left[4\pi r^2\frac{I_s(\theta_s,\phi_s)}{I_i(\theta_i,\phi_i)}\right] \\
&= 4\pi\left|\iint_{-\infty}^{+\infty}\mathbf{F}[(\theta_s,\phi_s|\theta_i,\phi_i),p,q]dpdq\right|^2.
\end{aligned} \tag{14b}$$

Second, there is the intensity distribution of the total field defined by

$$I_t(\mathbf{r}) = |\mathbf{E}_t(\mathbf{r})|^2 = |\mathbf{E}_b(\mathbf{r}) + \mathbf{E}_s(r)|^2, \tag{15}$$

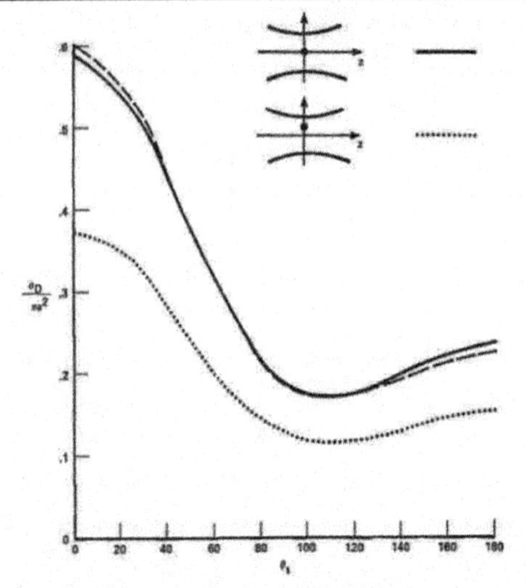

Fig. 4. Beam scattering patterns for a sphere of $n = 1.55$ and $ka = 1.0$. The dashed curve represents a single-plane wave incidence case and is included for comparison. The solid line represents a laser beam with spot size $W_0 = 5a$, and the particle is centered. The dotted curve represents the same below ($W_0 = 5a$), but the scatterer is off-centered to $\mathbf{d} = (2a, 0.5a, 0)$.

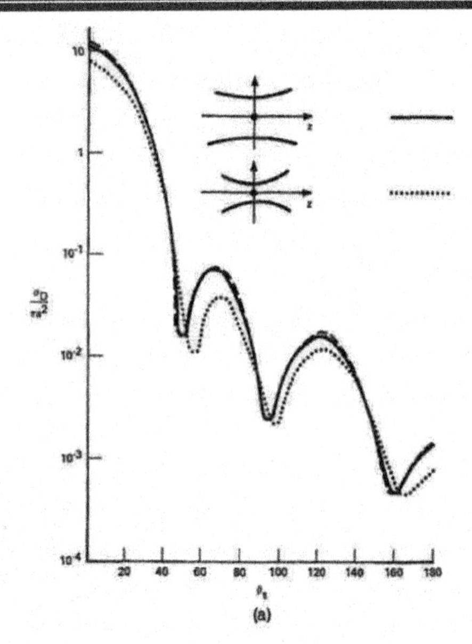

where $\mathbf{E}_b(\mathbf{r})$ is the incident beam field given by Eq. (1) or (2), and $\mathbf{E}_s(\mathbf{r})$ is the scattered field given by Eq. (13). This intensity distribution is normally calculated by assuming a finite distance for the scattering points.

III. Numerical Examples

Computations were done to obtain the normalized differential cross sections and the total intensity distributions for the two classes of particles, i.e., spheres and axisymmetric bodies. Different parameters for both the incident beam and the particle characteristics were used.

Parameters of the examples for spheres were chosen so that comparisons of the present exact results with the approximate results obtained in our previous paper[5] were possible. Figures 4 and 5 show the normalized differential cross-section patterns for such spheres for different beam parameters and for various positions of the scatterer. The same main characteristics as discovered in our earlier study[5] were obtained. It was found that decreasing the beam waist spot size produced a decrease and expansion in the scattering pattern. The expansion is in the direction of $\theta_s = \pm 180°$. Off-cen-

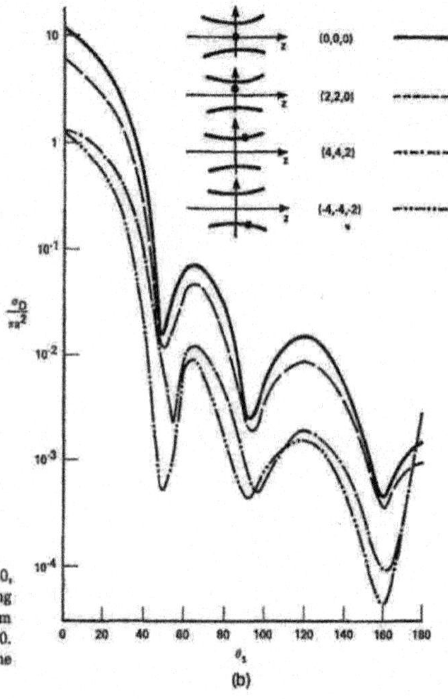

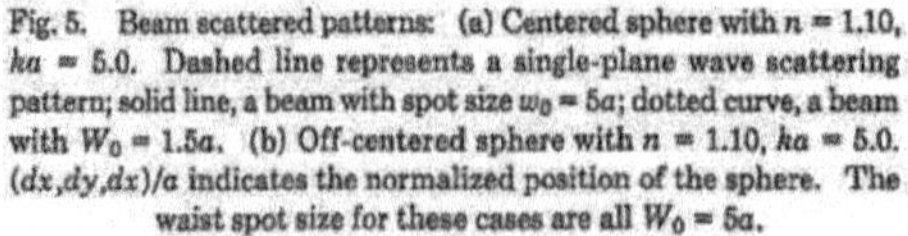
Fig. 5. Beam scattered patterns: (a) Centered sphere with $n = 1.10$, $ka = 5.0$. Dashed line represents a single-plane wave scattering pattern; solid line, a beam with spot size $w_0 = 5a$; dotted curve, a beam with $W_0 = 1.5a$. (b) Off-centered sphere with $n = 1.10$, $ka = 5.0$. $(dx, dy, dz)/a$ indicates the normalized position of the sphere. The waist spot size for these cases are all $W_0 = 5a$.

tering the particle caused an almost proportional decrease in the magnitude compared with the original pattern. However, one significant difference in the present exact results from the previous approximate results[5] is that the asymmetric patterns are obtained even for the cases where the particles are off-centered in the $z = 0$ plane. This was not observed in the approximate solution simply because of the assumptions of Rayleigh approximations on the field behavior within the particle. Moving the particle farther from the origin in the $z = 0$ plane increased the asymmetry, and the dips and peaks of the pattern showed shifts whose direction depended on the position of the scatterer. For example, moving the particle toward positive x values shifts the pattern toward $\theta_s = +180°$ in the first quadrant of the x-z plane which is the scattering plane. But these shifts are relatively minor for small displacements, and the asymmetry introduced in such cases is very small.

If an off-centered particle in the x-y plane is moved out of this plane in the direction of the z axis, the effect is an increase or a decrease in the magnitude of the scattering pattern depending on the original position of the particle. Another effect is a relatively small increase in the asymmetry of the pattern, which is expected, because now the particle sees a completely asymmetric incident field. The reason for the increase or decrease in the overall magnitude of the pattern is due to the nonuniform intensity distribution of the laser beam itself. It is found that the beam field magnitude or the intensity is not a monotonically decreasing function of z; that is, going away from the $z = 0$ plane does not necessarily mean a decreasing field amplitude. This is true only if the particle is moving close to the z axis. These points can be further clarified if the original beam expression of Eq. (1) is considered. The corresponding intensity distribution for the field is

$$I(r,z) = A_0^2 \left[\frac{W_0}{W(z)}\right] \exp[-2r^2/W^2(z)], \qquad (16)$$

where $A_0^2 = A_x^2 + A_y^2$ and $r^2 = x^2 + y^2$. The variation of this intensity with respect to z is given by

$$\frac{\partial}{\partial z} I(r,z) = A_0^2 \frac{2z}{z_0^2} \left[\frac{W_0}{W(z)}\right]^{1/2} \exp\left[\frac{2r^2}{W^2(z)}\right]\left[\frac{2r^2}{W^2(z)} - 1\right], \qquad (17)$$

which means that the laser beam intensity is an increasing function of z until

$$W(z) = \sqrt{2}r, \qquad (18)$$

and it started decreasing afterward. This is true only if r is kept constant in this process. Therefore, if the particle is originally off-centered to a value of r on the $z = 0$ plane and later moved to a nonzero value of z with the same r, the intensity of the incident radiation will be increasing until the beam spot size reaches the value given by Eq. (18). Figure 6 shows the form of the constant intensity contours of the beam for a fixed r-z plane. Note that Eq. (18), or Fig. 7, means that no intensity increases may be observed if the particle is originally located on the x-y plane at a position $r \leq W_0/\sqrt{2}$ and then moved out of this plane in the z direction. In this case the intensity will be decreasing.

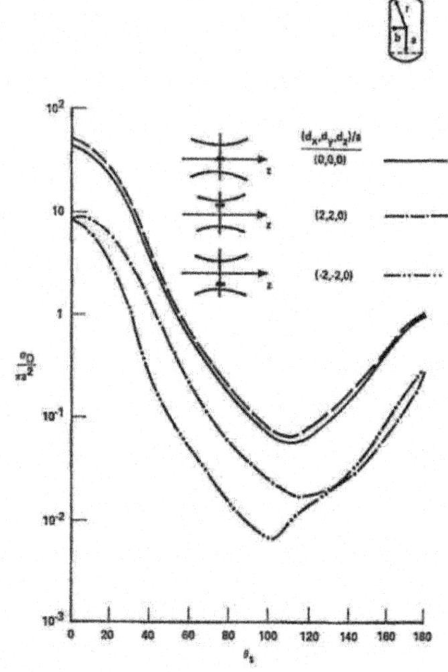

Fig. 6. Scattering patterns for a cylinder with spherical endcaps. The scatterer characteristics are: $\epsilon_r = 1.96$; $ka = 5.288$; $a/b = 1.73$; $r = 26$. The symmetry axis of the scatterer is along the z axis; that is, $(\theta_b,\phi_b) = (0,0)$. The dashed curve (without any dots) represents the single-plane wave scattering case, and the rest of the curves are for a Gaussian beam with a spot size $W_0 = 3a$. The displacement of the scatterer in each case is shown.

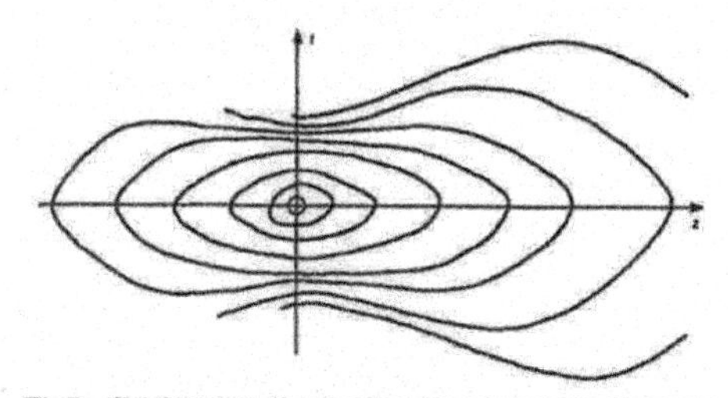

Fig. 7. Constant intensity contours for the beam fields. The center ones represent higher intensities. r and z axes may not be to the same scale. Note that for larger r values a scatterer sees increasing incident intensity as it moves along the z axis direction.

The facts discussed above are indeed verified by the computed results. For example, moving the particle of Fig. 5(b) from the $(2a,2a,0)$ point to the $(2a,2a,2a)$ point (which is not shown in the figure) did not cause significant change in the scattering pattern. But moving it from the $(4a,4a,0)$ to the $(4a,4a,2a)$ position

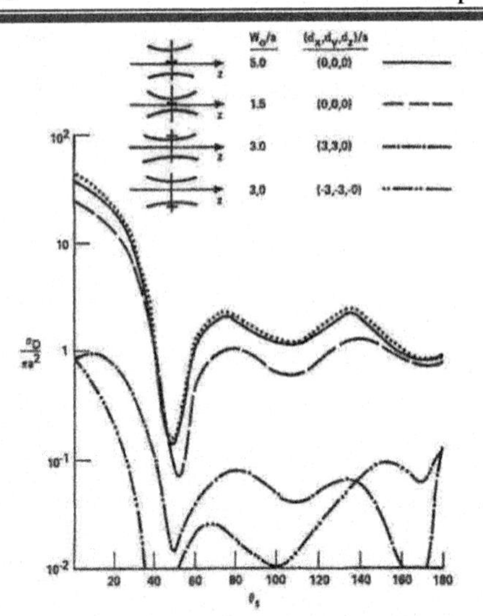

Fig. 8. Scattering patterns for oblate spheroids with $\epsilon_r = 5.0$, $ka = 4.309$, $a/b = 2.0$, where a and b are the semimajor and semiminor axes, respectively. The symmetry axis of the ellipsoid is along two directions; that is, $(\theta_b, \phi_b) = (0,0)$. The dotted curve represents the single-plane wave (in the z direction) scattering case, and the other curves are specified.

increased the pattern ~30% at the forward peak. Note that for the case of Fig. 5(b) the spot size is $5a$, and thus one can easily see the relationship

$$[(2a)^2 + (2a)^2] < \frac{(5a)^2}{2} < [(4a)^2 + (4a)^2], \qquad (19)$$

which verifies the condition of Eq. (18) and Fig. 7.

The scattering patterns for slightly off-centered particles are plotted for the positive half of the scattering plane because their asymmetry is very slight.

Comparison of the present exact results with the approximate results of our previous study[5] shows that the iterations used in the latter gave very accurate results. This is true if one neglects some of the asymmetry relations which could not be observed in the previous approximate results. On the average, deviations of these approximate results from the present exact results are ~5%. This difference decreases if the particle is placed close to origin and goes up to 25% for the cases in which the scatterer is away from the origin.

The next set of computations is done for axisymmetric nonspherical scatterers. Figures 6 and 8 give scattering patterns for oblate spheroids and cylinders for various beam parameters. These patterns are given in terms of the normalized differential cross section as defined by Eqs. (14). These examples do not introduce any new features to the main characteristics of the beam scattering. Decreasing the spot size pulls down the scattering pattern, and similar pattern shifts or expansions as discussed earlier are also observed. However, the latter feature is not as clear as in the case of spheres, because axisymmetric bodies normally have more complicated surface structure, and this prevents the occurrence of simple interference effects as discussed in our previous paper.[5]

Another set of computations is done for the total intensity patterns as defined in Eq. (15). Both the incident and scattered fields are now taken into consideration. These patterns correspond to experimentally measurable patterns for the total field. Figure 9 shows an example of this calculation for an oblate spheroid. Note that for small scattering angles one will be able to measure a slightly distorted incident Gaussian beam intensity form. For larger scattering angles, the incident beam intensity is very small, and thus the measured intensity approximates very closely the scattered field intensity pattern. However, for the cases involving situations where the incident field and the scattered field are of comparable magnitude, one is able to see a set of interference fringe patterns in the total field. This is due to the difference in the respective phases of the incident and the scattered fields. Note that decreasing the spot size of the beam increases the far-field beam angle, and thus one will see more fringing due to the larger range over which the incident and scattered fields are of comparable magnitude. For different scatterers, one may get a greater or lesser number of interference fringes in the total intensity pattern depending on the scattered field pattern and the properties of the incident laser beam.

IV. Conclusions

The exact solution of the focused laser beam scattering showed the same main characteristics as the approximate solutions which were obtained in our earlier work.[5] However, the asymmetry of the scattering pattern which is expected for an off-centered particle in the $z = 0$ plane can only be obtained by the present exact solution. The examples studied for axisymmetric nonspherical scatterers have the same characteristics except that for such scatterers off-centering gave enhanced asymmetry compared with the spherical scatterers.

The total field intensity patterns which are more useful for experimentalists showed that the measurable intensities for different ranges are as follows: (1) at small angles (within the far-field beam angle) the intensity of the laser beam itself; (2) for large angles (away from far-field beam angle) the intensity of the scattered fields alone; and (3) for the in-between cases a set of fringes due to the interference of the incident laser fields with the scattered fields.

It has been shown that use of the plane wave spectrum representation of the beam provides a simple way of dealing with beam scattering problems.

This approach can also be applied to scattering problems involving moving scatterer systems.[12]

This work was partly supported by the National Science Foundation and by the Environmental Protection Agency.

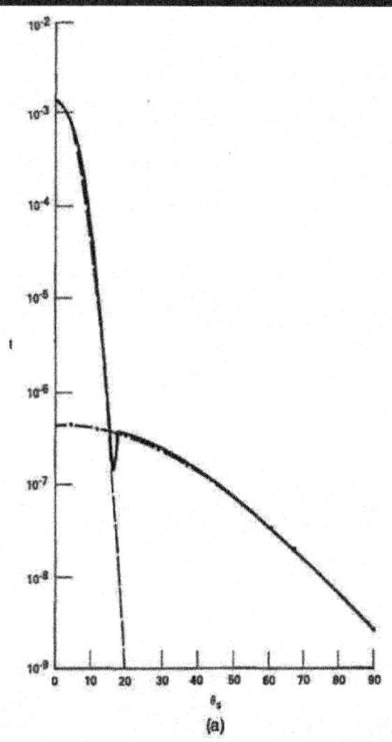

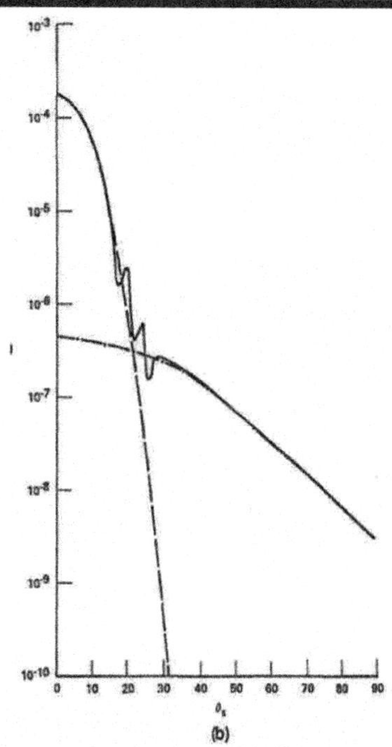

Fig. 9. Total field intensity patterns as defined by Eq. (15) in the text. The scatterer is a centered oblate spheroid with parameters ϵ_r = 1.44, ka = 3.0, a/b = 1.80, and its symmetry axis is along the z axis; that is, $(\theta_b,\phi_b) = (0,0)$. The intensity patterns are evaluated for r = 1 m. The incident beam has a linear polarization in the x direction. The total intensity is indicated by the solid line. The dashed line and the dashed line with dots indicate the incident beam intensity and the scattered intensity, respectively. The beam waist spot sizes are (a) $W_0 = 5a$ and (b) $W_0 = 3a$.

Appendix: Surface Integrals I, J, K, L and I', J', K', L' and Spherical Vector Harmonics

The surface integrals I, J, K, L and I', J', K', L' used in the scatterer matrix of Eq. (36) are given by the following relations:

$$I = \frac{k^2}{\pi}\int_s \hat{n}\cdot \mathbf{M}_\gamma^3(kr')\times \mathbf{M}_\mu^1(k'r')ds,$$

$$J = \frac{k^2}{\pi}\int_s \hat{n}\cdot \mathbf{M}_\gamma^3(kr')\times \mathbf{N}_\mu^1(k'r')ds,$$

$$K = \frac{k^2}{\pi}\int_s \hat{n}\cdot \mathbf{N}_\gamma^3(kr')\times \mathbf{M}_\mu^1(k'r')ds,$$

$$L = \frac{k^2}{\pi}\int_s n\cdot \mathbf{N}_\gamma^3(kr')\times \mathbf{M}^1(k'r')ds;$$

$$I' = \frac{k^2}{\pi}\int_s \hat{n}\cdot \mathbf{M}_\gamma^1(kr')\times \mathbf{M}_\mu^1(k'r')ds,$$

$$J' = \frac{k^2}{\pi}\int_s n\cdot \mathbf{M}_\gamma^1(kr')\times \mathbf{N}_\mu^1(k'r')ds,$$

$$K' = \frac{k^2}{\pi}\int_s \hat{n}\cdot \mathbf{N}_\gamma^1(kr')\times \mathbf{M}_\mu^1(k'r')ds,$$

$$L' = \frac{k^2}{\pi}\int_s \hat{n}\cdot \mathbf{N}_\gamma^1(kr')\times \mathbf{N}_\mu^1(k'r')ds,$$

where s indicates the surface of the scatterer. $\hat{n}$ is the unit vector normal to this surface,[2] and r' is the vector from the origin of the scatterer to the surface of it. The propagations constants are defined as $k = 2\pi/\lambda$, and $k' = \sqrt{\mu_r\epsilon_r}k \cdot \gamma$ and μ indicate the spherical harmonic indices σmn, where σ can be e (even) or o (odd). The spherical harmonics are given by

$$\mathbf{M}_{\sigma mn}^1(kr) = \sqrt{n(n+1)}\mathbf{C}_{\sigma mn}(\theta,\phi)j_n(kr),$$

$$\mathbf{N}_{\sigma mn}^1(kr) = n(n+1)\mathbf{P}_{\sigma mn}(\theta,\phi)j_n(kr) + \sqrt{n(n+1)}\mathbf{B}_{\sigma mn}(\theta,\phi)\frac{1}{kr}\frac{d}{dr}[rj_n(kr)],$$

$$\mathbf{M}_{\sigma mn}^3 = \sqrt{n(n+1)}\mathbf{C}_{\sigma mn}(\theta,\phi)h_n^{(1)}(kr),$$

$$\mathbf{N}_{\sigma mn}^3 = \sqrt{n(n+1)}\mathbf{P}_{\sigma mn}(\theta,\phi)h_n^{(1)}(kr) + \sqrt{n(n+1)}\mathbf{B}_{\sigma mn}(\theta,\phi)\frac{1}{kr}\frac{d}{dr}[rh_n^{(1)}(kr)],$$

where

$$\mathbf{P}_{\sigma mn}(\theta,\phi) = \hat{e}_r X_{\sigma nm}(\theta,\phi),$$

$$\mathbf{B}_{\sigma mn}(\theta,\phi) = \frac{1}{\sqrt{n(n+1)}}\nabla X_{\sigma mn}(\theta,\phi),$$

$$\mathbf{C}_{\sigma mn}(\theta,\phi) = \frac{1}{\sqrt{n(n+1)}} \nabla \times [\mathbf{r} X_{\sigma mn}(\theta,\phi)],$$

$$\mathbf{X}_{\sigma nm}(\theta,\phi) = P_n^m(\cos\theta) \begin{Bmatrix} \cos m\phi \\ \sin m\phi \end{Bmatrix} \quad \sigma = \begin{cases} \text{even,} \\ \text{odd} \end{cases}$$

$j_n(kr)$ and $h_n^{(1)}(kr)$ are the spherical Bessel and Hankel functions, respectively.

References

1. L. W. Casperson, C. Yeh, and W. F. Yeung, Appl. Opt. 16, 1104 (1977).
2. P. W. Barber and C. Yeh, Appl. Opt. 14, 2864 (1975).
3. P. C. Clemmow, *The Plane Wave Spectrum Representation of Electromagnetic Fields* (Pergamon, London, 1966).
4. L. W. Casperson and C. Yeh, Appl. Opt. 17, 1637 (1978).
5. S. Colak, C. Yeh, and L. W. Casperson, Appl. Opt. 18, 294 (1979).
6. J. E. Pearson, T. C. McGill, S. Kurtin, and A. Yariv, J. Opt. Soc. Am. 59, 1440 (1969).
7. N. Morita, T. Tanaka, T. Yamasaki, and Y. Nakanishi, IEEE Trans. Antennas Propag. AP-16, 724 (1968).
8. W. C. Tsai and R. J. Pogorzelski, J. Opt. Soc. Am. 65, 1457 (1975).
9. W. G. Tam, Appl. Opt. 16, 2016 (1977).
10. H. Chew, M. Kerker, and D. D. Cooke, Opt. Lett. 1, 138 (1977).
11. H. Chew, M. Kerker, and D. D. Cooke, Phys. Rev. A 16, 320 (1977).
12. S. Colak and C. Yeh, Appl. Opt. 19, 256 (1980).

Scattering of a focused beam by moving particles

S. Colak and C. Yeh

The scattering of a focused laser beam by moving particles is studied. A solution is given for a spherical scatterer that is small compared to the wavelength. Several numerical examples are given. It is concluded that for particles that have a short time of flight within the laser beam, the bandwidth and frequency spectrum of the scattered field is critically dependent on the laser beam shape. This fact is particularly important when information on the motion of small particles or macromolecules is to be inferred from the measured bandwidth and spectrum of the scattered field due to an incident wave that is not a plane wave.

I. Introduction

The use of lasers in scattering experiments provides us with the ability to focus a light beam into a small region. This enables us to select a single scattering body among an ensemble of scatterers.[1] It is recognized, however, that the scattering characteristics of a focused laser beam are substantially different from those of an incident plane wave.[2] New solutions of the scattering problem for the case of a sharply focused beam must therefore be found.

Single-particle scattering experiments may be classified in two general categories when the dynamic scattering characteristics are taken into consideration:

(1) The scattering particle is considered to be stationary.

(2) The scattering particle is moving.

The stationary case has been analyzed for a broad range of particles and beam characteristics.[3] Special treatment must be used when the particle is moving with respect to the incident beam (Fig. 1).

For the moving case with a focused beam, in addition to the expected simple Doppler shift in frequency for the scattered wave, one has to take into account the laser linewidth as well as the presence of a Doppler frequency width for the scattered field. This focused-beam situation is quite different from the case of the single monochromatic plane wave, in which the incident beam spectrum is simply a single frequency, while the scattered field spectrum is also a single Doppler-shifted frequency. In the case of a focused laser beam, the moving scatterer is expected to produce a scattered field pattern whose spectrum has a nonzero width. The purpose of this paper is to provide the properties of this spectrum function. Although our approach can be applied to particles of different shapes and composition, we shall only consider the case of a small moving sphere whose permittivity is either close to unity (the Rayleigh case) or infinity (the Thompson case).

II. Formulation of the Problem

In our previous papers[2,3] on focused laser beam scattering by stationary particles we had used the plane wave spectrum (PWS) representation of the incident laser beam:

$$\mathbf{E}_b(x,y,z,t) = \exp(i\omega_i t)\frac{b^2}{\pi}\iint_{-\infty}^{+\infty}\left(A_{0x}\hat{x} + A_{0y}\hat{y} + \frac{pA_{0x}+qA_{0y}}{s}\hat{z}\right) \times \exp[-b^2(p^2+q^2)]\exp[ik_0(px+qy-sz)]\,dp\,dq, \tag{1}$$

where $b = k_0W_0/2$, and $k_0 = 2\pi/\lambda_0 = \omega_i/C_0$. λ_0, ω_i, and W_0 are, respectively, the wave number, wavelength, radian frequency of oscillation, and the beam-waist spot size of the incident beam fields. A_{0x} and A_{0y} are the complex amplitudes of the x component and y component of the electric field vectors for the laser beam at the beam center. In an isotropic and homogeneous propagation medium, we have $s = (1-p^2-q^2)^{1/2}$, where p and q are the variables of the 2-D spatial frequency domain. p, q, and s can also be seen to be the direction cosines for the propagation vector of each component of the beam PWS.

To begin the solution, two coordinate systems are defined as the stationary-beam system and the moving-scatterer system. These will be denoted by O and O', respectively. Then a general form for the plane wave components of the beam, which is obtained from

The authors are with UCLA, Electrical Sciences & Engineering Department, Los Angeles, California 90024.

Received 10 May 1979.
0003-6935/80/020256-07$00.50/0.

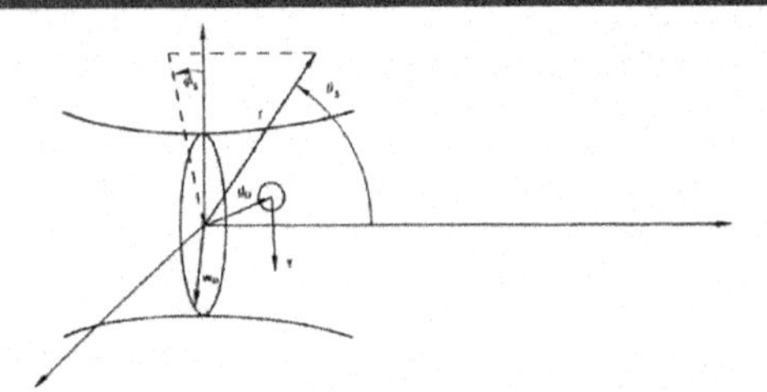

Fig. 1. Geometry of the problem: w_0 is the beam spot size; $\mathbf{v}$ and $\mathbf{d}_0$ are, respectively, the velocity and the zero-time position of the scatterer. The figure shows the $t = 0$ situation. $\mathbf{r}$ is the scattering direction and is defined by angles θ_s and ϕ_s.

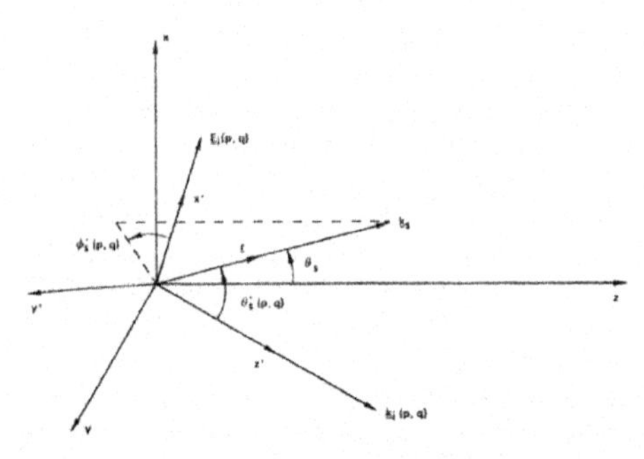

Fig. 2. Coordinates for the moving-scatterer frame. The scatterer-coordinate system (the primed one) is moving with a velocity $\mathbf{v}$ with respect to the stationary-beam system. x' and z' are chosen such that the incident electric field $\mathbf{E}_i(p,q)$ is along x' and the propagation vector $\mathbf{k}_i(p,q)$ is along z'. The scattering vector $\mathbf{k}_s$ makes a polar angle $\theta_s'(p,q)$ and an azimuthal angle $\phi_s'(p,q)$ with respect to the scatterer frame. Note that the intersection point of the dashed lines is in the $x'y'$ plane.

Eq. (1), will be defined in the stationary-beam system as follows (see Fig. 2):

$$\mathbf{E}_i(\mathbf{r},t,p,q) = \mathbf{A}_i(p,q)\exp\{i[\omega_i t + \mathbf{k}_i(p,q)\times\mathbf{r}]\}, \tag{2a}$$

where

$$\mathbf{A}_i(p,q) = \frac{b^2}{\pi}\left(A_{0x}\hat{x} + A_{0y}\hat{y} + \frac{pA_{0x}+qA_{0y}}{s}\hat{z}\right) \times \exp[-b^2(p^2+q^2)], \tag{2b}$$

$$\mathbf{k}_i(p,q) = k_0(p\hat{x} + q\hat{y} - s\hat{z}), \tag{2c}$$

$$s = (1 - p^2 - q^2)^{1/2}. \tag{2d}$$

The representation of this wave in the moving-scatterer system is found by a simple coordinate transformation. The result is

$$\mathbf{E}_i'(\mathbf{r}',t,p,q) = \mathbf{A}_i'(p,q)\exp\{i[\omega_i t + \mathbf{k}_i'(p,q) \times \mathbf{r}' + \xi(t,p,q)]\}, \tag{3a}$$

where

$$\mathbf{A}_i'(p,q) = [T]\mathbf{A}_i(p,q), \tag{3b}$$

$$\mathbf{k}_i'(p,q) = [T]\mathbf{k}_i(p,q), \tag{3c}$$

$$\xi(t,p,q) = \mathbf{k}_i(p,q)\times\mathbf{d}(t). \tag{3d}$$

The time-dependent displacement of the scatterer frame with respect to the beam system is defined by

$$\mathbf{d}(t) = \mathbf{d}_0 + \mathbf{v}t, \tag{4}$$

where $\mathbf{d}_0$ is the position of the scatterer at $t = 0$, and $\mathbf{v}$ is the velocity of the scatterer with respect to the beam system O. The matrix $[T]$ in Eq. (3) is introduced to include any possible coordinate rotations.

Let us assume that the particle is a small sphere with dielectric properties such that it falls into the region between the Rayleigh case ($n \approx 1$) and the Thompson case (perfectly conducting body). Shifrin[4] has solved the scattering problem for this type of stationary scatterer for an incident plane wave.

We shall now extend this method to treat the problem of moving scatterers. The moving-scatterer system is defined such that the incident field vector may lie in the $x'y'$ plane, and the propagation vector of the incident field will be in the z' direction. The transformation matrix in Eq. (3) takes the following form:

$$[T] = \begin{bmatrix} a_{11} & a_{12} & a_{13} \\ a_{21} & a_{22} & a_{23} \\ a_{31} & a_{32} & a_{33} \end{bmatrix}, \tag{5a}$$

where

$$\begin{bmatrix} a_{31} \\ a_{32} \\ a_{33} \end{bmatrix} = \begin{bmatrix} p \\ q \\ -s \end{bmatrix}, \tag{5b}$$

$$\begin{bmatrix} a_{11} \\ a_{12} \\ a_{13} \end{bmatrix} = \frac{1}{A_m}\begin{bmatrix} A_{0x} \\ A_{0y} \\ (pA_{0x} + qA_{0y})/s \end{bmatrix}, \tag{5c}$$

$$A_m = \left[A_{0x}^2 + A_{0y}^2 + \frac{(pA_{0x}+qA_{0y})^2}{s^2}\right]^{1/2}, \tag{5d}$$

$$\begin{bmatrix} a_{21} \\ a_{22} \\ a_{23} \end{bmatrix} = \begin{bmatrix} a_{32}a_{13} - a_{12}a_{33} \\ a_{33}a_{11} - a_{31}a_{13} \\ a_{31}a_{12} - a_{11}a_{32} \end{bmatrix}. \tag{5e}$$

Note that the matrix $[T]$ does not depend on the velocity of the particle. The particle velocity affects the fields through the phase term $\xi(t,p,q)$ as given by Eq. (3d). A similar velocity-dependent term will appear in the scattered field expressions [see Eq. (12b)]. Note also that the relativistic speeds are not considered in this study.

Using the above transformation one may obtain from Eq. (3) a simplified expression for the incident plane wave in the moving-scatterer frame O':

$$\mathbf{E}_i'(\mathbf{r}',t,p,q) = A_x(p,q)\hat{x} \times \exp\{i[\omega_i t - k_z'(p,q)z' + \xi(t,p,q)]\}, \tag{6a}$$

where

$$A_x(p,q) = \left[A_{0x}^2 + A_{0y}^2 + \frac{(pA_{0x} + qA_{0y})^2}{s^2}\right]^{1/2} \times \frac{b^2}{\pi} \exp[-b^2(p^2+q^2)], \quad (6b)$$

$$k_z'(p,q) = k_0 = \omega_i/c_0 = \omega_i(\epsilon_0\mu_0)^{1/2}, \quad (6c)$$

where ϵ_0 and μ_0 are the permittivity and the permeability of the homogeneous space surrounding the scatterer. Combining Eqs. (6c) and (6a) gives

$$\mathbf{E}_i'(\mathbf{r}',t,p,q) = A_x(p,q)\hat{x} \times \exp\{i[\omega_i t - k_0 z' + \xi(t,p,q)]\}, \quad (7)$$

which is a very simple representation of PWS components of the beam in the moving-scatterer frame. One notes that Eq. (7) has the same plane wave form as the incident field expression used by Shifrin.[4] Applying Shifrin's results directly to the present situation gives, for the case of a small spherical scatterer of radius a,

$$E_\theta^{s'}(\mathbf{r}',t,p,q) = \frac{\cos[\phi_s'(p,q)]A_x(p,q)\exp\{i[\omega_i t + \xi(t,p,q)]\}}{k_0}\frac{\exp(-ik_0 r')}{r'}(k_0 a)^3\left\{\frac{\gamma(u)}{2} + \cos[\theta_s'(p,q)]\right\}, \quad (8a)$$

$$E_\phi^{s'}(\mathbf{r}',t,p,q) = -\frac{\sin[\phi_s'(,q)]A_x(p,q)\exp\{i[\omega_i t + \xi(t,p,q)]\}}{k_0}\frac{\exp(-ik_0 r')}{r'}(k_0 a)^3\left\{1 + \frac{\gamma(u)}{2}\cos[\theta_s'(p,q)]\right\}, \quad (8b)$$

where

$$u = nk_0 a,$$

$$\gamma(u) = 1 + \frac{3}{u}\cot(u) - \frac{3}{u^2}, \quad (8c)$$

$E_\theta^{s'}$ and $E_\phi^{s'}$ are the transverse electric field components in the moving frame, and $\phi_s'(p,q)$ and $\theta_s'(p,q)$ are scattering angles defined by the scattering direction $\mathbf{r}_s'(p,q)$ as shown in Fig. 2. The p and q dependence of these terms comes from the fact that the scatterer system is rotated differently for each separate PWS component of the beam in such a way that the incident field expression may be reduced to the form given by Eq. (7). One notes that the $E_\theta^{s'}$ and $E_\phi^{s'}$ fields in Eq. (8) are defined in the scatterer-coordinate system O', which is rotated and shifted with respect to the stationary-beam system O. To obtain the expression for the complete beam spectrum, we shall first transform the fields $E_\theta^{s'}$ and $E_\phi^{s'}$ back to the beam-coordinate system, which is a stationary reference frame, and then apply the principle of superposition.

Let us assume that the propagation vector of the scattered field in the stationary-beam system is pointed along the direction

$$\hat{r}_s = \sin\theta_s\cos\phi_s\hat{x} + \sin\theta_s\sin\phi_s\hat{y} + \cos\theta_s\hat{z}, \quad (9)$$

so that $\mathbf{k}_s = k_0\hat{r}_s$. Transformation of Eq. (9) to the rotated-coordinate system O' can be achieved by using the $[T]$ matrix in Eq. (5). One has

$$\hat{r}_s'(p,q) = [T]\hat{r}_s$$

or

$$\hat{r}_s'(p,q) = \begin{bmatrix} r_{sx}'(p,q) \\ r_{sy}'p,q) \\ r_{sz}'(p,q) \end{bmatrix} = [T]\begin{bmatrix} \sin\theta_s\cos\phi_s \\ \sin\theta_s\sin\phi_s \\ \cos\theta_s \end{bmatrix}. \quad (10)$$

The scattering angles $\theta_s'(p,q)$ and $\phi_s'(p,q)$ in the rotated-scatterer system can be easily found from the above equation. They are

$$\theta_s'(p,q) = \cos^{-1}[r_{sz}'(p,q)] = \sin^{-1}[r_{xy}'(p,q)], \quad (11a)$$

$$\phi_s'(p,q) = \cos^{-1}\left[\frac{r_{sx}'(p,q)}{r_{xy}'(p,q)}\right] = \sin^{-1}\left[\frac{r_{sy}'(p,q)}{r_{xy}'(p,q)}\right], \quad (11b)$$

where

$$r_{xy}'(p,q) = \{[r_{sx}'(p,q)]^2 + [r_{sy}'(p,q)]^2\}^{1/2}. \quad (11c)$$

Using Eq. (8) one may obtain the Cartesian components of the scattered fields in the rotated-scatterer system.

$$\mathbf{E}_s(\mathbf{r},t,p,q) = [E_x(\mathbf{r},p,q,)\hat{x} + E_y(\mathbf{r},p,q)\hat{y} + E_z(\mathbf{r},p,q)\hat{z}]\frac{A_x(p,q)\exp\{i[\omega_i t + \xi(t,p,q)]\}}{k_0}\exp[i\eta(t,\mathbf{r})](k_0 a)^3\frac{\exp(-ik_0 r)}{r}, \quad (12a)$$

where

$$\eta(t,\mathbf{r}) = \mathbf{k}_s \times \mathbf{d}(t), \quad (12b)$$

The transformation of these Cartesian field components to the stationary beam system O is achieved by using the factor $[T]^{-1}$, which is the inverse of the coordinate transformation matrix given by Eq. (5). It can be seen that a new time-dependent phase-shift term appears. This term accounts for the Doppler shift in the scattered fields. This phase shift is only a function of the scattering angle and the particle velocity. The resultant scattered field in the beam system is

$$\begin{bmatrix} E_x(\mathbf{r},p,q) \\ E_y(\mathbf{r},p,q) \\ E_z(\mathbf{r},p,q) \end{bmatrix} = [T^{-1}]\begin{bmatrix} r_{sz}'\left(\frac{r_{sx}'}{r_{xy}'}\right)^2\alpha_\theta + \left(\frac{r_{sy}'}{r_{xy}'}\right)\alpha_\phi \\ \frac{r_{sz}'r_{sx}'r_{sy}'}{(r_{xy}')^2}\alpha_\theta - \frac{r_{sx}'r_{sy}'}{(r_{xy}')^2}\alpha_\phi \\ r_{sx}\alpha_\theta \end{bmatrix}, \quad (12c)$$

with

$$\alpha_\theta = \frac{\gamma(u)}{2} + r_{sz}', \quad (12d)$$

$$\alpha_\phi = 1 + \frac{\gamma(u)}{2}r_{sz}'. \quad (12e)$$

Equation (12) represents the scattered field due to a single plane wave in the stationary-beam coordinate frame O. The total scattered field due to the laser beam is found by superposition. Hence, the total scattered field expression is

$$\mathbf{E}_{ST}(\mathbf{r},t) = (k_0a)^3 \frac{\exp(-ik_0r)}{k_0r} \exp\{i[\omega_i t + \eta(t,\theta_s)]\}$$
$$\times \iint_{-\infty}^{+\infty} \mathbf{E}'_{sc}(\mathbf{r},p,q) \exp[i\xi(t,p,q)]dp\,dq, \tag{13a}$$

where

$$\mathbf{E}'_{sc}(\mathbf{r},p,q) = [E_x(\mathbf{r},p,q)\hat{x} + \mathbf{E}_y(\mathbf{r},p,q)\hat{y} + \mathbf{E}_z(\mathbf{r},p,q)\hat{z}]A_x(p,q). \tag{13b}$$

Assume that the time dependence of the displacement (or the motion) of the scatterer is given by Eq. (4). Using the relations for $\eta(t,\mathbf{r})$ and $\xi(t,p,q)$ as given by Eqs. (12) and (3), Eq. (13) becomes

$$\mathbf{E}_{ST}(\mathbf{r},t) = (k_0a)^3 \frac{\exp(ik_0r)}{k_0r} \exp[i(\omega_i t + \mathbf{k}_s\ \mathbf{V})t]$$
$$\times \iint_{-\infty}^{+\infty} \mathbf{E}_{sc}(\mathbf{r},p,q) \exp\{-i[\mathbf{k}_i(p,q)\ \mathbf{V}]t\}dp\,dq, \tag{14a}$$

where

$$\mathbf{E}_{sc}(\mathbf{r},p,q) = \mathbf{E}'_{sc}(\mathbf{r},p,q) \exp\{i\mathbf{d}_0 [\mathbf{k}_0 - \mathbf{k}_i(p,q)]\}. \tag{14b}$$

Scattering patterns for the present problem can now be obtained from the above relations for various scattering

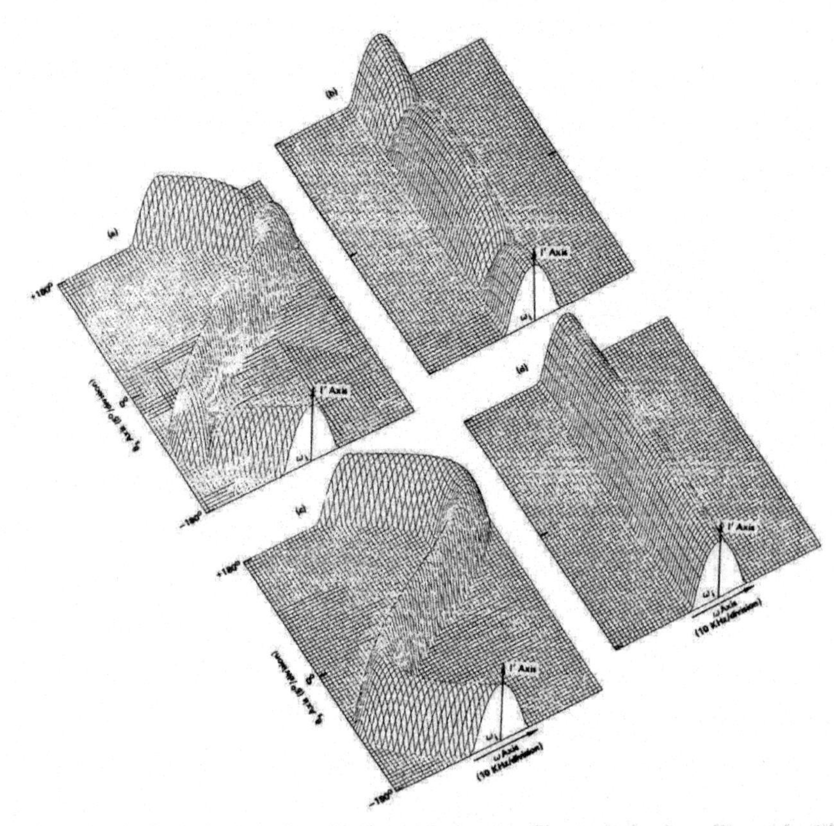

Fig. 3. Three-dimensional plots showing the effect of incident polarization at two different scattering planes. The spot size of the beam is 1 μm, and the velocity of particle is $\mathbf{v} = (10,0,0)$ cm/sec and $\mathbf{d}_0 = (0,0,0)$ μm. Other parameters are given in the text. (a) $A_{0x} = 1$, $A_{0y} = 0$, and $\phi = 0°$; (b) $A_{0x} = 1$, $A_{0y} = 0$, and $\phi_s = 90°$; (c) $A_{0x} = 0$, $A_{0y} = 1$, and $\phi_s = 0°$; (d) $A_{0x} = 0$, $A_{0y} = 1$, and $\phi_s = 90°$. The θ_s axis is 5°/div, and the ω axis is 10 kHz/div, and these values for plots are kept the same in Figs. 4 and 5. The incident frequency indicated in the figures is $\omega_i = 2\pi c/\lambda_i = 37.7 \times 10^{14}$ Hz.

angles θ_s and ϕ_s. Taking the Fourier transform of Eq. (14) gives

$$\mathbf{E}_{ST}(\mathbf{r},\omega) = (2\pi)^2(k_0a)^3\frac{\exp(-ik_0r)}{k_0r}\,\delta[\omega-(\omega_i+\mathbf{k}_s\mathbf{v})]*\left\{\iint_{-\infty}^{+\infty}\mathbf{E}_{sc}(\mathbf{r},p,q)\delta[\omega-\mathbf{k}_i(p,q)\mathbf{v}]\,dp\,dq\right\}, \tag{15}$$

where $*$ is used to indicate the convolution operation. As expected, when $\mathbf{v} = 0$, the above expression gives the scattering pattern as a function of the scattering angles θ_s and ϕ_s at a single oscillation frequency ω_i. On the other hand, when $\mathbf{v} \neq 0$ the integral on the right-hand side of Eq. (15) shows that the time-spectrum function of the scattered field is an integral sum of the impulse functions $\delta[\omega - \mathbf{k}_i(p,q)\mathbf{v}]$ with an envelope function $\mathbf{E}_{sc}(\mathbf{r},p,q)$. The convolution of this function with the first delta function $\delta[\omega - (\omega_i - \mathbf{k}_s\mathbf{v})]$ shifts the time-spectrum function to a center frequency

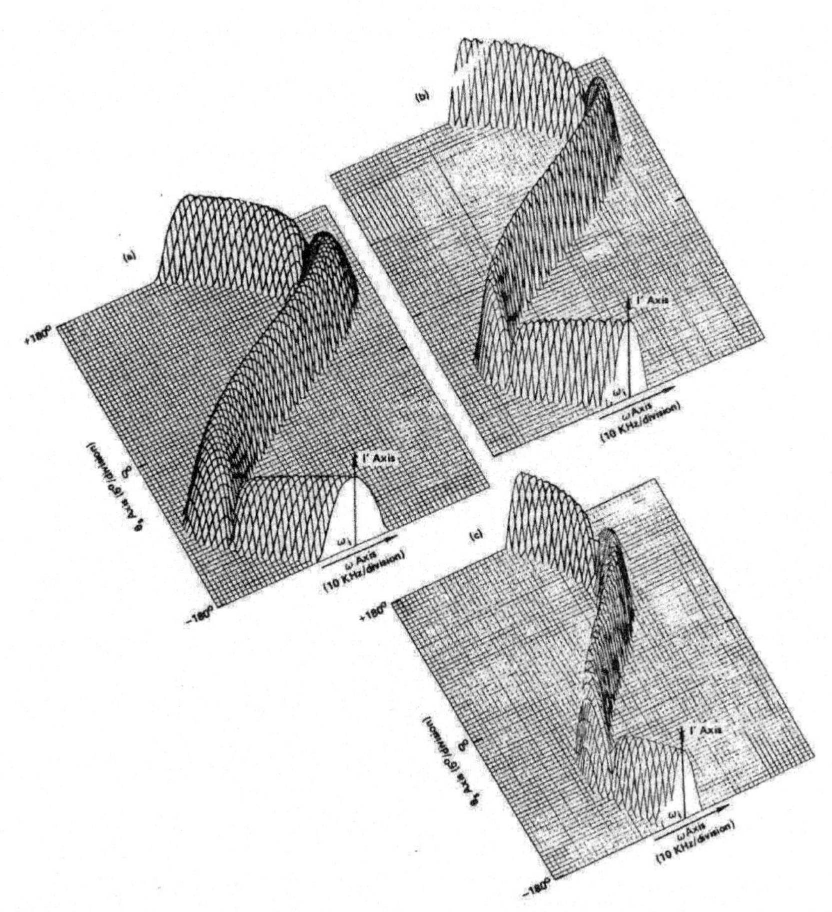

Fig. 4. Plots showing the effect of beam spot size and the speed of the scatterer. For these figures $A_{0x} = 1$, $A_{0y} = 0$, and the scattering plane is $\phi_s = 0$ plane and $\mathbf{d}_0 = (0,0,0)$ μm. (a) is the same as Fig. 3(a) and is included for comparison purposes. (b) $W_0 = 2$ μm, and $\mathbf{v} = (10,0,0)$ cm/sec. (c) $W_0 = 1$ μm, and $\mathbf{v} = (5,0,0)$ cm/sec. Other common parameters are given in the text and in earlier figures.

$$\omega_c = \omega_i + \mathbf{k}_s \cdot \mathbf{v}. \quad (16)$$

This means that at each different scattering direction, the scattered field possesses a time-frequency spectrum with a distinct center frequency whose value will depend on θ_s, ϕ_s, and $\mathbf{v}$ as well as a distinct envelope function that is a function of the parameters of the beam, the velocity of the scatterer, and the scattering direction.

The integral in Eq. (15) will give zero value when

$$\omega = \mathbf{k}_i(p,q) \cdot \mathbf{v} = k_0(pv_x + qv_y - dv_z). \quad (17)$$

This means that if the particle is moving with a velocity $\mathbf{v}$, a plane wave component of the PWS of the beam with the parameters p and q will generate a single scattered field frequency shift, given by above relation. So when performing numerical computations one may replace one of the integral variables (either p or q) by a relation obtained from Eq. (17). This procedure reduces the multiple integrations to a single integration.

III. Numerical Results and Conclusions

The spectrum of the scattered fields can be obtained from Eq. (15) by use of the Gaussian quadrature scheme. Typical results are displayed in Figs. 3–5. We have chosen the particle to be a spherical particle with radius $a = 0.05$ μm and index of refraction $n = 1.55$. The wavelength of the incident radiation is taken to be $\lambda = 0.5$ μm. The vertical axes in these figures represent the intensity of the scattered field, defined by

$$I' = r^2 \mathbf{E}_{ST}(\mathbf{r},\omega) \cdot \mathbf{E}^*_{ST}(\mathbf{r},\omega). \quad (18)$$

In the plots this quantity is shown on a logarithmic scale. The electric field amplitude of the incident beam is taken to be unity in all cases regardless of the polarization.

Figure 3 shows the effect of different incident polarization on the scattered field for a particle moving in the x direction. Since the particle is spherical and the beam

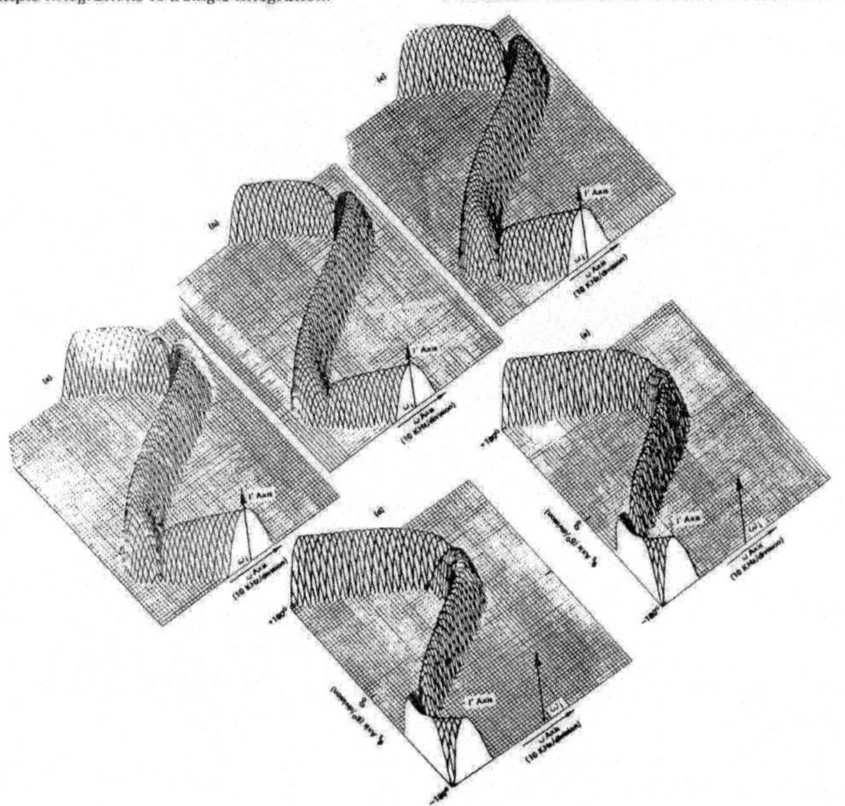

Fig. 5. Plots showing effects of particle velocity and the path it follows through the beam. For these figures $A_{0x} = 1$ and $A_{0y} = 0$, $W_0 = 1$ μm, and $|\mathbf{v}| = 10$ cm/sec although the path and direction of the motion change for each plot. (a) is the same as Fig. 3(a) and is included for reference purposes. (b) $\mathbf{v} = (10,0,0)$ cm/sec, $\mathbf{d}_0 = (0,1,0)$ μm; (c) $\mathbf{v} = (10,0,0)$ cm/sec, $\mathbf{d}_0 = (0,0,5)$ μm; (d) $\mathbf{v}$ (8,0,6) cm/sec, $\mathbf{d}_0 = (0,0,0)$ μm; (e) $\mathbf{v} = (8,0,6)$ cm/sec, $\mathbf{d}_0 = (0,0,3)$ μm.

has a circular cross section, these figures provide a general picture for the scattered field of a particle moving in the $z = 0$ plane. One notes that the bandwidth of the scattered field at a certain scattering angle θ_s is quite large. Its value may be comparable to the simple Doppler shift for the plane wave incident case. The S shape of the spectrum in Figs. 3(a) and 3(c) is due to the Doppler shift introduced to the scattered field by the motion of the scatterer. The decline in intensity shown in Figs. 3(a) and 3(b) near $\theta = \pm 90°$ is due to the fact that the incident light is polarized in the scattering plane. This behavior is not seen in Figs. 3(c) or 3(d), because in this case the polarization of the incident light is perpendicular to the scattering plane.

The effects of changing the speed of the particle and the spot size of the beam on the scattered field are shown in Fig. (4). The particle is moving in the x direction in the $z = 0$ plane, and the scattering plane is chosen to be the xz plane (or $\phi_s = 0$ plane). From Fig. 4(b) one notes that the effect of increasing the beam spot size is to introduce a decrease of the bandwidth of the spectrum function. Figure 4(c) shows the effect of decreasing the speed of the particle. Here, in addition to the bandwidth decrease, which is expected, one observes a decrease in the Doppler shift in the scattered fields and an increase of the intensity of the scattered field. This is expected because the time of flight of the particle within the beam spot size region increases, which in turn provides higher scattered energy.

The scattered spectrum functions for different velocities and different paths of the scatterer object are shown in Fig. 5.

One may easily conclude from the above specific example that a strongly focused beam alters significantly the bandwidth and frequency shift spectrum of the scattered field of a moving particle when compared to the case for a plane incident wave. This fact is particularly important when information on the motion of small particles or macromolecules is to be inferred from the measured bandwidth and spectrum of the scattered field due to an incident wave that is not a plane wave.[5,6]

References

1. L. W. Casperson, C. Yeh, and W. F. Yeung, Appl. Opt. **16**, 1104 (1977).
2. S. Colak, C. Yeh, and L. W. Casperson, Appl. Opt. **18**, 294 (1979).
3. C. Yeh, S. Colak, and P. W. Barber, "Scattering of Focused Beam by Arbitrarily Shaped Particles—Exact Solution," to be submitted to Appl. Opt.
4. K. S. Shifrin, *Scattering of Light in a Turbid Medium* (Nauka, Moscow, 1951); NASA Tech. Translation TTF-477, 1968, Chap. 4.
5. B. J. Berne and R. Pecora, *Dynamic Light Scattering* (Academic, New York, 1976).
6. C. T. O'Kanski, *Molecular Electrooptics: Theory and Methods* (Dekker, New York, 1976).

Reprinted from Journal of the Optical Society of America A, Vol. 1, page 359, April, 1984.

Matrix representations of the vector radiative-transfer theory for randomly distributed nonspherical particles

Akira Ishimaru

Department of Electrical Engineering, University of Washington, Seattle, Washington 98195

C. W. Yeh

Department of Electrical Engineering, University of California at Los Angeles, Los Angeles, California 90024

Received August 26, 1983; accepted December 18, 1983

General matrix representations of the vector radiative-transfer equation for randomly distributed nonspherical particles are given with compact representations of the extinction matrix and the Mueller matrix. The propagation of the coherent Stokes vector and the coherency matrix in such a medium is expressed in matrix form. The extinction matrix is related to the generalized optical theorem for partially polarized waves. The first-order scattering solution of the Stokes vector is given in matrix form, and discussions of the Fourier expansion of the equation of transfer and the limitation of the equation of transfer are given.

INTRODUCTION

In recent years, considerable attention has been paid to the problem of communication through various particles in the atmosphere, such as fog, ice particles, and snow.[1-8] These particles are often nonspherical and cause polarization-dependent attenuation, depolarization, and cross polarization. These polarization effects are also evident in lidar and radar detection and terrain scattering.[9-13]

For the complete polarization effects to be included in the radiative-transfer theory, the equation of transfer must be expressed in matrix form using the Stokes vector. Even though the vector radiative-transfer equation has been studied previously,[1-5,14] most of these studies were directed to spherical particles, which resulted in considerable simplification of the equation. For example, the extinction matrix becomes a scalar, and the Mueller matrix becomes a function of $\phi - \phi'$, as is discussed below. For nonspherical particles, the coherent field becomes depolarized. Even though these cross-polarization effects have been analyzed for nonspherical rain drops and ice particles,[6-8,15] we present a compact matrix representation of the propagation of the coherent field in terms of the Stokes vector. We also present a compact representation of the extinction and Mueller matrices using the direct product. The first-order scattering solution is then given by a matrix representation of the Stokes vector. These matrix representations should facilitate calculation of the propagation and scattering characteristics of the Stokes vector in nonspherical particles.

VECTOR RADIATIVE-TRANSFER EQUATION FOR PLANE-PARALLEL MEDIUM

Let us first consider the wave $[E'(\hat{s}')]$ with the vertical E_v' and the horizontal E_h' components incident upon a single particle in the direction of a unit vector $\hat{s}'$ ($\mu' = \cos\theta'$, ϕ') (Fig. 1). The scattered wave $[E(\hat{s})]$ with the vertical E_v and the horizontal E_h components in the direction $\hat{s}$ ($\mu = \cos\theta$, ϕ) at a distance R in the far zone of the particle is related to $[E'(\hat{s}')]$ by the 2×2 scattering amplitude matrix $[f(\hat{s}, \hat{s}')]$:

$$[E(\hat{s})] = \exp(ikR)/R[f(\hat{s}, \hat{s}')][E'(\hat{s}')], \qquad (1)$$

where k is the wave number and

$$[E(\hat{s})] = \begin{bmatrix} E_v(\hat{s}) \\ E_h(\hat{s}) \end{bmatrix}, \qquad [E'(\hat{s}')] = \begin{bmatrix} E_v'(\hat{s}') \\ E_h'(\hat{s}') \end{bmatrix},$$

$$[f(\hat{s}, \hat{s}')] = [f_{ij}(\hat{s}, \hat{s}')]. \qquad (2)$$

The scattering-amplitude matrix [Eqs. (2)] is a function of $\hat{s}$ and $\hat{s}'$ and should be expressed in the coordinate system attached to the medium when used with the radiative-transfer equation. In practice, however, the scattering amplitude of nonspherical particles is more conveniently expressed in the coordinate system attached to the particle, as this simplifies the numerical calculations. The transformation of the scattering amplitude from the representation in the particle coordinate to the medium coordinate can be done, in general, by using Euler's angle transformation.[16] Simple transformations can be used if the particle sizes are small compared with a wavelength.[12,13] These are not discussed in this paper. Here we assume that the scattering-amplitude matrix $[f]$ in the coordinate system attached to the medium is known.

If all the particles are identical and oriented in the same direction, the scattering amplitude $[f]$ is a deterministic function of $\hat{s}$ and $\hat{s}'$. However, in general, the particles have size distributions and random orientations, and the scattering-amplitude matrix, the extinction matrix, and the Mueller matrix to be used with the radiative-transfer equation should be the ensemble average of these quantities. Similarly, the Stokes vector should be the ensemble average, as is indicated below. For spherical particles, the scattering-amplitude matrix has been given in terms of the Mie solution.[2,3]

Let us define the Stokes vector $[I(\bar{r}, \hat{s})]$ at $\mathbf{r}$ pointed in the direction $\hat{s}$:

0740-3232/84/040359-06$02.00

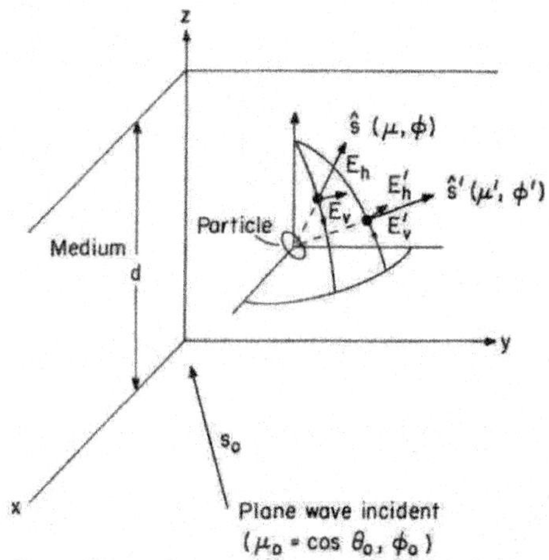

Fig. 1. Geometry for scattering-amplitude matrix $[f_{ij}(\hat{s}, \hat{s}')]$ in plane-parallel medium.

$$[I(\bar{r}, \hat{s})] = \begin{bmatrix} \langle E_v(\bar{r}, \hat{s})E_v^*(\bar{r}, \hat{s})\rangle \\ \langle E_h(\bar{r}, \hat{s})E_h^*(\bar{r}, \hat{s})\rangle \\ 2\,\mathrm{Re}\langle E_v(\bar{r}, \hat{s})E_h^*(\bar{r}, \hat{s})\rangle \\ 2\,\mathrm{Im}\langle E_v(\bar{r}, \hat{s})E_h^*(\bar{r}, \hat{s})\rangle \end{bmatrix}, \tag{3}$$

where Re and Im denote the real and imaginary parts, respectively, and the angle brackets denote the ensemble average. The vector equation of transfer can then be given as[1,2]

$$\frac{\mathrm{d}}{\mathrm{d}s}[I] = [\mathrm{A}][I] + \rho \int [S][I']\mathrm{d}\omega', \tag{4}$$

where $[I] = [I(\bar{r}, \hat{s})]$ and $[I'] = [I'(\bar{r}, \hat{s}')]$. The 4×4 matrix $[\mathrm{A}] = [\mathrm{A}(\hat{s})]$ is the extinction matrix, and $[S] = [S(\hat{s}, \hat{s}')]$ is the 4×4 Mueller matrix; ds is the elementary distance in the direction of $\hat{s}$, ρ is the number density (the number of particles per unit volume), and dω' is the elementary solid angle in the direction of $\hat{s}'$. If the particles have size distributions, $\rho[\ ]$ means $\int_0^\infty [\ ]N(D)\mathrm{d}D$, where $N(D)$ is the number of particles having the sizes between D and $D + \mathrm{d}D$. The extinction matrix $[\mathrm{A}]$ and the Mueller matrix $[S]$ can be expressed in terms of the scattering-amplitude matrix $[f]$, and these representations are given in the following sections.

If all the particles are identical and aligned in the same directions, the extinction matrix $[\mathrm{A}]$ and the Mueller matrix $[S]$ are deterministic functions of $\hat{s}$ and $\hat{s}'$. However, in general, the particles may be oriented with certain probability distributions. Then $[\mathrm{A}]$ and $[S]$ and $[I]$ are, in general, correlated. It is, however, reasonable to assume that the correlation at the same point is negligible, and we can approximate $\langle[\mathrm{A}][I]\rangle_p$ by $\langle[\mathrm{A}]\rangle_p\langle[I]\rangle_p$ and $\langle[S][I']\rangle_p$ by $\langle[S]\rangle_p\langle[I']\rangle_p$, where $\langle\ \rangle_p$ means the average over the orientation of the particles.

If a plane wave is incident upon the parallel-plane medium in the direction $\hat{s}_0(\mu_0 = \cos\theta_0, \phi_0)$, we can write the Stokes vector $[I]$ as a sum of the coherent intensity $[I_c]$ and the incoherent (or diffuse) intensity $[I_d]$:

$$[I(\bar{r}, \hat{s})] = [I_c(\bar{r}, \hat{s}_0)]\delta(\hat{s} - \hat{s}_0) + [I_d(\bar{r}, \hat{s})]. \tag{5}$$

The coherent intensity $[I_c]$ satisfies the following equation:

$$\frac{\mathrm{d}}{\mathrm{d}s}[I_c] = [\mathrm{A}][I_c]. \tag{6}$$

Therefore we get

$$\frac{\mathrm{d}}{\mathrm{d}s}[I_d] = [\mathrm{A}][I_d] + \rho \int [S][I'_d]\mathrm{d}\omega' + [I_i], \tag{7}$$

where $[I_i(r, \hat{s})]$ is the incident intensity matrix given by

$$[I_i(\bar{r}, \hat{s})] = \rho[S(\hat{s}, \hat{s}_0)][I_c(\bar{r}, \hat{s}_0)]. \tag{8}$$

The vector radiative-transfer equation (7) is most useful for plane-wave incidence. In the following sections, we give detailed discussions on the extinction matrix $[\mathrm{A}]$ and the Mueller matrix $[S]$ in terms of the scattering-amplitude matrix $[f]$. First we start with the discussion of the coherent field, as this clarifies the relationship between $[f]$ and $[\mathrm{A}]$.

PROPAGATION OF THE COHERENT FIELD

In general, the field $[E]$ consists of the coherent (average) field $[E_c]$ and the diffuse (incoherent) field $[E_d]$:

$$[E] = [E_c] + [E_d], \tag{9}$$

where $[E_c] = \langle[E]\rangle$ and $\langle[E_d]\rangle = 0$. The propagation characteristics of the coherent field can be expressed in terms of the field $[E_c]$, the Stokes vector $[I_c]$, or the coherency matrix $[J_c]$. We present these three representations in this section.

Field Representation

Consider the coherent field $[E_c]$ propagating in the direction $\hat{s}$ over the distance s. Omitting the factor $\exp(iks - i\omega t)$, the coherent field satisfies the following equation[3,4]:

$$\frac{\mathrm{d}}{\mathrm{d}s}[E_c] = [M][E_c]. \tag{10}$$

Here

$$[E_c] = \begin{bmatrix} \langle E_v\rangle \\ \langle E_h\rangle \end{bmatrix}, \qquad [M] = [M_{ij}], \qquad M_{ij} = \frac{i2\pi\rho}{k} f_{ij}(\hat{s}, \hat{s}).$$

Equation (10) is valid for a volume density much less than 0.1%,[17] and it also neglects the small longitudinal component of the electric field.

The solution to Eq. (10) is easily obtained by the matrix eigenvalue technique. The eigenvalues γ_1 and γ_2 are obtained by

$$\begin{vmatrix} M_{11} - \gamma & M_{12} \\ M_{21} & M_{22} - \gamma \end{vmatrix} = 0. \tag{11}$$

Let the eigenvectors corresponding to γ_1 and γ_2 be denoted by $[A_1]$ and $[A_2]$, respectively. Then we write

$$[E_c(s)] = C_1[A_1]\exp(\gamma_1 s) + C_2[A_2]\exp(\gamma_2 s)$$
$$= [A]\begin{bmatrix} \exp(\gamma_1 s) & 0 \\ 0 & \exp(\gamma_2 s) \end{bmatrix}\begin{bmatrix} C_1 \\ C_2 \end{bmatrix},$$
$$[A] = [[A_1][A_2]], \tag{12}$$

where $[A]$ is a 2×2 matrix whose first column is $[A_1]$ and second column is $[A_2]$. C_1 and C_2 are arbitrary constants to be determined by the boundary condition. In terms of M_{ij}, we get

$$\left.\begin{matrix}\gamma_1\\ \gamma_2\end{matrix}\right\} = \frac{1}{2}\{M_{11} + M_{22} \pm [(M_{11} - M_{22})^2 + 4M_{12}M_{21}]^{1/2}\},$$

$$[A] = \begin{bmatrix} 1 & A_{12} \\ A_{21} & 1 \end{bmatrix},$$

$$A_{21} = -\frac{M_{21}}{M_{22} - \gamma_1}, \qquad A_{12} = -\frac{M_{12}}{M_{11} - \gamma_2}. \tag{13}$$

Now we apply the boundary condition that $[E_c(o)]$ is given at $s = 0$:

$$[E_c(o)] = [A]\begin{bmatrix} C_1 \\ C_2 \end{bmatrix}. \tag{14}$$

From this, we get

$$\begin{bmatrix} C_1 \\ C_2 \end{bmatrix} = [A]^{-1}[E_c(o)]. \tag{15}$$

Substituting Eq. (15) into Eq. (12), we get

$$[E_c(s)] = [T(s)][E_c(o)], \tag{16}$$

where

$$[T(s)] = [A]\begin{bmatrix} \exp(\gamma_1 s) & 0 \\ 0 & \exp(\gamma_2 s) \end{bmatrix}[A]^{-1}$$
$$= [T_1]\exp(\gamma_1 s) + [T_2]\exp(\gamma_2 s),$$
$$[T_1] = [A]\begin{bmatrix} 1 & 0 \\ 0 & 0 \end{bmatrix}[A]^{-1},$$
$$[T_2] = [A]\begin{bmatrix} 0 & 0 \\ 0 & 1 \end{bmatrix}[A]^{-1}.$$

The relationship in Eq. (16) gives the complete cross-polarization effects on the coherent-field propagation. This solution is usually used for the cross polarization that is due to nonspherical rain droplets.[6]

Coherency Matrix Representation

The field representation in Eq. (16) can be transformed into the representation in terms of the coherency matrix.

The coherency matrix $[J]$ is defined by

$$[J] = [J_{ij}] = \begin{bmatrix} \langle E_v E_v^* \rangle & \langle E_v E_h^* \rangle \\ \langle E_h E_v^* \rangle & \langle E_h E_h^* \rangle \end{bmatrix}. \tag{17}$$

We write $[J]$ as a sum of the average $[J_c]$ and the diffuse $[J_d]$ components. Making use of Eq. (16), we relate the coherency matrix $[J_c(s)]$ at s to $[J_c(o)]$ at $s = 0$ as follows:

$$[J_c(s)] = [E_c(s)][\tilde{E}_c^*(s)]$$
$$= [T(s)][J_c(o)][\tilde{T}^*(s)]. \tag{18}$$

Here $[\tilde{E}_c]$ is the transpose of the matrix $[E_c]$.

Stokes Vector Representation

It is often convenient to express the propagation characteristics by using the Stokes vector. Let us start with the coherency matrix $[J]$ and rewrite it in the following column matrix form:

$$[D] = \begin{bmatrix} J_{11} \\ J_{12} \\ J_{21} \\ J_{22} \end{bmatrix}. \tag{19}$$

It is then easy to relate it to the Stokes vector $[I]$:

$$[I] = [P][D], \tag{20}$$

where

$$[P] = \begin{bmatrix} 1 & 0 & 0 & 0 \\ 0 & 0 & 0 & 1 \\ 0 & 1 & 1 & 0 \\ 0 & -i & i & 0 \end{bmatrix}.$$

It should be noted that $[J]$ is Hermitian. As a consequence, even though $[P]$ and $[D]$ are complex in Eq. (20), $[I]$ is always real.

We now describe the propagation of the matrix $[D_c]$, the coherent part of $[D]$. This can be done by using Eq. (16) and straightforward but laborious manipulations. The resultant formula for the propagation of $[D_c]$ can be compactly expressed as

$$[D_c(s)] = [T(s)] \otimes [T^*(s)][D_c(o)], \tag{21}$$

where $[A] \otimes [B]$ is the direct product[18] defined by

$$[A] \otimes [B] = \begin{bmatrix} A_{11}[B] & A_{12}[B] \\ A_{21}[B] & A_{22}[B] \end{bmatrix}$$
$$= \begin{bmatrix} A_{11}B_{11} & A_{11}B_{12} & A_{12}B_{11} & A_{12}B_{12} \\ A_{11}B_{21} & A_{11}B_{22} & A_{12}B_{21} & A_{12}B_{22} \\ A_{21}B_{11} & A_{21}B_{12} & A_{22}B_{11} & A_{22}B_{12} \\ A_{21}B_{21} & A_{21}B_{22} & A_{22}B_{21} & A_{22}B_{22} \end{bmatrix}. \tag{22}$$

In Eq. (21), it is understood that the direct product should be taken before other multiplications are performed. Using Eqs. (20) and (21), we get

$$[I_c(s)] = [Q(s)][I_c(o)], \tag{23}$$

where

$$[Q(s)] = [P][T] \otimes [T^*][P]^{-1}.$$

Equation (23) gives the Stokes vector for the coherent field at a distance s, given the Stokes vector at $s = 0$.

In view of Eq. (16), we can also write Eq. (23) as follows:

$$[Q(s)] = \sum_{j=1}^{4} [Q_j(s)]\exp(\lambda_j s). \tag{24}$$

Here,

$$[Q_1(s)] = [P][T_1] \otimes [T_1^*][P]^{-1},$$
$$[Q_2(s)] = [P][T_1] \otimes [T_2^*][P]^{-1},$$
$$[Q_3(s)] = [P][T_2] \otimes [T_1^*][P]^{-1},$$
$$[Q_4(s)] = [P][T_2] \otimes [T_2^*][P]^{-1},$$
$$\lambda_1 = \gamma_1 + \gamma_1^*, \qquad \lambda_2 = \gamma_1 + \gamma_2^*,$$
$$\lambda_3 = \gamma_1^* + \gamma_2, \qquad \lambda_4 = \gamma_2 + \gamma_2^*.$$

It is also possible to express the propagation characteristics of $[I_c]$ starting with Eq. (6) and using the eigenvalue technique. Letting λ_i and $[\beta_i]$ be the eigenvalue and the eigenvector for the extinction matrix $[A]$,

$$|[A] - \lambda_i[U]|[\beta_i] = 0, \quad (25)$$

where $[U]$ is the 4 × 4 unit matrix, we get

$$[I_c(s)] = [Q(s)][I_c(o)],$$
$$[Q(s)] = [\beta][\lambda][\beta]^{-1}, \quad (26)$$

where $[\beta] = [\beta_1\beta_2\beta_3\beta_4]$ and

$$[\lambda] = \begin{bmatrix} \exp(\lambda_1 s) & 0 & 0 & 0 \\ 0 & \exp(\lambda_2 s) & 0 & 0 \\ 0 & 0 & \exp(\lambda_3 s) & 0 \\ 0 & 0 & 0 & \exp(\lambda_4 s) \end{bmatrix}.$$

Equations (23) and (26) give the two alternative representations of the propagation characteristics of the Stokes vector. The eigenvalues λ_i in Eq. (25) for the extinction matrix $[A]$ should be identical with those appearing in Eq. (24).

EXTINCTION MATRIX, OPTICAL THEOREM, AND MUELLER MATRIX

It is tedious but straightforward to derive Eq. (6) from Eq. (10) and obtain in the process the extinction matrix $[A]$ given below:

$$[A] = \begin{bmatrix} 2\,\mathrm{Re}\,M_{11} & 0 & \mathrm{Re}\,M_{12} & \mathrm{Im}\,M_{12} \\ 0 & 2\,\mathrm{Re}\,M_{22} & \mathrm{Re}\,M_{21} & -\mathrm{Im}\,M_{21} \\ 2\,\mathrm{Re}\,M_{21} & 2\,\mathrm{Re}\,M_{12} & \mathrm{Re}(M_{11}+M_{22}) & -\mathrm{Im}(M_{11}-M_{22}) \\ -2\,\mathrm{Im}\,M_{21} & 2\,\mathrm{Im}\,M_{12} & \mathrm{Im}(M_{11}-M_{22}) & \mathrm{Re}(M_{11}+M_{22}) \end{bmatrix}. \quad (27)$$

It is also possible to express Eq. (27) in a compact form using the direct product. Toward this end, we use the following coherency matrix representations:

$$\frac{d}{ds}[J_c] = \frac{d}{ds}[E_c][\tilde{E}_c^*] + [E_c]\frac{d}{ds}[\tilde{E}_c^*]$$
$$= [M][J_c][U] + [U][J_c][\tilde{M}^*]. \quad (28)$$

From Eq. (28), we can easily get the alternative compact form

$$[A] = [P][[M] \otimes [U] + [U] \otimes [\tilde{M}^*]]. \quad (29)$$

Equation (27) also represents the generalization of the optical theorem for nonspherical particles. Note that for spherical particles $M_{11} = M_{22}$, and $M_{12} = M_{21} = 0$. Therefore, Eq. (27) reduces to $[A] = 2\,\mathrm{Re}\,M_{11}[U]$, where $2\,\mathrm{Re}\,M_{11} = -(4\pi\rho/k)\,\mathrm{Im}\,f_{11}(\hat{s},\hat{s}) = -\rho\sigma_t$, where σ_t is the extinction cross section of a single particle. Thus

$$[A][I] = -\rho\sigma_t[I] \quad (30)$$

for spherical particles, as expected.

Equation (27) is also consistent with the generalized vector optical theorem.[19,20] The sum of the first two rows of Eq. (27) represents the attenuation of the total intensity and is identified as $-\rho\sigma_{total}$, where σ_{total} is identical with the total cross section given in Eq. (1.78) of Ref. 19 for partially polarized waves.

The complete expression of the Mueller matrix $[S]$ is known[1] and is given by

$$[S] = \begin{bmatrix} |f_{11}|^2 & |f_{12}|^2 & \mathrm{Re}(f_{11}f_{12}^*) & -\mathrm{Im}(f_{11}f_{12}^*) \\ |f_{21}|^2 & |f_{22}|^2 & \mathrm{Re}(f_{21}f_{22}^*) & -\mathrm{Im}(f_{21}f_{22}^*) \\ 2\,\mathrm{Re}(f_{11}f_{21}^*) & 2\,\mathrm{Re}(f_{12}f_{22}^*) & \mathrm{Re}(f_{11}f_{22}^* + f_{12}f_{21}^*) & -\mathrm{Im}(f_{11}f_{22}^* - f_{12}f_{21}^*) \\ 2\,\mathrm{Im}(f_{11}f_{21}^*) & 2\,\mathrm{Im}(f_{12}f_{22}^*) & \mathrm{Im}(f_{11}f_{22}^* + f_{12}f_{21}^*) & \mathrm{Re}(f_{11}f_{22}^* - f_{12}f_{21}^*) \end{bmatrix}. \quad (31)$$

Here we present a compact expression of $[S]$ using the direct product. We start with Eq. (1), and, following the detailed procedure used in obtaining Eq. (23), we get the following:

$$[S(\hat{s},\hat{s}')] = [P][f(\hat{s},\hat{s}')] \otimes [f^*(\hat{s},\hat{s}')][P]^{-1}. \quad (32)$$

It is easy to show that Eqs. (31) and (32) are identical.

FIRST-ORDER SCATTERING SOLUTION FOR THE STOKES VECTOR

In the first-order scattering theory, the wave is assumed to propagate with the propagation constant of the coherent wave, to be scattered by the elementary volume dV of the medium once, and then to propagate to the observation point with the propagation constant of the coherent wave. The total scattered wave is then the sum of all the scattered waves from all points of the medium. The first-order scattering is applicable if the optical depth of the medium is much less than unity or if the medium is absorbing. For rain medium at 30 GHz, for example, the albedo is approximately 0.5, and the first-order solution is close to the complete radiative-transfer solution.[2]

By using the expression for the propagation of the Stokes vector, the first-order solution for the Stokes vector can be expressed in the following matrix form. Let us consider the plane-parallel medium (Fig. 2). The transmitted $[I_t(d,\hat{s})]$ and the backscattered $[I_b(o,\hat{s})]$ Stokes vectors are given by

$$[I_t(d,\hat{s})] = \int_0^d [Q_2][S(\hat{s},\hat{s}_0)][Q_1][I_0]\rho ds,$$

$$[I_b(o,\hat{s})] = \int_0^d [Q_3][S(\hat{s},\hat{s}_0)][Q_1][I_0]\rho ds, \quad (33)$$

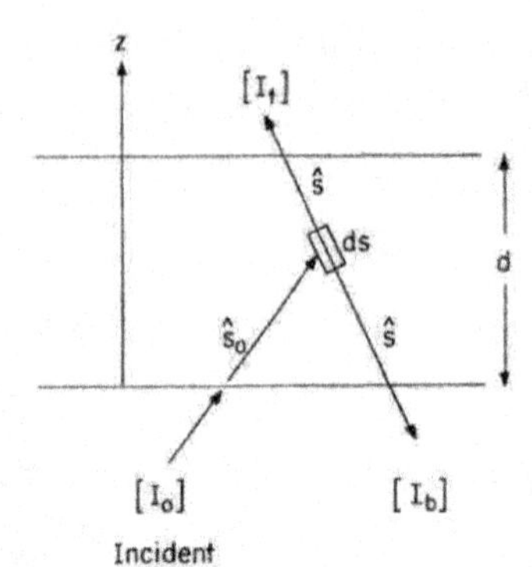

Fig. 2. First-order scattering for the transmitted $[I_t]$ and the backscattered $[I_b]$ Stokes vectors.

where

$$[I_0] = [I_0(\hat{s}_0)] \text{ is the incident Stokes vector,}$$

$$[Q_1] = \sum_{i=1}^{4} [Q_i(\hat{s}_0)]\exp\left(\lambda_i \frac{z}{\mu_0}\right),$$

$$[Q_2] = \sum_{j=1}^{4} [Q_j(\hat{s})]\exp\left(\lambda_j \frac{d-z}{\mu}\right),$$

$$[Q_3] = \sum_{j=1}^{4} [Q_j(\hat{s})]\exp\left(-\lambda_j \frac{z}{\mu}\right).$$

Note that the Stokes vector $[I_0]$ is incident upon the medium, propagates to ds represented by $[Q_1][I_0]$, is scattered by the medium in ds represented by $[S]$, and then propagates to the receiver represented by $[Q_2]$ or $[Q_3]$.

Performing the integrations in Eqs. (33), we get the final expressions

$$[I_t(d, \hat{s})] = \sum_{i=1}^{4} \sum_{j=1}^{4} \rho[Q_j(\hat{s})][S(\hat{s}, \hat{s}_0)][Q_i(\hat{s}_0)][G_{tij}][I_0],$$

$$[I_b(o, \hat{s})] = \sum_{i=1}^{4} \sum_{j=1}^{4} \rho[Q_j(\hat{s})][S(\hat{s}, \hat{s}_0)][Q_i(\hat{s}_0)][G_{bij}][I_0], \quad (34)$$

where

$$[G_{tij}] = \frac{\exp\left(\lambda_i \frac{d}{\mu_0}\right) - \exp\left(\lambda_j \frac{d}{\mu}\right)}{\frac{\lambda_i}{\mu_0} - \frac{\lambda_j}{\mu}} \frac{1}{\mu}, \qquad \mu > 0,$$

$$[G_{bij}] = \frac{\exp\left(\lambda_i \frac{d}{\mu_0} - \lambda_j \frac{d}{\mu}\right) - 1}{\frac{\lambda_i}{\mu_0} - \frac{\lambda_j}{\mu}} \frac{1}{(-\mu)}, \qquad \mu < 0.$$

Equations (34) are a compact matrix representation of the first-order scattering solution of the Stokes vector.

FOURIER EXPANSION OF VECTOR RADIATIVE-TRANSFER EQUATION

In our previous papers,[2,4,5] we described the complete solution of the equation of transfer (7) for the case of spherical particles. We expanded Eq. (7) in a Fourier series in ϕ. We then noted that, if the particles are spherical or if the particles are axially symmetric with the axis pointed in the z direction, then the extinction matrix $[\Lambda]$ becomes independent of ϕ and the Mueller matrix $[S]$ becomes a function of $\phi - \phi'$.

In these cases, all the coupling among the Fourier components disappears, and the Fourier components of the equation of transfer become independent of one another. We can then solve the equation for each Fourier component separately.[2,4,5]

In a general case of nonspherical particles, all Fourier components are coupled to one another, as indicated below, and detailed numerical studies have not been reported.

Let us expand Eq. (7) in Fourier series in ϕ as follows:

$$[I_d] = \sum_m [I]_m \exp(-jm\phi),$$

$$[\Lambda] = \sum_m [\Lambda]_m \exp(-jm\phi),$$

$$[S] = \sum_m \sum_{m'} [S]_{m,m'} \exp(-jm\phi + jm'\phi'),$$

$$[I_i] = \sum_m [I_i]_m \exp(-jm\phi), \quad (35)$$

where

$$[I]_m = [I]_{-m}^*, \qquad [\Lambda]_m = [\Lambda]_{-m}^*,$$
$$[I_i]_m = [I]_{-m}^*, \qquad [S]_{m,m'} = [S]_{-m,-m'}^*.$$

We then obtain

$$\mu \frac{d}{dz} [I]_m = \sum_{m'} [\Lambda]_{m-m'} [I]_{m'} + \sum_{m'} \int [S]_{m,m'} [I']_{m'} d\mu' + [I_i]_m. \quad (36)$$

Note that, for spherical particles and z-oriented axially symmetric particles, $[\Lambda]_m = 0$ for all $m \neq 0$. In this special case, $[S]$ becomes a single summation $\Sigma_m [S]_m \exp[-jm(\phi - \phi')]$.

CIRCULAR-POLARIZATION REPRESENTATION

Instead of the vertical and horizontal polarization E_v and E_h, we can also use a circular-polarization (CP) representation consisting of the right-handed E_+ and the left-handed E_- polarizations:

$$[E_{CP}] = \begin{bmatrix} E_+ \\ E_- \end{bmatrix} = [C_p] \begin{bmatrix} E_v \\ E_h \end{bmatrix},$$

$$[C_p] = \frac{1}{2^{1/2}} \begin{bmatrix} 1 & -i \\ 1 & +i \end{bmatrix}. \quad (37)$$

The corresponding Stokes vector $[I_{CP}]$ for the CP representation is related to $[I]$ by

$$[I_{CP}] = [Q_{CP}][I], \quad (38)$$

where

$$[I_{CP}] = \begin{bmatrix} I_c \\ I_+ \\ I_- \\ I_c^* \end{bmatrix} = \begin{bmatrix} \langle E_+E_-^* \rangle \\ \langle E_+E_+^* \rangle \\ \langle E_-E_-^* \rangle \\ \langle E_+^*E_- \rangle \end{bmatrix}, \qquad [Q_{CP}] = \frac{1}{2}\begin{bmatrix} 1 & -1 & -i & 0 \\ 1 & 1 & 0 & -1 \\ 1 & 1 & 0 & 1 \\ 1 & -1 & i & 0 \end{bmatrix}.$$

For example, by using $[I_{CP}]$, Eq. (23) can be rewritten as

$$[I_{CP}(S)] = [Q_{CP}][Q][Q_{CP}]^{-1}[I_{CP}(o)]. \tag{39}$$

LIMITATION OF THE RADIATIVE-TRANSFER THEORY

The radiative-transfer theory described in this paper is applicable when the particle density is approximately less than 1% in volume. If the particle sizes are less than a wavelength, the effective cross section decreases as the density increases, while the cross section increases slightly with the density if the particle sizes are greater than a wavelength. These effective cross sections must be taken into account for high-density cases.[17,21,22]

CONCLUSION

In this paper, we have presented a general matrix formulation of the vector radiative transfer in Eqs. (4) and (7). The propagation characteristics of the coherent field are expressed in terms of the field in Eq. (16), the coherency matrix in Eq. (18), and the Stokes vector in Eq. (23), (24), or (26). The extinction matrix, the optical theorem, and the Mueller matrix are given, and the general first-order scattering solution in Eqs. (34) of the propagation of the Stokes vector has been carried out. Included also are the discussion on Fourier expansion of the equation of transfer, a CP representation, and the limitations of the radiative transfer theory.

ACKNOWLEDGMENT

This research was supported by the U.S. Army Research Office.

REFERENCES

1. A. Ishimaru, *Wave Propagation and Scattering in Random Media* (Academic, New York, 1978), Vol. I.
2. A. Ishimaru and R. L.-T. Cheung, "Multiple scattering effects on wave propagation due to rain," Ann. Telecommun. **35**, 373–379 (1980).
3. A. Ishimaru and R. L.-T. Cheung, "Multiple-scattering effect on radiometric determination of rain attenuation at millimeter wavelengths," Radio Sci. **15**, 507–516 (1980).
4. A. Ishimaru, R. Woo, J. W. Armstrong, and D. Blackman, "Multiple scattering calculations of rain effects," Radio Sci. **17**, 1425–1433 (1982) (special issue on NASA propagation studies).
5. R. L.-T. Cheung and A. Ishimaru, "Transmission, backscattering and depolarization of waves in randomly distributed spherical particles," Appl. Opt. **21**, 3792–3798 (1982).
6. T. Oguchi, "Scattering from hydrometeors: a survey," Radio Sci. **16**, 691–730 (1981).
7. C. Yeh, R. Woo, J. W. Armstrong, and A. Ishimaru, "Scattering by single ice needles and plates at 30 GHz," Radio Sci. **17**, 1503–1510 (1982).
8. C. Yeh, R. Woo, J. W. Armstrong, and A. Ishimaru, "Scattering by Pruppacher-Pitter raindrops at 30 GHz," Radio Sci. **17**, 575–765 (1982).
9. W.-M. Boerner, A. K. Jordan, and I. W. Kay, "Introduction to the special issue on inverse methods in electromagnetics," IEEE Trans. Antennas Propag. **AP-29**, 185–191 (1981).
10. W.-M. Boerner, M. B. El-Arini, C.-Y. Chan, and P. M. Mastoris, "Polarization dependence in electromagnetic inverse problems," IEEE Trans. Antennas Propag. **AP-29**, 262–271 (1981).
11. J. R. Huynen, "Phenomenological theory of radar targets," Ph.D. Thesis (Technical University of Delft, Rotterdam, The Netherlands, 1970).
12. R. H. Lang, "Electromagnetic backscattering from a sparse distribution of lossy dielectric scatterers," Radio Sci. **16**, 15–30 (1981).
13. T. Tsang, M. C. Kubacsi, and J. A. Kong, "Radiative transfer theory for active remote sensing of a layer of small ellipsoidal scatterers," Radio Sci. **16**, 321–329 (1981).
14. S. Chandrasekhar, *Radiative Transfer* (Oxford U. Press, London, 1950).
15. T. Oguchi, "Electromagnetic wave propagation and scattering in rain and other hydrometeors," Proc. IEEE **71**, 1029–1078 (1983).
16. H. Goldstein, *Classical Mechanics* (Addison-Wesley, Reading, Mass., 1981).
17. A. Ishimaru and Y. Kuga, "Attenuation constant of a coherent field in a dense distribution of particles," J. Opt. Soc. Am. **72**, 1317–1320 (1982).
18. B. A. Robson, *Polarization Phenomena* (Clarendon, Oxford, 1974).
19. R. G. Newton, *Scattering Theory of Waves and Particles* (McGraw-Hill, New York, 1966).
20. M. A. Karam and A. K. Fung, "Vector forward scattering theorem," Radio Sci. **17**, 752–756 (1982).
21. V. N. Bringi, V. V. Varadan, and V. K. Varadan, "Coherent wave attenuation by a random distribution of particles," Radio Sci. **17**, 946–952 (1982).
22. L. Tsang, J. A. Kong, and T. Habashy, "Multiple scattering of acoustic waves by random distribution of discrete spherical scatterers with the quasicrystalline and Percus–Yevick approximation," J. Acoust. Soc. Am. **71**, 552–558 (1982).

Radio Science, Volume 19, Number 5, Pages 1356–1366, September–October 1984

Multiple scattering calculations for nonspherical particles based on the vector radiative transfer theory

A. Ishimaru[1]

Electrical Engineering Department, University of Washington, Seattle

D. Lesselier

Groupe d'Electromagnetisme, Laboratoire des Signaux et Systemes, CNRS-ESE, Gif-sur-Yvette, France

C. Yeh[1]

Electrical Engineering Department, University of California at Los Angeles, California

(Received December 13, 1983; revised April 30, 1984; accepted May 14, 1984.)

On the basis of the vector radiative transfer equation, multiple scattering calculations were performed for an obliquely incident linearly polarized plane wave upon a plane-parallel slab consisting of arrays of vertically oriented, uniformly distributed nonspherical spheroidal particles of identical size. Results are given in terms of the Stokes' parameters for the incoherent field. Owing to the symmetry of the particles and their orientation, decoupling of the Fourier expansion coefficients in the solution of the radiative transfer equation occurs. Each Fourier component satisfies a single equation which is then solved by the Gauss' quadrature and the eigenvalue-eigenvector technique. The behavior of the forward scattered incoherent field is investigated for low-loss as well as high-loss particles, for different densities and shapes, and for various angles of incidence. Comparison with the results from the approximate first-order theory shows good quantitative agreement for thin and sparcely populated layers.

1. INTRODUCTION

Successful multiple scattering calculations based on the vector radiative transfer theory have been performed for an ensemble of spherical particles recently [*Cheung and Ishimaru*, 1982; *Ishimaru et al.*, 1982]. Here, we shall consider the case for an ensemble of nonspherical particles (spheroids). Identical and uniformly distributed spheroids are assumed to be aligned along the z axis of a plane-parallel slab of thickness d. (The z axis is perpendicular to the slab interface.) A linearly polarized coherent plane wave is assumed to be incident obliquely upon this slab. The resultant incoherent field is calculated according to the vector radiative transfe. theory [*Chandrasekhar*, 1950; *Ishimaru*, 1978]. Solution of the integrodifferential equation of radiative transfer gives the Stokes vector (i.e., the incoherent intensity matrix) of this field as a function of the scattering amplitudes of a single particle [*Yeh et al.*, 1982*a*, *b*] and of the slab density and thickness. Due to the symmetry of the particles and their orientation, decoupling of the Fourier expansion coefficients in the solution of the radiative transfer equation occurs [*Ishimaru and Yeh*, 1984]. Each Fourier component satisfies a single equation which is then solved by the Gauss' quadrature and the eigenvalue-eigenvector technique. The behavior of the forward scattered incoherent field is investigated for low-loss as well as high-loss particles, for different densities and shapes and for various angles of incidence. Comparison will also be made with the result obtained according to the first-order expressions (A. Ishimaru et al., unpublished manuscript, 1984).

Other notable references on the scattering of single particles and on multiple scattering by an ensemble of spherical particles are *Mon*, [1982]; *Olsen*, [1982]; *Oguchi* [1983]; *Stuebing* [1983]; *Lang* [1981]; *Tsang et al.* [1981]; *Varadan et al.* [1983]; *Varadan and Varadan* [1980]; *Karam and Fung* [1983].

[1]Also at EMtec Engineering, Inc., Los Angeles, California.

Paper number 4S0736.
0048-6604/84/004S-0736$08.00

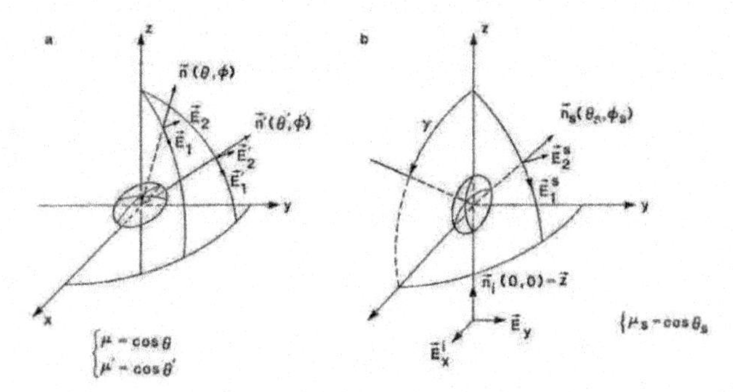

Fig. 1. Single-particle scattering geometry.

2. ANALYTICAL FORMULATION

Consider a layer of spheroidal particles which extends from $z = 0$ to $z = d$, as shown in Figure 1. We shall assume that both boundaries of the scattering medium are flat and that the particles are homogeneously distributed with a uniform particle density. These spheroidal particles are all aligned along the z axis and oriented such that the z axis is the axis of symmetry. A linearly polarized coherent plane wave is assumed to be incident along the direction $\mathbf{n}_i(\theta_i, \phi_i)$. Consider now the case of the scattering of plane wave by a single particle. Representing the vertical and horizontal components of the incident electric field by (E'_1, E'_2) in the incident direction $\mathbf{n}'(\theta', \phi')$ and the vertical and horizontal components of the scattered electric field by (E_1, E_2) in the direction $\mathbf{n}(\theta, \phi)$, one may define the following relationship according to *Van de Hulst* [1957] (see Figure 1*a*):

$$\begin{bmatrix} E_1 \\ E_2 \end{bmatrix} = \frac{e^{ikR}}{R} \begin{bmatrix} f_{11} & f_{12} \\ f_{21} & f_{22} \end{bmatrix} \begin{bmatrix} E'_1 \\ E'_2 \end{bmatrix} \tag{1}$$

where $f_{ij}(\phi, \phi'; \phi, \phi')$ are complex scattering amplitudes. For axially symmetric particles, the scattering amplitudes f_{11} and f_{22} are even functions of $(\phi - \phi')$ while the scattering amplitudes f_{12} and f_{21} are odd functions of $(\phi - \phi')$, and these functions are functions of $(\phi - \phi')$ only, i.e.,

$$f_{ij}(\theta, \theta'; \phi, \phi') = f_{ij}(\theta, \theta'; \phi - \phi') \tag{2}$$

The scattering functions $f_{ij}(\theta, \theta'; \phi - \phi')$ are also related to the particle scattering functions for a single particle $\tilde{f}_{ij}(\theta_s, \phi_s, \alpha = \theta')$ as follows (see Figure 1*b*):

$$[f_{ij}(\theta, \theta'; \phi - \phi')] = \pm[R(\beta)][\tilde{f}_{ij}(\theta_s, \phi_s, \alpha = \theta')] \tag{3}$$

where

$$[f_{ij}] = \begin{bmatrix} f_{11} & f_{12} \\ f_{21} & f_{22} \end{bmatrix} \qquad [\tilde{f}_{ij}] = \begin{bmatrix} \tilde{f}_{11} & \tilde{f}_{12} \\ \tilde{f}_{21} & \tilde{f}_{22} \end{bmatrix}$$

$$[R(\beta)] = \begin{bmatrix} \cos\beta & \sin\beta \\ -\sin\beta & \cos\beta \end{bmatrix}$$

$$\cos\theta_s = \cos\theta\cos\theta' + \sin\theta\sin\theta'\cos(\phi - \phi')$$

$$(0 \le \theta_s < \pi)$$

$$\cos\beta = [\sin\theta\cos\theta' - \sin\theta'\cos\theta\cos(\phi - \phi')]/\sin\theta_s$$

$$\cos\gamma_s = [\cos\theta'\sin\theta\cos(\phi - \phi') - \cos\theta\sin\theta']/\sin\theta_s$$

$$\phi_s = \gamma_s + \pi \tag{4}$$

The signs of γ_s and β are equal to the sign of $\sin(\phi - \phi')$. In equation (3) the upper sign (plus) refers to the case $\theta' = 0$ and the lower sign (minus) refers to the case $\theta' \neq 0$.

Let us consider the following special cases:

1. For a normally incident plane wave ($\theta' = 0$), one has

$$\theta_s = \theta \qquad \beta = 0 \qquad \phi_s = \phi - \phi'$$

$$\begin{aligned} [f_{ij}(\theta, 0, \phi - \phi')] &= [R(0)][\tilde{f}_{ij}(\theta, \phi - \phi', 0)] \\ &= [\tilde{f}_{ij}(\theta, \phi - \phi', 0)] \\ &= [\tilde{f}_{ij}(\theta, 0, 0)][R(\phi - \phi')] \end{aligned} \tag{5}$$

2. For the forward scattering case ($\theta = \theta'$, $\phi = \phi'$), one has

$$\theta_s = 0 \qquad \beta = 0 \qquad \phi_s = \pi$$
$$[f_{ij}(\theta', \theta', 0)] = \text{a diagonal matrix with } f_{ii} = \tilde{f}_{ii}(0, 0, \theta) \tag{6}$$

3. For the backward scattering case ($\theta + \theta' = \pi$,

$\phi - \phi' = \pi$), one has

$$\theta_s = \pi \qquad \beta = \pi \qquad \phi_s = \pi$$

$$[f_{ij}(\pi - \theta', \theta', \pi)] \tag{7}$$

$$= \text{a diagonal matrix with } f_{ii} = f_{ii}(\pi, \pi, \theta')$$

Consider now the modified Stokes' parameters (I_1, I_2, U, V): [*Ishimaru*, 1978]

$$I_1 = \langle E_1 E_1^* \rangle, \; I_2 = \langle E_2 E_2^* \rangle$$
$$U = 2\,\mathrm{Re}\,\langle E_1 E_2^* \rangle \qquad V = 2\,\mathrm{Im}\,\langle E_1 E_2^* \rangle \tag{8}$$

where E_1 and E_2 are the electric field components in the θ and ϕ directions, respectively (Figure 1), Re and Im denote the real part and the imaginary part, and the asterisk means the complex conjugate. In other words, E_1 and E_2 are, respectively, the vertical and the horizontal components of the electric field.

The electric field components (E_{c1}, E_{c2}) for the coherent wave which propagates in the direction of the incident wave (θ_i, $\phi_i = 0$), satisfy the following equation: [*Olsen*, 1982; *Ishimaru and Yeh*, 1984; *Oguchi*, 1983].

$$\frac{d}{ds}\begin{bmatrix} E_{c1} \\ E_{c2} \end{bmatrix} = i\lambda\rho \begin{bmatrix} f_{11} & f_{12} \\ f_{21} & f_{22} \end{bmatrix} \begin{bmatrix} E_{c1} \\ E_{c2} \end{bmatrix} \tag{9}$$

where $s = z/\mu_i$ and $\mu_i = \cos\theta_i$, and f_{ij} are the forward scattering amplitudes in the direction (θ_i, $\phi_i = 0$); λ is the wavelength and ρ is the particle density. Solving (9) gives

$$\begin{bmatrix} E_{c1}(z) \\ E_{c2}(z) \end{bmatrix} = \begin{bmatrix} \exp(i\lambda\rho \tilde{f}_{11}(0, 0, \theta_i)z/\mu_i) & 0 \\ 0 & \exp(i\lambda\rho \tilde{f}_{22}(0, 0, \theta_i)z/\mu_i) \end{bmatrix} \cdot \begin{bmatrix} E_{c1}(0) \\ E_{c2}(0) \end{bmatrix} \tag{10}$$

For an incident x-polarized (vertically polarized) or y-polarized (horizontally polarized) plane wave, the coherent intensity matrix $[I_c]$ at z and in the direction (θ_i, 0) are, respectively,

$$[I_c(z, \mu_i)] = \begin{bmatrix} \exp(\gamma_{11} z/\mu_i) \\ 0 \\ 0 \\ 0 \end{bmatrix} \tag{11}$$

or

$$[I_c(z, \mu_i)] = \begin{bmatrix} 0 \\ \exp(\gamma_{22} z/\mu_i) \\ 0 \\ 0 \end{bmatrix} \tag{12}$$

with

$$\gamma_{11} = -2\lambda\rho\,\mathrm{Im}\,[\tilde{f}_{11}(0, 0, \theta_i)]$$
$$\gamma_{22} = -2\lambda\rho\,\mathrm{Im}\,[\tilde{f}_{22}(0, 0, \theta_i)]$$

According to the radiative transfer theory, the intensity matrix $[I(z, \mu, \phi)]$ for the incoherent wave generated by multiple scattering effect in the direction ($\mu = \cos\theta$, ϕ) satisfies the following integrodifferential equation [*Ishimaru and Yeh*, 1984].

$$\mu \frac{d}{dz}[I(z, \mu, \phi)] = [E(\mu)][I(z, \mu, \phi)] + \int_{-1}^{+1}\int_0^{2\pi} [S(\mu, \mu', \phi - \phi')][I(z, \mu', \phi')]\, d\mu'\, d\phi' + [I_i(z, \mu, \phi)] \tag{13}$$

and the boundary conditions

$$[I(z = 0, \mu, \phi)] = 0 \qquad 0 \le \mu \le 1 \tag{14}$$
$$[I(z = d, \mu, \phi)] = 0 \qquad 0 \ge \mu \ge -1 \tag{15}$$

The various symbols in (13) are given as follows:

1. $[E]$ is a 4×4 extinction matrix which is composed of the forward scattering amplitudes f_{11} and f_{22} in the direction (μ, ϕ); i.e., $f_{ii}(\mu, \mu' = \mu, 0) = \tilde{f}_{ii}(0, 0, \mu)$, with $ii = 11$ or 22. So,

$$[E(\mu)] = \lambda\rho \begin{bmatrix} -2\,\mathrm{Im}\,(f_{11}) & 0 & 0 & 0 \\ 0 & -2\,\mathrm{Im}\,(f_{22}) & 0 & 0 \\ 0 & 0 & -\mathrm{Im}\,(f_{11} + f_{22}) & \mathrm{Re}\,(f_{11} - f_{22}) \\ 0 & 0 & 0 & -\mathrm{Im}\,(f_{11} + f_{22}) \end{bmatrix} \tag{16}$$

For spherical particles, $[E(\mu)] = -\lambda\rho\sigma_t[I]$ where $[I]$ is the 4×4 unit matrix and σ_t as the total scattering cross section for a sphere (i.e., $\sigma_t = 2\ \mathrm{Im}\ (f_{11}) = 2\ \mathrm{Im}\ (f_{22})$).

2. $[S]$ is a 4×4 scattering matrix which is composed of the scattering amplitudes, $f_{ij}(\mu, \mu'; \phi - \phi')$:

$$[S(\mu, \mu', \phi - \phi')] = \begin{bmatrix} [S_1(\mu, \mu', \phi - \phi')] & [S_2(\mu, \mu', \phi - \phi')] \\ [S_3(\mu, \mu', \phi - \phi')] & [S_4(\mu, \mu', \phi - \phi')] \end{bmatrix} \quad (17)$$

where $[S_1]$ and $[S_4]$ are even functions of $\phi - \phi'$ and are given by the following

$$[S_1] = \rho \begin{bmatrix} |f_{11}|^2 & |f_{12}|^2 \\ |f_{21}|^2 & |f_{22}|^2 \end{bmatrix} \quad (18)$$

$$[S_4] = \rho \begin{bmatrix} \mathrm{Re}\ (f_{11} f_{22}^* + f_{12} f_{21}^*) & -\mathrm{Im}\ (f_{11} f_{22}^* - f_{12} f_{21}^*) \\ \mathrm{Im}\ (f_{11} f_{22}^* + f_{12} f_{21}^*) & \mathrm{Re}\ (f_{11} f_{22}^* - f_{12} f_{21}^*) \end{bmatrix} \quad (19)$$

and $[S_2]$ and $[S_3]$ are odd functions of $\phi - \phi'$ and are given by the following

$$[S_2] = \rho \begin{bmatrix} \mathrm{Re}\ (f_{11} f_{12}^*) & -\mathrm{Im}\ (f_{11} f_{12}^*) \\ \mathrm{Im}\ (f_{11} f_{12}^*) & -\mathrm{Im}\ (f_{21} f_{22}^*) \end{bmatrix} \quad (20)$$

$$[S_3] = \rho \begin{bmatrix} \mathrm{Re}\ (f_{11} f_{21}^*) & \mathrm{Re}\ (f_{12} f_{22}^*) \\ \mathrm{Im}\ (f_{11} f_{21}^*) & \mathrm{Im}\ (f_{12} f_{22}^*) \end{bmatrix} \quad (21)$$

3. $[I_i]$ is a 4×1 source matrix which represents the incident coherent wave in the direction (μ, ϕ). For $\mu' = \mu_i$, $\phi' = 0$, one has

$$[I_i(z, \mu, \phi)] = [S(\mu, \mu_i, \phi)][I_c(z, \mu_i)] \quad (22)$$

Substituting (11) or (12) into (22) gives

$$[I_i(z, \mu, \phi)] = [F(\mu, \mu_i, \phi)] \exp\ (\gamma_{ii} z/\mu_i) \quad (23)$$

where the matrix $[F]$ has been partitioned into upper block, which contains elements F_1 and F_2 that are even functions of ϕ, and lower block, which contains elements F_3 and F_4 that are odd functions of ϕ, i.e.,

$$[F] = \begin{bmatrix} F_1 \\ F_2 \\ F_3 \\ F_4 \end{bmatrix}$$

For an incident plane wave which is polarized in the x direction, one has

$$\begin{bmatrix} F_1 \\ F_2 \end{bmatrix} = \begin{bmatrix} \rho|f_{11}|^2 \\ \rho|f_{21}|^2 \end{bmatrix} \quad (24)$$

$$\begin{bmatrix} F_3 \\ F_4 \end{bmatrix} = \begin{bmatrix} \rho 2\ \mathrm{Re}\ (f_{11} f_{21}^*) \\ \rho 2\ \mathrm{Im}\ (f_{11} f_{21}^*) \end{bmatrix} \quad (25)$$

with $\gamma_{ii} = \gamma_{11} = -2\lambda\rho\ \mathrm{Im}\ (\bar{f}_{11}(0, 0, \theta_i))$ and F_i and f_{ij} are functions of (μ, μ_i, ϕ); for an incident plane wave which is polarized in the y direction, one has

$$\begin{bmatrix} F_1 \\ F_2 \end{bmatrix} = \begin{bmatrix} \rho|f_{12}|^2 \\ \rho|f_{22}|^2 \end{bmatrix} \quad (26)$$

$$\begin{bmatrix} F_3 \\ F_4 \end{bmatrix} = \begin{bmatrix} \rho 2\ \mathrm{Re}\ (f_{12} f_{22}^*) \\ \rho 2\ \mathrm{Im}\ (f_{12} f_{22}^*) \end{bmatrix} \quad (27)$$

with $\gamma_{ii} = \gamma_{22} = -2\lambda\rho\ \mathrm{Im}\ (\bar{f}_{22}(0, 0, \theta_i))$.

Let us now consider the solution of (13). For the special situation stated above, because of the even-odd symmetry with respect to ϕ, one may expand the matrices $[I]$ and $[F]$ in terms of even-odd Fourier series as follows [*Cheung and Ishimaru*, 1982; *Ishimaru et al.*, 1982]:

$$[I(z, \mu, \phi)] = \begin{bmatrix} \sum_{m=0}^{\infty} I_{m1}(z, \mu) \cos m\phi \\ \sum_{m=0}^{\infty} I_{m2}(z, \mu) \cos m\phi \\ \sum_{m=1}^{\infty} U_m(z, \mu) \sin m\phi \\ \sum_{m=1}^{\infty} V_m(z, \mu) \sin m\phi \end{bmatrix} \quad (28)$$

and

$$[F(\mu, \mu_i, \phi)] = \begin{bmatrix} \sum_{m=0}^{\infty} F_{m1}(\mu, \mu_i) \cos m\phi \\ \sum_{m=0}^{\infty} F_{m2}(\mu, \mu_i) \cos m\phi \\ \sum_{m=1}^{\infty} F_{m3}(\mu, \mu_i) \sin m\phi \\ \sum_{m=1}^{\infty} F_{m4}(\mu, \mu_i) \sin m\phi \end{bmatrix} \quad (29)$$

So, upon substitution into (13), one has

$$\mu \frac{d}{ds} [I(z, \mu)]_m = [E(\mu)][I(z, \mu)]_m + \int_{-1}^{+1} [L(\mu, \mu')]_m [I(z, \mu')]_m\, d\mu' + [F(\mu, \mu_i)]_m \exp\ (\gamma_{ii} z/\mu_i) \quad (30)$$

with the boundary conditions

$$[I(0, \mu)]_m = 0 \qquad 0 \le \mu \le +1$$
$$[I(d, \mu)]_m = 0 \qquad 0 \ge \mu \ge -1 \quad (31)$$

The extinction matrix is given by (16) and

$$[I(z, \mu)]_m = \begin{bmatrix} I_{m1}(z, \mu) \\ I_{m2}(z, \mu) \\ U_m(z, \mu) \\ V_m(z, \mu) \end{bmatrix} \tag{32}$$

$$[L(\mu, \mu')]_m = \begin{bmatrix} [S_1(\mu, \mu')]_m & [S_2(\mu, \mu')]_m \\ -[S_3(\mu, \mu')]_m & [S_4(\mu, \mu')]_m \end{bmatrix} \tag{33}$$

$$[S_k(\mu, \mu')]_m = 2 \int_0^{\pi} [S(\mu, \mu', \phi - \phi')]_k p(m(\phi' - \phi))\, d(\phi' - \phi) \tag{34}$$

$$\begin{aligned} p(m(\phi' - \phi)) &= \cos m(\phi' - \phi) \quad \text{if } k = 1, 4 \\ &= \sin m(\phi' - \phi) \quad \text{if } k = 2, 3 \end{aligned} \tag{35}$$

$$[F(\mu, \mu_i)]_m = \begin{bmatrix} F_{m1}(\mu, \mu_i) \\ F_{m2}(\mu, \mu_i) \\ F_{m3}(\mu, \mu_i) \\ F_{m4}(\mu, \mu_i) \end{bmatrix} \tag{36}$$

$$F_{mk}(\mu, \mu_i) = \begin{Bmatrix} 1 & \text{if } m = 0 \\ 2 & \text{if } m \neq 0 \end{Bmatrix} \int_0^{\pi} F_k(\mu, \mu_i, \phi) q(m\phi)\, d\phi \tag{37}$$

$$\begin{aligned} q(m\phi) &= \cos m\phi \quad \text{if } k = 1, 2 \\ &= \sin m\phi \quad \text{if } k = 3, 4 \end{aligned} \tag{38}$$

$[S_k]$ are given by (18)–(21), and F_k are given by (24)–(27).

The integral in (30) can be approximated by the Gauss' quadrature technique [*Dahlquist and Björk*, 1974; *Ishimaru*, 1978] as follows:

$$\begin{aligned} \int_{-1}^{+1} [L(\mu, \mu')]_m [I(z, \mu')]_m\, d\mu' &= \sum_{k=1}^{N} a_k \{[L(\mu, \mu_k)]_m [I(z, \mu_k)]_m \\ &+ [L(\mu, \mu_{-k})]_m [I(z, \mu_{-k})]_m\} \end{aligned} \tag{39}$$

where $\mu_k (1 > \mu_1 > \mu \cdots > \mu_N > 0;\ \mu_{-k} = -\mu_k)$ are the roots of the Legendre polynomial of degree $2N$ and a_k are the Gauss' quadrature expansion coefficients. Substituting (39) into (30) and simplifying yields a set of K first-order differential equations at discrete value μ_k (with $\pm k = 1, 2, \cdots, N$; $K = 8N (m \neq 0)$ or $K = 4N (m = 0)$):

$$\frac{d}{dz}[\chi(z)]_m + [\xi]_m = [\chi_i]_m \exp(\gamma_{ii} z/\mu_i) \tag{40}$$

with

$$[\chi(z)]_m = \begin{bmatrix} [I(z, \mu_N)]_m \\ \vdots \\ [I(z, \mu_{-N})]_m \end{bmatrix} \tag{41}$$

$$[\chi_i]_m = \begin{bmatrix} [F(\mu_N, \mu_i)]_m \\ \vdots \\ [F(\mu_{-N}, \mu_i)]_m \end{bmatrix} \tag{42}$$

$$[\xi]_m = \begin{bmatrix} [\kappa(\mu_N, \mu_N)]_m \cdots [\kappa(\mu_N, \mu_{-N})]_m \\ [\kappa(\mu_{-N}, \mu_N)]_m \cdots [\kappa(\mu_{-N}, \mu_{-N})]_m \end{bmatrix} \tag{43}$$

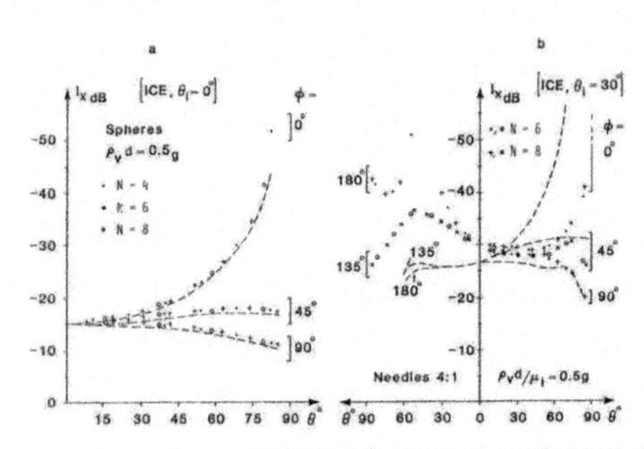

Fig. 2. I_x versus θ for various values of N. Ice spheres and ice needles were considered. First-order results are shown by dashed lines.

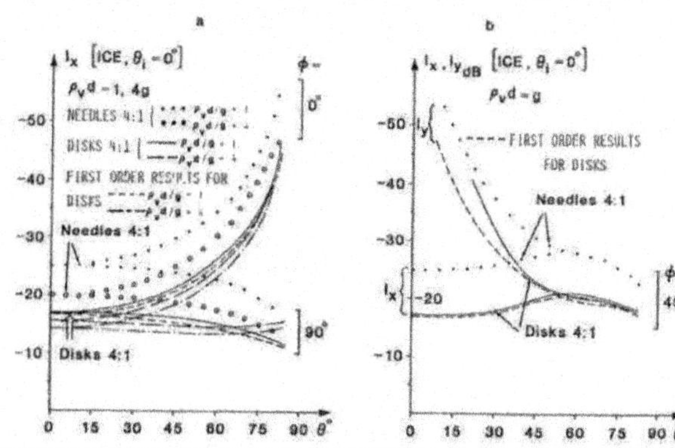

Fig. 3. Co- and cross-polarized intensities versus θ for various $\rho_v d/g$. First-order results for ice disks are also shown.

$$[\kappa(\mu_k, \mu_l)]_m = -\frac{1}{\mu_k}\{a_l[L(\mu_k, \mu_l)]_m - \delta_{k,l}[E(\mu_k)]\}$$

$$\delta_{k,l} = 1 \qquad (k = l)$$

$$\delta_{k,l} = 0 \qquad (k \neq l) \tag{44}$$

The boundary conditions now become

$$[I(0, \mu_k)]_m = 0$$

$$[I(d, \mu_{-k})]_m = 0$$

$$k = 1, 2, \cdots, N \tag{45}$$

The boundary-value problem (i.e., (40)–(45)) is now solved by the eigenvalue-eigenvector technique [*Ishimaru*, 1978; *Ishimaru et al.*, 1982] (see Appendix A). Knowing $[I]_m$, one can then compute the intensity matrix for the incoherent wave, $[I]$, from (28). One notes that for a normally incident plane wave, due to symmetry, only $[F]_0$ and $[F]_2$ terms are present. Hence only $[I]_0$ and $[I]_2$ survives; so the Fourier series summation for $[I]$ only requires the sum of two terms, the $m = 0$ and $m = 2$ terms. For obliquely incident case, one must sum over all m terms. Knowing $[I]$, which is the incoherent intensity matrix, one may easily obtain the copolarized incoherent intensity (I_x) and the cross-polarized incoherent intensity (I_y) for an x-polarized incident wave as follows [*Ishimaru et al.*, 1982]:

I_x(copolarized)

$$= \mu^2 \cos^2 \phi I_1 + I_2 \sin^2 \phi - U\mu \sin \phi \cos \phi$$

I_y(cross-polarized)

$$= \mu^2 \sin^2 \phi I_1 + I_2 \cos^2 \phi + U\mu \sin \phi \cos \phi \tag{46}$$

3. NUMERICAL RESULTS AND DISCUSSION

As illustrative examples, we have performed numerical computations according to the above technique for the following parameters.

High-loss particles:

oblate spheroidal (disk shape) smoke particles with 10 : 1, 4 : 1, and 1 : 1 flatness ratios (i.e., $a/b = 10, 4, 1$);
dielectric constant of particle $= 2 + i$;
size of particles: $a = 1.0$ mm;
wave number: $k = 0.63$/mm;

Low-loss particles:

oblate spheroidal (disk shape) as well as prolate spheroidal (needle shape) ice particles with 10 : 1, 4 : 1, and 1 : 1 flatness (or sharpness) ratios (i.e., $a/b = 10, 4, 1$);

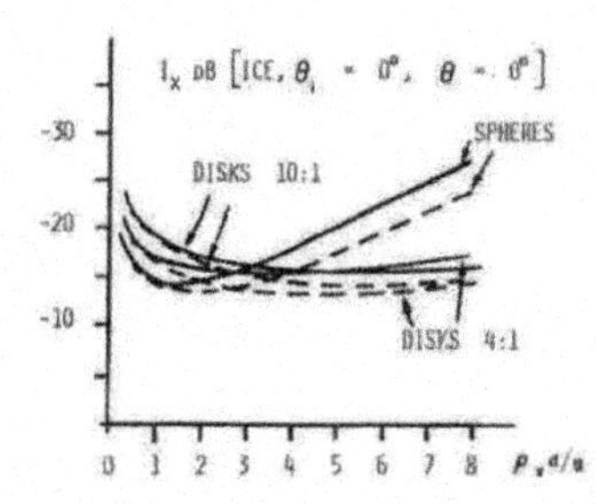

Fig. 4. Copolarized intensity versus $\rho_v d/g$ for ice disks and spheres. Solid curves are exact transport solutions and dashed curves are first-order results.

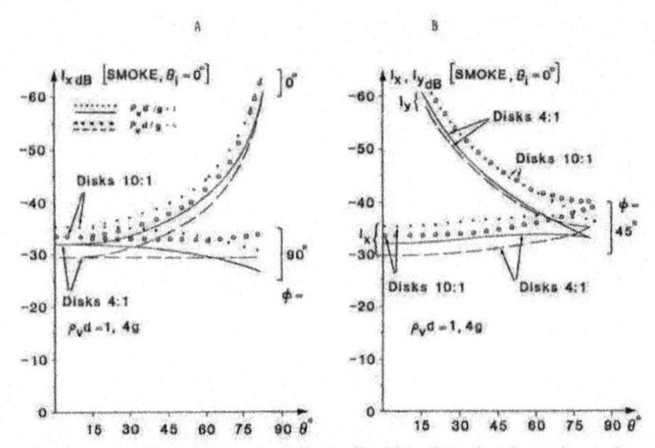

Fig. 5. Co- and cross-polarized intensities versus θ for smoke disks. Note that first-order results are identical to those shown.

dielectric constant of particle = $3.168 + 0.085i$;
size of particles: $a = 1.0$ mm;
wave number: $k = 0.63$/mm.

These particles are assumed to be uniformly distributed with particle number density ρ. The corresponding volume density is, therefore, $\rho_v = \rho\pi a^2 b/6$ for oblate particles and $\rho_v = \rho\pi ab^2/6$ for prolate particles. The thickness of the slab is d. The incident coherent plane wave is assumed to be polarized in the x direction and incident at the following angles (θ_i, $\phi_i = 0$): $\theta_i = 0°, 30°, 60°$. Incoherent intensities as a function of observation angle θ are computed for various values of $\rho_v d/\mu_i$, various angles of incidence, various particle shapes, and various particle orientations. Results are displayed in Figures 2–10. Plotted in these figures are also the results obtained according to the first-order expressions derived earlier (A. Ishimaru et al., unpublished manuscript, 1984). It is important to note that the optical depth given by $\tau_i = \gamma_{ii} d/\mu_i = -2\lambda\rho d \times \mathrm{Im}\,[\bar{f}_{11}(0, 0, \theta_i)]/\mu_i$ for the high-loss particles is different than that for low-loss particles (see Table 1). Hence different values of $\rho_v d/\mu_i = gL$, $L = 0.5$–8 are chosen for low-loss ice particles $g = 10^{-2}\pi/24$ than for high-loss particles $g = 10^{-3}\pi/48$.

Let us now discuss the computational aspect of this problem. Listed below are items of consequence with respect to the accuracy and convergence of our numerical solutions:

Subroutines from IMSL library were used to solve the complex linear system of equations for $[\chi]_m$ (equation (22)). Results were compared with those obtained according to Le Foll's algorithm for large

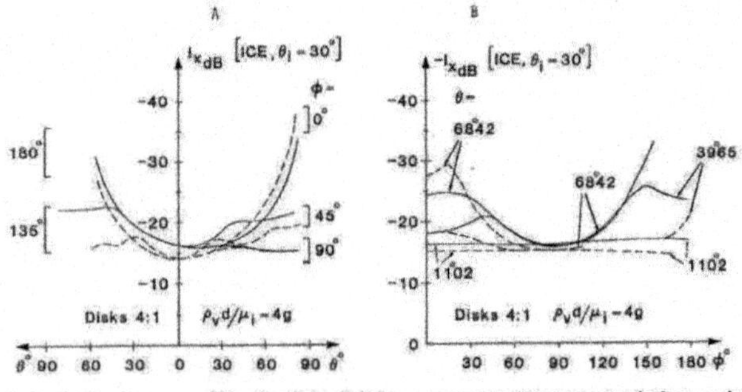

Fig. 6. Copolarized intensity versus θ for ice disks. Solid curves are exact transport solutions and dashed curves are first-order results.

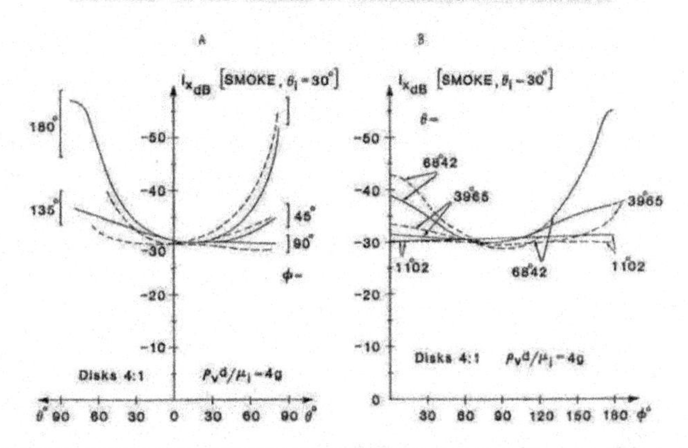

Fig. 7. Copolarized intensity versus θ for smoke disks. Solid curves are exact transport solutions and dashed curves are first-order results.

and ill-conditioned system of equations [*Bolomey et al.*, 1980]. No difference was found. The eigenvalues-eigenvectors of $[\xi]_m$ were also computed by the use of a subroutine from IMSL library. Simpson's rule was used to compute the integrals for the Fourier components for the scattering and source matrices. The required f_{ij} were obtained through the Lagrange's interpolation scheme applied to a set of tabulated values for $\bar{f}_{ij}$ given by *Barber and Yeh* [1975]. Adequate accuracy was obtained when the number of integration points was chosen to be 25. Doubling P doubled the computation time while less than 0.01% differences in decibels for the resultant intensities were found.

For the cases treated here, the Fourier series converges very rapidly even for obliquely incident wave. Only three terms ($m = 0, 1, 2$) are needed. Doubling the number of terms merely changes the resultant intensities by at most 0.3% in decibels while it increases the computation time by about 50%. Consequently, we have used at most seven terms to perform our computation.

To learn the number of terms N needed in the Gauss' quadrature technique to evaluate (39), we have computed the incoherent intensities using $N = 4, 6, 8$ for spherical particles and $N = 6, 8$ for nonspherical particles. Results are displayed in Figure 2. It is known that the choice of too small a value for N may lead to inaccurate description of varying intensities while the choice of too large a

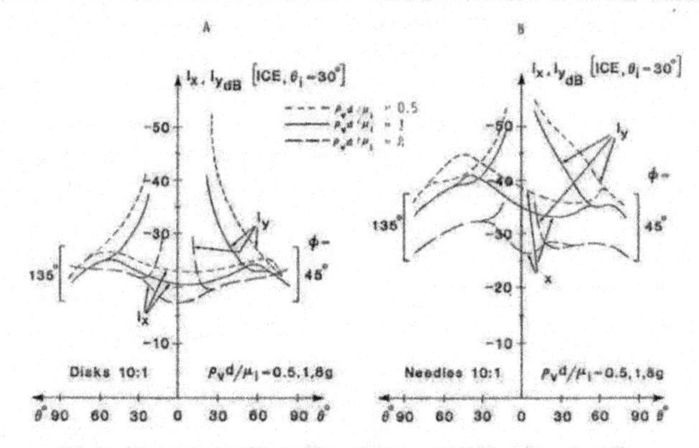

Fig. 8. Co- and cross-polarized intensities versus θ for ice disks and needles.

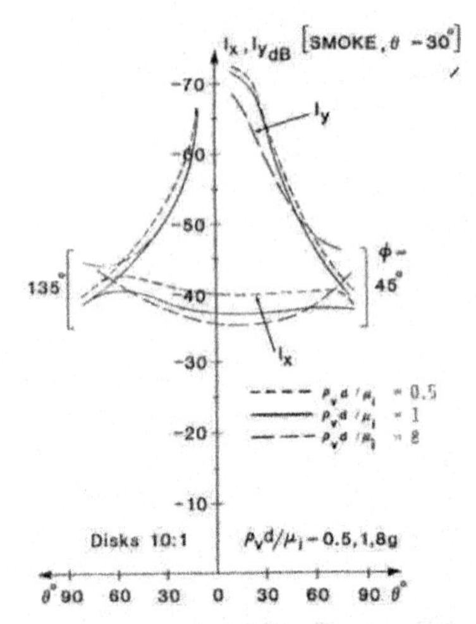

Fig. 9. Co- and cross-polarized intensities versus θ for smoke disks.

value for N may lead to unacceptable computer time and cost as well as other numerical problems associated with the manipulation of large matrices. According to Figure 2, no noticeable difference was observed for all the N cases that we tested. Hence $N = 6$, corresponding to $\theta_k = 11.02°$, 25.30°, 39.65°, 54.03°, 68.42°, 82.81°, and $\theta_{-k} = 180° - \theta_k$, were chosen for all subsequent calculations. The corresponding rank for the linear system of equations as well as the number of eigenvalues is then $K = 24$ (for $m = 0$) and $K = 48$ (for $m \neq 0$).

Because of the numerical constraints discussed above, there is a limitation on the number of figures of accuracy one may achieve for the eigenvalues. Unfortunately, for large optical thicknesses, small variations in the eigenvalues may cause very large changes in the values of the incoherent intensities. Hence there is a limit on the achievable accuracy of the incoherent intensities when the density is high and when optical depth is large.

Return now to the discussion of the behavior of the incoherent intensities for low-loss and high-loss nonspherical particles. The computed results are displayed in Figures 3–9. Plotted in these figures are also the first-order results. One observes the following:

1. For a given angle of incidence the behavior of the copolarized incoherent intensity I_x versus $\rho_v d$ curve remains relatively unaltered for all types of spherical or nonspherical particles. I_x increases as $\rho_v d$ increases and reaches a maximum and then decreases slowly as $\rho_v d$ further increases. The maximum is reached earlier for disks than needles. Similar behavior is also observed for the cross-polarized incoherent intensity I_y. Higher variations are observed for low-loss particles than for high-loss particles.

2. For the normal incident case, the shape of I_x versus θ curves remains the same for all types of spherical or nonspherical particles. In the $\phi = 0°$ plane, I_x decreases monotonically as θ increases from zero; in the $\phi = 90°$ plane, I_x remains relatively constant as θ increases. The fact that I_x is maximum at $\theta = 0°$ signifies the presence of the dipole scattering effect. Other observations are as follows: Varying $\rho_v d$ merely changes the amplitudes of I_x. For the same $\rho_v d$, fatter (more spherical) particles produce higher I_x. For the same optical depth, low-loss disks produce higher I_x (about 15 dB higher) than high-loss disks.

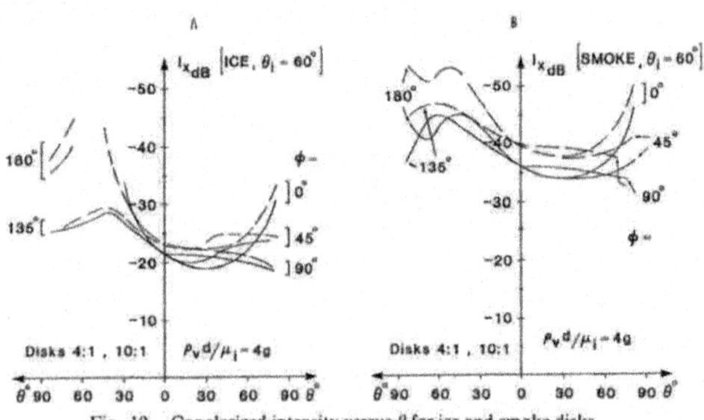

Fig. 10. Copolarized intensity versus θ for ice and smoke disks.

TABLE 1. Optical Depths τ_i of the Plane-Parallel Slab

	Incidence					
	$\theta_i = 0°$		$\theta_i = 30°$		$\theta_i = 60°$	
Particles	Low-Loss	High-Loss	Low-Loss	High-Loss	Low-Loss	High-Loss
Disks 10 : 1	0.700	1.182	0.544	0.965	0.245	0.534
Disks 4 : 1	1.110	1.015	0.898	0.866	0.476	0.559
Spheres	0.861	2.600	0.861	2.600	0.861	2.600
Needles 4 : 1	0.157	···	0.240	···	0.413	···
Needles 10 : 1	0.068	···	0.118	···	0.220	···

Volume density ρ_V, width d under normal or oblique incidence as functions of the particles' shape and material, for $\rho_V d/\cos\theta_i = gL$, $L = 4$, $g = 10^{-3}\pi/48$ (high-loss) or $10^{-2}\pi/24$ (low-loss); τ_i are proportional to L.

3. For the normal incident case, the cross-polarized intensity I_y increases sharply as ϕ increases from 0° and reaches a maximum when $\phi = 45°$. In the $\phi = 45°$ plane, I_y increases sharply as θ increases while the copolarized intensity I_x in this plane remains relatively constant. This behavior is true for high-loss as well as low-loss particles; the only difference is in the magnitudes of the intensities. For high-loss particles, I_y is lower than I_x except when θ is near 90°, where I_y is equal to I_x. For low-loss particles, I_y is very close to I_x for θ larger than only a few tenths of a degree when $\rho_v d$ is high.

4. For oblique incident case, as expected, I_x versus θ or I_y versus θ curves are no longer symmetric about $\theta = 0°$. The behavior of these curves in the $\phi = 0°$ plane is still relatively similar to the normal incident case, except the curves are tilted toward $\theta = \theta_i$. The shape factor for the particles appears to be more pronounced when the wave is obliquely incident; i.e., the I_x, I_y versus θ, ϕ curves are more asymmetrical for sharper needles or flatter disks.

A word about the validity of first-order result is in order. As one may observe from all these figures that the first-order theory does provide the general trend for all the cases considered. However, it is important to observe that very significant difference is noted when quantitative comparisons are made. The first-order formulas may yield values that are as much as 10 dB below or above those predicted by the more exact calculation given in this paper. Perhaps, we have demonstrated the importance of carrying out the more exact radiative transfer treatment for the multiple scattering problem even for relatively sparsely populated medium. At the same time, we have shown that if only an idea on the general behavior for the induced incoherent intensities is needed, the first-order approach is still attractive owing to its simplicity.

APPENDIX A

The solution of the K first-order differential equations with constant coefficients (equation (27), $K = 4N$ ($m = 0$) or $8N$ ($m \neq 0$)), which is

$$\frac{d}{dz}[\chi(z)] + [\xi][\chi(z)] = [\chi_i] \exp(\gamma_{ii} z/\mu_i) \quad \text{(A1)}$$

is the sum of a particular solution $[\chi_p(z)]$

$$\left[[\xi] + \frac{\gamma_{ii}}{\mu_i}[H]\right][\chi_p(z)] = [\chi_i] \exp(\gamma_{ii} z/\mu_i) \quad \text{(A2)}$$

with $[H]$ being the unit matrix, and of the complementary solution

$$[\chi_{cp}(z)] = \sum_{k=-K}^{+K} c_k[\beta_k]e^{-\lambda_k z} \quad \text{(A3)}$$

where the λ_k and $[\beta_k]$ are the eigenvalues and eigenvectors of the matrix $[\xi]$. The c_k are complex-valued coefficients that are determined from K boundary conditions at $z = 0$ and at $z = d$ (cf equation (45)). Due to the particular symmetry of the $[\xi]$ matrix the λ_k eigenvalues are composed of real-valued pairs (λ_k, $-\lambda_k$) and/or of complex-valued quadruplets (λ_k, $-\lambda_k$, λ_k^*, $-\lambda_k^*$), where an asterisk means the complex conjugate. One notes that the λ_k are proportional to ρ.

Acknowledgments. The authors wish to acknowledge with thanks the partial support received from the Army Research Office under the direction of Walter Flood. Doris C. Blackman of JPL has been of great help with the interpolation scheme.

REFERENCES

Barber, P., and C. Yeh, Scattering of electromagnetic waves by arbitrarily shaped dielectric bodies, *Appl. Opt.*, *14*(12), 2864–2872, 1975.

Bolomey, J. C., F. Hillaire, and D. Lesselier, Numerical study of thick antennas by means of the Albert and Synge integral equation, *Ann. Telecommun.*, *35*(5–6), 183–192, 1980.

Chandrasekhar, S., *Radiative Transfer*, 393 pp., Clarendon, Oxford, 1950.

Cheung, R. L.-T., and A. Ishimaru, Transmission, backscattering and depolarization of waves by randomly distributed spherical particles, *Appl. Opt.*, *21*(20), 3792–3798, 1982.

Dahlquist, G., and A Björk, *Numerical Methods*, *Ser. in Automat. Comput.*, vol. 1, translated from French by N. Anderson, 564 pp., Prentice-Hall, Englewood Cliffs, N. J., 1974.

Reprinted from Applied Optics, Vol. 23, page 4132, November 15, 1984

First-order multiple scattering theory for nonspherical particles

A. Ishimaru, C. Yeh, and D. Lesselier

Using the vector radiative transfer equation, we have obtained the first-order solution for the problem dealing with multiple scattering from an ensemble of nonspherical particles. Representative results for low-absorbent disks (oblate spheroids) as well as high-absorbent disks were obtained. In general, it was found that multiple scattering effects were more prominent for low-absorbent particles.

I. Introduction

It is known that the atmosphere contaminated by smoke or hydrometeors (ice or rain) has profound effects on the propagation of high-frequency electromagnetic energy, which in turn may degrade the performance of many modern communication or detection systems.[1,2] Smoke or hydrometeors may take on different shapes other than spheres,[3,4] and the close proximity of these particles may enhance multiple scattering effects.[5,6] It is, therefore, important to assess the relative importance of nonspherical vs spherical particle effects in the multiple scattering of electromagnetic waves by an ensemble of particles.

Various studies of the multiple scattering effects of hydrometeors on microwave propagation have been conducted.[5,7-10] While earlier studies dealt with scalar fields and first-order solutions,[7,8] more recent ones have considered circularly or linearly polarized incident waves and exact solutions to the equation of transfer.[5] In all the previous studies, only spherical particles were considered. The emphasis in this paper will be on the effects of nonspherical particles compared with spherical particles on the coherent as well as incoherent intensities produced by multiple scattering.

Our formulation for the incoherent intensities is based on the equation of transfer for the Stokes' parameters,[11] while the formulation for the coherent intensities is based on the classical approach of van de Hulst.[12] A linearly polarized wave is assumed to be obliquely incident on a plane-parallel medium containing the ensemble of particles. In principle, the equation of transfer can be solved exactly using the eigenvalue-eigenvector technique[11] with the scattering characteristics of the particles being calculated using the extended boundary condition method[13] and a given drop size and orientation distribution. Numerical computation based on this technique is being carried out, and results will be presented in another paper.[14] The main purpose of the present paper is to present the first-order solution to this problem dealing with nonspherical particles. It is believed that providing the first-order expressions for the incoherent intensities represents the necessary initial step toward the complete solution of the multiple scattering problem.

As specific examples, numerical results based on the first-order solution are given for the co-polarized and cross-polarized intensities as functions of observation angle and optical depth for two types of particle: those with high loss tangent (i.e., high absorbent) and those with low loss tangent (i.e., low absorbent). As expected, multiple scattering effects are more pronounced for ensemble of low absorbent particles than for ensembles of high absorbent particles.

A. Ishimaru is with University of Washington, Department of Electrical Engineering, Seattle, Washington 98195; C. Yeh is with University of California, Los Angeles, Electrical Engineering Department, Los Angeles, California 90024; and D. Lesselier is with CNRS-ESE, Laboratoire des Signaux et Systemes, Plateau du Moulon, 91190 Gif-sur-Yvette, France.

Received 3 November 1983.
0003-6935/84/224132-08$02.00/0.

II. Formulation of the Problem–The Transport Equation Approach

The transport theory approach will be used to deal with the multiple scattering problem.[15] The depolarization effects due to irregularly shaped particles as well as the multiple scattering effects will be included in the equation of radiative transfer.[11] Solution of this transport equation will provide information on the needed incoherent specific intensity of transmitted wave.

Let us introduce the modified Stokes parameters $(\mathbf{I}_1, \mathbf{I}_2, U, V)$ as follows[12]:

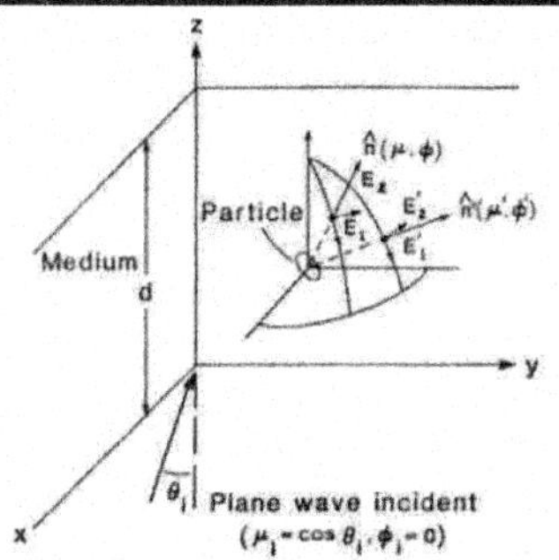

Fig. 1. Geometry showing particle medium ($0 \leq x \leq d$), the direction $(\mu_i, 0)$ of the incident plane wave, the direction (μ', ϕ') of the wave incident upon a particle, and the direction (μ, ϕ) of the scattered wave. E_1 and E_2 are the electric field components in the θ and ϕ directions, respectively.

$$\mathbf{I}_1 = \langle E_1 E_1^* \rangle, \mathbf{I}_2 = \langle E_2 E_2^* \rangle,$$
$$U = 2\,\mathrm{Re}\langle E_1 E_2^* \rangle, V = 2\,\mathrm{Im}\langle E_1 E_2^* \rangle, \tag{1}$$

where Re and Im denote the real part and the imaginary part, and the asterisk means the complex conjugate.

If a plane wave is incident upon the parallel–plane medium in the direction $\hat{n}_i(\mu_i = \cos\theta_i, \phi_i)$ (see Fig. 1), we can write the Stokes vector $(\mathbf{I})$ as a sum of the coherent intensity $[I_c]$ and the incoherent (diffuse) intensity $[I_d]$:

$$[\mathbf{I}(\mathbf{r}, \hat{n})] = [I_c(\mathbf{r}, \hat{n}_i)]\delta(\hat{n} - \hat{n}_i) + [I_d(\mathbf{r}, \hat{n})]. \tag{2}$$

The coherent intensity $[I_c]$ satisfies the following equation:

$$\frac{d}{dn}[I_c] = [\mathrm{A}][I_c], \tag{3}$$

while the incoherent intensity $[I_d]$ satisfies the following equation

$$\frac{d}{dn}[I_d] = [\mathrm{A}][I_d] + \rho \int [\mathrm{S}][I_d']d\omega' + [I_i] \tag{4}$$

with

$$[I_i(\mathbf{r}, \hat{n})] = \rho[\mathrm{S}(\hat{n}, \hat{n}_i)][\mathbf{I}_c(\mathbf{r}, \hat{n}_i)], \tag{5}$$

$\rho[\ ]$ means $\int_0^\infty [\ ]N(D)dD$, where $N(D)$ is the number of particles having the sizes between D and $D + dD$. The extinction matrix [A] and the Mueller matrix [S] can be expressed in terms of the scattering amplitude matrix [f]; dn is the elementary distance in the direction of $\hat{n}$, and $d\omega'$ is the elementary solid angle in the direction $\hat{n}'$.

The incident specific intensity matrix $[I_i]$ is the source function generated by the scattering of the reduced incident specific intensity $[I_{ri}]$ propagating in the direction of $\mu_i = \cos\theta_i$ and $\phi_i = 0$ and is given by

$$[I_i] = \int [\mathrm{S}][I_{ri}]d\omega', \tag{6}$$

where $[I_{ri}]$ is given by the incident Stokes vector $[\mathbf{I}_0]$

$$[I_{ri}] = [Q][\mathbf{I}_0], \tag{7}$$

and $[Q]$ is the 4×4 matrix.[11] If the incident wave with the total intensity of unity is polarized parallel to the x-z plane (vertical polarization), $[\mathbf{I}_0]$ is given by

$$[\mathbf{I}_0] = \begin{bmatrix} 1 \\ 0 \\ 0 \\ 0 \end{bmatrix}. \tag{8}$$

Similarly, if the incident wave is polarized perpendicular to the x-z plane (horizontal polarization), we have

$$[\mathbf{I}_0] = \begin{bmatrix} 0 \\ 1 \\ 0 \\ 0 \end{bmatrix}. \tag{9}$$

The boundary condition for the incoherent specific intensity (I_d) is

$$[I_d] = 0 \cos\theta > 0 (0 \leq \mu \leq 1) \text{ at } z = 0,$$
$$[I_d] = 0 \cos\theta < 0 (0 \geq \mu \geq -1) \text{ at } z = d. \tag{10}$$

The equation of transfer [Eq. (4)] and the boundary condition [Eq. (10)] constitute the complete mathematical formulation of the problem.[11] This is an integro-differential equation for the incoherent intensity, $[I_d]$ with five variables (z, μ, μ', ϕ, and ϕ').[15] Special solution of this equation has been given elsewhere.[14]

III. Solution for the Coherent Intensities

In general, the field $[E]$ consists of the coherent (average) field $[E_c]$ and the incoherent (diffuse) field $[E_d]$[16]:

$$[E] = [E_c] + [E_d], \tag{11}$$

where $[E_c] = \langle [E] \rangle$ and $\langle [E_d] \rangle = 0$. The coherent field $[E_c]$ propagating in the direction $\hat{n}$ over distance s satisfies the following equation:

$$\frac{d}{ds}[E_c] = [M][E_c]. \tag{12}$$

The factor $\exp(iks - i\omega t)$ has been suppressed. Here,

$$[E_c] = \begin{bmatrix} \langle E_1 \rangle \\ \langle E_2 \rangle \end{bmatrix}, \quad [M] = [M_{ij}],$$
$$M_{ij} = \frac{i2\pi\rho}{k} f_{ij}(\hat{n}, \hat{n}), \tag{13}$$

where ρ and k are, respectively, the number density and wave number, E_1 and E_2 are the electric field components in the θ and ϕ directions, respectively (see Fig. 1), and f_{ij} are the elements of the 2×2 scattering amplitude matrix $[f(\hat{n}, \hat{n}')]$[13,17]:

$$[E(\hat{n})] = \exp(ikR)/R[f(\hat{n}, \hat{n}')][E'(\hat{n}')]. \tag{14}$$

$[E(\hat{n})]$ and $[E'(\hat{n}')]$ are, respectively, the scattered wave from a single particle in the direction $\hat{n}$ and the incident wave upon this single particle in the direction $\hat{n}'$.

Let us now consider the specific case of an incoming wave which is a normally incident coherent plane wave, linearly polarized in the x direction. The particles are

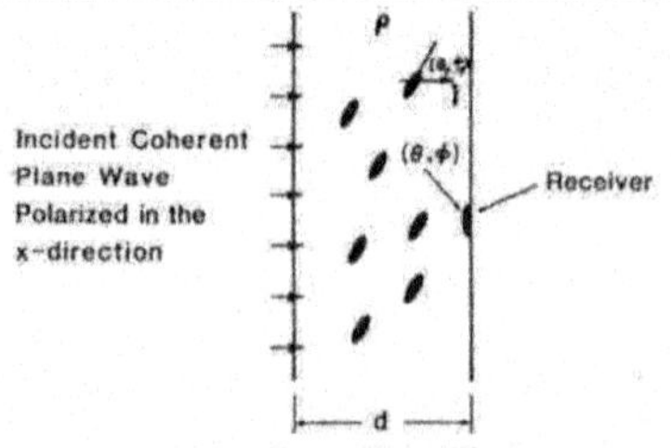

Fig. 2. Geometry of the problem. The axially symmetric particles are aligned in the (θ_p,ϕ_p) direction.

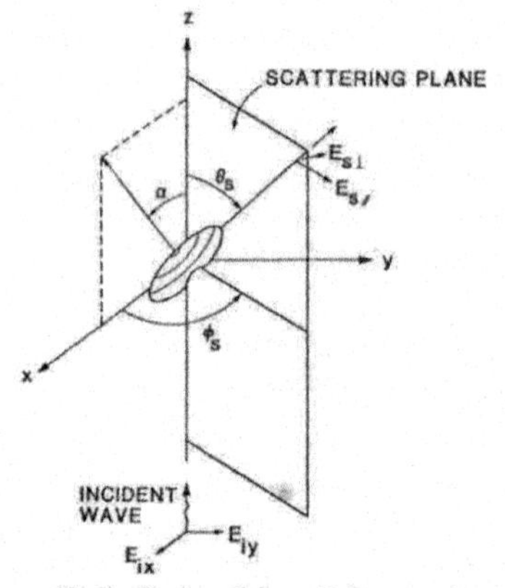

Fig. 3. Single particles scattering geometry.

axially symmetric and oriented in the same direction given by (θ_p,ϕ_p) (see Fig. 2). We are looking for the coherent intensity (I_{cx}) for the x-polarized (co-polarized) field and the coherent intensity (I_{cy}) for the y-polarized (cross-polarized) field at the receiver as a function of particle density. I_{cx} and I_{cy} are defined as follows:

$$I_{cx} = |\langle E_{cx}\rangle|^2 \quad I_{cy} = |\langle E_{cy}\rangle|^2, \tag{15}$$

where E_{cx} and E_{cy} are, respectively, the coherent field in the x direction and in the y direction.

First the incident wave $[E_0]$ expressed in the medium coordinate needs to be transformed to the coordinate system attached to the particle by a rotation of axis. This is accomplished by the multiplication by the rotation matrix $[R_p]$. The incident field is now expressed as (E_{ix},E_{iy}) in Fig. 3. These two components propagate with the propagation constants K_x and K_y to be shown in Eqs. (20) and (21). After propagating over distance d, these two components (E_{ix},E_{iy}) must be expressed in the medium coordinate. This is done by the rotation of axis given by $[R_p]^{-1}$. Therefore, the coherent field (E_x,E_y) at $z = d$ is given by $[T][E_0]$. The coherent intensities are

$$I_{cx} = |T_{11}|^2, \tag{16}$$

$$I_{cy} = |T_{12}|^2 \tag{17}$$

with

$$[T] = \begin{bmatrix} T_{11} & T_{12} \\ T_{21} & T_{22} \end{bmatrix} = \exp(iK_xd)[R_p]^{-1}\begin{bmatrix} 1 & 0 \\ 0 & 0 \end{bmatrix}[R_p] + \exp(iK_yd)[R_p]^{-1}\begin{bmatrix} 0 & 0 \\ 0 & 1 \end{bmatrix}[R_p], \tag{18}$$

$$[R_p] = \begin{bmatrix} \cos\phi_p & \sin\phi_p \\ -\sin\phi_p & \cos\phi_p \end{bmatrix}, \tag{19}$$

$$K_x = k + \frac{2\pi}{k}\rho[\bar{f}_{11}(0,0,\theta_p)], \tag{20}$$

$$K_y = k + \frac{2\pi}{k}\rho[\bar{f}_{22}(0,0,\theta_p)], \tag{21}$$

where k = free-space wave number,
d = thickness of the slab, and
$[R_p]^{-1}$ = inverse of $[R_p]$.

The functions $\bar{f}_{11}$ and $\bar{f}_{22}$ are given in the single particle scattering expression as follows[13,18]:

$$\begin{bmatrix} E_{s\parallel} \\ E_{s\perp} \end{bmatrix} = \frac{\exp(ikR)}{R}\begin{bmatrix} \bar{f}_{11}(\theta_s,\phi_s,\alpha) & \bar{f}_{12}(\theta_s,\phi_s,\alpha) \\ \bar{f}_{21}(\theta_s,\phi_s,\alpha) & \bar{f}_{22}(\theta_s,\phi_s,\alpha) \end{bmatrix}\begin{bmatrix} E_{ix} \\ E_{iy} \end{bmatrix},$$

where the symbols have been defined in Fig. 3. Equations (16) and (17) give the incoherent intensities in the multiple scattering problem for an incident coherent plane wave.

IV. First-Order Solution for the Incoherent Intensities

Using an iterative approach one may expand Eq. (4) as follows: Let

$$\underline{L} = \mu\frac{d}{dz} + [T], \tag{23}$$

$$\underline{S} = \int[S]d\omega'; \tag{24}$$

then from Eq. (2),

$$[I_d] = (\underline{L})^{-1}\underline{S}[I] + (\underline{L})^{-1}[I_i]. \tag{25}$$

First iteration yields

$$[I_d] = (\underline{L})^{-1}[I_i] + (\underline{L})^{-1}\underline{S}(\underline{L})^{-1}[I_i] + \dots. \tag{26}$$

The first term on the right-hand side of the equation represents the first-order solution of the specific incoherent intensities. The incoming wave is assumed to be a normally incident coherent plane wave linearly polarized in the x direction. The particles are oriented in the same direction given by (θ_p,ϕ_p) (see Fig. 2). We are looking for the incoherent intensity (I_{dx}) for the x-polarized (co-polarized) field and the incoherent intensity (I_{dy}) for the y-polarized (cross-polarized) field in the direction (θ,ϕ). The first-order solution for the incoherent intensities are given by the following expressions:

$$I_{dx} = \rho \left([A_{11}B_{11}C_{11}D_{11}][\overline{\alpha}] \begin{bmatrix} A_{11}^* \\ B_{11}^* \\ C_{11}^* \\ D_{11}^* \end{bmatrix} \right), \tag{27}$$

$$I_{dy} = \rho \left([A_{21}B_{21}C_{21}D_{21}][\overline{\alpha}] \begin{bmatrix} A_{21}^* \\ B_{21}^* \\ C_{21}^* \\ D_{21}^* \end{bmatrix} \right), \tag{28}$$

$$\left.\begin{aligned}
[A] &= \begin{bmatrix} A_{11} & A_{12} \\ A_{21} & A_{22} \end{bmatrix} = [\mathrm{Re}][T_1^s][F_e][T_1^0], \\
[B] &= \begin{bmatrix} B_{11} & B_{12} \\ B_{21} & B_{22} \end{bmatrix} = [\mathrm{Re}][T_1^s][F_e][T_2^0], \\
[C] &= \begin{bmatrix} C_{11} & C_{12} \\ C_{21} & C_{22} \end{bmatrix} = [\mathrm{Re}][T_2^s][F_e][T_1^0], \\
[D] &= \begin{bmatrix} D_{11} & D_{12} \\ D_{21} & D_{22} \end{bmatrix} = [\mathrm{Re}][T_2^s][F_e][T_2^0],
\end{aligned}\right. \tag{29}$$

$$[\mathrm{Re}] = \begin{bmatrix} \cos\theta\cos\phi & -\sin\phi \\ \cos\theta\sin\phi & \cos\phi \end{bmatrix}, \tag{30}$$

$$[T_1^s] = [R_q]^{-1} \begin{bmatrix} 1 & 0 \\ 0 & 0 \end{bmatrix} [R_q], \tag{31}$$

$$[T_2^s] = [R_q]^{-1} \begin{bmatrix} 0 & 0 \\ 0 & 1 \end{bmatrix} [R_q], \tag{32}$$

$$\left.\begin{aligned}
[T_1^0] &= \begin{bmatrix} 1 & 0 \\ 0 & 0 \end{bmatrix} [R_p], \\
[T_2^0] &= \begin{bmatrix} 0 & 0 \\ 0 & 1 \end{bmatrix} [R_p], \\
[R_q] &= \begin{bmatrix} \cos\phi_q & \sin\phi_q \\ -\sin\phi_q & \cos\phi_q \end{bmatrix}, \\
[R_p] &= \begin{bmatrix} \cos\phi_p & \sin\phi_p \\ -\sin\phi_p & \cos\phi_p \end{bmatrix},
\end{aligned}\right\} \tag{33}$$

$$[F_e] = \begin{bmatrix} \bar{f}_{11}(\theta,\phi-\phi_p,\theta_p) & \bar{f}_{12}(\theta,\phi-\phi_p,\theta_p) \\ \bar{f}_{21}(\theta,\phi-\phi_p,\theta_p) & \bar{f}_{22}(\theta,\phi-\phi_p\theta_p) \end{bmatrix}, \tag{34}$$

$$\alpha_{ij} = \exp(p_i + p_j^*) \left\{ \frac{\exp[(q_i + q_j^*)d] - 1}{\mu(q_i + q_j^*)} \right\}, \tag{35}$$

$$\left.\begin{aligned}
\mu &= \cos\theta, \\
p_1 &= iK_{s1}d/\mu, \\
p_2 &= p_1, \\
p_3 &= iK_{s2}d/\mu, \\
p_4 &= p_3, \\
q_1 &= i[K_x - (K_{s1}/\mu)], \\
q_2 &= i[K_y - (K_{s1}/\mu)], \\
q_3 &= i[K_x - (K_{s2}/\mu)], \\
q_4 &= i[K_y - (K_{s2}/\mu)], \\
K_{s1} &= k + \frac{2\pi}{k}\rho[\bar{f}_{11}(0,0,\theta_q)], \\
K_{s2} &= k + \frac{2\pi}{k}\rho[\bar{f}_{22}(0,0,\theta_q)],
\end{aligned}\right\} \tag{36}$$

and K_x and K_y are given by Eqs. (20) and (21), respectively. The angles θ_q and ϕ_q are given by the following expressions:

$$\cos\theta_q = \sin\theta\sin\theta_p\cos(\phi - \phi_p) + \cos\theta\cos\theta_p, \tag{37}$$

$$\cos\phi_q = \frac{1}{\sin\theta_q}[\sin\theta_p\cos\theta\cos(\phi - \phi_p) - \cos\theta_p\sin\theta]. \tag{38}$$

All the symbols used in these expressions have been defined earlier. The physical meaning of the expressions (27) and (28) is that $[T_1^0]$ and $[T_2^0]$ represent the rotation of the polarization axis from those of the incident wave to the principal axis of the particle, $[F_e]$ represents the scattering by the particle, and $[T_1^s]$ and $[T_2^s]$ represents the rotation of the polarization axis from those of the particles axis to those of the scattered wave. K_x and K_y are the propagation constants for the incident wave, and K_{s1} and K_{s2} are the propagation constants for the scattered wave, and α_{ij} represents the integrated propagation effects from $z = 0$ to d. This is the first-order solution for the incoherent intensities (I_{dx} and I_{dy}) in the multiple scattering problem for an incident coherent plane wave.

V. Numerical Results and Discussion

Before we display the numerical results, let us define the following quantities of interest:

CPI = the transmitted co-polarized coherent intensity for an incident x-polarized coherent wave [from Eq. (16)],
= $10\log_{10}[I_{cx}]$

XPI = the transmitted cross-polarized coherent intensity for an incident x-polarized coherent wave [from Eq. (17)],
= $10\log_{10}[I_{cy}]$,

CIDX = the transmitted co-polarized incoherent intensity for an incident x-polarized coherent wave [from Eq. (27)],
= $10\log_{10}[I_{dx}]$,

CIDY = the transmitted cross-polarized incoherent intensity for an incident x-polarized coherent wave [from Eq. (28)],
= $10\log_{10}[I_{dy}]$,

These quantities are functions of thickness of the slab, orientation-size-shape of the particles and observation angle (θ,ϕ).

As illustrative examples we have chosen the following cases:

a. High-Loss Case[18]: Oblate spheroidal smoke particles with 10:1, 4:1 and 1:1 flatness ratios (i.e., a/b = 10, 4, 1; Dielectric constant of particle = $2 + i$; Thickness of slab: d; Size of particles: $a = 0.1$ and 1.0 mm; Wave number: $k \simeq 0.6$/mm; Particle volume density: $\rho_v = \rho V = \rho\pi a^2 b/6$.

b. Low-Loss Case[19]: Oblate spheroidal ice particles with 10:1, 4:1, and 1:1 flatness ratios (i.e., a/b = 10, 4, 1); Dielectric constant of particle = $3.168 + 0.0085i$; Thickness of slab: d; Size of particles: a = 0.1 or 1.0 mm; Wave number: $k \simeq 6$/mm; Particle volume density: $\rho_v = \rho V = \rho\pi a^2 b/6$.

Calculations have been carried out for a normally incident plane wave polarized in the x direction according to Eqs. (16), (17), (27), and (28). Results are displayed in Figs. 4–13. The transmitted coherent intensities as a function of the product $\rho_v d$ of particle volume density times slab thickness for various particle sizes and particle orientations are shown in Figs. 4 and 5 for the high-loss case and in Figs. 6 and 7 for the low-loss case. The behavior of these intensities can be summarized in the following:

For the same amount of particle material per unit volume (ρ_v), the transmitted co-polarized intensity CPI is weaker for flatter disks for very small size particles.

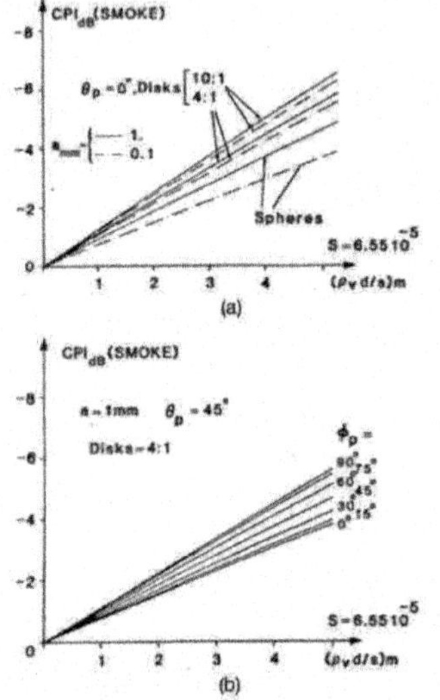

Fig. 4. Co-polarized coherent intensity CPI vs $\rho_v d$ in the case of smoke particles symmetrical or not about the incident plane: (a) small and large spheres and disks orientated along the direction of incidence. (b) large canted disks with various symmetry planes.

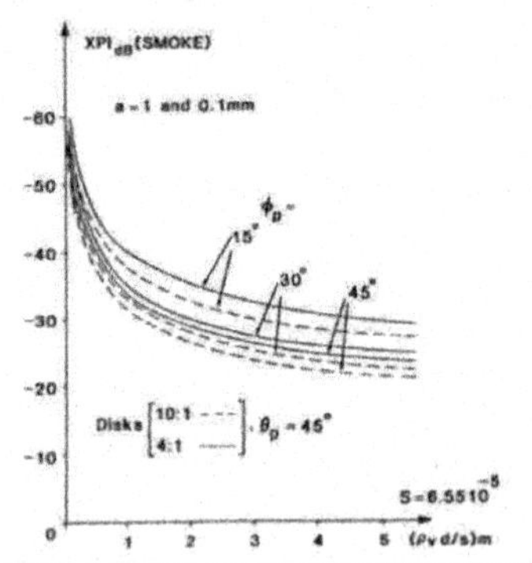

Fig. 5. Cross-polarized coherent intensity XPI vs $\rho_v d$ in the case of canted smoke particles not symmetrical about the incidence plane. Curves obtained for small ones and large ones are indistinguishable (differences <0.2 dB). XPI is the same for complementary ϕ_p.

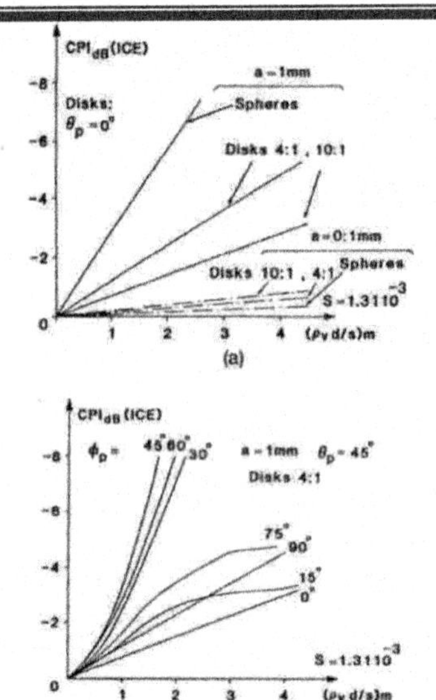

Fig. 6. Similar to Fig. 5 in the case of ice particles.

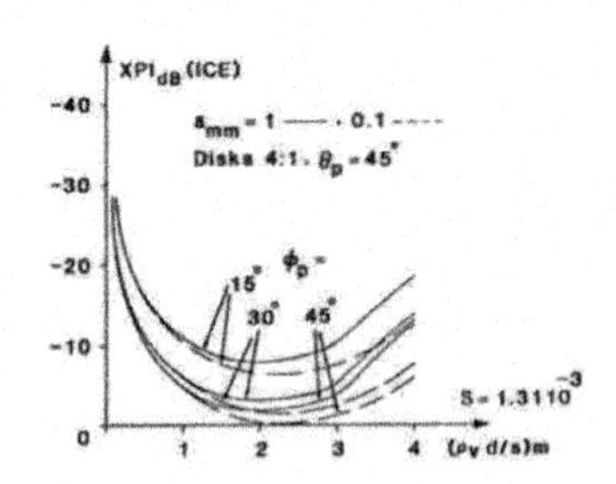

Fig. 7. Similar to Fig. 5 in the case of ice particles. Curves for small and large particles now differ.

This is because flatter disks correspond to a higher number of particles per unit volume (ρ) for a fixed size a and for a fixed ρ_v. Apparently the presence of more particles means higher attenuation for the co-polarized coherent wave.

For high-loss particles, CPI decreases linearly with respect to $\rho_v d$, while for low-loss particles, CPI may decrease somewhat faster than at a linear rate as $\rho_v d$ increases.

For a fixed $\rho_v d$, larger high-loss particles only increase slightly the attenuation of CPI. On the other hand, the orientation of the plates appears to have a much larger effect on the attenuation of CPI. This may be due to the generation of cross-polarized field for obliquely oriented disk particles.

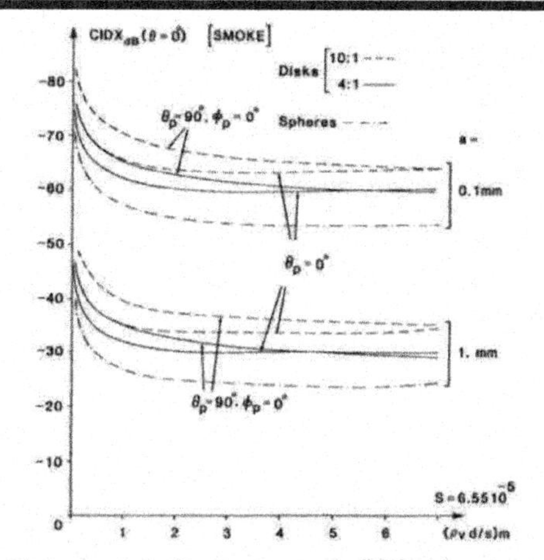

Fig. 8. Co-polarized incoherent intensity CIDX in the zenith direction vs $\rho_v d$ in the case of small and large smoke particles symmetrical about the incidence plane.

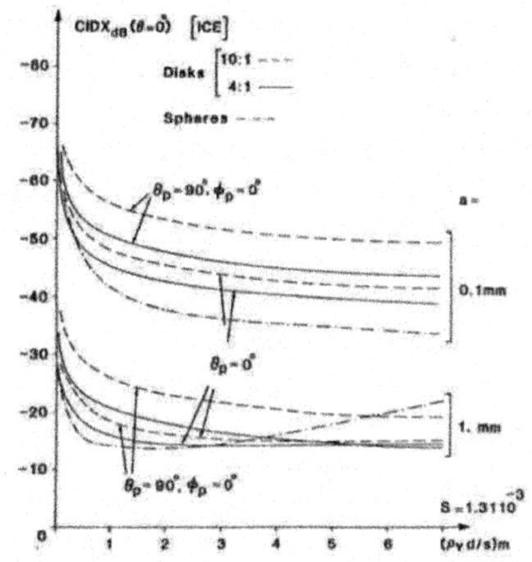

Fig. 9. Similar to Fig. 8 in the case of ice particles.

For high-loss particles, the cross-polarized coherent intensity XPI for obliquely oriented disks appears to increase and approaches a limit as $\rho_v d$ increases. (It is understood that no cross-polarized field is induced for spheres or for certain symmetrically oriented disks.) For low-loss particles, XPI increases and reaches a plateau and then decreases as $\rho_v d$ increases. In general, XPI for the low-loss case is much higher than that for the high-loss case (by almost 20 dB). This is because much stronger cross-polarized field is generated from low-loss particles.

The transmitted co-polarized incoherent intensities (CIDX) as a function of $\rho_v d$ for various sizes, shapes, and orientations are shown in Fig. 8 for the high-loss case and in Fig. 9 for the low-loss case. The behavior of these intensities can be summarized in the following:

The transmitted co-polarized incoherent intensity (CIDX) is very weak for small $\rho_v d$. As $\rho_v d$ increases, CIDX also increases until it reaches a plateau. For high- as well as low-loss particles and for a fixed $\rho_v d$, larger size particles produce higher CIDX and fatter bodies such as spheres also produce higher CIDX.

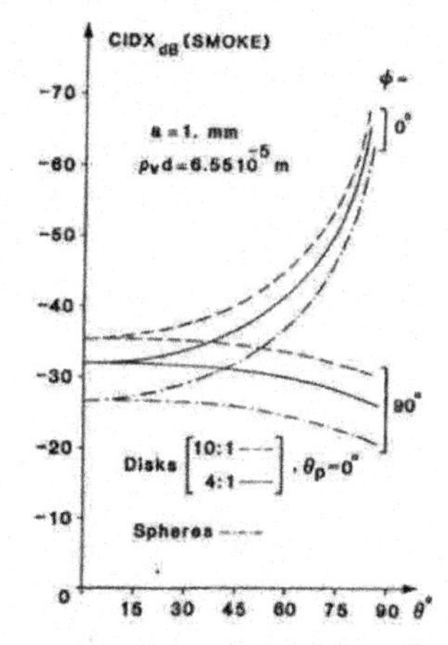

Fig. 10. At given $\rho_v d$, patterns of the co-polarized incoherent intensity CIDX vs the angle from zenith θ in the case of large smoke spheres and disks orientated along the direction of incidence.

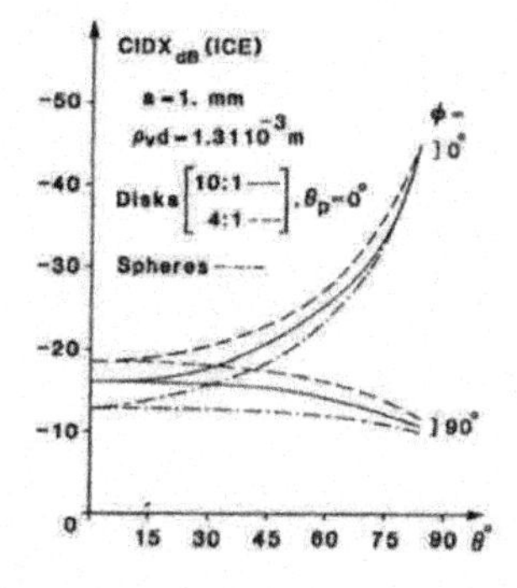

Fig. 11. Similar to Fig. 10 in the case of ice particles.

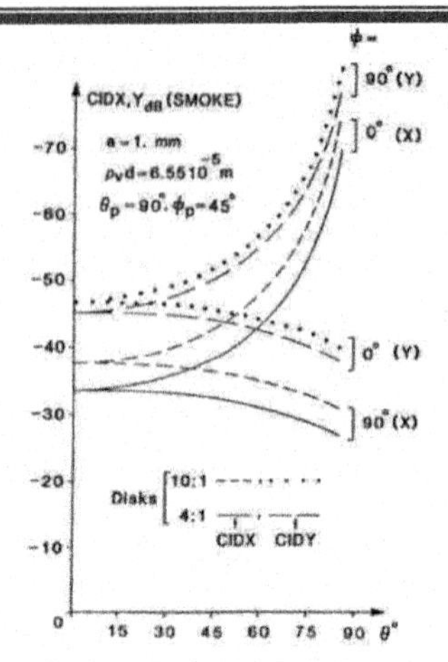

Fig. 12. At given $\rho_v d$, patterns of the co- and cross-polarized incoherent intensities CIDX and CIDY vs the angle from zenith θ in the case of large smoke disks not symmetrical about the incidence plane.

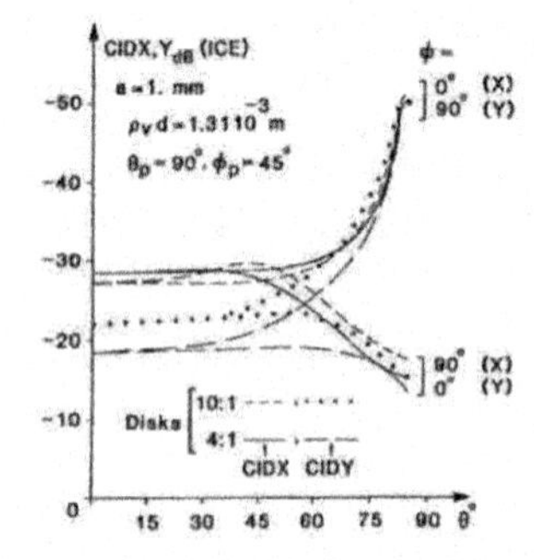

Fig. 13. Similar to Fig. 12 in the case of ice particles.

The shape of the CIDX vs $\rho_v d$ curve remains unaltered for different shapes of particle or for different particle orientations.

Figures 10–13 show the transmitted co-polarized incoherent intensities (CIDX) and/or the transmitted cross-polarized incoherent intensities (CIDY) as a function of the observation angle θ for a given set of parameters $(\rho_v d, \theta_p, \phi_p, a)$. One observes the following:

As θ changes from 0 to 90°, CIDX decreases in the observation plane $\phi = 0°$ and increases in the observation plane $\phi = 90°$. The behavior is similar for spheres or disks.

The shape of the CIDX vs θ curves remain relatively unchanged for the high- or for the low-loss particles; only the absolute magnitudes for CIDX are different.

The cross-polarized incoherent intensities (CIDY) only exist for obliquely positioned disks (i.e., $\theta_p = 90°$, $\phi_p = 45°$). For low values of $\rho_v d$ as θ changes from 0 to 90°, CIDY also decreases in the observation plane $\phi = 90°$. The behavior is similar for different disks. For high values of $\rho_v d$, the structure for the CIDY vs θ curve becomes more complicated.

The relative amplitudes of CIDX and CIDY are very different for high- and low-loss particles. For high-loss particles, CIDY is usually smaller than CIDX. For low-loss particles, CIDY is of the same order of magnitude as CIDX.

VI. Conclusion

First-order analytic expressions for the incoherent intensities as well as the coherent intensities were obtained for a layer of spherical as well as nonspherical particles. Cross-polarization effects for nonspherical particles were included. Using these expressions, numerical examples for low- and high-loss nonspherical particles were also presented. It should be noted that these approximate expressions are valid for a medium with an optical depth ≪1 or when it is highly absorbing.

When this work was done D. Lesselier was a visiting scholar at the UCLA Department of Electrical Engineering. This work was partly supported by U.S. Army Research Office, Research Triangle Park, N.C.

References

1. A. Ishimaru, *Wave Propagation and Scattering in Random Media, Vols. I and II* (Academic, New York, 1978).
2. The 1980, 1981 and 1983 CSL Scientific Conference on Obscuration and Aerosol Research, Aberdeen Proving Ground (U.S. Army), Md.; organized by Ed. Stuebing.
3. L. J. Ippolito, "Radio-Propagation for Space Communication Systems," Proc. IEEE 69, 697 (1981).
4. J. P. Mon, "Backward and Forward Scattering of Microwaves by Ice Particles: A Review," Radio Sci. 17, 953 (1982).
5. A. Ishimaru, R. Woo, J. W. Armstrong, and D. C. Blackman, "Multiple Scattering Calculations of Rain Effects," Radio Sci. 17, 1425 (1982).
6. L. Tsang, M. C. Kubacsi, and J. A. Kong, "Radiative Transfer Theory for Active Remote Sensing of a Layer of Small Ellipsoidal Scatterers," Radio Sci. 16, 321 (1981).
7. A. Ishimaru and R. L. T. Cheung, "Multiple Scattering on Wave Propagation due to Rain," Ann. Telecommun. 35, 373 (1980).
8. T. Oguchi, "Scattering from Hydrometeors: A Survey," Radio Sci. 16, 691 (1981).
9. R. L. Olsen, "A Review of Theories of Coherent Radio Wave Propagation through Precipitation Media of Randomly Oriented Scatterers, and the Role of Multiple Scattering," Radio Sci. 17, 913 (1982).
10. R. H. Lang, "Electromagnetic Backscattering from a Sparse Distribution of Lossy Dielectric Scatterers," Radio Sci. 16, 15 (1981).
11. A. Ishimaru and C. Yeh, "Matrix Representations of the Vector Radiative Transfer Theory for Randomly Distributed Nonspherical Particles," J. Opt. Soc. Am. A 1, 359 (1984).
12. H. C. van de Hulst, *Light Scattering by Small Particles* (Wiley New York, 1957).

13. P. Barber and C. Yeh, "Scattering of Electromagnetic Waves by Arbitrarily Shaped Dielectric Bodies," Appl. Opt. **14**, 2864 (1975).
14. A. Ishimaru, D. Lesselier, and C. Yeh, "Multiple Scattering Calculations for Non-Spherical Particles Based on the Vector Radiation Transfer Equation," Radio Sci. (1984) (to appear).
15. S. Chandrasekhar, *Radiative Transfer* (Clarendon, Oxford, 1950).
16. Z. Sekera, "Scattering Matrices and Reciprocity Relationships for Various Representations of the State of Polarization," J. Opt. Soc. Am. **56**, 1732 (1966).
17. V. V. Varadan and V. K. Varadan, Eds., "Acoustic Electromagnetic and Elastic Ware Scattering: Focus on the T Matrix Approach," (Pergamon, New York, 1980).
18. M. Kerker, *The Scattering of Light* (Academic, New York, 1969).
19. C. Yeh, R. Woo, J. W. Armstrong, and A. Ishimaru, "Scattering by Pruppacher-Pitter Raindrops at 30 GHz," Radio Sci. **17**, 757 (1982)).
20. C. Yeh, R. Woo, A. Ishimaru, and J. W. Armstrong, "Scattering by Single Ice Needles and Plates at 30 GHz," Radio Sci. **17**, 1503 (1982).

Reprinted from THE PHYSICAL REVIEW, Vol. 135, No. 5A, A1193–A1201, 31 August, 1964
Printed in U. S. A.

Perturbation Approach to the Diffraction of Electromagnetic Waves by Arbitrarily Shaped Dielectric Obstacles*

C. YEH
Electrical Engineering Department, University of Southern California, Los Angeles, California
(Received 6 March 1964; revised manuscript received 30 April 1964)

A perturbation method is developed to consider the problem of the diffraction of electromagnetic waves by an arbitrarily shaped dielectric obstacle whose boundary may be expressed in the general form, in spherical coordinates, $r_p=r_0[1+\delta f_1(\theta,\phi)+\delta^2 f_2(\theta,\phi)+\cdots]$ where r_0 is the radius of an unperturbed sphere and $f_n(\theta,\phi)$ are arbitrary, single-valued and analytic functions. δ is chosen such that

$$\sum_{n=1}^{\infty} |\delta^n f_n(\theta,\phi)| < 1,\ 0\leq\theta\leq\pi,\ 0\leq\phi\leq 2\pi.$$

Detailed analysis is carried out to the first order in δ. Procedures to obtain higher order terms are also indicated. The perturbation solutions are valid for the near zone region of the obstacle as well as for the far zone region and they are applicable for all frequencies. Possible applications of this perturbation technique to elementary-particle scattering problems and other electromagnetic scattering problems are noted.

I. INTRODUCTION

THE exact solution of the problem of the diffraction of electromagnetic waves by an obstacle of given shape and electromagnetic properties can be found only in a few cases.[1,2] For example, the diffraction of waves by a conducting or dielectric sphere, by dielectric coated spheres and by a perfectly conducting disk are the few three-dimensional problems that have been solved rigorously. The need for approximate methods to treat the more general cases of diffraction from arbitrarily sphaped obstacles is quite apparent. The variational principles[3,4] provide a very powerful tool in obtaining an approximate expression for the scattering cross section; but it is not possible to derive from the variational principles a description of the electromagnetic fields. Furthermore, the success of the variational approach depends to a great extent on the trial function. At low frequencies, the Rayleigh method is very useful.[5,6] However, the solutions of Laplace's equation are still required. At very high frequencies, the treatment of diffraction problems by geometric and physical optics techniques developed by Fock[7] and Keller[8] is very successful. An approximate or perturbation method in the medium frequency range still remains to be found.

* This work was supported by the Air Force Cambridge Research Laboratories.

[1] R. King and T. T. Wu, *The Scattering and Diffraction of Waves* (Harvard University Press, Cambridge, Massachusetts, 1959).
[2] C. J. Bouwkamp, Rept. Progr. Phys. 17, 35 (1954).
[3] P. M. Morse and H. Feshbach, *Methods of Theoretical Physics* (McGraw-Hill Book Company, Inc., New York, 1953).
[4] H. Levine and J. Schwinger, *Theory of Electromagnetic Waves* (Interscience Publications, Inc., New York, 1951).
[5] Lord Rayleigh, Phil. Mag. 44, 28 (1897).
[6] A. F. Stevenson, J. Appl. Phys. 24, 1134 (1953).
[7] V. A. Fock, J. Phys. (USSR) 10, 130 (1946); 10, 399 (1946); see also *Thirteen Papers by V. A. Fock*, edited by N. A. Logan (Antenna Laboratory, Air Force Cambridge Research Center, Bedford, Massachusetts, 1957).
[8] J. B. Keller, J. Opt. Soc. Am. 52, 102 (1962).

In this paper, the boundary perturbation technique[9] will be extended to consider the problem of diffraction of waves by a dielectric object with perturbed boundary. This perturbation method is based on a Taylor expansion of the boundary conditions at the perturbed boundary.[10] Since this approach attacks the complete boundary-value problem, the perturbation solution for the field components is valid for the near zone (i.e., near the obstacle) as well as for the far zone and is valid for all frequencies. Similar procedure has been used recently by Erma[11] in his treatment of the electrostatic problem for irregularly shaped conductors.

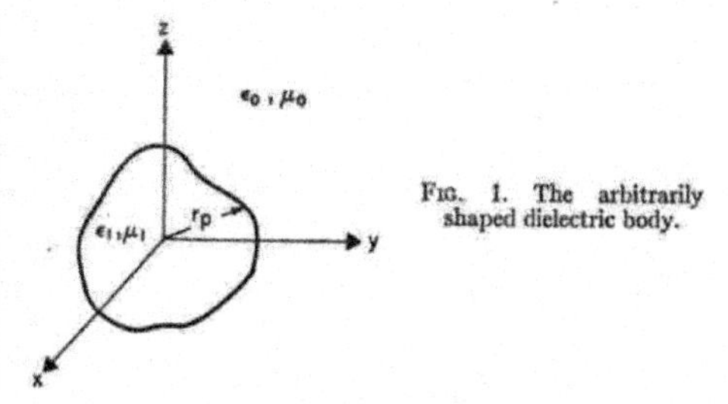

FIG. 1. The arbitrarily shaped dielectric body.

II. THE PERTURBATION SOLUTION

It is assumed that an arbitrarily shaped dielectric body which has a permittivity ϵ_1 and a permeability μ_1, is embedded in a homogeneous dielectric medium (ϵ_0,μ_0). The boundary of the dielectric body (Fig. 1) takes the shape of a perturbed sphere which may be expressed by the following equation

$$r_p=r_0(1+\delta f_1(\theta,\phi)+\delta^2 f_2(\theta,\phi)+\cdots), \tag{1a}$$

where r_0 is the radius of the unperturbed sphere, δ is a smallness parameter, and the $f_n(\theta,\phi)$ are arbitrary, single valued, continuous functions satisfying the conditions

$$f_n(\theta,0)=f_n(\theta,2\pi);\quad \sum_{n=1}^{\infty}|\delta^n f_n(\theta,\phi)|<1,\qquad 0\leq\theta\leq\pi,\quad 0\leq\phi\leq 2\pi. \tag{1b}$$

The spherical coordinates (r,θ,ϕ) are used.

Let the given exciting field (which need not necessarily be a plane wave) be denoted by $\mathbf{E}^{(i)}$, $\mathbf{H}^{(i)}$, the scattered field by $\mathbf{E}^{(s)}$, $\mathbf{H}^{(s)}$, and the field inside the dielectric body by $\mathbf{E}^{(t)}$, $\mathbf{H}^{(t)}$. The zeroth-order solution will be designated by a subscript 0, the first-order solution by subscript 1, etc. Hence, the resultant scattered fields and the resultant transmitted fields inside the body are respectively,

$$\begin{aligned}\mathbf{E}^{(s)}&=\mathbf{E}_0^{(s)}+\delta\mathbf{E}_1^{(s)}+\delta^2\mathbf{E}_2^{(s)}+\cdots,\\ \mathbf{H}^{(s)}&=\mathbf{H}_0^{(s)}+\delta\mathbf{H}_1^{(s)}+\delta^2\mathbf{H}_2^{(s)}+\cdots,\end{aligned} \tag{2}$$

and

$$\begin{aligned}\mathbf{E}^{(t)}&=\mathbf{E}_0^{(t)}+\delta\mathbf{E}_1^{(t)}+\delta^2\mathbf{E}_2^{(t)}+\cdots,\\ \mathbf{H}^{(t)}&=\mathbf{H}_0^{(t)}+\delta\mathbf{H}_1^{(t)}+\delta^2\mathbf{H}_2^{(t)}+\cdots.\end{aligned} \tag{3}$$

The higher order solutions are generated from the known zeroth-order solution; i.e., $\mathbf{E}^{(i)}$, $\mathbf{H}^{(i)}$, $\mathbf{E}_0^{(s)}$, $\mathbf{H}_0^{(s)}$, $\mathbf{E}_0^{(t)}$, and $\mathbf{H}_0^{(t)}$ are assumed known quantities. For the sake of clarity and simplicity, only the first-order solution will be carried out in detail. The higher order solution can be obtained in a similar fashion.

The boundary conditions require the continuity of tangential electric and magnetic fields at the boundary surface $r=r_p$:

$$\mathbf{n}\times[\mathbf{E}^{(i)}(r_p,\theta,\phi)+\mathbf{E}^{(s)}(r_p,\theta,\phi)]=\mathbf{n}\times\mathbf{E}^{(t)}(r_p,\theta,\phi), \tag{4}$$

$$\mathbf{n}\times[\mathbf{H}^{(i)}(r_p,\theta,\phi)+\mathbf{H}^{(s)}(r_p,\theta,\phi)]=\mathbf{n}\times\mathbf{H}^{(t)}(r_p,\theta,\phi), \tag{5}$$

where $\mathbf{n}$ is a unit vector outward normal to the boundary surface and can be written as

$$\mathbf{n}\simeq\mathbf{e}_r-\delta\frac{\partial f_1}{\partial\theta}\mathbf{e}_\theta-\delta\frac{1}{\sin\theta}\frac{\partial f_1}{\partial\phi}\mathbf{e}_\phi, \tag{6}$$

to the first order in δ in spherical coordinates. $\mathbf{e}_r$, $\mathbf{e}_\theta$, and $\mathbf{e}_\phi$ are respectively the unit vectors in r, θ, and ϕ directions. f_1 has been defined in Eq. (1). Carrying out the vector operations and expressing Eqs. (4) and (5) to the first order in δ in component form with the help of Eqs. (2) and (3), one obtains

$$\begin{aligned}\mathbf{e}_r:\quad &\delta(\partial f_1/\partial\theta)[E_\phi^{(i)}(r_p,\theta,\phi)+E_{0\phi}^{(s)}(r_p,\theta,\phi)]+\delta\frac{1}{\sin\theta}\frac{\partial f_1}{\partial\phi}[E_\theta^{(i)}(r_p,\theta,\phi)+E_{0\theta}^{(s)}(r_p,\theta,\phi)]\\ &=\delta\frac{\partial f_1}{\partial\theta}E_{0\phi}^{(t)}(r_p,\theta,\phi)+\delta\frac{1}{\sin\theta}\frac{\partial f_1}{\partial\phi}E_{0\theta}^{(t)}(r_p,\theta,\phi).\end{aligned} \tag{7}$$

$$\begin{aligned}\mathbf{e}_\theta:\quad &E_\phi^{(i)}(r_p,\theta,\phi)+E_{0\phi}^{(s)}(r_p,\theta,\phi)+\delta\left\{E_{1\phi}^{(s)}(r_p,\theta,\phi)-\frac{1}{\sin\theta}\frac{\partial f_1}{\partial\phi}[E_r^{(i)}(r_p,\theta,\phi)+E_r^{(s)}(r_p,\theta,\phi)]\right\}\\ &=E_{0\phi}^{(t)}(r_p,\theta,\phi)+\delta\left\{E_{1\phi}^{(t)}(r_p,\theta,\phi)-\frac{1}{\sin\theta}\frac{\partial f_1}{\partial\phi}E_r^{(t)}(r_p,\theta,\phi)\right\}.\end{aligned} \tag{8}$$

[9] P. M. Morse and H. Feshbach, J. Opt. Soc. Am. 52, 1052 (1962).
[10] See, for example, P. C. Clemmow and V. H. Weston, Proc. Roy. Soc. (London) A264, 246 (1961); C. J. Marcinkowki and L. B. Felsen, J. Res. Natl. Bur. Std. 66D, 699 (1962); 66D, 707 (1962).
[11] V. A. Erma, J. Math. Phys. 4, 1517 (1963).

$$\mathbf{e}_\phi:\ E_\theta^{(i)}(r_p,\theta,\phi)+E_{0\theta}^{(s)}(r_p,\theta,\phi)+\delta\{E_{1\theta}^{(s)}(r_p,\theta,\phi)+(\partial f_1/\partial\theta)[E_r^{(i)}(r_p,\theta,\phi)+E_{0r}^{(s)}(r_p,\theta,\phi)]\}$$
$$=E_{0\theta}^{(t)}(r_p,\theta,\phi)+\delta\{E_{1\theta}^{(t)}(r_p,\theta,\phi)+(\partial f_1/\partial\theta)E_{0r}^{(t)}(r_p,\theta,\phi)\}. \tag{9}$$

$$\mathbf{e}_r:\ \delta(\partial f_1/\partial\theta)[H_\phi^{(i)}(r_p,\theta,\phi)+H_{0\phi}^{(s)}(r_p,\theta,\phi)]+\delta\frac{1}{\sin\theta}\frac{\partial f_1}{\partial\phi}[H_\theta^{(i)}(r_p,\theta,\phi)+H_{0\theta}^{(s)}(r_p,\theta,\phi)]$$
$$=\delta\frac{\partial f_1}{\partial\theta}H_{0\phi}^{(t)}(r_p,\theta,\phi)+\delta\frac{1}{\sin\theta}\frac{\partial f_1}{\partial\phi}H_{0\theta}^{(t)}(r_p,\theta,\phi). \tag{10}$$

$$\mathbf{e}_\theta:\ H_\phi^{(i)}(r_p,\theta,\phi)+H_{0\phi}^{(s)}(r_p,\theta,\phi)+\delta\left\{H_{1\phi}^{(s)}(r_p,\theta,\phi)-\frac{1}{\sin\theta}\frac{\partial f_1}{\partial\phi}[H_r^{(i)}(r_p,\theta,\phi)+H_{0r}^{(s)}(r_p,\theta,\phi)]\right\}$$
$$=H_{0\phi}^{(t)}(r_p,\theta,\phi)+\delta\left\{H_{1\phi}^{(t)}(r_p,\theta,\phi)-\frac{1}{\sin\theta}\frac{\partial f_1}{\partial\phi}H_{0r}^{(t)}(r_p,\theta,\phi)\right\}. \tag{11}$$

$$\mathbf{e}_\phi:\ H_\theta^{(i)}(r_p,\theta,\phi)+H_{0\theta}^{(s)}(r_p,\theta,\phi)+\delta\{H_{1\theta}^{(s)}(r_p,\theta,\phi)+(\partial f_1/\partial\theta)[H_r^{(i)}(r_p,\theta,\phi)+H_{0r}^{(s)}(r_p,\theta,\phi)]\}$$
$$=H_{0\theta}^{(t)}(r_p,\theta,\phi)+\delta\{H_{1\theta}^{(t)}(r_p,\theta,\phi)+(\partial f_1/\partial\theta)H_{0r}^{(t)}(r_p,\theta,\phi)\}. \tag{12}$$

Equations (7) and (10) are satisfied by the zeroth-order solution. We now expand the above functions in Eqs. (8), (9), (11), and (12) to order δ in Taylor series about the unperturbed boundary $r=r_0$, obtaining

$$E_\phi^{(i)}(r_0,\theta,\phi)+E_{0\phi}^{(s)}(r_0,\theta,\phi)-E_{0\phi}^{(t)}(r_0,\theta,\phi)$$
$$=\delta\left\{\frac{1}{\sin\theta}\frac{\partial f_1}{\partial\phi}[E_r^{(i)}(r_0,\theta,\phi)+E_{0r}^{(s)}(r_0,\theta,\phi)]-E_{1\phi}^{(s)}(r_0,\theta,\phi)-r_0f_1[E_\phi^{(i)\prime}(r_0,\theta,\phi)+E_{0\phi}^{(s)\prime}(r_0,\theta,\phi)]\right\}$$
$$-\delta\left\{\frac{1}{\sin\theta}\frac{\partial f_1}{\partial\phi}E_{0r}^{(t)}(r_0,\theta,\phi)-E_{1\phi}^{(t)}(r_0,\theta,\phi)-r_0f_1E_{0\phi}^{(t)\prime}(r_0,\theta,\phi)\right\}, \tag{13}$$

$$E_\theta^{(i)}(r_0,\theta,\phi)+E_{0\theta}^{(s)}(r_0,\theta,\phi)-E_{0\theta}^{(t)}(r_0,\theta,\phi)$$
$$=-\delta\{E_{1\theta}^{(s)}(r_0,\theta,\phi)+(\partial f_1/\partial\theta)[E_r^{(i)}(r_0,\theta,\phi)+E_{0r}^{(s)}(r_0,\theta,\phi)]+r_0f_1[E_\theta^{(i)\prime}(r_0,\theta,\phi)+E_{0\theta}^{(s)\prime}(r_0,\theta,\phi)]\}$$
$$+\delta\{E_{1\theta}^{(t)}(r_0,\theta,\phi)+(\partial f_1/\partial\theta)E_{0r}^{(t)}(r_0,\theta,\phi)+r_0f_1E_{0\theta}^{(t)\prime}(r_0,\theta,\phi)\}, \tag{14}$$

$$H_\phi^{(i)}(r_0,\theta,\phi)+H_{0\phi}^{(s)}(r_0,\theta,\phi)-H_{0\phi}^{(t)}(r_0,\theta,\phi)$$
$$=\delta\left\{\frac{1}{\sin\theta}\frac{\partial f_1}{\partial\phi}[H_r^{(i)}(r_0,\theta,\phi)+H_{0r}^{(s)}(r_0,\theta,\phi)]-H_{1\phi}^{(s)}(r_0,\theta,\phi)-r_0f_1[H_\phi^{(i)\prime}(r_0,\theta,\phi)+H_{0\phi}^{(s)\prime}(r_0,\theta,\phi)]\right\}$$
$$-\delta\left\{\frac{1}{\sin\theta}\frac{\partial f_1}{\partial\phi}H_{0r}^{(t)}(r_0,\theta,\phi)-H_{1\phi}^{(t)}(r_0,\theta,\phi)-r_0f_1H_{0\phi}^{(t)\prime}(r_0,\theta,\phi)\right\}, \tag{15}$$

$$H_\theta^{(i)}(r_0,\theta,\phi)+H_{0\theta}^{(s)}(r_0,\theta,\phi)-H_{0\theta}^{(t)}(r_0,\theta,\phi)$$
$$=-\delta\{H_{1\theta}^{(s)}(r_0,\theta,\phi)+(\partial f_1/\partial\theta)[H_r^{(i)}(r_0,\theta,\phi)+H_{0r}^{(s)}(r_0,\theta,\phi)]+r_0f_1[H_\theta^{(i)\prime}(r_0,\theta,\phi)+H_{0\theta}^{(s)\prime}(r_0,\theta,\phi)]\}$$
$$+\delta\{H_{1\theta}^{(t)}(r_0,\theta,\phi)+(\partial f_1/\partial\theta)H_{0r}^{(t)}(r_0,\theta,\phi)+r_0f_1H_{0\theta}^{(t)\prime}(r_0,\theta,\phi)\}, \tag{16}$$

where the prime signified the derivative of the function with respect to r_0. The left-hand sides of the above equations are equal to zero by virtue of the zeroth-order solution. Hence, the right-hand sides of the above equations must vanish identically. Rearranging and combining Eqs. (13) and (14) gives

$$[E_{1\theta}^{(s)}(r_0,\theta,\phi)-E_{1\theta}^{(t)}(r_0,\theta,\phi)]\mathbf{e}_\theta+[E_{1\phi}^{(s)}(r_0,\theta,\phi)-E_{1\phi}^{(t)}(r_0,\theta,\phi)]\mathbf{e}_\phi=u_1(r_0,\theta,\phi)\mathbf{e}_\theta+u_2(r_0,\theta,\phi)\mathbf{e}_\phi, \tag{17}$$

and combining Eqs. (15) and (16) gives

$$[H_{1\theta}^{(s)}(r_0,\theta,\phi)-H_{1\theta}^{(t)}(r_0,\theta,\phi)]\mathbf{e}_\theta+[H_{1\phi}^{(s)}(r_0,\theta,\phi)-H_{1\phi}^{(t)}(r_0,\theta,\phi)]\mathbf{e}_\phi=v_1(r_0,\theta,\phi)\mathbf{e}_\theta+v_2(r_0,\theta,\phi)\mathbf{e}_\phi, \tag{18}$$

where

$$u_1(r_0,\theta,\phi)=(\partial f_1/\partial\theta)[E_{0r}^{(i)}(r_0,\theta,\phi)-E_r^{(t)}(r_0,\theta,\phi)-E_{0r}^{(s)}(r_0,\theta,\phi)]$$
$$+r_0f_1[E_{0\theta}^{(i)\prime}(r_0,\theta,\phi)-E_\theta^{(t)\prime}(r_0,\theta,\phi)-E_{0\theta}^{(s)\prime}(r_0,\theta,\phi)],$$

$$u_2(r_0,\theta,\phi)=\frac{1}{\sin\theta}\frac{\partial f_1}{\partial\phi}[E_r^{(t)}(r_0,\theta,\phi)+E_{0r}^{(s)}(r_0,\theta,\phi)-E_{0r}^{(i)}(r_0,\theta,\phi)]$$
$$+r_0f_1[E_{0\phi}^{(i)\prime}(r_0,\theta,\phi)-E_\phi^{(t)\prime}(r_0,\theta,\phi)-E_{0\phi}^{(s)\prime}(r_0,\theta,\phi)], \quad (19)$$

$$v_1(r_0,\theta,\phi)=(\partial f_1/\partial\theta)[H_{0r}^{(i)}(r_0,\theta,\phi)-H_r^{(t)}(r_0,\theta,\phi)-H_{0r}^{(s)}(r_0,\theta,\phi)]$$
$$+r_0f_1[H_{0\theta}^{(i)\prime}(r_0,\theta,\phi)-H_\theta^{(t)\prime}(r_0,\theta,\phi)-H_{0\theta}^{(s)\prime}(r_0,\theta,\phi)],$$

$$v_2(r_0,\theta,\phi)=\frac{1}{\sin\theta}\frac{\partial f_1}{\partial\phi}[H_r^{(t)}(r_0,\theta,\phi)+H_{0r}^{(s)}(r_0,\theta,\phi)-H_{0r}^{(i)}(r_0,\theta,\phi)]$$
$$+r_0f_1[H_{0\phi}^{(i)\prime}(r_0,\theta,\phi)-H_\phi^{(t)\prime}(r_0,\theta,\phi)-H_{0\phi}^{(s)\prime}(r_0,\theta,\phi)].$$

It is noted that the resultant fields given by Eqs. (2) and (3) must satisfy the wave equation. It is therefore clear that each term in Eqs. (2) and (3) must separately satisfy the wave equation. Consequently, the general expressions for $\mathbf{E}_1^{(s)}$, $\mathbf{H}_1^{(s)}$, $\mathbf{E}_1^{(t)}$, and $\mathbf{H}_1^{(t)}$, that are appropriate to the present problem, are[12]

$$\mathbf{E}_1^{(s)}=\sum_{m,n}A_{e,omn}\mathbf{M}_{e,omn}^{(s)}+B_{e,omn}\mathbf{N}_{e,omn}^{(s)}, \quad (20)$$

$$\mathbf{H}_1^{(s)}=\sum_{m,n}\frac{k_0}{i\omega\mu_0}(A_{e,omn}\mathbf{N}_{e,omn}^{(s)}+B_{e,omn}\mathbf{M}_{e,omn}^{(s)}), \quad (21)$$

$$\mathbf{E}_1^{(t)}=\sum_{m,n}C_{e,omn}\mathbf{M}_{e,omn}^{(t)}+D_{e,omn}\mathbf{N}_{e,omn}^{(t)}, \quad (22)$$

$$\mathbf{H}_1^{(t)}=\sum_{m,n}\frac{k_1}{i\omega\mu_1}(C_{e,omn}\mathbf{N}_{e,omn}^{(t)}+D_{e,omn}\mathbf{M}_{e,omn}^{(t)}), \quad (23)$$

where

$$\begin{aligned}
\mathbf{M}_{e,omn}^{(s)}&=h_n^{(1)}(k_0r)\mathbf{m}_{e,omn},\\
\mathbf{N}_{e,omn}^{(s)}&=\frac{1}{k_0r}h_n^{(1)}(k_0r)\mathbf{l}_{e,omn}+\frac{1}{k_0r}\frac{\partial}{\partial r}[rh_n^{(1)}(k_0r)](\mathbf{e}_r\times\mathbf{m}_{e,omn}),\\
\mathbf{M}_{e,omn}^{(t)}&=j_n(k_1r)\mathbf{m}_{e,omn},\\
\mathbf{N}_{e,omn}^{(t)}&=\frac{1}{k_1r}j_n(k_1r)\mathbf{l}_{e,omn}+\frac{1}{k_1r}\frac{\partial}{\partial r}[rj_n(k_1r)](\mathbf{e}_r\times\mathbf{m}_{e,omn}),
\end{aligned} \quad (24)$$

with

$$\begin{aligned}
\mathbf{m}_{e,omn}&=\mp\frac{mP_n^m(\cos\theta)}{\sin\theta}\,{}^{\sin}_{\cos}\,m\phi\,\mathbf{e}_\theta-\frac{\partial P_n^m(\cos\theta)}{\partial\theta}\,{}^{\cos}_{\sin}\,m\phi\,\mathbf{e}_\phi,\\
\mathbf{l}_{e,omn}&=n(n+1)P_n^m(\cos\theta)\,{}^{\cos}_{\sin}\,m\phi\,\mathbf{e}_r.
\end{aligned} \quad (25)$$

$h_n^{(1)}(k_0r)$ and $j_n(k_1r)$ are, respectively, spherical Hankel and spherical Bessel functions; $P_n^m(\cos\theta)$ are associated Legendre polynomials. $k_0^2=\omega^2\mu_0\epsilon_0$ and $k_1^2=\omega^2\mu_1\epsilon_1$. $A_{e,omn}$, $B_{e,omn}$, $C_{e,omn}$ and $D_{e,omn}$ are yet unknown arbitrary constants that can be determined from Eqs. (17) and (18) using the orthogonality properties of the angular

[12] J. A. Stratton, *Electromagnetic Theory* (McGraw-Hill Book Company, Inc., New York, 1941).

functions. Substituting Eqs. (20) through (23) into Eqs. (17) and (18), and making use of the following orthogonality relations

$$\mathbf{l}_{e,omn}\cdot\mathbf{m}_{e,omn}=0,\quad \mathbf{l}_{e,omn}\cdot(\mathbf{e}_r\times\mathbf{m}_{e,omn})=0,\quad \mathbf{m}_{e,omn}\cdot(\mathbf{e}_r\times\mathbf{m}_{e,omn})=0,$$

$$\int_0^\pi\int_0^{2\pi}(\mathbf{l}_{e,omn}\cdot\mathbf{l}_{e,om'n'})\sin\theta d\theta d\phi=\begin{cases}0, & \text{for } m\neq m',\, n\neq n'\\ \dfrac{2n^2(n+1)^2}{2n+1}\dfrac{(n+m)!}{(m-m)!}(1+\delta_{0m})\pi & \text{for } m=m',\, n=n',\end{cases}$$

$$\int_0^\pi\int_0^{2\pi}(\mathbf{m}_{e,omn}\cdot\mathbf{m}_{e,om'n'})\sin\theta d\theta d\phi=\int_0^\pi\int_0^{2\pi}[(\mathbf{e}_r\times\mathbf{m}_{e,omn})\cdot(\mathbf{e}_r\times\mathbf{m}_{e,om'n'})]\sin\theta d\theta d\phi$$
$$=\begin{cases}0, & \text{for } m\neq m',\, n\neq n'\\ \dfrac{2n(n+1)}{2n+1}\dfrac{(n+m)!}{(n-m)!}(1+\delta_{0m})\pi, & \text{for } m=m',\, n=n',\end{cases} \tag{26}$$

$$\delta_{0m}=\begin{cases}1, & m=0\\ 0, & m>0,\end{cases}$$

one obtains

$$A_{e,omn}h_n^{(1)}(k_0r_0)-C_{e,omn}j_n(k_1r_0)=\frac{1}{p_{mn}}\int_0^\pi\int_0^{2\pi}\mathbf{u}\cdot\mathbf{m}_{e,omn}\sin\theta d\theta d\phi, \tag{27}$$

$$B_{e,omn}\frac{1}{k_0r_0}\frac{\partial}{\partial r_0}[r_0h_n^{(1)}(k_0r_0)]-D_{e,omn}\frac{1}{k_1r_0}\frac{\partial}{\partial r_0}[r_0j_n(k_1r_0)]=\frac{1}{p_{mn}}\int_0^\pi\int_0^{2\pi}\mathbf{u}\cdot(\mathbf{e}_r\times\mathbf{m}_{e,omn})\sin\theta d\theta d\phi, \tag{28}$$

$$A_{e,omn}\frac{1}{i\omega\mu_0r_0}\frac{\partial}{\partial r_0}[r_0h_n^{(1)}(k_0r_0)]-C_{e,omn}\frac{1}{i\omega\mu_1r_0}\frac{\partial}{\partial r_0}[r_0j_n(k_1r_0)]=\frac{1}{p_{mn}}\int_0^\pi\int_0^{2\pi}\mathbf{v}\cdot(\mathbf{e}_r\times\mathbf{m}_{e,omn})\sin\theta d\theta d\phi, \tag{29}$$

$$B_{e,omn}\frac{k_0}{i\omega\mu_0}h_n^{(1)}(k_0r_0)-D_{e,omn}\frac{k_1}{i\omega\mu_1}j_n(k_1r_0)=\frac{1}{p_{mn}}\int_0^\pi\int_0^{2\pi}\mathbf{v}\cdot\mathbf{m}_{e,omn}\sin\theta d\theta d\phi, \tag{30}$$

with

$$\mathbf{u}=u_1(r_0,\theta,\phi)\mathbf{e}_\theta+u_2(r_0,\theta,\phi)\mathbf{e}_\phi, \tag{31}$$

$$\mathbf{v}=v_1(r_0,\theta,\phi)\mathbf{e}_\theta+v_2(r_0,\theta,\phi)\mathbf{e}_\phi, \tag{32}$$

$$p_{mn}=\frac{2n+1}{2n(n+1)}\frac{(n-m)!}{(n+m)!}\frac{1}{(1+\delta_{0m})\pi}. \tag{33}$$

u_1, u_2, v_1, and v_2 are given by Eq. (19). The coefficients $A_{e,omn}$, $B_{e,omn}$, $C_{e,omn}$, and $D_{e,omn}$ can be found readily from the above equations. Substituting these coefficients back to Eqs. (20) through (23) gives the first-order correction to the electromagnetic fields due to the departure of the boundary surface from a perfect sphere with radius r_0. Higher order corrections can be found successively in the same manner. It is interesting to note from the above analysis that, in general, the perturbed wave will have all components of electromagnetic fields even if the incident wave is a pure *TE* wave ($E_r^{(i)}=0$) or a pure *TM* wave ($H_r^{(i)}=0$).

Since the exact solution to the problem of the diffraction of electromagnetic waves by a three-dimensional dielectric obstacle other than a sphere is not available, it is therefore not possible to compare the result obtained by the above perturbation approach with a known one. However, as a partial check, the problem of the diffraction of a plane wave by a dielectric sphere of radius $r_0(1+\delta)$ was carried out in detail using the above derived formulas. Results are found to be in complete agreement with the solutions obtained by expanding the exact solutions to the first order in δ.

III. AN EXAMPLE: THE SCATTERING OF PLANE WAVES BY A DIELECTRIC SPHEROID

As a less trivial example of the application of the theory derived in Sec. II, the problem of the scattering of plane waves by a dielectric spheroid with small eccentricity will be considered. It is assumed that the incident plane wave with its electric vector polarized in the x direction is propagating in the direction of the negative z axis. The equation of a spheroidal surface is given by

$$r_p=r_0[(1-2\delta\sin^2\theta)^{1/2}], \tag{34}$$

where

$$\delta=[1-(r_0/(r_0+\Delta r_0))^2] \tag{35}$$

($\delta<0$: prolate spheroid; $\delta>0$: oblate spheroid), and $2r_0$ and $2(r_0+\Delta r_0)$ are the lengths of the two axes of the

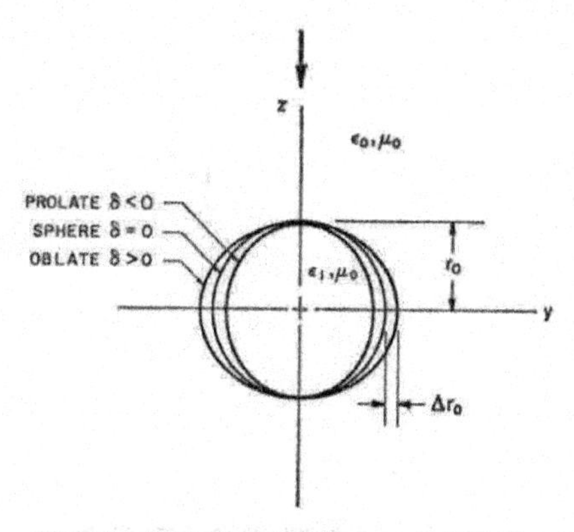

FIG. 2. The dielectric spheroid. The arrow indicates the direction of the incident wave.

spheroid. (See Fig. 2.) For small eccentricity, one has

$$r_p \simeq r_0[1+\delta\sin^2\theta]. \tag{36}$$

Comparing Eqs. (36) and (1a) gives

$$f_1(\theta,\phi)=\sin^2\theta. \tag{37}$$

The unperturbed solution to the problem of the scattering of plane waves by a dielectric sphere is well known[12]:

$$\mathbf{E}^{(i)}=\sum_{n=1}^{\infty}(-i)^n\frac{2n+1}{n(n+1)}(\mathbf{M}_{o1n}^{(1)}+i\mathbf{N}_{e1n}^{(1)}), \tag{38a}$$

$$\mathbf{H}^{(i)}=\sum_{n=1}^{\infty}\frac{k_0}{i\omega\mu_0}(-i)^n\frac{2n+1}{n(n+1)}(\mathbf{N}_{o1n}^{(1)}+i\mathbf{M}_{e1n}^{(1)}), \tag{38b}$$

$$\mathbf{E}_0^{(s)}=\sum_{n=1}^{\infty}(-i)^n\frac{2n+1}{n(n+1)}(a_n^s\mathbf{M}_{o1n}^{(s)}+ib_n^s\mathbf{N}_{e1n}^{(s)}), \tag{39a}$$

$$\mathbf{H}_0^{(s)}=\sum_{n=1}^{\infty}\frac{k_0}{i\omega\mu_0}(-i)^n\frac{2n+1}{n(n+1)}(a_n^s\mathbf{N}_{o1n}^{(s)}+ib_n^s\mathbf{M}_{e1n}^{(s)}), \tag{39b}$$

$$\mathbf{E}_0^{(t)}=\sum_{n=1}^{\infty}(-i)^n\frac{2n+1}{n(n+1)}(a_n^t\mathbf{M}_{o1n}^{(t)}+ib_n^t\mathbf{N}_{e1n}^{(t)}), \tag{40a}$$

$$\mathbf{H}_0^{(t)}=\sum_{n=1}^{\infty}\frac{k_0}{i\omega\mu_0}(-i)^n\frac{2n+1}{n(n+1)}\times(a_n^t\mathbf{N}_{e1n}^{(t)}+ib_n^t\mathbf{M}_{e1n}^{(t)}), \tag{40b}$$

with

$$a_n^s=\frac{\mu_0 j_n(k_0r_0)[k_1r_0j_n(k_1r_0)]'-\mu_1 j_n(k_1r_0)[k_0r_0j_n(k_0r_0)]'}{\mu_1 j_n(k_1r_0)[k_0r_0h_n^{(1)}(k_0r_0)]'-\mu_0 h_n^{(1)}(k_0r_0)[k_1r_0j_n(k_1r_0)]'}, \tag{41a}$$

$$b_n^s=\frac{(k_0/k_1)^2\mu_1 j_n(k_0r_0)[k_1r_0j_n(k_1r_0)]'-\mu_0 j_n(k_1r_0)[k_0r_0j_n(k_0r_0)]'}{\mu_0 j_n(k_1r_0)[k_0r_0h_n^{(1)}(k_0r_0)]'-\mu_1(k_0/k_1)^2h_n^{(1)}(k_0r_0)[k_1r_0j_n(k_1r_0)]'}, \tag{41b}$$

$$a_n^t=\frac{(-i)\mu_1/k_0r_0}{\mu_0 h_n^{(1)}(k_0r_0)[k_1r_0j_n(k_1r_0)]'-\mu_1 j_n(k_1r_0)[k_0r_0h_n^{(1)}(k_0r_0)]'}, \tag{42a}$$

$$b_n^t=\frac{(-i)\mu_1k_1r_0}{\mu_1(k_0r_0)^2h_n^{(1)}(k_0r_0)[k_1r_0j_n(k_1r_0)]'-\mu_0(k_1r_0)^2j_n(k_1r_0)[k_0r_0h_n^{(1)}(k_0r_0)]'}. \tag{42b}$$

$\mathbf{M}_{e,omn}^{(1)}$ and $\mathbf{N}_{e,omn}^{(1)}$ are obtained, respectively, by replacing $h_n^{(1)}(k_0r)$ by $j_n(k_0r)$ in $\mathbf{M}_{e,omn}^{(s)}$ and $\mathbf{N}_{e,omn}^{(s)}$. The prime in the above expressions denotes differentiation with respect to k_0r_0 or k_1r_0 as appropriate.

To find the first-order perturbation solution, we first substitute Eqs. (38) through (40) into Eq. (19) obtaining

$$u_1(r_0,\theta,\phi)=\sum_{p=1}^{\infty}(-i)^p\frac{2p+1}{p(p+1)}\left[P_p\frac{\partial f_1}{\partial\theta}(\mathbf{l}_{e1p}\cdot\mathbf{e}_r)+Q_pf_1(\mathbf{m}_{o1p}\cdot\mathbf{e}_\theta)+R_pf_1((\mathbf{e}_r\times\mathbf{m}_{e1p})\cdot\mathbf{e}_\theta)\right], \tag{43}$$

$$u_2(r_0,\theta,\phi)=\sum_{p=1}^{\infty}(-i)^p\frac{2p+1}{p(p+1)}[Q_pf_1(\mathbf{m}_{o1p}\cdot\mathbf{e}_\phi)+R_pf_1((\mathbf{e}_r\times\mathbf{m}_{e1p})\cdot\mathbf{e}_\phi)], \tag{44}$$

$$v_1(r_0,\theta,\phi)=\sum_{p=1}^{\infty}(-i)^p\frac{2p+1}{p(p+1)}\frac{k_0}{i\omega\mu_0}\left[S_p\frac{\partial f_1}{\partial\theta}(\mathbf{l}_{o1p}\cdot\mathbf{e}_r)+T_pf_1((\mathbf{e}_r\times\mathbf{m}_{o1p})\cdot\mathbf{e}_\theta)+U_pf_1(\mathbf{m}_{e1p}\cdot\mathbf{e}_\theta)\right], \tag{45}$$

$$v_2(r_0,\theta,\phi)=\sum_{p=1}^{\infty}(-i)^p\frac{2p+1}{p(p+1)}\frac{k_0}{i\omega\mu_0}[T_pf_1((\mathbf{e}_r\times\mathbf{m}_{o1p})\cdot\mathbf{e}_\phi)+U_pf_1(\mathbf{m}_{e1p}\cdot\mathbf{e}_\phi)], \tag{46}$$

where

$$P_p=(i/k_0r_0)[b_p^t(k_0/k_1)j_p(k_1r_0)-j_p(k_0r_0)-b_p^s h_p^{(1)}(k_0r_0)], \tag{47}$$

$$Q_p=a_p^t k_1r_0 j_p'(k_1r_0)-k_0r_0 j_p'(k_0r_0)-a_p^s k_0r_0 h_p^{(1)\prime}(k_0r_0), \tag{48}$$

$$R_p=ik_1r_0b_p^t[(1/k_1r_0)(k_1r_0j_p(k_1r_0))']'-ik_0r_0[(1/k_0r_0)(k_0r_0j_p(k_0r_0))']'-ik_0r_0b_p^s[(1/k_0r_0)(k_0r_0h_p^{(1)}(k_0r_0))']', \tag{49}$$

$$S_p=(k_1\mu_0/k_0\mu_1)(1/k_1r_0)a_p^t j_p(k_1r_0)-(1/k_0r_0)j_p(k_0r_0)-(1/k_0r_0)a_p^s h_p^{(1)}(k_0r_0), \tag{50}$$

$$T_p=(k_1\mu_0/k_0\mu_1)a_p^t k_1r_0[(1/k_1r_0)(k_1r_0j_p(k_1r_0))']'-k_0r_0[(1/k_0r_0)(k_0r_0j_p(k_0r_0))']' -a_p^s k_0r_0[(1/k_0r_0)(k_0r_0h_p^{(1)}(k_0r_0))']', \tag{51}$$

$$U_p=i(k_1\mu_0/k_0\mu_1)b_p^t k_1r_0 j_n'(k_1r_0)-ik_0r_0j_n'(k_0r_0)-ib_n^s k_0r_0h_n^{(1)\prime}(k_0r_0). \tag{52}$$

The expansion coefficients for the first-order perturbation fields are then found by putting expressions (43) through (46) into Eqs. (27) through (30) and carrying out the integration where possible. One has

$$\begin{aligned}
&A_{emn}=B_{omn}=C_{emn}=D_{omn}=0 \quad \text{for all } m \text{ and } n,\\
&A_{omn}=B_{emn}=C_{omn}=D_{emn}=0 \quad \text{for} \quad m\neq 1 \text{ and all } n,\\
&A_{o1n}=[\chi_{o1n}j_n(k_1r_0)-\alpha_{o1n}(k_1\mu_0/k_0\mu_1)(1/k_1r_0)(k_1r_0j_n(k_1r_0))']/\Gamma,\\
&B_{e1n}=[-\gamma_{e1n}(1/k_1r_0)(k_1r_0j_n(k_1r_0))'+\beta_{e1n}(k_1\mu_0/k_0\mu_1)j_n(k_1r_0)](k_0\mu_1/k_1\mu_0)/\Gamma,\\
&C_{o1n}=[\chi_{o1n}h_n^{(1)}(k_0r_0)-\alpha_{o1n}(1/k_0r_0)(k_0r_0h_n^{(1)}(k_0r_0))']/\Gamma,\\
&D_{e1n}=[-\beta_{e1n}h_n^{(1)}(k_0r_0)+\gamma_{e1n}(1/k_0r_0)(k_0r_0h_n^{(1)}(k_0r_0))'](k_0\mu_1/k_1\mu_0)/\Gamma,
\end{aligned} \tag{53}$$

with

$$\Gamma=(1/k_0r_0)(k_0r_0h_n^{(1)}(k_0r_0))'j_n(k_1r_0)-(k_1\mu_0/k_0\mu_1)(1/k_1r_0)(k_1r_0j_n(k_1r_0))'h_n^{(1)}(k_0r_0),$$

$$\begin{aligned}
\alpha_{o1n}&=\frac{\pi}{p_{1n}}\sum_{p=1}^{\infty}(-i)^p\frac{2p+1}{p(p+1)}[P_pJ_{11n}{}^p+Q_pJ_{21n}{}^p+R_pJ_{31n}{}^p],\\
\beta_{e1n}&=\frac{\pi}{p_{1n}}\sum_{p=1}^{\infty}(-i)^p\frac{2p+1}{p(p+1)}[P_pJ_{41n}{}^p+Q_pJ_{31n}{}^p+R_pJ_{21n}{}^p],\\
\gamma_{e1n}&=\frac{\pi}{p_{1n}}\sum_{p=1}^{\infty}(-i)^p\frac{2p+1}{p(p+1)}[S_pJ_{11n}{}^p+T_pJ_{31n}{}^p+U_pJ_{21n}{}^p],\\
\chi_{e1n}&=\frac{\pi}{p_{1n}}\sum_{p=1}^{\infty}(-i)^p\frac{2p+1}{p(p+1)}[S_pJ_{41n}{}^p+T_pJ_{21n}{}^p+U_pJ_{31n}{}^p],
\end{aligned} \tag{54}$$

where $J_{11n}{}^p$, $J_{21n}{}^p$, $J_{31n}{}^p$, and $J_{41n}{}^p$, which are definite integrals involving the associated Legendre functions, are given in the Appendix. Hence, the scattered fields correct to the first order in δ are

$$\mathbf{E}^{(s)}=\sum_{n=1}^{\infty}\left[(-i)^n\frac{2n+1}{n(n+1)}a_n^s+\delta A_{o1n}\right]\mathbf{M}_{o1n}{}^{(s)}+\left[i(-i)^n\frac{2n+1}{n(n+1)}b_n^s+\delta B_{e1n}\right]\mathbf{N}_{e1n}{}^{(s)}, \tag{55}$$

$$\mathbf{H}^{(s)}=\sum_{n=1}^{\infty}\frac{k_0}{i\omega\mu_0}\left\{\left[(-i)^n\frac{2n+1}{n(n+1)}a_n^s+\delta A_{o1n}\right]\mathbf{N}_{o1n}{}^{(s)}+\left[i(-i)^n\frac{2n+1}{n(n+1)}b_n^s+\delta B_{e1n}\right]\mathbf{M}_{e1n}{}^{(s)}\right\}. \tag{56}$$

Of particular interest is the far zone behavior of the scattered field. The radial component of the scattered field may be neglected at large r because of its rapid fall off compared to the θ or ϕ component. Consequently, the scattered field has the form of a spherically outgoing wave, i.e.,

$$\mathbf{E}^{(s)}\sim\frac{e^{ik_0r}}{k_0r}\sum_{n=1}^{\infty}(-i)^{n+1}\left\{\left[V_n\frac{P_n{}^1(\cos\theta)}{\sin\theta}-iW_n\frac{\partial}{\partial\theta}P_n{}^1(\cos\theta)\right]\cos\phi\mathbf{e}_\theta-\left[V_n\frac{\partial}{\partial\theta}P_n{}^1(\cos\theta)-iW_n\frac{P_n{}^1(\cos\theta)}{\sin\theta}\right]\sin\phi\mathbf{e}_\phi\right\}, \tag{57}$$

where

$$V_n=(-i)^n[(2n+1)/n(n+1)]a_n^s+\delta A_{e1n}, \quad (58)$$

$$W_n=i(-i)^n[(2n+1)/n(n+1)]b_n^s+\delta B_{e1n}. \quad (59)$$

Rewriting Eq. (57) gives

$$\mathbf{E}^{(s)}\sim E_\theta^{(s)}\mathbf{e}_\theta+E_\phi^{(s)}\mathbf{e}_\phi, \quad (60)$$

with

$$E_\theta^{(s)}=(e^{ik_0r}/k_0r)\cos\phi S_1(\theta), \quad (61)$$

$$E_\phi^{(s)}=-(e^{ik_0r}/k_0r)\sin\phi S_2(\theta), \quad (62)$$

where $S_1(\theta)$ and $S_2(\theta)$ are called the complex amplitudes of the scattered radiation for the two polarizations. The squares of the absolute values of $S_1(\theta)$ and $S_2(\theta)$ are called the intensities of the scattered radiation for the two polarizations; i.e.,

$$I_\theta=|S_1(\theta)|^2, \text{ and } I_\phi=|S_2(\theta)|^2. \quad (63)$$

The backscattering cross section or the radar cross section is also of interest. It is defined by

$$\sigma=\lim_{r\to\infty}4\pi r^2|E^s|^2/|E^i|^2|_{\theta=0^\circ}, \quad (64)$$

or from Eq. (57),

$$\sigma=\frac{4\pi}{k_0^2}\left|\sum_{n=1}^{\infty}(-i)^{n+1}\frac{n(n+1)}{2}(V_n-iW_n)\right|^2. \quad (65)$$

Simplifying gives

$$\sigma\simeq\frac{4\pi}{k_0^2}\Bigg\{\left|\sum_{n=1}^{\infty}(-1)^n\frac{2n+1}{2}(-i)(a_n^s+b_n^s)\right|^2 +\delta\Bigg[\left(\sum_{n=1}^{\infty}(-1)^n\frac{2n+1}{2}(-i)(a_n^s+b_n^s)\right)\times\left(\sum_{n=1}^{\infty}(-i)^{n+1}\frac{n(n+1)}{2}(A_{e1n}-iB_{e1n})\right)^* +\left(\sum_{n=1}^{\infty}(-1)^n\frac{2n+1}{2}(-i)(a_n^s+b_n^s)\right)^*\times\left(\sum_{n=1}^{\infty}(-i)^{n+1}\frac{n(n+1)}{2}(A_{e1n}-iB_{e1n})\right)\Bigg]\Bigg\}, \quad (66)$$

where the asterisk indicates the complex conjugate of the function. The first term on the right-hand side of the above equation represents the backscattering cross section of an unperturbed sphere, while the other term corresponds to the first-order correction due to small eccentricity.

To qualitatively illustrate how the solutions behave, numerical computations are carried out using the high-speed IBM-7094 computer. It is assumed that $(\epsilon_1/\epsilon_0)^{1/2}=1.33$ and $\mu_1/\mu_0=1.0$. The Bessel functions and associated Legendre functions are computed from available subroutines. The integrals in the Appendix are evaluated numerically by Simpson's rule. Five cases of the spheroidal shape are considered:

$$(r_0+\Delta r_0)/r_0=0.9, 0.95, 1.0, 1.05, 1.1.$$

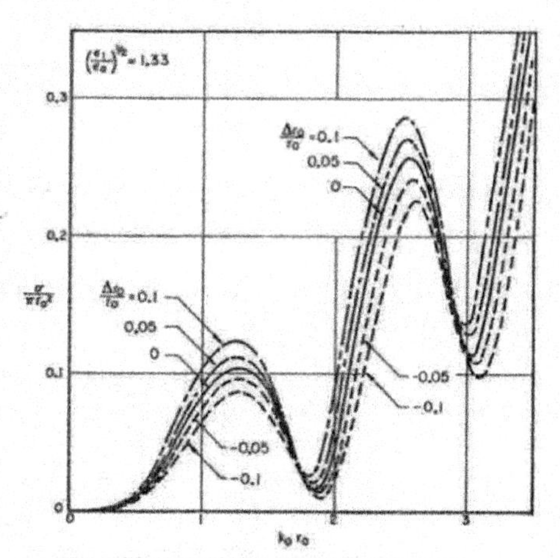

FIG. 3. The normalized backscattering cross sections for nose-on incidence.

The normalized backscattering cross sections $(\sigma/\pi r_0^2)$ as a function of k_0r_0 for $0\lesssim k_0r_0\lesssim 3.5$ for these five cases have been computed. Results are given in Fig. 3. Figure 4 shows the variation of the polarization of the scattered wave as a function of the polar angle θ with $k_0r_0=2$ for various spheroidal shapes. The polarization is often defined as

$$P=(I_\parallel-I_\perp)/(I_\parallel+I_\perp).$$

For the present case under consideration, $I_\parallel=I_\theta$ and $I_\perp=I_\phi$. It can be observed from these figures that, in general, polarization shows a greater sensitivity to the deformation of the spherical obstacle than does the normalized backscattering cross section.

It should be noted that although the numerical results given here are computed from the first-order solutions it is still expected that the results would be good approximation to the exact solutions for $|\Delta r_0/r_0|\lesssim 0.05$.

IV. CONCLUSIONS

The problem of the diffraction of electromagnetic waves by a dielectric body with perturbed boundary has been considered using the boundary perturbation technique. The solution is valid for the near zone (i.e., near the dielectric body) as well as for the far zone and is good for all frequencies. Since the perturbation solution satisfies Maxwell's equations, the boundary conditions, and the radiation condition for the scattered wave at infinity, hence it is unique. It should be noted that with slight modifications of Eqs. (20) through (23) the above derived results are also applicable for a

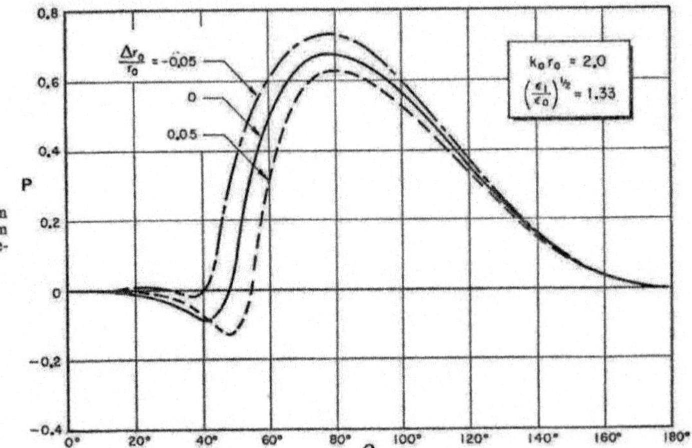

FIG. 4. The polarization of light scattered from dielectric spheroid for nose-on incidence.

radially inhomogeneous dielectric body with perturbed boundary.[13]

Further applications of this perturbation technique can be found in the scattering of electromagnetic or acoustic waves from hard or soft objects, in the scattering of x ray or of light by interstellar matter, and in elementary particle scattering theory.

ACKNOWLEDGMENTS

I wish to thank the reviewer for calling my attention to previous related work as represented by Ref. 10. I also would like to thank Mrs. G. Knudson of the Hughes Aircraft Company for providing the subroutines to compute Bessel functions. The use of the computing facilities at the Western Data Processing Center at UCLA is gratefully acknowledged.

[13] C. Yeh, Phys. Rev. 131, 2350 (1963).

APPENDIX

The definite integrals $J_{11n}{}^{p}$, $J_{21n}{}^{p}$, $J_{31n}{}^{p}$, and $J_{41n}{}^{p}$ are defined as follows:

$$J_{11n}{}^{p}=\int_0^{\pi} p(p+1)\frac{df_1}{d\theta}P_p{}^1P_n{}^1d\theta, \tag{A1}$$

$$J_{21n}{}^{p}=\int_0^{\pi} f_1\left[\frac{dP_n{}^1}{d\theta}\frac{dP_p{}^1}{d\theta}+\frac{P_p{}^1P_n{}^1}{\sin^2\theta}\right]\sin\theta d\theta, \tag{A2}$$

$$J_{31n}{}^{p}=\int_0^{\pi} f_1\left[P_p{}^1\frac{dP_n{}^1}{d\theta}+\frac{dP_p{}^1}{d\theta}P_n{}^1\right]d\theta, \tag{A3}$$

$$J_{41n}{}^{p}=\int_0^{\pi} p(p+1)\frac{df_1}{d\theta}P_p{}^1\frac{dP_n{}^1}{d\theta}\sin\theta d\theta, \tag{A4}$$

where $f_1=\sin^2\theta$ and $df_1/d\theta=\sin 2\theta$.

Reprinted from *Printed in U.S.A.*
JOURNAL OF MATHEMATICAL PHYSICS VOLUME 6, NUMBER 12 DECEMBER 1965

Perturbation Method in the Diffraction of Electromagnetic Waves by Arbitrarily Shaped Penetrable Obstacles*

C. YEH
Electrical Engineering Department, University of Southern California, Los Angeles, California
(Received 16 November 1964)

The perturbation technique which is based on a Taylor series expansion of the boundary conditions at the perturbed boundary is extended to consider the problem of the diffraction of waves by a dielectric object with perturbed boundary. Since this approach attacks the complete boundary-value problem, the result is valid for the near zone as well as for the far zone and is valid for all frequencies. In a way of illustration, the problem of the diffraction of electromagnetic waves by a dielectric cylinder with perturbed boundary is treated. A specific example on the scattering of plane waves by a dielectric elliptic cylinder with small eccentricity is given. Numerical results are also computed for this specific example and are compared with those obtained from the exact solution.

I. INTRODUCTION

EXACT solutions of boundary-value problems in the theory of electromagnetic wave diffraction are available only for certain specific bodies of relatively simple shape.[1,2] For example, the available exact solutions for cylindrical bodies without sharp edges are limited to those with circular, elliptic or parabolic cross-sections. The diffraction of waves by a conducting or dielectric sphere, by dielectric coated spheres and by a perfectly conducting disk are the few three dimensional problems that have been solved rigorously. The need for approximate methods to treat the more general cases of diffraction from arbitrarily shaped obstacles is quite apparent. The variational principles[3,4] provide a very powerful tool in obtaining approximate expression for the scattering cross section; but it is not possible to derive from the variational principles a description of the electromagnetic fields. Furthermore, the success of the variational approach depends to a great extent on the trial function. At low frequencies, the Rayleigh method[5,6] is very successful. However, the solutions of Laplace's equation are still required. At very high frequencies, the treatment of diffraction problems by geometric and physical optics techniques developed by Fock[7] and Keller[8] is very successful. An approximate or perturbation method in the medium frequency range still remains to be found.

In the present work the boundary perturbation technique[9,10] which is based on a Taylor expansion of the boundary conditions at the perturbed boundary will be extended to consider the problem of the diffraction of waves by a dielectric object with perturbed boundary. Since this approach attacks the complete boundary-value problem, the perturbation solution for the field components is valid for the near zone (i.e., near the obstacle) as well as for the far zone and is valid for all frequencies. In a way of illustration, the problem of the diffraction of electromagnetic waves by a dielectric cylinder with perturbed boundary will be treated. A specific example on the scattering of plane waves by a dielectric elliptic cylinder with small eccentricity will be given. The more involved case of the diffraction by a dielectric sphere with perturbed boundary can be solved in a similar manner.[11]

It is hoped that this perturbation approach will not only find applications in microwave and plasma physics but also in collision theory, acoustics, meteorology and astrophysics.[12]

II. THE PERTURBATION SOLUTION

It is assumed that an arbitrarily shaped dielectric cylinder which has a permittivity ϵ_1 and a per-

* Supported in part by the U. S. Naval Ordnance Test Station, Pasadena.

1 R. King and T. T. Wu, *The Scattering and Diffraction of Waves* (Harvard University Press, Cambridge, Massachusetts, 1959).
2 C. J. Bouwkamp, Repts. Progr. Phys. **17**, 35 (1954).
3 P. M. Morse and H. Feshback, *Methods of Theoretical Physics* (McGraw-Hill Book Company, Inc., New York, 1953).
4 H. Levine and J. Schwinger, *Theory of Electromagnetic Waves* (Interscience Publishers, Inc., New York, 1951).
5 Lord Rayleigh, Phil. Mag. **44**, 28 (1897).
6 A. F. Stevenson, J. Appl. Phys. **24**, 1134 (1953).
7 V. A. Fock, Phys. (USSR) **10**, 130, 399 (1946).
8 J. B. Keller, J. Opt. Soc. Am. **52**, 102 (1962).
9 Reference 3, p. 1052.
10 P. C. Clemmow and V. H. Weston, Proc. Roy. Soc. (London) **A264**, 246 (1961).
11 C. Yeh, Phys. Rev. **135**, A1193 (1964).
12 See H. C. van de Hulst's article in *Electromagnetic Scattering* (Pergamon Press, Ltd., Oxford, 1963).

meability μ_1, is embedded in a homogeneous dielectric medium (ϵ_0, μ_0). The boundary of the cross section of the dielectric cylinder (Fig. 1) takes the

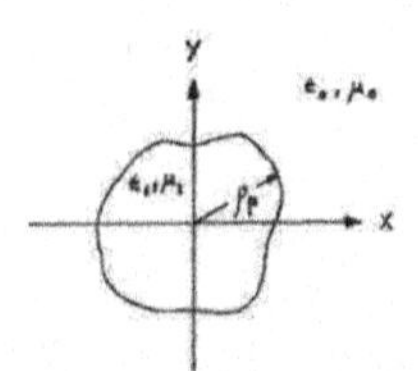

FIG. 1. The arbitrarily shaped penetrable cylinder.

shape of a perturbed circle which may be expressed by the following equation:

$$\rho_p = \rho_0[1 + \delta f_1(\phi) + \delta^2 f_2(\phi) + \cdots], \quad (1)$$

where ρ_0 is the radius of the unperturbed circle, δ is a smallness parameter, and $f_n(\phi)$ are arbitrary, single-valued, continuous functions satisfying the condition

$$\sum_{n=1}^{\infty} |\delta^n f_n(\phi)| < 1, \quad 0 \le \phi \le 2\pi.$$

The cylindrical coordinates (ρ, ϕ, z) are used.

Let the given exciting field (which need not necessarily be a plane wave) be denoted by $\mathbf{E}^{(i)}, \mathbf{H}^{(i)}$, the scattered field by $\mathbf{E}^{(s)}, \mathbf{H}^{(s)}$, and the field inside the dielectric body by $\mathbf{E}^{(t)}, \mathbf{H}^{(t)}$. The zeroth-order solution will be designated by a subscript 0, the first-order solution by subscript 1, etc. Hence, the resultant scattered fields and the resultant transmitted fields inside the body are, respectively,

$$\mathbf{E}^{(s)} = \mathbf{E}_0^{(s)} + \delta\mathbf{E}_1^{(s)} + \delta^2\mathbf{E}_2^{(s)} + \cdots,$$
$$\mathbf{H}^{(s)} = \mathbf{H}_0^{(s)} + \delta\mathbf{H}_1^{(s)} + \delta^2\mathbf{H}_2^{(s)} + \cdots, \quad (2)$$

and

$$\mathbf{E}^{(t)} = \mathbf{E}_0^{(t)} + \delta\mathbf{E}_1^{(t)} + \delta^2\mathbf{E}_2^{(t)} + \cdots,$$
$$\mathbf{H}^{(t)} = \mathbf{H}_0^{(t)} + \delta\mathbf{H}_1^{(t)} + \delta^2\mathbf{H}_2^{(t)} + \cdots. \quad (3)$$

The higher-order solutions are generated from known zeroth-order solution; i.e., $\mathbf{E}^{(i)}, \mathbf{H}^{(i)}, \mathbf{E}_0^{(s)}, \mathbf{H}_0^{(s)}, \mathbf{E}_0^{(t)}$, and $\mathbf{H}_0^{(t)}$ are assumed known qualities. For the sake of clarity and simplicity, only the first-order solution will be carried out in detail. The higher-order solution can be obtained in a similar fashion.

The boundary conditions require the continuity of tangential electric and magnetic fields at the boundary surface $\rho = \rho_p$:

$$\mathbf{n} \times [\mathbf{E}^{(i)}(\rho_p, \phi, z) + \mathbf{E}^{(s)}(\rho_p, \phi, z)] = \mathbf{n} \times \mathbf{E}^{(t)}(\rho_p, \phi, z), \quad (4)$$

$$\mathbf{n} \times [\mathbf{H}^{(i)}(\rho_p, \phi, z) + \mathbf{H}^{(s)}(\rho_p, \phi, z)] = \mathbf{n} \times \mathbf{H}^{(t)}(\rho_p, \phi, z), \quad (5)$$

where $\mathbf{n}$ is a unit vector outward normal to the boundary surface and can be written as

$$\mathbf{n} \simeq \mathbf{e}_\rho - \delta \frac{\partial f_1}{\partial \phi} \mathbf{e}_\phi \quad (6)$$

to the first order in δ in cylindrical coordinates. $\mathbf{e}_\rho$ and $\mathbf{e}_\phi$ are, respectively, the unit vectors in ρ and ϕ directions. $f_1(\phi)$ has been defined in Eq. (1). Carrying out the vector operations and expressing Eqs. (4) and (5) to the first order in δ in component form with the help of Eqs. (2) and (3), one obtains

$$\mathbf{e}_\rho: \quad \delta \frac{\partial f_1}{\partial \phi}[E_z^{(i)}(\rho_p, \phi, z) + E_{0z}^{(s)}(\rho_p, \phi, z)] = \delta \frac{\partial f_1}{\partial \phi} E_{0z}^{(t)}(\rho_p, \phi, z), \quad (7)$$

$$\mathbf{e}_\phi: \quad E_z^{(i)}(\rho_p, \phi, z) + E_{0z}^{(s)}(\rho_p, \phi, z) + \delta E_{1z}^{(s)}(\rho_p, \phi, z) = E_{0z}^{(t)}(\rho_p, \phi, z) + \delta E_{1z}^{(t)}(\rho_p, \phi, z), \quad (8)$$

$$\mathbf{e}_z: \quad E_\phi^{(i)}(\rho_p, \phi, z) + E_{0\phi}^{(s)}(\rho_p, \phi, z) + \delta\left\{E_{1\phi}^{(s)}(\rho_p,\phi,z) + \frac{\partial f_1}{\partial \phi}[E_\rho^{(i)}(\rho_p,\phi,z) + E_{0\rho}^{(s)}(\rho_p,\phi,z)]\right\} = E_{0\phi}^{(t)}(\rho_p,\phi,z) + \delta\left[E_{1\phi}^{(t)}(\rho_p,\phi,z) + \frac{\partial f_1}{\partial \phi} E_{0\rho}^{(t)}(\rho_p,\phi,z)\right], \quad (9)$$

$$\mathbf{e}_\rho: \quad \delta \frac{\partial f_1}{\partial \phi}[H_z^{(i)}(\rho_p, \phi, z) + H_{0z}^{(s)}(\rho_p, \phi, z)] = \delta \frac{\partial f_1}{\partial \phi} H_{0z}^{(t)}(\rho_p, \phi, z), \quad (10)$$

$$\mathbf{e}_\phi: \quad H_z^{(i)}(\rho_p, \phi, z) + H_{0z}^{(s)}(\rho_p, \phi, z) + \delta H_{1z}^{(s)}(\rho_p, \phi, z) = H_{0z}^{(t)}(\rho_p, \phi, z) + \delta H_{1z}^{(t)}(\rho_p, \phi, z), \quad (11)$$

$$\mathbf{e}_z: \quad H_\phi^{(i)}(\rho_p, \phi, z) + H_{0\phi}^{(s)}(\rho_p, \phi, z) + \delta\left\{H_{1\phi}^{(s)}(\rho_p,\phi,z) + \frac{\partial f_1}{\partial \phi}[H_\rho^{(i)}(\rho_p,\phi,z) + H_{0\rho}^{(s)}(\rho_p,\phi,z)]\right\} = H_{0\phi}^{(t)}(\rho_p,\phi,z) + \delta\left[H_{1\phi}^{(t)}(\rho_p,\phi,z) + \frac{\partial f_1}{\partial \phi} H_{0\rho}^{(t)}(\rho_p,\phi,z)\right]. \quad (12)$$

Equations (7) and (10) are satisfied by the zeroth-order solution. We now expand the above functions in Eqs. (8), (9), (11), and (12) to order δ in Taylor series about the unperturbed boundary $\rho = \rho_0$, obtaining

$$E_z^{(i)}(\rho_0, \phi, z) + E_{0z}^{(s)}(\rho_0, \phi, z) - E_{0z}^{(t)}(\rho_0, \phi, z) = \delta\{E_{1z}^{(t)}(\rho_0, \phi, z) - E_{1z}^{(s)}(\rho_0, \phi, z)$$

$$- \rho_0 f_1[E_z^{(i)\prime}(\rho_0, \phi, z) + E_{0z}^{(s)\prime}(\rho_0, \phi, z) - E_{0z}^{(t)\prime}(\rho_0, \phi, z)]\}, \tag{13}$$

$$E_\phi^{(i)}(\rho_0, \phi, z) + E_{0\phi}^{(s)}(\rho_0, \phi, z) - E_{0\phi}^{(t)}(\rho_0, \phi, z) = \delta\Big\{E_{1\phi}^{(t)}(\rho_0, \phi, z) - E_{1\phi}^{(s)}(\rho_0, \phi, z) - \rho_0 f_1[E_\phi^{(i)\prime}(\rho_0, \phi, z) + E_{0\phi}^{(s)\prime}(\rho_0, \phi, z) - E_{0\phi}^{(t)\prime}(\rho_0, \phi, z)] - \frac{\partial f_1}{\partial \phi}[E_\rho^{(i)}(\rho_0, \phi, z) + E_{0\rho}^{(s)}(\rho_0, \phi, z) - E_{0\rho}^{(t)}(\rho_0, \phi, z)]\Big\}, \tag{14}$$

$$H_z^{(i)}(\rho_0, \phi, z) + H_{0z}^{(s)}(\rho_0, \phi, z) - H_{0z}^{(t)}(\rho_0, \phi, z) = \delta\{H_{1z}^{(t)}(\rho_0, \phi, z) - H_{1z}^{(s)}(\rho_0, \phi, z) - \rho_0 f_1[H_z^{(i)\prime}(\rho_0, \phi, z) + H_{0z}^{(s)\prime}(\rho_0, \phi, z) - H_{0z}^{(t)\prime}(\rho_0, \phi, z)]\}, \tag{15}$$

$$H_\phi^{(i)}(\rho_0, \phi, z) + H_{0\phi}^{(s)}(\rho_0, \phi, z) - H_{0\phi}^{(t)}(\rho_0, \phi, z) = \delta\Big\{H_{1\phi}^{(t)}(\rho_0, \phi, z) - H_{1\phi}^{(s)}(\rho_0, \phi, z) - \rho_0 f_1[H_\phi^{(i)\prime}(\rho_0, \phi, z) + H_{0\phi}^{(s)\prime}(\rho_0, \phi, z) - H_{0\phi}^{(t)\prime}(\rho_0, \phi, z)] - \frac{\partial f_1}{\partial \phi}[H_\rho^{(i)}(\rho_0, \phi, z) + H_{0\rho}^{(s)}(\rho_0, \phi, z) - H_{0\rho}^{(t)}(\rho_0, \phi, z)]\Big\}, \tag{16}$$

where the prime signifies the derivative of the function with respect to ρ_0. The left-hand sides of the above equations are equal to zero by virtue of the zeroth-order solution. Hence, the right-hand sides of the above equations must vanish identically. Rearranging and combining Eqs. (13) and (14) gives

$$[E_{1z}^{(s)}(\rho_0, \phi, z) - E_{1z}^{(t)}(\rho_0, \phi, z)]\mathbf{e}_\phi + [E_{1\phi}^{(s)}(\rho_0, \phi, z) - E_{1\phi}^{(t)}(\rho_0, \phi, z)]\mathbf{e}_z = u_1(\rho_0, \phi, z)\mathbf{e}_\phi + u_2(\rho_0, \phi, z)\mathbf{e}_z, \tag{17}$$

and combining Eqs. (15) and (16) gives

$$[H_{1z}^{(s)}(\rho_0, \phi, z) - H_{1z}^{(t)}(\rho_0, \phi, z)]\mathbf{e}_\phi + [H_{1\phi}^{(s)}(\rho_0, \phi, z) - H_{1\phi}^{(t)}(\rho_0, \phi, z)]\mathbf{e}_z = v_1(\rho_0, \phi, z)\mathbf{e}_\phi + v_2(\rho_0, \phi, z)\mathbf{e}_z, \tag{18}$$

where

$$u_1(\rho_0, \phi, z) = \rho_0 f_1[E_{0z}^{(t)\prime}(\rho_0, \phi, z) - E_z^{(i)\prime}(\rho_0, \phi, z) - E_{0z}^{(s)\prime}(\rho_0, \phi, z)],$$

$$u_2(\rho_0, \phi, z) = \rho_0 f_1[E_{0\phi}^{(t)\prime}(\rho_0, \phi, z) - E_\phi^{(i)\prime}(\rho_0, \phi, z) - E_{0\phi}^{(s)\prime}(\rho_0, \phi, z)] + \frac{\partial f_1}{\partial \phi}[E_{0\rho}^{(t)}(\rho_0, \phi, z) - E_\rho^{(i)}(\rho_0, \phi, z) - E_{0\rho}^{(s)}(\rho_0, \phi, z)],$$

$$v_1(\rho_0, \phi, z) = \rho_0 f_1[H_{0z}^{(t)\prime}(\rho_0, \phi, z) - H_z^{(i)\prime}(\rho_0, \phi, z) - H_{0z}^{(s)\prime}(\rho_0, \phi, z)],$$

$$v_2(\rho_0, \phi, z) = \rho_0 f_1[H_{0\phi}^{(t)\prime}(\rho_0, \phi, z) - H_\phi^{(i)\prime}(\rho_0, \phi, z) - H_{0\phi}^{(s)\prime}(\rho_0, \phi, z)] + \frac{\partial f_1}{\partial \phi}[H_{0\rho}^{(t)}(\rho_0, \phi, z) - H_\rho^{(i)}(\rho_0, \phi, z) - H_{0\rho}^{(s)}(\rho_0, \phi, z)]. \tag{19}$$

It is noted that the resultant fields given by Eqs. (2) and (3) must satisfy the wave equation. It is therefore clear that each term in Eqs. (2) and (3) must separately satisfy the wave equation. Consequently, the general expressions for the longitudinal components of $\mathbf{E}_1^{(s)}$, $\mathbf{H}_1^{(s)}$, $\mathbf{E}_1^{(t)}$, and $\mathbf{H}_1^{(t)}$, that are appropriate to the present problem, are[13]

$$E_{1z}^{(s)} = \int_{-\infty}^{\infty} \sum_{n=-\infty}^{\infty} A_n H_n^{(1)}[(k_0^2 - h^2)^{\frac{1}{2}}\rho] e^{in\phi} e^{-ihz}\, dh, \tag{20}$$

$$H_{1z}^{(s)} = \int_{-\infty}^{\infty} \sum_{n=-\infty}^{\infty} B_n H_n^{(1)}[(k_0^2 - h^2)^{\frac{1}{2}}\rho] e^{in\phi} e^{-ihz}\, dh, \tag{21}$$

$$E_{1z}^{(t)} = \int_{-\infty}^{\infty} \sum_{n=-\infty}^{\infty} C_n J_n[(k_1^2 - h^2)^{\frac{1}{2}}\rho] e^{in\phi} e^{-ihz}\, dh, \tag{22}$$

$$H_{1z}^{(t)} = \int_{-\infty}^{\infty} \sum_{n=-\infty}^{\infty} D_n J_n[(k_1^2 - h^2)^{\frac{1}{2}}\rho] e^{in\phi} e^{-ihz}\, dh, \tag{23}$$

where $H_n^{(1)}[(k_0^2 - h^2)^{\frac{1}{2}}\rho]$ and $J_n[(k_1^2 - h^2)^{\frac{1}{2}}\rho]$ are, respectively, Hankel and Bessel functions, and $k_0^2 = \omega^2\mu_0\epsilon_0$ and $k_1^2 = \omega^2\mu_1\epsilon_1$. A_n, B_n, C_n, and D_n are yet unknown arbitrary constants that can be determined from Eqs. (17) and (18) using the orthogonality properties of the trigonomatric functions. The transverse components of $\mathbf{E}_1^{(s)}$, $\mathbf{H}_1^{(s)}$, $\mathbf{E}_1^{(t)}$, and $\mathbf{H}_1^{(t)}$ can be obtained from Maxwell's equations with the help of Eqs. (20) through (23).

Substituting the expressions for $\mathbf{E}_1^{(s)}$, $\mathbf{H}_1^{(s)}$, $\mathbf{E}_1^{(t)}$, and $\mathbf{H}_1^{(t)}$ into Eqs. (17) and (18), and making use of the orthogonality properties of the trigonometric functions, one obtains

$$A_n H_n^{(1)}[(k_0^2 - h^2)^{\frac{1}{2}}\rho_0] - C_n J_n[(k_1^2 - h^2)^{\frac{1}{2}}\rho_0] = \frac{1}{2\pi}\int_0^{2\pi} S_1(\rho_0, \phi) e^{-in\phi}\, d\phi, \tag{24}$$

$$A_n \frac{\partial}{\partial \rho_0} H_n^{(1)}[(k_0^2 - h^2)^{\frac{1}{2}}\rho_0] - C_n \frac{\mu_0}{\mu_1}\frac{\partial}{\partial \rho_0} J_n[(k_1^2 - h^2)^{\frac{1}{2}}\rho_0]$$

[13] J. A. Stratton, *Electromagnetic Theory* (McGraw-Hill Book Company, Inc., New York, 1941).

$$= \frac{-i\omega\mu_0}{2\pi} \int_0^{2\pi} T_2(\rho_0, \phi) e^{-in\phi} \, d\phi, \qquad (25)$$

$$B_n H_n^{(1)}[(k_0^2 - h^2)^{\frac{1}{2}} \rho_0] - D_n J_n[(k_1^2 - h^2)^{\frac{1}{2}} \rho_0] = \frac{1}{2\pi} \int_0^{2\pi} T_1(\rho_0, \phi) e^{-in\phi} \, d\phi, \qquad (26)$$

$$B_n \frac{\partial}{\partial \rho_0} H_n^{(1)}[(k_0^2 - h^2)^{\frac{1}{2}} \rho_0] - D_n \frac{\epsilon_0}{\epsilon_1} \frac{\partial}{\partial \rho_0} J_n[(k_1^2 - h^2)^{\frac{1}{2}} \rho_0] = \frac{i\omega\epsilon_0}{2\pi} \int_0^{2\pi} S_2(\rho_0, \phi) e^{-in\phi} \, d\phi, \qquad (27)$$

where S_1, S_2, T_1, and T_2 are defined as follows:

$$u_1(\rho_0, \phi, z) = \int_{-\infty}^{\infty} S_1(\rho_0, \phi) e^{-ihz} \, dh,$$

$$u_2(\rho_0, \phi, z) = \int_{-\infty}^{\infty} S_2(\rho_0, \phi) e^{-ihz} \, dh,$$

$$v_1(\rho_0, \phi, z) = \int_{-\infty}^{\infty} T_1(\rho_0, \phi) e^{-ihz} \, dh,$$

$$v_2(\rho_0, \phi, z) = \int_{-\infty}^{\infty} T_2(\rho_0, \phi) e^{-ihz} \, dh. \qquad (28)$$

u_1, u_2, v_1 and v_2 are given by Eq. (19). The coefficients A_n, B_n, C_n, and D_n can be found readily from the above equations. Substituting these coefficients back to Eqs. (20)–(23) gives the first-order correction to the electromagnetic fields due to the departure of the boundary surface from a perfect circular cylinder with radius ρ_0. Higher-order corrections can be found successively in the same manner. It is interesting to note that in general the perturbed wave will have all components of electromagnetic fields even if the incident wave is a pure TE wave ($E_z^{(i)} = 0$) or a pure TM wave ($H_z^{(i)} = 0$).

III. THE SCATTERING OF PLANE WAVES BY A DIELECTRIC ELLIPTIC CYLINDER

As an example of the application of the theory derived in Sec. II, the problem of the scattering of plane waves by a dielectric elliptic cylinder with small eccentricity will be considered. It is assumed that the incident plane wave with its electric vector polarized in the z direction is propagating in the direction of the positive x axis. The equation of an ellipse is given by

$$\rho_p = \frac{\rho_0}{[1 - 2\delta \sin^2 (\phi - \phi_0)]^{\frac{1}{2}}}, \qquad (29)$$

where

$$\delta = \frac{1}{2}\left[1 - \left(\frac{\rho_0}{\rho_0 + \Delta\rho_0}\right)^2\right], \qquad (30)$$

$2\rho_0$ and $2(\rho_0 + \Delta\rho_0)$ are the lengths of the two axes of the ellipse. ϕ_0 is the angle between the x axis and the major axis of the ellipse if $\delta < 0$, and it is the angle between the x axis and the minor axis of the ellipse if $\delta > 0$. For small eccentricity, one has

$$\rho_p \simeq \rho_0[1 + \delta \sin^2 (\phi - \phi_0)]. \qquad (31)$$

Comparing Eqs. (31) and (1) gives

$$f_1(\phi) = \sin^2 (\phi - \phi_0). \qquad (32)$$

The unperturbed solution to the problem of the scattering of normally incident plane (E) wave by a dielectric circular cylinder is well known:

$$\mathbf{E}^{(i)} = \sum_{n=-\infty}^{\infty} (i)^n J_n(k_0\rho) e^{in\phi} \mathbf{e}_z, \qquad (33a)$$

$$\mathbf{H}^{(i)} = \frac{1}{i\omega\mu_0} \sum_{n=-\infty}^{\infty} (i)^n \times \left[\frac{in}{\rho} J_n(k_0\rho)\mathbf{e}_\rho - \frac{d}{d\rho} J_n(k_0\rho)\mathbf{e}_\phi\right] e^{in\phi}, \qquad (33b)$$

$$\mathbf{E}_0^{(s)} = \sum_{n=-\infty}^{\infty} a_n (i)^n H_n^{(1)}(k_0\rho) e^{in\phi} \mathbf{e}_z, \qquad (34a)$$

$$\mathbf{H}_0^{(s)} = \frac{1}{i\omega\mu_0} \sum_{n=-\infty}^{\infty} a_n (i)^n \times \left[\frac{in}{\rho} H_n^{(1)}(k_0\rho)\mathbf{e}_\rho - \frac{d}{d\rho} H_n^{(1)}(k_0\rho)\mathbf{e}_\phi\right] e^{in\phi}, \qquad (34b)$$

$$\mathbf{E}_0^{(t)} = \sum_{n=-\infty}^{\infty} b_n (i)^n J_n(k_1\rho) e^{in\phi} \mathbf{e}_z, \qquad (35a)$$

$$\mathbf{H}_0^{(t)} = \frac{1}{i\omega\mu_1} \sum_{n=-\infty}^{\infty} b_n (i)^n \times \left[\frac{in}{\rho} J_n(k_1\rho)\mathbf{e}_\rho - \frac{d}{d\rho} J_n(k_1\rho)\mathbf{e}_\phi\right] e^{in\phi}, \qquad (35b)$$

with

$$a_n = \frac{J_n(k_1\rho_0)J_n'(k_0\rho_0) - (\epsilon_1\mu_0/\epsilon_0\mu_1)^{\frac{1}{2}} J_n(k_0\rho_0) J_n'(k_1\rho_0)}{(\epsilon_1\mu_0/\epsilon_0\mu_1)^{\frac{1}{2}} H_n^{(1)}(k_0\rho_0) J_n'(k_1\rho_0) - H_n^{(1)\prime}(k_0\rho_0) J_n(k_1\rho_0)}, \qquad (36a)$$

$$b_n = \frac{J_n'(k_0\rho_0) H_n^{(1)}(k_0\rho_0) - J_n(k_0\rho_0) H_n^{(1)\prime}(k_0\rho_0)}{(\epsilon_1\mu_0/\epsilon_0\mu_1)^{\frac{1}{2}} H_n^{(1)}(k_0\rho_0) J_n'(k_1\rho_0) - H_n^{(1)\prime}(k_0\rho_0) J_n(k_1\rho_0)}. \qquad (36b)$$

The prime in the above expressions denotes differentiation with respect to $k_0\rho_0$ or $k_1\rho_0$ as appropriate.

To find the first-order perturbation solution, we first substitute Eqs. (33) through (35) into Eq. (19)

obtaining

$$u_1(\rho_0, \phi) = \rho_0 f_1 \sum_{p=-\infty}^{\infty} \alpha_p e^{ip\phi},$$

$$u_2(\rho_0, \phi) = 0,$$

$$v_1(\rho_0, \phi) = 0,$$

$$v_2(\rho_0, \phi) = \rho_0 f_1 \sum_{p=-\infty}^{\infty} \beta_p e^{ip\phi} + \frac{\partial f_1}{\partial \phi} \sum_{p=-\infty}^{\infty} \gamma_p e^{ip\phi}, \tag{37}$$

where

$$\alpha_p = (i)^p [b_p k_1 J_p'(k_1\rho_0) - k_0 J_p'(k_0\rho_0) - a_p k_0 H_p^{(1)\prime}(k_0\rho_0)],$$

$$\beta_p = (i)^p \left[\frac{-b_p k_1^2}{i\omega\mu_1} J_p''(k_1\rho_0) + \frac{k_0^2}{i\omega\mu_0} J_p''(k_0\rho_0) + \frac{a_p k_0^2}{i\omega\mu_0} H_p^{(1)\prime\prime}(k_0\rho_0) \right],$$

$$\gamma_p = (i)^p \left[\frac{p b_p}{\omega\mu_1\rho_0} J_p(k_1\rho_0) - \frac{p}{\omega\mu_0\rho_0} J_p(k_0\rho_0) - \frac{p a_p}{\omega\mu_0\rho_0} H_p^{(1)}(k_0\rho_0) \right]. \tag{38}$$

The expansion coefficients for the first-order perturbation fields are then found by putting expressions (37) into Eqs. (24)–(27) and carrying out the integration involving the angular functions. One has

$$A_n H_n^{(1)}(k_0\rho_0) - C_n J_n(k_1\rho_0) = \chi_n,$$

$$A_n(\rho_0 k_0) H_n^{(1)\prime}(k_0\rho_0) - C_n(k_1\rho_0)\left(\frac{\mu_0}{\mu_1}\right) J_n'(k_1\rho_0) = \eta_n,$$

$$B_n = 0, \qquad D_n = 0, \tag{39}$$

where

$$\chi_n = \tfrac{1}{2}[\rho_0\alpha_n - \tfrac{1}{2}\rho_0\alpha_{n+2} e^{-2i\phi_0}],$$

$$\eta_n = -\tfrac{1}{2} i\omega\mu_0 [\rho_0^2 \beta_n - \tfrac{1}{2}\rho_0^2 \beta_{n+2} e^{-2i\phi_0} + i\rho_0 \gamma_{n+2} e^{2i\phi_0}]. \tag{40}$$

The following expressions have been used:

$$\int_0^{2\pi} \sin^2(\phi - \phi_0) e^{i(p-n)\phi}\, d\phi = \pi[\delta_{p,n} - \tfrac{1}{2} e^{-2i\phi_0} \delta_{2,(p-n)}], \tag{41}$$

$$\int_0^{2\pi} \sin^2(\phi - \phi_0) e^{i(p-n)\phi}\, d\phi = i\pi e^{2i\phi_0} \delta_{2,(p-n)}, \tag{42}$$

with

$$\delta_{r,m} = 0, \quad r \neq m,$$
$$= 1, \quad r = m.$$

Solving Eqs. (39) gives

$$A_n = \frac{\eta_n J_n(k_1\rho_0) - (k_1\rho_0)(\mu_0/\mu_1)\chi_n J_n'(k_1\rho_0)}{k_0\rho_0 H_n^{(1)\prime}(k_0\rho_0) J_n(k_1\rho_0) - (k_1\rho_0)(\mu_0/\mu_1) J_n'(k_1\rho_0) H_n^{(1)}(k_0\rho_0)}, \tag{43}$$

$$C_n = \frac{\chi_n(k_0\rho_0) H_n^{(1)\prime}(k_0\rho_0) - \eta_n H_n^{(1)}(k_0\rho_0)}{k_0\rho_0 H_n^{(1)}(k_0\rho_0) J_n(k_1\rho_0) - (k_1\rho_0)(\mu_0/\mu_1) J_n'(k_1\rho_0) H_n^{(1)}(k_0\rho_0)}. \tag{44}$$

Hence, the scattered fields correct to the first order in δ are

$$\mathbf{E}^{(s)} = \sum_{n=-\infty}^{\infty} [(i)^n a_n + \delta A_n] H_n^{(1)}(k_0\rho) e^{in\phi} \mathbf{e}_z, \tag{45}$$

$$\mathbf{H}^{(s)} = \frac{1}{i\omega\mu_0} \sum_{n=-\infty}^{\infty} [(i)^n a_n + \delta A_n] \times \left[\frac{in}{\rho} H_n^{(1)}(k_0\rho) \mathbf{e}_\rho - \frac{d}{d\rho} H_n^{(1)}(k_0\rho) \mathbf{e}_\phi \right] e^{in\phi}, \tag{46}$$

Of particular interest is the behavior of the backscattering cross section which is defined as the ratio of the total power scattered by a fictitious isotropic scatterer which scatters energy in all directions with intensity equal to that scattered directly back toward the source by the actual scattering object, to the incident power per unit area on the scatterer; i.e.,

$$\sigma_B^E = \lim_{\rho\to\infty} 2\pi\rho \frac{|E_z^{(s)}|^2}{|E_z^{(i)}|^2} \quad \text{at} \quad \phi = \pi, \tag{47}$$

or, using Eq. (45),

$$\sigma_B^E = \frac{4}{k_0} \left| \sum_{n=-\infty}^{\infty} [(i)^n a_n + \delta A_n] \exp i\left(\frac{n\pi}{2} - \frac{\pi}{4}\right) \right|^2. \tag{48}$$

Simplifying gives

$$\sigma_B^E = \frac{4}{k_0} \left\{ \left| \sum_{n=-\infty}^{\infty} (i)^n a_n \exp i\left(\frac{n\pi}{2} - \frac{\pi}{4}\right) \right|^2 + \delta \left[\left(\sum_{n=-\infty}^{\infty} (i)^n a_n \exp i\left(\frac{n\pi}{2} - \frac{\pi}{4}\right) \right) \times \left(\sum_{n=-\infty}^{\infty} A_n \exp i\left(\frac{n\pi}{2} - \frac{\pi}{4}\right) \right)^* + \left(\sum_{n=-\infty}^{\infty} (i)^n a_n \exp i\left(\frac{n\pi}{2} - \frac{\pi}{4}\right) \right)^* \times \left(\sum_{n=-\infty}^{\infty} A_n \exp i\left(\frac{n\pi}{2} - \frac{\pi}{4}\right) \right) \right] \right\}, \tag{49}$$

where the star above the series indicates the complex conjugate of the function. The first term on the right-hand side of the above equation represents the back-scattering cross section of an unperturbed circular cylinder, while the other term corresponds to the first-order correction due to small eccentricity.

To qualitatively illustrate how the solutions behave, the backscattering cross sections as a function of frequency are computed. Numerical computations are carried out using the high-speed IBM 7090 computer. It is assumed that $\epsilon_1/\epsilon_0 = 2.0$ and $\mu_1/\mu_0 = 1.0$. Two cases of perturbed cylindrical shape are considered: $\delta = 0.1, 0.05$. Different angles of incidence are used. Results are shown in Figs. 2 through 4. Numerical investigation shows that the first order perturbation solution should be good approximation to the exact solution for $|\delta| \leq 0.05$.

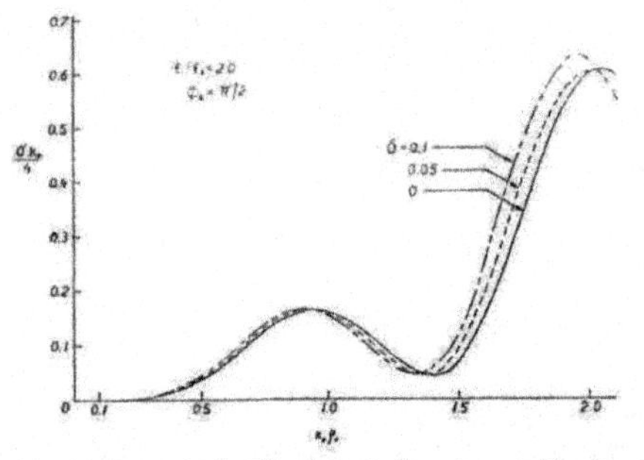

FIG. 2. The normalized back-scattering cross section for an elliptical dielectric cylinder. The direction of the incident wave is parallel to the major axis of the ellipse. $k_0\rho_0$ is the normalized semi-minor axis.

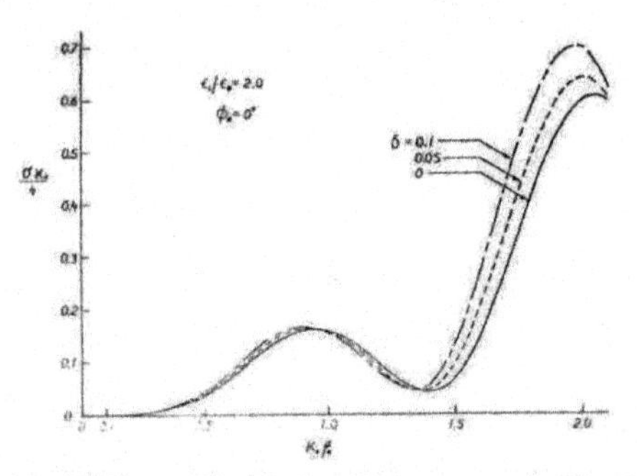

FIG. 3. The normalized back-scattering cross section for an elliptical dielectric cylinder. The direction of the incident wave is parallel to the minor axis of the ellipse. $k_0\rho_0$ is the normalized semi-minor axis.

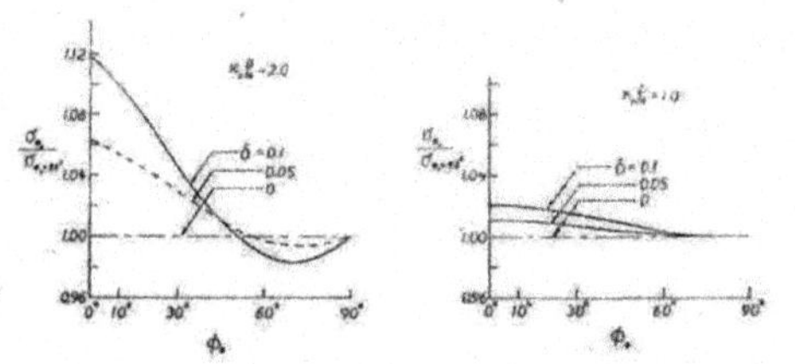

FIG. 4. Relative back-scattering cross section as a function of the direction of the incident wave.

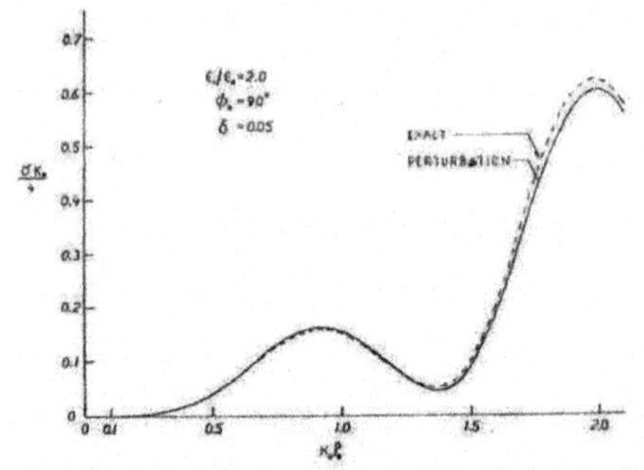

FIG. 5. Comparison between the normalized back-scattering cross section for an elliptical dielectric cylinder obtained from exact solution and that obtained according to the perturbation method.

It is always desirable to compare the results obtained from the approximate approach to available exact results. Numerical computations are therefore carried out from the exact formal solution of the diffraction of plane waves by dielectric elliptic cylinder. The exact solution is given in terms of infinite series of Mathieu functions.[14] The comparison between the exact results and the perturbation results is shown in Fig. 5. It can be seen that the agreement is very good. However it is expected that for flatter elliptical cross section, higher-order perturbation solutions must be included.

ACKNOWLEDGMENT

The use of the computing facilities at the Western Data Processing Center at University of California, Los Angeles is gratefully acknowledged.

[14] C. Yeh, J. Math. Phys. 4, 65 (1963).

4

GUIDED WAVES AND WAVEGUIDES

Understanding guided waves and designing guided wave structures (waveguides) are important aspects in electromagnetics. Basically, there are two frequency regions of interest for low-loss guided wave structures, one below 30 GHz and the other higher than 30 GHz. For frequencies below 30 GHz, mostly metal-based structures are used, and for frequencies above 30 GHz, increasing skin-depth losses in metal means that low-loss guiding structures must be made with low-loss dielectric material.

As early as the late 1950s and early 1960s, we realized that to attain high-data-rate transmission for long distances, guiding structures must be metal-free. This meant dielectric waveguides would be the prime candidates. Reviewing the literature showed that prior to 1960, other than dielectric slabs or circular rods, the important canonical elliptical dielectric waveguide had not been considered. In fact, the lack of theoretical treatment of dielectric guiding structures was apparent. We decided to set our goal to remedy this situation. Realization of this goal enabled us to be the pioneers in providing theoretical and/or experimental results in the following areas:

- Elliptical dielectric waveguides—fundamental solution
- TM wave propagation in sinusoidally stratified dielectric or plasma media—fundamental solution
- An efficient method to treat layered (or radially inhomogeneous) circular fibers (submitted for publication on Feb. 27, 1976). This method was then used by others to treat the simpler problems dealing with layered planar structures.
- Finite element method to treat arbitrarily shaped inhomogeneous dielectric waveguides (submitted for publication on Sept. 4, 1974). It has subsequently been developed into a commercial product.

- Beam propagation method applied to optical fibers (submitted for publication on Apr. 28, 1978). This pioneering work has subsequently been developed into a commercial product.
- TLM (transmission–line-matrix) method applied to dielectric or stripline waveguides
- Shepherding effects in fibers
- Low-loss terahertz waveguides
- Wavelength-division-multiplexed (WDM) solitons in fibers

We have compiled the above subjects into a book entitled *The Essence of Dielectric Waveguides* (Springer, New York, 2008).

Following are the abstracts of the papers included in this chapter.

4.1 Elliptical dielectric waveguides

4.1.1 Exact solutions. (Paper 4-1-1)

The problem of electromagnetic propagation along a cylinder of elliptical cross section is considered. Two infinite determinants representing the characteristic equations for the two types of hybrid waves (the ${}_eHE_{mn}$ and the ${}_oHE_{mn}$ waves) are derived. It is found that there exist two dominant waves that possess zero cutoff frequencies. The characteristic roots of these two dominant waves are computed for various values of eccentricities and relative dielectric constants. Theoretical propagation constants for the dominant wave are verified by experiments.

4.1.2 Attenuation calculation. (Paper 4-1-2)

The problem of the propagation of electromagnetic waves along a cylinder of elliptical cross section is considered. There exist two dominant waves that possess zero cutoff frequencies. The attenuation characteristics of these dominant waves are analyzed theoretically and experimentally. Good agreements are obtained. It is shown that one of the dominant waves, with its electric field parallel at the center of the cylinder to the minor axis of the elliptical cylinder, can propagate along a flat dielectric strip with much lower loss than can the HE_{11} wave along a circular cylinder having the identical cross-sectional area.

4.1.3 Weakly guiding case. (Paper 4-1-3)

Approximate and much simplified dispersion relations are obtained for the problem of optical wave propagation within weakly guiding elliptical fibers. The refractive index difference between the core and its cladding of weakly guiding optical fibers that are contenders for use as practical optical communication lines is very small; i.e., $(n_1/n_0 - 1) \ll 1$, where n_1 is the core index and n_0 is the cladding index. These greatly simplified dispersion relations are then used to calculate the propagation constants for several higher-order modes on an elliptical fiber.

4.1.4 Ferrite-filled elliptical waveguides. (Papers 4-1-4a, 4-1-4b)

This paper deals with electromagnetic wave propagation in a longitudinally magnetized, ferrite-filled elliptical waveguide. Using the wave equations and boundary conditions for a ferrite-filled waveguide of arbitrary cross-sectional shape, it is shown that for an elliptical cross section, the characteristic equations for the propagation constant take the form of even and odd infinite determinants. Solutions for the characteristic equations are obtained for various values of applied field strength, saturation magnetization, and waveguide ellipticity. The existence of three types of cutoff conditions is demonstrated.

Detailed analyses were also carried out for the Faraday effect of electromagnetic waves propagating in longitudinally magnetized, ferrite-filled elliptical waveguides. Few resemblances were found between the Faraday rotation effect in a circular waveguide and that in an elliptical waveguide. The circular waveguide features strong coupling between the dominant vertically and horizontally polarized modes, whereas in the elliptical ferrite-filled guide, very little coupling occurs for large eccentricities.

4.2 Wave guidance in periodic dielectric/plasma medium

4.2.1 Sinusoidally stratified dielectric medium. (Paper 4-2-1)

The problem of the propagation of TM waves in a sinusoidally stratified dielectric medium is considered. The propagation characteristics are determined from the stability diagram of the resultant Hill's equation. Numerical results show that the stability diagrams for Hill's equations and those for Mathieu's equation are quite different. Consequently, the dispersion properties of TM waves and TE waves in this stratified medium are also different. Detailed dispersion characteristics of TM waves in an infinite stratified medium and in waveguides filled with this stratified material are obtained.

4.2.2 Sinusoidally stratified plasma medium. (Paper 4-2-2)

The problem of the propagation of electromagnetic waves in a sinusoidally stratified plasma medium is treated analytically. The propagation characteristics of TE and TM waves are determined respectively from the characteristic equations of the resultant Mathieu and Hill equations. Detailed dispersion characteristics of TE and TM waves in an infinite stratified plasma medium and in waveguides filled longitudinally with this stratified dispersion material are given. It is found that although stop-band and pass-band structures exist for the ω-β diagrams of both TE and TM waves, detailed dispersion properties for TE and TM waves are quite different for most frequency ranges except when $[(\omega/\omega_{\nu 0})^2 - 1]^2 \gg \delta^2$, where ω is the frequency of the propagating wave, $\omega_{\nu 0}$ is the average plasma frequency of the inhomogeneous plasma medium, and δ is the amplitude of the sinusoidally varying term for the electron density profile $(0 \leq \delta \leq 1)$.

4.2.3 Periodic structure in integrated optics. (Paper 4-2-3)

Thin-film dielectric waveguides with a periodic refractive index, a periodic substrate, or periodic surface are studied. The field is determined from Maxwell's equations using Floquet's theorem. The Brillouin diagram and the interaction regions are investigated. The bandwidth and the attenuation coefficients of the interaction regions are given as a function of the optical wavelength. A number of applications in active and passive integrated systems are discussed.

4.2.4 Frequency selective coupler for integrated optics systems. (Paper 4-2-4)

A frequency selective coupler, which consists of two thin-film waveguides imbedded in a periodic medium, is studied using the Brillouin diagram. Detailed results for the relative bandwidth and the coupling factor are plotted as a function of normalized frequency for a representative case.

4.2.5 Stop bands for optical wave propagation in cholesteric liquid crystals. (Paper 4-2-5)

The stop-band characteristics in cholesteric-liquid-crystal half-space are investigated using the exact wave solution, the Floquet theorem, and the corresponding Brillouin diagram. The other half-space is filled with a uniform dielectric medium. By appropriate choice of dielectric constant of the uniform medium and the angle of incidence of the incoming plane wave, the stop band may split into two or three stop bands.

4.2.6 Mode conversion in periodically disturbed thin-film waveguides. (Paper 4-2-6)

Mode conversion in a periodically perturbed thin-film optical waveguide is studied in detail. Three different types of perturbations are considered: periodic index of refraction of the film, periodic index of refraction in the substrate, and periodic boundary. The applications in filters, mode converters, and distributed feedback lasers are discussed.

4.2.7 Transversely bounded DFB lasers. (Paper 4-2-7)

Bounded distributed-feedback (DFB) lasers are studied in detail. Threshold gain and field distribution for a number of configurations are derived and analyzed. More specifically, the thin-film guide, fiber, diffusion guide, and hollow channel with inhomogeneous-cladding DFB lasers are considered. Different-modes feedback and the effects of the transverse boundaries are included. A number of applications are also discussed.

4.2.8 Transients in a periodic slab: coupled waves approach. (Paper 4-2-8)

The reflection and transmission of rectangular and Gaussian pulses impinging on a periodically stratified slab are studied, and a number of examples are illustrated and analyzed. The coupled-waves approach is used to derive the reflection and transmission coefficient, then the fast Fourier transform (FFT) is used to illustrate the transient responses.

4.3 An efficient 4 x 4 matrix method to compute the propagation characteristics of radially stratified fibers. (Paper 4-3, submitted for publication on Feb. 27, 1976)

An efficient method is introduced in this paper to compute the dispersion characteristics as well as the Poynting flux distribution of radially stratified fibers.

Only 4 x 4 matrix operations were needed. Detailed results are given for several representative radially inhomogeneous fibers of practical interest.

4.4 Finite element approach

4.4.1 Arbitrarily shaped inhomogeneous optical fiber or integrated optical waveguides. (Paper 4-4-1)

Using the finite element technique, a numerical method is developed so that one may obtain the propagation characteristics of optical waves along guiding structures whose cores may be of arbitrary cross-sectional shape and whose material media may be inhomogeneous in more than one transverse direction. Several specific examples are given, and the results are compared with those obtained by other exact or approximate methods. Very close agreement was found. The method developed here can be easily applied to many important problems dealing with practical optical fiber or integrated optical waveguide whose cross-sectional index of refraction distribution may be quite arbitrary.

4.4.2 Single-mode optical waveguides. (Paper 4-4-2)

An efficient and powerful technique has been developed to treat the problem of wave propagation along arbitrarily shaped single-mode dielectric waveguides with inhomogeneous index variations in the cross-sectional plane. The technique is based on a modified finite-element method. Illustrative examples were given for (1) the triangular fiber guide, (2) the elliptical fiber guide, (3) the single-material fiber guide, (4) the rectangular fiber guide, (5) the embossed integrated optics guide, (6) the diffused channel guide, and (7) the optical stripline guide.

4.5 Beam propagation method (BPM) or scalar wave FFT method

4.5.1 Propagation of truncated Gaussian beams in multimode or single-mode fiber guides. (Paper 4-5-1, submitted for publication on Jan, 21, 1977)

The ability to predict the light-propagation characteristics in various practical multimode guiding structures is very important in optical fiber communications and in optical image transfer. The usual mode-by-mode analysis is impractical when the guiding structure is capable of supporting hundreds or thousands of modes. In this study, a numerical technique is described that is capable of providing useful data on the propagation characteristics of optical multimode guiding structures whose index of refraction variation may be quite arbitrary. As a specific example, the problem of infinite or truncated Gaussian beam propagation in a radially inhomogeneous fiber with parabolic index profile is solved. The numerical results for the infinite Gaussian beam case are compared with exact analytical data; they are in complete agreement. (Note that this method can also be applied to single-mode structures.)

4.5.2 Multimode or single-mode inhomogeneous fiber couplers. (Paper 4-5-2)

A numerical technique to obtain the wave behavior in tightly coupled multimode fibers with inhomogeneous indices is introduced in this paper. The specific problem of the coupling characteristics of two parallel multimode fibers whose index profile is parabolic is treated in detail. It was found that in spite of the fact that rather complicated coupling behavior is observed when multimodes exist, total guided power still exchanges among the fibers in a periodic manner, and the coupling length still increases monotonically as a function of the separation distance between the fibers. It has also been demonstrated that by simply specifying the index profiles of the coupling structure (provided that the profiles are slowly varying), the coupling characteristics can be generated with our technique. (Note that this technique can also be applied to single-mode fiber couplers.)

4.5.3 On multimode or single-mode fiber couplers, tapers, and horns. (Paper 4-5-3)

Using the scalar wave-FFT technique (or BPM), it is demonstrated that the propagation characteristics of light beams in many practical multimode or single-mode fiber structures such as fiber couplers, tapers, or horns can be obtained.

4.5.4 On weakly guiding single-mode optical waveguides. (Paper 4-5-4)

This paper presents a powerful technique to treat the problem of wave propagation along weakly guiding single-mode optical waveguides. The shape as well as the 3-D index variation of the guide may be quite arbitrary as long as the weakly guiding character is preserved. The technique is based on the solution of a scalar-wave equation of the forward-marching FFT or the BPM technique method. Excellent agreements were obtained with known solutions for various optical guiding structures. We have also demonstrated the capability of this technique to treat other exotic single-mode structures.

4.6 Nonlinear optical fibers

4.6.1 Pulse shepherding in nonlinear fiber optics. (Paper 4-6-1)

In a wavelength-division multiplexed fiber system, where pulses on different wavelength beams may co-propagate in a single-mode fiber, the cross-phase modulation (CPM) effects caused by the nonlinearity of the optical fiber are unavoidable. In other words, pulses on different-wavelength beams can interact with and affect each other through the intensity dependence of the refractive index of the fiber. Although CPM will not cause energy to be exchanged among the beams, the pulse shapes and locations on these beams can be altered significantly. This phenomenon makes possible the manipulation and control of pulses co-propagating on different-wavelength beams through the introduction of a shepherd pulse at a separate wavelength. How this can be accomplished is demonstrated in this paper.

4.6.2 Experimental verification of the pulse shepherding concept in dispersion-shifted single-mode fiber for bit-parallel wavelength links. (Paper 4-6-2)

A new way to dynamically control in-flight pulses by a co-propagating shepherd pulse in a WDM single-mode fiber system was proposed at the MPPOI '96 Conference. That system functionally resembles an optical fiber ribbon cable, except that all the bits pass on one fiberoptic waveguide. This single-fiber bit-parallel wavelength link can be used to extend the (speed x distance) product of emerging cluster computer networks, such as the MyriNet, SCI, Hippi-6400, and ShuffleNet. Here, we present the first experimental evidence that this pulse shepherding effect can be observed in a commercially available Corning DS (dispersion-shifted) fiber. Computer simulation results are first presented for the case observed in the laboratory setup. A discussion of the experiment setup and measurement procedures is given. Experimental results are then compared with computer-generated results. Excellent agreement is observed.

4.6.3 Enhanced pulse compression in a nonlinear fiber by wavelength division multiplexed optical pulse. (Paper 4-6-3)

A way to compress an optical pulse in a single-mode fiber is presented in this paper. By the use of the cross-phase modulation (CPM) caused by the nonlinearity of the optical fiber, a shepherd pulse propagating on a different-wavelength beam in a wavelength-division-multiplexed single-mode fiber system can be used to enhance the pulse compression of a co-propagating primary pulse. Although CPM will not cause energy to be exchanged among the beams, the pulse shapes on these beams can be altered significantly. For example, a 1 mW peak power 10 ps primary pulse on a given wavelength beam may be compressed by a factor of as much as 25 when a co-propagating 10 ps shepherd pulse of peak power of 49 mW on a different-wavelength beam is similarly compressed. Results of a systematic study on this effect are presented in this paper. Furthermore, even when the primary pulse on a given wavelength beam has a peak power of much less than 1 mW, it still can be compressed by the same compression factor as a co-propagating shepherd pulse of peak power much larger than 1 mW on a different-wavelength beam as it undergoes compression. Through CPM, co-propagating pulses on separate beams appear to share the nonlinear effect induced on any one of the pulses on separate beams.

4.6.4 An all-optical long-distance multi-Gbytes/s bit-parallel WDM single-fiber link. (Paper 4-6-4)

An all-optical long-distance (>30 km) bit-parallel WDM single-fiber link with 12 bit-parallel channels having 1 Gbyte/s capacity has been designed. That system functionally resembles an optical fiber ribbon cable, except that all bits pass on one fiberoptic waveguide. This single-fiber bit-parallel wavelength link can be used to extend the (speed x distance) product of emerging cluster computer networks such as MyriNet, SCI, Hippi-6400, and ShuffleNet. Here, the detailed design of this link using the commercially available Corning DS fiber is given. To demonstrate the viability of this link, two WDM channels at 1530 and 1545 nm carrying 1 ns pulses on each channel were sent through a single 25.2 km long Corning DS fiber. The walk-off was 200 ps, well within the allowable setup and hold time for the standard ECL logic, which is 350 ps for a bit period of 1 ns. This result implies that 30 bit-parallel beams spaced 1 nm apart between 1530 nm and 1560 nm, each carrying 1 Gbits/s, can be sent through a 25.2 km Corning DS fiber carrying information at a 30 Gb/s rate.

4.6.5 Generation of time-aligned picosecond pulses on wavelength-division-multiplexed beams in a nonlinear fiber. (Paper 4-6-5)

A fundamentally different way to generate time-aligned data pulses on wavelength-division-multiplexed (WDM) beams in a single-mode fiber is found. A large-amplitude pulse called a shepherd pulse is launched on one of the co-propagating beams (shepherd pulse is defined as a pulse that can affect other co-propagating pulses in a WDM format while maintaining its own propagating behavior). Initially, at the launching plane, no other pulse exists on any of the other co-propagating WDM beams. Due to the nonlinear cross-phase modulation effect, time-aligned pulses are generated on all other beams after a given fiber length. Both theoretical and experimental results are presented.

4.6.6 Existence of optical solitons on wavelength division multiplexed beams in a nonlinear fiber. (Paper 4-6-6)

A simple analytic expression for the initial fundamental optical solitons on wavelength-division-multiplexed (WDM) beams in a nonlinear fiber has been found. For an ideal fiber with no loss and uniform group-velocity dispersion (GVD) in the anomalous GVD region, the initial form is $[1+2(M-1)]^{-1/2}\text{sech}(\tau)$, where M is the number of WDM beams and τ is the normalized time. Computer simulation shows that these initial pulses on WDM beams in this fiber will propagate undistorted without change in their shapes for arbitrarily long distances. The discovery of the existence of solitons on WDM beams presents the ultimate goal for optical fiber communication on multiple-wavelength beams in a single fiber.

4.7 Optical waveguide theory. (Paper 4-7)

As optical fiber technology matures, complexity of optical waveguides and waveguide components also grows. Traditional techniques that may be used to analyze simple step-index circular fibers are no longer adequate. An assessment of several modern analytical/numerical techniques that have been used successfully to treat problems of practical interest is given. Illustrative numerical examples are also presented.

4.8 Dynamic three-dimensional TLM analysis of microstriplines on anisotropic substrate. (Paper 4-8)

The frequency-dependent propagation characteristics of a hybrid mode along microstriplines on anisotropic substrates are presented for the case in which the constitutive parameter tensors may be diagonalized. A generalization of the three-dimensional transmission-line matrix (TLM) numerical procedure is used to obtain results for the phase constant β, effective permittivity, ϵ_{eff}, and the dielectric impedance, Z, all as functions of the frequency and the shape ratio (w/h). Also shown are results for coupled microstrips on a sapphire substrate.

4.9 Modeling of star-shaped and parallel-wire carrier distribution systems. (Paper 4-9)

It is often desirable to represent carrier distribution systems by networks so analysis can be made to predict the behavior of these systems. Modeling of two canonical carrier distribution systems, the star-shaped system and the parallel-wire system, by networks was carried out in this paper. The representation is given in terms of lumped parameters when the lengths of the carrier lines are short and is given in terms of lumped and distributed parameters when the lengths of the carrier lines are long.

4.10 Leaky waves in a heteroepitaxial film. (Paper 4-10)

Theoretical as well as experimental investigations were carried out for the propagation of leaky waves along an important class of optical thin-film waveguides. For this type of guiding structure, which may consist of heteroepitaxial deposition of ZnS or ZnSe on GaAs, the dielectric constant of the thin film is less than that of the substrate. Consequently, only the leaky type of guided waves may exist. Theoretical results show that the attenuation constants of leaky modes, which may be TE or TM, are inversely proportional to $(thickness\ of\ layer)^3$ and directly proportional to $(wavelength)^2$, and that TM modes are more lossy than TE modes. The existence of these leaky modes is also demonstrated for three different thicknesses.

4.11 Terahertz and millimeter dielectric waveguides

4.11.1 Dielectric ribbon waveguide: an optimum configuration for ultra-low-loss millimeter/submillimeter dielectric waveguide. (Paper 4-11-1)

Dielectric ribbon waveguide supporting the ${}_{e}HE_{11}$.dominant mode can be made to yield an attenuation constant for this mode of less than 20 dB/m in the millimeter/submillimeter-wavelength range. The waveguide is made with a high dielectric-constant low-loss material such as alumina or sapphire. It takes the form of thin dielectric ribbon surrounded by lossless dry air. A detailed theoretical analysis of the attenuation and field extent characteristics for the low-loss dominant ${}_{e}HE_{11}$ mode along a ribbon dielectric waveguide was carried out using the exact finite-element technique as well as two approximate techniques. Analytical predictions were then verified by measurements on ribbon waveguides made with Rexolite using the highly sensitive cavity resonance method. Excellent agreement was found.

4.11.2 Communication at millimetre-submillimetre wavelengths using a ceramic ribbon. (Paper 4-11-2)

Following the discovery by Kao and Hockman that ultra-low-loss optical fibers could be made from pure silica through the elimination of impurities, the ability to guide signals effectively at optical wavelengths has been assured, but there remains an important region of the spectrum—from 30 to 3000 GHz (the millimetre-submillimetre band)—where low-loss waveguides are unknown. The main problem here in finding low-loss solids is no longer one of eliminating impurities but is due to the presence of intrinsic vibration absorption bands. The use of highly conducting materials is also precluded, owing to high skin-depth losses in this part of the spectrum. Here, we show that a combination of material and waveguide geometry can circumvent these difficulties. We adopt a ribbon-like structure with an aspect ratio of 10:1, fabricated from ceramic alumina (Coors 998 Alumina), and the resulting waveguide has an attenuation factor of less than 10 dB km^{-1} in the millimeter-submillimeter band. This attenuation is more than 100 times smaller than that of a typical ceramic (or other dielectric) circular rod waveguide and is sufficient for immediate application.

4.11.3 Low-loss terahertz ribbon waveguides. (Paper 4-11-3)

The submillimeter-wave or terahertz band (1 mm–100 μm wavelength) is one of the last unexplored frontiers in the electromagnetic spectrum. A major stumbling block hampering instrument deployment in this frequency range is the lack of a low-loss guiding structure equivalent to the optical fiber that is so prevalent at the visible wavelengths. The presence of strong inherent vibrational absorption bands in solids and the high skin-depth losses of conductors make the traditional stripline circuits, conventional dielectric lines, or metallic waveguides, which are common at microwave frequencies, much too lossy to use at THz bands. Even the modern surface plasmon polariton waveguides are much too lossy for long-distance transmission in the THz bands. We describe a concept for overcoming this drawback and describe a new family of ultra-low-loss ribbon-based waveguide structures and matching components for propagating single-mode THz signals. For straight runs, this ribbon-based waveguide can provide an attenuation constant that is more than 100 times less than that of conventional dielectric or metallic waveguides. Problems dealing with sufficient coupling of power into and out of the ribbon guide, achieving low-loss bends and branches, and forming THz circuit elements are discussed in detail. We note that active circuit elements can be integrated directly onto the ribbon structure (when it is made with semiconductor material) and that the absence of metallic structures in the ribbon guide provides the possibility of high power-carrying capability. It thus appears that this ribbon-based dielectric waveguide and associated components can be used as fundamental building blocks for a new generation of ultra-high-speed electronic integrated circuits of THz interconnects.

4.11.4 Attenuation measurement of very low loss dielectric waveguides by the cavity resonator method applicable in the millimeter/submillimeter wavelength range. (Paper 4-11-4)

A dielectric waveguide shorted at both ends is constructed as a cavity resonator. By measuring the Q of this cavity, one can determine the attenuation constant of the guided mode on this dielectric structure. The complex permittivity of the dielectric waveguide material can also be derived from the measurements. Measurements were made at *Ka*-band for dielectric waveguides constructed of nonpolar, low-loss polymers such as Teflon, polypropylene, polyethylene, polystyrene, and Rexolite.

4.12 Wave propagation in hot plasma waveguides with infinite magnetostatic fields. (Paper 4-12)

Solutions of the "dispersion relation" for an unbounded Maxwellian electron plasma, $k^2/k_D^2 = Z'(\omega/kv_{th})$, for ω real are given. Solutions of the corresponding relation for a circular waveguide filled with a hot plasma with an infinite magnetostatic field are also presented. It is shown that these solutions are of importance in evaluating the fields in such waveguides because they determine poles of the integrands that arise in the integral formulation of the problem. The integral solution for the field in the waveguide due to a general symmetric excitation at the wall of the waveguide is given.

4.13 Distribution networks and electrically controllable couplers for integrated optics. (Paper 4-13)

The power distribution as a function of propagation distance in a network of coupled optical waveguides is determined for several interesting cases. An electrically controllable coupler is proposed and analyzed in detail. High-efficiency coupling and decoupling between two optical guides can be accomplished with the use of an electro-optically generated dynamic channel of finite length located between the two guides.

4.14 Highly sensitive quantum well infrared photodetectors. (Paper 4-14)

A fundamentally new method for light coupling in quantum well infrared photodetectors that provides a tenfold improvement over an optimized grating coupler is presented. It is based on the prism-film coupler concept developed earlier for selective-mode coupling into integrated optical circuits. In this article, this concept is specifically used to turn the incident electrical field from one that is polarized parallel to the quantum well layer to one that is mostly perpendicular to the layer, thereby increasing dramatically the sensitivity of the quantum well infrared photodetector. Detailed sample design and its computer simulation results are given and discussed.

4.15 How does one induce leakage in an optical fiber link? (Paper 4-15)

Three nondestructive methods to induce the leakage of optical signal from optical fibers will be discussed: (1) the index-matching-fluid method, (2) the temperature method, and (3) the bending method. Experiments were performed for these cases. Results show that all three methods are effective in inducing leakage from plastic-clad fibers, while only the bending method is effective in glass-clad fibers.

Reprinted from JOURNAL OF APPLIED PHYSICS, Vol. 33, No. 11, 3235-3243, November, 1962

Elliptical Dielectric Waveguides*

C. YEH†
California Institute of Technology, Pasadena, California
(Received March 7, 1962)

The problem of electromagnetic wave propagation along a dielectric cylinder of elliptical cross section is considered. Two infinite determinants representing the characteristic equations for the two types of hybrid waves (the $_eHE_{mn}$ and the $_oHE_{mn}$ waves) are derived. It is found that there exists two dominant waves which possess zero cutoff frequencies. The characteristic roots of these two dominant waves are computed for various values of eccentricity and relative dielectric constant. Theoretical propagation constants for the dominant waves are verified by experiments.

I. INTRODUCTION

THE concept of guiding electromagnetic waves either along a single conducting wire with finite surface impedance or along a dielectric rod is not new. As early as 1899, Sommerfeld[1] conceived the idea of guiding a circularly symmetric TM wave along a conducting wire with small surface resistivity. In 1910, Hondros and Debye[2] demonstrated theoretically that it is possible to propagate a TM wave along a lossless dielectric cylinder. However, due to the large field extent outside the wire and the relatively high attenuation of this surface wave, the "open-wire" line remained a novelty for almost half a century. Recent developments in the generation and application of millimeter and sub-millimeter electromagnetic waves, the availability of very low loss dielectrics, and the development of fiber optics, have renewed interest in the surface-wave guides. There have appeared numerous papers and reports concerning various forms of surface-wave guides and the feasibility of these guides as practical transmission lines. The most recent investigations have been reported by Goubau,[3] Barlow and Karbowiak,[4] Sheibe, King, and Van Zieland,[5] and Roberts[6] on the single-wire line; and by Elsasser,[7] Chandler,[8] King,[9] Snitzer,[10] and Wiltse[11] on the circular dielectric rod, to mention only a few. A rather complete bibliography on the subject of surface-wave propagation is given in a paper by Harvey.[12]

In order that a dielectric rod may be a low loss surface-wave guiding device, one must choose a small value of ka, where k is the free-space wavenumber and a is the radius of the dielectric cylinder. In the millimeter wavelength range, the radius of the dielectric cylinder becomes inconveniently small. Fortunately, it has been found experimentally[13] that if a circular rod is flattened, (i.e., if a circular rod is rendered to an elliptical rod of the same cross-sectional area), the attenuation of the dominant mode may be reduced considerably, provided that the electric field of the dominant mode is parallel at the center of the rod to the minor axis of the elliptical rod. Furthermore, it is noted that so far there exists no satisfactory way of analyzing the problem of surface-wave propagation along a dielectric rod of elliptical cross section.[14] It is therefore the purpose of the present investigation to develop a method to analyze this problem theoretically, to examine in particular the propagation characteristics of the dominant modes, and to perform experiments to verify the analytic results. This paper presents some of the results of this investigation.

II. FORMULATION OF THE PROBLEM

It is assumed that an infinitely long dielectric cylinder of elliptical cross section is imbedded in an infinite dielectric medium of dielectric constant ϵ_0 and magnetic permeability μ_0. The cylinder has a dielectric constant ϵ_1 and a magnetic permeability μ_0. The conductivity in both media is assumed to be zero; and $\epsilon_1 > \epsilon_0$. We further assume that the exciting source is so far away that, in the region of interest, the surface wave dominates the radiated wave from the source.

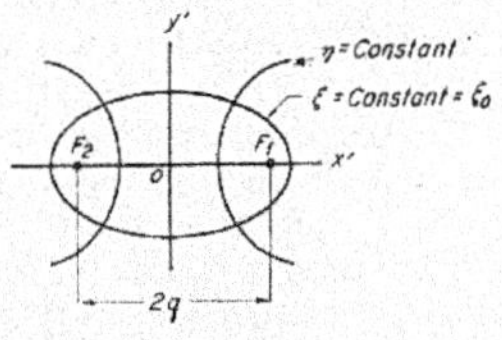

FIG. 1. Cross section of elliptical dielectric waveguide. F_1 and F_2 are the focuses of the ellipse. The distance between focuses is the focal distance $2q$.

* Supported by the Air Force Office of Scientific Research.
† Present address: Electrical Engineering Department, University of Southern California, Los Angeles, California.

1 A. Sommerfeld, Ann. Physik Chem. 67, 233 (1899).
2 D. Hondros and P. Debye, Ann. Physik 32, 465 (1910).
3 G. Goubau, J. Appl. Phys. 21, 119 (1950).
4 H. Barlow and A. Karbowiak, Proc. IEE 100, Pt III, 321 (1953).
5 E. Scheibe, B. King, and D. Van Zeeland, J. Appl. Phys. 25, 790 (1954).
6 T. Roberts, J. Appl. Phys. 24, 57 (1954).
7 W. Elsasser, J. Appl. Phys. 20, 1193 (1949).
8 C. Chandler, J. Appl. Phys. 20, 1188 (1949).
9 D. King, J. Appl. Phys. 23, 699 (1952).
10 E. Snitzer, Rept. No. 1, Research Center, American Optical Company, Southbridge, Massachusetts (1961).
11 J. Wiltse, I.R.E. Trans. on Microwave Theory Tech. MTT-7, 65 (1959).
12 A. Harvey, I.R.E. Trans. on Microwave Theory Tech. MTT-8, 30 (1960).
13 M. King and J. Wiltse, Rept. No. 1, Electrical Communication Inc., St. Petersburg, Florida (1960).
14 This problem had been attempted by Karbowiak (reference 15), and by King and Wiltse (reference 13). Most recently Lynbimov *et al.* (reference 16) also considered this problem.
15 A. Karbowiak, Brit. J. Appl. Phys. 5, 328 (1954).
16 L. Lynkinov, G. Veselov, and N. Bei, Radio Eng. and Electronics (U.S.S.R.) 6, 1668 (1961).

To analyze this problem, the elliptical cylinder coordinates (ξ,η,z), as shown in Fig. 1, are introduced. In terms of the rectangular coordinates (x',y',z'), the elliptical cylinder coordinates are defined by the following relations:

$$x'=q\cosh\xi\cos\eta,$$
$$y'=q\sinh\xi\sin\eta,$$
$$z'=z,$$
$$(0\leq\xi<\infty,\ 0\leq\eta\leq2\pi),\tag{1}$$

where q is the semifocal length of the ellipse. One of the confocal elliptic cylinders with $\xi=\xi_0$ is assumed to coincide with the boundary of the solid dielectric cylinder, and the z axis coincides with its longitudinal axis.

The harmonic time dependence of $e^{-i\omega t}$ for all field components is assumed. The rationalized mks system is used throughout this work. We shall confine our treatment to waves propagating along the positive z axis. In complex representation, these assumptions result in a multiplication of all wavefunctions by $e^{-i\omega t+i\beta z}$. The propagation constant β is to be determined from the boundary conditions.

The appropriate solutions of the wave equation for this problem are as follows:

For region 1 $(0\leq\xi\leq\xi_0)$

$$\Lambda_1=\begin{Bmatrix} Ce_n(\xi)ce_n(\eta) & \text{(even)} \\ Se_n(\xi)se_n(\eta) & \text{(odd)} \end{Bmatrix},\tag{2}$$

and for region 0 $(\xi_0\leq\xi<\infty)$

$$\Lambda_0=\begin{Bmatrix} Fek_n(\xi)ce_n^*(\eta) & \text{(even)} \\ Gek_n(\xi)se_n^*(\eta) & \text{(odd)} \end{Bmatrix},\tag{3}$$

where the abbreviations,

$$\begin{aligned} Ce_n(\xi)&=Ce_n(\xi,\gamma_1^2) & ce_n(\eta)&=ce_n(\eta,\gamma_1^2)\\ Se_n(\xi)&=Se_n(\xi,\gamma_1^2) & se_n(\eta)&=se_n(\eta,\gamma_1^2)\\ Fek_n(\xi)&=Fek_n(\xi,-\gamma_0^2) & ce_n^*(\eta)&=ce_n(\eta,-\gamma_0^2)\\ Gek_n(\xi)&=Gek_n(\xi,-\gamma_0^2) & se_n^*(\eta)&=se_n(\eta,-\gamma_0^2),\end{aligned}\tag{4}$$

have been used. γ_1^2 and $-\gamma_0^2$ represent, respectively, $(k_1^2-\beta^2)q^2/4$ and $(k_0^2-\beta^2)q^2/4$ with $k_1^2=\omega^2\mu\epsilon_1$ and $k_0^2=\omega^2\mu\epsilon_0$. Λ_1 may be either E_{z_1} or H_{z_1}; Λ_0 may be E_{z_0} or H_{z_0}.

The boundary conditions require the continuity of the tangential electric and magnetic field components, E_z, E_η, H_z, and H_η, across the boundary surface $\xi=\xi_0$.

It will be shown later (in Sec. III) that, in order to satisfy the above boundary conditions, E_z and H_z must both be present.[17] Hence, all modes on an elliptical dielectric cylinder are hybrid. Furthermore, since the angular Mathieu functions are not only functions of η, the angular coordinates, but also of the electrical properties of the medium in which they apply, the field components in both regions must be represented by an infinite product terms of Mathieu and modified Mathieu functions. Due to the asymmetry of the elliptical cylinder, it is possible to have two orientations for the field configurations. Thus, a hybrid wave on an elliptical dielectric rod will be designated by a prescript e or 0, indicating an even wave or an odd wave. The axial magnetic and electric field of an even wave are represented by even and odd Mathieu functions, respectively, and those of an odd wave by odd and even Mathieu functions, respectively. The letter HE is used to designate the hybrid wave. A double subscript (m,n) will also be employed; (m,n) denotes the order of wave which corresponds to the order (m,n) for an HE_{mn} wave on a circular dielectric cylinder when the eccentricity of the ellipse becomes zero.

[17] The circularly symmetric TE or TM wave on a circular dielectric rod is an exception.

III. THE DETERMINANTAL EQUATIONS

${}_e\mathrm{HE}_{mn}$ Wave

According to the definition given in the previous sections the most general expressions for the axial magnetic and electric fields of an ${}_e\mathrm{HE}_{mn}$ wave are: For region 1 $(0\leq\xi\leq\xi_0)$

$$H_{z_1}=\sum_{m=0}^{\infty}A_mCe_m(\xi)ce_m(\eta)e^{i\beta z}\tag{5a}$$

$$E_{z_1}=\sum_{m=1}^{\infty}B_mSe_m(\xi)se_m(\eta)e^{i\beta z},\tag{5b}$$

and for region 0 $(\xi_0\leq\xi<\infty)$

$$H_{z_0}=\sum_{r=0}^{\infty}L_rFek_r(\xi)ce_r^*(\eta)e^{i\beta z}\tag{6a}$$

$$E_{z_0}=\sum_{r=1}^{\infty}P_rGek_r(\xi)se_r^*(\eta)e^{i\beta z},\tag{6b}$$

where A_m, B_m, L_r, and P_r are the arbitrary constants. All transverse fields can be derived from Maxwell's equations.

Equating the tangential electric and magnetic fields at the boundary surface, $\xi=\xi_0$, using the expansions

$$ce_r^*(\eta)=\sum_{n=0}^{\infty}{}'\alpha_{r,n}ce_n(\eta),\quad se_r^*(\eta)=\sum_{n=1}^{\infty}{}'\beta_{r,n}se_n(\eta),$$

$$\frac{d}{d\eta}[ce_m(\eta)]=\sum_{n=1}^{\infty}{}'\chi_{m,n}se_n(\eta),$$

$$\frac{d}{d\eta}[se_m(\eta)]=\sum_{n=0}^{\infty}{}'\nu_{m,n}ce_n(\eta),\tag{7}$$

and applying the orthogonality relations of Mathieu

function, leads to

$$A_n a_n = \sum_{r=0}^{\infty}{}' L_r l_r \alpha_{r,n} \tag{8a}$$

$$B_n b_n = \sum_{r=1}^{\infty}{}' P_r p_r \beta_{r,n} \tag{8b}$$

$$\frac{\omega\epsilon_1}{\beta} B_n b_n' + \left[1+\frac{\gamma_1^2}{\gamma_0^2}\right] \sum_{r=1}^{\infty}{}' A_r a_r \chi_{r,n} = \left(-\frac{\gamma_1^2}{\gamma_0^2}\right)\frac{\omega\epsilon_0}{\beta} \sum_{r=1}^{\infty}{}' P_r p_r' \beta_{r,n} \tag{8c}$$

$$\frac{\omega\mu}{\beta} A_n a_n' - \left[1+\frac{\gamma_1^2}{\gamma_0^2}\right] \sum_{r=0}^{\infty}{}' B_r b_r \nu_{r,n} = \left(-\frac{\gamma_1^2}{\gamma_0^2}\right)\frac{\omega\mu}{\beta} \sum_{r=0}^{\infty}{}' L_r l_r' \alpha_{r,n}, \quad \begin{pmatrix} n=0,\,2,\,4\cdots \\ \text{or } n=1,\,3,\,5\cdots \end{pmatrix}, \tag{8d}$$

where the abbreviations

$$\begin{aligned} a_n &= Ce_n(\xi_0) & a_n' &= (d/d\xi_0)Ce_n(\xi_0) \\ b_n &= Se_n(\xi_0) & b_n' &= (d/d\xi_0)Se_n(\xi_0) \\ l_r &= Fek_r(\xi_0) & l_r' &= (d/d\xi_0)Fek_n(\xi_0) \\ p_r &= Gek_r(\xi_0) & p_r' &= (d/d\xi_0)Gek_n(\xi_0) \end{aligned} \tag{9}$$

have been used. $\alpha_{r,n}$, $\beta_{r,n}$, $\chi_{m,n}$, and $\nu_{m,n}$ are given in the Appendix. The prime over the summation sign indicates that odd or even integer values of r are to be taken according as n is odd or even. Simplifying Eqs. (8) and making the identifications

$$g_{m,n} = \left(1+\frac{\gamma_1^2}{\gamma_0^2}\right) l_m \sum_{r=1}^{\infty}{}' \chi_{r,n}\alpha_{m,r}$$

$$s_{m,n} = -\left(1+\frac{\gamma_1^2}{\gamma_0^2}\right) p_m \sum_{r=1}^{\infty}{}' \nu_{r,n}\beta_{m,r}$$

$$h_{m,n} = \frac{\omega\epsilon_1}{\beta}\frac{b_n'}{b_n} p_m \beta_{m,n} + \frac{\gamma_1^2}{\gamma_0^2}\frac{\omega\epsilon_0}{\beta} p_m' \beta_{m,n}$$

$$t_{m,n} = -\frac{\omega\mu}{\beta}\frac{a_n'}{a_n} l_m \alpha_{m,n} + \frac{\gamma_1^2}{\gamma_0^2}\frac{\omega\mu}{\beta} l_m' \alpha_{m,n}, \tag{10}$$

one obtains

$$\sum_{m=0}^{\infty}{}' \left[L_m g_{m,n} + P_m h_{m,n}\right] = 0 \tag{11a}$$

$$\sum_{m=0}^{\infty}{}' \left[L_m t_{m,n} + P_m s_{m,n}\right] = 0 \quad \begin{pmatrix} n=0,\,2,\,4\cdots \\ \text{or } n=1,\,3,\,5\cdots \end{pmatrix}. \tag{11b}$$

Equations (11) are two sets of infinite homogeneous linear algebraic equations in L_m and P_m. For a nontrivial solution the determinant of these equations must vanish. The roots of this infinite determinant provide the values from which the propagation constant β can be determined. For example, the infinite determinant for the $m=1$ mode is

$$\begin{vmatrix} g_{1,1} & h_{1,1} & g_{3,1} & h_{3,1} & g_{5,1} & h_{5,1} & \cdot & \cdot \\ t_{1,1} & s_{1,1} & t_{3,1} & s_{3,1} & t_{5,1} & s_{5,1} & \cdot & \cdot \\ g_{1,3} & h_{1,3} & g_{3,3} & h_{3,3} & g_{5,3} & h_{5,3} & \cdot & \cdot \\ t_{1,3} & s_{1,3} & t_{3,3} & s_{3,3} & t_{5,3} & s_{5,3} & \cdot & \cdot \\ g_{1,5} & h_{1,5} & g_{3,5} & h_{3,5} & g_{5,5} & h_{5,5} & \cdot & \cdot \\ t_{1,5} & s_{1,5} & t_{3,5} & s_{3,5} & t_{5,5} & s_{5,5} & \cdot & \cdot \\ \cdot & \cdot & \cdot & \cdot & \cdot & \cdot & \cdot & \cdot \\ \cdot & \cdot & \cdot & \cdot & \cdot & \cdot & \cdot & \cdot \end{vmatrix} = 0. \tag{12}$$

Due to the extreme complexity of this infinite determinant, the roots of this determinant can only be obtained numerically by the method of successive approximations.[18] This point will be discussed further in Sec. V. It was found numerically that the first root of $m=1$ mode was governed principally by the expression[19]

$$\begin{vmatrix} g_{1,1} & h_{1,1} \\ t_{1,1} & s_{1,1} \end{vmatrix} = 0 \tag{13}$$

as long as the elliptical cross section is not too flat (i.e., $\xi_0 > 0.5$).

${}_0\mathrm{HE}_{mn}$ Wave

The expressions for the axial magnetic and electric fields of an ${}_0\mathrm{HE}_{mn}$ wave are:
For region 1 ($0 \leq \xi \leq \xi_0$)

$$H_{z_1} = \sum_{m=1}^{\infty} C_m Se_m(\xi) se_m(\eta) e^{i\beta z} \tag{14a}$$

$$E_{z_1} = \sum_{m=0}^{\infty} D_m Ce_m(\xi) ce_m(\eta) e^{i\beta z}, \tag{14b}$$

and for region 0 ($\xi_0 \leq \xi < \infty$)

$$H_{z_0} = \sum_{r=1}^{\infty} G_r Gek_r(\xi) se_r^*(\eta) e^{i\beta z} \tag{15a}$$

$$E_{z_0} = \sum_{r=0}^{\infty} F_r Fek_r(\xi) ce_r^*(\eta) e^{i\beta z}, \tag{15b}$$

[18] L. Kantorovich and V. Krylov, *Approximate Methods of Higher Analysis* (Interscience Publishers, Inc., New York, 1958).
[19] For any other modes, say the mth mode, successive approximations should start from the factor

$$\begin{vmatrix} g_{m,m} & h_{m,m} \\ t_{m,m} & s_{m,m} \end{vmatrix} = 0.$$

where C_m, D_m, G_r, and F_r are the arbitrary constants. Upon matching the boundary conditions at $\xi=\xi_0$ and applying the similar mathematical operations as for the ${}_e\text{HE}_{mn}$ mode, one can easily obtain the characteristic equation for the ${}_0\text{HE}_{mn}$ wave

$$\begin{vmatrix} g_{1,1}^* & h_{1,1}^* & g_{3,1}^* & h_{3,1}^* & g_{5,1}^* & h_{5,1}^* & \cdot & \cdot \\ t_{1,1}^* & s_{1,1}^* & t_{3,1}^* & s_{3,1}^* & t_{5,1}^* & s_{5,1}^* & \cdot & \cdot \\ g_{1,3}^* & h_{1,3}^* & g_{3,3}^* & h_{3,3}^* & g_{5,3}^* & h_{5,3}^* & \cdot & \cdot \\ t_{1,3}^* & s_{1,3}^* & t_{3,3}^* & s_{3,3}^* & t_{5,3}^* & s_{5,3}^* & \cdot & \cdot \\ g_{1,5}^* & h_{1,5}^* & g_{3,5}^* & h_{3,5}^* & g_{5,5}^* & h_{5,5}^* & \cdot & \cdot \\ t_{1,5}^* & s_{1,5}^* & t_{3,5}^* & s_{3,5}^* & t_{5,5}^* & s_{5,5}^* & \cdot & \cdot \\ \cdot & \cdot & \cdot & \cdot & \cdot & \cdot & \cdot & \cdot \\ \cdot & \cdot & \cdot & \cdot & \cdot & \cdot & \cdot & \cdot \end{vmatrix} = 0, \qquad (16)$$

where

$$g_{m,n}^* = \left(1+\frac{\gamma_1^2}{\gamma_0^2}\right) p_m \sum_{r=1}^{\infty}{}' \nu_{r,n}\beta_{m,r}$$

$$s_{m,n}^* = -\left(1+\frac{\gamma_1^2}{\gamma_0^2}\right) l_m \sum_{r=1}^{\infty}{}' \chi_{r,n}\alpha_{m,r}$$

$$h_{m,n}^* = \frac{\omega\epsilon_1}{\beta}\frac{a_n'}{a_n} l_m \alpha_{m,n} + \frac{\gamma_1^2}{\gamma_0^2}\frac{\omega\epsilon_0}{\beta} l_m' \alpha_{m,n}$$

$$t_{m,n}^* = \frac{\omega\mu}{\beta}\frac{b_n'}{b_n} p_m \beta_{m,n} + \frac{\gamma_1^2}{\gamma_0^2}\frac{\omega\mu}{\beta} p_m' \beta_{m,n}. \qquad (17)$$

To simplify the notations, the following dimensionless qualities are introduced:

$$x^2 = q^2 \cosh^2\xi_0 (k_1^2-\beta^2) = 4\cosh^2\xi_0 \gamma_1^2$$

$$y^2 = -q^2 \cosh^2\xi_0 (k_0^2-\beta^2) = 4\cosh^2\xi_0 \gamma_0^2. \qquad (18)$$

Hence the infinite determinants are functions of x, y, ξ_0, and ϵ_1/ϵ_0 only.

It can easily be shown that the characteristic equations for the ${}_e\text{HE}_{mn}$ and ${}_0\text{HE}_{mn}$ waves degenerate to the well-known characteristic equation for the HE_{mn} wave as the elliptical cross section degenerates to a circular one.

IV. CUTOFF FREQUENCIES OF THE DOMINANT MODES

The dominant modes are defined as the modes having the lowest cutoff frequencies.

In order to have a guided wave, β^2, x^2, and y^2 must all be real and positive. If y^2 is negative and real, the expressions for the field components will indicate the presence of an outgoing radial wave at a large distance from the surface of the dielectric rod, which can only come from an infinitely long line-type source (in the z direction) located at some finite ξ. Such source has not been postulated in the formulation. In fact, the concern here is with the source-free problem. Thus, y^2 must be positive for all surface guided waves and consequently the lowest permissible value of y^2 is zero. The frequency corresponding to this value of y^2, called the cutoff frequency of the wave, is

$$\omega_{\text{cutoff}} = x/q \cosh\xi_0 [\mu\epsilon_0(\epsilon_1/\epsilon_0 - 1)]^{\frac{1}{2}}, \qquad (19)$$

where x corresponds to the root of the characteristic equation with $y^2=0$. Physically it means that below this cutoff frequency the structure can no longer support such a wave and thereby ceases to be a binding medium for this wave.

The approximate expressions of the radial Mathieu functions for small x and y are quite complicated and involved. The reader is referred to reference 20 for the derivation and the explicit forms of these expressions. It can be shown that for very small values of x and y,

$$\begin{aligned} \alpha_{r,n} \sim \beta_{r,n} &\sim 1 \quad \text{when } r=n \\ &\sim 0 \quad \text{when } r\neq n \end{aligned}$$

and

$$\begin{aligned} \nu_{m,n} \sim -\chi_{m,n} &\sim m \quad \text{when } m=n \\ &\sim 0 \quad \text{when } m\neq n. \end{aligned} \qquad (20)$$

Hence, the infinite determinant (12) for the even wave collapses and reduces approximately to

$$x^2 \approx \frac{4(m^2-1)[m(1+\epsilon_0/\epsilon_1)+Q_1(\epsilon_0/\epsilon_1)\tanh\xi_0+Q_2\coth\xi_0]}{(e^{2\xi_0}/\cosh^2\xi_0)[(m+1)+(m-1)e^{-4\xi_0}]}, \qquad (21)$$

when $m\geq 3$ (m is odd) and

$$x^2 \approx \frac{8[1+(\epsilon_0/\epsilon_1)(1+\tanh\xi_0)+\coth\xi_0]}{(-\epsilon_0/\epsilon_1)(e^{2\xi_0}/\cosh^2\xi_0)(3-2e^{-\xi_0})\ln(e^{\alpha}ye^{\xi_0}/2\cosh\xi_0)}, \qquad (22)$$

when $m=1$, where Q_1 and Q_2 are finite constants, and α is the Euler's constant.

Upon inspection of Eq. (21), one immediately concludes that the right-hand side of this equation is always positive and nonzero and is not necessarily small for all values of ξ_0 and ϵ_0/ϵ_1. Thus, the imposed small x

[20] C. Yeh, California Institute of Technology, Antenna Laboratory Tech. Rept. No. 27 (1962).

TABLE I. An illustration of the rate of convergence for the infinite determinant.

ξ_0	y	x (2×2)	x (4×4)	x (6×6)	x (8×8)
3.0	0.1	1.10	1.10	...	...
3.0	1.5	1.83	1.83	...	...
3.0	2.9	2.01	2.02	...	...
0.7	0.1	1.42	1.46	1.47	...
0.7	1.5	2.30	2.40	2.42	...
0.7	2.9	2.51	2.67	2.68	...
0.2	0.1	2.72	2.77	2.79	...
0.2	0.9	3.85	3.94	3.99	4.01
0.2	1.7	4.43	4.54	4.61	4.63

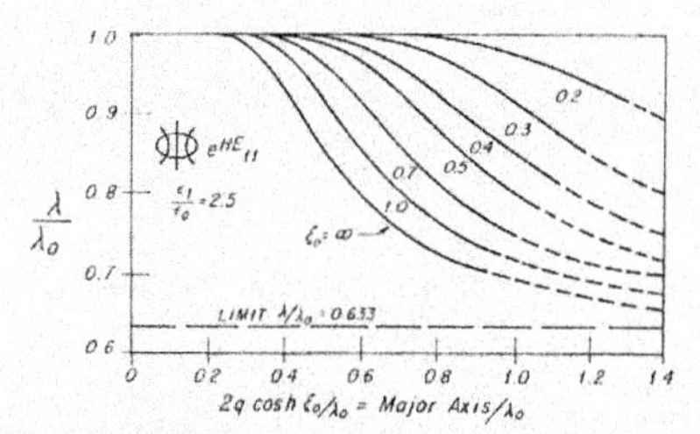

FIG. 3. Normalized guide wavelength λ/λ_0 as function of normalized major axis for the ${}_e\mathrm{HE}_{11}$ wave.

approximation is not valid. The same conclusion may be reached for $m\geq 0$ even waves. From Eq. (22), one notes that as y approaches zero, the right-hand side of this equation also approaches zero since the factor $\ln(e^{\alpha}ye^{\xi_0}/2\cosh\xi_0)$ approaches $-\infty$. In other words, as y approaches zero, x also becomes zero and the imposed small x approximation is valid. Therefore ${}_e\mathrm{HE}_{11}$ wave is the only even wave that possesses zero cutoff frequencies. It is also interesting to note [from Eq. (22)]

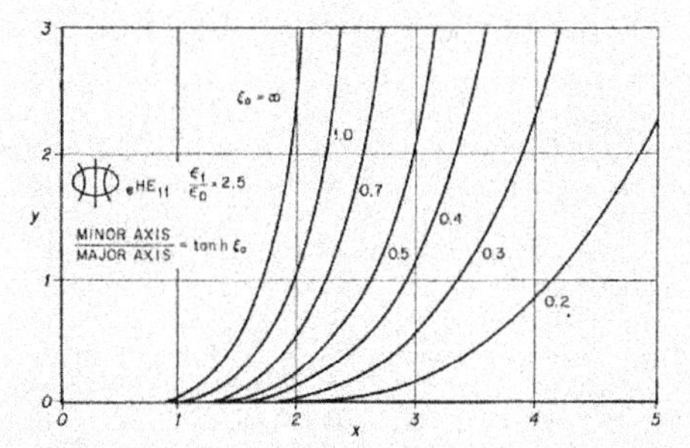

FIG. 2. Roots of the characteristic equation for the ${}_e\mathrm{HE}_{11}$ wave.

that as the elliptical cross section becomes flatter (i.e., $\xi_0\rightarrow 0$) the cutoff frequencies of the ${}_e\mathrm{HE}_{11}$ wave approaches zero slower. Carrying out the same analysis for the ${}_o\mathrm{HE}_{mn}$ wave, one obtains the similar conclusions. (i.e., ${}_o\mathrm{HE}_{11}$ wave is the only odd wave that has zero cutoff frequencies.) It is further conjectured here that the cutoff frequencies of all the other modes become higher for flatter elliptical dielectric cylinders.

V. SOLUTION OF THE DETERMINANTAL EQUATIONS FOR THE DOMINANT WAVES

The computations were carried out on a high-speed electronic computer, the IBM 7090.

It is known[21] that the periodic Mathieu functions can be expanded in terms of infinite series of trigonometric functions and that the corresponding radial Mathieu functions can be expanded in terms of infinite series of Bessel functions. The coefficients of expansion for a certain finite range of γ^2 have been tabulated by the National Bureau of Standards.[22] These coefficients were stored in the computer's memory. A three-point Lagrangian interpolation subroutine was used to interpolate the required coefficients from the stored values. The roots of the determinantal equations for the dominant modes were found by the successive approximation method.[18] It was found (numerically) that the infinite determinants converge rather rapidly within the present region of interests (i.e., $0\leq x\leq 5$ and $0\leq y\leq 3$). The rate of convergence for the infinite determinant can best be illustrated by the following example:

From Table I, for the ${}_e\mathrm{HE}_{11}$ wave with $\epsilon_1/\epsilon_0=2.5$, where the column heading (2×2), (4×4), etc. means that the values of x were obtained from a 2×2 determinant, a 4×4 determinant, etc., it can readily be seen that, as expected, the infinite determinant converges faster as ξ_0 gets larger (i.e., as the elliptical cross section approaches a circular one). Furthermore, for small values of y, the determinant also converges very fast. Same rate of convergence was observed for the odd dominant wave.

The Even Dominant Wave, ${}_e\mathrm{HE}_{11}$ Wave

The roots of the determinantal equation for the ${}_e\mathrm{HE}_{11}$ wave are shown in Fig. 2 for the case of $\epsilon_1/\epsilon_0=2.5$. Figure 3 gives the relation between the normalized guide wavelength λ/λ_0 and the normalized major axis (NMA) $2q\cosh\xi_0/\lambda_0$ for various values of ξ_0 with $\epsilon_1/\epsilon_0=2.5$. $\lambda_0=2\pi/[\omega(\mu\epsilon_0)^{\frac{1}{2}}]$ is the free-space wave-

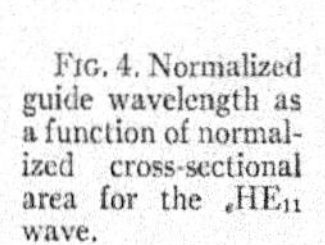

FIG. 4. Normalized guide wavelength as a function of normalized cross-sectional area for the ${}_e\mathrm{HE}_{11}$ wave.

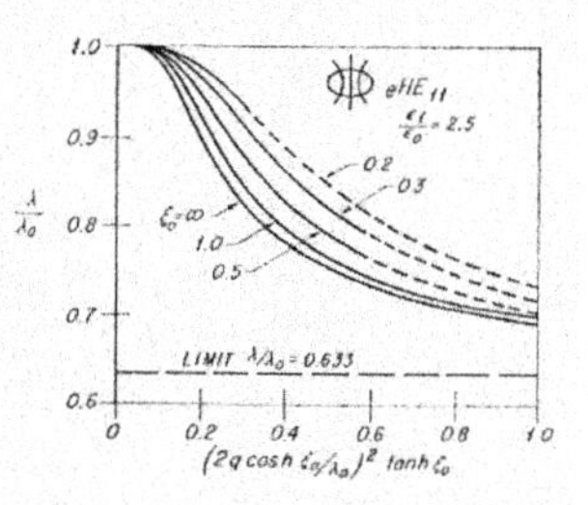

[21] N. McLachlan, *Theory and Application of Mathieu Functions* (Oxford University Press, Oxford, England, 1951).

[22] National Bureau of Standards, "*Tables Relating to Mathieu Functions* (Columbia University Press, New York, 1951).

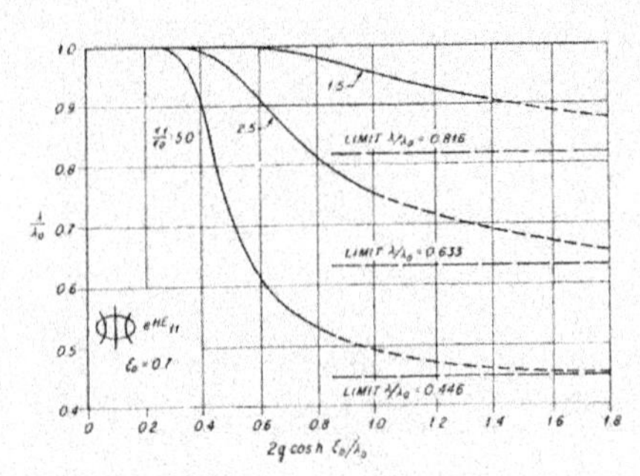

FIG. 5. Normalized guide wavelength of the ${}_eHE_{11}$ wave as a function of normalized major axis for various values of ϵ_1/ϵ_0.

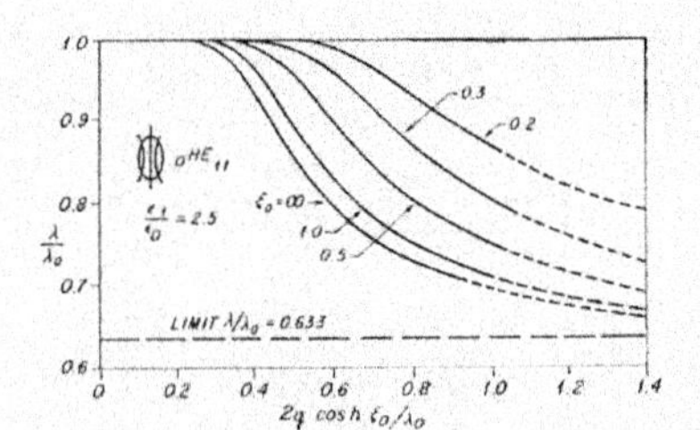

FIG. 7. Normalized guide wavelength as a function of normalized major axis for the ${}_oHE_{11}$ wave.

length. As expected, no cutoff frequency exists for this mode, and the cutoff frequency approaches zero slower as the elliptical cross section becomes flatter. For small values of NMA which corresponds to small values of y, the guide wavelength λ approaches the free-space wavelength λ_0, and the radial Mathieu functions describing the fields outside the dielectric rod decay very slowly; physically it means that the field strength of this wave falls off very slowly away from the rod and only a small part of the total energy is transported within the dielectric rod. For large values of NMA which corresponds to large values of y, the guide wavelength approaches asymptotically to the characteristic wavelength of the material $\lambda_M=\lambda_0/(\epsilon_1/\epsilon_0)^{\frac{1}{2}}$, and the radial Mathieu functions describing the fields outside the rod disappear very quickly and almost all the energy is transported within the dielectric rod.

Figure 4 shows curves of the normalized guide wavelength as functions of normalized cross-sectional area (NCSA) $(2q\cosh\xi_0/\lambda_0)^2\tanh\xi_0$ for various ξ_0 with $\epsilon_1/\epsilon_0=2.5$. For a fixed value of NCSA, λ/λ_0 is larger for flatter elliptical cross section. This behavior suggests that the field intensity is more concentrated in a circular rod. It is also noted that the flatter is the cross section, the smaller is the variation of λ/λ_0 as a function of NCSA.

The effect of the variation of relative dielectric constant ϵ_1/ϵ_0 on the propagation constant can be seen readily from Fig. 5, in which λ/λ_0 is plotted against NMA for various values of ϵ_1/ϵ_0 with $\xi_0=0.7$. As ϵ_1/ϵ_0 becomes larger, λ approaches to the characteristic wavelength of the rod material $\lambda_M=\lambda_0/(\epsilon_1/\epsilon_0)^{\frac{1}{2}}$ faster. This is the same behavior exhibited by the dominant wave propagating along a circular dielectric rod.

The Odd Dominant Wave, ${}_oHE_{11}$ Wave

The roots of the determinantal equation, the normalized guide wavelength as a function of NMA, and λ/λ_0 as a function of NCSA for the ${}_oHE_{11}$ wave are shown respectively in Figs. 6, 7, and 8. The general characteristics of the guide wavelength as a function of NMA, NCSA, ξ_0, and ϵ_1/ϵ_0 are very similar to those indicated by the even dominant wave. However, unlike the case for the ${}_eHE_{11}$ wave, it seems that the elliptical rod is a better binding geometry for the ${}_oHE_{11}$ wave than a circular rod, since for a fixed value of NCSA, λ/λ_0 is smaller for flatter cross section. The fact, that the curves for various values of ξ_0 are quite close to each other, suggests that the field intensity is quite uniform for this odd dominant wave in the dielectric rod.

It should be noted that the propagation characteristic of the ${}_eHE_{11}$ wave and the ${}_oHE_{11}$ wave pass smoothly to that of the circular HE_{11} wave as $\xi_0\rightarrow\infty$.

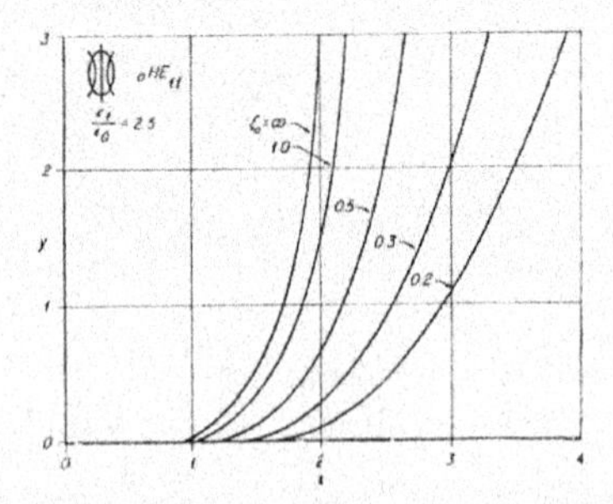

FIG. 6. Roots of the characteristic equation for the ${}_oHE_{11}$ wave.

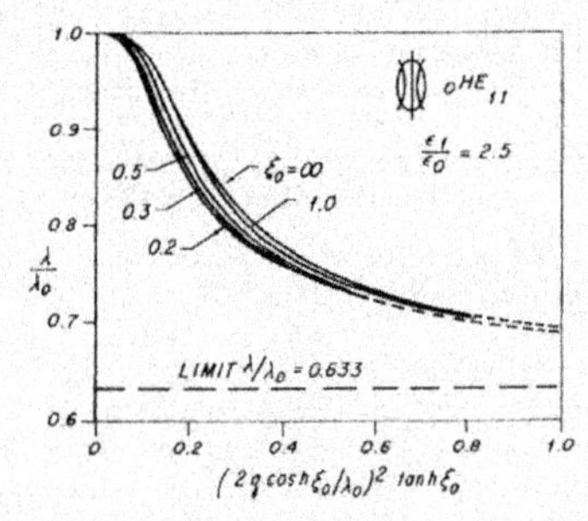

FIG. 8. Normalized guide wavelength as a function of normalized cross-sectional area.

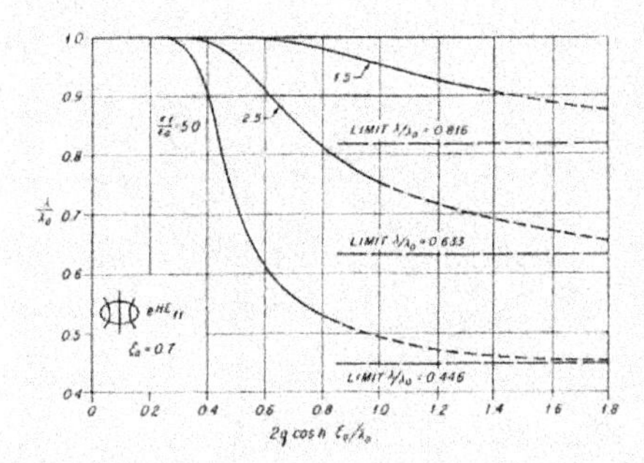

FIG. 5. Normalized guide wavelength of the $_e$HE$_{11}$ wave as a function of normalized major axis for various values of ϵ_1/ϵ_0.

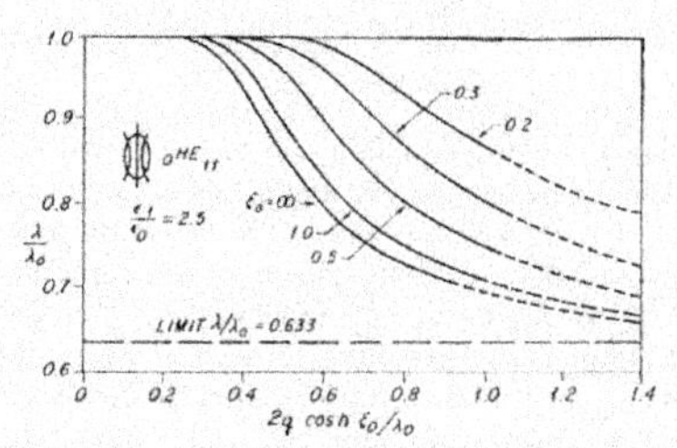

FIG. 7. Normalized guide wavelength as a function of normalized major axis for the $_o$HE$_{11}$ wave.

length. As expected, no cutoff frequency exists for this mode, and the cutoff frequency approaches zero slower as the elliptical cross section becomes flatter. For small values of NMA which corresponds to small values of y, the guide wavelength λ approaches the free-space wavelength λ_0, and the radial Mathieu functions describing the fields outside the dielectric rod decay very slowly; physically it means that the field strength of this wave falls off very slowly away from the rod and only a small part of the total energy is transported within the dielectric rod. For large values of NMA which corresponds to large values of y, the guide wavelength approaches asymptotically to the characteristic wavelength of the material $\lambda_M=\lambda_0/(\epsilon_1/\epsilon_0)^{\frac{1}{2}}$, and the radial Mathieu functions describing the fields outside the rod disappear very quickly and almost all the energy is transported within the dielectric rod.

Figure 4 shows curves of the normalized guide wavelength as functions of normalized cross-sectional area (NCSA) $(2q\cosh\xi_0/\lambda_0)^2\tanh\xi_0$ for various ξ_0 with $\epsilon_1/\epsilon_0=2.5$. For a fixed value of NCSA, λ/λ_0 is larger for flatter elliptical cross section. This behavior suggests that the field intensity is more concentrated in a circular rod. It is also noted that the flatter is the cross section, the smaller is the variation of λ/λ_0 as a function of NCSA.

The effect of the variation of relative dielectric constant ϵ_1/ϵ_0 on the propagation constant can be seen readily from Fig. 5, in which λ/λ_0 is plotted against NMA for various values of ϵ_1/ϵ_0 with $\xi_0=0.7$. As ϵ_1/ϵ_0 becomes larger, λ approaches to the characteristic wavelength of the rod material $\lambda_M=\lambda_0/(\epsilon_1/\epsilon_0)^{\frac{1}{2}}$ faster. This is the same behavior exhibited by the dominant wave propagating along a circular dielectric rod.

The Odd Dominant Wave, $_o$HE$_{11}$ Wave

The roots of the determinantal equation, the normalized guide wavelength as a function of NMA, and λ/λ_0 as a function of NCSA for the $_o$HE$_{11}$ wave are shown respectively in Figs. 6, 7, and 8. The general characteristics of the guide wavelength as a function of NMA, NCSA, ξ_0, and ϵ_1/ϵ_0 are very similar to those indicated by the even dominant wave. However, unlike the case for the $_e$HE$_{11}$ wave, it seems that the elliptical rod is a better binding geometry for the $_o$HE$_{11}$ wave than a circular rod, since for a fixed value of NCSA, λ/λ_0 is smaller for flatter cross section. The fact, that the curves for various values of ξ_0 are quite close to each other, suggests that the field intensity is quite uniform for this odd dominant wave in the dielectric rod.

It should be noted that the propagation characteristic of the $_e$HE$_{11}$ wave and the $_o$HE$_{11}$ wave pass smoothly to that of the circular HE$_{11}$ wave as $\xi_0\to\infty$.

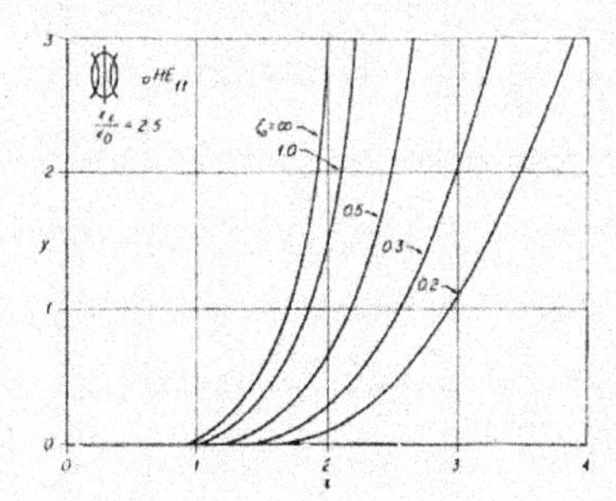

FIG. 6. Roots of the characteristic equation for the $_o$HE$_{11}$ wave.

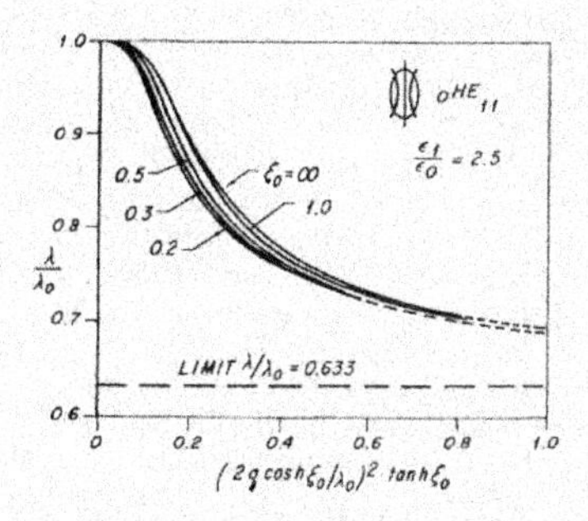

FIG. 8. Normalized guide wavelength as a function of normalized cross-sectional area.

VI. FIELD CONFIGURATIONS OF THE DOMINANT WAVE

In practice the field configurations are most quickly found by inspection of the mode function. It is found that the patterns of the electric and magnetic field lines are quite similar to those known in a hollow metallic guide. However, owing to the absence of the metallic shield around the dielectric, the field is no longer confined to the inner space. Furthermore, due to the absence of the conducting walls and therefore the absence of the conduction currents, all the electric and magnetic field lines must form closed loops.

Figures 9(a) and 9(b) show the transverse cross-sectional field distributions of the ${}_eHE_{11}$ wave and the ${}_0HE_{11}$ wave, respectively. Perspective views of the electric lines of force are shown in Figs. 10(a) and 10(b). The fact that the cross-sectional field configurations are similar to the corresponding dominant waves in the metal tube waveguides suggests a simple method of exciting these dominant dielectric waves.

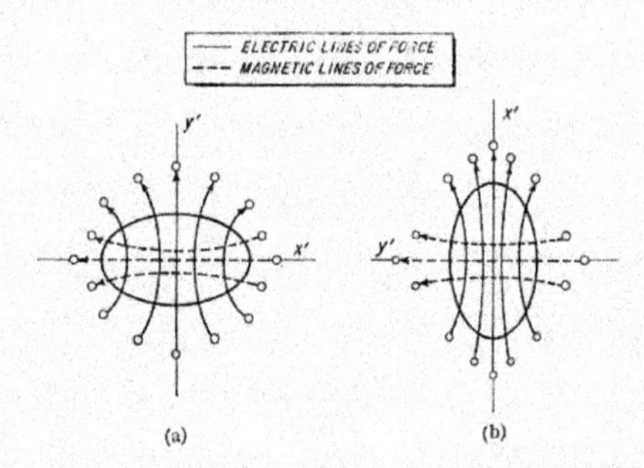

Fig. 9. (a) Cross-sectional field configuration of the ${}_eHE_{11}$ wave. (b) Cross-sectional field configuration of the ${}_0HE_{11}$ wave.

VII. EXPERIMENTAL RESULTS

To experimentally investigate the dominant electromagnetic wave propagation along an elliptical dielectric rod, the wavelength along the rod was measured using the apparatus shown schematically in Fig. 11. For the sake of convenience and simplicity, measurements were performed in the X-band frequency range. Standard X-band components were used: X-13 Varian reflex klystron and its power supply, isolator, attenuator, high Q circular-mode cavity resonator, rectangular metal waveguide, a slotted line section, and an HP standing wavemeter. Since a rectangular metal guide operating in the dominant TE_{10} mode has an electric field whose configuration is roughly similar to the transverse component of the electric field of the ${}_eHE_{11}$ wave or the ${}_0HE_{11}$ wave on the dielectric guide, the transfer of microwave energy could be made by simply inserting the dielectric rod longitudinally into the metal guide for a short distance. The orientation of the cross section depends upon whether ${}_eHE_{11}$ wave or ${}_0HE_{11}$ wave is desired. To improve matching and to minimize reflection, a flare pyramidal horn whose flare angle was adjusted for best energy transfer was connected to the metal guide. The dielectric rod was tapered to a point within the guide and after emerging from the metal guide the rod was tapered to whatever size was required for a given test. The other end of the dielectric rod was machined very flat to accommodate a good contact with a flat aluminum plate which was used as a good shorting device. The aluminum plate was made large enough to intercept practically all of the energy outside the dielectric rod. Since dielectric rods of elliptical cross section were not commercially available, they were machined from available, rectangular Lucite strips which were at least five and one-half feet long. The major axis and ξ_0 of these elliptical rods ranged from 1.5 to 0.5 in. and $\xi_0=\infty$ to $\xi_0=0.37$. The dielectric constant of these rods were found to vary between $\epsilon_1/\epsilon_0=2.5$ to $\epsilon_1/\epsilon_0=2.6$.

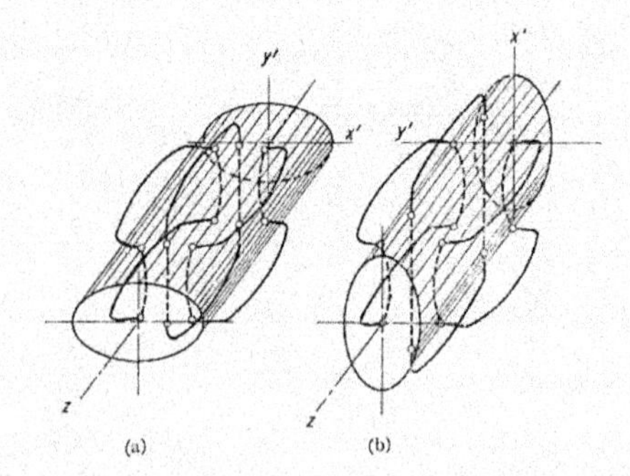

Fig. 10. (a) A sketch of the electric lines of force for the ${}_eHE_{11}$ wave. (b) A sketch of the electric lines of force for the ${}_0HE_{11}$ wave.

With the help of a transit and a level this whole experiment setup was aligned carefully. The dielectric rod must be very straight and its axis perpendicular to the shorting plate. To minimize sagging of some small

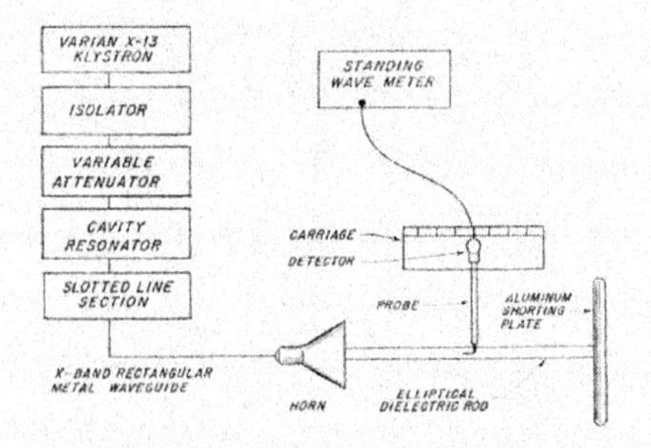

Fig. 11. Block diagram showing arrangement of components for determining guide wavelength of an elliptical dielectric waveguide.

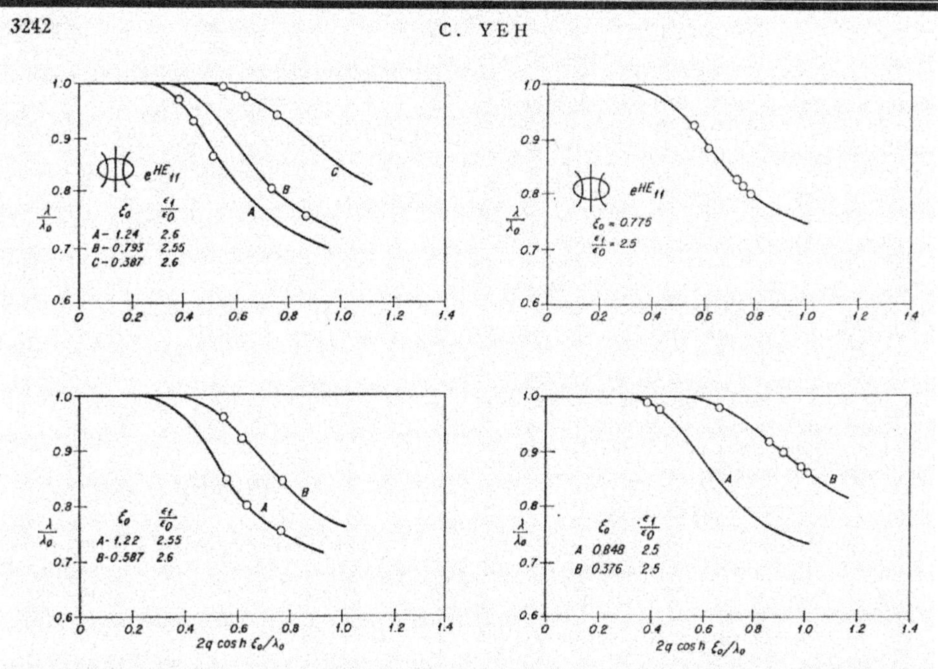

Fig. 12. Normalized guide wavelength as a function of normalized major axis for the $_e$HE$_{11}$ wave. Circles are experimental points.

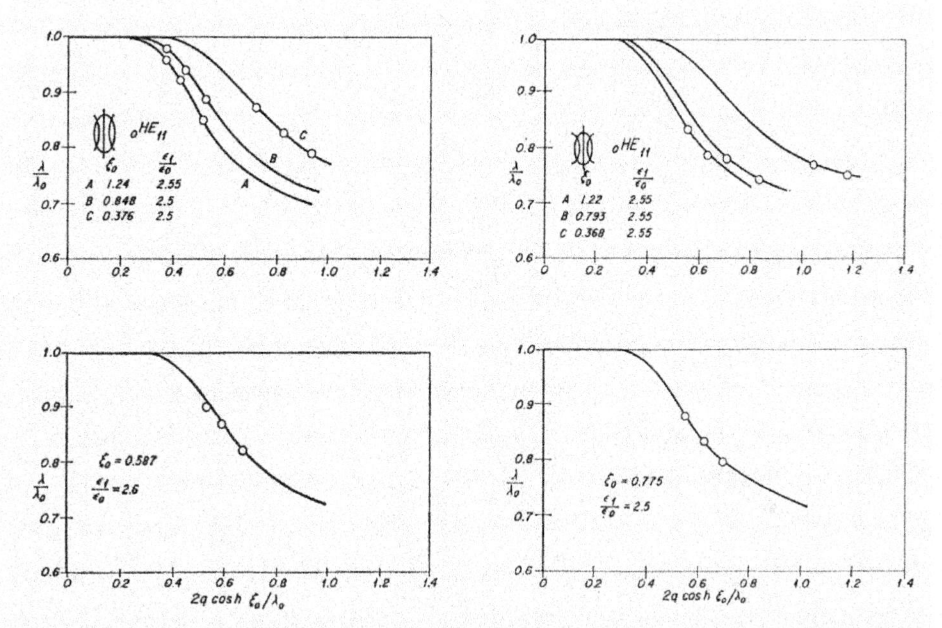

Fig. 13. Normalized guide wavelength as a function of normalized major axis for the $_o$HE$_{11}$ wave. Circles are experimental points.

or flat rods, very thin nylon threads were used to provide support. Although disturbances due to these threads were unavoidable, because of the sizes of rods used at this frequency range very little perturbation was observed.

The wavelengths were measured using the standing wave technique. The distance between nulls of the standing wave pattern that resulted from energy reflected from the shorted end of the dielectric guide was measured by a small electric probe. This distance is equal to $\lambda/2$, where λ is the guide wavelength.

The measured guide wavelength values for the two dominant waves together with the theoretical curves are summarized in Figs. 12 and 13. It can be seen that the analytic and experimental results are in very good agreement. Incidentally, the above experiment suggests a rather convenient way of measuring the dielectric constants of certain low loss dielectric materials.

ACKNOWLEDGMENTS

The author wishes to express his sincere thanks to Professor C. H. Papas for his stimulating criticisms and his encouragement throughout all phases of this investigation.

The author acknowledges the Western Data Processing Center at the University of California at Los Angeles for the use of the computing facilities.

APPENDIX. FORMULAS FOR $\alpha_{r,m}$, $\beta_{r,n}$, $\chi_{m,n}$ AND $\nu_{m,n}$

It can readily be shown from the theory of Mathieu functions that

$$\alpha_{r,n} = \int_0^{2\pi} ce_r^*(\eta) ce_n(\eta) d\eta \Big/ \int_0^{2\pi} ce_n^2(\eta) d\eta$$

$$\beta_{r,n} = \int_0^{2\pi} se_r^*(\eta) se_n(\eta) d\eta \Big/ \int_0^{2\pi} se_n^2(\eta) d\eta$$

$$\chi_{m,n} = \int_0^{2\pi} ce_m'(\eta) se_n(\eta) d\eta \Big/ \int_0^{2\pi} se_n^2(\eta) d\eta$$

$$\nu_{m,n} = \int_0^{2\pi} se_m'(\eta) ce_n(\eta) d\eta \Big/ \int_0^{2\pi} ce_n^2(\eta) d\eta,$$

where the prime signifies the derivative of the function with respect to its argument. The above integrals can easily be integrated using the series expansions of Mathieu functions in terms of trigonometric functions (see reference 21).

Attenuation in a Dielectric Elliptical Cylinder*

CAVOUR W. H. YEH†, MEMBER, IRE

Summary—The problem of the propagation of electromagnetic waves along a dielectric cylinder of elliptical cross section is considered. There exist two dominant waves which possess zero cutoff frequencies. The attenuation characteristics of these dominant waves are analyzed theoretically and experimentally. Good agreements are obtained. It is shown that one of the dominant waves, with its electric fields parallel at the center of the cylinder to the minor axis of the elliptical cylinder, can propagate along a flat dielectric strip with much lower loss than can the HE_{11} wave along a circular cylinder having the identical cross-sectional area.

Introduction

RECENT DEVELOPMENTS in the generation and application of millimeter and sub-millimeter electromagnetic waves, the availability of very low loss dielectrics, and the development of fiber optics, have renewed interest in the surface waveguides.

It is known that the attenuation factor of a dominant wave in a conventional metal tube waveguide becomes unbearably high at millimeter or sub-millimeter wavelength.[1] New ways must therefore be found to guide electromagnetic energy at very high frequencies. The use of a dielectric coated wire,[2] or the use of a dielectric rod[3-6] as a practical millimeter waveguide has been proposed. However, in order that the dielectric rod may be a low loss guiding device, one must choose a small value of ka where k is the free-space wave number and a is the radius of the dielectric cylinder.[3] In the millimeter wavelength range the radius of the cylinder becomes inconveniently small. Fortunately, it is found that if the circular rod is flattened, (*i.e.*, if a circular rod is deformed into an elliptical rod of the same cross-sectional area), the attenuation of the dominant wave may be reduced considerably, provided that the electric field of the dominant wave is parallel at the center of the cylinder to the minor axis of the elliptical rod. The problem of the propagation of electromagnetic wave along a circular dielectric cylinder has been considered by many authors.[3-5,7]

The problem of the propagation of electromagnetic waves along an elliptical dielectric cylinder (dielectric strip) has been presented elsewhere.[8] It is the purpose

* Received July 26, 1962; revised manuscript received November 13, 1962. This paper was supported by the Air Force Office of Scientific Research.

† Department, of Electrical Engineering University of Southern California, Los Angeles. Formerly with Antenna Laboratory, California Institute of Technology, Pasadena, Calif.

[1] It is noted that the attenuation factor of TE_{01} mode in a circular metal waveguide decreases as frequency increases. However, to avoid mode conversions (since TE_{01} mode is not the dominant wave), the circular waveguide must meet close tolerances. Hence, it is rather expensive to manufacture.

[2] G. Goubau, "Surface waves and their application to transmission lines," *J. Appl. Phys.*, vol. 21, pp. 1119–1120; November, 1950.

[3] C. H. Chandler, "An investigation of dielectric rod as waveguides," *J. Appl. Phys.*, vol. 20, pp. 1188–1192; December, 1949.

[4] W. M. Elsasser, "Attenuation in a dielectric circular rod," *J. Appl. Phys.*, vol. 20, pp. 1193–1196; December, 1949.

[5] D. D. King, "Dielectric image line," *J. Appl. Phys.*, vol. 23, pp. 699–700; June, 1952.

[6] A. F. Harvey, "Periodic and guiding structures at microwave frequencies," IRE TRANS. ON MICROWAVE THEORY AND TECHNIQUES, vol. MTT-8, pp. 30–61; January, 1960.

[7] J. R. Carson, S. P. Mead and S. A. Schelkunoff, "Hyperfrequency waveguides—mathematical theory," *Bell Sys. Tech. J.*, vol. 15, pp. 310–328; March, 1936.

[8] C. W. H. Yeh, "Electromagnetic Surface-Wave Propagation Along a Dielectric Cylinder of Elliptical Cross-Section," Antenna Lab., California Inst. Tech., Pasadena, Tech. Rept. No. 27; January, 1962. See also, "On the elliptical dielectric waveguides," *J. Appl. Phys.*, vol. 33, pp. 3235–3243; November, 1962.

of this paper to examine the attenuation characteristics of the dominant modes, and to perform experiments to verify the analytic results.

Analysis

Only the results of the theory of electromagnetic wave propagation along an elliptical dielectric cylinder that are required to calculate the attenuation characteristics of the dominant waves due to dielectric loss will be given here. It is known that two types of dominant wave are possible. One, with its electric field lines parallel at the center of the rod to the minor axis of the elliptical cross section, is called the even dominant wave, the $_eHE_{11}$ wave; the other, with its electric field lines parallel at the center of the rod to the major axis of the elliptical cross section, is called the odd dominant wave, the $_oHE_{11}$ wave. Both of these dominant waves have zero cutoff frequency.

The Even Dominant Wave, the $_eHE_{11}$ Wave

The longitudinal components of the $_eHE_{11}$ mode are,[8,9] inside an elliptical dielectric rod,

$$H_{z_1} = \sum_{m=1}^{\infty}{}' A_m Ce_m(\xi) ce_m(\eta)$$

$$E_{z_1} = \sum_{m=1}^{\infty}{}' B_m Se_m(\xi) se_m(\eta), \tag{1}$$

and, outside the rod,

$$H_{z_0} = \sum_{r=1}^{\infty}{}' L_r Fek_r(\xi) ce_r^*(\eta)$$

$$E_{z_0} = \sum_{r=1}^{\infty}{}' P_r Gek_r(\xi) se_r^*(\eta), \tag{2}$$

where the prime on the summation sign means that only odd integers are to be summed. The transverse field components can be obtained from Maxwell's equations using (1) and (2). A_m, B_m, L_r and P_r are arbitrary constants which are related by the boundary conditions. The common factor $e^{i\beta z - i\omega t}$, where β is the propagation constant and ω is the frequency of oscillation in free-space, has been omitted. The parameters x^2 and y^2 are related to the propagation constant β by the following relations:

$$x^2 = q^2 \cosh^2 \xi_0 (k_1^2 - \beta^2)$$

$$y^2 = -q^2 \cosh^2 \xi_0 (k_0^2 - \beta^2) \tag{3}$$

where q is the semifocal length of the ellipse, ξ_0 is the boundary of the elliptical rod in the elliptical cylinder coordinates (ξ, η, z), $k_1^2 = \omega^2 \mu \epsilon_1$, $k_0^2 = \omega^2 \mu \epsilon_0$, ϵ_1 and ϵ_0 are respectively the dielectric constant of the rod and of the free-space, and μ is the permeability of the rod and of the free-space.

[9] N. W. McLachlan, "Theory and Application of Mathieu Functions," University Press, Oxford, Eng.; 1951.

Application of the boundary conditions yields a determinantal equation for the $_eHE_{11}$ wave. This equation, which is a function of x, y, ξ_0 and ϵ_1/ϵ_0, is

$$\begin{vmatrix} g_{1,1} & h_{1,1} & g_{3,1} & h_{3,1} & g_{5,1} & h_{5,1} & \cdot & \cdot \\ t_{1,1} & s_{1,1} & t_{3,1} & s_{3,1} & t_{5,1} & s_{5,1} & \cdot & \cdot \\ g_{1,3} & h_{1,3} & g_{3,3} & h_{3,3} & g_{5,3} & h_{5,3} & \cdot & \cdot \\ t_{1,3} & s_{1,3} & t_{3,3} & s_{3,3} & t_{5,3} & s_{5,3} & \cdot & \cdot \\ g_{1,5} & h_{1,5} & g_{3,5} & h_{3,5} & g_{5,5} & h_{5,5} & \cdot & \cdot \\ t_{1,5} & s_{1,5} & t_{3,5} & s_{3,5} & t_{5,5} & s_{5,5} & \cdot & \cdot \\ \cdot & \cdot & \cdot & \cdot & \cdot & \cdot & \cdot & \cdot \\ \cdot & \cdot & \cdot & \cdot & \cdot & \cdot & \cdot & \cdot \end{vmatrix} = 0, \tag{4}$$

where

$$g_{m,n} = \left(1 + \frac{x^2}{y^2}\right) l_m \sum_{r=1}^{\infty}{}' \chi_{r,n} \alpha_{m,r}$$

$$s_{m,n} = -\left(1 + \frac{x^2}{y^2}\right) p_m \sum_{r=1}^{\infty}{}' \nu_{r,n} \beta_{m,r}$$

$$h_{m,n} = \frac{\omega\epsilon_1}{\beta} \frac{b_n'}{b_n} p_m \beta_{m,n} + \frac{x^2}{y^2} \frac{\omega\epsilon_0}{\beta} p_m' \beta_{m,n}$$

$$t_{m,n} = \frac{\omega\mu}{\beta} \frac{a_n'}{a_n} l_m \alpha_{m,n} + \frac{x^2}{y^2} \frac{\omega\mu}{\beta} l_m' \alpha_{m,n} \tag{5}$$

in which the abbreviations

$$\begin{aligned} a_m &= Ce_m(\xi_0) & a_m' &= Ce_m'(\xi_0) \\ b_m &= Se_m(\xi_0) & b_m' &= Se_m'(\xi_0) \\ l_m &= Fek_m(\xi_0) & l_m' &= Fek_m'(\xi_0) \\ p_m &= Gek_m(\xi_0) & p_m' &= Gek_m'(\xi_0) \end{aligned} \tag{6}$$

have been used. $\alpha_{r,n}$, $\beta_{r,n}$, $\chi_{m,n}$, and $\nu_{m,n}$ are given in Appendix I. The prime on the modified Mathieu functions signifies the derivative of the function with respect to ξ_0. The roots of this determinantal equation (4) give the relation between x^2 and y^2 as functions of ξ_0 and ϵ_1/ϵ_0. The normalized guide wavelength λ/λ_0, where $\lambda = 2\pi/\beta$, as functions of $(2q \cosh \xi_0)/\lambda_0$, ϵ_1/ϵ_0, and ξ_0 can easily be computed using (3).

The roots of the determinantal equation for the $_eHE_{11}$ mode have been computed for the case of $\epsilon_1/\epsilon_0 = 2.5$.[8] The relations between the normalized guide wavelength λ/λ_0 and the normalized major axis $(2q \cosh \xi_0)/\lambda_0$ for various values of ξ_0 with $\epsilon_1/\epsilon_0 = 2.5$ have also been obtained. Results are presented in Figs. 1 and 2.

It is well known that if the power loss (due to imperfect dielectric) per wavelength along the rod is small compared to the power flowing along the rod, the attenuation constant α can be calculated using the following formula:[4]

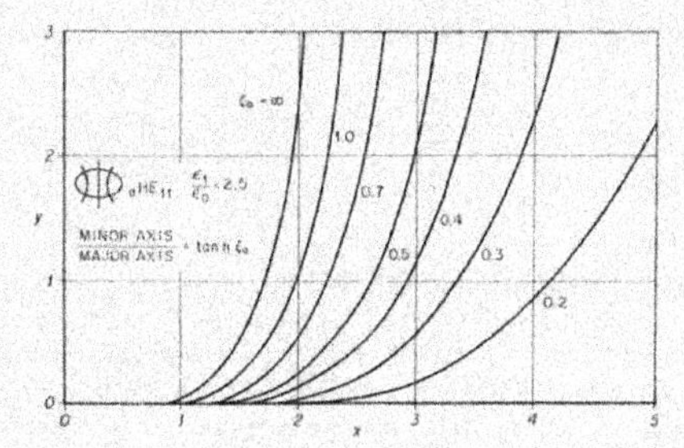

Fig. 1—Roots of the characteristic equation for the $_e$HE$_{11}$ wave.

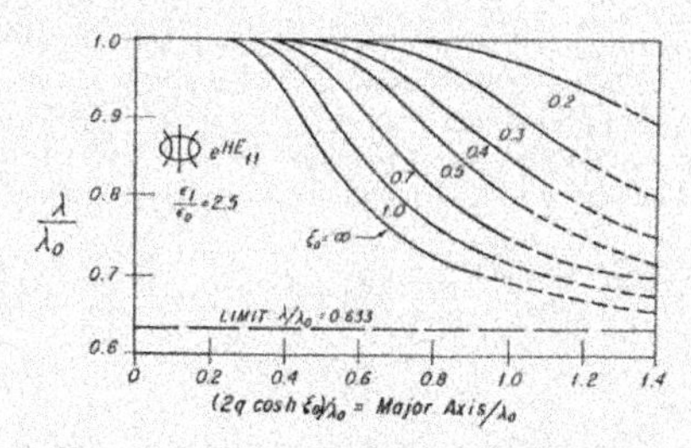

Fig. 2—Normalized guide wavelength λ/λ_0 as function of normalized major axis for the $_e$HE$_{11}$ wave.

$$\alpha = \frac{8.686}{2}\sigma_d\sqrt{\frac{\mu}{\epsilon_0}}\left|\frac{\int_{Ai} E\cdot E^* dA}{\frac{\mu}{\epsilon_0}\int_A \mathbf{e}_z\cdot(E_A\times H_A^*)dA}\right|$$

$$= 4.343\,\sigma_d\sqrt{\frac{\mu}{\epsilon_0}}\,R \text{ (db/ meter)} \tag{7}$$

where σ_d is the conductivity of the dielectric rod and $\mathbf{e}_z$ is the unit vector in the direction of propagation. A_i is the cross-sectional area of the dielectric rod and A is the total cross-sectional area of the guide. It is assumed that guide wavelength $\lambda = 2\pi/\beta$ is not affected by the presence of small dielectric loss, and that the mode functions in the case of small dielectric loss differ from those of the lossless case only by a multiplicative attenuation factor $e^{-\alpha z}$.

Substituting the appropriate expressions for the $_e$He$_{11}$ mode into (7) and simplifying gives

$$R = \frac{f_1}{f_2 + f_3} \tag{8}$$

where

$$f_1 = \mathbf{C}_1\left[\mathbf{C}_2(\text{A I}) + \left(\frac{B_1}{A_1}\right)^2\left(\frac{\epsilon_0}{\mu}\right)(\text{A II}) + 2\sqrt{\mathbf{C}_2}\left(\frac{B_1}{A_1}\right)\sqrt{\frac{\epsilon_0}{\mu}}(\text{A III})\right] + \left(\frac{B_1}{A_1}\right)^2\left(\frac{\epsilon_0}{\mu}\right)(\text{A IV}), \tag{9a}$$

$$f_2 = \mathbf{C}_1\left[\sqrt{\mathbf{C}_2}(\text{A I}) + \frac{\epsilon_1}{\epsilon_0}\sqrt{\mathbf{C}_2}\frac{\epsilon_0}{\mu}\left(\frac{B_1}{A_1}\right)^2(\text{A II}) + \sqrt{\frac{\epsilon_0}{\mu}}\frac{B_1}{A_1}\left(1 + \frac{\epsilon_1}{\epsilon_0}\mathbf{C}_2\right)(\text{A III})\right], \tag{9b}$$

$$f_3 = \frac{x^4}{y^4}\mathbf{C}_1\left[\left(\frac{L_1}{A_1}\right)^2\sqrt{\mathbf{C}_2}(\text{B I}) + \left(\frac{P_1}{A_1}\right)^2\sqrt{\mathbf{C}_2}\frac{\epsilon_0}{\mu}(\text{B II}) + \sqrt{\frac{\epsilon_0}{\mu}}\left(\frac{L_1}{A_1}\right)\left(\frac{P_1}{A_1}\right)(1 + \mathbf{C}_2)(\text{B III})\right] \tag{9c}$$

in which

$$\mathbf{C}_1 = \frac{x^2 + y^2\frac{\epsilon_1}{\epsilon_0}}{x^4\left(\frac{\epsilon_1}{\epsilon_0} - 1\right)}, \qquad \mathbf{C}_2 = \frac{x^2 + y^2}{x^2 + \frac{\epsilon_1}{\epsilon_0}y^2}$$

$$(\text{A I}) = \sum_n{}'\left(\frac{A_n}{A_1}\right)^2[R_{nn}{}^e I_{nn}{}^{e\prime} + R_{nn}{}^{e\prime} I_n{}^e] + \sum_n{}'\sum_{m\,(n\neq m)}{}'\left(\frac{A_n}{A_1}\right)\left(\frac{A_m}{A_1}\right)[R_{nm}{}^e I_{nm}{}^{e\prime}],$$

$$(\text{A II}) = \sum_n{}'\left(\frac{B_n}{B_1}\right)^2[R_{nn}{}^{0\prime} I_n{}^0 + R_{nn}{}^0 I_{nn}{}^{0\prime}] + \sum_n{}'\sum_{m\,(n\neq m)}{}'\left(\frac{B_n}{B_1}\right)\left(\frac{B_m}{B_1}\right)[R_{nm}{}^0 I_{nm}{}^{0\prime}],$$

$$(\text{A III}) = \sum_n{}'\sum_m{}'\left(\frac{A_n}{A_1}\right)\left(\frac{B_m}{B_1}\right)[T_{nm}{}^e J_{nm}{}^0 - T_{nm}{}^0 J_{nm}{}^e],$$

$$(\text{A IV}) = \sum_n{}'\sum_m{}'\left(\frac{B_n}{B_1}\right)\left(\frac{B_m}{B_1}\right)Q_{nm}{}^0,$$

$$(\text{B I}) = \sum_n{}'\left(\frac{L_n}{L_1}\right)^2[R_{nn}{}^e I_{nn}{}^{e\prime} + R_{nn}{}^{e\prime} I_n{}^e] + \sum_n{}'\sum_{m\,(n\neq m)}{}'\left(\frac{L_n}{L_1}\right)\left(\frac{L_m}{L_1}\right)[R_{nm}{}^e I_{nm}{}^{e\prime}],$$

$$(\text{B II}) = \sum_n{}'\left(\frac{P_n}{P_1}\right)^2[R_{nn}{}^{0\prime} I_n{}^0 + R_{nn}{}^0 I_{nn}{}^{0\prime}] + \sum_n{}'\sum_{m\,(n\neq m)}{}'\left(\frac{P_n}{P_1}\right)\left(\frac{P_m}{P_1}\right)[R_{nm}{}^0 I_{nm}{}^{0\prime}],$$

$$(\text{B III}) = \sum_n{}'\sum_m{}'\left(\frac{L_n}{L_1}\right)\left(\frac{P_m}{P_1}\right)[T_{nm}{}^e J_{nm}{}^0 - T_{nm}{}^0 J_{nm}{}^e]. \tag{10}$$

f_2 and f_3 represent the portion of the total transmitted power being carried inside and outside the dielectric cylinder, respectively. The R's, I's, J's, T's and Q's are given in Appendix I. The ratios between various arbitrary constants can be obtained from the following:

$$A_n a_n = \sum_{r=1}^{\infty}{}' L_r l_r \alpha_{r,n}, \tag{11a}$$

$$B_n b_n = \sum_{r=1}^{\infty}{}' P_r p_r \beta_{r,n}, \tag{11b}$$

$$\frac{\omega\epsilon_1}{\beta} B_n b_n' + \left[1+\frac{x^2}{y^2}\right] \sum_{r=1}^{\infty}{}' A_r a_r \chi_{r,n} = \left(-\frac{x^2}{y^2}\right)\frac{\omega\epsilon_0}{\beta}\sum_{r=1}^{\infty}{}' P_r p_r' \beta_{r,n} \tag{11c}$$

$$\frac{\omega\mu}{\beta} A_n a_n' - \left[1+\frac{x^2}{y^2}\right] \sum_{r=1}^{\infty}{}' B_r b_r \nu_{r,n} = \left(-\frac{x^2}{y^2}\right)\frac{\omega\mu}{\beta}\sum_{r=1}^{\infty}{}' L_r l_r' \alpha_{r,n}. \tag{11d}$$

$$(n = 1, 3, 5 \cdots)$$

These equations are obtained by matching the boundary conditions at $\xi=\xi_0$.[8] It is an easy matter to express one arbitrary constant in terms of another through algebraic manipulations of (11a)–(11d). It is found numerically that, for the $_e\mathrm{HE}_{11}$ wave,

$$1 \gg \left|\frac{A_3}{A_1}\right| \gg \left|\frac{A_5}{A_1}\right| \gg \cdots \gg \left|\frac{A_\infty}{A_1}\right|$$

$$1 \gg \left|\frac{B_3}{B_1}\right| \gg \left|\frac{B_5}{B_1}\right| \gg \cdots \gg \left|\frac{B_\infty}{B_1}\right|$$

$$1 \gg \left|\frac{L_3}{L_1}\right| \gg \left|\frac{L_5}{L_1}\right| \gg \cdots \gg \left|\frac{L_\infty}{L_1}\right|$$

$$1 \gg \left|\frac{P_3}{P_1}\right| \gg \left|\frac{P_5}{P_1}\right| \gg \cdots \gg \left|\frac{P_\infty}{P_1}\right| \tag{12}$$

for most cases investigated. Hence, the infinite series appearing in (10) coverge rather rapidly.

The Odd Dominant Wave, The $_o\mathrm{HE}_{11}$ Wave

The longitudinal components of the $_o\mathrm{HE}_{11}$ mode[8,9] inside an elliptical dielectric rod are

$$H_{z_1} = \sum_{m=1}^{\infty}{}' C_m Se_m(\xi) se_m(\eta) e^{i\beta z}$$

$$E_{z_1} = \sum_{m=1}^{\infty}{}' D_m Ce_m(\xi) ce_m(\eta) e^{i\beta z}, \tag{13}$$

and outside the rod

$$H_{z_0} = \sum_{r=1}^{\infty}{}' G_r Gek_r(\xi) se_r^*(\eta)$$

$$E_{z_0} = \sum_{r=1}^{\infty}{}' F_r Fek_r(\xi) ce_r^*(\eta) \tag{14}$$

where C_m, D_m, G_r and F_r are arbitrary constants. The characteristic equation for this mode can be obtained in a similar way as for the $_e\mathrm{HE}_{11}$ mode and will not be given here. Numerical solutions of the characteristic equation for the $_o\mathrm{HE}_{11}$ wave are presented in Figs. 3 and 4.

By a similar process as for the $_e\mathrm{HE}_{11}$ wave, it can be proven that the attenuation constant of the $_o\mathrm{HE}_{11}$ wave is as follows:

$$\alpha' = 4.343\, \sigma_d \sqrt{\frac{\mu}{\epsilon_0}}\, R' \text{ (db/meter)} \tag{15}$$

with

$$R' = \frac{f_1'}{f_2' + f_3'} \tag{16}$$

where f_1', f_2', and f_3' can be obtained by replacing, respectively, the odd modified Mathieu functions and the even modified Mathieu functions, A_m, B_m, L_r, P_r, $\alpha_{m,r}$, $\beta_{m,r}$, $\chi_{r,n}$, $\nu_{r,n}$, f_1, f_2, and f_3, in (9) by the even modified Mathieu functions and the odd modified Mathieu function, C_m, D_m, G_r, F_r, $\beta_{m,r}$, $\alpha_{m,r}$, $\nu_{r,n}$, $\chi_{r,n}$, f_1', f_2', and f_3'.

Numerical Results

The computations were carried out on a high speed electronic computer, the IBM 7090.

It is known that the periodic Mathieu functions can be expanded in terms of infinite series of trigonometric functions and that the corresponding radial Mathieu functions can be expanded in terms of infinite series of Bessel functions.[9] The coefficients of expansion for a certain finite range of x^2 or y^2 have been tabulated by the National Bureau of Standards.[10] These coefficients were stored in the computer's memory. A three-point Lagrangian interpolation sub-routine was used to interpolate the required coefficients from the stored values. The radial integrals in (8) and (16) are evaluated numerically using Simpson's rule. Numerical values of the infinite determinants are obtained by the the method of successive approximations.[11] Results of this very lengthy computation[12] are shown in Figs. 5–8.

[10] National Bureau of Standards, "Tables Relating to Mathieu Functions," Columbia University Press, New York, N. Y.; 1951.

[11] L. Kantorovich and V. Krylov, "Approximate Methods of Higher Analysis," Interscience Publishers, Inc., New York, N. Y.; 1958.

[12] It takes almost 25 minutes of continuous computation by the IBM 7090 to obtain each curve.

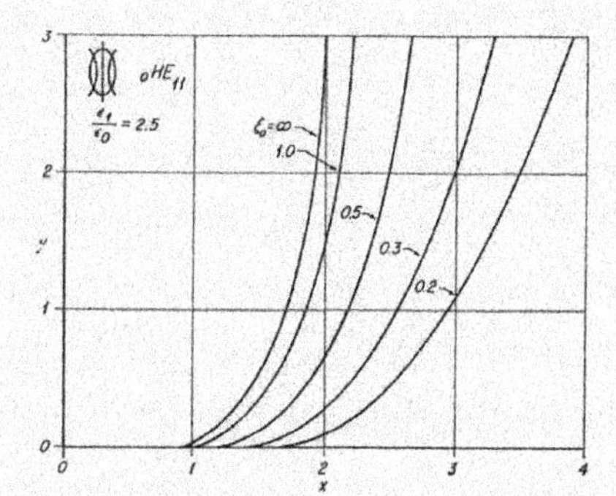

Fig. 3—Roots of the characteristic equation for the $_{o}HE_{11}$ wave.

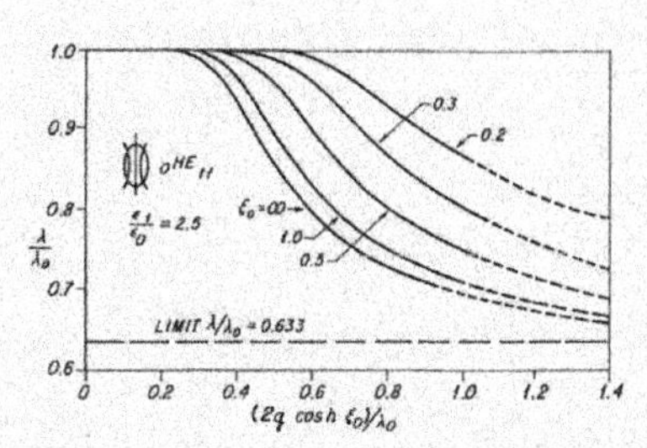

Fig. 4—Normalized guide wavelength λ/λ_0 as function of normalized major axis for the $_{o}HE_{11}$ wave.

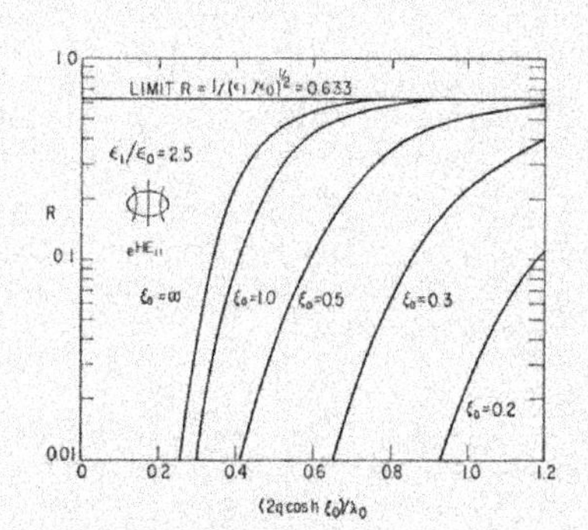

Fig. 5—Attenuation factor R for the $_{e}HE_{11}$ wave as a function of normalized major axis.

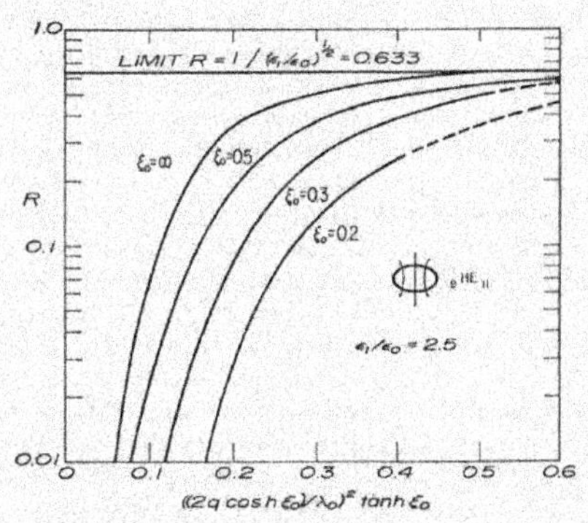

Fig. 6—Attenuation factor R for the $_{e}HE_{11}$ wave as a function of normalized cross-sectional area.

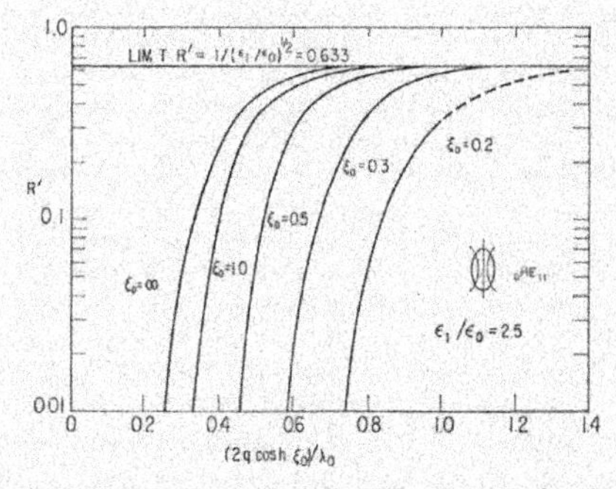

Fig. 7—Attenuation factor R' for the $_{o}HE_{11}$ wave as a function of normalized major axis.

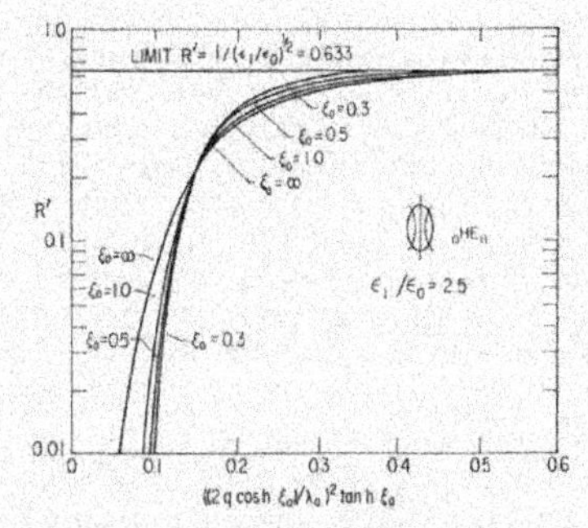

Fig. 8—Attenuation factor R' for the $_{e}HE_{11}$ wave as a function of normalized cross-sectional area.

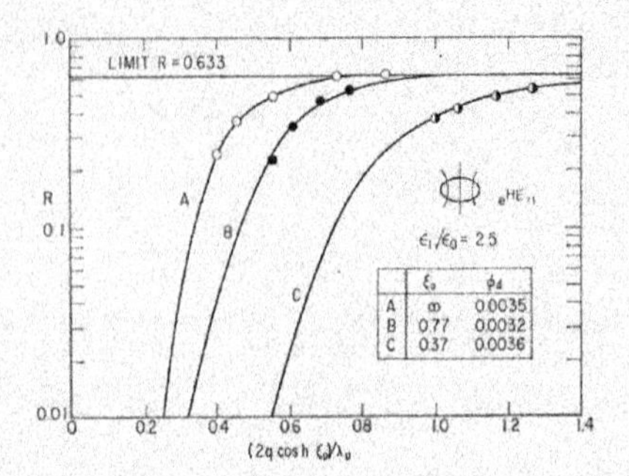

	ξ_0	ϕ_d
A	∞	0.0035
B	0.77	0.0032
C	0.37	0.0036

Fig. 9—R as a function of $(2q \cosh \xi_0)/\lambda_0$ for the $_e\text{HE}_{11}$ wave. Circles are experimental points.

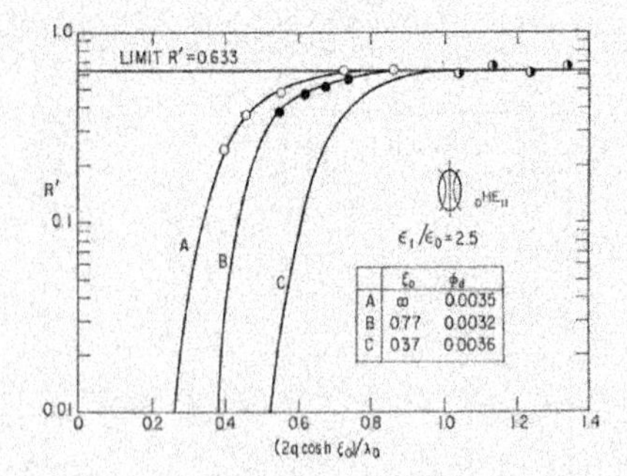

	ξ_0	ϕ_d
A	∞	0.0035
B	0.77	0.0032
C	0.37	0.0036

Fig. 10—R' as a function of $(2q \cosh \xi_0)/\lambda_0$ for the $_o\text{HE}_{11}$ wave. Circles are experimental points.

The $_e\text{HE}_{11}$ *Wave*

In Fig. 5 the value R, which is directly proportional to the attenuation constant α, for the $_e\text{HE}_{11}$ wave is plotted against the normalized major axis (NMA), $(2q \cosh \xi_0)/\lambda_0$, for various values of ξ_0 with $\epsilon_1/\epsilon_0 = 2.5$. For sufficiently large values of NMA, R tends toward the value $1/\sqrt{\epsilon_1/\epsilon_0}$, which is the attenuation factor for a plane wave in an infinite medium, for all values of ξ_0; for small enough values of NMA, R can be made as small as desired. This behavior is attributed to the fact that when NMA is sufficiently large almost all of the energy of the wave is transmitted inside the rod, and for small values of NMA most of the energy is transported outside the rod. It is also clear that R tends to the limit $1/\sqrt{\epsilon_1/\epsilon_0}$ much more slowly as ξ_0 gets smaller. Fig. 6 shows curves of the attenuation factor R as functions of the normalized cross-sectional area (NCSA) $((2q \cosh \xi_0)/\lambda_0)^2 \tanh \xi_0$, for various ξ_0 with $\epsilon_1/\epsilon_0 = 2.5$. For a fixed value of NCSA, R is smaller for flatter elliptical cross section. It is for this reason that an elliptical dielectric rod, supporting the $_e\text{HE}_{11}$ wave, is superior as a guiding device than a circular dielectric rod, supporting the HE_{11} wave and having the same cross section area as the elliptical rod. It is further noted that the variation of slopes with respect to NCSA is smaller for flatter rods in the low loss region. This means that a small imperfection in the dimensions of a flatter rod would induce a smaller change in the attenuation factor. Hence the manufacturing tolerances are not as stringent for flatter rods.

It should be noted that the attenuation characteristics of the $_e\text{HE}_{11}$ wave pass smoothly to those of the HE_{11} wave on a circular rod as $\xi_0 \to \infty$.

The $_o\text{HE}_{11}$ *Wave*

Figs. 7 and 8 show, respectively, the variation of R', which is directly proportional to the attenuation constant α', with respect to NMA and to NCSA for various ξ_0 with $\epsilon_1/\epsilon_0 = 2.5$. The general characteristics of the attenuation constant as a function of NMA or as a function of NCSA are very similar to those indicated by the even dominant wave. However, unlike the case for the $_e\text{HE}_{11}$ wave, it seems that the flatter elliptical cross-sectional rod supporting the $_o\text{HE}_{11}$ wave does not present a more favorable attenuation factor. The fact that the curves for various values of ξ_0 in Fig. 8 are quite close to each other suggests that the field intensity is quite uniform for this odd dominant wave in the dielectric rod. It is further noted that in the low loss region a small variation in NMA introduces a rather significant fluctuation in R'.

It is quite obvious that the $_e\text{HE}_{11}$ mode is more suitable than the $_o\text{HE}_{11}$ mode as a transmission mode on an elliptical dielectric rod.

Experimental Results

Attenuation measurements were made by probe, using the standard standing wave technique.[13] For the sake of convenience and simplicity, measurements were performed in the X-Band frequency range. The measured attenuation constants for the two dominant waves together with the theoretical curves are shown in Figs. 9 and 10. It can be seen that the analytic and experimental results are in very good agreement. Incidentally, it may be interesting to mention that since dielectric rods of elliptical cross section were not commercially available, they had to be machined from available rectangular lucite strips which were at least $5\frac{1}{2}$ ft long.

Conclusions

The results obtained in this research lead to the following conclusions:

1) The even dominant $_e\text{HE}_{11}$ wave propagating along a flat elliptical dielectric strip has considerably lower loss than the HE_{11} wave propagating along a circular dielectric cylinder having the same cross-sectional area as the strip.

[13] E. L. Ginzton, "Microwave Measurements," McGraw-Hill Book Company, Inc., New York, N. Y.; 1957.

2) The use of an elliptical cross-sectional rod minimizes deplorization effects, because the guide wavelengths differ for the even and odd modes. It is known that internal strain, nonuniform dimensions, and bends of the circular rod cause the HE_{11} wave to change polarization.
3) Since the attenuation constant of an $_eHE_{11}$ wave is a slower varying function of the dimensions for a strip than for a circular rod, wider dimensional tolerances are permitted. Hence, a strip is easier to fabricate.
4) A flat elliptical dielectric strip has a larger surface area than a circular rod of identical cross-sectional area, thus it would be easier to handle at very high frequencies, such as in the mm wavelength range.

The greatest attraction of a dielectric strip as a millimeter or infrared wave transmission line is in its simplicity of construction, its low cost and ease of manufacture, and its flexibility.

In conclusion, it is remarked that the usefulness of the dielectric strip as a practical electromagnetic wave guiding device depends greatly upon the field extent of the guided waves outside the dielectric strip. This problem will be discussed elsewhere.

APPENDIX I

Formulas for $\alpha_{r,n}$, $\beta_{r,n}$, $\chi_{m,n}$, *and* $\nu_{m,n}$

$$\alpha_{r,n} = \frac{\int_0^{2\pi} ce_r^*(\eta)ce_n(\eta)d\eta}{\int_0^{2\pi} ce_n^2(\eta)d\eta},$$

$$\beta_{r,n} = \frac{\int_0^{2\pi} se_r^*(\eta)se_n(\eta)d\eta}{\int_0^{2\pi} se_n^2(\eta)d\eta},$$

$$\chi_{m,n} = \frac{\int_0^{2\pi} ce_m'(\eta)se_n(\eta)d\eta}{\int_0^{2\pi} se_n^2(\eta)d\eta},$$

$$\nu_{m,n} = \frac{\int_0^{2\pi} se_m'(\eta)ce_n(\eta)d\eta}{\int_0^{2\pi} ce_n^2(\eta)d\eta},$$

where the prime signifies the derivative of the function with respect to η. The above integral can easily be integrated using the expansions of Mathieu functions in terms of the trigonometric functions.[9]

Angle Integrals Involving Mathieu Functions

$$I_n^e = \int_0^{2\pi} ce_n^2(\eta)d\eta$$

$$I_n^0 = \int_0^{2\pi} se_n^2(\eta)d\eta$$

$$I_{nm}^{e'} = \int_0^{2\pi} ce_n'(\eta)ce_m'(\eta)d\eta$$

$$I_{nm}^{0'} = \int_0^{2\pi} se_n'(\eta)se_m'(\eta)d\eta$$

$$J_{nm}^e = \int_0^{2\pi} ce_n(\eta)se_m'(\eta)d\eta$$

$$J_{nm}^0 = \int_0^{2\pi} ce_n'(\eta)se_m(\eta)d\eta$$

$$I_n^e = \int_0^{2\pi} ce_n^{*2}(\eta)d\eta$$

$$I_n^0 = \int_0^{2\pi} se_n^{*2}(\eta)d\eta$$

$$I_{nm}^{e'} = \int_0^{2\pi} ce_n^{*\prime}(\eta)ce_m^{*\prime}(\eta)d\eta$$

$$I_{nm}^{0'} = \int_0^{2\pi} se_n^{*\prime}(\eta)se_m^{*\prime}(\eta)d\eta$$

$$J_{nm}^e = \int_0^{2\pi} ce_n^*(\eta)se_m^{*\prime}(\eta)d\eta$$

$$J_{nm}^0 = \int_0^{2\pi} ce_n^{*\prime}(\eta)se_m^*(\eta)d\eta$$

where the prime signifies the derivative of the function with respect to η. Again the above integrals can easily be integrated analytically.[8]

Radial Integrals Involving Modified Mathieu Functions

All radial integrals are integrated numerically by Simpson's rule. Let us introduce the following normalized dimensionless variables:

$$z_1 = \frac{xe^{\xi}}{2\cosh\xi_0} \quad \text{and} \quad z_2 = \frac{ye^{\xi}}{2\cosh\xi_0}.$$

The limits of integration instead of being from 0 to ξ_0 and ξ_0 to ∞ will be from

$$\frac{x}{2\cosh\xi_0} \quad \text{to} \quad \frac{xe^{\xi_0}}{2\cosh\xi_0} \quad \text{and} \quad \frac{ye^{\xi_0}}{2\cosh\xi_0}$$

to ∞, respectively. Assuming

$$a_1 = \frac{x}{2\cosh\xi_0}, \quad a_2 = \frac{xe^{\xi_0}}{2\cosh\xi_0}, \quad b_1 = \frac{ye^{\xi_0}}{2\cosh\xi_0} \text{ and } b_2 = \infty,$$

we have the following radial integrals:

$$R_{nm}^{e} = \int_{a_1}^{a_2} Ce_n(z_1)Ce_m(z_1)\,\frac{dz_1}{z_1}$$

$$R_{nm}^{0} = \int_{a_1}^{a_2} Se_n(z_1)Se_m(z_1)\,\frac{dz_1}{z_1}$$

$$R_{nm}^{e'} = \int_{a_1}^{a_2} Ce_n'(z_1)Ce_m'(z_1)z_1dz_1$$

$$R_{nm}^{0'} = \int_{a_1}^{a_2} Se_n'(z_1)Se_m'(z_1)z_1dz_1$$

$$T_{nm}^{e} = \int_{a_1}^{a_2} Ce_n(z_1)Se_m'(z_1)dz_1$$

$$T_{nm}^{0} = \int_{a_1}^{a_2} Ce_n'(z_1)Se_m(z_1)dz_1$$

$$R_{nm}^{e} = \int_{b_1}^{b_2} Fek_n(z_2)Fek_m(z_2)\,\frac{dz_2}{z_2}$$

$$R_{nm}^{0} = \int_{b_1}^{b_2} Gek_n(z_2)Gek_m(z_2)\,\frac{dz_2}{z_2}$$

$$R_{nm}^{e'} = \int_{b_1}^{b_2} Fek_n'(z_2)Fek_m'(z_2)z_2dz_2$$

$$R_{nm}^{0'} = \int_{b_1}^{b_2} Gek_n'(z_2)Gek_m'(z_2)z_2dz_2$$

$$T_{nm}^{e} = \int_{b_1}^{b_2} Fek_n(z_2)Gek_m'(z_2)dz_2$$

$$T_{nm}^{0} = \int_{b_1}^{b_2} Fek_n'(z_2)Gek_m(z_2)dz_2$$

where the prime indicates the derivative of the function with respect to its argument.

Integrals Involving Mathieu and Modified Mathieu Functions

The following double integrals can easily be separated into two single integrals,[8] one involving only with the radial component ξ and the other involving only with the angular component η. The double integrals are

$$Q_{nm}^{0} = \int_0^{\xi_0}\int_0^{2\pi} Se_n(\xi)Se_m(\xi)se_n(\eta)se_m(\eta) \cdot [\sinh^2\xi + \sin^2\eta]\,\frac{d\eta d\xi}{\cosh^2\xi_0},$$

$$Q_{nm}^{e} = \int_0^{\xi_0}\int_0^{2\pi} Ce_n(\xi)Ce_m(\xi)ce_n(\eta)ce_m(\eta) \cdot [\sinh^2\xi + \sin^2\eta]\,\frac{d\eta d\xi}{\cosh^2\xi_0},$$

$$\mathcal{Q}_{nm}^{0} = \int_{\xi_0}^{\infty}\int_0^{2\pi} Gek_n(\xi)Gek_m(\xi)se_n^*(\eta)se_m^*(\eta) \cdot [\sinh^2\xi + \sin^2\eta]\,\frac{d\eta d\xi}{\cosh^2\xi_0},$$

$$\mathcal{Q}_{nm}^{e} = \int_{\xi_0}^{\infty}\int_0^{2\pi} Fek_n(\xi)Fek_m(\xi)ce_n^*(\eta)ce_m^*(\eta) \cdot [\sinh^2\xi + \sin^2\eta]\,\frac{d\eta d\xi}{\cosh^2\xi_0}.$$

ACKNOWLEDGMENT

The author wishes to express his sincere thanks to Prof. C. H. Papas for stimulating criticisms and encouragement throughout all phases of this investigation. The Western Data Processing Center at the University of California, Los Angeles, is also gratefully thanked for the author's use of its computing facilities.

Reprinted from IEEE TRANSACTIONS
ON *ANTENNAS AND PROPAGATION*
Volume AP-11, Number 2, March, 1963

PRINTED IN THE U.S.A.

Optical and Quantum Electronics 8 *(1976) 43–47*

Modes in weakly guiding elliptical optical fibres*

C. YEH
Electrical Sciences and Engineering Department, University of California, Los Angeles, California 90024 USA

Received 14 July 1975

Approximate and much simplified dispersion relations are obtained for the problem of optical wave propagation within weakly guiding elliptical fibres. The refractive index difference between the core and its cladding of weakly guiding optical fibres that are contenders for use as practical optical communication lines is very small; i.e., $(n_1/n_0 - 1) \ll 1$ where n_1 is the core index and n_0 is the cladding index. These greatly simplified dispersion relations are then used to calculate the propagation constants for several higher order modes on an elliptical optical fibre.

1. Introduction

The announcement of the attainment of glass fibres with attenuation below 20 dB km^{-1} initiated a major movement in the tele-communication communities throughout the world to study in earnest the implementation of optical fibres in high data rate communication systems. The basic configuration of an optical fibre transmission line consists of a circular cylindrical core of radius a and index of refraction n_1 coated with a cladding material of refractive index $n_0 (n_1 > n_0)$. The problem of the propagation characteristics of different modes in such a circular guiding structure has been studied extensively [2–5]. However, if the core of the fibre is somewhat deformed, say into an elliptical cross-sectional shape, a single mode in a circular fibre may split into two modes with different polarizations and different propagation velocities in the deformed elliptical fibres. These elliptical deformations will therefore introduce delay distortion to the transmitted signal in a single mode fibre and additional delay distortion in a multi-mode fibre. Although exact dispersion relations for various modes in terms of infinite determinants whose elements are given as functions of radial and angular Mathieu functions exist for the elliptical fibre, numerical computations have only been attempted for the dominant ${}_e\mathrm{HE}_{11}$ and ${}_o\mathrm{HE}_{11}$ modes [6]. Even for these lowest order modes, the task was very onerous. Approximate perturbation analysis have also been carried out for these dominant modes when the eccentricity of the deformed fibre is near unity [7, 8]. The need to have simplified formulas for the dispersion relations that are valid for all eccentricities and from which the propagation characteristics of higher order modes may be easily computed is quite apparent. Fortunately, the refractive index difference between the core and its cladding of optical fibres that are contenders for use as practical optical communication lines is very small; i.e., $(n_1/n_0 - 1) \ll 1$. Approximate analysis are carried out using this condition. It will be shown that very significant simplifications are achieved. The purpose of this paper is to show these simplified dispersion relations for higher order modes and to provide some computed results for these higher order modes. It should be noted that elliptical optical dielectric waveguides are good approximations to many guide configurations encountered in the study of integrated optical circuits and so have additional applications in their own right.

2. Exact dispersion relations for modes in elliptical fibres

Summarized in the following are the exact dispersion relations for guided modes in an elliptical fibre whose core material has a dielectric constant of ϵ_1 and whose cladding material has a dielectric constant

*Supported partly by NELC, San Diego; this paper was presented at 1974 URSI Electromagnetic Waves Conference, London.

of ϵ_0. To satisfy the boundary condition completely, both longitudinal electric and magnetic fields must be present; in other words only hybrid type (HE) modes may exist in elliptical fibres. Furthermore, due to the asymmetry of the elliptical fibre, two types of modes may exist: The even type (${}_e\mathrm{HE}_{mn}$) and the odd type (${}_o\mathrm{HE}_{mn}$) modes. The longitudinal fields are [6]:

Even modes (${}_e\mathrm{HE}_{mn}$)

$$\begin{aligned} {}_eH_z &= \sum_{m=0}^{\infty} {}_eA_m \mathrm{Ce}_m(\xi, \gamma_1^2)\mathrm{ce}_m(\eta, \gamma_1^2) && (\xi \leqslant \xi_0) \\ &= \sum_{m=0}^{\infty} {}_eL_m \mathrm{Fek}_m(\xi, |\gamma_0^2|)\mathrm{ce}_m(\eta, -\gamma_0^2) && (\xi \geqslant \xi_0) \\ {}_eE_z &= \sum_{m=1}^{\infty} {}_eB_m \mathrm{Se}_m(\xi, \gamma_1^2)\mathrm{se}_m(\eta, \gamma_1^2) && (\xi \leqslant \xi_0) \\ &= \sum_{m=1}^{\infty} {}_eP_m \mathrm{Gek}_m(\xi, |\gamma_0^2|)\mathrm{se}_m(\eta, -\gamma_0^2) && (\xi \geqslant \xi_0) \end{aligned} \tag{1}$$

Odd modes (${}_o\mathrm{HE}_{mn}$)

$$\begin{aligned} {}_oH_z &= \sum_{m=1}^{\infty} {}_oA_m \mathrm{Se}_m(\xi, \gamma_1^2)\mathrm{se}_m(\eta, \gamma_1^2) && (\xi \leqslant \xi_0) \\ &= \sum_{m=1}^{\infty} {}_oL_m \mathrm{Gek}_m(\xi, |\gamma_0^2|)\mathrm{se}_m(\eta, -\gamma_0^2) && (\xi \geqslant \xi_0) \\ {}_oE_z &= \sum_{m=0}^{\infty} {}_oB_m \mathrm{Ce}_m(\xi, \gamma_1^2)\mathrm{ce}_m(\eta, \gamma_1^2) && (\xi \leqslant \xi_0) \\ &= \sum_{m=0}^{\infty} {}_oP_m \mathrm{Fek}_m(\xi, |\gamma_0^2|)\mathrm{ce}_m(\eta, -\gamma_0^2) && (\xi \geqslant \xi_0) \end{aligned} \tag{2}$$

where ${}_{e,o}A_m$, ${}_{e,o}B_m$, ${}_{e,o}L_m$ and ${}_{e,o}P_m$ are arbitrary constants and

$$\gamma_1^2 = (\omega^2\mu\epsilon_1 - \beta^2)q^2/4, \tag{3}$$

$$\gamma_0^2 = (\beta^2 - \omega^2\mu\epsilon_0)q^2/4. \tag{4}$$

All field components are assumed to be multiplied by the factor $\exp(-i\omega t + i\beta z)$ which is suppressed throughout. β is the propagation constant, ω is the frequency of the guided wave, q is the semi-focal length of the elliptical fibre and ξ_0 is the boundary of the fibre in elliptical coordinates (ξ, η, z).

Matching the tangential electric and magnetic fields at the boundary $\xi = \xi_0$, making use of the orthogonality relations of Mathieu functions, and setting the determinant of the resultant algebraic equations to zero, one finally obtains the exact dispersion relations for the guided modes:

$$\begin{vmatrix} {}_{e,o}g_{1,1} & {}_{e,o}h_{1,1} & {}_{e,o}g_{3,1} & {}_{e,o}h_{3,1} & \cdot & \cdot \\ {}_{e,o}t_{1,1} & {}_{e,o}s_{1,1} & {}_{e,o}t_{3,1} & {}_{e,o}e_{3,1} & \cdot & \cdot \\ {}_{e,o}g_{1,3} & {}_{e,o}h_{1,3} & {}_{e,o}g_{3,3} & {}_{e,o}h_{3,3} & \cdot & \cdot \\ {}_{e,o}t_{1,3} & {}_{e,o}s_{1,3} & {}_{e,o}t_{3,3} & {}_{e,o}s_{3,3} & \cdot & \cdot \\ \cdot & \cdot & \cdot & \cdot & & \\ \cdot & \cdot & \cdot & \cdot & & \end{vmatrix} = 0 \tag{5}$$

and

$$\begin{vmatrix} {}_{e,o}g_{0,0} & {}_{e,o}h_{0,0} & {}_{e,o}g_{2,0} & {}_{e,o}h_{2,0} & \cdot & \cdot \\ {}_{e,o}t_{0,0} & {}_{e,o}s_{0,0} & {}_{e,o}t_{2,0} & {}_{e,o}s_{2,0} & \cdot & \cdot \\ {}_{e,o}g_{0,2} & {}_{e,o}h_{0,2} & {}_{e,o}g_{2,2} & {}_{e,o}h_{2,2} & \cdot & \cdot \\ {}_{e,o}t_{0,2} & {}_{e,o}s_{0,2} & {}_{e,o}t_{2,2} & {}_{e,o}h_{2,2} & \cdot & \cdot \\ \cdot & \cdot & \cdot & \cdot & \cdot & \cdot \\ \cdot & \cdot & \cdot & \cdot & \cdot & \cdot \end{vmatrix} = 0 \tag{6}$$

where

$$
\begin{aligned}
{}_e g_{m,n} &= \left(1+\frac{\gamma_1^2}{\gamma_0^2}\right)\mathrm{Fek}_m(\xi_0,|\gamma_0^2|)\sum_{r=1}^{\infty}{}' \chi_{r,n}\alpha_{m,r}\\
{}_e s_{m,n} &= -\left(1+\frac{\gamma_1^2}{\gamma_0^2}\right)\mathrm{Gek}_m(\xi_0,|\gamma_0^2|)\sum_{r=1}^{\infty}{}' \nu_{r,n}\beta_{m,r}\\
{}_e h_{m,n} &= \left[\frac{\omega\epsilon_1}{\beta}\frac{\mathrm{Se}_n'(\xi_0,\gamma_1^2)}{\mathrm{Se}_n(\xi_0,\gamma_1^2)}\mathrm{Gek}_m(\xi_0,|\gamma_0^2|)+\frac{\gamma_1^2}{\gamma_0^2}\frac{\omega\epsilon_0}{\beta}\mathrm{Gek}_m'(\xi_0,|\gamma_0^2|)\right]\beta_{m,n}\\
{}_e t_{m,n} &= \left[\frac{\omega\mu}{\beta}\frac{\mathrm{Ce}_n'(\xi_0,\gamma_1^2)}{\mathrm{Ce}_n(\xi_0,\gamma_1^2)}\mathrm{Fek}_m(\xi_0,|\gamma_0^2|)+\frac{\gamma_1^2}{\gamma_0^2}\frac{\omega\mu}{\beta}\mathrm{Fek}_m'(\xi_0,|\gamma_0^2|)\right]\alpha_{m,n}
\end{aligned}
\tag{7}
$$

with

$$
\begin{aligned}
\alpha_{r,n} &= \int_0^{2\pi}\mathrm{ce}_r(\eta,-\gamma_0^2)\mathrm{ce}_n(\eta,\gamma_1^2)\mathrm{d}\eta \Big/ \int_0^{2\pi}\mathrm{ce}_n^2(\eta,\gamma_1^2)\mathrm{d}\eta\\
\beta_{r,n} &= \int_0^{2\pi}\mathrm{se}_r(\eta,-\gamma_0^2)\mathrm{se}_n(\eta,\gamma_1^2)\mathrm{d}\eta \Big/ \int_0^{2\pi}\mathrm{se}_n^2(\eta,\gamma_1^2)\mathrm{d}\eta\\
\chi_{r,n} &= \int_0^{2\pi}\mathrm{ce}_r'(\eta,\gamma_1^2)\mathrm{se}_n(\eta,\gamma_1^2)\mathrm{d}\eta \Big/ \int_0^{2\pi}\mathrm{se}_n^2(\eta,\gamma_1^2)\mathrm{d}\eta\\
\nu_{r,n} &= \int_0^{2\pi}\mathrm{se}_r'(\eta,\gamma_1^2)\mathrm{ce}_n(\eta,\gamma_1^2)\mathrm{d}\eta \Big/ \int_0^{2\pi}\mathrm{ce}_n^2(\eta,\gamma_1^2)\mathrm{d}\eta
\end{aligned}
\tag{8}
$$

where the prime signifies the derivatives of the function with respect to ξ_0 or η as appropriate. Expressions for ${}_o g_{mn}$, ${}_o s_{mn}$, ${}_o h_{mn}$ and ${}_o t_{mn}$ are obtained by appropriately replacing Fek by Gek, Gek by Fek, Se by Ce, Ce by Se, $\chi_{r,m}$ by $\nu_{r,m}$, $\nu_{r,m}$ by $\chi_{r,m}$, $\alpha_{m,r}$ by $\beta_{m,r}$ and $\beta_{m,r}$ by $\alpha_{m,r}$ in the expressions for ${}_e g_{mn}$, ${}_e s_{mn}$, ${}_e h_{mn}$ and ${}_e t_{mn}$.

The propagation characteristics of modes in elliptical fibres can be computed from the exact dispersion relations given above. However, as pointed out in the introduction, the computation is extremely involved.

3. Approximate dispersion relations

Let us now define the following parameters:

$$U^2 = q^2(\cosh^2\xi_0)(k_1^2-\beta^2) = 4\cosh^2\xi_0\gamma_1^2 \tag{9}$$

$$W^2 = -q^2(\cosh^2\xi_0)(k_0^2-\beta^2) = 4\cosh^2\xi_0\gamma_0^2 \tag{10}$$

$$
\begin{aligned}
V^2 &= U^2+W^2 = q^2(\cosh^2\xi_0)\omega^2\mu\epsilon_1\left(1-\frac{\epsilon_0}{\epsilon_1}\right)\\
&= q^2(\cosh^2\xi_0)\omega^2\mu\epsilon_1\delta
\end{aligned}
\tag{11}
$$

$$\theta_p = \sqrt{\delta}\,\frac{U}{V} = \left[1-\frac{\beta^2}{k_1^2}\right]^{1/2}. \tag{12}$$

These parameters are similar to those defined by Snyder [9] in his treatment of the circular fibre case. In fact these parameters are identical to Snyder's when the elliptical cylinder degenerates to a circular one.

In a similar manner as that carried out by Snyder for the circular fibre case, we shall expand all functions in terms of the smallness parameter θ_p if $(n_1/n_0-1)\ll 1$. In other words when $\theta_p \ll 1$, considerable simplification results. Hence, we have

$$-\gamma_0^2 = \gamma_1^2 - \tfrac{1}{4}\omega^2\mu\epsilon_1\delta = \gamma_1^2 + 0(\theta_p^2) \tag{13}$$

$$\mathrm{ce}_r(\eta,-\gamma_0^2) = \mathrm{ce}_r(\eta,\gamma_1^2) + 0(\theta_p^2) \tag{14}$$

$$\mathrm{se}_r(\eta, -\gamma_0^2) = \mathrm{se}_r(\eta, \gamma_1^2) + 0(\theta_p^2) \tag{15}$$

$$\alpha_{r,m} = \left[\int_0^{2\pi} \mathrm{ce}_r(\eta, \gamma_1^2)\mathrm{ce}_n(\eta, \gamma_1^2)\mathrm{d}\eta \Big/ \int_0^{2\pi} \mathrm{ce}_n^2(\eta, \gamma_1^2)\mathrm{d}\eta\right] + 0(\theta_p^2) = C_{r,m}^{(1)}\Delta_{r,m} + 0(\theta_p^2) \tag{16}$$

$$\beta_{r,m} = \left[\int_0^{2\pi} \mathrm{se}_r(\eta, \gamma_1^2)\mathrm{se}_n(\eta, \gamma_1^2)\mathrm{d}\eta \Big/ \int_0^{2\pi} \mathrm{se}_n^2(\eta, \gamma_1^2)\mathrm{d}\eta\right] + 0(\theta_p^2) = C_{r,m}^{(2)}\Delta_{r,m} + 0(\theta_p^2) \tag{17}$$

$$1 + \frac{\gamma_1^2}{\gamma_0^2} = 0(\theta_p^2) \tag{18}$$

where $C_{r,n}^{(1),(2)}$ are known constants and $\Delta_{r,m}$ is the Kronecker delta which is zero when $r \neq n$ and is unity when $r = n$.

Substituting Equations 13–18 into Equation 7 gives

$$_{e,o}g_{m,n} = 0(\theta_p^2) \tag{19}$$

$$_{e,o}s_{m,n} = 0(\theta_p^2) \tag{20}$$

$$_{e,o}h_{m,n} = {}_{e,o}D_{m,n}^{(1)}\Delta_{m,n} + 0(\theta_p^2) \tag{21}$$

$$_{e,o}t_{m,n} = {}_{e,o}D_{m,n}^{(2)}\Delta_{m,n} + 0(\theta_p^2) \tag{22}$$

where $_{e,o}D_{m,n}^{(1),(2)}$ can be obtained readily from Equation 7.

Inserting Equations 19–22 into Equation 5 or 6, one may show that all off-diagonal terms are all of the order of θ_p^2. For example, from Equation 5 one has

$$\begin{vmatrix} 0(\theta_p^2) & {}_{e,o}h_{1,1} & 0(\theta_p^2) & 0(\theta_p^2) & \cdot & \cdot \\ {}_{e,o}t_{11} & 0(\theta_p^2) & 0(\theta_p^2) & 0(\theta_p^2) & \cdot & \cdot \\ 0(\theta_p^2) & 0(\theta_p^2) & 0(\theta_p^2) & {}_{e,o}h_{3,3} & \cdot & \cdot \\ 0(\theta_p^2) & 0(\theta_p^2) & {}_{e,o}t_{3,3} & 0(\theta_p^2) & \cdot & \cdot \\ \cdot & \cdot & \cdot & \cdot & \cdot & \cdot \\ \cdot & \cdot & \cdot & \cdot & \cdot & \cdot \end{vmatrix} = 0 \tag{23}$$

Ignoring terms of the order of θ_p^2, the zero order dispersion relations for $_e\mathrm{HE}_{mn}$ and for $_o\mathrm{HE}_{mn}$ modes are, respectively,

$$\frac{1}{\gamma_1^2}\frac{\mathrm{Ce}_m'(\xi_0, \gamma_1^2)}{\mathrm{Ce}_m(\xi_0, \gamma_1^2)} + \frac{1}{\gamma_0^2}\frac{\mathrm{Fek}_m'(\xi_0, |\gamma_0^2|)}{\mathrm{Fek}_m(\xi_0, |\gamma_0^2|)} = 0 \tag{24}$$

and

$$\frac{1}{\gamma_1^2}\frac{\mathrm{Se}_m'(\xi_0, \gamma_1^2)}{\mathrm{Se}_m(\xi_0, \gamma_1^2)} + \frac{1}{\gamma_0^2}\frac{\mathrm{Gek}_m'(\xi_0, |\gamma_0^2|)}{\mathrm{Gek}_m(\xi_0, |\gamma_0^2|)} = 0. \tag{25}$$

These are the simplified dispersion relations that we are seeking. These relations are valid if $(n_1/n_0 - 1) \ll 1$. It can be seen that the above Equations 24 and 25 are much simpler expressions for the dispersion relations than the infinite determinants as represented by Equations 5 and 6.

4. Numerical examples

To verify the accuracy of the approximate dispersion relations, numerical computations were carried out for several specific cases using first the exact dispersion relations and then the approximate ones. For the cases considered in which $n_1/n_0 \lesssim 1$, the approximate dispersion relations yield results that are within 2% of the exact values, and are hence indistinguishable from the exact results when plotted on Figs. 1 and 2. Although only $\xi_0 = 0.5$ cases were shown in these figures, we have also carried out computations for other ξ_0 values corresponding to various ellipticities of the elliptical cross-section. Again the results from the approximate dispersion relations are generally accurate to within 2% of the exact values.

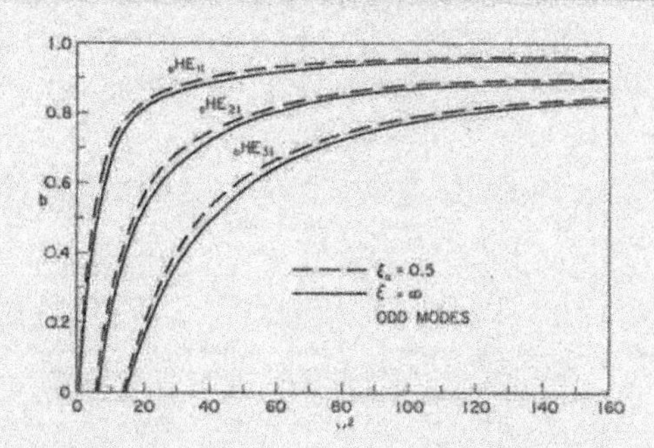

Figure 1 Normalized propagation constant, $b = [(\beta/k_0) - n]/(n_c - n)$, as a function of $V^2 = q^2(\cosh^2 \xi_o)\omega^2\mu\epsilon_1(1 - \epsilon_0/\epsilon_1)$ for the odd modes. n = cladding index and n_c = core index.

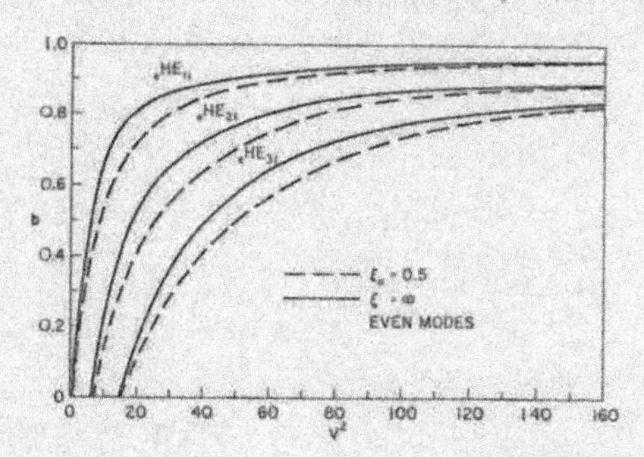

Figure 2 Normalized propagation constant, $b = [(\beta/k_0) - n]/(n_c - n)$, as a function of $V^2 = q^2(\cosh^2 \xi_o)\omega^2\mu\epsilon_1(1 - \epsilon_0/\epsilon_1)$ for the even modes. n = cladding index and n_c = core index.

References

1. F. P. KAPRON, D. B. KECK and R. D. MAURER, *Appl. Phys. Lett.* (1970) 423.
2. P. J. B. CLARRICOATS, 'Electromagnetic Aspects of Optical Waveguides,' Research Report, Dept. Elec. Eng., Queen Mary College, University of London (1973).
3. C. YEH, Advances in communication through light fibres, in 'Advances in Communication Systems, Theory and Application,' edited by A. VITERBI and A. V. BALAKRISHNAN, Academic Press, (1975).
4. D. MARCUSE, 'Light Transmission Optics,' Van Nostrand Reinhold Company, New York (1972).
5. N. S. KAPANY and J. J. BURKE, 'Optical Waveguides,' Academic Press, New York (1972).
6. C. YEH, *J. Appl. Phys.* 33 (1962) 3235.
7. W. O. SCHLOSSER, *Bell Sys. Tech. J.* 51 (1972) 487.
8. R. B. DYOTT and J. R. STERN, *Elec. Lett.* 7 (1971) 82.
9. A. W. SNYDER, *IEEE Trans. on Microwave Theory and Technique*, MTT-12, (1969) 1130; D. GLOGE, *Appl. Optics* 10 (1971) 2252.

Reprinted from JOURNAL OF APPLIED PHYSICS, Vol. 41, No. 8, 3210–3220, July 1970

Ferrite-Filled Elliptical Waveguides. I. Propagation Characteristics

W. E. SALMOND AND C. YEH

Electrical Sciences and Engineering Department, University of California, Los Angeles, California 90024

(Received 20 February 1970)

This paper deals with electromagnetic wave propagation in a longitudinally magnetized, ferrite-filled elliptical waveguide. Using the wave equations and boundary conditions for a ferrite-filled waveguide of arbitrary cross-sectional shape, it is shown that for an elliptical cross section the characteristic equations for the propagation constant take the form of even and odd infinite determinants. Solutions for the characteristic equations are obtained for various values of applied field strength, saturation magnetization, and waveguide ellipticity. The existence of three types of cutoff conditions is demonstrated.

I. INTRODUCTION

The success in the application of ferrite-filled wave guide devices as phase shifters or isolators in the microwave frequency region is well known. The theory of propagation in longitudinally magnetized, ferrite-filled waveguides was first considered by Gamo,[1] Van Trier,[2] Kales,[3] and Suhl and Walker.[4] Each of the above authors derived the wave equation and the electromagnetic field equations for the ferrite-filled circular guide. They deduced that the modes must be hybrid (TE, TM, or TEM modes were not possible) and that they fall into two groups, having either an $e^{+jn\phi}$ or $e^{-jn\phi}$ angular dependence, and with unequal propagation constants. It was found that for $n=+1$ the transverse field vectors of these modes were circularly polarized in opposite senses on the axis of the guide. Furthermore, it has been shown that the above situation differs somewhat from that of the infinite medium. The counterrotating circularly polarized modes do have unequal propagation constants, but for the case of a filled waveguide they have different field configurations as well; i.e., the two modes no longer combine to give a linearly polarized transverse field pattern which simply rotates as the wave advances in the direction of propagation. Various experimental techniques have been devised for making microwave measurements with longitudinally magnetized, circular ferrite-filled waveguides: Sakiotis and Chait,[5] Hogan,[6] and many others have been able to directly measure the Faraday rotation and nonreciprocal attenuation by using a rotatable probe; Lax and Button[7] show that the phase shift and attenuation of each of the circularly polarized components of the TE_{11} mode can be measured precisely by means of a microwave bridge circuit; Van Trier[8] uses a resonant cavity to measure the wavelengths of the two circularly polarized waves, from which he deduces the propagation constants and the Faraday rotation.

All previous rigorous treatment on the problem of wave propagation in a longitudinally magnetized ferrite-filled waveguide were carried out for a waveguide with circular cross section.[9] Due to the symmetry of the circular waveguide the solution for this problem may be expressed in closed form. It is of interest, however, to understand how the propagation characteristics of the guided wave are affected when this symmetry no longer exists. In other words, we wish to examine the propagation characteristics of a deformed ferrite-filled circular waveguide, such as an elliptical waveguide. It will be shown in this paper that the problem of a longitudinally magnetized ferrite-filled elliptical waveguide can be solved rigorously, although the solution is much more involved than that for the circular case. Significant differences between the propagation characteristics of a ferrite-filled elliptical waveguide and a ferrite-filled circular waveguide are found. It is expected that the results from the present analysis may also be applied to the case of a ferrite-filled rectangular waveguide.

II. FORMULATION OF THE PROBLEM

It is assumed that an infinitely long metallic waveguide of elliptical cross section with an impressed magnetostatic field H_0 in the $+z$ axial direction is filled completely with a lossless ferrite medium which may be characterized by a scalar permittivity ϵ and a tensor permeability

$$\boldsymbol{\mu}=\begin{bmatrix} \mu & -j\kappa & 0 \\ -j\kappa & \mu & 0 \\ 0 & 0 & \mu_z \end{bmatrix}, \tag{1a}$$

where

$$\mu=\mu_0+[\bar{\gamma}M_s\omega_0\mu_0/(\omega_0^2-\omega^2)], \tag{1b}$$

$$\kappa=-\bar{\gamma}M_s\omega\mu_0/(\omega_0^2-\omega^2), \tag{1c}$$

$$\mu_z=\mu_0, \tag{1d}$$

$\omega_0=\bar{\gamma}H_0$, the processional frequency. M_s is the saturation magnetization and $\bar{\gamma}$ is the ratio of the magnetic moment to the angular moment of an electron. A harmonic time dependence of $e^{j\omega t}$ for all electromagnetic field components is also assumed. We shall confine our treatment to waves propagating along the positive z axis (the axial direction). In complex representation, these assumptions result in a multiplication of all wavefunctions by $\exp(j\omega t - j\beta z)$. The propagation constant β is to be determined from the boundary conditions.

To analyze this problem, the elliptical cylinder coordinate system (ξ, η, z), as shown in Fig. 1, is introduced. In terms of rectangular coordinates (x', y', z') the elliptical cylinder coordinates are defined by the following relations:

$$x' = h\cosh\xi\cos\eta,$$

$$y' = h\sinh\xi\sin\eta,$$

$$z' = z, \qquad (0 \leq \xi < \infty, \quad 0 \leq \eta < 2\pi), \tag{2}$$

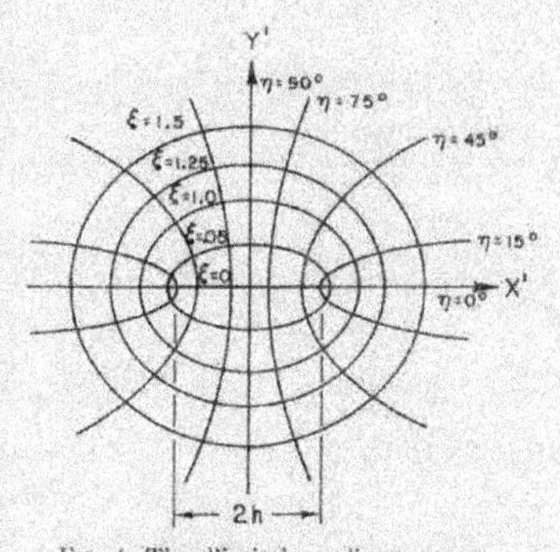

Fig. 1. The elliptical coordinate system.

where h is the semifocal length of the ellipse. One of the confocal elliptic cylinders $\xi = \xi_0$ is assumed to coincide with the boundary of the metallic waveguide, and the z axis coincides with its longitudinal axis. The solution of Maxwell's equations in a homogeneous anisotropic medium have been derived by Kales[9] and others.[10] For the case of a longitudinally magnetized ferrite-filled waveguide of arbitrary cross section, the axial and transverse electric and magnetic fields are given by

$$E_z = s_1 u_1 + s_2 u_2, \tag{3}$$

$$H_z = s_1[(s_1 - a)/b]u_1 + s_2[(s_2 - a)/b]u_2, \tag{4}$$

$$\mathbf{E}_t = -j\beta\nabla_t(u_1 + u_2) + (\mu/\kappa\beta)\mathbf{a}_z \times \nabla_t[(s_1 - a)u_1 + (s_2 - a)u_2], \tag{5}$$

$$\mathbf{H}_t = (1/\omega\kappa)\nabla_t[(\omega^2\mu\epsilon - \beta^2 - s_1)u_1 + (\omega^2\mu\epsilon - \beta^2 - s_2)u_2]. \tag{6}$$

Here $\mathbf{a}_z$ is a unit vector in the z direction and ∇_t is the transverse del operator. The field quantities u_1 and u_2 satisfy the wave equations

$$\nabla_t^2 u_1 + s_1 u_1 = 0,$$

$$\nabla_t^2 u_2 + s_2 u_2 = 0,$$

and

$$a = (\omega^2\mu\epsilon - \beta^2) - \omega^2\epsilon(\kappa^2/\mu),$$

$$b = j\mu_0\omega\kappa\beta/\mu,$$

$$c = (\mu_0/\mu)(\omega^2\mu\epsilon - \beta^2),$$

$$d = -j\omega\epsilon\kappa\beta/\mu,$$

$$s_{1,2} = [(a+c)/2] \pm \tfrac{1}{2}[(a+c)^2 - 4(ac - bd)]^{1/2}. \tag{7}$$

If we now specify the waveguide to be elliptical in cross section, the appropriate expressions for u_1 and u_2 are[11,12]

$$u_1 = \sum_{n=0}^{\infty} A_n Ce_n(\xi, q_1)ce_n(\eta, q_1) + \sum_{n=1}^{\infty} B_n Se_n(\xi, q_1)se_n(\eta, q_1), \tag{8}$$

$$u_2 = \sum_{r=0}^{\infty} C_r Ce_r(\xi, q_2)ce_r(\eta, q_2) + \sum_{r=1}^{\infty} D_r Se_r(\eta, q_2)se_r(\eta, q_2), \tag{9}$$

where $q_{1,2} = s_{1,2}h^2/4$. $Ce_n(\xi, q_{1,2})$ and $Se_n(\xi, q_{1,2})$ are, respectively, the even and the odd radial Mathieu functions; $ce_n(\eta, q_{1,2})$ and $se_n(\eta, q_{1,2})$ are, respectively, the even and the odd angular Mathieu functions. A_n, B_n, C_r, and D_r are the arbitrary constants.

III. THE CHARACTERISTIC EQUATIONS

Satisfying the boundary conditions which require that the tangential components of the electric field must be zero at the metallic wall, $\xi = \xi_0$, we have

$$s_1 \sum_{n=0}^{\infty} A_n Ce_n(\xi_0)ce_n(\eta) + s_2 \sum_{r=0}^{\infty} C_r Ce_r^*(\xi_0)ce_r^*(\eta) = 0, \tag{10}$$

$$s_1 \sum_{n=0}^{\infty} B_n Se_n(\xi_0)se_n(\eta) + s_2 \sum_{r=0}^{\infty} D_r Se_r^*(\xi_0)se_r^*(\eta) = 0, \tag{11}$$

$$-j\frac{\beta^2\kappa}{\mu}\left[\sum_{n=1}^{\infty} B_n Se_n(\xi_0)se_n'(\eta)+\sum_{r=1}^{\infty} D_r Se_r^*(\xi_0)se_r^{*\prime}(\eta)\right]$$
$$+(s_1-a)\sum_{n=0}^{\infty} A_n Ce_n'(\xi_0)ce_n(\eta)+(s_2-a)\sum_{r=0}^{\infty} C_r Ce_r^{*\prime}(\xi_0)ce_r^*(\eta)=0, \quad (12)$$

$$-j\frac{\beta^2\kappa}{\mu}\left[\sum_{n=0}^{\infty} A_n Ce_n(\xi_0)ce_n'(\eta)+\sum_{r=0}^{\infty} C_r Ce_r^*(\xi_0)ce_r^{*\prime}(\eta)\right]$$
$$+(s_1-a)\sum_{n=1}^{\infty} B_n Se_n'(\xi_0)se_n(\eta)+(s_2-a)\sum_{r=1}^{\infty} D_r Se_r^{*\prime}(\xi_0)se_r^*(\eta)=0, \quad (13)$$

where the abbreviations,

$$Ce_n(\xi_0)=Ce_n(\xi_0, q_1), \qquad ce_n(\eta)=ce_n(\eta, q_1),$$
$$Se_n(\xi_0)=Se_n(\xi_0, q_1), \qquad se_n(\eta)=se_n(\eta, q_1),$$
$$Ce_n^*(\xi_0)=Ce_n(\xi_0, q_2), \qquad ce_n^*(\eta)=ce_n(\eta, q_2),$$
$$Se_n^*(\xi_0)=Se_n(\xi_0, q_2), \qquad se_n^*(\eta)=se_n(\eta, q_2),$$

have been used. The prime indicates the derivative of the function with respect to ξ_0 or η, as appropriate.

Multiplying both sides of Eqs. (10) and (12) by $ce_n(\eta)$ and both sides of Eqs. (11) and (13) by $se_n(\eta)$, and integrating with respect to η from 0 to 2π leads to

$$A_n Ce_n(\xi_0)=-\frac{s_2}{s_1}\sum_{r=0}^{\infty}{}' C_r Ce_r^*(\xi_0)\alpha_{r,n}^*, \quad (14)$$

$$B_n Se_n(\xi_0)=-\frac{s_2}{s_1}\sum_{r=1}^{\infty}{}' D_r Se_r^*(\xi_0)\beta_{r,n}^*, \quad (15)$$

$$-j\frac{\beta^2\kappa}{\mu}\left[\sum_{r=1}^{\infty}{}' B_r Se_r(\xi_0)\nu_{r,n}+\sum_{r=1}^{\infty}{}' D_r Se_r^*(\xi_0)\nu_{r,n}^*\right]+(s_1-a)A_n Ce_n'(\xi_0)+(s_2-a)\sum_{r=0}^{\infty}{}' C_r Ce_r^{*\prime}(\xi_0)\alpha_{r,n}^*=0, \quad (16)$$

$$-j\frac{\beta^2\kappa}{\mu}\left[\sum_{r=0}^{\infty}{}' A_r Ce_r(\xi_0)\chi_{r,n}+\sum_{r=0}^{\infty}{}' C_r Ce_r^*(\xi_0)\chi_{r,n}^*\right]$$
$$+(s_1-a)B_n Se_n'(\xi_0)+(s_2-a)\sum_{r=1}^{\infty}{}' D_r Se_r^{*\prime}(\xi_0)\beta_{r,n}^*=0, \qquad (n=0, 2, 4, \cdots, \text{ or } n=1, 3, 5, \cdots), \quad (17)$$

where

$$\alpha_{r,n}^*=\int_0^{2\pi} ce_r^*(\eta)ce_n(\eta)d\eta \Big/ \int_0^{2\pi} ce_n^2(\eta)d\eta,$$

$$\beta_{r,n}^*=\int_0^{2\pi} se_r^*(\eta)se_n(\eta)d\eta \Big/ \int_0^{2\pi} se_n^2(\eta)d\eta,$$

$$\nu_{r,n}^*=\int_0^{2\pi} se_r^{*\prime}(\eta)ce_n(\eta)d\eta \Big/ \int_0^{2\pi} ce_n^2(\eta)d\eta,$$

$$\nu_{r,n}=\int_0^{2\pi} se_r'(\eta)ce_n(\eta)d\eta \Big/ \int_0^{2\pi} ce_n^2(\eta)d\eta,$$

$$\chi_{r,n}^*=\int_0^{2\pi} ce_r^{*\prime}(\eta)se_n(\eta)d\eta \Big/ \int_0^{2\pi} se_n^2(\eta)d\eta,$$

$$\chi_{r,n}=\int_0^{2\pi} ce_r'(\eta)se_n(\eta)d\eta \Big/ \int_0^{2\pi} se_n^2(\eta)d\eta. \quad (18)$$

The prime over the summation sign indicates that r takes on odd or even integer values depending upon whether

n is odd or even. Simplifying Eqs. (14)–(17) and making the substitutions

$$a_{m,n}=-j\frac{\beta^2\kappa}{\mu}Ce_m^*(\xi_0)\left[\chi_{m,n}^*-\frac{s_2}{s_1}\sum_{r=0}^{\infty}{}'\chi_{r,n}\alpha_{m,r}^*\right],$$

$$b_{m,n}=\beta_{m,n}^*\left[(s_2-a)Se_m^{*\prime}(\xi_0)-\frac{(s_1-a)s_2}{s_1}\frac{Se_n'(\xi_0)}{Se_n(\xi_0)}Se_m^*(\xi_0)\right],$$

$$c_{m,n}=\alpha_{m,n}^*\left[(s_2-a)Ce_m^{*\prime}(\xi_0)-\frac{(s_1-a)s_2}{s_1}\frac{Ce_n'(\xi_0)}{Ce_n(\xi_0)}Ce_m^*(\xi_0)\right],$$

$$d_{m,n}=-j\frac{\beta^2\kappa}{\mu}Se_m^*(\xi_0)\left[\nu_{m,n}^*-\frac{s_2}{s_1}\sum_{r=1}^{\infty}{}'\nu_{r,n}\beta_{m,r}^*\right], \tag{19}$$

one obtains

$$\sum_{m=0}^{\infty}{}'C_m a_{m,n}+\sum_{m=1}^{\infty}{}'D_m b_{m,n}=0, \tag{20a}$$

$$\sum_{m=0}^{\infty}{}'C_m c_{m,n}+\sum_{m=1}^{\infty}{}'D_m d_{m,n}=0, \qquad (n=0,2,4,\cdots,\text{ or } n=1,3,5,\cdots). \tag{20b}$$

Equations (20) are two sets of infinite, linear, homogeneous equations for C_m and D_m. For a nontrivial solution the determinant of the coefficients must vanish. The resulting expression is the characteristic equation, the roots of which determine the propagation constant β for the ferrite-filled elliptical waveguide. The infinite determinant for the odd modes ($m=1, 3, 5, \cdots$) is

$$\begin{vmatrix} a_{11} & b_{11} & a_{31} & b_{31} & a_{51} & b_{51} & \cdot & \cdot & \cdot \\ c_{11} & d_{11} & c_{31} & d_{31} & c_{51} & d_{51} & \cdot & \cdot & \cdot \\ a_{13} & b_{13} & a_{33} & b_{33} & a_{53} & b_{53} & \cdot & \cdot & \cdot \\ c_{13} & d_{13} & c_{33} & d_{33} & c_{53} & d_{53} & \cdot & \cdot & \cdot \\ a_{15} & b_{15} & a_{35} & b_{35} & a_{55} & b_{55} & \cdot & \cdot & \cdot \\ c_{15} & d_{15} & c_{35} & d_{35} & c_{55} & d_{55} & \cdot & \cdot & \cdot \\ \vdots & \vdots & \vdots & \vdots & \vdots & \vdots & \vdots & \vdots & \vdots \end{vmatrix}=0, \tag{21}$$

while the infinite determinant for the even modes ($m=0, 2, 4, \cdots$) is

$$\begin{vmatrix} c_{00} & c_{20} & d_{20} & c_{40} & d_{40} & \cdot & \cdot & \cdot \\ a_{02} & a_{22} & b_{22} & a_{42} & b_{42} & \cdot & \cdot & \cdot \\ c_{02} & c_{22} & d_{22} & c_{42} & d_{42} & \cdot & \cdot & \cdot \\ a_{04} & a_{24} & b_{24} & a_{44} & b_{44} & \cdot & \cdot & \cdot \\ c_{04} & c_{24} & d_{24} & c_{44} & d_{44} & \cdot & \cdot & \cdot \\ \vdots & \vdots & \vdots & \vdots & \vdots & \vdots & \vdots & \vdots \end{vmatrix}=0. \tag{22}$$

Equations (21) and (22) are extremely complex functions of β, and their solutions can be found only by the use of numerical methods.[12]

It can easily be shown that as the elliptical waveguide degenerates to a circular one (i.e., as $h\to 0$ and $\xi_0\to\infty$ such that $h\exp(\xi_0)/2\to r_0$, where r_0 is the radius of the degenerated circle) the above determinants reduce to the following expression

$$a_{m,m}d_{m,m}-b_{m,m}c_{m,m}=0,$$

which is the characteristic equation for the circular ferrite-filled waveguide,[4] as expected.

IV. CLASSIFICATION OF MODES AND CUTOFF CONDITIONS

In order to satisfy the boundary conditions, the modes in a ferrite-filled waveguide must be hybrid, requiring the presence of both an E_z and H_z,[4] but we have yet to decide exactly how these modes are to be identified. Eqs. (1b) and (1c) give the variation of μ and κ versus the applied magnetic field H_0, and it can be demonstrated that as $H_0\to\infty$, $\kappa\to 0$, and $\mu\to\mu_z$. We expect, then, that as H_0 becomes large, the propagation constants for the hybrid modes will be asymptotic to the modes in a dielectric-filled guide, where the permittivity of the dielectric equals that of the ferrite and the permeability of the dielectric is equal to μ_z. The terms "HE" mode and "EH" mode will thus be used to designate those hybrid modes which, in the limit as $H_0\to\infty$, correspond to the TE and TM modes in a filled waveguide containing unmagnetized ferrite.[13] This scheme of classification will be used in lieu of the usual method of identifying modes according to their form at cutoff,

TABLE I. Zeros of radial Mathieu functions and their derivatives.

ξ	$Ce_n(\xi, q)=0$		$Se_n(\xi, q)=0$		$Ce_n'(\xi, q)=0$		$Se_n'(\xi, q)=0$	
				First root				
	$n=0$	$n=1$	$n=2$	$n=1$	$n=0$	$n=1$	$n=2$	$n=1$
0.25	11.399	15.271	...	...	...	0.834	14.802	11.092
0.50	3.178	5.384	14.124	10.649	10.406	0.691	4.728	2.844
0.75	1.492	2.971	6.739	4.584	4.400	0.519	2.337	1.194
1.00	0.825	1.815	3.751	2.372	2.258	0.362	1.317	0.598
1.25	0.484	1.128	2.205	1.329	1.269	0.240	0.778	0.325
1.50	0.290	0.700	1.322	0.773	0.746	0.154	0.461	0.185
	$n=2$	$n=3$	$n=4$	$n=3$	$n=2$	$n=3$	$n=4$	$n=3$
0.25	19.983	25.618	...	...	2.831	...	...	...
0.50	8.400	12.271	23.239	18.308	2.312	4.820	10.585	7.291
0.75	5.090	7.862	12.735	9.450	1.696	3.468	6.171	3.992
1.00	3.251	5.121	7.662	5.515	1.148	2.284	3.771	2.379
1.25	2.060	3.246	4.681	3.327	0.736	1.428	2.306	1.443
1.50	1.283	2.007	2.855	2.020	0.457	0.875	1.404	0.877
				Second root				
	$n=0$	$n=1$	$n=2$	$n=1$	$n=0$	$n=1$	$n=2$	$n=1$
0.25	...	...	...	...	...	...	...	...
0.50	22.694	27.532	...	...	...	13.786	27.247	22.469
0.75	9.540	12.483	...	16.322	16.154	6.506	12.257	9.365
1.00	4.854	6.748	10.595	8.224	8.107	3.664	6.583	4.731
1.25	2.697	3.938	6.053	4.530	4.454	2.210	3.803	2.612
1.50	1.567	2.378	3.582	2.610	2.565	1.357	2.264	1.507
	$n=2$	$n=3$	$n=4$	$n=3$	$n=2$	$n=3$	$n=4$	$n=3$
0.25	...	...	...	...	...	...	...	...
0.50	21.250	...	...	...	...	22.700	...	...
0.75	15.985	...	29.168	24.326	9.225	12.611	19.734	15.704
1.00	4.756	11.900	16.580	13.375	5.554	7.934	11.525	8.846
1.25	5.511	7.421	9.950	7.861	3.475	5.041	6.993	5.268
1.50	3.407	4.630	6.054	4.734	2.168	3.144	4.265	3.188

[a] ... implies that the zero occurs for $q>25$.

because the latter method sometimes leads to ambiguity.

In accordance with the above scheme of mode classification, Suhl and Walker[4] have shown that the circular ferrite-filled waveguide will support the same hierarchy of modes as does the unmagnetized ferrite guide—a phenomenon which must also hold true for the elliptical case. Let us recall that the propagation constant β for an elliptical waveguide filled with unmagnetized ferrite can be obtained from the relation

$$4q_{np}'/h^2=\omega^2\mu\epsilon-\beta_{np}^2 \quad \text{for TE modes,} \qquad (23)$$

and from

$$4q_{np}/h^2=\omega^2\mu\epsilon-\beta_{np}^2 \quad \text{for TM modes,} \qquad (24)$$

where q_{np}' and q_{np} are, respectively, the roots of equations

$$\frac{\partial}{\partial\xi}\begin{pmatrix} Ce_n \\ Se_n \end{pmatrix}(\xi_0, q)=0 \qquad \text{(TE modes),} \qquad (25a)$$

and

$$\begin{pmatrix} Ce_n \\ Se_n \end{pmatrix}(\xi_0, q)=0 \qquad \text{(TM modes).} \qquad (25b)$$

The subscript p refers to the order of the zero and the subscript n refers to the order of the function. In addition, a prescript "e" or "o" must be specified in order to denote whether the mode is even or odd. For example, the solution of $Se_n(\xi_0, q_{np})=0$ yields the ${}_o\mathrm{TM}_{np}$ modes; the solution of $(\partial/\partial\xi)Ce_n(\xi_0, q_{np})=0$ yields the ${}_e\mathrm{TE}_{np}$ modes. The roots of Eq. (25) are tabulated in Table I. By using Table I and Eqs. (23) and (24), we can predict the order of mode propagation, along with the asymptotic value of β (as H_0 becomes very large), for all modes in the ferrite-filled elliptical waveguide.

As far as the cutoff frequencies for the various possible modes are concerned, let us again refer back to the treatment of ferrite-filled circular waveguides by

Suhl and Walker.[4] They show that cutoff can occur in three different ways. The first of these (Type 1) is similar to that which occurs in an air-filled waveguide—the propagation constant β goes to zero, and the field is represented as a superposition of plane waves propagating at right angles to the axis of the guide. The second type of cutoff (Type 2) occurs when $\mu_{\text{eff}}=\mu\pm\kappa=0$, and does so only for sufficiently small waveguide radius. In this case the field components E and B both vanish, leaving only a circularly polarized H vector rotating in planes perpendicular to the waveguide axis. Such a situation could not exist in a guide filled with an isotropic medium; it is possible here because of the vanishing of the effective value of permeability. The third type of cutoff (Type 3) occurs when μ (the main diagonal element of the permeability tensor) goes to zero. It is the most unusual of the three because it occurs without β becoming zero. Returning now to the examination of the cutoff conditions for the elliptical ferrite-filled guide, we note that for the cutoff condition termed "Type 1," $\beta=0$ and the scalar Helmholtz equations may be written as

$$\nabla_t^2E_z+aE_z=0, \qquad \nabla_t^2H_z+cH_z=0.$$

Similarly, the boundary conditions become

$$(\partial/\partial\xi)H_z(\xi,\eta)|_{\xi=\xi_0}=0, \qquad E_z(\xi,\eta)|_{\xi=\xi_0}=0.$$

Above equations indicate that for $\beta=0$ the solution can be separated into TE and TM modes, as in the case of a waveguide filled with an isotropic dielectric material. From Eqs. (23) and (24) we see that for a dielectric-filled elliptical guide, the square of the wave number is equal to $4q/h^2$, when β goes to zero. Therefore, in the case of a magnetized ferrite-filled elliptical waveguide the Type 1 cutoff conditions become

$$4q_{n,p}/h^2=a|_{\beta=0}=\omega^2\mu\epsilon-\omega^2\epsilon\kappa^2/\mu, \tag{26a}$$

$$4q_{n,p}'/h^2=c|_{\beta=0}=\omega^2\mu_z\epsilon, \tag{26b}$$

where the subscripts n and p refer to the TE and TM modes which exist in an isotropic ferrite-filled waveguide. If we make the substitutions

$$\bar{\omega}_0=\bar{\gamma}H_0/\omega \qquad \bar{\omega}_M=\bar{\gamma}M_s/\omega,$$

in the expressions for μ and κ, these may be rewritten as follows:

$$\mu=\mu_0\{1+[\bar{\omega}_M\bar{\omega}_0/(\bar{\omega}_0^2-1)]\}, \tag{27a}$$

$$\kappa=-\mu_0[\bar{\omega}_M/(\bar{\omega}_0^2-1)]. \tag{27b}$$

The cutoff conditions then take the form

$$4q_{n,p}/h^2=\omega^2\mu_0\epsilon[(\bar{\omega}_0^2-1)+2\bar{\omega}_M\bar{\omega}_0+\bar{\omega}_M^2/(\bar{\omega}_0^2-1)+\bar{\omega}_M\bar{\omega}_0], \tag{28a}$$

$$4q_{np}'/h^2=\omega^2\mu_z\epsilon. \tag{28b}$$

When the applied field is zero (i.e., for $\bar{\omega}_0=0$, $\bar{\omega}_M\neq0$), Eqs. (28a) and (28b) indicate, respectively, the frequencies above which the various HE and EH modes can propagate in the guide. For $\bar{\omega}_0\neq0$, however, (28b) no longer applies, and all Type 1 cutoffs, $(\omega_0)_{c0}=(\omega_0)_1$, for both EH and HE modes occur according to Eq. (28a).

In addition to the mode classifications discussed previously, one may—in a manner analogous to the infinite medium case—classify the modes as having either positive or negative rotation with respect to the applied field,[14] with effective permeabilities $\mu_+=\mu-\kappa$ and $\mu_-=\mu+\kappa$. A "Type 2" cutoff occurs when the effective permeability of the positive rotating mode equals zero. The expression for this cutoff is thus $\mu-\kappa=0$, or

$$1+[\bar{\omega}_M\bar{\omega}_0/(\bar{\omega}_0^2-1)]+[\bar{\omega}_M/(\bar{\omega}_0^2-1)]=0. \tag{29}$$

Solving for $\bar{\omega}_0$ gives

$$\bar{\omega}_0=-\tfrac{1}{2}\bar{\omega}_M\pm\tfrac{1}{2}[\bar{\omega}_M^2-4(\bar{\omega}_M-1)]^{1/2}=-\tfrac{1}{2}\bar{\omega}_M\pm\tfrac{1}{2}(\bar{\omega}_M-2). \tag{30}$$

Discarding the (+) sign we have

$$\bar{\omega}_0=(\bar{\omega}_0)_2=1-\bar{\omega}_M, \tag{31}$$

as the Type 2 cutoff condition.

A "Type 3" cutoff results when the value of μ, itself, becomes negative, although here, unlike the previous two cases, the value of γ is not zero. At the point $\mu=0$ the wavenumber $(s_2)^{1/2}$ tends to an infinite imaginary value, while $(s_1)^{1/2}$ remains finite and real. Setting μ equal to zero in Eq. (27a), we have

$$(\bar{\omega}_0^2-1+\bar{\omega}_M\bar{\omega}_0)/(\bar{\omega}_0-1)=0. \tag{32}$$

Solving for $\bar{\omega}_0$

$$(\bar{\omega}_0)_3=-(\bar{\omega}_M/2)+[(\bar{\omega}_M/2)^2]^{1/2}+1. \tag{33}$$

These three types of cutoffs and the modes which produce them will be illustrated in the following section.

V. SOLUTION OF THE CHARACTERISTIC EQUATIONS FOR THE DOMINANT MODES

It is known[11] that the angular Mathieu functions can be expanded in terms of an infinite series of trigonometric functions and that the corresponding radial Mathieu functions can be expanded in terms of an infinite series of Bessel functions. These series were employed in the present computation. The propagation constants (i.e., the roots of the transcendental characteristic equations) for the various modes were then obtained using the method of linear interpolation. It was found (numerically) that the infinite

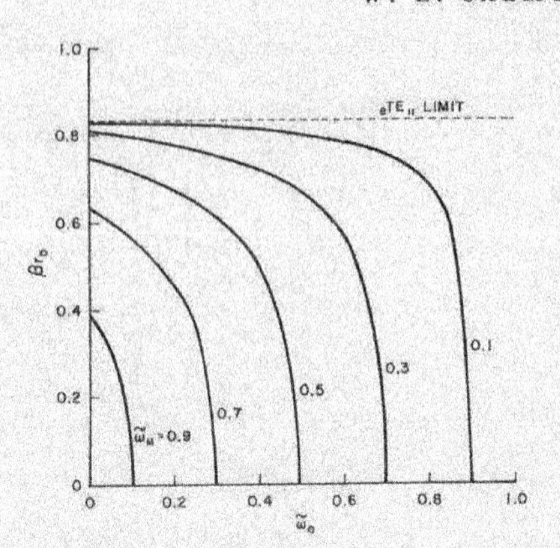

FIG. 2. Propagation constant vs applied magnetic field for the HE_{11} mode: $\xi_0=0.5$, $\bar{h}=0.1$, $\tilde{\omega}_0=\gamma H_0/\omega$, $\tilde{\omega}_M=\gamma M_s/\omega$, $r_0=h(\cosh\xi_0 \sinh\xi_0)^{1/2}$, ω=constant.

determinants converged quite rapidly, such that they could be approximated very well by determinants of finite order N; e.g., a 4×4 determinant yielded 3-place accuracy for all β's of the first azimuthal type. In general, it can be shown that the β's for the $m=k$ modes are governed principally by the expression

$$\begin{vmatrix} a_{k,k} & b_{k,k} \\ c_{k,k} & d_{k,k} \end{vmatrix} = 0.$$

All computations were performed on the IBM 360/75 computer. The propagation constants are analyzed as a function of the applied magnetic field, the saturation moment of the ferrite and the eccentricity of the waveguide cross section. The computations are limited to those modes having azimuthal mode number $m=1$.

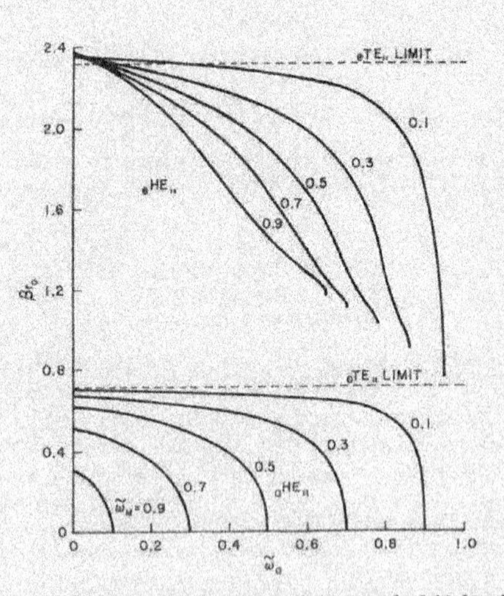

FIG. 3. Propagation constant vs applied magnetic field for the HE_{11} mode: $\xi_0=05$, $\bar{h}=0.175$.

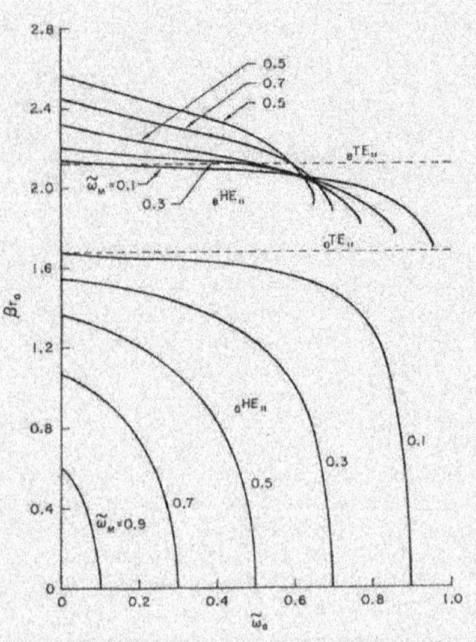

FIG. 4. Propagation constant vs applied magnetic field for the HE_{11} mode: $\xi_0=1.0$, $\bar{h}=0.1$.

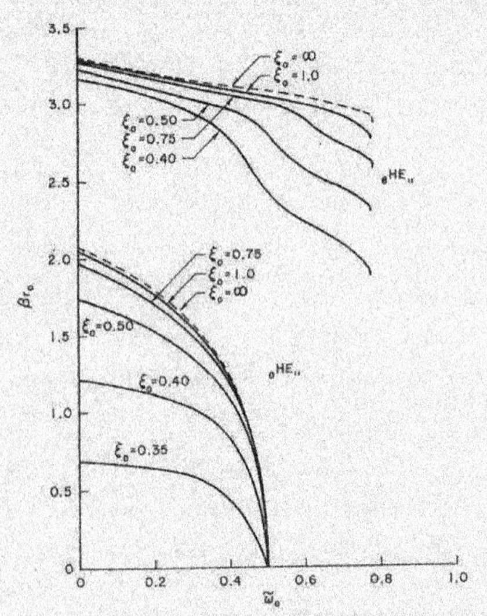

FIG. 5. Transition from elliptical to circular cross section for the HE_{11} modes: $\tilde{\omega}_M=0.5$, $\bar{A}_z=\pi/36$.

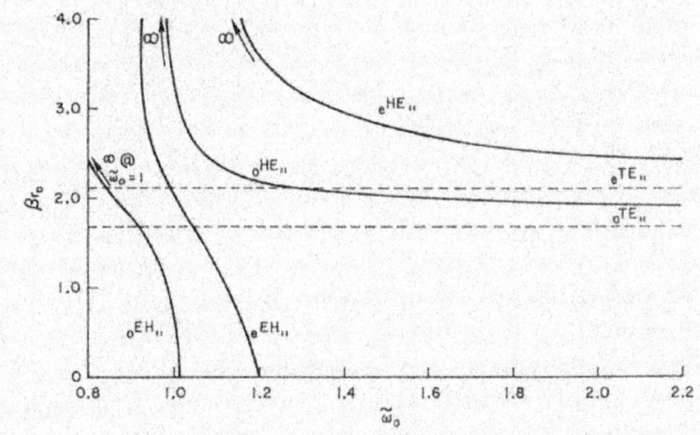

FIG. 6. Propagation constant vs applied magnetic field for the HE_{11} modes above resonance, plus incipient EH_{11} modes: $\xi_0 = 1.0$, $\tilde{h} = 0.1$, $\tilde{\omega}_M = 0.5$.

All results in this section assume a relative dielectric constant $\epsilon_r = \epsilon/\epsilon_0 = 10$ and a relative permeability $\mu_r = \mu_z/\mu_0 = 1.0$. The phase constant β has been normalized to the radius of a circular waveguide having the same cross-sectional area as the elliptical guide. The semifocal length h of the elliptical cross section has, in turn, been normalized to the wavelength of the source frequency; i.e.,

$$\tilde{h} = h/\lambda_0 = hk_0/2\pi, \tag{34}$$

where k_0 is the free space wave number. Using the above normalization for h, the type 1 cutoff conditions, Eqs. (28a) and (28b), become

$$\tilde{h} = (1/\pi)(q_{mn}'/\epsilon_r\mu_r)^{1/2}, \tag{35a}$$

$$\tilde{h} = (1/\pi)(q_{mn}/\epsilon_r\mu_e)^{1/2}, \tag{35b}$$

for the HE_{mn} and EH_{mn} modes, respectively, where

$$\mu_e = [(\tilde{\omega}_0^2 - 1) + 2\tilde{\omega}_M\tilde{\omega}_0 + \tilde{\omega}_M^2]/[(\tilde{\omega}_0^2 - 1) + \tilde{\omega}_M\tilde{\omega}_0]. \tag{36}$$

According to Eqs. (25a) and (25b), the first mode to propagate in the elliptical guide is the $_eHE_{11}$ mode. This occurs when

$$\pi^{-1}(_eq_{11}'/\epsilon_r\mu_r)^{1/2} < \tilde{h} < \pi^{-1}(_oq_{11}'/\epsilon_r\mu_r)^{1/2}. \tag{37}$$

In Fig. 2 the propagation constant for the $_eHE_{11}$ mode is plotted as a function of the normalized magnetic-field intensity, with the normalized saturation moment as a parameter. The waveguide cross section is characterized by $\xi_0 = 0.5$, $\tilde{h} = 0.1$. It can be seen that as the magnetic field is increased from zero, the propagation constant decreases toward zero and a type 2 cutoff at $\tilde{\omega}_0 = 1 - \tilde{\omega}_M$. The dashed line represents the value which βr_0 takes on in the limit as $\tilde{\omega}_0 \to \infty$, when the medium becomes isotropic and the $_eHE_{11}$ mode becomes the $_eTE_{11}$ mode.

For the range of $\tilde{h}$ described above the $_eHE_{11}$ mode is an elliptically polarized mode which rotates positively with respect to the direction of the applied field. Furthermore, for this range of $\tilde{h}$ the $_eHE_{11}$ mode is the only mode which propagates. Consequently, when $\tilde{h}$ satisfies the inequality (37) there can be no Faraday effect in the waveguide. The existence of a single mode without its orthogonal counterpart is one of the ways in which the elliptical waveguide distinguishes itself from the circular guide.

When $\tilde{h} = (1/\pi)(_oq_{11}'/\epsilon_r\mu_r)^{1/2}$ the $_oHE_{11}$ mode propagates and joins the $_eHE_{11}$ mode as the dominant pair in the ferrite-filled waveguide. At this point the $_eHE_{11}$ mode relinquishes its Type 2 cutoff to the $_oHE_{11}$ mode. The $_oHE_{11}$ mode is thus a positively rotating mode which cuts off at $\tilde{\omega}_0 = 1 - \tilde{\omega}_M$ with $\beta r_0 = 0$. The $_eHE_{11}$ mode, on the other hand, now becomes a negatively rotating mode which terminates at $\mu = 0$ with $\beta r_0 \neq 0$—a Type 3 cutoff. Figures 3 and 4 show the below resonance propagation constants as a function of $\tilde{\omega}_0$, $\tilde{\omega}_M$ and the ellipticity of the cross section.

Figure 5 illustrates the transition from elliptical to circular cross section for the HE_{11} modes in a waveguide having cross-sectional area $\tilde{A}_x = \pi\tilde{h}^2 \cosh\xi_0 \times \sinh\xi_0 = \pi/36$. When $\xi_0 \to \infty$ this corresponds to a circular waveguide of normalized radius $\tilde{r}_0 = 0.167$.

Figure 6 shows βr_0 for magnetic field intensities larger than or equal to the value required for resonance. At resonance the values of μ and k become infinite. Consequently, the value of βr_0 for the HE_{11} modes is infinite when $\tilde{\omega}_0 = 1$. As $\tilde{\omega}_0$ is increased above resonance the propagation constants decrease toward their respective nonmagnetic asymptotes.

In addition to presence of the HE_{11} modes, the resonance region is characterized by the propagation of an infinite number of higher-order modes. These are the so-called incipient modes—i.e., modes which are cut off in the unmagnetized condition because

$$\tilde{h} < (1/\pi)(q_{mp}/\epsilon_r\mu_e)^{1/2}.$$

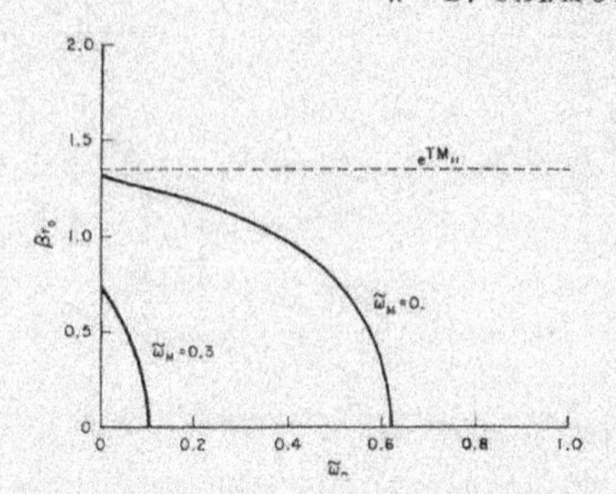

FIG. 7. Propagation constant vs applied magnetic field for the $_e$EH$_{11}$ mode: $\xi_0=1.0$, $\bar{h}=0.145$.

They propagate near resonance because of the large values of μ and k. The incipient modes exist in the region $\tilde{\omega}_0>\tilde{\omega}_0|_{\mu=0}$ and they all terminate with a Type 1 cutoff.[15] In Fig. 6 the HE$_{11}$ modes are accompanied by the incipient EH$_{11}$ modes, which also originate at $\tilde{\omega}_0=1$ with $\beta r_0=\infty$, and which cut off at succeedingly higher (or lower) values of $\tilde{\omega}_0$ as $\tilde{\omega}_M$ is increased (or decreased) within the range $0<\omega_M<1$.

The $_e$EH$_{11}$ and $_o$EH$_{11}$ modes make their appearance when

$$\bar{h}>(1/\pi)({}_eq_{11}/\epsilon_r\mu_{e0})^{1/2},$$

$$\bar{h}>(1/\pi)({}_oq_{11}/\epsilon_r\mu_{e0})^{1/2},$$

respectively, where μ_{e0} is the effective permeability of the EH$_{11}$ modes when $\tilde{\omega}_0=0$ and is given by

$$\mu_{e0}=\mu_e\,\tilde{\omega}_{0=0}=1-\tilde{\omega}_M^2. \tag{38}$$

Like the HE$_{11}$ modes, the even and odd EH$_{11}$ modes represent elliptically polarized waves which rotate in opposite directions at the axis of the guide; the $_o$EH$_{11}$ mode rotates in the positive sense, while the $_e$EH$_{11}$ modes rotates in the negative sense. Also, we see from (38) that the propagation of the EH$_{11}$ modes, and this applies to all EH modes, depends upon the saturation moment of the ferrite as well as the effective size of the waveguide. In contrast to the HE$_{mp}$ modes, which propagate when

$$\bar{h}>(1/\pi)(q_{mp}'/\epsilon_r\mu_r)^{1/2},$$

for all values of $\tilde{\omega}_M$, the EH$_{mp}$ modes exist for the case

$$\pi^{-1}(q_{mp}/\epsilon_r\mu_e)^{1/2}<\bar{h}<\pi^{-1}(q_{m.p+1}'/\epsilon_r\mu_r)^{1/2},$$

only for sufficiently small $\tilde{\omega}_M$. The preceding statement applies to the EH$_{11}$ modes for values of $\tilde{\omega}_0$ below resonance, or, more specifically, for values of $\tilde{\omega}_0<\tilde{\omega}_0|_{\mu=0}$. Above resonance the propagating EH modes behave in a manner similar to that of the HE modes. The above facts are demonstrated in Figs. 7–9.

For the range of $\bar{h}$ considered in Figs. 7 and 8, i.e., for

$$\pi^{-1}({}_eq_{11}/\epsilon_r\mu_e)^{1/2}<\bar{h}<\pi^{-1}({}_eq_{12}'/\epsilon_r\mu_r)^{1/2}, \tag{39}$$

all EH$_{11}$ cutoffs are of Type 1 and are in accordance with Eq. (28a). When $(\omega_0)_1$ is plotted as a function of $\tilde{\omega}_M$ and $q_{mp}=2(q_{mp})^{1/2}/\bar{h}$, the results appear as in Fig. 10. For example, consider the cutoff point of the EH$_{11}$ mode when $\bar{h}=0.165$, $\xi_0=1.0$, and $\tilde{\omega}_M=0.3$. From Table I, ${}_eq_{11}|_{\xi_0=1.0}=1.815$, which implies that $\bar{q}_{11}\approx16.4$. Figure 10 then yields the results $(\tilde{\omega}_0)_1\approx4.3$, which concurs with the cutoff point in Fig. 8. The cutoffs for the incipient EH modes were also found to correspond with Eq. (28a), although these have not been illustrated.

The transition from elliptical to circular cross section for the EH$_{11}$ modes is given in Fig. 11. Notice that for the EH$_{11}$ modes, convergence is not nearly as rapid as it was for the HE$_{11}$ modes (Fig. 5). This undoubtedly is a result of the increased complexity of the EH$_{11}$ mode pattern.

For all values of $\bar{h}$ within the range of the inequality (39) the character of the HE$_{11}$ and EH$_{11}$ modes remains essentially unchanged. However, when

$$\bar{h}=(1/\pi)({}_eq_{12}'/\epsilon_r\mu_r)^{1/2}$$

(the condition for propagation of the $_e$HE$_{12}$ mode),

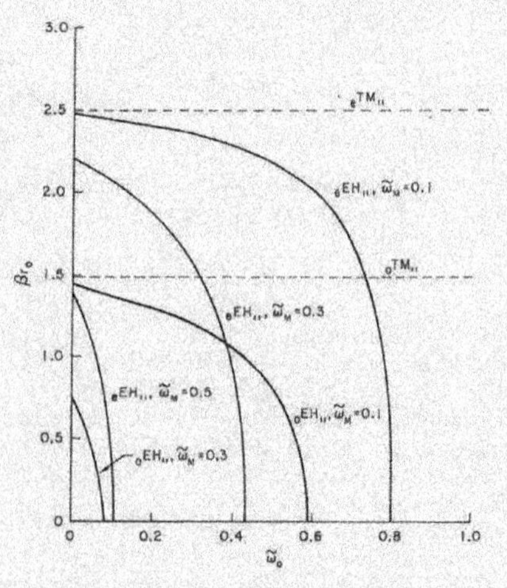

FIG. 8. Propagation constant vs applied magnetic field for the $_e$EH$_{11}$ and $_o$EH$_{11}$ modes: $\xi_0=1.0$, $\bar{h}=0.165$.

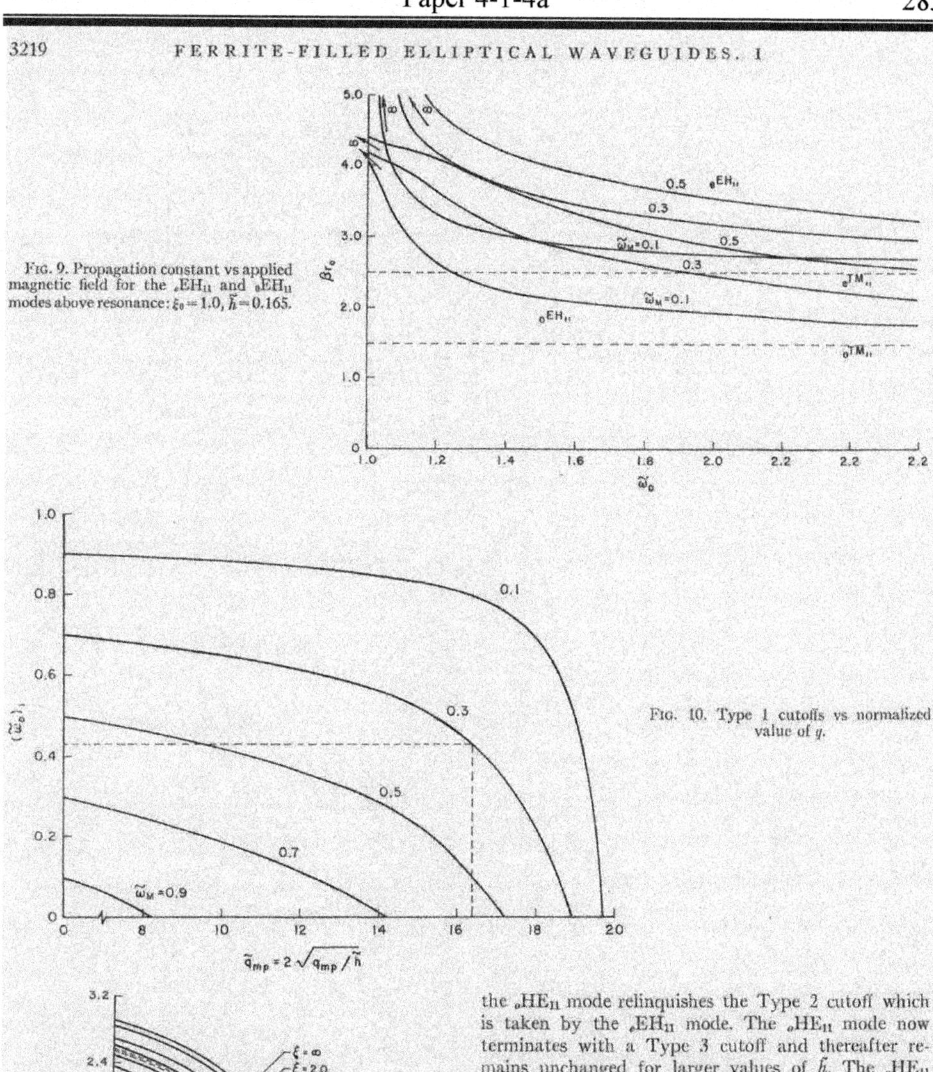

FIG. 9. Propagation constant vs applied magnetic field for the ${}_e\mathrm{EH}_{11}$ and ${}_o\mathrm{EH}_{11}$ modes above resonance: $\xi_0=1.0$, $\tilde{h}=0.165$.

FIG. 10. Type 1 cutoffs vs normalized value of q.

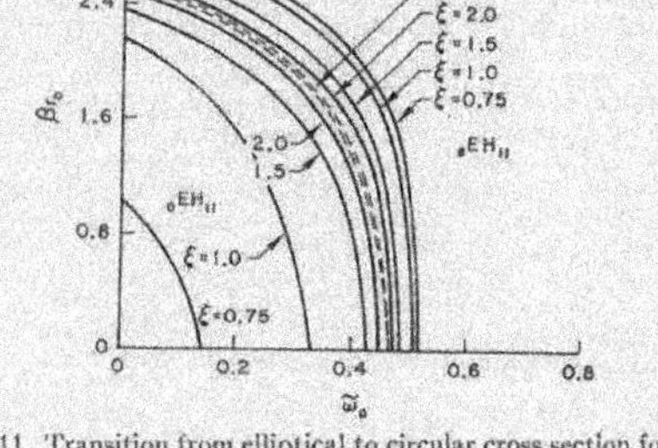

FIG. 11. Transition from elliptical to circular cross section for the EH_{11} modes: $\tilde{\omega}_M=0.3$, $\tilde{A}_z=\pi/17.5$.

the ${}_e\mathrm{HE}_{11}$ mode relinquishes the Type 2 cutoff which is taken by the ${}_e\mathrm{EH}_{11}$ mode. The ${}_o\mathrm{HE}_{11}$ mode now terminates with a Type 3 cutoff and thereafter remains unchanged for larger values of $\tilde{h}$. The ${}_e\mathrm{HE}_{11}$ mode then relinquishes its Type 1 cutoff to the newly-propagating ${}_e\mathrm{HE}_{12}$ mode. Here is an example of a mode (the ${}_e\mathrm{HE}_{12}$) which, in the limit of increasing magnetic field, becomes a normally propagating TE mode, but which has TM mode properties at cutoff. This provides justification for the method of mode classification described earlier. At the point where

$$\tilde{h}=(1/\pi)({}_oq_{12}'/\epsilon_r\mu_r)^{1/2},$$

the ${}_e\mathrm{EH}_{11}$ mode assumes a Type 3 cutoff and relinquishes its Type 2 cutoff to the ${}_o\mathrm{EH}_{11}$ mode. The

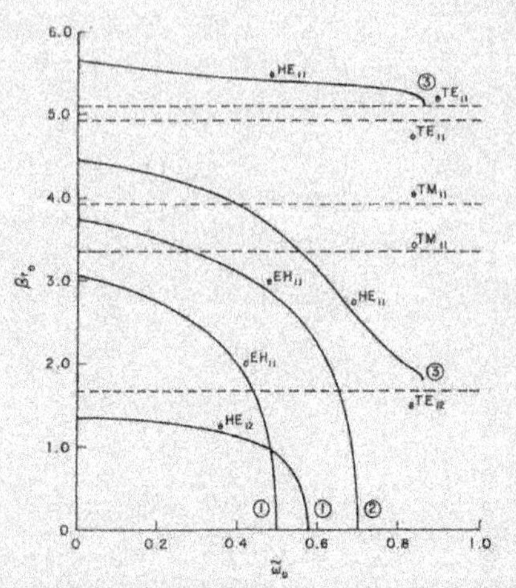

FIG. 12. Propagation constant vs applied magnetic field for the $_e$HE$_{11}$ and $_o$HE$_{11}$ modes, $_e$EH$_{11}$ and $_o$EH$_{11}$ modes, plus the $_e$HE$_{12}$ mode: $\xi_0=1.0$, $\tilde{h}=0.2$, $\tilde{\omega}_M=0.3$.

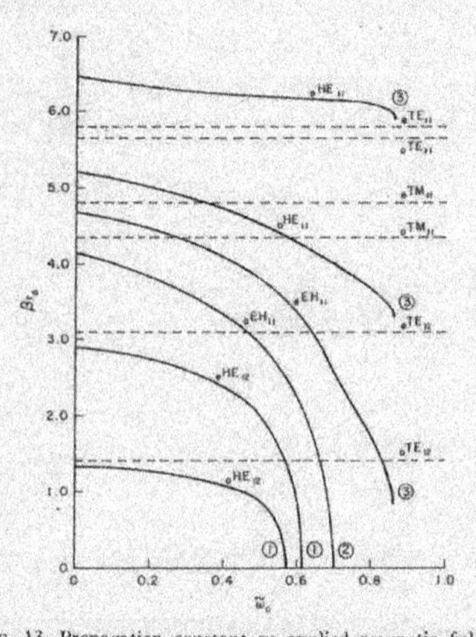

FIG. 13. Propagation constant vs applied magnetic field for the $_e$HE$_{11}$ and $_o$HE$_{11}$ modes, $_e$EH$_{11}$ and $_o$EH$_{11}$ modes, $_e$He$_{12}$ and $_o$HE$_{12}$ modes: $\xi_0=1.0$, $\tilde{h}=0.225$, $\tilde{\omega}_M=0.3$.

$_e$EH$_{11}$ mode, in turn, relinquishes its Type 1 cutoff to the $_o$HE$_{12}$ mode, which propagates for

$$\tilde{h}>(1/\pi)({}_oq_{12}'/\epsilon_r\mu_r)^{1/2}.$$

Figures 12 and 13 depict the mode sequences and cutoffs described above for values of $\tilde{\omega}_0$ below resonance. Above resonance the situation remains unchanged except for the addition of the propagating HE$_{12}$ modes.

From a knowledge of the propagation characteristics of the first few modes, the course of events for all higher order modes of the first azimuthal type, as well as for those having an azimuthal model number of 2 or greater, is easily understood. All EH$_{mp}$ modes initiate propagation with Type 1 cutoffs, which are passed on to the subsequently propagating HE$_{m,p+1}$ modes; the Type 2 cutoffs for the $_o$HE$_{mp}$ and the $_e$EH$_{mp}$ modes are relinquished upon propagation of the $_e$HE$_{m,p+1}$ and $_o$HE$_{m,p+1}$ modes, respectively.

It should be mentioned that the results of this section are consistent with those obtained by Suhl and Walker[4] for the circular ferrite-filled waveguide. The differences result, of course, from the ellipticity of the cross section. In the present analysis the even and odd hybrid modes initiate propagation for different values of $\tilde{h}$, whereas in the circular guide ($h\cosh\xi_0\to r_0$) both modes propagate simultaneously at the appropriate value of $\tilde{r}_0$. Thus, for the circular case, one finds that as the HE$_{12}$ modes begin to propagate, the $_e$EH$_{11}$ mode jumps directly to a Type 3 cutoff, while the $_o$EH$_{11}$ jumps directly to a Type 2 cutoff.

[1] H. Gamo, J. Phys. Soc. Japan **8**, 176 (1953).
[2] A. A. Van Trier, Appl. Sci. Res. **B3**, 305 (1953).
[3] M. L. Kales, J. Appl. Phys. **24**, 604 (1953).
[4] H. Suhl and L. R. Walker, Bell System Tech. J. **33**, 579, 939, 1133 (1954); Phys. Rev. **86**, 122 (1952).
[5] B. G. Sakiotis and H. N. Chait, Proc. IRE **41**, 87 (1953).
[6] C. L. Hogan, Bell System Tech. J. **31**, 22 (1952).
[7] B. Lax and K. J. Button, *Microwave Ferrites and Ferrimagnetics* (McGraw-Hill, New York, 1962).
[8] A. A. Van Trier, IRE Trans. Antennas Propagation **4**, 502 (1956).
[9] P. J. B. Clarricoats, *Microwave Ferrites* (Chapman and Hall, London, 1961).
[10] R. N. Ghose, *Microwave Circuit Theory and Analysis* (McGraw-Hill, New York, 1963).
[11] N. W. McLachlan, *Theory and Application of Mathieu Functions* (Dover, New York, 1964).
[12] C. Yeh, J. Appl. Phys. **33**, 3235 (1962).
[13] L. J. Chu, J. Appl. Phys. **9**, 583 (1938).

[14] A Wave with positive rotation will be defined as being polarized in the sense that would advance a right-hand screw in the direction of the applied field; a mode with negative rotation would advance a left-hand screw in the direction of the static field. Unlike the modes in the infinite ferrite medium, however, those in the filled wave guide are elliptically polarized instead of circularly polarized.

[15] Suhl and Walker[4] are rather unclear as to the nature and type of these incipient modes. They seem to indicate that both EH and HE incipient modes exist and that they both terminate with Type 1 cutoffs at different values of $\tilde{\omega}_0$. We found incipient modes of the EH type only. Those HE modes which propagate for the unmagnetized case, i.e., for $h>(1/\pi)(q'_{mp}/\epsilon_r\mu_r)^{1/2}$, will propagate in the incipient mode region and approach their isotropic asymptotes as $\omega_0\to\infty$. But when $\tilde{h}<(1/\pi)(q_{mp}/\epsilon_r\mu_r)^{1/2}$ there is no evidence of the existence of any HE modes in this region. The EH modes, on the other hand, have been found for

$$\tilde{h}<(1/\pi)(q_{mp}/\epsilon_r\mu_r)^{1/2},$$

or

$$\tilde{h}>(1/\pi)(q_{mp}/\epsilon_r\mu_e)^{1/2}.$$

Reprinted from JOURNAL OF APPLIED PHYSICS, Vol. 41, No. 8, 3221–3226, July 1970

Ferrite-Filled Elliptical Waveguides. II. Faraday Effects

W. E. SALMOND AND C. YEH
Electrical Sciences and Engineering Department, University of California, Los Angeles, California 90024
(Received 20 February 1970)

Detailed analyses were carried out for the Faraday effect of electromagnetic waves propagating in longitudinally magnetized ferrite-filled elliptical waveguides. Few resemblances were found between the Faraday rotation effect in a circular waveguide and that in an elliptical waveguide. The circular waveguide features strong coupling between the dominant vertically and horizontally polarized modes, whereas in the elliptical ferrite-filled guide very little mode coupling occurs for large eccentricities.

I. INTRODUCTION

A knowledge of the propagation constant, as given in the previous paper (I),[1] allows the calculation of the field quantities. The expressions for the electric and magnetic fields can be derived as infinite product series of angular and radial Mathieu functions. When evaluated at the waveguide axis, the fields are found to be elliptically polarized waves with the principal axes of the ellipse corresponding to the axis of guide itself. The field patterns for the dominant modes have been computed for several values of ξ_0, where the cross-sectional area remains constant. In the analysis of the Faraday effect, the rotation which takes place for the infinite medium is compared to that of the filled waveguide. The Faraday rotation which takes place at the axis of the elliptical guide is plotted as a function of the waveguide eccentricity for the case of small gyrotropy. The rotation of the entire cross section is plotted for two particular cases.

II. TRANSVERSE FIELD EXPRESSIONS

Using Eqs. (3)–(9) given in paper I, one may readily derive the transverse field expressions. They are

$$\sigma E_\xi = -j\beta(\lambda_1+\lambda_2) - (\mu/\beta\kappa)[(s_1-a)\lambda_3 + (s_2-a)\lambda_4], \tag{1a}$$

$$\sigma E_\eta = -j\beta(\lambda_3+\lambda_4) + (\mu/\beta\kappa)[(s_1-a)\lambda_1 + (s_2-a)\lambda_2], \tag{1b}$$

$$\sigma H_\xi = (1/\omega\kappa)[(\omega^2\mu\epsilon-\beta^2-s_1)\lambda_1 + (\omega^2\mu\epsilon-\beta^2-s_2)\lambda_2] + j\omega\epsilon(\lambda_3+\lambda_4), \tag{1c}$$

$$\sigma H_\eta = (1/\omega\kappa)[(\omega^2\mu\epsilon-\beta^2-s_1)\lambda_3 + (\omega^2\mu\epsilon-\beta^2-s_2)\lambda_4] - j\omega\epsilon(\lambda_1+\lambda_2), \tag{1d}$$

where

$$\lambda_1 = \partial u_1/\partial\xi = \sum_{m=0}^{\infty}{}' A_m Ce'_m(\xi, q_1) ce_m(\eta, q_1) + \sum_{m=1}^{\infty}{}' B_m Se'_m(\xi, q_1) se_m(\eta, q_1), \tag{2a}$$

$$\lambda_2 = \partial u_2/\partial\xi = \sum_{m=0}^{\infty}{}' C_m Ce'_m(\xi, q_2) ce_m(\eta, q_2) + \sum_{m=1}^{\infty}{}' D_m Se'_m(\xi, q_2) se_m(\eta, q_2), \tag{2b}$$

$$\lambda_3 = \partial u_1/\partial\eta = \sum_{m=0}^{\infty}{}' A_m Ce_m(\xi, q_1) ce'_m(\eta, q_1) + \sum_{m=1}^{\infty}{}' B_m Se_m(\xi, q_1) se'_m(\eta, q_1), \tag{2c}$$

$$\lambda_4 = \partial u_2/\partial\eta = \sum_{m=0}^{\infty}{}' C_m Ce_m(\xi, q_2) ce'_m(\eta, q_2) + \sum_{m=1}^{\infty}{}' D_m Se_m(\xi, q_2) se'_m(\eta, q_2). \tag{2d}$$

In order to evaluate the series for the field components, it remains to determine the constants A_m, B_m, C_m, and D_m. When the value of β is known, the latter two constants may be obtained from the characteristic equations

$$\sum_{m=0}^{\infty}{}' C_m a_{m,n} + \sum_{m=1}^{\infty}{}' D_m b_{m,n} = 0, \tag{3a}$$

$$\sum_{m=0}^{\infty}{}' C_m c_{m,n} + \sum_{m=1}^{\infty}{}' D_m d_{m,n} = 0. \tag{3b}$$

Using (3a) and (3b), we may arbitrarily assume a value for D_1 and solve for the remaining constants in terms D_1. Now if the infinite determinant in the characteristic equation is approximated by a determinant of order M, Eqs. (3) reduce to a set of $M-1$ simultaneous linear equations for C_m and D_m. Once these constants have been determined, A_m and B_m can be calculated; i.e.,

$$A_m = \frac{-1}{Ce_m(\xi_0, q_1)}\left(\frac{s_2}{s_1}\right)\sum_{r=1}^{\infty}{}' C_r Ce_r(\xi_0, q_2)\alpha^*_{r,m}, \tag{4a}$$

$$B_m = \frac{-1}{Se_m(\xi_0, q_1)}\left(\frac{s_2}{s_1}\right)\sum_{r=1}^{\infty}{}' D_r Se_r(\xi_0, q_2)\beta^*_{r,m}. \tag{4b}$$

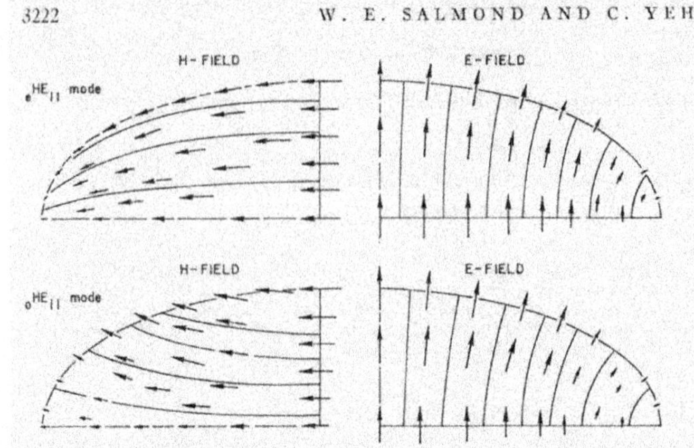

FIG. 1. Transverse field patterns for the $_e$He$_{11}$ and $_o$HE$_{11}$ modes: $\omega t=0$, $z=0$, $\xi_0=0.5$, $\bar{\omega}_M=\bar{\omega}_0=0.3$, $h=0.175$.

III. TRANSVERSE FIELD CONFIGURATIONS FOR DOMINANT MODES

Having solved for the constants A_m, B_m, C_m, and D_m, we may thus evaluate the **E** and **H** fields at any point (ξ, η) inside a filled elliptical waveguide of semifocal length h and eccentricity $e=1/\cosh\xi_0$. Let us first consider the fields on the axis of the guide. Here, as in paper *I*, the discussion shall be restricted to those modes which, in the limit of infinite applied field, become normally propagating TE_{mp} and TM_{mp} modes, where $m=1$ (for $m\neq 1$ the transverse fields vanish at the axis). By evaluating the transverse fields at $\xi=0$, $\eta=(\pi/2)$, one can readily show that $\mathbf{E}_t$ and $\mathbf{H}_t$ represent elliptically polarized waves on the axis of the guide, with the principal axes of the ellipse in the directions of $\mathbf{a}_\eta$ and $\mathbf{a}_\xi$, which, in this case, correspond to the $\mathbf{a}_x$ and $\mathbf{a}_y$ directions. Thus, $\mathbf{E}_t$ and $\mathbf{H}_t$ may be written in the form

$$\mathbf{E}_t=(E_\xi\mathbf{a}_y\pm jE_\eta\mathbf{a}_x)e^{j(\omega t-\beta z)}, \tag{5a}$$

$$\mathbf{H}_t=(H_\eta\mathbf{a}_x\mp jH_\xi\mathbf{a}_y)e^{j(\omega t-\beta z)}, \tag{5b}$$

where E_ξ, E_η, H_ξ, and H_η are real quantities. The $\pm$ sign indicates that the wave can have either positive or negative elliptical polarization, depending upon whether the propagating mode is odd or even. The corresponding $\mathbf{E}_t$ and $\mathbf{H}_t$ vectors will then rotate in opposite directions at the axis. Equations (5) represent time-average field quantities; the instantaneous field quantities are written according to

$$\mathcal{E}=\sqrt{2}\ \mathrm{Re}[\mathbf{E}e^{j\omega t}], \tag{6}$$

so for $z=0$

$$\mathcal{E}_t=\sqrt{2}\ [E_\xi \cos\omega t\mathbf{a}_y\mp E_\eta \sin\omega t\mathbf{a}_x], \tag{7a}$$

$$\mathcal{H}_t=\sqrt{2}\ [H_\eta \cos\omega t\mathbf{a}_x\pm H_\xi \sin\omega t\mathbf{a}_y]. \tag{7b}$$

Equations (7) demonstrate the fact that as ωt varies from 0 to 2π, the tip of the $\mathcal{E}_t$ (or $\mathcal{H}_t$) vector traces out an ellipse, the principal axes of which correspond to the x and y axes.

The preceding statements hold true only for the fields at the z axis. At other points in the cross section E_ξ, E_η, H_ξ, and H_η are complex quantities

$$E_\xi=|E_\xi|e^{j\phi_1} \qquad H_\xi=|H_\xi|e^{j\phi_3},$$

$$E_\eta=|E_\eta|e^{j\phi_2} \qquad H_\eta=|H_\eta|e^{j\phi_4},$$

with the instantaneous fields (at $z=0$) being written as

$$\mathcal{E}_t=\sqrt{2}\ [|E_\xi|\cos(\omega t+\phi_1)\mathbf{a}_\xi\mp|E_\eta|\sin(\omega t+\phi_2)\mathbf{a}_\eta], \tag{8a}$$

$$\mathcal{H}=\sqrt{2}\ [|H_\xi|\sin(\omega t+\phi_3)\mathbf{a}_\xi\mp|H_\eta|\cos(\omega t+\phi_4)\mathbf{a}_\eta]. \tag{8b}$$

Here it is obvious that the principal axes of the ellipse no longer correspond to the $\mathbf{a}_x$ and $\mathbf{a}_y$ directions.

Because the computation of the field quantities (which must be preceded by the computation of β and the constants $A_m, \cdots, D_m$) is a very involved process, only the transverse configurations for the dominant $_e$HE$_{11}$ and $_o$HE$_{11}$ modes will be considered. In performing the computations it was found that the series for the **E** and **H** fields converged quite satisfactorily for all $\xi<\xi_0$ when β was calculated from a 6×6 determinant. It is inherent in these calculations, though, that the accuracy in β be quite high—at least six significant figures.

Figures 1 and 2 show the typical transverse **E** and **H** field patterns of the $_e$HE$_{11}$ and $_o$HE$_{11}$ modes for two ferrite-filled waveguides having the same cross-sectional area but with different ellipticities. These patterns correspond to the case $\bar{\omega}_0=\bar{\omega}_M=0.3$ with $\omega t=0$. The direction of flow is indicated by the solid lines; the field strength is proportional to the length of the arrows. Figure 3 demonstrates that the two modes rotate in opposite directions with time.

In Fig. 1 the **E** fields for the $_e$HE$_{11}$ and $_o$HE$_{11}$ modes have been scaled in a ratio of 1 to 10, while the **H** fields are scaled in a ration of 1 to 3. In Fig. 2 a common normalization exists for the two modes. In Fig. 3, however, for the case $\xi_0=1$, $\omega t=\pi/2$, the **E** fields are scaled

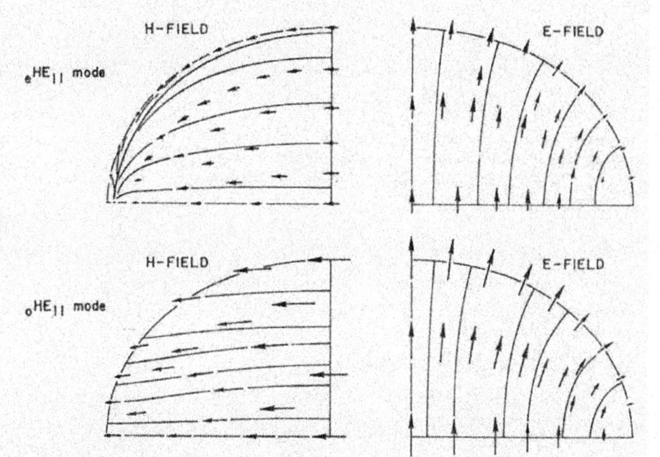

FIG. 2. Transverse field patterns for the $_e\mathrm{HE}_{11}$ and $_o\mathrm{HE}_{11}$ modes: $\omega t=0$, $z=0$, $\xi_0=1.0$, $\bar{\omega}_M=\bar{\omega}_0=0.3$, $h=0.10$.

in a ration of 4 to 1. If one considers the fields at the axis of the guide, the interpretation is as follows: the $_e\mathrm{HE}_{11}$ mode represents a negative elliptically polarized wave whose semimajor axis corresponds to the semiminor axis of the waveguide; the $_o\mathrm{HE}_{11}$ mode represents a positive elliptically polarized wave whose semimajor axis corresponds to that of the waveguide when $z=0$. In terms of Eq. (5a) we have

$$_e\mathbf{E}_t=({_eE_\xi}\mathbf{a}_\xi+j{_eE_\eta}a_\eta)\exp[j(\omega t-\beta_e z)], \quad (9a)$$

$$_o\mathbf{E}_t=({_oE_\xi}\mathbf{a}_\xi+j{_oE_\eta}\mathbf{a}_\eta)\exp[j(\omega t-\beta_o z)], \quad (9b)$$

where ${_eE_\xi}$, ${_eE_\eta}$, ${_oE_\xi}$, ${_oE_\eta}$ all have the same sign and

$$|{_eE_\xi}|>|{_eE_\eta}| \qquad |{_oE_\xi}|<|{_oE_\eta}|.$$

When the waveguide cross section becomes a circle

$$|{_eE_\xi}|=|{_eE_\eta}|=E_e \qquad |{_oE_\xi}|=|{_oE_\eta}|=E_0,$$

where the difference between E_e and E_0 is a function of the size of the guide and the gyrotropy of the medium.

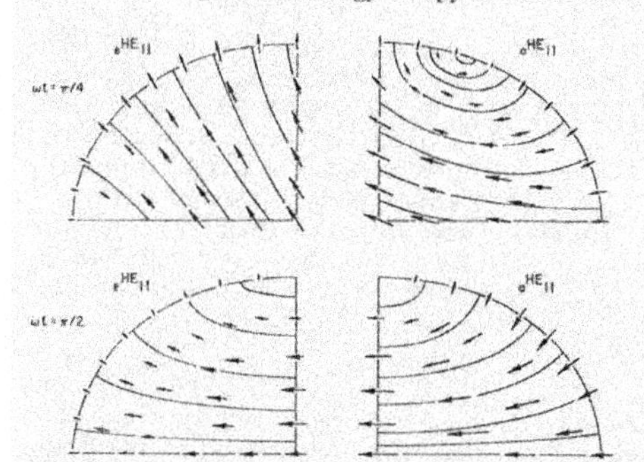

FIG. 3. Transverse field patterns (E-fields only) for the HE_{11} modes: $\xi_0=1.0$, $\bar{\omega}_M=\bar{\omega}_0=0.3$, $h=0.1$, $z=0$. $_e\mathrm{HE}_{11}$ pattern is mirror image of actual field configuration.

The results are in agreement with those of Suhl and Walker[2] as the cross section tends to a circle.

IV. FARADAY ROTATION IN ELLIPTICAL WAVEGUIDES

The Faraday rotation phenomenon is most clearly demonstrated for the case of an infinite gyromagnetic medium where the direction of propagation (assumed here to be the z direction) coincides with that of the applied magnetic field. It is known that the natural modes of propagation in an infinite ferrite medium are circularly polarized waves having equal amplitudes but rotating in opposite directions. These waves propagate with phase constants β_- and β_+ and may be written as

$$\mathbf{E}_-=(E_0/2)(\mathbf{a}_x+j\mathbf{a}_y)\exp(-j\beta_- z), \quad (10a)$$

$$\mathbf{E}_+=(E_0/2)(\mathbf{a}_x-j\mathbf{a}_y)\exp(-j\beta_+ z). \quad (10b)$$

Combining Eqs. (10a) and (10b) yields

$$\mathbf{E}_t=\mathbf{E}_++\mathbf{E}_-=E_0[\mathbf{a}_x\cos\tfrac{1}{2}(\beta_+-\beta_-)z-\mathbf{a}_y\sin\tfrac{1}{2}(\beta_+-\beta_-)z]\times\exp[-\tfrac{1}{2}(\beta_++\beta_-)z]. \quad (11)$$

The resultant wave $\mathbf{E}_t$ is thus a linearly polarized wave which has a propagation constant equal to the average of those of the right and left circularly polarized modes and whose plane of polarization has been rotated by an angle

$$\theta=\tan^{-1}(E_y/E_x)$$
$$=\tan^{-1}[-\tan\tfrac{1}{2}(\beta_+-\beta_-)z]$$
$$=-\tfrac{1}{2}(\beta_+-\beta_-)z, \quad (12)$$

with respect to the x axis.

If we constrain the magnetized ferrite medium to the region inside a perfectly conducting waveguide, the Faraday effect takes on a somewhat different form. As a result of the imposed boundary conditions the counterrotating modes are now elliptically polarized; i.e.,

$$\mathbf{E}_+ = (A\mathbf{a}_\xi + jB\mathbf{a}_\eta)\exp(-j\beta_- z), \quad (13a)$$

$$\mathbf{E}_- = (C\mathbf{a}_\xi + jD\mathbf{a}_\eta)\exp(-j\beta_+ z), \quad (13b)$$

and the resultant waveform—formerly a linearly polarized wave—is now an elliptically polarized wave. The field in a ferrite-filled elliptical waveguide resulting from (13a) and (13b) is

$$\mathbf{E}_t = [A\exp(-j\beta_- z) + C\exp(j\beta_+ z)]\mathbf{a}_\xi + j[B\exp(-j\beta_- z) + D\exp(j\beta_+ z)]\mathbf{a}_\eta. \quad (14)$$

Following the same procedure used for the infinite medium, we obtain

$$\mathbf{E}_t = \{[(A+C)\cos\theta + j(A-C)\sin\theta]\mathbf{a}_\xi + j[(B+D)\cos\theta + j(B-D)\sin\theta]\mathbf{a}_\eta\} \times \exp[-\tfrac{1}{2}(\beta_+ - \beta_-)z], \quad (15)$$

where $\theta = [(\beta_+ - \beta_-)/2]z$. If we make the substitutions

$$\hat{E}_\xi = (A+C)\cos\theta + j(A-C)\sin\theta = E_\xi \exp(j\phi_\xi), \quad (16a)$$

$$\hat{E}_\eta = (B+D)\cos\theta + j(B-D)\sin\theta = E_\eta \exp(j\phi_\eta), \quad (16b)$$

the previous expression for the transverse field becomes

$$\mathbf{E}_t = (\hat{E}_\xi \mathbf{a}_\xi + j\hat{E}_\eta \mathbf{a}_\eta)\exp[-\tfrac{1}{2}(\beta_+ - \beta_-)z]. \quad (17)$$

Equation (17) is easily recognized as an elliptically polarized wave whose major axis has been rotated by some angle τ with respect to the ξ direction. The angle τ will be called the rotation angle and is given by

$$\tau = \tfrac{1}{2}\tan^{-1}[2E_\xi E_\eta \cos(\phi_\eta - \phi_\xi)/(E_\xi^2 - E_\eta^2)]. \quad (18)$$

The variation of the rotation angle with θ (or, more precisely, with z) constitutes the Faraday rotation effect in a ferrite-filled waveguide. Unfortunately, the rotation angle, as defined above, is a rather complicated function of θ. For the sake of illustration, it turns out that the angle τ can be defined quite easily if Eq. (17) is expressed as a sum of two circularly polarized waves:

$$\mathbf{E}_t = \hat{E}_+(\mathbf{a}_\xi + j\mathbf{a}_\eta) + \hat{E}_-(\mathbf{a}_\xi - j\mathbf{a}_\eta), \quad (19)$$

where

$$\hat{E}_+ = (\hat{E}_\xi + \hat{E}_\eta)/2, \qquad \hat{E}_- = (\hat{E}_\xi - \hat{E}_\eta)/2, \quad (20)$$

and

$$\hat{E}_-/\hat{E}_+ = Re^{j\delta}. \quad (21)$$

If one adopts this procedure, it can easily be shown that the rotation angle is given by

$$\tau = \delta/2, \quad (22)$$

and that the axial ratio (the ratio of semimajor to semiminor axis) is given by $(1+R)/(1-R)$.

Let us now return to Eq. (15) to examine the waveguide Faraday effect. In Eq. (15) we have not, as yet, specified whether the quantities A, B, C, and D are real or complex. In general they are complex with unequal amplitudes, but we shall begin by considering the special case of $A = B$, $C = -D$, where A and C are real. This corresponds to the situation which exists at the axis of a circular ferrite-filled waveguide. Dividing (15) into circularly polarized waves with complex amplitudes $\hat{E}_+$ and $\hat{E}_-$, we have

$$\hat{E}_+ + \hat{E}_- = (A+C)\cos\theta + j(A-C)\sin\theta, \quad (23a)$$

$$\hat{E}_+ - \hat{E}_- = (A-C)\cos\theta + j(A+C)\sin\theta, \quad (23b)$$

or

$$\hat{E}_+ = A(\cos\theta + j\sin\theta) = A\exp(j\theta), \quad (24a)$$

$$\hat{E}_- = C(\cos\theta - j\sin\theta) = C\exp(-j\theta), \quad (24b)$$

which yields

$$\tau = \delta/2 = (-2\theta)/2 = -[(\beta_+ - \beta_-)/2]z, \quad (25)$$

the identical result obtained for the case of the infinite medium. Hence, the Faraday effect at the center of a ferrite-filled circular waveguide consists of the uniform rotation of the principal axes of an elliptically polarized wave, the axial ratio of which is constant and equal to $(A+C)/(A-C)$. For small ferrite gyrotropy this ratio can become quite large, making the situation very similar to that of the infinite medium.

At the axis of the elliptical ferrite-filled guide, A, B, C, and D are still real quantities but their amplitudes are all unequal; i.e.,

$$\hat{E}_+ + \hat{E}_- = (A+C)\cos\theta + j(A-C)\sin\theta, \quad (26a)$$

$$\hat{E}_+ - \hat{E}_- = (B+D)\cos\theta + j(B-D)\sin\theta, \quad (26b)$$

or

$$2\hat{E}_+ = \{(A+C+B+D)^2\cos^2\theta + (A+B-C-D)^2\sin^2\theta\}^{1/2} \times \exp\left[j\tan^{-1}\left(\frac{A+B-C-D}{A+B+C+D}\tan\theta\right)\right], \quad (27a)$$

$$2\hat{E}_- = \{(A+C-B-D)^2\cos^2\theta + (A-B-C+D)^2\sin^2\theta\}^{1/2} \times \exp\left[j\tan^{-1}\left(\frac{A-B-C+D}{A+C-B-D}\tan\theta\right)\right], \quad (27b)$$

$$\tau = \frac{1}{2}\left\{\tan^{-1}\left(\frac{A-B-C+D}{A+C-B-D}\tan\theta\right) - \tan^{-1}\left(\frac{A+B-C-D}{A+B+C+D}\tan\theta\right)\right\}. \quad (28)$$

Therefore, we see that on the axis of the elliptical guide, neither is the rotation rate uniform nor is the axial ratio constant. Both are, instead, functions of the angle θ and of the ellipticities of the component waves.

The Faraday rotation at the axis of the ferrite-filled waveguide is illustrated in Fig. 4. Here the rotation angle τ (measured clockwise from the y-axis) is plotted as a function of

$$\theta(\xi_0) = \{[\beta_+(\xi_0) - \beta_-(\xi_0)]/2\}z,$$

for two cases of small gyrotropy[3] and for various values of waveguide ellipticity. In all cases the waveguide cross section is constant and θ always varies between

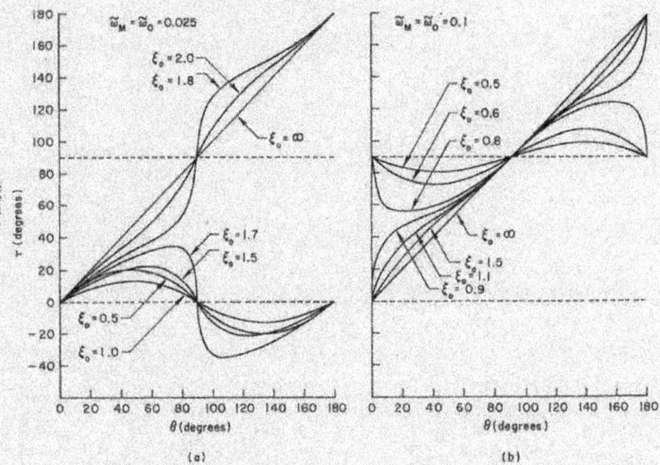

FIG. 4. Faraday rotation for the $_eHE_{11}$ and $_oHE_{11}$ modes at the axis of an elliptical ferrite-filled waveguide of normalized cross-section area $A_z = \pi/36$.

0 and π. It is significant that for most cases of large ellipticity the major axis of the polarization ellipse does not make a complete rotation. It merely oscillates about its orientation at $z=0$ when θ varies from 0 to 180°. As the elliptical cross section becomes more circular the angle of departure from the $z=0$ position increases until ξ_0 becomes large enough to allow a rotation of 180°. Notice, though, that it is possible to have a 180° rotation even when the elliptical cross section is very flat. In this situation, however, the axial ratio of the polarization ellipse is very small (close to unity), resulting in a polarization ellipse that is very nearly circular. This is in contrast to the circular guide where the axial ratio is large and the polarization ellipse is highly elliptical. Fig. 5 shows the axial ratio $(1+R)/(1-R)$ plotted as a function of $\xi_0/(1+\xi_0)$ for $\bar{\omega}_M = \bar{\omega}_0 = 0.025$.

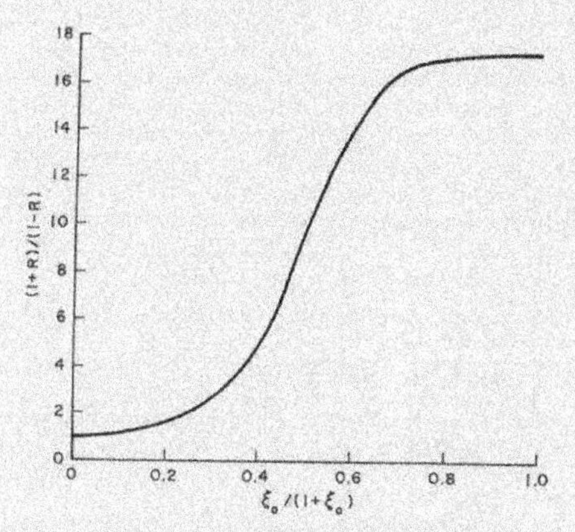

FIG. 5. Axial ratio of polarization ellipse versus waveguide ellipticity: $A_z = \pi/36$, $\theta = 0°$, $\bar{\omega}_M = \bar{\omega}_0 = 0.025$.

Figures 6 and 7 illustrate the Faraday rotation for one quadrant of the waveguide cross section for cases of small gyrotropy. The arrows point in the direction of the major axis of the polarization ellipse; their length is proportional to the field strength in this direction. Notice that, as the circular waveguide ($\xi_0 = \infty$) is

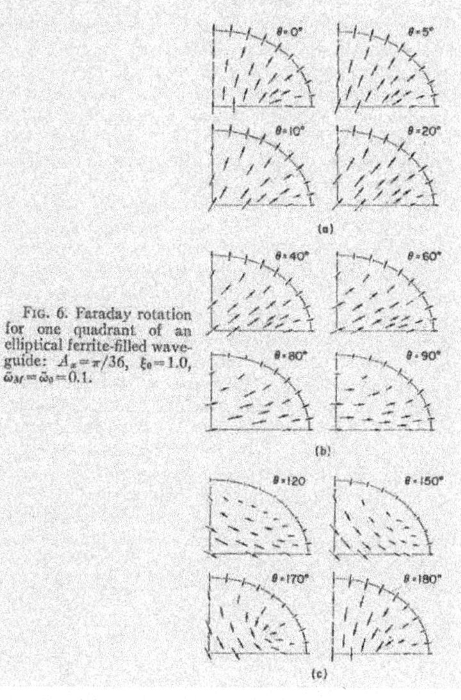

FIG. 6. Faraday rotation for one quadrant of an elliptical ferrite-filled waveguide: $A_z = \pi/36$, $\xi_0 = 1.0$, $\bar{\omega}_M = \bar{\omega}_0 = 0.1$.

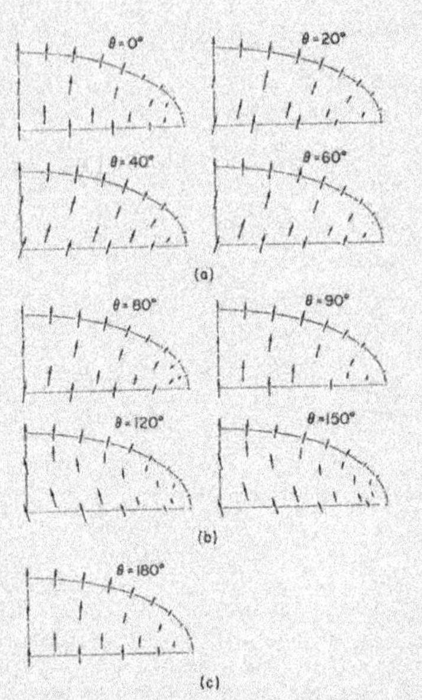

FIG. 7. Faraday rotation for one quadrant of an elliptical ferrite-filled waveguide: $A_z=\pi/36$, $\xi_0=0.5$, $\bar{\omega}_M=\bar{\omega}_0=0.025$.

slightly flattened (Fig. 6), a portion of the cross section rotates in the opposite direction of the pattern as a whole. This portion initially rotates in accordance with the rest of the pattern, but beyond a certain value of θ the direction of rotation is reversed. As the eccentricity becomes larger this effect begins to characterize the entire cross section—to the point where rotation no longer takes place (Fig. 7).

The above results can be clarified if one considers the coupled mode point of view for propagation in gyromagnetic media. That is, the behavior of the ferrite-loaded region may be described in terms of a coupling introduced by the magnetized ferrite between the two modes which are orthogonal in the absence of any gyrotropy. It is known from the theory of distributed mode coupling[4] that two modes propagating with the same phase velocity can completely transfer energy between one another if a uniformly distributed coupling exists between them. Such is the case for the uniformly magnetized infinite ferrite medium, where we may identify vertically and horizontally polarized modes as the dominant modes in the medium. A wave which, at some given point, is vertically polarized will decrease in amplitude as it travels further through the medium, and the horizontally polarized wave will increase in amplitude. The situation is much the same in a circular ferrite-filled waveguide; coupling occurs between the normally degenerate modes with $\cos n\phi$ and $\sin n\phi$ variations, although the presence of boundary conditions in the form of the waveguide walls prevents a complete interchange of energy between these modes. The result is an elliptically polarized wave whose major axis is alternately vertically and horizontally polarized as the wave travels down the guide.

The elliptical ferrite-filled guide differs from the above two cases in the respect that the dominant vertically and horizontally polarized modes (${}_eTE_{11}$ and ${}_oTE_{11}$) are no longer degenerate. As the elliptical cross section becomes flat, a significantly large difference exists between the phase constants of the dominant modes. This results in a correspondingly large decrease in mode coupling, such that Faraday rotation (in its familiar form) can no longer exist, or at least to the point where it becomes insignificant.

V. CONCLUSIONS

In summary we may state that with respect to mode types and propagation characteristics there exist both notable differences and great similarities between the elliptical and circular ferrite-filled waveguides. The major differences result from the nondegeneracy of the vertically and horizontally polarized mode pairs in the elliptical guide. These nondegeneracies provide a bandwidth for single-mode propagation. In addition, the ellipticity of the cross section was seen to affect the order of mode propagation, as well as the hierarchy of cutoff conditions. On the other hand the character of the $\bar{\omega}_0-\beta$ propagation curves and cutoff types is similar to that of the circular guide.

In considering the Faraday effect, however, there is little comparison between the elliptical and circular waveguides. Here, instead, one expects to find a close resemblance between the elliptical and rectangular ferritefilled guides. The circular waveguide features strong coupling between the dominant vertically and horizontally polarized modes, whereas in the elliptical ferrite-filled guide very little mode coupling occurs for large eccentricities. The similarities of mode patterns for the dominant modes for the isotropically-filled elliptical and rectangular waveguides indicate that in a longitudinally magnetized, ferrite-filled rectangular waveguide, the coupling between the dominant modes, and, hence, the Faraday rotation should compare very closely with that found in the elliptical guide.

[1] W. E. Salmond and C. Yeh, J. Appl. Phys. **41**, 3210 (1970) (preceding paper).

[2] H. Suhl and L. R. Walker, Bell System Tech. J. **33**, 579 (1954); **33**, 959 (1954); **33**, 1133 (1954).

[3] All Faraday rotation devices operate in the range of small applied fields. Typically H_0 is chosen such that the ferrite is below saturation.

[4] S. E. Miller, Bell System Tech. J. **33**, 661 (1954).

Reprinted from IEEE TRANSACTIONS
ON MICROWAVE THEORY AND TECHNIQUES
Volume MTT-13, Number 3, May, 1965
Pp. 297-302

Transverse Magnetic Wave Propagation in Sinusoidally Stratified Dielectric Media

C. YEH, MEMBER, IEEE, K. F. CASEY, AND Z. A. KAPRIELIAN, MEMBER, IEEE

***Abstract*—The problem of the propagation of TM waves in a sinusoidally stratified dielectric medium is considered. The propagation characteristics are determined from the stability diagram of the resultant Hill's equation. Numerical results show that the stability diagrams for Hill's equation and those for Mathieu's equation are quite different. Consequently, the dispersion properties of TM waves and TE waves in this stratified medium are also different. Detailed dispersion characteristics of TM waves in an infinite stratified medium and in waveguides filled longitudinally with this stratified material are obtained.**

Introduction

THE PROBLEM OF electromagnetic wave propagation in a sinusoidally varying dielectric medium is not only of interest from a theoretical point of view but also possesses many possible applications [1], [2]. For example, a section of waveguide filled with this type of inhomogeneous dielectric may be used as a band-pass filter in the mm or in the optical range. The use of an ultrasonic standing wave as a modulating device for certain pressure sensitive media, such as carbon disulfide, pentane, or nitric acid at optical frequencies to achieve a sinusoidally varying dielectric medium may be proposed. Other applications, such as the study of acoustically modulated plasma column and the analysis of sinusoidally modulated dielectric slab antenna, have also been proposed. Furthermore, the results should be very useful in the study of wave propagation in solids [3].

It can be shown [2] that two types of waves, propagating in the direction of the dielectric inhomogeneity, may exist: one with its electric vector transverse to the direction of propagation called a TE wave, and the other with its magnetic vector transverse to the direction of propagation, called a TM wave. The resultant differential equations for TE waves and TM waves are, respectively, the Mathieu and the Hill differential equations [4], [5]. The simpler case of the propagation of TE waves in a sinusoidally stratified dielectric medium has been considered most recently by Tamir, Wang and Oliner [1], and discussed briefly by Yeh and Kaprielian [2]. The purpose of this paper is to consider the problem of the propagation of TM waves in such an inhomogeneous medium. Since the solution of a Hill equation is required, it is expected that the results will be rather in-

Manuscript received November 5, 1964; revised December 30, 1964. The work in this paper was supported by the Technical Advisory Committee of the Joint Services Electronics Program.
The authors are with the Dept. of Electrical Engineering, University of Southern California, Los Angeles, Calif.

volved. It is found that the propagation characteristics of TM waves are quite different from those of TE waves for large dielectric variations.

Formulation of the Problem

It is assumed that the inhomogeneous dielectric medium under consideration fills the entire space and possesses a relative permittivity

$$\frac{\epsilon(z)}{\epsilon_0} = A\left(1 - \delta \cos \frac{2\pi z}{d}\right) \tag{1}$$

and a relative permeability

$$\frac{\mu}{\mu_0} = 1 \tag{2}$$

in the (x, y, z) rectangular coordinates. ϵ_0 and μ_0 are, respectively, the free space permittivity and permeability. A and δ are known positive constants. Furthermore, $0 \leq \delta < 1$. d denotes the period of the sinusoidal variation (Fig. 1).

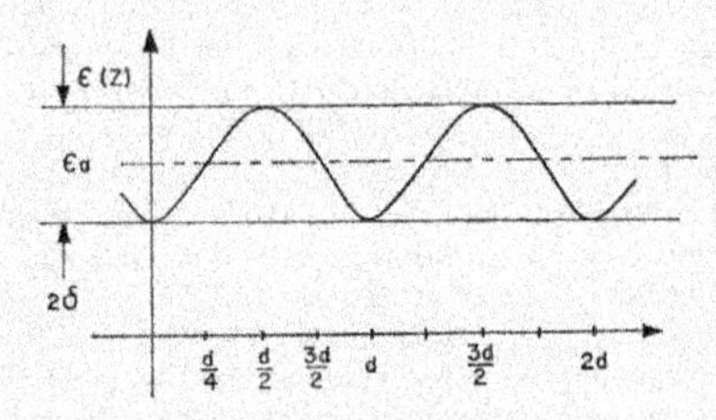

Fig. 1. Variation of permittivity as a function of longitudinal distance.

The source-free vector wave equations in this medium are:

$$\nabla \times \nabla \times E - k_0^2(\epsilon(z)/\epsilon_0)E = 0 \tag{3}$$

$$\nabla \times \nabla \times H - \frac{\nabla\epsilon(z)}{\epsilon(z)} \times \nabla \times H - k_0^2(\epsilon(z)/\epsilon_0)H = 0 \tag{4}$$

where E and H are, respectively, the electric and magnetic field vectors, $k_0^2 = \omega^2\mu_0\epsilon_0$, and a time dependence $e^{-i\omega t}$ is assumed. It can be shown that all field components in this medium can be obtained from the scalar quantities $\Phi(x, y, z)$ and $\Psi(x, y, z)$ as follows [2]:

$$E^{(m)} = \nabla \times [\Phi(x, y, z)e_z] \tag{5}$$

$$H^{(m)} = \frac{-i}{\omega\mu_0} \nabla \times \nabla \times [\Phi(x, y, z)e_z] \tag{6}$$

for transverse electric waves; and

$$H^{(e)} = \nabla \times [\Psi(x, y, z)e_z] \tag{7}$$

$$E^{(e)} = \frac{i}{\omega\epsilon(z)} \nabla \times \nabla \times [\Psi(x, y, z)e_z] \tag{8}$$

for transverse magnetic waves. e_z is the unit vector in the z direction. Upon substituting (5) into (3), and (7) into (4), carrying out the vector operations and separating variables in rectangular coordinates, one obtains

$$\Phi(x, y, z) = \begin{Bmatrix}\sin\\ \cos\end{Bmatrix}(sx)\begin{Bmatrix}\sin\\ \cos\end{Bmatrix}(wy)\left\{U^{(1),(2)}(z)\right\} \tag{9}$$

and

$$\Psi(x, y, z) = \begin{Bmatrix}\sin\\ \cos\end{Bmatrix}(px)\begin{Bmatrix}\sin\\ \cos\end{Bmatrix}(qy)\left\{V^{(1),(2)}(z)\right\} \tag{10}$$

where s, w, p, and q are separation constants. $U^{(1),(2)}(z)$ and $V^{(1),(2)}(z)$ satisfy, respectively, the differential equations

$$\left\{\frac{d^2}{dz^2} + [k_0^2(\epsilon(z)/\epsilon_0) - s^2 - w^2]\right\} U^{(1),(2)}(z) = 0 \tag{11}$$

and

$$\left\{\frac{d^2}{dz^2} - \left(\frac{d\epsilon(z)}{dz}\right)\frac{1}{\epsilon(z)}\frac{d}{dz} + [k_0^2(\epsilon(z)/\epsilon_0) - p^2 - q^2]\right\} V^{(1),(2)}(z) = 0. \tag{12}$$

Since we are only concerned with the propagation of transverse magnetic waves in this inhomogeneous medium, the transverse electric waves will not be considered further. Putting (1) into (12), introducing the dimensionless variable $\xi = \pi z/d$, and making the substitution

$$V^{(1),(2)}(z) = (1 - \delta \cos 2\xi)^{1/2} W^{(1),(2)}(\xi) \tag{13}$$

gives

$$\left[\frac{d^2}{d\xi^2} + \lambda(\xi)\right] W^{(1),(2)}(\xi) = 0 \tag{14}$$

where

$$\lambda(\xi) = \frac{2\delta \cos 2\xi}{1 - \delta \cos 2\xi} - \frac{3\delta^2 \sin^2 2\xi}{(1 - \delta \cos 2\xi)^2} + \left(\frac{k_0 d}{\pi}\right)^2 \cdot \left\{A - A\delta \cos 2\xi - \left[\left(\frac{p}{k_0}\right)^2 + \left(\frac{q}{k_0}\right)^2\right]\right\}. \tag{15}$$

It is noted that since $\lambda(\xi)$ is an even periodic function, it may be expanded in the Fourier cosine series

$$\lambda(\xi) = \theta_0 + 2\sum_{n=1}^{\infty} \theta_n \cos 2n\xi \tag{16}$$

in which

$$\theta_0 = \left(\frac{k_0 d}{\pi}\right)^2\left[A - \left(\frac{p}{k_0}\right)^2 - \left(\frac{q}{k_0}\right)^2\right] - \left[\frac{1}{\sqrt{1 - \delta^2}} - 1\right] \tag{17a}$$

$$\theta_1 = -\frac{\delta}{2}\left(\frac{k_0 d}{\pi}\right)^2 A + \frac{4b^3 - 2b}{b^2 - 1} \tag{17b}$$

$$\theta_n = \frac{(3n + 1)b^{n+2} - (3n - 1)b^n}{b^2 - 1} \quad (n \geq 2) \tag{17c}$$

with

$$b = \frac{1}{\delta} - \frac{1}{\delta}\sqrt{1-\delta^2}. \qquad (17d)$$

The above series converges absolutely for $0 \leq \delta < 1$. Substituting (16) into (14), one obtains

$$\left[\frac{d^2}{d\xi^2} + \theta_0 + 2\sum_{n=1}^{\infty}\theta_n \cos 2n\xi\right] W^{(1),(2)}(\xi) = 0 \qquad (18)$$

which is the general form of Hill's equation [4]-[6]. It is known that two types of solutions for the Hill equation exist: one called the stable type, and the other called the unstable type. In order to have propagating waves in the z direction, only the stable type is allowed.

Solutions of Hill's Equation

With the help of Floquet's Theorem [6] concerning wave propagation in periodic media, the solutions of Hill's equation can be expressed in the following form:

$$W^{(1),(2)}(\xi) = e^{\pm i\beta\xi} \sum_{n=-\infty}^{\infty} C_n(\beta) e^{\pm 2in\xi} \qquad (19)$$

where β and $C_n(\beta)$ are yet unknown coefficients. After substituting (19) into (18), and simplifying, one obtains the following recursion relations:

$$-(\beta + 2n)^2 C_n + \sum_{m=-\infty}^{\infty} \theta_m C_{n-m} = 0 \qquad (20)$$

$$(n = \cdots -2, -1, 0, 1, 2, \cdots)$$

with $\theta_{-m} = \theta_m$. The above is a set of an infinite number of homogeneous linear algebraic equations in C_n. For a nontrivial solution to exist the determinant of these equations must vanish. This equation is called the characteristic equation of the Hill equation. Using an ingenious method described in Morse and Feshbach [6], it is possible to simplify this characteristic equation to give

$$\sin^2 \frac{\pi\beta}{2} = \Delta(0) \sin^2 \frac{\pi\sqrt{\theta_0}}{2} \qquad (21)$$

where $\Delta(0)$ is the determinant of the matrix $[M]$, whose elements are

$$M_{mm} = 1$$

$$M_{mn} = \frac{-\theta_{m-n}}{4m^2 - \theta_0} \qquad m \neq n. \qquad (22)$$

The characteristic number β can be obtained from (21).

Real values of β yield stable solutions to Hill's equation, while complex values of β produce unstable solutions. Physically speaking the stable solutions correspond to modulated propagating waves, and the unstable solutions correspond to damped or growing waves. (The fields for the growing waves do not satisfy the radiation condition at infinity, hence they must be omitted.)

Numerical computation has been carried out for (21). The values for the infinite determinant $\Delta(0)$ were obtained by the successive approximation method [7]. In other words, computations were carried out for a 3×3 determinant, a 4×4 determinant, a 5×5 determinant, etc., until the desired accuracy was reached. It was found (numerically) that the infinite determinant converges quite rapidly within the present region of interests. At no time was any determinant greater than 7×7 required to achieve an accuracy of three significant figures.

Results of the computation are given in terms of a "stability diagram," which is customary in the study of Hill-type equations. Figures 2 and 3 show, respectively, the "stability diagram" for the cases $\delta = 0.25$ and $\delta = 0.4$. The unshaded areas in these figures are the "stable regions" wherein β is purely real; the shaded areas are the "unstable regions" wherein β is complex. It is noted that the value of β in the unshaded regions is bounded by

$$m \leq |\beta| \leq m + 1 \quad (m = 0, 1, 2, \cdots) \qquad (23)$$

so that the value of m may be used to label the appropriate regions as shown in Figs. 2 and 3.

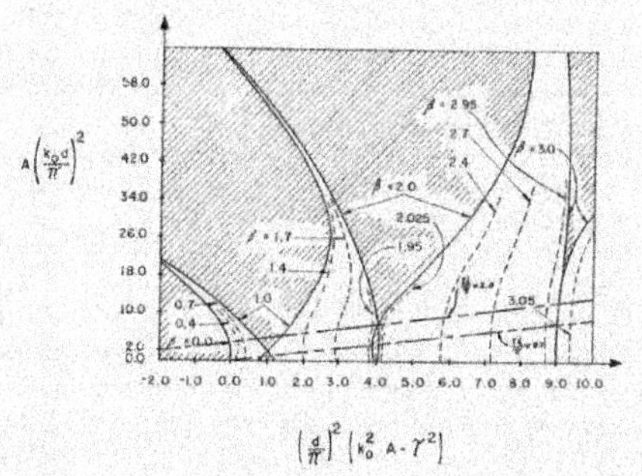

Fig. 2. Stability chart for Hill's equation with $\delta = 0.25$. Unstable regions are shaded. Family of straight lines represents (28) for various values of $\gamma d/\pi$.

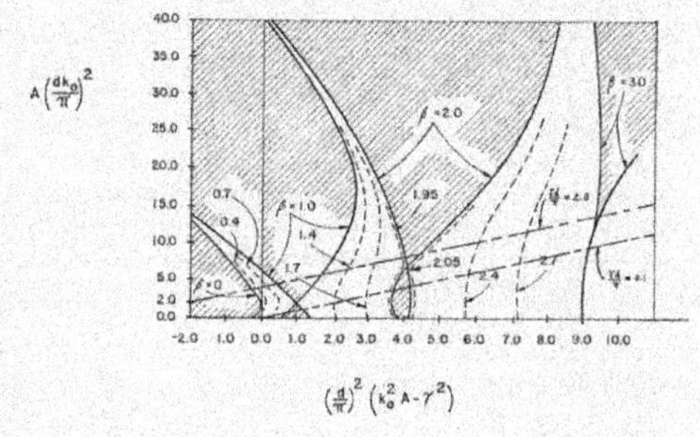

Fig. 3. Stability chart for Hill's equation with $\delta = 0.4$. Unstable regions are shaded. Family of straight lines represent (28) for various values of $\gamma d/\pi$.

It is interesting to note the differences between the stability diagram for Mathieu equation and those given in Figs. 2 and 3. Unlike the Mathieu case, curves separating the stable and unstable regions do not necessarily meet at the abscissa. As a matter of fact, in some instances they cross over each other (as can be seen from these figures) near the point $\theta_0=4.0$. As δ increases the overlapped region becomes larger.

Propagation Characteristics of TM Waves

A. Infinite region filled with sinusoidally stratified dielectric

The transverse magnetic field components of a TM wave in an infinite medium filled with sinusoidally stratified dielectric can be obtained from (7) and (10):

$$H_x^{(e)} = \sum_{n=-\infty}^{\infty} iqC_n e^{ipx} e^{iqy} e^{i(\beta+2n)\pi z/d}\left(1 - \delta\cos\frac{2\pi z}{d}\right)^{1/2} \quad (24)$$

$$H_y^{(e)} = \sum_{n=-\infty}^{\infty} -ipC_n e^{ipx} e^{iqy} e^{i(\beta+2n)\pi z/d} \cdot\left(1 - \delta\cos\frac{2\pi z}{d}\right)^{1/2} \quad (25)$$

where the coefficients C_n can be determined from (20) in terms of C_0. C_0 is obtained from a normalization condition. All electric field components may be found from Maxwell's equations.

Unlike the case of a TM wave propagating in an infinite homogeneous medium in which β is simply related to p and q by the following:

$$\beta^2 = k^2 - \gamma^2$$

with $\gamma^2=p^2+q^2$ and $k^2=\omega^2\mu\epsilon$ where μ and ϵ are the permeability and permittivity of the homogeneous medium, the propagation constant β for the inhomogeneous case is related to p and q through the stability diagrams given by Figs. 2 and 3. Real values of β as a function of real values of γ for a fixed value of A, δ, and k_0d are shown in Figs. 4 through 6. It is recalled that complex values of β indicate the presence of damped waves (i.e., nonpropagating waves). p and q are taken to be real. The unshaded regions in these figures indicate the regions in which β is real (i.e., regions in which propagating waves may exist). One notes from these figures that for very small values of k_0d, say $k_0d<0.2$, as long as $\gamma^2<k_0^2A$, β is always real. However as k_0d increases, there exist regions in which β is complex even though $\gamma^2<k_0^2A$. The presence of these stop band and pass band regions is characteristic of wave propagation in periodic structures [3].

B. Waveguide filled longitudinally with sinusoidally stratified dielectric

It is assumed that a rectangular waveguide of dimension h_1 and h_2 is filled completely with an inhomogeneous dielectric medium, which varies sinusoidally in the longitudinal direction. The general expressions for the transverse magnetic field components of a TM wave are:

$$H_x^{(e)} = \sum_{m=1}^{\infty}\sum_{r=1}^{\infty}\sum_{n=-\infty}^{\infty} C_n^{m,r}\sin\frac{m\pi x}{h_1}\cos\frac{r\pi y}{h_2}e^{i(\beta^{m,r}+2n)\pi z/d} \cdot\frac{r\pi}{h_2}\left(1-\delta\cos\frac{2\pi z}{d}\right)^{1/2} \quad (26)$$

$$H_y^{(e)} = \sum_{m=1}^{\infty}\sum_{r=1}^{\infty}\sum_{n=-\infty}^{\infty} C_n^{m,r}\cos\frac{m\pi x}{h_1}\sin\frac{r\pi y}{h_2}e^{i(\beta^{m,r}+2n)\pi z/d} \cdot\left[-\left(\frac{m\pi}{h_1}\right)\right]\left(1-\delta\cos\frac{2\pi z}{d}\right)^{1/2} \quad (27)$$

where $C_0^{m,r}$ are arbitrary constants, and $C_n^{m,r}$ can be determined from (20) in terms of $C_0^{m,r}$. Expressions for the electric field components can easily be derived from Maxwell's equations.

To obtain the dispersion curves for this case, one combines (17a) and (17b)

$$\theta_0 + \frac{2\theta_1}{\delta} = -\left(\frac{\gamma d}{\pi}\right)^2 - \left[\frac{1}{\sqrt{1-\delta^2}} - 1\right] + \frac{4b^3-2b}{b^2-1} \quad (28)$$

where b is given by (17d) and

$$\gamma^2 = \left(\frac{m\pi}{h_1}\right)^2 + \left(\frac{r\pi}{h_2}\right)^2. \quad (29)$$

Expression (28) is the equation of a family of straight lines as shown in Figs. 2 and 3. For given values of $\gamma d/\pi$ and δ one can then obtain the $\omega-\beta$ diagram from these figures. The $\omega-\beta$ diagrams are given for $\gamma d/\pi=0.1$, 2.0, 5.0, and $\delta=0.25$, 0.40 in Figs. 7 through 12. The pass band and stop band characteristics can clearly be seen. It is interesting to note that the first pass band starts at a frequency which is lower than the cutoff frequency for an identical waveguide filled with homogeneous dielectric material, which has a dielectric constant equal to the average value of the inhomogeneous dielectric.

The $\omega-\beta$ diagrams given by Figs. 7 through 12 are also applicable for circular waveguide except γ^2 is given by $(\Gamma_{mr}/\rho_0)^2$, where ρ_0 is the radius of the circular waveguide and Γ_{mr} are the roots of the equation

$$J_m(\Gamma_{mr}) = 0.$$

References

[1] Tamir, T., H. C. Wang, and A. A. Oliner, Wave propagation in sinusoidally stratified dielectric media, *IEEE Trans. on Microwave Theory and Techniques*, vol MTT-12, May 1964, pp 323–335.
[2] Yeh, C., and Z. A. Kaprielian, On inhomogeneously filled waveguides, USCEC Rept 84-206, Electrical Engineering Dept., University of Southern California, Los Angeles, 1963.
[3] Brillouin, L., *Wave Propagation in Periodic Structures*. New York: Dover, 1953.
[4] McLachlan, N. W., *Theory and Application of Mathieu Functions*. Oxford, England: University Press, 1951.
[5] Meixner, J., and F. W. Schäfke, *Mathieusche Funktionen und Sphäroidfunktionen*. Berlin, Germany: Springer-Verlag, 1954.
[6] Morse, P. M., and H. Feshbach, *Methods of Theoretical Physics*. New York: McGraw-Hill, 1953.
[7] Kantorovich, L., and V. Krylov, *Approximate Methods of Higher Analysis*. New York: Interscience Publishers, Inc., 1958.

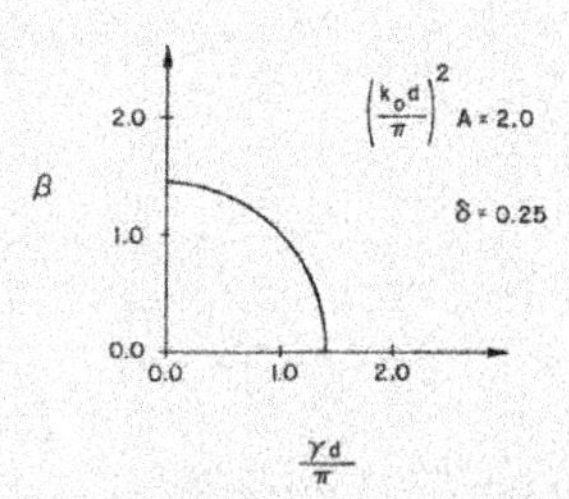

Fig. 4. The propagation constant β as a function of $\gamma d/\pi$.

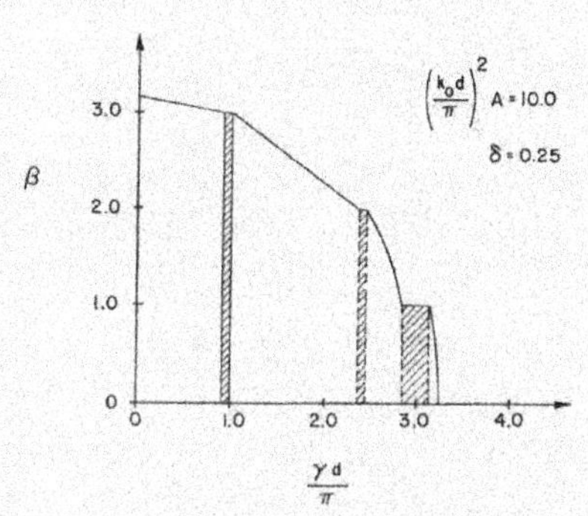

Fig. 5. The propagation constant β as a function of $\gamma d/\pi$. Unstable regions are shaded.

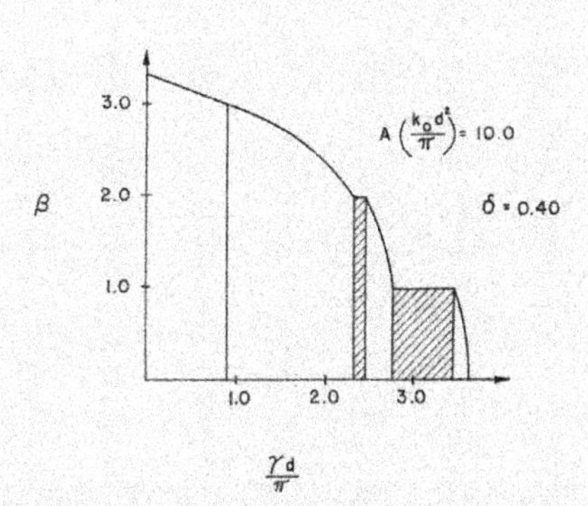

Fig. 6. The propagation constant β as a function of $\gamma d/\pi$. Unstable regions are shaded.

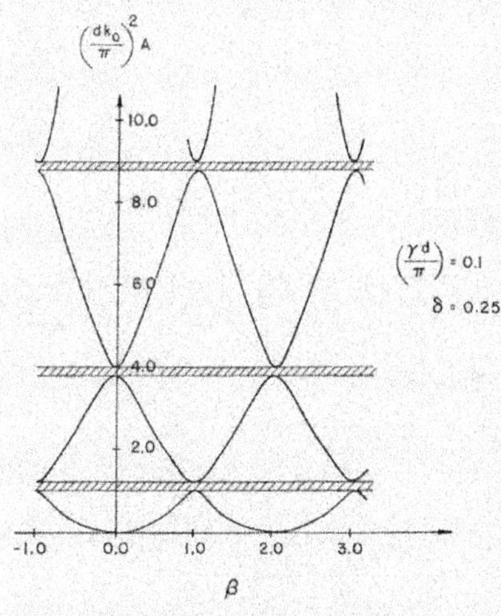

Fig. 7. Frequency as a function of β with $\delta = 0.25$. Stop bands are shaded.

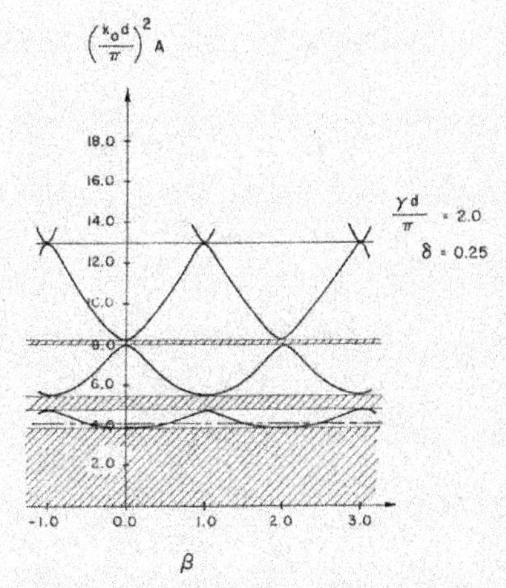

Fig. 8. Frequency as a function of β with $\delta = 0.25$. Stop bands are shaded.

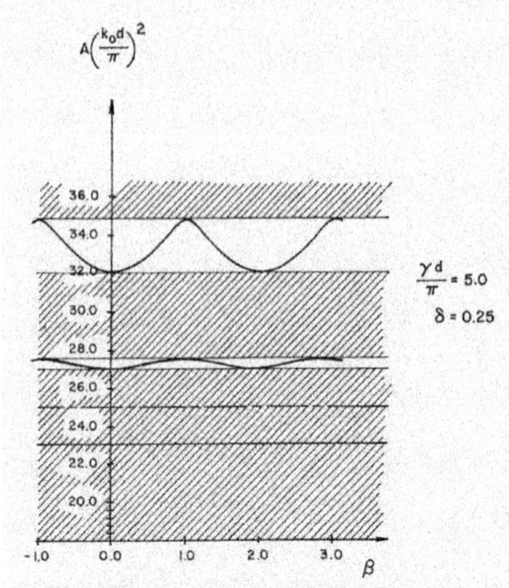

Fig. 9. Frequency as a function of β with $\delta = 0.25$. Stop bands are shaded. The dot-dash line represents the cutoff frequency of a waveguide filled with a homogeneous dielectric medium with $\epsilon = \epsilon_a$. A very narrow stable region near $A(k_0 d/\pi)^2 = 23.0$ is indicated by a solid line.

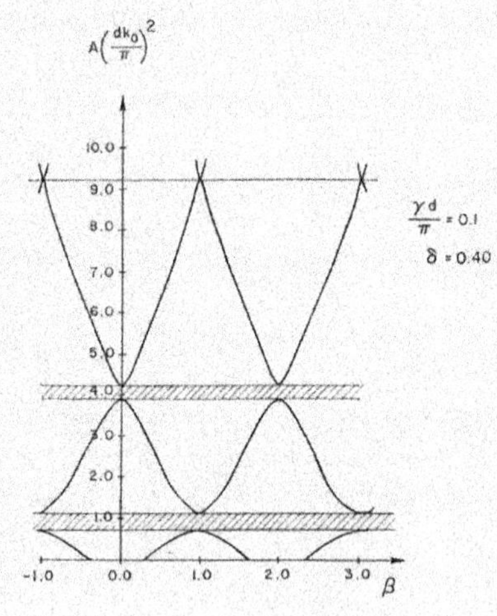

Fig. 10. Frequency as a function of β with $\delta = 0.4$. Stop bands are shaded.

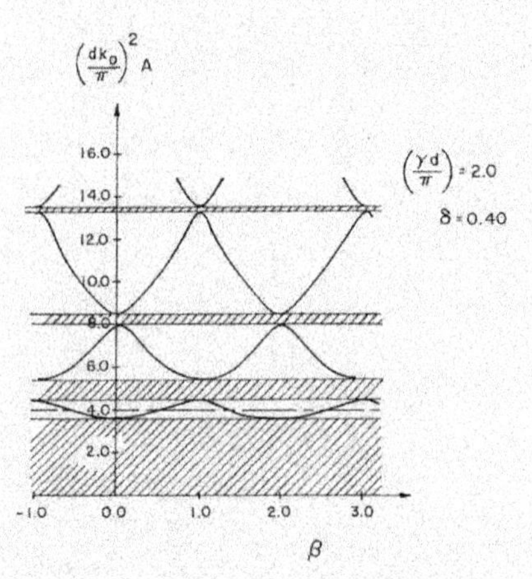

Fig. 11. Frequency as a function of β with $\delta = 0.4$. Stop bands are shaded. The dot-dash line represents the cutoff frequency of a waveguide filled with a homogeneous dielectric medium with $\epsilon = \epsilon_a$.

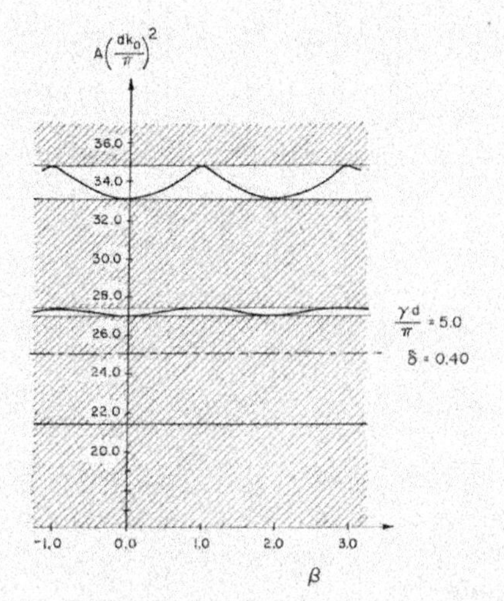

Fig. 12. Frequency as a function of β with $\delta = 0.4$. Stop bands are shaded. The dot-dash line represents the cutoff frequency of a waveguide filled with a homogeneous dielectric medium with $\epsilon = \epsilon_a$. A very narrow stable region near $A(dk_0/\pi)^2 = 21$ is indicated by a solid line.

Reprinted from
JOURNAL OF MATHEMATICAL PHYSICS VOLUME 10, NUMBER 5 MAY 1969

Wave Propagation in Sinusoidally Stratified Plasma Media*

K. F. CASEY AND J. R. MATTHES
Electrical Engineering Department, Air Force Institute of Technology, Dayton, Ohio
AND
C. YEH
Department of Engineering, University of California, Los Angeles, California

(Received 1 August 1968)

The problem of the propagation of electromagnetic waves in a sinusoidally stratified plasma media is treated analytically. The propagation characteristics of TE and TM waves are determined, respectively, from the characteristic equations of the resultant Mathieu and Hill equations. Detailed dispersion characteristics of TE and TM waves in an infinite stratified plasma medium and in waveguides filled longitudinally with this stratified dispersion material are given. It is found that, although the stop-band and pass-band structures exist for the ω–β diagrams of both TE and TM waves, detailed dispersion properties for TE and TM waves are quite different for most frequency ranges except when $[(\omega/\omega_{p0})^2 - 1]^2 \gg \delta^2$, where ω is the frequency of the propagating waves, ω_{p0} is the average plasma frequency of the inhomogeneous plasma medium, and δ is the amplitude of the sinusoidally varying term for the electron-density profile $(0 \leq \delta \leq 1)$.

I. INTRODUCTION

The problem of electromagnetic wave propagation in a sinusoidally stratified medium is not only of interest from a theoretical point of view but also possesses many possible applications. For example, a section of waveguide filled with this type of inhomogeneous dielectric may be used as a band-pass filter in the millimeter-wave or optical region. The use of an ultrasonic standing wave as a modulating device for certain pressure-sensitive media, such as carbon disulfide, pentane, or nitric acid at optical frequencies to achieve a sinusoidally varying dielectric medium may be proposed.[1] Other applications of wave interactions in periodic media, such as the deflection of laser beams,[2] the measurement of acoustic properties in crystals,[3] and Čerenkov radiation in periodically stratified media,[4] can also be found. In recent years, the problem of wave propagation in sinusoidally stratified dielectric nondispersive media has been considered in detail by several authors.[5] However, the results are not applicable for dispersive media.

The purpose of this investigation is to consider the propagation characteristics of waves in a dispersive sinusoidally varying medium. Specifically, the dispersive medium is assumed to be an inhomogeneous cold plasma with a sinusoidally varying free electron density profile.

Two types of waves may exist: one with its electric vector transverse to the direction of the inhomogeneity, called a TE wave, and the other with its magnetic vector transverse to the direction of inhomogeneity, called a TM wave. Both types of waves will be treated. Detailed dispersion characteristics of TE and TM waves in an infinite stratified plasma medium and in waveguides filled longitudinally with this inhomogeneous plasma will be presented.

It is hoped that these results will be useful in the diagnostics of plasmas[6] and in the study of wave propagation in solids.[7]

II. FORMULATION OF THE PROBLEM

It is assumed that the inhomogeneous plasma medium under consideration fills the entire space and possesses a plasma frequency

$$\omega_p^2(z) = n(z)e^2/m\epsilon_0, \tag{1}$$

where z is the axial coordinate, $n(z)$ is the free electron density, e is the electron charge, m is the electron mass, and ϵ_0 is the free-space permittivity.

The source-free vector wave equations in this medium are

$$\nabla \times \nabla \times \mathbf{E} - k_0^2(\epsilon(z)/\epsilon_0)\mathbf{E} = 0, \tag{2}$$

$$\nabla \times \nabla \times \mathbf{H} - \frac{\nabla\epsilon(z)}{\epsilon(z)} \times \nabla \times \mathbf{H} - k_0^2\left(\frac{\epsilon(z)}{\epsilon_0}\right)\mathbf{H} = 0, \tag{3}$$

* Supported partly by the Office of Naval Research and partly by the National Science Foundation.

[1] C. Yeh and Z. A. Kaprielian, "On Inhomogeneously Filled Waveguides," USCEC Dept. 84-206, Electrical Engineering Department, University of Southern California, Los Angeles, 1963.

[2] M. G. Cohn and E. I. Gordon, Bell System Tech. J. **44**, 693 (1965).

[3] H. H. Parker, E. F. Kelly, and D. I. Bulef, Appl. Phys. Letters **5**, 7 (1964).

[4] K. F. Casey, C. Yeh, and Z. A. Kaprielian, Phys. Rev. **140**, B768 (1965).

[5] T. Tamir, H. C. Wang, and A. A. Oliner, IEEE Trans. Microwave Theory Tech., MT **12**, 323 (1964).

[6] M. A. Heald and C. B. Wharton, *Plasma Diagnostics with Microwaves* (John Wiley & Sons, Inc., New York, 1965).

[7] L. Brillouin, *Wave Propagation in Periodic Structures* (Dover Publications, Inc., New York, 1953).

where **E** and **H** are, respectively, the electric and magnetic field vectors, $k_0^2 = \omega^2\mu_0\epsilon_0$, and a time dependence $e^{-i\omega t}$ is assumed. μ_0 is the free-space permeability. The dielectric permittivity of the inhomogeneous cold plasma is related to the plasma frequency by the following relation:

$$\epsilon(z) = \epsilon_0[1 - \omega_p^2(z)/\omega^2]. \tag{4}$$

It can be shown that all field components in this medium can be obtained from the scalar quantities $\Phi(x, y, z)$ and $\Psi(x, y, z)$ as follows[1]:

$$\mathbf{E}^{(m)} = \nabla \times [\Phi(x, y, z)\mathbf{e}_z], \tag{5}$$

$$\mathbf{H}^{(m)} = (-i/\omega\mu_0)\nabla \times \nabla \times [\Phi(x, y, z)\mathbf{e}_z], \tag{6}$$

for transverse electric waves, and

$$\mathbf{H}^{(e)} = \nabla \times [\Psi(x, y, z)\mathbf{e}_z], \tag{7}$$

$$\mathbf{E}^{(e)} = (i/\omega\epsilon(z))\nabla \times \nabla \times [\Psi(x, y, z)\mathbf{e}_z], \tag{8}$$

for transverse magnetic waves. $\mathbf{e}_z$ is the unit vector in the z direction. Upon substituting (5) into (3) and (7) into (4), carrying out the vector operations, and separating variables in rectangular coordinates, one obtains

$$\Phi(x, y, z) = \begin{Bmatrix}\sin\\ \cos\end{Bmatrix}(sx)\begin{Bmatrix}\sin\\ \cos\end{Bmatrix}(wy)\Bigg\}U^{(1),(2)}(z), \tag{9}$$

$$\Psi(x, y, z) = \begin{Bmatrix}\sin\\ \cos\end{Bmatrix}(px)\begin{Bmatrix}\sin\\ \cos\end{Bmatrix}(qy)\Bigg\}V^{(1),(2)}(z), \tag{10}$$

where s, w, p, and q are separation constants. $U^{(1),(2)}(z)$ and $V^{(1),(2)}(z)$ satisfy, respectively, the differential equations

$$\left\{\frac{d^2}{dz^2} + \left[k_0^2\left(\frac{\epsilon(z)}{\epsilon_0}\right) - s^2 - w^2\right]\right\}U^{(1),(2)}(z) = 0 \tag{11}$$

and

$$\left\{\frac{d^2}{dz^2} - \left(\frac{d\epsilon(z)}{dz}\right)\frac{1}{\epsilon(z)}\frac{d}{dz} + \left[k_0^2\left(\frac{\epsilon(z)}{\epsilon_0}\right) - p^2 - q^2\right]\right\} \times V^{(1),(2)}(z) = 0. \tag{12}$$

Introducing the dimensionless variable $\xi = \pi z/d$, where d has the dimension of length and will be defined later, it can be shown that Eqs. (11) and (12) can be put into the following form:

$$\left[\frac{d^2}{d\xi^2} + \lambda(\xi)\right]W^{(1),(2)}(\xi) = 0, \tag{13}$$

where

$$\lambda(\xi) = \frac{d^2}{\pi^2}\left[k_0\left(\frac{\epsilon(\xi)}{\epsilon_0}\right) - s^2 - w^2\right] \tag{14}$$

if

$$W^{(1),(2)}(\xi) = U^{(1),(2)}(\xi), \tag{15}$$

and

$$\lambda(\xi) = \frac{1}{\epsilon^2(\xi)}\left\{[\epsilon^{\frac{1}{2}}(\xi)]'' + \frac{\epsilon'(\xi)}{\epsilon(\xi)}[\epsilon^{\frac{1}{2}}(\xi)]' + \frac{d^2}{\pi^2}\left[k_0\left(\frac{\epsilon(\xi)}{\epsilon_0}\right) - p^2 - q^2\right]\epsilon^{\frac{1}{2}}(\xi)\right\} \tag{16}$$

if

$$W^{(1),(2)}(\xi) = \epsilon^{-\frac{1}{2}}(\xi)V^{(1),(2)}(\xi). \tag{17}$$

The primes indicate the derivative of the function with respect to ξ.

III. THE SINUSOIDALLY STRATIFIED PLASMA MEDIUM

If the free electron density distribution is assumed to have a sinusoidal stratification as follows:

$$n(z) = n_0\left(1 - \delta\cos\frac{2\pi z}{d}\right), \tag{18}$$

where n_0 is the average electron density, δ is a known constant with $0 \leq \delta \leq 1$, and d denotes the period of the sinusoidal variation, then the dielectric constant of this plasma medium is

$$\epsilon(\xi) = \epsilon_0\left\{\left(1 - \frac{\omega_{p0}^2}{\omega^2}\right) + \frac{\omega_{p0}^2}{\omega^2}\delta\cos 2\xi\right\} \tag{19}$$

and

$$\omega_{p0}^2 = n_0 e^2/m\epsilon_0.$$

Substituting Eq. (19) into (14) gives

$$\lambda^{\mathrm{TE}}(\xi) = \left(\frac{k_0 d}{\pi}\right)^2 - \left(\frac{k_{p0} d}{\pi}\right)^2 - \left(\frac{sd}{\pi}\right)^2 - \left(\frac{wd}{\pi}\right)^2 + \left(\frac{k_{p0} d}{\pi}\right)^2\delta\cos 2\xi, \tag{20}$$

with

$$k_{p0}^2 = \omega_{p0}^2\mu_0\epsilon_0,$$

for the transverse electric wave; substituting Eq. (19) into (16) gives

$$\lambda^{\mathrm{TM}}(\xi) = -\frac{2(\omega_{p0}/\omega)^2\delta\cos 2\xi}{1 - (\omega_{p0}/\omega)^2(1 - \delta\cos 2\xi)} - \frac{3(\omega_{p0}/\omega)^2\delta^2\sin^2 2\xi}{[1 - (\omega_{p0}/\omega)^2(1 - \delta\cos 2\xi)]^2} + \left(\frac{k_0 d}{\pi}\right)^2\left[1 - \left(\frac{\omega_{p0}}{\omega}\right)^2(1 - \delta\cos 2\xi)\right] - \left(\frac{pd}{\pi}\right)^2 - \left(\frac{qd}{\pi}\right)^2, \tag{21}$$

for the transverse magnetic wave. Since, according to Eqs. (20) and (21), $\lambda^{\mathrm{TE}}(\xi)$ or $\lambda^{\mathrm{TM}}(\xi)$ is an even periodic function, it can, therefore, be represented by a

Fourier cosine series

$$\lambda^{TE,TM}(\xi) = \theta_0^{TE,TM} + 2\sum_{n=1}^{\infty}\theta_n^{TE,TM}\cos 2n\xi, \quad (22)$$

where

$$\theta_0^{TE} = \left(\frac{k_0 d}{\pi}\right)^2 - \left(\frac{k_{p0} d}{\pi}\right)^2 - \left(\frac{sd}{\pi}\right)^2 - \left(\frac{wd}{\pi}\right)^2, \quad (23a)$$

$$\theta_1^{TE} = \tfrac{1}{2}\delta(k_{p0}d/\pi)^2, \quad (23b)$$

$$\theta_n^{TE} = 0, \quad n \geq 2, \quad (23c)$$

and

$$\theta_0^{TM} = \left(\frac{k_0 d}{\pi}\right)^2 - \left(\frac{k_{p0} d}{\pi}\right)^2 - \left(\frac{pd}{\pi}\right)^2 - \left(\frac{qd}{\pi}\right)^2 - \left[\frac{1}{(1-\Lambda^2)^{\frac{1}{2}}} - 1\right], \quad (24a)$$

$$\theta_1^{TM} = \tfrac{1}{2}\delta\left(\frac{k_{p0} d}{\pi}\right)^2 + \frac{4b^3 - 2b}{b^2 - 1}, \quad (24b)$$

$$\theta_n^{TM} = \frac{(3n+1)b^{n+2} - (3n-1)b^n}{b^2 - 1}, \quad n \geq 2, \quad (24c)$$

with

$$b = \Lambda^{-1} - \Lambda^{-1}(1 - \Lambda^2)^{\frac{1}{2}}, \quad (24d)$$

$$\Lambda = \delta/(1 - \omega^2/\omega_{p0}^2). \quad (24e)$$

It should be noted that the above Fourier series representation for the TM case converges absolutely when

$$0 \leq |\delta/(1 - \omega^2/\omega_{p0}^2)| < 1 \quad (25)$$

with $0 \leq \delta \leq 1$.

IV. SOLUTIONS OF HILL'S EQUATION

Substituting (22) into (13), one obtains

$$\left[\frac{d^2}{d\xi^2} + \theta_0^{TE,TM} + 2\sum_{n=1}^{\infty}\theta_n^{TE,TM}\cos 2n\xi\right]W^{(1),(2)}(\xi) = 0 \quad (26)$$

which is the general form of Hill's equation.[8,9] It is known that two types of solutions of Hill's equations exist: one called the stable type and the other called the unstable type. In order to have propagating waves in the z direction, only the stable type is allowed.

With the help of Floquet's Theorem,[7–9] the solutions of Hill's equation can be expressed in the following form:

$$W^{(1),(2)}(\xi) = e^{\pm i\nu\xi}\sum_{n=-\infty}^{\infty} C_n(\nu)e^{\pm 2in\xi}, \quad (27)$$

where ν, the characteristic exponent, and $C_n(\nu)$ are as yet unknown. After substituting (27) into (26) and simplifying, one obtains the following recursion relations:

$$-(\nu + 2n)^2 C_n + \sum_{m=-\infty}^{\infty}\theta_m^{TE,TM}C_{n-m} = 0,$$

$$n = \cdots -2, -1, 0, 1, 2, \cdots, \quad (28)$$

with $\theta_{-m}^{TE,TM} = \theta_m^{TE,TM}$. It is understood that $\nu = \nu^{TE}$ and $C_n = C_n^{TE}$ when θ_m^{TE} are used, while $\nu = \nu^{TM}$ and $C_n = C_n^{TM}$ when θ_m^{TM} are used. Equation (28) is a set of an infinite number of homogeneous linear algebraic equations in C_n. For a nontrivial solution to exist the characteristic number ν and the coefficients θ_m must satisfy the characteristic equation of the Hill equation[9]:

$$\sin^2\frac{\pi\nu}{2} = \Delta(0)\sin^2\frac{\pi(\theta_0^{TE,TM})^{\frac{1}{2}}}{2}. \quad (29)$$

$\Delta(0)$ is the determinant of the matrix $[M]$ whose elements are

$$M_{mm} = 1,$$

$$M_{mn} = \frac{-\theta_{m-n}^{TE,TM}}{4m^2 - \theta_0^{TE,TM}}, \quad m \neq n. \quad (30)$$

The characteristic number ν can be obtained from (29).

Real values of ν yield stable solutions to Hill's equation, while complex values of ν produce unstable solutions. Physically speaking the stable solutions correspond to modulated propagating waves, and the unstable solutions correspond to damped or growing waves. For the present problem, the fields for the growing waves do not satisfy the radiation condition at infinity, hence they must be omitted.

Numerical computation has been carried out for (29). The values for the infinite determinant $\Delta(0)$ were obtained by the successive approximation method.[10] In other words, computations were carried out for a 3×3 determinant, a 5×5 determinant, etc., until the desired accuracy was reached. It was found (numerically) that the infinite determinant converges quite rapidly within the present region of interest. For example, for small values of δ and k_{p0}, such that $\delta \leq 0.1$ and $(k_{p0}d/\pi)^2 \leq 0.5$, at no time was any determinant greater than 7×7 required to achieve an accuracy of three significant figures. For large values of δ and k_{p0}, no determinants of order greater than 15×15 were required to obtain the desired accuracy.

V. PROPAGATION CHARACTERISTICS OF TE WAVES

Returning now to the problem of obtaining the propagation characteristics of waves in a sinusoidally

[8] J. Meixner and F. W. Schäfke, *Mathieusche Funktionen und Sphäroidfunktionen* (Springer-Verlag, Berlin, 1954).

[9] P. M. Morse and H. Feshbach, *Methods of Theoretical Physics* (McGraw-Hill Book Co., Inc., New York, 1953).

[10] L. Kantorovich and V. Krylov, *Approximate Methods of Higher Analysis* (Interscience Publishers, New York, 1958).

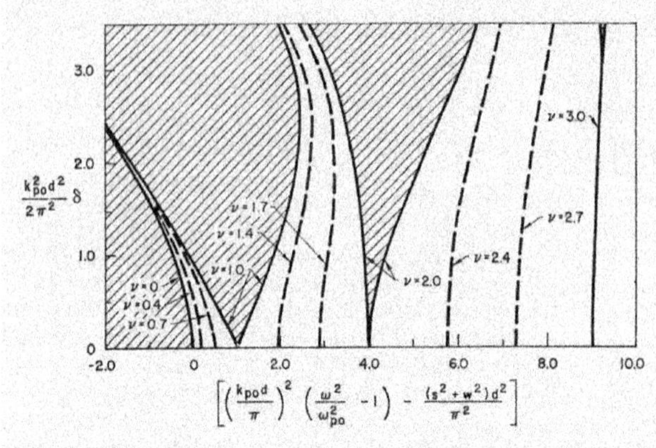

FIG. 1. Stability chart for Mathieu's equation from which the propagation characteristics of TE waves may be determined. Unstable regions are shaded.

stratified plasma medium, we note, upon substituting Eq. (22) for TE waves into Eq. (13), that the resultant differential equation for TE waves is the Mathieu equation, a special case of Hill's equation. It is customary to express the results of computation for the characteristic exponents of Mathieu functions in terms of a "stability diagram."[9,11] Figure 1 shows the stability diagram which was obtained from previously tabulated values for the characteristic exponents of Mathieu functions. The unshaded areas are the "stable regions" wherein ν is purely real; the shaded areas are the "unstable regions" wherein ν is complex.

A. Infinite Region Filled With Sinusoidally Stratified Plasma

The transverse electric field components of a TE wave in an infinite medium filled with sinusoidally stratified plasma can be obtained from Eqs. (5) and (6):

$$E_x^{TE}=\sum_{n=-\infty}^{\infty} iwC_n^{TE}e^{isx}e^{iwy}e^{i(\nu+2n)\pi z/d}, \quad (31)$$

$$E_y^{TE}=\sum_{n=-\infty}^{\infty} -isC_n^{TE}e^{isx}e^{iwy}e^{i(\nu+2n)\pi z/d}, \quad (32)$$

where the coefficients C_n^{TE} can be determined from Eq. (28) in terms of C_0^{TE}. The propagation constant β is related to ν by the equation $\beta=\nu\pi/d$. C_0^{TE} is obtained from a normalization condition. All magnetic field components may be found from Maxwell's equations.

Unlike the case of a TE wave propagating in an infinite homogeneous plasma in which β is simply related to s and w by the following:

$$\beta^2=\omega^2\mu_0\epsilon_0(1-\omega_p^2/\omega^2)-(s^2+w^2), \quad (33)$$

where ω_p is the plasma frequency of the homogeneous plasma, the propagation constant β for the inhomogeneous case is related to s and w through the stability diagram given by Fig. 1. Real values of β as a function of real values of (s^2+w^2) for fixed values of δ, $k_0 d$, and $k_{p0}d$ are shown in Fig. 2. It is recalled that complex values of β indicate the presence of damped waves (i.e., nonpropagating waves). s and w are taken to be real. The unshaded regions in these figures indicate the regions in which β is real (i.e., regions in which propagating waves may exist). Equation (33) is also plotted in Fig. 2 with $\omega_p=\omega_{p0}$. As one can see, the propagation characteristics of waves in the inhomogeneous plasma are significantly modified from those in the homogeneous plasma. The presence of stop-band and pass-band regions in Fig. 2 is characteristic of wave propagation in periodic structures.

B. Waveguide Filled Longitudinally With Sinusoidally Stratified Plasma

It is assumed that a rectangular waveguide of dimensions h_1 and h_2 is filled completely with an

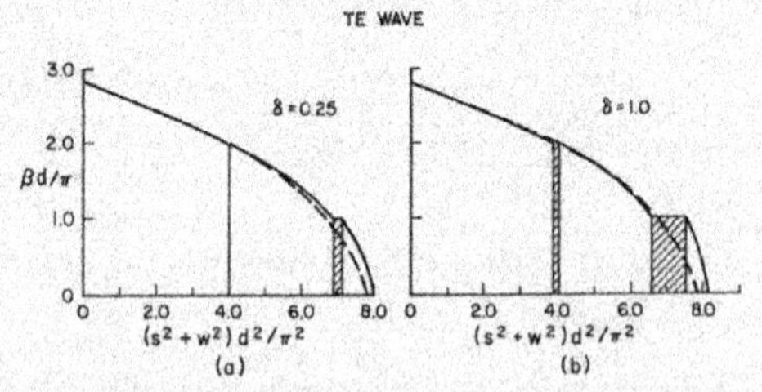

FIG. 2. The propagation constant $\beta d/\pi$ as a function of $(s^2+w^2)d^2/\pi^2$ with $(k_{p0}d/\pi^2)=1.0$ and $(k_0 d/\pi)^2=9.0$. Stop bands are shaded. The dashed line represents the behavior of the propagation constant for an equivalent homogeneous plasma medium with $\omega_p=\omega_{p0}$ (TE wave).

[11] T. Tamir, Math. Computation 16, 100 (1962).

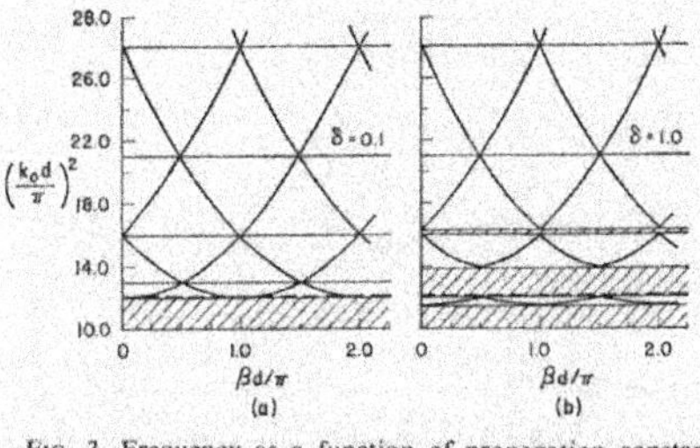

FIG. 3. Frequency as a function of propagation constant with $(k_{p0}d/\pi)^2 = 2.0$ and $(\gamma d/\pi)^2 = 10.0$. Stopbands are shaded. Dashed line indicates the cutoff frequency for a waveguide filled with an equivalent homogeneous plasma medium with $\omega_p = \omega_{p0}$ (TE wave).

inhomogeneous plasma medium, which varies sinusoidally in the longitudinal direction. The general expressions for the transverse electric field components of a TE wave are

$$E_x^{\rm TE} = \sum_{m=1}^{\infty}\sum_{r=1}^{\infty}\sum_{n=-\infty}^{\infty} - C_n^{\rm TE}\frac{r\pi}{h_2} \times \cos\frac{m\pi x}{h_1}\sin\frac{r\pi y}{h_2}e^{i(\nu+2n)\pi z/d}, \quad (34)$$

$$E_y^{\rm TE} = \sum_{m=1}^{\infty}\sum_{r=1}^{\infty}\sum_{n=-\infty}^{\infty} C_n^{\rm TE}\frac{m\pi}{h_1} \times \sin\frac{m\pi x}{h_1}\cos\frac{r\pi y}{h_2}e^{i(\nu+2n)\pi z/d}, \quad (35)$$

with $\beta = \nu\pi/d$. The $C_n^{\rm TE}$ are arbitrary constants and $C_n^{\rm TE}$ can be determined from Eq. (28) in terms of $C_0^{\rm TE}$. Expressions for the magnetic field components can easily be derived from Maxwell's equations.

The dispersion relations expressed in terms of the ω–β diagrams can be found from Fig. 1 for fixed values of δ, $k_{p0}d$, and $s^2 + w^2$, where

$$(\gamma^{\rm TE})^2 = s^2 + w^2 = \left(\frac{m\pi}{h_1}\right)^2 + \left(\frac{r\pi}{h_2}\right)^2. \quad (36)$$

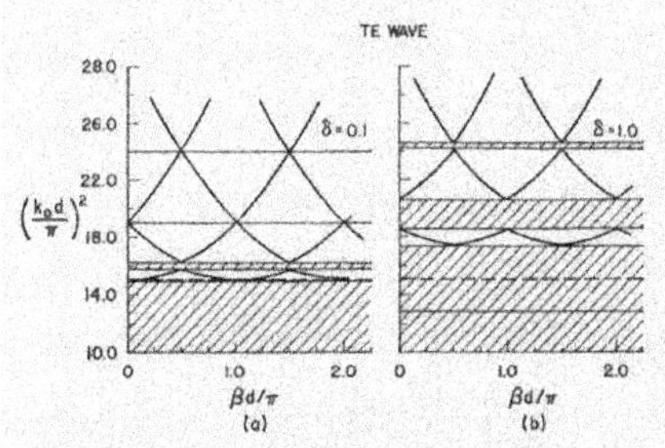

FIG. 4. Frequency as a function of propagation constant with $(k_{p0}d/\pi)^2 = 5.0$ and $(\gamma d/\pi)^2 = 10.0$. Stop bands are shaded. Dashed line indicates the cutoff frequency for a waveguide filled with an equivalent homogeneous plasma medium with $\omega_p = \omega_{p0}$ (TE wave).

The ω–β diagrams are given for $\delta = 0.1$, 1.0, $(k_{p0}d/\pi)^2 = 2.0$, 5.0 and $(s^2 + w^2)(d/\pi)^2 = 5.0$, 10.0 in Figs. 3 and 4. The pass-band and stop-band characteristics can clearly be seen. Furthermore, an increase in δ results in a corresponding increase in the bandwidth of the stop bands. Also shown in these figures is the cutoff frequency for an identical waveguide filled with homogeneous plasma which has a plasma frequency equal to ω_{p0}. It is noted that the cutoff frequency for the equivalent homogeneous case is above the cutoff frequency of the first pass band for the inhomogeneous case.

VI. PROPAGATION CHARACTERISTICS OF TM WAVES

Unlike the TE case, the TM case requires the solution of the more complex Hill equation. It is more straightforward to obtain the propagation characteristics of TM waves directly from Eq. (29) than to use the stability diagram, as described earlier for TE waves.

It is recalled that, in order that Eq. (12) for TM waves may be put in the form of the Hill equation, a very important limitation on the ratio ω/ω_{p0} must be satisfied, i.e., from Eq. (25),

$$0 \le \frac{1}{|1 - \omega^2/\omega_{p0}^2|} < \frac{1}{\delta} \quad (37)$$

with $0 \le \delta \le 1$. A sketch of the above relation is given in Fig. 5. The shaded area indicates the forbidden region in which the relation (37) is not satisfied. It was found numerically that, for moderate values of $(p^2 + q^2)(d/\pi)^2$, the lower cutoff frequency of the first pass band occurs at frequencies considerably above ω_{p0}.

A. Infinite Region Filled With Sinusoidally Stratified Plasma

The transverse magnetic-field components of a TM wave in an infinite medium filled with sinusoidally stratified plasma can be obtained from Eqs. (7)

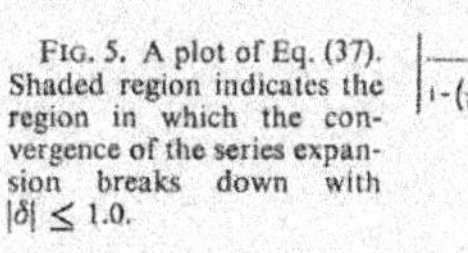
FIG. 5. A plot of Eq. (37). Shaded region indicates the region in which the convergence of the series expansion breaks down with $|\delta| \le 1.0$.

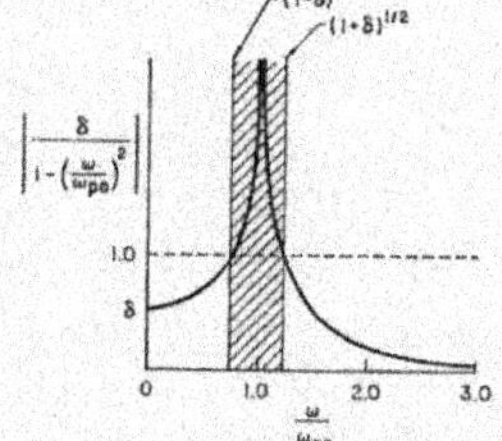

and (8):

$$H_x^{\rm TM} = \sum_{n=-\infty}^{\infty} iqC_n^{\rm TM} e^{ipx} e^{iqy} e^{i(\nu+2n)\pi z/d} \times \left[1 - \left(\frac{\omega_{p0}}{\omega}\right)^2 \left(1 - \delta \cos \frac{2\pi z}{d}\right)\right]^{\frac{1}{2}}, \quad (38)$$

$$H_y^{\rm TM} = \sum_{n=-\infty}^{\infty} - ipC_n^{\rm TM} e^{ipx} e^{iqy} e^{i(\nu+2n)\pi z/d} \times \left[1 - \left(\frac{\omega_{p0}}{\omega}\right)^2 \left(1 - \delta \cos \frac{2\pi z}{d}\right)\right]^{\frac{1}{2}}, \quad (39)$$

with $\beta = \nu\pi/d$. The coefficients $C_n^{\rm TM}$ can be determined from Eq. (28) in terms of $C_0^{\rm TM}$. The electric field components can be found from Maxwell's equations.

Real values of β as a function of real values of $(p^2 + q^2)$ for fixed values of δ, k_0d, and $k_{p0}d$ are shown in Fig. 6. Again, the pass-band and stop-band features are apparent in this figure. Results are rather similar to those given in Fig. 2 for TE wave, although quantitatively the results are different.

B. Waveguide Filled Longitudinally With Sinusoidally Stratified Plasma

The general expressions for the transverse magnetic-field components of a TM wave in a rectangular waveguide of dimensions h_1 and h_2 filled longitudinally with an inhomogeneous plasma medium are

$$H_x^{\rm TM} = \sum_{m=1}^{\infty} \sum_{r=1}^{\infty} \sum_{n=-\infty}^{\infty} C_n^{\rm TM} \sin \frac{m\pi x}{h_1} \cos \frac{r\pi y}{h_2} e^{-(\nu+2n)\pi z/d} \times \frac{r\pi}{h_2}\left[1 - \left(\frac{\omega_{p0}}{\omega}\right)^2 \left(1 - \delta \cos \frac{2\pi z}{d}\right)\right]^{\frac{1}{2}}, \quad (40)$$

$$H_y^{\rm TM} = \sum_{m=1}^{\infty} \sum_{r=1}^{\infty} \sum_{n=-\infty}^{\infty} C_n^{\rm TM} \cos \frac{m\pi x}{h_1} \sin \frac{r\pi y}{h_2} e^{i(\nu+2n)\pi z/d} \times \left(-\frac{m\pi}{h_1}\right)\left[1 - \left(\frac{\omega_{p0}}{\omega}\right)^2 \left(1 - \delta \cos \frac{2\pi z}{d}\right)\right]^{\frac{1}{2}}, \quad (41)$$

where the coefficients $C_n^{\rm TM}$ can be determined from Eq. (28) in terms of $C_0^{\rm TM}$. The electric-field components can be found from Maxwell's equations.

Solving Eq. (29) for fixed values of δ, $k_{p0}d$, and $(p^2 + q^2)$, where

$$(\gamma^{\rm TM})^2 = p^2 + q^2 = \left(\frac{m\pi}{h_1}\right)^2 + \left(\frac{r\pi}{h_2}\right)^2,$$

yields the propagation constant β as a function of ω as shown in Figs. 7 and 8. Again, the propagation characteristics of TM and TE waves are similar. However, the bandwidths, center frequencies of the stop bands, and the waveguide cutoff frequencies vary considerably for the two modes. Also noted from the definitions (23) and (24) is that $\theta_n^{\rm TM} \simeq \theta_n^{\rm TE}$ provided that

$$\frac{\delta^2}{[(\omega/\omega_{p0})^2 - 1]^2} \to 0.$$

In other words, if $[(\omega/\omega_{p0})^2 - 1]^2 \gg \delta^2$, the propagation characteristics of TE and TM waves are almost the same. Otherwise, the results for TE and TM waves will be quite different. For example, in the lower frequency ranges, the bandwidths of the stop bands for the TM waves are greater than the bandwidths of

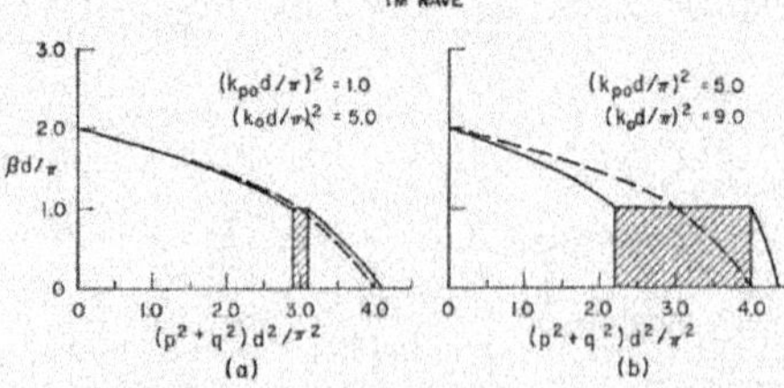

FIG. 6. The propagation constant $\beta d/\pi$ as a function of $(p^2 + q^2) d^2/\pi^2$ with $\delta = 0.25$. Stop bands are shaded. The dashed line represents the behavior of the propagation constant for an equivalent homogeneous plasma medium with $\omega_p = \omega_{p0}$ (TM wave).

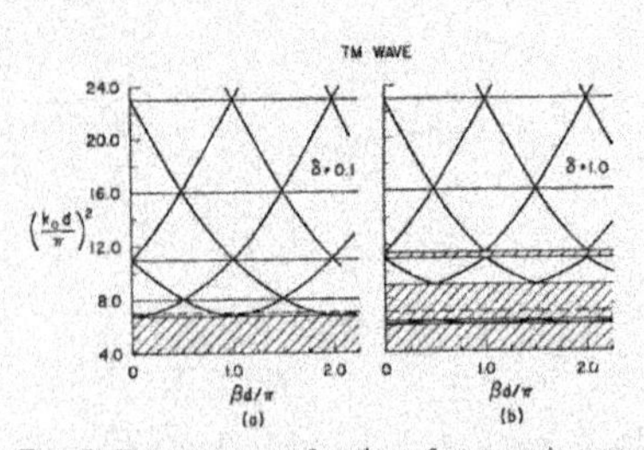

FIG. 7. Frequency as a function of propagation constant with $(k_{p0}d/\pi)^2 = 2.0$ and $(\gamma d/\pi)^2 = 5.0$. Stop bands are shaded. Dashed line indicates the cutoff frequency for a waveguide filled with an equivalent homogeneous plasma with $\omega_p = \omega_{p0}$ (TM wave).

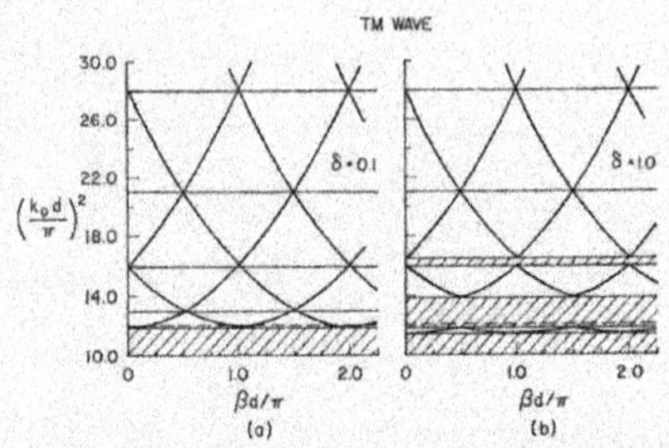

FIG. 8. Frequency as a function of propagation constant with $(k_{p0}d/\pi)^2 = 2.0$ and $(\gamma d/\pi)^2 = 10.0$. Stop bands are shaded. Dashed line indicates the cutoff frequency for a waveguide filled with an equivalent homogeneous plasma with $\omega_p = \omega_{p0}$ (TM wave).

the corresponding stop bands for the TE waves, and the waveguide cutoff frequencies for the TM waves are lower than those of the TE waves. (Compare Fig. 8 with Fig. 3.) As the frequency is increased beyond the second pass band, the bandwidths of the stop bands are approximately the same.

The ω–β diagrams given by Figs. 3 and 4 for TE waves and Figs. 7 and 8 for TM waves are also applicable for waveguides of arbitrary cross-section. For example, in a circular waveguide of radius ρ_0, $\gamma^{TE,TM}$ is given by $\Gamma_{mr}^{TE,TM}/\rho_0$, where $\Gamma_{mr}^{TE,TM}$ are, respectively, the rth roots of the equations

$$J'_m(\Gamma_{mr}^{TE}) = 0$$

and

$$J_m(\Gamma_{mr}^{TM}) = 0.$$

Periodic structures in integrated optics*

C. Elachi and C. Yeh

Electrical Sciences and Engineering Department, University of California, Los Angeles, California 90024
and
Jet Propulsion Laboratory, California Institute of Technology, Pasadena, California 91109
(Received 11 January 1973; in final form 12 March 1973)

Thin-film dielectric waveguides with a periodic refractive index, a periodic substrate, or periodic surface are studied. The field is determined from Maxwell's equations using Floquet's theorem. The Brillouin diagram and the interaction regions are investigated. The bandwidth and the attenuation coefficients of the interaction regions are given as a function of the optical wavelength. A number of applications in active and passive integrated optics systems are discussed.

I. INTRODUCTION

The recent introduction of the concept of integrated optics[1-3] has stimulated a great deal of interest in the design of thin-film miniaturized optical systems. In the last three years, many experiments have been performed in light generation, waveguiding, coupling, deflection, parametric interactions, and others in thin-film structures.[1-4]

Periodic dielectric structures can be used in many integrated optical devices. The periodicity could be generated either by chemical processes, by ion milling, by volume or surface acoustic waves, by electro-optical periodic systems, or by photodimer memories.[5-9] Such structures could be used in DFB (distributed feed-back) lasers,[9,10] integrated optics filter, frequency-selective couplers, and phase-matchable nonlinear interactions.[8]

In this paper we shall study the waveguiding properties of three important types of periodic structures: periodically inhomogeneous thin-film waveguide [Fig. 1(a)], periodically inhomogeneous substrate guide [Fig. 1(b)], and thin-film waveguide with periodic surfaces (periodic thickness) [Fig. 1(c)].

We shall first obtain the general solutions of the wave equation in sinusoidally stratified media. The solutions will then be applied to three specific periodic structure problems as mentioned earlier. Many possible applications of this type of periodic structure to passive and active integrated optics systems will be discussed.

II. SOLUTION OF THE WAVE EQUATION IN PERIODICALLY STRATIFIED DIELECTRIC MEDIA

The source-free wave vector equations in a medium whose permittivity is a function of z, the axial coordinate, are[11]

$$\nabla\times\nabla\times\mathbf{E}-k_0^2[\epsilon(z)/\epsilon_0]\mathbf{E}=0, \qquad (1)$$

$$\nabla\times\nabla\times\mathbf{H}-[\nabla\epsilon(z)/\epsilon(z)]\times\nabla\times\mathbf{H}-k_0^2[\epsilon(z)/\epsilon_0]\mathbf{H}=0, \qquad (2)$$

where $k_0=\omega(\mu_0\epsilon_0)^{1/2}$, ϵ_0 and μ_0 are, respectively, the free-space permittivity and permeability, $\mathbf{E}$ and $\mathbf{H}$ are, respectively, the electric and magnetic field vectors, a time dependence $\exp(-i\omega t)$ is assumed and suppressed throughout, and $\epsilon(z)$ is the dielectric permittivity of the inhomogeneous medium. Let $\epsilon(z)$ be a periodically varying function of z:

$$\epsilon(z)=\epsilon[1+\eta f(Kz)],$$

where η and K are known constants and $f(\xi)$ is a periodic function. The solution of the wave equation consists of an infinite number of space harmonics (Floquet form). For the TE wave, we can write (for propagation in the xz plane)

$$\mathbf{E}^{(TE)}=\mathbf{e}_y\sum_{n=-\infty}^{n=+\infty}C_n^{TE}\begin{pmatrix}\sin(s_nx)\\ \cos(s_nx)\end{pmatrix}\exp(i\kappa_nz); \qquad (3)$$

and for the TM wave,

$$\mathbf{H}^{(TM)}=\mathbf{e}_y\sum_{n=-\infty}^{n=+\infty}C_n^{TM}\begin{pmatrix}\sin(p_nx)\\ \cos(p_nx)\end{pmatrix}\exp(i\kappa_n'z); \qquad (4)$$

where η is an integer, s_n and p_n are separation constants, C_n^{TE} and C_n^{TM} are the amplitudes of the space harmonics, and κ_n and κ_n' are the propagation constants

$$\kappa_n=\kappa+nK,$$

$$\kappa_n'=\kappa'+nK,$$

where κ and κ' may be determined from the dispersion relation and the boundary conditions of various problems that we shall consider.

Replacing $\mathbf{E}$ and $\mathbf{H}$ by their Floquet form in Eqs. (1) and (2), replacing $\epsilon(z)$ by its exponential Fourrier series, and equating the terms with the same z dependence, we obtain the following set of infinite equations:

$$(\kappa_n^2+s_n^2)C_n^{TE}-\frac{k_0^2}{\epsilon_0}\sum_{m=-\infty}^{m=+\infty}a_{n\mp m}C_m^{TE}=0, \qquad (5)$$

$$(\kappa_n'^2+p_n^2)C_n^{TM}-\frac{k_0^2}{\epsilon_0}\sum_{m=-\infty}^{m=+\infty}L_{nm}C_m^{TM}=0, \qquad (6)$$

$$L_{nm}=a_{n\mp m}\pm\frac{\epsilon_0K}{k_0^2}\sum_{l=-\infty}^{l=+\infty}la_la'_{n-l\mp m},$$

where a_j and a_j' are the Fourier series coefficients of $\epsilon(z)$ and $1/\epsilon(z)$. The solution of the above equations gives the relative amplitudes C_n^{TE}/C_0^{TE} and C_n^{TM}/C_0^{TM}. C_0^{TE} and C_0^{TM} can be determined from the source condition (or the excitation condition).[12] The nontriviality condition gives the dispersion relation.

III. THIN-FILM LONGITUDINALLY INHOMOGENEOUS OPTICAL WAVEGUIDE

In this section we shall consider the problem of the propagation of a transverse electric wave (TE) along a thin-film optical waveguide where the permittivity varies as [Fig. 1(a)]

$$\epsilon(z)=\epsilon_1[1+\eta\cos Kz],$$

and where the thickness is $2L$. This waveguide is sub-

FIG. 1. Geometry of the periodic structures. In (d), $X(z)$ is the surface impedance.

merged in a uniform medium of dielectric constant ϵ_2 with $\epsilon_2 < \epsilon_1$.

Substituting Eq. (3) into Eq. (1) gives the electric field component of the TE wave $[\mathbf{E}(x,z) = E_y(x,z)\mathbf{e}_y]$ within the inhomogeneous layer:

$$E_y = \sum_{n=-\infty}^{n=+\infty} A_n u_n(x) \exp(i\kappa_n z); \tag{7}$$

where $u_n(x) = \cos(s_n x)$ (for even mode) or $\sin(s_n x)$ (for odd mode), and A_n are yet unknown amplitudes of the space harmonics. Since the fields are independent of y, all other components of the electric field are zero. In the surrounding medium, we have for the guided waves

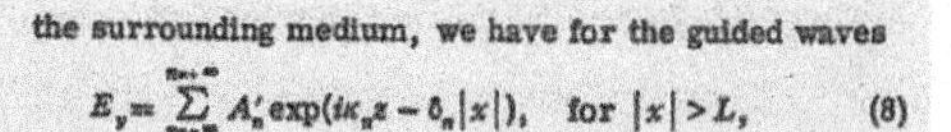

$$E_y = \sum_{n=-\infty}^{n=+\infty} A'_n \exp(i\kappa_n z - \delta_n |x|), \quad \text{for } |x| > L, \tag{8}$$

where A'_n are yet unknown amplitudes of the space harmonics and δ_n is the transverse wave number. The magnetic field components may be found from Maxwell's equations.

The boundary conditions at $|x| = L$ demand the continuity of the tangential electric and magnetic field. Hence, we have

$$\begin{aligned} A_n u_n(L) &= A'_n \exp(-\delta_n L), \\ A_n \frac{du_n}{dx}\bigg|_{x=L} &= -A'_n \delta_n \exp(-\delta_n L). \end{aligned} \tag{9}$$

Simplifying,

$$\begin{aligned} \delta_n &= s_n \tan s_n L \qquad \text{(even modes)} \\ &= s_n(-\cot a n s_n L) \quad \text{(odd modes)}. \end{aligned} \tag{10}$$

From the wave equation, one has

$$\kappa_n^2 - \delta_n^2 = \mu_0 \epsilon_2 \omega^2 \tag{11}$$

and

$$D_n A_n u_n(x) + A_{n+1} u_{n+1}(x) + A_{n-1} u_{n-1}(x) = 0, \tag{12}$$

where

$$D_n = \frac{2}{\eta}\left(1 - \frac{s_n^2 + \kappa_n^2}{\mu_0 \epsilon_1 \omega^2}\right).$$

To simplify the notation, we did not include the index corresponding to different possible waveguide modes. Equation (12) cannot be satisfied for all x, but as we will only study the interaction regions of identical modes in the limit η small, the $u_n(x)$ will be the same for the two phase-matched space harmonics.

When $\eta = 0$, Eq. (12) gives $\eta D_n = 0$; hence,

$$\kappa_n^2 + s_n^2 = \mu_0 \epsilon_1 \omega^2 \tag{13}$$

and the resulting Brillouin diagram $\omega(\kappa)$ consists of an infinite number of identical curves centered at $\kappa = -nK$. Each curve is identical to the well-known Brillouin curve of a homogeneous dielectric waveguide (Fig. 2). For $\eta = 0$, only the central curve has physical meaning. Since we are mainly interested in the bounded wave, the following condition must be satisfied:

$$\kappa_n^2 > \mu_0 \epsilon_2 \omega^2, \tag{14}$$

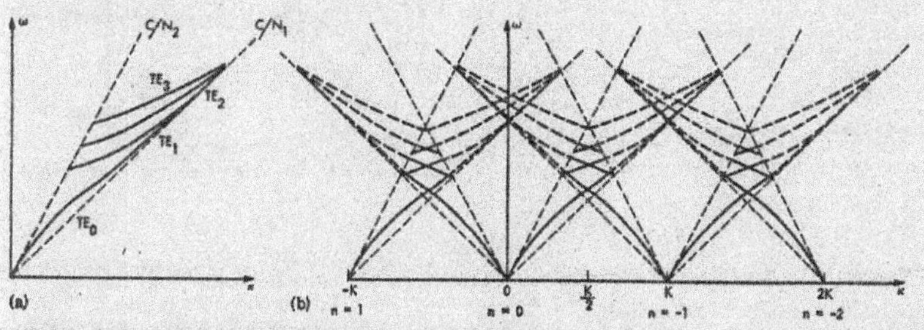

FIG. 2. (a) ω–κ diagram (Brillouin diagram) for a typical dielectric waveguide. (b) ω–κ diagram for a typical periodic dielectric waveguide.

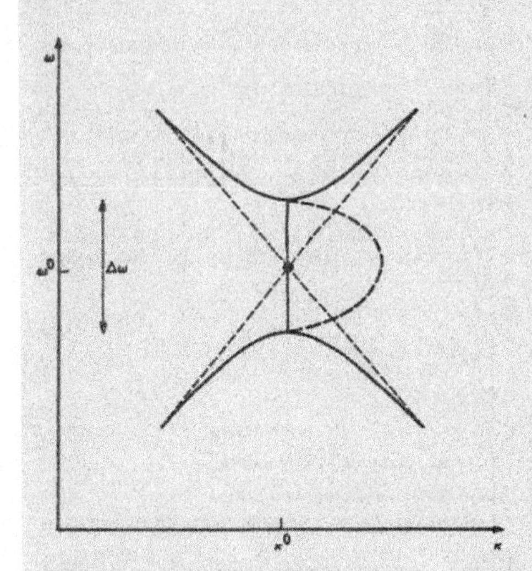

FIG. 3. ω-κ diagram near an interaction region. The solution for κ is complex. The dashed curve is the imaginary part of κ and the solid curve is the real part of κ.

and the corresponding region is shown in Fig. 2. It shows that only successive space harmonics intersect in the bounded region.[13]

For $\eta \neq 0$, strong coupling occurs at the intersection points leading to stop-band interactions (complex κ) between identical modes of successive space harmonics. The behavior of κ near those points requires the solution of the nontriviality condition of the system of Eq. (12). For small values of η, a first-order Taylor series development which takes into account only the two interacting space harmonics can be used to determine the solution in the interaction regions between successive space harmonics.

This paper is confined to the study of stop-band interactions. The codirectional passive interactions are not studied because, in most practical cases, the refractive indices of the film and the substrate are not very different from each other, and these interactions occur in the radiation region of the Brillouin diagram.

In order that interactions between harmonics of the mth mode may occur, the following condition must be satisfied:

$$\frac{K}{2} > \kappa_{\text{cutoff}} \text{ (of the } m\text{th mode)} = \frac{N_2}{(N_1^2 - N_2^2)^{1/2}} \frac{m\pi}{2L}.$$

This implies that

$$\Lambda = \frac{2\pi}{K} < \frac{2L}{m}\left(\frac{N_1^2}{N_2^2} - 1\right)^{1/2}, \tag{15}$$

where $N_1^2 = \epsilon_1/\epsilon_0$ and $N_2^2 = \epsilon_2/\epsilon_0$.

Consider now the interaction point $(\omega^0, \kappa^0 = \frac{1}{2}K)$ between the space harmonics $n = 0$ and $n = -1$. Near that point we can write

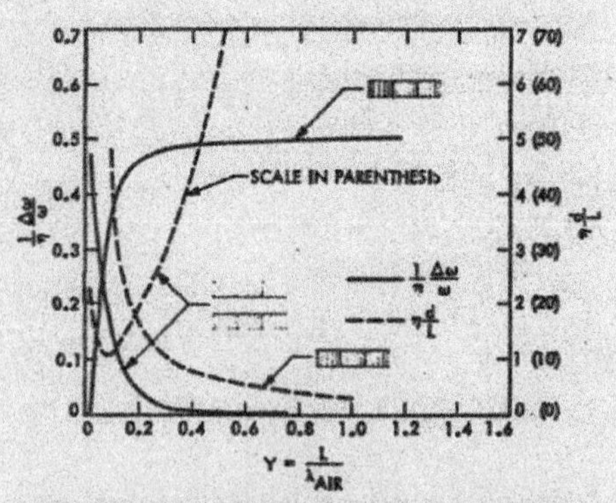

FIG. 4. Plots of the relative width $(1/\eta)\Delta\omega/\omega$ and attenuation factor $\eta d/L$ for the first interaction region of the main even mode as a function of L/λ_0. λ_0 is wavelength in vacuum. The cases of a periodic waveguide and a periodic substrate are plotted. The waveguide index is $N_1 = 2$ and the substrate index is $N_2 = 1.5$.

$$\kappa = \kappa^0(1 + \eta g) \quad \text{and} \quad \omega = \omega^0(1 + \eta f),$$

and $u_0 \approx u_{-1}$. The system of Eqs. (12) reduces to

$$\left.\begin{array}{l} D_0 A_0 + A_{-1} = 0 \\ D_{-1} A_{-1} + A_0 = 0 \end{array}\right\} \Rightarrow D_0 D_{-1} = 1. \tag{16}$$

Applying the Taylor series development to Eqs. (10), (11), and (16), we get

$$\alpha^2 f^2 - g^2 = \beta^2, \tag{17}$$

where

$$\alpha = \left(\frac{2\omega^0}{Kc}\right)^2 \left(\frac{N_2^2 + qN_1^2}{1 + q}\right),$$

$$\beta = \left(\frac{N_1 \omega^0}{Kc}\right)^2 \left(\frac{q}{1 + q}\right),$$

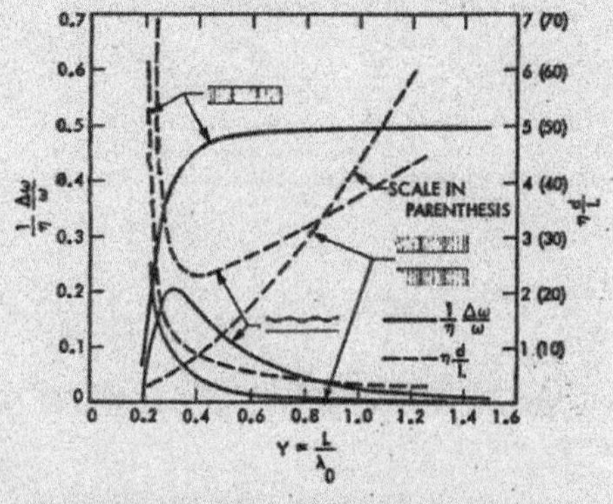

FIG. 5. Plots of the relative width $(1/\eta)\Delta\omega/\omega$ and attenuation factor $\eta d/L$ for the first interaction region of the first odd mode as a function of L/λ_0. λ_0 is wavelength in vacuum. The cases of a periodic waveguide, periodic surface, and a periodic substrate are plotted. The waveguide index is $N_1 = 2$ and the substrate index is $N_2 = 1.5$.

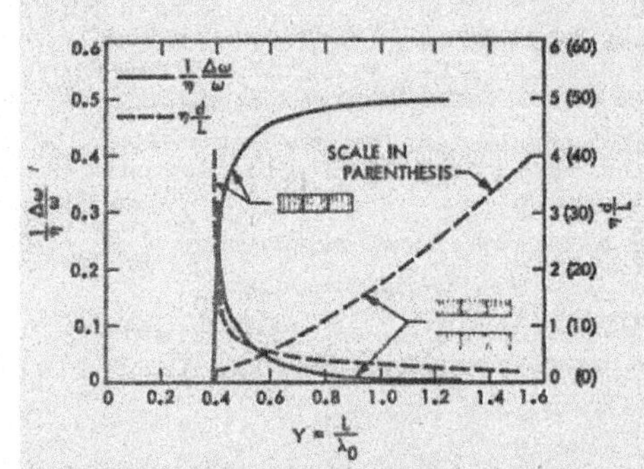

FIG. 6. Plots of the relative width $(1/\eta)\Delta\omega/\omega$ and attenuation factor $\eta d/L$ for the first interaction region of the second even mode as a function of L/λ_0. λ_0 is wavelength in vacuum. The cases of a periodic waveguide and a periodic substrate are plotted. The waveguide index is $N_1=2$ and the substrate index is $N_2=1.5$.

$$q=\left(\frac{\delta^0}{s^0}\right)^2\left(1\pm\frac{s^0L}{\sin s^0L\cos s^0L}\right),$$

where the ± sign corresponds to the even(+) and odd(−) modes, and the values of δ^0 and s^0 correspond to ω^0 and κ^0. Equation (16) corresponds to a hyperbola for $|f| \geq \beta/\alpha$ and an ellipse (g imaginary) for $|f| < \beta/\alpha$, as sketched in Fig. 3. The two characteristic parameters of interest are the relative bandwidth of the interaction $\Delta\omega/\omega=2\eta\beta/\alpha$, and the maximum of imaginary (κ) $=\eta K\beta/2$ and its inverse d. In Figs. 4–6 we plot these two parameters as a function of L/λ, where λ corresponds to the free-space wavelength for the first two even modes and the first odd mode. In all these cases, the interaction is most efficient ($d/L\to 0$) when the energy of the wave is mostly confined within the waveguide (i.e., $L/\lambda\to\infty$).

IV. HOMOGENEOUS THIN-FILM WAVEGUIDE SURROUNDED BY A LONGITUDINALLY PERIODIC MEDIUM

Let us now treat the problem of a TE wave propagating along a homogeneous thin film immersed in an inhomogeneous medium whose dielectric constant is given by

$$\epsilon_2(z)=\epsilon_2(1+\eta'\cos Kz).$$

The dielectric constant of the film is ϵ_1 [see Fig. 1(b)]. This problem may be solved in a manner similar to the previous case. The corresponding systems of equations to be solved are Eq. (10) and

$$s_n^2+\kappa_n^2=\mu_0\epsilon_1\omega^2, \quad (18)$$

$$D_n'A_n'\exp(-\delta_n x)+A_{n+1}'\exp(-\delta_{n+1}x)+A_{n-1}'\exp(-\delta_{n-1}x)=0, \quad (19)$$

where

$$D_n'=\frac{2}{\eta'}\left(1-\frac{\kappa_n^2-\delta_n^2}{\mu_0\epsilon_2\omega^2}\right).$$

The limitations on the solution of Eq. (19) are the same as for Eq. (12). The Brillouin diagram is the same as in Fig. 2 and the solution in the interaction region for small η' is

$$\alpha^2f^2-g^2=\beta'^2,$$

where

$$\beta'=(N_2\omega^0/Kc)^2(1+q)^{-1}.$$

The corresponding characteristic parameters are given in Table I and plotted in Figs. 4–6. Unlike the previous case, we note that mode interaction occurs most efficiently near the cutoff where most of the energy is in the substrate. The mode interaction characteristics for these two structures are equivalent if η and η' satisfy the following relation:

$$\eta\beta=\eta'\beta'\Rightarrow\eta/\eta'=\beta'/\beta.$$

This ratio is plotted in Fig. 7.

V. OPTICAL GUIDE WITH SINUSOIDALLY VARYING SURFACE

The third problem we shall consider is the case of a TE wave propagating along an optical thin-film guide whose surface is a sinusoidal function of the longitudinal coordinate [Fig. 1(c)]; i.e.,

$$L(z)=L[1+\eta''\cos Kz].$$

The substrate is assumed to be a perfect conductor and the upper half-space has a dielectric constant $\epsilon_2<\epsilon_1$. We assume $\eta''KL\ll 1$, so that the Rayleigh assumption[14] for scattering from periodic surfaces is valid.

Owing to the boundary periodicity, the field is the sum of an infinite number of space harmonics (Floquet's theorem). Therefore, the field expression can be written as follows (we are mainly interested in the bounded wave):

$$E=\sum_{n=-\infty}^{n=+\infty}A_n\sin(s_nx)\exp(i\kappa_nz), \quad \text{for } 0<x\leq L(z)$$

$$=\sum_{n=-\infty}^{n=+\infty}A_n'\exp(-\delta_n|x|+i\kappa_nz), \quad \text{for } x\geq L(z),$$

where A_n and A_n' are the amplitudes of the space harmonics. The above expressions satisfy the boundary

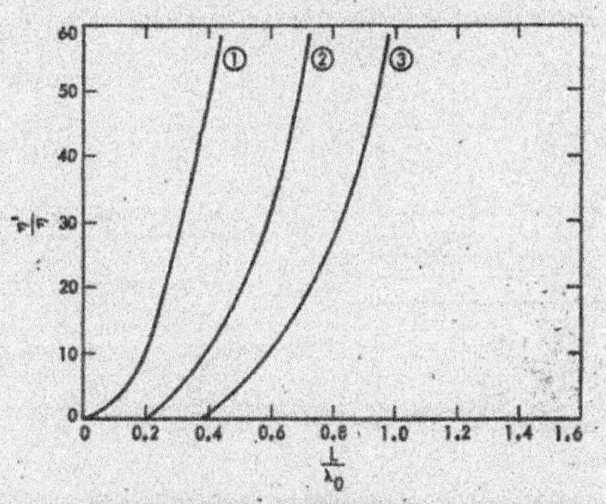

FIG. 7. Equivalence ratio η'/η between the periodic substrate and periodic waveguide configuration. Curve 1 corresponds to the basic even mode. Curve 2 corresponds to the first odd mode. Curve 3 corresponds to the second even mode.

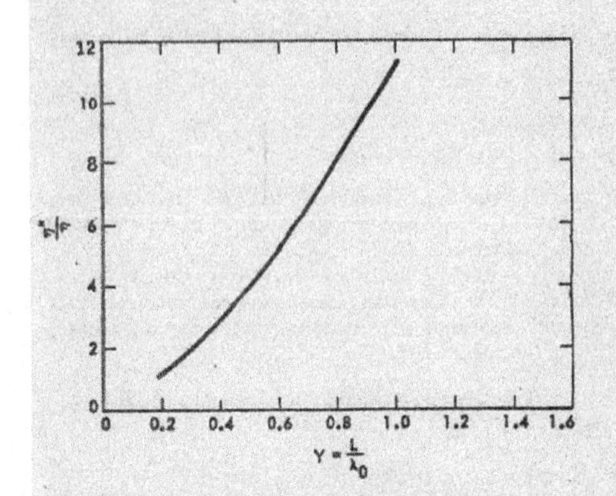

FIG. 8. Equivalence ratio η''/η between the periodic thickness and periodic waveguide index configuration for the first odd mode.

condition at $x=0$. From Maxwell's equations we may obtain the following relations:

$$\kappa_n^2 + s_n^2 = \mu_0\varepsilon_1\omega^2, \quad (20)$$

$$\kappa_n^2 - \delta_n^2 = \mu_0\varepsilon_2\omega^2. \quad (21)$$

The boundary conditions at $x=L(z)=L(1+\eta''\cos Kz)$ are (i) tangential **E** must be continuous; and (ii)

$$(\mathbf{H}-\mathbf{H}')\times\mathbf{e}=0, \quad (22)$$

where **e** is a unit vector normal to the surface $=(\mathbf{e}_x + \eta'' KL\sin Kz\,\mathbf{e}_z)/(1+\eta''^2K^2L^2\sin^2 Kz)^{1/2}$ and H and H' are the magnetic field on both sides of the boundary.

From condition (i) and the field expression we get, after applying a first-order Taylor series development ($\eta'' KL \ll 1$),

$$\sum_n A_n(\sin s_n L + \eta'' s_n L\cos Kz\cos s_n L)\exp(i\kappa_n z) = \sum_n A_n'(1-\eta''\delta_n L\cos Kz)\exp(i\kappa_n z).$$

Writing $\cos Kz$ in an exponential form and equating the terms with the same longitudinal wave vector, we get

$$(a_n/s_n)\tan s_n L + \tfrac{1}{2}\eta'' L(a_{n+1}+a_{n-1}) - (b_n/\delta_n) + \tfrac{1}{2}\eta'' L(b_{n+1}+b_{n-1}) = 0, \quad (23)$$

where $a_n = s_n A_n\cos(s_n L)$ and $b_n = \delta_n A_n'\exp(-\delta_n L)$. Similarly, from Eq. (22) we find

$$a_n - \tfrac{1}{2}\eta'' L(C_{n+1}a_{n+1} + C_{n-1}a_{n-1}) + b_n + \tfrac{1}{2}\eta'' L(B_{n+1}b_{n+1} + B_{n-1}b_{n-1}) = 0, \quad (24)$$

where

$$C_n = [(K\kappa_n + s_n^2)/s_n]\tan s_n L, \quad B_n = (K\kappa_n - \delta_n^2)/\delta_n.$$

The infinite systems of Eqs. (23) and (24) can be written in a matrix form:

$$\|M\|\cdot|a| = \|N\|\cdot|b|,$$

$$\|P\|\cdot|a| = \|Q\|\cdot|b|,$$

or

$$(\|M\| - \|N\|\cdot\|Q\|^{-1}\cdot\|P\|)\cdot|a| = 0, \quad (25)$$

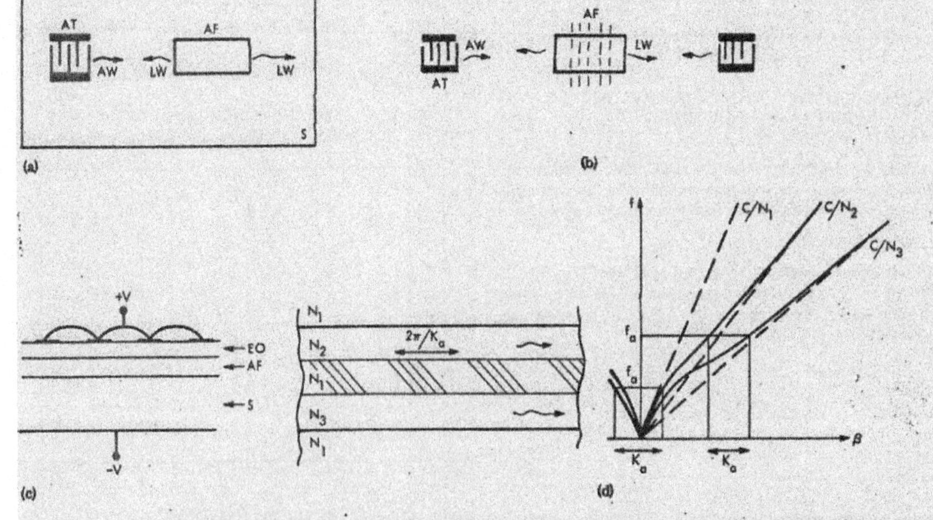

FIG. 9. Some optical systems using periodic structures: (a) DFB laser using a propagating acoustic wave; (b) DFB pulsing laser using a stationary acoustic wave; (c) DFB laser using an electro-optic periodic structure; (d) frequency-selective coupler with the corresponding Brillouin diagram. AT—acoustic transducer; AF—amplifying film; AW—acoustic wave; LW—light wave; EO—electro-optic film, S—substrate.

TABLE I. Characteristic parameters with $Y=\omega L/2\pi c$.

Structure	$\frac{1}{\eta}\frac{\Delta\omega}{\omega}$	$\frac{\eta d}{L}$	$\frac{\Delta\omega}{\omega}\frac{d}{L}$
periodic waveguide index	$\frac{qN_1^2}{2(N_2^2+qN_1^2)}$	$\frac{L}{\Lambda}\frac{1+q}{q}\frac{1}{\pi N_1^2 Y^2}$	$\frac{L}{2\Lambda}\frac{1+q}{N_2^2+qN_1^2}\frac{1}{\pi Y^2}$
periodic substrate index	$\frac{N_1^2}{2(N_2^2+qN_1^2)}$	$\frac{L}{\Lambda}(1+q)\frac{1}{\pi N_2^2 Y^2}$	same as above
periodic waveguide thickness	$(\delta^0 L)^2\frac{N_1^2-N_2^2}{N_2^2+qN_1^2}$	$\frac{K}{\delta^0}\frac{1+q}{(N_1^2-N_2^2)(2\pi Y)^2}$	same as above

where the elements of the matrices $\|M\|$, $\|N\|$, $\|P\|$, and $\|Q\|$ can be easily determined from Eqs. (23) and (24). The notriviality condition of Eq. (25), with Eqs. (20) and (21), gives the dispersion relation $\kappa(\omega)$ and the relative amplitudes a_n/a_0 and b_n/a_0.

For $\eta''=0$, Eqs. (23) and (24) reduce to

$$(a_n/s_n)\tan s_n L = b_n/\delta_n,$$

$$a_n = -b_n,$$

or

$$\delta_n = -s_n \cot s_n L. \qquad (26)$$

The corresponding Brillouin diagram is the same as for the previous cases (odd modes only).

For small $\eta''\neq 0$, a first-order Taylor series development gives

$$\alpha^2 f^2 - g^2 = \beta''^2,$$

where

$$\beta'' = \frac{2\delta^0 L}{1+q}(N_1^2 - N_2^2)\left(\frac{\omega^0}{Kc}\right)^2.$$

The two characteristic parameters $\Delta\omega/\omega$ and d/L are given in Table I and plotted in Fig. 6 for the first mode with $N_2=1.5$ and $N_1=2$. We see that for $\omega\to\infty$ ($L/\lambda\to\infty$), the periodicity has little effect ($d/L\to\infty$, $\Delta\omega/\omega\to 0$) because the wave energy is mostly confined inside the waveguide and is not near the boundary. The boundary periodicity has also little effect on the propagating waves in the limit $\omega\to\omega_{\text{cutoff}}$ when most of the energy is in the substrate. The boundary perturbation is most effective for $L/\lambda\approx 0.35$, where $d/L=2.3/\eta$ and $\Delta\omega/\omega=0.2\eta$.

In Fig. 8 we plotted the equivalence ratio η''/η between the surface periodic case and the inside index periodic case.

Another interesting structure which can be treated in a similar way is shown in Fig. 1(d)—the dielectric waveguide which has a periodic boundary impedance

$$Z = iX_0[1+\eta\cos Kz].$$

Such an impedance can be generated by using a grooved substrate. This impedance can be controllable (slightly), for matching purposes, by filling the substrate grooves with an electro-optic material whose index (and, therefore, the effective groove depth) can be controlled by an applied electric voltage.

VI. DISCUSSIONS AND POSSIBLE APPLICATIONS

Periodic structures can be of two types: (i) permanent, where the periodicity in the refractive index, gain coefficient (in the case of an active medium), and boundary of the optical medium is permanent (chemical, doping, ion implementation, or others). This type of structure was used in DFB lasers[10,15] and in couplers. (ii) Nonpermanent or dynamically controllable, where the periodicity can be easily written and erased (photodimer memories,[6] electro-optical periodic structures), or is dynamically controllable (beam interference,[7] surface or bulk acoustic waves). We shall now discuss the application of some of these structures in active and passive integrated optics systems.

Periodic structures in amplifying films are used in DFB lasers.[10,15,16] Kogelnik and Shank[10] showed that very small changes (10^{-4}—10^{-5}) in the refractive index of the guiding film are sufficient to generate oscillation in a DFB laser. Using the results of Sec. V and Fig. 8, we see that this is equivalent to boundary rippling of the order of 10—20 Å depending on the guide parameters. This surface perturbation could be obtained by a surface acoustic wave [Fig. 9(a)]. The lasing wavelength of the DFB laser is given by the Bragg condition (neglecting the acoustic wave motion):

$$\lambda_0 = 2N_1\Lambda/m,$$

where m is an integer. Therefore, the acoustic frequency falls in the range of few GHz. This domain of the acoustic spectrum is of great interest in integrated acoustics and it is presently possible to fabricate such surface-wave devices for operating frequencies up to 3 GHz.[17] In Fig. 9(b), a second acoustic transducer, at 180° from the first one, would generate a standing acoustic wave leading to a DFB laser pulsing at twice the acoustic frequency. The lasing frequency can be tuned by changing the acoustic frequency.

The electro-optic effect can also be used in thin-film structures.[1,18] If a spatial periodic electric voltage is applied to an electro-optic material, the resulting periodicity generates the DFB effect. The electro-optic material can be used as a substrate or as a matrix for the lasing material [Fig. 9(c)]. Index change of the order of 10^{-4} can be easily achieved with only a few volts. Periodic structures using the electro-optic effect have also been used for wave modulation.[19]

In passive systems, periodic structures can be used for distributed selective coupling between two closely spaced waveguides.[20] They have already been used for the coupling of a free wave to a waveguide. Let us now consider the system shown in Fig. 9(d). It can be easily seen that there is no evanescent coupling for the fundamental mode at all frequencies since, for a given f, β is not the same in both guides, and the periodic structure would only couple a signal of frequency f_0, even if the waveguides may carry a large number of signals with different frequencies. To obtain optimum coupling, the parameters must be adequately chosen. Such a system can be designed to yield selective coupling at two or more specific frequencies, if we make adequate use of the different modes of the waveguides and the guide dimensions. If the periodic structure is of the dynamic type, the frequencies can be varied. The frequency-selective coupler can be used as the basic unit of an optical multiplexer to select specific channels from a large number of communication channels in a thin-film guide.

Another application is the use of the periodic structures as integrated optical filters. For an 0.8-μ film with a surface rippling of 40 Å and an optical wavelength of $\lambda_0 = 0.8\ \mu$, the attenuation factor could reach 14 dB/mm (power).

In conclusion, we note that periodic structures have characteristics which can be used in many active and passive integrated-optics systems. The stop-band pass-band property can be used to design integrated optics filters. Attenuations of the order of tens of dB/mm can be easily obtained. The feedback property of periodic structures can be used in active systems to design DFB lasers. Different configurations can be used: index periodicity in the film, index periodicity of the substrate, or film-thickness periodicity. The presence of space harmonics with different phase velocities can also be used for phase-matched nonlinear interactions.

*Partly supported by the Naval Electronics Laboratory Center and partly supported by NASA under Contract No. NAS7-100.

[1]P. K. Tien, Appl. Opt. 10, 2395 (1971).
[2]S. E. Miller, Bell Syst. Tech. J. 48, 2059 (1969).
[3]R. Shubert and J. H. Harris, IEEE Trans. Microwave Theory Tech. 16, 1048 (1968).
[4]J. L. Altman, *Microwave Circuits* (Nostrand, Princeton, N.J., 1964).
[5]P. K. Tien, Appl. Opt. 11, 205 (1972).
[6]R. L. Fork, K. R. German, and E. A. Chandross, Appl. Phys. Lett. 20, 139 (1972).
[7]R. V. Pole, S. E. Miller, J. H. Harris, and P. K. Tien, Appl. Opt. 11, 1675 (1972).
[8]S. Somekh and A. Yariv, Appl. Phys. Lett. 21, 140 (1972).
[9]J. E. Bjorkholm and C. V. Shank, Appl. Phys. Lett. 20, 3061 (1972).
[10]H. Kogelnik and C. V. Shank, J. Appl. Phys. 43, 2327 (1972).
[11]C. Yeh, K. F. Casey, and Z. A. Kaprielian, IEEE Trans. Microwave Theory Tech. 13, 297 (1965).
[12]C. Elachi, IEEE Trans. Antennas Propag. 20, 280 (1972).
[13]E. S. Cassedy, Proc. IEEE 112, 269 (1965).
[14]Lord Rayleigh, *The Theory of Sound* (Dover, New York, 1945), Vol. II.
[15]D. P. Schinke, R. G. Smith, E. G. Spencer, and M. F. Galvin, Appl. Phys. Lett. 21, 494 (1972).
[16]K. O. Hill and A. Watanabe, Opt. Commun. 5, 389 (1972).
[17]R. M. White, Proc. IEEE 58, 1238 (1970).
[18]D. J. Channin, Appl. Phys. Lett. 19, 128 (1971).
[19]D. P. Gia Russo and J. H. Harris, Appl. Opt. 12, 2786 (1971).
[20]C. Elachi and C. Yeh, Optics Comm. (to be published).

FREQUENCY SELFCTIVE COUPLER FOR INTEGRATED OPTICS SYSTEMS*

C. ELACHI
University of California, Los Angeles, California 90024, USA

and

C. YEH
California Institute of Technology, Jet Propulsion Laboratory, Pasadena, California 91103, USA

Received 15 January 1973

A frequency selective coupler which consists of two thin film waveguides imbedded in a periodic medium is studied using the Brillouin diagram. Detailed results for the relative bandwidth and the coupling factor are plotted as a function of normalized frequency for a representative case.

Many techniques have been used to couple a light beam to a thin film waveguide such as the prism coupler, rating coupler and Bragg coupler (for references see ref. [1]). Many theoretical studies also have been carried out to investigate the properties of these structures (for references see ref. [2]). Coupling between two integrated optics waveguides using the evanescent wave was studied by Marcatili [3] and reported by Ihaya et al. [4]. In this communication, we shall consider a frequency selective coupler which consists of two thin films of different index of refraction, N_1 and N_3, imbedded in a periodic medium (fig. 1). Although this study is limited to the basic TE even mode, similar results may also be obtained for all other modes by the same technique.

Let us assume that the refractive index of the substrate is

$$N_2(z) = N_2(1 + \tfrac{1}{2}\eta \cos Kz). \tag{1}$$

When $\eta = 0$, no coupling is possible because at each frequency $\omega/2\pi$, the longitudinal wavevectors in the two waveguides are different (see fig. 1a). For $\eta \neq 0$, phase matching would then occur at two frequencies, $\omega_c/2\pi$ and $\omega_c'/2\pi$ as shown in fig. 1b.

Because of the spatial periodicity of the medium, the field must be expressed in terms of an infinite number of space harmonics as follows:

$$E = \sum_{\eta=-\infty}^{\eta=+\infty} [A_n \exp(ik_n x) + B_n \exp(-ik_n x)] \exp(i\kappa_n z), \tag{2}$$

where $\kappa_n = \kappa + nK$ are longitudinal wave vectors, A_n, B_n are constants which can be determined from the boundary and source conditions, and k_n is the transverse wave vector:

$$\begin{aligned} k_n &= k_{1n} \text{ in medium 1,} \\ &= k_{3n} \text{ in medium 3,} \\ &= i\delta_n \text{ in medium 2 (evanescent wave).} \end{aligned}$$

From the wave equation, the wave vectors are related by the following relations:

$$\kappa_n^2 + k_{1n}^2 = (N_1 \omega/c)^2, \tag{3}$$

$$\kappa_n^2 + k_{3n}^2 = (N_3 \omega/c)^2, \tag{4}$$

* Partly supported by the Naval Electronics Laboratory Center and partly supported by NASA under Contract NAS 7–100.

Volume 7, number 3 OPTICS COMMUNICATIONS March 1973

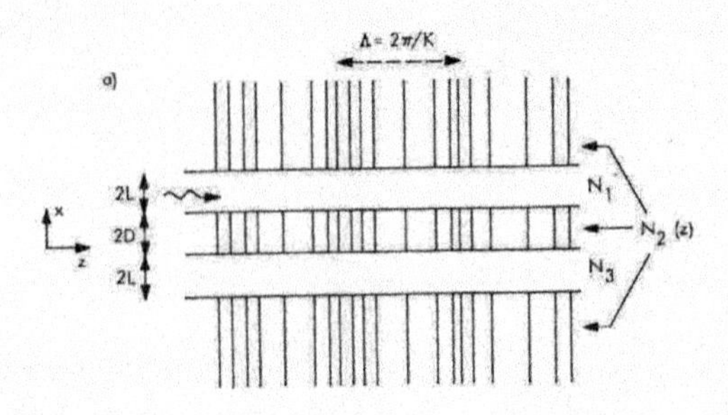

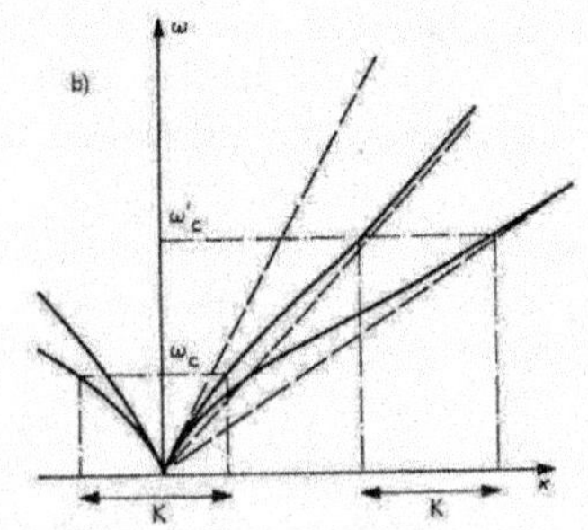

Fig. 1. (a) Geometry. The two thin waveguides of thickness $2L$ and refractive index N_1 and N_3 are imbedded in a periodic dielectric of refractive index $N_2(z) = N_2[1 + \frac{1}{2}\eta \cos Kz]$. The separation between the two guides is $2D$. (b) Brillouin diagram of the two waveguides when $\eta = 0$. No evanescent wave coupling is possible. When $\eta \neq 0$, only the frequencies $\omega_c/2\pi$ and $\omega_c'/2\pi$ can be coupl

$$\det \|\mathbf{M}\| = 0, \tag{5}$$

where c is the speed of light in vacuum. Eq. (5) is the dispersion equation in a space-periodic medium [6, 7] and the matrix $\|\mathbf{M}\|$ is a tridiagonal matrix with elements:

$$M_{n,n} = D_n = \frac{2}{\eta}\left[1 - \frac{\kappa_n^2 - \delta_n^2}{(N_2\omega/c)^2}\right],$$

$$M_{n,n+1} = 1,$$

$$M_{n,m} = 0, \text{ otherwise}.$$

Satisfying the boundary conditions gives

$$[R_{21}^2 \exp(2ik_{1n}L) - \exp(-2ik_{1n}L)]$$
$$\times [R_{23}^2 \exp(2ik_{3n}L) - \exp(-2ik_{3n}L)]$$
$$= -4R_{21}R_{23}\sin(2k_{1n}L)\sin(2k_{3n}L)\exp(-4\delta_n D), \tag{6}$$

where

$$R_{21} = \frac{k_{1n} - i\delta_n}{k_{1n} + i\delta_n}, \qquad R_{23} = \frac{k_{3n} - i\delta_n}{k_{3n} + i\delta_n},$$

$2L$ = waveguide thickness,

$2D$ = coupling region thickness.

For $D \to \infty$, eq. (6) reduces to $R_{21} = \pm \exp(-2ik_{1n}L)$ and $R_{23} = \pm \exp(-2ik_{3n}L)$ which are the well-known relations for two uncoupled dielectric waveguides.

The system of equations (3)–(6) can be solved numerically to determine the dispersion equation $\kappa(\omega)$, but the main characteristics of the solution can be determined from the limit of $\eta \to 0$ and large D. In this limit, the Brillouin diagram consists of periodic cells each identical to the diagram of the homogenous case (see fig. 2). For $\eta \neq 0$, strong coupling occurs between the different space-harmonics at their intersection points. There are two types of coupling.

(1) Co-directional (fig. 2a, point 5), where the energy is periodically interchanged between the two space harmonics. For the case where N_1, N_2 and N_3 are not very different from each other, this coupling occurs in the radiation region.

(2) Contra-directional (fig. 2a) where the energy is transferred from one space harmonic to the other. At points 1 and 2, the coupling is between harmonics of the same waveguide. At points 3 and 4 the coupling is between the harmonics of the two guides leading to energy transfer. Near these points, the solution for κ is complex and the corresponding Brillouin diagram is shown in fig. 2b.

The solution for κ in the coupling region 3 was determined in the case of η small using a first order Taylo series development which takes into account only the two interacting harmonics at that point. In fig. 3a we plot the relative bandwidth $\Delta\omega/\omega$ of the coupling region as a function of L/λ_0 for $D/L = 1$ and different

Volume 7, number 3 OPTICS COMMUNICATIONS March 1973

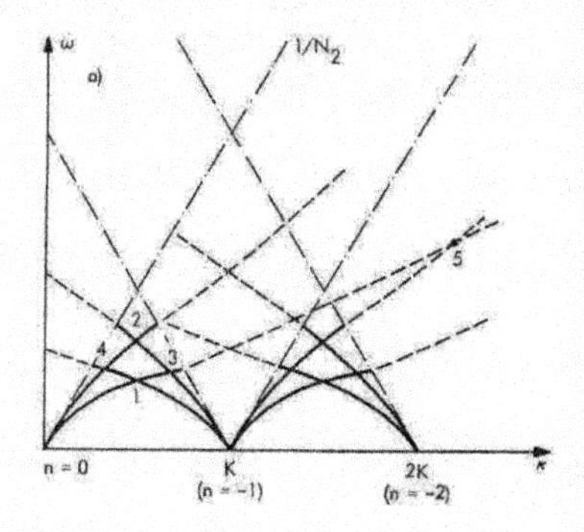

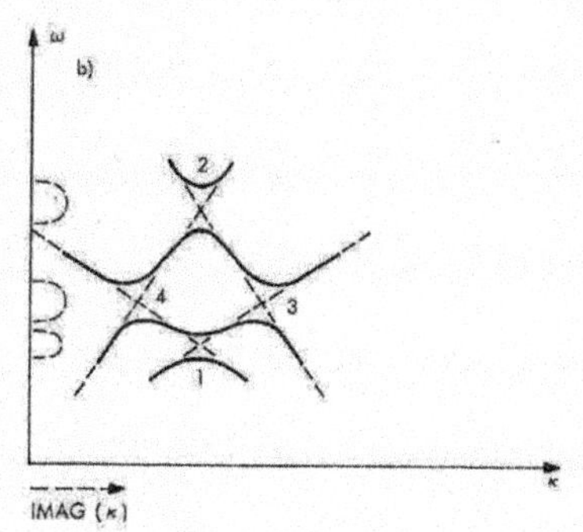

Fig. 2. (a) Brillouin diagram of the periodic coupler. The dashed part corresponds to the radiation region. The interaction between the different space harmonics can be of the co-directional type (point 5) or contra-directional type (points 1–4). (b) Detailed diagram of the contra-directional interaction region. The dashed curves on the left correspond to imaginary κ.

values of the index of refraction, where λ_0 is the wavelength in vacuum. Each value of λ_0 corresponds to a specific value of K. The coupling bandwidth goes to zero for $L/\lambda_0 \to \infty$ or D/L large, as is expected, because of the exponential nature of the evanescent wave.

In fig. 3b we plot $\eta d/L$ as a function of L/λ for $D/L = 1$, where $d = (\text{maximum of imaginary } \kappa)^{-1}$ is the distance at which the amplitude of the incident wave is attenuated by a factor e = 2.72 ..., its energy being transferred to the second waveguide. For each value of D/L, there is an optimum value of L/λ_0 which corresponds to the strongest coupling (smallest d). In fact, for $L/\lambda_0 \to \infty$, there is no evanescent wave, and the coupling disappears, and for $L/\lambda_0 \to 0$, the energy of the evanescent wave is spread over a large region leading to a small energy density, and therefore weak coupling.

As an example, for $\lambda_0 = 10\ \mu$, $D = L = 2\ \mu$ and $\eta = 1 \times 10^{-3}$, then $d = 500\ \mu$ (Curve 2 in fig. 3b) which corresponds to an energy coupling of 17 dB/mm.

The coupler studied in this paper can be used as a basic unit in an optical multiplexer. The selective coupling frequency can, if desired, be dynamically tuned if one or more of the media (waveguides or substrate) are made of electro-optical material such that their index of refraction can be changed by applying an electrical voltage.

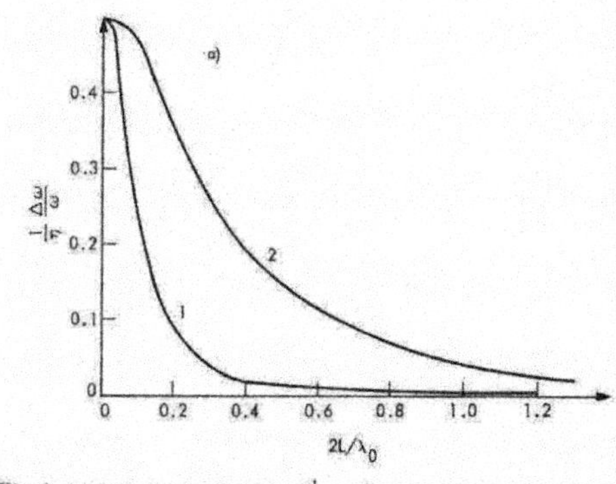

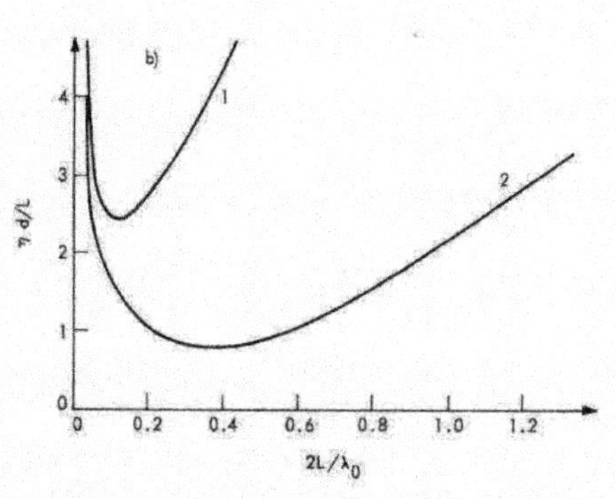

Fig. 3. (a) Relative bandwidth $\eta^{-1}\ \Delta\omega/\omega$ and (b) coupling characteristic length $\eta\ d/L$ as a function of $2L/\lambda_0$ where λ_0 is the wavelength in vacuum. Curves 1 correspond to $N_1 = 2$, $N_2 = 1.5$ and $N_3 = 2.5$. Curves 2 correspond to $N_1 = 1.5$, $N_2 = 1.4$ and $N_3 = 1.6$.

References

[1] P.I. Tien, Appl. Opt. 10 (1971) 2395.
[2] J.H. Harris et al., Appl. Opt. 11 (1972) 2234.
[3] E.A.J. Marcatilli, Bell System Tech. J. 48 (1969) 2071.
[4] A. Ihaya et al., Proc. IEEE (1972) 470.
[5] S.E. Miller, Bell System Tech. J. 48 (1969) 2189.
[6] C. Elachi and C. Yeh, to be published.
[7] C. Elachi, IEEE Trans. AP-20 (1972) 334.

Reprinted from:
JOURNAL OF THE OPTICAL SOCIETY OF AMERICA VOLUME 63, NUMBER 7 JULY 1973

Stop bands for optical wave propagation in cholesteric liquid crystals*

C. Elachi and C. Yeh
Electrical Sciences and Engineering Department, University of California, Los Angeles, California 90024
Jet Propulsion Laboratory, California Institute of Technology, Pasadena, California 91103
(Received 17 January 1973)

The stop-band characteristics of waves in cholesteric-liquid-crystal half-space are investigated using the exact wave solution, the Floquet theorem, and the corresponding Brillouin diagram. The other half-space is filled with a uniform dielectric medium. By appropriate choice of the dielectric constant of the uniform medium and the angle of incidence of the incoming plane wave, the stop band may split into two or three stop bands.

Index Heading: Cholesteric liquid crystals.

Cholesteric liquid crystals can be represented by a structure consisting of molecules arranged in thin anisotropic layers, with the successive layers rotated through a small angle, leading to a spiral configuration.[1,2] This configuration gives unique optical properties that can be used in many optical applications. These crystals have a very strong rotatory factor: 60 000°/mm vs 300°/mm for ordinary organic liquids. The periodicity of the structures leads to pass-band and stop-band characteristics for waves propagating through such structures; therefore they can provide a wide variety of optical-filtering functions.[3] Furthermore, the fact that the pitch of the spiral configuration is a function of temperature, pressure, added chemicals, etc., may permit numerous other applications.[4,5] Many authors have studied the propagation of optical waves in cholesteric liquid crystals,[6–11] using various configurations and various methods of solutions. In this paper, we shall consider the case of an obliquely incident wave upon a half-space of cholesteric liquids (Fig. 1). The Brillouin diagram[12,13] will be extensively used to obtain the properties of waves in the stop-band regions (relative bandwidth and attenuation factor) as functions of the incident angles.

Cholesteric liquid crystals (CLC) can be characterized by a dielectric tensor

$$\varepsilon(z)=\begin{Vmatrix}\bar{\epsilon}+\eta\cos Kz & \eta\sin Kz & 0\\ \eta\sin Kz & \bar{\epsilon}-\eta\cos Kz & 0\\ 0 & 0 & \epsilon_3\end{Vmatrix}, \quad (1)$$

where $K=4\pi/p$. This dielectric tensor can be represented by an ellipsoid of principal axes ϵ_3 (parallel to z), $\bar{\epsilon}+\eta$ and $\bar{\epsilon}-\eta$. The ellipsoid spirals around the z axis with a pitch p. ϵ_3, η, and $\bar{\epsilon}$ are assumed to be known constants. Owing to the symmetry of an ellipsoid, the CLC can be considered as a periodic medium of period $p/2$. The value of p is usually of the order of 1 μm (fraction of to a few microns).

We assume that the electric vector of an incident wave has the form

$$\mathbf{E}^{(i)}=\mathbf{E}e^{ik'(\sin\theta x+\cos\theta z)},$$

where $k'^2=\mu_0\epsilon'\omega^2$, ϵ' is the dielectric constant of the uniform medium. θ is the angle of incidence, as shown in Fig. 1. The spatial periodicity of the CLC medium demands that the field within the liquid crystal must be represented by an infinite series of spatial harmonics (according to the Floquet theorem).[14] Hence,

$$\mathbf{E}^{(\mathrm{CLC})}=\sum_{n=-\infty}^{n=+\infty}\mathbf{E}_n e^{i(k'\sin\theta_i x+i\kappa_n z)}, \quad (2)$$

where $\mathbf{E}_n$ are arbitrary constants, $\kappa_n=\kappa+nK$, and κ is to be determined from Maxwell's equations. Putting the expressions for ε and $\mathbf{E}^{(\mathrm{CLC})}$ in Maxwell's equations, and equating the terms that have the same z dependence, we obtained

$$D_nE_{xn}+\eta[E_{x,n+1}+E_{x,n-1}+iE_{y,n+1}-iE_{y,n-1}]=0, \quad (3)$$

$$D_n'E_{yn}-\eta[E_{y,n+1}+E_{y,n-1}+iE_{x,n+1}-iE_{x,n-1}]=0, \quad (4)$$

$$E_{zn}=C_nE_{xn},$$

where

$$D_n=2\bar{\epsilon}-2\kappa_n^2\Big/\left(\frac{\omega^2}{c^2}-\frac{k'^2\sin^2\theta}{\epsilon_3}\right),$$

$$D_n'=2\left(\bar{\epsilon}-\frac{k'^2\sin^2\theta+\kappa_n^2}{\omega^2/c^2}\right),$$

$$C_n=-\frac{k'\kappa_n\sin\theta}{(\omega^2/c^2)\epsilon_3-k'^2\sin^2\theta},$$

and E_{xn}, E_{yn}, E_{zn} are the (x,y,z) components of each harmonic vector $\mathbf{E}_n$ and c is the speed of light. Applying the nontriviality condition to Eqs. (3) and (4) gives the dispersion relation for κ. Relative amplitudes

$$E_{xn}/E_{x0},\quad E_{yn}/E_{x0},\quad E_{zn}/E_{x0}$$

may also be obtained from Eqs. (3) and (4); E_{x0} can be found from the boundary conditions.

The dispersion relation as well as the relative amplitudes may be obtained numerically. However, a great deal of information concerning the propagation constant κ can be obtained by considering the case in which $\eta\rightarrow 0$. When $\eta=0$, we have D_n, $D_n'=0$. Hence from

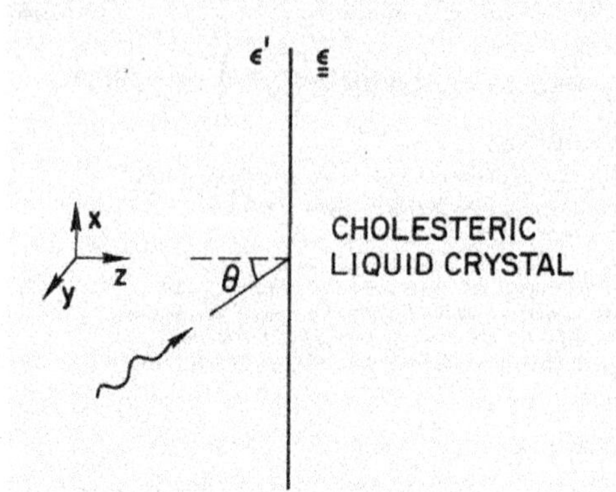

FIG. 1. Geometry of the problem.

$D_n=0$, we obtained

$$\kappa_n=\pm\alpha_1\omega/c, \tag{5}$$

and from $D_n'=0$, we obtained

$$\kappa_n=\pm\alpha_2\omega/c, \tag{6}$$

where

$$\alpha_1=\sqrt{\bar{\epsilon}}[1-(\epsilon'/\epsilon_3)\sin^2\theta]^{\frac{1}{2}}$$

and

$$\alpha_2=\sqrt{\bar{\epsilon}}(1-(\epsilon'/\bar{\epsilon})\sin^2\theta)^{\frac{1}{2}}.$$

So the corresponding Brillouin diagram consists of a set of four lines repeated periodically for $\kappa=-nK$ (Fig. 2). When $\eta=0$, only the set corresponding to $n=0$ has any physical meaning.

When $\eta\neq 0$, strong interaction occurs near the intersection points, due to the phase matching between different space harmonics, leading to stop-band-type interaction, i.e., κ is complex (see Fig. 2). For small η, and considering only the interactions between successive harmonics n and $n-1$, we can apply the first-order Taylor-series expansion to Eqs. (3) and (4). Near each intersection point (ω^0,κ^0), we may write

$$\kappa=\kappa^0(1+\eta a),\quad \omega=\omega^0(1+\eta b),$$

where ηa and ηb would correspond to the relative change in the wave vector and the frequency near the interaction point. Replacing κ and ω by their Taylor-series expression in Eqs. (3) and (4) we found (see Fig. 3)

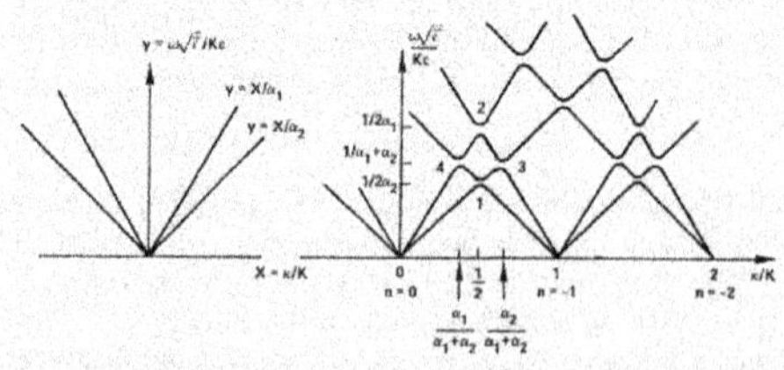

FIG. 2. Brillouin diagrams for waves in cholesteric liquid crystals.

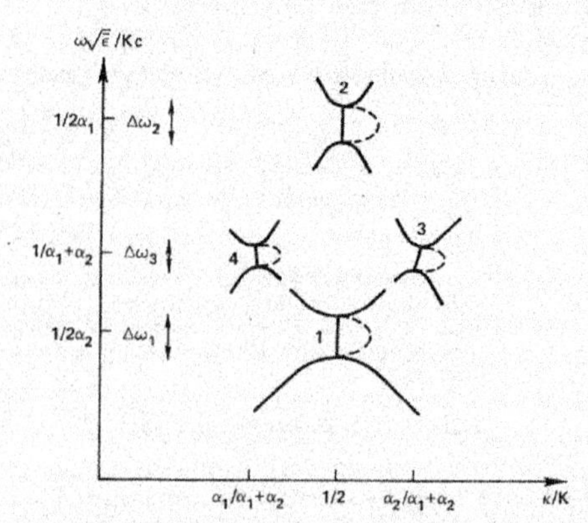

FIG. 3. Sketch of the Brillouin diagram near an interaction region. The dashed line corresponds to the imaginary part of κ. The continuous line corresponds to the real part of κ.

at intersection point 1 $\quad b^2-a^2=1/(4\bar{\epsilon}\alpha_2^2)^2, \quad (7)$

at intersection point 2 $\quad b^2-a^2=1/(4\bar{\epsilon})^2, \quad (8)$

at intersection point 3

$$(b-a)[b+(\alpha_2/\alpha_1)a]=1/(4\bar{\epsilon}\alpha_2)^2, \tag{9}$$

at intersection point 4

$$(b-a)[b+(\alpha_1/\alpha_2)a]=1/(4\bar{\epsilon}\alpha_2)^2. \tag{10}$$

The corresponding Brillouin diagram is sketched in Fig. 3. Two important characteristic parameters (relative bandwidth $\Delta\omega/\omega$ and the maximum value of the imaginary part of κ) are given in Table I.

When $\theta=0$, $\alpha_1=\alpha_2$ and the four intersection points degenerate to only a single point; hence, only one stop band exists. As θ increases from zero, the four intersection points may separate from each other. Hence, the existence of three stop bands. However, the widths of

TABLE I. Characteristic parameters.

Intersection point	$\Delta\omega/\omega$	$\Delta\kappa/\kappa$
1	$\eta/[2\alpha_2^2\bar{\epsilon}]$	$\eta/[4\alpha_2^2\bar{\epsilon}]$
2	$\eta/[2\bar{\epsilon}]$	$\eta/[4\bar{\epsilon}]$
3	$\eta\left(\frac{\alpha_1}{\alpha_2}\right)^{\frac{1}{2}}\Big/[\bar{\epsilon}(\alpha_1+\alpha_2)]$	$\eta\left(\frac{\alpha_1}{\alpha_2}\right)^{\frac{1}{2}}\Big/[4\bar{\epsilon}(\alpha_2)]$
4	$\eta\left(\frac{\alpha_2}{\alpha_1}\right)^{\frac{1}{2}}\Big/[\bar{\epsilon}(\alpha_1+\alpha_2)]$	$\eta\left(\frac{\alpha_2}{\alpha_1}\right)^{\frac{1}{2}}\Big/[4\bar{\epsilon}(\alpha_2)]$

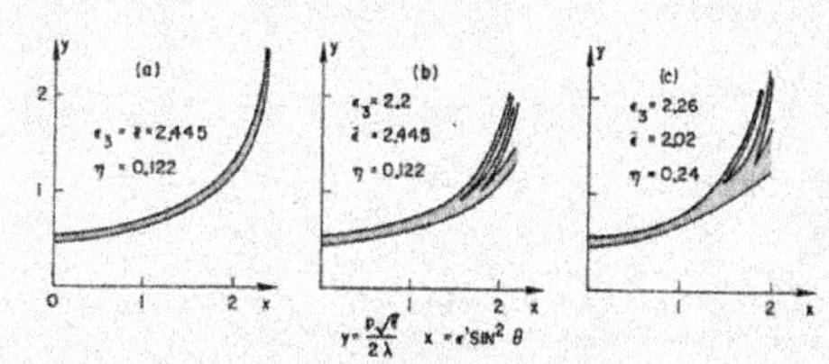

FIG. 4. Interaction region (dark area) as a function of $\epsilon' \sin^2\theta$ for various values of η, $\bar{\epsilon}$, ϵ_3. The y and x are defined in the figure and do not correspond to the spatial coordinates.

the intersection regions are large relative to the separation, so overlapping occurs. Consequently, only one stop band exists. As θ becomes larger, two or three stop bands may exist, provided that we choose ϵ' appropriately. These possibilities are shown in Fig. 4, where we plotted the width of the stop band as a function of $\epsilon' \sin^2\theta$ for different values of $\bar{\epsilon}$, η, and ϵ_3. In Fig. 4(a), the interaction regions never separate; so only one stop band exists for all values of θ. In Fig. 4(b), if $\epsilon'<1.48$, one stop band exists for all values of θ. If $1.48<\epsilon'<1.72$, two stop bands exist for large θ. If $\epsilon'>1.72$, there is one stop band for relatively small values of θ, two for intermediate θ, and three for large θ. In the preparation of Fig. 4, we assumed that $\eta<0.24$; therefore the error of the Taylor-series expansion is less than 1% [error $\sim(\eta/\bar{\epsilon},\epsilon_3)^2 \simeq 0.01$]. This simple use of the Brillouin diagram clarifies the reason for the triplet structure first computed numerically in Ref. 7.

Let us now determine the separation points analytically for small values of η. It is easy to show that regions 1, 3, and 4 would overlap if (see Fig. 3)

$$\frac{1}{2}\left[\left|\frac{\Delta\omega}{\omega}\right|_1+\left|\frac{\Delta\omega}{\omega}\right|_3\right]>\frac{1}{\alpha_1+\alpha_2}-\frac{1}{2\alpha_2},$$

whereas regions 2, 3, and 4 would overlap if

$$\frac{1}{2}\left[\left|\frac{\Delta\omega}{\omega}\right|_1+\left|\frac{\Delta\omega}{\omega}\right|_3\right]>\frac{1}{2\alpha_1}-\frac{1}{\alpha_1+\alpha_2}.$$

These relations are obtained for the case $\alpha_1<\alpha_2$. For the case $\alpha_2<\alpha_1$, the two parameters α_1 and α_2 have to be interchanged.

At normal incidence, the stop band occurs at $\omega=cK/4\sqrt{\bar{\epsilon}}$ (i.e., $p=\lambda/\sqrt{\bar{\epsilon}}$) and has a bandwidth $\Delta\omega/\omega=\Delta\lambda/\lambda=\eta/2\bar{\epsilon}$. The maximum value of the imaginary part of κ, which represents the attenuation constant, can be determined from the relations in Table I.

The results of this paper can be easily verified experimentally by measuring the oblique reflection coefficient of a liquid-crystal film as a function of the incidence angle. Such an experiment requires adequate choice of the incident wavelength and incident medium. As shown in Fig. 4(b), for $p\sqrt{\bar{\epsilon}}/2\lambda>1$, and $\epsilon'\geq 2$, the reflectance would have three peaks which correspond to the three stop bands.

REFERENCES

*Partly supported by the Naval Electronics Laboratory Center and partly supported by NASA under contract No. NAS 7-100.

[1] H. de Vries, Acta Crystallogr. **4**, 219 (1951).

[2] G. H. Conners, J. Opt. Soc. Am. **58**, 875 (1968).

[3] J. Adams, W. Haas, and J. Dailey, J. Appl. Phys. **42**, 4096 (1971).

[4] Groupe des Cristaux Liquides, La Recherche (Paris) **2**, 435 (1971).

[5] *Liquid Crystals and Their Applications*, edited by T. Kallard (Optosonic Press, New York, 1970).

[6] J. L. Fergason, Mol. Cryst. Liq. Cryst. **1**, 293 (1966).

[7] D. Taupin, J. Phys. (Paris) **30**, 4 (1969).

[8] D. W. Berreman and T. J. Scheffer, Phys. Rev. Lett. **25**, 577 (1970).

[9] E. I. Kats, Zh. Eksp. Teor. Fiz. **59**, 1854 (1970) [Sov. Phys.-JETP **32**, 1004 (1970)].

[10] R. Dreher, G. Meier, and A. Saupe, Mol. Cryst. Liq. Cryst. **13**, 17 (1971).

[11] A. S. Marathay, Opt. Commun. **3**, 369 (1971).

[12] T. Tamir, H. C. Wang, and A. A. Oliner, IEEE Trans. Antennas Propag. **12**, 323 (1964).

[13] C. Elachi, IEEE Trans. Antennas Propag. **20**, 534 (1972).

[14] C. Yeh, K. F. Casey, and Z. A. Kaprielian, IEEE Trans. Microwave Theory Tech. **13**, 297 (1965).

Mode conversion in periodically disturbed thin-film waveguides*

C. Elachi and C. Yeh

Jet Propulsion Laboratory, California Institute of Technology, Pasadena, California 91103

and

Electrical Sciences and Engineering Department, University of California, Los Angeles, California 90024

(Received 4 February 1974)

Mode conversion in a periodically perturbed thin-film optical waveguide is studied in detail. Three different types of perturbations are considered: periodic index of refraction of the film, periodic index of refraction of the substrate, and periodic boundary. The applications in filters, mode converters, and distributed feedback lasers are discussed.

I. INTRODUCTION

Periodic structures have a large field of applications in active and passive optical thin-film structures. Their stop-band passband characteristic can be used for distributed feedback,[1-5] filtering,[4,5] and coupling.[6] Their space-harmonic characteristic can be mainly used for phase-matched nonlinear interactions[7,8] and for coupling to drifting electrons in a thin-film semiconductor.

In a recent paper,[4] the authors have studied the coupling between identical modes in different types of periodic optical thin-film waveguides (surface periodicity, waveguide index periodicity, and substrate periodicity). In this paper we extend our previous work to the case of coupling between nonidentical modes and we evaluate the efficiency of the mode conversion and its use for mode generation, filtering, and distributed feedback. The last application consisting of the use of a higher-order mode to carry the feedback function in a DFB (distributed feedback) laser operating on a lower forward mode. As we will show in this paper, this scheme is sometimes more efficient than in the case where the feedback and direct waves are in the same mode. We also show that in a number of cases, the coupling coefficient (for mode conversion, feedback, or filtering) reaches a maximum in a specific frequency region thus allowing optimization of the system.

The approach used in this paper is slightly different than in Ref. 4; however, for the sake of continuity and to avoid repetitions, our previous notations and a number of results and figures will be used here. Our study is limited to the case of TE waves, but the approach can be applied in a straightforward manner, for TM waves.

We shall study three important types of periodic structures [Fig. 2 upper right-hand corner, or Figs. 1(a) and 1(b) Ref. 4]: periodically inhomogeneous thin-film guides, periodically inhomogeneous substrate guide, and thin-film waveguide with period surfaces. All these structures are technologically feasible.[1-3]

II. WAVE SOLUTION AND BRILLOUIN DIAGRAM

Electromagnetic waves can be guided in a structure which consists of a thin-film dielectric of relative permittivity ϵ_1 imbedded in a medium of refractive permittivity $\epsilon_2 < \epsilon_1$. For a transverse electric (TE) guided wave the field expression is[4,9]

$$\mathbf{E} = \mathbf{e}_y C u(x) \exp(i\kappa z), \tag{1a}$$

$$u(x) = \cos(sx)/\cos(sL) \quad \text{(even modes, } |x| \leq L)$$
$$= \sin(sx)/\sin(sL) \quad \text{(odd modes, } |x| \leq L)$$
$$= \exp(\delta L - \delta|x|) \quad \text{(even modes, } |x| \geq L)$$
$$= \text{sign}(x)\exp(\delta L - \delta|x|) \quad \text{(odd modes, } |x| \geq L), \tag{1b}$$

where $2L$ is the waveguide thickness, C is the field at the the boundary, and s, δ, and κ are the components of the wave vector. These wave vectors are related to the frequency $\omega/2\pi$ by the disperson relations

$$\kappa^2 + s^2 = \epsilon_1 k^2, \tag{2a}$$

$$\kappa^2 - \delta^2 = \epsilon_2 k^2, \tag{2b}$$

$$\delta = s\ \tan(sL) \quad \text{(even modes)}$$

$$= s[-\cot(sL)] \quad \text{(odd modes)}, \tag{2c}$$

where $k = \omega/c$. The above relations have multiple solutions which correspond to the different modes.

If the optical guide is periodic, the field would consist of an infinite number of space harmonics[10] of longitudinal wave vectors

$$\kappa_{pn} = \kappa_p + nK, \tag{3}$$

where p is the mode index, n is the space-harmonic index, $K = 2\pi/\Lambda$ where Λ is the perturbation period, and κ_p is to be determined. The corresponding Brillouin diagram consists of an infinite number of subdiagrams, each identical to the diagram of a homogeneous thin-film waveguide (Fig. 2, Ref. 4). Strong phase-matched coupling occurs at the intersection points between different harmonics, leading to reflection or mode conversion. Two types of coupling could occur.

(i) Codirectional where the group velocities of the two coupled harmonics are parallel [Fig. 3(a)]. In this case the energy is transferred back and forth between the two harmonics.

(ii) Contradirectional, where the group velocities are antiparallel [Fig. 3(b)]. In this case, there is a one-way energy transfer.

III. MODE CONVERSION IN A PERIODICALLY INHOMOGENEOUS GUIDE

Let us consider the case of a periodically inhomogeneous guide imbedded in a homogeneous substrate, where

$$\epsilon_1(z) = \epsilon_1[1 + \eta\cos(Kz)], \quad \text{with } \eta \ll 1 \tag{4}$$

$$\epsilon_2 = \text{const}$$

Without any loss of generality, we consider the interaction between the $n=0$ space harmonic of the pth mode and the neighboring $n=\pm 1$ space harmonics of the qth mode. The phase-matching condition is

$$\kappa_p \pm \kappa_q = K. \quad (5)$$

The plus and minus signs correspond, respectively, to the contradirectional and codirectional interaction. Let ω_{pq} be the frequency at which this condition is satisfied. For η small, all the other space harmonics can be neglected.

To understand and formulate the coupling mechanism, let us consider the pth mode wave of frequency $\omega=\omega_{pq}+\Delta\omega$. The corresponding electric field can be written

$$E_{po} = C_{po} u'_{po}(x)\exp(i\kappa'_p z), \quad (6)$$

where $\kappa'_p=\kappa_p+\Delta\kappa$, and $u'_{po}(x)$ is given by Eq. 1(b) with s and δ replaced by $s'_p=s_p+\Delta s_p$ and $\delta'_p=\delta_p+\Delta\delta_p$, respectively. The terms κ_p, s_p, and δ_p correspond to the homogeneous guide, and $\Delta\kappa$, Δs_p, and $\Delta\delta_p$ are small perturbations caused by the inhomogenety. Due to the presence of the periodic component in the dielectric constant [Eq. (3)] a spatially periodic convection current J_c is generated:

$$\begin{aligned} J_c &= -i\omega\eta\epsilon_0\epsilon_1\cos(Kz)E_{po} \\ &= -i\omega\eta\tfrac{1}{2}(\epsilon_0\epsilon_1)C_{po}u'_{po}(x)h(x) \\ &\quad\times\{\exp[i(\kappa'_p+K)z]+\exp[i(\kappa'_p-K)z]\} \end{aligned} \quad (7)$$

where $h(x)=1$ for $|x|<L$ and $h(x)=0$ for $|x|>L$. This current will excite the two neighboring space harmonics. Let us consider the case of contradirectional longitudinal phase matching. Then a backward qth mode wave $E_{q,-1}$ will be excited. However to determine the effective excitation current for the new qth mode, the current J_c has to be expanded as a function of the transverse modes:

$$u'_{po}(x)h(x)=\textstyle\sum_j a'_{pj}u_j(x) \quad (8)$$

and only the term a'_{pj} has to be taken into account. This corresponds to transverse phase matching. The expansion coefficients a'_{pj} are given by

$$\begin{aligned} a'_{pj} &= [\textstyle\int_{-\infty}^{+\infty} u'_p(x)h(x)u^*_j(x)\,dx][\int_{-\infty}^{+\infty} u_j(x)u^*_j(x)\,dx]^{-1} \\ &= [\textstyle\int_{-L}^{+L} u'_p(x)u^*_j(x)\,dx][\int_{-\infty}^{+\infty} u_j(x)u^*_j(x)\,dx]. \end{aligned} \quad (9)$$

Then the corresponding wave equation is

$$\left(\frac{\partial^2}{\partial x^2}+\frac{\partial^2}{\partial z^2}+\epsilon_1\frac{\omega^2}{c^2}\right)E_{q,-1}$$
$$= -\eta\frac{\omega^2}{2}\frac{\epsilon_1}{c}C_{po}a'_{pq}u_q(x)\exp[i(\kappa'_p-K)z] \quad (10)$$

for $|x|<L$, and for $|x|>L$ we replace ϵ_1 by ϵ_2 in the parentheses. Replacing $E_{q,-1}$ by an expression similar to Eq. (6), we obtain

$$[(s_q+\Delta s_q)^2+(\kappa_q+\Delta\kappa)^2-\epsilon_1\omega^2/c^2]C_{q,-1}u'_q(x)$$
$$=\eta\frac{\omega^2}{2}\frac{\epsilon_1}{c^2}C_{po}a'_{pq}u_q(x)\exp[i(\kappa'_p+\kappa'_q-K)z] \quad (11)$$

for $|x|<L$ and

$$[-(\delta_q+\Delta\delta_q)^2+(\kappa_q+\Delta\kappa)^2-\epsilon_2\omega^2/c^2]C_{q,-1}u'_q(x)$$
$$=\eta\frac{\omega^2}{2}\frac{\epsilon_1}{c^2}C_{po}a'_{pq}u_q(x)\exp[i(\kappa'_p+\kappa'_q-K)z] \quad (12)$$

for $|x|>L$. Using the dispersion relation for the unperturbed case [Eq. (2)], neglecting second-order terms (i.e., small term $\times u'_q=$ small term $\times u_q$), and expressing Δs_q and $\Delta\delta_q$ as function of $\Delta\kappa_q$ and $\Delta\omega$ by differentiating Eqs. (2a)–(2c). Equations (12) and (13) reduce to

$$\left(-\frac{\Delta\kappa}{\kappa_q}+B_q\frac{\Delta\omega}{\omega_{pq}}\right)C_{q,-1}=\eta\frac{\epsilon_1}{4}\left(\frac{\omega}{c\kappa_q}\right)^2 a_{pq}C_{p,0}, \quad (13)$$

where, for the even modes

$$B_q=\left(\frac{\omega_{pq}}{c\kappa_q}\right)^2\left(\epsilon_1\frac{\delta_qL+\sin^2(s_qL)}{\delta_qL+1}+\epsilon_2\frac{\cos^2(s_qL)}{\delta_qL+1}\right),$$

$$a_{pq}=-(\textstyle\int_{-L}^{+L}u_pu^*_q)(\int_{-\infty}^{+\infty}u_qu^*_q)^{-1}$$

$$=\frac{2\delta_q\cos^2(s_qL)}{(\delta_p+\delta_q)(1+\delta_qL)},\quad p\neq q$$

$$=\frac{\delta_pL+\sin^2(s_pL)}{1+\delta_pL},\qquad p=q.$$

For the odd modes, the sines and cosines must be exchanged.

At the same time, the qth mode generates a convection current which excites the pth mode; therefore the coefficients $C_{q,-1}$ and $C_{p,0}$ are also related by

$$\left(-\frac{\Delta\kappa}{\kappa_p}+B_p\frac{\Delta\omega}{\omega}\right)C_{p,0}=\eta\frac{\epsilon_1}{4}\left(\frac{\omega}{c\kappa_p}\right)^2 a_{qp}C_{q,-1}, \quad (14)$$

where B_p and a_{qp} are equivalent to B_q and a_{pq} with p and q interchanged. Thus to satisfy Eqs. (14) and (15), we must have

$$X=(\alpha_p-\alpha_q)Y\pm[(\alpha_p+\alpha_q)^2Y^2-\eta^2\gamma^2_{pq}]^{1/2}, \quad (15)$$

where

$$X^+=\Delta\kappa_p/K;\quad X^-=\Delta\kappa_q/K;\quad Y=\Delta\omega/\omega_{pq};$$

$$\alpha_i=(|\kappa_i|/2K)B_i\quad \text{with } i=p,q;$$

$$\gamma_{pq}=\frac{\epsilon_1}{4}\left(\frac{\omega}{c}\right)^2\frac{1}{K}\left(\frac{a_{pq}a_{qp}}{|\kappa_p\kappa_q|}\right)^{1/2};$$

$$\kappa_q=-|\kappa_q|.$$

The term X corresponds to the normalized change of the longitudinal wave vector when the operating frequency is equal to $\omega_{pq}(1+Y)$. The characteristic of the above solution is that the longitudinal wave vector is complex in a frequency band (called stop band) centered at ω_{pq} and of total width

$$\Omega=2\eta\omega_{pq}\gamma_{pq}/(\alpha_p+\alpha_q) \quad (16)$$

and the imaginary component of the wave vector has a maximum equal to

$$M=\eta K\gamma_{pq}. \quad (17)$$

The corresponding Brillouin diagram is shown in Fig. 1(b). In Figs. 2 and 3 we plotted the two parameters Ω/ω_{pq} and $1/M\lambda$ as a function of L/λ for mode conversion (or mode coupling) from the forward basic mode to

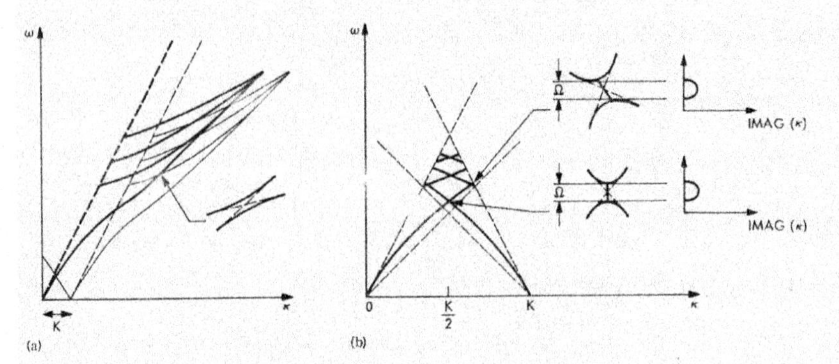

FIG. 1. Interaction regions between two space harmonics; (a) codirectional interaction, (b) contradirectional symmetric and nonsymmetric interaction.

the backward basic and second even ($p=2$) modes. λ is the wavelength in vacuum.

As we would expect from physical reasoning, the value of $1/M\lambda$ (i.e., attenuation or coupling length) is large near cutoff because most of the energy is in the substrate where no coupling occurs. As the frequency increases, more optical energy is enclosed into the periodic guide leading to stronger coupling (i.e., $1/M\lambda$ smaller). This trend continues for the 0-0 modes coupling. However, for the 0-2 modes coupling, after a certain optimal frequency, $1/M\lambda$ starts increasing again because, at high frequencies, the overlap term a_{pq} goes to zero, leading to weaker coupling. The coupling bandwidth behaves in a reverse manner. It should be pointed out that no coupling occurs between an even and an odd mode because a_{pq} vanishes.

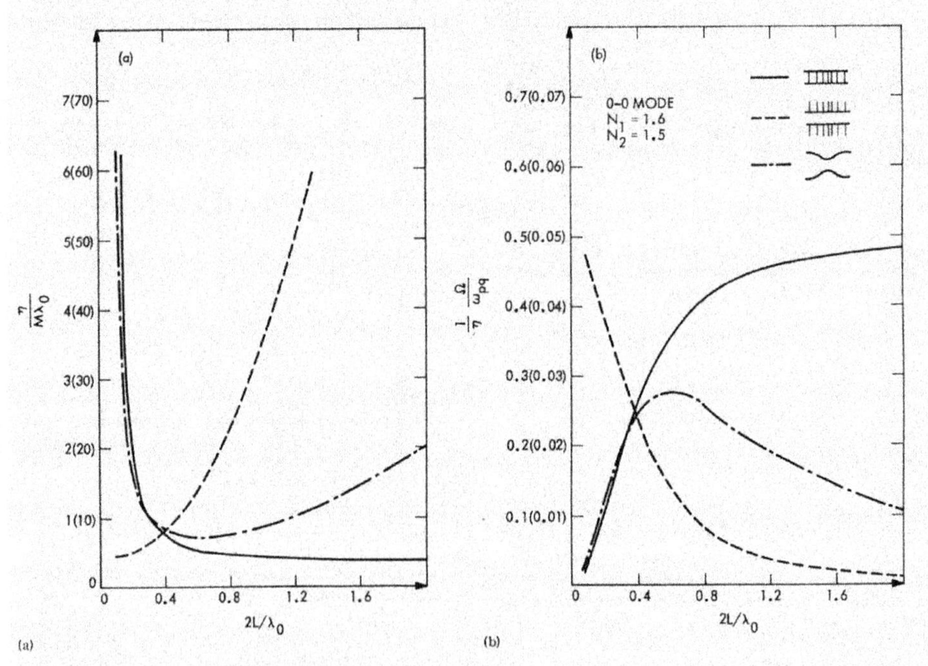

FIG. 2. Normalized interaction length and bandwidth for the 0-0 modes interaction. The scale in parenthesis corresponds to the surface corrugation case.

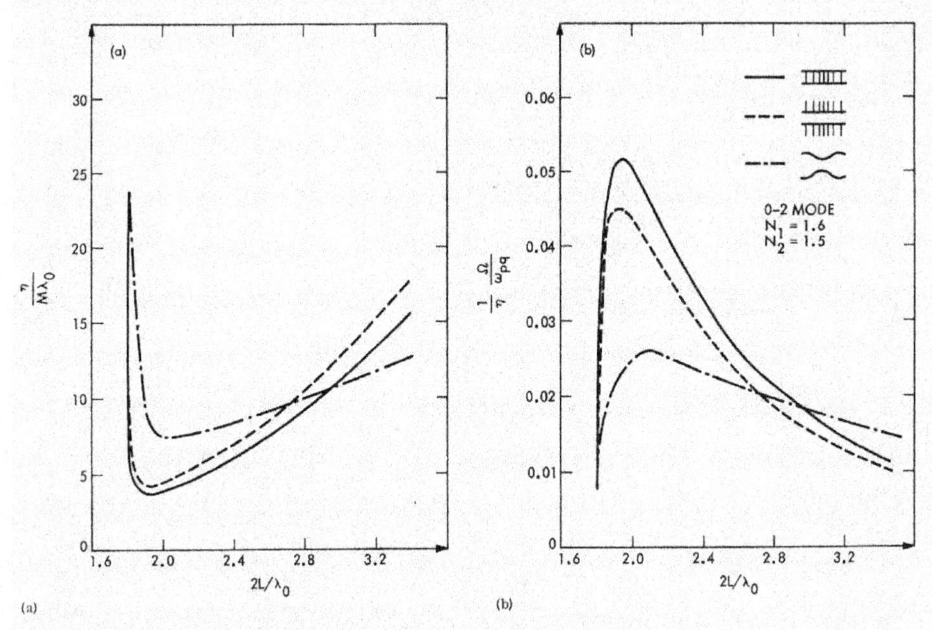

(a) (b)

FIG. 3. Normalized interaction length and bandwidth for the 0-2 modes interaction.

In the case of codirectional interaction, the solution for X is

$$X=(\alpha_p+\alpha_q)Y\pm[(\alpha_p-\alpha_q)^2Y^2+\eta^2\gamma_{pq}^2]^{1/2} \tag{18}$$

and the corresponding diagram is shown in Fig. 3(a). In this case there is no stop band, but the energy is transferred back and forth between the two modes. The characteristic energy transfer length T_{pq} is given by

$$T_{pq}=\pi/2M$$

which can be determined from Figs. 2 and 3.

It should be mentioned that, in the case $p=q$, the above results are identical to the ones derived in Ref. 4 using a direct wave equation solution in a periodic medium.

IV. MODE CONVERSION IN A HOMOGENEOUS WAVEGUIDE WITH A PERIODIC SUBSTRATE

The same method used above can also be applied to the case where the waveguide is homogeneous and the substrate has a periodic dielectric constant. In this case, the source convection current is in the substrate, and Eqs. (16) and (19) are valid with γ_{pq} replaced by

$$\Gamma_{pq}=\frac{\epsilon_2}{4}\left(\frac{\omega}{c}\right)^2\frac{1}{K}\left(\frac{b_{pq}b_{qp}}{|K_pK_q|}\right)^{1/2}, \tag{19}$$

where

$$b_{pq}=\left(\int_{-\infty}^{-L}u_pu_q^*+\int_{L}^{+\infty}u_pu_q^*\right)\left(\int_{-\infty}^{+\infty}u_qu_q^*\right)^{-1}$$

$$=\frac{2\delta_q\cos^2(s_qL)}{(\delta_p+\delta_q)(1+\delta_qL)}.$$

We remark that for $p\neq q$

$$\Gamma_{pq}=(\epsilon_2/\epsilon_1)\gamma_{pq}. \tag{20}$$

This is an important result because it means that for ϵ_2 very close to ϵ_1, the coupling between different modes does not depend appreciably on the fact that the periodicity is in the guide or the substrate. In Figs. 2 and 3 we plotted the corresponding values of Ω/ω_{pq} and $1/M\lambda$. For the 0-0 coupling, $1/M\lambda$ increases with the frequency because the energy tends to concentrate into the guide leading to weak coupling. In the case of the 0-2 interaction, the coupling is also weak (i.e., $1/M\lambda$ large) near the mode 2 cutoff because its energy is very thinly spread into the substrate, and the overlap integral b_{pq} with the 0 mode is very small.

AREA S SURFACE CURRENT x z ϵ_2 ϵ_1 ℓ ≡

(a) (b)

FIG. 4. A surface perturbation could be replaced by a surface current.

V. MODE CONVERSION IN A HOMOGENEOUS GUIDE WITH A PERIODIC BOUNDARY

A similar approach can be used to study the case where the boundary between the waveguide and the substrate is periodically perturbed. Marcuse[9] determined the energy losses in such a guide by considering the boundary region as a thin inhomogeneous layer. In a previous paper[4] the authors determined the coupling between identical modes by an exact solution. Here we will determine the coupling strength and bandwidth by using an excitation surface current to represent the surface perturbation.

Let us consider the pth mode wave of frequency $\omega = \omega_{pq} + \Delta\omega$. From Maxwell's equations we know that (see Fig. 4)

$$\int_l \mathbf{H} = -i\epsilon_0\epsilon_1\omega \int_s \mathbf{E}. \qquad (21)$$

The surface perturbation is equivalent to a surface current, $\mathbf{J}_s$, on a straight boundary, which is related to the fields by

$$\int_l \mathbf{H} = -i\epsilon_0\epsilon_2\omega \int_s \mathbf{E} + \int_s \mathbf{J}_s. \qquad (22)$$

The above two equations imply that

$$\int_s \mathbf{J}_s = -i\epsilon_0(\epsilon_1 - \epsilon_2)\omega \int_s \mathbf{E} \Rightarrow$$

$$J_s\Delta z = -i\epsilon_0(\epsilon_1 - \epsilon_2)\omega\Delta x\Delta z E|_{x=L}$$

which gives the value of J_s [using $\Delta x = \eta L \cos(Kz)$]

$$J_s = -i\epsilon_0\eta L(\epsilon_1 - \epsilon_2)\omega \cos(Kz) C_{p0} \exp(i\kappa_p' z). \qquad (23)$$

This surface current could be phase matched with the neighboring space harmonics. For the case of contradirectional phase matching [Eq. (8)] a backward qth mode is excited. The boundary condition for this new mode is

$$H_z(x=L^+) - H_z(x=L^-) = J_{s1}, \qquad (24)$$

where J_{s1} is the component in phase with the generated wave. The above equation gives

$$[s_q' \tan(s_q' L) - \delta_q'] C_{q,-1} = \tfrac{1}{2}\eta(\epsilon_1 - \epsilon_2) L (\omega_{pq}/c) C_{p0} \qquad (25)$$

for the even modes. For the odd modes, the tangent must be replaced by minus cotangent. At the same time, the qth mode excites the pth mode, leading a relation similar to Eq. (25) (with p and q exchanged), which then implies that

$$[s_q' \tan(s_q' L) - \delta_q'][s_p' \tan(s_p' L) - \delta_p'] = [\tfrac{1}{2}\eta(\epsilon_1 - \epsilon_2) L]^2 (\omega_{pq}/c)^4. \qquad (26)$$

This relation replaces the boundary condition [Eq. (2c)] of the homogeneous guide, and couples the two modes through the perturbation of the boundary. The other dispersion relations [Eqs. (2a) and (2b)] are still valid.

Replacing s', δ', and κ' by their expressions and following the same method as previously we find that Eq. (16) is still valid with γ_{pq} replaced by

$$\gamma_{pq}' = \frac{\epsilon_1 - \epsilon_2}{2}\left(\frac{\omega}{c}\right)^2 \frac{1}{K}\left(\frac{d_p d_q}{|\kappa_p \kappa_q|}\right)^{1/2}, \qquad (27)$$

where

$$d_i = \frac{\delta_i L \cos^2(s_i L)}{1 + \delta_i L}, \quad i = p, q.$$

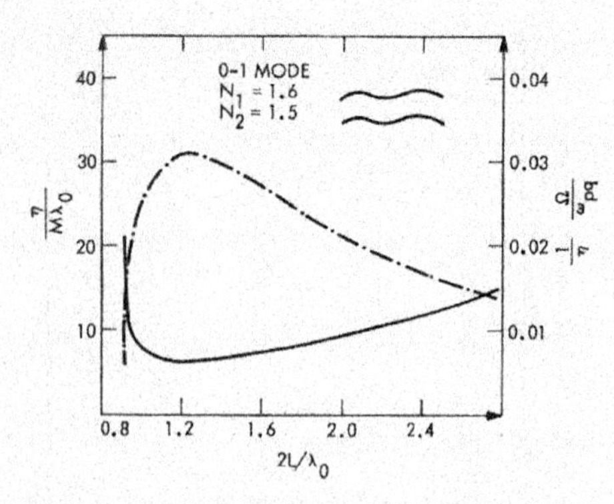

FIG. 5. Normalized interaction length and bandwidth for the 0-1 modes interaction in an antisymmetrically corrugated guide.

For an odd mode, the cosine should be replaced by sine. This result is identical to the one derived by the authors[4] using the exact Floquet solution.

In a symmetrically perturbed waveguide only even-even and odd-odd couplings occur. Even-odd couplings occur in the antisymmetric case. If only one boundary is perturbed, then the above results are still valid with η replaced by $\frac{1}{2}\eta$.

In Figs. 2, 3, and 5 we plotted $\eta/M\lambda$ and $\Omega/\eta\omega_{pq}$ for the 0-0, 0-1, and 0-2 modes coupling. We see that near cutoff, where most of the energy is spread in the substrate, the coupling is weak. The same happens at high frequency when most of the energy is inside the guide and the field is small at the boundary. The strongest coupling (and therefore feefback or filtering) occurs at an optimum frequency somewhere inbetween.

VI. APPLICATIONS AND CONCLUSION

The results of this paper can be applied to the design of integrated optic filters and distributed feedback lasers. The terms η/M_{pq} and Ω_{pq} describe the coupling, filtering, and feedback efficiency. The coupling coefficient can be used to determine the gain threshold of a thin-film distributed feedback laser.[11] It is clear that for an optimum design the operating frequency has to be selected in a well-specified region. It should be pointed out that the product

$$\frac{\Omega}{\omega_{pq}} \frac{1}{M\lambda} = \frac{2}{\lambda(\alpha_p + \alpha_q)} \qquad (28)$$

is the same for all three structures studied and depends only on the properties of the unperturbed guide and the operating frequency. This implies that if a structure is designed for a large coupling coefficient, then its bandwidth would be very narrow and vice versa.

Even though this paper was limited to the case of thin-film waveguides, the same methods could be used to study other types of structures: diffuse waveguides for DFB lasers and filters, diffuse capillary waveguides for CO_2 lasers, and fiber waveguides.

Finally, it should be mentioned that the above results could be easily generalized to any type of periodicity which can be expanded in a Fourier series.[12] In that case, η has to be replaced by the coefficient of the Fourier component used for phase matching.

ACKNOWLEDGMENT

The authors would like to thank Dr. C.H. Papas and Dr. G. Evans for many helpful discussions.

*Partly supported by NASA under contract No. NAS7-100 and partly by NELC.

[1]H. Kogelnik and C.V. Shank, J. Appl. Phys. 43, 2327 (1972).
[2]K.O. Hill and A. Watanabe, Opt. Commun. 5, 289 (1972).
[3]D.P. Schinke, R.G. Smith, E.G. Spencer, and M.F. Gavin, Appl. Phys. Lett. 21, 494 (1972).
[4]C. Elachi and C. Yeh, J. Appl. Phys. 44, 3146 (1973).
[5]A. Yariv, IEEE J. Quantum Electron. QE-9, 919 (1973).
[6]C. Elachi and C. Yeh, Opt. Commun. 7, 201 (1973).
[7]N. Bloembergen and A.J. Sievers, Appl. Phys. Lett. 17, 483 (1970).
[8]S. Somekh and A. Yariv, Appl. Phys. Lett. 21, 140 (1972).
[9]D. Marcuse, *Light Transmission Optics* (Van Nostrand, New York, 1973).
[10]N. Brillouin, *Wave propagation in periodic structures* (Dover, New York, 1953).
[11]C. Elachi, G. Evans, and C. Yeh, Integrated Optics Conference, Optical Society of America, New Orleans, 1974 (unpublished).
[12]C. Elachi, IEEE Trans. Antenna Prop. AP-20, 534 (1972).

Reprinted from:
JOURNAL OF THE OPTICAL SOCIETY OF AMERICA VOLUME 65, NUMBER 4 APRIL 1975

Transversely bounded DFB lasers*

Charles Elachi and Gary Evans

Jet Propulsion Laboratory, California Institute of Technology, Pasadena, California 91103

Cavour Yeh

University of California, Los Angeles, California 90024
(Received 29 June 1974)

Bounded distributed-feedback (DFB) lasers are studied in detail. Threshold gain and field distribution for a number of configurations are derived and analyzed. More specifically, the thin-film guide, fiber, diffusion guide, and hollow channel with inhomogeneous-cladding DFB lasers are considered. Optimum points exist and must be used in DFB laser design. Different-modes feedback and the effects of the transverse boundaries are included. A number of applications are also discussed.

Index Headings: Lasers; Resonant modes.

Great interest has been recently generated in the use of distributed-feedback (DFB) cavities in laser systems. The first DFB laser was developed by Kogelnik and Shank[1,2] in 1971; since then, a number of scientists have reported the development of different types of DFB lasers.[3-8] In most cases, interest was mainly directed toward thin-film integrated-optics sources. However, the use of the distributed-feedback cavity in other types of lasers can have major advantages because it would eliminate the need for mirrors, which are one of the major problems in the design of high-power lasers (because of mirror burning) or ultra-high-frequency lasers (because of low mirror reflectance).

In this paper, we study the interaction of electromagnetic waves with transversely bounded, active, periodic structures that can be used for DFB lasers. We study the coupling between different guided modes and take into account their effective gain. The laser-action threshold gain and the longitudinal field distribution are derived and discussed for a number of different laser configurations. We will specifically consider thin-film lasers, diffuse-guide lasers, fiber lasers, and capillary lasers with inhomogeneous cladding (Fig. 1), and we will show that there are optimum designs for which the gain required for oscillation is at a minimum. Periodic-index and surface-perturbation distributed feedback are covered. Finally, we discuss the applications of the mentioned structures, and, more generally, of the distributed-feedback concept in the fields of planar integrated optics, optic-fiber communication, high-power lasers, vuv lasers, and possibly x-ray lasers.

To avoid excessive mathematical derivations, we limit our study to the case of TE waves for both planar and circular configurations; however, the approach is directly applicable to TM, HE, and EH modes. An $e^{-i\omega t}$ time dependence is assumed.

I. THRESHOLD AND LONGITUDINAL FIELD DISTRIBUTION

Kogelnik and Shank[2] studied the threshold condition and longitudinal field distribution in transversely unbounded DFB lasers in which the forward and backward waves have identical gains and wave vectors. In this section, we extend their results to the case in which the two interacting waves are different. This case is usually encountered when coupling occurs between different guided modes.

In principle, a periodic structure supports an infinite number of space harmonics.[9,10] However, for the case of small periodic perturbations, the coupled-mode theory can be applied near the Bragg frequencies, where only the two phase-matched interacting waves have significant amplitudes. The first-order Bragg phase-matching condition for feedback is

$$\beta_p + \beta_q = 2\pi/\Lambda, \tag{1}$$

where β_p and β_q are the longitudinal wave vectors of the coupled pth and qth modes, and Λ is the period of the perturbation. Let ω_{pq} be the corresponding frequency.

If the operating frequency ω is very close to ω_{pq} (i.e., $\omega = \omega_{pq} + \Delta\omega$), then the coupled-waves equations for the forward pth mode wave $F_p(z)e^{i\beta_p z}$ and backward qth mode wave $B_q(z)e^{-i\beta_q z}$ are[2]

$$\begin{aligned} +\frac{dF_p}{dz} - (C_p g + i\Delta\beta_p)F_p &= i\chi_{pq}B_q\,, \\ -\frac{dB_q}{dz} - (C_q g + i\Delta\beta_q)B_q &= i\chi_{pq}F_p\,, \end{aligned} \tag{2}$$

where C_p and C_q are the gain-efficiency coefficients, g is the medium gain (see Sec. III), χ_{pq} is the coupling coefficient that takes into account the transverse phase matching (see Sec. II), and $\Delta\beta_p$ and $\Delta\beta_q$ are related to $\Delta\omega$ by

$$\begin{aligned} \Delta\beta_p &= \psi_p\Delta\omega = \left.\frac{\partial\beta_p}{\partial\omega}\right|_{\omega_{pq}}\Delta\omega\,, \\ \Delta\beta_q &= \psi_q\Delta\omega = \left.\frac{\partial\beta_q}{\partial\omega}\right|_{\omega_{pq}}\Delta\omega\,, \end{aligned} \tag{3}$$

where ψ_p and ψ_q are inversely proportional to the slopes of the Brillouin diagram (i.e., group velocity) of the unperturbed guide at ω_{pq}. The values of ψ_i are given in Table I for all of the configurations considered.

The general solution of the coupled-waves equations (2) is of the form

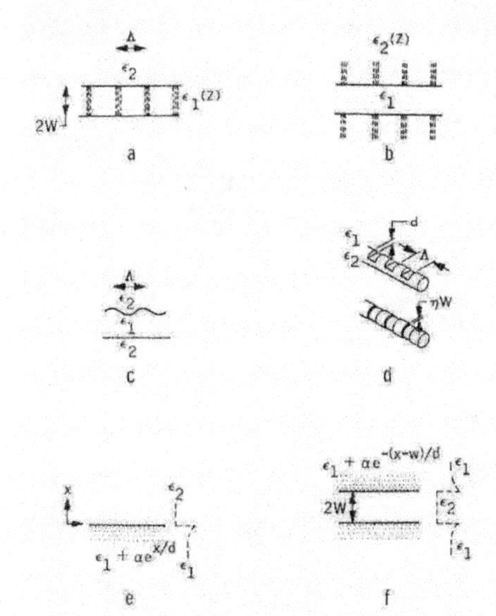

FIG. 1. Bounded DFB laser structures studied. (a) Thin-film DFB laser with a periodic guide index. (b) Thin-film DFB laser with a periodic substrate index. (c) Thin-film DFB laser with a periodic surface corrugation. The cases of one-boundary and both-boundaries corrugation were considered. (d) DFB fiber laser. The upper configuration can be generated by particle-beam machining of a fiber. The lower configuration is harder to achieve; however, it can be treated analytically and is equivalent to the upper one (see text). (e) Diffuse DFB laser. The distributed feedback can be generated by corrugating the surface, and guiding is achieved because of the inhomogeneity. (f) Diffuse-capillary DFB laser. Only the planar structure is considered. In all the following figures, the real dielectric constants are $\epsilon_1=(3.6)^2$ and $\epsilon_2=(3.5)^2$ for the thin-film strucutre; $\epsilon_1=(1.5)^2$ and $\epsilon_2=1$ for the diffuse structures; and $\epsilon_1=(1.5)^2$ and $\epsilon_2=(1.4)^2$ for the fiber structure.

$$F_p=f_1e^{\gamma z}+f_2e^{\gamma' z},$$
$$B_q=b_1e^{\gamma z}+b_2e^{\gamma' z}, \tag{4}$$

and, as the DFB laser is self-oscillating, the boundary conditions

$$F_p(-L/2)=B_q(L/2)=0\;, \tag{5}$$

where L is the laser length, must be satisfied in the case of no end reflections. Replacing F_p and B_q in Eq. (2) by their expressions in Eq. (4), and using the boundary conditions (5), we find

$$\gamma,\gamma'=\frac{C_p-C_q}{2}g+i\frac{\psi_p-\psi_q}{2}\Delta\omega \pm\left[\left(\frac{C_p+C_q}{2}g+i\frac{\psi_p+\psi_q}{2}\Delta\omega\right)^2+\chi_{pq}^2\right]^{1/2}, \tag{6}$$

$$\frac{\gamma-\gamma'}{2}=\pm i\chi_{pq}\sinh\left(\frac{\gamma-\gamma'}{2}L\right), \tag{7}$$

$$F_p=2f_1\sinh\left[\frac{\gamma-\gamma'}{2}\left(z+\frac{L}{2}\right)\right]\exp\left(\frac{\gamma+\gamma'}{2}z-\frac{\gamma-\gamma'}{4}L\right), \tag{8}$$

$$B_q=\pm 2f_1\sinh\left[\frac{\gamma-\gamma'}{2}\left(z-\frac{L}{2}\right)\right]\exp\left(\frac{\gamma+\gamma'}{2}z-\frac{\gamma-\gamma'}{4}L\right). \tag{9}$$

Equations (6) and (7) relate the coupling coefficient, threshold gain, and laser length. Equations (8) and (9) give the longitudinal field distribution. We remark that

$$D=\frac{\gamma-\gamma'}{2}=\left[\left(\frac{C_p+C_q}{2}g+i\frac{\psi_p+\psi_q}{2}\Delta\omega\right)^2+\chi_{pq}^2\right]^{1/2}, \tag{10}$$

$$S=\frac{\gamma+\gamma'}{2}=\frac{C_p-C_q}{2}g+i\frac{\psi_p-\psi_q}{2}\Delta\omega\,. \tag{11}$$

In the case of coupling between two identical modes (i.e., $p=q$), we get $\gamma'=-\gamma$ and

$$\gamma=[(C_pg+i\psi_p\Delta\omega)^2+\chi_{pq}^2]^{1/2}. \tag{12}$$

This equation was derived by Kogelnik and Shank.[2] The threshold condition [Eq. (7)] in the case of $p\neq q$ depends on $\frac{1}{2}(C_p+C_q)g$ and $\frac{1}{2}(\psi_p+\psi_q)\Delta\omega$. Thus the results of Kogelnik and Shank are applicable to the average effective gain and average $\Delta\beta$. However, this is not true for the longitudinal average irradiance distribution which, on the surface of the guide, is proportional to

$$\begin{aligned} I(z)&=\langle[F_p(z)+B_q(z)][F_p(z)+B_q(z)]^*\rangle\\ &\quad+\langle[F_q(z)+B_p(z)][F_q(z)+B_p(z)]^*\rangle\\ &=A\cosh[(C_p-C_q)g\,z]\left\{\left|\sinh\left(Dz+D\frac{L}{2}\right)\right|^2\right.\\ &\quad\left.+\left|\sinh\left(Dz-D\frac{L}{2}\right)\right|^2\right\}, \end{aligned} \tag{13}$$

where A is a normalized parameter that is a function of the laser power and the transverse distribution of the excited modes. The sum in Eq. (13) occurs because both the p and q modes are excited in both the forward and backward directions. Thus, usually there are two independent sets of coupled waves: forward p coupled to a backward q, and a backward p coupled to a forward q.

The solution for Eq. (7) is multivalued, leading to different longitudinal modes.[2] These modes correspond to the laser-cavity spectrum.

II. COUPLING COEFFICIENT

The coupling coefficient χ_{pq} can be derived by solving exactly the wave equation, taking into account all of the space harmonics.[11,12] However, in the case of a small periodic disturbance, a simple perturbation method[13,14] gives the same results as the exact solution.[11]

Let us consider a guided wave propagating in the unperturbed guide. The normalized electric field can be expressed in the form

$$E(x,z)=u(x)e^{i\beta z}, \tag{14}$$

where $u(x)$ is the transverse distribution, which is com-

TABLE I. Parameters F, ψ, and χ for the different configurations studied.

GUIDE	PERIODICITY	F	ψ	χ_{pq}
SLAB (ϵ_2, ϵ_1, 2w)	GUIDE INDEX	*EVEN MODES: $\tan^2(sw)\left[1+\frac{2sw}{\sin(2sw)}\right]$ *ODD MODES: $-\mathrm{cotan}^2(sw)\left[1-\frac{2sw}{\sin(2sw)}\right]$	$\frac{k}{\beta c}(\epsilon_1 A+\epsilon_2 B)$ *EVEN MODES: $A=\frac{\delta w+\sin^2(sw)}{1+\delta w}$ $B=\frac{\cos^2(sw)}{1+\delta w}$ *ODD MODES: $\sin^2\to\cos^2$ $\cos^2\to\sin^2$	$\eta\frac{\epsilon_1}{4}k^2(P_{pq}P_{qp})^{1/2}$ $P_{ij}=2\frac{\delta_i}{\beta_i}\frac{\delta_j}{\delta_i+\delta_j}$ FOR $i\neq j$ $P_{ij}=A_i/\beta_i$ FOR $i=j$
	SUBSTRATE INDEX			$\eta\frac{\epsilon_2}{4}k^2(P_{pq}P_{qp})^{1/2}$ $P_{ij}=2\frac{\delta_i}{\beta_i}\frac{\delta_j}{\delta_i+\delta_j}$
	BOUNDARY			$\eta\frac{\epsilon_1-\epsilon_2}{2}k^2w(P_pP_q)^{1/2}$ $P_i=\beta_i/\delta_i B_i$
FIBER (ϵ_2, ϵ_1, 2w)	BOUNDARY	$\frac{\delta}{s}\left[\frac{J_0/J_1+sw(J_0/J_1)'}{K_0/K_1+\delta w(K_0/K_1)'}\right]$	$\frac{k}{\beta c}\left(\frac{\epsilon_1 A+\epsilon_2 B}{A+B}\right)$ WHERE $A=-\delta\left[\delta w\frac{J_1'K_0}{J_1K_1}+\frac{J_0}{J_1}-sw\right]$ $B=s\left[sw\frac{K_1'J_0}{J_1K_1}+\frac{K_0}{K_1}-\delta w\right]$	AS JUST ABOVE WITH $P=\frac{\beta J_0}{\delta J_1}\left[sw\frac{K_1'}{K_1}+\delta w\frac{J_1'}{J_1}-\frac{s^2+\delta^2}{s\delta}\right]$
DIFFUSE (ϵ_2, $\epsilon_1+ae^{x/d}$)	BOUNDARY	$\gamma\delta dR/\nu$ WHERE: $\gamma=2kd\sqrt{a}$ $R=\frac{\partial}{\partial\nu}\left(\frac{J_{\nu-1}-J_{\nu+1}}{J_\nu}\right)$	$\left[\frac{\sqrt{a}}{c}(1+2\delta d)\frac{J_\nu'}{J_\nu}+\gamma\frac{\sqrt{a}}{c}\frac{J_\nu''}{J_\nu}-\epsilon_2\frac{k}{c\delta}\frac{\partial J_\nu}{\partial\nu}\right]/P\to$	AS ABOVE WITH $P=\frac{2\beta d}{\nu}(\gamma+2\delta d)\frac{\partial J_\nu/\partial\nu}{J_\nu}+\beta/\delta$
CAPILLARY WITH DIFFUSE CLADDING (2w, ϵ_2, $\epsilon_1+ae^{\frac{x+w}{d}}$)	BOUNDARY	*EVEN MODES: $\gamma\delta dR/\nu\,(\tanh\delta w+\delta w\,\mathrm{sech}^2\delta w)$ * ODD MODES: $\tanh\to\mathrm{cotanh}$ $\mathrm{sech}^2\to\mathrm{cosech}^2$	$\left[\frac{\sqrt{a}}{c}(1+2\delta d\tanh\delta w)\frac{J_\nu'}{J_\nu}+\gamma\frac{\sqrt{a}}{c}\frac{J_\nu''}{J_\nu}-\epsilon_2\frac{k}{c\delta}\frac{\partial J_\nu}{\partial\nu}\left\{\tanh\delta w-\epsilon_2\frac{kw}{c}\mathrm{sech}^2\delta w\right\}\right]/P\to$	AS ABOVE WITH $P=\frac{2\beta d}{\nu}\left[\gamma+2\delta d\tanh\delta w\right]\frac{\partial J_\nu/\partial\nu}{J_\nu}+\frac{\beta}{\delta}(\tanh\delta w+\delta w\,\mathrm{sech}^2\delta w)$

pletely determined by the guide configuration, the transverse wave vectors in the guide s, and the cladding (or substrate) δ. The wave-vector components β, s, and δ are related to the operating frequency by the dispersion relations

$$s^2+\beta^2=\epsilon_1 k^2, \tag{15}$$

$$-\delta^2+\beta^2=\epsilon_2 k^2, \tag{16}$$

$$sf(sw)=\delta h(\delta w), \tag{17}$$

where ϵ_1, ϵ_2 are the relative dielectric constants of the guide and the cladding, $2W$ is the width (or diameter) of the guide, and $u(x)$, $f(\xi)$, and $h(\xi)$ are given in Appendix A. The dispersion relation has multiple solutions which correspond to the different guided modes. For the detailed study of the dispersion, field distributions and properties of the structures studied, the reader is referred to Appendix A and Refs. 13 and 15–17.

If any of the parameters ϵ_1, ϵ_2, w or the boundary is periodically perturbed so that

$$(\epsilon_1,\epsilon_2,w)'=(\epsilon_1,\epsilon_2,w)[1+\eta\cos(2\pi z/\Lambda)] \tag{18}$$

or x (boundary) $=\eta'\cos(2\pi z/\Lambda)$ (for the $\frac{1}{2}$-space diffuse guide), with $\eta\ll 1$, (or $\eta'\ll\lambda$), then a pth mode wave $u_p(x)e^{i\beta_p z}$ would generate a displacement current J_c or surface current J_s of different longitudinal wave vector (see Appendix B), i.e.,

$$J_{c,s}\sim\eta\begin{pmatrix}\epsilon_1 & u_p(x)\\ \epsilon_2 & u_p(x)\\ w & u_p(w)\end{pmatrix}e^{i\beta_p z}\cos\left(\frac{2\pi}{\Lambda}z\right)$$

$$\sim\exp\left[i\left(\beta_p+\frac{2\pi}{\Lambda}\right)z\right]+\exp\left[\left(\beta_p-\frac{2\pi}{\Lambda}z\right)\right]. \tag{19}$$

Any of the two components of the perturbation current could be longitudinally phase matched with, and therefore be a source current for, a backward mode q that satisfies the condition

$$\beta_p\pm\beta_q=2\pi/\Lambda. \tag{20}$$

The (+) sign corresponds to contradirectional (or feedback) coupling and the (−) sign corresponds to the codirectional coupling. Only the feedback case is considered in the rest of the paper.

Owing to the presence of the source current, the wave-vector components, s, δ, and β will be perturbed and have to be written as $s'=s+\Delta s$, $\delta'=\delta+\Delta\delta$, and $\beta'=\beta+\Delta\beta$. In the case of guide-index perturbation, the new wave-vector components still satisfy Eqs. (16) and (17), however, Eq. (15) will have to be replaced by a new equation that takes into account the perturbing current.[14] In the case of cladding perturbation, only Eq. (16) must be changed; in the case of surface perturbation, only Eq. (17) must be changed. Solving for $\Delta\beta$, we get the coupling coefficient

$$\chi_{pq}=\mathrm{Im}(\Delta\beta).$$

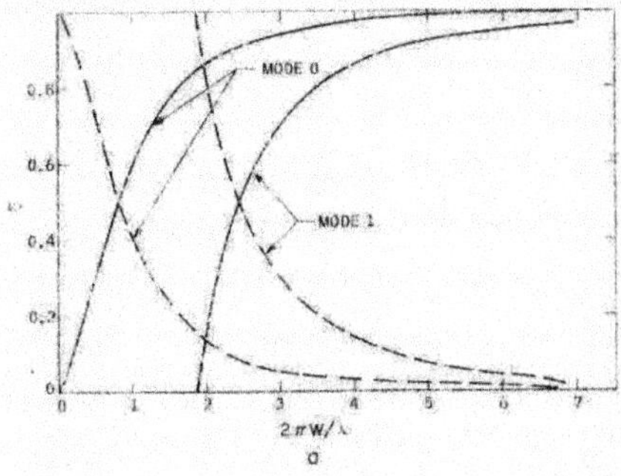

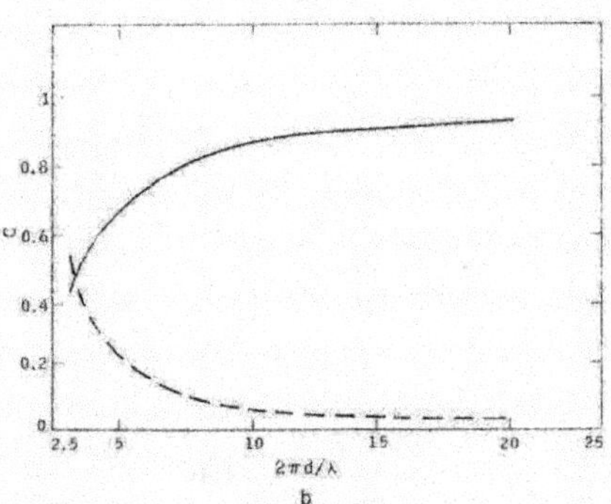

FIG. 2. Effective gain coefficient. (a) Case of a thin film with active guide (continuous line) or substrate (dashed line). (b) Case of mode 0 in a diffuse guide with active homogeneous (dashed line) or inhomogeneous (continuous line) half-space.

The details of this approach can be found in Ref. 14.

In Table I, we give the expressions for the coupling coefficients of the configurations studied. The corresponding plots are discussed in Sec. V.

The approach outlined in the foregoing is valid for ϵ and η complex; thus both index perturbation and gain perturbation are covered.

III. EFFECTIVE GAIN COEFFICIENT

If the guide or the cladding is an active medium that has gain coefficient g, the effective gain coefficient would then be equal to Cg. This is because the optical energy is never completely confined to the guide or the cladding. The coefficient C can be determined by using complex dielectric constants and solving the dispersion relations [Eqs. (15)–(17)] for β which is now complex. In the case of low gain, a first-order Taylor expansion gives

$$C=\epsilon_1^{1/2}\,(k/\beta)\,\frac{F}{1+F} \tag{21}$$

for an active guide, and

$$C=\epsilon_2^{1/2}\,(k/\beta)\,\frac{1}{1+F} \tag{22}$$

for an active cladding, where

$$F=\frac{\delta}{s}\,\frac{f+s\,df/ds}{h+\delta\,dh/d\delta}\,. \tag{23}$$

The expressions of F are given in Table I.

In the case of an active slab or fiber (TE mode) guide, the coefficient can be determined from simple physical reasoning as

$$C=\frac{P_i}{P_o+P_i}\,\frac{1}{\cos\phi}\,, \tag{24}$$

where P_i and P_o are, respectively, the power inside and outside the guide, and ϕ is the angle between the

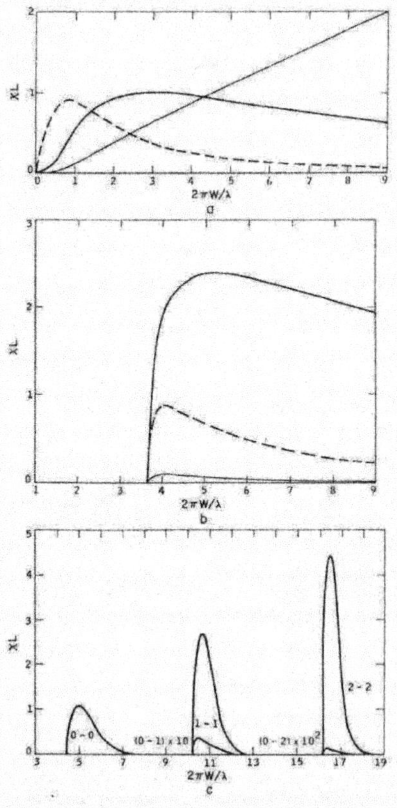

FIG. 3. Coupling coefficient. (a) 0–0 mode coupling in a thin-film waveguide. (b) 0–2 mode coupling. The dashed line corresponds to a periodic substrate for $\eta L/w=2.5$. The heavy line corresponds to a periodic boundary for $\eta L/w=25$. The light line corresponds to a periodic guide index for $\eta L/w=0.25$. The light line levels off at large values of $2\pi w/\lambda$. (c) Different modes coupling in a fiber for $\eta L/w=4$.

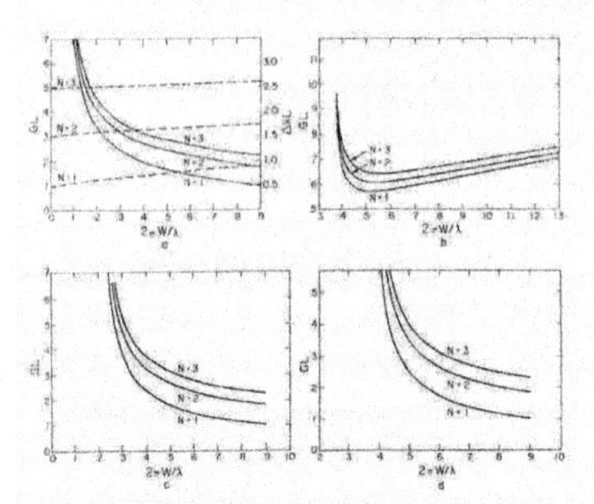

FIG. 4. Threshold-gain coefficient (continuous line) and longitudinal spectrum or frequency mismatch ΔkL (dashed line) for a thin-film DFB laser with periodic guide index and an active guide operating at different modes coupling (or feed back). (a) 0–0 coupling, (b) 0–2 coupling, (c) 1–1 coupling, (d) 2–2 coupling. The different values of N correspond to the different longitudinal modes. $\eta L/w = 0.25$.

optical ray and the z axis. Both Eqs. (23) and (24) give the same value for C. Several plots for different cases are given in Fig. 2.

IV. HIGHER-ORDER INTERACTIONS

High-order Bragg coupling can also be used for feed back. In this case the phase-matching condition is

$$\beta_p + \beta_q = n2\pi/\Lambda, \tag{25}$$

where n is an integer. This could be achieved in two different ways.

If the perturbation is not sinusoidal but a periodic function that can be written in a Fourier-series form, then the nth component can be used for phase-matched coupling and all of the results of this paper are valid with η replaced by the nth Fourier coefficient.

Even if the perturbation is sinusoidal, higher-order coupling exists. However, in this case the coupling coefficient is proportional to

$$\chi \sim \eta^n . \tag{26}$$

V. NUMERICAL RESULTS AND APPLICATIONS

All four configurations shown in Fig. 1 were studied in detail. In the case of a thin slab, we considered the distributed feed back due to index sinusoidal periodicity in the guide and the substrate; for all four configurations, we considered the case of sinusoidal boundary perturbation. The major characteristics of interest are (i) the threshold-gain curves as functions of the operating wavelength for different longitudinal modes, (ii) the longitudinal modes spectrum, and (iii) the longitudinal field distribution. First, we will discuss the behavior of the gain coefficient and coupling coefficient. We then discuss in detail the thin-film case, and finally present in a more-condensed way the properties of the other structures, with emphasis on practical applications. For simplicity, all of the structures will be divided into two regions. In the cases of Fig. 1(a) to 1(d), the central region will be called "guide" and the outside region "cladding." In the case of Fig. 1(e) the lower half-space will be called "guide" and the upper region "cladding." In the case of Fig. 1(f), the central region will be called "channel" and the outer region "guide." This notation is arranged so that at high frequency all of the energy is in the guide region. This is evident in the case of the slab and the fiber. In the case of the inhomogeneous structures, the energy tends to concentrate in the high-index region next to the boundary but still in the guide region.

Gain and coupling coefficients

The simplest characteristic that could be generalized is the behavior of the gain coefficient C_q. In the case of an active guide, C_q increases as the frequency increases away from cutoff because more energy is concentrated into the guide. In the case of an active cladding or channel, C_q decreases down to 0 as the frequency increases. This behavior is clearly seen in Fig. 2.

In the case of a slab with guide-index periodicity, the coupling starts at zero, near cutoff, because most of the energy is in the substrate, and then increases with frequency. If the coupled modes are identical, this coefficient levels off when almost all of the energy is inside the guide. However, if the coupled modes are not identical, the coupling coefficient reaches a peak and then goes to zero because the overlap integral vanishes at high frequency (i.e., the distributions inside the guide are orthogonal). This behavior is clearly seen in Figs. 3(a) and 3(b). This type of periodic structure couples only modes of the same symmetry (even–even or odd–odd).

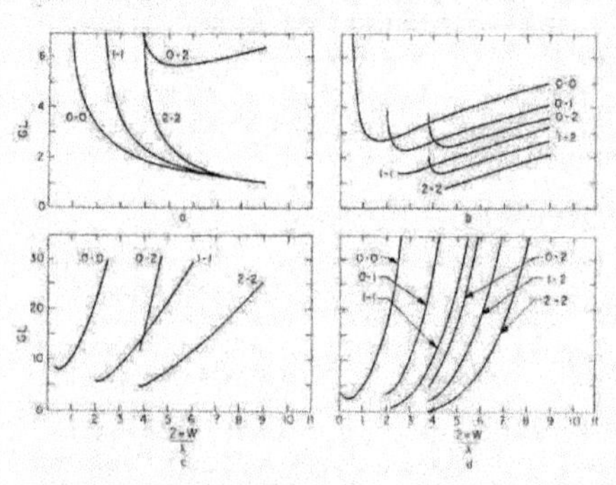

FIG. 5. Threshold-gain coefficient for a thin-film DFB laser. (a) Active guide and periodic guide index, $\eta L/w = 0.25$; (b) active guide and periodic substrate index, $\eta L/w = 2.5$; (c) active substrate and periodic guide index, $\eta L/w = 0.25$; (d) active substrate and periodic substrate index, $\eta L/w = 2.5$.

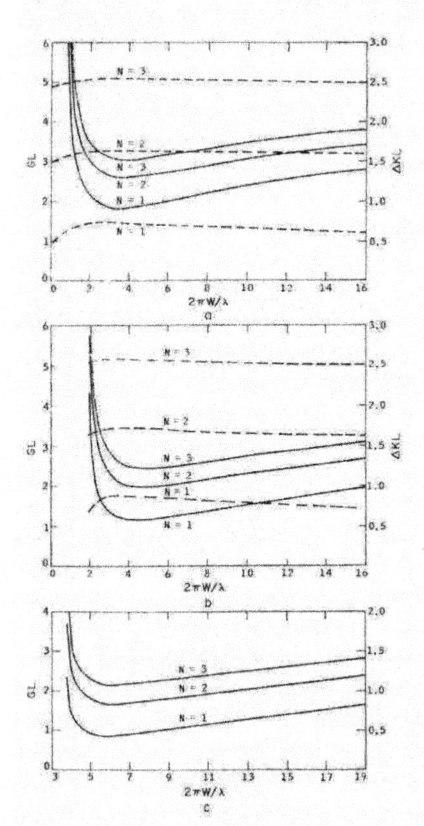

FIG. 6. Threshold-gain coefficient (continuous line) and longitudinal spectrum (dashed line) for a thin-film DFB laser with surface corrugations. N corresponds to the different longitudinal spectra and $\eta L/w = 25$. (a) 0–0 mode coupling, (b) 0–1 mode coupling (nonsymmetric corrugation), (c) 0–2 coupling.

In the case of a slab with a substrate periodic index, the coupling starts at zero, increases to a maximum, then decreases to zero when most of the energy is confined inside the guide [Figs. 3(a) and 3(b)].

In the case of surface corrugation, the coupling is small near cutoff because most of the energy is spread in the cladding. At first, the coupling increases with the frequency. In the case of the slab and fiber, Fig. 3, it reaches a maximum then decreases because the energy gets into the guide and the field at the surface becomes very small. In the case of the diffuse guide, the energy tends to concentrate into the high-index region that is next to the perturbed boundary. Thus, the coupling continues to increase until, evidently at very high frequency, ηw becomes greater than λ and our theory is no longer valid (the Rayleigh condition[18] is no longer satisfied.) This behavior is shown clearly in Fig. 9.

A symmetric surface corrugation will couple only even-even modes or odd–odd. An antisymmetric corrugation would couple only even–odd modes. If only one boundary is corrugated, then all types of coupling are possible. In this last case, the values of η have to be doubled in the figures.

Slab-waveguide DFB laser

In Figs. 4–6, we plotted the threshold gain GL required for oscillation and the frequency mismatch ($\Delta kL = \Delta\omega L/c$) as a function of the operating wavelength, for a number of cases and for different longitudinal modes. Some of these curves show that there are optimum regions where the gain is at a minimum. These regions would correspond to an optimum design. To illustrate, let us consider the case of a surface-corrugated DFB laser with gain in the guide and 0–0 mode coupling [Fig. 6(a)]. For $\lambda = 0.9$ μm, $2w = 1.2$ μm, $L = 1$ mm, and $\eta = 1.5\times10^{-2}$, the threshold-gain coefficients for the first three longitudinal modes are, respectively,

$$G_1 = 20 \text{ cm}^{-1},$$

$$G_2 = 27 \text{ cm}^{-1},$$

$$G_3 = 30 \text{ cm}^{-1}.$$

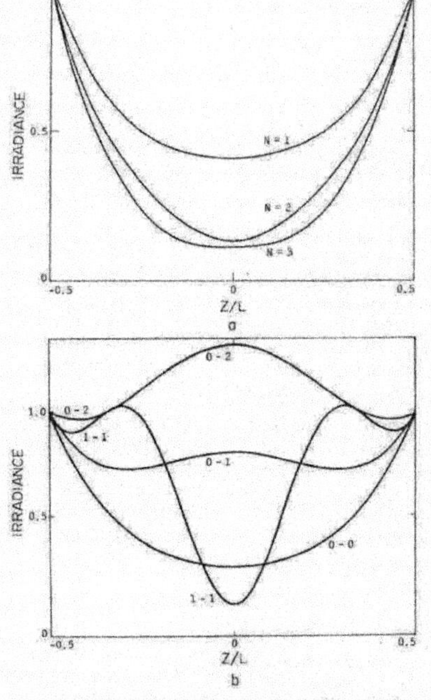

FIG. 7. Field distribution along a thin-film DFB laser with surface corrugation. (a) 0–0 mode coupling, $2\pi w/\lambda = 3.5$, and different values of N. (b) $N = 1$, $2\pi w/\lambda = 6$, and different modes coupling.

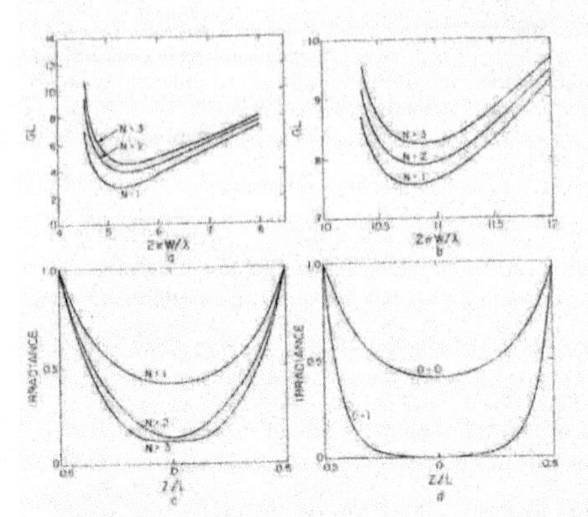

FIG. 8. Threshold-gain coefficient [(a) and (b)] and field distribution [(c) and (d)] for a DFB fiber laser. (a) 0–0 mode coupling, (b) 0–1 mode coupling, (c) 0–0 mode coupling, (d) $N=1$. $\eta L/w = 4$.

If we consider the 0–1 modes coupling (i.e., adequately change the grating period), the threshold-gain coefficients decrease to

$$G_1 = 12 \ \text{cm}^{-1},$$

$$G_2 = 20 \ \text{cm}^{-1},$$

$$G_3 = 25 \ \text{cm}^{-1}.$$

For the 0–2 modes coupling, the threshold gains are even lower. These gains are well within the limit of available active materials (i.e., semiconductors, dye).

It is clear from the curves in Figs. 4–6 that the selection of the DFB laser parameter is critical. The first longitudinal mode does not oscillate exactly at the Bragg frequency. This is because the reflected wave is shifted by $\pi/2$ relative to the orginal wave, and the double reflected wave is out of phase with the original one. This occurs only for the cases of boundary perturbation or real-index perturbation.

In Fig. 7, we plotted the longitudinal irradiance distribution in the case of 0–0 and 0–1 coupling in a surface-corrugated guide at different frequencies. The irradiance varies appreciably as a function of z, suggesting that, in the case of above-threshold laser operation, the field distribution has to be accounted for in the original coupled equations, leading to a nonlinear system of equations.

Fiber DFB laser

The configuration for a fiber DFB laser that we considered is that shown in Fig. 1 (d). The upper configuration can be achieved by particle-beam machining or other techniques. For angularly symmetric TE modes, this configuration is equivalent to the lower configuration with groove depth equal to

$$\eta w = \frac{d}{\pi}\left(\frac{d}{2w}\right)^{1/2},$$

which can be derived by Fourier-series expansion.

In Fig. 8, we plotted the threshold gain and field distribution for a fiber DFB laser. The general behavior is similar to the case of a slab. To illustrate, let us consider the case in which $d/2w = 0.01$, $\eta = 6\times10^{-4}$, $L/w = 6600$, and $\lambda/w = 1.2$. For $\lambda = 1$ μm, this corresponds to $w = 0.83$ μm, $d = 0.016$ μm, and $L = 5.5$ mm. The corresponding threshold gain for the 0–0 mode coupling is [Fig. 8(a)]

$$GL \simeq 3 - \text{total gain} = 25 \text{ dB}$$

for the first longitudinal mode. This gain can be achieved with many active materials (for instance dye or neodymium doping). A larger value of $d/2w$ would lead to even lower threshold gain. Such a laser could

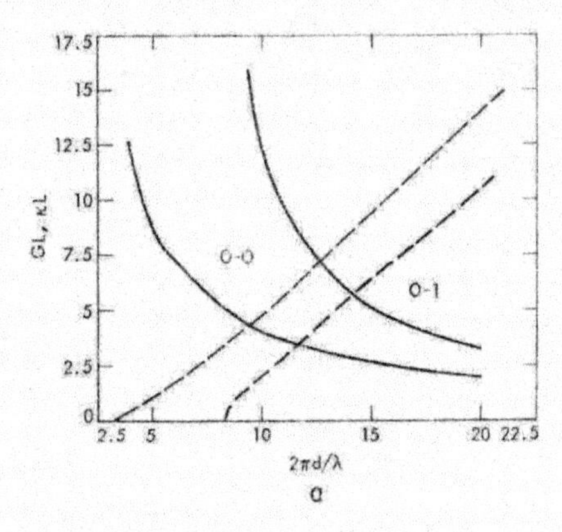

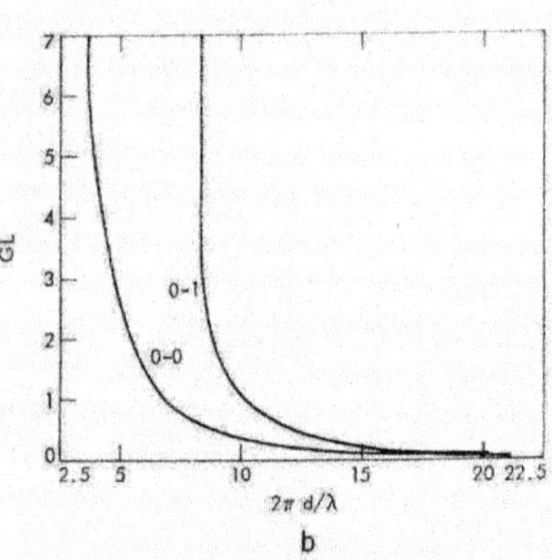

FIG. 9. Threshold-gain coefficient (continuous line) and coupling coefficient (dashed line) for half-space diffuse guide. (a) Gain in the homogeneous upper half-space. (b) Gain in the diffuse lower half-space. $\eta L/d^2 = 75$, $\alpha = 0.1$.

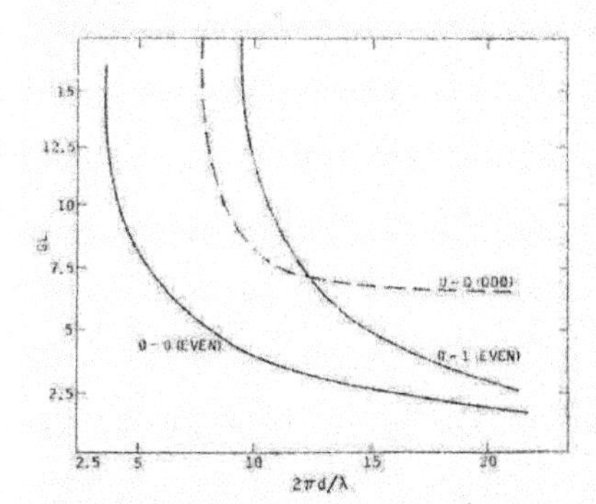

FIG. 10. Threshold-gain coefficient for a diffuse-capillary DFB laser. $\alpha = 0.1$ (continuous line) and $\alpha = 0.025$ (dashed line). $\eta L/d^2 = 75$.

be used as the light source in an optical-network system thus eliminating the need of coupling from a planar source to a circular guide.

Diffuse and channel DFB laser

In Figs. 9 and 10 we plotted the threshold gain for the diffuse guide Fig. 1 (c) and channel guide Fig. 1 (f) for a number of cases. The diffuse guide is attractive because it is very simple to construct. The channel guide can be used in capillary gas lasers, with the advantage that the cladding inhomogeneity allows waveguiding; however, it has the disadvantage that the active medium interacts with the evanescent wave only.

To illustrate, let us consider the case of a diffuse-guide DFB laser. For $\lambda = 0.63\ \mu m$, $d = 1\ \mu m$, $\eta = 0.015\ \mu m$, the threshold gain required for the case in which the inhomogeneous half-space is active is

$$GL = 0.4 \quad \text{(for 0–0 mode coupling)},$$

$$GL = 1.5 \quad \text{(for 0–1 mode coupling)}.$$

These gains are well below the limits of many active materials (dye, semiconductors). If we consider the case of a capillary DFB laser with diffuse cladding, then for $\lambda = 10\ \mu m$, $d = 33\ \mu m$, $2w = 20\ \mu m$ and $\eta = 0.5\ \mu m$ we get

$$GL = 1.8.$$

VI. SUMMARY AND CONCLUSION

We studied the application of the DFB concept to a number of waveguide lasers. An optimum choice of the waveguide parameters and the feedback modes could minimize the gain needed for oscillation. In most cases, an optimum region was shown to exist. This optimum region cannot be derived by a treatment of the unbounded case.

We have also shown that in all of the configurations studied, threshold oscillation can be achieved by use of very reasonable values for the gain and geometric dimensions. Thus, the DFB approach seems to be feasible, not only for thin-film lasers but for a wide variety of lasers in the optical regions. This concept was even suggested for x-ray lasers[19] using zeolite crystals as a DFB capillary-type cavity and a gas, highly excited by optical or electron-beam pumping, as the active medium.

The details of the analysis presented in this paper can be found in Ref. 20.

ACKNOWLEDGMENT

We would like to thank Professor C. H. Papas for many helpful discussions.

APPENDIX A: DISPERSION RELATIONS FOR TE MODES

For a thin-film dielectric slab (Fig. 1), the transverse field distribution is[17]

$$u(x) = \begin{cases} \cos(sx)/\cos(sw), & \text{even modes} \\ \sin(sx)/\sin(sw), & \text{odd modes} \end{cases} \quad |x| < w$$
$$\qquad \exp(-\delta|x| + \delta w), \quad x > w$$

and the dispersion relations are

$$s^2 + \beta^2 = \epsilon_1 k^2, \tag{A1}$$

$$-\delta^2 + \beta^2 = \epsilon_2 k^2, \tag{A2}$$

$$\delta = s \begin{cases} \tan(sw) & \text{(even)} \\ -\cot(sw) & \text{(odd)}. \end{cases} \tag{A3}$$

For a fiber (Fig. 1) the transverse distribution of the longitudinal magnetic field (for TE waves) is given by[17]

$$u(r) = \begin{cases} e^{i\nu\phi} J_\nu(sr)/J_\nu(sw) \\ e^{i\nu\phi} K_\nu(\delta r)/K_\nu(\delta w) \end{cases}$$

and the dispersion relation for $\nu = 0$ are Eq. (A1), Eq. (A2), and

$$s\frac{J_0(sw)}{J_1(sw)} = -\delta\frac{K_0(\delta w)}{K_1(\delta w)}, \tag{A4}$$

where J_ν and K_ν are Bessel functions.

For a diffuse half-space (Fig. 1) the transverse field distribution is[16]

$$u(x) = \begin{cases} e^{-\delta x}, & x > 0 \\ J_\nu(2\sqrt{\alpha}\, kd\, e^{x/2d})/J_\nu(2\sqrt{\alpha}\, kd), & x < 0 \end{cases}$$

and the dispersion relations are

$$-\delta^2 + \beta^2 = \epsilon_2 k^2, \tag{A5}$$

$$-(\nu/2d)^2 + \beta^2 = \epsilon_1 k^2, \tag{A6}$$

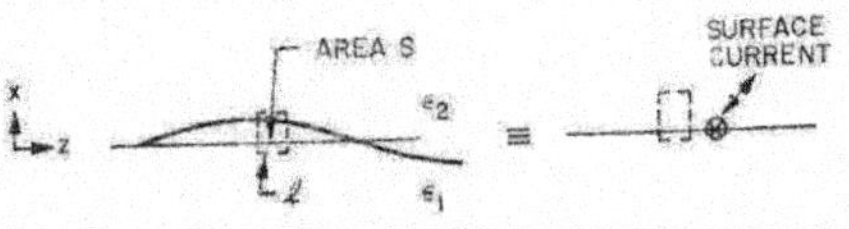

FIG. 11. Contour of integration that gives the surface current equivalent to the surface perturbation.

$$-\sqrt{\alpha}\, kd\, \frac{J_\nu'(2\sqrt{\alpha}\, kd)}{J_\nu(2\sqrt{\alpha}\, kd)} = \delta d \,. \qquad \text{(A7)}$$

For a hollow slab imbedded in an inhomogeneous substrate, we get

$$u(x) = \begin{cases} \left.\begin{array}{ll} \cosh(\delta x)/\cosh(\delta w) & \text{(even modes)} \\ \sinh(\delta x)/\sinh(\delta w) & \text{(odd modes)} \end{array}\right\}, & |x| < w \\ J_\nu\left(2\sqrt{\alpha}\, kd \exp\left(-\dfrac{|x| - w}{2d}\right)\right)\Big/ J_\nu(2\sqrt{\alpha}\, kd)\,, & |x| > w \end{cases}$$

and the dispersion relations are Eq. (A5), Eq. (A6), and

$$-\sqrt{\alpha}\, kd\, \frac{J_\nu'(2\sqrt{\alpha}\, kd)}{J_\nu(2\sqrt{\alpha}\, kd)} = \delta d \begin{cases} \tanh(\delta w) & \text{(even modes)} \\ \coth(\delta w) & \text{(odd modes).} \end{cases} \qquad \text{(A8)}$$

The expressions of the functions f and h used in the text are evident from Eqs. (A3), (A4), (A7), and (A8).

APPENDIX B: EQUIVALENT SOURCE CURRENTS

Ampere's equation in a dielectric is

$$\nabla \times \vec{H} = -i\omega\epsilon\vec{E} + \vec{J}\,.$$

If the dielectric is perturbed by a value $\Delta\epsilon \ll \epsilon$, then

$$\nabla \times \vec{H} = -i\omega\epsilon\vec{E} - i\omega\Delta\epsilon\vec{E} + \vec{J} = -i\omega\epsilon\vec{E} + \vec{J}_e + \vec{J}\,,$$

where $\vec{J}_e = -i\omega\Delta\epsilon\vec{E}$ is an equivalent displacement current. In the case of surface perturbation, a contour integral near the boundary (see Fig. 11) gives an equivalent surface current

$$\vec{J}_s = -i\epsilon_0(\epsilon_1 - \epsilon_2)\,\eta w\omega \cos(Kz)\, C_p\, e^{i\beta_p z}\,.$$

For a periodic perturbation, both J_e and J_s would be periodic and would consist of an infinite number of terms,

$$J_{e,s} = \sum_n J_n \exp\left[i\left(\beta_p + n\frac{2\pi}{\Lambda}\right)z\right]\,.$$

In the case of sinusoidal periodicity, then only J_1 and J_{-1} are different from zero.

*Partly supported by NASA Contract No. NAS7-100 and partly by NELC.

[1] H. Kogelnik and C. V. Shank, Appl. Phys. Lett. 18, 152 (71).
[2] H. Kogelnik and C. V. Shank, J. Appl. Phys. 43, 2327 (72).
[3] C. V. Shank, J. E. Bjorkholm, and H. Kogelnik, Appl. Phys. Lett. 18, 395 (71).
[4] K. O. Hill and A. Watanabe, Opt. Commun. 5, 289 (72).
[5] D. P. Schinke, R. G. Smith, E. G. Spencer, and M. F. Gavin, Appl. Phys. Lett. 21, 494 (72).
[6] J. E. Bjorkholm and C. V. Shank, Appl. Phys. Lett. 20, 3061 (72).
[7] J. E. Bjorkholm, T. P. Sosnowski, and C. V. Shank, Appl. Phys. Lett. 22, 132 (73).
[8] M. Nakamura, A. Yariv, H. W. Yen, S. Somekh, and H. L. Garvin, Appl. Phys. Lett. 22, 515 (73).
[9] N. Brillouin, *Wave Propagation in Periodic Structures* (Dover, New York, 1954).
[10] T. Tamir, H. C. Wang, and A. A. Oliner, IEEE Trans. MTT-12, 323 (64).
[11] C. Elachi and C. Yeh, J. Appl. Phys. 44, 3146 (73).
[12] R. E. DeWaines and W. F. Hall, Appl. Phys. Lett. 23, 28 (73).
[13] D. Marcuse, IEEE Trans. QE-81, 61 (72).
[14] C. Elachi and C. Yeh, J. Appl. Phys. 45, 3494 (1974).
[15] N. S. Kapany and J. J. Burke, *Optical Waveguides* (Academic, New York, 1972).
[16] E. H. Conwell, Appl. Phys. Lett. 23, 328 (73).
[17] D. Marcuse, *Light Transmission Optics* (Van Nostrand, New York, 1973).
[18] Lord Rayleigh, *The Theory of Sound, Vol. II* (Dover, New York, 1945).
[19] C. Elachi, G. Evans, and F. Grunthaner, Appl. Opt. 14, 14 (1975).
[20] G. Evans, PhD thesis (California Institute of Technology, 1975).

352 IEEE TRANSACTIONS ON ANTENNAS AND PROPAGATION, VOL. AP-23, NO. 3, MAY 1975

Transients in a Periodic Slab: Coupled Waves Approach

CHARLES ELACHI, MEMBER, IEEE, DWIGHT L. JAGGARD, STUDENT MEMBER, IEEE AND C. YEH, MEMBER, IEEE

Abstract—**The reflection and transmission of rectangular and Gaussian pulses impinging on a periodically stratified slab are studied and a number of examples are illustrated and analyzed. The coupled waves approach is used to derive the reflection and transmission coefficient, then the fast Fourier transform is used to illustrate the transient responses.**

I. Introduction

Most of the work in the field of electromagnetic wave propagation in periodic structures has been directed toward continuous monochromatic waves [1]–[6]. However, it is of importance to study the transient response of these structures to finite pulses. In this paper we study the reflection and transmission of a rectangular pulse and a Gaussian pulse impinging on a periodic structure of finite length l. In our analysis, we consider the general case of transversely bounded structures where the coupled waves could be different modes and therefore have different longitudinal wave-vectors [6]. For simplicity, and with no loss of generality, the numerical examples will correspond to the special case of transversely unbounded structures. The coupled wave approach [7]–[9] is used, and we discuss briefly its limitations relative to the more general Floquet approach [1]–[5]. There is no limitation on the type of periodicity as long as the coupling coefficient χ is small enough to satisfy the coupled wave approach conditions.

II. Transfer Functions

Let us consider a wave $F(z)U_p(x)e^{i\beta_p z}$ propagating in a transversely bounded periodic structure. The index p represents the guided mode, β_p is its longitudinal wave number, and $U_p(x)$ is its transverse dependence. This wave can be effectively coupled to a backward q mode wave $B(z)U_q(x)e^{-i\beta_q z}$ if the phase matching condition

$$\beta_p + \beta_q = \frac{2\pi}{\Lambda} \tag{1}$$

is satisfied, where Λ is the period of the structure (Fig. 1). Let $\omega_{pq}/(2\pi)$ be the frequency at which phase matching occurs. The coupled waves equations are well known to be [7]–[9]:

$$\frac{dF}{dz} - i\delta_p F = i\chi_{pq} B$$

Manuscript received July 29, 1974; revised December 30, 1974. This work was sponsored in part by NASA under Contract NAS7-100 and in part by NELC.
C. Elachi and D. L. Jaggard are with the California Institute of Technology, Jet Propulsion Laboratory, Pasadena, Calif. 91103.
C. Yeh is with the Electrical Sciences and Engineering Department, University of California, Los Angeles, Calif. 90024.

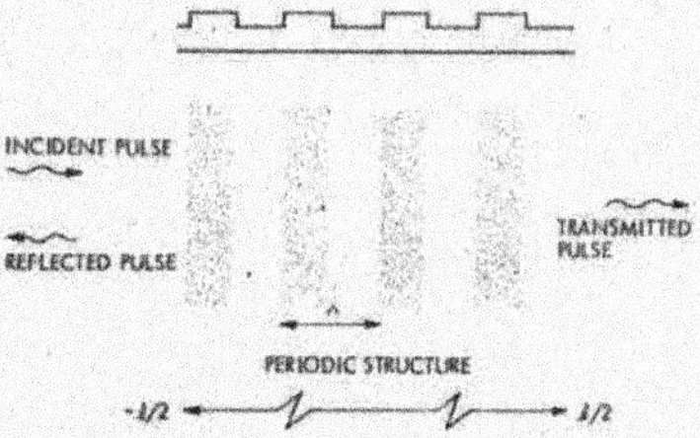

Fig. 1. Configuration. l is thickness of periodic slab.

$$\frac{dB}{dz} + i\delta_q B = -i\chi_{pq} F \tag{2}$$

where χ_{pq} is the coupling coefficient which depends on the transverse functions $U_p(x)$ and $U_q(x)$ [6], δ_p and δ_q are the wave numbers' mismatches and they are related to the frequency mismatch by

$$\delta_p = \psi_p \,\Delta\omega = \left.\frac{\partial \beta_p}{\partial \omega}\right|_{\omega_{pq}} \Delta\omega$$

$$\delta_q = \psi_q \,\Delta\omega = \left.\frac{\partial \beta_q}{\partial \omega}\right|_{\omega_{pq}} \Delta\omega$$

$$\Delta\omega = \omega - \omega_{pq}$$

$$\omega = \text{angular frequency.} \tag{3}$$

ψ_p and ψ_q are the slopes of the dispersion curves of the guided waves. If l is the length of the coupling region, then the longitudinal boundary conditions are (see Fig. 1)

$$F\left(-\frac{l}{2}\right) = 1 \qquad F\left(\frac{l}{2}\right) = T \tag{4}$$

$$B\left(-\frac{l}{2}\right) = R \qquad B\left(\frac{l}{2}\right) = 0 \tag{5}$$

where T and R are the transmission and reflection coefficients. The solution for the coupled wave equation (2) with the above boundary conditions gives

$$R = \frac{i\chi_{pq}}{D \coth Dl - i\dfrac{\delta_p + \delta_q}{2}} \tag{6}$$

$$T = \frac{D e^{-i(\delta_q - \delta_p)l}}{D \cosh Dl - i\dfrac{\delta_p + \delta_q}{2} \sinh Dl} \tag{7}$$

Printed in U.S.A. Annals No. 505AP008

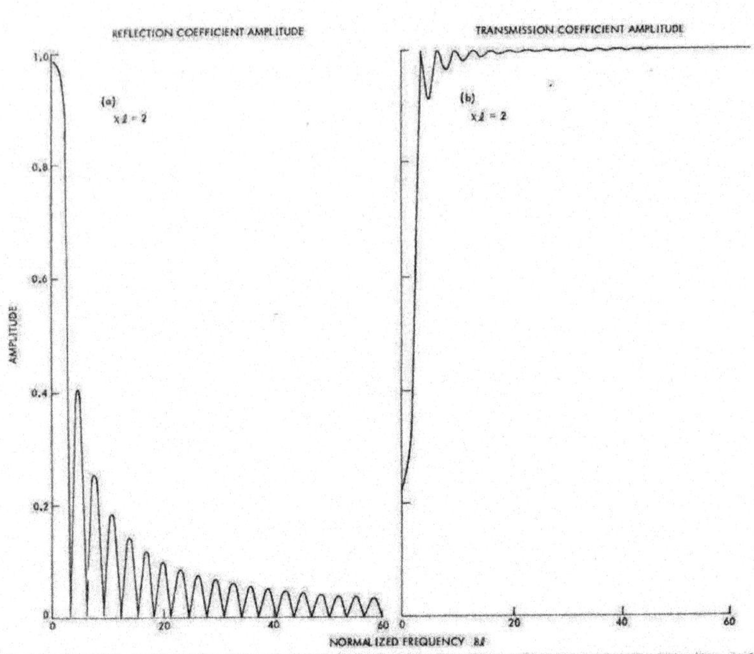

Fig. 2. Magnitude of reflection and transmission coefficients as function of normalized frequency mismatch: $\delta l = \Delta\omega l/c$ for fixed coupling: $\chi l = 2$.

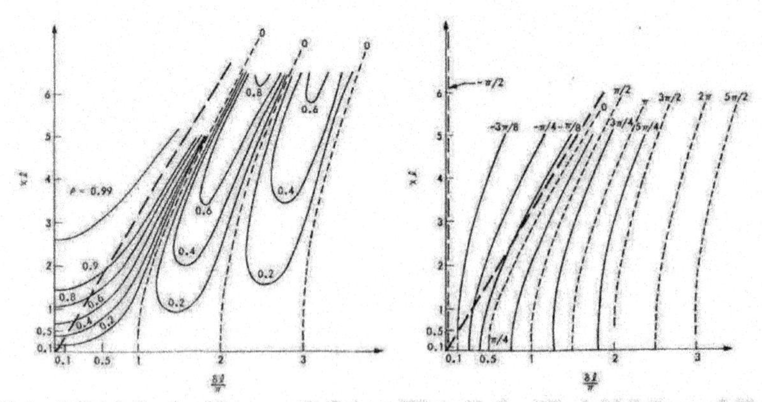

Fig. 3. Equimagnitude and equiphase curves of reflection coefficient as function of δl and χl. Reflection curve in Fig. 2 corresponds to horizontal line ($\chi l = 2$) across graph.

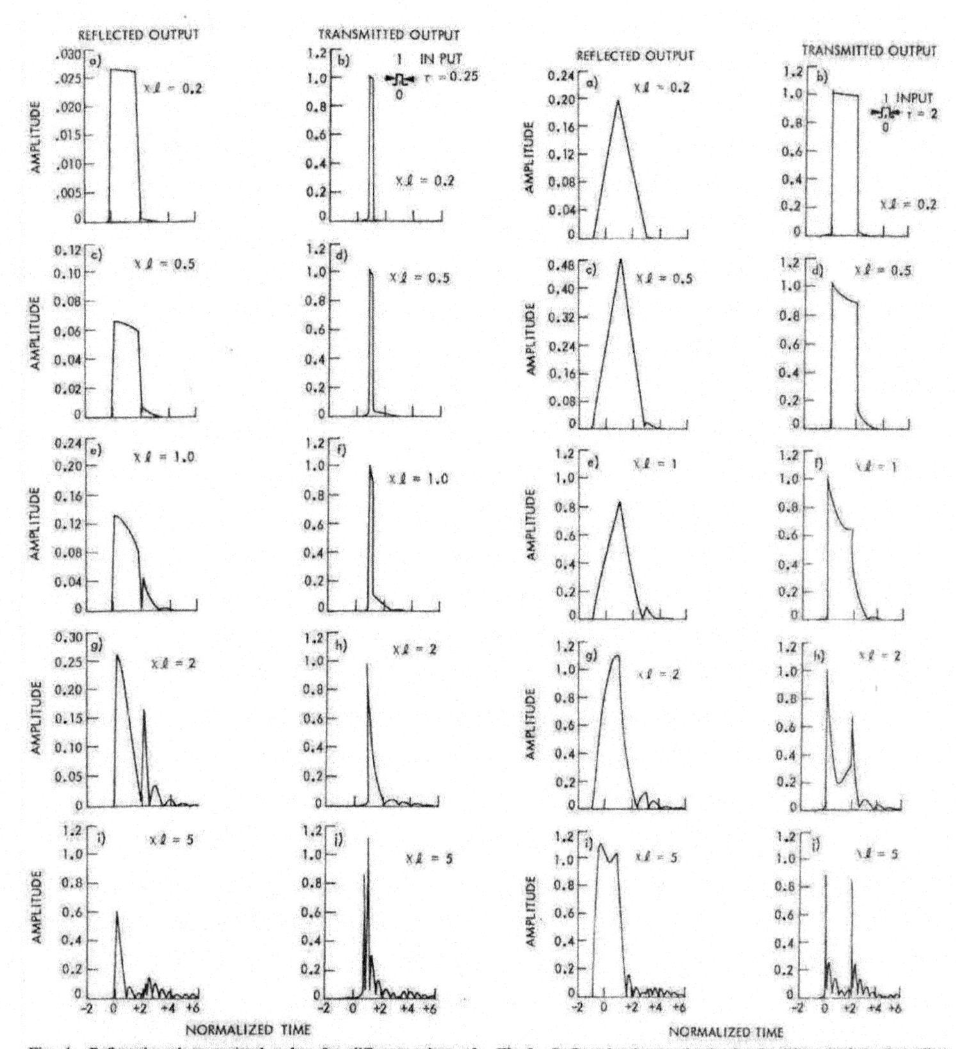

Fig. 4. Reflected and transmitted pulses for different values of coupling coefficient. Incident rectangular pulse has length of 0.25 time units and carrier frequency ω_0 at Bragg condition. Remark difference in vertical scales.

Fig. 5. Reflected and transmitted pulses for different values of coupling coefficient. Incident rectangular pulse has length of 2.0 time units and carrier frequency ω_0 at Bragg condition. Remark difference in vertical scales.

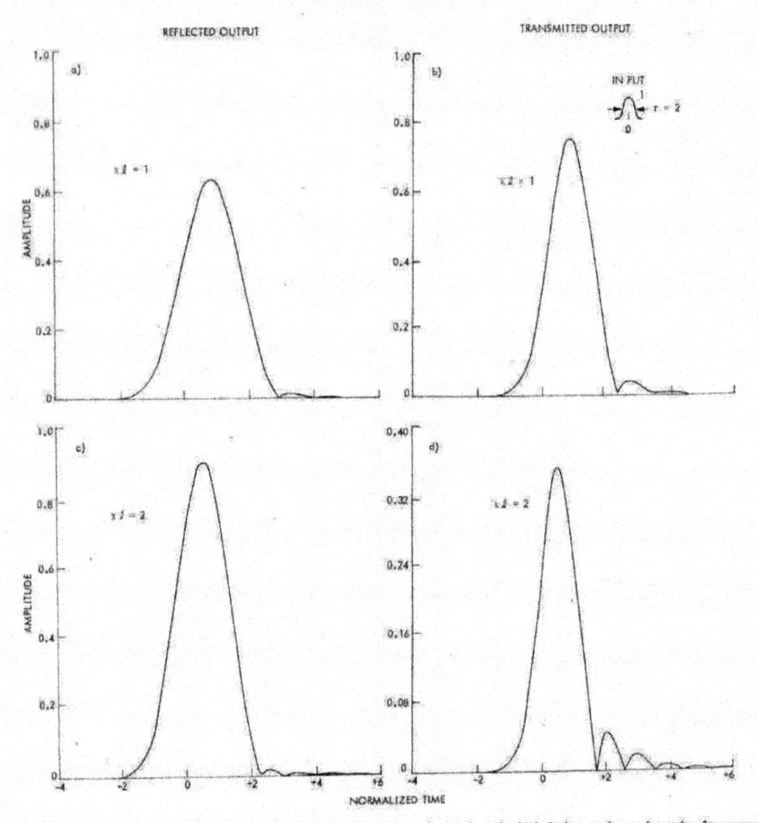

Fig. 6. Reflected and transmitted pulses for case of incident Gaussian pulse of width 2 time units and carrier frequency ω_0. Remark difference in vertical scales.

where

$$D = \left[\chi_{pq}^2 - \left(\frac{\delta_p + \delta_q}{2}\right)^2\right]^{1/2}.$$

If the two coupled modes are identical, then $\delta_p = \delta_q = \delta$, and

$$R = \frac{i\chi}{D \coth Dl - i\delta} \tag{8}$$

$$T = \frac{D}{D \cosh Dl - i\delta \sinh Dl} \tag{9}$$

$$D = (\chi^2 - \delta^2)^{1/2}.$$

From (6) and (7) we can derive the characteristics of the reflection and transmission coefficients of the periodic structure:

1) maximum reflection

$$R_M = i \tanh(\chi_{pq} l); \tag{10}$$

2) minimum transmission

$$T_m = 1/\cosh(\chi_{pq} l); \tag{11}$$

3) $R = 0$ and $T = 1$ for

$$\delta_p + \delta_q = 2[\chi_{pq}^2 + (n\pi/l)^2]^{1/2}, \quad n = 1, 2, \cdots \tag{12}$$

the corresponding phase of R is equal to $(2n - 1)\pi/2$;

4) $|R|^2 + |T|^2 = 1$.

In Fig. 2 are shown the magnitudes of the reflection and transmission coefficients given in (8) and (9) as functions of δl for a given value of χl. Fig. 3 shows the equimagnitude and equiphase curves of the reflection coefficient plotted as a function of χl and δl. Fig. 2 corresponds to a horizontal line ($\chi l = 2$) across Fig. 3. If the two coupled modes are different (i.e., $p \neq q$), the curves in Figs. 2 and 3 are still valid if we replace δ by $(\delta_p + \delta_q)/2$ and χ by χ_{pq}.

III. Transient Response

The reflection and transmission of a pulse with a relatively narrow spectrum near ω_{pq} can now be derived from the

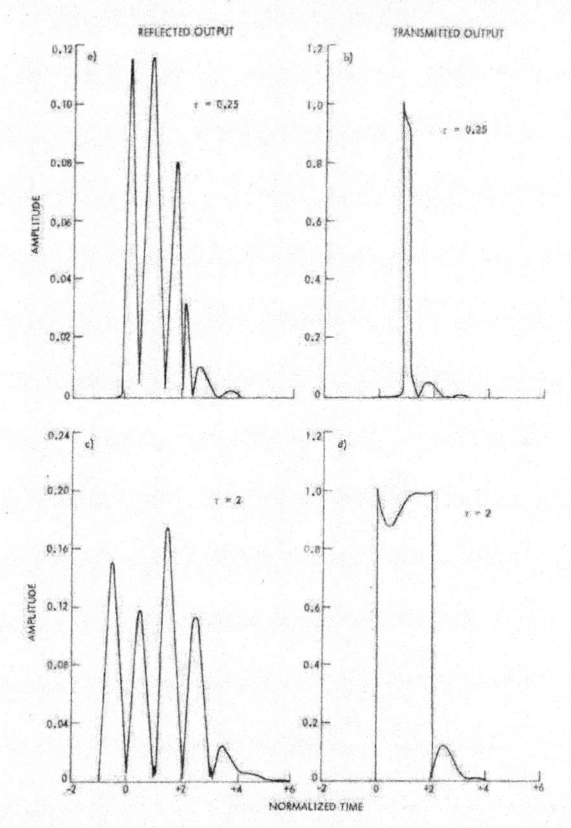

Fig. 7. Reflected and transmitted pulses for case of incident rectangular pulse of width 0.25 time units and 2.0 time units and carrier frequency ω_1 at first zero of reflection coefficient. Remark difference in vertical scales. $\chi l = 1$.

reflection and transmission coefficients of each spectral component. For an incident pulse $g(t)$ of spectrum $G(\omega)$

$$G(\omega) = \int_{-\infty}^{+\infty} g(t)e^{i\omega t}\, dt \tag{13}$$

the reflected pulse is

$$r(t) = \frac{1}{2\pi}\int_{-\infty}^{+\infty} G(\omega)R(\omega)e^{-i\omega t}\, d\omega \tag{14}$$

and the transmitted pulse is

$$p(t) = \frac{1}{2\pi}\int_{-\infty}^{+\infty} G(\omega)T(\omega)e^{-i\omega t}\, d\omega. \tag{15}$$

Numerical inversion has been used by a number of authors [10], [11], [and others] to obtain time response to radiation, propagation, and scattering problems by convolution of transforms. The method used in this paper was the Cooley–Tukey fast Fourier transform (FFT) [12], with $2^{11} = 2048$ samples in the FFT to calculate $r(t)$ and $p(t)$.

We studied the reflection and transmission of several rectangular pulses of carrier frequency ω_0, ω_1, and ω_2, where ω_0 is the Bragg frequency, and ω_1 and ω_2 correspond to the first two zeros of the reflection coefficient. The normalized pulse lengths τ were chosen to be 0.25 and 2.0 time units, where each time unit corresponds to the transit time of the pulse across the slab of length l.

Fig. 4 displays the envelopes of the reflected and transmitted pulses which result from an incident pulse of length $\tau = 0.25$, with a carrier frequency exactly at phase match ($\omega = \omega_0$). The different cases correspond to different values of the coupling coefficient χ. In all the illustrations, the time reference is such that $t = 0$ corresponds to the instant when the center of the incident pulse is at the first boundary. For weak coupling, the reflected pulse is spread over 2.25 time units (Figs. 4(a) and (b)) because the energy is reflected from the successive stratifications and the echo from the last layer would take a round trip time of 2 units. Since the successive reflections are relatively weak ($\chi l \ll 1$), the reflected pulse is quasi-rectangular, and the transmitted

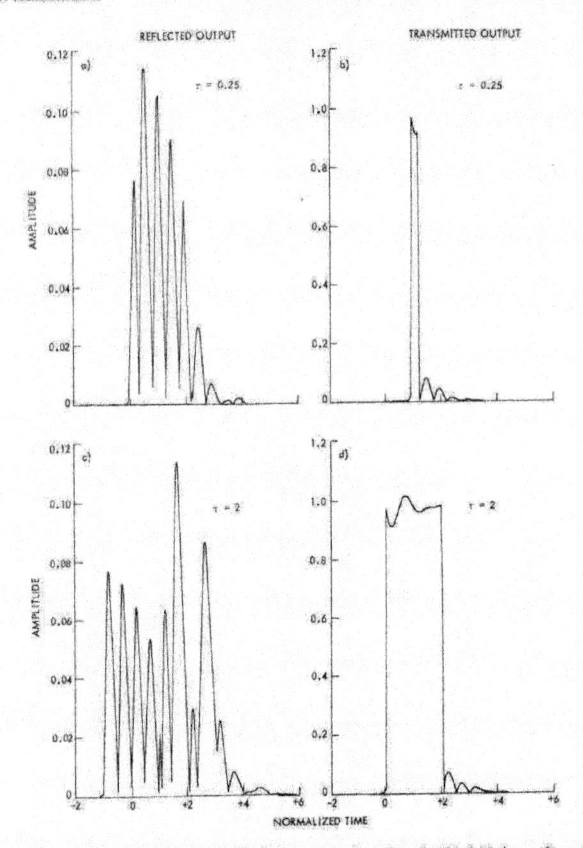

Fig. 8. Reflected and transmitted pulses for case of incident rectangular pulse of width 0.25 time units and 2.0 time units and carrier frequency ω_2 at second zero of reflection coefficient. Remark difference in vertical scales. $\chi l = 1$.

pulse is similar to the incident pulse with a time delay. As the coupling is increased (i.e., stronger reflections from each layer), multiple interference plays an appreciable role which leads to transients over a longer time period. The reflected pulse reaches a maximum at time $t = 0.25$ when the whole incident pulse is just inside the slab. The subsequent fall-off in the reflected pulse amplitude is due to the fact that a large portion of the signal has already been reflected back by the first few stratifications. This fall-off is faster for stronger coupling (Figs. 4(e), (g), and (i)). Additionally, a complicated interference tail of multiple subpulses is present. For strong coupling, the transmitted pulse consists mainly of two narrow peaks, 0.25 time units apart, which correspond to differentiation of the incident pulse.

In Fig. 5, the same sequence is repeated for the case of a rectangular pulse of length $\tau = 2$. For weak coupling (Fig. 5(a)), the reflected pulse is similar to the autocorrelation of the incident pulse. This is due to the similarity of the input pulse spectrum and the reflection coefficient in the frequency domain, and does not hold for other pulse widths or pulses in general. For strong coupling, the transmitted pulse is again similar to the absolute derivative of the incoming pulse, with the characteristic interference echoes. It should be pointed out that the reflected echo might have an amplitude larger than 1 (Figs. 5(g) and (i)) due to favorable coherent addition of successive reflections.

Fig. 6 displays the reflected and transmitted pulses for a Gaussian incident pulse of width $\tau = 2$ (width is taken at the $1/e$ values). Since a Gaussian pulse spectrum contains smaller high frequency components, the reflected and transmitted pulses are grossly similar to the incident pulse. However, if the incident pulse is narrowed to $\tau = 0.25$ (not shown here), the results are somewhat similar to those obtained for a rectangular pulse of the same width.

Finally, a number of curves were generated for the case where the carrier frequency is shifted to the first (Fig. 7) or second (Fig. 8) zero of the reflection coefficient. In these cases, the reflected pulse consists of a number of interference pulses which are different from the incident one. We remark that for $\tau = 2$, the width of each main pulse is approximately 1 time unit (in Fig. 7) and $\frac{1}{2}$ time unit (in Fig. 8).

IV. Limitations and Applications

The preceding results were based on the coupled wave approach which only accounts for the main forward and backward waves and neglects the presence of higher harmonics. This is valid as long as the coupling is weak enough such that the wave amplitude does not change much over a wavelength or similarly over a structure period, and the input bandwidth is not too large. Otherwise the Floquet approach [1]–[5], which accounts for all space-harmonics, has to be used. This approach, even though it is simple to formulate, is relatively involved numerically and is left for future work.

Periodic structures, in different forms, have played an important role in many fields: microwave tubes, filters, antennas, and solid state theory. Recently their application in the emerging field of optical communication and integrated optics [5] has attracted appreciable attention. As it is very likely that optical communication will be in a digital form, the results of this paper are very relevant. For instance, a periodic structure can be used as a differentiator for edge detection (Fig. 5(j)) or multiple pulse generator (Fig. 7). The results can also be applied to many fields like microwave filters, optical filters, and sounders of subsurface layers.

References

[1] N. Brillouin, *Wave Propagation in Periodic Structures.* New York: Dover, 1953.
[2] T. Tamir, H. C. Wang, and A. A. Oliner, "Wave propagation in sinusoidally stratified dielectric media," *IEEE Trans. Microwave Theory Tech.*, vol. MTT-12, pp. 323–335, May 1964.
[3] K. F. Casey, J. R. Mathis, and C. Yeh, "Wave propagation in sinusoidally stratified plasma," *J. Math. Phys.*, vol. 10, pp. 891–987, 1969.
[4] T. Tamir and H. C. Wang, "Scattering of electromagnetic waves by a sinusoidally stratified half-space," *Can. J. Phys.*, vol. 44, pp. 2073–2093, 1966.
[5] C. Elachi and C. Yeh, "Periodic structures in integrated optics," *J. Appl. Phys.*, vol. 44, pp. 3146–3152, 1973.
[6] ——, "Mode conversion in periodically distributed thin film waveguides," *J. Appl. Phys.*, vol. 45, p. 3494, 1974.
[7] W. H. Louisell, *Coupled mode and parametric electronics.* New York: Wiley, 1960.
[8] D. Marcuse, *Light Transmission Optics.* New York: Van Nostrand, 1973.
[9] H. Kogelnik and C. V. Shank, "Coupled wave theory of distributed feedback lasers," *J. Appl. Phys.*, vol. 43, pp. 2327–2335, 1972.
[10] P. S. Ray and J. J. Stephens, "Far-field transient backscattering by ice spheres," *Radio Sci.*, vol. 9, pp. 43–55, 1974.
[11] D. A. Hill and J. R. Wait, "The transient electromagnetic response of a spherical shell of arbitrary thickness," *Radio Sci.*, vol. 7, pp. 911–935, 1972.
[12] J. W. Cooley, P. A. Lewis and P. D. Welch, *The Fast Fourier Transform Algorithm.* Yorktown Heights, N.Y.: IBM Watson Research Center, 1969.

Reprinted from **APPLIED OPTICS**, Vol. 16, page 483, February 1977

Computing the propagation characteristics of radially stratified fibers: an efficient method

C. Yeh and G. Lindgren

An efficient method is introduced in this paper to compute the dispersion characteristics as well as the Poynting flux distribution of radially stratified fibers. Only 4 × 4 matrix operations were needed. Detailed results are given for several representative radially inhomogeneous fibers of practical interest.

I. Introduction

It has become increasingly clear that radially inhomogeneous fibers must be used to maximize the information carrying capacities of the optical fiber transmission system. Efficient analytical means must be found to predict the propagation characteristics of various modes in such inhomogeneous fibers. It is known that unlike the homogeneous fiber case, the wave equation governing the fields for radially inhomogeneous fibers consists of two coupled second-order differential equations whose solutions are usually very difficult to obtain. Using the numerical integration technique, Dil and Blok[1] and later Vassell[1] solved these equations for a fiber with radial parabolic dielectric profile. Yip and Ahmew,[2] using the same technique, solved the problem of a cladded fiber with a radial parabolic index profile. However, this method of solution is very time consuming, hence very expensive. Furthermore to achieve computational efficiency, their consideration is restricted to the square law distribution of permittivity wherein a power series expansion may be used to represent the field variation as a function of the radial coordinate.

A useful approximate approach for an analytical solution of electromagnetic problems involving a radially inhomogeneous column is to subdivide it into thin homogeneous layers and to solve an easier problem in each layer. The fields in each layer are expanded in appropriate eigenfunctions and the expansion coefficients determined by matching boundary conditions. However, this straightforward approach becomes much too tedious, and the number of simultaneous equations to be solved tends to be prohibitively large as the number of layers increases.[3] Using this technique, the number of layers that may be used is quickly limited by the capacity of a modern computer. It is therefore quite apparent that a different approach must be taken. The purpose of the present investigation is to seek a simple way to solve the problem of guided waves in inhomogeneous fibers without the forementioned difficulties and limitations. (Exhaustive comparison with other techniques in terms of computer time is not possible or warranted, but typically the computer time requirement for the present technique using five layers is three times less than that used in Ref. 3 and ten times less than that used in Refs. 1 and 2.) We shall show that by appropriately manipulating the simultaneous equations obtained according to the homogeneous layers approach, we shall only deal with 4 × 4 type matrix operations. Hence the constraint on the number of layers used to approximate the inhomogeneous profiles can be eliminated. Furthermore, the required results can be obtained very quickly (efficient usage of computer time) with this technique. We shall apply this method to determine the dispersion characteristics and the Poynting flux distribution of several radially inhomogeneous fibers of practical interests.

II. Formulation of the Problem

Without loss of generality we may assume that the expressions for the field components of all modes are multiplied by the factor $\exp(in\theta + i\beta z - i\omega t)$, which will be suppressed throughout.

Dividing the fiber guide into $m + 1$ region, as shown in Fig. 1, we may write the expressions for the tangential fields in these regions as follows:
in region 1,

$$\begin{bmatrix} E_z^{(1)} \\ \eta H_z^{(1)} \\ \rho E_\theta^{(1)} \\ \eta\rho H_\theta^{(1)} \end{bmatrix} = \begin{bmatrix} c_1(r) & 0 & 0 & 0 \\ 0 & d_1(r) & 0 & 0 \\ e_1(r) & f_1(r) & 0 & 0 \\ q_1(r) & h_1(r) & 0 & 0 \end{bmatrix} \begin{bmatrix} C_1 \\ D_1 \\ 0 \\ 0 \end{bmatrix}; \qquad (1)$$

in region $m (m > 1)$,

When this work was done both authors were with University of California, Electrical Science & Engineering Department, Los Angeles, California 90024. G. Lindgren is now with Hughes Aircraft Company, Canoga Park, California 00000.
Received 27 February 1976.

$$\begin{bmatrix} E_z^{(m)} \\ \eta H_z^{(m)} \\ \rho E_\theta^{(m)} \\ \eta\rho H_\theta^{(m)} \end{bmatrix} = \begin{bmatrix} c_m(r) & 0 & c_m'(r) & 0 \\ 0 & d_m(r) & 0 & d_m'(r) \\ e_m(r) & f_m(r) & e_m'(r) & f_m'(r) \\ g_m(r) & h_m(r) & g_m'(r) & h_m'(r) \end{bmatrix} \begin{bmatrix} C_m \\ D_m \\ C_m' \\ D_m' \end{bmatrix}; \quad (2)$$

in region $m+1$ (the outermost region),

$$\begin{bmatrix} E_z^{(m+1)} \\ \eta H_z^{(m+1)} \\ \rho E_\theta^{(m+1)} \\ \eta\rho H_\theta^{(m+1)} \end{bmatrix} = \begin{bmatrix} 0 & 0 & s(r) & 0 \\ 0 & 0 & 0 & \tau(r) \\ 0 & 0 & u(r) & v(r) \\ 0 & 0 & w(r) & \chi(r) \end{bmatrix} \begin{bmatrix} 0 \\ 0 \\ G \\ F \end{bmatrix}; \quad (3)$$

where

$$c_m(r) = J_n(p_m r), \qquad c_m'(r) = N_n(p_m r),$$
$$d_m(r) = iJ_n(p_m r), \qquad d_m'(r) = N_n(p_m r)i,$$
$$e_m(r) = -\frac{\beta n k_o}{p_m^2} J_n(p_m r), \qquad e_m'(r) = -\frac{\beta n k_o}{p_m^2} N_n(p_m r),$$
$$f_m(r) = -\frac{ik_o}{p_m} k_o r J_n'(p_m r)i, \qquad f_m'(r) = \frac{-ik_o}{p_m} k_o r N_n'(p_m r)i,$$
$$g_m(r) = i\frac{\epsilon_m}{\epsilon_o}\frac{k_o}{p_m} k_o r J_n'(p_m r), \qquad g_m'(r) = \frac{\epsilon_m}{\epsilon_o}\frac{k_o}{p_m} k_o r N_n'(p_m r),$$
$$h_m(r) = -\frac{n\beta k_o}{p_m^2} \times J_n(pr)i, \qquad h_m'(r) = \frac{-n\beta k_o}{p_m} N_n(p_m r)i,$$
$$s(r) = K_n(qr),$$
$$\tau(r) = K_n(qr),$$
$$u(r) = \frac{\beta n k_o}{q^2} K_n(qr),$$
$$v(r) = \frac{ik_o}{q} k_o r K_n'(qr)i,$$
$$w(r) = -i\frac{\epsilon_{m+1}}{\epsilon_o}\frac{k_o}{q} \times k_o r K_n'(qr),$$
$$\chi(r) = \frac{\beta n k_o}{q^2} K_n(qr)i,$$
$$p_m^2 = \omega^2\mu_o\epsilon_m - \beta^2,$$
$$k_m^2 = \omega^2\mu_o\epsilon_m, \qquad k_o = \omega(\mu_o\epsilon_o)^{1/2},$$
$$q^2 = \beta^2 - \omega^2\mu_o\epsilon_{m+1}, \qquad \eta = [(\mu_o)/(\epsilon_o)]^{1/2}, \quad (4)$$

and $C_1, D_1, \ldots, C_m, D_m, C_m', D_m' \ldots G, F$ are arbitrary constants. It has been assumed that within each region the permittivity is a constant.

Matching the tangential electric and magnetic fields at the bounding surfaces, i.e., $r = r_1, r_2, \ldots r_m$ (see Fig. 1), gives

$$\begin{bmatrix} \bar{c}_1 & 0 & 0 & 0 \\ 0 & 0 & \bar{d}_1 & 0 \\ \bar{e}_1 & 0 & \bar{f}_1 & 0 \\ \bar{g}_1 & 0 & \bar{h}_1 & 0 \end{bmatrix} \begin{bmatrix} C_1 \\ 0 \\ D_1 \\ 0 \end{bmatrix} = \begin{bmatrix} c_2 & c_2' & 0 & 0 \\ 0 & 0 & d_2 & d_2' \\ e_2 & e_2' & f_2 & f_2' \\ g_2 & g_2' & h_2 & h_2' \end{bmatrix} \begin{bmatrix} C_2 \\ C_2' \\ D_2 \\ D_2' \end{bmatrix}$$

$$\begin{bmatrix} \bar{c}_2 & \bar{c}_2' & 0 & 0 \\ 0 & 0 & \bar{d}_2 & \bar{d}_2' \\ \bar{e}_2 & \bar{e}_2' & \bar{f}_2 & \bar{f}_2' \\ \bar{g}_2 & \bar{g}_2' & \bar{h}_2 & \bar{h}_2' \end{bmatrix} \begin{bmatrix} C_2 \\ C_2' \\ D_2 \\ D_2' \end{bmatrix} = \begin{bmatrix} c_3 & c_3' & 0 & 0 \\ 0 & 0 & d_3 & d_3' \\ e_3 & e_3' & f_3 & f_3' \\ g_3 & g_3' & h_3 & h_3' \end{bmatrix} \begin{bmatrix} C_3 \\ C_3' \\ D_3 \\ D_3' \end{bmatrix}$$

$$\begin{bmatrix} \bar{c}_{m-1} & \bar{c}_{m-1}' & 0 & 0 \\ 0 & 0 & \bar{d}_{m-1} & \bar{d}_{m-1}' \\ \bar{e}_{m-1} & \bar{e}_{m-1}' & \bar{f}_{m-1} & \bar{f}_{m-1}' \\ \bar{g}_{m-1} & \bar{g}_{m-1}' & \bar{h}_{m-1} & \bar{h}_{m-1}' \end{bmatrix} \begin{bmatrix} C_{m-1} \\ C_{m-1}' \\ D_{m-1} \\ D_{m-1}' \end{bmatrix} = \begin{bmatrix} c_m & c_m' & 0 & 0 \\ 0 & 0 & d_m & d_m' \\ e_m & e_m' & f_m & f_m' \\ g_m & g_m' & h_m & h_m' \end{bmatrix} \begin{bmatrix} C_m \\ C_m' \\ D_m \\ D_m' \end{bmatrix}$$

$$\begin{bmatrix} \bar{c}_m & \bar{c}_m' & 0 & 0 \\ 0 & 0 & \bar{d}_m & \bar{d}_m' \\ \bar{e}_m & \bar{e}_m' & \bar{f}_m & \bar{f}_m' \\ \bar{g}_m & \bar{g}_m' & \bar{h}_m & \bar{h}_m' \end{bmatrix} \begin{bmatrix} C_m \\ C_m' \\ D_m \\ D_m' \end{bmatrix} = \begin{bmatrix} \bar{s} & 0 & 0 & 0 \\ 0 & 0 & \bar{\tau} & 0 \\ \bar{u} & 0 & \bar{v} & 0 \\ \bar{w} & 0 & \bar{\chi} & 0 \end{bmatrix} \begin{bmatrix} E \\ 0 \\ F \\ 0 \end{bmatrix}, \quad (5)$$

where the bar signifies that the functions $\bar{c}_n, \bar{c}_n' \ldots \bar{h}_n, \bar{h}_n'$ are evaluated at the surface $r = r_n$. The functions $c_n, c_n' \ldots h_n, h_n'$ are evaluated at the surface $r = r_{n-1}$. It is clear that if one sets the determinant of the above simultaneous equations to zero, one obtains in a straightforward manner the dispersion relation for the inhomogeneous fiber problem. However, one notes that the size of the resultant determinant depends directly on the number of layers that we use. The unattractive numerical problems associated with a very large size matrix are well known. Fortunately, for the inhomogeneous fiber problem, a way can be found to avoid working with this large-size matrix.

Realizing the fact that we may express the array

$$\begin{bmatrix} C_2 \\ C_2' \\ D_2 \\ D_2' \end{bmatrix}$$

in terms of

$$\begin{bmatrix} C_3 \\ C_3' \\ D_3 \\ D_3' \end{bmatrix}$$

etc., one may write down the following chain equation:

$$\mathbf{M}_1 \begin{bmatrix} C_1 \\ 0 \\ D_1 \\ 0 \end{bmatrix} = M_2 \mathbf{M}_2^{-1} M_3 \mathbf{M}_3^{-1} \ldots M_m \mathbf{M}_m^{-1} M_{m+1} \begin{bmatrix} E \\ 0 \\ F \\ 0 \end{bmatrix}, \quad (6)$$

where

$$M_1 = \begin{bmatrix} c_1 & 0 & 0 & 0 \\ 0 & 0 & d_1 & 0 \\ e_1 & 0 & f_1 & 0 \\ g_1 & - & h_1 & 0 \end{bmatrix},$$

$$M_m = \begin{bmatrix} c_m & c_m' & 0 & 0 \\ 0 & 0 & d_m & d_m' \\ e_m & e_m' & f_m & f_m' \\ g_m & g_m' & h_m & h_m' \end{bmatrix},$$

$$\mathrm{M}_m^{-1} = \begin{bmatrix} c_m & c_m' & 0 & 0 \\ 0 & 0 & d_m & d_m' \\ e_m & e_m' & f_m & f_m' \\ g_m & g_m' & h_m & h_m' \end{bmatrix}^{-1},$$

$$M_{m+1} = \begin{bmatrix} s & 0 & 0 & 0 \\ 0 & 0 & \tau & 0 \\ u & 0 & v & 0 \\ w & 0 & \chi & 0 \end{bmatrix}.$$

Hence, we are now only dealing with matrix of size 4 × 4. This rearrangement eliminates the need to compute large matrix. The size of our present matrix is independent of the number of layers that we use. Hence, we may use as many layers as we wish to achieve the desired accuracy. Rewriting Eq. (6) gives

$$\begin{bmatrix} c_1 & 0 & -M_{11} & -M_{13} \\ 0 & d_1 & -M_{21} & -M_{23} \\ e_1 & f_1 & -M_{31} & -M_{33} \\ g_1 & h_1 & -M_{41} & -M_{43} \end{bmatrix} \begin{bmatrix} C_1 \\ D_1 \\ E \\ F \end{bmatrix} = 0, \tag{7}$$

where M_{mn} are elements of the matrix M with

$$M = M_2 \mathrm{M}_2^{-1} M_3 \mathrm{M}_3^{-1} \ldots M_m \mathrm{M}_m^{-1} M_{m+1}.$$

Setting the determinant of Eq. (7) to zero, one obtains the dispersion relation from which the propagation constants of various modes along an inhomogeneous fiber guide may be found. In other words, the propagation constants of various modes correspond to the roots of the dispersion relation. Given all known constants such as the frequency, the size of the fiber, the dielectric constant, thicknesses of the layers, etc., the roots can be found by the well-known Newton's method.

III. Numerical Results

The above technique is used to obtain the dispersion characteristics and Poynting flux distribution of several radially inhomogeneous fibers of practical interest. The refractive index profiles of these fibers are shown in Fig. 2. For the present computation, the refractive index is assumed to be independent of the frequency of operation. Figures 3 and 4 give, respectively, the normalized propagation constant $\beta_N = \beta/k_0$ vs the normalized frequency k_0a curves for several lower order modes of a homogeneous fiber and of an inhomogeneous with parabolic index variation. From a comparison of Fig. 3 to Fig. 4, one notices that the excursion of β/k_0 from the cutoff value to far above cutoff value occurs over a much larger range of k_0a when the profile is parabolic rather than homogeneous. This agrees with the minimal dispersion predictions provided by geometrical optics. The other interesting feature of the parabolic profile is that a number of the modes have merged or possess nearly degenerate dispersion curves.

The doughnut profile was studied in two forms, which are shown in Fig. 5. Figure 6 presents their respective dispersion diagrams. One notices from the curves that the dispersion is intermediate between the homogeneous core and the parabolic core and further that the mode merging is not complete. Profile (a) is less dispersive than (b) which follows since $\langle n \rangle_{nA}$ of (a) is less than that of (b).

Finally, the $HE_{1,1}$ mode dispersion characteristics of the three profiles (step, parabolic, and doughnut) are compared in Fig. 7. This shows that for a given radius and Δn, a guide is less dispersive as the profile departs from a step profile.

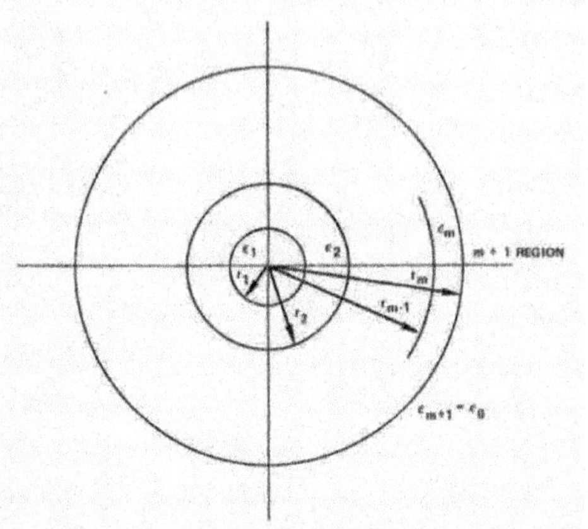

Fig. 1. Geometry of the inhomogeneous fiber.

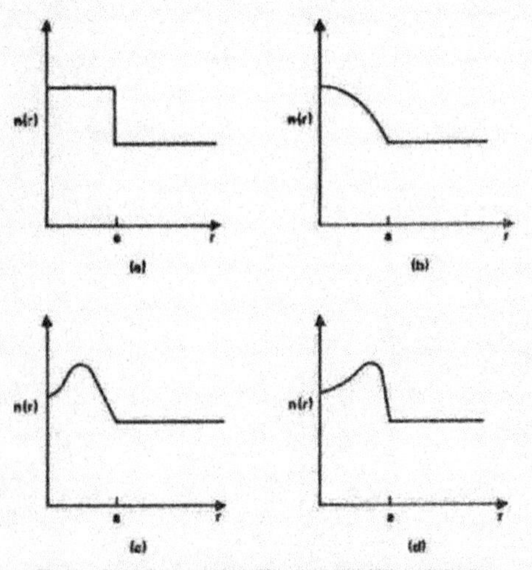

Fig. 2. Refractive-index profiles of several fibers of interest.

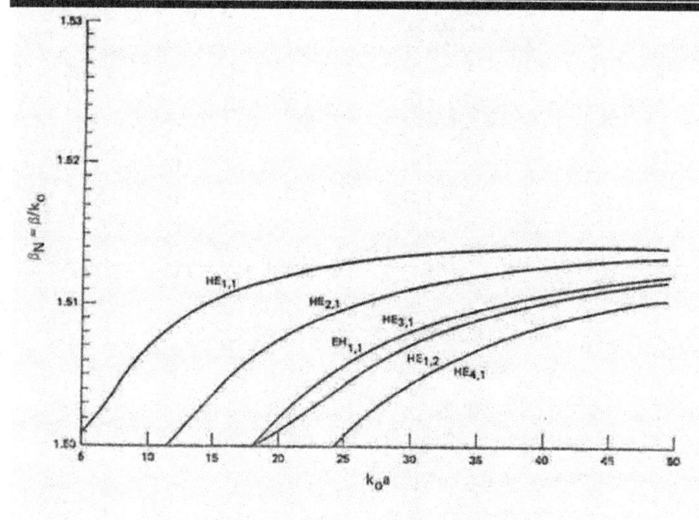

Fig. 3. Dispersion curves for homogeneous core fiber with $n_1 = 1.515$ and $n_2 = 1.50$. a is the radius of the core.

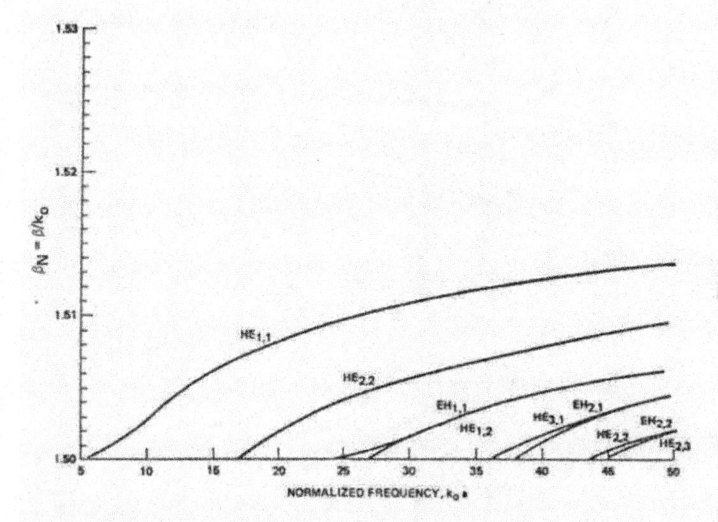

Fig. 4. Dispersion curves for parabolic core (five-layer) fiber with $n_{MAX} = 1.515$ and $n_2 = 1.50$.

The Poynting flux in the mth region of a fiber is given by

$$S_z(m) = \frac{1}{2} \mathrm{Re}\,[E_r(m)H\phi(m)^* - E_\phi(m)H_r(m)^*]. \qquad (8)$$

The Poynting flux (energy density distribution) was computed for several representative profiles given in Fig. 2. The results are presented primarily at two operating conditions, near cutoff and far above cutoff, since the flux confinement is intermediate for other operating conditions.

Figures 8–12 show the behavior of some homogeneous core modes along with the percentage of power carried inside and outside the core region. The Poynting flux diagrams are shown throughout as normalized to the peak intensity of the flux and as a function of r along outward rays of peak intensity.

Upon examination of the Poynting flux diagrams, one will observe that the Poynting flux S_z has a slight discontinuity at $r = a$. This discontinuity arises simply because the E_r component of the field is not continuous across the boundary. Similarly, for the inhomogeneous core profiles, S_z is discontinuous at each stratification boundary. Figures 13 and 14 show S_z for the parabolic

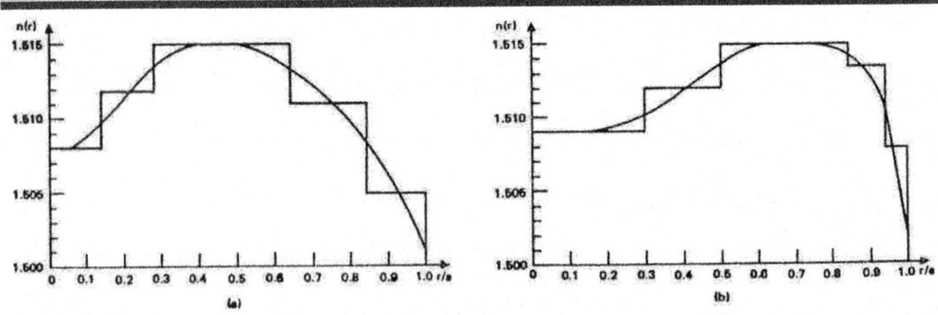

Fig. 5. Staircase approximations of two doughnut refractive-index profiles.

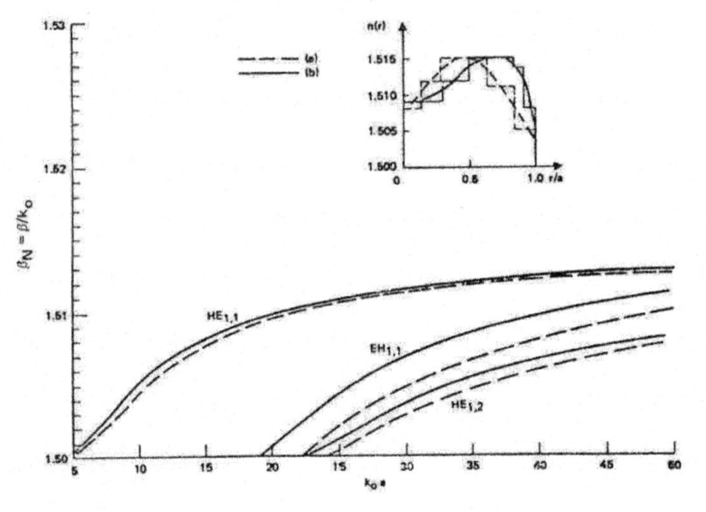

Fig. 6. Comparison of dispersion curves for two fibers with different doughnut refractive-index profiles.

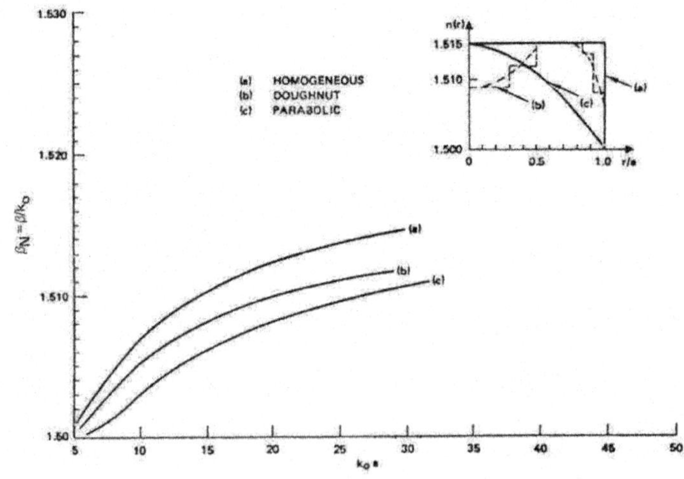

Fig. 7. Comparison of $HE_{1,1}$ mode dispersion curves for homogeneous, doughnut, and parabolic refractive-index profile fibers.

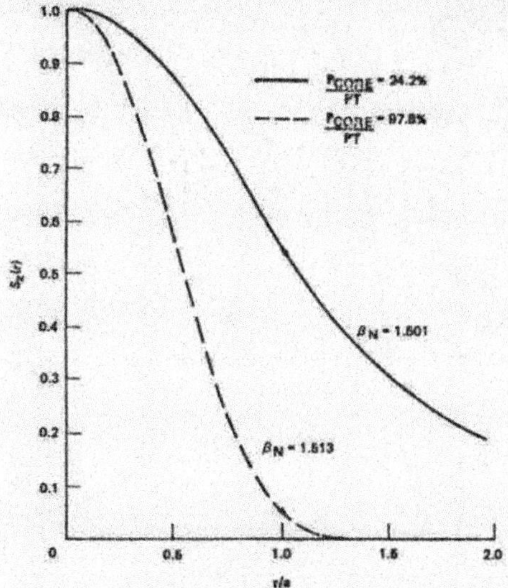

Fig. 8. Poynting flux characteristics of $HE_{1,1}$ mode. Fiber profile is homogeneous, $n_1 = 1.515$ and $n_2 = 1.500$.

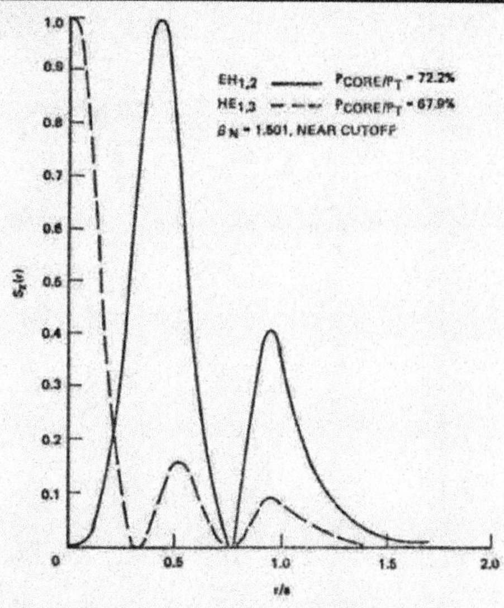

Fig. 9. Poynting flux characteristics of $EH_{1,2}HE_{1,3}$ modes. Fiber profile is homogeneous, $n_1 = 1.515$ and $n_2 = 1.500$.

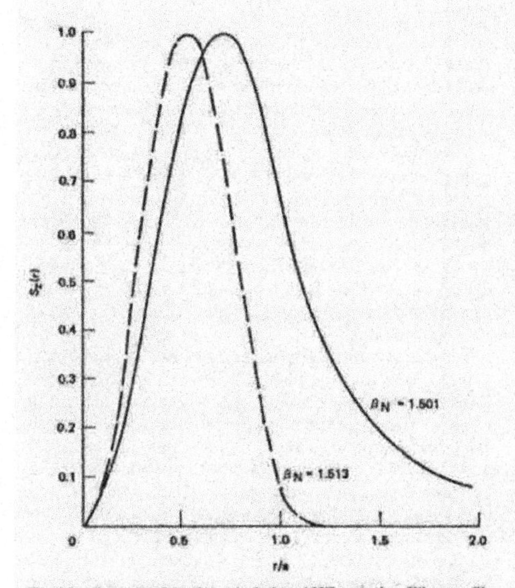

Fig. 10. Poynting flux characteristics of $HE_{2,1}$ mode. Fiber profile is homogeneous, $n_1 = 1.515$ and $n_2 = 1.500$.

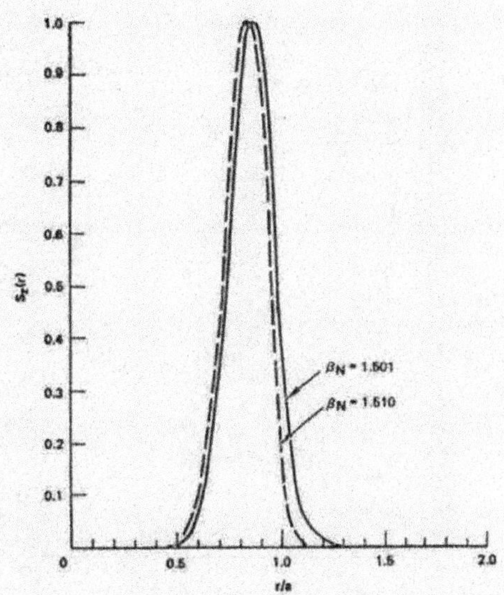

Fig. 11. Poynting flux characteristics of $HE_{10,1}$ mode. Fiber profile is homogeneous, $n_1 = 1.515$ and $n_2 = 1.500$.

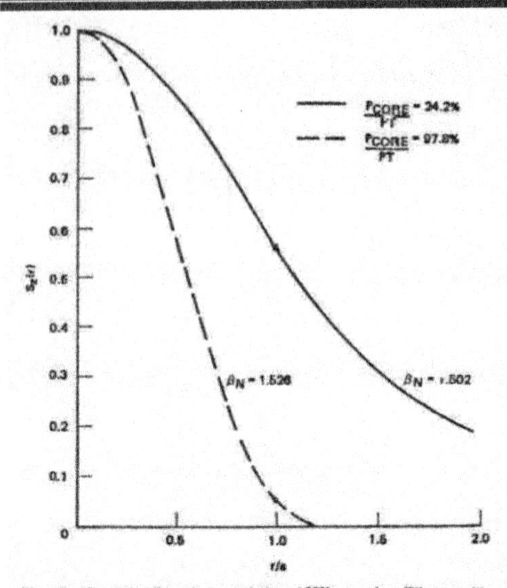

Fig. 12. Poynting flux characteristics of $HE_{1,1}$ mode. Fiber profile is homogeneous, $n_1 = 1.53$ and $n_2 = 1.500$.

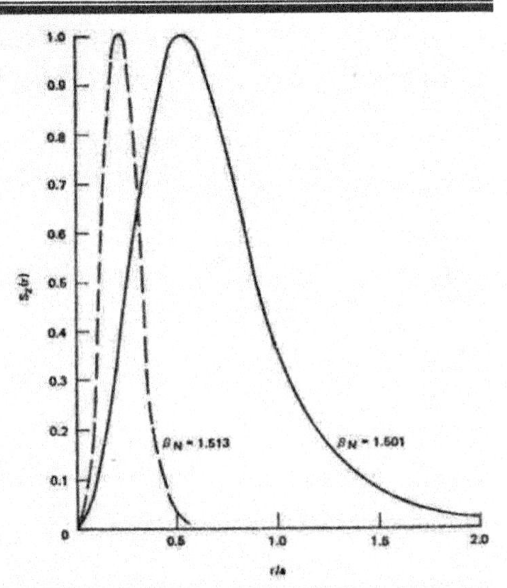

Fig. 14. Poynting flux characteristics of $HE_{2,1}$ mode. Fiber profile is parabolic (five layer), $n_{MAX} = 1.515$ and $n_2 = 1.500$.

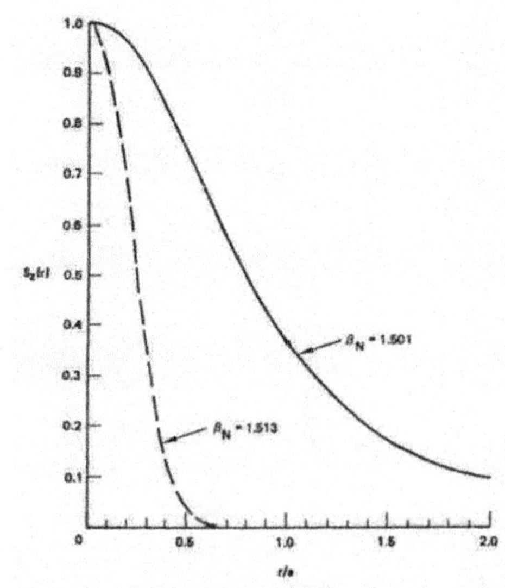

Fig. 13. Poynting flux characteristics of $HE_{1,1}$ mode. Fiber profile is parabolic (five layer), $n_{MAX} = 1.515$ and $n_2 = 1.500$.

core inhomogeneity; Figs. 15 and 16 show S_z for the doughnut profile. In these diagrams, the discontinuity is so slight that it is difficult to detect on the graphs; this is just what one would expect if the medium were continuously inhomogeneous.

The interesting feature of the inhomogeneous core guides is that the Poynting flux concentrates around the region of highest refractive index as the frequency is increased. In the case of the doughnut profiles at far above cutoff, the middle core region takes on the behavior of a slab guide where the peak refractive index serves as the core, and the layers inside and outside appear as the equivalent of a slab with upper and lower cladding regions.

Finally, a comparison was made between the homogeneous core and the parabolic core fiber with respect to radius and field confinement when used in a single mode application; the basis for the comparison was the $HE_{1,1}$ mode operated at $\beta_N = 1.507$, where $n_1 = 1.515$ and $n_2 = 1.500$. Figure 17 shows a rescaled parabolic core dispersion curve superimposed upon the homogeneous core $HE_{1,1}$ dispersion curve. The parabolic core fiber has a radius that is 1.66 times larger than the homogeneous core fiber. This is significant since it lends itself to improved launching efficiency and easier splicing between fibers. Figure 18 shows that the relative field confinement of the $r = 1.66a$ parabolic profile

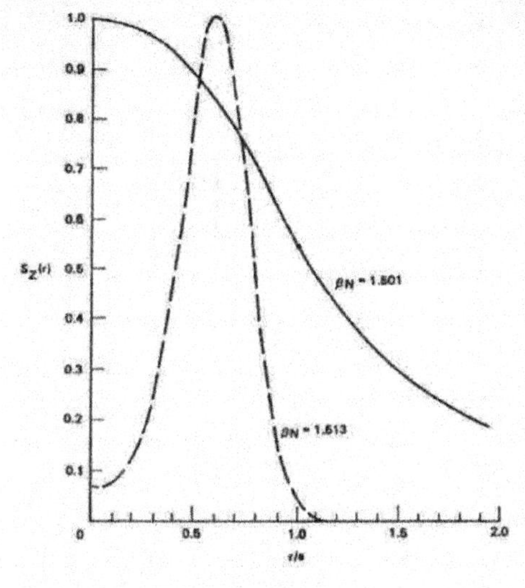

Fig. 15. Poynting flux characteristics of $HE_{1,1}$ mode. Fiber profile is doughnut (five layer), $n_{MAX} = 1.515$ and $n_2 = 1.500$.

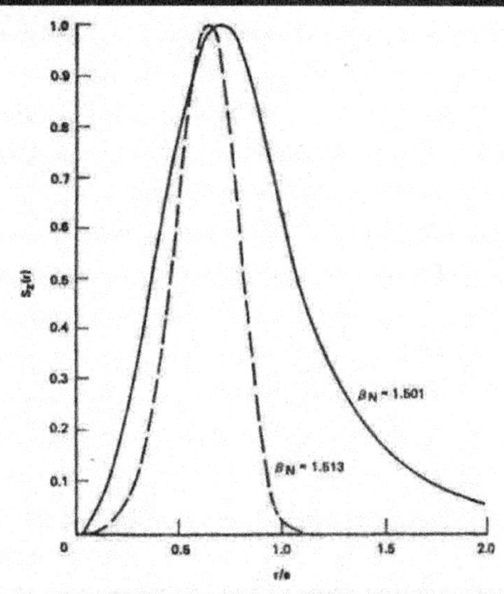

Fig. 16. Poynting flux characteristics of $HE_{2,1}$ mode. Fiber profile is doughnut (five layer), $n_{MAX} = 1.515$ and $n_2 = 1.500$.

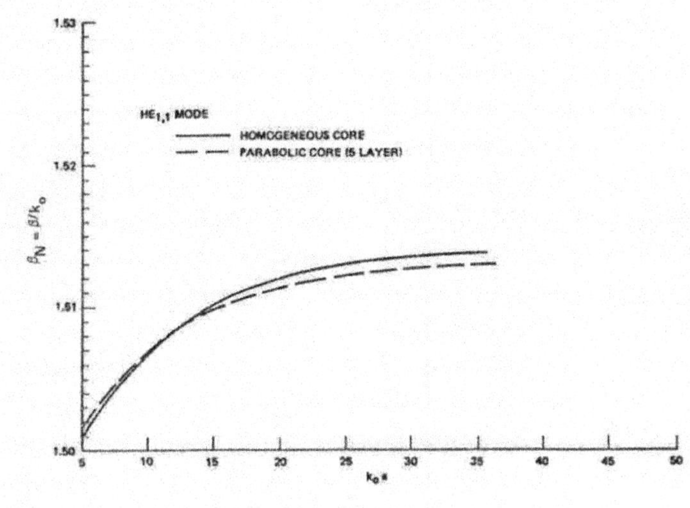

Fig. 17. Comparison of dispersion curves for homogeneous parabolic (five-layer) refractive-index profile fibers. Radius of parabolic core fiber is 1.66 times larger than homogeneous core radius.

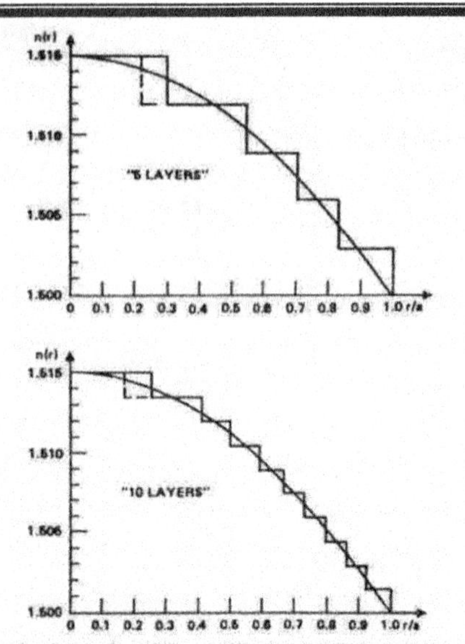

Fig. 18. Comparison of Poynting flux characteristics for homogeneous and parabolic (five-layer) core fibers. Radius of parabolic core fiber is 1.66 times larger than homogeneous core radius.

guide matches very well to a step profile guide of radius $r = a$.

In terms of dispersion, the refractive index profile has a pronounced effect on the guide dispersion characteristic. Of the profiles studied, the parabolic profile guide demonstrates the least dispersion. The factors that minimize pulse distortion in parabolic profile guide can be deduced from the generated dispersion curves. When one compares parabolic profile guide with step profile guide, the important differences are that for a parabolic profile guide (1) each mode is much less dispersive, (2) many modes have degenerate dispersion curves, and (3) for any given core size a and index difference Δn, the total number of propagating modes has been reduced, and these modes are mostly degenerate.

In terms of field confinement, the parabolic profile guide has a desirable property: when operating at single mode the parabolic profile guide with a core radius 1.66 times larger than a step profile guide has the same spot size as that of a step profile guide. This comparison was made on the basis that both guides are operating at the same frequency and wavenumber ($\beta_N = 1.507$, where $1.500 < \beta_N < 1.515$). If one connected the two guides with mismatched core sizes, one may expect efficient transfer of energy at the junction since the spot sizes, frequency, and wavenumbers all match. And finally, a comparison of the guide dispersions when operated under the above conditions indicates that the parabolic guide is still less dispersive than the step guide.

The doughnut profile guide was also studied in terms of dispersion and field confinement. The dispersion characteristics are intermediate to a step and parabolic profile guide, and the modes are not fully degenerate although the dispersion curves do show movement toward degeneracy. Hence, one would expect that multimode pulse distortion would also be intermediate to the other two types of guide. The doughnut guide has interesting focusing properties that confirm one's geometrical optics model of doughnut guidance. Even the fundamental $HE_{1,1}$ mode at high frequencies focuses onto the high index region as if the high index region were a slab guide. Thus, the doughnut guide has interesting possibilities as a transition from fiber to slab guide and an n-port power divider. For instance, one could join fiber of any profile to doughnut profile guide. Then, allow the doughnut guide to become very multimode by an increase in its diameter. When sufficiently multimode, the doughnut profile guide high-index region can be divided into azimuthal segments, each segment corresponding to a slab waveguide. The azimuthal segments can correspond to n-ports of a power divider with equal or unequal division.

And finally, it is interesting to note some general properties of fiber power flux density S_z. First, to evaluate $S_z(r)$, one must use accurate values for ω-β solutions; otherwise, large errors in S_z occur at the layer boundaries. Second, near cutoff, the percentage of power carried in the core vs that in the cladding does not approach zero for all modes; for instance, the $HE_{1,3}$ mode carries 69% of the power in the core even near cutoff. Third, far above cutoff, almost all the power is carried in the core, which confirms one's geometrical

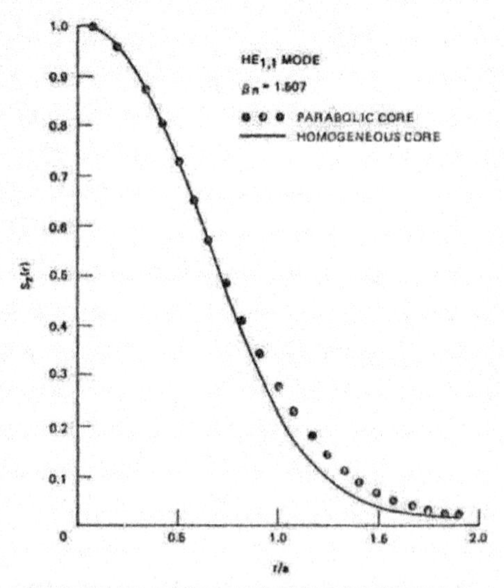

Fig. 19. Staircase approximations of parabolic refractive-index profile.

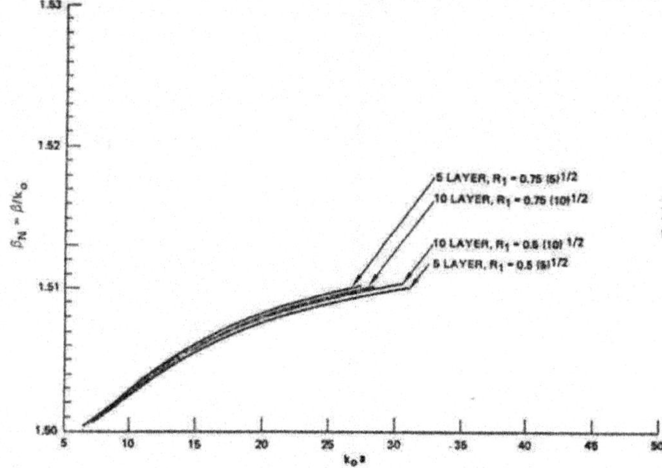

Fig. 20. Dispersion curves for $HE_{1,1}$ mode of parabolic refractive-index profile fiber approximated by five and ten layers.

Table I. Illustration of Error in $S_z(r)$ due to Error in $(\beta_N, k_o a)$ and P

β_N	$k_o a$	$\Delta k_o a$ Error	$S_z(r=a^-)$	$S_z(r=a^+)$	P
1.501	5.13593	0.0	.54888	.54438	-.99886
		0.5	.47937	.68664	-.996*
		1.0	.41087	.82311	-.996*
1.507	10.19825	0.0	.21555	.21770	-.99399
		0.5	.17623	.22763	-.996*
		1.0	.14052	.23554	-.996*
1.513	26.13023	0.0	.04209	.04231	-.99376
		0.5	.03343	.04215	-.996*
		1.0	.02581	.04171	-.996*

* P was fixed at -.996

Fiber Parameters: Homogeneous core, $n_1 = 1.515$, $n_2 = 1.500$

Dispersion Curve: $HE_{1,1}$ mode

optics model of fiber. And fourth, for a given mode and fiber profile, there is a relationship between R_β,

$$R_\beta = \frac{\beta_N - n_2}{n_1 - n_2}; \; n_2 < \beta_N < n_1,$$

and the percentage of power carried in the core region that is independent of Δn. For example, the $HE_{1,1}$ mode operated at $R_\beta = 0.0666$ has 24.2% the power in the core whether Δn is 1% or 2%.

It may be worthwhile to mention that to obtain the pulse dispersion characteristics of an optical fiber, the group velocities of various modes must be computed. Detailed discussion of this important pulse dispersion problem for an inhomogeneous fiber will be given in another paper that is under preparation.

IV. Discussion of Computational Problems

In the study of the parabolic profile, the question arises: how many layers of stratification are needed, and how should they be distributed? Figure 19 shows the distributions that were selected to determine an answer. Figure 20 demonstrates the results. Clearly, the more layers of stratification one uses, the better the accuracy. On the other hand, as the number of layers is increased, the computer cost increases faster than linearly. Hence, five layers have been used to demonstrate the dispersion characteristics in terms of a reasonable compromise between accuracy and cost. The accuracy could be improved by the use of an unequal distribution of the stratification, that is, use large increments where the profile is flat and small increments where the profile is steep. Excellent accuracy could be achieved with ten layers as shown by Fig. 20 wherein ten layer stratification was relatively insensitive to a minor shift in a layer boundary location. However, the ten layer solutions were three times costlier to obtain than the five layer solutions. On the other hand, a four layer stratification showed a significantly different dispersion characteristic than did the five layer model. Hence five layer stratification was selected as the minimal number of layers that would achieve respectable accuracy at the lowest possible cost.

The cost associated with the use of more layers can be reduced by a decrease in the size of the search performed in (β_N, k_0a) space. This can be achieved by the definition of a small search neighborhood around the five layer (β_N, k_0a) solution.

Besides the cost of additional layers, further layering of the profile causes other computational problems. For any propagating mode, β_N is always takes on a value $n_i \leq \beta_N \leq n_{i+1}$. As the number of layers is increased, $\beta_N \simeq n_i$ or $\beta_N \simeq n_{i+1}$. Whenever this occurs, the radial wavenumber in the n_i region is approximately zero. This causes a computational problem in evaluation of the Bessel functions for the n_i region. However, this can be overcome through the use of small argument Bessel functions.

The computation of S_z is quite an involved process; in view of this, an investigation was undertaken to determine if one could easily generate S_z plots by the use of approximate values of (β_N, k_0a) such as could be taken from a dispersion diagram and by taking advantage of the properties of P (for HE modes, $P \simeq -1$, for EH modes, $P \simeq +1$). The value P is defined as follows: $P = [(\omega\mu)/\beta][(H_z)/(E_z)]$. The study of a homogeneous core $S_z(r)$ equation indicated that accurate values of P and (β_N, k_0a) are required; otherwise, erroneously large discontinuities in $S_z(r)$ occur at $r = a$. The results are summarized in Table I.

This work was supported in part by the U.S. Office of Naval Research.

References

1. J. G. Dil and H. Blok, Opto-electronics 5, 415 (1973); M. O. Vassell, Opto-electronics 6, 271 (1974).
2. G. L. Yip and Y. H. Ahmex, in *Proc. URSI Symposium of Electromagnetic Wave Theory* (1974), p. 272.
3. P. J. B. Clarricoats and K. B. Chan, Electron. Lett. 6, 694 (1970).

Arbitrarily shaped inhomogeneous optical fiber or integrated optical waveguides*

C. Yeh, S. B. Dong, and W. Oliver

School of Engineering and Applied Science, University of California, Los Angeles, California 90024
(Received 5 September 1974)

Using the finite element technique, a numerical method is devloped so that one may obtain the propagation characteristics of optical waves along guiding structures whose cores may be of arbitrary cross-sectional shape and whose material media may be inhomogeneous in more than one transverse direction. Several specific examples are given and the results are compared with those obtained by other exact or approximate methods. Very close agreement was found. The method developed here can easily be applied to many important problems dealing with practical optical fiber or integrated optical waveguides whose cross-sectional index of refraction distribution may be quite arbitrary.
PACS numbers: 84.40.V, 84.40.E, 42.80.L, 42.80.M

I. INTRODUCTION

Since the attainment of glass fibers with attenuation below 3 dB/km, communication via optical fibers is no longer a dream but a reality. Understanding of the propagation characteristics of optical signals along guiding structures depends on the availability of analytic or numerical solutions of the problems. Analytic solutions are known for only a few light-guiding structures with simple geometrical shapes, such as circular cylinders,[1] elliptical cylinders,[2] or planar layers.[3] Approximate solutions are also available for homogeneous rectangular cylinders provided that the refractive index of the core region is only slightly higher than that of the surrounding region.[4] The need to have a method that is capable of providing the solutions to the problem of waveguiding along practical guiding structures, which may be inhomogeneous optical fibers with noncircular core cross-sections or may be inhomogeneous channel or embossed integrated optical waveguides, is quite apparent.

Significant advances in recent years in the successful application of the finite element method to the complicated structural and continuum mechanics problems[5] provided the impetus for us to search for the possible application of the finite element technique to problems involving optical waveguiding. This paper describes our successful attempt in finding the solution to the problem of wave propagation along arbitrarily shaped inhomogeneous optical waveguides using the finite element technique.

II. FORMULATION OF THE PROBLEM

An arbitrarily shaped inhomogeneous optical waveguide is immersed in several dielectrics as shown in Fig. 1. The optical guide is assumed to be uniform along its longitudinal z axis. The dielectric inhomogeneity of the guide is a function of the transverse coordinates, i.e., $\epsilon_1=\epsilon_1(x,y)$, where ϵ_1 is the dielectric permittivity of the core region. The magnetic permeabilities of all regions are taken to be identical and equal to the free-space value μ_0. It is assumed that all fields outside the region bounded by the dashed boundary have decayed sufficiently that for all practical purposes no guided energy exists outside this region. (See Fig. 1.)

The governing equations for the longitudinal fields $(E_z^{(p)}, H_z^{(p)})$ of a guided wave propagating along the z direction are[6]

$$(\nabla_t + k_p^2)\begin{Bmatrix}E_z^{(p)}\\ H_z^{(p)}\end{Bmatrix}=0, \tag{1}$$

where ∇_t is the transverse del operator,

$$k_p^2=(\omega/c)^2(\epsilon_p/\epsilon_0)-\beta^2, \tag{2}$$

and the superscript (p) as well as the subscript p represent the pth region of the guide. The cross section of the guide is divided into P regions. In each of these regions, the medium is assumed to be uniform having a constant dielectric permittivity ϵ_p. The factor $\exp(-i\omega t+i\beta z)$ is assumed for all field components and suppressed throughout; ω is the frequency and β is the propagation constant. At the interfaces of the regions appropriate boundary conditions must be satisfied by the fields, i.e., the tangential electric and magnetic fields which are $E_z^{(p)}, H_z^{(p)}$,

$$\tau_p\left[\gamma\left(\frac{\epsilon_0}{\mu_0}\right)^{1/2}\frac{\partial E_z^{(p)}}{\partial s}-\frac{\partial H_z^{(p)}}{\partial n}\right],$$

and

$$\tau_p\left[\frac{\epsilon_p}{\epsilon_0}\left(\frac{\epsilon_0}{\mu_0}\right)^{1/2}\frac{1}{\gamma}\frac{\partial E_z^{(p)}}{\partial n}+\frac{\partial H_z^{(p)}}{\partial s}\right],$$

with

$$\tau_p=\frac{\gamma^2-1}{\gamma^2-\epsilon_p/\epsilon_0}, \tag{3}$$

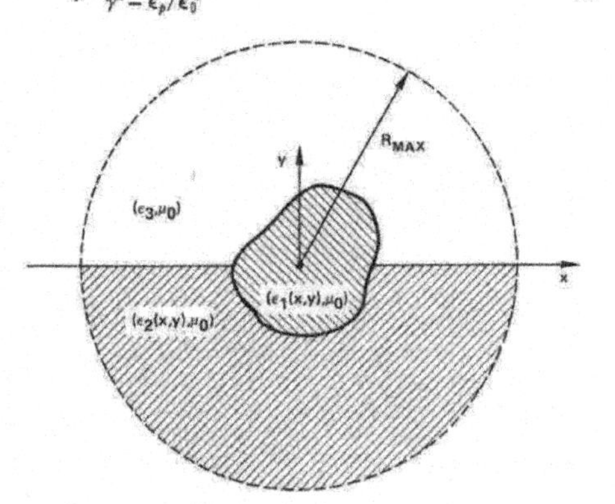

FIG. 1. Geometry of the problem.

$$\gamma = \beta c/\omega, \tag{4}$$

must be continuous from one region to another. The symbols s and n are, repectively, the tangential and normal directions along an interface and $\mathbf{n}\times\mathbf{s} = \hat{e}_z$, where $\hat{e}_z$ is the unit vector along the z axis. All transverse field components are derivable from the longitudinal electric and magnetic fields as follows:

$$E_x^{(p)} = \frac{i\omega\mu_0}{k_p^2}\left[\frac{\partial H_z^{(p)}}{\partial z} + \left(\frac{\epsilon_0}{\mu_0}\right)^{1/2}\gamma\frac{\partial E_z^{(p)}}{\partial x}\right],$$

$$E_y^{(p)} = \frac{i\omega\mu_0}{k_p^2}\left[-\frac{\partial H_z^{(p)}}{\partial x} + \left(\frac{\epsilon_0}{\mu_0}\right)^{1/2}\gamma\frac{\partial E_z^{(p)}}{\partial y}\right], \tag{5}$$

$$H_x^{(p)} = \frac{i\omega\epsilon_0}{k_p^2}\left[-\frac{\epsilon_p}{\epsilon_0}\frac{\partial E_z^{(p)}}{\partial y} + \gamma\left(\frac{\mu_0}{\epsilon_0}\right)^{1/2}\frac{\partial H_z^{(p)}}{\partial x}\right],$$

$$H_y^{(p)} = \frac{i\omega\epsilon_0}{k_p^2}\left[\frac{\epsilon_p}{\epsilon_0}\frac{\partial E_z^{(p)}}{\partial x} + \gamma\left(\frac{\mu_0}{\epsilon_0}\right)^{1/2}\frac{\partial H_z^{(p)}}{\partial y}\right].$$

According to the well-known Euler theorem, which states that if the integral

$$I(E_z^{(p)}, H_z^{(p)}) = \sum_{p=1}^{p}\int\int_{S_p} f_p\left(x, y, E_z^{(p)}, H_z^{(p)}, \frac{\partial E_z^{(p)}}{\partial x}, \frac{\partial H_z^{(p)}}{\partial x}, \frac{\partial E_z^{(p)}}{\partial y}, \frac{\partial H_z^{(p)}}{\partial y}\right)dx\,dy \tag{6}$$

is to be minimized, then the necessary and sufficient condition for this minimum to be reached is that the unknown functions $E_z^{(p)}(x, y)$ and $H_z^{(p)}(x, y)$ should satisfy the following differential equations:

$$\frac{\partial}{\partial x}\left(\frac{\partial f_p}{\partial(\partial E_z^{(p)}/\partial x)}\right) + \frac{\partial}{\partial y}\left(\frac{\partial f_p}{\partial(\partial E_z^{(p)}/\partial y)}\right) - \frac{\partial f_p}{\partial(E_z^{(p)})} = 0,$$

$$\frac{\partial}{\partial x}\left(\frac{\partial f_p}{\partial(\partial H_z^{(p)}/\partial x)}\right) + \frac{\partial}{\partial y}\left(\frac{\partial f_p}{\partial(\partial H_z^{(p)}/\partial y)}\right) - \frac{\partial f_p}{\partial(H_z^{(p)})} = 0 \tag{7}$$

within the same region, provided $(E_z^{(p)}, H_z^{(p)})$ satisfy the same boundary conditions in both cases. One can show simply that the equivalent formulation to that of Eq. (1) is the requirement that the surface integral given below and taken over the whole region, should be minimized[7]

$$I = \sum_{p=1}^{p}\int\int_{S_p}\left(\tau_p|\nabla H_z^{(p)}|^2 + \gamma^2\tau_p\frac{\epsilon_p}{\epsilon_0}\left|\frac{1}{\gamma}\left(\frac{\epsilon_0}{\mu_0}\right)^{1/2}\nabla E_z^{(p)}\right|^2 + 2\tau_p\hat{e}_z\cdot\left[\frac{1}{\gamma}\left(\frac{\epsilon_0}{\mu_0}\right)^{1/2}\nabla E_z^{(p)}\times\nabla H_z^{(p)}\right] - (\omega/c)^2\left\{(H_z^{(p)})^2 + \gamma^2\frac{\epsilon_p}{\epsilon_0}\left[\frac{1}{\gamma}\left(\frac{\epsilon_0}{\mu_0}\right)^{1/2}E_z^{(p)}\right]^2\right\}\right)dx\,dy, \tag{8}$$

with $(E_z^{(p)}, H_z^{(p)})$ obeying the same boundary conditions. Hence the Euler equations of the variational statement are

$$\delta I = \delta\sum_{p=1}^{p}\int\int_{S_p}\left(\tau_p|\nabla H_z^{(p)}| + \gamma^2\tau_p\frac{\epsilon_p}{\epsilon_0}\left|\frac{1}{\gamma}\left(\frac{\epsilon_0}{\mu_0}\right)^{1/2}\nabla E_z^{(p)}\right|^2 + 2\tau_p\hat{e}_z\cdot\left[\frac{1}{\gamma}\left(\frac{\epsilon_0}{\mu_0}\right)^{1/2}\nabla E_z^{(p)}\times\nabla H_z^{(p)}\right] - \left(\frac{\omega}{c}\right)^2\left\{(H_z^{(p)})^2 + \gamma^2\frac{\epsilon_p}{\epsilon_0}\left[\frac{1}{\gamma}\left(\frac{\epsilon_0}{\mu_0}\right)^{1/2}E_z^{(p)}\right]^2\right\}\right)dx\,dy = 0. \tag{9}$$

III. THE FINITE ELEMENT APPROXIMATION

In the finite element approximation, the primary dependent variables are replaced by a system of discretized variables over the domain of consideration. Therefore, the initial step is a discretization of the original domain into many subregions. For the present analysis, there are a number of regions in the composite cross section of the waveguide for which the permittivity is distinct. Each of these regions is discretized into a number of smaller triangular subregions interconnected at a finite number of points called nodes. Appropriate relationships can then be developed to represent the waveguide characteristics in all triangular subregions. These relationships are assembled into a system of algebraic equations governing the entire cross section.

Equation (9) provides the basis for generating the elemental waveguide characteristics. Let N_p denote the number of triangular subregions in a uniform dielectric region S_p and let $I_{p(n)}$ denote the integral in Eq. (9) for the nth triangle in S_p. Then Eq. (9) takes the slightly modified form

$$\delta I = \delta\sum_{p=1}^{p}\left(\sum_{n}^{N_p} I_{p(n)}\right) = \delta\sum_{p=1}^{p}\left\{\sum_{n}^{N_p}\int\int_{S_{p(n)}}\left[\tau_p|\nabla\mathcal{H}_z^{(p)}|^2 + \gamma^2\tau_p\frac{\epsilon_p}{\epsilon_0}|\nabla\mathcal{E}_z^{(p)}|^2 + 2\tau_p\gamma^2\hat{e}_z\cdot(\nabla\mathcal{E}_z^{(p)}\times\nabla\mathcal{H}_z^{(p)}) - \left(\frac{\omega}{c}\right)^2[(\mathcal{H}_z^{(p)})^2 + \gamma^2\frac{\epsilon_p}{\epsilon_0}(\mathcal{E}_z^{(p)})^2]\right]dS_{p(n)}\right\} = 0, \tag{10}$$

with

$$\mathcal{H}_z^{(p)} = H_z^{(p)},$$

$$\mathcal{E}_z^{(p)} = \frac{1}{\gamma}\left(\frac{\epsilon_0}{\mu_0}\right)^{1/2}E_z^{(p)},$$

where $S_{p(n)}$ is the triangular subregion of the nth element in regions S_p. In the present development approximate longitudinal electric and magnetic fields, which are linear over its triangular region, will be expressed in terms of their values at the node. These approximate fields will be sufficient to assume interelement continuity. It is noted that although higher-order approximations for these fields can be taken, requiring a consideration of more generalized coordinates, it was not pursued here because the linear approximation, when used in conjunction with the efficient algebraic eigensolution technique, provides more than sufficient accuracy in the analysis.

To facilitate the development, let triangular coordinates ξ_j $(j = 1, 2, 3)$ be introduced as shown in Fig. 2. The coordinates ξ_j of a point $P = P(\xi_1, \xi_2, \xi_3)$ are defined as follows:

$$\xi_j = A_j/A, \tag{11}$$

where A_j is the area of the subtriangle subtending the jth vertex, and A is the total triangular area. Since $A = A_1 + A_2 + A_3$ so ξ_j is constrained by the relation

$$\xi_1 + \xi_2 + \xi_3 = 1. \tag{12}$$

Introducing the notations

$$\{\xi^*\} = \begin{bmatrix}\xi_1\\ \xi_2\\ \xi_3\end{bmatrix} \quad\text{and}\quad \{\xi^*\}^T = (\xi_1, \xi_2, \xi_3), \tag{13}$$

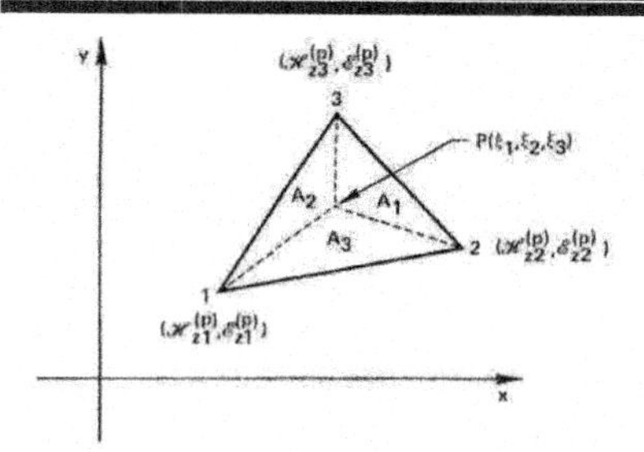

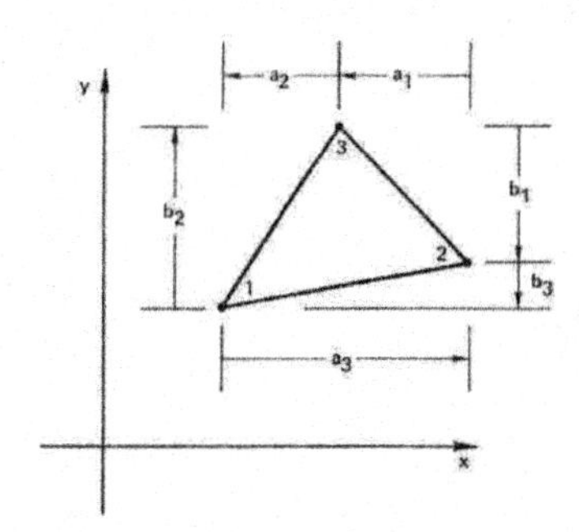

FIG. 2. The triangular coordinates.

and assuming (x_j, y_j) as the Cartesian coordinates of the jth vertex of the triangle, one can show that $\{\xi^*\}$ is related to the Cartesian coordinates (x, y) though the relation

$$\{\xi^*\}=\frac{1}{2A}\begin{bmatrix}2A_{23} & b_1 & a_1\\ 2A_{31} & b_2 & a_2\\ 2A_{12} & b_3 & a_3\end{bmatrix}\begin{bmatrix}1\\ x\\ y\end{bmatrix}, \tag{14}$$

where

$$\begin{aligned}&A_{mn}=x_m y_n - x_n y_m \quad (m, n=1,\ 2,\ 3),\\ &b_1=y_2-y_3,\quad b_2=y_3-y_1,\quad b_3=y_1-y_2,\\ &a_1=x_3-x_2,\quad a_2=x_1-x_3,\quad a_3=x_2-x_1.\end{aligned} \tag{15}$$

Denoting $\mathcal{H}_{zm}^{(p)}$ and $\mathcal{E}_{zm}^{(p)}$ $(m=1,\ 2,\ 3)$ as the vertex values of the normalized longitudinal magnetic and electric fields, the approximate linear fields within a triangle can be stated in terms of the following interpolation formula:

$$\begin{aligned}&\mathcal{H}_z^{(p)}(x, y)=\{\mathcal{H}_{zm}^{(p)}\}^T\{\xi^*\},\\ &\mathcal{E}_z^{(p)}(x, y)=\{\mathcal{E}_{zm}^{(p)}\}^T\{\xi^*\},\end{aligned} \tag{16}$$

where

$$\begin{aligned}&\{\mathcal{H}_{zm}^{(p)}\}^T=(\mathcal{H}_{z1}^{(p)}, \mathcal{H}_{z2}^{(p)}, \mathcal{H}_{z3}^{(p)}),\\ &\{\mathcal{E}_{zm}^{(p)}\}^T=(\mathcal{E}_{z1}^{(p)}, \mathcal{E}_{z2}^{(p)}, \mathcal{E}_{z3}^{(p)}).\end{aligned} \tag{17}$$

The following derivatives of the field variables will be needed in subsequent derivations:

$$\begin{aligned}&\frac{\partial \mathcal{H}_z^{(p)}}{\partial x}=\{\mathcal{H}_{zm}^{(p)}\}^T\left\{\frac{\partial \xi^*}{\partial x}\right\},\\ &\frac{\partial \mathcal{H}_z^{(p)}}{\partial y}=\{\mathcal{H}_{zm}^{(p)}\}^T\frac{\partial \xi^*}{\partial y},\\ &\frac{\partial \mathcal{E}_z^{(p)}}{\partial x}=\{\mathcal{E}_{zm}^{(p)}\}^T\frac{\partial \xi^*}{\partial x},\\ &\frac{\partial \mathcal{E}_z^{(p)}}{\partial y}=\{\mathcal{E}_{zm}^{(p)}\}^T\frac{\partial \xi^*}{\partial y},\end{aligned} \tag{18}$$

where $\partial\xi^*/\partial x$ and $\partial\xi^*/\partial y$ can be obtained directly from Eq. (14); they are

$$\left\{\frac{\partial \xi^*}{\partial x}\right\}=\frac{1}{2A}\begin{bmatrix}b_1\\ b_2\\ b_3\end{bmatrix},\quad \left\{\frac{\partial \xi^*}{\partial y}\right\}=\frac{1}{2A}\begin{bmatrix}a_1\\ a_2\\ a_3\end{bmatrix}. \tag{19}$$

The various terms in Eq. (10) can now be expanded in terms of the nodal values. For example,

$$\int\int_{S_{p(n)}} \tau_p |\nabla \mathcal{H}_z^{(p)}|^2 dS_{p(n)}=\tau_p\{\mathcal{H}_{zm}^{(p)}\}^T[I_1]\{\mathcal{H}_{zm}^{(p)}\},$$

$$\int\int_{S_{p(n)}} \gamma^2\tau_p\frac{\epsilon_p}{\epsilon_0}|\nabla \mathcal{E}_z^{(p)}|^2 dS_{p(n)}=\gamma^2\tau_p\frac{\epsilon_p}{\epsilon_0}\{\mathcal{E}_{zm}^{(p)}\}^T[I_1]\{\mathcal{E}_{zm}^{(p)}\},$$

$$\int\int_{S_{p(n)}} 2\tau_p\gamma^2\hat{e}_z\cdot[\nabla\mathcal{E}_z^{(p)}\times\nabla\mathcal{H}_z^{(p)}]dS_{p(n)}$$

$$=2\tau_p\gamma^2[\{\mathcal{E}_{zm}^{(p)}\}^T[I_2]\{\mathcal{H}_{zm}^{(p)}\}-\{\mathcal{H}_{zm}^{(p)}\}^T[I_2]\{\mathcal{E}_{zm}^{(p)}\}],$$

$$-\left(\frac{\omega}{c}\right)^2\int\int_{S_{p(n)}}[\mathcal{H}_z^{(p)}]^2 dS_{p(n)}=-\left(\frac{\omega}{c}\right)^2\{\mathcal{H}_{zm}^{(p)}\}^T[I_3]\{\mathcal{H}_{zm}^{(p)}\},$$

$$-\left(\frac{\omega}{c}\right)^2\int\int_{S_{p(n)}}\gamma^2\frac{\epsilon_p}{\epsilon_0}[\mathcal{E}_z^{(p)}]^2 dS_{p(n)}$$

$$=-\left(\frac{\omega}{c}\right)^2\gamma^2\frac{\epsilon_p}{\epsilon_0}\{\mathcal{E}_{zm}^{(p)}\}^T[I_3]\{\mathcal{E}_{zm}^{(p)}\}, \tag{20}$$

with

$$\begin{aligned}&[I_1]=\int\int_{S_{p(n)}}\left[\left\{\frac{\partial\xi^*}{\partial x}\right\}\left\{\frac{\partial\xi^*}{\partial x}\right\}^T+\left\{\frac{\partial\xi^*}{\partial y}\right\}\left\{\frac{\partial\xi^*}{\partial y}\right\}^T\right]dS_{p(n)},\\ &[I_2]=\int\int_{S_{p(n)}}\left[\left\{\frac{\partial\xi^*}{\partial x}\right\}\left\{\frac{\partial\xi^*}{\partial y}\right\}^T-\left\{\frac{\partial\xi^*}{\partial y}\right\}\left\{\frac{\partial\xi^*}{\partial x}\right\}^T\right]dS_{p(n)},\\ &[I_3]=\int\int_{S_{p(n)}}[\{\xi^*\}\{\xi^*\}^T]dS_{p(n)}.\end{aligned} \tag{21}$$

These integrals can be readily evaluated. Evaluating the expression $I_{p(n)}$ in Eq. (10), using the expansions in Eq. (20), gives

$$I_{p(n)}=\{\theta_n\}^T[A_n]\{\theta_n\}-(\omega^2/c^2)\{\theta_n\}^T[B_n]\{\theta_n\}, \tag{22}$$

where $\{\theta_n\}$ is an ordered array of nodal values of $\mathcal{H}_z^{(p)}$ and $\mathcal{E}_z^{(p)}$ and

$$\{\theta_n\}^T=\{\mathcal{H}_{z1}^{(p)}, \mathcal{E}_{z1}^{(p)}, \mathcal{H}_{z2}^{(p)}, \mathcal{E}_{z2}^{(p)}, \mathcal{H}_{z3}^{(p)}, \mathcal{E}_{z3}^{(p)}\},$$

and $[A_n]$ and $[B_n]$ are symmetric matrices comprising the integrals in Eqs. (20) and (21). Summing the contributions of all triangles over all the regions of the guide yields

$$I=\{\Theta\}^T[\mathbf{A}]\{\Theta\}-(\omega/c)^2\{\Theta\}^T[\mathbf{B}]\{\Theta\}, \tag{23}$$

where $\{\Theta\}$ is an ordered array of the longitudinal electromagnetic nodal variables and $[\mathbf{A}]$ and $[\mathbf{B}]$ are waveguide matrices. Hence, the variation of I in Eq. (23) gives the following algebraic eigenvalue problem:

$$[\mathbf{A}]\{\Theta\}=k_0^2[\mathbf{B}]\{\Theta\}, \tag{24}$$

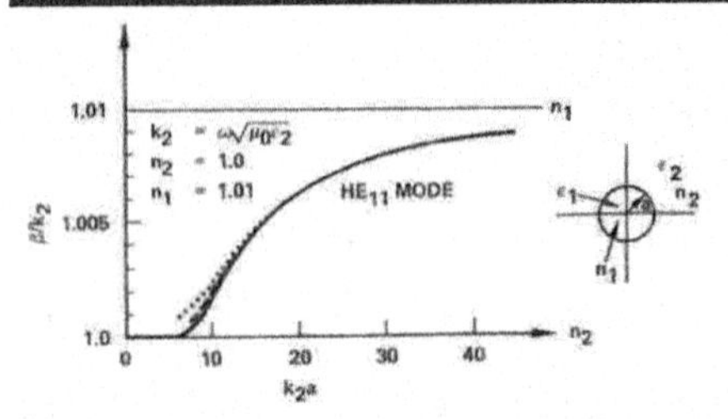

FIG. 3. Dispersion relation of a circular fiber. Solid line, exact. Dashed line, 412 modes, 733 elements in 90° sector; $R_{max}/2 = 40$. Dotted line, 152 nodes, 262 elements in 180° sector, $R_{max}/a = 10$.

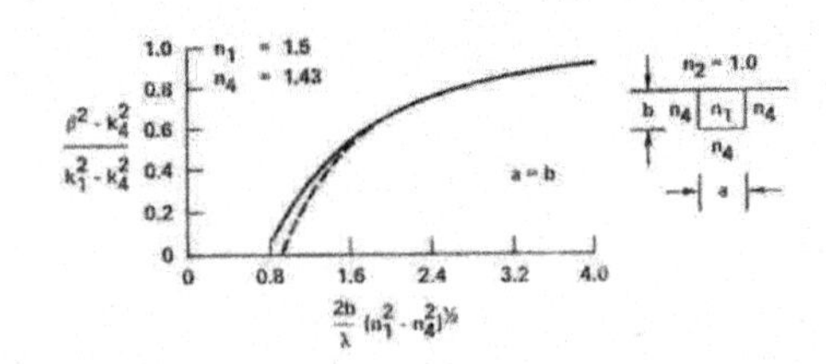

FIG. 5. Dispersion relation of an integrated optical waveguide. Solid line, final element result. Dashed line, Marcatili's approximate results.

with

$$k_0 = \omega/c.$$

Solution of this eigenvalue problem will provide the required results on the propagation constants of various modes on a particular guide.

IV. APPLICATIONS TO SEVERAL OPTICAL WAVEGUIDE PROBLEMS

This finite elements method of solution will now be applied to several practical problems in integrated and fiber optics. It is important to recognize that the success of this technique depends intimately on how quickly and efficiently one may obtain the solution of the eigenvalue problem, i.e., Eq. (24). We shall make use here of a very efficient eigensolution technique,[8,9] developed originally for mechanical systems. It permits a group of eigenvalues and eigenvectors to be extracted from the algebraic eigensystem in the frequency range of interest. The essence of the method is the reduction of the rank of the original algebraic eigensystem by means of a suitable subset of generalized coordinates. The eigenvalue analysis is conducted in the reduced space. An iteration scheme is augmented in order to achieve convergence. Convergence is assured because the entire process is merely an extension of the Stodola-Vianello method applied simultaneously to a group of eigenvectors rather than to only one at a time. In order to translate

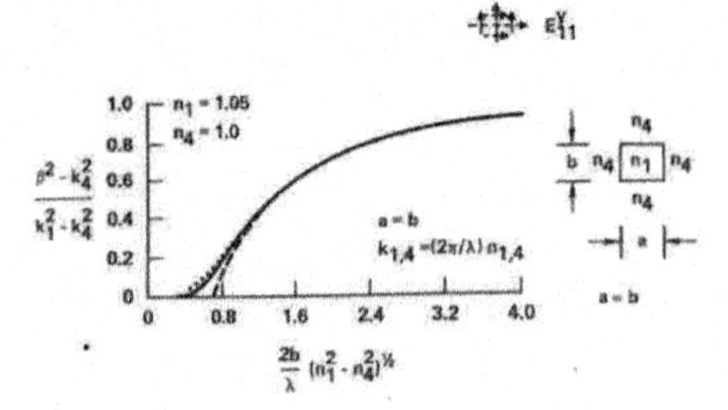

FIG. 4. Dispersion relation of a rectangular fiber. Solid line, Goell's result. Dashed line, Maracatili's approximate result. Dotted line, finite element result.

the eigenvalue problem posed by Eq. (24) to the frequency range of interest, it is only necessary to regard the eigenvalue parameter k_0^2 as being composed of two parts:

$$k_0^2 = k_c^2 + (\Delta k)^2, \tag{25}$$

where k_c^2 is the point of interest in the frequency range and $(\Delta k)^2$ is the incremental part which together with k_c^2 constitutes the time eigenvalue. Substituting Eq. (25) into Eq. (24) and rearranging terms yields

$$[\mathbf{A}^*]\{\Theta\} = (\Delta k)^2[\mathbf{B}]\{\Theta\}, \tag{26}$$

where

$$[\mathbf{A}^*] = [\mathbf{A}] - k_c^2[\mathbf{B}]. \tag{27}$$

The algebraic eigensystem (26) can be solved by the same technique. In this case, the method will yield a subset of eigenvalues $(\Delta k)^2$ whose moduli are the smallest in the translated problem. Once they have been found, the true eigenvalues are recovered by means of Eq. (25).

In the present computer code, ten reduced coordinates were used and convergence of the two eigenvalues with the smallest moduli was sought before the iteration process was considered completed. By varying the dimensionless propagation constant γ over the range of interest, it was possible to trace the frequency spectra for the various modes in a very efficient computational manner.

Three specific optical waveguides were considered: (a) The important uniform-core circular fiber guide of radius a with core index $n_1 = 1.01$ and cladding index $n_2 = 1.0$; (b) a square dielectric guide of size $a \times a$ with uniform core of index $n_1 = 1.05$ and uniform cladding of index $n_4 = 1.0$; (c) an imbedded channel integrated optical circuit guide with index $n_1 = 1.5$ whose core width is a and whose core height is a. This guide is imbedded in a substrate with index $n_4 = 1.43$. The upper of the structure is in a medium with index $n_2 = 1.0$. Formal exact solution is known for case (a);[1] Goell's circular harmonics approximate solution[10] is available for case (b); no exact numerical solution is known for case (c). Marcatili's approximate solutions are available for cases (b) and (c).[4] Comparison will be made of the results obtained according to the present finite element technique with those obtained previously by other means as mentioned above.

Figure 3 shows the normalized propagation constant β/k_2 for the dominant HE_{11} mode as a function of the normalized frequency k_2a with $k_2 = n_2\omega/c$. Three curves were shown: the solid curve gives the exact solution: the dotted curve shows the case in which the active region (the region within the dash boundary in Fig. 1) has been divided into 524 triangular elements and $R_{max}/a = 10$, where R_{max} in the radius of the dashed boundary circle and a is the core radius; the dashed curve shows the case in which the active region has been divided into 2932 triangular elements and $R_{max}/a = 40$. It can be seen easily that since a larger number of elements and a larger active region were used to obtain the dashed curve, the computed results should be closer to the true (actual) dispersion curve which was given by the solid curve. It is also expected that near the cutoff region, the field extends a large distance away from the core region, i.e., the active region is quite large. Therefore, we expect the results obtained according to the finite element technique to require a much larger active region to achieve the desired accuracy when the operating frequency is near the cutoff frequency of the mode. Far away from the cutoff region the field intensity is concentrated near or within the core region; therefore very accurate results are easily obtainable.

To see how well the present finite element technique can be used to solve optical waveguide problems whose exact analytic solutions are not yet available, two specific integrated optical circuits problems [cases (b) and (c)] were considered. The dispersion relations for the dominant E^y_{11} mode were obtained.[4] Results are shown in Fig. 4 for case (b) and in Fig. 5 for case (c). In these figures the normalized propagation constant $(\beta^2 - k_4^2)/(k_1^2 - k_4^2)$ is plotted against the normalized frequency $2b(n_1^2 - n_4^2)^{1/2}/\lambda$, where $k_{1,4} = (2\pi/\lambda)n_{1,4}$ and λ is the free-space wavelength. It can be seen that very good agreements are obtained among the three different methods for case (b) and among the two different methods for case (c).

In conclusion we would like to remark that the usefulness of the finite element technique is certainly not to reproduce the results which may be obtained using various approximation techniques such as Marcatili's or Goell's method, but to provide a straightforward way to solve problems whose solutions are not obtainable using other known methods and whose solutions are of great practical interest. We are referring to problems in which the cross-sectional core geometries may be quite arbitrary and the core or the surrounding medium may be nonuniform. These problems are of great practical importance since in practice it is very difficult, if not impossible, to carefully control the core shape and the uniformity of the guiding medium, in making various integrated optical circuit (IOC) guides.

*Work partly supported by NELC, San Diego, Calif.

[1]E. Snitzer, J. Opt. Soc. Am. 51, 491 (1961).
[2]C. Yeh, J. Appl. Phys. 33, 3235 (1962).
[3]N.S. Kapany and J.J. Burke, *Optical Waveguides* (Academic, New York, 1973); D. Marcuse, *Theory of Dielectric Optical Waveguides* (Academic, New York, 1974).
[4]E.A.J. Marcatili, Bell Syst. Tech. J. 48, 2071 (1969).
[5]*The Mathematics of Finite Elements and Applications*, edited by J.R. Whiteman (Academic, New York, 1973); O.C. Zienkiewicz and C.K. Cheung, *The Finite Element Method in Structural and Continuum Mechanics* (McGraw-Hill, New York, 1967).
[6]C. Yeh, in *Advances in Communication Systems—Theory and Applications*, edited by A.J. Viterbi and A.V. Balakrishnan (Academic, New York, to be published).
[7]S. Ahmed and P. Daly, Proc. Inst. Electr. Eng. 116, 1661 (1969).
[8]S.B. Dong, J.A. Wolf, Jr., and F.E. Peterson, Int. J. Numerical Methods Eng. 4, 155 (1972).
[9]J.A. Wolf, Jr. and S.B. Dong, SAE Trans. 80, 2619 (1971).
[10]J.E. Goell, Bell Syst. Tech. J. 48, 2133 (1969).

Single-mode optical waveguides

C. Yeh, K. Ha, S. B. Dong, and W. P. Brown

An efficient and powerful technique has been developed to treat the problem of wave propagation along arbitrarily shaped single-mode dielectric waveguides with inhomogeneous index variations in the cross-sectional plane. This technique is based on a modified finite-element method. Illustrative examples were given for the following guides: (a) the triangular fiber guide; (b) the elliptical fiber guide; (c) the single material fiber guide; (d) the rectangular fiber guide; (e) the embossed integrated optics guide; (f) the diffused channel guide; (g) the optical stripline guide.

I. Introduction

It has become increasingly clear that single strand optical fiber cables will eventually replace copper-based coaxial cables or twisted wire pairs as high-data rate communication lines. The basic configuration of an optical fiber consists of a circular cylindrical core of radius a and index of refraction n_1, coated with a cladded material of refractive index $n_0(n_1 > n_0)$. The complete analytic solution of the problem on the propagation characteristics of different modes in such a circular homogeneous core guiding structure was first given by Carson *et al.*[1] as early as 1936. Snitzer[2] in 1961 re-solved this problem and applied the result to the case of light propagation along a homogeneous core circular fiber. The first complete analysis of noncircular elliptical fiber was given by Yeh[3] in 1962. It was not until 1969 that Goell[4] presented his circular-harmonic computer analysis of a rectangular homogeneous dielectric waveguide, and Marcatili[5] presented his approximate analysis of the rectangular homogeneous dielectric structures. One notes, however, that Goell's technique can only be used efficiently on guide shapes that are close to a circle, while Marcatili's technique is most reliable for rectangular guides when the fields are mostly contained within the core region. Both techniques can only be applied to homogeneous core guides.

W. P. Brown is with Hughes Research Laboratories, Malibu, California 90268; the other authors are with University of California at Los Angeles, Electrical Sciences & Engineering Department, Los Angeles, California 90024.

Received 14 September 1978.
0003-6935/79/101490-15$00.50/0.

Significant progress has also been made in solving problems dealing with radially inhomogeneous circular fibers. A survey on this subject is included in a paper by Yeh and Lindgren.[6] It is generally accepted that no simple and satisfactory means exists in analyzing the difficult problem of waveguiding in arbitrarily shaped inhomogeneous fibers. The purpose of this paper is to specifically address this problem. It will be shown that when a modified finite-element technique was applied to this type of problem, the propagation constants, as well as the field intensity distributions for the dominant mode in guides with various different cross-sectional shapes and inhomogeneous material composition, can be readily obtained. A modified finite-element approach using a decaying parameter to model the field behavior far from the core region enables us to obtain much better results with the same number of elements near the cutoff region where the fields extend relatively far from the core region.

Specific waveguide shapes that we have considered are: (a) the triangular fiber guide; (b) the elliptical fiber guide; (c) the single material fiber guide; (d) the rectangular fiber guide; (e) the embossed integrated optics guide; (f) the diffused channel integrated optical circuit guide; and (g) the optical stripline guide. The dominant modes in these guides were treated. We have also applied this technique successfully to treat the important and well-known radially inhomogeneous circular fiber guide problem.[7]

II. Modified Finite Element Approach

In a succinct manner, we shall describe in the following our modified finite element approach to the solution of the fiber and IOC problems.

Consider an arbitrary shaped inhomogeneous optical waveguide. The inhomogeneity over its cross section can be approximated by a model consisting of a number of regions, each with uniform dielectric permittivity ϵ. For the present analysis, let the magnetic permeabilities of all regions be the same as free space $\mu = \mu_o$. In a typical region (say the pth), the governing equations for the longitudinal fields $[E_z^{(p)}, H_z^{(p)}]$ of a guided wave propagating along the z-direction are:

$$(\nabla_t^2 + k_p^2)\begin{bmatrix} E_z^{(p)} \\ H_z^{(p)} \end{bmatrix} = 0, \quad (1)$$

where ∇_t^2 is the transverse Laplacian operator, and k_p^2 is given by

$$k_p^2 = (\omega/c)^2(\epsilon_p/\epsilon_o) - \beta^2, \quad (2)$$

with ϵ_p as the dielectric permittivity. The harmonic factor $\exp[i(-\omega t + \beta z)]$, where ω is the frequency and β the propagation constant, appears in all field components and will be omitted in our discussion. Interface continuity between two contiguous regions (say the pth and qth) requires that

$$E_z^{(p)} = E_z^{(q)},$$

$$H_z^{(p)} = H_z^{(q)},$$

$$\tau_p \left[\gamma\left(\frac{\epsilon_o}{\mu_o}\right)^{1/2} \frac{\partial E_z^{(p)}}{\partial s} - \frac{\partial H_z^{(p)}}{\partial n}\right] = \tau_q \left[\gamma\left(\frac{\epsilon_o}{\mu_o}\right)^{1/2} \frac{\partial E_z^{(q)}}{\partial s} - \frac{\partial H_z^{(q)}}{\partial n}\right], \quad (3)$$

$$\tau_p \left[\frac{\epsilon_p}{\epsilon_o \gamma}\left(\frac{\epsilon_o}{\mu_o}\right)^{1/2} \frac{\partial E_z^{(p)}}{\partial n} + \frac{\partial H_z^{(p)}}{\partial s}\right] = \tau_q \left[\frac{\epsilon_q}{\epsilon_o \gamma}\left(\frac{\epsilon_o}{\mu_o}\right)^{1/2} \frac{\partial E_z^{(q)}}{\partial n} + \frac{\partial H_z^{(q)}}{\partial s}\right],$$

along their common boundary, where s and n refer to tangential and normal directions, respectively, with $n \times s = e_z$ defining the unit normal along the z-direction. In Eq. (3), τ_i and γ are given by

$$\tau_i = (\gamma^2 - 1)/(\gamma^2 - \epsilon_i/\epsilon_o),$$

$$\gamma = (\beta c)/\omega. \quad (4)$$

All transverse field components can be derived from the longitudinal electric and magnetic fields by the equations

$$E_x^{(p)} = \frac{i\omega\mu_o}{k_p^2}\left[\frac{\partial H_z^{(p)}}{\partial y} + \left(\frac{\epsilon_o}{\mu_o}\right)^{1/2} \gamma \frac{\partial E_z^{(p)}}{\partial x}\right],$$

$$E_y^{(p)} = \frac{i\omega\mu_o}{k_p^2}\left[-\frac{\partial H_z^{(p)}}{\partial x} + \left(\frac{\epsilon_o}{\mu_o}\right)^{1/2} \gamma \frac{\partial E_z^{(p)}}{\partial y}\right], \quad (5)$$

$$H_x^{(p)} = \frac{i\omega\epsilon_o}{k_p^2}\left[-\frac{\epsilon_p}{\epsilon_o}\frac{\partial E_z^{(p)}}{\partial y} + \left(\frac{\mu_o}{\epsilon_o}\right)^{1/2} \gamma \frac{\partial H_z^{(p)}}{\partial x}\right],$$

$$H_y^{(p)} = \frac{i\omega\epsilon_o}{k_p^2}\left[\frac{\epsilon_p}{\epsilon_o}\frac{\partial E_z^{(p)}}{\partial x} + \left(\frac{\mu_o}{\epsilon_o}\right)^{1/2} \gamma \frac{\partial H_z^{(p)}}{\partial y}\right].$$

The variational principle

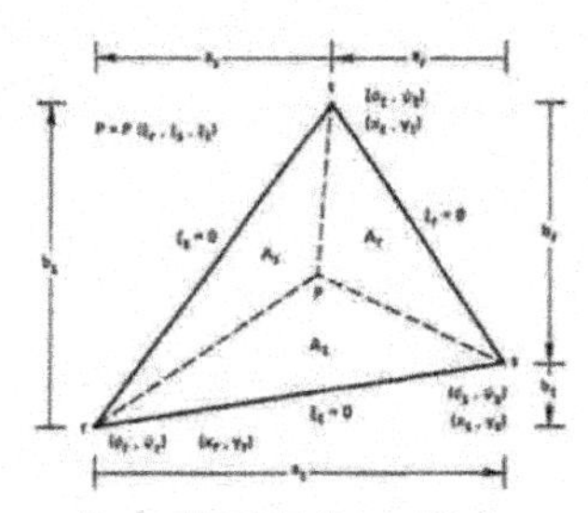

Fig. 1. The triangular coordinates.

$$\delta I = 0, \quad (6)$$

where

$$I = \sum_{p=1} I_p = \sum_{p=1} \iint \left(\tau_p \left|\nabla H_z^{(p)}\right|^2 + \gamma^2 \tau_p \frac{\epsilon_p}{\epsilon_o}\left|\frac{1}{\gamma}\left(\frac{\epsilon_o}{\mu_o}\right)^{1/2} \nabla E_z^{(p)}\right|^2 + 2\tau_p \hat{e}_z \cdot \left[\frac{1}{\gamma}\left(\frac{\epsilon_o}{\mu_o}\right)^{1/2} \nabla E_z^{(p)} \times \nabla H_z^{(p)}\right] - \left(\frac{\omega}{c}\right)^2 \left\{[H_z^{(p)}]^2 + \gamma^2 \frac{\epsilon_p}{\epsilon_o}\left[\frac{1}{\gamma}\left(\frac{\epsilon_o}{\mu_o}\right)^{1/2} E_z^{(p)}\right]^2\right\}\right) dxdy \quad (7)$$

yields as its Euler equations and natural boundary conditions the governing Eqs. (1) and continuity conditions (3) for all regions comprising the waveguide's cross section.

In what follows, scalar potential functions ϕ and ψ will be used, and they are related to the longitudinal fields by

$$\phi = H_z; \quad \psi = [(\omega\epsilon_o)/\beta]E_z. \quad (8)$$

The finite element method will be used to determine the propagation characteristics in optical fibers. This method is especially suited for problems involving complicated geometries and variable index of refraction. The bases for generating the finite element equations are Eq. (6) with functional (7). The essence of this technique will now be summarized.

The initial step in the finite element analysis is the discretization of the waveguide's cross section and its surroundings into a large number of subregions or elements. A uniform dielectric region within an inhomogeneous waveguide should not be confused with a finite element subregion. It is possible to model a dielectric region by as many elements as needed in order to properly represent the physical behavior. Herein, triangular or quadrilateral elements will be used. The triangular element is more fundamental, because two triangles can be assembled into a quadrilateral. Therefore, only the elemental properties for a triangle will be developed.

In the finite element discretization, the potential ϕ_j and ψ_j at the element's vertices or nodes will be considered as the primary dependent variables of the problem. Since the description of the element geometry and functions pertains to a triangle, the triangular coordinates will be introduced. Consider a triangle with nodes r, s, t as shown in Fig. 1. A point P within the triangle has coordinates $P = P(\xi_r, \xi_s, \xi_t)$, where triangular coordinates ξ_j have the values

$$\xi_j = A_j/A \quad (j = r,s,t) \quad (9)$$

with A_j as the subtriangular area subtending the jth node, and A is the total area of the triangle. These triangular coordinates are constrained by

$$\xi_r + \xi_s + \xi_t = 1. \quad (10)$$

The transformation between triangular and Cartesian coordinates is given by

$$\{\xi\} = \begin{bmatrix} \xi_r \\ \xi_s \\ \xi_t \end{bmatrix} = \frac{1}{2A} \begin{bmatrix} 2A_{st} & b_r & a_r \\ 2A_{tr} & b_s & a_s \\ 2A_{rs} & b_t & a_t \end{bmatrix} \begin{bmatrix} 1 \\ x \\ y \end{bmatrix}, \tag{11}$$

where

$$\begin{aligned} &A_{mn} = x_m y_n - x_n y_m \quad (m,n = r,s,t), \\ &a_r = x_t - x_s; \quad a_s = x_r - x_t; \quad a_t = x_s - x_r, \\ &b_r = y_s - y_t; \quad b_s = y_t - y_r; \quad b_t = y_r - y_s, \end{aligned} \tag{12}$$

where (x_j,y_j) $(j = r,s,t)$ are the cartesian components of the nodes.

For the present formulation, let potential functions $\phi_p(\xi_j)$ and $\psi_p(\xi_j)$ be interpolated linearly over each triangular cross section. This can be expressed as

$$\phi_p(\xi_j) = \langle \phi_r, \phi_s, \phi_t \rangle \begin{bmatrix} \xi_r \\ \xi_s \\ \xi_t \end{bmatrix} = \langle \phi_p \rangle \{\xi\}, \tag{13}$$

$$\psi_p(\xi_j) = \langle \psi_r, \psi_s, \psi_t \rangle \begin{bmatrix} \xi_r \\ \xi_s \\ \xi_t \end{bmatrix} = \langle \psi_p \rangle \{\xi\}, \tag{14}$$

where ϕ_j, $\psi_j (j = r,s,t)$ are the nodal values of the potentials.

Cartesian derivatives of $\phi_p(\xi_j)$ and $\psi_p(\xi_j)$ are needed in functional (7), and they can be obtained from Eqs. (13) and (14) by means of chain rule differentiation using Eq. (11). The Cartesian derivatives of $\{\xi\}$ are

$$\frac{\partial}{\partial x}\{\xi\} = \frac{1}{2A} \begin{bmatrix} b_r \\ b_s \\ b_t \end{bmatrix} = \{\mathbf{b}\}, \tag{15}$$

$$\frac{\partial}{\partial y}\{\xi\} = \frac{1}{2A} \begin{bmatrix} a_r \\ a_s \\ a_t \end{bmatrix} = \{\mathbf{a}\}, \tag{16}$$

so that the derivatives of ϕ and ψ have the forms

$$\begin{aligned} &\partial\phi/\partial x = \langle \phi_p \rangle^T \{\mathbf{b}\}; \quad \partial\phi/\partial y = \langle \phi_p \rangle^T \{\mathbf{a}\}, \\ &\partial\psi/\partial x = \langle \psi_p \rangle^T \{\mathbf{b}\}; \quad \partial\psi/\partial y = \langle \psi_p \rangle^T \{\mathbf{a}\}. \end{aligned} \tag{17}$$

Substituting the appropriate terms into Eq. (7) and integrating over the element's area yield the following equation for I_p:

$$I_p = \langle \theta_p \rangle^T [A_p] \{\theta_p\} - \Gamma \langle \theta_p \rangle^T [B_p] \{\theta_p\}, \tag{18}$$

where Γ is the eigenvalue parameter defined by

$$\begin{aligned} &\Gamma = (\omega/c)^2 (1 - \bar{\beta}^2), \\ &\bar{\beta} = \beta c/\omega, \end{aligned} \tag{19}$$

and $\langle \theta_p \rangle$ is the assembled array of nodal potentials:

$$\langle \theta_p \rangle^T = \{\phi_r, \psi_r, \phi_s, \psi_s, \phi_t, \psi_t\}. \tag{20}$$

The matrices $[A_p]$ and $[B_p]$ are given by

$$[A_p] = \begin{bmatrix} \frac{\tau_p}{4A}(b_r^2 + a_r^2) & 0 & \frac{\tau_p}{4A}(b_r b_s + a_r a_s) & \frac{\tau_p \bar{\beta}^2}{4A}(b_s a_r - b_r a_s) & \frac{\tau_p}{4A}(b_r b_t + a_r a_t) & \frac{\tau_p \bar{\beta}^2}{4A}(b_t a_r - b_r a_t) \\ & \frac{\tau_p \bar{\beta}^2 \epsilon_p}{4A}(b_r^2 + a_r^2) & \frac{\tau_p \bar{\beta}^2}{4A}(b_r a_s - b_s a_r) & \frac{\tau_p \bar{\beta}^2 \epsilon_p}{4A}(b_r b_s + a_r a_s) & \frac{\tau_p \bar{\beta}^2}{4A}(b_r a_t - b_t a_r) & \frac{\tau_p \bar{\beta} \epsilon_p}{4A}(b_r b_t + a_r a_t) \\ & & \frac{\tau_p}{4A}(b_s^2 + a_s^2) & 0 & \frac{\tau_p}{4A}(b_s b_t + a_s a_t) & \frac{\tau_p \bar{\beta}^2}{4A}(b_t a_s - b_s a_t) \\ & & & \frac{\tau_p \bar{\beta}^2 \epsilon_p}{4A}(b_s^2 + a_s^2) & \frac{\tau \bar{\beta}^2}{4A}(b_s a_t - b_t a_s) & \frac{\tau_p \bar{\beta}^2 \epsilon_p}{4A}(b_s b_t + a_s a_t) \\ & & & & \frac{\tau_p}{4A}(b_t^2 + a_t^2) & 0 \\ & \text{symmetric} & & & & \frac{\tau_p \bar{\beta}^2 \epsilon_p}{4A}(b_t^2 + a_t^2) \end{bmatrix} \tag{21}$$

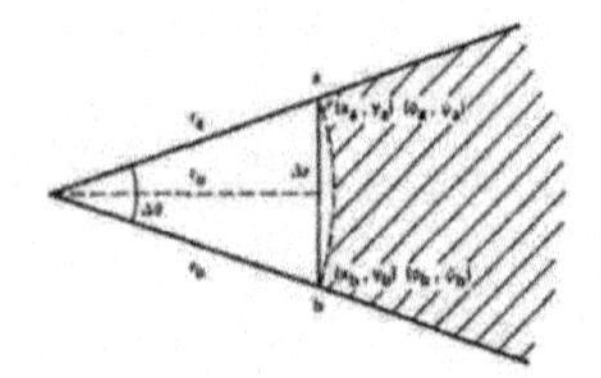

Fig. 2. A generalized special element.

$$[B_p] = \begin{bmatrix} \frac{1}{6}A & 0 & \frac{1}{12}A & 0 & \frac{1}{12}A & 0 \\ & \frac{1}{6}A\bar{\beta}^2\epsilon_p & 0 & \frac{1}{12}A\bar{\beta}^2\epsilon_p & 0 & \frac{1}{12}A\bar{\beta}^2\epsilon_p \\ & & \frac{1}{6}A & 0 & \frac{1}{12}A & 0 \\ & & & \frac{1}{6}A\bar{\beta}^2\epsilon_p & 0 & \frac{1}{12}A\bar{\beta}^2\epsilon_p \\ & & & & \frac{1}{6}A & 0 \\ & \text{symmetric} & & & & \frac{1}{6}A\bar{\beta}^2\epsilon_p \end{bmatrix} \quad (22)$$

Both E_z and H_z fields exist in the medium (or media) surrounding the optical waveguide and must be represented in the finite element model. These fields extend to infinity, and their decay is exponential. One method for modeling the surrounding is with sufficiently large-size elements. At the termination of the finite element model, prescribe that the fields ϕ and ψ vanish identically. If this boundary is sufficiently far away, the fields within the waveguide will not be significantly influenced by this approximation.[7]

However, a better method of modeling that will reduce the number of degrees-of-freedom in the problem is by means of a boundary element, where the exponential decay is incorporated into the approximation of the fields. (This method is called the modified finite-element approach.) Consider such a boundary element as shown in Fig. 2. Let the fields vary linearly in the azimuthal direction and exponential with the radial distance from the origin, which should be the central point of the waveguide. These fields may be stated as

$$\phi_p(r,\theta) = \exp(-\alpha r)\left(\phi_a + \frac{\phi_b - \phi_a}{\Delta\theta}\theta\right);$$

$$\psi_p(r,\theta) = \exp(-\alpha r)\left(\psi_a + \frac{\psi_b - \psi_a}{\Delta\theta}\theta\right), \quad (23)$$

where

$$\Delta\theta = (\Delta r)/r_o$$

$$r_o = (r_a + r_b)/2 \quad \Delta r = [(x_a - x_b)^2 + (y_a - y_b)^2]^{1/2}, \quad (24)$$

$$r_a = (x_a^2 + y_a^2)^{1/2} \quad r_b = (x_b^2 + y_b^2)^{1/2}$$

where α will be assigned. The value for α is chosen to yield the most accurate eigenfrequency (see later discussion). Carrying out the same process as that for obtaining I_p for the triangle, the following equation for the boundary element results:

$$I_{pb} = \langle\theta_{pb}\rangle^T[A_{pb}]\{\theta_{pb}\} - \Gamma\langle\theta_{pb}\rangle^T[B_{pb}]\{\theta_{pb}\}, \quad (25)$$

where $\langle\theta_{pb}\rangle$ contains the following nodal potentials:

$$\langle\theta_{pb}\rangle^T = \langle\phi_a,\psi_a,\phi_b,\psi_b\rangle. \quad (26)$$

The elements in $[A_{pb}]$ and $[B_{pb}]$ are

$$[A_{pb}] = \begin{bmatrix} \tau_p\left[\frac{\Delta\theta}{12}(2\alpha r_o + 1)\exp(-2\alpha r_o) + \frac{1}{\Delta\theta}E_1(2\alpha r_o)\right] & 0 & \tau_p\left[\frac{\Delta\theta}{24}(2\alpha r_o + 1)\exp(-2\alpha r_o) - \frac{1}{\Delta\theta}E_1(2\alpha r_o)\right] & \frac{1}{2}\tau_p\bar{\beta}^2\exp(-2\alpha r_o) \\ & \bar{\beta}^2\epsilon_p\tau_p\left[\frac{\Delta\theta}{12}(2\alpha r_o + 1)\exp(-2\alpha r_o) + \frac{1}{\Delta\theta}E_1(2\alpha r_o)\right] & -\frac{1}{2}\tau_p\bar{\beta}^2\exp(-2\alpha r_o) & \bar{\beta}^2\epsilon_p\tau_p\left[\frac{\Delta\theta}{24}(2\alpha r_o + 1) + \frac{1}{\Delta\theta}E_1(2\alpha r_o)\right] \\ & & \tau_p\left[\frac{\Delta\theta}{12}(2\alpha r_o + 1)\exp(-2\alpha r_o) - \frac{1}{\Delta\theta}E_1(2\alpha r_o)\right] & 0 \\ \text{symmetric} & & & \bar{\beta}^2\epsilon_p\tau_p\left[\frac{\Delta\theta}{24}(2\alpha r_o + 1)\exp(-2\alpha r_o) - \frac{1}{\Delta\theta}E_1(2\alpha r_o)\right] \end{bmatrix} \quad (27)$$

where

$$E_1 = \int_{r_o}^{\infty} \frac{1}{r} \exp(-2\alpha r) dr,$$

$$[B_{pb}] = \begin{bmatrix} \frac{1}{12}\frac{\Delta\theta}{\alpha^2}(2\alpha r_o + 1)\exp(-2\alpha r_o) & 0 & \frac{1}{24}\frac{\Delta\theta}{\alpha^2}(2\alpha r_o + 1)\exp(-2\alpha r_o) & 0 \\ & \frac{1}{12}\bar{\beta}^2\epsilon_p\frac{\Delta\theta}{\alpha^2}(2\alpha r_o + 1)\exp(-2\alpha r_o) & 0 & \frac{1}{24}\bar{\beta}^2\epsilon_p\frac{\Delta\theta}{\alpha^2}(2\alpha r_o + 1)\exp(-2\alpha r_o) \\ & & \frac{1}{12}\frac{\Delta\theta}{\alpha^2}(2\alpha r_o + 1)\exp(-2\alpha r_o) & 0 \\ & & & \frac{1}{12}\bar{\beta}^2\epsilon_p\frac{\Delta\theta}{\alpha^2}(2\alpha r_o + 1)\exp(-2\alpha r_o) \end{bmatrix} \quad (28)$$

Summing the contributions I_p of all the triangles and I_{pb} of all the boundary elements yields the following equation for I:

$$I = \langle\theta\rangle^T[A]|\theta| - \Gamma\langle\theta\rangle^T[B]|\theta|, \quad (29)$$

where $\langle\theta\rangle$ is an ordered array of the longitudinal electromagnetic nodal variables, and $[A]$ and $[B]$ are the waveguide matrices. Variation of Eq. (29) with respect to the nodal variables leads to the following algebraic eigenvalue problem:

$$[A]|\theta| = \Gamma[B]|\theta|. \quad (30)$$

Algebraic eigenvalue problem (30) is solved by means of an efficient direct-iterative technique described in Ref. 8. The essence of this method is a reduction of the rank of the algebraic eigensystem with a suitably chosen set of reduced generalized coordinates. The eigenvalue problem is solved in the reduced space. This process is iterated until convergence of the eigenvalues is achieved. Its convergence is assured because the method is akin to the power method. However, in this case it is applied to a group of trial vectors simultaneously instead of only one vector in the classical power method approach. This method is extremely efficient.

The best value for α is determined huristically. In other words, since the value of α determines how fast the field decays away from the guiding structure, the best value for α will be the value which yields a decaying solution that matches closest the actual solution. This value of α will also give the most accurate eigenfrequency. As an illustration, we shall consider the following situation: Given a circular fiber with core index, $n_1 = 1.01$ and cladding index $n_2 = 1.0$, for $\bar{\beta} = 1.0004$, $R_o/a = 14.0$ and 157 quadrilateral elements in one quadrant of the fiber, we shall plot the eigenfrequency k_2a as a function of α as shown in Fig. 3. It can be seen from Fig. 3 that the normalized eigenfrequency varies greatly with respect to α. This is true particularly for small values of α. As α increases, the normalized eigenfrequency also increases until a maximum value is reached. For α larger than 0.2, the normalized eigenfrequency ceases to vary. This is because all fields decay abruptly from the dashed boundary, which is equivalent to imposing the boundary conditions $E_z = 0$ and $H_z = 0$ at the dashed boundary. Hence the contribution from these special elements is zero for $\alpha > 0.2$, and the normalized eigenfrequency is not affected by the change of α for $\alpha > 0.2$. Now, the best value of α is the value for which the most accurate eigenfrequency is obtained. For the present case, the best value of α is given by the point where the eigenfrequency achieves the highest value because it comes closest to the actual value. This actual value is obtained using very fine grid sizes and many elements so that accurate field description is achieved.

III. Accuracy of the Finite-Element Approach

To study the accuracy of the finite-element approach described above, we shall compare our results with those obtained according to other exact or approximate

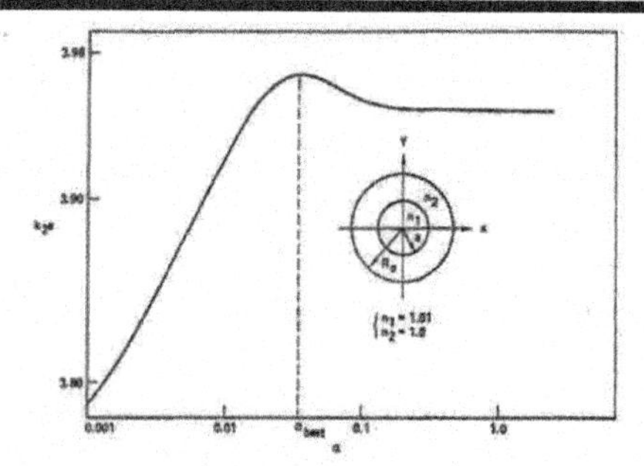

Fig. 3. Variation of eigenfrequency as a function of α for $\widetilde{\beta}$ = 1.0004.

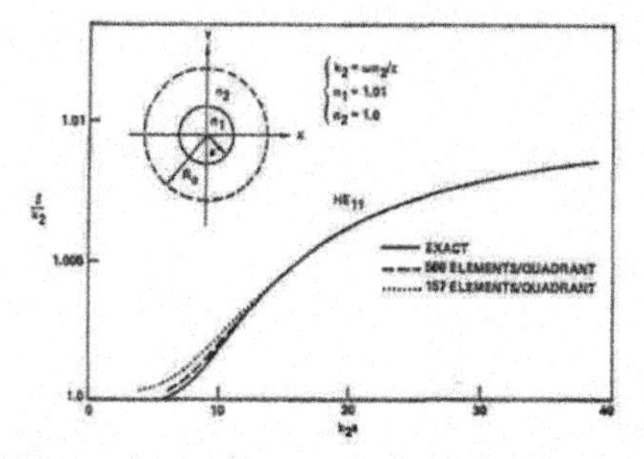

Fig. 4. Comparison of finite element approach results with the exact results for the dispersion characteristics of the HE_{11} mode in a step-index fiber.

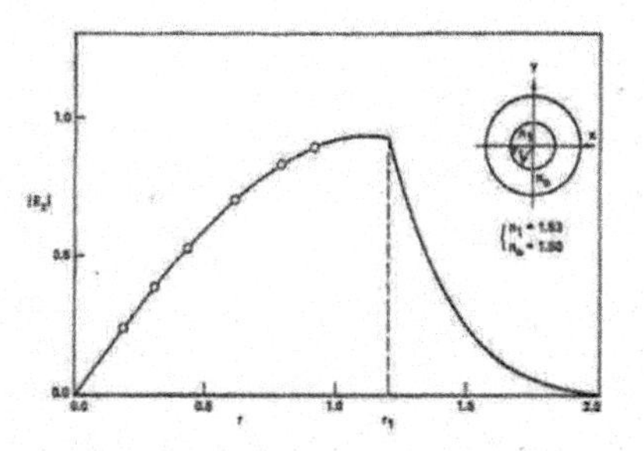

Fig. 5. Comparison of the finite element results for the normalized field component $|E_z|$ of the HE_{11} mode for a step-index fiber with the exact Bessel function $J_1(\mu r)$. The solid line represents finite element results. The circles are calculated according to the Bessel function.

methods. Specifically we shall consider the comparison of the following five types of waveguides: (1) circular step-index fiber; (2) circular graded-index fiber; (3) rectangular dielectric waveguides; (4) channel waveguides; (5) strip-line type optical waveguides. The confidence we gain from this study will enable us to use this finite-element technique to treat other guiding structures, which have not been analyzed elsewhere.

A. Circular Step-Index Fiber

Exact modal solution exists for the circular step-index fiber.[1,2] Comparison of our finite-element results with the exact results can be readily made. In Fig. 3 we have displayed three dispersion curves for the dominant HE_{11} mode: the solid curve is for the exact modal result, the dotted curve is obtained according to our finite-element approach using 157 quadrilateral elements within one quadrant of the dashed boundary circle with $R_{max}/a = 14.0$, where R_{max} is the radius of the dashed circle, and a is the core radius as shown in Fig. 3, and the dashed curve is also obtained according to the finite-element approach using 588 quadrilateral elements within a quadrant of the dashed boundary circle with $R_{max}/a = 42.0$. Because of the presence of the two degrees of symmetry, one along the x axis and the other along the y axis, only one quadrant of the circular cross section needs be used. Imposing the boundary condition of either $H_z = 0$ or $E_z = 0$ on the x axis or y axis yields different modes of interest.

It can be seen from Fig. 4 that, as expected, the accuracy of the finite-element technique improves with a larger number of elements, and better agreement with exact result is found at higher frequencies. One also realizes that at low frequencies, the field extends a large distance from the core region, so a large value of R_o/a as well as a larger number of elements are needed to achieve accurate results. Reasons for the discrepancies between the exact curve and the finite-element curves are summarized as follows:

(a) Insufficient number of elements were used, especially at lower frequencies.

(b) The element's boundary does not fit the curved boundary at the core–cladding interface.

(c) The trial functions, which are linear within each element, do not fit the exact solution.

(d) Round-off errors are present.

(e) Residual errors caused by the premature termination of the iterations used to evaluate the eigenvalues.

It should be concluded from the above consideration that the finite-element technique can be used successfully to obtain the dispersion characteristics of circular step-index fibers.

Realizing the fact that the field component E_z in the homogeneous core region of a step-index fiber is proportional to the Bessel function $J_1(\gamma r)$, one may compare the amplitude of E_z in the core region with the exact analytic function. Excellent agreement as shown in Fig. 5 was found.

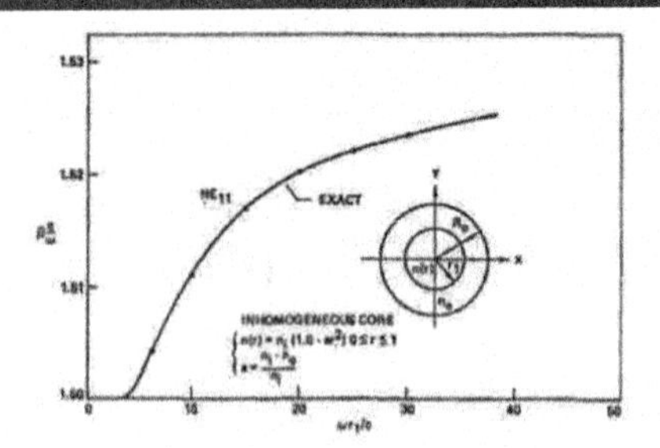

Fig. 6. Comparison of the finite element approach results with the exact Yeh and Lindgren's results for the dispersion characteristics of the HE_{11} mode in a graded-index fiber. Data are finite element results.

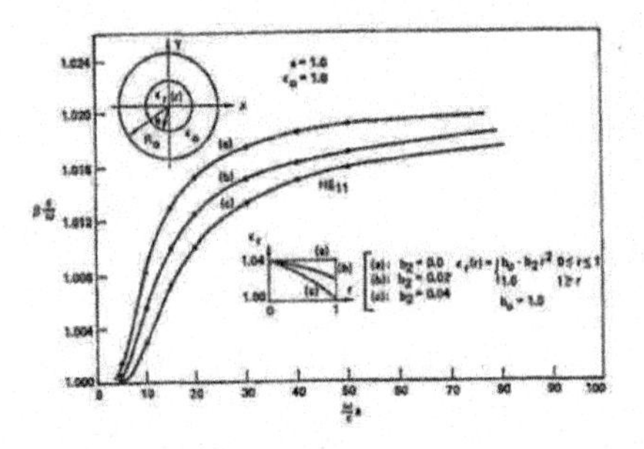

Fig. 7. Comparison of finite element approach results with the exact Dil and Blok's results for the dispersion characteristics of the HE_{11} mode in a graded-index fiber. Data are finite element results.

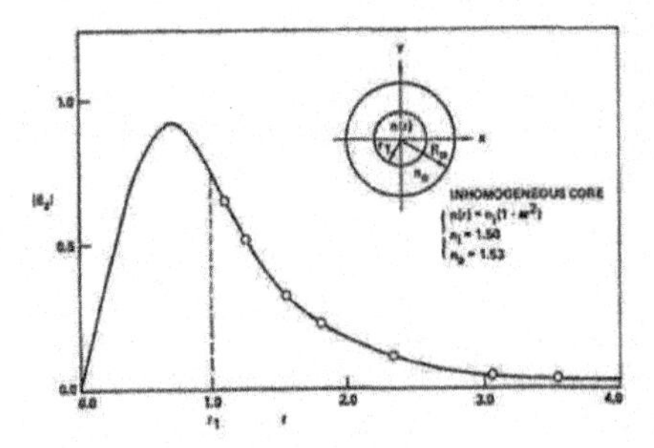

Fig. 8. Comparison of the finite element results for the normalized field component $|E_z|$ of the HE_{11} mode for a graded-index fiber with the exact modified Bessel function $K_1(\gamma r)$. The solid line represents the finite element results. The circles are calculated according to the modified Bessel function.

B. Circular Graded-Index Fiber

Because of the enhanced information carrying capacity of this family of fibers, they have been studied extensively. At least two effective techniques have been perfected to treat this type of radially inhomogeneous fibers.[6,9] Using 588 quadrilateral elements for one quadrant of the circular inhomogeneous fiber with core index

$$n(r) = n_i(1 - ar^2), \quad n_i = 1.53, \quad n_0 = 1.50, \quad a = (n_i - n_0)/n_i,$$

and cladding index $n_o = 1.50$, the finite element approach was used to obtain the dispersion curve. Results are shown in Fig. 6. In the same figure the results obtained according to Yeh and Lindgren's staircase (10-layer) approach were also plotted. The two curves are indistinguishable from each other.

Further comparison was carried out with Dil and Blok's differential equation approach. The results are shown in Fig. 7. Again no noticeable differences can be seen.

Recognizing the fact that the field component E_z in the homogeneous cladding region of an inhomogeneous core fiber is proportional to the analytic function $K_1(\gamma r)$, one may compare the amplitudes of E_z in the cladding region with the exact analytic function. Excellent agreement as shown in Fig. 8 was found.

One may conclude again from the above investigation that the finite element approach with the given elements does offer very accurate results for the radially inhomogeneous fibers.

C. Rectangular Dielectric Waveguides

Goell,[4] using the circular harmonic expansion technique, and Marcatili,[5] using approximate techniques, obtained the dispersion characteristics of rectangular dielectric waveguides. We have also solved the same problem using the finite element method. Results are shown in Fig. 9. Again excellent agreement was found. Our result matches better with Goell's result.

D. Channel Waveguides

Only Marcatili's approximate results are available for this type of waveguide.[5] Figure 10 shows the comparison between our finite element results and Marcatili's. The agreement is very good at high frequencies. The two curves deviate from each other at lower frequencies. Since the finite element approach provides better coverage of the field outside the core region, we may conclude that the finite element results are more accurate.

E. Optical Stripline Guide

The optical stripline is a planar waveguide with a strip of slightly lower index on a higher index thin film.[10] An advantage of this guide is a relaxation of the stringent requirement for smoothness of waveguide side walls because a small portion of the field strength impinges on the side walls of the strip.

Furuta *et al.*[11] obtained the dispersion curve of this guide using the effective index method. Marcatili also gave an approximate solution for this problem.[12] Re-

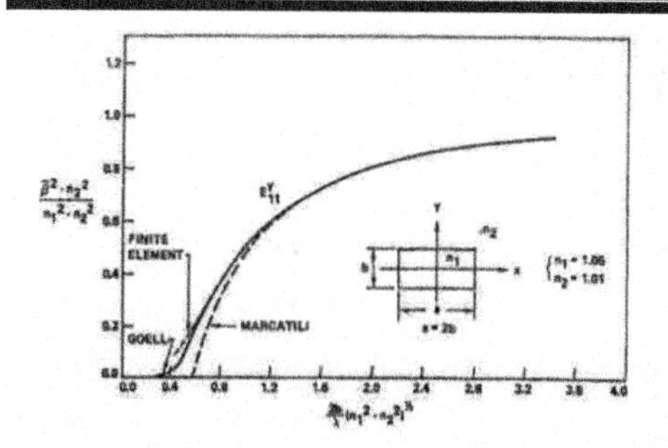

Fig. 9. Comparison of the finite element results for the HE_{11} mode with Goell's results and with Marcatili's approximate results for the rectangular fiber guide.

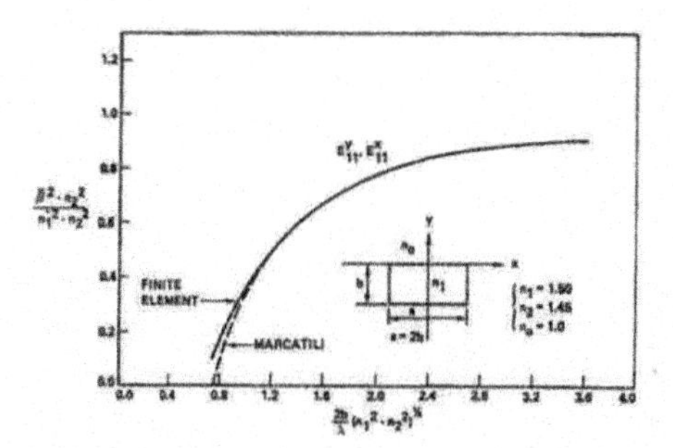

Fig. 10. Comparison of the finite element results for the lowest order mode with Marcatili's approximate results for the channel waveguide.

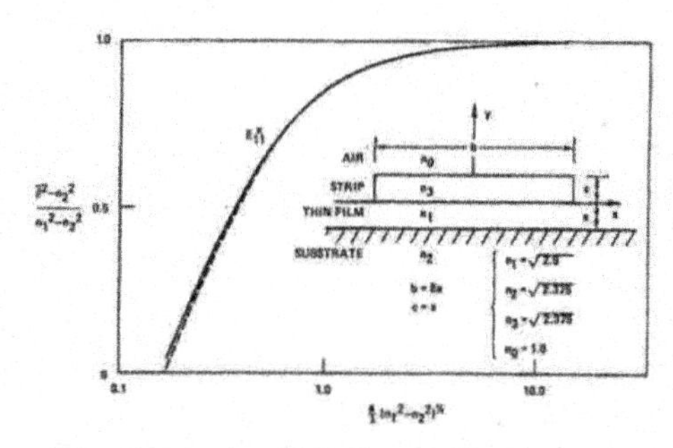

Fig. 11. Dispersion curves for the dominant E^x_{11} mode in an optical stripline waveguide. The solid line represents results found according to the finite-element method, while the dashed line represents results found according to the vector variational method.

cently, Ohtaka *et al.*[13] analyzed this problem using the vector variational method.

The cross-sectional geometry of the optical stripline is given in Fig. 11. It also shows the refractive-index distribution where n_1, n_2, n_3, and n_4 are the refractive index of thin film, substrate, strip, and air, respectively. These refractive indices satisfy one of the following relations for the stripline:

$$n_1 > n_2 \geq n_3 > n_4, \qquad n_1 > n_3 \geq n_2 > n_4. \tag{31}$$

In the present study, the numerical data of the refractive indices are chosen to be $n_1 = (2.5)^{1/2}$, $n_2 = n_3 = (2.375)^{1/2}$, and $n_4 = 1.0$. This refractive-index distribution corresponds to the case where the thin film and the substrate are glass and the ambient is air.

The fundamental modes of this stripline guide are the E^y_{11} mode and E^x_{11} mode. The principal transverse field components of the E^y_{11} mode are E_y and H_x, and those of the E^x_{11} mode are E_x and H_y. The present analysis involves only solving the E^x_{11} mode.

The finite element model employs 900 elements and 928 nodes in one-half of the cross section because of the symmetry on the y axis. The dispersion curve for the E^x_{11} mode is computed with the boundary condition, and $E_z = 0$ on the y axis. This curve is plotted in Fig. 11 where a semilog graph is used. The solid curve is the result of the finite element method, and the dotted one is that of the vector variational method. $[(\beta^2 - n_2^2)/(n_1^2 - n_2^2)]$ vs $(a/\lambda)(n_1^2 - n_2^2)^{1/2}$ is plotted where λ is the free space wavelength, and a is the thickness of thin film. These two curves are very close to each other. Therefore, the finite element method confirms the result of the vector variational method.

Let us now compare our numerical results with those of the effective index method, Marcatili's method, and the variational method. Table I gives the numerical results of the E^x_{11} mode obtained by the finite element, variational, effective index, and Marcatili's methods for the optical stripline shown in Fig. 11.

Table I. Numerical Examples for the Optical Stripline

		$\bar{\beta}^2 - n_2^2/n_1^2 - n_2^2$			
Mode	$a/\lambda(n_1^2 - n_2^2)^{1/2}$	Finite element	Vari.[13]	E.I.[11]	Marcatili[12]
E^x_{11}	0.63	0.724	0.720	0.725	0.716
	0.25	0.270	0.256	0.278	0.167

The normalized frequencies obtained by the effective index method and the variational method are in good agreement with those obtained by the finite element method. Marcatili's results are not in good agreement with those of the finite element method, the variational method, and the effective index method, especially near the cutoff frequency. This was already shown for the rectangular fiber in which the finite element solution is closer to Goell's computer solution than Marcatili's, especially in the neighborhood of cutoff frequency.

The next step is to check the mode shape for E^x_{11} and compare it with the variational method since the other two analytical approximate methods cannot be used to obtain the field intensity distribution. The variational

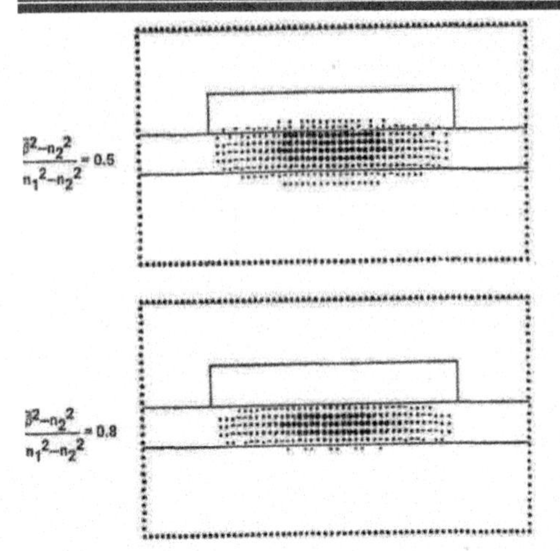

Fig. 12. Gray-scale plots for the power intensity distributions of the dominant E^y_{11} mode in an optical stripline waveguide at two different frequencies.

method produces the intensity distribution of the transverse magnetic field, which is almost the same shape as the power intensity plot generated by the finite element method.

Figure 12 shows the power intensity distribution on the gray-scale taken at the normalized propagation constant = 0.5 and 0.8. This figure is almost identical with those obtained by the variational method. In Fig. 12, the intensity distribution is well-confined in the x-direction by the dielectric strip because the width b of the strip is larger compared with the thin film thickness a and the wavelength λ. It is also confined in the y-direction by the thin film. Therefore, the optical stripline minimizes scattering loss and undesired mode conversion caused by imperfections of the etched side walls of the optical waveguide.

Having provided the above rather convincing demonstration of the versatility and the degree of accuracy of the finite element method as applied to optical waveguides, we can now confidently apply this technique to various arbitrarily shaped and inhomogeneous optical guides. As a rule of thumb, approximately 200 elements per quadrant are needed to achieve the desired three-figure accuracy for fields that are 95% confined within the core region and for the first three lower order modes.

IV. Numerical Examples

A. Triangular Fiber Guides

The finite element program for optical waveguides has been tested for a variety of problems including homogeneous and inhomogeneous circular fibers and homogeneous rectangular integrated optical waveguides. Excellent agreement has been obtained with the exact or approximate results when available.

The propagation characteristics of the dominant mode in a triangular optical fiber guide are obtained. It is assumed that this guide has an equilateral triangle core region with a refractive index $n_1 = 1.5085$ and a uniform surrounding region with a refractive index $n_2 = 1.5$.

The finite element model employs 689 elements and 659 nodes and requires 1500K storage on the computer if all nodes are used. It was necessary to devise a special subroutine to input the rectangular coordinate positions of all the nodes, especially for the nodes at the boundary.

Finite elements of the rectangular type, which is the special kind of quadrilateral element, are used. Some triangular elements are also used to match the boundary of the triangular core region. The finite element in the surrounding medium increases in size with increasing distance from the origin. The subdivisions in the core region are uniform.

Since this guide has a symmetry line along the y axis, the total field distribution needs to be calculated for only one-half of the guide. The boundary condition $H_z = 0$ is used on this line to obtain the dispersion curve for the E^y_{11} mode.

The normalized propagation constant is plotted in Fig. 13 against the normalized frequency, $[(2b)/\lambda](n_1^2 - n_2^2)^{1/2}$, where b is the height of this guide. The curve is plotted from $(\bar{\beta}^2 - \bar{n}_2^2)/(n_1^2 - n_2^2) = 0.1$ since the active region below this point is too large to be contained by the present finite element model.

To determine whether this curve represents the E^y_{11} mode, the power intensity is plotted on the gray-scale. Figure 14 shows the intensity distribution for three different values of the normalized propagation constant. These figures closely match the anticipated intensity distribution.

The core region in Figure 14 resembles an isosceles triangle rather than the original equilateral triangle because the computer printer increment is larger in the vertical direction than in the horizontal direction. This intensity test shows that the dispersion curve in Fig. 13 represents that of the E^y_{11} mode.

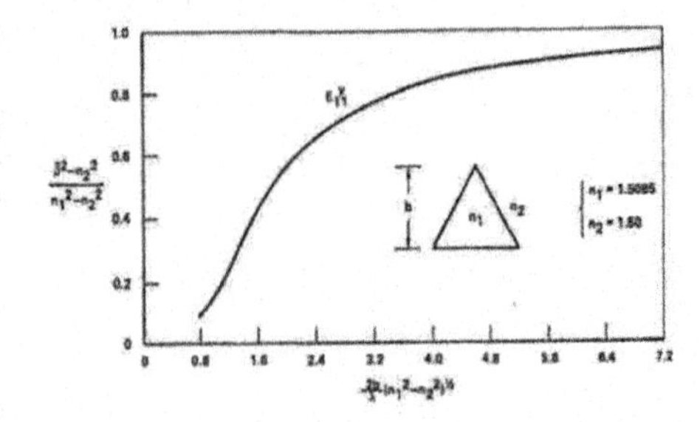

Fig. 13. Dispersion curve for the dominant mode in a triangular fiber.

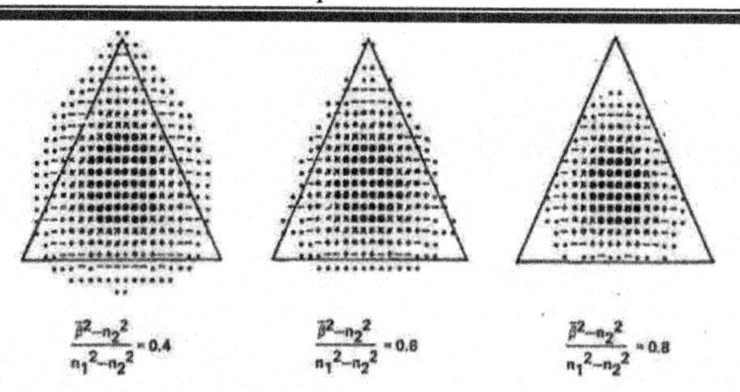

Fig. 14. Gray-scale plots for the power intensity distributions of the dominant mode in a triangular fiber at three different frequencies.

B. Elliptical Fiber Guide

This fiber consists of a uniform elliptical core region with a refractive index $n_1 = 1.5085$ and a uniform cladding region with a refractive index $n_2 = 1.5$. The aspect ratio of this guide b/a is 2, where a and b are the length of the major and minor axes, respectively.

The finite element model employs 615 elements and 658 nodal points. Only one-quarter of the entire cross section is covered by these elements, since this guide has two symmetry lines, one along the x axis and the other along the y axis. Most elements are of the rectangular type. Some triangular elements are also used to fit the elliptical core boundary.

With the same boundary conditions on the symmetry lines, the dispersion curve of the E_{11}^y mode is obtained. Figure 15 shows the dispersion where the propagation constant $(\beta^2 - n_2^2)/(n_1^2 - n_2^2)$ is plotted as a function of $[(2b)/\lambda](n_1^2 - n_2^2)^{1/2}$.

The power intensity is also plotted on the gray scale. Figure 16 shows the intensity distribution for the normalized propagation constants 0.2, 0.6, and 0.8. This figure fits the anticipated behavior of the Poynting vector exactly, i.e., the gray intensity is darkest at the center of the core region, and the field extends farther into the cladding medium near the cutoff frequency.

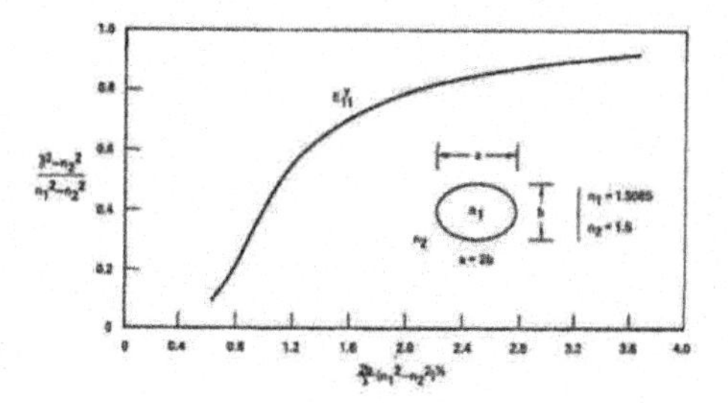

Fig. 15. Dispersion curve for the dominant $_eHE_{11}$ mode in an elliptical fiber.

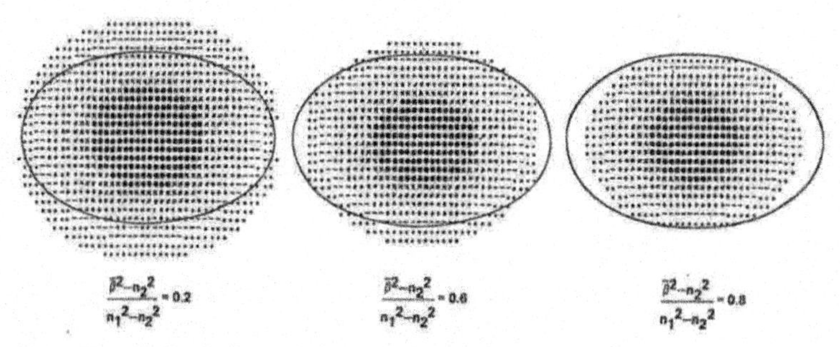

Fig. 16. Gray-scale plots for the power intensity distributions of the dominant $_eHE_{11}$ mode in an elliptical fiber at three different frequencies.

The results for elliptical fiber have practical importance since circular fibers commonly used in optical communication systems are easily deformed to various elliptical shapes. The present finite element technique can then be readily applied to these elliptical fiber problems having different aspect ratios.

C. Single Material Fiber Guide

Typically, optical fibers are constructed with a central glass core surrounded by a glass cadding with a slightly lower refractive index. The single material fiber is created by a structural form that uses only a single low-loss material.

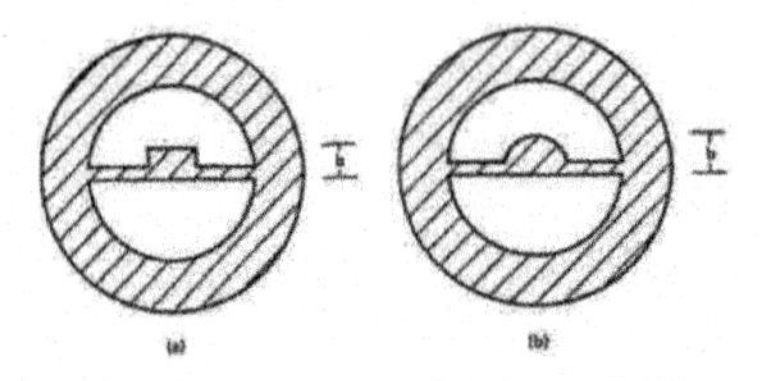

Fig. 17. Cross-sectional views of single material fibers.

Figure 17 shows cross-sectional views of two possible forms for the single material (SM) fiber. The guided energy is concentrated primarily in the central enlargement. The fields decay exponentially outward from the central enlargement. The guided wave fields at the outside cylinder can be made negligibly small if spacing between the central enlargement and outer cylinder is sufficiently large.

Only the SM fiber with cylindrical core is investigated. The complete dimensions of the SM fiber used in the present study are given in Fig. 18. The slab has a thickness of $(5/23)b$, where b is the total height of the central enlargement. It is also assumed that this guide is made of glass material with a refractive index $n_1 = 1.5$. The refractive index outside the central enlargement is equal to 1.0. The finite element model of this guide has 913 elements and 969 nodes. These elements are employed in one-half of the cross section due to the symmetry line along the y axis. The dispersion curve for the E_{11}^{y} mode is plotted in Fig. 18. In Fig. 19 the intensity distributions are plotted on the gray scale. Some errors appear in these figures, especially near the sharp edge. In order to correct these errors, the present finite element model must be modified to allocate more fine subdivisions near this area. Since these errors only

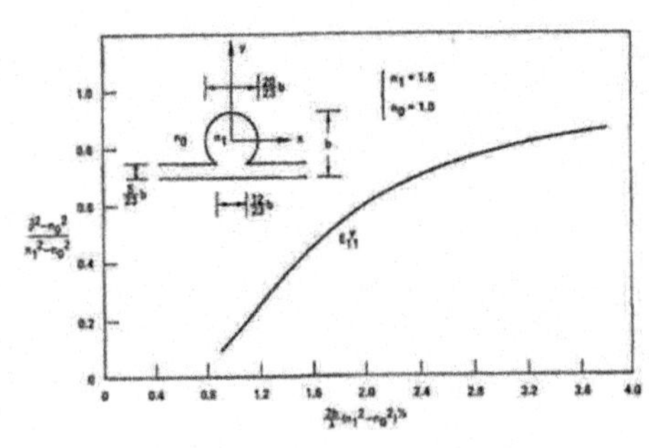

Fig. 18. Dispersion curve for the dominant mode in a single material fiber.

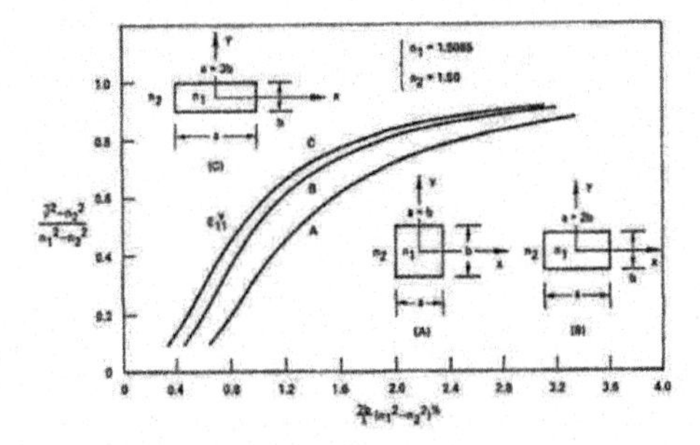

Fig. 20. Dispersion curves for the dominant ${}_eHE_{11}$ mode in rectangular fibers with different aspect ratios.

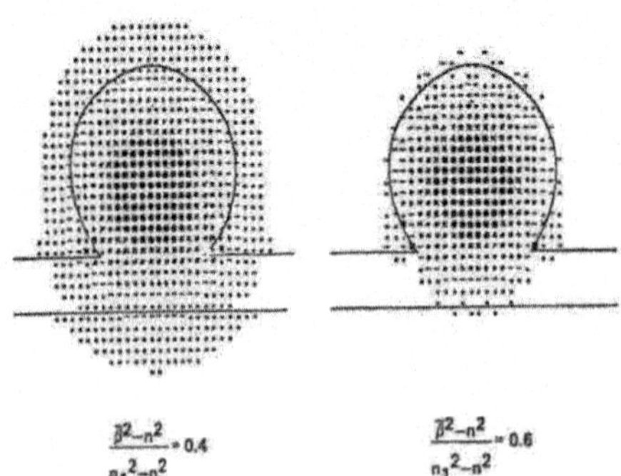

Fig. 19. Gray-scale plots for the power intensity distributions of the dominant mode in a single material fiber at three different frequencies.

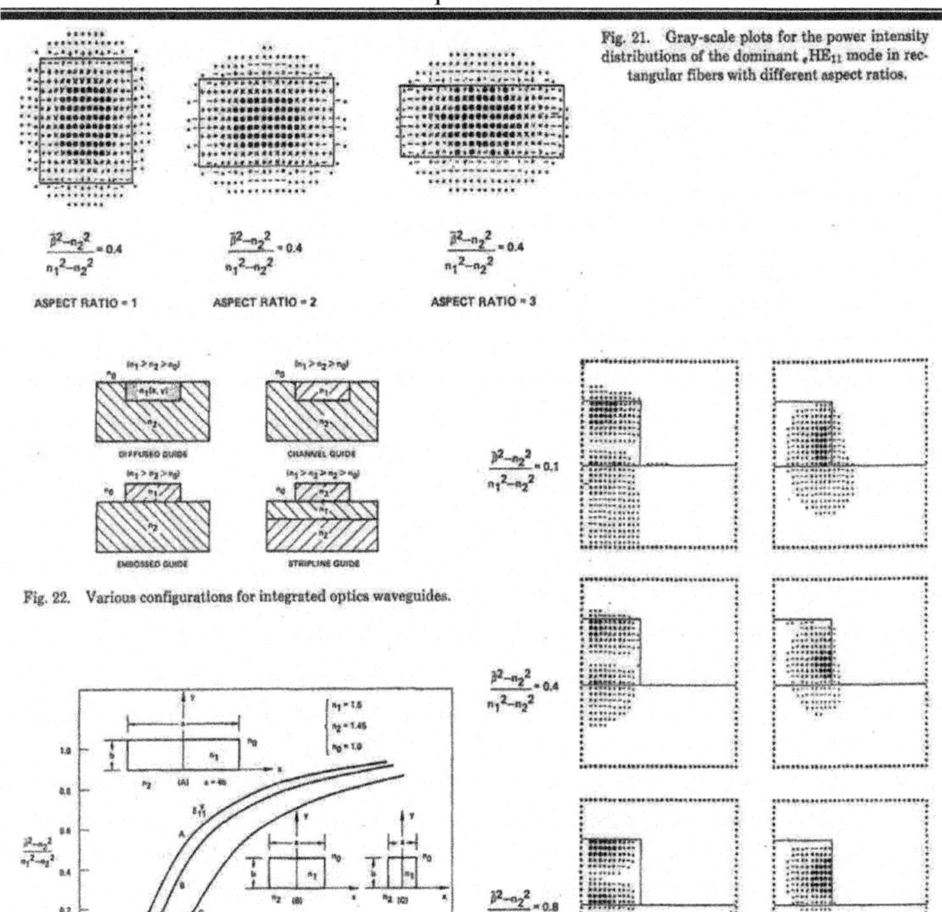

Fig. 21. Gray-scale plots for the power intensity distributions of the dominant $_eHE_{11}$ mode in rectangular fibers with different aspect ratios.

Fig. 22. Various configurations for integrated optics waveguides.

Fig. 23. Dispersion curves for the dominant E^y_{11} mode in three embossed rectangular waveguides.

Fig. 24. Gray-scale plots for the magnitude distribution of E_x and H_z for the dominant E^y_{11} mode in embossed rectangular waveguides at three different frequencies.

occur in a few elements out of several hundred elements, the dispersion curve can be still considered accurate. These pictures are taken at three different values of the normalized propagation constant. The smaller this constant, the farther the fields extend out from the circular region. The gray-scale plot also shows that the fields do not extend far toward the outer cylinder, even near the cutoff; the main portion of the power intensity is well-confined within the central enlargement. Therefore, our assumption that the outer cylinder can be neglected is reasonable.

D. Rectangular Fiber Guide

Rectangular fibers with different aspect ratios are investigated. These ratios are 1, 2, and 3. The refractive indices in the core and cladding regions are 1.5085 and 1.5, respectively. The finite element model uses 450 elements and 464 nodal points in one-quarter of the cross section. The fundamental E^y_{11} mode is investigated with the appropriate boundary conditions on the symmetry lines. Figure 20 shows the dispersion curves for this mode. Curves (A), (B), and (C) correspond to the aspect ratio 1, 2, and 3, respectively. In Fig. 21 the intensity distribution is given for each aspect ratio. These pictures are shown for the normalized propagation constant of 0.4. The intensity figures obtained by the finite element technique are very similar to Goell's results. The intensity distribution for the square core shown in Fig. 21 resembles that of the rectangular core, due to different computer printer increments in the vertical and horizontal directions.

E. Embossed Integrated Optics Guide

The waveguide commonly used for integrated optical circuitry has a rectangular core configuration. This guide consists of a rectangular core cross section, surrounded by several dielectrics of smaller refractive indices. Figure 22 shows the cross-sectional view of rectangular waveguides commonly employed for integrated optics. These are called diffused, channel, embossed, and stripline guide, respectively.

In this section, the embossed waveguide for several different aspect ratios is investigated. This guide has an embossed rectangular region with a refractive index $n_1 = 1.5$ and a uniform substrate region with a refractive index $n_2 = 1.45$, which is in contact with a third dielectric, air. The finite element model employs 900 elements and 928 nodes in one-half of the entire cross section. This embossed guide supports two dominant hybrid modes, E^y_{11} and E^x_{11}. The present investigation concentrates on the E^y_{11} mode. The dispersion curve for the E^y_{11} mode can be obtained with the boundary condition, $H_z = 0$ on the y axis. In Fig. 23, the dispersion curves for the E^y_{11} mode are plotted for different aspect ratios. Curves (A), (B), and (C) correspond to the aspect ratio 1, 2, and 4, respectively. These curves are plotted for $(\beta^2 - n_2^2)/(n_1^2 - n_2^2)$ vs $[(2b)/\lambda](n_1^2 - n_2^2)^{1/2}$, where b is the height of the embossed region. The larger the aspect ratio, the higher will be the position of the dispersion curve. The gray-scale plots of E_z and H_z are shown in Fig. 24 where only one-half of the cross section appears. These pictures represent the case for which the aspect ratio is 2. Very similar figures are also obtained for different aspect ratios. Figure 24 is taken at the normalized propagation constant of 0.1, 0.4, and 0.8, respectively, where the smaller value is closer to the cutoff region. The E_z and H_z fields are expected to extend into the substrate for small propagation constant and concentrate in the embossed region for large propagation constant. The dark region in Fig. 24 gradually shrinks as the propagation constant increases. This confirms the anticipated behavior of the E_z and H_z fields.

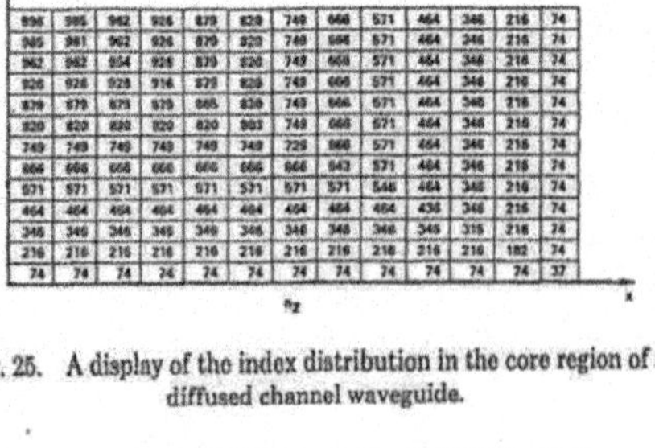

F. Diffused Channel Guide

The dielectric diffused channel waveguide is an attractive guide for integrated optical circuitry because this guide can be fabricated by a masked diffusion process, already a well-established technique in the production of integrated electronic circuits. The feasibility of producing efficient high-speed electrooptic modulators and switches has been demonstrated for this type of guide.

Several theoretical solutions are available for planar diffused waveguides.[14] More recently, a variational method using a parabolic cylinder function[15] has been used to calculate dispersion curves for diffused channel waveguides.

The diffused channel waveguides have a rectangular core region surrounded by a uniform substrate with slightly smaller refractive index. The diffusion process is assumed to be limited to the core region. The ratio of mask gap a to diffusion depth b is 2.

In the present analysis, we have chosen two different types of the refractive index variation, $n_1(x,y)$, depending on the diffusion process. The first is unidirectional diffusion, i.e., diffusion along the y axis only, and the second is circular diffusion, i.e., diffusion along the y and x axes.

Although the refractive index of the diffused guide fabricated by masked diffusion consists of a complementary error function, a parabolic form of the refractive index is used. The refractive index chosen in unidirectional diffusion

$$n_1(x,y) = n_2 + \frac{-(n_{1m} - n_2)}{b^2}(y^2 - b^2), \quad -b \le y \le 0, \quad (32)$$

where b is the diffusion depth. n_{1m} is the maximum refractive index at the surface, and n_2 is the substrate refractive index. The index variation described by Eq. (32) has a maximum value at the surface and decreases parabolically along the y axis to the lower refractive index of the substrate.

The refractive index for the circular diffusion is described as

$$n_1(x,y) = n_2 + \frac{-(n_{1m} - n_2)}{L^2}(x^2 + y^2 - L^2),$$

$$-(a/2) \le x \le a/2 - b \le y \le 0, \quad (33)$$

where

$$L = (b^2 + x^2)^{1/2} \text{ when } y \ge x,$$

$$L = [(a/2)^2 + y^2]^{1/2} \text{ when } x < y,$$

where a and b are the width and depth of the core region, respectively. L is the length of the straight line from the origin to the core boundary which intersects a point $P(x,y)$. If rectangular coordinates of $P(x,y)$

y

n_0

996	985	962	926	879	820	749	666	571	464	346	216	74
985	981	962	926	879	820	749	666	571	464	346	216	74
962	962	954	925	879	820	749	666	571	464	346	216	74
926	926	925	916	879	820	749	666	571	464	346	216	74
879	879	879	879	865	820	749	666	571	464	346	216	74
820	820	820	820	820	803	749	666	571	464	346	216	74
749	749	749	749	749	749	729	666	571	464	346	216	74
666	666	666	666	666	666	666	643	571	464	346	216	74
571	571	571	571	571	571	571	571	546	464	346	216	74
464	464	464	464	464	464	464	464	464	436	346	216	74
346	346	346	346	346	346	346	346	346	346	315	216	74
216	216	216	216	216	216	216	216	216	216	216	182	74
74	74	74	74	74	74	74	74	74	74	74	74	37

n_2 x

Fig. 25. A display of the index distribution in the core region of a diffused channel waveguide.

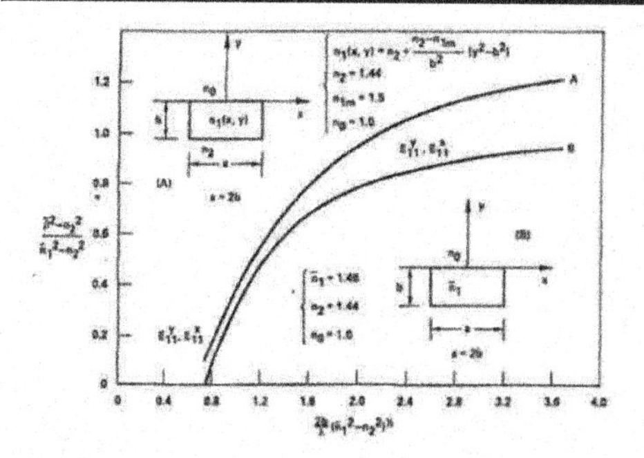

Fig. 26. Dispersion curves for the dominant E^y_{11} or E^x_{11} mode in a 1-D diffused channel waveguide and a uniform channel waveguide.

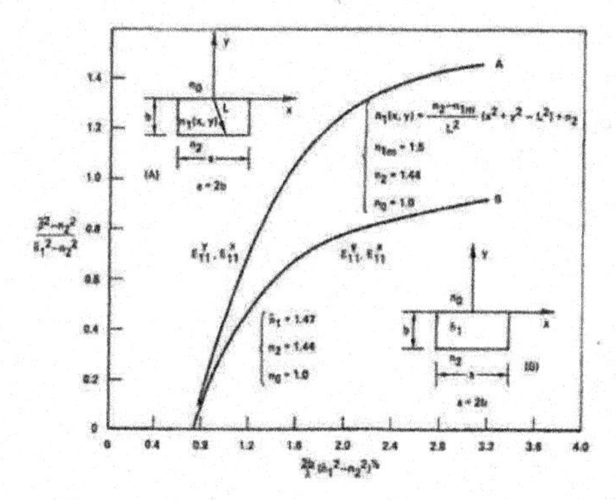

Fig. 27. Dispersion curves for the dominant E^y_{11} or E^x_{11} mode in a 2-D diffused channel waveguide and in a uniform channel waveguide.

satisfy $y \geq x$ or $y < x$, L becomes $(b^2 + x^2)^{1/2}$ or $[(a/2)^2 + y^2]^{1/2}$, respectively. The maximum length of L is $[(a/2)^2 + b^2]^{1/2}$ when $P(x,y)$ is located on the line which connects the origin and the corner of the core region, ($x = b/2$, $y = b$).

n_{1m} is the maximum refractive index at the origin, and n_2 is the substrate refractive index. The refractive index is maximum at the origin and decreases as $1/r^2$ along the x axis and y axis to the substrate index. This index variation can be realized in practice when the source of diffusion is located only at the origin.

Figure 25 shows the finite element model in the core region with the complete variation of the refractive index described by Eq. (33). This figure only shows one-half of the core region due to the symmetry along the y axis. In Fig. 25, the finite element model uses 169 square elements for the core region. Each element has its own refractive index calculated by averaging x and y coordinates of nodal points and substituting these coordinates into Eq. (32).

The refractive index can be also obtained from the number shown in Fig. 16 as

$$n_1(x,y) = n_2 + (n_{1m} - n_2) \times [N/(1000)], \qquad (34)$$

where N is the number given in Fig. 25.

The numbers in Fig. 25 are obtained by computing the refractive index described by Eq. (33) and normalized to 1000. Figure 25 clearly shows that the refractive index is maximum at the origin and decreases as $1/r^2$ to the lower substrate index. This figure is merely a simple pictorial representation of the refractive-index variation of Eq. (33). In the present analysis, we have chosen n_{1m} and n_2 to be 1.5 and 1.44, respectively, for both types of diffusion.

A total of 900 rectangular elements have been used for both types of diffusion. As previously mentioned, the core region was subdivided into 169 square elements of the same size. The element area increases proportionally to its distance from the origin for the substrate region.

The fundamental modes of this guide are the E^y_{11} and E^x_{11} mode. The boundary condition for the E^y_{11} mode on the y axis is $H_z = 0$. That for the E^x_{11} mode is $E_z = 0$. Only the E^y_{11} mode is investigated here.

The dispersion curve of the E^y_{11} mode for the unidirectional diffusion is given in Fig. 26. The curve for the circular diffusion is shown in Fig. 27, where $[(2b)/\lambda](\bar{n}_1^2 - n_2^2)^{1/2}$ is the normalized frequency, $(\beta^2 - n_2^2)/(\bar{n}_1^2 - n_2^2)$ is the normalized propagation constant, and $\bar{n}_1$ is the average value of the refractive index in the diffused core region. The averaged refractive index $\bar{n}_1$, is determined by the following equation:

$$\bar{n}_1 = \frac{1}{A}\int_0^b \int_{-\frac{a}{2}}^{\frac{a}{2}} n_1(x,y)dxdy, \qquad (35)$$

where A is the total area of the core region. Substituting Eqs. (32) and (33) into Eq. (35), $\bar{n}_1$ is computed as 1.48 for the unidirectional diffusion and 1.47 for the circular diffusion.

Curve (A) in Figs. 26 and 27 is the finite element solution for the diffused channel guide, and curve (B) is Marcatili's result for the uniform channel guides with the core refractive index $\bar{n}_1$. Curve (A) in both figures shows that the normalized propagation constant can be greater than 1.0 since this coordinate consists of $\bar{n}_1$ instead of n_1.

To determine whether curve (A) represents the E^y_{11} mode, the longitudinal electric and magnetic fields are sketched on the gray scale. The gray-scale plots for E_z and H_z are given in Fig. 28, where only one-half of the guide cross section is plotted. These figures are for the case of the circular diffusion. Similar figures are also obtained for the unidirectional diffusion. Figure 28 is plotted for normalized propagation constants of 0.2, 0.8, and 1.4, respectively.

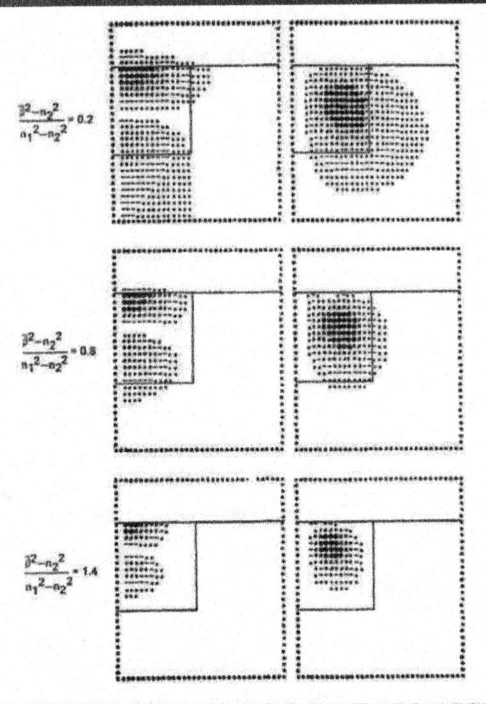

Fig. 28. Gray-scale plots for the magnitude distributions of E_x and H_x for the dominant E_{11}^y mode in a 2-D diffused channel waveguide at three different frequencies.

IV. Conclusion

We have demonstrated in this paper the versatility of the finite-element technique in the treatment of wave propagation problems dealing with various fiber optics waveguides and integrated optics waveguides. Not only are the propagation constants obtainable by this technique, but the field distributions are also readily available. Only when the operating frequencies are close to the cutoff frequency of the mode under consideration is the limitation of the finite-element technique approached. For these cases, mode field extends greatly beyond the waveguide core, and many more elements are needed to represent accurately the field region. Consequently, the finite computer machine size became the limitation. For example, at frequencies for which the normalized propagation constant $(\beta^2 - n_2^2)/(n_1^2 - n_2^2)$ is below 0.1, it takes more than 4 million bites of active memory core (the limit of our IBM 650/91 computer) to achieve the desired accuracy of 2 decimal places. Fortunately, these low-frequency cases do not usually correspond to the practical cases of interest.

With this technique, systematic studies of different noncircular and/or nonuniform fiber or integrated optics guides can now be carried out.

This work was supported in part by the U. S. Office of Naval Research.

References

1. J. Carson, S. P. Mead, and S. A. Schelkunoff, Bell Syst. Tech. J. 15, 310 (1936).
2. E. Snitzer, J. Opt. Soc. Am. 51, 491 (1961).
3. C. Yeh, J. Appl. Phys. 33, 3235 (1962).
4. J. Goell, Bell Syst. Techn. J. 48, 2133 (1969).
5. E. Marcatili, Bell Syst. Tech. J. 48, 2071 (1969).
6. C. Yeh and G. Lindgren, Appl. Opt. 16, 483 (1977).
7. C. Yeh, S. Dong, and W. Oliver, J. Appl. Phys. 46, 2125 (1975).
8. S. Dong *et al.*, Int. J. Numer. Methods Eng. 4, 155 (1972).
9. J. Dil and H. Blok, Opto-Electronics 5, 415 (1973).
10. F. Blum, D. Shaw, and W. C. Holton, Appl. Phys. Lett. 25, 116 (1974).
11. H. Furuta *et al.*, Appl. Opt. 13, 322 (1974).
12. E. Marcatili, Bell Syst. Tech. J. 53, 645 (1974).
13. M. Ohtaka *et al.*, IEEE J. Quantum Electron. **QE-12**, 378 (1976).
14. G. Hocker and W. K. Burns, IEEE J. Quantum Electron. **QE-11**, 270 (1975).
15. H. Taylor, IEEE J. Quantum Electron. **QE-12**, 748 (1976).

Propagation of truncated Gaussian beams in multimode fiber guides

C. Yeh, L. Casperson, and B. Szejn
Electrical Sciences and Engineering Department, University of California, Los Angeles, California 90024
(Received 21 January 1977)

The ability to predict the light propagation characteristics in various practical multimode guiding structures is very important in optical fiber communications and in optical image transfer. The usual mode-by-mode analysis is impractical when the guiding structure is capable of supporting hundreds or thousands of modes. In this study a numerical technique is described which is capable of providing useful data on the propagation characteristics of optical multimode guiding structures whose index of refraction variation may be quite arbitrary. As a specific example, the problem of infinite or truncated Gaussian beam propagation in a radially inhomogeneous fiber with parabolic index profile is solved. The numerical results for the infinite Gaussian beam case are compared with exact analytical data and they are in complete agreement.

I. INTRODUCTION

The merits of optical fibers as communication lines are well recognized, and a considerable amount of theoretical and experimental work has been carried out on the guiding characteristics of these devices.[1] However, in order to obtain analytical results, various approximations must be used.[2-4] For small single-mode fibers (or fibers that can support only a few low-order modes) the usual assumption is that the guiding region has a purely radial index of refraction variation. With this assumption, the electromagnetic fields can be obtained from Maxwell's equations either by dividing the radially inhomogeneous region into many stratified layers of uniform index[4] or by numerically integrating the resultant radial differential equations.[3] Recently, using the finite-element approach, we have been able to solve problems involving fibers of arbitrary shape and arbitrary index variation in the transverse directions.[5] Although this is a powerful approach when dealing with single-mode fibers, it is inefficient and costly (in terms of computer time) to obtain results for multimode fibers.

For large multimode fibers, the fields can be obtained using either the scalar wave equation or geometrical optics.[6] Since a more complete description of the wave behavior is obtainable from the scalar wave approach, it is therefore preferred. Again, the purely radial index variation is usually assumed. Analytical results have been obtained for some specific index profiles such as the parabolic or the fourth-order radial variation. It is well known that practical limitations may cause actual fibers to have index profiles that deviate (sometimes considerably) from the ideal situations described above. Furthermore, small random inhomogeneities may also exist due to frozen-in thermal fluctuations or a random distribution of ion concentration superimposed on a smoothly varying profile.[7] The deviations from an ideal refractive-index distribution cause mode conversion of an incident beam and deformation of the optical waveforms during propagation through the glass fiber,[8] while the presence of random inhomogeneities causes spatial fluctuations of the beam amplitude and phase.[9] Since the above-mentioned imperfections are present in most multimode fibers, a means of predicting the propagation behavior of light beams guided by these fibers is desirable. The purpose of this paper is to describe an efficient numerical approach to the solution of this problem. As an illustration, detailed computations will be carried out for the case of infinite or truncated Gaussian beams propagating in a radially inhomogeneous fiber with a parabolic index profile.

II. FORMULATION OF THE PROBLEM

It is well known that the solution of the exact electromagnetic equations for a spatially inhomogeneous medium is a formidable problem. Fortunately, there are some approximations that can be made to simplify this task. First, the light wavelengths of interest are much shorter than the inhomogeneity scale length. This enables us to neglect polarization effects. Hence, the optical field can be derived from a scalar $u(\mathbf{x}, z)$ that satisfies the reduced wave equation[10]

$$[\nabla^2 + k^2 n^2(\mathbf{x}, z)]\, u(\mathbf{x}, z) = 0\,, \tag{1}$$

where k is the wave number $2\pi/\lambda$ and λ is the laser wavelength. $n(\mathbf{x}, z)$ is the spatially inhomogeneous re-

 0030-3941/78/6807-0989$00.50

fractive index of the medium. Second, if we write u as the product of a factor e^{ikn_0z} that accounts for the rapid change in the phase of u along the direction of propagation and a complex amplitude $A(\mathbf{x}, z)$, a further simplification of the calculational problem results. One then has

$$\left(i2kn_0\frac{\partial}{\partial z}+\nabla_T^2+k^2[n^2(\mathbf{x},z)-n_0^2]\right)A(\mathbf{x},z)=-\frac{\partial^2 A(\mathbf{x},z)}{\partial z^2}, \quad (2)$$

where ∇_T^2 is the transverse Laplacian $[\partial^2/\partial x^2+\partial^2/\partial y^2]$, and n_0 is a given constant which represents the refractive index of some uniform medium. At laser wavelengths the complex amplitude $A(\mathbf{x})$ varies much more rapidly transverse to the direction of propagation than it does along the direction of propagation. This enables us to make the paraxial approximation, wherein the term on the right-hand side of Eq. (2) is neglected (in the Russian literature this is called the parabolic approximation). So the complex amplitude now satisfies

$$\left(i2kn_0\frac{\partial}{\partial z}+\frac{\partial^2}{\partial x^2}+\frac{\partial^2}{\partial y^2}+k^2[n^2(\mathbf{x},z)-n_0^2]\right)A(\mathbf{x},z)=0. \quad (3)$$

In addition to Eq. (3), the complex amplitude satisfies an initial condition on the fiber end, at $z=0$,

$$A(\mathbf{x},0)=u(\mathbf{x},0) \quad (4)$$

and the boundary condition

$$A(\pm\infty,z)=0. \quad (5)$$

If a truncated Gaussian beam is focused on one end of the optical guide, then

$$u(x,y,0)=u_0\exp(-r^2/w^2) \quad \text{for } 0\le r\le b,$$
$$=0 \quad \text{for } r>b, \quad (6)$$

where $r^2=x^2+y^2$, w is the spot size of the beam, and b is the radius of the truncated beam at $z=0$.

III. THE NUMERICAL APPROACH

We propose to solve Eq. (3) numerically in the following manner: Let us write $A(\mathbf{x}, z)$ in the form

$$A(\mathbf{x},z)=\exp[\Gamma(\mathbf{x},z)]\,v(\mathbf{x},z), \quad (7)$$

where $\Gamma(\mathbf{x}, z)$ is a phase function associated with the medium inhomogeneities

$$\Gamma(\mathbf{x},z)=\frac{ik}{2n_0}\int_{z_0}^{z}[n^2(x,y,z')-n_0^2]\,dz'. \quad (8)$$

The modified complex amplitude $v(\mathbf{x}, z)$ then satisfies the equation

$$i2kn_0\frac{\partial}{\partial z}v(\mathbf{x},z)+e^{-\Gamma}\nabla_T^2[e^{\Gamma}v(\mathbf{x},z)]=0. \quad (9)$$

Although Eq. (9) does not look any easier to solve than Eq. (3), it is easier to solve numerically, because, for sufficiently small increments in the z direction and an appropriately chosen lower limit in the integral in Eq. (8), the value of $v(x, y, z+\Delta z)$ can be obtained to a good approximation by solving the simpler equation

$$\left(i2kn_0\frac{\partial}{\partial z}+\nabla_T^2\right)v(\mathbf{x},z)=0, \quad (10)$$

with the initial condition

$$v(x,y,0)=u(x,y,0). \quad (11)$$

Physically, these equations approximate the propagation in the inhomogeneous medium by a two-step process at each z increment. First, we propagate the field $u(\mathbf{x}, z)$ at z to $z+\Delta z$ assuming that the intervening space is homogeneous. The effect of the inhomogeneities between z and $z+\Delta z$ is then accounted for by multiplying this solution by the phase factor $\exp(\Gamma)$.

In this research we have solved Eq. (10) by the fast Fourier-transform technique. Replacing the Laplacian by its finite-difference equivalent but still retaining the z derivative, the solution of Eq. (10) can be expressed in the form

$$v(m,n,z)=\sum_{m',n'=0}^{N-1}V(m',n',z)\exp\left(\frac{i2\pi}{N}(mm'+nn')\right), \quad (12)$$

where $x=m\Delta x$, $y=n\Delta x$ and $V(m', n', z)$ satisfies

$$\left(i2kn_0\frac{\partial}{\partial z}+\frac{f(m',n')}{(\Delta x)^2}\right)V(m',n',z)=0, \quad (13)$$

with the initial condition

$$V(m',n',z_i)=\frac{1}{N^2}\sum_{m,n=0}^{N-1}v(m,n,z_i)\exp[\Gamma(m,n,z_i)]\times\exp\left(-\frac{i2\pi}{N}(m'm+n'n)\right). \quad (14)$$

The function $f(m', n')$ is determined by the difference approximation used to represent the Laplacian in Eq. (10). For example, if $\nabla_T^2 v$ is approximated by the simple central difference expression

$$\nabla_T^2 v=[1/(\Delta x)^2][v(m+1,n,z)-2v(m,n,z)+v(m-1,n,z)+v(m,n+1,z)-2v(m,n,z)+v(m,n-1,z)], \quad (15)$$

then $f(m', n')$ is

$$f(m',n')=-4[\sin^2(\pi m'/N)+\sin^2(\pi n'/N)]. \quad (16)$$

Note that the series in Eq. (12) is simply the discrete Fourier transform of the function $V(m', n', z)$ and thus can be evaluated numerically for a given $V(m', n', z)$ by a fast Fourier transform algorithm. Furthermore, the function $V(m', n', z)$ is readily determined from Eq. (13) as

$$V(m',n',z_i)=V(m',n'z_i)\exp\left(\frac{-if(m',n')}{2k(\Delta x)^2 n_0}(z-z_i)\right), \quad (17)$$

where $V(m', n', z_i)$ is given by the series in Eq. (14), which can also be evaluated by a fast Fourier transform algorithm. To summarize, we will step from z to $z+\Delta z$ as follows: (1) Take the inverse discrete Fourier transform of $u(m, n, z)=\exp[\Gamma(m, n, z)]\,v(m, n, z)$ by means of an inverse fast Fourier transform algorithm; (2) multiply the result by $\exp[-if(m', n')\Delta z/2k(\Delta x)^2 n_0]$ and take the discrete Fourier transform with a fast Fourier transform algorithm; and (3) multiply the result by $\exp[\Gamma(m, n, z+\Delta z)]$ to yield $u(m, n, z+\Delta z)$. This process is repeated until we have reached the desired z plane.

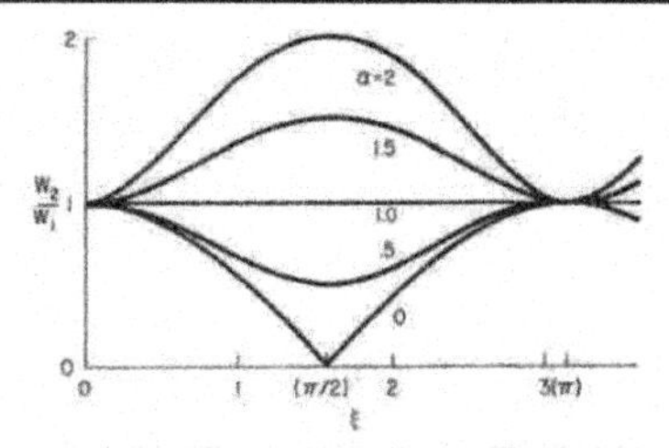

FIG. 1. Normalized beam waist w_2/w_1 as a function of normalized axial distance ξ along the fiber axis for various α. w_1 and w_2 are, respectively, the beam waist at the input and that along the fiber axis.

IV. AN ILLUSTRATIVE EXAMPLE—GAUSSIAN BEAM PROPAGATION IN A RADIALLY INHOMOGENEOUS FIBER

The numerical technique outlined earlier can be used to solve many interesting and important problems in fibers and integrated optics. As an illustration, the problem of beam propagation in an optical guide whose cross-sectional index variation is a function of the radial coordinates will be considered.

It has been well recognized that analytic solutions for the infinite Gaussian beam propagation problems exist only for certain specific radial index profiles. Even for these cases, the solutions are often involved and cumbersome to use. By simply specifying the transverse index variation $n(x, y)$, we can readily calculate the propagation characteristics of an incident beam in this medium according to the scheme discussed in Sec. III. Some of the relevant questions that can be answered by this kind of calculations are the following: (1) How well will this medium confine and focus the beam? (2) What is the focal distance? (3) How does the beam profile of an initial Gaussian beam change as it propagates in a medium with an index variation $n(x, y)$?

To demonstrate the usefulness of our numerical approach we consider the specific case of an infinite Gaussian beam propagating in a radially inhomogeneous fiber that possesses a parabolic radial index profile as follows:

$$n(x, y) = n_0 - \tfrac{1}{2} n_2 r^2 , \qquad (18)$$

where $r^2 = x^2 + y^2$, r is the radial coordinate, n_0 is the index at $r=0$, and n_2 is a given constant with dimension $(1/\mathrm{m})^2$. Inserting this parabolic index profile into our computer code and assuming an initial Gaussian beam of the form

$$u(x, y, 0) = \exp(-r^2/w^2) , \qquad (19)$$

one may obtain the required propagation characteristics of the beam. The results are summarized in Fig. 1, in which the normalized beam waist w_2/w_1, where w_2 is the beam waist along the fiber axis and w_1 is the beam waist of the input beam, is plotted against ξ, the normalized axial distance along the fiber axis for various values of α. The coefficients ξ and α are defined as follows:

$$\xi = z\left(\frac{n_2}{n_0}\right)^{1/2} , \quad n_0 \alpha = \frac{\lambda}{\pi w_1^2}\left(\frac{n_0}{n_2}\right)^{1/2} , \qquad (20)$$

where λ is the free-space wavelength of the beam. The results shown in Fig. 1 are indistinguishable from the exact analytic solutions.[6] To illustrate the intensity distributions of the beam as it propagates down the fiber guide, Fig. 2 is introduced for the case in which

$$\begin{aligned} &n_0 = 2.0 , \\ &n_2/n_0 = 8.0 \times 10^{-9}\ (\mu\mathrm{m})^{-2} , \\ &= 70.7\ \mu\mathrm{m} , \\ &\lambda = 0.8\ \mu\mathrm{m} , \\ &\Delta z = 878.1\ \mu\mathrm{m} , \\ &z_{\min} = 17562\ \mu\mathrm{m} , \end{aligned} \qquad (21)$$

where Δz is the step size of our numerical technique and $z_{\min}$ is the distance from the beam entrance to the plane at which the beam spot-size is at its minimum value.

To further demonstrate our numerical approach, we consider the case of a truncated Gaussian beam propagating in a radially inhomogeneous fiber whose index profile is given by Eq. (18). The beam at the entrance of the fiber takes on the form given by Eq. (6) with $u_0 = 1$. Results of our computation with the constants given in Eqs. (21) and $b = 75\ \mu\mathrm{m}$ are shown in Fig. 3. It can be seen that rather unusual intensity distributions are observed as the truncated Gaussian beam propagates down the fiber, and smoothing of the beam shape also

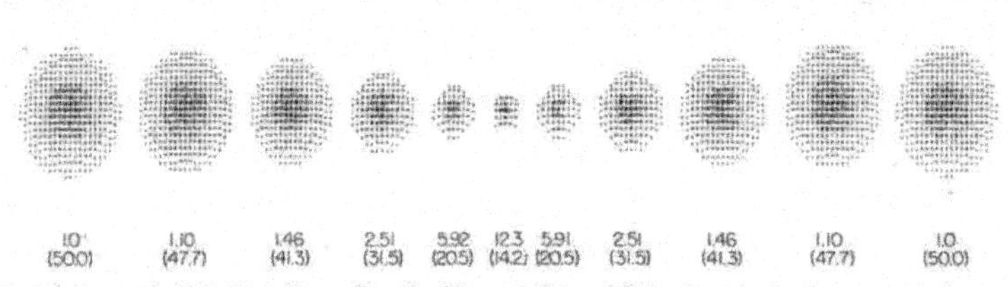

FIG. 2. Selected grey-scale intensity patterns along the fiber axis for an infinite Gaussian beam propagating in a radially inhomogeneous fiber with parabolic index profile. The values in the brackets represent the beam waists in μm of the beam while the other unbracketed values represent the highest intensity values for these patterns. Note that the Gaussian profile of the beam is retained throughout the propagation path. For the chosen parameters given in the text, it takes a 35124-μm-long axial distance to complete one cycle as shown.

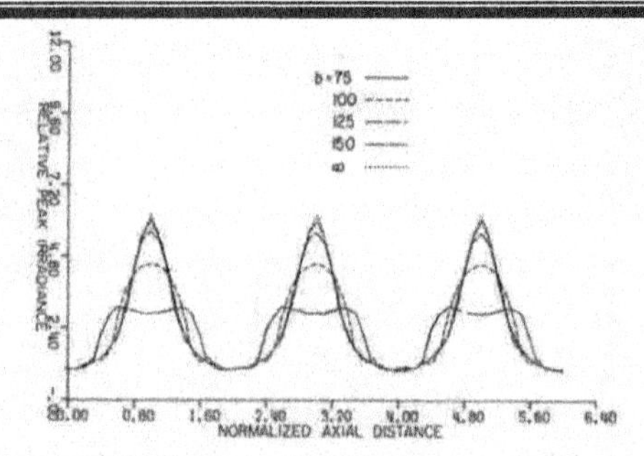

FIG. 8. Relative peak irradiance as a function of normalized axial distance $2\xi/\pi$ for various values of beam truncation radius (b) in μm. The coefficient for the parabolic index profile is chosen to be $n_2/n_1 = 4\times10^{-9}$ $(\mu m)^{-2}$.

tion of the normalized axial distance (ξ) for various values of beam truncation radius (b). The three sets correspond to the three values of n_2/n_0, which were chosen to be 8×10^{-9}, 4×10^{-9}, 8×10^{-10} $(\mu m)^{-2}$. It can be observed from these figures that truncation of an incident Gaussian beam alters significantly the way the beam waist changes as the beam propagates down the fiber. To see how the peak intensity of the beam varies as it propagates down the fiber corresponding to the several cases given in Figs. 4–6, we introduce Figs. 7–9.

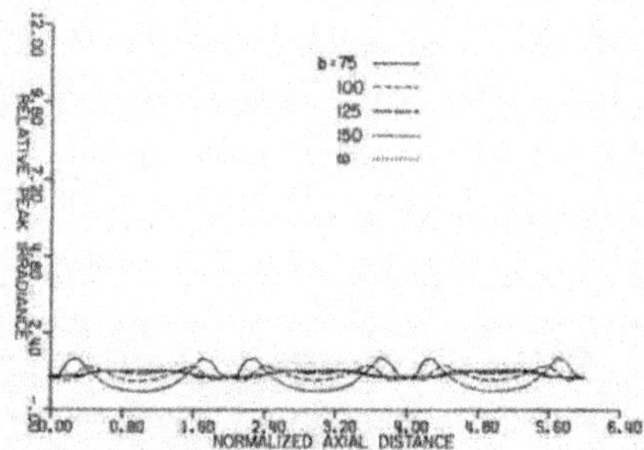

FIG. 9. Relative peak irradiance as a function of normalized axial distance $2\xi/\pi$ for various values of beam truncation radius (b) in μm. The coefficient for the parabolic index profile is chosen to be $n_2/n_1 = 8\times10^{-10}$ $(\mu m)^{-2}$.

Truncation of an incident Gaussian beam clearly affects greatly the behavior of peak beam intensity along the fiber. Results of these plots reinforce the importance of our technique since the results show that one cannot extrapolate readily from the infinite Gaussian beam case where analytic results are available to the truncated Gaussian beam case.

V. CONCLUSION

We have successfully demonstrated that the FFT technique can be used efficiently to obtain the light-beam propagation characteristics in inhomogeneous multimode fibers. It is now possible by the same methods to treat the intriguing problems[11] of (a) the power transfer between neighboring multimode guides, (b) the branching of power as a multimode guide splits into two or more guides, and (c) the guiding characteristics of beams in tapered multimode guides.

ACKNOWLEDGMENTS

The authors wish to express their sincere gratitude to Dr. W. P. Brown, Jr. of Hughes Research Laboratories, Malibu, Calif. for introducing the FFT technique to them.

*Partly supported by the Office of Naval Research and partly supported by the National Science Foundation.

[1]S. E. Miller, E. A. J. Marcatili, and T. Li, Proc. IEEE 61, 1703 (1973).

[2]E. A. J. Marcatili, Bell. Syst. Tech. J. 48, 2071 (1969); J. E. Goell, *ibid.*, 48, 2133 (1969).

[3]I. G. Dil and H. Blok, Opto-Electronics 5, 415 (1973); P. J. B. Clarricoats and K. B. Chan, Electronics Lett. 6, 694 (1970).

[4]C. Yeh and G. Lindgren, Appl. Opt. 16, (1977) (to be published).

[5]C. Yeh, S. B. Dong, and W. Oliver, J. Appl. Phys. 46, 2125 (1975).

[6]L. W. Casperson, Appl. Opt. 12, 2434 (1973).

[7]H. Kita, I. Kotano, T. Uchida, and M. Furukawa, J. Am. Ceram. Soc. 54, 321 (1971).

[8]W. A. Gambling and H. Matsumura, Opto-Electronics 5, 429 (1973).

[9]V. I. Tatarski, *Wave Propagation in a Turbulent Medium* (McGraw-Hill, New York, 1961).

[10]J. W. Goodman, *Introduction to Fourier Optics* (McGraw-Hill, New York, 1968).

[11]M. K. Barnoski, M. D. Rourke, S. M. Jensen, and H. R. Friedrick, "Coupling Components for Single Optical Fibers," presented at IEDM meeting, Washington, D. C., December, 1976 (unpublished).

Reprinted from APPLIED OPTICS, Vol. 18, page 489, February 15, 1979

Multimode inhomogeneous fiber couplers

C. Yeh, W. P. Brown, and R. Szejn

A numerical technique to obtain the wave behavior in tightly coupled multimode fibers with inhomogeneous indices is introduced in this paper. The specific problem of the coupling characteristics of two parallel multimode fibers whose index profile is parabolic is treated in detail. It was found that in spite of the fact that rather complicated coupling behavior is observed when multimodes exist, total guide power still exchanges among the fibers in a periodic manner, and the coupling length still increases monotonically as a function of the separation distance between the fibers. It has also been demonstrated that by simply specifying the index profiles of the coupling structure (provided that the profiles are slowly varying), the coupling characteristics can be generated with our technique.

I. Introduction

The achievement of low-loss high-bandwidth transmission has made the optical fiber waveguide the leading contender as the transmission medium for a variety of future systems ranging in length from meters to kilometers. This has strongly stimulated efforts in supporting technologies, such as fiber preparation and cabling, long life solid state sources, and high performance receivers. As a result of considerable progress in these areas, utilization of optical fibers in military systems and commercial systems appears imminent.

There are currently strong trends in system design toward microminiaturization digital processing and system level integration in order to achieve smaller size, weight, and power consumption, along with lower cost and improved reliability. These trends naturally point to data bus multiplexing, i.e, the interconnection of a number of spatially distributed terminals via fiber optic waveguide cables. The key component for any fiber data bus system is the fiber coupler. Since multimode single-strand fiber is used as a communication link in a data bus system, multimode fiber couplers must be designed and used.

To design any coupler properly for a multimode single-fiber data bus system, detailed analysis of waveguide propagation should be carried out. Existing techniques based on the finite elements method,[1] the coupled mode theory,[2] or the geometrical optics method[3] are often inadequate or inefficient. Although the finite elements method is a very powerful approach when dealing with single-mode fibers or couplers, it is very costly (in terms of computer time) to obtain any results for multimode structures. Similarly, coupled mode theory has been used quite successfully in predicting the coupling efficiencies of single-mode uniform-core fiber structures, but when used for multimode uniform-core structures, many gross assumptions must be made in order to obtain any results.[4] Although the geometrical optics method using the ray-tracing technique may yield zero-order results for multimode structures, it is too crude to preduce the wave behavior of light signals in couplers. Recently, Arnaud[4] developed a technique based on the Cook adiabatic coupler principle[5] to treat the multimode coupler problem. He claims that "it may be used to couple two optical fibers because the dimensions are not critical; only slowness is required." However, the validity of this approach has not been established. In this paper we shall present a technique (discussed in the next section) which will yield accurate results for this multimode inhomogeneous fiber coupler problem without making *a priori* assumptions on the relevance of the dimensions of the structure and the slowness of the coupling. It is noted that analysis based on the scalar wave approximation has been very successful in predicting the properties of many optical devices whose characteristic dimensions are on the order of many wavelengths.[5] The scalar wave approach is therefore preferred.

In the following we shall first present the formulation and the numerical approach in solving the multimode coupler problem. Then specific results on the two fiber coupler problems will be given.

C. Yeh is with University of California, Electrical Science & Engineering Department, Los Angeles, California 90024; the other authors are with Hughes Research Laboratories, Malibu, California 90265.
Received 28 April 1978.
0003-6935/79/040489-07$00.50/0.

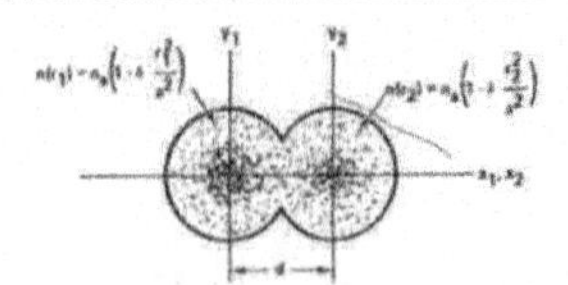

Fig. 1. The fiber coupler.

II. Formulation of the Problem

The geometry of a multimode inhomogeneous fiber coupler is shown in Fig. 1. Two graded-index fibers with index variation given by

$$n(r_{1,2}) = n_a\left(1 - \delta\frac{r_{1,2}^2}{a^2}\right),$$

where $r_{1,2}$ are the radial coordinates of 1 and 2 fibers, respectively, and n_a, δ, a are all known constants, are fused together as shown. The separation distance between the centers of the fibers is d. A Gaussian beam representable by

$$u(x,y) = u_o \exp\left\{\left[-\left(x + \frac{d}{2}\right)^2 - y^2\right]/w^2\right\},$$

where $u(x,y)$ is the scalar wave function of the beam, and u_o, w are given constants, is incident on one of the fibers. We wish to learn how this beam evolves as it propagates down the coupled structure. In other words, the coupling distance and the beam shape will be obtained.

It is well known that solving the exact electromagnetic equations for a spatially inhomogeneous medium is a formidable problem. Multimode couplers can be approximated by structures with a spatially nonuniform dielectric medium. Fortunately, there are some approximations that can be made to simplify this task.[6] First, the light wavelengths of interest are much shorter than the inhomogeneity scale length. This enables us to neglect polarization effects. Hence, the optical field can be derived from a scalar $u(x,z)$ that satisfied the reduced wave equation[7]

$$[\nabla^2 + k^2n^2(\mathbf{x},z)]u(\mathbf{x},z) = 0, \tag{1}$$

where k is the wavenumber $2\pi/\lambda$, λ is the laser wavelength, and $n(\mathbf{x},z)$ is the spatially inhomogeneous refractive index of the medium. Second, if we write u as the product of a factor $\exp(ikn_oz)$ that accounts for the rapid change in the phase of u along the direction of propagation and a complex amplitude $A(\mathbf{x},z)$, a further simplification of the calculational problem results:

$$\left\{i2kn_o\frac{\partial}{\partial z} + \nabla_T^2 + k^2\left[n^2(\mathbf{x},z) - n_0^2\right]\right\}A(\mathbf{x},z) = -\frac{\partial^2 A(\mathbf{x},z)}{\partial z^2}, \tag{2}$$

where ∇_T^2 is the transverse Laplacian $\partial^2/\partial x^2 + \partial^2/\partial y^2$, and n_o is a given constant which represents the refractive index of some uniform medium. At laser wavelengths, the complex amplitude $A(\mathbf{x})$ varies much more rapidly transverse to the direction of propagation than it does along the direction of propagation. This enables us to make the paraxial approximation and neglect the term on the right side of Eq. (2). (In the Russian literature this is called the parabolic approximation.) So, the complex amplitude now satisfies

$$\left\{i2kn_o\frac{\partial}{\partial z} + \frac{\partial^2}{\partial x^2} + \frac{\partial^2}{\partial y^2} + k^2\left[n^2(\mathbf{x},z) - n_0^2\right]\right\}A(\mathbf{x},z) = 0. \tag{3}$$

In addition to Eq. (3), the complex amplitude satisfies an initial condition on the fiber end; at $z = 0$,

$$A(\mathbf{x},0) = u(\mathbf{x},0), \tag{4}$$

and the boundary condition is

$$A(\pm\infty,z) = 0. \tag{5}$$

If a truncated Gaussian beam is focused on one end of the optical guide,

$$\begin{aligned} u(x,y,o) &= u_o \exp(-r^2/w^2) \quad &\text{for } 0 \le r \le b \\ &= 0 &\text{for } r > b, \end{aligned} \tag{6}$$

where $r^2 = x^2 + y^2$, w is the spot size of the beam, and b is the radius of the truncated beam at $z = 0$.

III. Numerical Approach

We shall solve Eq. (3) numerically in the following manner. Let us write $A(\mathbf{x},z)$ in the form

$$A(\mathbf{x},z) = \exp[\Gamma(\mathbf{x},z)]v(\mathbf{x},z), \tag{7}$$

where $\Gamma(\mathbf{x},z)$ is a phase function associated with the medium inhomogeneities

$$\Gamma(\mathbf{x},z) = \frac{ik}{2n_o}\int_{z_o}^{z}[n^2(x,y,z') - n_0^2]dz'. \tag{8}$$

The modified complex amplitude $v(\mathbf{x},z)$ then satisfies the equation

$$i2kn_o\frac{\partial}{\partial z}v(\mathbf{x},z) + \exp(-\Gamma)\nabla_T^2[\exp(\Gamma)v(\mathbf{x},z)] = 0. \tag{9}$$

Although Eq. (9) does not look any easier to solve than Eq. (3), it is easier to solve numerically because, for sufficiently small increments in the z direction and an appropriately chosen lower limit in the integral in Eq. (8), the value of $v(x,y,z + \Delta z)$ can be obtained to a good approximation by solving the simpler equation

$$\left(i2kn_o\frac{\partial}{\partial z} + \nabla_T^2\right)v(\mathbf{x},z) = 0 \tag{10}$$

with the initial condition

$$v(x,y,0) = u(x,y,0). \tag{11}$$

Physically, these equations approximate the propagation in the inhomogeneous medium by a two-step process at each z increment. First, we propagate the field $u(\mathbf{x},z)$ at z to $z + \Delta z$ assuming that the intervening space is homogeneous. The effect of the inhomogeneities between z and $z + \Delta z$ is then accounted for by multiplying this solution by the phase factor $\exp(\Gamma)$.

Equation (10) is solved by the fast Fourier transform technique. Replacing the Laplacian by its finite difference equivalent but still retaining the z derivative, the solution of Eq. (10) can be expressed in the form

$$v(m,n,z) = \sum_{m',n'=0}^{N-1}\sum V(m',n',z)\exp\left[\frac{i2\pi}{N}(mm'+nn')\right], \tag{12}$$

where $x = m\Delta x$, $y = n\Delta x$, and $V(m',n',z)$ satisfy

$$\left[i2kn_o\frac{\partial}{\partial z} + \frac{f(m',n')}{(\Delta x)^2}\right]V(m',n',z) = 0 \tag{13}$$

with the initial conditions

$$V(m',n',z_i) = \frac{1}{N^2}\sum_{m,n=0}^{N-1}\sum v(m,n,z_i)\exp[\Gamma(m,n,z_i)] \times \exp\left[-\frac{i2\pi}{N}(m'm+n'n)\right]. \tag{14}$$

The function $f(m',n')$ is determined by the difference approximation used to represent the Laplacian in Eq. (10). For example, if $\nabla_T^2 v$ is approximated by the simple central difference equation

$$\nabla_T^2 v = \frac{1}{(\Delta x)^2}[v(m+1,n,z) - 2v(m,n,z) + v(m-1,n,z) + v(m,n+1,z) - 2v(m,n,z) + v(m,n-1,z)], \tag{15}$$

$f(m',n')$ is

$$f(m',n') = -4\left[\sin^2\left(\frac{\pi m'}{N}\right) + \sin^2\left(\frac{\pi n'}{N}\right)\right]. \tag{16}$$

Note that the series in Eq. (12) is simply the discrete Fourier transform of the function $V(m',n',z)$ and thus can be evaluated numerically for a given $V(m',n',z)$ by a fast Fourier transform algorithm. Furthermore, the function $V(m',n',z)$ is readily determined from Eq. (13) as

$$V(m',n',z) = V(m',n',z_i)\exp\left[\frac{-if(m',n')}{2k(\Delta x)^2 n_o}(z-z_i)\right], \tag{17}$$

where $V(m',n',z_i)$ is given by the series in Eq. (14), which can also be evaluated by a fast Fourier transform algorithm. To summarize, we will step from z to $z + \Delta z$ as follows: (1) take the inverse discrete Fourier transform of $u(m,n,z) = [\exp\Gamma(m,n,z)]v(m,n,z)$ by means of an inverse fast Fourier transform algorithm; (2) multiply the result by $\exp[-if(m',n')\Delta z/2k(\Delta x)^2 n_o]$ and take the discrete Fourier transform with a fast Fourier transform algorithm; and (3) multiply the result by $\exp[\Gamma(m,n,z+\Delta z)]$ to yield $u(m,n,z+\Delta z)$. This process is repeated until we have reached the desired z plane.

IV. Results and Discussion

Before we proceed with the numerical computations, it should be recalled that we are dealing completely with total field quantities and not with the modes. In other words, we are interested in how the total field evolves as it propagates down the guiding structure; we are not interested in how each mode propagates. Nevertheless, it is recognized that the total field may be decomposed into a set of orthonormal guided modes. For example, an incident Gaussian beam with a given beamwidth w may excite many modes in a parabolic-index-profile fiber when $\alpha > 1$ or when $\alpha < 1$ with

$$\alpha = (2\lambda a)/[\pi n_a w^2(2\delta)^{1/2}]$$

or may excite only one mode when $\alpha = 1$. Only the $\alpha = 1$ single mode will propagate down the parabolic-index-profile guide without experiencing the focusing and defocusing effects. For $\alpha \neq 1$ cases, the input Gaussian beam will experience focusing and defocusing effects. Another way to interpret the above phenomenon is that multimodes with different propagation constants are excited by the input beam when $\alpha \neq 1$, while only single mode is excited when $\alpha = 1$. The stronger are the focusing–defocusing effects, the higher is the content of different modes. Using this reasoning, one may expect that if two identical parabolic-index fibers were placed side by side with each other and if the beamwidth were so chosen that $\alpha = 1$ is obtained, one would expect the guided power to interchange between the two fibers in a periodic manner with a distinct single coupling length. On the other hand, if α departs from unity, many modes are generated and many beat coupling lengths occur, so the guided power will no longer interchange among the guides in a simple manner. For the $\alpha \ll 1$ or $\alpha \gg 1$ case, even larger numbers of modes are excited. Depending upon how the incident energy is distributed among the excited modes the back and forth (periodic) power exchange phenomenon could still

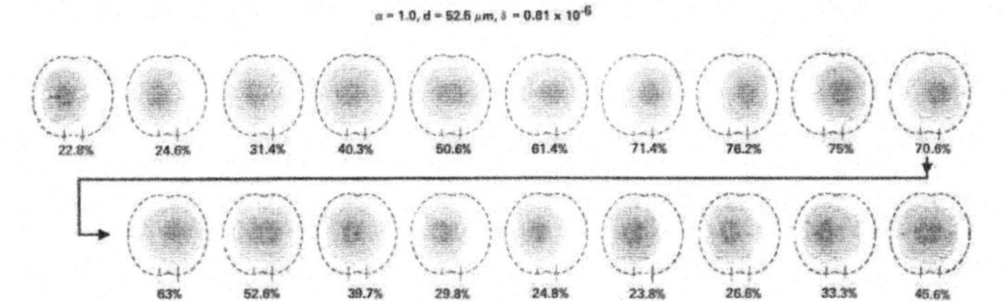

Fig. 2. A cross-sectional view of the power density distribution for the guided wave. Note the power exchange phenomenon as the wave propagates along the coupled structure. The percent value indicates the percent of total power contained in the right-hand half of the structure (i.e., fiber 2). The distance between each frame is 24,682 μm.

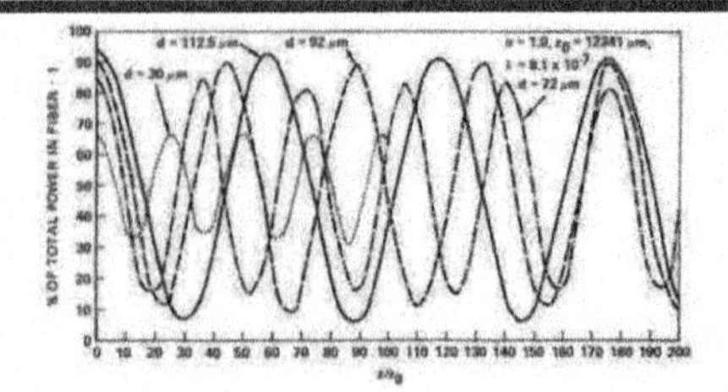

Fig. 3. Percent of total power in the left-hand half of the structure (i.e., fiber 1) as a function of the axial distance along the coupling structure. This is the single-mode coupling case.

prevail if the major amount of incident energy is distributed in several low-order modes. Only when the incident energy is distributed evenly among all modes and only in the limit of infinite number of modes will the input power be split in a 50–50 manner between the two fibers.[4]

Using the numerical technique described above, we can now study quantitatively the multimode fiber coupler problem instead of merely understanding this coupler problem intuitively in a heuristic manner.

A. $\alpha = 1$ Case

This is the single-mode case. Choosing the parameters

$$n_a = 2.0, \quad \lambda = 0.8\ \mu\text{m},$$
$$\delta = 0.81 \times 10^{-6}, \quad w = 100\ \mu\text{m},$$
$$a = 50\ \mu\text{m},$$

one has $\alpha = 1$. Shown in Fig. 2 is a typical picture of how a Gaussian beam evolves as it propagates down a coupling structure made with two graded-index fibers. One clearly sees the back-and-forth transfer of guided power from one guide to the other in a rather complicated manner. When all the power is transferred from one fiber to the other, the original Gaussian beam profile is recovered provided that the separation of the fibers is relatively large. Small separation of the fiber tends to distort the parabolic-index profile so that the input Gaussian beam can no longer be considered as the single orthonormal mode of the guide composed of half of the coupling structure. Even though periodic power exchange can still be observed, the beam pattern tends to be somewhat distorted.

By integrating over the cross section of one of the fibers, one may obtain the amount of guided power contained in that fiber. This is shown in Fig. 3 in which the total guided power in fiber 1 is plotted against the normalized longitudinal distance z/z_0 along the fiber for various separation distances d. z_0 is a normalizing constant equal to 12,341 μm for the specific example under consideration. It should be noted that the spot size of the incident Gaussian beam has been kept constant for the various fiber separations. For example, in the separation distance d = 112.5-μm case, about 7.14% of the incident beam power exists in the fiber 2 portion of the coupling structure, while 92.86% of the incident power is in fiber 1. When the separation distance d = 30 μm, about 33.3% of the incident beam power exists in the fiber 2 portion of the coupling structure, while 66.6% of the incident power is in fiber 1. For this near single-mode case, one may observe from Fig. 3 that the total guided power in fibers 1 and 2 exchanges in a periodic sinusoidal manner and that the coupling length which is defined as the distance for which maximum guided power is transferred from one guide to the other becomes shorter as the separation distance between the axes of the guides is shortened. At a separation distance of 112.5 μm, it takes about 37 cm for the guided power to exchange among the fibers, while at a separation distance of 30 μm it takes only 14.8 cm.

B. $\alpha = 0.4$ Case

Turning our attention to the slightly multimode case with parameters

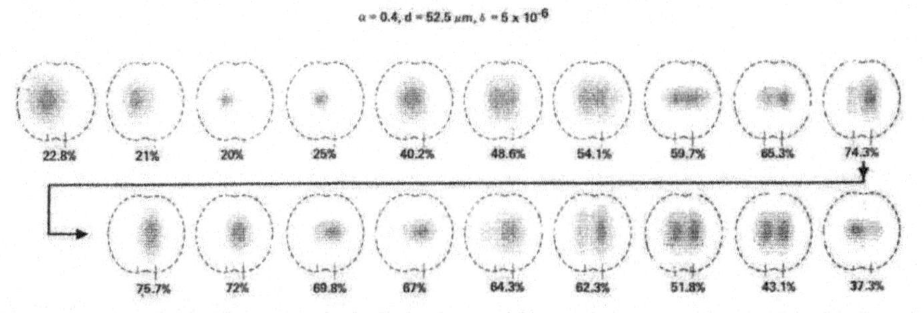

Fig. 4. A cross-sectional view of the power density distribution for the guided wave. Note the power exchange phenomenon as the wave propagates along the coupled structure. The percent value indicates the percent of total power contained in the right-hand half of the structure (i.e., fiber 2). The distance between each frame is 9934 μm.

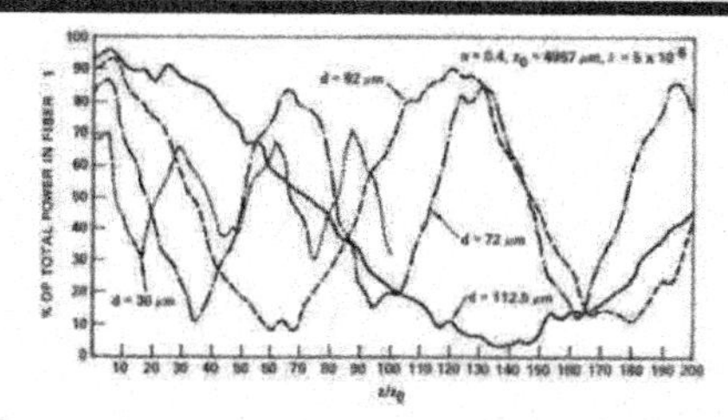

Fig. 5. Percent of total power in the left-hand half of the structure (i.e., fiber 1) as a function of the axial distance along the coupling structure. This is the multimode coupling case.

$$n_a = 2.0, \quad \lambda = 0.8\ \mu\text{m},$$
$$\delta = 5 \times 10^{-6}, \quad w = 100\ \mu\text{m},$$
$$a = 50\ \mu\text{m},$$

one obtains $\alpha = 0.4$. This means that even though the dominant power is carried by the dominant mode, other higher order modes were also excited. These higher order modes contribute to the focusing–defocusing character of the beam. It is anticipated that the coupling effects will no longer be uniform along the longitudinal distance and stronger coupling exists when the beam is defocused. Coupling length can no longer be obtained unambiguously due to the presence of higher order mode coupling and thus the presence of beat coupling lengths. As an illustration, the power exchange phenomenon for this slightly multimode case is shown in Fig. 4. Because of the complex wave interaction that is taking place in this coupling structure, the beam can no longer retain the Gaussian shape as it propagates down the structure. It is nevertheless of interest to note that when the maximum amount of power is transferred to the other fiber a Gaussian beam shape (although somewhat deformed) is recovered.

In Fig. 5, we have plotted the total amount of power carried in fiber 1 as a function of the normalized longitudinal distance z/z_0 with $z_0 = 4967\ \mu$m for various separation distances. It is seen that unlike the single-mode $\alpha = 1$ case, the power exchange among the fibers no longer varies sinusoidally as a function of the longitudinal distance due to the presence of multimode coupling. However, it appears that the periodic power exchange phenomenon is still retained. It is also noted that initial power transfer originates from fiber 2 to fiber 1 rather than the situation demonstrated for the $\alpha = 1$ case. It is still true that the coupling distance becomes shorter as the fiber separation becomes closer. One may also deduce from Figs. 4 and 5 that the distortion of the beam is much more pronounced when the fiber separation distance is small and when multimodes exist. From Fig. 5, it appears that the coupling length for the separation distance $d = 112.5$-μm case is about 69.5 cm, while it is 28.8 cm for the separation distance $d = 30$-μm case. The fact that the coupling length is longer for the multimode $\alpha = 0.4$ case than the comparable situation for the single-mode $\alpha = 1$ case is somewhat deceiving. This is because the index profiles are different for these two cases. If the same index profile were used, the coupling length for the multimode case would be shorter than that for the single-mode case.

C. $\alpha < 0.4$ Case

Using the parameters

$$n_a = 2.0, \quad \lambda = 0.8\ \mu\text{m},$$
$$\delta = 10^{-5}, \quad w = 100\ \mu\text{m},$$
$$a = 50\ \mu\text{m},$$

one has $\alpha = 0.285$. Based on the previous discussion one concludes that when an incident beam is very sharply focused within the fiber, many higher order modes have been generated. It is recognized that maximum power transfer (coupling) occurs when the beamwaist is the largest. Since the incident beam is so sharply focused, within a given longitudinal distance the

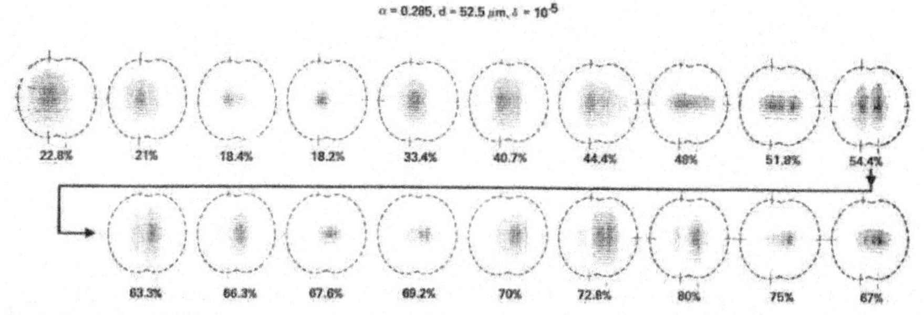

Fig. 6. A cross-sectional view of the power density distribution for the guided wave. Note the power exchange phenomenon as the wave propagates along the coupled structure. The percent value indicates the percent of total power contained in the right-hand half of the structure (i.e., fiber 2). The distance between each frame is 7024 μm.

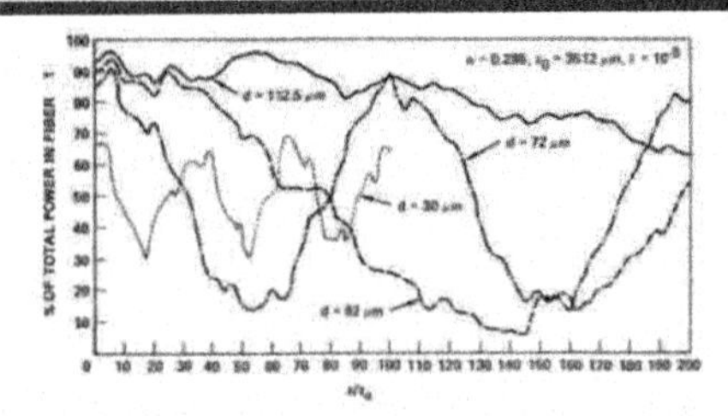

Fig. 7. Percent of total power in the left-hand half of the structure (i.e., fiber 1) as a function of the axial distance along the coupling structure. This is the multimode coupling case.

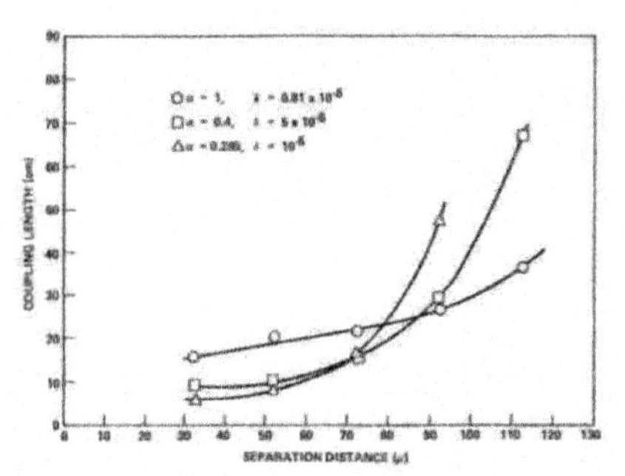

Fig. 8. Coupling length as a function of separation distance for various single-mode and multimode cases for the fiber coupler.

beam went through many cycles of focusing and defocusing, which implies that there exist many occasions when maximum power transfer takes place. Shown in Fig. 6 is a figure of several cycles of maximum power transfer. Synchronous focusing and defocusing of the beams in both fibers can be observed. In other words maximum power transfer tends to occur among similar beams in neighboring fibers even for the multimode case. For the chosen parameters, the incident Gaussian beam excited many modes in an unequal (amplitude-wise) fashion. It is therefore expected that the guided power would not be divided equally among the two fibers.[4] The back-and-forth power exchange phenomenon should still be present as one may see from Fig. 6. This type of wave coupling behavior highlights the importance of the beam characteristics of the incident beam. Not all modes are excited equally by an incident Gaussian beam. Hence, one cannot usually expect the split of the guided beam energy in a 50–50 manner even for highly multimoded fibers.[8] In fact, it seems most likely that the back-and-forth power exchange behavior will prevail even for highly multimoded fiber couplers. This fact adds to the complexity in the proper design of multimode fiber couplers. It also means that the simple intuitive design idea, which assumes that the incident guided beam energy in a multimode guide is split in a 50–50 manner among two coupled multimode fibers, is not easily justified.

In Fig. 7, we have plotted the total amount of power carried in fiber 1 as a function of the normalized longitudinal distance z/z_0 with $z_0 = 3512\ \mu$m for various separation distances. It is seen that the curves are even more rugged than those in Fig. 5 for the $\alpha = 0.4$ case. This behavior is again attributable to the presence of many higher order modes. The outstanding feature of these curves is that periodic power exchange still prevails between the fibers.

V. Conclusions

Based on the above calculations one may summarize the results on the coupling length as a function of the separation distance for various values of α in Fig. 8. It is seen that at a small separation distance for which the parabolic index profile of each fiber is substantially distorted, the coupling length for the highly multimoded guide is shorter than that for the single-mode guide. At a large separation distance for which the parabolic index profile of each fiber is preserved, the coupling length for the highly multimoded guide appears longer than that for the single-mode guide. However, this observation is somewhat deceiving because the index profile for the multimode case was not the same as that for the single-mode case. When the same index profiles are used, the beam for the single-mode case occupies only a very small region in the center of the guide, while the beam for the multimode case expands (defocuses) and contracts (focuses) periodically so that stronger coupling occurs when the beam expands. Hence, the coupling length for the multimode case is shorter. It is of interest to compare the multimode coupler results with the exact single-mode coupling results for two slab guides (Fig. 9). The dominant similarity is that for both cases the coupling length is a monotonically increasing function of the separation distance.

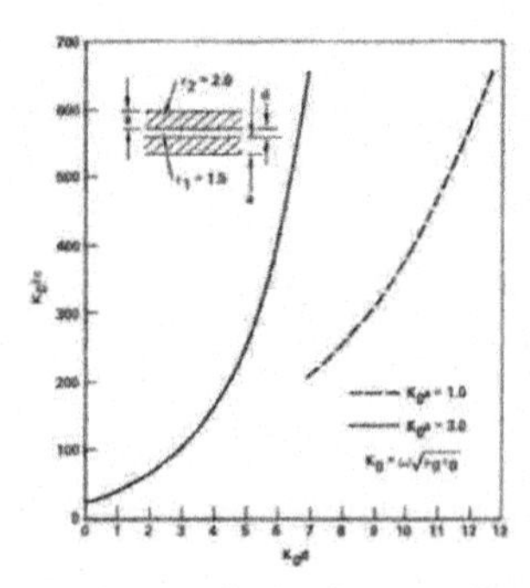

Fig. 9. Coupling length as a function of separation distance for the single-mode coupling case for two parallel slab waveguides.

Several general observations may be made on the multimode parabolic index fiber coupler that are consistent with our calculated results:

(a) For an on-axis Gaussian beam, periodic power exchange between two graded-index fibers always prevails.

(b) The coupling lengths are usually shorter when operating under multimode condition than single-mode condition.

(c) Coupling characteristics are critically dependent on the initial beam characteristics. Hence, the beam characteristics at the entrance of the fiber coupler are very important in order to predict accurately the coupling behavior of the coupler.

This research was partly supported by the Office of Naval Research and the National Science Foundation.

References

1. C. Yeh, S. B. Dong, and W. Oliver, J. Appl. Phys. **46**, 2125 (1975).
2. D. Marcuse, Bell Syst. Tech. J. **50**, 1791 (1971).
3. A. H. Cherin and E. J. Murphy, Bell Syst. Tech. J. **54**, 17 (1975).
4. J. A. Arnaud, Bell Syst. Tech. J. **54**, 1431 (1975).
5. J. S. Cook, Bell Syst. Tech. J. **34**, 807 (1955).
6. C. Yeh, L. Casperson, and B. Szejn, J. Opt. Soc. Am., to appear (1978).
7. J. W. Goodman, *Introduction to Fourier Optics* (McGraw-Hill, New York, 1968).
8. K. Ogawa, Bell Syst. Tech. J. *56*,729 (1977).

On multimode optical fiber couplers, tapers, and horns

C. YEH

EMtec Engineering, Inc.
Los Angeles, Calif. 90024

Using the scalar wave-FFT technique, it is demonstrated that the propagation characteristics of light beams in many practical multimode fiber structures such as fiber couplers, tapers, or horns can be obtained.

*This research was supported by the Electronic Systems Division of RADC under Contract No. F19628–78–C–0206.

Currently, there are strong trends in system design toward microminiturization digital processing and system level integration in order to achieve smaller size, weight, and power consumption, along with lower cost and improved reliability. These trends naturally point to data-bus multiplexing, i.e. the interconnection of a number of spatially distributed terminals via fiber-optic waveguide cables. The key components for any fiber data-bus system are the fiber couplers, branches, and tapers. Since multimode single-strand fiber is used as a communication link in a data-bus system, multimode fiber components must be designed and used.

To design these components properly for a multimode single-fiber data-bus system, quantitative understanding of the wave propagation phenomenon in these structures must first be achieved. Existing analytic techniques based on the finite-elements method,[1] the coupled-mode theory,[2] or the geometrical-optics method[3] are usually inadequate. Although the finite-elements method is a very powerful approach when dealing with single-mode fibers or couplers, it is very inefficient and costly (in terms of computer time) to obtain any results for multimode structures. Similarly, coupled-mode theory has been used quite successfully in predicting the coupling efficiencies of single-mode structures but cannot be used for multimode structures. Although the geometrical-optics method using the ray-tracing technique may yield zero-order results for multimode structures, it is too crude to predict the wave behavior of light signals in couplers, tapers, and branches. It has been shown that, if certain limiting conditions (e.g. small index differences between core and cladding regions and gentle index profile variations) are satisfied, the scalar wave approximation will yield accurate results for multimode fiber structures.[4] These limiting conditions are usually met by most practical multimode fiber structures.

The mathematical approach

It is well known that solving the exact electromagnetic equations for a spatially inhomogeneous medium is a formidable problem. Multimode couplers and mode converters can be approximated by structures with a spatially nonuniform dielectric medium. Fortunately, there are some approximations that can be made to simplify this task. First, the light wavelengths of interest must be shorter than the inhomogeneity scale length. This enables us to neglect polarization effects. Hence, the optical field can be derived from a scalar $u(x,z)$ that satisfies the reduced wave equation[5,6]

$$[\nabla^2+k^2n^2(x,z)]u(x,z)=0 \tag{1}$$

where k is the wave number $2\pi/\lambda$, λ is the laser wavelength, and $n(x,z)$ is the spatially inhomogeneous refractive index of the medium. Second, if we write u as the product of a factor e^{ikn_0z} that accounts for the rapid change in the phase of u along the direction of propagation and complex amplitude $A(x,z)$, a further simplification of the calculation problem results:

$$\left[i2kn_0\frac{\partial}{\partial z}+\nabla_T^2+k^2\left(n^2(\underline{x},z)-n_0^2\right)\right] A(\underline{x},z)=-\frac{\partial^2 A(x,z)}{\partial z^2} \tag{2}$$

where ∇_T^2 is the transverse Laplacian $\partial^2/\partial x^2+\partial^2/\partial y^2$ and n_0 is a given constant which represents the refractive index of some uniform medium. At laser wavelengths, the complex amplitude $A(\underline{x})$ varies much more rapidly transverse to the direction of propagation than it does along the direction of propagation. This enables us to make the paraxial approximation and neglect the term on the right of Eq. (2) (in the Russian literature this is called the parabolic approximation). So, the complex amplitude now satisfies

$$\left[i2kn_0\frac{\partial}{\partial z}+\frac{\partial^2}{\partial x^2}+\frac{\partial^2}{\partial y^2}+k^2\left(n^2(\underline{x},z)-n_0^2\right)\right] A(\underline{x},z)=0 \tag{3}$$

In addition to Eq. (3), the complex amplitude satisfies an initial condition on the fiber end; at $z=0$,

$$A(\underline{x},0)=u(\underline{x},0) \tag{4}$$

and the boundary condition is

$$A(\pm\infty,z)=0 \tag{5}$$

If a truncated Gaussian beam is focused on one end of the optical guide, then

$$u(x,y,0)=u_0\exp(-r^2/w^2) \quad \text{for } 0\le r\le b$$
$$=0 \qquad \text{for } r>b \tag{6}$$

where $r^2=x^2+y^2$, w is the spot size of the beam, and b is the radius of the truncated beam at $z=0$.

We have solved Eq. (3) numerically in the following manner. Let us write $A(\underline{x},z)$ in the form

$$A(\underline{x},z)=\exp[\Gamma(\underline{x},z)]v(x,z) \tag{7}$$

where $\Gamma(\underline{x},z)$ is a phase function associated with the medium inhomogeneities

$$\Gamma(x,z)=\frac{ik}{2n_0}\int_{z_0}^{z}\left[n^2(\underline{x},y,z')-n_0^2\right]dz' \tag{8}$$

The modified complex amplitude $v(x,z)$ then satisfies the equation

$$i2kn_0\frac{\partial}{\partial z}v(\underline{x},z)+e^{-\Gamma}\nabla_T^2[e^{\Gamma}v(\underline{x},z)]=0 \tag{9}$$

Although Eq. (9) does not look any easier to solve than Eq. (3), it is easier to solve numerically because, for sufficiently small increments in the z direction and an appropriately chosen lower limit in the integral in Eq. (8), the value of $v(x,y,z+\Delta z)$ can be obtained to a good approximation by solving the simpler equation

$$\left[i2kn_0 \frac{\partial}{\partial z} + \nabla_T^2 \right] v(\underline{x},z)=0 \tag{10}$$

with the initial condition

$$v(x,y,0)=u(x,y,0) \tag{11}$$

Physically, these equations approximate the propagation in the inhomogeneous medium by a two-step process at each z increment. First, we propagate the field $u(x,z)$ at z to $z+\Delta z$ assuming that the intervening space is homogeneous. The effect of the inhomogeneities between z and $z+\Delta z$ is then accounted for by multiplying this solution by the phase factor $\exp\Gamma$.

In this research we have solved Eq. (10) by the fast Fourier transform technique. Replacing the Laplacian by its finite difference equivalent but still retaining the z derivative, the solution of Eq. (10) can be expressed in the form

$$v(m,n,z)= \sum_{m',n'=0}^{N-1}\sum V(m',n',z) \exp\left[\frac{i2\pi}{N} (mm'+nn') \right] \tag{12}$$

where $x=m\Delta x$, $y=n\Delta x$, and $V(m',n',z)$ satisfy

$$\left[i2kn_0 \frac{\partial}{\partial z} + \frac{f(m',n')}{(\Delta x)^2} \right] V(m',n',z)=0 \tag{13}$$

with the initial conditions

$$V(m',n',z_i)= \frac{1}{N^2} \sum_{m,n=0}^{N-1}\sum V(m,n,z_i) \exp[\Gamma(m,n,z_i)]$$

$$\exp\left[-\frac{i2\pi}{N}(m'm+n'n) \right] \tag{14}$$

The function $f(m',n')$ is determined by the difference approximation used to represent the Laplacian in Eq. (10). For example, if $\nabla_T^2 v$ is approximated by the simple central difference expression

$$\begin{aligned}\nabla_T^2 v = \frac{1}{(\Delta x)^2}[&v(m+1,n,z)-2v(m,n,z)\\ &+v(m-1,n,z)+v(m,n+1,z)\\ &-2v(m,n,z)+v(m,n-1,z)]\end{aligned} \tag{15}$$

then $f(m',n')$ is

$$f(m',n')=-4\left[\sin^2\left(\frac{\pi m'}{N}\right) + \sin^2\left(\frac{\pi n'}{N}\right) \right] \tag{16}$$

Note that the series in Eq. (12) is simply the discrete Fourier transform of the function $V(m',n',z)$, and thus can be evaluated numerically for a given $V(m',n',z)$ by a fast Fourier transform algorithm. Furthermore, the function $V(m',n',z)$ is readily determined from Eq. (13) as

$$V(m',n',z)=V(m',n'z_i)\exp\left[\frac{-if(m',n')}{2k(\Delta x)^2n_0}(z-z_i)\right] \tag{17}$$

where $V(m',n'z_1)$ is given by the series in Eq. (14), which can also be evaluated by a fast Fourier transform algorithm. To summarize, we will step from z to $z+\Delta z$ as follows: (1) take the inverse discrete Fourier transform of $u(m,n,z)=[\exp\Gamma(m,n,z)]\cdot v(m,n,z)$ by means of an inverse fast Fourier transform algorithm; (2) multiply the result by $\exp[-if(m',n')\Delta z/2k(\Delta x)^2n_0]$ and take the discrete Fourier transform with a fast Fourier transform algorithm; and (3) multiply the result by $\exp[\Gamma(m,n,z+\Delta z)]$ to yield $u(m,n,z+z)$. This process is repeated until we have reached the desired z plane.

Using this algorithm we have been successful in obtaining many meaningful results for various multimode structures. These results are summarized in the following section.

Multimode fiber components

Before we proceed with the presentation of the numerical results, it should be recalled that we are dealing completely with total field quantities and not with the modes. In other words, we are interested in how the total field evolves as it propagates down the guiding structure; we are not interested in how each mode propagates. Nevertheless, it is recognized that the total field may be decomposed into a set of orthonormal guided modes. For example, an incident Gaussian beam with a given beamwidth w may excite many modes in a parabolic-index-profile fiber when $\alpha>1$ or when $\alpha<1$ with

$$\alpha=(2\lambda a)/[\pi n_a w^2(2\delta)^{1/2}]$$

or may excite only one mode when $\alpha=1$. Only the $\alpha=1$ single mode will propagate down the parabolic-index-profile guide without experiencing the focusing and defocusing effects. For $\alpha\neq1$ cases, the input Gaussian beam will experience focusing and defocusing effects. Another way to interpret the above phenomenon is that multimodes with different propagation constants are excited by the input beam when $\alpha\neq1$, while only single mode is excited when $\alpha=1$. The stronger the focusing-defocusing effects, the higher the content of different modes.

Dual fiber couplers

The geometry of a multimode inhomogeneous fiber coupler is shown in Fig. 1. Two graded-index fibers with index variation given by

$$n(r_{1,2})=n_a\left(1-\delta\frac{r^2_{1,2}}{a^2_{1,2}}\right)$$

where $r_{1,2}$ are the radial coordinates of 1 and 2 fibers, respectively, and n_a, δ, a are all known constants, fused together as shown. The separation distance between the centers of the fibers is d. A Gaussian beam representable by

$$u(x,y)=u_0\exp\left\{\left[-\left(x+\frac{d}{2}\right)^2-y^2\right]/w^2\right\}$$

where $u(x,y)$ is the scalar wave function of the beam and u_0, w are given constants, is incident on one of the fibers. We wish to learn how this beam evolves as it propagates down the coupled structure. In other words, the coupling distance and the beam shape will be obtained.

Results of our investigation on the multimode dual-fiber couplers are shown in Figs. 2 to 4. It is seen that if two identical parabolic-index fibers were placed side by side and if the beamwidth were so chosen that $\alpha=1$ is obtained, one would expect the guided power to interchange between the two fibers in a periodic manner with a distinct single coupling length. On the other hand, if α departs from unity, many modes are generated and many beat coupling lengths occur, so the guided power will no longer interchange among the guides in a simple manner. For the $\alpha<<1$ or $\alpha>>1$ case, even larger numbers of modes are excited. Depending on how the incident energy is distributed among the excited modes, the back and forth (periodic) power exchange phenomenon could still prevail if the major amount of incident energy is distributed in several low-order modes. Only when the incident energy is distributed evenly among all modes and only in the limit of infinite number of modes will the input power be split in a 50-50 manner between the two fibers.

An example of the evolution of field intensity distribution in the two-parallel-fiber coupler is given in Fig. 5.

Fiber tapers

Fiber tapers have been commonly used to interconnect two optical waveguides of different cross-sectional sizes or shapes. The effects of the transition on the guided fields are shown in Figs. 6 to 8. It can be seen that there are very profound effects on the guided fields. Extensive mode conversion may be present even for gently varying tapers.

Fiber horns

Numerical computations to obtain the field behavior in a multimode fiber-horn structure have also been carried out using the technique outlined earlier. The cross-sectional index profiles are specified for each step in the direction of wave propagation. Shown in Figs. 9 to 11 are plots of beam waist as a function of a longitudinal distance for various incident beam sizes.

Conclusions

We have successfully developed a computational technique, based on the scalar wave-FFT method, to treat problems dealing with various multimode fiber components, such as fiber couplers, fiber tapers or horns, fiber branches, or mode converters. The only significant restriction to keep in mind is that index variations of the structure under consideration must be gentle.[8] This consideration is normally satisfied in most practical situations. By simply specifying the index profile at each longitudinal z step and knowing the initial beam shape, one may generate the propagation characteristics of the beam as it propagates down the structure.

Using this technique, we were able to trace the evolution of a given beam as it propagates down a given multimode graded-index fiber structure. If the multimode structure were a multimode coupler made with two or more paralleled graded-index fibers in close proximity, we could predict the coupling distances as well as the power distribution of the beam in such coupler; if the structure were a tapered or a flared multimode fiber, we could learn whether a given tapered structure or a given horn structure could still confine or guide a given beam.

With the help of this newly developed technique, one may obtain the propagation characteristics of beams in realistic multimode fiber structures, such as data-bus-type multichannel couplers, tapered or flared transition joints, and multimode fibers with nonideal parabolic-index profiles.

Acknowledgment

The author thanks L. Eyges and A. Yang, Rome Air Development Center, Hanscom AFB, Mass., for continuing interest and support.

References

[1]C. Yeh, S. B. Dong, and W. Oliver, *J. Appl. Phys.*, **46**, 2125 (1975).
[2]D. Marcues, *BSTJ*, **50**, 1791 (1971).
[3]A. H. Cherin and E. J. Murphy, *BSTJ*, **54**, 17 (1975).
[4]C. Yeh, "Optical Waveguide Theory," *IEEE Trans. Circuits Syst.*, CAS–26, 1011 (1979).
[5]J. W. Goodman, Introduction to Fourier Optics. McGraw-Hill, New York (1968).
[6]C. Yeh, "Scalar-Wave Approach for Inhomogeneous Fiber Problems," to be published in *Applied Physics Letters*. Solution of the single-mode homogeneous dielectric waveguide problem based on a scalar-wave equation has been obtained by L. Eyges, P. Gianino, and P. Wintersteiner ("Modes of Dielectric Waveguides on Arbitrary Cross-Sectional Shape," to be published in *Applied Optics*).

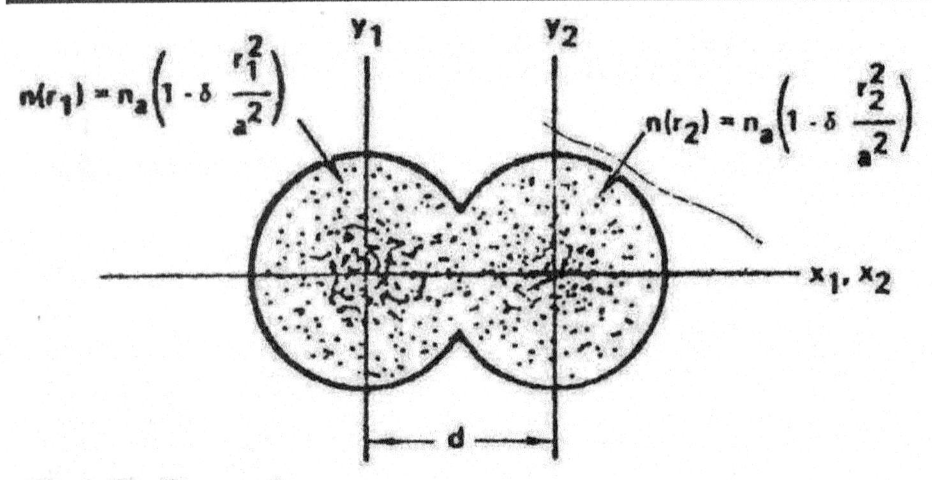

Fig. 1. The fiber coupler.

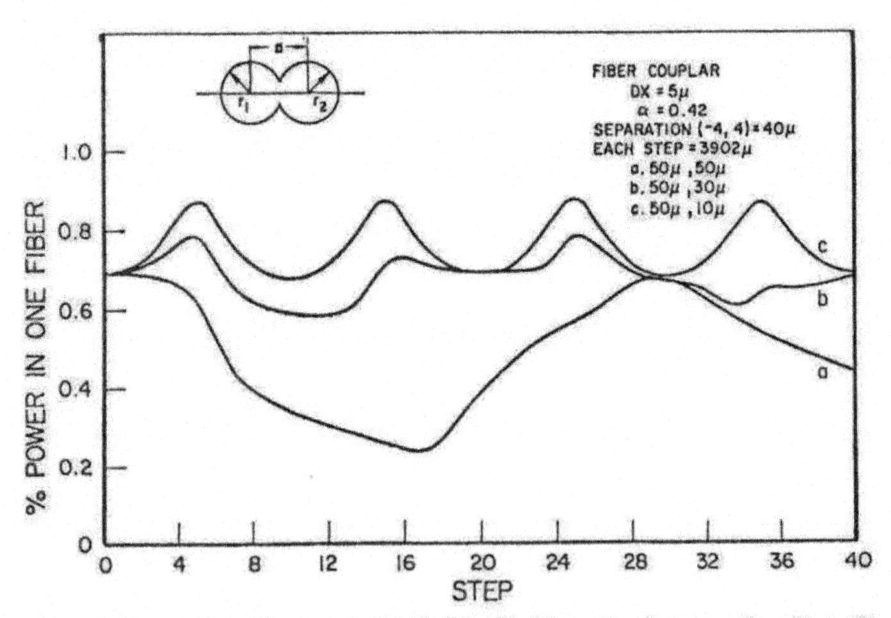

Fig. 2. Percent of total power in the left half of the structure as a function of the axial distance along the coupling structure. The separation is 40 μm.

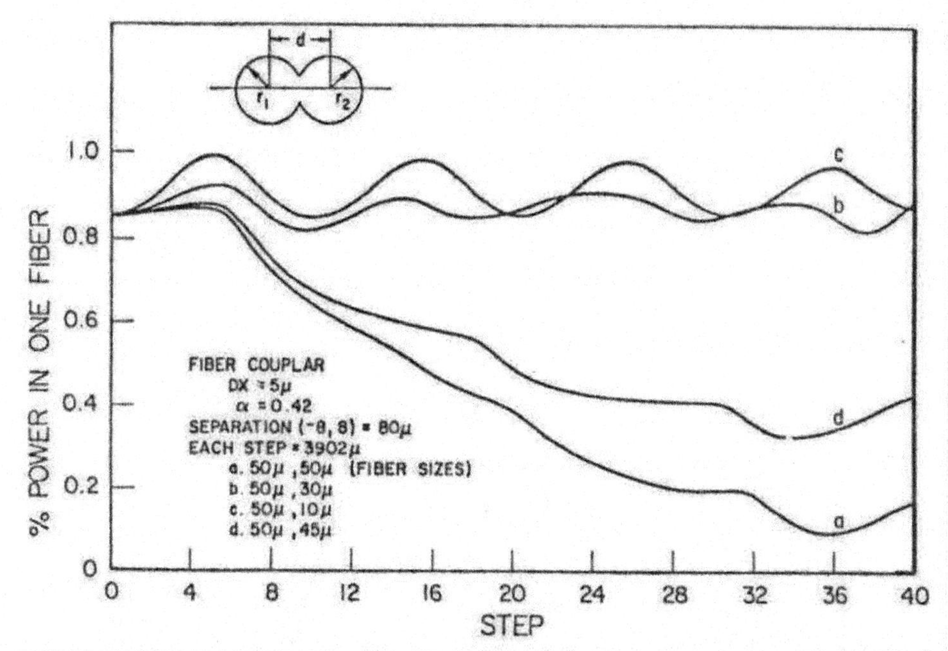

Fig. 3. Percent of total power in the left half of the structure as a function of the axial distance along the coupling structure. The separation is 80 μm.

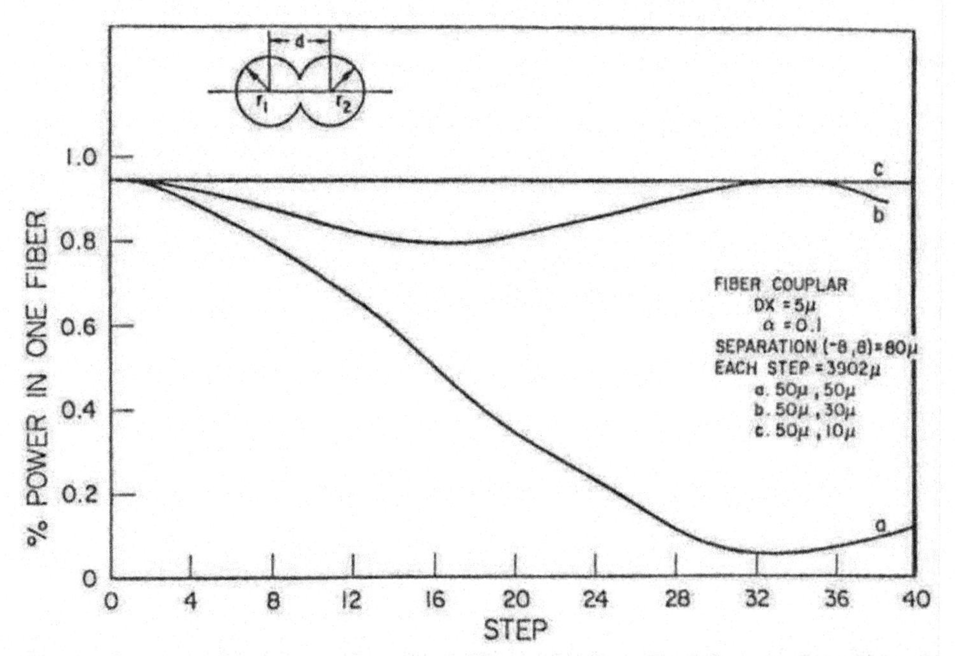

Fig. 4. Percent of total power in the left half of the structure as a function of the axial distance along the coupling structure. The separation is 80 μm.

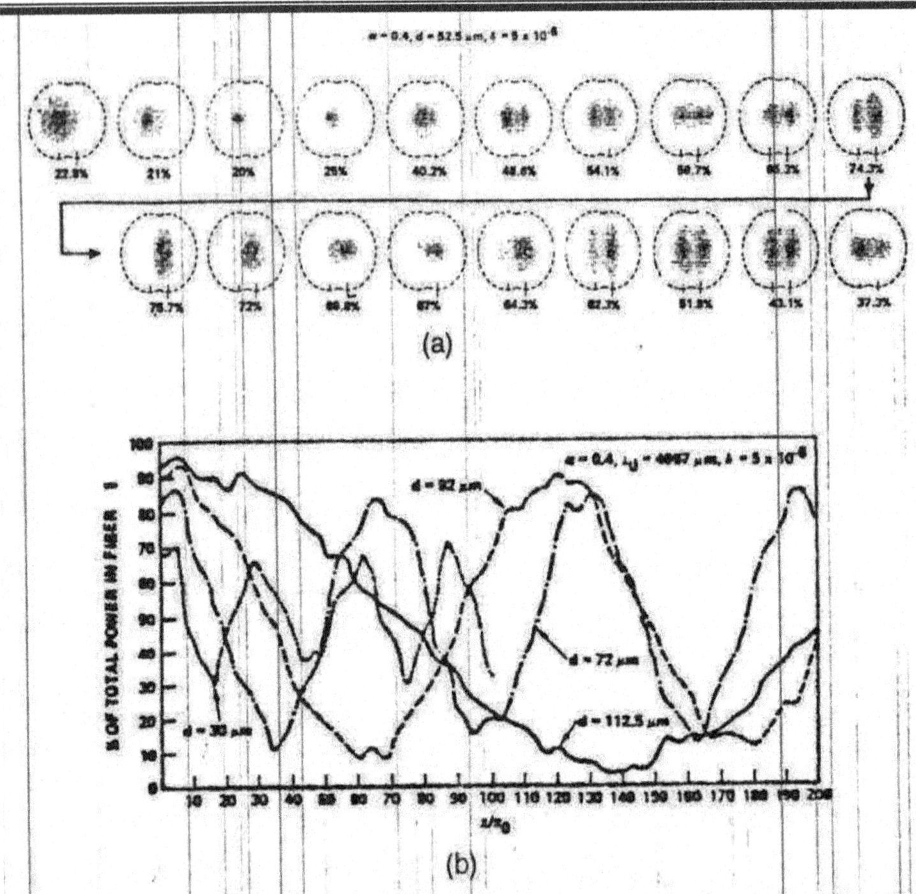

Fig. 5. (*a*) A cross-sectional view of the power intensity distribution for the guided wave. Note the power exchange phenomenon as the wave propagates along the coupled structure. The percent value indicates the percent of the total power contained in the right half of the structure (i.e. fiber 2). The distance between each frame is 9934 μm. (*b*) Percent of total power in the left half of the structure (i.e. fiber 1) as a function of the axial distance along the coupling structure. This is the multimode coupling case.

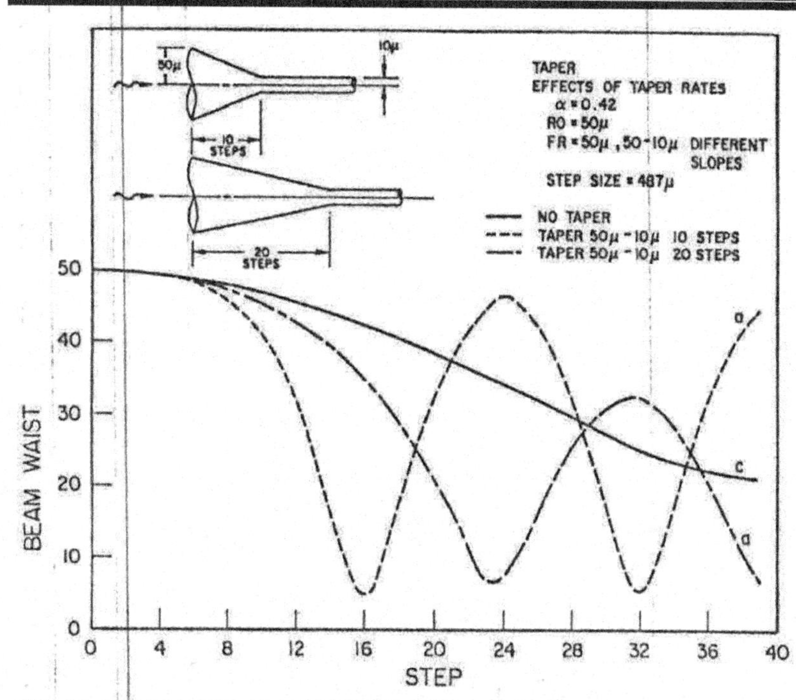

Fig. 6. Beam waist as a function of the axial distance.

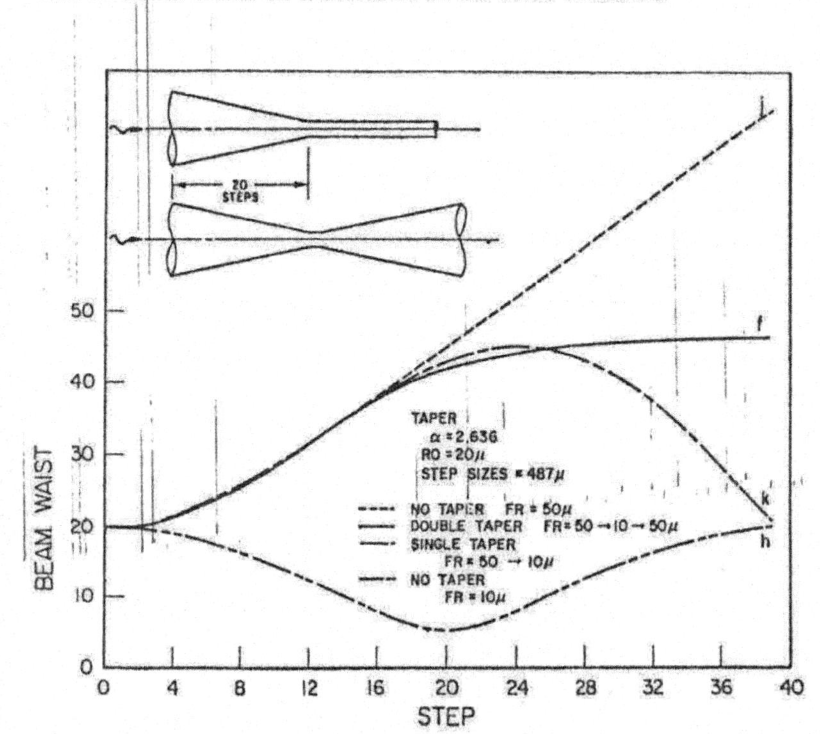

Fig. 7. Beam waist as a function of the axial distance.

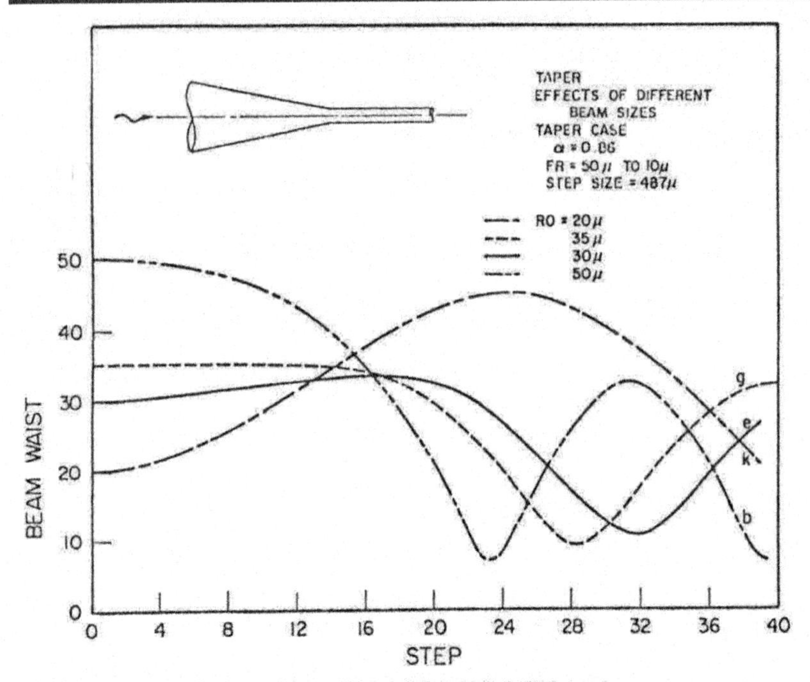

Fig. 8. Beam waist as a function of the axial distance.

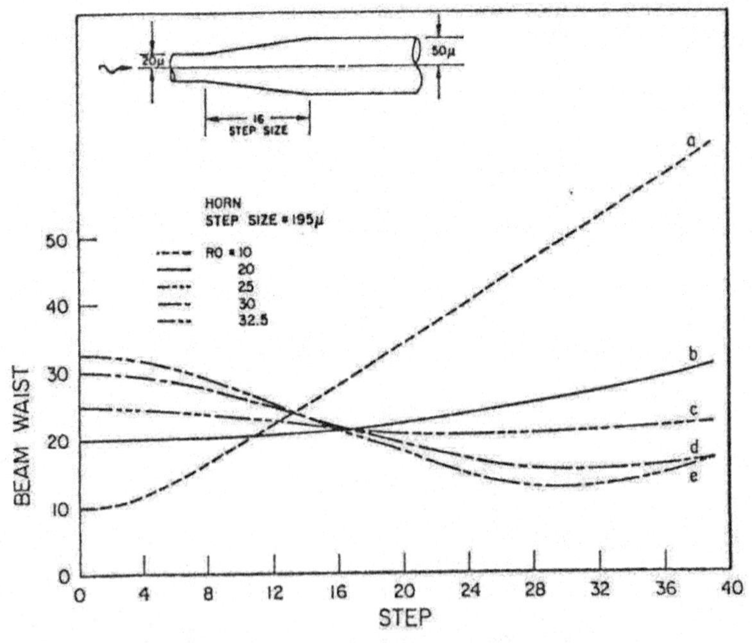

Fig. 9. Beam waist as a function of the axial distance.

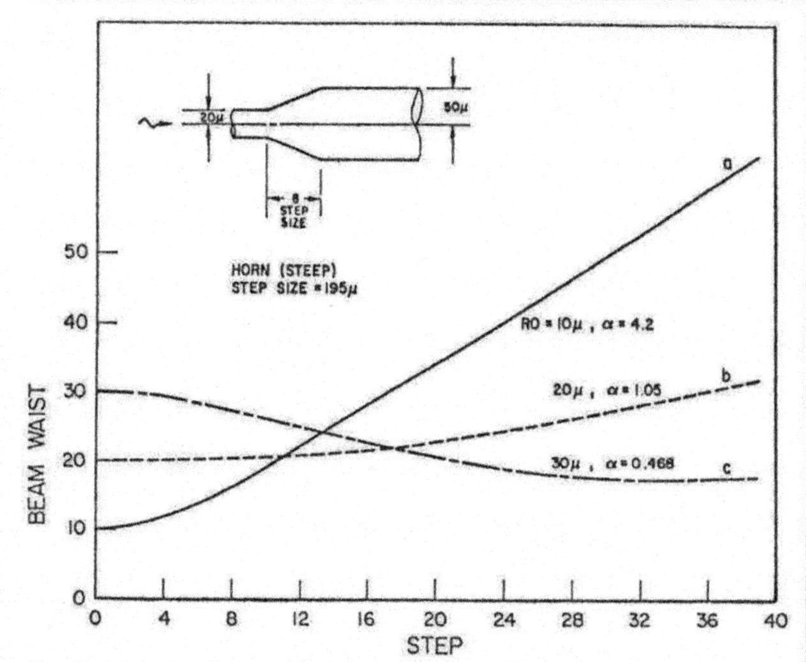

Fig. 10. Beam waist as a function of the axial distance.

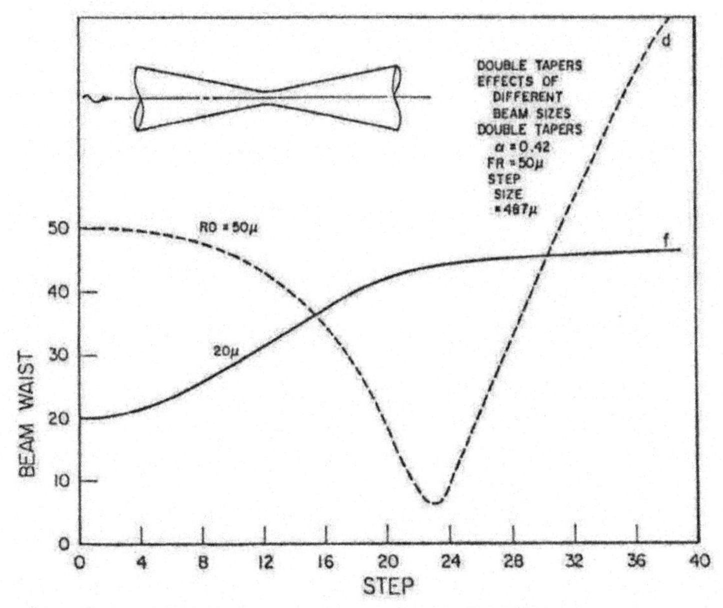

Fig. 11. Beam waist as a function of the axial distance.

JOURNAL OF LIGHTWAVE TECHNOLOGY, VOL. LT-3, NO. 1, FEBRUARY 1985 199

On Weakly Guiding Single-Mode Optical Waveguides

CAVOUR YEH, SENIOR MEMBER, IEEE, AND FARZIN MANSHADI

Abstract—This paper presents a powerful technique to treat the problem of wave propagation along weakly guiding single-mode optical waveguides. The shape as well as the 3-D index variation of the guide may be quite arbitrary as along as the weakly guiding character is preserved. The technique is based on the solution of a scalar-wave equation by the forward-marching fast Fourier transform (FFT) method. Excellent agreements were obtained, with known solutions, for various optical guiding structures. We have also demonstrated the capability of this technique to treat other exotic single-mode structures.

I. Introduction

Although several techniques have been successfully applied to the analysis of homogeneous optical waveguide structures [1], until relatively recently a simple, straightforward, and versatile technique for the treatment of arbitrarily shaped, inhomogeneous optical waveguides was apparently lacking. This paper will address this problem. We will show that the scalar-wave fast-Fourier-transform marching technique (SW-FFT) may be readily adapted to obtain the needed characteristics of most practical (i.e., weakly guiding) arbitrarily shaped inhomogeneous optical waveguides [2], [3].

A few words about the development of SW-FFT technique is in order. This technique was first used by Yeh *et al.* [2] in 1977 to solve the problem of the propagation of truncated Gaussian beams in a fiber guide. Feit and Fleck [6] subsequently used a similar technique to treat the problem of circular fiber waveguides. Yeh and his colleagues published a series of papers [3]-[5] describing the application of this technique to various multimode and single-mode fiber structures such as fiber couplers, fiber horns, and branches.

Starting with a brief summary of the SW-FFT technique in conjunction with a spectral analysis of the spatial configuration of the field, we shall first compare our results for several simple fiber structures such as, step-index circular fiber, graded-index circular fiber, elliptical fiber and triangular fiber, with those obtained using other known techniques [7]-[9]. Then, results will be presented for fiber structures such as circular fiber with arbitrary index variation (other than radial) and fiber with longitudinal index variation. The principal purpose of this paper is to show that our SW-FFT technique can be used to solve problems which cannot be handled easily by other approximate or numerical techniques and that the field solution of a scalar-wave equation for an arbitrarily shaped inhomogeneous fiber is easily obtainable.

Manuscript received March 20, 1984; revised August 20, 1984.
The authors are with the Electrical Engineering Department, University of California, Los Angeles, CA 90024.

II. The SW-FFT Approach

It has been universally accepted that the full set of Maxwell equations resulting in the vector-wave equation must be used to treat waveguides supporting single mode. This requirement confines the analytical treatment to only a few simple structures [1]. Fortunately, most practical low-loss optical waveguides fall into the category of the weakly guiding structures for which the variations in refractive-index of the guiding region are very small [10]. Accounts of weakly guding waveguides has been given by Snyder and Love [10], by Gloge [17], and by Marcuse [18]. For the square of index profile given by the following expression:

$$n^2(\boldsymbol{x}) = n_0^2 [1 - 2\Delta f(\boldsymbol{x})] \tag{1}$$

where n_0 is a given constant, Δ is a profile height parameter and $f(x)$ is the functional variation of the profile, the weakly guiding approximation assumes that $\Delta \ll 1$. Under this approximation wave solutions for these weakly guiding structures may be found from the solutions of the scalar-wave equation [3], [10]. Physically it means that:

modes are nearly TEM waves;

spatial dependence of transverse field is governed by the scalar-wave equation; and

polarization properties of the guiding structure, as manifested by the $\nabla_t \ln n^2$ in the vector-wave equation where ∇_t is the transverse del operator, are ignored.

Snyder has derived polarization corrections to the scalar propagation constant using the perturbation methd [10]

$$\delta\beta_m \simeq \frac{a(2\Delta)^{3/2}}{2V} \frac{\int_A (\nabla_t \cdot \boldsymbol{E}_{t_m}) \boldsymbol{E}_{t_m} \cdot \nabla_t f(\boldsymbol{x})\, dA}{\int_A |\boldsymbol{E}_{t_m}|^2\, dA} \tag{2}$$

with

$$\boldsymbol{E}_{t_m} = E_{x_m} \boldsymbol{e}_x + E_{y_m} \boldsymbol{e}_y \tag{3}$$

$$E_{x_m} \sim \phi_m(\boldsymbol{x}) \exp [j(\beta_m + \delta\beta_{x_m}) z] \tag{4}$$

$$E_{y_m} \sim \phi_m(\boldsymbol{x}) \exp [j(\beta_m + \delta\beta_{y_m}) z] \tag{5}$$

$$[\nabla_t^2 + (k^2 n^2(\boldsymbol{x}) - \beta_m^2)]\, \phi_m(\boldsymbol{x}) = 0 \tag{6}$$

and

$k \quad \frac{2\pi}{\lambda}$, the free-space wavelength is λ,

β_m scalar-wave propagation constant for mth mode,
$\boldsymbol{x}$ transverse position vector,
$\boldsymbol{e}_x, \boldsymbol{e}_y$ unit vectors parallel to x, y Cartesian axis, (7)
V $= k\, a\, n_0 (2\Delta)^{1/2}$,
ϕ_m transverse scalar-wave function for mth mode,
a core radius or half-width,
m mth mode, and
A infinite cross section.

The SW-FFT approach starts with the 3-D scalar-wave equation [2]

$$[\nabla^2 + k^2 n^2(\boldsymbol{x}, z)]\, u(\boldsymbol{x}, z) = 0 \tag{8}$$

where ∇ is the 3-D del operator and u is the total scalar field which is related to the vector electric field $\boldsymbol{E}(\boldsymbol{x}, z)$ as follows

$$\boldsymbol{E}(\boldsymbol{x}, z) = \boldsymbol{e}_p u(\boldsymbol{x}, z). \tag{9}$$

Here, $\boldsymbol{e}_p$ is a unit vector in the direction of the initial polarization of the wave. The scalar-wave equation, (8), with the boundary condition on the initial surface, and the radiation condition at infinity completely specifies $u(\boldsymbol{x}, z)$, from which one may obtain the electromagnetic field vectors $\boldsymbol{E}$ and $\boldsymbol{H}$ [2]. The evolution of the scalar-wave function $u(\boldsymbol{x}, z)$ as the wave propagates along z, the direction of propagation, can be found according to the forward-marching fast Fourier transform technique (the SW-FFT approach) which has been described in detail in our 1978 paper [2]. Several important characteristics of this technique are summarized in the following.

To obtain the scalar-wave equation (see (8)), the term $\nabla(\epsilon^{-1}\nabla\epsilon \cdot \boldsymbol{E})$ where $\epsilon = n^2(\boldsymbol{x}, z)$ in the vector-wave equation is discarded. This means that the polarization properties of the guiding structure are ignored. Hence, the waveguide is properly defined as the weakly guiding structure.

Paraxial approximation (or the parabolic approximation) was used.

In contrast with the usual modal approach which deals with one mode at a time, this technique provides the evolution of the total field as it propagates down the guiding structure. It is recognized, however, that the total field may be decomposed into a set of orthonormal guided modes.

The forward-marching characteristic of this technique implies that backward propagating waves are not allowed to exist (or are ignored); i.e., no reflection along the direction of propagation is taken into consideration.

This SW-FFT technique will now be applied to the problem of wave propagation along a number of circular or noncircular inhomogeneous and weakly guiding dielectric structures to yield information on the total field. Knowing the complete description of the total field one may derive the propagation characteristics for each normal mode on this given structure [11].

III. Modal Characteristics of the Field

All of the information necessary for a complete description of the field in modes can be extracted from the total scalar-wave function obtained according to the technique described in the previous section. Take the fact that the complex field amplitude $u(x, y, z)$ can be expressed as a superposition of orthonormal mode eigenfunctions as follows:

$$u(x, y, z) = \sum_m A_m \phi_m(x, y) \exp(-i\beta_m z) \tag{10}$$

where m is the mode number, β_m are the mode eigenvalues (or propagation constants), and A_m are the mode amplitudes which are determined by a given incident field $u_0(x, y, z)$. The mode propagation constants β_m can be determined from a computation of the correlation function [11]

$$P(z) = \iint_A u^*(x, y, 0)\, u(x, y, z)\, dx\, dy,$$

$$A = \text{cross-sectional area}. \tag{11}$$

If use is made of (10) and the orthogonality of the mode eigenfunctions $u_m(x, y)$, (11) reduces to

$$P(z) = \sum_m |A_m|^2 \exp(-i\beta_m z). \tag{12}$$

The Fourier transform of (12) with respect to z is

$$P(\beta) = \sum_m |A_m|^2\, \delta(\beta - \beta_m) \tag{13}$$

which suggests that the calculated spectrum of $P(z)$ will display a series of resonances with maxima at $\beta = \beta_m$ and peak values proportional to the mode weight coefficients

$$w_m = |A_m|^2. \tag{14}$$

In practice only a finite record of $P(z)$ is available and that record must be enhanced with the multiplication by a window function $w(z)$ before Fourier Transform is computed [12]. The resulting resonances in the spectrum $P(\beta)$, after multiplication by a window function, will thus exhibit a finite width and shape that is characteristic of the record length Z and the window function $w(z)$. A sample curve of $P(\beta)$ for a waveguide with a ten propagation mode is shown in Fig. 1. Since, in general, the resonant peaks do not coincide with the sampled values of β, errors will result in the values of w_m and β_m obtained form the maxima in the sampled data set for $P(\beta)$. For example, the maximum uncertainty in β_m so determined will be

$$\Delta\beta_m = \frac{1}{2}\Delta\beta = \frac{\pi}{Z} \tag{15}$$

where $\Delta\beta$ is the sampling interval in $P(\beta)$. It is possible to reduce the uncertainties in β_m and w_m by lengthening Z, but it is far more efficient to first fit the correct lineshape function intrinsic to $w(z)$ to the sampled values of $P(\beta)$ that are closest to the resonance under consideration and to determine these β_m and w_m from the resulting lineshape fit. For the window function

$$w(z) = 1 - \cos\frac{2\pi z}{Z}, \quad z \leqslant Z$$
$$= 0, \quad z > Z \tag{16}$$

used here, the normalized line shape function corresponding to

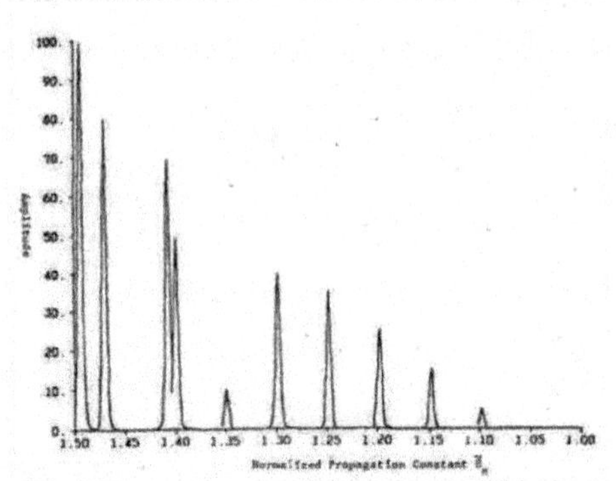

Fig. 1. Modal power spectrum for dielectric waveguide with ten propagation modes.

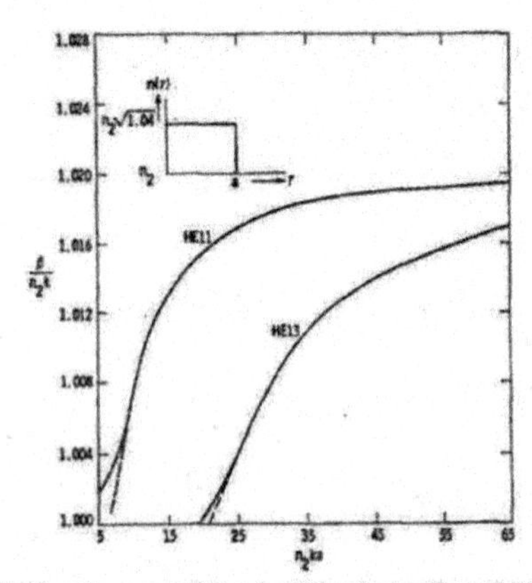

Fig. 2. Dispersion curves for a step-index circular fiber. Solid line: Dil and Block's results. Dashed line: scalar-wave approach results.

the harmonic z-dependence $\exp(-i\beta_m z)$ is

$$(\beta-\beta_m)=\frac{1}{Z}\int_0^Z \exp\,[i(\beta-\beta_m)z]\;w(z)\,dz$$
$$=\frac{\exp\{i[(\beta-\beta_m)Z+2\pi]\}-1}{i(\beta-\beta_m)Z}$$
$$-\mathcal{L}\frac{1}{2}\left[\frac{\exp\{i[(\beta-\beta_m)Z+2\pi]\}-1}{i[(\beta-\beta_m)Z+2\pi]}+\frac{\exp\{i[(\beta-\beta_m)Z-2\pi]\}-1}{i[(\beta-\beta_m)Z-2\pi]}\right]. \quad (17)$$

Therefore, it is possible to represent $P(\beta)$ over the range of β that corresponds to guided modes as

$$P(\beta)=\sum_m w_m \mathcal{L}(\beta-\beta_m). \quad (18)$$

By fitting $(\beta-\beta_m)$ to sampled values of $P(\beta)$ it is possible to compute extremely accurate solution for w_m and β_m.

IV. Accuracy of the Scalar-Wave Approach

To study the accuracy of the scalar-wave approach described above, we shall compare our results with those obtained according to other exact or approximate methods. Specifically we shall compare the results for the following waveguides: 1) the step-index circular fiber; 2) the graded-index circular fiber; 3) the rectangular fiber; 4) the elliptical fiber; 5) the triangular fiber; and 6) the diffused-channel rectangular waveguide.

A. The Step-Index Circular Fiber

The step-index circular fiber consists of a dielectric cylinder with homogeneous refractive index equal to n_1, surrounded by a cladding layer with a smaller index n_2. Due to pulse widening effects when two or more modes are propagating, this waveguide is usually used for single-mode operations. The cladding layer not only is used for mechanical support but also for selecting the n_1/n_2 ratio to be close to unity so as to allow for large core size even under single-mode operating conditions.

The exact solution for the propagation characteristics of the step-index fiber is developed by Dil and Block [8]. For their computation, the refractive index was assumed to be

$$n(r)=\begin{cases} n_2\sqrt{1.04}, & \text{for } 0\leqslant r\leqslant a \\ n_2, & \text{for } r>a. \end{cases} \quad (19)$$

The dispersion characteristics of this fiber is also computed by the scalar-wave approach described in the previous section. In Fig. 2, the dispersion curves for the two modes are displayed where the solid line is the result reported by Dil and Block and the darkened line is the result obtained by the scalar-wave approach. As can be seen the two methods produce almost identical results everywhere except near the cutoff region of the modes. This is due to the fact that near the cutoff region the field is more loosely bound to the core region.

B. Graded-Index Circular Fiber

A graded-index circular fiber is the most commonly used optical fiber today. This radially inhomogeneous dielectric waveguide has been developed for use as a multimode waveguide in optical communication systems. It is shown that the parabolic refractive-index variation, that is usually used with this type of fiber, will equalize the group velocities of all the propagating mode. Therefore, the spreading of a pulse due to delay differences between the propagating modes is minimized.

This type of optical fiber is also analyzed by Dil and Block and the exact solution for the modal characteristics are obtained. Additionally, the finite element solution is presented by Yeh *et al.* [13] which yields the exact result. In Fig. 3, the dispersion curves for the first five lower order modes of a graded-index circular fiber are shown when the refractive index is assumed to be

$$n(r)=\begin{cases} n_2\sqrt{1.04-0.04\left(\dfrac{r}{a}\right)^2}, & \text{for } 0\leqslant r\leqslant a \\ n_2, & \text{for } r>a. \end{cases} \quad (20)$$

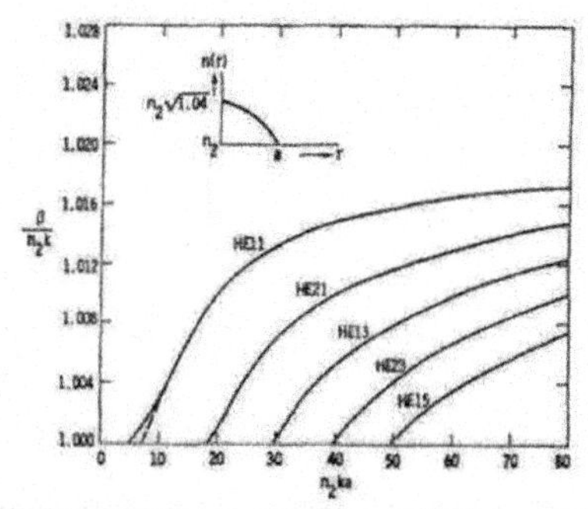

Fig. 3. Dispersion curves for a graded-index circular fiber. Solid line: Dil and Block's and finite element method's results. Dashed line: scalar-wave approach results.

In this figure, the solid line is the result reported by Dil and Block and Yeh's finite element method and the dashed line is the one generated by the scalar-wave approach. Again, it can be observed that the scalar-wave approach produces accurate results everywhere except at lower frequency range near the cutoff region of the dominant mode.

C. A Rectangular Fiber

A rectangular optical fiber is a dielectric rod, with rectangular cross section, surrounded by a medium with a refractive index less than the index of the rod. Several solutions for this optical fiber are available such as, Goell's circular harmonic approximate solution [9], Marcatili's approximate solution [14] and the finite element method solution [13]. Goell's technique is based on the expansion of the fields in terms of a series of circular harmonics, that is, Bessel and modified Bessel functions multiplied by trigonometric functions. Then the fields inside the waveguide are matched to those outside the core at the boundaries to obtain equations which are solved numerically. Marcatili's approximate technique is based on solving the Maxwell equations in close form but with many simplifying assumptions. The finite element method solves the vector-wave equation by the use of a variational expression of the wave propagation equation. This method yields the exact solution for the modal characteristics of the wave.

To compare the results obtained by the scalar-wave approach to the ones presented by Goell and Marcatili, a rectangular dielectric waveguide with core refractive index equal to 1.05 surrounded by a medium of index 1.01 is considered here. The aspect ratio of the guide a/b is assumed to be 2, where a and b represent the width and the height of the rectangular core, respectively. The dispersion curve for this problem is shown in Fig. 4, for the dominant modes E_{11}^{y} and E_{11}^{x}, which are almost degenerate for small core and cladding index differences. In Fig. 4, the normalized propagation constant

$$\frac{\bar{\beta}^2 - n_2^2}{n_1^2 - n_2^2}$$

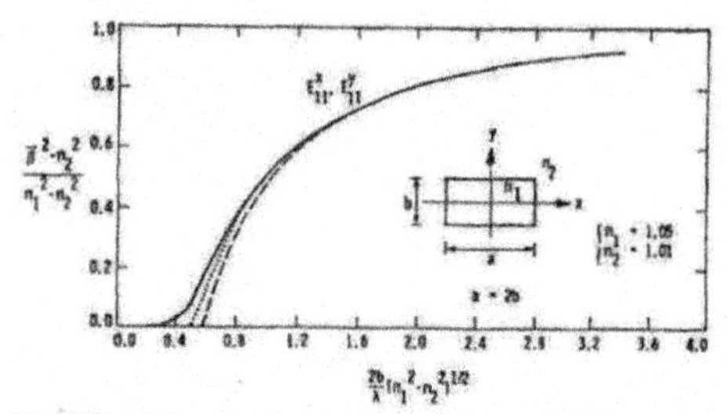

Fig. 4. Dispersion curve for a rectangular fiber. Solid line: Goell's result. Dash line: Marcatili's approximate result. Dotted line: scalar-wave approach result.

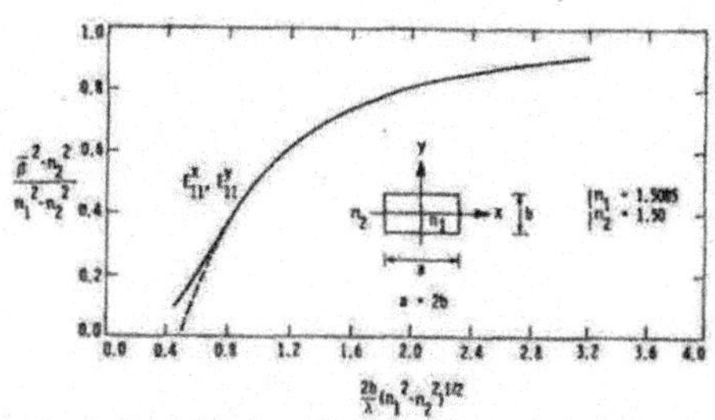

Fig. 5. Dispersion curves for a rectangular fiber. Solid line: finite element method result. Dashed line: scalar-wave approach result.

is plotted against the normalized frequency

$$\frac{2b}{\lambda}(n_1^2 - n_2^2)^{1/2}$$

where $\bar{\beta} = \beta/k$, β is the propagation constant, and k and λ are defined in previous sections. The solid curve is obtained using Goell's computer solution of the boundary value problem. The dashed line is the Marcatili's approximate solution. The dotted line is the scalar-wave approach solution. Goell's results are more reliable than those given by Marcatili's method, since the latter solution was obtained by assuming simple field distribution in the core region to match the fields along the four sides of the rectangular core. Therefore, comparing the curves of Fig. 4, we conclude that the scalar-wave approach is superior to Marcatili's approximate method.

To compare the results obtained by scalar-wave approximation to those from the finite element method, a rectangular waveguide with core refractive index of 1.5085 and cladding index of 1.50 is considered. The dispersion curves for this problem are given in Fig. 5. The solid line shows the results from the finite element method and the dashed line shows the results obtained by the scalar-wave approach. As can be observed, the two methods agree very closely everywhere except near the cutoff, where we expect the scalar-wave approach to be inaccurate.

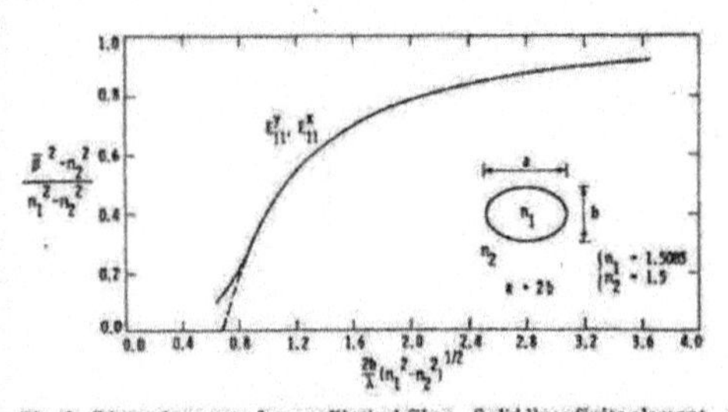

Fig. 6. Dispersion curve for an elliptical fiber. Solid line: finite element method result. Dashed line: scalar-wave approach result.

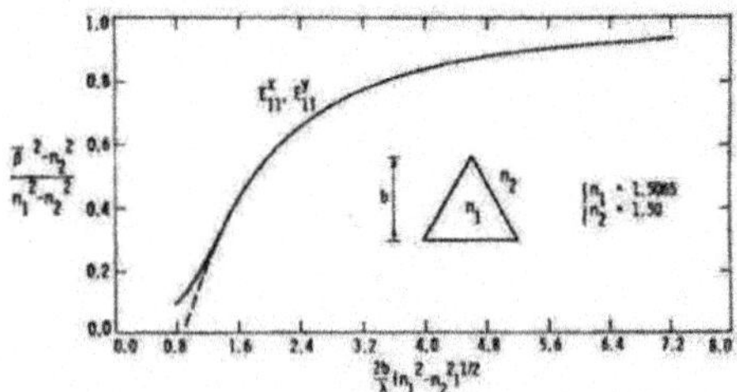

Fig. 7. Dispersion curve for a triangular fiber. Solid line: finite element result. Dashed line: scalar-wave approach result.

D. An Elliptical Fiber

An elliptical optical fiber is a dielectric rod with elliptical cross section and refractive index n_1, surrounded by a medium of smaller refractive index n_2. This optical fiber has practical importance since circular fibers commonly used in optical communication systems are easily deformed to elliptical shapes. This type of optical fiber was first analyzed by Yeh [7], [15] who showed that for this class of dielectric waveguides all modes are hybrid modes. Later, Yeh *et al.* [13] computed the dispersion characteristics of elliptical fibers by the finite element method. The scalar-wave approach can also be used to analyze this type of optical fibers.

Consider an elliptical optical fiber with core index $n_1 = 1.5085$ and cladding index $n_2 = 1.5$. The aspect ratio of the guide b/a is assumed to be 2, where a and b are the length of the major and minor axis of the core, respectively. The dispersion curve of the dominant modes for this problem is given in Fig. 6. The solid line shows the result obtained from the finite element method and the dashed line represents the result computed by the scalar-wave approach. As can be seen the two methods agree very closely almost everywhere except near the cutoff region.

E. A Triangular Fiber

Consider a triangular optical fiber with the core refractive index $n_1 = 1.5085$, and the cladding refractive index $n_2 = 1.50$. Assume the fiber to have an equilateral triangular core region with a height equal to b. The dispersion curve of the dominant modes are given in Fig. 7. The solid line shows the results from the finite element method [13], while the dashed line shows the results obtained from the scalar-wave approach. Similar to previous problems the two methods are in close agreement everywhere except near the cutoff region, where we expect the finite element technique to yield more accurate values.

F. A Diffused-Channel Rectangular Waveguide

A diffused-channel rectangular waveguide consists of a dielectric rod with rectangular cross section and refractive index n_1, embedded in another dielectric of slightly smaller index $n_2 = n_1(1 - \Delta)$; all surrounded by a medium which could be air or a dielectric medium with index n_0. For this class of optical waveguides the Marcatili's approximate solution [14] as well as the finite element method solution [13] are available.

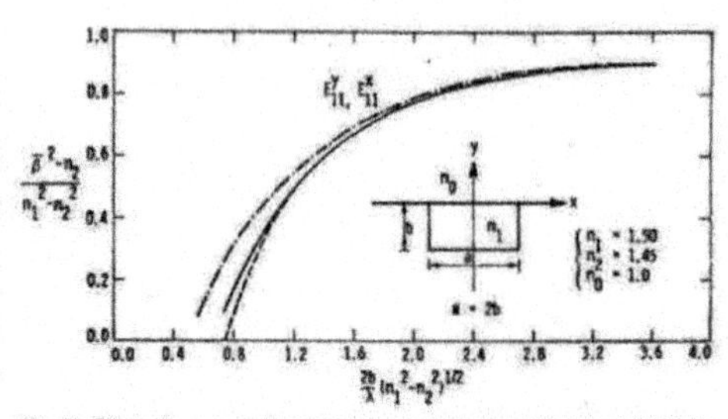

Fig. 8. Dispersion curve for a uniform channel waveguide. Solid line: finite element result. Dashed line: Marcatili's approximate result. Dotted-dashed line: scalar-wave approach results.

To compare the solution of the scalar-wave approach with the solution obtained from the methods mentioned above, consider a diffused-channel waveguide with $n_1 = 1.5$, $n_2 = 1.45$, and $n_0 = 1$. The aspect ratio of the guide a/b is assumed to be 2, where a is the width and b is the height of the channel. The dispersion curves of the dominant modes for this problem are given in Fig. 8. In this figure, the solid line is the finite element result, the dashed line is the Marcatili's approximate result, and the dotted-dashed line shows the scalar-wave approach solution. As can be seen, for this problem the scalar-wave approach does not yield accurate results even for the range away from the cutoff region. This is because of the breakdown of the scalar-wave approximation due to large index differences for this type of problem. In general the scalar-wave approach does not yield accurate results for problems where the index difference is larger than 10 percent. Fortunately, for most practical cases the index difference is only a few percent which makes them suitable for treatment by the method described here.

V. Numerical Examples

In the previous section it was demonstrated that, within its limitation, the scalar-wave approach is a versatile and accurate method applicable to optical-fiber waveguides. Now we can

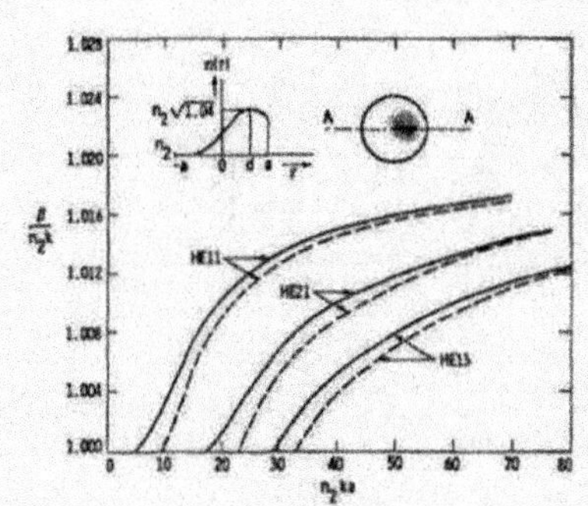

Fig. 9. Dispersion curves for a graded-index circular fiber with shifted index. Solid line: results for $d = 0$. Dashed line: results for $d = a/2$.

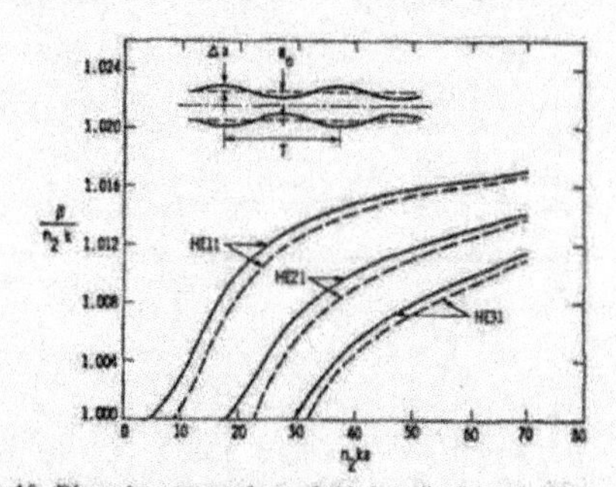

Fig. 10. Dispersion curves of a graded-index circular optical fiber with sinusoidal taper along the axis. Solid line: $\Delta a = 0$. Dashed line: $\Delta a = a_0/2$.

confidently apply this technique to more complicated inhomogeneous optical waveguides.

A. A Non-Axisymmetric Graded-Index Fiber

Consider a graded-index circular fiber, or the type discussed above, but with an index profile with its axis of symmetry shifted to one side by d. The geometry of this problem as well as its dispersion curves for the three lowest order modes are given in Fig. 9. In this figure the normalized propagation constant β/n_2k is shown versus the normalized frequency n_2ka. The solid line is for the case where $d = 0$, and the dashed line is for $d = a/2$. It must be noted that the variation of the dispersion characteristics of this guide does not depend on d in a linear fashion, but that it changes much faster as d is increased from zero to a. The example discussed here envisions one of the possible ways that a graded optical fiber can be distorted due to fabrication errors.

B. Optical Fiber With Sinusoidal Index Variation Along Its Axis

The scalar-wave approach can be applied to three-dimensional problems the same way as it is applied to two-dimensional problems without significant additional difficulty. To illustrate a numerical example, consider a graded-index circular fiber with sinusoidally varying radius, and, therefore, refractive index, along its axis (z-axis). The fiber radius is assumed to have the following functional form

$$a(z) = a_0 + \Delta a \cos\left[\frac{2\pi}{T} z\right] \tag{21}$$

where a_0 is the mean, Δa is the taper amplitude, and T is the period of the variation. It should be mentioned that Δa and T must be chosen in such a way that changes in the refractive index, from one propagation step to the next, is smooth. This would guarantee that the SW-FFT approach remains valid for this type of problem. However, it should be noted that using the forward-marching SW-FFT approach precludes the consideration of any backward wave excited by the longitudinal index variation. In other words, the propagation constant of the wave is always real and there exist no bandpass characteristics for the sinusoidally varying guide [16].

Dispersion curves for the first three higher order modes of a sinusoidally varying optical fiber are given in Fig. 10. The solid lines show the results for $\Delta a = 0$, and the dashed lines present the solution for $a = a_0/2$. The refractive index is assumed to be formulated as in (20) with $n_2 = 1.45$, and a as given in (21). It can be observed that the normalized propagation constant β/n_2k, has decreased as Δa is increased. This means that the effective refractive index of the optical fiber decreases as Δa is increased.

VI. Conclusion

We have successfully demonstrated that the SW-FFT technique can be used efficiently to obtain the complete description of the total field propagating along a general shaped, weakly guiding dielectric waveguide with arbitrary index variations. Knowledge of the evolution of the total field may be used to yield readily the propagation characteristics of different modes. Comparisons have also been made with other known results. Excellent agreements were obtained. As long as the premise of weakly guiding condition [10], [17], [18] is not violated, this SW-FFT technique is indeed a very powerful and efficient method to predict the propagation characteristics of guided field along any dielectric fiber or integrated optical circuit guide. Equation (2) can be used to obtain polarization corrections to the scalar propagation constant. We wish to emphasize that the present technique is capable of providing solutions to problem that could not be solved readily by other means. We are in the process of extending this technique to treat other important practical problems such as single-mode couplers and dielectric horns or branches.

References

[1] C. Yeh, "Optical waveguide theory," *IEEE Trans. Circuits Syst.*, vol. CAS-26, pp. 1011-1019, 1979.

[2] C. Yeh, L. Casperson, and B. Szejn, "Propagation of truncated Gaussian beams in multi-mode fiber guides," *J. Opt. Soc. Amer.*, vol. 68, pp. 989-993, 1978.

[3] C. Yeh, L. Casperson, and W. P. Brown, "Scalar wave approach

JOURNAL OF LIGHTWAVE TECHNOLOGY, VOL. LT-3, NO. 1, FEBRUARY 1985 205

for single-mode inhomogeneous fiber problems," *Appl. Phys. Lett.*, vol. 34, pp. 460–462, 1979.
[4] C. Yeh, W. P. Brown, and R. Szejn, "Multi-mode inhomogeneous fiber couplers," *Appl. Opt.*, vol. 18, pp. 489–498, 1979.
[5] C. Yeh, "On multi-mode optical fiber couplers, tapers, and horns," in *Physics of Fiber Optics, Advances in Ceramics*, vol. 2, Bernard Bendow and S. S. Mitra, Ed. The Amer. Ceramic Soc., 1981.
[6] M. D. Feit and J. A. Fleck, Jr., "Light propagation in graded-index optical fibers," *Appl. Opt.*, vol. 17, pp. 3990–3998, 1978.
[7] C. Yeh, "Elliptical dielectric waveguides," *J. Appl. Phys.*, vol. 33, pp. 3235–3243, 1962.
[8] J. G. Dil and H. Blok, "Propagation of electromagnetics surface waves in a radially inhomogeneous optical waveguides," *Opto-Elec.*, vol. 5, pp. 415–428, 1973.
[9] J. E. Goell, "A circular harmonic computer analysis of rectangular dielectric waveguide," *Bell Syst. Tech. J.*, vol. 48, pp. 2133–2160, 1969.
[10] A. W. Snyder and J. D. Love, *Optical Waveguide Theory*. London, England: Chapman and Hall, 1983.
[11] M. D. Feit and J. A. Fleck, Jr., "Computation of mode properties in optical fiber waveguides by a propagating beam method," *Appl. Opt.*, vol. 19, pp. 1154–1164, 1980.
[12] F. J. Harris, "On the use of windows for harmonic analysis with the discrete Fourier transform," *Proc. IEEE*, vol. 66, pp. 51–63, 1978.
[13] C. Yeh, K. Ha, S. B. Dong, and W. P. Brown, "Single-mode optical waveguides," *Appl. Opt.*, vol. 18, pp. 1490–1504, 1979.
[14] E. A. J. Marcatili, "Dielectric rectangular waveguide and directional coupler for integrated optics," *Bell Syst. Tech. J.*, vol. 48, pp. 2071–2102, 1969.
[15] C. Yeh, "Modes in weakly guiding elliptical optical fibers," *Opt. Quantum Electron.*, vol. 8, pp. 43–47, 1976.
[16] C. Elachi and C. Yeh, "Periodic structures in integrated optics," *J. Appl. Phys.*, vol. 44, pp. 3146–3154, 1973.
[17] D. Gloge, "Weakly guiding fibers," *Appl. Opt.*, vol. 10, pp. 2252–2258, 1971.
[18] D. Marcuse, "Theory of dielectric optical waveguides." New York: Academic, 1974.

Cavour Yeh (S'56-M'63-SM'82) was born in Nanking, China, on August 11, 1936. He received the B.S., M.S., and Ph.D. degrees in electrical engineering from the California Institute of Technology, Pasadena, CA, in 1957, 1958, and 1962, respectively.

He is presently professor of electrical engineering at the University of California at Los Angeles (U.C.L.A). He joined U.C.L.A. in 1967 after serving on the faculty of U.S.C. from 1962 to 1967. His current areas of research interest are optical and millimeter-wave guiding structures, gigabit-rate fiber-optic local area network, and scattering of electromagnetic waves by penetrable irregularly shaped objects.

Dr. Yeh is a member of Eta Kappa Nu, Sigma Xi, and the Optical Society of America.

*

Farzin Manshadi was born in Yazd, Iran, on April 22, 1950. He received the B.S. degree from the Arya Mehr University of Technology, Tehran, Iran, in electrical engineering in 1972. He came to the United States in 1974 and received the M.S. degree and Ph.D. degree in electrical engineering, majoring in electromagnetics, from the University of California at Los Angeles, CA, in 1976 and 1983, respectively.

From 1972 to 1974 he was with the Iranian National Radio and Television Organization. He has been with the Jet Propulsion Laboratory, Pasadena, CA, since 1981, where he is currently a member of Technical staff in Radio Engineering and Microwave Subsystems Section. He has been involved in development of techniques for control of large space antennas based on their RF characteristics and presently is responsible for design and development of Passive Microwave devices as well as a new generation of corrugated feed horns for NASA's Deep Space Network.

Pulse shepherding in nonlinear fiber optics

C. Yeh[a] and L. Bergman
Jet Propulsion Laboratory, MS 525-3660, California Institute of Technology, 4800 Oak Grove Drive, Pasadena, California 91109

(Received 29 April 1996; accepted for publication 7 June 1996)

In a wavelength division multiplexed fiber system, where pulses on different wavelength beams may copropagate in a single mode fiber, the cross-phase-modulation (CPM) effects caused by the nonlinearity of the optical fiber are unavoidable. In other words, pulses on different wavelength beams can interact with and affect each other through the intensity dependence of the refractive index of the fiber. Although CPM will not cause energy to be exchanged among the beams, the pulse shapes and locations on these beams can be altered significantly. This phenomenon makes possible the manipulation and control of pulses copropagating on different wavelength beams through the introduction of a shepherd pulse at a separate wavelength. How this can be accomplished is demonstrated in this paper. © *1996 American Institute of Physics.* [S0021-8979(96)03218-5]

I. INTRODUCTION

The successful design of dispersion-shifted and dispersion-flattened optical fibers having low dispersion over a relatively large wavelength range 1.3–1.6 μm (Ref. 1) enhances the viability of a multichannel wavelength division multiplexed (WDM) system.[2] All channels will experience similar low dispersion. This design is achieved through the use of multiple cladding layers.[3] Therefore, it is conceivable that this type of multiple-cladding-layers design technique may be used to custom design the desired dispersion characteristics.[4] For example, by minimizing the chromatic dispersion over a wavelength band in which WDM channels are assigned, group–velocity mismatch for these channels may be eliminated, resulting in the desirable simultaneous arrival of signals in these WDM channels. Time aligned, simultaneous arrival of WDM bit pulses is very important for a new class of bit parallel wavelength (BPW) link system used in high speed (>10 Gbyte/s) single fiber computer buses.[2]

In spite of the intrinsically small value of the nonlinearity coefficient in fused silica, because of low loss and long interaction length, the nonlinear effects in optical fibers made with fused silica cannot be ignored even at relatively low power levels.[5] This nonlinear phenomenon in fibers has been used successfully to generate optical solitons,[6] to compress optical pulses,[7] to produce timing maintenance in optical communications systems,[8] to transfer energy from a pump wave to a Stokes wave through the Raman gain effect,[9] to transfer energy from a pump wave to a counterpropagating Stokes wave through the Brillouin gain effect,[10] and to produce four-wave mixing.[11] Now, we wish to add one more application, the shepherding effect.

In a WDM system, the cross-phase-modulation (CPM) effects,[12,13] caused by the nonlinearity of the optical fiber are unavoidable. These CPM effects occur when two or more optical beams copropagate simultaneously and affect each other through the intensity dependence of the refractive index. This CPM phenomenon can be used to produce an interesting pulse shepherding effect. The purpose of this paper is to report this effect and to describe how it may be utilized to align the arrival time of pulses that are otherwise misaligned.

II. FORMULATION OF THE PROBLEM

The fundamental equations governing M copropagating waves in a nonlinear fiber including the CPM phenomenon are the coupled nonlinear Schrodinger equations[14]

$$\frac{\partial A_j}{\partial z}+\frac{1}{v_{gj}}\frac{\partial A_j}{\partial t}+\frac{1}{2}\alpha_j A_j = \frac{1}{2}\beta_{2j}\frac{\partial^2 A_j}{\partial t^2}-\gamma_j\left(|A_j|^2+2\sum_{m\neq j}^{M}|A_m|^2\right)A_j \quad (j=1,2,3,\ldots,M). \tag{1}$$

Here, for the jth wave, $A_j(z,t)$ is the slowly varying amplitude of the wave, v_{gj}, the group velocity, β_{2j}, the dispersion coefficient ($\beta_{2j} = dv_{gj}^{-1}/d\omega$), α_j, the absorption coefficient,

$$\gamma_j=\frac{n_2\omega_j}{cA_{\text{eff}}}, \tag{2}$$

is the nonlinear index coefficient with A_{eff} as the effective core area and $n_2 = 3.2\times10^{-16}$ cm^2/W for silica fibers, ω_j is the carrier frequency of the jth wave, c is the speed of light, and z is the direction of propagation along the fiber.

Introducing the normalizing coefficients

$$\tau=\frac{t-(z/v_{g1})}{T_0},\quad d_{1j}=(v_{g1}-v_{gj})/v_{g1}v_{gj},\quad \xi=z/L_{D1},\quad L_{D1}=T_0^2/|\beta_{21}|, \tag{3}$$

and setting

$$u_j(\tau,\xi)=(A_j(z,t)/\sqrt{P_{0j}})\times\exp(\alpha_j L_{D1}\xi/2), \tag{4}$$

$$L_{NLj}=1/(\gamma_j P_{0j}),\qquad L_{Dj}=T_0^2/|\beta_{2j}|, \tag{5}$$

gives

[a]Electronic mail: cavour@solstice.jpl.nasa.gov

 0021-8979/96/80(6)/3174/5/$10.00

$$i\frac{\partial u_j}{\partial \xi}=\frac{\mathrm{sgn}(\beta_{2j})L_{D1}}{2L_{Dj}}\frac{\partial^2 u_j}{\partial \tau^2}-i\frac{d_{1j}}{T_0}L_{D1}\frac{\partial u_j}{\partial \tau}-\frac{L_{D1}}{L_{NLj}}$$

$$\times\left[\exp(-\alpha_j L_{D1}\xi)|u_j|^2\right.$$

$$\left.+2\sum_{m\neq j}^{M}\exp(-\alpha_m L_{D1}\xi)|u_m|^2\right]u_j$$

$$(j=1,2,3,\ldots M). \quad (6)$$

Here, T_0 is the pulse width, P_{0j} is the incident optical power of the jth beam, and d_{1j}, the walk-off parameter between beam 1 and beam j, describes how fast a given pulse in beam j passes through the pulse in beam 1. In other words, the walk-off length is

$$L_{W(1j)}=T_0/|d_{1j}|. \quad (7)$$

So, $L_{W(1j)}$ is the distance for which the faster moving pulse (say, in beam j) completely walked through the slower moving pulse in beam 1. The nonlinear interaction between these two optical pulses ceases to occur after a distance $L_{W(1j)}$. For CPM to take effect significantly, the group–velocity mismatch must be held to near zero.

It is also noted from Eq. (6) that the summation term in the bracket representing the CPM effect is twice as effective as the self-phase-modulation (SPM) effect for the same intensity. This means that the nonlinear effect of the fiber medium on a beam may be enhanced by the copropagation of another beam with the same group velocity.

III. NUMERICAL SOLUTION

Equation (6) is a set of simultaneous coupled nonlinear Schrodinger equations that may be solved numerically by the split-step Fourier method. This method was used successfully earlier to solve the problem of beam propagation in complex fiber structures, such as fiber couplers,[15] and to solve the thermal blooming problem for high energy laser beams.[16] According to this method, the solutions may be advanced first using only the nonlinear part of Eq. (6). Next the solutions are allowed to advance using only the linear part of Eq. (6). This forward stepping process is repeated until the desired destination is reached. The Fourier transform is accomplished numerically via the well-known fast Fourier transform technique.

With this approach, the evolution of all the pulses on all the copropagating WDM beams as they propagate down the fiber may be obtained. It was through these numerical computations that we discovered the interesting pulse shepherding as well as the beam compression effects.[17]

Consider now the evolution of two single pulses on two copropagating beams whose operating wavelengths are separated by $\Delta\lambda=4$ nm. For this case, the four-wave-mixing effect is negligible. It is further assumed that the signal carrying pulses on each beam are separated by sufficiently large time intervals so that no interaction among succeeding pulses on the same beam occurs. The physical parameters chosen for the simulation correspond to an actual system that is of interest,

L =length of fiber=50 km
β_2 =dispersion coefficient=-1.61 ps^2/km
λ_1 =operating wavelength of beam No. 1=1.55 μm
λ_2 =operating wavelength of beam No. 2=1.546 μm
γ =nonlinear index coefficient=20W^{-1} km^{-1}
P_0 =incident power of each beam=1 mW
α =attenuation or absorption of each beam in fiber =0.2 dB/km
v_g =group velocity of the beam=2.051147×10^8m/s
d_{1j} =walk-off parameter between beam No. 1 and beam No. $j=v_{g1}-v_{gj}=0$ (no walk off)
T_0 =pulse width=10 ps.

With these values, the dispersion length L_D or the nonlinear length L_{NL}, which provides the length scale over which the dispersive or nonlinear effects for pulses on a single beam become important for pulse evolution along a fiber of length L, is

$$L_D=62 \text{ km}$$

or

$$L_{NL}=50 \text{ km}.$$

It is noted that in an idealized situation of zero fiber attenuation, the soliton propagation condition for a single beam results when

$$N^2=L_D/L_{NL},$$

and N is an integer. For multiple interacting beams, there is no condition under which solitons may exist even if the fiber is lossless.

IV. DISCUSSION OF THE RESULTS: DYNAMIC CONTROL OF IN-FLIGHT PULSES WITH A SHEPHERD PULSE

A. Without shepherd pulse

Shown in Fig. 1(a) is the evolution of two Gaussian pulses on two different wavelength beams as they propagate in this single mode fiber. These pulses are initially offset by 1/2 pulse width. It is seen that the pulses are affected by each other. Because of the nonlinear SPM and CPM effects, the pulses tend to attract each other. They appear to congregate towards a region of higher induced index of refraction. The leading pulse is pulled back while the trailing pulse is pushed forward so that these pulses tend to align with each other.

This observation is consistent with the earlier discovery of the self-focusing effect[18] where the induced higher refraction index region caused by a higher beam intensity tends to "attract" the propagating optical wave, resulting in the "focusing" of this optical wave. It is also consistent with the concept used to confine a thermally bloomed high-energy laser beam, where multiple surrounding beams are used to create an index environment in which the central main beam tends to expand less due to the lowering of the surrounding index of refraction caused by the heating from the surrounding beams.[16] It is also consistent with the "dragging effect" that occurs in weakly birefringent fibers.[19]

It is expected, however, that when the two copropagating pulses on two separate wavelength beams are separated by a

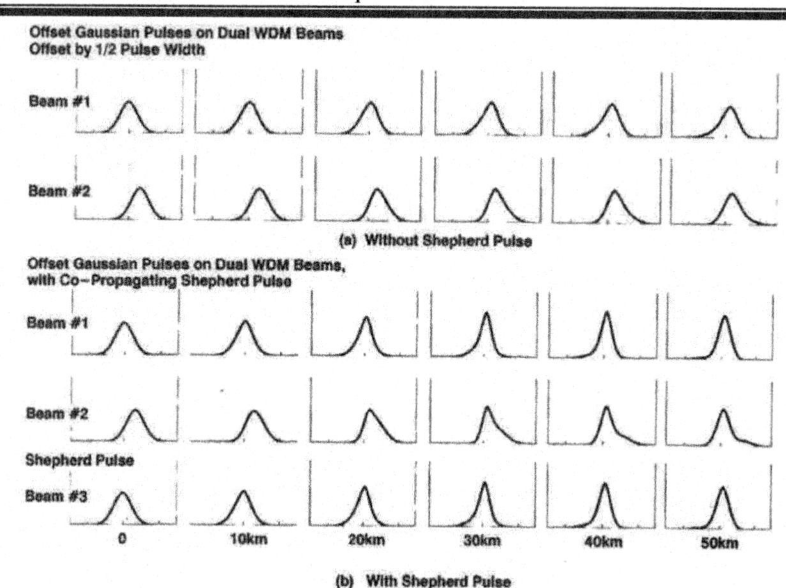

FIG. 1. Evolution of two Gaussian pulses on two WDM beams, separated by 1/2 pulse width: (a) without shepherd pulse on the third beam, and (b) with shepherd pulse on the third beam.

sufficiently large distance, these two pulses will not interact with each other. This fact is demonstrated in Fig. 2(a), where the two copropagating pulses are separated by one pulse width. Each pulse propagates independently as if it were not aware of the presence of the other pulse. It, thus, appears that once these pulses are launched in this manner, the separation of these pulses cannot be altered, except through the introduction of a shepherd pulse as shown in the next section.

B. With shepherd pulse

It will be shown that the widely separated pulses on beam No. 1 and beam No. 2, as shown in Fig. 2(a), can be brought significantly closer to each other by the launching of another pulse on a separate, 1.542 μm wavelength beam with the proper magnitude and at the proper time. This pulse is called the shepherd pulse because of its shepherding behavior on the other pulses. In other words, it is possible to pull back the leading pulse and at the same time to push forward the trailing pulse to achieve near pulse alignment. This is shown in Fig. 2(b).

The magnitude, the shape, and the location of the shepherd pulse, all contribute to the eventual success of this scheme to align these copropagating pulses. The fundamental phenomena that govern this scheme are the SPM, CPM, and group-velocity dispersion (GVD).[12] Computer simulation shows that a lower magnitude shepherd pulse does not possess sufficient attractive strength to pull the shepherded pulses together. For example, a magnitude 1 shepherd pulse, $\exp(-0.5\tau^2)$, situated in the middle of the shepherded pulses, can only bring these pulses 10% closer to each other, while a magnitude 2 shepherd pulse, $2\exp(-0.5\tau^2)$, similarly situated, can almost align these pulses [see Fig. 2(b)]. It does not follow, however, that an even higher magnitude shepherd pulse can bring the shepherded pulses together sooner, because a magnitude 3 shepherd pulse's tremendous "pull" on the shepherded pulses tends to break up these pulses through the introduction of higher oscillations. There is a limit as to how strong the shepherd pulse can be.

Broadening the shepherd pulse, i.e., using a $2\times\exp(-0.05\tau^2)$ pulse, only sharpens the shepherded pulses due to an increased apparent medium nonlinearity. The use of this broadened shepherd pulse can only bring the shepherded pulses 30% closer to each other, a far cry from the alignment achieved by the sharper shepherd pulse of $2\exp(-0.5\tau^2)$.

The next step, perhaps, is to use two shepherd pulses on two different wavelength beams to further enhance the shepherding effect. One, the $2\exp(-0.5\tau^2)$ shepherd pulse, on beam No. 3 at 1.542 μm may be used to pull the two shepherded pulses together, and the other, the $2\times\exp(-0.5\tau^2)$ shepherd pulse, on beam No. 4 at 1.538 μm may be used to sharpen the two shepherded pulses. This simulation is done. It is discovered that the added strength of the two shepherd pulses tends to break up the shepherded pulses into several oscillating pulses, an undesirable phenomenon.

The above computer simulation shows that there exists an optimum shepherd pulse with a certain magnitude, pulse

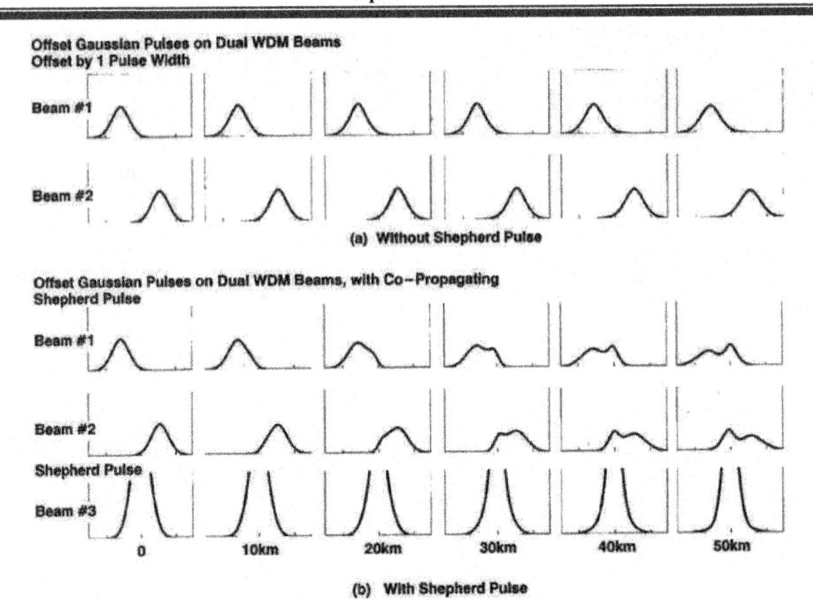

FIG. 2. Evolution of two Gaussian pulses on two WDM beams, separated by 1 pulse width: (a) without shepherd pulse on the third beam, and (b) with shepherd pulse on the third beam.

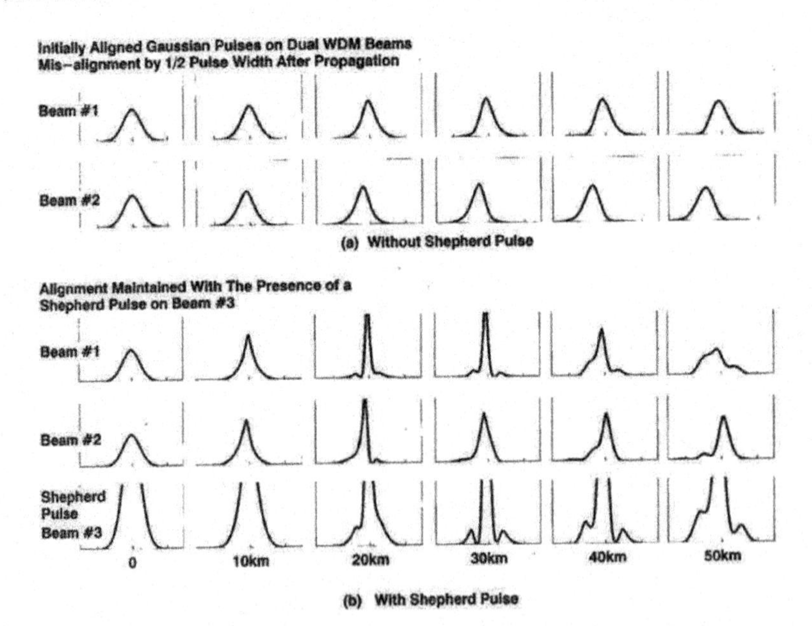

FIG. 3. Evolution of two initially aligned Gaussian pulses on two WDM beams: (a) after propagation, separation occurs for pulses on beam No. 1 and beam No. 2 without shepherd pulse on the third beam, and (b) alignment maintained for pulses on beam No. 1 and beam No. 2 with shepherd pulse on the third beam.

width, pulse shape, and location with respect to the shepherded pulses that can provide the best alignment for these pulses. For the example with the physical parameters given here, the optimum shepherd pulse appears to be the 2 exp $(-0.5\tau^2)$ pulse situated between the two shepherded pulses.

For the case of reduced separation of the pulses to be shepherded, as for the case shown in Fig. 1, a much more dramatic demonstration on the successful alignment achievable by a well designed shepherd pulse can be seen in Fig. 1(b). Here, the Gaussian pulses on beam No. 1 and on beam No. 2 are offset by 1/2 pulse width. A Gaussian shepherd pulse of unity magnitude, aligned with the pulse on beam No. 1, is introduced on beam No. 3 whose wavelength is 1.542 μm. This wavelength is 4 nm from beam No. 2 and 8 nm from beam No. 1, assuring that the four wave mixing effect is negligible. It is observed that this shepherd pulse is capable of achieving excellent alignment of the wayward pulse (pulse on beam No. 2) with the reference pulse (pulse on beam No. 1).

Another case demonstrating the effectiveness of a shepherd pulse to control and align the shepherded pulses is shown in Fig. 3. Here, two Gaussian pulses on two different wavelength beams with wavelengths of 1.55 and 1.546 μm, originating in an aligned position as shown in Fig. 3(a), begin to separate from each other because of a slight difference in the group velocities for these two beams. Without the presence of a shepherd pulse, these beams will be approximately 1/2 pulse width apart at 50 km downstream as can be seen from Fig. 3(a). With the shepherd pulse of $2 \times \exp(-0.5\tau^2)$ on a third beam with wavelength 1.542 μm, originally aligned with the two shepherded pulses and propagating at the same velocity as the pulse on beam No. 1, at 50 km downstream, the shepherded pulses are still aligned as shown in Fig. 3(b).

V. CONCLUSIONS

Demanding "fast" walk off of copropagating beams from each other in order to avoid any deleterious walk-off effect among the beams and to minimize the interaction among these beams due to the nonlinear behavior of the fiber medium, is unnecessary. On the contrary, it is found that, by requiring as little walk off as possible, the "shepherding" effect among the various beams may be used to "herd" them together resulting in the desirable characteristic of simultaneous arrival of copropagating beams in a BPW (bit-parallel wavelength) system.[2]

What has been demonstrated here is that through the introduction of a shepherd pulse on a separate wavelength beam, it is possible to *dynamically* manipulate, control, and reshape pulses on copropagating beams. This dynamic control feature from a shepherd pulse is a unique one.

ACKNOWLEDGMENTS

The research described in this paper was performed by the Center for Space Microelectronics Technology, Jet Propulsion Laboratory, California Institute of Technology, and was sponsored by the Ballistic Missile Defense Organization, Office of Innovative Science and Technology, through an agreement with the National Aeronautics and Space Administration. One of the authors, C.Y. wishes to thank Dr. W. P. Brown for his help in writing the soliton program.

[1] L. G. Cohen, W. L. Mammel, and S. J. Jang, Electron. Lett. **18**, 1023 (1982); B. J. Ainslie and C. R. Day, J. Lightwave Technol. **LT-4**, 967 (1986).
[2] L. A. Bergman, A. J. Mendez, and L. S. Lome, SPIE Crit. Rev. **CR62**, 210 (1996).
[3] C. Yeh and G. Lindgren, Appl. Opt. **16**, 483 (1977).
[4] H. H. Kuehl, J. Opt. Soc. Am. B **5**, 709 (1988).
[5] E. P. Ippen, in *Laser Applications to Optics and Spectroscopy*, edited by S. F. Jacobs, M. Sargent III, J. F. Scott, and M. O. Scally (Addison-Wesley, Reading, MA, 1975), Vol. 2, Chap. 6.
[6] A. Hasegawa and F. D. Tappert, Appl. Phys. Lett. **23**, 142 (1973); L. F. Mollenauer, R. H. Stolen, and J. P. Gordon, Phys. Rev. Lett. **45**, 1095 (1980).
[7] L. F. Mollenauer, R. H. Stolen, J. P. Gordon, and W. J. Tomlinson, Opt. Lett. **8**, 289 (1983).
[8] J. P. Gordon and H. A. Haus, Opt. Lett. **11**, 665 (1986); D. Woods, J. Lightwave Technol. **8**, 1097 (1990); Y. Kodama and A. Hasegawa, Opt. Lett. **17**, 31 (1992); L. F. Mollenauer, J. P. Gordon, and S. G. Evangelides, *ibid.* **17**, 1575 (1992); J. D. Moores, W. S. Wong, and H. A. Haus, Opt. Commun. **113**, 153 (1994).
[9] R. H. Stolen and E. P. Ippen, Appl. Phys. Lett. **22**, 276 (1973); V. A. Vysloukh and V. N. Serkin, Pis'ma Zh. Eksp. Teor. Fiz. (JETP Lett.) **38**, 170 (1983); E. M. Dianov, A. Ya. Karasik, P. V. Mamyshev, A. M. Prokhorov, M. F. Stel'makh, and A. A. Fomichev, *ibid.* **41**, 294 (1985).
[10] E. P. Ippen and R. H. Stolen, Appl. Phys. Lett. **21**, 539 (1972); N. A. Olsson and J. P. van der Ziel, Appl. Phys. Lett. **48**, 1329 (1986).
[11] R. H. Stolen, J. E. Bjorkholm, and A. Ashkin, Appl. Phys. Lett. **24**, 308 (1974).
[12] G. P. Agrawal, *Nonlinear Fiber Optics* (Academic, New York, 1989).
[13] *Optical Solitons—Theory and Experiment, Cambridge Studies in Modern Optics 10*, edited by J. R. Taylor (Cambridge University Press, Cambridge, 1992).
[14] G. P. Agrawal, Phys. Rev. Lett. **59**, 880 (1987).
[15] C. Yeh, W. P. Brown, and R. Szejn, Appl. Opt. **18**, 489 (1979).
[16] C. Yeh, J. E. Pearson, and W. P. Brown, Appl. Opt. **15**, 2913 (1976).
[17] The beam compression effect due to the presence of a shepherd pulse will be discussed in a paper elsewhere.
[18] F. Shimizu, Phys. Rev. Lett. **19**, 1097 (1967).
[19] C. R. Menyuk, J. Opt. Soc. Am. B **5**, 392 (1988).

Experimental Verification of the Pulse Shepherding Concept in Dispersion-Shifted Single-Mode Fiber for Bit-Parallel Wavelenght Links

L. Bergman, J. Morookian, C. Yeh,
and S. Monacos

Reprint

Proceedings of the

**International Conference
Massively Parallel Processing
Using Optical Interconnections**

Montreal, Canada
June 22-24, 1997

Washington ♦ Los Alamitos ♦ Brussels ♦ Tokyo

PUBLICATIONS OFFICE, 10662 Los Vaqueros Circle, P.O. Box 3014, Los Alamitos, CA 90720-1314 USA

Experimental Verification of the Pulse Shepherding Concept in Dispersion–Shifted Single–Mode Fiber for Bit–Parallel Wavelength Links

L. Bergman, J. Morookian, C. Yeh and S. Monacos
Jet Propulsion Laboratory
California Institute of Technology
Pasadena, California 91109

Abstract

A new way to dynamically control in–flight pulses by a co–propagating shepherd pulse in a wavelength division multiplexed (WDM) single–mode fiber system was proposed at the MPPOI '96 Conference. That system functionally resembles an optical fiber ribbon cable, except that all the bits pass on one fiber optic waveguide. This single fiber bit parallel wavelength link can be used to extend the (speed x distance) product of emerging cluster computer networks, such as, the MyriNet, SCI, Hippi–6400, ShuffleNet, etc. Here, we shall present the first experimental evidence that this pulse shepherding effect can be observed in a commercially available Corning DS (dispersion–shifted) fiber. Computer simulation results will first be presented for the case observed in the laboratory setup. A discussion of the experiment setup and measurement procedures will be given. Experimental results will then be compared with computer generated results. Excellent agreement is observed. Future experiments dealing with the shepherding effect among more than two co–propagating pulses will be performed.

I. Introduction

The concept of using a shepherd pulse to promote time-alignment of co-propagating pulses in a bit-parallel wavelength division multiplexed system [1] for a single-mode fiber was presented at the MPPOI '96 Conference [2]. The proposed concept is based on the cross phase modulation (CPM) effects [3] caused by the nonlinearity of the optical fiber in a wavelength division multiplexed (WDM) system. These CPM effects occur when two or more optical beams co-propagate simultaneously and affect each other through the intensity dependence of the refractive index. This CPM phenomenon can be used to produce an interesting pulse shepherding effect to align the arrival time of pulses which are otherwise misaligned. This same CPM effect can also be used to generate time-aligned co-propagating pulses on different wavelength beams.

An example of the pulse shepherding effect [4] is shown below:

Let us assume that two gaussian pulses on two different wavelength beams with wavelengths of 1.55 μm and 1.546 μm, originating in an aligned position as shown in Fig. 1(a), begin to separate from each other due to slight difference in the group velocities for these two beams. Without the presence of a shepherd pulse, these beams will be approximately 1/2 pulsewidth apart at 50 km downstream as can be seen from Fig. 1(a). With the shepherd pulse of $2 \exp(-0.5 \tau^2)$ on a third beam with wavelength 1.542 μm, originally aligned with the two shepherded pulses and propagating at the same velocity as the pulse on beam #1, at 50 km downstream, the shepherded pulses are still aligned as shown in Fig. 1(b).

What this means is that through the introduction of a shepherd pulse on a separate wavelength beam, it is possible to dynamically manipulate, control and reshape pulses on co-propagating beams in a WDM system. This dynamic control feature from a shepherd pulse will enable the eventual construction of a time-aligned bit-parallel wavelength link as an interconnect with exceptionally high speed, low latency, simplified electronics interface (with no speed bottleneck), and extensibility to all-optical packet networks.

Here, we shall present the first experimental result showing the existence of this pulse shepherding effect.

0-8186-7974-3/97 $10.00 © 1997 IEEE

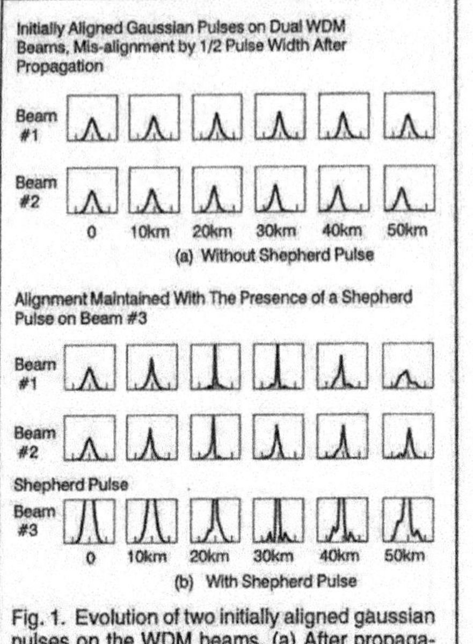

Fig. 1. Evolution of two initially aligned gaussian pulses on the WDM beams. (a) After propagation, separation occurs for pulses on beam #1 and beam #2 without shepherd pulse on the third beam.(b) Alignment maintained for pulses on beam #1 and beam #2 with shepherd pulse on the third beam.

A review of the theoretical background will first be given. Computer simulation results for the case corresponding to that observed in the laboratory will be presented. Experimental setup and measurement procedures will then be discussed. Measured results will be compared with computer results. Evidence of the shepherding effect will be presented. Finally, a discussion on future experiments dealing with the shepherding effect among more than two co-propagating pulses will be given.

II. A Review of the Theoretical Foundation

The fundamental equations governing M numbers of co-propagating waves in a nonlinear fiber including the CPM phenomenon are the coupled nonlinear Schrodinger equations [3,4]:

$$\frac{\partial A_j}{\partial z} + \frac{1}{v_{gj}}\frac{\partial A_j}{\partial t} + \frac{1}{2}\alpha A_j = \frac{1}{2}\beta_{2j}\frac{\partial^2 A_j}{\partial t^2}$$

$$-\gamma_j \left(|A_j|^2 + 2\sum_{m\neq j}^{M} |A_m|^2 \right) A_j$$

$$(j = 1,2,3, \ldots\ldots M) \quad (1)$$

Here, for the jth wave, $Aj(z,t)$ is the slowly-varying amplitude of the wave, v_{gj}, the group velocity, β_{2j}, the dispersion coefficient ($\beta_{2j} = dv_{gj}^{-1}/d\omega$), α_j, the absorption coefficient,

$$\gamma_j = \frac{n_2\omega_j}{c\, A_{eff}} \quad (2)$$

is the nonlinear index coefficient with A_{eff} as the effective core area and $n_2 = 3.2 = 10^{-16}$ cm^2/W for silica fibers, ω_j is the carrier frequency of the jth wave, c is the speed of light, and z is the direction of propagation along the fiber.

Introducing the normalizing coefficients

$$\tau = \frac{t - (z/v_{g1}}{T_0}$$

$$d_{1j} = (v_{g1} - v_{gj})/v_{g1}v_{gj},$$

$$\xi = z/L_{D1},$$

$$L_{D1} = T_0^2/|\beta_{21}|, \quad (3)$$

and setting

$$u_j(\tau,\xi) = (A_j(z,t)/\sqrt{P_{0j}})\ \exp(\alpha_j L_{D1}\xi/2) \quad (4)$$

$$L_{NLj} = 1/(\gamma_j P_{0j})$$

$$L_{Dj} = T_0^2/|\beta_{2j}| \quad (5)$$

gives

$$i\frac{\partial u_j}{\partial \xi} = \frac{sgn(\beta_{2j})L_{D1}}{2L_{Dj}}\frac{\partial^2 u_j}{\partial \tau^2}$$

$$- i\frac{d_{1j}}{T_0}L_{D1}\frac{\partial u_j}{\partial \tau}$$

$$-\frac{L_{D1}}{L_{NLj}}\ [\ \exp(-\alpha_j L_{D1}\xi)\ |u_j|^2$$

$$+ 2\sum_{m\neq j}^{M} exp\ (-\alpha_m L_{D1}\xi)\ |u_m|^2]\ u_j$$

$(j = 1,2,3, \ldots M)$ (6)

Here, T_0 is the pulse width, P_{0j} is the incident optical power of the jth beam, and d_{lj}, the walk-off parameter between beam l and beam j, describes how fast a given pulse in beam j passes through the pulse in beam l. In other words, the walk-off length is

$$L_{w(lj)} = T_0 \,/\, |d_{lj}|. \quad (7)$$

So, $L_{W(1j)}$ is the distance for which the faster moving pulse (say, in beam j) completely walked through the slower moving pulse in beam 1. The nonlinear interaction between these two optical pulses ceases to occur after a distance $L_{W(1j)}$. For cross-phase-modulation (CPM) to take effect significantly, the group-velocity mismatch must be held to near zero.

It is also noted from Eq. (6) that the summation term in the bracket representing the cross-phase-modulation (CPM) effect is twice as effective as the self-phase-modulation (SPM) effect for the same intensity. This means that the nonlinear effect of the fiber medium on a beam may be enhanced by the co-propagation of another beam with the same group velocity.

Equation (6) is a set of simultaneous coupled nonlinear Schrodinger equations which may be solved numerically by the split-step Fourier method, which was used successfully earlier to solve the problem of beam propagation in complex fiber structures, such as, the fiber couplers [5], and to solve the thermal blooming problem for high energy laser beams [6]. According to this method, the solutions may be advanced first using only the nonlinear part of the equations. And then the solutions are allowed to advance using only the linear part of Eq. (6). This forward stepping process is repeated over and over again until the desired destination is reached. The Fourier transform is accomplished numerically via the well-known Fast Fourier Transform Technique.

III. Computer Simulation Results

Based on the above numerical technique, computer simulation is carried out for the case corresponding to that performed in the laboratory.

Two beams with wavelength separated by 5 nm (nano–meter) are launched into a single mode fiber: One beam carries a 20 ps gaussian pulse while the other beam carries a 200 ps gaussian pulse to simulate the cw signal in the experiment. The evolution of the two pulses on these two co-propagating beams is the focus of our simulation. It is noted that the four wave mixing effect is negligible for this case. Let us label the initial 20 ps gaussian pulse carried by one of the beam as the shepherd (S) pulse and the other 200 ps pulse on the other beam as the primary (P) pulse. The parameters that we use for the simulation are:

L = length of fiber = 50 km
β_2 = dispersion coefficient = 2 ps^2/km
λ_1 = operating wavelength of beam #1 = 1.55 µm
λ_2 = operating wavelength of beam #2 = 1.545 µm
γ = nonlinear index coefficient = 20 W^{-1}km^{-1}
α = attenuation or absorption of each beam in fiber = 0.2 dB/km
v_g = group velocity of the beam = 2.051147×10^8 m/sec
d_{12} = walk-off parameter between beam #1 and beam #2 < 1 ps/km
T_0 = pulse width = 20 ps.

Shown in Fig. 2 is the evolution of these two pulses on two different wavelength beams as they propagate in this single mode fiber. Since both pulses are operating in the positive dispersion region, i.e., the dispersion coefficient β_2 is positive, neither pulse will undergo pulse–compression. Since the dispersion coefficient is quite small, for the distance considered, neither pulse will experience significant pulse-broadening.

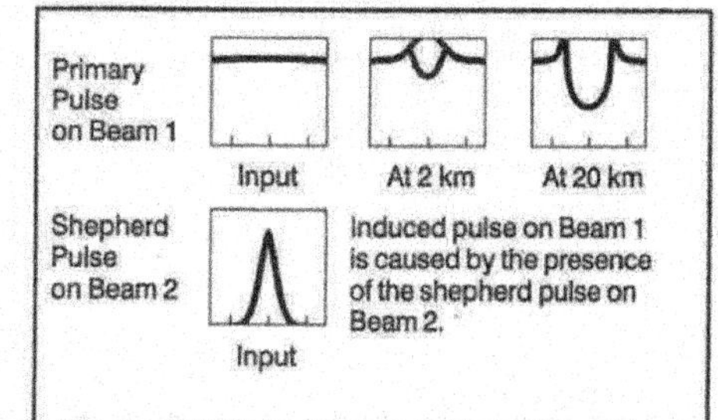

Fig. 2. Computer simulation results for the case of a 20 ps pulse (the shepherd pulse) co-propagating with a 200 ps pulse (approximating a cw primary pulse) in a 20 km dispersion shifted fiber with negligible walkoff. A dip on the 200 ps pulse appeared at the end of the fiber indicating the presence of the shepherding effect.

One notes that in the absence of the shepherding effect (i.e., the CPM effect) these pulses will propagate independent of each other. However, due to the presence of the shepherding effect, very significant changes are observed on the 200 ps primary pulse. On that primary pulse, a dip appears at the location which is aligned with the 20 ps shepherd pulse. This dip appears to grow deeper and broader as both pulses propagate down the fiber, eventually reaching the shape of an inverted 20 ps gaussian pulse. This inverted guassian pulse is superposed over the 200 ps primary pulse. This inverted gaussian pulse on a long plateau looks very much like a dark soliton pulse. Also noted is a narrow rim around this dip. Due to the averaging technique used in the experimental measurement, this narrow rim will not appear in the measured picture of the induced primary pulse; only a dip will appear in the picture.

The effect of the small walk-off is to shift the induced inverted pulse on the 200 ps primary beam slightly. The shepherding effect also skews slightly the symmetry of both the shepherd pulse and the induced inverted pulse on the primary pulse.

This very distinctive feature of an induced inverted pulse on a broad primary pulse which is clearly caused by the shepherding effect has been used to experimentally verify the existence of the shepherding effect.

IV. Experimental Setup and Procedures

A schematic block diagram of the experimental setup is shown in Fig. 3. The pulse source is an Erbium Doped Fiber Ring Laser (EDFRL), producing a 100 MHz train of pulses 20 ps in length at a wavelength near 1551 nm. This Erbium Ring pulse is named the shepherd pulse, operating at peak power of higher than 200 mW. The primary source is a DFB laser diode at 1545 nm operated under a dc bias well above threshold. This cw output from the primary laser diode source is about 1 mW which is amplified through an Erbium Doped Fiber Amplifier (EDFA) to around 33 mW.

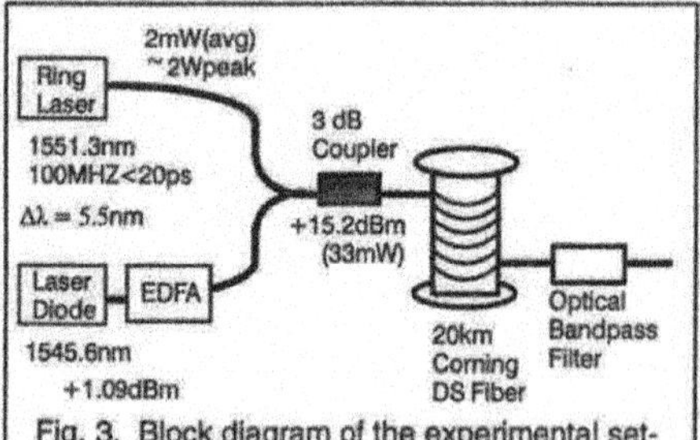

Fig. 3. Block diagram of the experimental set-up to detect the shepherding effect.

As shown in Fig. 3, signals from these two sources of two different wavelengths are then combined using a 2 to 1 fiber coupler. The combined output is sent through a 20 km spool of Corning DS fiber. At the output end of the fiber, an optical bandpass filter is used to reject the pulse signal from the ring laser. The signal from the laser diode is detected and viewed on an oscilloscope. A picture of this output is shown in Fig. 4. A dip on the cw signal is observed indicating the presence of the shepherding effect as predicted by the computer simulation result.

This is the very first time that this shepherding effect has been observed. This experiment also shows that for the length of fiber that we used, i.e., 2 km long, the walk-off effect of this commercially available Corning DS fiber [7] is less than 1 ps/km.

V. Discussion and Future Research

The pictures shown in Fig. 4 clearly demonstrate not only the existence of an induced inverted pulse which can only come about because of the shepherding effect but also the growth of this induced pulse as the interaction distance grows longer as predicted by our computer simulation.

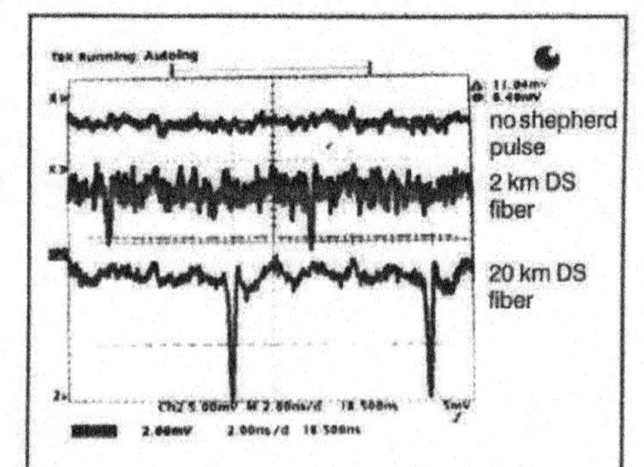

Fig. 4. Picture of the output of the cw primary source. The first line represents the output of the cw primary source signal without the presence of the shepherd pulse. The second line represents the output of the primary pulse with the presence of the shepherd pulse for a 2 km long Corning DS fiber. The third line represents the output of the primary pulse with the presence of the shepherd pulse for a 20 km long Corning DS fiber. A dip is seen indicating the successful interaction of the shepherd pulse with the primary signal.

That this shepherding effect is observable in a commercially available Corning SMF–DS fiber [7] (a dispersion shifted fiber) is worth noting. This means we are now in a position to perform further experiments corresponding to the cases produced by computer simulations without waiting for the production of an idealized fiber.

From a practical point of view, it is worth noting that this single fiber bit parallel wavelength link with shepherding pulse(s) can be used to extend greatly the (speed x distance) product of emerging cluster computer networks, such as, the MyriNet, SCI, Hippi–6400, ShuffleNet, etc. The distance may exceed many kilometers, a distance much beyond the capability of fiber ribbons.

Acknowledgements

The research described in this paper was performed by the Center for Space Microelectronics Technology, Jet Propulsion Laboratory, California Institute of Technology, and was sponsored by the Ballistic Missile Defense Organization, Office of Innovative Science and Technology, through an agreement with the National Aeronautics and Space Administration.

References

[1] L. A. Bergman, A. J. Mendez, and L. S. Lome, "Bit–parallel wavelength links for high performance computer networks", in SPIE Critical Review of Optical Science and Technology, *Optoelectronic Interconnects and Packaging*, edited by Ray T. Chen and Peter S. Cuilfoyle, vol. CR62, p.p. 210–226, (1996).

[2] L. A. Bergman and C. Yeh, "Dynamic alignment of pulses in bit-parallel wavelength links using a shepherd pulse in nonlinear fibers for massively parallel processing computer networks", Presented at the Third International Conference on Massively ParallelProcessingUsingOpticalInterconnections (MPPPOI'96), Maui, Hawaii, October 27–29, 1996

[3] G. P. Agrawal, *Nonlinear Fiber Optics*, Academic Press, New York (1989); J. R. Taylor, Ed., *Optical Solitons – Theory and Experiment*, Cambridge Studies in Modern Optics 10, Cambridge University Press, Cambridge (1992).

[4] C. Yeh and L. A. Bergman, J. Appl. Phys. 80, 3174 (1996).

[5] C. Yeh, W. P. Brown, and R. Szejn, Appl. Opt. 18, 489 (1979).

[6] C. Yeh, J. E. Pearson, and W. P. Brown, Appl. Opt. 15, 2913 (1976).

[7] "Single-mode dispersion", MM26, Opto-Electronics Group, Corning Inc., Corning, NY 14831, (1/96)

PHYSICAL REVIEW E VOLUME 57, NUMBER 2 FEBRUARY 1998

Enhanced pulse compression in a nonlinear fiber by a wavelength division multiplexed optical pulse

C. Yeh and L. Bergman
Jet Propulsion Laboratory, California Institute of Technology, Pasadena, California 91109
(Received 2 July 1997)

A way to compress an optical pulse in a single-mode fiber is presented in this paper. By the use of the cross-phase modulation (CPM) effect caused by the nonlinearity of the optical fiber, a shepherd pulse propagating on a different wavelength beam in a wavelength division multiplexed single-mode fiber system can be used to enhance the pulse compression of a copropagating primary pulse. Although CPM will not cause energy to be exchanged among the beams, the pulse shapes on these beams can be altered significantly. For example, a 1-mW peak power 10-ps primary pulse on a given wavelength beam may be compressed by a factor of as much as 25 when a copropagating 10-ps shepherd pulse of peak power of 49 mW on a different wavelength beam is similarly compressed. Results of a systematic study on this effect are presented in this paper. Furthermore, even when the primary pulse on a given wavelength beam has a peak power of much less than 1 mW, it can still be compressed by the same compression factor as a copropagating shepherd pulse of peak power much larger than 1 mW on a different wavelength beam as it undergoes compression. Through CPM, copropagating pulses on separate beams appear to share the nonlinear effect induced on any one of the pulses on separate beams. [S1063-651X(98)12902-1]

PACS number(s): 42.81.Dp

I. INTRODUCTION

In spite of the intrinsically small value of the nonlinearity coefficient in fused silica, due to low loss and long interaction length, the nonlinear effects in optical fibers made with fused silica cannot be ignored even at relatively low power levels [1]. This nonlinear phenomenon in fibers has been used successfully to generate optical solitons [2], to compress optical pulses [3], to transfer energy from a pump wave to a Stokes wave through the Raman gain effect [4], to transfer energy from a pump wave to a counterpropagating Stokes wave through the Brillouin gain effect [5], to produce four-wave mixing [6], and to dynamically shepherd pulses [7].

In a wavelength division multiplexed (WDM) system, the cross-phase modulation (CPM) effects [8,9] caused by the nonlinearity of the optical fiber are unavoidable. These CPM effects occur when two or more optical beams copropagate simultaneously, and effect each other through the intensity dependence of the refractive index. This CPM phenomenon can be used to produce an interesting pulse shepherding effect to align the arrival time of pulses which are otherwise misaligned. This same CPM effect can also be used to produce a highly compressed pulse on a different wavelength beam.

The usual soliton-effect compressor [3,10–13], which makes use of higher-order solitons supported by fiber as a result of interplay between self-phase modulation (SPM) and anomalous group-velocity dispersion (GVD), is well known. It is found here that the interplay between CPM and GVD may also provide similar pulse compression effects. The significant difference is that pulse compression can take place for pulses on a different wavelength beam. This means that the high power pulse on one wavelength beam may be used to provide high compression to a low power pulse on another wavelength beam. The purpose of this paper is to provide detailed simulation results on this type of pulse compression technique.

II. FORMULATION OF THE PROBLEM

The fundamental equations governing M numbers of copropagating waves in a nonlinear fiber including the CPM phenomenon are the coupled nonlinear Schrödinger equations [7,14]

$$\frac{\partial A_j}{\partial z}+\frac{1}{v_{gj}}\frac{\partial A_j}{\partial t}+\frac{1}{2}\alpha_j A_j$$

$$=\frac{i}{2}\beta_2\frac{\partial^2 A_j}{\partial t^2}-\gamma_j\left(|A_j|^2+2\sum_{m\neq j}^{M}|A_m|^2\right)A_j$$

$$(j=1,2,3,\ldots,M) \tag{1}$$

Here, for the jth wave, $A_j(z,t)$ is the slowly varying amplitude of the wave, v_{gj} the group velocity, β_{2j} the dispersion coefficient $(\beta_{2j}=dv_{gj}^{-1}/d\omega)$, α_j the absorption coefficient, and

$$\gamma_j=\frac{n_2\omega_j}{cA_{\rm eff}} \tag{2}$$

is the nonlinear index coefficient, with $A_{\rm eff}$ as the effective core area and $n_2=3.2\times10^{-16}\ {\rm cm}^2/{\rm W}$ for silica fibers, ω_j is the carrier frequency of the jth wave, c is the speed of light, and z is the direction of propagation along the fiber.

Introducing the normalizing coefficients

$$\tau = \frac{t-(z/v_{g1})}{T_0},$$

$$d_{1j} = (v_{g1} - v_{gj})/v_{g1}v_{gj},$$

$$\xi = z/L_{D1},$$

$$L_{D1} = T_0^2/|\beta_{21}|, \quad (3)$$

and setting

$$u_j(\tau,\xi) = (A_j(z,t)/\sqrt{P_{0j}})\exp(\alpha_j L_{D1}\xi/2), \quad (4)$$

$$L_{NLj} = 1/(\gamma_j P_{0j}),$$

$$L_{Dj} = T_0^2/|\beta_{2j}| \quad (5)$$

gives

$$i\frac{\partial u_j}{\partial \xi} = \frac{\text{sgn}(\beta_{2j})L_{D1}}{2L_{Dj}}\frac{\partial^2 u_j}{\partial \tau^2} - i\frac{d_{1j}}{T_0}L_{D1}\frac{\partial u_j}{\partial \tau} - \frac{L_{D1}}{L_{NLj}}\left[\exp(-\alpha_j L_{D1}\xi)|u_j|^2 + 2\sum_{m\neq j}^{M}\exp(-\alpha_m L_{D1}\xi)|u_m|^2\right]u_j$$

$$(j = 1,2,3,\ldots,M). \quad (6)$$

Here T_0 is the pulse width, P_{0j} is the incident optical power of the jth beam, and d_{1j}, the walk-off parameter between beam 1 and beam j, describes how fast a given pulse in beam j passes through the pulse in beam 1. In other words, the walk-off length is

$$L_{W(1j)} = T_0/|d_{1j}|. \quad (7)$$

So $L_{W(1j)}$ is the distance for which the faster moving pulse (say, in beam j) completely walked through the slower moving pulse in beam 1. The nonlinear interaction between these two optical pulses ceases to occur after a distance $L_{W(1j)}$. For CPM to take effect significantly, the group-velocity mismatch must be held to near zero.

It is also noted from Eq. (6) that the summation term in the bracket representing the CPM effect is twice as effective as the SPM effect for the same intensity. This means that the nonlinear effect of the fiber medium on a beam may be enhanced by the copropagation of another beam with the same group velocity.

III. NUMERICAL SOLUTION

Equation (6) is a set of simultaneous coupled nonlinear Schrödinger equations which may be solved numerically by the split-step Fourier method, which was used successfully earlier to solve the problem of beam propagation in complex fiber structures, such as the fiber couplers [15], and to solve the thermal blooming problem for high-energy laser beams [16]. According to this method, the solutions may be advanced first using only the nonlinear part of the equations. Then, the solutions are allowed to advance using only the linear part of Eq. (6). This forward stepping process is repeated over and over again until the desired destination is reached. The Fourier transform is accomplished numerically via the well-known fast Fourier transform technique. Due to the large dynamic range of the pulse width, a mesh size of 2048 with $\Delta\tau = 0.01$ was used.

Using the above approach, the evolution of all the pulses on all the copropagating WDM beams as they propagate down the fiber may be obtained. It was through these numerical computations that we discovered the interesting pulse shepherding and beam compression effects [7]. As expected, these effects only exist when group-velocity mismatch for the interested beams is negligible. In other words, there is no walk-off [7,14] among the interested beams. This can be accomplished through proper tailoring of the dispersion characteristics of a single-mode fiber [17].

Now consider the evolution of two single soliton pulses on two copropagating beams whose operating wavelengths are separated by $\Delta\lambda > 4$ nm. For this case, the four wave mixing effect is negligible. Let us label the first pulse as the primary (P) pulse and the second pulse as the shepherd (S) pulse. The soliton number N_j for the pulse on the jth beam is defined as

$$N_j^2 = L_{Dj}/L_{NLj}.$$

Furthermore, we assume that there is negligible walk-off, i.e.,

$$d_{1j} = (\text{walk-off parameter between beam No. 1 and beam } j) = v_{g1} - v_{gj} = 0,$$

and there is no loss, i.e.,

$$\alpha_j = (\text{attenuation or absorption of beam } j \text{ in fiber}) = 0.$$

The neglect of fiber loss is justified since the fiber lengths typically employed are only a small fraction of the absorption length ($\alpha_j L \ll 1$). Strictly speaking, for multiple interacting beams, there is no condition under which solitons may exist even if the fiber is lossless. However, numerical simulation shows that significant pulse compression still exists for these interacting pulses.

IV. DISCUSSION OF THE RESULTS

(i) *Shepherd and primary pulses are all in the anomalous dispersion region.* For solitons propagating on a single beam in silica fibers, pulse compression is experienced when N, the soliton order, is larger than 1 [8]. This effect is due to the interaction of self-phase modulation and anomalous group-velocity dispersion during propagation. When two aligned pulses, one called the primary pulse and the other called the shepherd pulse, on two different wavelength beams copropagate in a single-mode silica fiber, compression of both pulses occurs due to the interaction of cross-phase modulation of these two pulses and anomalous GVD during propagation.

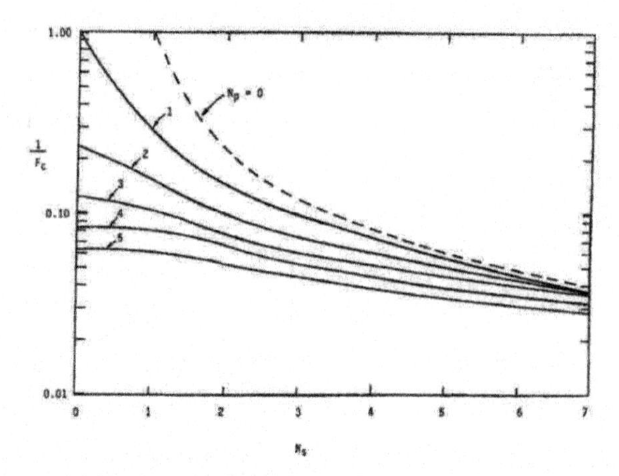

FIG. 1. Compression factor (F_c) for various soliton values (N_p) of a primary pulse (P) as a function of soliton values (N_s) of a copropagating shepherd pulse (S). The compression factor for the primary pulse is the same as the compression factor for the shepherd pulse. The initial pulse width for the primary pulse and that for the shepherd pulse are identical. The compression factor F_c is defined as the ratio between the full width at half maximum for the initial uncompressed pulse and that for the final compressed pulse.

A. Initial pulse widths are identical

Computer simulation results are shown in Figs. 1–4 for copropagating pulses with identical initial pulse width. Both pulses are in the anomalous GVD regime. In Fig. 1 the maximum amount of compression experienced by both pulses, the primary (P) pulse and the shepherd (S) pulse, are plotted against the soliton order N_s for the shepherd pulse for various cases of the primary pulse with the soliton order N_p. The amount of compression is expressed by the compression factor F_c, which is defined as [3]

$$F_c = T_{\mathrm{FWHM}} / T_{\mathrm{COMP}},$$

where the subscript FWHM means the full width at half

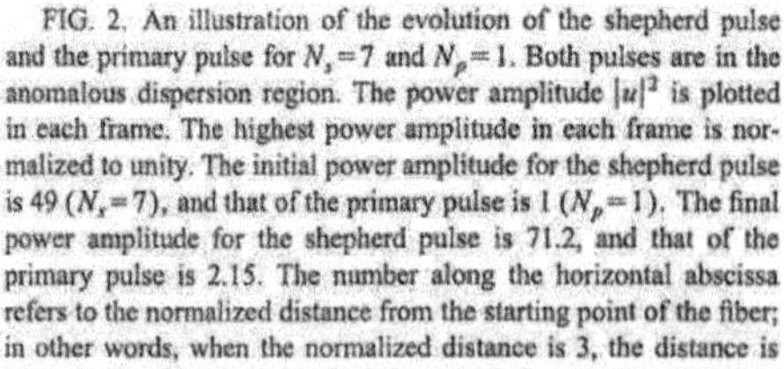

FIG. 2. An illustration of the evolution of the shepherd pulse and the primary pulse for $N_s = 7$ and $N_p = 1$. Both pulses are in the anomalous dispersion region. The power amplitude $|u|^2$ is plotted in each frame. The highest power amplitude in each frame is normalized to unity. The initial power amplitude for the shepherd pulse is 49 ($N_s = 7$), and that of the primary pulse is 1 ($N_p = 1$). The final power amplitude for the shepherd pulse is 71.2, and that of the primary pulse is 2.15. The number along the horizontal abscissa refers to the normalized distance from the starting point of the fiber; in other words, when the normalized distance is 3, the distance is $3(z_{\mathrm{opt}}/z_0)z_0/5$, where $z_0 = (\pi/2)L_{Ds}$, and L_{Ds} is the dispersion length of the shepherd pulse. z_{opt} is the optimum fiber length in km for the shepherd pulse when it experiences maximum pulse compression. Note that both pulses with different initial soliton numbers are similarly compressed, and the degree of compression for both pulses is higher than that experienced by each pulse when propagating alone. The dispersion coefficients β_{2s} and β_{2p} have units of (ps^2/km). All other numbers in the figure are dimensionless.

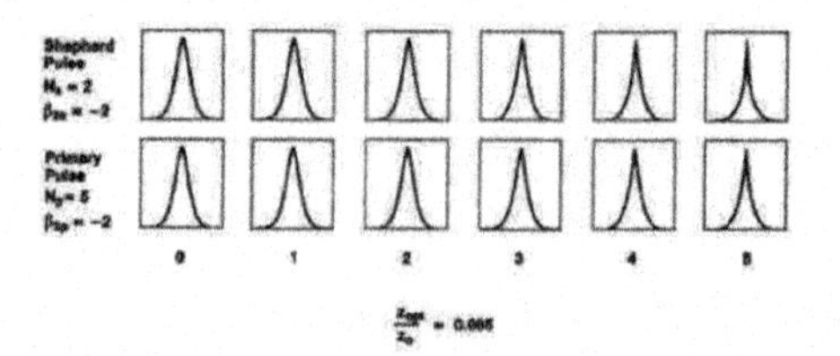

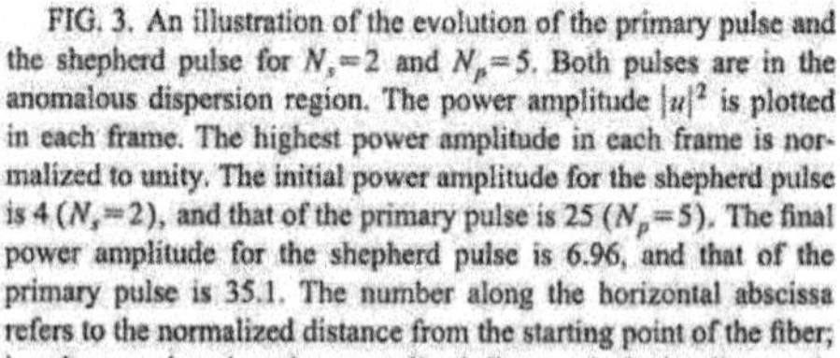

FIG. 3. An illustration of the evolution of the primary pulse and the shepherd pulse for $N_s = 2$ and $N_p = 5$. Both pulses are in the anomalous dispersion region. The power amplitude $|u|^2$ is plotted in each frame. The highest power amplitude in each frame is normalized to unity. The initial power amplitude for the shepherd pulse is 4 ($N_s = 2$), and that of the primary pulse is 25 ($N_p = 5$). The final power amplitude for the shepherd pulse is 6.96, and that of the primary pulse is 35.1. The number along the horizontal abscissa refers to the normalized distance from the starting point of the fiber; in other words, when the normalized distance is 3, the distance is $3(z_{\mathrm{opt}}/z_0)z_0/5$, where $z_0 = (\pi/2)L_{Ds}$, and L_{Ds} is the dispersion length of the shepherd pulse. z_{opt} is the optimum fiber length in km for the shepherd pulse when it experiences maximum pulse compression. Note that both pulses with different initial soliton numbers are similarly compressed, and that the degree of compression for both pulses is higher than that experienced by each pulse when propagating alone. The dispersion coefficients β_{2s} and β_{2p} have units of (ps^2/km). All other numbers in the figure are dimensionless.

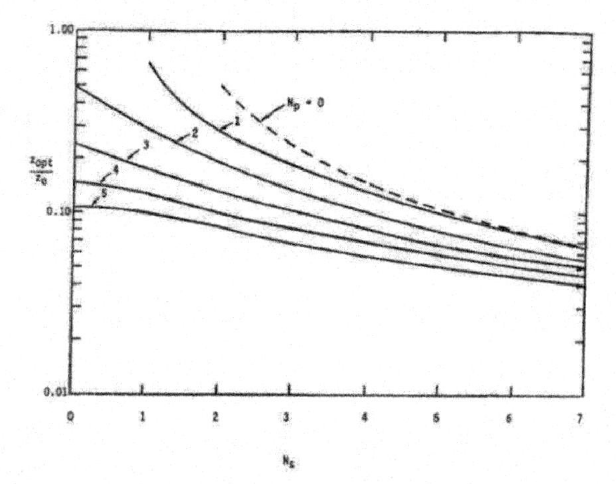

FIG. 4. Normalized optimum fiber length as a function of N_s for various fixed values of N_p. $z_0=(\pi/2)L_{Ds}$, and L_{Ds} is the dispersion length of the shepherd pulse. z_{opt} is the optimum fiber length in km for the shepherd pulse when it experiences maximum pulse compression.

maximum of the pulse and the subscript COMP means the FWHM of the compressed pulse. It is seen that, in the absence of the shepherd pulse, i.e., $N_s=0$, the primary pulse undergoes the well-known soliton compression process for a single soliton pulse for soliton number $N>1$. As expected, the primary pulse retains its shape when $N_p=1$. But, when a copropagating shepherd pulse is present, both pulses undergo the same compression even if N_p is not equal to N_s or if $N_s<1$ or if $N_s\ll 1$. Furthermore, the amount of compression is always larger than that achievable by a single stand-alone pulse.

For $N_s>N_p$, the shepherd pulse helps to compress the primary pulse further, especially when soliton number for the primary pulse is near unity. For example, as N_s varies from 1 to 7, the pulse width of the $N_p=1$ primary pulse can be compressed by the shepherd pulse by a factor of 27, while the pulse width of the $N_p=2$ primary pulse will be compressed by a factor of 7. For an $N_p=5$ primary pulse, its pulse width will be reduced by a factor of only 2.2 as N_s varies from 1 to 7. In other words, the weaker the intensity of the primary pulse the more its pulse width will be compressed by the presence of a copropagating high intensity shepherd pulse. Figure 2 gives an illustration of the evolution of the pulse shapes of the primary and shepherd pulses for the case where $N_s=7$ and $N_p=1$.

For $N_s<N_p$, the shepherd pulse still helps to compress the primary pulse further, but the effect is much more moderate. For example, as N_s varies from 0 to 2, the pulse width of the $N_p=2$ primary pulse is compressed by a factor of 2.4, while the pulse width of the $N_p=5$ primary pulse will be compressed by a factor of only 2 as N_s varies from 0 to 5. This means that to effectively enhance the pulse compression of a primary pulse, a higher intensity shepherd pulse must be used. Figure 3 shows the evolution of the pulse shapes of the primary and shepherd pulses for the case where $N_s=2$ and $N_p=5$.

It is known that a single pulse with $N<1$, no pulse compression will occur. Hence a $N_p<1$ primary pulse traveling alone, or a $N_s<1$ shepherd pulse traveling alone, will not experience any pulse compression. This is no longer true when these pulses copropagate in the fiber. Even when $N_p+N_s<1$, a slight pulse compression may still be observed for both the primary and secondary pulses. This is caused by the nonlinearity of the fiber medium. One also notes that when $N_p\ll 1$ and $N_s>1$, pulse compression will be experienced by both the primary and shepherd pulses. The same degree of pulse compression will occur on the primary pulse even when $N_p\ll 1$. The degree of pulse compression for the primary or shepherd pulse is governed by the $N_s>1$ shepherd pulse.

Figure 4 shows the normalized optimum fiber length z_{opt}/z_{0p} for the primary pulse as a function of N_s for various fixed values of N_p, where z_{opt} is the optimum fiber length in km for the primary or shepherd pulse when it experiences maximum pulse compression and $z_{0p}=(\pi/2)L_{Dp}$. Here L_{Dp} is dispersion length for the primary pulse defined in Eq. (5). It is of interest to note that z_{opt} for the primary pulse occurs at the same location or very near the same location as that for the shepherd pulse. This means that the maximum pulse compression for the primary pulse and that for the shepherd pulse occur at the same location and at the same time. For high values of N_s, this normalized optimum fiber length can be much smaller than unity, indicating that the maximum pulse compression could occur at a length many times smaller than the dispersion length. Using, as an example, the physical parameters

$$\beta_2=(\text{dispersion coefficient})=-2.0\ \text{ps}^2/\text{km},$$

$$\lambda_1=(\text{operating wavelength of beam No. 1})=1.552\ \mu\text{m},$$

$$\lambda_2=(\text{operating wavelength of beam No. 2})=1.548\ \mu\text{m},$$

γ=(nonlinear index coefficient)=20 W^{-1} km^{-1}

P_0=(incident power of each beam)=1 mW,

α=(attenuation or absorption of each beam in fiber)

=0 dB/km,

v_g=(group velocity of the beam=2.051 147×10^8 m/s),

d_{1j}=(walk-off parameter between beam No. 1 and beam j)

$=v_{g1}-v_{gj}=0$ (no walk-off),

T_0=(pulse width)=10 ps,

one has

$$L_{Dp}=50 \text{ km}.$$

Take the case of $N_p=5$ and $N_s=7$, one finds $z_{opt}/z_{0p}=0.04$ from Fig. 4. This means that maximum pulse compression can occur in a fiber with length of only 2.0 km long. For higher values of N_p and/or N_s, this length can be made even shorter.

B. Initial pulse widths are not identical

We also investigated the case where the pulse width of the primary pulse and that of the shepherd pulse are not identical. Let us consider the case where a primary pulse has an initial intensity of $N_p=1$, and a shepherd pulse has an initial intensity of $N_s=9$. It was assumed that the pulse width of the shepherd pulse is varied from the same to several times (3–5 times) wider than that of the primary pulse. Our computer simulation shows that the primary pulse is similarly compressed for all the above cases. In other words, varying the pulse width of the shepherd pulse does not appear to affect the minimum pulse width achievable for the primary pulse, although the distance required to gain this minimum pulse width for the primary is increased as the pulse width of the shepherd pulse is increased. The amount of pulse compression for the primary pulse is governed by the intensity of the accompanying shepherd pulse. It is observed that, for the broad shepherd pulse, only the central portion of the shepherd pulse that overlaps the primary pulse is significantly affected and undergoes compression.

This simulation shows that the broader shepherd pulse with high intensity appears to enhance (or increase) the strength of the nonlinear coefficient of the fiber medium for the primary pulse, so as to enhance the pulse compression effect experienced by the primary pulse. This means that there is a way to increase the nonlinear effect of the medium dynamically through the addition of a broad, high intensity shepherd pulse. The amount of enhancement and the duration are controlled by the intensity and the pulse width of the shepherd pulse. The nonlinear effect of the medium is transferred to the primary pulse through the CPM effect.

Let us now investigate the case where the intensity of the narrow shepherd pulse is much higher than that of the broad primary pulse. In this simulation, the initial intensity of the narrow shepherd pulse is taken to be $N_s=9$, and that of the broad primary pulse is $N_p<1$. Both pulses undergo compression. The degree of compression is mostly governed by the high intensity narrow shepherd pulse. For example, at the maximum compression distance, the shepherd pulse is compressed by a factor of approximately 16, while a narrow pulse with the same compressed pulse width as that of the shepherd pulse appears to have been generated on top of the broad small intensity primary pulse which appears as the pedestal for the narrow pulse.

It is noted here that what has been described above has practical significance. This scheme provides a practical pure optical way of generating very narrow bits on different wavelength streams for the bit-parallel data format.

(ii) *The shepherd pulse is in the normal dispersion region and the primary pulse is in the anomalous dispersion regime.* It is known that pulse compression of a single pulse in a fiber occurs because of the interaction of the nonlinear effect and the anomalous GVD effect [8]. This interaction also gives birth to the possible existence of a soliton pulse with $N=1$. The above simulation results show that when a shepherd pulse is added as a copropagating companion primary pulse, enhancement of pulse compression of the primary pulse is observed. It is of interest to learn if this pulse compression enhancement of the primary pulse still exists if the shepherd pulse is launched on a beam whose wavelength falls in the normal GVD regime. This computer experiment has been carried out. In this experiment, N_p is set to unity with $\beta_{2p}=-2$, while N_s is set to 9 with $\beta_{2s}=+2$. It is expected that without the shepherd pulse, the primary pulse is a soliton pulse which will retain its shape without pulse compression or pulse spreading as it propagates down the fiber. Also, without the primary pulse, the high amplitude shepherd pulse in the normal dispersion regime is expected to propagate without experiencing pulse compression. When both of these pulses copropagate on two separate beams, the pulse shepherding effect is observed, but no pulse compression is observed.

If N_p and N_s are both set equal to 9, the high amplitude of the primary pulse in the anomalous dispersion regime produces large pulse compression, but the degree of pulse compression (i.e., the narrowness of the compressed pulse) is not influenced by the presence of the high amplitude shepherd pulse in the normal dispersion regime. On the other hand, a very significant dip appears in the center of the shepherd pulse in the normal dispersion regime, breaking the original single shepherd pulse into two pulses. This is very different than the case where both primary and shepherd pulse are in the anomalous dispersion region. There both pulses undergo compression.

(iii) *The shepherd pulse and primary pulses are all in the normal dispersion region.* When both shepherd and primary pulses are in the normal dispersion region, no pulse compression occurs. Pulses tend to congregate toward the region of higher induced index of refraction.

Summary of the above discussion. The interaction between two separate pulses copropagating on two different wavelength beams in a single-mode fiber is studied in detail. It is shown that the cross-phase modulation effect can be used effectively to provide another way to generate pulse compression in the anomalous disper-

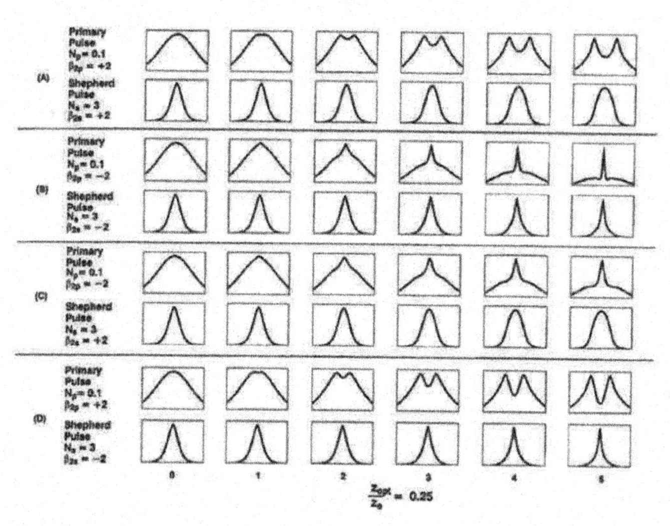

FIG. 5. Evolution of two propagating pulses in various different dispersion regions. The initial pulse amplitude of the primary pulse (pulse 1) is $N_p=0.1$, and the initial pulse amplitude of the shepherd pulse (pulse 2) is $N_s=3$. The initial pulse width of primary pulse (pulse 1) is three times the initial pulse width of the shepherd pulse (pulse 2). (A) Primary pulse 1 and shepherd pulse 2 are both in the normal dispersion region ($\beta_2=+2$). (B) Primary pulse 1 and shepherd pulse 2 are both in the anomalous dispersion region ($\beta_2=-2$). (C) Primary pulse 1 is in the anomalous dispersion region ($\beta_2=-2$), and shepherd pulse 2 is in the normal dispersion region ($\beta_2=+2$). (D) Primary pulse 1 is in the normal dispersion region ($\beta_2=+2$), and shepherd pulse 2 is in the anomalous dispersion region ($\beta_2=-2$). The power amplitude $|u|^2$ is plotted in each frame. The highest power amplitude in each frame is normalized to unity. The initial power amplitude for the shepherd pulse is 9 ($N_s=3$), and that of the primary pulse is 0.01 ($N_p=0.1$). The final power amplitudes for the shepherd pulse are (A)=6.59, (B)=14.8, (C)=6.59, and (D)=14.8 and those of the primary pulse are (A)=0.0116, (B)=0.0317, (C)=0.0195, and (D)=0.0121. The number along the horizontal abscissa refers to the normalized distance from the starting point of the fiber; in other words, when the normalized distance is 3, the distance is $3(z_{opt}/z_0)z_0/5$, where $z_0=(\pi/2)L_{Ds}$, and L_{Ds} is the dispersion length of the shepherd pulse. z_{opt} is the optimum fiber length in km for the shepherd pulse when it experiences maximum pulse compression. The dispersion coefficients β_{2s} and β_{2p} have units of (ps^2/km). All other numbers in the figure are dimensionless.

sion region of a single-mode fiber. Due to the nonlinearity of the fiber medium, a slight pulse compression still occurs when the sum of the soliton numbers for the two beams is less than unity.

A more complex interaction is observed when one of the pulses is propagating in the normal dispersion region. The pulse in the normal dispersion region is seen to be broken up by the compression of the high soliton number pulse in the anomalous dispersion region. It also appears that if the pulse in the normal dispersion region is very broad compared with the high intensity narrow pulse in the anomalous dispersion region, a dark solitonlike pulse can be generated on top of the broad pulse in the normal dispersion region, while the pulse in the anomalous dispersion region undergoes the usual pulse compression. Figure 5 is introduced to illustrate the evolution of the two propagating pulses when they exist in various different combinations of the dispersion regions.

It should be noted that the dispersion region in which the beam resides (i.e., where the beam wavelength resides) is all important in determining the behavior of the pulse on that beam even in the presence of a copropagating pulse on a different wavelength beam. The copropagating shepherd pulse, through the cross-phase modulation effect due to the Kerr index nonlinearity, provides an additional phase retardation to the primary pulse as it travels down the fiber. In other words, an additional frequency chirp (in addition to that caused by self-phase modulation) is added to the primary pulse by the copropagating shepherd pulse.

This "chirped" primary pulse is acted upon by the fiber's dispersion to yield the expected behavior. For example, if the primary pulse is on a beam whose wavelength is in the anomalous dispersion region (negative GVD region) and if the chirp caused by self- and cross-modulation effects is high enough, the leading half of the pulse containing the lowered frequencies will be retarded, while the trailing half, containing the higher frequencies, will be advanced, and the primary pulse will tend to collapse upon itself resulting in pulse narrowing or pulse compression [see Figs. 5(B) and 5(C)].

On the other hand, if the primary pulse is on a beam whose wavelength is in the normal dispersion region (positive GVD region), the presence of a copropagating shepherd pulse on a different wavelength beam induces a dark-soliton-like behavior for the primary pulse, confirming the fact that the dispersive region in which the wavelength of the beam resides determines the propagation characteristic of that pulse. In contrast with the bright soliton case, a dark soliton possesses a nontrivial phase profile which is a function of time, resulting in a rapid dip in the intensity of a broad pulse [see Figs. 5(A) and 5(D)].

An investigation was also carried out for the interaction of pulses on more than two beams. As many as ten simultaneously propagating pulses on ten separate beams, with one carrying the shepherd pulse, were used. It was found that a single large amplitude shepherd pulse could similarly and simultaneously affect the other nine small amplitude pulses. The evolution of each of the small amplitude pulses depended mainly on the interaction of that pulse with the large amplitude shepherd pulse according to the manner discussed above for the two beam interaction case. Through CPM, copropagating pulses on separate beams appear to share the nonlinear effect induced on any one of the pulses on separate beams.

This investigation shows that for a wavelength division multiplexed (WDM) system, one shepherd pulse can cause the compression of all the other wavelength pulses, thereby improving their pulse widths as well as the separation of different pulses. Furthermore, since the longer wavelength pulses are compressed at rate different from the shorter wavelength pulses, one may conceivably give all pulses the same time width, which may make detection and discrimination easier to accomplish.

V. CONCLUSION

A way to compress bright or dark pulse is found. The nonlinear cross-phase modulation (CPM) effect is used to accomplish this on two or more copropagating pulses on two or more wavelength division multiplexed (WDM) beams in a single-mode fiber. Numerical simulation shows that the effectiveness of compression is similar to that displayed by a single higher-order soliton pulse propagating in a single beam. That this CPM effect can be used to compress pulses whose amplitudes are much less than unity (the traditional soliton number for a single beam) as long as a copropagating pulse on a WDM beam undergoes compression, should be noted.

ACKNOWLEDGMENTS

The research described in this paper was performed by the Center for Space Microelectronics Technology, Jet Propulsion Laboratory, California Institute of Technology, and was sponsored by the Ballistic Missile Defense Organization, Office of Innovative Science and Technology through an agreement with the National Aeronautics and Space Administration.

[1] E. P. Ippen, in *Laser Applications to Optics and Spectroscopy*, edited by S. F. Jacobs, M. Sargent III, J. F. Scott, and M. O. Scally (Addison-Wesley, Reading, MA 1975), Vol. 2, Chap. 6.

[2] A. Hasegawa and F. D. Tappert, Appl. Phys. Lett. **23**, 142 (1973); L. F. Mollenauer, R. H. Stolen, and J. P. Gordon, Phys. Rev. Lett. **45**, 1095 (1980).

[3] L. F. Mollenauer, R. H. Stolen, J. P. Gordon, and W. J. Tomlinson, Opt. Lett. **8**, 289 (1983).

[4] R. H. Stolen and E. P. Ippen, Appl. Phys. Lett. **22**, 276 (1973); V. A. Vysloukh and V. N. Serkin, Pisma Zh. Eksp. Teor. Fiz. **38**, 170 (1983) [JETP Lett. **38**, 199 (1983)]; E. M. Dianov, A. Ya. Karasik, P. V. Mamyshev, A. M. Prokhorov, M. F. Stel'makh, and A. A. Fomichev, *ibid.* **41**, 294 (1985) [*ibid.* **41**, 294 (1985)].

[5] E. P. Ippen and R. H. Stolen, Appl. Phys. Lett. **21**, 539 (1972); N. A. Olsson and J. P. van der Ziel, *ibid.* **48**, 1329 (1986).

[6] R. H. Stolen, J. E. Bjorkholm, and A. Ashkin, Appl. Phys. Lett. **24**, 308 (1974).

[7] C. Yeh and L. Bergman, J. Appl. Phys. **80**, 3175 (1996).

[8] G. P. Agrawal, *Nonlinear Fiber Optics* (Academic, New York, 1989).

[9] *Optical Solitons—Theory and Experiment*, edited by J. R. Taylor, Cambridge Studies in Modern Optics Vol. 10 (Cambridge University Press, Cambridge, 1992).

[10] R. A. Fisher, P. L. Kelley, and T. K. Gustafson, Appl. Phys. Lett. **14**, 140 (1969).

[11] H. Nakatsuka, D. Grischkowsky, and A. C. Balant, Phys. Rev. Lett. **47**, 910 (1981); R. L. Fork, C. H. Brito Cruz, P. C. Becker, and C. V. Shank, Opt. Lett. **12**, 483 (1987).

[12] W. J. Tomlinson, R. H. Stolen, and C. V. Shank, J. Opt. Soc. Am. B **1**, 139 (1984); W. J. Tomlinson and W. H. Knox, *ibid.* **4**, 1404 (1987).

[13] D. Mestdagh, Appl. Opt. **26**, 5234 (1987).

[14] G. P. Agrawal, Phys. Rev. Lett. **59**, 880 (1987).

[15] C. Yeh, W. P. Brown, and R. Szejn, Appl. Opt. **18**, 489 (1979).

[16] C. Yeh, J. E. Pearson, and W. P. Brown, Appl. Opt. **15**, 2913 (1976).

[17] *Single-Mode Dispersion*, Report No. MM26, Opto-Electronics Group, Corning Inc., Corning, NY, 1996; L. G. Cohen, W. L. Mammel, and S. J. Jang, Electron. Lett. **18**, 1023 (1982); B. J. Ainslie and C. R. Day, J. Lightwave Technol. **LT-4**, 967 (1986).

JOURNAL OF LIGHTWAVE TECHNOLOGY, VOL. 16, NO. 9, SEPTEMBER 1998 1577

An All-Optical Long-Distance Multi-Gbytes/s Bit-Parallel WDM Single-Fiber Link

L. Bergman, J. Morookian, and C. Yeh, *Fellow, IEEE, Fellow, OSA*

Abstract—An all-optical long-distance (>30 km) bit-parallel wavelength division multiplexed (WDM) single-fiber link with 12 bit-parallel channels having 1 Gbyte/s capacity has been designed. That system functionally resembles an optical fiber ribbon cable, except that all the bits pass on one fiber-optic waveguide. This single-fiber bit parallel wavelength link can be used to extend the (speed $\times$ distance) product of emerging cluster computer networks, such as, the MyriNet, SCI, Hippi-6400, ShuffleNet, etc. Here, the detailed design of this link using the commercially available Corning DS (dispersion-shifted) fiber is given. To demonstrate the viability of this link, two WDM channels at wavelengths 1530 and 1545 nm carrying 1 ns pulses on each channel were sent through a single 25.2-km long Corning DS fiber. The walkoff was 200 ps, well within the allowable setup and hold time for the standard ECL logic which is 350 ps for a bit period of 1 ns. This result implies that 30 bit-parallel beams spaced 1 nm apart between 1530–1560 nm, each carrying 1 Gbits/s signal, can be sent through a 25.2-km Corning DS fiber carrying information at a 30 Gb/s rate.

***Index Terms*—Optical fiber communication, optical propagation in nonlinear media, optical pulses, optical solitons, optical waveguides, single-mode fiber, wavelength division multiplexing.**

I. Introduction

UNLIKE the usual wavelength division multiplexed (WDM) format where input parallel pulses are first converted into a series of single pulses which are then launched on different wavelength beams into a single-mode fiber, the bit-parallel (BP) WDM format was proposed [1], [2]. Under this BP-WDM format, no parallel to serial conversion of the input signal is necessary, parallel pulses are launched simultaneously on different wavelength beams. Time alignment of the pulses for a given signal byte is very important.

There exists a competing non-WDM approach to transmit parallel bits—the fiber optic ribbon approach—where parallel bits are sent through corresponding parallel fibers in a ribbon format. However, it is very difficult to maintain time alignment of the parallel pulses due to practical difficulty in manufacturing identical uniform fibers. Furthermore, it is known that computer vendors would like to apply the same technology to increase the bandwidth of campus network in support cluster computing, and to provide salable external I/O networks for clusters of massively parallel processor (MPP) supercomputers (i.e., multiple network channels connected to one machine). Cluster computing is expected to gain greater importance in the near future as users tap the latent unused computer cycles of company workstations (sometimes in off hours) to work on large problems, rather than buying a specific supercomputer. In DoD applications, it would enable high-performance computers to be deployed in embedded systems. For high-performance computing environments, clusters of MPP supercomputers can also be envisioned. This concept elevates the cluster computing model to a new level. In this case, not only is high bandwidth and low latency required, but now interchannel message synchronization also becomes important among the parallel network channels entering the machine—especially if all machines are tightly coupled together to work on one large problem. In the limit, the aggregate bandwidth required to interconnect two large MPP supercomputers approaches the bisection bandwidth of the internal communication network of the machine. For example, in a 2^n-node hypercube interconnected MPP architecture, there would be up to 2^{n-1} links between each half of the machine (e.g., 1024 processor nodes would have 512 links at 200 Mbytes/s per link, or 102 Gbytes/s total). The need for a single media parallel interconnect is apparent. Thus, the single-fiber WDM format of transmitting parallel bits rather than a fiber ribbon format may be the media of choice.

The purpose of this paper is, first, to present the detailed design of a long distance (32 km) all-optical bit-parallel WDM single-fiber link with 12 bit-parallel channels having 1 Gbyte/s capacity using available components and fiber. The speed-distance product for this link is 32 Gbytes/s-km while the maximum speed-distance product for fiber ribbon is less than 100 Mbytes/s-km.

Then, to demonstrate the viability of this link, two WDM channels at wavelengths 1530 and 1545 nm carrying 1 ns pulses on each channel were sent through a single 25.2-km long Corning DS fiber. The walkoff was 200 ps, well within the allowable setup and hold time for the standard ECL logic which is 350 ps for a bit period of 1 ns.

Manuscript received October 13, 1997. This work was supported by the Ballistic Missile Defense Organization, Office of Innovative Science and Technology, through an agreement with the National Aeronautics and Space Administration. This work was done by the Center for Space Microelectronics Technology, Jet Propulsion Laboratory, California Institute of Technology.

The authors are with the Jet Propulsion Laboratory, California Institute of Technology, Pasadena, CA 91109 USA.

Publisher Item Identifier S 0733-8724(98)05903-9.

II. A Review of the Theoretical Foundation

This section provides the theoretical foundation for the wave propagation of parallel pulses on different wavelength beams in a linear/nonlinear fiber.

The fundamental equations governing M numbers of co-propagating waves in a linear/nonlinear fiber including the

CPM phenomenon are the coupled nonlinear Schrodinger equations [3], [4]

$$\frac{\partial A_j}{\partial z} + \frac{1}{v_{gj}}\frac{\partial A_j}{\partial t} + \frac{1}{2}\alpha_j A_j = \frac{1}{2}\beta_{2j}\frac{\partial^2 A_j}{\partial t^2} - \gamma_j\left(|A_j|^2 + 2\sum_{m\neq j}^{M}|A_m|^2\right)A_j \quad (j = 1,2,3,\cdots M). \tag{1}$$

Here, for the jth wave, $A_j(z,t)$ is the slowly varying amplitude of the wave v_{gj}, the group velocity β_{2j}, the dispersion coefficient ($\beta_{2j} = dv_{gj}^{-1}/d\omega$), α_j, the absorption coefficient, and

$$\gamma_j = \frac{n_2\omega_j}{cA_{\text{eff}}} \tag{2}$$

is the nonlinear index coefficient with A_{eff} as the effective core area and $n_2 = 3.2 \times 10^{-16}$ cm^2/W for silica fibers, ω_j is the carrier frequency of the jth wave, c is the speed of light, and z is the direction of propagation along the fiber. (For a linear fiber, the nonlinear coefficient γ_j is zero resulting in the decoupling of copropagating beams, i.e., each beam propagates independently of all other copropagating beams.)

Introducing the normalizing coefficients

$$\begin{aligned}\tau &= \frac{t-(z/v_{g1})}{T_0}\\ d_{1j} &= (v_{g1}-v_{gj})/v_{g1}v_{gj}\\ \xi &= z/L_{D1}\\ L_{D1} &= T_0^2/|\beta_{21}|\end{aligned} \tag{3}$$

and setting

$$u_j(\tau,\xi) = (A_j(z,t)/\sqrt{P_{0j}})\exp(\alpha_j L_{D1}\xi/2) \tag{4}$$

$$\begin{aligned}L_{NLj} &= 1/(\gamma_j P_{0j})\\ L_{Dj} &= T_0^2/|\beta_{2j}|\end{aligned} \tag{5}$$

gives

$$i\frac{\partial u_j}{\partial \xi} = \frac{\text{sgn}(\beta_{2j})L_{D1}}{2L_{Dj}}\frac{\partial^2 u_j}{\partial \tau^2} - i\frac{d_{1j}}{T_0}L_{D1}\frac{\partial u_j}{\partial \tau} - \frac{L_{D1}}{L_{NLj}}\left[\exp(-\alpha_j L_{D1}\xi)|u_j|^2 + 2\sum_{m\neq j}^{M}\exp(-\alpha_m L_{D1}\xi)|u_m|^2\right]u_j \quad (j = 1,2,3,\cdots,M). \tag{6}$$

Here, T_0 is the pulse width, P_{0j} is the incident optical power of the jth beam, and d_{1j}, the walk-off parameter between beam 1 and beam j, describes how fast a given pulse in beam j passes through the pulse in beam 1. In other words, the walk-off length is

$$L_{W(1j)} = T_0/|d_{1j}|. \tag{7}$$

So, $L_{W(1j)}$ is the distance for which the faster moving pulse (say, in beam j) completely walked through the slower moving pulse in beam 1. The nonlinear interaction between these two optical pulses ceases to occur after a distance $L_{W(1j)}$. For cross-phase modulation (CPM) to take effect significantly, the group-velocity mismatch must be held to near zero.

It is also noted from (6) that the summation term in the bracket representing the cross-phase modulation (CPM) effect is twice as effective as the self phase modulation (SPM) effect for the same intensity. This means that the nonlinear effect of the fiber medium on a beam may be enhanced by the copropagation of another beam with the same group velocity.

Equation (6) is a set of simultaneous coupled nonlinear Schrodinger equations which may be solved numerically by the split-step Fourier method, which was used successfully earlier to solve the problem of beam propagation in complex fiber structures, such as, the fiber couplers [5], and to solve the thermal blooming problem for high-energy laser beams [6]. According to this method, the solutions may be advanced first using only the nonlinear part of the equations. And then the solutions are allowed to advance using only the linear part of (6). This forward stepping process is repeated over and over again until the desired destination is reached. The Fourier transform is accomplished numerically via the well-known fast Fourier transform Technique.

This nonlinear interaction of copropagating beams, for short, high-intensity pulses, is the subject of intense research. Some of the preliminary results have been published [4].

III. Elements of a 12-Bit Parallel WDM System

Let us now return to the design of our BP-WDM system. Due to the relatively broad pulsewidths (~1 ns) and low power levels of the data pulses, nonlinear interaction of copropagating pulses can be consider to be negligible [8]. It is expected that 12 separate beams will be used. Anticipating the use of erbium amplifier, beam separation among these 12 beams must be limited by the useful bandwidth of the erbium amplifier which is from 1535 to 1560 nm. Hence, separation between neighboring beams must be less than 25/12 = 2.08 or 2 nm. A block diagram of the link is shown in Fig. 1.

A. The Transmitter

The transmitter of the system consists of 12 discreet distributed-feedback laser diodes [9] and a 16-to-1 fiber coupler. Each laser element is selected to fall within the erbium gain bandwidth at a preselected $\Delta\lambda$ from its neighbors. To minimize system cost, the lasers are directly modulated with NRZ data at a rate up to 1 Gb/s each, for an aggregate of 1 Gbyte/s. The timing of the bits in any word are aligned at the input to the fiber link by adjusting the phase of the laser drive signal for each bit using conventional electrical delay components. The optical power coupled into the fiber arms at the input to the 16-to-1 coupler is about 0 dBm (i.e., 1 mW).

B. The Single-Mode Fiber

Corning DS fiber is chosen to be the single-mode fiber for this system because of its desirable dispersion characteristics [7]. The dispersion characteristics of this fiber is shown in Fig. 2. It is seen that for the wavelength range of interest (1535–1560 nm), the dispersion coefficient, $|\beta_2|$, is around 2

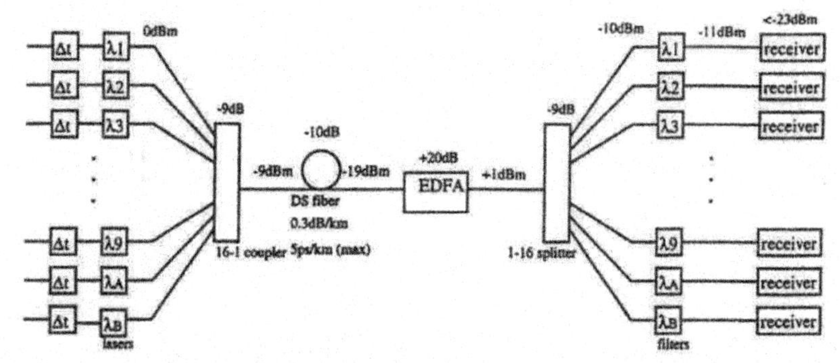

Fig. 1. Block diagram for an all-optical 12 channel bit-parallel WDM single-fiber system.

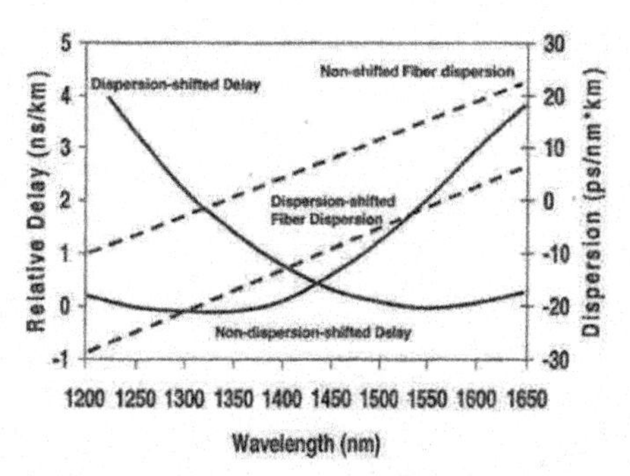

Fig. 2. Dispersion characteristics of Corning DS fiber.

ps^2/km. The difference of group velocities as a function of the wavelength of the beams have been measured and are displayed in Fig. 3. It is seen that the maximum difference in group velocity over the wavelength of interest is 5 ps/km. An erbium-doped fiber amplifier (EDFA) is used to boost the power at the receiver.

C. The Receiver

The receiver of the system consists of a 1-to-16 fiber splitter, 12 optical bandpass filters, and 12 fiber-optic receivers.

IV. Design Considerations

A. Wavelength Spacing Consideration

A 1 Gbyte/s, each bit path must have a minimum bandwidth of 2 GHz to reproduce the data. In estimating the spread of the optical spectrum of each laser element, a 4 GHz bandwidth will be assumed. The spread of each element's spectrum $\Delta\lambda$ is then 0.032 nm for a 4 GHz bandwidth which is well within the 2 nm beam separation between neighboring beams. It should be noted that any spectral broadening of the pulse due to chirp or other factors will be much less than the 2 nm beam separation that has been used for our system. Furthermore, the 2 nm beam separation also lessens the demand on the optical bandpass filters used to separate the WDM beams at the receiver end.

B. Skew and Walk-Off Consideration

At 1 Gb/s, the bit period is approximately 1 ns. For the worst case, the setup and hold time for standard ECL logic is 350 ps. This means that there is a leeway of (1000-350)/2 = 325 ps in which the pulses may drift away from each other. If one limits the skew or walkoff to half of 325 ps, then the maximum length of fiber which can be used is 160/5 = 32 km.

C. Loss Consideration

For a maximum length of 32 km, it is clear that an EDFA will be needed to increase the power at the receiver. As indicated in Fig. 1, a gain of 20 dB via the EDFA will provide a gain margin of more than 12 dB at the receiver.

V. Experimental Demonstration of a Two Wavelengths BP-WDM System

The experimental setup is shown in Fig. 4. Two beams from two laser diodes whose wavelengths are 1530 and 1545 nm, are modulated by nano-second size pulses. These beams whose spectral shapes are displayed in Fig. 5, are coupled simultaneously into a Corning DS fiber. A picture of the pulses on these two beams before they were launched into the fiber is shown in Fig. 6. It is seen that these nano second size pulses were well aligned at the entrance of the fiber link.

The spool of Corning DS fiber used for our experimental link was 25.2 km long. The output was displayed in Fig. 7. One can readily measure the shift or the walkoff between these pulses—it was 200 ps or 6 ps/km. This result is consistent with our previous measurement displayed in Fig. 3. There, the walkoff was measured between a tunable ring laser and a 1545-nm laser diode.

It is noted that the experimentally measured walkoff of 200 ps for this two wavelength BP-WDM demonstration is well within the allowable setup and hold time for the standard ECL logic which is 350 ps for a bit period of 1 ns.

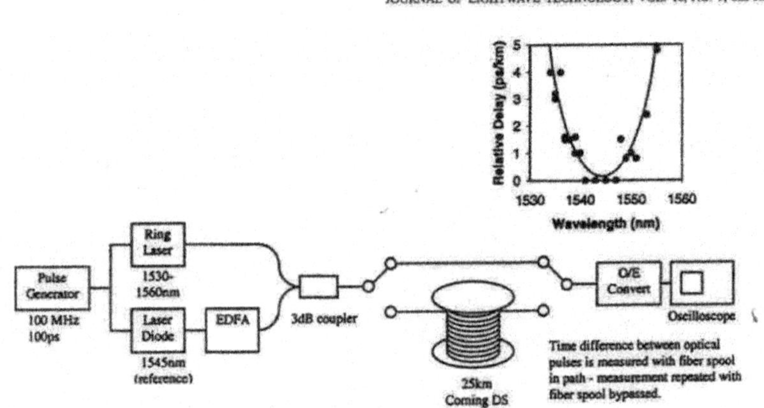

Fig. 3. Measured group velocity differences for different wavelength beams. The sources are a tunable ring laser and a DFB laser diode at 1545 nm.

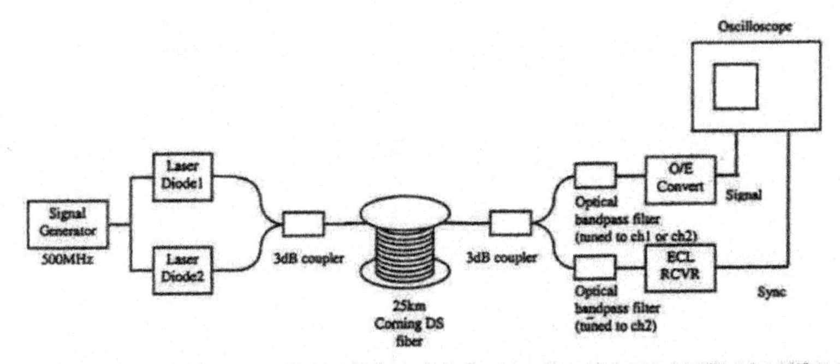

Fig. 4. The experimental setup for the measurement of bit-parallel nanosecond pulses propagating on two beams at 1530 and at 1545 nm in a 25.2-km long Corning DS fiber.

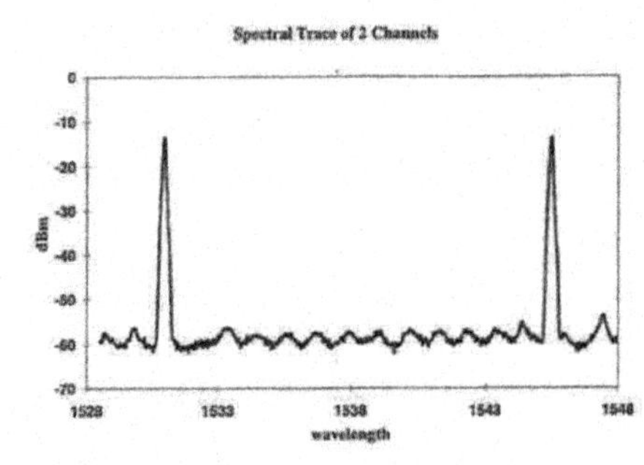

Fig. 5. Spectral shapes for the two copropagating beams.

VI. Conclusion

This paper shows that, using available components and fiber, one can design a 32-km BP-WDM single-fiber data link with 1 Gbyte/s capacity. This is an all-optical link with byte-wide optical path which can bypass any electrical bottleneck. A demonstration link can be readily built between JPL and Caltech for HIPPI 6400 using this design.

It was also shown through an actual experiment that nanosecond size pulses on two BP-WDM beams at 1530 and 1545 nm can be successfully transmitted through a 25.2-km long Corning DS fiber with acceptable walkoff which is well within the allowable setup and hold time of standard ECL logic circuits. As can be seen from Fig. 3 that the maximum walkoff between any beams located within the wavelength range of 1530 and 1560 nm is 200 ps. This result implies that 30 bit-parallel beams spaced 1 nm apart from 1530 to 1560 nm, each carrying 1 Gb/s signal, can be sent through a 25.2-km Corning DS fiber at an information rate of 30 Gb/s. This means that the speed-distance product for this link is about 94 Gbytes/s-km, a number way beyond the best that fiber ribbon can offer.

An all-optical adaptive bit alignment scheme for future ultrahigh-capacity BP-WDM link with 10-ps bit pulses is being studied using our newly developed shepherding pulse technique [4].

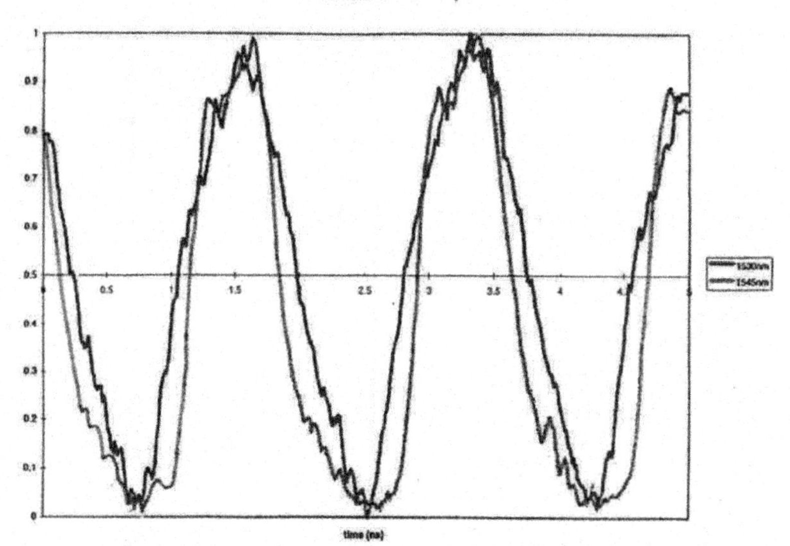

Fig. 6. A picture of the nanosecond pulses on the two copropagating beams before entering the fiber. These pulses are very well aligned at the entrance to the fiber link.

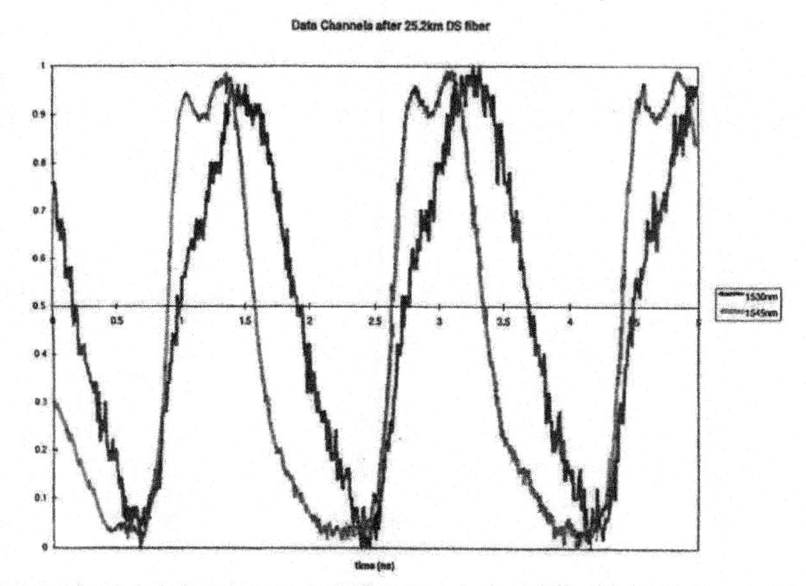

Fig. 7. A picture of the nanosecond pulses on the two copropagating beams after passing through 25.2 km of the fiber. The alignment of the two pulses is shifted 200 ps at the output of the fiber link. This shift represents the walkoff among these different wavelength beams.

ACKNOWLEDGMENT

The authors wish to thank Dr. L. Lome of BMDO for his encouragement and stimulating discussions.

REFERENCES

[1] L. A. Bergman, A. J. Mendez, and L. S. Lome, "Bit-parallel wavelength links for high performance computer networks," in *Proc. SPIE Critical Review of Optical Science and Technology, Optoelectronic Interconnects and Packaging*, Ray T. Chen and Peter S. Cuilfoyle, Eds., 1996, vol. CR62, pp. 210–226.

[2] L. A. Bergman and C. Yeh, "Dynamic alignment of pulses in bit-parallel wavelength links using a shepherd pulse in nonlinear fibers for massively parallel processing computer networks," presented at the Third Int. Conf. Massively Parallel Processing Using Optical Interconnections (MPPPOI'96), Maui, Hawaii, Oct. 27–29, 1996.

[3] G. P. Agrawal, *Nonlinear Fiber Optics*. New York: Academic, 1989; see also, J. R. Taylor, Ed., "Optical solitons—Theory and experiment," in *Cambridge Studies in Modern Optics 10*. Cambridge, U.K.: Cambridge University Press, 1992.

[4] C. Yeh and L. A. Bergman, "Pulse shepherding in nonlinear fiber optics," *J. Appl. Phys.*, vol. 80, p. 3174, 1996; see also, C. Yeh and L. A. Bergman, "Enchanced pulse compression in a nonlinear fiber by a wavelength division multiplexed optical pulse," *Phys. Rev. E*, vol. 57, pp. 2398–2404, 1998; C. Yeh, L. Bergman, J. Morookian, and S. Monacos, "Generation of time-aligned picosecond pulses on wavelength division multiplexed beams in a nonlinear fiber," *Phys. Rev. E*, vol. 57, pp. 6135–6139, May 1998.

[5] C. Yeh, W. P. Brown, and R. Szejn, "Multimode inhomogeneous fiber couplers'," *Appl. Opt.*, vol. 18, pp. 489–495, 1979.

[6] C. Yeh, J. E. Pearson, and W. P. Brown, "Enhanced focal-plane irradiance in the presence of thermal blooming," *Appl. Opt.*, vol. 15, pp. 2913–2916, 1976.

[7] Corning Inc., "Single-mode dispersion," *MM26, Opto-Electronics Group*. Corning, NY, Jan. 1996.

[8] A. E. Willner, L. G. Kazovsky, and S. Benedetto, "*Advanced Optical Communication Systems*. Norwood, MA: Artech House, 1996.

[9] L. Davis, M. G. Young, and S. Forouhar, "Wavelength control in mode-locked Lasers for WDM Applications (Invited Paper)," in *Proc. SPIE's Photon. West*, San Jose, CA, Jan. 1998; see also L. Davis, M. G. Young, and S. Forouhar, "Mode-locked lasers for WDM applications," presented at IEEE/LEOS Summer Topical Meeting on WDM Components Technol., Montreal, P.Q., Canada, Aug. 1997; see also, L. Davis, J. Singletery, M. G. Young, T. A. Vang, M. A. Mazed, and S. Forouhar, "20 GHz mode-locked lasers for WDM applications (Invited Paper)," in *Proc. SPIE's Photon. West*, San Jose, CA, Feb. 1997.

[10] G. Jeong and J. W. Goodman, *J. Lightwave Technol.*, vol. 14, p. 655, 1996.

L. Bergman received the M.S. degree from the California Institute of Technology (Caltech), Pasadena, and the Ph.D. degree from Chalmers University of Technology, Gothenburg, Sweden, both in electrical engineering.

For the past 25 years, he has worked at the Jet Propulsion Laboratory (JPL), Pasadena, CA, in the area of fiber-optic local area networks, terabit all-optical computer networks, fiber-optic sensors, and optical interconnections for computers. He has authored over 80 papers in the fields of fiber optics and high-speed communications, received four patents, and lectured at local universities. In 1984, he was awarded the first place prize at the European Conference on Optical Communication (ECOC), Stuttgart, Germany, for a new system approach that achieves 5 Gbits/s data rates for ring LAN's. In 1993, he earned the Technology and Applications Program (TAP) Directorate Exceptional Service Award for sustained research contributions to the fields of fiber-optic network and supercomputer optics, motion picture, and data processing industries as well as nearby universities in the areas of electrooptic system design, telecommunications, and computer science. He presently is the Deputy Manager of the Information and Computing Technologies Research Section, and is also the Project Engineer for the JPL Supercomputer Center.

Dr. Bergman is a member of Tau Beta Pi, Eta Kappa Nu, Phi Kappa Phi, and Sigma Xi.

J. Morookian received the B.S. degree in electrical engineering in 1991 from the University of Southern California (USC), Los Angeles.

He subsequently assumed a full-time position at the Jet Propulsion Laboratory (JPL), California Institute of Technology, Pasadena, in May 1991. During his four graduate years as a National Merit Scholar at USC, he worked at the Center for Laser Studies, furthering research in optical memory systems, power-by-light applications, and optical code division multiple access (CDMA). As a member of the High-Speed Optical Systems Group at JPL, he has worked on several optical networking projects, including an optical CDMA scheme; fempto-second laser pulse source; high-speed time division multiplexing (TDM), and high-speed wavelength division multiplexing (WDM). He has coauthored several papers on these subjects.

Mr. Morookian is a member of the USC Engineering Honors Group, Alpha Lambda Delta, and Tau Beta Pi.

C. Yeh (S'56–M'63–SM'82–F'85) received the B.S., M.S. and Ph.D. degrees in electrical engineering from the California Institute of Technology (Caltech), Pasadena, in 1957, 1958, and 1962, respectively.

In 1962, he joined the University of Southern California, Los Angeles, as an Assistant Professor of Electrical Engineering and became an Associate Professor in 1967. He moved to the University of California at Los Angeles (UCLA) in 1967 as an Associate Professor of Electrical Engineering and became a Professor in 1972. Throughout more than 30 years of his professional career, he was a consultant to many industrial companies, such as the Hughes Research Laboratories, The Dikewood Corporation, the Aerospace Corporation, etc. Starting in 1992, he left UCLA and has been a Consulting Engineer at the Jet Propulsion Laboratory, Pasadena, CA. He has published more than 120 papers in fiber optics and applied electromagnetic waves, where many of his publications are widely cited. Examples of his key research are propagation of wavelength division multiplexed soliton pulses in a nonlinear fiber, ceramic ribbon waveguide—an ultralow-loss (less than 5 dB/km) millimeter/submillimeter dielectric waveguide, a random-access protocol for unidirectional ultrahigh-speed (multigigabit rate) optical fiber network, single-mode optical waveguides by the vector finite-element method, propagation of optical waves in an arbitrarily shaped fiber, fiber couplers, or integrated optical circuit by the scalar beam propagation method, scattering of a single submicron particle by focused laser beams, scattering of electromagentic waves by arbitrarily shaped dielectric bodies, reflection and transmission of electromagnetic waves by a relativistically moving dielectric slab or halfspace, diffraction of waves by an elliptical or parabolic dielectric cylinder, and elliptical dielectric waveguides or optical fibers, etc.

Dr. Yeh is a member of Eta Kappa Nu and Sigma XI and a Fellow of the Optical Society of America (OSA).

PHYSICAL REVIEW E VOLUME 57, NUMBER 5 MAY 1998

Generation of time-aligned picosecond pulses on wavelength-division-multiplexed beams in a nonlinear fiber

C. Yeh, L. Bergman, J. Morookian, and S. Monacos
Jet Propulsion Laboratory, California Institute of Technology, Pasadena, California 91109
(Received 24 July 1997; revised manuscript received 30 January 1998)

A fundamentally different way to generate time-aligned data pulses on wavelength-division-multiplexed (WDM) beams in a single-mode fiber is found. A large-amplitude pulse called a shepherd pulse is launched on one of the copropagating beams (shepherd pulse is defined as a pulse that can affect other copropagating pulses in a WDM format while maintaining its own propagation behavior). Initially at the launching plane no other pulse exists on any of the other copropagating wavelength-division-multiplexed beams. Due to the nonlinear cross-phase modulation effect, time-aligned pulses are generated on all other beams after a given fiber length. Both theoretical and experimental results are presented. [S1063-651X(98)09105-3]

PACS number(s): 42.81.−i, 42.65.Re

I. INTRODUCTION

The difficulty in the generation of time-aligned pulses in the picosecond range on wavelength-division-multiplexed (WDM) beams is well recognized. Yet, these time-aligned pulses are the backbone for the future ultra-high-speed bit-parallel communication system [1,2]. A way to generate these pulses is described here.

In spite of the intrinsically small value of the nonlinearity coefficient in fused silica, due to low loss and long interaction length, the nonlinear effects in optical fibers made with fused silica cannot be ignored even at relatively low power levels [3]. This nonlinear phenomenon in fibers has been used successfully to generate optical solitons [4], to compress optical pulses [5], to transfer energy from a pump wave to a Stokes wave through the Raman gain effect [6], to transfer energy from a pump wave to a counterpropagating Stokes wave through the Brillouin gain effect [7], to produce four-wave mixing [8], to dynamically shepherd pulses [9], and to enhance pulse compression [10]. Now, we wish to add one more: the generation of time-aligned pulses.

In a wavelength-division-multiplexed fiber system, the cross-phase modulation (CPM) effects [11,12] caused by the nonlinearity of the optical fiber are unavoidable. These CPM effects occur when two or more optical beams copropagate simultaneously and affect each other through the intensity dependence of the refractive index. This CPM phenomenon is used to generate time-aligned data pulses.

II. THEORETICAL FOUNDATION

The fundamental equations governing M numbers of copropagating waves in a nonlinear fiber including the CPM phenomenon are the coupled nonlinear Schrödinger equations [9–11]:

$$i\frac{\partial u_j}{\partial \xi}=\frac{\operatorname{sgn}(\beta_{2j})L_{D1}}{2L_{Dj}}\frac{\partial^2 u_j}{\partial \tau^2}-i\frac{d_{1j}}{T_0}L_{D1}\frac{\partial u_j}{\partial \tau}-\frac{L_{D1}}{L_{NLj}}\left[\exp(-\alpha_j L_{D1}\xi)|u_j|^2+2\sum_{m\neq j}^{M}\exp(-\alpha_m L_{D1}\xi)|u_m|^2\right]u_j$$

$$(j=1,2,3,\ldots,M). \quad (1)$$

Here, for the jth wave, u_j is the normalized slowly varying amplitude of the wave, β_{2j} the dispersion coefficient ($\beta_{2j}=dv_{gj}^{-1}/d\omega$), v_{gj} the group velocity, α_j the absorption coefficient, γ_j the nonlinear index coefficient, $d_{1j}=(v_{g1}-v_{gj})/v_{g1}v_{gj}$, $L_{NLj}=1/(\gamma_j P_{0j})$, $L_{Dj}=T_0^2/|\beta_{2j}|$, $\tau=[t-(z/v_{g1})]/T_0$, and $\xi=z/L_{D1}$, where z is the direction of propagation along the fiber and t is the time coordinate. Also, T_0 is the pulse width, P_{0j} is the incident optical power of the jth beam, and d_{1j}, the walk-off parameter between beam 1 and beam j, describes how fast a given pulse in beam j passes through the pulse in beam 1. In other words, the walk-off length is $L_{W(1j)}=T_0/|d_{1j}|$. So, $L_{W(1j)}$ is the distance for which the faster moving pulse (say, in beam j) completely walked through the slower moving pulse in beam 1. The nonlinear interaction between these two optical pulses ceases to occur after a distance $L_{W(1j)}$. For cross-phase modulation to take effect significantly, the group-velocity mismatch must be held to near zero.

It is also noted from Eq. (1) that the summation term in the bracket representing the cross-phase modulation effect is twice as effective as the self-phase modulation effect for the same intensity. This means that the nonlinear effect of the fiber medium on a beam may be enhanced by the copropagation of another beam with the same group velocity.

Equation (1) is a set of simultaneous coupled nonlinear Schrödinger equations that may be solved numerically by the

1063-651X/98/57(5)/6135(5)/$15.00

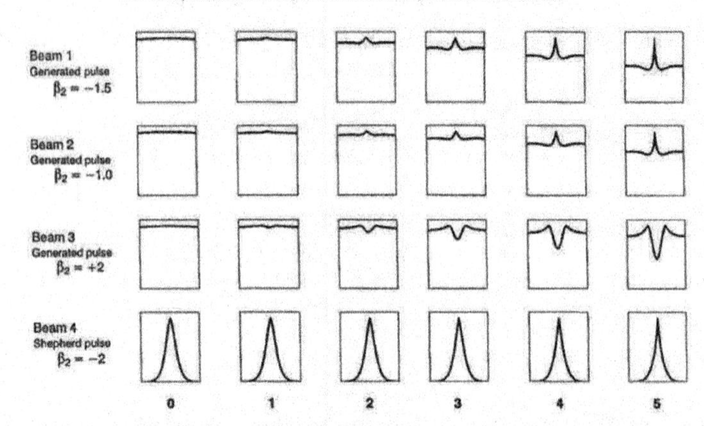

FIG. 1. Pulse evolution picture for the generation of three simultaneous pulses on three beams with separate wavelengths by a large-amplitude shepherd pulse on the fourth beam. For beam 1, $\beta_2=-1.5$, for beam 2, $\beta_2=-1.0$, for beam 3, $\beta_2=+2.0$, and for beam 4, the shepherd pulse beam, $\beta_2=-2.0$. The effect of different values of the dispersion coefficient on the induced pulses can be seen. In the negative β_2 region (the anomalous group-velocity dispersion region of the fiber), the induced pulses are "bright" pulses, and in the positive β_2 region (the normal group-velocity dispersion region of the fiber) the induced pulse is a "dark" pulse. The higher the $|\beta_2|$ value, the higher the amplitude of the induced pulse. (This case does not correspond to the case considered in Figs. 3, 4, and 5.)

split-step Fourier method, which was used successfully earlier to solve the problem of beam propagation in complex fiber structures, such as the fiber couplers [13], and to solve the thermal blooming problem for high-energy laser beams [14]. According to this method, first, the solutions may be advanced using only the nonlinear part of the equations. Then the solutions are allowed to advance using only the linear part of Eq. (1). This forward stepping process is repeated over and over again until the desired destination is reached. The Fourier transform is accomplished numerically via the well-known fast-Fourier-transform technique.

III. HOW TO GENERATE TIME-ALIGNED PULSES

A high-power, picosecond pulse, called the shepherd pulse, is launched on a given beam. A number of low-power beams that are selected based on the wavelength-division-multiplexed format are launched without any signal pulses into a single-mode nonlinear fiber. These beams copropagate with the beam carrying the shepherd pulse in this fiber. It will be shown, first through numerical simulation results, and then through experimental measurements, that time-aligned pulses will appear on these low-power WDM beams. The nonlinear cross-phase modulation effect in a single-mode fiber is instrumental in the generation of these time-aligned pulses. It is also required that the "walk-off" among all the beams be kept at a minimum acceptable value. A more detailed discussion on this requirement will be given later.

IV. COMPUTER SIMULATION RESULTS

Computer-simulation results for the generation of three simultaneous pulses on three beams with separate wavelengths by a large amplitude shepherd pulse on the fourth beam are shown in Fig. 1. It is assumed that wavelength separation among the beams is larger than 5 nm and the shepherd pulse has a pulse width of 60 ps. Due to this wide wavelength separation of the beams as well as the width of the shepherd pulse, the four-wave-mixing effect is negligible for this case. Other parameters are: the length of fiber $L=20$ km, the dispersion coefficient $|\beta_2|=2$ ps^2/km, the operating wavelength of beam 1 $\lambda_1=1.55$ μm, the operating wavelength of beam 2 $\lambda_2=1.545$ μm, the operating wavelength of beam 3 $\lambda_3=1.535$ μm, the operating wavelength of beam 4 $\lambda_4=1.555$ μm, the nonlinear index coefficient $\gamma=20$ W^{-1}km^{-1}, the attenuation or absorption of each beam in fiber $\alpha=0.2$ dB/km, the group velocity of the beam $v_g=2.051\,147\times10^8$ m/sec, the walk-off parameter between the slowest beam and the fastest beam $d_{12}<3$ ps/km, and the pulse width $T_0=60$ ps. It has been assumed that the dispersion coefficients for these beams are: for beam 1, $\beta_2=-1.5$, for beam 2, $\beta_2=-1.0$, for beam 3, $\beta_2=+2.0$, and for beam 4, the shepherd pulse beam, $\beta_2=-2.0$ The effect of different values of the dispersion coefficient on the induced pulses can be seen from the resultant data. In the negative β_2 region (the anomalous group-velocity dispersion region), the induced pulses are "bright" pulses and, in the positive β_2 region (the normal group-velocity dispersion region), the induced pulse is a "dark" pulse, i.e., a dip. The higher the $|\beta_2|$ value, the higher the amplitude of the induced pulse. For shepherd pulse having a large amplitude, say, $N_S=5$, where N_S is the soliton number for the shepherd pulse, the fiber length at which the shepherd pulse experiences maximum pulse compression is $0.105L_{DS}$ where L_{DS} is the dispersion length for the shepherd pulse. This is also the length at which maximum amplitude for the induced pulse is generated. The evolution of these generated pulses from 0 to this length is shown in Fig. 1.

Our simulation shows that large walk-off among the beams, i.e., walk-off larger than a full shepherd pulse width within the dispersion length for the shepherd pulse, would

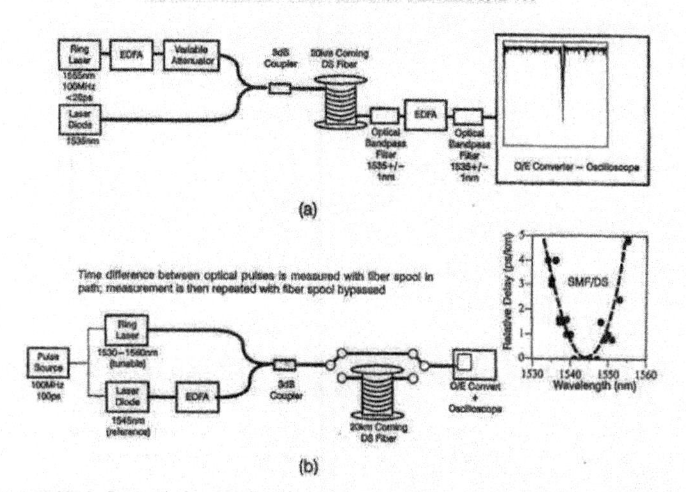

FIG. 2. (a) A schematic block diagram for the experimental setup to measure the generated pulses. (b) A schematic block diagram for the experimental setup to measure the "walk-off" characteristics of the Corning DS fiber. The maximum walk-off for the wavelength range 1535–1560 nm is less than 4 ps/km.

destroy the capability of the shepherd pulse to generate time-aligned pulses on the copropagating primary beams. It is for this reason that the selection of a proper fiber is of utmost importance. The following experiment will show that this demand, although rather stringent, can still be satisfied.

V. EXPERIMENTAL SETUP AND RESULTS

Schematic block diagrams of two experimental setups are shown in Fig. 2. The pulse source is a tunable erbium-doped fiber-ring laser, producing a 100-MHz train of pulses 60 ps in width between the wavelength range of 1530–1560 nm. This erbium ring pulse is named the shepherd pulse, operating at a peak power of higher than 200 mW. The primary sources are distribution feedback laser diodes at 1535, 1540, 1545, and 1557 nm, operated under a dc bias well above threshold. This cw output from the primary laser diode source is about 1 mW, which is amplified through an erbium-doped fiber amplifier to around 33 mW.

A. Measurement of fiber characteristics

As indicated in Sec. IV, the selection of a proper fiber is of great importance in the successful generation of time-aligned pulses on copropagating WDM beams. The Corning dispersion shifted (DS) fiber [15] is chosen to be the single-mode fiber for the experiment because of its desirable dispersion and walk-off characteristics. To learn quantitatively the behavior of this fiber, the walk-off characteristics of this fiber are measured and displayed in Fig. 3. It is seen that for the wavelength range of interest (1535–1560 nm), the dispersion coefficient β_2 varies between +2 and −2 ps^2/km. The difference of group velocities as a function of the wavelength of the beams varies between 0 and 4 ps/km. An erbium-doped fiber amplifier (EDFA) is used to boost the power at the receiver.

B. Generation of time-aligned pulses on copropagating WDM beams

Two sets of experiments were performed: The first set dealt with the generation of "dark" pulses on two or three semiconductor laser sources by a shepherd pulse on the ring laser; the second set dealt with the generation of "bright" pulses on a semiconductor laser source by a shepherd pulse on the ring laser. Signals from these sources (one or more signal sources from semiconductor lasers and one shepherd source from a ring laser) of different wavelengths are combined using two 2-to-1 fiber couplers. The combined output is sent through a 20-km spool of Corning DS fiber. At the output end of the fiber, an optical bandpass filter is used to reject the shepherd pulse signal from the ring laser. The signal from each laser diode is detected and viewed on an oscilloscope. A picture of this output is shown in Fig. 2(a). A dip (dark pulse) or a rise (bright pulse) on the cw signal from a laser diode indicates the presence of a generated pulse as predicted by the computer-simulation result.

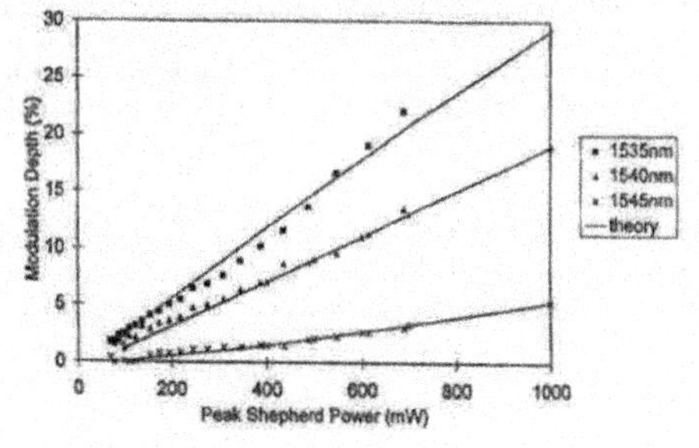

FIG. 3. Measured and computed modulation depths of the generated "dark" pulses on three different wavelength beams as a function of the shepherd-pulse peak power.

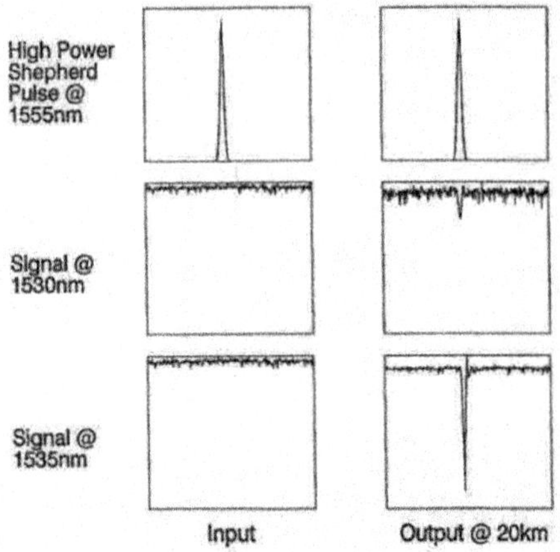

FIG. 4. Oscilloscope picture for the generated pulses on two different wavelength primary beams by a copropagating shepherd pulse on the third beam. "Dark" pulses are generated on the diode laser beams, since the operating wavelengths of these primary beams fall into the normal group-velocity dispersion region of the fiber.

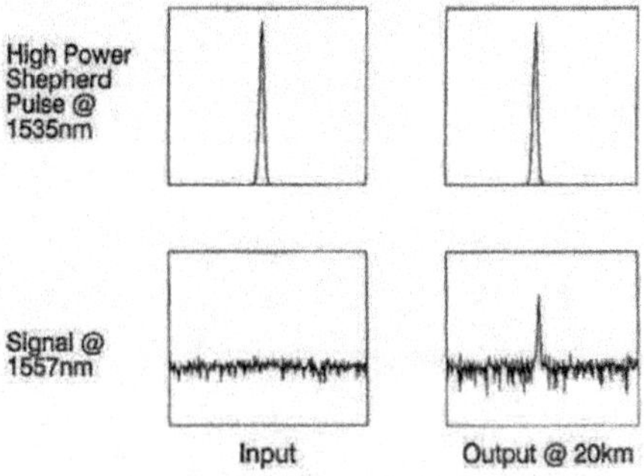

FIG. 5. Oscilloscope picture for the generated pulse on a primary beam by a copropagating shepherd pulse on the second beam. The "bright" pulse is generated on the diode laser beam, since the operating wavelength of the primary beam falls into the anomalous group-velocity dispersion region of the fiber.

1. Set 1: Generation of "dark" pulses

Systematic measurements are made for pulses generated on two or three primary cw beams from laser-diode sources due to the presence of a shepherd pulse on a beam from the ring laser. Figure 4 shows the input and output pulses on the two primary cw beams with wavelengths 1530 and 1535 nm, and on the ring laser beam operating at 1555 nm. It is seen that there were no pulses on the two primary beams at the input and there was a large shepherd pulse on the ring laser beam at the input. After passing through the fiber, there appeared three time-aligned copropagating output pulses on all three beams. As predicted by the theory, two dark pulses were generated on the primary diode laser beams because the operating wavelengths fell in the normal group-velocity dispersion region of the fiber and one was the shepherd pulse on the ring laser beam.

Shown in Fig. 3 is a plot of modulation depths on the cw beams as a function of the shepherd pulse peak power. Solid lines represent the computer-simulation results and the data points represent the measured results. Very close agreement is observed. It is noted that walk-off between the shepherd pulse and the generated primary pulses is less than 1 ps/km. For a fiber length of 20 km, the maximum pulse misalignment is 20 ps or 1/3 pulse width without the pulse-shepherding effect. It is expected that the presence of the unavoidable pulse-shepherding effect will diminish the pulse misalignment to a negligible level, as observed in the experimental results.

2. Set 2: Generation of "bright" pulses

To verify the theoretical prediction that bright pulses can be generated if the operating wavelength of the primary laser diode falls in the anomalous group-velocity dispersion region of the fiber, the following experiment was performed: Two beams, one from the ring laser, carrying the shepherd pulse at 1535 nm, and the other from the diode laser, carrying no pulse at 1557 nm, were combined and sent through the Corning DS fiber. The input and output pulses on these beams are displayed in Fig. 5. As predicted by the theory, a bright pulse is generated on the diode laser beam because its operating wavelength falls in the anomalous group-velocity dispersion region of the fiber.

VI. CONCLUSIONS

The above experiments show that through the use of a shepherd pulse on a copropagating wavelength-division-multiplexed beam, a simple way is found to generate time-aligned pulses on the other different wavelength beams in a nonlinear fiber. This was accomplished experimentally using a Corning single-mode-fiber–dispersion-shifted fiber. It should be noted that the success of this technique depends on the condition that the amount of walk-off or drifting between the large-amplitude shepherd pulse and the generated pulses must be less than half of the pulse width of the shepherd pulse for the entire pulse-generation interaction length of fiber. It should be noted that these generated pulses are "stable" pulses in the sense that they remain even after the shepherd pulse is channeled away. Successful generation of these time-aligned pulses on WDM beams is crucial to the realization of the future ultrahigh data-rate bit-parallel wavelength-division-multiplexed single-fiber transmission system.

ACKNOWLEDGMENTS

The research described in this paper was performed by the Center for Space Microelectronics Technology, Jet Propulsion Laboratory, California Institute of Technology, and was sponsored by the Ballistic Missile Defense Organization, Office of Innovative Science and Technology, through an agreement with the National Aeronautics and Space Administration.

[1] L. A. Bergman, A. J. Mendez, and L. S. Lome, in *Optoelectronic Interconnects and Packaging*, Vol. CR62 of *SPIE Critical Review of Optical Science and Technology*, edited by Ray T. Chen and Peter S. Cuilfoyle (SPIE, Bellingham, WA, 1996), pp. 210–226.

[2] L. A. Bergman and C. Yeh, in *Proceedings of the Third International Conference on Massively Parallel Processing Using Optical Interconnections (MPPOI'96), Maui, Hawaii, 1996*, edited by E. Schenfeld (IEEE Computer Society Press, New York, 1996).

[3] E. P. Ippen, in *Laser Applications to Optics and Spectroscopy*, edited by S. F. Jacobs, M. Sargent III, J. F. Scott, and M. O. Scully (Addison-Wesley, Reading, MA, 1975), Vol. 2, Chap. 6.

[4] A. Hasegawa and F. D. Tappert, Appl. Phys. Lett. **23**, 142 (1973); L. F. Mollenauer, R. H. Stolen, and J. P. Gordon, Phys. Rev. Lett. **45**, 1095 (1980).

[5] L. F. Mollenauer, R. H. Stolen, J. P. Gordon, and W. J. Tomlinson, Opt. Lett. **8**, 289 (1983).

[6] R. H. Stolen and E. P. Ippen, Appl. Phys. Lett. **22**, 276 (1973); V. A. Vysloukh and V. N. Serkin, Pis'ma Zh. Eksp. Teor. Fiz. **38**, 170 (1983) [JETP Lett. **38**, 199 (1983)]; E. M. Dianov, A. Ya. Karasik, P. V. Mamyshev, A. M. Prokhorov, M. F. Stel'makh, and A. A. Fomichev, *ibid*. **41**, 294 (1985) [*ibid*. **41**, 294 (1985)].

[7] E. P. Ippen and R. H. Stolen, Appl. Phys. Lett. **21**, 539 (1972); N. A. Olsson and J. P. van der Ziel, *ibid*. **48**, 1329 (1986).

[8] R. H. Stolen, J. E. Bjorkholm, and A. Ashkin, Appl. Phys. Lett. **24**, 308 (1974).

[9] C. Yeh and L. Bergman, J. Appl. Phys. **80**, 3175 (1996).

[10] C. Yeh and L. Bergman, Phys. Rev. E **57**, 2398 (1998).

[11] G. P. Agrawal, *Nonlinear Fiber Optics* (Academic Press, New York, 1989); A. Hasagawa and Y. Kodama, *Solitons in Optical Communications, Oxford Series in Optical and Imaging Sciences* (Clarendon, London, 1995), Vol. 7; *Optical Solitons–Theory and Experiment, Cambridge Studies in Modern Optics 10*, edited by J. R. Taylor (Cambridge University Press, Cambridge, 1992); M. Islam, *Ultrafast Fiber Switching Devices and Systems, Cambridge Studies in Modern Optics* (Cambridge University Press, London, 1992), Vol. 12.

[12] L. Wang and C. C. Yang, Opt. Lett. **15**, 474 (1990); V. V. Aftanasyev, Yu. S. Kivshar, V. V. Konotop, and V. N. Serkin, *ibid*. **14**, 805 (1989); S. Trillo, S. Wabnitz, E. M. Wright, and G. I. Stegeman, *ibid*. **13**, 871 (1989).

[13] C. Yeh, W. P. Brown, and R. Szejn, Appl. Opt. **18**, 489 (1979).

[14] C. Yeh, J. E. Pearson, and W. P. Brown, Appl. Opt. **15**, 2913 (1976).

[15] Opto-Electronics Group, Corning Incorporated Report No. MM26, 1996 (unpublished); L. G. Cohen, W. L. Mammel, and S. J. Jang, Electron. Lett. **18**, 1023 (1982); B. J. Ainslie and C. R. Day, J. Lightwave Technol. **LT-4**, 967 (1986).

PHYSICAL REVIEW E VOLUME 60, NUMBER 2 AUGUST 1999

Existence of optical solitons on wavelength division multiplexed beams in a nonlinear fiber

C. Yeh and L. A. Bergman
Jet Propulsion Laboratory, California Institute of Technology, 4800 Oak Grove Drive, Pasadena, California 91109
(Received 9 March 1999)

A simple analytic expression for the initial fundamental optical solitons on wavelength division multiplexed (WDM) beams in a nonlinear fiber has been found. For an ideal fiber with no loss and uniform group-velocity dispersion (GVD) in the anomalous GVD region, the initial form is $[1+2(M-1)]^{-1/2}\operatorname{sech}(\tau)$, where M is the number of WDM beams and τ is the normalized time. Computer simulation shows that these initial pulses on WDM beams in this fiber will propagate undistorted without change in their shapes for arbitrarily long distances. The discovery of the existence of solitons on WDM beams presents the ultimate goal for optical fiber communication on multiple wavelength beams in a single fiber. [S1063-651X(99)06208-X]

PACS number(s): 42.81.Dp, 42.65.Tg

I. INTRODUCTION

The discovery in 1973 that an optical soliton [1] on a single wavelength beam can exist in fiber is one of the most significant events since the perfection of low-loss optical fiber communication. This means that, in principle, data pulses may be transmitted in a fiber without degradation forever. This soliton discovery sets the ultimate goal for optical fiber communication on a single wavelength beam.

Another most significant event is the development of wavelength division multiplexed (WDM) transmission in a single mode fiber [2]. This means that multiple beams of different wavelengths, each carrying its own data load, can propagate simultaneously in a single mode fiber. This WDM technique provides dramatic increase in the bandwidth of a fiber. However, due to the presence of complex nonlinear interaction between copropagating pulses on different wavelength beams, it is no longer certain that WDM solitons can exist.

The existence of solitons is a blissful event in nature. It is a marvel that the delicate balance between the dispersion effect and the nonlinear effect can allow a specially shaped optical pulse to propagate in the fiber without degradation. This is called a temporal soliton [1]. It is an equal marvel that the delicate balance between the diffraction effect and the nonlinear effect can also allow a specially shaped pulse to propagate in a planar waveguide or array waveguides without degradation. This is called a spatial soliton [3]. They occur only on a single wavelength beam.

When beams with different wavelengths copropagate in a single mode fiber, such as in the wavelength division multiplexed case [2], interaction of pulses on different beams via the nonlinear cross phase modulation (CPM) effect (the Kerr effect) is usually instrumental in destroying the integrity of solitons on these wavelength multiplexed beams. Other recent applications of CPM effect in fiber have been reported [4].

The purpose of this paper is to show that temporal solitons can exist on WDM beams in a single fiber under appropriate conditions. The existence of these solitons critically depends on the presence of the nonlinear cross phase modulation effect of the WDM beams. Just as in the earlier single-beam soliton case, this discovery, sets the ultimate goal for optical fiber communication on WDM beams.

II. THE FUNDAMENTAL EQUATIONS

The fundamental equations governing M numbers of copropagating waves in a nonlinear fiber including the CPM phenomenon are the coupled nonlinear Schrödinger equations [5]:

$$\frac{\partial A_j}{\partial z}+\frac{1}{v_{gj}}\frac{\partial A_j}{\partial t}+\frac{1}{2}\alpha_j A_j = \frac{i}{2}\beta_{2j}\frac{\partial^2 A_j}{\partial t^2}-\gamma_j\left(|A_j|^2+2\sum_{m\neq j}^{M}|A_m|^2\right)A_j \quad (j=1,2,3,\ldots,M). \tag{1}$$

Here, for the jth wave, $A_j(z,t)$ is the slowly varying amplitude of the wave, v_{gj} the group velocity, β_{2j} the dispersion coefficient ($\beta_{2j}=dv_{gj}^{-1}/d\omega$), α_j the absorption coefficient, and

$$\gamma_j=\frac{n_2\omega_j}{cA_{\text{eff}}} \tag{2}$$

is the nonlinear index coefficient with A_{eff} as the effective core area and $n_2=3.2\times10^{-16}\,\text{cm}^2/\text{W}$ for silica fibers, ω_j is the carrier frequency of the jth wave, c is the speed of light, and z is the direction of propagation along the fiber.

Introducing the normalizing coefficients

$$\tau=\frac{t-(z/v_{g1})}{T_0}, \quad d_{1j}=(v_{g1}-v_{gj})/v_{g1}v_{gj}, \quad \xi=z/L_{D1}, \quad L_{D1}=T_0^2/|\beta_{21}|, \tag{3}$$

and setting

$$u_j(\tau,\xi)=[A_j(z,t)/\sqrt{P_{0j}}]\exp(\alpha_j L_{D1}\xi/2), \tag{4}$$

$$L_{nlj} = 1/(\gamma_j P_{0j}),$$

$$L_{Dj} = T_0^2/|\beta_{2j}| \qquad (5)$$

gives

$$i\frac{\partial u_j}{\partial \xi} = \frac{\text{sgn}(\beta_{2j})L_{D1}}{2L_{Dj}}\frac{\partial^2 u_j}{\partial \tau^2} - i\frac{d_{1j}}{T_0}L_{D1}\frac{\partial u_j}{\partial \tau} - \frac{L_{D1}}{L_{nlj}}\left(\exp(-\alpha_j L_{D1}\xi)|u_j|^2 + 2\sum_{m\neq j}^{M}\exp(-\alpha_m L_{D1}\xi)|u_m|^2\right)u_j$$

$$(j = 1,2,3,\dots,M). \qquad (6)$$

Here, T_0 is the pulse width, P_{0j} is the incident optical power of the jth beam, and d_{1j}, the walk-off parameter between beam 1 and beam j, describes how fast a given pulse in beam j passes through the pulse in beam 1. In other words, the walk-off length is

$$L_{W(1j)} = T_0/|d_{1j}|. \qquad (7)$$

So, $L_{W(1j)}$ is the distance for which the faster moving pulse (say, in beam j) completely walked through the slower moving pulse in beam 1. The nonlinear interaction between these two optical pulses ceases to occur after a distance $L_{W(1j)}$. For cross phase modulation to take effect significantly, the group-velocity mismatch must be held to near zero.

Finding the analytic solution of Eq. (6), which is a set of simultaneous coupled nonlinear Schrödinger equations, is a formidable task. However, it may be solved numerically by the split-step Fourier method, which was used successfully earlier to solve the problem of beam propagation in complex fiber structures, such as the fiber couplers, and to solve the thermal blooming problem for high energy laser beams [6]. According to this method, the solutions may be advanced first using only the nonlinear part of the equations, and then the solutions are allowed to advance using only the linear part of Eq. (6). This forward stepping process is repeated over and over again until the desired destination is reached. The Fourier transform is accomplished numerically via the well-known fast Fourier transform technique.

III. SOLITON ON A SINGLE BEAM

It is well known that, for an idealized fiber with no loss, an optical soliton on a single wavelength beam takes the initial form [1,5]

$$u(0,\tau) = N\,\text{sech}(\tau), \qquad (8)$$

where N is the soliton magnitude and

$$N^2 = \frac{L_D}{L_{nl}}. \qquad (9)$$

It is also known that the single-beam soliton equation is

$$i\frac{\partial u}{\partial \xi} = -\frac{1}{2}\frac{\partial^2 u}{\partial \tau^2} - N^2[|u|^2]u_j. \qquad (10)$$

Here, the dispersion length L_D and the nonlinear length L_{nl} are defined earlier in Eq. (6). In the case of anomalous group-velocity dispersion (GVD) for a soliton, $\text{sgn}(\beta_2) = -1$. For the fundamental soliton case, $N=1$. This means that when an initial pulse with pulse shape given by Eq. (8) with an amplitude of unity is launched inside an ideal lossless fiber, the pulse will retain its hyperbolic secant shape without degradation for arbitrarily long distances. One notes that the delicate balance between the dispersion effect represented by L_D and the nonlinear self-phase modulation effect represented by L_{nl} occurs at $N=1$ for the fundamental soliton. The nonlinear effect on a pulse for a single wavelength beam is embodied in L_{nl}, while the dispersion effect on the pulse is embodied in L_D.

IV. SOLITONS ON WAVELENGTH DIVISION MULTIPLEXED BEAMS

It is of interest to learn whether solitons exist on WDM beams in a fiber. Starting with an idealized fiber which is lossless (i.e., $\alpha_j = 0$ for all beams) and which possesses uniform group-velocity dispersion (i.e., $v_{gj} = v_g$ for all beams) within the wavelength range under investigation, the equations governing the propagation characteristics of signal pulses are

$$i\frac{\partial u_j}{\partial \xi} = -\frac{1}{2}\frac{\partial^2 u_j}{\partial \tau^2} - \frac{L_D}{L_{nl}} \times \left(|u_j|^2 + 2\sum_{m\neq j}^{M}|u_m|^2\right)u_j$$

$$(j = 1,2,3,\dots,M). \qquad (11)$$

The anomalous GVD case in which $\text{sgn}(\beta_{2j}) = -1$ is considered. It is seen from the above equation that the summation term representing the cross phase modulation effect is twice as effective as the self-phase modulation (SPM) effect for the same intensity. This observation also provides the idea that cross phase modulation may be used in conjunction with self-phase modulation on the WDM pulses to counteract the GVD effect, thus producing WDM solitons. Comparing the bracketed terms in Eqs. (10) and (11) shows that if one chooses the correct amplitudes for the initial pulses on WDM beams and retains the hyperbolic secant pulse form, it may be possible to construct a set of initial pulses which will propagate in the same manner as the single soliton pulse case, i.e., undistorted and without change in shape for arbitrarily long distances. Let us choose the initial pulses as follows:

$$u_j(0,\tau) = [1 + 2(M-1)]^{-1/2}\,\text{sech}(\tau) \quad (j = 1,2,3,\dots,M), \qquad (12)$$

where M is the number of WDM beams.

Using these initial pulse forms numerical simulation was carried out to solve Eq. (11). The split-step Fourier method was used. The fiber parameters used for the simulation are L, length of fiber equal to 1000 km; β_2, dispersion coefficient, equal to -2 ps^2/km; γ, nonlinear index coefficient, equal to 20 W^{-1} km^{-1}; T_0, pulse width, equal to 10 ps; L_D $=50$ km; and $L_{nl}=50$ km. Four cases with $M=1,2,3,4$ were treated. The $M=1$ case corresponds to the well-known single soliton case; here, the amplitude for the fundamental soliton is 1. For the two-beam case, the amplitude is $(3)^{-1/2}=0.57735$. For the three-beam case, it is $(5)^{-1/2}$ $=0.4472136$. For the four-beam case, it is $(7)^{-1/2}$ $=0.37796447$. It is noted that the amplitude of the fundamental solitons on WDM multibeams becomes successively smaller as the number of beams is increased. This is because the nonlinear effect becomes more pronounced when more beams are present. Numerical simulation shows that after propagating 1000 km through this fiber the original pulse shape for all these WDM pulses remains unchanged. It thus appears that the initial forms chosen for the pulses on WDM beams are the correct soliton forms for WDM beams.

V. CONCLUSION

The existence of optical solitons on wavelength division multiplexed beams in a fiber is not only of fundamental interest but also has enormous implications in the field of optical fiber communications. It is conceivable that multiple terabits of information can be sent through a single fiber in the bit-parallel wavelength division multiplexed format [2] without degradation.

ACKNOWLEDGMENTS

The authors wish to thank Dr. L. Lome of BMDO for his encouragement and support. The research described in this paper was performed by the Center for Space Microelectronics Technology, Jet Propulsion Laboratory, California Institute of Technology, and was sponsored by the Ballistic Missile Defense Organization, Office of Innovative Science and Technology, through an agreement with the National Aeronautics and Space Administration.

[1] A. Hasegawa and F. D. Tappert, Appl. Phys. Lett. **23**, 142 (1973); L. F. Mollenauer, R. H. Stolen, and J. P. Gordon, Phys. Rev. Lett. **45**, 1095 (1980).

[2] L. A. Bergman, A. J. Mendez, and L. S. Lome, SPIE Crit. Rev. **CR62**, 210 (1996); N. Edagawa, K. Mochizuki, and Y. Iwamoto, Electron. Lett. **23**, 196 (1987); L. F. Mollenauer, R. H. Stolen, and M. N. Islam, Opt. Lett. **10**, 229 (1985); A. Hasegawa, Appl. Opt. **23**, 3302 (1984).

[3] H. S. Eisenberg, Y. Silberberg, R. Morandotti, A. R. Boyd, and J. S. Aitchison, Phys. Rev. Lett. **81**, 3383 (1998).

[4] C. Yeh and L. Bergman, Phys. Rev. E **57**, 2398 (1998); C. Yeh, L. Bergman, J. Morookian, and S. Monacos, *ibid.* **57**, 6135 (1998); C. Yeh and L. Bergman, J. Appl. Phys. **80**, 3175 (1996).

[5] G. P. Agrawal, *Nonlinear Fiber Optics* (Academic, New York, 1989); *Optical Solitons—Theory and Experiment*, edited by J. R. Taylor, Cambridge Studies in Modern Optics 10 (Cambridge University Press, Cambridge, England, 1992).

[6] C. Yeh, W. P. Brown, and R. Szejn, Appl. Opt. **18**, 489 (1979); C. Yeh, J. E. Pearson, and W. P. Brown, *ibid.* **15**, 2913 (1976).

IEEE TRANSACTIONS ON CIRCUITS AND SYSTEMS, VOL. CAS-26, NO. 12, DECEMBER 1979 1011

Optical Waveguide Theory

C. W. YEH, MEMBER, IEEE

Invited Paper

Abstract—As optical fiber technology matures, complexity of optical waveguides and waveguide components also grows. Traditional techniques which may be used to analyze simple step-index circular fibers are no longer adequate. An assessment of several modern analytical/numerical techniques which have been used successfully to treat problems of practical interests is given. Illustrative numerical examples are also presented.

I. Introduction

AFTER being nurtured through the critical research and development stage in the early 1970's, optical fiber technology has finally come of age in the late 1970's. It is only a matter of time before the use of optical fiber links as practical communication lines becomes widespread. In order to properly design and use an optical fiber link, the propagation characteristics and field distributions of the propagating modes in the optical waveguide must be known. Knowledge of the propagation constants of guided modes as a function of frequency provides information on the bandwidth capability of the fiber guide under consideration while knowledge of the field distributions of guided modes provides information on how to couple light energy efficiently into and out of the fiber guide. Furthermore, this knowledge on the propagating modes also provides the starting point for many treatments on the effects of bending, tapering, surface irregularities, scattering centers, or mode conversions.

The purpose of this article is to assess several modern analytical/numerical techniques which have been used successfully in obtaining the propagation characteristics and field distributions of guided modes in optical waveguides. Some historical background information on the analysis of optical (dielectric) waveguides will first be given. Then the pros and cons as well as the limitations of various available analytical/numerical techniques will be discussed. Finally, presentation will be given on several selected promising modern techniques together with illustrations.

II. Historical Background

The concept of guiding light (electromagnetic waves) in a dielectric fiber is not new. Hondros and Debye, in 1910, showed analytically that a circularly symmetric transverse-magnetic (TM) mode can be guided by a dielectric cylinder with dielectric constant ϵ_1, situated in free space with dielectric constant ϵ_0 ($\epsilon_1 > \epsilon_0$). The existence of this wave was demonstrated experimentally by Zahn and Rüter and Schriever in 1915 [2]. The complete treatment of all guided modes that can be supported by a dielectric cylinder in free-space was carried out by Carson, Mead, and Schelkunoff in 1936 [3]. They were the first ones to show that all noncircularly symmetric modes in a circular fiber are hybrid modes (i.e., longitudinal electric and magnetic fields must both be present for asymmetric modes), that only one mode, the lowest order hybrid mode HE_{11}, has zero cutoff frequency, and that all other modes have finite cutoff frequencies below which they cease to exist. Numerical results for the propagation constants of several lower order modes in circular fiber and the experimental verification were carried out by Elsasser [4] and Chandler in 1949 [5]. Snitzer [6], in 1961, resolved the circular dielectric waveguide problem and applied the result to the case of light propagation along a homogeneous core circular fiber. The first complete analysis of noncircular elliptical fiber was given by Yeh in 1962 [7]. He showed that all modes must be of the hybrid type in a noncircular fiber and that there exists two dominant modes which possess zero cutoff frequencies. It was not until 1969, that Goell [8] presented his circular-harmonic computer analysis of a rectangular homogeneous dielectric waveguide and Marcatili [9] presented his approximate analysis of the rectangular homogeneous dielectric structures. Recognizing the fact that the index difference between the inner core region and the outer cladding region of an optical fiber is quite small (of the order of a few percent), Snyder [10] and Gloge [11] presented simplified approximate expressions for the propagation parameters in a circular fiber in 1969 and in 1971, respectively. Significant progress has also been made in solving problems dealing with radially inhomogeneous circular fibers. A survey on this subject is included in a paper by Yeh and Lindgren [12]. As the optical fiber technology matures, in the late 1970's, demand for more sophisticated and versatile analytical/numerical techniques to solve problems dealing with arbitrarily shaped, inhomogeneous dielectric waveguides and their associated components also grows. Recent successful development of the scalar wave—fast Fourier transform (FFT) technique [13], [14], the finite element technique [15] and the extended boundary condition technique [16] will certainly expand our ability in predicting and understanding the wave behavior in these complex guiding structures.

Manuscript received May 31, 1979. This work was supported in part by ONR.

The author is with the Department of Electrical Sciences and Engineering, University of California, Los Angeles, CA 90024.

III. Discussion of Available Analytic/Numerical Techniques

The governing equation for the guided-wave field of a fiber structure is the vector-wave equation:

$$\nabla\times\nabla\times E-\omega^2\mu\epsilon(r)E=0 \tag{1}$$

where ω and E are, respectively, the frequency and the electric-field vector of the guided wave, while μ and $\epsilon(r)$ are, respectively, the permeability and the permittivity of the guiding medium. The propagation characteristics of the guided wave are obtained by requiring that the guided-wave field must satisfy the proper boundary conditions at the interface of two different media (i.e., tangential electric- and magnetic-field vectors must be continuous across the boundary) and the radiation condition for field that extends to infinity [17].

When the guided wave propagates along a perfect straight-line path, one may assume that every component of the electromagnetic wave may be represented in the form:

$$f(u,v)e^{-i\beta z}e^{i\omega t} \tag{2}$$

in which z is chosen as the propagation direction and u,v are generalized orthogonal coordinates in a transverse plane. β is the propagation constant and ω is the frequency of the wave. Under this assumption, the transverse-field components in homogenous isotropic medium (ϵ,μ) are

$$E_u=\frac{-i}{\gamma^2}\left(\frac{\beta}{h_1}\frac{\partial E_z}{\partial u}+\frac{\omega\mu}{h_2}\frac{\partial H_z}{\partial v}\right) \tag{3}$$

$$E_v=\frac{-i}{\gamma^2}\left(\frac{\beta}{h_2}\frac{\partial E_z}{\partial v}-\frac{\omega\mu}{h_1}\frac{\partial H_z}{\partial u}\right) \tag{4}$$

$$H_u=\frac{-i}{\gamma^2}\left(\frac{\beta}{h_1}\frac{\partial H_z}{\partial u}-\frac{\omega\epsilon}{h_2}\frac{\partial H_z}{\partial v}\right) \tag{5}$$

$$H_v=\frac{-i}{\gamma^2}\left(\frac{\beta}{h_2}\frac{\partial H_z}{\partial v}+\frac{\omega\epsilon}{h_1}\frac{\partial E_z}{\partial u}\right) \tag{6}$$

with

$$\gamma^2=k^2-\beta^2 \tag{7}$$

and

$$k^2=\omega^2\mu\epsilon \tag{8}$$

where ϵ is the permittivity of the medium and μ is the permeability of the medium, and the longitudinal-field components satisfy the following equation:

$$\left[\frac{1}{h_1h_2}\left(\frac{\partial}{\partial u}\frac{h_2}{h_1}\frac{\partial}{\partial u}+\frac{\partial}{\partial v}\frac{h_1}{h_2}\frac{\partial}{\partial v}\right)+(k^2-\beta^2)\right]\left\{\begin{matrix}E_z\\H_z\end{matrix}\right\}=0 \tag{9}$$

where h_1 and h_2 are the metric coefficients for the orthogonal curvilinear coordinates. Only discrete values of β will satisfy the boundary conditions. These allowed β values are called eigenvalues, and corresponding to these eigenvalues are the eigenfunctions. Each eigenvalue β corresponds to the propagation constant of a certain guided mode. It is pointed out here that TM guided modes refer to waves having $H_z=0$, TE guided modes having $E_z=0$, HE or EH guided modes (hybrid modes) having all field components not equal to zero.

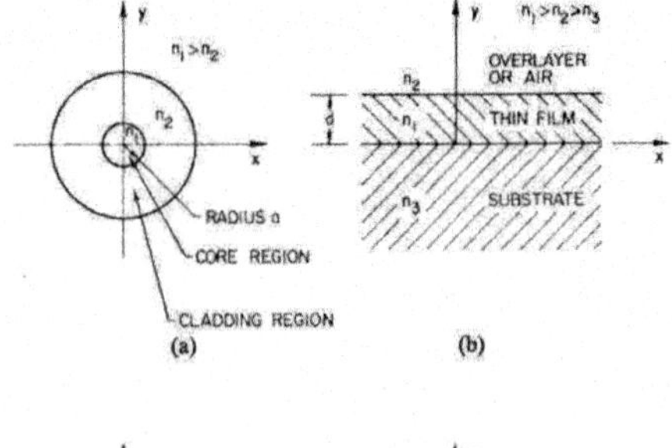

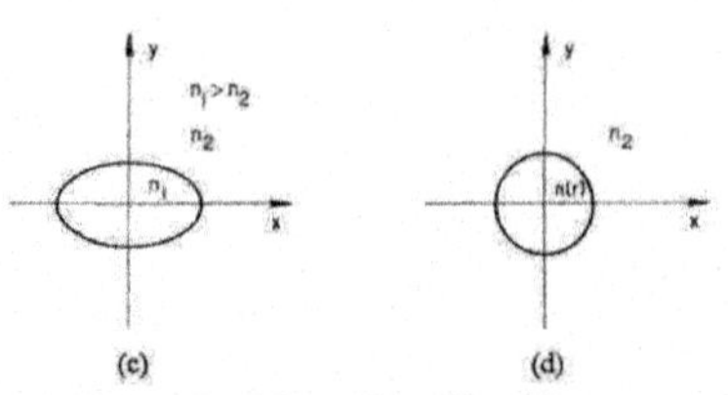

Fig. 1. Cross-sectional views of the light guiding structures that could be treated by the separation of variable technique. (a) Step-index circular fiber. (b) Slab guide. (c) Step-index elliptical fiber. (d) Graded-index circular fiber.

Two general types of fiber structures are of practical interest: the multimode fiber structures and the single-mode fiber structures. The multimode structures are capable of supporting many guided modes (say >10 modes) while the single-mode structures may support only a few lower order guided modes (say <10 modes). In the following, we shall divide our discussion of available analytic/numerical techniques into these two general categories:

A. The Single-Mode Case

1) Separation of Variables Method: For homogeneous medium, or some special inhomogeneous medium, the vector-wave equation is separable in three coordinate systems: the rectangular coordinates, the cylindrical coordinates and the spherical coordinates. The rectangular coordinates are particularly suited for slab-type guiding structures (see Fig. 1) and the cylindrical coordinates are suited for the step-index circular cylindrical fibers or radially graded index circular cylindrical fibers or step-index elliptical cylindrical fibers. (see Fig. 1). These geometrics are of great practical importance

Basically, the separation of variables method starts with the appropriate eigensolutions of the wave equations in the core and the cladding regions of the guiding structure. Then, by matching the boundary conditions at the interface, which fits the contour of one of the coordinate surfaces, one may obtain a dispersion relation from which the propagation constant β behavior may be found. Typically, the circular cylindrical fiber structure can support a family of circularly symmetric transverse-electric TE_{om} or transverse-magnetic TM_{om} modes (whose fields are independent of the azimuthal coordinate) and a family of hybrid HE_{nm} or EH_{nm} modes. The subscripts n and m denote, respectively, the number of cyclic variation with the azimuthal coordinate and the mth root of the dispersion relation which is obtained by satisfying the appropriate boundary conditions. The symbol HE refers to mode with the ratio $(\mu_0\omega/\beta)(H_z/E_z) = -1$ far from the cutoff frequency while the symbol EH refers to modes with the ratio $(\mu_0\omega/\beta)(H_z/E_z) = +1$ far from the cutoff frequency.

The pros and cons of this method are summarized as follows.

Pros:

- conceptually simple to use; analytic expressions are available,
- results are exact and applicable to small as well as large index differences for the core and cladding regions.

Cons:

- may be applied to only a small family of geometries such as the step-index circular cylindrical fibers, the radially graded index circular cylindrical fibers, the step-index elliptical cylindrical fibers, the slab dielectric guides,
- dispersion relations are transcendental functions, hence difficult to use.

2) Circular Harmonics Expansion Method: This method was first used by Goell [8] in his treatment of the step-index rectangular fibers. His computer analysis is based on an expansion of the guided electromagnetic field in terms of a series of circular harmonics. For example, the expressions for the longitudinal electric and magnetic fields in the core and cladding regions are, respectively,

$$E_z^{(\text{core})} = \sum_{n=-\infty}^{\infty} A_n J_n(\gamma_1 r) e^{in\theta} e^{-i\beta z + i\omega t} \quad (10)$$

$$H_z^{(\text{core})} = \sum_{n=-\infty}^{\infty} B_n J_n(\gamma_1 r) e^{in\theta} e^{-i\beta z + i\omega t} \quad (11)$$

and

$$E_z^{(\text{cladding})} = \sum_{n=-\infty}^{\infty} C_n K_n(\gamma_2 r) e^{in\theta} e^{-i\beta z + i\omega t} \quad (12)$$

$$H_z^{(\text{cladding})} = \sum_{n=-\infty}^{\infty} D_n K_n(\gamma_2 r) e^{in\theta} e^{-i\beta z + i\omega t} \quad (13)$$

where

$$\gamma_1 = (k_1^2 - \beta^2)^{1/2}, \quad \gamma_2 = (\beta^2 - k_2^2)^{1/2}$$
$$k_1^2 = \omega^2 \mu \epsilon_1, \quad k_2^2 = \omega^2 \mu \epsilon_2$$

ϵ_1 and ϵ_2, are, respectively, the core dielectric constant and the cladding dielectric constant. A_n, B_n, C_n, and D_n are arbitrary constants. J_n and K_n are the nth-order Bessel

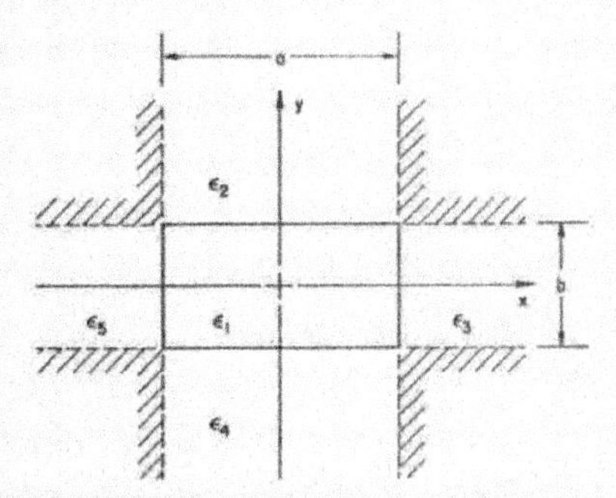

Fig. 2. Geometry of the rectangular dielectric guide immersed in different dielectrics.

functions and modified Bessel functions, respectively. All transverse fields can be found from (3)–(6). Point matching of the tangential electric and magnetic fields on the boundary of the core and cladding regions yields a set of linear algebraic equations containing the unknown coefficients A_n, B_n, C_n, and D_n. Setting the determinant of this set of linear equations to zero gives a dispersion relation from which the propagation constants of various modes may be computed. The size of the determinant depends on the number of points that were used for matching.

Pros:

- may be used to treat noncircular fibers with arbitrary core/cladding index difference.

Cons:

- convergence of the results is not guaranteed for shapes that deviate significantly from the circular fiber shape; i.e., good for square fiber but not good for rectangular fibers,
- only applicable to step-index fibers; i.e., core and cladding regions must contain uniform dielectrics,
- this is a purely numerical approach; demand on computer (time and money) may be excessive if many matching points were used.

3) Marcatili's Approximate Approach: Recognizing the fact that most of the guided energy is confined within the dielectric core region for a wide range of parameters and very little energy is guided in the corner regions of a rectangular dielectric guide, Marcatilli [9] formulated an approximate solution to the problem of waveguiding by rectangular dielectric structures by ignoring the matching of fields along the edges of the shaded area in Fig. 2. By matching the tangential electric and magnetic fields only along the four sides of region 1, and assuming that the field components in region 1 vary sinusoidally in the x and y directions, those in 2 and 4 vary sinusoidally along x and exponentially along y, and those in regions 3 and 5 vary sinusoidally along y and exponentially along x, one may obtain a dispersion relation from which the propagation constants of various modes may be calculated.

1014 IEEE TRANSACTIONS ON CIRCUITS AND SYSTEMS, VOL. CAS-26, NO. 12, DECEMBER 1979

Pros:

-resultant mathematical expressions are easy to manipulate since only simple sinusoidal and exponential functions have been used,
- since analytic expressions of the fields were obtained, they provide easy insight and physical interpretation of the results,
- may be used to treat any rectangular shaped homogeneous core optical guides,
- results are quite accurate as long as the guided energy is mostly confined within the core region.

Cons:

- only applicable to step-index rectangular structures,
- results are unreliable near the cutoff region of the mode, where the guided fields are not necessarily confined within the core region,
- this is an heuristic approach; regions of validity of the results are unknown.

4) Finite-Element Approach: This is the single most powerful technique [15] developed in recent years to treat the single-mode problems dealing with arbitrarily shaped dielectric waveguides with inhomogeneous index variations in the cross-sectional plane. The governing longitudinal fields of the guided wave are first expressed as a functional as follows:

$$
\begin{aligned}
I &= \sum_{p=1} I_p \\
&= \sum_{p=1} \iint \Bigg\{ \tau_p |\nabla H_z^{(p)}|^2 + \gamma^2 \tau_p \frac{\epsilon_p}{\epsilon_0} \left| \frac{1}{\gamma} \left(\frac{\epsilon_0}{\mu_0} \right)^{1/2} \nabla E_z^{(p)} \right|^2 \\
&\quad + 2\tau_p \hat{e}_z \cdot \left[\frac{1}{\gamma} \left(\frac{\epsilon_0}{\mu_0} \right)^{1/2} \nabla E_z^{(p)} \cdot \nabla H_z^{(p)} \right] \\
&\quad - \left(\frac{\omega}{c} \right)^2 \left[(H_z^{(p)})^2 + \gamma^2 \frac{\epsilon_p}{\epsilon_0} \left[\frac{1}{\gamma} \left(\frac{\epsilon_0}{\mu_0} \right)^{1/2} E_z^{(p)} \right]^2 \right] \Bigg\} dx\, dy
\end{aligned}
\tag{14}
$$

where

$$
\gamma = \frac{\beta c}{\omega}, \quad \tau_p = \frac{\gamma^2 - 1}{\gamma^2 - \epsilon_p / \epsilon_0}
$$

the symbol p represents the pth region when one divides the guiding structure into many appropriate regions. Minimizing the above surface integral over the whole region is equivalent to satisfying the wave equation (1) and the boundary conditions for E_z and H_z. In the finite element approximation, the primary dependent variables are replaced by a system of discretized variables over the domain of consideration. Therefore, the initial step is a discretization of the original domain into many subregions. For the present analysis, there are a number of regions in the composite cross section of the waveguide for which the permittivity is distinct. Each of these regions is discretized into a number of smaller triangular subregions interconnected at a finite number of points called nodes. Appropriate relationships can then be developed to represent the waveguide characteristics in all triangular subregions. These relationships are assembled into a system of algebraic equations governing the entire cross section. Taking the variation of these equations with respect to the nodal variable leads to an algebraic eigenvalue problem from which the propagation constant for a certain mode may be determined.

Pros:

- may be used to treat any arbitrarily shaped, inhomogeneous dielectric guides,
- numerical results can be generated very efficiently,
- results are based on the exact Maxwell equations.

Cons:

- no analytic expression is available.

5) Extended Boundary Condition Method: For arbitrarily shaped step-index fibers, integral representations for the longitudinal electric and magnetic fields can be derived which satisfy the appropriate wave equations and all the necessary boundary conditions. By expanding the longitudinal fields in terms of a complete set of circular harmonics and by making use of analytic continuation technique, one may reduce the integral representations to a set of linear algebraic equations [16]. Setting its determinant to zero and finding its roots yields the propagation constants.

Pros:

- may be used to treat any arbitrarily shaped, step-index fibers,
- numerical results can be generated very efficiently,
- results are based on the exact Maxwell equations.

Cons:

- this technique cannot be used to treat inhomogeneous index structure.

6) Scalar Wave–FFT Approach: All of the above techniques were based on the exact vector-wave equations. Hence they were valid for large as well as small-index differences between core and cladding regions. One recognizes, however, that for most practical optical waveguides the index difference between the core and cladding regions are only of the order of a few percentages [13], [14]. It has been shown recently that if certain limiting conditions (such as small-index differences and gentle-index profiles) are satisfied [18], the scaler-wave approximation will yield valid results for single-mode structures. These limiting conditions are usually satisfied by many practical fiber or integrated optics structures.

Starting with the scaler-wave equation and making use of the paraxial approximation, one may develop a propagation code based on FFT technique to trace the evolution of the transverse field as it propagates down a fiber structure. This technique not only provides information on the field distribution, it also yields information on the propagation constant of the dominant mode.

Pros:

- may be used to treat arbitrarily shaped, inhomogeneous fiber structures such as fiber couplers, horns or tapers,
- efficient computational time.

Cons:

- index variation must be gentle and index differences must be small.

Knowing the pros and cons of the above techniques in their treatment of single-mode structures, one may choose the most efficient method in dealing with the problem at hand. In Section IV, several examples of contemporary interest will be given.

B. The Multimode Case

In principle, all of the single-mode techniques discussed earlier may be used to deal with the multimode structures. Contribution from each mode may be summed according to the principle of superposition to yield the correct result for the multimode case. However, this approach may be extremely cumbersome and sometimes unworkable when hundreds or even thousands of propagating modes are supportable by the multimode structure. Hence, the treatment of multimode structures based on the modal concept, perhaps should be altered. In the following, we shall discuss two approaches which are not based on this modal concept:

1) Geometrical Optics Approach: Because of its conceptual simplicity, the ray-tracing geometrical optics technique has always been of importance whenever optics problems are encountered. By tracing the ray paths according to the theory of geometrical optics, one is able to calculate the evolution of an input optical signal as it propagates down the waveguide. It can be seen, however, that this technique could become extremely laborious and that, since diffraction phenomenon was not taken into account, the result may also be rather inaccurate.

Pros:

- calculation involves elementary concepts and procedures.

Cons:

- calculations may be tedious,
- results may be grossly inaccurate due to over simplification of the propagation phenomenon.

2) Scalar Wave–FFT Approach: This technique was discussed earlier. It is seen that this scalar wave–FFT approach [13], [14] does not depend on the decomposition of fields into individual eigenmodes. It deals completely with total field quantities and not with individual modes. In other words, the evolution of the total field is calculated according to the scalar-wave equation as it propagates down the guiding structure. Since the wave nature of the field is taken into consideration, diffraction phenomenon is automatically included. Hence, this approach provides an accurate description of the guided wave in a multimode structure.

Pros:

- decomposition of field into modes is not necessary for multimode structures,
- provide accurate wave description of the propagating field,
- may be used to treat complex arbitrarily shaped, inhomogeneous multimode structures such as: multistrand fiber couplers, fibers with general index profiles, horns, tapers and branches, etc.

Cons:

- index variation must be gentle and index differences must be small,
- depolarization phenomenon is ignored.

IV. Selected Examples

According to the previous discussions, the most promising new techniques in dealing with general shaped optical waveguide structures having inhomogeneous index profiles are the finite-element method and the scalar wave –FFT method. In the following, we shall provide several illustrative examples of these techniques. In order to demonstrate the versatility of these techniques, we have purposely chosen examples concerning rather unusual guiding structures.

A. Examples Based on the Finite-Element Method

It has been mentioned earlier that the finite-element method is a very powerful technique. It can be used to solve single-mode problems dealing with guiding structures whose cores may be of arbitrary cross-sectional shape and whose material media may be inhomogeneous in more than one transverse direction. Since this technique is based on the exact Maxwell's equations, its results are valid for large or small index variations.

1) The Optical Stripline Guide: The optical stripline is a planar waveguide with a strip of slightly lower index on a higher index thin film. An advantage of this guide is a relaxation of the stringent requirement for smoothness of waveguide side walls because a small portion of the field strength impinges on the side walls of the strip.

The cross-sectional geometry of the optical stripline is given in Fig. 3. It also shows the refractive-index distribution where n_1, n_2, n_3, and n_0 are the refractive index of thin film, substrate, strip, and air, respectively. These refractive indexes satisfy one of the following relations for the stripline:

$$n_1 > n_2 > n_3 > n_0$$
$$n_1 > n_3 > n_2 > n_0.$$

1016 IEEE TRANSACTIONS ON CIRCUITS AND SYSTEMS, VOL. CAS-26, NO. 12, DECEMBER 1979

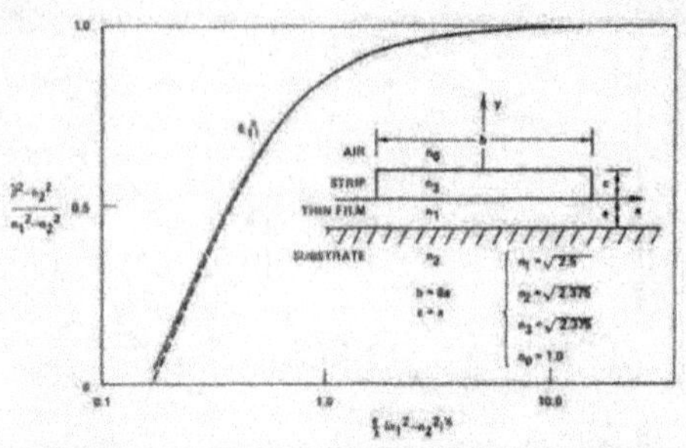

Fig. 3. Dispersion curves for the dominant E_{11}^{y} mode in an optical stripline waveguide. Solid line represents results found according to the finite-element method, while the dashed line represents results found according to the vector-variational method.

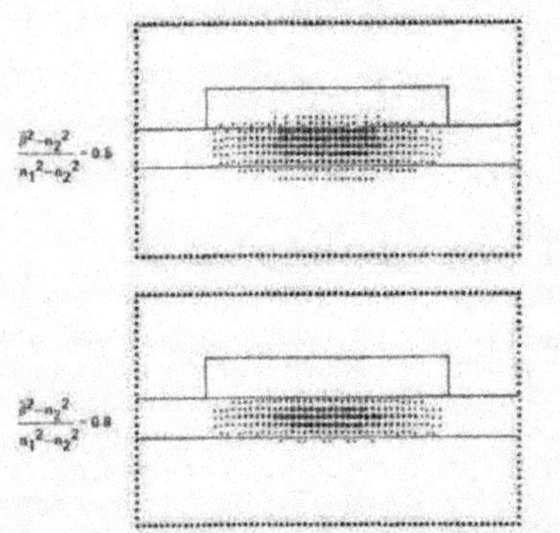

Fig. 4. Grey-scale plots for the power intensity distributions of the dominant E_{11}^{y} mode in an optical stripline waveguide at two different frequencies.

In the present example, the numerical data of the refractive indexes is chosen to be $n_1=\sqrt{2.5}$, $n_2=n_3=\sqrt{2.375}$, and $n_0=1.0$. This refractive-index distribution corresponds to the case where the thin film and the substrate are glass and the ambient is air.

The fundamental modes of this stripline guide are the E_{11}^{y} mode and E_{11}^{x} mode. The principal transverse-field components of the E_{11}^{y} mode are E_y and H_x and those of the E_{11}^{x} mode are E_x and H_y. The present analysis involves only solving the E_{11}^{x} mode.

The finite element model employs 900 elements and 928 nodes in one-half of the cross section because of the symmetry on the y axis. The dispersion curve for the E_{11}^{x} mode is computed with the boundary condition, $E_z=0$ on the y axis. The curve is plotted in Fig. 3.

Fig. 4 shows the power intensity distribution on the grey-scale taken at the normalized propagation constant $=0.5$ and 0.8. It is seen that the intensity distribution is well confined in the x direction by the dielectric strip because the width b of the strip is larger compared with

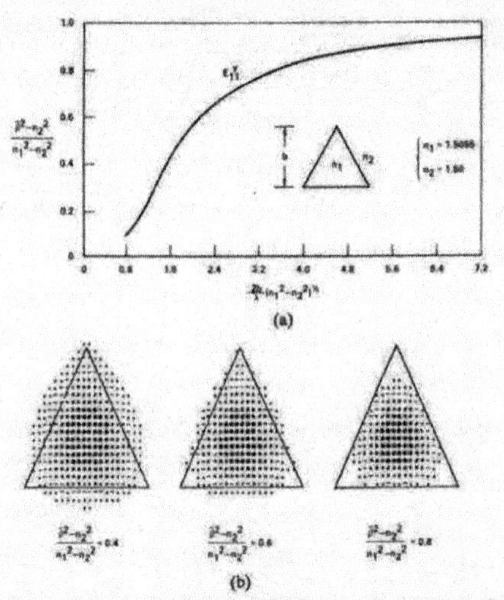

Fig. 5. (a) Dispersion curve for the dominant mode in a triangular fiber. (b) Grey-scale plots for the power intensity distributions of the dominant mode in a triangular fiber at three different frequencies.

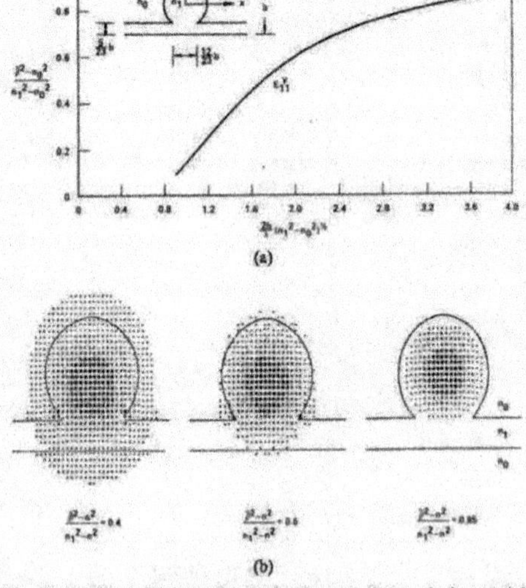

Fig. 6. (a) Dispersion curve for the dominant mode in a single material fiber. (b) Grey-scale plots for the power intensity distributions of the dominant mode in a single material fiber at three different frequencies.

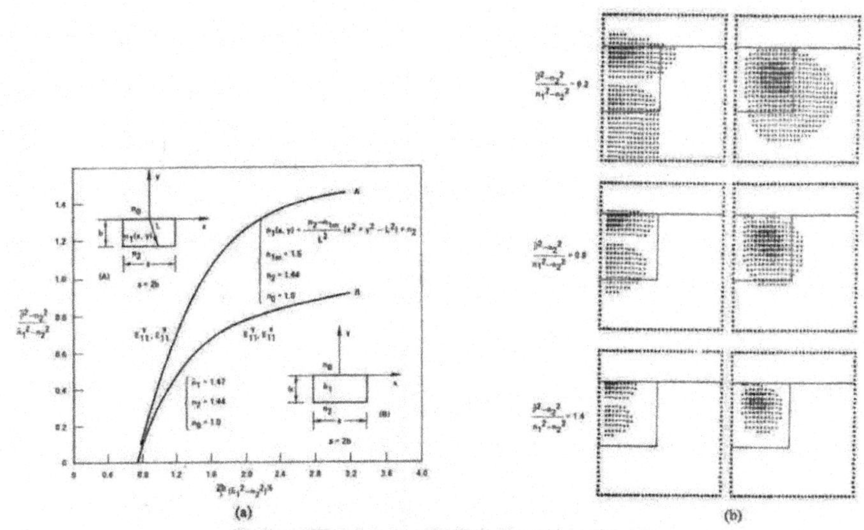

Fig. 7. (a) Dispersion curves for the dominant E^y_{11} or E^x_{11} mode in a two-dimensional diffused channel waveguide and in a uniform channel waveguide. (b) Grey-scale plots for the magnitude distributions of E_z and H_z for the dominant E^y_{11} mode in a two-dimensional diffused channel waveguide at three different frequencies.

the thin film thickness a and the wavelength λ. It is also confined in the y direction by the thin film. Therefore, the optical stripline minimizes scattering loss and undesired mode conversion caused by imperfections of the etched side walls of the optical waveguide.

2) Triangular Fiber Guide: Shown in Fig. 5 are the propagation characteristics and the power distribution of the dominant mode in a triangular optical fiber guide. These results were obtained by the finite element method employing 689 elements and 659 nodes.

3) Single Material Fiber Guide: Typically, optical fibers are constructed with a central glass core surrounded by a glass cladding with a slightly lower refractive index. The single material fiber is created by a structural form that uses only a single low-loss material. The guided energy is concentrated primarily in the central enlargement. The fields decay exponentially outward from the central enlargement. The guided-wave fields at the outside cylinder can be made negligibly small if spacing between the central enlargement and outer cylinder is sufficiently large. The propagation characteristics and the intensity distributions of the dominant mode are shown in Fig. 6.

4) The Diffused Channel Guide: To illustrate that the finite-element technique can just as easily be used to treat inhomogeneous structures as homogeneous ones, the dominant-mode propagation characteristics and its magnitude distributions of the longitudinal fields for the diffused channel guide were obtained and displayed in Fig. 7.

B. Examples Based on the Scalar-Wave–FFT Method

Multimode optical structures are of great practical importance, As noted earlier, the only efficient method which incorporates the wave nature of the guided energy in multimode structure is the scalar wave–FFT method. The only significant restriction to keep in mind is that index variations of the structure under consideration must be gentle.

1) Truncated Gaussian-Beam Propagation in Multimode Inhomogeneous Fiber Guide: It is recognized that analytic solutions for Gaussian-beam propagation problems exist only for certain specific radial-index profiles. Even for these cases, the solutions are often involved and cumbersome to use. By simply specifying the transverse-index variation of the guiding structure, we can readily calculate the propagation characteristics of an incident beam in this medium using the scalar wave–FFT technique. Shown in Fig. 8 is the evolution of intensity patterns of an incident truncated Gaussian beam as it propagates down a multimode fiber structure with parabolic-index profile.

2) Multimode Inhomogeneous Fiber Couplers: To further demonstrate the versatility and the power of this scalar wave–FFT technique, the exceedingly complex coupling problem dealing with two neighboring multimode graded index fibers was treated. Shown in Fig. 9 are the evolution

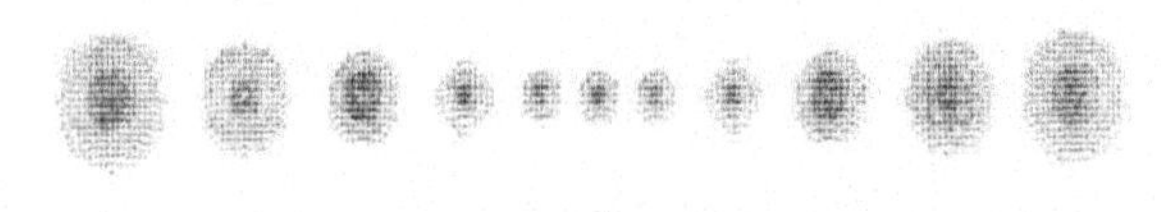

Fig. 8. Selected grey-scale intensity patterns along the fiber axis for a truncated Gaussian beam propagating in a radially inhomogeneous fiber with parabolic index profile. The values in the brackets represent the beam waists in micrometers of the beam while the other unbracketed values represent the highest intensity values for these patterns. Note that the truncated Gaussian profile of the beam is not preserved along the propagation path. At some points along the path, the intensity at the center of the beam takes a dip. Owing to the presence of the diffraction effects, the beam is not completely symmetrical about a certain "focal point" where the beam achieves the smallest beam waist.

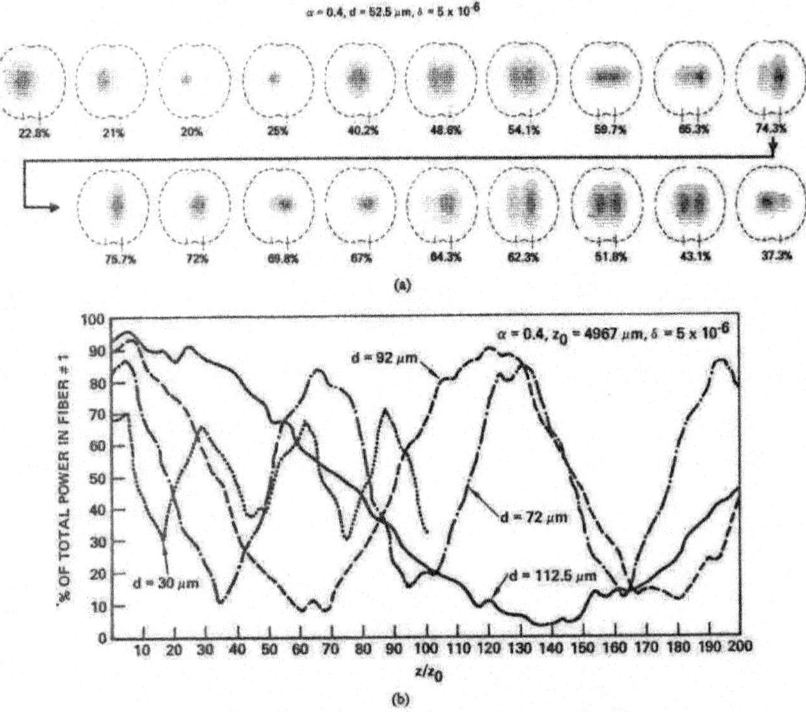

Fig. 9. (a) A cross-sectional view of the power density distribution for the guided wave. Note the power exchange phenomenon as the wave propagates along the coupled structure. The percent value indicates the percent of total power contained in the right-hand half of the structure (i.e., fiber 2). The distance between each frame is 9934 μm. (b) Percent of total power in the left-hand half of the structure (i.e., fiber 1) as a function of the axial distance along the coupling structure. This is the multimode coupling case.

of the power density distribution for the guided beam as it travels down the coupled structure and the percent of total power in one-half of the structure as a function of the axial distance.

V. Conclusions

This is a rather subjective review of the available analytic/numerical techniques to treat problems dealing with guided waves in optical waveguide structures. It is unavoidable to conclude from the above discussion tha significant advances in the theoretical treatment of optica guided wave problems in the last decade have been in the numerical/computational areas. This is not to minimize the traditional thought that analytic expressions for simplified/idealized structures provide a great deal of insight into many physical phenomenon of fundamental importance. However, from the engineering standpoint, it is not sufficient just to understand the physical phenomenon of wave interaction taking place in an idealized structure but rather, one must be able to generate solid quantitative data for practical complex guiding structures. Furthermore, the trend toward data bus multiplexing [19] (the interconnection of a number of spatially distributed terminals via fiber optic waveguide cables) as well as integration of optical circuits for data processing [20] (integrated optics) point to the development of optical waveguide components such as couplers, branches, tapers, and integrated optical circuits. The only viable means of analyzing these complex structures is to make use of the highly versatile techniques using computers.

References

[1] D. Handros and P. Debye, *Ann. der Physik*, vol. 32, p. 465, 1910.
[2] H. Zahn, *Ann. der Physik*, vol. 49, p. 907, 1916.
[3] J. K. Carson, S. P. Mead, and S. A. Schelkunoff, *Bell Syst. Tech. J.*, vol. 15, p. 310, 1936.
[4] W. Elsasser, *J. Appl. Phys.*, vol. 20, p. 1193, 1949.
[5] C. Chandler, *J. Appl. Phys.*, vol. 20, p. 1188, 1949.
[6] E. Snitzer, *J. Opt. Soc. Amer.*, vol. 51, p. 491, 1961.
[7] C. Yeh, *J. Appl. Phys.*, vol. 33, p. 3235, 1962.
[8] J. E. Goell, *Bell Syst. Tech. J.*, vol. 48, p. 2133, 1969.
[9] E. A. J. Marcatili, *Bell Syst. Tech. J.*, vol. 48, p. 2071, 1969.
[10] A. W. Snyder, *IEEE Trans. Microwave Theory Tech.*, vol. MTT-17, p. 1138, 1969.
[11] D. Gloge, *Appl. Opt.*, vol. 10, p. 2442, 1971.
[12] C. Yeh and G. Lindgren, *Appl. Opt.*, vol. 16, p. 483, 1977.
[13] C. Yeh, L. Casperson, and B. Szejn, *J. Opt. Soc. Amer.*, vol. 68, 1978.
[14] C. Yeh, W. P. Brown, and R. Szejn, *Appl. Opt.*, vol. 18, p. 489, 1979.
[15] C. Yeh, K. Ha, S. B. Dong, and W. P. Brown, *Appl. Opt.*, vol. 18, p. 1596, 1979.
[16] L. Eyges, P. Gianino, and P. Wintersteiner, "Modes of dielectric waveguides of arbitrary cross-sectional shapes," *Appl. Opt.*, to be published.
[17] C. Yeh, "Advances in communication through light fibers," in *Advances in Communication Systems*. vol. 4, Theory and Applications, New York: Academic, 1975.
[18] C. Yeh, L. Casperson, and W. P. Brown, *Appl. Phys. Lett.*, vol. 19, p. 456, 1979.
[19] H. Taylor, *Appl. Opt.* vol. 17, p. 1493, 1978.
[20] D. Marcuse, *Integrated Optics*. NY: IEEE Press, 1972.

C. W. Yeh (S'56–M'63), photo and biography not available at time of publication.

IEEE TRANSACTIONS ON MICROWAVE THEORY AND TECHNIQUES, VOL. MTT-33, NO. 9, SEPTEMBER 1985 789

Dynamic Three-Dimensional TLM Analysis of Microstriplines on Anisotropic Substrate

G. E. MARIKI AND C. YEH, FELLOW, IEEE

Abstract—The frequency-dependent propagation characteristics of a hybrid mode along microstriplines on anisotropic substrates are presented for the case where the constitutive parameter tensors may be diagonalized. A generalization of the three-dimensional transmission-line-matrix (TLM) numerical procedure is used to obtain results for the phase constant β, effective permittivity ϵ_{eff}, and the characteristic impedance Z, all as functions of frequency and the shape ratio (w/h). Also shown are results for coupled microstrips on a sapphire substrate.

I. Introduction

Although the work reported in this paper was carried out several years ago (but unpublished) [1], recent interest in high-frequency (millimeter and submillimeter wave-length) microstrip circuits has prompted us to publish our results [2]. It appears that the transmission-line-matrix (TLM) approach that we used to solve the many problems dealing with enclosed microstrips with anisotropic and/or inhomogeneous substrates is rapidly becoming a very acceptable and viable way of dealing with these problems in spite of its demand for large computer memory and time [3], [4]. This is because of the enormous decrease in computational cost as well as the realization of the simplicity and versatility of the TLM approach as compared with other available numerical or analytical means [5]–[7].

Most of the original work on microstrip was based on a TEM approximation mainly because the resultant formulation is vastly simplified, and the solutions obtained agree closely with experimental results in the low-frequency range (below X-band). That this is so is exemplified by the pioneering works of Wheeler [8], using a conformal mapping technique, and Silvester [9] who applied a Green's function formulation. As the need arose for hybrid integrated circuits operating at frequencies as high as 20 GHz and above, the TEM solutions were no longer satisfactory since dispersion, which is significant at high frequencies, is ignored under a TEM approximation. Several techniques have been applied to determine the dispersion properties of microstrip. Among these were Getsinger's empirical formulation [10] and the TLM method used by Akhtarzad and Johns [11], [12]. The above-mentioned authors have presented results for the phase constant β and/or effective permittivity ϵ_{eff} as functions of frequency. Their results were strictly for isotropic substrates.

One of the practical design difficulties of using isotropic substrates, such as alumina, is the significant variation in the dielectric permittivity from different manufacturers or even from batch to batch from the same manufacturer. This essentially means that repeated measurements of the dielectric permittivity are required for accurate design of microstrip circuits. The use of anisotropic substrates with stable electrical properties, such as sapphire, alleviates this difficulty although it introduces a new problem in that new techniques have to be developed to analyze microstriplines on anisotropic substrates. The TEM approach was again used by Owens, Aitken, and Edwards [13] and Alexopoulos *et al.* [14] to determine the quasi-static properties of microstriplines on anisotropic substrates with a diagonal dielectric tensor $\underline{\epsilon}$. Measurements of the dispersion characteristics of microstrip on sapphire substrates were also reported by Edwards and Owens [15]. The TEM solutions mentioned above suffer from the same limitations as those developed for isotropic substrates. The need for an efficient technique to determine the frequency-dependent (dynamic) propagation characteristics of microstriplines on anisotropic substrates is therefore apparent. This was the motivation behind our work which was completed in 1978. Since then a number of publications have appeared describing various techniques [5]–[7].

In the following, a description of the modified TLM technique applied to the enclosed microstrip on anisotropic substrate problems will be given. Numerical results with a discussion on the accuracy of the TLM technique will then be presented. Also shown are further examples showing the application to microstrip on periodically varying substrates as well as the coupled microstrips on anisotropic substrate.

II. The Transmission-Line-Matrix Technique

The geometry of the shielded microstripline is shown in Fig. 1. It consists of a conductive metallic strip of width w and zero thickness placed on a dielectric substrate of thickness h and a box-type enclosure of a perfectly conducting material. The thickness of the metallic strip is assumed to be vanishingly small and the properties of the

Manuscript received January 23, 1985; revised May 1, 1985.

The authors are with the Electrical Engineering Department, University of California at Los Angeles, Los Angeles, CA 90024.

0018-9480/85/0900-0789$01.00 ©1985 IEEE

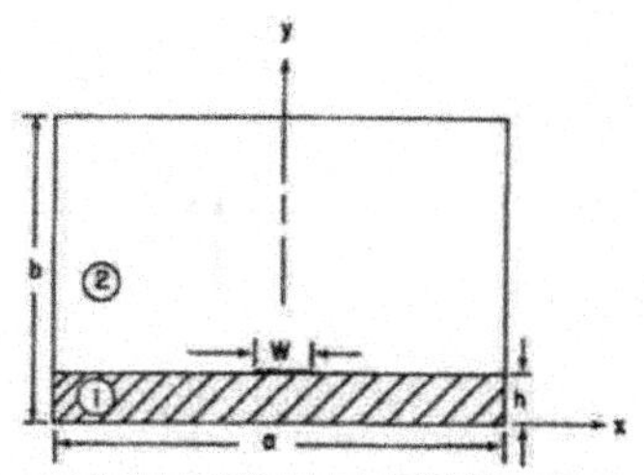

Fig. 1. Cross-sectional geometry of a single microstripline.

substrate are characterized by diagonal tensors of rank two, i.e., a dyadics, for the permittivity $\underline{\underline{\epsilon}}$, permeability $\underline{\underline{\mu}}$, and conductivity $\underline{\underline{\sigma}}$. The tensor elements are, in general, functions of the cross-sectional coordinates x and y in a rectangular coordinate system. Thus

$$\underline{\underline{\epsilon}} = \epsilon_0 \begin{bmatrix} \epsilon_{xx}(x, y) & 0 & 0 \\ 0 & \epsilon_{yy}(x, y) & 0 \\ 0 & 0 & \epsilon_{zz}(x, y) \end{bmatrix} \tag{1}$$

$$\underline{\underline{\mu}} = \mu_0 \begin{bmatrix} \mu_{xx}(x, y) & 0 & 0 \\ 0 & \mu_{yy}(x, y) & 0 \\ 0 & 0 & \mu_{zz}(x, y) \end{bmatrix} \tag{2}$$

$$\underline{\underline{\sigma}} = \begin{bmatrix} \sigma_{xx}(x, y) & 0 & 0 \\ 0 & \sigma_{yy}(x, y) & 0 \\ 0 & 0 & \sigma_{zz}(x, y) \end{bmatrix} \tag{3}$$

where ϵ_0 and μ_0 are the free-space permittivity and permeability, respectively. Maxwell equations in component form are

$$\frac{\partial E_z}{\partial y} - \frac{\partial E_y}{\partial z} = -\mu_0 \mu_{xx}(x, y) \frac{\partial H_x}{\partial t} \tag{4}$$

$$\frac{\partial E_x}{\partial z} - \frac{\partial E_z}{\partial x} = -\mu_0 \mu_{yy}(x, y) \frac{\partial H_y}{\partial t} \tag{5}$$

$$\frac{\partial E_y}{\partial x} - \frac{\partial E_x}{\partial y} = -\mu_0 \mu_{zz}(x, y) \frac{\partial H_z}{\partial t} \tag{6}$$

$$\frac{\partial H_z}{\partial y} - \frac{\partial H_y}{\partial z} = (\sigma_{xx}(x, y) + \epsilon_0 \epsilon_{xx}(x, y)) \frac{\partial}{\partial t} E_x \tag{7}$$

$$\frac{\partial H_x}{\partial z} - \frac{\partial H_z}{\partial x} = (\sigma_{yy}(x, y) + \epsilon_0 \epsilon_{yy}(x, y)) \frac{\partial}{\partial t} E_y \tag{8}$$

$$\frac{\partial H_y}{\partial x} - \frac{\partial H_x}{\partial y} = (\sigma_{zz}(x, y) + \epsilon_0 \epsilon_{zz}(x, y)) \frac{\partial}{\partial t} E_z. \tag{9}$$

The TLM technique will be used to solve the above set of equations together with the appropriate boundary conditions. We shall assume that the guided wave is propagating in the z-direction

$$E, H \sim e^{-j\beta z}$$

where β is the propagation constant.

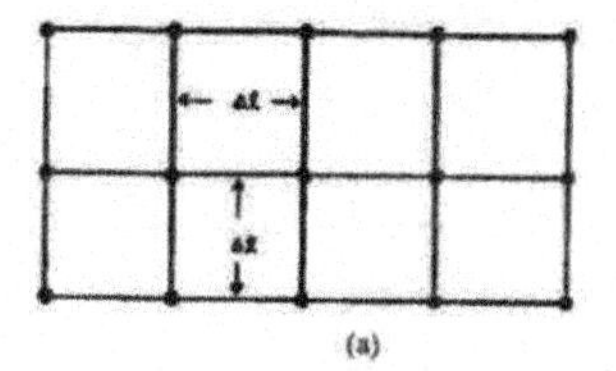

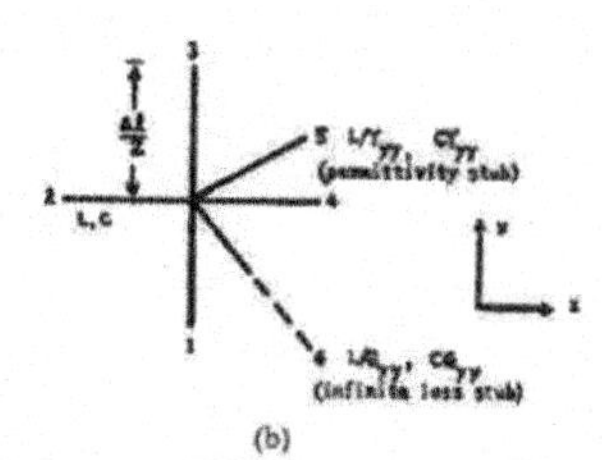

Fig. 2. (a) Shunt-connected TEM lines. (b) A generalized shunt node.

A. The Shunt Node

Fig. 2(a) shows how a number of ideal lossless two-wire transmission lines can be connected to form a two-dimensional transmission-line matrix. Shunt nodes are formed where the lines cross and these present impedance discontinuities for waves propagating along the lines. The internodal separation Δl is uniform throughout the matrix. Note that for clarity in Fig. 2(a), single lines are used to represent a transmission-line pair. A generalized shunt node is depicted in Fig. 2(b). Here an open-circuited shunt stub of length $\Delta l/2$ and normalized characteristic admittance Y_{yy} (normalized with respect to the characteristic impedance of the main line) is attached to the node. This is called the permittivity stub. Also shown in Fig. 2(b) is an infinite loss stub of normalized characteristic admittance G_{yy}. Hence, one notices that a pulse injected into the loss stub from the node will be completely lost. A lumped parameter equivalent network for an elementary matrix section is shown in Fig. 3. In this equivalent network, the short section of an open-circuited transmission line (the permittivity stub) is represented by a shunt admittance of value $(jY_{yy}/\sqrt{L/C})\tan(\omega\Delta l/2c)$. If $\omega\Delta l/2c = \pi\Delta l/\lambda \ll 1$, then $\tan(\omega\Delta l/2c) \simeq (\omega\Delta l/2)\sqrt{LC}$. Therefore, the admittance of the open-circuited stub is $\simeq j\omega CY_{yy}\Delta l/2$. This can be recognized as a capacitive admittance. Hence, the total capacitance at the node is

$$C' = 2C\left(1 + \frac{Y_{yy}}{4}\Delta l\right). \tag{10}$$

The infinite loss stub of normalized characteristic admittance G_{yy} is represented by a lumped conductance of

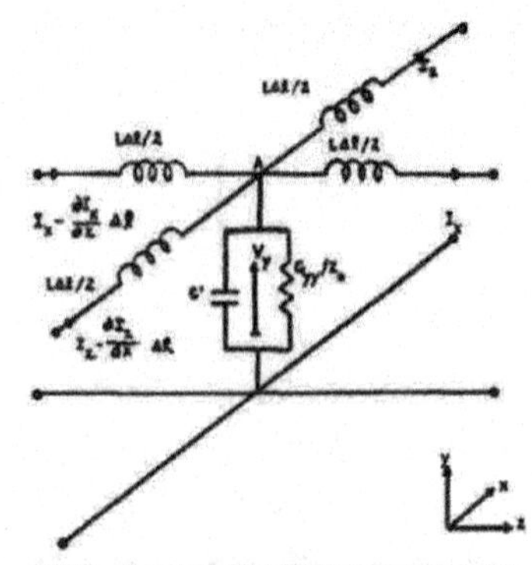

Fig. 3. Shunt node lumped network representation.

magnitude

$$G_{yy}\sqrt{\frac{C}{L}} = \frac{G_{yy}}{Z_0}$$

where

$$Z_0 = \sqrt{\frac{L}{C}}. \quad (11)$$

Application of Kirchoff's Current Law at node A gives (see Fig. 3)

$$\frac{\partial I_x}{\partial z} - \frac{\partial I_z}{\partial x} = \left[\frac{G_{yy}}{Z_0\,\Delta l} + 2C\left(1 + \frac{Y_{yy}}{4}\right)\frac{\partial}{\partial t}\right]V_y. \quad (12)$$

The above analysis was carried out for a shunt node in the $x-z$ plane. A similar analysis for a shunt node in the $x-y$ and $y-z$ planes leads to, respectively,

$$\frac{\partial I_y}{\partial x} - \frac{\partial I_x}{\partial y} = \left(\frac{G_{zz}}{Z_0\,\Delta l} + 2C\left(1 + \frac{Y_{zz}}{4}\right)\frac{\partial}{\partial t}\right)V_z \quad (13)$$

and

$$\frac{\partial I_z}{\partial_y} - \frac{\partial I_y}{\partial z} = \left(\frac{G_{xx}}{Z_0\,\Delta l} + 2C\left(1 + \frac{Y_{xx}}{4}\right)\frac{\partial}{\partial t}\right)V_x. \quad (14)$$

One notes that the subscripts for the currents do not correspond to the coordinate direction along which the current flows; the subscripts are assigned at the series node (see Fig. 4) where a series node in, say, the $y-z$ plane has common node current I_x, etc. The choice of the subscripts makes it easier to identify the line equations with Maxwell's equations. The use of stubs, each with a different characteristic admittance, allows the TLM model to represent anisotropic media correctly.

From (7) and (14), the following equivalences between the TLM equation and Maxwell's equations can be identified:

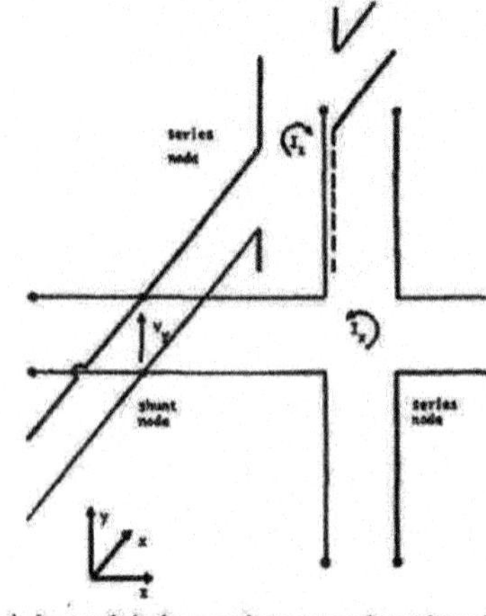

Fig. 4. A shunt node in the $x-z$ plane connected to series nodes in the $y-z$ and $x-y$ planes.

$$H_z = I_z \quad (15)$$

$$H_y = I_y \quad (16)$$

$$\sigma_{xx} = \frac{G_{xx}}{Z_0\,\Delta l} \quad (17)$$

$$\epsilon_0 = 2C \quad (18)$$

$$\epsilon_{xx} = \frac{4 + Y_{xx}}{4} \quad (19)$$

$$E_x = V_x. \quad (20)$$

Similarly, from (8) and (12), we have

$$H_x = I_x \quad (21)$$

$$E_y = V_y \quad (22)$$

$$\sigma_{yy} = \frac{G_{yy}}{Z_0\,\Delta l} \quad (23)$$

$$\epsilon_{yy} = \frac{4 + Y_{yy}}{4}. \quad (24)$$

Equations (9) and (13) yield the following equivalences:

$$E_z = V_z \quad (25)$$

$$\sigma_{zz} = \frac{G_{zz}}{Z_0\,\Delta l} \quad (26)$$

$$\epsilon_{zz} = \frac{4 + Y_{zz}}{4}. \quad (27)$$

From the above analysis, it is concluded that half of Maxwell's equations can be fully accounted for by three shunt nodes oriented in the $x-y$, $y-z$, and $x-z$ planes. The remaining half of Maxwell's equations will be satisfied by the series node.

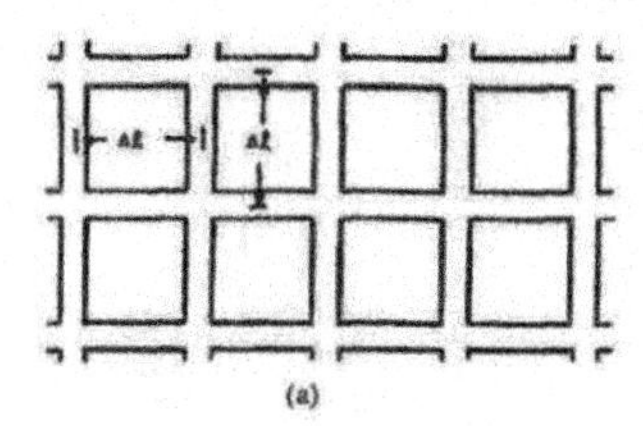

(a)

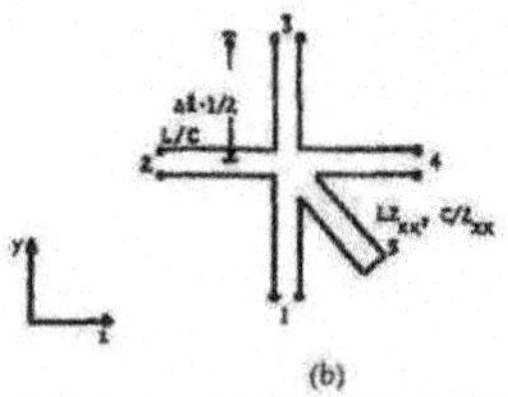

(b)

Fig. 5. (a) Series-connected TEM lines. (b) A generalized series node.

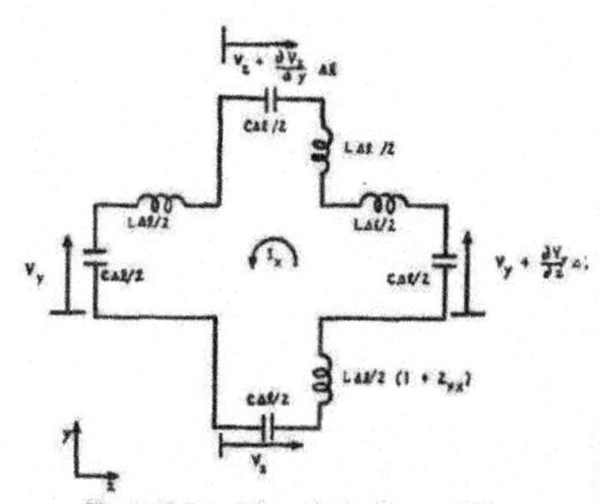

Fig. 6. Series node lumped network representation.

B. The Series Node

A series-connected two-dimensional transmission-line matrix is shown in Fig. 5(a). As in the shunt-connected TLM, the lines are ideal and lossless and the uniform mesh parameter is Δl. Fig. 5(b) shows a generalized series node which is equipped with a short-circuited stub called the permeability stub. The equations satisfied by the series node can be derived using the lumped parameter representation shown in Fig. 6. The input impedance of the short-circuited transmission line is

$$Z_{in} = jZ_{xx}\sqrt{\frac{L}{C}}\tan(\omega\Delta l/2c) \approx j\omega LZ_{xx}\Delta l/2. \quad (28)$$

The validity of (28) is the same as was stipulated for the short open-circuited stub associated with the shunt node, namely,

$$\frac{\omega\Delta l}{2c} = \frac{\pi\Delta l}{\lambda} \ll 1. \quad (29)$$

Equation (28) represents an inductive impedance, the magnitude of the inductance being

$$L' = \left(\frac{Z_{xx}}{2}\Delta l\right)L. \quad (30)$$

Application of Kirchoff's Voltage Law around the loop at the series node (Fig. 6) gives

$$\frac{\partial V_z}{\partial y} - \frac{\partial V_y}{\partial z} = 2L\left(1+\frac{Z_{xx}}{4}\right)\frac{\partial I_x}{\partial t}. \quad (31)$$

A series node in the $x-y$ and $x-z$ planes will, respectively, satisfy the following equations:

$$\frac{\partial V_y}{\partial x} - \frac{\partial V_x}{\partial y} = 2L\left(1+\frac{Z_{zz}}{4}\right)\frac{\partial I_z}{\partial t} \quad (32)$$

and

$$\frac{\partial V_x}{\partial z} - \frac{\partial V_z}{\partial x} = 2L\left(1+\frac{Z_{yy}}{4}\right)\frac{\partial I_y}{\partial t}. \quad (33)$$

The following additional equivalences complete the TLM modeling of Maxwell's equations:

$$\mu_0 \equiv 2L \quad (34)$$

$$\mu_{xx} \equiv \frac{4+Z_{xx}}{4} \quad (35)$$

$$\mu_{yy} \equiv \frac{4+Z_{yy}}{4} \quad (36)$$

$$\mu_{zz} \equiv \frac{4+Z_{zz}}{4}. \quad (37)$$

We have demonstrated that the use of three shunt and three series nodes oriented in the $x-y$, $x-z$, and $y-z$ planes enables all of Maxwell's equations for an inhomogeneous anisotropic medium to be accounted for properly. What remains is to devise a way of interconnecting the six nodes to form a three-dimensional node in space. One notices that no restriction on the spatial distribution or magnitudes of $\underline{\epsilon}$, $\underline{\mu}$, and $\underline{g}$ has been placed.

C. The Three-Dimensional Node

Fig. 7 shows the construction of a three-dimensional node capable of representing three-dimensional space and satisfying Maxwell's equations for inhomogenous anisotropic media. The three-dimensional node is an interconnection of three shunt nodes and three series nodes structured in such a manner that there is one shunt node and one series node in each coordinate plane. The use of these six nodes results in a model that properly accounts for all of Maxwell's field equations. Again, in Fig. 7, a single line is used to represent a transmission-line pair. The nodes are named to correspond to the field quantity they represent.

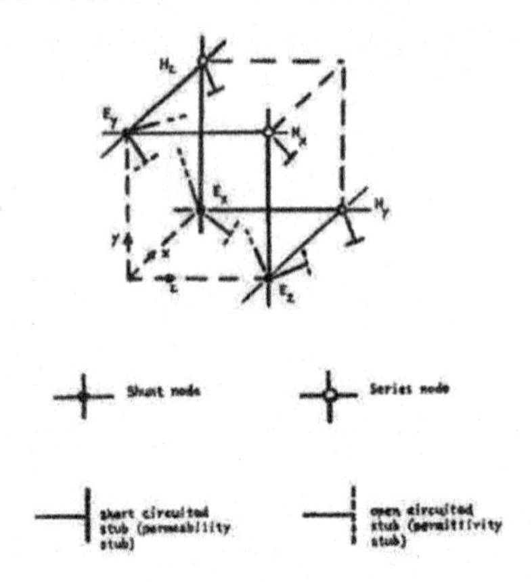

Fig. 7. A three-dimensional node.

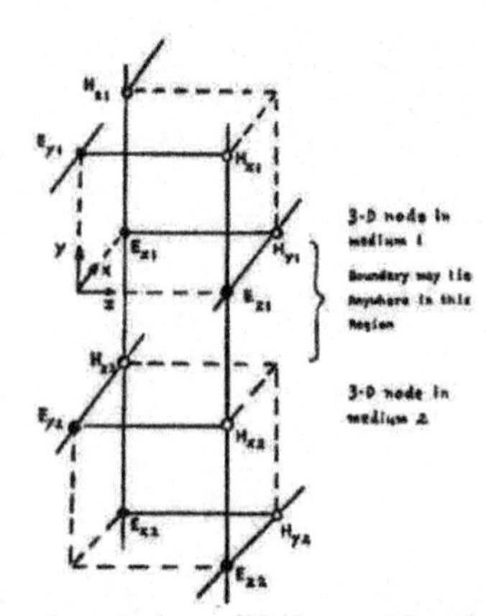

Fig. 8. Continuity of tangential fields across a dielectric boundary.

Thus, the common voltage at shunt node E_x corresponds to the x component of the electric field. The common current at series node H_x corresponds to the x component of the magnetic field, and so on.

To represent a three-dimensional propagation space, a number of these three-dimensional nodes are connected to form a three-dimensional mesh network. Maxwell's equations are thus satisfied at each three-dimensional node.

D. Boundary Conditions in the TLM Model

1) Electric and Magnetic Walls: In the plane of an electric wall, the tangential electric field must vanish. Similarly, in the plane of a magnetic wall, the tangential magnetic field must be zero. Since the corresponding quantities for the electric and magnetic field in the TLM model are the voltage and current in the transmission lines, electric and magnetic walls can be easily simulated in the TLM model by short-circuiting and open-circuiting the nodes, respectively. For example, to set E_x and E_y equal to zero in a particular plane, all shunt nodes E_x and E_y lying in that plane are shorted. To set, say, H_y and H_z equal to zero in some plane, the series nodes H_y and H_z in that plane are simply open-circuited.

2) Dielectric Boundary: The continuity of tangential electric and magnetic fields across a dielectric/dielectric boundary is automatically satisfied in the TLM model when the three-dimensional nodes are joined up by elementary sections of ideal transmission lines. For example, for a dielectric/dielectric boundary in the $x-z$ plane as shown in Fig. 8, since the common voltages at the shunt nodes correspond to the electric field and the common currents at the series nodes correspond to the magnetic field, the following equations valid for a transmission-line element joining the nodes on either side of the boundary are applicable:

$$E_{z1} = E_{z2} + \frac{\partial E_{z2}}{\partial y}\Delta l \tag{38}$$

$$E_{x1} = E_{x2} + \frac{\partial E_{x2}}{\partial y}\Delta l \tag{39}$$

$$H_{x1} = H_{x2} + \frac{\partial H_{x2}}{\partial y}\Delta l \tag{40}$$

$$H_{z1} = H_{z2} + \frac{\partial H_{z2}}{\partial y}\Delta l. \tag{41}$$

Since the voltage and current in the transmission line are smooth functions of position along the line, the continuity of the tangential fields across a boundary placed in between the nodes is assured.

E. The Numerical Procedure

1) Series and Shunt Nodes Scattering Matrices: In the TLM method, the numerical procedure involves determination of the impulse response of the network. Delta-function impulses are introduced at various locations in the matrix and these travel along the ideal transmission lines at the speed of light before being scattered at the nodes. Any of the six field components may be excited initially by specifying initial impulses at the appropriate nodes. Likewise, the response for any of the field components may be monitored by recording the pulses that pass through the relevant nodes.

The shunt and series nodes represent impedance discontinuities to the traveling pulses. From Figs. 2(b) and 3(b), the voltage scattering matrices for the shunt and series

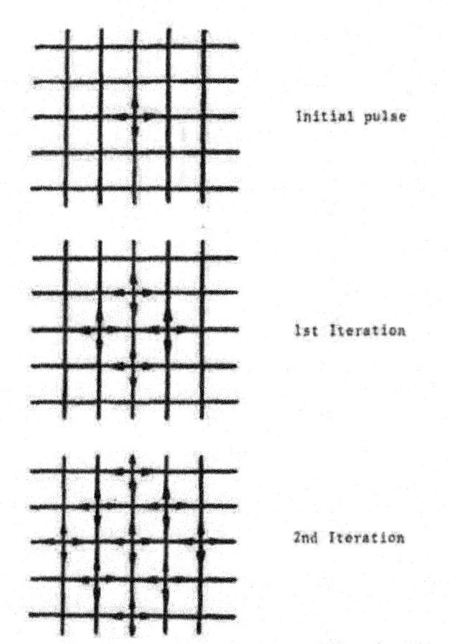

Fig. 9. Propagation of a point source on a two-dimensional homogeneous TLM model.

nodes are

$$[S]_{sh}=\frac{2}{Y}\begin{bmatrix}1&1&1&1&Y_{jj}\\1&1&1&1&Y_{jj}\\1&1&1&1&Y_{jj}\\1&1&1&1&Y_{jj}\\1&1&1&1&Y_{jj}\end{bmatrix}-[I] \tag{42}$$

where

$$Y=4+Y_{jj}+G_{jj} \tag{43}$$

$[I]$ is the unit matrix and *sh* stands for shunt. The subscripts *jj* may be *xx*, *yy*, or *zz* as appropriate. For the series node, we have

$$[S]_s=\frac{2}{Z}\begin{bmatrix}-1&1&1&-1&-1\\1&-1&-1&1&1\\1&-1&-1&1&1\\1&1&1&-1&-1\\-Z_{jj}&Z_{jj}&Z_{jj}&-Z_{jj}&-Z_{jj}\end{bmatrix}+[I] \tag{44}$$

where

$$Z=4+Z_{jj}. \tag{45}$$

Thus, an impulse impinging on a shunt node would be scattered in accordance with (42) and an impulse incident on a series node would be scattered following (44).

2) Pulse Propagation in the TLM Model: The propagation of pulses in the TLM model is illustrated in Fig. 9, where the first two iterations following an initial excitation pulse in a two-dimensional shunt-connected transmission-line matrix are shown. For simplicity, we assume free-space propagation, in which case there are no open-circuited shunt stubs connected to the nodes. The distance between the nodes is Δl and the time interval between iterations is $t_0=\Delta l/c$ since the individual pulses travel at the speed of light. Upon impinging on a node, the pulse is scattered into the four coordinate directions in accordance with (42).

Propagation in a three-dimensional model can be visualized in a similar manner. In this case, the pulses are scattered at the shunt nodes as well as the series nodes.

3) Form of Output and Accuracy of Results: The output response function consists of a train of impulses of varying magnitude in the time domain separated by a time interval $\Delta l/c$. Thus, the theoretical frequency response obtained by taking the Fourier transform of the output response consists of a series of delta functions in the frequency domain corresponding to the modal resonant frequencies of the cavity for which a solution exists. Truncation of the output impulse response function (due to practical reasons) causes a spreading of the solution delta function into $\sin x/x$ type curves.

Let $V_{out}(t)$ be the output impulse function taken for N iterations of the matrix, i.e., the function starts at time $t=0$ and finishes at time $t=N\Delta l/c$. $V_{out}(t)$ may be regarded as an impulse function $V_\infty(t)$, extending to time = infinity multiplied by a rectangular time function $V_p(t)$ of unit height and width $N\Delta l/c$ as follows:

$$V_{out}(t)=V_\infty(t)\times V_p(t) \tag{46}$$

where

$$V_p(t)=\begin{cases}1, & 0\leqslant t\leqslant \dfrac{N\Delta l}{c}\\ 0, & \text{elsewhere}\end{cases}. \tag{47}$$

If the Fourier transform of $V_{out}(t)$, $V_\infty(t)$, and $V_p(t)$ are $S_{out}(f)$, $S_\infty(f)$, and $S_p(f)$, respectively, then the Fourier transform of (46) is given by the convolution of $S_\infty(f)$ and $S_p(f)$, i.e.,

$$S_{out}(f)=\int_{-\infty}^{\infty}S_\infty(\alpha)S_p(f-\alpha)\,d\alpha \tag{48}$$

and

$$S_p(f)=\frac{N\Delta l}{c}\frac{\sin\frac{\pi N\Delta lf}{c}}{\frac{\pi N\Delta lf}{c}}e^{-j(\pi N\Delta lf/c)}. \tag{49}$$

Equations (47) and (48) indicate that $S_p(f)$, a curve of $\sin x/c$ form, is placed in each of the positions of the delta functions of the exact response $S_\infty(f)$, in both the positive and negative frequency planes. The accuracy of the result depends on the number of iterations N, since the greater N, the sharper the maximum peak of the curve. The accuracy is also affected by interference from the tail regions of the reflected solutions in the negative frequency plane as well as from the tail regions of neighboring solution points corresponding to other modes of propagation.

III. Accuracy of the TLM Approach

To study the accuracy of the TLM approach described above, we shall compare our results with those obtained using quasi-static approximations. Specifically, we shall consider the comparison of the single microstrip problem for which quasi-static solutions have been presented [13], [14]. The confidence we gain from this study will enable us to use this TLM technique to treat other microstrip problems and present results which have not been analyzed elsewhere.

Quasi-static solutions for the single microstrip problem have been computed by Owens *et al.* [13] and Alexopoulos *et al.* [14]. Owens *et al.* have applied the method of finite differences to compute the capacitance per unit length C. Knowing the value of C, they were able to compute the low-frequency effective dielectric constant ϵ_{e0}. An analytical expression for an equivalent isotropic dielectric constant ϵ_{req} was developed. The implication is that a microstripline on an isotropic substrate with a dielectric constant of ϵ_{req} would exhibit the same electrical behavior as a microstripline on sapphire. The expression given by Owens *et al.* is [13]

$$\epsilon_{req} = 12.0 - \frac{1.21}{1 + 0.39(\log(10w/h))^2}. \tag{50}$$

It must be emphasized that this equation is applicable to sapphire substrates only. Alexopoulos *et al.* approached the problem of a microstripline on an anisotropic substrate again from a quasi-static basis [14]. Their method is based on a Green's function formulation to compute the static capacitance of the line. Their results are presented in terms of the variation of the phase velocity with w/h.

Experimental measurements on the dispersion characteristics of microstrip on sapphire have been reported by Edwards and Owens [15]. In attempting to fit Getsinger's dispersion formula to their experimental results, they found an empirical formula for Getsinger's G factor for sapphire substrates.

The following scheme was adopted to verify the TLM solutions:

1) ϵ_{e0} and z_0 were determined from the TLM dispersion curves.
2) ϵ_{req} was computed from (50).
3) Getsinger's dispersion formula was used to determine ϵ_{eff} and this was compared with the TLM solutions.

Single crystal sapphire is a uniaxial crystal characterized by a dielectric tensor

$$\underline{\epsilon} = \epsilon_0 \begin{bmatrix} \epsilon_{xx} & 0 & 0 \\ 0 & \epsilon_{yy} & 0 \\ 0 & 0 & \epsilon_{zz} \end{bmatrix} \tag{51}$$

with $\epsilon_{xx} = \epsilon_{zz} = 9.4$ and $\epsilon_{yy} = 11.6$. In writing (51), we have assumed that the crystal is oriented such that the y-axis is parallel to the optical axis. When used in microstrip circuits, sapphire substrates are usually cut with their plane surfaces perpendicular to the optical axis so that the material is constant everywhere in the plane of the substrate. A propagating wave along the microstripline is then not subjected to a change in permittivity at bends or corners in the line.

The geometry of the problem is shown in Fig. 1. The fundamental mode of propagation has even symmetry about the y-axis, i.e., at $x = 0$. Hence, the boundary conditions specified in the input data of the TLM computer program are:

1) $E_x = 0$ and $E_z = 0$ along $y = 0$ and $y = b$.
2) $E_y = 0$ and $E_z = 0$ along $x = a$.
3) $H_y = 0$ and $H_z = 0$ along $x = 0$.
4) $E_x = 0$ and $E_z = 0$ for $y = h$ and $-w/2 \leq x \leq w/2$.

The dielectric material was assumed to be lossless and so were the walls of the enclosure. Hence, to satisfy the above boundary conditions, infinite conductivity was specified along all electric walls, and even-mode symmetry about the y-axis was imposed thus making this axis a magnetic wall. Initially, the minimum number of nodes (corresponding to $h = \Delta l$) were used to obtain the dispersion curves for $w/h = 3,5$. Dispersion analysis by the TLM technique involves resonating a section of the transmission line by placing shorting planes along the axis of propagation (the z-axis in this case), such that the images of the line in the shorting planes appear to be continuations of the structure. Each separation of the shorting planes then equals half of the guided wavelength for the fundamental mode at the frequency given by the resonant frequency of the cavity. If the distance between the shorting planes is $2L$, the phase constant is given by $\beta = \pi/2L$. For TEM waves, β is a linearly increasing function of frequency since the phase velocity is constant and uniform throughout the medium.

In Figs. 10 and 11, the TLM results for the dispersion curves depicting the phase constant as a function of frequency for $w/h = 3$ and $w/h = 5$ are shown. In both cases, the substrate height was $h = \Delta l$ and $b/h = 6$. Also shown in the same figures are the quasi-static solutions of Alexopoulos [14] and Getsinger's dynamic solutions computed as outlined above [10].

In executing the TLM program, 1000 iterations were used to ensure convergence of the solution. This figure on the number of iterations was arrived at after several runs of the entire program for iterations between 200 and 2000. Beyond 1000 iterations, the change in the resonant frequency of the cavity was less than 1.0 percent. For iterations less than 400, the TLM results were rather unstable, varying by as much as ± 10 percent from the convergent solution. This is because the peak of the $\sin x/x$ computer output response curve is not clearly defined for a low number of iterations (N). Also, the proximity of neighboring solution points corresponding to higher order even symmetry modes results in a larger error when N is small.

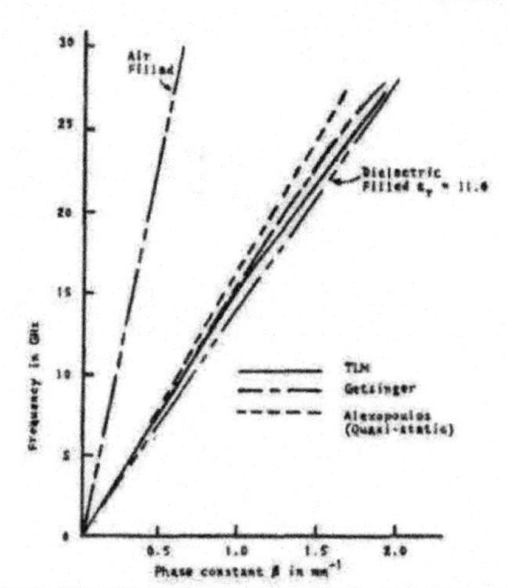

Fig. 10. Dispersion diagram for single microstrip on sapphire substrate $w/h = 3.0$, $h = \Delta l$.

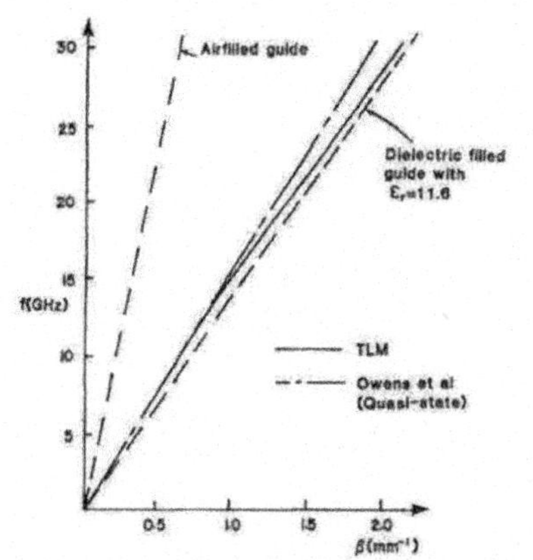

Fig. 11. Dispersion curves for the phase constant β ($w/h = 5.0$, $h = \Delta l$). Single microstrip on sapphire substrate.

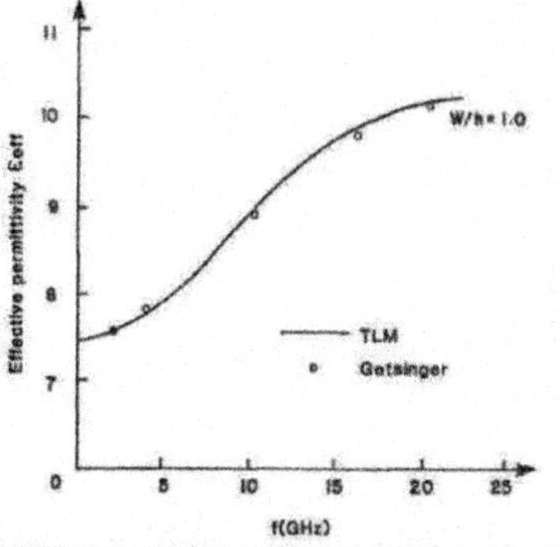

Fig. 12. Effective permittivity versus frequency ($w/h = 1.0$, $h = 3\Delta l$). Single microstrip on sapphire substrate.

To enhance the fundamental-mode field configuration, E_y was excited at all the nodes lying directly below the strip, and E_x was excited along the edge of the strip to provide the correct biasing for the field configuration of the fundamental mode. This choice was made again after a detailed study of the convergence of the results with respect to the manner in which the network was excited. Convergence was much faster if the approximate field distribution was specified at the onset and the network excited to enhance this mode. We note that the use of symmetry greatly reduces the storage requirements; this fact was fully exploited throughout the TLM analyses for the various problems treated in this paper.

Even though a very small number of nodes were used to describe the geometry of Fig. 1, the TLM solution was in very good agreement with Getsinger's dynamic solution (within 7 percent) and the low-frequency results agreed entirely with the quasi-static results due to Alexopoulos (see Figs. 10 and 11). Also shown in the same figures are the solutions for the air-filled and dielectric-filled microstrip.

Increasing the number of nodes should improve the accuracy of the TLM results. This was investigated by using two mesh points for the substrate, i.e., putting $h = 2\Delta l$. As before, E_y was excited at all nodes lying under the strip and E_x at all nodes along the edge of the strip. Agreement with Getsinger's result was improved from about 7 percent to better than 2 percent. The TLM results for this case are shown in Fig. 10 and 11. The quasi-static solutions obtained by Owens *et al.* using the method of finite differences are also shown for comparison at low frequencies [13]. Fig. 12 shows the variation of the effective permittivity with frequency for the case $w/h = 1$. Here, $h = 3\Delta l$ and agreement with Getsinger's result was extremely good—better than 0.5 percent for the most part.

A special feature of the TLM technique is the potential ability to determine the relative magnitudes of all the six electromagnetic-field components at a particular frequency for a specific mode of propagation, thus enabling a full hybrid-mode solution to be obtained with each run of the computer program. Defining the characteristic impedance of the microstripline as $Z_0 = E_y/H_x$ for a node lying directly below the center of the strip, we computed the frequency dependence of the characteristic impedance of the line.

The computation of the characteristic impedance was carried out simultaneously with the determination of the dispersion curves presented earlier, i.e., the voltage at the shunt node E_y and the current at series node H_x are recorded at the end of each iteration. The results are shown

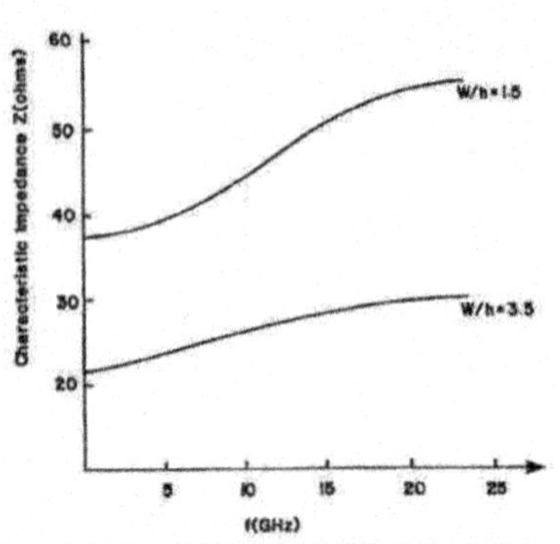

Fig. 13. Characteristic impedance versus frequency. Single microstrip on sapphire substrate with $h = 1.0$ mm.

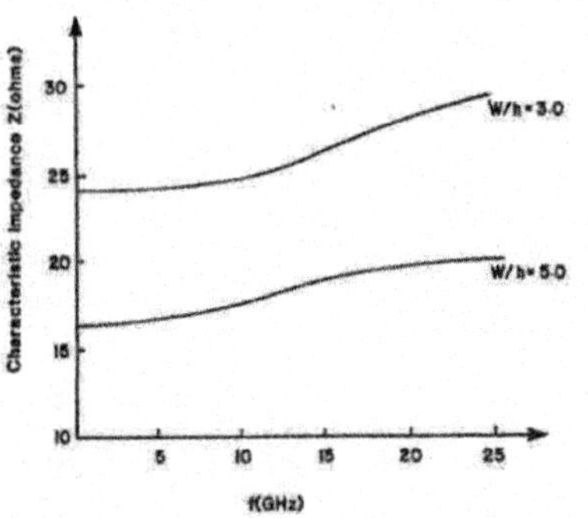

Fig. 14. Characteristic impedance versus frequency. Single microstrip on sapphire substrate with $h = 0.5$ mm.

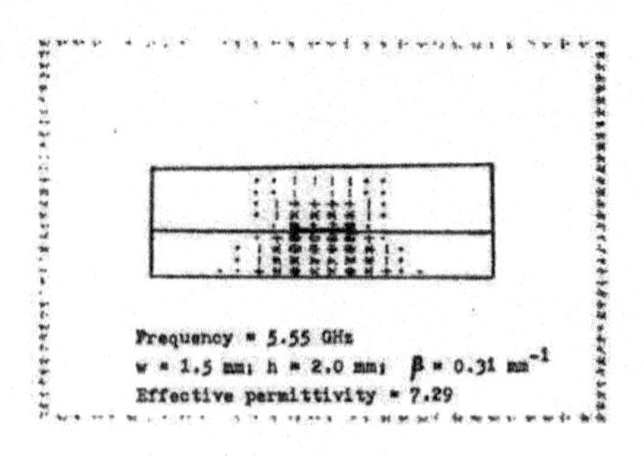

GREY-SCALE CHARACTERS AND RANGES

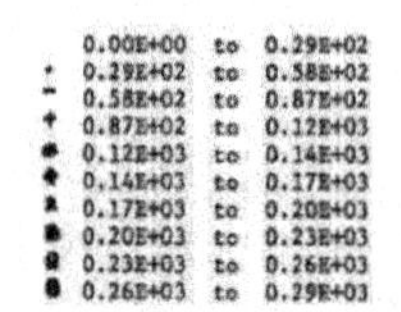

	0.00E+00	to	0.29E+02
·	0.29E+02	to	0.58E+02
-	0.58E+02	to	0.87E+02
+	0.87E+02	to	0.12E+03
[illegible]	0.12E+03	to	0.14E+03
[illegible]	0.14E+03	to	0.17E+03
[illegible]	0.17E+03	to	0.20E+03
[illegible]	0.20E+03	to	0.23E+03
[illegible]	0.23E+03	to	0.26E+03
[illegible]	0.26E+03	to	0.29E+03

Fig. 15. Electric-field (E_y) distribution for a single microstrip on sapphire substrate.

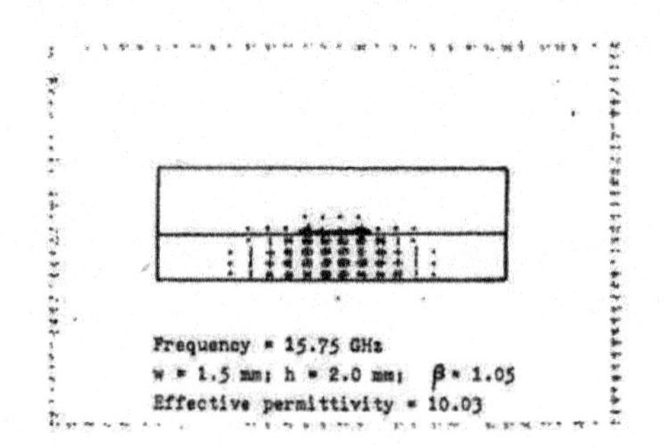

GREY SCALE CHARACTERS AND RANGES

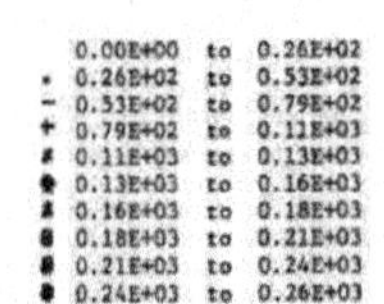

	0.00E+00	to	0.26E+02
·	0.26E+02	to	0.53E+02
-	0.53E+02	to	0.79E+02
+	0.79E+02	to	0.11E+03
[illegible]	0.11E+03	to	0.13E+03
[illegible]	0.13E+03	to	0.16E+03
[illegible]	0.16E+03	to	0.18E+03
[illegible]	0.18E+03	to	0.21E+03
[illegible]	0.21E+03	to	0.24E+03
[illegible]	0.24E+03	to	0.26E+03

Fig. 16. Electric-field (E_y) distribution for a single microstrip on sapphire substrate.

in Figs. 13 and 14. The effective width of the microstrip line decreases with increasing frequency. Since the line impedance Z_0 is inversely proportional to the effective width w_{eff}, Z_0 therefore increases as frequency is increased.

As a further demonstration of the flexibility and generality of the TLM procedure, a gray-scale plot of the transverse field distribution was obtained for low and high frequencies. The nonhomogeneous field distribution is evident in Figs. 15 and 16, where E_y is plotted for single microstripline on sapphire with a strip-width to substrate-height ratio of 0.75. In order to see the details of the field distribution, more mesh points are needed than would be required in a dispersion analysis. In the above figures, four mesh points were used to describe the substrate, i.e., $h = 4\Delta l$, six mesh points were used for the free space above the strip. This was found to be adequate at a high frequency since most of the field was then concentrated in the substrate. At the lower frequency, the results were slightly modified by the upper wall of the enclosure, resulting in a flattening of the field distribution at the top. This effect was judged to be tolerable since the field intensity at the wall had decayed to about a tenth of its peak value.

IV. Numerical Examples

To demonstrate the versatility of the TLM method in solving different microstrip problems, we shall provide the following additional numerical examples.

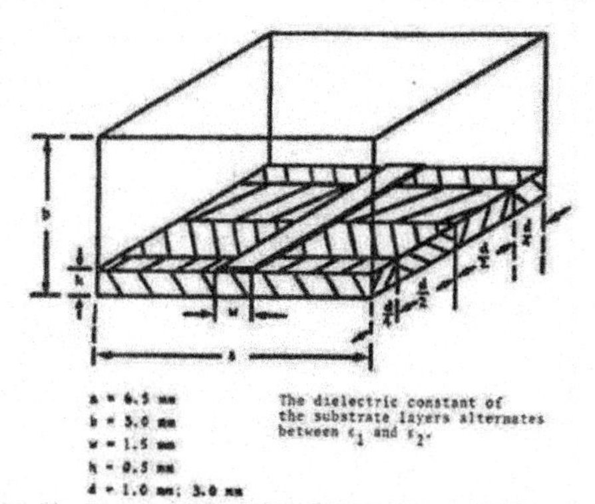

Fig. 17. Geometry of microstrip on substrate with periodically stratified index of refraction.

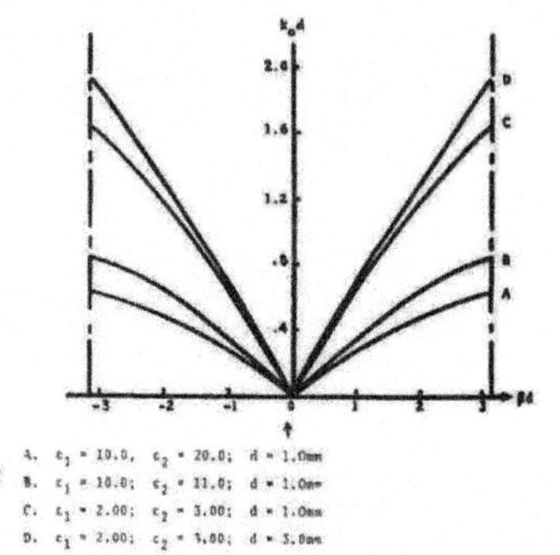

Fig. 18. $\omega-\beta$ diagram for microstrip on periodically stratified index of refraction.

A. Microstrip on Substrate with Periodically Stratified Dielectrics

Shown in Fig. 17 is a microstripline placed on a substrate made up of alternate layers of isotropic dielectric material of relative permittivities ϵ_1 and ϵ_2. The length of the period cell is d. Note that the two end layers in the cavity are of width $d/4$ so that the images of the structure in the end shorting planes appear to be continuations of the structure.

According to Floquet's theorem, an infinite set of spatial harmonics exists for guided waves along a periodic structure. These spatial harmonics must be present simultaneously in order that the total field may satisfy the boundary conditions. The eigenvalue equation for β for a periodic structure will always yield solutions $\beta_n = \beta + 2n\pi/d$, in addition to the fundamental solution. These other possible solutions are clearly the propagation constants of the spatial harmonics. A complete ω-β diagram thus exhibits $k_0 d$ as a periodic function of βd, that is, the βd curve is continued periodically outside the range

$$-\pi \leqslant \beta d \geqslant \pi.$$

The principal values of βd are plotted in Fig. 18 for various values of ϵ_1 and ϵ_2. The length of the unit cell in curves 1, 2, and 3 is $d = 1.0$ mm. The cutoff frequency for the low-frequency passband is given by the value of $k_0 d$ when $\beta d = \pm\pi$. Note that Fig. 18 shows only the first passband of the periodic structure. Examination of the diagram shows that the cutoff frequency may be reduced by increasing ϵ_1 or ϵ_2. Although the value of $k_0 d$ at cutoff was somewhat increased when the magnitude of d was tripled, the overall effect of increasing d was to reduce the cutoff frequency since $k_0 d$ did not triple at cutoff. In general, one may deduce the following conclusions: 1) To increase the upper cutoff frequency, one should lower the dielectric constant of the material. The phase velocity of the wave is lowered by reducing the length of the periodic cell. Hence, by controlling the width of the cell, it is possible to adjust the phase velocity of the wave. 2) Microstrip on a substrate with a periodically stratified index of refraction exhibits the slow-wave and filtering properties common to all periodic waveguiding structures.

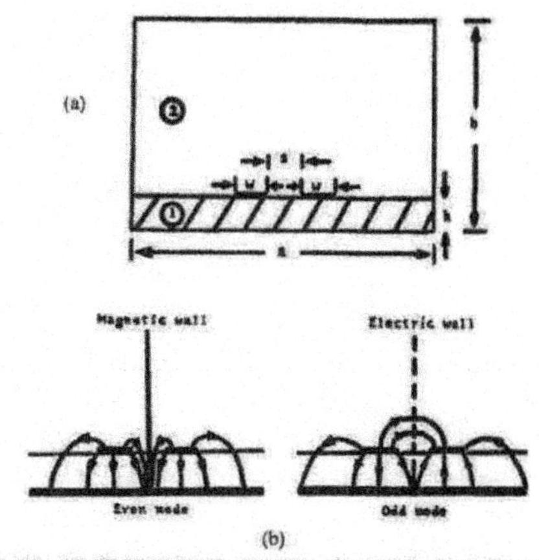

Fig. 19. (a) Cross-sectional geometry of coupled microstrip pair. (b) Electric flux lines for the two fundamental modes of two coupled microstrips.

B. Coupled Microstriplines on Sapphire

The frequency-dependent even- and odd-mode phase constants were determined for the structure of Fig. 19

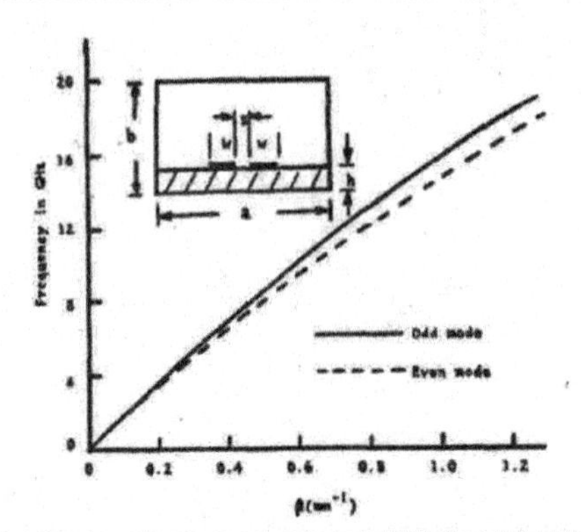

Fig. 20. Dispersion diagram for even and odd fundamental modes of edge-coupled microstrip.

using sapphire substrate as an example. Also computed was the frequency-dependent coupling length for the same structure. The results are depicted in Fig. 20. Although a b/h ratio of only 3 was used, the results obtained by moving the upper boundary farther away were not significantly different. The number of iterations of the computer program was kept at 1000 to minimize truncation errors.

V. Conclusions

The simplicity of the TLM technique in solving problems dealing with complex waveguiding structures should be noted. Although we have demonstrated only the solutions of microstrip problems, the technique can be easily extended to solve waveguiding problems for integrated optical circuits (IOC) or fiber optics [16]. The guiding medium may be anisotropic and inhomogeneous.

References

[1] G. E. Mariki, "Analysis of microstrip lines on inhomogeneous anisotropic substrates," Ph.D. dissertation, University of California, Los Angeles, June 1978.
[2] A. S. Kariotis, "The millimeter wave market: An overview," *Microwave J.*, vol. 27, no. 7, p. 24, 1984.
[3] D. Choi and W. J. K. Hoefer, "A scalar TLM model for three-dimensional wave simulation," *IEEE Trans. Microwave Theory Tech.*, 1984.
[4] T. Kitazawa and Y. Hayashi, "Propagation characteristics—strip lines with multi-layered anisotropic media," *IEEE Trans. Microwave Theory Tech.*, vol. MTT-31, pp. 429–433, 1983.
[5] A-M. A. El-Sherbiny, "Hybrid mode analysis of microstrip lines on anisotropic substrates," *IEEE Trans. Microwave Theory Tech.*, vol. MTT-29, p. 1261, 1981. N. Yoshida, I. Fukai, and J. Fuknoka, "Application of Bergeron's method to anisotropic media," *Trans. IECE Japan*, vol. J64B, pp. 1242–1249, 1981.
[6] H. Lee and V. K. Tripathi, "Spectral domain analysis of frequency dependent propagation characteristics of planar structures on uniaxial medium," *IEEE Trans. Microwade Theory Tech.*, vol. MTT-30, p. 1188, 1982. M. Horno, "Quasistatic characteristics of covered coupled microstrips on anisotropic substrates: Spectral and variational analysis," *IEEE Trans. Microwave Theory Tech.*, vol. MTT-30, p. 1888, 1982.
[7] R. Mittra and T. Itoh, "A new technique for the analysis of the dispersion characteristics of microstrip lines," *IEEE Trans. Microwave Theory Tech.*, vol. MTT-19, pp. 47–56, Jan. 1971. T. Itoh and R. Mittra, "Spectral domain approach for calculating the dispersion characteristics of microstrip lines," *IEEE Trans. Microwave Theory Tech.*, vol. MTT-21, pp. 496–499, July 1973.
[8] H. A. Wheeler, "Transmission-line properties of parallel strips separated by a dielectric sheet," *IEEE Trans. Microwave Theory Tech.*, vol. MTT-13, pp. 172–185, Mar. 1965.
[9] P. Silvester, "TEM wave properties of microstrip transmission lines," *Proc. Inst. Elec. Eng.*, vol. 115, pp. 43–48, Jan. 1968.
[10] W. J. Getsinger, "Microstrip dispersion model," *IEEE Trans. Microwave Theory Tech.*, vol. MTT-21, pp. 34–39, 1973.
[11] S. Akhtarzad and P. B. Johns, "The solution of Maxwell's equations in three space dimensions and time by the TLM method," *Proc. Inst. Elec. Eng.*, vol. 122, 1975.
[12] S. Akhtarzad and P. B. Johns, "Generalized elements for the TLM method of numerical analysis," *Proc. Inst. Elec. Eng.*, vol. 122, 1975.
[13] R. P. Owens *et al.*, "Quasi-static characteristics of microstrip on anisotropic sapphire substrates," *IEEE Trans. Microwave Theory Tech.*, vol. MTT-24, pp. 499–505, Aug. 1976.
[14] N. G. Alexopoulos, S. Kerner, and C. M. Krowne, "Dispersionless coupled microstrip over fused silica-like anisotropic substrates," *Electron. Lett.*, vol. 12, pp.579–580, Oct. 1976. N. G. Alexopoulos *et al.* "On the characteristics of single and coupled microstrip on anisotropic substrates," *IEEE Trans. Microwave Theory Tech.*, vol. MTT-26, June 1978.
[15] T. Edwards and R. P. Owens, "2–18 GHz dispersion measurements on 10–100 ohm microstrip lines on sapphire," *IEEE Trans. Microwave Theory Tech.*, vol. MTT-24, pp. 506–513, Aug. 1976.
[16] C. Yeh, K. Ha, and S. B. Dong, "Single mode optical waveguides," *Appl. Opt.*, vol. 18, p. 1490, 1979.

G. E. Mariki was born in Moshi, Tanzania, on December 1, 1948. He received the B.S. degree in electrical engineering from the University of London (Imperial College), England, in 1972, the M.S. degree in telecommunication systems from the University of Essex, England, in 1973, and the Ph.D. degree in electrical engineering from the University of California at Los Angeles (UCLA) in 1978.

From 1973–1975, he was a Lecturer in the Department of Electrical Engineering at the University of Dar es Salaam, Tanzania. From 1978–1980, he was an Assistant Professor of Electrical Engineering at the Carnegie-Mellon University, Pittsburgh. Dr. Mariki's present address is unknown.

C. Yeh (S'56–M'63–SM'82–F'85) was born in Nanking, China, on August 11, 1936. He received the B.S., M.S., and Ph.D. degrees in electrical engineering from the California Institute of Technology, Pasadena, CA, in 1957, 1958, and 1962, respectively.

He is presently Professor of electrical engineering at the University of California at Los Angeles (UCLA). He joined UCLA in 1967 after serving on the faculty of USC from 1962 to 1967. His current areas of research interest are optical and millimeter-wave guiding structures, gigabit-rate fiber-optic local area network, and scattering of electromagnetic waves by penetrable irregularly shaped objects.

Dr. Yeh is a member of Eta Kappa Nu, Sigma Xi, and the Optical Society of America.

Modeling of Star-shaped and Parallel-wire Carrier Distribution Systems

by C. YEH,* F. MANSHADI *and* G. MARIKI

Electrical Sciences and Engineering Department
University of California
Los Angeles, CA. 90024, U.S.A.
and W. P. BROWN
Hughes Research Laboratories
Malibu, CA 90265, U.S.A.

ABSTRACT: *It is often desirable to represent carrier distribution systems by networks so that analysis can be made to predict the behavior of these systems. Modeling of two canonical carrier distribution systems, the star-shaped system and the parallel-wire system by networks was carried out in this paper. The representation is given in terms of lumped parameters when the lengths of the carrier lines are short and it is given in terms of lumped and distributed parameters when the lengths of the carrier lines are long.*

I. Introduction

To analyze exactly the guiding characteristic of carrier waves along multi-conductor over-head power lines and to develop a complete circuit model for the guiding structure, Maxwell's equations with the appropriate boundary conditions at the interface such as, the earth–air interface, the wire junctions and the wire conductor–air interface, the radiation conditions for the open-wire lines, and the appropriate source and termination conditions, must be solved. This is obviously a Herculean task. Fortunately, the frequencies of the carrier waves of intersect are limited within a band from 1 kHz to 500 kHz. At these frequencies, the size of conductors, the spacings between the conductors, and the heights of conductors above the earth's surface are much smaller than the free-space wavelengths of the carrier waves. Hence a low-frequency approximation of Maxwell's equations may be used. Furthermore, as an *a priori* assumption, one may consider the radiation problem and the transmission problem as two independent problems. [This assumption can be justified by the use of the perturbation analysis (**1**)].

For the radiation problem, it is assumed that a set of known propagating currents representing the carrier wave exists on straight conductors. These currents will excite TM radiation waves and Zenneck-type surface waves (**2**). Excitation of these waves may be considered as a source of (small) power loss by the carrier waves.

For the transmission problem it is assumed that the dominant wave is a TEM wave with the propagation velocity near that of light in vacuum. Hence the

* Consultant: Hughes Research Laboratories, 3011 Malibu Canyon Road, Malibu, CA 90265, U.S.A.

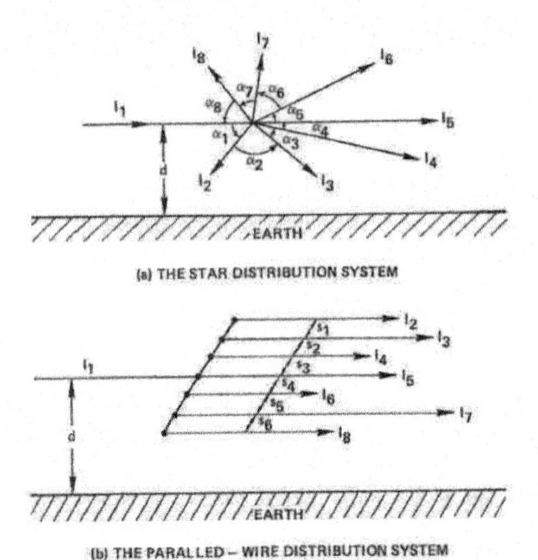

FIG. 1. Two basic types of carrier distribution system

field distributions and configurations are governed by the static Maxwell's equations. The effects of finite conductivity and non-unity dielectric constant of the earth as well as the non-zero skin depth of the conductors at the carrier frequencies can be taken into account by the inclusion of correction terms in the distributed series impedance and the distributed shunt admittances representation of the transmission line (**3**). The carrier current and voltage distribution of a multi-conductor power line can then be found by solving a set of coupled transmission line equations with the given initial conditions. Or, if the typical lengths of the transmission lines are short (say, <30 miles), a lumped parameter network representation of the carrier distribution system would be adequate. The voltage and current for the carrier signals can be easily calculated from this network model.

Two basic types of carrier distribution system as shown in Fig. 1 may be envisioned: (a) The star-shaped distribution system and (b) the parallel distribution system. The star-shaped system may approximate the situation when the carrier signal originating from a sub-station is dispersed in all directions in a star format while the parallel-wire format occurs when the signal is fed along parallel streets.

II. Calculation of Mutual Inductances

A. *Derivation of Basic Formula for Star-shaped Distribution System*

The basic star-shaped distribution system is initially assumed to be a number of thin wires connected as shown in Fig. 1(a) above a perfectly conducting earth which acts as return for the carrier currents in the wires. To construct a network model for this type of system, the mutual inductance among the wires must be taken into consideration. The canonical solution for the mutual inductance of two intersecting wires in a plane which is parallel to the conducting earth and is at a height of d meters above the earth will first be obtained. [See Fig. 2(a)].

Given a current i_b flowing in wire ⓑ as shown in Fig. 2(a), the vector potential is given by (4)

$$\mathbf{A}_b = \mathbf{e}_x \frac{\mu}{4\pi}\left[\int_0^{l_b} \frac{i_b\,dx'}{\sqrt{[(x-x')^2+(z-d)^2]}} - \int_0^{l_b} \frac{i_b\,dx'}{\sqrt{[(x-x')^2+y^2+(z+d)^2]}}\right]$$

$$= \mathbf{e}_x \frac{i_b\mu}{4\pi}\left[\ln\left[\frac{-x+\sqrt{[x^2+y^2+(z+d)^2]}}{-x+\sqrt{x^2+y^2+(z-d)^2]}}\right]\right.$$

$$\left. - \ln\left[\frac{(l_b-x)+\sqrt{[l_b-x)^2+y^2+(z+d)^2]}}{(l_b-x)+\sqrt{[(l_b-x)^2+y^2+(z-d)^2]}}\right]\right] \quad (1)$$

where $\mathbf{e}_x$ is a unit vector in the x direction, and l_b is the length of wire ⓑ. The mutual inductance between wire ⓐ and wire ⓑ is

$$M_{ab} = \frac{1}{i_b}\int_{ⓐ} \mathbf{A}_b \cdot \mathrm{d}l_a, \quad (2)$$

with

$$\mathrm{d}l_a = \mathrm{d}x \cos\alpha\ \mathbf{e}_x + \mathrm{d}_y \sin\alpha\ \mathbf{e}_y,$$

where α is the angle between wire ⓐ and wire ⓑ at the intersecting point, and $\mathbf{e}_y$ is a unit vector in the y direction. Substituting (1) into (2) and simplifying gives

$$M_{ab} = \frac{\mu \cos\alpha}{2\pi}\left\{\int_0^{l_a \cos\alpha}\left[\ln\left(\frac{-x+\sqrt{[1+\tan^2\alpha)x^2+4d^2]}}{-x+\sqrt{[(1+\tan^2\alpha)x^2]}}\right)\right.\right.$$

$$\left.\left. - \ln\left(\frac{(l_b-x)+\sqrt{[(l_b-x)^2+x^2\tan^2\alpha+4d^2]}}{(l_b-x)+\sqrt{[(l_b-x)^2+x^2\tan^2\alpha]}}\right)\right]\mathrm{d}x\right\}. \quad (3)$$

Performing the integration in the above equation yields the mutual inductance between 2 wires as shown in Fig. 2(a). These integrals can be easily computed numerically with the help of a computer if all the constants are given.

Referring now back to the star-shaped system as shown in Fig. 1(a) and assuming $d=20$ m, $\alpha=\alpha_1=\alpha_2\ldots=\alpha_8=\pi/4$ and $l_b\to\infty$, one has

$$M_{12} = \frac{\mu\sqrt{2}}{4\pi}\int_0^{l_a/\sqrt{2}} \ln\left(\frac{-x+\sqrt{(2x^2+4d^2)}}{-x+\sqrt{(2x^2)}}\right)\mathrm{d}x \quad (4)$$

$$M_{13} = 0 \quad (5)$$

$$M_{14} = \frac{\mu\sqrt{2}}{4\pi}\int_0^{l_a/\sqrt{2}} \ln\left(\frac{x+\sqrt{(2x^2+4d^2)}}{x+\sqrt{(2x^2)}}\right)\mathrm{d}x \quad (6)$$

$$M_{15} = \frac{\mu}{2\pi}\left[l_a \ln\left(\frac{l_a+\sqrt{(l_a+4d^2)}}{2l_a}\right) - \sqrt{(l_a^2+4d^2)} + (l_a+2d)\right] \quad (7)$$

$$M_{16} = M_{14} \quad (8)$$

$$M_{17} = M_{13} = 0 \quad (9)$$

$$M_{18} = M_{12}. \quad (10)$$

Mutual inductances between wire 2 and other wires, etc. can be calculated in a similar manner as given above. Hence, the results will not be repeated here.

The above expressions have been evaluated numerically with $d=20$ meters and for various values of $l_a/\sqrt{2}$. The results are tabulated in Table 1. It can be seen that, as expected, the mutual inductance approaches a certain constant as $l_a \to \infty$. This is because the flux lines are linked most closely at the intersections of the wires. It should be noted that M_{23}, $M_{24} \ldots M_{28}$ etc. can be calculated in a similar manner as shown from the general expression Eq. (3).

B. *Derivation of Basic Formula for Parallel Distribution System*

The basic parallel wire distribution system is assumed to be a number of thin wires connected as shown in Fig. 1(b) above a perfectly conducting earth which acts as return for the carrier currents in the wires. To construct a network model for this type of system, the mutual inductances among the wires must be taken into consideration. The canonical solution for the mutual inductances of two parallel wires in a plane which is parallel to the conducting earth and is at a height of d meters above the earth will first be obtained. [See Fig. 2(b)].

The mutual inductance between wire ⓐ and wire ⓑ as shown in Fig. 2(b) can be calculated in a similar manner as given in the above section, that is

$$M_{ab} = \frac{\mu}{2\pi}\left\{ (l_a + l_b)\ln\left[\frac{(l_a+l_b)+\sqrt{[(l_a+l_b)^2+s^2]}\sqrt{(4d^2+s^2)}}{s(l_a+l_b)+\sqrt{[(l_a+l_b)^2+4d^2+s^2]}}\right]\right.$$

$$+ l_b \ln\left[\frac{(l_b+\sqrt{(l_b^2+4d^2+s^2)})s}{\sqrt{(4d^2+s^2)}((l_b+\sqrt{[l_b^2+s^2]}))}\right]$$

$$+ l_a \ln\left[\frac{(l_a+\sqrt{(l_a^2+4d^2+s^2)})s}{\sqrt{(4d^2+s^2)}(l_a+\sqrt{l_a^2+s^2}))}\right]$$

$$-\sqrt{[(l_a+l_b)^2+s^2]}+\sqrt{[(l_a+l_b)^2+4d^2+s^2]}$$

$$+\sqrt{(l_b^2+s^2)}-\sqrt{(l_b^2+4d^2+s^2)}$$

$$+\sqrt{(l_a^2+s^2)}-\sqrt{(l_a^2+4d^2+s^2)}$$

$$\left. -s+\sqrt{(4d^2+s^2)}\right\} \tag{11}$$

where l_a and l_b are the lengths of the wire ⓐ and wire ⓑ, respectively. For the special case of $l_b=\infty$, the mutual inductance M_{ab} is given by the expression

$$M_{ab} = \frac{\mu}{2\pi}\left\{ l_a \ln\left[\frac{l_a+\sqrt{(l_a^2+4d^2+s^2)}}{l_a+\sqrt{(l_a^2+s^2)}}\right]+\sqrt{(l_a^2+s^2)}\right.$$

$$\left. -\sqrt{(l_a^2+4d^2+s^2)}-s+\sqrt{(4d^2+s^2)}\right\}. \tag{12}$$

As a numerical example, let $l_1=300$ meters, $l_2=l_3=\ldots=l_8=\infty$, $d=20$ meters, $s_1=s_2=s_3\ldots=s_6=100$ meters. [Refer to Fig. 1(b).] One has

$$M_{12}=M_{18}=0.310\ \mu\text{H}$$

$$M_{13}=M_{17}=0.568\ \mu\text{H}$$

$$M_{14}=M_{16}=1.28\ \mu\text{H}$$

$$M_{15}=7.73\ \mu\text{H}.$$

The above results are still valid if $l_2, l_3\ldots l_8$ are not very long as long as l_2, $l_3\ldots l_3$ are much longer than l_1, s_1, $s_2\ldots s_6$.

The mutual inductance between two infinitely long parallel wires spaced s apart and at a height of d above a perfectly conducting earth is

$$M'_{ab}=\frac{\mu}{2\pi}\ln\left[\frac{4d^2+s^2}{s^2}\right]\text{(H/m)}. \tag{13}$$

For the above numerical example, one has

$$M'_{23}=M'_{34}=M'_{45}=M'_{56}=M'_{67}=M'_{78}=0.0297\ \mu\text{H/m}$$

$$M'_{24}=M'_{35}=M'_{46}=M'_{57}=M'_{68}=0.00784\ \mu\text{H/m}$$

$$M'_{25}=M'_{56}=M'_{47}=M'_{58}=0.00352\ \mu\text{H/m}$$

$$M'_{26}=M'_{37}=M'_{48}=0.00199\ \mu\text{H/m}$$

$$M'_{27}=M'_{38}=0.00128\ \mu\text{H/m}$$

$$M'_{28}=0.000887\ \mu\text{H/m}.$$

III. Equivalent Network Representation of Star-shaped or Parallel-wire Carrier Distribution System

Now we are in a position to construct equivalent networks for these two distribution systems. Using these equivalent networks one can quickly calculate the voltages or the currents for the carrier waves at the various loads. Depending upon the lengths of the carrier lines, it is sometimes necessary to represent the distribution systems by lumped elements as well as distributed parameters. For lengths much shorter than the free space wavelength of the carrier signal, representation by lumped elements will suffice.

A. *Lumped Parameter Representation*

Referring again back to Fig. 1(a), one sees that if l_1, $l_2\ldots l_8$ are short, say, less than 30 miles, one may represent this star-shaped distribution system by lumped parameters as follows:

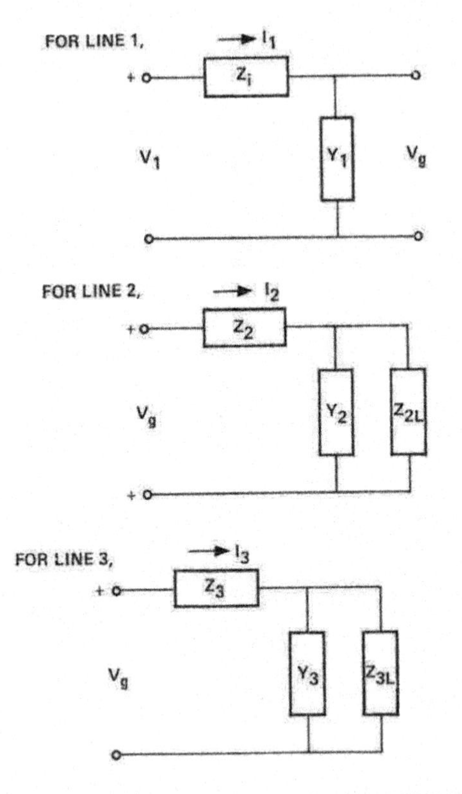

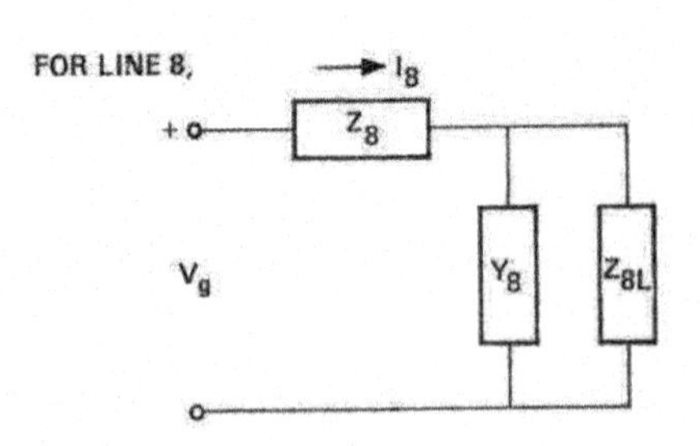

where

$$Z_n = R_n + j\omega L_{nT}$$

$$Y_n = \frac{1}{R_{ng}} + j\omega C_{nT}$$

ω = frequency of the carrier wave (14)

$n = 1, 2, \ldots 8.$

The load impedances of the lines are represented by Z_{nL} $(n=1,2\ldots8)$. The mutual impedances between Z_1, $Z_2\ldots Z_8$ are given by M_{nm} $(n=1,2\ldots8,\ m=1,2\ldots8,\ n\neq m)$ which can be calculated according to the formation given earlier. The symbols R_n, L_{nT}, R_{ng} and C_{nT} have the same meaning as those given by Hedman (**3**).

Assuming that V_1 is given, one can calculate all the currents $I_1,\ldots I_8$ and V_g from Eq. (15) shown on the following page.

It can be seen that the above equivalent network can also be good representation of the parallel-wire distribution system provided that s, the distance between parallel wire, is small. It is quite elementary to include the effects of finite léngth of the cross bar; for example, the section of the cross bar between two parallel wires can be represented by the following net-work:

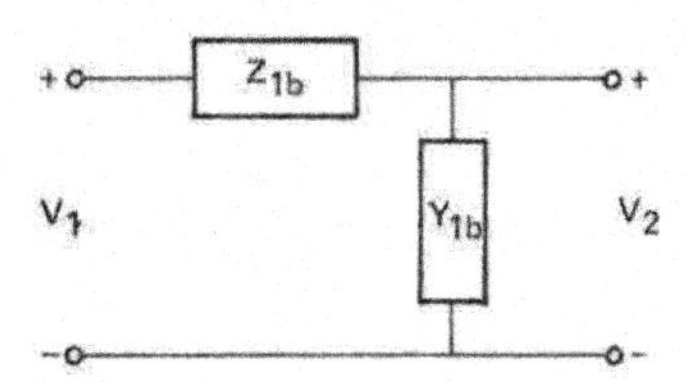

where Z_{1b} and Y_{1b} are the impedance and admittance of the cross bar and its image.

B. *Lumped and Distributed Parameters Representation*

Referring to Fig. 1, note that if l_2, $l_3\ldots l_8$ are long, it would not be satisfactory to represent the distribution system by lumped parameters only. Both lumped and distributed parameters must be used. In the following, we shall present the equivalent circuits for the carrier distribution system which include lumped as well as distributed parameters:

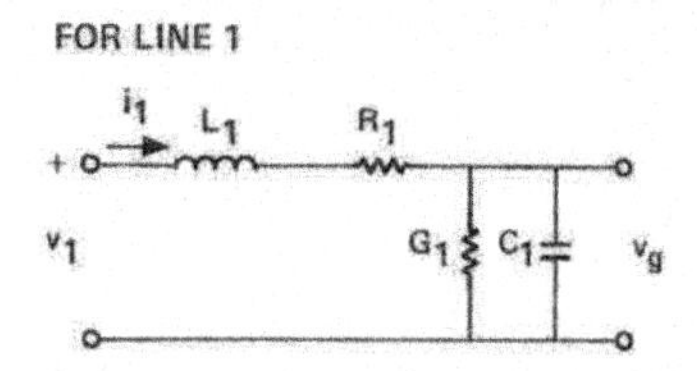

FOR LINE 2

$$
\begin{bmatrix}
Z_1 & j\omega M_{12} & j\omega M_{13} & j\omega M_{14} & j\omega M_{15} & j\omega M_{16} & j\omega M_{17} & j\omega M_{18} & -1 \\
j\omega M_{21} & Z_2+\dfrac{1}{\dfrac{1}{Z_{2L}}+Y_2} & j\omega M_{23} & j\omega M_{24} & j\omega M_{25} & j\omega M_{26} & j\omega M_{27} & j\omega M_{28} & -1 \\
j\omega M_{31} & j\omega M_{32} & Z_3+\dfrac{1}{\dfrac{1}{Z_{3L}}+Y_3} & j\omega M_{34} & j\omega M_{35} & j\omega M_{36} & j\omega M_{37} & j\omega M_{38} & -1 \\
j\omega M_{41} & j\omega M_{42} & j\omega M_{43} & Z_4+\dfrac{1}{\dfrac{1}{Z_{4L}}+Y_4} & j\omega M_{45} & j\upsilon M_{46} & j\omega M_{47} & j\omega M_{48} & -1 \\
j\omega M_{51} & j\omega M_{52} & j\omega M_{53} & j\omega M_{54} & Z_5+\dfrac{1}{\dfrac{1}{Z_{5L}}+Y_5} & j\omega M_{56} & j\omega M_{57} & j\omega M_{58} & -1 \\
\cdot & \cdot & \cdot & \cdot & \cdot & \cdot & \cdot & \cdot & \cdot \\
j\omega M_{81} & j\omega M_{82} & j\omega M_{83} & j\omega M_{84} & j\omega M_{85} & j\omega M_{86} & j\omega M_{87} & Z_8+\dfrac{1}{\dfrac{1}{Z_{8L}}+Y_8} & -1 \\
-1 & 1 & 1 & 1 & 1 & 1 & 1 & 1 & \dfrac{1}{Y_1}
\end{bmatrix}
\begin{bmatrix} I_1 \\ I_2 \\ I_3 \\ I_4 \\ I_5 \\ I_8 \\ V_g \end{bmatrix}
=
\begin{bmatrix} V_1 \\ 0 \\ 0 \\ 0 \\ 0 \\ 0 \\ \cdot \\ \cdot \\ \cdot \\ 0 \end{bmatrix}
\tag{15}
$$

$$
V_n = I_n \frac{1}{\dfrac{1}{Z_{nL}} + Y_n}
$$

$$
n = 2, 3, \ldots 8.
$$

C. Yeh, F. Manshadi, G. Mariki and W. P. Brown

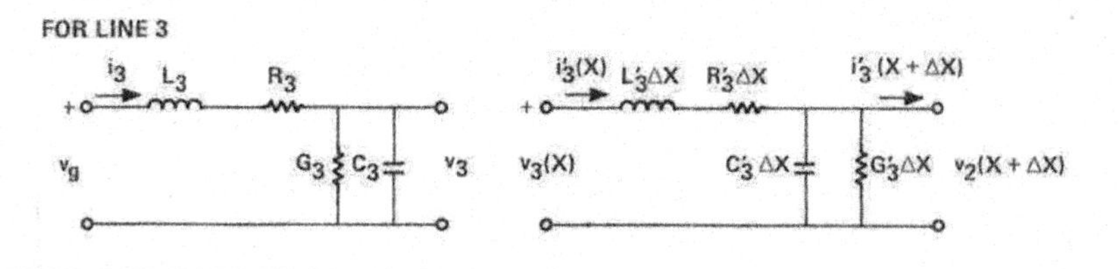

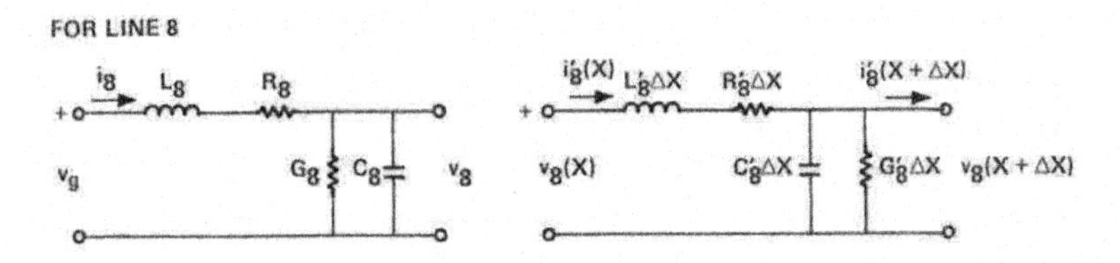

The lumped mutual impedances between $L_1, L_2 \ldots L_8$ are given by $M_{nm}(n = 1, 2, 3 \ldots, m = 1, 2, 3 \ldots, n \neq m)$ which can be calculated according to the formulas given in Section II. The distributed mutual impedances between $L'_2, L'_3 \ldots L'_8$ are given by M'_{nm} $(n = 2, 3, 4 \ldots 8, m = 2, 3, 4 \ldots 8, n \neq m)$. $L_1, L_2 \ldots L_8$ and $R_1, R_2 \ldots R_8$ are, respectively, the self-inductances and the self-resistances of the lines, while $C_1, C_2, \ldots C_8$ and $G_1, G_2, \ldots G_8$ are, respectively, the capacitances and leakage conductances of the lines. The prime refers to the distributed parameters.

Given $v_1(t)$, solving the following set of equation derived from the above network representations, will yield the required information on the currents at various loads:

For lumped-parameter networks,

$$v_g = v_1 + L_1 \frac{\partial i_1}{\partial t} + R_1 i_1 + M_{12} \frac{\partial i_2}{\partial t} + M_{13} \frac{\partial i_3}{\partial t} + \ldots + M_{18} \frac{\partial i_8}{\partial t}$$

$$v_2 = v_g + L_2 \frac{\partial i_2}{\partial t} + R_2 i_2 + M_{21} \frac{\partial i_1}{\partial t} + M_{23} \frac{\partial i_3}{\partial t} + \ldots + M_{28} \frac{\partial i_8}{\partial t}$$

$$v_3 = v_g + L_3 \frac{\partial i_3}{\partial t} + R_3 i_3 + M_{31} \frac{\partial i_1}{\partial t}$$

$$+ M_{32} \frac{\partial i_2}{\partial t} + M_{34} \frac{\partial i_4}{\partial t} + \ldots + M_{38} \frac{\partial i_8}{\partial t} \tag{16}$$

$$v_8 = v_g + L_8 \frac{\partial i_8}{\partial t} + R_8 i_8 + M_{81} \frac{\partial i_1}{\partial t} + M_{82} \frac{\partial i_2}{\partial t} + \ldots + M_{87} \frac{\partial i_7}{\partial t}$$

For distributed-parameter networks,

$$i_1 = C_1 \frac{\partial v_8}{\partial t} + G_1 v_8 + i_2 + i_3 + \ldots + i_8$$

$$\frac{\partial v_2}{\partial x} = -L_2' \frac{\partial i_2'}{\partial t} - R_2' i_2' - M_{23}' \frac{\partial i_3'}{\partial t} - M_{24}' \frac{\partial i_4'}{\partial t} - \ldots - M_{28}' \frac{\partial i_8'}{\partial t}$$

$$\frac{\partial i_2'}{\partial x} = -C_2' \frac{\partial v_2}{\partial t} - G_2' v_2 \tag{17}$$

$$\frac{\partial v_3}{\partial x} = -L_3' \frac{\partial i_3'}{\partial t} - R_3' i_3' - M_{32}' \frac{\partial i_2'}{\partial t} - M_{34}' \frac{\partial i_4'}{\partial t} - \ldots - M_{38}' \frac{\partial i_8'}{\partial t}$$

$$\frac{\partial i_3'}{\partial x} = -C_3' \frac{\partial v_3}{\partial t} - G_3' v_3$$

$$\frac{\partial v_8}{\partial x} = -L_8' \frac{\partial i_8'}{\partial t} - R_8' i_8' - M_{82}' \frac{\partial i_2'}{\partial t} - M_{83}' \frac{\partial i_3'}{\partial t} - \ldots - M_{87}' \frac{\partial i_7'}{\partial t}$$

$$\frac{\partial i_8'}{\partial x} = -C_8' \frac{\partial v_8}{\partial t} - G_8' v_8.$$

The connecting equations are:

$$i_2 = i_2' + C_2 \frac{\partial v_2}{\partial t} + G_2 v_2$$

$$i_3 = i_3' + C_3 \frac{\partial v_3}{\partial t} + G_3 v_3 \tag{18}$$

$$i_8 = i_8' + C_8 \frac{\partial v_8}{\partial t} + G_8 v_8.$$

The above sets of equations can be solved with the appropriate initial and boundary conditions.

IV. Discussion

We have just provided a method to analyze the behavior of two canonical types of carrier distribution systems. Other more realistic and more complex distribution systems can be broken down into many parts which can usually be identified by either one of the two canonical types analyzed in this paper. These parts can be connected together using the conditions of continuity of current and voltage. The total network can then be analyzed using the usual network analysis to provide the needed information on the currents or voltages at the loads.

It should be noted that earth conduction effects and the skin effect of the wires can be taken into account by the use of Carson–Wise earth correction terms as demonstrated by Hedman (**3**), Perz (**5**) and others (**6**). In other words, referring back to Eq. (14), the earth correction terms may be included in Z_n

and Y_n in a straightforward manner (**3**, **5**, **6**). Furthermore, if necessary, the radiation effects of the wires may also be included as a resistance correction term for the real part of Z_n. This resistance correction term is simply the radiation resistance of that section of the transmission line (**2**).

V. Acknowledgment

We wish to thank Dr. Lian of the Hughes Research Laboratories for suggesting this problem.

References

(**1**) J. R. Carson, "The Rigorous and Approximate Theorem of Electrical Transmission Along Wires", *Bell. Sys. Tech. J.*, Vol. 7, p. 11, 1928.

(**2**) R. E. Collin, "Field Theory of Guided Waves", McGraw-Hill Book Co., New York, 1960.

(**3**) D. H. Hedman, "Propagation on Overhead Transmission Lines I, II". *IEEE Trans. on Power Apparatus and Systems*, Vol. 84, p. 200, 1965.

(**4**) W. R. Smythe, "Static and Dynamic Electricity", McGraw-Hill Book Co., New York, 1950.

(**5**) M. C. Perz, "A Method of Analysis of Power Line Carrier Problems on Three-Phase Lines". *IEEE Trans. on Power Apparatus and Systems*, Vol. 85, p. 686, 1964.

(**6**) K. D. Tran and J. Robert, "New study of Parameters of EHV Multiconductor Transmission Lines with Earth Return". *IEEE Trans. on Power Apparatus and Systems*, Vol. 90, p. 452, 1971.

Leaky waves in a heteroepitaxial film

D. B. Hall

Naval Electronics Laboratory Center, Dan Diego, California 92152

C. Yeh

Electrical Sciences and Engineering Department, University of California, Los Angeles, California 90024

(Received 9 November 1972)

Theoretical as well as experimental investigations were carried out for the propagation of leaky waves along an important class of optical thin-film waveguides. For this type of guiding structure which may consist of heteroepitaxial deposition of ZnS or ZnSe on GaAs, the dielectric constant of the thin film is less than that of the substrate. Consequently, only the leaky type of guided waves may exist. Theoretical results show that the attenuation constants of leaky modes, which may be TE or TM, are inversely proportional to (thickness of layer)^3 and directly proportional to (wavelength)^2, and that TM modes are more lossy than TE modes. The existence of these leaky modes has also been demonstrated for three different thicknesses.

INTRODUCTION

The development of low-loss fibers in recent years has spurred renewed interest in the study of optical-fiber communication lines as well as the possible construction of optical integrated circuits. One of the very basic configurations for optical integrated circuits is the deposition of a layer of material of thickness $2a$ and dielectric constant ϵ_2 on a substrate whose dielectric constant is ϵ_3 (see Fig. 1). Optical waves may be supported along and within this layer. It is well known from electromagnetic theory that in order that this type of structure may support a surface guided wave, the dielectric constant of the three different regions must be such that $\epsilon_1 < \epsilon_2$ and $\epsilon_3 < \epsilon_2$.[1-3] ϵ_1 is the dielectric constant of the region above the thin layer. However, in many important physical situations such as the deposition of ZnS ($\epsilon/\epsilon_0 = 5.48$) or ZnSe ($\epsilon/\epsilon_0 = 6.66$) on GaAs ($\epsilon/\epsilon_0 \approx 15$) the dielectric constant of the substrate is higher than that of the layer. Hence, the ordinary surface guided wave cannot exist in this structure. The purpose of this paper is to investigate the problem of wave propagation along a structure where $\epsilon_1 < \epsilon_2 < \epsilon_3$. It will be shown that when the thickness of the layer is large compared with a wavelength of light, low-loss leaky modes may exist. Experimental results verifying the existence of these modes will also be shown.

LEAKY MODES ALONG AN ASYMMETRIC DIELECTRIC GUIDE

Two types of leaky modes may exist along an asymmetric dielectric guide: TE and TM leaky modes. The TE leaky modes have a single component of electric field E_y and magnetic field components H_x and H_z, while the TM leaky modes have a single component of magnetic field H_y and electric field components E_x and E_z. Due to the asymmetry of the three-layer structure, in general, the modes may not be separated into even and odd modes.

The electric field component of the TE leaky modes which propagate in the $+z$ direction takes the form

$$E_y = A\exp[i(k_1x + \gamma z - \omega t)], \qquad x = 2a \qquad (1a)$$

$$E_y = (B\cos k_2x + C\sin k_2x)\exp[i(\gamma z - \omega t)], \quad 2a \geq x \geq 0 \qquad (1b)$$

$$E_y = D\exp[i(-k_3x + \gamma z - \omega t)], \qquad x \leq 0 \qquad (1c)$$

where the transverse wave numbers k_1, k_2, and k_3 are defined in the following relations:

$$k_1^2 = (\epsilon_1/\epsilon_2)k^2 - \gamma^2, \qquad (2a)$$

$$k_2^2 = k^2 - \gamma^2, \qquad (2b)$$

$$k_3^2 = (\epsilon_3/\epsilon_2)k^2 - \gamma^2 \qquad (2c)$$

with

$$k^2 = \omega^2\mu_0\epsilon_2. \qquad (3)$$

Since $\epsilon_1 < \epsilon_2 < \epsilon_3$, no TE surface-wave guided modes are possible. All TE modes are leaky in nature, i.e., the propagation constant γ for these modes are complex with $\mathrm{Re}(\gamma) > 0$ and $\mathrm{Im}(\gamma) > 0$. A, B, C, and D are arbitrary constants. The tangential component of the magnetic fields can be found from Maxwell's equations:

$$H_z = A(k_1/\omega\mu_0)\exp[i(k_1x + \gamma z - \omega t)], \qquad x \geq 2a \qquad (4a)$$

$$H_z = -(i/\omega\mu_0)(-Bk_2\sin k_2x + Ck_2\cos k_2x) \times \exp[i(\gamma z - \omega t)], \qquad 2a \geq x \geq 0 \qquad (4b)$$

$$H_z = -D(k_3/\omega\mu_0)\exp[i(-k_3x + \gamma z - \omega t)], \qquad x \leq 0. \qquad (4c)$$

Matching the tangential electric and magnetic fields at the boundaries $x = 0$, $x = 2a$, we obtain

$$A\exp(2ik_1a) - B\cos(2k_2a) - C\sin(2k_2a) = 0, \qquad (5a)$$

$$k_2A\exp(2ik_1a) - iBk_2\sin(2k_2a) + ik_2C\cos(2k_2a) = 0, \qquad (5b)$$

$$B - D = 0, \qquad (5c)$$

$$ik_2C - k_3D = 0. \qquad (5d)$$

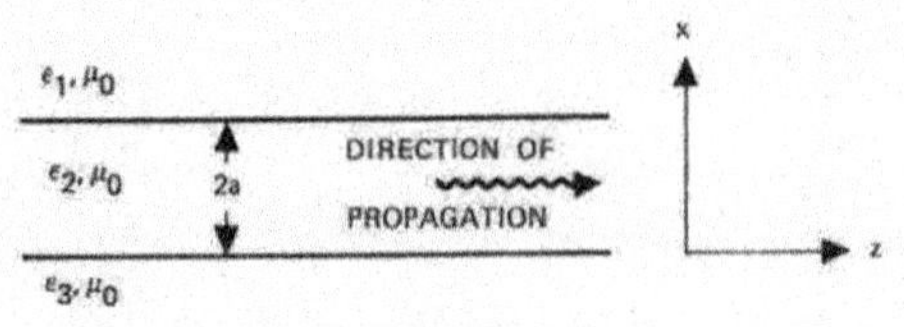

FIG. 1. Geometry of the leaky waveguide $\epsilon_1 < \epsilon_2 < \epsilon_3$.

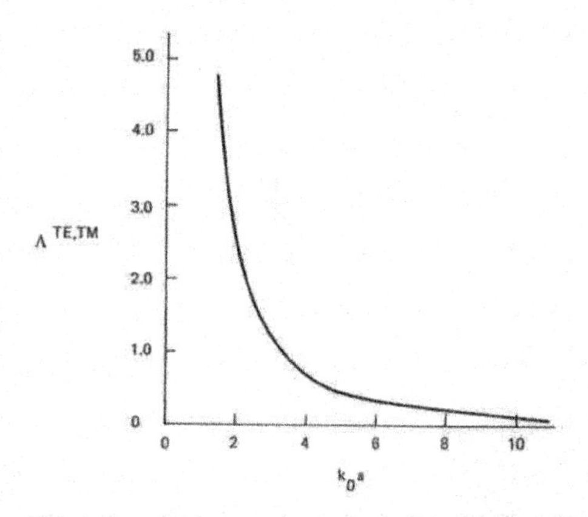

FIG. 2. Normalized attenuation constant in dB as a function of k_0a.

Setting the determinant of the above linear simultaneous algebraic equations to zero gives the characteristic equation for the TE leaky modes:

$$\tan(2k_2a) = i(k_1k_2 + k_3k_2)/(k_2^2 + k_1k_3). \tag{6}$$

The analysis for the TM leaky modes is similar to the one for the TE leaky modes. H_y and E_z, the tangential electric and magnetic fields for the TM leaky modes, are matched at the boundaries $x=0$ and $x=2a$. The determinant for the resulting linear simultaneous algebraic equations is set to zero and gives the characteristic equation for the TM leaky modes:

$$\tan(2k_2a) = i(\epsilon_2\epsilon_3k_1k_2 + \epsilon_1\epsilon_2k_3k_2)/(\epsilon_1\epsilon_3k_2^2 + \epsilon_2^2k_1k_3). \tag{7}$$

One notes that the characteristic equations for the TE and TM leaky modes possess the same form except the right-hand side is different.

APPROXIMATE SOLUTIONS OF THE CHARACTERISTIC EQUATIONS

It can be shown that when $\epsilon_1 < \epsilon_2 < \epsilon_3$, no real roots for γ, the propagation constant, may be found from the characteristic equations, i.e., no surface-wave modes may exist on such structures. All roots for γ will be complex. In other words, waves "guided" along this structure will be attenuated in the direction of propagation. We are interested in the relatively low loss modes. It is expected that when $ka \to \infty$, the low-loss wave will approach the case of a plane wave propagating in a medium with dielectric constant ϵ_2 and permeability μ_0. Hence, making the initial approximation

$$\gamma \approx k, \tag{8}$$

we have from Eqs. (2a)–(2c) that

$$k_1 \approx ik(1-\epsilon_1/\epsilon_2)^{1/2}, \tag{9}$$

$$k_3 \approx k(\epsilon_3/\epsilon_2 - 1)^{1/2}, \tag{10}$$

and

$$ka \gg k_2a. \tag{11}$$

Using these approximations, Eq. (6) for the TE leaky modes becomes

$$\tan(2k_2a) \approx i(k_2/k_1 + k_2/k_3) \tag{12}$$

and Eq. (7) for the TM leaky modes becomes

$$\tan(2k_2a) \approx i(\epsilon_1k_2/\epsilon_2k_1 + \epsilon_3k_2/\epsilon_2k_3). \tag{13}$$

The right-hand sides of Eqs. (12) and (13) are complex numbers much smaller in magnitude than unity. Hence, applying the perturbation technique we obtain for the TE modes

$$2k_2a \approx (n+1)\pi(1 + \theta_r/2ka + i\theta_i/2ka), \tag{14}$$

where $n = 0, 1, 2, \ldots$ is the mode order and

$$\theta_i \equiv (\epsilon_3/\epsilon_2 - 1)^{-1/2}, \tag{15}$$

$$\theta_r \equiv (1 - \epsilon_1/\epsilon_2)^{-1/2}. \tag{16}$$

The corresponding characteristic equation for the TM modes is

$$2k_2a \approx (n+1)\pi\left[1 + \frac{\epsilon_1}{\epsilon_2}\frac{\theta_r}{2ka} + i\frac{\epsilon_3}{\epsilon_2}\frac{\theta_i}{2ka}\right]. \tag{17}$$

Substitution of Eqs. (14)–(17) into Eq. (2b) gives

$$\gamma_{TE} = k[1 - (k_2/k)^2]^{1/2} \approx k[1 - \tfrac{1}{2}(k_2/k)^2]$$
$$\approx k\left[1 - \frac{(n+1)^2\pi^2}{8k^2a^2}\left(1 + \frac{\theta_r}{ka} + i\frac{\theta_i}{ka}\right)\right] \tag{18}$$

for the TE modes and

$$\gamma_{TM} \approx k\left[1 - \frac{(n+1)^2\pi^2}{8k^2a^2}\left(1 + \frac{\epsilon_1}{\epsilon_2}\frac{\theta_r}{ka} + i\frac{\epsilon_3}{\epsilon_2}\frac{\theta_i}{ka}\right)\right] \tag{19}$$

for the TM modes. Remembering that $ka \gg 1$, the phase constants and rates of power attenuation for the TE and TM modes are, to first order,

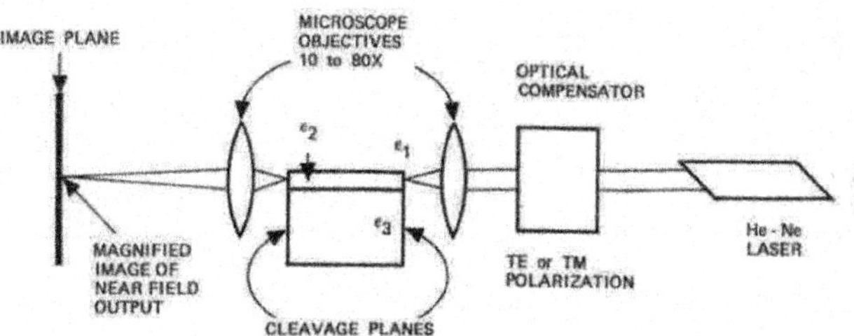

FIG. 3. Experimental setup for observing the leaky waveguide modes.

(a) 2 μm THICK LAYER

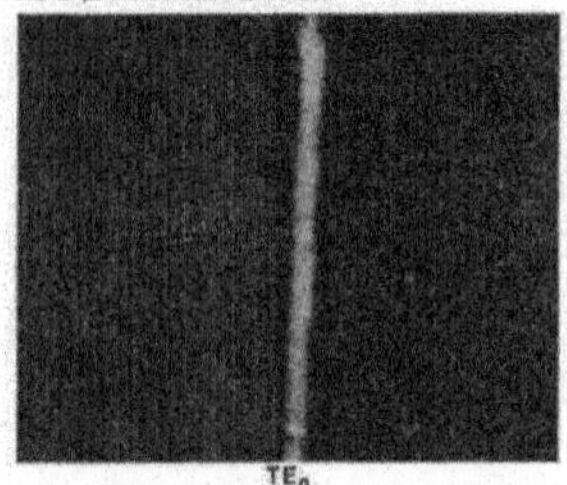

(b) 4 μm THICK LAYER

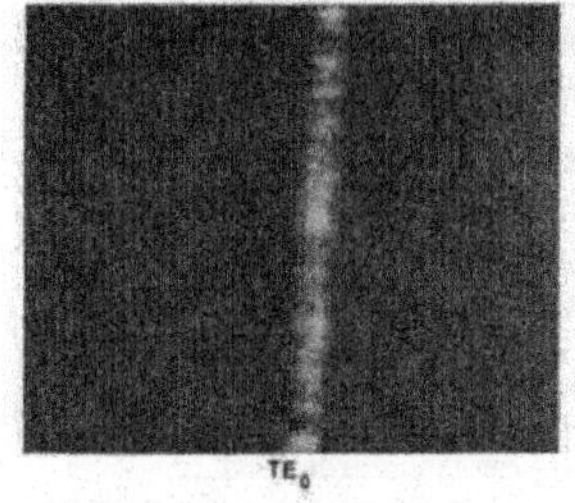

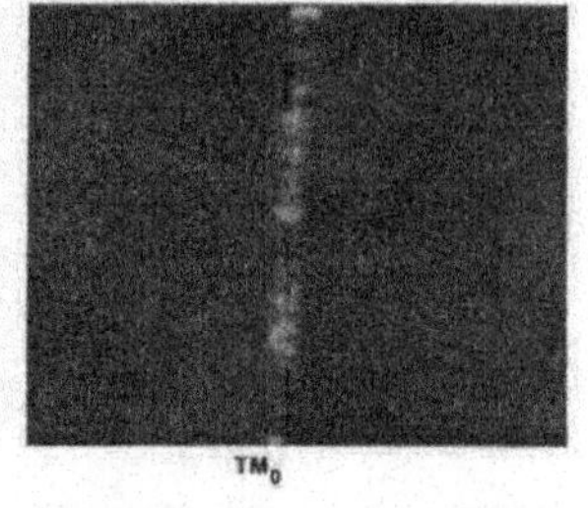

(c) 14 μm THICK LAYER

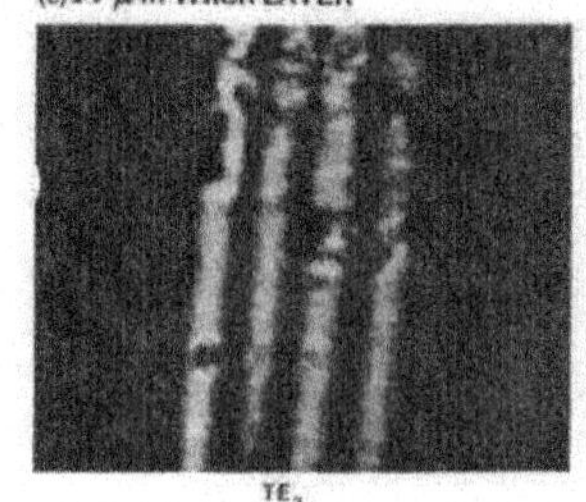

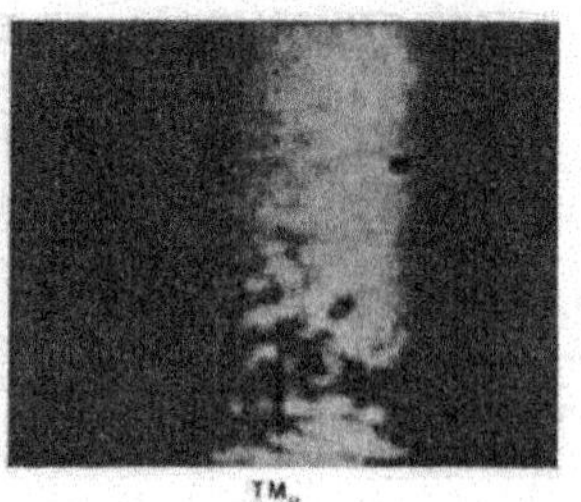

FIG. 4. Photographs of mode profiles.

$$\beta_{TE}=\mathrm{Re}(\gamma_{TE})\approx k\,[1-\pi^2(n+1)^2/8k^2a^2], \quad (20)$$

$$\alpha_{TE}=2\mathrm{Im}(\gamma_{TE})\approx k\,\pi^2(n+1)^2\theta_i/4k^3a^3, \quad (21)$$

$$\beta_{TM}=\mathrm{Re}(\gamma_{TM})\approx k[1-\pi^2(n+1)^2/8k^2a^2], \quad (22)$$

$$\alpha_{TM}=2\mathrm{Im}(\gamma_{TM})\approx k[\pi^2(n+1)^2\theta_i/4k^3a^3](\epsilon_3/\epsilon_2). \quad (23)$$

Upon comparison of Eq. (20) with Eq. (22) and Eq. (21) with Eq. (23), one notes that the frequency dependence and the guide-width dependence for the attenuation constants of TE and TM modes of the same order are identical. Only the magnitudes of the attenuation constants for these modes are different; $\alpha_{TM}=(\epsilon_3/\epsilon_2)\alpha_{TE}$. Also noted is the fact that the phase constants for these modes are identical.

Rearranging Eqs. (21) and (23) gives

$$\Delta^{TE}=\frac{\epsilon_2}{\epsilon_0}\frac{\alpha_{TE}a}{(n+1)^2\theta_i}=4.34\frac{\pi^2}{4k_0^2a^2}\ \text{(dB)}, \quad (24)$$

$$\Delta^{TM}=\frac{\epsilon_2}{\epsilon_0}\frac{\alpha_{TM}a}{(n+1)^2\theta_i}\frac{\epsilon_2}{\epsilon_3}=4.34\frac{\pi^2}{4k_0^2a^2}\ \text{(dB)}, \quad (25)$$

where $k_0=2\pi/\lambda_0$. λ_0 is the free-space wavelength and Δ^{TE} and Δ^{TM} are called, respectively, the normalized attenuation constants for the TE and TM modes. To illustrate how the normalized attenuation constant varies as a function of k_0a, Fig. 2 is introduced. It can be seen that substantial attenuation results for even moderate values of k_0a. As expected, when the thickness of the slab becomes very large, the guided-wave approaches the plane-wave case and the attenuation constant becomes very small.

EXPERIMENTAL VERIFICATION OF THE LEAKY MODES

The TE and TM leaky modes along a structure with $\epsilon_1 < \epsilon_2 < \epsilon_3$ were investigated using the experimental setup shown in Fig. 3. Light from a HeNe laser at $\lambda_0=6328$ Å was focused on the (110) cleaved edge of a ZnSe epitaxial layer with optical dielectric constant $\epsilon_2/\epsilon_0=6.66$ sandwiched between air with $\epsilon_1/\epsilon_0=1$ and a GaAs substrate with $\epsilon_3/\epsilon_0\approx 15$.[4,5] (At 6328 Å GaAs has a complex dielectric constant with an imaginary part less than 10% of the real part[5]; this should not effect greatly the analysis in the previous section where the imaginary part was ignored). The near-field output from the ZnSe layer was magnified and projected on a screen where it was photographed. Three different structures were constructed and tested. (a) ZnSe layer of thickness $2a \approx 2\ \mu$ and length of 2.2 mm. (b) ZnSe layer of thickness $2a \approx 4\ \mu$ and length of 8.6 mm. (c) ZnSe layer of thickness $2a \approx 14\ \mu$ and length of 11.5 mm. At $\lambda_0 = 6328$ Å, the first structure supported only the lowest TE_0 mode; losses were too high for the TM_0 mode to propagate. The second structure supported the TE_0 mode and, with much higher losses, the TM_0 mode, and the third structure supported TE and TM modes with mode orders observed from zero to six. Representative modes profiles from the three guides are shown in Fig. 4.

As predicted by our theoretical analysis, the attenuation for the TM mode is significantly higher than for the TE mode. It is therefore not surprising to note that guiding of the TM mode was not observed for the first structure. It was also observed that the amount of light passing through the guiding structure is critically dependent upon the thickness of the layer as anticipated from our theoretical work.

In conclusion we state that we have successfully demonstrated the existence of these leaky modes. It is hoped that our theoretical results may be used to obtain the attenuation constants of the leaky modes on these increasingly important IOC structures.

ACKNOWLEDGMENT

One of us (C.Y.) wishes to thank Dr. Don Albares and his group for their kind hospitality during his stay at NELC last summer.

[1]R. Collins, *Field Theory of Guided Waves* (McGraw-Hill, New York, 1961).
[2]R. L. Fork, K. R. German, and E. A. Chandross, Appl. Phys. Lett. **20**, 139 (1972).
[3]J. J. Burke, Appl. Opt. **9**, 2444 (1970).
[4]D. T. F. Marple, J. Appl. Phys. **35**, 539 (1964).
[5]H. R. Philipp and H. Enrenreich, Phys. Rev. **129**, 1550 (1963).

IEEE TRANSACTIONS ON MICROWAVE THEORY AND TECHNIQUES, VOL. 38, NO. 6, JUNE 1990 691

Dielectric Ribbon Waveguide: An Optimum Configuration for Ultra-Low-Loss Millimeter/Submillimeter Dielectric Waveguide

C. YEH, FELLOW, IEEE, FRED I. SHIMABUKURO, MEMBER, IEEE, AND J. CHU, STUDENT MEMBER, IEEE

Abstract —Dielectric ribbon waveguide supporting the ${}_eHE_{11}$ dominant mode can be made to yield an attenuation constant for this mode of less than 20 dB/km in the millimeter/submillimeter-wavelength range. The waveguide is made with a high-dielectric-constant, low-loss material such as alumina or sapphire. It takes the form of thin dielectric ribbon surrounded by lossless dry air. A detailed theoretical analysis of the attenuation and field extent characteristics for the low-loss dominant ${}_eHE_{11}$ mode along a ribbon dielectric waveguide was carried out using the exact finite-element technique as well as two approximate techniques. Analytical predictions were then verified by measurements on ribbon guides made with Rexolite using the highly sensitive cavity resonator method. Excellent agreement was found.

I. Introduction

THE PHENOMENAL success of the dielectric fiber as an ultra-low-loss optical waveguide has enticed us to reconsider the viability of the dielectric rod as a low-loss millimeter/submillimeter (mm/sub-mm) waveguide. A survey of commercially available materials shows that two classes of material may be excellent candidates as low-loss dielectric materials for mm/sub-mm wave applications [1]–[6]: (1) crystalline material such as quartz, alumina, and sapphire and (2) nonpolar polymers such as poly-4-methylpentene-1 (TPX), PTFE (Teflon), polyethylene (LDPE), and polypropylene. Another way to minimize the attenuation constant for the guided wave along a dielectric structure is to use special waveguide configurations. This paper will first provide a brief survey of commercially available low-loss dielectric material and highlight possible ways to reduce the loss factor. Then we will focus our attention on identifying low-loss configurations. It is shown that a properly configured waveguide can support the dominant mode with a loss factor as much as 50–100 times below that for an equivalent circular dielectric waveguide. A loss factor of less than 20 dB/km can be realized with presently available material. This waveguide takes the form of a thin dielectric ribbon surrounded by lossless dry air. Theoretical analyses have been carried out based on three approaches: the slab approach [7], Marcatili's approach [8], and the exact finite-element approach [9]. Experimental verification of selected cases has also been carried out using the unique ultra-high-Q dielectric waveguide cavity resonator apparatus that we developed [10]. Our investigation shows that it is feasible to design a long-distance mm/sub-mm wave communication line with losses approaching 20 dB/km using the dielectric ribbon waveguide made with commercially available low-loss, high-dielectric-constant material.

Manuscript received December 8, 1988; revised February 18, 1990. This work was supported in part by the UCLA–TRW MICRO Program and by the U.S. Air Force under Contract F04701-87-C-0088.

C. Yeh and J. Chu are with the Electrical Engineering Department, UCLA, Los Angeles, CA 90024-1594.
F. I. Shimabukuro is with the Aerospace Corporation, P.O. Box 92957, Los Angeles, CA 90009.

IEEE Log Number 9035405.

II. A Discussion on Low-Loss Dielectric Material

A series of very detailed measurements in the mm/sub-mm wavelength range on the dielectric constants and loss tangents of groups of promising low-loss materials has been performed by the MIT "Mag-Lab" group in recent years. Results of their findings were summarized in a very comprehensive paper by Afsar and Button [1]. Afsar [3] also presented his measured results on several very low loss nonpolar polymers. A sample list of two types of commercially available low-loss materials is given in Table I. It can be seen that the polymer material in general has a much lower dielectric constant than the crystalline material. The best loss tangent is of the order of 10^{-4}. Using a nominal dielectric constant of 2.0, the attenuation constant for plane wave in this bulk material is 1.3 dB/m at 100 GHz, which is already better than the 2.4 dB/m loss for conventional metallic waveguides at this frequency. The attenuation constant for plane wave is calculated from the following equation [11]:

$$\alpha = 8.686\left(\pi\sqrt{\epsilon_r}/\lambda_0\right)\tan\delta \quad (\text{dB/m}). \tag{1}$$

Here ϵ_r is the relative dielectric constant, λ_0 is the free-space wavelength, and $\tan\delta$ is the loss tangent. According to (1), it appears that in addition to requiring as small a loss tangent as possible, a lower dielectric constant is also helpful in achieving lower loss. Hence, flexible nonpolar polymers such as LDPE (polyethylene) and

692 IEEE TRANSACTIONS ON MICROWAVE THEORY AND TECHNIQUES, VOL. 38, NO. 6, JUNE 1990

TABLE I

Crystalline Material [4]–[6]	Dielectric constant	Loss tangent
ZnS (at 100 GHz)	8.4	2×10^{-3}
Alumina (at 10 GHz)	9.7	2×10^{-4}
Sapphire (at 100 GHz)	9.3–11.7	4×10^{-4}
Quartz (at 100 GHz)	3.8–4.8	5×10^{-4}
KRS-5 (at 94.75 GHz)	30.5	1.9×10^{-2}
KRS-6 (at 94.75 GHz)	28.5	2.3×10^{-2}
$LiNbO_3$ (at 94.75 GHz)	6.7	8×10^{-3}
Polymer Material [1]–[3]	Dielectric constant	Loss tangent
Teflon (at 100 GHz) (PTFE)	2.07	2×10^{-4}
Rexolite (at 10 GHz)	2.55	1×10^{-3}
RT-Duroid 5880 (at 10 GHz)	2.2	9×10^{-4}
Polyethylene (at 100 GHz) (LDPE)	2.306	3×10^{-4}
TPX (Poly-4 methylpentene-1) (at 100 GHz)	2.071	6×10^{-4}
Polypropylene (at 100 GHz)	2.261	7×10^{-4}

PTFE (Teflon) may be good choices for making low-loss mm/sub-mm waveguides. However, this conclusion can be deceiving because it is based purely on the low-loss property of the waveguide material; i.e., only bulk material loss is considered and the effect of waveguide configuration on losses has not been included. If the configuration effect is taken into account, material with lower dielectric constant may not offer the advantage of lower attenuation constant as indicated by (1). (Detailed consideration will be given to this in Section III.)

One way to construct low-loss waveguide material is to use an artificial dielectric [12], [13]. The artificial dielectric may be composed of alternate longitudinal layers of low-loss, high-dielectric-constant material such as quartz and air. One may also interpret this approach as a way of altering the configuration of guided wave structure, which will be addressed in the next section. The artificial dielectric material may also be constructed with small particles (size $\ll \lambda_0$) of low-loss, high-dielectric-constant material such as small quartz or sapphire spheres suspended in air. For example, the dielectric constant and loss tangent of sapphire at 100 GHz are 10 and 4×10^{-4}, respectively. Roughly speaking, if a volume contains 10% sapphire powder (small spheres), the equivalent dielectric constant would be approximately 2, and the equivalent loss tangent could be reduced to 4×10^{-5}. Thus high-dielectric-constant, low-loss crystalline materials are excellent candidates for the construction of artificial low-loss dielectric materials.

Returning to our discussion of polymers, it is known that the molecules that make up a typical molecular crystal are bound together by strong valence forces and are held in their correct places in the lattice by much weaker van der Waals forces. The major absorption mechanisms in nonpolar polymers in the mm/sub-mm region are [3]:

1) resonances of normal modes of macromolecular helices;
2) absorption spectrum of impurities such as catalyst residues, antioxidants, ionic impurities, plasticizers, catalyst residues, and other additives;
3) amorphous behavior of polymers.

Recognizing the mechanisms that are generally responsible for the absorption spectra of these polymers in the mm/sub-mm/far-infrared region, one may selectively improve the loss characteristics of a given polymer. For example, impurities may be controlled and reduced; other ions may be introduced to stiffen the long chain molecules so that the chain-twisting vibrations in the mm-wave frequencies may be dampened, reducing the absorptions at these frequencies; and the amorphous characteristics of the polymer can be altered by using different cooling/melting processes.

It will be shown in the following section that a high-dielectric-constant material with low loss tangent is the preferred material for the construction of specially configured ultra-low-loss waveguides.

III. Low-Loss Configuration Study

Reducing the loss tangent of the bulk material will certainly improve the attenuation constant of a guided wave along a dielectric waveguide made with such material. It appears that other major factors that can influence the attenuation characteristics of guided wave along a dielectric structure are the size and shape of the waveguide. The attenuation constant for a dielectric waveguide with arbitrary cross-sectional shape is given by the following expression [11]:

$$\alpha = 8.686\pi(1/\lambda_0)(L_1 + L_0) \quad (\mathrm{dB/m}) \tag{2}$$

where

$$L_{1,0} = (\epsilon_{1,0}) \tan\delta_{1,0} R_{1,0} \tag{3}$$

$$R_{1,0} = \left| \frac{\int_{A_{1,0}} (E_{1,0}\cdot E_{1,0}^*)\, dA}{\sqrt{\frac{\mu}{\epsilon}}\left[\int_{A_1} e_z\cdot(E_1\times H_1^*)\, dA + \int_{A_0} e_z\cdot(E_0\times H_0^*)\, dA\right]} \right|. \tag{4}$$

Here, the subscripts 1 and 0 refer, respectively, to the core region and the cladding region of the guide, $\epsilon_{1,0}$ and $\tan\delta_{1,0}$ are, respectively, the relative dielectric constant and the loss tangent of the dielectric material, λ_0 is the free-space wavelength, ϵ and μ are, respectively, the

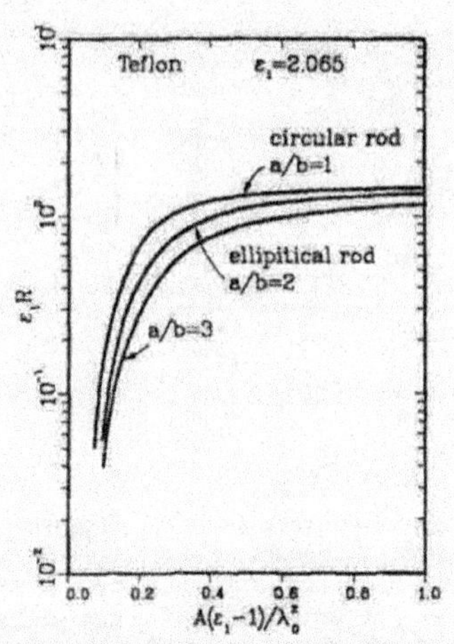

Fig. 1. Configuration loss factor $\epsilon_1 R$ as a function of normalized frquency for an elliptical Teflon rod supporting the dominant ${}_eHE_{11}$ mode. Here A is the cross-sectional area, λ_0 is the free-space wavelength, a is the semimajor axis of the elliptical rod, and b is the semiminor axis. Note that flatter rod yields a smaller configuration loss factor for the same cross-sectional area.

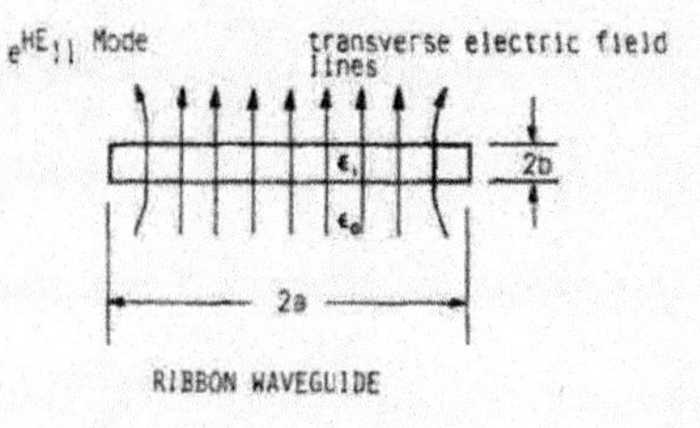

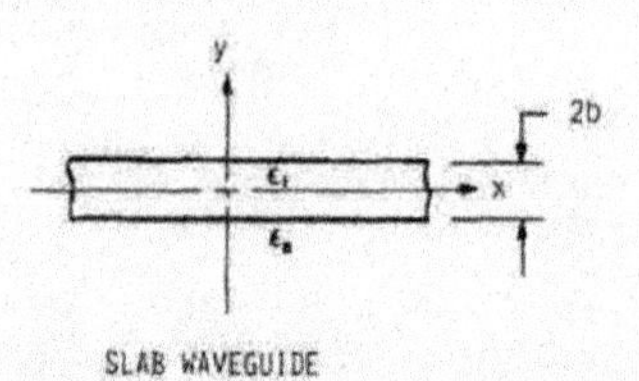

Fig. 2. Cross-sectional geometries for ribbon waveguide and slab waveguide.

permittivity and permeability of free space, e_z is the unit vector in the direction of propagation, A_1 and A_0 are, respectively, the cross-sectional areas of the core and the cladding region, and E and H are the electric and magnetic field vectors of the guided mode under consideration.

If the core and cladding regions contain similar dielectric material, as in the case of optical fiber waveguide, the attenuation constant α will be relatively insensitive to the geometry of the guide because $(R_1 + R_0)$ will be insensitive to the geometry of the guide. For this case, the attenuation of the wave is determined totally by the loss tangent of the material, and the guide configuration is unimportant. On the other hand, if the cladding region (region 0) contains low-loss dry air or is a vacuum, then the loss factor $\epsilon_1 R_1$, which is sensitive to the guide configuration and the frequency of operation, will play an important role in determining the attenuation constant of the mode guided by the dielectric structure. This loss factor $\epsilon_1 R_1$ could vary from a very small value to $\sqrt{\epsilon_1}$, which is the case for a plane wave propagating in a dielectric medium with relative dielectric constant ϵ_1. So, for a given operating frequency, the smaller the factor $\epsilon_1 R_1$, the more desirable the configuration. As an example, the $\epsilon_1 R_1$ factors for an elliptical Teflon dielectric waveguide supporting the dominant ${}_eHE_{11}$ mode as a function of the normalized cross-sectional area for different (major axis)/(minor axis) ratios are given in Fig. 1. It is seen that a mere flattening of a circular dielectric rod along the maximum intensity of the electric field lines for the dominant ${}_eHE_{11}$ mode can improve the $\epsilon_1 R_1$ factor (hence, α) by a factor of 2 or more [11]. It appears that flattening the circular dielectric rod tends to redistribute and spread out the electric field intensities in such a way that the factor $\int_{A_1}(E_1 \cdot E_1^*)\,dA$ in (4) (hence, α) is substantially reduced. This deduction leads us to the conclusion that a very flat elliptical cylinder or simply a thin ribbon may be an extremely attractive low-loss configuration.

A. Wave Guidance Along a Dielectric Slab

Let us now turn our attention to the problem of wave guidance by an infinite flat plate as shown in Fig. 2. Two types of dominant modes may exist on this structure: the TM mode (with E_y, E_z, H_x), which is the low-loss mode, and the TE mode (with H_y, H_z, E_x), which is the high-loss mode. The field components for these modes are given below [7], [12]. (Since only the symmetric modes are considered, the following field expressions, applicable in the region $y > 0$, are used.)

TM Wave: In region 1 (the core region),

$$E_y^{(1)} = \frac{-j\beta}{p_1} B \cos p_1 y$$
$$E_z^{(1)} = B \sin p_1 y$$
$$H_x^{(1)} = \frac{-j\omega\epsilon_1\epsilon}{p_1} B \cos p_1 y \tag{5}$$

and in region 0 (the cladding region),

$$E_y^{(0)} = \frac{-j\beta}{p_0} Ce^{-p_0 y}$$
$$E_z^{(0)} = Ce^{-p_0 y}$$
$$H_x^{(0)} = \frac{-j\omega\epsilon_0\epsilon}{p_0} Ce^{-p_0 y} \tag{6}$$

with

$$k_0^2 = \omega^2 \mu \epsilon_0 \epsilon \qquad k_1^2 = \omega^2 \mu \epsilon_1 \epsilon$$

$$p_0^2 = \beta^2 - k_0^2 \text{ and } p_1^2 = k_1^2 - \beta^2.$$

TE Wave: In region 1 (the core region),

$$H_y^{(1)} = \frac{j\beta D}{p_1} \cos p_1 y$$

$$H_z^{(1)} = D \sin p_1 y$$

$$E_x^{(1)} = \frac{j\omega\mu D}{p_1} \cos p_1 y \tag{7}$$

and in region 0 (the cladding region),

$$H_y^{(0)} = \frac{j\beta}{p_0} F e^{-p_0 y}$$

$$H_z^{(0)} = F e^{-p_0 y}$$

$$E_x^{(0)} = \frac{j\omega\mu}{p_0} F e^{-p_0 y}. \tag{8}$$

We have assumed that the expressions for the field components of all modes are multiplied by the factor $\exp(j\omega t - j\beta z)$, which will be suppressed throughout. Here β and ω are, respectively, the propagation constant and angular frequency of the wave, and z is the direction of propagation of the wave. Matching the tangential electric and magnetic fields at the boundary surface $y = b$ yields the dispersion relations for the symmetric TM and TE modes, from which the $\omega - \beta$ characteristics for these modes may be found. The dispersion relations and the ratios of unknown coefficients are as follows.

TM Wave:

$$\tan p_1 b = \frac{\epsilon_1}{\epsilon_0} \frac{p_0 b}{p_1 b} \qquad p_1^2 + p_0^2 = k_0^2 \left(\frac{\epsilon_1}{\epsilon_0} - 1 \right)$$

$$\frac{B}{C} = \frac{e^{-p_0 b}}{\sin p_1 b}. \tag{9}$$

TE Wave:

$$\tan p_1 b = \frac{p_0 b}{p_1 b} \qquad p_1^2 + p_0^2 = k_0^2 \left(\frac{\epsilon_1}{\epsilon_0} - 1 \right)$$

$$\frac{D}{F} = \frac{e^{-p_0 b}}{\sin p_1 b}. \tag{10}$$

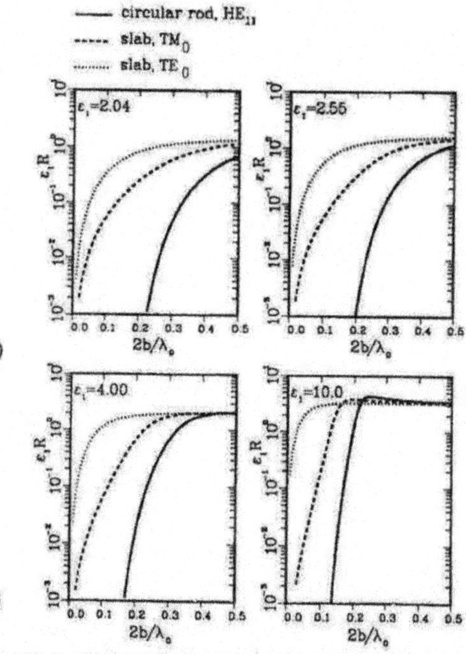

Fig. 3. Configuration loss factor $\epsilon_1 R$ as a function of normalized frequency for a dielectric slab of thickness $2b$ supporting the dominant TE_0 and dominant TM_0 modes and for a circular rod of radius b supporting the dominant HE_{11} mode for various dielectric materials.

Referring to (4), one may also calculate the configuration loss factor R as follows:

$$R^{(TM)} = \frac{(p_0 b)^3}{2(p_1 b)(k_0 b)(\beta b)} \cdot \frac{\left[2p_1 b \left(1 + \left(\frac{\beta b}{p_1 b} \right)^2 \right) - \left(1 - \left(\frac{\beta b}{p_1 b} \right)^2 \right) \sin 2p_1 b \right]}{\left[\frac{1}{2} \left(\frac{p_0 b}{p_1 b} \right)^3 \frac{\epsilon_1}{\epsilon_0} (2p_1 b + \sin 2p_1 b) + \sin^2 p_1 b \right]} \tag{11}$$

$$R^{(TE)} = \frac{1}{2} \frac{k_0 b}{\beta b} \left(\frac{p_0 b}{p_1 b} \right)^3 \cdot \frac{(2p_1 b + \sin 2p_1 b)}{\left[\left(\frac{p_0 b}{p_1 b} \right)^3 \frac{1}{2} (2p_1 b + \sin 2p_1 b) + \sin^2 p_1 b \right]}. \tag{12}$$

As expected, one can easily show the limiting case for $R^{(TE)}$ and $R^{(TM)}$ as $(2b/\lambda_0) \to \infty$; it is

$$R^{(TE)} = R^{(TM)} \to 1/\sqrt{\epsilon_1}.$$

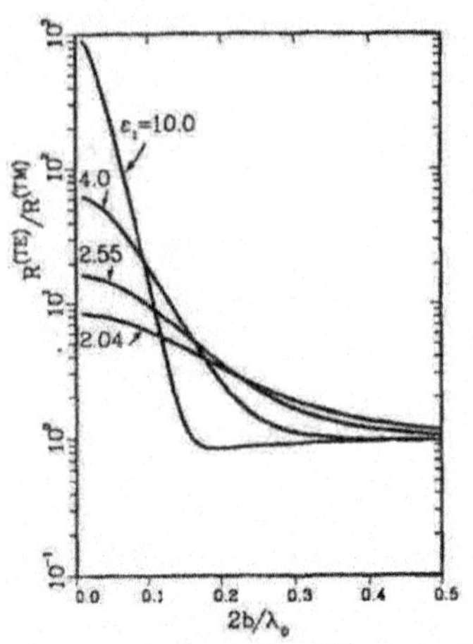

Fig. 4. Ratio of the configuration loss factor for TE_0 and TM_0 versus the normalized frequency for various dielectric materials. Note that the effect of higher dielectric constant material on the ratio is much more pronounced.

Although $R^{(TM)}$ and $R^{(TE)}$ approach the same limit as $2b/\lambda_0$ approaches ∞, the behavior of $R^{(TM)}$ versus $2b/\lambda_0$ and that of $R^{(TE)}$ versus $2b/\lambda_0$ are very different. Fig. 3 gives plots of $\epsilon_1 R^{(TM)}, \epsilon_1 R^{(TE)}$ versus $2b/\lambda_0$ for different values of ϵ_1. The normalized coefficient $R^{(TM),(TE)}\epsilon_1$ is used because it is proportional to the attenuation constant $\alpha^{(TM),(TE)}$. For a given $2b/\lambda_0$ and ϵ_1, $\epsilon_1 R^{(TM)}$, in general, is significantly lower than $\epsilon_1 R^{(TE)}$, indicating that the dominant TM_0 mode is the low-loss mode. The ratios of $[\epsilon_1 R^{(TE)}]/[\epsilon_1 R^{(TM)}]$ versus $2b/\lambda_0$ for three values of ϵ_1 are shown in Fig. 4. It is of interest to note that the ratio is higher for higher ϵ_1. For example, at the nominal operating frequency of $2b/\lambda_0 = 0.1$, when $\epsilon_1 = 2.04$ (Teflon), the ratio $\epsilon_1 R^{(TE)}/\epsilon_1 R^{(TM)}$ is 6; when $\epsilon_1 = 2.55$ (Rexolite), the ratio is 10; and when $\epsilon_1 = 4$ (quartz) the ratio is 19, indicating that the loss factor for the TM_0 mode is 19 times smaller than that for the TE_0 mode for material with a higher dielectric constant. This fact appears to suggest that high-dielectric-constant material with low loss tangent is the most desirable material for low-loss dielectric waveguides.

Since the field guided along a dielectric waveguide without cladding material extends into the region beyond the dielectric core, it is of interest to learn the relationship between the loss factor $\epsilon_1 R$ and the field extent beyond the core region. In Fig. 5, the loss factor $\epsilon_1 R$ is plotted against the normalized field extent beyond the core surface, expressed by the distance d from the core surface at which the power density of the guided mode has decayed to $1/e$ of its value at the core surface divided by the free-space wavelength. So, given $\epsilon_1 R$ and ϵ_1, one may obtain the normalized distance from the core surface at which the guided power has decayed to its $1/e$ value at

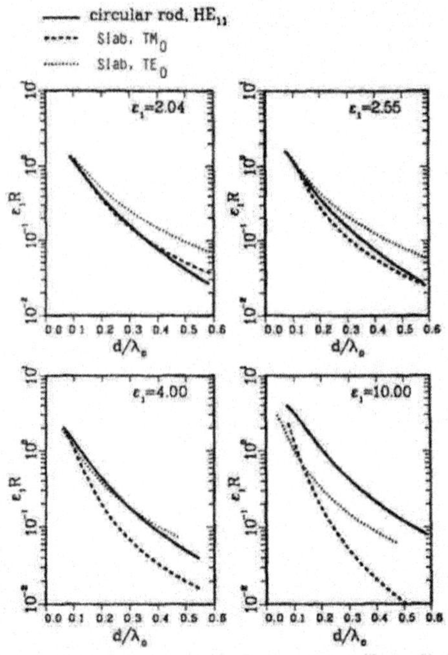

Fig. 5. Configuration loss factor $\epsilon_1 R$ as a function of normalized power decaying distance d/λ_0. Here d is the $1/e$ power decaying distance from the surface of the slab or rod, as appropriate. Note that for high-dielectric-constant materials, the power decaying distance is much shorter for a slab guide than that for a circular rod guide for the same configuration loss factor.

the core surface. For example, if $\epsilon_1 R$ is 0.1 and $\epsilon_1 = 4.0$ (quartz), $d/\lambda_0 = 0.26$ for the low-loss TM_0 mode on a slab, $d/\lambda_0 = 0.38$ for the HE_{11} mode on a circular rod, and $d/\lambda_0 = 0.41$ for the high-loss TE mode on a slab.

B. Wave Guidance Along a Dielectric Ribbon

Recognizing the fact that the dominant ${}_eHE_{11}$ mode guided along a flat dielectric ribbon with aspect ratio greater than, say, 10 must behave similarly to the dominant TM_0 mode guided along a dielectric slab and that the only significant differences must be due to the fringing fields at the edges of the flat ribbon guide, one may make use of the slab results to obtain the approximate results for the ${}_{e,o}HE_{11}$ modes along a dielectric ribbon guide with high aspect ratio. It is also expected that at very low frequencies, because of the large field extent around the guiding structure, implying that the field pattern for an infinite plate would be substantially different from that for a ribbon structure, the loss factor behavior for infinite plate and ribbon will also be very different. This very low frequency region would not be the region of interest because the mode is too loosely guided for any

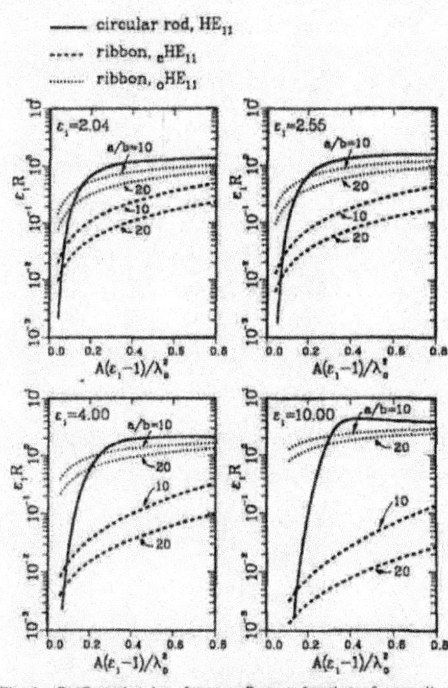

Fig. 6. Configuration loss factor $\epsilon_1 R$ as a function of normalized frequency for a dielectric ribbon with width $2a$ and thickness $2b$ for various dielectric materials. This calculation is based on the slab approximation for which all fields external as well as internal are confined within a width of $2a$. Note, for a given normalized frequency, the dramatic difference between the configuration loss factor for a ribbon supporting the low-loss TM wave and that for a circular rod supporting the HE_{11} mode, especially for higher dielectric constant material.

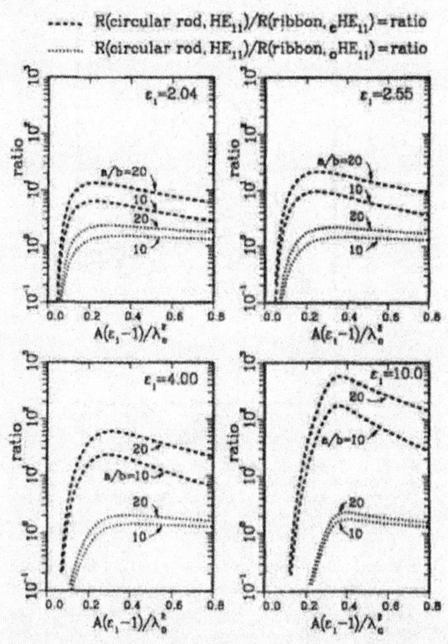

Fig. 7. Ratio of the configuration loss factor for circular rod and ribbon versus the normalized frequency for ribbon of various aspect ratios and for different dielectric materials. Note that for $\epsilon_1 = 10$ and $a/b = 20$, the ratio at $A(\epsilon_1 - 1)/\lambda_0^2 = 0.35$ is as high as 600.

practical applications. The loss factor $\epsilon_1 R$ is plotted against the normalized area, $A(\epsilon_1 - 1)/\lambda_0^2$, where A is the cross-sectional area of the guide, in Fig. 6. It can be seen that there is a dramatic difference between the loss factors for the ribbon TM ($_eHE_{11}$) mode, the ribbon TE ($_oHE_{11}$) mode, and the circular rod HE_{11} mode for the same normalized area. The loss factor for flat ribbon guide supporting the $_eHE_{11}$ mode could be as much as 100 times smaller than that for a circular rod guide supporting the HE_{11} mode. Furthermore, for a rather broad region of normalized area, the loss factor, $\epsilon_1 R$, is reasonably flat for the ribbon guide while it is rather steep for a circular rod guide, indicating that the ribbon guide possesses rather stable low-loss behavior for any possible fluctuation in operating frequencies or cross-sectional area changes.

Another way of demonstrating the advantage of ribbon guide over circular rod guide is shown in Fig. 7, where the loss factor ratio, $\epsilon_1 R$ (for circular rod)/$\epsilon_1 R$ (for ribbon), is plotted against the normalized cross-sectional area. Again, a dramatic difference is seen. For example, from Fig. 7 with $\epsilon_1 = 10$ (sapphire), if the normalized area is 0.35, the ratio could be as high as 400 for a ribbon with aspect ratio of 20:1, indicating that the loss factor for a ribbon guide could be as much as 400 times less than that for a circular rod. These curves also show the advantage of using high-dielectric-constant and low-loss-tangent material to construct the waveguide structure. Fig. 8 demonstrates this.

A major concern of any open guiding structure is the field extent outside the core region. The fact that the loss factor for a flat ribbon can be made so much smaller than that for an equivalent circular rod is primarily due to the spreading of the guided power in the lossless outer (non-core) region. But the distinguishing feature of a ribbon guide is its expanding surface area, which enables the guided mode to attach to it. This feature is very much unlike the case for the circular rod, which possesses very minimal surface area; hence, its guided mode (in the low-loss region) tends to be loosely attached to the guide and can easily detach itself and become a radiated wave. How rapidly the power density of the guided mode decays from the core surface is shown in Fig. 9 for the four low-loss guiding materials. In this figure, the normalized distance, d/λ_0, is plotted against the normalized area.

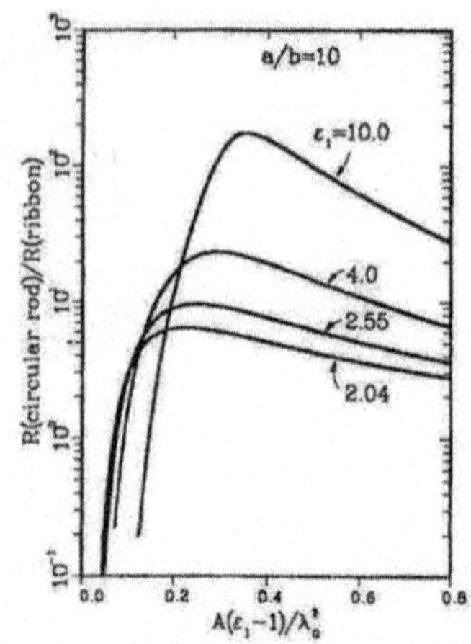

Fig. 8. Ratio of the configuration loss factor for circular rod and ribbon versus the normalized frequency for ribbon with an aspect ratio of 10 for various dielectric materials. Note the dramatic increase of the ratio as ϵ_1 is increased.

Here, d is the distance from the core surface at which the power density of the guided mode has decayed to $1/e$ of its value at the core surface. For a typical operating range $(0.2 < A(\epsilon_1-1)/\lambda_0^2 < 1.0)$, the normalized field extent, d/λ_0, is less than 0.5. In other words, it is safe to conclude that most of the guided power is confined within a region whose outer boundary is situated at least one free-space wavelength away from the core boundary. For millimeter or submillimeter operation, this requirement is easily accommodated.

IV. Theoretical Verifications

In the previous section, the solution for a plane slab is used to form the solution for a ribbon with high aspect ratio. Further refinement of the slab solution can be obtained using an approximate approach developed by Marcatili [8]. He formulated an approximate solution to the problem of wave guidance by rectangular dielectric structures by ignoring the matching of fields along the corners of the rectangular dielectric structure. By matching the tangential electric and magnetic fields along the four sides of the rectangular core region, and assuming that the field components in the core region vary sinusoidally in the two transverse directions along its major and minor axes (those in upper and lower regions outside the core varying sinusoidally along the direction for the major axis and decaying exponentially along the direction for the minor axis, and those in the left and right regions outside the core varying sinusoidally along the direction for the minor axis and decaying exponentially along the direction for the major axis), one may obtain a dispersion relation from which the propagation constants of various modes may be calculated. For the low-loss TM wave, the propagation constant β can be found by solving the following equations:

$$2ak_x = \pi - 2\tan^{-1}\left[k_x/(k_1^2-k_0^2-k_x^2)^{1/2}\right]$$
$$2bk_y = \pi - 2\tan^{-1}\left[(k_y/\epsilon_1)/(k_1^2-k_0^2-k_y^2)^{1/2}\right]$$
$$\beta^2 = k_1^2 - k_x^2 - k_y^2. \quad (13)$$

Here, $2b$ and $2a$ are, respectively, the height and the width of the ribbon guide; $k_1^2 = \omega^2\mu\epsilon_1\epsilon$ and $k_0^2 = \omega^2\mu\epsilon_0\epsilon$. The configuration loss factor $R_{\text{Marcatili}}^{(\text{TM})}$ is

$$R_{\text{Marcatili}}^{(\text{TM})} = |I|/\left[\sqrt{\mu/\epsilon}\,|I_p|\right] \quad (14)$$

with

$$\begin{aligned} I = {} & (\omega\epsilon_1\beta)^{-2}ab\Big\{(k_xk_y)^2[1-\text{sinc}(2k_xa)] \\ & \cdot[1-\text{sinc}(2k_yb)] \\ & +(k_1^2-k_y^2)^2[1+\text{sinc}(2k_xa)][1+\text{sinc}(2k_yb)] \\ & +(\beta k_y)^2[1+\text{sinc}(2k_xa)][1-\text{sinc}(2k_yb)]\Big\} \end{aligned}$$

$$\begin{aligned} I_p = {} & (k_1^2-k_y^2)ab(\omega\epsilon_1\beta)^{-1}[1+\text{sinc}(2k_xa)] \\ & \cdot[1+\text{sinc}(2k_yb)] \\ & +(k_1^2-k_y^2)a(\omega\epsilon_0\beta)^{-1} \\ & \cdot(k_1^2-k_0^2-k_y^2)^{-1/2}[1+\text{sinc}(2k_xa)]\cos^2(k_yb) \\ & +(k_0^2-k_y^2)b(\omega\epsilon_0\beta)^{-1}(k_1^2-k_0^2-k_x^2)^{-1/2} \\ & \cdot\cos^2(k_xa)[1+\text{sinc}(2k_yb)]. \end{aligned}$$

To compare the configuration loss factors $R_{\text{Marcatili}}^{(\text{TM})}$ with $R^{(\text{TM})}$ (according to the slab model), Fig. 10 is introduced. In this figure the normalized configuration loss factors for both models are plotted against the normalized area. It is seen that, for a normalized frequency larger than 0.1 and $b/a > 3$, the loss factors for both models are quite close to each other, indicating that the slab model approximation approaches Marcatili's approximation. One also observes that as the aspect ratio increases for the rectangular dielectric waveguide, as expected, the loss factor based on Marcatili's approximation approaches that based on the slab model. The fact that Marcatili's curves are above the slab model curves is worth noting; it means that the nonuniform distribution of the electric field intensity within the rectangular core region tends to increase the configuration loss factor. This conclusion is consistent with our previous conjecture that achieving a uniform distribution of the electric field intensity within the core region promotes the low-loss behavior of the guided mode. Hence, flat ribbon with high aspect ratio appears to be the optimal configuration in achieving a low loss factor.

The above results, shown in Fig. 10, reaffirm the validity of the slab model in providing a good theoretical guideline for designing ultra-low-loss ribbon dielectric waveguides.

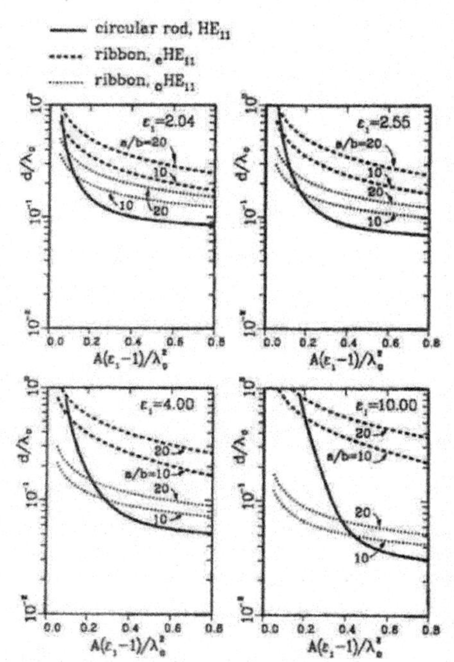

Fig. 9. Normalized power decaying distance d/λ_0 as a function of normalized frequency for various ribbon aspect ratios and for different dielectric materials. Here d is the $1/e$ power decaying distance from the surface of the ribbon or rod. Note that for all regions of interest (i.e., $0.2 < A(\epsilon_1 - 1)\lambda_0^2 < 0.8$), d is less than one free-space wavelength.

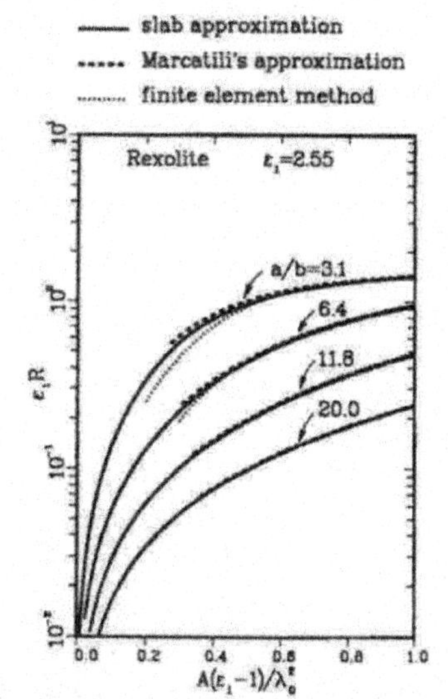

Fig. 10. Configuration loss factor versus normalized frequency for the dielectric ribbon waveguide. Results are obtained according to two approximate methods: the slab approximation and Marcatili's approximation, and one exact method: the finite element method. Within the region of interest, i.e., $0.3 < A(\epsilon_1 - 1)/\lambda_0^2 < 2.0$, results from approximate methods agree very closely with those from the exact method for flat ribbon with $a/b > 10$, where $2a$ is the width and $2b$ is the thickness of the ribbon; only when $a/b < 6.4$ are the differences noticeable. This graph shows that the approximation approaches can be used with confidence to predict the configuration loss factor behavior for thin ribbons.

To further validate the slab model, an exact approach based on the solution of Maxwell's equations by the finite element approach is used to calculate the configuration loss factor. According to this finite element approach [9], the governing longitudinal fields of the guided wave are first expressed as a functional as follows:

$$I = \sum_{p=1} I_p$$

$$= \sum_{p=1} \iint \Bigg\{ \tau_p |\nabla H_z^{(p)}|^2 + \gamma^2 \tau_p \epsilon_p \left| \frac{1}{\gamma} \left(\frac{\epsilon_0}{\mu} \right)^{1/2} \nabla E_z^{(p)} \right|^2 + 2\gamma^2 \tau_p \hat{e}_z \cdot \left[\frac{1}{\gamma} \left(\frac{\epsilon_0}{\mu} \right)^{1/2} \nabla E_z^{(p)} \times \nabla H_z^{(p)} \right] - \left(\frac{\omega}{c} \right)^2 (\gamma^2 - 1) \left[\left(H_z^{(p)} \right)^2 + \gamma^2 \frac{\epsilon_p}{\epsilon_0} \cdot \left[\frac{1}{\gamma} \left(\frac{\epsilon_0}{\mu} \right)^{1/2} E_z^{(p)} \right]^2 \right] \Bigg\} \, dx\, dy \tag{15}$$

where

$$\gamma = \frac{\beta c}{\omega} \qquad \tau_p = \frac{\gamma^2 - 1}{\gamma^2 - \epsilon_p}.$$

Here ϵ_p is the dielectric constant in the pth region, $\hat{e}_z$ is a unit vector in the z direction, and c is the speed of light in vacuum. The symbol p represents the pth region when one divides the guiding structure into appropriate regions. Minimizing the above surface integral over the whole region is equivalent to satisfying the wave equation and the boundary conditions for E_z and H_z. In the finite element approximation, the primary dependent variables are replaced by a system of discretized variables over the domain of consideration. Therefore, the initial step is a discretization of the original domain into many subregions. For the present analysis, there are a number of regions in the composite cross section of the waveguide for which the permittivity is distinct. Each of these regions is discretized into a number of smaller triangular subregions interconnected at a finite number of points, called nodes. Appropriate relationships can then be developed to represent the waveguide characteristics in all triangular subregions. These relationships are assembled into a system of algebraic equations governing the entire cross section. Taking the variation of these equations with respect to the nodal variable leads to an algebraic eigenvalue problem from which the propagation constant for a certain mode may be determined. The longitudinal electric field, $E_z^{(p)}$, and the longitudinal magnetic field, $H_z^{(p)}$, in each subdivided pth region are also generated in this formalism. All transverse fields in the pth region can subsequently be produced from the longitudinal fields. A complete knowledge of the fields can be used to generate the configuration loss factor according to (4). Results are also shown in Fig. 10, where the configuration loss factors for four rectangular ribbon guides with aspect ratios of 3.1, 6.4, 11.8, and 20 are plotted as a function of their normalized areas. In the same figure, results based on the slab model, as well as on Marcatili's approximation, are also given. It is seen that when the aspect ratio is 3.1 the curve based on the exact analysis is substantially below those based on Marcatili's method or on the slab model. As expected, however, the agreement is better for higher frequencies and for rectangular guides with higher aspect ratios. In fact, one may conclude from the above illustration that, for ribbon guide with large aspect ratios ((height/width) > 5) and for the frequency region [area($\epsilon_1 - 1$)/(free-space wavelength)^2] > 0.3, the configuration loss factor calculated according to Marcatili's method or the slab model gives an extremely good approximation to the true value and may be readily used to design low-loss ribbon dielectric waveguides.

V. Experimental Verifications

This exceptionally low loss behavior of the dielectric ribbon waveguide supporting the dominant ${}_eHE_{11}$ mode will now be verified by measurements. A newly designed dielectric waveguide cavity resonator which is capable of supporting the dominant mode is used [10]. A schematic diagram of the measurement setup is shown in Fig. 11.

A dielectric rod resonant cavity consists of a dielectric waveguide of length d terminated at its ends by sufficiently large, flat, highly reflecting plates that are perpendicular to the axis of the guide. Microwave energy is coupled into and out of the resonator through small coupling holes at both ends of the cavity. For best results, the holes are dimensioned such that they are beyond cutoff. At resonance, the length of the cavity, L, must be $m\lambda_g/2$ (m an integer), where λ_g is the guide wavelength of the particular mode under consideration. By measuring the resonant frequency of the cavity, one may obtain the guide wavelength of that particular guided mode in the dielectric waveguide. The propagation constant, β, of that mode is related to λ_g and v_p, the phase velocity, as follows:

$$\beta = \frac{2\pi}{\lambda_g} = \frac{\omega}{v_p}. \tag{16}$$

The Q of a resonator is indicative of the energy storage capability of a structure relative to the associated energy

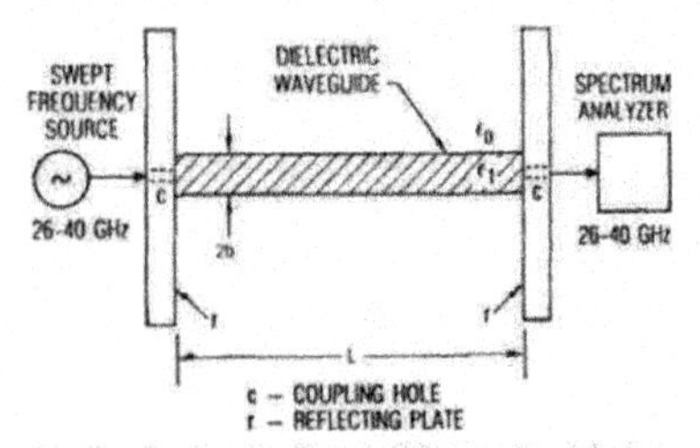

Fig. 11. A schematic diagram of the experimental setup.

TABLE II
Rexolite Strip Waveguides ($\epsilon_1 = 2.55$, $\tan\delta = 0.9\times10^{-3}$)

	2a (cm)	2b (cm)	a/b	L (cm)	Area (cm^2)
WG1	0.767	0.251	3.1	20.32	0.193
WG2	1.072	0.167	6.4	20.32	0.180
WG3	1.40	0.14	10.0	60.96	0.195

dissipation arising from various loss mechanisms, such as those due to the imperfection of the dielectric material and the finite conductivity of the end plates. The common definition for Q is applicable to the dielectric rod resonator and is given by

$$Q = \omega\frac{\overline{W}}{\overline{P}} \tag{17}$$

where ω is the angular frequency of oscillation, $\overline{W}$ is the total time-averaged energy stored, and $\overline{P}$ is the average power loss. Three Rexolite dielectric strip waveguides were fabricated and placed in a parallel-plate resonator. The dimensions of these waveguides are listed in Table II. The measurement procedure was described in a previous paper [10]. The coupling was such that only the dominant mode was excited, and the primary loss mechanism in the resonator was the dielectric loss. A swept signal was coupled into the cavity, and the output took the form of a series of narrow resonances; the resonant frequencies and half power bandwidths were measured with the spectrum analyzer. At each resonance the Q is given by

$$Q_m = \frac{f_m}{\Delta f_m} \tag{18}$$

where f_m is the mth resonance and Δf_m is the half-power bandwidth at that resonance. Plots of the measured Q's for these waveguides are shown in Fig. 12. As explained in [10] the primary loss mechanism in this measurement configuration is the dielectric loss, and the measured Q is the dielectric Q. For this case the relation between α and Q is [10], [14]

$$\alpha = (v_p/v_g)\beta/(2Q) \tag{19}$$

and the measured $\epsilon_1 R$ is given by

$$\epsilon_1 R = (v_p/v_g)(\beta/Q)/\left(\omega\sqrt{\mu\epsilon_0}\tan\delta\right) \tag{20}$$

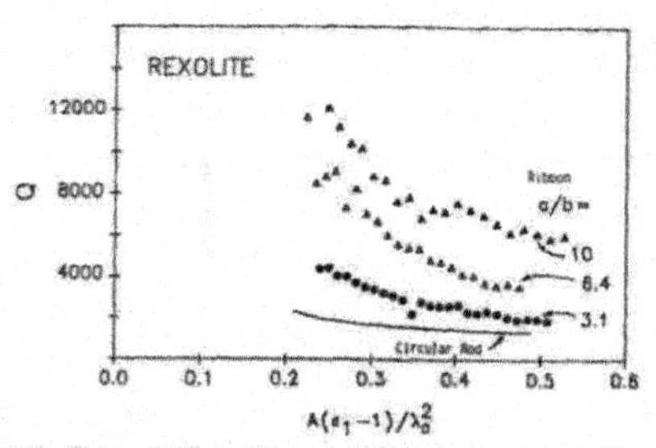

Fig. 12. Measured Q as a function of frequency for the four Rexolite waveguides. The dielectric constant and loss tangent of Rexolite are, respectively, 2.55 and 0.9×10^{-3}. Only the low-loss $_eHE_{11}$ mode is supported by the structure.

where β, v_p, and v_g were measured as described in [10] and $\tan\delta$ is the value previously determined for Rexolite. A plot of the external power density distribution across the three waveguides using an electric probe is shown in Fig. 13. The height of the probe was positioned such that the power level was 10 dB below the level at the surface of the waveguide. At this level the Q was not significantly affected by the presence of the probe. Plots of the external power density decay away from the surface of the waveguide are shown in Fig. 14, along with the calculated values. Plots of the measured $\epsilon_1 R$'s for the Rexolite rectangular waveguides are shown in Fig. 15 along with the calculated values for these waveguide dimensions. Excellent agreement was found for all three samples used in our experiment.

VI. Conclusion

This investigation shows that, by using a high-aspect-ratio ($a/b>10$) dielectric ribbon waveguide made with high-dielectric-constant ($\epsilon_1>9$) low-loss ($\tan\delta_1\approx10^{-4}$) material, it is possible to design an ultra-low-loss mm/sub-mm wave transmission system with the following features [15]:

- Extremely low attenuation constant for the dominant guided mode—the attenuation constant can be made lower than 10–20 dB/km in the mm/sub-mm wavelength range.
- This low-loss waveguide structure can be made with *known* low-loss dielectric material such as alumina, quartz, or sapphire—no major breakthrough in research on low-loss materials is needed to achieve the target of less than 10–20 dB/km for the attenuation constant.
- The guide can be made flexible; i.e., it can turn corners.
- The guide is economical and easy to manufacture.
- The guide is EMP resistant.
- Unlike metallic structure, this guide presents a relatively low scattering profile.

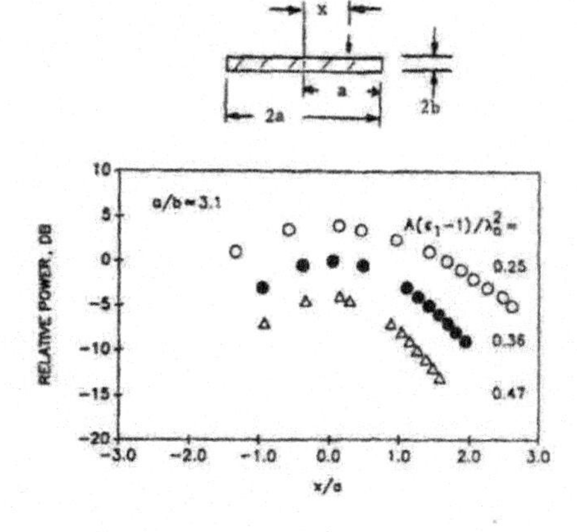

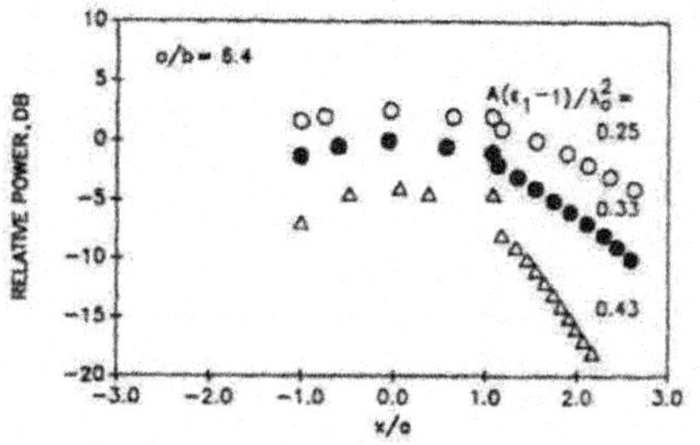

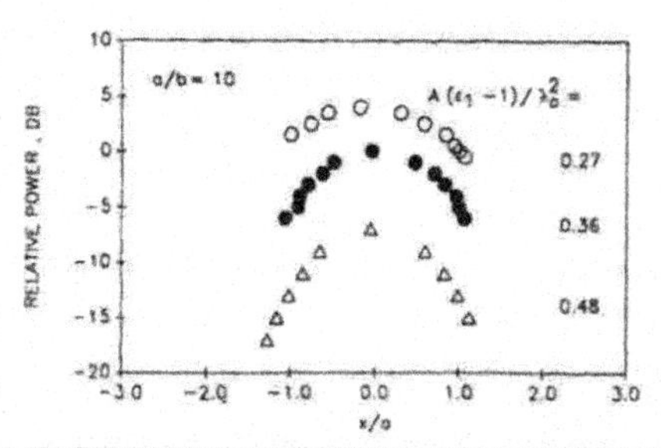

Fig. 13. External power density distribution for the guided mode for three Rexolite waveguide samples.

- It is easy to couple power into and out of the guiding structure.
- Using photolithographic techniques, circuits can be conveniently etched on waveguide surface.

Realization of our ultra-low-loss dielectric waveguide will encourage further development in the perfection of a new class of low-loss dielectric waveguides and components for use in the mm/sub-mm wavelength range.

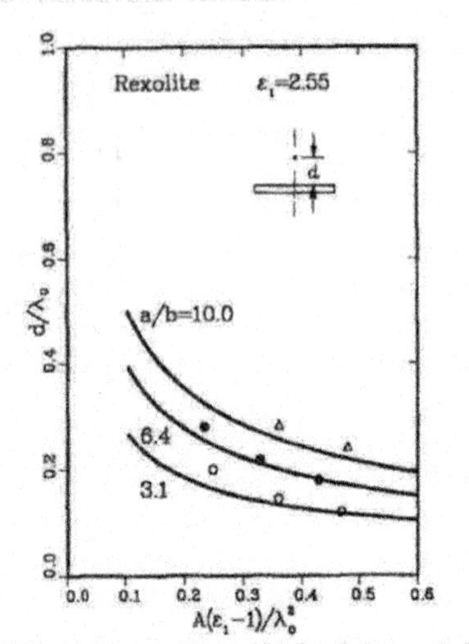

Fig. 14. External power density as a function of the normalized distance s/λ_0; s is the distance away from the surface of the guide. Solid lines are calculated results.

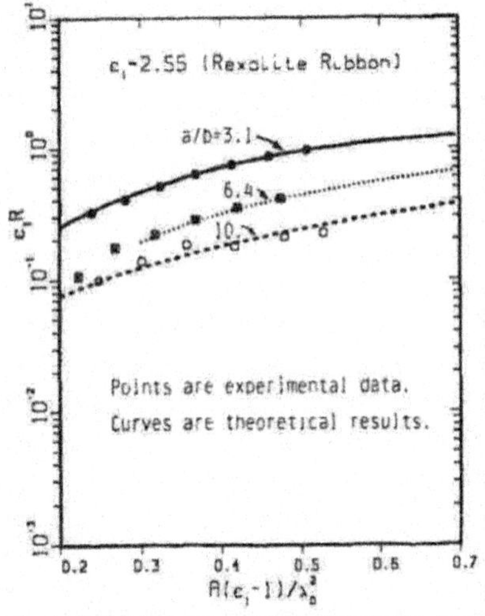

Fig. 15. Comparison between calculated configuration loss factor and measured data for dielectric ribbon waveguide with three different aspect ratios.

ACKNOWLEDGMENT

The authors express their gratitude to H. B. Dyson for being primarily responsible for obtaining the measurement data. They also would like to thank S. L. Johns for assisting in the measurements and plotting some of the data and G. G. Berry for fabricating the dielectric waveguides. The authors at UCLA wish to thank Dr. J. Hamada and Dr. B. Wong for their enthusiastic support of the UCLA–TRW MICRO Program.

REFERENCES

[1] M. N. Afsar and K. J. Button, "Millimeter-wave dielectric measurement of materials," *Proc. IEEE*, vol. 73, pp. 131–153 1985.
[2] R. Birch, J. D. Dromey, and J. Lisurf, "The optical constants of some common low-loss polymers between 4 and 40 cm^{-1}," *Infrared Physics*, vol. 21, pp. 225–228, 1981.
[3] M. N. Afsar, "Precision dielectric measurements of nonpolar polymers in the millimeter wavelength range," *IEEE Trans. Microwave Theory Tech.*, vol. MTT-33, pp. 1410–1415, 1985.
[4] J. R. Birch and T. J. Parker, *Infrared and Millimeter Waves*, vol. 2, K. J. Button, Ed. New York: Academic Press, 1979.
[5] W. B. Bridges, "Low loss flexible dielectric waveguide for millimeter wave transmission and its application to devices," California Institute of Technology, Report No. SRO-005-1 and No. SRO-005-2, 1979–1982.
[6] W. B. Bridges, M. B. Kline, and E. Schweig, *IEEE Trans. Microwave Theory Tech.*, vol. MTT-30, pp. 286–292, 1982.
[7] S. Ramo, J. R. Whinnery, and T. VanDuzer, *Fields and Waves in Communication Electronics*, 2nd ed. New York: Wiley, 1984.
[8] E. A. J. Marcatili, *Bell Syst. Tech. J.*, vol. 48, p. 2071, 1969.
[9] C. Yeh, K. Ha, S. B. Dong, and W. P. Brown, "Single-mode optical waveguides," *Appl. Opt.*, vol. 18, pp. 1490–1504, 1979.
[10] F. I. Shimabukuro and C. Yeh, "Attenuation measurement of very low loss dielectric waveguides by the cavity resonator method applicable to millimeter/submillimeter wavelength range," *IEEE Trans. Microwave Theory Tech.*, vol. 36, pp. 1160–1167, 1988.
[11] C. Yeh, "Attenuation in a dielectric elliptical cylinder," *IEEE Trans. Antennas Propagat.*, vol. AP-11, pp. 177–184, 1963.
[12] R. E. Collin, *Field Theory of Guided Waves*. New York: McGraw-Hill, 1960.
[13] C. Yeh, "On single-mode polarization preserving multi-layered optical fiber," *J. Electromagn. Waves and Appl.*, vol. 2, pp. 379–390, 1988.
[14] C. Yeh, "A relation between α and Q," *Proc. IRE*, vol. 50, p. 2143, 1962.
[15] C. Yeh, F. I. Shimabukuro, and J. Chu, *Appl. Phys. Lett.*, to be published.

C. Yeh (S'56–M'63–SM'82–F'85) was born in Nanking, China, on August 11, 1936. He received the B.S., M.S., and Ph.D. degrees in electrical engineering from the California Institute of Technology, Pasadena, in 1957, 1958, and 1962, respectively.

He is presently Professor of Electrical Engineering at the University of California at Los Angeles (UCLA). He joined UCLA in 1967 after serving on the faculty of USC from 1962 to 1967. His current areas of research interest are optical and millimeter-wave guiding structures, gigabit-rate fiber-optic local area networks, and scattering of electromagnetic waves by penetrable, irregularly shaped objects.

Dr. Yeh is a fellow of the Optical Society of America and a member of Eta Kappa Nu and Sigma Xi.

Fred I. Shimabukuro (S'55–M'56) was born in Honolulu, HI, on September 3, 1932. He received the B.S. and M.S. degrees in electrical engineering from the Massachusetts Institute of Technology, Cambridge, in 1955 and 1956, respectively, and the Ph.D. degree from the California Institute of Technology, Pasadena, in 1962.

He worked with the Hughes Aircraft Company from 1956 to 1958, and since 1962 has been at the Aerospace Corporation, El Segundo, CA. His current research activity is in millimeter- and submillimeter-wave technology.

Dr. Shimabukuro is a member of Sigma Xi.

J. Chu (S'80) was born in Taiwan on November 14, 1959. He received the B.S. degree in engineering from UCLA in 1982, and the M.S. degree in electrical engineering from the California Institute of Technology, Pasadena, in 1983. He is presently completing the requirements for the Ph.D. degree in electrical engineering at UCLA. His current research activity is in millimeter-wave guiding structures and in quantum electronics.

Communication at millimetre–submillimetre wavelengths using a ceramic ribbon

C. Yeh*, F. Shimabukuro*†, P. Stanton*, V. Jamnejad*, W. Imbriale* & F. Manshadi*

* *Jet Propulsion Laboratory, California Institute of Technology, 4800 Oak Grove Drive, Pasadena, California 91109, USA*
† *The Aerospace Corp., El Segundo, California 90009, USA*

Following the discovery by Kao and Hockman[1–3] that ultra-low-loss optical fibres could be made from pure silica through the elimination of impurities, the ability to guide signals effectively at optical wavelengths has been assured. But there remains an important region of the spectrum—from 30 to 3,000 GHz (the millimetre–submillimetre band)—where low-loss waveguides are unknown. The main problem here in finding low-loss solids is no longer one of eliminating impurities, but is due to the presence of intrinsic vibration absorption bands[4–6]. And the use of highly conducting materials is also precluded owing to high skin-depth losses[7,8] in this part of the spectrum. Here we show that a combination of material and waveguide geometry can circumvent these difficulties. We adopt a ribbon-like structure with an aspect ratio of 10:1, fabricated from ceramic alumina (Coors' 998 Alumina), and the resulting waveguide has an attenuation factor of less than 10 dB km^{-1} in the millimetre–submillimetre band. This attenuation is more than 100 times smaller than that of a

typical ceramic (or other dielectric) circular rod waveguide and is sufficient for immediate application.

Examination of the fundamental equation[7,8,9–11] governing the attenuation constant (α) of a dominant mode guided by a simple solid dielectric waveguide surrounded by lossless dry air shows that it is dependent on the material loss factor ($\tan\delta_1$) and the dielectric constant (ϵ_1) of the dielectric material and the geometrical size and shape of the guiding structure, described by the geometrical loss factor[10,11] ($\epsilon_1 R$), all of which are related to the attenuation constant (in $\mathrm{dB\,m^{-1}}$) as follows:

$$\alpha = 8.686\pi(1/\lambda_0)(\epsilon_1 R\tan\delta_1) \quad (1)$$

where λ_0 is the free-space wavelength. The geometrical loss factor is given by:

$$\epsilon_1 R = \frac{\epsilon_1 \int_A (\mathbf{E}_1 \cdot \mathbf{E}_1^*)dA}{377[\int_A \mathbf{e}_z \cdot (\mathbf{E}_1 \times \mathbf{H}_1^*)dA + \int_{A_0} \mathbf{e}_z \cdot (\mathbf{E}_0 \times \mathbf{H}_0^*)dA]} \quad (2)$$

where $\mathbf{e}_z$ is the unit vector in the direction of propagation, A and A_0 are, respectively, the cross-sectional areas of the core and the surrounding region and $(\mathbf{E}_1,\mathbf{H}_1)$ and $(\mathbf{E}_0,\mathbf{H}_0)$ are, respectively, the modal electric and magnetic field vectors of the guided mode in the core region and in the surrounding region. As the material loss factor and the dielectric constant of a solid are fixed, the only way to reduce the attenuation constant is to find the proper cross-sectional geometry of the waveguide. After performing a systematic normal-mode study on a variety of geometries, we have shown[12–14] that a ribbon-shaped guide made with low-loss, high-dielectric-constant ceramic material, such as alumina, can yield an attenuation constant for the dominant TM (transverse magnetic)-like mode of less than $0.005\ \mathrm{dB\,m^{-1}}$. (Two dominant modes with no cut-off frequency can be supported by this ceramic ribbon structure[9]: a TE (transverse electric)-like dominant mode with most of its electric field aligned

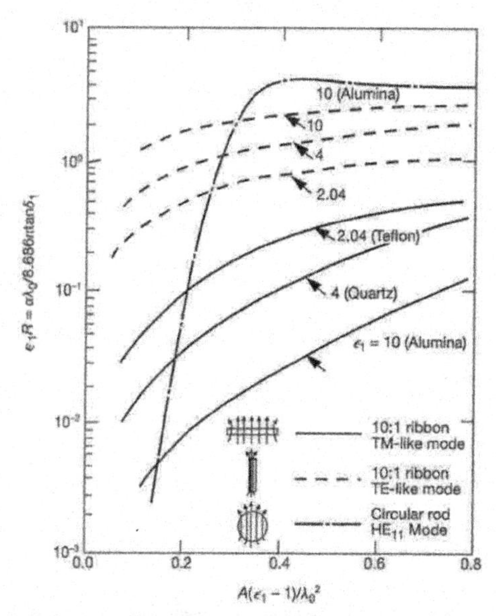

Figure 1 The geometrical loss factor $\epsilon_1 R$ as a function of the normalized cross-sectional area $A(\epsilon_1-1)/\lambda_0^2$. Here, A is the cross-sectional area of the waveguide, ϵ_1 is the relative dielectric constant of the dielectric guide and λ_0 is the free-space wavelength. Dielectric ribbons with an aspect ratio of 10 and an alumina circular rod are considered. For the ribbon case, the geometrical loss factors for the dominant TM-like (low-loss) and TE-like (high-loss) modes are obtained for three different dielectric materials: alumina with $\epsilon_1 = 10$, quartz with $\epsilon_1 = 4$ and teflon with $\epsilon_1 = 2.04$. The case for the alumina circular rod supporting the dominant HE_{11} mode is shown for comparison purposes. The alumina ribbon supporting the TM-like mode provides the most dramatic reduction in the geometrical loss factor when compared with that for the alumina circular rod. Sketches of transverse electric field lines for the TE-like, TM-like and HE_{11} modes are also shown.

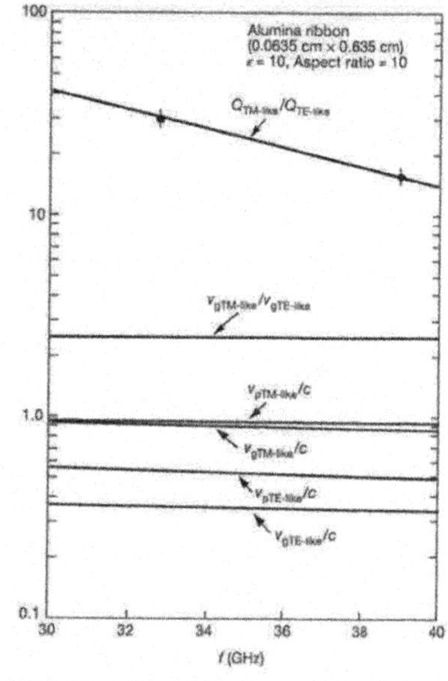

Figure 2 Ratios of $Q_{TM\text{-}like}/Q_{TE\text{-}like}$, $v_{gTM\text{-}like}/v_{gTE\text{-}like}$, $v_{pTM\text{-}like}/c$, $v_{gTM\text{-}like}/c$, $v_{pTE\text{-}like}/c$ and $v_{gTE\text{-}like}/c$ versus the frequency f (in GHz). Values taken for a 10:1 aspect ratio alumina ribbon with dimensions 0.0635 cm × 0.635 cm and dielectric constant $\epsilon_1 = 10$, supporting the TM-like mode or the TE-like mode. Here, $Q_{TM\text{-}like}$, $v_{gTM\text{-}like}$, $v_{pTM\text{-}like}$ and $Q_{TE\text{-}like}$, $v_{gTE\text{-}like}$, $v_{pTE\text{-}like}$ are the resonant Q, the group velocity and the phase velocity of the TM-like and TE-like modes, respectively. The velocity of light in vacuum is c. The ratio $Q_{TM\text{-}like}/Q_{TE\text{-}like}$ can vary from 42 at 30 GHz to 14 at 40 GHz. This means that, for a low-loss alumina ribbon with a loss tangent of 0.00005, if $Q_{TE\text{-}like}$ is measured as 22,760 at 30 GHz (this Q value is well within the measurement capability of our apparatus), then $Q_{TM\text{-}like}$ must be 955,900 (this Q value is well beyond the measurement capability of any known room-temperature resonant-cavity apparatus). To ensure that this measurement capability is not exceeded, so that $Q_{TE\text{-}like}$ and $Q_{TM\text{-}like}$ may both be measured directly in order to verify the calculated ratio ($Q_{TM\text{-}like}/Q_{TE\text{-}like}$), high-loss alumina ribbon made by coating the low-loss alumina ribbon with a thin layer of dried Indian ink is used. Measured results are shown as data points and calculated results are shown as curves. Excellent agreement is found. A photograph of the alumina-ribbon waveguide resonator is also shown.

Table 1 Measured Q, attenuation constant and loss tangent for ceramic waveguides

Batch number	Frequency (GHz)	$Q_{TE\text{-like}}$	$Q_{TM\text{-like}}$	α_{TM} (dB m^{-1})	$\tan\delta_1$ ($\times 10^{-4}$)
1	38.60	3,860 ± 300	64,700 ± 4,800	0.062 ± 0.005	2.8 ± 0.21
1	32.80	3,920 ± 290	121,700 ± 9,000	0.026 ± 0.002	2.8 ± 0.21
2	38.89	6,480 ± 490	103,700 ± 7,800	0.035 ± 0.003	1.59 ± 0.12
2	32.98	7,233 ± 540	216,990 ± 1,6300	0.014 ± 0.001	1.59 ± 0.12
3	38.96	10,948 ± 450	10,948 ± 450	1.45 ± 0.09	1.0 ± 0.04
3	30.03	11,117 ± 500	11,117 ± 500	1.17 ± 0.08	1.0 ± 0.04

Three batches of alumina material samples were obtained from Coors Ceramic Company (Golden, Colorado, USA). Batch 1, made from Coors' Superstrate 996 S20-71 (99.6%, Hirel, Thin Film Substrate), contains ribbons with dimensions 0.0635 cm × 0.635 cm × 11.43 cm. Batch 2, made from Coors' extruded 998 Alumina (99.8% alumina) rectangular rod, contains ribbons with dimensions 0.0635 cm × 0.635 cm × 91 cm. Batch 3 is Coors' extruded 998 Alumina (99.8% alumina) circular rod with dimensions 0.244 cm (diameter) × 91 cm (length). Numerous repeated measurements are made on these samples at various frequencies within the frequency band from 30–40 GHz, using the waveguide resonator technique. As Batch 3 contains a 91-cm-long circular alumina rod, the guiding property of the dominant mode is independent of the orientation of the transverse electric field. Thus, $Q_{TE\text{-like}} = Q_{TM\text{-like}} = Q_{HE}$ for the circular rod[8]. Here Q_{HE} is the Q value for the HE_{11} mode on a circular dielectric rod.

parallel to the major axis of the ribbon and a TM-like dominant mode with most of its electric field aligned parallel to the minor axis of the ribbon, as sketched in Fig. 1.) As a comparison, at an operating-frequency band of around 100 GHz, the attenuation constant for the Teflon dielectric waveguide is 1.3 dB m^{-1}, for the usual metallic rectangular waveguide is 2.4 dB m^{-1} and for the microstripline is 3 dB m^{-1} (ref. 15). This remarkable low-loss behaviour of a ceramic ribbon guiding the dominant TM-like mode, as well as the loss behaviour of a ceramic ribbon guiding the TE-like mode, is shown in Fig. 1. A suitable choice of configuration and dielectric constant can significantly reduce the geometrical loss factor for the TM-like mode.

We draw three conclusions from our analytical investigation:

(1) Much lower geometrical loss factors are obtained for high-aspect-ratio ribbon waveguide with high dielectric constant. For example, when the normalized cross-sectional area, $A(\epsilon_1 - 1)/\lambda_0^2$, is 0.4, the geometrical loss factor (as well as the attenuation factor) for this ribbon supporting the dominant TM-like mode is about 140 times smaller than that for a circular rod with the same cross-sectional area supporting the dominant HE_{11} mode.

(2) In the low-loss region, that is, $\epsilon_1 R < 0.05$, the geometrical loss factor curve for the 10:1 ribbon with dielectric constant $\epsilon_1 = 10$ supporting the TM-like mode is much flatter than that for the circular rod supporting the HE_{11} mode. This indicates that the geometrical loss factor for the ribbon is insensitive to small deviations of the normalized cross-sectional area of the ribbon, whereas the geometrical loss factor for the circular rod is very sensitive to size changes in the rod. This means that TM-like mode on the ribbon is a very stable mode, not easily disturbed by any geometrical imperfections.

(3) Separation of the geometrical loss curves for the TE-like and TM-like modes becomes larger for larger dielectric constant of the guiding ribbon. Furthermore, there is a definite relationship between the geometrical loss curves for the two modes. These facts are very significant, because they can be used to devise a fundamentally new way to measure the super-low-loss characteristics of TM-like mode guided along ceramic ribbon.

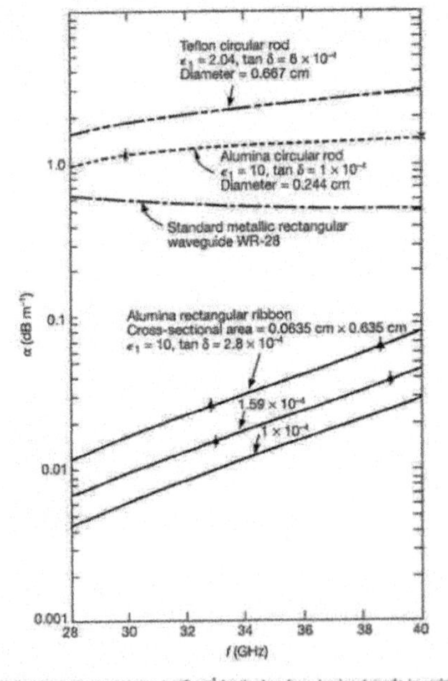

Figure 3 Attenuation constant α in dB m^{-1} for the low-loss dominant mode in various guiding structures versus frequency f (in GHz) in the K_a band (26.5–40 GHz). Measured results are shown as data points and theoretical results are shown as curves. Excellent agreement is found.

We used the cavity resonator method[16] to verify experimentally the very low attenuation constant of the ceramic ribbon waveguide. By measuring the Q of the alumina-ribbon resonator as shown in Fig. 2, we obtained the attenuation constant of the dielectric waveguide α (in dB m^{-1}) according to the formula[17]:

$$\alpha = 8.686(v_p/v_g)(\beta/2Q) = (8.686\pi/Q)(v_g/c)(1/\lambda_g) \qquad (3)$$

where β is the propagation constant of the mode under consideration, and v_p and v_g are the phase and the group velocity, respectively, of that mode. Several factors affect the accuracy and sensitivity of this technique: the alignment of the dielectric waveguide with the coupling holes, the alignment of the parallel plates, the uniformity or the straightness of the dielectric waveguide, the coupling or radiation losses and the metallic wall losses of the plates. Previous work[16] shows that the maximum Q that can be reliably measured by this technique in the K_a band (26.5–40 GHz) is approximately 30,000. This limits the smallest value of the measured attenuation constant to around 0.1 dB m^{-1}, whereas we expect the ceramic ribbon to exhibit a value around a tenth of that.

To circumvent this difficulty, we developed an indirect way to measure this ultra-low attenuation. As seen from equations (1–3), there is a definite relationship between the attenuation constants or Q values for the TM-like and TE-like modes on a dielectric ribbon:

$$\begin{aligned} r_\alpha &= (\alpha_{TE\text{-like}}/\alpha_{TM\text{-like}}) = (Q_{TM\text{-like}}/Q_{TE\text{-like}})(v_{gTM\text{-like}}/v_{gTE\text{-like}}) \\ &= (\epsilon_1 R)_{TE\text{-like}}/(\epsilon_1 R)_{TM\text{-like}} \end{aligned} \qquad (4)$$

where $v_{gTE\text{-like}}$ and $v_{gTM\text{-like}}$ are the group velocity of the TE-like and TM-like modes, respectively, and $Q_{TE\text{-like}}$ and $Q_{TM\text{-like}}$ are the Q values for the TE-like and TM-like modes, respectively. Measured results are displayed in Fig. 2 as data points and calculated theoretical results are given as curves. Excellent agreement between measured values and theoretical values affirms the correctness of the derived theoretical ratio ($Q_{TM\text{-like}}/Q_{TE\text{-like}}$). This relationship can now be used reliably to obtain $Q_{TM\text{-like}}$ when $Q_{TE\text{-like}}$ is known.

Using this technique, we measured the Q and the attenuation constant for the TM-like mode on ultra-low-loss alumina ribbon. Measured results are summarized in Table 1. The uncertainty in the measured value is discussed above. The measured data points as well as the calculated results are displayed in Fig. 3. Excellent agreement between the experimental data and theoretical results can be seen. The importance of the geometry of the guide is apparent from these results. At 30 GHz, the alumina rectangular ribbon with aspect ratio of 10 and a loss tangent of 0.000159 can support a low-loss TM-like mode with an attenuation constant at 0.0098 dB m^{-1}, which is 120 times less than that for the dominant mode on an alumina circular rod with similar cross-sectional area and a loss tangent of 0.0001. It is also 61 times less than that for the dominant mode in a standard metallic rectangular waveguide (WR-28). Significant improvement to less than 0.006 dB m^{-1} can easily be obtained for the ribbon if the same alumina material that was used for the circular rod is used.

To demonstrate the viability of the alumina ribbon as an actual transmission line, we have performed the following experiment. Shown in Fig. 4a are two horns, one transmitting and one receiving, separated by a free-space distance of 86 cm. A 120-ps pulse is emitted from the transmitting horn and is received by the receiving horn as pulse B after traversing through this free-space distance of 86 cm. We performed another experiment, where a 91-cm-long ceramic-ribbon waveguide with cross-sectional area of 0.0635 cm × 0.635 cm was inserted between the horns. Specially designed transitions are used to maximize launching and receiving efficiencies. A launching (or receiving) efficiency of 84% (or loss of less than 0.825 dB) at 39.86 GHz has been measured for an exponential launching horn. The same 120-ps pulse is now sent through this ceramic-ribbon waveguide structure. The received pulse is labelled pulse A. The received pulses for these two cases are displayed in Fig. 4a. The received signal through the ceramic-ribbon waveguide is at least 21 dB greater than that through free space, providing clear evidence that the signal can easily be guided by the ceramic ribbon.

The ceramic-ribbon waveguide is an open structure surrounded by dry air. How to support such a structure is an important consideration. One supporting structure appears to be most promising—support made with plastic fishing-line, that is, nylon thread. (See upper photograph in Fig. 4.) Thin nylon (low dielectric constant) threads, spaced at least 10 cm apart, are strung across wooden rails separated by 5 cm (far enough apart so that the exterior guiding field at the ribbon edges has decayed to negligible value). Ceramic-ribbon waveguide can simply be laid on top of the nylon threads along the middle of the rails. The nylon threads can easily support the ceramic ribbon. Any perturbation caused by the nylon-thread support on the propagation characteristics of the guided TM-like mode on the ceramic ribbon waveguide is not detectable.

Another practical problem is how to join sections of ceramic-ribbon waveguides. We have discovered that a 'shiplap' joint may be used to provide a strong bond between two ends of ceramic ribbon. A picture of the shiplap joint is shown in Fig. 4. A quarter-inch length of the jointing ribbon end is ground to a thickness of 0.0317 cm, which is half the original thickness of the ribbon. The jointing end of the other ceramic ribbon is prepared similarly. The ends are lapped together, aligned and then glued with 'super glue', resulting in a strong bond. We found no measurable loss attributable to the joint. Finally, to further understand how power is being guided along a high-dielectric-constant (ϵ_1 = 10) thin ribbon structure that can provide the very low attenuation constant for the dominant mode, see Fig. 4b.

Just as the first 20-dB km^{-1} optical fibre made in the late 1960s produced a revolution in optical communication[1-3], so may the attainment of 10 dB km^{-1} ceramic ribbon provide an opening to communications at 30–3000 GHz. Some immediate applications in this frequency band may be: transporting signals for Jet Propulsion

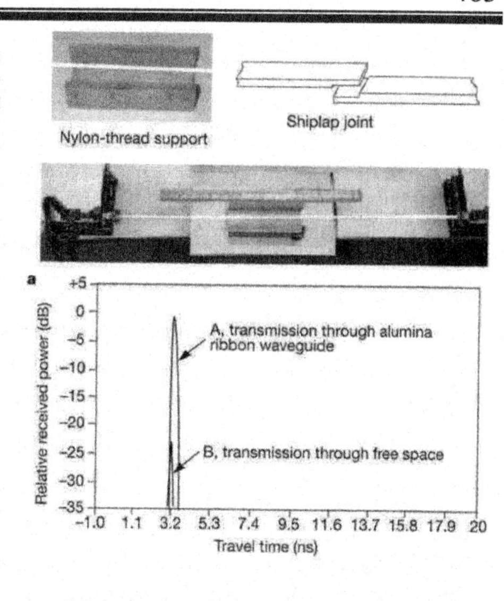

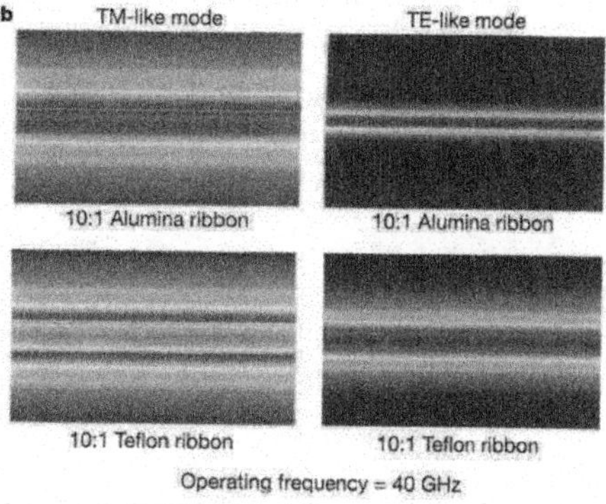

Figure 4 Transmission by ceramic ribbon and its power distribution. **a**, Comparison between the received pulses for a 120-ps pulse sent through horns linked by the alumina-ribbon waveguide (pulse A) and that linked by free-space (pulse B). The picture of a 91-cm-long alumina-ribbon waveguide with cross-sectional area of 0.0635 cm × 0.635 cm with launching and receiving horns is shown. The slight delay of the arrival of pulse A indicates that, owing to wave guidance by the alumina ribbon, pulse A is being guided by the ribbon and is propagating at the group velocity of the TM-like mode on this structure. This guided group velocity is slower than c, the free-space group velocity. To prevent sagging, the ceramic waveguide is supported by taut, widely spaced nylon threads. Inset, to make a long distance transmission line, sections of ceramic ribbon may be joined together in a 'shiplap' manner with 'super glue'. No measurable loss was noted due to the nylon-thread support or the joins. **b**, Normalized power-intensity distribution for the dominant normal modes (TM-like mode and TE-like mode) on a 10:1 dielectric ribbon structure. The highest power intensity is indicated in red and the lowest power intensity is indicated in blue. The cross-sectional sizes are chosen for single-mode operation at 40 GHz. Unlike the traditional dielectric waveguide case, the distinguishing feature here is that there is a dip in the power intensity for the TM-like mode within the thin alumina guiding ribbon.

Laboratory's deep space network and on spacecraft, transporting signals for detectors, sensors and phase array antennas or carrying long distance signals, to name a few. It is conceivable that ceramic-ribbon waveguide may be the backbone of future communication systems in this frequency band. □

Received 3 June 1999; accepted 31 January 2000.

1. Kao, K. C. & Hockman, G. A. Dielectric fiber surface waveguides for optical frequencies. *Proc. IEE* **133**, 1151–1158 (1966).
2. Agrawal, G. P. *Fiber Optic Communication Systems* (Wiley Series in Microwave and Optical Engineering, New York, 1997).
3. Marcuse, D. *Light Transmission Optics* (Van Nostrand-Reinhold, New York, 1972).
4. Afsar, M. N. & Button, K. J. Millimeter-wave dielectric measurement of materials. *Proc. IEEE* **73**, 131–153 (1985).
5. Birch, R., Dromey, J. D. & Lisurf, J. The optical constants of some common low-loss polymers between 4 and 40 cm^{-1}. *Infrared Phys.* **21**, 225–228 (1981).
6. Afsar, M. N. Precision dielectric measurements of nonpolar polymers in millimeter wavelength range. *IEEE Trans. Microwave Theor. Tech.* **33**, 1410–1415 (1985).
7. Yeh, C. in *American Institute of Physics Handbook* (ed. Gray, D. E.) 3rd edn (McGraw Hill, New York, 1972).
8. Ramo, S., Whinnery, J. R. & Van Duzer, T. *Fields and Waves in Communication Electronics* 2nd edn (Wiley, New York, 1984).
9. Yeh, C. Elliptical dielectric waveguides. *J. Appl. Phys.* **33**, 3235–3243 (1962).
10. Yeh, C. Attenuation in a dielectric elliptical cylinder. *IEEE Trans. Antenna Propag.* **11**, 177–184 (1963).
11. Yeh, C., Shimabukuro, F. I. & Chu, J. Ultra-low-loss dielectric ribbon waveguide for millimeter/submillimeter waves. *Appl. Phys. Lett.* **54**, 1183–1185 (1989).
12. Yeh, C., Ha, K., Dong, S. B. & Brown, W. P. Single-mode optical waveguides. *Appl. Opt.* **18**, 1490–1504 (1979).
13. Taflove, A. *Computational Electrodynamics, the Finite-Difference Time-Domain Method* (Artech House, Norwood, MA, 1995).
14. Yeh, C., Casperson, L. & Szejn, B. Propagation of truncated gaussian beams in multimode fiber guides. *J. Opt. Soc. Am.* **68**, 989–993 (1978).
15. Koul, S. K. *Millimeter Wave and Optical Dielectric Integrated Guides and Circuits* (Wiley Series in Microwave and Optical Engineering, New York, 1997).
16. Shimabukuro, F. I. & Yeh, C. Attenuation measurement of very low loss dielectric waveguides by the cavity resonator method applicable in the millimeter/submillimeter wavelength range. *IEEE Trans. Microwave Theor. Tech.* **36**, 1160–1166 (1988).
17. Yeh, C. A relation between α and Q. *Proc. Inst. Radio Eng.* **50**, 2145 (1962).

Acknowledgements

We thank C. Stelzried, A. Bhanji, D. Rascoe and M. Gatti of JPL for their encouragement and support. Expert assistance from M. Ostrander in machining, R. Cirillo in experimental measurements and C. Copeland in graphic works is appreciated. The research was carried out at the Jet Propulsion Laboratory, California Institute of Technology, under a contract with NASA.

Correspondence and requests for materials should be addressed to C. Y. (e-mail: cavour.yeh@jpl.nasa.gov).

Low-loss terahertz ribbon waveguides

Cavour Yeh, Fred Shimabukuro, and Peter H. Siegel

The submillimeter wave or terahertz (THz) band (1 mm–100 μm) is one of the last unexplored frontiers in the electromagnetic spectrum. A major stumbling block hampering instrument deployment in this frequency regime is the lack of a low-loss guiding structure equivalent to the optical fiber that is so prevalent at the visible wavelengths. The presence of strong inherent vibrational absorption bands in solids and the high skin-depth losses of conductors make the traditional microstripline circuits, conventional dielectric lines, or metallic waveguides, which are common at microwave frequencies, much too lossy to be used in the THz bands. Even the modern surface plasmon polariton waveguides are much too lossy for long-distance transmission in the THz bands. We describe a concept for overcoming this drawback and describe a new family of ultra-low-loss ribbon-based guide structures and matching components for propagating single-mode THz signals. For straight runs this ribbon-based waveguide can provide an attenuation constant that is more than 100 times less than that of a conventional dielectric or metallic waveguide. Problems dealing with efficient coupling of power into and out of the ribbon guide, achieving low-loss bends and branches, and forming THz circuit elements are discussed in detail. One notes that active circuit elements can be integrated directly onto the ribbon structure (when it is made with semiconductor material) and that the absence of metallic structures in the ribbon guide provides the possibility of high-power carrying capability. It thus appears that this ribbon-based dielectric waveguide and associated components can be used as fundamental building blocks for a new generation of ultra-high-speed electronic integrated circuits or THz interconnects. © 2005 Optical Society of America

OCIS code: 230.7370, 230.7390, 230.7400, 130.2790.

1. Introduction

There exists a frequency "gap" between 300 GHz and 10 THz that has not been significantly exploited due to a lack of both sources of submillimeter-wave power and an efficient low-loss circuit interconnect and guide structure.[1–3] The lack of an efficient low-loss circuit interconnect has limited both the complexity and the scale of THz components. It has meant that different components (sources, detectors, amplifiers, antennas, etc.) must either be placed in direct proximity to one another or they must be coupled by free space through bulky, fixed path, hard to align, and often lossy optical elements.[3] Even chip-to-chip interconnections at THz frequencies represent a serious problem at the present time.[2]

C. Yeh (evepanda@aol.com) and F. Shimabukuro (shimas@worldnet.att.net) are with California Advanced Studies, 826 5th Street, Suite 3, Santa Monica, California 90403. P. H. Siegel, as well as C. Yeh, is with the Jet Propulsion Laboratory, California Institute of Technology, 4800 Oak Grove Drive, Pasadena, California 91109.

Received 3 March 2005; revised manuscript received 22 April 2005; accepted 6 May 2005.
0003-6935/05/285937-10$15.00/0

The realization of low-loss connections for electronic circuits operating in the frequency range from 300 GHz to 10 THz has been an unsolved problem for many years. Because of the presence of inherent vibrational absorption bands in solids, the elimination of impurities is no longer the solution for finding low-loss solids in this frequency range.[4–10] High skin-depth loss at these frequencies also eliminates the use of highly conducting materials to form metallic waveguides or coaxial lines.[11] It thus appears that continuously searching for ultra-low-loss solids or waveguide materials in this band is not likely to bear fruit.[12,13] Despite these formidable constraints, attempts have been made to transmit THz signals through oversized circular and/or rectangular metallic waveguides, through plastic ribbons with their associated high attenuation constants, and through the use of a surface plasmon polariton on a metallic wire with its obligatory high ohmic losses.[14–18] These guides cease to operate when path lengths of longer than a few centimeters are needed. In this paper we describe a way to guide THz signals with significantly lower loss. It is based on the use of specially designed low-loss dielectric ribbons that we discovered[19] with an attenuation constant that is as much as 100 times less than that of a conventional waveguide. The ribbon can be made with selected high dielectric constant materials (e.g., alumina, silicon, InP, GaAs, etc.) and can be used as the fundamental building blocks for a new generation of ultra-high-speed electronic integrated circuits or THz interconnects.[1-3]

Here we describe the fundamentals of this family of fully functional ribbon-based low-loss THz guiding systems[19-22] including the analysis of an efficient input–output coupling mechanism, a design that achieves low-loss bends and branches, and a means of forming THz active circuit elements. Numerical simulation studies were carried out between 30 GHz and 3 THz. One notes that with the dielectric ribbon all dimensions scale with free-space wavelength λ_0 and attenuation losses scale as λ_0^{-1}. The fact that active circuit elements may be integrated directly onto the planar ribbon structure and that the absence of metallic structures in the ribbon guide provides the possibility of high-power carrying capability should be noted. Just as the low-loss optical fiber has become the backbone of high-speed communication lines replacing copper-based transmission lines,[4-10] so may this low-loss dielectric ribbon circuit become the backbone of ultra-high-speed circuits replacing metal-based microstrip lines or waveguides.

2. Background

Because the ribbon concept that was discovered earlier[19] is not well known, we shall provide a brief description here. From the theory of wave propagation along a dielectric waveguide, the attenuation constant α of a dielectric waveguide surrounded by dry air is given by the following formula[19-22]:

$$\alpha = \frac{8.686\ \pi\varepsilon_1 R \tan\delta_1}{\lambda_0}, \tag{1}$$

where the geometric loss factor is defined as $\varepsilon_1 R$. It is given by

$$\varepsilon_1 R = \frac{\varepsilon_1 \int_{A_1} (\mathbf{E}_1 \cdot \mathbf{E}_1^*) dA}{(\mu/\varepsilon)^{1/2} \left[\int_{A_1} \mathbf{e}_z \cdot (\mathbf{E}_1 \times \mathbf{H}_1^*) dA + \int_{A_0} \mathbf{e}_z \cdot (\mathbf{E}_0 \times \mathbf{H}_0^*) dA \right]}. \tag{2}$$

Subscripts 1 and 0 denote quantities in the core and cladding regions, respectively. In Eq. (2), ε_1 and $\tan\delta_1$ are the relative dielectric constant and the loss tangent of the dielectric core material, respectively, μ and ε_0 are the permeability and permittivity of free space, respectively, λ_0 is the free-space wavelength in meters, and $\mathbf{e}_z$ is the unit vector in the direction of propagation. A_1 and A_0 are the cross-sectional areas of the core and the cladding region, respectively, and

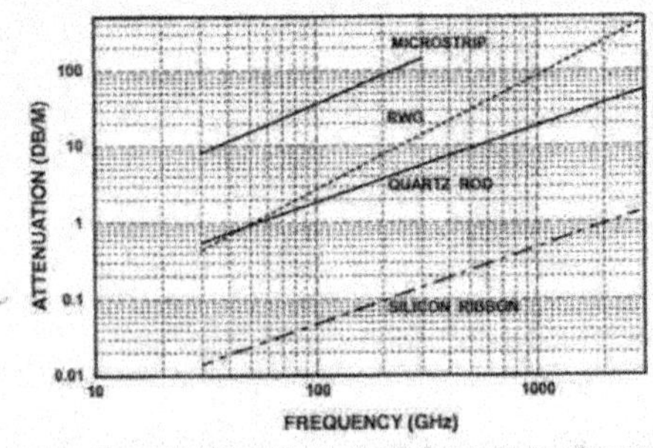

Fig. 1. Typical performance comparison between several conventional waveguide structures and the high dielectric constant (Si) ribbon waveguide for the frequency range from 30 GHz to 3 THz. Note that the waveguide losses of typical conventional waveguides can be as much as 100 times larger than those of the ribbon waveguide in this spectrum.

$(\mathbf{E}_1, \mathbf{H}_1)$ and $(\mathbf{E}_0, \mathbf{H}_0)$ are the modal electric and magnetic field vectors of the guided mode in the core region and in the cladding region, respectively. Examination of the fundamental equation (1) governing the attenuation constant of a dominant mode guided by a simple solid dielectric waveguide surrounded by lossless dry air shows that the attenuation is dependent on the loss factor, the dielectric constant of the dielectric material, and the geometric shape of the guiding structure.[19-22] Since the material loss factor and the dielectric constant of a solid are fixed, the only way to reduce the attenuation constant is to find the most appropriate cross-sectional geometry of the waveguide. A systematic study on a variety of geometries shows that a ribbon-shaped guide made with a low-loss, high dielectric constant material, such as alumina, can yield an attenuation constant for the dominant TM-like mode of less than 0.005 dB/m in the K_a band (28–40 GHz) (Ref. 19) and of less than 0.5 dB/m in the terahertz band (2–4 THz).

The significance of the development of the ribbon-based transmission system can be seen in Fig. 1. There the attenuation constant in dB/m is plotted as a function of operating frequency from 30 to 3000 GHz for several commonly used single-mode traditional waveguides and for the ribbon waveguide. These single-mode traditional waveguides are the

rectangular metallic waveguide, the circular dielectric guide with the same cross-sectional area as the dielectric ribbon, and the microstrip line. It is seen that at low frequencies (around 30 GHz), the rectangular metallic waveguide, quartz circular rod, and microstrip line are acceptable guides having attenuation constants ranging from 0.4 to 8 dB/m. The high loss of the microstrip line is tolerated since only short lengths are normally used. On the other hand, the silicon dielectric ribbon exhibits an attenuation constant of only 0.014 dB/m. As operating frequency f increases, attenuation also increases. For the rectangular metallic waveguide and the microstrip line, the attenuation increases at a rate proportional to approximately $f^{3/2}$. For the quartz circular rod and the silicon dielectric ribbon, it increases at a rate proportional to approximately f. At 300 GHz the attenuation for the microstrip line has reached approximately 150 dB/m (an onerous figure indeed), while that for the rectangular metallic waveguide is 15 dB/m and the quartz circular rod has a loss of 5.5 dB/m. At 300 GHz the attenuation for the silicon dielectric ribbon is approximately 0.15 dB/m, a figure significantly below that for the rectangular metallic waveguide at one tenth the frequency. At 3 THz the attenuation for the rectangular metallic waveguide has risen to more than 400 dB/m while that for the quartz circular rod has risen to more than 50 dB/m. These are extremely unattractive figures, rendering these waveguides impractical. Of course, the microstrip line is also impractical. It appears that the
only remaining candidate at 3 THz is the silicon or other low-loss high dielectric constant ribbon. There the attenuation for the ribbon is approximately 1.5 dB/m, which is a workable figure, and in fact the only viable guiding medium that has been proposed to date for cw THz signals.

Let us now turn to the important issues that involve utilizing the ribbon waveguide in actual circuits. These issues include how to transport THz signals around bends (corners) without excessive radiation loss, how to efficiently couple THz power to and from conventional waveguides and microstrip structures, and how to design specific circuit elements. Typical low-loss THz circuits or interconnects may contain not only the basic straight sections of low-loss dielectric ribbons but also the other necessary circuit components, such as couplers, bends and curves, branches and combiners, and filters. Many of these passive circuit components require a tightly confined mode. We will show that this is accomplished by using a polymer-coated high dielectric constant ribbon structure. Known numerical methods[21-27] will be used to analyze these components and to generate the desired results.

3. Ribbon Structures and Circuits

A. Polymer-Coated Alumina Ribbon Waveguide

From the discussion in Section 2, one expects that very low-loss guiding of terahertz band signals can be obtained with a ribbon-shaped guiding structure made with a moderately low-loss high dielectric constant material. However, since more than 90% of the guided power is carried in the lossless dry air region outside the ribbon structure, it is expected that a significant amount of guided power will be radiated and lost when the guiding structure encounters sharp curves or corners. To remedy this problem, a variant of the described ribbon guide is proposed wherein a thick polymer coating is added to the dielectric substrate in the vicinity of the bend or radiating region. A sketch of the structure is shown in Fig. 2. The polymer coating is introduced to provide tighter confinement of the guided power near the high dielectric constant ribbon. When a 10:1 aspect ratio, high dielectric constant ribbon ($0.625\ \lambda_0 \times 0.0625\ \lambda_0$) is coated on both sides with a 1/4 λ_0 thick Teflon layer, more than 90% of the guided power for the dominant TM-like mode is contained within the boundary of the coated waveguide. This means that this coated ribbon can guide the dominant mode around corners, can divide or combine the power into or out of multiple waveguides, or can keep the mode confined in regions where radiation is likely. The calculated attenuation constant for the dominant TM-like mode in the Teflon-coated alumina ribbon is approximately 0.2 dB/mm at 3 THz (or 0.02 dB/free-space wavelength), an acceptable value for short-distance propagation. It thus appears that the Teflon-coated alumina ribbon can be used (with appropriate input–output matching) as an intermediate section for the design of all the components mentioned above. For long straight-run distances, a bare high dielectric constant ribbon with an attenuation constant of 0.0006 dB/mm at 3 THz should still be used.

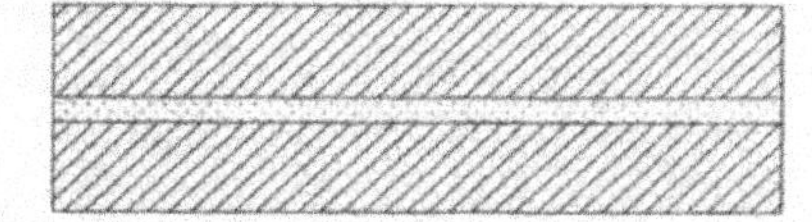

Fig. 2. Longitudinal cross-sectional geometry of a polymer-coated high dielectric constant ribbon. The thickness and the width of the high dielectric constant ribbon are, respectively, approximately $0.0635\ \lambda_0$ and $0.635\ \lambda_0$. The thickness of the polymer coating is approximately $0.25\ \lambda_0$ and the width is approximately $0.635\ \lambda_0$. The dielectric constant of the ribbon is 10 while that of the polymer is 2.04.

A summary of the analyses and uses of the polymer-coated ribbon waveguide follows:

(a) The distribution of the guided power (Poynting's vector) for the dominant ${}_eHE_{11}$ mode (TM-like mode) as a function of the distance away from the center major axis of the coated guide is shown in Fig. 3. Several cases with various Teflon-coating thicknesses are shown. It is seen that there exists a large discontinuity in the power distribution at the boundary between the low dielectric constant Teflon medium and the high dielectric constant alumina. The reason is that the normal displacement vector (D) is

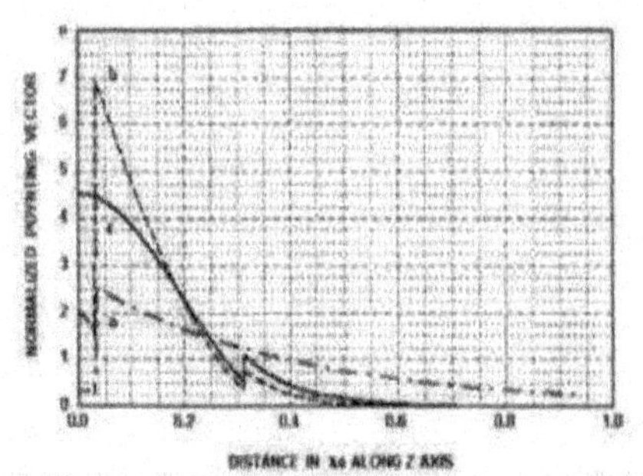

ig. 3. Normalized power intensity in a polymer-coated high dielectric constant ribbon supporting the dominant TM-like mode. hree cases are shown. For case (a) the guiding structure is a bare .635 λ_0 (width) × 0.06235 λ_0 (height) high dielectric constant ribon surrounded by dry air; for case (b) it is the same high dielectric onstant ribbon coated on both (height) sides of the ribbon with a ayer of Teflon 0.25 λ_0 thick; for case (c) it is a plain Teflon ribbon ith dimensions of 0.635 λ_0 (width) × 0.6235 λ_0 (height) surounded by dry air. Here $\epsilon_2 = 2.06$ and $\epsilon_1 = 10$ are, respectively, he dielectric constants of Teflon and high dielectric constant maerial, and λ_0 is the free-space wavelength. The distributions of uided power for the three cases are as follows: For case (a) 1.04% s in the high dielectric constant material and 98.96% is in the air egion; for case (b) 5.75% is in the high dielectric constant material, 7.8% is in the Teflon material, and 6.45% is in the air region; for ase (c) 69.23% is in the Teflon material and 10.77% is in the air egion. Note that for case (b) more than 93% of the guided power is ontained within the coated waveguide structure.

ontinuous at the boundary and therefore the normal lectric vector has a large discontinuity at the boundry when the dielectric constant differs greatly at the nterface, resulting in a large discontinuity in the istribution of the Poynting's vector (power density) cross the boundary. Specifically, the normal electric vector undergoes a jump at the boundary that equal to the ratio of the inner to outer relative dieletric constants of the two dielectrics involved. For th thin high dielectric constant ribbon case, the pow density within the high dielectric constant region quite small while the power density just outside th high dielectric constant boundary is very large. Th distinctive hole-in-the-middle power distribution f this high dielectric constant ribbon must be consi ered in the design of an efficient coupler for this ril bon structure. When a thin layer (of around 1/4 λ_0) the polymer is coated on the high dielectric constar ribbon, more than 90% of the guided power is cor tained within the coated guiding structure. This fa is very important because this means that this coate ribbon can turn corners without suffering significar radiation losses.

(b) A plot of the relative intensity envelope of th transverse E field in the upper half plane for a Teflor coated high dielectric constant waveguide is shown i Fig. 4. The top row shows the cross-sectional view the field and the bottom row shows the side view the field distribution in the direction of propagatior The alumina is 0.6 λ_0 wide and 0.06 λ_0 thick. Thre cases are considered: In case (a) there is no polyme coat; in case (b) the Teflon coat is 0.1 λ_0 thick; and i case (c) the Teflon coat is 0.26 λ_0 thick. Notice that th field is more confined to the structure for thicke coats. Most of the guided power is confined within th coated guiding structure when the Teflon thickness i about 1/4 λ_0. When the thickness is reduced to zer most of the power is confined within a distance of 1 λ from the surface of the high dielectric constant ribbo in free space. Although there is a dip in the trans verse field in the high dielectric constant materia the high dielectric constant ribbon is very thin an the dip may not be clearly seen in Fig. 4.

(c) Possible applications are as follows. A polyme

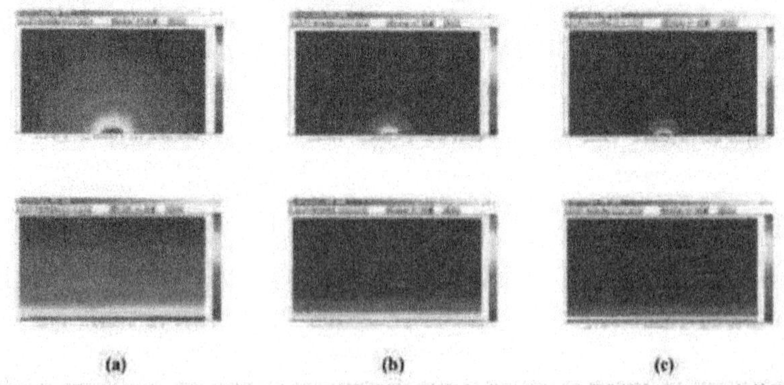

ig. 4. Relative intensity (red is high and blue is low intensity) envelope of the transverse E field in the upper half plane for eflon-coated high dielectric constant waveguide. The top row shows the cross-sectional view of the field and the bottom row shows the sid ew of the field distribution in the direction of propagation. A sketch of the structure is shown in Fig. 2. The high dielectric constant ribbo 0.6 λ_0 wide and 0.06 λ_0 thick. In (a) there is no polymer coat, in (b) the Teflon coat is 0.1 λ_0 thick, and in (c) the Teflon coat is 0.26 λ ick. Note that the field is more confined to the structure for thicker coats.

coated high dielectric constant ribbon waveguide may be used: (1) as a transition region for (a) turning corners, (b) for making waveguide splits, (c) for shielding the guided wave from extraneous obstacles or interference, and (d) for providing efficient excitation of a desired mode. (Illustrative designs are given in the following sections); and (2) as a fundamental circuit element for designing low-pass, high-pass, or bandpass filters, couplers, patch antennas, array antenna elements, etc. Various high dielectric constant semiconductor materials such as GaAs, InP, or Si may be used to construct ribbon or polymer-coated ribbon waveguides. The additional advantage is that active elements may now be constructed from and/or on these ribbon or coated-ribbon waveguides, enabling the natural integration of passive and/or active elements and waveguiding structures. Such low-loss transmission media do not exist at this time for common commercial circuits at millimeter wavelengths, submillimeter wavelengths, or at THz frequencies. Components that would benefit from the use of Si, GaAsG, or InP ribbon waveguides include amplifiers, filters, up–down converters, oscillators, radiating antennas, phase shifters, millimeter wave monolithic integrated circuits (MMIC), or THz interconnects. As the frequency increases, the semiconductor-composed ribbon waveguide circuits become more desirable and can provide a whole new category of rf components and subsystems. They also solve major interconnect problems that now limit multichip MMICs at millimeter wavelengths.[6,7] Furthermore, they also provide high-power carrying capabilities.

B. Efficient Couplers for the Ribbon Waveguides

Three conditions must be satisfied to achieve efficient coupling of power into and out of a dielectric ribbon waveguide through the use of a transition.[28] (1) Impedance matching: The wave impedance of an incident wave (mode) must be as close as possible to that of the guided mode (TM-like mode) on the dielectric ribbon waveguide (Minimizing Fresnel loss belongs in this category.) This means that the transition between guides with highly dissimilar impedances must be very gradual. (2) Field matching: The transverse-field configuration (pattern) of the incident wave (mode) must be as close as possible to that of the guided mode (TM-like mode) on the dielectric ribbon waveguide. The presence of dissimilar transverse-field lines will induce radiated waves or higher-order modes. (3) Phase velocity matching: The phase velocity of the incident wave (mode) must be as close as possible to that of the guided mode (TM-like mode) on the dielectric ribbon waveguide. This means that the transition between guides with different phase velocities must be very gentle. It is perhaps worthwhile to mention that there exists another way of transferring power from one waveguide to another by the coupled-mode approach. According to the coupled-mode perturbation theory, power can be transferred from one guide to another if the modes on these guides possess a similar phase velocity and if

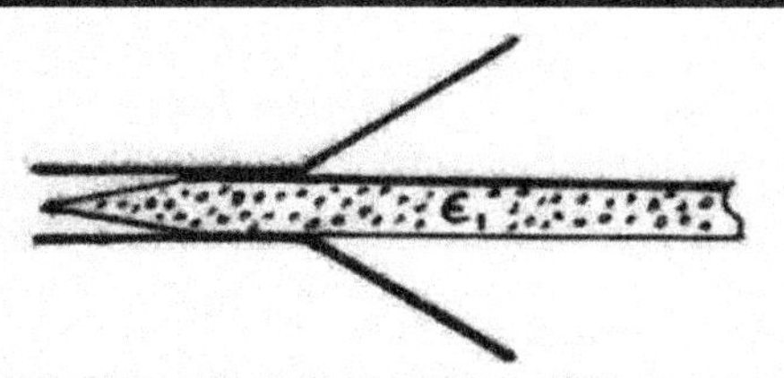

Fig. 5. Rectangular metal waveguide to low dielectric constant ribbon waveguide transition.

the proximity of the guides does not significantly affect the mode pattern of the guided wave on each guide. This concept has been used successfully to design traveling wave tubes, optical fiber couplers, and integrated optical planar waveguide couplers. The same concept can certainly be used here to design ribbon waveguide couplers.

1. *Transition between a Rectangular Metallic Waveguide and a High Dielectric Constant Ribbon Waveguide: Excitation of a Wave on a High Dielectric Constant Ribbon Guide*

It is of paramount importance to be able to input power to the low-loss ribbon waveguide with a minimum of excitation loss to achieve an overall low-loss transport medium for terahertz signals. A conventional way of transferring power from a metallic waveguide to a low dielectric constant polymer waveguide is by use of a transition horn.[28] The cross-sectional size of the polymer guide is normally larger than that of the metallic waveguide. To provide a good impedance match and a good field pattern match, the polymer dielectric waveguide is inserted into the metallic waveguide, fully filling its cross section, and the inserted end of the polymer guide is gently tapered to a point. Most of the guided power is already contained within the polymer at the mouth of the metallic waveguide. Mismatching of the transverse field for the polymer-filled metallic waveguide and that of the polymer dielectric waveguide will excite a radiated wave, which will detract from the guided power. To remedy this situation a metallic transition horn is placed at the end of the polymer-filled metallic waveguide (Fig. 5). In this way a very gentle transition of the field inside the polymer-filled metallic waveguide to that of the pure polymer waveguide occurs. The guided power inside a polymer-filled metallic waveguide is fully launched onto the polymer guide, generating very little lost radiated power. A coupling efficiency of as high as 98% has been achieved.[28]

The situation is quite different for making a transition from a rectangular metallic waveguide to a high dielectric constant ribbon guide. Two salient features stand out: (1) The transverse electric field pattern for the ribbon guide has a significant dip inside the high dielectric constant ribbon while that for the rectangular metallic waveguide does not. (2) Due to the severe discontinuity of the transverse

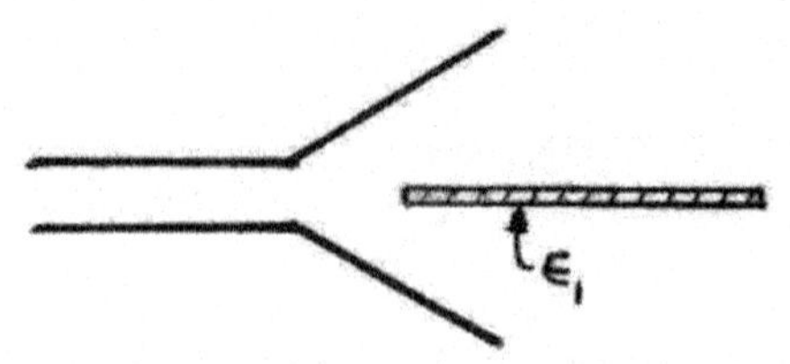

Fig. 6. Rectangular metal waveguide to high dielectric constant ribbon waveguide transition.

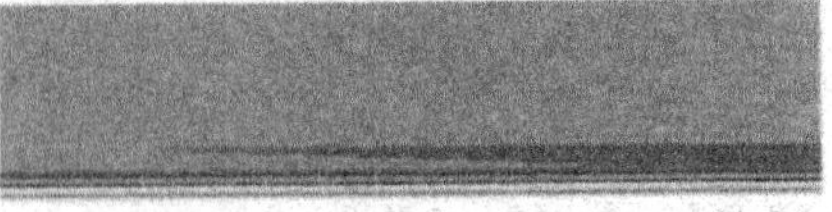

Fig. 8. Side view of the top half of transitions from a high dielectric constant ribbon to a polymer-coated high dielectric constant ribbon using an inverse taper. The high dielectric constant ribbon is 0.6 λ_0 wide and 0.06 λ_0 thick. The polymer is Teflon (0.6 λ_0 wide) and the final coating is 0.26 λ_0 thick. The transition length is 6 λ_0.

electric field at the surface of a high dielectric constant ribbon, tapering the thickness of the ribbon can cause significant perturbation to the guided wave, thus producing a large radiation loss. This means that the conventional way of making the transition for a rectangular metallic waveguide to a polymer dielectric waveguide cannot be used for the high dielectric constant ribbon guide. Here, to minimize impedance mismatch, the high dielectric constant ribbon should not be inserted into the metallic rectangular waveguide. Instead it will be placed partially into the mouth of the horn and only its width will be tapered in the transition region. (See Fig. 6.) Note that tapering the width of the ribbon would not cause a large disturbance to the transverse electric field due to the tangential field continuity condition. Despite the field mismatch in the region inside the high dielectric constant ribbon, strong coupling of the electric field on the surface of the high dielectric constant ribbon with the relatively uniform transverse electric field of the horn and good matching of the phase velocity for these fields can provide an excellent condition for the efficient launching of the desired guided wave on the high dielectric constant ribbon guide. Using this transition method a measured value of 0.35 dB mismatch loss was attained at the K_a band.[19]

2. *Transition between the High Dielectric Constant Ribbon Waveguide and the Strongly Guided Polymer-Coated High Dielectric Constant Ribbon Waveguide*

To show the compatibility between the high dielectric constant ribbon and the (low dielectric constant) polymer-coated high dielectric constant ribbon, we have studied the transition problem involving these structures. The transition from the low-loss high dielectric constant ribbon to a tightly guided polymer-coated high dielectric constant ribbon can be accomplished by tapering the polymer coating. Two types of tapering have been studied: the first is to linearly taper the polymer thickness from the surface of the high dielectric constant ribbon to the desired coating thickness (Fig. 7), and the second type of tapering is done by inverting the first linear taper (Fig. 8). Both tapering sections are 6 λ_0 long. Figure 8 shows a sketch of a new transition section that will gently capture and guide the surface wave on the high dielectric constant ribbon and couple it into the strongly guided polymer-coated ribbon. Since the key to avoiding the generation of a radiated wave is to perturb the guided field as little as possible, the transition (an inverted polymer dielectric taper) is initiated in a region of weak field strength. Its influence is gradually increased through the thickening of the polymer layer, eventually reaching the desired coating thickness. Computer simulations using the finite-difference time domain (FDTD) program[27] illustrate how the incident guided wave evolves as it propagates through this tapered transition (Figs. 9 and 10). The numerical results show that the insertion loss for the first linear taper is 0.3 dB while that for the second inverted taper is 0.22 dB. These are acceptable figures if only a couple of transition regions occur along the guide, indicating good compatibility between the high dielectric constant ribbon and the Teflon-coated high dielectric constant ribbon. According to the reciprocity theorem, the insertion loss for the inverse taper (polymer coated to uncoated ribbon) should be identical. Lower-loss transitions, including step-compensated versions, are under investigation.

Fig. 7. Side view of the top half of transitions from a high dielectric constant ribbon to a polymer-coated high dielectric constant ribbon using a gradual taper. The high dielectric constant ribbon is 0.6 λ_0 wide and 0.06 λ_0 thick. The polymer is Teflon (0.6 λ_0 wide and 0.26 λ_0 thick). The transition length is 6 λ_0.

Fig. 9. For the structure shown in Fig. 7 the FDTD simulation is obtained at 3 THz for the envelope of the transverse E field of the TM-like mode, which propagates from the bare high dielectric constant ribbon through the transition to the coated high dielectric constant waveguide. The direction of propagation is to the right. The color intensity scale shows the relative intensity of the transverse field at the center of the guide. The insertion loss of this transition is 0.3 dB.

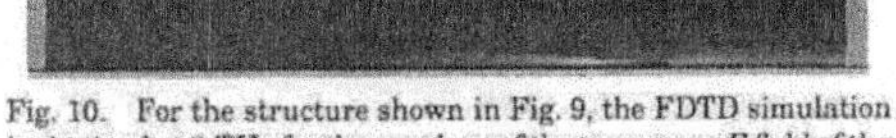

Fig. 10. For the structure shown in Fig. 9, the FDTD simulation is obtained at 3 THz for the envelope of the transverse E field of the TM-like mode that propagates from the bare high dielectric constant ribbon through the transition to the coated high dielectric constant waveguide. The direction of propagation is to the right. The color intensity scale shows the relative intensity of the transverse field at the center of the ribbon guide. The insertion loss of this transition is 0.22 dB.

3. Step-Index Transition from High Dielectric Constant Ribbon to Polymer-Coated High Dielectric Constant Ribbon

Strictly speaking, a step-index transition is a discontinuity, not a transition. It is nevertheless an important circuit element because it can be used to design filters for the ribbon-based circuits provided that radiation loss is not excessive.[12,13] A picture of a step-index transition is shown in Fig. 11(a). Figure 11(b) shows the envelope of the E field perpendicular to the ribbon surface as the guided wave propagates through this transition from a bare high dielectric constant ribbon to a polymer-coated high dielectric constant ribbon and back to a bare high dielectric constant ribbon. A standing wave is clearly seen at the input end of this transition. Although a significant reflective component is evident, it is encouraging to note from the S parameter[11–13,20] calculation that the radiation loss is not excessive. The insertion radiation loss is approximately 0.5 dB. Therefore, the step index can be used as a circuit component, as in a filter design.[12,13] Note that this design is limited to single-mode waveguides. Although the ribbon structure and the polymer-coated ribbon structure are operating in the single-mode region, these are still open surface-wave structures, implying the possible presence of a radiated wave. Therefore, limiting the excitation of the radiated wave is necessary to ensure the successful application of this design approach.

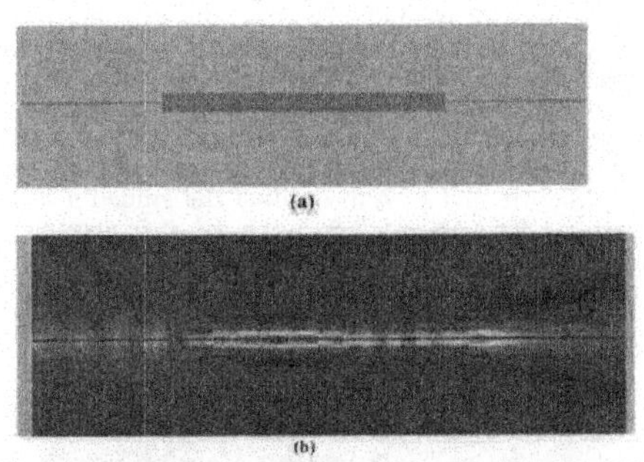

Fig. 11. (a) Side view sketch of a step-index transition from high dielectric constant ribbon to a polymer-coated high dielectric constant ribbon. The high dielectric constant ribbon is 0.6 λ_0 wide and 0.06 λ_0 thick. The top and bottom polymer coats are each 0.6 λ_0 wide and 0.26 λ_0 thick. (b) Side view of the envelope of the transverse E field for the TM-like mode propagating through the structure shown in (a). The polymer is Teflon. The simulation at 3 THz shows that for this step-index transition there is significant reflection but little radiation loss. This means that the step-index transition can be used to design circuits, such as filters, on the high dielectric constant ribbon waveguide. The insertion radiation loss is around 0.5 dB.

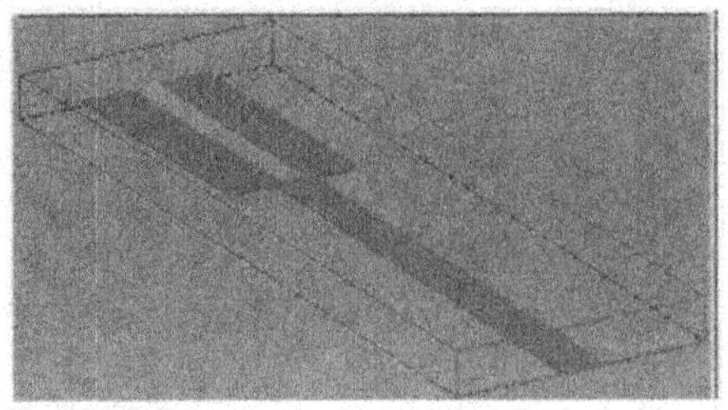

Fig. 12. Sketch of the butt-jointed transition from microstrip line to polymer-coated high dielectric constant waveguide on a ground plane.

The combination of the transition section with the polymer-coated high dielectric constant guide having a highly confined field can be used to build other components such as power splitters and combiners, waveguide twists, couplers, isolators, or Faraday rotators. The waveguide structure can be supported by ultrathin dielectric wires (as long as their diameter is small compared to a wavelength) without any significant added losses.

4. Transition between Microstrip Line and High Dielectric Constant Ribbon

Microstrip lines of several forms (strip line, coplanar waveguide, microstrip, etc.) have been the backbone for wave propagation on planar microwave circuits for many decades. Despite their high-loss characteristics at submillimeter wavelengths, they are still being used due to a lack of alternative structures. To couple the available microstrip line geometries with the new low-loss ribbon structures, it is important to find an efficient and readily fabricated transition. For interchip or off-chip use the simplest transition is to directly "butt-joint" the microstrip line with the ribbon waveguide. Because of inherent mismatches, this gives an unacceptable mismatch. However, if a butt joint is made between the microstrip and a polymer-coated ribbon the situation is not nearly as bad. A sketch of such a transition (for analysis purposes) is shown in Fig. 12. Most of the transverse electric field for the microstrip is confined between its top conductor and its bottom conducting ground plane, and for the polymer-coated alumina ribbon over a ground plane the transverse electric field of the dominant mode is mostly confined within the polymer layer. The similarity of these transverse electric field patterns enables one to butt-joint the polymer microstrip line to the polymer-coated high dielectric constant ribbon on a ground plane and to expect excellent field

Fig. 13. Side view of the envelope of the transverse E field for the TM-like mode propagating through the structure shown in Fig. 12. The microstrip line structure is on the left with a 0.22 λ_0 thick polyethylene substrate on a ground plane and a 0.32 λ_0 wide metal microstrip line conductor. The polymer-coated high dielectric constant ribbon waveguide is on the right, where the ribbon is 0.6 λ_0 wide, the high dielectric constant material is 0.03 λ_0 thick, and the polymer coat is polyethylene (0.2 λ_0 thick) on top of the high dielectric constant material, also placed on the ground plane. The microstrip line is 4 λ_0 long and the coated polymer ribbon is 8 λ_0 long. The transmission loss from left to right is 0.35 dB.

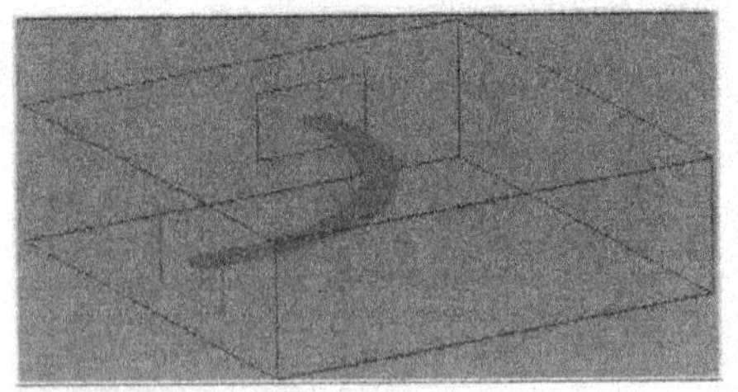

Fig. 14. A 90° alumina ribbon bend. The ribbon is 0.6 λ_0 wide with a 1 λ_0 input and output straight section and a 4 λ_0 inside radius bend.

matching for a low-loss transition. In this simulation a polyethylene substrate is used in the microstrip line and a polyethylene coating is used on the high dielectric constant ribbon. These dielectrics were chosen to get a closer impedance and phase velocity match. Computer simulation of the envelope of the transverse E field as it propagates from the microstrip line into the polymer-coated high dielectric constant ribbon is shown in Fig. 13. A slight standing-wave pattern is seen in the simulation. This can be further reduced by optimizing the structural dimensions. A total loss (reflection and radiation) of 0.24 dB was found for this transition.

If the microstrip line dielectric is made with a high dielectric constant material rather than with polyethylene, the simple butt-joint technique does not work well. There is a now substantial difference between the transverse electric field for the microstrip and that for the high dielectric constant ribbon: The highest concentration of the transverse electric field resides in the high dielectric constant substrate between the conducting surfaces of the microstrip line while the highest field concentration resides just outside the surface of the high dielectric constant ribbon with a substantial dip inside the high dielectric constant material. It is therefore prudent not to butt-joint the two ends of these different waveguides. In this situation an additional transition section must be employed. From the experimental results at the K_u band using an alumina (high dielectric constant) ribbon partially inserted into a horn, the following microstrip line-to-coated high dielectric constant ribbon can be envisioned. One must first transform the high concentration of the transverse electric field inside the high dielectric constant substrate of the microstrip line to a transverse field confined between two conducting planes filled with air. Tapering the width of the high dielectric constant substrate to a sharp edge and simultaneously widening the upper conducting microstrip line accomplishes this. This section of the transition basically transforms the high dielectric constant material-filled microstrip line to an air-filled parallel-plate transmission line. The upper and lower plates of this transmission line are flared into a horn structure. The high dielectric constant ribbon is then inserted part of the way into this flared region to accommodate efficient transfer of the guided wave. It should be noted that, since the metal surface forms an integral part of microstrip line and since the loss for metallic material increases dramatically as operating frequency increases, the above transition consideration is only meaningful for frequencies of much less than 100 GHz and not for the THz band. Further investigation of this problem is in progress.

C. Bends and Corners

Since the highest concentration of guided field resides just outside and near the surface of the high dielectric constant ribbon waveguide, the ribbon waveguide is a true surface waveguide. As such the external field can easily "slide" off the guiding structure and transform into the radiated field when the ribbon guide bends (Fig. 14). This is clearly seen in Fig. 15 where the behavior of the dominant electric field is plotted as the ribbon guide turns a 90° corner with a radius of curvature of 4 λ_0. Almost all of the guided field is radiated. To develop a low-loss bend, it is necessary first to transition the surface-guided, low-loss high dielectric constant ribbon to the strongly guided

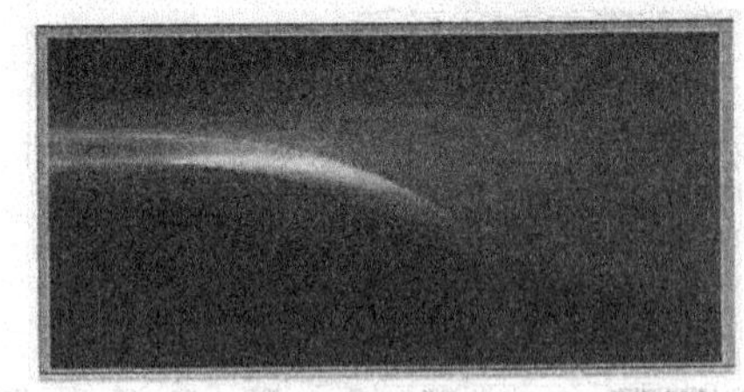

Fig. 15. Top view of the envelope of the transverse E field for the TM-like mode, just outside the high dielectric constant ribbon, as it propagates around the bend shown in Fig. 14. Almost all of the initial guided wave is lost to radiation.

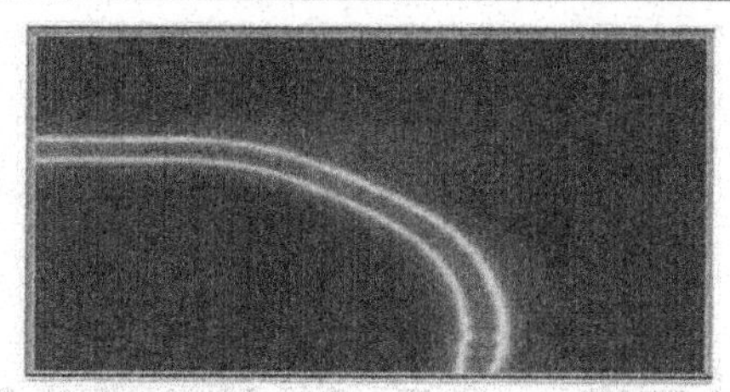

Fig. 16. Top view of the envelope of the transverse E field for the TM-like mode, just inside the Teflon–air boundary, for the high dielectric constant ribbon bend shown in Fig. 12 with a Teflon coat (0.26 λ_0 thick) on either side of the alumina. Almost all of the initial guided wave is transmitted through the bend. The radiation loss is less than 0.1 dB.

polymer-coated high dielectric constant ribbon that was discussed earlier. Basically the low-loss bends are bends using a polymer-coated high dielectric constant ribbon waveguide. Computer simulation of the guided field around a 90° bend having a radius of curvature of 4 λ_0 on the polymer-coated ribbon was carried out. This is displayed in Fig. 16. It is seen that since most of the guided field is already confined within the polymer-coated guide prior to turning the corner, very little guided field is transformed into a radiated field when the polymer-coated guide turns a corner. The total 90° cornering loss is found to be less than 0.2 dB.

4. Conclusion

We have shown how the lack of a viable low-loss transmission system in the THz spectrum may be overcome with the use of newly developed ribbon-based waveguides[19] and transitions. The propensity of the field to lie outside the guide substrate was shown to be problematic due to loose fundamental mode coupling and hence increased radiation around bends, transitions, or discontinuities. This problem has been overcome by adding a low dielectric constant polymer coating to the top and bottom of the ribbon guide in the immediate vicinity of the radiating regions. The extra dielectric loss introduced by the coating is minimized by keeping the coated length short (<10 wavelengths). Several new transitions into and out of the coated and uncoated ribbon guide were proposed and analyzed. These gave mismatch losses below 0.25 dB per transition, which is acceptable for small numbers of transition regions within a circuit. Lower-loss transitions are being investigated for more complex circuits that might contain many discontinuous regions. The same polymer-coated ribbon structures that are used to confine the dominant mode around bends can be used to make low-loss transitions into and out of a low dielectric constant microstrip line or can be used as filter elements on a straight ribbon guide. Plans for realizing a ribbon guide at THz frequencies are ongoing, with silicon being one of the materials of choice for operation at these wavelengths. Although this paper presents the basis and first realization to our knowledge of ribbon guide circuit elements for terahertz applications, there are still material and construction issues that must be solved before these components can find their way into practical circuits. One of the first applications is to build flexible sensor and/or source heads for some of the terahertz imaging instruments that are beginning to emerge.[29] At this time to our knowledge no other low-loss guide media exists at these wavelengths. We hope that the concepts presented here will lead to many more component and transition realizations and to practical use of the ribbon guide in the near future.

Just as the low-loss optical fiber has become the backbone of high-speed communication links, replacing copper-based transmission lines, we hope that the low-loss dielectric ribbon guide can become the backbone of ultra-high-speed circuits, replacing metal-based microstrip lines or waveguides. Armed with this low-loss THz interconnect media, one may find applications in medical diagnostics, radar, mechanically flexible sensors and sources, short- or long-distance communication links, MMIC interconnects, and perhaps even in micromachined on-chip transmission-line networks for ultrafast computers and/or processors.

5. Appendix A: Comments on the Use of Pure Numerical Methods

The power of using a pure numerical approach in solving many problems that cannot be solved by analytical means is apparent. Nevertheless, it must be noted that results obtained by numerical techniques are only as good as the understanding of the problem that is usually guided by the analytic solution of a canonical problem. Here the canonical problem is wave propagation on an elliptical dielectric waveguide, which was solved many years ago by Yeh.[21,22] Associated with this problem is the problem of wave propagation on a circular dielectric waveguide, which was solved even earlier by Carson, *et al.*[30] Some of the major findings that are applicable to the ribbon waveguide are as follows[21,22]:

(a) Only hybrid modes containing all six field components can exist on an elliptical dielectric waveguide or on an alumina ribbon waveguide.

(b) The dominant modes are also hybrid modes, one with its dominant electric field parallel to the major axis of the ribbon and one with its dominant electric field parallel to the minor axis of the ribbon.

(c) No evanescent mode can exist on the ribbon waveguide. Below the cutoff of a higher-order mode, that mode simply no longer exists.

(d) Unless the excitation field matches exactly the total modal field of modes that can exist on the dielectric structure, unguided radiation fields will be launched and lost.

(e) Any disturbance or deviation (such as curves, bending, twisting, narrowing or broadening, blemishes, etc.) that occurs to a perfectly straight dielectric structure will generate radiated waves and/or other higher-order guided waves if they can exist on this structure. No evanescent modes are generated because they do not exist on a dielectric waveguide.

(f) Characteristic impedance for a given propagating mode used in transmission lines, microstrip lines, or metallic waveguides is usually defined as the ratio of the transverse electric field to the transverse magnetic field of that mode.[12,13,21,22] When this definition is used for the hybrid mode on an elliptical or ribbon dielectric waveguide, the characteristic impedance becomes dependent on the transverse spatial coordinates of the ribbon. In the traditional microwave circuit design, the characteristic impedance of a transmission circuit must have a value that is independent of any spatial coordinates.[11,12,20-22] This ambiguity must be resolved before the conventional circuit techniques can be used to analyze the ribbon waveguides based on the conventional transmission-line theory.

This research was supported in part by the National Science Foundation through grant DMI-0214150 and in part by the National Institutes of Health through grant 1-R2-EB004203-01.

References

1. J. Mullins, "Using unusable frequencies," IEEE Spectrum **39**, 22–23 (2002).
2. D. van der Weide, "Applications and outlook for electronic terahertz technology," Opt. Photon. News **14**, 48–53 (2003).
3. P. H. Siegel, "Terahertz technology," IEEE Trans. Microwave Theory Tech. **MTT-50**, 910–928 (2002).
4. M. N. Afsar and K. J. Button, "Millimeter-wave dielectric measurements of materials," Proc. IEEE **73**, 131–153 (1985).
5. R. Birch, J. D. Dromey, and J. Lisurf, "The optical constants of some common low-loss polymers between 4 and 40 cm^{-1}," Infrared Phys. **21**, 225–228 (1981).
6. M. N. Afsar, "Precision dielectric measurements of nonpolar polymers in the millimeter wavelength range," IEEE Trans. Microwave Theory Tech. **MTT-33**, 1410–1415 (1985).
7. J. W. Lamb, "Miscellaneous data on materials for millimetre and submillimetre optics," Int. J. Infrared Millim. Waves **17**, 1997–2034 (1996).
8. K. C. Kao and G. A. Hockman, "Dielectric fiber surface waveguides for optical frequencies," IEE Proc. Optoelectron. **133**, 1151–1158 (1966).
9. G. P. Agrawal, *Fiber Optic Communication Systems*, Wiley Series in Microwave and Optical Engineering (Wiley, New York, 1997).
10. D. Marcuse, *Light Transmission Optics* (Van Nostrand-Reinhold, New York, 1972).
11. S. Ramo, J. R. Whinnery, and T. Van Duzer, *Fields and Waves in Communication Electronics*, 2nd ed. (Wiley, New York, 1984).
12. S. K. Koul, *Millimeter Wave and Optical Dielectric Integrated Guides and Circuits*, (Wiley Series in Microwave and Optical Engineering, (Wiley, New York, 1997).
13. T. C. Edwards, *Foundations for Microstrip Circuit Design* (Wiley, New York, 1981).
14. G. Gallot, S. P. Jamison, R. W. McGowan, and D. Grischkowsky, "Terahertz waveguides," J. Opt. Soc. Am. B **17**, 851–863 (2000).
15. J.-F. Roux, F. Aquistapace, F. Garet, L. Duvillaret, and J.-L. Coutaz, "Grating-assisted coupling of terahertz waves into a dielectric waveguide studied by terahertz time-domain spectroscopy," Appl. Opt. **41**, 6507–6513 (2002).
16. G. L. Carr, M. C. Martin, W. C. McKinney, K. Jordan, G. R. Neill, and G. P. Williams, "High-power terahertz radiation from relativistic electrons," Nature **420**, 153–156 (2002).
17. R. Mendis and D. Grischkowsky, "Plastic ribbon THz waveguides," J. Appl. Phys. **88**, 4449–4451 (2000).
18. K. Wang and D. M. Mittleman, "Metal wires for terahertz waveguiding," Nature **432**, 376–379 (2004).
19. C. Yeh, F. Shimabukuro, P. Stanton, V. Jamnejad, W. Imbriale, and A. F. Manshadi, "Communication at millimetre-submillimetre wavelengths using ceramic ribbon," Nature **404**, 584–588 (2000).
20. C. Yeh, "Dynamic Fields," in *American Institute of Physics Handbook*, 3rd ed., D. E. Gray, ed. (McGraw-Hill, New York, 1972).
21. C. Yeh, "Elliptical dielectric waveguides," J. Appl. Phys. **33**, 3235–3243 (1962).
22. C. Yeh, "Attenuation in a dielectric elliptical cylinder," IEEE Trans. Antennas Propag. **AP-11**, 177–184 (1963).
23. C. Yeh, K. Ha, S. B. Dong, and W. P. Brown, "Single-mode optical waveguides," Appl. Opt. **18**, 1490–1504 (1979).
24. A. Taflove and S. C. Hagness, *Computational Electrodynamics: The Finite-Difference Time-Domain Method*, 2nd ed. (Artech House, Norwood, Mass., 2000).
25. C. Yeh, L. Casperson, and B. Szejn, "Propagation of truncated Gaussian beams in multimode fiber guides," J. Opt. Soc. Am. **68**, 989–993 (1978).
26. K. S. Yee, "Numerical solution of initial boundary value problems involving Maxwell's equation in isotropic media," IEEE Trans. Antennas Propag. **AP-14**, 302–307 (1966).
27. QuickWave-3D FDTD Software, QWED Sp.z o.o., ul. Zwyciezcow 3 4/2, 03-938 Warszawa, Poland.
28. W. Schlosser and H. G. Unger, "Partially filled waveguides and surface waveguides in rectangular cross section," in *Advances in Microwaves*, L. Young, ed. (Academic, New York, 1966).
29. P. H. Siegel, S. E. Fraser, W. Grundfest, C. Yeh, and F. Shimabukuro, "Flexible Ribbon Guide for In-Vivo and Hand-Held THz Imaging," proposal to NIH PAR-03-075, Technology Development for Biomedical Applications R21, September 2003.
30. J. K. Carson, S. P. Mead, and S. A. Schelkunoff, "Cylindrical dielectric waveguide," Bell Syst. Tech. J. **15**, 310 (1936).

Attenuation Measurement of Very Low Loss Dielectric Waveguides by the Cavity Resonator Method Applicable in the Millimeter/Submillimeter Wavelength Range

FRED I. SHIMABUKURO, MEMBER, IEEE, AND C. YEH, FELLOW, IEEE

Abstract —A dielectric waveguide shorted at both ends is constructed as a cavity resonator. By measuring the Q of this cavity, one can determine the attenuation constant of the guided mode on this dielectric structure. The complex permittivity of the dielectric waveguide material can also be derived from the measurements. Measurements were made at Ka-band for dielectric waveguides constructed of nonpolar, low-loss polymers such as Teflon, polypropylene, polyethylene, polystyrene, and rexolite.

I. Introduction

BY USING A specially configured dielectric rod made from low-loss, nonpolar polymers, one can construct millimeter/submillimeter dielectric waveguides supporting the dominant mode with a very small attenuation coefficient. To verify experimentally the low-loss characteristics of such waveguides, an accurate measurement scheme must be devised. A logical solution is the construction of a cavity consisting of a length of a dielectric rod waveguide supporting the mode of interest, with parallel shorting plates at both ends [1]. At a resonant frequency of such a cavity, the guide wavelength, λ_g, is obtained from the cavity spacing, and the attenuation constant, α, can be obtained from the measured Q. This cavity method also provides an accurate determination of the dielectric properties of the waveguide material.

This paper will first describe the theoretical foundation for this cavity technique. Then, a detailed discussion and derivation of the relationship between α and Q are given. Finally, experimental results for several low-loss dielectric materials are presented.

II. Theoretical Foundation

The geometry of a dielectric rod resonator, including a schematic of the measurement system, is shown in Fig. 1. The signals are coupled in and out of the resonator through

Manuscript received August 13, 1987; revised January 30, 1988. This work was supported by the U.S. Air Force under Contract FO4701-87-C-0088 and by the UCLA–TRW Micro Program.

F. I. Shimabukuro is with the Aerospace Corporation, P.O. Box 92957, Los Angeles, CA 90009.

C. Yeh is with the Electrical Engineering Department, University of California at Los Angeles, Los Angeles, CA 90024.

IEEE Log Number 8821626.

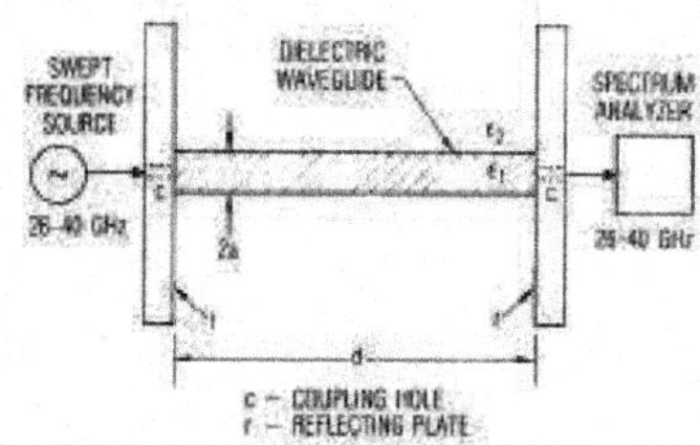

Fig. 1. The schematic of the dielectric waveguide cavity resonator, including measurement setup.

small coupling holes in the center of the reflecting plates. For a circular step-index dielectric rod, the HE_{11} mode is the dominant guided mode for this dielectric waveguide [2], [3]. The longitudinal fields of this HE_{11} mode resonating between two shorting, parallel plates are, inside the core region ($\rho < a$),

$$E_{zi} = AJ_1(u\rho)\sin\phi\cos\beta z \tag{1}$$

$$H_{zi} = BJ_1(u\rho)\cos\phi\sin\beta z \tag{2}$$

$$u^2 = k_1^2 - \beta^2 \qquad k_1^2 = \omega^2\mu\epsilon_1 \tag{3}$$

and

$$\beta = \frac{m\pi}{d}, \qquad m = 1,2,3\cdots.$$

Outside the core region ($\rho > a$), they are

$$E_{zo} = CK_1(w\rho)\sin\phi\cos\beta z \tag{4}$$

$$H_{zo} = DK_1(w\rho)\cos\phi\sin\beta z \tag{5}$$

with

$$w^2 = \beta^2 - k_2^2 \qquad k_2^2 = \omega^2\mu\epsilon_2. \tag{6}$$

In the previous equations, A, B, C, and D are arbitrary constants, $J_1(u\rho)$ is the Bessel function, $K_1(w\rho)$ is the modified Bessel function, a is the radius of the dielectric rod, d is the spacing between the shorting plates, ϵ_1 and ϵ_2 are the permittivities of the regions inside and outside the

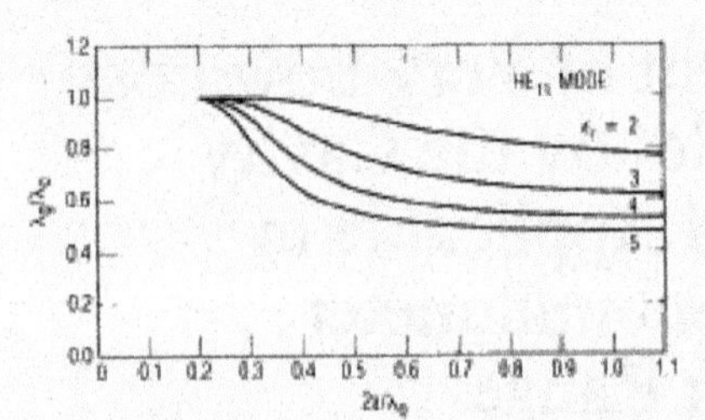

Fig. 2. Dispersion of the HE_{11} mode of a rod dielectric waveguide of radius a. The solution is given as a plot of the normalized guide wavelength as a function of normalized rod diameter. λ_0 is the free-space wavelength.

core, respectively, ω is the angular frequency of the resonant mode, and $\mu = \mu_0$ is the permeability of free space. In this study, the region outside the core is free space, and $\epsilon_2 = \epsilon_0$. It is noted that all transverse fields $(E_\phi, E_\rho, H_\phi, H_\rho)$ may be derived from the longitudinal fields, E_z and H_z [4]. By satisfying the boundary conditions at $\rho = a$, the following dispersion relation is obtained:

$$\left[\frac{J_1'(ua)}{uJ_1(ua)} + \frac{K_1'(wa)}{wK_1(wa)}\right]\left[\frac{k_1^2 J_1'(ua)}{uJ_1(ua)} + \frac{k_0^2 K_1'(wa)}{wK_1(wa)}\right] = \left(\frac{\beta}{a}\right)^2\left[\frac{1}{u^2} + \frac{1}{w^2}\right]^2. \quad (7)$$

The solution of this dispersion relation will yield the guide wavelength, λ_g, of the cavity for the HE_{11} mode for given a, d, ϵ_1/ϵ_0, μ_0, and ω. Results for various values of ϵ_r $(= \epsilon_1/\epsilon_0)$ are shown in Fig. 2.

III. Ultrahigh Q Dielectric Rod Resonant Cavity

As shown in Fig. 1, a dielectric rod resonant cavity consists of a dielectric waveguide of length d terminated at its ends by sufficiently large, flat, and highly reflecting plates that are perpendicular to the axis of the guide. Microwave energy is coupled into and out of the resonator through small coupling holes at both ends of the cavity. For best results, the holes are dimensioned such that they are beyond cutoff. At resonance, the length of the cavity, d, must be $m\lambda_g/2$ (m an integer), where λ_g is the guide wavelength of the particular mode under consideration. By measuring the resonant frequency of the cavity, one may obtain the guide wavelength of that particular guided mode in the dielectric waveguide. The propagation constant, β, of that mode is related to λ_g and v_p, the phase velocity, as follows:

$$\beta = \frac{2\pi}{\lambda_g} = \frac{\omega}{v_p}. \quad (8)$$

The Q of a resonator is indicative of the energy storage capability of a structure relative to the associated energy dissipation arising from various loss mechanisms, such as those due to the imperfection of the dielectric material and the finite conductivity of the end plates. The common definition for Q is applicable to the dielectric rod resonator and is given by

$$Q = \omega\frac{\overline{W}}{\overline{P}} \quad (9)$$

where ω is the angular frequency of oscillation, $\overline{W}$ is the total time-averaged energy stored, and $\overline{P}$ is the average power loss.

For the case under study, with carefully machined dielectric rods and proper cavity alignment, the time-averaged power dissipation $\overline{P}$ consists of two parts, the power loss due to the dielectric rod and that due to the metal end walls, namely,

$$\overline{P} = \overline{P}_{\text{dielectric}} + \overline{P}_{\text{wall}}.$$

The power dissipation due to the dielectric rod is given by

$$\overline{P}_{\text{dielectric}} = \frac{1}{2}\sigma_d \int_0^d \int_{A_d} (E_1 \cdot E_1^*)\, dA\, dz \quad (10)$$

where E_1 is the electric field within the dielectric rod, σ_d is the conductivity of the dielectric, A_d is the cross-sectional area of the dielectric rod, and the asterisk denotes the complex conjugate. The losses due to both end walls are given by

$$\overline{P}_{\text{wall}} = 2\left(\frac{R_s}{2}\right)\int_{A_w} (H_t \cdot H_t^*)\, dA \quad (11)$$

where $R_s = \sqrt{\omega\mu/2\sigma_r}$, the wall surface resistivity, σ_r is the conductivity of the reflector material, and H_t is the tangential component of the magnetic field along the metal wall. Here, A_w is the area of each conducting wall. There is also a loss due to the coupling hole, but, as in this experiment, the coupling can be made sufficiently small, such that the primary wall losses can be considered to be the ohmic wall losses:

$$\overline{W} = 2\overline{W}_m = 2\overline{W}_e = \mu\int_V (H \cdot H^*)\, dV = \epsilon\int_V (E \cdot E^*)\, dV \quad (12)$$

where V is the total volume of the cavity, $\overline{W}_m$ and $\overline{W}_e$ are the time-averaged magnetic and electric energies, respectively, and H and E are the total fields. Equation (9)–(12) can be rearranged to obtain

$$\frac{1}{Q} = \frac{\overline{P}}{\omega\overline{W}} = \frac{\overline{P}_{\text{dielectric}}}{\omega\overline{W}} + \frac{\overline{P}_{\text{wall}}}{\omega\overline{W}} = \frac{1}{Q_d} + \frac{1}{Q_w}. \quad (13)$$

The term Q_d is the Q factor of the cavity if the end plates were perfectly conducting, and Q_w is the Q factor of the cavity if the dielectric were perfect. From (13) we have

$$Q_d = \frac{\omega\overline{W}}{\overline{P}_{\text{dielectric}}} = \frac{1}{2\tan\delta\frac{\epsilon_1}{\epsilon_0}}\frac{C_T}{C_D} \quad (14)$$

$$Q_W = \frac{\omega\overline{W}}{\overline{P}_{\text{wall}}} = \frac{d}{2\delta_s}\frac{C_T}{C_W} \quad (15)$$

1162 IEEE TRANSACTIONS ON MICROWAVE THEORY AND TECHNIQUES, VOL. 36, NO. 7, JULY 1988

where $\tan\delta$ ($=\sigma/\omega\epsilon_1$) is the loss tangent of the dielectric rod, and δ_r ($=2R_s/\omega\mu$) is the skin depth of the metallic end plates. The ratios C_T/C_D and C_T/C_W are dimensionless quantities involving integrals of the fields.

It is noted that Q_d is independent of the length of the cavity whereas Q_w is proportional to the length. For a long cavity, $Q_w \gg Q_d$, and $Q = Q_d$. By measuring the Q of the cavity with $Q_w \gg Q_d$, one can obtain the attenuation constant α of the given mode.

In 1944 Davidson and Simmonds [5] derived a relation between the Q of a cavity composed of a uniform transmission line with short-circuiting ends, and the attenuation constant α of such a transmission line. Later, in 1950, Barlow and Cullen [6] rederived this relation. These authors showed that this relation is quite general and is applicable to uniform metal tube waveguides with arbitrary cross section. Since then, one of the standard techniques for the measurement of the attenuation constant α is the use of the cavity method.[1] This method offers an excellent way of measuring the attenuation constant of the guide when the loss is quite small. Later on this method was generalized and applied to open waveguides, such as the single wire line, the dielectric cylinder guide, and associated guides, by various authors, e.g., [1], [7].

However, it should be remembered that the formula by Davidson and Simmonds and Barlow and Cullen is derived under the assumption that there exists a single equivalent transmission line for the mode under consideration. This assumption is true for a pure TE, TM or TEM mode, but it is not clear that such a single equivalent transmission line exists for the hybrid waves. This suspicion arises from the fact that a) the TE and TM waves are intimately coupled to each other, and b) the characteristic impedance defined by Schelkunoff [8] is not constant with respect to the transverse coordinates. It is, therefore, very difficult to conceive the possibility that there exists a single equivalent transmission line for this hybrid mode; at best the hybrid wave may be represented by a set of transmission lines coupled tightly with one another. Hence, the formula by Davidson and Simmonds and Barlow and Cullen is *not* applicable to the hybrid wave.[2]

A more general relation between Q and α can be obtained without using the transmission line equivalent circuit, provided that α is very small compared with β. The propagation constant of a guided wave with small attenuation constant at ω is

$$\Gamma(\omega)=\alpha(\omega)+j\beta(\omega), \qquad j=\sqrt{-1}. \tag{16}$$

It can be shown that for a waveguide placed between reflecting parallel plates, with minuscule coupling to external circuits,

$$P_t \sim \frac{1}{\left|1-r^2\exp(-2\Gamma d)\right|^2} \tag{17}$$

where P_t is the power transmission of the resonator, r is the reflection coefficient at each wall, Γ is the propagation constant, given in (16), and d is the distance between the reflecting plates. At the half-power transmission points,

$$P_t(\omega=\omega_0)=2P_t(\omega=\omega_0\pm\Delta\omega) \tag{18}$$

and

$$\beta=\beta_0+\Delta\beta$$
$$\alpha=\alpha_0+\Delta\alpha. \tag{19}$$

For the case $r=1$ and $\alpha d \ll 1$, and using (17)–(19), one gets

$$\Delta\beta=\alpha. \tag{20}$$

Since

$$\Delta\beta=\frac{\partial\beta}{\partial\omega}\Delta\omega \qquad v_p=\frac{\omega}{\beta} \qquad v_g=\frac{\partial\omega}{\partial\beta}$$

and

$$Q=\frac{\omega_0}{2\Delta\omega}$$

we finally arrive at the relation

$$\alpha=\frac{\omega_0}{2Qv_g}=\frac{v_p}{v_g}\frac{\beta}{2Q}. \tag{21}$$

This is the general relation that we are seeking. This result was also obtained by Yeh [9] using an alternative approach. Substituting the values of v_p/v_g for TE, TM or TEM into (21), one gets the relations derived by Davidson *et al.* For the TM or TE mode,

$$\frac{v_p}{v_g}=\frac{1}{1-\left(\dfrac{\lambda}{\lambda_c}\right)^2} \qquad \alpha=\frac{1}{1-\left(\dfrac{\lambda}{\lambda_c}\right)^2}\frac{\beta}{2Q}$$

and for the TEM mode,

$$v_p/v_g=1 \qquad \alpha=\beta/2Q$$

where λ_c is the cutoff wavelength.

The group and phase velocity of the dominant modes can be obtained easily from the $\omega-\beta$ diagram. A sketch of the $\omega-\beta$ diagram for the propagating modes is shown in Fig. 3. It can be seen that at low frequencies or small β's, $v_p \approx v_g$ and again at very high frequencies or large β's, $v_p \approx v_g$. Therefore, the relation $\alpha=\beta/2Q$ is applicable only at very low frequencies or at very high frequencies.

Returning now to the problem of measuring the attenuation constant of very low loss dielectric waveguides, one notes that using the dielectric waveguide cavity technique, a Q of the order of 30000 can readily be measured. At the higher frequencies this value of Q corresponds to a loss tangent of the order of 10^{-5}. A schematic of the experiment is shown in Fig. 1. A dielectric rod waveguide is

[1] The procedures of this method in general are the following: Short-circuit the uniform transmission line under consideration at both ends and measure the Q of such a resonator. From the knowledge of the measured Q and other constants, such as the cutoff frequency of the guide and the frequency of oscillation, it is an easy matter to obtain from the formula derived by these authors.

[2] But several investigators, apparently unaware of this restriction, used this formula in their investigations of the hybrid wave.

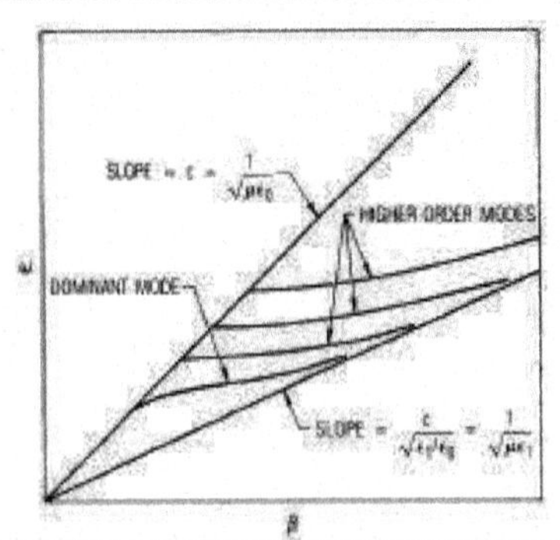

Fig. 3. Sketch of the $\omega-\beta$ diagram for several propagating modes along a dielectric waveguide.

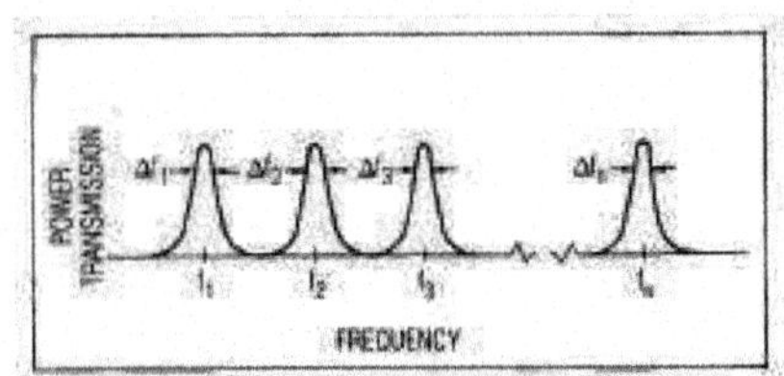

Fig. 4. Power output of a swept input signal through a dielectric waveguide in a parallel-plate resonator.

placed in a parallel-plate cavity, and a swept signal frequency is transmitted through the waveguide cavity and detected by a spectrum analyzer. The signals are coupled through very small holes in the circular gold plated reflectors. The plates are large enough (6 in. diameter) such that the fields beyond the plate diameter are insignificant. The output is a series of narrow transmission resonances at $f_1, f_2, \cdots, f_m$ with half-power bandwidths, $\Delta f_1, \Delta f_2, \cdots, \Delta f_m$, respectively (see Fig. 4). At each resonant frequency the guide wavelength is given by

$$\lambda_{gm} = \frac{2d}{m} \tag{22}$$

and the Q by

$$Q_m = \frac{f_m}{\Delta f_m} \tag{23}$$

where d is the length of the waveguide and m is the mth resonance. The integer m is the number of guide half-wavelengths at a particular resonant frequency. From a, the dielectric rod radius, the spacing d, the guide wavelength λ_g, and the number m, the relative dielectric constant, $\epsilon_r = \epsilon_1/\epsilon_0$, can be determined at the different frequencies using the solutions of (7).

With careful alignment of the waveguide and the shorting plates the primary loss mechanisms to be considered are the wall losses and the dielectric loss. From previous

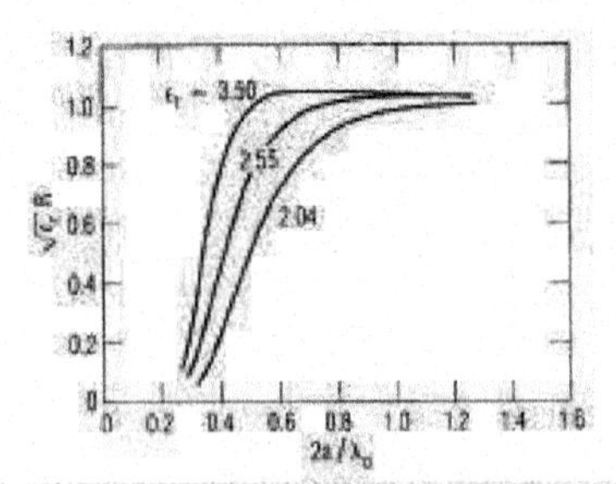

Fig. 5. Plots of the attenuation factor R in a circular dielectric waveguide of radius a for different relative permittivities.

discussion,

$$\frac{1}{Q_m} = \frac{1}{Q_d} + \frac{1}{Q_w} \tag{24}$$

where Q_m is the measured Q of the mth mode, recalling that Q_d is independent of cavity length, whereas Q_w is proportional to the cavity length. For the different dielectric waveguides used in this study, the calculated Q_w ranges from 18000 to 21000 d, where d is the length in cm. Experimentally, the effect of the wall losses, whether due to the coupling or to the ohmic dissipation, on the cavity Q could not be detected; therefore,

$$Q_w \gg Q_d. \tag{25}$$

The measurement verification of (25) will be discussed in the next section.

The general relation between Q and α for a short-circuited low-loss waveguide given in (21) is rewritten as

$$\alpha = 8.686 \frac{v_p}{v_g} \frac{\beta}{2Q} \quad \text{dB/m} \tag{26}$$

where $v_p = \omega/\beta$ and $v_g = d\omega/d\beta$. It has been shown that, for a dielectric rod waveguide [10],

$$\alpha = 4.343\omega\sqrt{\mu\epsilon_0} \tan\delta\epsilon_r R \tag{27}$$

where

$$R = \left| \frac{\int_{A_d} (E_1 \cdot E_1^*)\, dA}{\sqrt{\frac{\mu}{\epsilon_0}} \int_A e_z \cdot (E \times H^*)\, dA} \right|. \tag{28}$$

As before $\epsilon_r = \epsilon_1/\epsilon_0$. A_d is the cross-sectional area of the core region of the dielectric waveguide. A is the total cross-sectional area. E_1 is the electric field within the dielectric rod, e_z is the unit vector along the direction of propagation, and E and H are the total fields. The quantity R is a frequency-dependent geometrical factor which can be computed. The loss tangent can be obtained

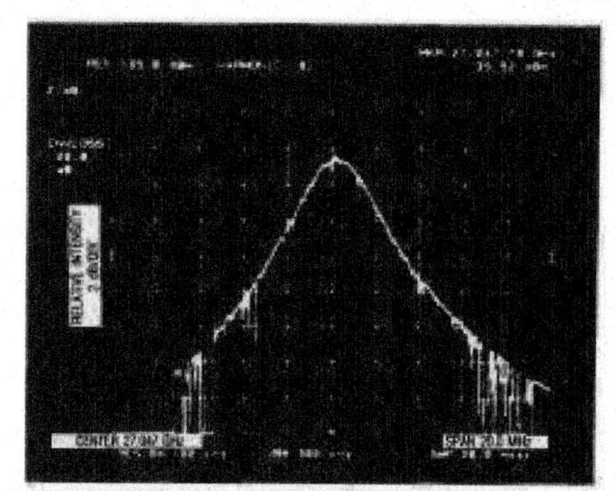

Fig. 6. Photograph of the output on the spectrum analyzer through the dielectric waveguide in the parallel-plate cavity at a transmission resonance.

by combining (26) and (27):

$$\tan\delta = \frac{\dfrac{v_p}{v_g}\dfrac{\beta}{Q}}{\omega\sqrt{\mu\epsilon_0\epsilon_r}R}. \qquad (29)$$

For a circular dielectric waveguide, one can calculate R for different values of ϵ_r. This is shown in Fig. 5. Hence, by measuring the Q of a dielectric rod in a parallel-plate resonator, the loss tangent of the dielectric and the attenuation constant for the corresponding mode can be obtained. This scheme provides an extremely accurate way of measuring the electrical properties (ϵ_r, $\tan\delta$) of ultra-low-loss dielectrics as well as the low attenuation constant for a dielectric waveguide supporing the dominant mode.

III. Experimental Results

Circular dielectric rod waveguides were made of Teflon, rexolite, polystyrene, polyethylene, and polypropylene. The diameters ranged from 0.4 to 0.63 cm, and the lengths from 15.2 to 20.3 cm. These waveguides were placed in a parallel-plate resonator. A swept frequency signal at Ka-band (26.5–40 GHz) was coupled into the resonator and the output was detected by a spectrum analyzer. The input and output coupling was done through a small hole (1.5 mm diam.) in an iris in WR-28 waveguide. With this coupling, only the HE_{11} dominant mode was excited. This was verified by mapping the fields outside the dielectric waveguide with an electric probe. A sample measurement of the transmission resonance on the spectrum analyzer is shown in Fig. 6, for a Teflon rod waveguide.

At each resonance the Q is measured. The results are shown in Fig. 7. Because m is known to be an integer it can be readily determined by measuring the guide wavelength approximately with a probe. Once m is known, the guide wavelengths at the various resonant frequencies are accurately determined, and the ω–β diagram can be generated, and α can be determined from (26). In this investigation the dielectric waveguides had a circular cross section and the following procedure was utilized. Once the guide wavelength and the waveguide dimensions were known, ϵ_r was determined from (7). Assuming the value of ϵ_r for Teflon in Table I, v_p and v_g can be calculated from (7), for a rod diameter of 0.635 cm. The comparison between the calculated and measured values of v_p and v_g is shown in Fig. 8. The measured group velocity was obtained by assuming a linear relation between adjacent measured values on the ω–β curve. The attenuation coefficient was calculated from (26) and $\tan\delta$ was obtained from (29). For the circular waveguide, the field

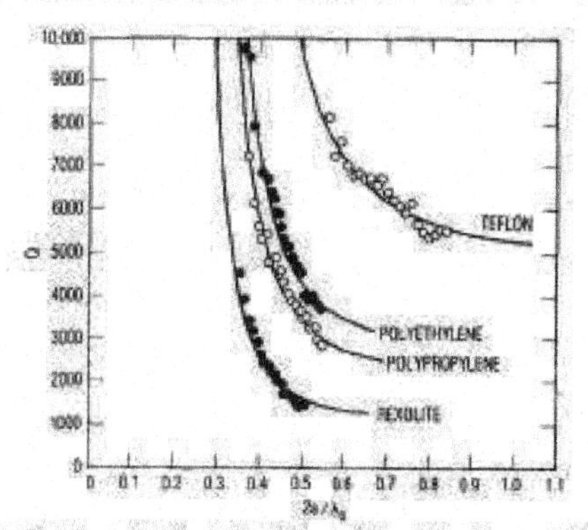

Fig. 7. Measured Q's of the different circular dielectric waveguides. The solid line is the theoretical Q_d curve using the permittivities given in Table I.

TABLE I
Measured Relative Permittivities and Loss Tangents, Ka-Band

Material	Estimates with Standard Error ϵ_r	$10^3 \tan\delta$
Teflon	2.0422 ± 0.0006	0.217 ± 0.006
Polypropylene	2.261 ± 0.001	0.50 ± 0.03
Polyethylene	2.302 ± 0.003	0.38 ± 0.02
Polystyrene	2.542 ± 0.001	0.87 ± 0.07
Rexolite	2.548 ± 0.001	0.89 ± 0.07

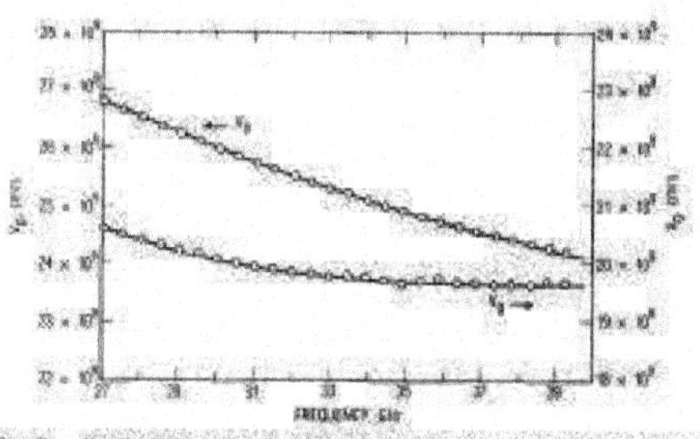

Fig. 8. Comparison of measured and calculated group and phase velocities for a Teflon rod waveguide of diameter 0.635 cm. The solid lines are calculated and the measurements are indicated by circles.

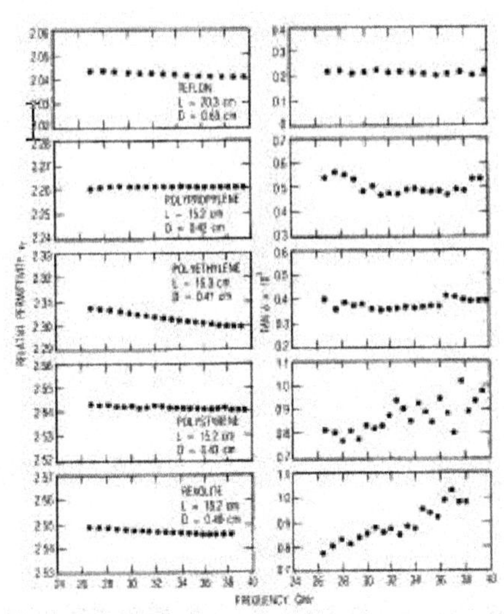

Fig. 9. Derived values of ϵ_r and $\tan\delta$ from the measurements for different dielectric materials.

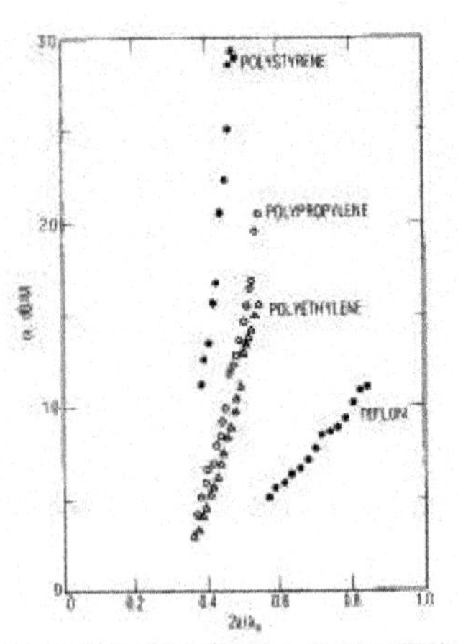

Fig. 10. Measured attenuation coefficients for the different dielectric waveguides corresponding to Fig. 8. Polystyrene and rexolite have similar attenuation characteristics.

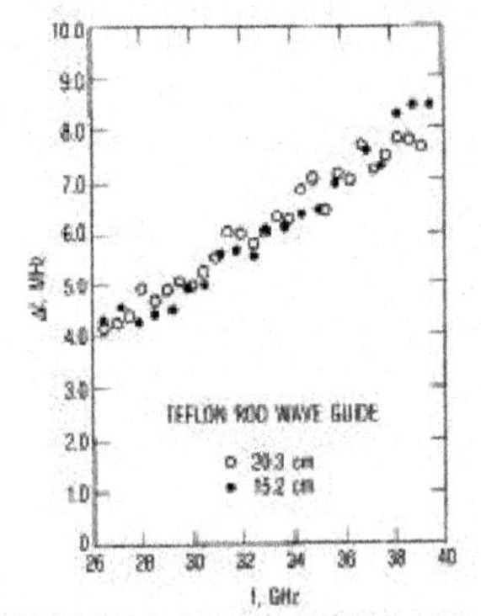

Fig. 11. Plot of half-power transmission bandwidths at the different resonances for two different lengths (6 in. and 8 in.) for circular Teflon waveguides.

configurations were known, and $\tan\delta$ was also calculated from (14), giving the same results as (29).

The measured relative permittivities and loss tangents at different resonant frequencies for the materials above are shown in Fig. 9. The average values with the corresponding standard deviations are given in Table I. A brief discussion, including references, of alternate methods used to determine the complex permittivities of materials at the millimeter wavelengths has been given in [11] and [12]. The corresponding attenuation coefficients for these dielectric waveguides are shown in Fig. 10. In Fig. 11 are shown plots of the half-power bandwidths at the different resonances for two lengths of 0.635-cm-diameter Teflon waveguide. The plot indicates that the measured Q's are primarily due to the dielectric losses. If the wall losses were significant, the Q's of the shorter length waveguide would have been noticeably lower at the lower frequencies and the derived loss tangents in Fig. 9 would have been noticeably higher. As a further check on the coupling effects, the insertion losses of the resonator system with a Teflon waveguide were measured at resonances near 27, 33, and 39 GHz. The measured insertion losses were −71 dB, −63 dB, and −51 dB, respectively, at these three frequencies.

It is clear that for low-loss performance in circular dielectric waveguides, one should use small-diameter rods made from material with small relative permittivity and loss tangent. At the *Ka*-band the attenuation in a dielectric rod waveguide for small $2a/\lambda_0$ can be less than that of a conventional rectangular metallic waveguide. Because the surface resistivity of metals is proportional to the square root of frequency, the losses of metallic waveguides increase with frequency relative to that of a dielectric waveguide. This is shown in Fig. 12. The attenuation coefficients of different silver rectangular waveguides and of circular Teflon waveguides at the indicated frequencies are plotted in the figure. The assumption is that for the Teflon rod, $2a/\lambda_0 = 0.4$ at the indicated frequencies. Since the attenuation coefficient of the dielectric waveguide can be further reduced by using other than a circular cross section, dielectrics show promise as viable guiding structures at the millimeter and submillimeter wavelengths.

To summarize, a resonator method applicable at the millimeter and submillimeter wavelengths which can accurately measure the attenuation coefficient of ultra-low-loss

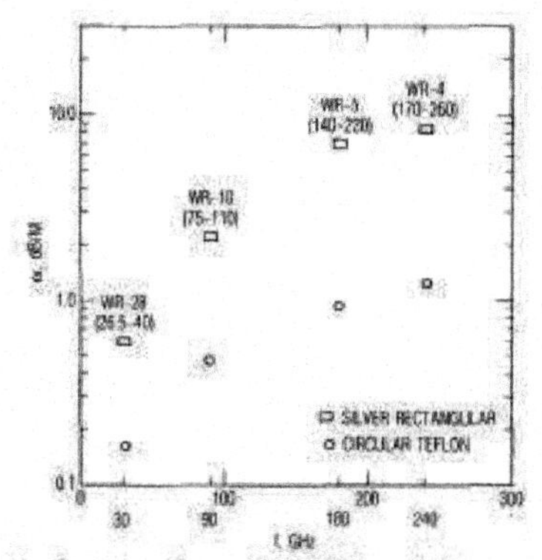

Fig. 12. Comparison of attenuation coefficients of silver rectangular and Teflon circular waveguides at the indicated frequencies. The waveguide range of the designated metal waveguides are shown in parentheses. For the dielectric waveguide, it is assumed that $2a/\lambda_0 = 0.4$ at the indicated frequencies.

dielectric waveguides has been described. In addition, the complex permittivity of the dielectric material of the waveguide can be derived. Since the fields are confined close to the dielectric core, long resonators can be conveniently implemented, permitting accurate measurements of α, ϵ_r, and $\tan\delta$.

Acknowledgment

The authors wish to thank H. B. Dyson for his invaluable help in setting up the experiment and making the measurements, and G. G. Berry for fabricating the Fabry–Perot plates and dielectric waveguides. One of the authors (C. Y.) wishes to thank Dr. J. Hamada and Dr. B. Wong for their enthusiastic support of the UCLA–TRW MICRO Program.

References

[1] C. Chandler, "Investigation of dielectric rod as waveguides," *J. Appl. Phys.*, vol. 20, pp. 1188–1192, 1949.

[2] J. R. Carson, S. P. Mead, and S. A. Schelkunoff, "Hyper-frequency waveguides—Mathematical theory," *Bell Syst. Tech. J.*, vol. 15, pp. 310–333, 1936.

[3] C. Yeh, "Advances in communication through light fibers," in *Advances in Communication Systems*, vol. 4, A. Viterbi, Ed. New York: Academic Press, 1975.

[4] S. Ramo, J. R. Whinnery, and T. Van Duzer, *Fields and Waves in Communication Electronics*, 2nd ed. New York: Wiley, 1984.

[5] C. F. Davidson and J. C. Simmonds, "Cylindrical cavity resonators," *Wireless Eng.*, vol. 31, pp. 420–424, 1944.

[6] H. M. Barlow and A. L. Cullen, *Microwave Measurements*, London: Constable and Co., 1950.

[7] D. D. King and S. P. Schlesinger, "Losses in dielectric image lines," *IRE Trans. Microwave Theory Tech.*, vol. MTT-5, pp. 31–35, 1957.

[8] S. A. Schelkunoff, "The impedance concept and its application to problems of reflection refraction, shielding and power absorption," *Bell Syst. Tech. J.*, vol. 17, pp. 17–48, 1938.

[9] C. Yeh, "A relation between α and Q," *Proc. IRE*, vol. 50, p. 2143, 1962.

[10] C. Yeh, "Attenuation in a dielectric elliptical cylinder," *IEEE Trans. Antennas Propagat.*, vol. AP-11, pp. 177–184, 1963.

[11] M. N. Afsar, "Dielectric measurements of millimeter wave materials," *IEEE Trans. Microwave Theory Tech.*, vol. MTT-32, pp. 1598–1609, 1984.

[12] F. I. Shimabukuro, S. Lazar, M. R. Chernick, and H. B. Dyson, "A quasi-optical method for measuring the complex permittivity of materials," *IEEE Trans. Microwave Theory Tech.*, vol. MTT-32, pp. 659–665, 1984.

Fred I. Shimabukuro (M'56) was born in Honolulu, HI, on September 3, 1932. He received the B.S. and M.S. degrees in electrical engineering from M.I.T., Cambridge, MA, in 1955 and 1956, respectively, and the Ph.D degree from the California Institute of Technology, Pasadena, CA, in 1962.

He worked at the Hughes Aircraft Company from 1956 to 1958, and since 1962 has been employed at the Aerospace Corporation in El Segundo, CA. His current research activity is in millimeter and submillimeter wave technology.

Dr. Shimabukuro is a member of Sigma Xi.

C. Yeh (S'56–M'63–SM'82–F'85) was born in Nanking, China, on August 11, 1936. He received the B.S., M.S., and Ph.D. degrees in electrical engineering from the California Institute of Technology, Pasadena, CA, in 1957, 1958, and 1962, respectively.

He is presently Professor of Electrical Engineering at the University of California at Los Angeles (UCLA). He joined UCLA in 1967 after serving on the faculty of USC from 1962 to 1967. His current areas of research interest are optical and millimeter-wave guiding structures, gigabit-rate fiber-optic local area networks, and scattering of electromagnetic waves by penetrable, irregularly shaped objects.

Dr. Yeh is a member of Eta Kappa Nu, Sigma Xi, and the Optical Society of America.

THE PHYSICS OF FLUIDS VOLUME 8, NUMBER 4 APRIL 1965

Wave Propagation in Hot Plasma Waveguides with Infinite Magnetostatic Fields

H. H. Kuehl, G. E. Stewart, and C. Yeh
Electrical Engineering Department, University of Southern California, Los Angeles, California
(Received 8 September 1964; final manuscript received 24 November 1964)

Solutions of the "dispersion relation" for an unbounded Maxwellian electron plasma, $2k^2/k_D^2 = Z'(\omega/kv_{th})$ for ω real are given. Solutions of the corresponding relation for a circular waveguide filled with a hot plasma with an infinite magnetostatic field are also presented. It is shown that these solutions are of importance in evaluating the fields in such waveguides since they determine the poles of the integrands which arise in the integral formulation of the problem. The integral solution for the field in the waveguide due to a general symmetric excitation at the wall of the waveguide is given.

I. INTRODUCTION

In recent years there has been much interest in guided waves which propagate in conducting cylindrical waveguides containing plasmas. In general, except for stability considerations,[1] previous studies have been concentrated on cold plasmas for which the dielectric constant is a relatively simple function of frequency. For example, in a circular waveguide completely filled with a cold plasma in an infinite axial magnetostatic field it can be shown[2] that the H-modes are unaffected by the plasma whereas the E-modes satisfy the dispersion relation

$$k^2 = k_0^2 - \frac{(p_{nm}/b)^2}{1 - (\omega_p^2/\omega^2)}, \qquad (1)$$

where k is the wavenumber, ω is the angular frequency, ω_p is the angular plasma frequency, b is the radius of waveguide, p_{nm} is the mth root of the nth order Bessel function of the first kind, and $k_0 = \omega/c$, where c is the velocity of light. Equation (1) is valid if the phase velocity ω/k of the wave is much greater than the thermal velocity of the electrons, the so-called cold plasma approximation. However, as ω approaches ω_p, (1) predicts that k becomes large so the phase velocity can become comparable to the thermal velocity, violating the cold plasma approximation. In this region it is possible for a large number of electrons to stay in phase with the wave for appreciable times, producing, in the linear approximation, the phenomenon of Landau damping and also a change in the phase velocity from that of the cold plasma theory. Therefore it is the purpose of this paper to examine the waves which may propagate in the axial direction of a plasma-filled circular waveguide without neglect of thermal velocities of the plasma particles. The reason for the choice of the completely-filled guide rather than the more realistic partially-filled guide is its relative mathematical simplicity. Many of the salient features of wave propagation in the completely filled guide will, however, be similar to the case of the partially filled guide. It will be assumed that an infinite axial magnetostatic field is present although the results will apply approximately to cases in which the cyclotron frequency is much greater than the other characteristic frequencies. Collision losses have also been neglected in the analysis. This is consistent with the examination of highly ionized, hot plasmas since the relaxation frequencies associated with collisions between electrons and ions fall off rapidly with increasing temperatures.

II. PLASMA OSCILLATIONS IN AN UNBOUNDED PLASMA

It is expected that for large wavenumbers, the modes in a plasma waveguide approach those in an unbounded plasma since for small wavelengths the boundaries become unimportant. For an infinite magnetic field, the case of interest here, it is well known that it is possible for transverse plane waves and longitudinal plasma oscillations to propagate along the magnetic field. The transverse waves are not affected by the plasma, propagating with the velocity of light. Longitudinal plasma oscillations have been the subject of extensive study[3,4] although solutions have been obtained under the assumption that the wavenumber k is real and the frequency ω is complex. In many cases of experimental interest, however, ω is real. It is this class of solutions which are considered here since they correspond to the more usual excitation conditions in waveguides.

[1] E. R. Harrison, Proc. Phys. Soc. (London) 79, 317 (1962).
[2] A. W. Trivelpiece and R. W. Gould, J. Appl. Phys. 30, 1784 (1959).
[3] L. Landau, J. Phys. (USSR) 10, 25 (1946).
[4] J. D. Jackson, J. Nucl. Energy, Pt. C 1, 171 (1960).

In the linear approximation, the dispersion relation for longitudinal oscillations in a Maxwellian electron plasma can be written[3-5]

$$k^2/k_D^2 = \tfrac{1}{2}Z'(\omega/kv_{th}) \qquad (2)$$

where the ion motion has been neglected. In (2), k is the wavenumber, v_{th} is the thermal velocity given by $v_{th} = (2\kappa T/m)^{\frac{1}{2}}$, where κ is Boltzmann's constant, T is the electron temperature, m is the electron mass, and k_D is the Debye wavenumber given by $k_D = \sqrt{2}\,\omega_p/v_{th}$, where ω_p is the plasma frequency. Z is the plasma dispersion function[6] and the prime denotes differentiation of Z with respect to its argument. The derivation of (2) has generally been presented in terms of an initial value problem in which k is assumed real and (2) yields the complex values of ω which persist at large time. However, for a plasma excited by a harmonically oscillating source at a given real value of the frequency ω, the solutions for k satisfying (2) are also of importance. For example, Landau[3] has shown that, for an oscillating electric field impressed at the boundary of a semi-infinite plasma (which is equivalent to a nonintercepting grid carrying an oscillating charge in an unbounded plasma), the electric field can be expressed in terms of two integrals in the k plane in which the roots of (2) determine poles of one of the integrands. Both integrands also have poles at the roots of the equation obtained by replacing $Z(\omega/kv_{th})$ by $\bar{Z}(\omega/kv_{th})$ in (2) where $\bar{Z}(z)$ is given by

$$\bar{Z}(z) = Z(z) - 2i\pi^{\frac{1}{2}}\exp(-z^2). \qquad (3)$$

However, it can be shown that the roots of the equation so obtained are simply the negative of the roots of (2) so it is not necessary to consider them separately. Moreover, because the poles determined by the roots of (2) and the analogous equation with $Z \rightarrow \bar{Z}$ are dependent on the plasma properties and not on the particular form of the harmonically varying excitation, they occur in the integrands of integrals arising in the study of arbitrary types of excitation. To evaluate such integrals, it is helpful to know the positions of the poles of the integrands in the k plane. Indeed, in the Landau problem the residue contribution of the least damped pole predominates to large distances from the excitation if $\omega \approx \omega_p$. Fried and Gould[5] give a convenient method for solving (2) for k real and ω complex. A similar method, discussed below, can be used for the present case in which ω is real and k is complex.

Letting $\zeta = \omega/kv_{th} = (\omega k_D)/(\sqrt{2}\,\omega_p k)$, (2) can be rewritten in the form

$$\omega^2/\omega_p^2 = \zeta^2 Z'(\zeta). \qquad (4)$$

Since ω/ω_p is real, the imaginary part of (4) gives

$$0 = \operatorname{Im}[\zeta^2 Z'(\zeta)], \qquad (5)$$

which does not involve ω/ω_p explicitly and determines a locus of points in the ζ plane. At any of these points the corresponding value of ω^2/ω_p^2 is obtained from the real part of (4) which is

$$\omega^2/\omega_p^2 = \operatorname{Re}[\zeta^2 Z'(\zeta)] \qquad (6)$$

and k/k_D is obtained from the definition of ζ given by

$$k/k_D = \omega/(\sqrt{2}\omega_p\zeta). \qquad (7)$$

Curves in quadrant IV of the $\zeta = x + iy$ plane obtained from (5) with the aid of a digital computer are shown in Fig. 1. There is actually an infinite number of branches but only the first few are shown. It should be noted that the entire imaginary axis is a solution of (5) since $\operatorname{Im} Z'(iy) = 0$. Equation (5) also yields curves in quadrant III of the ζ plane which are not shown because they are simply mirror images about the imaginary axis of the solutions in quadrant IV. These curves are somewhat similar to those for the case of real k and complex ω which are derived from the equation[5]

$$0 = \operatorname{Im}[Z'(\zeta)], \qquad (8)$$

but of course differ quantitatively because of the factor ζ^2 in (5). Only the portions of the curves of Fig. 1 for which $\operatorname{Re}[\zeta^2 Z'(\zeta)]$ is positive are valid

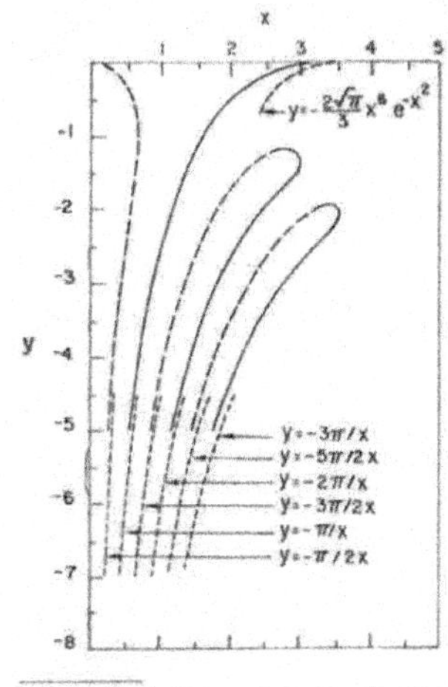

FIG. 1. Locus of points in the $\zeta = \omega/kv_{th}$ plane of solutions of (6) for ω real. Asymptotic solutions are also shown.

[5] B. D. Fried and R. W. Gould, Phys. Fluids 4, 139 (1961).

[6] B. D. Fried and S. D. Conte, *The Plasma Dispersion Function* (Academic Press Inc., New York, 1961).

higher-order root becomes equal to that of this root for frequencies not much greater than ω_p.

The approximate equation for k_I/k_D obtained from (9) and (13) is

$$\frac{k_I}{k_D} \approx \tfrac{3}{2}(\tfrac{1}{2}\pi)^{\frac{1}{2}}(\omega^2/\omega_p^2 - 1)^{-2} \exp[-\tfrac{3}{2}(\omega^2/\omega_p^2 - 1)], \quad (14)$$

which is valid for $|\omega/k| \gg v_{th}$. This is also shown in Fig. 2 and is a good approximation only very near ω_p.

III. THE PLASMA-FILLED WAVEGUIDE

In order to treat the case of a hot plasma in an infinite magnetostatic field bounded by conducting metal walls, it is necessary to solve the Vlasov and Maxwell equations in cylindrical coordinates. The Vlasov equation for the electron distribution function F is

$$\frac{\partial F}{\partial t} + \mathbf{v}\cdot\nabla F - \frac{e}{m}[\mathbf{E} + \mathbf{v}\times\mathbf{B}]\cdot\nabla_v F = 0. \quad (15)$$

Since the case of an infinite axial magnetostatic field given by

$$\mathbf{B}_0 = \hat{z}B_0, \qquad B_0 \to \infty \quad (16)$$

is to be examined, F is expanded as follows:

$$F = F^{(0)} + F^{(1)}/B_0 + F^{(2)}/B_0^2 + \cdots. \quad (17)$$

Substituting (16) and (17) into (15), one obtains to lowest order in B_0

$$(\mathbf{v}\times\mathbf{B}_0)\cdot\nabla_v F^{(0)} = B_0\left(v_y\frac{\partial F^{(0)}}{\partial v_x} - v_x\frac{\partial F^{(0)}}{\partial v_y}\right) = 0. \quad (18)$$

Therefore, the v_x and v_y dependence of $F^{(0)}$ is only through the variable $(v_x^2 + v_y^2)$. To next order in B_0, (15) yields

$$\frac{\partial F^{(0)}}{\partial t} + \mathbf{v}\cdot\nabla F^{(0)} - \frac{e}{m}[\mathbf{E} + \mathbf{v}\times\mathbf{B}]\cdot\nabla_v F^{(0)} - \frac{e}{B_0 m}(\mathbf{v}\times\mathbf{B}_0)\cdot\nabla_v F^{(1)} = 0, \quad (19)$$

where $\mathbf{B}$ is any magnetic field other than $\mathbf{B}_0$. Letting

$$f = \iint F^{(0)}\, dv_x\, dv_y, \quad (20)$$

integrating (19) over v_x and v_y, and using the fact that $F^{(0)}$ is an even function of v_x and v_y, one obtains

$$\frac{\partial f}{\partial t} + v_z\frac{\partial f}{\partial z} - \frac{e}{m}E_z\frac{\partial f}{\partial v_z} = 0. \quad (21)$$

As $B_0 \to \infty$, $F \to F^{(0)}$ so (20) gives

$$f \to \iint F\, dv_x\, dv_y, \qquad B_0 \to \infty. \quad (22)$$

Thus, (21) is the exact equation which the electron distribution function (integrated over v_x and v_y) satisfies in an infinite z-directed magnetic field. Linearizing (21) by letting $f = f_0 + f_1$, one obtains for f_1

$$\frac{\partial f_1}{\partial t} + v_z\frac{\partial f_1}{\partial z} - \frac{e}{m}E_z\frac{\partial f_0}{\partial v_z} = 0. \quad (23)$$

Equation (23) together with Maxwell's equations can be solved for the filled waveguide under the condition that the exciting electric field is "turned on" at $t = 0$ so that all quantities can be Laplace transformed in time. Assuming all first order quantities are zero at $t = 0$, the Laplace-transformed Maxwell equations can be combined to give

$$\nabla^2\mathbf{E}(\mathbf{r},\omega) + \omega^2\mu_0\epsilon_0\mathbf{E}(\mathbf{r},\omega) = -i\omega\mu_0\mathbf{J}(\mathbf{r},\omega) - \frac{i}{\omega\epsilon_0}\nabla\nabla\cdot\mathbf{J}(\mathbf{r},\omega), \quad (24)$$

where $\mathbf{E}(\mathbf{r},\omega)$ is the transformed electric field given by

$$\mathbf{E}(\mathbf{r},\omega) = \int_0^\infty \mathbf{E}(\mathbf{r},t)\,e^{i\omega t}\,dt \quad (25)$$

and $\mathbf{J}(\mathbf{r},\omega)$ is the transformed current density. Since F is an even function of v_x and v_y, the current density has only a z component given by

$$J_z(\mathbf{r},t) = -en_0\int_{-\infty}^{\infty} f_1 v_z\, dv_z, \quad (26)$$

where n_0 is the zero-order electron density. Thus the z component of (24) is given by

$$\nabla^2 E_z(\mathbf{r},\omega) + \omega^2\mu_0\epsilon_0 E_z(\mathbf{r},\omega) = -i\omega\mu_0 J_z(\mathbf{r},\omega) - \frac{i}{\omega\epsilon_0}\frac{\partial^2}{\partial z^2}J_z(\mathbf{r},\omega). \quad (27)$$

Carrying out the Fourier transform of (27) with respect to z, one obtains

$$[\nabla_T^2 + k_0^2 - k^2]E_z(x,y,k,\omega) = -\frac{i}{\omega\epsilon_0}(k_0^2 - k^2)J_z(x,y,k,\omega), \quad (28)$$

where the transformed field is given by

$$E_z(x,y,k,\omega) = \frac{1}{(2\pi)^{\frac{1}{2}}}\int_{-\infty}^{\infty} E_z(\mathbf{r},\omega)\,e^{-ikz}\,dz \quad (29)$$

and similarly for $J_z(x,y,k,\omega)$. In (28), ∇_T^2 is the transverse Laplacian and $k_0^2 = \omega^2\mu_0\epsilon_0$. By carrying out the Fourier and Laplace transforms of (23) and

(26) and inserting the results into (28), one obtains

$$\left\{\nabla_T^2 + (k_0^2 - k^2)\left[1 - \frac{\omega_p^2}{k^2}\int_{-\infty}^{\infty}\frac{\partial f_0/\partial v_z}{v_z - \omega/k}\,dv_z\right]\right\} \cdot E_z(x, y, k, \omega) = 0. \quad (30)$$

Letting

$$K = 1 - \frac{\omega_p^2}{k^2}\int_{-\infty}^{\infty}\frac{\partial f_0/\partial v_z}{v_z - \omega/k}\,dv_z, \quad (31)$$

Eq. (30) can be written

$$[\nabla_T^2 + (k_0^2 - k^2)K]E_z(x, y, k, \omega) = 0. \quad (32)$$

For a waveguide of circular cross section, (32) must be solved in circular cylindrical coordinates, r, ϕ, and z. The solution of (32) which is finite on the axis is

$$E_z(r, \phi, k, \omega) = AJ_n\{[K(k_0^2 - k^2)]^{\frac{1}{2}}r\}\, e^{\pm in\phi}, \quad (33)$$

where A is an arbitrary constant and J_n is the nth order Bessel function of the first kind. The arbitrary constant A is determined by the excitation. We assume a harmonically varying ϕ-independent axial electric field impressed at the wall $r = b$ of the guide localized in the vicinity of $z = 0$. Thus

$$E_z(r = b, z, t) = E_0(z)\, e^{-i\omega_0 t}, \qquad t > 0, \quad (34)$$

where ω_0 is assumed to have a small positive imaginary part. The Fourier–Laplace transform yields

$$E_z(r = b, k, \omega) = iE_0(k)/(\omega - \omega_0), \quad (35)$$

where $E_0(k)$ is the Fourier transform of $E_0(z)$. Equating (33) and (35) at $r = b$ determines the arbitrary constant A so that (33) becomes

$$E_z(r, k, \omega) = \frac{iE_0(k)J_0\{[K(k_0^2 - k^2)]^{\frac{1}{2}}r\}}{(\omega - \omega_0)J_0\{[K(k_0^2 - k^2)]^{\frac{1}{2}}b\}}. \quad (36)$$

At large time the inverse Laplace transform can be carried out since only the contribution of the pole at $\omega = \omega_0$ will remain because the other contribution will damp in time. Thus for long times after the excitation has been turned on, one obtains

$$E_z(r, k, t) = E_0(k)\frac{J_0\{[K(k_0^2 - k^2)]^{\frac{1}{2}}r\}}{J_0\{[K(k_0^2 - k^2)]^{\frac{1}{2}}b\}}\, e^{-i\omega_0 t}, \quad (37)$$

where K and k_0 are evaluated at $\omega = \omega_0$. Since ω_0 has a small positive imaginary part, it is evident that $\text{Im}\,(\omega_0/k) < 0$ when $k < 0$ and $\text{Im}\,(\omega_0/k) > 0$ when $k > 0$. Moreover, the integral in (31) defines a different function of ω/k for $\text{Im}\,(\omega/k) < 0$ than for $\text{Im}\,(\omega/k) > 0$. Therefore, when the inverse Fourier transform of (37) is carried out by integrating along the entire real k axis, the integration is split into two parts as follows:

$$E_z(r, z, t) = \frac{e^{-i\omega_0 t}}{(2\pi)^{\frac{1}{2}}}\left[\int_{-\infty}^{0} E_0(k)\frac{J_0\{[K^-(k_0^2 - k^2)]^{\frac{1}{2}}r\}}{J_0\{[K^-(k_0^2 - k^2)]^{\frac{1}{2}}b\}}\, e^{ikz}\, dk + \int_0^{\infty} E_0(k)\frac{J_0\{[K^+(k_0^2 - k^2)]^{\frac{1}{2}}r\}}{J_0\{[K^+(k_0^2 - k^2)]^{\frac{1}{2}}b\}}\, e^{ikz}\, dk\right], \quad (38)$$

where K^- and K^+ are the functions obtained from (31) when $\text{Im}\,(\omega_0/k) < 0$ and $\text{Im}\,(\omega_0/k) > 0$, respectively. For a Maxwellian, these functions, when analytically continued throughout the complex plane, are given by[6]

$$K^+ = 1 - (k_D^2/2k^2)Z'(\omega/kv_{th}), \quad (39)$$

$$K^- = 1 - (k_D^2/2k^2)\bar{Z}'(\omega/kv_{th}), \quad (40)$$

where $\bar{Z}$ is given by (3). Using (39) and (40) in (38) and noting that Z' and $\bar{Z}'$ are entire functions, it is valid to let ω_0 in (38) be purely real and this is henceforth assumed.

For a given $E_0(k)$ and large z, it is possible under certain conditions to evaluate (38) asymptotically in a manner similar to that used by Landau.[2] It is of importance to determine the poles of the integrands occurring in (38) which are given by

$$K^+(k_0^2 - k^2) = (p_{0m}/b)^2, \quad (41)$$

$$K^-(k_0^2 - k^2) = (p_{0m}/b)^2, \quad (42)$$

where p_{0m} is the mth root of the zero-order Bessel function. Since the roots of (42) are just the negative of the roots of (41), only (41) is considered. Equation (41) yields two types of solutions; those with $|\omega_0/k| \gtrsim c$ for which $K^+ \approx 1 - \omega_p^2/\omega_0^2$ (if $\text{Im}\,k < 0$) so that (1) is valid, and those with $|\omega_0/k| \ll c$. For the latter solutions it is justified to neglect k_0^2 with respect to k^2 in (41). With this approximation, (41) becomes

$$K^+ = -(p_{0m}/kb)^2 \quad (43)$$

from which it is evident that for $(p_{0m}/|k|\,b)^2 \ll 1$, (43) becomes $K^+ \approx 0$ which is identical to (2). Thus, as is expected, for small wavelengths the boundaries are unimportant, the roots becoming identical to those of unbounded plasma oscillations.

Equation (43) can be solved in a manner similar to that discussed in Sec. II for the unbounded plasma. Dropping the subscript on ω_0 and letting $\zeta = \omega/kv_{th}$, (43) can be rewritten in the form

$$\omega^2/\omega_p^2 = \zeta^2[Z'(\zeta) - 2(p_{0m}/k_D b)^2]. \quad (44)$$

Comparison of (44) with (4) shows that the effect of the finite boundaries is to subtract the constant term $2(p_{0m}/k_D b)^2$ from Z'. In hot plasmas for which $k_D b \approx 1$, this term is important over a wide range of

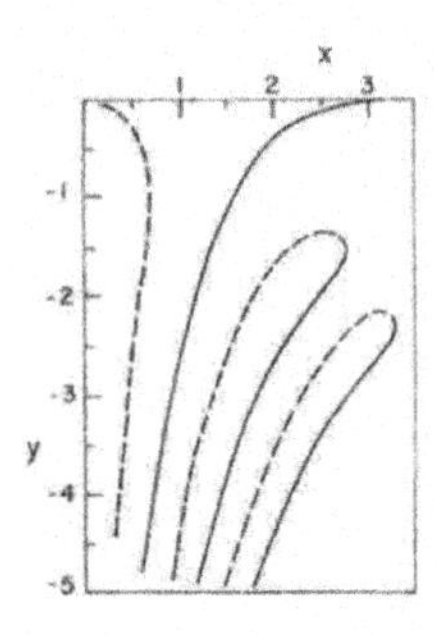

FIG. 4. Locus of points in the $\zeta = \omega/kv_{th}$ plane of solutions of (45) for a plasma-filled circular waveguide with $2(p_{0m}/k_Db)^2 = 0.1$.

frequencies but in lower temperature plasmas where $k_Db \gg 1$, it is important only where $Z'(\zeta)$ is small which is generally where $|\zeta| = |\omega/kv_{th}| \gg 1$, corresponding to the cold plasma regime. The imaginary and real parts of (44) are

$$0 = \operatorname{Im}\{\zeta^2[Z'(\zeta) - 2(p_{0m}/k_Db)^2]\}, \qquad (45)$$

$$\omega^2/\omega_p^2 = \operatorname{Re}\{\zeta^2[Z'(\zeta) - 2(p_{0m}/k_Db)^2]\}. \qquad (46)$$

Equation (45) determines a locus of points in the ζ plane from which ω/ω_p and k/k_D are obtained from (46) and (7). Figure 4 shows the curves determined from (45) for $2(p_{0m}/k_Db)^2 = 0.1$. For $m = 1$, $p_{01} = 2.405$, which corresponds to $k_Db = 10.76$. From the asymptotic expansion of Z, the branch of Fig. 4

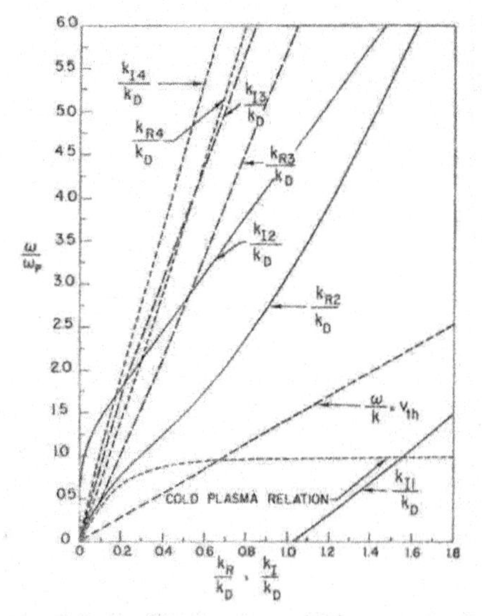

FIG. 5. Real and imaginary parts of k/k_D vs ω/ω_p for the first four roots of (44) for a plasma-filled waveguide with $2(p_{0m}/k_Db)^2 = 0.1$. The cold plasma relation is also shown.

which approaches the real axis for large x is given approximately by

$$y \approx -(\tfrac{2}{3}\pi^{\frac{1}{2}})x^6 \exp(-x^2)[1 + (\tfrac{4}{3}x^4)(\tfrac{3}{2}p_{0m}/k_Db)^2]^{-1} \qquad (47)$$

for large x, which can be combined with (1) to obtain an approximate damping constant analogous to (14). Below the $-45°$ line, $y = -x$, the curves become identical to those of the unbounded plasma, Fig. 1, since there $Z'(\zeta)$ becomes large, making the term $2(p_{0m}/k_Db)^2$ in (45) unimportant. The entire imaginary axis is again a solution of (45). Figure 5 shows the curves of ω/ω_p vs k_R/k_D and k_I/k_D for this case. Only the first four roots of (44) for $2(p_{0m}/k_Db)^2 = 0.1$ are shown in Fig. 5. The root labeled k_{I1}/k_D corresponds to the negative imaginary axis of Fig. 4. At $\omega = 0$, $k_{I1}/k_D = [1 + (p_{0m}/k_Db)^2]^{\frac{1}{2}}$. Not shown is the root corresponding to the positive imaginary axis of Fig. 4. This root is purely imaginary and negative with the limiting values

$$k_I/k_D = -[1 + (p_{0m}/k_Db)^2]^{\frac{1}{2}}$$

at $\omega = 0$, and $k_I/k_D = 0$ at $\omega/\omega_p = [1 + (cp_{0m}/\omega_pb)^2]^{\frac{1}{2}}$. Also shown is the cold plasma relation (1) which the real part of the first root approaches at long wavelengths although it is a poor approximation for these parameters. Comparison of Fig. 5 with Fig. 2 shows that the boundaries affect the second root greatly in the large wavelength domain, the damping being increased significantly from the unbounded case because of the decreased phase velocity. The higher order roots of Fig. 3 are not greatly affected by the boundaries.

Figure 6 shows curves of the first three roots for $2(p_{0m}/k_Db)^2 = 0.01$, which for $p_{01} = 2.405$ corresponds to $k_Db = 34.02$. Because the Debye length is smaller than that of Fig. 5, the cold plasma relation (1) is a better approximation to the first root at long wavelengths.

IV. CONCLUSIONS

In Sec. II has been outlined a method for the extraction of the roots of the dispersion relation for electron plasma oscillations in an unbounded plasma for real ω and complex k. For a Maxwellian velocity distribution, the dispersion relation yields an infinite number of roots although for frequencies near the plasma frequency, the higher-order roots are more highly damped. The presence of the higher-order roots is quite sensitive to the zero-order distribution function. For example, the resonance-shaped zero-order distribution function

$$f_0(v) = (v_{th}/\pi)(v_{th}^2 + v^2)^{-1} \qquad (48)$$

yields the dispersion relation

$$\omega = \omega_p - ikv_{th} \tag{49}$$

and there are no other roots.

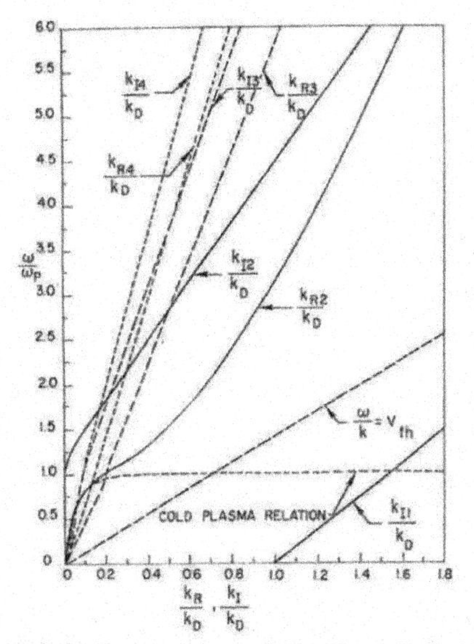

FIG. 6. Real and imaginary parts of k/k_D vs ω/ω_p for the first four roots of (44) for a plasma-filled waveguide with $2(p_{0m}/k_Db)^2 = 0.01$. The cold plasma relation is also shown.

For the Maxwellian distribution, the computed curves of the first root of the dispersion relation, Fig. 2, show that, except for $\omega \approx \omega_p$, the hydrodynamic solution (13) is a poor approximation.

Solutions for the plasma-filled waveguide are presented in Sec. III. From (44), it is evident that the boundaries are of importance if $2(p_{0m}/k_Db)^2$ is comparable in magnitude to $Z'(\zeta)$. For hot and/or low-density plasmas, these two terms can be comparable over a wide frequency range whereas in lower-temperature and/or dense plasmas, the term $2(p_{0m}/k_Db)^2$ is small and is of importance only when $Z'(\zeta)$ is small, the cold plasma regime.

Comparison of the unbounded plasma curves, Figs. 2 and 3, with those for finite boundaries, Fig. 5, shows that the second root is most profoundly affected by the boundaries, with an appreciable increase in spatial damping for low frequencies because of the decreased phase velocity. The higher-order roots are influenced less by the boundaries because they arise from the lower branches of Fig. 4 where the magnitude of $Z'(\zeta)$ is generally larger than for the upper branch, making the term $2(p_{0m}/k_Db)^2$ of less importance.

ACKNOWLEDGMENT

This work was supported by the Joint Services Electronics Program (U. S. Army, U. S. Navy, and U.S. Air Force) under Grant No. AF-AFOSR-496-64.

Reprinted from APPLIED OPTICS, Vol. 13, page 1372, June 1974

Distribution Networks and Electrically Controllable Couplers for Integrated Optics

C. Elachi and C. Yeh

The power distribution as a function of propagation distance in a network of coupled optical waveguides is determined for several interesting cases. An electrically controllable coupler is proposed and analyzed in detail. High efficiency coupling and decoupling between two optical guides can be accomplished with the use of an electrooptically generated dynamic channel, of finite length, located in between the two guides.

I. Introduction

Recently, a number of researchers[1,2] have reported the development of thin film and channel waveguide optical couplers for use in the emerging field of integrated optics. Applications of these couplers in optical networks, modulators, and multiplexers/demultiplexers, would be drastically increased if the coupling is dynamically controllable by an electric signal. Electrooptic substrates can be used to control the coupling coefficient between two waveguides, but such a scheme would be inefficient because of the upper limitation on the change of the refractive index of existing materials that could be achieved with reasonable voltage. In this communication, we study the power distributions as a function of the distance from the input plane in a network of N parallel guides. Then we discuss a number of functions that could be achieved using coupled optical waveguides, and we will study in detail a scheme for an electrically controllable coupler.

II. Symmetric and Nonsymmetric Optical Networks

Let us consider N identical optical waveguides with K being the coupling coefficient between two neighboring guides. The field E_n in the nth guide is determined by the system of equations:

$$(dE_n)/(dz) = -iKE_{n+1} - iKE_{n-1} \text{ for } 2 \leqslant n \leqslant N-1,$$

$$(dE_1)/(dz) = -iKE_2,$$

$$(dE_N)/(dz) = -iKE_{N-1},$$

where we have neglected the direct coupling between nonneighboring guides. If the input light is fed into the mth guide, the normalized initial condition is

$$E_n(0) = \begin{cases} 1 \text{ for } n = m, \\ 0 \text{ for } n \neq m. \end{cases}$$

The above system of equations can be written in a matrix form as follows:

$$(d/(d\xi)E = \mathbf{M}\cdot E \qquad (1)$$

$$E(0) = c,$$

where $\xi = Kz$ and

$$E = \begin{pmatrix} E_1 \\ E_2 \\ \vdots \\ E_N \end{pmatrix} \qquad c = \begin{pmatrix} 0 \\ \vdots \\ 1 \\ \vdots \\ 0 \end{pmatrix} \leftarrow m\text{th element}$$

$$\mathbf{M} = -i \begin{pmatrix} 0 & 1 & 0 & \dots & 0 \\ 1 & 0 & 1 & \dots & 0 \\ 0 & 1 & 0 & \ddots & 0 \\ & & \ddots & \ddots & 1 \\ 0 & & & 1 & 0 \end{pmatrix}$$

Equation (1) is a well known differential equation[3] that can be solved by determining the eigenvalues and eigenvectors of the matrix $\mathbf{M}$.

For $N \to \infty$, the solution of Eq. (1) is the well known Bessel functions:

$$E_n = (-i)^{|n-m|}J_{|n-m|}(2\xi).$$

For N finite, the solution can be determined in a straightforward manner with a digital computer. In Fig. 1 we present the power $P_n = E_nE_n^*$ for a number of configurations ($N = 2$, 3, and 5) with different input conditions. These various configurations can be used to perform a number of functions in optical networks. Some of the possible applications are shown in Fig. 2 and discussed below.

C. Elachi is with the California Institute of Technology, Pasadena, California 91109; C. Yeh is with the University of California, Los Angeles, California 90026.
Received 21 September 1973.

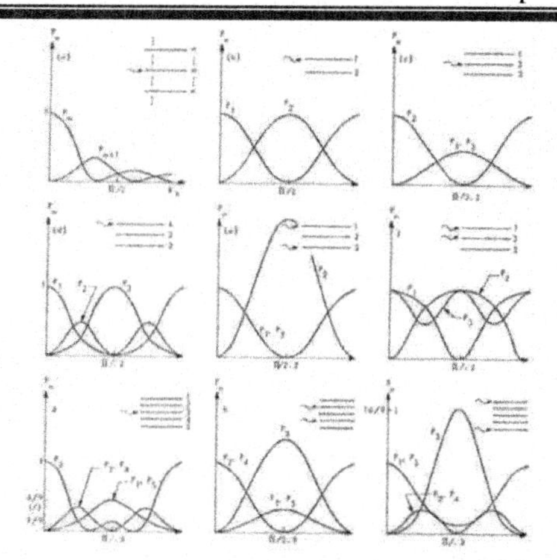

Fig. 1. Power distribution in coupled optical networks. P_n is the power in the nth guide as a function of the propagation distance z. K is the coupling constant between neighboring guides. Nonneighboring guides are assumed to be uncoupled. For the case where there is more than one input, these inputs are assumed to be in phase.

Figure 2(a) shows an energy transfer function configuration. Complete transfer occurs if the coupling length is an odd integer of $\pi/2K$ [see Fig. 1(b)]. Figure 2(b) shows an energy divider configuration. The input energy is equally and completely divided between the two outputs if the coupling length is $\pi/2K\sqrt{2}$ [see Fig. 1(c)]. An elementary ligic system is shown in Fig. 2(c). The two inputs A and B are in phase, and the output is given by the truth table. Figures 2(d) and 2(e) correspond to energy transfer from two inputs to two outputs, where the useful information is the amplitude or the frequency of the signal (assuming that the two frequencies ω_1 and ω_2 are not very different so that K is approximately the same for both). In Figs. 2(f) and 2(g) we present a possible configuration for an electrically controllable coupler or switch that is discussed in the next section.

III. Electrically Controllable Coupler

The basic configuration for an electrically controllable coupler is shown in Figs. 2(f) and 2(h). The two permanent channel guides are imbedded at the surface of an electrooptic substrate. They can be formed by proton bombardment,[4] ion implantation, diffusion, or other techniques. The two guides are located such that the direct coupling is very weak. In the region between the two guides, a third channel of finite length is dynamically generated by applying a voltage to the two electrodes shown in the figure. The resulting electric field generates a local change in the refractive index. The feasibility of such an electrooptically generated channel guide was recently reported by Channin.[5] This controllable channel plays the role of a bridge between the two permanent guides. The electrodes should be located such that the cross section of the dynamic guide is similar to the cross section of the two permanent guides.

The power distribution as a function of the propagation distance in a three guides system, where the energy is fed in the first guide, is shown in Fig. 1(d) and is given by [from Eq. (1)]:

$$P_1(z) = \frac{1}{4}[\cos(K\sqrt{2}z) + 1]^2,$$

$$P_2(z) = \frac{1}{2}[\sin(K\sqrt{2}z)]^2,$$

$$P_3(z) = \frac{1}{4}[\cos(K\sqrt{2}z) - 1]^2,$$

where we assumed that the direct coupling coefficient K' between the two permanent guides is $\ll K$. Complete energy transfer between the two permanent guides occurs if the controllable channel has a length $L = \pi/K\sqrt{2}$.

In the absence of the bridge channel, the power distribution in the two permanent guides is

$$P_1'(z) = \cos^2(K'z),$$

$$P_2'(z) = \sin^2(K'z),$$

and the energy transferred over the length L is

$$\Delta P = \sin^2[(\pi/\sqrt{2})(K'/K)].$$

Therefore the dynamic efficiency of the coupler can be defined as

$$\eta = 1 - \Delta P = \cos^2[(\pi/\sqrt{2})(K'/K)].$$

The analytic expression of the coupling coefficients was derived by Marcatilli[6] as

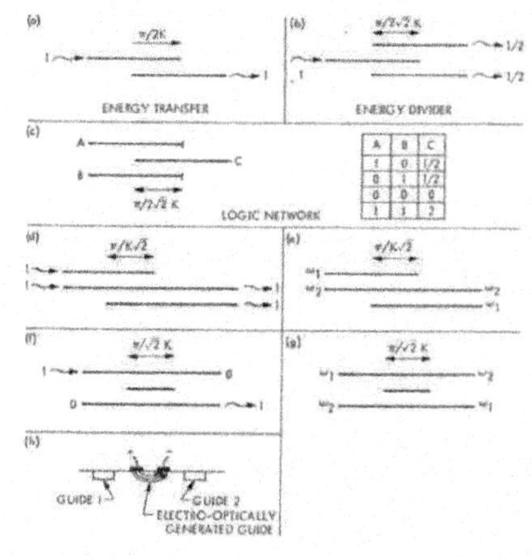

Fig. 2. Different configurations of optical network that could be used in energy transfer, energy distribution, controlled switching (see text).

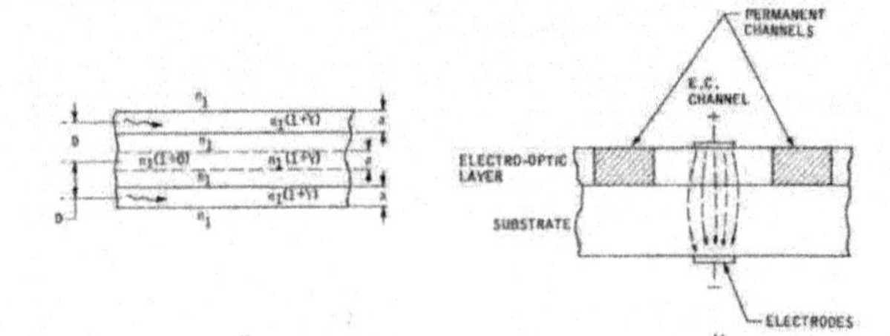

Fig. 3. (a) Simplified (ECC) for thin film guides. γ is the percentage change of the index of refraction in the permanent guides. γ is also taken as the percentage change due to the electrooptic effect. (b) Another possible configuration for an ECC.

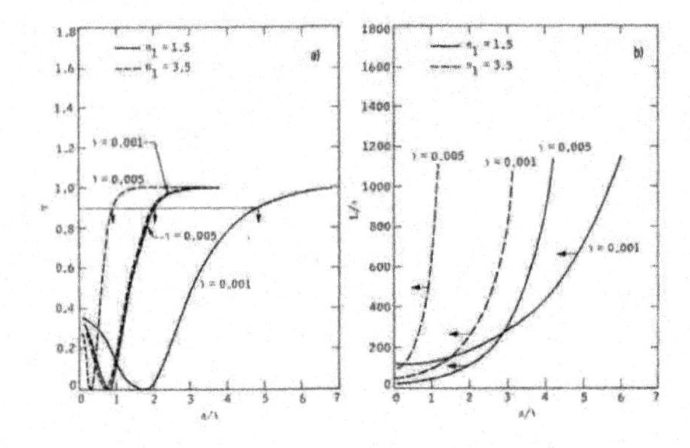

Fig. 4. Dynamic efficiency and effective length of the ECC shown in Fig. 3(a), as a function of a/λ, for different values of n_1 and γ. The value of D/a is taken equal to 1.5.

$$K = [(2s^2\delta)/(s^2 + \delta^2)]\ \{\exp[-\delta(D - a)]\}/(\kappa a)$$

and

$$K' = [(2s^2\delta)/(s^2 + \delta^2)]\ \{\exp[-\delta(2D - a)]\}/(\kappa a),$$

where a is the width of each channel, D is the distance between the center lines, κ and s are, respectively, the propagation constants along and perpendicular to the propagation direction in the coupler plane, and δ is the exponential full off between the guides. The above expressions were derived for well-confined modes, but they may be used as a good approximation in the general use. Using the above expressions of K and K' we can express L and η as

$$L = [\pi/(2\sqrt{2})][(s^2 + \delta^2)/(s^2\delta)]\kappa a\ \exp[\delta(D - a)],$$

$$\eta = \cos^2[(\pi/\sqrt{2})\ \exp(-\delta D)].$$

We carried out a numerical study of the simplified coupler scheme shown in Fig. 3(a), where the guides

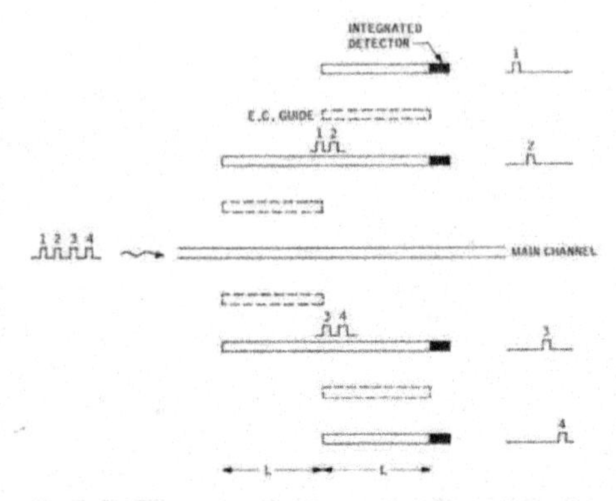

Fig. 5. Possible configuration for a 4-channel optical demultiplexer. The dashed guides are electrically controllable.

are thin film layers. n_1 is the index of refraction of the substrate, and γ is the percentage increase generated when a voltage is applied. γ is also taken as the percentage index increase in the permanent guides. In Fig. 4 we plotted η and L/λ as a function of a/λ for a fixed value of D/a, where λ is the optical wavelength in vacuum. It is clear that high efficiency ($\eta \geq 90\%$) is possible at high frequencies, and larger values of n_1 or γ lead to a wider region of high efficiency. On the other hand, the value of L/λ increases with a/λ. To illustrate, let us choose $n_1 = 3.5$, $\gamma = 0.001$, and $\lambda = 1.15\ \mu$. For an efficiency of 90%,

$$a = 2.3\ \mu, D = 3.45\ \mu, \text{and } L = 304\ \mu.$$

If we want to increase the efficiency to 99%,

$$a = 3.4\ \mu, D = 5.1\ \mu, \text{and } L = 920\ \mu.$$

For shorter wavelengths, the above numbers are proportionally smaller. In Fig. 3(b) another scheme for an electrically controllable coupler is shown.

IV. Conclusion

The above results show that some simple function and efficient dynamic switching are possible over relatively short distances. The dynamic coupler may play a central role in complex optical networks and multiplexers/demultiplexers. A scheme for a 4-channel multiplexer/demultiplexer is shown in Fig. 5.

C. Elachi is also with the University of California at Los Angeles and the Jet Propulsion Laboratory. This research was funded partly by NECL.

References

1. A. Ihaya, H. Furuta, and H. Noda, Proc. IEEE 60, 470 (1972).
2. S. Somekh, E. Garmire, A. Yariv, H. L. Garvin, and R. G. Hunsperger, Appl. Phys. Lett. 22, 46 (1973).
3. R. Bellman, *Introduction to Matrix Algebra* (McGraw-Hill, New York, 1970).
4. E. Garmire, H. Stoll, A. Yariv, and R. G. Hunsperger, Appl. Phys. Lett. 21, 87 (1972).
5. D. J. Channin, Appl. Phys. Lett. 19, 128 (1971).
6. E. A. J. Marcatilli, Bell Syst. Tech. J. 48, 2071 (1969).
7. H. F. Taylor, J. Appl. Phys., 44, 3257 (1973).

JOURNAL OF APPLIED PHYSICS VOLUME 86, NUMBER 5 1 SEPTEMBER 1999

Highly sensitive quantum well infrared photodetectors

Tom Cwik[a] and C. Yeh
Jet Propulsion Laboratory, California Institute of Technology, 4800 Oak Grove Drive, Pasadena, California 91109

(Received 19 January 1999; accepted for publication 28 May 1999)

A fundamentally new method for light coupling in quantum well infrared photodetectors that provides a ten-fold improvement over an optimized grating coupler is presented. It is based on the prism-film coupler concept developed earlier for selective mode coupling into integrated optical circuits. In this article this concept is specifically used to turn the incident electric field from one that is polarized parallel to the quantum well layer to one that is mostly perpendicular to the layer, thereby increasing dramatically the sensitivity of the quantum well infrared photodetector. Detailed sample design and its computer simulation results are given and discussed. © *1999 American Institute of Physics.* [S0021-8979(99)03517-3]

I. INTRODUCTION

A limitation of quantum well infrared photodetectors is that they can only detect (absorb) infrared light whose electric field is polarized in a direction which is perpendicular to the plane of the quantum well layers.[1] The incident infrared light usually takes the form of a plane wave whose electric field is polarized in a direction that is parallel to the plane of the quantum well layers. Therefore, much effort has been focused on ways to induce the generation of electric fields having the desired polarization.

Historically, a direct approach for light coupling is to orient the quantum well infrared photodetector at a 45° angle to the incident infrared radiation. The incident electric field will have a component ($1/\sqrt{2}$ of the incident electric vector) that can produce absorption of photons, and any additional scattering due to the mesa structure can enhance this absorption.[2] When the photodetectors are used in focal plane arrays, the photodetector array is oriented perpendicular to the scene to be imaged, the direction of propagation is perpendicular to the photodetector, and the electric vector is parallel to the quantum well layers. No absorption is possible in this orientation. To produce absorption in this focal plane array application, different structures have been developed. One structure uses a random surface at the base of the mesa structure to scatter the incident radiation into the correct polarization for absorption. Substantial enhancement in quantum well infrared photodetector responsivity compared to the 45° incidence structure has been reported.[3] To improve upon the random surface grating, a variety of grating couplers has been developed.[4,5] These couplers consist of one- or two-dimensional periodic profiles designed into the base of the mesa structure to convert the normally incident radiation to waves propagating parallel to the quantum well layers. For metallic profiles, an efficient grating design minimizes the energy reflected into the (0, 0) harmonic, while maximizing the energy coupled into the propagating $(\pm 1, \pm 1)$ harmonics that have a component of the electric field perpendicular to the quantum well layers. If the design operates just at the point of propagation of these first harmonics, the propagation vector is exactly parallel to the quantum well layers and maximum coupling can be achieved. This theoretical performance is achieved for infinite gratings, and is modified for finite sized gratings and when truncation effects and scattering due to the finite mesa size are considered. The performance of these grating structures can be enhanced to reduce the dark current associated with the quantum well structure, by patterning the multiquantum well layers, removing some of the wells.[6] To date the optimized grating coupled quantum well infrared photodetector offers the best performance.

The purpose of this article is to show a new way to improve performance and that a ten-fold improvement over an optimized grating coupler is attainable. This new way is based on the prism-film coupler concept developed earlier for selective mode coupling into integrated optical circuits. In this article, this concept is specifically used to turn the incident electric field from one that is polarized parallel to the quantum well layer to one that is mostly perpendicular to the layer, thereby increasing dramatically the sensitivity of the quantum well infrared photodetector. A detailed description on how this is accomplished will first be given. Computer simulations were carried out for several specific geometries corresponding to achievable practical designs. Figure of merits for these cases as well as that for an optimized 50 μm grating for the band between 14 and 16 μm will be given. Finally, we shall discuss the viability of the new design as a practical quantum well infrared photodetector device.

II. THE EVANESCENT-WAVE PRISM-FILM COUPLER CONCEPT

To efficiently couple light into dielectric waveguides and excite a given mode for integrated optic applications, evanescent-wave prism couplers were developed 30 years ago.[7,8] By feeding a light beam into a film through a broad surface of the film, one avoids the difficult problem of focusing a light beam through a rough film edge. Since the film and the prism are coupled over a length of many optical

[a]Electronic mail: cwik@jpl.nasa.gov

2780 J. Appl. Phys., Vol. 86, No. 5, 1 September 1999

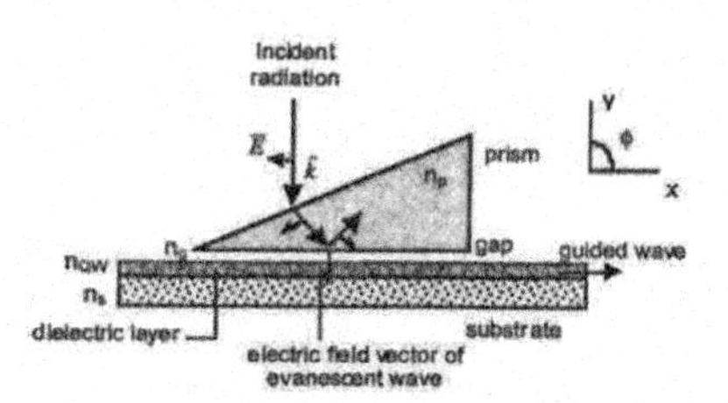

FIG. 1. Geometry of prism coupler for integrated optics application. Solid lines indicate direction of propagation; smaller arrows indicate electric field vector. The electric vector of the evanescent wave is polarized vertically. The indices of refraction n_p, n_g, n_{QW}, and n_s are of the prism, gap, guided wave, and substrate layers, respectively.

wavelengths, energy is transferred continuously between them as waves propagate over the coupling region and as shown in Ref. 7, very high coupling efficiency can be obtained. In this article our goal is to change the polarization of an incident wave from a direction which is parallel to the quantum well layers to a direction which is normal to the layers. It will be shown later that this evanescent-wave prism film coupler is capable of providing this transformation.

Shown in Fig. 1 is an evanescent-wave prism-film coupler. A plane wave with the electric field polarized parallel to the substrate layers is incident on the prism. It enters the prism and is totally reflected at or near the critical angle at the base of the prism. Because of the total reflection, the field in the prism is a standing wave and the field beyond the base of the prism is an exponentially decaying evanescent field that decreases rapidly away from the prism surface. As seen from Fig. 1, the prism, placed on top of the quantum well layers, maintains a small uniform air gap between the base surface of the prism and the top surface of the layers. The evanescent field below the prism then penetrates into the layers and excites a surface wave into the layers. When the gap thickness t and relative indices are tuned appropriately, up to 80% of the incident radiation can be coupled into a mode within the dielectric layer.[7] For the critical angle to be reached, $n_p<n_g$, and for a guided wave to be supported in the dielectric layer, $n_{QW}>n_g$, n_s where the indices of refraction n_p, n_g, n_{QW}, and n_s are of the prism, gap, guided wave and substrate layers, respectively. The prism is truncated at some distance along the path of coupling to eliminate energy being coupled back into the prism along the direction of propagation.

Furthermore, one may note from Fig. 1 that the electric field component of the evanescent field below the prism is now normal to the bottom surface of the prism. This means that the electric field component of the incident wave has been successfully transformed to the direction needed by the quantum well infrared photodetector layers for efficient absorption.

It is noted that several factors govern the response of this quantum well infrared photodetector. They include the angle of the prism, the index of the prism, the indices and the thickness of the quantum well layers, the separation of the air gap, and the length of the coupling region. Proper adjustment of these factors produces the desired design.

III. ANALYSIS

The earlier description of the evanescent-wave coupler is based on an optical modal analysis since the size of the prism thin-film coupler is electrically large. In this article, the infrared coupler is much smaller in size, operating in the resonant-length regime. Therefore, a direct solution of Maxwell's equations throughout the device is necessary. A method that couples a finite element solution of Maxwell's equations inside the device with an integral equation representation of the fields on the boundary of the finite element mesh is used. The integral equation includes the radiation boundary condition to accurately truncate the computational domain. By coupling the two field representations at the surface, a system of equations that describes the fields throughout the device is found.

A complete description of the numerical approach is found in Ref. 9. The analysis considers a two-dimensional scattering geometry consisting of an interior and exterior regions bounded by a surface. The surface S, taken to lie on the boundary of the device, divides the geometry into interior region V and exterior region Ω. Volume V can be inhomogeneous, containing lossy material and perfect conductors that can lie partly on the surface S, therefore modeling the device materials accurately. The magnetic field is polarized in the z direction and is produced by an incident magnetic field

$$H^{\rm inc}(\rho)=\hat{z}e^{jk(x\,{\rm Cos}(\phi)+y\,{\rm Sin}(\phi))}e^{j\omega t} \tag{1}$$

due to a source in the exterior region Ω. The incident magnetic field is denoted $H^{\rm inc}$, k is the wavenumber, ϕ is the polar angle (see Fig. 1), and the time dependence is $e^{j\omega t}$.

A. The wave and integral equations

In the inhomogeneous region V, the wave equation for the total magnetic field $H=\hat{z}H_z$

$$\nabla\times\frac{1}{\epsilon_r(\rho)}\nabla\times H(\rho)-\mu_r(\rho)k^2H(\rho)=0 \quad \rho\in V \tag{2}$$

is used; ρ is the two-dimensional position vector. Both ϵ_r and μ_r are functions of position and may be complex, i.e., $\epsilon_r=\epsilon'-j\epsilon''$ and $\mu_r=\mu'-j\mu''$, to account for dielectric and magnetic loss. In the homogenous exterior region Ω, the surface magnetic field integral equation[10]

$$H(\rho)=2H^{\rm inc}(\rho)+\frac{jk}{2}\oint_{\partial v}H_{\rm tan}(\rho')H_1^{(2)}(k|\rho-\rho'|)\times{\rm Cos}(\hat{n}',\rho-\rho')dl'+\frac{\omega\epsilon}{2}\oint_{\partial v}E_{\rm tan}(\rho')\times H_0^{(2)}(k|\rho-\rho'|)dl' \quad \rho,\rho'\in S \tag{3}$$

is used; $H_{\rm tan}=\hat{n}\times H^+$ and $E_{\rm tan}=\hat{n}\times E^+$ are the tangential field components just exterior to the surface ∂V. $H_0^{(2)}$ and $H_1^{(2)}$ are the outgoing first and second order Hankel functions; $(\hat{n}',\rho$

J. Appl. Phys., Vol. 86, No. 5, 1 September 1999

$-\rho'$) denotes the angle between source and observation points ρ and ρ', respectively, and principal value integrations are implied.

B. Boundary condition enforcement

With an equation written for the fields interior to the volume, and one also written for the fields on the surface that satisfies the radiation condition, it is only necessary to enforce boundary conditions to fully satisfy Maxwell's equations. Boundary conditions must be satisfied at all material boundaries, at infinity, and at the boundary surface of the two solutions (S). At material boundaries, which only reside within V, boundary conditions are satisfied in the finite element solution by a proper choice of elements, or by explicitly zeroing tangential electric field components to model perfectly conducting surfaces. As noted earlier, in the integral Eq. (3) the Sommerfeld radiation condition is naturally enforced by the outgoing Green's function. The only boundary conditions remaining need be satisfied at the surface S. Equating tangential electric field components gives

$$E_{\tan}(\rho)+\frac{1}{j\omega\epsilon}\hat{n}\times\nabla\times H(\rho)=0 \quad \rho\in S \tag{4}$$

on the surface S. If part of the surface encloses a perfect conductor, $H(\rho)$ within the conductor is identically zero, and the usual boundary condition $E_{\tan}=0$ applies. Similarly, equating tangential magnetic field components gives

$$H_{\tan}(\rho)-\hat{n}\times H(\rho)=0 \quad \rho\in S \tag{5}$$

on the surface S.

C. Discretization

The wave Eq. (2) and integral Eq. (3), as well as boundary conditions (4) and (5), are combined into three equations. This system of equations is discretized and solved numerically yielding the magnetic fields in the interior region and tangential magnetic and electric fields on the surface S.

For a finite element solution of the wave Eq. (2) is dotted against a testing function $T=\hat{z}T_z$, integrated over the volume, and integrated again by parts yielding the weak form

$$\int_V\left[\frac{1}{\epsilon_r}\nabla T_z\cdot\nabla H_z-\mu_r k^2 T_z\cdot H_z\right]dv$$

$$+\int_{\partial V}\frac{1}{\epsilon_r}T\cdot[\hat{n}\times\nabla\times \underset{\sim}{H}]ds \tag{6}$$

The boundary condition on tangential components of the electric field [Eq. (4)] is combined into this equation by noting that in the contour integral in Eq. (6), $\hat{n}\times\nabla\times H=j\omega\epsilon E_{\tan}$; therefore, substituting Eq. (4) into this integral gives

$$\int_V\left[\frac{1}{\epsilon_r}\nabla T_z\cdot\nabla H_z-\mu_r k^2 T_z\cdot H_z\right]dv-j\omega\epsilon_0\int_{\partial V}T_z E_{\tan}ds \tag{7}$$

for the first equation of the system ($E_{\tan}=\hat{t}E_{\tan}$, $\hat{t}$ is the surface tangent). The second equation is found by enforcing Eq. (5) in a weak sense, i.e., dotting Eq. (5) against a testing function U and integrating

$$\int_{\partial V}U\cdot[\hat{n}\times H(\rho)-H_{\tan}]ds=0. \tag{8}$$

The third and final equation of the system is found from a standard moment solution to Eq. (3), dotting the equation against a testing function W and integrating along the contour S. This equation will not be written explicitly, the reader is referred to the method of moments literature (e.g., Ref. 11) for further details.

Equations (7), (8), and (3) tested against W, are discretized, forming the system of matrix equations

$$\begin{bmatrix} K & C & 0\\ C^t & 0 & T\\ 0 & G_e & G_h\end{bmatrix}\begin{bmatrix} H\\ E_{\tan}\\ H_{\tan}\end{bmatrix}=\begin{bmatrix}0\\0\\V^{\rm inc}\end{bmatrix}, \tag{9}$$

where K is the square sparse matrix resulting from the volume integral of Eq. (7); its order is the number of nodes in the finite element grid. C and T are coupling matrices found from evaluating Eq. (8) (C^t denotes the transpose) and G_e and G_h are moment method matrices found from the discretization of Eq. (3). The solution of the linear system [Eq. (9)] produces the magnetic field H everywhere inside the device, and the tangential fields E and H on the surface. Numerical convergence of the solution is confirmed by successively finer discretizations of the finite element mesh. Accuracy studies and comparisons to measurements for a range of objects can be found in Ref. 9.

IV. COMPUTER SIMULATION RESULTS

A finite element mesh is generated over the prism coupler structure. A two-dimensional simulation (a cut in the x-y plane) is performed, solving for the z component of the magnetic field, H_z, due to the incident plane wave. Since it is the y component of the electric field, E_y that induces absorption of photons in the quantum well layers, the electric field is calculated from the magnetic field using an interpolation method.[12] To measure the effectiveness of the absorption of the incident radiation, the magnitude squared of E_y is calculated and integrated over the quantum well layer. This is the figure of merit that is used to compare with other methods. The figure of merit is defined as

$$F=\int\int|E_y(x,y)|^2dxdy, \tag{10}$$

where the integral is over the active region.

For the quantum well infrared photodetector coupler, the substrate layer shown in Fig. 1 is replaced by a metal layer and the active region is placed directly below the prism (Fig. 2). The dimensions and materials used in the design are shown in the table in Fig. 2 for optimized absorption at 15 μm. For this design, the quantum well layer is 50 μm by 1.875 μm. The prism angle θ, based on the optical design

2782 J. Appl. Phys., Vol. 86, No. 5, 1 September 1999

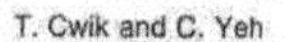

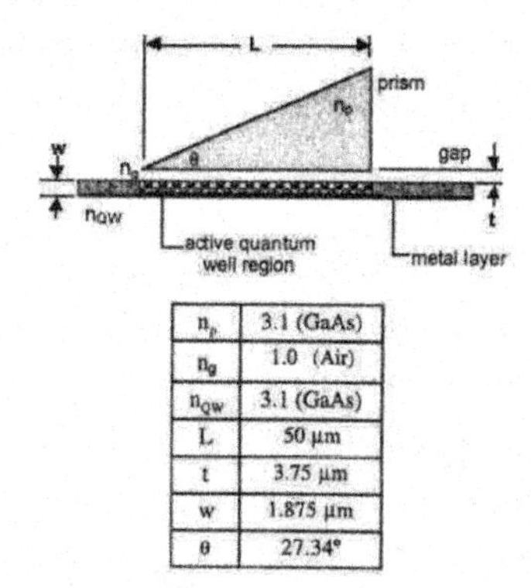

n_p	3.1 (GaAs)
n_g	1.0 (Air)
n_{QW}	3.1 (GaAs)
L	50 μm
t	3.75 μm
w	1.875 μm
θ	27.34°

FIG. 2. Geometry of prism coupler for light coupling application. The geometry and variables are shown in the figure with the design values shown in the table at the bottom. The prism length is L, the prism angle is θ, the gap and active gallium arsenide thicknesses are t and w, respectively. The indices of refraction n_p, n_g, and n_{QW} are of the prism, gap, and quantum well layers, respectively.

outlined in Sec. II is chosen to provide a totally reflected wave when the incident plane wave reaches the bottom surface of the prism.

Shown in Fig. 3 is a plot of the figure of merit as a function of wavelength for the prism coupler shown in Fig. 2. The results show several peaks in the figure of merit indicating that very efficient absorption occurs at those wavelengths. The presence of these peaks is not surprising since the quantum well layer waveguide is truncated (terminated) at both ends forming a surface waveguide resonator with specific resonant frequencies. The layer is truncated to both contain the energy for increased absorption and to isolate one element from another when this structure is used as a single pixel in a focal plane array. The multiple resonances though were not found to be dependent on a single dimension of the structure. Rather they depend upon a combination of the prism angle, gap thickness, and thickness of the quantum well substrate. Also noted is the fact that the absorption peaks are not uniform in magnitude due to the frequency dependence of the prism-waveguide coupling effect. For the chosen parameters, the highest absorption peak occurs at 14.95 μm. What is surprising is the height of this peak; it is ten times higher that the highest valley where the highest valley is already comparable to the best figure of merit of an optimized grating coupler. The wavelength location of this peak can be adjusted by changing the prism angle, the thickness of the air gap, or the thickness of the quantum well substrate. The ultrasensitivity achievable by this prism-coupled quantum well infrared photodetector device is its most distinguishing feature. As a comparison, plotted in Fig. 3 is the figure of merit for an optimized grating structure consisting of a 50 μm mesa with ten grating periods and an active quantum well region of size 50 μm by 3.75 μm in cross section. It is seen that at the resonant wavelengths, the figure of merit for the prism-coupled quantum well infrared photodetector is far greater than that for the grating structure.

To further understand the prism coupled quantum well infrared photodetector, the distribution of the magnitude of E_y inside the prism, the air gap and the quantum well layer is obtained and displayed in Figs. 4–6 for three different wavelengths (14.950, 15.150, and 14.267 μm). When the incident radiation is at 14.950 μm, according to Fig. 3, the figure of merit is at a maximum indicating that most of the incident

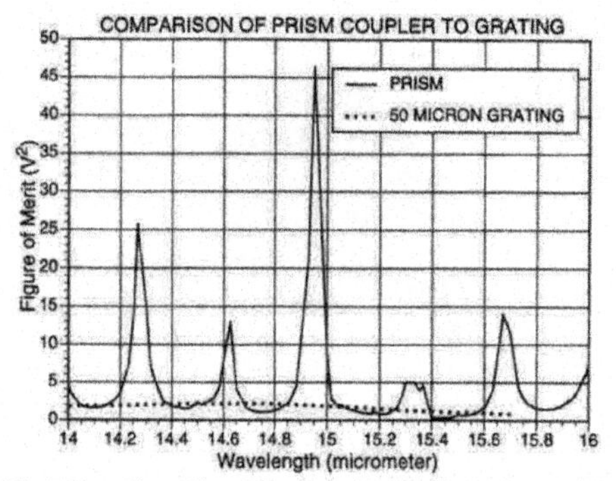

FIG. 3. Comparison of figures of merit for the prism coupler shown in Fig. 2 to that of an optimized 50 μm grating coupler. The figure of merit is defined in Eq. (10).

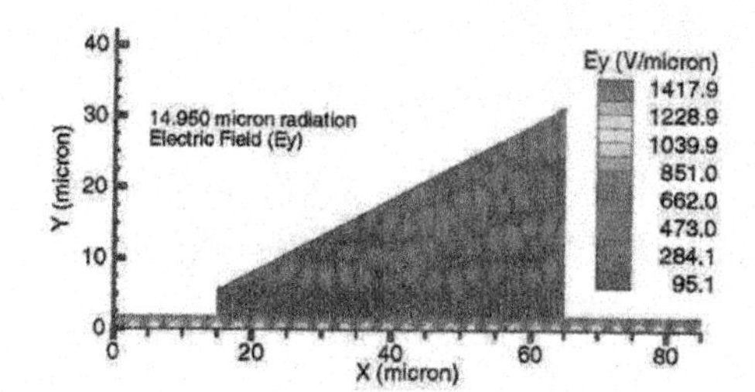

FIG. 4. Field display at 14.950 μm radiation. The quantum well active region is directly below the prism. The wavelength of operation is at a relative peak in the figure of merit in Fig. 3.

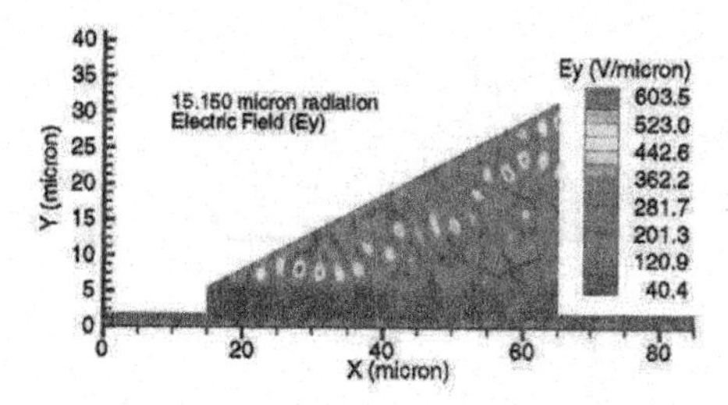

FIG. 5. Field display at 15.150 μm radiation. This is at a null of the figure of merit indicating little coupling. Note the scale is also lower on this plot than on Figs. 4 and 6. The quantum well active region is directly below the prism.

J. Appl. Phys., Vol. 86, No. 5, 1 September 1999

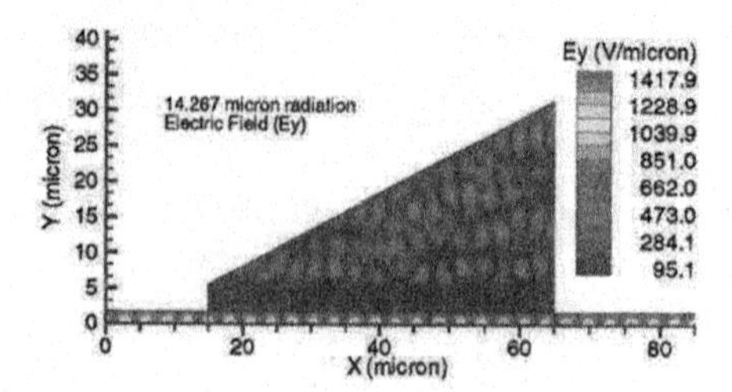

FIG. 6. Field display at 14.267 μm radiation. This is a relative maximum of the figure of merit. The quantum well active region is directly below the prism.

parallel polarized field has been transformed to vertically polarized field inside the quantum well layer. Hence, as expected, Fig. 4 shows concentration of high magnitude E_y field (red) inside the quantum well layer. When the incident radiation is at 15.150 μm, according to Fig. 3, the figure of merit is at a minimum indicating that only a limited amount of the incident parallel polarized field has been transformed to vertically polarized field inside the quantum well layer. Hence, only low magnitude E_y (blue) is shown inside the (active) quantum well layer, while higher magnitude E_y (green) is found inside the (inactive) prism. The next figure is for incident radiation at 14.267 μm, another relative peak in the figure of merit in Fig. 3, and a wavelength indicating good absorption by the quantum well layer. Figure 6 shows this situation, where, as expected, concentrations of high magnitude E_y field (red) inside the quantum well layer are found. It is seen that an excellent transformation of parallel polarized incident radiation to concentrations of vertically polarized radiation inside the quantum well layer can occur using this prism coupler technique.

V. A MONOLITHIC DESIGN

The earlier design can be limiting due to the necessity of an air-gap layer between the prism and the active gallium arsenide layer. Fabrication difficulties as well as practical

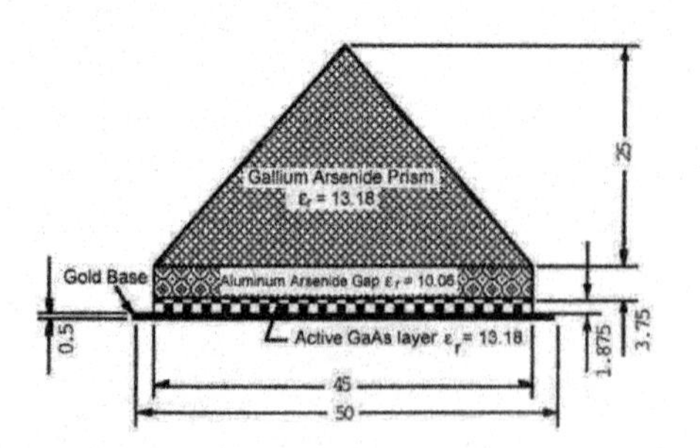

FIG. 7. Geometry of prism coupler with aluminum arsenide gap layer. Relative permittivity values used in calculations are shown for the materials. Dimensions are in micrometers.

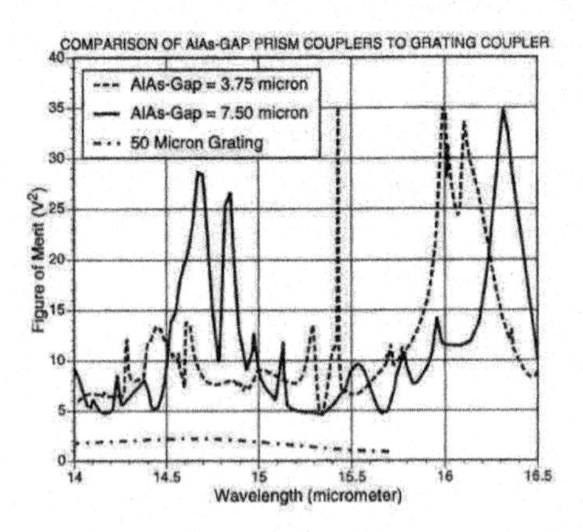

FIG. 8. Comparison of figures of merit for aluminum arsenide-gap prism couplers shown in Fig. 7 with optimized grating coupler. Two different gap thicknesses are shown.

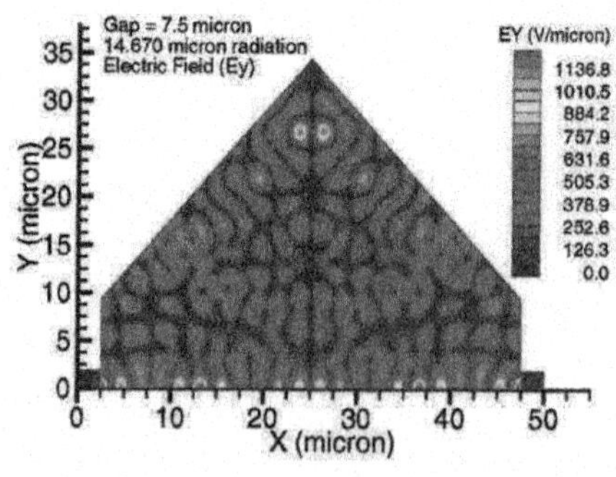

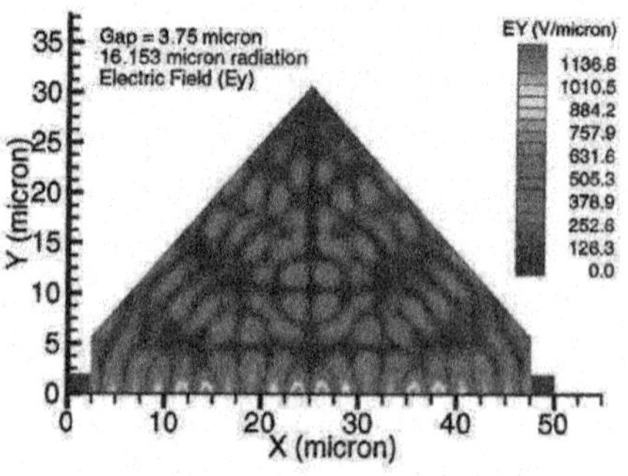

FIG. 9. Electric field plots (E_y) for AlAs-gap prisms of different gap thicknesses. The wavelengths of operation correspond to peak figures of merit indicated in Fig. 8.

2784 J. Appl. Phys., Vol. 86, No. 5, 1 September 1999 T. Cwik and C. Yeh

lifecycle limitations of the air-gap layer when the sensor is used in a space environment can limit its usefulness. To remedy this, a design consisting of gallium arsenide for the prism, aluminum gallium arsenide for the gap region and active gallium arsenide for the quantum well layer is used. The quantum well layer is placed on top of a gold base. This design allows monolithic fabrication since two materials with relatively uniform lattice constants and the greatest differences in their index of refraction are used.[13] To produce a more compact design, the right-angle prism was changed to a pyramid design, thus limiting its height. Dimensions and material parameters are shown in Fig. 7. Because of the choice of materials as well as the resonant size of the device, the performance of this prism coupled quantum well infrared photodetector must be calculated using the exact wave equation approach discussed in Sec. III rather than the more intuitive ray optics approach discussed in Sec. II. In spite of the replacement of the air gap by a layer of aluminum gallium arsenide, substantial coupling of the desired polarized field will still take place resulting in dramatic increase in the sensitivity of this photodetector at certain wavelengths.

Shown in Fig. 8 is the figure of merit for designs with two different gap widths compared to the grating coupler design. It is noted that nearly uniformly periodic resonant character of the figure of merit has disappeared and the overall level is higher than the air-gap design. It is clear that sharp resonances can be produced from the air-gap optical design, while these are smoothed when the higher index aluminum gallium arsenide gap is used and the prism angle is reduced from the optimum optical design of 76° to the 48° design in Fig. 7. Similarly a resonance does not exist at 15 μm for this nonoptimal design.

Shown in Fig. 9 are electric field plots throughout the prism coupler for the two different aluminum arsenide-gap sizes. The wavelengths of operation for each simulation correspond to a peak energy level in Fig. 8. In comparison to similar figures for the air-gap prism (Figs. 4 and 6), electric fields in this aluminum arsenide-gap prism are not as well trapped in the active region. This is due to the lack of confinement in the gallium arsenide base region which acts as a waveguide structure when bounded by the air-gap above. As in the air-gap prism though, a discontinuity in the active region to the left and right of the base is considered essential to force trapping of the energy in the active region. This trapping of the energy is thought to form the resonant peaks in the energy coupling curves and when used in an active array also eliminates energy leaking from one pixel to the next as it propagates along the active region.

VI. CONCLUSION

A new way to dramatically increase the sensitivity of quantum well infrared photodetectors has been found. It is based on the evanescent-wave prism coupler originally developed for integrated optics applications. In this article, it is used to change the polarization of an incident plane wave to a desired polarization that enhances absorption thereby improving greatly the sensitivity of this quantum well infrared photodetector. A specific design that allows monolithic fabrication has been proposed and its performance has been obtained by numerical simulation. A ten-fold increase in the sensitivity over an optimized grating coupler at a given wavelength is demonstrated. By choosing the appropriate parameters such as the gap separation, the angle of the prism and the thickness of the quantum layer, it is possible to tune the maximum sensitivity of the detector to a given wavelength. It is believed that a broadband detector may also be designed using the same evanescent-wave prism coupler concept. Here, a detuning mechanism must be introduced. This research is presently being pursued. Finally, it should be noted that the design given in this article is aimed towards what can be implemented in practice.

ACKNOWLEDGMENTS

The work described in this publication was carried out by the Jet Propulsion Laboratory, California Institute of Technology under a contract with the National Aeronautics and Space Administration. The supercomputer used in this investigation was provided by funding from the NASA Offices of Earth Science, Aeronautics, and Space Science. Part of the research reported here was performed using the HP SPP-2000 operated by the Center for Advanced Computing Research at Caltech; access to this facility was provided by the California Institute of Technology.

[1] B. F. Levine, J. Appl. Phys. **74**, R1 (1993).
[2] S. D. Gunapala *et al.*, IEEE Device **44**, 45 (1997).
[3] G. Sarusi, B. F. Levine, S. J. Pearton, K. M. S. Bandara, and R. E. Leibenguth, Appl. Phys. Lett. **64**, 960 (1994).
[4] Y. C. Wang and S. S. Li, J. Appl. Phys. **74**, 2192 (1993).
[5] J. Y. Andersson and L. Lundqvist, J. Appl. Phys. **71**, 3600 (1992).
[6] T. R. Schimert *et al.*, Appl. Phys. Lett. **68**, 2846 (1996).
[7] P. K. Tien, R. Ulrich, and R. J. Martin, Appl. Phys. Lett. **14**, 291 (1969).
[8] R. Shubert, and J. Harris, IEEE Trans. Antennas Propag. **16**, 1048 (1968).
[9] T. Cwik, IEEE Trans. Antennas Propag. **40**, 1496 (1992).
[10] A. Poggio and E. Miller, *Computer Techniques for Electromagnetics* (Pergamon, London, 1973), Chap. 4.
[11] A. Glisson and D. Wilton, IEEE Trans. Antennas Propag. **28**, 593 (1980).
[12] A. Borgioli and T. Cwik, IEEE AP-S International Symposium and URSI Radio Sci. Meet. Atlanta, Georgia, 1998.
[13] *Properties of Aluminum Gallium Arsenide*, edited by S. Adochi (INSPEC, IEE, London, 1993), Chap. 5.

AGARD-CP-219

AGARD-CP-219

ADVISORY GROUP FOR AEROSPACE RESEARCH & DEVELOPMENT

7 RUE ANCELLE 92200 NEUILLY SUR SEINE FRANCE

AGARD CONFERENCE PROCEEDINGS No. 219

Optical Fibres, Integrated Optics and Their Military Applications

Edited by
Dr.H.Hodara

Published May 1977
Hanford House, London.

NORTH ATLANTIC TREATY ORGANIZATION

HOW DOES ONE INDUCE LEAKAGE IN AN OPTICAL FIBER LINK?

C. Yeh and A. Johnston
Electrical Sciences and Engineering Dept.
University of California
Los Angeles, California 90024

SUMMARY

Three non-destructive methods to induce the leakage of optical signal from optical fibers will be discussed: (1) The index-matching-fluid method, (2) The temperature method, (3) The bending method. Experiments were performed for these cases. Results show that all three methods are effective in inducing leakage from plastic clad fibers while only the bending method is effective for glass-clad fibers.

INTRODUCTION

It has almost been taken for granted that, because optical fiber transmission line is free of RF leakage and cross-talk problems, it can be used readily as a secure information link. Furthermore, the intrinsic crack propagation characteristic of glass prevents cutting into the glass core to gain access to the optical signal without severing it. This talk is an attempt to discuss several non-destructive ways of tapping an optical fiber link. We are interpreting the word tapping in its broadest sense; i.e., any scheme that would induce the leakage of light from the fiber is interpreted as a viable means of tapping the fiber.

Two families of low-loss multi-mode fiber exist: (a) The plastic-clad fiber family, (b) The glass-clad fiber family which includes the graded-index fibers. The plastic-clad fiber usually consist of a glass core and a plastic sheath as its cladding. The cladding of a small section (say, 5 cm.) of a fiber link (say, longer than 20 m.) can be easily stripped off without noticably affecting the strenght of the transmitted signal. On the other hand the cladding of the glass-clad fiber is usually intimately bonded to the glass core in such a way that removing the glass cladding will usually damage the core structure. Different ways must therefore be devised to tap these fibers. Three non-destructive methods may be used to induce the leakage of optical signal from these fibers: (1) The index-matching-fluid method, (2) The temperature method, (3) The bending method.

THE INDEX-MATCHING-FLUID METHOD

The index-matching-fluid method is particularly suited for tapping plastic clad fiber. After stripping off the plastic cladding the bare glass-core section is immersed in an index-matching fluid whose index of refraction may be so chosen that only a small controlled amount of light is allowed to leak out of the core of the multi-mode fiber.

According to the theory of guided waves along optical fibers,[1] the number of modes N that a fiber may carry can be estimated from a very simple relation:

$$N = \tfrac{1}{2}\left(\frac{\omega a}{c}\right)^2 n_1^2\left(1 - \frac{n_2^2}{n_1^2}\right) \qquad (1)$$

$$\left(n_1 - n_2\right)/n_2 \ll 1$$

where $\frac{\omega}{c} = k$ is the free-space wave number, a is the radius of the fiber core and n_1 and n_2 are respectively the indices of refraction for the core and for cladding which is the index-matching-fluid. Therefore the number of modes that can be supported by the fiber guide is directly proportional to the difference of the square of indices of refraction. In other words, when the cladding index approaches the core index, the number of propagating modes approaches a small value. By finely adjusting the value for $(n_1 - n_2)$, the number of propagating modes may be adjusted. If the modes excitation condition is such that all modes are equally excited, one may derive an approximate expression for the fractional power carried in the cladding region. It is

$$\frac{\text{power transmitted in the cladding}}{\text{total transmitted power}} \approx \frac{8}{3V} \qquad (2)$$

$$\text{with} \quad V = \left(\frac{\omega a}{c}\right) n_1 \left(1 - \frac{n_2^2}{n_1^2}\right)^{1/2} \quad \text{and} \quad (n_1 - n_2)/n_2 \ll 1.$$

By adjusting the cladding index, (in the present case, it is the index-matching-fluid) one may control the amount of power carried in the cladding region which can then be tapped off very easily. Relations (1) and (2) have been plotted in Fig. 1.

To verify the above observation, measurements were carried out. A schematic diagram of the experimental setup is shown in Fig. 2. Results of our measurements are shown in Fig. 1. This experiment shows that one may induce a controlled amount of leakage for the plastic-clad fiber link with this method.

For the glass-clad fiber, the index of the glass core or that of the glass cladding varies as a function of temperature in a similar fashion such that the NA of the fiber remains unchanged.[2] Hence, no leakage of light occurs as the temperature changes. As far as the plastic clad fiber is concerned, since the index of the glass core and that of the plastic cladding vary as a function of temperature in a dissimilar manner, the NA of the fiber is significantly changed as the temperature varies. Hence, leakage of light occurs as the temperature changes.

To confirm the above observation, measurements were carried out. The schematic diagram of the experimental setup is similar to that shown in Fig. 2, except the segment where we placed the index-matching fluid is replaced by a dewar which provides a controlled temperature enviornment for a length of fiber. To make the desired measurements it is not necessary to strip the cladding from the fiber as was done for the previous method. Detailed experimental data are shown in Fig. 3 for the plastic-clad fiber and for the glass-clad fiber. As expected no leakage of light is detectable for the glass-clad fiber as the temperature changes. On the other hand, very significant leakage of light occurs for the plastic-clad fiber when the temperature varies through a critical region. Apparently a phase transition for the plastic cladding occurs in this temperature range so that the index for the plastic changes significantly.

So far, it appears that glass-clad fiber is immune to our attempts to induce leakage. It will be shown, however, that the next attempt will be successful.

THE BENDING METHOD

By bending a fiber, radiation or cladding modes may be induced due to the sharp curvature of the fiber. According to a simplified analysis,[3] the fraction of modes lost in a bent step-index type fiber is $2a/(R\delta)$ where a is the core radius, $\delta = 1 - (n_2/n_1)^2$, and R is the curvature radius. The curvature loss as a function of the curvature for the step-index fiber is sketched in Fig. 4. It can be seen that very large curvature loss can be induced.

Measurements were performed using the basic experimental set-up as sketched in Fig. 2. The fiber under examination was tightly wound tem times around a post with predetermined radius of curvature. By using posts of different radius, we can adjust the bending radius of the fiber. Results of our measurements are shown in Fig. 5. It can be seen that, as expected, leakage from the fiber line is very significant when the radius of curvature reaches a certain critical value. One notes that the bending method to induce leakage is equally effective when applied to the glass-clad fiber as to the plastic-clad fiber.

CONCLUSIONS

The fact that the above methods provide "non-destructive" and "recoverable" ways of inducing leakage in an optical fiber is worth noting. We have also shown experimentally that when the distrubances caused by the above schemes were removed, the transmission characteristics of the fiber link returned to normal. The security implication of the above experiments is clear. It appears that the only way to insure that no leakage of information had occured is to monitor continuously the power level of the received signal. Techniques to achieve this will not be discussed here.

REFERENCES

1. C. Yeh, "Advances in Communication Through Light Fibers in Advances in Communication Systems," Vol. 4, Theory and Applications, Academic Press, New York (1975); D. Gloge, Appl. Opt. 10, 2253 (1971).

2. C. J. Parker and W. A. Popov, Appl. Opt. 10, 2137 (1971).

3. D. Gloge, Appl. Opt. 11, 2506 (1972).

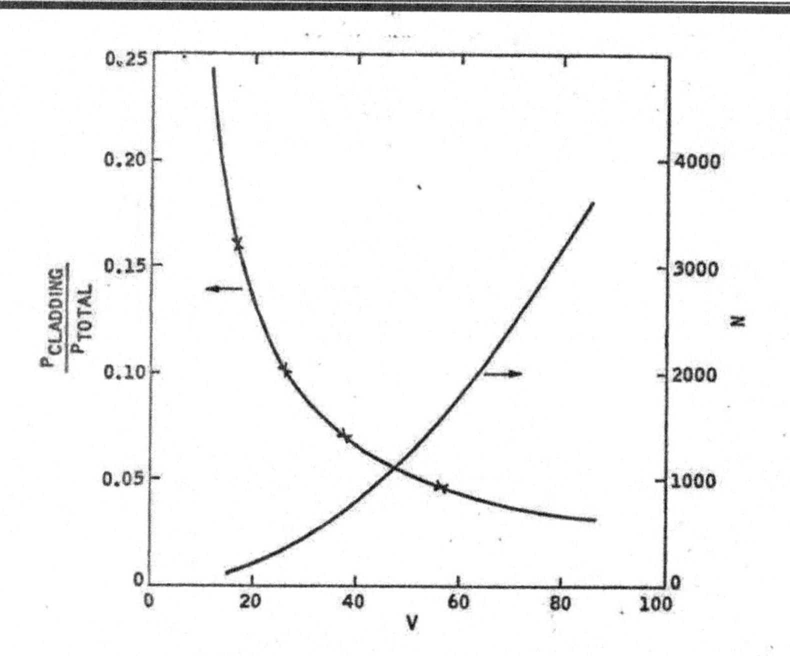

Figure 1 Normalized power in the cladding and the number of modes (N) as a function of $V = k_1a(1-(n_2/n_1)^2)^{1/2}$. The crosses are experimental points. The arrows indicate the ordinates for the curves.

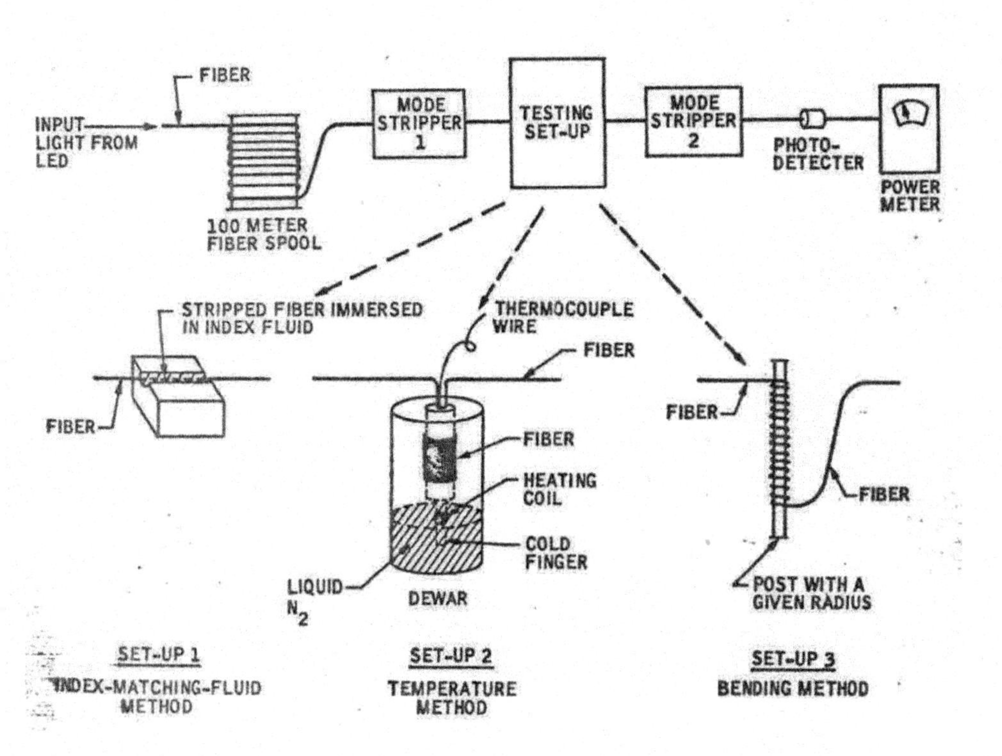

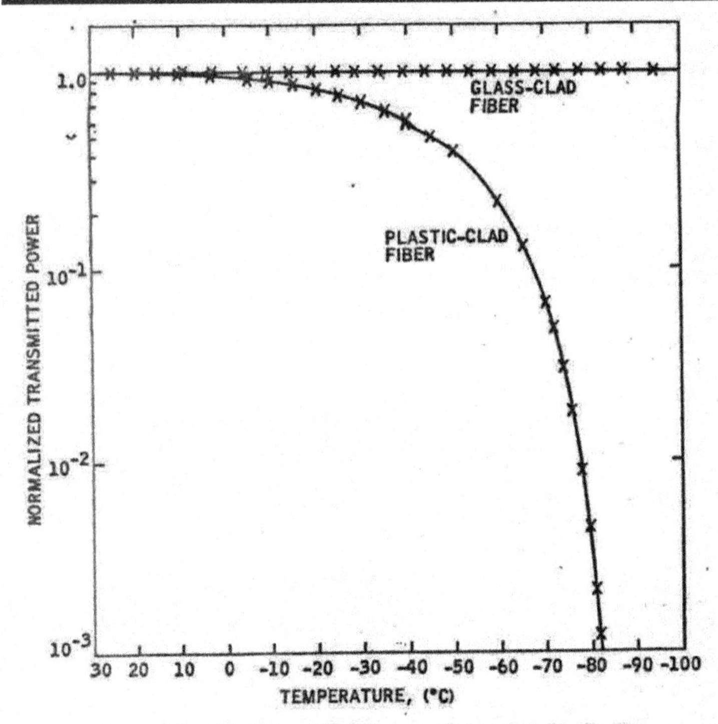

Figure 3 Normalized Transmitted Power vs. Temperature for Plastic-Clad Fiber and Glass-Clad Fiber.

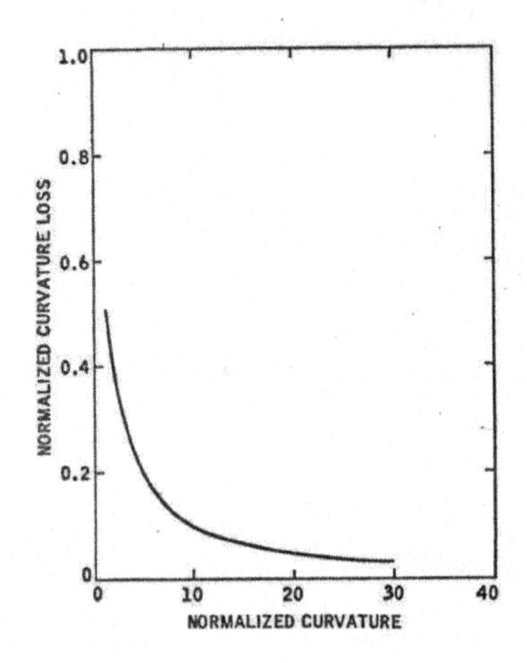

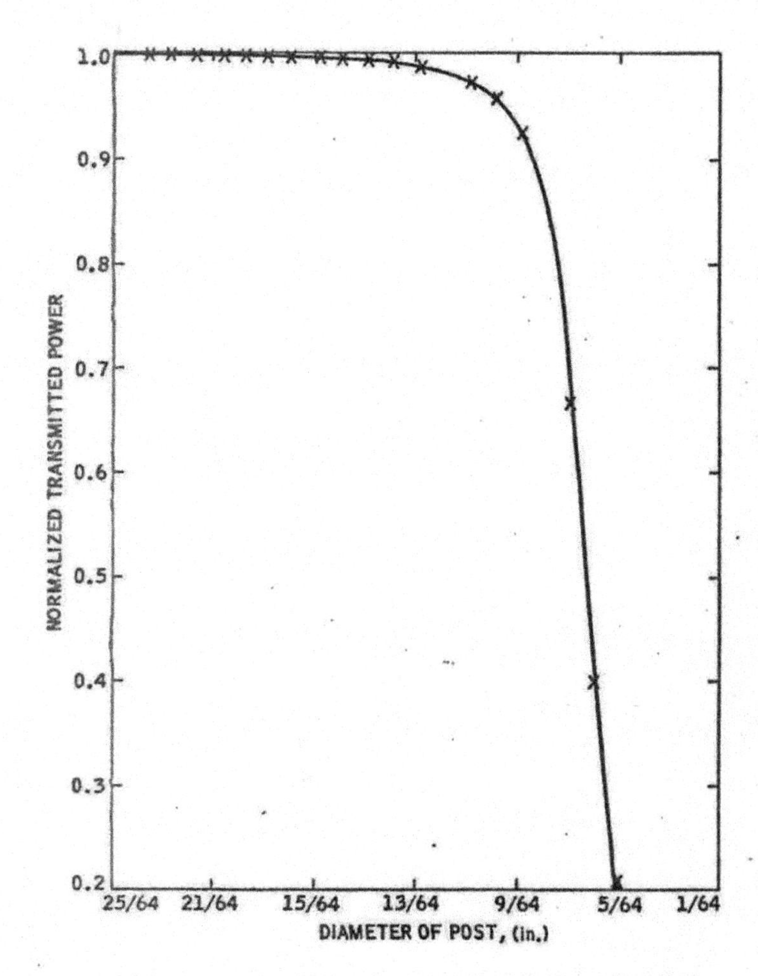

Figure 5 Normalized Transmitted Power vs. Curvature for Glass-Clad Fiber. A length of 1 meter is wound around the post.

5

ANTENNAS, RADIATION, AND PROPAGATION

By the 1960s, most of the theories on fundamental radiators, such as dipole, linear, spiral, horn, dish, aperture, and array antennas in air, were well established. Still unknown were the effects on radiated fields when dielectrics or plasma sheath surrounded the antenna; therefore, we concentrated our effort in solving canonical coated-antenna problems.

The appearance of lasers meant that we should study the propagation characteristics of laser beams in the atmosphere, which is a turbulent as well as nonlinear medium. The possibility of using optical laser beams to communicate at high data rate with spacecraft or with other planets prompted us to investigate the problem.

Abstracts of the papers in this chapter are provided below.

5.1 Antennas and radiators

5.1.1 On the dielectric coated prolate spheroidal antenna. (Paper 5-1-1)

The particular problem studied in this paper is the unsymmetrically fed dielectric (or plasma) coated linear (or prolate spheroidal) antenna. The exact analytical solution is obtained. The presence of this dielectric sheath, which may introduce multiple reflections within the sheath, is expected to modify quite considerably the radiation characteristics and the distribution of the current densities on the conducting surface. Because of the involved boundary conditions for this problem, it is almost impossible to use the integral equation method or the biconical antenna approach. Fortunately, the mathematical difficulties can be overcome by using the Fourier-Láme method. It is noted that the presence of the dielectric coating introduces further complications in matching the boundary conditions, which are not present in the uncoated case, since the angular spheroidal functions are not only functions of η, the angular coordinates, but also of the electrical properties of the medium in which they apply. This difficulty may be resolved by means of the orthogonality relations of the spheroidal functions, however. It is for this reason that the mode concept as introduced by Chu and Stratton in discussing the uncoated spheroidal antenna must be modified for the present problem. Also noted are the facts that coated prolate spheroids are good approximations to the bodies of missiles reentering the earth's atmosphere and so have a physical counterpart in their own right, and that the problem of end effects, which are always present when other configurations are used to approximate this problem, is nonexistent.

5.1.2 Radiation from an axially slotted cylinder coated with an inhomogeneous dielectric sheath. (Paper 5-1-2)

The expressions for the radiated fields of an axially slotted infinite cylinder coated with a radially inhomogeneous dielectric sheath are obtained. The two-dimensional vector wave equation in this inhomogeneous medium is separated using the vector wave function of Hansen and Stratton. The problem of finding the solution of Maxwell's equations in such an inhomogeneous medium reduces to the solution of two ordinary differential equations. Numerical computations are carried out for a specific variation of the dielectric sheath, i.e., $\epsilon(r) = \epsilon_0 \alpha / k_0 r$, where α is a constant, $k_0^2 = \omega^2 \mu \epsilon_0$, and ϵ_0 and μ are respectively the free-space dielectric constant and free-space permeability. Radiation patterns are plotted for various values of $k_0 a$, $k_0 b$, and α, where a and b are respectively the inner and outer radius of the sheath. These patterns are also compared with the homogeneous sheath problem; the dielectric constant of the homogeneous sheath is taken to be the average value of the homogeneous sheath. Results are discussed.

5.1.3 Dyadic Green's function for a radially inhomogeneous spherical medium. (Paper 5-1-3)

The general expression for the dyadic Green's function in a radially inhomogeneous dielectric medium is obtained. It is then applied to the problem of radiation from an electric dipole in such a medium. Possible applications to electromagnetic scattering problems and to elementary particle scattering problems are noted.

5.1.4 External field produced by an arbitrary slot on a ribbon coated with a penetrable sheath. (Paper 5-1-4)

Expressions are derived for the external electromagnetic field produced by an arbitrary slot on the conducting surface of a dielectric-coated ribbon. It is assumed that the electric field tangential to the slot is a prescribed function. Boundary conditions are satisfied using orthogonality characteristics of Mathieu functions. The integrals in the formal solutions are evaluated by the saddle point method for the far-zone fields. These far-zone field solutions are then specialized to thin circumferential and axial slots. Illustrative numerical computations for the far-zone field of a thin axial slot are carried out.

5.1.5 An application of Sommerfeld's complex-order wavefunctions to an antenna problem. (Paper 5-1-5)

Using the orthogonality relations of Sommerfeld's complex-order wavefunctions, the exact solution for the problem of electromagnetic radiation from a circularly symmetric slot on the conducting surface of a dielectric-coated cone is obtained. The results are valid for the near-zone as well as for the far-zone region, and they are applicable for arbitrary-angle cones. It is noted that the technique used to solve this problem may be applied to similar types of problems involving conical structure, such as the diffraction of waves by a dielectric-coated, spherically tipped cone.

5.1.6 Electromagnetic radiation from an arbitrary slot on a conducting cylinder coated with a uniform cold plasma sheath with an axial static magnetic field. (Paper 5-1-6)

The expressions for the radiated fields of an arbitrary slot on a conducting cylinder coated with a cold plasma sheath with an axial static magnetic field are obtained. It is well known from the plane-wave analysis that the presence of a static magnetic field has a considerable effect on the transmission characteristics of the signal through the plasma sheath. The principal concern of this paper is to consider the effects of a static magnetic field on the radiation pattern of a plasma-coated slot antenna. Extensive numerical computations are carried out for the special case of an infinite axial slot. It is found that if the plasma frequency is moderate, a small axial static magnetic field could change significantly the radiation patterns of the slotted antenna. Furthermore, owing to the anisotropy of the plasma medium, the radiation patterns are asymmetric.

5.1.7 Radiation through an inhomogeneous, magnetized plasma sheath. (Paper 5-1-7)

In this communication, we comment on the predicted radiation characteristics for an axially slotted cylinder that is covered with a magnetized, inhomogeneous plasma sheath. Parabolic electron density distributions, ranging from nearly constant value that drops abruptly to zero at the edge to a steadily decreasing function that smoothly becomes zero at the edge have been assumed. Radiation patterns from such slotted, sheath-clad cylinders were computed for a wide range of principal parameters: frequency, magnetic field intensity, plasma density, and profile shape. In general, at low frequencies (sheath thickness much smaller than a wavelength), the shape of the radiation pattern was not significantly affected by the presence of the sheath. Because the sheath became increasingly transparent at high frequencies, the patterns were also relatively independent of the sheath parameters and approached the patterns from a slotted cylinder without coating. Significant pattern changes were noticed primarily at intermediate frequencies, between the "high" and "low" frequency ranges. For example, magnetic effects caused the patterns to be asymmetric; however, it was discovered that profile shape, the parameter of most interest to us in this correspondence, exerted only a very minor influence on the pattern shapes. Furthermore, such a profile effect was noticeable only under relatively severe conditions within the sheath, e.g., thick sheaths, very dense plasma, and resonances.

5.2 Propagation

5.2.1 Enhanced focal-plane irradiance in the presence of thermal blooming. (Paper 5-2-1)

The use of multiple transmitter beams is shown to significantly increase the peak focal-plane irradiance that can be achieved in the presence of thermal blooming. Computer simulation studies of the beam propagation problem show an increase by more than a factor of 2 in the irradiance of a single beam and an increase by a factor of 9 when three coherent beams are focused on the same target spot. Preliminary experimental results with three mutually incoherent, nonoverlapping beams are in qualitative agreement with the computer simulation.

5.2.2 Propagation of laser beams having an on-axis null in the presence of thermal blooming. (Paper 5-2-2)

The propagation of focused beams having an on-axis null is considered in the presence of thermal blooming. Two cases are treated: (a) beam profiles that have an irradiance zero in the beam center at the focal plane as well as the transmitter aperture and (b) beam profiles that have an on-axis irradiance null only at the transmitter. It is demonstrated that none of the beam profiles considered in case (a) has a meaningful advantage over a Gaussian beam profile. Some of the case (b) profiles do produce a larger bloomed irradiance in the focal plane, particularly when the intensity distribution is very uniform in its nonzero regions. Addition of a simple central obscuration to a "filled" irradiance distribution is found to have no advantage, however, for the cases considered.

5.2.3 Earth-to-deep-space optical communications system with adaptive tilt and scintillation correction by use of near-Earth relay mirrors. (Paper 5-2-3)

Performance of an Earth-to-deep-space optical telecommunications system is degraded by distortion of the beam as it propagates through the turbulent atmosphere. Conventional approaches to correcting distortions, based on natural or artificial guide stars, have practical difficulties or are not adequate for correction of distortions, which are important for Earth-to-deep-space optical links. A beam-relay approach that overcomes these difficulties is discussed. A downward-directed laser near an orbiting relay mirror provides a reference for atmospheric correction. Adaptive optics at the ground station compensate the uplink beam so that after it passes through the atmosphere uplink, propagation effects are removed. The orbiting mirror then directs the corrected beam to the distant spacecraft.

ON THE DIELECTRIC COATED PROLATE SPHEROIDAL ANTENNA

By Cavour W. H. Yeh

1. Introduction. One of the most important and most basic type of radiating structures is the linear antenna. There are at least three methods for attacking this problem. They are the integral equation method, the biconical antenna method, and the Fourier-Láme method. The integral equation technique was first developed by Hallén[1] who used it to analyze the problem of thin antennas. More recently the problem of determining the current and impedance of an unsymmetrically driven cylinder antenna was formulated by King,[2] who made use of an integral equation which he solved by the method of successive approximations. In the middle 1930's Schelkunoff[3] proposed the use of a thin biconical antenna model to analyze the thin layer antenna. The biconical model has the advantages that the problem may be formulated in spherical coordinates, and that results are given in terms of well-known spherical wave functions. In the Fourier-Láme method the wave equation is separated in coordinates of the prolate spheroid, and the solutions appear as an infinite series of spheroidal functions. Since a very thin spheroid matches almost exactly the shape of a linear antenna, and since the spheroidal system (with the spherical as its special case) is the only "separable" system* that contains coordinate surface of finite dimensions, the spheroidal antenna has long received considerable attention. However, due to the complexity of spheroidal functions, not until 1941 did Chu and Stratton[4] succeed in obtaining and interpreting the complete solution in the case of forced oscillations. The merits and the short comings of these three methods of approach are discussed very thoroughly in a book by Schelkunoff.[3] The principal objection in using the prolate spheroidal functions was that these functions had not been tabulated and in order to find numerical values, very laborious computations had to be carried out. However, since the publication of rather elaborate tables of spheroidal functions by Stratton, Morse, Chu, Little and Corbató,[5] this objection no longer exists. It is, therefore, feasible and meaningful to express solutions in terms of spheroidal functions.

The particular problem studied in this paper is the unsymmetrically fed, dielectric (or plasma) coated linear (or prolate spheroidal) antenna. The exact analytic solution is obtained. The presence of this dielectric sheath which may introduce multiple reflections within the sheath, is expected to modify quite considerably the radiation characteristics and the distribution of the current densities on the conducting surface. Due to the involved boundary conditions for this problem, it is almost impossible to use the integral equation method or the biconical antenna approach. Fortunately, the mathematical difficulties can be overcome by using the Fourier-Láme method. It is noted that the presence

* "Separable" system means that in this coordinate system the wave equations are separable.

of the dielectric coating introduces further complications in matching the boundary conditions, which are not present in the uncoated case, since the angular spheroidal functions are not only functions of η, the angular coordinates, but also of the electrical properties of the medium in which they apply. However, this difficulty may be resolved by means of the orthogonality relations of the spheroidal functions. It is for this reason that the mode concept as introduced by Chu and Stratton[4] in discussing the uncoated spheroidal antenna, must be modified for the present problem. Also noted are the facts that coated prolate spheroids are good approximations to the bodies of missiles reentering the earth's atmosphere and so have a physical counterpart in their own right, and that the problem of end effects which is always present when other configurations are used to approximate this problem is non-existent.

2. Formulation of the Problem. To analyze this problem, the prolate spheroidal coordinates (ξ, η, ϕ), as shown in Fig. 1, are introduced. In terms of the spherical coordinates (R, θ, ϕ), the spheroidal coordinates are defined by the following relations:

$$
\begin{aligned}
&R = q(\xi^2 + \eta^2 - 1)^{\frac{1}{2}}, \qquad \cos\theta = \xi\eta/(\xi^2 + \eta^2 - 1)^{\frac{1}{2}} \\
&\phi = \phi, \qquad (\xi \geqq 1, -1 \leqq \eta \leqq 1, 0 \leqq \phi \leqq 2\pi)
\end{aligned}
\tag{1}
$$

where q is the semi-focal length of the spheroid. The contour surface of constant ξ are confocal spheroids, and those of constant η are confocal hyperboloids. The outer boundary of the dielectric coated conducting antenna is assumed to coincide with one of the confocal spheroids with $\xi = \xi_1$; the inner boundary is assumed to coincide with one of the confocal spheroids with $\xi = \xi_0$. It will be assumed that this dielectric coated spheroid is embedded in a homogeneous perfect dielectric medium (ϵ_0, μ; $\sigma_0 = 0$), and that the applied electric field intensity is circularly symmetric about the major axis and linearly polarized in a direction parallel to this axis. The dielectric coating has a permittivity of ϵ_1, a per-

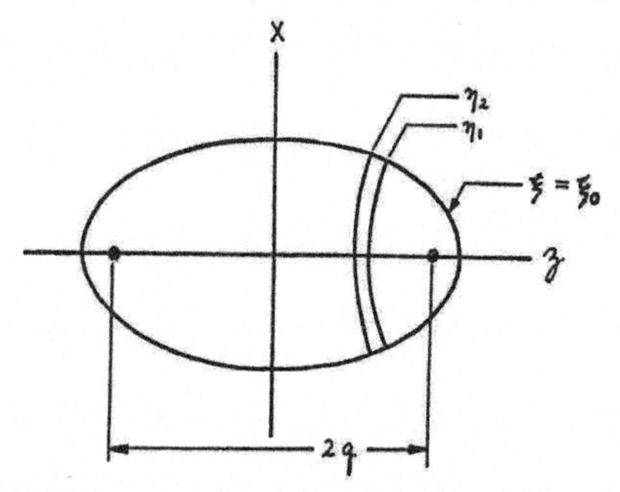

FIG. 1. The prolate spheroidal coordinate system.

meability μ, and a conductivity of zero. A possible solution of the wave equation for a source-free homogeneous perfect dielectric medium is then $q[(\xi^2 - 1)(1 - \eta^2)]^{\frac{1}{2}} R(\xi)\, \Theta(\eta) e^{-i\omega t}$, where R and Θ satisfy the differential equation

$$(\xi^2 - 1)\frac{d^2R}{d\xi^2} + 4\xi\frac{dR}{d\xi} - (c - q^2k^2\xi^2)R = 0 \tag{2}$$

$$(1 - \eta^2)\frac{d^2\Theta}{d\eta^2} - 4\eta\frac{d\Theta}{d\eta} + (c - q^2k^2\eta^2)\Theta = 0 \tag{3}$$

in which c is the separation constant and $k^2 = \omega^2\mu\epsilon$. It appears that both R and Θ satisfy the same differential equation

$$(1 - z^2)\frac{d^2W}{dz^2} - 4z\frac{dW}{dz} + (c - q^2k^2z^2)W = 0. \tag{4}$$

The solutions of this equation have been discussed quite thoroughly by Chu and Stratton.[4] These solutions are called the spheroidal wave functions.

It is expected that all field components remain finite at the poles $\eta = \pm 1$; thus, the angular function will be represented by the angular spheroidal function of the first kind $\mathrm{Se}_{1,n}^1(qk, \eta)$ with the sequence in n according to increasing values of c. (This periodic spheroidal function of the first kind exists only for certain characteristic values of c.) The solutions of (2) corresponding to the angular functions $\mathrm{Se}_{1,n}^1(qk, \eta)$ having the same characteristic values of c, are the radial spheroidal functions of the first kind $\mathrm{Re}_{1,n}^1(qk, \xi)$ and the radial spheroidal functions of the second kind $\mathrm{Re}_{1,n}^2(qk, \xi)$.

The proper choice of these radial functions to represent the electromagnetic fields depends upon the boundary conditions. All field components must be finite in all regions (i.e. the region within the dielectric shell and the region outside the dielectric coated spheroid). In addition all field components for the radiated wave must satisfy Sommerfeld's radiation condition at infinity. Consequently the appropriate solutions of the wave equation for the region within the dielectric shell are

$$q[(\xi^2 - 1)(1 - \eta^2)]^{\frac{1}{2}} \left\{ \begin{matrix} \mathrm{Re}_{1,n}^1(qk_1, \xi) \\ \mathrm{Re}_{1,n}^2(qk_1, \xi) \end{matrix} \right\} \mathrm{Se}_{1,n}^1(qk_1, \eta), \tag{5}$$

and those for the radiated wave are

$$q[(\xi^2 - 1)(1 - \eta^2)]^{\frac{1}{2}} \left\{ \begin{matrix} \mathrm{Re}_{1,n}^3(qk_0, \xi) \\ \mathrm{Re}_{1,n}^4(qk_0, \xi) \end{matrix} \right\} \mathrm{Se}_{1,n}^1(qk_0, \eta), \tag{6}$$

where

$$\mathrm{Re}_{1,n}^3(qk_0, \xi) = \mathrm{Re}_{1,n}^1(qk_0, \xi) + i\,\mathrm{Re}_{1,n}^2(qk_0, \xi) \tag{7}$$

$$\mathrm{Re}_{1,n}^4(qk_0, \xi) = \mathrm{Re}_{1,n}^1(qk_0, \xi) - i\,\mathrm{Re}_{1,n}^2(qk_0, \xi) \tag{8}$$

and $k_0^2 = \omega^2\mu\epsilon_0$, $k_1^2 = \omega^2\mu\epsilon_1$. As $qk_0\xi \to \infty$ we find that

$$\mathrm{Re}^{3}_{1,n}(qk_0\,,\xi) \to -\frac{1}{(qk_0\,\xi)^2}\,e^{-i(qk_0\xi-\frac{1}{2}n\pi)} \tag{9}$$

$$\mathrm{Re}^{4}_{1,n}(qk_0\,,\xi) \to -\frac{1}{(qk_0\,\xi)^2}\,e^{i(qk_0\xi-\frac{1}{2}n\pi)}. \tag{10}$$

The complete steady-state field expressions may be obtained from Maxwell's equations: For the region within the dielectric sheath

$$H^i_\phi = q[(\xi^2-1)(1-\eta^2)]^{\frac{1}{2}}e^{-i\omega t}\sum_{n=0}^{\infty}[A_n\,\mathrm{Re}^{1*}_n(\xi)+B_n\,\mathrm{Re}^{2*}_n(\xi)]\mathrm{Se}^{1*}_n(\eta) \tag{11}$$

$$E^i_\xi = \frac{i}{\omega\epsilon_1}\left(\frac{\xi^2-1}{\xi^2-\eta^2}\right)^{\frac{1}{2}}e^{-i\omega t} \cdot\sum_{n=0}^{\infty}[A_n\,\mathrm{Re}^{1*}_n(\xi)+B_n\,\mathrm{Re}^{2*}_n(\xi)]\,\frac{d}{d\eta}\,[(1-\eta^2)\mathrm{Se}^{1*}_n(\eta)] \tag{12}$$

$$E^i_\eta = \frac{-i}{\omega\epsilon_1}\left(\frac{1-\eta^2}{\xi^2-\eta^2}\right)e^{-i\omega t}\sum_{n=0}^{\infty}\left\{A_n\,\frac{d}{d\xi}\,[(\xi^2-1)\mathrm{Re}^{1*}_n(\xi)] + B_n\,\frac{d}{d\xi}\,[(\xi^2-1)\,\mathrm{Re}^{2*}_n(\xi)]\right\}\mathrm{Se}^{1*}_n(\eta), \tag{13}$$

and for the regio n outside the dielectric coated spheroid

$$H^r_\phi = q[(\xi^2-1)(1-\eta^2)]^{\frac{1}{2}}e^{-i\omega t}\sum_{n=0}^{\infty}[C_n\,\mathrm{Re}^3_n(\xi)]\,\mathrm{Se}^1_n(\eta) \tag{14}$$

$$E^r_\xi = \frac{i}{\omega\epsilon_0}\left(\frac{\xi^2-1}{\xi^2-\eta^2}\right)^{\frac{1}{2}}e^{-i\omega t}\sum_{n=0}^{\infty}[C_n\,\mathrm{Re}^3_n(\xi)]\,\frac{d}{d\eta}\,[(1-\eta^2)\mathrm{Se}^1_n(\eta)] \tag{15}$$

$$E^r_\eta = \frac{-i}{\omega\epsilon_0}\left(\frac{1-\eta^2}{\xi^2-\eta^2}\right)^{\frac{1}{2}}e^{-i\omega t}\sum_{n=0}^{\infty}\left\{C_n\,\frac{d}{d\xi}\,[(\xi^2-1)\,\mathrm{Re}^3_n(\xi)]\right\}\mathrm{Se}^1_n(\eta). \tag{16}$$

A_n, B_n, and C_n are arbitrary constants which are to be determined by the boundary conditions and by the distribution of applied e.m.f. The following abbreviations have been used:

$$\mathrm{Re}^{1*}_n(\xi) = \mathrm{Re}^1_{1,n}(qk_1\,,\xi), \qquad \mathrm{Re}^{2*}_n(\xi) = \mathrm{Re}^2_{1,n}(qk_1\,,\xi),$$

$$\mathrm{Re}^3_n(\xi) = \mathrm{Re}^3_{1,n}(qk_0\,,\xi),$$

$$\mathrm{Se}^{1*}_n(\eta) = \mathrm{Se}^1_{1,n}(qk_1\,,\eta), \qquad \mathrm{Se}^1_n(\eta) = \mathrm{Se}^1_{1,n}(qk_0\,,\eta).$$

3. The Mathematical Solution. The boundary conditions require the continuity of the tangential electric and magnetic fields at the boundary surface, $\xi = \xi_1$. On the conducting surface, $\xi = \xi_0$, the tangential electric field must be zero everywhere except across the gap where it is equal to the applied field. Let the applied field be defined by

$$E^{\text{app.}}_\eta = E_0G(\eta)e^{-i\omega t} \tag{17}$$

where $G(\eta)$ is defined as follows:

$$G(\eta) = \begin{cases} 1 & \text{for} \quad \eta_1 < \eta < \eta_2 \\ 0 & \text{for} \quad -1 < \eta < \eta_2, \quad \text{and} \quad \eta_1 < \eta < 1. \end{cases} \tag{18}$$

$|\eta_1 - \eta_2|$ is the gap width (See Fig. 1). Expanding the applied field in terms of spheroidal functions gives

$$E_\eta^{\text{app.}} = + \frac{1}{i\omega\epsilon_1}\left[\frac{1-\eta^2}{\xi_0^2 - \eta^2}\right]^{\frac{1}{2}} \sum_{l=0}^{\infty} P_l \operatorname{Se}_l^{1*}(\eta) e^{-i\omega t} \tag{19}$$

where

$$P_l = + \frac{i\omega\epsilon_1}{N_l} \int_{\eta_2}^{\eta_1} E_0[(\xi_0^2 - \eta^2)(1-\eta^2)]^{\frac{1}{2}} \operatorname{Se}_l^{1*}(\eta)\, d\eta \tag{20}$$

in which the normalizing factor is

$$N_l = \int_{-1}^{1} (1-\eta^2)[\operatorname{Se}_l^{1*}(\eta)]^2\, d\eta. \tag{21}$$

Expression (19) represents $E_\eta^{\text{app.}}$ for all values of η between -1 and 1.

Upon matching the tangential magnetic and electric fields at $\xi = \xi_1$, one obtains

$$\begin{aligned} \textstyle\sum_{n=0}^{\infty} [A_n \operatorname{Re}_n^{1*}(\xi_1) + B_n \operatorname{Re}_n^{2*}(\xi_1)] \operatorname{Se}_n^{1*}(\eta) \\ = \textstyle\sum_{n=0}^{\infty} [C_n \operatorname{Re}_n^{3}(\xi_1)] \operatorname{Se}_n^{1}(\eta), \end{aligned} \tag{22}$$

$$\begin{aligned} \frac{\epsilon_0}{\epsilon_1}\sum_{n=0}^{\infty}\left\{A_n \frac{d}{d\xi_1}[(\xi_1^2-1)\operatorname{Re}_n^{1*}(\xi_1)] + B_n \frac{d}{d\xi_1}[(\xi_1^2-1)\operatorname{Re}_n^{2*}(\xi_1)]\right\}\operatorname{Se}_n^{1*}(\eta) \\ = \sum_{n=0}^{\infty} C_n \frac{d}{d\xi_1}[(\xi_1^2-1)\operatorname{Re}_n^{3}(\xi_1)]\operatorname{Se}_n^{1}(\eta). \end{aligned} \tag{23}$$

It is noted that in contrast with the spherical and circular cylindrical case the angular functions in the spheroidal case are functions not only of the angular component but also of the characteristics of the medium. Consequently, the summation signs and the angular functions in the above equations may not be omitted. However, it will be shown that this difficulty may be overcome by the use of the orthogonality properties of the angular spheroidal function. Substituting the expansion

$$\operatorname{Se}_n^{1*}(\eta) = \textstyle\sum_{m=0}^{\infty} \alpha_{n,m} \operatorname{Se}_m^{1}(\eta) \tag{24}$$

into equations (22) and (23), and applying the orthogonality relations of the spheroidal function, leads to the following expressions:

$$\textstyle\sum_{n=0}^{\infty} (A_n a_n + B_n b_n)\alpha_{n,m} = C_m c_m \tag{25}$$

$$\frac{\epsilon_0}{\epsilon_1}\sum_{n=0}^{\infty} (A_n a_n' + B_n b_n')\alpha_{n,m} = C_m c_m' \qquad \begin{pmatrix} m = 0, 2, 4, \cdots \\ \text{or } m = 1, 3, 5 \cdots \end{pmatrix} \tag{26}$$

where the abbreviations

$$a_n = \mathrm{Re}_n^{1*}(\xi_1), \qquad a_n' = \frac{d}{d\xi_1}[(\xi_1^2 - 1)\,\mathrm{Re}_n^{1*}(\xi_1)],$$

$$b_n = \mathrm{Re}_n^{2*}(\xi_1), \qquad b_n' = \frac{d}{d\xi_1}[(\xi_1^2 - 1)\,\mathrm{Re}_n^{2*}(\xi_1)], \tag{27}$$

$$c_n = \mathrm{Re}_n^{3}(\xi_1), \qquad c_n' = \frac{d}{d\xi_1}[(\xi_1^2 - 1)\,\mathrm{Re}_n^{3}(\xi_1)],$$

have been used. $\alpha_{n,m}$ is given in the appendix. Expressing B_n in terms of A_n gives (in matrix notation)

$$B_n = R_{n,m}^{-1} D_m \tag{28}$$

where $R_{n,m}^{-1}$ is the inverse of the matrix

$$[R_{m,n}] = \left[\left(\frac{\epsilon_0}{\epsilon_1} b_n' c_m - b_n c_m'\right)\alpha_{n,m}\right],$$

and D_m is a column matrix

$$\left[\sum_{r=0}^{\infty} A_r \left(a_r c_m' - \frac{\epsilon_0}{\epsilon_1} a_r' c_m\right)\alpha_{r,m}\right].$$

Equation (28) can also be written in the following form:

$$B_n = \textstyle\sum_{r=0}^{\infty} g_{n,r} A_r \tag{29}$$

where $g_{n,r}$ are obtained by the use of equation (28).

At the surface of the conducting spheroid, $\xi = \xi_0$, expressions (13) and (19) must be identically equal for all values of η between -1 and 1. Therefore,

$$A_n \frac{d}{d\xi_0}[(\xi_0^2 - 1)\,\mathrm{Re}_n^{1*}(\xi_0)] + B_n \frac{d}{d\xi_0}[(\xi_0^2 - 1)\,\mathrm{Re}_n^{2*}(\xi_0)] = P_n \tag{30}$$

where P_n is given by equation (20). Making the identification

$$d_n = \frac{d}{d\xi_0}[(\xi_0^2 - 1)\,\mathrm{Re}_n^{1*}(\xi_0)], \quad f_n = \frac{d}{d\xi_0}[(\xi_0^2 - 1)\,\mathrm{Re}_n^{2*}(\xi_0)] \tag{31}$$

and substituting equation (29) into (30), one finds

$$A_n d_n + f_n \textstyle\sum_{r=0}^{\infty} g_{n,r} A_r = P_n . \tag{32}$$

Solving for A_n gives

$$A_n = Q_{n,m}^{-1} P_m \tag{33}$$

where $Q_{n,m}^{-1}$ is the inverse of the matrix

$$[Q_{m,n}] = [(d_n \delta_{nm} + g_{n,m} f_n)], \tag{34}$$

and $\delta_{n,m}$ is the Kronecker delta which is equal to zero when $n \neq m$ and is equal to unity when $n = m$. P_m is a column matrix. With the knowledge of A_n and B_n, the coefficient C_n can easily be computed using either equation (25) or (26).

The fields in the dielectric shell and in the free-space are now completely determined. At large distances from the spheroid the asymptotic expression for $\mathrm{Re}_n^3(\xi)$ leads to

$$H_\phi^r = \left(\frac{\epsilon_0}{\mu}\right)^{\frac{1}{2}} E_\eta^r = -\left(\frac{1}{q\xi k_0^2}\right) e^{-i(\omega t - k_0 q\xi)} \sum_{n=0}^{\infty} (-i)^n C_n \,\mathrm{Se}_n^1(\eta). \qquad (35)$$

The distance from the origin to the point of observation approaches $q\xi$. The radiation pattern is determined by the sum of the angular functions.

In their discussion of the forced oscillations of a prolate spheroid, Chu and Stratton[4] introduced the concept of modes. They defined the modes as characterized by the index l which is the expansion index for the applied field. (See equation (19).) l could also very well be the expansion index for the radiated field in their problem, since there was no cross coupling between the expansion coefficients for the applied field and the expansion coefficients for the radiated field. However, the presence of a dielectric sheath around the spheroid introduces coupling between these coefficients. Each expansion coefficient for the radiated field is coupled with *all* expansion coefficients for the applied field.* (See equations (25), (29), and (33).) Thus one may state that all modes are coupled to each other through the dielectric sheath. It is therefore noted that for the present problem one must be more specific in using the term "modes." For example, one may speak of the "mode of oscillation" using n as the index of the mode, where n is the expansion index for the radiated field. (See equation (35).)

4. Conclusions. By the use of the orthogonality properties of the spheroidal functions, the exact solution for the electromagnetic field excited by a dielectric coated prolate spheroidal antenna is obtained. It can be seen that because of the involved coupling between the expansion coefficients for the radiated field and those for the applied field, the field expressions are considerably more complicated than those for the un-coated case. Consequently, the concept of "modes" as discussed by Chu and Stratton[4] must be modified, and must be more specific. Numerical computations are now underway using the available tables on spheroidal functions prepared by Stratton, Morse, Chu, Little, and Corbató,[5] and using the high-speed computer, the IBM 7090. A somewhat complete presentation of the numerical results is almost certainly complicated and lengthy and must therefore be reserved for future communications.

It is noted here that (25), (28), and (33) give (as they should) the correct values of the C_n, B_n, and A_n in the special case when the spheroid degenerates to a sphere. For this special case, $q \to 0$, $\xi \to \infty$ such that $q\xi \to r$, where r is the radius of a sphere. Thus, equations (25), (28), and (33) reduce respectively to

* An exception of this occurs when the spheroid degenerates to a sphere, or when the dielectric sheath disappears.

$$(A_n a_n + B_n b_n) = C_n c_n \tag{36}$$

$$B_n = \frac{A_n\left(a_n\,c_n' - \frac{\epsilon_0}{\epsilon_1}\,a_n'\,c_n\right)}{\left(\frac{\epsilon_0}{\epsilon_1}\,b_n'\,c_n - b_n\,c_n'\right)} = g_{n,n}\,A_n \tag{37}$$

and

$$A_n = \frac{P_n}{(d_n + h_{n,n}\,f_n)} \tag{38}$$

where

$$a_n = \frac{1}{k_1\,r_1}\,j_{n+1}(k_1\,r_1), \quad a_n' = \frac{1}{qk_1}\,[k_1\,r_1\,j_n(k_1\,r_1) - (n+1)j_{n+1}(k_1\,r_1)],$$

$$b_n = \frac{1}{k_1\,r_1}\,n_{n+1}(k_1\,r_1), \quad b_n' = \frac{1}{qk_1}\,[k_1\,r_1\,n_n(k_1\,r_1) - (n+1)n_{n+1}(k_1\,r_1)],$$

$$c_n = \frac{1}{k_0\,r_1}\,h_{n+1}^{(1)}(k_0\,r_1), \quad c_n' = \frac{1}{qk_0}\,[k_0\,r_1\,h_n^{(1)}(k_0\,r_1) - (n+1)h_{n+1}^{(1)}(k_0\,r_1)],$$

$$d_n = \frac{1}{qk_1}\,[k_1\,r_0\,j_n(k_1\,r_0) - (n+1)j_{n+1}(k_1\,r_0)],$$

$$f_n = \frac{1}{qk_1}\,[k_1\,r_0\,n_n(k_1\,r_0) - (n+1)n_{n+1}(k_1\,r_0)],$$

$$P_n = \frac{i\omega\epsilon_1}{N_n}\int_{\eta_2}^{\eta_1} E_0\,\xi_0\,P_{n+1}^1(\eta)\,d\eta, \quad N_n = \int_{-1}^{1}[P_{n+1}^1(\eta)]^2\,d\eta.$$

j_n, n_n, $h_n^{(1)}$, and P_{n+1}^1 are respectively the spherical Bessel function, the spherical Neumann function, the spherical Hankel function, and the associated Legendre polynomial. r_0 and r_1 are the radii of the inner and outer spheres. Expressions (36), (37), and (38) are then the solution for a coated spherical antenna.

In conclusion we shall remark that the technique used in obtaining the solution for this problem is also applicable to similar type of problems, such as the diffraction of waves by a thin penetrable (dielectric or plasma) spheroid, the diffraction of waves by a penetrable ribbon (elliptical cylinder) and other associated acoustical problems.

The author wishes to express his appreciation to Dr. H. H. Kuehl for reading this paper, and to thank Miss Linda LaBella for typing this manuscript. Support for this study by the Air Force Cambridge Research Laboratories is gratefully acknowledged.

5. Appendix. *Formula for* $\alpha_{n,m}$. Multiplying both sides of (24) by

$$\mathrm{Se}_m^1(\eta)(1-\eta^2),$$

integrating with respect to η from -1 to $+1$, and using the orthogonality relation for the spheroidal function, one obtains

$$\alpha_{n,m} = \frac{1}{M_m}\int_{-1}^{+1} \mathrm{Se}_n^{1*}(\eta)\,\mathrm{Se}_m^1(\eta)(1-\eta^2)\,d\eta$$

where

$$M_m = \int_{-1}^{+1} [\mathrm{Se}_m^1(\eta)]^2(1-\eta^2)\,d\eta$$

It is known that the angular spheroidal function may be expanded in terms of associated Legendre's functions:

$$(1-\eta^2)^{\frac{1}{2}}\,\mathrm{Se}_m^{1*}(\eta) = {\sum}'^{\infty}_{l=0,1,2\cdots} d_l^{m*} P_{l+1}^1(\eta)$$

$$(1-\eta^2)^{\frac{1}{2}}\,\mathrm{Se}_m^{1}(\eta) = {\sum}'^{\infty}_{l=0,1,2\cdots} d_l^{m} P_{l+1}^1(\eta)$$

where d_n^{m*} and d_n^m are the expansion coefficients[4] which have been tabulated,[5] and the prime over the summation sign indicates that odd or even integer values of l are to be taken according as m is odd or even. The integrals can now be evaluated with the help of the orthogonality characteristics of associated Legendre function:

$$\int_{-1}^{+1} [\mathrm{Se}_m^1(\eta)]^2(1-\eta^2) = 2\sum_{l=0,1,2}^{\infty}{}' [d_l^m]^2\frac{(l+2)!}{l!(2l+3)}$$

$$\int_{-1}^{+1} \mathrm{Se}_n^{1*}(\eta)\,\mathrm{Se}_m^1(\eta)\,(1-\eta^2)\,d\eta = 2\sum_{l=0,1,2}^{\infty}{}' d_l^{n*}\,d_l^m\frac{(l+2)!}{l!(2l+3)}.$$

Thus,

$$\alpha_{n,m} = \frac{\displaystyle\sum_{l=0,1,2}^{\infty}{}' d_l^{n*}\,d_l^m\frac{(l+2)!}{l!(2l+3)}}{\displaystyle\sum_{l=0,1,2}^{\infty}{}' [d_l^m]^2\frac{(l+2)!}{l!(2l+3)}}.$$

[It should be noted that throughout this paper the double subscripts (n, m), (m, n), (n, r), or (r, m) must *both* be even or odd (for example, let us consider the subscripts (n, m), if n is odd m must also be odd, when n is even, m must also be even). The above characteristics are due to the properties of the spheroidal functions.]

REFERENCES

1. E. Hallén, "Über die elektrischen Schwingungen in drahtförmigen Leitern," Uppsala Universitets Årsskrift, no. 1; 1930.
2. R. King, "Asymmetrically driven antennas and the sleeve dipole," Technical Report No. 93, Craft Lab., Harvard University; 1949.
3. S. A. Schelknoff, "Advanced Antenna Theory," John Wiley and Sons, New York; 1952.
4. L. J. Chu and J. A. Stratton, "Forced Oscillation of a Prolate Spheroid," Jour. of

Appl. Phys. **12**, pp. 241–248; 1941. L. J. Chu and J. A. Stratton, "Elliptic and Spheroidal Wave Functions," Jour. of Math. and Phys., Vol. **XX**, No. 3; 1941.

5. J. A. Stratton, P. M. Morse, L. J. Chu, J. D. C. Little, and F. J. Corbató, "Spheroidal Wave Functions," Published jointly by the Technology Press of M.I.T. and John Wiley and Sons, Inc., N.Y., 1956.

Electrical Engineering Department
University of Southern California
Los Angeles, California

(Received June 13, 1962)

Radiation from an axially slotted cylinder coated with an inhomogeneous dielectric sheath

C. YEH and Z. A. KAPRIELIAN
Electrical Engineering Department, University of Southern California, Los Angeles, California

MS. received 9th April 1963, *in revised form 6th June* 1963

The expressions for the radiated fields of an axially slotted infinite cylinder coated with a radially inhomogeneous dielectric sheath are obtained. The two-dimensional vector wave equation in this inhomogeneous medium is separated using the vector wave function method of Hansen and Stratton. The problem of finding the solution of Maxwell's equation in such an inhomogeneous medium reduces to the solution of two ordinary differential equations. Numerical computations are carried out for a specific variation of the dielectric sheath, i.e. $\epsilon(r) = \epsilon_0\alpha/k_0 r$, where α is a constant, $k_0 = \omega^2\mu\epsilon_0$, and ϵ_0 and μ are respectively the free-space dielectric constant and the free-space permeability. Radiation patterns are plotted for various values of k_0a, k_0b and α, where a and b are respectively the inner and the outer radius of the sheath. These patterns are also compared with the homogeneous sheath problem; the dielectric constant of the homogeneous sheath is taken to be the average value of the inhomogeneous sheath. Results are discussed.

1. Introduction

As a space vehicle re-enters the atmosphere a plasma sheath surrounding the vehicle is generated. It is well known that the sheath is inhomogeneous. However, in order that this problem may be rendered to theoretical analyses, most investigators made the assumption that the plasma sheath is homogeneous. It is the purpose of this paper to investigate the validity of this assumption. To bring out the effects on the radiation characteristics of a slotted antenna* on the vehicle due to the inhomogeneity of the sheath and to avoid unnecessary complicated mathematical operations, the following idealized problem is analysed. An infinitely long cylinder coated with a radially inhomogeneous dielectric sheath is excited by an axial slot across which is maintained a constant distribution of electric field. The analytical expressions for the radiated fields are given. Numerical computations are carried out for a specific variation of the dielectric constant, i.e. $\epsilon(kr) = \epsilon_0/kr$, where $k^2 = \omega^2\mu\epsilon_0/\alpha^2$, ϵ_0 is the free-space dielectric constant, α is a constant and r is the radial component in the cylindrical coordinates. With this choice of dielectric variation the wave functions can be expressed in terms of familiar functions (i.e. Bessel functions of non-integer order). The computed radiation patterns are then compared with those for an identical structure coated with a homogeneous dielectric sheath whose dielectric constant equals the average value of the inhomogeneous dielectric. Results are discussed.

The slotted-cylinder antenna has been considered by many investigators (Silver and Saunders 1950, Wait 1959). In most of the previous theoretical work the slot is assumed to be cut on a circular or an elliptical cylinder of perfect conductivity and infinite length. Recently, Wait and Mientka (1957) solved the problem of a slotted-cylinder antenna coated with a homogeneous dielectric sheath. In the present work the vector wave-function method of Hansen (1935) and Stratton (1941) will be extended to represent the fields in a radially stratified cylindrical medium without any restriction upon the exact nature of the dielectric constant. According to this method all field components in this medium can be obtained from the scalar quantity $\Phi(r, \theta)$ or $\Psi(r, \theta)$ as follows:

$$\left.\begin{aligned} \mathbf{E} &= \nabla \times \{\Phi(r, \theta)\mathbf{e}_r\} \\ \mathbf{H} &= \frac{-i}{\omega\mu_0} \nabla \times \nabla \times \{\Phi(r, \theta)\mathbf{e}_r\} \end{aligned}\right\} \quad (1)$$

for TE waves; and

$$\left.\begin{aligned} \mathbf{H} &= \nabla \times \{\Psi(r, \theta)\mathbf{e}_r\} \\ \mathbf{E} &= \frac{+i}{\omega\epsilon(r)} \nabla \times \nabla\{\Psi(r, \theta)\mathbf{e}_r\} \end{aligned}\right\} \quad (2)$$

for TM waves. Since an infinitely long axial slot is assumed, all field components are then independent of the axial components. $\mathbf{e}_r$ is the unit vector in the r direction and the harmonic time dependence $e^{-i\omega t}$ is assumed and suppressed throughout. The above formulation assures the fulfilment of the divergence conditions in Maxwell's equations. It is noted that μ_0 may also be a function of r in this formulation. The solutions for $\Phi(r, \theta)$ and $\Psi(r, \theta)$ may be obtained respectively by substituting equations (1) and (2) into Maxwell's equations. Hence

$$\Phi(r, \theta) = \begin{Bmatrix} U_n^{(1)}(r) \\ U_n^{(2)}(r) \end{Bmatrix} e^{\pm in\theta} \quad (3)$$

and

$$\Psi(r, \theta) = \begin{Bmatrix} V_n^{(1)}(r) \\ V_n^{(2)}(r) \end{Bmatrix} e^{\pm in\theta} \quad (4)$$

where $U_n^{(1),(2)}(r)$ and $V_n^{(1),(2)}(r)$ satisfy respectively the differential equations

$$\left\{\frac{d^2}{dr^2} - \frac{1}{r}\frac{d}{dr} + \left(\omega^2\mu_0\epsilon(kr) + \frac{1-n^2}{r^2}\right)\right\}U_n^{(1),(2)} = 0 \quad (5)$$

and

* Slotted antennae are used extensively on high-velocity vehicles because there are no protruding obstacles to introduce extra drag.

$$\left[\frac{d^2}{dr^2}-\left\{1+\frac{r}{\epsilon(kr)}\frac{d\epsilon(kr)}{dr}\right\}\left(\frac{1}{r}\frac{d}{dr}\right)+\left\{\omega^2\mu_0\epsilon(kr)+\frac{1}{r\epsilon(kr)}\frac{d\epsilon(kr)}{dr}+\frac{1-n^2}{r^2}\right\}\right]V_n^{(1),(2)}=0. \tag{6}$$

2. Formal solution

To formulate the problem, we consider the electromagnetic field radiation from a perfectly conducting infinite cylinder of radius a coated with an inhomogeneous concentric dielectric sheath that has an outer radius b, and fed by an axial slot across which is maintained a constant distribution of electric field defined by

$$\mathbf{E}=\mathbf{e}_\theta E_0 G(\theta)\,e^{-i\omega t} \tag{7}$$

where $G(\theta)$ is

$$G(\theta)=\left\{\begin{matrix}1 & \text{for} -\tfrac{1}{2}\theta_0<\theta<\tfrac{1}{2}\theta_0\\ 0 & \text{for}\ \tfrac{1}{2}\theta_0<\theta<-\tfrac{1}{2}\theta_0\end{matrix}\right\} \tag{8}$$

and $\mathbf{e}_\theta$ is the unit vector in the θ direction (see figure 1). Since only circumferential currents can be generated by such a source, the radial and circumferential components of the magnetic vector $\mathbf{H}$ and the axial components of the electric vector $\mathbf{E}$ are identically zero. Thus only TM waves are generated. The most general expressions for H_z and E_θ for $r>b$ are the following:

$$H_z^{\text{rad}}=\sum_{n=-\infty}^{\infty}C_n\mathrm{H}_n^{(1)}(k_0r)\,e^{in\theta} \tag{9}$$

and

$$E_\theta^{\text{rad}}=\sum_{n=-\infty}^{\infty}C_n(-i)\left(\frac{\mu_0}{\epsilon_0}\right)^{1/2}\mathrm{H}_n^{(1)\prime}(k_0r)e^{in\theta} \tag{10}$$

where $k_0=2\pi/\lambda_0$, λ_0 being the free-space wavelength and where $\mathrm{H}_n^{(1)}(k_0r)$ are the Hankel functions of the first kind and order n, the prime indicates the derivative of the function with respect to its argument, and C_n are arbitrary constants to be determined. The expressions for H_z and E_θ for $a<r<b$ must be of the form

$$H_z^{\text{sheath}}=\sum_{n=-\infty}^{\infty}\frac{1}{kr}\{A_nV_n^{(1)}(kr)+B_nV_n^{(2)}(kr)\}e^{in\theta} \tag{11}$$

$$E_\theta^{\text{sheath}}=\sum_{n=-\infty}^{\infty}\frac{-ik}{\omega\epsilon(kr)}\left[A_n\frac{d}{d(kr)}\left\{\frac{V_n^{(1)}(kr)}{kr}\right\}+B_n\frac{d}{d(kr)}\left\{\frac{V_n^{(2)}(kr)}{kr}\right\}\right]e^{in\theta} \tag{12}$$

where A_n and B_n are arbitrary constants as yet unknown, and $V_n^{(1),(2)}(kr)$ satisfy the differential equation (6).

To obtain an analytical expression for the applied field as given by (7), it is expanded in a complex Fourier series; that is,

$$E_\theta^{\text{app.}}=E_0G(\theta)=\frac{E_0}{\pi}\sum_{n=-\infty}^{\infty}\frac{\sin(n\tfrac{1}{2}\theta_0)}{n}e^{in\theta}. \tag{13}$$

This infinite series represents the E_θ field on the conducting cylinder for all values of θ between $-\pi$ and π.

Satisfying the appropriate boundary conditions at $r=a$ and $r=b$ (i.e. the tangential electric and magnetic field must be continuous at $r=b$, and $E_\theta^{\text{app.}}$ and E_θ^{sheath} must be identical at $r=a$) gives the following relations:

$$A_n=\frac{E_0 i\omega\epsilon(kr)}{n\pi k}\,\frac{\sin(n\tfrac{1}{2}\theta_0)}{\dfrac{d}{d(ka)}\left\{\dfrac{V_n^{(1)}(ka)}{ka}\right\}+P_n(kb)\dfrac{d}{d(ka)}\left\{\dfrac{V_n^{(2)}(ka)}{ka}\right\}} \tag{14}$$

$$B_n=A_nP_n(kb) \tag{15}$$

$$C_n=A_nQ_n(kb) \tag{16}$$

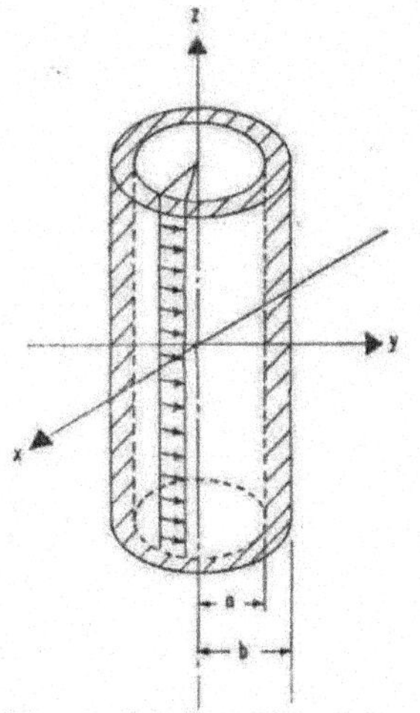

Figure 1. Geometrical configuration.

where

$$P_n(kb)=\frac{\mathrm{H}_n^{(1)}(k_0b)\dfrac{d}{d(kb)}\left\{\dfrac{V_n^{(1)}(kb)}{kb}\right\}-\dfrac{\omega\epsilon(kb)}{k}\left(\dfrac{\mu_0}{\epsilon_0}\right)^{1/2}\left\{\dfrac{V_n^{(1)}(kb)}{kb}\right\}\mathrm{H}_n^{(1)\prime}(k_0b)}{\dfrac{\omega\epsilon(kb)}{k}\left(\dfrac{\mu_0}{\epsilon_0}\right)^{1/2}\left\{\dfrac{V_n^{(2)}(kb)}{kb}\right\}\mathrm{H}_n^{(1)\prime}(k_0b)-\dfrac{d}{d(kb)}\left\{\dfrac{V_n^{(2)}(kb)}{kb}\right\}\mathrm{H}_n^{(1)}(k_0b)} \tag{17}$$

$$Q_n(kb)=\frac{-\left\{\dfrac{V_n^{(1)}(kb)}{kb}\right\}\dfrac{d}{d(kb)}\left\{\dfrac{V_n^{(2)}(kb)}{kb}\right\}+\left\{\dfrac{V_n^{(2)}(kb)}{kb}\right\}\dfrac{d}{d(kb)}\left\{\dfrac{V_n^{(1)}(kb)}{kb}\right\}}{\dfrac{\omega\epsilon(kb)}{k}\left(\dfrac{\mu_0}{\epsilon_0}\right)^{1/2}\left\{\dfrac{V_n^{(2)}(kb)}{kb}\right\}\mathrm{H}_n^{(1)\prime}(kb)-\dfrac{d}{d(kb)}\left\{\dfrac{V_n^{(2)}(kb)}{kb}\right\}\mathrm{H}_n^{(1)}(k_0b)}. \tag{18}$$

The formal solution for this problem is obtained when one substitutes equations (14) through (16) into equations (9) through (12).

At large distances from the cylinder (i.e. $k_0 r \gg 1$), the asymptotic expression for the derivative of the Hankel function

$$\frac{d}{d(k_0 r)}\{H_n^{(1)}(k_0 r)\} \to i\left(\frac{2}{\pi k_0 r}\right)^{1/2} \exp i\left(k_0 r - \frac{2n+1}{4}\pi\right) \quad (19)$$

is applicable. Substituting (14), (16) and (19) into (10) gives the expression

$$E_\theta^{\mathrm{rad}} = i\frac{E_0 \omega \epsilon(ka)}{\pi k}\left(\frac{2\mu_0}{\pi k_0 r \epsilon_0}\right)^{1/2} \exp i(k_0 r - \tfrac{1}{4}\pi)$$
$$\times \sum_{n=-\infty}^{\infty}\left[\frac{\sin(\tfrac{1}{2}n\theta_0) Q_n(kb) \exp in(\theta - \tfrac{1}{2}\pi)}{n\left\{\frac{d}{d(ka)}\left\{\frac{V_n^{(1)}(ka)}{ka}\right\} + P_n(kb)\frac{d}{d(ka)}\left\{\frac{V_n^{(2)}(ka)}{ka}\right\}\right\}}\right]. \quad (20)$$

The absolute value of the above equation represents the radiation pattern of the radiating structure.

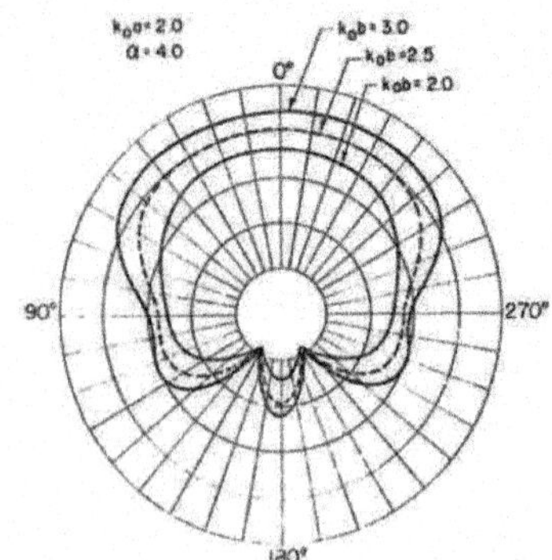

Figure 3. Radiation patterns as a function of sheath thickness.

3. A specific example

To get an idea of how the solutions behave, numerical computations are carried out for a specific variation of the dielectric constant. The inhomogeneous dielectric is taken to be

$$\epsilon(kr) = \frac{\epsilon_0}{kr} \quad (21)$$

where $k^2 = \omega^2\mu\epsilon_0/\alpha^2$ and α is a constant. Putting (21) into (6), one notes that the solutions may be expressed in terms of well-known functions, i.e.

$$V_n^{(1),(2)}(kr) = (kr)^{1/2}\begin{bmatrix} J_{(1+4n^2)^{1/2}}\left\{\frac{2k_0}{k}(kr)^{1/2}\right\} \\ J_{-(1+4n^2)^{1/2}}\left\{\frac{2k_0}{k}(kr)^{1/2}\right\}\end{bmatrix} \quad (22)$$

where $J_{\pm\nu}(p)$ are the Bessel functions of order $\pm\nu$ and argument p.

Numerical results for the radiation pattern are presented in figures 2–9. In the figures, $|E_\theta^{\mathrm{rad}}|$ is plotted against θ in polar form for various values of k_0a, k_0b, and α with $\theta_0 = 0{\cdot}1$ rad.

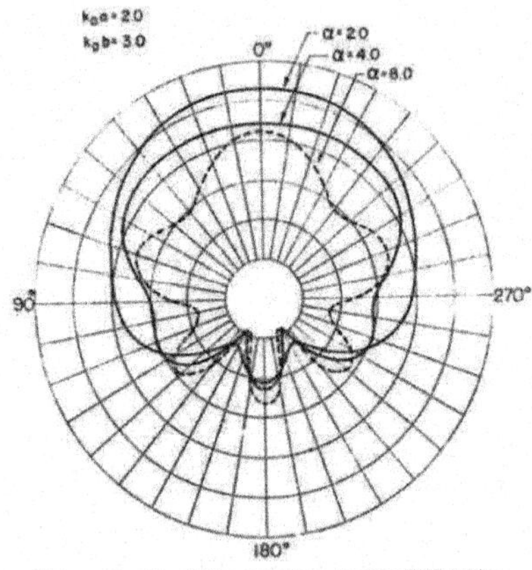

Figure 4. Radiation patterns as a function of α.

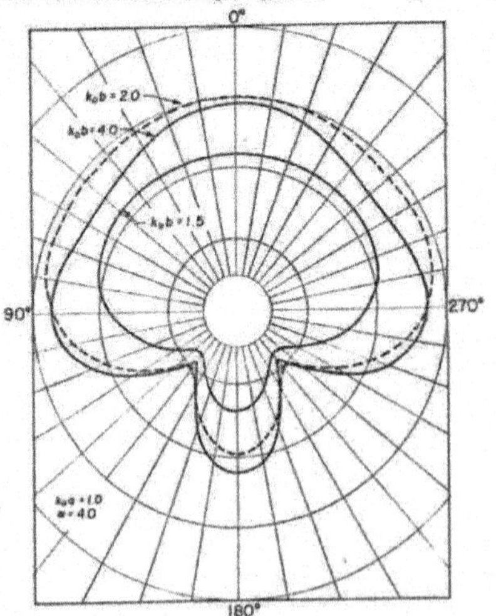

Figure 2. Radiation patterns as a function of sheath thickness.

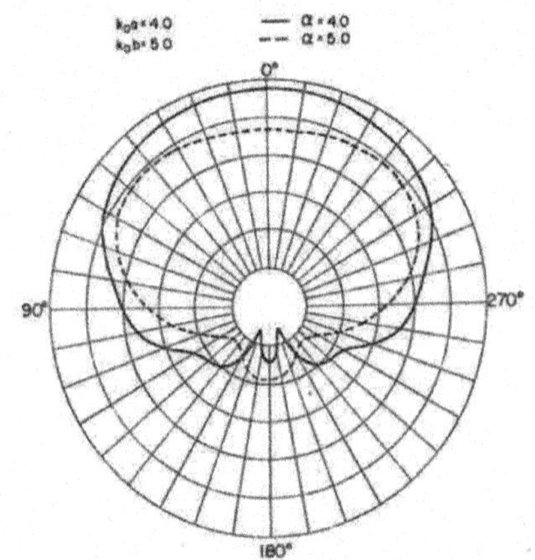

Figure 5. Radiation patterns as a function of α.

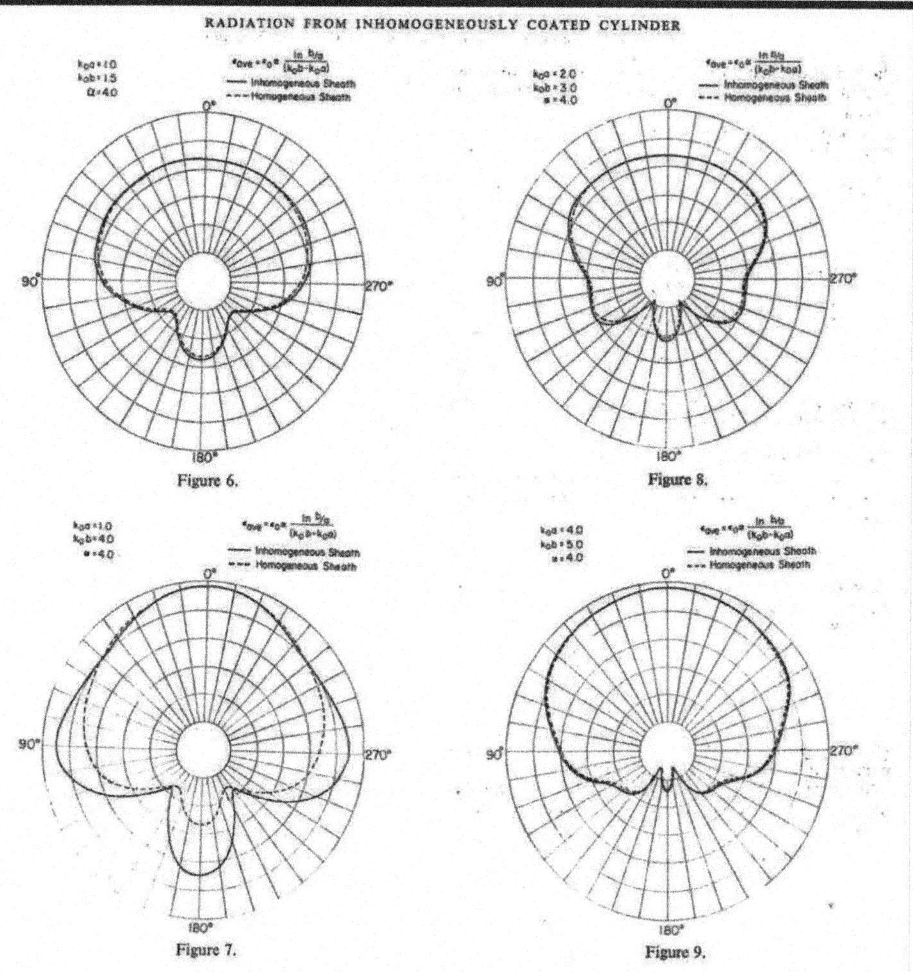

Figures 6, 7, 8, 9. A comparison between the radiation pattern for a homogeneous dielectric sheath with that for an inhomogeneous sheath.

It should be noted that each pattern is individually normalized. Consequently, direct comparison of relative amplitudes from one figure to another is not possible. It is noted that the series in equation (20) converges rather rapidly for small values of k_0a and k_0b. For $k_0a < 1$ and $k_0b < 2$, only four terms are required in order to get a three significant figure accuracy; for $k_0a < 4$ and $k_0b < 5$, eight terms are required to give the same accuracy. Figures 2 and 3 show the variation of the radiation pattern with respect to the variation of the sheath's thickness. It can be seen that the thicker the sheath the wider is the spread of the pattern towards the rear of the cylinder. This is probably due to the fact that the 'ray path' (optically speaking) within the sheath is bent more towards the rear of the cylinder for thicker sheath. The presence of a peripheral standing wave is quite evident. It seems that the presence of the sheath tends to enhance the standing-wave pattern. Changing the thickness of the sheath effectively alters the effective dielectric constant of the sheath; hence, the period of the standing-wave pattern is somewhat altered. It is also clear that the peripheral surface wave is guided, with considerable leakage, around the cylinder.

Varying the constant α changes essentially the effective dielectric constant of the sheath. Thus, the period of the peripheral standing wave is altered. This effect can be observed clearly from figures 4 and 5.

To compare the radiation pattern of a cylinder coated with a homogeneous dielectric sheath with that of an identical cylinder coated with an inhomogeneous dielectric sheath, figures 6–9 are introduced. The dielectric constant of the homogeneous sheath is taken to be the average value of the inhomogeneous sheath. It can be seen that if the thickness of the sheath is moderate, very little differences are observed as far as the normalized shapes of these radiation patterns are concerned. The only significant difference is that the radiation pattern for the inhomogeneous case spreads out more towards the rear of the cylinder. This is the expected behaviour from the 'ray path' point of view of the problem. As the thickness of the sheath becomes larger, the discrepancy becomes greater as illustrated by figure 7. This increased discrepancy can also be explained by the 'ray path' theory.

In conclusion, it is noted that for this particular numerical example the shapes of the radiation patterns for the inhomogeneous case and for the homogeneous case are very similar as long as the sheath is not very thick (i.e. $k_0b - k_0a < 1$). However the relative amplitudes of these patterns for these two cases may be quite different. It has been shown elsewhere (Yeh and Kaprielian 1963) that the backscattering cross section of a cylinder coated with a homogeneous sheath is quite different from that of one coated with an inhomogeneous sheath. It should be remarked that the formulation of this problem is applicable to plasma-coated cylinders with inhomogeneous plasma-density distribution in the radial direction. This problem is applicable to the re-entry problem.

Acknowledgments

This work was supported by the U.S. Air Force Cambridge Research Laboratory.

References

HANSEN, W. W., 1935, *Phys. Rev.*, **47**, 139.

SILVER, S., and SAUNDERS, W. K., 1950, *J. Appl. Phys.*, **21**, 745.

STRATTON, J. A., 1941, *Electromagnetic Theory* (New York: McGraw-Hill).

WAIT, J. R., 1959, *Electromagnetic Radiation from Cylindrical Structures* (New York: Pergamon Press).

WAIT, J. R., and MIENTKA, W., 1957, *J. Res. Nat. Bur. Stand.*, **58**, 287.

YEH, C., and KAPRIELIAN, Z. A., 1963, *Canad. J. Phys.*, **41**, 143.

UNWIN BROTHERS LIMITED, WOKING AND LONDON

Dyadic Green's Function for a Radially Inhomogeneous Spherical Medium*

C. YEH
Electrical Engineering Department, University of Southern California, Los Angeles, California
(Received 29 March 1963)

The general expression for the dyadic Green's function in a radially inhomogeneous dielectric medium is obtained. It is then applied to the problem of radiation from an electric dipole in such a medium. Possible applications to electromagnetic scattering problems and to elementary particle scattering problems are noted.

I. INTRODUCTION

IN a recent article by Wyatt,[1] the problem of scattering of electromagnetic plane waves by an inhomogeneous spherically symmetric object was considered. He formulated it as a boundary-value problem; i.e., appropriate expressions for the field components are obtained for the region within the inhomogeneous spherical particle and for the homogeneous region outside the particle. The expansion coefficients in the series expansions of the interior and scattered fields are then determined by satisfying the proper boundary conditions at the interface of the two regions. However, it is noted,[2] that in many instances there may not be any distinct boundary separating the inhomogeneous and homogeneous regions. For example, we have the problems of the scattering of soft x rays and light by large molecules and the scattering of microwaves by lenses made of artificial dielectrics, and the analogous problems in elementary particle scattering theory in which the potential is never discontinuous.

It is, therefore, the purpose of the present paper to consider the problem of electromagnetic wave propagation in a continuously inhomogeneous spherically symmetric medium. The vector wave equations in the radially stratified medium will be separated in spherical coordinates by the method of Hansen and Stratton.[3] The dyadic Green's function in such a medium will be derived. The total field from a dipole source in this medium is then obtained. In the conclusions, several possible applications of the results are pointed out. It may be interesting to mention that Schwinger[4] and Morse and Feshbach[5] advocate that the introduction of a dyadic Green's function by means of which the vector wave equation satisfied by $\mathbf{E}$ or $\mathbf{H}$ can be integrated presents the most elegant way of dealing with many electromagnetic problems.

* Supported by the Air Force Cambridge Research Laboratories.
[1] P. J. Wyatt, Phys. Rev. 127, 1837 (1962).
[2] L. I. Schiff, J. Opt. Soc. Am. 52, 140 (1962).
[3] W. W. Hansen, Phys. Rev. 47, 139 (1935); J. A. Stratton, *Electromagnetic Theory* (McGraw-Hill Book Company, Inc., New York, 1941).
[4] J. Schwinger, Comm. Pure Appl. Math. 3, 355 (1950).
[5] P. M. Morse and H. Feshbach, *Methods of Theoretical Physics* (McGraw-Hill Book Company, Inc., New York, 1953).

II. SOLUTIONS OF VECTOR WAVE EQUATIONS

Maxwell's equations in a radially stratified medium take the form

$$\nabla\times\mathbf{H}=\mathbf{J}-i\omega\epsilon(r)\mathbf{E}, \tag{1}$$

$$\nabla\times\mathbf{E}=i\omega\mu_0\mathbf{H}. \tag{2}$$

$\mathbf{E}$ and $\mathbf{H}$ are electric and magnetic-field vectors, $\mathbf{J}$ is the current density, $\epsilon(r)$ is the inhomogeneous dielectric constant and μ_0 is taken to be the free-space permeability. A time dependence of $e^{-i\omega t}$ is assumed. Combining Eqs. (1) and (2) gives

$$\nabla\times\nabla\times\mathbf{E}-\omega^2\mu_0\epsilon(r)\mathbf{E}=i\omega\mu_0\mathbf{J}. \tag{3}$$

The dyadic Green's function $\mathbf{\Gamma}(r,r')$ satisfies the following equation:

$$\nabla\times\nabla\times\mathbf{\Gamma}(\mathbf{r},\mathbf{r}')-\omega^2\mu_0\epsilon(r)\mathbf{\Gamma}(\mathbf{r},\mathbf{r}')=\mathbf{I}\delta(\mathbf{r}-\mathbf{r}'), \tag{4}$$

where $\mathbf{I}$ is the unit dyadic and $\delta(\mathbf{r}-\mathbf{r}')$ is a delta function. It can be shown with the help of the vector Green's theorem that if $\mathbf{\Gamma}(\mathbf{r},\mathbf{r}')$ and $\mathbf{J}(\mathbf{r}')$ are known, $\mathbf{E}(\mathbf{r})$ can be found by the relation

$$\mathbf{E}(\mathbf{r})=i\omega\mu_0\int_{v'}\mathbf{\Gamma}(\mathbf{r},\mathbf{r}')\cdot\mathbf{J}(\mathbf{r}')dv', \tag{5}$$

where the integration is performed over the volume v' containing the source currents. It is known[5] that $\mathbf{\Gamma}(\mathbf{r},\mathbf{r}')$ may be expanded in terms of the eigenfunctions of the following vector wave equations:

$$\nabla\times\nabla\times\mathbf{E}-\omega^2\mu_0\epsilon(r)\mathbf{E}=0, \tag{6}$$

$$\nabla\times\nabla\times\mathbf{H}-[\nabla\epsilon(r)/\epsilon(r)]\times\nabla\times\mathbf{H}-\omega^2\mu_0\epsilon(r)\mathbf{H}=0. \tag{7}$$

Hence, the solutions for these equations will be our concern in this section.

According to the vector wave-function method of Hansen and Stratton,[3] the above equations can be reduced to two scalar wave equations by separating the fields into two linearly independent fields; viz., the transverse electric (TE) and the transverse magnetic (TM) fields.[6] Since $\epsilon(r)$ $\mathbf{E}$ and $\mathbf{H}$ are solenoidal vectors, the field components can be derived from the scalar

[6] See also C. T. Tai, Appl. Sci. Res. Sec. B 7, 113 (1958).

quantities $\Phi(r,\theta,\phi)$ and $\Psi(r,\theta,\phi)$ as follows:

$$\mathbf{E}^{(m)}=\nabla\times(\Phi(r,\theta,\phi)\mathbf{e}_r), \quad (8)$$

$$\mathbf{H}^{(m)}=(1/i\omega\mu_0)\nabla\times\nabla\times(\Phi(r,\theta,\phi)\mathbf{e}_r) \quad (9)$$

for TE waves, and

$$\mathbf{H}^{(e)}=\nabla\times(\Psi(r,\theta,\phi)\mathbf{e}_r), \quad (10)$$

$$\mathbf{E}^{(e)}=(i/\omega\epsilon(r))\nabla\times\nabla\times(\Psi(r,\theta,\phi)\mathbf{e}_r) \quad (11)$$

for TM waves. $\mathbf{e}_r$ is the unit vector in the radial direction and the superscripts (m) and (e) denote TE and TM waves, respectively. The above formulation assures the fulfillment of the divergence conditions in Maxwell's equations. The solutions for $\Phi(r,\theta,\phi)$ or $\Psi(r,\theta,\phi)$ can be obtained, respectively, by substituting Eq. (8) into (6) or (10) into (7), carrying out the vector operations and separating the variables. One has

$$\Phi(r,\theta,\phi)=\begin{Bmatrix}U_n^{(1)}(r)\\U_n^{(2)}(r)\end{Bmatrix}\begin{Bmatrix}P_n^m(\cos\theta)\\Q_n^m(\cos\theta)\end{Bmatrix}\begin{Bmatrix}\sin m\phi\\\cos m\phi\end{Bmatrix} \quad (12)$$

and

$$\Psi(r,\theta,\phi)=\begin{Bmatrix}V_n^{(1)}(r)\\V_n^{(2)}(r)\end{Bmatrix}\begin{Bmatrix}P_n^m(\cos\theta)\\Q_n^m(\cos\theta)\end{Bmatrix}\begin{Bmatrix}\sin m\phi\\\cos m\phi\end{Bmatrix}, \quad (13)$$

where $P_n^m(\cos\theta)$ and $Q_n^m(\cos\theta)$ are the associated Legendre's polynomial and $U_n^{(1),(2)}(r)$ and $V_n^{(1),(2)}(r)$ satisfy, respectively, the differential equations

$$\left[\frac{d^2}{dr^2}+\left(\omega^2\mu_0\epsilon(r)-\frac{n(n+1)}{r^2}\right)\right]U_n^{(1),(2)}(r)=0 \quad (14)$$

and

$$\left[\frac{d^2}{dr^2}-\frac{1}{\epsilon(r)}\frac{d\epsilon(r)}{dr}\frac{d}{dr}+\left(\omega^2\mu_0\epsilon(r)-\frac{n(n+1)}{r^2}\right)\right]V_n^{(1),(2)}(r)=0. \quad (15)$$

The solutions of these differential equations depend upon the dielectric variation $\epsilon(r)$. For instance, $\epsilon(r)=\epsilon_0$, a constant, the solutions are the spherical Bessel functions multiplied by r. More will be said about these equations in the conclusions.

III. DERIVATION OF THE DYADIC GREEN'S FUNCTION

Returning now to the problem of deriving the proper dyadic Green's function in a radially inhomogeneous spherical medium, we note that the appropriate dyadic Green's function must (a) be a solution of Eq. (4), (b) satisfy Sommerfeld's radiation condition, and (c) be finite in the source-free region. Conditions (b) and (c) are satisfied if we expand the dyadic Green's function in terms of the eigenfunctions of the wave equations, i.e.,

$$\mathbf{\Gamma}(\mathbf{r},\mathbf{r}')=\sum_m\sum_n A_{e,omn}^{(m)}\mathbf{E}_{e,omn}^{(m)(3)}(r,\theta,\phi)\mathbf{E}_{e,omn}^{(m)(1)}(r',\theta',\phi')+A_{e,omn}^{(e)}\mathbf{E}_{e,omn}^{(e)(3)}(r,\theta,\phi)\mathbf{E}_{e,omn}^{(e)(1)}(r',\theta',\phi'), \quad (16)$$

for the region $r>r'$, and

$$\mathbf{\Gamma}(\mathbf{r},\mathbf{r}')=\sum_m\sum_n A_{e,omn}^{(m)}\mathbf{E}_{e,omn}^{(m)(1)}(r,\theta,\phi)\mathbf{E}_{e,omn}^{(m)(3)}(r',\theta',\phi')+A_{e,omn}^{(e)}\mathbf{E}_{e,omn}^{(e)(1)}(r,\theta,\phi)\mathbf{E}_{e,omn}^{(e)(3)}(r',\theta',\phi') \quad (17)$$

for the region $r<r'$. The following abbreviations have been used:

$$\mathbf{E}_{e,omn}^{(m)(p)}=\nabla\times(\Phi_{e,omn}^{(p)}\mathbf{e}_r), \quad (18)$$

$$\mathbf{E}_{e,omn}^{(e)(p)}=\frac{i}{\omega\epsilon(r)}\nabla\times\nabla\times(\Psi_{e,omn}^{(p)}\mathbf{e}_r), \quad (p=1,3) \quad (19)$$

where

$$\Phi_{e,omn}^{(p)}=U_n^{(p)}(r)P_n^m(\cos\theta)\begin{matrix}\cos\\ \sin\end{matrix}m\phi, \quad (20)$$

$$\Psi_{e,omn}^{(p)}=V_n^{(p)}(r)P_n^m(\cos\theta)\begin{matrix}\cos\\ \sin\end{matrix}m\phi. \quad (p=1,3) \quad (21)$$

$U_n^{(p)}(r)$ and $V_n^{(p)}(r)$ are solutions of Eqs. (14) and (15), respectively. The superscript p indicates the nature of the required solutions. $p=1$ denotes the standing wave solution which corresponds to the spherical Bessel function $j_n(\omega(\mu\epsilon_0)^{1/2}r)$ multiplied by r when the dielectric variation $\epsilon(r)$ becomes a constant ϵ_0; $p=3$ denotes the traveling wave solution which corresponds to the spherical Hankel function $h_n^{(1)}(\omega(\mu\epsilon_0)^{1/2}r)$ multiplied by r when the dielectric variation $\epsilon(r)$ becomes a constant ϵ_0. $A_{e,omn}^{(m)}$ and $A_{e,omn}^{(e)}$ are arbitrary constants that are to be determined by satisfying Eq. (4).

Let us premultiply Eq. (4) by $\mathbf{e}_\theta$ which is the unit vector in the θ direction, integrate with respect to r from $r=r'-\delta$ to $r=r'+\delta$ and make $\delta\to 0$. The result is

$$\left\{-\frac{\partial}{\partial r}[r\mathbf{e}_\theta\cdot\mathbf{\Gamma}(\mathbf{r},\mathbf{r}')]+\frac{\partial}{\partial\theta}[\mathbf{e}_r\cdot\mathbf{\Gamma}(\mathbf{r},\mathbf{r}')]\right\}\Bigg|_{r=r'-0}^{r=r'+0}=\mathbf{e}_\theta\frac{\delta(\theta-\theta')\delta(\phi-\phi')}{r'\sin\theta}, \quad (22)$$

where $\mathbf{e}_r$ is the unit vector in the r direction. Substituting Eqs. (16) and (17) into (22) gives

$$\sum_m \sum_n A_{e,omn}{}^{(m)} \left\{ -\frac{\partial}{\partial r'}[r'\mathbf{e}_\theta \cdot \mathbf{E}_{e,omn}{}^{(m)(3)}(r',\theta,\phi)]\mathbf{E}_{e,omn}{}^{(m)(1)}(r',\theta',\phi') \right.$$

$$\left. + \frac{\partial}{\partial r'}[r'\mathbf{e}_\theta \cdot \mathbf{E}_{e,omn}{}^{(m)(1)}(r',\theta,\phi)]\mathbf{E}_{e,omn}{}^{(m)(3)}(r',\theta',\phi') \right\}$$

$$+ A_{e,omn}{}^{(e)} \left\{ \left[-\frac{\partial}{\partial r'}[r'\mathbf{e}_\theta \cdot \mathbf{E}_{e,omn}{}^{(e)(3)}(r',\theta,\phi) + \frac{\partial}{\partial \theta}[\mathbf{e}_{r'} \cdot \mathbf{E}_{e,omn}{}^{(e)(3)}(r',\theta,\phi)] \right] \mathbf{E}_{e,omn}{}^{(e)(1)}(r',\theta',\phi') \right.$$

$$\left. - \left[-\frac{\partial}{\partial r'}[r'\mathbf{e}_\theta \cdot \mathbf{E}_{e,omn}{}^{(e)(1)}(r',\theta,\phi)] + \frac{\partial}{\partial \theta}[\mathbf{e}_{r'} \cdot \mathbf{E}_{e,omn}{}^{(e)(1)}(r',\theta,\phi)] \right] \mathbf{E}_{e,omn}{}^{(e)(3)}(r',\theta',\phi') \right\} = \mathbf{e}_\theta \frac{\delta(\theta-\theta')\delta(\phi-\phi')}{r'\sin\theta}. \quad (23)$$

Multiplying both sides of the above equation by $\sin\theta' \mathbf{E}_{e,omn}{}^{(m)(1)}(r',\theta',\phi')$, integrating over θ' from $\theta'=0$ to $\theta'=\pi$ and ϕ' from $\phi'=0$ to $\phi'=2\pi$ and making use of the orthogonality relations[5]

$$\int_0^{2\pi}\int_0^{\pi} \mathbf{E}_{e,omn}{}^{(m)(p)}(r',\theta',\phi') \cdot \mathbf{E}_{e,omn}{}^{(m)(q)}(r',\theta',\phi') \sin\theta' d\theta' d\phi' = \delta_{mm'}\delta_{nn'}(1+\delta_{om}) \times \frac{2\pi n(n+1)}{2n+1} \frac{(n+m)!}{(n-m)!} \frac{U_n{}^{(p)}(r')}{r'} \frac{U_n{}^{(q)}(r')}{r'}, \quad (24)$$

$$\int_0^{2\pi}\int_0^{\pi} \mathbf{E}_{e,omn}{}^{(m)(p)}(r',\theta',\phi') \cdot \mathbf{E}_{e,omn}{}^{(e)(q)}(r',\theta',\phi') \sin\theta' d\theta' d\phi' = 0, \quad (25)$$

where $\delta_{mn'}$, $\delta_{nn'}$, and δ_{om} are the Kronecker deltas, and the Wronskian relation

$$U_n{}^{(3)}(r')\frac{d}{dr'}U_n{}^{(1)}(r') - U_n{}^{(1)}(r')\frac{d}{dr}U_n{}^{(3)}(r') = \frac{1}{ik_0}, \quad (26)$$

where $k_0{}^2 = \omega\mu_0\epsilon_0$, one obtains

$$A_{e,omn}{}^{(m)} = ik_0 \frac{(2n+1)}{2\pi(1+\delta_{om})n(n+1)} \frac{(n-m)!}{(n+m)!}. \quad (27)$$

Multiplying both sides of Eq. (23) by $\sin\theta' \mathbf{E}_{e,omn}{}^{(e)(1)}(r',\theta',\phi')$, integrating over θ' from $\theta'=0$ to $\theta'=\pi$ and ϕ' from $\phi'=0$ to $\phi'=2\pi$, and making use of the orthogonality relations (25) and

$$\int_0^{2\pi}\int_0^{2\pi} \mathbf{E}_{e,omn}{}^{(e)(p)}(r',\theta',\phi') \cdot \mathbf{E}_{e,omn}{}^{(e)(q)}(r',\theta',\phi') \sin\theta' d\theta' d\phi' = \frac{\pi(1+\delta_{mo})}{\omega^2\epsilon^2(r')} \frac{2n(n+1)}{(2n+1)} \frac{(n+m)!}{(n-m)!} \delta_{mn'}\delta_{nn'} \times \left[\frac{n(n+1)}{r'^4} V_n{}^{(p)}(r')V_n{}^{(q)}(r') + \frac{1}{r'^2}\left(\frac{d}{dr'}V_n{}^{(p)}(r')\right)\left(\frac{d}{dr'}V_n{}^{(q)}(r')\right) \right] \quad (28)$$

and the Wronskian relation

$$V_n{}^{(3)}(r')\frac{d}{dr'}V_n{}^{(1)}(r') - V_n{}^{(1)}(r')\frac{d}{dr'}V_n{}^{(3)}(r') = \frac{1}{ik_0}\frac{\epsilon(r')}{\epsilon_0}, \quad (29)$$

one obtains

$$A_{e,omn}{}^{(e)} = ik_0 \frac{(2n+1)}{2\pi(1+\delta_{om})n(n+1)} \frac{(n-m)!}{(n+m)!}\left(\frac{\epsilon_0}{\mu_0}\right). \quad (30)$$

It is noted that, in deriving the Wronskian relations using the asymptotic representations for the radial functions, the assumption $\epsilon(r) \to \epsilon_0$ as $r \to \infty$ has been used. Substituting Eqs. (27) and (30) into (16) and (17) gives the required dyadic Green's function in a radially stratified spherical medium.

IV. RADIATION FROM AND ELECTRIC DIPOLE

The field from an electric dipole located at (x',y',z') can now be found directly from the dyadic Green's function through the relation

$$\mathbf{E}(\mathbf{r})=\omega^2\mu_0\mathbf{\Gamma}(\mathbf{r},\mathbf{r}')\cdot\mathbf{p}(\mathbf{r}'), \tag{31}$$

where $\mathbf{p}$ is the dipole moment. As a specific example, let us assume that the electric dipole, which has a dipole moment p_x, is pointed in the x direction and located at $r'=a$, $\theta'=\pi$, $\phi'=0$. The electromagnetic field due to the dipole in this radially stratified medium is given by

$$\mathbf{E}^{>}=\frac{ik_0^2p_x}{4\pi\epsilon_0}\sum_{n=0}^{\infty}\frac{2n+1}{n(n+1)}(-1)^n\frac{1}{a}\left\{U_n^{(1)}(a)\mathbf{E}_{o1n}^{(m)(3)}(r,\theta,\phi)+\frac{i}{\omega\mu_0}\frac{\epsilon_0}{\epsilon(a)}\left[\frac{d}{da}V_n^{(1)}(a)\right]\mathbf{E}_{e1n}^{(e)(3)}(r,\theta,\phi)\right\}, \tag{32}$$

$$\mathbf{H}^{>}=(1/i\omega\mu_0)\nabla\times\mathbf{E}^{>} \tag{33}$$

for $r\geq a$, and

$$\mathbf{E}^{<}=\frac{ik_0^2p_x}{4\pi\epsilon_0}\sum_{n=1}^{\infty}\frac{2n+1}{n(n+1)}(-1)^n\frac{1}{a}\left\{U_n^{(3)}(a)\mathbf{E}_{o1n}^{(m)(1)}(r,\theta,\phi)+\frac{i}{\omega\mu_0}\frac{\epsilon_0}{\epsilon(a)}\left[\frac{d}{da}V_n^{(3)}(a)\right]\mathbf{E}_{e1n}^{(e)(1)}(r,\theta,\phi)\right\}, \tag{34}$$

$$\mathbf{H}^{<}=(1/i\omega\mu_0)\nabla\times\mathbf{E}^{<} \tag{35}$$

for $r\leq a$, where

$$\mathbf{E}_{o1n}^{(m)(p)}(r,\theta,\phi)=\nabla\times[U_n^{(p)}(r)P_n^1(\cos\theta)\sin\phi\,\mathbf{e}_r], \tag{36}$$

$$\mathbf{E}_{e1n}^{(e)(p)}(r,\theta,\phi)=(1/\omega\epsilon(r))\nabla\times\nabla\times[V_n^{(p)}(r)P_n^1(\cos\theta)\cos\phi\,\mathbf{e}_r],\quad(p=1,3). \tag{37}$$

For the special case of $\epsilon(r)=\epsilon_0$, Eqs. (32) through (37) give (as they should) the correct expressions for the electromagnetic fields of a dipole in the inhomogeneous free-space.[6] This is because

$$U_n^{(1)}(r)=V_n^{(1)}(r)=k_0rj_n(k_0r), \tag{38}$$

$$U_n^{(3)}(r)=V_n^{(3)}(r)=k_0rh_n^{(1)}(k_0r) \tag{39}$$

in a homogeneous medium.

V. CONCLUSIONS

The general expression for the dyadic Green's function in a radially continuously varying spherically symmetric dielectric medium is obtained. The result is expressed in terms of the associated Legendre polynomials, trigonometric functions, and the radial functions which are the appropriate solutions of two differential equations. These radial functions depend, of course, on the specific dielectric variations. It should be noted that the task of finding these proper radial functions is by no means trivial. However, in some instances the solutions may be expressed in terms of some well-investigated function, such as the hypergeometric functions or the confluent hypergeometric functions.[7] For example, if $\epsilon(r)=\epsilon_0(1+\alpha r^{-1})$ where α is a constant, the solutions of Eq. (14) are the well-known Coulomb wave functions.[5]

In a recent paper by Schiff,[2] he advocates that in electromagnetic diffraction theory, much more attention has been devoted in the past to scattering regions that are piecewise homogeneous than to scatterers that have continuously variable dielectric constant, conductivity, etc. In order that one may consider successfully the analog between elementary particle scattering theory in which the potential is never discontinuous and the electromagnetic problem, the problem of wave propagation in a continuously inhomogeneous medium must be considered.[1,8] It is hoped that the problem considered here will provide a useful beginning for this very involved problem.

Further applications of the dyadic Green's function derived in this paper can be found in the scattering of soft x rays from various macromolecules and viruses which have characteristic diffuse surfaces, the scattering of electromagnetic waves by a plasmoid, or the scattering of infrared radiation by small inhomogeneous particles.

[7] E. T. Whittaker and G. N. Watson, *Modern Analysis* (Cambridge University Press, London, 1948).

[8] H. Überall, Phys. Rev. 128, 2429 (1962); L. I. Schiff, *ibid.* 103, 443 (1956).

EXTERNAL FIELD PRODUCED BY AN ARBITRARY SLOT ON A RIBBON COATED WITH A PENETRABLE SHEATH*)

by C. YEH

Department of Electrical Engineering, University of Southern California, Los Angeles, California, U.S.A.

Summary

Expressions are derived for the external electromagnetic field produced by an arbitrary slot on the conducting surface of a dielectric coated ribbon. It is assumed that the electric field tangential to the slot is a prescribed function. Boundary conditions are satisfied using the orthogonality characteristics of Mathieu functions. The integrals in the formal solution are evaluated by the saddle point method for the far zone fields. These far zone field solutions are then specialized to thin circumferential and axial slots. Illustrative numerical computations for the far zone field of a thin axial slot are carried out.

§ 1. *Introduction.* Slot antennas have been extensively utilized as microwave radiating systems. They are particularly useful for high-velocity vehicles, since there is no protruding obstacle to introduce extra drag.

The theoretical analysis of the radiation characteristics of slot antennas is very difficult except for cases where the slot is located on structures of simple geometric shapes**). Sinclair[1]), Papas[2]), Silver and Saunders[3]) and others[4]) have considered the problem of slot radiators on the circular cylinder, while Stratton and Chu[5]), Bailin and Silver[6]), etc. have attacked the exterior problems of slots on spheroids, spheres and cones. The solution for the radiated

---

*) This work was supported by Air Force Cambridge Research Laboratories, Office of Aerospace Research, U.S.A.

**) Even for these simplified cases, the interior and exterior fields must be treated separately. Hence, for the exterior problem, the fields across the slot are assumed known quantities.

fields produced by a slot on the elliptical cylinder has also been obtained by Sinclair[7]), Wong[8]) and Wait[9]).

When dielectric sheath or a plasma sheath surrounds the slot radiators, it is quite safe to assume that the radiated fields will be effected. The problems of the dielectric (or plasma) coated spheres or circular cylinders have been analyzed by several authors[10]) [11]). Those problems present no new mathematical difficulties and can be analyzed in a similar manner as for the uncoated ones. However, this is no longer true for the dielectric coated elliptic cylinders or ribbons, since the angular Mathieu functions are not only functions of η, the angular coordinates, but also of the electrical properties of the medium in which they apply. Fortunately, this difficulty may be overcome by means of the orthogonality relations of the Mathieu functions.

It is the purpose of this paper to derive expressions for the fields produced by a slot of arbitrary shape on the conducting surface of a dielectric coated elliptic cylinder of infinite length. The tangential components of the electric field in the slot are assumed to be prescribed. The integrals in the formal solution are then evaluated approximately for the far zone fields by a direct application of the saddle point method. These far zone field solutions are then specialized to thin circumferential and axial slots. Illustrative numerical computations, for the far zone field of the thin axial slot at the center of a dielectric coated ribbon, were carried out.

§ 2. *Formulating of the problem.* To analyze this problem, the elliptical cylinder coordinates (ξ, η, z), as shown in fig. 1, are introduced. In terms of the rectangular coordinates (x, y, z) the elliptical cylinder coordinates are defined by the following relations:

$$\begin{aligned} x &= q \cosh \xi \cos \eta, \\ y &= q \sinh \xi \sin \eta, \\ z &= z \\ &(0 \leq \xi < \infty,\ 0 \leq \eta \leq 2\pi), \end{aligned} \tag{1}$$

where q is the semi-focal length of the ellipse. The contour surfaces of constant ξ are confocal elliptic cylinders, and those of constant η are confocal hyperbolic cylinders. The outer boundary of the dielectric-coated conducting elliptic cylinder is assumed to coincide with one of the confocal elliptic cylinders with $\xi = \xi_1$; the inner boundary is assumed to coincide with one of the confocal elliptic cylinders with $\xi = \xi_0$. The electrical constants of the coating are ε_1 and μ and those of the homogeneous (air) space outside are ε_0 and μ.

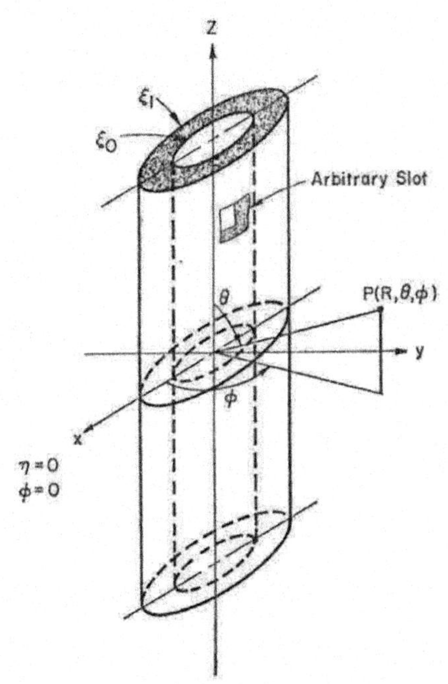

Fig. 1. Geometrical parameters.

A slot of arbitrary shape is located on the conducting cylinder bounded in the axial direction by the planes $z = z_1$ and $z = z_2$ and confined circumferentially by the curves $\eta_1(z)$ and $\eta_2(z)$. The tangential electric field in the slot will in general have components in both the η and z directions. The prescribed electric field in the slot is defined in the following manner:

$$E_\eta(\xi_0, \eta, z) = E_\eta^0(\eta, z), \; E_z(\xi_0, \eta, z) = E_z^0(\eta, z) \tag{2}$$

for $\eta_1 \leq \eta \leq \eta_2$ and $z_1 \leq z \leq z_2$, whereas

$$E_\eta(\xi_0, \eta, z) = 0, \; E_z(\xi_0, \eta, z) = 0 \tag{3}$$

for points (ξ_0, η, z) outside the slot.

A solution of Maxwell's equation will now be constructed which assumes the prescribed value of the tangential electric field on the conducting cylinder and gives rise to outgoing waves at infinity.

The field in a source-free homogeneous region can be generally represented as a superposition of a complete set of TE and TM cylindrical waves[12] as follows:

$$E_{z(1,0)} = \int_{-\infty}^{\infty} \sum_{m=0}^{\infty} (k_{(1,0)}^2 - \beta^2)\, A_m^{(1,0)}(\xi,\eta,\beta)\, e^{i(\beta z - \omega t)}\, d\beta, \tag{4}$$

$$H_{z(1,0)} = \int_{-\infty}^{\infty} \sum_{m=0}^{\infty} (k_{(1,0)}^2 - \beta^2)\, A_m^{+(1,0)}(\xi,\eta,\beta)\, e^{i(\beta z - \omega t)}\, d\beta, \tag{5}$$

$$E_{\xi(1,0)} = \int_{-\infty}^{\infty} \sum_{m=0}^{\infty} \left[\frac{i\beta}{p}\frac{\partial A_m^{(1,0)}}{\partial \xi} + \frac{i\mu\omega}{p}\frac{\partial A_m^{+(1,0)}}{\partial \eta}\right] e^{i(\beta z - \omega t)}\, d\beta, \tag{6}$$

$$H_{\xi(1,0)} = \int_{-\infty}^{\infty} \sum_{m=0}^{\infty} \left[-\frac{i\omega\varepsilon_{(1,0)}}{p}\frac{\partial A_m^{(1,0)}}{\partial \eta} + \frac{i\beta}{p}\frac{\partial A_m^{+(1,0)}}{\partial \eta}\right] e^{i(\beta z - \omega t)}\, d\beta, \tag{7}$$

$$E_{\eta(1,0)} = \int_{-\infty}^{\infty} \sum_{m=0}^{\infty} \left[\frac{i\beta}{p}\frac{\partial A_m^{(1,0)}}{\partial \eta} - \frac{i\omega\mu}{p}\frac{\partial A_m^{+(1,0)}}{\partial \xi}\right] e^{i(\beta z - \omega t)}\, d\beta, \tag{8}$$

$$H_{\eta(1,0)} = \int_{-\infty}^{\infty} \sum_{m=0}^{\infty} \left[\frac{-i\omega\varepsilon_{(1,0)}}{p}\frac{\partial A_m^{(1,0)}}{\partial \xi} - \frac{i\beta}{p}\frac{\partial A_m^{+(1,0)}}{\partial \eta}\right] e^{i(\beta z - \omega t)}\, d\beta, \tag{9}$$

where the symbol (1,0) is used to designate region 1 or 0 (the region within the dielectric shell or the free-space region).

$p = q\,(\sinh^2 \xi + \sin^2 \eta)^{\frac{1}{2}}$, $k_1^2 = \omega^2 \mu \varepsilon_1$, $k_0^2 = \omega^2 \mu \varepsilon_0$, and $A_m^{(1,0)}$ or $A_m^{+(1,0)}$ satisfy the wave equation

$$\frac{\partial^2 A_m^{(1,0)}}{\partial \xi^2} + \frac{\partial^2 A_m^{(1,0)}}{\partial \eta^2} + [q^2(k_{(1,0)}^2 - \beta^2)(\sinh^2\xi + \sin^2\eta)]\, A_m^{(1,0)} = 0. \tag{10}$$

The appropriate required expressions for $A_m^{(1,0)}$ are[13]

$$A_m^{(1)} = [A_m^e\, Ce_m(\xi, \gamma_1^2) + B_m^e\, Fey_m(\xi, \gamma_1^2)]\, ce_m(\eta, \gamma_1^2) + \\ + [A_m^0\, Se_m(\xi, \gamma_1^2) + B_m^0\, Gey_m(\xi, \gamma_1^2)]\, se_m(\eta, \gamma_1^2), \tag{11}$$

$$A_m^{(0)} = [F_m^e\, Me_m^{(1)}(\xi, \gamma_0^2)]\, ce_m(\eta, \gamma_0^2) + [F_m^0\, Ne_m^{(1)}(\xi, \gamma_0^2)]\, se_m(\eta, \gamma_0^2), \tag{12}$$

where $A_m^{(e,0)}$, $B_m^{(e,0)}$ and $F_m^{(e,0)}$ are arbitrary constants, and

$\gamma_1^2 = q^2(k_1^2 - \beta^2)$, $\gamma_0^2 = q^2(k_0^2 - \beta^2)$. $ce_m(\eta, \gamma^2)$ and $se_m(\eta, \gamma^2)$ are the even and odd angular Mathieu functions, and

$$\begin{Bmatrix} Ce_m\ (\xi, \gamma^2) \\ Fey_m(\xi, \gamma^2) \end{Bmatrix} \quad \text{and} \quad \begin{Bmatrix} Se_m\ (\xi, \gamma^2) \\ Gey_m(\xi, \gamma^2) \end{Bmatrix}$$

are the even and odd radial Mathieu functions. $Me_m^{(1)}(\xi, \gamma^2)$ and $Ne_m^{(1)}(\xi, \gamma^2)$ are defined as

$$\begin{aligned} Me_m^{(1)}(\xi, \gamma^2) &= Ce_m(\xi, \gamma^2) + i\, Fey_m(\xi, \gamma^2), \\ Ne_m^{(1)}(\xi, \gamma^2) &= Se_m(\xi, \gamma^2) + i\, Gey_m(\xi, \gamma^2). \end{aligned} \tag{13}$$

The appropriate expressions for $A_m^{+(1,0)}$ are of the same form as $A_m^{(1,0)}$ with arbitrary constants $C_m^{(e,0)}$, $D_m^{(e,0)}$ and $G_m^{(e,0)}$ instead of $A_m^{(e,0)}$, $B_m^{(e,0)}$ and $F_m^{(e,0)}$.

The formal solution of the problem is obtained when the coefficients $A_m^{(e,0)}$, $B_m^{(e,0)}$, $C_m^{(e,0)}$, $D_m^{(e,0)}$, $F_m^{(e,0)}$, $G_m^{(e,0)}$ are expressed in terms of the specified tangential electric field on the surface of the elliptic cylinder.

§ 3. *The formal solution.* The boundary conditions require the continuity of the tangential electric and magnetic fields at the boundary surface $\xi = \xi_1$. On the conducting surface, $\xi = \xi_0$, the tangential electric field must be zero everywhere except across the slot where it is equal to the applied field. It is possible to express the tangential field on the conducting cylinder as a combined azimuthal Fourier series and axial Fourier integral:

$$E_\eta(\xi_0, \eta, z) = \frac{1}{2\pi} \sum_{m=0}^{\infty} \frac{1}{p} \int_{-\infty}^{\infty} [W_m^e\, ce_m(\eta, \gamma_1^2) + W_m^0\, se_m(\eta, \gamma_1^2)]\, e^{i\beta z}\, d\beta, \tag{14a}$$

$$E_z(\xi_0, \eta, z) = \frac{1}{2\pi} \sum_{m=0}^{\infty} \int_{-\infty}^{\infty} [U_m^e\, ce_m(\eta, \gamma_1^2) + U_m^0\, se_m(\eta, \gamma_1^2)]\, e^{i\beta z}\, d\beta, \tag{14b}$$

where

$$W_m^e = \int_{z_1}^{z_2} \frac{1}{N_m^e} \int_{\eta_1(z')}^{\eta_2(z')} E_\eta^0(\eta', z')\, ce_m(\eta', \gamma_1^2)\, p'\, e^{-i\beta z'}\, d\eta'\, dz',$$

$$W_m^0 = \int_{z_1}^{z_2} \frac{1}{N_m^0} \int_{\eta_1(z')}^{\eta_2(z')} E_\eta^0(\eta', z',)\, se_m(\eta', \gamma_1^2)\, p'\, e^{-i\beta z'}\, d\eta'\, dz',$$

$$U_m^e = \int_{z_1}^{z_2} \frac{1}{N_m^e} \int_{\eta_1(z')}^{\eta_2(z')} E_z^0(\eta', z')\, ce_m(\eta', \gamma_1^2)\, e^{-i\beta z'}\, d\eta'\, dz',$$

$$U_m^0 = \int_{z_1}^{z_2} \frac{1}{N_m^0} \int_{\eta_1(z')}^{\eta_2(z')} E_z^0(\eta', z')\, se_m(\eta', \gamma_1^2)\, e^{-i\beta z'}\, d\eta'\, dz'.$$

$$N_m^e = \int_0^{2\pi} ce_m^2(\eta, \gamma_1^2)\, d\eta, \quad N_m^0 = \int_0^{2\pi} se_m^2(\eta, \gamma_1^2)\, d\eta, \tag{15}$$

and $p' = q(\sinh^2 \xi_0 + \sin^2\eta')^{\frac{1}{2}}$. E_η^0 and E_z^0 are the applied fields across the slot.

Upon matching the tangential magnetic and electric fields at $\xi = \xi_1$, one obtains

$$\sum_{m=0}^{\infty} (k_1^2 - \beta^2)\, A_m^{(1)}(\xi_1, \eta, \beta) = \sum_{m=0}^{\infty} (k_0^2 - \beta^2)\, A_m^{(0)}(\xi_1, \eta, \beta), \tag{16}$$

$$\sum_{m=0}^{\infty} (k_1^2 - \beta^2)\, A_m^{+(1)}(\xi_1, \eta, \beta) = \sum_{m=0}^{\infty} (k_0^2 - \beta^2)\, A_m^{+(0)}(\xi_1, \eta, \beta), \tag{17}$$

$$\sum_{m=0}^{\infty} \frac{\partial A_m^{+(1)}}{\partial \xi_1} = \sum_{m=0}^{\infty} \left\{ \frac{\beta}{\omega\mu} \left[\left(\frac{k_0^2 - \beta^2}{k_1^2 - \beta^2} \right) - 1 \right] \frac{\partial A_m^{(0)}}{\partial \eta} + \frac{\partial A_m^{+(0)}}{\partial \xi_1} \right\}, \tag{18}$$

$$\sum_{m=0}^{\infty} \left(\frac{\varepsilon_1}{\varepsilon_0} \frac{\partial A_m^{(1)}}{\partial \xi_1} \right) = \sum_{m=0}^{\infty} \left\{ \frac{\partial A_m^{(0)}}{\partial \xi_1} + \frac{\beta}{\omega\varepsilon_0} \left[1 - \left(\frac{k_0^2 - \beta^2}{k_1^2 - \beta^2} \right) \right] \frac{\partial A_m^{+(0)}}{\partial \eta} \right\}. \tag{19}$$

On the surface of the conducting elliptical cylinder, $\xi = \xi_0$, the assumed field (given by equations (4) through (9)) must be identically equal to the applied field (given by equations (14)) for all values of z and η. Therefore,

$$\sum_{m=0}^{\infty} (k_1^2 - \beta^2)\, A_m^{(1)}(\xi_0, \eta, \beta) = \frac{1}{2\pi} \sum_{m=0}^{\infty} [U_m^e\, ce_m(\eta, \gamma_1^2) + U_m^0\, se_m(\eta, \gamma_1^2)], \tag{20}$$

$$\sum_{m=0}^{\infty} \left[i\beta \frac{\partial A_m^{(1)}(\xi_0, \eta, \beta)}{\partial \eta} - i\omega\mu \frac{\partial A_m^{+(1)}(\xi_0, \eta, \beta)}{\partial \xi_0} \right] = \frac{1}{2\pi} \sum_{m=0}^{\infty} [W_m^e\, ce_m(\eta, \gamma_1^2) + W_m^0\, se_m(\eta, \gamma_1^2)]. \tag{21}$$

It is noted that in contrast with the circular cylinder and spherical cases the angular functions in the elliptical cylinder case are functions not only of the angular coordinate but also of the characteristics of the medium. Consequently, the summation signs and the angular functions in the above equations may not be omitted. However, it will be shown that this difficulty may be overcome by the use of the orthogonality properties of Mathieu functions[13]. Substituting the expansions

$$ce_m(\eta, \gamma_1^2) = \sum_{n=0}^{\infty}{}' \alpha_{m,n}\, ce_n(\eta, \gamma_0^2), \tag{22a}$$

$$se_m(\eta, \gamma_1^2) = \sum_{n=1}^{\infty}{}' \beta_{m,n}\, se_n(\eta, \gamma_0^2), \tag{22b}$$

$$\frac{\mathrm{d}}{\mathrm{d}\eta}[ce_m(\eta, \gamma_0^2)] = \sum_{n=1}^{\infty}{}' \gamma_{m,n}\, se_n(\eta, \gamma_0^2), \tag{22c}$$

$$\frac{\mathrm{d}}{\mathrm{d}\eta}[se_m(\eta, \gamma_0^2)] = \sum_{n=1}^{\infty}{}' \chi_{m,n}\, ce_n(\eta, \gamma_0^2), \tag{22d}$$

$$\frac{\mathrm{d}}{\mathrm{d}\eta}[ce_m(\eta, \gamma_1^2)] = \sum_{n=1}^{\infty}{}' \tau_{m,n}\, se_n(\eta, \gamma_1^2), \tag{22e}$$

$$\frac{\mathrm{d}}{\mathrm{d}\eta}[se_m(\eta, \gamma_1^2)] = \sum_{n=1}^{\infty}{}' K_{m,n}\, ce_n(\eta, \gamma_1^2), \tag{22f}$$

into equations (16) through (21), and applying the orthogonality relations of Mathieu functions, leads to the following expressions:

$$\frac{1}{h^2}\sum_{m=0}^{\infty}{}' (A_m^e a_m^e + B_m^e b_m^e)\, \alpha_{mn} = F_n^e f_n^e, \tag{23}$$

$$\frac{1}{h^2}\sum_{m=1}^{\infty}{}' (A_m^0 a_m^0 + B_m^0 b_m^0)\, \beta_{mn} = F_n^0 f_n^0, \tag{24}$$

$$\frac{1}{h^2}\sum_{m=0}^{\infty}{}' (C_m^e a_m^e + D_m^e b_m^e)\, \alpha_{mn} = G_n^e f_n^e, \tag{25}$$

$$\frac{1}{h^2}\sum_{m=1}^{\infty}{}' (C_m^0 a_m^0 + D_m^0 b_m^0)\, \beta_{mn} = G_n^0 f_n^0, \tag{26}$$

$$\sum_{m=0}^{\infty}{}' (C_m^e a_m^{e'} + D_m^e b_m^{e'})\, \alpha_{mn} = G_n^e f_n^{e'} + \sum_{m=1}^{\infty}{}' \frac{\beta}{\omega\mu}(h^2-1)\, F_m^0 f_m^0 \chi_{mn}, \tag{27}$$

$$\sum_{m=1}^{\infty}{}' (C_m^0 a_m^{0'} + D_m^0 b_m^{0'}) \beta_{mn} = G_n^0 f_n^{0'} + \sum_{m=1}^{\infty}{}' \frac{\beta}{\omega\mu} (h^2 - 1) F_m^e f_m^e \gamma_{mn}, \quad (28)$$

$$\sum_{m=0}^{\infty}{}' \frac{\varepsilon_1}{\varepsilon_0}(A_m^e a_m^{e'} + B_m^e b_m^{e'}) \alpha_{mn} = F_n^e f_n^{e'} + \sum_{m=1}^{\infty}{}' \frac{\beta}{\omega\varepsilon_0}(1-h^2) G_m^0 f_m^0 \chi_{mn}, \quad (29)$$

$$\sum_{m=1}^{\infty}{}' \frac{\varepsilon_1}{\varepsilon_0}(A_m^0 a_m^{0'} + B_m^0 b_m^{0'}) \beta_{mn} = F_n^0 f_n^{0'} + \sum_{m=1}^{\infty}{}' \frac{\beta}{\omega\varepsilon_0}(1-h^2) G_m^e f_m^e \gamma_{mn}, \quad (30)$$

$$(k_1^2 - \beta^2)(A_n^e c_n^e + B_n^e d_n^e) = \frac{1}{2\pi} U_n^e, \quad (31)$$

$$(k_1^2 - \beta^2)(A_n^0 c_n^0 + B_n^0 d_n^0) = \frac{1}{2\pi} U_n^0, \quad (32)$$

$$\sum_{m=0}^{\infty}{}' i\beta(A_m^e c_m^e + B_m^e d_m^e) \tau_{mn} - i\omega\mu(C_n^0 c_n^{0'} + D_n^0 d_n^{0'}) = \frac{1}{2\pi} W_n^0, \quad (33)$$

$$\sum_{m=1}^{\infty}{}' i\beta(A_m^0 c_m^0 + B_m^0 d_m^0) K_{mn} - i\omega\mu(C_n^e c_n^{e'} + D_n^e d_n^{e'}) = \frac{1}{2\pi} W_n^e, \quad (34)$$

$$(n = 0, 2, 4 \ldots \quad \text{or} \quad n = 1, 3, 5 \ldots)$$

where the abbreviations

$$a_m^e = Ce_m(\xi_1, \gamma_1^2), \qquad a_m^{e'} = \frac{d}{d\xi_1} [Ce_m(\xi_1, \gamma_1^2)],$$

$$a_m^0 = Se_m(\xi_1, \gamma_1^2), \qquad a_m^{0'} = \frac{d}{d\xi_1} [Se_m(\xi_1, \gamma_1^2)],$$

$$b_m^e = Fey_m(\xi_1, \gamma_1^2), \qquad b_m^{e'} = \frac{d}{d\xi_1} [Fey_m(\xi_1, \gamma_1^2)],$$

$$b_m^0 = Gey_m(\xi_1, \gamma_1^2), \qquad b_m^{0'} = \frac{d}{d\xi_1} [Gey_m(\xi_1, \gamma_1^2)],$$

$$c_m^e = Ce_m(\xi_0, \gamma_1^2), \qquad c_m^{e'} = \frac{d}{d\xi_0} [Ce_m(\xi_0, \gamma_1^2)],$$

$$c_m^0 = Se_m(\xi_0, \gamma_1^2), \qquad c_m^{0'} = \frac{d}{d\xi_0} [Se_m(\xi_0, \gamma_1^2)],$$

$$d_m^e = Fey_m(\xi_0, \gamma_1^2), \qquad d_m^{e'} = \frac{d}{d\xi_0} [Fey_m(\xi_0, \gamma_1^2)],$$

$$d_m^0 = Gey_m(\xi_0, \gamma_1^2), \qquad d_m^{0'} = \frac{d}{d\xi_0}\,[Gey_m(\xi_0, \gamma_1^2)],$$

$$f_m^e = Me_m^{(1)}(\xi_1, \gamma_0^2), \qquad f_m^{e'} = \frac{d}{d\xi_1}\,[Me_m^{(1)}(\xi_1, \gamma_0^2)],$$

$$f_m^0 = Ne_m^{(1)}(\xi_1, \gamma_0^2), \qquad f_m^{0'} = \frac{d}{d\xi_1}\,[Ne_m^{(1)}(\xi_1, \gamma_0^2)],$$

$$h^2 = (k_0^2 - \beta^2)/(k_1^2 - \beta^2), \tag{35}$$

have been used. α_{mn}, β_{mn}, γ_{mn}, χ_{mn}, τ_{mn}, and K_{mn} are given in the appendix. The prime on the summation sign means that when n is odd, the above series are summed over all odd values of m; and when n is even, the series are summed over all even values of m. The unknown coefficients $A_m^{e,0}$, $B_m^{e,0}$, $C_m^{e,0}$, $D_m^{e,0}$, $F_m^{e,0}$ and $G_m^{e,0}$ can now be obtained in terms of the known coefficients $U_n^{e,0}$ and $W_n^{e,0}$ from equations (23) through (34). The process is quite straight forward although the algebraic operations may be quite lengthy. For example, combining equations (24), (25), (27), (30), (32) and (34) gives

$$\sum_{m=0}^{\infty}{}' C_m^e s_{mn}^e + A_m^0 t_{mn}^0 = Q_n^0, \tag{36}$$

$$\sum_{m=0}^{\infty}{}' C_m^e u_{mn}^e + A_m^0 v_{mn}^0 = P_n^0 \tag{37}$$

$$(n = 0, 2, 4 \ldots, \quad \text{or} \quad n = 1, 3, 5 \ldots),$$

and combining equations (23), (26), (28), (29), (31) and (33) gives

$$\sum_{m=0}^{\infty}{}' C_m^0 s_{mn}^0 + A_m^e t_{mn}^e = Q_n^e, \tag{38}$$

$$\sum_{m=0}^{\infty}{}' C_m^0 u_{mn}^0 + A_m^e v_{mn}^e = P_n^e \tag{39}$$

$$(n = 0, 2, 4 \ldots, \quad \text{or} \quad n = 1, 3, 5 \ldots),$$

where

$$s_{mn}^e = \alpha_{mn}\left\{a_m^{e'} - \frac{f_n^{e'}}{f_n^e}\frac{1}{h^2}\,a_m^e - c_m^{e'}\frac{b_m^{e'}}{d_m^{e'}} + \frac{f_n^{e'}}{f_n^e}\frac{1}{h^2}\frac{b_m^e}{d_m^{e'}}\,c_m^{e'}\right\},$$

$$s_{mn}^0 = \beta_{mn}\left\{a_m^{0'} - \frac{f_n^{0'}}{f_n^0}\frac{1}{h^2}\,a_m^0 - c_m^{0'}\frac{b_m^{0'}}{d_m^{0'}} + \frac{f_n^{0'}}{f_n^0}\frac{1}{h^2}\frac{b_m^0}{d_m^{0'}}\,c_m^{0'}\right\},$$

$$t^e_{mn} = \frac{\beta}{\omega\mu}(1-h^2)\frac{1}{h^2}\left[a^e_m - \frac{b^e_m}{d^e_m}c^e_m\right]\sum_{r=0}^{\infty}{}' \alpha_{mr}\gamma_{rn},$$

$$t^0_{mn} = \frac{\beta}{\omega\mu}(1-h^2)\frac{1}{h^2}\left[a^0_m - \frac{b^0_m}{d^0_m}c^0_m\right]\sum_{r=0}^{\infty}{}' \beta_{mr}\chi_{rn},$$

$$u^e_{mn} = \frac{\beta}{\omega\varepsilon_0}(h^2-1)\frac{1}{h^2}\left[a^e_m - \frac{b^e_m}{d^{e'}_m}c^{e'}_m\right]\sum_{r=0}^{\infty}{}' \alpha_{mr}\gamma_{rn},$$

$$u^0_{mn} = \frac{\beta}{\omega\epsilon_0}(h^2-1)\frac{1}{h^2}\left[a^0_m - \frac{b^0_m}{d^{0'}_m}c^{0'}_m\right]\sum_{r=0}^{\infty}{}' \beta_{mr}\chi_{rn},$$

$$v^e_{mn} = \alpha_{mn}\left\{\frac{\varepsilon_1}{\varepsilon_0}a^{e'}_m - \frac{\varepsilon_1}{\varepsilon_0}\frac{b^{e'}_m}{d^e_m}c^e_m - \frac{1}{h^2}\frac{f^{e'}_n}{f^e_n}a^e_m + \frac{1}{h^2}\frac{f^{e'}_n}{f^e_n}\frac{b^e_m}{d^e_m}c^e_m\right\},$$

$$v^0_{mn} = \beta_{mn}\left\{\frac{\varepsilon_1}{\varepsilon_0}a^{0'}_m - \frac{\varepsilon_1}{\varepsilon_0}\frac{b^{0'}_m}{d^0_m}c^0_m - \frac{1}{h^2}\frac{f^{0'}_n}{f^0_n}a^0_m + \frac{1}{h^2}\frac{f^{0'}_n}{f^0_n}\frac{b^0_m}{d^0_m}c^0_m\right\},$$

$$Q^e_n = \sum_{m=0}^{\infty}{}'\left\{\frac{p^0_m}{d^{0'}_m}\beta_{mn}\left[\frac{1}{h^2}\frac{f^{0'}_n}{f^0_n}b^0_m - b^{0'}_m\right] + \right.$$
$$\left. + \frac{\beta}{\omega\mu}(h^2-1)\frac{1}{h^2}\sum_{r=0}^{\infty}{}'\frac{b^e_r}{d^e_r}q^e_r\alpha_{rm}\gamma_{mn}\right\},$$

$$Q^0_n = \sum_{m=0}^{\infty}{}'\left\{\frac{p^e_m}{d^{e'}_m}\alpha_{mn}\left[\frac{1}{h^2}\frac{f^{e'}_n}{f^e_n}b^e_m - b^{e'}_m\right] + \right.$$
$$\left. + \frac{\beta}{\omega\mu}(h^2-1)\frac{1}{h^2}\sum_{r=0}^{\infty}{}'\frac{b^0_r}{d^0_r}q^0_r\beta_{rm}\chi_{mn}\right\},$$

$$P^e_n = \sum_{m=0}^{\infty}{}'\left\{\frac{q^e_m}{d^e_m}\alpha_{mn}\left[\frac{1}{h^2}\frac{f^{e'}_n}{f^e_n}b^e_m - \frac{\varepsilon_1}{\varepsilon_0}b^{e'}_m\right] + \right.$$
$$\left. + \frac{\beta}{\omega\varepsilon_0}(1-h^2)\frac{1}{h^2}\sum_{r=0}^{\infty}{}'\frac{b^0_r}{d^{0'}_r}p^0_r\beta_{rm}\chi_{mn}\right\},$$

$$P^0_n = \sum_{m=0}^{\infty}{}'\left\{\frac{q^0_m}{d^0_m}\beta_{mn}\left[\frac{1}{h^2}\frac{f^{0'}_n}{f^0_n}b^0_m - \frac{\varepsilon_1}{\varepsilon_0}b^{0'}_m\right] + \right.$$
$$\left. + \frac{\beta}{\omega\varepsilon_0}(1-h^2)\frac{1}{h^2}\sum_{r=0}^{\infty}{}'\frac{b^e_r}{d^{e'}_r}p^e_r\alpha_{rm}\gamma_{mn}\right\},$$

$$p^e_l = \left(\frac{\beta}{\omega\mu}\frac{1}{2\pi(k_i^2-\beta^2)}\sum_{r=0}^{\infty}{}' U^0_r K_{rl}\right) + \frac{i}{2\pi\omega\mu}W^e_l,$$

$$p_l^0 = \left(\frac{\beta}{\omega\mu}\,\frac{1}{2\pi(k_1^2-\beta^2)}\,\sum_{r=0}^{\infty}{}' \,U_r^e \tau_{rl}\right) + \frac{i}{2\pi\omega\mu}\,W_l^0,$$

$$q_l^e = \frac{1}{2\pi(k_1^2-\beta^2)}\,U_l^0,$$

$$q_l^0 = \frac{1}{2\pi(k_1^2-\beta^2)}\,U_l^e. \tag{40}$$

The coefficients A_m^0 and C_m^e, and A_m^e and C_m^0 can be obtained readily from equations (36) and (37), and equations (38) and (39) respectively. The coefficients B_m^e, B_m^0, D_m^e, D_m^0, F_n^e, F_n^0, G_n^e, and G_n^0 can be found respectively from equations (31), (32), (34), (33), (23), (24), (25) and (26). The roots of the determinant of eqs. (36) and (37) provide the propagation constants of a set of surface waves along a dielectric coated elliptical conductor. Due to the asymmetry of the elliptical cylinder, it is possible to have two orientations for the field configurations. The propagation constants of the other set of surface waves are obtained from the roots of the determinant of eqs. (38) and (39). Finding the roots of these determinants in itself may prove to be a rather worthy project. The solution provides the propagation characteristics of the guided waves along an elliptical Goubau line[14].

Substituting the expressions for these coefficients back into eqs. (4) through (9), one obtains the formal solution of the problem. The integration with respect to β is not readily carried out for the general case. It is possible, however, if certain simplifying assumptions are made that an approximate evaluation can be carried out. For example, if the fields are to be observed at a large distance from the cylinder the integration with respect to β can be carried out by the saddle point method as will be illustrated in the next section.

§ 4. *The far-zone field.* The integrals, in the free-space region, to be evaluated are of the following form:

$$I_n^{r,0} = \int_{-\infty}^{\infty} T_n^{e,0}(\beta) \begin{Bmatrix} Me_n^{(1)} \\ Ne_n^{(1)} \end{Bmatrix} (\xi, \gamma_0^2)\, e^{i\beta z}\, d\beta, \tag{41}$$

where the dependence on the characteristics of the coating and the excitation parameters are included in the factor $T_n^{e,0}(\beta)$. It is noted that $T_n^{e,0}(\beta)$ are slowly varying functions compared with

$$\begin{Bmatrix} Me_n^{(1)} \\ Ne_n^{(1)} \end{Bmatrix} (\xi, \gamma_0^2) \, e^{i\beta z}.$$

The integration of β is along the real axis from $-\infty$ to $+\infty$ and the path of integration is indented above the branch point at $\beta = k_0$ and below the branch point at $\beta = -k_0$. Since $\xi \gg 1$ in the far field, the asymptotic expansion of the radial Mathieu functions, which are

$$\begin{aligned} Me_n^{(1)}(\xi, \gamma_0^2) &\simeq w_n^e(\beta)(2/\pi\,\rho\sqrt{k_0^2-\beta^2})^{\frac{1}{2}}\, e^{i[\rho\sqrt{k_0^2-\beta^2}-(2n+1/4)\pi]}, \\ Ne_n^{(1)}(\xi, \gamma_0^2) &\simeq w_n^0(\beta)(2/\pi\,\rho\sqrt{k_0^2-\beta^2})^{\frac{1}{2}}\, e^{i[\rho\sqrt{k_0^2-\beta^2}-(2n+1/4)\pi]}, \end{aligned} \tag{42}$$

where

$$\begin{aligned} w_n^e(\beta) &= p_n(-1)^n, \\ w_n^0(\beta) &= s_n(-1)^n, \end{aligned}$$

may be employed. The constants p_n and s_n are defined in [13]. ρ is the radial component in the circular cylindrical coordinates, (ρ, ϕ, z). Substituting (42) into (41) gives

$$I_n^{e,0} = \int_{-\infty}^{\infty} T_n^{e,0}(\beta) w_n^{e,0}(\beta)(2/\pi\,\rho\sqrt{k_0^2-\beta^2})^{\frac{1}{2}}\, e^{i[\rho\sqrt{k_0^2-\beta^2}-(2n+1/4)\pi]}\, e^{i\beta z}\, d\beta. \tag{43}$$

This integral can be evaluated immediately by using Van der Waerden's modified saddle point method[15] and thus eliminating the usual deformation of the contour to the path of steepest descent. Hence, (43) becomes

$$I_n^{e,0} \simeq (2\pi)^{\frac{1}{2}}\,(-i)\, e^{ik_0R}\, R^{-1}\, e^{-in\pi/2}\, T_n^{e,0}(k_0\cos\theta)\, w_n^{e,0}(k_0\cos\theta), \tag{44}$$

where

$$R = (\rho^2 + z^2)^{\frac{1}{2}},\ \theta = \tan^{-1}\rho/z,$$

and where terms containing R^{-2}, R^{-3}, etc. have been neglected. The spherical coordinates, (R, θ, ϕ), have been employed. The integrals of the form

$$J_n^{e,0} = \int_{-\infty}^{\infty} S_n^{e,0}(\beta) \frac{\partial}{\partial\xi} \begin{Bmatrix} Me_n^{(1)} \\ Ne_n^{(1)} \end{Bmatrix} (\xi, \gamma_0^2) \, e^{i\beta z}\, d\beta \tag{45}$$

can also be evaluated in a similar way. The result is

$$J_n^{e,0} \simeq (2\pi)^{\frac{1}{2}} k_0 \sin\theta \, e^{ik_0R} R^{-1} \rho \, e^{-in\pi/2} S_n^{e,0}(k_0 \cos\theta) \, w_n^{e,0}(k_0 \cos\theta). \quad (46)$$

The components of E_ξ and E_z in the R-direction that are of the order of magnitude R^{-1} cancel one another; the same occurs for the magnetic field. The radiation field is thus transverse to the R-direction and, in general, has non-zero components:

$$\begin{aligned} E_\theta &\simeq E_\xi \cos\theta - E_z \sin\theta, \\ E_\phi &\simeq E_\eta, \\ H_\theta &= -(\varepsilon_0/\mu)^{\frac{1}{2}} E_\phi, \\ H_\phi &= (\varepsilon_0/\mu)^{\frac{1}{2}} E_\theta. \end{aligned} \quad (47)$$

The far-zone radiation field of the arbitrary slot on the dielectric coated elliptic cylinder is now completely specified in terms of the prescribed tangential electric field. The remaining task is to carry out the integrations of $W_m^{e,0}$ and $U_m^{e,0}$ with respect to η' and z'.

§ 5. *Illustrative examples.* Even after making the far-zone approximations, the expressions for the fields are still quite cumbersome. The situations are simplified somewhat if the excitation source consists of a circumferential slot or a thin axial slot. The solutions of these special cases can be readily obtained from the general expressions and will be illustrated as follows:

1. The uniformly excited circumferential slot. The slot runs completely around the cylinder and has a height $2w$ which is very much smaller than the circumference. For convenience we shall take $z_1 = -w$, $z_2 = w$. It is also noted that $\eta_1(z') = 0$, $\eta_2(z') = 2\pi$. The tangential electric field in the slot has only a z component and is independent of η; thus

$$E_\eta^0(\eta', z') = 0; \; E_z^0(\eta', z') = E_0. \quad (48)$$

According to equation (15),

$$W_m^{e,0} = 0, \quad (49)$$

$$\begin{aligned} U_m^e &= 2E_0 w \, (\sin\beta \, w/\beta w) \left[\frac{1}{N_0^e} \int_0^{2\pi} ce_0(\eta', \gamma_1^2) \, d\eta' \right] \quad (m = 0), \\ &= 0 \quad (m \neq 0), \\ U_m^0 &= 0, \end{aligned} \quad (50)$$

and according to equation (40),

$$p_i^e = 0, \quad q_i^e = 0. \tag{51}$$

Hence, all coefficients are present. The field expressions are not much simplified.

2. The thin axial slot. It is assumed that a thin axial slot of width w is parallel to the axis of the cylinder. The length of the slot is $2l$. Across the slot is assumed a voltage distribution of $V(z)$. Thus, the tangential electric field in the slot has only an η component; i.e.,

$$E_\eta^0(\eta', z') = \frac{V(z')}{w} \delta(\eta' - \eta_0); \quad E_z^0(\eta', z') = 0, \tag{52}$$

where $\delta(\eta' - \eta_0)$ is such that

$$\int_0^{2\pi} \delta(\eta' - \eta_0)\, p'\, \mathrm{d}\eta' = 1 \qquad (\eta_1 < \eta_0 < \eta_2). \tag{53}$$

η_0 is the angular location of the slot on the cylinder. From equation (15), one gets

$$U_m^{e,0} = 0,$$

$$W_m^e = \frac{1}{N_m^e} \int_{-l}^{l} \int_0^{2\pi} \frac{V(z')}{w} \delta(\eta' - \eta_0)\, ce_m(\eta', \gamma_1^2)\, p'\, \mathrm{e}^{-i\beta z'}\, \mathrm{d}\eta'\, \mathrm{d}z'$$

$$= \frac{1}{N_m^e} ce_m(\eta_0, \gamma_1^2) \left[\int_{-l}^{l} \frac{V(z')}{w} \mathrm{e}^{-i\beta z'}\, \mathrm{d}z' \right],$$

$$W_m^0 = \frac{1}{N_m^0} se_m(\eta_0, \gamma_1^2) \left[\int_{-l}^{l} \frac{V(z')}{w} \mathrm{e}^{-i\beta z'}\, \mathrm{d}z' \right]. \tag{54}$$

Hence, according to equation (40),

$$q_i^e = q_i^0 = 0, \quad p_i^{e,0} = \frac{i}{2\pi\omega\mu} W_i^{e,0}, \tag{55}$$

and all coefficients are present. If the slot is sufficiently narrow, the voltage $V(z')$ may be approximated by the expression

$$V(z') = V_0 \frac{\sin k_0(l - |z'|)}{\sin k_0 l}, \tag{56}$$

where V_0 is the voltage at the center of the slot, the integral in (54) can be evaluated:

$$\int_{-l}^{l} \frac{V(z')}{w} e^{-i\beta z'} dz' = \frac{2V_0}{w} \frac{(\cos \beta l - \cos k_0 l)}{k_0(1 - \beta^2/k_0^2) \sin k_0 l}, \tag{57}$$

with $\beta = k_0 \cos \theta$ in the far zone.

§ 6. *Conclusions.* By the use of the orthogonality properties of the Mathieu functions, the exact solution for the electromagnetic field produced by an arbitrary slot on the surface of a conducting elliptic cylinder coated with a dielectric (or plasma) sheath is

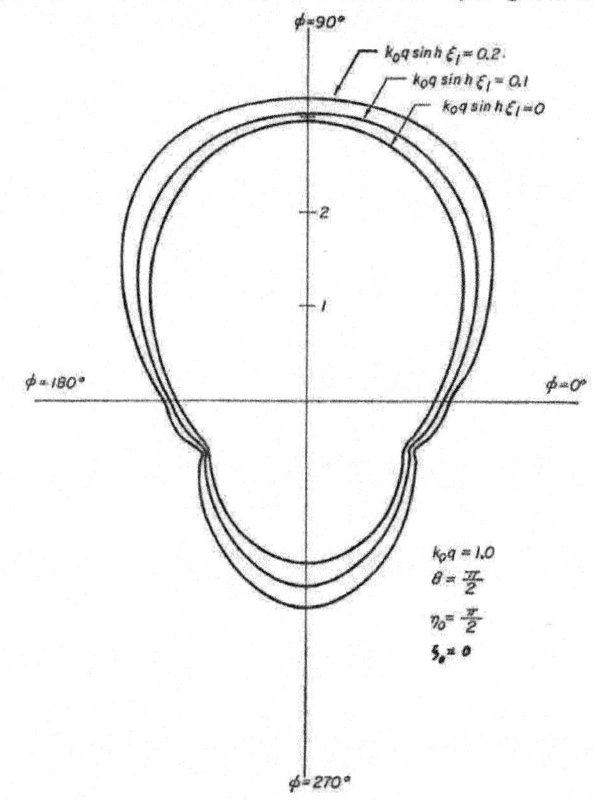

Fig. 2. Radiation patterns in the equatorial plane of a thin axial slot in the centre of a dielectric coated strip of width $2q$.

obtained. It can be seen that because of the involved coupling between the expansion coefficients for the radiated field and those for the applied field, the field expressions are considerably more complicated than those for the uncoated ones. Initial numerical computations are carried out using the available tables on the angular and radial Mathieu functions[16]), and using the IBM 7090 high-speed computor. Results show that the infinite series for the radiated fields converge very rapidly for small values of k_0q and for thin dielectric coating. For example, the relative radiated field $|E_{\phi_0}|$ for the axial slot at the center of a dielectric coated ribbon*) ($\eta_0 = \pi/2$) is computed as a function of ϕ for the equatorial plane ($\theta = \pi/2$). It is found that only the first four or five terms of the

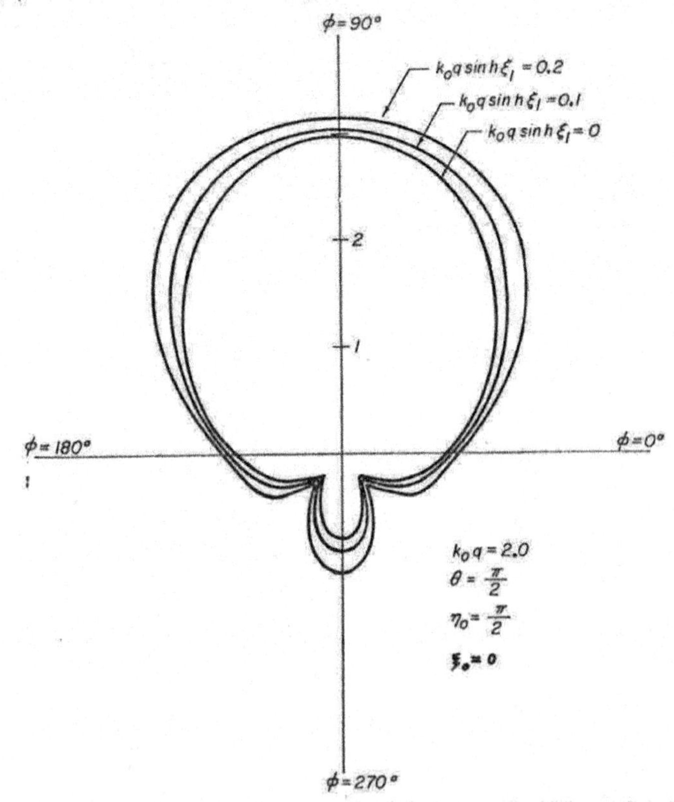

Fig. 3. Radiation patterns in the equatorial plane of a thin axial slot in the center of a dielectric coated strip of width $2q$.

infinite series are needed as long as $k_0q \cosh \xi_0$ is less than 4 and the thickness of the dielectric coating, $k_0q(\sinh \xi_1 - \sinh \xi_0)$, does not exceed 0.2. Results of the computation are given in figures 2 and 3. Two cases were considered: One with $k_0q = 1$,

*) The computational work becomes somewhat simpler when a dielectric coated ribbon is considered.

k_0q sinh $\xi_1 = 0$, 0.1, 0.2; the other with $k_0q = 2$, k_0q sinh $\xi_1 = 0$, 0.1, 0.2. It can be seen that the results approach smoothly those for the un-coated ribbon radiator as the thickness of the dielectric coating approaches zero [9]. It is interesting to note that the presence of the dielectric sheath tends to spread out the radiation pattern, and to diffract more field around to the rear of the ribbon. Like the case of a dielectric coated circular cylinder, the dielectric coating on a ribbon seems to enhance the ripples in the curves. The evidence of trapped circumferential surface waves with considerable leakage is also quite clear.

It is interesting to note that the mathematical technique used to satisfy the boundary conditions may prove worthy when one considers the case of an anisotropic plasma coated elliptical cylinder or other related problems.

Appendix. Applying the orthogonality characteristics of Mathieu functions [13] to equations (22), one obtains

$$\alpha_{mn} = \int_0^{2\pi} ce_m(\eta, \gamma_1^2)\, ce_n(\eta, \gamma_1^2)\, d\eta \,/ \int_0^{2\pi} ce_n^2(\eta, \gamma_0^2)\, d\eta,$$

$$\beta_{mn} = \int_0^{2\pi} se_m(\eta, \gamma_1^2)\, se_n(\eta, \gamma_0^2)\, d\eta \,/ \int_0^{2\pi} se_n^2(\eta, \gamma_0^2)\, d\eta,$$

$$\gamma_{mn} = \int_0^{2\pi} ce_m'(\eta, \gamma_0^2)\, se_n(\eta, \gamma_0^2)\, d\eta \,/ \int_0^{2\pi} se_n^2(\eta, \gamma_0^2)\, d\eta,$$

$$\chi_{mn} = \int_0^{2\pi} se_m'(\eta, \gamma_0^2)\, ce_n(\eta, \gamma_0^2)\, d\eta \,/ \int_0^{2\pi} ce_n^2(\eta, \gamma_0^2)\, d\eta,$$

$$\tau_{mn} = \int_0^{2\pi} ce_m'(\eta, \gamma_1^2)\, se_n(\eta. \gamma_1^2)\, d\eta \,/ \int_0^{2\pi} se_n^2(\eta, \gamma_1^2)\, d\eta,$$

$$K_{mn} = \int_0^{2\pi} se_m'(\eta, \gamma_1^2)\, ce_n(\eta, \gamma_1^2)\, d\eta \,/ \int_0^{2\pi} ce_n^2(\eta, \gamma_1^2)\, d\eta,$$

where the prime signifies the derivative of the function with respect to η. The above integrals can be easily integrated using the trigonometric expansions for the periodic Mathieu functions [13].

Acknowledgements. I would like to thank Miss Linda LaBella for typing this manuscript. The use of the computing facilities at the Western Data Processing Center at U.C.L.A.is gratefully acknowledged.
Received 11th December, 1962.

REFERENCES

1) Sinclair, G., Proc. Inst. Radio Engrs. **36** (1948) 1487.
2) Papas, C. H., J. Math. Phys. **28** (1950) 227.
3) Silver, S. and W. K. Saunders, J. Appl. Phys. **21** (1950) 153.
4) Pistolkors, A. A., J. Tech. Phys. U.S.S.R. **17** (1947) 377.
5) Chu, L. J. and J. A. Stratton, J. Appl. Phys. **12** (1941) 236, 241.
6) Bailin, L. L. and S. Silver, IRE Trans. on Antennas and Propagation AP-4 (1956) 5.
7) Sinclair, G., Proc. Inst. Radio Engrs. **33** (1951) 660.
8) Wong, J. Y., Proc. Inst. Tadio Engrs. **41** (1953) 1172.
9) Wait, J. R., J. Appl. Phys. **26** (1955) 458.
10) Wait, J. R., and W. Mientka, J. of Research N. B. S. **58** (1957) 287.
11) Scharfman, H. and D. D. King, Proc. Inst. Radio Engrs. **42** (1954) 854.
12) Stratton, J. A., Electromagnetic Theory, McGraw-Hill Book Comp. Inc., New York, 1941. p. 354.
13) McLachlan, N. W., Theory and Application of Mathieu Functions, University Press, Oxford, 1951.
14) Goubau, G., J. Appl. Phys. **21** (1950) 1119.
15) Van der Waerden, B. L., Appl. sci. Research **B2** (1951) 43.
16) Tables Relating to Matieu Functions, Columbia University Press, New York, 1951.

JOURNAL OF MATHEMATICAL PHYSICS VOLUME 5, NUMBER 3 MARCH 1964

An Application of Sommerfeld's Complex-Order Wavefunctions to an Antenna Problem*

C. Yeh
Electrical Engineering Department, University of Southern California, Los Angeles, California
(Received 27 February 1963; and in revised form 1 November 1963)

Using the orthogonality relations of Sommerfeld's complex-order wavefunctions, the exact solution for the problem of electromagnetic radiation from a circularly symmetric slot on the conducting surface of a dielectric-coated cone is obtained. The results are valid for the near-zone region as well as for the far-zone region, and they are applicable for arbitrary-angle cones. It is noted that the technique used to solve this problem may be applied to similar types of problems involving conical structure, such as the diffraction of waves by a dielectric-coated, spherically tipped cone.

I. INTRODUCTION

THE problems of scattering of waves by a perfectly conducting conical obstacle or radiation from such a structure have been considered by many authors.[1-5] The exact mathematical solution to the problem of the diffraction of waves by a finite, perfectly conducting cone has recently been obtained by Northover.[6] However, the corresponding solution for the diffraction by or radiation from a dielectric-coated, semi-infinite conical structure has not been found. It is the purpose of this paper to present the exact solution of the radiation from this dielectric-coated structure. It is shown that certain mathematical difficulties can be overcome by the use of Sommerfeld's complex-order wavefunctions[7] and their orthogonality properties.

II. FORMULATION OF THE PROBLEM

To analyze this problem, the spherical coordinates (r, θ, ϕ) are used. The geometry of this conical structure is shown in Fig. 1. The vertex of the cone is taken to coincide with the origin of the spherical polar coordinates. To eliminate the singularity at the vertex, a small perfectly conducting spherical boss of radius a, with its center at the origin, is situated at the tip of the cone. The outer boundary of the dielectric-coated cone is assumed to coincide with $\theta = \theta_1$; the inner boundary is assumed to coincide with $\theta = \theta_0$. The dielectric coating has a permittivity of ϵ_1, a permeability μ, and a conductivity of zero. It is assumed that this radiating structure is embedded in a homogeneous, perfect dielectric medium $(\epsilon_0, \mu; \sigma_0 = 0)$, and that the applied electric-field intensity across the slot, which is located on the perfectly conducting conical surface $\theta = \theta_0$, is circularly symmetric about the axis of the cone and linearly polarized in the radial direction.

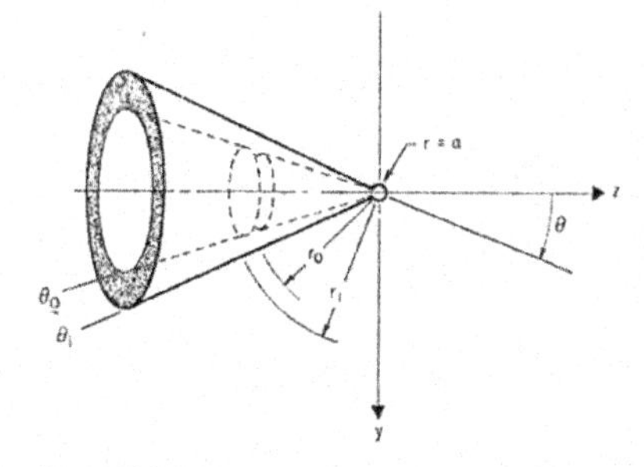

Fig. 1 The dielectric coated spherically tipped cone.

Due to the symmetrical characteristics of this problem, all components of the electromagnetic field are independent of the azimuthal angle ϕ. For a TM wave, the nonvanishing components are E_r, E_θ, and H_ϕ. The wave equation in spherical coordinates takes the form

$$\frac{\partial^2}{\partial r^2}(rH_\phi) + \frac{1}{r}\frac{\partial}{\partial \theta}\left[\frac{1}{\sin\theta}\frac{\partial}{\partial\theta}(H_\phi \sin\theta)\right] + k^2 r H_\phi = 0, \qquad (1)$$

where $k^2 = \omega^2\mu\epsilon$ and the steady-state time depend-

* This work was supported by the Air Force Cambridge Research Laboratories.

[1] H. S. Carslaw, Math Ann. **75**, 133 (1914).

[2] W. W. Hansen and L. I. Schiff, "Theoretical Study of Electromagnetic Waves Scattered from Shaped Metal Surfaces," Quarterly Progress Rept. No. 4, Stanford University Microwave Laboratory, September, 1948.

[3] L. B. Felsen, J. Appl. Phys. **26**, 138 (1955).

[4] K. M. Siegel, R. F. Goodrich, and V. H. Weston, Appl. Sci. Res. **B8**, 8 (1959).

[5] J. B. Keller, "Back Scattering from a Finite Cone," New York University, Courant Institute of Mathematical Sciences, Division of Electromagnetic Research, Res. Rept. No. EM-127, (1959).

[6] F. H. Northover, Quart. J. Mech. Appl. Math. **15**, 1 (1962).

[7] A. Sommerfeld, *Partial Differential Equations* (Academic Press Inc., New York, 1949).

ence $e^{-i\omega t}$ has been assumed. Setting

$$H_\phi = i\omega\epsilon(\partial u/\partial\theta) \tag{2}$$

in Eq. (1) gives

$$(\nabla^2 + k^2)u(r, \theta) = 0. \tag{3}$$

A possible solution of Eq. (3) is then $u(r, \theta) = R(r)\Theta(\theta)$, where R and Θ satisfy the differential equations

$$d/dr\,[r^2(dR/dr)] + (k^2r^2 - c)R = 0, \tag{4}$$

$$d/d\theta\,[\sin\theta(d\Theta/d\theta)] + c\sin\theta\Theta = 0, \tag{5}$$

in which c is the separation constant. If one chooses $c = \nu(\nu + 1)$ where ν may be a complex number, the solutions of Eq. (5) are the Legendre functions

$$\begin{Bmatrix} P_\nu(\cos\theta) \\ Q_\nu(\cos\theta) \end{Bmatrix}.$$

The corresponding solutions of Eq. (4) are the spherical Hankel functions

$$\begin{Bmatrix} h_\nu^{(1)}(kr) \\ h_\nu^{(2)}(kr) \end{Bmatrix}.$$

The proper choice of these functions to represent the electromagnetic fields depends upon the boundary conditions. All field components must be finite in all regions (i.e., the region within the dielectric sheath and the region outside the sheath). In addition, all field components for the radiated wave must satisfy Sommerfeld's radiation condition at infinity. Consequently, the appropriate solution for the region inside the dielectric sheath is

$$u_s(r, \theta) = \sum_\nu h_\nu^{(1)}(k_1 r) \times [A_\nu P_\nu(\cos\theta) + B_\nu Q_\nu(\cos\theta)], \tag{6}$$

and that for the radiated wave,

$$u_r(r, \theta) = \sum_{\nu'} G_{\nu'} h_{\nu'}^{(1)}(k_0 r) P_{\nu'}(\cos\theta), \tag{7}$$

where $k_1^2 = \omega^2\mu\epsilon_1$ and $k_0^2 = \omega^2\mu\epsilon_0$. A_ν, B_ν, and $G_{\nu'}$ are arbitrary constants to be determined by the boundary conditions. The summation is over all values of ν which are determined by the boundary condition on the spherical boss at the tip of the cone; i.e., the tangential electric field must vanish on the perfectly conducting spherical boss:

$$\partial/\partial r\,[rh_\nu^{(1)}(k_1 r)]|_{r=a} = 0, \tag{8}$$

$$\partial/\partial r\,[rh_{\nu'}^{(1)}(k_0 r)]|_{r=a} = 0. \tag{9}$$

The roots of ν from Eqs. (8) and (9) will be designated, respectively, by ν_n and ν_n'. It should be noted that $h_\nu^{(1)}(k_1 r)$ or $h_{\nu'}^{(1)}(k_0 r)$ are orthogonal over the range $k_1 a$ to ∞ or $k_0 a$ to ∞, respectively. (The proof is given in Appendix A.) It is because of this orthogonality property of these radial functions that they are so useful for the conical problems. The orthogonality characteristic of the Hankel functions with complex order was first investigated by Sommerfeld.[7] Hence these functions are also called Sommerfeld's complex-order wavefunctions.[8]

III. THE MATHEMATICAL SOLUTION

The boundary conditions require the continuity of the tangential electric and magnetic fields at the boundary surface $\theta = \theta_1$. On the conducting surface $\theta = \theta_0$, the tangential electric field must be zero everywhere except across the gap where it is equal to the applied field. Let the applied field be defined by

$$E_r^{\text{app}} = E_0 d(r) e^{-i\omega t}, \tag{10}$$

where $d(r)$ is defined as follows:

$$d(r) = \begin{cases} 1 & \text{for } r_0 < r < r_1, \\ 0 & \text{for } a < r < r_0, \text{ and } r > r_1. \end{cases} \tag{11}$$

$|r_1 - r_0|$ is the gap width (see Fig. 1). Expanding the applied field in terms of Sommerfeld's complex-order wavefunctions, gives

$$E_r^{\text{app}} = \frac{1}{r}\sum_{\nu_n} L_{\nu_n}\nu_n(\nu_n + 1)h_{\nu_n}^{(1)}(k_1 r)P_{\nu_n}(\cos\theta_0) \times e^{-i\omega t}, \tag{12}$$

where

$$L_{\nu_n} = \frac{1}{\nu_n(\nu_n + 1)P_{\nu_n}(\cos\theta_0)N_{\nu_n}(k_1 a)} \times \int_{r_0}^{r_1} E_0 r h_{\nu_n}^{(1)}(k_1 r)\, d(k_1 r), \tag{13}$$

in which the normalizing factor is

$$N_{\nu_n}(k_1 a) = \int_{k_1 a}^{\infty} [h_{\nu_n}^{(1)}(k_1 r)]^2\, d(k_1 r). \tag{14}$$

Expression (12) represents E_r^{app} for all values of r between a and ∞.

Upon matching the tangential magnetic and electric fields at $\theta = \theta_1$, one obtains

$$i\omega\epsilon_0 \sum_{\nu_m'} G_{\nu_m'} h_{\nu_m'}^{(1)}(k_0 r)\left[\frac{d}{d\theta_1}P_{\nu_m'}(\cos\theta_1)\right] = i\omega\epsilon_1 \sum_{\nu_n} h_{\nu_n}^{(1)}(k_1 r)\left[A_{\nu_n}\frac{d}{d\theta_1}P_{\nu_n}(\cos\theta_1) + B_{\nu_n}\frac{d}{d\theta_1}Q_{\nu_n}(\cos\theta_1)\right], \tag{15}$$

[8] C. H. Papas, J. Math. and Phys. 33, 269 (1954).

$$\sum_{\nu_m'} G_{\nu_m'}\nu_m'(\nu_m'+1)h_{\nu_m'}^{(1)}(k_0r)P_{\nu_m'}(\cos\theta_1) = \sum_{\nu_n}\nu_n(\nu_n+1)h_{\nu_n}^{(1)}(k_1r) \times [A_{\nu_n}P_{\nu_n}(\cos\theta_1)+B_{\nu_n}Q_{\nu_n}(\cos\theta_1)]. \quad (16)$$

It is noted that, in contrast with the spherical and circular cylindrical boundary-value problems, the boundary conditions cannot be satisfied by equating each term of the series expansion. For the present case, the above equations must be satisfied for all values of r from $r = a$ to $r = \infty$. Consequently, the orthogonality properties of the radial function must be utilized to overcome the difficulty. Substituting the expansion

$$h_{\nu_n}^{(1)}(k_1r) = \sum_{\nu_m'} \alpha_{\nu_n,\nu_m'} h_{\nu_m'}^{(1)}(k_0r) \quad (17)$$

into Eqs. (15) and (16), and applying the orthogonality relations of the radial function, leads to the following expressions:

$$\frac{\epsilon_0}{\epsilon_1} G_{\nu_m'} g'_{\nu_m'} = \sum_{\nu_n}(A_{\nu_n}a'_{\nu_n} + B_{\nu_n}b'_{\nu_n})\alpha_{\nu_n,\nu_m'}, \quad (18)$$

$$G_{\nu_m'}\nu_m'(\nu_m'+1)g_{\nu_m'} = \sum_{\nu_n}(A_{\nu_n}a_{\nu_n} + B_{\nu_n}b_{\nu_n}) \times \nu_n(\nu_n+1)\alpha_{\nu_n,\nu_m'} \quad (\nu_m' = \nu_0', \nu_1', \nu_2', \cdots), \quad (19)$$

where the abbreviations

$$a_{\nu_n} = P_{\nu_n}(\cos\theta_1), \quad a'_{\nu_n} = (d/d\theta_1)P_{\nu_n}(\cos\theta_1),$$
$$b_{\nu_n} = Q_{\nu_n}(\cos\theta_1), \quad b'_{\nu_n} = (d/d\theta_1)Q_{\nu_n}(\cos\theta_1),$$
$$g_{\nu_m'} = P_{\nu_m'}(\cos\theta_1), \quad g'_{\nu_m'} = (d/d\theta_1)P_{\nu_m'}(\cos\theta_1) \quad (20)$$

have been used. $\alpha_{\nu_n,\nu_m'}$ is given in Appendix B. Expressing B_{ν_n} in terms of A_{ν_n} gives (in matrix notation)

$$B_{\nu_n} = R^{-1}_{\nu_n,\nu_m'} D_{\nu_m'}, \quad (21)$$

where $R^{-1}_{\nu_n,\nu_m'}$ is the inverse of the matrix

$$[R_{\nu_m',\nu_n}] = \left[\left(\frac{\epsilon_1}{\epsilon_0} b'_{\nu_n} g_{\nu_m'}\nu_m'(\nu_m'+1) - b_{\nu_n}\nu_n(\nu_n+1)g'_{\nu_m'}\right)\alpha_{\nu_n,\nu_m'}\right],$$

and $D_{\nu_m'}$ is a column matrix

$$\left[\sum_{\nu_s} A_{\nu_s}\left(a_{\nu_s}\nu_s(\nu_s+1)g'_{\nu_m'} - \frac{\epsilon_1}{\epsilon_0} a'_{\nu_s} g_{\nu_m'}\nu_m'(\nu_m'+1)\right)\alpha_{\nu_s,\nu_m'}\right].$$

Equation (21) can also be written in the form

$$B_{\nu_n} = \sum_{\nu_s} h_{\nu_n,\nu_s} A_{\nu_s}, \quad (22)$$

where h_{ν_n,ν_s} are obtained using Eq. (21).

At the surface of the conducting cone, $\theta = \theta_0$, E_r in the dielectric sheath and Eq. (12) must be identically equal for all values of r between a and ∞. Therefore,

$$A_{\nu_n}P_{\nu_n}(\cos\theta_0) + B_{\nu_n}Q_{\nu_n}(\cos\theta_0) = L_{\nu_n}P_{\nu_n}(\cos\theta_0), \quad (23)$$

where L_{ν_n} is given by Eq. (13). Making the identification

$$d_{\nu_n} = P_{\nu_n}(\cos\theta_0), \quad f_{\nu_n} = Q_{\nu_n}(\cos\theta_0), \quad (24)$$

and substituting Eq. (22) into Eq. (23), one finds

$$A_{\nu_n}d_{\nu_n} + f_{\nu_n}\sum_{\nu_s} h_{\nu_n,\nu_s}A_{\nu_s} = L_{\nu_n}d_{\nu_n}. \quad (25)$$

Solving for A_{ν_n} gives

$$A_{\nu_n} = [Q^{-1}_{\nu_s,\nu_n}][L_{\nu_n}d_{\nu_n}], \quad (26)$$

where $Q^{-1}_{\nu_s,\nu_n}$ is the inverse of the matrix

$$[Q_{\nu_m,\nu_n}] = [(d_{\nu_n}\delta_{\nu_n,\nu_m} + h_{\nu_n,\nu_m}f_{\nu_n})], \quad (27)$$

and δ_{ν_n,ν_m} is the Kronecker delta which is equal to zero when $\nu_n \neq \nu_m$, and is equal to unity when $\nu_n = \nu_m$. $[L_{\nu_n}d_{\nu_n}]$ is a column matrix. With the knowledge of A_{ν_n} and B_{ν_n}, the coefficient $G_{\nu_m'}$ can easily be computed using either Eq. (18) or Eq. (19).

The electromagnetic fields in the dielectric shell and in the free space are now completely determined. At large distances from the radiating source, the asymptotic expressions for $h_{\nu'}^{(1)}(k_0r)$ which is

$$(e^{ik_0r}/k_0r)e^{-i(\nu'+1)\frac{1}{2}\pi},$$

leads to

$$H_\phi^r = (i\omega\epsilon_0 e^{ik_0r}/k_0r) \times \sum_{\nu'} G_{\nu'}\left[\frac{\partial}{\partial\theta}P_{\nu'}(\cos\theta)\right]e^{-i(\nu'+1)\frac{1}{2}\pi}. \quad (28)$$

IV. NUMERICAL COMPUTATIONS

The influence of the presence of a dielectric sheath upon the electromagnetic field radiated from a spherically tipped cone can now be computed. However, it is noted that the computation is by no means trivial since the required Sommerfeld's complex-order wavefunctions have not been tabulated; only certain limiting values are known at present. Hence, the task of tabulating these complex-order wavefunctions was first carried out for $kr \leq 10$ and $|\nu| \leq 10$. This tabulation is given elsewhere.[9]

In a recent article by Keller, Rubinow, and

[9] C. Yeh, "Tabulation of Complex Order Spherical Hankel Functions," USCEC Rept., Electrical Engineering Dept., University of Southern California, (June, 1963).

Goldstein,[10] the complex zero $\nu_n(z)$, $n = 1, 2 \cdots$ of $H_\nu^{(1)}(z)$, $dH_\nu^{(1)}(z)/dz$ and $dH_\nu^{(1)}(z)/dz + iZH_\nu^{(1)}(z)$ were investigated. They obtained approximate expressions for ν_n for small and large values of z by using appropriate approximate representations for Bessel functions. Also given were the numerical solutions of ν_n $n = 1, 2, 3, 4, 5$ for $H_\nu^{(1)}(z)$ and $dH_\nu^{(1)}(z)/dz$, with $0.01 \leq z \leq 7$. Similar procedures as those given by Keller, Rubinow, and Goldstein (KRG) will be used to find the roots of Eqs. (8) and (9). However it will be pointed out that the approximate expressions of ν_n for $H_\nu^{(1)}(z)$ and $dH_\nu^{(1)}(z)/dz$ given here for $z \ll 1$ are better approximations than those given by KRG. Starting with the expression

$$H_\nu^{(1)}(z) = (J_{-\nu}(z) - J_\nu(z)e^{-i\nu\pi})/i \sin \nu\pi \tag{29}$$

with

$$J_{\pm\nu}(z) = \sum_{m=0}^{\infty} \frac{(-1)^m(\frac{1}{2}z)^{\pm\nu+2m}}{m!\,\Gamma(\pm\nu + 1 + m)}, \tag{30}$$

and making the appropriate approximation for $0 < z \ll 1$, one has

$$-\sum_{m=1}^{\infty} \frac{\zeta(2m+1)}{(2m+1)} \nu^{2m+1} + (\tfrac{1}{2}i\pi - \gamma - \log \tfrac{1}{2}z)\nu - in\pi + O(z^2) = 0, \tag{31}$$

where γ is the Euler's constant and ζ is the Riemann zeta function. The condition $H_\nu^{(1)}(z) = 0$ has been used to obtain Eq. (31). It should be noted that Eq. (31) is identical to Eq. (9) in KRG's paper. However instead of expanding ν as a power series in $[\log \frac{1}{2}z]^{-1}$ as given by KRG, successive-approximation method is used here to solve for ν. We have

$$\nu_n = \frac{1}{[\frac{1}{2}i\pi - \gamma - \log \frac{1}{2}z]} \left\{ in\pi + \tfrac{1}{3}[\zeta(3)] \times \left[\frac{in\pi}{(\frac{1}{2}i\pi - \gamma - \log \frac{1}{2}z)}\right]^3 + O(z^2) \right\} \quad z \ll 1. \tag{32}$$

This expression is different from KRG's Eq. (10). To get an idea of the improvement, a numerical example is carried out. For $z = 0.01$, KRG's Eq. (10) yields $\mathrm{Re}\,\nu_1 = 0.205$, $\mathrm{Im}\,\nu_1 = 0.613$, while Eq. (32) gives $\mathrm{Re}\,\nu_1 = 0.180$, $\mathrm{Im}\,\nu_1 = 0.593$. The numerical solution is $\mathrm{Re}\,\nu_1 = 0.184$, $\mathrm{Im}\,\nu_1 = 0.592$.

In the same way, when $z \ll 1$, we have

$$\nu_n = \frac{1}{(\frac{1}{2}i\pi - \gamma - \log \frac{1}{2}z)} \left\{ i\pi(n - \tfrac{1}{2}) + \tfrac{1}{3}[\zeta(3)]\left[\frac{i\pi(n-\frac{1}{2})}{(\frac{1}{2}i\pi - \gamma - \log \frac{1}{2}z)}\right]^3 + O(z^2) \right\} \tag{33}$$

[10] J. B. Keller, S. I. Rubinow, and M. Goldstein, J. Math. Phys. 4, 829 (1963).

for the roots of $dH_\nu^{(1)}(z)/dz = 0$. This Eq. (33) is also different from KRG's Eq. (16).

Returning now to the problem of finding the roots of Eq. (8) or (9), we note that Eq. (8) or (9) can be simplified to give

$$H_\mu^{(1)}(v) + 2v(d/dv)H_\mu^{(1)}(v) = 0, \tag{34}$$

where $\mu = \nu + \frac{1}{2}$ and v may be k_0a or k_1a. When $v \ll 1$, the following expression can be obtained using Eqs. (29) and (30):

$$\log \frac{\Gamma(\mu+1)}{\Gamma(-\mu+1)} = \log\left(\frac{1+2\mu}{1-2\mu}\right) + 2n\pi i + 2\mu \log \tfrac{1}{2}v - i\mu\pi + O(v^2). \tag{35}$$

Substituting the following series:

$$\log \frac{\Gamma(\mu+1)}{\Gamma(-\mu+1)} = -2\gamma\mu - 2\sum_{m=1}^{\infty} \frac{\zeta(2m+1)}{2m+1} \mu^{2m+1}, \tag{36}$$

$$\log\left(\frac{1+2\mu}{1-2\mu}\right) = 2\sum_{m=1}^{\infty} \frac{\mu^{2m+1}}{2m+1} 2^{2m+1} \tag{37}$$

into Eq. (35) and simplifying, gives

$$\mu(2 \log \tfrac{1}{2}v - i\pi) = -2n\pi i - 2\gamma\mu - 2\sum_{m=1}^{\infty} \frac{2^{2m+1} + \zeta(2m+1)}{2m+1} \mu^{2m+1} + O(u^2). \tag{38}$$

When $v \ll 1$, using the method of successive approximation, one obtains

$$\mu_n = \frac{1}{\frac{1}{2}i\pi - \gamma - \log \frac{1}{2}v} \left\{ in\pi + \tfrac{1}{3}[8 + \zeta(3)] \times \left[\frac{1}{\frac{1}{2}i\pi - \gamma - \log \frac{1}{2}v}\right]^3 + O(v^2) \right\}. \tag{39}$$

It is noted that for very small values of v, Eq. (39) approaches Eq. (32). Hence the roots for Eq. (34) approaches the roots for $H_\mu^{(1)}(v) = 0$ when v is very small.

Numerical solutions of Eq. (34) were carried out for moderate values of $v (0.1 < v < 1.0)$ using the IBM 7090 computer. It is found that the roots μ_n given by Eq. (39) are within 15% of those given by Eq. (34) provided that $v \leq 0.1$ and $n \leq 5$, and the roots computed from the simplified Eq. (35) are within 3% of those obtained from Eq. (34) when $v \leq 0.1$. As a matter of fact, the roots from the simplified Eq. (35) are in agreement within 5% of those obtained from Eq. (34) even when $v \leq 0.3$. It should be noted that the roots are symmetric about the origin since $H_{-\mu}^{(1)}(v) = e^{-i\mu\pi}H_\mu^{(1)}(v)$.[10]

For large values of v and $|\nu_n|$, the approximate expression for the roots of Eq. (34) can be derived with the help of Debye's expansion for the Bessel functions.[10,11] Since we are concerned with the problem of a cone with a small spherical boss at the tip, the roots of Eq. (34) for large value of v will not be considered.

Since the tables for complex-order Legendre functions are also not available, numerical values are computed from the series expansions for the Legendre functions.[12] To compute $\alpha_{\nu_n,\nu_m'}$ from Eq. (B1) (see Appendix B), one must evaluate the integral

$$I = \int_{k_0 a}^{\infty} h_{\nu_n}^{(1)}(k_1 r) h_{\nu_m'}^{(1)}(k_0 r)\, d(k_0 r). \quad (40)$$

This integral may be separated into two parts as follows:

$$I = I_1 + I_2, \quad (41)$$

where

$$I_1 = \int_{k_0 a}^{\xi} h_{\nu_n}^{(1)}(k_1 r) h_{\nu_m'}^{(1)}(k_0 r)\, d(k_0 r), \quad (42)$$

$$I_2 = \int_{\xi}^{\infty} h_{\nu_n}^{(1)}(k_1 r) h_{\nu_m'}^{(1)}(k_0 r)\, d(k_0 r). \quad (43)$$

The integral I_2 may be evaluated approximately using the asymptotic expression for the spherical Hankel function,

$$h_\nu^{(1)}(kr) \approx (1/kr) e^{ikr} e^{-i(\nu+1)\frac{1}{2}\pi}, \quad (44)$$

provided that $kr \gg |\nu|$. ξ is chosen such that $\xi \gg |\nu_n|$ and $|\nu_m'|$. Substituting Eq. (44) into Eq. (43) and carrying out the integration gives

$$I_2 \approx \frac{k_0}{k_1} e^{-i(\nu_n+\nu_m'+2)\frac{1}{2}\pi} \left\{ \frac{1}{\xi} \cos(\eta\xi) - \eta[\tfrac{1}{2}\pi - \mathrm{Si}(\eta\xi)] + i\left[\frac{1}{\xi}\sin(\eta\xi) - \eta\,\mathrm{Ci}(\eta\xi)\right] \right\}, \quad (45)$$

with

$$\eta = 1 + k_1/k_0.$$

$\mathrm{Si}(x)$ and $\mathrm{Ci}(x)$ are, respectively, Sine integral and Cosine integral,[13] i.e.,

$$\mathrm{Si}(x) = \tfrac{1}{2}\pi - \int_x^{\infty} \frac{\sin u}{u}\, du,$$

$$\mathrm{Ci}(x) = \int_{\infty}^{x} \frac{\cos u}{u}\, du. \quad (46)$$

The integral I_1 cannot be integrated analytically and must therefore be evaluated numerically. Simpson's rule was used. It is found from the computations that the coefficients $\alpha_{\nu_n,\nu_m'}$ converge quite rapidly for small values of $\chi = k_1/k_0 - 1$; only the first few terms of the infinite series [Eq. (17)] are needed as long as χ is less than 3.

To qualitatively illustrate how the solutions behave, the radiation patterns of $|H_\phi^i|$ given by Eq. (28) are computed. It is assumed that the conical structure with $\theta_0 = 165°$ (i.e., a cone angle of 30°) is excited by a delta slot located at $k_0 r_1$. Two cases are considered: one with $k_0 r_1 = \pi$, and the other with $k_0 r_1 = 3\pi$. The assumption of a delta-slot source simplifies considerably the expression for L_{ν_n} in Eq. (13); we have

$$L_{\nu_n} = \frac{E_0 r_1 k_1 h_{\nu_n}^{(1)}(k_1 r_1)}{\nu_n(\nu_n+1) P_{\nu_n}(\cos\theta_0) N_{\nu_n}(k_1 a)}.$$

Three cases of the thickness of the dielectric coating with $\epsilon_1/\epsilon_0 = 2.0$ are considered: $\theta_1 = 165°$, $\theta_1 = 162.5°$, and $\theta_1 = 160°$. The spherical boss at the tip of the cone has a radius of $k_0 a = 0.1$. The roots ν_n and ν_m' are computed, respectively, from Eq. (8) with $k_1 a = 0.1414$, and from Eq. (9) with $k_0 a = 0.1$. A total of 20 roots each were found for ν_n and ν_m'; i.e., $n = \pm 1, \pm 2, \cdots \pm 10$ and $m = \pm 1, \pm 2, \cdots \pm 10$. Results for the computed radiation patterns are shown in Figs. 2 and 3. Numerical investigation

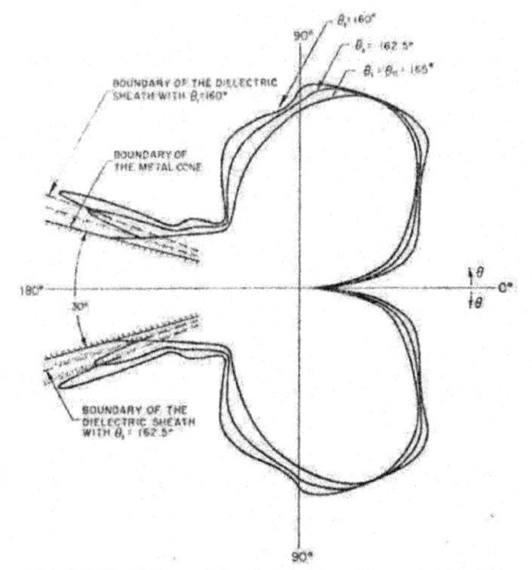

FIG. 2. Radiation patterns for dielectric coated cone excited by axially symmetric circumferential slot. Thickness of the dielectric sheath is indicated by θ_1.

[11] W. Franz, Z. Naturforsch. 9A, 705 (1954).

[12] Bateman manuscript project, *Higher Transcendental Functions*, edited by A. Erdély (McGraw-Hill Book Company, Inc., New York, 1953), Vol. 1.

[13] E. Jahnke and F. Emde, *Tables of Functions*, (Dover Publications, Inc., New York, 1945).

shows that the infinite series representing these expansion coefficients for the radiated wave converge quite rapidly for small values of k_1a and k_1r_1. Upon comparison of the radiation pattern for the uncoated, spherically tipped cone (i.e., the $\theta_1 = 165°$ case in Fig. 2) with the pattern given by Bailin and Silver[14] for the uncoated cone, one notes that if the spherical boss is small (i.e., $ka \ll 1$), and if $k_0r_1 > 1$, the presence of the spherical tip alters only slightly the radiation patterns for the cone. Only the pattern in the forward direction is somewhat enlarged; this is due to the induced currents on the spherical boss. The presence of the dielectric sheath tends to spread out the radiation pattern, and to diffract more field to the rear of the cone; the radiated field in the forward direction was not altered significantly. It can also be observed that the dielectric coating seems to enhance the ripples in the curves.

It is remarked here that the formal exact solutions given in Sec. III are valid for the near zone (i.e., near the conical structure) as well as for the far zone.

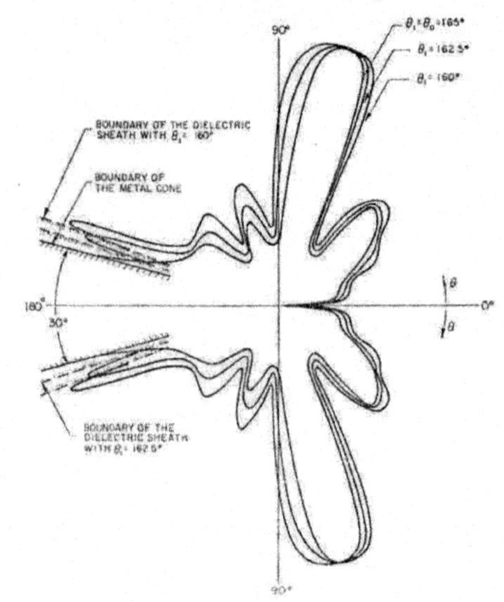

FIG. 3. Radiation patterns for dielectric coated cone excited by axially symmetric circumferential slot. Thickness of the dielectric sheath is indicated by θ_1.

APPENDIX A. ORTHOGONALITY CHARACTERISTICS OF $h_\nu^{(1)}(kr)$

To show that the radial functions $h_\nu^{(1)}(kr)$ are orthogonal over the range $kr = ka$ to $kr = \infty$,[7] one notes that for any ν, say, ν_n,

$$(kr)(d^2/d(kr)^2)(krh_{\nu_n}^{(1)}(kr)) + ((kr)^2 - \nu_n(\nu_n + 1))h_{\nu_n}^{(1)}(kr) = 0, \quad \text{(A1)}$$

and for any other value of ν, say, ν_m,

$$(kr)(d^2/d(kr)^2)(krh_{\nu_m}^{(1)}(kr)) + ((kr)^2 - \nu_m(\nu_m + 1))h_{\nu_m}^{(1)}(kr) = 0. \quad \text{(A2)}$$

Multiplying the first equation by $h_{\nu_m}^{(1)}(kr)$ and the second by $h_{\nu_n}^{(1)}(kr)$ and integrating the difference from $kr = ka$ to $kr = \infty$, one gets

$$[\nu_m(\nu_m + 1) - \nu_n(\nu_n + 1)] \int_{ka}^{\infty} h_{\nu_n}^{(1)}(kr)h_{\nu_m}^{(1)}(kr)\, d(kr)$$
$$= \int_{ka}^{\infty} \left[krh_{\nu_n}^{(1)}(kr) \frac{d^2}{d(kr)^2}(krh_{\nu_m}^{(1)}(kr)) - krh_{\nu_m}^{(1)}(kr) \frac{d^2}{d(kr)^2}(krh_{\nu_n}^{(1)}(kr)) \right] d(kr). \quad \text{(A3)}$$

Integrating the above by parts gives

$$[\nu_m(\nu_m + 1) - \nu_n(\nu_n + 1)] \int_{ka}^{\infty} h_{\nu_n}^{(1)}(kr)h_{\nu_m}^{(1)}(kr)\, d(kr)$$
$$= krh_{\nu_n}^{(1)}(kr) \frac{d}{d(kr)}(krh_{\nu_m}^{(1)}(kr)) - krh_{\nu_m}^{(1)}(kr) \frac{d}{d(kr)}(krh_{\nu_n}^{(1)}(kr)) \Big|_{ka}^{\infty}. \quad \text{(A4)}$$

The terms on the right-hand side of the equal sign are zero by virtue of the boundary condition (8) or (9) and the asymptotic behavior of $h_{\nu_n}^{(1)}(kr)$. Hence,

$$\int_{ka}^{\infty} h_{\nu_n}^{(1)}(kr)h_{\nu_m}^{(1)}(kr)\, d(kr) = \delta_{\nu_n,\nu_m} N_{\nu_n}(ka), \quad \text{(A5)}$$

where δ_{ν_n,ν_m} is the Kronecker delta and $N_{\nu_n}(ka)$ is a normalization factor which can be obtained from Eq. (A4) by an application of de l'Hospital's rule for the limit $\nu_n \to \nu_m$. Substituting the relations

$$\Delta\nu_m = \nu_n - \nu_m,$$

and

$$krh_{\nu_n}^{(1)}(kr) = krh_{\nu_m}^{(1)}(kr) + (\partial/\partial\nu_m)(krh_{\nu_m}^{(1)}(kr))\Delta\nu_m$$

into Eq. (A4) and using the boundary condition

$$[\partial/\partial(ka)][kah_{\nu_m}^{(1)}(ka)] = 0,$$

one obtains the normalization factor

$$N_{\nu_n}(ka) = \int_{ka}^{\infty} [h_{\nu_n}^{(1)}(kr)]^2\, d(kr)$$
$$= \frac{1}{2\nu_n + 1} \left\{ kah_{\nu_n}^{(1)}(ka) \frac{\partial^2}{\partial(ka)\,\partial(\nu_n)} [kah_{\nu_n}^{(1)}(ka)] \right\}. \quad \text{(A6)}$$

[14] L. L. Bailin and S. Silver, IRE Trans. Antennas Propagation 4, 5 (1956).

APPENDIX B. FORMULA FOR $\alpha_{\nu_n \nu_m'}$

Multiplying both sides of (17) by $h_{\nu_m'}^{(1)}(k_0 r)$, integrating with respect to $k_0 r$ from $k_0 a$ to ∞, and using the orthogonality relation for the radial function (A5), one obtains

$$\alpha_{\nu_n, \nu_m'} = \frac{1}{M_{\nu_m'}} \int_{k_0 a}^{\infty} h_{\nu_n}^{(1)}(k_1 r) h_{\nu_m'}^{(1)}(k_0 r)\, d(k_0 r), \qquad \text{(B1)}$$

where

$$M_{\nu_m'} = \int_{k_0 a}^{\infty} [h_{\nu_m'}(k_0 r)]^2\, d(k_0 r)$$

$$= \frac{1}{2\nu_m' + 1} \left\{ k_0 a h_{\nu_m'}^{(1)}(k_0 a) \frac{\partial^2}{\partial(k_0 a)\, \partial(\nu_m')} [k_0 a h_{\nu_m'}^{(1)}(k_0 a)] \right\}. \qquad \text{(B2)}$$

ELECTROMAGNETIC RADIATION FROM AN ARBITRARY SLOT ON A CONDUCTING CYLINDER COATED WITH A UNIFORM COLD PLASMA SHEATH WITH AN AXIAL STATIC MAGNETIC FIELD*

C. YEH

Electrical Engineering Department, University of Southern California, Los Angeles, California

Received October 29, 1963

ABSTRACT

The expressions for the radiated fields of an arbitrary slot on a conducting cylinder coated with a cold plasma sheath with an axial static magnetic field are obtained. It is well known from the plane-wave analysis that the presence of a static magnetic field has a considerable effect on the transmission characteristics of the signal through the plasma sheath. The principal concern of this paper is to consider the effects of a static magnetic field on the radiation pattern of a plasma-coated slot antenna. Extensive numerical computations are carried out for the special case of an infinite axial slot. It is found that, if the plasma frequency is moderate, a small axial static magnetic field could change significantly the radiation patterns of the slotted antenna. Furthermore, owing to the anisotropy of the plasma medium, the radiation patterns are asymmetric.

I. INTRODUCTION

As a space vehicle reenters the atmosphere, a plasma sheath surrounding the vehicle is generated. It is well known that the presence of such a plasma sheath interrupts the telemetering of necessary data as a result of the diminished signal. The possibility of the elimination of the reentry radio blackout using an impressed static magnetic field has been discussed by many investigators (Hodara *et al.* 1960; Dirsa 1960; Harley and Tyras 1961; Hodara 1961). However, most previous work was carried out assuming a planar geometry (Hodara *et al.* 1960; Dirsa 1960; Harley and Tyras 1961; Hodara 1961; Miller 1962; Hodara and Cohn 1962; Allis *et al.* 1963). The purpose of this paper is to analyze the radiation characteristics of an arbitrary slot on the conducting surface of a circular cylinder coated with a cold lossless plasma with an impressed static magnetic field in the axial direction. The presence of such a static magnetic field changes significantly the electromagnetic properties of the plasma. Because of the anisotropic characteristics of the plasma sheath, it is expected that the radiation pattern of a symmetrical slot will be asymmetrical. Numerical computations were carried out for the special case of an infinitely long axial slot for various values of normalized plasma- and gyro-frequencies.

II. FORMULATION OF THE PROBLEM

The geometry of the present problem is shown in Fig. 1. The plasma surrounding the cylinder is anisotropic by virtue of an impressed magnetostatic field in the axial direction. It is assumed that the plasma may be adequately described in terms of an average macroscopic dielectric tensor. This dielectric tensor of the cold anisotropic plasma with the magnetostatic field in the positive z direction is given by (Allis *et al.* 1963)

*This study was supported by the Air Force Cambridge Research Laboratories.

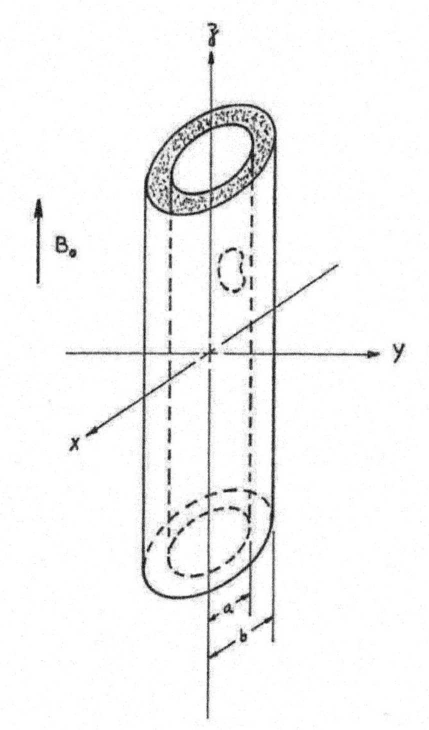

FIG. 1. Geometrical configuration.

$$\mathbf{\varepsilon} = \begin{bmatrix} \epsilon_1 & -i\epsilon_2 & 0 \\ i\epsilon_2 & \epsilon_1 & 0 \\ 0 & 0 & \epsilon_3 \end{bmatrix} \tag{1}$$

in which

$$\begin{aligned} \epsilon_1 &= \epsilon_0 \left(1 - \frac{\omega_p^2}{\omega^2 - \omega_g^2}\right), \\ \epsilon_2 &= \epsilon_0 \left(\frac{\omega_p^2 \, \omega_g/\omega}{\omega^2 - \omega_g^2}\right), \\ \epsilon_3 &= \epsilon_0 \left(1 - \frac{\omega_p^2}{\omega^2}\right), \end{aligned} \tag{2}$$

where ω_p and ω_g are the angular electron plasma- and electron gyro-frequencies, respectively, ϵ_0 is the dielectric constant of free space, and ω is the angular frequency. A time dependence of $e^{i\omega t}$ is assumed. It is further assumed that the applied fields across the arbitrary slot on the conducting cylinder are specified. Hence, the tangential electric field can be written as a combined azimuthal Fourier series and axial Fourier integral as follows (Wait 1959):

(3) $$E_z(a, \theta, z) = \int_{-\infty}^{\infty} \sum_{n=-\infty}^{\infty} U_n e^{-in\theta} e^{-i\beta z} d\beta,$$

(4) $$E_\theta(a, \theta, z) = \int_{-\infty}^{\infty} \sum_{n=-\infty}^{\infty} W_n e^{-in\theta} e^{-i\beta z} d\beta,$$

with

(5) $$U_n = \frac{1}{4\pi^2} \int_{z_1}^{z_2} \int_{\theta_1(z')}^{\theta_2(z')} E_{z\,\mathrm{slot}}(a, \theta', z') e^{in\theta'} e^{i\beta z'} d\theta' dz',$$

(6) $$W_n = \frac{1}{4\pi^2} \int_{z_1}^{z_2} \int_{\theta_1(z')}^{\theta_2(z')} E_{\theta\,\mathrm{slot}}(a, \theta', z') e^{in\theta'} e^{i\beta z'} d\theta' dz'.$$

where the slot is considered to be bounded by $\theta_1 \leqslant \theta' \leqslant \theta_2$ and $z_1 < z' < z_2$ in terms of the primed coordinates, and $E_{z\mathrm{slot}}$ and $E_{\theta\mathrm{slot}}$ are the specified tangential electric fields across the slot.

The solutions of Maxwell's equations in an anisotropic homogeneous medium in the cylindrical coordinates are well known (Allis *et al.* 1963). Hence, the proper expressions for the axial electric and magnetic field for the region within the plasma sheath are

(7) $$E_z^i = \int_{-\infty}^{\infty} \sum_{n=-\infty}^{\infty} \Phi_{zn}^{(i)}(r) e^{-in\theta} e^{-i\beta z} d\beta,$$

(8) $$H_z^i = \int_{-\infty}^{\infty} \sum_{n=-\infty}^{\infty} \psi_{zn}^{(i)}(r) e^{-in\theta} e^{-i\beta z} d\beta,$$

with

(9) $$\Phi_{zn}^{(i)}(r) = \alpha_1[A_n J_n(T_1 r) + B_n Y_n(T_1 r)] + \alpha_2[C_n J_n(T_2 r) + D_n Y_n(T_2 r)],$$

(10) $$\psi_{zn}^{(i)}(r) = [A_n J_n(T_1 r) + B_n Y_n(T_1 r)] + [C_n J_n(T_2 r) + D_n Y_n(T_2 r)],$$

where

(11) $$T_1^2 = \frac{i\omega\mu_0}{id + \alpha_1 c} \quad \text{and} \quad T_2^2 = \frac{i\omega\mu_0}{id + \alpha_2 c}.$$

α_1 and α_2 are the roots of the algebraic equation

(12) $$\alpha^2 + i\alpha\left(\frac{d}{c} + \frac{s}{c}\frac{\mu_0}{\epsilon_3}\right) + \frac{\mu_0}{\epsilon_3} = 0,$$

and

(13) $$c = \frac{\beta\gamma_2}{\gamma_1^2 - \gamma_2^2}, \qquad d = -\frac{\omega\mu_0\gamma_1}{\gamma_1^2 - \gamma_2^2}, \qquad s = \frac{\omega\epsilon_1\gamma_1 + \omega\epsilon_2\gamma_2}{\gamma_1^2 - \gamma_2^2},$$

$$\gamma_1^2 = (\beta^2 - \omega^2\mu_0\epsilon_1)^2, \qquad \gamma_2^2 = (\omega^2\mu_0\epsilon_2)^2.$$

μ_0 is the permeability of free space, and $J_n(x)$ and $Y_n(x)$ are respectively the Bessel function and the Neumann function of order n and argument x. A_n, B_n, C_n, and D_n are the arbitrary constants that are to be determined from the boundary conditions. It is noted that all transverse components of the electromagnetic field in this anisotropic medium can easily be obtained from Maxwell's equations by differentiation. The required expressions for the axial field components for free space (i.e., $r \geqslant b$) in the cylindrical coordinates are also well known. They are

$$E_z^{(r)} = \int_{-\infty}^{\infty} \sum_{n=-\infty}^{\infty} \Phi_{zn}^{(r)}(r) e^{-in\theta} e^{-i\beta z} dz, \tag{14}$$

$$H_z^{(r)} = \int_{-\infty}^{\infty} \sum_{n=-\infty}^{\infty} \psi_{zn}^{(r)}(r) e^{-in\theta} e^{-i\beta z} dz, \tag{15}$$

where

$$\Phi_{zn}^{(r)} = F_n H_n^{(2)}(ur), \tag{16}$$

$$\psi_{zn}^{(r)} = G_n H_n^{(2)}(ur), \tag{17}$$

with $u^2 = k_0^2 - \beta^2$ and $k_0^2 = \omega^2 \mu_0 \epsilon_0$. The Hankel function $H_n^{(2)}(ur)$ of the second kind of order n and argument ur is used to assure the proper asymptotic behavior of the radiated waves for large values of r. F_n and G_n are arbitrary constants. Again the transverse components of the field can be found from Maxwell's equations.

III. THE FORMAL SOLUTIONS

The boundary conditions require the continuity of the tangential electric and magnetic fields at the boundary surface, $r = b$. On the conducting surface, $r = a$, the tangential electric field must be zero everywhere except across the slot, where it is equal to the applied field. Hence, we have

$$\begin{bmatrix} a_1 & b_1 & c_1 & d_1 & f_1 & 0 \\ a_2 & b_2 & c_2 & d_2 & 0 & g_2 \\ a_3 & b_3 & c_3 & d_3 & f_3 & g_3 \\ a_4 & b_4 & c_4 & d_4 & f_4 & g_4 \\ a_5 & b_5 & c_5 & d_5 & 0 & 0 \\ a_6 & b_6 & c_6 & d_6 & 0 & 0 \end{bmatrix} \begin{bmatrix} A_n \\ B_n \\ C_n \\ D_n \\ F_n \\ G_n \end{bmatrix} = \begin{bmatrix} 0 \\ 0 \\ 0 \\ 0 \\ U_n \\ W_n \end{bmatrix}, \tag{18}$$

where

$$a_1 = \alpha_1 J_n(T_1 b), \tag{19}$$

$$a_2 = J_n(T_1 b),$$

$$a_3 = \frac{k_0^2 - \beta^2}{\gamma_1^2 - \gamma_2^2} \left[\gamma_1 \frac{n}{b} \beta \alpha_1 J_n(T_1 b) + \gamma_1 \beta \alpha_1 T_1 J_n'(T_1 b) - i\omega\mu_0 \gamma_1 T_1 J_n'(T_1 b) - i\omega\mu_0 \gamma_2 \frac{n}{b} J_n(T_1 b) \right],$$

$$a_4 = -\frac{k_0^2 - \beta^2}{\gamma_1^2 - \gamma_2^2} \left[\frac{n}{b} \gamma_1 \beta J_n(T_1 b) + \beta \gamma_1 T_1 J_n'(T_1 b) \right.$$

$$+ (i\omega\epsilon_1\gamma_1 + i\omega\epsilon_2\gamma_2)\alpha_1 T_1 J_n'(T_1 b) + \left(i\omega\epsilon_2\gamma_1 \frac{n}{b} + i\omega\epsilon_1\gamma_2 \frac{n}{b}\right) \alpha_1 J_n(T_1 b)\Bigg],$$

$$a_5 = \alpha_1 J_n(T_1 a),$$

$$a_6 = \frac{1}{\gamma_1^2 - \gamma_2^2}\Bigg[\gamma_1 \frac{n}{a} \beta\alpha_1 J_n(T_1 a) + \gamma_2\beta\alpha_1 T_1 J_n'(T_1 a) - i\omega\mu_0\gamma_1 T_1 J_n'(T_1 a) - i\omega\mu_0\gamma_2 \frac{n}{a} J_n(T_1 a)\Bigg];$$

$$b_1 = \alpha_1 Y_n(T_1 b),$$

$$b_2 = Y_n(T_1 b),$$

$$b_3 = \frac{k_0^2 - \beta^2}{\gamma_1^2 - \gamma_2^2}\Bigg[\gamma_1 \frac{n}{b} \beta\alpha_1 Y_n(T_1 b) + \gamma_2\beta\alpha_1 T_1 Y_n'(T_1 b) - i\omega\mu_0\gamma_1 T_1 Y_n'(T_1 b) - i\omega\mu_0\gamma_2 \frac{n}{b} Y_n(T_1 b)\Bigg],$$

$$b_4 = -\frac{k_0^2 - \beta^2}{\gamma_1^2 - \gamma_2^2}\Bigg[\frac{n}{b} \gamma_1\beta Y_n(T_1 b) + \beta\gamma_1 T_1 Y_n'(T_1 b) + (i\omega\epsilon_1\gamma_1 + i\omega\epsilon_2\gamma_2)\alpha_1 T_1 Y_n'(T_1 b) + \left(i\omega\epsilon_2\gamma_1 \frac{n}{b} + i\omega\epsilon_1\gamma_2 \frac{n}{b}\right) \alpha_1 Y_n(T_1 b)\Bigg],$$

$$b_5 = \alpha_1 Y_n(T_1 a),$$

$$b_6 = \frac{1}{\gamma_1^2 - \gamma_2^2}\Bigg[\gamma_1 \frac{n}{a} \beta\alpha_1 Y_n(T_1 a) + \gamma_2\beta\alpha_1 T_1 Y_n'(T_1 a) - i\omega\mu_0\gamma_1 T_1 Y_n'(T_1 a) - i\omega\mu_0\gamma_2 \frac{n}{a} Y_n(T_1 a)\Bigg];$$

$$c_1 = \alpha_2 J_n(T_2 b),$$

$$c_2 = J_n(T_2 b),$$

$$c_3 = \frac{k_0^2 - \beta^2}{\gamma_1^2 - \gamma_2^2}\Bigg[\gamma_1 \frac{n}{b} \beta\alpha_2 J_n(T_2 b) + \gamma_2\beta\alpha_2 T_2 J_n'(T_2 b) - i\omega\mu_0\gamma_1 T_2 J_n'(T_2 b) - i\omega\mu_0\gamma_2 \frac{n}{b} J_n(T_2 b)\Bigg],$$

$$c_4 = -\frac{k_0^2 - \beta^2}{\gamma_1^2 - \gamma_2^2}\Bigg[\frac{n}{b} \gamma_1\beta J_n(T_2 b) + \beta\gamma_1 T_2 J_n'(T_2 b) + (i\omega\epsilon_1\gamma_1 + i\omega\epsilon_2\gamma_2)\alpha_2 T_2 J_n'(T_2 b) + \left(i\omega\epsilon_2\gamma_1 \frac{n}{b} + i\omega\epsilon_1\gamma_2 \frac{n}{b}\right) \alpha_2 J_n(T_2 b)\Bigg],$$

$$c_5 = \alpha_2 J_n(T_2 a),$$

$$c_6 = \frac{1}{\gamma_1^2 - \gamma_2^2}\Bigg[\gamma_1 \frac{n}{a} \beta\alpha_2 J_n(T_2 a) + \gamma_2\beta\alpha_2 T_2 J_n'(T_2 a)$$

$$- i\omega\mu_0\gamma_1 T_2 J_n'(T_2a) - i\omega\mu_0\gamma_2 \frac{n}{a} J_n(T_2a)\Big];$$

$$d_1 = \alpha_2 Y_n(T_2b),$$

$$d_2 = Y_n(T_2b),$$

$$d_3 = \frac{k_0^2 - \beta^2}{\gamma_1^2 - \gamma_2^2}\Big[\gamma_1 \frac{n}{b} \beta\alpha_2 Y_n(T_2b) + \gamma_2\beta\alpha_2 T_2 Y_n'(T_2b)$$

$$- i\omega\mu_0\gamma_1 T_2 Y_n'(T_2b) - i\omega\mu_0\gamma_2 \frac{n}{b} Y_n(T_2b)\Big],$$

$$d_4 = -\frac{k_0^2 - \beta^2}{\gamma_1^2 - \gamma_2^2}\Big[\frac{n}{b}\gamma_1\beta Y_n(T_2b) + \beta\gamma_1 T_2 Y_n'(T_2b)$$

$$+ (i\omega\epsilon_1\gamma_1 + i\omega\epsilon_2\gamma_2)\alpha_2 T_2 Y_n'(T_2b) + \left(i\omega\epsilon_2\gamma_1 \frac{n}{b} + i\omega\epsilon_1\gamma_2 \frac{n}{b}\right)\alpha_2 Y_n(T_2b)\Big];$$

$$d_5 = \alpha_2 Y_n(T_2a),$$

$$d_6 = \frac{1}{\gamma_1^2 - \gamma_2^2}\Big[\gamma_1 \frac{n}{a} \beta\alpha_2 Y_n(T_2a) + \gamma_2\beta\alpha_2 T_2 Y_n'(T_2a)$$

$$- i\omega\mu_0\gamma_1 T_2 Y_n'(T_2a) - i\omega\mu_0\gamma_2 \frac{n}{a} Y_n(T_2a)\Big];$$

$$f_1 = H_n^{(2)}(ub),$$

$$f_3 = -\frac{n}{b}\beta H_n^{(2)}(ub),$$

$$f_4 = -i\omega\epsilon_0 u H_n^{(2)\prime}(ub);$$

$$g_2 = -H_n^{(2)}(ub),$$

$$g_3 = -i\omega\mu_0 u H_n^{(2)\prime}(ub),$$

$$g_4 = -\frac{n}{b}\beta H_n^{(2)}(ub).$$

In these expressions, the prime signifies the derivative of the function with respect to its argument. It is noted that the roots of the determinant of equation (18) provide the propagation constants of surface waves along a conducting cylinder coated with an anisotropic plasma sheath. Solving equation (18) gives the expressions for the arbitrary constants, A_n, B_n, C_n, D_n, F_n, and G_n, in terms of U_n and W_n. These constants when substituted back into equations (7), (8), (14), and (15) constitute the formal solution of the problem. The integration with respect to β in these equations cannot be carried out readily for the general case. However, it is possible to carry out an approximate evaluation, if certain simplifying assumptions are made. For instance, if the fields are to be observed at large distances from the cylinder, the integration with respect to β may be carried out by the saddle-point method, as will be illustrated in the next section.

IV. THE RADIATION FIELD

The integrals, in the free-space region, to be evaluated are of the following form:

$$I_n = \int_{-\infty}^{\infty} S_n(\beta) H_n^{(2)}(ur) e^{-i\beta z} d\beta, \tag{20}$$

where the dependence of the characteristics of the coated cylinder and the excitation parameters are included in the parameter $S_n(\beta)$. It is noted that since $S_n(\beta)$ is a slowly varying function compared with

$$H^{(2)}(\sqrt{k_0^2 - \beta^2}\, r) e^{-i\beta z}$$

for $k_0 r \gg 1$, the integral can be evaluated approximately by the saddle-point method (Van der Waerden 1951; Knop 1961). The integration of β is along the real axes from $-\infty$ to $+\infty$, and the path of integration is indented above the branch point at $\beta = k_0$ and below the branch point at $\beta = -k_0$. In the far zone, $k_0 r \gg 1$, the asymptotic expression for the Hankel function may be used, i.e.,

$$H_n^{(2)}(ur) \sim \sqrt{\frac{2}{\pi ur}} \exp\left[-i\left(ur - \frac{2n+1}{4}\pi\right)\right]. \tag{21}$$

Substituting equation (21) into equation (20) gives

$$I_n \approx \int_{-\infty}^{\infty} \sqrt{\frac{2}{\pi ur}} S_n(\beta) \exp\left[-i\left(ur - \frac{2n+1}{4}\pi + \beta z\right)\right] d\beta. \tag{22}$$

This integral can be evaluated immediately by applying the saddle-point integration method. The result is

$$I_n \approx \sqrt{2\pi}\, i \exp[-i(k_0 R - \tfrac{1}{2} n\pi)]\, S_n(k_0 \cos\theta)/R, \tag{23}$$

where $R = (r^2 + z^2)^{\frac{1}{2}}$ and $\theta = \tan^{-1}(r/z)$. The terms containing $1/R^2$, $1/R^3$, etc. in the above equation have been neglected. The integral of the form

$$P_n = \int_{-\infty}^{\infty} T_n(\beta) \frac{d}{dr} [H_n^{(2)}(ur)]\, e^{-i\beta z} d\beta \tag{24}$$

can also be evaluated in a similar way. To the same order of approximation as equation (23), one gets

$$P_n \approx \sqrt{2\pi}\, k_0 \exp[-i(k_0 R - \tfrac{1}{2} n\pi)] \sin\theta\, T_n(k_0\cos\theta)/R. \tag{25}$$

The far-zone fields in terms of spherical coordinates (R, θ, ϕ) are then given by

$$\begin{aligned} E_\theta &\approx E_r \cos\theta - E_z \sin\theta, & H_\theta &\approx H_r \cos\theta - H_z \sin\theta, \\ E_\phi &\approx E_\theta, & H_\phi &\approx H_\theta, \\ E_R &\approx E_r \sin\theta + E_z \cos\theta & H_R &\approx H_r \sin\theta - H_z \cos\theta \\ &\approx 0, & &\approx 0. \end{aligned} \tag{26}$$

It is noted that when the terms containing R^{-2}, R^{-3}, etc. are neglected, the radiation fields are transverse. The far-zone radiation fields are now completely specified in terms of the prescribed tangential electric fields.

V. A NUMERICAL EXAMPLE

To get an idea of how the solution behaves, numerical computations are carried out using an IBM 7090 computer for a specific case of the prescribed tangential electric fields. It is assumed that the slot is an infinitely long narrow axial slot across which is maintained a constant distribution of electric field defined by

$$E = e_\theta E_0 G(\theta), \tag{27}$$

where

$$G(\theta) = \begin{cases} 1 & \text{for } -\tfrac{1}{2}\theta_0 \leqslant \theta \leqslant \tfrac{1}{2}\theta_0 \\ 0 & \text{for } \tfrac{1}{2}\theta_0 < \theta < -\tfrac{1}{2}\theta_0 , \end{cases} \tag{28}$$

where $a\theta_0$ is the slot width. Referring to equations (5) and (6), one notes that, for the present case,

$$U_n = 0, \tag{29}$$

$$W_n = (E_0/n\pi)\sin(n\theta_0/2). \tag{30}$$

Hence, the θ component of the radiated electric field is

$$E_\theta^{(r)} = \sum_{n=-\infty}^{\infty} L_n H_n^{(2)\prime}(k_0 r) e^{-in\theta}, \tag{31}$$

where

$$L_n = \frac{E_0}{n\pi}\sin(\tfrac{1}{2}n\theta_0)\frac{J_n(T_1 b)Q_n(T_1 b) - Y_n(T_1 b)Q_n(T_1 b)}{H_n^{(2)\prime}(k_0 b)[Y_n(T_1 b)P_n(T_1 a) - J_n(T_1 b)Q_n(T_1 a)] + R_n}, \tag{32}$$

with

$$R_n = \frac{\epsilon_0\epsilon_1 T_1}{k_0}\frac{H_n^{(2)}(k_0 b)}{(\epsilon_2^{\,2} - \epsilon_1^{\,2})}[Q_n(T_1 b)P_n(T_1 a) - P_n(T_1 b)Q_n(T_1 a)], \tag{33}$$

$$P_n(T_1 r) = J_n'(T_1 r) + n\frac{\epsilon_2}{\epsilon_1}\frac{1}{T_1 r}J_n(T_1 r), \tag{34}$$

$$Q_n(T_1 r) = Y_n'(T_1 r) + n\frac{\epsilon_2}{\epsilon_1}\frac{1}{T_1 r}Y_n(T_1 r), \tag{35}$$

$$T_1^{\,2} = \frac{\epsilon_1^{\,2} - \epsilon_2^{\,2}}{\epsilon_1\epsilon_0}k_0^{\,2}. \tag{36}$$

The prime signifies the derivative of the function with respect to its argument. It should be noted that T_1 may be real or purely imaginary depending upon whether T_1^2 is positive or negative. If T_1 is real, the θ component of the radiated electric field is properly represented by equations (31) through (36); if T_1 is imaginary, the Bessel function or Neumann function with the argument T_1r must be replaced by the proper modified Bessel functions, $I_n(|T_1|r)$ or $K_n(|T_1|r)$.

At large distances from the cylinder (i.e., $k_0r \gg 1$), the asymptotic expression for the derivative of the Hankel function

$$H_n^{(2)\prime}(k_0r) \to -i\left(\frac{2}{\pi k_0 r}\right)^{\frac{1}{2}} \exp\left[-i\left(k_0r - \frac{2n+1}{4}\pi\right)\right] \tag{37}$$

is applicable. Substituting (37) into (31) gives

$$E_\theta^{(r)} = -i\left(\frac{2}{\pi k_0 r}\right)^{\frac{1}{2}} \exp[-i(k_0r - \tfrac{1}{4}\pi)] \sum_{n=-\infty}^{\infty} L_n \exp[-in(\theta - \tfrac{1}{2}\pi)]. \tag{38}$$

The absolute value of the above equation represents the radiation pattern of this radiating structure. It should be noted that for the general case (i.e. $\omega_g \neq 0$), $L_{-n} \neq L_n(-1)^n$. Hence, the radiation patterns are not symmetrical with respect to $\theta = 0$. When the applied static magnetic field is zero, the plasma sheath behaves like an isotropic dielectric medium with a permittivity of $\epsilon_0[1 - \omega_p^2/\omega^2]$. Then $L_{-n} = L_n(-1)^n$ and the asymmetric characteristic of the radiation patterns disappears.

Numerical results for the radiation pattern are presented in Figs. 2 through 8. In these figures, $|E_\theta^{(r)}|$ is plotted against θ in polar form for various values of k_0a, k_0b, ω_p/ω, and ω_g/ω with $\theta_0 = 0.1$ rad. Figures 2–4 show the variation of the radiation pattern with respect to the variation of the normalized gyrofrequency.

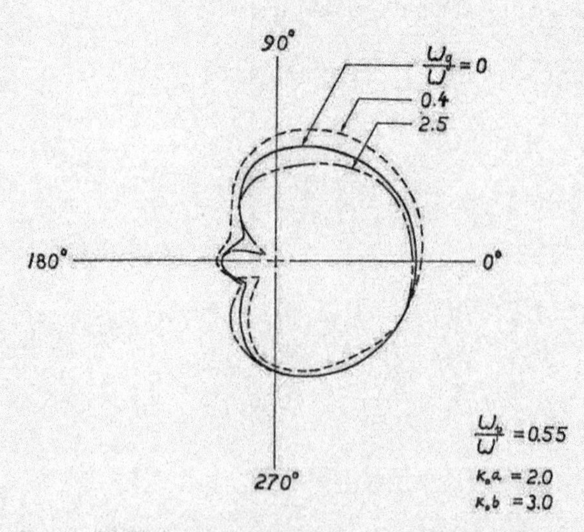

FIG. 2. Radiation patterns as a function of gyrofrequencies.

The effects of the presence of a static magnetic field and the asymmetrical characteristics of the radiation patterns due to anisotropy of the plasma sheath are clearly evident. When the normalized plasma frequency is small, i.e., when the plasma density is low, the use of a static magnetic field does not change the radiation characteristics of the slot antenna significantly, as can be seen from

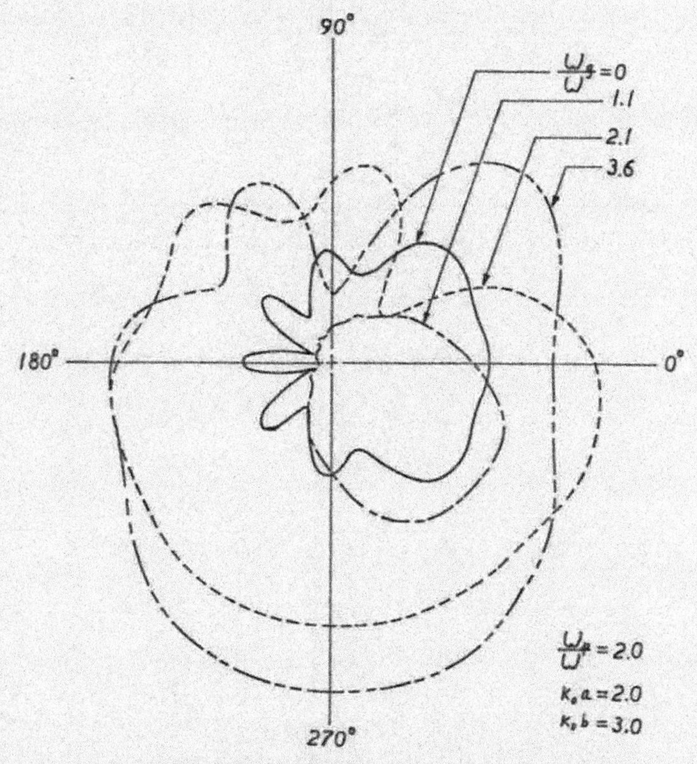

FIG. 3. Radiation patterns as a function of gyrofrequencies.

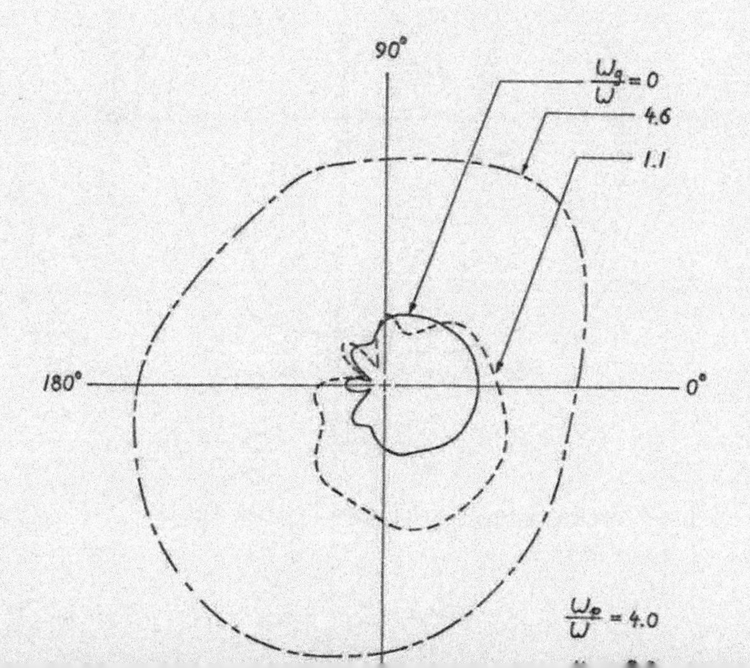

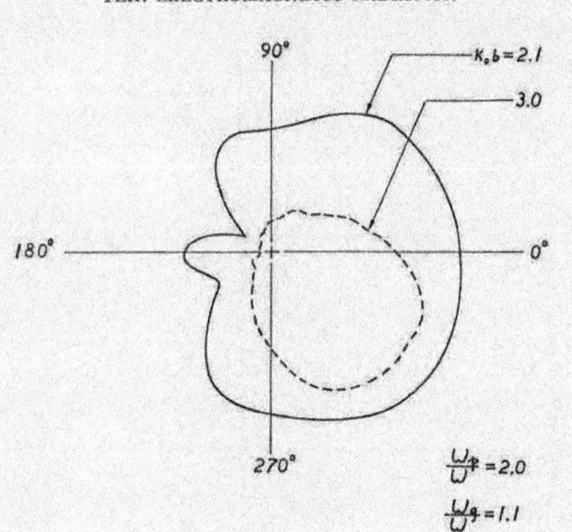

FIG. 5. Radiation patterns as a function of sheath thickness.

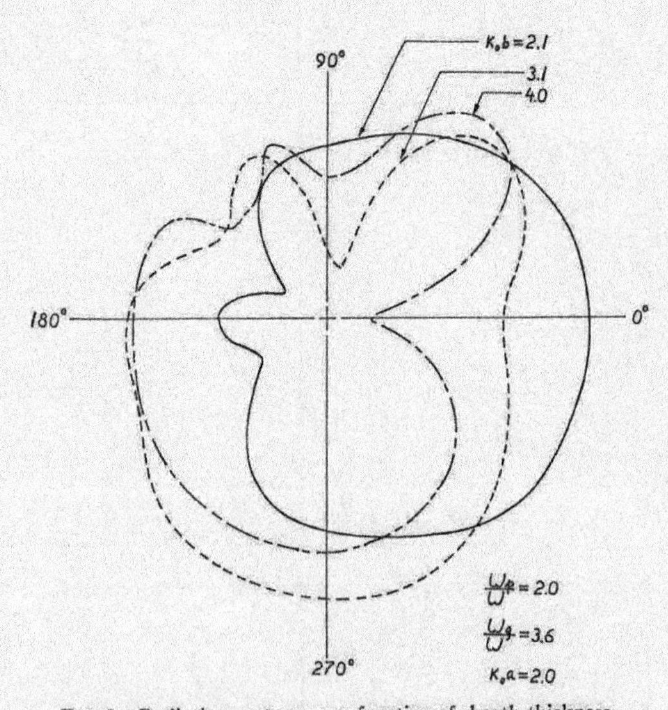

FIG. 6. Radiation patterns as a function of sheath thickness.

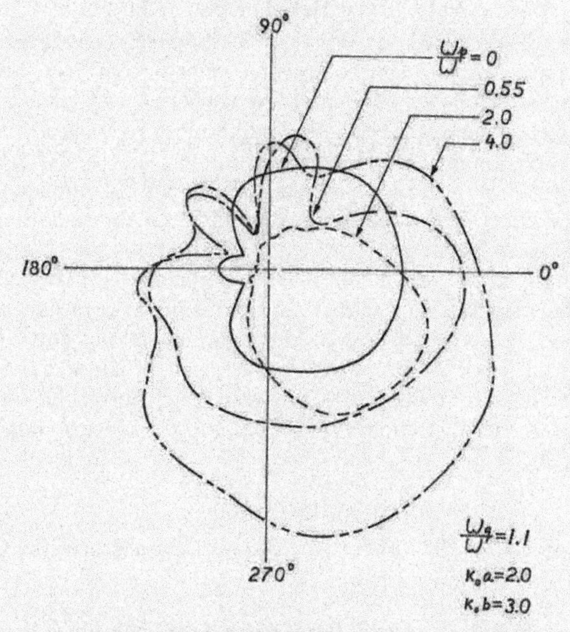

FIG. 7. Radiation patterns as a function of plasma frequencies.

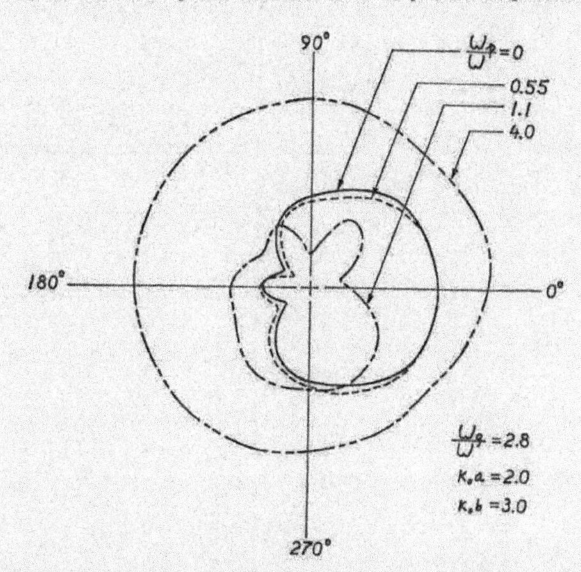

FIG. 8. Radiation patterns as a function of plasma frequencies.

Fig. 2. This is because the equivalent dielectric constant of the sheath is quite close to the free-space permittivity. However, when the plasma density is increased (as shown in Figs. 3 and 4), even the presence of a small static magnetic field changes significantly the radiation patterns of the slot. One can also observe the presence of peripheral standing waves from these figures.

Increasing the thickness of the sheath enhances the effects of the presence of the plasma sheath as well as the effects of the applied static magnetic field on the radiation patterns. This can be observed clearly from Figs. 5 and 6. As expected, when the sheath is very thin and the plasma density is moderate, the radiation pattern approaches that of a slot radiating in free space.

To obtain the variation of the radiation patterns as a function of the normalized plasma density, Figs. 7 and 8 are introduced. For a fixed value of the static magnetic field, the effect of this static magnetic field on the radiation pattern increases as the plasma frequency is increased.

In conclusion, it is noted that if the plasma frequency is moderate, even a small applied static magnetic field could alter significantly the radiation pattern of the slot antenna. The presence of a static magnetic field not only affects the radiation pattern but also the transmission coefficient of the signal. This was demonstrated by Hodara (1960, 1961), who analyzed the problem of transmission of a plane electromagnetic wave through a plane plasma sheath with a static magnetic field.

ACKNOWLEDGMENT

The use of the computing facilities at the Western Data Processing Center at U.C.L.A. is gratefully acknowledged.

REFERENCES

ALLIS, W. P., BUCHBAUM, S. J., and BERS, A. 1963. Waves in plasmas (John Wiley and Sons, Inc., New York).
DIRSA, E. F. 1960. Proc. I.R.E. **48**, 703.
HARLEY, T. P. and TYRAS, G. 1961. Proc. I.R.E. **49**, 1822.
HODARA, H. 1961. Proc. I.R.E. **49**, 1825.
HODARA, H. and COHN, G. I. 1962. I.R.E., Trans. Antennas Prop. **AP-10**, 581.
HODARA, H., ROEMER, H. R., and COHN, G. I. 1960. A New Approach to Space Communications, Proc. XIth Intern. Astronautical Conf., Stockholm, Sweden.
KNOP, C. M. 1961. I.R.E., Trans. Antennas Prop. **AP-9**, 535.
MILLER, G. F., III. 1962. Phys. Fluids, **5**, 899.
VAN DER WAERDEN, B. L. 1951. Appl. Sci. Research Sect. B, **2**, 33.
WAIT, J. R. 1959. Electromagnetic radiation from cylindrical structures (Pergamon Press, New York).

Radiation Through an Inhomogeneous, Magnetized Plasma Sheath

In a recent article[1] the solution to the wave equation in an axially magnetized, radially inhomogeneous plasma medium was presented by the present authors. This solution has been applied to the scattering of a plane H wave from magnetized, inhomogeneous plasma columns[1,2] and from cylinders which are covered with a magnetized, inhomogeneous plasma sheath.[3]

In this communication we would like to comment on the predicted radiation characteristics for an axially slotted cylinder which is covered with a magnetized, inhomogeneous plasma sheath (Fig. 1). Parabolic electron density distributions, ranging from a nearly constant value which drops abruptly to zero at the edge to a steadily decreasing function which smoothly becomes zero at the edge (Fig. 2), have been assumed.

Radiation patterns from such slotted, sheath-clad cylinders were computed for a wide range of the principal parameters: frequency, magnetic field intensity, plasma density, and profile shape. In general, at low frequencies (sheath thickness much smaller than a wavelength) the shape of the radiation pattern was not significantly affected by the presence of the sheath. Since the sheath became increasingly transparent at high frequencies, the patterns were also relatively independent of the sheath parameters and approached the patterns from a slotted cylinder without a coating.

Significant pattern changes were noticed primarily at intermediate frequencies, between the "high" and "low" frequency ranges. For example, magnetic effects caused the pattern to be asymmetric.[4] However, it was discovered that profile shape, the parameter of most interest to us in this correspondence, exerted only a very minor influence on the pattern shapes.[5] Furthermore, such a profile-effect was noticeable only under relatively severe conditions within the sheath, e.g. thick sheaths, very dense plasma, resonance, etc.

Figure 3 is presented to illustrate the influence of electron profile upon the radiation pattern. It has been assumed that

$$\omega a/c = 2.95; \quad eBa/mc = 3.0;$$

$$\sqrt{\frac{Ne^2}{m\epsilon_0}}\,\frac{a}{c} = 0.9,$$

where c is the velocity of light in a vacuum and N is the axial electron density for profile #2.

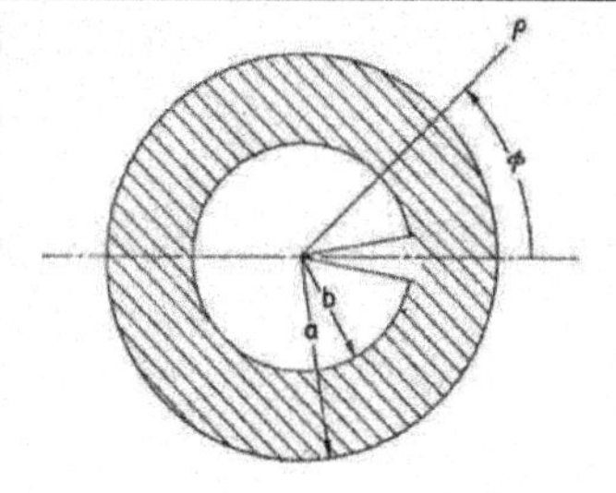

Fig. 1. Cross section of infinite slotted cylinder with plasma sheath.

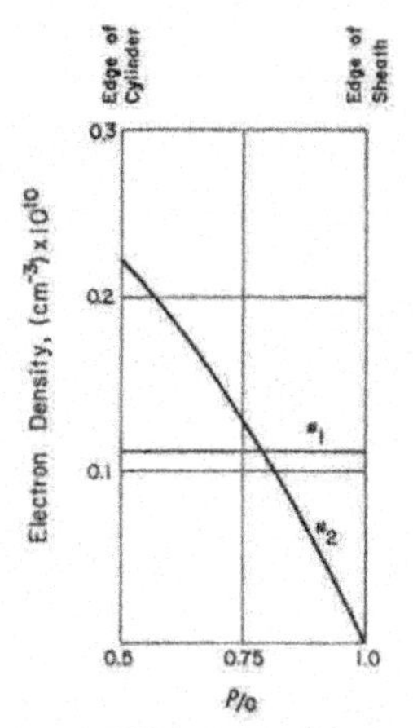

Fig. 2. Electron density profiles.

Fig. 3. Radiation patterns ($\omega a/c = 2.95$; $eBa/mc = 3.0$; $\sqrt{(Ne^2/m\epsilon_0)}(a/c) = 0.9$).

A specific numerical example for these parameters would be: $a = 10$ cm; $b = 5$ cm; a free-space wavelength of 21.3 cm (1.41 GHz); magnetic flux density, B, of 0.51 kG, giving rise to an upper-hybrid resonant frequency[1] near 1.43 GHz. The two electron densities, normalized to contain equal numbers of electrons, are plotted in Fig. 2. This example has been chosen because conditions within the sheath are rather extreme: the sheath is thick, and the signal frequency is near the upper-hybrid resonance.

The two radiation patterns in Fig. 3, produced by the two different profiles but otherwise equal conditions, are not exceptionally different. The forward lobe for profile #1 is enhanced. However, the remainder of the patterns, including the deep null near 270°, is rather similar. As the frequency moves away from the resonance, or as the sheath becomes thinner, the two patterns rapidly coalesce. (It should be noted that these remarks apply only to the *shape* of the pattern. Relative magnitudes cannot be directly compared inasmuch as equal slot voltage, rather than equal transmitter voltage, has been assumed.)

It is concluded that, for the class of problems considered, profile shape does not seriously influence the shape of the radiation pattern, even under such relatively extreme conditions as those above.

C. Yeh
Dept. of Elec. Engrg.
University of Southern California
Los Angeles, Calif.
W. V. T. Rusch
Crawford Hill Lab.
Bell Telephone Labs., Inc.
Holmdel, N. J.

Manuscript received September 6, 1966.

[1] C. Yeh and W. V. T. Rusch, "Interaction of microwaves with an inhomogeneous and anisotropic plasma column," *J. Appl. Phys.*, vol. 36, pp. 2302–2306, July 1965.

[2] C. Yeh and W. V. T. Rusch, "Effects of electron density distribution on the backscattering from an anisotropic plasma cylinder," *Electronics Letters*, vol. 1, p. 25, March 1965.

[3] W. V. T. Rusch and C. Yeh, "Scattering of microwave by a cylinder coated with an inhomogeneous and anisotropic sheath," *IEEE Trans. on Antennas and Propagation*, submitted for publication.

[4] C. Yeh, "Electromagnetic radiation from an arbitrary slot on a conducting cylinder coated with a uniform cold plasma sheath with an axial static magnetic field," *Can. J. Phys.*, vol. 42, pp. 1369–1381, July 1964.

[5] C. T. Swift, "Radiation patterns of a slotted cylinder antenna in the presence of an inhomogeneous lossy plasma," *IEEE Trans. on Antennas and Propagation*, vol. AP-12, pp. 728–738, November 1964.

Reprinted from IEEE TRANSACTIONS ON *ANTENNAS AND PROPAGATION*
Volume AP-15, Number 2, March, 1967
p. 328

Reprinted from APPLIED OPTICS, Vol. 15, page 2913, November 1976

Enhanced focal-plane irradiance in the presence of thermal blooming

C. Yeh, J. E. Pearson, and W. P. Brown, Jr.

The use of multiple transmitter beams is shown to significantly increase the peak focal-plane irradiance that can be achieved in the presence of thermal blooming. Computer simulation studies of the beam propagation problem show over a factor of 2 increase in the irradiance of a single beam and a factor of 9 increase when three coherent beams are focused on the same target spot. Preliminary experimental results with three mutually noncoherent, nonoverlapping beams are in qualitative agreement with the computer simulation.

It is well known that nonlinear thermal blooming imposes a severe constraint on the amount of cw focal-plane irradiance that can be achieved when high power optical beams are propagated in the atmosphere. For instance, the initially circularly symmetric irradiance distribution of a Gaussian beam distorts and spreads into a crescent-shaped distribution,[1,2] and there is an optimum transmitter power P_{opt} above which the peak irradiance decreases.[3,4] The peak irradiance at the focal point is reduced because the area of the crescent is much larger than the area of a diffraction-limited spot, and there is an optimum power because the strength of the beam-induced negative thermal lens is power-dependent. Several schemes have been proposed to compensate for this effect with varying degrees of success.[1,4-8] These schemes can generally be classified into two basic categories: (1) phase compensation[1,4-6] and (2) irradiance tailoring.[7,8] If the initial phase front of a focused transmitted beam is properly chosen, some of the thermal blooming phase distortions are removed, thus enhancing the focal-point irradiance. By tailoring the irradiance distribution of the transmitted beam, correctable phase changes along the propagation path can be achieved so again enhancement of the irradiance at the focal plane can be observed.

The purpose of this paper is to describe a new irradiance tailoring technique that can produce high beam irradiance at the target plane. The technique uses multiple transmitted beams placed so the outer beams steer or guide the inner beam (or beams). The effect is similar to a leaky waveguide[9] and to the guiding that occurs in hollow waveguides such as those used by waveguide CO_2 lasers.[10]

When a nonuniform optical beam (e.g., with a Gaussian profile) is propagated in an absorbing medium, the beam produces an intensity-dependent index of refraction that is lower in the hotter regions of the beam. This type of index profile is characteristic of negative lenses and acts to defocus or spread the beam, producing the phenomenon known as thermal blooming. The case of greatest interest is one for which the medium moves transversely to the beam (a wind exists). In this case the most pronounced blooming effect is in the direction perpendicular to the wind direction, since the wind carries the heated (lower index) medium out of the optical beam in the other direction. The net effect is similar to a thick, distorted, cylindrical lens. By creating low index regions above and below a central beam, as shown in Fig. 1(a), the rate of beam divergence tends to be decreased, as illustrated in Fig. 1(b). The diagram shows only three beams, but more can be used; the more beams, the more effective the confinement of the central beam. For simplicity this paper will treat only the three-beam case, and all beams will have truncated Gaussian profiles.

The most general three-beam case is shown in Fig. 2. The spacing of the three beams at the transmitter and their relative focal spot locations can all be varied to optimize the irradiance of the central beam at the focal plane.[11] The three beams can be mutually noncoherent (different wavelengths) or spatially separated enough not to interact, or they can be coherent and focused to one spot.

Figure 3(a) shows the bloomed focal-plane irradiance distribution for a single beam. Figure 3(b) shows the improvement for the case of three noncoherent beams of differing wavelength as found by computer simulation of the nonlinear propagation problem.[12] The improvement is about a factor of 2 in peak irradiance.

When this work was done, all authors were with Hughes Research Laboratories, Malibu, California 90265. The permanent address of C. Yeh is University of California, Department of Electrical Engineering, Los Angeles, California 90024.

Received 19 November 1975.

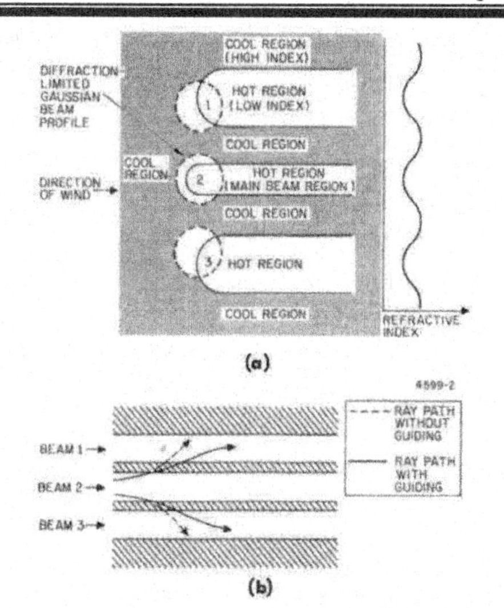

Fig. 1. (a) Effect of multiple Gaussian beams (noninteracting) on atmospheric index of refraction when atmosphere is absorbing; (b) schematic of how the index profile of the three beams in (a) create a guiding effect on the central beam.

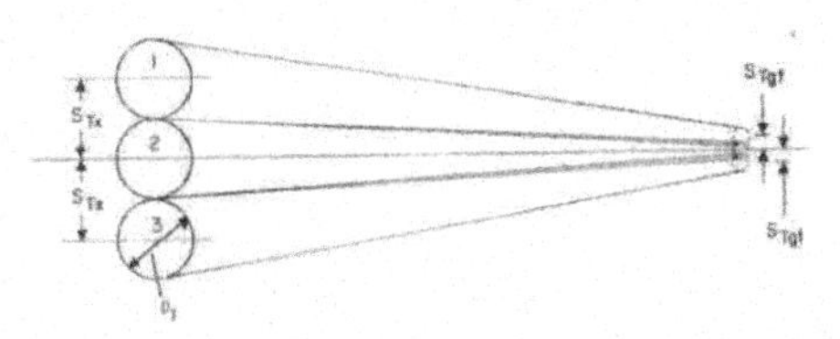

Fig. 2. Most general three-beam case. The separation S_{Tx} can be obtained by physically separating the beams at the transmitter, and the separation S_{Tgt} can be obtained by appropriately tilting beams 1 and 3. Beams 1, 2, and 3 may all be coherent beams or may have different wavelengths. The initial diameter of each transmitted beam is D_T.

Significantly better results are obtained, however, if all three beams are coherent with each other; in effect, a three-beam phased array is used. This case is shown in Fig. 3(c) for $S_{Tx} = D_T$ = transmitter diameter and $S_{Tgt} = 0$ (see Fig. 2). The computer simulation predicts a factor of 9 greater irradiance than with a single beam containing the optimum power.

We have also performed a simple experiment to demonstrate the multiple beam effect. A 1-W argon laser was split into three equal intensity beams, which were focused through an absorbing liquid cell (cell transmission = 0.41). The cell pathlength was 20 cm; and the liquid, iodine dissolved in methanol, flowed transversely to the optical path to insure convection-dominated heat transfer as assumed in the computer calculations. The beams out of the cell were recollimated at a small diameter and allowed to expand by diffraction to produce an image of the focal-plane irradiance distribution of just the central beam.

The results of this experiment are shown in Fig. 4; the pictures on the left side of Fig. 4(a) and 4(b) are black and white photographs of the far-field intensity distribution of the center beam taken off a color TV monitor. The black and white video signal has been sliced into eight levels, each of which is displayed as a different color on the TV monitor. The pictures on the right of

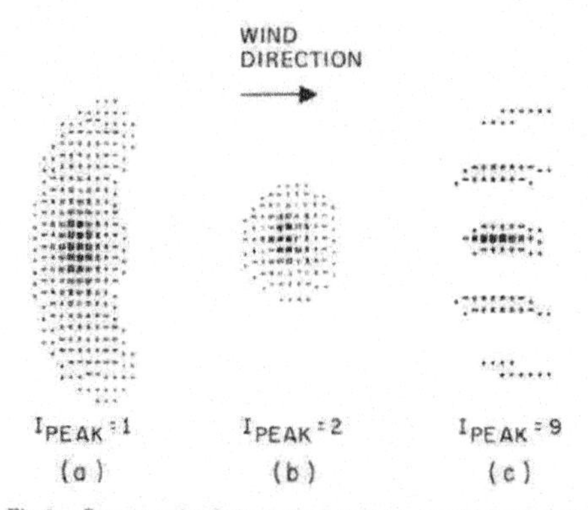

Fig. 3. Computer simulation target irradiance profiles for thermal blooming: (a) thermally bloomed single beam; (b) central beam irradiance only, when two auxiliary beams of different wavelengths are used as in Figs. 1 and 2; (c) total target irradiance (all three beams) when the three beams are coherent.

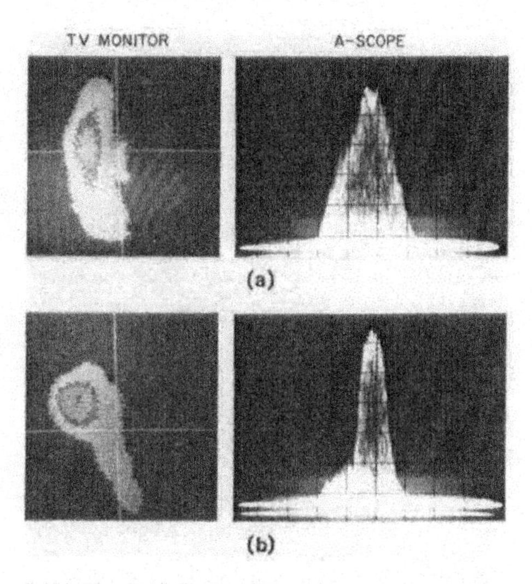

Fig. 4. Experimentally observed far-field irradiance patterns of the central beam in a three-beam array: (a) outside, guiding beams not present; (b) guiding beams turned on; the peak target irradiance increases by 1.8 dB.

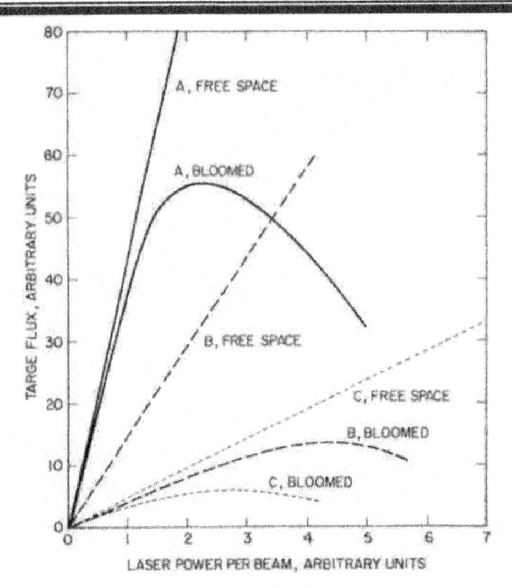

Fig. 5. Summary of computer simulation of three-beam propagation with blooming for 10.6-μm propagation in the atmosphere. A slew rate of 20 mrad/sec is also used: *A*—three-beam, coherent array; *B*—three noninteracting Gaussian beams overlapping at target; *C*—single Gaussian beam.

the figure are A-scope presentations (profiles) of the same intensity distributions. The peak irradiance increases by 1.8 dB when the two outer guiding beams are turned on. The improved beam formation is evident from the pictures. As also observed in the computer simulation, the improvement in the central beam focal-plane irradiance is very sensitive to the location of the guiding beams in both the transmitter and focal planes.

From the data in Figs. 3 and 4, we conclude that confinement of the central beam is indeed produced by the outer two beams. Not only is the peak irradiance for the central guided beam increased as compared with that of the unguided case, but the beam pattern is also improved. The agreement between theory and experiment for beam profiles (Figs. 2 and 3) is qualitatively quite good. The 1.8-dB improvement in peak irradiance observed experimentally is also in reasonably good agreement with the theoretical value of 3 dB, considering the difficulty of determining and maintaining the experimental beam alignment and pointing and of setting the blooming medium parameters so the theoretical propagation conditions are well matched by the experiment.

When three coherent beams are focused at the same focal plane point [Fig. 3(c)], the resultant beam pattern is basically the interference pattern for three separate beams focused to the same spot at the focal plane. In other words, thermal blooming degrades these three beams independently. The use of three separate beams focused on the same point did not introduce larger thermal blooming effects. In fact, we believe that if the location and size of the three beams at the transmitter are optimized (this could even be done adaptively), even higher focal plane irradiance might be achieved by optimizing the beam guiding effect. Of course, the use of more than three beams will produce even higher irradiance. Figure 5 summarizes the quantitative results of our computer simulation runs for 10.6-μm propagation of one and three beams.

Since the maximum peak focal-plane irradiance is known to depend on the transmitter diameter,[12] it is reasonable to ask whether the enhanced focal-plane irradiances shown in Figs. 3–5 are merely due to a larger effective transmitter aperture. The answer is yes, but only in part. Increasing the over-all transmitter diameter but keeping a single, Gaussian-profile beam will increase the focal-plane irradiance for three reasons: (1) the optimum transmitter power is increased; (2) the focal spot size is reduced; and (3) the strongest lens effects occur closer to the focal plane (and thus have a shorter lever arm for defocusing the beam). The multiple beam technique, however, does not increase the optimum transmitter power *per beam* (see Fig. 5) but does greatly increase the peak irradiance at a fixed power in a single beam. The optimum *total* transmitted power is also increased by a factor of 3.

Tripling the transmitter diameter will increase the optimum total transmitter power by only a factor[3] of $(3)^{0.5}$, but the maximum achievable target irradiance is increased by[3] $(3)^{2.5} = 15.6$. Since this is 1.7 times larger than the factor of 9 shown in Fig. 4, it appears to be better to increase the transmit aperture rather than to use multiple beams. The advantage may be illusory, however. As the transmit aperture is increased, the degrading effects of atmospheric turbulence increase rapidly enough to offset much of the blooming reduction. The smaller individual multiple beams will be less affected by turbulence. In addition, practical consideration such as optical element size and pointer/tracker system designs may make the use of multiple small beams a much more attractive approach than the use of larger, single beams to achieve high target irradiance in the presence of blooming. The multiple beam technique may also be coupled with coherent optical adaptive techniques (COAT)[1,5,6] to further remove blooming distortions, as well as to compensate for turbulence distortions. Work is now in progress to explore this possibility as well as to investigate other promising and simple transmitter intensity profiles.[8]

The authors wish to thank T. R. O'Meara and V. Evtuhov for valuable discussions. Critical comments by W. B. Bridges are gratefully acknowledged.

This work sponsored in part by the Air Force Systems Command's Rome Air Development Center, Griffiss AFB, N.Y.

References

1. W. P. Brown, Jr., and J. E. Pearson, "Multidither COAT Compensation for Thermal Blooming and Turbulence: Experimental and Computer Simulations Results (U)," presented at First DoD Conference on High Energy Laser Technology, San Diego, Calif., October 1974; J. E. Pearson, "COAT Measurements and Analysis," RADC-TR-75-101, May 1975, available from NTIS or RADC.
2. F. G. Gebhardt and D. C. Smith, IEEE J. Quantum Electron. **QE-7**, 63 (1971).
3. E. H. Tahken and D. M. Cordroy, Appl. Opt. **13**, 2753 (1974).
4. L. C. Bradley and J. Herrman, Appl. Opt. **13**, 331 (1974).
5. W. B. Bridges and J. E. Pearson, Appl. Phys. Lett. **26**, 539 (1975).
6. J. E. Pearson, W. B. Bridges, S. Hansen, T. A. Nussmeier, and M. E. Pedinoff, Appl. Opt. **15**, 611, (1976).
7. J. Wallace, I. Itzkam, and J. Camm, J. Opt. Soc. Am. **64**, 1123 (1974).
8. J. E. Pearson, C. Yeh, and W. P. Brown, Jr., to be published in J. Opt. Soc. Am. **66**, No. 11 (Nov. 1976).
9. D. B. Hall and C. Yeh, J. Appl. Phys. **44**, 2271 (1973).
10. R. L. Abrams and W. B. Bridges, IEEE J. Quantum Electron. **QE-9**, 940 (1973).
11. This optimization could even be done automatically using techniques similar to those in Ref. 1, 4, and 6.
12. W. P. Brown, Jr., unpublished.

Propagation of laser beams having an on-axis null in the presence of thermal blooming*

J. E. Pearson, C. Yeh,† and W. P. Brown, Jr.
Hughes Research Laboratories, 3011 Malibu Canyon Road, Malibu, California 90265
(Received 4 December 1975; revision received 19 July 1976)

The propagation of focused beams having an on-axis irradiance null is considered in the presence of thermal blooming. Two cases are treated: (a) beam profiles that have an irradiance zero in the beam center at the focal plane as well as the transmitter aperture; (b) beam profiles that have an on-axis irradiance null only at the transmitter. It is demonstrated that none of the beam profiles considered in case (a) has a meaningful advantage over a Gaussian beam profile. Some of the case (b) profiles do produce a larger bloomed irradiance in the focal plane, particularly when the initial intensity distribution is very uniform in its nonzero regions. Addition of a simple central obscuration to a "filled" irradiance distribution is found to have no advantage, however, for the cases considered.

I. INTRODUCTION

When an optical laser beam is propagated through an absorbing medium with a negative refractive index temperature coefficient ($dn/dT < 0$), the phenomenon of thermal blooming[1,2] occurs. In the presence of thermal blooming, the propagation medium acts like a negative lens that defocuses, spreads, and distorts the optical beam. This power-dependent phenomenon can be particularly harmful when the optical system is attempting to produce maximum focal plane irradiance.

It has been suggested that the use of a transmitted beam that has an annular intensity profile may provide higher focal plane irradiance in the presence of thermal blooming[3] than that produced by a Gaussian beam. The physical reasoning behind this suggestion is straightforward. A Gaussian intensity profile produces a quadratic refractive index variation in the medium that is minimum at the beam center where most of the optical power occurs. With an annular beam that has an on-axis null, the negative lens induced in the medium has a larger refractive index on-axis than off-axis. The result is a positive lens for light that is close to the beam axis, and this lens tends to counteract some of the off-axis negative lens beam spreading as the optical beam propagates.

This article presents the results of an investigation aimed at quantifying how much thermal blooming distortions are reduced by the use of laser beam that have an on-axis intensity zero. Such beams can be produced either by apodization (truncation), or by use of laser resonators that have annular gain profiles,[4] or by many high-energy laser devices. The investigations were performed by computer simulation of the propagation problem. The computer code[5] solves the coupled differential equations that describe both the medium dynamics and the diffraction propagation of a focused laser beam. Convection-dominated heat transfer produced by a transverse wind is assumed and the quantity of interest is the peak focal plane irradiance.

Two types of annular or "on-axis-null" intensity profiles are treated. Section II discusses those that have an intensity zero in the beam center, both in the focal plane (far-field) as well as at the transmitter (near-field) and thus are free-space propagation modes. Section III discusses those profiles that have a zero in the beam center only at the initial transmitter plane. Due to diffraction, the initial on-axis irradiance is not zero at the focal plane. An appropriate choice of the initial profile can indeed increase the maximum focal plane irradiance. The standard of comparison in all cases is the focal plane irradiance produced by a Gaussian beam that is truncated at the 10% intensity radius.

II. "ON-AXIS-NULL" IRRADIANCE PROFILE AT THE TRANSMITTING APERTURE (NEAR-FIELD) AND AT THE FOCAL PLANE (FAR-FIELD)

In this section we consider cases where the irradiance null in the middle of the beam is retained as the focused beam is propagated to the target. To first demonstrate that an appropriately chosen transmitter irradiance distribution can indeed be focused to produce an on-axis-null focal-plane irradiance distribution, we start with a transmitter aperture field given by[4]

$$E(r_1, \theta_1) = A_m f_m(r_1)\cos(m\theta_1), \quad (m = 0, 1, 2, \ldots), \tag{1}$$

where A_m is the maximum field magnitude, $f_m(r_1)$ may be a continuous or a discontinuous function of r_1, and (r_1, θ_1) are the transverse polar coordinates of the transmitting aperture. From Fresnel diffraction formulas we obtain the field u at the focal plane:

$$u(x_0, y_0, z) = \frac{e^{jkz}}{j\lambda z}\exp\left(j\frac{k}{2z}(x_0^2+y_0^2)\right) \times \int_{-\infty}^{\infty}\int A(r_1, \theta_1)\exp\left(-j\frac{2\pi}{\lambda z}(x_0x_1+y_0y_1)\right)dx_1\,dy_1, \tag{2}$$

where z is the focal distance, λ the free-space wavelength, $k = 2\pi/\lambda$, (x_0, y_0) are the transverse rectangular coordinates at the focal plane, and (x_1, y_1) are the transverse rectangular coordinates at the transmitting aperture plane. Changing Eq. (2) into polar coordinates and carrying out the angular integration, we find

$$u(r_0, \theta_0, z) = \frac{e^{jkz}}{j\lambda z}\exp\left(j\frac{k}{2z}r_0^2\right)\int_0^{\infty} f_m(r_1)r_1\,dr_1 \times \int_0^{2\pi}\cos(m\theta)\exp\left(-j\frac{2\pi}{\lambda z}r_1r_0\cos(\theta_1-\theta_0)\right)d\theta_1$$
$$= \frac{e^{jkz}}{j\lambda z}\exp\left(j\frac{k}{2z}r_0^2\right)2\pi\cos(m\theta_0)e^{-j(m\pi/2)}I_m(r_0), \tag{3}$$

where

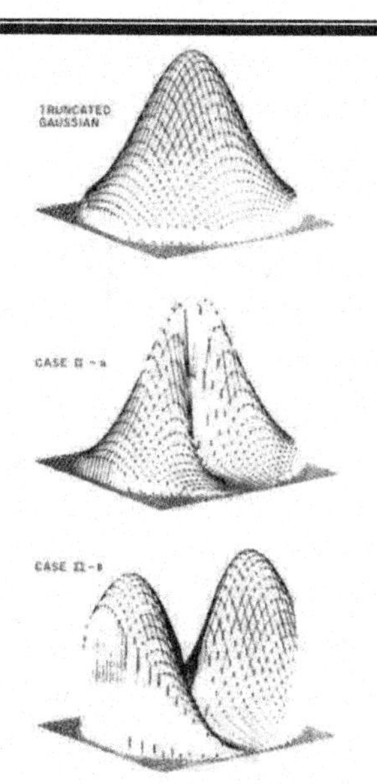

FIG. 1. Three-dimensional plots of transmitter irradiance profiles that retain their shape for free-space propagation. The truncated Gaussian beam is used as a comparison reference for all other beams considered. The two "on-axis-null" profiles shown here also have an on-axis null in the focal plane.

$$I_m(r_0)=\int_0^{\infty} A_m f_m(r_1) J_m\left(\frac{2\pi}{\lambda z} r_1 r_0\right) r_1\, dr_1 . \quad (4)$$

The on-axis irradiance occurs at $r_0 = 0$ and is given by

$$|u(0, \theta_0, z)|^2 = \left(\frac{2\pi \cos(m\theta_0)}{\lambda z}\right)^2 I_m^2(0) . \quad (5)$$

Since $J_m(0) = 0$ when $m \neq 0$, Eq. (4) indicates that $I_m \equiv 0$ for $m \neq 0$. An on-axis-null focal plane irradiance is thus obtained for all $m \neq 0$. When $m = 0$, however, the on-axis focal plane irradiance may take on any value, including zero, depending upon the initial radial dependence of u.

The particular transmitter irradiance profiles that we have chosen to study are given in Eqs. (6) and (7) and are illustrated in Fig. 1. The truncated Gaussian beam we are using for comparison purposes is also shown in Fig. 1 and defined in Eq. (8):

Case (II-a).

$$|u|^2 = A_1^2 \exp(-r_1^2/\rho_0^2) \cos^2\theta_1 \quad \text{for } r_1 \leq a_0 ,$$
$$= 0 \quad \text{for } r_1 > a_0 ; \quad (6)$$

Case (II-b).

$$|u|^2 = A_2^2 \exp\left[-\left(\frac{r_1 - r_m}{\rho_0}\right)^2 \frac{1}{\rho^2}\right] \cos^2\theta_1$$
$$\text{for } b_0 \leq r_1 \leq a_0 \text{ and } b_0 < r_m < a_0 , \quad (7)$$
$$= 0 \quad \text{otherwise;}$$

Trucated Gaussian.

$$|u|^2 = A_3^2 \exp(-r_1^2/\rho_0^2) \quad \text{for } r_1 \leq a_0 ,$$
$$= 0 \quad \text{otherwise} . \quad (8)$$

For these beams, the first two of which can be produced by coaxial-resonator, annular-gain lasers,[4] ρ_0 is the e-folding intensity radius of the Gaussian beam in Eq. (8), ρ is a constant less than or equal to 1, r_m is the adjustable maximum intensity radius, and b_0 and a_0 are the inner and outer radii of a coaxial laser resonator. The beams in Eqs. (6) and (7) are assumed oriented in such a way that the $\theta_1 = \pi/2$ axis is parallel to the transverse wind direction. This orientation produces minimum blooming since the heated air from one main transmitter lobe does not pass through the region heated by the other main lobe. The field amplitudes A_1, A_2, and A_3 are chosen so that the total transmitted power is constant. The expressions used to obtain these amplitudes are derived in the Appendix.

Figure 2 compares the initial transmitter irradiance distributions of the three beams to the focal plane ir-

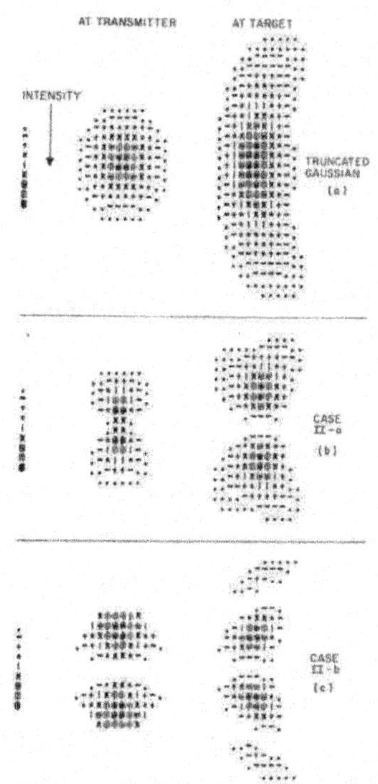

FIG. 2. Pseudogray scale plots of transmitter and focal plane irradiance distributions for the three beams in Fig. 1. A moderate amount of thermal blooming is present in the propagation path.

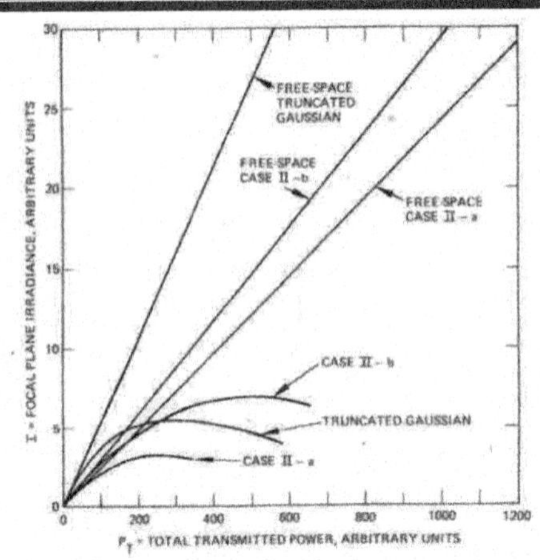

FIG. 3. Peak focal plane irradiance versus total transmitter power for the same three cases as in Figs. 1 and 2. The axis units are normalized values as discussed in the Appendix.

radiance distributions for a representative transmitter power. The parameters necessary to specify the blooming strength are given in the Appendix. For all the computational results presented here, we have chosen values for the various constants to correspond to the obscuration ratio in some laser pointer/tracker systems now in use, and also to give the most uniform initial intensity distribution; we have found that thermal blooming is minimized for uniform initial intensity profiles. The values used in our calculations are $a_0/\rho_0 = 1.52$, $a_0/b_0 = 7$, $r_m/a_0 = \frac{2}{3}$, and $\rho = 1$. Note that in the absence of blooming effects, the focal plane distributions would look like the transmitter distributions for each case.

The peak focal plane irradiance as a function of total transmitted laser power for these three beam cases is given in Fig. 3. Both the irradiance and the transmitter power are plotted in normalized units as discussed in the Appendix. The maximum focal plane irradiance for case (II-b) is a factor of 1.2 larger than that obtained with the truncated Gaussian beam. The higher irradiance is obtained by increasing the total laser power at the aperture plane by a factor of 2.75. Case (II-a) yields lower peak focal plane irradiance even with higher total laser power. Hence no significant improvement is obtained with transmitting intensity profiles tailored according to cases (II-a) or (II-b).

III. "ON-AXIS-NULL" INTENSITY PROFILE ONLY AT THE TRANSMITTING APERTURE (NEAR-FIELD)

We now treat cases for which the on-axis intensity null occurs only at the transmitting aperture. In other words, under thermal blooming conditions, the cooler region in the middle of the beam does not exist throughout the focused propagation path to the focal plane. Four specific aperture intensity profiles are considered and compared to the truncated Gaussian in Eq. (8):

Case (III-a).

$$|u|^2 = A_3^2 (r_1/\rho_0) \exp^4[-(r_1^2/\rho_0^2)] \quad \text{for } r_1 < a_0, \qquad (9)$$
$$= 0 \quad \text{for } r_1 > a_0;$$

Case (III-b).

$$|u|^2 = A_4^2 \exp\left[-\left(\frac{r_1 - r_m}{\rho_0}\right)^2 \frac{1}{\rho^2}\right] \quad \text{for } b_0 < r_1 < a_0, \qquad (10)$$
$$= 0 \quad \text{otherwise};$$

Case (III-c).

$$|u|^2 = A_5^2 \quad \text{for } b_0 < r_1 < a_0, \qquad (11)$$
$$= 0 \quad \text{otherwise};$$

Case (III-d).

$$|u|^2 = A_6^2 \exp[-(r_1^2/\rho_0^2)] \quad \text{for } b_0 < r_1 < a_0,$$
$$= 0 \quad \text{otherwise}.$$

The beam amplitudes A_3, A_4, A_5, A_6 are chosen to keep the total transmitter power constant. The calculations

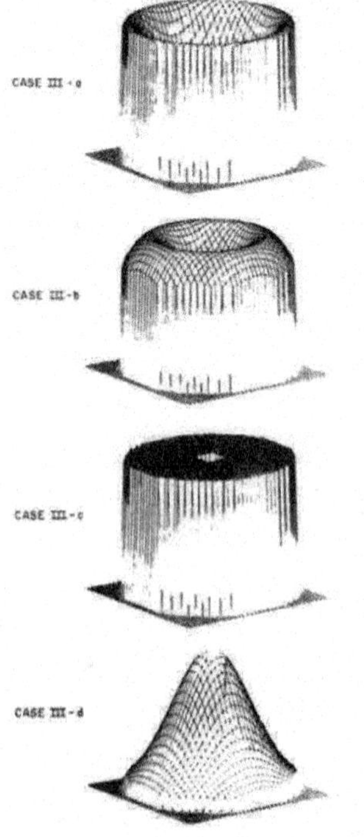

FIG. 4. Three-dimensional plots of transmitter irradiance profiles that have an on-axis null initially that fills in as the beams propagate. Case (III-a) is a mixture of cylindrical modes TEM_{00} and TEM_{01} oscillating in phase composition. Case (III-b) is similar to Case (II-b), except $|u|^2$ has no θ dependence. Case (III-c) is a uniform annular beam. Case (III-d) is an annularly truncated Gaussian beam.

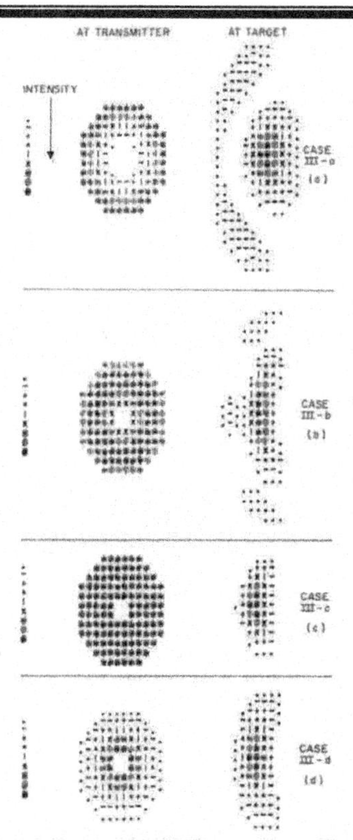

FIG. 5. Pseudogray scale plots of transmitter and focal plane irradiance distributions for the three beams in Fig. 4. A moderate amount of thermal blooming is present in the propagation path.

for these amplitudes are given in the Appendix. Computer-generated plots of the initial beam profiles are illustrated in Fig. 4.

The transmitter aperture intensity profiles as well as the corresponding focal plane intensity profiles for these cases are displayed in Fig. 5 for a representative laser power level. Figure 6 compares the normalized peak focal plane irradiance as a function of normalized total input transmitting power for these four cases to that for the truncated Gaussian case. Unlike the cases treated in Sec. II, significantly larger peak focal plane irradiance is achieved for two of the four cases considered as compared with that for the truncated solid Gaussian. Case (III-d) yields a peak focal plane irradiance which is the same as that for the truncated Gaussian beam.

The intensity profile for case (III-a) is obtained by combining the TEM_{00} and TEM^{*}_{10} modes of a cylindrical cavity oscillating in phase opposition Laguerre-Gaussian modes[6,7]. The peak focal plane irradiance achieved with this profile is almost twice that of the truncated solid Gaussian beam with a total laser power roughly 2.5 times that for the truncated solid Gaussian beam. Other ways to increase focal plane irradiance by this amount such as coherent optical adaptive techniques (COAT)[8] may be more practical and efficient if this type of mode is difficult to produce in practice in high-power lasers.

In cases (III-b) and (III-c), the peak focal plane irradiance is also about twice the value for the truncated solid Gaussian beam, but as shown in Fig. 6, this increase is obtained at about 90% of the total input laser power of the truncated solid Gaussian beam. A very uniform intensity distribution is thus desirable for reducing blooming, but may be very difficult to obtain in practice.

IV. CONCLUSION

These results lead us to conclude that many special laser modes or intensity distributions that have an on-axis null or minimum do not offer any significant advantages for reducing thermal blooming when the goal is to achieve maximum focal plane irradiance for a given transmitter power. In particular, the laser modes considered here that produce an on-axis null in both the near-field and far-field have no advantages over intensity distributions that have the null only in the near-field of the transmitter. Significantly less blooming does occur, however, for certain annular circularly symmetric laser modes and for initial irradiance profiles that are very uniform in the nonzero regions (a uniform annulus, for example).

One question that is frequently asked is whether the "holes" introduced into beams by unstable resonators should be eliminated or whether they may provide some advantage. Although we have not considered a

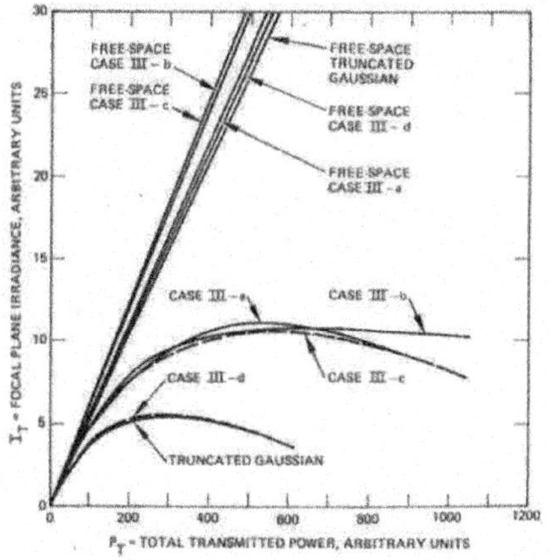

FIG. 6. Peak focal plane irradiance versus total transmitter power for the four beams in Figs. 5 and 6. The axis units are normalized as discussed in the Appendix.

sufficient number of cases to make a definite conclusion, our studies to date of filled beams and of beams with a central hole or obscuration indicate that the hole makes very little difference, at least as far as thermal blooming is concerned. When the goal is to minimize thermal blooming distortions so that maximum focal-plane irradiance is achieved, then it is much more profitable to force the initial beam intensity profile to be as uniform as possible in its nonzero regions rather than attempt to produce or eliminate central holes in the beam. Increasing the initial transmitted beam diameter is also quite effective in reducing blooming distortions, particularly when coupled with adaptive optical techniques.[9]

Since it may be impractical in many high-power laser applications to achieve either sufficiently large apertures or to produce the uniform or special intensity distributions that minimize thermal blooming, other irradiance-tailoring techniques[10,11] or phase-front-tailoring techniques[8,12] may be more useful. Additional work is now in progress in both of these areas.

ACKNOWLEDGMENT

We thank R. M. Szejn for his able assistance with the computer programs employed in this work.

APPENDIX

In this Appendix we derive the normalizations used in the text for the focal plane irradiance and for the field amplitudes for the various beams in terms of the total laser transmitter power. The total transmitted power is given by

$$P_T = \int_{A_T} |E|^2 dA_T = \frac{4\pi}{m+1}\rho_0^2 A_m^2 F_1 \frac{a_0}{\rho_0}, \tag{A1}$$

where A_T is the aperture area, E is the field amplitude as defined in Eq. (1), and F_1 is a constant given by

$$F_1(a_0/\rho_0) = \int_0^{a_0/\sqrt{2}\rho_0} f_m^2(\sqrt{2}\rho_0 x) x\, dx. \tag{A2}$$

The maximum aperture dimension is a_0 and ρ_0 is the e-folding intensity radius of the Gaussian beam in Eq. (8).

The focal plane irradiance as plotted in Figs. 3 and 6 is found from Eqs. (3) and (4):

$$|u(r_0, \theta_0, z)|^2 = \left(\frac{2\pi\cos(m\theta_0)}{\lambda z}\right)^2 I_m^2(r_0, z), \tag{A3}$$

with I_m defined by Eq. (4). We are interested in the maximum focal plane irradiance, which will occur at $\theta_0 = 0$ and at some r_0, depending on the initial irradiance distribution. Using $\theta_0 = 0$ and combining Eqs. (A2), (A3), and (4), the irradiance can be written

$$|u|^2 = 4\pi(m+1)\left(\frac{\rho_0}{\lambda z}\right)^2 e^{-\alpha z} \frac{F_2^2(a_0/\rho_0, \alpha)P_T}{F_1(a_0/\rho_0^2)}, \tag{A4}$$

where α is the propagation medium absorption, F_1 is defined in Eq. (A2), and F_2 is given by

$$F_2(a_0/\rho_0, \beta) = \int_0^{a_0/\sqrt{2}\rho_0} f_m(\sqrt{2}\rho_0 x)\, J_m(\beta x) x\, dx \tag{A5}$$

TABLE AI. Values for peak focal plane irradiance calculations.

Case	m	F_1	(F_2)max	(β)max
Inf. Gaus.	0	0.25	0.50	0
Trunc. Gaus.	0	0.225	0.3421	0
II-a	1	0.225	0.167	2.50
II-b	1	0.508	0.279	2.31
III-a	0	0.0509	0.161	0
III-b	0	0.508	0.536	0
III-c	0	0.566	0.566	0
III-d	0	0.214	0.331	0

and

$$\beta = 2\pi\sqrt{2}\rho_0 r_0/\lambda z. \tag{A6}$$

We have evaluated the integral function F_2 in Eq. (A5) for each of the 7 radial f_m functions discussed in the text [Eqs. (6)–(12)] to find the maximum value of F_2 as a function of the normalized parameter ρ. The values $a_0/\rho_0 = 1.52$, $a_0/b_0 = 7$, $r_m/a_0 = \frac{2}{3}$, and $\rho = 1$ were used for this evaluation. Table AI gives the results of this analysis. To obtain numerical results, we chose $\lambda = 10.6\ \mu m$, $z = 2$ km, and $\rho_0 = 23$ cm (corresponding to $a_0 = 35$ cm). The irradiance plotted as the ordinate axes in Figs. 3 and 6 is thus given by

$$|u|^2 = 0.0592(m+1)(F_2^2/F_1)(\mathrm{S.R.})P_T \tag{A7}$$

in units of power per square centimeter, with (S.R.) being the strehl ratio determined by the thermal blooming computer code (S.R. = 1 for free-space propagation). For the blooming calculations, the following parameters were taken:

absorption $= 2\times 10^{-4}\ m^{-1}$,

transverse wind velocity $= 10$ m/s,

beam slew rate $= 20$ mrad.

*This work sponsored in part by the Air Force Systems Command's Rome Air Development Center, Griffiss AFB, NY.

†Permanent address: Electrical Engineering Department, University of California at Los Angeles, Los Angeles, Calif. 90024.

[1]J. P. Gordon, C. C. Leite, R. S. Moore, S. P. S. Porto, and J. R. Whinnery, J. Appl. Phys. 36, 3 (1965).
[2]A concise review of this topic for gaseous media is presented in J. N. Hayes, P. B. Ulrich, and A. H. Aitken, Appl. Opt. 11, 257 (1972).
[3]P. B. Ulrich, J. Opt. Soc. Am. 64, 549 (1974).
[4]L. W. Casperson and M. S. Shekhani, Appl. Opt. 14, 1653 (1975).
[5]W. P. Brown, Jr. (unpublished).
[6]G. D. Boyd and J. P. Gordon, Bell Syst. Tech. J. 40, 489 (1961).
[7]G. Goubau and F. Schwering, IRE Trans. Antennas and Propag. AP-9, 248 (1961).
[8]W. B. Bridges and J. E. Pearson, Appl. Phys. Lett. 26, 539 (1975).
[9]W. P. Brown, Jr (unpublished).
[10]J. Wallace, I. Itzkam, and J. Camm, J. Opt. Soc. Am. 64, 1123 (1974).
[11]C. Yeh, J. E. Pearson, and W. P. Brown, Jr., "Enhanced Target Irradiance in the Presence of Thermal Blooming," Appl. Opt. (to be published).
[12]L. C. Bradley and J. Herrman, Appl. Opt. 13, 331 (1974).

July 15, 1998 / Vol. 23, No. 14 / OPTICS LETTERS 1087

Earth-to-deep-space optical communications system with adaptive tilt and scintillation correction by use of near-Earth relay mirrors

J. W. Armstrong, C. Yeh, and K. E. Wilson

Jet Propulsion Laboratory, California Institute of Technology, 4800 Oak Grove Drive, Pasadena, California 91109

Received April 28, 1998

Performance of an Earth-to-deep-space optical telecommunications system is degraded by distortion of the beam as it propagates through the turbulent atmosphere. Conventional approaches to correcting distortions, based on natural or artificial guide stars, have practical difficulties or are not adequate for correction of distortions, which are important for Earth-to-deep-space optical links. A beam-relay approach that overcomes these difficulties is discussed. A downward-directed laser near an orbiting relay mirror provides a reference for atmospheric correction. Adaptive optics at the ground station compensate the uplink beam so that after it passes through the atmosphere uplink propagation effects are removed. The orbiting mirror then directs the corrected beam to the distant spacecraft. © 1998 Optical Society of America

OCIS codes: 060.4510, 010.1080, 010.1300.

Optical telecommunications will be the next technological step in wideband Earth-to-deep-space communication. However, propagation of an optical beam through the irregular atmosphere significantly distorts the signal, requiring correction schemes for deep-space communications. Adaptive optics have been used successfully to correct some propagation effects in astronomical[1] and "Star Wars" (Ref. 2) applications. These correction schemes require an optical reference source that is very near the target direction. For example, artificial guide stars (AGS's) have been used successfully for correction of distortions, which are important for astronomical imaging.[3]

For optical telecommunications, adaptive optics will be necessary for high-data-rate downlinks and coherent applications. For the uplink, low-order optical distortions, especially the tilt, or transverse gradient of the phase, are of particular importance. If the tilt is too large, the beam is not pointed at the target, and the link degrades or is lost completely. The situation is particularly acute for the uplink in very long-distance communication (e.g., Earth to Pluto), in which nearly diffraction-limited beams are required for adequate signal strength when the signal reaches the spacecraft.

Atmospheric tilt can be quantified[4,5] by the phase-structure function $D_\phi(r)$, which is the mean-square phase difference at a transverse separation r in the receiving plane. For isotropic Kolmogorov turbulence $D_\phi(r) = 6.88(r/r_0)^{5/3} = (r/b_{\text{coh}})^{5/3}$, where r_0 is the Fried parameter, over a wide range of scales. A typical refractive tilt is $(\lambda/2\pi)\text{grad}(\phi) \sim \lambda/(2\pi b_{\text{coh}})$. Even at infrared wavelengths, a 1-m-class telescope typically has a diameter $\gg r_0$ and without compensation has tilt errors that can be large compared with the diffraction limit. In this Letter we outline difficulties with conventional schemes for atmospheric-tilt correction when they are applied to Earth-to-deep-space optical telecommunications systems and describe a new approach that can overcome these problems.

Several methods have been suggested to correct for propagation through the atmosphere, particularly for astronomical imaging. These methods involve natural or artificial reference sources near the desired uplink direction. These methods are still under development, and their final performance will be application dependent. AGS-based correction methods applied to the uplink optical telecommunications problem (as distinguished from the imaging problem) present some difficulties. We outline these difficulties below to contrast these approaches with the solution that we propose.

The straightforward way to measure and correct tropospheric tilt and scintillation is to use a natural star[6] nearby the desired uplink direction as a reference source. However, the star must be bright (to get enough photons to sense the tilt in the ~1–100 ms before it changes significantly[7]) and near the desired direction (to ensure that the distortion is correlated between the reference star and the uplink beam direction). The latter requirement can be stringent when high correlation of the two optical paths is necessary. Bright stars near a desired target direction are not generally available, so a natural-guide-star scheme is not practical for operational high-availability systems.

An alternative to using natural guide stars for astronomical imaging is to form a monochromatic AGS, produced by a laser that is colocated with the telescope. Such a system cannot be used to measure tilt, however, because atmospheric tilt is common to both the laser beam that is propagating up through the atmosphere and the backscattered light from the AGS. The result is that tilt cannot be sensed or corrected by such a system.

More-elaborate artificial polychromatic guide-star systems, which exploit dispersion in the refractive index of air between the ultraviolet and the infrared, have been proposed for tilt measurement. To break the symmetry of the upward- and the downward-going paths, mesospheric sodium is excited with laser beams

0146-9592/98/141087-03$15.00/0

1088 OPTICS LETTERS / Vol. 23, No. 14 / July 15, 1998

at 589 and 569 nm. When the atoms relax, they fluoresce at wavelengths of 0.33–2.3 μm. The dispersion of the refractive index causes the tilt to be different at different wavelengths; observation of the angular offset of the AGS's at different wavelengths allows the tilt to be measured.[8] This approach is under development; a challenge for a practical implementation is generating sufficient laser power to allow for diurnal and seasonal variations in the sodium column density.

Two or more monochromatic AGS's produced by laser transmitters that are physically separated from the main telescope have been proposed. In suitable circumstances the tilt in the target direction can be deduced. The technical difficulties with a practical implementation of this idea have been discussed.[9]

An optical downlink from the distant spacecraft itself cannot be used as an AGS reference because the spacecraft's downlink will be too weak to drive an adaptive system, and, in general, aberration will cause the downlink photons to arrive from the wrong direction. The downlink and uplink directions are separated by $2\Delta v/c$, where Δv is relative transverse velocity, ≈30 km/s of Earth's orbital speed. Thus a typical aberration angle will be ≈200 μrad, far larger than the isoplanatic angle for observations in the visible (but comparable with the isoplanatic angle in the near infrared[10]). We propose a fundamentally different way to communicate with a distant spacecraft from the ground in the presence of atmospheric distortions that avoids the above-discussed difficulties.

Since optical propagation in interplanetary space is distortionless, a diffraction-limited optical beam can be sent from a point outside the Earth's atmosphere to the spacecraft without error. This diffraction-limited beam could be formed either by a transmitter above the atmosphere or by a mirror that redirects a beam coming through the atmosphere from the ground.

The idea behind this system is that a relay mirror is located in Earth orbit, above the distorting atmosphere. A reference beam, produced by a small onboard laser, provides the information that a ground-based adaptive optics system needs to correct the distortions and deliver a diffraction-limited beam to the orbiting relay mirror. The reference source is slightly separated from the relay mirror to allow for aberration caused by relay–ground-station relative motion. The relay mirror directs the corrected uplink communications beam to the distant spacecraft. Figure 1 is a drawing of this system in its simplest form (one reflection off a relay mirror); for some geometries intermediate relay to a second orbiting mirror may be useful. Unlike other atmospheric-compensation approaches, the beam-relay method offers one the ability to monitor in space the degree of compensation achieved on the uplink. By placing optical detectors in the interstices between mirror segments, one can monitor and map the optical power distribution across the aperture. These data could be relayed to the ground station along with metrology data on the mirror segments. With appropriate registration of the detectors, they could provide real-time information on the irradiance pattern at the relay mirror and of the beam directed to the spacecraft.

The beam-relay system proposed here solves the principal problem: The reference source for the adaptive optics is above the atmosphere, the tilt is estimated from a wave that makes only one passage through the troposphere, and adaptive optics can be used to produce a diffraction-limited communications beam that can be aimed by the relay mirror. Some system design considerations are discussed below.

The reference must be bright enough for rapid (approximately millisecond) tilt and higher-order corrections. This brightness can be accomplished with a modest-power laser and small optics.[11] The signal-to-noise ratio (SNR) for the adaptive optics system will be limited by some combination of sky background and receiver noise. The shot-noise-limited receiver SNR is $\eta P_R/(2h\nu B)$, where η is the quantum efficiency, P_R is the received power, h is Planck's constant, ν is the photon frequency, and B is the receiver bandwidth. The following is an approximate calculation for the longest Earth-relay distance that we might contemplate: If light from a 1-mW laser at $\lambda \sim 0.5$ μm on a spacecraft in geosynchronous orbit were transmitted to the Earth through ~10-cm optics, a 1-m telescope on the ground would collect $P_R \approx 30$ nW, giving shot-noise SNR of ≈70 dB for $B \approx 1$ kHz. (Daytime SNR will be reduced by sky background noise. Even for observations at 0.5 μm, with a 100-μrad field of view and a filter bandwidth $\Delta\lambda \approx 0.1$ nm, and when one is looking ≈10° or farther from the Sun, the sky background would contribute only ~1% of the power received from the spacecraft reference, however.) Thus modest-power reference sources provide an adequate SNR for the adaptive optics system.

The reflector need not be a single mirror (there may be practical reasons to synthesize the aperture from smaller segments.) The relay mirror (or mirror segments) could be flat. System engineering tradeoffs (e.g., size of reflector or number of segments versus distance to the orbiting mirror, complexity of tracking low-Earth orbiters, number of relays required) can

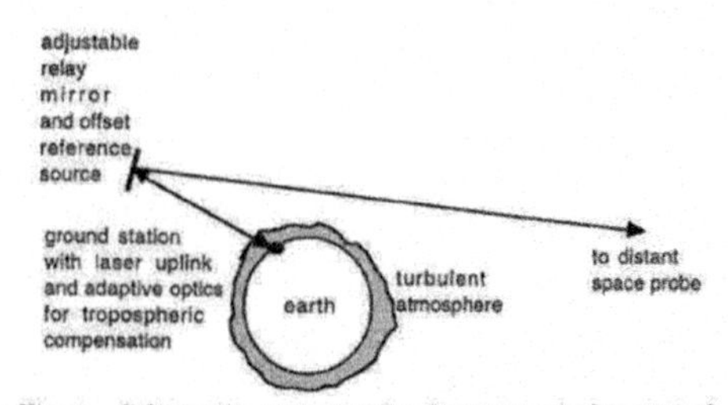

Fig. 1. Schematic, not-to-scale diagram of the optical telecommunications beam-relay system. The beam from a laser that is slightly offset from the relay mirror (compensating for aberration) propagates down to the Earth, giving a reference for the adaptive optics at the ground station. The atmospherically compensated uplink communications beam is transmitted to the orbiting relay mirror and then directed to the distant deep-space probe.

be made. Suppose, for example, that the relay were in relatively low orbit with a typical ground-station-to-relay distance of ~1000 km. A diffraction-limited beam from a 1-m aperture with $\lambda/D \sim 0.5 \times 10^{-6}$ rad propagating 10^6 m to the relay would have a spot size of ≈1.5 m. Restricting operations with a given relay spacecraft to intervals when the angle of incidence on the mirror is ≲45° means that the mirror would have to be ≈2 m in size. (Moving the relay to geosynchronous orbit would require fewer relay spacecraft and might have operational advantages but would require a larger structure in space. A smaller propagation distance to the relay is an attractive way to minimize mirror size.)

If the relay is in near-Earth orbit, aberration is $2\Delta v/c \lesssim 8$ μrad (here Δv is set by relay orbital speed and Earth rotation speed). Thus the reference beacon should be offset from the mirror by ≲8 m. (A value of 8 μrad is comparable with the isoplanatic angle in the visible and much smaller than the isoplanatic angle in the infrared, so phase variations on the reference beam will be highly correlated with those on the communications beam.) For a near-Earth relay spacecraft, the transverse velocity will change as it passes over the ground station. Thus one must take steps to produce the reference beam at the appropriate (time-dependent) position with respect to the mirror. (If the relay were in geosynchronous orbit, the reference source might be on a separate small spacecraft ≈600 m from the mirror.)

The uplink communications beam must have relatively high power to communicate throughout the solar system. Part of this power will be absorbed by the atmosphere, producing temperature gradients that will contribute additional refractive tilt. The extent of heating depends on beam power and atmospheric opacity at the uplink wavelength. For a 1-kW uplink at ≈1 μm on a 1-m telescope, beam heating may induce tilts that are nonnegligible for deep-space telecommunications applications. Any beam-heating tilt variations will, however, largely also be observed by the downlink reference beam and thus are correctable by the adaptive optics system.

The relay mirror is a spaceborne structure with a characteristic width of ≳2 m, depending on the orbit chosen. To relay the uplink communications beam accurately to a distant spacecraft, the relay mirror should be oriented in space to better than 10^{-6} rad (only the control points defining the mirror-surface positions, not the whole structure, need to be oriented to this accuracy). The spacecraft target position would be determined within a reference frame of relatively bright stars. After these guide stars are acquired, the structure would then be oriented to the correct position for beam relay; the desired orientation would also have to be continuously controlled as the ground-station–relay-spacecraft geometry changed. In support of near-future optical interferometry and astrometry in space, there is interest in assembly and control of space structures that are comparable in size with what is proposed here. Building a beam-relay spacecraft with the required capability and at an acceptable cost will clearly benefit from these and other technological developments (vibration isolation, thermal control, laser metrology to measure positions of optics).[12]

Uncompensated propagation through the Earth's turbulent atmosphere fundamentally degrades optical communications with distant spacecraft. Scintillation and tilt of the uplink wave are of particular concern. We have summarized guide-star methods to correct tilt and scintillation in the optical communications problem and discussed a fundamentally different approach, beam relay, which can provide essentially perfect tilt and scintillation correction. This system employs an orbiting relay mirror with a bright laser reference source used by ground adaptive optics for tilt and scintillation correction of the communications beam. This atmospherically compensated communications beam is transmitted to the relay mirror and from there directed to a distant spacecraft. The system requires a space platform that provides accurate, dynamic beam aiming via a relatively large (≳2-m) flat mirror surface. General requirements for such a system were addressed. The requirements are stringent but not inconsistent with current and near-future technologies.

We thank Wilbur P. Brown of the University of New Mexico and Phillips Laboratory, Albuquerque, New Mexico, for stimulating comments. The research described here was carried out at the Jet Propulsion Laboratory, California Institute of Technology, under a contract with NASA.

References

1. R. Q. Fugate, D. L. Fried, G. A. Ameer, B. R. Boeke, S. L. Browne, P. H. Roberts, R. E. Ruane, G. A. Tyler, and L. M. Wopat, Nature (London) **353**, 144 (1991).
2. R. Braham, IEEE Spectrum **34**, 26 (1997); also see the other articles in that issue.
3. C. E. Max, S. S. Olivier, H. W. Friedman, J. An, K. Avicola, B. V. Beeman, H. D. Bissinger, J. M. Brase, G. V. Erbert, D. T. Gavel, K. Kanz, M. C. Liu, B. Macintosh, K. P. Neeb, J. Patience, and K. E. Waltjen, Science **277**, 1649 (1997).
4. I. Ishimaru, *Wave Propagation and Scattering in Random Media* (Academic, New York, 1978).
5. B. J. Rickett, Annu. Rev. Astron. Astrophys. **28**, 561 (1990).
6. S. S. Olivier and D. T. Gavel, J. Opt. Soc. Am. A **11**, 368 (1994).
7. G. Tyler, J. Opt. Soc. Am. A **11**, 358 (1994).
8. R. Foy, A. Migus, F. Biraben, G. Grynberg, P. R. McCullough, and M. Tallon, Astron. Astrophys. Suppl. **111**, 569 (1995).
9. R. Ragazzoni, Astron. Astrophys. **305**, L13 (1996).
10. M. C. Roggermann, B. M. Welsh, and R. Q. Fugate, Rev. Modern Phys. **69**, 437 (1997).
11. A. H. Greenaway, Proc. SPIE **1494**, 386 (1991).
12. Committee on Advanced Space Technology (D. Hastings, chair), National Research Council, *Space Technology for the New Century* (National Academy Press, Washington, D.C., 1998).

6

MOVING MEDIUM

Investigation on the interaction of electromagnetic waves with moving dielectric/plasma structures was started in the mid-1960s. The goal was to learn the effect of the movement of these structures on the reflection, scattering, guidance, and/or radiation of electromagnetic waves. The theory of special relativity, Lorentz transformation, and Minkowski transformation must be used to solve these problems.

Einstein's special theory of relativity is given below,

- *Postulate 1*: When properly formulated, the laws of physics are invariant to a transformation from one reference system to another moving with a linear, uniform relative velocity.
- *Postulate 2*: The velocity of propagation of an electromagnetic disturbance in free space is a universal constant, c, which is independent of the reference system.

Based on these postulates, the Lorentz transformations between an inertial frame, $S(r,t)$, and another inertial frame, $S'(r',t')$, which is moving at a uniform velocity, v, with respect to S, can be written in the general form

$$\boldsymbol{r}' = \boldsymbol{r} - \gamma \boldsymbol{v} t + (\gamma - 1)\frac{\boldsymbol{r}\cdot\boldsymbol{v}}{v^2} v$$

$$t' = \gamma(t - \frac{\boldsymbol{r}\cdot\boldsymbol{v}}{c^2})$$

where

$$\gamma = (1 - \beta^2)^{-1/2}, \quad \beta = \frac{v}{c}$$

$$\boldsymbol{r} = x\boldsymbol{e}_x + y\boldsymbol{e}_y + z\boldsymbol{e}_z$$

$$c = velocity\ of\ light\ in\ vacuum$$

and $\boldsymbol{e}_x$, $\boldsymbol{e}_y$, $\boldsymbol{e}_z$ are the unit vectors in the x, y, and z directions. To assure the covariance of the Maxwell equations between S and S' systems, the following Minkowski transformations for the field vectors must be used:

$$\boldsymbol{E}' = \gamma(\boldsymbol{E} + \boldsymbol{v} \times \boldsymbol{B}) + (1 - \gamma)\frac{\boldsymbol{E} \cdot \boldsymbol{v}}{v^2}\boldsymbol{v}$$

$$\boldsymbol{B}' = \gamma\left(\boldsymbol{B} - \frac{1}{c^2}\,\boldsymbol{v} \times \boldsymbol{E}\right) + (1 - \gamma)\frac{\boldsymbol{B} \cdot \boldsymbol{v}}{v^2}\boldsymbol{v}$$

$$\boldsymbol{D}' = \gamma\left(\boldsymbol{D} + \frac{1}{c^2}\,\boldsymbol{v} \times \boldsymbol{H}\right) + (1 - \gamma)\frac{\boldsymbol{D} \cdot \boldsymbol{v}}{v^2}\boldsymbol{v}$$

$$\boldsymbol{H}' = \gamma(\boldsymbol{H} - \boldsymbol{v} \times \boldsymbol{D}) + (1 - \gamma)\frac{\boldsymbol{H} \cdot \boldsymbol{v}}{v^2}\boldsymbol{v}$$

where $(\boldsymbol{E}, \boldsymbol{H}, \boldsymbol{D}, \boldsymbol{B})$ are the electromagnetic fields in the S system and $(\boldsymbol{E}', \boldsymbol{H}', \boldsymbol{D}', \boldsymbol{B}')$ are the electromagnetic fields in the S' system. Another consequence of these postulates is the birth of the principle of phase invariance, that is:

$$-\boldsymbol{k}' \cdot \boldsymbol{r} + \omega' t' = -\boldsymbol{k} \cdot \boldsymbol{r} + \omega t$$

or, using the Lorentz transforms, we obtain

$$\boldsymbol{k}' = \boldsymbol{k} - \gamma\frac{\boldsymbol{v}\omega}{c^2} + (\gamma - 1)\frac{\boldsymbol{k} \cdot \boldsymbol{v}}{v^2}\boldsymbol{v}$$

$$\omega' = \gamma(\omega - \boldsymbol{v} \cdot \boldsymbol{k})$$

Here, $(\omega, \boldsymbol{k})$ are the frequency and wave vector in the S system, and $(\omega', \boldsymbol{k}')$ are the frequency and wave vector in the S' system. Using these transformations, we were able to solve the problems discussed in the following papers.

6.1 Moving dielectric half-space or slabs

6.1.1 Reflection and transmission of electromagnetic waves by a moving dielectric medium. (Paper 6-1-1)

The reflection and transmission of electromagnetic waves by a moving dielectric medium are investigated theoretically, and the reflection and transmission coefficients are determined. Two cases of movement are considered: (a) The dielectric medium moves parallel to the interface;(b) the dielectric medium moves perpendicular to the interface.

Various interesting features concerning the variation of the reflection and transmission coefficients, angles of reflection and transmission, and frequencies of the reflected and transmitted wave, as a function of the velocity of the moving medium, are discussed.

6.1.2 Brewster angle for a dielectric medium moving at relativistic speed. (Paper 6-1-2)

Several interesting features concerning the reflection of obliquely incident electromagnetic waves from a dielectric half-space moving at relativistic velocities are discussed. Two cases of the movement are considered: (a) The dielectric medium moves parallel to the interface; (b) the dielectric medium moves perpendicular to the interface. Both polarizations of the incident wave are considered: (a) The electric field of the incident field lies in the plane of incidence; (b) the electric field lies in a plane that is normal to the plane of incidence. The variation of the Brewster angle as a function of the velocity of the medium is examined in particular. Also considered are the energy relations of this moving-boundary problem.

6.1.3 Reflection and transmission of electromagnetic waves by a moving dielectric slab. (Paper 6-1-3)

The reflection and transmission of electromagnetic waves by a moving dielectric slab are investigated theoretically, and the reflection and transmission coefficients are determined. Two cases are considered: (a) The dielectric slab moves parallel to the interface; (b) the dielectric slab moves perpendicular to the interface. Various interesting features concerning the variation of the reflection and transmission coefficients, angles of reflection and transmission, and frequencies of the reflected and transmitted wave, as a function of the velocity of the moving medium, are discussed.

6.1.4 Reflection and transmission of electromagnetic waves by a moving dielectric slab. II. parallel polarization. (Paper 6-1-4)

The reflection and transmission of a plane wave, with the electric vector polarized in the plane of incidence, by a moving dielectric slab, are investigated theoretically. Two cases of the movement are considered: (a) The dielectric slab moves parallel to the interface; (b) the dielectric slab moves perpendicular to the interface. It is shown that, in general, the reflection and transmission coefficients for an incident plane wave with its electric vector polarized in the plane of incidence are different from those for an incident plane wave with its electric vector polarized normal to the plane of incidence, except for case (b) for normally incident waves. Detailed results on the reflection and transmission coefficients for case (a) for normally incident waves are given and discussed.

6.1.5 Reflection and transmission of electromagnetic waves by a moving plasma medium. (Paper 6-1-5)

The reflection and transmission of electromagnetic waves by a moving dispersive dielectric half-space or slab are investigated theoretically. The dispersive dielectric medium is assumed to be a cold plasma medium. Two cases of the medium are considered: (a) The plasma medium moves parallel to the interface; (b) the plasma medium moves perpendicular to the interface. It is interesting to note that for the case (a), the reflected and transmitted fields for an incident-plane E wave are independent from the movement of the plasma medium. Detailed results on the reflection and transmission coefficients for case (b) are given and discussed. In the present analysis, the electric vector of the incident wave is assumed to be polarized in a direction that is normal to the plane of incidence.

6.1.6 Reflection and transmission of electromagnetic waves by a moving plasma medium. II. parallel polarizations. (Paper 6-1-6)

The reflection and transmission of electromagnetic waves by a moving dispersive dielectric half-space or slab are investigated theoretically. The dispersive medium is assumed to be a cold plasma medium. Two cases of the movement are considered: (a) The plasma medium moves parallel to the interface; (b) the plasma medium moves perpendicular to the interface. The electric vector of the incident wave is assumed to be polarized in the plane of incidence (parallel polarization). Unlike the normal polarization case, which was treated earlier, the reflected and transmitted fields for the present case are functions of the velocity of the moving plasma medium for both cases, (a) and (b). An illustrative example is given for case (a) at normal incidence.

6.1.7 Reflection from a dielectric-coated moving mirror. (Paper 6-1-7)

The effects of a dielectric coating on a moving mirror upon the characteristics of the reflected waves are investigated. Both polarizations of the incident waves are considered. Two cases of the movement are considered: (a) The coated mirror moves parallel to the interface; (b) the coated mirror moves perpendicular to the interface. Various interesting features concerning the variation of the reflection coefficient, angle of reflection, phase shift, and frequency of the reflected wave, as a function of the velocity of the moving coated mirror, are discussed.

6.2 Guided wave on moving structure

6.2.1 Propagation along moving dielectric wave guides. (Paper 6-2-1)

The problem of the propagation of modes along a moving dielectric interface is considered. Two types of dielectric structures are considered in particular: a moving dielectric slab and a moving dielectric circular cylinder. The characteristic equations for the modes along these two types of structures are derived, and a detailed discussion concerning these characteristic equations is also given. As a specific example, numerical results for the guide wavelength of the dominant TE wave along a moving dielectric slab as a function of frequency for various values of v_z/c, where v_z is the velocity of the moving dielectric material and c is the velocity of light in vacuum, are presented. It is found that if $v_z/c > (\epsilon_1/\epsilon_0)^{-1/2}$, the moving dielectric structure can support a forward wave as well as a backward wave, and there also exists a high-frequency cutoff for the dominant mode under consideration.

6.2.2 Wave propagation on a moving plasma column. (Paper 6-2-2)

This article presents the results of the problem for the propagation of surface waves along a moving plasma column.

6. 3 Scattering by moving plasma column or sheath

6.3.1 Scattering [of] obliquely incident microwaves by a moving plasma column. (Paper 6-3-1)

The problem of the interaction of obliquely incident microwaves with a plasma column moving uniformly in the axial direction is treated analytically. An arbitrary polarization is assumed. Two methods of solving this problem are presented. Extensive numerical results for the scattered energy in the backward and broadside directions, and the angular distribution of the scattered energy are obtained for various interesting ranges of the parameters involved. It is found that cross-polarized field components are induced even at normal incidence when the plasma medium is moving with respect to the observer and that cross-polarized field components disappear at an incident angle $\theta = sin^{-1}(v_z/c)$, where v_z is the velocity of the moving plasma and c is the velocity of light in vacuum, when the plasma is imbedded in free space.

6.3.2 Diffraction of waves by a conducting cylinder coated with a moving plasma sheath. (Paper 6-3-2)

The scattering of plane electromagnetic waves by a perfectly conducting cylinder coated with a moving dielectric or plasma sheath is investigated theoretically. The homogeneous sheath is assumed to be moving in the axial direction with a uniform velocity v_z with respect to the conducting cylinder. Solutions of this problem are obtained by making use of the special theory of relativity, the covariance of Maxwell's equations, and the Lorentz transformations. Results are given in terms of the radiation patterns of the scattered fields. A unique feature concerning mode coupling between the incident wave and the scattered wave is found. Even at normal incidence for $v_z \neq 0$, an incident E wave or H wave will produce a scattered wave that contains both E wave and H waves. Detailed discussions are presented.

6.4 Radiation from an aperture in a conducting cylinder coated with a moving plasma sheath. (Paper 6-4)

The electromagnetic radiation from an aperture on a conducting cylinder coated with a moving isotropic plasma sheath is considered. Numerical results are presented to illustrate the radiation patterns as a function of sheath velocity and plasma frequency for the circumferential slot and axial slot apertures. It is found for the circumferential slot aperture that the radiation is enhanced in the direction of the sheath motion when the plasma is over-dense and that relatively little change occurs when the sheath is under-dense. For the axial slot, it is found that an electromagnetic field is radiated whose polarization is normal to that of the field radiated under stationary conditions, in addition to a field of the usual polarization. Significant alterations of the radiation patterns from their form when the sheath is stationary can occur at relatively small velocities if the wave frequency is near the plasma frequency.

6.5 A proposed method of shifting the frequency of light waves. (Paper 6-5)

A simple method of shifting the frequency of light waves with precision, based on the Doppler principle, is proposed. The proposed system consists of two reflecting mirrors. One mirror is assumed to be moving with a velocity v while the other mirror is assumed to be stationary. Practical limitations such as mirror loss and beam spread, are discussed. A numerical example is given.

Reprinted from JOURNAL OF APPLIED PHYSICS, Vol. 36, No. 11, 3513–3517, November 1965

Reflection and Transmission of Electromagnetic Waves by a Moving Dielectric Medium*

C. YEH
Electrical Engineering Department, University of Southern California, Los Angeles, California
(Received 3 May 1965)

The reflection and transmission of electromagnetic waves by a moving dielectric medium are investigated theoretically, and the reflection and transmission coefficients are determined. Two cases of the movement are considered: (a) The dielectric medium moves parallel to the interface. (b) The dielectric medium moves perpendicular to the interface.

Various interesting features concerning the variation of the reflection and transmission coefficients, angles of reflection and transmission, and the frequencies of the reflected and transmitted wave, as a function of the velocity of the moving medium, are discussed.

I. INTRODUCTION

THE problem of the reflection of plane electromagnetic waves by a perfectly reflecting moving mirror has been discussed many years ago by various authors.[1,2] Sommerfeld gave a rather comprehensive treatment of this problem in his book.[2] However, it is somewhat surprising to learn that the basic problem of the reflection and transmission of plane waves by a uniformly moving dielectric half-space has not been given.[3] The purpose of this paper is to present the solution to this important problem. The principle of phase invariance is used to treat this problem. Several interesting features concerning the variation of the reflection and transmission coefficients, the angles of reflection and transmission, and the frequencies of the reflected and transmitted waves, as a function of the velocity of the moving medium are discussed.

II. THE FORMAL SOLUTION

The geometry of this problem is shown in Fig. 1. It is assumed that a moving homogeneous dielectric medium having a permittivity of ϵ_1, a permeability of μ_0, and a conductivity of zero, occupies half of space ($z<0$), while the other half of space is filled by empty free-space (ϵ_0,μ_0). The boundary between the two media is thus a plane, $z=0$. The motion of the moving medium is taken to be uniform (i.e., independent of position and time). Two cases of the movement are considered: (a) The dielectric medium moves parallel to the interface in the x direction with a constant velocity v_x. (b) The dielectirc medium moves perpendicular to the interface in the z direction with a constant velocity v_z. Finally, the incident wave in the free-space region is assumed to be plane with a harmonic time dependence. The case for an incident E wave is analyzed in detail.

In the observer's system S which is stationary with respect to the free-space the incident plane wave takes the form

$$E_y^{(i)}=E_0e^{i(k_xx+k_zz)}e^{-i\omega t}, \tag{1}$$

$$B_y^{(i)}=0, \tag{2}$$

where E_0 and ω are, respectively, the amplitude and the frequency of the incident wave, $k_x=k_0\sin\theta_0$, $k_z=-k_0\cos\theta_0$, and $k_0=\omega(\mu_0\epsilon_0)^{\frac{1}{2}}$. θ_0 is the angle between the propagation vector and the positive z axis in the x–z plane. All other field components can be obtained from Maxwell's equations.

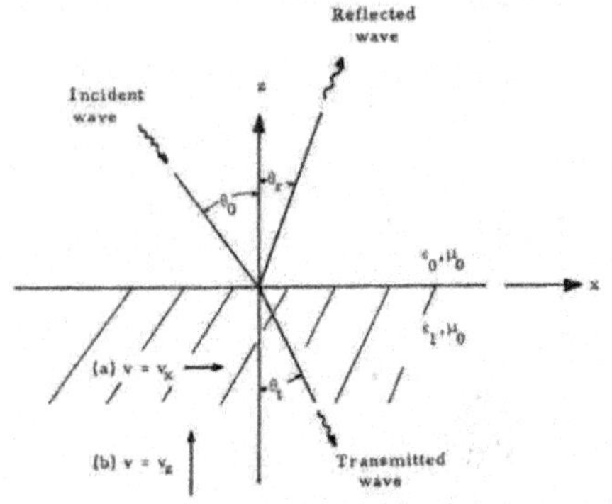

FIG. 1. The geometry of the problem.

* Work supported by the Technical Advisory Committee of the Joint Services Electronics Program and by the U. S. Naval Ordnance Test Station

[1] W. Pauli, *Theory of Relativity* (Pergamon Press, Inc., New York, 1958).

[2] A. Sommerfeld, *Optik* (Akademische Verlagsgesellschaft, Leipzig, 1959), 2nd ed.

[3] (a) As far as the writer is aware this problem has not been treated in as complete a fashion as that given here. Most recently, C. T. Tai treated the problem of the reflection by a dielectric medium moving in a direction transverse to the direction of an incident wave under the condition that the velocity of motion is very small compared to the velocity of light (Antenna Laboratory Report No. 1691-7, Ohio State University). Through private communications, I learned that he has recently also treated the case described in his report without the limitation on the speed of the dielectric medium. However, he did not treat the case in which the dielectric medium is moving towards or away from an incident wave. It appears that this case happens to be the more interesting one in terms of reflection and transmission coefficients as well as the angle of reflection and transmission. In his work he did not make use of the phase invariance principle; rather he made use of Maxwell's equations for a moving dielectric medium as viewed from a stationary system. Due to the simplicity of the procedures described here, the phase invariance approach is preferred for the present problem. (b) The corresponding problem of the reflection and transmission of sound by a medium moving parallel to the interface has been given by J. B. Keller [J. Acoust. Soc. Am. 27, 1044 (1955)].

In the moving system S' which is stationary with respect to the uniformly moving dielectric half-space, the incident plane wave takes the form

$$E_{y'}^{(i)'}=E_0'e^{i(k_x'x'+k_z'z')}e^{-i\omega't'}, \tag{3}$$

$$B_{y'}^{(i)'}=0, \tag{4}$$

where k_x', k_z', ω', and E_0' are related to combinations of k_x, k_z, ω, and E_0 by Eqs. (15) when the dielectric half-space is moving in the positive x direction and by Eqs. (18) when the dielectric half-space is moving in the positive z direction. x', z', and t' are related to combinations of x, z, and t by the Lorentz transformations. The reflected wave and the transmitted wave must, respectively, have the form

$$E_{y'}^{(r)'}=A_r'e^{i(k_x'x'-k_z'z')}e^{-i\omega't'}, \tag{5}$$

$$B_{y'}^{(r)'}=0, \tag{6}$$

and

$$E_{y'}^{(t)'}=G_t'\exp\{i[k_x'x'-(\omega'^2\mu_0\epsilon_1-k_x'^2)^{\frac{1}{2}}z']\}e^{-i\omega't'}, \tag{7}$$

$$B_{y'}^{(t)'}=0, \tag{8}$$

where A_r' and G_t' are arbitrary constants to be determined by the boundary conditions. Matching the tangential electric and magnetic fields at the boundary $z'=0$, one obtains

$$A_r'=E_0'\frac{(\omega'^2\mu_0\epsilon_1-k_x'^2)^{\frac{1}{2}}+k_z'}{k_z'-(\omega'^2\mu_0\epsilon_1-k_x'^2)^{\frac{1}{2}}}, \tag{9}$$

$$G_t'=E_0'\frac{2k_z'}{k_z'-(\omega'^2\mu_0\epsilon_1-k_x'^2)^{\frac{1}{2}}}. \tag{10}$$

In the observer's system S, the reflected wave and the transmitted wave take the following forms: For the reflected wave

$$E_y^{(r)}=A_re^{i(k_x^{(r)}x-k_z^{(r)}z)}e^{-i\omega^{(r)}t}, \tag{11}$$

$$B_y^{(r)}=0; \tag{12}$$

for the transmitted wave

$$E_y^{(t)}=G_te^{i(k_x^{(t)}x+k_z^{(t)}z)}e^{-i\omega^{(t)}t}, \tag{13}$$

$$B_y^{(t)}=0. \tag{14}$$

A_r, $k_x^{(r)}$, $k_z^{(r)}$, $\omega^{(r)}$, G_t, $k_x^{(t)}$, $k_z^{(t)}$, and $\omega^{(t)}$ are related to combinations of A_r', k_x', k_z', ω', and G_t' by Eqs. (16) when the dielectric half-space is moving in the x direction and by Eqs. (19) when the dielectric half-space is moving in the z direction.

Case (a): $v=v_xe_x$

Let us consider the case in which the dielectric half-space is moving at a uniform velocity v_x in the x direction. Making use of the covariance of Maxwell's equations and the phase invariance of a uniform plane wave, we have the following transformations [referring to Eq. (3)]:

$$\begin{aligned}
k_x'&=\gamma_x[k_x-(\omega v_x/c^2)],\\
k_z'&=k_z,\\
\omega'&=\gamma_x(\omega-v_xk_x),\\
E_0'&=\gamma_xE_0(1-v_xk_x/\omega),
\end{aligned} \tag{15}$$

where $\gamma_x=(1-v_x^2/c^2)^{-\frac{1}{2}}$ and c is the velocity of light. Viewing from the observer's system S, we have [referring to Eqs. (11) and (13)],

$$\begin{aligned}
\omega^{(r)}&=\omega^{(t)}=\gamma_x(\omega'+v_xk_x'),\\
k_x^{(r)}&=k_x^{(t)}=\gamma_x[k_x'+(v_x\omega'/c^2)],\\
k_z^{(r)}&=k_z',\\
k_z^{(t)}&=-[\omega'^2\mu_0\epsilon_1-k_x'^2]^{\frac{1}{2}},\\
A_r&=\gamma_xA_r'(1+v_xk_x'/\omega'),\\
G_t&=\gamma_xG_t'(1+v_xk_x'/\omega').
\end{aligned} \tag{16}$$

Substituting Eq. (15) into (16) gives

$$\omega^{(r)}=\omega^{(t)}=\omega, \tag{17a}$$

$$k_x^{(r)}=k_x^{(t)}=k_x=k_0\sin\theta_0, \tag{17b}$$

$$k_z^{(r)}=k_z=-k_0\cos\theta_0, \tag{17c}$$

$$k_z^{(t)}=-\gamma_xk_0\left[\frac{\epsilon_1}{\epsilon_0}(1-\beta_x\sin\theta_0)^2-(\sin\theta_0-\beta_x)^2\right]^{\frac{1}{2}}, \tag{17d}$$

$$A_r=E_0\frac{\cos\theta_0+(k_z^{(t)}/k_0)}{\cos\theta_0-(k_z^{(t)}/k_0)}, \tag{17e}$$

$$G_t=E_0\frac{2\cos\theta_0}{\cos\theta_0-(k_z^{(t)}/k_0)}, \tag{17f}$$

where $\beta_x=v_x/c$. It is interesting to note that there exists no Doppler shift in frequency for the reflected and the transmitted waves. Furthermore, according to Eqs. (17b) and (17c) the familiar law concerning the equality of the angle of incidence and the angle of reflection is preserved. The same observation has been made concerning the moving mirror problem. In that case, the mirror is assumed to be a perfectly conducting wall. On the other hand, the angle of refraction, defined as $\theta_t=\tan^{-1}|k_x^{(t)}/k_z^{(t)}|$, is a function of the velocity of the dielectric medium as well as a function of the permittivity of the dielectric medium. As v_x approaches the velocity of light c, θ_t approaches zero.

Case (b): $v=v_ze_z$

Now let us consider the case in which the dielectric half-space is moving at a uniform velocity v_z in the positive z direction. Again making use of the covariance of Maxwell's equations and the phase invariance of a uniform plane wave, we obtained the following transformations:

$$\omega' = \gamma_z\omega(1 - v_z k_z/\omega),$$
$$k_x' = k_x,$$
$$k_z = \gamma_z[k_z - (\omega v_z/c^2)],$$
$$E_0' = \gamma_z E_0(1 - v_z k_z/\omega), \quad (18)$$

where $\gamma_z = 1/[1 - (v_z/c)^2]^{\frac{1}{2}}$. Viewing from the observer's system S, we have

$$\omega^{(r)} = \gamma_z(\omega' - v_z k_z'),$$
$$k_x^{(r)} = k_x^{(t)} = k_x',$$
$$k_z^{(r)} = -\gamma_z[-k_z' + (\omega' v_z/c^2)],$$
$$\omega^{(t)} = \gamma_z[\omega' - v_z(\omega'^2\mu_0\epsilon_1 - k_x'^2)^{\frac{1}{2}}],$$
$$k_z^{(t)} = \gamma_z[(v_z\omega'/c^2) - (\omega'^2\mu_0\epsilon_1 - k_x'^2)^{\frac{1}{2}}],$$
$$A_r = \gamma_z A_r'(1 - v_z k_z'/\omega'),$$
$$G_t = \gamma_z G_t'[1 - (v_z/\omega)(\omega'^2\mu_0\epsilon_1 - k_x'^2)^{\frac{1}{2}}]. \quad (19)$$

Substituting Eq. (18) into (19) gives

$$\omega^{(r)} = \omega\gamma_z^2[(1+\beta_z^2) + 2\beta_z\cos\theta_0], \quad (20a)$$
$$k_x^{(r)} = k_x = k_0\sin\theta_0, \quad (20b)$$
$$k_z^{(r)} = -k_0\gamma_z^2[2\beta_z + \cos\theta_0(1+\beta_z^2)], \quad (20c)$$
$$\omega^{(t)} = \omega\gamma_z^2[(1+\beta_z\cos\theta_0) - \beta_z Q], \quad (20d)$$
$$k_z^{(t)} = k_0\gamma_z^2[\beta_z(1+\beta_z\cos\theta_0) - Q], \quad (20e)$$
$$A_r = E_0\frac{\omega^{(r)}}{\omega}\left[\frac{(\cos\theta_0+\beta_z) - Q}{(\cos\theta_0+\beta_z) + Q}\right], \quad (20f)$$
$$G_t = E_0\left[\frac{2(\omega^{(t)}/\omega)(\cos\theta_0+\beta_z)}{(\cos\theta_0+\beta_z)+Q}\right], \quad (20g)$$

where

$$\beta_z = v_z/c \quad \text{and} \quad Q = [(\epsilon_1/\epsilon_0)(1+\beta_z\cos\theta_0)^2 - \sin^2\theta_0(1-\beta_z^2)]^{\frac{1}{2}}. \quad (20h)$$

Unlike case (a), there exists Doppler shift in frequency for the reflected wave as well as for the transmitted wave. And the amount of frequency shifts are different for the reflected wave and for the transmitted wave. The frequency shift for the reflected wave is independent of the permittivity of the dielectric medium, while the frequency shift for the transmitted wave is a function of ϵ_1. Also noted from Eqs. (20b) and (20c) is the fact that the angle of reflection $\theta_r = \tan^{-1}|k_x^{(r)}/k_z^{(r)}|$ is a function of the velocity of the medium. Consequently, the angle of reflection is no longer equal to the angle of incidence. When the velocity of the moving dielectric medium approaches the speed of light the angle of reflection approaches zero, and becomes almost independent of the angle of incidence, i.e., the propagation vector of the reflected wave is normal to the dielectric surface. Similar conclusions may be obtained for the propagation vector of the transmitted wave except the propagation vector is pointing in the other direction.

III. THE REFLECTION AND TRANSMISSION COEFFICIENTS

The reflection coefficient and the transmission coefficient are defined, respectively, by the relations

$$R = \mathbf{n}\cdot\mathbf{S}_r/\mathbf{n}\cdot\mathbf{S}_i \quad (21)$$

and

$$T = \mathbf{n}\cdot\mathbf{S}_t/\mathbf{n}\cdot\mathbf{S}_i, \quad (22)$$

where $\mathbf{n}$ is the unit vector normal to the interface and

$$\mathbf{S}_i = \tfrac{1}{2}(\mathbf{E}^{(i)} \times \mathbf{H}^{*(i)}), \quad (23)$$
$$\mathbf{S}_r = \tfrac{1}{2}(\mathbf{E}^{(r)} \times \mathbf{H}^{*(r)}), \quad (24)$$
$$\mathbf{S}_t = \tfrac{1}{2}(\mathbf{E}^{(t)} \times \mathbf{H}^{*(t)}). \quad (25)$$

The * signifies the complex conjugate of the function. It can also be shown that $R+T=1$.

The reflection and transmission coefficients for an incident E wave are, respectively,

$$R^{(E)} = (A_r/E_0)^2[\cos\theta^{(r)}/\cos\theta_0], \quad (26)$$
$$T^{(E)} = (G_t/E_0)^2[\epsilon^{(t)}/\epsilon_0]^{\frac{1}{2}}(\cos\theta^{(t)}/\cos\theta_0), \quad (27)$$

where

$$\cos\theta^{(r)} = 1/[1 + (k_x^{(r)}/k_z^{(r)})^2]^{\frac{1}{2}},$$
$$\cos\theta^{(t)} = 1/[1 + (k_x^{(t)}/k_z^{(t)})^2]^{\frac{1}{2}},$$
$$(\epsilon^{(t)}/\epsilon_0)^{\frac{1}{2}} = [\omega^{(t)}(\mu_0\epsilon_0)^{\frac{1}{2}}]^{-1}(k_x^{(t)2} + k_z^{(t)2})^{\frac{1}{2}}. \quad (28)$$

$k_x^{(r)}$, $k_z^{(r)}$, $\omega^{(t)}$, $k_x^{(t)}$, and $k_z^{(t)}$ are given by Eqs. (17) when the dielectric half-space is moving uniformly in the position x direction, and they are given by Eqs. (20) when the dielectric half-space is moving uniformly in the positive z direction. Simplifying Eqs. (26) and (27), one has for $\mathbf{v} = \mathbf{v}_x\mathbf{e}_x$,

$$R_x^{(E)} = \left[\frac{\cos\theta_0 + (k_z^{(t)}/k_0)_x}{\cos\theta_0 - (k_z^{(t)}/k_0)_x}\right]^2, \quad (29)$$

$$T_x^{(E)} = \frac{-4\cos\theta_0(k_z^{(t)}/k_0)_x}{[\cos\theta_0 - (k_z^{(t)}/k_0)_x]^2}, \quad (30)$$

with

$$(k_z^{(t)}/k_0)_x = -\gamma_x[(\epsilon_1/\epsilon_0)(1-\beta_x\sin\theta_0)^2 - (\sin\theta_0 - \beta_x)^2]^{\frac{1}{2}},$$

and for $\mathbf{v} = \mathbf{v}_z\mathbf{e}_z$

$$R_z^{(E)} = -\left(\frac{\omega^{(r)}}{\omega}\right)^2\left[\frac{(\cos\theta_0+\beta_z) - Q}{(\cos\theta_0+\beta_z)+Q}\right]^2 \times \frac{(k_z^{(r)}/k_0)_z}{[(k_z^{(r)}/k_0)_z^2 + \sin^2\theta_0]^{\frac{1}{2}}}\frac{1}{\cos\theta_0}, \quad (31)$$

$$T_z^{(E)} = -\left(\frac{(k_z^{(t)}/k_0)_z}{\cos\theta_0}\right)\frac{4(\cos\theta_0+\beta_z)^2(\omega^{(t)}/\omega)_z}{[(\cos\theta_0+\beta_z)+Q]^2}, \quad (32)$$

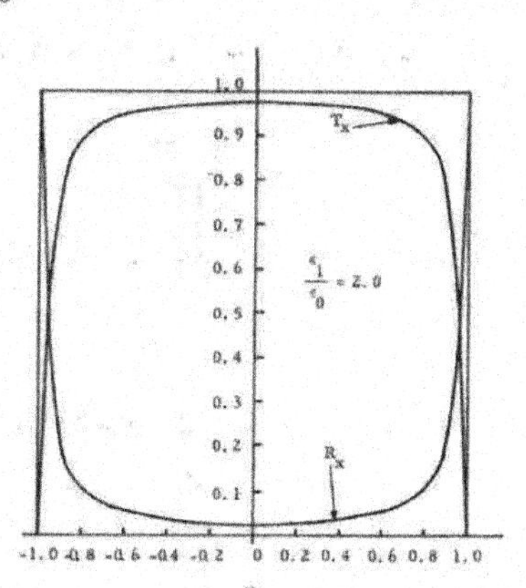

FIG. 2. The reflection and transmission coefficients for $v=v_x$.

with

$$(\omega^{(r)}/\omega)_z=\gamma_z^2[(1+\beta_z^2)+2\beta_z\cos\theta_0],$$

$$(k_z^{(r)}/k_0)_z=-\gamma_z^2[2\beta_z+\cos\theta_0(1+\beta_z^2)],$$

$$Q=[(\epsilon_1/\epsilon_0)(1+\beta_z\cos\theta_0)^2-\sin^2\theta_0(1-\beta_z^2)]^{\frac{1}{2}},$$

$$(\omega^{(t)}/\omega)_z=\gamma_z^2[(1+\beta\cos\theta_0)-\beta_z Q],$$

$$(k_z^{(t)}/k_0)_z=\gamma_z^2[\beta_z(1+\beta_z\cos\theta_0)-Q].$$

It can be seen from the above equations that the reflection and the transmission coefficients are rather complicated functions of the angle of incidence, the velocity, and the dielectric constant of the medium. To have a qualitative idea of how the reflection and the transmission coefficients vary as a function of the velocity of the moving medium, we now consider the limiting case of normal incidence. At normal incidence, i.e., $\theta_0=0$, Eqs. (29) through (32) reduce to

$$R_z^{(E)}=\left[\frac{1-\gamma_z[(\epsilon_1/\epsilon_0)-\beta_z^2]^{\frac{1}{2}}}{1+\gamma_z[(\epsilon_1/\epsilon_0)-\beta_z^2]^{\frac{1}{2}}}\right]^2, \tag{33}$$

$$T_z^{(E)}=\frac{4\gamma_z[(\epsilon_1/\epsilon_0)-\beta_z^2]^{\frac{1}{2}}}{\{1+\gamma_z[(\epsilon_1/\epsilon_0)-\beta_z^2]^{\frac{1}{2}}\}^2}, \tag{34}$$

$$R_x^{(E)}=\left[\left(\frac{1+\beta_x}{1-\beta_x}\right)\left(\frac{1-(\epsilon_1/\epsilon_0)^{\frac{1}{2}}}{1+(\epsilon_1/\epsilon_0)^{\frac{1}{2}}}\right)\right]^2, \tag{35}$$

$$T_x^{(E)}=\frac{4[(\epsilon_1/\epsilon_0)^{\frac{1}{2}}\beta_x-1][\beta_x-(\epsilon_1/\epsilon_0)^{\frac{1}{2}}]}{(1-\beta_x)^2[1+(\epsilon_1/\epsilon_0)^{\frac{1}{2}}]^2}. \tag{36}$$

One can readily show that $R_z^{(E)}+T_z^{(E)}=1$ and $R_x^{(E)}+T_x^{(E)}=1$ for all velocities.

Equations (33) through (36) are plotted in Figs. 2 and 3 for $\epsilon_1/\epsilon_0=2.0$. In these figures the reflection coefficient and the transmission coefficient are plotted as a function of the velocity of the medium. Figure 2 shows that the reflection coefficient increases monotonically to unity as v_x approaches the velocity of light in vacuum while the transmission coefficient decreases monotonically from $4(\epsilon_1/\epsilon_0)^{\frac{1}{2}}/[1+(\epsilon_1/\epsilon_0)^{\frac{1}{2}}]^2$ to zero as v_x varies from 0 to c, the velocity of light in vacuum. In other words, $v_x\rightarrow c$, most of the energy is reflected and nothing is transmitted. The dielectric medium for $v_x=c$ case acts like a perfectly conducting wall. Also noted is the fact that the T_x vs v_x/c and the R_x vs v_x/c curves are symmetrical with respect to the axis $v_x/c=0$ as expected.

Figure 3 shows that the reflection coefficient also increases monotonically as v_z/c increases and the transmission coefficient also decreases monotonically as v_z/c increases. However, in the present cases R_z increases without bounds and T_z decreases without bounds as v_z/c approaches unity. At $v_z/c=(\epsilon_0/\epsilon_1)^{\frac{1}{2}}$ which is the Čerenkov threshold velocity, $R_z=1$ and $T_z=0$. In other words when the dielectric half-space is moving towards the incident plane wave at the phase velocity of light in the stationary dielectric medium all the incident energy is reflected and no energy is transmitted into the dielectric medium. On the other hand, if the dielectric medium is moving away from the incident

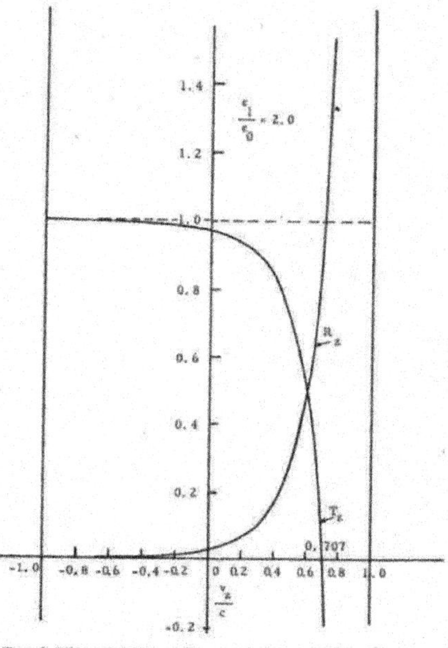

FIG. 3. The reflection and transmission coefficients for $v=v_z$.

wave, the reflection coefficient decreases monotonically to zero while the transmission coefficient increases monotonically to unity as $-v_z/c$ approaches unity.

It is noted that the formulas for the reflection and the transmission coefficients may also be obtained by using the Fresnel formulas.[4]

[4] C. Møller, *The Theory of Relativity* (Oxford University Press, London, 1957), p. 209.

Similar techniques may be used to obtain the reflection and transmission coefficients of an incident H plane wave.

ACKNOWLEDGMENTS

I wish to thank Professor C. T. Tai for sending me his memo describing the topics to be presented in an URSI spring meeting, 1965, at Washington, D. C.

Reprinted from JOURNAL OF APPLIED PHYSICS, Vol. 38, No. 13, 5194–5200, December 1967

Brewster Angle for a Dielectric Medium Moving at Relativistic Speed*

C. YEH†
Department of Engineering, University of California, Los Angeles, California
(Received 28 April 1967; in final form 14 August 1967)

Several interesting features concerning the reflection of obliquely incident electromagnetic waves from a dielectric half-space moving at relativistic velocities are discussed. Two cases of the movement are considered: (a) The dielectric medium moves parallel to the interface. (b) The dielectric medium moves perpendicular to the interface. Both polarizations of the incident wave are considered: (a) The electric field of the incident wave lies in plane of incidence. (b) The electric field of the incident wave lies in a plane which is normal to the plane of incidence. The variation of the Brewster angle as a function of the velocity of the medium is examined in particular. Also considered in detail are the energy relations of this moving boundary problem.

The purpose of this paper is to present several interesting features concerning the reflection of obliquely incident electromagnetic waves from a dielectric halfspace moving at relativistic velocities. In particular, the variation of the Brewster angle as a function of the velocity of the medium is examined.

Making use of the Lorentz transformations, the covariance of Maxwell's equations and the phase invariance of a uniform plane wave,[1] we may derive the following expressions for the reflection coefficients of a moving dielectric half-space[2]:

$$R_{\parallel,\perp}^{(x)}=[(m_{\parallel,\perp}\cos\theta_0-P)/(m_{\parallel,\perp}\cos\theta_0+P)]^2 \quad (1)$$

$$R_{\parallel,\perp}^{(z)}=(\omega^{(r)}/\omega)^2\{[m_{\parallel,\perp}(\cos\theta_0+\beta_z)-Q]/[m_{\parallel,\perp}(\cos\theta_0+\beta_z)+Q]\}^2\cos\theta^{(r)}/\cos\theta_0 \quad (2)$$

with

$$P=\gamma_x[(\epsilon_1/\epsilon_0)(1-\beta_x\sin\theta_0)^2-(\sin\theta_0-\beta_x)^2]^{1/2}, \quad (3)$$

$$Q=[(\epsilon_1/\epsilon_0)(1+\beta_z\cos\theta_0)^2-(1-\beta_z^2)\sin^2\theta_0]^{1/2}, \quad (4)$$

$$\omega^{(r)}/\omega=\gamma_z^2[(1+\beta_z^2)+2\beta_z\cos\theta_0], \quad (5)$$

$$\cos\theta^{(r)}=\gamma_z^2[2\beta_z+(1+\beta_z^2)\cos\theta_0]/\{\gamma_z^4[2\beta_z+(1+\beta_z^2)\cos\theta_0]^2+\sin^2\theta_0\}^{1/2},$$

$$\gamma_x=(1-\beta_x^2)^{-1/2}, \qquad \gamma_z=(1-\beta_z^2)^{-1/2},$$

$$\beta_x=v_x/c, \qquad \beta_z=v_z/c,$$

$$m_{\parallel}=\epsilon_1/\epsilon_0, \qquad m_{\perp}=1. \quad (6)$$

* This work was supported by the National Science Foundation.
† Formerly with the Electrical Engineering Department, University of Southern California, Los Angeles, California.
[1] C. Møller, *The Theory of Relativity* (Oxford University Press, New York, 1967).
[2] It is noted that Eqs. (1) and (2) cannot be found in the literature, although the problem of the reflection of electromagnetic waves by a moving dielectric half-space is one of the basic problems in electrodynamics of moving media. The case of the reflection of an E wave was considered earlier [C. Yeh, J. Appl. Phys. **36**, 3513 (1965)]. Most recently the problem of the total reflection of waves at the interface of moving media was considered by T. Shiozawa and N. Kumagai [Proc. IEEE **55**, 1243 (1967)].

$R_{\|,\perp}^{(x)}$ and $R_{\|,\perp}^{(z)}$ are, respectively, the reflection coefficient for the case in which the dielectric half-space is moving parallel to the interface with velocity v_x and for the case in which the dielectric half-space is moving normal to the interface with velocity v_z. (See Fig. 1). The subscript $\|$ signifies that the electric vector of the incident plane wave lies in the plane of incidence, while the subscript $\perp$ indicates that the electric vector of the incident plane wave lies in a plane which is normal to the plane of incidence. θ_0 and $\theta^{(r)}$ are, respectively, the angle of incidence and the angle of reflection. The dielectric half-space is assumed to be characterized by a permittivity of ϵ_1 and a permeability of μ_0. The other half of the space is filled by empty free space (ϵ_0, μ_0). c is the velocity of light in free space.

Case A: DIELECTRIC HALF-SPACE MOVING PARALLEL TO THE INTERFACE

It can be shown for this case that the frequency of the reflected wave and the incident wave remain the same

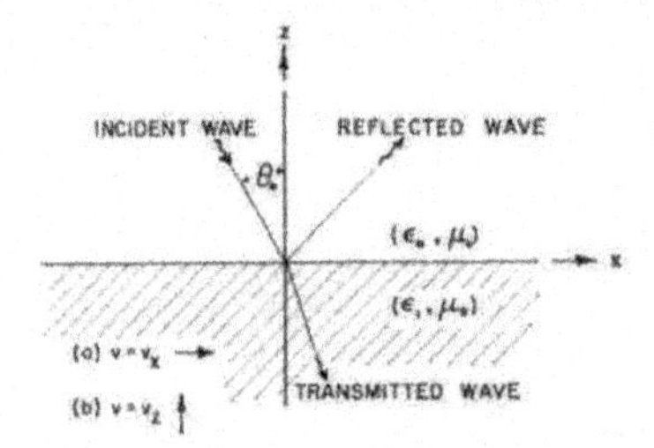

FIG. 1. Geometry of the problem.

and that the angle of reflection is the same as the angle of incidence. This is true for both polarizations of the incident wave. According to Eq. (1), we noted that the reflection coefficient is affected by the movement of the dielectrics. To have a qualitative idea of how the reflection coefficient varies as functions of v_x and θ_0, Figs. 2 and 3 are introduced. In these figures, the reflection coefficient is plotted against the angle of incidence for various values of β_x. A relative dielectric constant (ϵ_1/ϵ_0) of 2.0 is assumed.

From Fig. 2 we note that for $\beta_x>0$, as the angle of incidence θ_0 changes from 0°, $R_\perp^{(x)}$ decreases very slowly until a minimum is reached, and then increases to unity as θ_0 approaches 90°. For $\beta_x<0$, $R_\perp^{(x)}$ increases monotonically to unity as θ_0 changes from 0° to 90°. The movement of the dielectric does not seem to affect significantly the behavior of $R_\perp^{(x)}$ as a function of θ_0. The same can not be said about the behavior of $R_\|^{(x)}$ as a function of θ_0 for various β_x as shown in Fig. 3.

Fig. 3 shows that as β_x is increased from 0, $R_\|^{(x)}$ at $\theta=0°$ decreases until $\beta_x=\beta_0=[\epsilon_1/\epsilon_0/(1+\epsilon_1/\epsilon_0)]^{1/2}$ at

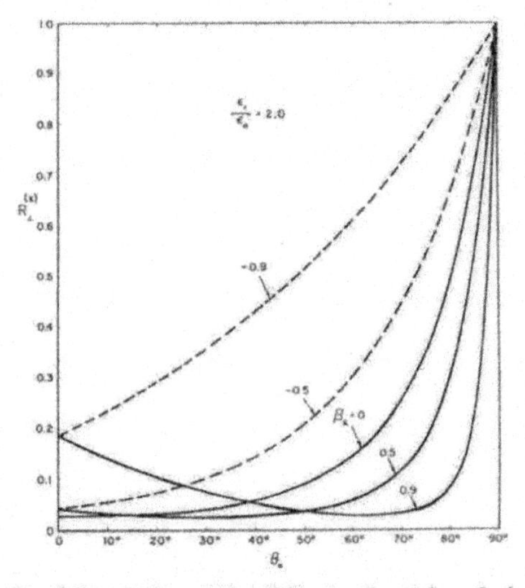

FIG. 2. The reflection coefficient $R_\perp^{(x)}$ as functions of the angle of incidence for various values of β_x.

which point $R_\|^{(x)}$ at $\theta_0=0°$ is zero. As β_x varies from β_0 to 1, $R_\|^{(x)}$ at $\theta_0=0°$ increases monotonically from 0 to 1. As expected, $R_\|^{(x)}$ at $\theta_0=0°$ is an even function of β_x. The general characters of the $R_\|^{(x)}$-vs-θ_0 curve for the stationary dielectric half-space case are still retained for the moving dielectric half-space case for

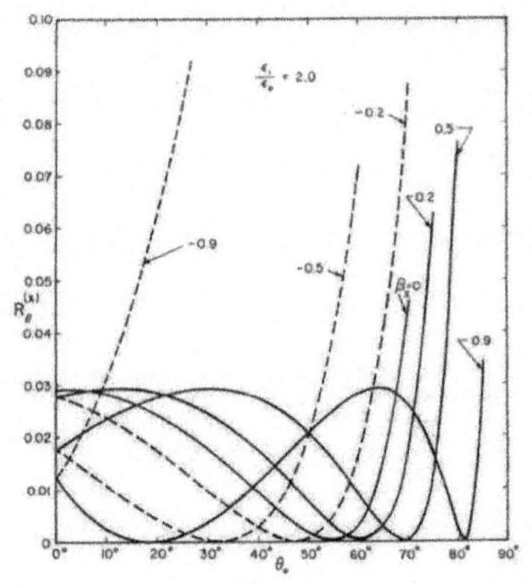

FIG. 3. The reflection coefficient $R_\|^{(x)}$ as functions of the angle of incidence for various values of β_x.

FIG. 4. The Brewster angle $\theta_{\parallel}^{(x)}$ as functions of β_x.

$-\beta_0>\beta_x>\beta_0$ (β_0 has been given earlier), i.e., $R_{\parallel}^{(x)}$ varies from a finite value at $\theta_0=0$ to 0 at a certain incident angle and then increases to unity as θ_0 approaches 90°. For $1\geq\beta_x>\beta_0$, there exist two values of θ_0 at which $R_{\parallel}^{(x)}=0$. For $-1<\beta_x<-\beta_0$, $R_{\parallel}^{(x)}$ is a monotonically increasing function of θ_0. The angle at which $R_{\parallel}^{(x)}=0$ is called the Brewster angle. From Eq. (1), we can find the relationship between the Brewster angle, the relative dielectric constant, and the velocity of the moving medium:

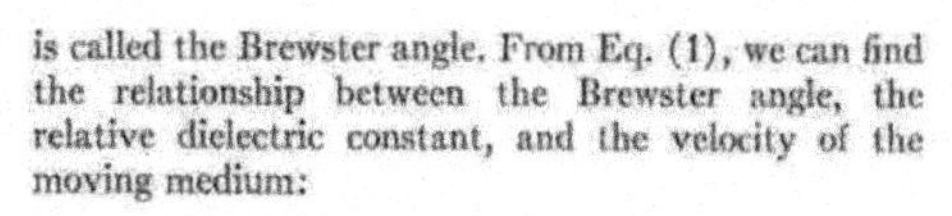

$$[(\epsilon_1/\epsilon_0+1)-\beta_x^2(\epsilon_1/\epsilon_0)]\sin^2\theta_{\parallel}^{(x)}-2\beta_x\sin\theta_{\parallel}^{(x)}+[\beta_x^2(\epsilon_1/\epsilon_0+1)-\epsilon_1/\epsilon_0]=0, \quad (7)$$

where $\theta_{\parallel}^{(x)}$ is the Brewster angle for the case in which the medium is moving parallel to the interface. For very small values of β_x, we have from Eq. (7)

$$\sin\theta_{\parallel}^{(x)}=[(\epsilon_1/\epsilon_0)/(1+\epsilon_1/\epsilon_0)]^{1/2}+[1/(1+\epsilon_1/\epsilon_0)]\beta_x+O(\beta_x^2). \quad (8)$$

Here, the parallel movement of the dielectric half space has a first-order effect on the Brewster angles. Solution of Eq. (7) for arbitrary values of β_x is given in Fig. 4 with $\epsilon_1/\epsilon_0=2.0$.

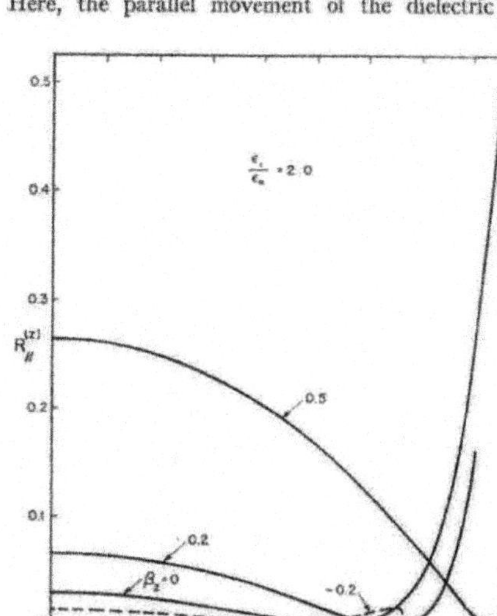

FIG. 6. The reflection coefficient $R_{\parallel}^{(x)}$ as functions of the angle of incidence for various values of β_x.

According to Fig. 4, we observed that the Brewster angle increases from $\tan^{-1}(\epsilon_1/\epsilon_0)^{1/2}$ to 90°, as β_x changes from 0 to 1.0, and that another Brewster angle appears at $\beta_x=\beta_0$, where β_0 is given earlier, and varies from 0° to 90° as β_x increases from β_0 to 1. In other words, unlike the stationary dielectric half-space case, there exist two Brewster angles for $\beta_0\leq\beta_x<1.0$. This effect is due to the variation of the value of the equivalent dielectric constant as functions of the angle of incidence of the plane wave and the velocity of the movement, as viewed by a stationary observer. As

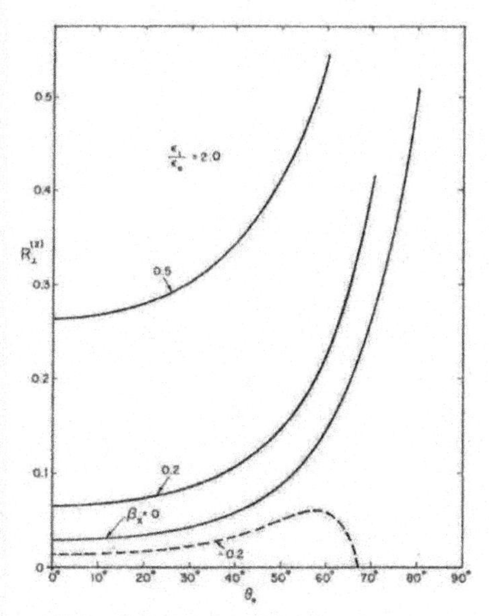

FIG. 5. The reflection coefficient $R_{\perp}^{(x)}$ as functions of the angle of incidence for various values of β_x.

β_z changes from 0 to $-\beta_0$, the Brewster angle varies from $\tan^{-1}(\epsilon_1/\epsilon_0)^{1/2}$ to 0°. There is no Brewster angle for $-1<\beta_z<-\beta_0$.

CASE B: DIELECTRIC HALF-SPACE MOVING NORMAL TO THE INTERFACE

In contrast with case (A), the frequency of the reflected wave as well as the angle of reflection are functions of the velocity of the moving dielectric half-space as can be seen from Eqs. (5) and (6). Furthermore, the frequency of the reflected wave and the angle of reflection are independent of the dielectric constant of the moving medium. To show how the reflection coefficient varies as functions of v_z and θ_0, we introduce Figs. 5 and 6. Again a relative dielectric constant (ϵ_1/ϵ_0) of 2.0 is assumed.

Figure 5 exhibits a set of $R_\perp^{(z)}$-vs-θ_0 curves for various values of β_z. The reflection coefficient $R_\perp^{(z)}$ increases without bounds from

$$R_0^{(z)}=(\{[(1+\beta_z)/(1-\beta_z)]\times\{[1-(\epsilon_1/\epsilon_0)^{1/2}]/[1+(\epsilon_1/\epsilon_0)^{1/2}]\})^2,$$

as θ_0 varies from 0° to 90° provided that $0<\beta_z\leq 1$. The fact that the reflection coefficient could be greater than unity is worth noting. It means that the reflected energy can be more than the energy of the incident wave. Apparently there is energy transfer from the moving dielectric to the reflected wave. Also noted is the fact that the angle of reflection is always less than the angle of incidence for $\beta_z>0$.

As the dielectric medium recedes from the incident wave, (i.e., $\beta_z<0$), $R_\perp^{(z)}$ increases from $R_0^{(z)}$ at $\theta_0=0°$ and then drops to 0 at

$$\theta_0=\theta_{0c}=\cos^{-1}[2|\beta_z|/(1+\beta_z^2)] \quad \text{for} \quad \beta_z<0;$$

beyond θ_{0c}, the reflected wave is evanescent. This is because at this critical incident angle (θ_{0c}), the angle of reflection is 90°; beyond this θ_{0c}, the angle of reflection is greater than 90°. As β_z decreases from 0, θ_{0c} also decreases. The existence of θ_{0c} for the moving medium is quite unique. It appears that there is no analogous counterpart for the case of the reflection from a stationary dielectric medium.

Figure 6 shows that, for $\beta_z>0$, $R_{||}^{(z)}$ changes from $R_0^{(z)}$ at $\theta_0=0$ to zero at a certain incident angle called $\theta_{||}^{(z)}$ and then increases without bounds, as θ_0 approaches 90°. For $\beta_z<0$, $R_{||}^{(z)}$ decreases from $R_0^{(z)}$ at $\theta_0=0$ to zero at $\theta_0=\theta_{||}^{(z)}$ and then increases as θ_0 increases from $\theta_{||}^{(z)}$. Then when θ_0 approaches θ_{0c}, $R_{||}^{(z)}$ again drops down to zero; beyond θ_{0c}, the reflected wave is evanescent.

The angle, $\theta_{||}^{(z)}$, is called the Brewster angle. From Eq. (2), we may obtain an equation relating $\theta_{||}^{(z)}$

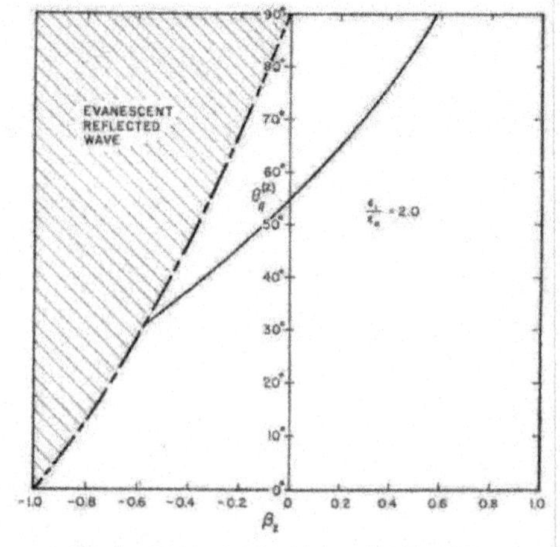

FIG. 7. The Brewster angle $\theta_{||}^{(z)}$ as functions of β_z.

with β_z

$$[(\epsilon_1/\epsilon_0+1)-\beta_z^2]\cos^2\theta_{||}^{(z)}+2\beta_z(\epsilon_1/\epsilon_0)\cos\theta_{||}^{(z)}+[(\epsilon_1/\epsilon_0+1)\beta_z^2-1]=0. \quad (9)$$

For very small values of β_z, Eq. (9) gives

$$\cos\theta_{||}^{(z)}=\left(\frac{1}{1+\epsilon_1/\epsilon_0}\right)^{1/2}-\left(\frac{\epsilon_1/\epsilon_0}{1+\epsilon_1/\epsilon_0}\right)\beta_z+O(\beta_z^2). \quad (10)$$

Hence, the normal movement of the dielectric half-space has a first-order effect on the Brewster angles. Figure 7 is introduced to show the response of $\theta_{||}^{(z)}$ as a function of β_z with $\epsilon_1/\epsilon_0=2.0$. Figure 7 also shows the θ_{0c}-vs-β_z curve. The reflected waves, with parameters which fall in the region to the left of this θ_{0c}-vs-β_z curve, are evanescent. Consequently, we note from Fig. 7 that the Brewster angle exists only if $-(1+\epsilon_1/\epsilon_0)^{-1/2}\leq\beta_z\leq(1+\epsilon_1/\epsilon_0)^{-1/2}$. For $1\geq\beta_z\geq(1+\epsilon_1/\epsilon_0)^{-1/2}$, there is no Brewster angle. For $-1<\beta_z<-(1+\epsilon_1/\epsilon_0)^{-1/2}$, the existence of Brewster angles according to Eq. (9) is meaningless, since the reflected waves are evanescent.

Because of the unique behavior of $R_{||,\perp}^{(z)}$, it is of interest at this point to investigate the energy relations for the case in which the dielectric half-space is moving normal to the interface. To determine the energy transfer at the interface, let us first imagine a small volume (a pill box), whose dimension normal to the interface is smaller than its other dimension by an order of magnitude, to be placed so that one of its larger surfaces lies in the vacuum and the other in the moving dielectric; both are parallel to the interface. The net time averaged Poynting's Vector, $\mathbf{S}_{||,\perp}^{(net)}$, which is the net energy per second coming out of the small volume in a direction that is normal to the interface, is defined as follows:

$$\mathbf{S}_{\parallel,\perp}^{(\mathrm{net})}\cdot\mathbf{e}_z=(-\mathbf{S}_{\parallel,\perp}^{(i)}+\mathbf{S}_{\parallel,\perp}^{(r)}+\mathbf{S}_{\parallel,\perp}^{(t)})\cdot\mathbf{e}_z, \quad (11)$$

where $\mathbf{S}_{\parallel,\perp}^{(i)}$, $\mathbf{S}_{\parallel,\perp}^{(r)}$, and $\mathbf{S}_{\parallel,\perp}^{(t)}$ are, respectively, the Poynting's vector for the incident wave, the reflected wave, and the transmitted wave[3]:

$$\mathbf{S}_{\parallel}^{(i)}\cdot\mathbf{e}_z=-\tfrac{1}{2}E_x^{(i)\parallel}H_y^{*(i)\parallel} \quad (12)$$

$$\mathbf{S}_{\perp}^{(i)}\cdot\mathbf{e}_z=\tfrac{1}{2}E_y^{(i)\perp}H_x^{*(i)\perp} \quad (13)$$

$$\mathbf{S}_{\parallel}^{(r)}\cdot\mathbf{e}_z=\tfrac{1}{2}E_x^{(r)\parallel}H_y^{*(r)\parallel} \quad (14)$$

$$\mathbf{S}_{\perp}^{(r)}\cdot\mathbf{e}_z=-\tfrac{1}{2}E_y^{(r)\perp}H_x^{*(r)\perp} \quad (15)$$

$$\mathbf{S}_{\parallel}^{(t)}\cdot\mathbf{e}_z=-\tfrac{1}{2}E_x^{(t)\parallel}H_y^{*(t)\parallel} \quad (16)$$

$$\mathbf{S}_{\perp}^{(t)}\cdot\mathbf{e}_z=\tfrac{1}{2}E_y^{(t)\perp}H_x^{*(t)\perp}. \quad (17)$$

The * signifies the complex conjugate of the function. All the field components in Eqs. (12)–(17) are given in Appendix A. Substituting Eqs. (12)–(17) into Eq. (11) gives

$$\mathbf{S}_{\parallel,\perp}^{(\mathrm{net})}\cdot\mathbf{e}_z=W_{\parallel,\perp}\Bigg(-\cos\theta_0+[2\beta_z+\cos\theta_0(1+\beta_z^2)/(1+\beta_z^2)+2\beta\cos\theta_0](A_r^{\parallel,\perp})^2 +\left\{\beta_z(1-\epsilon_1/\epsilon_0)+\frac{[1-(\epsilon_1/\epsilon_0)\beta_z^2][Q-\beta_z(1+\beta_z\cos\theta_0)]}{1+\beta_z\cos\theta_0-\beta_zQ}\right\}(\gamma_z^2/m_{\parallel,\perp})(G_t^{\parallel,\perp})^2\Bigg), \quad (18)$$

where $W_\parallel=H_0^2/2c\epsilon_0$ and $W_\perp=E_0^2/2c\mu_0$. E_0 and H_0 are, respectively, the amplitudes of the incident E wave and the incident H wave. $A_r^{\parallel,\perp}$ and $G_t^{\parallel,\perp}$ are given in Appendix A[4]. It is noted from Eq. (18) that in general $\mathbf{S}_{\parallel,\perp}^{(\mathrm{net})}\cdot\mathbf{e}_z$ is not zero, except when $\beta_z=0$. Positive or negative $\mathbf{S}_{\parallel,\perp}^{(\mathrm{net})}\cdot\mathbf{e}_z$ implies, respectively, that the sum of the radiated energy of the reflected wave and the transmitted wave is greater or less than the radiated energy of the incident wave.

Let us now examine the rate of change of stored energy due to the movement of the dielectric interface within the small pill-box that we introduced earlier. The stored electric and magnetic energy per unit volume is defined as follows[3]:

$$\text{Total stored energy per unit volume within the pill box}=\tfrac{1}{4}\mathbf{E}\cdot\mathbf{D}^*+\tfrac{1}{4}\mathbf{H}\cdot\mathbf{B}^*. \quad (19)$$

E, **D**, **H**, **B** are the electromagnetic fields within the pill-box. Using the above definition, one may obtain the rate of change of stored energy due to the movement of the dielectric interface:

$$U_{\parallel,\perp}=v_z[(\tfrac{1}{4}\mathbf{E}^{(t)\parallel,\perp}\cdot\mathbf{D}^{(t)\parallel,\perp}+\tfrac{1}{4}\mathbf{H}^{(t)\parallel,\perp}\cdot\mathbf{B}^{(t)\parallel,\perp})-(\tfrac{1}{4}\mathbf{E}^{(i)\parallel,\perp}\cdot\mathbf{D}^{(i)\parallel,\perp}+\tfrac{1}{4}\mathbf{H}^{(i)\parallel,\perp}\cdot\mathbf{B}^{(i)\parallel,\perp}) -(\tfrac{1}{4}\mathbf{E}^{(r)\parallel,\perp}\cdot\mathbf{D}^{(r)\parallel,\perp}+\tfrac{1}{4}\mathbf{H}^{(r)\parallel,\perp}\cdot\mathbf{B}^{(r)\parallel,\perp})]. \quad (20)$$

Simplifying Eq. (20) gives

$$U_{\parallel,\perp}=V_{\parallel,\perp}\beta_z\Bigg((\gamma_z^2/m_{\parallel,\perp})\left\{(\epsilon_1/\epsilon_0-\beta_z^2)+\beta_z(\epsilon_1/\epsilon_0-1)\frac{[Q-\beta_z(1+\beta_z\cos\theta_0)]}{[(1+\beta_z\cos\theta_0)-\beta_zQ]}\right\}(G_t^{\parallel,\perp})^2-1-(A_r^{\parallel,\perp})^2\Bigg), \quad (21)$$

with $V_\parallel=\frac{1}{2}\mu_0H_0^2c$ and $V_\perp=\frac{1}{2}\epsilon_0E_0^2c$.

According to the principle of conservation of energy, we conclude that the sum of the net rate of change of radiated energy per unit area, $\mathbf{S}_{\parallel,\perp}^{(\mathrm{net})}\cdot\mathbf{e}_z$ and the rate of change of stored energy per unit area, $U_{\parallel,\perp}$ must be equal to the rate of change of the mechanical energy per unit area, $M_{\parallel,\perp}$, in order to keep the moving dielectric half-space moving at a uniform velocity, v_z. Hence, we have

$$M_{\parallel,\perp}=\mathbf{S}_{\parallel,\perp}^{(\mathrm{net})}\cdot\mathbf{e}_z+U_{\parallel,\perp}. \quad (22)$$

It is noted that $M_{\parallel,\perp}$ may be positive or negative. Positive $M_{\parallel,\perp}$ means that mechanical energy must be added to the moving structure to assure uniform motion, while negative $M_{\parallel,\perp}$ means that mechanical energy must be carried away from the moving structure to assure uniform motion.

The fact that the quantity $M_{\parallel,\perp}$ is indeed the mechanical energy can be demonstrated by the application of the concept of stress tensors.[3] As an example, let us consider the $\perp$ case. The radiation pressure exerted on the inter-

[3] J. A. Stratton, *Electromagnetic Theory* (McGraw-Hill Book Co., Inc., New York 1941).

[4] The transmission coefficient $T_t^{(E)}$ given in Ref. 2 should read as follows:

$$T_t^{(E)}=\gamma_z^2\left\{\beta_z(1-\epsilon_1/\epsilon_0)+\frac{[1-(\epsilon_1/\epsilon_0)\beta_z^2][Q-\beta_z(1+\beta_z\cos\theta_0)]}{1+\beta_z\cos\theta_0-\beta_zQ}\right\}(G_t^{\perp})^2.$$

I wish to thank the reviewer for his contribution to the paragraph on stress tensor.

face in the S' system in which the dielectric medium is stationary, is

$$F_z'^{(\perp)}=+\epsilon_0 E_0'^2\cos^2\theta'\,2(\epsilon_1/\epsilon_0-1)/[\cos\theta'+(\epsilon_1/\epsilon_0-\sin^2\theta')^{1/2}]^2, \tag{23}$$

where

$$E_0'=E_0\gamma_z(1+\beta_z\cos\theta_0),$$
$$\cos\theta'=(\cos\theta_0+\beta_z)/(1+\beta_z\cos\theta_0),$$
$$\sin\theta'=(1-\cos^2\theta')^{1/2}.$$

Transforming the above expression back to the S system, we have

$$F_z^{(\perp)}=\gamma_z F_z'^{(\perp)}=-\epsilon_0 E_0^2\gamma_z^3(1+\beta_z\cos\theta_0)^2(\cos\theta_0+\beta_z)^2 2(\epsilon_1/\epsilon_0-1)/[(\cos\theta_0+\beta_z)+Q]^2 \tag{24}$$

By comparing with Eq. (22), we can show that $M_\perp=v_zF_z^{(\perp)}$. Hence, $M_\perp$ is indeed the mechanical energy.

To illustrate how the rate of change of mechanical energy per unit area ($M_{\|,\perp}$) varies as a function of β_z, Figs. 8 and 9 are introduced. Figure 8 gives a set of $M_\perp/\frac{1}{2}\epsilon_0E_0^2c$-vs-$\beta_z$ curves for two values of the angle of incidence for the $\perp$ polarization case, while Fig. 9 gives a set of $M_\|/\frac{1}{2}\epsilon_0H_0^2c$-vs-$\beta_z$ curves for the $\|$ polarization case. Both Figs. 8 and 9 show that when the dielectric half-space is moving towards the incident wave, mechanical energy must be carried away from the moving dielectric half-space. In other words the dielectric half-space is being sucked towards the incident wave for positive β_z. Furthermore, it appears that as β_z increases from 0 to +1, the normalized rate of change of mechanical energy per unit area is monotonically decreasing function of β_z.

According to Figs. 8 and 9, as the dielectric half-space moves away from the incident wave, mechanical energy must be supplied to the dielectric half-space in order to maintain a uniform velocity. As β_z varies from 0 to −1, the normalized rate of change of mechanical energy per unit area increases and reaches a maximum value and then decreases to zero as β_z approaches −1 for the normal incidence case. For oblique incidence case, similar behavior as that discussed above is noted except $M_{\|,\perp}/V_{\|,\perp}$ reaches zero at $\beta_z=-\cos\theta_0$. It should be noted that in Figs. 8 and 9, the scale for positive values of $M_{\|,\perp}/V_{\|,\perp}$ is different than that for negative values of $M_{\|,\perp}/V_{\|,\perp}$.

In conclusion, we observe that the movement of the dielectric half-space introduces many features concerning the reflected wave and the energy relations that are significantly different from those for a stationary dielectric half-space case.

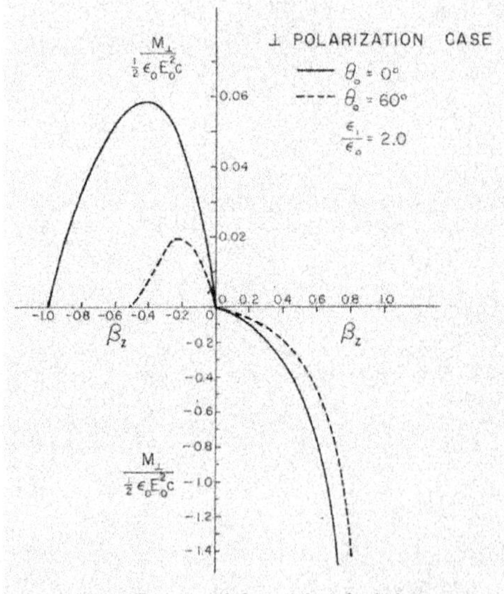

FIG. 8. Normalized rate of change of mechanical energy per unit area $M_\perp/\frac{1}{2}\epsilon_0E_0^2c$ as functions of β_z for various values of θ_0 for the $\perp$ polarization case. Note that the vertical scale is not uniform.

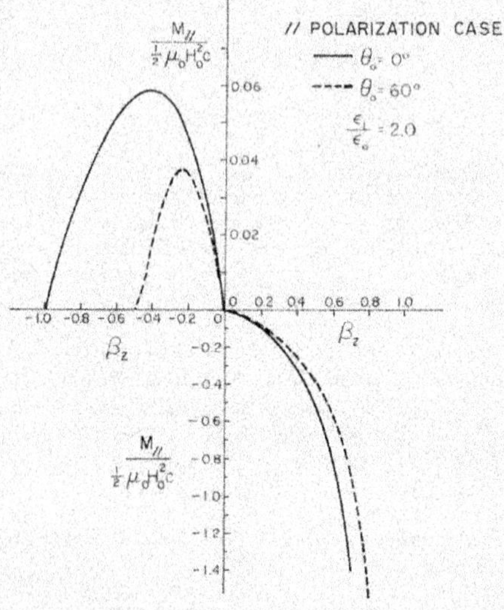

FIG. 9. Normalized rate of change of mechanical energy per unit area $M_\|/\frac{1}{2}\mu_0H_0^2c$ as functions of β_z for various values of θ_0 for the $\|$ polarization case. Note that the vertical scale is not uniform.

APPENDIX

The electromagnetic fields for the $\perp$ polarization case are as follows[2,4]:

For the incident wave,

$$E_y^{(i)\perp}=E_0\exp[i(k_xx+k_zz)]\exp(-i\omega t),$$
$$H_x^{(i)\perp}=-(k_z/\omega\mu_0)E_y^{(i)\perp},$$
$$H_z^{(i)\perp}=(k_x/\omega\mu_0)E_y^{(i)\perp},$$
$$D_y^{(i)\perp}=\epsilon_0E_y^{(i)\perp}; \tag{A1}$$

for the reflected wave,

$$E_y^{(r)\perp}=E_0A_r^{\perp}\exp[i(k_x^{(r)}x-k_z^{(r)}z)]\exp(-i\omega^{(r)}t),$$
$$H_x^{(r)\perp}=(k_z^{(r)}/\omega^{(r)}\mu_0)E_y^{(r)\perp},$$
$$H_z^{(r)\perp}=(k_x^{(r)}/\omega^{(r)}\mu_0)E_y^{(r)\perp},$$
$$D_y^{(r)\perp}=\epsilon_0E_y^{(r)\perp}; \tag{A2}$$

for the transmitted wave,

$$E_y^{(t)\perp}=E_0G_t^{\perp}\exp[i(k_x^{(t)}x+k_z^{(t)}z)]\exp(-i\omega^{(t)}t),$$
$$H_x^{(t)\perp}=(\gamma_z^2/\mu_0c)[\beta_z(1-\epsilon_1/\epsilon_0)-[k_z^{(t)}c/\omega^{(t)}]\times(1-\beta_z^2\epsilon_1/\epsilon_0)]E_y^{(t)\perp},$$
$$H_z^{(t)\perp}=(k_x^{(t)}/\omega^{(t)}\mu_0)E_y^{(t)\perp},$$
$$D_y^{(t)\perp}=(\gamma_z^2/\mu_0c^2)[(\epsilon_1/\epsilon_0-\beta_z^2)-(k_z^{(t)}c\beta/\omega^{(t)})\times(\epsilon_1/\epsilon_0-1)]E_y^{(t)\perp}, \tag{A3}$$

and

$$k_x=k_x^{(r)}=k_x^{(t)}=k_0\sin\theta_0=(\omega/c)\sin\theta_0,$$
$$k_z=-k_0\cos\theta_0,$$
$$k_z^{(r)}=-k_0\gamma_z^2[2\beta_z+\cos\theta_0(1+\beta_z^2)],$$
$$k_z^{(t)}=k_0\gamma_z^2[\beta_z(1+\beta_z\cos\theta_0)-Q],$$
$$\omega^{(r)}=\omega\gamma_z^2[(1+\beta_z^2)+2\beta_z\cos\theta_0],$$
$$\omega^{(t)}=\omega\gamma_z^2[(1+\beta_z\cos\theta_0)-\beta_zQ],$$
$$A_r^{\perp}=(\omega^{(r)}/\omega)\{[(\cos\theta_0+\beta_z)-Q]/[(\cos\theta_0+\beta_z)+Q]\},$$
$$G_t^{\perp}=2(\omega^{(t)}/\omega)(\cos\theta_0+\beta_z)/[(\cos\theta_0+\beta_z)+Q]. \tag{A4}$$

Q is given by Eq. (4).

The electromagnetic fields for the $\parallel$ polarization case are as follows:

For the incident wave,

$$H_y^{(i)\parallel}=H_0\exp[i(k_xx+k_zz)]\exp(-i\omega t),$$
$$E_x^{(i)\parallel}=(k_z/\omega\epsilon_0)H_y^{(i)\parallel},$$
$$E_z^{(i)\parallel}=(k_x/\omega\epsilon_0)H_y^{(i)\parallel}$$
$$B_y^{(i)\parallel}=\mu_0H_y^{(i)\parallel}; \tag{A5}$$

for the reflected wave,

$$H_y^{(r)\parallel}=H_0A_r^{\parallel}\exp[i(k_x^{(r)}x-k_z^{(r)}z)]\exp(-i\omega^{(r)}t),$$
$$E_x^{(r)\parallel}=-(k_z^{(r)}/\omega^{(r)}\epsilon_0)H_y^{(r)\parallel},$$
$$E_z^{(r)\parallel}=-(k_x^{(r)}/\omega^{(r)}\epsilon_0)H_y^{(r)\parallel},$$
$$B_y^{(r)\parallel}=\mu_0H_y^{(r)\parallel}; \tag{A6}$$

for the transmitted wave,

$$H_y^{(t)\parallel}=H_0G_t^{\parallel}\exp[i(k_x^{(t)}x+k_z^{(t)}z)]\exp(-i\omega^{(r)}t),$$
$$E_x^{(t)\parallel}=(\gamma_z^2/\epsilon_1c)\{\beta_z(\epsilon_1/\epsilon_0-1)+(k_z^{(t)}c/\omega^{(t)}\times[1-(\epsilon_1/\epsilon_0)\beta_z^2]\}H_y^{(t)\parallel},$$
$$E_z^{(t)\parallel}=-[k_x^{(t)}/\omega^{(t)}\epsilon_1]H_y^{(t)\parallel},$$
$$B_y^{(t)\parallel}=(\gamma_z^2/\epsilon_1c^2)[(\epsilon_1/\epsilon_0-\beta_z^2)+(k_z^{(t)}c/\omega^{(t)})\beta_z\times(1-\epsilon_1/\epsilon_0)]H_y^{(t)\parallel}; \tag{A7}$$

and

$$A_r^{\parallel}=(\omega^{(r)}/\omega)\left[\frac{(\epsilon_1/\epsilon_0)(\cos\theta_0+\beta_z)-Q}{(\epsilon_1/\epsilon_0)(\cos\theta_0+\beta_z)+Q}\right],$$
$$G_t^{\parallel}=\frac{2(\omega^{(t)}/\omega)(\epsilon_1/\epsilon_0)(\cos\theta_0+\beta_z)}{(\epsilon_1/\epsilon_0)(\cos\theta_0+\beta_z)+Q}. \tag{A8}$$

Reprinted from THE PHYSICAL REVIEW, Vol. 144, No. 2, 665-669, 15 April 1966
Printed in U. S. A.

Reflection and Transmission of Electromagnetic Waves by a Moving Dielectric Slab*

C. YEH
Electrical Engineering Department, University of Southern California, Los Angeles, California

AND

K. F. CASEY
Electrical Engineering Department, Air Force Institute of Technology, Wright Patterson Air Force Base, Ohio

(Received 1 November 1965)

The reflection and transmission of electromagnetic waves by a moving dielectric slab are investigated theoretically and the reflection and transmission coefficients are determined. Two cases of the movement are considered: (a) the dielectric slab moves parallel to the interface; (b) the dielectric slab moves perpendicular to the interface. Various interesting features concerning the variation of the reflection and transmission coefficients, angles of reflection and transmission, and the frequencies of the reflected and transmitted wave, as a function of the velocity of the moving medium, are discussed.

I. INTRODUCTION

THE effects of a perfectly reflecting moving boundary upon an incident plane electromagnetic wave were discussed many years ago by various authors.[1-3] The formula for the equivalent index of refraction of a dielectric medium moving at a uniform velocity with respect to a reference frame S, as viewed from the reference frame S, was first derived by Fresnel.[1-3] The well-known Fresnel formula was then verified experimentally by Fizeau. Sommerfeld gave a rather comprehensive treatment of these interesting problems in his book.[3] However, it is somewhat surprising to learn that the problem of the reflection and transmission of plane waves by a uniformly moving dielectric slab has not been treated.[4] The purpose of this paper is to present the solution to this important problem. The result shows that there exists no Doppler shift in frequency for the transmitted wave due to the movement of the slab. Furthermore, the sum of the reflection coefficient and the transmission coefficient is not unity in general. Discussion of these features as well as several other interesting features concerning the variation of the reflection and transmission coefficients, the angles of reflection and transmission, and the frequency of the reflected waves, as a function of the velocity of the moving medium will be given.

* Supported by the Naval Ordnance Test Station.

[1] W. Pauli, *Theory of Relativity* (Pergamon Press, Inc., New York, 1958).

[2] C. Møller, *The Theory of Relativity* (Oxford University Press, New York, 1952).

[3] A. Sommerfeld, *Optik* (Akademische Verlagsgesellschaft, Leipzig, 1959), 2nd ed.

[4] Most recently, Tai treated the problem of reflection by a dielectric half-space moving in a direction transverse to the direction of an incident wave. [C. T. Tai, Antenna Laboratory Report No. 1691-7, Ohio State University, 1964 (unpublished); oral presentation of the 1965 Spring URSI meeting in Washington, D.C.] The case in which the dielectric half-space is moving towards or away from an incident wave has been given by C. Yeh, J. Appl. Phys. 36, 3513 (1965).

II. THE FORMAL SOLUTION

The geometry of this problem is shown in Fig. 1. A homogeneous dielectric slab having a permittivity of ϵ_1, a permeability of μ_0, and a conductivity of zero, is assumed to occupy the space $d \geq z' \geq 0$ in the S' system which is stationary with respect to the slab. The region outside the dielectric slab is filled by empty free space (ϵ_0, μ_0). It is assumed that the dielectric slab may move in the following directions: (a) The slab moves parallel to the interface in the x direction with a constant velocity v_x. (b) The slab moves perpendicular to the interface in the z direction with a constant velocity v_z. Finally, the incident wave in the free-space region is assumed to be plane with a harmonic time dependence. Only the case for an incident E wave will be analyzed in detail.

In the observer's system S which is stationary with respect to the free space, the incident plane wave takes the form

$$E_y^{(i)} = E_0 e^{i(k_x x - k_z z)} e^{-i\omega t}, \tag{1}$$

$$B_y^{(i)} = 0, \tag{2}$$

where E_0 and ω are, respectively, the amplitude and the frequency of the incident wave, $k_x = k_0 \sin\theta_0$, $k_z = k_0 \cos\theta_0$, and $k_0 = \omega(\mu_0 \epsilon_0)^{1/2}$. θ_0 is the angle between the propagation vector and the positive z axis in the x-z plane.

In the moving system S' which is stationary with respect to the uniformly moving dielectric slab, the incident plane wave can be represented by the expressions

$$E_{y'}^{(i)'} = E_0' e^{i(k_x' x' - k_z' z')} e^{-i\omega' t'}, \tag{3}$$

$$B_{y'}^{(i)'} = 0, \tag{4}$$

where k_x', k_z', ω', and E_0' are related to combinations of k_x, k_z, ω, and E_0 by Eqs. (17) when the dielectric slab is moving in the positive x direction and by Eqs. (21) when the dielectric slab is moving in the positive z direction. x', z', and t' are related to combinations of x, z, and t by the Lorentz transformations. The reflected wave, the wave within the slab, and the transmitted wave must, respectively, have the form

$$E_{y'}^{(r)'} = A_r' e^{i(k_x' x' + k_z' z')} e^{-i\omega' t'}, \tag{5}$$

$$B_{y'}^{(r)'} = 0, \tag{6}$$

$$E_{y'}^{(p)'} = \{B_p' \exp[-i(\omega'^2 \mu_0 \epsilon_1 - k_x'^2)^{1/2} z'] + C_p' \exp[i(\omega'^2 \mu_0 \epsilon_1 - k_x'^2)^{1/2} z']\} \times e^{ik_x' x'} e^{-i\omega' t'}, \tag{7}$$

$$B_{y'}^{(p)'} = 0, \tag{8}$$

and

$$E_{y'}^{(t)'} = G_t' e^{i(k_x' x' - k_z' z')} e^{-i\omega' t'}, \tag{9}$$

$$B_{y'}^{(t)'} = 0. \tag{10}$$

A_r', B_p', C_p', and G_t' are arbitrary constants to be determined by the boundary conditions. Of interest are the reflected-wave coefficient A_r' and the transmitted-wave coefficient G_t'. Matching the tangential electric and magnetic fields at the boundaries $z' = d$ and $z' = 0$, one obtains

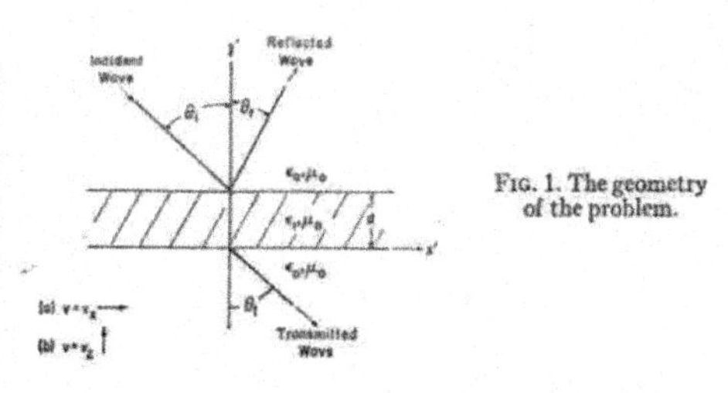

Fig. 1. The geometry of the problem.

$$A_r' = \frac{iE_0'(\xi'^2 - k_z'^2) e^{-2ik_z' d} \sin\xi' d}{2\xi' k_z' \cos\xi' d - i(k_z'^2 + \xi'^2) \sin\xi' d}, \tag{11}$$

$$G_t' = \frac{2E_0' \xi' k_z' e^{-ik_z' d}}{2\xi' k_z' \cos\xi' d - i(k_z'^2 + \xi'^2) \sin\xi' d}, \tag{12}$$

where $\xi' = (\omega'^2 \mu_0 \epsilon_1 - k_x'^2)^{1/2}$. In the observer's system S, the reflected wave and the transmitted wave take the following forms:

For the reflected wave

$$E_y^{(r)} = A_r \exp[i(k_x^{(r)} x + k_z^{(r)} z)] \exp[-i\omega^{(r)} t], \tag{13}$$

$$B_y^{(r)} = 0; \tag{14}$$

for the transmitted wave

$$E_y^{(t)} = G_t \exp[i(k_x^{(t)} x - k_z^{(t)} z)] \exp[-i\omega^{(t)} t], \tag{15}$$

$$B_y^{(t)} = 0. \tag{16}$$

A_r, $k_x^{(r)}$, $k_z^{(r)}$, $\omega^{(r)}$, G_t, $k_x^{(t)}$, $k_z^{(t)}$, and $\omega^{(t)}$ are related to combinations of A_r', k_x', k_z', ω', and G_t' by Eqs. (18) when the dielectric slab is moving in the x direction, and by Eqs. (22) when the dielectric slab is moving in the z direction.

Case (a): $\mathbf{v} = v_x \mathbf{e}_x$

Let us consider the case in which the dielectric slab is moving at a uniform velocity v_x in the x direction. Making use of the covariance of Maxwell's equations and the phase invariance of a uniform plane wave, we have the following transformations [referring to Eq. (3)]:

$$\begin{aligned} k_x' &= \gamma_x[k_x - \omega v_x/c^2] = \gamma_x k_0[\sin\theta_0 - \beta_x], \\ k_z' &= k_z = k_0 \cos\theta_0, \\ \omega' &= \gamma_x(\omega - v_x k_x) = \gamma_x \omega[1 - \beta_x \sin\theta_0], \\ \xi' &= \gamma_x k_0[(\epsilon_1/\epsilon_0)(1 - \beta_x \sin\theta_0)^2 - (\sin\theta_0 - \beta_x)^2]^{1/2}, \\ E_0' &= \gamma_x E_0(1 - v_x k_x/\omega) = \gamma_x E_0[1 - \beta_x \sin\theta_0], \end{aligned} \tag{17}$$

where $\gamma_x = 1/(1-\beta_x^2)^{1/2}$, $\beta_x = v_x/c$, and c is the velocity of light. Viewing from the observer's system S, we have [referring to Eqs. (5)–(10)],

$$
\begin{aligned}
&\omega^{(r)}=\omega^{(t)}=\gamma_x(\omega'+v_xk_x'),\\
&k_x^{(r)}=k_x^{(t)}=\gamma_x(k_x'+v_x\omega'/c^2),\\
&k_z^{(r)}=k_z^{(t)}=k_z',\\
&A_r=\gamma_xA_r'(1+v_xk_x'/\omega'),\\
&G_t=\gamma_xG_t'(1+v_xk_x'/\omega').
\end{aligned}
\tag{18}
$$

Substituting Eq. (17) into (18) gives

$$
\begin{aligned}
&\omega^{(r)}=\omega^{(t)}=\omega,\\
&k_x^{(r)}=k_x^{(t)}=k_x=k_0\sin\theta_0,\\
&k_z^{(r)}=k_z^{(t)}=k_z=k_0\cos\theta_0,\\
&A_r=E_0\left\{\frac{i(\epsilon_1/\epsilon_0-1)(1-\beta_x\sin\theta_0)^2\gamma_x^2\sin(\eta_xk_0d)\exp(-2ik_0d\cos\theta_0)}{2\eta_x\cos\theta_0\cos(\eta_xk_0d)-i(\eta_x^2+\cos^2\theta_0)\sin(\eta_xk_0d)}\right\},\\
&G_t=E_0\left\{\frac{2\eta_x\cos\theta_0\exp(-ik_0d\cos\theta_0)}{2\eta_x\cos\theta_0\cos(\eta_xk_0d)-i(\eta_x^2+\cos^2\theta_0)\sin(\eta_xk_0d)}\right\},
\end{aligned}
\tag{19}
$$

where

$$
\eta_x=\gamma_x[(1-\beta_x\sin\theta_0)^2(\epsilon_1/\epsilon_0)-(\sin\theta_0-\beta_x)^2]^{1/2}. \tag{20}
$$

According to Eqs. (19), we note that there exists no Doppler shift in frequency for the reflected and the transmitted waves. Furthermore, the familiar law concerning the equality of the angle of incidence and the angle of reflection is preserved. On the other hand, the coefficients for the reflected and the transmitted waves are affected by the transverse motion of the slab.

Case (b): $\mathbf{v}=v_z\mathbf{e}_z$

It is assumed that the dielectric slab is moving at a uniform velocity v_z in the positive z direction. Again making use of the covariance of Maxwell's equations and the phase invariance of a uniform plane wave, we obtain the following transformations:

$$
\begin{aligned}
&\omega'=\gamma_z(\omega+v_zk_z)=\gamma_z\omega(1+\beta_z\cos\theta_0), && k_x'=k_x=k_0\sin\theta_0,\\
&k_z'=\gamma_z(k_z+\omega v_z/c^2)=\gamma_zk_0(\cos\theta_0+\beta_z), && \xi'=k_0[\gamma_z^2(\epsilon_1/\epsilon_0)(1+\beta_z\cos\theta_0)^2-\sin^2\theta_0]^{1/2},\\
&E_0'=\gamma_zE_0(1+v_zk_z/\omega)=\gamma_zE_0(1+\beta_z\cos\theta_0),
\end{aligned}
\tag{21}
$$

where $\gamma_z=1/(1-\beta_z^2)^{1/2}$ and $\beta_z=v_z/c$. Viewing from the observer's system S, we have

$$
\begin{aligned}
&\omega^{(r)}=\gamma_z(\omega'+v_zk_z'),\quad k_x^{(r)}=k_x^{(t)}=k_x', && k_z^{(r)}=\gamma_z(k_z'+\omega'v_z/c^2),\\
&\omega^{(t)}=\gamma_z(\omega'\mp v_zk_z'),\quad k_z^{(t)}=\gamma_z(k_z'-\omega'v_z/c^2), && A_r=\gamma_zA_r'(1+v_zk_z'/\omega'),\\
&G_t=\gamma_zG_t'(1\mp v_zk_z'/\omega').
\end{aligned}
\tag{22}
$$

Substituting Eq. (21) into (22) gives

$$
\begin{aligned}
&\omega^{(t)}=\omega,\\
&\omega^{(r)}=\omega\gamma_z^2[(1+\beta_z^2)+2\beta_z\cos\theta_0],\\
&k_x^{(r)}=k_x^{(t)}=k_0\sin\theta_0,\\
&k_z^{(r)}=k_0\gamma_z^2[2\beta_z+\cos\theta_0(1+\beta_z^2)],\\
&k_z^{(t)}=k_0\cos\theta_0,\\
&A_r=E_0\frac{i\gamma_z^4(\epsilon_1/\epsilon_0-1)(1+\beta_z\cos\theta_0)^2[(1+\beta_z^2)+2\beta_z\cos\theta_0]\sin(k_0\eta_zd)\exp[-2i\gamma_zk_0d(\cos\theta_0+\beta_z)]}{2\gamma_z\eta_z(\cos\theta_0+\beta_z)\cos(\eta_zk_0d)-i[\eta_z^2+\gamma_z^2(\cos\theta_0+\beta_z)^2]\sin(\eta_zk_0d)},\\
&G_t=E_0\frac{2\gamma_z(\cos\theta_0+\beta_z)\eta_z\exp[-i\gamma_zk_0d(\cos\theta_0+\beta_z)]}{2\gamma_z\eta_z(\cos\theta_0+\beta_z)\cos(\eta_zk_0d)-i[\eta_z^2+\gamma_z^2(\cos\theta_0+\beta_z)^2]\sin(\eta_zk_0d)},
\end{aligned}
\tag{23}
$$

with

$$\eta_z=[\gamma_z^2(\epsilon_1/\epsilon_0)(1+\beta_z\cos\theta_0)^2-\sin^2\theta_0]^{1/2}. \quad (24)$$

Unlike case (a), there exists a Doppler shift in frequency for the reflected wave. But there is no frequency shift for the transmitted wave. The frequency shift for the reflected wave is independent of the permittivity of the slab and depends only on the slab velocity and the angle of incidence. The angle of reflection ($\theta_r=\tan^{-1}|k_z^{(r)}/k_x^{(r)}|$) is no longer equal to the angle of incidence However, the transmitted wave still propagates in the same direction as the incident wave The coefficients for the reflected and the transmitted waves are affected by the motion of the slab.

III. THE REFLECTION AND TRANSMISSION COEFFICIENTS

The reflection coefficient and the transmission coefficient are defined, respectively, by the relations

$$R=\mathbf{n}\cdot\mathbf{S}_r/\mathbf{n}\cdot\mathbf{S}_i, \quad (25)$$

and

$$T=\mathbf{n}\cdot\mathbf{S}_t/\mathbf{n}\cdot\mathbf{S}_i, \quad (26)$$

where $\mathbf{n}$ is the unit vector normal to the interface and

$$S_i=\tfrac{1}{2}(\mathbf{E}^{(i)}\times\mathbf{H}^{*(i)}), \quad (27)$$

$$S_r=\tfrac{1}{2}(\mathbf{E}^{(r)}\times\mathbf{H}^{*(r)}), \quad (28)$$

$$S_t=\tfrac{1}{2}(\mathbf{E}^{(t)}\times\mathbf{H}^{*(t)}). \quad (29)$$

The asterisk signifies the complex conjugate of the function. Simplifying Eqs. (25) and (26) gives

$$R=|A_r/E_0|^2\cos\theta^{(r)}/\cos\theta_0, \quad (30)$$

$$T=|G_t/E_0|^2\cos\theta^{(t)}/\cos\theta_0, \quad (31)$$

where

$$\cos\theta^{(r)}=k_z^{(r)}/[k_x^{(r)2}+k_z^{(r)2}]^{1/2}, \quad (32)$$

$$\cos\theta^{(t)}=k_z^{(t)}/[k_x^{(t)2}+k_z^{(t)2}]^{1/2}, \quad (33)$$

$k_x^{(r)}$, $k_z^{(r)}$, $k_x^{(t)}$, and $k_z^{(t)}$ are given by Eqs. (19) when the dielectric slab is moving uniformly in the positive x direction, and they are given by Eqs. (23) when the dielectric slab is moving uniformly in the positive z direction. Making the appropriate substitutions into Eqs. (30) and (31), one has, for the case $\mathbf{v}=v_x\mathbf{e}_x$,

$$R_x=\frac{\gamma_x^4(\epsilon_1/\epsilon_0-1)^2(1-\beta_x\sin\theta_0)^4\sin^2(\eta_xk_0d)}{4\eta_x^2\cos^2\theta_0\cos^2(\eta_xk_0d)+(\eta_x^2+\cos^2\theta_0)^2\sin^2(\eta_xk_0d)}, \quad (34)$$

and

$$T_x=1-R_x; \quad (35)$$

for the case $\mathbf{v}=v_z\mathbf{e}_z$,

$$R_z=\frac{\chi\gamma_z^4(\epsilon_1/\epsilon_0-1)^2(1+\beta_z\cos\theta_0)^4\sin^2(\eta_zk_0d)}{4\gamma_z^2\eta_z^2(\cos\theta_0+\beta_z)^2\cos^2(\eta_zk_0d)+[\eta_z^2+\gamma_z^2(\cos\theta_0+\beta_z)^2]^2\sin^2(\eta_zk_0d)}, \quad (36)$$

and

$$T_z=\frac{4\gamma_z^2\eta_z^2(\cos\theta_0+\beta_z)^2}{4\gamma_z^2\eta_z^2(\cos\theta_0+\beta_z)^2\cos^2(\eta_zk_0d)+[\eta_z^2+\gamma_z^2(\cos\theta_0+\beta_z)^2]^2\sin^2(\eta_zk_0d)}, \quad (37)$$

where

$$\chi=\frac{\gamma_z^4[2\beta_z+(1+\beta_z^2)\cos\theta_0](1+2\beta_z\cos\theta_0+\beta_z^2)^2}{\cos\theta_0\{\sin^2\theta_0+\gamma_z^4[2\beta_z+(1+\beta_z^2)\cos\theta_0]^2\}^{1/2}}. \quad (38)$$

It is interesting to note that for the case $\mathbf{v}=v_z\mathbf{e}_z$, ($v_z\neq0$), $R_z+T_z\neq1$. This is because of energy transfer from the moving slab to the reflected wave. A similar situation also occurs in the case of a perfectly reflecting mirror moving normal to its surface. In that case, the power density of the reflected wave in the direction normal to the surface is greater than that of the incident wave.

It can be seen from Eqs. (34)–(38) that the reflection and the transmission coefficients are rather complicated functions of the angle of incidence, the velocity, the thickness, and the dielectric constant of the slab. To have a qualitative idea of how the reflection and the transmission coefficients vary as a function of the velocity of the

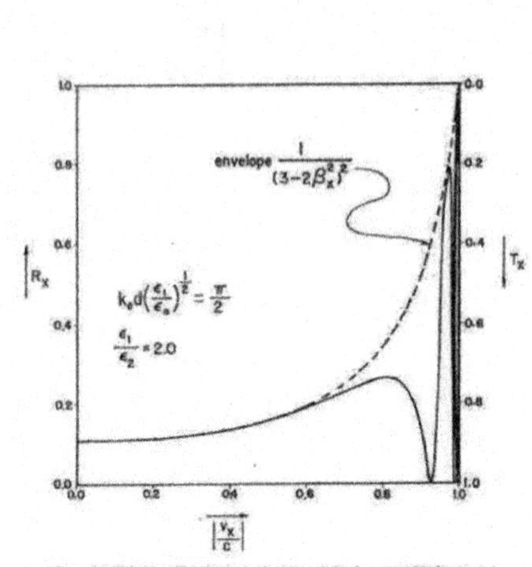

FIG. 2. The reflection and transmission coefficients as functions of $|v_x/c|$ for normal incidence.

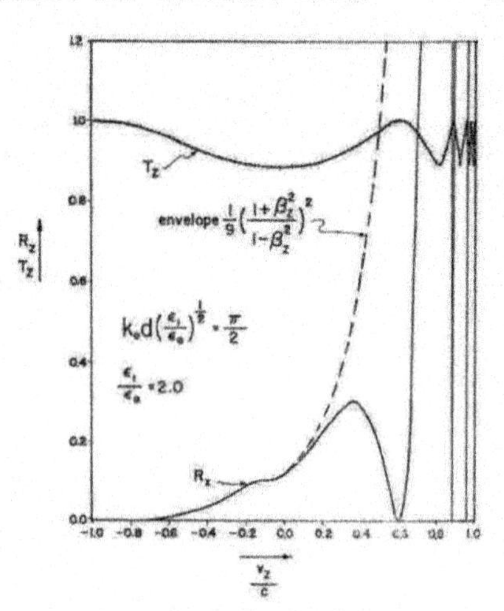

FIG. 3. The reflection and transmission coefficients as functions of v_z/c for normal incidence.

moving slab, we shall consider the limiting case of normal incidence. At normal incidence, i.e., $\theta_0=0$, Eqs. (34)–(38) reduce to

$$R_x=\frac{\gamma_x^4\left(\frac{\epsilon_1}{\epsilon_0}-1\right)^2}{4\gamma_x^2(\epsilon_1/\epsilon_0-\beta_x^2)\cot^2[\gamma_x k_0 d(\epsilon_1/\epsilon_0-\beta_x^2)^{1/2}]+[\gamma_x^2(\epsilon_1/\epsilon_0-\beta_x^2)+1]^2}, \tag{39}$$

$$T_x=1-R_x, \tag{40}$$

$$R_z=\left(\frac{1+\beta_z}{1-\beta_z}\right)^2\frac{1\left(\frac{\epsilon_1}{\epsilon_0}-1\right)^2}{4(\epsilon_1/\epsilon_0)\cot^2[\gamma_z k_0 d(\epsilon_1/\epsilon_0)^{1/2}(1+\beta_z)]+((\epsilon_1/\epsilon_0)+1))^2}, \tag{41}$$

$$T_z=\frac{4(\epsilon_1/\epsilon_0)/\sin^2[\gamma_z k_0 d(\epsilon_1/\epsilon_0)^{1/2}(1+\beta_z)]}{4(\epsilon_1/\epsilon_0)\cot^2[\gamma_z k_0 d(\epsilon_1/\epsilon_0)(1+\beta_z)]+((\epsilon_1/\epsilon_0)+1)^2}. \tag{42}$$

Equations (39)–(42) are plotted in Figs. 2 and 3. In these figures the reflection coefficient and the transmission coefficient are plotted as a function of the velocity of the moving slab. It is assumed that $\epsilon_1/\epsilon_0=2.0$ and $k_0d(\epsilon_1/\epsilon_0)^{1/2}=\pi/2$. Figure 2 shows that the reflection coefficient and the transmission coefficient oscillate more and more rapidly as β_x approaches unity. The oscillations are caused by the rapid change of the equivalent electrical thickness of the slab with velocity as viewed from the S system. Inspection of Eqs. (39) and (40) shows that R_x and T_x are even functions of β_x, as expected.

Figure 3 shows that the reflection coefficient also oscillates more and more rapidly as β_z varies from -1 and $+1$ and its envelope increases monotonically without bounds from zero. On the other hand, one notes from the same figure that the transmission coefficient oscillates between $4/(1+\epsilon_0/\epsilon_1)^2$ and 1 provided that $k_0d(\epsilon_1/\epsilon_0)^{1/2}=\pi/2$, and the frequency of oscillation increases as β_z changes from -1 to 1. The fact that the reflection coefficient can be greater than 1 is worth noting. It means that the reflected energy can be more than the energy of the incident wave as far as the observer who is stationary with respect to the S system is concerned. Apparently, there is energy transfer from the moving slab to the reflected wave. It is also interesting to note that as far as an observer who is stationary with respect to the S system is concerned, the frequency of the transmitted wave suffers no frequency shift due to the constant motion of the dielectric slab.

Reprinted from THE PHYSICAL REVIEW, Vol. 167, No. 3, 875–877, 15 March 1968
Printed in U. S. A.

Reflection and Transmission of Electromagnetic Waves by a Moving Dielectric Slab. II. Parallel Polarization*

C. YEH
Department of Engineering, University of California, Los Angeles, California 90024
(Received 1 November 1967)

The reflection and transmission of a plane wave, with its electric vector polarized in the plane of incidence, by a moving dielectric slab are investigated theoretically. Two cases of the movement are considered: (a) the dielectric slab moves parallel to the interface, (b) the dielectric slab moves perpendicular to the interface. It is shown that, in general, the reflection and transmission coefficients for an incident plane wave with its electric vector polarized in the plane of incidence are different from those for an incident plane wave with its electric vector polarized normal to the plane of incidence, except for case (b) for normally incident waves. Detailed results on the reflection and transmission coefficients for case (a) for normally incident waves are given and discussed.

IN a previous article,[1] the problem of the reflection and transmission of a plane electromagnetic wave by a moving dielectric slab was considered. Various interesting features concerning the variation of the reflection and transmission coefficients, the angles of reflection and transmission, and the frequencies of the reflected and transmitted wave as a function of the velocity of the moving medium, were observed. However, only the case in which the electric vector of the incident wave is polarized normal to the plane of incidence (perpendicular polarization) was considered. The purpose of this work is to present the solution for the other polarization; i.e., the case in which the electric vector of the incident wave is polarized in the plane of incidence (parallel polarization) will be considered. It is found that the reflection and transmission coefficients are significantly different for the two polarizations.

A harmonic plane wave in the free-space regions with its electric vector polarized in the plane of incidence is assumed to be incident upon a moving dielectric slab of thickness d. (See Fig. 1 in I.) In the observer's system S the incident plane wave is

$$H_y^{(i)}=H_0 e^{i(k_x x-k_z z)}e^{-i\omega t}, \tag{1}$$

$$D_y^{(i)}=0, \tag{2}$$

where H_0 and ω are, respectively, the amplitude and the frequency of the incident wave, $k_x=k_0\sin\theta_0$, $k_z=k_0\cos\theta_0$, and $k_0=\omega(\mu_0\epsilon_0)^{1/2}$. θ_0 is the angle between the propagation vector and the positive z axis in the x-z plane. The reflected wave and the transmitted wave, in the observer's system S, take the following forms:

For the reflected wave,

$$H_y^{(r)}=A_r e^{i[k_x^{(r)}x+k_z^{(r)}z]}e^{-i\omega^{(r)}t}, \tag{3}$$

$$D_y^{(r)}=0; \tag{4}$$

for the transmitted wave

$$H_y^{(t)}=G_t e^{i[k_x^{(t)}x-k_z^{(t)}z]}e^{-i\omega^{(t)}t}, \tag{5}$$

$$D_y^{(t)}=0. \tag{6}$$

The values of A_r and G_t are given later. It can be shown that $k_x^{(r)}$, $k_x^{(t)}$, $k_z^{(r)}$, $k_z^{(t)}$, $\omega^{(r)}$, and $\omega^{(t)}$ are the same as those given in I. In other words, the angle of reflection, the angle of transmission, and the frequencies of the reflected and transmitted waves are the same for both polarizations. Making use of the principle of phase invariance of plane waves, the covariance of Maxwell's equations, and the Lorentz transformation,

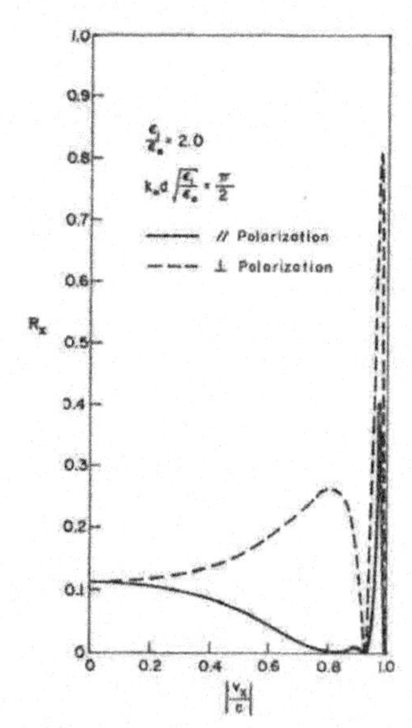

FIG. 1. The reflection coefficients R_x as functions of $|v_x/c|$ for normal incidence. (Note that $T_x=1-R_x$.)

* Supported by the National Science Foundation.
[1] C. Yeh and K. F. Casey, Phys. Rev. 144, 665 (1966); hereafter referred to as I.

and satisfying the boundary conditions, one obtains the following relations[1,2]:

(a) If the slab is moving uniformly with a velocity v_x in the positive x direction,

$$A_r = H_0 \frac{ie^{-2ik_0 d\cos\theta_0}\{[\eta_x(\epsilon_0/\epsilon_1)]^2 - \cos^2\theta_0\}\sin(k_0 d\eta_x)}{2(\epsilon_0/\epsilon_1)\eta_x \cos\theta_0 \cos(k_0 d\eta_x) - i\{[(\epsilon_0/\epsilon_1)\eta_x]^2 + \cos^2\theta_0\}\sin(k_0 d\eta_x)}, \tag{7a}$$

$$G_t = H_0 \frac{2(\epsilon_0/\epsilon_1)\eta_x \cos\theta_0 e^{-ik_0 d\cos\theta_0}}{2(\epsilon_0/\epsilon_1)\eta_x \cos\theta_0 \cos(k_0 d\eta_x) - i\{[(\epsilon_0/\epsilon_1)\eta_x]^2 + \cos^2\theta_0\}\sin(k_0 d\eta_x)}, \tag{7b}$$

with

$$\begin{aligned} \eta_x &= \gamma_x[(1-\beta_x \sin\theta_0)^2(\epsilon_1/\epsilon_0) - (\sin\theta_0 - \beta_x)^2]^{1/2}, \\ \gamma_x &= (1-\beta_x^2)^{-1/2}, \\ \beta_x &= v_x/c, \\ c &= \text{the velocity of light in vacuum.} \end{aligned} \tag{7c}$$

(b) If the slab is moving uniformly with a velocity v_z in the positive z direction,

$$A_r = H_0 \frac{i\gamma_z^2(1+2\beta_z \cos\theta_0 + \beta_z^2)e^{-2ik_0 d\gamma_z(\cos\theta_0+\beta_z)}\{[\eta_z(\epsilon_0/\epsilon_1)]^2 - \gamma_z^2(\cos\theta_0+\beta_z)^2\}\sin(k_0 d\eta_z)}{2(\epsilon_0/\epsilon_1)\eta_z\gamma_z(\cos\theta_0+\beta_z)\cos(\eta_z k_0 d) - i\{[\eta_z(\epsilon_0/\epsilon_1)]^2 + \gamma_z^2(\cos\theta_0+\beta_z)^2\}\sin(k_0 d\eta_z)}, \tag{8a}$$

$$G_t = H_0 \frac{2(\epsilon_0/\epsilon_1)\eta_z\gamma_z(\cos\theta_0+\beta_z)e^{-ik_0 d\gamma_z(\cos\theta_0+\beta_z)}}{2(\epsilon_0/\epsilon_1)\eta_z\gamma_z(\cos\theta_0+\beta_z)\cos(\eta_z k_0 d) - i\{[\eta_z(\epsilon_0/\epsilon_1)]^2 + \gamma_z^2(\cos\theta_0+\beta_z)^2\}\sin(k_0 d\eta_z)}, \tag{8b}$$

with

$$\eta_z = [\gamma_z^2(\epsilon_1/\epsilon_0)(1+\beta_z \cos\theta_0)^2 - \sin^2\theta_0]^{1/2}, \quad \gamma_z = (1-\beta_z^2)^{-1/2}, \quad \beta_z = v_z/c. \tag{8c}$$

It is noted that the coefficients of the reflected and transmitted waves are significantly different for the two different polarizations of an incident plane wave.

The reflection and transmission coefficients for the parallel polarization cases are, respectively,

$$R_{x,z} = (A_r A_r^*/H_0^2)p_z \tag{9}$$

and

$$T_{x,z} = G_t G_t^*/H_0^2, \tag{10}$$

where A_r and G_t are given by Eq. (7) and $p_z = 1$ when the dielectric slab is moving in the x direction; and A_r and G_t are given by Eq. (8) and

$$p_z = \frac{2\beta_z + \cos\theta_0(1+\beta_z^2)}{[(1+\beta_z^2)+2\beta_z \cos\theta_0]\cos\theta_0},$$

when the dielectric slab is moving uniformly in the positive z direction. Simplifying Eqs. (9) and (10), one has for $\mathbf{v} = v_x\mathbf{e}_x$

$$R_x = \frac{\{[\eta_x(\epsilon_0/\epsilon_1)] - \cos^2\theta_0\}^2 \sin^2(k_0 d\eta_x)}{4[(\epsilon_0/\epsilon_1)\eta_x]^2 \cos^2\theta_0 \cos^2(k_0 d\eta_x) + \{[(\epsilon_0/\epsilon_1)\eta_x]^2 + \cos^2\theta_0\}^2 \sin^2(\eta_x k_0 d)}, \tag{11}$$

$$T_x = 1 - R_x, \tag{12}$$

and for $\mathbf{v} = v_z\mathbf{e}_z$

$$R_z = \chi \frac{\gamma_z^4[1+2\beta_z \cos\theta_0 + \beta_z^2]^2\{[\eta_z(\epsilon_0/\epsilon_1)]^2 - \gamma_z^2(\cos\theta_0+\beta_z)^2\}^2 \sin^2(k_0 d\eta_z)}{4[(\epsilon_0/\epsilon_1)\eta_z]^2\gamma_z^2(\cos\theta_0+\beta_z)^2 \cos^2(\eta_z k_0 d) + \{[\eta_z(\epsilon_0/\epsilon_1)]^2 + \gamma_z^2(\cos\theta_0+\beta_z)^2\}^2 \sin^2(\beta_z k_0 d)}, \tag{13}$$

$$T_z = \frac{4(\epsilon_0/\epsilon_1)^2\eta_z^2\gamma_z^2(\cos\theta_0+\beta_z)^2}{4[(\epsilon_0/\epsilon_1)\eta_z]^2\gamma_z^2(\cos\theta_0+\beta_z)^2 \cos^2(\eta_z k_0 d) + \{[\eta_z(\epsilon_0/\epsilon_1)]^2 + \gamma_z^2(\cos\theta_0+\theta_z)^2\}^2 \sin^2(\eta_z k_0 d)}, \tag{14}$$

$$\chi = \frac{\gamma_z^2[2\beta_z + \cos\theta_0(1+\beta_z^2)]}{\cos\theta_0\{\gamma_z^4[2\beta_z + \cos\theta_0(1+\beta_z^2)]^2 + \sin^2\theta_0\}^{1/2}}. \tag{15}$$

[2] C. Möller, *The Theory of Relativity* (Oxford University Press, London, 1957).

To have a qualitative idea of how the reflection and transmission coefficients vary as a function of the velocity of the moving medium, we shall consider the limiting case of normal incidence. At normal incidence, i.e., $\theta_0=0$, Eqs. (11–14) reduce to

$$R_x=\left(\frac{\epsilon_0}{\epsilon_1}\right)^2\gamma_x^2\frac{[(\epsilon_1/\epsilon_0)-1]^2\{\beta_x^2[(\epsilon_1/\epsilon_0)+1]^2-(\epsilon_1/\epsilon_0)\}^2}{4[(\epsilon_1/\epsilon_0)-\beta_x^2]\cos^2(k_0d\eta_x^0)+(\epsilon_0/\epsilon_1)^2\gamma_x^2\{\epsilon_1/\epsilon_0[1+(\epsilon_1/\epsilon_0)]-\beta_x^2[1+(\epsilon_1/\epsilon_0)^2]\}^2}, \tag{16}$$

$$T_x=1-R_x, \tag{17}$$

$$R_z=\left(\frac{1+\beta_z}{1-\beta_z}\right)^2\frac{[(\epsilon_0/\epsilon_1)-1]^2\sin^2(k_0d\eta_z^0)}{4(\epsilon_0/\epsilon_1)\cos^2(k_0d\eta_z^0)+[(\epsilon_0/\epsilon_1)+1]^2\sin^2(k_0d\eta_z^0)}, \tag{18}$$

$$T_z=\frac{4(\epsilon_0/\epsilon_1)}{4(\epsilon_0/\epsilon_1)\cos^2(k_0d\eta_z^0)+[(\epsilon_0/\epsilon_1)+1]^2\sin^2(k_0d\eta_z^0)}, \tag{19}$$

with

$$\eta_x^0=\left[\frac{(\epsilon_1/\epsilon_0)-\beta_x^2}{1-\beta_x^2}\right]^{1/2}; \tag{20}$$

$$\eta_z^0=\left[\frac{\epsilon_1}{\epsilon_0}\left(\frac{1+\beta_z}{1-\beta_z}\right)\right]. \tag{21}$$

It is interesting to note that when the dielectric slab is moving in the z direction the reflection and transmission coefficients for a normally incident wave with parallel polarization are identical to those for a normally incident wave with perpendicular polarization [i.e., Eq. (18) is the same as Eq. (41) in I and Eq. (19) is the same as Eq. (42) in I].[3] On the other hand, when the dielectric slab is moving in the x direction the reflection and transmission coefficients for a normally incident wave are quite different for the two different polarizations.

Equation (15) is plotted in Fig. 1. The reflection coefficient is plotted as a function of the velocity of the moving slab. It is assumed that $\epsilon_1/\epsilon_0=2.0$ and $k_0d(\epsilon_1/\epsilon_0)^{1/2}=\frac{1}{2}\pi$. It can be seen from Fig. 1 that as β_x increases, the reflection coefficient for the parallel polarization case decreases monotonically until

$$\beta_x=\left[\frac{\epsilon_1/\epsilon_0}{1+(\epsilon_1/\epsilon_0)}\right]^{1/2};$$

at this velocity the reflection coefficient is zero and the transmission coefficient is unity. As β_x increases further, the oscillatory behavior of the reflection coefficient can be observed. This is because of the change of the electrical thickness of the slab as β_x varies. The reflection coefficient becomes zero at

$$\beta_x=\left[\frac{(n\pi/k_0d)^2-\epsilon_1/\epsilon_0}{(n\pi/k_0d)^2-1}\right]^{1/2}$$

for integer values of n and for $\beta_x<1$. At $\beta_x=1$, the reflection coefficient is unity, i.e., all the incident energy is reflected. For the sake of comparison, the reflection coefficient for the perpendicular polarization case is also plotted as a function of β_x in Fig. 1.

In conclusion, one observes that the characteristics of the reflection and transmission coefficients for an incident plane wave with its electric vector polarized in the plane of incidence, as a function of the velocity of the slab, are significantly different from those for an incident plane wave with its electric vector polarized normal to the plane of incidence. Even for normally incident plane waves, the reflection coefficients for the two different polarizations are different except when the slab is moving in a direction which is normal to the interface.

[3] Note that the numerator of Eqs. (39) and (40) in Ref. 1 should be multiplied by $(\epsilon_1/\epsilon_0-1)^2$, and that the right-hand side of Eq. (38) in I should be multiplied by γ_x^2.

Reprinted from JOURNAL OF APPLIED PHYSICS, Vol. 37, No. 8, 3079–3082, July 1966

Reflection and Transmission of Electromagnetic Waves by a Moving Plasma Medium*

C. YEH
Electrical Engineering Department, University of Southern California, Los Angeles, California
(Received 24 January 1966)

The reflection and transmission of electromagnetic waves by a moving dispersive dielectric half-space or slab are investigated theoretically. The dispersive dielectric medium is assumed to be a cold plasma medium. Two cases of the movement are considered: (a) The plasma medium moves parallel to the interface. (b) The plasma medium moves perpendicular to the interface. It is interesting to note that, for case (a), the reflected and transmitted fields for an incident plane E wave are independent to the movement of the plasma medium. For case (b), the reflected wave and the transmitted wave are functions of the velocity of the moving plasma medium. Detailed results on the reflection and transmission coefficients for case (b) are given and discussed. In the present analysis, the electric vector of the incident wave is assumed to be polarized in a direction that is normal to the plane of incidence.

IN a previous article,[1] the problem of the reflection and transmission of electromagnetic waves by a moving dielectric medium was considered. Various interesting features concerning the variation of the reflection and transmission coefficients, angles of reflection and transmission, and the frequencies of the reflected and transmitted wave as a function of the velocity of the moving medium were observed. However, only nondispersive dielectric medium was considered. If the medium is assumed to be dispersive, further interesting features concerning the reflected wave and the transmitted wave are noted. The purpose of this article is to present results of the problem of the reflection and transmission of waves by a moving dispersive dielectric half-space or slab. Specifically, the dispersive medium is assumed to be a homogeneous cold plasma. The permittivity of a cold plasma is

$$\epsilon_p = \epsilon_0[1-(\omega_p^2/\omega'^2)], \quad (1)$$

where ϵ_0 is the free-space permittivity, ω_p is the plasma frequency, and ω' is the frequency in the moving system S' which is stationary with respect to the uniformly moving medium.

A harmonic plane wave in the free-space region with its electric vector polarized in the y direction, is assumed to be incident upon a moving semi-infinite plasma half-space or upon a moving plasma slab of thickness d (see Fig. 1). In the observer's system S

* Supported partly by the Joint Services Electronics Program and partly by the Naval Ordnance Test Station.
[1] C. Yeh, J. Appl. Phys. 36, 3513 (1965).

which is stationary with respect to the free space the incident plane wave is

$$E_y^{(i)} = E_0 e^{i(k_x x + k_z z)} e^{-i\omega t}, \tag{2}$$

$$B_y^{(i)} = 0, \tag{3}$$

where E_0 and ω are, respectively, the amplitude and the frequency of the incident wave $k_x = k_0 \sin\theta_0$, $k_z = -k_0 \cos\theta_0$, and $k_0 = \omega(\mu_0\epsilon_0)^{1/2}$. θ_0 is the angle between the propagation vector and the positive z axis in the x–z plane. The reflected wave and the transmitted wave, in the observer's system S, takes the following forms: For the reflected wave

$$E_y^{(r)} = A_r \exp[i(k_x^{(r)}x - k_z^{(r)}z)] \exp[-i\omega^{(r)}t], \tag{4}$$

$$B_y^{(r)} = 0; \tag{5}$$

for the transmitted wave

$$E_y^{(t)} = G_t \exp[i(k_x^{(t)}x - k_z^{(t)}z)] \exp[-i\omega^{(t)}t], \tag{6}$$

$$B_y^{(t)} = 0. \tag{7}$$

A_r, G_t, $k_x^{(r)}$, $k_x^{(t)}$, $k_z^{(r)}$, $k_z^{(t)}$, $\omega^{(r)}$, and $\omega^{(t)}$ are given later. Making use of the principle of phase invariance of plane waves, the covariance of Maxwell's equations and the Lorentz transformations, and satisfying the boundary conditions, one obtains the following relations[1-3]:

(a) For a moving half-space, if it is moving uniformly with a velocity v_x in the positive x direction,

$$\omega^{(r)} = \omega^{(t)} = \omega, \tag{8a}$$

$$k_x^{(r)} = k_x^{(t)} = k_x = k_0 \sin\theta_0, \tag{8b}$$

$$k_z^{(r)} = k_z = -k_0 \cos\theta_0, \tag{8c}$$

$$k_z^{(t)} = -k_0[\cos^2\theta_0 - (\omega_p^2/\omega^2)]^{1/2}, \tag{8d}$$

$$A_r = E_0 \frac{\cos\theta_0 + (k_z^{(t)}/k_0)}{\cos\theta_0 - (k_z^{(t)}/k_0)}, \tag{8e}$$

$$G_t = E_0 \frac{2\cos\theta_0}{\cos\theta_0 - (k_z^{(t)}/k_0)}; \tag{8f}$$

if the half-space is moving uniformly with a velocity v_z in the positive z direction,

$$\omega^{(r)} = \omega\gamma_z^2[(1+\beta_z^2) + 2\beta_z \cos\theta_0], \tag{9a}$$

$$k_x^{(r)} = k_x = k_0 \sin\theta_0, \tag{9b}$$

$$k_z^{(r)} = -k_0\gamma_z^2[2\beta_z + \cos\theta_0(1+\beta_z^2)], \tag{9c}$$

$$\omega^{(t)} = \omega\gamma_z^2[(1+\beta_z\cos\theta_0) - \beta_z Q], \tag{9d}$$

$$k_z^{(t)} = k_0\gamma_z^2[\beta_z(1+\beta_z\cos\theta_0) - Q], \tag{9e}$$

$$A_r = E_0 \frac{\omega^{(r)}}{\omega}\left[\frac{(\cos\theta_0+\beta_z) - Q}{(\cos\theta_0+\beta_z) + Q}\right], \tag{9f}$$

$$G_t = E_0\left[\frac{2(\omega^{(t)}/\omega)(\cos\theta_0+\beta_z)}{(\cos\theta_0+\beta_z) + Q}\right], \tag{9g}$$

with $\beta_z = v_z/c$, $\gamma_z = 1/(1-\beta_z^2)^{1/2}$, and

$$Q = (1/\gamma_z)[\gamma_z^2(1+\beta_z\cos\theta_0)^2 - (\omega_p^2/\omega^2) - \sin^2\theta_0]^{1/2}.$$

(b) For a moving slab of thickness d, if it is moving uniformly with a velocity v_x in the positive x direction,

$$\omega^{(r)} = \omega^{(t)} = \omega, \tag{10a}$$

$$k_x^{(r)} = k_x^{(t)} = k_x = k_0 \sin\theta_0, \tag{10b}$$

$$k_z^{(r)} = k_z^{(t)} = k_z = -k_0 \cos\theta_0, \tag{10c}$$

$$A_r = E_0\left\{\frac{-i(\omega_p/\omega)^2 \sin(\eta_z k_0 d)\exp(-2ik_0 d\cos\theta_0)}{2\eta_z \cos\theta_0 \cos(\eta_z k_0 d) - i(\eta_z^2 + \cos^2\theta_0)\sin(\eta_z k_0 d)}\right\}, \tag{10d}$$

$$G_t = E_0\left\{\frac{2\eta_z \cos\theta_0 \exp(-ik_0 d\cos\theta_0)}{2\eta_z \cos\theta_0 \cos(\eta_z k_0 d) - i(\eta_z^2 + \cos^2\theta_0)\sin(\eta_z k_0 d)}\right\}, \tag{10e}$$

with

$$\eta_z = [\cos^2\theta_0 - \omega_p^2/\omega^2]^{1/2}; \tag{10f}$$

if the slab is moving uniformly with a velocity v_z in the positive z direction,

$$\omega^{(r)} = \omega\gamma_z^2[(1+\beta_z^2) + 2\beta_z\cos\theta_0], \tag{11a}$$

$$k_x^{(r)} = k_x^{(t)} = k_0 \sin\theta_0, \tag{11b}$$

$$k_z^{(r)} = -k_0\gamma_z^2[2\beta_z + \cos\theta_0(1+\beta_z^2)], \tag{11c}$$

$$\omega^{(t)} = \omega, \tag{11d}$$

[2] W. Pauli, *Theory of Relativity* (Pergamon Press, New York, 1958).
[3] A. Sommerfeld, *Optik* (Akademische Verlagsgesellschaft, Leipzig, 1959), 2nd. ed.

$$k_z^{(t)} = -k_0 \cos\theta_0, \tag{11e}$$

$$A_r = E_0 \frac{-i\gamma_z^2(\omega_p^2/\omega^2)[(1+\beta_z^2)+2\beta_z\cos\theta_0]\sin(\eta_z k_0 d)\exp[-2i\gamma_z k_0 d(\cos\theta_0+\beta_z)]}{2\gamma_z\eta_z(\cos\theta_0+\beta_z)\cos(\eta_z k_0 d)-i[\eta_z^2+\gamma_z^2(\cos\theta_0+\beta_z)^2]\sin(\eta_z k_0 d)}, \tag{11f}$$

$$G_t = E_0 \frac{2\gamma_z\eta_z(\cos\theta_0+\beta_z)\exp[-i\gamma_z k_0 d(\cos\theta_0+\beta_z)]}{2\gamma_z\eta_z(\cos\theta_0+\beta_z)\cos(\eta_z k_0 d)-i[\eta_z^2+\gamma_z^2(\cos\theta_0+\beta_z)^2]\sin(\eta_z k_0 d)}, \tag{11g}$$

with

$$\eta_z = [\gamma_z^2(1+\beta_z\cos\theta_0)^2 - (\omega_p^2/\omega^2) - \sin^2\theta_0]^{\frac{1}{2}}. \tag{11h}$$

It is rather surprising to note from Eqs. (8) and (10) that all reflected and transmitted fields for both the plasma half-space case and the plasma slab case are independent to the velocity of the plasma medium which is moving in a direction parallel to the interface.

Of special interest are the reflection coefficient for the half-space case and the reflection and transmission coefficient for the slab case. The reflection coefficient and the transmission coefficient are defined, respectively, by the relations

$$R = \mathbf{n}\cdot\mathbf{S}_r/\mathbf{n}\cdot\mathbf{S}_i, \tag{12}$$

and

$$T = \mathbf{n}\cdot\mathbf{S}_t/\mathbf{n}\cdot\mathbf{S}_i, \tag{13}$$

where $\mathbf{n}$ is the unit vector normal to the interface and

$$\mathbf{S}_i = \tfrac{1}{2}(\mathbf{E}^{(i)}\times\mathbf{H}^{*(i)}), \tag{14}$$

$$\mathbf{S}_r = \tfrac{1}{2}(\mathbf{E}^{(r)}\times\mathbf{H}^{*(r)}), \tag{15}$$

$$\mathbf{S}_t = \tfrac{1}{2}(\mathbf{E}^{(t)}\times\mathbf{H}^{*(t)}). \tag{16}$$

The * signifies the complex conjugate of the function. The more interesting case of a plasma medium moving in a direction normal to its interface is now considered in detail. The reflection coefficient for a plasma half-space moving in the z direction is[4]

$$R_z^{(HS)} = (A_r A_r^*/E_0^2)(k_z^{(r)}/k_0\cos\theta_0)(\omega/\omega^{(r)}), \tag{17}$$

where A_r, $k_z^{(r)}$, and $\omega^{(r)}$ are given in Eq. (9). The reflection and transmission coefficient for a plasma slab moving in the z direction are, respectively,

$$R_z^{(S)} = (A_r A_r^*/E_0^2)(k_z^{(r)}/k_0\cos\theta_0)(\omega/\omega^{(r)}) \tag{18}$$

and

$$T_z^{(S)} = G_t G_t^*/E_0^2, \tag{19}$$

where A_r, $k_z^{(r)}$, $\omega^{(r)}$, and G_t are given in Eq. (11). It can be seen from the above equations that the reflection and transmission coefficients are rather complicated functions of the angle of incidence, the velocity, and the plasma frequencies of the medium. To have a qualitative idea of how the reflection and transmission coefficients vary as a function of the velocity of the moving medium, we shall consider the limiting case of normal incidence.

Figure 2 shows the reflection coefficient $R_z^{(HS)}$ for the plasma half-space case as a function of β_z for various values of ω_p/ω. Unlike the nondispersive dielectric half-space case, the reflection coefficient no longer increases monotonically as β_z increases for all values of ω_p/ω. For the case $\omega_p/\omega>1$, $R_z^{(HS)}$ reaches a maximum as β_z increases from -1.0 and then decays to the value $(\omega_p/\omega)^4$. Furthermore, if $\omega_p/\omega>1$ and $\beta_z>1$, $R_z^{(HS)}$ is always greater than unity. In other words, when the plasma half-space is moving towards the incident wave, the reflected energy is greater than the

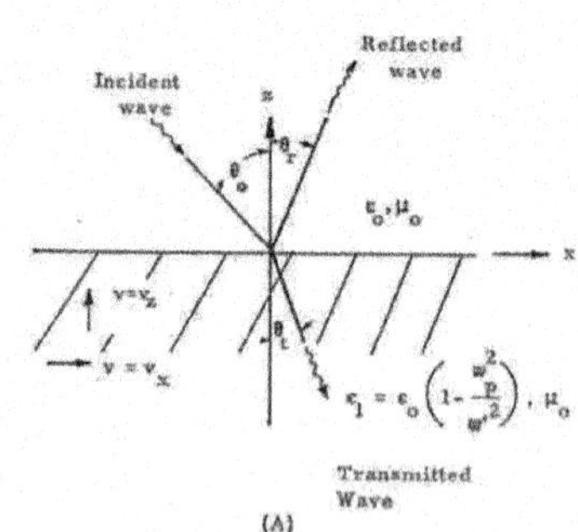

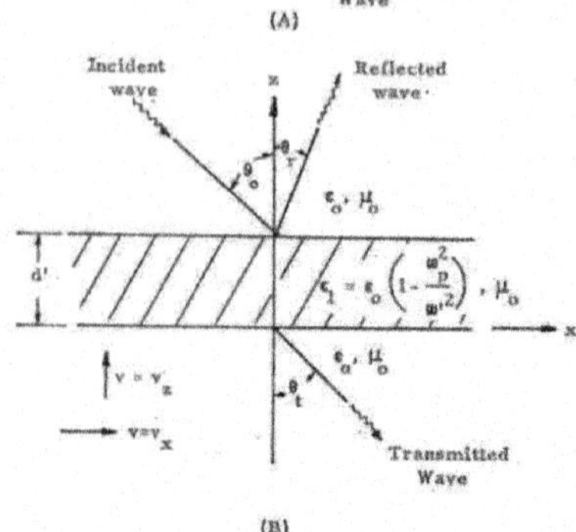

FIG. 1. Geometry of the problem. (A) Moving plasma half-space, (B) moving plasma slab.

[4] The definitions given by Eqs. (12) and (13) for the reflection and transmission coefficients are only meaningful for harmonic time-dependent fields (i.e., all fields will have $e^{-i\Omega t}$ dependence where Ω is real). Although the incident wave as viewed from the S system has harmonic time dependence, the frequency of the transmitted wave in the moving plasma medium as viewed from the s system [i.e., $\omega^{(t)}$ in Eq. (9d)], in general, may be complex. Consequently, the transmission coefficient given by Eq. (13) is not applicable to the case in which the plasma half-space is moving towards or away from the incident wave.

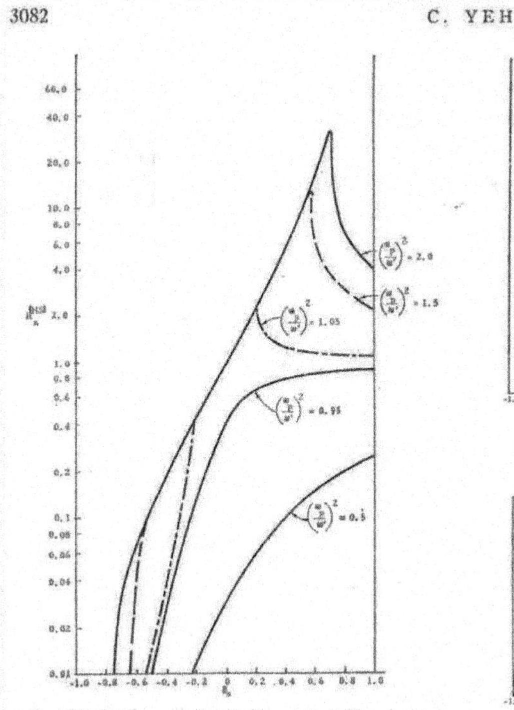

Fig. 2. Reflection coefficient for the moving half-space case.

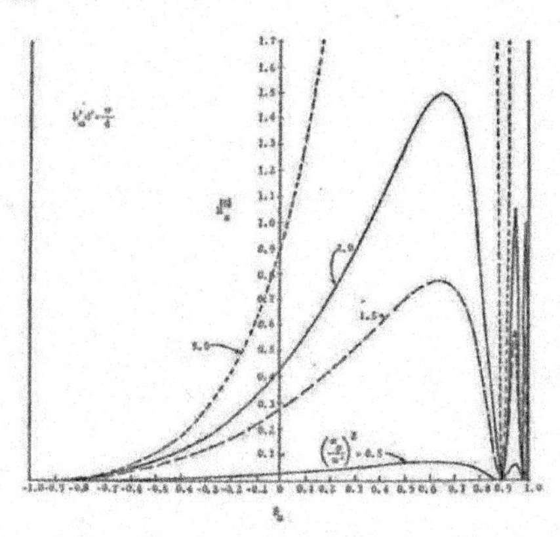

Fig. 3. Reflection coefficient for the moving slab case.

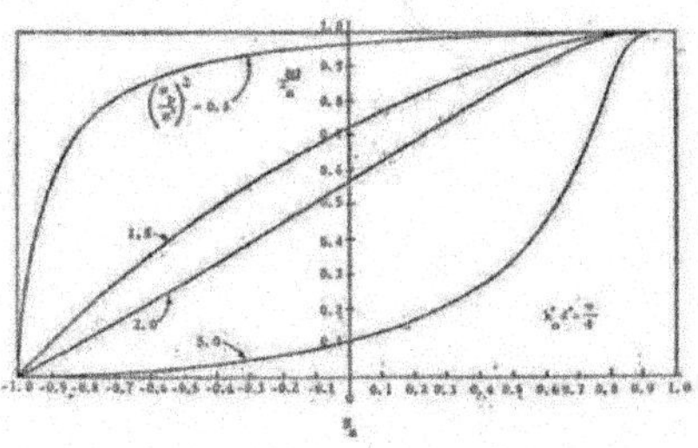

Fig. 4. Transmission coefficient for the moving slab case.

energy of the incident wave. Another interesting feature of the $\omega_p/\omega>1$ case is that a portion of the $R_z^{(HS)}$ vs β_z curve is identical for different values of ω_p/ω. For example, the portion of the curve between $\beta_z\approx-0.5$ and $\beta_z\approx0.55$ is the same for $(\omega_p/\omega)^2=1.5$ case and for $(\omega_p/\omega)^2>1.5$ case. If $\omega_p/\omega<1$, $R_z^{(HS)}$ increases monotonically from zero to $(\omega_p/\omega)^4$ as β_z varies from -1 to $+1$.

The reflection coefficient $R_z^{(S)}$ and the transmission coefficient $T_z^{(S)}$ for a moving plasma slab are plotted, respectively, in Figs. 3 and 4 as a function of the velocity of the slab. It is assumed that $k_0d=\pi/4$, where d is the thickness of the slab in the reference frame which is at rest with the plasma slab and k_0 is the free-space wave number. The reflection coefficient as well as the transmission coefficient decrease monotonically to zero for all values of ω_p/ω, as β_z changes from 0 to -1. As the velocity of the slab moving towards the incident wave increases, i.e., as β_z increases from 0, the reflection coefficient increases and reaches a maximum and then starts to oscillate between zero and certain constant. The envelope of the oscillating function is a decaying function; it approaches $(\omega_p/\omega)^4$ as β_z approaches $+1$. The maximum value of $R_z^{(S)}$ increases as the plasma density increases. The transmission coefficient increases monotonically as β_z increases from 0 to about 0.9. As β_z increases from 0.9 to 1.0, very small ripples are observed as $T_z^{(S)}$ approaches 1.0. Also noted is the fact that $T_z^{(S)}$ approaches unity faster as the plasma becomes less dense. It is further noted that $R_z^{(S)}+T_z^{(S)}=1$ at $\beta_z=0$, as expected. However if $\beta_z\neq0$, then $R_z^{(S)}+T_z^{(S)}\neq1$ in general.

In conclusion, one observes that the characteristics of the reflection and transmission coefficients for a dispersive medium as a function of the velocity of the medium are significantly different from those for a nondispersive medium.

Reprinted from JOURNAL OF APPLIED PHYSICS, Vol. 38, No. 7, 2871–2873, June 1967

Reflection and Transmission of Electromagnetic Waves by a Moving Plasma Medium. II. Parallel Polarizations*

C. YEH†
School of Engineering, University of California, Los Angeles, California
(Received 16 February 1967)

The reflection and transmission of electromagnetic waves by a moving dispersive dielectric half-space or slab are investigated theoretically. The dispersive medium is assumed to be a cold plasma medium. Two cases of the movement are considered: (a) the plasma medium moves parallel to the interface; (b) the plasma medium moves perpendicular to the interface. The electric vector of the incident wave is assumed to be polarized in the plane of incident (parallel polarization). Unlike the normal polarization case which was treated earlier, the reflected and transmitted fields for the present case are functions of the velocity of the moving plasma medium for both cases (a) and (b). An illustrative numerical example is given for case (a) at normal incidence.

In a recent article,[1] the problem of the reflection and transmission by a moving plasma medium of electromagnetic wave with its electric vector polarized in a direction that is normal to the plane of incidence was considered. Various interesting features concerning the variation of the reflection and transmission coefficients, angles of reflection and transmission, and the frequencies of the reflected and transmitted wave as a function of the velocity of the moving plasma were observed. In particular, it was found that if the plasma medium is moving parallel to the interface, the reflected and transmitted fields for this polarization are independent of the movement of the plasma.

The purpose of this work is to treat the problem of the reflection and transmission of waves with the other polarization, i.e., with the electric vector polarized in the plane of incidence, by a moving plasma half-space or slab. It will be shown that significant differences are found in the characteristics of the reflection and transmission coefficients for the two polarizations.

A harmonic plane wave in the free-space region with its magnetic vector polarized in the y direction is assumed to be incident upon a moving semi-infinite plasma half-space or upon a moving plasma slab of thickness d (see Fig. 1 in I). The moving medium is assumed to be a homogeneous cold plasma medium. In the observer's system S which is stationary with respect to the free space, the incident plane wave is

$$H_y^{(i)}=H_0\exp[i(k_xx+k_zz)]\exp(-i\omega t), \quad (1)$$

$$D_y^{(i)}=0, \quad (2)$$

where H_0 and ω are, respectively, the amplitude and the frequency of the incident wave, $k_x=k_0\sin\theta_0$, $k_z=-k_0\cos\theta_0$, and $k_0=\omega(\mu_0\epsilon_0)^{1/2}$. θ_0 is the angle of incidence. The reflected wave and the transmitted wave, in the observer's system S, take the following forms:

for the reflected wave,

$$H_y^{(r)}=C_r\exp[i(k_x^{(r)}x-k_z^{(r)}z)]\exp[-i\omega^{(r)}t], \quad (3)$$

$$D_y^{(r)}=0; \quad (4)$$

for the transmitted wave,

$$H_y^{(t)}=K_t\exp[i(k_x^{(t)}x-k_z^{(t)}z)]\exp[-i\omega^{(t)}t], \quad (5)$$

$$D_y^{(t)}=0. \quad (6)$$

C_r and K_t will be given later. It can be shown that

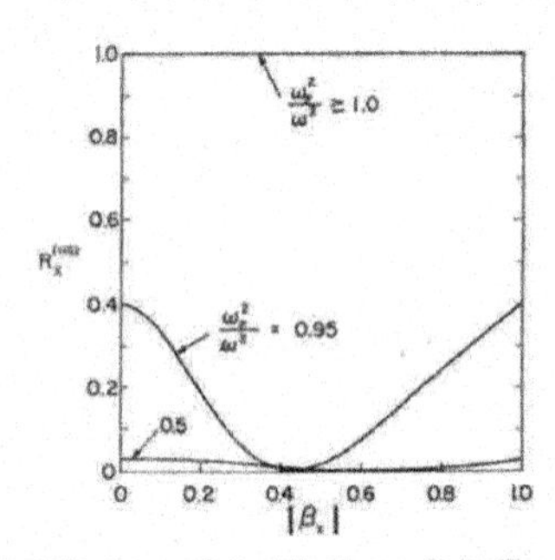

FIG. 1. Reflection coefficient for the moving-half-space case.

$k_x^{(r)}$, $k_x^{(t)}$, $k_z^{(r)}$, $k_z^{(t)}$, $\omega^{(r)}$, and $\omega^{(t)}$ are the same as those given in I. In other words, the angle of reflection, the angle of transmission, and the frequencies of the reflected and transmitted waves are the same for both polarizations. Making use of the principle of phase invariance of plane waves, the covariance of Maxwell's equations and the Lorentz transformations, and satisfying the boundary conditions, one obtains the following relations[1,2]:

(a) For a moving plasma half-space, if it is moving

* Supported by the National Science Foundation.
† Formerly with the Electrical Engineering Department, University of Southern California, Los Angeles, Calif.
[1] C. Yeh, J. Appl. Phys. 37, 3079 (1966); hereafter referred to as I.
[2] A. Sommerfeld, *Optik* (Akademische Verlagsgesellschaft, Leipzig, 1954), 2nd ed.

uniformly with a velocity v_x in the positive x direction,

$$C_r = H_0(\cos\theta_0 - m_x)/(\cos\theta_0 + m_x), \tag{7a}$$

$$K_t = H_0 2\cos\theta_0/(\cos\theta_0 + m_x), \tag{7b}$$

$$m_x = \left(\cos^2\theta_0 - \frac{\omega_p^2}{\omega^2}\right)^{1/2} \bigg/ \left[1 - \frac{\omega_p^2}{\omega^2}\frac{1}{\gamma_x^2(1-\beta_x\sin\theta_0)^2}\right], \tag{7c}$$

with $\beta_x = v_x/c$, $\gamma_x = 1/(1-\beta_x^2)^{1/2}$, and ω_p is the plasma frequency; if the half-space is moving uniformly with a velocity v_z in the positive z direction,

$$C_r = H_0(\omega^{(r)}/\omega)[(m_z - Q)/(m_z + Q)], \tag{8a}$$

$$K_t = H_0(\omega^{(t)}/\omega)[2m_z/(m_z + Q)], \tag{8b}$$

with $\beta_z = v_z/c$, $\gamma_z = 1/(1-\beta_z^2)^{1/2}$,

$$\omega^{(r)}/\omega = \gamma_z^2[(1+\beta_z^2) + 2\beta_z\cos\theta_0], \tag{8c}$$

$$\omega^{(t)}/\omega = \gamma_z^2[(1+\beta_z\cos\theta_0) - \beta_z Q], \tag{8d}$$

$$m_z = (\cos\theta_0 + \beta_z)\left[1 - \frac{\omega_p^2}{\omega^2}\frac{1}{\gamma_z^2(1+\beta_z\cos\theta_0)^2}\right], \tag{8e}$$

and

$$Q = [(\cos\theta_0 + \beta_z)^2 - (1-\beta_z^2)(\omega_p^2/\omega^2)]^{1/2}. \tag{8f}$$

(b) For a moving slab of thickness d, if it is moving uniformly with a velocity v_x in the positive x direction,

$$C_r = H_0 \frac{i(m_x^2 - \cos^2\theta_0)\sin(k_0 d\eta_x)\exp(-2ik_0 d\cos\theta_0)}{2m_x\cos\theta_0\cos(k_0 d\eta_x) - i(m_x^2 + \cos^2\theta_0)\sin(k_0 d\eta_x)}, \tag{9a}$$

$$K_t = H_0 \frac{2m_x\cos\theta_0\exp(-ik_0 d\cos\theta_0)}{2m_x\cos\theta_0\cos(k_0 d\eta_x) - i(m_x^2 + \cos^2\theta_0)\sin(k_0 d\eta_x)}, \tag{9b}$$

$$\eta_x = [\cos^2\theta_0 - (\omega_p^2/\omega^2)]^{1/2}, \tag{9c}$$

where m_x is given by Eq. (7c); if the slab is moving uniformly with a velocity v_z in the positive z direction,

$$C_r = H_0 \frac{i(\omega^{(r)}/\omega)(Q^2 - m_z^2)\sin(k_0 d\gamma_z Q)\exp[-2ik_0 d\gamma_z(\cos\theta_0 + \beta_z)]}{2m_z Q\cos(k_0 d\gamma_z Q) - i(Q^2 + m_z^2)\sin(k_0 d\gamma Q)}, \tag{10a}$$

$$K_t = H_0 \frac{2m_z Q\exp[-ik_0 d\gamma_z(\cos\theta_0 + \beta_z)]}{2m_z Q\cos(k_0 d\gamma_z Q) - i(Q^2 + m_z^2)\sin(k_0 d\gamma_z Q)}, \tag{10b}$$

where $\omega^{(r)}/\omega$, m_z, and Q, are given, respectively, by Eqs. (8c), (8e), and (8f).

In contrast with the case treated in I, we note from Eqs. (7) and (9) that for an incident wave with its electric vector polarized in the plane of incidence, the reflected and transmitted fields are functions of the velocity of the plasma medium which is moving in a direction parallel to the interface.

The reflection coefficient for a moving plasma half-space is

$$R_{x,z}^{(HS)} = (C_r C_r^*/H_0^2)p_{x,z}, \tag{11}$$

where C_r is given by Eq. (7a) and $p_x = 1$ when the medium is moving in the x direction; and C_r is given by Eq. (8a) and

$$p_z = \frac{2\beta_z + \cos\theta_0(1+\beta_z^2)}{[(1+\beta_z^2) + 2\beta_z\cos\theta_0]\cos\theta_0} \tag{12}$$

when the medium is moving in the z direction. The * signifies the complex conjugate of the function. The reflection and transmission coefficient for a moving plasma slab are, respectively,

$$R_{x,z}^{(S)} = (C_r C_r^*/H_0^2)p_{x,z}, \tag{13}$$

and

$$T_{x,z}^{(S)} = K_t K_t^*/H_0^2, \tag{14}$$

where C_r and K_t are given in Eq. (9) and $p_x = 1$ when the slab is moving in the x direction; and C_r and K_t are given in Eq. (1) and p_z is given by Eq. (12) when the slab is moving in the z direction.

To obtain a qualitative idea of how the reflection and transmission coefficients vary as a function of the velocity of the moving medium, we consider the limiting case of normal incidence. At normal incidence, the reflection and transmission coefficients for both polarizations of the incident wave remain the same when the plasma medium is moving in the z direction. Hence,

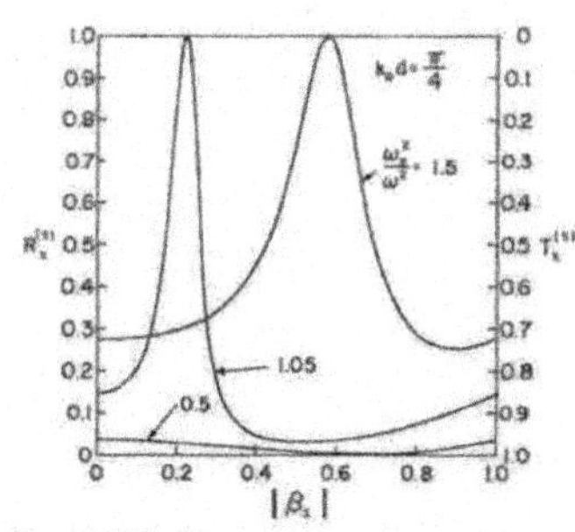

FIG. 2. Reflection and transmission coefficients for the moving-slab case.

numerical examples are given only for the case when the plasma medium is moving in the x direction.

Figure 1 shows the reflection coefficient $R_x^{(HS)}$ for the plasma half-space case as a function of β_x for various values of ω_p/ω. As expected, $R_x^{(HS)}$ is an even function of β_x. For the case $\omega_p/\omega \geq 1$, $R_x^{(HS)}$ is independent of β_x and is equal to unity. However, according to Eq. (7), the phase of the reflected wave would still be a function of β_x. If $\omega_p/\omega < 1$, $R_x^{(HS)}$ decreases from $R_x^{(HS)}(0)$ to 0 with

$$R_x^{(HS)}(0) = \left[\frac{1-(1-\omega_p^2/\omega^2)^{1/2}}{1+(1-\omega_p^2/\omega^2)^{1/2}}\right]^2,$$

as $|\beta_x|$ varies from 0 to β_{x0} where

$$\beta_{x0} = \left[\left(1-\frac{\omega_p^2}{\omega^2}\right)^{1/2} - \left(1-\frac{\omega_p^2}{\omega^2}\right)\right]^{1/2} \Big/ \left(\frac{\omega_p}{\omega}\right);$$

then $R_x^{(HS)}$ increases from 0 to $R_x^{(HS)}(0)$ as $|\beta_x|$ varies from β_{x0} to 1. Hence, the movement of the plasma medium only causes a reduction of the reflected energy. As ω_p/ω decreases, the plasma medium becomes more and more transparent to the incident wave and $R_x^{(HS)}$ approaches zero.

The reflection coefficient $R_x^{(S)}$ and the transmission coefficient $T_x^{(S)}$ for a moving plasma slab are plotted in Fig. 2 as a function of β_x for various values of ω_p/ω. The relation $k_0 d = \pi/4$ is assumed where d is the thickness of the slab in the reference frame which is at rest with the plasma slab and k_0 is the free-space wavenumber. It is noted that $R_x^{(S)} + T_x^{(S)} = 1$ for all values of β_x. For $\omega_p/\omega < 1$ and $k_0 d < [1-(\omega_p^2/\omega^2)]^{-1/2}\pi/2$, the general behavior of $R_x^{(S)}$ vs β_x curve resembles the behavior of $R_x^{(HS)}$ vs β_x curves as shown in Fig. 1. Most of the incident energy is transmitted through the slab when $\omega_p/\omega < 1$. For $\omega_p/\omega > 1$, $R_x^{(S)}$ increases from $R_x^{(S)}(0)$ which is the reflection coefficient for a stationary slab to unity as β_x increases from 0 to β_{x1} where

$$\beta_{x1} = (\omega_p^2/\omega^2 - 1)^{1/2}(\omega_p/\omega)^{-1}.$$

As β_x increases from β_{x1}, $R_x^{(S)}$ decreases and drops below the value $R_x^{(S)}(0)$ and then increases and approaches $R_x^{(S)}(0)$ when $\beta_x = 1$. The fact that $R_x^{(S)}$ is unity at β_{x1} is worth noting; it means that all the incident energy is reflected when the slab is moving at a speed of β_{x1} regardless of the thickness of the slab as long as $d > 0$.

In conclusion, one observes that the movement of the plasma medium not only affects the reflection and transmission coefficients of an incident wave, but also affects different polarizations of the incident wave in a distinctively different manner.

Reprinted from JOURNAL OF THE OPTICAL SOCIETY OF AMERICA, Vol. 57, No. 5, 657–661, May 1967
Printed in U. S. A.

Reflection from a Dielectric-Coated Moving Mirror*

C. YEH
Electrical Engineering Department, University of Southern California, Los Angeles, California 90007
(Recieved 22 December 1966)

The effects of a dielectric coating on a moving mirror upon the characteristics of the reflected wave are investigated. Both polarizations of incident plane waves are considered. Two cases of the movement are assumed: (a) the coated mirror moves parallel to the interface; (b) the coated mirror moves perpendicular to the interface. Various interesting features concerning the variation of the reflection coefficient, angle of reflection, the phase shift and the frequency of the reflected wave, as a function of the velocity of the moving coated mirror, are discussed.

INDEX HEADINGS: Reflection; Dielectric layers; Mirrors; Reflectance.

THE classic problem of the reflection of a plane electromagnetic wave by a perfectly reflecting moving mirror has been discussed some years ago by various authors.[1–3] Many interesting features concerning the variation of the reflectance, the angle of reflection, and the frequency of the reflected wave, as a function of the velocity of the moving mirror were obtained. For example, when the mirror is moving parallel to its interface the reflectance, the angle of reflection, and the frequency of the reflected wave are exactly the same as those for the reflected wave from a stationary mirror. In other words, it is not possible to detect the transverse motion of a mirror from the reflected wave. However, when the mirror is moving in a direction which is normal to the interface, the reflectance, the angle of reflection, and the frequency of the reflected wave are all functions of v_z/c, where v_z is the velocity of the moving mirror and c is the velocity of light in vacuum. For a normally incident wave, as v_z/c varies from 0 to -1, (i.e., the mirror is receding from the incident wave), the reflectance decreases monotonically from 1 to 0, and the frequency of the reflected wave goes from ω to 0 where ω is the frequency of the incident wave. As v_z/c increases from 0 to $+1$, (i.e., the mirror is moving towards the incident wave), the reflectance increases monotonically without bound from unity, the angle of reflection changes from the zero-velocity value to 0°, and the frequency of the reflected wave increases from ω to ∞. The fact that the reflectance can be greater than unity is worth noting. It means that the reflected flux is greater than the incident flux. Apparently, power is transferred from the moving mirror to the reflected wave. A scheme for amplifying electromagnetic waves can be based on this.

The purpose of the present work is to investigate the effects of a dielectric coating on the moving mirror upon the characteristics of the reflected wave. Results concerning the reflectance, the angle of reflection, and the phase shift and the frequency of the reflected wave, as functions of the velocity of the moving dielectric coated mirror are given in this paper.

FORMAL SOLUTION

The geometry of this problem is shown in Fig. 1. It is assumed that, in the S' system which is stationary with respect to the moving dielectric-coated mirror, a homogeneous dielectric sheath having a permittivity of ϵ_1, a permeability of μ_0, and a conductivity of zero, occupies the space $0 \leq z' \leq d$ while a perfectly conducting half-space mirror occupies the space $z' \leq 0$. The region outside the dielectric-coated mirror is filled by empty free space (ϵ_0, μ_0). The coated mirror may move in the following directions: (a) parallel to the interface in the x direction with a constant velocity v_x, (b) perpendicular to the interface in the z direction with a constant velocity v_z. Finally, the incident wave in the free-space region is assumed to be plane with a harmonic time dependence. Two types of incident waves

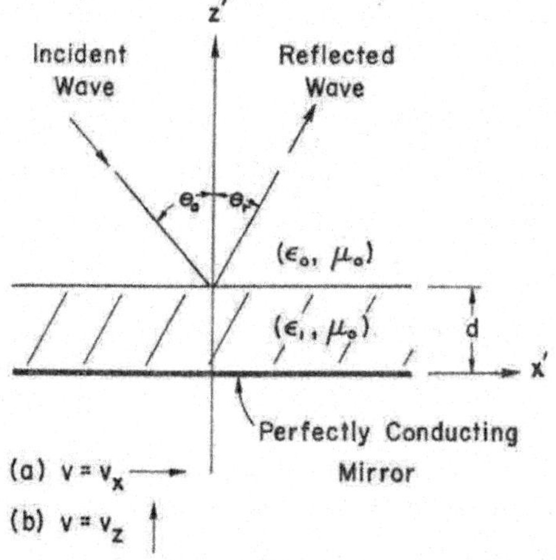

FIG. 1. Dielectric-coated moving mirror.

* Work supported partly by the National Science Foundation.
[1] W. Pauli, *Theory of Relativity* (Pergamon Press, Inc., New York, 1958).
[2] C. Møller, *The Theory of Relativity* (Oxford University Press, New York, 1952).
[3] A. Sommerfeld, *Optik* (Akademische Verlagsgesellschaft, Leipzig, 1954), 2nd ed.

are possible. One, called an E wave, is defined by $H_y=0$; the other, called an H wave, is defined by $E_y=0$.

E Wave

In the observer's system S, which is stationary with respect to the free space, an incident E wave takes the form

$$E_y^{(i)}=E_0 \exp[i(k_x x-k_z z)]e^{-i\omega t} \tag{1}$$

$$B_y^{(i)}=0, \tag{2}$$

where E_0 and ω are, respectively, the amplitude and the frequency of the incident wave, $k_x=k_0\sin\theta_0$, $k_z=k_0\cos\theta_0$, and $k_0=\omega(\mu_0\epsilon_0)^{\frac{1}{2}}$, θ_0 is the angle between the propagation vector and the positive z axis in the x-z plane.

In the moving system S', which is stationary with respect to the uniformly moving dielectric-coated mirror, the incident plane wave can be represented by the expressions

$$E_y^{(i)\prime}=E_0' \exp[i(k_x'x'-k_z'z')]e^{-i\omega' t'} \tag{3}$$

$$B_y^{(i)\prime}=0, \tag{4}$$

where k_x', k_z', ω', and E_0' are related to combinations of k_x, k_z, ω, and E_0 by Eqs. (12) when the coated mirror is moving in the positive x direction and by Eqs. (15) when the coated mirror is moving in the positive z direction. x', z', and t' are related to combinations of x, z, and t by the Lorentz transformations. The reflected wave and the wave within the dielectric sheath must, respectively, have the form

$$E_y^{(r)\prime}=A_r' \exp[i(k_x'x'+k_z'z')]e^{-i\omega' t'}, \tag{5}$$

$$B_y^{(r)\prime}=0, \tag{6}$$

and

$$E_y^{(p)\prime}=B_y' \sin[(\omega'^2\mu_0\epsilon_1-k_x'^2)^{\frac{1}{2}}z']e^{ik_x'x'}e^{-i\omega' t'}, \tag{7}$$

$$B_y^{(p)\prime}=0. \tag{8}$$

A_r' and B_y' are arbitrary constants to be determined by the boundary conditions. Matching the tangential electric and magnetic fields at the boundary $z'=d$, we obtain

$$A_r'=E_0'e^{-2ik_z'd}\left[\frac{i\sin(\eta'd)-(\eta'/k_z')\cos(\eta'd)}{i\sin(\eta'd)+(\eta'/k_z')\cos(\eta'd)}\right], \tag{9}$$

where $\eta'=(\omega'^2\mu_0\epsilon_1-k_x'^2)^{\frac{1}{2}}$. In the observer's system S, the reflected wave takes the following form

$$E_y^{(r)}=A_r \exp[i(k_x^{(r)}x+k_z^{(r)}z)]e^{-i\omega^{(r)}t} \tag{10}$$

$$B_y^{(r)}=0. \tag{11}$$

A_r, $k_x^{(r)}$, $k_z^{(r)}$, and $\omega^{(r)}$ are related to combinations of A_r', k_x', k_z', and ω' by Eqs. (13) when the dielectric-coated mirror is moving in the x direction, and by Eq. (16) when the mirror is moving in the z direction.

Case (a): $\mathbf{v}=v_x\mathbf{e}_x$

Let us consider the case in which the dielectric-coated mirror is moving at a uniform velocity v_x in the x direction. Making use of the covariance of Maxwell's equations and the phase invariance of a uniform plane wave, we have the following transformations

$$\begin{aligned}
k_x'&=\gamma_x k_0[\sin\theta_0-\beta_x]\\
k_z'&=k_0\cos\theta_0\\
\omega'&=\gamma_x\omega[1-\beta_x\sin\theta_0]\\
\eta'&=\gamma_x k_0[(\epsilon_1/\epsilon_0)(1-\beta_x\sin\theta_0)^2-(\sin\theta_0-\beta_x)^2]^{\frac{1}{2}}\\
E_0'&=\gamma_x E_0[1-\beta_x\sin\theta_0],
\end{aligned} \tag{12}$$

where $\gamma_x=1/(1-\beta_x^2)^{\frac{1}{2}}$, $\beta_x=v_x/c$, and c is the velocity of light in vacuum. Viewing from the observer's system S, we have

$$\begin{aligned}
\omega^{(r)}&=\gamma_x(\omega'+v_x k_x')\\
k_x^{(r)}&=\gamma_x(k_x'+v_x\omega'/c^2)\\
k_z^{(r)}&=k_z'\\
A_r&=\gamma_x A_r'(1+v_x k_x'/\omega').
\end{aligned} \tag{13}$$

Combining Eqs. (12) and (13) gives

$$\omega^{(r)}=\omega \tag{14a}$$

$$k_x^{(r)}=k_0\sin\theta_0 \tag{14b}$$

$$k_z^{(r)}=k_0\cos\theta_0 \tag{14c}$$

$$A_r=[-E_0]\exp[-i\Phi_x^{(E)}] \tag{14d}$$

with

$$\Phi_x^{(E)}=2k_0 d\cos\theta_0+2\tan^{-1}[(\cos\theta_0/\eta_x)\tan(\eta_x k_0 d)] \tag{14e}$$

$$\eta_x=\gamma_x[(1-\beta_x\sin\theta_0)^2(\epsilon_1/\epsilon_0)-(\sin\theta_0-\beta_x)^2]^{\frac{1}{2}}. \tag{14f}$$

According to Eqs. (14), we note that the reflected wave from a dielectric-coated mirror moving in the x direction is very similar to that from an uncoated mirror. The only difference is in the phase of the reflected wave. No phase shift is observed for the reflected wave from an uncoated moving mirror, while the phase shift for the reflected wave from the dielectric-coated mirror is a function of the speed of the movement, the dielectric constant and the thickness of the sheath. A numerical example is presented later for the phase shift as a function of β_x.

Case (b): $\mathbf{v}_z=\mathbf{e}_z$

Assume that the coated mirror is moving at a uniform velocity v_z in the positive z direction. Again making use of the covariance of Maxwell's equations and the phase invariance of a uniform plane wave, we obtain the following transformations

$$\begin{aligned}
\omega'&=\gamma_z\omega(1+\beta_z\cos\theta_0)\\
k_x'&=k_0\sin\theta_0\\
k_z'&=\gamma_z k_0(\cos\theta_0+\beta_z)\\
\eta'&=k_0[\gamma_z^2(\epsilon_1/\epsilon_0)(1+\beta_z\cos\theta_0)^2-\sin^2\theta_0]^{\frac{1}{2}}\\
E_0'&=\gamma_z E_0(1+\beta_z\cos\theta_0),
\end{aligned} \tag{15}$$

where $\gamma_z=1/(1-\beta_z^2)^{\frac{1}{2}}$ and $\beta_z=v_z/c$. Viewing the mirror from the observer's system S, we have

$$\begin{aligned}\omega^{(r)}&=\gamma_z(\omega'+v_zk_z')\\k_x^{(r)}&=k_x'\\k_z^{(r)}&=\gamma_z(k_z'+\omega'v_z/c^2)\\A_r&=\gamma_zA_r'(1+v_zk_z'/\omega').\end{aligned}\tag{16}$$

Combining Eqs. (15) and (16) gives

$$\omega^{(r)}=\omega\gamma_z^2[(1+\beta_z^2)+2\beta_z\cos\theta_0]\tag{17a}$$

$$k_x^{(r)}=k_0\sin\theta_0\tag{17b}$$

$$k_z^{(r)}=k_0\gamma_z^2[2\beta_z+\cos\theta_0(1+\beta_z^2)]\tag{17c}$$

$$A_r=[-E_0]\gamma_z^2[1+2\beta_z\cos\theta_0+\beta_z^2]\times\exp[-i\Phi_z^{(E)}],\tag{17d}$$

with

$$\Phi_z^{(E)}=2k_0d\gamma_z(\cos\theta_0+\beta_z)+2\tan^{-1}\{[\gamma_z(\cos\theta_0+\beta_z)/\eta_z]\tan(\eta_zk_0d)\}\tag{17e}$$

$$\eta_z=[\gamma_z^2(\epsilon_1/\epsilon_0)(1+\beta_z\cos\theta_0)^2-\sin^2\theta_0]^{\frac{1}{2}}.\tag{17f}$$

Again, the reflected wave from the coated mirror moving in the z direction is very similar to that from an uncoated mirror. The difference is in the phase of the reflected wave. The fact that the amplitude of the reflected wave, for the present case also, can be greater than the amplitude of the incident wave for $v_z/c>0$ is still evident. The frequency shift for the reflected wave is independent of the permittivity of the dielectric sheath and depends only on the mirror velocity and the angle of incidence. The angle of reflection is no longer equal to the angle of incidence.

H Wave

Similar analysis may be carried out for an incident H wave. Only the results are given here.

In the observer's system S, which is stationary with respect to the free space, an incident H wave takes the form

$$H_y^{(i)}=H_0\exp[i(k_xx-k_zz)]e^{-i\omega t}\tag{18}$$

$$D_y^{(i)}=0,\tag{19}$$

where H_0 is the amplitude of the incident wave. The reflected wave in the S system is then

$$H_y^{(r)}=C_r\exp[i(k_x^{(r)}x+k_z^{(r)}z)]e^{-i\omega^{(r)}t}\tag{20}$$

$$D_y^{(r)}=0,\tag{21}$$

where

$$C_r=H_0\exp[-i\Phi_x^{(H)}]\tag{22}$$

$$\Phi_x^{(H)}=2k_0d\cos\theta_0-2\tan^{-1}[(\epsilon_0/\epsilon_1)(\eta_x/\cos\theta_0)\tan(\eta_xk_0d)]\tag{23}$$

and η_x, $k_x^{(r)}$, $k_z^{(r)}$, and $\omega^{(r)}$ are given in Eq. (14), when the dielectric-coated mirror is moving in the x direction;

$$C_r=H_0\gamma_z^2[1+2\beta_z\cos\theta_0+\beta_z^2]\exp[-i\Phi_z^{(H)}]\tag{24}$$

$$\Phi_z^{(H)}=2k_0d\gamma_z(\cos\theta_0+\beta_z)-2\tan^{-1}\{(\epsilon_0/\epsilon_1)[\eta_z/\gamma_z(\cos\theta_0+\beta_z)]\tan(\eta_zk_0d)\}\tag{25}$$

and η_z, $k_x^{(r)}$, $k_z^{(r)}$, and $\omega^{(r)}$ are given in Eq. (17), when the coated mirror is moving in the z direction.

Again the only difference between the reflected wave from the coated and that from uncoated mirror for an incident H wave is in the phase shift.

The reflectances for both polarizations of the incident wave are also of interest. They are

$$R_x=1,\tag{26}$$

$$R_z=\frac{\gamma_z^6[1+2\beta_z\cos\theta_0+\beta_z^2]^2[2\beta_z+\cos\theta_0(1+\beta_z^2)]}{\cos\theta_0\{\sin^2\theta_0+\gamma_z^4[2\beta_z+\cos\theta_0(1+\beta_z^2)]^2\}^{\frac{1}{2}}},\tag{27}$$

where R_x and R_z are, respectively, the reflectance for the case in which the dielectric-coated mirror is moving in the x direction and for the case in which the coated

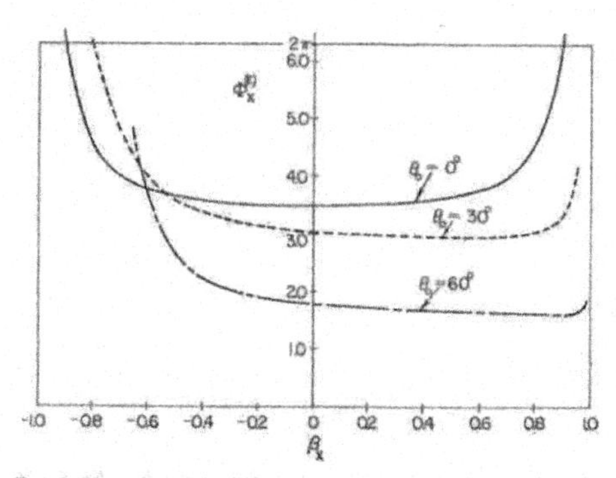

FIG. 3. Phase function of the reflected E wave as a function of β_x for various angles of incidence. $K_0d=\pi/4$; $\epsilon_1/\epsilon_0=2.0$.

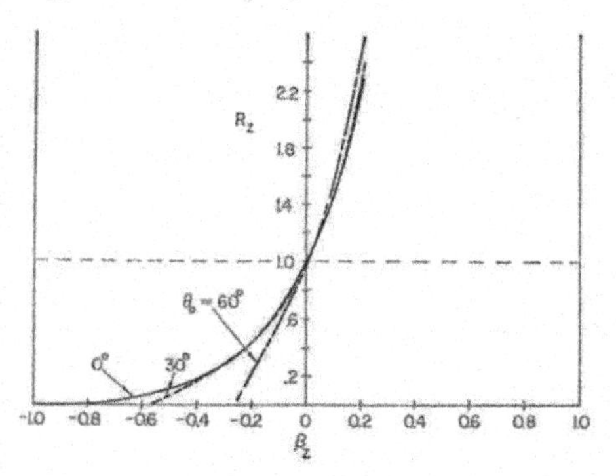

FIG. 2. Reflectance as a function of β_z for various angles of incidence.

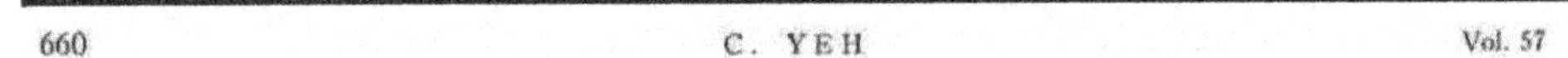

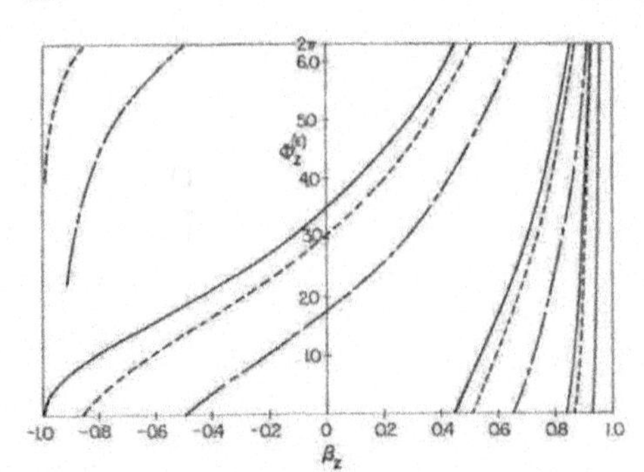

FIG. 4. Phase function of the reflected E wave as a function of β_z for various angles of incidence: 0° ———, 30° - - - -, 60° -——-. $K_0d=\pi/4$, $\epsilon_1/\epsilon_0=2.0$.

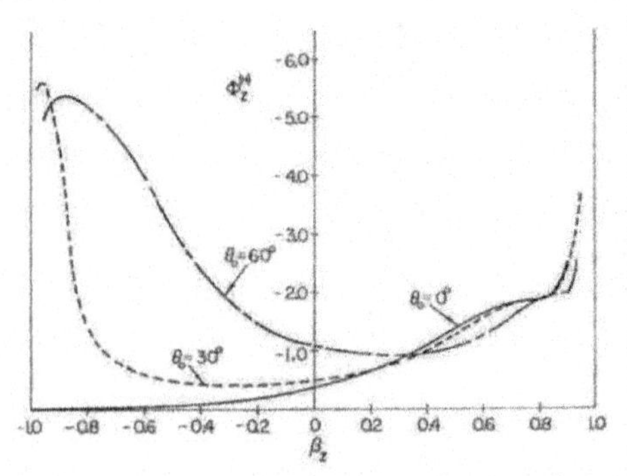

FIG. 6. Phase function of the reflected H wave as a function of β_z for various angles of incidence.

mirror is moving in the z direction. The reflectances for the two polarizations are identical. It is interesting to note that the reflectances are independent of the dielectric coating. In other words, the reflectance for a coated mirror and that for an uncoated mirror are indistinguishable even if the mirror is moving.

NUMERICAL EXAMPLES

To give a qualitative idea of how the reflectance and the phase of the reflected wave vary as a function of the velocity of the moving coated mirror for various angles of incidence, Figs. 2 through 6 are introduced. It is assumed that $\epsilon_1/\epsilon_0=2.0$ and $k_0d=\pi/4$. The reflectance R_z as a function of β_z is shown, in Fig. 2. That R_z can be greater than unity for all values of θ_0, the angle of incidence, is quite evident. As β_z increases from 0 to +1, R_z increases monotonically from 1 to ∞; as β_z decreases from 0, R_z decreases monotonically from 1 to 0. For oblique incidence, the reflectance becomes zero when the coated mirror is moving away at a speed of $\beta_z=|(\sin\theta_0-1)/\cos\theta_0|$; at this speed the angle of reflection is $\pi/2$. For $\beta_z<|(\sin\theta_0-1)/\cos\theta_0|$, no energy is reflected.

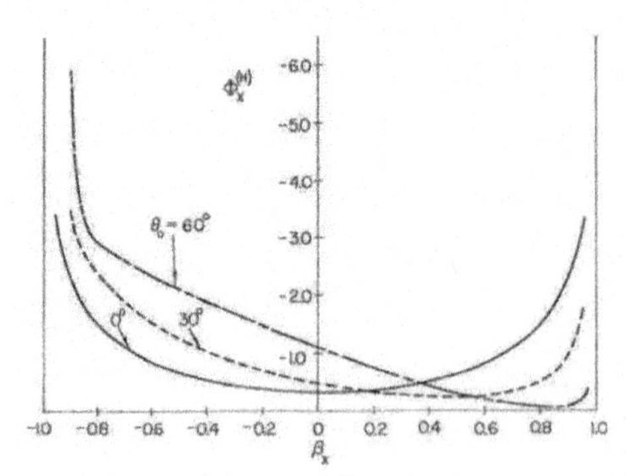

FIG. 5. Phase function of the reflected H wave as a function of β_z for various angles of incidence.

Figure 3 shows that the phase function $\Phi_z^{(E)}$ for an incident E wave is a monotonically increasing function as $|\beta_z|$ varies from 0 to unity. This is probably due to the increase of the equivalent dielectric constant of the sheath as $|\beta_z|$ is increased. For a normally incident E wave, $\Phi_z^{(E)}$ is symmetric about $\beta_z=0$, and $\Phi_z^{(E)}$ is relatively constant for $0<|\beta_z|<0.6$. As the angle of incidence, θ_0, increases from 0° which is the normal incidence case, the $\Phi_z^{(E)}$ vs β_z curve shifts towards the positive part of β_z, and the curve is no longer symmetric about $\beta_z=0$ for $\theta_0\neq 0°$.

For an incident E wave, the phase function $\Phi_x^{(E)}$ increases monotonically as β_x varies from -1 to $+1$ as shown in Fig. 4. The rate of change of $\Phi_x^{(E)}$ increases as β_x increases from 0. As expected, the $\Phi_x^{(E)}$ vs β_x curve is asymmetric about $\beta_x=0$. For an obliquely incident E wave, the behavior of the $\Phi_x^{(E)}$ vs β_x curves is very similar to the $\theta_0=0°$ case. However, since there is no reflected energy for $\beta_x<[(\sin\theta_0-1)/\cos\theta_0]$, according to Fig. 2, the portion of the $\Phi_x^{(E)}$ vs β_x curve within the region $-1<\beta_x<[(\sin\theta_0-1)/\cos\theta_0]$ is meaningless.

The phase function of an incident H wave was similarly computed. The results are shown in Figs. 5 and 6. The characteristics of the $\Phi_x^{(H)}$ vs β_x curves, shown in Fig. 5, are very similar to those shown in Fig. 3 for an incident E wave except that the phase function $\Phi_x^{(H)}$ becomes more negative as $|\beta_x|$ increases for normally incident H wave. Again the variation of $\Phi_x^{(H)}$, as a function of β_x, is due to the change of the effective dielectric constant of the sheath as a function of β_x. Figure 6 shows that the phase function $\Phi_z^{(H)}$ for a normally incident H wave is still a monotonically increasing function of β_z. For an obliquely incident H wave, the behavior of $\Phi_z^{(H)}$ as a function of β_z is somewhat more erractic.

In conclusion, we note that the reflectances, the angle of reflection, and the frequency of the reflected wave for both polarizations of the incident wave are not affected by the coating of a perfectly conducting mirror with a sheath of lossless nondispersive dielectric material; only the phase of the reflected wave is affected by the dielectric coating.

Reprinted from JOURNAL OF THE OPTICAL SOCIETY OF AMERICA, Vol. 58, No. 6, 767–770, June 1968
Printed in U. S. A.

Propagation along Moving Dielectric Wave Guides*

C. YEH
Department of Engineering, University of California, Los Angeles, California 90024
(Received 1 November 1967)

The problem of the propagation of modes along a moving dielectric interface is considered. Two types of dielectric structures were considered in particular: a moving dielectric slab and a moving dielectric circular cylinder. The characteristic equations for modes along these two types of structures are derived and a detailed discussion concerning these characteristic equations is also given. As a specific example, numerical results for the guide wavelength of the dominant TE wave along a moving dielectric slab as a function of frequency for various values of v_z/c, where v_z is the velocity of the moving dielectric material and c is the velocity of light in vacuum, were presented. It is found that if $v_z/c > (\epsilon_1/\epsilon_0)^{-\frac{1}{2}}$, the moving dielectric structure can support a forward wave as well as a backward wave and there also exists a high-frequency cutoff for the dominant mode under consideration.

INDEX HEADINGS: Electrodynamics; Mode propagation; Wave guides; Dielectric layers.

THERE has been renewed interest in and emphasis on the theory of the electrodynamics of moving media in recent years. The work of Boffi,[1] Fano–Chu–Adler,[2] Tai,[3] Elliott,[4] and Chu–Haus–Penfield[5] are a few notable examples. This interest is, perhaps, prompted by the desire of students of electromagnetic theory to understand more deeply the relation of the special theory of relativity and electrodynamics. Many interesting and perhaps unexpected results are obtained as a consequence of imposing the relativity principle. Important applications of the electrodynamics of moving media have also been developed lately. The problem of the radiation of a dipole immersed in a moving dielectric medium was studied by Lee and Papas.[6] Collier and Tai[7] and others[8] considered the propagation of waves in a waveguide filled with a dielectric medium which is moving uniformly in the axial direction. The propagation characteristics of electromagnetic waves in moving anisotropic plasma have also been investigated.[9] The Fresnel reflection coefficients for a moving dielectric half space were obtained by Yeh.[10] Compton[11] and Tai[12] recently gave an expression for the time-dependent Green's function for electromagnetic waves in moving media. In all of these problems, very interesting results were obtained.

* Work supported by the National Science Foundation.
[1] L. V. Boffi, "Electrodynamics of Moving Media," Ph.D. dissertation, Mass. Inst. of Tech., Cambridge (1958).
[2] R. M. Fano, L. J. Chu, and R. B. Adler, *Electromagnetic Fields, Energy and Forces* (John Wiley & Sons, Inc., New York, 1960), p. 315.
[3] C. T. Tai, Proc. IEEE **52**, 685 (1964).
[4] R. S. Elliott, *Electromagnetics* (McGraw-Hill Book Co., New York, 1966), pp. 98–272.
[5] L. J. Chu, H. A. Haus, and P. Penfield, Jr., Proc. IEEE **54**, 920 (1966).
[6] K. S. Lee and C. H. Papas, J. Math. Phys. **5**, 1668 (1964).
[7] J. R. Collier and C. T. Tai, IEEE Trans. Microwave Theory Techniques **MTT-13**, 441 (1965).
[8] H. Fujioka, K. Yoshida, and N. Kumagai, IEEE Trans. Microwave Theory Techniques **MTT-15**, 265 (1967); P. Daly, **MTT-15**, 274 (1967).
[9] C. T. Tai, Radio Sci. **69D**, 401 (1965).
[10] C. Yeh, J. Appl. Phys. **36**, 3513 (1965).
[11] R. T. Compton, Jr., J. Math. Phys. **7**, 2145 (1966).
[12] C. T. Tai, J. Math. Phys. **8**, 646 (1967).

The purpose of this paper is to present another important application of the electrodynamics of moving media, namely, the problem of the propagation of surface waves along a moving dielectric interface. The solution of this problem is an initial and necessary step towards the solution of the problem of the radiation from a source in the presence of a moving dielectric interface. Two kinds of dielectric structures will be considered: a plane dielectric sheet and a circular dielectric cylinder. The isotropic dielectric material having a permittivity of ϵ_1, a permeability of μ_0, and a conductivity of zero is assumed to be moving in a direction which is parallel to the direction of propagation of the surface waves.

MODE PROPAGATION ALONG MOVING DIELECTRIC SLABS

Two types of modes may be guided along a moving dielectric slab: TE and TM modes. The TE modes have a single component of electric field E_y and magnetic-field components H_x and H_z, while the TM modes have a single component of magnetic field H_y and electric-field components E_x and E_z. The dielectric slab is assumed to have a thickness of $2d$.

A. TE Modes

In the moving system S' which is stationary with respect to the uniformly moving dielectric slab, the electric-field component of the TE modes takes the form[13]

$$E'_{y'}=\begin{cases} A'\exp\{-(p'|x'|-d)-j\beta'z'+j\omega't'\} & |x'|\geq d \\ A'\sec(h'd)\cos(h'x')\exp(-j\beta'z'+j\omega't') & |x'|\leq d \end{cases} \quad (1)$$

for the even solution (i.e., $E'_{y'}$ is symmetric with respect

[13] R. E. Collins, *Field Theory of Guided Waves* (McGraw-Hill Book Co., New York, 1960), pp. 470–474.

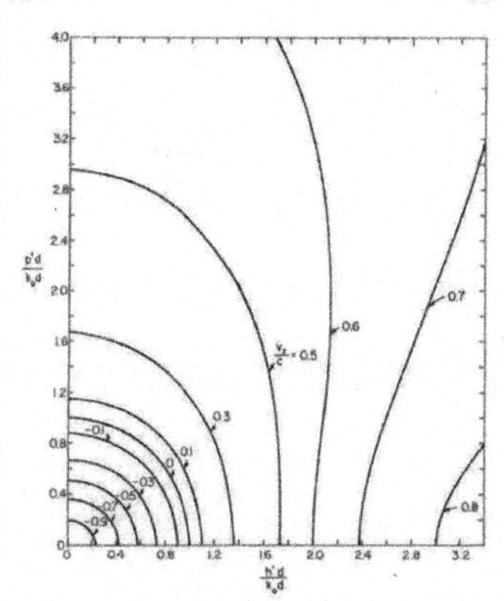

FIG. 1. $(h'd/k_0d)$ vs $(p'd/k_0d)$ for various v_s/c with $\epsilon_1/\epsilon_0=2.0$; i.e., a plot of Eq. (23).

to the plane $x'=0$) and the form

$$E'_{y'}=\begin{cases} A'\exp\{-p'(x'-d)-j\beta'z'+j\omega't'\} & x'\geq d \\ A'\csc(h'd)\sin(h'x')\exp(-j\beta'z'+j\omega't') & |x'|\leq d \\ -A'\exp\{p'(x'+d)-j\beta'z'+j\omega't'\} & x'\leq -d \end{cases} \quad (2)$$

for the odd solution (i.e., $E'_{y'}$ is asymmetric with respect to the plane $x'=0$). It is noted that

$$(\beta'd)^2=\omega'^2\mu_0\epsilon_1 d^2-(h'd)^2 \quad (3)$$

$$(\beta'd)^2=\omega'^2\mu_0\epsilon_0 d^2+(p'd)^2. \quad (4)$$

Matching the tangential electric and magnetic fields at the boundaries, $x'=\pm d$, we obtain

$$p'd=h'd\tanh'd \quad (5)$$

for the even solution, and

$$p'd=-h'd\coth'd \quad (6)$$

for the odd solution. Combining Eqs. (3) and (4) gives

$$(p'd)^2+(h'd)^2=[(\epsilon_1/\epsilon_0)-1]\omega'^2\mu_0\epsilon_0 d^2. \quad (7)$$

Making use of Minkowski's transformation theory and the Lorentz transformations,[14] we have the following expressions for the electric-field component of the TE modes in the observer's system S:

For the even solution

$$E_y=\begin{cases} A\exp\{-(p'|x|-d)-j\beta_z+j\omega t\} & |x'|\geq d \\ A\sec(h'd)\cos(h'x)\exp(-j\beta z+j\omega t) & |x'|\leq d \end{cases} \quad (8)$$

with

$$p'd=h'd\tanh'd \quad (9)$$

$$(p'd)^2+(h'd)^2=\left(\frac{\epsilon_1}{\epsilon_0}-1\right)(k_0d^2)\gamma^2 \times\left[1+\frac{v_s}{c}\frac{1}{k_0d}(k_0^2d^2+p'^2d^2)^{\frac{1}{2}}\right]^2 \quad (10)$$

$$(\beta d)^2=(k_0d)^2+(p'd)^2, \quad (11)$$

where $k_0=\omega(\mu_0\epsilon_0)^{\frac{1}{2}}$, c is the speed of light in vacuum, $\gamma^2=1/[1-(v_s/c)^2]$, A is an arbitrary constant, and v_s is the velocity of the moving dielectric slab.

For the odd solution

$$E_y=\begin{cases} A\exp\{-p'(x-d)-j\beta z+j\omega t\} & x\geq d \\ A\csc(h'd)\sin(h'x)\exp(-j\beta z+j\omega t) & |x|\leq d \\ -A\exp\{p'(x+d)-j\beta z+j\omega t\} & x\leq -d \end{cases} \quad (12)$$

with

$$p'd=-h'd\coth'd \quad (13)$$

as well as Eqs. (10) and (11). The propagation constant βd for the even solution or for the odd solution can be obtained, respectively, by solving Eqs. (9)–(11), or Eqs. (10), (11), and (13).

B. TM Modes

Similar analysis may be carried out for TM modes. In the observer's system S, the magnetic-field component of the TM modes takes the following form[13]:

For the even solution

$$H_y=\begin{cases} B\cos(h'x)\exp(-j\beta z+j\omega t) & |x|\leq d \\ B\cos(h'd)\exp\{-p'(|x|-d)-j\beta z+j\omega t\} & |x|\geq d \end{cases} \quad (14)$$

with

$$(\epsilon_1/\epsilon_0)p'd=h'd\tan(h'd) \quad (15)$$

$$(p'd)^2+(h'd)^2=\left(\frac{\epsilon_1}{\epsilon_0}-1\right)(k_0d)^2\gamma^2 \times\left[1+\frac{v_s}{c}\frac{1}{k_0d}(k_0^2d^2+p'^2d^2)^{\frac{1}{2}}\right]^2 \quad (16)$$

$$(\beta d)^2=(k_0d)^2+(p'd)^2, \quad (17)$$

[14] A. Sommerfeld, *Electrodynamics* (Academic Press Inc., New York, 1952), p. 120.

where B is an arbitrary constant; for the odd solution

$$H_y = \begin{cases} B \exp\{-p'(x-d) - j\beta z + j\omega t\} & x \geq d \\ B \csc(h'd) \sin(h'x) \exp(-j\beta z + j\omega t) & -d \leq x \leq d \\ -B \exp\{p'(x+d) - j\beta z + j\omega t\} & x \leq -d, \end{cases} \tag{18}$$

with

$$(\epsilon_1/\epsilon_0)p'd = -h'd \cot(h'd) \tag{19}$$

as well as Eqs. (16) and (17). Again the propagation constant βd for the even solution or for the odd solution can be obtained, respectively, by solving Eqs. (15)–(17), or Eqs. (16), (17), and (19).

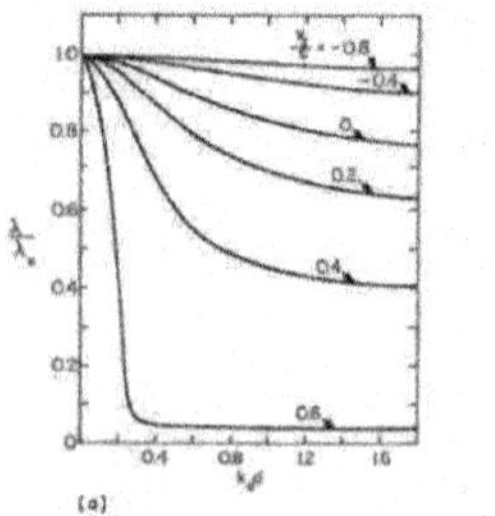
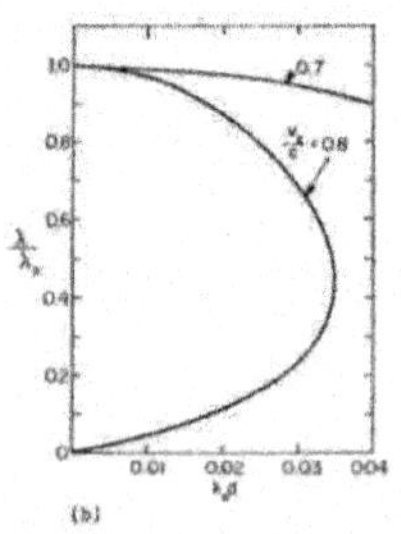

FIG. 2. Normalized guide wavelength λ/λ_0 as a function of k_0d for the dominant TE even mode along a moving-dielectric slab with thickness $2d$ and $\epsilon_1/\epsilon_0 = 2.0$. Note that the scale for k_0d in (b) is expanded.

MODE PROPAGATION ALONG A MOVING DIELECTRIC CYLINDER

Unlike the dielectric-slab case, the modes cannot, in general, be separated into TE or TM modes. All modes with angular dependence are a combination of a TE and TM mode, and are classified as hybrid HE modes. This is still true when the dielectric cylinder moves in the axial direction. In the observer's system S, the axial components of the electric and magnetic fields are

$$E_z = \begin{cases} f_m^{(1)}(r) \exp(-jm\theta - j\beta z + j\omega t) & r < a \\ f_m^{(2)}(r) \exp(-jm\theta - j\beta z + j\omega t) & r > a \end{cases} \tag{20}$$

$$H_z = \begin{cases} g_m^{(1)}(r) \exp(-jm\theta - j\beta z + j\omega t) & r < a \\ g_m^{(2)}(r) \exp(-jm\theta - j\beta z + j\omega t), & r > a, \end{cases} \tag{21}$$

where $f_m^{(1),(2)}(r)$ and $g_m^{(1),(2)}(r)$ are appropriate combinations of Bessel or modified Bessel functions. Using a similar technique, as described for the moving dielectric-slab case, we may obtain the characteristic equation for the moving-dielectric-cylinder case

$$\left[\frac{1}{h'd}\frac{J_m'(h'd)}{J_m(h'd)} + \frac{1}{p'd}\frac{K_m'(p'd)}{K_m(p'd)}\right] \times \left[\frac{1}{h'd}\frac{J_m'(h'd)}{J_m(h'd)} + \frac{\epsilon_0}{\epsilon_1}\frac{1}{p'd}\frac{K_m'(p'd)}{K_m(p'd)}\right] = m^2 \frac{[(h'd)^2 + (p'd)^2][(p'd)^2 + (h'd)^2\epsilon_0/\epsilon_1]}{(h'd)^4(p'd)^4} \tag{22}$$

$$(p'd)^2 + (h'd)^2 = \left(\frac{\epsilon_1}{\epsilon_0} - 1\right)(k_0d)^2\gamma^2 \times \left[1 + \frac{v_z}{c}\frac{1}{k_0d}(k_0^2d^2 + p'^2d^2)^{\frac{1}{2}}\right]^2 \tag{23}$$

$$(\beta d)^2 = (k_0d)^2 + (p'd)^2. \tag{24}$$

Here d represents the radius of the dielectric cylinder and $J_m(\xi)$ and $K_m(\xi)$ are, respectively, the Bessel function and the modified Bessel function of order m and argument ξ. The prime on the Bessel function of the modified Bessel function indicates the derivative of the function with respect to its argument.

It is of interest to note from the above analyses for both kinds of mode structure that as far as the characteristic equations are concerned, the velocity of the moving dielectric enters into only one of the characteristic equations [i.e., Eqs. (16) or (23)] and that equation is the same for mode propagation along a moving-dielectric slab or along a moving-dielectric cylinder. Consequently, the computation of the characteristic roots for the modes along a moving-dielectric structure is greatly simplified. For example, we suppose that we wish to find the propagation constant β of a hybrid HE_{mn} mode for a moving-dielectric rod. Solution of the transcendental Eq. (22) provides a plot of $p'd$ vs $h'd$ for fixed values of m and ϵ_1/ϵ_0. Solution of the algebraic Eq. (23) provides another plot of $p'd$ vs $h'd$ for fixed values of k_0d, ϵ_1/δ_0, and v_z/c. The intersection of these two plots gives the root for Eqs. (22) and (23) for fixed values of k_0d, ϵ_1/ϵ_0, and v_z/c. Substituting the value for $p'd$ corresponding to the root into Eq. (24) gives the propagation constant βd. It is important to note that the much more difficult plot for the transcendental Eq. (22) is independent of the velocity of the dielectric rod while the simpler plot for the algebraic Eq. (23), which is also valid for the dielectric-slab problem, is a function of v_z/c. Hence the solution for the transcendental equation for the stationary-dielectric problem may be used directly for the moving-dielectric problem.

A set of $(p'd/k_0d)$ vs. $(h'd/k_0d)$ curves computed from Eq. (23) for various values of v_z/c with $\epsilon_1/\epsilon_0 = 2.0$, is given in Fig. 1. This set of curves is independent of k_0d and is symmetrical about $(p'd/k_0d)$ axis as well as $(h'd/k_0d)$ axis. At $v_z/c = 0$, as expected, the curve is a circular arc. For $v_z/c < 0$, i.e., when the dielectric material is moving in a direction which is opposite to the direction of propagation of the mode, the curves look like elliptical arcs. As v_z/c approaches -1, the size of the arc approaches 0. As v_z/c increases from 0, the size of the arc grows. For $v_z > 0.5$ the curves no longer look like the elliptical arcs. When $p'd/k_0d = 0$, these curves intercept the $h'd/k_0d$ axis at $[(\epsilon_1/\epsilon_0 - 1)(1 + v_z/c)/(1 - v_z/c)]^{1/2}$. When $h'd/k_0d = 0$ and for $v_z/c \leq (\epsilon_0/\epsilon_1)^{1/2}$, these curves intercept the $p'd/k_0d$ axis at $[(\epsilon_1/\epsilon_0 - 1) \times (1 - (v_z/c)^2)]^{1/2}/(1 - (v_z/c)[\epsilon_1/\epsilon_0]^{1/2})$. For $v_z/c > (\epsilon_0/\epsilon_1)^{1/2}$, the curves will not intercept the $p'd/k_0d$ axis and the asymptotes of these curves are $(h'd/k_0d) = \pm[((\epsilon_1/\epsilon_0)(v_z/c) - 1)/(1 - (v_z/c)^2)]^{1/2}(p'd/k_0d)$.

To illustrate how the normalized guide wavelength λ/λ_0 for the dominant $TE^{(\text{even})}$ mode along a moving-dielectric slab of thickness $2d$, varies as a function of k_0d for various values of v_z/c, Fig. 2 is introduced. It is assumed that $\epsilon_1/\epsilon_0 = 2.0$. At $v_z/c = 0$, as expected, λ/λ_0 is unity for $k_0d = 0$ and λ/λ_0 approaches asymptotically to $[\epsilon_1/\epsilon_0]^{-1/2}$ as k_0d becomes infinite. However, when $v_z/c < 0$, λ/λ_0 varies from 1 to $(1 - (v_z/c)[\epsilon_1/\epsilon_0]^{1/2})/([\epsilon_1/\epsilon_0]^{1/2} - v_z/c)$ which is larger than $[\epsilon_1/\epsilon_0]^{-1/2}$ for negative values of v_z/c. When $v_z/c = -1$, $\lambda/\lambda_0 = 1$ for all values of k_0d. In other words, when the dielectric medium is moving opposite to the direction of mode propagation, the guide wavelength of the mode becomes closer to the free-space wavelength or the mode becomes less attached to the guiding structure. On the other hand, when the dielectric medium is moving in the same direction as the mode (i.e., $v_z/c > 0$), the guide wavelength approaches $(1 - (v_z/c)[\epsilon_1/\epsilon_0]^{1/2})([\epsilon_1/\epsilon_0]^{1/2} - v_z/c)$ which is smaller than $[\epsilon_1/\epsilon_0]^{1/2}$ as k_0d increases or the mode clings closer to the guiding structure. When $v_z/c = [\epsilon_1/\epsilon_0]^{-1/2}$, λ/λ_0 varies from 1 to 0 as k_0d varies from 0 to ∞. As v_z/c goes beyond $[\epsilon_1/\epsilon_0]^{-1/2}$, the characteristics of the λ/λ_0 vs k_0d curve are altered significantly as can be seen in Fig. 2(b). For example, when $v_z/c = 0.8$, the guiding structure can support not only the usual forward mode but also a backward mode within the region $0 \leq k_0d < 0.035$. Beyond $k_0d = 0.035$, the structure can no longer support this dominant mode; i.e., there exists a high-frequency cutoff for this mode. This is a unique feature of the moving nondispersive dielectric structure.

We note from this example that the movement of the dielectric medium introduces many interesting features which cannot be found for the stationary-dielectric problem. It is expected that these unique features will also be present for mode propagation along a moving-dielectric rod.

Reprinted from JOURNAL OF APPLIED PHYSICS, Vol. 39, No. 13, 6112-6113, December 1968

Wave Propagation on a Moving Plasma Column*

C. YEH
Department of Engineering, University of California, Los Angeles, California 90024
(Received 6 May 1968; in final form 16 August 1968)

Investigations on the interaction of electromagnetic waves with moving media produced many interesting results which cound not be found had the media stayed stationary. For example, the radiation pattern of a source immersed in a moving dielectric medium is found to be shifted towards the direction of the movement[1-3]; and if the medium is moving at a velocity higher than the phase velocity of the electromagnetic wave in the medium, all the radiated fields will be compressed into a conical region. The Fresnel reflection coefficient of a plane wave incident upon a dielectric half space, which is moving towards the incident wave, can be greater than unity[4]; and the angle of reflection is no longer equal to the angle of incidence. Owing to the axial movement of plasmas in a plasma column, an axial electric field for the scattered wave is produced, although the electric field of a normally incident plane wave is polarized in the transverse direction.[5] These are only a few recent examples on the applications of the electrodynamics of moving media. The purpose of the present article is to present the results of the problem for the propagation of surface waves along a moving plasma column.[6]

A circular plasma column of radius a is assumed to be surrounded by a free-space region. The plasma medium having a permittivity of

$$\epsilon_p=\epsilon_0[1-(\omega_p^2/\omega'^2)] \tag{1}$$

is assumed to be moving uniformly with a velocity v_z in the axial direction (i.e., the z direction), which is parallel to the direction of propagation of the surface waves. In Eq. (1), ϵ_0 is the free-space permittivity, ω_p is the plasma frequency, and ω' is the frequency in the moving system S′ which is stationary with respect to the plasma medium. We are interested in the propagation characteristics of surface waves along this structure. It can be shown that in the S′ system the plasma structure can support two types of surface modes: a TM circularly symmetric mode and a hybrid HE angularly dependent mode. The axial components of electric and magnetic fields in the moving system S′ are

$$\begin{aligned}E_z'&=A_m'I_m(p'r')\exp(-im\theta'-i\beta'z'+i\omega't') && r<a\\ &=B_m'K_m(q'r')\exp(-im\theta'-\beta'z'+i\omega't') && r>a\end{aligned} \tag{2}$$

$$\begin{aligned}H_z'&=C_m'I_m(p'r')\exp(-im\theta'-i\beta'z'+i\omega't') && r<a\\ &=D_m'K_m(q'r')\exp(-im\theta'-i\beta'z'+i\omega't') && r>a\end{aligned} \tag{3}$$

with

$$(\beta'a)^2=(p'a)^2+(\omega'a/c)^2[1-(\omega_p^2/\omega'^2)], \tag{4}$$

$$(\beta'a)^2=(q'a)^2+(\omega'a/c)^2, \tag{5}$$

where I_m and K_m are modified Bessel functions and A_m', B_m', C_m', and D_m' are arbitrary constants. Matching the tangential electric and magnetic fields at the boundary $r'=a$, one obtains

$$\begin{aligned}&\{[K_m'(q'a)/q'aK_m(q'a)]-[I_m'(p'a)/p'aI_m(p'a)]\}\\ &\times\{[K_m'(q'a)/q'aK_m(q'a)]-[1-(\omega_p^2/\omega'^2)][I_m'(p'a)/p'aI_m(p'a)]\}\\ &=m^2[(p'^2a^2-q'^2a^2)/(p'a)^4(q'a)^4]\\ &\times\{(p'a)^2-[1-(\omega_p^2/\omega'^2)](q'a)^2\},\end{aligned} \tag{6}$$

where the prime on the modified Bessel function indicates the derivative of the function with respect to its argument.

Making use of Lorentz transformations,[7] we have

$$\beta=\gamma[\beta'+\omega'(v_z/c^2)], \tag{7a}$$

$$\omega=\gamma(\omega'+\beta'v_z), \tag{7b}$$

$$(\beta a)^2=\{(p'a)^2+(a^2/c^2)[\gamma^2(\omega-\beta v_z)^2-\omega_p^2]\}/\{\gamma^2[1-(\omega/\beta)(v_z/c^2)]^2\} \tag{7c}$$

$$(\beta a)^2=(q'a)^2+(\omega a/c)^2. \tag{7d}$$

The dispersion relation is still given by Eq. (6) with

$$\omega_p^2/\omega'^2=\frac{\omega_p^2/\omega^2}{\gamma^2(1-(v_z/c)\{1+[q'^2a^2/(\omega_p^2a^2/c^2)](\omega_p^2/\omega^2)\}^{1/2})^2}, \tag{8}$$

$$p'^2a^2=q'^2a^2+(\omega_p^2a^2/c^2), \tag{9}$$

$$\beta^2a^2=q'^2a^2+(\omega^2/\omega_p^2)(\omega_p^2a^2/c^2), \tag{10}$$

where β is the propagation constant of the surface waves.

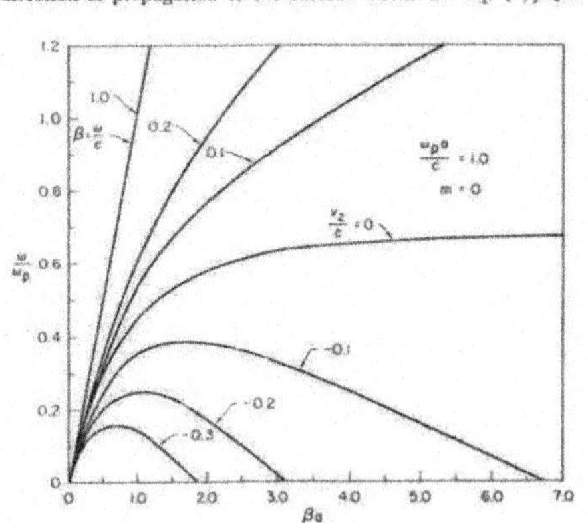

FIG. 1. The ω-β diagram for the circularly symmetric TM mode ($m=0$) propagating along a moving plasma column.

It is of interest to investigate analytically the properties of the dispersion relation for the extreme values of βa. At cutoff, $q'^2a^2=0$; in other words, the surface wave is very loosely bounded to the guiding structure near cutoff and below the cutoff frequency of a certain mode, the guiding structure can no longer support this mode. By substituting $q'a\to 0$ into Eqs. (6), (8), (9), and (10), one can show that there exists no cutoff frequency for the circularly symmetric TM mode ($m=0$) and for the lowest-order hybrid mode ($m=1$). On the other hand, there exists cutoff frequencies for the higher-order hybrid modes ($m\geqq 2$). The cutoff frequencies are governed by the following equation:

$$(\omega/\omega_p)^2_{\text{cutoff}}\cong\{[1+(v_z/c)/1-(v_z/c)]\}(\omega/\omega_p)^2_{\text{cutoff for } v_z=0 \text{ case}}$$

and

$$(\omega/\omega_p)^2_{\text{cutoff for } v_z=0 \text{ case}}=\{2+[PI_m(P)/(m-1)I_{m-1}(P)]\}^{-1}$$

with

$$P=\omega_p a/c.$$

It can be seen that when the plasma is moving in the direction of wave propagation the cutoff frequencies for the moving plasma case are higher than those for the stationary plasma case and when the plasma is moving in the opposite direction of wave propagation the cutoff frequencies for the moving plasma case are lower than those for the stationay plasma case.

Another interesting property for the guided wave along a moving plasma column can be obtained from the dispersion relation. Unlike the stationary plasma waveguide case,[8] in which all guided modes have an upper cutoff frequency, i.e., $\omega=\omega_p/\sqrt{2}$, all guided modes propagating along a moving plasma possess no upper cutoff frequency provided that the plasma medium is moving in the same direction of the propagating modes. When the plasma medium is moving in the opposite direction of the propagating modes, backward wave appears as the propropagation constant βa increases and there exists a maximum value for βa. This point can best be indicated by the following numerical examples.

To obtain the propagation constant βa of a particular surface wave mode, m, it is necessary to first solve the transcendental Eq. (6) with Eqs. (8) and (9) for the root $q'a$ for fixed values of ω/ω_p and $\omega_p a/c$ and then substitute the resultant values for $q'a$, ω/ω_p, and $\omega_p a/c$ into Eq. (10). Figures 1 and 2 are introduced, respectively, to illustrate hcw the propagation constants βa for the dominant circularly symmetric TM mode ($m=0$) and for the dipolar hybrid HE mode ($m=1$), vary as a function of ω/ω_p for various values of v_z/c with $\omega_p a/c=1.0$. It can be seen from these figures that the general behaviors of the ω/ω_p vs βa curves for the $m=0$ mode and the $m=1$ mode are quite similar. At $v_z/c=0$, as expected, βa is zero for $\omega/\omega_p=0$ and βa becomes infinite as ω/ω_p approaches asymptotically $1/\sqrt{2}$. However, when $v_z/c>0$, the high frequency cutoff for $m=0$, 1 modes disappears; βa increases without bounds as ω/ω_p increases. When $v_z/c=1$, β becomes ω/c. In other words, when the plasma medium is moving in the direction of wave propagation, the propagation constant becomes closer to the free-space propagation constant or the surface wave becomes less attached to the plasma column. On the other hand, when the plasma medium is moving opposite

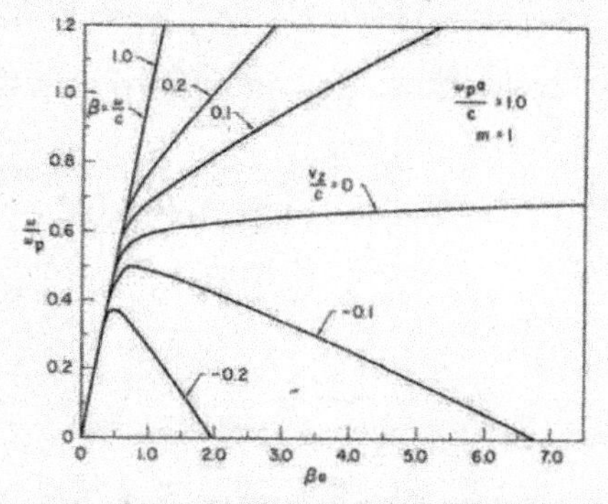

FIG. 2. The ω-β diagram for the lowest-order hybrid HE mode ($m=1$) propagating along a moving plasma column.

to the direction of propagation of the surface wave (i.e., $v_z/c<0$), in addition to the presence of a high-frequency cutoff, it is noted that the guiding structure can not only support a forward wave but also a backward wave. Furthermore, as v_z/c decreases from 0 the high-frequency cutoff decreases and when $v_z/c=1.0$, no surface wave can exist along this structure. One also notes that the phase velocity for the backward wave is less than that for the forward wave and that below a certain frequency the group velocity for the backward wave is almost independent of frequency and approaches the speed of the moving plasma. The value for $\omega_p a/c$ is so chosen that the appearance of backward wave is due solely to the movement of the plasma medium and not due to the intrinsic property of a stationary plasma column as a backward wave structure.

In conclusion, we note from this specific example that the movement of the plasma medium introduces many interesting features which cannot be found for the stationary plasma problem. It is expected that these unique features will also be present for the surface wave propagation along structures of different geometrical shapes, such as slabs. Furthermore, the case of the wave propagation along a moving plasma column with an impressed dc magnetic field in the axial direction can also be treated in a similar manner.

* Supported by the National Science Foundation.

[1] C. T. Tai, J. Math. Phys. 8, 646 (1967).
[2] R. T. Compton, Jr., J. Math. Phys. 7, 2145 (1966).
[3] K. S. Lee and C. H. Papas, J. Math. Phys. 5, 1668 (1964).
[4] C. Yeh, J. Appl. Phys. 36, 3513 (1965).
[5] A. M. Messiaen and P. E. Vandenplas, Phys. Rev. 149, 131 (1966).
[6] A qualitative discussion of the propagation characteristics of surface space-charge wave ($m=0$ mode) for nonrelativistically drifting plasma under the quasistatic approximations was given by A. W. Trivelpiece [Tech. Rept. 7, California Institute of Technology, Pasadena, Calif. (1958)].
[7] W. Pauli, *Theory of Relativity* (Pergamon Press, Inc., New York, 1958).
[8] A. W. Trivelpiece and R. W. Gould, J. Appl. Phys. 30, 1784 (1959).

Reprinted from JOURNAL OF APPLIED PHYSICS, Vol. 40, No. 13, 5066–5075, December 1969

Scattering Obliquely Incident Microwaves by a Moving Plasma Column

C. YEH
Electrical Sciences and Engineering Department, University of California, Los Angeles, California 90024
(Received 23 October 1968; in final form 2 May 1969)

The problem of the interaction of obliquely incident microwaves with a plasma column which is moving uniformly in the axial direction is treated analytically. An arbitrary polarization for the incident plane wave is assumed. Two methods in solving this problem are presented. Extensive numerical results for the scattered energy in the backward and broadside directions and the angular distribution of the scattered energy are obtained for various interesting ranges of the parameters involved. It is found that cross-polarized field components are induced even at normal incidence when the plasma medium is moving with respect to the observer and that cross-polarized field components disappear at an incident angle $\theta_0 = \sin^{-1}(v_z/c)$, where v_z is the velocity of the moving plasma and c is the speed of light in vacuum when the plasma column is imbedded in free space.

I. INTRODUCTION

The problem of the scattering of microwaves by a plasma column has been intensively studied by various authors in recent years.[1] Vandenplas[2] in a recent book summarized the investigations on electron waves and resonances in bounded plasmas, which were carried out within the past 10 years. The scattering characteristics of waves by a homogeneous,[3] or radially inhomogeneous,[4] or anisotropic,[5] or radially inhomogeneous and anisotropic[6] cold plasma column have been quite thoroughly investigated. A preliminary study of the problem of the scattering of a normally incident plane wave whose electric field is polarized in the transverse direction by a cold plasma column with a small axial drift was carried out by Messian and Vandenplas.[7] They found that the slight axial drift of the plasma column is responsible for the coupling between an incident H wave and a scattered E wave at normal incidence. As pointed out by Messian and Vandenplas, this phenomenon can be considered as the plasma high-frequency analog of the static field induced by Roentgen–Eichenwald[8] current. The purpose of the present analysis is to treat the more general problem of the scattering of an obliquely incident plane wave with arbitrary polarizations (i.e., containing either E wave or H wave or both) by a cold plasma column which is moving axially with a uniform velocity v_z (which may be very large) with respect to an observer in the laboratory. It is known that for a stationary plasma cylinder an obliquely incident E or H wave will excite both E and H scattered wave except at normal incidence.[9] It will be shown here that when the plasma cylinder is moving in an axial direction in free space both E and H scattered wave will be excited for an obliquely or normally incident E or H wave, but when the angle of incidence $\theta_0 = \sin^{-1} v_z/c$, where c is the velocity of light in vacuum, an obliquely incident E wave or H wave will excite, respectively, only a scattered E wave or H wave. Two methods in solving the moving plasma column problem are illustrated. Detailed scattering characteristics are also discussed.

It is curious to note that in the past five years a great deal of work on the reflection and refraction of waves by various moving penetrable half-spaces or slabs was carried out by various authors[10–13] in an attempt to understand more deeply the problem of the interaction of electromagnetic waves with moving media. Solution of the present problem certainly provides further information on the diffraction of waves by a finite (resonant) size obstacle containing moving medium.

II. FORMULATION OF THE PROBLEM

An infinite plasma cylinder of radius a immersed in free space is coaxial with the z axis (Fig. 1). It shall be assumed that the plasma medium is moving in the axial z direction at a uniform velocity v_z. In the rest frame of the plasma (the primed frame), the plasma medium may be characterized by a permittivity ϵ and permeability μ_0, with

$$\epsilon = \epsilon_0[1-(\omega_p^2/\omega'^2)],$$

where ω_p is the plasma frequency, ω' is the frequency of the wave in the primed frame, and ϵ_0 is the free-space permittivity. A plane wave is assumed to impinge upon this plasma structure. We are interested in the scattering characteristics of the scattered wave.

---

[1] M. A. Heald and C. B. Wharton, *Plasma Diagnostics with Microwaves* (John Wiley & Sons, Inc., New York, 1965).
[2] P. E. Vandenplas, *Electron Waves and Resonances in Bounded Plasmas* (John Wiley & Sons, Inc., New York, 1968).
[3] H. Shapiro, Antenna Lab. Tech. Rep., No. 11, California Institute of Technology (1957); A. Dattner, Ericsson Tech. 2, 309 (1957).
[4] F. A. Albini and R. G. Jahn, J. Appl. Phys. **32**, 75 (1961).
[5] P. M. Platzman and H. T. Ozaki, J. Appl. Phys. **31**, 1597 (1960).
[6] C. Yeh and W. V. T. Rusch, J. Appl. Phys. **36**, 2302 (1965).
[7] A. M. Messian and P. E. Vandenplas, Phys. Rev. **149**, 131 (1966).
[8] W. C. Roentgen, Ann. Phys. **35**, 264 (1888); A. Eichenwald, Ann. Phys. **11**, 421 (1903).
[9] H. Wilhelmsson, Trans. Chalmers Univ. Technol. Gothenburg 155 (1954); 206 (1958).
[10] C. Yeh, J. Appl. Phys. **36**, 3513 (1965); **38**, 5194 (1967).
[11] C. T. Tai, Antenna Lab. Rep. No. 1691-7, Ohio State University (1964).
[12] H. Fujioka, F. Nibei, and N. Kumagai, J. Appl. Phys. **39**, 2161 (1968).
[13] C. Yeh and K. F. Casey, Phys. Rev. **144**, 665 (1966).

There are two ways of solving the present problem. The first method is to solve the problem in a frame (the primed frame) in which the plasma medium is at rest.[14] One then transforms the solution back to the observer's frame (the unprimed frame) in which the medium is moving by the use of the Lorentz transformations and the covariance of Maxwell's equations. The other method[11,15] is to solve the problem directly in the observer's frame in which the medium is moving by making use of the constitutive relations for fields in the moving medium and the boundary conditions for relatively moving media. The first method is particularly suited for the case in which the region outside the moving plasma is the free-space medium (ϵ_0, μ_0). On the other hand, if the region surrounding the moving plasma is some dielectric medium (such as glass) with constitutive parameters $(\epsilon_1 \neq \epsilon_0, \mu_0)$, then the second method should be used. Because of the simplicity of the first method, we shall therefore make use of it in solving the present problem. Solution for the more general case of a moving plasma surrounded by glass wall is also obtained by the use of the second method and given in the Appendix.

The case for an incident plane wave with its electric vector polarized in the axial direction (i.e., an E wave) will first be analyzed. In the observer's system S, the axial components of the incident plane wave in free space takes the form

$$E_z^{(i)} = E_0 \cos\theta_0 \exp(-ik_0 \cos\theta_0 y + ik_0 \sin\theta_0 z) \exp(-i\omega t) \tag{1}$$

$$H_z^{(i)} = 0, \tag{2}$$

where E_0 and ω are, respectively, the amplitude and the frequency of the incident wave and $k_0 = \omega(\mu_0\epsilon_0)^{1/2}$. θ_0 is the angle between the propagation vector and the positive y axis in the y–z plane.

In the moving system S' which is stationary with respect to the uniformly moving plasma medium, the incident plane wave takes the form

$$E_{z'}^{(i)'} = E_0' \cos\theta' \exp(-ik_0' \cos\theta' y' + ik_0' \sin\theta' z') \exp(-i\omega' t')$$
$$= F_E' \sum_{n=-\infty}^{\infty} (-1)^n J_n(k_0' r' \cos\theta') \exp(in\phi') \tag{3}$$

$$H_{z'}^{(i)'} = 0, \tag{4}$$

where

$$\omega' = \gamma_z \omega[1 - \beta_z \sin\theta_0] \tag{5a}$$

$$\gamma_z = (1 - \beta_z^2)^{-1/2}, \qquad \beta_z = v_z/c,$$

c = speed of light in vacuum

$$k_0' \cos\theta' = k_0 \cos\theta_0 \tag{5b}$$

$$\sin\theta' = (\sin\theta_0 - \beta_z)/(1 - \beta_z \sin\theta_0) \tag{5c}$$

$$k_0' = \omega'(\mu_0\epsilon_0)^{1/2} = \gamma_z k_0[1 - \beta_z \sin\theta_0] \tag{5d}$$

$$E_0' = \gamma_z E_0[1 - \beta_z \sin\theta_0] \tag{5e}$$

$$F_E' = E_0' \cos\theta' \exp(ik_0' \sin\theta' z' - i\omega' t')$$
$$= E_0 \cos\theta_0 \exp(ik_0 \sin\theta_0 z - i\omega t). \tag{5f}$$

The above expansions are obtained by making use of the principle of phase invariance of plane waves, the Lorentz transformations, and the covariance of Maxwell's equations. $J_n(p)$ is the Bessel function of order n and argument p. A polar coordinate system (r', ϕ', z') is introduced. The scattered wave and the penetrated wave in the column must, respectively, have the form[16]

$$E_{z'}^{(s)'} = F_E' \sum_{n=-\infty}^{\infty} (-1)^n A_n' H_n^{(1)}(k_0' r' \cos\theta') \exp(in\phi') \tag{6}$$

$$H_{z'}^{(s)'} = F_E' \sum_{n=-\infty}^{\infty} (-1)^n B_n' i\left(\frac{\epsilon_0}{\mu_0}\right)^{1/2} \times H_n^{(1)}(k_0' r' \cos\theta') \exp(in\phi') \tag{7}$$

and

$$E_{z'}^{(p)'} = F_E' \sum_{n=-\infty}^{\infty} (-1)^n C_n' J_n(\lambda' r') \exp(in\phi') \tag{8}$$

$$H_{z'}^{(p)'} = F_E' \sum_{n=-\infty}^{\infty} (-1)^n D_n' i\left(\frac{\epsilon_0}{\mu_0}\right)^{1/2} J_n(\lambda' r') \exp(in\phi') \tag{9}$$

with

$$\lambda' = k_0'[(\mu_1\epsilon_1/\mu_0\epsilon_0) - \sin^2\theta']^{1/2}, \tag{10}$$

where (μ_1, ϵ_1) characterizes the electromagnetic property of the column in the S' system and $H_n^{(1)}(k_0' r' \cos\theta')$ is the Hankel function. A_n', B_n', C_n', and D_n' are yet unknown arbitrary constants to be determined according to the appropriate boundary conditions.

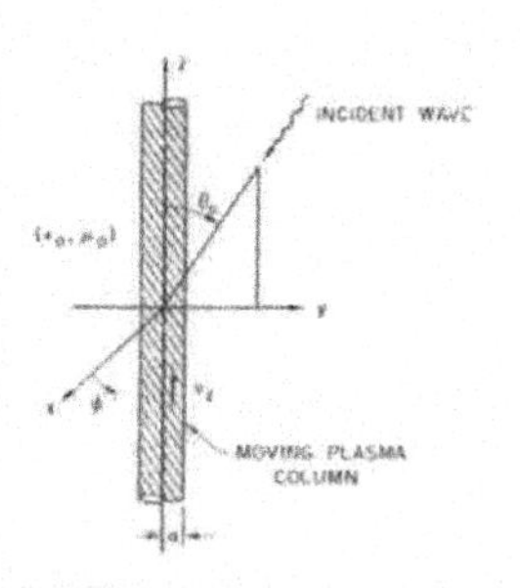

FIG. 1. The geometry of the problem.

[14] A. Sommerfeld, *Optik* (Akademische Verlagsgesellschaft, Leipzig, 1959), 2nd ed.

[15] C. T. Tai, Proc. IEEE 52, 685, 307 (1964).

[16] It is understood that if λ' is imaginary, the Bessel function $J_n(\lambda' r')$ should be replaced by the modified Bessel function $I_n(|\lambda'| r')$.

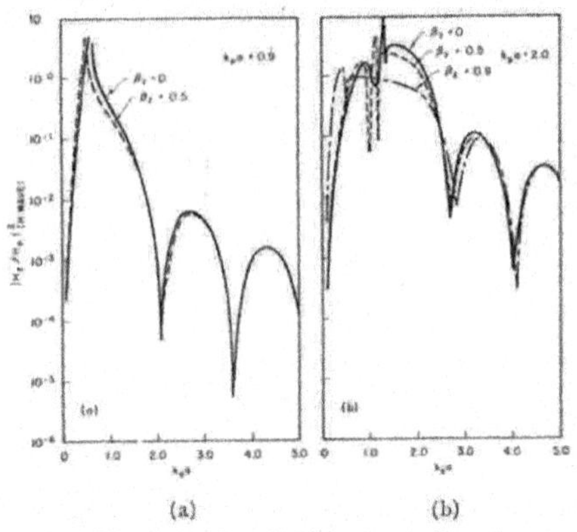

(a) (b)

FIG. 2. The backscattered energy of an incident H wave ($\theta_0=0°$, $\phi=90°$).

III. FORMAL SOLUTIONS

Satisfying the boundary conditions in the S' system, which require the continuity of the tangential electric and magnetic fields at the boundary surface $r'=a$, gives the following equation from which the unknown coefficients A_n', B_n', C_n', and D_n' can be obtained:

$$\begin{bmatrix} a_{11} & a_{12} & a_{13} & a_{14} \\ a_{21} & a_{22} & a_{23} & a_{24} \\ a_{31} & a_{32} & a_{33} & a_{34} \\ a_{41} & a_{42} & a_{43} & a_{44} \end{bmatrix} \begin{bmatrix} A_n' \\ B_n' \\ C_n' \\ D_n' \end{bmatrix} = \begin{bmatrix} b_1 \\ b_2 \\ b_3 \\ b_4 \end{bmatrix} \tag{11}$$

$$a_{11}=H_n^{(1)}(k_0'a\cos\theta') \qquad b_1=-J_n(k_0'a\cos\theta')$$
$$a_{12}=0 \qquad b_2=0$$
$$a_{13}=-J_n(\lambda'a) \qquad b_3=\sin\theta' nJ_n(k_0'a\cos\theta')$$
$$a_{14}=0 \qquad b_4=-k_0'a\cos\theta' J_n'(k_0'a\cos\theta')$$
$$a_{21}=0$$
$$a_{22}=H_n^{(1)}(k_0'a\cos\theta')$$
$$a_{23}=0$$
$$a_{24}=-J_n(\lambda'a)$$
$$a_{31}=-n\sin\theta' H_n^{(1)}(k_0'a\cos\theta')$$
$$a_{32}=k_0'a\cos\theta' H_n^{(1)\prime}(k_0'a\cos\theta')$$
$$a_{33}=(k_0'/\lambda')^2\cos^2\theta' nJ_n(\lambda'a)\sin\theta'$$
$$a_{34}=-(k_0'/\lambda')^2\cos^2\theta'\lambda' bJ_n(\lambda'a)(\mu_1/\mu_0)$$
$$a_{41}=k_0'a\cos\theta' H_n^{(1)\prime}(k_0'a\cos\theta')$$
$$a_{42}=-\sin\theta' nH_n^{(1)}(k_0'a\cos\theta')$$
$$a_{43}=-(k_0'/\lambda')^2\cos^2\theta'(\epsilon_1/\epsilon_0)\lambda' aJ_n'(\lambda'a)$$
$$a_{44}=(k_0'/\lambda')^2\cos^2\theta'\sin\theta' nJ_n(\lambda'a). \tag{12}$$

This is the formal solution for the problem of the scattering of a stationary dielectric cylinder by an obliquely incident plane E wave in the S' system. It is noted that the scattered wave as well as the penetrated wave contains both E and H waves although only an E wave is incident upon the cylinder. If the incident plane wave is an H wave, the above results are still applicable provided that we replace $\mathbf{E}'$ by $\mathbf{H}'$ and $\mathbf{H}'$ by $-\mathbf{E}'$, ϵ by μ and μ by ϵ, throughout.[17]

In the observer's system S, the field components of the scattered wave are

$$E_z^{(s)}=E_{z'}^{(s)\prime} \tag{13a}$$
$$H_z^{(s)}=H_{z'}^{(s)\prime} \tag{13b}$$
$$E_\phi^{(s)}=\gamma_z[E_{\phi'}^{(s)\prime}-v_z\mu_0H_{r'}^{(s)\prime}] \tag{13c}$$
$$H_\phi^{(s)}=\gamma_z[H_{\phi'}^{(s)\prime}+v_z\epsilon_0E_{r'}^{(s)\prime}] \tag{13d}$$
$$E_r^{(s)}=\gamma_z[E_{r'}^{(s)\prime}+v_z\epsilon_0H_{\phi'}^{(s)\prime}] \tag{13e}$$
$$H_r^{(s)}=\gamma_z[H_{r'}^{(s)\prime}-v_z\mu_0E_{\phi'}^{(s)\prime}]. \tag{13f}$$

Upon inspection of the above expressions, one notes that even at normal incidence ($\theta_0=0°$) when $v_z\neq0$, an incident E or H wave will produce a scattered wave which contains both E and H waves. This is a rather unique feature concerning the coupling between the incident wave and the scattered wave, which is only present when the plasma column is moving. A possible experimental set up to verify this phenomenon is discussed in Sec. IV.

At large distances from the cylinder, the asymptotic expression for the Hankel function

$$H_n^{(1)}(k_0r\cos\theta_0)\to(2/\pi k_0r\cos\theta_0)^{1/2}\times\exp\{i[k_0r\cos\theta_0-(2n+1)\pi/4]\}$$

is applicable provided that $k_0r\cos\theta_0\gg1$ and $k_0r\cos\theta_0\gg n$. Using the above equation, we obtain the following expressions for the far-zone scattered fields in the S system:

for incident E wave

$$\left|\frac{E_z^{(s)}}{E_0}\right|_{(E\text{ wave})}\sim\left|\sum_{n=-\infty}^{\infty}(-1)^nA_n'\exp[in(\phi-\tfrac{1}{2}\pi)]\right|, \tag{14}$$

$$\left|\frac{H_z^{(s)}}{E_0(\epsilon_0/\mu_0)^{1/2}}\right|_{(E\text{ wave})}\sim\left|\sum_{n=-\infty}^{\infty}(-1)^nB_n'\exp[in(\phi-\tfrac{1}{2}\pi)]\right|, \tag{15}$$

[17] J. R. Wait, Can. J. Phys. 33, 189 (1955).

for incident H wave

$$\left|\frac{H_z^{(s)}}{H_0}\right|_{(H\text{ wave})}\sim\left|\sum_{n=-\infty}^{\infty}(-1)^n A_n'\exp[in(\phi-\tfrac{1}{2}\pi)]\right|,\tag{16}$$

$$\left|\frac{E_z^{(s)}}{H_0(\mu_0/\epsilon_0)^{1/2}}\right|_{(H\text{ wave})}\sim\left|\sum_{n=-\infty}^{\infty}(-1)^n B_n'\exp[in(\phi-\tfrac{1}{2}\pi)]\right|.\tag{17}$$

It can be shown from Eq. (11) that $A_{-n}=A_n$, $B_{-n}=-B_n$, $C_{-n}=C_n$, and $D_{-n}=-D_n$ with

$$J_{-n}(p)=(-1)^nJ_n(p)\quad\text{and}\quad Y_{-n}(p)=(-1)^nY_n(p).$$

Making use of this relation Eqs. (14)–(17) may be written in the following forms:

$$\left|\frac{E_z^{(s)}}{E_0}\right|_{(E\text{ wave})}\sim\left|A_0'+2\sum_{n=1}^{\infty}(-1)^nA_n'\cos n(\phi-\tfrac{1}{2}\pi)\right|\tag{18}$$

$$\left|\frac{H_z^{(s)}}{E_0(\epsilon_0/\mu_0)^{1/2}}\right|_{(E\text{ wave})}\sim\left|B_0'+2i\sum_{n=1}^{\infty}(-1)^nB_n'\sin n(\phi-\tfrac{1}{2}\pi)\right|\tag{19}$$

$$\left|\frac{H_z^{s}}{H_0}\right|_{(H\text{ wave})}\sim\left|A_0'+2\sum_{n=1}^{\infty}(-1)^nA_n'\cos n(\phi-\tfrac{1}{2}\pi)\right|\tag{20}$$

$$\left|\frac{E_z}{H_0(\mu_0/\epsilon_0)^{1/2}}\right|_{(H\text{ wave})}\sim\left|B_0'+2i\sum_{n=1}^{\infty}(-1)^nB_n'\sin n(\phi-\tfrac{1}{2}\pi)\right|.\tag{21}$$

It is noted that the cross-polarized fields given by Eqs. (19) and (21) are zero at $\phi=\frac{1}{2}\pi$ or $\frac{3}{2}\pi$, the backward or forward scattering direction. This is because B_0' in the above equations is zero. Furthermore, one also notes from Eq. (5c) that at $\sin\theta_0=\beta_z$, θ' is zero. Substituting this value of θ_0 into Eq. (12) yields $B_n'=0$ and $D_n'=0$. In other words, as far as the scattered wave is concerned, the axial movement of the plasma column is similar to having a plane wave incident at an angle $\theta_0=\sin^{-1}\beta_z$. Therefore, if there is a plane wave incident at an angle θ_0, if the plasma column is moving at β_z, and if $\theta_0=\sin^{-1}\beta_z$, then the scattered wave will contain no cross-polarized components.

IV. NUMERICAL RESULTS

To have a qualitative idea of how the scattered field behaves as a function of the velocity of the moving medium, numerical computations are carried out. In the S system, the permittivity and the permeability of a cold plasma medium are, respectively,

$$\epsilon_1/\epsilon_0=1-(\omega_p^2/\omega^2)[1/\gamma_z^2(1-\beta_z\sin\theta_0)^2]\tag{22}$$

$$\mu_1/\mu_0=1.$$

Substituting these expressions into Eq. (10) gives

$$\lambda'=k_0[\cos^2\theta_0-(\omega_p^2/\omega^2)]^{1/2}\tag{23}$$

which is independent of the movement of the column.

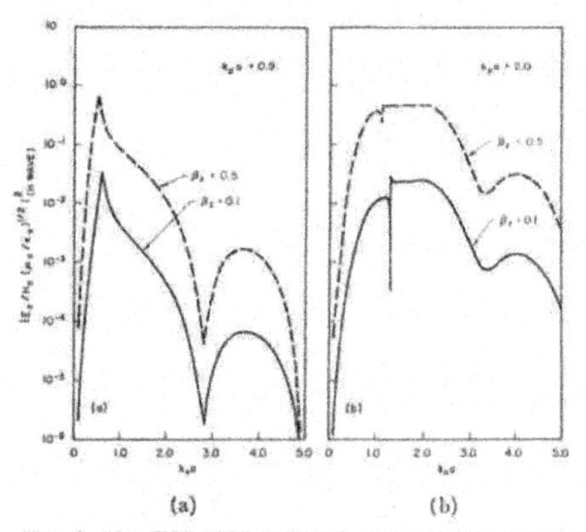

FIG. 3. The differential scattered energy in the broadside direction for the cross-polarized component for an incident H wave ($\theta_0=0°$, $\phi=0°$).

A. Incident H Wave

Considerations will first be given to the case in which the H field of an incident plane wave is parallel to the axis of the cylinder. The effects of the movement of the plasma column on the backscattered energy ($|H_s/H_0|^2_{\phi=90°}$) as a function of k_0a for various plasma densities are shown in Figs. 2(a) and (b). The independent variable for these curves is k_0a, and the variable parameter is k_pa which is defined by

$$k_pa=\omega_p(\mu_0\epsilon_0)^{1/2}a.$$

It is well known that at $\beta_z=0$ and for small value of k_pa, the backscattered energy in Fig. 2(a) consists primarily of a series of monotonically decreasing "geometrical" resonances upon which is superimposed a single resonance of the "dipolar" type[18] which occurs at $\omega\approx\omega_p/\sqrt{2}$. The dominant effect of increasing β_z when k_pa is small is a slight shift of the "dipolar" resonance. This shift of the "dipolar" frequency has also been predicted according to the theory of wave propagation for the HE_{11} mode along a moving plasma column.[19] The effect of the movement of the plasma medium upon the backscattered energy is more pronounced for denser plasma

---

[18] T. H. Stix, *The Theory of Plasma Waves* (McGraw-Hill Book Co., New York, 1962).
[19] C. Yeh, J. Appl. Phys. 39, 6112 (1968).

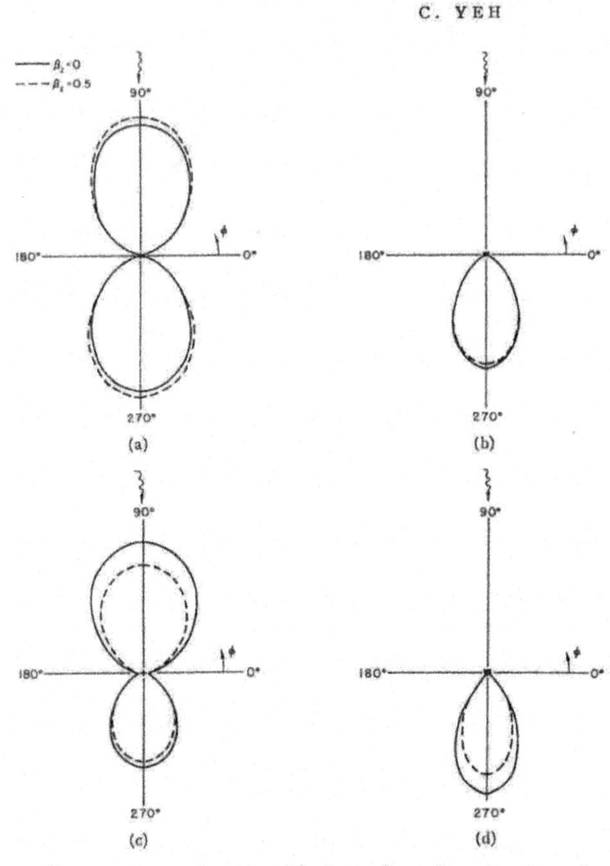

FIG. 4. Polar diagrams for $|H_z/H_0|^2$ for an incident H wave ($\theta_0=0°$). (a) $k_pa=0.9$, $k_0a=0.5$; scale ratio=2.0. (b) $k_pa=0.9$, $k_0a=2.0$; scale ratio=1.0. (c) $k_pa=2.0$, $k_0a=0.5$, scale ratio=2. (d) $k_pa=2.0$, $k_0a=2.5$; scale ratio=1.0.

medium as can be seen from Fig. 2(b). Again, the major variances occur near the "dipolar" resonance frequency. The presence of standing waves in the plasma column near this dipolar frequency is quite evident.

Since the cross-polarized component for the scattered wave is zero at the forward and backward directions, we shall therefore consider the scattered energy for the cross-polarized component in the broadside direction, i.e., at 90° from the backscattering direction. Figures 3(a) and (b) give the variation in scattered energy for the cross-polarized field ($|E_z/H_0(\mu_0/\epsilon_0)^{1/2}|^2_{\phi=0°}$) for various β_z and k_pa. Again, the "dipolar" type of resonance is evident in these figures. It is also clear from these figures that the magnitude of scattered energy for the cross-polarized field is directly proportional to β_z.

To investigate the angular distribution of scattered energy, Figs. 4 and 5 are introduced. From these figures one notes that the angular distribution of the scattered energy is very insensitive to the movement of the plasma column although the existence of the cross-polarized component of the scattered field for a normally incident plane wave depends intimately upon the movement of the plasma column. According to Figs. 4(a) and (c) we see that a dipole pattern for the scattered H_z field prevails with its major lobes pointed in the forward and backward directions when the frequency of the incident wave is below the plasma frequency. For frequencies above the plasma frequencies a single forward lobe dominates the scattering pattern for H_z field as shown in Figs. 4(b) and (d). A dipole pattern also prevails for the cross-polarized component E_z when $\omega<\omega_p$ (the overdense case), except the major lobes are pointed in the broadside directions (i.e., $\phi=0°$ and 180°) [see Figs. 5(a) and (c)]. As the frequency of the incident wave is increased above the plasma frequency (the underdense case) the two major lobes tend to tilt towards the forward direction as shown in Figs. 5(c) and (d).

B. Incident E Wave

We shall now consider the case in which the E field of an incident plane wave is parallel to the axis of the

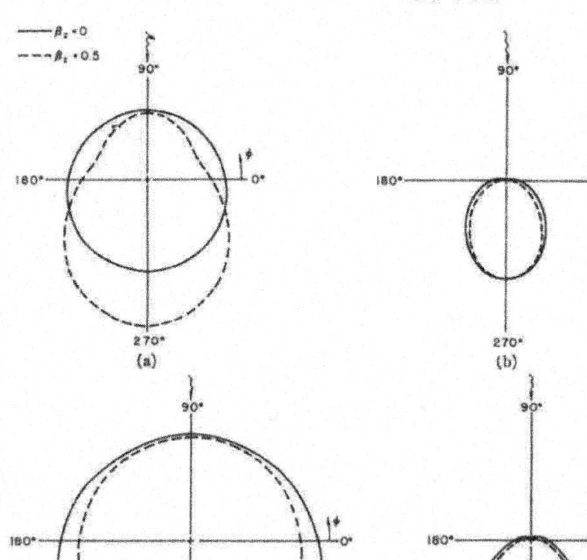

FIG. 8. Polar diagrams for $|E_z/E_0|^2$ for an incident E wave ($\theta_0 = 0°$). (a) $k_p a = 0.9$, $k_0 a = 0.5$; scale ratio = 1.0. (b) $k_p a = 0.5$, $k_0 a = 2.0$; scale ratio = 1.0. (c) $k_p a = 2.0$, $k_0 a = 0.5$; scale ratio = 1.0. (d) $k_p a = 2.0$, $k_0 a = 2.5$; scale ratio = 1.0.

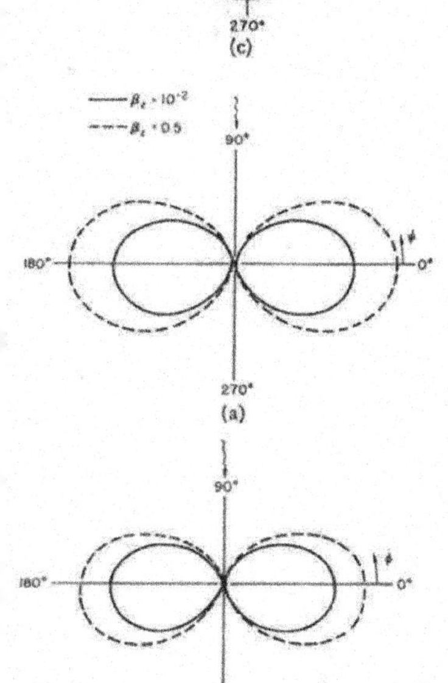

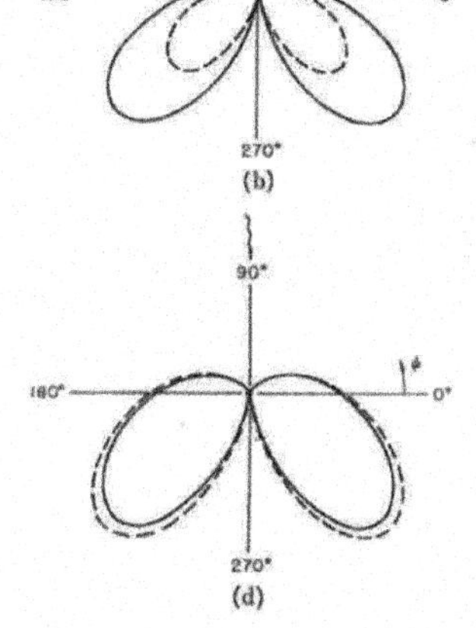

FIG. 9. Polar diagrams for $|H_z/E_0(\mu_0/\epsilon_0)^{1/2}|^2$ for an incident E wave ($\theta_0 = 0°$). (a) $k_p a = 0.9$, $k_0 a = 0.5$; scale ratio = 4×10^3. (b) $k_p a = 0.9$, $k_0 a = 2.0$; scale ratio = 4×10^3. (c) $k_p a = 2.0$, $k_0 a = 0.5$; scale ratio = 4×10^3. (d) $k_p a = 2.0$, $k_0 a = 2.5$; scale ratio = 2×10^3.

is a much more sensitive function of β_z for an incident E wave than for an incident H wave. One may again attribute this behavior to the presence of dipolar type of resonance when $\beta_z \neq 0$ for the incident E wave case.

Since the cross-polarized component for the scattered wave for an incident E wave is again zero at the forward and backward directions, we shall therefore consider the scattered energy for the cross-polarized component in the broadside direction. Figures 7(a) and (b) give the variation in scattered energy for the cross-polarized field $(|H_z/E_0(\epsilon_0/\mu_0)^{1/2}|^2_{\phi=0°})$ for various β_z and k_pa. Of course, the dipolar type of resonance is evident in these figures. Furthermore, the magnitude of the scattered energy for the cross-polarized field increases with increasing β_z.

Figures 8 and 9 are introduced to show the angular distribution of scattered energy for a normally incident E wave. A polar plot for $|E_z/E_0|^2$ vs ϕ for $\theta_0 = 0°$ is given in Fig. 8. For frequencies below the plasma frequency, the radiation patterns for $|E_z/E_0|^2$ are rather insensitive to ϕ; the patterns tilt slightly towards the forward direction as shown in Figs. 8(a) and (c). For frequencies above the plasma frequency, the radiation patterns for $|E_z/E_0|^2$ show a single forward lobe as shown in Figs. 8(b) and (d). Upon comparison of Figs. 9 and 5, we see that the angular distribution for $|H_z/E_0(\epsilon_0/\mu_0)^{1/2}|^2$ for an incident E wave is very similar to that for $|E_z/H_0(\mu_0/\epsilon_0)^{1/2}|^2$ for an incident H wave. For $\omega < \omega_p$, a dipole pattern with its main lobes pointed in the broadside direction prevails for the cross-polarized component H_z. Again, for $\omega > \omega_p$ the two major lobes tend to tilt towards the forward direction as shown in Figs. 9(c) and (d). From these figures one may again conclude that the angular distribution of the scattered energy for an incident E wave is not very sensitive to the movement of the plasma column although the existence of the cross-polarized component of the scattered field for a normally incident plane wave depends intimately upon the movement of the plasma column.

Computations were also carried out for obliquely incident E or H wave. However, the results will not be included here since the shapes of the curves are not significantly effected by the variation in θ_0. Of course, the response of the scattered field for obliquely incident wave is no longer symmetric about $\beta_z = 0$.

The most significant aspect of this problem that is revealed by the present consideration is the presence of cross-polarized component of the scattered field for a normally incident plane wave due to the movement of the plasma medium. This phenomenon may be verified quite easily in a laboratory experiment. A conceivable laboratory setup is shown in Fig. 10. The incident H wave may originate from a rectangular horn A supporting the TE_{10} mode and oriented with its electric field lines parallel to the x axis. According to the above numerical analysis, maximum scattered cross-polarized field will occur at $\phi = 0°$ or $180°$ if the frequency of the incident wave is below the plasma frequency. Hence, the receiving rectangular horn B operating in the TE_{10} mode should be located at $\phi = 0°$ or $180°$ and oriented in such a way that its electric field lines will be parallel to the z axis. If the frequency of the incident wave is above the plasma frequency, the receiving horn should be located somewhere towards the forward scattering direction according to the computed results. For example, according to Fig. 5(b) when $k_pa = 0.9$, $k_0a = 2.0$, the receiving horn should be located near $\phi \simeq 320°$ in order to receive maximum cross-polarized field $|E_z|$.

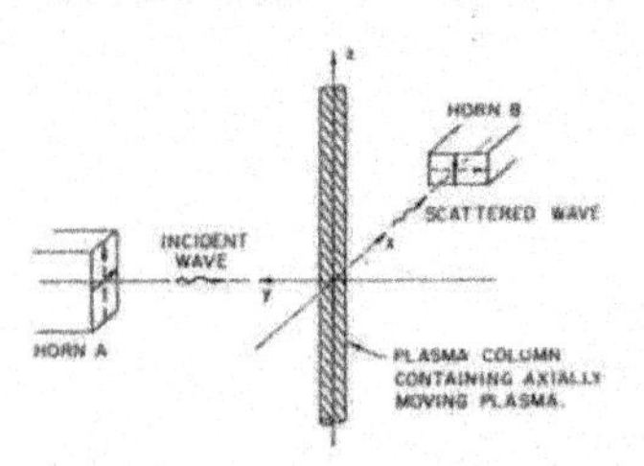

FIG. 10. A proposed experimental setup for the detection of the induced cross-polarized wave. Solid arrows indicate electric field lines. Dotted arrows indicate magnetic field lines.

For the case of an incident E wave, horn B may be used as a sending horn and horn A may then be used as a receiving horn. In this case the cross-polarized field measured by horn A is the $|H_z|$ component.

APPENDIX

A cylindrical plasma column of radius a is surrounded by a coaxial glass sheath of outer radius b. The plasma medium is moving in the axial (z-) direction with a velocity v_z. A plane wave with its electric vector polarized in the axial direction (i.e., an E wave) is assumed to be incident upon this cylindrical structure. We shall solve this problem directly in the observer's frame in which the plasma medium is moving, by making use of the constitutive relations for fields in the moving medium and the boundary conditions for relatively moving media. The constitutive relations and the wave equation for fields in the axial moving plasma medium are given in the following[30]:

$$\mathbf{D}_t^{(p)} = \alpha\epsilon'\mathbf{E}_t^{(p)} + (\delta/c)\,\mathbf{e}_z \times \mathbf{H}_t^{(p)} \qquad \text{(A1a)}$$

$$D_z^{(p)} = \epsilon' E_z^{(p)} \qquad \text{(A1b)}$$

$$\mathbf{B}_t^{(p)} = \alpha\mu'\mathbf{H}_t^{(p)} - (\delta/c)\,\mathbf{e}_z \times \mathbf{E}_t^{(p)} \qquad \text{(A1c)}$$

$$B_z^{(p)} = \mu' H_z^{(p)} \qquad \text{(A1d)}$$

and

$$(\nabla_t^2 + K^2) E_z^{(p)} = 0 \qquad \text{(A2a)}$$

$$(\nabla_t^2 + K^2) H_z^{(p)} = 0 \qquad \text{(A2b)}$$

[30] J. A. Kong and D. K. Cheng, IEEE Trans. Microwave Theory Tech. MTT-16, 99 (1968).

with

$$K^2=(1/\alpha)(k^2\alpha^2-d^2)$$

$$d=k_z+(\omega\delta/c)$$

$$\alpha=(1-\beta^2)/(1-n^2\beta^2)$$

$$\delta=\beta(n^2-1)/(1-n^2\beta^2)$$

$$n^2=\mu'\epsilon'c^2$$

$$k^2=\omega^2\mu'\epsilon'$$

$$\epsilon'=\epsilon_0[1-(\omega_p^2/\omega'^2)],\qquad \mu'=\mu_0$$

$$\omega'=(\omega-v_sk_z)(1-\beta^2)^{-1/2}$$

$$\beta=v_s/c$$

and

$$\mathbf{E}_t^{(p)}=[-id/(d^2-\alpha^2k^2)][\nabla_tE_z^{(p)}+(\omega\mu'\alpha/d)\nabla_t\times(H_z^{(p)}\mathbf{e}_z)] \quad \text{(A3a)}$$

$$\mathbf{H}_t^{(p)}=[-id/(d^2-\alpha^2k^2)][\nabla_tH_z^{(p)}-(\omega\epsilon'\alpha/d)\nabla_t\times(E_z^{(p)}\mathbf{e}_z)]. \quad \text{(A3b)}$$

The subscript "t" denotes field quantities transverse to the z direction, and ∇_t is the usual transverse del operator. The factor $\exp(ik_zz)\exp(-i\omega t)$ has been assumed and suppressed throughout for all field components in Eqs. (A1)–(A3).

The axial components of an incident plane E wave in free space (ϵ_0, μ_0) take the form

$$E_z^{(i)}=F_z\sum_{n=-\infty}^{\infty}(-1)^nJ_n(K_0r)\exp(in\phi) \quad \text{(A4a)}$$

$$H_z^{(i)}=0 \quad \text{(A4b)}$$

with

$$F_z=E_0\cos\theta_0$$

$$k_z=k_0\sin\theta_0$$

$$K_0^2=k_0^2-k_z^2$$

$$k_0^2=\omega^2\mu_0\epsilon_0.$$

The axial components for the scattered wave in free space are

$$E_z^{(s)}=F_z\sum_{n=-\infty}^{\infty}(-1)^nA_1^{(n)}H_n^{(1)}(K_0r)\exp(in\phi) \quad \text{(A5a)}$$

$$H_z^{(s)}=F_z\sum_{n=-\infty}^{\infty}(-1)^nA_2^{(n)}H_n^{(1)}(K_0r)\exp(in\phi). \quad \text{(A5b)}$$

$J_n(K_0r)$ and $H_n^{(1)}(K_0r)$ are, respectively, the Bessel and Hankel functions. All transverse components for the incident and scattered wave may be found from the following equations:

$$\mathbf{E}_t=[-ik_z/(k_z^2-k_0^2)]\{\nabla_tE_z+(\omega\mu_0/k_z)\nabla_t\times(H_z\mathbf{e}_z)\}, \quad \text{(A6a)}$$

$$\mathbf{H}_t=[-ik_z/(k_z^2-k_0^2)]\{\nabla_tH_z-(\omega\epsilon_0/k_z)\nabla_t\times(E_z\mathbf{e}_z)\}. \quad \text{(A6b)}$$

The axial components of the wave in the glass sheath (ϵ_1, μ_1) are

$$E_z^{(g)}=F_z\sum_{n=-\infty}^{\infty}(-1)^n[A_3^{(n)}J_n(K_1r)+A_4^{(n)}N_n(K_1r)]\exp(in\phi) \quad \text{(A7a)}$$

$$H_z^{(g)}=F_z\sum_{n=-\infty}^{\infty}(-1)^n[A_5^{(n)}J_n(K_1r)+A_6^{(n)}N_n(K_1r)]\exp(in\phi) \quad \text{(A7b)}$$

with $K_1^2=k_1^2-k_z^2$ and $k_1^2=\omega^2\mu_1\epsilon_1$. $N_n(K_1r)$ is the Neumann function. All transverse components for the wave in the glass sheath may be found from Eqs. (A6), provided that we replace k_0, ϵ_0, μ_0 by k_1, ϵ_1, μ_1, respectively, and we make use of Eqs. (A7). The axial components of the penetrated wave within the plasma column are found from Eqs. (A2):

$$E_z^{(p)}=F_z\sum_{n=-\infty}^{\infty}(-1)^nA_7^{(n)}J_n(Kr)\exp(in\phi) \quad \text{(A8a)}$$

$$H_z^{(p)}=F_z\sum_{n=-\infty}^{\infty}(-1)^nA_8^{(n)}J_n(Kr)\exp(in\phi). \quad \text{(A8b)}$$

The transverse fields may be obtained from Eqs. (A3). The unknown coefficients $A_1^{(n)}, A_2^{(n)}\cdots A_8^{(n)}$ are to be found from the following boundary conditions:

at $r=b$,

$$E_z^{(i)}+E_z^{(s)}=E_z^{(g)} \quad \text{(A9a)}$$

$$H_z^{(i)}+H_z^{(s)}=H_z^{(g)} \quad \text{(A9b)}$$

$$E_\phi^{(i)}+E_\phi^{(s)}=E_\phi^{(g)} \quad \text{(A9c)}$$

$$H_\phi^{(i)}+H_\phi^{(s)}=H_\phi^{(g)}; \quad \text{(A9d)}$$

at $r=a$,

$$E_z^{(p)}=E_z^{(g)} \quad \text{(A10a)}$$

$$H_z^{(p)}=H_z^{(g)} \quad \text{(A10b)}$$

$$E_\phi^{(p)}(1+\beta\delta)+\alpha\mu'v_sH_r^{(p)}=E_\phi^{(g)}+\mu_1v_sH_r^{(g)} \quad \text{(A10c)}$$

$$H_\phi^{(p)}(1+\beta\delta)-\alpha\epsilon'v_sE_r^{(p)}=H_\phi^{(g)}-\epsilon_1v_sE_r^{(g)}. \quad \text{(A10d)}$$

Satisfying the above boundary conditions gives the following set of algebraic equations from which we may solve for the unknown coefficients:

$$\sum_{l=1}^{8}a_{ml}A_l^{(n)}=\xi_m \qquad (m=1,2\cdots 8), \quad \text{(A11)}$$

where

$$a_{11}=H_n^{(1)}(K_0b),\qquad \xi_1=-J_n(K_0b)$$

$$a_{13}=-J_n(K_1b)$$

$$a_{14}=-N_n(K_1b)$$

$$a_{22}=H_n^{(1)}(K_0b)$$

$$a_{25}=-J_n(K_1b)$$

$$a_{26}=-N_n(K_1b)$$

$$a_{31}=(in/k_0b)H_n^{(1)}(K_0b),\qquad \xi_3=-(in/k_0b)J_n(K_0b)$$

$$a_{32}=-(\omega\mu_0K_0/k_0k_z)H_n^{(1)\prime}(K_0b)$$

$$a_{33}=-(K_0^2/K_1^2)(in/k_0b)J_n(K_1b)$$

$$a_{34}=-(K_0^2/K_1^2)(in/k_0b)N_n(K_1b)$$

$$a_{35}=(K_0^2/K_1)(\omega\mu_1/k_0k_z)J_n'(K_1b)$$

$$a_{36}=(K_0^2/K_1)(\omega\mu_1/k_0k_z)N_n'(K_1b)$$

$$a_{41}=(\omega\epsilon_0K_0/k_0k_z)H_n^{(1)\prime}(K_0b)$$

$$\xi_4=-(\omega\epsilon_0K_0/k_0k_z)J_n'(K_0b)$$

$$a_{42}=(in/k_0b)H_n^{(1)}(K_0b)$$

$$a_{43}=-(K_0^2\omega\epsilon_1/K_1k_0k_z)J_n'(K_1b)$$

$$a_{44}=-(K_0^2\omega\epsilon_1/K_1k_0k_z)N_n'(K_1b)$$

$$a_{45}=-(K_0^2/K_1^2)(in/k_0b)J_n(K_1b)$$

$$a_{46}=-(K_0^2/K_1^2)(in/k_0b)N_n(K_1b)$$

$$a_{53}=J_n(K_1a)$$

$$a_{54}=N_n(K_1a)$$

$$a_{57}=-J_n(Ka)$$

$$a_{65}=J_n(K_1a)$$

$$a_{66}=N_n(K_1a)$$

$$a_{68}=-J_n(Ka)$$

$$a_{73}=(in/k_0a)J_n(K_1a)[1-\mu_1v_z(\omega\epsilon_1/k_z)]$$

$$a_{74}=(in/k_0a)N_n(K_1a)[1-\mu_1v_z(\omega\epsilon_1/k_z)]$$

$$a_{75}=(K_1/k_0)J_n'(K_1a)[\mu_1v_z-(\omega\mu_1/k_z)]$$

$$a_{76}=(K_1/k_0)N_n'(K_1a)[\mu_1v_z-(\omega\mu_1/k_z)]$$

$$a_{77}=-\frac{d}{k_z}\frac{k_z^2-k_1^2}{d^2-\alpha^2k^2}\frac{in}{k_0a}J_n(Ka)\left[(1+\beta\delta)-\frac{\omega\mu'\epsilon'v_z\alpha^2}{d}\right]$$

$$a_{78}=-\frac{d}{k_z}\frac{k_z^2-k_1^2}{d^2-\alpha^2k^2}\frac{K}{k_0}J_n'(Ka)\left[\alpha\mu'v_z-(1+\beta\delta)\frac{\omega\mu'\alpha}{d}\right]$$

$$a_{83}=(K_1/k_0)J_n'(K_1a)[(\omega\epsilon_1/k_z)-\epsilon_1v_z]$$

$$a_{84}=(K_1/k_0)N_n'(K_1a)[(\omega\epsilon_1/k_z)-\epsilon_1v_z]$$

$$a_{85}=(in/k_0a)J_n(K_1a)[1-(\omega\epsilon_1\mu_1v_z/k_z)]$$

$$a_{86}=(in/k_0a)N_n(K_1a)[1-(\omega\epsilon_1\mu_1v_z/k_z)]$$

$$a_{87}=-\frac{d}{k_z}\frac{k_z^2-k_1^2}{d^2-\alpha^2k^2}\frac{K}{k_0}J_n'(Ka)\left[(1+\beta\delta)\frac{\omega\epsilon'\alpha}{d}-\alpha\epsilon'v_z\right]$$

$$a_{88}=-\frac{d}{k_z}\frac{k_z^2-k_1^2}{d^2-\alpha^2k^2}\frac{in}{k_0a}J_n(Ka)\left[(1+\beta\delta)-\frac{\omega\mu'\epsilon'v_z\alpha^2}{d}\right].$$

All other a_{ml} and ξ_m not listed above are zero. This completes the solution for the problem for an incident plane E wave. If the incident plane wave is an H wave, the above results are still applicable provided that we replace **E** by **H** and **H** by −**E**, ϵ by μ, and μ by ϵ, throughout.

It can easily be shown that if we let $b\to a$ or $\epsilon_1\to\epsilon_0$ and $\mu_1\to\mu_0$, the solution given in the Appendix approaches that given in the text, i.e., Eqs. (11)–(13). Many features concerning the scattered wave, such as the production of the scattered E and H wave by a normally incident E or H wave and the fact that the induced cross-polarized fields are zero at the backward and forward scattering direction, are still retained for the case of the glass-enclosed plasma column.

Reprinted from:

JOURNAL OF MATHEMATICAL PHYSICS VOLUME 11, NUMBER 1 JANUARY 1970

Diffraction of Waves by a Conducting Cylinder Coated with a Moving Plasma Sheath*

C. YEH
Department of Engineering, University of California, Los Angeles, California

(Received 4 March 1969)

The scattering of plane electromagnetic waves by a perfectly conducting cylinder coated with a moving dielectric or plasma sheath is investigated theoretically. The homogeneous sheath is assumed to be moving in the axial direction with a uniform velocity v_z with respect to the conducting cylinder. Solutions of this problem are obtained by making use of the special theory of relativity, the covariance of Maxwell's equations, and the Lorentz transformations. Results are given in terms of the radiation patterns of the scattered fields. A rather unique feature concerning mode coupling between the incident wave and the scattered wave is found. Even at normal incidence for $v_z \neq 0$, an incident E wave or H wave will produce a scattered wave which contains both E and H waves. Detailed discussions are presented.

I. INTRODUCTION

In an attempt to understand the problem of the interaction of electromagnetic waves with moving penetrable medium, a great deal of work on the reflection and refraction of waves by various moving penetrable media has been carried out in recent years.[1-3] Many interesting and sometimes unexpected results are obtained. However, the problem of the diffraction of waves by a finite (resonant) size obstacle containing moving medium has not been considered. The purpose of this paper is to treat this problem. Specifically, the problem of the scattering of electromagnetic waves by a conducting cylinder coated with a dielectric or plasma sheath which is moving axially with a uniform velocity v_z is solved. This problem is not only of interest from a theoretical point of view but also has an important application, i.e., the understanding of the re-entry problem. It is well known that the plasma surrounding a re-entry vehicle streams pass the conducting core and that the vehicle is moving with respect to an observer.

Solutions of this problem are obtained by making use of the special theory of relativity, the covariance of Maxwell's equations, and the Lorentz transformations.[4] Several interesting features concerning the radiation patterns and the magnitude of the scattered waves as a function of the velocity of the moving medium are discussed.

II. FORMULATION OF THE PROBLEM

The geometry of this problem is shown in Fig. 1. It is assumed that an infinite, perfectly conducting cylinder of radius a, surrounded by a homogeneous moving plasma sheath of thickness $(b - a)$, is immersed in free-space (ϵ_0, μ_0). The plasma sheath is moving in the axial direction with respect to the conducting cylinder at a uniform velocity v_z. The incident wave in the free-space region is assumed to be plane with a harmonic time dependence. The case for an incident E wave is analyzed in detail.

In the observer's system S, which is stationary with respect to the conducting cylinder, the axial components of the incident plane wave in free space takes the form

$$E_z^{(i)} = E_0 \cos\theta_0 \exp(-ik_0\cos\theta_0 y + ik_0 \sin\theta_0 z) \times \exp(-i\omega t), \quad (1)$$

$$H_z^{(i)} = 0, \quad (2)$$

where E_0 and ω are, respectively, the amplitude and the frequency of the incident wave and $k_0 = \omega(\mu_0\epsilon_0)^{\frac{1}{2}}$. θ_0 is the angle between the propagation vector and the positive y axis in the y-z plane.

In the moving system S', which is stationary with respect to the uniformly moving plasma sheath, the

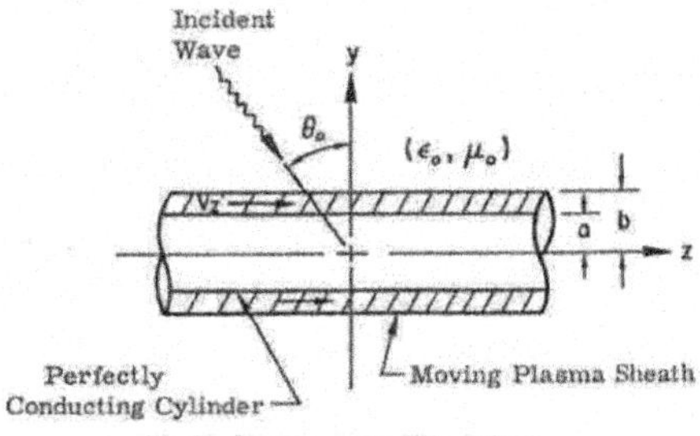

FIG. 1. The geometry of the problem.

* Supported by the National Science Foundation.

[1] H. Fujioka, F. Nihei, and N. Kumagai, J. Appl. Phys. **39**, 2161 (1968).
[2] C. Yeh, J. Appl. Phys. **38**, 5194 (1967).
[3] V. P. Pyati, J. Appl. Phys. **38**, 652 (1967).
[4] C. Møller, *The Theory of Relativity* (Oxford University Press, London, 1957).

incident plane wave takes the form

$$E_z^{(i)\prime} = E_0' \cos\theta' \exp(-ik_0' \cos\theta' y' + ik_0' \sin\theta' z') \times \exp(-i\omega' t')$$
$$= F_E' \sum_{n=-\infty}^{\infty} (-1)^n J_n(k_0' r' \cos\theta') e^{in\phi'}, \quad (3)$$

$$H_z^{(i)\prime} = 0, \quad (4)$$

where

$$\omega' = \gamma_z \omega(1 - \beta_z \sin\theta_0), \quad (5a)$$
$$\gamma_z = (1 - \beta_z^2)^{-\frac{1}{2}}, \quad \beta_z = v_z/c,$$
$$c = \text{speed of light in vacuum},$$
$$k_0' \cos\theta' = k_0 \cos\theta_0, \quad (5b)$$
$$\sin\theta' = (\sin\theta_0 - \beta_z)/(1 - \beta_z \sin\theta_0), \quad (5c)$$
$$k_0' = \omega'(\mu_0\epsilon_0)^{\frac{1}{2}} = \gamma_z k_0(1 - \beta_z \sin\theta_0), \quad (5d)$$
$$E_0' = \gamma_z E_0(1 - \beta_z \sin\theta_0), \quad (5e)$$
$$F_E' = E_0' \cos\theta' \exp(ik_0' \sin\theta' z' - i\omega' t')$$
$$= E_0 \cos\theta_0 \exp(ik_0 \sin\theta_0 z - i\omega t). \quad (5f)$$

The above expansions are obtained by making use of the principle of phase invariance of plane waves, the Lorentz transformations, and the covariance of Maxwell's equations. $J_n(p)$ is the Bessel function of order n and argument p. A polar coordinate system (r', ϕ', z') is introduced. The scattered wave and the penetrated wave in the sheath must have the form[5]

$$E_z^{(s)\prime} = F_E' \sum_{n=-\infty}^{\infty} (-1)^n A_n' H_n^{(1)}(k_0' r' \cos\theta') e^{in\phi'}, \quad (6)$$

$$H_z^{(s)\prime} = F_E' \sum_{n=-\infty}^{\infty} (-1)^n B_n' i(\epsilon_0/\mu_0)^{\frac{1}{2}} H_n^{(1)}(k_0' r' \cos\theta') e^{in\phi'} \quad (7)$$

and

$$E_z^{(p)\prime} = F_E' \sum_{n=-\infty}^{\infty} (-1)^n C_n' P_n(\lambda' r') e^{in\phi'}, \quad (8)$$

$$H_z^{(p)\prime} = F_E' \sum_{n=-\infty}^{\infty} (-1)^n D_n' i(\epsilon_0/\mu_0)^{\frac{1}{2}} Q_n(\lambda' r') e^{in\phi'}, \quad (9)$$

respectively, with

$$P_n(\lambda' r') = J_n(\lambda' r') - [J_n(\lambda' a)/N_n(\lambda' a)] N_n(\lambda, r'), \quad (10)$$

$$Q_n(\lambda' r') = J_n(\lambda' r') - \left(\frac{dJ_n(\lambda' a)}{d(\lambda' a)} \Big/ \frac{dN_n(\lambda' a)}{d(\lambda' a)}\right) N_n(\lambda' r'), \quad (11)$$

$$\lambda' = k_0'(\mu_1\epsilon_1/\mu_0\epsilon_0 - \sin^2\theta')^{\frac{1}{2}}, \quad (12)$$

[5] J. R. Wait, *Electromagnetic Radiation from Cylindrical Structures* (Pergamon Press, Inc., New York, 1959).

where (μ_1, ϵ_1) characterizes the electromagnetic property of the sheath in the S' system and $H_n^{(1)}(k_0' r' \cos\theta')$ is the Hankel function. A_n', B_n', C_n', and D_n' are as yet unknown arbitrary constants to be determined according to the appropriate boundary conditions.

III. FORMAL SOLUTIONS

Satisfying the boundary conditions in the S' system, which requires the continuity of the tangential electric and magnetic fields at the boundary surface $r' = b$, gives the following equation from which the unknown coefficients A_n', B_n', C_n', and D_n' can be obtained:

$$\begin{bmatrix} a_{11} & a_{12} & a_{13} & a_{14} \\ a_{21} & a_{22} & a_{23} & a_{24} \\ a_{31} & a_{32} & a_{33} & a_{34} \\ a_{41} & a_{42} & a_{43} & a_{44} \end{bmatrix} \begin{bmatrix} A_n' \\ B_n' \\ C_n' \\ D_n' \end{bmatrix} = \begin{bmatrix} b_1 \\ b_2 \\ b_3 \\ b_4 \end{bmatrix}, \quad (13)$$

where

$$\begin{aligned}
a_{11} &= H_n^{(1)}(k_0' b \cos\theta'), \\
a_{12} &= 0, \\
a_{13} &= -P_n(\lambda' b), \\
a_{14} &= 0, \\
a_{21} &= 0, \\
a_{22} &= H_n^{(1)}(k_0' b \cos\theta'), \\
a_{23} &= 0, \\
a_{24} &= -Q_n(\lambda' b), \\
a_{31} &= -n \sin\theta' H_n^{(1)}(k_0' b \cos\theta'), \\
a_{32} &= k_0' b \cos\theta' H_n^{(1)\prime}(k_0' b \cos\theta'), \\
a_{33} &= (k_0'/\lambda')^2 \cos^2\theta' n P_n(\lambda' b) \sin\theta', \\
a_{34} &= -(k_0'/\lambda')^2 \cos^2\theta' \lambda' b Q_n(\lambda' b)(\mu_1/\mu_0), \\
a_{41} &= k_0' b \cos\theta' H_n^{(1)\prime}(k_0' b \cos\theta'), \\
a_{42} &= -\sin\theta' n H_n^{(1)}(k_0' b \cos\theta'), \\
a_{43} &= -(k_0'/\lambda')^2 \cos^2\theta' (\epsilon_1/\epsilon_0) \lambda' b P_n'(\lambda' b), \\
a_{44} &= (k_0'/\lambda')^2 \cos^2\theta' \sin\theta' n Q_n(\lambda' b),
\end{aligned} \quad (14)$$

and

$$\begin{aligned}
b_1 &= -J_n(k_0' b \cos\theta'), \\
b_2 &= 0, \\
b_3 &= \sin\theta' n J_n(k_0' b \cos\theta'), \\
b_4 &= -k_0' b \cos\theta' J_n'(k_0' b \cos\theta').
\end{aligned}$$

This is the formal solution for the problem of the scattering of a stationary dielectric coated cylinder by an obliquely incident plane E wave in the S' system. It is noted that the scattered wave as well as the penetrated wave contain both E and H waves, although only an E wave is incident upon the coated cylinder. If the incident wave is an H wave, the above results are still applicable provided that we replace

E' by H' and H' by $-E'$, ϵ by μ and μ by ϵ, Q_n by P_n and P_n by Q_n, throughout.

In the observer's system S, the field components of the scattered wave are

$$E_z^{(s)} = E_{z'}^{(s)'}, \tag{15}$$

$$H_z^{(s)} = H_{z'}^{(s)'}, \tag{16}$$

$$E_\phi^{(s)} = \gamma_z(E_{\phi'}^{(s)'} - v_z\mu_0 H_{r'}^{(s)'}), \tag{17}$$

$$H_\phi^{(s)} = \gamma_z(H_{\phi'}^{(s')} + v_z\epsilon_0 E_{r'}^{(s)'}), \tag{18}$$

$$E_r^{(s)} = \gamma_z(E_{r'}^{(s)'} + v_z\epsilon_0 H_{\phi'}^{(s)'}), \tag{19}$$

$$H_r^{(s)} = \gamma_z(H_{r'}^{(s)'} - v_z\mu_0 E_{\phi'}^{(s)'}). \tag{20}$$

Upon inspection of the above expressions, one notes that, even at normal incidence ($\theta_0 = 0°$) when $v_z \neq 0$, an incident E or H wave will produce a scattered wave which contains both E and H waves. This is a rather unique feature concerning the coupling between the incident wave and the scattered wave, which is only present when the sheath is moving. It is also worthwhile to point out that the above results are equally valid when the perfectly conducting center core is moving with respect to the plasma sheath or when the perfectly conducting center core is stationary with respect to the plasma sheath. This is because the boundary conditions remain unchanged and time independent whether the perfectly conducting cylinder is moving or not, so long as the movement is parallel to the interface.

At large distances from the cylinder, the asymptotic expression for the Hankel function

$$H_n^{(1)}(k_0 r\cos\theta_0) \to \left(\frac{2}{\pi k_0 r\cos\theta_0}\right)^{\frac{1}{2}} e^{i[k_0 r\cos\theta_0 - \frac{1}{4}(2n+1)\pi]}$$

is applicable provided that $k_0 r\cos\theta_0 \gg 1$ and $k_0 r\cos\theta_0 \gg n$. Using the above equation, we obtain the following expressions for the far-zone scattered fields in the S system:

for incident E wave,

$$\left|\frac{E_z^{(s)}}{E_0}\right|_{E\text{ wave}} \sim \left|\sum_{n=-\infty}^{\infty}(-1)^n A_n' e^{in(\phi-\frac{1}{2}\pi)}\right|, \tag{21}$$

$$\left|\frac{H_z^{(s)}}{E_0(\epsilon_0/\mu_0)^{\frac{1}{2}}}\right|_{E\text{ wave}} \sim \left|\sum_{n=-\infty}^{\infty}(-1)^n B_n' e^{in(\phi-\frac{1}{2}\pi)}\right|; \tag{22}$$

for incident H wave,

$$\left|\frac{H_z^{(s)}}{H_0}\right|_{H\text{ wave}} \sim \left|\sum_{n=-\infty}^{\infty}(-1)^n A_n' e^{in(\phi-\frac{1}{2}\pi)}\right|, \tag{23}$$

$$\left|\frac{E_z^{(s)}}{H_0(\mu_0/\epsilon_0)^{\frac{1}{2}}}\right|_{H\text{ ewav}} \sim \left|\sum_{n=-\infty}^{\infty}(-1)^n B_n' e^{in(\phi-\frac{1}{2}\pi)}\right|. \tag{24}$$

IV. DISCUSSION OF THE RESULTS

To have a qualitative idea of how the scattered fields behave as a function of the velocity of the moving medium, numerical computations are carried out for the moving-plasma-sheath case. The permittivity and permeability of a cold plasma medium in the S' system are, respectively,

$$\epsilon_1/\epsilon_0 = 1 - \omega_p^2/\omega'^2,$$
$$\mu_1/\mu_0 = 1. \tag{25}$$

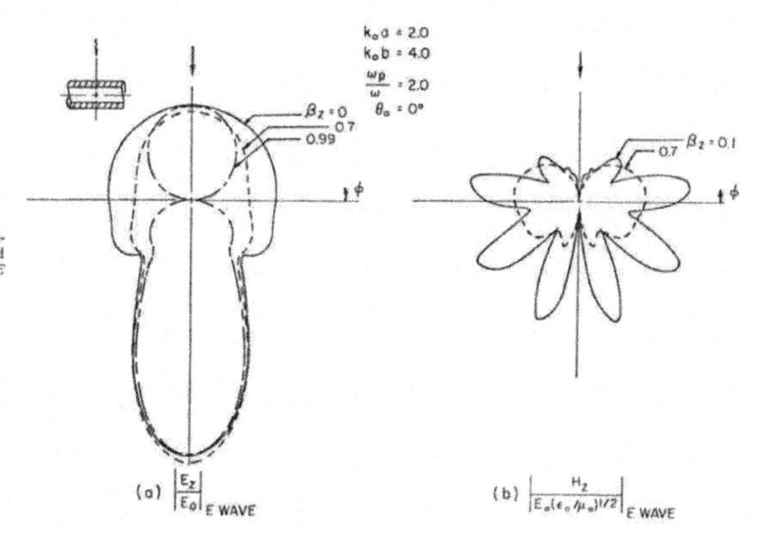

Fig. 2. Radiation patterns of the scattered waves for an incident E wave with $\theta_0 = 0°$.

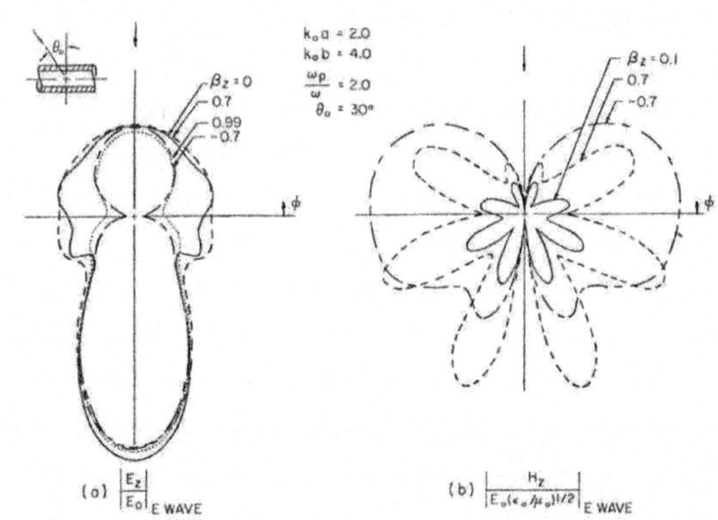

FIG. 3. Radiation patterns of the scattered waves for an incident E wave with $\theta_0 = 30°$.

In the S system, they are

$$\epsilon_1/\epsilon_0 = 1 - (\omega_p^2/\omega^2)[\gamma_z^2(1 - \beta_z \sin\theta_0)^2]^{-1},$$
$$\mu_1/\mu_0 = 1, \tag{26}$$

respectively. Substituting these expressions into Eq. (12) gives

$$\lambda' = k_0(\cos^2\theta_0 - \omega_p^2/\omega^2)^{\frac{1}{2}}, \tag{27}$$

which is independent of the movement of the sheath. Radiation patterns of the scattered waves [i.e., Eqs. (21)–(24)] are obtained for various values of θ_0 and v_z/c with $k_0a = 2.0$, $k_0b = 4.0$, and $\omega_p/\omega = 2.0$. In Figs. 2 and 3, the radiation patterns are plotted for various values of β_z and θ_0. Two angles of incidence, $\theta_0 = 0°, 30°$, are considered. It is noted that only representative patterns were shown in these figures. As can be seen from Fig. 2(a), the forward main lobe for the radiation patterns of $|E_z/E_0|_{E\text{ wave}}$ remain relatively unchanged as $|\beta_z|$ increases. On the other hand, $|E_z/E_0|_{E\text{ wave}}$ changes quite significantly in other directions; as β_z increases from 0, nulls appear in the $\phi = 0°$ and $180°$ directions. The fact that the movement of the sheath introduces coupling between an incident E wave with the scattered H wave even at normal incidence can best be seen from Fig. 2(b). As β_z increases from 0, a multilobe radiation pattern for $|H_z/E_0(\epsilon_0/\mu_0)^{\frac{1}{2}}|$ is produced; for higher values of β_z, the radiation pattern becomes basically a two-lobe structure. For all values of β_z, there exist two nulls in the forward and backward directions for

$$|H_z/E_0(\epsilon_0/\mu_0)^{\frac{1}{2}}|_{E\text{ wave}},$$

while two main lobes exist for $|E_z/E_0|_{E\text{ wave}}$ in the forward and backward directions. Similar radiation patterns are obtained for the $\theta_0 = 30°$ case as shown in Fig. 3. The general behavior of these patterns as a function of β is very similar to that for the $\theta_0 = 0°$ case. The only major difference is that at $\beta_z = 0$ the pattern for $|H_z/E_0(\epsilon_0/\mu_0)^{\frac{1}{2}}|_{E\text{ wave}}$ is not zero.

Not only are significant variations for the radiation

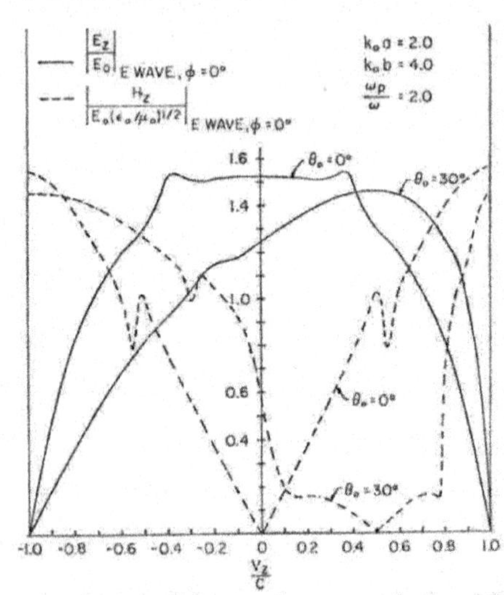

FIG. 4. Magnitude of the scattered waves as a function of the velocity of the plasma sheath for an incident E wave.

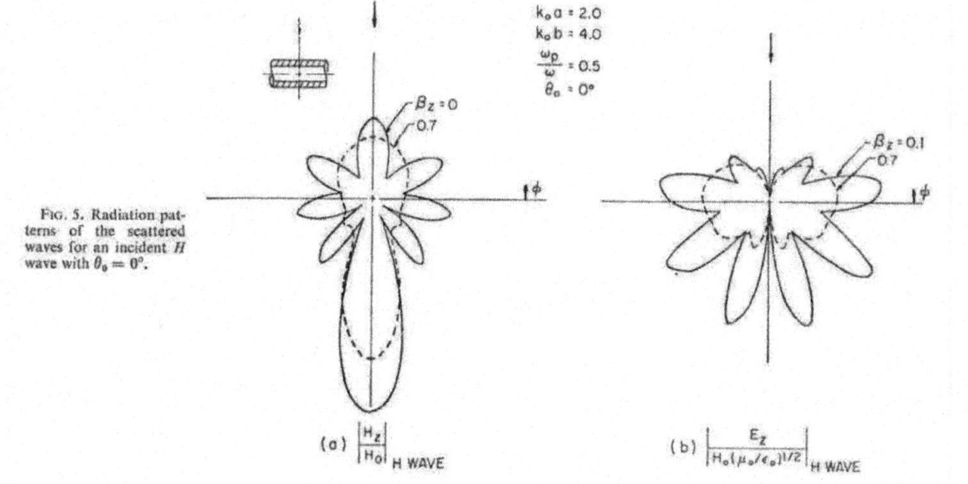

FIG. 5. Radiation patterns of the scattered waves for an incident H wave with $\theta_0 = 0°$.

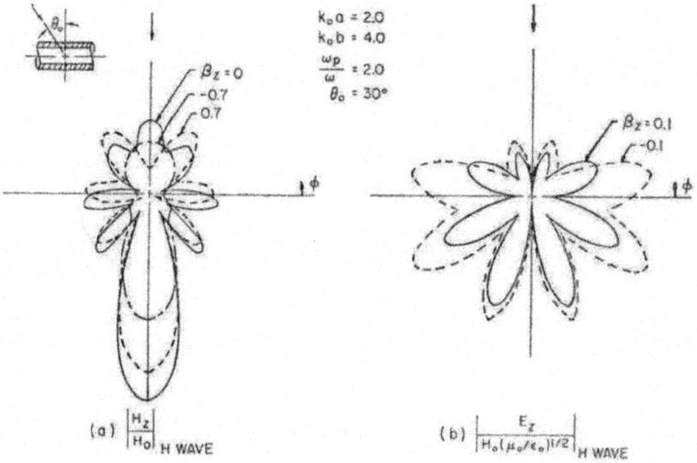

FIG. 6. Radiation patterns of the scattered waves for an incident H wave with $\theta_0 = 30°$.

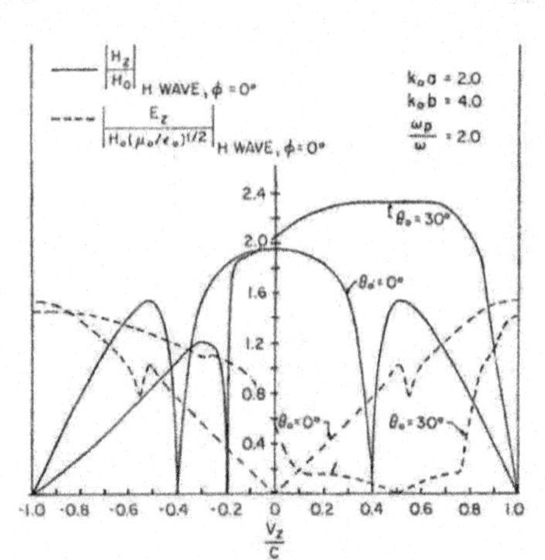

FIG. 7. Magnitude of the scattered waves as a function of the velocity of the plasma sheath for an incident H wave.

patterns observed for the scattered wave as β_z and θ_0 vary, but also for the magnitude of the scattered wave. Figure 4 is introduced to indicate the variation of $|E_z/E_0|_{E\text{ wave}}$ and $|H_z/E_0(\epsilon_0/\mu_0)^{\frac{1}{2}}|_{E\text{ wave}}$ at $\phi = 0°$ as a function of the velocity of the moving sheath. As $|v_z| \rightarrow c$, the magnitude of the scattered E wave at $\phi = 0°$ approaches zero and the magnitude of the scattered H wave at $\phi = 0°$ approaches a certain constant value. So the mode-coupling phenomenon appears to be most prominent at $\phi = 0°$ as $|\beta_z| \rightarrow 1$. As expected, the magnitude as well as the radiation patterns for the scattered wave are not symmetric with respect to β_z for $\theta_0 \neq 0°$.

Similar computations were carried out for an incident H wave. Results are shown in Figs. 5–7. It appears that the radiation patterns for an incident H wave are affected more dominantly by the movement of the sheath than those for an incident E wave. Again, coupling exists between an incident H wave with the scattered E wave at normal incidence. Computations were also carried out for $\omega_p/\omega = 0.5$, the under-dense case. Similar results as those discussed above were obtained. Since the plasma medium at $\omega_p/\omega = 0.5$ is rather transparent to the incident wave, the scattered fields $|E_z/E_0|_{E\text{ wave}}$ or $|H_z/H_0|_{H\text{ wave}}$ are not very sensitive to the movement of the plasma medium. Hence, the results for the $\omega_p/\omega = 0.5$ case are not included here.

IEEE TRANSACTIONS ON ANTENNAS AND PROPAGATION, VOL. AP-17, NO. 6, NOVEMBER 1969 757

Radiation from an Aperture in a Conducting Cylinder Coated with a Moving Plasma Sheath

KENDALL F. CASEY, MEMBER, IEEE, AND CAVOUR YEH

Abstract—The electromagnetic radiation from an aperture on a conducting cylinder coated with a moving isotropic plasma sheath is considered. Numerical results are presented to illustrate the radiation patterns as functions of sheath velocity and plasma frequency for the circumferential slot and axial slot apertures. It is found for the circumferential slot aperture that the radiation is enhanced in the direction of sheath motion when the plasma is overdense and that relatively little change occurs when the sheath is underdense. For the axial slot, it is found that an electromagnetic field is radiated whose polarization is normal to that of the field radiated under stationary conditions, in addition to a field of the usual polarization. Significant alterations of radiation patterns from their form when the sheath is stationary can occur at relatively small velocities if the wave frequency is near the plasma frequency.

I. Introduction

The problem of determining the electromagnetic field radiated by an aperture in a conducting cylinder surrounded by a plasma sheath has been intensively studied in recent years. Various types of stationary sheaths have been discussed by Knop [1] (homogeneous, isotropic), Swift [2] (inhomogeneous, isotropic), Yeh [3] (homogeneous, anisotropic), and Yeh and Rusch [4] (inhomogeneous, anisotropic), to name but a few. The purpose of this paper is to consider the effect of motion of an isotropic, homogeneous plasma sheath on the radiation pattern of an aperture in a conducting cylinder. Particular attention will be given to the special cases of circumferential slot and longitudinal slot apertures.

It is known [5] that the motion of a slab of dielectric material has a significant effect on the reflection and transmission coefficients of plane waves incident upon it. When the motion of the slab is in the direction parallel to its surfaces, the angles of incidence, reflection, and transmission are all equal, as they are when the slab is stationary, but the amplitudes of the reflected and transmitted fields are strong functions of the velocity of the slab. Thus it is to be expected that the radiation pattern of an antenna radiating through a moving slab or cylindrical sheath will be significantly altered by the motion.

In Section II of this paper, we present the formal solution to the problem of radiation from an arbitrary aperture in a conducting cylinder coated with an isotropic plasma layer moving in the axial direction. In Sections III and IV, this formal solution is specialized to the cases of circumferential and axial slot apertures. Numerically computed radiation patterns are presented and discussed. The results are summarized in Section V.

Manuscript received October 24, 1968; revised May 13, 1969. This work was supported by NSF.

K. F. Casey is with the Department of Electrical Engineering, Air Force Institute of Technology, Wright-Patterson AFB, Dayton, Ohio 45433.

C. Yeh is with the Department of Engineering, University of California, Los Angeles, Calif. 90024.

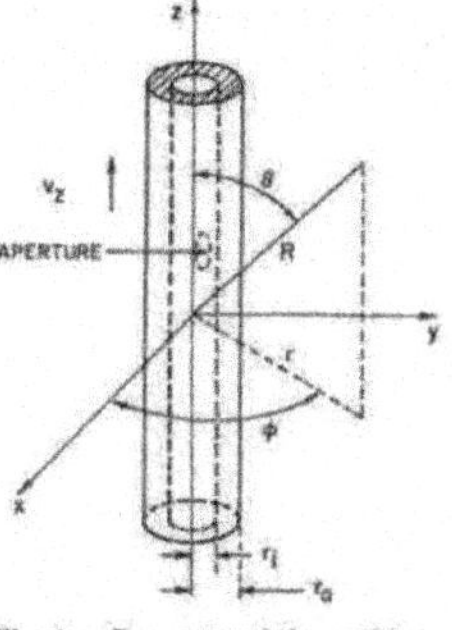

Fig. 1. Geometry of the problem.

II. Formulation and Formal Solution

The geometry of the problem is shown in Fig. 1. An infinitely long conducting circular cylinder at $r = r_i$ in the cylindrical coordinates (r,ϕ,z) is surrounded by a moving plasma sheath of outer radius r_0. The velocity of the sheath is $\mathbf{v} = v_z\mathbf{a}_z$, parallel to the axis of the cylinder. An aperture, over which the tangential components of the electric field are known, lies in the surface of the conducting cylinder.

The permittivity of the plasma sheath in its rest frame (the S' system) is

$$\epsilon' = \epsilon_0(1 - \omega_p^2/\omega'^2) \tag{1}$$

where ϵ_0 denotes the permittivity of free space, ω_p the plasma frequency, and ω' the wave frequency in the S' system. The time dependence of the fields in the S system, in which the conducting cylinder is at rest, is assumed to be $\exp(i\omega t)$. The permeability of the sheath in its rest frame is denoted μ' and is taken equal to μ_0, the permeability of free space. The region outside the sheath is free space.

In the S system, the constitutive relations in the moving plasma sheath are, denoting field quantities in the sheath with the superscript s [6],

$$\mathbf{D}_t^s = a\epsilon'\mathbf{E}_t^s + (b/c)\mathbf{a}_z \times \mathbf{H}_t^s \tag{2a}$$

$$D_z^s = \epsilon' E_z^s \tag{2b}$$

$$\mathbf{B}_t^s = a\mu'\mathbf{H}_t^s - (b/c)\mathbf{a}_z \times \mathbf{E}_t^s \tag{3a}$$

$$B_z^s = \mu' H_z^s \tag{3b}$$

in which

$$a = (1 - \beta^2)/(1 - n^2\beta^2) \quad (4)$$

$$b = \beta(n^2 - 1)/(1 - n^2\beta^2) \quad (5)$$

$$\beta = v_s/c \quad (6)$$

$$n^2 = \mu'\epsilon'c^2 \quad (7)$$

where c is the speed of light in free space. t subscripts denote field quantities transverse to the z direction.

Let the field quantities be expressed in terms of their Fourier transforms as follows:

$$\mathbf{E}^s = \mathbf{E}_t^s + E_z^s\mathbf{a}_z = \int_{-\infty}^{\infty} \exp(-ik_z z)(\hat{\mathbf{E}}_t^s + \hat{E}_z^s\mathbf{a}_z)\, dk_z \quad (8)$$

and

$$\mathbf{H}^s = \mathbf{H}_t^s + H_z^s\mathbf{a}_z = \int_{-\infty}^{\infty} \exp(-ik_z z)(\hat{\mathbf{H}}_t + \hat{H}_z\mathbf{a}_z)\, dk_z. \quad (9)$$

Then $\hat{\mathbf{E}}_t^s$ and $\hat{\mathbf{H}}_t^s$ are given in terms of $\hat{E}_z^s$ and $\hat{H}_z^s$ by [6]

$$\hat{\mathbf{E}}_t^s = \frac{id}{d^2 - a^2k^2}\left(\nabla_t\hat{E}_z^s + \frac{\omega\mu' a}{d}\nabla_t \times \hat{H}_z^s\mathbf{a}_z\right) \quad (10)$$

$$\hat{\mathbf{H}}_t^s = \frac{id}{d^2 - a^2k^2}\left(\nabla_t\hat{H}_z^s - \frac{\omega\epsilon' a}{d}\nabla_t \times \hat{E}_z^s\mathbf{a}_z\right) \quad (11)$$

in which

$$d = k_z + \omega b/c \quad (12)$$

$$k^2 = \omega^2\mu'\epsilon' \quad (13)$$

and ∇_t is the usual transverse del operator in rectangular coordinates

$$\nabla_t = \mathbf{a}_x\frac{\partial}{\partial x} + \mathbf{a}_y\frac{\partial}{\partial y}. \quad (14)$$

$\hat{E}_z^s$ and $\hat{H}_z^s$ satisfy the equations

$$(\nabla_t^2 + K^2)\hat{E}_z^s = 0 \quad (15)$$

$$(\nabla_t^2 + K^2)\hat{H}_z^s = 0 \quad (16)$$

with

$$K^2 = (1/a)(k^2a^2 - d^2).^1 \quad (17)$$

ω', the frequency in the rest frame of the sheath, is given by

$$\omega' = (\omega - v_s k_z)(1 - \beta^2)^{-1/2}. \quad (18)$$

In the free space surrounding the sheathed cylinder, we have, denoting field quantities with the superscript r:

$$\mathbf{D}^r = \epsilon_0\mathbf{E}^r \quad (19a)$$

$$\mathbf{B}^r = \mu_0\mathbf{H}^r \quad (19b)$$

$$\mathbf{E}^r = \mathbf{E}_t^r + E_z^r\mathbf{a}_z = \int_{-\infty}^{\infty} \exp(-ik_z z)(\hat{\mathbf{E}}_t^r + \hat{E}_z^r\mathbf{a}_z)\, dk_z \quad (20)$$

[1] It is tedious but straightforward to show that when ϵ' is given by (1) and ω' by (18), K^2 reduces to $k_0^2(1 - \omega_p^2/\omega^2) - k_z^2$, which is independent of β.

$$\mathbf{H}^r = \mathbf{H}_t^r + H_z^r\mathbf{a}_z = \int_{-\infty}^{\infty} \exp(-ik_z z)(\hat{\mathbf{H}}_t^r + \hat{H}_z^r\mathbf{a}_z)\, dk_z \quad (21)$$

$$\hat{\mathbf{E}}_t^r = [ik_z/(k_z^2 - k_0^2)][\nabla_t\hat{E}_z^r + (\omega\mu_0/k_z)\nabla_t \times \hat{H}_z^r\mathbf{a}_z] \quad (22)$$

$$\hat{\mathbf{H}}_t^r = [ik_z/(k_z^2 - k_0^2)][\nabla_t\hat{H}_z^r - (\omega\epsilon_0/k_z)\nabla_t \times \hat{E}_z^r\mathbf{a}_z] \quad (23)$$

with

$$k_0^2 = \omega^2\mu_0\epsilon_0 \quad (24)$$

$$(\nabla_t^2 + K_0^2)\hat{E}_z^r = 0 \quad (25)$$

$$(\nabla_t^2 + K_0^2)\hat{H}_z^r = 0 \quad (26)$$

$$K_0^2 = k_0^2 - k_z^2. \quad (27)$$

At the boundary between the sheath and free space, the tangential components of the vectors $\mathbf{E} + \mathbf{v} \times \mathbf{B}$ and $\mathbf{H} - \mathbf{v} \times \mathbf{D}$ must be continuous. Thus at $r = r_0$

$$\hat{E}_z^s = \hat{E}_z^r \quad (28)$$

$$\hat{H}_z^s = \hat{H}_z^r \quad (29)$$

$$\hat{E}_\phi^s(1 + \beta b) + a\mu' v_s\hat{H}_r^s = \hat{E}_\phi^r + \mu_0 v_s\hat{H}_r^r \quad (30)$$

$$\hat{H}_\phi^s(1 + \beta b) - a\epsilon' v_s\hat{E}_r^s = \hat{H}_\phi^r - \epsilon_0 v_s\hat{E}_r^r. \quad (31)$$

Further, at $r = r_i$, the tangential components of the electric field must vanish except for points in the aperture. Denoting the tangential field components in the aperture E_z^a and E_ϕ^a, we have

$$\hat{E}_z^s = \hat{E}_z^a \quad (32)$$

$$\hat{E}_\phi^s = \hat{E}_\phi^a \quad (33)$$

at $r = r_i$, where

$$E_z^a = \int_{-\infty}^{\infty} \exp(-ik_z z)\hat{E}_z^a\, dk_z \quad (34)$$

and

$$E_\phi^a = \int_{-\infty}^{\infty} \exp(-ik_z z)\hat{E}_\phi^a\, dk_z. \quad (35)$$

The solutions of (15), (16), (25), and (26) appropriate to the problem under consideration are

$$\hat{E}_z^s = \sum_{n=-\infty}^{\infty} \exp(-in\phi)[A_n(k_z)J_n(Kr) + B_n(k_z)Y_n(Kr)] \quad (36)$$

$$\hat{H}_z^s = \sum_{n=-\infty}^{\infty} \exp(-in\phi)[C_n(k_z)J_n(Kr) + D_n(k_z)Y_n(Kr)] \quad (37)$$

$$\hat{E}_z^r = \sum_{n=-\infty}^{\infty} \exp(-in\phi)F_n(k_z)H_n^{(2)}(K_0r) \quad (38)$$

$$\hat{H}_z^r = \sum_{n=-\infty}^{\infty} \exp(-in\phi)G_n(k_z)H_n^{(2)}(K_0r). \quad (39)$$

TABLE I
ELEMENTS OF MATRIX OF COEFFICIENTS

p	a_p	b_p	c_p	d_p	f_p	g_p
1	$J_n(Kr_0)$	$Y_n(Kr_0)$	0	0	$-H_n^{(2)}(K_0r_0)$	0
2	0	0	$J_n(Kr_0)$	$Y_n(Kr_0)$	0	$-H_n^{(2)}(K_0r_0)$
3	$\frac{-in}{K^2r_0} J_n(Kr_0) \cdot(\beta k_0 - k_z)$	$\frac{-in}{K^2r_0} Y_n(Kr_0) \cdot(\beta k_0 - k_z)$	$\frac{\mu'}{K} J_n'(Kr_0) \cdot(\omega - v_z k_z)$	$\frac{\mu'}{K} Y_n'(Kr_0) \cdot(\omega - v_z k_z)$	$\frac{in}{K_0^2r_0} H_n^{(2)}(K_0r_0) \cdot(\beta k_0 - k_z)$	$\frac{-\mu_0}{K_0} H_n^{(2)\prime}(K_0r_0) \cdot(\omega - v_z k_z)$
4	$\frac{-\epsilon'}{K} J_n'(Kr_0) \cdot(\omega - v_z k_z)$	$\frac{-\epsilon'}{K} Y_n'(Kr_0) \cdot(\omega - v_z k_z)$	$\frac{-in}{K^2r_0} J_n(Kr_0) \cdot(\beta k_0 - k_z)$	$\frac{-in}{K^2r_0} Y_n(Kr_0) \cdot(\beta k_0 - k_z)$	$\frac{\epsilon_0}{K_0} H_n^{(2)\prime}(K_0r_0) \cdot(\omega - v_z k_z)$	$\frac{in}{K_0^2r_0} H_n^{(2)}(K_0r_0) \cdot(\beta k_z - k_z)$
5	$J_n(Kr_i)$	$Y_n(Kr_i)$	0	0	0	0
6	$\frac{-nd}{K^2r_ia} J_n(Kr_i)$	$\frac{-nd}{K^2r_ia} Y_n(Kr_i)$	$\frac{i\omega\mu'}{K} J_n'(Kr_i)$	$\frac{i\omega\mu'}{K} Y_n'(Kr_i)$	0	0

$J_n(\cdot)$ and $Y_n(\cdot)$ denote the Bessel and Neumann functions of order n, and $H_n^{(2)}(\cdot)$ denotes the Hankel function of the second kind of order n. A_n, B_n, C_n, D_n, F_n, and G_n are constants to be determined by the boundary conditions. It is noted that the forms chosen for $\hat{E}_z^r$ and $\hat{H}_z^r$ guarantee that the radiation condition at $r = \infty$ is satisfied.

Let $\hat{E}_z^a$ and $\hat{E}_\phi^a$ be written as follows:

$$\hat{E}_z^a = \sum_{n=-\infty}^{\infty} U_n \exp(-in\phi) \tag{40}$$

$$\hat{E}_\phi^a = \sum_{n=-\infty}^{\infty} W_n \exp(-in\phi) \tag{41}$$

in which

$$U_n = (1/2\pi)^2 \int_{ap}\int E_z^a \exp(in\phi' + ik_z z')\, d\phi'\, dz' \tag{42}$$

$$W_n = (1/2\pi)^2 \int_{ap}\int E_\phi^a \exp(in\phi' + ik_z z')\, d\phi'\, dz' \tag{43}$$

where the integrations are taken over the aperture. Then applying the boundary conditions (28)–(33), the following set of equations in the coefficients $A_n - G_n$ is obtained:

$$\begin{bmatrix} a_1 & b_1 & 0 & 0 & f_1 & 0 \\ 0 & 0 & c_2 & d_2 & 0 & g_2 \\ a_3 & b_3 & c_3 & d_3 & f_3 & g_3 \\ a_4 & b_4 & c_4 & d_4 & f_4 & g_4 \\ a_5 & b_5 & 0 & 0 & 0 & 0 \\ a_6 & b_6 & c_6 & d_6 & 0 & 0 \end{bmatrix} \begin{bmatrix} A_n \\ B_n \\ C_n \\ D_n \\ F_n \\ G_n \end{bmatrix} = \begin{bmatrix} 0 \\ 0 \\ 0 \\ 0 \\ U_n \\ W_n \end{bmatrix}. \tag{44}$$

The elements of the matrix of coefficients in (44) are given in Table I. The primes denote differentiation with respect to the argument. Equation (44) constitutes the formal solution to the problem, since when the coefficients $A_n - G_n$ are found, all field components in both regions of the problem may be determined.

The integrals with respect to k_z may easily be evaluated for the radiation field outside the sheath by the usual saddle-point technique. The integrals which appear are of the form

$$M_n = \int_{-\infty}^{\infty} S_n(k_z) H_n^{(2)}(\sqrt{k_0^2 - k_z^2}\, r) \exp(-ik_z z)\, dk_z \tag{45}$$

and

$$N_n = \int_{-\infty}^{\infty} T_n(k_z)(d/dr) H_n^{(2)}(\sqrt{k_0^2 - k_z^2}\, r) \exp(-ik_z z)\, dk_z \tag{46}$$

whose asymptotic values are, respectively,

$$M_{na} = 2i \exp[-i(k_0 R - \tfrac{1}{2}n\pi)] S_n(k_0 \cos\theta)/R \tag{47}$$

and

$$N_{na} = 2k_0 \exp[-i(k_0 R - \tfrac{1}{2}n\pi)] T_n(k_0 \cos\theta) \sin\theta/R \tag{48}$$

in the spherical coordinates (R,θ,ϕ).

In Sections III and IV, these results are applied to the uniformly excited circumferential slot and axial slot apertures.

III. THE CIRCUMFERENTIAL SLOT

In this section it is assumed that

$$E_z^a = \delta(z) \tag{49}$$

$$E_\phi^a = 0 \tag{50}$$

so that the aperture is a very thin circumferential slot across which a voltage $\exp(i\omega t)$ is applied. For this case,

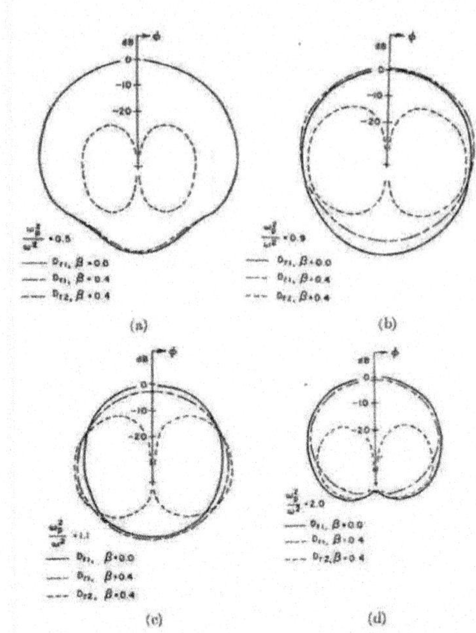

Fig. 4. Radiation patterns of axially slotted cylinder for $k_0 r_i = 2.0$, $k_0 r_0 = 3.0$; $\omega_p^2/\omega^2 = 0.5, 0.9, 1.1, 2.0$; $\beta = 0.0, 0.4$.

increases further, the shape of the pattern changes little from that for $\beta = 0.2$. When $\omega_p^2/\omega^2 = 1.1$, a lobe appears near $\theta = \pi$ and moves in the direction of decreasing θ, weakening as β increases. For values of β greater than 0.2, the pattern changes relatively little, the dip in the pattern weakening and moving toward $\theta = 0$. As in the case $\omega_p^2/\omega^2 = 0.9$, the patterns for $\beta \neq 0$ can differ from that for $\beta = 0$ by 30 dB or more, even for relatively small values of β.

IV. The Axial Slot

In this section, it is assumed that

$$E_z^a = 0 \tag{55}$$

$$E_\phi^a = (1/r_i)\delta(\phi) \tag{56}$$

so that the aperture is a very thin slot cut longitudinally in the cylinder, across which a voltage $\exp(i\omega t)$ is applied. The excitation, boundary conditions, and fields are independent of z.

When $\beta = 0$, the only field components present are E_ϕ, E_r, and H_z; however, when $\beta \neq 0$, all six field components are present. Accordingly, we consider two radiation patterns:

$$D_{r1} = 2\pi r\eta_0 \,|\, H_z^r \,|^2 = \frac{8\eta_0}{K_0\pi} \left| \sum_{n=-\infty}^{\infty} \exp(-in\phi) i^{-n} G_n(0) \right|^2 \tag{57}$$

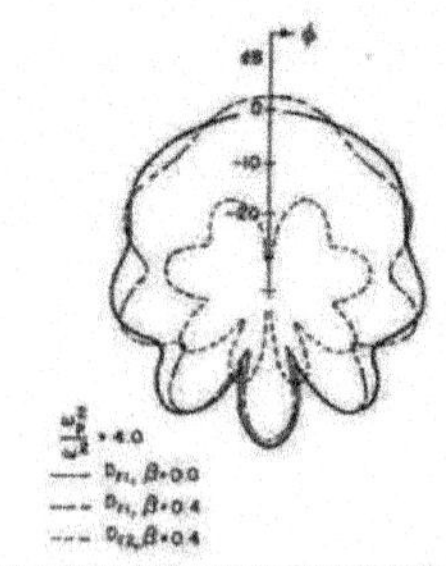

Fig. 5. Radiation patterns of axially slotted cylinder for $k_0 r_i = 2.0$, $k_0 r_0 = 3.0$; $\omega_p^2/\omega^2 = 4.0$; $\beta = 0.0, 0.4$.

and

$$D_{r2} = \frac{2\pi r}{\eta_0} \,|\, E_z^r \,|^2 = \frac{8}{K_0 \eta_0 \pi} \left| \sum_{n=-\infty}^{\infty} \exp(-in\phi) i^{-n} F_n(0) \right|^2 \tag{58}$$

in which D_{r1} is the pattern which is present for $\beta = 0$; the "cross-polarized" pattern D_{r2} is present only when $\beta \neq 0$. $F_n(0)$ and $G_n(0)$ are given by

$$F_n(0) = \frac{\{-1/[(k_0 r_0)(K r_0) r_i]\} F_a H_n^{(2)}(k_0 r_0) \cdot (\eta\beta/\pi^2)(1 - k_0^2/K^2)}{D} \tag{59}$$

and

$$G_n(0) = \frac{[1/i\pi^2\eta_0(k_0 r_0) r_i] \cdot [F_a(k_0/K) H_n^{(2)\prime}(k_0 r_0) - F_d(k^2/K^2) H_n^{(2)}(k_0 r_0)]}{D} \tag{60}$$

in which

$$\begin{aligned} D = {} & [(K/k_0) F_b H_n^{(2)\prime}(k_0 r_0) + F_c H_n^{(2)}(k_0 r_0)] \\ & \cdot [F_a(k_0/K) H_n^{(2)\prime}(k_0 r_0) - F_d(k^2/K^2) H_n^{(2)}(k_0 r_0)] \\ & - (\eta\beta/k_0 r_0)^2 (1 - k_0^2/K^2)^2 F_a F_b [H_n^{(2)}(k_0 r_0)]^2 \end{aligned} \tag{61}$$

with

$$F_a = J_n(K r_i) Y_n(K r_0) - J_n(K r_0) Y_n(K r_i) \tag{62}$$

$$F_b = J_n(K r_0) Y_n'(K r_i) - J_n'(K r_i) Y_n(K r_0) \tag{63}$$

$$F_c = J_n'(K r_i) Y_n'(K r_0) - J_n'(K r_0) Y_n'(K r_i) \tag{64}$$

$$F_d = J_n(K r_i) Y_n'(K r_0) - J_n'(K r_0) Y_n(K r_i). \tag{65}$$

When β vanishes, $F_n(0)$ vanishes and $G_n(0)$ becomes simply

$$G_n(0)\,|_{\beta=0} = [i\pi^2\eta_0 r_i(k_0 r_0)]^{-1} \cdot [(K/k_0) F_b H_n^{(2)\prime}(k_0 r_0) + F_c H_n^{(2)}(k_0 r_0)]^{-1}. \tag{66}$$

Numerical computation of these radiation patterns has been carried out for $k_0 r_i = 2.0$, $k_0 r_0 = 3.0$, and various values of ω_p^2/ω^2 and β. The results are shown in Figs.

TABLE II

Amount in Decibels
(peak values of D_{r1} exceed those of D_{r2})

ω_p^2/ω^2 \ β	0.1	0.2	0.3	0.4	0.5	0.6	0.7	0.8	0.9
0.25	38.3	32.3	28.8	26.4	24.6	23.1	21.9	20.9	20.1
0.50	29.7	23.8	20.6	18.5	16.9	15.7	14.7	14.0	13.5
0.90	13.7	8.9	7.0	6.4	6.4	6.5	6.8	7.2	7.6
1.1	15.6	8.9	4.5	2.2	2.0	3.0	5.1	5.1	6.0
1.5	23.8	17.5	13.6	10.4	7.7	5.5	4.3	4.1	4.8
2.0	26.2	20.1	16.4	13.5	11.0	8.7	6.7	5.3	5.0

4 and 5 and in Table II. The zero-decibel level in the figures is that for $\phi = 0$ and $\beta = 0$.

Radiation patterns for the cases $\omega_p^2/\omega^2 = 0.5, 0.9, 1.1$, and 2.0 and $\beta = 0.0$ and 0.4 are shown in Fig. 4. For these values of ω_p^2/ω^2 as well as smaller values for which the patterns are not shown, the patterns are uncomplicated functions of ϕ. It is evident from these figures that the motion of the sheath has a relatively minor effect on the pattern D_{r1} and that this small effect becomes more pronounced as ω_p^2/ω^2 is increased. More noticeable is the introduction of the cross-polarized field whose pattern is D_{r2} as β is increased from zero. This cross-polarized field radiates most strongly off-broadside and vanishes for $\phi = 0$ or π. One will notice in Fig. 4 in the patterns for $\beta = 0.4$ that D_{r2} is largest in comparison with D_{r1} for the values of ω_p^2/ω^2 nearest unity. That this is so for other values of β is illustrated in Table II, in which are listed for various values of ω_p^2/ω^2 and β the amounts in decibels by which the peak values of D_{r1} exceed those of D_{r2}. It is evident from the table that even for small values of β, D_{r2} may be significant with respect to D_{r1} if ω_p^2/ω^2 is near unity.

Radiation patterns for the case $\omega_p^2/\omega^2 = 4.0$, $\beta = 0.0$ and 0.4 are shown in Fig. 5. The effect of the motion on the pattern D_{r1} is to cause the lobes in the pattern for $\beta = 0$ to shift toward $\phi = \pi$ and to decrease in magnitude. For larger values of β than are shown, this trend continues until the lobes are concentrated near $\phi = \pi$ and are of extremely small magnitude. The remainder of the pattern D_{r1} is smooth, with a peak value at $\phi = 0$ near 0 dB for $\beta = 0.8$. The pattern D_{r2} has maxima where D_{r1} has minima, and for larger values of β the lobes in D_{r2} move toward $\phi = \pi$ and decrease in magnitude until the pattern D_{r2} is a smooth function of ϕ, with peak values near -10 dB at $\phi = \pm 50°$ for $\beta = 0.8$.

V. Summary

The formal solution to the problem of determining the radiation from an aperture in a conducting cylinder surrounded by a moving isotropic plasma sheath has been found. Numerical results illustrating the behavior of the radiation patterns of circumferential and axial slots were presented. It was found that the motion of the sheath causes changes in the shape of the radiation pattern of a circumferential slot and introduces a cross-polarized electromagnetic field into the radiation from an axial slot. In both cases, it was found that when the wave frequency and plasma frequency are nearly equal, significant pattern alteration can occur as a result of the sheath motion even if the speed is relatively small.

Acknowledgment

The authors wish to express their thanks to R. Kahler for his assistance in the preparation of the figures.

References

[1] C. M. Knop, "The radiation fields from a circumferential slot on a metal cylinder coated with a lossy dielectric," *IRE Trans. Antennas and Propagation*, vol. AP-9, pp. 535–545, November 1961.
[2] C. T. Swift, "Radiation patterns of a slotted-cylinder antenna in the presence of an inhomogeneous lossy plasma," *IEEE Trans. Antennas and Propagation*, vol. AP-12, pp. 728–738, November 1964.
[3] C. Yeh, "Electromagnetic radiation from an arbitrary slot on a conducting cylinder coated with a uniform cold plasma sheath with an axial static magnetic field," *Can. J. Phys.*, vol. 42, pp. 1369–1381, July 1964.
[4] C. Yeh and W. V. T. Rusch, "Radiation through an inhomogeneous magnetized plasma sheath," *IEEE Trans. Antennas and Propagation* (Communications), vol. AP-15, p. 328, March 1967.
[5] C. Yeh and K. F. Casey, "Reflection and transmission of electromagnetic waves by a moving dielectric slab," *Phys. Rev.*, vol. 144, pp. 665–669, April 15, 1966.
[6] J. A. Kong and D. K. Cheng, "On guided waves in moving anisotropic media," *IEEE Trans. Microwave Theory and Techniques*, vol. MTT-16, pp. 99–103, February 1968.
[7] G. Meltz, "Radiation from a homogeneous plasma-covered slotted cylinder," Sperry-Rand Research Center, Rept. SRRC RR-64-23, March 1964.

Reprinted by permission from IEEE TRANSACTIONS ON ANTENNAS AND PROPAGATION
Vol. AP-17, No. 6, November 1969
pp. 757-762

Reprinted from

Volume 9, Number 5 APPLIED PHYSICS LETTERS 1 September 1966

A PROPOSED METHOD OF SHIFTING THE FREQUENCY OF LIGHT WAVES*

C. Yeh
Electrical Engineering Department
University of Southern California
Los Angeles, California
(Received 27 June 1966)

A simple method of shifting the frequency of light waves with precision, based on the Doppler principle, is proposed. The proposed system consists of two reflecting mirrors. One mirror is assumed to be moving with a velocity v while the other mirror is assumed to be stationary. Practical limitations such as mirror loss, beam spread, etc., are discussed. A numerical example is also given.

The difficulty of tuning the frequency of a laser beam is well known.[1] The purpose of this Letter is to propose a simple way of shifting the frequency of light waves with precision. The proposed method is based on the Doppler principle.[2] It is known that the frequency of the reflected wave from a perfectly reflecting mirror moving towards a normally incident plane wave is shifted according to the following equation

$$\omega^{(r)} = \omega_0 \left(\frac{1+\beta}{1-\beta}\right) \tag{1}$$

where $\omega^{(r)}$ and ω_0 are, respectively, the frequency for the reflected wave and for the incident wave, and $\beta = v/c$ in which c is the velocity of light in vacuum and v is the velocity of the moving mirror.

The proposed system is shown in Fig. 1. Mirror A is moving with a velocity v, and mirror B is assumed to be stationary. The distance between mirror A and B is l(cm) at $t = 0$. Mirror A travels a distance of d(cm) in t_d sec (i.e., $t_d = d/v$.) In t_d sec, the plane electromagnetic wave could have travelled a distance of b(cm) $= ct_d$. Let us assume that the mean distance between mirror A and mirror B during this time period t_d sec, is $l - d/2$. Then the number of bounces that the plane wave makes within the time period t_d is

$$n = \frac{b}{2\left(l-\frac{d}{2}\right)} = \frac{ct_d}{2\left(l-\frac{d}{2}\right)} = \frac{c}{v}\,\frac{d}{2\left(l-\frac{d}{2}\right)}\,.$$

The resultant frequency after n bounces is

$$\omega^{(r)} = \omega_0\left(\frac{1+\beta}{1-\beta}\right)^n_{n\beta\ll 1} \simeq \omega_0(1+2n\beta) = \omega_0 + \frac{d}{\left(l-\frac{d}{2}\right)}\,\omega_0 = \omega_0 + \Delta\omega\,. \tag{2}$$

Figure 2 is introduced to summarize the relationship between the frequency shift ($\Delta\omega/\omega_0$), the number of bounces (n), the length ratio (d/l), and the velocity of the moving mirror.

In order that the above proposed technique of shifting the frequency of light waves will be practical, the following physical limitations must be discussed:

(*a*) Mirror Loss. The number of bounces is limited by the mirror reflectivity. If R_0 is the reflection coefficient of the mirror, then the equivalent reflection coefficient, after n round trips, is R_0^{2n} where the factor 2 is due to the presence of two mirrors (i.e., mirrors A and B). It can be seen that for larger n, the reflectance of the mirror is very critical. For a typical mirror with reflectance of 99.7% (ref. 3), the resultant signal, after 1000 round trips, is 52 dB down from the initial signal.

(*b*) Pulse Width. According to Eq. (2), we have

$$\frac{\Delta\omega}{\omega_0} = \frac{d}{\left(l-\frac{d}{2}\right)}\,. \tag{3}$$

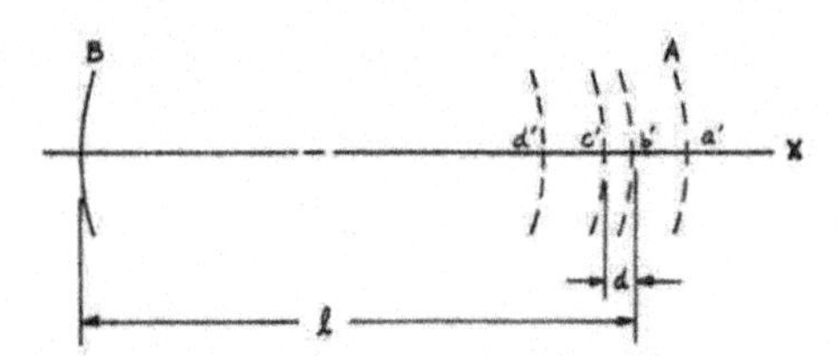

Fig. 1. Schematic diagram of the moving mirror system. Mirror B is fixed. Mirror A oscillates between position a' and d'. Mirror A accelerates from position a' to position b', coasts with a constant velocity v from position b' to position c', decelerates from position c' to position d', and then returns to position a'. Incident light pulse strikes mirror A when it is located at position b' moving towards position c'.

*Supported by the Joint Services Electronics Program.

Volume 9, Number 5 APPLIED PHYSICS LETTERS 1 September 1966

Knowing the frequency shift ($\Delta\omega/\omega_0$) and the initial length of the cavity (l), d is then determined. d is the distance travelled by the moving mirror A. The pulse width of an incident wave must be less than or equal to $t_d/2$ where $t_d = d/v$ and v is the velocity of the moving mirror. Since it is most likely that in practice mirror A would undergo oscillatory type of motion, the next incident pulse must be synchronized such that the beginning of this pulse will bounce off mirror A when the mirror is located at b' and moving towards c'. (See Fig. 1.)

(c) Mode Matching and Cross-Talk. Similar techniques, as used in the design of the folded optical delay line[4] to overcome the beam spread problem as well as the problem of minimizing the cross-talk between the original intense beam and the weaker output beam, may be used. Detailed descriptions of the techniques have been given in ref. 4. The movement of mirror A will only enhance beam separation, hence minimizing cross-talk. The problem of inlet and exit of the light beam may also be handled in a similar manner as the folded optical delay line case.[4]

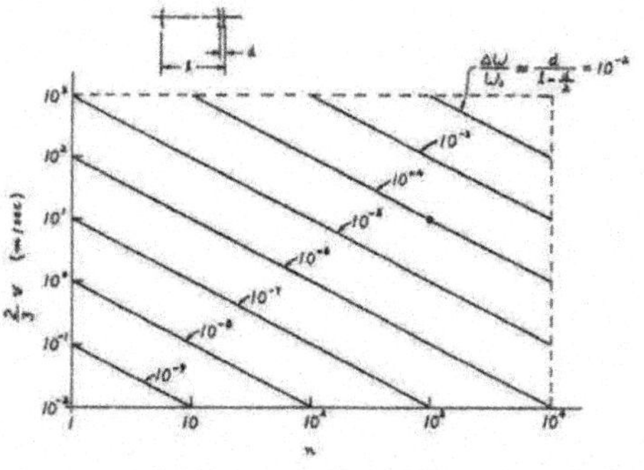

Fig. 2. The required velocity of the moving mirror (v) as a function of the required number of bounces (n) of the input light wave to achieve the necessary frequency shift ($\Delta\omega/\omega_0$). The upper bound on v, as shown by the dashed line, is arbitrarily taken to be the maximum velocity that may be obtained in the laboratory. The upper bound on n, as shown by the dashed line, is determined mainly by the reflectivity of the mirrors. It is arbitrarily chosen to be 10^4. After 10^4 bounces, the signal is diminished by −520 dB provided that the reflectivity of the mirror is 99.7%. The dot indicates the operating point for the numerical example given in the text.

(d) Vibration of the Moving Mirror. It is expected that the frequency of vibration of the moving mirror will be much less than $1/t_d$; therefore during the interaction period (t_d), the vibration of mirror A will not matter.

(e) Frequency Shift for Obliquely Incident Waves. In order to have proper beam separation, the direction of the incident beam must not be exactly normal to the mirror surface. The resultant frequency after n bounces is then[5]

$$\omega^{(r)} = \omega_0 \left(\frac{1+2\beta\cos\theta_0+\beta^2}{1-\beta^2}\right)^n_{n\beta\ll 1} \approx \omega_0(1+2n\beta\cos\theta_0),$$

where θ_0 is the angle of incidence.

In spite of the above limitations, it is believed that the proposed technique of shifting the frequency of light waves is workable. To have an idea of the order of magnitude of various quantities discussed above, the following numerical example is introduced:

- l (Initial mirror separation) = 1 m
- ω_0 (Incident laser beam frequency) = $2\pi \times 3/5 \times 10^{15}$ cps (~5000 Å)
- $\Delta\omega$ (Desired frequency shift) = $2\pi \times 3/5 \times 10^{11}$ cps (~5 mm)
- v (Velocity of the moving mirror) = 15 m/sec
- d (Distance travelled by the moving mirror with constant v) ≈ .1 mm. Maximum pulse width ≈ 1/30 msec
- n (Number of round trips) = 1000
- R (Resultant signal strength with mirror reflectivity of 99.7%) ≈ −52 dB.

It is noted that for the present example the required velocity of the moving mirror is very slow. A factor of 10 increase in speed can easily be obtained; increasing the speed by a factor of 10 means that the desired frequency shift can also be increased by a factor of 10 provided that the number of round trips remains the same.

[1] K. Tomiyasu, "Laser Bibliography, III," *IEEE J. Quantum Electronics* **QE-2**, No. 6, 1966.
[2] C. H. Papas, *Theory of Electromagnetic Wave Propagation*, (McGraw-Hill Book Company, New York, (1965).
[3] D. L. Perry, *Appl. Opt.* **4**, 987 (1965).
[4] D. R. Herriott and H. J. Schulte, *Appl. Opt.* **4**, 883 (1965).
[5] C. Yeh, *J. Appl. Phys.* **36**, 3513 (1965).

7

CLASSICAL RADIATION FROM MOVING PARTICLES

Electromagnetic radiation can be generated classically or quantum mechanically. Radiation can be generated classically in one of five ways, when a charged particle: (1) accelerates or decelerates (Bremsstrahlung), (2) moves along a curved path at a constant velocity (cyclotron radiation), (3) moves at a constant velocity that is faster than the phase velocity of light in the medium (Čerenkov radiation), (4) moves at a uniform velocity along an uneven surface (Smith-Purcell radiation), or (5) moves through two media with different electrical properties (transition radiation).

We are interested in the classical radiation that can be generated when a charged particle moves inside a nonuniform dielectric medium or moves near or through a dielectric body. Problems were treated in the following papers.

7.1 Čerenkov radiation in inhomogeneous periodic media. (Paper 7-1)

The formal exact solution to the problem of the radiation of a charged particle traveling with a constant velocity in a periodically inhomogeneous medium is obtained. As a specific example, the case with a sinusoidally varying dielectric profile is treated in detail. Results of the computation are summarized in two graphs from which information concerning threshold velocity for a particular mode, the emission angles for various radiating modes, and the cutoff frequency for certain modes can be found. Unlike the case of Čerenkov radiation in a homogeneous medium, there exist radiating modes in this inhomogeneous dielectric case even when the velocity of the charged particle is below the threshold Čerenkov velocity. A formal expression for the radiation spectrum is also given. Approximate expressions for the radiated fields and for the radiation spectrum are obtained when the variation of the permittivity is small. Results are discussed and interpreted.

7.2 Transition radiation in a periodically stratified plasma. (Paper 7-2)

The solution to the problem of determining the transition radiation emitted when a charged particle moves uniformly in a periodically stratified cold isotropic plasma is obtained. The electromagnetic field may be expressed in terms of an infinite number of normal modes. The conditions under which some of these modes are radiative are discussed, and expressions for the field components and the energy spectral density are obtained. It is found that radiation may be emitted for arbitrarily small particle velocities and that the strongest emission takes place in a frequency band just above the average plasma frequency. Numerical results are presented to illustrate the behavior of the spectra of the various radiative modes as the frequency and plasma parameters are varied.

7.3 Diffraction radiation from a charged particle moving through a penetrable sphere. (Paper 7-3)

The formal exact solution to the problem of the radiation of a charged particle traveling with a constant velocity through a dielectric sphere is obtained. The electromagnetic field may be expressed in terms of an infinite number of normal modes. It is found that radiation may be emitted for arbitrarily small particle velocities. For larger spheres (i.e., $ka \gg 1$ when k is the wave number and a is the radius of the sphere), the radiated field is predominantly Čerenkov-type radiation when the particle is above the Čerenkov threshold velocity. For small spheres ($ka \ll 1$) and low particle velocities, the radiation is shown to be mainly of the transition type. Numerical results are presented to illustrate the behavior of the spectra of the various lower-order radiative modes as the velocity of the particle is varied.

7.4 On the Čerenkov threshold associated with synchrotron radiation in a dielectric medium. (Paper 7-4)

A formal exact expression for the radiation field of a charged particle executing a circular orbit in a dispersive dielectric medium is obtained by means of a dyadic Green's function approach. The field is expressed as an infinite sum of vector spherical harmonics. The radiated power P is computed numerically, and a set of universal curves of $nP/q^2\Omega^2$ vs $n\beta$ is obtained where q is the charge on the particle, n is the index of refraction of the medium, Ω is the orbital angular velocity of the particle, and $\beta = v/c$, where v is the particle velocity and c is the speed of light in vacuum. The Čerenkov threshold phenomenon is shown to be manifest primarily in the high-frequency portion of the radiated spectrum. For an index refraction of unity, the result is shown (at low velocities) to be identical with both

Larmor's formula for the radiated power and Schwinger's result for the angular distribution of the radiation.

7.5 Synchrotron-diffraction radiation spectra in the presence of a penetrable sphere. (Paper 7-5)

The radiation from a charged particle orbiting both inside and outside a dielectric sphere is expressed as a series of vector spherical harmonics. The results of numerical computation of the radiated spectra for various situations are presented and discussed. When the particle orbits far outside the sphere, the interaction between the two is weak and the radiation is of the synchrotron type. The radiated power is shown to decrease with increasing sphere size in this situation; however, when the particle orbits close to or inside the sphere, the interaction is very strong; this being characterized by the appearance of resonances of the sphere in the radiated spectrum. The resonances are shown to be responsible for the power in a particular harmonic being radiated almost entirely into a single spherical harmonic in some instances.

Reprinted from THE PHYSICAL REVIEW, Vol. 140, No. 3B, B768–B775, 8 November 1965
Printed in U. S. A.

Čerenkov Radiation in Inhomogeneous Periodic Media*

K. F. CASEY,† C. YEH, AND Z. A. KAPRIELIAN
Electrical Engineering Department, University of Southern California, Los Angeles, California
(Received 4 June 1965)

The formal exact solution to the problem of the radiation of a charged particle traveling with a constant velocity in a periodically inhomogeneous medium is obtained. As a specific example, the case with a sinusoidally varying dielectric profile is treated in detail. Results of the computation are summarized in two graphs from which information concerning threshold velocity for a particular mode, the emission angles for various radiating modes, and the cutoff frequency for a certain mode can be found. Unlike the case of Čerenkov radiation in a homogeneous medium, there exist radiating modes in this inhomogeneous-dielectric case even when the velocity of the charged particle is below the threshold Čerenkov velocity. A formal expression for the radiation spectrum is also given. Approximate expressions for the radiated fields and for the radiation spectrum are obtained when the variation of the permittivity is small. Results are discussed and interpreted.

I. INTRODUCTION

IF the velocity of electrons traveling in a dielectric medium is higher than the phase velocity of light in the medium, radiation is observed. This is the well-known Čerenkov effect.[1] The theoretical analysis of the Čerenkov effect was first obtained by Frank and Tamm,[2] who treated the problem of the radiation from an electron moving uniformly in a homogeneous dielectric medium. Extension of their analysis to anisotropic and dispersive media has been carried out by various authors.[3] The problem of the emission from a particle traversing a piecewise homogeneous dielectric medium

* This work was supported by the Joint Services Electronics Program (U. S. Army, U. S. Navy, and U. S. Air Force) under Grant No. AF-AFOSR-496-65.

† Present address: U. S. Air Force Institute of Technology, Wright-Patterson Air Force Base, Dayton, Ohio.

[1] P. A. Čerenkov, Phys. Rev. 52, 378 (1937).

[2] I. M. Frank and I. Tamm, Dokl. Akad. Nauk. S.S.S.R. 14, 109 (1937).

[3] J. V. Jelley, *Čerenkov Radiation and its Applications* (Pergamon Press, New York, 1958).

was considered by Fainberg and Khiznyak[4] and by Garybyan.[5] However, very little work has been carried out on the problem of the radiation from an electron moving uniformly in a continuously inhomogeneous periodic dielectric medium. It is expected that if the wavelength of the emitted radiation is much smaller than the period of the inhomogeneity, the Čerenkov radiation would depend to a great extent upon the permittivity in the immediate neighborhood of the particle, which may be assumed nearly constant. The WKB method was used by Ter-Mikaelyan[6] to investigate this problem. On the other hand, if the wavelength of the emitted radiation is much greater than the period of the dielectric variation, the Čerenkov radiation will depend primarily upon the average value of the permittivity of the medium. At wavelength comparable to the period of the medium, the behavior of the radiation is not readily apparent.

It is therefore the purpose of this paper to treat this problem. A formal solution will be obtained for the spectral density of the radiation emitted by a charged-particle traveling uniformly in a dielectric medium which is periodically and continuously inhomogeneous in the direction along the particle path. Detailed analysis is carried out for a specific dielectric variation, i.e.,

$$\epsilon(z)=\epsilon_0[1-\delta\cos(2\pi z/p)],$$

where ϵ_0 is the average value of the dielectric constant, δ is the magnitude of the variation and p is the periodicity of the variation. If the magnitude of the dielectric variation is small, approximate analytic results can be found. It is noted that owing to the inhomogeneity of the medium, there exists not only Čerenkov-type radiation by also transition-type radiation as well.

II. FORMULATION OF THE PROBLEM

It is assumed that an inhomogeneous dielectric medium fills the entire space and possesses a relative permittivity

$$\epsilon(z)/\epsilon_0=\epsilon(z+p)/\epsilon_0, \tag{1}$$

a relative permeability

$$\mu/\mu_0=1, \tag{2}$$

and a conductivity $\sigma=0$, in the (x,y,z) rectangular coordinate system. p is the periodicity of the dielectric variation, ϵ_0 and μ_0 are, respectively, the free-space permeability and $\epsilon(z)$ is an analytic function. A charged particle is moving in the z direction through this medium at a constant speed v. Denoting the charge by q, one has for the current density $\mathbf{J}$ due to the passage of this particle

$$\mathbf{J}=(q/2\pi\rho)\delta(\rho)\delta(t-z/v)\mathbf{e}_z, \tag{3}$$

where $\mathbf{e}_z$ is a unit vector in the z direction, $\rho^2=x^2+y^2$, and δ is the Dirac delta function. Maxwell's equations for the present situation are

$$\nabla\times\mathbf{E}=-\mu\partial\mathbf{H}/\partial t \tag{4}$$

$$\nabla\times\mathbf{H}=\epsilon(z)\frac{\partial\mathbf{E}}{\partial t}+\frac{q}{2\pi\rho}\delta(\rho)\delta\left(t-\frac{z}{v}\right)\mathbf{e}_z, \tag{5}$$

where $\mathbf{E}$ and $\mathbf{H}$ are, respectively, the electric and magnetic fields. Taking the Fourier transformation with respect to time of Eqs. (4) and (5) gives

$$\nabla\times\mathcal{E}=i\omega\mu\mathcal{H} \tag{6}$$

$$\nabla\times\mathcal{H}=-i\omega\epsilon(z)\mathcal{E}+(q/4\pi^2\rho)\delta(\rho)e^{i\omega z/v}\mathbf{e}_z. \tag{7}$$

$\mathcal{E}$ and $\mathcal{H}$ denote the Fourier transforms of $\mathbf{E}$ and $\mathbf{H}$, respectively; they are related to $\mathbf{E}$ and $\mathbf{H}$ by

$$\mathbf{E}=\int_{-\infty}^{\infty}\mathcal{E}e^{-i\omega t}d\omega, \tag{8}$$

$$\mathbf{H}=\int_{-\infty}^{\infty}\mathcal{H}e^{-i\omega t}d\omega. \tag{9}$$

To solve Eqs. (6) and (7) for the region $\rho>0$, let us introduce a scalar function $\psi(\rho,z)$ as follows:

$$\mathcal{H}=\nabla\times[\psi(\rho,z)\mathbf{e}_z], \tag{10}$$

$$\mathcal{E}=\frac{i}{\omega\epsilon(z)}\nabla\times\nabla\times[\psi(\rho,z)\mathbf{e}_z]. \tag{11}$$

$\psi(\rho,z)$ satisfies the following partial differential equation:

$$\frac{1}{\rho}\frac{\partial}{\partial\rho}\left[\rho\frac{\partial\psi}{\partial\rho}\right]-\frac{1}{\epsilon}\frac{d\epsilon}{dz}\frac{d\psi}{dz}+\frac{\partial^2\psi}{\partial z^2}+\omega^2\mu\rho\psi=0. \tag{12}$$

This method of solving the vector wave equation for an inhomogeneous medium is a modified version of the vector wave-function method of Hansen[7] and Stratton.[8] ψ must also satisfy the following boundary conditions:

(a) the Sommerfeld radiation condition at $\rho=\infty$,
(b) Ampere's law at $\rho=0$, i.e.,

$$\lim_{\rho\to 0}2\pi\rho\frac{\partial\psi}{\partial\rho}=-\frac{q}{2\pi}e^{i\omega z/v}. \tag{13}$$

III. FORMAL SOLUTION FOR ψ

The appropriate solution for ψ can be obtained from Eq. (12) using the method of separation of variables:

$$\psi(\rho,z)=H_0^{(1)}(\gamma\rho)Z(z), \tag{14}$$

[4] Ya. Fainberg and N. Khiznyak, Zh. Eksperim. i Teor. Fiz. 32, 883 (1957) [English transl.: Soviet Phys.—JETP 5, 720 (1957)].
[5] G. Garybyan, Zh. Eksperim. i Teor. Fiz. 35, 1435 (1958) [English transl.: Soviet Phys.—JETP 8, 1003 (1959)].
[6] M. Ter-Mikaelyan, Dokl. Akad. Nauk SSSR 134, 318 (1960) [English transl.: Soviet Phys.—Doklady 5, 1015 (1960)].
[7] W. W. Hansen, Phys. Rev. 47, 139 (1935).
[8] J. A. Stratton, *Electromagnetic Theory* (McGraw-Hill Book Company, Inc., New York, 1941).

where $H_0^{(1)}(\gamma\rho)$ is the Hankel function of the first kind of order zero, γ^2 is a separation constant, and $Z(z)$ satisfies the following differential equation:

$$\frac{d^2Z}{dz^2}-\frac{1}{\epsilon}\frac{d\epsilon}{dz}\frac{dZ}{dz}+(\omega^2\mu_0\epsilon-\gamma^2)Z=0. \qquad (15)$$

In order that the radiation condition at infinity can be satisfied, both the real and imaginary parts of γ must be positive.

The solution of Eq. (15) depends, of course, on the dielectric variation, $\epsilon(z)$. Making the substitution

$$\xi=\pi z/p, \qquad (16)$$

$$Z(\xi)=\epsilon^{1/2}(\xi)u(\xi), \qquad (17)$$

one has

$$\frac{d^2u}{d\xi^2}+\left[\frac{1}{2\epsilon}\frac{d^2\epsilon}{d\xi^2}-\frac{3}{4\epsilon^2}\left(\frac{d\epsilon}{d\xi}\right)^2+(\omega^2\mu_0\epsilon-\gamma^2)\frac{p^2}{\pi^2}\right]u=0. \qquad (18)$$

If one assumes that $\epsilon(\xi)$ is an even-periodic function of period π, the function in the brackets in Eq. (18) may be represented in terms of a Fourier cosine series. Hence, (18) becomes

$$\frac{d^2u}{d\xi^2}+[\theta_0+2\sum_{n=1}^{\infty}\theta_n\cos 2n\xi]u=0 \qquad (19)$$

which is the canonical form of Hill's equation. The formal solutions of Eq. (19) can be obtained with the help of Floquet's theorem.[9,10] They are

$$u(\xi)=he_\beta^{(1,2)}(\xi)=e^{\pm i\beta\xi}\sum_{n=-\infty}^{\infty}b_n(\beta)e^{\pm 2ni\xi}, \qquad (20)$$

where the superscripts (1) and (2) refer, respectively, to the + and the − signs on the right-hand side of this equation. β are the roots of the following characteristic equation:

$$\sin^2(\tfrac{1}{2}\pi\beta)=\Delta(0)\sin^2(\tfrac{1}{2}\pi\sqrt{\theta_0}) \qquad (21)$$

in which $\Delta(0)$ is an infinite determinant whose elements are

$$\Delta(0)_{mm}=1,$$

$$\Delta(0)_{mn}=\theta_{m-n}/(\theta_0-4m^2) \quad (m\neq n). \qquad (22)$$

The coefficients $b_n(\beta)$ are determined from the recurrence relation

$$-(\beta+2n)^2b_n(\beta)+\sum_{m=-\infty}^{\infty}\theta_m b_{n-m}(\beta)=0. \qquad (23)$$

The most general expression for ψ that satisfies the radiation condition at $\rho=\infty$ is

$$\psi=\epsilon^{1/2}(\xi)\int_{-\infty}^{\infty}F(\beta)he_\beta^{(1)}(\xi)H_0^{(1)}(\gamma\rho)d\beta, \qquad (24)$$

where $F(\beta)$ is an arbitrary function of β. Substituting Eq. (24) into the required boundary condition (13) at $\rho=0$ gives,

$$\epsilon^{1/2}(z)\int_{-\infty}^{\infty}\left(\frac{4}{i}\right)F(\beta)he_\beta^{(1)}(\pi z/p)d\beta=\frac{q}{2\pi}e^{i\omega z/v}. \qquad (25)$$

Division of Eq. (25) by $(4/i)\epsilon^{1/2}(z)e^{i\omega z/v}$ yields

$$\int_{-\infty}^{\infty}F(\beta)he_\beta^{(1)}(\pi z/p)e^{-i\omega z/v}d\beta=\frac{iq}{8\pi}\epsilon^{-1/2}(z). \qquad (26)$$

Since the right-hand side of (26) is periodic with period p, the left-hand side must also be periodic with period p. Hence, substitution of $(z+p)$ for z in (26) yields

$$\int_{-\infty}^{\infty}F(\beta)he_\beta^{(1)}\left(\frac{\pi z}{p}\right)e^{i\beta\pi-i\omega z/v-i\omega p/v}d\beta=\frac{iq}{8\pi}\epsilon^{-1/2}(z). \qquad (27)$$

Subtraction of (26) from (27) yields

$$\int_{-\infty}^{\infty}F(\beta)he_\beta^{(1)}\left(\frac{\pi z}{p}\right)e^{-i\omega z/v}[e^{i\beta\pi-i\omega p/v}-1]d\beta=0. \qquad (28)$$

In order that linear independence of the Hill functions be preserved, the factor in square brackets in (28) must be zero. This requires that

$$\beta=(\omega p/v\pi)+2l, \qquad (29)$$

where l is any integer. Hence, Eq. (24) becomes

$$\psi=\sum_{l=-\infty}^{\infty}\epsilon^{1/2}(z)F_lH_0^{(1)}(\gamma_l^\alpha\rho)he_{\alpha+2l}^{(1)}(\pi z/p), \qquad (30)$$

where $\alpha=\omega p/v\pi$, and γ_l^α denotes the value of γ for which $\beta=\alpha+2l$. F_l can be determined from the relation

$$\frac{iq}{8\pi}\epsilon^{-1/2}(z)e^{i\omega z/v}=\sum_{l=-\infty}^{\infty}F_lhe_{\alpha+2l}^{(1)}\left(\frac{\pi z}{p}\right) \qquad (31)$$

with the help of the orthogonality properties of Hill's functions.[9,10] We have

$$F_l=\frac{\pi}{pC_l}\int_0^p\frac{iq}{8\pi}\epsilon^{-1/2}(z)he_{2l+\alpha}^{(1)*}\left(\frac{\pi z}{p}\right)e^{i\omega z/v}dz, \qquad (32)$$

where C_l is a normalization factor which is defined

$$\int_0^\pi|he_{2l+\alpha}^{(1)}(\xi)|^2d\xi=C_l. \qquad (33)$$

Equation (30) constitutes the formal solution for ψ.

[9] P. M. Morse and H. Feshbach, *Methods of Theoretical Physics* (McGraw-Hill Book Company, Inc., New York, 1953).

[10] N. W. McLachlan, *Theory and Application of Mathieu Functions* (Oxford University Press, Oxford, 1951).

III. THE RADIATION SPECTRUM

We shall now obtain a formal expression for the energy radiated by the particle per unit length and per unit frequency interval. The total energy W radiated by the particle is equal to

$$W=\int_{t=-\infty}^{\infty} P dt, \tag{34}$$

where P denotes the radiated power. By Poynting's theorem,[8] the power radiated into a cylinder of radius ρ coaxial with the track of the particle and of length $2L$ is

$$P=-2\pi\rho\int_{z=-L}^{L} dz\{E_z(\rho,z,t)H_\phi(\rho,z,t)\}. \tag{35}$$

Expressing $E_z(\rho,z,t)$ and $H_\phi(\rho,z,t)$ in terms of their Fourier transforms and substituting (35) into (34) yields

$$W=-2\pi\rho\int_{t=-\infty}^{\infty} dt\int_{z=-L}^{L} dz\int_{\omega=-\infty}^{\infty} d\omega\int_{\omega'=-\infty}^{\infty} d\omega' \times[E_z(\rho,z,\omega)H_\phi(\rho,z,\omega')e^{it(\omega+\omega')}]. \tag{36}$$

Interchanging the order of integration gives

$$W=-4\pi^2\rho\int_{z=-L}^{L} dz\int_{\omega=-\infty}^{\infty} d\omega \times[\mathcal{E}_z(\rho,z,\omega)\mathcal{H}_\phi(\rho,z,-\omega)]. \tag{37}$$

Since $E_z(\rho,z,t)$ and $H_\varphi(\rho,z,t)$ are real, it follows that

$$W=-4\pi^2\rho\int_{z=-L}^{L} dz\int_{\omega=0}^{\infty} d\omega[\mathcal{E}_z(\rho,z,\omega)\mathcal{H}_\phi^*(\rho,z,\omega) +\mathcal{E}_z^*(\rho,z,\omega)\mathcal{H}_\phi(\rho,z,\omega)]. \tag{38}$$

The energy radiated per unit path length and per unit frequency interval $d^2W/dLd\omega$ is given by

$$\frac{d^2W}{dLd\omega}=-4\pi^2\rho\frac{1}{2L}\int_{z=-L}^{L} dz[\mathcal{E}_z(\rho,z,\omega)\mathcal{H}_\phi^*(\rho,z,\omega) +\mathcal{E}_z^*(\rho,z,\omega)\mathcal{H}_\phi(\rho,z,\omega)]. \tag{39}$$

Inserting the relations

$$\mathcal{E}_z(\rho,z,\omega)=\frac{1}{i\omega\epsilon\rho}\frac{\partial}{\partial\rho}\left(\rho\frac{\partial\psi}{\partial\rho}\right), \tag{40}$$

$$\mathcal{H}_\phi(\rho,z,\omega)=-\partial\psi/\partial\rho \tag{41}$$

into (39), where ψ is given in Eq. (30), and utilizing the large-argument asymptotic expression for the Hankel functions

$$\lim_{X\to\infty} H_0^{(1)}(X)=\left(\frac{2}{\pi X}\right)^{1/2} e^{i X-i(\pi/4)}, \tag{42}$$

one has

$$\frac{d^2W}{dLd\omega}=\frac{8\pi}{\omega L}\sum_{l=-\infty}^{\infty}\sum_{m=-\infty}^{\infty} F_lF_m^*(\gamma_l^\alpha)^{3/2} (\gamma_m^{\alpha*})^{1/2}\exp[i(\gamma_l^\alpha-\gamma_m^{\alpha*})\rho] \times\int_{-L}^{L} he_{\alpha+2l}\left(\frac{\pi z}{p}\right)he_{\alpha+2m}^*\left(\frac{\pi z}{p}\right)dz. \tag{43}$$

Utilizing the orthogonality properties of the Hill functions and carrying out the integral in Eq. (43) gives

$$\frac{d^2W}{dLd\omega}=\frac{16}{\omega}\sum_{l=-\infty}^{\infty}|F_l|^2C_l(\gamma_l^\alpha)^{3/2}(\gamma_l^{\alpha*})^{1/2} \times\exp[i(\gamma_l^\alpha-\gamma_l^{\alpha*})\rho] \tag{44}$$

If γ_l^α is imaginary, the corresponding term in the series becomes zero at large values of ρ, so one has finally

$$\frac{d^2W}{dLd\omega}=\frac{16}{\omega}\sum_l|F_l|^2C_l(\gamma_l^\alpha)^2, \tag{45}$$

where the summation is taken over those values of l for which γ_l^α is real. Equation (45) constitutes the formal solution for the spectral density of the radiation.

V. CLASSIFICATION OF THE RADIATION

Owing to the inhomogeneity of the dielectric medium, the radiation given off by a charged particle moving uniformly through this medium contains not only Čerenkov-type radiation but also transition-type radiation. In order to classify these types of radiation, let us first consider the degenerate case of the radiation in a homogeneous medium. For this degenerate case, the Hill functions, $he_\beta^{(1)}(\xi)$, reduce to the exponential functions, i.e.,

$$he_\beta^{(1)}(\xi)\to e^{i\beta\xi}, \tag{46}$$

where

$$\beta=\sqrt{\theta_0}=(\omega^2\mu_0\epsilon-\gamma^2)(p/\pi)^2. \tag{47}$$

The coefficient F_l given by Eq. (32) reduces to

$$F_l=\frac{1}{p}\left(\frac{iq}{8\pi}\right)\frac{1}{\sqrt{\epsilon}}\int_0^p e^{2li\pi z/p}dz \tag{48}$$

$$\begin{aligned}&=0 && \text{for } l\neq 0\\ &=iq/8\pi\sqrt{\epsilon} && \text{for } l=0.\end{aligned} \tag{49}$$

Hence, the wave function ψ given by Eq. (30) becomes

$$\psi=\frac{iq}{8\pi}e^{i\omega z/v}H_0^{(1)}(\gamma_0^\alpha\rho), \tag{50}$$

where

$$(\gamma_0^\alpha)^2=\omega^2\mu_0\epsilon-\omega^2/v^2. \tag{51}$$

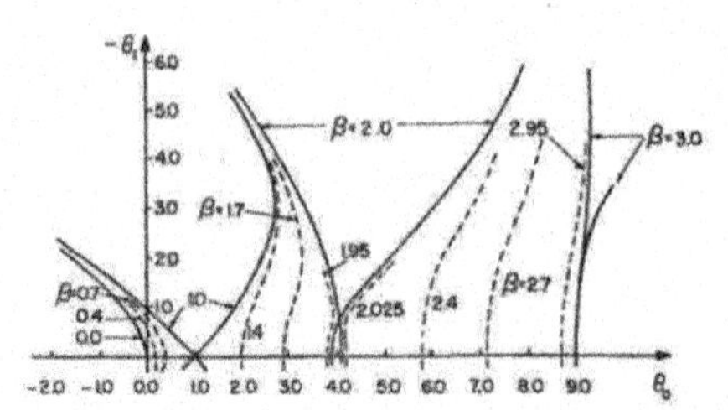

FIG. 1. Stability diagram for Hill's equation with $\delta=0.25$.

And, the energy radiated per unit path length and per unit frequency interval is then given by

$$\frac{d^2W}{dLd\omega}=\frac{q^2}{4\pi\epsilon\omega}(\gamma_0{}^a)^2 \quad \text{for} \quad \gamma_0{}^a \text{ real}$$
$$=0 \quad \text{for} \quad \gamma_0{}^a \text{ imaginary}. \tag{52}$$

Therefore, only the $l=0$ term survives as the inhomogeneous medium degenerates to a homogeneous one. This $l=0$ term corresponds to the Čerenkov radiation term. It is defined that the $l=0$ term in Eq. (30) will be designated as the Čerenkov radiation term and the $l\neq0$ terms will be designated as the transition radiation terms. Consequently, the $l=0$ term in Eq. (45) will be designated as the radiated energy spectrum due to the Čerenkov effect while the $l\neq0$ terms in Eq. (45) will be designated as the radiated energy spectrum due to the transition effect.

The angles of emission of the radiation can be readily obtained from Eq. (30) and the expression for the Hill functions, Eq. (20). The radiation is composed of an infinite number of cylindrical waves emitted at angles

$$\theta_{ln}=\arctan\left[\frac{\gamma_l{}^a}{(\beta_l+2n)\pi/p}\right] \tag{53}$$

with respect to the positive z-axis, where n and l are any integers and β_l is given by Eq. (29). It will be noted that for any value of l for which $\gamma_l{}^a$ is real, radiation is emitted at an infinite number of angles ranging from zero, when n is very large and positive, to π, when n is very large and negative. One would

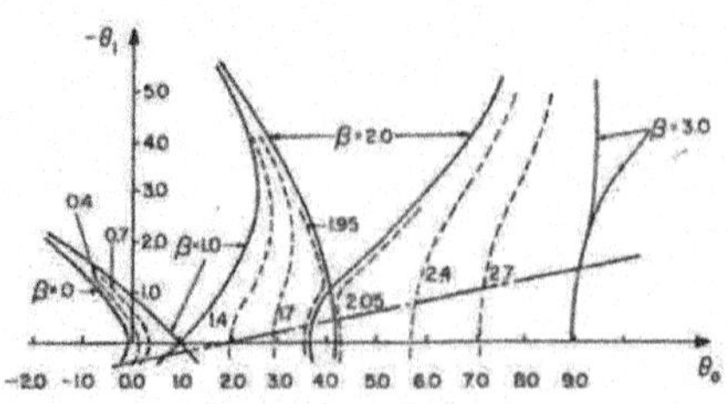

FIG. 2. Stability diagram for Hill's equation with $\delta=0.4$. The $\gamma_l{}^a=0$ curve is given by the dot-dashed line.

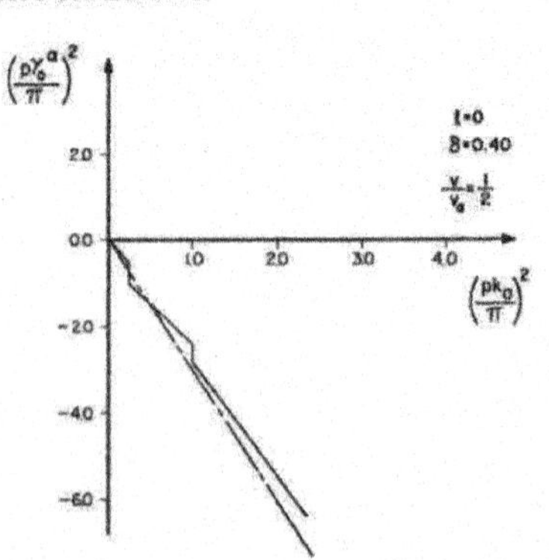

FIG. 3. Transverse separation constant $(p\gamma_0/\pi)^2$ as a function of frequency $(pk_0/\pi)^2$ for Čerenkov mode ($l=0$) well below the threshold velocity. The dot-dashed line indicates the curve for the homogeneous case. No radiation is emitted.

expect that the angle at which the dominant Čerenkov radiation will be emitted is the angle θ_{00}:

$$\theta_{00}=\arctan\left(\frac{\gamma_0{}^a}{\beta_0\pi/p}\right),$$
$$=\arctan\left(\frac{v\gamma_0{}^a}{\omega}\right), \tag{54}$$

since this is the angle at which the Čerenkov radiation is emitted in a homogeneous medium.

VI. AN EXAMPLE: THE DIELECTRIC PROFILE $\epsilon(z)=\epsilon_a[1-\delta\cos(2\pi z/p)]$

In order to obtain quantitative results from the formal solution, it is necessary to assume a specific dielectric profile. Let us consider the following dielectric variation:

$$\epsilon(z)=\epsilon_a[1-\delta\cos(2\pi z/p)], \tag{55}$$

where ϵ_a is the average value of the permittivity and δ gives the relative amplitude of the variation. Furthermore, $0\leq\delta<1$. Substituting Eq. (55) into Eq. (18) gives the coefficients θ_0 and θ_n in Eq. (19). They are

$$\theta_0=\left(\frac{p}{\pi}\right)^2(\omega^2\mu_0\epsilon_a-\gamma_l{}^{a2})-\left(\frac{1}{(1-\delta^2)^{1/2}}-1\right), \tag{56}$$

$$\theta_1=-\frac{\delta}{2}\left(\frac{p}{\pi}\right)^2\omega^2\mu_0\epsilon_a+\frac{4C_1{}^3-2C_1}{C_1{}^2-1}, \tag{57}$$

$$\theta_n=\frac{(3n+1)C_1{}^{n+2}-(3n-1)C_1{}^n}{C_1{}^2-1}, \quad (n\geq2), \tag{58}$$

in which

$$C_1=(1/\delta)-(1/\delta)(1-\delta^2)^{1/2}. \tag{59}$$

For a given δ, depending upon the values of θ_0 and θ_1, Eq. (19) would yield solutions that are stable or unstable.[11] In order to satisfy Sommerfeld's radiation condition only the stable solutions are allowed. For a given value of δ, a diagram giving the stable and unstable regions can be constructed. Two such stability diagrams for $\delta=0.25$ and $\delta=0.40$ are given in Figs. 1 and 2. These diagrams provide information concerning the threshold velocity for each mode (i.e., for each l) and the angles of emission for various radiating components.

To illustrate how one may obtain this information from this diagram, let us consider the following procedures. Combining Eqs. (56) and (57) and eliminating $\omega^2\mu_0\epsilon_a p^2/\pi^2(k_a^2p^2/\pi^2)$ gives

$$-\theta_1=\frac{\delta}{2}\theta_0+\left(\frac{p\gamma_l^a}{\pi}\right)^2\frac{\delta}{2}+\left[\frac{\delta}{2}\left(\frac{1}{(1-\delta^2)^{1/2}}-1\right)-\frac{4C_1^3-2C_1}{C_1^2-1}\right], \quad (60)$$

where C_1 is given by Eq. (59). For a fixed value of δ expression (60) gives a family of straight lines corresponding to various values of $(p\gamma_l^a/\pi)^2$. The line for which $(p\gamma_l^a/\pi)^2=0$ is drawn in Fig. 2. Above this line, $(p\gamma_l^2/\pi)^2$ is positive corresponding to radiation, while below this line, it is negative corresponding to radially evanescent fields and therefore no radiation is emitted. For a given velocity of the charged particle (v/v_a), for a given frequency (k_ap/π) and for a given mode (l), one may compute β from Eq. (29). $-\theta_1$ can also be computed knowing (k_ap/π). The intersection between the lines $\beta=$ constant and $-\theta_1=$ constant provides the point from which one may obtain the value for θ_0.

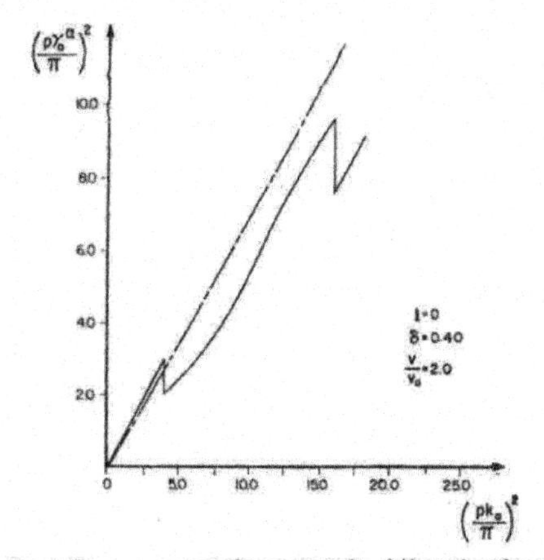

FIG. 4. Transverse separation constant $(p\gamma_0/\pi)^2$ as a function of frequency $(pk_a/\pi)^2$ for Čerenkov mode ($l=0$) above the threshold velocity. Dot-dashed line indicates the curve for the homogeneous case. Radiation is emitted at all frequencies.

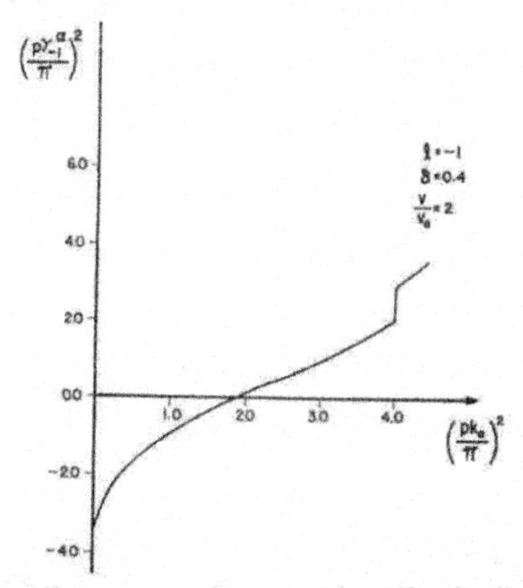

FIG. 5. Transverse separation constant $(p\gamma_{-1}/\pi)^2$ as a function of frequency $(pk_a/\pi)^2$ for $l=-1$ transition radiation mode above the Čerenkov threshold velocity. Radiation is emitted when $(pk_a/\pi)^2$ is greater than 1.9.

Hence the value of $(p\gamma_l^a/\pi)^2$ can be computed from Eq. (56). The sign of $(p\gamma_l^a/\pi)^2$ provides the information whether this particular mode radiates or not. As a specific example, let $\delta=0.4$, $pk_a/\pi=2.0$, $v/v_a=1.414$, $l=0$; we then have $\beta=1.414$, and $-\theta_1=0.4$. The intersection of $\beta=1.414$ and $-\theta_1=0.4$ yields the value for θ_0 which is 2.05. Using Eq. (56) we have $(p\gamma_l^a/\pi)^2=2.87$, which is positive, hence radiation does occur for this particular case. Knowing $(p\gamma_l^a/\pi)^2$, it is a simple matter to compute the emission angles from Eq. (53). They are for $n=0$, $\theta_{00}=50.2°$; $n=1$, $\theta_{01}=26.1°$; $n=2$, $\theta_{02}=17.4°$, etc.

It is noted from Eq. (29) that no matter how small the velocity of the charged particle is, β can always be adjusted using l to give modes which would radiate.

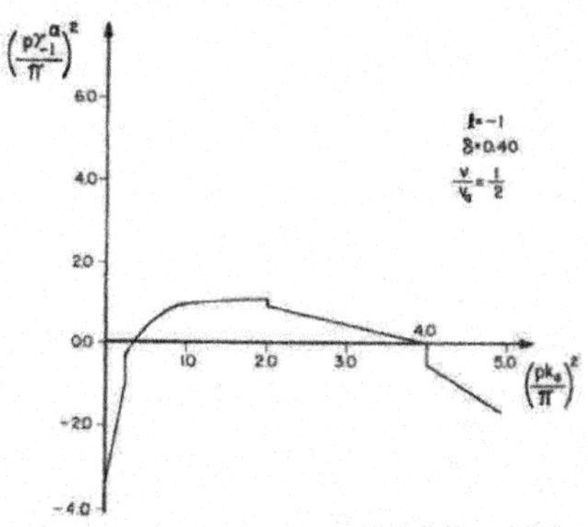

FIG. 6. Transverse separation constant $(p\gamma_{-1}/\pi)^2$ as a function of frequency $(pk_a/\pi)^2$ for $l=-1$ transition radiation mode below the Čerenkov threshold. Radiation is emitted only in the frequency range $0.35<(pk_a/\pi)^2<4.0$.

[11] Stable solutions refer to solutions which possess real values of β while the unstable solutions refer to solutions which possess complex values of β. [See Eq. (20).]

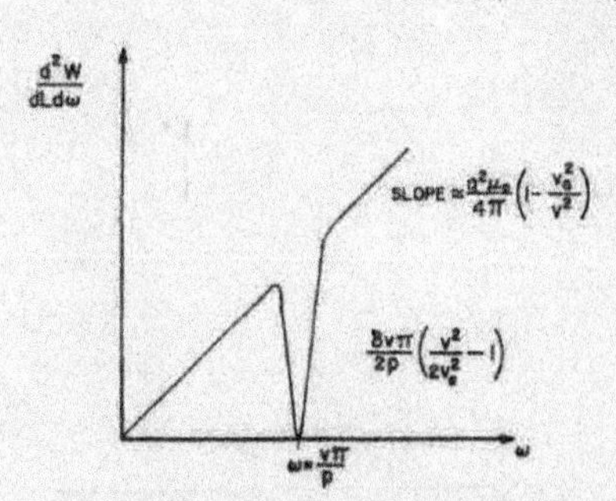

FIG. 7. Spectral density of Čerenkov radiation ($l=0$ mode) above the Čerenkov threshold velocity.

This is because l may take on positive as well as negative integer values. This conclusion is in contrast with the case of Čerenkov radiation in a homogeneous medium in which radiation will take place only if a certain threshold velocity is reached. This observation may be understood by the fact that transition radiation occurs for any velocities.

Figures 3 through 6 display the functional variation of $(p\gamma_l^a/\pi)^2$ with respect to $(pk_a/\pi)^2$ for various values of l and the ratio v/v_a when $\delta=0.40$. The fact that $(p\gamma_l^a/\pi)^2$ can be positive even if $v<v_a$ is apparent in Fig. 6.

VII. APPROXIMATE SOLUTIONS FOR SMALL δ

If the amplitude of the dielectric variation given by Eq. (55) is very small, i.e., if $\delta\ll 1$, it is possible to derive approximate analytic expressions for the scalar wave function ψ as well as the energy radiated per unit path length and per unit frequency interval $d^2W/dLd\omega$. Retaining terms correct to the first order in δ, we have

$$\begin{aligned}
\theta_0&=(p/\pi)^2(\omega^2\mu\epsilon_a-\gamma_l^{a2}),\\
\theta_1&=\delta[1-(p^2/2\pi^2)\omega^2\mu\epsilon_a],\\
\theta_n&=0\quad(n\geq 2),\\
b_{-1}(\beta)/b_0(\beta)&=-\theta_1/[\theta_0-(\beta-2)^2],\\
b_1(\beta)/b_0(\beta)&=-\theta_1/[\theta_0-(\beta+2)^2],\\
b_n(\beta)/b_0(\beta)&=0\quad(|n|\geq 2),\\
C_l&=\pi|b_0(\alpha+2l)|^2,\\
\beta&=\sqrt{\theta_0}.
\end{aligned}\tag{61}$$

The Hill's function then reduces to

$$he_\beta^{(1)}(\pi z/p)=e^{i\omega z/v}e^{i2l\pi z/p}[b_{-1}(\beta)e^{-2\pi iz/p}+b_0(\beta)+b_1(\beta)e^{2\pi iz/p}].\tag{62}$$

Substituting Eq. (62) into Eq. (32) and carrying out the integration gives

$$\begin{aligned}
F_{-1}&\approx\frac{iq}{8\pi\epsilon_a}\left\{\frac{b_1(\alpha-2)}{[b_0(\alpha-2)]^2}+\frac{\delta}{4b_0(\alpha-2)}\right\},\\
F_0&\approx\frac{iq}{8\pi\epsilon_a}\left\{\frac{1}{b_0(\alpha)}\right\},\\
F_1&\approx\frac{iq}{8\pi\epsilon_a}\left\{\frac{b_{-1}(\alpha+2)}{[b_0(\alpha+4)]^2}+\frac{\delta}{4b_0(\alpha+2)}\right\},\\
F_{\pm l}&\approx 0,\quad l>1.
\end{aligned}\tag{63}$$

Inserting Eq. (63) into Eq. (30), one obtains the approximate expression for the scalar wave function:

$$\begin{aligned}
\psi\approx\frac{iq}{8\pi}e^{i\omega z/v}\Bigg(&H_0^{(1)}\left\{\left[k_a^2-\left(\frac{\omega}{v}-\frac{2\pi}{p}\right)^2\right]^{1/2}\rho\right\}e^{-2i\pi z/p}\frac{\delta}{4}\left[1+\frac{1-\frac{1}{2}(pk_a/\pi)^2}{(\omega p/v\pi)-1}\right]\\
&+H_0^{(1)}\left[\left(k_a^2-\frac{\omega^2}{v^2}\right)^{1/2}\rho\right]\left[1-e^{-2\pi iz/p}\frac{\delta}{4}\left(1+\frac{1-\frac{1}{2}(pk_a/\pi)^2}{(\omega p/v\pi)-1}\right)-e^{2\pi iz/p}\frac{\delta}{4}\left(1-\frac{1-\frac{1}{2}(pk_a/\pi)^2}{(\omega p/v\pi)+1}\right)\right]\\
&+H_0^{(1)}\left\{\left[k_a^2-\left(\frac{\omega}{v}+\frac{2\pi}{p}\right)^2\right]^{1/2}\rho\right\}e^{2i\pi z/p}\frac{\delta}{4}\left[1-\frac{1-\frac{1}{2}(pk_a/\pi)^2}{(\omega p/v\pi)+1}\right]\Bigg).
\end{aligned}\tag{64}$$

It is noted that as δ approaches zero, ψ reduces to the expression for the homogeneous case. As p approaches zero, ψ also reduces to the same expression for the homogeneous case. This is because as p approaches zero, even if δ is finite, the medium becomes macroscopically homogeneous with an average permittivity ϵ_a. Equation (64) also shows that to the first order in δ, the threshold velocity for the Čerenkov component ($l=0$) is $1/(\mu_0\epsilon_a)^{1/2}$ which is identical to the threshold velocity for the homogeneous case. The threshold conditions for the $l=\pm 1$ components of the field are, respectively,

$$k_a^2-[(\omega/v)\pm(2\pi/p)]^2=0.\tag{65}$$

It is also possible from the threshold conditions, Eq. (65), to obtain the frequency ranges for the $l=\pm 1$ component radiation. For $v>1/(\mu\epsilon_a)^{1/2}$, we have

$$\omega\geq 2\pi v/p[v(\mu\epsilon_a)^{1/2}+1]\tag{66}$$

for the $l=-1$ component and

$$\omega\geq 2\pi v/p[v(\mu\epsilon_a)^{1/2}-1]\tag{67}$$

for the $l=+1$ component. When $v<1/(\mu\epsilon_a)^{1/2}$, we have

$$2\pi v/p[v(\mu\epsilon_a)^{1/2}+1]\leq\omega\leq 2\pi v/p[1-v(\mu\epsilon_a)^{1/2}]\tag{68}$$

for the $l=-1$ component. Although radiation occurs

below the usual Čerenkov threshold velocity, the spectral density of the radiation for the $l=-1$ component is quite small. (It is of the order of δ^2.) This fact will be shown below.

An approximate expression for the spectral density can be obtained with the help of Eqs. (61) through (64). Substituting these equations into expression (45) and carrying out the algebraic manipulation gives

$$\frac{d^2W}{dLd\omega}\approx\frac{q^2}{4\pi\epsilon_a\omega}$$
$$\times\Bigg\{(\gamma_{-1}^{a})^2\frac{[f^2-2\delta f(\alpha-1)+\delta^2(\alpha-1)^2](\alpha-3)^2}{f^2[(\alpha-1)^2+(\alpha-3)^2]+16(\alpha-1)^2(\alpha-3)^2}$$
$$+(\gamma_0^{a})^2\frac{\left[16(\alpha^2-1)\left(1+\frac{3\delta^2}{8}\right)+\delta f\right](\alpha^2-1)}{f^2[(\alpha-1)^2+(\alpha+1)^2]+16(\alpha^2-1)^2}$$
$$+(\gamma_1^{a})^2\frac{[f^2+2\delta f(\alpha+1)+\delta^2(\alpha+1)^2](\alpha+3)^2}{f^2[(\alpha+1)^2+(\alpha+3)^2]+16(\alpha+1)^2(\alpha+3)^2}\Bigg\}\quad(69)$$

in which γ_{-1}^{a}, γ_0^{a}, and γ_1^{a} must be real,

$$f=\frac{\delta}{2}\left(\frac{pk_a}{\pi}\right)^2-\delta$$

and

$$\omega p/v\pi=\alpha.$$

If any of these γ_l^{a} in Eq. (69) is imaginary, the term containing this γ_l^{a} should be set to zero.

The behavior of the $l=0$ (Čerenkov) term in Eq. (69) is very similar to that of the spectral density of Čerenkov radiation in a homogeneous medium, i.e.,

$$\left.\frac{d^2W}{dLd\omega}\right|_{l=0\text{ term}}\approx\frac{q^2\mu\omega}{4\pi}\left(1-\frac{v_a^2}{v^2}\right)\quad(70)$$

with $v_a=1/(\mu\epsilon_a)^{1/2}$, except in the vicinity of $\omega p/v\pi=1$, When $\omega p/v\pi=1$, this $l=0$ term is zero. A sketch of the spectral density for the $l=0$ term as a function of frequency is given in Fig. 7.

The $l=+1$ (transition) term in Eq. (69) is of the order of δ^2 over the whole frequency spectrum. The $l=-1$ (transition) term is also of the order of δ^2 over the whole frequency spectrum except around $\omega p/v\pi=1$. At this point, the $l=-1$ term is of the order of unity and has the value

$$\left.\frac{d^2W}{dLd\omega}\right|_{l=-1\text{ term at }\omega p/v\pi=1}\approx\frac{q^2\mu\omega_0}{4\pi}\left(1-\frac{v_a^2}{v^2}\right).$$

In other words, a peak occurs for the $l=-1$ term at a point where a null occurs for the $l=0$ term. A sketch of the spectral density for the $l=-1$ term as a function of frequency is also given in Fig. 7.

It can be shown from the approximate expression for the scalar wave function, Eq. (64), that at frequencies other than $\omega p/v\pi=1$, the phase for the $l=0$ Čerenkov radiation is stationary at the angle

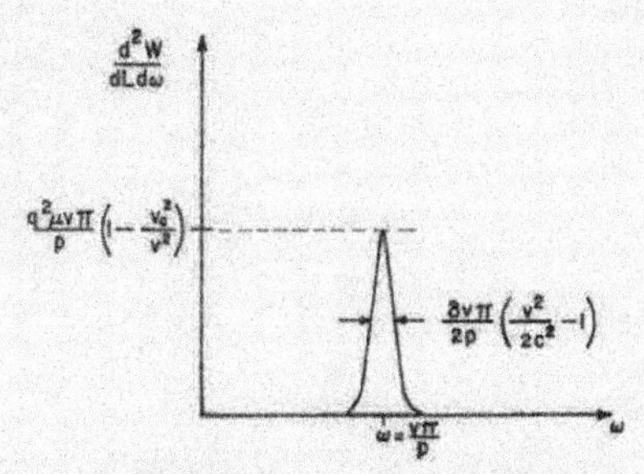

FIG. 8. Spectral density of $l=-1$ transition radiation mode above the Čerenkov threshold velocity.

$$\theta|_{l=0}\approx\arctan\left(\frac{v^2}{v_a^2}-1\right)^{1/2}\quad(71)$$

which is simply the Čerenkov radiation angle in a homogeneous medium. On the other hand, around the frequency $\omega p/v\pi=1$ the phase for the $l=-1$ transition radiation is stationary at the angle

$$\theta|_{l=-1}\approx\arctan\left[-\left(\frac{v^2}{v_a^2}-1\right)^{1/2}\right]\quad(72)$$

which is $(\pi-\theta_c)$, where θ_c is the Čerenkov angle. Thus the peak $l=-1$ transition radiation is emitted in the backward direction while the $l=0$ Čerenkov radiation is emitted in the forward direction.

One way of interpreting the above result qualitatively may be given in terms of the Bragg reflection condition. Above the Čerenkov threshold velocity, the charged particle emits radiation at an angle θ_c with respect to the direction of travel of the particle. The radiated wave then undergoes multiple reflections due to the striations of the medium. Bragg reflections occur when the following condition is satisfied:

$$n\lambda=2p\cos\theta_c,\quad(73)$$

where n are positive integers,

$$\lambda=2\pi v_a/\omega\quad(74)$$

and

$$\cos\theta_c=v_a/v.\quad(75)$$

Hence, the Bragg condition becomes

$$\omega p/v\pi=n.\quad(76)$$

When $n=1$, we have the condition described earlier. Therefore according to this argument one would expect a null in the Čerenkov radiation spectrum and a corresponding peak in the transition radiation spectrum when $\omega p/v\pi=1$. The Bragg-reflection analysis indicates that one might expect other peaks and nulls in the various spectra when $\omega p/v\pi$ takes on integer value other than unity.

PHYSICAL REVIEW A VOLUME 2, NUMBER 3 SEPTEMBER 1970

Transition Radiation in a Periodically Stratified Plasma

K. F. Casey*
Electrical Engineering Department, Air Force Institute of Technology, Wright-Patterson Air Force Base, Ohio 45433
and
C. Yeh
Department of Electrical Sciences and Engineering, University of California, Los Angeles, California 90024
(Received 25 February 1970)

The solution to the problem of determining the transition radiation emitted when a charged particle moves uniformly in a periodically stratified cold, isotropic plasma is obtained. The electromagnetic field may be expressed in terms of an infinite number of normal modes. The conditions under which some of these modes are radiative are discussed, and expressions for the field components and the energy spectral density are obtained. It is found that radiation may be emitted for arbitrarily small particle velocities and that the strongest emission takes place in a frequency band just above the average plasma frequency. Numerical results are presented to illustrate the behavior of the spectra of the various radiative modes as the frequency and plasma parameters are varied.

I. INTRODUCTION

When a charged particle moves with constant speed through an inhomogeneous medium, there are two (macroscopic) mechanisms by which radiation may be emitted. Radiation of the Čerenkov[1,2] type is expected if the particle moves close to or through a region in which the phase velocity of light is less than the speed of the particle. For radiation of this type to occur, the speed of the particle must be greater than the smallest phase velocity encountered. Transition radiation,[3-5] on the other hand, may be expected to occur at any particle speed. As the charged particle moves uniformly in the continuously inhomogeneous medium, its images will not, in general, be in uniform motion, but will be accelerated. Transition radiation may be thought of as being emitted by these accelerated image charges. Since the nonuniform motion of the images will occur even if the moving charged particle is traveling slowly, there is no velocity threshold for transition radiation.

In a previous paper,[6] the emission of Čerenkov and transition radiation by a charged particle moving uniformly in a periodically stratified nondispersive dielectric medium was considered. It was found that the electromagnetic field excited by the passage of the charged particle may be expressed in terms of an infinite number of normal modes. Each of these normal modes was a modulated cylindrical wave, propagating in the direction of motion of the particle at a phase speed equal to the velocity of the particle. Some of these modes were also propagating in the outward direction, away from the track of the particle. Furthermore, the threshold velocity, cutoff frequency, and emission angles for each mode were also found. However, these results are not applicable to dispersive media.

The purpose of this investigation is to consider the radiation characteristics of a charged particle moving uniformly in a dispersive sinusoidally stratified medium. Specifically, the dispersive medium is assumed to be an inhomogeneous cold, isotropic plasma with a sinusoidally varying free-electron density profile. Since the permittivity of the medium is everywhere less than that of free space, only transition-type radiation is expected. Further, the emission is expected to occur over a restricted frequency range: At very low frequencies the plasma is opaque, while at very high frequencies the permittivity is sensibly identical to that of free space, and the inhomogeneity of the plasma, which causes the transition radiation to occur, disappears. In Sec. II–VI, the formal solution to this problem will be obtained. Some approximate results for the frequency ranges over which radiation is expected are obtained and compared with exact numerical data. Finally, numerical results are presented to illustrate the behavior of the spectral density of the emitted radiation as the parameters of the problem are varied.

II. FORMULATION OF PROBLEM AND FORMAL SOLUTION

The geometry of the problem is shown in Fig. 1. It is assumed that the inhomogeneous, isotropic cold plasma medium under consideration fills the entire space. The medium is characterized by a plasma frequency ω_p whose square varies with the axial coordinate z of the circular-cylindrical coordinates (ρ, ϕ, z) as follows:

$$\omega_p^2(z) = \omega_{p0}^2[1+\Delta \cos(2\pi z/d)] . \tag{1}$$

ω_{p0}^2 denotes the average squared plasma frequency, Δ the modulation index ($|\Delta| < 1$), and d the period of the variation. The permittivity of the medium ϵ_p is then given by

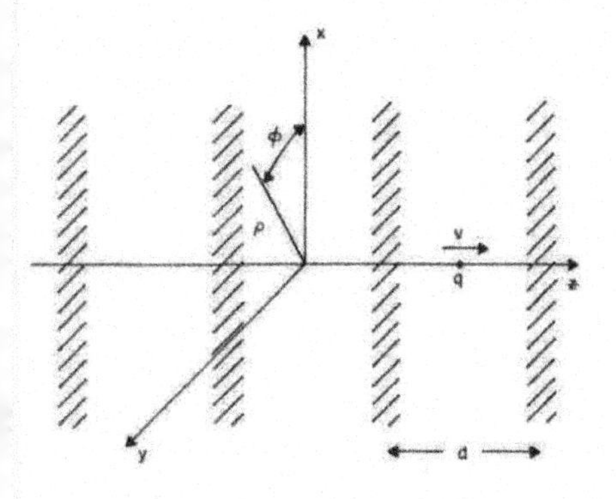

FIG. 1 Geometry of the problem.

$$\epsilon_p(z) = \epsilon_0(1-\omega_{p0}^2/\omega^2)[1-\delta \cos(2\pi z/d)] , \tag{2}$$

in which ϵ_0 denotes the free-space permittivity, ω the frequency of a wave whose time dependence is taken to be $e^{-i\omega t}$, and δ the modulation index of the permittivity:

$$\delta = \frac{\Delta}{\omega^2/\omega_{p0}^2 - 1} . \tag{3}$$

The permeability of the medium is taken to be μ_0, the free-space permeability; the effects of collisions are ignored.

A particle of charge q moves with constant speed v along the z axis. The motion of the charge excites an axisymmetric TM electromagnetic field whose electric and magnetic vectors may be expressed in terms of a scalar potential function ψ as follows, for $\rho > 0$:

$$\vec{\mathcal{E}} = \int_{-\infty}^{\infty} -(1/i\omega\epsilon)(\nabla\times\nabla\times\psi\vec{a}_z)e^{-i\omega t}d\omega , \tag{4}$$

$$\vec{\mathcal{H}} = \int_{-\infty}^{\infty} (\nabla\times\psi\vec{a}_z)e^{-i\omega t}d\omega . \tag{5}$$

$\vec{a}_z$ denotes the unit vector in the z direction. The scalar function ψ satisfies the partial differential equation

$$\frac{1}{\rho}\frac{\partial}{\partial\rho}\left(\rho\frac{\partial\psi}{\partial\rho}\right)+\frac{\partial^2\psi}{\partial z^2}-\frac{1}{\epsilon_p}\frac{d\epsilon_p}{dz}\frac{\partial\psi}{\partial z}+\omega^2\mu_0\epsilon_p\psi=0 \tag{6}$$

subject to the following boundary conditions: (a) the Sommerfeld radiation condition at $\rho=\infty$ and (b) Ampere's law at $\rho=0$, i.e.,

$$\lim_{\rho\to 0}\left(-2\pi\rho\frac{\partial\psi}{\partial\rho}\right)=\frac{q}{2\pi}e^{i\omega z/v} . \tag{7}$$

The energy spectral density of the emitted radiation, which is the quantity of principal interest in the present paper, is easily found in terms of the scalar potential function ψ. Letting $\mathcal{U}$ denote the average energy radiated per meter and per radian per second, one easily obtains the result

$$\mathcal{U} = \lim_{\substack{\rho\to\infty \\ L\to\infty}} \frac{8\pi^2}{\omega L}\int_{-L/2}^{L/2}\frac{1}{\epsilon_p}\,\mathrm{Im}\left[\frac{\partial}{\partial\rho}\left(\rho\frac{\partial\psi}{\partial\rho}\right)\frac{\partial\psi^*}{\partial\rho}\right]dz . \tag{8}$$

The asterisk (*) denotes complex conjugation and Im[ ] denotes the imaginary part.

The problem now becomes one of determining the scalar potential function ψ. The elementary-product solutions ψ_e of Eq. (6) which satisfy the Sommerfeld radiation condition at $\rho=\infty$ are readily shown to be

$$\psi_e = \epsilon_p^{1/2}(z)H_0^{(1)}(\gamma\rho)z_\gamma(z) . \tag{9}$$

$H_0^{(1)}(\ )$ denotes the Hankel function of the first kind of order zero, γ is a separation constant, and $z_\gamma(z)$ is that solution of the differential equation

$$\frac{d^2 z_r}{dz^2}+\left[\frac{1}{2\epsilon_p}\frac{d^2\epsilon_p}{dz^2}-\frac{3}{4\epsilon_p^2}\left(\frac{d\epsilon_p}{dz}\right)^2+\omega^2\mu_0\epsilon_p-\gamma^2\right]z_r=0\ , \tag{10}$$

which reduces, when $\Delta\to 0$, to $\exp[iz(\omega^2\mu_0\epsilon_p-\gamma^2)^{1/2}]$. Defining

$$\zeta=\pi z/d\ , \tag{11}$$

$$u(\zeta)=z_r(z)\ , \tag{12}$$

$$\epsilon_p(\zeta)=\epsilon_0(1-\omega_{p0}^2/\omega^2)(1-\delta\cos 2\zeta)\ , \tag{13}$$

one obtains the equation for $u(\zeta)$:

$$\frac{d^2u}{d\zeta^2}+\left[\frac{1}{2\epsilon_p}\frac{d^2\epsilon_p}{d\zeta^2}-\frac{3}{4\epsilon_p^2}\left(\frac{d\epsilon_p}{d\zeta}\right)^2+\left(\frac{d}{\pi}\right)^2(\omega^2\mu_0\epsilon_p-\gamma^2)\right]u=0\ . \tag{14}$$

Because of the periodicity of ϵ_p and the resulting periodicity of the function in square brackets in Eq. (14), the function $u(\zeta)$ may be written in the Floquet form[7]

$$u(\zeta)=e^{i\beta\zeta}\phi(\beta;\zeta)\ , \tag{15}$$

in which β is the so-called "characteristic exponent", and $\phi(\beta;\zeta)$ is a periodic function of ζ, having the period π. The relation of β to the separation constant γ and the other parameters of the problem will be clarified presently. Using the form of Eq. (15), the following expression for the elementary-product solution ψ_e is obtained:

$$\psi_e=\epsilon^{1/2}(z)H_0^{(1)}[\gamma(\beta)\rho]e^{i\beta\pi z/d}\phi(\beta;\pi z/d)\ . \tag{16}$$

The dependence of γ upon β has been indicated.

Now let the ψ_e of Eq. (16) be superposed to match the boundary condition of Ampere's law at $\rho=0$. Letting curly brackets here denote a linear superposition of elementary functions ψ_e, one obtains from Eqs. (7) and (16):

$$\lim_{\rho\to 0} -2\pi\rho\{\epsilon^{1/2}(z)e^{i\beta\pi z/d}\times\phi(\beta;\pi z/d)\gamma(\beta)H_0^{(1)\prime}[\gamma(\beta)\rho]\}=(q/2\pi)e^{i\omega z/v}\ , \tag{17}$$

in which the prime denotes differentiation with respect to the argument. Taking the limit in Eq. (17) yields the result

$$\{e^{i\beta\pi z/d}\phi(\beta;\pi z/d)\}=\left(iq/8\pi\epsilon^{1/2}(z)\right)e^{i\omega z/v}\ . \tag{18}$$

Now, the right-hand side of Eq. (18) has the form of a propagating wave multiplied by a periodic function of z. Those functions on the left-hand side which take this form are those for which

$$\beta=(\omega/v)(d/\pi)+2l\ , \tag{19}$$

in which l is any integer. Defining

$$\alpha=(\omega/v)d/\pi\ , \tag{20}$$

one obtains the general solution ψ, given by

$$\psi=\epsilon^{1/2}(z)\sum_{l=-\infty}^{\infty}F_l H_0^{(1)}(\gamma_l\rho)\times e^{i(\alpha+2l)(\pi z/d)}\phi[\alpha+2l;(\pi z/d)]\ , \tag{21}$$

in which $\gamma_l\equiv\gamma(\alpha+2l)$ and the coefficients F_l are as yet undetermined.

Equation (18) representing the Ampere's-law condition at $\rho=0$ becomes

$$\frac{iq}{8\pi\epsilon^{1/2}(z)}e^{i\alpha\pi z/d}=\sum_{l=-\infty}^{\infty}F_l e^{i(\alpha+2l)}\times\phi[\alpha+2l;(\pi z/d)]\ . \tag{22}$$

This equation may be used to obtain the coefficients F_l by virtue of the fact that the functions $e^{i(\alpha+2l)\zeta}\phi(\alpha+2l;\zeta)$ possess the orthogonality property

$$\int_0^\pi e^{i(\alpha+2l)\zeta}\phi(\alpha+2l;\zeta)e^{-i(\alpha+2m)\zeta}\times\phi(\alpha+2m;-\zeta)d\zeta=C_l\delta_{lm}\ , \tag{23}$$

where C_l is a normalization factor and δ_{lm} is the Kronecker-δ symbol. Using Eqs. (23) and (22), one obtains

$$F_l=(iq/8\pi C_l)\int_0^\pi\epsilon^{-1/2}(\zeta')\times e^{-2il\zeta'}\phi(\alpha+2l;-\zeta')d\zeta'\ . \tag{24}$$

To obtain the energy spectral density in terms of the formal solution for ψ given in Eq. (21), one substitutes that expression into Eq. (8) and performs the operations implied there. There results

$$\mathfrak{U}=\sum_l\mathfrak{U}_l=(16/\omega)\sum_l|F_l|^2C_l\gamma_l^2\ , \tag{25}$$

in which the summation is taken over those values of l for which $\gamma_l^2>0$. Equation (25) is the desired result, expressed in terms of the solutions of Eq. (14). In Sec. III, these solutions will be obtained and applied to the evaluation of the spectral density $\mathfrak{U}$.

III. SOLUTION OF THE HILL EQUATION

Returning now to the solution of Eq. (14), we note that the function in square brackets is even and periodic in ζ with period π as a result of the even periodicity of $\epsilon(\zeta)$. Thus, it may be written as a Fourier cosine series; Eq. (14) becomes

$$\frac{d^2u}{d\zeta^2}+\left(\theta_0+2\sum_{n=1}^{\infty}\theta_n\cos 2n\zeta\right)u=0\ , \tag{26}$$

which is the canonical form for Hill's equation.[7] In the event that the series S, defined by

$$S=\sum_{n=1}^{\infty}\theta_n\ , \tag{27}$$

is absolutely convergent, the solutions to Eq. (26)

may be readily found.

When $\epsilon_p(\zeta)$ is given by Eq. (13), the coefficients θ_n are

$$\theta_0 = \left(\frac{k_0 d}{\pi}\right)^2 - \left(\frac{k_{p0} d}{\pi}\right)^2 - \left(\frac{\gamma d}{\pi}\right)^2 + 1 - (1-\delta^2)^{-1/2} , \quad (28)$$

$$\theta_1 = -\frac{\Delta}{2}\left(\frac{k_{p0} d}{\pi}\right)^2 + \frac{4b_1^3 - 2b_1}{b_1^2 - 1} , \quad (29)$$

$$\theta_n = \frac{(3n+1)b_1^{n+2} - (3n-1)b_1^n}{b_1^2 - 1} , \ (n \geq 2) \quad (30)$$

in which

$$k_0 = \omega(\mu_0 \epsilon_0)^{1/2} , \quad (31)$$

$$k_{p0} = \omega_{p0}(\mu_0 \epsilon_0)^{1/2} , \quad (32)$$

$$b_1 = (1/\delta)[1 - (1-\delta^2)^{1/2}] . \quad (33)$$

The series S is absolutely convergent if $|\delta| < 1$, i.e., if

$$|\omega/\omega_{p0}| < (1 - |\Delta|)^{1/2} \quad (34)$$

or

$$|\omega/\omega_{p0}| > (1 + |\Delta|)^{1/2} . \quad (35)$$

It will be noted that the range of frequencies for which S is not absolutely convergent is that range for which ϵ_p is zero for some values of z. Numerical results to be discussed later indicate that radiation is emitted only for frequencies which satisfy Eq. (35); thus the absolute convergence of S is assumed in what follows.

The function $u(\zeta)$ can be written in the Floquet form[7]

$$u(\zeta) = \sum_{n=-\infty}^{\infty} b_n(\beta) e^{i(2n+\beta)\zeta} . \quad (36)$$

Substituting Eq. (36) into Eq. (26), one obtains a set of recurrence relations

$$[\theta_0 - (\beta + 2n)^2] b_n(\beta) + \sum_{m=-\infty}^{\infty} \theta_{|m|} b_{n-m}(\beta) = 0 , \quad (37)$$

for any integer n. For a nontrivial solution to exist, the characteristic exponent β and the coefficient θ_n must be related by the characteristic equation

$$\sin^2(\tfrac{1}{2}\pi\beta) = \Delta(0) \sin^2(\tfrac{1}{2}\pi)(\theta_0)^{1/2} , \quad (38)$$

in which $\Delta(0)$ is an infinite determinant whose elements are

$$\Delta(0)_{m,m} = 1 , \quad (39)$$

$$\Delta(0)_{m,n} = \theta_{|m-n|}/(\theta_0 - 4m^2) \quad (m \neq n) . \quad (40)$$

The convergence of the infinite determinant $\Delta(0)$ is assured if S is absolutely convergent. If the characteristic equation is satisfied, Eqs. (37) may be solved for ratios $b_n(\beta)/b_0(\beta)$. One therefore obtains the appropriate solution for Hill's equation.

Now the spectral density of the radiation may be expressed in terms of the coefficients b_n. Let

$$D_k(\delta) = (1/\pi)\int_0^{\pi} (1 - \delta\cos 2\zeta)^{-1/2} e^{-2ik\zeta}\, d\zeta . \quad (41)$$

Then from Eqs. (24) and (36), it is easy to show that

$$F_l = \frac{iq}{8\pi C_l[\epsilon_0(1-\omega_{p0}^2/\omega^2)]^{1/2}} \times \sum_{n=-\infty}^{\infty} b_n(\alpha + 2l) D_{n+l}(\delta) . \quad (42)$$

From Eqs. (23) and (36), one obtains

$$C_l = \pi \sum_{n=-\infty}^{\infty} b_n^2(\alpha + 2l) . \quad (43)$$

Thus $\mathfrak{U}_l$, the spectral density of the lth radiating mode, is given by

$$\mathfrak{U}_l = \frac{q^2}{4p}\left(\frac{\mu_0}{\epsilon_0}\right)^{1/2}\left(\frac{\gamma_l d}{\pi}\right)^2\left(\frac{k_0 d}{\pi}\right)^{-1}\left(1 - \frac{\omega_{p0}^2}{\omega^2}\right)^{-1} \times \left|\sum_{n=-\infty}^{\infty} b_n(\alpha+2l) D_{n+l}(\delta)\right|^2 \Big/ \sum_{n=-\infty}^{\infty} b_n^2(\alpha + 2l)$$

$$\equiv (q^2/4p)(\mu_0/\epsilon_0)^{1/2}\mathfrak{U}_l^n \quad (44)$$

if $\gamma_l^2 > 0$, and zero otherwise. $\mathfrak{U}_l^n$ denotes the normalized energy spectral density of the lth mode.

IV. RADIATION THRESHOLDS

To determine whether or not a given mode is radiative, one must solve the characteristic equation (38) for γ_l as a function of $\beta = \alpha + 2l$. Since the characteristic equation is transcendental in the variable γ, numerical techniques are necessary in general. However, if Δ, the modulation index of the free-electron density in the plasma, is very small, some approximate results may be obtained.

In the limit as Δ(or δ) approaches zero, the coefficients θ_n all approach zero except θ_0, which becomes

$$\theta_0|_{\delta=0} = \left(\frac{k_0 d}{\pi}\right)^2 - \left(\frac{k_{p0} d}{\pi}\right)^2 - \left(\frac{\gamma d}{\pi}\right)^2 , \quad (45)$$

and the characteristic equation becomes

$$\theta_0 = \beta^2 = (\alpha + 2l)^2 . \quad (46)$$

Upon elimination of θ_0 from Eqs. (45) and (46), we obtain

$$\left(\frac{\gamma_l d}{\pi}\right)^2 = \left(1 - \frac{c^2}{v^2}\right)\left(\frac{k_0 d}{\pi}\right)^2 - 4l\frac{c}{v}\left(\frac{k_0 d}{\pi}\right) - \left(\frac{k_{p0} d}{\pi}\right)^2 - 4l^2 . \quad (47)$$

Setting $(\gamma_l d/\pi)^2 = 0$ and solving for $k_0 d/\pi$ (the normalized cutoff frequencies) yields the result that $(\gamma_l d/\pi)^2 > 0$ if

$$\text{(a)} \quad \frac{k_{p0} d}{\pi} < \frac{2|l|}{(c^2/v^2 - 1)^{1/2}} \equiv \nu_l , \quad (48)$$

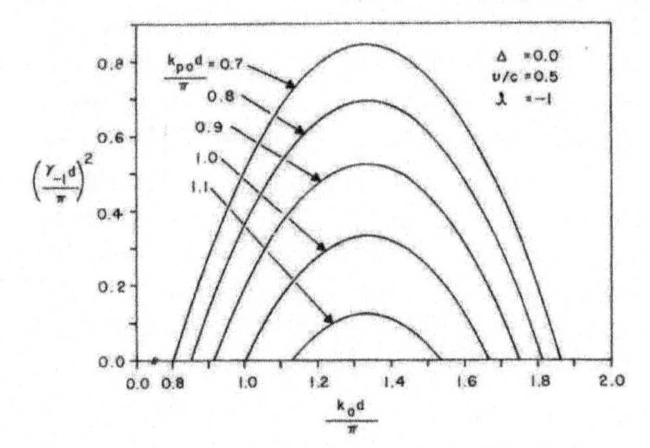

FIG. 2. Normalized transverse separation constant $(\gamma_{-1}d/\pi)^2$ versus k_0d/π in homogeneous limit $\Delta=0$. Values of $(\gamma_{-2}d/\pi)^2$ versus k_0d/π for $\Delta=0$ are obtained by multiplying $k_{p0}d/\pi$ and k_0d/π by 2, $(\lambda_{-1}d/\pi)^2$ by 4.

(b) $l<0$, (49)

(c) $$\frac{2|l|}{(c^2/v^2-1)}\left[\frac{c}{v}-\left(1-\frac{(k_{p0}d/\pi)^2}{y_l^2}\right)^{1/2}\right]<k_0d/\pi$$
$$<\frac{2|l|}{(c^2/v^2-1)}\left[\frac{c}{v}+\left(1-\frac{(k_{p0}d/\pi)^2}{y_l^2}\right)^{1/2}\right]. \quad (50)$$

Thus the lth mode ($l<0$) is radiative over the range given in Eq. (50) if the normalized average plasma frequency $k_{p0}d/\pi$ is less than a critical value which depends on l and v/c. When $v/c \ll 1$, this critical value is

$$y_l = 2|l|v/c \ . \quad (51)$$

It is apparent that if the mode $l=-|L|$ can be radiative, so also can be the modes $l=-|L|-1$, $-|L|-2, \ldots$. Further, it is clear that no matter what nonzero values are assigned to $k_{p0}d/\pi$ and v/c, it is possible to find a large negative value of l such that Eq. (48) is satisfied. Thus, radiation is emitted for arbitrarily small v or large $k_{p0}d/\pi$.

Values of $(\gamma_l d/\pi)^2$ as a function of normalized frequency k_0d/π for $v/c=0.5$, $l=-1$ and -2, $\Delta=0$ [cf. Eq. (47)], and $\Delta=0.4$, and various values of $k_{p0}d/\pi$, are shown in Figs. 2–4. It will be noted that when $k_{p0}d/\pi$ is relatively small, the curves for $\Delta=0$ and $\Delta=0.4$ are very similar, except for the discontinuities which occur when $\alpha=(c/v)\times(k_0d/\pi)$ takes on certain integer values. This occurs because the right-hand side of the characteristic equation is double valued when $\beta=\alpha+2l$ is a nonzero integer. As $k_{p0}d/\pi$ is increased, the deviation of the $\Delta=0.4$ curves from parabolic shape increases. This is a consequence of the fact that

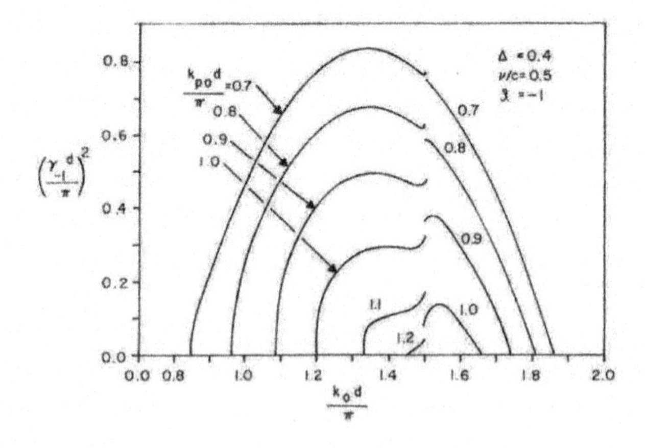

FIG. 3. Normalized transverse separation constant $(\gamma_{-1}d/\pi)^2$ versus k_0d/π for $\Delta=0.4$.

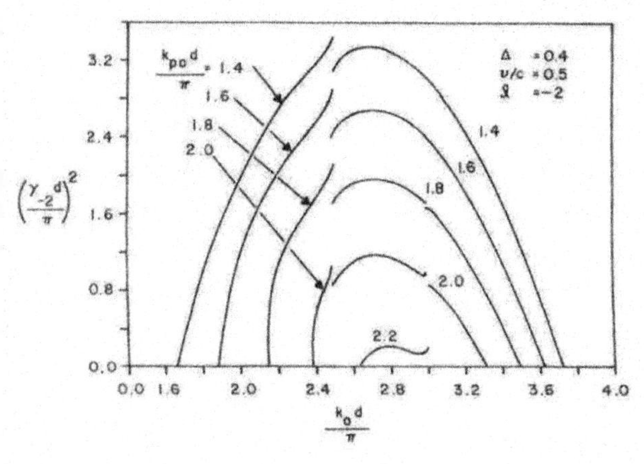

FIG. 4. Normalized transverse separation constant $(\gamma_{-2}d/\pi)^2$ versus $k_0 d/\pi$ for $\Delta=0.4$.

for fixed Δ and $\omega>\omega_{p0}$, an increase of $k_{p0}d/\pi$ increases δ, the inhomogeneity in permittivity. Finally, it will be noted that the low-frequency cutoffs in the curves for $\Delta=0.4$ occur at frequencies which satisfy Eq. (35), as has been noted earlier.

V. RADIATION SPECTRA

The normalized energy spectral density $\mathcal{U}_l^n$ *of the lth mode is defined as follows* [cf. Eq. (44)]:

$$\mathcal{U}_l^n=\left(\frac{\gamma_l d}{\pi}\right)^2\left(\frac{k_0 d}{\pi}\right)^{-1}\left(1-\frac{\omega_{p0}^2}{\omega^2}\right)^{-1}$$

$$\times\left|\sum_{n=-\infty}^{\infty} b_n(\alpha+2l)D_{n+l}(\delta)\right|^2\Big/\sum_{n=-\infty}^{\infty} b_n^2(\alpha+2l)\ . \tag{52}$$

The coefficients $D_k(\delta)$ are given by [cf. Eq. (41)]

$$D_k(\delta)=(\tfrac{1}{2}\delta)^{|k|}\sum_{l=0}^{\infty}(\tfrac{1}{2}\delta)^{2l}\Gamma_l^{|k|}\ , \tag{53}$$

in which

$$\Gamma_l^{|k|}=\frac{(4l+2|k|)!\,2^{-4l-2|k|}}{l!\,(l+|k|)!\,(2l+|k|)!}\ . \tag{54}$$

The coefficients $b_n(\beta)$ are obtained from the recurrence relations of Eq. (37) by writing those equations in the following matrix form:

$$\begin{bmatrix}\cdots & \cdots & \cdots & \cdots & \cdots\\ \cdots & \theta_0-(\beta-2)^2 & \theta_1 & \theta_2 & \cdots\\ \cdots & \theta_1 & \theta_0-\beta^2 & \theta_1 & \cdots\\ \cdots & \theta_2 & \theta_1 & \theta_0-(\beta+2)^2 & \cdots\\ \cdots & \cdots & \cdots & \cdots & \cdots\end{bmatrix}\begin{bmatrix}\vdots\\ b_{-1}(\beta)\\ b_0(\beta)\\ b_1(\beta)\\ \vdots\end{bmatrix}=0\ . \tag{55}$$

Setting the determinant of the matrix of coefficients equal to zero yields the characteristic equation (38). Rearranging the set of equations (55) and deleting the central equation ($n=0$), one obtains a set of equations for the ratios $b_n(\beta)/b_0(\beta)$ as follows:

$$\begin{bmatrix}\cdots & \cdots & \cdots & \cdots\\ \cdots & \theta_0-(\beta-2)^2 & \theta_2 & \cdots\\ \cdots & \theta_2 & \theta_0-(\beta+2)^2 & \cdots\\ \cdots & \cdots & \cdots & \cdots\end{bmatrix}\begin{bmatrix}\vdots\\ b_{-1}(\beta)/b_0(\beta)\\ b_1(\beta)/b_0(\beta)\\ \vdots\end{bmatrix}=-\begin{bmatrix}\vdots\\ \theta_1\\ \theta_1\\ \vdots\end{bmatrix}\ . \tag{56}$$

The system of equations (56) is inverted to yield the ratios $b_n(\beta)/b_0(\beta)$ by Cramer's rule. θ_0 and β are, of course, related by the characteristic equation.

By inverting the system of equations (56) and evaluating the coefficients $D_k(\delta)$, one may calculate $\mathfrak{U}_l^n$ for any set of parameters. Numerical calculation of $\mathfrak{U}_l^n$ as a function of (k_0d/π) has been carried out for $\Delta = 0.4$, $v/c = 0.5$, $l = -1$ and -2, and various values of $k_{p0}d/\pi$. The ranges of normalized frequency over which the spectra are nonzero are given in Figs. 3 and 4.

The results of the calculation for the $l = -1$ mode are shown in Figs. 5 and 6. As has been pointed out earlier, the lower limit of the frequency band over which radiation is emitted satisfies Eq. (35), so that energy is radiated only when the plasma is everywhere transparent. The upper limit of the band effectively occurs slightly above $k_0d/\pi = 1.5$, although $(\gamma_{-1}d/\pi)^2$ is positive for larger values of k_0d/π for most of the values of $k_{p0}d/\pi$ considered (cf. Fig. 3). For values of $k_{p0}d/\pi$ smaller than those for which curves are shown (e.g., $k_{p0}d/\pi = 0.5$), the normalized spectral density $\mathfrak{U}_{-1}^n$ is everywhere below 0.001, even at $k_0d/\pi = 1.5$. For values of $k_{p0}d/\pi$ much larger than those for which curves are shown, the $l = -1$ mode is nonradiative.

Spectral density curves for the $l = -2$ mode are

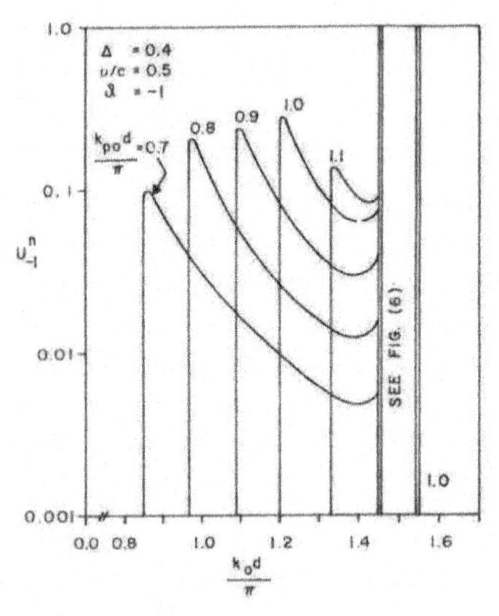

Fig. 5. Normalized energy spectral density $\mathfrak{U}_{-1}^n$ versus k_0d/π; $\Delta = 0.4$.

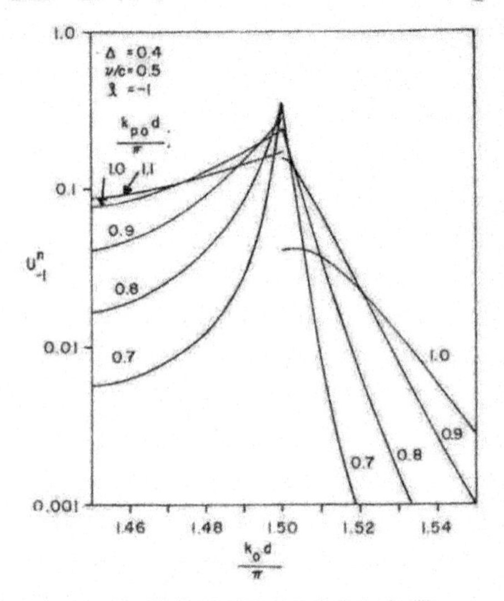

FIG. 6. $\mathfrak{U}_{-1}^n$ versus k_0d/π for $\Delta = 0.4$: detail.

shown in Figs. 7 and 8. For values of $k_{p0}d/\pi$ much smaller than those for which curves are shown, the spectral density is everywhere very small. The principal difference between the $l = -1$ and $l = -2$ curves is that, for the range of parameters included, the $l = -2$ curves have two discontinuities, while the $l = -1$ curves have but one. This is simply a result of the fact that the frequency range involved for the $l = -2$ modes is larger and that $\alpha = c/v(k_0d/\pi) - 4$ takes a nonzero integer value at $k_0d/\pi = 2.5$ and 3.0. The low-frequency cutoff again satisfies Eq. (35), and the effective high-frequency cutoff occurs at $k_0d/\pi = 3.0$, even though $(\gamma_{-2}d/\pi)^2$ is nonzero for higher frequencies, for most of the values of $k_{p0}d/\pi$ considered (cf. Fig. 4).

For the values of $k_{p0}d/\pi$ considered, the next $(l = -3)$ mode has a very small spectral density over its radiative frequency range. When $k_{p0}d/\pi$ increases to, say, 2.4, this mode emits a substantial amount of energy. Thus the pattern of the emission emerges: most of the radiation occurs in a frequency band beginning slightly above the average plasma frequency; this energy is carried by the "dominant" (i.e., smallest $|l|$) mode. Strictly speaking, energy is radiated at all frequencies above the low-frequency cutoff, but most of the emission is very weak indeed.

VI. SUMMARY

The formal solution for the radiation emitted by

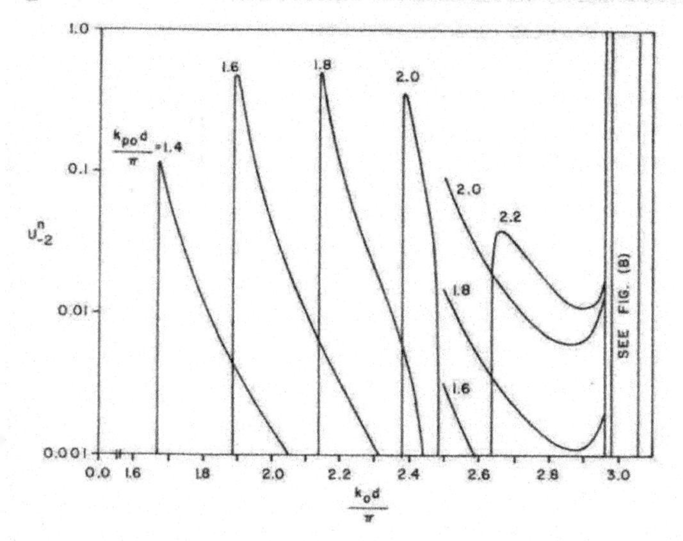

FIG. 7. Normalized energy spectral density $\mathfrak{U}_{-2}^{n}$ versus $k_0 d/\pi$; $\Delta = 0.4$.

a charged particle traversing a periodically stratified plasma has been obtained. It has been shown that transition radiation is emitted for any particle speed or average electron density of the plasma. The emission is most pronounced in a band of frequencies beginning slightly above the average plasma frequency.

It is important to note that the individual modes excited by the passage of the charged particle through the plasma are not coupled, but are excited independently of each other [cf. Eqs. (20) and (24)]. As a consequence, the modes also radiate independently of each other [cf. Eq. (25)]. This is in contrast to problems of the Smith-Purcell[8,9] type, where the various "space harmonics" (analogous to the modes in the present problem) are coupled by the boundary conditions. Thus certain types of resonances found in Smith-Purcell radiation[9] are not found in the present problem and allied problems. The shape of the spectral density curves is determined solely by the coupling between the souce current and the Hill function for the mode in question.

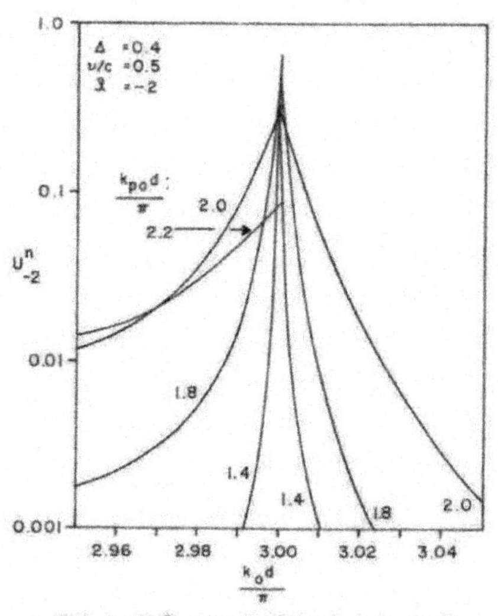

FIG. 8. $\mathfrak{U}_{-2}^{n}$ versus $k_0 d/\pi$ for $\Delta = 0.4$: detail.

---

*Present address: Dep't. of Electrical Engineering, Kansas State University, Manhattan, Kansas 66502.

[1] P. A. Čerenkov, Phys. Rev. 52, 378 (1937).
[2] I. M. Frank and I. Tamm, Dokl. Akad. Nauk SSSR 14, 109 (1937).
[3] V. L. Ginzburg and I. M. Frank, Zh. Eksperim. i Teor. Fiz. 16, 15 (1946).
[4] G. Garybyan, Zh. Eksperim. i Teor. Fiz. 33, 1403 (1957) [Soviet Phys. JETP 6, 1079 (1958)].
[5] A. Amatuni and N. Korkhmazyan, Zh. Eksperim. i Teor. Fiz. 39, 1011 (1960) [Soviet Phys. JETP 12, 703 (1961)].

[6]K. F. Casey, C. Yeh, and Z. A. Kaprielian, Phys. Rev. 140, B768 (1965).

[7]E. T. Whittaker and G. N. Watson, *A Course of Modern Analysis* (Cambridge U. P., New York, 1963).

[8]S. J. Smith and E. M. Purcell, Phys. Rev. 92, 1069 (1953).

[9]A. Hessel, Can. J. Phys. 42, 1196 (1964).

PHYSICAL REVIEW A VOLUME 8, NUMBER 1 JULY 1973

Diffraction Radiation from a Charged Particle Moving through a Penetrable Sphere*

R. Pogorzelski and C. Yeh

Electrical Sciences and Engineering Department, University of California, Los Angeles, California 90024

(Received 12 July 1972)

The formal exact solution to the problem of the radiation of a charged particle traveling with a constant velocity through a dielectric sphere is obtained. The electromagnetic field may be expressed in terms of an infinite number of spherical normal modes. It is found that radiation may be emitted for arbitrarily small particle velocities. For larger spheres (i.e., $ka \gg 1$ when k is the wave number and a is the radius of the sphere) the radiated field is predominantly Čerenkov-type radiation when the velocity of the particle is above the Čerenkov threshold velocity. For small spheres ($ka \gg 1$) and low particle velocities the radiation is shown to be mainly of transition type. Numerical results are presented to illustrate the behavior of the spectra of the various lower-order radiative modes as the velocity of the particle is varied.

I. INTRODUCTION

When a charged particle moves with constant velocity through or by an obstacle, there are three (macroscopic) mechanisms by which radiation may be emitted. Radiation of the Čerenkov type[1,2] is expected if the particle moves along or through a region in which the phase velocity of light is less than the speed of the particle. For radiation of this type to occur, the speed of the particle must be greater than the smallest phase velocity encountered. Transition radiation,[3-6] and diffraction radiation,[7-9] on the other hand, may be expected to occur at any particle speed. Transition radiation which occurs when the particle passes from one electrical medium to another, and diffraction radiation which occurs when the particle moves in the vicinity of a localized inhomogeneity in a medium, may be thought of as being emitted by the accelerated motion of the induced image charges. Since the nonuniform motion of the images will occur even if the moving charged particle is traveling slowly, there is no velocity threshold for transition or diffraction radiation.

Most previous studies[8] on diffraction radiation were carried out for perfectly conducting (impenetrable) bodies such as conducting half-planes, screens, or gratings, open ends of metallic waveguides, or conducting spheres. The important problem of diffraction radiation due to the presence of dielectric (penetrable) bodies has not been considered. It is expected that, owing to the presence of multiple reflections within the penetrable obstacles, the radiation characteristics for a uniformly moving charged particle passing by or through such penetrable obstacles will be quite different from those for the impenetrable case. An important feature of diffraction radiation is that the sources in general excite a continuous spectrum of frequencies; hence, it is essential that exact solutions to the problems of diffraction radiation be obtained. In this paper we shall treat the problem of diffraction radiation from a uniformly moving charged particle passing through a dielectric sphere. Exact solutions are obtained by expanding the incident fields due to the moving charge in terms of spherical harmonics and by matching the incident field and diffracted field with the interior field at the boundary of the dielectric sphere. In other words, the electromagnetic field excited by the passage of the charged particle may be expressed in terms of an infinite number of normal modes.[10] Numerical computations were carried out for several low order modes for a wide frequency spectrum. Detailed discussion of the results is given. Several important approximate solutions are also presented.

II. FORMULATION OF PROBLEM

It is assumed that a charge q moving in free space with velocity $\vec{v}$ along the positive z axis is incident upon a dielectric sphere of radius a (see Fig. 1). The dielectric sphere possesses a relative permittivity ϵ_1/ϵ_0 and a relative permeability $\mu/\mu_0=1$. ϵ_0 and μ_0 are, respectively, the free-space permittivity and permeability.

Maxwell's equations for the persent situation are

$$\nabla\times\vec{H}=\frac{\partial\vec{D}}{\partial t}+q\vec{v}\delta(\vec{r}-\vec{v}t) , \tag{1}$$

$$\nabla\times\vec{E}=-\mu_0\frac{\partial\vec{H}}{\partial t} , \tag{2}$$

$$\nabla\cdot\vec{D}=q\delta(\vec{r}-\vec{v}t) , \tag{3}$$

$$\nabla\cdot\vec{H}=0 , \tag{4}$$

where $\vec{E}$ and $\vec{H}$ represent the electric and magnetic fields and $\vec{r}$ is the radial vector. For the region $r>a$, $\vec{D}=\epsilon_0\vec{E}$, and for the region $r<a$, $\vec{D}=\epsilon_1\vec{E}$. Naturally, spherical coordinates will be used.

The charge density can then be written

$$\rho=q\delta(\vec{r}-\vec{v}t)=\frac{q}{r\sin\theta}\delta(r\sin\theta)\delta(r\cos\theta-vt) \tag{5}$$

and the corresponding current density is

$$\vec{J}(\vec{r},t)=\frac{q\vec{v}}{r\sin\theta}\delta(r\sin\theta)\delta(r\cos\theta-vt) . \tag{6}$$

Taking the Fourier transforms of Eqs. (1) and (2)

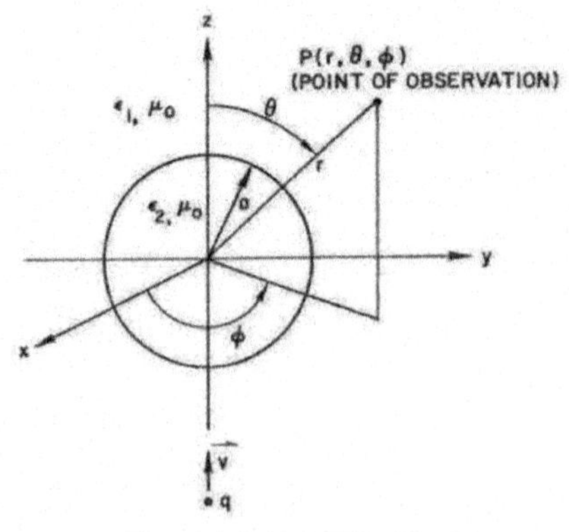

FIG. 1. Geometry of the problem.

with respect to time gives

$$\nabla\times\vec{E}_\omega=i\omega\mu_0\vec{H}_\omega , \tag{7}$$

$$\nabla\times\vec{H}_\omega=-i\omega\vec{D}_\omega+\frac{q}{r\sin\theta}\delta(r\sin\theta)e^{i(\omega/v)r\cos\theta}\cos\theta\hat{a}_r ; \tag{8}$$

$\vec{E}_\omega$, $\vec{D}_\omega$, and $\vec{H}_\omega$ denote the Fourier transforms of $\vec{E}$, $\vec{D}$, and $\vec{H}$, respectively; they are related to $\vec{E}$, $\vec{D}$, and $\vec{H}$ by

$$\vec{E}=\int_{-\infty}^{\infty}\vec{E}_\omega e^{-i\omega t}(d\omega/2\pi) , \tag{9}$$

$$\vec{D}=\int_{-\infty}^{\infty}\vec{D}_\omega e^{-i\omega t}(d\omega/2\pi) , \tag{10}$$

$$\vec{H}=\int_{-\infty}^{\infty}\vec{H}_\omega e^{-i\omega t}(d\omega/2\pi) . \tag{11}$$

We wish to find the appropriate expressions for $\vec{E}_\omega$, $\vec{D}_\omega$, and $\vec{H}_\omega$ from Eqs. (7) and (8) subject to the satisfaction of the boundary conditions at the surface of the dielectric sphere which require the continuity of tangential electric and magnetic fields and subject to the Sommerfeld radiation condition at $r=\infty$.

III. FORMAL SOLUTION

The most elegant and useful way of solving diffraction problems with sources is to use the dyadic-Green's-function method.[11] Combining Eqs. (7) and (8) gives

$$\nabla\times\nabla\times\vec{E}-\omega^2\mu_0\epsilon\vec{E}=i\omega\mu_0\vec{J}_\omega ,$$

where

$$\vec{J}_\omega=\frac{q\delta(r\sin\theta)}{r\sin\theta}e^{i(\omega/v)r\cos\theta}\cos\theta\hat{a}_r . \tag{12}$$

In free space, $\epsilon=\epsilon_0$. The free-space dyadic Green's function $\overleftrightarrow{G}_0(\vec{r}|\vec{r}')$ satisfies the following equation:

$$\nabla\times\nabla\times\overleftrightarrow{G}_0(\vec{r}|\vec{r}')-\omega^2\mu_0\epsilon_0\overleftrightarrow{G}_0(\vec{r}|\vec{r}')=\overleftrightarrow{I}\delta(\vec{r}-\vec{r}') , \tag{13}$$

where $\overleftrightarrow{I}$ is the unit dyadic and $\delta(\vec{r}-\vec{r}')$ is a δ function. One may show with the help of the vector Green's theorem that if $\overleftrightarrow{G}_0(\vec{r}|\vec{r}')$ and $\vec{J}_\omega(\vec{r}')$ are known, $\vec{E}_\omega(\vec{r})$ can be found by the relation

$$\vec{E}_\omega(\vec{r})=i\omega\mu_0\int_{V'}\overleftrightarrow{G}_0(\vec{r}|\vec{r}')\cdot\vec{J}_\omega(\vec{r}')dV' , \tag{14}$$

where the integration is performed over the volume V' containing the source currents. $\overleftrightarrow{G}_0(\vec{r}|\vec{r}')$ may be expanded in terms of the eigenfunctions of the vector wave equation. In spherical coordinates, we have

$$\overleftrightarrow{G}_0(\vec{r}|\vec{r}')=\frac{ik}{4\pi}\sum_{l=1}^{\infty}\frac{2l+1}{l(l+1)}\times\begin{cases}\vec{M}_l^{(3)}(kr)\vec{M}_l^{(1)}(kr')+\vec{N}_l^{(3)}(kr)\vec{N}_l^{(1)}(kr') ; & r>r'\\ \vec{M}_l^{(1)}(kr)\vec{M}_l^{(3)}(kr')+\vec{N}_l^{(1)}(kr)\vec{N}_l^{(3)}(kr') ; & r<r'\end{cases}$$

where

$$\vec{M}_l^{(1)\atop(3)}(kr) = -\begin{Bmatrix} j_l(kr) \\ h_l^{(1)}(kr)\end{Bmatrix} \frac{\partial}{\partial\theta} P_l(\cos\theta)\hat{a}_\phi \ ,$$

$$\vec{N}_l^{(1)\atop(3)} = \frac{l(l+1)}{kr}\begin{Bmatrix} j_l(kr) \\ h_l^{(1)}(kr)\end{Bmatrix} P_l(\cos\theta)\hat{a}_r + \frac{1}{kr}\frac{\partial}{\partial r}\left(r\begin{Bmatrix} j_l(kr) \\ h_l^{(1)}(kr)\end{Bmatrix}\right)\frac{\partial}{\partial\theta} P_l(\cos\theta)\hat{a}_\theta \ , \tag{15}$$

where $k=\omega(\mu_0\epsilon_0)^{1/2}$ and $\hat{a}_r$, $\hat{a}_\theta$, and $\hat{a}_\phi$ are the spherical coordinate unit vectors. The function $j_l(kr)$ is the spherical Bessel function of the first kind and the function $h_l^{(1)}(kr)$ is the spherical Hankel function of the first kind. $P_l(\cos\theta)$ are Legendre polynomials. Owing to the symmetry of the present problem, only the $m=0$ (cylindrically symmetric) terms were required.

The presence of the dielectric sphere introduces an additional scattered field. Hence, by superposition, the total dyadic Green's function consists of the unperturbed infinite-space dyadic Green's function $\vec{G}_0(\bar{r}|\bar{r}')$ plus the contribution due to the scattered field.

For $r'>a$, we have

$$\vec{G}(\bar{r}|\bar{r}') = \vec{G}_0(\bar{r}|\bar{r}')\big|_{r>r', k=k_1} + \vec{G}_s^{(11)}(\bar{r}|\bar{r}') \ , \quad r>r'$$

$$\vec{G}(\bar{r}|\bar{r}') = \vec{G}_0(\bar{r}|\bar{r}')\big|_{r<r', k=k_1} + \vec{G}_s^{(11)}(\bar{r}|\bar{r}') \ , \quad a<r<r'$$

$$\vec{G}(\bar{r}|\bar{r}') = \vec{G}_s^{(21)}(\bar{r}|\bar{r}') \ , \quad r<a$$

where

$$\vec{G}_s^{(11)}(\bar{r}|\bar{r}') = \frac{ik_1}{4\pi}\sum_{l=1}^{\infty}\frac{2l+1}{l(l+1)}\left[A_{1l}\vec{M}_l^{(3)}(k_1r)\vec{M}_l^{(3)}(k_1r') + B_{1l}\vec{N}_l^{(3)}(k_1r)\vec{N}_l^{(3)}(k_1r')\right], \tag{16}$$

$$\vec{G}_s^{(21)}(\bar{r}|\bar{r}') = \frac{ik_1}{4\pi}\sum_{l=1}^{\infty}\frac{2l+1}{l(l+1)}\left[C_{1l}\vec{M}_l^{(1)}(k_2r)\vec{M}_l^{(3)}(k_1r') + D_{1l}\vec{N}_l^{(1)}(k_2r)\vec{N}_l^{(3)}(k_1r')\right],$$

and for $r'<a$, we have

$$\vec{G}(\bar{r}|\bar{r}') = \vec{G}_s^{(12)}(\bar{r}|\bar{r}'), \quad r>a$$

$$\vec{G}(\bar{r}|\bar{r}') = \vec{G}_0(\bar{r}|\bar{r}')\big|_{r'<r, k=k_2} + \vec{G}_s^{(22)}(\bar{r}|\bar{r}') \ , \quad r'<r<a$$

$$\vec{G}(\bar{r}|\bar{r}') = \vec{G}_0(\bar{r}|\bar{r}')\big|_{r<r', k=k_2} + \vec{G}_s^{(22)}(\bar{r}|\bar{r}') \ , \quad r<r'$$

where

$$\vec{G}_s^{(12)}(\bar{r}|\bar{r}') = \frac{ik_2}{4\pi}\sum_{l=1}^{\infty}\frac{2l+1}{l(l+1)}\left[A_{2l}\vec{M}_l^{(3)}(k_1r)\vec{M}_l^{(1)}(k_2r') + B_{2l}\vec{N}^{(3)}(k_1r)\vec{N}_l^{(1)}(k_2r')\right], \tag{17}$$

$$\vec{G}_s^{(22)}(\bar{r}|\bar{r}') = \frac{ik_2}{4\pi}\sum_{l=1}^{\infty}\frac{2l+1}{l(l+1)}\left[C_{2l}\vec{M}_l^{(1)}(k_2r)\vec{M}_l^{(1)}(k_2r') + D_{2l}\vec{N}_l^{(1)}(k_2r)\vec{N}_l^{(1)}(k_2r')\right].$$

The subscript s designates the scattered part of the dyadic Green's function. A_{1l}, B_{1l}, C_{1l}, D_{1l}, A_{2l}, B_{2l}, C_{2l}, and D_{2l} are arbitrary constants that are to be determined from the boundary conditions on the surface of the dielectric sphere at $r=a$. The constant $k_1=\omega(\mu_0\epsilon_1)^{1/2}$ and the constant $k_2=\omega(\mu_0\epsilon_2)^{1/2}$, where ϵ_1 and ϵ_2 are, respectively, the dielectric constant outside the sphere $(r>a)$ and that inside the sphere $(r\leq a)$.

It is noted that the transformed current source extends from $r=0$ to $r=\infty$ for $\theta=0$ and $\theta=\pi$. Therefore, the dyadic Green's function must be represented by two separate expressions; one for $r'>a$ and the other for $r'<a$.

In order to match the tangential electric and magnetic fields at $r=a$, we require the continuity of

$$\hat{a}_r\times\vec{G} \tag{18}$$

and

$$(1/\mu)\hat{a}_r\times(\nabla\times\vec{G}) \ , \tag{19}$$

where $\hat{a}_r$ is the unit vector in the radial direction.

$$j_l(k_1a)+A_{1l}h_l^{(1)}(k_1a) = C_{1l}j_l(k_2a) \ ,$$

$$[k_1a\,j_l(k_1a)]' + A_{1l}[k_1ah_l^{(1)}(k_1a)]' = C_{1l}[k_2a\,j_l(k_2a)]' \ , \tag{20}$$

$$k_1j_l(k_1a)+B_{1l}k_1h_l^{(1)}(k_1a) = D_{1l}k_2j_l(k_2a) \ ,$$

$$\frac{1}{k_1a}[k_1a\,j_l(k_1a)]' + \frac{B_{1l}}{k_1a}[k_1ah_l^{(1)}(k_1a)]' = \frac{D_{1l}}{k_2a}[k_2a\,j_l(k_2a)]' \ , \tag{21}$$

$$h_l^{(1)}(k_2a)+C_{2l}j_l(k_2a) = A_{2l}h_l^{(1)}(k_1a) \ ,$$

$$[k_2ah_l^{(1)}(k_2a)]' + C_{2l}[k_2a\,j_l(k_2a)]' = A_{2l}[k_1ah_l^{(1)}(k_1a)]', \tag{22}$$

$$k_2h_l^{(1)}(k_2a)+D_{2l}k_2j_l(k_2a) = B_{2l}k_1h_l^{(1)}(k_1a) \ ,$$

$$\frac{1}{k_2a}[k_2ah_l^{(1)}(k_2a)]' + \frac{D_{2l}}{k_2a}[k_2a\,j_l(k_2a)]' = \frac{B_{2l}}{k_1a}[k_1ah_l^{(1)}(k_1a)]' \ . \tag{23}$$

The coefficients A_{1l}, B_{1l}, C_{1l}, D_{1l}, A_{2l}, B_{2l}, C_{2l}, and D_{2l} may be obtained from the above equations. If we are only interested in the far-zone fields, r will never be less than r'. Thus for $r'>a$,

$$\vec{G}(\bar{r}|\bar{r}') = \frac{ik_1}{4\pi}\sum_{l=1}^{\infty}\frac{2l+1}{l(l+1)}\vec{N}_l^{(3)}(k_1r)\left[\vec{N}_l^{(1)}(k_1r') + B_{1l}\vec{N}_l^{(3)}(k_1r')\right] \ , \tag{24}$$

and for $r' < a$,

$$\overleftrightarrow{G}(\bar{r}|\bar{r}') = \frac{ik_2}{4\pi} \sum_{l=1}^{\infty} \frac{2l+1}{l(l+1)} B_{2l} \bar{N}_l^{(3)}(k_1 r) \bar{N}_l^{(1)}(k_2 r') \quad . \tag{25}$$

The electric field is then

$$\bar{E}_\omega(\bar{r}) = -\frac{\omega \mu_0 q}{4\pi} \sum_{l=1}^{\infty} (2l+1) \Bigg[B_{2l} \int_0^a \frac{j_l(k_2 r')}{r'} [e^{i(\omega/v)r'} - (-1)^l e^{-i(\omega/v)r'}] dr' + \int_a^{\infty} \left(\frac{j_l(k_1 r') + B_{1l} h_l^{(1)}(k_1 r')}{r'} \right) [e^{i(\omega/v)r'} - (-1)^l e^{-i(\omega/v)r'}] dr' \Bigg] \bar{N}_l^{(3)}(k_1 r) \; , \tag{26}$$

where B_{1l} and B_{2l} can be found from Eqs. (20)–(23). Making the appropriate normalization, we have

$$\bar{E}_\omega(\bar{r}) = -\left(\frac{\mu_0 q \omega}{4\pi}\right) \sum_{l=1}^{\infty} (2l+1) \Bigg[B_{2l} \int_0^{x_2} \frac{j_l(x)}{x} [e^{i\alpha_2 x} - (-1)^l e^{-i\alpha_2 x}] dx + \int_{x_1}^{\infty} \left(\frac{j_l(x) + B_{1l} h_l^{(1)}(x)}{x} \right) [e^{i\alpha_1 x} - (-1)^l e^{-i\alpha_1 x}] dx \Bigg] \bar{N}_l^{(3)}(k_1 r) \; , \tag{27}$$

where $x_1 = k_1 a$, $x_2 = k_2 a$, $\alpha_1 = 1/n_1\beta$, $\alpha_2 = 1/n_2\beta$, $\beta = v/c$, $x' = k_1 r'$, $x = k_2 r'$, $n_1 = (\epsilon_1/\epsilon_0)^{1/2}$, and $n_2 = (\epsilon_2/\epsilon_0)^{1/2}$.

Equation (27) constitutes the formal solution for the transformed electric field. In Figs. 2–4 a unitless quantity $|A_l|/\omega$ is plotted as a function of $k_1 a$ to indicate the character of this result. A_l is the mode amplitude defined by

$$\bar{E}_\omega(\bar{r}) = \frac{\mu_0}{4\pi} q \sum_{l=1}^{\infty} A_l \bar{N}_l^{(3)}(k_1 r) \quad .$$

Thus A_l contains all of the l and ω dependence of the coefficient of $\bar{N}_l^{(3)}$.

IV. SEVERAL SPECIAL CASES

When the charged particle moves at a slow speed such that $n_2\beta < 1$, i.e., below the "Čerenkov limit," one may profitably expand the solution in powers of β. When l is odd we obtain, by integration by parts,

$$\begin{aligned} \bar{E}(\bar{r}) = &-\left(\frac{\mu_0 q\omega}{4\pi}\right) \sum_{l=1}^{\infty} (2l+1) \left[\left(\frac{\sin\alpha_2 x_2}{\alpha_2 x_2} \right) j_l(x_2) B_{2l} - [j_l(x_1) + B_{1l} h_l^{(1)}(x_1)] \left(\frac{\sin\alpha_1 x_1}{\alpha_1 x_1} \right) \right] \bar{N}_l^{(3)}(k_1 r) \\ &-\left(\frac{\mu_0 q\omega}{4\pi}\right) \sum_{l=1}^{\infty} (2l+1) \Bigg[\left(\frac{\cos\alpha_2 x_2}{(\alpha_2 x_2)^2} \right) [(l-1) j_l(x_2) - x_2 j_{l+1}(x_2)] \\ &- \frac{\cos\alpha_1 x_1}{(\alpha_1 x_1)^2} \left((l-1)[j_l(x_1) + B_{1l} h_l^{(1)}(x_1)] - x_1 [j_{l+1}(x_1) + B_{1l} h_{l+1}^{(1)}(x_1)] \right) \Bigg] \bar{N}_l(k_1 r) + O(\beta^3) \quad . \end{aligned} \tag{28}$$

When l is even,

$$\begin{aligned} \bar{E}(\bar{r}) = &\left(\frac{\mu_0 q\omega}{4\pi}\right) \sum_{l=1}^{\infty} (2l+1) \left[\left(\frac{\cos\alpha_2 x_2}{\alpha_2 x_2} \right) j_l(x_2) B_{2l} - [j_l(x_1) + B_{1l} h_l^{(1)}(x_1)] \left(\frac{\cos\alpha_1 x_1}{\alpha_1 x_1} \right) \right] \bar{N}_l^{(3)}(k_1 r) \\ &-\left(\frac{\mu_0 q\omega}{4\pi}\right) \sum_{l=1}^{\infty} (2l+1) \Bigg[\frac{\sin\alpha_2 x_2}{(\alpha_2 x_2)^2} [(l-1) j_l(x_2) - x_2 j_{l+1}(x_2)] \\ &- \frac{\sin\alpha_1 x_1}{(\alpha_1 x_1)^2} \left((l-1)[j_l(x_1) + B_{1l} h_l^{(1)}(x_1)] - x_1 [j_{l+1}(x_1) + B_{1l} h_{l+1}^{(1)}(x_1)] \right) \Bigg] \bar{N}_l^{(3)}(k_1 r) + O(\beta^3) \quad . \end{aligned} \tag{29}$$

Considering only the lowest-order terms $[O(\beta)]$ and noting that $\alpha_1 x_1 = \alpha_2 x_2$, we obtain for l odd,

$$\bar{E}_l(\bar{r}) = \left(\frac{\mu_0 q\omega(2l+1)}{2\pi} \right) \left[F_l(x_1) j_l(x_1) - F_l(x_2) \left(\frac{h_l^{(1)}(x_2)}{h_l^{(1)}(x_1)} \right) j_l(x_2) \right] \left(\frac{\sin\alpha_1 x_1}{\alpha_1 x_1} \right) \bar{N}_l^{(3)}(k_1 r) \tag{30}$$

and for l even,

$$\bar{E}_l(\bar{r}) = \left(\frac{\mu_0 q\omega(2l+1)}{2\pi i} \right) \left[F_l(x_1) j_l(x_1) - F_l(x_2) \left(\frac{h_l^{(1)}(x_2)}{h_l^{(1)}(x_1)} \right) j_l(x_2) \right] \left(\frac{\cos\alpha_1 x_1}{\alpha_1 x_1} \right) \bar{N}_l^{(3)}(k_1 r) \; , \tag{31}$$

where

$$F_l(x)=\left(\frac{[xh_l^{(1)}(x)]'}{xh_l^{(1)}(x)}-\frac{[xj_l(x)]'}{xj_l(x)}\right)$$
$$\times\left[\frac{[x_1h_l^{(1)}(x_1)]'}{x_1h_l^{(1)}(x_1)}-\left(\frac{x_1}{x_2}\right)\frac{[x_2j_l(x_2)]'}{x_2j_l(x_2)}\right]^{-1} .$$

One notes that the velocity enters these expressions only through the factors $\cos\alpha_1x_1/\alpha_1x_1$ and $\sin\alpha_1x_1/\alpha_1x_1$. When β is small, these factors oscillate rapidly as functions of x_1 compared with the other factors in $\vec{E}_l(\vec{r})$ and they are easily discernible in Fig. 2 as the rapid oscillations beneath the envelope. This lowest-order approximation is in excellent agreement with the results of numerical evaluation of (2.7) for $\beta=0.1$.

As β is increased this approximation becomes poorer and above the Čerenkov velocity it is totally useless. The effect of exceeding the Čerenkov velocity may be observed by analyzing the behavior of $\vec{E}(\vec{r})$ for large k_1r, i.e., large x_1. We start with Eq. (27) and note that the second integral approaches zero as x_1 approaches infinity and only the first integral remains. As x_2 approaches infinity the upper limit of this integral becomes infinite and the radiation is then due entirely to the Čerenkov effect. Consider now the velocity-dependent part of Eq. (27) which is

$$I_l=\int_0^{\infty}\frac{j_l(x)}{x}\left[e^{i\alpha_2x}-(-l)^le^{-i\alpha_2x}\right]dx \ , \qquad (32)$$

where

$$\alpha_2=\frac{1}{n_2\beta} .$$

For the $l=1$ mode, we have

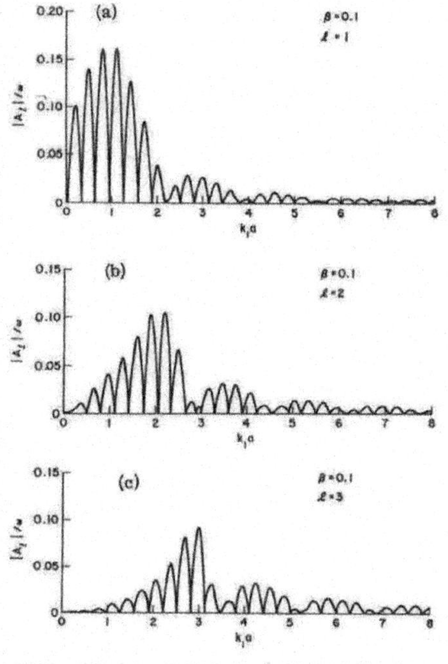

FIG. 2. Scattering coefficient as a function of k_1a for low particle velocity.

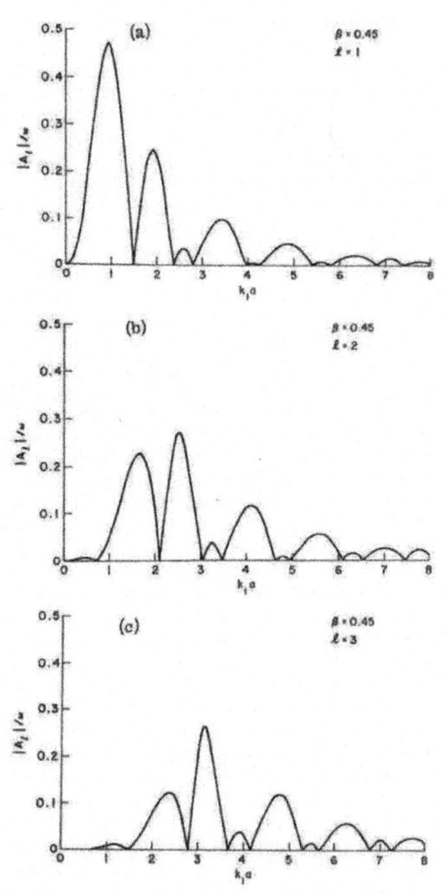

FIG. 3. Scattering coefficient as a function of k_1a for particle velocity below the Čerenkov threshold.

$$I_1 = 2\int_0^\infty \frac{j_1(x)}{x}\cos(\alpha_2 x)\,dx \quad . \tag{33}$$

Substituting the expansion $j_1(x) = \sin x/x^2 - \cos x/x$ in Eq. (33) and integrating by parts yields

$$I_1 = (1-\alpha_2^2)\int_0^\infty (\sin x \cos\alpha_2 x/x)\,dx \quad . \tag{34}$$

Since

$$\int_0^\infty \frac{\sin x \cos\alpha_2 x}{x}\,dx = \begin{cases} 0 & \text{for } |\alpha_2| > 1 \ , \\ \tfrac{1}{2}\pi & \text{for } |\alpha_2| < 1 \ , \end{cases} \tag{35}$$

one has

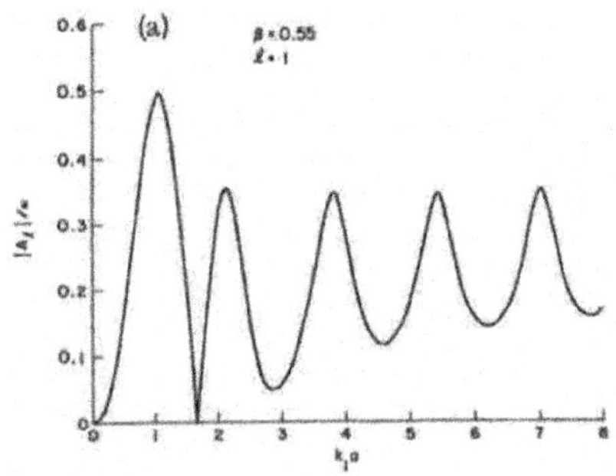

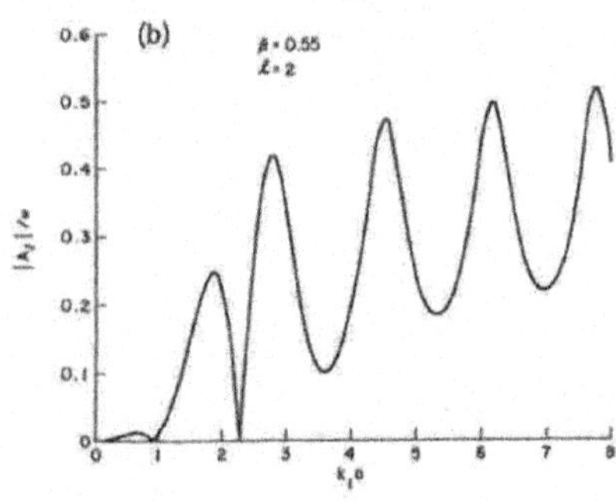

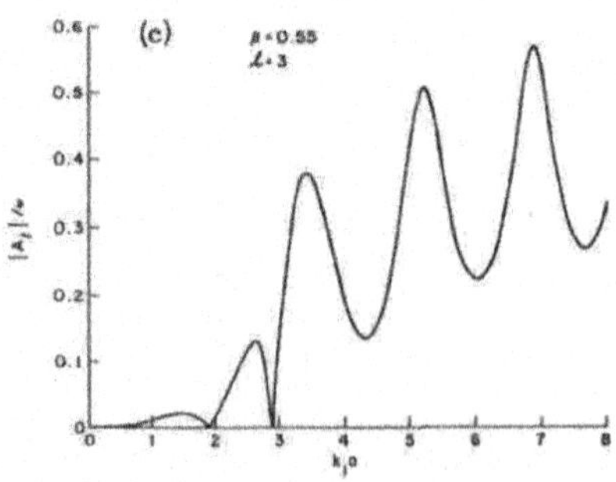

FIG. 4. Scattering coefficient as a function of k_1a for particle velocity above the Čerenkov threshold.

$$I_1 = \begin{cases} 0 & \text{for } \beta < 1/n_2 \ , \\ \tfrac{1}{2}(1-\alpha^2)\pi & \text{for } \beta > 1/n_2 \ , \end{cases} \tag{36}$$

which clearly exhibits the Čerenkov threshold characteristic. Carrying out a similar procedure for the $l = 2$ mode gives

$$I_2 = \cdot \begin{cases} 0 & \text{for } \beta < 1/n_2 \ , \\ \alpha(1-\alpha^2)\tfrac{1}{2}\pi i & \text{for } \beta > 1/n_2 \ . \end{cases} \tag{37}$$

The question now arises as to whether or not I_l exhibits threshold behavior for all l. This can be demonstrated by making use of the following recursion relation:

$$j_p(x) = [(2p-1)/x]j_{p-1}(x) - j_{p-2}(x) \quad . \tag{38}$$

Using (38), we obtain

$$\begin{aligned} I_p &= \int_0^\infty \frac{j_p(x)}{x}[e^{i\alpha_2 x} - (-1)^p e^{-i\alpha_2 x}]\,dx \\ &= \int_0^\infty \frac{(2p-1)j_{p-1}(x)}{x^2}[e^{i\alpha_2 x} - (-1)^p e^{-i\alpha_2 x}]\,dx \\ &\quad - \int_0^\infty \frac{j_{p-2}(x)}{x}[e^{i\alpha_2 x} - (-1)^p e^{-i\alpha_2 x}]\,dx \\ &= i\int (2p-1)I_{p-1}\,d\alpha_2 - I_{p-2} + C \ , \end{aligned} \tag{39}$$

where C is a constant of integration. Now, if I_{p-1} and I_{p-2} are zero for $\alpha_2 > 1$, I_p remains constant for $1 < \alpha_2 < \infty$. The Riemann–Lebesgue lemma indicates that $I_p(\alpha_2) \to 0$ as $\alpha_2 \to \infty$ so I_p is zero for $1 < \alpha_2 < \infty$. Similarly, if I_{p-1} and I_{p-2} are zero for $\alpha_2 < -1$, I_p is zero for $\alpha_2 < -1$. Hence I_p may be written

$$I_p(\alpha_2) = i(2p-1)\int_{-1}^{\alpha_2} I_{p-1}(\alpha)\,d\alpha - I_{p-2}(\alpha_2) \tag{40}$$

and it exhibits the Čerenkov threshold, i.e., $I_p = 0$ for $|\alpha_2| > 1$. We have shown that if I_{p-1} and I_{p-2} exhibit the threshold, then I_p does also and we have shown that I_1 and I_2 exhibit threshold behavior. Thus, by induction $I_l(\alpha) = 0$ for $|\alpha| > 1$, which was to be demonstrated. This means that $\vec{E}(\vec{r})$ itself exhibits the Čerenkov threshold for large k_1a. This threshold behavior is evident in Figs. 3 and 4 which show the result of numerical computation of $\vec{E}_\omega(\vec{r})$ below $n_2\beta = 1$ ($\beta = 0.45$) and above $n_2\beta = 1$ ($\beta = 0.55$); $n_2 = 2$. The oscillations in the amplitude as a function of k_1a at large k_1a and $\beta > 1/n_2$ are due to the oscillatory behavior of the coefficient B_{2l} which becomes

$$2e^{i(x_2 - x_1)}\left\{\left(1 + \frac{x_2}{x_1}\right)\left[1 + (-1)^l\left(\frac{x_2 - x_1}{x_2 + x_1}\right)e^{2ix_2}\right]\right\}^{-1} \tag{41}$$

as x_1 and x_2 become very large. In this limit $\vec{E}_\omega(\vec{r})$ may be written in the following form:

$$\vec{E}_\omega(\vec{r}) \to -\left(\frac{\mu_0 q\omega}{4\pi}\right)\left(\frac{x_1}{x_2}\right)\sum_{l=1}^{\infty}(2l+1)\left(\frac{2x_2}{x_1+x_2}\,e^{i(x_2-x_1)}\right) \times\left\{\sum_{k=0}^{\infty}\left[-(-1)^l\left(\frac{x_2-x_1}{x_2+x_1}\right)e^{2ix_2}\right]^k\right\} I_l \vec{N}_l^{(3)}(k_1 r) \quad (42)$$

This form suggests the following process: A wave is generated by the charge moving through the material inside the sphere. Each of its component spherical modes then undergoes multiple reflections from the inner surface of the sphere represented by the factor in curly brackets { }. The factor e^{2ix_2} accounts for the phase shift in propagating from the center of the sphere to the surface and back. At each reflection a certain amount of radiation is transmitted represented by the factor $[2x_2/(x_1+x_2)]e^{i(x_2-x_1)}$. Note that this factor appears only once per term of the series while the reflection factor appears k times. This is because a wave generated within the sphere and reflected k times from the inside of the surface requires only one transmission through the surface in order to be radiated away. The factor $e^{i(x_2-x_1)}$ is necessary to match the $e^{ik_2 r}/r$ internal wave to the $e^{ik_1 r}/r$ external wave at $r=a$.

The other limiting case studied is the small $k_1 a$ or Rayleigh limit for small β. Using the asymptotic expressions for Bessel functions of small arguments we find that

$$\vec{E}(\vec{r}) \to \left[3\left(\frac{\mu_0 q\omega}{2\pi}\right)\beta\left(\frac{n_2^2-1}{n_2^2+2}\right)\sin\left(\frac{k_1 a}{\beta}\right)\right]\vec{N}_1^{(3)}(k_1 r) \quad (43)$$

This exhibits the rapid oscillatory behavior shown in Figs. 2–4 by virtue of the $\sin(k_1 a/\beta)$ factor. Expression (43) is in essence the ordinary dipole-term spectrum encountered in Rayleigh scattering but the spectrum is multiplied by

$$\sin\frac{k_1 a}{\beta} = \sin\omega\frac{a}{v} = \frac{1}{2i}\left(e^{i\omega(a/v)} - e^{-i\omega(a/v)}\right) \quad (44)$$

This corresponds, in the time domain, to time delay and a time advance of a/v seconds in the normal dipole response. That is, the radiation is emitted when the particle crosses the surface of the sphere and is termed transition radiation.

V. DISCUSSION OF NUMERICAL RESULTS AND CONCLUSION

Expression (27) was evaluated numerically using four-point Gaussian quadrature over x intervals of 0.1. The results are depicted in Figs. 2–4. Figure 2 shows the mode amplitude $|A_l|$ for $l=1$, 2, and 3, respectively, when the particle velocity is low, i.e., $\beta=0.1$. These results agree closely with those obtained using the low-velocity approximate results (30) and (31). From these equations we see that the rapid oscillatory behavior of the curves stems from the $\sin\alpha_1 x_1/\alpha_1 x_1$ and $\cos\alpha_1 x_1/\alpha_1 x_1$ factors, the envelope being determined by the rather complicated expression in square brackets. That is, the oscillatory behavior is determined by the particle velocity while the envelope is determined by the size and index of refraction of the dielectric sphere.

As the particle velocity is increases the rapid oscillations becomes less rapid as indicated in Fig. 3. Here the velocity is $0.45c$ which is just below the Čerenkov velocity c/n_2 for the sphere. In this case it is difficult to distinguish the two oscillatory factors, that involving the velocity and that determined by the sphere. They are nearly equal in oscillation rate.

The most striking change, however, occurs when one exceeds the Čerenkov velocity, c/n_2. This case is illustrated in Fig. 4. Here the velocity is $0.55c$, whereas the Čerenkov velocity for $n_2=2$ is $0.5c$. Note that now the amplitude no longer decreases to zero as ka approaches infinity. This is a manifestation of the production of Čerenkov radiation within the sphere. It has an infinite bandwidth because dispersion has been neglected. In reality, as the frequency approaches infinity the index of refraction of the material of which the sphere is made approaches unity; thus causing the Čerenkov radiation to approach zero in the high-frequency limit. The oscillations determined by the size and index of refraction of the sphere as explained previously [see Eqs. (30) and (31)] are now the more rapid ones. The particle velocity determines the slower oscillation [$\sim\sin(x_1/0.55)$] discernible in Fig. 4. Thus, above the Čerenkov velocity the sphere parameters and particle velocity switch roles in their determination of the slow and rapid oscillation rates compared to that below the Čerenkov velocity.

In essence, then, the radiation which is generated when a charged particle passes through a penetrable sphere is primarily of Čerenkov type for a large sphere and a high particle velocity and primarily of transition (dipole) type for small spheres and low particle velocity.

---

*Partly supported by the National Science Foundation.

[1] P. A. Čerenkov, Phys. Rev. 52, 378 (1937).
[2] I. M. Frank and I. Tamm, Dokl. Akad. Nauk SSSR 14, 109 (1937).
[3] V. L. Ginzburg and I. M. Frank, Zh. Eksp. Teor. Fiz. 16, 15 (1946).

[4]G. Garibyan, Zh. Eksp. Teor. Fiz. **33**, 1403 (1957) [Sov. Phys.-JETP **6**, 1079 (1958)].
[5]F. G. Bass and V. M. Yakovenko, Usp. Fiz. Nauk **86**, 189 1965) [Sov. Phys.-Usp. **8**, 420 (1965)].
[6]K. F. Casey and C. Yeh, Phys. Rev. A **2**, 810 (1970).
[7]A. P. Kazantsev and G. I. Snadutovich, Dokl. Akad. Nauk SSSR **147**, 74 (1962) [Sov. Phys.-Dokl. **7**, 990 (1963)].
[8]B. M. Bolotovskii and G. V. Voskresenskii, Usp. Fiz. Nauk **88**, 209 (1966) [Sov. Phys.-Usp. **9**, 73 (1966)].
[9]S. J. Smith and E. M. Purcell, Phys. Rev. **92**, 1069 (1953).
[10]J. A. Stratton, *Electromagnetic Theory* (McGraw-Hill, New York, 1941).
[11]C. T. Tai, *Dyadic Green's Functions in Electromagnetic Theory* (International Textbook, San Francisco, Calif., 1971).

On the Čerenkov threshold associated with synchrotron radiation in a dielectric medium*

R. J. Pogorzelski[†] and C. Yeh

Electrical Sciences and Engineering Department, University of California, Los Angeles, California 90024

K. F. Casey

Department of Electrical Engineering, Kansas State University, Manhattan, Kansas 66504

(Received 18 February 1974; in final form 13 May 1974)

A formal exact expression for the radiation field of a charged particle executing a circular orbit in a dispersive dielectric medium is obtained by means of a dyadic Green's function approach. The field is expressed as an infinite sum of vector spherical harmonics. The radiated power P is computed numerically and a set of universal curves of $nP/q^2\Omega^2$ vs $n\beta$ is obtained where q is the charge on the particle, n is the index of refraction of the medium, Ω is the orbital angular velocity of the particle, and $\beta = v/c$, where v is the particle velocity and c is the speed of light in vacuum. The Čerenkov threshold phenomenon is shown to be manifest primarily in the high-frequency portion of the radiated spectrum. For an index of refraction of unity, the result is shown (at low velocities) to be identical with both Larmor's formula for the radiated power and Schwinger's result for the angular distribution of the radiation.

INTRODUCTION

It is well known that a charged particle moving at superluminal velocity in a straight line through a dielectric medium will radiate electromagnetic energy known as Čerenkov radiation.[1,2] Where the velocity is subluminal no radiation will occur. Thus, the radiation exhibits a "threshold" at the velocity of light in the medium. In an earlier publication[3] we demonstrated how the Čerenkov threshold came about in the solution expressed in terms of a Green's function expanded in spherical vector harmonics in the case of a straight particle path. It was shown that the velocity dependence of the radiation was contained in the factor

$$\int_0^\infty [j_l(x)/x][\exp(i\alpha x) - (-1)^l \exp(-i\alpha x)]\,dx,$$

where $\alpha = (n\beta)^{-1}$, $\beta = v/c$, $n = \epsilon_r^{1/2}$ (the relative permittivity of the medium), and l is the order of the spherical harmonic θ dependence. This integral is zero for $n\beta < 1$ and nonzero for $n\beta > 1$ regardless of the index l. Thus, the threshold occurs in *each term* of the spherical vector harmonic series. The purpose of this paper is to discuss the Čerenkov threshold phenomenon in the case of a particle executing a circular motion in a homogeneous dielectric medium. Regardless of the velocity, such a particle radiates synchrotron radiation.[4] However, above the velocity of light in the medium some manifestation of a Čerenkov effect should be discernable. The problem is formulated by means of the free-space dyadic Green's function[5] expanded in vector spherical harmonics[6] and appropriately modified to account for a homogeneous dielectric medium. We use this expansion to arrive at a result which explicitly shows the relative significance of each multipole comprising the radiation field. This gives an indication of the angular dependence of the radiation to be expected at each harmonic without detailed computation. The result of this formulation is an expression for the radiated energy as a function of the particle velocity and the index of refraction of the medium. The expression takes the form of a series of spherical radiative mode contributions at each harmonic of the orbital frequency of the charged particle. The series is computed for each of several harmonics at a number of values of n, where n is the index of refraction of the medium and β is the ratio of the particle velocity to the velocity of light in vacuum. Temporal dispersion must be included to yield a finite result. The computational results are plotted and the Čerenkov effect is manifest. Moreover, universal curves are obtained by plotting the total radiated energy multiplied by the index of refraction and divided by the square of the orbital angular velocity and the square of the charge as a function of $n\beta$.

FORMULATION

The geometry of the problem is shown in Fig. 1. A particle of charge q orbits the z axis at radius r_0 and angular velocity Ω. The medium is a homogeneous dielectric with an index of refraction $n = (\epsilon/\epsilon_0)^{1/2}$, where ϵ is the permittivity of the dielectric and ϵ_0 is that of free space. The charge density may be written

$$\rho = q\delta(\mathbf{r} - \mathbf{v}t)$$
$$= (q/r_0^2)\delta(r - r_0)\delta(\cos\theta)\delta(\phi - \Omega t). \tag{1}$$

The corresponding current density is

$$\mathbf{J} = q\mathbf{v}\delta(\mathbf{r} - \mathbf{v}t)$$
$$= (q\Omega/r_0)\delta(r - r_0)\delta(\cos\theta)\delta(\phi - \Omega t)\hat{a}_\phi, \tag{2}$$

where $\hat{a}_\phi$ is a unit vector in the ϕ direction. This cur-

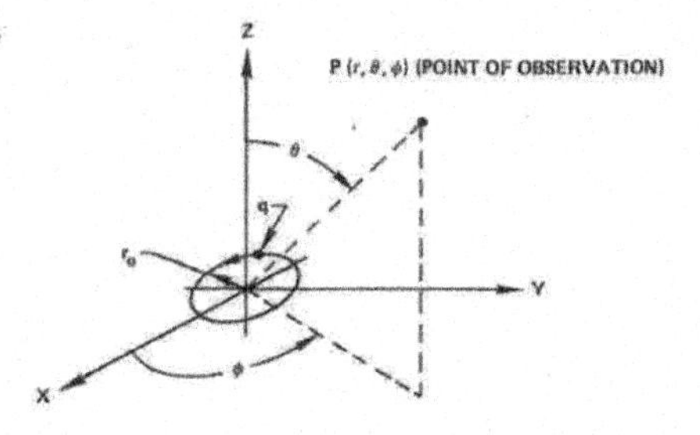

FIG. 1. Geometry of the problem.

rent density may be expressed in terms of its Fourier spectrum as

$$\mathbf{J}=\int_{-\infty}^{\infty}\mathbf{J}_{\omega}\exp(-i\omega t)\,\frac{d\omega}{2\pi}, \tag{3}$$

where

$$\mathbf{J}_{\omega}=(q/r_0)\,\delta(r-r_0)\,\delta(\cos\theta)\exp[i(\omega/\Omega)t]\hat{a}_{\phi}. \tag{4}$$

Thus we wish to solve for $\mathbf{E}_{\omega}$ by using the inhomogeneous vector wave equation

$$\nabla\times\nabla\times\mathbf{E}_{\omega}-\omega^2\mu_0\epsilon\mathbf{E}_{\omega}=i\omega\mu_0\mathbf{J}_{\omega}, \tag{5}$$

where $\mathbf{J}_{\omega}$ is given by Eq. (4) and $\mathbf{E}_{\omega}$ is the Fourier spectrum of the electric field intensity $\mathbf{E}$.

SOLUTION OF THE PROBLEM

The vector wave equation, Eq. (5), will be solved formally by means of the appropriate dyadic Green's function. In free space, $\epsilon=\epsilon_0$. The free-space dyadic Green's function $\mathbf{G}_0(r\,|\,r')$ satisfies the following equation:

$$\nabla\times\nabla\times\mathbf{G}_0(\mathbf{r}\,|\,\mathbf{r}')-\omega^2\mu_0\epsilon_0\mathbf{G}_0(\mathbf{r}\,|\,\mathbf{r}')=\mathbf{I}\delta(\mathbf{r}-\mathbf{r}'), \tag{6}$$

where $\mathbf{I}$ is the unit dyadic. One may show with the help of the vector Green's theorem that if $\mathbf{G}_0(\mathbf{r}\,|\,\mathbf{r}')$ and $\mathbf{J}_{\omega}(\mathbf{r}')$ are known, then $\mathbf{E}_{\omega}(\mathbf{r})$ can be found by using the relation

$$\mathbf{E}_{\omega}(\mathbf{r})=i\omega\mu_0\int_{v'}\mathbf{G}_0(\mathbf{r}\,|\,\mathbf{r}')\cdot\mathbf{J}_{\omega}(\mathbf{r}')\,dv', \tag{7}$$

where v' is the volume occupied by the current density $\mathbf{J}_{\omega}$. $\mathbf{G}_0(\mathbf{r}\,|\,\mathbf{r}')$ may be expanded in terms of the eigenfunctions of the vector wave equation. In spherical coordinates we have

$$\mathbf{G}_0(\mathbf{r}\,|\,\mathbf{r}')=\frac{ik}{4\pi}\sum_{l=1}^{\infty}\sum_{m=0}^{l}(2-\delta_0)\frac{2l+1}{l(l+1)}\frac{(l-m)!}{(l+m)!}$$
$$\times\begin{cases}\mathbf{M}_{elm}^{(3)}(kr)\mathbf{M}_{elm}^{(1)}(kr')+\mathbf{N}_{elm}^{(3)}(kr)\mathbf{N}_{elm}^{(1)}(kr')+\mathbf{M}_{0lm}^{(3)}(kr)\mathbf{M}_{0lm}^{(1)}(kr')+\mathbf{N}_{0lm}^{(3)}(kr)\mathbf{N}_{0lm}^{(1)}(kr') & r>r',\\ \mathbf{M}_{elm}^{(1)}(kr)\mathbf{M}_{elm}^{(3)}(kr')+\mathbf{N}_{elm}^{(1)}(kr)\mathbf{N}_{elm}^{(3)}(kr')+\mathbf{M}_{0lm}^{(1)}(kr)\mathbf{M}_{0lm}^{(3)}(kr')+\mathbf{N}_{0lm}^{(1)}(kr)\mathbf{N}_{0lm}^{(3)}(kr') & r<r',\end{cases}$$

where

$$\mathbf{M}_{elm}^{(1)}(kr)=-\frac{m}{\sin\theta}j_l(kr)P_l^m(\cos\theta)\sin(m\phi)\hat{a}_{\theta}-j_l(kr)\frac{\partial}{\partial\theta}P_l^m(\cos\theta)\cos(m\phi)\hat{a}_{\phi},$$

$$\mathbf{M}_{0lm}^{(1)}(kr)=+\frac{m}{\sin\theta}j_l(kr)P_l^m(\cos\theta)\cos(m\phi)\hat{a}_{\theta}-j_l(kr)\frac{\partial}{\partial\theta}P_l^m(\cos\theta)\sin(m\phi)\hat{a}_{\phi},$$

$$\mathbf{N}_{elm}^{(1)}(kr)=\frac{l(l+1)}{kr}j_l(kr)P_l^m(\cos\theta)\cos(m\phi)\hat{a}_r+\frac{1}{kr}\frac{\partial}{\partial r}[rj_l(kr)]\left(\frac{\partial}{\partial\theta}P_l^m(\cos\theta)\cos(m\phi)\hat{a}_{\theta}-\frac{m}{\sin\theta}P_l^m(\cos\theta)\sin(m\phi)\hat{a}_{\phi}\right),$$

$$\mathbf{N}_{0lm}^{(1)}(kr)=\frac{l(l+1)}{kr}j_l(kr)P_l^m(\cos\theta)\sin(m\phi)\hat{a}_r+\frac{1}{kr}\frac{\partial}{\partial r}[rj_l(kr)]\left(\frac{\partial}{\partial\theta}P_l^m(\cos\theta)\sin(m\phi)\hat{a}_{\theta}+\frac{m}{\sin\theta}P_l^m(\cos\theta)\cos(m\phi)\hat{a}_{\phi}\right), \tag{8}$$

and $\delta_0=0$ for $m\neq0$ and $\delta_0=1$ for $m=0$. The superscript (3) indicates that j_l is to be replaced by $h_l^{(1)}$, the spherical Hankel function of the first kind. Substituting Eqs. (4) and (8) into Eq. (7) and carrying out part of the integration, we find that the electric field for $r>r_0$ is given by

$$\mathbf{E}_{\omega}(\mathbf{r})=\frac{\omega\mu_0 qkr_0}{2\pi}\sum_{lm}\frac{2l+1}{l(l+1)}\frac{(l-m)!}{(l+m)!}\Bigg\{j_l(kr_0)\frac{\partial P_l^m}{\partial\theta}\bigg|_{\theta=\pi/2}\int_{-\infty}^{\infty}\cos(m\phi)\exp\left[i\left(\frac{\omega}{\Omega}\right)\phi\right]d\phi\,\mathbf{M}_{elm}^{(3)}$$
$$+j_l(kr_0)\frac{\partial P_l^m}{\partial\theta}\bigg|_{\theta=\pi/2}\int_{-\infty}^{\infty}\sin(m\phi)\exp\left[i\left(\frac{\omega}{\Omega}\right)\phi\right]d\phi\,\mathbf{M}_{0lm}^{(3)}+\frac{1}{kr_0}\frac{\partial}{\partial r}[rj_l(kr)]\bigg|_{r=r_0}mP_l^m(0)\int_{-\infty}^{\infty}\sin(m\phi)\exp\left[i\left(\frac{\omega}{\Omega}\right)\phi\right]d\phi\,\mathbf{N}_{elm}^{(3)}$$
$$-\frac{1}{kr_0}\frac{\partial}{\partial r}[rj_l(kr)]\bigg|_{r=r_0}mP_l^m(0)\int_{-\infty}^{\infty}\cos(m\phi)\exp\left[i\left(\frac{\omega}{\Omega}\right)\phi\right]d\phi\,\mathbf{N}_{0lm}^{(3)}\Bigg\}. \tag{9}$$

Now,

$$P_{\nu}^{\mu}(0)=2^{\mu}\pi^{-1/2}\cos\left[\tfrac{1}{2}\pi(\nu-\mu)\right]\frac{\Gamma[\frac{1}{2}(\nu+\mu+1)]}{\Gamma[\frac{1}{2}(\nu+\mu+2)]} \tag{10}$$

and

$$\frac{\partial P_{\nu}^{\mu}(x)}{\partial x}\bigg|_{x=0}=2^{\mu+1}\pi^{-1/2}\sin\left[\tfrac{1}{2}\pi(\nu+\mu)\right]\frac{\Gamma[\frac{1}{2}(\nu+\mu+2)]}{\Gamma[\frac{1}{2}(\nu+\mu+1)]}. \tag{11}$$

Therefore, performing the integrations on ϕ, we obtain

$$\mathbf{E}_{\omega}(\mathbf{r})=\frac{\omega\mu_0 qkr_0}{2\pi}\sum_{lm}\frac{2l+1}{l(l+1)}\frac{(l-m)!}{(l+m)!}$$

$$\times\left\{j_l(kr_0)2^{m+1}\pi^{-1/2}\sin[\tfrac{1}{2}\pi(l+m)]\frac{\Gamma[\frac{1}{2}(l+m+2)]}{\Gamma[\frac{1}{2}(l-m+1)]}\left\langle\left[\frac{1}{2}\delta\left(m+\frac{\omega}{\Omega}\right)+\frac{1}{2}\delta\left(m-\frac{\omega}{\Omega}\right)\right]\mathbf{M}^{(3)}_{elm}+\left[\frac{1}{2i}\delta\left(m+\frac{\omega}{\Omega}\right)\right.\right.\right.$$

$$\left.\left.-\frac{1}{2i}\delta\left(m-\frac{\omega}{\Omega}\right)\right]\mathbf{M}^{(3)}_{0lm}\right\rangle+\frac{1}{kr_0}\frac{\partial}{\partial r}[rj_l(kr)]\Big|_{r=r_0}m2^m\pi^{-1/2}\cos[\tfrac{1}{2}\pi(l-m)]\frac{\Gamma[\frac{1}{2}(l+m+1)]}{\Gamma[\frac{1}{2}(l-m+2)]}$$

$$\left.\times\left\langle\left[\frac{1}{2i}\delta\left(m+\frac{\omega}{\Omega}\right)-\frac{1}{2i}\delta\left(m-\frac{\omega}{\Omega}\right)\right]\mathbf{N}^{(3)}_{elm}-\left[\frac{1}{2}\delta\left(m+\frac{\omega}{\Omega}\right)+\frac{1}{2}\delta\left(m-\frac{\omega}{\Omega}\right)\right]\mathbf{N}^{(3)}_{0lm}\right\rangle\right\}. \tag{12}$$

This is the formal solution of Eq. (5) for the Fourier transform electric field radiated by the orbiting charged particle.

DISCUSSION

We define Q_{1l} and Q_{2l} such that,

$$\mathbf{E}_\omega(\mathbf{r})=q\sum_{lm}Q_{1l}\left\{\left[\frac{1}{2}\delta\left(m+\frac{\omega}{\Omega}\right)+\frac{1}{2}\delta\left(m-\frac{\omega}{\Omega}\right)\right]\mathbf{M}^{(3)}_{elm}+\left[\frac{1}{2i}\delta\left(m+\frac{\omega}{\Omega}\right)-\frac{1}{2i}\delta\left(m-\frac{\omega}{\Omega}\right)\right]\mathbf{M}^{(3)}_{0lm}\right\}$$

$$+Q_{2l}\left\{\left[\frac{1}{2i}\delta\left(m+\frac{\omega}{\Omega}\right)-\frac{1}{2i}\delta\left(m-\frac{\omega}{\Omega}\right)\right]\mathbf{N}^{(3)}_{elm}-\left[\frac{1}{2}\delta\left(m+\frac{\omega}{\Omega}\right)+\frac{1}{2}\delta\left(m-\frac{\omega}{\Omega}\right)\right]\mathbf{N}^{(3)}_{0lm}\right\}$$

$$=\sum_{lm}\mathbf{E}_{lm}\,. \tag{13}$$

Since the spherical vector harmonics $\mathbf{M}$ and $\mathbf{N}$ are orthogonal functions over 4π sr of solid angle Ω, it can easily be shown that the total power radiated is the sum of the powers P_{lm} radiated into each of the spherical modes. The power radiated in the even and odd lmth modes combined, P_{lm}, is given by

$$P_{lm}=\lim_{r\to\infty}\tfrac{1}{2}\mathrm{Re}\int_{4\pi}\mathbf{E}_{lm}\times\mathbf{H}^*_{lm}\cdot\hat{a}_r r^2\,d\Omega$$

$$=\frac{2\pi q^2}{\omega k_0\mu_0 n}\frac{l(l+1)}{2l+1}\frac{(l+m)!}{(l-m)!}(Q^2_{1l}+Q^2_{2l})\Omega^2, \tag{14}$$

where

$$k_0^2=\omega^2\mu_0\epsilon_0 \quad\text{and}\quad \mathbf{H}_{lm}=\frac{1}{i\omega\mu_0}\nabla\times\mathbf{E}_{lm}\,.$$

Noting that $kr_0=mn\beta$, where $\beta=v/c$, we find that a plot of $\sum_{l=m}^{\infty}nP_{lm}/q^2\Omega^2$) as a function of $n\beta$ is a universal curve representing the total radiated power at frequency $m\Omega$ for any charge, any index of refraction, any radius, and any velocity. We therefore define a quantity P_{Tj} such that

$$P_{Tj}=\sum_{m=1}^{j}\sum_{l=m}^{\infty}P_{lm}\,, \tag{15}$$

i.e., P_{Tj} is the total power radiated into the first j harmonics of Ω. Figure 2 shows $nP_{Tj}/q^2\Omega^2$ as a function of $n\beta$ for values of j from 1 to 15. Note that

$$P_{Tp}-P_{Tp-1}=\sum_{l=p}^{\infty}P_{lp}\,, \tag{16}$$

i.e., the space between the $(p-1)$th and pth curves represents the power radiated at $\omega=p\Omega$.

As an example, we consider a particle of charge 10^{-19} C executing an orbit of radius 2 cm at an angular frequency of 10^{10} rad/sec in a medium whose dispersion curve is shown in Fig. 3. The shape of this curve between $\omega=10$ and 11 Ω is irrelevant. The power radiated into the first ten harmonics can be read directly from Fig. 2(b) at $n=r_0\Omega/c=2.4\times0.02\times10^{10}/3\times10^{8}=1.6$ on the $j=10$ curve. Thus we find that

$$nP_{T10}/q^2\Omega^2=1642\ js/C^2$$

or

$$P_{T10}=0.687\times10^{-15}\ \mathrm{W}. \tag{17}$$

Now, above $\omega=10\ \Omega$, $n=1$ and β remains at 0.667, i.e., $n\beta=0.667$. One can easily see from Fig. 2(a) that the contribution from harmonics above 10 Ω at $n\beta=0.667$ is negligible and the total radiated power is about $\frac{2}{3}\times10^{-15}$ W. Had the index of refraction dropped to unity between $\omega=\Omega$ and $\omega=2\Omega$, the total power radiated would have been

$$P=P_{T1}\big|_{n\beta=1.6}+(P_T\big|_{n\beta=0.667}-P_{T1}\big|_{n\beta=0.667})$$

$$=8.2\times10^{-18}+(33-7.6)\times10^{-18}\approx0.34\times10^{-16}\ \mathrm{W}. \tag{18}$$

The quantity in parentheses is the power radiated into the second and higher harmonics while P_{T1} is radiated into the fundamental.

Figure 2 may be used for any $n(\omega)$ by treating each harmonic m individually and by using the n appropriate to the harmonic in question; i.e., $n(m\Omega)$ when determining $n\beta$ and the power. Finally, we note that for $\beta\ll1$ and $n=1$ the curve of Fig. 2(a) is represented by

$$P/q^2\Omega^2=(\mu_0/\epsilon_0)^{1/2}(\beta^2/6\pi)\approx20\beta^2 \tag{19}$$

which is identical with Larmor's well-known result.[7]

Schwinger has studied the radiation from a charged particle accelerating in vacuum and used a circular path as an example.[8] His result, appropriately generalized to account for nonunity index of refraction, is

$$W_m=\frac{\Omega q^2 m}{4\pi r_0\epsilon_0 n^2}\left[2n^2\beta^2J'_{2m}(2mn\beta)+(n^2\beta^2-1)\int_0^{2mn\beta}J_{2m}(x)\,dx\right],$$

(Ref. 8')

where W_m is the power radiated at $\omega=m\Omega$ and J_{2m} and J'_{2m} are the cylindrical Bessel function of order $2m$ and its derivative, respectively. While it has not been

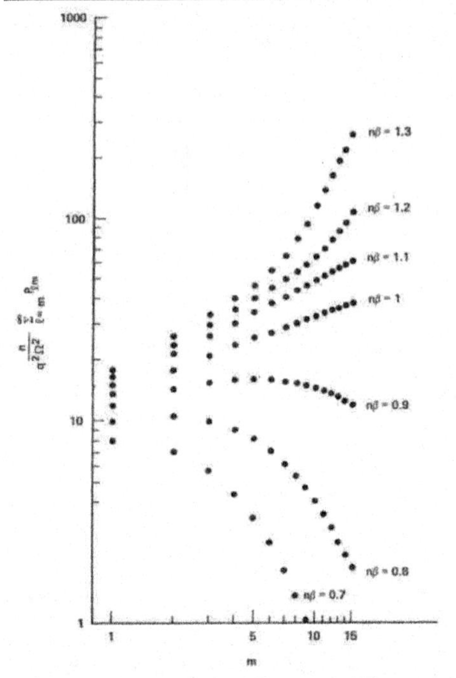

FIG. 4. Amplitudes of the first 15 harmonics of Ω for several values of $n\beta$.

either as

$$j_{l-1}(kr_0) - (l/kr_0)j_l(kr_0) \tag{22a}$$

or as

$$-j_{l+1}(kr_0) + [(l+1)/kr_0]j_l(kr_0). \tag{22b}$$

For large l, where the transition from small to significant is most pronounced, $l+1 \approx l-1 \approx l$ and the term is significant for $n\beta > (l-1)/l \approx (l+1)/l \approx 1$. Thus, for large l and for $m=l$ (large m also) a rather pronounced threshold, the Čerenkov threshold, is manifest. Since it is for large m that this effect appears and since $\omega = m\Omega$, we should see a pronounced Čerenkov threshold in the high-frequency portion of the radiated spectrum. Figure 4 shows the amplitude of the first 15 harmonics of Ω for several values of $n\beta$. Note that for $n\beta \gg 1$, the high-frequency harmonics exceed the low-frequency harmonics in amplitude. Thus, the sum on m is divergent and the total radiated power becomes infinite. This is a manifestation of a pronounced Čerenkov threshold at high frequencies and a less pronounced one at low frequencies. Of course, in fact, the total power is finite because at sufficiently high frequencies the index of refraction becomes unity and the sum on m converges (for $\beta < 1$). Finally, we note that at a given value of m (given harmonic of Ω) the relative significance of the terms for various l values is determined primarily by $j_l(kr_0) = j_l(nm\beta)$. Thus, this formulation points out that the most significant multipoles are those corresponding to $m < l < m(n\beta)$ for $n\beta > 1$ and, of course, $l=m$ for $n\beta < 1$.

Based on the above discussion we may make several observations about the angular distribution of the radiation without detailed computation. First, for $n\beta < 0.1$, at each harmonic $m\Omega$ the power is essentially all radiated with an angular dependence

$$\left(\frac{\partial P_m^m(\cos\theta)}{\partial\theta}\right)^2 + \frac{m^2}{\sin^2\theta}[P_m^m(\cos\theta)]^2;$$

i.e., for small $n\beta$

$$\sum_{l=m}^{\infty} P_{lm} \approx P_{mm} \approx \frac{q^2\Omega\beta}{2\pi r_0\epsilon_0 n}\left(\frac{mn\beta}{2}\right)^{2m}\frac{1}{[(m-1)!]^2}$$

$$\times(\sin\theta)^{2m-2}(1+\cos^2\theta),$$

where we have used the fact that $P_m^m(\cos\theta) = (2m)!\sin^m\theta/2^m m!$. This becomes identical with the low-velocity limit of Schwinger's vacuum result when n is put equal to unity. Note that in this limit the angular dependence is independent of the index of refraction. For $0.1 < n\beta < 1$, the dominant term is still the one for $l=m$, but the higher l values gradually appear as $n\beta$ increases and the angular distribution becomes more complicated. For $n\beta$ slightly larger than unity a result similar to that for low velocity is obtained at the lower frequencies of the radiated spectrum. Suppose $n\beta = 1.1$. Then, for $m < 10$ the dominant term is still the one for $l=m$. However, for $m=10$, the $l=11$ term has become significant along with the $l=10$ term. Thus, the higher harmonics have more complicated angular distributions from the point of view of their expression in terms of a coherent sum over l.

In general, for large values of $n\beta$ or for high-order harmonics or both, the angular distribution involves a coherent sum of many terms, i.e., many l values, and this formulation leaves no alternative but to sum the series to obtain this angular distribution. However, it does point out which l values are significant and under what circumstances one term predominates. That is, for low values of $n\beta$ or low-order harmonics or both, one term of the sum over l predominates. In this circumstance the total radiation pattern is a simple incoherent sum of the $(\partial P_m^m/\partial\theta)^2 + (m^2/\sin^2\theta)(P_m^m)^2$ power distribution of each harmonic, $m\Omega$; i.e.,

$$P(\theta) \approx \sum_{m=1}^{\infty}\frac{q^2\Omega^2}{\omega k_0\mu_0 n}(Q_{1m}^2 + Q_{2m}^2)\left[\left(\frac{\partial P_m^m}{\partial\theta}\right)^2 + \frac{m^2}{\sin^2\theta}(P_m^m)^2\right]$$

and, of course, Q_{1m} and Q_{2m} rapidly approach zero for as m approaches infinity if $n\beta < 1$. For $n\beta > 1$ the infinite series diverges but a finite result is obtained by terminating the sum when the index of refraction becomes unity as discussed previously.

*Partly supported by a National Science Foundation Grant.
†Present address: Department of Electrical Engineering, University of Mississippi, University, Miss. 38677.
[1] P.A. Čerenkov, Phys. Rev. 52, 378 (1937).
[2] I.M. Frank and I. Tamm, Dokl. Akad. Nauk SSSR 14, 109 (1937).
[3] R.J. Pogorzelski and C. Yeh, Phys. Rev. A 8, 137 (1973).
[4] See, for example, J.D. Jackson, *Classical Electrodynamics* (Wiley, New York, 1962), pp. 481–8.
[5] C.T. Tai, *Dyadic Green's Functions in Electromagnetic Theory* (International Textbook, San Francisco, 1971).
[6] J.A. Stratton, *Electromagnetic Theory* (McGraw-Hill, New York, 1941).
[7] In Ref. 4, p. 469.
[8] J. Schwinger, Phys. Rev. 75, 1912 (1949).
[8'] We are indebted to the reviewer for pointing out this generalization of Schwinger's result.
[9] K. Kitao, Prog. Theor. Phys. 23, 759 (1960).
[10] See for example, C.H. Papas, *Theory of Electromagnetic Wave Propagation* (McGraw-Hill, New York, 1965), p. 34

Synchrotron-diffraction radiation spectra in the presence of a penetrable sphere*

R. J. Pogorzelski† and C. Yeh

Electrical Sciences and Engineering Department, University of California, Los Angeles, California 90024
(Received 23 May 1974; in final form 24 July 1974)

The radiation from a charged particle orbiting both inside and outside a dielectric sphere is expressed as a series of vector spherical harmonics. The results of numerical computation of the radiated spectra for various situations are presented and discussed. When the particle orbits far outside the sphere, the interaction between the two is weak and the radiation is of the synchrotron type. The radiated power is shown to decrease with increasing sphere size in this situation. However, when the particle orbits close to or inside the sphere, the interaction is very strong; this being characterized by the appearance of resonances of the sphere in the radiated spectrum. The resonances are shown to be responsible for the power in a particular harmonic being radiated almost entirely into a single spherical harmonic in some instances.

INTRODUCTION

Most previous studies on diffraction radiation[1-3] were carried out for perfectly conducting (impenetrable) bodies such as conducting half-plane, screens, or gratings, open ends of metallic waveguides, or conducting spheres. Very little work has been carried out for the corresponding problems of diffraction radiation due to the presence of penetrable (dielectric) bodies.[4] This paper deals with the case of a charged particle orbiting concentrically around a dielectric sphere. We wish to learn to what extent the synchrotron radiation is affected by the presence of a penetrable sphere. It is expected that due to the presence of multiple reflections within the sphere and diffraction phenomena, the radiation characteristics for this charged particle will be significantly modified. It should be noted that since an essential feature of diffraction radiation is that the sources excite a spectrum of frequencies, it is important that exact solution (valid for all frequency ranges) to the problems of diffraction radiation be obtained.

FORMULATION

Figure 1 illustrates the physical situation with which we will concern ourselves. A point charge q orbits the z axis at radius r_0 and angular velocity Ω. Also present is a dielectric sphere of radius a, index of refraction $n=(\epsilon_1/\epsilon_0)^{1/2}$, and permeability μ_0 centered at the origin of coordinates. The figure shows the case where a is less than r_0, but the $r_0 < a$ case is also considered in this work. We now express the charge density ρ in the form

$$\rho = q(\mathbf{r}-\mathbf{v}t) = (q/r_0^2)\,\delta(r-r_0)\,\delta(\cos\theta)\,\delta(\phi-\Omega t). \tag{1}$$

The current density $\mathbf{J}$ which corresponds to this is

$$\mathbf{J} = q\mathbf{v}\,\delta(\mathbf{r}-\mathbf{v}t) = (q\Omega/r_0)\,\delta(r-r_0)\,\delta(\cos\theta)\delta(\phi-\Omega t)\hat{a}_\phi, \tag{2}$$

where $\hat{a}_\phi$ is a unit vector in the ϕ direction. Now, this current density may be expressed in terms of its Fourier spectrum $\mathbf{J}_\omega$ as

$$\mathbf{J} = \int_{-\infty}^{\infty} \mathbf{J}_\omega \exp(-i\omega t)\, d\omega/2\pi, \tag{3}$$

where

$$\mathbf{J}_\omega = (q/r_0)\,\delta(r-r_0)\,\delta(\cos\theta)\exp[i(\omega/\Omega)t]\,\hat{a}_\phi. \tag{4}$$

Thus we reduce the problem to solving for $\mathbf{E}_\omega$ the inhomogeneous vector wave equation

$$\nabla\times\nabla\times\mathbf{E}_\omega - \omega^2\mu_0\epsilon\mathbf{E}_\omega = i\omega\mu_0\mathbf{J}_\omega, \tag{5}$$

where $\epsilon=\epsilon_1$ for $r<a$ and $\epsilon=\epsilon_0$ for $r>a$, $\mathbf{J}_\omega$ is given by (4), and $\mathbf{E}_\omega$ is the Fourier spectrum of the electric field intensity $\mathbf{E}$.

SOLUTION OF THE PROBLEM

The above vector wave equation can be solved with the aid of the appropriate dyadic Green's function.[5] In free space $\epsilon=\epsilon_0$. The free-space dyadic Green's function $\mathbf{G}_0(\mathbf{r}\,|\,\mathbf{r}')$ satisfies the equation

$$\nabla\times\nabla\times\mathbf{G}_0(\mathbf{r}\,|\,\mathbf{r}') - \omega^2\mu_0\epsilon_0\mathbf{G}_0(\mathbf{r}\,|\,\mathbf{r}') = \mathbf{I}\,\delta(\mathbf{r}-\mathbf{r}'), \tag{6}$$

where $\mathbf{I}$ is the unit dyadic. Application of the vector Green's theorem leads to the result

$$\mathbf{E}_\omega(\mathbf{r}) = i\omega\mu_0\int_{V'} \mathbf{G}_0(\mathbf{r}\,|\,\mathbf{r}')\cdot\mathbf{J}_\omega(\mathbf{r}')\,dV', \tag{7}$$

where V' is the volume occupied by the current density $\mathbf{J}_\omega\cdot\mathbf{G}_0(\mathbf{r}\,|\,\mathbf{r}')$ may be expanded in terms of the eigenfunctions of the vector wave equation. In spherical coordinates such an expansion takes the form

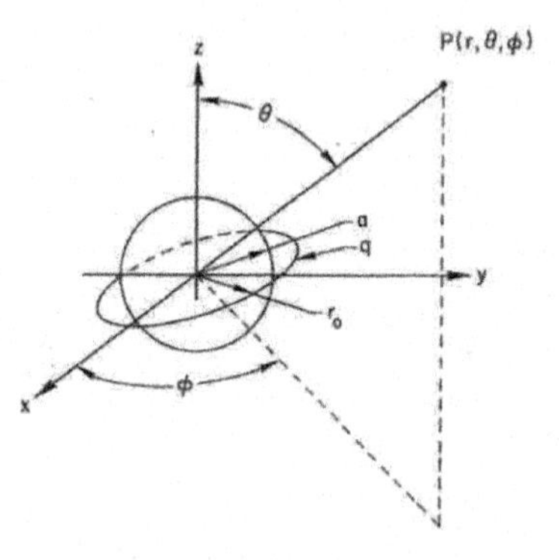

FIG. 1. Charge q orbiting a dielectric sphere.

$$\mathbf{G}_0(\mathbf{r}|\mathbf{r}') = \frac{ik}{4\pi}\sum_{l=1}^{\infty}\sum_{m=0}^{l}(2-\delta_0)\frac{2l+1}{l(l+1)}\frac{(l-m)!}{(l+m)!}\times[\mathbf{M}_{elm}^{(3)}(kr)\mathbf{M}_{elm}^{(1)}(kr')+\mathbf{N}_{elm}^{(3)}(kr)\mathbf{N}_{elm}^{(1)}(kr')+\mathbf{M}_{olm}^{(3)}(kr)\mathbf{M}_{olm}^{(1)}(kr')$$

$$+\mathbf{N}_{olm}^{(3)}(kr)\mathbf{N}_{olm}^{(1)}(kr')], \qquad r>r'$$

$$=\frac{ik}{4\pi}\sum_{l=1}^{\infty}\sum_{m=0}^{l}(2-\delta_0)\frac{2l+1}{l(l+1)}\frac{(l-m)!}{(l+m)!}$$

$$\times[\mathbf{M}_{elm}^{(1)}(kr)\mathbf{M}_{elm}^{(3)}(kr')+\mathbf{N}_{elm}^{(1)}(kr)\mathbf{N}_{elm}^{(3)}(kr')+\mathbf{M}_{olm}^{(1)}(kr)\mathbf{M}_{olm}^{(3)}(kr')+\mathbf{N}_{olm}^{(1)}(kr)\mathbf{N}_{olm}^{(3)}(kr')], \qquad r<r'$$

where

$$\mathbf{M}_{elm}^{(1)}(kr) = -\frac{m}{\sin\theta}j_l(kr)P_l^m(\cos\theta)\sin(m\phi)\hat{a}_\theta - j_l(kr)\frac{\partial}{\partial\theta}P_l^m(\cos\theta)\cos(m\phi)\hat{a}_\phi,$$

$$\mathbf{M}_{olm}^{(1)}(kr) = +\frac{m}{\sin\theta}j_l(kr)P_l^m(\cos\theta)\cos(m\phi)\hat{a}_\theta - j_l(kr)\frac{\partial}{\partial\theta}P_l^m(\cos\theta)\sin(m\phi)\hat{a}_\phi,$$

$$\mathbf{N}_{elm}^{(1)}(kr) = \frac{l(l+1)}{kr}j_l(kr)P_l^m(\cos\theta)\cos(m\phi)\hat{a}_r + \frac{1}{kr}\frac{\partial}{\partial r}[rj_l(kr)]\left(\frac{\partial}{\partial\theta}P_l^m(\cos\theta)\cos(m\phi)\hat{a}_\theta - \frac{m}{\sin\theta}P_l^m(\cos\theta)\sin(m\phi)\hat{a}_\phi\right),$$

$$\mathbf{N}_{olm}^{(1)}(kr) = \frac{l(l+1)}{kr}j_l(kr)P_l^m(\cos\theta)\sin(m\phi)\hat{a}_r + \frac{1}{kr}\frac{\partial}{\partial r}[rj_l(kr)]\left(\frac{\partial}{\partial\theta}P_l^m(\cos\theta)\sin(m\phi)\hat{a}_\theta + \frac{m}{\sin\theta}P_l^m(\cos\theta)\cos(m\phi)\hat{a}_\phi\right), \quad (8)$$

$k=\omega(\mu_0\epsilon_0)^{1/2}$ and $\delta_0=0$ for $m\neq 0$ and $\delta_0=1$ for $m=0$. The superscript (3) indicates that the spherical Bessel function j_l is to be replaced by $h_l^{(1)}$, the spherical Hankel function of the first kind.

The presence of the dielectric sphere introduces an additional scattered field. Hence, by superposition, the total dyadic Green's function consists of the unperturbed infinite-space dyadic Green's function $\mathbf{G}_0(\mathbf{r}|\mathbf{r}')$ plus the contribution due to the scattered field.

For $r'>a$, we have

$$\mathbf{G}(\mathbf{r}|\mathbf{r}') = \mathbf{G}_0(\mathbf{r}|\mathbf{r}')|_{r>r'} + \mathbf{G}_s^{(11)}(\mathbf{r}|\mathbf{r}'), \qquad r>r'$$

$$= \mathbf{G}_0(\mathbf{r}|\mathbf{r}')|_{r<r'} + \mathbf{G}_s^{(11)}(\mathbf{r}|\mathbf{r}'), \qquad a<r<r'$$

$$= \mathbf{G}_s^{(21)}(\mathbf{r}|\mathbf{r}'), \qquad r<a$$

where

$$\mathbf{G}_s^{(11)}(\mathbf{r}|\mathbf{r}') = \frac{ik}{4\pi}\sum_{l=1}^{\infty}\frac{2l+1}{l(l+1)}[A_{1l}\mathbf{M}_l^{(3)}(kr)\mathbf{M}_l^{(3)}(kr')$$

$$+B_{1l}\mathbf{N}_l^{(3)}(kr)\mathbf{N}_l^{(3)}(kr')],$$

$$\mathbf{G}_s^{(21)}(\mathbf{r}|\mathbf{r}') = \frac{ik}{4\pi}\sum_{l=1}^{\infty}\frac{2l+1}{l(l+1)}[C_{1l}\mathbf{M}_l^{(1)}(k_1r)\mathbf{M}_l^{(3)}(kr')$$

$$+D_{1l}\mathbf{N}_l^{(1)}(k_1r)\mathbf{N}_l^{(3)}(kr')], \quad (9)$$

and for $r'<a$, we have

$$\mathbf{G}(\mathbf{r}|\mathbf{r}') = \mathbf{G}_s^{(12)}(\mathbf{r}|\mathbf{r}'), \qquad r>a$$

$$= \mathbf{G}_0(\mathbf{r}|\mathbf{r}')|_{\substack{r'<r\\k=k_1}} + \mathbf{G}_s^{(22)}(\mathbf{r}|\mathbf{r}'), \qquad r'<r<a$$

$$= \mathbf{G}_0(\mathbf{r}|\mathbf{r}')|_{\substack{r<r'\\k=k_1}} + \mathbf{G}_s^{(22)}(\mathbf{r}|\mathbf{r}'), \qquad r<r'$$

where

$$\mathbf{G}_s^{(12)}(\mathbf{r}|\mathbf{r}') = \frac{ik_1}{4\pi}\sum_{l=1}^{\infty}\frac{2l+1}{l(l+1)}[A_{2l}\mathbf{M}_l^{(3)}(kr)\mathbf{M}_l^{(1)}(k_1r')$$

$$+B_{2l}\mathbf{N}^{(3)}(kr)\mathbf{N}_l^{(1)}(k_1r')],$$

$$\mathbf{G}_s^{(22)}(\mathbf{r}|\mathbf{r}') = \frac{ik_1}{4\pi}\sum_{l=1}^{\infty}\frac{2l+1}{l(l+1)}[C_{2l}\mathbf{M}_l^{(1)}(k_1r)\mathbf{M}_l^{(1)}(k_1r')$$

$$+D_{2l}\mathbf{N}_l^{(1)}(k_1r)\mathbf{N}_l^{(1)}(k_1r')]. \quad (10)$$

The constant $k=\omega(\mu_0\epsilon_0)^{1/2}$ and the constant $k_1=\omega(\mu_0\epsilon_1)^{1/2}$, where ϵ_0 and ϵ_1 are, respectively, the dielectric constant outside the sphere $(r>a)$ and that inside the sphere $(r>a)$. The subscript s designates the scattered part of the dyadic Green's function. $A_{1l}, B_{1l}, C_{1l}, D_{1l}, A_{2l}, B_{2l}, C_{2l}$, and D_{2l} are arbitrary constants that are to be determined from the boundary conditions on the surface of the dielectric sphere at $r=a$. That is, we require continuity of

$$\hat{a}_r\times\mathbf{G} \quad (11)$$

and

$$(1/\mu)\hat{a}_r\times(\nabla\times\mathbf{G}), \quad (12)$$

where $\hat{a}_r$ is a unit vector in the radial direction. Since we are only interested in the far zone fields, r will never be less than r' and similarly r will never be less than a. Thus, only the A's and B's need be obtained as the C's and D's play no role. The boundary conditions (11) and (12) then yield

$$A_{1l} = -\frac{f_l(k_2a)-f_l(k_1a)}{f_l(k_2a)-g_l(k_1a)}\frac{j_l(k_1a)}{h_l^{(1)}(k_1a)}, \quad (13)$$

$$B_{1l} = -\frac{k_2^2f_l(k_1a)-k_1^2f_l(k_2a)}{k_2^2g_l(k_1a)-k_1^2f_l(k_2a)}\frac{j_l(k_1a)}{h_l^{(1)}(k_1a)}, \quad (14)$$

$$A_{2l} = \frac{f_l(k_2a)-g_l(k_2a)}{f_l(k_2a)-g_l(k_1a)}\frac{h_l^{(1)}(k_2a)}{h_l^{(1)}(k_1a)}, \quad (15)$$

$$B_{2l} = \frac{k_1k_2f_l(k_2a)-k_1k_2g_l(k_2a)}{k_1^2f_l(k_2a)-k_2^2g_l(k_1a)}\frac{h_l^{(1)}(k_2a)}{h_l^{(1)}(k_1a)}, \quad (16)$$

where,

$$f_l(x) = \frac{1}{j_l(x)}\frac{d}{dx}[xj_l(x)],$$

$$g_l(x) = \frac{1}{h_l^{(1)}(x)}\frac{d}{dx}[xh_l^{(1)}(x)].$$

Now, making use of Eq. (7) and recalling that $\mathbf{J}_\omega$ is given by (4) we find that for $r_0>a$,

$$\mathbf{E}_\omega = \omega\mu_0 q k r_0 \sum_{lm} \frac{2l+1}{l(l+1)}\frac{(l-m)!}{(l+m)!}\Big[\big[j_l(kr_0)+A_{1l}h_l^{(1)}(kr_0)\big]\frac{\partial P_l^m}{\partial\theta}\Big|_{\theta=\pi/2}\int_{-\infty}^{\infty}\cos(m\phi)\exp\Big(i\frac{\omega}{\Omega}\phi\Big)\,d\phi\,\mathbf{M}_{elm}^{(3)}$$

$$+\big[j_l(kr_0)+A_{1l}h_l^{(1)}(kr_0)\big]\frac{\partial P_l^m}{\partial\theta}\Big|_{\theta=\pi/2}\int_{-\infty}^{\infty}\sin(m\phi)\exp\Big(i\frac{\omega}{\Omega}\phi\Big)\,d\phi\,\mathbf{M}_{olm}^{(3)}$$

$$+\frac{1}{kr}\frac{\partial}{\partial r}r\big[j_l(kr)+B_{1l}h_l^{(1)}(kr)\big]_{r=r_0} m\,P_l^m(0)\int_{-\infty}^{\infty}\sin(m\phi)\exp\Big(i\frac{\omega}{\Omega}\phi\Big)\,d\phi\,N_{elm}^{(3)}$$

$$-\frac{1}{kr}\frac{\partial}{\partial r}r\big[j_l(kr)+B_{1l}h_l^{(1)}(kr)\big]_{r=r_0} m\,P_l^m(0)\int_{-\infty}^{\infty}\cos(m\phi)\exp\Big(i\frac{\omega}{\Omega}\phi\Big)\,d\phi\,N_{olm}^{(3)}\ . \tag{17}$$

However,

$$P_\nu^\mu(0)=2^\mu\pi^{-1/2}\cos\Big(\frac{\pi}{2}(\nu-\mu)\Big)\frac{\Gamma[\frac12(\nu+\mu+1)]}{\Gamma[\frac12(\nu-\mu+2)]}$$

and,

$$\frac{\partial P_\nu^\mu(x)}{\partial x}\Big|_{x=0}=2^{\mu+1}\pi^{-1/2}\sin\Big(\frac{\pi}{2}(\nu+\mu)\Big)\frac{\Gamma[\frac12(\nu+\mu+2)]}{\Gamma[\frac12(\nu-\mu+1)]}.$$

Thus, performing the integrations on ϕ we obtain for $r_0 > a$,

$$\mathbf{E}_\omega=\omega\mu_0 q k r_0\sum_{lm}\frac{2l+1}{l(l+1)}\frac{(l-m)!}{(l+m)!}\big[j_l(kr_0)+A_{1l}h_l^{(1)}(kr_0)\big]2^{m+1}\pi^{-1/2}\sin\Big(\frac{\pi}{2}(l+m)\Big)\frac{\Gamma[\frac12(l+m+2)]}{\Gamma[\frac12(l-m+1)]}$$

$$\times\Big\langle\Big[\frac12\delta\Big(m+\frac{\omega}{\Omega}\Big)+\frac12\delta\Big(m-\frac{\omega}{\Omega}\Big)\Big]\mathbf{M}_{elm}^{(3)}+\Big[\frac{1}{2i}\delta\Big(m+\frac{\omega}{\Omega}\Big)-\frac{1}{2i}\delta\Big(m-\frac{\omega}{\Omega}\Big)\Big]\mathbf{M}_{olm}^{(3)}\Big\rangle$$

$$+\frac{1}{kr}\frac{\partial}{\partial r}r\big[j_l(kr)+B_{1l}h_l^{(1)}(kr)\big]_{r=r_0} m\,2^m\pi^{-1/2}\cos\Big[\frac{\pi}{2}(l-m)\Big]\frac{\Gamma[\frac12(l+m+1)]}{\Gamma[\frac12(l-m+2)]}\Big\langle\Big[\frac{1}{2i}\delta\Big(m+\frac{\omega}{\Omega}\Big)-\frac{1}{2i}\delta\Big(m-\frac{\omega}{\Omega}\Big)\Big]\mathbf{N}_{elm}^{(3)}$$

$$+\Big[\frac12\delta\Big(m+\frac{\omega}{\Omega}\Big)+\frac12\delta\Big(m-\frac{\omega}{\Omega}\Big)\Big]\mathbf{N}_{olm}^{(3)}\Big\rangle\ . \tag{18}$$

Similarly, for $r_0 < a$,

$$\mathbf{E}_\omega=\omega\mu_0 q k r_0\sum_{lm}\frac{2l+1}{l(l+1)}\frac{(l-m)!}{(l+m)!}\Big\{A_{2l}j_l(kr_0)2^{m+1}\pi^{-1/2}\sin\Big(\frac{\pi}{2}(l+m)\Big)\frac{\Gamma[\frac12(l+m+2)]}{\Gamma[\frac12(l-m+1)]}$$

$$\times\Big\langle\Big[\frac12\delta\Big(m+\frac{\omega}{\Omega}\Big)+\frac12\delta\Big(m-\frac{\omega}{\Omega}\Big)\Big]\mathbf{M}_{elm}^{(3)}+\Big[\frac{1}{2i}\delta\Big(m+\frac{\omega}{\Omega}\Big)-\frac{1}{2i}\delta\Big(m-\frac{\omega}{\Omega}\Big)\Big]\mathbf{M}_{olm}^{(3)}\Big\rangle$$

$$+\frac{B_{2l}}{kr}\frac{\partial}{\partial r}r j_l(kr)\Big|_{r=r_0} m\,2^m\pi^{-1/2}\cos\Big(\frac{\pi}{2}(l-m)\Big)\frac{\Gamma[\frac12(l+m+1)]}{\Gamma[\frac12(l-m+2)]}\Big\langle\Big[\frac{1}{2i}\delta\Big(m+\frac{\omega}{\Omega}\Big)-\frac{1}{2i}\delta\Big(m-\frac{\omega}{\Omega}\Big)\Big]\mathbf{N}_{elm}^{(3)}$$

$$+\Big[\frac12\delta\Big(m+\frac{\omega}{\Omega}\Big)+\frac12\delta\Big(m-\frac{\omega}{\Omega}\Big)\Big]\mathbf{N}_{olm}^{(3)}\Big\rangle\Big\}, \tag{19}$$

where $k_1=\omega(\mu_0\epsilon_1)^{1/2}$.

These represent the formal solution of Eq. (5) for the Fourier transform of the radiated electric field.

DISCUSSION

We define Q_{1l}, Q_{2l}, and $\mathbf{E}_{lm}$ such that

$$\mathbf{E}_\omega(r)=q\sum_{lm}Q_{1l}\Big\{\Big[\frac12\delta\Big(m+\frac{\omega}{\Omega}\Big)+\frac12\delta\Big(m-\frac{\omega}{\Omega}\Big)\Big]\mathbf{M}_{elm}^{(3)}+\Big[\frac{1}{2i}\delta\Big(m+\frac{\omega}{\Omega}\Big)-\frac{1}{2i}\delta\Big(m-\frac{\omega}{\Omega}\Big)\Big]\mathbf{M}_{olm}^{(3)}\Big\}$$

$$+Q_{2l}\Big\{\Big[\frac{1}{2i}\delta\Big(m+\frac{\omega}{\Omega}\Big)-\frac{1}{2i}\delta\Big(m-\frac{\omega}{\Omega}\Big)\Big]\mathbf{N}_{elm}^{(3)}-\Big[\frac12\delta\Big(m+\frac{\omega}{\Omega}\Big)+\frac12\delta\Big(m-\frac{\omega}{\Omega}\Big)\Big]\mathbf{N}_{olm}^{(3)}\Big\}$$

$$=\sum_{lm}\mathbf{E}_{lm}. \tag{20}$$

Since the spherical vector harmonics $\mathbf{M}_{elm}^{(3)}$, $\mathbf{M}_{olm}^{(3)}$, $\mathbf{N}_{elm}^{(3)}$ and $\mathbf{N}_{olm}^{(3)}$ are orthogonal functions over 4π steradians of solid angle Ω, it can easily be shown that the total power radiated is the sum of the powers P_{lm} radiated into each of the spherical modes. The power radiated in the even and odd lmth modes combined, P_{lm}, is given by

$$P_{lm}=\lim_{r\to\infty}\tfrac12\,\mathrm{Re}\int_{4\pi}\mathbf{E}_{lm}\times\mathbf{H}_{lm}^*\cdot\hat{a}_r r^2\,d\Omega$$

$$=\frac{2\pi q^2}{k\mu_0\omega}\frac{l(l+1)}{2l+1}\frac{(l+m)!}{(l-m)!}\big[Q_{1l}^2+Q_{2l}^2\big]\Omega^2, \tag{21}$$

where $k=\omega(\mu_0\epsilon_0)^{1/2}$ and $\mathbf{H}_{lm}=(i\omega\mu_0)^{-1}\nabla\times\mathbf{E}_{lm}$. Performing the summation on l from m to ∞ for each m we

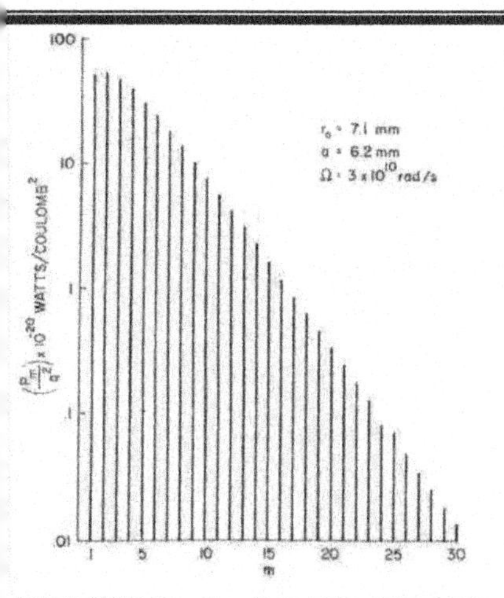

FIG. 2. Radiated spectrum for particle orbiting outside a dielectric sphere.

obtain the power radiated into each harmonic $m\Omega$:

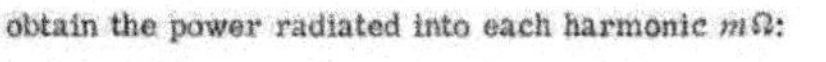

$$P_m = \sum_{l=m}^{\infty} P_{lm}; \tag{22}$$

that is, the power P_m is radiated at $\omega = m\Omega$. Figures 2–7 show P_m as a function of m for various values of a with r_0 fixed, while Figs. 8–12 show P_m for various values of r_0 with a fixed. In all cases $\Omega = 3\times10^{10}$ rad/s and the index of refraction of the sphere is 2.

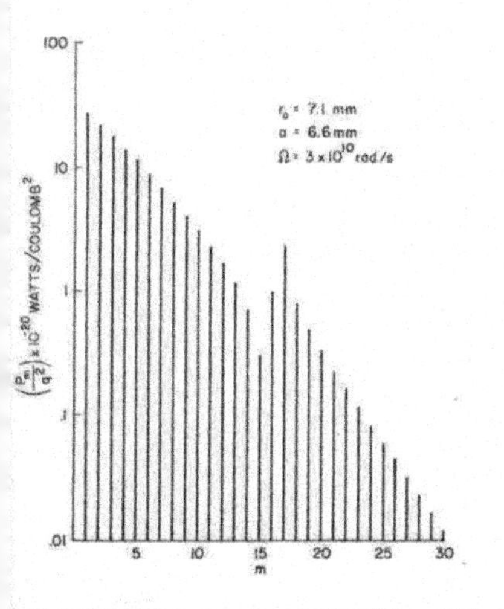

FIG. 3. Radiated spectrum for particle orbiting outside a dielectric sphere.

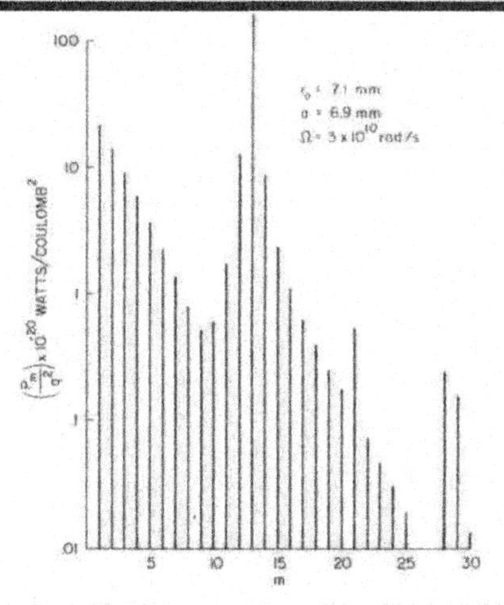

FIG. 4. Radiated spectrum for particle orbiting outside a dielectric sphere.

Referring to the sequence of spectra for fixed r_0 $= 7.1$ mm we note that for $a = 6.2$ mm the spectrum is essentially a synchrotron spectrum except for a slight resonance deviation in the neighborhood of the 24th and 25th harmonics (see Fig. 2). As a progresses from 6.2 through 6.6, to 6.9 mm (see Figs. 3 and 4) we see this resonance become stronger because the surface of the dielectric sphere comes closer and closer to the charged particle orbit, thus increasing the interaction.

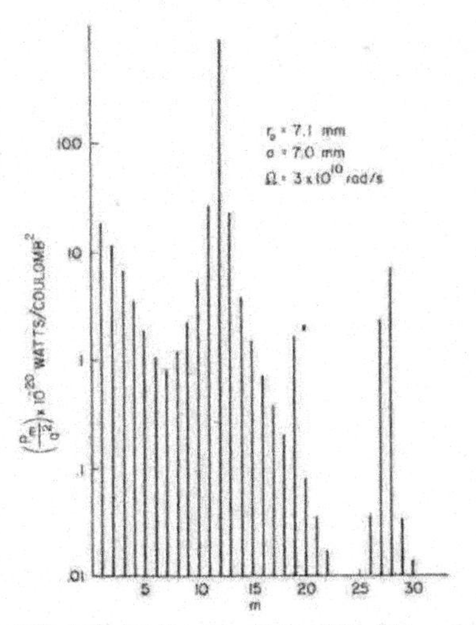

FIG. 5. Radiated spectrum for particle orbiting outside a dielectric sphere.

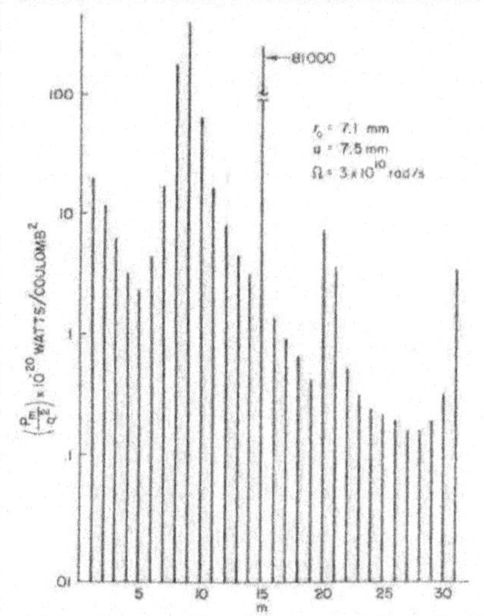

FIG. 6. Radiated spectrum for particle orbiting inside a dielectric sphere.

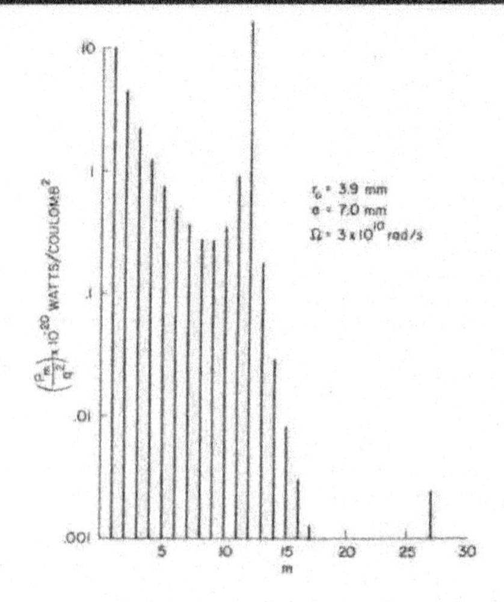

FIG. 8. Radiated spectrum for particle orbiting inside a dielectric sphere.

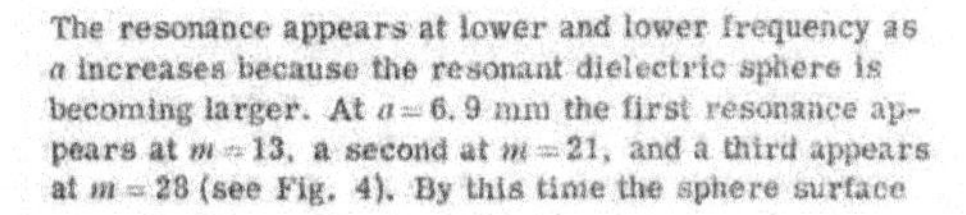

The resonance appears at lower and lower frequency as a increases because the resonant dielectric sphere is becoming larger. At $a = 6.9$ mm the first resonance appears at $m = 13$, a second at $m = 21$, and a third appears at $m = 28$ (see Fig. 4). By this time the sphere surface is only 0.2 mm away from the orbit and the interaction is very strong. At $a = 7$ mm the sphere is 0.1 mm inside the orbit and the resonances are even stronger. Proceeding further to $a = 7.5$ mm (see Fig. 6) whence the particle is enveloped by the sphere, we see a very

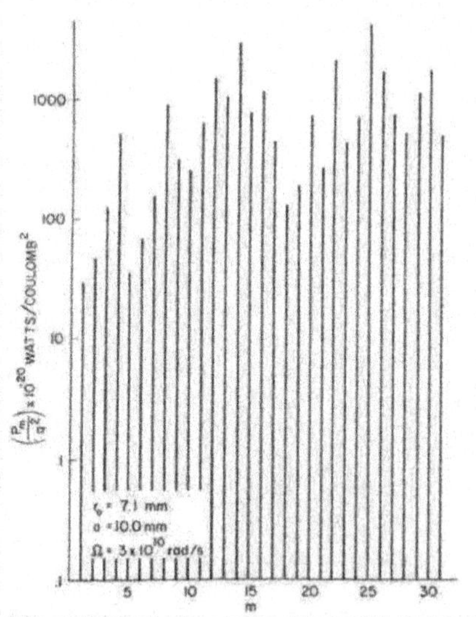

FIG. 7. Radiated spectrum for particle orbiting inside a dielectric sphere.

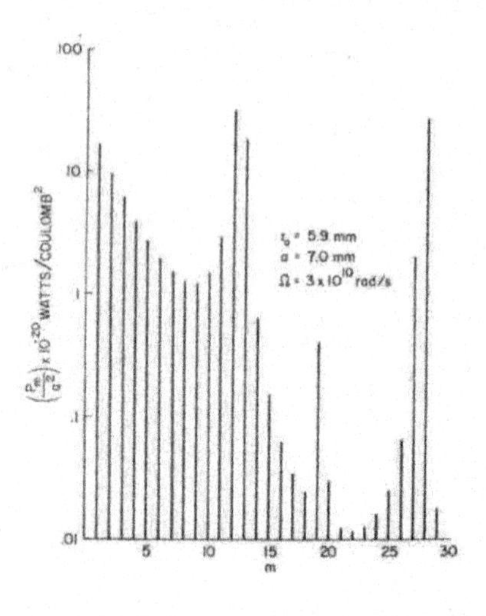

FIG. 9. Radiated spectrum for particle orbiting inside a dielectric sphere.

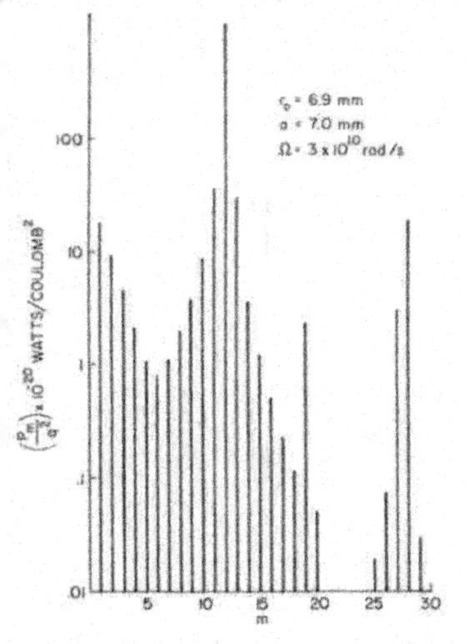

FIG. 10. Radiated spectrum for particle orbiting inside a dielectric sphere.

strong interaction in which the same resonances are visible together with a fourth at $m=31$ and the second one at $m=15$ now dominates the others. Figure 7 shows the spectrum for $a=10.0$ mm. Here the resonances have become so strong and so close together as to render the spectrum extremely complex.

Fixing a at 7 mm and varying r_0 we arrive at the spectra shown in Figs. 8–12. In Figs. 8–10 the orbit is inside the sphere and we observe that as the orbit size increases the significance of the higher harmonics increases. This is due to an increase in the particle velocity.[8] In Figs. 11 and 12 the orbit is outside the sphere and the further the orbit is from the sphere surface the weaker the interaction and consequently the less pronounced the resonances. Again the higher harmonics become more significant as the particle velocity increases. We note also that since the sphere size is fixed, the resonances always appear at the same frequencies.

The resonances discussed above occur because of minima in the denominators of the quantities A_{1l}, B_{1l}, A_{2l}, and B_{2l}. It is reasonable to anticipate that each minimum occurs in one particular term of the summation on l for the harmonic in question. Indeed we find that upon the occurrence of a very dominant resonance such as that at the 15th harmonic in the case where $r_0=7.1$ mm and $a=7.5$ mm nearly all of the energy is radiated into a particular spherical mode; in this case the $l=18$, $m=15$ mode. The other modes at this frequency contain only one part in 10^6 of the energy. Another somewhat less pronounced example is the first resonance in the spectra for $r_0=7.1$ mm. For $a=7.5$ mm it occurs for $l=9$ and $m=9$ and for $a=7$ mm it occurs for $l=12$ and $m=12$. Thus the spherical mode in which the largest energy is radiated varies as the sphere size changes even though we are in a sense dealing with the "same" resonance; that is, we can follow the shift of this resonance as a is varied. Of course, in the series of spectra for fixed a (Figs. 8–12) the resonant mode order is fixed; that is, it is independent of r_0 and in the case of the first resonance it is always $l=12$, $m=12$. This is because A_{1l}, B_{1l}, A_{2l}, and B_{2l} are independent of r_0.

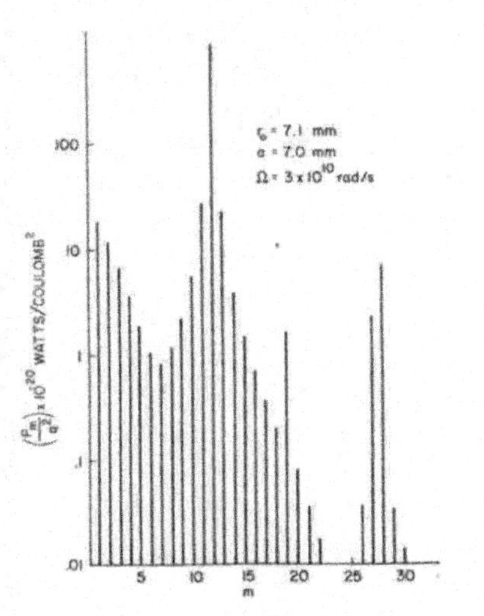

FIG. 11. Radiated spectrum for particle orbiting outside a dielectric sphere.

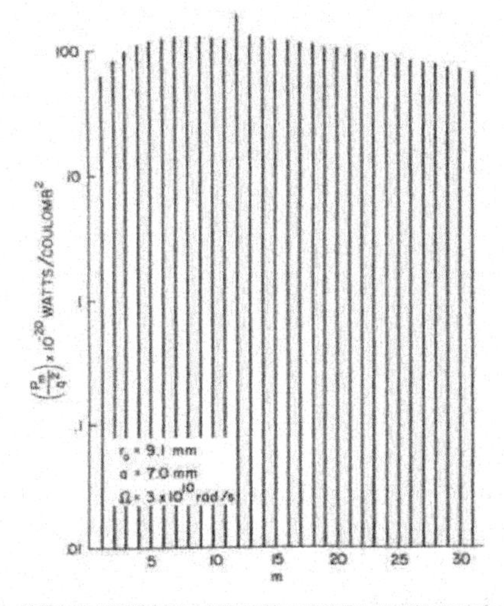

FIG. 12. Radiated spectrum for particle orbiting outside a dielectric sphere.

We note finally that though many resonances occur for the various values of l and m, only those for which l is not too much larger than m are significant. That is, were a resonance to occur for $l=m+12$ it would never be noticed in the spectrum (without considerable effort) because the mode itself in the absence of resonance is many orders of magnitude smaller than the $l=m$ mode.

CONCLUSION

The radiation from a charged particle orbiting both inside of and outside of a dielectric sphere has been expressed as a sum of vector spherical harmonics specified by indices l and m. The frequency of a given harmonic is $m\Omega$, where Ω is the orbital angular frequency. When the particle orbits close to or inside of the sphere, resonances of the sphere were found to appear in the radiated spectrum. At a resonance, one term of the expansion on l for the resonant harmonic m predominates and the radiation at $\omega=m\Omega$ is radiated for the most part into a single spherical harmonic. In the absence of a resonance effect the dominant l value is $l=m$ and in many cases the resonance is not strong enough to change this fact.

*Partly supported by a grant from the National Science Foundation.
†Present address: Department of Electrical Engineering, University of Mississippi, University, Miss. 38677.
[1]A. P. Kanzantsev and G. I. Snadutovich, Sov. Phys. Dokl. 7, 990 (1963).
[2]B. M. Bolotovskii and G. V. Voskresenskii, Sov. Phys. Usp. 9, 73 (1966).
[3]S. J. Smith and E. M. Purcell, Phys. Rev. **92**, 1069 (1953).
[4]R. Pogorzelski and C. Yeh, Phys. Rev. A **8**, 137 (1973).
[5]C. T. Tai, *Dyadic Green's Functions in Electromagnetic Theory* (International Textbook, San Francisco, 1971).
[6]R. Pogorzelski, C. Yeh, and K. Casey, J. Appl. Phys. 45, 5251 (1974).

8

HIGH-DATA–RATE FIBER NETWORKS AND LOW-NOISE FIBEROPTICS RECEIVER

By the late 1970s and early 1980s, most practical concerns for point-to-point single-mode fiber links, such as low-loss fibers, low-loss fiber splices, fiber connectors/splices, efficient couplers, high-data-rate laser sources, and detectors, had been solved. The next step was to use optical fibers for high-data-rate links for data nodes (stations), i.e., the formation of fiber optics local area network (FOLAN) or wide area network (WAN). To operate these networks efficiently, appropriate protocols and network architectures had to be developed, providing

- high-bandwidth efficiency,
- bounded delay,
- high fault tolerance,
- random access.

We undertook a research program to achieve this goal. We also considered the low-noise fiberoptics receiver problem. The following papers provide the results of our research.

8.1 RATO-Net: a random-access protocol for unidirectional ultra-high-speed optical fiber network. (Paper 8-1)

The local network medium is a pair of unidirectional fiber-optic busses to which stations are connected via passive taps. For this configuration, a new random-access protocol called RATO (random access, time-out) is presented in this paper. This RATO provides random access, fairness, and bounded delay access to all stations and is particularly suited for ultra-high-speed transmission when the performance of the popular Ethernet becomes unattractive. Simplicity and ease of hardware implementation of RATO under ultra-high-speed environment is emphasized because the only control requirements are the sensing of activity in the bus and a fixed time delay between consecutive transmissions from the same station. Simulation and performance of RATO with other schemes have been carried out. In ultra-high-speed wide-area networks, RATO outperforms all these other schemes. RATO was tested in a three-station pilot network: Excellent agreement was found between predicted and measured performance.

8.2 U-Net: a unidirectional fiber bus network. (Paper 8-2)

To satisfy the growing demands of local area community users, new architectures are required that provide very high bandwidth and satisfy real-time delay constraints. Fiber-optics networks appear to be a good choice for meeting these requirements. This paper describes a fiber-optics bus architecture, U-Net, which is based on a token protocol, offers high bandwidth efficiency, guarantees bounded delay, and is highly fault tolerant. U-Net performance compares very favorably with other fiber bus networks.

8.3 Token-based protocols for high-speed optical-fiber networks. (Paper 8-3)

The local network medium is a pair of unidirectional fiber-optic busses to which stations are connected via passive taps. For this configuration, we present several protocols that provide round-robin bounded delay access to all stations and are particularly suited to high-speed transmission. The common characteristic of the protocols is the use of the *token* as the synchronizing event to schedule the transmission. The token may be explicit (as in U-Net) or implicit (as in Tokenless Net). It may be used all the time, or it may be used simply to resolve collisions (as in Buzz-Net). The protocols are shown to be cost effective at very high (bandwidth x length) products that are the unique characteristic of high-speed single-mode fiber networks. Furthermore, they are robust to failure because of the passive interfaces and the totally distributed control. The implementation of these protocols on fiber-optic busses is also discussed in the paper.

8.4 Interconnection of fiber optics local area networks. (Paper 8-4)

The purpose of this paper is twofold: (1) to review the need for and the requirements of fiber optics local area networks (FOLAN) interconnects, and (2) to present Tree-Net, a novel, multilevel, interconnected FOLAN architecture. We first review the basic FOLAN architectures and their requirements. Then we discuss the problem of FOLAN interconnection and review some interconnect alternatives. Finally, we introduce Tree-Net as an example of interconnect schemes and evaluate its performance characteristics.

8.5 BUZZ-NET: a hybrid random access/virtual token local network. (Paper 8-5)

3uzz-net is a local network supported by a pair of unidirectional busses, to which stations are connected via passive interfaces. The access protocol is a hybrid that combines random access and virtual token features. More precisely, the network operates in random access mode at light load and in virtual token mode at heavy load. Buzz-net can find applications in fiber-optics networks, which are intrinsically unidirectional. This paper describes the protocol and compares its performance to that of other unidirectional schemes.

8.6 Low-noise fiber optics receiver with super-beta bipolar transistors. (Paper 8-6)

A low-noise wideband optical fiber receiver has been successfully designed using super-beta bipolar transistors (BJTs) at the front end. Even with *commercially* available super-beta devices, which are not optimized for our application, the obtainable input sensitivity for medium- and high-bandwidth optical receivers is comparable or superior to the best FET design. To demonstrate this concept, a 10 MHz analog receiver was built with a super-beta BJT at the input stage. This receiver achieved an expected average input noise current density of less than 0.4 $pA/\sqrt{Hz}$ over the full bandwidth for a transresistence of 500 kΩ. Detailed design procedures are given. The noise characteristics of a 50 MHz receiver using super-beta BJTs are also obtained.

78 JOURNAL OF LIGHTWAVE TECHNOLOGY, VOL. 8, NO. 1, JANUARY 1990

RATO-Net: A Random-Access Protocol for Unidirectional Ultra-High-Speed Optical Fiber Network

CAVOUR YEH, FELLOW, IEEE, MAURICE LIN, MEMBER, IEEE, MARIO GERLA, AND PAULO RODRIGUES

Abstract—The local network medium is a pair of unidirectional fiber optic busses to which stations are conneted via passive taps. For this configuration, a new random-access protocol called RATO (random-access, time-out) is presented in this paper. This RATO provides random-access, fairness, and bounded delay access to all stations, and is particularly suited for ultra-high-speed transmission when the performance of the popular Ethernet becomes unattractive. Simplicity and ease of hardware implementation of RATO under ultra-high-speed environment is emphasized because the only control requirements are the sensing of activity in the bus and a fixed time delay between consecutive transmissions from the same station. Simulation and performance comparison of RATO with other schemes have been carried out. In ultra-high-speed wide-area networks, RATO outperforms all these other schemes. RATO was tested in a three station pilot network: Excellent agreement was found between predicted and measured performance.

I. Introduction

THE NEED for high speed integrated fiber optics local area networks (FOLAN) is becoming apparent in many applications [1]–[13]. In the commercial sector an example is the large corporation wishing to combine all of its communications requirements on a fiber cable which spans the entire building or plant. The system may include thousands of telephones, hundreds of terminals, several video channels used for video conferencing, security monitoring, education, etc. This plant may be connected to other plants within the same metropolitan area via a metropolitan area network. Furthermore, cross country connections may be established via a satellite microwave or fiber WAN (wide area network) [14], [15].

Other civilian FOLAN applications include air traffic control and factory automation. The air traffic control center of the future will be characterized by a distributed processing environment [16]. Radar processors, general purpose processor, controllers' workstations, etc., will be connected by a high speed local network carrying radar data, command and control information, and digitized voice and video (real time) signals. Considering that a large ATC center may have several hundred workstations, it is easy to understand why second generation FOLAN's will be required to meet ATC communications needs. Automation of the factory plant has been propagating very rapidly in recent years, to become one of the most important applications of local networking, so much so that a manufactoring automation protocol (MAP) standard has been established for factory local networks (FLN's) [17]. Although the current standard is based on broad-band cable implementations, the FOLAN alternative is actively investigated for its light weight, ease of installation, EMI robustness, and high bandwidth.

Another example of service integration, this time in the defense sector, is the battlefield system [1], [8]. The basic battlefield system envisioned here consists of several outposts and observation stations (manned or unmanned) interconnected with each other and with command posts, and distributed over a radius of several miles, different media (multimedia environment). Radio (RF) communications will be used for mobile units (autos, tanks, airplanes, etc.) Satellite communications will be necessary to interconnect distant battlefield areas. The optical medium will provide the tightly meshed local interconnection among outposts and command posts. It will also provide a much higher bandwidth (up to the gigabit range) as compared with the megabit-per-second bandwidth typically available via RF and satellite.

Thus, the high speed requirement is the natural consequence of two facts: the user population growth and the increasing bandwidth of individual user connections (both data and image). For instance, the uncompressed, digitized, color video signal occupies a bandwidth of 80 Mbit/s and higher. Also, high volume disk-to-disk transfers may reach a rate of several megabits per second [18].

The design of very high speed FOLAN's poses two types of challenges. First, as speed increases, the ratio of propagation time over packet transmission time increases, thus making the carrier sense multiple access-collision detect (CSMA-CD) protocol [19], [20] generally used in Ethernet-type networks, inadequate for FOLAN's. Thus, new protocols must be developed. Secondly, the implementation of the optic/electric interface clearly becomes

Manuscript received July 5, 1988; revised March 7, 1989. This work was partially supported by the NSF and by ARO.
C. Yeh and M. Lin are with the Electrical Engineering Department, University of California, Los Angeles, Los Angeles, CA 90024.
M. Gerla is with the Computer Science Department, University of California, Los Angeles, Los Angeles, CA 90024.
P. Rodrigues was with the Computer Science Department, University of California, Los Angeles, Los Angeles, CA 90024. He is now with NCE, Federal University of Rio de Janeiro, 20001 Rio de Janeiro, Brazil.
IEEE Log Number 8928607.

a delicate task at gigabit speed [5], [6], [21], [22]. The high speed requires a careful coordination between protocol design and interface design. Namely, the protocol must be designed so that the amount of processing required at gigabit speed is minimal to keep the gigabit logic simple. The less time-critical functions should be performed off-line.

The integration of data with voice and video also poses a challenge in the design of the protocol. In fact, while data can be freely buffered and flow controlled, voice and real time video must be delivered to destination with tight delay constraints, otherwise unacceptable degradation will occur. This suggests that voice and video connections be handled in a special way. Namely, a connection should be accepted only if sufficient bandwidth is available. Once accepted, the connection should be guaranteed the required bandwidth [23]–[26].

In this paper, we will focus on the analysis, simulation, and experimentation of a new protocol for a dual unidirectional bus structure which possesses the desirable random access and guaranteed delay features. Furthermore, because of its simplicity, this protocol is particularly suited for ultra-high-speed fiber network systems. The simplicity stems from the fact that the only control requirements for this protocol are the sensing of activity in the bus and a fixed time delay between consecutive transmissions from the same station.

II. Statement of the Problem

One of the very unique features of high-speed (gigabit-per-second) FOLAN's is that end-to-end propagation time could be larger than the packet transmission time [24]. In copper based LAN's (e.g., Ethernet), on the other hand, because of the relatively low bandwidth (megabit-per-second) the propagation time is almost always shorter than the packet transmission time. It is therefore of interest to develop, analyze, and test new random-access protocols which can operate efficiently also in the FOLAN environment.

Another unique characteristic of high-speed FOLAN's is the potential mismatch between network speed and terminal speed; namely the optical portion of the FOLAN could be operating at a gigabit-per-second rate while the local terminal may be operating at a much lower rate (say the kilobits-per-second or megabits-per-second range). The interface portion of FOLAN will play a central role in the successful implementation of high-speed FOLAN's.

The proposed architecture is the dual bus configuration shown in Fig. 1. In this topology there are only two connection points per station per bus, and expansion of the network is easily done at both ends. In the proposed protocol, one has the option to perform unidirectional or bidirectional transmissions. (i.e., each packet is transmitted only on one bus, or on both busses) Unidirectional transmission requires a session setup phase in order to discover the physical location of the destination. On the other hand, bidirectionality, though wasting some of the bandwidth, avoids the setup phase and allows direct addressing by name (independent of physical station address). That is, processes can be moved about in the network without other processes needing to be aware of their current physical location. Addressing by name can be achieved by using a word associative buffer in each bus interface to hold the names of the processes resident in the attached components. Session connections can be established in hardware without intervention of higher level protocols.

For ultrahigh speed FOLAN, it is desirable to emphasize the simplicity in the implementation of a given protocol [5], [6]. The random access time-out (RATO) protocol that we developed is simple to implement because the only control requirements are the sensing of activity in the bus, and the enforcement of a fixed time delay between consecutive transmissions from the same station. RATO transmissions are controlled separately in each direction. If bidirectionality is required, a packet can be queued for independent transmission in opposite directions. However, when a session is established between processes residing in different stations, the processes may be able to determine their location relative to each other during the set-up phase, and consequently, stations may attempt to transmit only in a single direction. We further assume that the receiver is able to detect a packet even when the packet is immediately preceded by some truncated transmission.

When a station has a packet to transmit, it performs the following steps:

1) The station senses the bus. If the bus is busy, it defers until the bus is idle.
2) The station starts transmitting the packet. If a collision with an upstream transmission occurs, the station aborts its packet and repeats step 1. Otherwise, step 3 is performed next. Observe that the incoming transmission gets only corrupted in its first d seconds, where d is the station reaction delay. Typically, $d \ll$ preamble length. Thus, the packet preamble guarantees data integrity and allows reliable packet reception at downstream stations.
3) The station observes a time out of T_0 s before it considers a new packet for transmission. If the transmission queue is empty after the elapsed T_0 s, the station goes idle until a new packet arrives. Then the station performs step 1 again.

Detailed performance analysis and simulation of this RATO-net have been carried out and compared with other more conventional schemes (i.e., token, ethernet, etc.) They will be discussed in the following sections.

III. Performance Analysis

For this RATO protocol time-out T_0 is critical to provide fair access to all stations in both directions. We determine T_0 such that all stations have a chance for a successful transmission in a finite time.

Consider the left-to-right (L-to-R) bus in Fig. 1. Due to transmission deferral, downstream stations are

80 JOURNAL OF LIGHTWAVE TECHNOLOGY, VOL. 8, NO. 1, JANUARY 1990

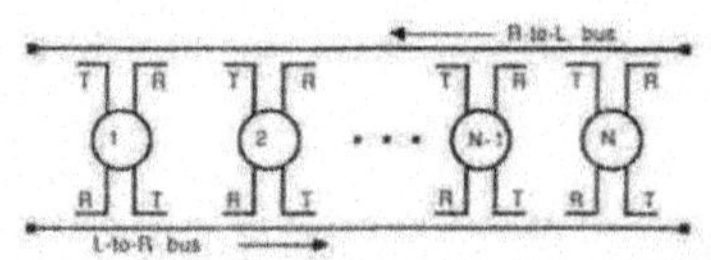

Fig. 1. The unidirectional bus topology—The RATO topology.

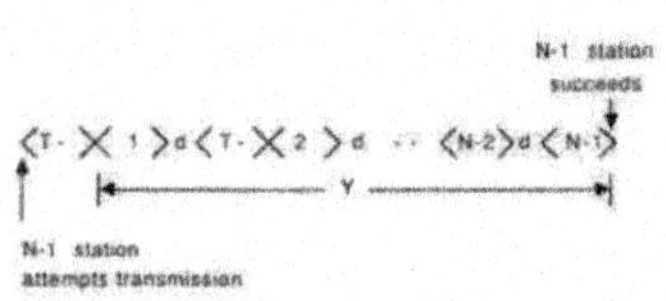

Fig. 2. Worst case insertion delay for RATO.

preempted by upstream transmissions. Therefore, the worst case condition for the insertion of a packet occurs at station $N - 1$ (station N only transmits on the R-to-L bus). Let us investigate the worst case for station $N - 1$ trying to transmit to station N. Assume that station $N - 1$ detects the bus idle and starts transmitting. After $T - \epsilon$, where T is one packet transmission time and ϵ is very small, the transmission is almost completed but a collision with a transmission from station 1 occurs. Station $N - 1$ defers and attempts again when the bus is idle (within d s of reaction delay). When the transmission is almost completed a collision from station 2 now occurs. Collisions from other stations follow this pattern until station $N - 1$ finally succeeds after transmission from station $N - 2$. The sequence of events as seen by an observer at station N is depicted in Fig. 2, where a worst case collision is represented by $\langle T \geq \epsilon \rangle$, $\langle i \rangle$ is a successful transmission of duration T by station i, and $\langle d \rangle$ is a station reaction delay.

From the figure we get:

$$Y = (N-1)T + (N-3)(T-\epsilon) + (N-2)d, \quad N > 2. \tag{1}$$

To provide a finite insertion delay to station $N - 1$ we must guarantee that the next transmission by station 1 (also applicable to other stations) does not occur before T_0 s where T_0 is given by

$$T_0 \geq \lim_{\epsilon \to 0} (Y - T).$$

Hence

$$T_0 \geq (2N-5)T + (N-2)d, \quad N > 2. \tag{2}$$

For $N \gg 1$ and $T \gg d$ we have $T_0 \geq 2NT$.

It is clear that all the other stations, not only station 1, must also be subjected to the same constraint.

Assuming that $N > 2$ and that $T_0 = (2N - 5)T + (N - 2)d$, let us evaluate RATO's performance in terms of throughput and delay performance.

A. Throughput Analysis

In the following analysis, we assume that data and preamble transmission times of each packet are constant and equal to T_r and T_p, respectively. Here $T = T_r + T_p$.

1) Light Load Analysis: At light load a new packet always arrives after the old packet has been serviced. Hence, the network is at a steady state, where output packet rate equals input packet rate. Thus, with each station having a packet arrival rate λ packets/second and with total i active stations on the bus, the throughput S, in the steady state is then given by

$$S(i) = i\lambda T_r.$$

Noted that the most downstream station cannot transmit packets on the same bus; therefore, the maximum throughput is given by

$$S_{\max} = \lambda(N-1)T_r.$$

2) Heavy Load Analysis: A station at heavy load is defined as a packet arrival occurs immediately after the buffer of the station becomes available. At heavy load, time out T_0 forces transmissions to be clustered together in rounds starting every $T_0 + T$ s. A round is depicted in Fig. 3.

From Fig. 3 the bus throughput when i stations are at heavy load is:

$$S(i) = \frac{iT_r}{T_0 + T} = \frac{i}{(N-2)} \frac{T_r}{(2T + d)}, \quad \text{for } i \leq N-1. \tag{3}$$

Channel capacity or maximum throughput S is obtained when $i = N - 1$. The maximum throughput is independent of τ (the end-to-end delay) and approaches 0.50 as $T_r \gg T_p$ and $N \gg 1$. This relationship implies that packet length, T_r must be large to provide a good throughput, especially since the preamble must increase with transmission speed. Independence from τ makes it possible to cover large distances and still maintain acceptable throughput. It is noted that the 50-percent capacity of this scheme is due to the penalty one must accept in order to guarantee the random access character of this protocol. Note also the fact that bus throughput is entirely independent of end-to-end length and channel speed, i.e., throughput is not a function of end-to-end propagation delay and packet transmission delay.

Other proposed protocols such as Fastnet, Expressnet, U-net, Buzz-net all exhibit certain degrees of dependency on network length and transmission rate due to their use of explicit or implicit token. In those schemes, a token, whether explicit or implicit, is normally used to define the boundary of a round robin cycle. By using a token, the round robin cycle approximately equals the sum of the total transmission time of a packet train and the token latency. At low transmission rate, the token latency is very small compared to the total transmission time of a train of packets. As a result, the overhead of using a token is

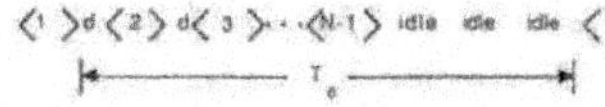

⟨1⟩ d ⟨2⟩ d ⟨3⟩ . . . ⟨N-1⟩ idle idle idle ⟨1⟩ . .

T_e

Fig. 3. Heavy load round in RATO.

almost negligible. However, that will not be the case when the transmission rate is increased and/or the network geographic size is lengthened. The transmission time of a packet is inversely proportional to the bit rate while the token latency is proportional to the network length and is independent of the transmission rate. Consequently, the ratio of the token latency over the cycle time which is the sum of token latency and total transmission time is no longer small. Performance of this network then deteriorates drastically. Hence, it is clear that a protocol whose performance depends on the physical length and bit rate is inherently designed for low bit rate transmissions and/or short distance communications. On the other hand, a protocol, such as RATO, whose performance is truly immune to both propagation and transmission delay, is more suitable for high bandwidth network such as an optical fiber network.

B. Delay Analysis

In order to simplify the analysis, we will make use of certain assumptions. It will be shown by simulation study that our approximate analysis still provides excellent results.

The assumptions are: a) the packet arrival of each station is a Poisson process with the same average rate λ packets per second, b) each station possesses one buffer which implies that any packet that arrives before the buffer has a chance to empty its content, will be lost, and that a new packet will be accepted only if the buffer is empty upon its arrival. Since RATO protocol requires each station to exercise a predetermined time-out period after each successful transmission, an accepted packet may have to stay in the buffer without being processed until the time-out period is over. So, the service facility of each station may be modeled by a two-stage sever as shown in Fig. 4. In this figure, the large oval boundary represents the service facility of a station. At any instant, at most one packet can be kept inside the facility. When a new packet enters the facility, it moves into the first stage where the service time ranges from 0 to T_0 s. Upon departure from stage one, it proceeds immediately to the second stage where the service time is at least T. Clearly, the total time that a packet spent in the facility is a random variable which is the sum of two independent random variables. Hence, the average service time in the facility is the sum of the average time a packet spent in stage one and in stage two.

The calculation of average service time at stage one $\overline{W}$ is straightforward. Because of the memoryless property of Poisson process, time considered for averaging starts from the moment that the buffer becomes available to infinity. If the arrival time t of a packet is less than T_0 s, the packet must wait $T_0 - t$ s. If the packet arrives after T_0 s,

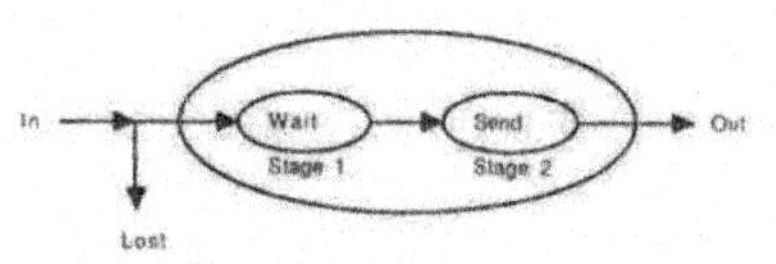

Fig. 4. Two-stage server facility.

it then spends 0 s at the stage one. Therefore, $\overline{W}$ is calculated by

$$\overline{W} = \int_0^{T_0} (T_0 - t)\lambda e^{-\lambda t}\, dt = T_0 - \frac{1}{\lambda}(1 - e^{-\lambda T_0}). \tag{4}$$

The average service time in stage two equals the sum of one packet transmission time T and the average deferral delay $\overline{dd}$. The exact analysis of the average deferral delay is very complicated. To simplify the analysis, it is assumed that a packet from a given station has an equal risk to collide with that from any of its upstream stations. Let p be the collision probability. To find the average deferral delay caused by any upstream station, we need to know the average time spent by any upstream station on the transmission of one packet. In the case of no collision, transmission of one packet takes exactly T s. However, it takes more than T s if a collision occurs. With collision, total time spent on the transmitting of one packet equals the sum of wasted time and the T s. The wasted time can range from ϵ seconds to $T - \epsilon$ s, where ϵ is a negligible small number. Thus, the time for transmission of one packet takes from T s to $2T - \epsilon$ s. Since it is randomly distributed, the average time spent on one packet's transmission is $\frac{3}{2}T$ s. The average deferral delay imposed upon a downstream station by any upstream station is $(3/2)pT$ s, if the fruitless transmission of the downstream station aborted by the first arriving upstream packet is considered belonging to the upper most station, which does not suffer from any deferral delay. For the ith station, there are $i - 1$ upstream stations; therefore, the average deferral delay d_i is given by

$$d_i = \tfrac{3}{2}pT(i - 1). \tag{5}$$

The average deferral delay over $N - 1$ stations is thus

$$\overline{dd} = \frac{1}{N-1}\sum_{i=1}^{N-1} d_i = \frac{N-2}{2} \times \frac{3}{2}pT. \tag{6}$$

Since p is the probability that a downstream packet may collide with an upstream packet, this is equivalent to the probability that an upstream station transmits a packet during a "vulnerable period." The "vulnerable period" is shown in the Fig. 5. Duration of the vulnerable period is equal to twice as long as a packet's transmission time T and is the same for every upstream station. So p is the probability that an upstream station has a packet ready to be sent and transmits the packet during the vulnerable period. The event that an upstream station has a packet ready

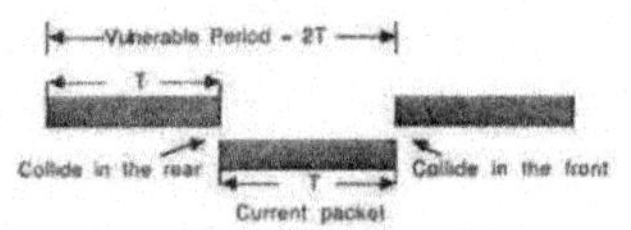

Fig. 5. Length of a vulnerable period.

to be sent and the event that the upstream transmission is taken place during the vulnerable period are independent events. There, p is the product of the probabilities of the occurrence of the events. The probability that an upstream station has a packet ready to be sent consists of the probability that a packet emerges from a timeout and the probability that a packet arrives after the timeout. A packet emerges from a time out means that the station was at a time out stage when the packet arrived and was not at the stage now. That is, the time-out-over event has occurred. The probability that an arrival occurs between $0 < t < T_0$ is

$$\text{Prob}\,[0 < t < T_0] = 1 - e^{-\lambda T_0} \tag{7}$$

where t is the packet arrival time.

To find the probability that the station is no longer at a time-out state now, let us consider the following problem. Given a packet arrival occurs at time 0 s, the occurrence of time-out-over event will come between time 0 s and T_0 s. Let W be the time that the event occurs. What is the probability that the event occurs after $\overline{W}$ s? Clearly, W is a random variable whose value can be any number between 0 and T_0. Since the average waiting time of all packets is $\overline{W}$ s, the probability that the station currently is not a a time-out stage is given by the probability that value of the random variable W is greater than the average waiting time $\overline{W}$. Assuming that the occurrence of the time-out-over event is totally random; thus, the distribution function of the random variable W is a uniform distribution. The probability is therefore given by

$$\text{Prob}\,[W > \overline{W}] = \frac{T_0 - \overline{W}}{T_0}$$

and the probability that a packet emerges from a time out is given by

$$(1 - e^{-\lambda T_0})\left(1 - \frac{\overline{W}}{T_0}\right) \tag{8}$$

where $\overline{W}$ is given by (4).

The probability that a packet arrives after the time out is given by

$$\text{Prob}\,[t > T_0] = e^{-\lambda T_0}. \tag{9}$$

From (8) and (9), the probability that an upstream station has a packet ready to be sent is obtained by

$$(1 - e^{-\lambda T_0})\left(1 - \frac{\overline{W}}{T_0}\right) + e^{-\lambda T_0}. \tag{10}$$

The probability that an upstream station transmits during the vulnerable period is given by

$$\text{Prob}\,[\text{an upstream station transmits during } 2T] = 1 - e^{-\mu 2T} \tag{11}$$

where μ is the average packet service rate per station. Except the discarded packets, operation of the bus is at a steady state. That is, from the bus point of view the output packet rate equals the input packet rate, or

$$\mu = \lambda_0 \tag{12}$$

where λ_0 is the effective packet arrival rate, which is the rate of packets accepted by the buffer of a station.

Since packets which arrive while the buffer is not available are discarded, the effective packet arrival rate λ_0 is obtained by multiplying the actual arrival rate λ by the acceptance fraction. So, the effective arrival rate or the average packet service rate is given by

$$\lambda_0 = \mu = \frac{\lambda}{\lambda(\overline{W} + \overline{dd} + T) + 1}. \tag{13}$$

Using (10), (11), and (13), the collision probability p is given by

$$p = \left((1 - e^{-\lambda T_0})\left(1 - \frac{\overline{W}}{T_0}\right) + e^{-\lambda T_0}\right) \cdot (1 - e^{-2\lambda_0 T}). \tag{14}$$

Substituting (13) and (14) into (6), the average deferral delay $\overline{dd}$ can be found by numerical methods.

Finally, the total average service time per packet $\overline{X}$ is given by

$$\overline{X} = \overline{W} + T + \overline{dd}. \tag{15}$$

The insertion delay (ID) which excludes the packet transmission time, is

$$\text{ID} = \overline{W} + \overline{dd}. \tag{16}$$

Since the output packet rate per station is μ, throughput of the bus, in bits per second, can be easily found and is given by

$$S = \mu(N-1)T_r = \frac{\lambda}{\overline{X}\lambda + 1}(N-1)T_r. \tag{17}$$

Equations (16) and (17) are used later to compare with the simulation data. The results are surprisingly good with the simulation results.

Delay performance is measured in terms of the insertion delay (ID). At light load the bus is usually idle and very few collisions take place. With light traffic the insertion delay is negligible if packet arrival time is greater than T_0. In case of multipacket messages, the ID for the first packet is 0 and is T_0 for the other packets of the same message.

At heavy load, the insertion delay (IDH) always equals T_0. This is because "phasing" of the transmission at the various stations takes place. However, the worst case for ID (MID) occurs at intermediate load when station $N-1$ (assuming left to right transmissions) tries to transmit after a T_0-s wait and finds the bus idle. When the transmission is almost completed, a collision occurs with a transmission from station 1. Then station $N-1$ attempts again and suffers a collision with station 2. Collisions with the other stations follow this pattern until station $N-1$ finally succeeds after the transmission from station $N-2$. The above pattern was also actually assumed as the worst case in calculating T_0. Therefore, from Fig. 2 and remembering that station $N-1$ has already waited T_0 at the beginning of the events, MID $= 2T_0 + T$. For $N >> 1$ and $T >> d$ we find MID $= 4NT$. This worst case result is approximately twice the value of IDH. Of course, the events leading to the worst case are very unlikely, and MID should neither affect the average delay nor the delay distribution.

We have given a lower bound on the value of T_0 to guarantee fair access and bounded delays. A major drawback is the dependency of T_0 on the product NT. If T_0 is set to its minimum acceptable value, then a new station insertion should be followed by a correspondent increase in T_0. In case of station deletion, T_0 should be decreased, to avoid wasted bandwidth and unnecessary delay.

To summarize, a very simple random access protocol with time-out control (RATO) has been described. The protocol uses time out T_0 as its only control and relies on deferral to upstream transmissions. Due to its simplicity, RATO implementation cost would be most economical.

IV. Simulator

In this section the development of a RATO simulator is described. The simulator can provide not only the average performance of the network as a whole but also the performance of separate stations [28]. Although RATO protocol is designed for dual unidirectional buses, it is adequate to study its performance on one bus. Hence, simulation is carried out for a single unidirectional bus. In the simulator design the discrete event scheduling technique is used to model the network. And all the statistical data can be obtained from the occurrences of events. There are eight events and five states in the simulator program which was written in C and run on a Pyramid computer. List of events and their schedulings are shown in Fig. 6(a) and the state diagram is shown in Fig. 6(b).

In the simulator design each station contains only one buffer. Each station is allowed to send only one packet at a time, and the packet size always equals to the buffer size which is the same for every station. Furthermore, no random line noise is considered. Consequently, once a packet is completely sent, that packet is assumed to have reached its destination intact. As a matter of fact, the only way that a packet can be lost is due to the transmission deferral to the upstream packet trains. Again, as in the analytic model, packet arrival for every station is a Poisson process.

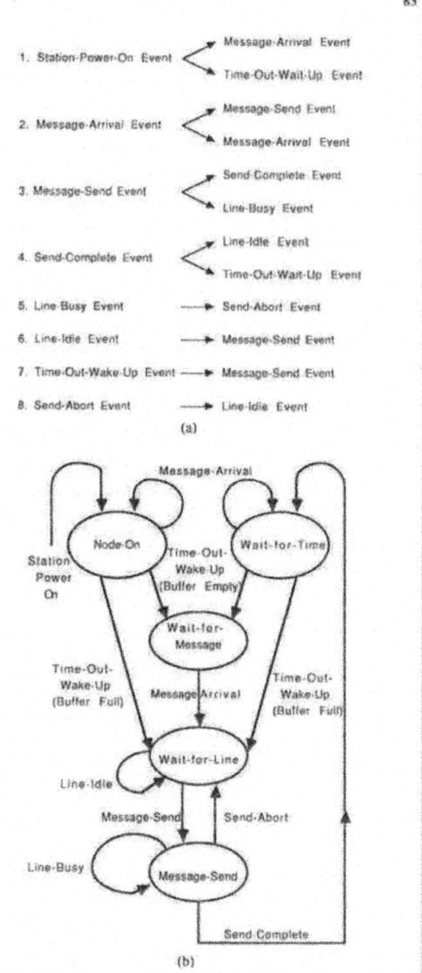

Fig. 6. (a) Events and schedulings. (b) State diagram of the simulator.

V. Specific Examples and Performance Comparison with Other Known Networks

To guarantee the random access property of RATO, the maximum throughput of the RATO network must be limited to 50 percent. But, the throughput of RATO is independent of propagation delay and transmission delay, so, unlike other protocols, the performance of RATO is not affected by the network geographic size and transmission rate. The throughput of RATO is compared with six other protocol schemes [12], [19], [24], [29], [31] in Figs. 7 and 8 in which throughput is plotted against the number of active stations for various protocols. Heavy load con-

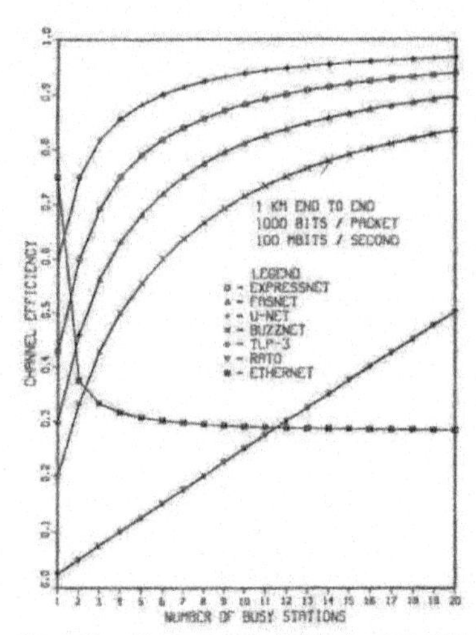

Fig. 7. Throughput comparison for 1-km length network.

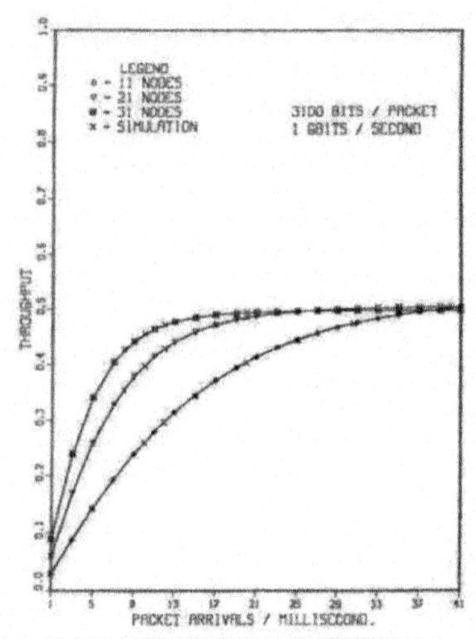

Fig. 9. Average throughput of RATO protocol.

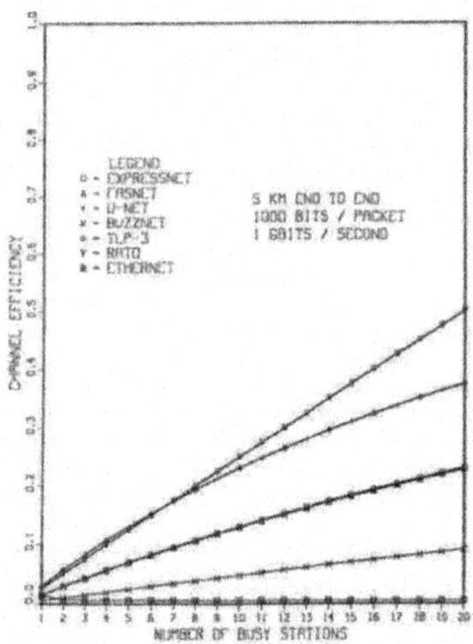

Fig. 8. Throughput comparison for 5-km length network.

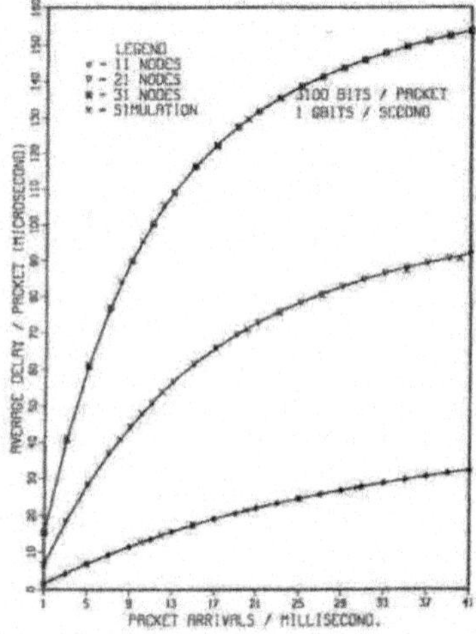

Fig. 10. Average insertion delay of RATO protocol.

dition is assumed for each station. In Fig. 7, the maximum network length is assumed to be 1 km, packet size is assumed to be 1000 bits, the transmission rate is taken to be 100 Mbit/s, and total number of stations connected to the network is 21. At low transmission rate it appears that RATO's throughput is the worst among all seven protocol schemes. As the network length and the transmission rate are increased to 5 km and 1 Gbit/s, respectively, Fig. 8 shows that the performance of RATO is the best among all seven schemes. This is because the performance of all the other protocols degrades drastically for wide area, high-bit-rate networks while that of RATO remains unchanged.

In Figs. 9 and 10, the analytic results are compared with the results obtained through simulation. Throughput is plotted against input packet arrival rate for various number of station nodes. As the number of stations connected to the bus is increased the network throughput

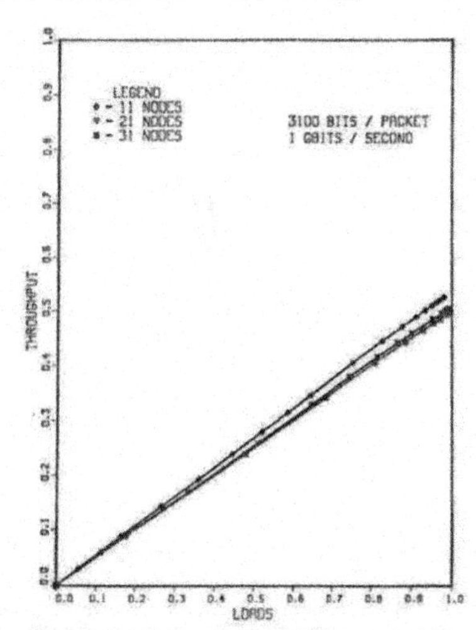

Fig. 11. Average throughput of RATO versus protocol.

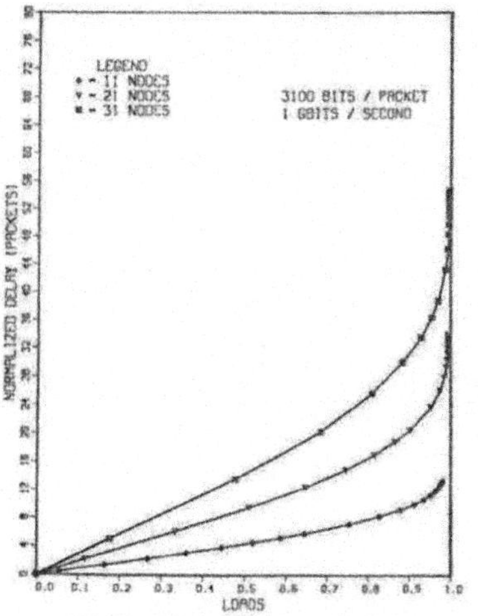

Fig. 12. Delay of RATO versus loads.

reaches the heavy load condition faster. In Fig. 10, the average delay per packet versus input packet arrival rate is shown. The delay is clearly proportional to the number of stations connected to the bus. Very close agreement is found between analytic results and simulated data, verifying the validity of the approximations used in deriving the analytic results.

Since the packet rate alone cannot reflect the amount of network load, let us introduce a load factor ρ which is defined as follows:

$$\rho = \frac{\text{Total packets accepted by the network per unit time}}{\text{Maximum packets serviced by the network per unit time (channel capacity)}}.$$

In this definition, the packet counts do not include those that are lost due to lack of available buffer. Hence, we have

$$\rho = \lambda_0(T_0 + T) \tag{18}$$

where λ_0 is the effective packet arrival rate, T_0 is the time-out period, and T is the single packet transmission time including the preamble overhead.

The throughput of (17) now reads

$$S = \rho \frac{(N - 1)T_r}{T_0 + T}. \tag{19}$$

Throughput S in (19) is plotted against the load factor ρ in Fig. 11 for different number of station nodes. As expected the maximum throughput is about 50 percent at the maximum load factor of unity. For the same given load, smaller number of station nodes N gives higher throughput. The variation becomes negligible when N becomes large.

The normalized insertion delay as a function of load is shown in Fig. 12. Here, the normalization constant is the transmission time for one packet. Because the load is a function of both insertion delay and packet arrival rate, no simple expression can be found for insertion delay as a function of load alone. Since both insertion delay and load are function of packet arrival rate, by varying the arrival rate one can obtain the relation between insertion delay and ρ. The results are shown in Fig. 12. One notes that the insertion delay is proportional to the total number of stations connected to the network, the delay is modest when the load is not close to the heavy load condition and the insertion delay is bounded even at load equal to 1.

Because we assume that each station has only one buffer, some packets will be lost due to nonavailability of buffer. Therefore, the knowledge on how the offered load and the network load are related is very important. Similar to the definition for the load factor, the offered load is defined as the ratio of the actual packet arrival rate to the maximum packet output rate. The expression for the offered load G is

$$G = \lambda(T_0 + T). \tag{20}$$

Results for the offered load versus network load are shown in Fig. 13. It is seen that network load is almost the same as the offered load when the load is around 50 percent and below. However, as the network load approaches 75 per-

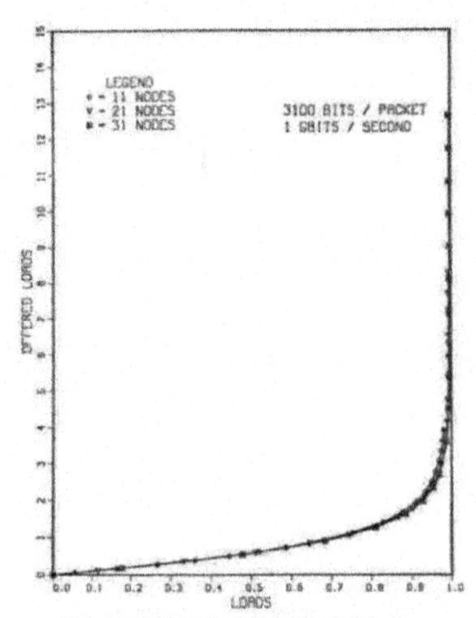

Fig. 13. Offered loads of RATO versus loads.

cent, the offered load approaches 100 percent. This implies that 25 percent of the requests for service will be denied due to the lack of buffer. When the offered load is greater than 100 percent, the network load is also increased, but more packets are lost.

VI. RATO Experiments

Having developed and analyzed a new protocol (RATO) which is particularly suited for high-speed FOLAN, it would be of interest to build an experimental network with RATO protocol to verify some of the analytical results. To simulate the high-speed FOLAN condition our experimental RATO network must possess the unique feature that end-to-end propagation time would be larger than the packet transmission time. Furthermore a buffer will be built into the interface of each station to simulate the condition of potential mismatched between network speed and terminal speed. A three-station experimental setup, based on a single unidirectional bus configuration with random access time-out protocol network, is built (see Fig. 14). (An operational RATO network would require a dual unidirectional bus configuration as shown in Fig. 1 in order that any station may reach any other stations.) The three local stations are three IBM-PC's. The needed interfaces with buffer has been designed and built. In the laboratory environment the spacing between stations was set to be much less than 1 km. To simulate the condition encountered in very high-speed gigabit-per-second rate FOLAN's for which the propagation delay is much longer than packet length (time), the delay between sending of the packet from one station and receiving of the packet at the next station is artificially created by electronic means. The generated delay may be set to be anywhere from 0.1 to 15 s, which is appropriate for our experiment that was performed using a baud rate of 1200 Bd. Each IBM-PC is connected to the unidirectional FOLAN through an interface called terminal network connector (TNC). A block diagram of the interface TNC is shown in Fig. 15. The TNC consists of eight modules: a central controller, two buffers, an encoder, a decoder, a line sensor, a transmitter, and a receiver. The central controller module has a microprocessor and memories [28]. Our RATO protocol is programmed into the microprocessor. Central controller also performs the handshaking operation for data transfer between external devices, such as the terminal computer and buffers in the TNC. Two separate buffers, the transmitting buffer and the receiving buffer, are used in the TNC. Buffers are used to match the operating speed of a peripheral device and that of the fiber network. Packets in the transmitting buffer are kept there until they are allowed to be sent without collision. The input data stream coming into the transmitting buffer may take the form of parallel streams or serial stream while the output data stream must take the form of a serial stream. On the other hand, the receiving buffer may accept data in a serial stream while the output data from it may take the form of parallel streams or serial stream.

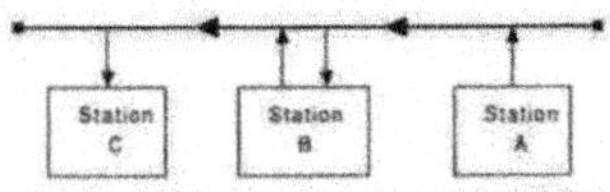

Fig. 14. A three-station bus network for RATO.

To send a packet to the fiber network, the NRZ signal from the transmitting buffer in the interface is first converted into the NRZI line format by the encoder. The converted signal then passes through a switch located in the transmitter module in order to reach the input of the light source. This switch is used for transmission abortion when collision occurs. The encoded signal then switches ON/OFF (i.e., modulates) the light source to achieve electrooptical conversion and modulation. The optical pulses emitted from the light source are coupled into an optical fiber line which guides the optical pulses to the next station.

When receiving a packet from the optical fiber, a receiver converts the optical signal into an amplified electrical signal. The amplified signal is sent to both a line sensor module and a clock recovery module. Upon receiving the signal, the line sensor module immediately puts out a line busy signal which opens a switch to allow the bypass signal to go through without being interrupted. The line sensor module continuously sends out line busy signal until the line is detected idle. At the same time, the clock recovery module extracts clock signal from the incoming data stream and sends the clock signal to the decoder and the receiving buffer. The decoder converts the received NRZI line code into the NRZ standard binary digital format and stores the converted data in the receiving buffer for further processing.

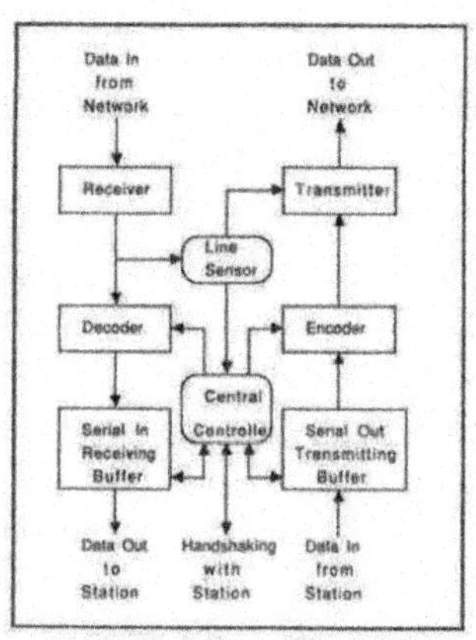

Fig. 15. Block diagram of the interface.

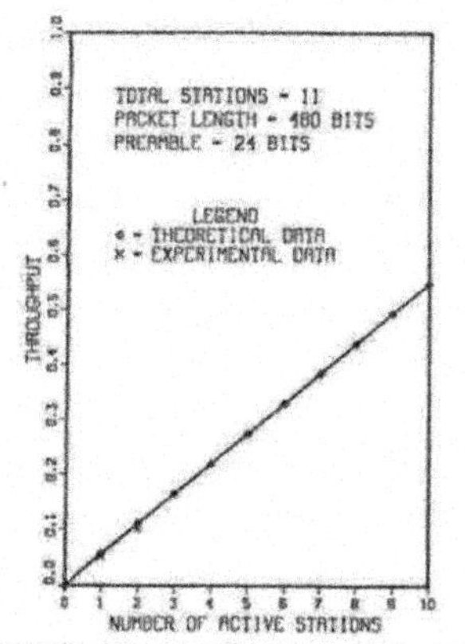

Fig. 16. Comparison of experimental results with theoretical data for throughput versus active stations for RATO protocol.

An important part of our interface design is the time out mechanism. Because time out duration depends on number of stations connected to the network, it is necessary to have a programmable time out system. Also, the time out period should be very long compared to a bit transmission line. A software programmable time out system is used. For the present RATO system, the shortest time out period can be set to be 100 ms and the longest time period can be set to be 15 s.

An experiment is carried out to measure throughput as a function of the number of active stations and throughput as a function of packet length. Measured results are compared with calculated results in Figs. 16 and 17.

In Fig. 16 the packet length is 480 bits, which contains 40 bytes of data, and 20 bytes of header. The preamble field is 24 bits, and the station reaction delay is 22 bits. The station reaction delay is the time between the activity that appears on the line and the activity known to the station. The reaction delay is the sum of the delay in detecting "line idle" and delay in detecting "line busy." In our design, the longest delay in detecting line idle takes 20 bits. Detection of line busy can be done much quicker using a 2-bit delay. Hence, the time-out period is equal to 8964 bits. Since the baud rate is 1200 Bd, the time-out period is translated into 7.47 s. Because the smallest unit in our software resolution is 0.1 s, the time-out period is then set to be 7.5 s. With only three stations available, we are able to measure the throughputs of one or two active stations. So, in our experiment, the average time required to deliver one packet from one active station is 7.94 s while it is 3.97 s from two active stations. The corresponding channel efficiencies are 0.05 and 0.1, respectively. Measured results are compared with calculated results in Fig. 16. The channel efficiency as a function of different packet lengths is shown in Fig. 17. One notes that efficiency increases slightly when the packet length is increased. This is because overhead length which

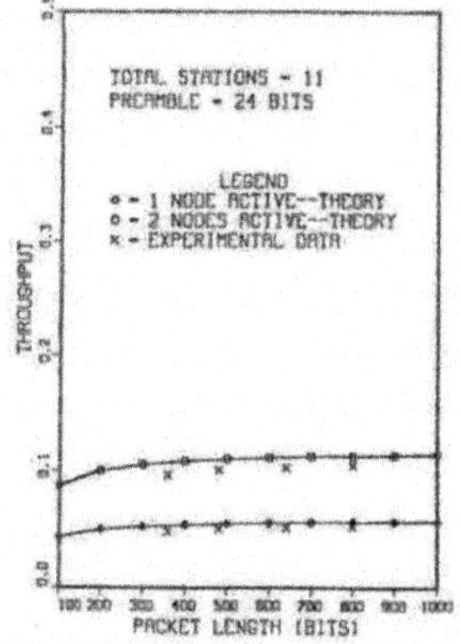

Fig. 17. Comparison of experimental results with theoretical data for throughput versus packet lengths for RATO protocols.

includes preamble field and station reaction delay, is a relatively small fixed quantity. In general, the throughput is quite insensitive to the variation of packet length. Excellent agreement between calculated results and measured results is seen from these figures.

VII. Conclusion

A new random access easy-to-implement protocol RATO has been developed and tested for a dual unidirectional bus structure made with optical fibers. Because it allows the packet transmission time to be shorter than the end-to-end propagation time, this protocol is particularly suited for ultra-high-bandwidth FOLAN's. In addition to having the feature of guaranteed delay, RATO also assures fairness among all stations. Performance comparison with other known protocol schemes, such as ethernet, fastnet, etc., shows that at ultra-high-bit-rate transmission for wide-area networks, RATO out-performs all these protocols. Simulation study was also carried out for RATO; very close agreement was found among results found from simulation and those calculated from analytical formulas. To test the operation of RATO, a three-station hardware-implemented RATO net was built and experimental measurements were performed. Excellent agreement was found between measured data and calculated results.

The basic RATO scheme may be modified to accommodate "burst-mode" operation, or to increase the overall channel efficiency. These modified RATO schemes will be analyzed and discussed in a subsequent paper. Expansion of the RATO type network as well as the possible support of voice/video connections will also be discussed there.

References

[1] L. U. Dworkin, and J. R. Christian, "Army fiber optic program: An update," *IEEE Trans. Commun.*, vol. COM-26, no. 7, pp. 999–1006, 1978.

[2] G. Lutes, "Optical fiber applications in the NASA deep space network," *Fiber Optic Technol.*, pp. 115–121, Sept. 1982.

[3] T. Minami *et al.*, "A 200-Mbit/s multiservice optical LAN using synchronous TDM loop structure," *IEEE J. Selected Areas Commun.*, no. 6, pp. 849–858, Nov. 1985.

[4] M. M. Nassehi *et al.*, "Fiber optic configurations for local area networks," *IEEE J. Selected Areas Commun.*, no. 6, pp. 941–949, Nov. 1985.

[5] R. V. Schmidt and H. Kogelnik, *Appl. Phys. Lett.*, vol. 28, p. 503, 1976.

[6] O. G. Ramer, C. Mohr, J. Pikulski, *IEEE J. Quantum Electron.*, vol. QE-18, p. 1772, 1982.

[7] J. Pingry, "Local area networks in fiber," presented at IFOC, Summer 1982.

[8] W. Vlasak and G. Pfister, "A fiber optic local area network solution for tactical command and control systems," presented at SPIE Meet., San Diego, CA, 1983.

[9] R. V. Schmidt *et al.*, "Fibernet II: A fiber optic ethernet," *IEEE J. Selected Areas Commun.*, vol. SAC-1, pp. 702–711, Nov. 1983.

[10] E. G. Rawson and R. M. Metcalfe, "Fibernet: Multimode optical fiber for local computer networks," *IEEE Trans. Commun.*, vol. COM-26, no. 7, pp. 983–990, July 1978.

[11] E. Y. Rocher, "Applications of monomode fiber to local networks," *IEEE J. Selected Areas Commun.*, vol. SAC-3, no. 6, pp. 897–907, Nov. 1985.

[12] J. P. Limb and C. Flores, "Description of Fasnet—A unidirectional local-area communications network," *Bell Syst. Tech. J.*, vol. 61, no. 7, pp. 1413–1440, Sept. 1982.

[13] M. Gerla, P. Rodrigues, and C. Yeh, "U-Net: A unidirectional fiber optics bus network," in *FOCLAN Conf. Proc.* (Las Vegas, NV), Sept. 1984, pp. 295–299.

[14] N. Abramson, "Packet switching with satellites," in *Proc. AFIPS Conf.* (Montvale, NJ), 1973, vol. 42, pp. 695–702.

[15] D. T. W. Sze, "A metropolitan area network," *IEEE J. Selected Areas Commun.*, vol. SAC-3, no. 6, pp. 815–824, Nov. 1985.

[16] A. G. Zellweger, "The next generation air traffic control automation system: Designing machines for people," *J. Telecommun. Networks*, vol. 4, no. 1, pp. 8–19, 1985.

[17] M. A. Kaminski, Jr., "Protocols for communicating in the factory," *IEEE Spectrum*, vol. 23, no. 4, pp. 56–62, Apr. 1986.

[18] T. Lissack and B. Maglaris, "Digital switching in local area networks," *IEEE Commun. Mag.*, vol. 21, no. 3, pp. 26–37, May 1983.

[19] DEC, INTEL, and XEROX, "The Ethernet, a local area network, data link layer and physical layer specification," available from DEC Inc., Intel Inc. and Xerox Inc., Tech. Rep. version 2.0, Nov. 1982.

[20] J. Shoch and J. Hupp, "Measured performance of an Ethernet local network," Xerox Corp., PARC Rep. CSL-80-02, Feb. 1980.

[21] H. F. Taylor, "Guided wave electrooptic devices for logic and computation," *Appl. Opt.*, vol. 17, p. 1493, 1978.

[22] F. J. Leonberger, C. E. Woodward, and D. L. Spears, "Design and development of a high-speed electrooptic A/D converter," *IEEE Trans. Circuits Syst.*, vol. CAS-26, p. 1125, 1979.

[23] M. Gerla, "Routing and flow control in ISDN," in *ICCC Conf. Proc.* (Munich, W. Germany), pp. 643–647, Sept. 1986.

[24] M. Gerla, P. Rodrigues, and C. Yeh, "Token based protocols for high speed optical fiber networks," *J. Lightwave Technol.*, vol. LT-3, no. 3, pp. 449–466, June 1985.

[25] M. Gerla, P. Rodrigues, and C. Yeh, "BUZZ-NET: A hybrid random access/virtual token local network," in *Proc. Globecom 1983* (San Diego, CA), pp. 1509–1513, Dec. 1983.

[26] M. Gerla, P. Rodrigues, and C. Yeh, "U-Net: A unidirection fiber bus network," in *Proc. of FOC/LAN 1984*, (Las Vegas, NV), Sept. 1984, pp. 295–299.

[27] L. Kleinrock, *Queueing Systems*, vol. 1, *Theory*. New York: Wiley, 1975.

[28] D. Ferrari, *Computer Systems Performance Evaluation*. Englewood Cliffs, NJ: Prentice-Hall, 1978.

[29] L. Fratta, F. Borgonovo, and F. A. Tobagi, "The EXPRESS-NET: A local area communication network integrating voice and data," in *Proc. Int. Conf. Performance of Data Communication Systems and Their Applications* (Paris, France), Sept. 1981, pp. 77–88.

[30] J. Uffenbeck, *Microcomputers and Microprocessors*. Englewood Cliffs, NJ: Prentice-Hall, 1985.

[31] J. W. Reedy and J. R. Jones, "Methods of collision detection in fiber optic CSMA/CD networks," *IEEE J. Selected Areas Commun.*, vol. SAC-3, no. 6, pp. 890–896, Nov. 1985.

Cavour Yeh (S'56–M'63–SM'82–F'85) was born in Nanking, China, on August 11, 1936. He received the B.S., M.S., and Ph.D. degrees in electrical engineering from the California Institute of Technology, Pasadena, CA, in 1957, 1958, and 1962, respectively.

He is presently professor of electrical engineering at the University of California at Los Angeles (UCLA). He joined UCLA in 1967 after serving on the faculty of U.S.C. from 1962 to 1967. His current areas of research interest are optical and millimeter-wave guiding structures, gigabit-rate fiber-optic local area network, and scattering of electromagnetic waves by penetrable irregularly shaped objects.

Dr. Yeh is a member of Eta Kappa Nu and Sigma Xi, and a Fellow of the Optical Society of America.

Maurice Lin received the B.S. degree in physics from National Central University, Taiwan, in 1974, the M.S. degree in physics from University of Kansas in 1979, the M.S. degree in computer science from Georgia Institute of Technology in 1980, and the Ph.D. degree in electrical engineering from UCLA in 1988.

From 1980 to 1983 he was with Gearhart Industries, Ft. Worth, TX, as a Software Engineer. He joined Bellcore, Red Bank, NJ, as Member of the Technical Staff in 1988. His main research interests include the design of distributed computer communication systems and networks and quantum electronics.

Mario Gerla (M'75) received the graduate degree in engineering from the Politecnico di Milano, in 1966, and the M.S. and Ph.D. degrees in engineering from UCLA in 1970 and 1973, respectively.

From 1973 to 1976 he was with network Analysis Corporation, New York, NY, where he was involved in several computer network design projects for both government and industry. From 1976 to 1977 he was with Tran Telecommunications, Los Angeles, CA, where he participated in the development of an integrated packet and circuit network. Since 1977, he has been on the Faculty of the Computer Science Department at UCLA. His research interests include the design, performance evaluation, and control of distributed computer communication systems and networks.

Paulo Rodrigues received the M.Sc. degree from COPPE, Rio de Janeiro, Brazil, in 1977, and the Ph.D. degree from UCLA in 1984, all in computer science.

He has been a Consulting Computer Analyst and Professor at the Federal University of Rio de Janeiro (UFJR), Brazil, since 1975, where he participated actively in the development of microsystems and peripheral devices. In 1978 he was director of EBC, a Brazilian computer industry, and in 1979 he joined the Ph.D. program at UCLA where he worked as a Research Assistant in high speed fiber optics local area network research.

He is presently in the research division of the Computer Center (NCE) and UFRJ. His current technical interests are in design and performance analysis of computer networks, local area networking supporting integrated services, interconnection of computer networks and distributed operating systems.

FOC-LAN 84

U-NET: A UNIDIRECTIONAL FIBER BUS NETWORK

M. Gerla, P. Rodrigues, C. Yeh

UCLA, School of Engineering
Los Angeles, CA 90024

ABSTRACT

To satisfy the growing demands of local area communications users, new architectures are required which provide very high bandwidth and satisfy real time delay constraints. Fiber optics networks appear to be a good choice for meeting these requirements. This paper describes a fiber optics bus architecture, U-Net, which is based on a token protocol, offers high bandwidth efficiency, guarantees bounded delay, and is highly fault tolerant. U-Net performance compares very favorably with other fiber bus networks.

1. INTRODUCTION

Recent years have witnessed a rapid growth in local area communications needs corresponding to increasing sophistication of users and the emergence of new applications such as real-time voice and video, high speed printers, graphics, etc. These environments require systems able to handle very high throughput among users several kilometers apart, while satisfying delay constraints which become severe when real-time traffic is involved. The answer to these demands is a suitable choice of transmission medium, topology, and access protocol.

Among the various transmission media, optical fiber appears to be an extremely promising technology. Fiber offers high bandwidth, immunity against electrical and magnetic interference and protection against signal leakage. It is easy to install due to its light weight and small size, and has lower attenuation than coaxial cable.

Fiber optics topologies can be configured in three basic ways: ring, star and bus. Ring implementations require active repeaters and substantial logic working at the channel speed (e.g. address recognition and flag setting) at each station.

Their cost and reliability, in spite of some fail-safe node proposals [ALBA 82], may constrain the use of ring interfaces at very high speed. Furthermore, special procedures are needed to recover from token loss or duplication and to avoid non-recognized packets circulating in the ring endlessly.

This research was supported in part by NSF under contract No. ECS-80-20300 and in part by Hewlett Packard and the State of California under a UC-MICRO grant. Mr. Rodrigues was also supported by the Brazilian Research Council (CNPq) under contract 200.123/79 and by the Federal University of Rio de Janeiro (UFRJ).

Star topologies can be built by connecting several stations via an optical coupler. Star configured networks have been implemented using either passive [RAWS 74] or active [RAWS 82] components. Both of these networks use CSMA-CD access protocol, which was successfully experimented in ETHERNET. However, CSMA-CD performance becomes very poor when the end-to-end propagation delay on the bus is comparable to (or larger than) the packet transmission time [TOBA 80]. Consequently, the effort required to push the technology to higher channel bandwidth is rewarded by an only marginal improvement in network throughput.

The third alternative is the bus. Since optical fibers are unidirectional in nature a unidirectional bus system (UBS) must be used. The following surveys a number of unidirectional bus architectures and provides a detailed description of U-Net.

2. UNIDIRECTIONAL BUS SYSTEMS

In a unidirectional bus signals propagate in only one direction. Broadcast communications are achieved by using either two separate busses with signals propagating in opposite directions (twin bus implementation), or one bus folded to visit all stations twice.

If two separate unidirectional busses are used, two independent transmitters and receivers (one for each bus) are required for each interface (see Fig. 1). If a folded bus is used (as in Fig. 2 and 3), one section of the bus is used for transmission and another is used for reception. Therefore, only one transmitter and one receiver are required. In addition, a sensor is usually installed in front of the transmitter to permit carrier sensing. As we shall show, carrier sensing is an essential feature in UBS protocols. In a twin bus implementation, carrier sensing is provided directly by the receiver, making a separate sensor unnecessary.

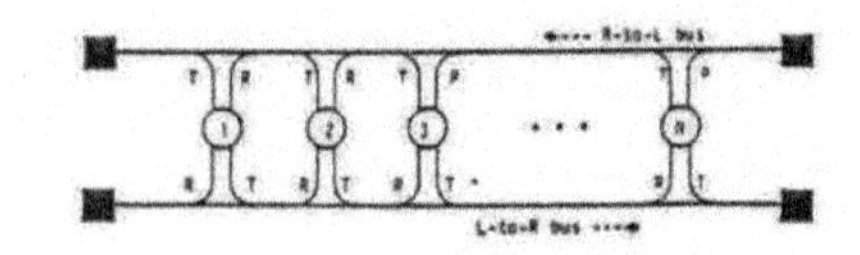

Fig. 1 - Twin Bus Topology

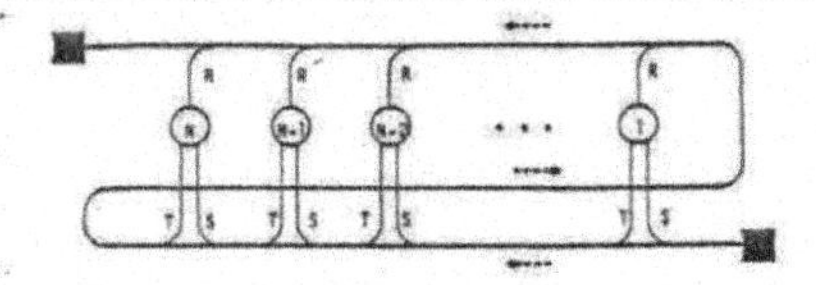

Fig. 2 - UBS Folded Topology

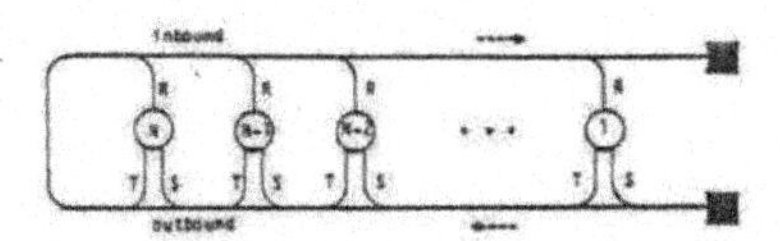

Fig. 3 - C-Net and D-net Topology

In recent years, extensive research has been conducted on UBS local area networks, and number of efficient protocols have been proposed. Express-net [FRAT 81], based on a folded unidirectional bus (Fig. 2), achieves conflict-free and bounded delay packet transmission by means of a round-robin oriented access protocol. C-Net [MARS 81], based on topology shown in Fig. 3, achieves high throughput efficiency and bounded delay by employing a protocol which combines implicit token and random access schemes. D-Net [TSEN 83] also uses the topology shown in Fig. 3. In D-Net, the right most station has the responsibility for generating tokens.

Several schemes have been proposed for the twin bus topology. Fasnet [LIMB 82] is a synchronous slotted network with end stations (the right-most and the left-most stations) responsible for slot generation and bit synchronization. DCR Net [TAKA 83] employs a deterministic contention resolution scheme. The normal mode of operation is random access CSMA-CD. Once a collision occurs, it is resolved using an implicit token passing scheme. Buzz-Net [GERL 83] proposes a similar hybrid mode of operation (random and token passing).

The latter two schemes have an advantage over Fasnet because they are able to establish a token passing round (after collision) without prior knowledge of the end stations. End stations are dynamically elected before each token passing round, clearly improving robustness and ability to withstand station failure. However, the alternation between random access mode and token mode and the presence of collisions introduce large variances in delay and reduce channel efficiency.

To overcome this problem, a token-less protocol [RODR 84] was developed. This protocol permits operation in token mode without prior knowledge of end stations. This is accomplished by "electing" end stations at each round. No explicit token must be exchanged between end stations to support the token mode of operation (hence the name of the protocol); rather, scheduling and synchronization are provided by presence or absence of carrier. There are no collisions in this protocol since station election is performed at each round, not triggered by collisions as in DCR Net or Buzz-Net.

Unfortunately, the election of end stations at each round introduces a fixed overhead in the form of a dead interval equal to a round trip delay at the end of each round. If the network configuration is fairly stable (station failures and disconnections or new station insertions are rather unfrequent), this overhead can be reduced by performing the end station election only at initialization and possibly after a configuration change, and by requiring the end stations to remember their "state" and exchange a token. These procedures are the essence of the U-Net protocol described below.

3. U-NET PROTOCOL

U-Net (Unidirectional Network) is a local network consisting of two unidirectional busses. Stations are connected to the busses via passive taps (see Fig. 1). Each tap includes a receiver and a transmitter.

The receiver detects presence/absence of carrier. When carrier is present, the receiver attempts to acquire bit synchronization from the preamble. After acquisition, the receiver copies bus data into private memory.

The transmitter sends a preamble followed by the data packet after it has received the go-ahead by the access protocol. If the station senses carrier coming from upstream while transmitting, it aborts its own transmission and tries again following the incoming data.

We assume a reaction delay of d seconds between the time a station senses end of carrier on the bus and the time it can start transmission on the same bus. Likewise, there is a d second delay between the sensing of carrier coming from an upstream station and the interruption of an ongoing transmission.

The above functions are common to all UBS interfaces. Actual UBS protocols differ from each other in the way they use these basic functions to provide access scheduling and synchronization.

The U-Net protocol consists of two procedures. The first procedure, described in this section, defines access to the bus after the end stations have been elected and the token mode has been established. The second procedure, introduced in the next section, defines the election of end stations at network initialization and/or network configuration change.

The following describes the token mode of operation used in U-Net. The two end stations are defined as L (left) and R (right). Protocol operation can be viewed as a sequence of cycles. Each cycle is initiated by one end station, for example, R station. R sends a special bit pattern, called token, on the R-to-L bus. This token is followed by a data packet from R (if R has data to send).

Each station continuously monitors both busses for a token. Once the token is heard on a bus (henceforth referred to as the token bus), the station is allowed to transmit one packet on both busses. More precisely, immediately after hearing the token, the station begins transmitting the preamble on the token bus. If, after an interval d from the beginning of its transmission, the station does not hear conflict on the bus (conflict may occur if an upstream station on the token bus is also attempting to transmit), it proceeds transmitting the preamble on the token bus as well as on the reverse bus (i.e., the bus in the opposite direction). If conflict is detected (i.e., the station hears another preamble coming in from upstream while it is transmitting its own), the station aborts its transmission on the token bus and does not even attempt to transmit on the reverse bus. The station will restart transmission after the oncoming packet has passed. This procedure is called "probing" the token bus.

On the token bus, packets are appended to the token in the same way as cars in a train are appended to a locomotive. Each station has the chance to transmit in the train, and can transmit at most one packet. Packets on the bus are separated by gaps of size d. On the reverse bus, a similar train is also formed. However, packets are not preceded by a token; rather they are separated by larger gaps than the packets on the token bus. The size of the gap between two packets on the reverse bus is equal to twice the propagation delay between the two sending stations, plus $2d$. Fig. 4 shows the space-time diagram for a possible sequence of packets on the token bus and on the reverse bus. A snapshot of the system is also shown.

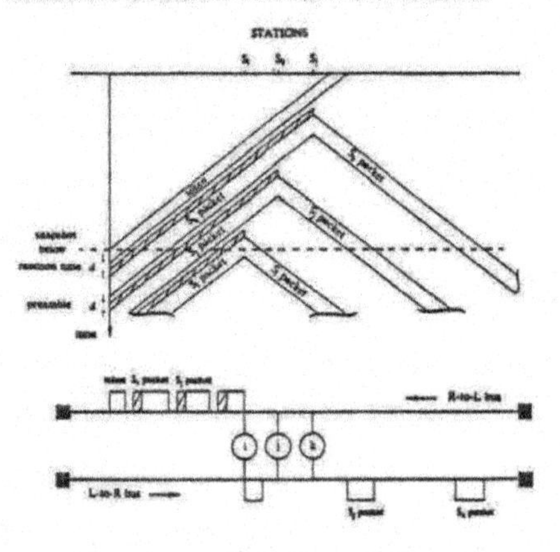

Fig. 4 - Space-Time Diagram and Snapshot

Another difference between the token bus and the reverse bus is that on the token bus the initial d seconds of the preamble may be damaged by conflicts. In fact, if the train carries N packets, the first K bits of the preamble (where K = dC, and C = bus speed) in the first packet correspond to the superimposition of N-1 preambles. The preamble must be large enough to allow bit sync to be acquired despite initial garbage.

It is important to note that each packet transmission is heard by all stations exactly once. Assuming the R-to-L bus is the token bus (see Fig.1), the packet transmitted by station i is received by station i+1, i+2, ... , and N on the token bus, and by station i-1, i-2, ..., 1 on the reverse bus. The transmission mode is implicitly a broadcast mode; specific knowledge of the destination station is unnecessary to properly route the packet.

The cycle terminates when the train terminates, i.e., when all the stations, including L and R, have had the opportunity to send their packets. The L station detects the end of the train from the absence of carrier for more than 2d seconds at the end of a packet (or token). After detecting the end of the train and (possibly) transmitting its own packet, the L station declares the cycle closed and starts a new cycle in the reverse direction by injecting a token in the reverse bus, which becomes the new token bus. The operation is the same with the roles of token bus and reverse bus interchanged.

4. END STATION ELECTION

U-Net is equipped with a dynamic procedure for electing end-stations. This procedure provides automatic recovery from station failure and from token loss, without operator intervention. It also permits smooth insertion of new stations in the system.

R is defined as the round trip propagation delay on the fiber cable. T_{MAX} is the maximum size packet transmission time. t_0 is the time required by an end station to "turn around" the token (read it from one bus and inject it onto the other bus).

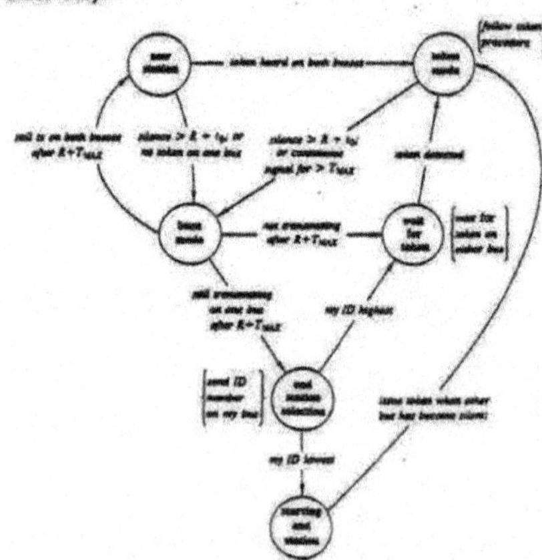

Fig. 5 - U-Net End Station Election State Diagram

Next, some observations. During normal token made operation there are short gaps between packets within each train, and larger gaps between trains. The distance between gaps is $\leq T_{MAX}$, by definition. If a continuous data stream of duration $> T_{MAX}$ is detected, it is interpreted as an anomaly. This property is exploited in the election procedure. As a second observation, the maximum duration of a silence gap at a station (the time during which both busses are sensed idle) during token mode operation is $R + t_0$. A larger silence gap denotes an abnormal situation (e.g. a failure).

The following describes the end station election procedure. During this procedure each station moves through the states shown in Fig. 5.

During normal operation each established (as opposed to new entering) station is found in the *token mode* state. Operation in this state was described in Sect. 3. From this state a station moves to the *buzz mode* state if it observes a silence gap $\geq R + t_0$, or it senses continuous signal for an interval $> T_{MAX}$.

In the *buzz mode* state a station issues a buzz tone on both busses. As a possible implementation, this buzz tone could consist of a preamble repeated continuously without gaps. During *buzz mode* a station defers to upstream stations by aborting its buzz tone when a buzz tone arrives from upstream.

After an interval $R + T_{MAX}$ from the time the first station entered *buzz mode*, all stations are necessarily in *buzz mode*. At this point, there are three possible conditions in which a station can be found:

(a) the station has deferred on both busses. In this case, the station is an intermediate station (i.e. not an end station.) It moves thus to the *wait for token* state. In this state, the station remains silent, awaiting for the token.

(b) the station is still transmitting on one bus (and has deferred on the other because a busy tone was detected or the bus is busy). The station is an end station and moves to the *end station selection* state, where one of the two end stations is selected to start the token cycle.

(c) the station is transmitting on both busses. This implies that there is only one station on the bus! The station moves to the *new station* state (to be defined later).

In the *end station selection* state the newly elected end stations must decide which starts first. To accomplish this, each station replaces the buzz tone with a pattern consisting of its ID number repeated over and over. The elected end stations compare the ID numbers. The high ID number station moves to the *wait for token* state. The low ID number station moves to the *starting end station* state, waiting for the reverse bus to become idle. It then issues a token and moves to the *token mode* state.

Upon hearing the token, all other stations move from the *wait for token* state to the *token mode* state.

A new entering station finds itself initially in the *new station* state. From this state, it must detect the token on both busses before moving to the *token* state. If a token is heard twice on the same bus, but not on the other bus, the station is the new end station. Thus, it moves to *buzz mode* to trigger a new election. Likewise, the station moves to *buzz mode* if a silence gap $> R + t_0$ is detected. This may happen at system initialization.

The election procedure may appear somewhat elaborate, but it is quite efficient. It requires approximately $3R + T_{MAX} + t_0$ to recover from failures. Typically, this is in the order of fractions of a millisecond for channel speeds over 100Mbps. The procedure is robust to any sort of failure. Even failures that occur during the recovery procedure are detected and recovered from.

5. PERFORMANCE ANALYSIS

It is of interest to compare the performance of U-Net with that of other bus protocols, namely, Tokenless Protocol (TLP), Express-net, and CSMA-CD. The performance measures considered here are network utilization at heavy load and insertion delay as a function of load. Insertion delay is defined as the time from the instant the packet is first considered for transmission until the beginning of its successful transmission.

Under equilibrium conditions, network utilization is defined as the ratio between the time in a round spent for packet transmission and the round duration, given that N stations are active and always transmittings in each round. Under these conditions, network utilization can be expressed analytically as a function of the number of active stations N and of the ratio $a = \tau/T$, where τ = end-to-end propagation delay and T = average packet transmission time [RODR 84]. Results for N = 20 are plotted in Fig. 6. Note that three different versions of Tokenless Protocol, namely, TPL-1, TPL-2 and TPL-3, are investigated [RODR 84]. The best performance is shown by TLP-3 and U-Net. TLP-1 and TLP-2 perform slightly worse than Express-net but quite better than CSMA-CD. CSMA-CD performance is negligibly affected by the number of active stations. For all the other protocols performance improves as the number of active stations increase.

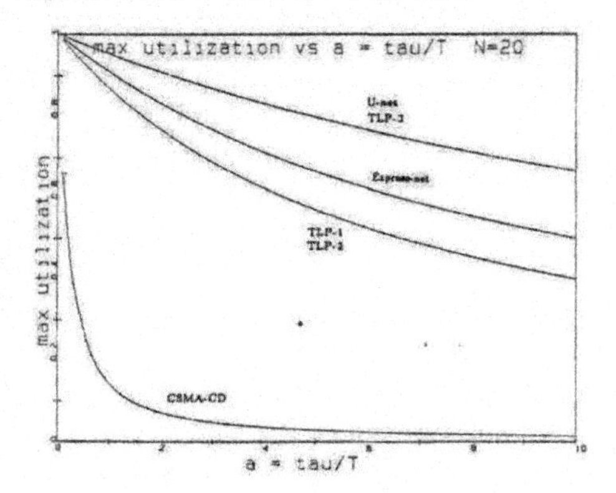

Fig. 6 - Utilization vs a = τ/T

In evaluating delay performance we assume that packets are generated at each station according to a Poisson process with rate λ and that stations have infinite buffer. Delay performance is measured in terms of the insertion delay (ID). Analytical expressions for ID at light and heavy load conditions can be derived, whereas results for intermediate load are obtained by simulation [RODR 84].

Simulation results for a 15 station network, average packet length = 1000 bits, and τ = 5 μs are reported in Fig. 7. Note that again U-Net and TLP-3 outperform all other schemes.

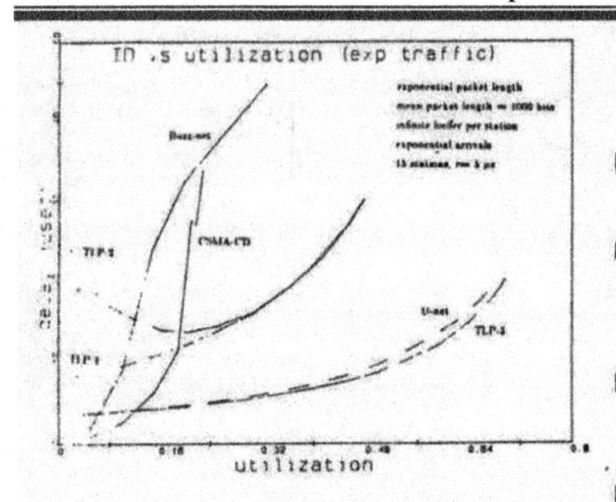

Fig. 7 - Average Insertion Delay vs Utilization

6. CONCLUSIONS

This paper describes a protocol suitable for the operation of a high speed twin fiber bus architecture. Operation is based on the token scheme, and therefore permits very high channel utilization and provides bounded delay. The protocol exhibits the best throughput and delay performance within the class of known twin bus protocols. A totally distributed procedure for end station election guarantees protection from station failure and token loss.

References

[ALBA 82] Albanese, A. "Fail-Safe Nodes for Lightguide Digital Networks", *Bell System Technical Journal*, Vol. 61, No. 2, February 1982, pp. 247-256.

[FRAT 81] Fratta, L., F. Borgonovo, and F. A. Tobagi, "The EXPRESS-NET: A Local Area Communication Network Integrating Voice and Data", in *Proceedings International Conference on Performance of Data Communication Systems and Their Applications*, Paris, France, September 1981.

[GERL 83] Gerla, M., P. Rodrigues, and C. Yeh, "Buzz-net: A Hybrid Random Access/Virtual Token Local Network", in *Proceedings Globecom '83*, S. Diego, CA. December 1983.

[LIMB 82] Limb, J. O. and C. Flores, "Description of Fasnet - A Unidirectional Local-Area Communications Network", *The Bell System Technical Journal*, Vol. 61, No. 7, September 1982, pp. 1413-1440.

[MARS 81] Marsan, M. A. and G. Albertengo, "C-Net: A Local Broadcast Communications Network Architecture", Internal Report, Politecnico di Torino, 1981.

[RAWS 74] Rawson, E. G., "Optical Fiber for Local Computer Networks", in *Proceedings Digest Topical Meeting on Optical Fiber Communications*, Washington, D. C., March 1974.

[RAWS 82] Rawson, E. G., and R. V. Schmidt, "FIBERNET II: An ETHERNET-Compatible Fiber Optic Local Area Network", in *Proceedings 82 Local Net*, Los Angeles, CA. 1982, pp. 42-46.

[RODR 84] Rodrigues, P.A., L. Fratta, and M. Gerla, "Tokenless Protocol for Fiber Optics Local Networks", *ICC Conference Proceedings*, Amsterdam, May 1984.

[TAKA 83] Takagi, A., S. Yamada, and S. Sugawara, "CSMA/CD With Deterministic Contention Resolution", *IEEE Journal on Selected Areas in Communications*, Vol. SAC-1, No. 5, November 1983.

[TOBA 80] Tobagi, F. A., and V. B. Hunt, "Performance Analysis of Carrier Sense Multiple Access with Collision Detection", *Computer Network*, Vol. 4, No. 5, October 1980, pp. 245-259.

[TSEN 83] Tseng, C. W. and B. U. Chen, "D-Net, A New Scheme for High Data Rate Optical Local Area Networks", *IEEE Journal on Selected Areas in Communications*, Vol. SAC-1, No. 3, April 1983.

JOURNAL OF LIGHTWAVE TECHNOLOGY, VOL. LT-3, NO. 3, JUNE 1985 449

Token-Based Protocols for High-Speed Optical-Fiber Networks

MARIO GERLA, MEMBER, IEEE, PAULO RODRIGUES, AND C. W. YEH, FELLOW, IEEE

Abstract—The local network medium is a pair of unidirectional fiber-optic busses to which stations are connected via passive taps. For this configuration, we present several protocols which provide round-robin, bounded delay access to all stations, and are particularly suited for high-speed transmission. The common characteristic of the protocols is the use of the *token* as the synchronizing event to schedule transmission. The token may be explicit (as in U-Net) or implicit (as in Tokenless Net). It may be used all the time, or it may be used simply to resolve collisions (as in Buzz-Net). The protocols are shown to be cost effective at very high (bandwidth) x (length) products that are the unique characteristic of high-speed single-mode fiber networks. Furthermore, they are robust to failures because of the passive interfaces and the totally distributed control. The implementation of these protocols on fiber-optic busses is also discussed in the paper.

I. Introduction

RECENT years have witnessed a rapid growth in local area communications needs corresponding to the increasing sophistication of users and the emergence of new applications such as real-time voice and video, high-speed printers, graphics, etc. These environments require systems able to handle very high throughput among users several kilometers apart, while satisfying delay constraints which become very tight when real-time traffic is involved. The answer to these demands is a suitable choice of transmission medium, topology, and access protocol.

Among the various transmission media, optical-fiber appears to be the most promising technology. Fiber offers an extremely high bandwidth x length product, up to 10 GHz $\times$ km for a single-mode operation. That is, a 10-GHz signal transmitted over 1-km of fiber without intermediate regeneration can be properly received using state-of-the-art transmitters and receivers. Other attractive features of the optical fiber include immunity against electromagnetic interference and protection against signal leakage. Fiber is easy to install due to its light weight and small size, and has lower attenuation than coaxial cable.

Fiber-optics topologies can be configured in three basic ways: ring, star, and bus. Ring implementations require

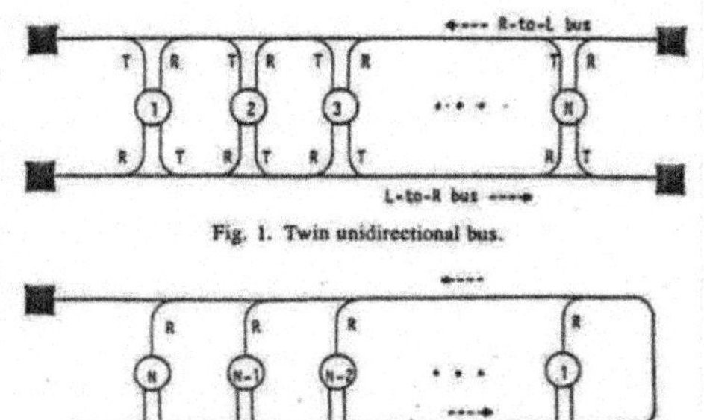

Fig. 1. Twin unidirectional bus.

Fig. 2. Folded unidirectional bus (Express-Net).

active repeaters and substantial logic working at channel speed (e.g., addrress recognition and flag setting) at each station. Their cost and reliability, in spite of some fail-safe node proposals [1], may set a limit to the use of ring interfaces at very high speed. Furthermore, special procedures are needed to recover from token loss or duplication and to avoid nonrecognized packets circulating endlessly in the ring.

Star topologies can be built by connecting several stations via a star coupler. Star configured networks have been implemented using either passive [9] or active [10] components. Most star networks use the CSMA-CD access protocol, which was successfully experimented in ETHERNET. However, CSMA-CD performance becomes very poor when the end-to-end propagation delay on the bus is comparable to (or larger than) the packet transmission time [16]. Consequently, the effort required to push the technology to higher channel bandwidth is rewarded only by a marginal improvement in network throughput.

The third alternative is the linear bus to which stations are coupled via passive interfaces. Since optical couplers are intrinsically unidirectional, a unidirectional bus system (UBS) must be used.

In a unidirectional bus signals propagate in only one direction. Thus full connectivity among stations is achieved by using either two separate busses with signals propagating in opposite directions (twin bus implementation), or one bus folded to visit all stations twice.

Manuscript received October 16, 1984; revised December 28, 1984. This work was supported by the following agencies: NSF-ECS-80-20300, NSF-INT-80-23740, MICRO (with Hewlett Packard) HD 831017, MICRO (with TTI), and the Brazilian Research Council, 200.123/79.

M. Gerla and C. W. Yeh are with the School of Engineering, UCLA, Los Angeles, CA 90024.

P. Rodrigues was with the School of Engineering, UCLA, Los Angeles, CA 90024. He is now with the Federal University of Rio de Janeiro, Brasil.

0733-8724/85/0600-0449$01.00 © 1985 IEEE

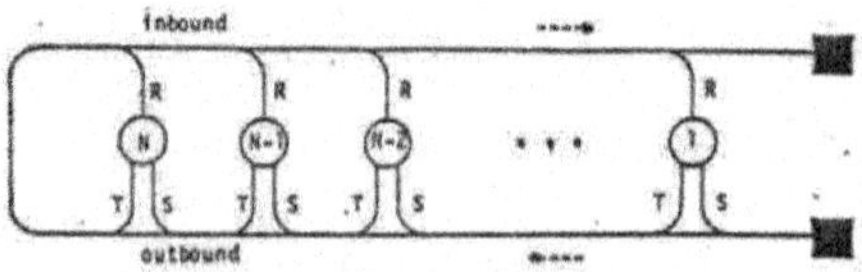

inbound
outbound

Fig. 3. Folded unidirectional bus (D-Net, C-Net, etc.).

If two separate pairs of unidirectional busses are used, two independent transmitters and receivers (one for each bus) are required for each interface (see Fig. 1). If a folded bus is used (as in Figs. 2 and 3), one section of the bus is used for transmission and another is used for reception. Therefore, only one transmitter and one receiver are required. In addition, a sensor is usually installed in front of the transmitter to permit carrier sensing. As we shall see, carrier sensing is an essential feature in UBS protocols. In a twin bus implementation, carrier sensing is provided directly by the receiver, making a separate sensor unnecessary.

In recent years, extensive research has been conducted on UBS local area networks, and several efficient protocols have been proposed. Express-net [2], based on a folded unidirectional bus (Fig. 2), achieves conflict-free transmissions and bounded delay by means of a round-robin oriented access protocol. C-Net [8], based on the topology shown in Fig. 3, achieves high throughput efficiency and bounded delay by employing a protocol which combines implicit token and random access schemes. D-Net [18] also uses the topology shown in Fig. 3. In D-Net, the right most station has the responsibility for generating explicit tokens, while in the first two schemes the implicit token (or, better, the synchronizing event) can be generated by any station.

Several schemes have been proposed for the twin bus topology. Fasnet [6] is a synchronous slotted network with end stations (the right-most and the left-most station) responsible for slot generation and bit synchronization. DCR Net [14] employs a deterministic contention resolution scheme. The normal mode of operation is random access CSMA-CD. Once a collision occurs, it is resolved using an implicit token passing scheme.

In this paper, we will focus on three protocols for twin unidirectional bus systems which were recently developed at UCLA, namely, U-Net, Buzz-Net, and Token-less Net. The protocols are described in detail, and their performance is compared to that of similar systems. Implementation aspects regarding the optical/electrical interfaces, the optical coupler ratio's, and the dynamic range of the receivers are also addressed.

II. U-Net Protocol

A. Basic Operation

Unidirectional Network (U-Net) is a local network consisting of two unidirectional busses [5]. Stations are connected to the busses via passive taps (see Fig. 1). Each tap includes a receiver and a transmitter.

The receiver detects presence/absence of carrier. When carrier is present, the receiver attempts to acquire bit synchronization from the preamble. After acquisition, the receiver copies bus data into private memory.

The transmitter sends a preamble followed by the data packet after it has received the go-ahead by the access protocol. If the station senses carrier coming from upstream while transmitting, it aborts its own transmission and tries again at the end of the incoming packet.

We assume a reaction delay of d seconds between the time a station senses end of carrier on the bus and the time it can start transmission on the same bus. Likewise, there is a d second delay between the sensing of carrier coming from an upstream station and the interruption of an ongoing transmission.

As we shall see, the above functions are common to all UBS interfaces. Actual UBS protocols differ from each other in the way they use these basic functions to provide access scheduling and synchronization.

The U-Net protocol consists of two procedures. The first procedure, described in this section, defines access to the bus after the end stations have been elected and the token mode has been established. The second procedure, introduced in the next section, defines the election of end stations at network initialization and/or network configuration change.

The following describes the token mode of operation used in U-Net. The two end stations are defined as L (left) and R (right). Protocol operation can be viewed as a sequence of cycles. Each cycle is initiated by one end station, for example, R station. R sends a special bit pattern, called token, on the R-to-L bus. This token is followed by a data packet from R (if R has data to send). Token and packet are separated by a gap on the order of d seconds.

Each station continuously monitors both busses for a token. Once the token is heard on a bus (henceforth referred to as the token bus), the station is allowed to transmit one packet on both busses. More precisely, immediately after hearing the token, the station begins transmitting the preamble on the token bus. If, after an interval d from the beginning of its transmission, the station does not hear conflict on the bus (conflict may occur if an upstream station on the token bus is also attempting to transmit), it proceeds transmitting the preamble on the token bus as well as on the reverse bus (i.e., the bus in the opposite direction). If conflict is detected (i.e., the station hears another preamble coming in from upstream while it is transmitting its own), the station aborts its transmission on the token bus and does not even attempt to transmit on the reverse bus. The station will restart transmission after the oncoming packet has passed by. This procedure is called "probing" the token bus.

On the token bus, packets are appended to the token in the same way as cars in a train are appended to a locomotive. Each station has the chance to transmit in the train, and can transmit at most one packet. Packets on the bus are separated by gaps of size d. On the reverse bus, a similar train is also formed. However, packets are not preceded by a token; rather they are separated by larger gaps than the packets on the token bus. The size of the gap

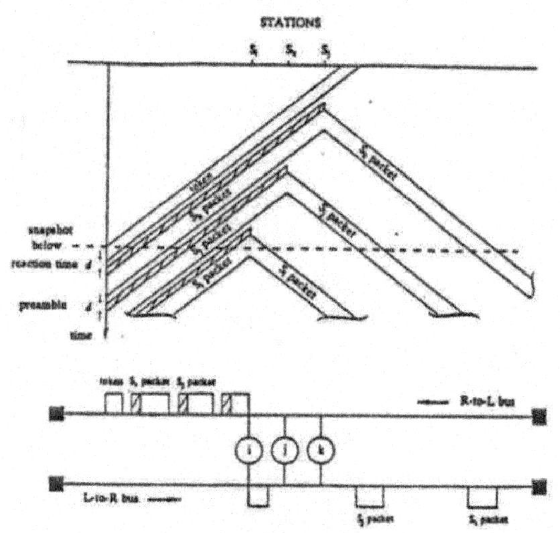

Fig. 4. Space-time diagram for U-Net.

between two packets on the reverse bus is equal to twice the propagation delay between the two sending stations, 'us $2d$. Fig. 4 shows the space-time diagram for a pos-..ole sequence of packets on the token bus and on the reverse bus. Each point in the diagram represents the position of the packet in one of the busses at a given time. A snapshot of the system is also shown.

Another difference between the token bus and the reverse bus is that on the token bus the initial d seconds of the preamble may be damaged by conflicts. In fact, if the train carries N packets, the first kbits of the preamble (where $K = dC$, and C = bus speed) in the first packet correspond to the superimposition of $N - 1$ preambles. The preamble must be large enough to allow bit sync to be acquired despite initial garbage. To simplify preamble acquisition, the beginning of the packet could be simply masked off by the receiver.

It is important to note that each packet transmission is heard by all stations exactly once. Assuming the R-to-L bus is the token bus (see Fig. 1), the packet transmitted by station i is received by station $i + 1, i + 2, \cdots$, and N on the token bus, and by station $i - 1, i - 2, \cdots, 1$ on the reverse bus. The transmission mode is implicitly a broadcast mode; specific knowledge of the destination station is unnecessary to properly route the packet.

The cycle terminates when the train terminates, i.e., when all the stations, including L and R, have had the portunity to send their packets. The L station detects ..e end of the train from the absence of carrier for more than $2d$ s at the end of a packet (or token). After detecting the end of the train and (possibly) transmitting its own packet, the L station declares the cycle closed and starts a new cycle in the reverse direction by injecting a token in the reverse bus, which becomes the new token bus. The operation is the same with the roles of token bus and reverse bus interchanged.

B. End Station Election

U-Net is equipped with a dynamic procedure for electing end-stations. This procedure provides automatic recovery from station failure from token loss, without operator intervention. It also permits smooth insertion of new stations in the system.

R is defined as the round trip propagation delay on the fiber cable. T_{MAX} is the maximum size packet transmission time. t_0 is the time required by an end station to "turn around" the token (read if from one bus and inject it onto the other bus).

Next, some observations. During normal token made operation there are short gaps between packets within each train, and larger gaps between trains. The distance between gaps is $\leq T_{MAX}$, by definition. If a continuous data stream of duration $> T_{MAX}$ is detected, it is interpreted as an anomaly. This property is exploited in the election procedure. As a second observation, the maximum duration of a silence gap at a station (the time during which both busses are sensed idle) during token mode operation is $R + t_0$. A larger silence gap denotes an abnormal situation (e.g., a failure or token loss).

The following describes the end station election procedure. During this procedure each station moves through the states shown in the state diagram in Fig. 5.

During normal operation each established (as opposed to new entering) station is found in the *token mode* state. Operation in this state was described in Section II-A. From this state a station moves to the *buzz mode* state if it ob-

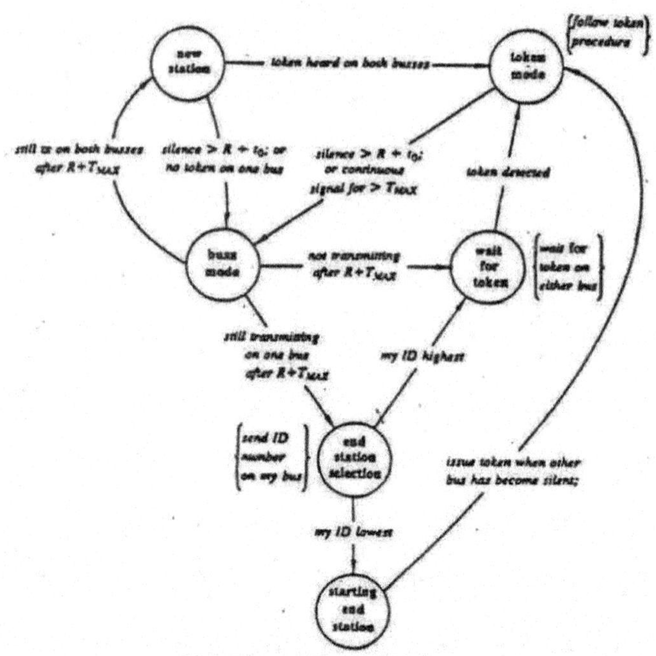

Fig. 5. U-Net end station election state diagram.

serves a silence gap $\geq R + t_0$, or it senses continuous signal for an interval $> T_{MAX}$.

In the *buzz mode* state a station issues a buzz tone on both busses. As a possible implementation, this buzz tone could consist of a preamble repeated continuously without gaps. During *buzz mode* a station defers to upstream stations by aborting its buzz tone when a buzz tone arrives from upstream.

After an interval $R + T_{MAX}$ from the time the first station entered *buzz mode*, all stations are necessarily in *buzz mode*. At this point, there are three possible conditions in which a station can be found.

a) The station has deferred on both buses. In this case, the station is an intermediate station (i.e., not an end station.) It moves thus to the *wait for token* state. In this state, the station remains silent, awaiting for the token.

b) The station is still transmitting on one bus (and has deferred on the other because a busy tone was detected or the bus is busy). The station is an end station and moves to the *end station selection* state, where one of the two end stations is selected to start the token cycle.

c) The station is transmitting on both busses. This implies that there is only one station on the bus! The station moves to the *new station* state (to be defined later).

In the *end station selection* state the newly elected end stations must decide which one should start the token cycle. To accomplish this, each station replaces the buzz tone with its ID number. The elected end stations compare the respective ID numbers. Using the rule that low ID starts the cycle, the high ID station moves to the *wait for token* state, and the low ID station moves to the *starting end station* state, waiting for the reverse bus to become idle. It then issues a token and moves to the *token mode* state.

Upon hearing the token, all other stations move from the *wait for token* state to the *token mode* state.

A new entering station finds itself initially in the *new station* state. From this state, it must detect the token on both busses before moving to the *token* state. If a token is heard twice on the same bus, but not on the other bus, the station is the new end station. Thus, it moves to *buzz mode* to trigger a new selection. Likewise, the station moves to *buzz mode* if a silence gap $> R + t_0$ is detected. This may happen at system initialization.

The election procedure may appear somewhat elaborate, but it is quite efficient. It requires approximately an interval $3R + T_{MAX} + t_0$ to recover from failures. Typically, this is in the order of fractions of a millisecond for channel speeds over 100 Mbps. The procedure is robust to any sort of failure. Even failures that occur during the recovery procedure are detected and recovered from.

III. Buzz-Net: A Hybrid Random Access/Implicit Token Protocol

One of the drawbacks of U-Net is the latency delay incurred by a station waiting for the token to come by. This delay is particularly annoying if there is only one station sending traffic (say, a large file) while all the other stations are idle. In this case U-Net forces the transmitting station to send only one packet at a time for each round-trip interval—a fairly inefficient proposition at very high bandwidth x length products.

This inefficiency can be overcome using a hybrid random access/implicit token protocol, which operates in random access mode when the system is lightly loaded (or a single station in transmitting) and reverts to implicit token mode when multiple source traffic builds up. This scheme we called *Buzz-Net*, because of the "buzz" pattern used to synchronize the system and drive it to token mode [3].

A. The Buzz-Net Algorithm

The network can operate in either of two modes: random access, or controlled access mode. Initially, a station starts in the *idle* state of the random access mode (see the state diagram in Fig. 6). When a packet arrives, the station moves to the *backlogged* state. From this state, transmission of the packet is attempted in random access mode.

a) If both busses are sensed idle, the station moves to the *Random Access Transmission* state. In this state, packet transmission immediately begins on both busses (it is assumed that the sender does not know the relative position of the destination on the bus).
b) If one bus is idle and one is busy, the station moves to the *Wait for EOC* state. Here, the station waits for end-of-carrier (EOC) on the busy bus.
c) If both busses are sensed busy or a buzz pattern is sensed, the station moves to the *Buzz-I* state, which is part of the controlled access procedure.

In the *Random Access Transmission* state the station proceeds transmitting on both busses. If, while transmitting, it is interfered by an upstream station (that is, it hears a Begin-of-Transmission (BOT) on one of the busses) it aborts its transmission and moves to *Buzz-I state*. The upstream transmission is allowed to proceed intact. If the transmission is successfully completed, the station moves *Idle* state.

In the *Wait for EOC* state, when *EOC* is sensed, the station moves to *Random Access Transmission* state. If, while in *Wait* state, the station senses a buzz pattern or it senses both busses busy, it moves to *Buzz-I* state.

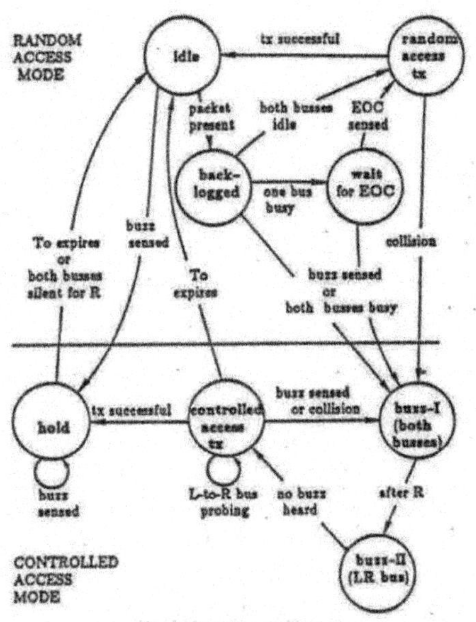

Fig. 6. Buzz-Net state diagram.

While in the random access mode a station with several packets ready for transmission may attempt to send them all in a single train, cycling between *Backlogged* and *Random Access Transmission* states, and thus capturing the channel and locking out the other stations. To avoid capture, a minimum interpacket gap must be observed between any consecutive packet transmissions. This minimum gap, on the order of a station reaction time interval d, allows downstream stations in the *Wait for EOC* state to detect *EOC* inside a train and, upon collision, force the system to controlled access mode, thus breaking capture.

In *Buzz-I* state a station transmits the buzz pattern on both busses (deferring, of course, to upstream transmissions) for R seconds, where R = round-trip delay. Because of deferrals, a station in the buzz state may actually buzz the busses only intermittently (or it may not buzz them at all). After R seconds, the station moves to *Buzz-II* state.

In *Buzz-II* state the station buzzes only the Left-to-Right bus, deferring as usual to upstream stations, until it hears no more buzzing on either bus. At this point, the station moves to *Controlled Access Transmission* state. The intermediate *Buzz-II* state guarantees that the leftmost (and *only* the leftmost) station starts the controlled access cycle when all the right-to-left buzzing has ceased.

In *Controlled Access Transmission* state each station is allowed to transmit its backlogged packet, and to move to *Hold* state thereafter. Controlled mode transmission is carried out much in the same way as in U-Net. As already discussed in U-Net, the left-to-right bus must be probed before transmission. That is, a station waits for the left-to-right bus to become free. Then it probes this bus by starting transmission of the preamble on it. If it does not hear upstream interference within a reaction time interval d, it will then proceed to transmit the remaining preamble also on the right-to-left bus, followed by the data packet. If interference is sensed, the station aborts its transmission and retries when the left-to-right bus is free again (i.e., after *EOC* is sensed). If a buzz is heard, the station moves back to *Buzz-I* state.

Clearly, in the controlled access mode, a "train" of packets is formed from left to right. Backlogged stations are allowed to append their packets to the train in a left to right order. At the end, all stations end up in the *Hold* state.

A time-out T_o from *Controlled Access Transmission* state to *Idle* state is provided to prevent a new station entering the system from locking up other stations finding themselves in *Controlled Access Transmission* state. A time-out T_o is also provided for the *Hold* state for similar reasons. A more detailed description of the recovery procedures when a new station joins the network is given in Section III-C below.

B. Buzz Signal Implementations

The buzz signal is a signal (or event) clearly distinguishable from regular packet flow. If the preamble pattern is uniquely distinguishable even when embeded in other data, then a possible buzz implementation consists of sending a prolonged preamble pattern. In this case, bit stuffing may be required to maintain data transparency within each packet.

An alternative buzz implementation which does not require bit stuffing consists of enforcing a minimum gap ΔT between any two consecutive packets on the bus. ΔT should be large enough so that a station in buzz mode can fill the gap with a burst of (arbitrary) data. The buzz implementation consists of sending one (or more, for reliability purposes) short burst(s) of unmodulated carrier where the length of a burst is less than the smallest packet size, but large enough to be safely detected by a station. A station recognizes the buzz condition when it detects on either bus the presence of a burst shorter than the minimum packet length. This scheme provides a faster detection of the buzz signal.

In general any method which permits some form of in-band signaling is a feasible buzzing method. The best method will most likely depend on interface implementation considerations, and may vary from application to application.

C. New Stations Joining The Network

A newly activated station may join the network at any time. In some cases, the joining process occurs transparently. In other situations, activity of the new station forces a transient phase which adds extra delay to the transmissions in process. However, whatever the case, the new st tion does not cause permanent disruption of network o eration, and the access algorithm automatically absor the external interference.

If the network is operating in random access mode, t new station is absorbed transparently. If, on the oth hand, the network is in controlled access mode the ne station will either move to *Buzz-I* state and participate a new buzzing phase, or it will successfully transmit packet after both busses are sensed idle. The first situati occurs because a buzz is detected by the new station, its transmission collides with a transmission of a bac logged station. In the worst case (stations 1 and N parti ipating in the new buzzing phase), an extra $2R$ interv may be necessary before transmissions are resumed. T second situation develops when the new station senses t right-to-left bus busy due to packet transmission by t next backlogged station situated upstream on that bu Upon detecting end of transmission, the new station tran mits on both busses and moves to *Idle* state. Other bac logged stations do not perceive this intrusion and beha normally.

Theoretically, there is still a chance that the new statio will keep transmitting in random access mode after reac ing the *Idle* state. In this case, any station that has pr viously moved to *Hold* state will eventually time out ar move back to *Idle*. Nevertheless, if these stations do n have any packet to transmit, the new station will contin to lock the remaining stations in controlled access mod The time out T_o in the *Controlled Access Transmissio* state prevents this capture effect.

As we have shown, the new station joins the set of acti stations gracefully. The extra delay added to controll access mode delay is in the best case 0, between 0 and 2 in most cases, and of the order of T_o in very unlikely wor case situations.

IV. Token-Less Protocol

The Buzz-net protocol is quite successful in eliminati token latency at light load. When load becomes substa tial, however, the delay performance degrades, and in fa becomes worse than the performance observed in U-Ne due to the overhead caused by the continuous switchi between random access and controlled access mode.

The attempt to retain the efficiency of U-Net at hea loads and, at the same time reduce token latency a eliminate the need for special token initialization proc dures, led to the creation of the *Token-less* protocol [11 This protocol, as the name indicates, does not use an e plicit token. Instead, it supports a round robin access mod (of the type seen in U-Net) by means of a synchronizi event which can be viewed as an "implicit" token. T synchronizing event is represented by the ceasing of a tivity on one of the busses. This event when detected a thorizes a station to "probe" the other bus and append i packet at the end of the "train" that has been formin on that bus. The details of the protocol are described below.

A. Principles Of Operation

In a tokenless network each station is connected to each bus with two passive taps, a *receiver* tap and a *transmit* tap. Through the receiver tap a station can receive all packets flowing in that bus and can monitor channel activity. More specifically it can observe presence or absence of activity (i.e., signal) and detect events as End of Activity EOA and Beginning of Activity BOA. As usual, the detection of these events occurs within the delay d which is the station reaction time.

It is also assumed that when a station detects a BOA event on either bus it stops transmission if any. That is, a station engaged in transmission always defers to an upstream transmission by aborting its own. The upstream transmission proceeds with only the first bits of the preamble corrupted no matter how many downstream stations are attempting to transmit. This guarantees that if the preamble is long enough a packet which has been completely transmitted by a station is correctly received by all (downstream) stations. We assume that an interface is able to corrrectly receive a packet even when this packet is immediately preceded by some truncated transmission. The underlying assumption is that the beginning-or-packet flag cannot be replicated within the packet data nor is contained in the activity signal (to be described below). Flags can be implemented as reserved bit patterns (in which case bit stuffing is required to preserve data transparency) or as code violations on the bit encoding level.

The transmit tap is used by a station to transmit either (data) packets or activity signal. As we shall see later, the purpose of the activity signal is to keep the downstream part of the channel busy. Its implementation (modulated or ummodulated carrier, random bits, continuous sequence of 1's, etc.) can be chosen according to the low level encoding utilized for transmission on the channel.

The goal of the protocol described in the following section is to guarantee collision-free transmissions among all stations with backlogged packets and to achieve good throughput/delay performance. Furthermore the need for special packets (e.g., token) is avoided and no central control is introduced. These characteristics are achieved by having EOA's events propagating alternatively in the two busses. One advantage of controlling the channel only through EOA events is to provide simple, reliable and low cost implemention even at very high speed. Another advantage, as we shall see later, includes easy implementation of initialization and recovery procedures for the protocols.

The EOA events can be seen as virtual tokens which allow the stations to transmit their packets in a round-robin fashion. These EOA events, detected simply by sensing the channel, propagate from left to right in one bus and from right to left in the other, thus minimizing the silent gap between the end of a round, and the beginning of the next.

B. The Basic Token-Less Protocol

The protocol basically consists of three procedures. The first procedure enables a station to recognize when it is its turn to transmit in a round. The second procedure enables a station to determine whether it is an extreme (left most or right most) active station. The third procedure enables a station which has been just powered-on to synchronize with other active stations, if any, or to initialize the round-robin cycle in an empty net. Part of this procedure also allows for recovery from failures caused by stations transmitting at the wrong time because of detection of false events.

There are different parameters and options which may be chosen when specifying the full protocol and in the following we will present in details three different implementations. Each one has some distinctive advantages depending on the traffic environment. Before getting into the details of the different implementations, we present the common foundation of the various versions of the Token-Less protocol.

A station, say S_i with a backlogged packet waits for the first of two events: EOA on either channel or time-out ND (Network Dead).

If an EOA (A) on channel A occurs first S_i starts transmitting activity signal on channel A. If BOA(A); occurs, stop transmission and wait for next EOA(A), otherwise after time-out d, start packet transmission on both channels. When packet transmission is completed S_i keeps transmitting activity signal on channel $\bar{A}$ until either a BOA($\bar{A}$) is detected or time-out Extreme Station (ES) occurs. If BOA($\bar{A}$) is detected then $A \leftarrow \bar{A}$ and repeat the above procedure. Otherwise if ES is reached, station S_i realizes that it is an extreme station. S_i then undertakes the *round restart procedure*.

If time-out ND occurs (meaning that no other station is active in the network) *Si* undertakes the *Initialization procedure*.

The flow diagram of the Basic Token-Less protocol is shown in Fig. 7.

C. Variations On The Basic Theme

As previously mentioned there are several ways to specify *Round Restart* and *Initialization* procedures and to choose the parameters ND and ES. Some of these are described in the following.

The state diagram shown in Fig. 8 defines the operation of TLP-1 (Token-Less protocol, Version 1). Explanation of the notation used in state diagrams is given in Table I. The states in the left side of the figure represent the *Initialization* procedure. A station is in these states only when it is powered-on. The time-out ND is set equal to a round-trip delay R. If the time-out ND is not reached, either because one channel is sensed busy or a BOA is detected, the station leaves the initialization state and reaches the WFT state. The *Round Restart* procedure is performed in state ES. In this state, the station transmits activity signal

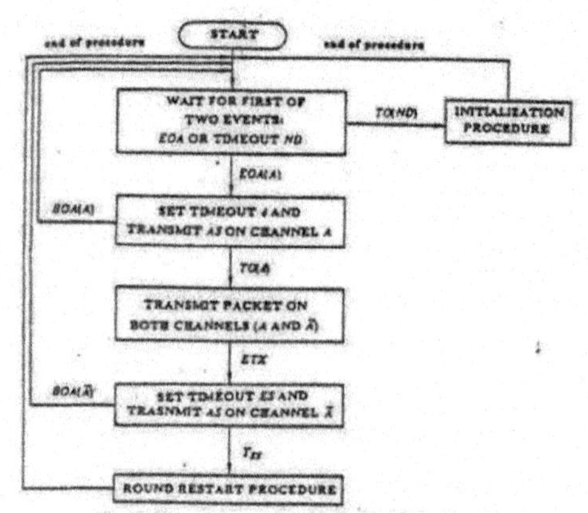

Fig. 7. Flow diagram of the basic token-less protocol.

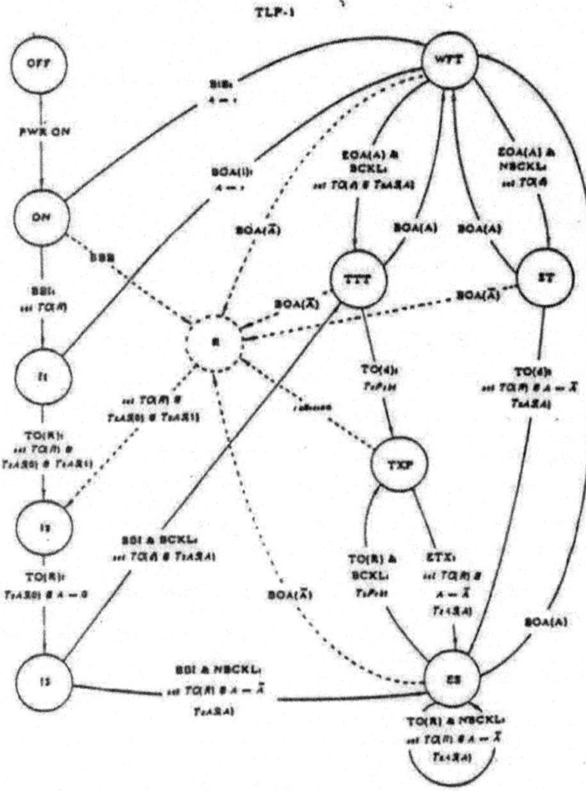

Fig. 8. State diagram of TLP-1.

TABLE I

TRANSITION:	"EVENTS: ACTIONS"	
DESCRIPTION OF EVENTS		
BiB	=	bus i busy
BBI	=	both busses idle
BBB	=	both busses busy
BCKL	=	backlogged packet
NBCKL	=	no backlogged packet
TxPCKT	=	transmit packet on both busses
ETX	=	end of packet transmission
TxAS	=	transmit activity signal
TxAS(A)	=	transmit activity signal on bus A
BOA(A)	=	beginning of activity on bus A
EOA(A)	=	end of activity on bus A
TO(x)	=	time-out of length x
Collision	=	detection of activity during packet transmission

in the direction opposite to the one where the virtual token is propagating. If the activity transmission lasts $ES = R$ s the station knows it is an extreme station and thus behaves accordingly. The dotted portion of the state diagram of Fig. 8 represents the recovery procedure which is undertaken any time an illegal event occurs.

An example of the operation of TLP-1 for a network with 10 stations is given in the space-time diagram shown in Fig. 9. The time intervals A, B, C, D, and E represent rounds. In round A the virtual token propagates from left to right. Stations 1, 3, 4, 5, 7, 8, 9, and 10 are powered-on, and stations 1, 7, and 8 have backlogged packets. In the next round, B, the virtual token propagates from right to left. A new station, station 6, is powered-on, and stations 6, 4, and 1 transmit a packet, and so on.

In TLP-1 a powered-on station always performs some activity on the channel even if it has no packets to transmit. This implies that the virtual token at each round goes back and forth between the two extreme powered-on stations. Thus any station can transmit a packet within a round. If traffic load is unbalanced and only a few stations are actually sending traffic this mode of operation introduces an unnecessary delay due to the fact that the virtual token must sweep the entire bus, rather than the section of the bus containing the stations involved in transmissions.

This inconvenience is eliminated in TLP-2 where the virtual token sweeps only the stations which have a packet to transmit. This is achieved, as shown in the state diagram of Fig. 10, by forcing a station with no backlogged packets back to the *Idle* state. All the other parameters are the same as in TPL-1. The behavior of TLP-2 for the same traffic pattern as in the previous example is shown in the space-time diagram of Fig. 11.

A substantial contribution to the overhead in both TLP-1 and TLP-2 is given by the time-out R between rounds. One way to reduce this delay is to set two time-outs ES_1 and ES_2 at each station equal to the round-trip delays from the station to either end of the busses. The appropriate time-out (ES_1 or ES_2) is then used when performing the *Round Restart* procedure. This method, however, introduces some additional complications when the network is installed and requires ES parameter updating at each station if extension are added to the bus.

Another method which is solely based on the protocol and not on the physical layout of the network is to take advantage of the fact that, in TLP-1, the extreme stations do not change very frequently. Thus, if a station is the right most station, say, in a round, it will very likely be again the right most station in the next round. Therefore, as soon as it has been granted access to the channel in one round it starts right away a new round in the opposite direction. This scheme, called TLP-3, eliminates the R seconds gap between rounds and works properly until a new station external to the present extreme stations is powered-on. In this case a collision is generated and the initialization procedure is invoked to restore the correct operation with the new powered-on station as an extreme station. Details of the algorithm TLP-3 are given in the state diagram of Fig. 12 where a flag $E(A)$ is used to signal if a station is an extreme one or not. The space-time diagram in Fig. 13 shows how this version works for the same example previously considered.

V. Performance Analysis

In this section we evaluate the performance of the various access protocols introduced so far and compare it with that of two other schemes that have been extensively studied, namely, Express Net and CSMA-CD [17].

The following performance measures are considered in this analysis:

1) bus utilization $U(i)$ at heavy load, defined as the net bus utilization when i stations are active and have infinite packet backlog;
2) average insertion delay defined as the interval between the time when the packet moves to the head of the transmitting queue, and the time when successful transmission begins. Note that insertion delay is equivalent to queueing delay when there is only one buffer per station. We distinguish between insertion delay at light load (IDL), and insertion delay at heavy load (IDH).

The above measures, albeit simple, provide us with useful criteria to decide whether a bus protocol is suitable or not for a given application. For example, interactive and real time applications are particularly sensitive to insertion delays. Batch data transfer on the other hand is mostly affected by bus throughput efficiency.

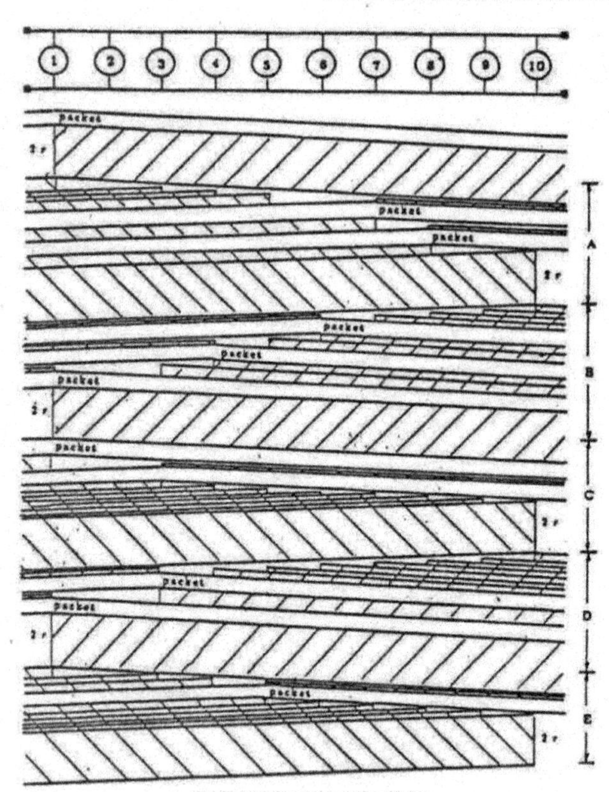

Fig. 9. Space-time diagram for TLP-1.

In our analysis, we assume that stations are uniformly spaced on the bus. For simplicity we also assume that detection time d is much smaller than the end-to-end propagation delay τ, and that the preamble is much shorter than the packet.

A. Bus Utilization

We begin by evaluating bus utilization $U(N)$ at heavy load. For token based protocols under equilibrium conditions, $U(N)$ is given by the ratio between the time in a round spent for packet transmissions and the round duration, given that N stations are active and always transmitting in each round. For CSMA-CD (which is not a token based protocol), the utilization will be computed separately.

The total packet transmission time during a round is the same for all schemes and is given by NT (where T = packet transmission time) since there are N backlogged stations and each station transmits in each round.

As for the round duration d, this varies from scheme to scheme. In U-Net we have

$$D_{\text{U-Net}} = NT + \tau.$$

In Buzz-Net, under heavy load conditions, the active stations will always conflict again at the end of a controlled phase. Therefore, the activity in the network is a succession of cycles where active stations are served in a round-robin way, lowest numbered stations first. The diagram in Fig. 14 portrays the cyclic pattern when all N stations are active. From Fig. 14 we can see that

$$D_{\text{Buzz-Net}} = NT + 6\tau$$

In TLP-1 and TLP-2, we have

$$D_{\text{TLP-1,2}} = NT + 3\tau.$$

In TLP-3 we have the same results as in U-Net, namely,

$$D_{\text{TLP-3}} = NT + \tau$$

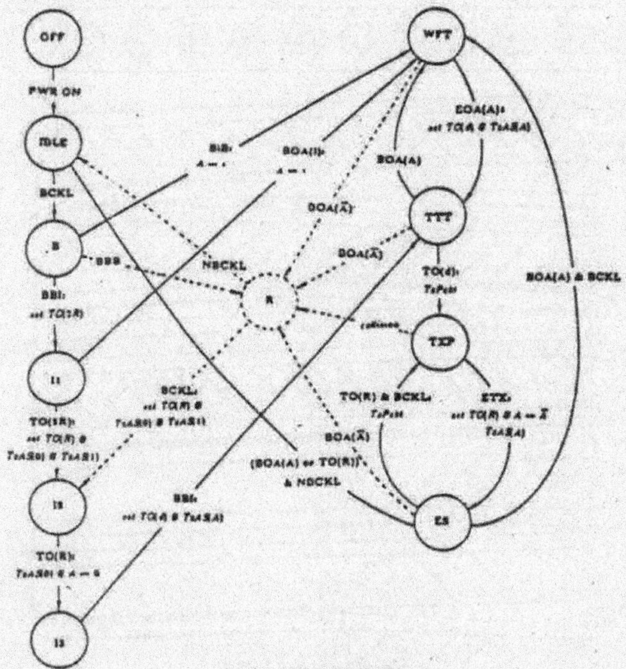

Fig. 10. State diagram of TLP-2.

Taking now the ratios of packet transmission time during the round and round time we obtain the following expressions for bus utilization

$$\text{U-NET: } \frac{1}{1 + a/N}$$

$$\text{Buzz-NET: } \frac{1}{1 + 6a/N}$$

$$\text{TLP-1,2: } \frac{1}{1 + 3a/N}$$

$$\text{TLP-3: } \frac{1}{1 + a/N}$$

$$\text{Express-Net: } \frac{1}{1 + 2a/N}$$

where $a = \tau/T$.

For CSMA-CD, bus utilization can be approximated by 7]

$$U = \frac{1}{1 + 5a}, \quad \text{for } a < 0.5$$

$$U = 1/7a, \quad \text{for } a > 0.5.$$

Note that CSMA-CD maximum utilization is not affected by the number of stations, as long as this number is fairly large.

The results are reported as a function of a in Fig. 15 for $N = 15$. The best performance is shown by TLP-3 and U-Net. TLP-1 and TLP-2 perform slightly worse than Express Net, but better than Buzz-Net. CSMA-CD performs very poorly for larger a, as expected.

It should be pointed out, however, that there is one important case in which Buzz-Net out performs all other token protocols. This is the case of single station with heavy backlog. In Buzz-Net, the single station can transmit an uninterrupted sequence of packets thus yielding $U = 1$. In the other schemes, subsequent packets from a single station are spaced by at least one round-trip delay.

B. Insertion Delay at Light Load

For light load conditions we assume that all stations are powered-on, and that the probability of two or more stations transmitting in a cycle is negligible.

It should be noticed that for U-Net and TLP protocols the token latency varies with the station position on the bus. It is minimum at the center, and it is maximum at the extremes. Correspondingly, th...

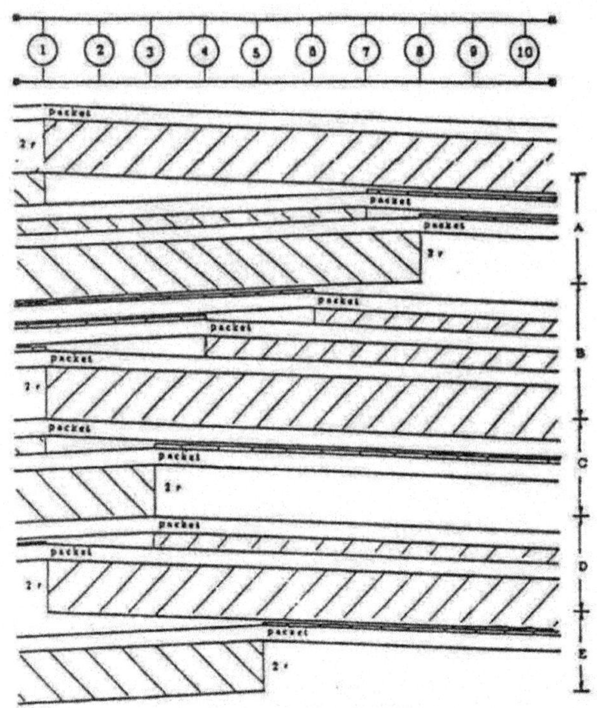

Fig. 11. Space-time diagram for TLP-2.

of IDL for each network. In the following, we report both values.

The values of average insertion delay for the various schemes are shown below:

U-Net	$\frac{\tau}{2} \leq \text{IDL} \leq \tau,$
Buzz-Net	$\text{IDL} = 0,$
TLP-1	$\frac{3}{2}\tau \leq \text{IDL} \leq \frac{5}{3}\tau,$
TLP-2	$\text{IDL} = 6\tau,$
TLP-3	$\frac{\tau}{2} \leq \text{IDL} \leq \tau,$
Express-Net	$\text{IDL} = \tau,$
CSMA-CD	$\text{IDL} = 0.$

We note that the best performance is obtained with Buzz-Net and CSMA-CD, as expected. We also note that TLP-2 displays a very high insertion delay. This is due to the fact that at very light load, the protocol must be reinitialized for each transmission. This requires on the order of three round-trip delays (i.e., 6τ), as can be seen from the diagram in Fig. 10.

C. Insertion Delay at Heavy Load

In heavy-load conditions we assume that a station has always a packet to transmit. The insertion delay is then obtained by computing the time required for the token to come back to the station, given that all the intermediate stations will transmit whenever they get the token. Again, an asymmetry is noticed in U-Net and TLP, in that each station will alternatively see a long cycle and a short cycle. However, one discovers that the two cycle times average out so that the insertion delay is the same for all stations.

The expressions for insertion delay at heavy load follow:

U-Net	$\text{IDH} = NT + \tau,$
Buzz-Net	$\text{IDH} = NT + 6\tau,$
TLP-1, 2	$\text{IDH} = NT + 3\tau,$
TLP-3	$\text{IDH} = NT + \tau,$
Express-Net	$\text{IDH} = NT + 2\tau,$
CSMA-CD	IDH = Unbounded.

From the results we notice that the delays of all the token based schemes are bounded—a key property due to the round robin scheduling. In contrast, the delay of the random access scheme, CSMA-CD, is unbounded.

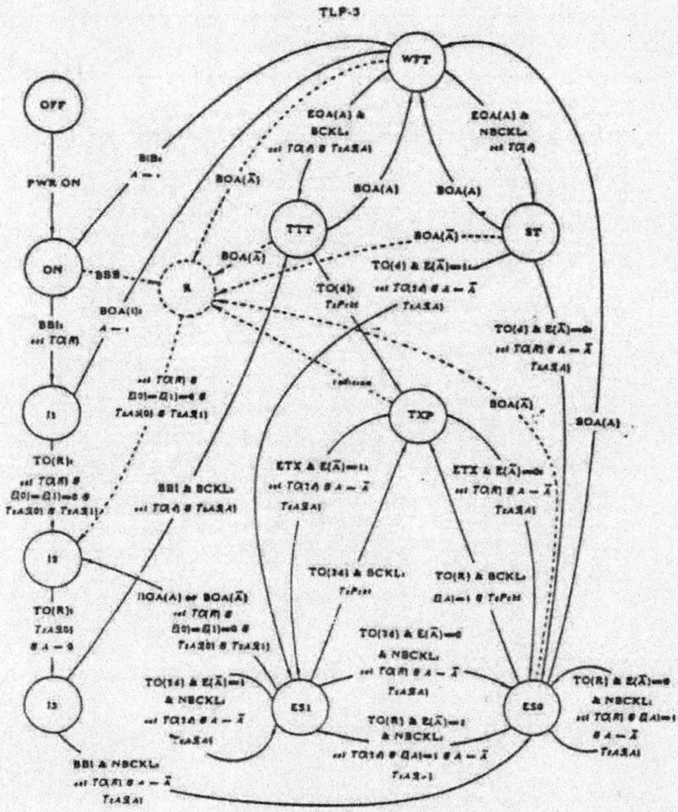

Fig. 12. State diagram of TLP-3.

Fig. 16 summarizes delay and utilization results in a ingle diagram, where delay is shown as a function of utiization for various access schemes. The network under tudy had the following characteristics:

$N = 15$
$\tau = 5\ \mu s$
$T = 1\ \mu s$ (*i.e.*, packet length = 1000 bits at 1 Gbps, or 100 bits at 100 Mbps)
Poisson arrivals (heavy load)
Exponential packet length distribution
Infinite buffers at each station.

The data points in Fig. 16 were obtained via simulation. xact analytic results are available only for light and heavy tilization. Thus approximate analytic models must be deeloped for intermediate values of utilization. One can erify that simulation results are in good agreement with he analytic results reported earlier for light and maximum tilization.

VI. Fiber Implementation Considerations

The various classes of token based protocols described in this paper can be applied to any unidirectional bus system regardless of the medium used (twisted pair, coaxial cable or optical fiber). However, the full potential of these protocols can be exploited only with optical fiber, for the following reasons:

The token based protocols permit to operate efficiently at much higher data rates than conventional carrier sense protocols. These higher data rates (above 500 Mbps, say) are economically achievable only with optical fibers.

The protocols permit efficient operation over large distances (more precisely, at very high Bandwidth × Length products). Only optical fiber permits the operation over several miles without intermediate repeaters. Cable requires signal

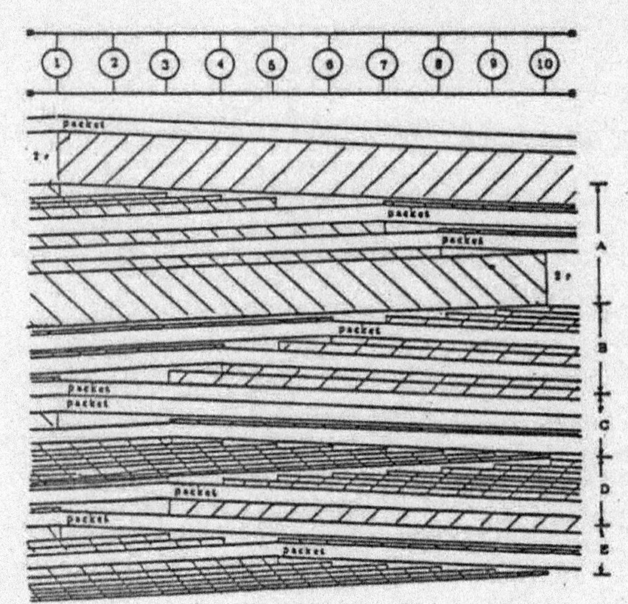

Fig. 13. Space-time diagram for TLP-3.

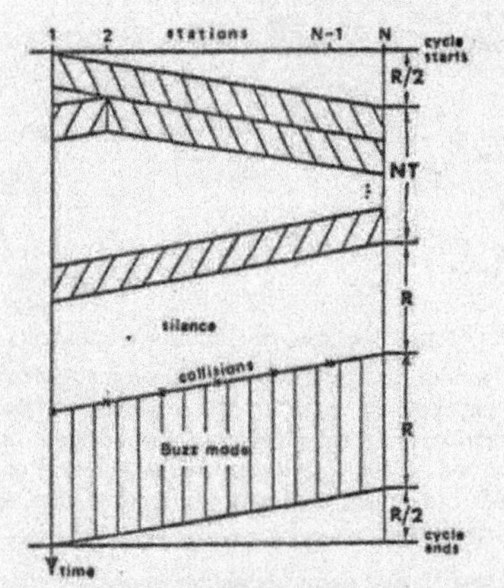

Fig. 14. Buzz-Net space-time diagram at heavy load.

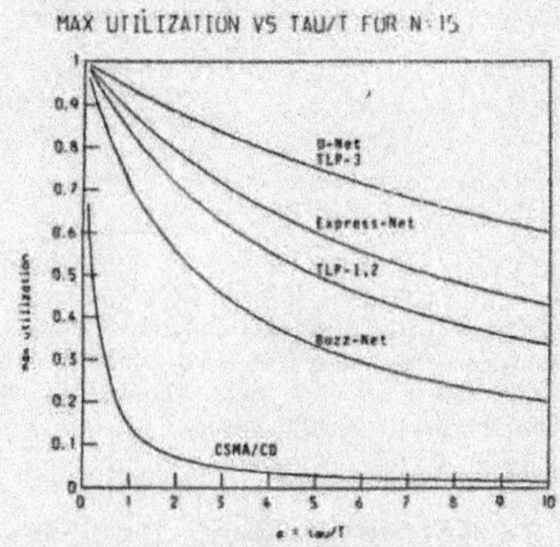

Fig. 15. Maximum utilization versus a.

defying the goal of totally, fault passive tolerant bus design.

It is therefore apparent that the full advantages of the token-based protocols can be obtained only in fiber-optics media. It is then appropriate to ask, whether the fiber implementation poses technical problems which are not present in cable implementations. Basically, there are three ... which the cable bus design differs from the optical bus design, namely:

optical/electrical interface,
optical coupler ratios and power budget; and
transmit and receive power calibration.

These issues are discussed in the following sections.

A. Optical/Electrical Interface

At low speeds, the design of the interface is very straightforward: The baseband electric signal, instead of

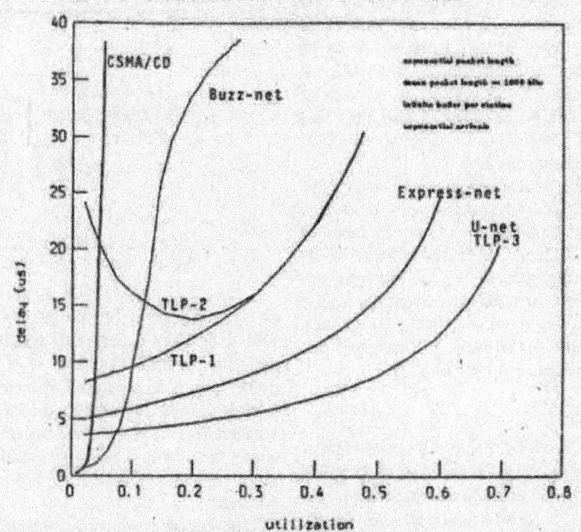

Fig. 16. Insertion delay versus utilization.

being injected directly in the cable, modulates a laser or an LED; on the receive side, the baseband electric signal is obtained from the p-i-n diode. The electronics is the same as in cable implementations.

The challenge arises when we operate at speeds above 100 Mbps. While the optical components can operate very efficiently at higher speeds, the electronic components show serious limitations. Without entering in details which are beyond the scope of this paper, we mention the fact that optical components can be used to compensate for the "inadequacy" of electronic components. On the one hand, optical logic can be used to perform the most time critical functions of the interface, such as beginning and end of carrier detection, preamble acquisition, clock acquisition, address recognition, etc. On the other hand, passive optical delay lines permit serial to parallel conversion of the bit stream traveling on the bus. Namely, this stream can be subdivided into N substreams which are out-of-phase by exactly one bit among each other. Thus, data can be written to (and read from) the bus at a speed which is N times less than the speed of the bus.

Akin to the optical delay line is the concept of "fiber buffer loop" proposed in [15]. The fiber loop provides a circulating buffer for the data in transfer between a low-speed computer device and a high-speed optical bus. Optical switches control the connection of loop to bus and of loop to computer device. By properly clocking the switches, data can be "strobed" in and out of the optical loop at a period which is multiples of the loop delay. A reduction in speed of several orders of magnitude can thus be obtained.

One should also note that the serial to parallel conversion solution is more effective in token busses than in token rings. In the bus situation, in fact, the only function that must be carried out serially, at bus speed, is the detection of beginning and end of carrier. This function is very straightforward to implement, even at extremely high speeds. All the other functions can be carried out through serial-to-parallel conversion. In the ring configuration, on the other hand, each interface must perform several logical functions on line, at bus speed, (e.g., clock acquisition, address recognition, etc.). This implies that the ring interface is generally more complex to implement than the bus interface at very high speeds.

B. Optical Coupling

One well-known drawback of optical receivers is the fact that they require a certain amount of optical power in order to operate properly (while electrical receivers can be built with very high input impedance so that power absorption from the line is minimal). Consequently, optical taps introduce a much higher power loss than electrical taps, posing a severe constraint on the number of stations that can be connected to the bus.

In a fiber bus, optical taps are generally implemented with biconical couplers, which allow to couple a fraction C of optical power from one fiber to the other. In addition to the "coupling loss" C, we must also account for an "excess loss" β corresponding to the fraction of optical power irradiated in the air at the junction.

Assuming that in the system under consideration the transmitters have maximum output power P_T and the receivers can detect reliably a minimum power P_S, then the ratio P_T/P_S is defined as the power margin M for the system. Clearly, for proper system operation, the sum of all tap and attenuator losses (in dB) along the bus must be less than M. One important problem in fiber bus design is therefore the determination of the maximum number of taps (i.e., stations) that can be installed on the bus. This number generally will depend on the coupling ratio C as well as on excess loss and attenuation.

Recent advances in single-mode fiber technology have driven excess loss below 0.1 dB and attenuation below 0.2 dB/km. Furthermore, the coupling ratio C can be tuned to match any desired value. Thus, the key parameter in the optimization is the coupling ratio C. Several different optimization problems can be formulated, depending on the constraints posed on C. The most common optimization assumes that C is the same for all taps. In this case, one easily finds that the optimal C is given by [13]

$$C = 1/(N - 1)$$

where N is the number of stations. From the above expression, one can then compute the max value of N for given power margin M and for given excess and attenuation loss per tap, γ. For example, for $M = 45$ dB and $\gamma = 0.2$ dB one finds $N_{MAX} = 24$ [12].

A more sophisticated optimization involves the individual adjustment of the ratio C for each tap. If this procedure is followed, one finds that for $M = 45$ dB and $\gamma = 0.2$ dB up to $N_{MAX} = 50$ stations can be installed on a bus [12]. Even better results can be obtained if within each interface the coupling ratio's of transmit and receive taps are individually optimized. In this case, the maximum number of stations becomes $N_{MAX} = 61$ for the above stated parameters [12].

C. Power Calibration

For proper signal detection at the p-i-n diode, the received power level should be approximately constant, independent of the transmitting station. In other words, the dynamic range allowed by the receiver is very limited, for proper operation.

If all the stations on the bus transmitted at the maximum power P_T, clearly the above condition would not be met. In fact, the signal received from a distant station may be as much as 40 dB lower than the signal received from a near station. Even worse, since in the token protocol the first few bits of the preamble may be the superposition of signals from several stations, the large disparity in received powers may seriously impair the effectiveness of the AGC (automatic gain control) mechanisms. There is therefore a need to calibrate the output power of the transmitters along the bus so that a receiver sees the same power, regardless of the station that is transmitting. This calibration can be carried out with a variable optical attenuator placed immediately after the laser.

With transmitter calibration in place, a given receiver will receive the same signal level regardless of the position

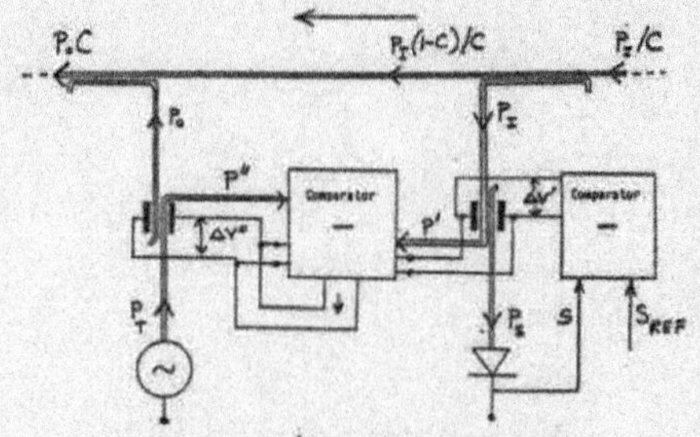

Fig. 17. Adaptive power calibration mechanism.

of the transmitter on the bus. However, the signal level will vary from receiver to receiver. For implementation and maintenance purposes it is convenient that the received signal level be the same at all stations. This would permit the replacement or the exchange of receiver units without requiring a cumbersome adjustment of operating levels each time. Again, the simplest solution to this problem is to place an optical variable attenuator in front of the receiver.

The receive and transmit attenuators can be adjusted at network initialization. One of the problems with this static adjustment approach, however, is the fact that whenever the system configuration is changed (e.g., a new tap is inserted, or the cable extended, etc.) a readjustment of the attenuators is required. To overcome this problem, an adaptive adjustment procedure must be used.

A possible adaptive adjustment scheme for the U-Net protocol is shown in Fig. 17. It is based on closed loop control. The optical attenuators are implemented with parallel optical guides with coupling ratio variable from zero to one depending on the voltage applied. We assume that the function of coupling ratio versus voltage is known for each attenuator. We let P_I be the incoming power from the bus and P_0 the outgoing power to the bus.

First, we must reduce the input power from P_I to P_s so that the detector works at the target operating point and provides the desired output level S_{REF}. This can be accomplished by comparing the actual output S with the reference S_{REF} and driving the attenuator with the difference signal $\Delta V'$.

Next, we must regulate the output P_0 so that this output generates on the bus the same signal strength as any other upstream station. From Fig. 17 we note that the following relationship must be satisfied (neglecting excess losses):

$$P_0 C = P_I(1 - C)^2/C.$$

Or

$$P_0 = P_I(1 - C)^2/C^2.$$

P_0 can be computed from P'' and $\Delta V''$ since the coupling vs voltage function of the attenuator is known. Likewise, P_I can be computed from P' and $\Delta V'$. The measurements of P', P'', $\Delta V'$ and $\Delta V''$ are fed to a comparator node which computes P_I and P_0 and generates a voltage $\Delta V''$ proportional to the value D given by

$$D = P_0 - P_I(1 - C)^2/C^2.$$

It should be noted that traffic on the U-Net bus is bursty, consisting of trains of packets separated by silence intervals. Thus, signal level and power measurements are also bursty. Since close loop controls require continuous inputs, the parameters S, P', and P'' in Fig. 17 are measured during active periods only and are averaged over an appropriate time window.

For the end station (in Fig. 17, the right most station) the closed-loop control is disabled. More precisely, the output power P_0 is set to P_T, to maximize the power margin. On the receive side, the input attenuator is set so that the coupling ratio $C = 1$, that is, the receiver is fully open and is ready to detect new upstream stations. The same setting is used also for new entering stations, and for system initialization.

VII. Conclusions and Future Research Directions

We have described in this paper three different token-based protocols designed for twin fiber bus networks. These protocols are suitable for high bandwidth x length products. They provide bounded delay at heavy load, thus permitting the integration of data and real time traffic (voice, video, etc.) on the same network. The logic needed t line speed is kept simple. In fact, the protocols are driven solely by the detection of activity on the channels. The completely distributed control and the passive nature of the interface guarantee robust, fail-safe operation.

The various versions of the protocols offer different performance characteristics. The choice of the protocol should depend ultimately on the application at hand.

One of the well-known drawbacks of fiber bus networks is the limitation on the number of stations that can be connected to the bus due to tap insertion loss. In this respect, twin bus architectures offer a slight advantage over "folded" bus architectures (such as Express-Net and D-Net) in that the latter need to support twice as many taps as the former on a single bus segment. Nevertheless, the maximum number of stations that can be accomodated on a twin bus is in the order of 20 to 30 in the best case. This severely limits the use of the fiber bus in applications with large number of devices. In order to overcome this problem, we are currently pursuing at UCLA the investigation of multilevel network architectures which permit to interconnect several busses, thus expanding the number of station supported, and yet maintain the high bandwidth and fault tolerance properties of the original structure.

Another key issue in local networks supporting integrated data, voice, and video traffic is the ability to guarantee bandwidth to real time (i.e., voice and video) users. Token-based protocols appear to be ideally suited for bandwidth allocation and control. Work is now in progress on the incorporation of such features in the access protocols.

References

[1] A. Albanese, "Fail-safe nodes for lightguide digital networks," *Bell System Tech. J.*, vol. 61, no. 2, pp. 247–256, Feb. 1982.

[2] L. Fratta, F. Borgonovo, and F. A. Tobagi, "The EXPRESS-NET: A local area communication network integrating voice and data," in *Proc. Int. Conf. Performance Data Commun. Syst. Their Appl.* (Paris, France), Sept. 1981.

[3] M. Gerla, P. Rodrigues, and C. Yeh, "Buzz-Net: A hybrid random access/virtual token network," in *Proc. Globecom '83* (San Diego, CA), Dec. 1983.

[4] M. Gerla, C. Yeh, and P. Rodrigues, "A token protocol for high speed fiber optics local networks," in *Proc. Opt. Fiber Commun. Conf.* (New Orleans, LA), Feb. 1983.

[5] M. Gerla, P. Rodrigues, C. Yeh, "U-Net: A unidirectional fiber bus network," FOC-LAN Conf. (Las Vegas, NV), Sept. 1984.

[6] J. O. Limb, and C. Flores, "Description of Fasnet—A unidirectional local-area communications network," *Bell Syst. Tech. J.*, vol. 61, no. 7, pp. 1413–1440, Sept. 1982.

[7] G. Lutes, "Optical fiber applications in the NASA deep space network," *Fiberoptic Technol.*, pp. 115–121, Sept. 1982.

[8] M. A. Marsan and G. Albertengo, "C-Net: A local broadcast communications network architecture," Politecnico di Torino, Int. Rep., 1981.

[9] E. G. Rawson, "Optical fiber for local computer networks," in *Proc. Dig. Top. Meeting Optical Fiber Commun.* (Washington, DC), March 1974.

[10] E. G. Rawson and R. V. Schmidt, "FIBERNET II: An ETHERNET-compatible fiber optic local area network," in *Proc. 82 Local Net* (Los Angeles, CA), 1982, pp. 42–46.

[11] P. A. Rodrigues, L. Fratta, M. Gerla, "Token-less protocols for fiber optics local area networks," in *ICC'84 Conf. Proc.* (Amsterdam, The Netherlands), May 1984.

[12] P. A. Rodrigues, "Access protocols for high speed fiber optics local networks," Ph.D. dissertation, UCLA, Computer Sci. Dept., Dec. 1984.

[13] R. V. Schmidt, *et al.*, "Fibernet II: A fiber optic ethernet," *IEEE J. Select. Areas Comm.*, vol. SAC-1, pp. 702–711, Nov. 1983.

[14] A. Takagi, S. Yamada, and S. Sugawara, "CSMA/CD with deterministic contention resolution," *IEEE J. Selected Areas Commun.*, vol. SAC-1, Nov. 1983.

[15] H. F. Taylor, "Technology and design considerations for a very-high-speed fiber optic data bus," *IEEE Select. Areas Commun.*, vol. SAC-1, 1981.

[16] F. A. Tobagi and V. B. Hunt, "Performance analysis of carrier sense multiple access with collision detection," *Computer Network*, vol. 4, pp. 245–259, Oct. 1980.

[17] M. N. Fine and F. A. Tobagi, "Performance of unidirectional broadcast local area networks: EXPRESS-NET and FASNET," *IEEE Trans. Comput.*, Dec. 1984.

[18] C. Tseng and B. Chen, "D-Net: A new scheme for high data rate optical local area networks," in *IEEE J. Selected Areas Commun.*, vol. SAC-1, Apr. 1983.

[19] M. V. Wilkes and D. J. Wheeler, "The Cambridge digital communication ring," in *Proc. Local Area Comm. Network Symp.*, pp. 47–61, May 1974.

*

Mario Gerla (M'75) received the graduate degree in engineering from the Politecnico di Milano, in 1966, and the M.S. and Ph.D. degrees in engineering from UCLA in 1970 and 1973, respectively.

From 1973 to 1976 he was with Network Analysis Corporation, New York, NY, where he was involved in several computer network design projects for both government and industry. From 1976 to 1977 he was with Tran Telecommunications, Los Angeles, CA, where he participated in the development of an integrated packet and circuit network. Since 1977, he has been on the Faculty of the Computer Science Department at UCLA. His research interests include the design, performance evaluation, and control of distributed computer communication systems and networks.

Paulo Rogrigues received the M.Sc. degree from COPPE, Rio de Janeiro, Brazil, in 1977, and the Ph.D. degree from UCLA in 1984, all in computer science.

He has been a Consulting Computer Analyst and Professor at the Federal University of Rio de Janeiro (UFJR), Brazil, since 1975, where he participated actively in the development of microsystems and peripheral devices. In 1978 he was director of EBC, a Brazilian computer industry, and in 1979 he joined the Ph.D. program at UCLA where he worked as a Research Assistant in high speed fiber optics local area network research.

He is presently in the research division of the Computer Center (NCE) and UFRJ. His current technical interests are in design and performance analysis of computer networks, local area networking supporting integrated services, interconnection of computer networks and distributed operating systems.

Cavour Yeh (S'56-M'63-SM'82) was born in Nanking, China, on August 11, 1936. He received the B.S., M.S., and Ph.D. degrees in electrical engineering from the California Institute of Technology, Pasadena, CA, in 1957, 1958, and 1962, respectively.

He is presently professor of electrical engineering at the University of California at Los Angeles (U.C.L.A). He joined U.C.L.A. in 1967 after serving on the faculty of U.S.C. from 1962 to 1967. His current areas of research interest are optical and millimeter-wave guiding structures, gigabit-rate fiberoptic local area network, and scattering of electromagnetic waves by penetrable irregularly shaped objects.

Dr. Yeh is a member of Eta Kappa Nu, Sigma Xi, and the Optical Society of America.

Interconnection of Fiber Optics Local Area Networks

Mario Gerla
Cavour Yeh

School of Engineering and
Applied Sciences
UCLA, Los Angeles, California 90024-1600

Abstract

The purpose of this paper is two-fold:

(a) To review the need for and the requirements of FOLAN interconnects; and

(b) To present Tree-Net, a novel, multilevel, interconnected FOLAN architecture.

We first review the basic FOLAN architectures and their requirements. Then, we discuss the problem of FOLAN interconnection and review some interconnect alternatives. Finally, we introduce Tree-Net, as an example of interconnect scheme, and evaluate its performance characteristics.

1. Introduction

Fiber Optics Local Area Networks (FOLAN's) are becoming increasingly popular for applications involving the exchange of very high data rates [PERS85]. The applications range from the transfer of large files, to the handling of real time control data, to the integration of data, voice and video services. It is to this last application (i.e. integration of services) that the high bandwidth of the FOLAN is ideally suited: therefore, much of the current research on FOLAN's is directed to providing an environment supportive of both real time traffic (voice and video) as well as the more traditional computer traffic (interactive, file transfers etc.) [PERS85].

As the interest in FOLAN's grows, so does the size of the systems to be connected via FOLAN's. One important example is the distribution of integrated services to a metropolitan area via a fiber optics MAN (Metropolitan Area Network). For many reasons (cost, reliability, traffic pattern, geographic layout, etc.) it is not practical to use a single network to span several hundred stations. A preferred solution is to develop smaller networks, and to interconnect them together via passive or active devices (e.g. star couplers, repeaters, bridges, gateways etc.)

While the interconnection of wide area networks (WANs) and conventional (i.e. non-fiber) local area networks (LANs) is a mature field, the interconnection of FOLAN's presents new challenges. First, one deals with much higher speeds (up to the Gbps range); this makes the problem of buffering data at the gateways a very delicate one. Secondly, the presence of real time traffic (i.e. voice and video) requires small delays and small delay variances so that packets can be delivered to the destination at regular intervals. Thus, WAN and LAN interconnect solutions cannot be easily adopted in FOLANs.

The purpose of this paper is two-fold:

(a) Analyze the requirements of FOLAN interconnects, and survey the interconnection alternatives.

(b) Present Tree-Net, a novel, multilevel, interconnected FOLAN architecture.

We first review the basic FOLAN architectures and their requirements. Then, we discuss the problem of FOLAN interconnection and review some interconnect alternatives. Finally, we introduce Tree-Net, as an example of interconnect scheme, and evaluate its performance characteristics.

2. Basic FOLAN Architectures

Various topologies and protocols have been proposed for FOLAN implementation, and many have actually been implemented. The topologies include: ring, bus, star, tree, and mesh; the protocols range from token to random access to store-and-forward queueing [STAL84].

Key requirements for a FOLAN are: throughput efficiency, low delay variance, guaranteed bandwidth (for real time traffic), fault tolerance/reliability and expandability. We will briefly survey the main FOLAN architecture alternatives in view of such requirements.

Currently, the most popular FOLAN architecture is the token ring, shown in Fig. 1 [BERG85], [FDDI84]. In the token ring, stations are connected to the fiber via active interfaces, namely, shift registers which permit the station to inspect a few bits on the fly and determine what needs to be done (e.g. copy the current packet to own memory; transmit new packet etc.). Access to the ring is controlled by the circulation of a token, to which stations append their packets. The transmitter is responsible for removing its packet after a full revolution [STAL84]. Since ring interfaces are active and can regenerate the signal, the ring can be viewed as a concatenation of unidirectional, point-to-point channels. The point-to-point fiber optic channel technology is well established; thus the token ring implementation does not pose major difficulties from the fiber stand-point. Furthermore, the token protocol provides Round Robin access to the network, thus permitting to control delays, to maintain priorities and to allocate bandwidth. A convincing proof of the success of the FOLAN ring alternative is the fact that the first FOLAN architecture proposed for standardization, FDDI (Fiber Distributed Data Interface) is indeed a token ring [FDDI84].

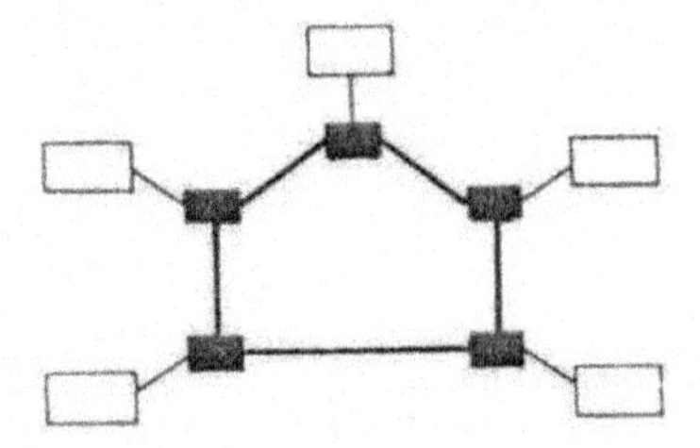

Fig. 1 Ring Topology

The ring itself, however, has some limitations. One drawback is the maintenance of clock synchronization among all stations. Another problem is the fact that packet header inspection must be carried out at line speed, on-the-fly, while the packet is transiting through the shift register. This may pose limitations on the speed at which the ring can operate. Finally, the presence of active interfaces makes the ring very vulnerable to failures. If an interface fails, the entire ring is down. All ring implementations, of course, have addressed the reliability and fault tolerance problem with great diligence. In the IBM implementation, for example, a central by-pass switch permits to detect and amputate faulty interfaces. In FDDI, two independent, counter rotating rings are used to enhance reliability. Still the fact remains that it does not seem practical to build rings with more than a few hundred stations for reliability reasons, for bandwidth limitations posed by active interface electronics, and for the fact that the ring topology cannot effectively cover a large geographical area.

Another topology which naturally lends itself to FOLAN implementations is the star topology (see Fig. 2). One of the first experimental FOLAN's, Fibernet, was a star [RAWS78]. The heart of the star FOLAN is the star coupler, which uniformly distributes the signal from one input to all the outputs. Each station is connected to one input and one output. The medium is a full broadcast medium: whatever a station transmits, all others will hear. The typical access protocol is Carrier Sense Multiple Access (CSMA): a station transmits whenever it senses the channel free. If two stations collide (because they transmitted at the same time), they try again later according to a well define procedure (e.g. random time-out).

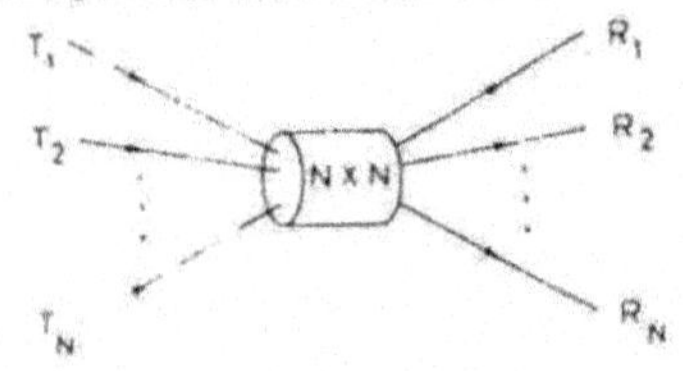

Fig. 2 Star Topology

The most popular star coupler available on the market is the biconical star coupler, obtained by fusing the various fibers together. One problem of this coupler is the fact that it is not suitable for single mode operation. To operate in single mode, one must use a "modular" star obtained by paralleling and cascading several binary couplers [MARH84]. This latter implementation is much more costly, however, than the biconical coupler. Furthermore, it does not allow implementation to an "active" star, with intermediate amplification of the signal in the star, before distribution to the outputs.

The major limitation, however, in using the star for serving a large user population over an extended geographical area rests on the CSMA protocol. This protocol becomes very inefficient when round trip propagation delay is larger than packet transmission time, a typical situation in very high speed MAN's. Other protocols could be used for the star in order to remove this inefficiency and provide, at the same time, guaranteed delay and bandwidth [KUNI82]. In fact, one such protocol will be described later in this paper. However, the general conclusion is that a single star net is not suitable to support a large user population.

Bus architectures have also been proposed for FOLANs, in part because the bus is one of the most popular topologies for conventional LANs, and also because of its fault tolerant features. In the bus, in fact, stations are connected to the cable via passive interfaces; the failure of one interface does not cause the entire bus to fail. Fiber optics bus architectures normally consist of one or two unidirectional busses to which stations are connected with passive couplers [GERL85], [TOBA83], [TSEN83]. Some typical topologies are shown in Fig. 3. The protocols commonly proposed for these networks are either implicit token protocols, or hybrid random access/token protocols. In the implicit token protocol, one station generates a token and the backlogged stations append their packets to the token in a Round Robin fashion. Thus, a train of packets forms after the token [GERL85]. In the hybrid protocol, stations access the channel in a random access mode during light load. When the load builds up, and therefore collisions occur, the system reverts to token mode [GERL87]. The weak point of bus FOLANs is the power loss at the optical couplers. With perfectly tuned coupling ratio's up to 40 or 50 stations can be supported [GERL85]. Still, this is clearly not sufficient for metropolitan area coverage.

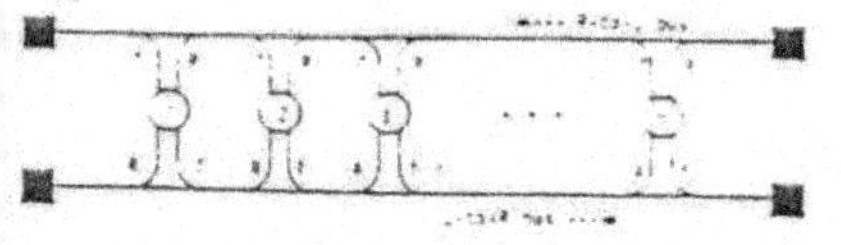

Twin Bus Topology

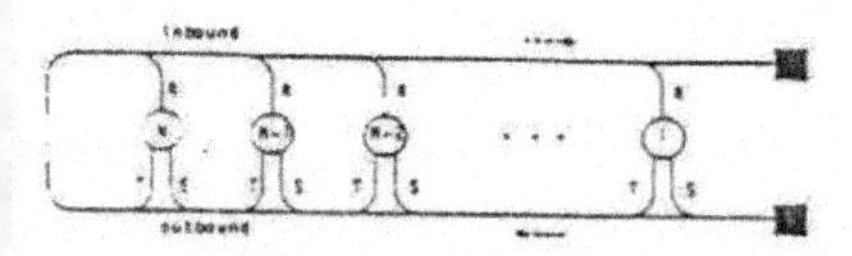

Folded Bus Topology (C-Net)

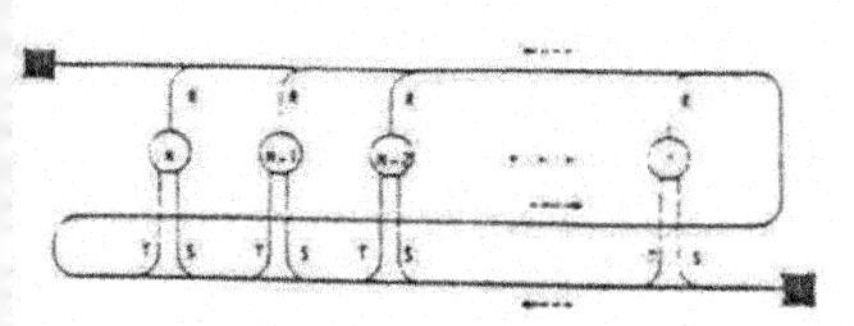

Folded Bus Topology (Express-Net)

Fig. 3 Bus Topologies

Tree topologies have received less attention than rings, busses and stars. One tree FOLAN implementation, however, has been reported, namely, Hubnet [LEE83]. The Hubnet topology shown in Fig. 4 is the superposition of two directional, rooted trees: the selection tree; and, the broadcast tree. User stations are connected at the leaves of the tree. The access to the network is essentially random access with capture. Namely, a packet attempts to capture a path to the root in the selection tree. If it succeeds, it then broadcasts to all stations on the broadcast tree. When two packets arrive at the same internal node in the selection tree, the node logic selects the first packet (i.e. the first packet captures the node) and forwards it on-the-fly to the next node higher up. The second packet is rejected and must re-try after time-out.

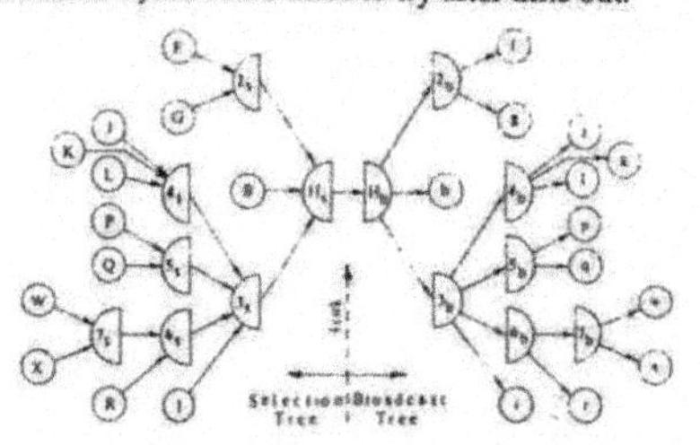

Fig. 4 Hubnet Topology (from [LEE83])

Weak points of Hubnet application to metropolitan network environments are: the inability to provide guaranteed delay and low delay variance at heavy load; and, the presence of active elements at the internal nodes (for packet selection and signal regeneration) which weakens network reliability. For either reason, it is not practical to build very large Hubnets. Some of these problems may be overcome by interconnecting smaller Hubnets via gateways [LEE83].

In comparing the features offered by the various FOLAN architectures, we note some common points. First, all schemes are designed to minimize the processing (if any) at intermediate nodes. This implies that no filtering (based on destination addresses) is performed within the network, so that the packet must be broadcast to all destinations. Along the same lines, to minimize nodal processing overhead, no store-and-forward buffering is performed at internal nodes.

Differences are noted in the access protocols. Ring and bus offer bounded delay and guaranteed bandwidth (because of the token protocol), thus are suitable for integration of data, voice, and video services. Star and tree are typically operated in a random access mode, thus cannot be directly applied to service integration. However, with proper design of the protocols, even star and tree topologies can be used in integrated services networks, as shown later in this paper.

One common limitation shared by all the proposed FOLAN architectures is the inability to provide adequate coverage of large user populations over metropolitan areas. This limitation stems from power loss (bus, star), inadequate reliability (ring, tree) and inadequate bandwidth (all). This limitation can be overcome by interconnecting these basic FOLANs into multilevel structures. The following section identifies the requirements for FOLAN interconnection and surveys some proposed interconnect schemes.

3. FOLAN Interconnection

FOLANs can be interconnected in many different ways. Here, we will examine two schemes, namely: repeaters and gateways.

A repeater essentially receives the signal from one network, it regenerates it and retransmits it to other networks. The connection between networks is thus a "physical" level connection, with reference to the OSI 7 layer model [GREE82]. The use of repeaters to strengthen the signal is helpful in FOLAN architectures limited by power loss in the passive taps (e.g. bus) or in the star coupler (e.g. star). For example, it is possible to increase the number of stations supported by a bus by cascading several busses via repeaters. Likewise, one may increase the station population in a star by using multilevel stars with intermediate stage amplification. Clearly, repeaters introduce single points of failures; thus redundancy must be provided.

While repeaters prove cost effective for moderate expansions of otherwise power limited FOLAN's, they cannot provide an adequate solution to the large population, metropolitan geography requirement for the following reasons:

- The traffic on each link grows linearly with network size (since each packet is re-broadcast over the entire network).
- If a round robin type scheme (e.g. token) is used to provide guaranteed delay and bandwidth, the delay latency increases linearly with the number of stations.
- The topological structure induced by the use of repeaters (e.g. cascaded busses) may not provide effective geographical coverage.

To obtain larger interconnect systems, gateways must be used. The gateway implements a packet level interconnection (as opposed to the physical level connection of the repeater). Namely, it performs packet buffering, routing and flow control. By inspecting the packet destination address, the gateway performs traffic screening, i.e. it routes a packet only to the intended destination network rather than broadcasting it to all. This screening and filtering function greatly alleviates the overload problem experienced when packets are broadcast over the entire network. Also, the "locality" behavior of the traffic, i.e. the fact that typically most of the traffic is local within the single FOLAN, helps reducing internet traffic.

Yet, the FOLAN interconnection via gateways is a challenging problem, in many respects more difficult than the gateway interconnection of conventional LANs because of the very high fiber speed, the integrated services implications and the larger geographical span. Key requirements for an effective interconnection are:

- Streamlined packet protocol in the gateway (to minimize processing overhead)
- Very simple routing and flow control protocols in the interconnection network (i.e. the high level network which connects the gateways)
- Bandwidth and delay guarantee (for integrated services support)
- Growth flexibility
- Fault tolerance

The interconnection network can be implemented with any of the previously mentioned schemes (ring, bus, tree, star). Furthermore, multilevel interconnection structures can be used. Thus, the number of possibilities is practically unlimited. Instead of attempting a taxonomy, we provide below a few examples of actual (proposed or implemented) FOLAN interconnect methods.

The conventional method for interconnecting rings is the ring hierarchy (see Fig. 5). An early proposal to the IEEE 802.6 committee for MAN standards consisted, in fact, in multiple ring layers [SZE85]. A key issue in ring interconnection is buffering and flow control at the gateway [BUX85]. Other potential problems in a multilevel network are the saturation of the ring at the top of the hierarchy, and the poor reliability of the single connected topology.

To remedy, in part, to the above problems, an embedded tree topology can be used. Fig. 6 shows such a topology proposed for the interconnection of busses [YEA87]. Note that the multiple paths enhence reliability and alleviate the problem of root saturation because of load sharing.

To further remove the danger of channel saturation in the interconnection network, a mesh topology has been proposed. An example of mesh topology is given in Fig. 7 where a "Manhattan Grid" is used to interconnect several user nets (busses) together [YEA87]. The main advantages of a mesh topology are the distribution of the traffic over many trunks, and the enhanced reliability. A drawback, however, is the increased complexity of the nodes, which must perform store-and-forward buffering, switching and routing [ALBA87].

In the balance of this paper, we present Tree-Net, an example of interconnected FOLAN architecture. The lowest level of Tree-Net consists of linear busses. The usses are interconnected by a tree (in a repeater-like ashion) supporting full broadcasting across all usses. Several Tree-Nets can then be interconnected ia gateways.

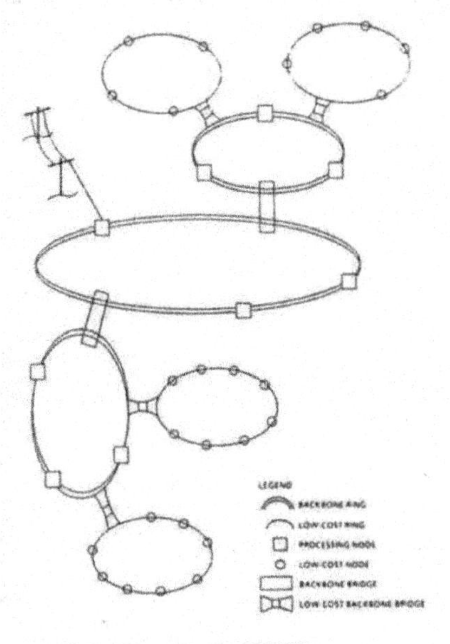

Fig. 5 Ring Hierarchy (from [SZE85])

4. Tree-Net: An Interconnected FOLAN Architecture

4.1. The Concept

The Tree-Net architecture here proposed can be viewed as a two level architecture, where the high level is a tree, and the low level is a linear bus (see Fig. 8). Stations are connected to the linear bus via passive taps; the tree itself is built using passive couplers.

In the simplest implementation of Tree-Net, no active components are present on the network path connecting any two stations, thus protecting the system from active component failures. This also implies that the signal transmitted by one station is broadcast to all other stations, with no filtering nor store-and-forward processing at any intermediate node (i.e. gateway). This permits to operate the network at very high aggregate data rate, without suffering of the bandwidth limitations imposed by the gateways.

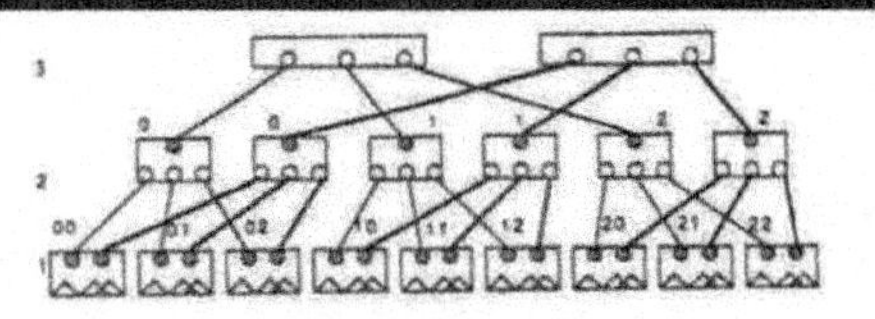

Fig. 6 Embedded Tree

Passive taps, full broadcasting and very high aggregate date rates are well advertised advantages of linear bus networks (e.g. Express-Net, U-Net, etc.)[GERL85]. While sharing these advantages Tree-Net offers additional features which permit to overcome some of the traditional drawbacks of the linear bus architecture. First, Tree-Net extends the number of stations that can be supported by an order of magnitude (from tens to hundreds, say). Secondly, the tree topology is better suited to cover a large geographical area (campus, industrial park, metropolitan area, etc.) than a linear topology. Thirdly, in the tree structure, the problem of transmitter and receiver calibration is simpler than in the linear bus. Finally, the optical couplers used to build the tree are simple 3dB couplers, while the couplers in the linear bus may need to be "tuned" depending on their position on the bus (i.e. the coupling ratio is adjusted to optimize the power budget) [RODR 84].

All the above advantages do not come, of course, for free. Tree-Net has also some drawbacks, most notably, an increased latency delay. In the following sections we review the components and protocols of Tree-Net, evaluate its performance and propose several extensions to the basic scheme.

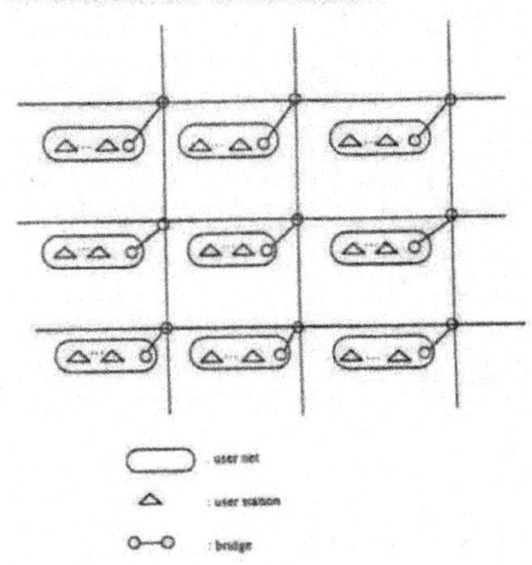

Fig. 7 Manhattan Grid

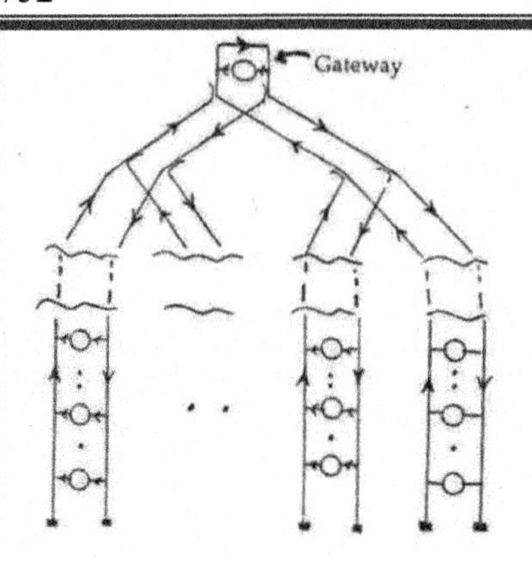

Fig. 8 Tree-Net Topology

4.2. The Bus Component

The basic building block of Tree-Net is a folded bus to which stations are connected via passive, directional taps (see Fig. 9). This architecture has been extensively studied, and several protocols have been reported in the literature. For our application, we assume that the access protocol is a "token" protocol. That is, the "end station" (see Fig. 9) starts a transmission cycle by issuing a token. Upon detecting the token (through the "sensor" port), downstream stations with a backlog of packets to send will append one packet to the token [TSEN83].

If there is more than one backlogged station, a "train" of packets will form after the token. Each station must then be able to locate the end of the train in order to attach its packet to it. This function is accomplished using a "probing" technique: namely upon sensing the end of a packet (or token) transmission, the backlogged station will start transmitting the preamble of its packet. If, while transmitting, the station hears a packet coming from upstream, it immediately aborts its transmission, "defers" to the incoming packet and tries again at the end of it. The result of this "collision" is damage of a few bits in the preamble. The preamble should be long enough so that proper synch acquisition is not compromised by the collision. More details on probing are found in [GERL85].

The token and the train of packets move from the transmit bus to the receive bus, where each station inspects the address of each packet. If the packet address matches its own address, the station copies the packet into its memory. On the receive bus, the end station looks for the end of the train. Upon sensing it, it issues a new token on the transmit bus, starting a new cycle.

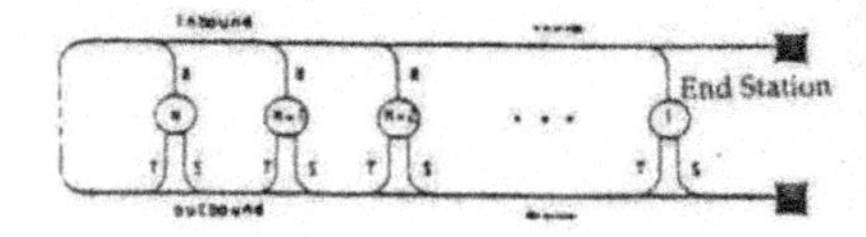

Fig. 9 Folded Bus Topology

4.3. The Tree

Two bus segments can be combined in "parallel" using the scheme shown in Fig. 10. Essentially, an "extension" bus is connected to the first bus via two couplers. The original token protocol can be easily extended to handle the two parallel bus configuration. Namely, "leaf" station A in branch A (see Fig. 10) starts the token cycle for branch A. (A-cycle). During the A-cycle, all backlogged stations in A can transmit, and both stations in A and B receive. At the end of the A-train, leaf station B starts the B-cycle, picking up the transmissions from the B-branch. Thus, the operation consists of an alternation of A and B cycles.

By applying the "parallel" combination process recursively, we obtain a binary tree structure, where the leaves correspond to linear bus segments (Fig. 8). The token protocol easily extends to the tree structure.

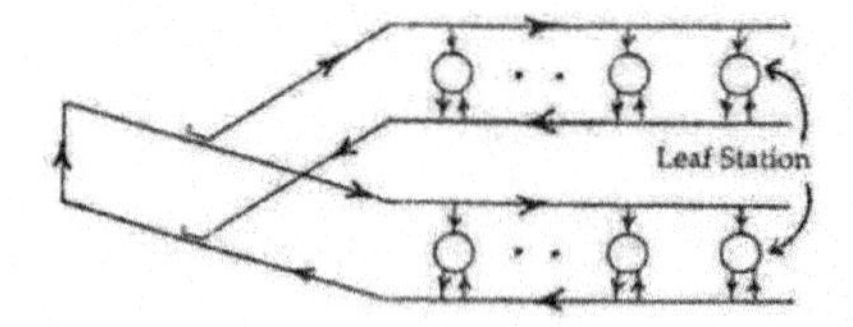

Fig. 10 Parallel Merge

The branches (i.e. leaves) take turns in issuing the token, according to a predefined order. A packet issued by a station is broadcast to all other stations. Thus, the tree can be viewed as a repeater-type interconnection of busses, as discussed in Section 3.

In our model, stations are connected to the tree only at the leaves. The scheme could be generalized by connecting stations also to internal links. For network access, the "internal" station is associated with the token cycle of one of the leaves in its sub-tree; that is, it is allowed to transmit at the end of the train from that specific leaf. As a special case, the station may be placed at the root of the tree: this is actually the position reserved for the gateway station (see Fig. 8). In this case, to reduce internet delays, the gateway is alowed to attach its packet at the end of a train from any leaf.

n this paper, for the sake of simplicity we will assume that stations can be connected only at the leaves except for the gateway station). Furthermore, we will assume that the binary tree is a full tree, that is, all the leaves are at the bottom level of the tree, and he bottom level is full.

A simple inspection of Fig. 8 reveals that from the functional standpoint, the tree could be replaced by a star. The star would actually provide power savings with respect to the tree and would in fact permit to support more stations. If the star is implemented with a biconical star coupler, however, it would preclude the use of single mode fibers as discussed in Section 2. This may be a serious limitation in very high speed FOLAN's operating at gigabit/sec. speeds. In order to accommodate single mode operations, a modular star based on 2 x 2 couplers may be used [MARH 84]. The modular star, however, requires many more couplers than the tree. A good compromise may be to use a modular star at the root, and to connect several sub-trees to the star. This option will be considered in a later section.

The actual Tree-Net layout may vary depending on the requirements of the specific application. A typical configuration for metropolitan area coverages is shown in Fig. 11. Here, we see that sets of internal tree nodes (i.e. couplers) may be grouped together in "distribution centers". At the root, a cluster may be created in the "central office" (this cluster may actually be replaced by a modular star, as later discussed).

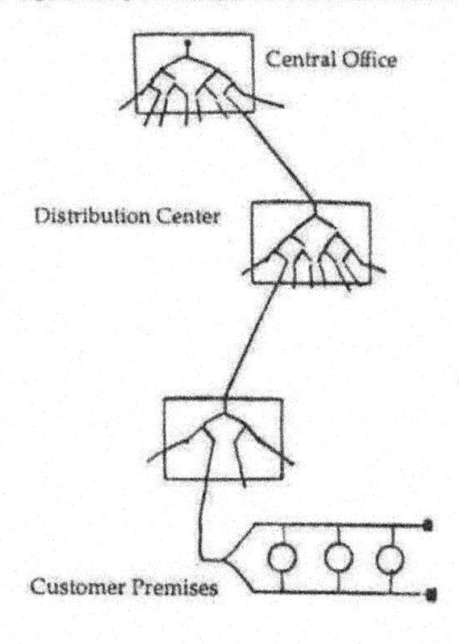

Fig. 11 Tree-Net Distribution System

One easily detects the analogy between the Tree-Net layout and the local distribution plant in the telephone network.

4.4. Maximum Number of Stations

An important figure of merit of a FOLAN architecture with passive taps is the maximum number of stations that can be connected. In fact, light detectors use much more signal power than their electronic counterparts (there is no such thing as a high impedance light sensor), so that the number of passive taps that can be cascaded on a linear bus or on a tree path is fairly limited. As we shall see, the tree structure offers substantial advantages over the linear bus in this regard. Still, even in the tree, the total number of stations is fairly modest unless intermediate amplification is used. Here again, the tree permits much more cost-effective amplification strategies than the bus.

To determine the maximum number of stations supported we must require that the signal transmitted by the light source (laser or LED) is properly detected at the end of the longest network path by the light detector.

First, some definitions and assumptions are in order. Let M be the power margin (in dB), defined as the difference between the laser output power (in dBm) and the minimum power detectable at the light detector (in dBm). Typically, M is on the order of 40 dB. The power reduction at a tree coupler is assumed to be 4 dB (3dB for power splitting, and 1dB for connector excess loss - a fairly conservative estimate). For the bus segment, tap coupling ratios may be individually adjusted to maximize station connectivity. Thus, the ratios vary from tap to tap, and depend on the total number of stations [GERL85]. To simplify the analysis, we assume that each bus tap causes a power reduction of 3dB (2dB for power spilling, and 1dB for excess loss).

Let L be the number of levels in the tree (see Fig. 12); let K be the number of stations on each bus segment. If no intermediate amplification is provided, the following inequality must be satisfied:

$$4L + 3K \leq M/2$$

or

$$K \leq M/6 - 4/3\,L \tag{1}$$

Noting that the total number of stations N is given by:

$$N = 2^L K \tag{2}$$

we have (by substituting (1) into (2):

$$N \leq 2^L (M/6 - 4/3\,L) \quad (3)$$

For M = 40, (3) becomes:

$$N = \leq 2^L (6.66 - 1.33\,L) \quad (4)$$

From (4), the maximum number of stations is plotted in Fig. 13, as a function of number of tree levels. We note that the maximum, N = 16, is obtained with L = 2 and k = 4, or L = 3 and K = 2, or, L = 4 and K = 1. For L = 0, i.e. single bus structure, the maximum is N = 6. Thus, the tree structure improves the situation (from 6 to 16), although the maximum number of supported stations is still too low for metropolitan area applications.

Station connectivity can be greatly improved by introducing an amplification stage at the root of the tree. In this case, the following inequality must be satisfied:

$$4\,L + 3\,K \leq M$$

Following the same steps as before we obtain (for M = 40 dB):

$$N \leq 2^L (13.33 - 1.33L) \quad (5)$$

The maximum number of stations for this case is plotted in Fig. 14. The maximum, N = 512, is obtained for L = 9 and K = 1; or L = 8 and K = 2 or, for L = 7 and k = 4. For L = 0, we find the maximum achievable with a single bus structure and with intermediate amplification namely, N = 13. The comparison with the previous result shows that amplification is much more effective (in terms of increasing the maximum number of stations) in Tree-Net than in the single bus structure.

In examining the above results, we note that the optimum is achieved with the pure tree structure (i.e. each leaf of the binary tree supports exactly one station). This result, however, is somewhat misleading because it does not take into account the delay performance. In the next section we address the delay issue and show that it pays to cluster stations on bus segments in order to reduce delays.

4.5. Access Delay

First, we define latency delay as the time interval between two successive token visits at the same station, in zero load conditions. Recalling that in Tree-Net each leaf is individually "polled" by the token, and letting:

d = average round trip propagation delay between root and leaf.

p = token processing and generation time (at the leaf station),

we obtain the following expression for latency delay LD:

$$LD = (d + p)\, 2^L$$

Average cycle time T in non-zero load conditions is related to latency delay as follows:

$$T = LD / (1 - u)$$

where u is the network load factor.

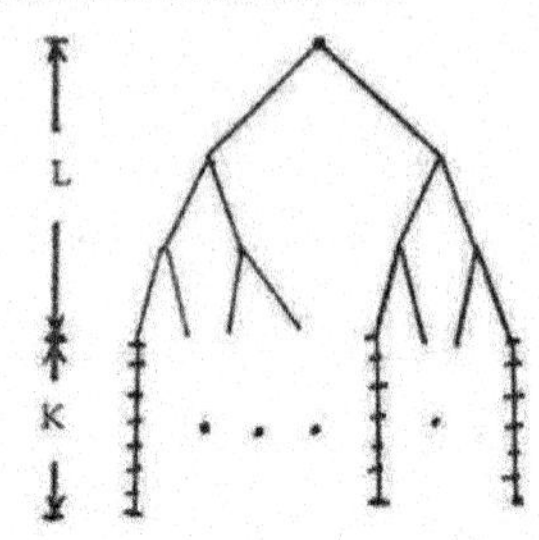

Fig. 12 Tree-Net Levels

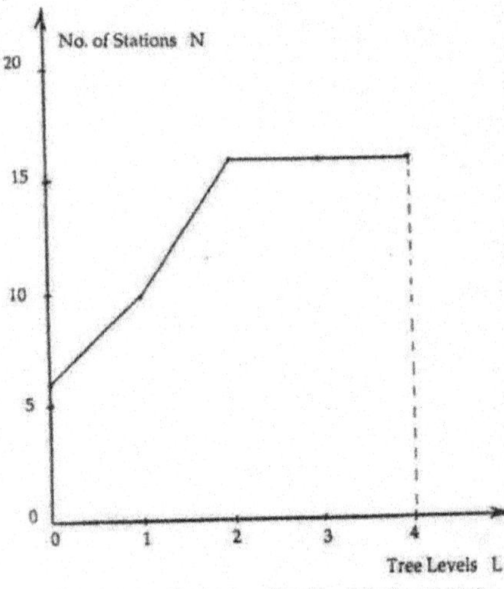

Fig. 13 No. of Stations vs Tree Levels (w/o repeater)

we have (by substituting (1) into (2):

$$N \leq 2^L(M/6 - 4/3\,L) \quad (3)$$

For M = 40, (3) becomes:

$$N = \leq 2^L(6.66 - 1.33\,L) \quad (4)$$

From (4), the maximum number of stations is plotted in Fig. 13, as a function of number of tree levels. We note that the maximum, N = 16, is obtained with L = 2 and k = 4, or L = 3 and K = 2, or, L = 4 and K = 1. For L = 0, i.e. single bus structure, the maximum is N = 6. Thus, the tree structure improves the situation (from 6 to 16), although the maximum number of supported stations is still too low for metropolitan area applications.

Station connectivity can be greatly improved by introducing an amplification stage at the root of the tree. In this case, the following inequality must be satisfied:

$$4\,L + 3\,K \leq M$$

Following the same steps as before we obtain (for M = 40 dB):

$$N \leq 2^L(13.33 - 1.33L) \quad (5)$$

The maximum number of stations for this case is plotted in Fig. 14. The maximum, N = 512, is obtained for L = 9 and K = 1; or L = 8 and K = 2 or, for L = 7 and k = 4. For L = 0, we find the maximum achievable with a single bus structure and with intermediate amplification namely, N = 13. The comparison with the previous result shows that amplification is much more effective (in terms of increasing the maximum number of stations) in Tree-Net than in the single bus structure.

In examining the above results, we note that the optimum is achieved with the pure tree structure (i.e. each leaf of the binary tree supports exactly one station). This result, however, is somewhat misleading because it does not take into account the delay performance. In the next section we address the delay issue and show that it pays to cluster stations on bus segments in order to reduce delays.

4.5. Access Delay

First, we define latency delay as the time interval between two successive token visits at the same station, in zero load conditions. Recalling that in Tree-Net each leaf is individually "polled" by the token, and letting:

d = average round trip propagation delay between root and leaf.

p = token processing and generation time (at the leaf station),

we obtain the following expression for latency delay LD:

$$LD = (d + p)\,2^L$$

Average cycle time T in non-zero load conditions is related to latency delay as follows:

$$T = LD\,/\,(1 - u)$$

where u is the network load factor.

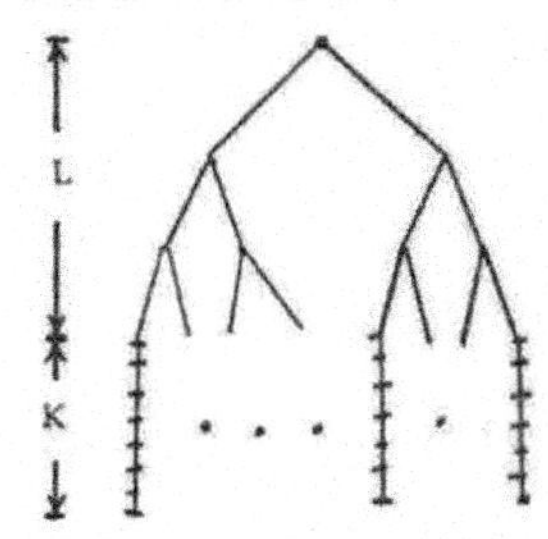

Fig. 12 Tree-Net Levels

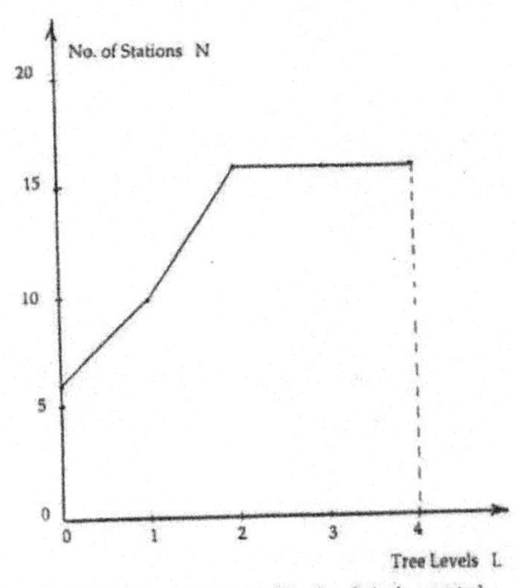

Fig. 13 No. of Stations vs Tree Levels (w/o repeater)

Access delay in Tree-Net may be defined as the interval between the time when a packet is ready for transmission in the station buffer (i.e. it is the first packet in the transmit queue), and the time when it is actually transmitted. Average access delay is clearly bounded by average cycle time.

For a MAN with a 5Km radius and operating at 1Gbps rate, reasonable values for the delay expression parameters are:

$$d = 50\mu sec$$

$$p = 1\mu sec$$

$$u = .5$$

Thus, $T \cong (.1)\, 2^L$ m sec. If the application supported by the MAN requires an access delay ≤ 2 m sec, then we must have:

$$2^L \leq 20$$

or

$$L \leq 4$$

Using the value L = 4 in expression (5) of the previous section, we find that the maximum number of stations supported is N = 128 (i.e. 16 bus segments, with 8 stations each).

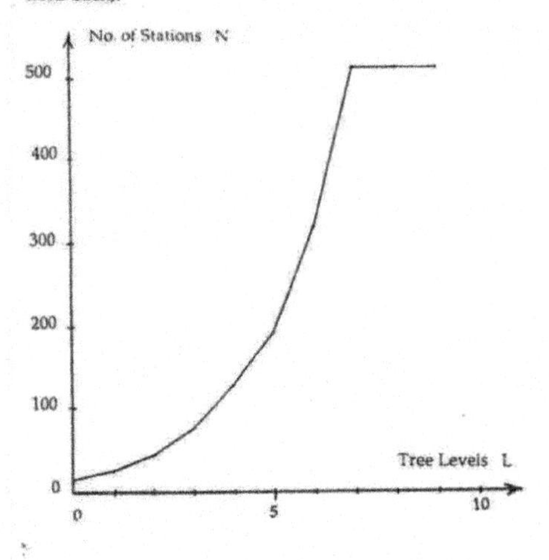

Fig. 14 No. of Stations vs Tree Levels (with repeater)

4.6. Gateway Interconnections

A Tree-Net can be connected to other Tree-Nets or to a wide area network (e.g. ISDN backbone) via a gateway. The gateway can be conveniently installed at the root of the tree (see Fig. 8). Its function is to inspect every packet coming up from the leaves and to extract and forward to other gateways the packets with a "foreign" address. Likewise, the gateway injects in Tree-Net the packets received from other networks.

It is important to note that the gateway, because of its privileged position in the tree, can append packets to the end of any "train" coming from any of the leaves. Thus, external packets can be transmitted on Tree-Net with very little latency delay, without waiting for an entire "polling" cycle to be completed. This feature clearly alleviates the buffer management requirements at the gateway.

The coverage of a typical metropolitan area may require several Tree-Nets. The Tree-Net gateways can be interconnected to each other and to WAN (Wide Area Network) gateways in various ways. Fig. 5 depicts a FOLAN ring interconnection but many other options are available as discussed in Section 3. The choice will depend on the geographical proximity of the gateways (in the limit they may all be located in the same central office), the internet traffic requirements, and the reliability and fault tolerance requirements.

Given the very high speed of Tree-Nets (on the order of the gigabit per second), the gateway must be able to handle an extremely high packet rate (up to the megapacket/sec). To this end, protocol processing must be streamlined to the extreme. A fast packet switch implementation based on the banyan switching fabric may be attractive [TURN86]. In this implementation, virtual circuits are maintained for all user sessions, and packets are routed according to virtual circuit ID numbers. Buffer flow control simply consists of dropping packets when buffers are full (more sophisticated controls are not possible at such high packet rates).

A common cause of congestion at gateways is the fact that the destination network cannot accept packets as fast as the origin networks generate them. Fortunately, in Tree-Net "foreign" packets have higher priority over local packets, so that queues will not form at the gateway-to-Tree-Net interface; rather, packets will queue in the origin station waiting for their turn to be transmitted. Once transmitted, the packets are essentially granted a congestion free path all the way to destination.

A policy which systematically favors foreign over local traffic may actually cause unfairness: a Tree-Net flooded with foreign traffic directed to it, may not be able to deliver its own originated traffic. To avoid this problem, flow control must be exercised at connection set-up time. That is, the traffic generated (in both directions) by a user session is estimated before the connection is established. The impact of this new traffic on origin, destination and interconnection network is evaluated. The new connection is accepted only if there is sufficient residual bandwidth (in the statistical sense) on the path.

4.7. Transmit Power Calibration

In a fiber optics broadcast network with passive taps the paths between different transmitter/receiver pairs may have different attenuation characteristics. Thus, if the transmit power level is the same for all transmitters, a receiver will receive packets from different sources at different power levels. This situation (often referred to as "dynamic range") makes proper reception difficult because the receiver cannot adjust to rapid fluctuations in input power. It is then necessary to "calibrate" the transmit power at each station, so that each detector receives the same power level from any sending station.

In Tree-Net, calibration can be accomplished with the following procedure. A station is chosen as a reference station (possible, the leaf station on the longest path from the root). This station periodically issues a token (as part of the access protocol). Each station will compare the power level of the token received from the reference station, with the power of the echo of its own transmission. It will dynamically adjust its transmit level until its echo power equals the reference power. A mechanism to implement this dynamic calibration was described in [GERL85].

It should be clear that the above procedure achieves our objective. In fact, if the reference and echo signal levels are the same at an arbitrary station, they must be identical also at the root of the tree. Therefore, they must be identical everywhere in the network. Furthermore, since each station calibrates its level to that of the reference station, the levels of any transmit stations will be identical at any arbitrary point in the network.

4.8. Fault Recovery Procedures

The ability to recover from faults is an important requirement for a network supporting many diverse applications. In Tree-Net, proper operation must be preserved even after failures of stations, gateway and amplifier. We review the various failure modes and present recovery schemes.

First, we consider the failure of the leaf station on a bus. Recall that the leaf station is responsible for issuing the token after detecting the end of train from the preceding bus (in the pre-established order). If the leaf station fails, no token is issued and the entire network fails. One way to avoid this problem is to modify the bus topology as shown in Fig. 15. In this modified topology, each station acts as if it were the leaf station, that is, upon detecting the end of train, it issues the token. Clearly, given the topology, station 2 (in Fig. 15) as it starts transmitting the token, will hear the token from station 1, and therefore it will immediately abort its transmission. This solution is inspired to the scheme used in Express-Net [TOBA83].

Another possible solution (which will work with the unmodified topology) consists of implementing staggered time-outs on the stations along the bus. Each station starts a time-out upon detecting the end-of-train; and, it issues a token if the time-out expires before a token has been heard on the bus. Since the time-outs are properly staggered, only one station will transmit the token. When the leaf station comes back up, it is automatically reinserted in the cycle.

Next, we consider the case in which all the stations on a bus segment (say, bus n in the polling sequence) have failed. Let T be the round trip delay from a leaf station to the root of Tree-Net and back. Bus n+1, i.e. the bus following bus n in the polling sequence upon noticing a silence period larger than T during which no token is heard, will assume that bus n has failed and issue a token, thus bypassing bus n.

Gateway and amplifier failures are handled by making these resources redundant. When the main unit fails, the backup is switched in. Monitoring and maintenance of these units is made easier by the fact that they are generally located in the Central Office (in the case of a Metropolitan Area Network) rather than on customer premises.

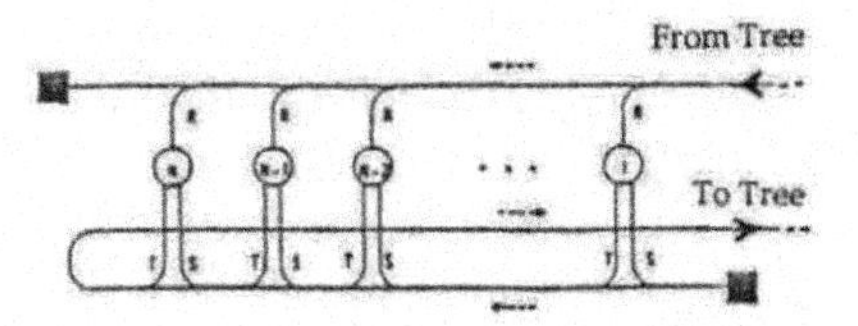

Fig. 15 Expressnet Type Bus Topology

4.9. Extensions of the Basic Concept

In Section 4 we have shown that the maximum number of stations supported by Tree-Net is about 500, using off-the-shelf technology and assuming amplification of the optical signal at the root. Here, we explore extensions to the basic Tree-Net structure which permit us to increase the number of stations.

The first modification consists of replacing the bus segments at the bottom of Tree-Net with "tree segments" (see Fig. 16). The stations, instead of being aligned along the bus as in Fig. 16a are now placed at the leaves of a tree segment as in Fig. 16b. The stations are also connected by a "control wire", a low bandwidth, unidirectional coaxial cable providing access control signaling [NASS85].

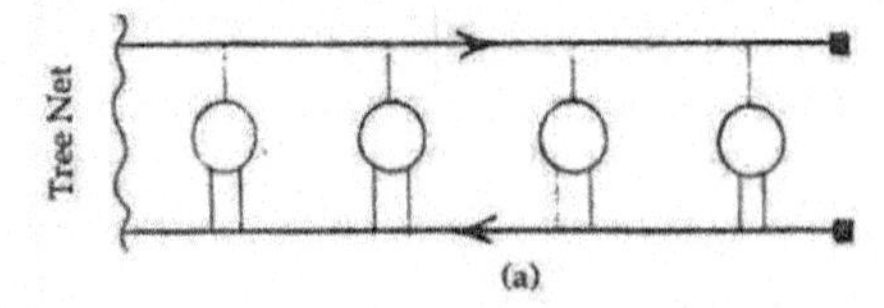

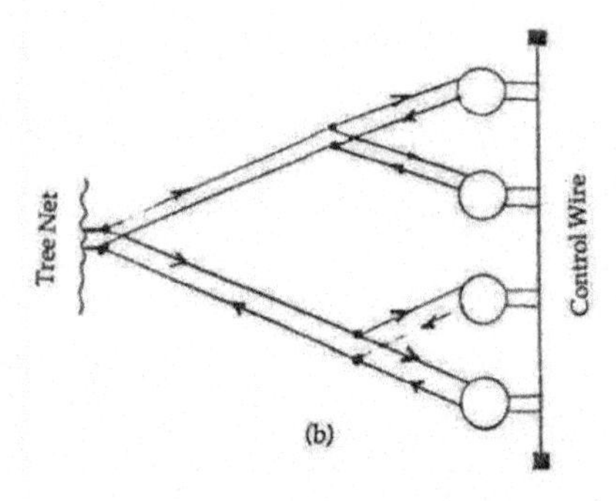

Fig. 16 Tree Segment Scheme

Access control in the tree segment works as follows. A station with packet (or token) to transmit, turns on its signal on the control wire. The leftmost station (station A in Fig. 16b) is responsible for issuing the token, so it always turns on the control signal. Upon detecting the end of the train generated by the previous segment, station A transmits the token followed by a packet (if any), and it drops the signal on the control wire. The next station along the control wire is now allowed to transmit its packet. After doing so, it will drop the signal on the control wire, enabling the following station to go, and so on until all the backlogged stations have transmitted their packets.

To protect the tree segment from leftmost station failure, a station will automatically take the role of leftmost station if it does not hear a control signal on the wire. Namely, it turns on its control signal and generates a token when required.

Note that the tree-segment can modularly replace the bus-segment in the network. In fact, the two types may actually co-exist in the same network.

The advantage of the tree-segment is that of reducing tap insertion loss with respect to the bus segment. Referring to the assumptions in Section 4, we find that the maximum number of stations grows from 16 to 32 (without amplification at the root); and, from 512 to 1024, with amplification.

Although this improvement may not seem large enough to justify considering the tree segment solution, there are other features that make this solution quite attractive. Referring to Section 4.4, one recalls that N = 512 was achieved with a bus segment with only two stations on it (K = 2). This in turns led to very high latency (as discussed in Section 5). To keep delay within an acceptable range, one is forced then to reduce N well below 512.

In the tree-segment solution, the choice of cluster size (i.e. the number of stations per segment) does not affect the total number of stations. Thus, the size can be properly chosen to suit the delay constraints.

Another very effective way of improving the total number of stations consists of interconnecting several Tree-Nets with a passive, modular star [MARH84], as shown in Fig. 17. Recall that the number of stages in the modular star is $\lg_2 P$, where P is the number of ports [MARH84]. Each stage introduces a 4db attenuation. Using the assumptions in Section 4.4, a system with a 4 stage star and with 3 level Tree-Nets can support 128 stations without requiring any amplification. This is a major improvement with respect to the 16 stations in a pure Tree-Net!

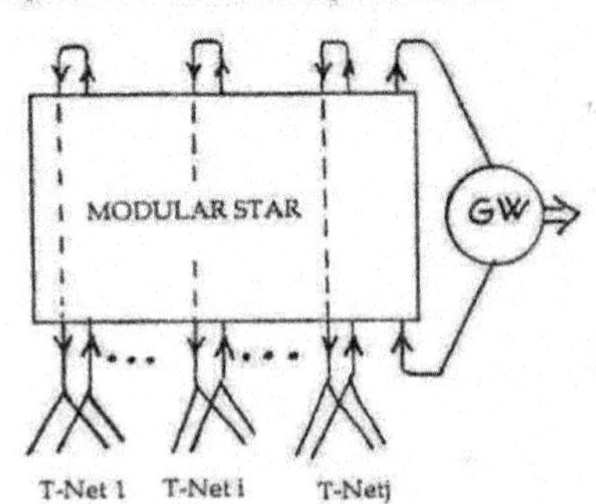

Fig. 17 Star Interconnection

If amplification is used, the modular star permits a dramatic increase in the total number of stations. Assuming that amplifiers are used both at the inputs and the outputs of the star, a configuration with a 10 stage star and 10-level Tree-Nets becomes feasible (from the power budget standpoint). This configuration supports up to 1,000,000 stations. Probably, such a large system is not practical for other reasons (the delay latency may be excessive; 5,000 couplers are required in the star; and, even the gigabit/sec bandwidth may not be sufficient to suport such a large population). However, a scaled-down version of hybrid star/tree architecture can be properly engineered and provide a very cost effective solution to the metropolitan coverage problem.

One may note that calibration can be carried out in the star/tree network in the same way as in Tree-Net, using a reference station. As for the interconnection of the star/tree network with other networks, this can be accomplished via a gateway connected across the star (see Fig. 17).

So far, we only have discussed topological modifications to the basic Tree-Net structure leaving the access protocol unaltered. One may also consider using different access protocols. A critical limitation in Tree-Net is latency delay, which is caused by the fact that the token must visit all the bus segments in sequence. Latency delay clearly limits the total number of stations that can be supported, especially when the geographical area to be covered is quite large, as in a MAN application. It seems appropriate, therefore, to explore access protocols other than the token scheme. We are currently investigating random access protocols as well as hybrid random access/reservation protocols for the Tree-Net topology.

5. Conclusions

In this paper we have focussed on the needs for interconnection of FOLANs. The principal application was the deployment of MANs (Metropolitan Area Networks) operating at gigabit/sec speed and above, and integrating data, voice, video services. We have shown that the existing basic FOLAN architectures cannot be stretched to satisfy high speed MAN requirements; and, have pointed to interconnection as a possible solution.

Among the interconnection alternatives considered were the repeater (at the physical level) and the gateway (at the packet, or network level). The following key requirements for MAN interconnection were identified: throughput efficiency; delay guarantee; fault tolerance; and expandability.

The Tree-Net architecture was then introduced as an example of cost effective FOLAN interconnection. Tree-Net has two levels of interconnection. At the low level, FOLAN busses are connected by a tree/star/repeater system. At the high level, gateways connect several Tree-Nets via a mesh topology.

The proposed architecture satisfies the basic requirements for MAN interconnection. Throughput efficiency is obtained through the use of an implicit token protocol within Tree-Net, and of fast packet switching and mesh topology at the higher level. Delay guarantee derives from the use of virtual circuits for real time traffic; bandwidth preallocation to real time connections at call set up time; and, built in priority for internet traffic. Fault tolerance follows from the fact that all network components in Tree-Net are passive, except for repeaters and gateways, which are redundant and are installed in a Central Office where extensive monitoring and maintenance facilities are presumed. Expandability within the tree topology is easily obtained by replacing a leaf with a subtree. At a higher level, new Tree-Nets can be deployed and connected to the internet mesh via gateways.

Work is currently in progress on refining and extending several aspects of Tree-Net. A hybrid random access/token mode scheme is being designed to overcome token latency. The trade-offs of star/tree combinations are investigated. Flow control and bandwidth allocation protocols for the internet are defined, with special attention to fairness issues. Various internet connection alternatives are investigated including the integration of fast packet switching and time division slot switching. Finally, the suitability of the Tree-Net concept for large FOLAN systems other than the MAN (e.g. automated factory; interconnection of existing FOLANs etc.) is investigated.

Acknowledgements

This research was supported by the State of California and Pacific Bell (through a MICRO grant); by NSF (INT 85-16798 and ECS 80-20300), and; by the US ARO (DAAG 29-85-0101).

References

[ALBA87] Albanese, A. and L. Fratta, "Integration of Services and Multi-user Groups Over Interconnected LAN's", *International Symposium on Data Communication Systems and Their Performance*, Rio de Janeiro, June '87.

[BERG85] Bergman, L. A. and S. T. Eng, "A Synchronous Fiber Optic Ring LAN for Multi-Gigabit/s Mixed Traffic Communications", *JSAC*, Nov. 1985.

[BUX85] Bux, W. and D. Grillo, "Flow Control in LANs of Interconnected Token Rings", *Trans. Comm.*, Oct. '85.

[FDDI84] FDDI Token Ring, ANSI X 3T9/84, Oct. 84.

[GERL85] Gerla, M. et al, "Token Based Protocols for High Speed FOLANs", *Journal of Lightwave Tech.*, June '85.

BUZZ-NET : a hybrid random access/virtual token local network

M. Gerla
P. Rodrigues
C. Yeh

UCLA - School of Engineering
Los Angeles, CA 90024

ABSTRACT

Buzz-net is a local network supported by a pair of unidirectional busses, to which stations are connected via passive interfaces. The access protocol is a hybrid which combines random access and virtual token features. More precisely, the network operates in random access mode at light load and virtual token mode at heavy load. Buzz-net can find applications in fiber optics networks, which are intrinsically unidirectional. This paper describes the protocol and compares its performance to that of other unidirectional bus schemes.

1 INTRODUCTION

Buzz-net is a local area communications network supported by a pair of unidirectional busses to which stations are connected via passive interfaces. Each station is connected to each bus with two taps, a *receive* tap and a *transmit* tap (see Fig.1). The access protocol is a hybrid random access/virtual token protocol. In principle, Buzz-net behaves as a random access network at light load. If there is an upsurge of traffic, all stations switch from random access to controlled access mode. The synchronizing event for this transition is a special "buzz" pattern emitted on the bus (hence the name of Buzz-net). In the controlled mode all backlogged stations take turns in transmitting one packet. When the controlled cycle is completed, random access mode is resumed.

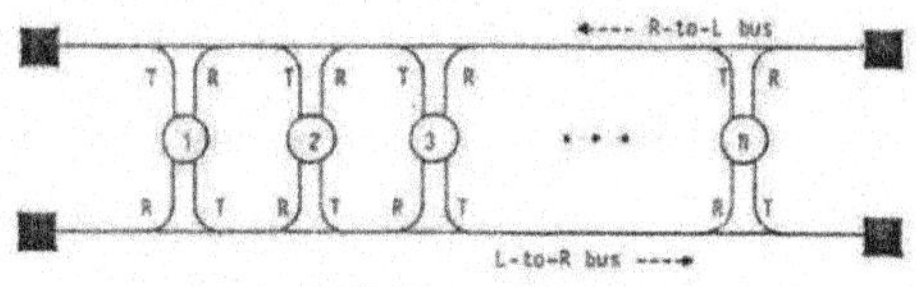

Fig.1 - Buzz-net topology

This research was supported by NSF contract No. ECS-80-20300. Mr.Rodrigues was also supported by the Brazilian Research Council (CNPq) under contract 200.123/79 and by the Federal University of Rio de Janeiro.

The main goal in the design of Buzz-net was to develop a local network that can: yield high throughput efficiency, provide bounded insertion delay, operate in fiber optic environment, run under totally distributed control, survive to processor failures, and allow automatic station insertions/removals.

Several local network architectures can be used in a fiber optics environment, namely, ring, star, and unidirectional bus. However, the ring presents fault tolerance problems in that a single failure may cause the entire system to fail. Furthermore, special procedures must be implemented to recover from token loss (in a token ring) and to remove undeliverable packets. The star architecture generally employs unslotted Aloha techniques and therefore exhibits unbounded delay and poor throughput at high bandwidth x length products.

Based on the above considerations, the unidirectional bus architecture with passive interfaces was selected. Currently, one of the main limitations of the passive bus architecture is the tap insertion loss. Recent advances in fiber optic technology, however, indicate that passive taps could be installed in the field with fractions of db loss. This will permit to support up to one hundred stations on the same bus without repeaters.

Within the family of unidirectional bus architectures we may distinguish two classes: the token (or virtual token) schemes, and the random access schemes. In the first class we mention Express-net [Frat81], D-net [Tsen82], and U-net [Gerl83]. Basically, all of these token schemes can provide good performance in a local fiber optics network environment. Each one of them, however, has some drawbacks. For example, the "folded" topology in Express-net and D-net causes higher attenuation than the twin bus topology since the signal must traverse twice as many taps. In D-net and Fasnet the network fails if the token generator(s) fail. In all schemes a token latency proportional to the end to end propagation delay is suffered at packet insertion. This translates into throughput degradation if only one station has data to send and can transmit only one packet per token.

In the random access family the most popular scheme is CSMA-CD [Metc76]. Although this scheme was initially developed for bidirectional busses, it can be extended to twin unidirectional busses. CSMA-CD eliminates token latency and provides high throughput to a single sending station. However, it shows throughput degradation, unbounded delays, and capture problems in heavy load, multistation situations.

Based on the above trade-offs, the "best of all worlds" appears to be a hybrid random access/token architecture. One such architecture, MAP, was proposed in [Mars81]. That architecture eliminated the latency problem, but did not resolve the single station throughput problem. Furthermore, the folded topology still caused an undesirable extra attenuation in the signal.

Buzz-net, described in this paper, appears to be a more viable hybrid architecture in that it combines many of the advantages of token and random access schemes without suffering of their limitations.

2 THE ALGORITHM

The network can operate in either of two states: random access and controlled access. Initially, a station starts in the *idle* state of the random access mode (see Fig.2). When a packet arrives, the station moves to the *backlogged* state. From this state, transmission of the packet is attempted in random access mode. Namely:

a. if both busses are sensed idle, the station moves to *Random Access Transmission* state. In this state, packet transmission immediately begins on both busses (it is assumed that the sender does not know the relative position of the destination on the bus).

b. if one bus is idle and one is busy, the station moves to *Wait for EOC* state. Here, the station waits for *EOC* (End-of-Carrier) on the busy bus.

c. if both busses are sensed busy or a buzz pattern is sensed, the station moves to the *Buzz-I* state, which is part of the controlled access procedure.

In the *Random Access Transmission* state the station proceeds transmitting on both busses. If, while transmitting, it is interfered by an upstream station (that is, it hears a BOT, Begin-of-Transmission, on one of the busses) it aborts its transmission and moves to *Buzz-I state*. The upstream transmission is allowed to proceed intact. If the transmission is successfully completed, the station moves to *Idle* state.

In the *Wait for EOC* state, when *EOC* is sensed, the station moves to *Random Access Transmission* state. If, while in *Wait* state, the station senses a buzz pattern or it senses both busses busy, it moves to *Buzz-I* state.

While in the random access mode a station with several packets ready for transmission may attempt to send them all in a single train, cycling between *Backlogged* and *Random Access Transmission* states, and thus capturing the channel and locking out the other stations. To avoid capture, a minimum interpacket gap

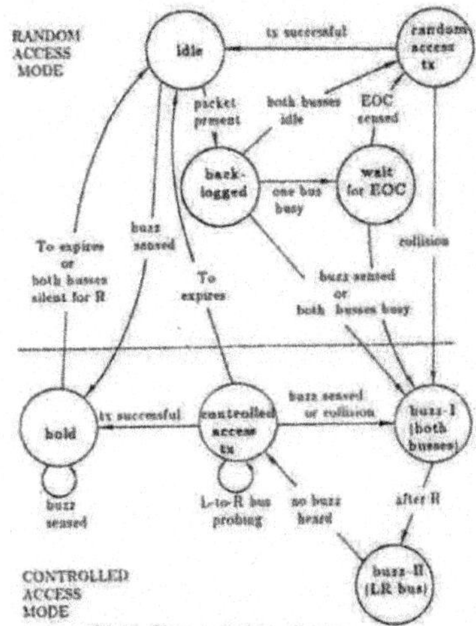

Fig.2 - Buzz-net state diagram

must be observed between any consecutive packet transmissions. This minimum gap, on the order of a station reaction time interval (= delay between detection of *EOC* on the bus and issue of *BOT* by the station), allows downstream stations in the *Wait for EOC* state to detect *EOC* inside a train and, upon collision, force the system to controlled access mode, thus breaking capture.

In *Buzz-I* state a station transmits the buzz pattern on both busses (deferring, of course, to upstream transmissions) for R seconds, where R = round trip delay. Because of deferrals, a station in the buzz state may actually buzz the busses only intermittently (or it may not buzz them at all). After *R* seconds, the station moves to *Buzz-II* state.

In *Buzz-II* state the station buzzes only the Left-to-Right bus, deferring as usual to upstream stations, until it hears no more buzzing on either bus. At this point, the station moves to *Controlled Access Transmission* state. The intermediate *Buzz-II* state guarantees that the leftmost (and *only* the leftmost) station starts the controlled access cycle when all the Right-to-Left buzzing has ceased.

In *Controlled Access Transmission* state each station is allowed to transmit its backlogged packet, and to move to *Hold* state thereafter. Controlled mode transmission is carried out much in the same way as in token networks, except that the Left-to-Right bus must be probed before transmission. That is, a station waits for the Left-to-Right bus to become free. Then it probes this bus by starting transmission of the preamble on it. If it does not hear upstream interference within a reaction time interval, it will then proceed to transmit the remaining preamble also on the Right-to-Left bus, followed by the data packet. If interference is sensed, the station aborts its transmission and retries when the

Left-to-Right bus is free again (i.e. after EOC is sensed). If a buzz is heard, the station moves back to *Buzz-I* state.

Clearly, in the controlled access mode, a "train" of packets is formed from left to right. Backlogged stations are allowed to append their packets to the train in a left to right order. At the end, all stations end up in the *Hold* state.

A time-out To from *Controlled Access Transmission* state to *Idle* state is provided to prevent a new station entering the system from locking up other stations in *Controlled Access Transmission* state. A time-out To is also provided for the *Hold* state for similar reasons. A more detailed description of the recovery procedures when a new station joins the network is given in section 4.

3 BUZZ SIGNAL IMPLEMENTATIONS

The buzz signal is a signal (or event) clearly distinguishable from regular packet flow. If the preamble pattern is uniquely distinguishable even when embedded in other data, then a simple buzz implementation consists of sending a prolonged preamble pattern. Bit stuffing may be required to maintain data transparency.

An alternative buzz implementation which does not require bit stuffing consists of enforcing a minimum gap ΔT between any two consecutive packets on the bus. ΔT is large enough so that a station in buzz mode can fill the gap with a burst of (arbitrary) data. The buzz implementation consists of sending one (or more,for reliability purposes) short burst(s) of unmodulated carrier where the length of a burst is less than the smallest packet size, but large enough to be safely detected by a station. A station recognizes the buzz condition when it detects on either bus the presence of a burst shorter than the minimum packet length. This scheme provides a faster detection of the buzz signal.

In general any method which permits some form of out-of-band signalling is a feasible buzzing method. The best method will most likely depend on interface implementation considerations, and may vary from application to application.

4 NEW STATIONS JOINING THE NETWORK

A newly activated station may join the network at any time. In some cases, the joining process occurs transparently. In other situations, activity of the new station forces a transient phase which adds extra delay to the transmissions in process. However, whatever the case, the new station does not cause permanent disruption of network operation, and the access algorithm automatically absorbs the external interference.

If the network is operating in random mode, the new station is absorbed transparently. If, on the other hand, the network is in controlled access mode the new station will either move to *Buzz-I* state and participate in a new buzzing phase, or it will successfully transmit its packet after both busses are sensed idle. The first situation occurs because a buzz is detected by the new station, or its transmission collides with a transmission of a backlogged station. In the worst case (stations 1 and N participating in the new buzzing phase), an extra $2R$ seconds may be necessary before transmissions are resumed. The second situation develops when the new station senses the right-to-left bus busy due to packet transmission by the next backlogged station situated upstream on that bus. Upon detecting end of transmission, the new station transmits on both busses and move to *Idle* state. Other backlogged stations do not perceive this intrusion and behave normally.

Nevertheless, there is a chance that the new station will keep transmitting in random mode after reaching the *Idle* state. In this case, any station that has previously moved to *Hold* state will eventually time out and move back to *Idle*. Nevertheless, if these stations do not have any packet to transmit, the new station will continue to lock the remaining stations in controlled access mode. The time out To in the *Controlled Access Transmission* state prevents this capture effect.

As shown, the new station joins that active stations gracefully. The extra delay added to controlled access mode delay is in the best case 0, between 0 and $2R$ in most cases, and of the order of To in very unlikely worst case situations.

5 PERFORMANCE EVALUATION

The following performance measures are used to evaluate Buzz-net and compare it with other unidirectional bus schemes:

1. Average insertion delay ID, defined as the interval between the time when the packet moves to the head of the transmitting queue and the time when successful transmission begins. Note that insertion delay is equivalent to queueing delay when threre is only one buffer per station. Insertion delay is evaluated as a function of number of active stations and of offered load. The average is over all stations and over time.

2. Maximum insertion delay MID, over all stations and over time. MID is a function of number of stations and offered load.

3. Heavy load bus utilization $S(i)$, defined as the net bus utilization when i stations are active and have infinite backlog.

The above measures, albeit simple, provide us with useful criteria to decide whether a bus protocol is suitable or not for a given application. For example, interactive and real time applications are particularly sensitive to average and maximum insertion delays. Batch data transfer is mostly affected by bus throughput efficiency.

In our analysis, we assume that stations are uniformly spaced on the bus. Thus, if N stations are present, the propagation delay between two adjacent stations is $s = D/(N-1)$, where D is the end-to-end delay. For simplicity we also assume that detection time δ is much smaller than the round trip delay R, and that the preamble is much shorter than the packet.

We begin by evaluating the bus utilization of Buzz-net at heavy load. Under heavy load conditions the active stations will always conflict again at the end of a controlled phase. Therefore, the activity in the network is a succession of cycles where active stations are served in a Round Robin way, lowest numbered stations first. The diagram in Fig.3 portrays the cyclic pattern when all N stations are active. From Fig.3 we can see that :

$$cycle = NT + 3R + 2a$$

where $T =$ packet transmission time, $T > 2a$. The inequality implies that no packets are successfully transmitted during random mode. The utilization is thus given by:

$$S(N) = \frac{NT}{NT + 3R + 2a}$$

When only i stations are active among N, and as we are interested in evaluating performance under fair conditions, we disregard the packets transmitted during random mode by the rightmost backlogged station. The worst case for the utilization occurs for the following set of active stations: $\{1,2,...,i\text{-}1,N\}$. For this situation and under the fair assumption, $S(i)$ is given by:

$$S(i) = \frac{iT}{iT + 3R + 2(N-i+1)a}, \text{ for } i > 1 \quad (1)$$

If only one station is active, that is $i = 1$, the station can transmit in random access mode since no collisions occur. Thus, we have $S(1) = 1$.

Average insertion delay tends to zero as the offered load goes to zero. In very light load, in fact, a station can immediately transmit, with negligible probability of collision. At heavy load, average insertion delay is closely related to utilization S. Namely, if i is the number of active stations:

$$ID(i) = iT/S(i) - T$$

For intermediate load values, the average insertion delay cannot be evaluated analytically since the lengths of random access and controlled access cycles are now random variables very difficult to characterize. Simulation was used to obtain intermediate load values. The results are shown in Fig.4. If all N stations are active, maximum insertion delay is equal to:

$$MID(N) = (N-1)T + 3R + 2a \text{ , at heavy load}$$

We now compare Buzz-net performance to that of other unidirectional bus schemes. First, we consider the token based schemes, namely, Express-net, D-net, and Fasnet. For Express-net [Frat81] and D-net [Tsen82] the throughput and delay expressions are identical for both schemes and, under the previous simplifying assumptions, are given by the following formulas:

$$S(i) = \frac{iT}{iT + R}, \text{for } i > 1 \quad (2)$$

$$ID(i) = \begin{cases} R/2 \text{ , at light load} \\ (i-1)T + R \text{ , at heavy load} \end{cases}$$

$$MID(i) = (i-1)T + R$$

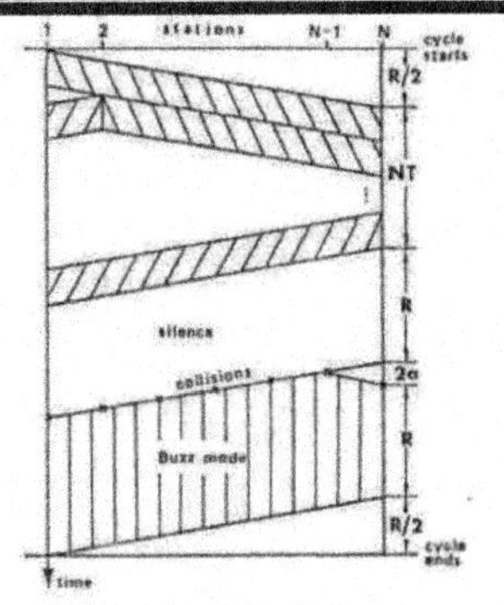

Fig.3 - Heavy load cycle diagram

In Fasnet, following the derivations in [Limb82], and neglecting any overhead in the transmission slot, the expressions are given by:

$$S(i) = \frac{iT}{(i+1)T + R}, \text{ for } i \geq 1 \quad (3)$$

$$ID = \begin{cases} R/2 + T/2 \text{ , at light load} \\ iT + R \text{ , at heavy load} \end{cases}$$

$$MID(i) = iT + R$$

Comparing (1) and (2) we can see that Express-net and D-net have slightly better utilization than Buzz-net for all values of i, except $i = 1$. For the case where $NT >> 3R$, all three nets have same asymptotic throughput of 1. If we compare the utilization of Buzz-net and Fasnet, we can see that Buzz-net performs better for $i = 1$ and for the cases where $T > 3R - 2(i-2)a$, $i > 1$.

If we compare the delay at light load we see that all nets perform worse than Buzz-net. As for the maximum insertion delays, the ranking is the same as for bus utilizations.

One important advantage of Buzz-net over all token networks is its very high utilization in the case of single station with heavy backlog. In Buzz-net, the single station can transmit an uninterrupted sequence of packets achieving the maximum possible throughput on the busses. In the other schemes, subsequent packets from a single station are spaced by at least one round trip delay.

If we now compare Buzz-net with CSMA-CD, we find that their performance is similar in very light load: for both, insertion delay tends to zero. The performance is the same also for the single active station with heavy backlog, where both Buzz-net and CSMA-CD achieve maximum bus utilization.

As load and number of stations increase, however, several problems arise in CSMA-CD. First, throughput degrades dramatically because of repeated collisions. Secondly, maximum insertion delay becomes unbounded.

Low-Noise Fiber Optics Receiver with Super-Beta Bipolar Transistors

Siegfried G. Knorr,* Osman Kaldirim,* and C. Yeh*

A low-noise wideband optical fiber receiver has been successfully designed using super-beta bipolar transistors (BJTs) at the front end. Even with commercially available super-beta devices, which are not optimized for our application, the obtainable input sensitivity for medium- and high-bandwidth optical receivers is comparable or superior to the best FET design. To demonstrate this concept, a 10-MHz analog receiver was built with a super-beta BJT at the input stage. This receiver achieved an expected average input noise current density of less than 0.4 pA/$\sqrt{\text{Hz}}$ over the full bandwidth for a transresistance of 500 kΩ. Detailed design procedures are given in this paper. The noise characteristics of a 50-MHz receiver using super-beta BJTs are also obtained.

The design of suitable fiber optics receivers has reached relative maturity,[1,2] even though requirements differ significantly from the more common wideband amplifier configurations that operate optimally with a given, relatively low, source impedance. Fiber optics receiver inputs are essentially driven from a current source, i.e., from a

* Electrical Sciences and Engineering Department, School of Engineering and Applied Science, University of California, Los Angeles, California.

Fiber and Integrated Optics, Volume 1, Number 4
0146-8030/78/0531-0369$02.00/0

reversed biased avalanche photodiode (APD) or a reversed biased PIN photodiode. In such an application, the total effective input capacitance must be kept at a minimum for a given sensitivity–bandwidth combination if a low-noise amplifier performance is to be achieved. In spite of existing low-noise receiver designs, it is important to investigate the problem of how to improve the design of low-noise fiber optics receiver modules further. We shall show that by properly designing the input amplifier device (the front end) and by using super-beta BJTs, we may simultaneously achieve a −56-dBm minimum detectable power (MPD) for a 10-MHz bandwidth and a maximum signal-to-noise ratio (SNR) of 55 dB for PIN-type photodiodes. Higher sensitivities may be obtained by using APDs instead of PIN-type photodiodes. Also, if desirable, the maximum SNR of the design example given here may be improved by lowering the effective input sensitivity.

We will discuss the merits of super-beta BJTs over FETs. Then, starting with the basic design considerations, we will give a detailed numerical example for a 10-MHz bandwidth receiver. Finally, we will compare the measured noise performance of our receiver with the calculated results. Excellent agreements were found between the measured and calculated data.

Comparison of the Merits of FET and BJT as Front-End Components

Since the gate current of FETs can be very small, which implies that the associated noise current is also very small, FETs are presently used in several commercially available optical receivers. On the other hand, since the conventional BJTs have common-emitter current gains in the order of $\beta_0 = 50$ to 300, the base bias current, and the resulting current noise, can be appreciable for typical collector current levels of 100 μA to 1 mA. It has been pointed out by Roko[3] that to reduce the base current noise component, the BJT must be biased at very low collector currents. Therefore, at such bias conditions, the g_m/C ratio for a BJT is quite poor compared with that of a FET device. At higher frequencies, the frequency-dependent noise component will be the dominant noise source for both the FET and the BJT. However, for wideband receiver systems, the FET exhibits a larger frequency-dependent noise component than that of the BJT. The basic reason is that BJTs have more favorable g_m/C ratios for typical bias points, as required for high-frequency operation.[4] As a result, the BJT may be superior in very wideband receivers, even though its base current noise component is higher at low frequencies than the gate current noise component of the FET. It is possible, however, that the *average* overall noise current density can be lower in a BJT for very wideband receivers if it is biased at a relatively high collector current.[4]

The underlying philosophy for choosing a two-stage shunt–shunt feedback amplifier configuration (Fig. 1) is that the magnitude and phase response, and therefore the transient response, can be predicted by design. The amplifier can be made unconditionally stable for a given feedback resistance, and, more importantly, it requires no compensation to achieve stability. The immediate benefit is that the amplifier is not slew-rate limited, that is, the small-signal and large-signal responses are essentially identical. Because the amplifier exhibits a large loop gain, the effective distortion, which can be kept very small, is usually determined by the output amplifier stage driving the load resistance.

Yet another advantage of the two-stage feedback amplifier is its predictable dc-output level. This will be important if the amplifier is to be dc-coupled. Contrary to FETs, the base-emitter bias potential V_{BE} of a BJT is a well-defined quantity, and its temperature coefficient is typically -2 mV/°C. For large loop gains, the change in V_{BE} will appear at the output of the feedback amplifier (collector of Q_5). If a more stable dc-output level is required, Q_1 may be replaced by a differential configuration.

Let us now consider the contribution of the various noise sources for a BJT and FET optical receiver design. In the following equations, we have

C_{tot} = total effective input capacitance of the input device

$$C_{tot}(\text{BJT}) = C_{in} + C_{DI} \tag{1a}$$

$$C_{tot}(\text{FET}) = C_{gs(\text{eff})} + C_{gd(\text{eff})} + C_{DI} \tag{1b}$$

C_{in} = total effective input capacitance of the BJT

C_{DI} = effective photodetector capacitance, including stray capacitance.

$$T_{tot}(\text{BJT}) = (r_{\pi}//R_S)C_{tot}(\text{BJT}) \tag{2}$$

= effective input time constant of the BJT amplifier stage.

$T_{tot}(\text{FET}) = R_S C_{tot}(\text{FET})$ (3)
= effective input time constant of the FET amplifier stage.

$V_T = kT/q$

$k = 8.62 \times 10^{-5}$ ev/°K (Boltzmann constant)

$q = 1.6 \times 10^{-19}$ As (electronic charge)

T = temperature in degrees Kelvin

R_S = source resistance of the input amplifier stage

g_m = transconductance

$g_m(\text{BJT}) = I_C/V_T$

I_C = collector bias current

r_b' = total effective base bulk resistance of BJT

$g_m(\text{FET}) = \dfrac{I_{DSS}}{V_P}\left(1 - \dfrac{V_{GS}}{V_P}\right)$ in the saturation region

I_{DSS} = drain current for the gate-to-source voltage if $V_{GS} = OV$

V_P = pinch-off voltage of FET

$r_\pi = \beta_0 (V_T/I_C)$

β_0 = common-emitter current gain of BJT at low frequencies

$\omega = 2\pi f$

P = constant (to account for FET device imperfections)

The frequency-dependent mean-square input noise current spectral density for a BJT in the common-emitter configuration is:

$$\frac{\overline{i_i^2}}{\Delta f}(\text{BJT}) = 2q\frac{I_C}{\beta_0^2}\left[\left(1 + \frac{r_\pi}{R_S}\right)^2 (1 + \omega^2 T_{tot}^2)\right] \tag{4a}$$

Since $\omega^2 T_{tot}^2 \gg 1$ for higher ω, Eq. (4a) reduces to

$$\frac{\overline{i_i^2}}{\Delta f}(\text{BJT}) \approx \left(\frac{\omega C_{tot}}{g_m}\right)^2 2qI_C \tag{4b}$$

or

$$\frac{i_i^2}{\Delta f}(\text{BJT}) \approx \frac{2q}{I_C}(\omega C_{tot} V_T)^2 \tag{4c}$$

The corresponding frequency-dependent relationship for a FET device, due to channel resistance in the common-source configuration, is:

$$\frac{\overline{i_i^2}}{\Delta f}(\text{FET}) = \frac{4kT}{R_S}\frac{0.7P}{g_m R_S}(1+\omega^2 T_{tot}^2) \tag{5a}$$

Again, with $\omega^2 T_{tot}^2 \gg 1$

$$\frac{\overline{i_i^2}}{\Delta f}(\text{FET}) \approx 0.7P\frac{4kT}{g_m}(\omega C_{tot})^2 \tag{5b}$$

Usually, for commercially available devices, $1<P<4$. To compare the frequency-dependent noise current densities for the two devices, we shall assume that (1) the total input capacitance of the BJT and the FET design is the same, since the capacitance is partly masked by the fixed photodiode capacitance C_{DI}, and (2) $P=1$, i.e., there are no imperfections in the FET device. Taking the ratio of Eqs. (4b) and (5b), one can show that the mean-square noise current density for a BJT device is 1.4 times less than that for a FET device for the same transconductance. Furthermore, the BJT device offers a number of advantages over the FET other than the inherent lower noise current densities that can be achieved at higher frequencies. These are: (1) BJT devices are easily biased and dc-coupled to subsequent amplifier stages, (2) super-beta BJTs exhibit very low $1/f$-noise components, and (3) the wideband noise performance of a BJT device is very close to the theoretical value.[5]

Design Considerations for the BJT Device

The basic BJT amplifier configuration is shown in Fig. 1. Q_1 is a super-beta BJT, and Q_2 acts as a grounded base stage, to reduce the Miller effect that is necessary for this circuit. Q_3 buffers the load resistance R_1 and reduces the effective source resistance of the differential amplifier Q_4, Q_5. The differential amplifier provides extra loop gain and assures the proper phase for the feedback signal. Q_5 also acts as a grounded base stage, so that the Miller effect for its load resistance R_5 is minimized. Although the usable gain of the differential amplifier configuration is only one-half of a single-ended stage (output is taken only from one output, i.e., from Q_5) this scheme is still preferable because of the simplicity in biasing. The source resistance R_S of Q_1 is equivalent to the transresistance of the amplifier at large loop

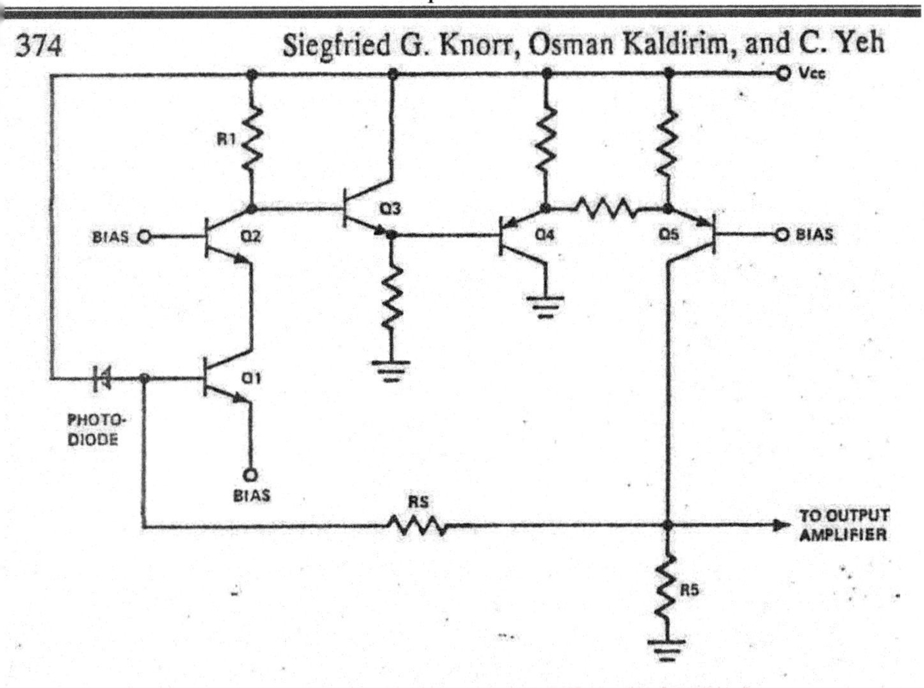

Fig. 1. Basic Circuit Configuration of BJT Fiber Optics Receiver.

gains, and R_S is chosen to suit desirable sensitivity–bandwidth combinations.

The equivalent input mean-square noise current spectral density of the amplifier may be approximated by

$$\frac{\overline{i_{iT}^2}}{\Delta f} \approx \frac{\overline{i_{i1}^2}}{\Delta f} + \frac{\overline{i_{i2}^2}}{\Delta f} + \frac{\overline{i_{i3}^2}}{\Delta f} + \frac{\overline{i_{i4}^2}}{\Delta f} \tag{6a}$$

$$\frac{\overline{i_{i1}^2}}{\Delta f} = \frac{4kT}{R_S} \tag{6b}$$

= noise component due to source resistance

$$\frac{\overline{i_{i2}^2}}{\Delta f} = 2qI_{B1} \tag{6c}$$

= noise component due to dc base current at the bias point

$$\frac{\overline{i_{i3}^2}}{\Delta f} \approx 2q\frac{I_{C2}}{\beta_0^2}\left[\left(1+\frac{r_{\pi 1}}{R_S}\right)^2\left(1+\omega^2 T_{\text{tot}}^2\right)\right] \tag{6d}$$

= noise component due to dc collector current at the bias point

$$\frac{\overline{i_{i4}^2}}{\Delta f} \approx \frac{4kT}{R_S^2}\left(r_{b1}' + \frac{1}{2g_{m1}}\right) \tag{6e}$$

= noise component due to base bulk resistance

In Eqs. (6d) and (6e) it has been assumed that $r_{b1}' \ll r_{\pi 1}$, which holds up well for this design. The noise component given by Eq. (6e) is usually negligible for large values of R_S. Equation (6a) does not include any additional noise components caused by the load resistance R_1 and the base bias current I_{B3} of Q_3. When properly designed, these noise components become negligible compared with other noise sources. For example, if

$$I_{B5} < 5 I_{C1} \tag{7}$$

and

$$\frac{4kT}{R_1} < 5(2qI_{C1}) \tag{8}$$

then their noise current contributions are less than 20% of the total noise current. It should be noted that if the succeeding gain stage adds significantly to the equivalent input noise current, a shunt-feedback amplifier configuration should be used instead of Q_2 and Q_3, as proposed by Hullett et al.[6] However, the shunt-feedback amplifier has a less desirable transient response and the obtainable bandwidth is reduced by the current gain A_i. In summary, we should aim for as large a transresistance as possible for the first amplifier stage, so that succeeding gain stages will add only an insignificant amount of noise to the equivalent input noise current.

Typical Numerical Example: Design of a 10-MHz Bandwidth Receiver

A super-beta input stage will now be designed with the following characteristics: (1) a bandwidth of $f_A = 10$ MHz, (2) a transresistance of $R_S = 500$ kΩ, to achieve a high sensitivity. The complete circuit diagram is shown in Fig. 2. The selected super-beta transistor for this application is a monolithic device by RCA, type CA 3095E. For the bias condition $I_{C1} = 200\ \mu$A, $\beta_0 = 1{,}500$ and $C_{in} = 3.0$ pF. The photodiode is a Hewlett-Packard device, type HP 5082-07, and with $C_{DI} = 4$ pF, $C_{tot} = 7$ pF. A single power supply of +12 V is used. The total linear dynamic range of the output amplifier is 1.0 V_{pp} (volts peak to peak) into a load resistance of $R_L = 50\ \Omega$.
To reduce wideband noise, a roll-off at

$$f_B = \sqrt{2} f_A \tag{9}$$

is added in the output amplifier, where $f_A = 10$ MHz is the overall

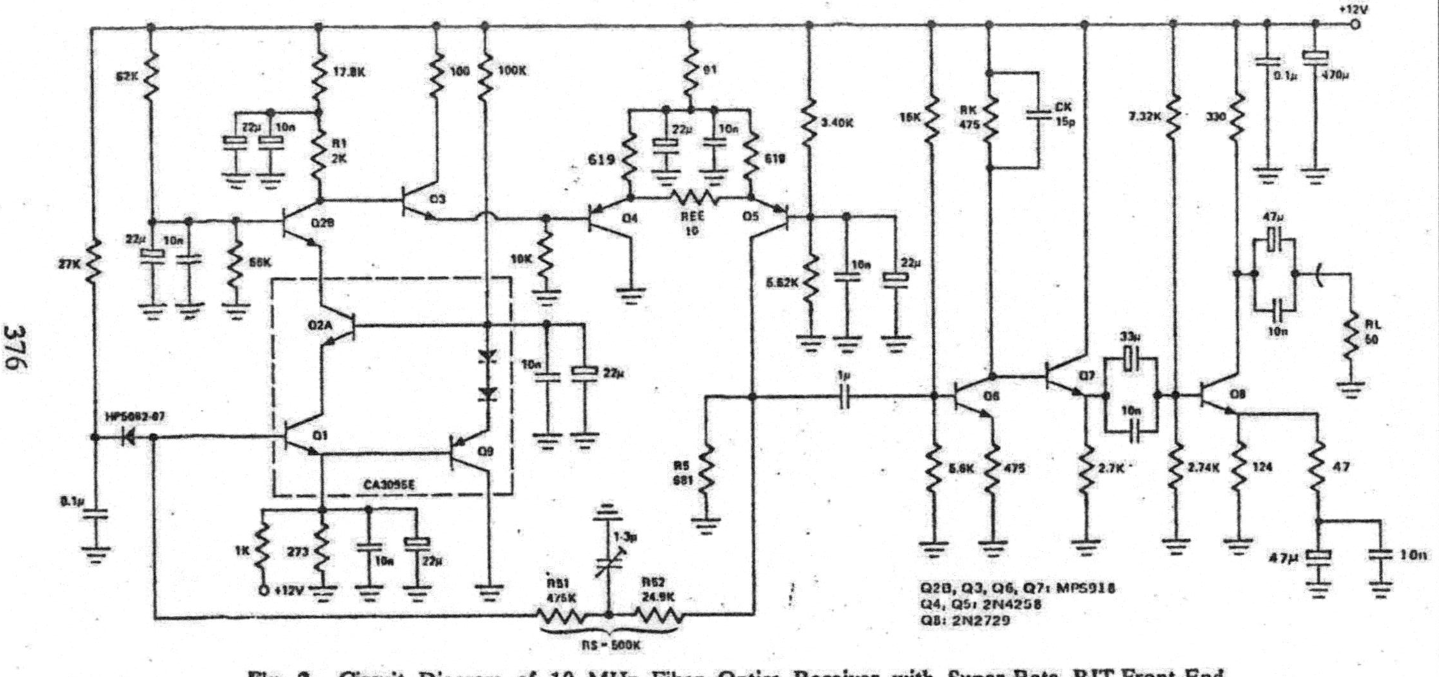

Fig. 2. Circuit Diagram of 10 MHz Fiber Optics Receiver with Super-Beta BJT Front-End.

bandwidth. The closed-loop bandwidth of the amplifier input stage must be

$$f_C = \sqrt{2} f_A \tag{10}$$

Hence,

$$f_A \approx \frac{1}{\sqrt{1/f_B^2 + 1/f_C^2}} \tag{11}$$

for a relatively "well-behaved" transient response, that is, one with little overshoot and ringing. Since the transient response of the amplifier can be controlled in this design, one can always assure a "well-behaved" transient response.

The input feedback amplifier stage of Fig. 1 has three important time constants T_1, T_2, and T_3:

$$T_1 = (r_\pi // R_S) C_{tot} \tag{12}$$

$$T_2 \approx R_1 C_2 \tag{13}$$

$$T_3 \approx R_5 C_3 \tag{14}$$

where $C_2 = 3$ pF and $C_3 = 5$ pF are the effective capacitances at the load resistors R_1 and R_5, respectively. In Eqs. (13) and (14) it has been assumed that the input resistances of Q_3 and Q_6 are negligible.

With the well-known feedback relationship, the closed-loop transimpedance is

$$Z_C(s) = \frac{Z_O(s)}{1 + M(s)} \tag{15a}$$

and

$$M(s) \approx \frac{1}{R_S} Z_O(s) \tag{15b}$$

where $Z_O(s)$ is the open-loop transimpedance and $M(s)$ is the loop gain of the feedback amplifier. If

$$\frac{1}{2\pi T_1} \ll f_A \tag{16}$$

which holds up well in this design, the low-frequency loop gain $M(O) \gg 1$, and both $Z_O(s)$ and $M(s)$ are well described by a single-pole frequency response $f_1 = 1/(2\pi T_1)$ within the useful passband f_A of the amplifier. Equation (15a) may then be approximated by

$$Z_C(s) \approx R_S \tag{15c}$$

In other words, Eq. (15c) ignores any contributions of the two higher frequency poles $f_2 = 1/(2\pi T_2)$ and $f_3 = 1/(2\pi T_3)$ to interfere with the major pole roll-off $f_1 = 1/(2\pi T_1)$ within the passband f_A of the feedback amplifier.

The required low-frequency loop gain $M(O)$ is then simply determined by Eqs. (10) and (12),

$$M(O) \approx f_C(2\pi T_1) \tag{17}$$

With

$$r_{\pi 1} = V_T/I_{C1} \tag{18a}$$

and

$$V_T = kT/q \approx 26 \text{ mV} \tag{18b}$$

at room temperature, $T_1 \approx 0.98$ μsec and $M(O) \approx 87$. The open-loop transresistance from Eq. (15b) is $Z_O(O) \approx 43.5$ MΩ, which may be equated to the individual voltage gains as follows:

$$Z_O(O) = R_{in(eff)} A_{V1} A_{V2} \tag{19}$$

$$R_{in(eff)} = r_{\pi 1} // R_S \tag{20}$$

$$A_{V1} \approx g_{m1} R_1 \tag{21}$$

$$A_{V2} \approx \frac{1}{2} \frac{g_{m5} R_5}{\left(1 + g_{m5} \frac{R_{EE}}{2}\right)} \tag{22}$$

The upper limits of R_1 and R_5 are determined by Eqs. (13) and (14) and by the desired phase margin of the open-loop transimpedance

$$Z_O(s) = \frac{Z_O(O)}{(1+sT_1)(1+sT_2)(1+sT_3)} \tag{23}$$

If the amplifier requires a phase margin of $\phi_m = 45°$, which results in a maximum flat amplitude response, the total phase contribution of T_2 and T_3 at f_C cannot exceed $\phi_x = 45°$, since T_1 is dominant and already contributes nearly to 90° phase shift at f_C; that is,

$$T_1 \gg T_2, T_3 \tag{24}$$

or

$$\phi_m|_{fc} = 180° + \phi_t|_{fc} = 45° \tag{25}$$

in which

$$\phi_i|_{fc} = -\arctan \sum_{i=1}^{i=3} \omega_C T_i \tag{26}$$

A suitable choice is $R_1 = 2\ \text{k}\Omega$ and $R_5 = 680\ \Omega$. From Eqs. (13) and (14) $T_2 = 6$ nsec and $T_3 = 3.4$ nsec. With $T_1 = 0.98\ \mu$sec, $\phi_i|_{fc} \approx -134.2°$ and $\phi_m \approx 45.8°$.

The voltage gain A_{V2} may now be calculated from Eqs. (20) and (21). One has $A_{V2} = 20.25$. What remains is to find a suitable selection of g_{m5} and R_{EE}, as dictated by Eq. (22). The signal swing across R_5 must be at least ± 0.5 V, assuming that the output amplifier gain $A_{VO} = 1$. The voltage drop across R_5 was chosen to be 3.3 V. Although this value is higher than necessary, it was chosen to accommodate large dc voltages originating from the photodiode for high input signal levels and low modulation indexes without cutting Q_5 off. With $I_{C5} \approx V_{R5}/R5$, g_{m5} and R_{EE} may be computed. The remaining resistor values to be determined are for dc bias setup, and their calculations are straightforward.

It should be noted that any distortion from the differential amplifier Q_4, Q_5 is reduced by the loop gain $M(s)$ of the amplifier. Since $M(O) \approx 87$, the effective distortion will be very small for the feedback amplifier configuration. For most practical applications, the dominant source of distortion is to be found in the output amplifier. Its distortion terms can be kept small by increasing the quiescent bias current for a given signal swing and by adding emitter degeneration. Both techniques have been employed in the present circuit.

The added roll-off in the output amplifier at $f_B = 14.1$ MHz consists of R_K and C_K plus the effective capacitance C_{Ka} at the collector of Q_6;

$$f_B = \frac{1}{2\pi R_K (C_K + C_{Ka})} \tag{27}$$

Using Eqs. (6a) to (6e), the input mean-square noise current density for an ideal "brickwall" response with an upper frequency bandwidth of $f_A = B$ can be obtained:

$$\left.\frac{\overline{i_{iT}^2}}{\Delta f}\right|_B \approx \frac{\overline{i_{i1}^2}}{\Delta f} + \frac{\overline{i_{i2}^2}}{\Delta f} + \frac{\overline{i_{i3}^2}}{\Delta f} + \frac{\overline{i_{i4}^2}}{\Delta f}$$

With $r_{b1}'=1{,}000\,\Omega$,

$$\left.\frac{\overline{i_{iT}^2}}{\Delta f}\right|_B=(3.32+4.3+21+0.007)\,10^{-26}\ \mathrm{A^2/Hz}$$

$$\left.\frac{\overline{i_{iT}}}{\Delta f}\right|_B=5.35\times10^{-13}\ \mathrm{A/\sqrt{Hz}}$$

As indicated before, the noise contribution due to base bulk resistance r_{b1}' is negligible for a high value of source resistance R_S. The total mean-square input noise is given by

$$\overline{i_{iT}^2}=\int_0^B\frac{\overline{i_{iT}^2}}{\Delta f}\,df \tag{28a}$$

Performing the integration

$$\overline{i_{iT}^2}\approx\left[\frac{4kT}{R_s}+2qI_{b1}+2q\frac{I_{C2}}{\beta_0^2}\left(1+\frac{r_{\pi1}}{R_S}\right)^2+\frac{4kT}{R_S^2}\left(r_{b1}'+\frac{1}{2g_{m1}}\right)\right]B$$
$$+\left[\frac{2qI_{C2}}{\beta_0^2}\left(1+\frac{r_{\pi1}}{R_S}\right)^2(2\pi BC_{tot})^2\right]\frac{B}{3} \tag{28b}$$

Taking $B=10^7$ and assuming an ideal "brickwall" filter response,

$$\overline{i_{iT}^2}\approx(3.32\times10^{-26}+4.3\times10^{-26}+6\times10^{-29}+7.1\times10^{-29})B$$
$$+[21\times10^{-26}]\frac{B}{3}$$
$$\overline{i_{iT}^2}=14.6\times10^{-19}\ \mathrm{A^2/Hz}$$

Put another way, the effective total input noise current is

$$\overline{i_{iT}}=1.2\,\mathrm{nA}\qquad\text{for}\quad B=10\,\mathrm{MHz}.$$

From the above results, we can see that the frequency-dependent noise source is the dominant one, followed by the base current noise component of the super-beta BJT. The advantage of a large current gain now becomes apparent. If we compare its noise contribution to the source resistance R_S, the base current produces as much noise as a 386-kΩ resistor. All three major noise components are of the same order, that is, Q_1 is biased at optimum base current.[4] However, if the bandwidth were increased, the frequency-dependent noise source would become dominant at higher frequencies.

The average value of the noise current spectral density over a bandwidth B is:

$$\frac{\overline{i_{iT}}}{\Delta f}(\mathrm{av}) = \frac{1}{\sqrt{B}}\overline{i_{iT}} \tag{29}$$

The noise equivalent power

$$\mathrm{NEP} = \frac{1}{R}\frac{\overline{i_{iT}}}{\Delta f}(\mathrm{av}) \tag{30}$$

where $R \approx 0.5$ A/W, the typical photodiode responsivity.
Finally,

$$\mathrm{MDP} = \mathrm{NEP}\sqrt{B} \tag{31}$$

$$\mathrm{MDP}\,(\mathrm{dBm}) = 10\log_{10}\frac{\mathrm{MDP}}{1\ \mathrm{mW}} \tag{32}$$

For the present numerical example, the average value of the noise current spectral density over a bandwidth of $B = 10$ MHz is about 0.38 pA/$\sqrt{\mathrm{Hz}}$, NEP$\approx 0.76 \times 10^{-12}$ W/$\sqrt{\mathrm{Hz}}$, and MDP$\approx$2.4 nW($-$56.2 dBm).

The total mean-square output noise voltage for a given bandwidth B ("brickwall" filter response) is

$$v_{oT}^2 = \int_0^B Z_C^2(s) A_{VO}^2 \frac{\overline{i_{iT}^2}}{\Delta f}\, df \tag{33}$$

$$v_{oT}^2 \approx R_S^2 A_{VO}^2 \int_0^B \frac{i_{iT}}{\Delta f}\, df \tag{34}$$

In Eq. (34) it has been assumed that (a) $Z_C(s) \approx R_S$ and (b) $A_{VO} \approx$ constant for a bandwidth of $B = 10$ MHz. Assumption (a) can be tested by using Eqs. (15) and (23). This assumption greatly simplifies to evaluate the integral of Eq. (33), and the resulting error is small provided $Z_C(s)$ is about constant to the upper band edge frequency $f_A \equiv B$ of the amplifier. In this two-stage feedback amplifier configuration, the condition $Z_C(s) \approx R_S$ up to $f = B$ can readily be met. Assumption (b) can be met by providing a larger bandwidth f_0 for the output amplifier section; for example, $f_0 \geq 2 f_C$.

Using the result of Eq. (28b), and with an output amplifier voltage gain of $A_{VO} = 1$,

$$\overline{v_{oT}} \approx 0.6\ \mathrm{mV\ (rms)}$$

The linear dynamic range of the output amplifier is 1 V_{pp} or 0.354

V_{rms}, so that the obtainable signal-to-noise ratio is

$$\text{SNR} = 20 \log_{10} \frac{\text{max. linear output voltage swing in rms}}{\text{total output noise voltage in rms}} \tag{35}$$

$$\text{SNR} = 55.4 \text{ dB}$$

The obtainable value of SNR can be improved if one is willing to trade the SNR for input sensitivity. For example, if $A_{VO} = 0.5$ instead of $A_{VO} = 1$, the SNR can be increased to 61.4 dB. However, the input sensitivity decreases by the same amount, from $\overline{i_{iT}} = 1.2$ nA to $\overline{i_{iT}} = 2.4$ nA.

Measured Performance of the Receiver Module

A 10-MHz bandwidth PIN-diode optical receiver module was built according to the above design. The noise performance of this module is shown in Fig. 3, where the measured values of the input noise current density are presented as a function of frequency. The measured values are within 2% of the theoretically predicted ones, but the difference cannot be seen in Fig. 3 due to the finite resolution of the vertical scale. The results also confirm that even for moderate bandwidths, the super-beta BJT can be a very suitable low-noise device for optical communication receivers.

Finally, in Fig. 4, the transient response of the optical receiver is shown. As predicted from the magnitude-phase response, there will be some finite overshoot and ringing in the transient response for a feedback amplifier with a 45° phase margin. The overshoot may be reduced drastically by allowing for more phase margin. The maximum flat amplitude response (45° phase margin) will yield ideally 4.3% of overshoot, whereas the maximum flat envelope delay response (60° phase margin) will yield 0.43% of overshoot. These numbers actually hold true only for an ideal two-pole transfer function. In our design, there is one major pole at $f_1 = 1/2\pi T_1$, and a number of very high-frequency transmission zeroes. There will be, therefore, some finite interaction with the two high-frequency poles of interest, $f_2 = 1/2\pi T_2$ and $f_3 = 1/2\pi T_3$. For completeness, a narrow pulse response for a pulse width of 70 n sec is shown in Fig. 5.

For higher bandwidth optical receiver systems, the super-beta BJT offers advantages similar to those in a FET design. This can be demonstrated theoretically for a receiver with $B = 50$ MHz bandwidth.

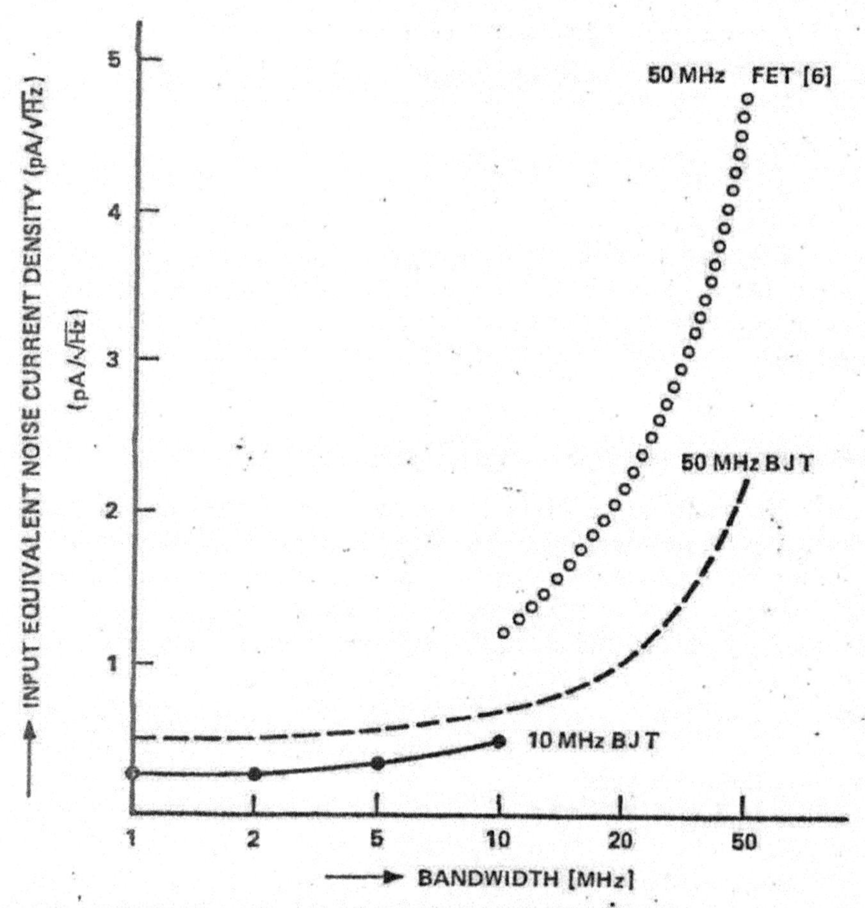

Fig. 3. Equivalent Input Current Noise Density against Bandwidth.
Straight Line: Computed and Measured Values of 10 MHz Super-Beta BJT Fiber Optics Receiver. The dots represent the measured values.
Dashed Line: Computed Values of 50 MHz Super-Beta BJT Fiber Optics Receiver.
Dotted Line: Computed Values of 50 MHz FET Fiber Optics Receiver [6].

For this example, it is advisable to increase the collector current of Q_1 from 200 μA to 500 μA in order to provide additional loop gain. The immediate effect of the collector current increase is the increase of C_{tot} from 7 pF to 10.5 pF. Also, the differential amplifier Q_4, Q_5 for a 50-MHz receiver should be replaced by a single amplifier stage, which increases the loop gain by a factor of 2. The source resistance R_S is decreased from 500 kΩ to 100 kΩ, which helps to reduce the loop gain requirement. Furthermore, $\beta_0 = 1{,}300$ for $I_C = 500$ μA. Under these

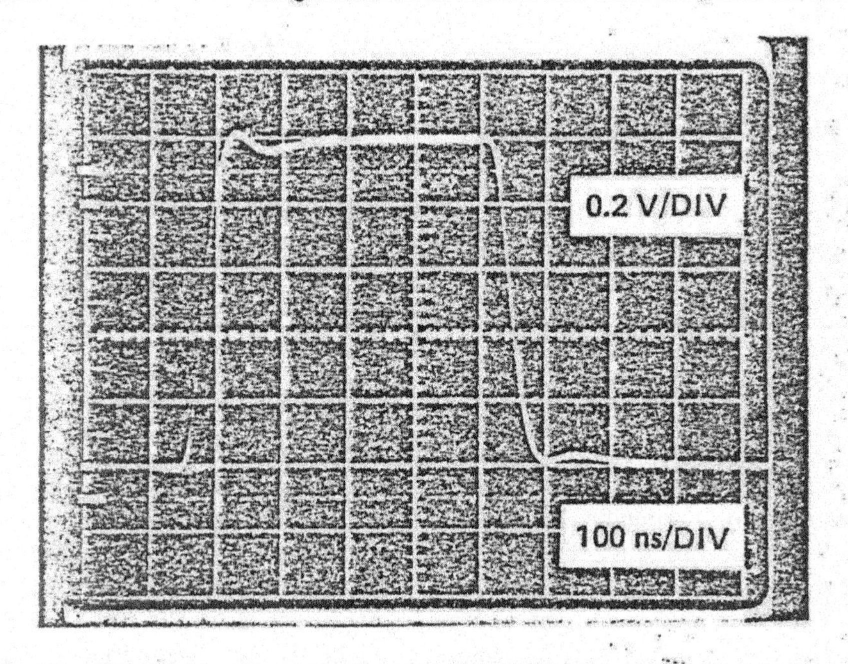

Fig. 4. Transient Pulse Response of BJT Fiber Optics Receiver.

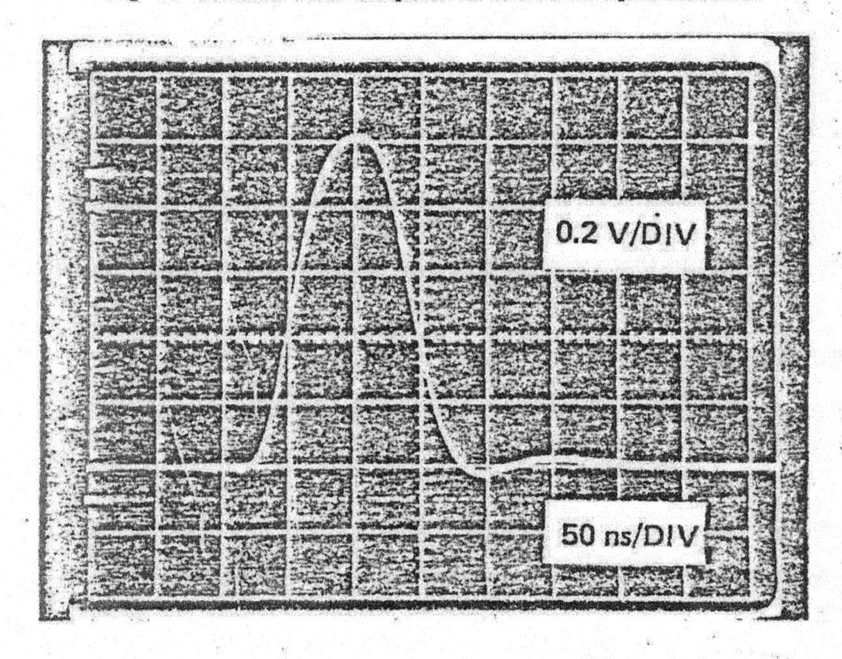

Fig. 5. Narrow Pulse Width (70 ns) Transient Response of BJT Fiber Optics Receiver.

conditions, the average noise current density over 50-MHz bandwidth is

$$1.36\ \mathrm{pA}/\sqrt{\mathrm{Hz}} \qquad \mathrm{NEP} \approx 2.8 \times 10^{-12}\ \mathrm{W}/\sqrt{\mathrm{Hz}}$$

$$\mathrm{MDP} \approx 19.2\ \mathrm{nW}\ (-47.2\ \mathrm{dBm})$$

The computed input noise current spectral density for the 50-MHz receiver is also shown in Fig. 3, by the dashed line. Comparing this curve to another analog receiver design using a high-performance JFET[6] (dotted line) shows that the super-beta BJT is superior in performance.

Conclusion

An optical communication receiver, with a super-beta BJT, has been designed for a specific bandwidth and input sensitivity. We have demonstrated the feasibility of building a low-noise PIN-diode optical receiver, using a super-beta BJT, that yields comparable or superior noise performance compared with a FET front-end design. At 10-MHz bandwidth, the MDP for the super-beta BJT design with a PIN-type photodiode is −56 dBM, and the maximum SNR (limited by nonlinearity) is 55.4 dB. With an APD, which is a lower capacitance device and possesses avalanche current gain, an even better MDP and maximum SNR can be expected. The optical receiver presented here possesses a predictable transient response, and the receiver requires no compensation to obtain stability. Excellent large signal behavior, stable operating points, and very low distortion are additional features inherent in this design. If present super-beta BJTs are optimized for this application it is estimated that even better noise performance can be obtained.

Acknowledgments

This research was partly supported by the Office of Naval Research and by the UCLA Academic Senate Grant No. 3235.

References

1. C. Yeh, "Advances in Communication Through Light Fibers," *Advances in Communication Systems: (Theory and Applications)* (Academic Press, New York, 1975), vol. 4; I. Jacobs and S. E. Miller, "Optical Transmission of Voice and Data," *IEEE Spectrum* 14 (2): 32–41 (1977).

2. S. D. Personick, "Receiver Design for Digital Fiber Optic Communication Systems, I & II," *Bell System Tech. J.* 52: 843–886 (1973).
3. G. H. S. Rokos, "Optical Detection Using Photodiodes," *Opto-electronics* 5: 351–366 (1973).
4. J. E. Goell, "Input Amplifiers for Optical PMC Receivers," *Bell System Tech. J.* 53: 1771–1793 (1974).
5. RCA Solid State Databook on Linear Integrated Circuits and MOS Devices, SSD-201B, pp. 189–198 (1974).
6. J. L. Hullet and T. V. Muoi, "A Feedback Receiver Amplifier for Optical Transmission Systems," *IEEE Trans. Commun.*, Com-24, no. 10, pp. 1180–1185 (October 1976).

9

\COUSTIC WAVES

lathematically speaking, acoustic waves satisfy the scalar wave equation while lectromagnetic waves satisfy the vector wave equation. One analytic way of olving the vector wave equation is to reduce it to a scalar wave equation. onsequently, the techniques used to analytically solve many electromagnetic ave problems can be used to solve a number of acoustic wave problems. In this hapter, we present the mathematical solutions of a few acoustic wave problems:

- Scattering by elliptical fluid cylinders
- Diffraction by penetrable oblate spheroids
- Scattering by penetrable prolate spheroids
- Scattering by liquid-coated prolate spheroids
- Diffraction of sound waves by a moving fluid cylinder
- Reflection and transmission of sound waves by a moving fluid layer
- Finite-element approach to solve acoustic waveguide problems

he abstracts for the papers in this chapter follow.

.1 Elliptical cylindrical liquid lens. (Paper 9-1)

he problem of focusing a pressure wave by a thick, elliptical cylindrical liquid ens has been treated in a rigorous manner. The size of the lens is on the order of r slightly greater than the wavelength of the incoming acoustic signal. Detailed omputation has been carried out for a representative case in which a liquid lens ith an index of refraction of 1.8 and a density of 1900 kg/m^3 is assumed to be nmersed in seawater. It is shown that a slight deformation of the boundary surface f a circular liquid lens results in significant changes in the location of the focal egion and that an elliptical lens may provide stronger focusing than an equivalent ircular lens. This research also points out the necessity of pursuing an nvestigation on thin lenses where the focal regions are located outside the lenses.

9.2 The diffraction of sound waves by penetrable disks. (Paper 9-2)

The exact solution of the diffraction of plane sound waves by an acousticall
penetrable oblate spheroid is obtained. Results are given in terms of spheroida
functions. It is found that each expansion coefficient of the scattered or transmitte
wave is coupled to all coefficients of the series expansion for the incident wav
except when the spheroid degenerates into a sphere. Decoupling of the coefficien
for certain special cases is discussed.

9.3 Scattering of acoustic waves by a penetrable prolate spheroid. I. liqui prolate spheroid (Paper 9-3)

This is the first of a series of reports on the scattering of waves by a penetrabl
prolate spheroid. In this letter, the exact solution of the scattering of acoustic wave
by a liquid prolate spheroid is obtained. Results are given in terms of spheroida
functions. It is found that each expansion coefficient of the scattered or transmitte
wave is coupled to all coefficients of the series expansion for the incident wave
except when the spheroid degenerates into a sphere. Detailed numerica
computations for the radiation patterns of the scattered wave are carried out for
specific example. It is found that, in general, the radiation patterns for the scattere
waves are more directional and possess more side lobes as the prolate spheroi
becomes more elongated.

9.4 Scattering by liquid-coated prolate spheroids. (Paper 9-4)

The exact solution of the diffraction of plane sound waves by a rigid prolat
spheroid coated with a confocal sheath of penetrable acoustic material is obtaine
Results are given in terms of spheroidal functions. Detailed numerica
computations for the radiation patterns of the scattered wave are carried out fc
certain acoustic parameters. It is found that the coating alters significantly th
scattering characteristics of an incident wave. Furthermore, because of th
coupling between each expansion coefficient for the scattered wave with a
expansion coefficients for the incident wave for the case of the coated spheroi
case, the Watson transformation technique cannot be used directly to yiel
expressions that are applicable at high frequencies.

.5 A further note on the reflection and transmission of sound waves by a noving fluid layer. (Paper 9-5)

n a previous article, the problem of the reflection and transmission of sound waves y a moving fluid layer was considered. The boundary conditions for the interface etween two stationary media were used, i.e., the pressure and the normal omponent of the velocity are assumed to be continuous at the interface; however, t the interface of two relatively moving media, the boundary conditions must be leduced from the requirements that the pressure must be continuous and the vortex heet must be a stream surface common to the flows in the two media. Consequently, the purpose of this paper is to present the solution for the moving luid problem using the more appropriate boundary conditions for moving media.

.6 Diffraction of sound waves by a moving fluid cylinder. (Paper 9-6)

The solution of the diffraction of sound waves by an axially moving fluid cylinder s obtained. Results indicate that there is no Doppler frequency shift for the cattered wave and that when the angle of incidence is normal to the cylinder axis, he scattered wave is independent of the movement of the medium. Numerical computations for the magnitudes as well as the radiation patterns of the scattered vave as a function of the velocity of the moving medium are also carried out for 'arious incident angles. It is found that the magnitudes and the radiation patterns of the scattered wave are quite sensitive to the variation in the angle of incidence even for slow relative movement of the fluid media.

).7 Triangular ridge acoustic waveguide and coupler. (Paper 9-7)

Jsing a more efficient computational scheme in solving the finite-element pproach to the problem of triangular ridge acoustic waveguides, we have nvestigated the higher-order modes in this single-material structure. The coupling characteristics of two neighboring triangular guides are also studied. Specific umerical results and discussions are given.

Reprinted from: The Journal of the Acoustical Society of America

Elliptical cylindrical liquid lens*

C. Yeh
Electrical Sciences and Engineering Department, University of California, Los Angeles, California 90024
(Received 12 January 1972)

The problem of focusing a pressure wave by a thick, elliptical cylindrical liquid lens has been treated in a rigorous manner. The size of the lens is of the order of or slightly greater than the wavelength of the incoming acoustic signal. Detailed computation has been carried out for a representative case in which a liquid lens with an index of refraction of 1.8 and a density of 1900 kg/m^3 is assumed to be immersed in seawater. It is shown that a slight deformation of the boundary surface of a circular liquid lens results in significant changes in the location of the focal region and that an elliptical lens may provide stronger focusing than an equivalent circular lens. This research also points out the necessity of pursuing an investigation on thin lenses where the focal regions are located outside the lenses.

Subject Classification: 13.6, 13.11t.

INTRODUCTION

The fact that liquid lenses may be used to enhance the characteristics of a wide variety of sonar systems has been generally recognized in recent years. The focusing properties of spherical homogeneous liquid lenses have been investigated experimentally by Toulis[1] and theoretically by Boyles.[2] Folds and Brown[3] also demonstrated the feasibility of using a circular cylindrical homogeneous liquid lens to focus incoming acoustic signals. More recently investigation of spherical lenses whose density distribution is inhomogeneous (such as the Luneburg-type lenses) has been initiated by Boyles.[4] Many interesting results were obtained. However, the desired inhomogeneity in any liquid lenses is very difficult to achieve in practice.

Our group is concerned basically with the problems associated with the noncircular homogeneous lenses. A noncircular lens with homogeneous density is much easier to manufacture than a circular inhomogeneous lens. Furthermore, it is also important to learn the consequences of deforming a circular lens. To initiate this program we shall first consider the focusing characteristics of an elliptical homogeneous liquid cylindrical lens.

When the characteristic size of the elliptical lens is much less than the wavelength of the acoustic signal (the low-frequency case), Rayleigh's solution[5] may be used. The pressure field within the low-frequency lens is quite uniform. When the characteristic size of the lens is much larger than the wavelength (the high-frequency case), the ray-optics approach[6] may be used to predict the focusing behavior although extreme care must be taken to distinguish the focusing characteristics of the intensity field and those of the pressure field. For the intermediate case, the full-wave solution to this problem must be obtained. Acoustic lenses, which will be used in future sonar systems, fall into this intermediate case. Consequently, we will be concerned chiefly with the rigorous solution of this problem. Owing to the complexity of the problem, a complete numerical presentation is not economical and will not be attempted. Numerical computations are therefore carried out for a representative case. Conclusions on the first phase of our work on elliptical-cylindrical liquid lenses are given in Sec. IV.

I. WAVE THEORY FOR ELLIPTICAL LIQUID LENSES

To analyze this problem, the elliptical cylindrical coordinates (ξ,η,z), as shown in Fig. 1, are introduced. In terms of the rectangular coordinates (x,y,z), the elliptical cylindrical coordinates are defined by the following relations:

$$x = h\cosh\xi\cos\eta,\quad y = h\sinh\xi\sin\eta,\quad z = z,$$
$$(0 \leq \xi < \infty,\quad 0 \leq \eta \leq 2\pi), \tag{1}$$

where h is the semifocal length of the ellipse. The contour surfaces of constant ξ are confocal elliptical cylinders, and those of constant η are confocal hyper-

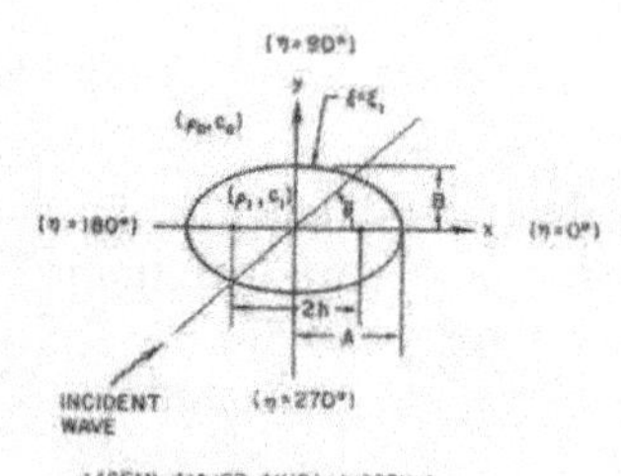

FIG. 1. Elliptical liquid lens. $2h$ is the focal length, ρ is the density of the fluid, and c is the sound speed. The arrow indicates the direction of incident wave.

bolic cylinders. The boundary of the liquid lens is assumed to coincide with one of the confocal elliptical cylinders with $\xi=\xi_1$. The liquid lens, composed of a nonabsorbing refractive liquid (i.e., a nonabsorbing liquid whose sound velocity is less than that of water), which has a density ρ and a sound velocity c_1, is immersed in water that is nonabsorbing and infinite in extent. Let ρ_0 denote the density of the water and c_0 the velocity of sound in the water. A harmonic-plane acoustic wave of angular frequency ω and acoustic pressure P_0 is assumed to be impinging upon the lens at an angle θ with the positive x axis (see Fig. 1). The incident wave p_i gives rise to an internal pressure wave p_p and to an external scattered wave p_s. For the present problem, it is the internal wave p_p that is to be determined.

The formal time-harmonic solutions of the source-free scalar wave equation, which is

$$[\nabla^2+(\omega/c)]p=0, \qquad (2)$$

where p is the scalar pressure, ω is the frequency, and c is the speed of sound in the acoustic medium of interest, in terms of Mathieu functions are[7]

$$\left\{\begin{matrix} Ce_n \\ Fey_n \end{matrix}\right. (\xi,q)\left.\vphantom{\begin{matrix} Ce_n \\ Fey_n \end{matrix}}\right\} ce_n(\eta,q), \quad \left\{\begin{matrix} Se_n \\ Gey_n \end{matrix}\right. (\xi,q)\left.\vphantom{\begin{matrix} Se_n \\ Gey_n \end{matrix}}\right\} se_n(\eta,q),$$

where $q=(\omega h/2c)^2$, $Ce_n(\xi,q)$, and $Fey_n(\xi,q)$ are the two independent even radial Mathieu functions corresponding to the even angular Mathieu function $ce_n(\eta,q)$ and $Se_n(\xi,q)$ and $Gey_n(\xi,q)$ are the two independent odd radial Mathieu functions corresponding to the odd angular Mathieu function $se_n(\eta,q)$. A time dependence of $e^{-i\omega t}$ is assumed and suppressed throughout.

Expressing in terms of Mathieu functions, we have, for the incident wave,

$$p_1=P_0\exp i(k_0x\cos\theta+k_0y\sin\theta)$$

$$=2P_0\sum_{n=0}^{\infty}\left[\frac{1}{p_{2n}}Ce_{2n}(\xi)ce_{2n}(\eta)ce_{2n}(\theta) + \frac{i}{p_{2n+1}}Ce_{2n+1}(\xi)ce_{2n+1}(\eta)ce_{2n+1}(\theta) + \frac{1}{s_{2n+2}}Se_{2n+2}(\xi)se_{2n+2}(\eta)se_{2n+2}(\theta) + \frac{i}{s_{2n+1}}Se_{2n+1}(\xi)se_{2n+1}(\eta)se_{2n+1}(\theta)\right]; \qquad (3)$$

for the scattered wave,

$$p_s=2P_0\sum_{n=0}^{\infty}\left[B_{2n}^{c}\frac{1}{p_{2n}}Me_{2n}^{(1)}(\xi)ce_{2n}(\eta)ce_{2n}(\theta) + B^{c}_{2n+1}\frac{i}{p_{2n+1}}Me_{2n+1}^{(1)}(\xi)ce_{2n+1}(\eta)ce_{2n+1}(\theta) + B^{s}_{2n+2}\frac{1}{s_{2n+2}}Ne_{2n+2}^{(1)}(\xi)se_{2n+2}(\eta)se_{2n+2}(\theta) + B_{2n+1}^{s}\frac{i}{s_{2n+1}}Ne_{2n+1}^{(1)}(\xi)se_{2n+1}(\eta)se_{2n+1}(\theta)\right]; \qquad (4)$$

and for the penetrated internal wave

$$p_p=2P_0\sum_{n=0}^{\infty}\left[A^{c}_{2n}\frac{1}{p^*_{2n}}Ce^*_{2n}(\xi)ce^*_{2n}(\eta) + A^{c}_{2n+1}\frac{i}{p^*_{2n+1}}Ce^*_{2n+1}(\xi)ce^*_{2n+1}(\eta) + A^{s}_{2n+2}\frac{1}{s^*_{2n+2}}Se^*_{2n+2}(\xi)se^*_{2n+2}(\eta) + A^{s}_{2n+1}\frac{i}{s^*_{2n+1}}Se^*_{2n+1}(\xi)se^*_{2n+1}(\eta)\right], \qquad (5)$$

where the abbreviations

$$\begin{matrix} Ce_n \\ Se_n \end{matrix}(\xi)=\begin{matrix} Ce_n \\ Se_n \end{matrix}(\xi,q_0), \quad \begin{matrix} ce_n \\ se_n \end{matrix}(\eta)=\begin{matrix} ce_n \\ se_n \end{matrix}(\eta,q_0),$$

$$\begin{matrix} Ce_n^* \\ Se_n^* \end{matrix}(\xi)=\begin{matrix} Ce_n \\ Se_n \end{matrix}(\xi,q_1), \quad \begin{matrix} ce_n^* \\ se_n^* \end{matrix}(\eta)=\begin{matrix} ce_n \\ se_n \end{matrix}(\eta,q_1),$$

have been used. Also

$$Me_n^{(1)}(\xi)=Ce_n(\xi,q_0)+iFey_n(\xi,q_0),$$

$$Ne_n^{(1)}(\xi)=Se_n(\xi,q_0)+iGey_n(\xi,q_0),$$

and $q_{0,1}=(\omega h/2c_{0,1})^2$. The p_n, s_n and p_n^*, s_n^* are normalization constants that are functions of q_0 and q_1, respectively. The coefficients $A_n^{c,s}$ and $B_n^{c,s}$ are to be found by applying the boundary conditions which require the continuity of the acoustic pressure and the continuity of the normal components of the particle velocity at the boundary surface of the liquid lens, $\xi=\xi_1$:

$$p_i+p_s=p_p|_{\xi=\xi_1}, \qquad (6)$$

$$\frac{1}{\rho_0}\left(\frac{\partial p_i}{\partial\xi}+\frac{\partial p_s}{\partial\xi}\right)=\frac{1}{\rho_1}\frac{\partial p_p}{\partial\xi}\bigg|_{\xi=\xi_1}, \qquad (7)$$

where ρ_0 and ρ_1 are, respectively, the density of the fluid surrounding the lens and that of the liquid lens.

Substituting Eqs. 3–5 into Eqs. 6 and 7 gives

$$\sum_{n=0}^{\infty}\Bigg\{\frac{1}{p_{2n}}[Ce_{2n}(\xi_1)+B^c{}_{2n}Me_{2n}{}^{(1)}(\xi_1)]ce_{2n}(\eta)ce_{2n}(\theta)+\frac{i}{p_{2n+1}}[Ce_{2n+1}(\xi_1)+B^c{}_{2n+1}Me^{(1)}{}_{2n+1}(\xi_1)]ce_{2n+1}(\eta)ce_{2n+1}(\theta)$$
$$+\frac{1}{s_{2n+2}}[Se_{2n+2}(\xi_1)+B^s{}_{2n+2}Ne^{(1)}{}_{2n+2}(\xi_1)]se_{2n+2}(\eta)se_{2n+2}(\theta)$$
$$+\frac{i}{s_{2n+1}}[Se_{2n+1}(\xi_1)+B^s{}_{2n+1}Ne^{(1)}{}_{2n+1}(\xi_1)]se_{2n+1}(\eta)se_{2n+1}(\theta)\Bigg\}$$
$$=\sum_{n=0}^{\infty}\Bigg\{\frac{A^c{}_{2n}}{p^*{}_{2n}}Ce^*{}_{2n}(\xi_1)ce^*{}_{2n}(\eta)+\frac{i}{p^*{}_{2n+1}}Ce^*{}_{2n+1}(\xi_1)ce^*{}_{2n+1}(\eta)A^c{}_{2n+1}$$
$$+\frac{A^s{}_{2n+2}}{s^*{}_{2n+2}}Se^*{}_{2n+2}(\xi_1)se^*{}_{2n+2}(\eta)+\frac{iA^s{}_{2n+1}}{s^*{}_{2n+1}}Se^*{}_{2n+1}(\xi_1)se^*{}_{2n+1}(\eta)\Bigg\} \quad (8)$$

and

$$\sum_{n=0}^{\infty}\Bigg\{\frac{1}{p_{2n}}[Ce'{}_{2n}(\xi_1)+B^c{}_{2n}Me^{(1)\prime}{}_{2n}(\xi_1)]ce_{2n}(\eta)ce_{2n}(\theta)+\frac{i}{p_{2n+1}}[Ce'{}_{2n+1}(\xi_1)+B^c{}_{2n+1}Me^{(1)\prime}{}_{2n+1}(\xi_1)]ce_{2n+1}(\eta)ce_{2n+1}(\theta)$$
$$+\frac{1}{s_{2n+2}}[Se'{}_{2n+2}(\xi_1)+B^s{}_{2n+2}Ne^{(1)\prime}{}_{2n+2}(\xi_1)]se_{2n+2}(\eta)se_{2n+2}(\theta)$$
$$+\frac{i}{s_{2n+1}}[Se'{}_{2n+1}(\xi_1)+B^s{}_{2n+1}Ne^{(1)\prime}{}_{2n+1}(\xi_1)]se_{2n+1}(\eta)se_{2n+1}(\theta)\Bigg\}$$
$$=\frac{\rho_0}{\rho_1}\sum_{n=0}^{\infty}\Bigg\{\frac{A^c{}_{2n}}{p^*{}_{2n}}Ce^{*\prime}{}_{2n}(\xi_1)ce^*{}_{2n}(\eta)+\frac{iA^c{}_{2n+1}}{p^*{}_{2n+1}}Ce^{*\prime}{}_{2n+1}(\xi_1)ce^*{}_{2n+1}(\eta)$$
$$+\frac{A^s{}_{2n+2}}{s^*{}_{2n+2}}Se^{*\prime}{}_{2n+2}(\xi_1)se^*{}_{2n+2}(\eta)+\frac{iA^s{}_{2n+1}}{s^*{}_{2n+1}}Se^{*\prime}{}_{2n+1}(\xi_1)se^*{}_{2n+1}(\eta)\Bigg\}, \quad (9)$$

where the prime on the radial Mathieu function denotes the derivatives of the function with respect to ξ_1. Multiplying both sides of Eqs. 8 and 9 by $ce_r(\eta)$ or $se_r(\eta)$, respectively, integrating with respect to η from 0 to 2π, making use of the orthogonality relations for angular Mathieu functions, and combining the resultant equations appropriately by eliminating the coefficients $B_r{}^{c,s}$, we finally obtain the following expressions:

$$\sum_{m=0}^{\infty} A^c{}_{2m}a^c{}_{2n,2m}=i\frac{2}{\pi}ce_{2n}(\theta), \quad (10)$$

$$\sum_{m=0}^{\infty} A^c{}_{2m+1}a^c{}_{2n+1,2m+1}=i\frac{2}{\pi}ce_{2n+1}(\theta), \quad (11)$$

$$\sum_{m=0}^{\infty} A^s{}_{2m+2}a^s{}_{2n+2,2m+2}=i\frac{2}{\pi}se_{2n+2}(\theta), \quad (12)$$

$$\sum_{m=0}^{\infty} A^s{}_{2m+1}a^s{}_{2n+2,2m+1}=i\frac{2}{\pi}se_{2n+1}(\theta), \quad (13)$$

where ($n=0, 1, 2, \cdots$), with

$$a^c{}_{2n,2m}=\frac{\alpha_{2n,2m}}{p_{2n}p^*{}_{2m}}\left[Ce^*{}_{2m}(\xi_1)Me_{2n}{}^{(1)\prime}(\xi_1)-\frac{\rho_0}{\rho_1}Ce^{*\prime}{}_{2m}(\xi_1)Me_{2n}{}^{(1)}(\xi_1)\right], \quad (14)$$

$$a^c{}_{2n+1,2m+1}=\frac{\alpha_{2n+1,2m+1}}{p_{2n+1}p^*{}_{2m+1}}\left[Ce^*{}_{2m+1}(\xi_1)Me^{(1)\prime}{}_{2n+1}(\xi_1)-\frac{\rho_0}{\rho_1}Ce^{*\prime}{}_{2m+1}(\xi_1)Me^{(1)}{}_{2n+1}(\xi_1)\right], \quad (15)$$

$$a^s{}_{2n+2,2m+2}=\frac{\beta_{2n+2,2m+2}}{s_{2n+2}s^*{}_{2m+2}}\left[Se^*{}_{2m+2}(\xi_1)Ne^{(1)\prime}{}_{2n+2}(\xi_1)-\frac{\rho_0}{\rho_1}Se^{*\prime}{}_{2m+2}(\xi_1)Ne^{(1)}{}_{2n+2}(\xi_1)\right], \quad (16)$$

$$a^s{}_{2n+1,2m+1}=\frac{\beta_{2n+1,2m+1}}{s_{2n+1}s^*{}_{2m+1}}\left[Se^*{}_{2m+1}(\xi_1)Ne^{(1)\prime}{}_{2n+1}(\xi_1)-\frac{\rho_0}{\rho_1}Se^{*\prime}{}_{2m+1}(\xi_1)Ne^{(1)}{}_{2n+1}(\xi_1)\right], \quad (17)$$

where

$$\alpha_{r,s}=\int_0^{2\pi} ce_r(\eta)ce_s{}^*(\eta)d\eta \Big/ \int_0^{2\pi} ce_r{}^2(\eta)d\eta \quad (18)$$

and

$$\beta_{r,s}=\int_0^{2\pi} se_r(\eta)se_s{}^*(\eta)d\eta \Big/ \int_0^{2\pi} se_r{}^2(\eta)d\eta. \quad (19)$$

he Wronskian relations for radial Mathieu functions given in Appendix A) have been used in the derivation f Eqs. 10–13.

The formal solution for the liquid elliptical lens roblem is obtained when one solves the simultaneous lgebraic equations, Eqs. 10–13, for the expansion oefficients for the penetrated pressure field, $A_r^{c,s}$. he expansion coefficients for the scattered pressure eld $B_r^{c,s}$ are related to the coefficient $A_r^{c,s}$ by the ollowing equations:

$$B^c{}_{2n}=\left\{-\frac{1}{p_{2n}}Ce_{2n}(\xi_1)+\sum_{m=0}^{\infty}\left[A^c{}_{2m}\frac{1}{p^*{}_{2m}}Ce^*{}_{2m}(\xi_1)\frac{\alpha_{2n,2m}}{ce_{2n}(\theta)}\right]\right\}\Big/\left[\frac{1}{p_{2n}}Me_{2n}{}^{(1)}(\xi_1)\right],\quad(20)$$

$$B^c{}_{2n+1}=\left\{-\frac{1}{p_{2n+1}}Ce_{2n+1}(\xi_1)+\sum_{m=0}^{\infty}\left[A^c{}_{2m+1}\frac{1}{p^*{}_{2m+1}}Ce^*{}_{2m+1}(\xi_1)\frac{\alpha_{2n+1,2m+1}}{ce_{2n+1}(\theta)}\right]\right\}\Big/\left[\frac{1}{p_{2n+1}}Me^{(1)}{}_{2n+1}(\xi_1)\right],\quad(21)$$

$$B^s{}_{2n+2}=\left\{-\frac{1}{s_{2n+2}}Se_{2n+2}(\xi_1)+\sum_{m=0}^{\infty}\left[A^s{}_{2m+2}\frac{1}{s^*{}_{2m+2}}Se^*{}_{2m+2}(\xi_1)\frac{\beta_{2n+2,2m+2}}{se_{2n+2}(\theta)}\right]\right\}\Big/\left[\frac{1}{s_{2n+2}}Ne^{(1)}{}_{2n+2}(\xi_1)\right],\quad(22)$$

$$B^s{}_{2n+1}=\left\{-\frac{1}{s_{2n+1}}Se_{2n+1}(\xi_1)+\sum_{m=0}^{\infty}\left[A^s{}_{2m+1}\frac{1}{s^*{}_{2m+1}}Se^*{}_{2m+1}(\xi_1)\frac{\beta_{2n+1,2m+1}}{se_{2n+1}(\theta)}\right]\right\}\Big/\left[\frac{1}{s_{2n+1}}Ne^{(1)}{}_{2n+1}(\xi_1)\right].\quad(23)$$

Substituting the results for the expansion coefficients $A_r^{c,s}$ into Eq. 5 gives the required penetrated pressure wave within the liquid lens.

A pressure-sensititive hydrophone with a suitable detector may be used to measure the root-mean-square pressure. Writing

$$p_p=p_p{}^r+ip_p{}^i,\quad(24)$$

where $p_p{}^r$ and $p_p{}^i$ are real, we have

$$|p_p|=(p_p{}^{r2}+p_p{}^{i2})^{\frac{1}{2}}\quad(25)$$

for the root-mean-square pressure within the lens. It is this quantity that we compute numerically.

II. DEGENERATE CASE OF THE CIRCULAR CYLINDRICAL LIQUID LENS

It is seen that the solution for the elliptical liquid lens is more involved than that for the circular cylindrical lens problem. This is because of the cross coupling between the expansion coefficients for the incident field and the expansion coefficients for the scattered and penetrated fields. That is, each expansion coefficient for the scattered or penetrated pressure field is coupled with all expansion coefficients for the incident pressure field.

Making use of the degenerate forms of the Mathieu functions (see Appendix B), one obtains the following degenerate expressions for the incident, scattered, and penetrated waves:

$$p_i=P_0J_0(k_0r)+2P_0\sum_{n=1}^{\infty}(-1)^nJ_{2n}(k_0r)\cos2n\eta\cos2n\theta+2P_0\sum_{n=0}^{\infty}\{(-1)^{n+1}J_{2n+2}(k_0r)\sin(2n+2)\eta\sin(2n+2)\theta$$
$$+i(-1)^nJ_{2n+1}(k_0r)[\cos(2n+2)\eta\cos(2n+2)\theta+\sin(2n+2)\eta\sin(2n+2)\theta]\}$$
$$=P_0J_0(k_0r)+2P_0\sum_{n=1}^{\infty}(i)^nJ_n(k_0r)\cos n(\eta-\theta),\quad(26)$$

$$p_s=P_0B_0{}^cH_0{}^{(1)}(k_0r)+2P_0\sum_{n=1}^{\infty}(-1)^nB^c{}_{2n}H^{(1)}{}_{2n}(k_0r)\cos2n\eta\cos2n\theta$$
$$+2P_0\sum_{n=0}^{\infty}\{(-1)^{n+1}B^s{}_{2n+2}H^{(1)}{}_{2n+2}(k_0r)\sin(2n+2)\eta\sin(2n+2)\theta$$
$$+i(-1)^nH^{(1)}{}_{2n+1}(k_0r)[B^c{}_{2n+1}\cos(2n+2)\eta\cos(2n+2)\theta+B^s{}_{2n+1}\sin(2n+2)\eta\sin(2n+2)\theta]\}$$
$$=P_0B_0H_0{}^{(1)}(k_0r)+2P_0\sum_{n=1}^{\infty}(i)^nB_nH_n{}^{(1)}(k_0r)\cos n(\eta-\theta),\quad(27)$$

$$p_p=\sqrt{2}P_0A_0^eJ_0(k_1r)+2P_0\sum_{n=1}^{\infty}(-1)^nA^e_{2n}J_{2n}(k_1r)\cos 2n\eta+2P_0\sum_{n=0}^{\infty}\{(-1)^{n+1}A^e_{2n+2}J_{2n+2}(k_1r)\sin(2n+2)\eta$$
$$+i(-1)^nJ_{2n+1}(k_1r)[A^e_{2n+1}\cos(2n+2)\eta+A^s_{2n+1}\sin(2n+2)\eta]\}$$
$$=P_0A_0J_0(k_1r)+2P_0\sum_{n=1}^{\infty}(i)^nA_nJ_n(k_1r)\cos n(\eta-\theta), \tag{28}$$

with

$$A_0=\sqrt{2}A_0^e=i\frac{2}{\pi}\bigg/\left\{k_0aJ_0(k_1a)H_0^{(1)\prime}(k_0a)-\frac{\rho_0}{\rho_1}k_1aJ_0'(k_1a)H_0^{(1)}(k_0a)\right\},$$

$$A_n=A_n^e/(\cos n\theta)=A_n^s/(\sin n\theta)$$
$$=i\frac{2}{\pi}\bigg/\left\{k_0aJ_n(k_1a)H_n^{(1)\prime}(k_0a)-\frac{\rho_0}{\rho_1}k_1aJ_n'(k_1a)H_n^{(1)}(k_0a)\right\},\quad (n\geq 1)$$

$$B_0=[-J_0(k_0a)+A_0J_0(k_1a)]/H_0^{(1)}(k_0a),$$
$$B_n=[-J_n(k_0a)+A_nJ_n(k_1a)]/H_n^{(1)}(k_0a),\qquad (n\geq 1). \tag{29}$$

Here a = radius of the degenerated circular cylinder. The prime indicates the derivative of the function with respect to its argument. The above results correspond to the well-known solutions for the scattering of waves by a circular, fluid cylinder.[8]

IV. NUMERICAL RESULTS AND DISCUSSIONS

The task of obtaining numerical results from the formal expressions for the pressure field within an elliptical lens is many orders of magnitude more difficult than that for the circular cylindrical lens problem. In the first place, Mathieu functions have not been adequately tabulated. (Perhaps, because of the number of variables associated with each angular Mathieu function or with each radial Mathieu function, the complete tabulation can never be justified.) In the second place, the programming of Mathieu functions from available expressions is not straightforward. Many numerical tricks must be incorporated in the computer program in order that the desired accuracy may be achieved. This is particularly true when q is large. It is known that the periodic angular Mathieu functions can be expanded in terms of infinite series of trigonometric functions and that the corresponding radial Mathieu functions can be expanded in terms of infinite series of Bessel functions or products of Bessel functions. It is found that computations for the $Ce_n(\xi,q)$ or $Se_n(\xi,q)$ radial Mathieu functions should be calculated from the expressions involving infinite series of Bessel functions while the computations for the $Fey_n(\xi,q)$ and $Gey_n(\xi,q)$ radial Mathieu functions should be calculated from the expressions involving infinite series of products of Bessel functions. To obtain the expansion coefficients we must first calculate the characteristic numbers according to the Bouwkamp–Blanch variational technique.[9]

Another added complexity for the present problem as compared with the circular cylindrical lens case or

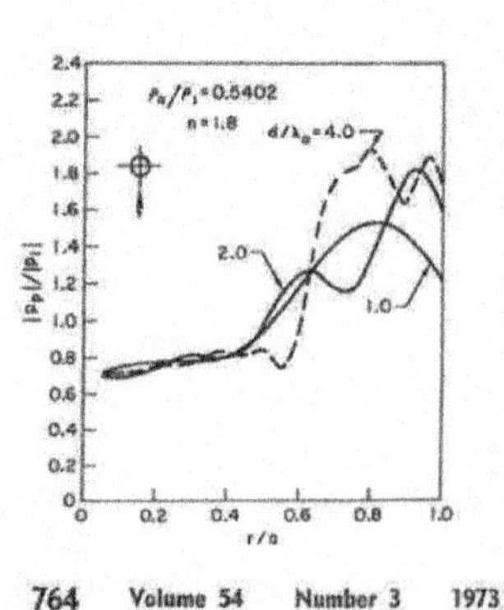

FIG. 2. Pressure distribution within a circular cylinder fluid lens along the forward direction. r is the distance from the origin to the point of observation and a is the radius of the lens. d is the diameter of the lens and λ_0 is the wavelength of the incident wave in the surrounding medium. n is the index of refraction of the lens.

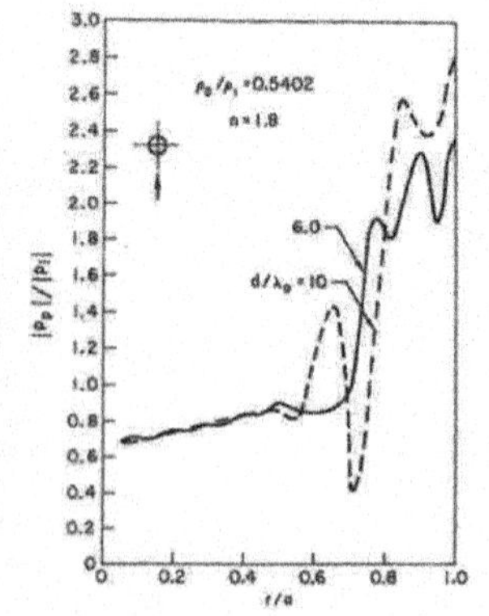

FIG. 3. Pressure distribution within a circular cylinder fluid lens along the forward direction.

with the perfectly reflecting elliptical cylinder case is the coupling of each expansion coefficient for the scattered or penetrated pressure field with all expansion coefficients for the incoming pressure field.[10] Consequently, in order to find n number of expansion coefficients (such as A_n^e from Eq. 10) it is necessary to carry out computations for a $m\times m$ matrix, where $m\geq n$. The value for m is determined by the successive approximation method; i.e., computations were carried out for a $n\times n$ matrix, a $(n+1)\times(n+1)$ matrix, etc., until the desired accuracy for the predetermined n-expansion coefficients was reached. Referring back to Eqs. 10–19, one notes that the coupling coefficients are governed by $\alpha_{r,s}$ and $\beta_{r,s}$. It can be seen that $\alpha_{r,s}$ or $\beta_{r,s}$ approaches zero slower as $|r-s|$ increases for larger differences of $|q_1-q_0|$. In other words, stronger coupling between the expansion coefficients for the penetrated pressure field and the expansion coefficients for the incident pressure field results when the value $|q_1-q_0|$ increases or when the elliptical cross section becomes flatter. For the cases considered here, $|q_1-q_0|$ may be quite large and the coupling could be quite strong.

Although the computer program prepared for this problem is valid for arbitrary values of q_0, q_1, ξ_1, θ, and ρ_0/ρ_1, from an economic (computer cost) as well as realistic point of view we shall consider one representative case: The lens is assumed to be immersed in seawater. As a standard velocity, we use $c_0=1500$ m/sec. This velocity corresponds to the velocity of surface seawater having a temperature of 13°C and a salinity of 35 parts per thousand. The density of seawater having this pressure, temperature, and salinity is $\rho_0=1026.4$ kg/m³. The liquid lens contains a particular refracting liquid called Fluorolube $(CF_2CFCl)_x$, which has a density $\rho_1=1900$ kg/m³ and a index of refraction

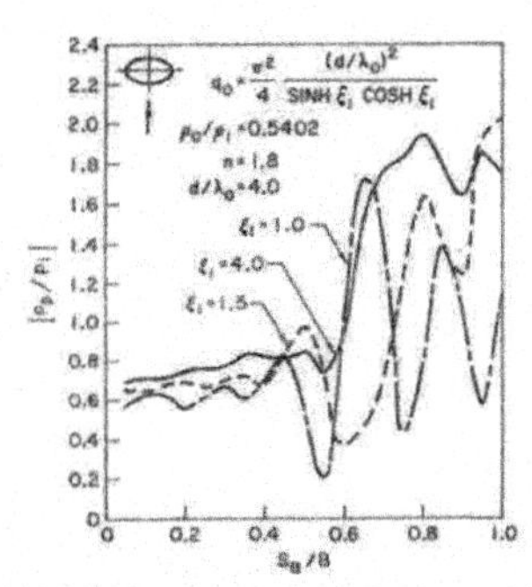

FIG. 4. Pressure distribution within an elliptical liquid lens along the forward direction (the minor axis). The cross-sectional area has been kept constant. d is the diameter of an equivalent circular cylinder with the same cross-sectional area as the elliptical cylinder. The incident wave is along the minor axis of the elliptical lens. S_B is the distance from the origin to the point of observation within the lens along the minor axis. B is the semiminor axis.

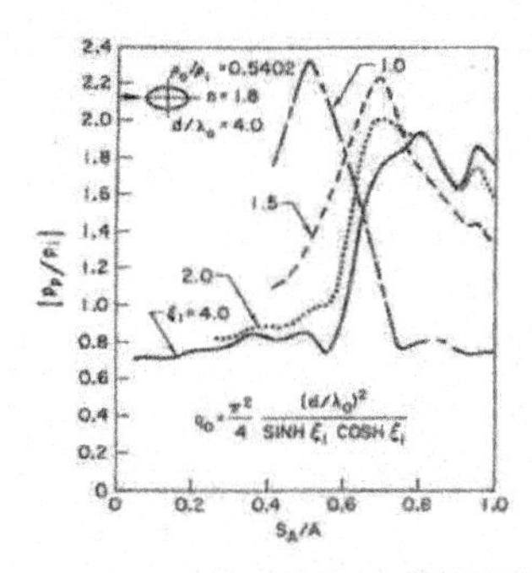

FIG. 5. Pressure distribution within an elliptical liquid lens along the forward direction (the major axis). The cross-sectional area has been kept constant. The incident wave is along the major axis of the elliptical lens. S_A is the distance from the origin to the point of observation within the lens along the major axis. A is the semimajor axis. Note that the focusing of the pressure amplitude is stronger for a flatter ellipse.

$n=c_0/c_1=1.8$, where c_1 is the sound speed in Fluorolube at 20°C.

A. Pressure Fields in a Circular Cylindrical Lens

In order to compare the results for an elliptical cylindrical lens with those for a circular cylindrical lens, we shall include here some pertinent results for the circular cylinder case. These results have not appeared in the literature. Of primary interest will be the pressure distributions along the forward direction of the incident wave. Knowledge of the pressure distribution will enable us to locate the position of highest pressure. Computations were carried out using the formulas given in Sec. II. Figures 2 and 3 show the variation of the relative root-mean-square pressure $|p_p|/|p_i|$ as a function of r/a for various values of d/λ_0. Here r is the distance measured from the center of the lens to the point of interest along the forward direction of the incident wave (see Fig. 1), d and a are, respectively, the diameter and radius of the circular lens, and λ_0 is the wavelength of the incident wave in the surrounding seawater. It can be seen that the location of maximum pressure is a rather complex function of d/λ_0. Owing to the interference of many diffracted rays with each other, it is very difficult to predict any meaningful results according to the simplified geometrical-optics technique for the d/λ_0 values that are under consideration. This observation points out the necessity in treating the medium-frequency-lens problem according to the wave theory as was carried out in this work. The fact that within the lens there exists a location (or locations) where the pressure field is higher than other locations and where $|p_p|/|p_i|>1$ should be noted. It means that a medium-frequency liquid cylindrical thick lens may serve as an effective device to collect an incoming acoustic signal.

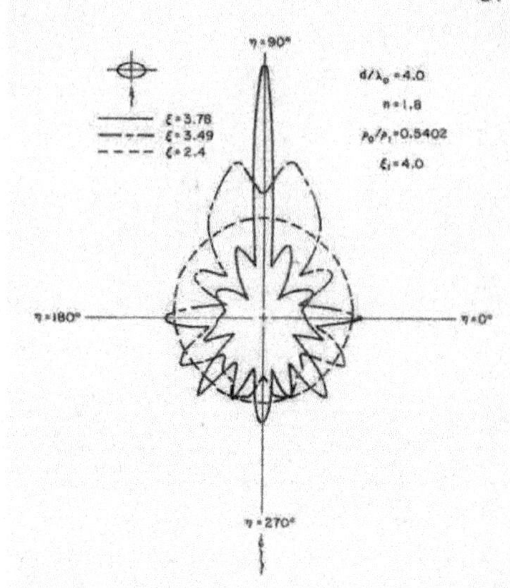

Fig. 6. Polar pressure distribution within a medium-size ($d/\lambda_0=4.0$) lens with $\xi_1=4.0$. The arrow indicates the direction of the incident wave. ξ represents the boundary of various ellipses at which the pressure magnitudes were computed.

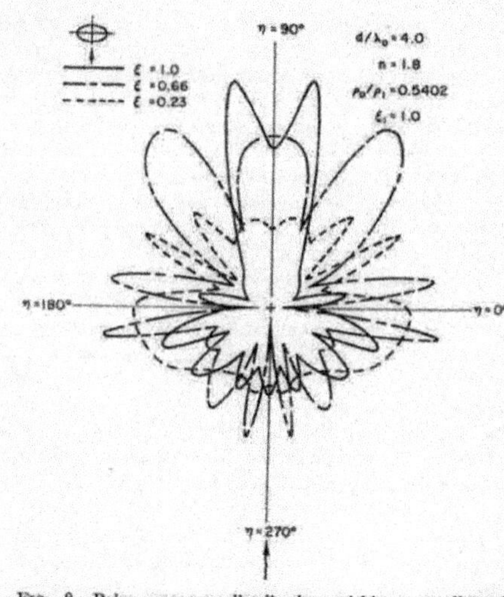

Fig. 8. Polar pressure distribution within a medium-size elliptical liquid lens with $\xi_1=1.0$. Incident wave is along the minor axis.

As the frequency of the incident wave increases or as the size of the lens increases (i.e., as d/λ_0 increases) the pressure distribution within the lens fluctuates more and more rapidly. Similar behavior has been observed be Boyles in his treatment of spherical liquid lens.[2]

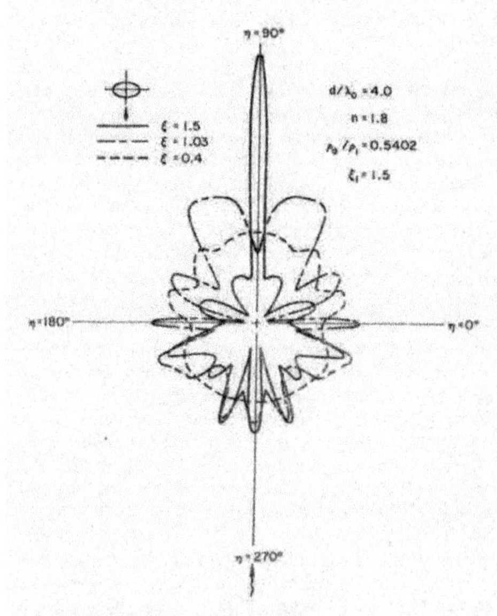

Fig. 7. Polar pressure distribution within a medium-size elliptical liquid lens with $\xi_1=1.5$. Incident wave is along the minor axis. ξ represents the boundary of various ellipses at which the pressure magnitudes were computed.

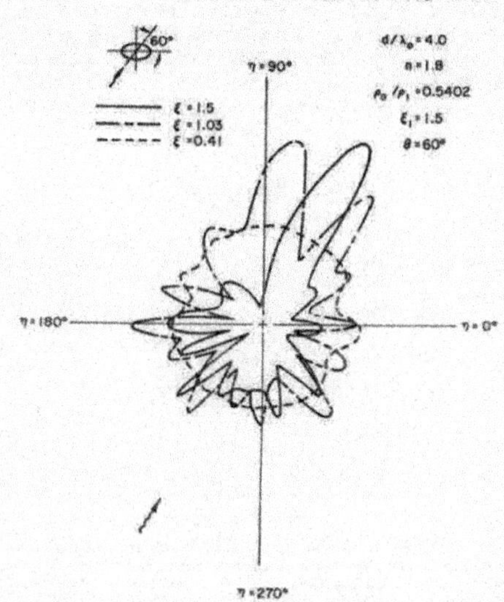

Fig. 9. Polar pressure distribution within a medium-size elliptical liquid lens with $\xi_1=1.5$. The direction of the incident wave is 60° from the major axis.

Having considered the circular liquid lens problems, we are now in a position to treat the elliptical liquid lens problem. Specifically, we would like to observe the change of pressure distribution as the circular cylinder is flattened.

B. Pressure Fields in an Elliptical Cylindrical Lens

It is well known that the diffracted fields are intimately related to the shapes of the diffracting objects. By deforming the circular boundary of a circular cylindrical lens, one expects to alter significantly the diffracted fields within the lens. Figure 4 gives the variation of $|p|/|p_i|$ within a cylindrical lens along the forward minor axis for three different shapes of elliptical cross section ($\xi_1=4.0$, 1.5, and 1.0). The cross-sectional area of these different shapes remains unaltered. The symbol d/λ_0 given in the figure represents the diameter/wavelength of an equivalent circular cylinder with the same cross-sectional area as the elliptical cylinder under consideration. In other words, by equating the cross-sectional area for an ellipse with that of an equivalent circle, one has

$$d/\lambda_0=(2/\pi)(q_0 \cosh\xi_1 \sinh\xi_1)^{\frac{1}{2}}.$$

The incident pressure wave is assumed to be propagating along the minor axis. It is recalled that ξ_1 is related to the ratio of major axis to minor axis by the relation major axis/minor axis$=\cosh\xi_1/\sinh\xi_1$. For the given examples, we have major axis/minor axis$=1.00067$, 1.1048, 1.313, corresponding to $\xi_1=4.0$, 1.5, and 1.0, respectively. One notes that even a small change in the cross-sectional shape of the lens introduces rather significant variations of the pressure field within the lens. The locations of maximum pressure is particularly sensitive to the shape changes. As the cylinder is flattened, the magnitude of the pressure field fluctuates more and more rapidly. On the other hand, the field near the origin is rather uniform and is not very sensitive to the deformation of the boundary.

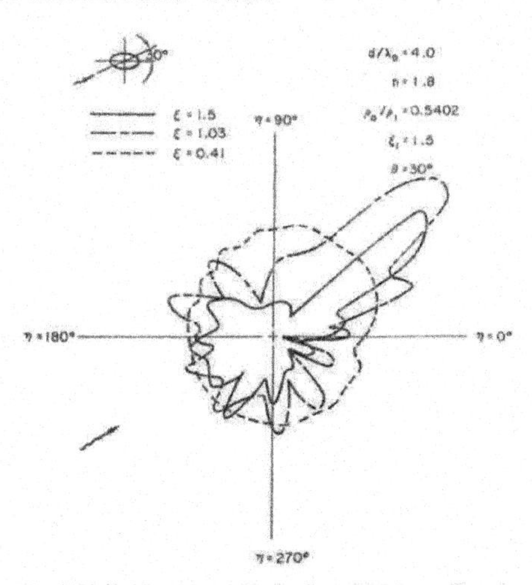

FIG. 10. Polar pressure distribution within a medium-size elliptical liquid lens with $\xi_1=1.5$. The direction of the incident wave is 30° from the major axis.

By rotating the lens 90° with respect to the direction of the incident wave, one may obtain the pressure distribution along the major axis when the direction of the incident wave is parallel to the major axis of the lens. It can be seen from Fig. 5 that the maximum magnitude of the pressure field within an elliptical lens is higher than that within a circular lens. In other words, the focusing power of a circular lens can be improved by simply deforming the circular cross section into an elliptical one.

To find out how the pressure field varies as one rotates around the central axis of the cylinder, Figs. 6–8 are

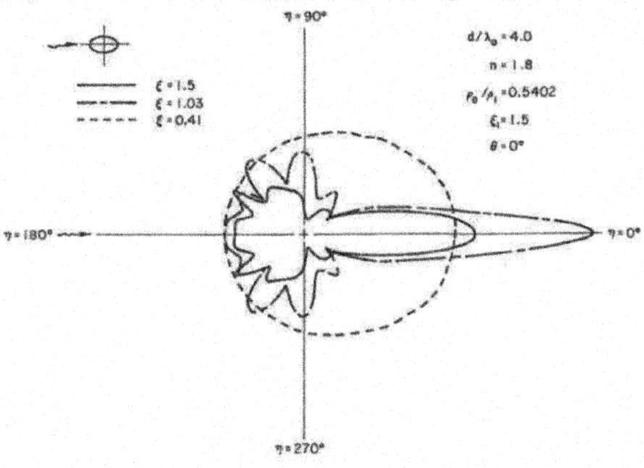

FIG. 11. Polar pressure distribution within a medium-size elliptical liquid lens with $\xi_1=1.5$.

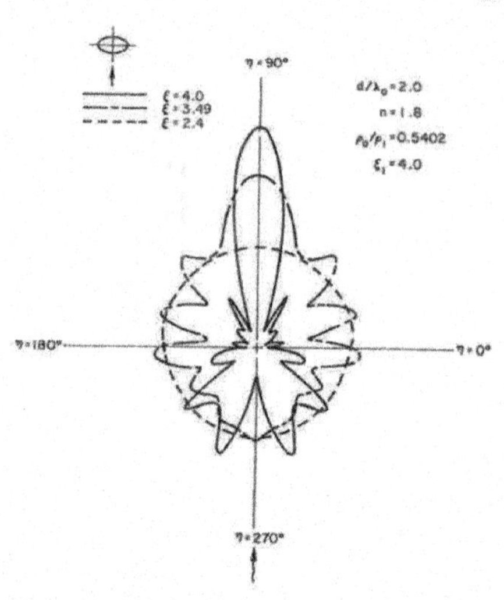

FIG. 12. Polar pressure distribution within a small-size ($d/\lambda_0=2.0$) elliptical liquid lens with $\xi_1=4.0$. The incident wave is along the minor axis.

introduced (with $d/\lambda_0=4.0$). In these figures the arrow indicates the direction of the incident wave. As expected, the fields become more directional as one moves

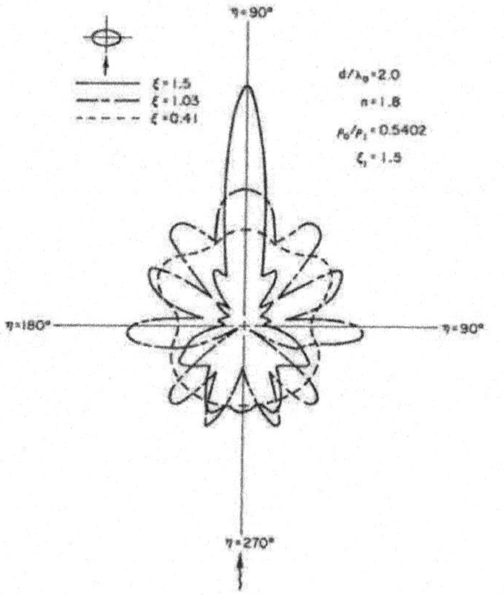

FIG. 13. Polar pressure distribution within a small-size elliptical liquid lens with $\xi_1=1.5$. The incident wave is along the minor axis.

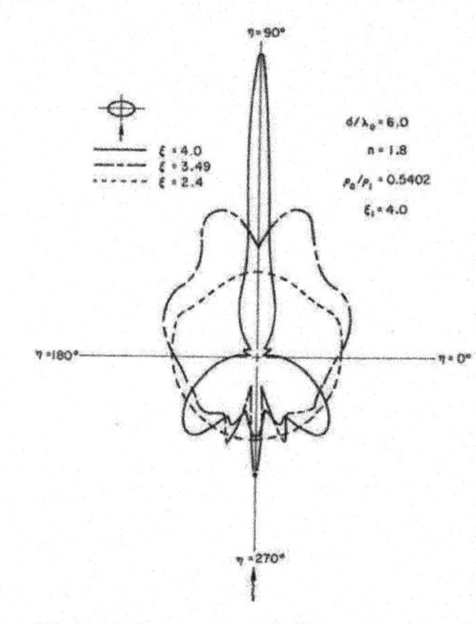

FIG. 14. Polar pressure distribution within a large-size ($d/\lambda_0=6.0$) elliptical liquid lens with $\xi_1=4.0$. The incident wave is along the minor axis.

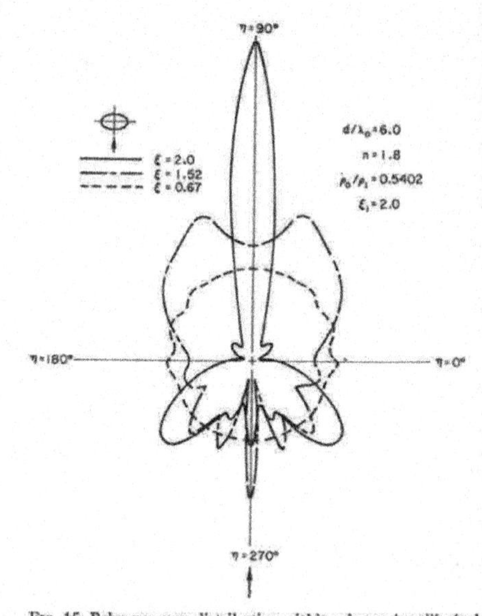

FIG. 15. Polar pressure distribution within a large-size elliptical liquid lens with $\xi_1=2.0$. The incident wave is along the minor axis.

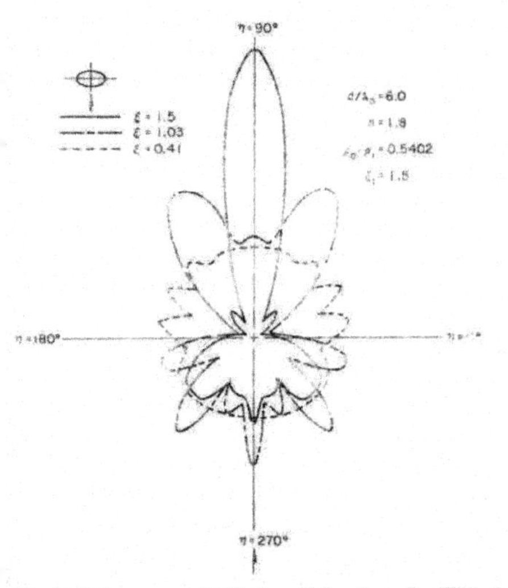

FIG. 16. Polar pressure distribution within a large-size elliptical liquid lens with $\xi_1=1.5$. The incident wave is along the minor axis.

away from the origin. A slight improvement of the beam width is achieved when the cross section is slightly deformed, i.e., when ξ_1 varies from 4.0 to 1.5. Further flattening of the cross section results in the degradation of the main lobe and the appearance of two major side lobes as shown in Fig. 8. It is speculated that the pressure field of a flat elliptical lens will converge at a point outside the lens. This means that a flat elliptical lens may be used as a thin acoustic lens.

It is of interest to investigate the dependence of the diffracted field within the elliptical lens as the direction of the incident wave is varied (see Figs. 9–11). Asymmetrical patterns are observed when the incident wave is not along the major or the minor axis. Since the deformation from a circular cylinder for the cases considered in these figures is not too great, the main lobe remains to be pointed in the forward direction of the incident wave.

According to Figs. 2 and 3, a higher concentration of the pressure field is achieved when larger liquid lenses are used. Computations are therefore carried out for different values of d/λ_0. Results are shown in Figs. 12–16. It can be observed that for small d/λ_0, the pressure fields are less sensitive to a deformation of the boundary (Compare Fig. 12 with Fig. 13).

V. CONCLUSIONS

Computations were carried out for a specific elliptical liquid lens made of Fluorolube. The lens is assumed to be immersed in seawater. Although many of the results, as given in the previous section, have been obtained, but because of the number of variables involved in this problem, such as the size of the lens, the ellipticity, the direction of the incident wave, the location of the point of interest, the relative density of the lens and its index of refraction, much more work remains to be carried out. Completion of this work is mainly dependent upon the availability of computer time and the interpretation of the numerical results rather than the development of new programs. The purpose of this report is to provide some new results obtained for the elliptical liquid lens. Based on the available numerical results several observations and conclusions can be made.

(1) Slight deformation of the boundary surface of a circular liquid lens results in significant changes in the location of maximum pressure magnitude. For a resonant-size thick liquid lens the change in pressure distribution cannot be predicted simply by an application of ray-optics technique. A full-wave treatment as was carried out in this paper must be used.

(2) More than 10% gain in the pressure amplitude may be realized for the specific case considered here by simply flattening a circular liquid lens into an elliptical liquid lens. The gain is achieved when the wave is incident along the major axis of the elliptical lens. It is anticipated that much more gain may be obtained for larger liquid lenses. This conjecture remains to be proven by further computed results.

(3) An elliptical lens may be used to shape the beam of a receiving microphone. Due to the noncircular nature of the elliptical lens asymmetrical beams may be achieved.

(4) One of the most interesting byproducts by the present investigation of thick elliptical lenses is to point out the need to study *thin* elliptical lenses. It appears that concentration of pressure amplitudes may be obtained outside of a thin elliptical lens when the incident wave is directed along the minor axis of the lens. A thin lens may be placed in front of a transducer to focus incoming pressure waves. Practically no work has been carried out on thin lenses for acoustic waves. This should be a very rich area for future research. We intend to at least follow through the work on the thin elliptical liquid lens.

APPENDIX A: WRONSKIAN RELATIONS FOR RADIAL MATHIEU FUNCTIONS

In the following we shall list the Wronskian relations for the radial Mathieu functions:

$$Ce_m(\xi,q)Fey_m'(\xi,q)-Ce_m'(\xi,q)Fey_m(\xi,q)=(2/\pi)p_m^2(q),$$
$$Se_m(\xi,q)Gey_m'(\xi,q)-Se_m'(\xi,q)Gey_m(\xi,q)=(2/\pi)s_m^2(q),$$

where p_m and s_m are jointing factors as given in McLachlan.[7]

APPENDIX B: DEGENERATE FORMS OF MATHIEU FUNCTIONS

As the ellipse tends to a circle $h\to 0$, $q\to 0$, $\xi\to\infty$, such that $h\cosh\xi = h\sinh\xi = \frac{1}{2}he^{\xi}\to r$ and $qe^{\xi}\to kr$ with $k=\omega/c$, the Mathieu functions become[7]

$$ce_m(\eta,q)\to \cos m\eta, \quad (m\geq 1),$$

$$\to 1/\sqrt{2}, \quad (m=0),$$

$$se_m(\eta,q)\to \sin m\eta, \quad (m\geq 1),$$

$$Ce_{2n}(\xi,q)/p_{2n}\to(-1)^n J_{2n}(kr),$$

$$Ce_{2n+1}(\xi,q)/p_{2n+1}\to(-1)^n J_{2n+1}(kr),$$

$$Se_{2n+2}(\xi,q)/s_{2n+2}\to(-1)^{n+1} J_{2n+2}(kr),$$

$$Se_{2n+1}(\xi,q)/s_{2n+1}\to(-1)^n J_{2n+1}(kr),$$

$$Me_{2n}(\xi,q)/p_{2n}\to(-1)^n H_{2n}^{(1)}(kr),$$

$$Me_{2n+1}(\xi,q)/p_{2n+1}\to(-1)^n H^{(1)}{}_{2n+1}(kr),$$

$$Ne_{2n+2}(\xi,q)/s_{2n+2}\to(-1)^{n+1} H^{(1)}{}_{2n+2}(kr),$$

$$Ne_{2n+1}(\xi,q)/s_{2n+1}\to(-1)^n H^{(1)}{}_{2n+1}(kr),$$

$$\frac{\partial}{\partial\xi}\{M_m(\xi,q)\}\to kr\frac{d}{d(kr)}\{R_m(kr)\},$$

where $M_m(\xi,q)$ may be any normalized radial Mathie[u] functions, such as $Ce_{2m}(\xi,q)/p_{2m}$, and $R_m(kr)$ may b[e] its corresponding degenerate form, such as (-1)[...] $J_{2n}(kr)$.

The degenerate forms of the functions $\alpha_{r,s}$ and $\beta_{r,s}$ ar[e]

$$\alpha_{r,s}\to 1, \quad (r=s),$$

$$\to 0, \quad (r\neq s),$$

$$\beta_{r,s}\to 1, \quad (r=s),$$

$$\to 0, \quad (r\neq s).$$

*Supported by the Office of Naval Research.

[1] W. J. Toulis, J. Acoust. Soc. Am. **35**, 286 (1963).
[2] C. A. Boyles, J. Acoust. Soc. Am. **38**, 393 (1965).
[3] D. L. Folds and D. H. Brown, J. Acoust. Soc. Am. **43**, 560 (1968).
[4] C. A. Boyles, J. Acoust. Soc. Am. **45**, 356 (1969).
[5] J. W. Strutt, *Theory of Sound* (Dover, New York, 1945).
[6] R. K. Luneburg, *Mathematical Theory of Optics* (U. of California Press, Berkeley and Los Angeles, 1964).
[7] N. McLachlan, *Theory and Application of Mathieu Functions* (Oxford U. P., New York, 1951). We shall follow the notations for Mathieu functions used by McLachlan.
[8] P. M. Morse and K. U. Ingard, *Theoretical Acoustics* (McGraw–Hill, New York, 1968).
[9] G. Blanch, J. Math. Phys. **25**, 1 (1946); C. J. Bouwkamp, J. Math. Phys. **26**, 79 (1947).
[10] Similar behavior was observed in the treatment of the liquid coated prolate spheroid problem [C. Yeh, J. Acoust. Soc. Am. **46**, 797 (1969)].

Sonderabdruck aus Annalen der Physik 7. Folge · Bd. 13 · Heft 1–2 · 1964
VERLAG VON JOHANN AMBROSIUS BARTH IN LEIPZIG
Printed in Germany

The Diffraction of Sound Waves by Penetrable Disks[1])

By C. Yeh

With 1 figure

Abstract

The exact solution of the diffraction of plane sound waves by an acoustically penetrable oblate spheroid is obtained. Results are given in terms of spheroidal functions. It is found that each expansion coefficient of the scattered or transmitted wave is coupled to all coefficients of the series expansion for the incident wave except when the spheroid degenerates to a sphere. Decoupling of the coefficients for certain special cases are discussed.

I. Introduction

The problem of the scattering of sound waves by rigid disks or by circular apertures was first treated rigorously by Bouwkamp[2]) in 1941 using the oblate spheroidal wave functions. Spence[3]) apparently unaware of Bouwkamp's work, also considered this problem in a similar fashion in 1948. Subsequent treatments and numerical calculations have been given by Spence[3]), Leitner[4]), and Meixner and Fritze[5]). The exact solution of the corresponding problem of the diffraction of electromagnetic waves by a perfectly conducting circular disk was obtained by Meixner and Andrejewski[6]) and by Flammer[7]) in 1950 and 1953 respectively. However, the solution for the diffraction of sound (scalar) waves by penetrable disks has not been found. The purpose of this paper is to present the exact solution of this problem. It is shown that certain mathematical difficulties can be overcome by separating the scalar wave equation in the oblate spheroidal coordinates, and by applying the orthogonality properties of the spheroidal functions. It is noted that since the angular oblate spheroidal functions are not only functions of η, the angular coordinates, but also of the acoustical properties of the medium in which they apply, each expansion coefficient for the scattered or transmitted wave is coupled to all coeffi-

[1]) This study was supported by the Air Force Cambridge Research Laboratories.
[2]) C. J. Bouwkamp, "Theoretische en numerieke behandeling van be buiging door een ronde opening", Diss. Grongingen, Groningen-Batavia, 1941; also, J. Math. and Phys. **26**, 79 (1947).
[3]) R. D. Spence, J. Acoust. Soc. Amer. **20**, 380 (1948); J. Acoust. Soc. Amer. **21**, 98 (1948).
[4]) A. Leitner, J. Acoust. Soc. Amer. **21**, 331 (1949).
[5]) J. Meixner and U. Fritze, Z. angew. Physik **1**, 535 (1949).
[6]) J. Meixner and W. Andrejewski, Ann. Physik **7**, 157 (1950).
[7]) C. Flammer, J. Appl. Phys. **24**, 1218 (1953); J. Appl. Phys. **24**, 1224 (1953).

cients of the series expansion for the incident wave. Also noted is the fact that acoustically penetrable oblate spheroids are good approximations to many physical objects.

II. Formulation of the problem

To analyze this problem, the oblate spheroidal coordinates $(\xi, \eta, \varnothing)$, as shown in Fig. 1, are introduced. In terms of the spherical coordinates $(R, \theta, \varnothing)$, the spheroidal coordinates are defined by the following relations:

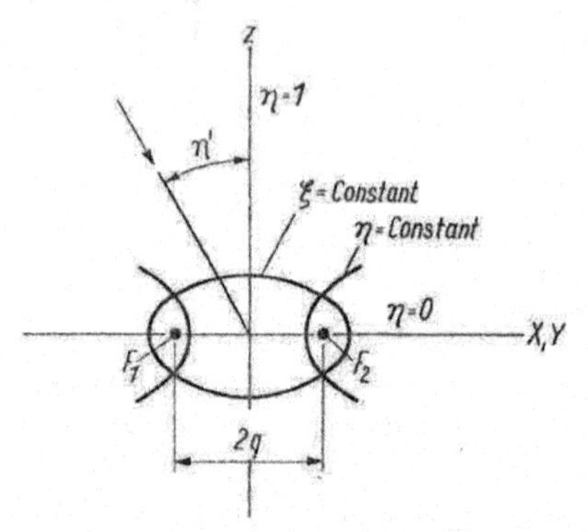

Fig. 1. The Oblate Spheroidal Coordinates. F_1 and F_2 are the foci. The distance between foci is the focal distance $2q$. The arrow indicates the direction of the incident wave

$$\begin{aligned} R &= q(\xi^2 - \eta^2 + 1)^{1/2} \\ \cos\theta &= \xi\eta/(\xi^2 - \eta^2 + 1)^{1/2} \\ \varnothing &= \varnothing \\ (0 \leq \xi &< \infty, -1 \leq \eta \leq 1, 0 \leq \varnothing \leq 2\pi) \end{aligned} \tag{1}$$

where q is the semi-focal length of the spheroid. The contour surface of constant ξ are confocal spheroids, and those of constant η are confocal hyperboloids. The boundary of the scattering obstacle is assumed to coincide with one of the confocal spheroids with $\xi = \xi_0$. The obstacle, having a density of ϱ_1 is assumed to be embedded in a medium having a density of ϱ_0. A possible solution of the scalar wave equation for a source-free homogeneous medium is then $R(\xi)\,\Theta(\eta)\,\Phi(\varnothing)\,e^{-i\omega t}$, where R, Θ and Φ satisfy respectively the differential equations

$$(\xi^2 + 1)\frac{d^2R}{d\xi^2} + 2\xi\frac{dR}{d\xi} - \left(A - q^2k^2\xi^2 - \frac{m^2}{\xi^2+1}\right)R = 0 \tag{2}$$

$$(1 - \eta^2)\frac{d^2\Theta}{d\eta^2} - 2\eta\frac{d\Theta}{d\eta} + \left(A + q^2k^2\eta^2 - \frac{m^2}{1-\eta^2}\right)\Theta = 0 \tag{3}$$

$$\frac{d^2\Phi}{d\varnothing^2} + m^2\Phi = 0 \tag{4}$$

in which A and m are the separation constants and $k^2 = \omega^2/c^2$, where c is the speed of sound in the medium. It appears that both R and Θ satisfy the same differential equation[8])

$$(1 - z^2)\frac{d^2W}{dz^2} - 2z\frac{dW}{dz} + \left(A - q^2k^2z^2 - \frac{m^2}{1-z^2}\right)W = 0. \tag{5}$$

The solutions of this equation have been discussed quite thoroughly by Chu and Stratton[9]), Meixner and Schäfke[10]), and Flammer[11]). These solutions are called the spheroidal wave functions.

---

[8]) We follow the notations adopted by Chu and Stratton; see also C. Flammer, "Spheroidal Wave Functions", Stanford University Press, Stanford, California, USA (1957). Note that the index (m,l) used here as well as in Flammer's book corresponds to the index $(m, l-m)$ used in Chu and Stratton's book.

[9]) L. J. Chu and J. A. Stratton, J. Math. and Phys. **20**, 259 (1941).

[10]) J. Meixner and F. W. Schäfke, "Mathieusche Funktionen und Sphäroidfunktionen", Springer-Verlag, Berlin, 1954.

[11]) C. Flammer, "Spheroidal Wave Functions", Stanford University Press, Stanford, California, 1957.

It is expected that the wave function remains finite at the poles $\eta = \pm 1$; thus the angular function will be represented by the angular spheroidal function of the first kind $S_{m,l}(-iqk, \eta)$ with the sequence in l according to increasing values of A. (This periodic spheroidal function of the first kind exists only for certain characteristic values of A.) The solutions of (2) corresponding to the angular function $S_{m,l}(-iqk, \eta)$, having the same characteristic values of A, are the radial spheroidal function of the first kind $R^{(1)}_{m,l}(-iqk, i\xi)$ and the radial spheroidal function of the second kind $R^{(2)}_{m,l}(-iqk, i\xi)$.

The proper choice of these radial functions to represent the wave function depends upon the boundary conditions. The wave function must be finite in all regions. In addition the scattered wave function must satisfy the radiation condition at infinity. Consequently the appropriate solutions of the wave equation for the region within the scattering obstacle are

$$R^{(1)}_{m,l}(-iqk_1, i\xi)\, S_{m,l}(-iqk_1, \eta) \begin{Bmatrix} \cos \\ \sin \end{Bmatrix} m\,\varnothing \tag{6}$$

and those for the scattered or radiated wave are

$$R^{(3)}_{m,l}(-iqk_0, i\xi)\, S_{m,l}(-iqk_0, \eta) \begin{Bmatrix} \cos \\ \sin \end{Bmatrix} m\,\varnothing, \tag{7}$$

where

$$R^{(3)}_{m,l}(-iqk_0, i\xi) = R^{(1)}_{m,l}(-iqk_0, i\xi) + i\, R^{(2)}_{m,l}(-iqk_0, i\xi)$$

and $k_0^2 = \frac{\omega^2}{c_0^2}$, $k_1^2 = \frac{\omega^2}{c_1^2}$. c_1 and c_0 are respectively the speed of sound in the obstacle and outside the obstacle. As $qk_0\xi \to \infty$, we find that

$$R^{(3)}_{m,l}(-iqk_0, i\xi) \to \frac{1}{q\,k_0\xi}\, e^{i\left(qk_0\xi - \frac{l+1}{2}\pi\right)}. \tag{8}$$

III. Scattering of a plane Wave

A plane pressure wave is assumed to be incident upon a penetrable oblate spheroidal disk at a direction specified by the polar angle θ' and the azimuthal angle $\varnothing'$. (See Fig. 1.) It can be shown that the incident plane pressure wave, having an intensity of γ_0, may be given by[11])

$$\begin{aligned} p_i = 2P_0 \sum_{l=0}^{\infty} \sum_{m=0}^{\infty} (i)^l \frac{2-\delta_m^0}{N_{m,l}} \\ \times S_{m,l}(-iqk_0, \eta)\, S_{m,l}(-iqk_0, \eta')\, R^{(1)}_{m,l}(-iqk_0, i\xi) \cos m(\varnothing - \varnothing') \end{aligned} \tag{9}$$

where $P_0 = \sqrt{2\varrho_0 c_0 \gamma_0}$, $\eta' = \cos\theta'$, $N_{m,l} = \int_{-1}^{1} S^2_{m,l}(-iqk_0, \eta)\, d\eta$, and δ_m^0 is the Kronecker Delta. $N_{m,l}$ is given in the Appendix. $\varnothing' = 0$ may be used without loss of generality. Referring to (6) and (7), one notes that the scattered wave and the transmitted wave inside the spheroid must be of the form

$$\begin{aligned} p_s = 2P_0 \sum_{l=0}^{\infty} \sum_{m=0}^{\infty} (i)^l \frac{2-\delta_m^0}{N_{m,l}} \\ \times A_{m,l}\, S_{m,l}(-iqk_0, \eta)\, S_{m,l}(-iqk_0, \eta')\, R^{(3)}_{m,l}(-iqk_0, i\xi) \cos m\varnothing \end{aligned} \tag{10}$$

and

$$p_t = 2P_0 \sum_{l=0}^{\infty} \sum_{m=0}^{\infty} (i)^l \frac{2-\delta_m^0}{N_{m,l}} \times B_{m,l} S_{m,l}(-iqk_1, \eta) S_{m,l}(-iqk_0, \eta') R_{m,l}^{(1)}(-iqk_1, i\xi) \cos m\varnothing \tag{11}$$

where $A_{m,l}$ and $B_{m,l}$ are arbitrary unknown coefficients that can be determined by applying the boundary conditions.

The boundary conditions require both that the pressure variation be continuous and that the normal components of particle velocity be continuous at the boundary. The particle velocity u is related to the pressure variation p by the relation[12])

$$u = \frac{1}{i\omega\varrho} \nabla p \tag{12}$$

where ϱ is the density of the medium. Matching the boundary conditions at $\xi = \xi_0$ gives

$$\begin{aligned} &\sum_{l=0}^{\infty} \sum_{m=0}^{\infty} (i)^l \frac{2-\delta_m^0}{N_{m,l}} [R_{m,l}^{(1)}(-iqk_0, i\xi_0) + A_{m,l} R_{m,l}^{(3)}(-iqk_0, i\xi_0)] \\ &\qquad \times S_{m,l}(-iqk_0, \eta) S_{m,l}(-iqk_0, \eta') \cos m\varnothing \\ &= \sum_{l=0}^{\infty} \sum_{m=0}^{\infty} (i)^l \frac{2-\delta_m^0}{N_{m,l}} B_{m,l} R_{m,l}^{(1)}(-iqk_1, i\xi_0) \\ &\qquad \times S_{m,l}(-iqk_1, \eta) S_{m,l}(-iqk_0, \eta) \cos m\varnothing, \end{aligned} \tag{13}$$

$$\begin{aligned} &\sum_{l=0}^{\infty} \sum_{m=0}^{\infty} (i)^l \frac{2-\delta_m^0}{N_{m,l}} \left[\frac{d}{d\xi_0} R_{m,l}^{(1)}(-iqk_0, i\xi_0) + A_{m,l} \frac{d}{d\xi_0} R_{m,l}^{(3)}(-iqk_0, i\xi_0)\right] \\ &\qquad \times S_{m,l}(-iqk_0, \eta) S_{m,l}(-iqk_0, \eta) \cos m\varnothing \\ &= \frac{\varrho_0}{\varrho_1} \sum_{l=0}^{\infty} \sum_{m=0}^{\infty} (i)^l \frac{2-\delta_m^0}{N_{m,l}} B_{m,l} \frac{d}{d\xi_0} R_{m,l}^{(1)}(-iqk_1, i\xi_0) \\ &\qquad \times S_{m,l}(-iqk_1, \eta) S_{m,l}(-iqk_0, \eta') \cos m\varnothing. \end{aligned} \tag{14}$$

It is noted that in contrast with the spherical and circular cylindrical case the polar angular functions in the spheroidal case are functions not only of the angular component η but also of the characteristics of the medium. Consequently, the summation signs on l and the angular spheroidal functions in the above equations may not be omitted. However, it will be shown that this difficulty may be overcome by the use of the orthogonality properties of the angular spheroidal function. Substituting the expansion

$$S_{m,l}(-iqk_0, \eta) = \sum_{n=0}^{\infty} \alpha_{l,n}^{(m)} S_{m,n}(-iqk_1, \eta) \tag{15}$$

---

[12]) P. M. Morse, "Vibration and Sound", McGraw-Hill Book Company, New York, 1948.

nto equations (13) and (14), and applying the orthogonality relations of the pheroidal functions, one obtains

$$\sum_{l=0}^{\infty} \frac{(i)^l}{N_{m,l}} [R_{m,l}^{(1)}(-iqk_0, i\xi_0) + A_{m,l} R_{m,l}^{(3)}(-iqk_0, i\xi_0)] S_{m,l}(-iqk_0, \eta') \alpha_{l,n}^{(m)} \\ = B_{m,n} R_{m,n}^{(1)}(-iqk_1, i\xi_0) S_{m,n}(-iqk_0, \eta'), \frac{(i)^n}{N_{m,n}}, \tag{16}$$

$$\sum_{l=0}^{\infty} \frac{(i)^l}{N_{m,l}} \left[\frac{d}{d\xi_0} R_{m,l}^{(1)}(-iqk_0, i\xi_0) + A_{m,l} \frac{d}{d\xi_0} R_{m,l}^{(3)}(-iqk_0, i\xi_0)\right] S_{m,l}(-iqk_0, \eta') \alpha_{l,n}^{(m)} \\ = \frac{\varrho_0}{\varrho_1} B_{m,n} \frac{d}{d\xi_0} R_{m,n}^{(1)}(-iqk_1, i\xi_0) S_{m,n}(-iqk_0 \cdot \eta') \frac{(i)^n}{N_{m,n}}, \tag{17}$$

$$(n = 0,1,2 \ldots .).$$

$\alpha_{l,n}^{(m)}$ is given in the Appendix. Solving (16) and (17) for $A_{m,l}$ gives (in matrix notation)

$$A_{m,l} = [Q_{l,n}^{(m)}]^{-1} D_{m,n} \tag{18}$$

where $[Q_{l,n}^{(m)}]^{-1}$ is the inverse of the square matrix

$$\left[\frac{(i)^l}{N_{m,l}} S_{m,l}(-iqk_0, \eta') \alpha_{l,n}^{(m)} \left[R_{m,l}^{(3)}(-iqk_0, i\xi_0) \right.\right. \\ \left.\left. - \frac{\varrho_1}{\varrho_0} \frac{R_{m,n}^{(1)}(-iqk_1, i\xi_0)}{\frac{d}{d\xi_0} R_{m,n}^{(1)}(-iqk_0, i\xi_0)} \frac{d}{d\xi_0} R_{m,l}^{(3)}(-iqk_0, \xi_0)\right]\right].$$

The index of this square matrix is (l, n). $D_{m,n}$ is a column matrix in n

$$\left[(-1) \sum_{l=0}^{\infty} \frac{(i)^l}{N_{m,l}} S_{m,l}(-iqk_0, \eta') \alpha_{l,n}^{(m)} \left[R_{m,l}^{(1)}(-iqk_0, i\xi_0) \right.\right. \\ \left.\left. - \frac{\varrho_1}{\varrho_0} \frac{R_{m,n}^{(1)}(-iqk_1, i\xi_0)}{\frac{d}{d\xi_0} R_{m,n}^{(1)}(-iqk_1, i\xi_0)} \frac{d}{d\xi_0} R_{m,l}^{(1)}(-iqk_0, i\xi_0)\right]\right].$$

The expansion coefficients for the transmitted wave, $B_{m,n}$, can easily be obtained using equations (18) and (16).

The expressions for the scattered wave and the transmitted wave are now completely determined. At large distances from the spheroid the asymptotic expression for $R_{m,l}^{(3)}(-iqk_0, i\xi)$ leads to

$$p_s \approx 2 P_0 \frac{(-i)}{qk_0\xi} e^{-i(\omega t - qk_0\xi)} \sum_{l=0}^{\infty} \sum_{m=0}^{\infty} \frac{2 - \delta_m^0}{N_{m,l}} A_{m,l} \\ \times S_{m,l}(-iqk_0, \eta) S_{m,l}(-iqk_0, \eta') \cos m\Phi. \tag{19}$$

The radiation pattern of the scattered wave is determined by the expression

$$|g(\eta, \Phi)| = \frac{1}{k_0} \left| \sum_{l=0}^{\infty} \sum_{m=0}^{\infty} \frac{2(2 - \delta_m^0)}{N_{m,l}} A_{m,l} \right. \\ \left. \times S_{m,l}(-iqk_0, \eta) S_{m,l}(-iqk_0, \eta') \cos m\Phi \right|, \tag{20}$$

where the vertical bars indicate that the absolute value is to be taken. The relative total scattering cross section is defined as

$$\sigma_r = \frac{\int_0^{2\pi}\int_0^{\pi} |g(\cos\Theta, \emptyset)|^2 \sin\theta \, d\theta \, d\emptyset}{2S} \tag{21}$$

where S is the area of the geometrical shadow

$$S = \pi q^2[(\xi_0^2+1)/(\xi_0^2+\eta'^2)]^{1/2}[\xi_0^2(1-\eta'^2)+(\xi_0^2+1)\eta'^2]. \tag{22}$$

In other words,

$$\sigma_r = \frac{1}{2S}\left[\frac{1}{k_0^2}\sum_{l=0}^{\infty}\sum_{m=0}^{\infty}\frac{8(2-\delta_m^0)}{N_{m,l}} A_{m,l}A^*_{m,l}S^2_{m,l}(-iqk_0,\eta')\right] \tag{23}$$

where $A^*_{m,l}$ indicates the complex conjugate of $A_{m,l}$.

IV. Serveal special cases

The expression for the expansion coefficient $A_{m,l}$ may be somewhat simplified for the following special cases:

a) Small eccentricity

If the shape of the oblate spheroidal obstacle approximates that of a perturbed sphere, i. e., $q \ll 1$ and $\xi_0 \gg 1$, only the zeroth order term in equations (16) and (17) need be retained. We have

$$\begin{aligned}
&\alpha_{l,n}^{(m)} \approx 0 && l \neq n\\
&\alpha_{l,l}^{(m)} \approx d^{ml}_{l-m}(-iqk_1)/d^{ml}_{l-m}(-iqk_0) && l = n\\
&R^{(1)}_{m,l}(-iqk_0, i\xi_0) \approx j_l(qk_0\xi_0)\\
&R^{(3)}_{m,l}(-iqk_0, i\xi_0) \approx h_l^{(1)}(qk_0\xi_0)\\
&R^{(1)}_{m,l}(-iqk_1, i\xi_0) \approx j_l(qk_1\xi_0) && (24)\\
&\frac{d}{d\xi_0}R^{(1)}_{m,l}(-iqk_0, i\xi_0) \approx qk_0 j_l'(qk_0\xi_0)\\
&\frac{d}{d\xi_0}R^{(3)}_{m,l}(-iqk_0, i\xi_0) \approx qk_0 h_l^{(1)\prime}(qk_0\xi_0)\\
&\frac{d}{d\xi_0}R^{(1)}_{m,l}(-iqk_1, i\xi_0) \approx qk_1 j_l'(qk_1\xi_0)\\
&S_{m,l}(-iqk_0, \eta') \approx d^{ml}_{l-m}(-iqk_0)P_l^m(\eta)'
\end{aligned}$$

where j_l, $h_l^{(1)}$ and P_l^m are respectively the spherical Bessel function, the spherical Hankel function and the associated Legendre polynomial. $d^{ml}_{l-m}(-iqk_0)$ or $d^{ml}_{l-m}(-iqk_1)$ are the expansion coefficients for the angular spheroidal functions. The prime indicates the derivative of the function with respect to its argument. Putting equation (24) into (16) and (17) gives the expression for $A_{m,l}$ for an almost degenerate spheroid:

$$A_{m,l} = \frac{\frac{c_0\,\varrho_0}{c_1\,\varrho_1} j_l'(qk_1\xi_0)\,j_l(qk_0\xi_0) - j_l(qk_1\xi_0)\,j_l'(qk_0\xi_0)}{j_l(qk_1\xi_0)\,h_l^{(1)\prime}(qk_0\xi_0) - \frac{c_0\,\varrho_0}{c_1\,\varrho_1} j_l'(qk_1\xi_0)\,h_l^{(1)}(qk_0\xi_0)}. \tag{25}$$

This is the same expression given by Anderson[13]) for the expansion coefficients for the scattered wave for the problem of diffraction of a plane pressure wave by a penetrable spherical obstacle. provided that we set $qk_1\xi_0 \to k_1 a$ and $qk_0\xi_0 \to k_0 a$ where a is the radius of the degenerated sphere.

b) Long Wavelength

If the operating frequencies are such that $qk_0 \ll 1$, $qk_1\xi_0 \ll 1$, $qk_0 \ll 1$, and $qk_1 \ll 1$, certain simplifications are again possible. According to the theory of spheroidal functions[9])[10])[11]), we find that for small values of qk_0 and qk_1,

$$S_{m,l}(-iqk_0, \eta) \approx d^{ml}_{l-m}(-iqk_0)\, P^m_l(\eta), \tag{26}$$

$$S_{m,l}(-iqk_1, \eta) \approx d^{ml}_{l-m}(-iqk_1)\, P^m_l(\eta), \tag{27}$$

such that

$$\begin{aligned} \alpha^{(m)}_{l,n} &\approx 0 && \text{for } l \neq n \\ &\approx \alpha^{(m)}_{l,l} \approx d^{ml}_{l-m}(-iqk_1)/d^{ml}_{l-m}(-iqk_0) && \text{for } l = n \end{aligned} \tag{28}$$

Hence, the infinite summation signs in equations (16) and (17) may be ignored and the summation index l may be replaced by n. The expansion coefficients for the spheroidal functions, $d^{ml}_{l-m}(-iqk_1)$ or $d^{ml}_{l-m}(-iqk_0)$ for small values of qk_1 and qk_0 may be expanded in power series of qk_1 or qk_0. The coefficients for the power series have been tabulated by Flammer[11]). The arbitrary constants, $A_{m,n}$ and $B_{m,n}$, can be obtaind eeasily from the simplified equations mentioned above.

c) $k_1/k_0 \approx 1$

If the density of the scattering obstacle is very close to that of the surrounding medium, the approximation

$$\begin{aligned} \alpha^{(m)}_{l,n} &\approx 0 && l \neq n \\ &\approx \alpha^{(m)}_{l,l} \approx d^{ml}_{l-m}(-iqk_1)/d^{ml}_{l-m}(-iqk_0) && l = n \end{aligned} \tag{29}$$

may again be used. One notes that, with the above approximation, the two infinite sets of equations, (16) and (17), may be decoupled to give the following set of equations:

$$[R^{(1)}_{m,n}(-iqk_0, i\xi_0) + A_{m,n} R^{(3)}_{m,n}(-iqk_0, i\xi_0)]\,\alpha^{(m)}_{n,n} = B_{m,n} R^{(1)}_{m,n}(-iqk_1, i\xi_0) \tag{30}$$

$$\begin{aligned} &\left[\frac{d}{d\xi_0} R^{(1)}_{m,n}(-iqk_0, i\xi_0) + A_{m,n} \frac{d}{d\xi_0} R^{(3)}_{m,n}(-iqk_0, i\xi_0)\right] \alpha^{(m)}_{n,n} \\ &\quad = B_{m,n} \frac{\varrho_0}{\varrho_1} \frac{d}{d\xi_0} R^{(1)}_{m,n}(-iqk_1, i\xi_0) \end{aligned} \tag{31}$$

The arbitrary constants $A_{m,n}$ and $B_{m,n}$ can easily be obtained from the above equations.

---

[13]) V. C. Anderson, J. Acoust. Soc. Amer. **22**, 426 (1950).

V. Conclusions

By the use of the orthogonality properties of the spheroidal functions, the exact solution for the scattering of a plane pressure wave by a penetrable disk is obtained. It is noted that unlike the case for sphere or circular cylinder, each expansion coefficient for the scattered or transmitted wave is coupled with all expansion coefficients for the incident wave. This characteristic is also found in the problem of diffraction of electromagnetic wave by an elliptical dielectric cylinder[14]. Numerical computations can be carried out using the available tables on spheroidal functions prepared by Stratton, Morse, Chu, Little, and Corbató[15] or by Flammer[11]. Preliminary results show that the infinite series converge rather rapidly for small values of qk_0; only the first few terms of the series are needed as long as $k_0 q$ is less than 5.

It is remarked that the technique used in obtaining the solution for this problem is also applicable to similar type of problems, such as the diffraction of waves by a penetrable prolate spheroid or by a rigid strip coated with some accoustically penetrable material and other associated electromagnetic problems.

Appendix

Formulae for $\alpha_{l,n}^{(m)}$ and $N_{m,l}$

Multiplying both sides of (15) by $S_{m,n}(-iqk_1, \eta)$, integrating with respect to η from -1 to $+1$, and using the orthogonality relation for the spheroidal function, one obtains

$$\alpha_{l,n}^{(m)} = \frac{1}{M_{m,n}} \int_{-1}^{+1} S_{m,l}(-iqk_0, \eta)\, S_{m,n}(-iqk_1, \eta)\, d\eta$$

where

$$M_{m,n} = \int_{-1}^{+1} [S_{m,n}(-iqk_1, \eta)^2]\, d\eta.$$

It is known that the angular spheroidal function may be expanded in terms of associated Legendre polynomials[9] [10] [11]:

$$S_{m,l}(-iqk_0, \eta) = \sum_{r=0,1}^{\infty}{}' d_r^{m,l}(-iqk_0)\, P_{m+r}^m(\eta)$$

$$S_{m,l}(-iqk_1, \eta) = \sum_{r=0,1}^{\infty}{}' d_r^{m,l}(-iqk_1)\, P_{m+r}^m(\eta)$$

where $d_r^{m,l}(-iqk_0)$ and $d_r^{m,l}(-iqk_1)$ are the expansion coefficients which have been tabulated[11] [15], and the prime over the summation sign indicates that odd or even integer values of r are to be taken according as $l-m$ is odd or even. The integrals can now be evaluated with the help of the orthogonality characteri-

---

[14] C. Yeh, J. Math. Phys. 4, 65 (1963).

[15] J. A. Stratton, P. M. Morse, L. J. Chu, J. D. C. Little and F. J. Corbató, "Spheroidal Wave Functions", Technology Press of M. I. T. and John Wiley and Sons, Inc., New York, 1956.

stics of associated Legendre functions[16]):

$$\int_{-1}^{+1} S_{m,l}(-iqk_0, \eta) S_{m,n}(-iqk_1, \eta) d\eta$$

$$= \sum_{r=0,1}^{\infty}{}' d_r^{m,l}(-iqk_0) d_r^{m,n}(-iqk_1) \frac{2}{2m+2r+1} \frac{(r+2m)!}{r!}$$

$$\int_{-1}^{+1} [S_{m,n}(-iqk_1, \eta)]^2 d\eta$$

$$= \sum_{r=0,1}^{\infty}{}' [d_r^{m,n}(-iqk_1)]^2 \frac{2}{2m+2r+1} \frac{(r+2m)!}{r!}.$$

Thus,

$$\alpha_{l,n}^{(m)} = \frac{\sum_{r=0,1}^{\infty}{}' d_r^{m,l}(-iqk_0) d_r^{m,n}(-iqk_1) \frac{2}{2m+2r+1} \frac{(r+2m)!}{r!}}{\sum_{r=0,1}^{\infty}{}' [d_r^{m,n}(-iqk_1)]^2 \frac{2}{2m+2r+1} \frac{(r+2m)!}{r!}}.$$

Similarily, one gets

$$N_{m,l} = \int_{-1}^{+1} [S_{m,l}(-iqk_0, \eta)]^2 d\eta$$

$$= \sum_{r=0,1}^{\infty}{}' [d_r^{m,l}(-iqk_0)]^2 \frac{2}{2m+2r+1} \frac{(r+2m)!}{r!}.$$

[16]) W. Magnus and F. Oberhettinger, "Formulas and Theorems for the Functions of Mathematical Physics", Chelsea Publishing Company, New York, 1954.

Los Angeles, California. Electrical Engineering Department University of Southern California.

Bei der Redaktion eingegangen am 27. Mai 1963.

Reprinted from THE JOURNAL OF THE ACOUSTICAL SOCIETY OF AMERICA, Vol. 42, No. 2, 518–521, August 1967

Scattering of Acoustic Waves by a Penetrable Prolate Spheroid. I. Liquid Prolate Spheroid*

C. YEH

Electrical Engineering Department, University of Southern California, Los Angeles, California 90007

This is the first of a series of reports on the scattering of waves by a penetrable prolate spheroid. In this letter, the exact solution of the scattering of acoustic waves by a liquid prolate spheroid is obtained. Results are given in terms of the spheroidal functions. It is found that each expansion coefficient of the scattered or transmitted wave is coupled to all coefficients of the series expansion for the incident wave, except when the spheroid degenerates to a sphere. Detailed numerical computations for the radiation patterns of the scattered wave are carried out for a sphecific example. It is found that, in general, the radiation patterns for the scattered waves are more directional and possess more side lobes as the prolate spheroid becomes more elongated.

DUE TO THE EXTREMELY HIGH ATTENUATION CHARACTERISTICS OF electromagnetic waves in sea water,[1] it appears that the use of acoustic waves for underwater detection is still one of the best ways. In order to obtain the required information concerning th scattering object, it is necessary to identify the returned signal In other words, one would like to be able to "see" underwater. I is well known that the scattered wave from an underwater objec is not only a function of the composition of the scattering objec but also of its physical shape.[2] However, in order that this probler may be rendered to theoretical analysis, most investigators mak the assumption that the scattering object is either an elasti sphere, an elastic circular cylinder, or a rigid spheroid.[3]

It has become painfully clear that the scattering characteristic of a body of spherical shape and a body of spheroidal shape ar ry different. It is not possible to extrapolate from the solution : a spherically shaped elastic body any information concerning e scattering by a spheroidally shaped elastic body. The major ficulty encountered in solving the elastic spheroid problem is e inseparability of vector wave equations in spheroidal co- dinates. Furthermore, owing to the complexity of spheroidal nctions and the lack of appropriate expressions suitable for nu- erical computation, very laborious computation had to be rried out in order to obtain numerical values.

Because of the importance of the spheroid problem with regard the scattering by nonspheroidal bodies, as well as its relevance the scattering by various physical objects, such as, submarines, ales, etc., the task of solving this problem has been undertaken. s an initial step towards the solution of diffraction of waves by a neral penetrable prolate spheroid such as the elastic prolate heroid, it is proposed that the problem of the diffraction by a uid prolate spheroid be considered. The liquid prolate spheroid the limiting case of an elastic spheroid with zero shear modulus. he wave-penetrable characteristics are still retained for the liquid heroid case. Since the scalar-wave equation governing the liquid heroid problem is separable in the prolate spheroid coordinates, act solution can be found. The results from the liquid case con- itute a check on the results for the elastic case as the shear modu- s approaches zero. Furthermore, the liquid spheroids are good proximations to bodies of many physical objects underwater, d so, have a physical counterpart in their own right.

The problem of the diffraction of acoustic waves by a rigid olate spheroid was considered many years ago.[4] Recently, solu- ons have been found for the scattering of sound by an impene- able prolate spheroid with surface-impedance boundary condi- ons.[5] However, very few numerical results were given. The cor- sponding problem for the diffraction of waves by a penetrable olate spheroid has not been considered. It is expected that, due the presence of multiple reflections within the spheroid, the attering characteristics for the penetrable spheroid will be quite fferent from those for an impenetrable one. Further complica- ons are introduced in matching the boundary conditions that re not present in the impenetrable case since the angular prolate heroidal functions are not only function of η, the angular co- rdinates, but also of the acoustical properties of the medium in hich they apply.

A harmonic-plane pressure wave in an acoustic medium, which characterized by (ρ_0, c_0), is assumed to be incident upon a pene- able prolate spheroid whose acoustic medium is characterized y (ρ_1, c_1) (see Fig. 1); ρ and c are, respectively, the density and the ound speed of the medium. The scattered wave can be repre- ented by the following series[6]:

$$p_{scat}=2P_0\sum_{m=0}^{\infty}\sum_{n=m}^{\infty} i^n\frac{(2-\delta_{0m})}{N_{mn}}A_{mn}S_{mn}(h_0,\eta_0)\times S_{mn}(h_0,\eta)R_{mn}^{(3)}(h_0,\xi)\cos m\Phi, \quad (1)$$

where

$$N_{mn}=\int_{-1}^{1}S_{mn}^2(h_0,\eta)d\eta, \delta_{0m}$$

s the Kronecker delta, $\eta=\cos\theta_0$, and θ_0 is the angle of incidence, $h_0=\omega q/c_0$, where ω is the frequency of the incident wave and q the emifocal length of the prolate spheroid. A time dependence of $e^{-i\omega t}$ is assumed and suppressed throughout. The notation for the pheroidal functions adopted by Flammer[7] is used. P_0 is the ampli- ude of the incident-plane wave. A_{mn} is determined from the oundary conditions. We have

$$\sum_{n=m}^{\infty}{}' i^n\beta_{ln}^{(m)}\gamma_{ln}^{(m)}A_{mn}=\sum_{n=m}^{\infty}{}' i^n\gamma_{ln}^{(m)r}\beta_{ln}^{(m)}, \quad (2)$$

$$(m=0,1,2,\cdots)$$
$$(l=m, m+1,\cdots)$$

with

$$\beta_{ln}^{(m)}=[(2-\delta_{0m})/N_{mn}]\alpha_{ln}^{(m)}S_{mn}(h_0,\eta_0),$$
$$\gamma_{ln}^{(m)}=\gamma_{ln}^{(m)r}+i\gamma_{ln}^{(m)i},$$

$$\gamma_{ln}^{(m)r}=\frac{\rho_0}{\rho_1}R_{mn}^{(1)}(h_0,\eta_0)\frac{d}{d\xi_0}R_{ml}^{(1)}(h_1,\xi_0)-R_{ml}^{(1)}(h_1,\xi_0)\frac{d}{d\xi_0}R_{mn}^{(1)}(h_0,\xi_0),$$

$$\gamma_{ln}^{(m)i}=\frac{\rho_0}{\rho_1}R_{mn}^{(2)}(h_0,\xi_0)\frac{d}{d\xi_0}R_{ml}^{(1)}(h_1,\xi_0)-R_{ml}^{(1)}(h_1,\xi_0)\frac{d}{\xi_0}R_{mn}^{(2)}(h_0,\xi_0), \quad (3)$$

$$\alpha_{ln}^{(m)}=\sum_{r=0,1}^{\infty}\frac{(r+2m)!}{(2r+2m+1)r!}d_r^{mn}(h_0)d_r^{ml}(h_1)\bigg/\sum_{r=0,1}^{\infty}\frac{(r+2m)!}{(2r+2m+1)r!}[d_r^{ml}(h_1)]^2,$$

where d_r^{mn} and d_r^{ml} are the expansion coefficients of angular-spheroidal function. In Eq. 3, even values of r are summed over when $(n-m)$ is even, and odd values of r are summed over when $(n-m)$ is odd. Furthermore, if l is odd n must be odd, and if l is even n must also be even. The prime on the summation sign means that when l is odd, the series in Eq. 2 are summed over all odd values of n, and when l is even, the series are summed over all even values of n.

Of special interest are the far-zone field and the radiation pattern of the scattered wave. From Eq. 2, we have

$$p_{scat}(\text{FAR-ZONE})\simeq(e^{ih_0\xi}/h_0\xi)P_0F(\eta,\Phi), \quad (4)$$

where

$$F(\eta,\Phi)=\frac{2}{i}\sum_{m=0}^{\infty}\sum_{n=m}^{\infty}\frac{2-\delta_{0m}}{N_{mn}}A_{mn}S_{mn}(h_0,\eta_0)S_{mn}(h_0,\eta)\cos m\Phi. \quad (5)$$

It is interesting to note, that unlike the case for a fluid sphere, or the case for a perfectly conducting prolate spheroid, each expansion coefficient of the scattered or transmitted wave for the liquid prolate spheroid is coupled to all coefficients of the series expansion for the incident wave. This characteristic is also found in the solution for the diffraction of waves by a dielectric-elliptical cylinder.[8] To illustrate qualitatively how the solution behaves, the radiation patterns of the scattered wave are computed. Numerical computations are carried out using the newly compiled tables on the radial prolate spheroidal functions.[9] However, the computation is by no means trivial, even with the help of modern high-speed computers. The complexity of the numerical computation can best be appreciated by noting that it is necessary to evaluate

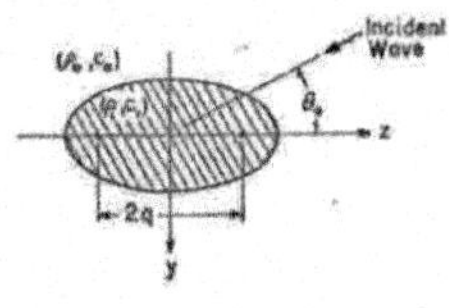

$$x=\frac{q}{2}[(1-\eta^2)(\xi^2-1)]^{1/2}\cos\phi$$
$$y=\frac{q}{2}[(1-\eta^2)(\xi^2-1)]^{1/2}\sin\phi$$
$$z=\frac{q}{2}\eta\xi$$
$$(-1\le\eta\le1,\ 1\le\xi<\infty,\ 0\le\phi\le2\pi)$$

FIG. 1. The geometry of the problem. (ξ,η,Φ) are the prolate spheroidal coordinates and q is the semifocal length.

a matrix of order $n\times n$, m times, in order to obtain the coefficient A_{mn} from Eq. 2. Each element in the matrix contains a combination of radial prolate spheroidal functions that are expanded in terms of infinite series of spherical Bessel functions. The evaluation of the infinite matrix, Eq. 2, was carried out by the successive-approximation method.[10] In other words, computations were carried out for a 4×4 matrix, a 5×5 matrix, a 6×6 matrix, etc., until the desired accuracy was reached. It was found (numerically) that the infinite matrix converges quite rapidly for the parameters that we considered below. At no time was any matrix greater than 10×10 (with $m=0, 1, \cdots, 6$) required to achieve an accuracy of three significant figures.

In order to bring out the effects of varying the geometric shape of the scattering object upon the radiation patterns of the scattered wave, it is assumed that the normalized volume of the scattering object remains constant, while ξ_0, or the eccentricity of th prolate spheroid, varies.[11] The normalized volume v of a prola spheroid is $(4\pi/3)h_0^3\xi_0(\xi_0^2-1)$. Two cases of v are considered: or with $v=4\pi/3$, and the other with $v=40\pi/3$. Three values of are used: 100 (sphere case), 1.05, and 1.006. It is assumed th $c_1=c_0$ and $\rho_0/\rho_1=2.0$. Results of the computation are shown Figs. 2 and 3. Figure 2 gives four diagrams indicating the radiatio patterns of the scattered wave for four different angles of incidenc with $v=4\pi/3$. In each of the diagrams in Fig. 2, three differer ξ_0, representing three different eccentricities, are used. As ex pected, owing to the symmetrical properties of the scatterin prolate spheroid, the radiation patterns are symmetrical abou $\theta=90°$ axis for an incident wave in the $\theta_0=90°$ direction, and a symmetrical about $\theta=0°$ axis for an incident wave in the $\theta_0=0$ direction. As the spheroid becomes more elongated, i.e., as

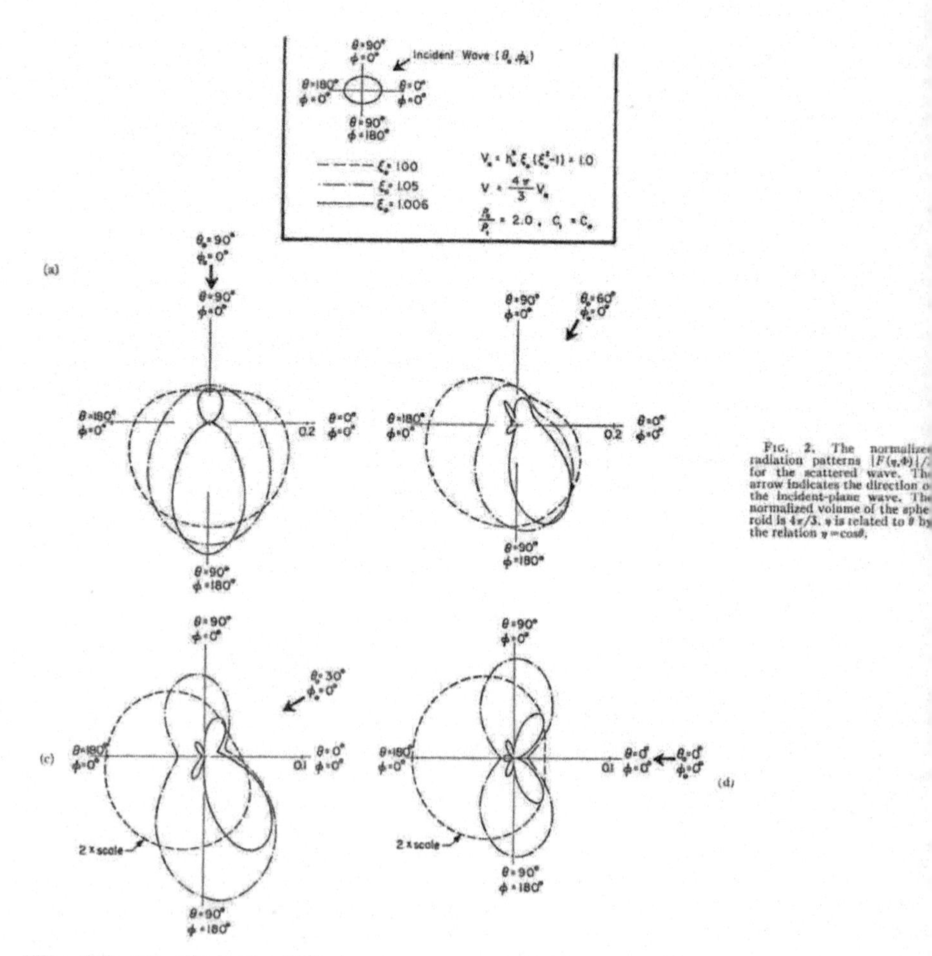

FIG. 2. The normalize radiation patterns $|F(\eta,\Phi)|/$ for the scattered wave. Th arrow indicates the direction o the incident-plane wave. The normalized volume of the sphe roid is $4\pi/3$. η is related to θ by the relation $\eta=\cos\theta$.

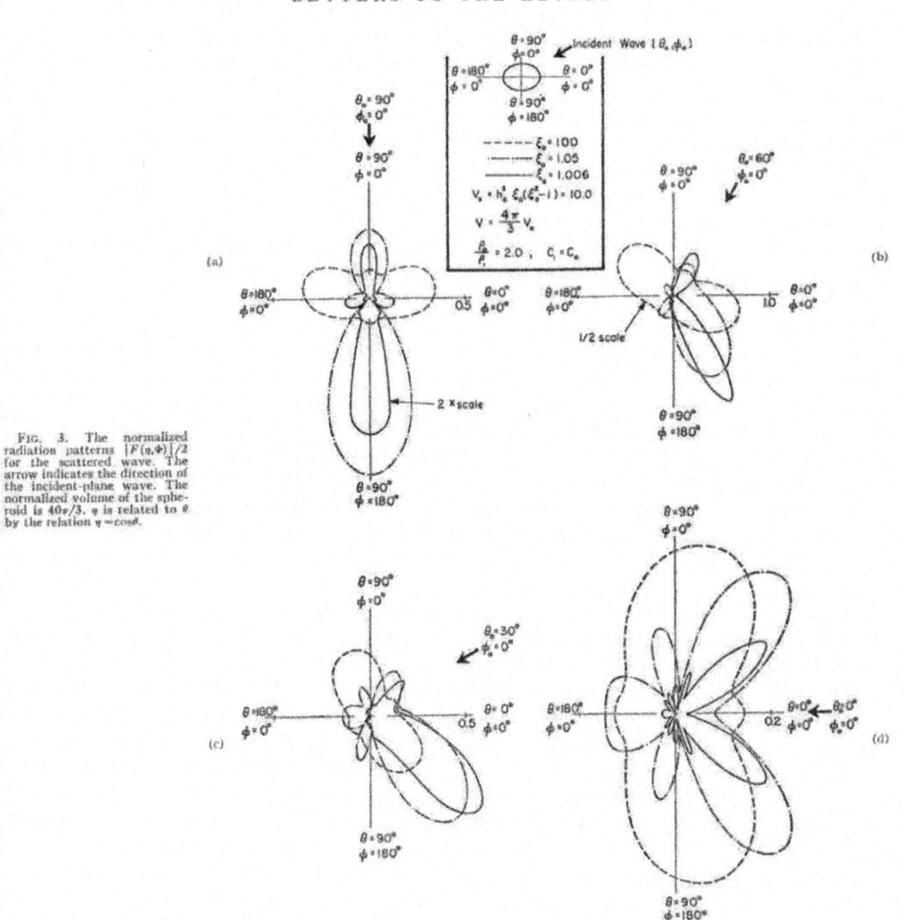

FIG. 3. The normalized radiation patterns $|F(\eta,\phi)|/2$ for the scattered wave. The arrow indicates the direction of the incident-plane wave. The normalized volume of the spheroid is $40\pi/3$. η is related to θ by the relation $\eta = \cos\theta$.

changes from 100 to 1.006, the radiation pattern becomes more directional and the multilobe structure becomes more prominent. For obliquely incident plane wave, as the spheroid becomes more elongated, the radiation pattern becomes more asymmetric about the axis of the incident wave; the multilobe characteristics still remain for elongated spheroids.

Figure 3 gives the radiation patterns for a larger spheroid, i.e., $v = 40\pi/3$. The discussion given above for Fig. 2 still applies here, except that in Fig. 3 more side lobes are evident. Furthermore, the main lobes for the elongated spheroid, as shown in Fig. 3, are more directional.

In conclusion, one notes that, in general, changing the spherical scattering body to an elongated spheroidal scattering body causes the radiation pattern for the scattered wave to be more directional and to have more side lobes. As stated in the introduction of this paper, the solution of the scattering of sound waves by a liquid prolate spheroid is just an initial step towards the solution for many problems involving penetrable prolate spheroids. The next step is to consider the elastic prolate spheroid problem.

* Supported by the Office of Naval Research.

[1] L. Batchelder, Proc. IEEE 53, 1310–1319 (1965).
[2] P. M. Morse, *Vibration and Sound* (McGraw-Hill Book Co., New York, 1948).
[3] W. P. Mason, Ed., *Physical Acoustics* (Academic Press Inc., New York, 1964).
[4] R. D. Spence and S. J. Granger, J. Acoust. Soc. Am. 23, 701–706 (1951).
[5] A. Silbiger, J. Acoust. Soc. Am. 35, 564–567 (1963).
[6] The formal solution to the diffraction of waves by a penetrable oblate spheroid has been given earlier [C. Yeh, Ann. Physik 13, 53–61 (1964)]. However, no numerical results were presented.
[7] C. Flammer, *Spheroidal Wave Functions* (Stanford University Press, Stanford, Calif. 1957).
[8] C. Yeh, J. Math. Phys. 4, 65–71 (1963).
[9] C. Chang and C. Yeh, "The Radial Prolate Spheroidal Functions," USCEE Rept. 166, Elec. Eng. Dept., Univ. of Southern California, Los Angles, Calif. (1966), pp. 524.
[10] L. Kantorovich and V. Krylov, *Approximate Methods of Higher Analysis* (Interscience Publishers, Inc., New York, 1958).
[11] The ratio of major axis/minor axis of a prolate spheroid is related to ξ_0 by the following formula: major axis/minor axis $= \xi_0/(\xi_0^2-1)^{\frac{1}{2}}$.

Reprinted from

THE JOURNAL OF THE ACOUSTICAL SOCIETY OF AMERICA

Volume 46 Number 3 (Part 2) SEPTEMBER 1969

Received 28 February 1969

13.6; 11.2

Scattering by Liquid-Coated Prolate Spheroids

C. Yeh

School of Engineering and Applied Science, University of California, Los Angeles, California 90024

The exact solution of the diffraction of plane sound waves by a rigid prolate spheroid coated with a confocal sheath of penetrable acoustic material is obtained. Results are given in terms of spheroidal functions. Detailed numerical computations for the radiation patterns of the scattered wave are carried out for certain acoustic parameters. It is found that the coating alters quite significantly the scattering characteristics of an incident wave. Furthermore, because of the coupling between each expansion coefficient for the scattered wave with all expansion coefficients for the incident wave for the coated spheroid case, the Watson transformation technique cannot be used directly to yield expressions that are applicable at high frequencies.

INTRODUCTION

Most previous theoretical analyses[1-6] were carried out with the assumption that the scattering objects were either elastic or fluid spheres, elastic or fluid cylinders, or rigid or soft spheroids. Because of the importance of the penetrable-spheroid problem with regard to the scattering by nonspherical bodies, the task of solving this problem was initiated. In a previous paper,[7] the problem of the scattering of waves by a liquid prolate spheroid was solved. Many interesting results concerning the radiation patterns of the scattered waves were found.

The purpose of the present work is to extend previously developed techniques in solving the liquid-prolate-spheriod problem to the problem of the scattering by a rigid prolate spheroid coated with a confocal sheath of penetrable acoustic material. The presence of even a very thin layer of acoustic material upon a rigid spheroid seems to alter quite drastically the scattering characteristics of an incident wave. It is further noted that the results obtained according to the present exact treatment will be very useful to those who are developing approximate or numerical techniques to solve the problems involving coated nonspherical obstacles.

[1] P. M. Morse and K. U. Ingard, *Theoretical Acoustics* (McGraw-Hill Book Co., New York, 1968).

[2] W. P. Mason, Ed., *Physical Acoustics* (Academic Press Inc., New York, 1964).

[3] R. D. Spence and S. J. Granger, J. Acoust. Soc. Amer. 23, 701–706 (1951).

[4] J. J. Faran, J. Acoust. Soc. Amer. 23, 405–418 (1951).

[5] R. Hickling, J. Acoust. Soc. Amer. 34, 1582–1586 (1962).

[6] R. D. Doolittle and H. Überall, J. Acoust. Soc. Amer. 43, 1–14 (1968).

[7] C. Yeh, J. Acoust. Soc. Amer. 42, 518 (L) (1967).

I. FORMULATION OF THE PROBLEM

To analyze this problem, the prolate-spheroidal coordinates (ξ,η,ϕ), as shown in Fig. 1, are introduced. In terms of the spherical coordinates (R,θ,ϕ), the spheroidal coordinates are defined by the following relations:

$$R=q(\xi^2+\eta^2-1)^{\frac{1}{2}},$$
$$\cos\theta=\xi\eta/(\xi^2+\eta^2-1)^{\frac{1}{2}}, \tag{1}$$
$$\phi=\phi,\quad (\xi\geq 1,\ -1\leq\eta\leq 1,\ 0\leq\phi\leq 2\pi),$$

where q is the semifocal length of the spheroid. The contour surface of constant ξ are confocal spheroids, and those of constant η are confocal hyperboloids. The outer boundary of a rigid prolate spheroid coated with a sheath of acoustic material characterized by (ρ_1,c_1) is assumed to coincide with one of the confocal spheroids with $\xi=\xi_0$; the inner rigid boundary is assumed to coincide with one of the confocal spheroids with $\xi=\xi_1$. It is further assumed that this coated spheroid is embedded in a homogeneous acoustic medium charac-

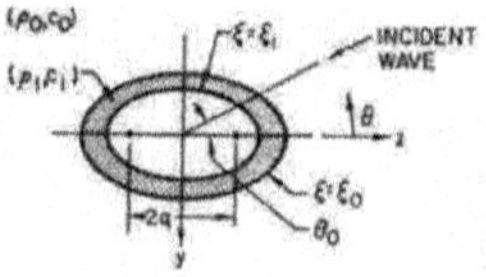

(ρ_0,c_0) (ρ_1,c_1) $\xi=\xi_1$ INCIDENT WAVE θ z $\xi=\xi_0$ θ_0 $2q$ y

Fig. 1. The coated prolate spheroid. ρ: Density. C: Speed of sound. q: Semifocal length.

rized by (ρ_0, c_0). ρ and c are, respectively, the density nd the sound speed of the medium.

The formal time-harmonic solutions of the source-ee scalar wave equation, which is

$$[\nabla^2+(\omega/c)^2]p=0, \quad (2)$$

here p is the scalar pressure, ω is the frequency, and is the speed of sound in the acoustic medium of terest, in terms of the spheroidal functions are[8]

$$\{R_{mn}^{(1),(2)}(h,\xi)\}\{S_{mn}(h,\eta)\}\begin{Bmatrix}\cos\\ \sin\end{Bmatrix} m\phi, \quad (3)$$

here $h=\omega q/c$, $R_{mn}^{(1),(2)}(h,\xi)$ are the radial prolate-pheroidal functions and $S_{mn}(h,\eta)$ are the angular rolate-spheroidal function. A time dependence of $e^{-i\omega t}$ is assumed and suppressed throughout.

A harmonic plane pressure wave having an amplitude f P_0 is assumed to be incident upon the coated prolate pheroid (see Fig. 1). Expressing in terms of the sphe-oidal functions, we have, for the incident wave,

$$p_i=2P_0\sum_{m=0}^{\infty}\sum_{n=m}^{\infty} i^n\frac{(2-\delta_{0m})}{N_{mn}}S_{mn}(h_0,\eta_0)$$

$$\times\{S_{mn}(h_0,\eta)R_{mn}^{(1)}(h_0,\xi)\cos m\phi\}, \quad (4)$$

or the scattered wave,

$$p_s=2P_0\sum_{m=0}^{\infty}\sum_{n=m}^{\infty} i^n\frac{(2-\delta_{0m})}{N_{mn}}A_{mn}S_{mn}(h_0,\eta_0)$$

$$\times\{S_{mn}(h_0,\eta)R_{mn}^{(3)}(h_0,\xi)\}\cos m\phi, \quad (5)$$

and for the penetrated wave within the sheath,

$$p_t=2P_0\sum_{m=0}^{\infty}\sum_{n=m}^{\infty} B_{mn}S_{mn}(h_1,\eta)$$

$$\times\{R_{mn}^{(1)}(h_1,\xi)-C_{mn}R_{mn}^{(2)}(h_1,\xi)\}\cos m\phi, \quad (6)$$

where $h_{0,1}=\omega q/c_{0,1}$,

$$N_{mn}=\int_{-1}^{1} S_{mn}^2(h_0,\eta)d\eta,$$

δ_{0m} is the Kronecker delta, $\eta_0=\cos\theta_0$, and θ_0 is the angle of incidence. A_{mn}, B_{mn}, and C_{mn} are arbitrary constants that are to be determined by the boundary conditions.

II. MATHEMATICAL SOLUTION

The boundary conditions require the continuity of the pressure and the normal components of the particle velocity at the boundary surface of the two fluids, $\xi=\xi_0$. On the rigid surface $\xi=\xi_1$, the particle velocity normal to the boundary must be zero. In other words,

$$p_i+p_s=p_t|_{\xi=\xi_0}, \quad (7)$$

$$\frac{1}{\rho_0}\left(\frac{\partial p_i}{\partial\xi}+\frac{\partial p_s}{\partial\xi}\right)=\frac{1}{\rho_1}\frac{\partial p_t}{\partial\xi}\bigg|_{\xi=\xi_0}, \quad (8)$$

$$\partial p_t/\partial\xi=0|_{\xi=\xi_1}. \quad (9)$$

Substituting Eqs. 4–6 into Eqs. 7–9 gives

$$\sum_n i^n\frac{(2-\delta_{0m})}{N_{mn}}S_{mn}(h_0,\eta_0)S_{mn}(h_0,\eta)[R_{mn}^{(1)}(h_0,\xi_0)$$

$$+A_{mn}R_{mn}^{(3)}(h_0,\xi_0)]=\sum_n B_{mn}S_{mn}(h_1,\eta)$$

$$\times\left[R_{mn}^{(1)}(h_1,\xi_0)-\frac{R_{mn}^{(1)\prime}(h_1,\xi_1)}{R_{mn}^{(2)\prime}(h_1,\xi_1)}R_{mn}^{(2)}(h_1,\xi_0)\right], \quad (10)$$

$$\sum_n i^n\frac{(2-\delta_{0m})}{N_{mn}}S_{mn}(h_0,\eta_0)S_{mn}(h_0,\eta)[R_{mn}^{(1)\prime}(h_0,\xi_0)$$

$$+A_{mn}R_{mn}^{(3)\prime}(h_0,\xi_0)]=\sum_n\frac{\rho_0}{\rho_1}B_{mn}S_{mn}(h_1,\eta)$$

$$\times\left[R_{mn}^{(1)\prime}(h_1,\xi_0)-\frac{R_{mn}^{(1)\prime}(h_1,\xi_1)}{R_{mn}^{(2)\prime}(h_1,\xi_1)}R_{mn}^{(2)\prime}(h_1,\xi_0)\right],$$

$$(m=0,1,\cdots), \quad (11)$$

where the prime on the radial prolate-spheroidal function denotes the derivative of the function with respect to ξ_0 or ξ_1, as appropriate. Multiplying both sides of Eqs. 10 and 11 by $S_{ml}(h_1,\eta)$ and integrating with respect to η from -1 to $+1$ gives

$$\sum_n i^n\frac{(2-\delta_{0m})}{N_{mn}}S_{mn}(h_0,\eta_0)\alpha_{ln}^{(m)}$$

$$\times[R_{mn}^{(1)}(h_0,\xi_0)+A_{mn}R_{mn}^{(3)}(h_0,\xi_0)]$$

$$=B_{ml}\left[R_{ml}^{(1)}(h_1,\xi_0)-\frac{R_{ml}^{(1)\prime}(h_1,\xi_1)}{R_{ml}^{(2)\prime}(h_1,\xi_1)}R_{ml}^{(2)}(h_1,\xi_0)\right], \quad (12)$$

$$\sum_n{}' i^n\frac{(2-\delta_{0m})}{N_{mn}}S_{mn}(h_0,\eta_0)\alpha_{ln}^{(m)}$$

$$\times[R_{mn}^{(1)\prime}(h_0,\xi_0)+A_{mn}R_{mn}^{(3)\prime}(h_0,\xi_0)]$$

$$=B_{ml}\frac{\rho_0}{\rho_1}\left[R_{ml}^{(1)\prime}(h_1,\xi_0)-\frac{R_{ml}^{(1)\prime}(h_1,\xi_1)}{R_{ml}^{(2)\prime}(h_1,\xi_1)}R_{ml}^{(2)\prime}(h_1,\xi_0)\right],$$

$$(m=0,1,2,\cdots),\ (l=m,m+1,\cdots), \quad (13)$$

[8] C. Flammer, *Spheroidal Wave Functions* (Stanford University Press, Stanford, Calif., 1957).

with

$$\alpha_{ln}^{(m)}=\int_{-1}^{+1} S_{mn}(h_0,\eta)S_{ml}(h_1,\eta)d\eta \Big/ \int_{-1}^{+1} S_{ml}^2(h_1,\eta)d\eta$$

$$=\sum_{r=0,1}^{\infty}{}' \frac{(r+2m)!}{(2r+2m+1)r!}d_r^{mn}(h_0)d_r^{ml}(h_1) \Big/ \sum_{r=0,1}^{\infty}{}' \frac{(r+2m)!}{(2r+2m+1)r!}[d_r^{ml}(h_1)]^2, \quad (14)$$

where d_r^{mn} and d_r^{ml} are the expansion coefficients of angular spheroidal function. The prime on the summation signs in Eqs. 12 and 13 means that l is odd, the series are summed over all odd values of n, and when l is even, the series are summed over all even values of n. In Eq. 14, even values of r are summed over when $(n-m)$ is even and odd values of r are summed over when $(n-m)$ is odd. Furthermore, if l is odd, n must also be odd; and if l is even, n must also be even.

Combining Eqs. 12 and 13 gives

$$\sum_n{}' P_{ln}^{(m)}A_{mn}=S_l^{(m)},$$

$$(m=0,\ 1,\ \cdots),\ (l=m,\ m+1,\ \cdots), \quad (15)$$

where

$$P_{ln}^{(m)}=\left[\left(\gamma_{ln}^{(m)r}\cos\frac{n\pi}{2}-\gamma_{ln}^{(m)i}\sin\frac{n\pi}{2}\right)+i\left(\gamma_{ln}^{(m)i}\cos\frac{n\pi}{2}+\gamma_{ln}^{(m)r}\sin\frac{n\pi}{2}\right)\right]\beta_{ln}^{(m)},$$

$$S_l^{(m)}=-\sum_n{}' i^n\gamma_{ln}^{(m)r}\beta_{ln}^{(m)},$$

with

$$\beta_{ln}^{(m)}=[(2-\delta_{0m})/N_{mn}]S_{mn}(h_0,\eta_0)\alpha_{ln}^{(m)},$$

$$\gamma_{ln}^{(m)r,i}=\frac{\rho_0}{\rho_1}R_{mn}^{(1),(2)}(h_0,\xi_0)\left[R_{ml}^{(1)\prime}(h_1,\xi_0)-\frac{R_{ml}^{(1)\prime}(h_1,\xi_1)}{R_{ml}^{(2)\prime}(h_1,\xi_1)}R_{ml}^{(2)\prime}(h_1,\xi_0)\right]$$

$$-R_{mn}^{(1),(2)\prime}(h_0,\xi_0)\left[R_{ml}^{(1)}(h_1,\xi_0)-\frac{R_{ml}^{(1)\prime}(h_1,\xi_1)}{R_{ml}^{(2)\prime}(h_1,\xi_1)}R_{ml}^{(2)}(h_1,\xi_0)\right].$$

The expansion coefficient A_{mn} for the scattered wave can now be obtained from Eq. 15.

It is seen that the solution for the coated spheroidal problem is much more involved than that for the rigid spheroid problem. This is because of the cross coupling between the expansion coefficients for the incident field and the expansion coefficients for the scattered and penetrated field. That is, each expansion coefficient for the scattered or penetrated pressure field is coupled with all expansion coefficients for the incident pressure field. It can be shown by the use of Eq. 14 that decoupling occurs when $h_0\rightarrow h_1$. As $h_0\rightarrow h_1$, $\alpha_{ln}^{(m)}\rightarrow\alpha_{ln}^{(m)}\delta_{ln}$, where δ_{ln} is the Kronecker delta, and the summation sign in Eq. 15 disappears. Equation 15 then becomes $P_{nn}^{(m)}A_{mn}=S_n^{(m)}$, and the expansion coefficient for the scattered field is $A_{mn}=S_n^{(m)}/P_{nn}^{(m)}$. The condition $h_0\rightarrow h_1$ is satisfied when the coated spheroid degenerates into a coated sphere, or when the sound speed of the sheath material approaches that of the surrounding medium, or when the frequency of the incident wave is very low such that $h_1\rightarrow 0$ and $h_0\rightarrow 0$ and $S_{mn}(h_{0,1},\eta)\rightarrow d_{nm}^{mn}(h_{0,1})P_n^m(\eta)$. Decoupling also occurs when the thickness of the coated sheath is zero—i.e., as $\xi_0\rightarrow\xi_1$, $A_{mn}=-R_{mn}^{(1)\prime}(h_0,\xi_0)/R_{mn}^{(3)\prime}(h_0,\xi_0)$, which is the solution for the rigid spheroid case.[3]

Of special interest are the far-zone field and the radiation pattern of the scattered wave. We have

$$p_s(\text{far-zone})\approx(e^{ih_0\xi}/h_0\xi)P_0F(\eta,\phi),$$

where

$$F(\eta,\phi)=\frac{2}{i}\sum_{m=0}^{\infty}\sum_{n=m}^{\infty}\frac{2-\delta_{0m}}{N_{mn}}A_{mn}S_{mn}(h_0,\eta_0)\times S_{mn}(h_0,\eta)\cos m\phi.$$

The radiation pattern of the scattered wave is defined as follows:

$$|f(\gamma,\phi)|=(c_0/\omega)|F(\eta,\phi)|.$$

This completes the derivation of the fundamental formulas involved in the diffraction of waves by a coated ridge prolate spheroid.

III. NUMERICAL EXAMPLES AND CONCLUSIONS

By the use of the orthogonality properties of the spheroidal functions, the exact solution for the scattering of a plane pressure wave by a rigid prolate spheroid coated with a penetrable sheath is obtained. It is noted that, since the angular spheroidal functions are not only functions of η (the angular coordinates) but also of the acoustical properties of the medium in which they apply, the resultant expression for each of the expansion coefficient for the scattered wave is a very complicated function of all orders of radial spheroidal functions. The task of obtaining numerical results for the present coated problem is many orders of magnitude more difficult than for the uncoated rigid spheroid problem. Nevertheless, to illustrate qualitatively how the formal solutions behave, the radiation patterns of the scattered pressure wave are computed. Numerical computations are carried out using the newly compiled Tables on the radial prolate spheroidal functions,[9] and

[9] C. Yeh and C. Chang, "The Radial Prolate Spheroidal Functions," USCEE Rep. 166, Elec. Eng. Dep., Univ. of Southern California, Los Angeles, Calif. (1966).

C. YEH

ing the IBM 7090 computer. The evaluations of the finite matrix, Eq. 15, were carried out by the succes-ve-approximation method; i.e., computations were arried out for a 4×4 matrix, a 5×5 matrix, a 6×6 atrix, etc., until the desired accuracy was reached. It as found (numerically) that the infinite matrix con-erges quite rapidly for the parameters that we con-dered below. At no time was any matrix greater than 2×12 (with $m=0, 1, \cdots 8$) required to achieve an ccuracy of three significant figures.

As a representative example, a density ratio of $/\rho_1=2.0$ and a sound-speed ratio of $c_0/c_1=2.0$ (or $/h_1=2.0$) were chosen. It is recalled that the subscript or 1 refers, respectively, to the surrounding medium the sheath medium. To check against known results a coated rigid sphere, two radiation patterns for a ear spherical prolate spheroid were given in Fig. 2. The ratio of major axis/minor axis is $\xi_1/(\xi_1^2-1)^{\frac{1}{2}}$ $100/(10^4-1)^{\frac{1}{2}}=1.00005$.) These radiation patterns re indistinguishable from those obtained using the phere model.

The radiation patterns for more elongated prolate pheroid ($\xi_1=1.01$) are given in Figs. 3 and 4 for two ncidence angles. The arrow indicates the direction of ncidence of the plane wave. Results for the uncoated ase ($\xi_{0,1}=1.01$) also check with available known results or the rigid spheroid case.[3] It is interesting to note that he presence of a penetrable sheath modifies quite gnificantly the radiation patterns. As a matter of fact, o resemblances between the radiation patterns can be ound when the normalized sheath thickness $h_0(\cosh\xi_0$ $-\cosh\xi_1)$, is $1.0\times(\cosh 1.04-\cosh 1.01)=0.0364$ at $=0°$. This result indicates that a great deal of scatter-ng occurs already at the outer boundary of the coated pheroid. Therefore, one cannot assume that the ominant contribution for the scattered wave is due to he center rigid core even though the thickness of the coated sheath is very thin. This result underlines the importance of the rigorous treatment for the coated-spheroid problem. From Figs. 3 and 4, one also notes that the radiation pattern for the scattered wave becomes more directional as the thickness of the sheath increases and that the main lobe direction of the patterns changes from a direction towards the incident wave to that away from the incident wave. For an obliquely incident wave, as the spheroid becomes more elongated, the radiation pattern becomes more asymmetric about the axis of the incident wave.

As the frequency of the incident pressure wave is increased, multilobe structures for the radiation pat-

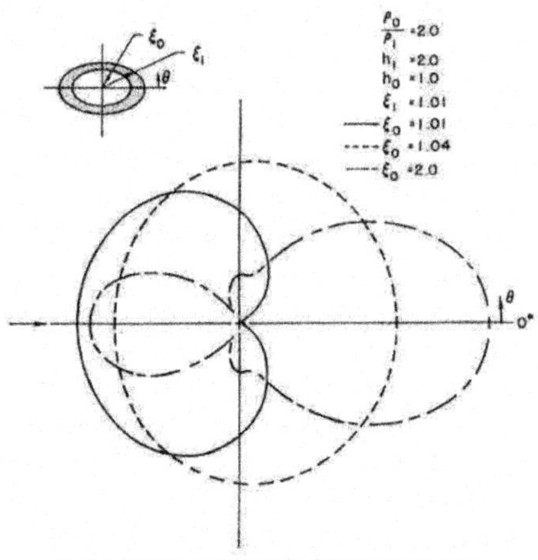

FIG. 3. Radiation patterns for the scattered waves.

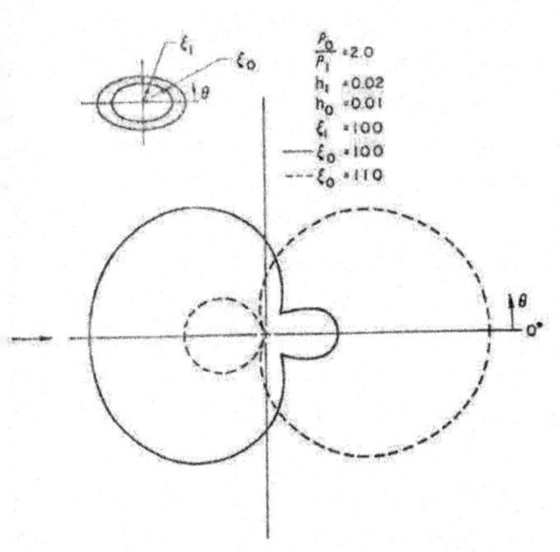

FIG. 2. Radiation patterns for the scattered waves.

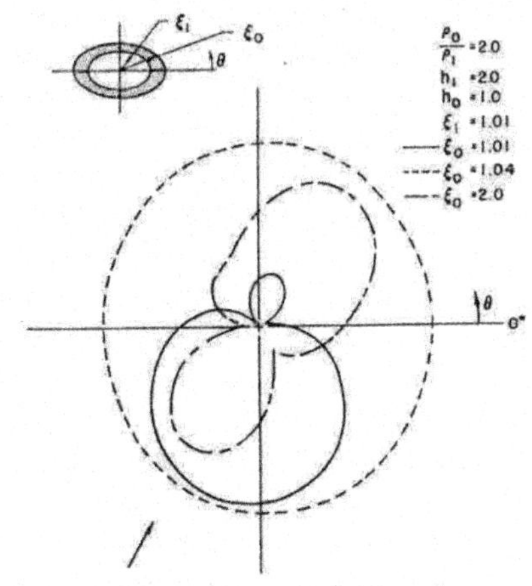

FIG. 4. Radiation patterns for the scattered waves.

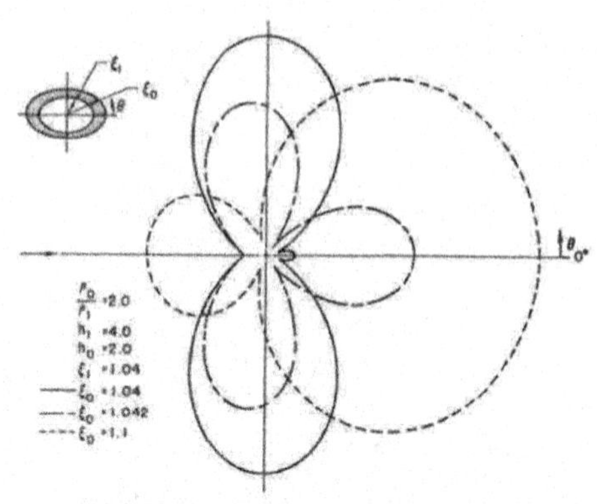

FIG. 5. Radiation patterns for the scattered waves.

terns are found. This effect can be seen from Figs. 5–7 in which three different directions of incidence for the plane wave are assumed. Again, we find that the presence of a thin coating alters quite drastically the scattering characteristics of the wave; i.e., not only is the direction of the main lobe changed significantly, more side lobes also appear.

The fact that the geometry of the scatterer alters significantly the radiation characteristics of the scattered wave should perhaps be emphasized here again. This is quite evident upon inspection of Figs. 2–7. A much more complex structure for the radiation pattern is found for an obliquely incident wave and for thicker coating.

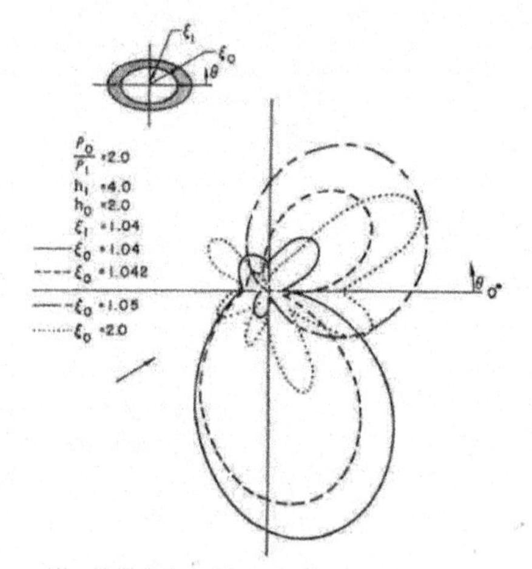

FIG. 6. Radiation patterns for the scattered waves.

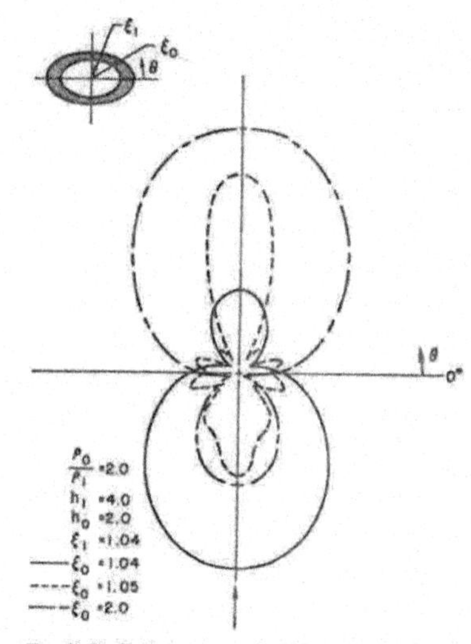

FIG. 7. Radiation patterns for the scattered waves.

It is well known that, as the frequency or the size of the scattering obstacle increases, the series representing the scattered wave converges more and more slowly—i.e., more and more terms are needed to yield the correct numerical result. One classic way to overcome this difficulty is the use of the Watson transformation technique. The Watson transformation transforms the harmonic series as given by Eq. 5 for the scattered wave into a residue series that converges very fast for high frequencies. This technique has been applied recently to the study of scattering of waves by large infinite cylinders.[6] However, because of the coupling between the expansion coefficients discussed above for the coated, rigid prolate-spheroid case, the Watson transformation[10] technique *cannot* be used directly here to yield expressions that are applicable at high frequencies. This feature for the present problem is certainly a significant departure from that for the problems involving spheres, circular cylinders or rigid spheroids. Hence, even at very high frequencies, computations for qualitative results must still be carried out from the given series solution.

[10] G. N. Watson, Proc. Roy. Soc. (London) A95, 83–99, 546–563 (1919).

Reprinted from THE JOURNAL OF THE ACOUSTICAL SOCIETY OF AMERICA, Vol. 43, No. 6, 1454–1455, June 1968

Printed in U. S. A.

A Further Note on the Reflection and Transmission of Sound Waves by a Moving Fluid Layer

C. YEH
Department of Engineering, University of California, Los Angeles, California 90024

The reflection and transmission of sound waves by a moving fluid layer are treated theoretically using the more proper boundary conditions, and the reflection and transmission coefficients are determined. A specific numerical example is also given.

IN A PREVIOUS ARTICLE,[1] THE PROBLEM OF THE REFLECTION AND transmission of sound waves by a moving fluid layer was considered. The boundary conditions for the interface between two stationary media were used.[2,3] (That is, the pressure and the normal component of the velocity are assumed to be continuous at the interface.) However, at the interface of two relatively moving media, the boundary conditions must be deduced from the requirements that the pressure must be continuous and the vortex sheet must be a stream surface common to the flows in the two media.[4,5] Consequently, the purpose of this note is to present the solution for the moving fluid problem using the more appropriate boundary conditions for moving media.

The geometry of the problem is shown in Fig. 1. It is now assumed that fluids in all three regions (Regions A, B, and C) are moving at three different speeds (say, V_0, V_1, and V_2). This assumption is more general than the ones used in the previous work.[1]

The incident and reflected pressure waves in Region A, the pressure wave in Region B, and the transmitted pressure wave in Region C take, respectively, the form

$$p_i^A = e^{ik_0(x\sin\theta_0 + y\cos\theta_0)} e^{-ik_0(V_0\sin\theta_0 + c_0)t}, \quad (1)$$

$$p_r^A = Re^{ik_0(x\sin\theta_0 - y\cos\theta_0)} e^{-ik_0(V_0\sin\theta_0 + c_0)t}, \quad (2)$$

$$p_1^B = (\alpha e^{ik_1\cos\theta_1 y} + \beta e^{-ik_1\cos\theta_1 y}) e^{ik_1\sin\theta_1 x} e^{-ik_1(V_1\sin\theta_1 + c_1)t}, \quad (3)$$

$$p_t^C = Te^{ik_2(x\sin\theta_2 + y\cos\theta_2)} e^{-ik_2(V_2\sin\theta_2 + c_2)t}, \quad (4)$$

where the amplitudes R and T, wavenumbers k_1 and k_2, and the

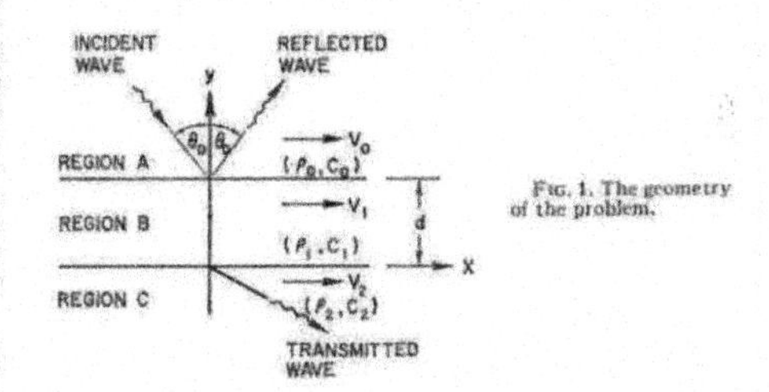

FIG. 1. The geometry of the problem.

wave angles θ_1 and θ_2 are to be determined. θ_0 is the angle of incidence. Satisfying the requirement that the vortex sheets separating the three media must have a definite phase velocity and wavenumber with respect to the x axis,[5] gives

$$k_0\sin\theta_0 = k_1\sin\theta_1 = k_2\sin\theta_2, \quad (5)$$

$$k_0(V_0\sin\theta_0 + c_0) = k_1(V_1\sin\theta_1 + c_1) = k_2(V_2\sin\theta_2 + c_2). \quad (6)$$

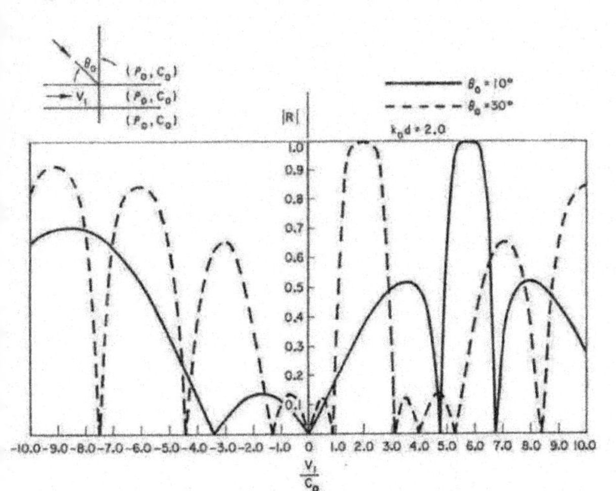

FIG. 2. The reflection coefficient as a function of V_1/c_0 for various angles of incidence.

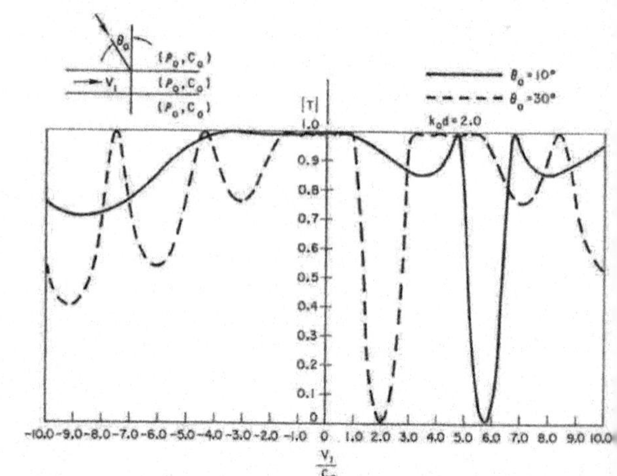

FIG. 3. The transmission coefficient as a function of V_1/c_0 for various angles of incidence.

From Eqs. 5 and 6, one obtains the wave angle of the transmitted wave,

$$\theta_2 = \sin^{-1}\{c_2 \sin\theta_0 / [(V_0 - V_2)\sin\theta_0 + c_0]\}, \tag{7}$$

which is independent of the movement of the layer.

Applying the boundary conditions,[4,5] which require the continuity of the pressure and the vortex sheet at the boundaries $y=d$ and $y=0$, gives

$$p_i^A + p_r^A = p_t^B|_{y=d}, \tag{8}$$

$$\frac{\sin\theta_0}{k_0\rho_0 c_0^2}\left(\frac{\partial p_i^A}{\partial y} + \frac{\partial p_r^A}{\partial y}\right) = \frac{\sin\theta_1}{k_1\rho_1 c_1^2}\frac{\partial p_t^B}{\partial y}\bigg|_{y=d}, \tag{9}$$

$$p_t^B = p_t^C|_{y=0}, \tag{10}$$

$$\frac{\sin\theta_1}{k_1\rho_1 c_1^2}\frac{\partial p_t^B}{\partial y} = \frac{\sin\theta_2}{k_2\rho_2 c_2^2}\frac{\partial p_t^C}{\partial y}\bigg|_{y=0}. \tag{11}$$

Substituting Eqs. 1–4 into the above equations and solving for R and T, one obtains

$$R = e^{2ik_0 d\cos\theta_0}\frac{[(1-\Gamma_1\Gamma_2)\cos(k_1 d\cos\theta_1) - i(\Gamma_1-\Gamma_2)\sin(k_1 d\cos\theta_1)]}{[(1+\Gamma_1\Gamma_2)\cos(k_1 d\cos\theta_1) + i(\Gamma_1+\Gamma_2)\sin(k_1 d\cos\theta_1)]}, \tag{12}$$

$$T = \frac{2e^{ik_0 d\cos\theta_0}}{[(1+\Gamma_1\Gamma_2)\cos(k_1 d\cos\theta_1) + i(\Gamma_1+\Gamma_2)\sin(k_1 d\cos\theta_1)]}, \tag{13}$$

where

$$\Gamma_1 = \rho_0 c_0^2 \sin 2\theta_1 / \rho_1 c_1^2 \sin 2\theta_0, \tag{14}$$

$$\Gamma_2 = \rho_1 c_1^2 \sin 2\theta_2 / \rho_2 c_2^2 \sin 2\theta_1. \tag{15}$$

θ_1 can be obtained by solving Eqs. 5 and 6. Again, as noted in the previous paper,[1] there exists no Doppler shift in frequency for the reflected and transmitted waves, and the angle of reflection is identical to the angle of incidence. It can be shown easily that when $d=0$, R and T given by Eqs. 12 and 13, respectively, reduce to Eqs. 3.4a and 3.5a given by Miles[5] for the moving half-space case.

To illustrate how the reflection and the transmission coefficients vary as a function of the velocity of the moving medium, numerical results for the following limiting case are presented:

$$\rho_0 = \rho_1 = \rho_2,\ c_0 = c_1 = c_2,\ V_0 = V_2 = 0. \tag{16}$$

This case is chosen so that the effect of the movement of the material layer can be brought out, since under this assumption, $R=0$ and $T=1$, when $V_1/c_0=0$. The reflection coefficient (R) and the transmission coefficient (T) are plotted in Figs. 2 and 3 as a function of the velocity of the moving layer for $k_0 d = 2.0$ and for two angles of incidence, $\theta_0 = 10°, 30°$. As can be seen from Figs. 2 and 3, the oscillatory behaviors of the $|R|$-vs-V_1/c_0 curves and the $|T|$-vs-V_1/c_0 curves are somewhat similar to those presented in the previous paper.[1] $|R|$ and $|T|$ are bounded between 0 and 1.

It should be emphasized here that the above results are valid only if the thickness of the viscous and thermal mixing region is small as compared with the wavelength.

Acknowledgment: I wish to thank Professor H. S. Ribner for calling my attention to the correct boundary conditions.

[1] C. Yeh, J. Acoust. Soc. Am. 41, 817–821 (1967).
[2] J. B. Keller, J. Acoust. Soc. Am. 27, 1044–1047 (1955).
[3] P. Franken and U. Ingard, J. Acoust. Soc. Am. 28, 126–127 (1956).
[4] H. S. Ribner, J. Acoust. Soc. Am. 29, 435–441 (1957).
[5] J. W. Miles, J. Acoust. Soc. Am. 29, 226–228 (1957).

Reprinted from THE JOURNAL OF THE ACOUSTICAL SOCIETY OF AMERICA, Vol. 44, No. 5, 1216–1219, November 1968

Diffraction of Sound Waves by a Moving Fluid Cylinder*

C. YEH

Department of Engineering, University of California, Los Angeles, California 90024

The solution of the diffraction of sound waves by an axially moving fluid cylinder is obtained. Results indicate that there is no Doppler frequency shift for the scattered wave and when the angle of incidence is normal to the cylinder axis, the scattered wave is independent of the movement of the medium. Numerical computations for the magnitudes as well as the radiation patterns of the scattered wave as a function of the velocity of the moving medium were also carried out for various incident angles. It is found that the magnitudes and the radiation patterns of the scattered wave are quite sensitive to the variation in the angle of incidence even for slow relative movement of the fluid media.

INTRODUCTION

THE problem of the scattering of acoustic waves by a cylindrical column containing homogeneous or inhomogeneous medium has been of continuing interest for many years.[1,2] Morse and Ungard gave a rather comprehensive treatment of this problem in their recent book.[1] However, in most previous work the medium is assumed to be stationary with respect to the source as well as to the observer. If the material medium of cylindrical shape moves at a uniform speed in the axial direction with respect to an observer, it is expected that the scattering characteristics of an incident sound wave will be affected significantly. The purpose of this paper is to present the solution to this problem. Several features concerning the radiation patterns and the magnitude of the scattered waves, as a function of the velocity of the moving medium are discussed. Result also shows that it is possible to obtain the velocity of the moving column by measuring the radiation pattern or the magnitude of the scattered wave as a function of the angle of incidence.

The related problems involving the reflection and transmission of sound waves by moving half-space or by moving layer have been treated by various authors.[3–5]

I. FORMULATION OF THE PROBLEM

The geometry of this problem is shown in Fig. 1. A circular column of homogeneous material characterized by (ρ_1,c_1) is assumed to occupy the space $r\leq a$. The region outside the cylinder $r>a$ is occupied by homogeneous material characterized by (ρ_0,c_0). The symbol ρ and c signify, respectively, the density and the speed of sound in the medium. It is assumed that the observer and the signal source are stationary with respect to each other. The circular cylinder and the surrounding medium are assumed, respectively, to be moving with respect to the observer at a constant speed of V_1 and V_0 in the x direction.

An incident harmonic plane pressure wave propagating in a direction that makes an angle θ_0 with the positive y axis (see Fig. 1), takes the form

$$p_i = e^{ik_0 x \sin\theta_0} e^{ik_0 y \cos\theta_0} e^{-ik_0(V_0 \sin\theta_0 + c_0)t}, \tag{1}$$

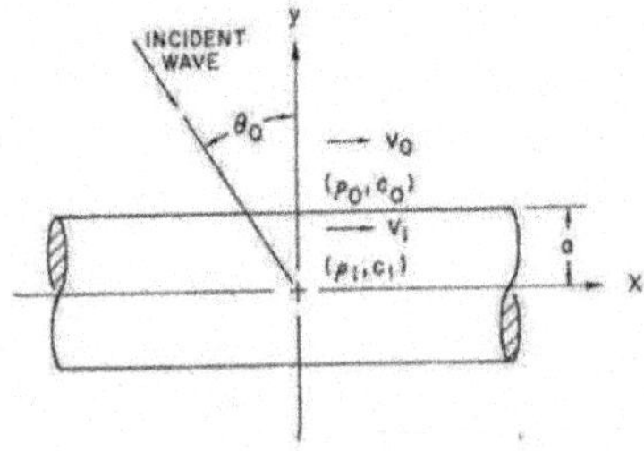

FIG. 1. The geometry of the problem.

* Supported by the Office of Naval Research.
[1] P. M. Morse and K. N. Ingard, *Theoretical Acoustics* (McGraw-Hill Book Co., New York, 1968).
[2] W. P. Mason, Ed., *Physical Acoustics* (Academic Press Inc., New York, 1964).
[3] J. W. Miles, J. Acoust. Soc. Amer. 29, 226–228 (1957).
[4] H. S. Ribner, J. Acoust. Soc. Amer. 29, 435–441 (1957).
[5] C. Yeh, J. Acoust. Soc. Amer. 43, 1454–1455 (L) (1968).

where k_0 is the wavenumber. The coordinates (y,z,x) are at rest with respect to the observer. Equation 1 is obtained by using a Galilean transformation that provides the transformation from a system that is at rest with respect to the moving fluid to a system that is at rest with respect to the observer. Making use of the Fourier–Bessel expansion theorem for a plane wave,[6] Eq. 1 can be expressed in terms of the cylindrical coordinates (r,ϕ,x) as follows:

$$p_i = e^{ik_0x\sin\theta_0}e^{-ik_0(V_0\sin\theta_0+c_0)t}\sum_{n=-\infty}^{\infty} i^n J_n(k_0r\cos\theta_0)e^{-in\phi}, \quad (2)$$

where $J_n(k_0r\cos\theta_0)$ is the Bessel function of order n and argument $k_0r\cos\theta_0$.

The scattered pressure wave and the pressure wave penetrated into the moving cylindrical column must, respectively, have the form

$$p_s = e^{ik_0x\sin\theta_0}e^{-ik_0(V_0\sin\theta_0+c_0)t} \times \sum_{n=-\infty}^{\infty} i^n A_n H_n^{(1)}(k_0r\cos\theta_0)e^{-in\phi} \quad (3)$$

and

$$p_t = e^{ik_1x\sin\theta_1}e^{-ik_1(V_1\sin\theta_1+c_1)t} \times \sum_{n=-\infty}^{\infty} i^n B_n J_n(k_1r\cos\theta_1)e^{-in\phi}, \quad (4)$$

where A_n and B_n are arbitrary constants and the wavenumber k_1, and wave angle θ_1 are to be determined. $H_n^{(1)}(k_0r\cos\theta_0)$ is the Hankel function of order n and argument $k_0r\cos\theta_0$, representing outgoing waves. It is noted that Eq. 3 satisfies the radiation condition.

II. FORMAL SOLUTION

Satisfying the requirement that the vortex sheet separating the two moving media must have a definite phase velocity and wavenumber with respect to the x axis,[3] gives

$$k_0\sin\theta_0 = k_1\sin\theta_1, \quad (5)$$

$$k_0(V_0\sin\theta_0+c_0) = k_1(V_1\sin\theta_1+c_1). \quad (6)$$

Since the boundary separating the two media is time independent, there exists no frequency shift for the scattered wave due to the axial movement of the column as far as an observer in the (y,z,x) system is concerned. Solving Eqs. 5 and 6 gives the wave angle

$$\theta_1 = \sin^{-1}\{c_1\sin\theta_0/[(V_0-V_1)\sin\theta_0+c_0]\}. \quad (7)$$

The approximate boundary conditions at $r=a$ are deduced from the facts that the pressure must be continuous and the vortex sheet must be a stream surface common to the flows in the two media.[3,4] Applying the boundary conditions at $r=a$ gives

$$p_i+p_s = p_t|_{r=a}, \quad (8)$$

$$\frac{\sin\theta_0}{k_0\rho_0c_0^2}\left(\frac{\partial p_i}{\partial r}+\frac{\partial p_s}{\partial r}\right) = \frac{\sin\theta_1}{k_1\rho_1c_1^2}\frac{\partial p_t}{\partial r}\bigg|_{r=a}. \quad (9)$$

Substituting Eqs. 2–4 into the above equations, one obtains

$$J_n(k_0a\cos\theta_0)+A_nH_n^{(1)}(k_0a\cos\theta_0) = B_nJ_n(k_1a\cos\theta_1), \quad (10)$$

$$J_n'(k_0a\cos\theta_0)+A_nH_n^{(1)\prime}(k_0a\cos\theta_0) = \frac{\sin2\theta_1}{\sin2\theta_0}\frac{\rho_0c_0^2}{\rho_1c_1^2}B_nJ_n'(k_1a\cos\theta_1), \quad (11)$$

where the prime signifies the derivative of the function with respect to its argument. Solving for the unknown coefficients A_n and B_n yields

$$A_n = \frac{J_n(k_1a\cos\theta_1)J_n'(k_0a\cos\theta_0)-\Gamma J_n(k_0a\cos\theta_0)J_n'(k_1a\cos\theta_1)}{\Gamma J_n'(k_1a\cos\theta_1)H_n^{(1)}(k_0a\cos\theta_0)-J_n(k_1a\cos\theta_1)H_n^{(1)\prime}(k_0a\cos\theta_0)}, \quad (12)$$

$$B_n = \frac{J_n'(k_0a\cos\theta_0)H_n^{(1)}(k_0a\cos\theta_0)-J_n(k_0a\cos\theta_0)H_n^{(1)\prime}(k_0a\cos\theta_0)}{\Gamma J_n'(k_1a\cos\theta_1)H_n^{(1)}(k_0a\cos\theta_0)-J_n(k_1a\cos\theta_1)H_n^{(1)\prime}(k_0a\cos\theta_0)}, \quad (13)$$

with

$$\Gamma = \frac{\sin\theta_1\cos\theta_1}{\sin\theta_0\cos\theta_0}\frac{\rho_0c_0^2}{\rho_1c_1^2}. \quad (14)$$

θ_1 is given by Eq. 7. It is noted that at normal incidence, $\theta_0=0$, the coefficients for the scattered wave and the penetrated wave (i.e., A_n and B_n) are independent of the velocity of the moving column.

[6] J. A. Stratton, *Electromagnetic Theory* (McGraw-Hill Book Co., New York, 1941), p. 371.

At large distances from the moving cylinder, the asymptotic expression for the Hankel function

$$H_n^{(1)}(k_0r\cos\theta_0) \to (2/\pi k_0r\cos\theta_0)^{\frac{1}{2}}e^{i[k_0r\cos\theta_0-(2n+1)\pi/4]} \quad (15)$$

is applicable provided that $k_0r\cos\theta_0 \gg 1$ and $k_0r\cos\theta_0 \gg n$. Using the above equation, we obtain the expression for the far-zone scattered field

$$p_s \simeq e^{i(k_0 z \sin\theta_0 + k_0 r \cos\theta_0 - \pi/4)} e^{-ik_0(V_0 \sin\theta_0 + c_0)t} \times (2/\pi k_0 r \cos\theta_0)^{\frac{1}{2}} \sum_{n=-\infty}^{\infty} A_n e^{-in\phi}. \quad (16)$$

This completes the derivation of the formulas involved in the scattering of waves by moving fluid cylinder.

III. DISCUSSION OF THE RESULTS

To have a qualitative idea of how the scattered pressures vary as a function of the velocity of the moving medium, we now consider the following limiting case: It is assumed that $\rho_0 = \rho_1$, $c_0 = c_1$, and $V_0 = 0$. Hence,

$$k_1 a \cos\theta_1 = k_0 a \{[1 - (V_1/c_0) \sin\theta_0]^2 - \sin^2\theta_0\}^{\frac{1}{2}}. \quad (17)$$

This assumption is so chosen as to bring out the effect of the movement of the material column, since under this assumption $A_n = 0$ and $B_n = 1$ when $V_1/c_0 = 0$. One also notes from Eq. 17 that $k_1 a \cos\theta_1$ is purely imaginary for $1/\sin\theta_0 - 1 < V_1/c_0 < 1/\sin\theta_0 + 1$ and $k_1 a \cos\theta_1$ is purely real for $-\infty < V_1/c_0 < 1/\sin\theta_0 - 1$ and $1/\sin\theta_0 + 1 < V_1/c_0 < \infty$. When $k_1 a \cos\theta_1$ is imaginary, the Bessel function $J_n(k_1 a \cos\theta_1)$ must be replaced by the modified Bessel function $I_n(|k_1 a \cos\theta_1|)$.

Numerical computation was carried out for the radiation pattern of the scattered pressure wave for various parameters of $k_0 a$, θ_0, and V_1/c_0. The radiation pattern of the scattered wave is defined as follows:

$$|p_s| \simeq \left| \sum_{n=-\infty}^{\infty} A_n e^{-in\phi} \right|, \quad (18)$$

where A_n is given by Eq. 12. In Figs. 2 and 3, the normalized radiation patterns are plotted for various values of V_1/c_0 with $k_0 a = 1.0$. Two angles of incidence, $\theta_0 = 10°$, $30°$, are assumed. It is noted that only representative patterns were shown in these figures. When the incident angle is 10° and $0 < V_1/c_0 < 4.0$, a dipole pattern, as shown in Fig. 2 for the case $V_1/c_0 = 0.5$, appears. As V_1/c_0 increases, the back lobe structure begins to diminish as can be seen for the case $V_1/c_0 = 5.0$. As V_1/c_0 increases further to 10.0, the dipole pattern is again recovered. When the medium is moving in the other direction, i.e., when $V_1/c_0 < 0$, for small negative values of V_1/c_0, again a two-lobe pattern (a main lobe in the forward direction and a minor lobe in the backward direction) is present. As $-V_1/c_0$ increases, the back lobe starts to shrink as can be seen for the case $V_1/c_0 = -5.0$. As $-V_1/c_0$ increases further, the radiation pattern becomes an omnidirectional one.

Significant changes for the radiation patterns of the scattered pressure wave are observed when the angle of incident wave changes to 30°. A two-lobe structure as shown in Fig. 3 for $V_1/c_0 = 0.5$ case is still present for $0 < V_1/c_0 < 1.0$, although the minor lobe is much smaller than the major lobe. As V_1/c_0 increases from 1, the pattern approaches an omnidirectional one as

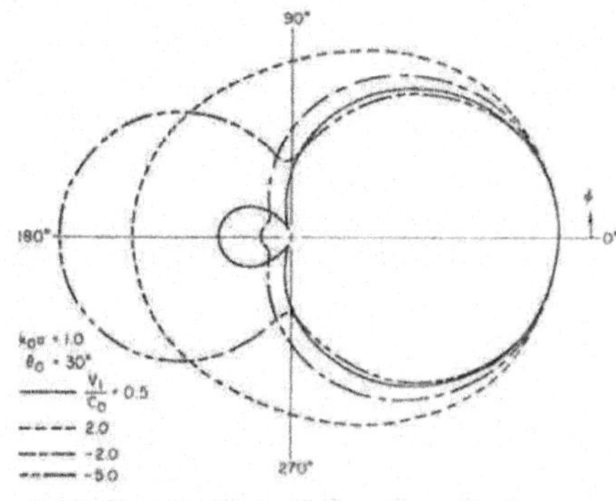

FIG. 3. The normalized radiation patterns for the scattered wave. $\theta_0 = 30°$.

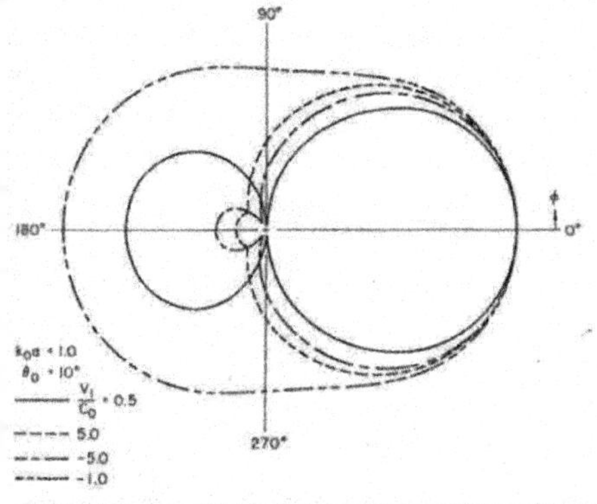

FIG. 2. The normalized radiation patterns for the scattered wave. $\theta = 10°$.

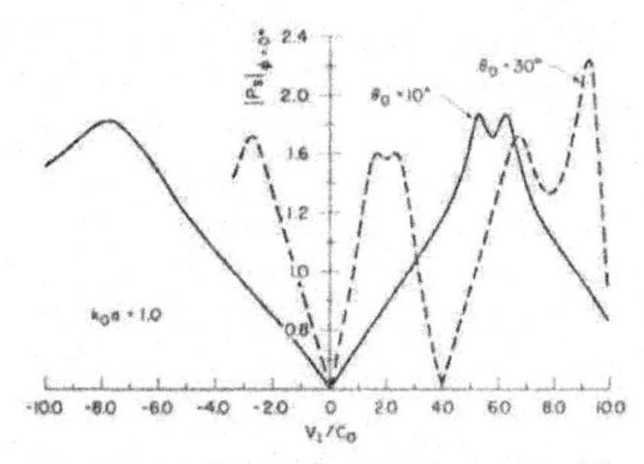

FIG. 4. The magnitude of the scattered wave at $\phi = 0°$ as a function of V_1/c_0 for various angles of incidence.

shown for $V_1/c_0=2.0$ case. If V_1/c_0 increases still further to 10.0, a multilobe structure for the radiation pattern is obtained. As V_1/c_0 decreases from -0.5 to -2.0 and then to -5.0, the pattern changes from a two-lobe structure to a single-lobe structure and then to another two-lobe structure.

Not only are significant variations for the radiation patterns for the scattered wave observed as V_1/c_0 and θ_0 vary, but also the magnitude of the scattered wave $|p_s|$. Figure 4 is introduced to indicate the variation of $|p_s|$ at $\phi=0°$ as a function of the velocity of the moving cylinder for $k_0a=1.0$ and for two angles of incidence, $\theta_0=10°$, $30°$. For very low velocities, i.e., $|V_1/c_0|\ll 1$, $|p_s|$ at $\phi=0°$ is a monotonically increasing function of V_1/c_0 for all angles of incidence except $\theta_0=0$. As $|V_1/c_0|$ increases, the magnitude of the scattered wave at $\phi=0°$ becomes an oscillatory functi of V_1/c_0. As expected, $|p_s|$ as well as the radiati patterns are not symmetric with respect to $V_1/c_0=$

In conclusion, one notes that the magnitude of t scattered signal as well as its radiation pattern a rather sensitive functions of the angle of inciden even for small values of V_1/c_0. Hence, it is possible obtain the velocity of the moving column by measuri the response of $|p_s|$ as a function of θ_0. It should emphasized here that the above results are valid on if the thickness of the viscous and thermal mixing regi is small as compared with the wavelength.

ACKNOWLEDGMENT

This work was supported by the Office of Nav Research.

Triangular ridge acoustic waveguide and coupler

W. A. Oliver,[a] S. B. Dong, and C.Yeh

School of Engineering and Applied Science, University of California, Los Angeles, California 90024

(Received 14 April 1979; accepted for publication 27 June 1980)

Using a more efficient computational scheme in solving the finite element approach to the problem of triangular ridge acoustic waveguides, we have investigated the higher-order modes in this single-material structure. The coupling characteristics of two neighboring triangular guides were also studied. Specific numerical results and discussions are given.

PACS numbers: 43.35.Pt

INTRODUCTION

In recent years several papers[1] have appeared discussing the prospects for applying surface acoustic wave (SAW) technology to the broad area of microwave circuitry and signal processing. As the elastic wavelength is five orders of magnitude smaller than the electromagnetic wavelength at the same frequency, one can envisage microwave acoustic circuits that are correspondingly smaller than their electromagnetic counterparts. For a SAW propagating on a piezoelectric medium, there is a traveling quasistatic electric field accompanying the elastic wave that decays away from the surface both in the elastic medium and in the air above. This exterior field provides the opportunity for a noncontacting interaction with the surface wave.

A fundamental prerequisite for the realization of the microwave acoustic potential is the development of such basic components as waveguides and directional couplers. Waldron[2] has argued that any such device requires a strongly confined wave. However, as the acoustic wavelength at microwave frequencies is so small (on the order of a few microns), the guiding scheme must include a massive substrate simply to provide mechanical support. A large amount of theoretical and experimental work has been conducted on a variety of schemes to overcome the natural diffraction of a surface wave. A sufficient condition for energy localization, i.e., guidance, is the reduction of the phase velocity of the wave in the guided region below that of the unperturbed Rayleigh surface wave. The method by which this is achieved may be broadly classified as physiographic or topographic. In the former category the phase velocity reduction is effected by a change in the physical properties of the guiding region relative to the substrate material. This is most commonly obtained with a thin-film overlay, such as gold on fused quartz. In the topographic case, guidance results from a local perturbation in the geometry of the surface such as a ridge or channel.

Energy coupling between two acoustic waveguides is achieved merely by running the guides close together. If two identical guides become aligned in parallel over a distance defined as the coupling length then, in principle, all the energy in the signal in one guide is transferred to the adjacent guide. This oscillation, or beating, results from the fact that the symmetric and antisymmetric wave components of the double-guide structure travel at different phase velocities. The coupling length L_c is related inversely to the difference between these two propagation constants (wavenumbers):

$$L_c = \pi/(\beta_{\text{antisym}} - \beta_{\text{sym}}). \quad (1)$$

Research on the double-guide coupler was pioneered by Tiersten[3] who developed a weak-coupling model for two parallel thin-film guides. Adkins and Hughes[4] subsequently altered and extended the model for thin-film guides to include strong coupling and obtained good experimental agreement.

The purpose of this paper is to present the results of our work on the triangular ridge guide and coupler. This triangular ridge structure provides strong guidance as well as low dispersion.

I. TRIANGULAR RIDGE GUIDE

Topographic waveguides possess the important practical advantage of consisting of a single material. The search for compatible materials at microwave frequencies is a major drawback of thin-film guides. Lagasse[5] has theoretically investigated ridge guides of various trapezoidal and triangular shapes with differing aspect ratios on an infinite substrate. Employing the finite element method of numerical analysis, he conducted a systematic search for a ridge shape offering strong guidance with low dispersion. His search was made difficult by the fact that these two features counteract each other, thus requiring a careful compromise. The result of his search was the selection of the triangular ridge guide with height/width ratio (H/W) equal to 0.80. This choice was found to offer low dispersion while still maintaining strong energy confinement within the ridge.

Using a finer finite element modeling mesh and a computationally more efficient program than Lagasse, we have investigated the higher-order modes, the modal displacement patterns, the energy confinement, and the effect of substrate depth. It was found for the ridge with H/W of 0.80 that a depth of 7 H simulates an infinite substrate in so far as the guided ridge mode is concerned. For normalized propagation constants (βW) below about 2.0, the ridge mode is no longer guided but reverts to a surface wave mode characterized by the particular substrate geometry employed.

For the ridge of H/W=0.80, the dispersion curves for the two lowest symmetric and antisymmetric modes

---

[a]Currently enrolled in the Graduate School of Management, UCLA, Los Angeles, CA 90024.

propagating in duralumin, an isotropic material, are plotted in Fig. 1, where V is the ridge wave phase velocity, V_R the Rayleigh surface wave velocity for the material, ω the radian frequency, and W the ridge base width. It is observed that the lowest antisymmetric mode has a comparatively broad frequency span of low dispersion in which the phase velocity is below that of the Rayleigh surface wave velocity for the material. Investigation of the modal displacement pattern reveals that the energy of this mode is strongly confined to the ridge (more so as frequency increases) and that the motion is characterized predominately by a lateral displacement.

The triangular ridge guide is seen to be ably suited to function as a low dispersion waveguide. Let us now proceed further and consider its incorporation in a double-guide structure capable of functioning as a directional coupler. It is recognized from the start that the feature of strong energy confinement conflicts with that of energy leakage, which is necessary for coupling between the guides. It is anticipated, therefore, that some design compromises will be necessary. The finite element method of approximate analysis will be employed in the study of these design compromises.

II. THE FINITE ELEMENT FORMULATION

The finite element method (FEM) is a numerical approximation technique based on the piecewise discretization of the body under study into many subregions. In each of these subregions, or elements, interpolation functions are defined to represent the dependent variables. These variables are to be determined in terms of their values at discrete points (nodes). Although it originated in the structural mechanics realm, the FEM has been increasingly applied to the solution of electromagnetics problems[6] as it can handle in a straightforward manner arbitrary boundaries and inhomogeneous and anisotropic media.

The utilization of the FEM requires the selection of the proper variational principle for the given problem, the expression of the functional involved in terms of approximate assumed field functions, and the minimization of this functional to obtain a set of governing equations. The proper variational principle for problems in dynamics is Hamilton's principle. The discretization of the body into subregions, such as triangles or quads for two-dimensional studies, will establish node points at which the unknown field values will be determined. The details of this formulation can be found in Ref. 7, a study which also accounted for piezoelectric coupling. Only the essence of this technique as it is related to the current problem is presented here.

For z-directed propagation down an elastic waveguide of arbitrary cross section, solutions for the displacement components take the form

$$\begin{aligned} u(x,y,z,t) &= [u_I(x,y) + ju_{II}(x,y)]e^{j(\omega t - \beta z)}, \\ v(x,y,z,t) &= [v_I(x,y) + jv_{II}(x,y)]e^{j(\omega t - \beta z)}, \\ w(x,y,z,t) &= [w_I(x,y) + jw_{II}(x,y)]e^{j(\omega t - \beta z)}, \end{aligned} \tag{2}$$

where β is the propagation constant or wavenumber. The two-wave complex representation is necessary in order to satisfy the wave equation for the most general case. This allows that the phase of the displacement components may vary over the cross section of the guide. Under special degenerate conditions, such as orthotropic or higher crystal symmetry, the second part of the wave (u_{II}, v_{II}, w_{II}) is identical to the first and may be omitted. The fields given by these expressions lead to energy terms which are positive-definite so that real frequencies can be obtained.

Because the z variation is already explicitly assumed, the finite element discretization occurs only in the transverse plane of the waveguide. The displacement fields will be approximated by a linear interpolation function defined over each of the planar triangular elements into which the cross section has been discretized. Smaller elements are used in areas in which more rapid spacial field variations are expected.

With the use of natural triangular coordinates, L_1, L_2, L_3 (refer to Fig. 2; any FEM book can be consulted about this coordinate system[8]), the expressions for the transverse dependence of the displacement fields

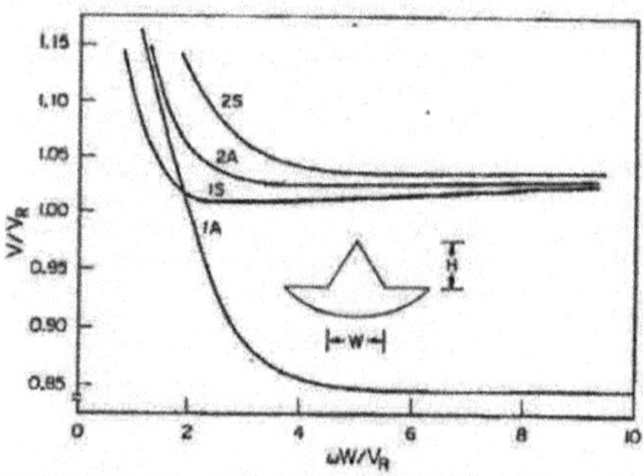

FIG. 1. Dispersion of first two symmetric and antisymmetric modes. Duralumin, H/W=0.80, substrate depth=7H, V_R=2920 m/s.

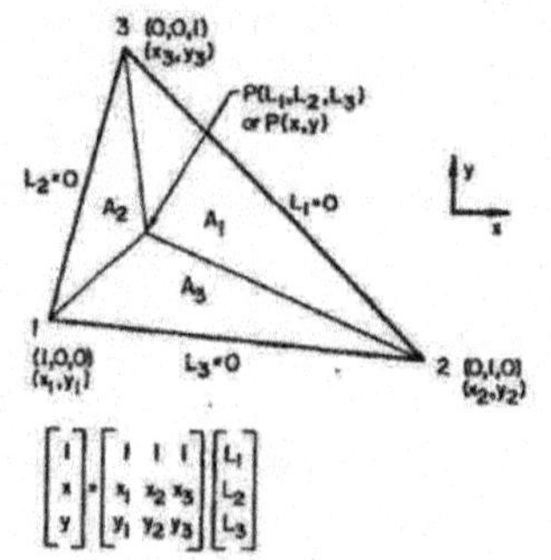

FIG. 2. The triangular coordinates.

$$\begin{bmatrix} 1 \\ x \\ y \end{bmatrix} = \begin{bmatrix} 1 & 1 & 1 \\ x_1 & x_2 & x_3 \\ y_1 & y_2 & y_3 \end{bmatrix} \begin{bmatrix} L_1 \\ L_2 \\ L_3 \end{bmatrix}$$

are

$$u_I(x,y)=L_1(x,y)U_{I1}+L_2(x,y)U_{I2}+L_3(x,y)U_{I3}$$
$$v_I(x,y)=L_1(x,y)V_{I1}+L_2(x,y)V_{I2}+L_3(x,y)V_{I3} \quad (3)$$
$$w_I(x,y)=L_1(x,y)W_{I1}+L_2(x,y)W_{I2}+L_3(x,y)W_{I3}$$

or $u_I=\{L\}^T\{U_I\}$, etc. in the matrix notation. The $\{U_I\}$ are the unknown values of the u displacement component at the corner nodes of the triangular element and $\{L\}$ is the linear interpolation function expressed in terms of natural coordinates. There is a similar interpolation for u_{II}, v_{II}, w_{II}. To simplify the notation, the I, II subscripts will now be dropped and the indicial notation, u_i, will be adopted for the three components, i.e., $u_1=u$, $u_2=v$, and $u_3=w$.

The functional for Hamilton's principle is the Lagrangian, $\mathfrak{L}$, defined as $\mathfrak{L}=T-V$, where T is the kinetic energy and V is the potential (or elastic strain) energy. The potential energy for a particular element is obtained by integrating the strain-energy density over the cross-sectional area, s_I, and one wavelength, λ:

$$V=\frac{1}{2}\int_0^\lambda\int_0^{s_I} S_{ij}C_{ijkl}S_{kl}\,ds_I\,dz, \quad (4)$$

where the S_{ij} are the Cartesian components of the strain tensor, given by

$$S_{ij}=\frac{1}{2}\left(\frac{\partial}{\partial x_j}u_i+\frac{\partial}{\partial x_i}u_j\right), \quad x_i=x,y,z \quad (5)$$

and the C_{ijkl} are the elastic moduli for the element. The corresponding kinetic energy expression is

$$T=\frac{1}{2}\int_0^\lambda\int_0^{s_I}\rho\dot{u}_i\dot{u}_i\,ds_I\,dz, \quad (6)$$

where ρ is the mass density of the element and the overdot denotes the derivative with respect to time. After differentiating the assumed displacement field, its substitution into the potential and kinetic energy expressions leads to forms given in the matrix notation by

$$V=\tfrac{1}{2}\{r\}^T[k]\{r\},$$
$$T=\tfrac{1}{2}\{\dot{r}\}^T[m]\{\dot{r}\}, \quad (7)$$

where $\{r\}$ is the ordered set of nodal values for the triangle and $[k]$ and $[m]$ are the corresponding "stiffness" and "mass" matrices.

The expressions for the potential and kinetic energies of the entire cross section are found by summation over all the triangles.

$$L=\Sigma(T-V)=\tfrac{1}{2}\{\dot{U}\}^T[M]\{\dot{U}\}-\tfrac{1}{2}\{U\}^T[K]\{U\}, \quad (8)$$

where these terms apply to the entire cross section. Hamilton's principle applied to the functional now leads to the equation

$$[K]\{U\}+[M]\{\ddot{U}\}=0. \quad (9)$$

For the assumed simple harmonic motion expressed in Eq. (2), we arrive at the following algebraic eigenvalue problem:

$$\{[K]-\Omega^2[M]\}\{U\}=0, \quad (10)$$

where Ω is the normalized eigenvalue, $\{U\}$ the eigenvector representing the displacement components for all the nodes, and $[K]$ and $[M]$ the assembled stiffness and mass matrices. The wavenumber term, β, occurs in $[K]$.

Equation (10) is solved by an efficient iterative eigensolution technique, described in detail in Ref. 9. The advantage of this solution method is an appreciable reduction of computational effort without sacrificing accurate modeling capability. The iterative portion of the procedure gives rapid convergence simultaneously for a set of eigenvalues and easily handles repeated eigenvalues. The iteration begins with ten trial eigenvectors and terminates when a preassigned convergence tolerance has been achieved by, say, the five lowest eigenvalues. Computational efficiency is also obtained by taking advantage of matrix symmetry and by the judicious numbering of the nodes so as to achieve a highly banded (i.e., sparse) matrix.

III. RIDGE COUPLER ANALYSIS

For a given double-guide coupler configuration, coupling would seem likely to increase as a result of any of the following changes:

(i) reduction in frequency, ω, because the wave becomes less tightly bound to the ridge, thus penetrating deeper into the substrate.

(ii) reduction in gap separation, C, because the wave decays exponentially away from the ridge.

(iii) reduction in ridge height, H, because the wave will be more loosely bound to the ridge.

The finite element method computer program was exercised to determine the coupling effect of each of these changes. The lowest symmetric and antisymmetric dispersion curves for a pair of duralumin guides of H/W =0.80 and C/W=0.50 are plotted in Fig. 3. Also plotted is the coupling length, which is derived from these curves according to Eq. (1).

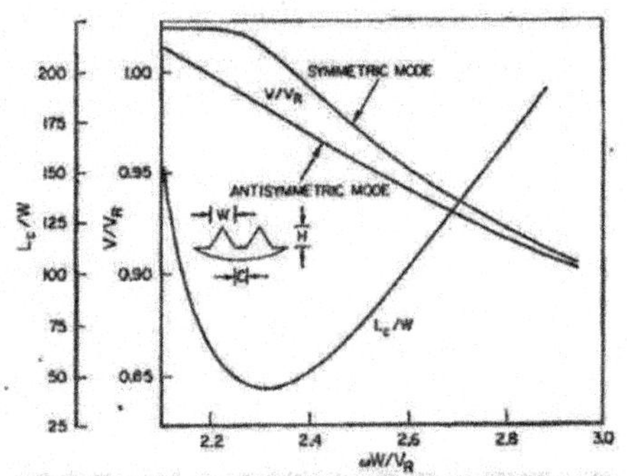

FIG. 3. Dispersion and coupling length of lowest double-guide mode. Duralumin, H/W=0.80, C/W=0.50, substrate depth =20 H.

A. Reduction in frequency

The plot of coupling length versus frequency (Fig. 3) reveals the existence of a frequency of minimum coupling length, i.e., maximum coupling. At lower frequencies the coupling steadily decreases. Experience with electromagnetic directional couplers would not lead one to expect this minimum. A similar effect for a rectangular gold strip on fused quartz was theoretically predicted and experimentally observed by Adkins and Hughes.[4] Without elaboration they characterized this effect as typifying "strong coupling."

An explanation for the existence of a frequency of maximum coupling may be found in the dispersion curves of Fig. 4. The curve representing the lowest symmetric mode of the double-guide structure cannot cross over that of the next lowest symmetric mode. The stability of the conservative system imposes this uniqueness constraint. The constraint forces a flattening of the symmetric mode curve and thus prevents coupling from increasing indefinitely as frequency is lowered. It is interesting to note that at the point at which the symmetric dispersion curves nearly intersect, the modal displacement patterns do "carry across," i.e., the modes rather abruptly exchange patterns.

B. Reduction in gap separation

The coupling curves for the guide of H/W=0.8 are plotted in Fig. 5 for a variety of gap separations. There is approximately a 20% decrease in minimum coupling length as the gap narrows from C/W =1.0 to C/W =0.125. There is likewise approximately a 10% shift to higher frequencies of this minimum value. This figure demonstrates that coupling is much more sensitive to frequency than to the gap separation. The frequency sensitivity of this coupler indicates that it would operate with a much lower bandwidth than for the single ridge case. This feature could be exploited in filter applications.

C. Reduction in ridge height

The results of the height reduction study for a separation of C/W=1.0 are presented in Table I. The inverse relationship observed between height and coupling length is contrary to that initially expected. This behavior may be understood by investigating the dispersion curves and the modal displacement patterns. The coupling length L_c is, according to Eq. (1), inversely related to the difference between the lowest symmetric and antisymmetric mode propagation constants. As the ridge height is reduced, these two dispersion curves move closer together, thus accounting for the increased coupling length. In the limit as the height approaches zero, L_c would become infinite simply because these two curves would coincide as they converge to an unguided surface wave.

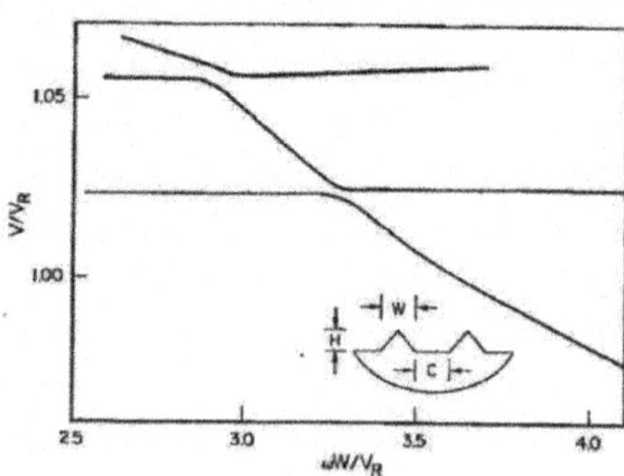

FIG. 4 Dispersion of first three symmetric double-guide modes. Duralumin, H/W=0.60, C/W=1.00, substrate depth =20 H.

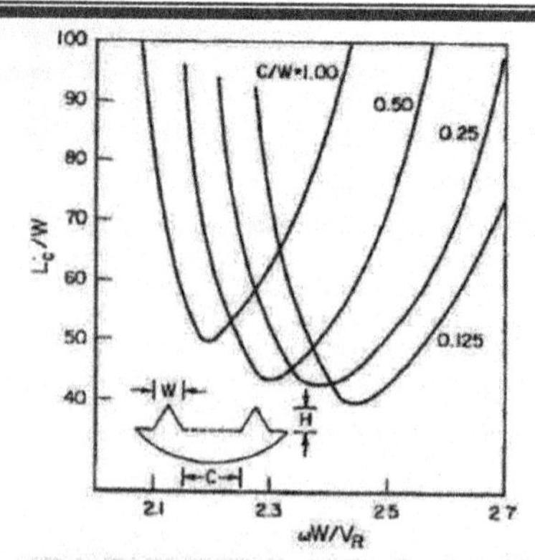

FIG. 5. Coupling lengths for various gap separations. Duralumin, H/W=0.80, substrate depth=20 H.

IV. CONCLUDING REMARKS

In the investigation of the double-guide ridge coupler, the effects of a number of important design parameters have been considered. Strong coupling between the two guides will be reflected in a low value for the coupling length, L_c. This value was observed to decrease not only for a reduction in gap separation, as expected, but also for an increase in ridge height. As frequency is lowered, L_c will not decrease indefinitely but rather to a minimum value, following which it again increases.

Plots such as the one in Fig. 5 are particularly useful for designing a directional coupler. Having specified the ridge aspect ratio and normalized operating frequency, one can use this figure as a design curve to determine the normalized gap separation for maximum coupling. Such design curves are important not only to optimally determine various parameters, but also to

TABLE I. Height reduction study (C/W=1.0).

H/W	Approx. minimum coupling length, L_c/W	Normalized frequency, $\omega W/V_R$
0.80	50	2.2
0.60	90	3.25
0.50	125	4.5
0.40	4000	7.2

bserve the sensitivity of the coupler to small changes n these parameters.

The present analysis has focused on the coupling be- ween two identical parallel guides. In the process of etermining how to maximize coupling, one also can earn how to minimize it. This information is impor- ant when it is necessary to isolate a guide from inter- erence with any other nearby guides. By designing to ninimize coupling, one can determine how closely acked together and for what distance parallel guide uns can be made.

The triangular ridge guide was studied because of the xcellent compromise it offers between strong guidance nd low dispersion. If one is willing to sacrifice the atter feature in order to obtain better coupling, then ther geometries may be considered. Because it can asily handle arbitrary geometries, the finite element nethod is ideally suited to generate design curves for arious topographic configurations.

R. M. White, "Surface Elastic Waves," Proc. IEEE 58, 1238–1276 (1970). E. Stern, "Microsound Components, Circuits, and Applications," IEEE Trans. Microwave Theory Tech. 17, 835–844 (1969). P. E. Lagasse, I. M. Mason and E. A. Ash, "Acoustic Surface Waveguides—Analysis and Assessment," IEEE Trans. Sonics Ultrason. 20, 143–154 (1973).
[2]R. A. Waldron, "Mode Spectrum of a Microsound Waveguide Consisting of an Isotropic Rectangular Overlay on a Perfectly Rigid Substrate," IEEE Trans. Sonics Ultrason. 18, 8–16 (1971).
[3]H. F. Tiersten, "Elastic Surface Waves Guided by Thin Films," J. Appl. Phys. 40, 770–789 (1969).
[4]L. R. Adkins and A. J. Hughes, "Elastic Surface Waves Guided by Thin Films: Gold on Fused Quartz," IEEE Trans. Microwave Theory Tech. 17, 904–911 (1969).
[5]P. E. Lagasse, "Higher-Order Finite Element Analysis of Topographic Guides Supporting Elastic Surface Waves," J. Acoust. Soc. Am. 53, 1116–1122 (1973).
[6]C. Yeh, S. B. Dong, and W. A. Oliver, "Arbitrarily Shaped Inhomogeneous Optical Fiber or Integrated Optical Waveguides," J. Appl. Phys. 46, 2125–2129 (1975).
[7]W. A. Oliver, "Finite Element Analysis of the Piezoelectric Topographic Waveguide," Ph.D. dissertation, School of Engineering and Applied Science, University of California, Los Angeles, 1974.
[8]C. S. Desai and J. F. Abel, *Introduction to the Finite Element Method* (Van Nostrand Reinhold, New York, 1972).
[9]S. B. Dong, J. A. Wolf, Jr., and F. E. Peterson, "On a Direct-Iterative Eigensolution Technique," Int. J. Numer. Methods Eng. 4, 155–161 (1972).

CPSIA information can be obtained at www.ICGtesting.com
Printed in the USA
BVOW06*0022260516

449360BV00003B/3/P